...ation provider it was only natural that **CCH** has developed a way to deliver ...ion that you need, quicker, efficiently, and in a format that you want. On the ...e you will be able to obtain and update OSHA regulations as soon as the gov- ...publishes them in the *Federal Register*. This will allow you to keep your *OSHA* ...*s Book* current, longer. So go to the website; check often; receive free safety tips; ... forms, checklists, and posters; stay current with the latest developments; and ...ber to stay safe!

...new home for all your safety needs can be found at:

**www.safety.cch.com**

**...nformation on further action.** The new CCH Safety Website (*www.safety.cch.com*), CCH EMPLOYMENT SAFETY AND HEALTH GUIDE (weekly), or HUMAN RESOURCES MANAGEMENT—OSHA COMPLIANCE (monthly) should be consulted for further court and administrative actions, as well as amendments made after August 13, 2002. Reports for these publications update the OSHA standards, which are reproduced in the EMPLOYMENT SAFETY AND HEALTH GUIDE in their entirety. The CD and On-Line versions of the CCH EMPLOYMENT SAFETY AND HEALTH GUIDE also contain the text of the standards for states that have OSHA-approved state plans, selected environmental safety laws, and the complete library of Occupational Safety and Health Review Commission decisions. The casebook series OCCUPATIONAL HEALTH AND SAFETY DECISIONS is issued as an adjunct to the EMPLOYMENT SAFETY AND HEALTH GUIDE. The CCH SAFETY MANAGEMENT HANDBOOK provides pratical information for an effective worker safety and health program, including laws, regulations, OSHA standards, and model programs.

*August 2002*

**CCH** INCORPORATED



# OSHA Standards for the Construction Industry

**(29 CFR PART 1926)**
**With Amendments as of August 13, 2002**

**Promulgated by the Occupational Safety and Health Administration United States Department of Labor**

**CCH Editorial Staff Publication**

**CCH** INCORPORATED
Chicago

Publisher: Catherine Wolfe
Executive Editor: Pamela K. Carron
Managing Editor: Jeff Reinholtz J.D.
Coordinating Editor: Steven E. Avadek
Production Coordinator: Theresa J. Jensen
Cover Design: Tom Mullen, Patrick Gallagher
Interior Design: Erika E. Lenz
Layout: Publications Design



ISBN 0-8080-0831-5

4025 W. Peterson Ave.
Chicago, IL 60646-6085
1 800 248 3248
www.safety.cch.com



# About This Boo

This book contains the occupational safety and he industry promulgated by the Occupational Saf effective August 13, 2002. Although this is produces in full the text of regulations containe 29 of the Code of Federal Regulations (CFR) F 1903 regulations governing inspections, ci lations covering occupational injury an 1990 carcinogen policy and model standard and Health Act of 1970 (the "General Duty" c

**Excerpts from Part 1910.** Throughout the construc ences to corresponding sections of Part 1910—the ge convenience, the editors of this book have inserted the co try section/appendix immediately following each cross-refer industry sections have a green background to differentiate t industry text. A finding table for all Part 1910 excerpts follows t this book.

**Recent Amendments.** A number of amendments, additions, and other OSHA standards have been implemented since the last edition of this book in February 2002:

—As part of OSHA's continuing efforts to determine which standards are incons duplicative, outdated, or in need of being rewritten in plain language, the agency revised the following standards: §1926.750, Steel Erection; and §1904, Recording an Reporting Occupational Injuries and Illnesses.

**Steel Erection Standard.** OSHA has issued a new standard on steel erection, §1926.750. The standard was developed through a negotiated rulemaking process by members of the Steel Erection Negotiated Rulemaking Advisory Committee (SENRAC), established by OSHA. The standard was originally scheduled to go into effect on July 18, 2001. However, OSHA chose to change the effective date to January 18, 2002, to provide the agency additional time to explain the new standard to the affected industry.

**Recordkeeping Standard.** OSHA has revised §1904—Recording and Reporting Occupational Injuries and Illnesses, to simplify and improve the overall system used to track work-related injuries and illnesses. Significant changes in the standard include a question/answer format and simplified forms. In addition, the revised standard promotes employee involvement in the recordkeeping process, clarifies provisions related to restricted work, provides more flexibility for using computers to meet recordkeeping requirements, and includes additional industries on the list of exempted employers. The majority of the revised recordkeeping standard took effect on January 1, 2002. However, three amendments will not take effect until January 1, 2003. Those amendments address the recording criteria for occupational hearing loss, the definition of "musculoskeletal disorder" (MSD), and the requirement that employers check the MSD column on the OSHA log. Furthermore, OSHA has replaced the old recordkeeping forms with three new forms: 300; 300A; and 301. For your convenience these forms are reproduced at the end of this book and labeled §1904 Forms.

**Product enhancements.** As part of **CCH** INCORPORATED's ongoing efforts to improve the quality of our products, this book has been redesigned. In response to customer requests, the editors of this book have chosen a new format to improve readability and overall appearance of the publication. Significant changes from previous editions include implementation of a new color scheme to highlight important material and a newly-designed cover. In addition, sections have been tabbed throughout the book, each section now includes a full citation line in color for easy reference, the index has been reformatted and now includes page numbers, graphics have been separated from text by a color border; and the reproduction of the new recordkeeping forms. Any questions or suggestions for further enhancements to this standards book may be directed to CCH Health and Human Resources customer service team members at 1-800-449-9525.

**—New Safety Website.** To meet our customers' needs the editors at **CCH** INCORPORATED have created a new website dedicated solely for the safety professional. The website contains forms, checklists, posters, relevant topical information, and much more... As a **CCH** subscriber, you are entitled to the most current OSHA information available. As a leading

# Table of Contents

## Subpart P—Excavations



## Subpart Q—Concrete and Masonry Construction



## Subpart R—Steel Erection



## Subpart S—Underground Construction, Caissons, Cofferdams and Compressed Air

## Subpart T—Demolition



## Subpart U—Blasting and the Use of Explosives



## Subpart V—Power Transmission and Distribution



## Subpart W—Rollover Protective Structures; Overhead Protection



## Subpart X—Stairways and Ladders

## 29 CFR Part 1990—Identification, Classification, and Regulation of Potential Occupational Carcinogens



## Subject Index For 29 CFR 1926, Sec. 5, 1903, 1904, and 1990



## Forms

# Excerpts from 1910—CCH's "In-Place Reporting"

As a value-added feature, the editors of this book have created an "in-place reporting" system that supplements the construction standards' cross-references with text of general industry standards where applicable. In-place reporting inserts the complete text of each general industry section/appendix immediately following each cross-reference. All incorporated general industry sections have a green background to differentiate them from the construction industry text. CCH's in-place reporting system ensures that employers in the construction industry have access to all of the OSHA requirements that apply to a construction employer.

For your convenience, the table below lists all of the general industry excerpts inserted in-place throughout this book.

## Cross References to the General Industry Standards Contained in the Construction Industry Standards Book

**Authority:** Section 107, Contract Work Hours and Safety Standards Act (Construction Safety Act) (40 U.S.C. 333); secs. 4, 6, 8, Occupational Safety and Health Act of 1970 (29 U.S.C. 653, 655, 657); Secretary of Labor's Order 12-71 (36 FR 8754), 8-76 (41 FR 25059), 1-90 (55 FR 9033),), or 6-96 (62 FR 111), as applicable; 29 CFR part 1911.

### §1926.1 Purpose and scope.

**1926.1(a)** This part sets forth the safety and health standards promulgated by the Secretary of Labor under section 107 of the Contract Work Hours and Safety Standards Act. The standards are published in subpart C of this part and following subparts.

**1926.1(b)** Subpart B of this part contains statements of general policy and interpretations of section 107 of the Contract Work Hours and Safety Standards Act having general applicability.

### §1926.2 Variances from safety and health standards.

**1926.2(a)** Variances from standards which are, or may be, published in this part may be granted under the same circumstances whereunder variances may be granted under section 6(b)(A) or 6(d) of the Williams-Steiger Occupational Safety and Health Act of 1970 (29 U.S.C. 65). The procedures for the granting of variances and for related relief under this part are those published in part 1905 of this title.

**1926.2(b)** Any requests for variances under this section shall also be considered requests for variances under the Williams-Steiger Occupational Safety and Health Act of 1970, and any requests for variances under Williams-Steiger Occupational Safety and Health Act with respect to construction safety or health standards shall be considered to be also variances under the Construction Safety Act. Any variance from a construction safety or health standard which is contained in this part and which is incorporated by reference in part 1910 of this title shall be deemed a variance from the standard under both the Construction Safety Act and the Williams-Steiger Occupational Safety and Health Act of 1970.

### §1926.3 Inspections—right of entry.

**1926.3(a)** It shall be a condition of each contract which is subject to section 107 of the Contract Work Hours and Safety Standards Act that the Secretary of Labor or any authorized representative shall have a right of entry to any site of contract performance for the following purposes:

**1926.3(a)(1)** To inspect or investigate the matter of compliance with the safety and health standards contained in subpart C of this part and following subparts; and

**1926.3(a)(2)** To carry out the duties of the Secretary under section 107(b) of the Act.

**1926.3(b)** For the purpose of carrying out his investigative duties under the Act, the Secretary of Labor may, by agreement, use with or without reimbursement the services, personnel, and facilities of any State or Federal agency. Any agreements with States under this section shall be similar to those provided for under the Walsh-Healey Public Contracts Act under 41 CFR part 50-205.

### §1926.4 Rules of practice for administrative adjudications for enforcement of safety and health standards.

**1926.4(a)** The rules of practice for administrative adjudications for the enforcement of the safety and health standards contained in subpart C of this part and the following subparts shall be the same as those published in part 6 of this title with respect to safety and health violations of the Service Contract Act of 1965 (69 Stat. 1035), except as provided in paragraph (b) of this section.

**1926.4(b)** In the case of debarment, the findings required by section 107(d) of the Act shall be made by the hearing examiner or the Assistant Secretary of Labor for Occupational Safety and Health, as the case may be. Whenever, as provided in section 107(d)(2), a contractor requests termination of debarment before the end of the 3-year period prescribed in that section, the request shall be filed in writing with the Assistant Secretary of Labor for Occupational Safety and Health who shall publish a notice in the FEDERAL REGISTER that the request has been received and afford interested persons an opportunity to be heard upon the request, and thereafter the provisions of part 6 of this title shall apply with respect to prehearing conferences, hearings and related matters, and decisions and orders.

### §1926.5 OMB control numbers under the Paperwork Reduction Act.

The following sections or paragraphs each contain a collection of information requirement which has been approved by the Office of Management and Budget under the control number listed.

| 29 CFR citation | OMB control No. |
|---|---|
| 1926.33 | 1218-0065 |
| 1926.50 | 1218-0093 |
| 1926.52 | 1218-0048 |
| 1926.53 | 1218-0103 |
| 1926.59 | 1218-0072 |
| 1926.60 | 1218-0183 |
| 1926.62 | 1218-0189 |
| 1926.64 | 1218-0200 |
| 1926.65 | 1218-0202 |
| 1926.103 | 1218-0099 |
| 1926.152 | 1218-0210 |
| 1926.200 | 1218-0132 |
| 1926.250 | 1218-0093 |
| 1926.251 | 1218-0233 |
| 1926.403 | 1218-0130 |
| 1926.404 | 1218-0130 |
| 1926.405 | 1218-0130 |
| 1926.407 | 1218-0130 |
| 1926.408 | 1218-0130 |
| 1926.453(a)(2) | 1218-0216 |
| 1926.502 | 1218-0197 |
| 1926.503 | 1218-0197 |
| 1926.550(a)(1) | 1218-0115 |
| 1926.550(a)(2) | 1218-0115 |
| 1926.550(a)(4) | 1218-0115 |
| 1926.550(a)(6) | 1218-0113 |
| 1926.550(a)(11) | 1218-0054 |
| 1926.550(a)(16) | 1218-0115 |
| 1926.550(b)(2) | 1218-0232 |
| 1926.550(g) | 1218-0151 |
| 1926.552 | 1218-0231 |
| 1926.652 | 1218-0137 |
| 1926.703 | 1218-0095 |
| 1926.800 | 1218-0067 |
| 1926.803 | 1218-0067 |
| 1926.900 | 1218-0217 |
| 1926.903 | 1218-0227 |
| 1926.1001 | 1218-0210 |
| 1926.1002 | 1218-0210 |
| 1926.1080 | 1218-0069 |
| 1926.1081 | 1218-0069 |
| 1926.1083 | 1218-0069 |
| 1926.1090 | 1218-0069 |
| 1926.1091 | 1218-0069 |
| 1926.1101 | 1218-0134 |
| 1926.1103 | 1218-0085 |
| 1926.1104 | 1218-0084 |
| 1926.1106 | 1218-0086 |
| 1926.1107 | 1218-0083 |
| 1926.1108 | 1218-0087 |
| 1926.1109 | 1218-0089 |

| 29 CFR citation | OMB control No. |
|---|---|
| 1926.1110 | 1218-0082 |
| 1926.1111 | 1218-0090 |
| 1926.1112 | 1218-0080 |
| 1926.1113 | 1218-0079 |
| 1926.1114 | 1218-0088 |
| 1926.1115 | 1218-0044 |
| 1926.1116 | 1218-0081 |
| 1926.1117 | 1218-0010 |
| 1926.1118 | 1218-0104 |
| 1926.1127 | 1218-0186 |
| 1926.1128 | 1218-0129 |
| 1926.1129 | 1218-0128 |
| 1926.1144 | 1218-0101 |
| 1926.1145 | 1218-0126 |
| 1926.1147 | 1218-0108 |
| 1926.1148 | 1218-0145 |

[61 FR 5509, Feb. 13, 1996, as amended at 63 FR 3814, Jan. 27, 1998; 63 FR 13340, Mar. 19, 1998; 63 FR 17094, Apr. 8, 1998; 64 FR 18810, Apr. 16, 1999]

**Authority:** Sec. 107, Contract Work Hours and Safety Standards Act (Construction Safety Act) (40 U.S.C. 333).

### §1926.10 Scope of subpart.

**1926.10(a)** This subpart contains the general rules of the Secretary of Labor interpreting and applying the construction safety and health provisions of section 107 of the Contract Work Hours and Safety Standards Act (83 Stat. 96). Section 107 requires as a condition of each contract which is entered into under legislation subject to Reorganization Plan Number 14 of 1950 (64 Stat. 1267), and which is for construction, alteration, and/or repair, including painting and decorating, that no contractor or subcontractor contracting for any part of the contract work shall require any laborer or mechanic employed in the performance of the contract to work in surroundings or under working conditions which are unsanitary, hazardous, or dangerous to his health or safety, as determined under construction safety and health standards promulgated by the Secretary by regulation.

### §1926.11 Coverage under section 103 of the act distinguished.

**1926.11(a)** ***Coverage under section 103.*** It is important to note that the coverage of section 107 differs from that for the overtime requirements of the Contract Work Hours and Safety Standards Act. The application of the overtime requirements is governed by section 103, which subject to specific exemptions, includes: (1) Federal contracts requiring or involving the employment of laborers or mechanics (thus including, but not limited to, contracts for construction), and (2) contracts assisted in whole or in part by Federal loans, grants, or guarantees under any statute "providing wage standards for such work." The statutes "providing wage standards for such work" include statutes for construction which require the payment of minimum wages in accordance with prevailing wage findings by the Secretary of Labor in accordance with the Davis-Bacon Act. A provision to section 103 excludes from the overtime requirements work where the Federal assistance is only in the form of a loan guarantee or insurance.

**1926.11(b)** ***Coverage under section 107.*** To be covered by section 107 of the Contract Work Hours and Safety Standards Act, a contract must be one which (1) is entered into under a statute that is subject to Reorganization Plan No. 14 of 1950 (64 Stat. 1267); and (2) is for "construction, alteration, and/or repair, including painting and decorating."

### §1926.12 Reorganization Plan No. 14 of 1950.

**1926.12(a)** ***General provisions.*** Reorganization Plan No. 14 of 1950 relates to the prescribing by the Secretary of Labor of "appropriate standards, regulations, and procedures" with respect to the enforcement of labor standards under Federal and federally assisted contracts which are subject to various statutes subject to the Plan. The rules of the Secretary of Labor implementing the Plan are published in part 5 of this title. Briefly, the statutes subject to the Plan include the Davis-Bacon Act, including its extension to Federal-aid highway legislation subject to 23 U.S.C. 113, and other statutes subject to the Plan by its original terms, statutes by which the Plan is expressly applied, such as the Contract Work Hours Standards Act by virtue of section 104(d) thereof.

**1926.12(b)** ***The Plan.*** (1) The statutes subject to Reorganization Plan No. 14 of 1950 are cited and briefly described in the remaining paragraphs of this section. These descriptions are general in nature and not intended to convey the full scope of the work to be performed under each statute. The individual statutes should be resorted to for a more detailed scope of the work.

**1926.12(b)(2)** *Federal-Aid Highway Acts.* The provisions codified in 23 U.S.C. 113 apply to the initial construction, reconstruction, or improvement work performed by contractors or subcontractors on highway projects on the Federal-aid systems, the primary and secondary, as well as their extensions in urban areas, and the Interstate System, authorized under the highway laws providing for the expenditure of Federal funds upon the Federal-aid system. As cited in 41 Op. A.G. 488, 496, the Attorney General ruled that the Federal-Aid Highway Acts are subject to Reorganization Plan No. 14 of 1950.

**1926.12(b)(3)** *National Housing Act (12 U.S.C. 1713, 1715a, 1715e, 1715k, 1715l(d)(3) and (4), 1715v, 1715w, 1715x, 1743, 1747, 1748, 1748h-2, 1750g, 1715l(h)(1), 1715z(j)(1), 1715z-1, 1715y(d), Subchapter 1x-A and 1x-B, 1715z-7).* This act covers construction which is financed with assistance by the Federal Government through programs of loan and mortgage insurance for the following purposes:

**1926.12(b)(3)(i)** Rental Housing—Section 1713 provides mortgage and insurance on rental housing of eight or more units and on mobile-home courts.

**1926.12(b)(3)(ii)** Section 1715a—Repealed.

**1926.12(b)(3)(iii)** Cooperative Housing—Section 1715e authorizes mortgage insurance on cooperative housing of five or more units as well as supplementary loans for improvement of repair or resale of memberships.

**1926.12(b)(3)(iv)** Urban Renewal Housing—Section 1715k provides mortgage insurance on single family or multifamily housing in approved urban renewal areas.

**1926.12(b)(3)(v)** Low or Moderate Income Housing—Section 1715L(d) (3) and (4) insures mortgages on low-cost single family or multifamily housing.

**1926.12(b)(3)(vi)** Housing for Elderly—Section 1715v provides mortgage insurance on rental housing for elderly or handicapped persons.

**1926.12(b)(3)(vii)** Nursing Homes—Section 1715w authorizes mortgage insurance on nursing home facilities and major equipment.

**1926.12(b)(3)(viii)** Experimental Housing—Section 1715x provides mortgage insurance on single family or multifamily housing with experimental design of materials.

**1926.12(b)(3)(ix)** War Housing Insurance—Section 1743 not active.

**1926.12(b)(3)(x)** Yield Insurance—Section 1747 insures investment returns on multifamily housing.

**1926.12(b)(3)(xi)** Armed Services Housing—Section 1748b to assist in relieving acute shortage and urgent need for family housing at or in areas adjacent to military installations.

**1926.12(b)(3)(xii)** Defense Housing for Impacted Areas—Section 1748h-2 provides mortgage insurance on single family or multifamily housing for sale or rent primarily to military or civilian personnel of the Armed Services, National Aeronautics and Space Administration, or Atomic Energy Commission.

**1926.12(b)(3)(xiii)** Defense Rental Housing—Section 1750g provides for mortgage insurance in critical defense housing areas.

**1926.12(b)(3)(xiv)** Rehabilitation—Section 1715L (h)(1) provides mortgage insurance for nonprofit organizations to finance the purchase and rehabilitation of deteriorating or substandard housing for subsequent resale to low-income home purchasers. There must be located on the property five or more single family dwellings of detached, semidetached, or row construction.

**1926.12(b)(3)(xv)** Homeowner Assistance—Section 1715Z(j)(1) authorizes mortgage insurance to nonprofit organizations or public bodies or agencies executed to finance sale of individual dwellings to lower income individuals or families. Also includes the rehabilitation of such housing if it is deteriorating or substandard for subsequent resale to lower income home purchasers.

**1926.12(b)(3)(xvi)** Rental Housing Assistance—Section 1715Z-1 authorizes mortgage insurance and interest reduction payments on behalf of owners of rental housing projects designed for occupancy by lower income families. Payments are also authorized for certain State or locally aided projects.

**1926.12(b)(3)(xvii)** Condominium Housing—Section 1715y(d) provides mortgage insurance on property purchased for the development of building sites. This includes waterlines and water supply installations, sewer lines and sewage disposal installations, steam, gas, and electrical lines and installations, roads, streets, curbs, gutters, sidewalks, storm drainage facilities, and other installations or work.

**1926.12(b)(3)(xviii)** Group Medical Practice Facilities—Subchapter LX-B authorizes mortgage insurance for the financing of construction and equipment, of facilities for group practice of medicine, optometry, or dentistry.

**1926.12(b)(3)(xix)** Nonprofit Hospitals—1715z-7 authorizes mortgage insurance to cover new and rehabilitated hospitals, including initial equipment.

**1926.12(b)(4)** *Hospital Survey and Construction Act, as amended by the Hospital and Medical Facilities Amendments of 1964 (42 U.S.C. 291e).* The provisions of this Act cover construction contracts made by State or local authorities or private institutions under Federal grant-in-aid programs for the construction of hospitals and other medical facilities.

**1926.12(b)(5)** *Federal Airport Act* (*49 U.S.C. 1114(b)*). The act provides grant-in-aid funds for airport construction limited to general site preparation runways, taxiways, aprons, lighting appurtenant thereto, and fire, rescue, and maintenance buildings. The act excludes construction intended for use as a public parking facility for passenger automobiles and the cost of construction of any part of an airport building except such of those buildings or parts of buildings to house facilities or activities directly related to the safety of persons at the airport.

**1926.12(b)(6)** *Housing Act of 1949* (*42 U.S.C. 1459*). Construction contracts awarded by local authorities financed with the assistance of loans and grants from the Federal Government. The construction programs are for slum clearance and urban renewal which includes rehabilitation grants, neighborhood development programs, neighborhood renewal plans, community renewal, demolition projects, and assistance for blighted areas. See the Housing Act of 1964, paragraph (b)(21) of this section, concerning financial assistance for low-rent housing for domestic farm labor.

**1926.12(b)(7)** *School Survey and Construction Act of 1950* (*20 U.S.C. 636*). This act provides for a Federal grant-in-aid program to assist in the construction of schools in federally affected areas.

**1926.12(b)(8)** *Defense Housing & Community Facilities & Services Act of 1951* (*42 U.S.C. 1592i*). Inactive Program.

**1926.12(b)(9)** *United States Housing Act of 1937* (*42 U.S.C. 1416*). This statute covers the construction of low-rent public housing and slum clearance projects awarded by local authorities. These projects are financed with the assistance of loans and grants from the Federal Government. The slum clearance is the demolition and removal of buildings from any slum area to be used for a low-rent housing project.

**1926.12(b)(10)** *Federal Civil Defense Act of 1950* (*50 U.S.C. App. 2281*). This act provides for Federal assistance to the several States and their political subdivisions in the field of civil defense which includes procurement, construction, leasing, or renovating of materials and facilities.

**1926.12(b)(11)** *Delaware River Basin Compact* (*sec. 15.1, 75 Stat. 714*). This joint resolution creates, by intergovernmental compact between the United States, Delaware, New Jersey, New York, and Pennsylvania, a regional agency for planning, conservation, utilization, development, management and control of the water and related sources of the Delaware River.

**1926.12(b)(12)** *Cooperative Research Act* (*20 U.S.C. 332a(c)*). This act provides Federal grants to a university, college, or other appropriate public or nonprofit private agency or institution for part or all of the cost of constructing a facility for research or for research and related purposes. Research and related purposes means research, research training, surveys, or demonstrations in the field of education, or the dissemination of information derived therefrom, or all of such activities, including (but without limitation) experimental schools, except that such term does not include research, research training, surveys, or demonstrations in the field of sectarian instruction or the dissemination of information derived therefrom. Construction includes new buildings, and the acquisition, expansion, remodeling, replacement, and alteration of existing buildings and the equipping of new buildings and existing buildings.

**1926.12(b)(13)** *Health Professions Educational Assistance Act of 1963* (*42 U.S.C. 292d (c)(4), 293a(c)(5)*). The provisions of this act provide for grants to assist public and nonprofit medical, dental, and similar schools for the construction, expansion, or renovation of teaching facilities.

**1926.12(b)(14)** *Mental Retardation Facilities Construction Act* (*42 U.S.C. 295(a)(2)(D), 2662(5), 2675(a)(5)*). This act authorizes Federal financial assistance in the construction of centers for research on mental retardation and related aspects of human development, of university-affiliated facilities for the mentally retarded and of facilities for the mentally retarded.

**1926.12(b)(15)** *Community Mental Health Centers Act* (*42 U.S.C. 2685(a)(5)*). This act authorizes Federal grants for the construction of public and other nonprofit community mental health centers.

**1926.12(b)(16)** *Higher Education Facilities Act of 1963* (*20 U.S.C. 753*). This act authorizes the grant or loan of Federal funds to assist public and other nonprofit institutions of higher education in financing the construction, rehabilitation, or improvement of academic and related facilities in undergraduate and graduate schools.

**1926.12(b)(17)** *Vocational Educational Act of 1963* (*20 U.S.C. 35f*). This act provides for Federal grants to the various States for construction of area vocational education school facilities.

**1926.12(b)(18)** *Library Services and Construction Act* (*20 U.S.C. 355e(a)(4)*). This act provides for Federal assistance to the various States for the construction of public libraries.

**1926.12(b)(19)** *Urban Mass Transportation Act of 1954* (*49 U.S.C. 1609*). This act provides for grants and loans to assist States and local public bodies and agencies thereof in financing the acquisition, construction, reconstruction, and improvement of facilities and equipment for use, by operation or lease or otherwise, in mass transportation service in urban areas and in coordinating such service with highway and other transportation in such areas.

**1926.12(b)(20)** *Economic Opportunity Act of 1964* (*42 U.S.C. 2947*). This act covers construction which is financed with assistance of the Federal Government for the following purposes:

**1926.12(b)(20)(i)** Authorizes Federal assistance for construction of projects, buildings and works which will provide young men and women in rural and urban residential centers with education, vocational training, and useful work experience (Title I).

**1926.12(b)(20)(ii)** Authorizes financial assistance for construction work planned and carried out at the community level for antipoverty programs (Title II):

**1926.12(b)(20)(ii)(*a*)** Authorizes loans to low income rural families by assisting them to acquire or improve real estate or reduce encumbrances or erect improvements thereon, and to participate in cooperative associations and/or to finance nonagricultural enterprises which will enable such families to supplement their income (Title III);

**1926.12(b)(20)(ii)(*b*)** Authorizes loans to local cooperative associations furnishing essential processing, purchasing, or marketing services, supplies, or facilities predominantly to low-income rural families (Title III);

**1926.12(b)(20)(ii)(c)** Authorizes financial assistance to States, political subdivisions of States, public and nonprofit agencies, institutions, organizations, farm associations, or individuals in establishing housing, sanitation, education, and child day-care programs for migrants and other seasonally employed agricultural employees and their families (Title III).

**1926.12(b)(20)(iii)** Authorizes loans or guarantees loans to small businesses for construction work (Title IV).

**1926.12(b)(20)(iv)** Authorizes the payment of the cost of experimental, pilot, or demonstration projects to foster State programs providing construction work experience or training for unemployed fathers and needy people (Title V).

**1926.12(b)(21)** *Housing Act of 1964* (*42 U.S.C. 1486(f); 42 U.S.C. 1452b(e)*). Provides financial assistance for low-rent housing for domestic farm labor. The Act further provides for loans, through public or private agencies, where feasible, to owners or tenants of property in urban renewal areas to finance rehabilitation required to conform the property to applicable code requirements or carry out the objectives of the urban renewal plan for the area.

**1926.12(b)(22)** *The Commercial Fisheries Research and Development Act of 1964* (*16 U.S.C. 779e(b)*). This Act authorizes financial assistance to State agencies for construction projects designed for the research and development of the commercial fisheries resources of the Nation.

**1926.12(b)(23)** *The Nurse Training Act of 1964* (*42 U.S.C. 296a(b)(5)*). This act provides for grants to assist in the construction of new facilities for collegiate, associate degree, and diploma schools of nursing, or replacement or rehabilitation of existing facilities of such schools.

**1926.12(b)(24)** *Elementary and Secondary Education Act of 1965* (*20 U.S.C. 241i, 848*). The purpose of the act is to provide financial assistance to local educational agencies serving areas with concentrations of children from low-income families for construction in connection with the expansion or improvement of their educational programs.

**1926.12(b)(25)** *Federal Water Pollution Control Act, as amended by the Water Quality Act of 1965* (*3 U.S.C. 466e(g)*). Provides for financial assistance to States or municipalities for construction of facilities in connection with the prevention and control of water pollution. This includes projects that will control the discharge into any waters of untreated or inadequately treated sewage.

**1926.12(b)(26)** *Appalachian Regional Development Act of 1965* (*40 U.S.C. App. 402*). Authorizes Federal assistance in the construction of an Appalachian development highway system; construction of multicounty demonstration health facilities, hospitals, regional health, diagnostic and treatment centers, and other facilities for health; seal and fill voids in abandoned mines and to rehabilitate strip mine areas; construction of school facilities for vocational education; and to assist in construction of sewage treatment works.

**1926.12(b)(27)** *National Technical Institute for the Deaf Act* (*20 U.S.C. 684(b)(5)*). Provides for financial assistance for institutions of higher education for the establishment, construction, including equipment and operation, of a National Institution for the Deaf.

**1926.12(b)(28)** *Housing Act of 1959 (12 U.S.C. 1701(q)(c)(3))*. This act authorizes loans to nonprofit corporations to be used for the construction of housing and related facilities for elderly families. Also, the provisions of the act provide for rehabilitation, alteration, conversion or improvement of existing structures which are otherwise inadequate for proposed dwellings used by such families.

**1926.12(b)(29)** *College Housing Act of 1950, as amended (12 U.S.C. 1749a(f))*. This act provides for Federal loans to assist educational institutions in providing housing and other educational facilities for students and faculties.

**1926.12(b)(30)** *Housing and Urban Development Act of 1965 (42 U.S.C. 1500c-3, 3107)*. This act provides for Federal assistance for the following purposes:

**1926.12(b)(30)(i)** Grants to States and local public bodies to assist in any construction work to be carried out under the open-space land and urban beautification provisions contained therein. It provides for parks and recreation areas, conservation of land and other natural resources, and historical and scenic purposes.

**1926.12(b)(30)(ii)** Grants to local public bodies and agencies to finance specific projects for basic public water facilities (including works for the storage, treatment, purification, and distribution of water), and for basic public sewer facilities (other than "treatment works" as defined in the Federal Water Pollution Control Act).

**1926.12(b)(30)(iii)** Grants to any local public body or agency to assist in financing neighborhood facilities. These facilities must be necessary for carrying out a program of health, recreational, social, or similar community service and located so as to be available for the use of the area's low or moderate income residents.

**1926.12(b)(31)** *National Foundation on the Arts and the Humanities Act of 1965 (20 U.S.C. 954(k))*. The act establishes the "National Foundation on the Arts and the Humanities" which may provide matching grants to groups (nonprofit organizations and State and other public organizations) and to individuals engaged in creative and performing arts for the entire range of artistic activity, including construction of necessary facilities.

**1926.12(b)(32)** *Public Works and Economic Development Act of 1965 (42 U.S.C. 3222)*. This act provides for Federal assistance for the following purposes:

**1926.12(b)(32)(i)** Grants for the acquisition or development of land or improvements for public works or development facility usage in redevelopment areas. It authorizes loans to assist in financing the purchase or development of land for public works which will assist in the creation of long-term employment opportunities in the area.

**1926.12(b)(32)(ii)** Loans for the purchase or development of land and facilities (including machinery and equipment) for industrial or commercial usage within redevelopment areas; guarantee of loans for working capital made to private borrowers by private lending institutions in connection with direct loan projects; and to contract to pay to, or on behalf of, business entities locating in redevelopment areas, a portion of the interest costs which they incur in financing their expansions from private sources.

**1926.12(b)(32)(iii)** Loans and grants to create economic development centers within designated county economic development districts.

**1926.12(b)(33)** *High-Speed Ground Transportation Study (40 U.S.C. 1636(b))*. This act provides for financial assistance for construction activities in connection with research and development of different forms of high-speed ground transportation and demonstration projects relating to intercity rail passenger service.

**1926.12(b)(34)** *Heart Disease, Cancer and Stroke Amendments of 1965 (42 U.S.C. 299(b)(4))*. This act provides for grants to public or nonprofit private universities, medical schools, research, institutions, hospitals, and other public and nonprofit agencies and institutions, or associations thereof to assist in construction and equipment of facilities in connection with research, training, demonstration of patient care, diagnostic and treatment related to heart disease, cancer, stroke, and other major diseases.

**1926.12(b)(35)** *Mental Retardation Facilities and Community Mental Health Centers Construction Act Amendments of 1965 (20 U.S.C. 618(g))*. These provisions provide for grants to institutions of higher education for construction of facilities for research or for research and related purposes relating to education for mentally retarded, hard of hearing, deaf, speech impaired, visually handicapped, seriously emotionally disturbed, crippled, or other health impaired children who by reason thereof require special education.

**1926.12(b)(36)** *Vocational Rehabilitation Act Amendments of 1965 (29 U.S.C. 41a(b)(4))*. This act authorizes grants to assist in meeting the costs of construction of public or other nonprofit workshops and rehabilitation facilities.

**1926.12(b)(37)** *Clean Air and Solid Waste Disposal Acts (42 U.S.C. 3256)*. This act provides for financial assistance to public (Federal, State, interstate, or local) authorities, agencies, and institutions, private agencies and institutions, and individuals in the construction of facilities for solid-waste disposal. The term construction includes the installation of initial equipment.

**1926.12(b)(38)** *Medical Library Assistance Act of 1965 (42 U.S.C. 280b-3(b)(3))*. This act provides for grants to public or private non-profit agencies or institutions for the cost of construction of medical library facilities.

**1926.12(b)(39)** *Veterans Nursing Home Care Act (38 U.S.C. 5035(a)(8))*. The construction industry health and safety standards do not apply to this act since it is not subject to Reorganization Plan No. 14 of 1950.

**1926.12(b)(40)** *National Capital Transportation Act of 1965 (40 U.S.C. 682(b)(4))*. This act provides for Federal assistance to the National Capital Transportation Agency for construction of a rail rapid transit system and related facilities for the Nation's Capital.

**1926.12(b)(41)** *Alaska Centennial—1967 (80 Stat. 82)*. The program under this legislation has expired.

**1926.12(b)(42)** *Model Secondary School for the Deaf Act (80 Stat. 1028)*. This act provides for funds to establish and operate, including construction and initial equipment of new buildings, expansion, remodeling, and alteration of existing buildings and equipment thereof, a model secondary school for the deaf to serve the residents of the District of Columbia and nearby States.

**1926.12(b)(43)** *Allied Health Professions Personnel Training Act of 1966 (42 U.S.C. 295h(b)(2)(E))*. This act provides for grants to assist in the construction of new facilities for training centers for allied health professions, or replacement or rehabilitation of existing facilities for such centers.

**1926.12(b)(44)** *Demonstration Cities and Metropolitan Development Act of 1966 (42 U.S.C. 3310; 12 U.S.C. 1715c; 42 U.S.C. 1416)*. This act provides for Federal assistance for the following purposes:

**1926.12(b)(44)(i)** Grants to assist in the construction, rehabilitation, alteration, or repair of residential property only if such residential property is designed for residential use for eight or more families to enable city demonstration agencies to carry out comprehensive city demonstration programs (42 U.S.C. 3310).

**1926.12(b)(44)(ii)** Amends the National Housing Act (12 U.S.C. 1715c) and the Housing Act of 1937 (42 U.S.C. 1416). See these acts for coverage.

**1926.12(b)(45)** *Air Quality Act of 1967 (42 U.S.C. 1857j-3)*. This act provides for Federal assistance to public or nonprofit agencies, institutions, and organizations and to individuals, and contracts with public or private agencies, institutions, or persons for construction of research and development facilities and demonstration plants relating to the application of preventing or controlling discharges into the air of various types of pollutants.

**1926.12(b)(46)** *Elementary and Secondary Education Amendments of 1967 (Title VII—Bilingual Education Act) (20 U.S.C. 880b-6)*. This act provides for Federal assistance to local educational agencies or to an institution of higher education applying jointly with a local educational agency for minor remodeling projects in connection with bilingual education programs to meet the special needs of children with limited English-speaking ability in the United States.

**1926.12(b)(47)** *Vocational Rehabilitation Amendments of 1967 (29 U.S.C. 42a(c)(3))*. This act authorizes Federal assistance to any public or nonprofit private agency or organization for the construction of a center for vocational rehabilitation of handicapped individuals who are both deaf and blind which shall be known as the National Center for Deaf-Blind Youths and Adults. Construction includes new buildings and expansion, remodeling, alteration and renovation of existing buildings, and initial equipment of such new, newly acquired, expanded, remodeled, altered, or renovated buildings.

**1926.12(b)(48)** *National Visitor Center Facilities Act of 1968 (40 U.S.C. 808)*. This act authorizes agreements and leases with the owner of property in the District of Columbia known as Union Station for the use of all or a part of such property for a national visitor center to be known as the National Visitor Center. The agreements and leases shall provide for such alterations of the Union Station Building as necessary to provide adequate facilities for visitors. They also provide for the construction of a parking facility, including necessary approaches and ramps.

**1926.12(b)(49)** *Juvenile Delinquency Prevention and Control Act of 1968 (42 U.S.C. 3843)*. This act provides for Federal grants to State, county, municipal, or other public agency or combination thereof for the construction of facilities to be used in connection with rehabilitation services for the diagnosis, treatment, and rehabilitation of delinquent youths and youths in danger of becoming delinquent.

**1926.12(b)(50)** *Housing and Urban Development Act of 1968 (including New Communities Act of 1968) (42 U.S.C. 3909)*. This act provides for Federal assistance for the following purposes:

**1926.12(b)(50)(i)** Guarantees, and commitments to guarantee, the bonds, debentures, notes, and other obligations issued by new community developers to help finance new community development projects.

**1926.12(b)(50)(ii)** Amends section 212(a) of the National Housing Act, adding section 236 for "Rental Housing for Lower Income Families" and section 242 "Mortgage Insurance for Nonprofit Hospitals" thereto.

**1926.12(b)(51)** *Public Health Service Act Amendment (Alcoholic and Narcotic Addict Rehabilitation Amendments of 1968) (42 U.S.C. 2681, et seq.)*. This act provides for grants to a public and nonprofit private agency or organization for construction projects consisting of any facilities (including post-hospitalization treatment facilities for the prevention and treatment of alcoholism or treatment of narcotic addicts.)

**1926.12(b)(52)** *Vocational Education Amendments of 1968 (20 U.S.C. 1246)*. This act provides for grants to States for the construction of area vocational education school facilities. The act further provides grants to public educational agencies, organizations, or institutions for construction of residential schools to provide vocational education for the purpose of demonstrating the feasibility and desirability of such schools. The act still further provides grants to State boards, to colleges and universities, to public educational agencies, organizations or institutions to reduce the cost of borrowing funds for the construction of residential schools and dormitories.

**1926.12(b)(53)** *Postal Reorganization Act (39 U.S.C. 410(d)(2))*. This Act provides for construction, modification, alteration, repair, and other improvements of postal facilities located in leased buildings.

**1926.12(b)(54)** *Airport and Airway Development Act of 1970 (Pub. L. 91-258, section 52(b)(7))*. This Act provides for Federal financial assistance to States and localities for the construction, improvement, or repair of public airports.

**1926.12(b)(55)** (i) *Public Law 91-230.* This Act provides for federal financial assistance to institutions of higher learning for the construction of a National Center on Educational Media and Materials for the Handicapped. The program under this statute expires on July 1, 1971. Public Law 91-230, section 662(1).

**1926.12(b)(55)(ii)** *Education of the Handicapped Act (20 U.S.C. 12326, 1404(a))*. This Act provides for financial assistance to States for construction, expansion, remodeling, or alteration of facilities for the education of handicapped children at the preschool, elementary school, and secondary school levels.

**1926.12(b)(56)** *Housing and Urban Development Act of 1970 (Pub. L. 91-609, section 707(b))*. This Act provides for grants to States and local public agencies to help finance the development of open-space or other land in urban areas for open-space uses. This Act becomes effective on July 1, 1971.

**1926.12(b)(57)** *Developmental Disabilities Services and Facilities Construction Amendments of 1970 (Pub. L. 91-517, section 135(a)(5))*. This Act authorizes grants to States for construction of facilities for the provision of services to persons with developmental disabilities who are unable to pay for such services.

**1926.12(b)(58)** *Rail Passenger Service Act of 1970 (Pub. L. 91-518, section 405(d))*. This statute provides that the National Railroad Passenger Corporation may construct physical facilities necessary to intercity rail passenger operations within the basic national rail passenger system designated by the Secretary of Transportation.

**1926.12(c)** ***VA and FHA housing.*** In the course of the legislative development of section 107, it was recognized that section 107 would not apply to housing construction for which insurance was issued by the Federal Housing Authority and Veterans' Administration for individual home ownership. Concerning construction under the National Housing Act, Reorganization Plan No. 14 of 1950 applies to construction which is subject to the minimum wage requirements of section 212(a) thereof (12 U.S.C. 1715c).

## §1926.13 Interpretation of statutory terms.

**1926.13(a)** The terms *construction, alteration,* and *repair* used in section 107 of the Act are also used in section 1 of the Davis-Bacon Act (40 U.S.C. 276a), providing minimum wage protection on Federal construction contracts, and section 1 of the Miller Act (40 U.S.C. 270a), providing performance and payment bond protection on Federal construction contracts. Similarly, the terms *contractor* and *subcontractor* are used in those statutes, as well as in Copeland (Anti-Kickback) Act (40 U.S.C. 276c) and the Contract Work Hours and Safety Standards Act itself, which apply concurrently with the Miller Act and the Davis-Bacon Act on Federal construction contracts and also apply to most federally assisted construction contracts. The use of the same or identical terms in these statutes which apply concurrently with section 107 of the Act have considerable precedential value in ascertaining the coverage of section 107.

**1926.13(b)** It should be noted that section 1 of the Davis-Bacon Act limits minimum wage protection to laborers and mechanics "employed directly" upon the "site of the work." There is no comparable limitation in section 107 of the Act. Section 107 expressly requires as a self-executing condition of each covered contract that no contractor or subcontractor shall require "any laborer or mechanic employed in the performance of the contract to work in surroundings or under working conditions which are unsanitary, hazardous, or dangerous to his health or safety" as these health and safety standards are applied in the rules of the Secretary of Labor.

**1926.13(c)** The term *subcontractor* under section 107 is considered to mean a person who agrees to perform any part of the labor or material requirements of a contract for construction, alteration or repair. Cf. MacEvoy Co. v. United States, 322 U.S. 102, 108-9 (1944). A person who undertakes to perform a portion of a contract involving the furnishing of supplies or materials will be considered a "subcontractor" under this part and section 107 if the work in question involves the performance of construction work and is to be performed: (1) Directly on or near the construction site, or (2) by the employer for the specific project on a customized basis. Thus, a supplier of materials which will become an integral part of the construction is a "subcontractor" if the supplier fabricates or assembles the goods or materials in question specifically for the construction project and the work involved may be said to be construction activity. If the goods or materials in question are ordinarily sold to other customers from regular inventory, the supplier is not a "subcontractor." Generally, the furnishing of prestressed concrete beams and prestressed structural steel would be considered manufacturing; therefore a supplier of such materials would not be considered a "subcontractor." An example of material supplied "for the specific project on a customized basis" as that phrase is used in this section would be ventilating ducts, fabricated in a shop away from the construction job site and specifically cut for the project according to design specifications. On the other hand, if a contractor buys standard size nails from a foundry, the foundry would not be a covered "subcontractor." Ordinarily a contract for the supplying of construction equipment to a contractor would not, in and of itself, be considered a "subcontractor" for purposes of this part.

## §1926.14 Federal contract for "mixed" types of performance.

**1926.14(a)** It is the intent of the Congress to provide safety and health protection of Federal, federally financed, or federally assisted construction. See, for example, H. Report No. 91-241, 91st Cong., first session, p. 1 (1969). Thus, it is clear that when a Federal contract calls for mixed types of performance, such as both manufacturing and construction, section 107 would apply to the construction. By its express terms, section 107 applies to a contract which is "for construction, alteration, and/or repair." Such a contract is not required to be exclusively for such services. The application of the section is not limited to contracts which permit an overall characterization as "construction contracts." The text of section 107 is not so limited.

**1926.14(b)** When the mixed types of performances include both construction and manufacturing, see also §1926.15(b) concerning the relationship between the Walsh-Healey Public Contracts Act and section 107.

## §1926.15 Relationship to the Service Contract Act; Walsh-Healey Public Contracts Act.

**1926.15(a)** A contract for "construction" is one for nonpersonal service. See, e.g., 41 CFR 1-1.208. Section 2(e) of the Service Contract Act of 1965 requires as a condition of every Federal contract (and bid specification therefor) exceeding $2,500, the "principal purpose" of which is to furnish services to the United States through the use of "service employees," that certain safety and health standards be met. See 29 CFR part 1925, which contains the Department rules concerning these standards. Section 7 of the Service Contract Act provides that the Act shall not apply to "any contract of the United States or District of Columbia for construction, alteration, and/or repair, including painting and decorating of public buildings or public works." It is clear from the legislative history of section 107 that no gaps in coverage between the two statutes are intended.

**1926.15(b)** The Walsh-Healey Public Contracts Act requires that contracts entered into by any Federal agency for the manufacture or furnishing of materials, supplies, articles, and equipment in any amount exceeding $10,000 must contain, among other provisions, a requirement that "no part of such contract will be performed nor will any of the materials, supplies, articles or equipment to be manufactured or furnished under said contract be manufactured or fabricated in any plants, factories, buildings, or surroundings or under working conditions which are unsanitary or hazardous or dangerous to the health and safety of employees engaged in the performance of said contract." The rules of the Secretary concerning these standards are published in 41 CFR part 50-204, and express the Secretary of Labor's interpretation and application of section 1(e) of the Walsh-Healey Public Contracts Act to certain particular working conditions. None of the described working conditions are intended to deal with construction activities, although such activities may conceivably be a part of a contract which is subject to the Walsh-Healey Public Contracts Act. Nevertheless, such activities remain subject to the general statutory duty prescribed by section 1(e). Section 103(b) of the Contract Work Hours and Safety Standards Act provides, among other things, that the Act shall not apply to any work required to be done in accordance with the provisions of the Walsh-Healey Public Contracts Act.

## §1926.16 Rules of construction.

**1926.16(a)** The prime contractor and any subcontractors may make their own arrangements with respect to obligations which might be more appropriately treated on a jobsite basis rather than individually. Thus, for example, the prime contractor and his subcontractors may wish to make an express agreement that the prime contractor or one of the subcontractors will provide all required first-aid or toilet facilities, thus relieving the subcontractors from the actual, but not any legal, responsibility (or, as the case may be, relieving the other subcontractors from this responsibility). In no case shall the prime contractor be relieved of overall responsibility for compliance with the requirements of this part for all work to be performed under the contract.

**1926.16(b)** By contracting for full performance of a contract subject to section 107 of the Act, the prime contractor assumes all obligations prescribed as employer obligations under the standards contained in this part, whether or not he subcontracts any part of the work.

**1926.16(c)** To the extent that a subcontractor of any tier agrees to perform any part of the contract, he also assumes responsibility for complying with the standards in this part with respect to that part. Thus, the prime contractor assumes the entire responsibility under the contract and the subcontractor assumes responsibility with respect to his portion of the work. With respect to subcontracted work, the prime contractor and any subcontractor or subcontractors shall be deemed to have joint responsibility.

**1926.16(d)** Where joint responsibility exists, both the prime contractor and his subcontractor or subcontractors, regardless of tier, shall be considered subject to the enforcement provisions of the Act.

NO.767 P001/002

# ORDER FORM

## STANDARDS BOOKS

GENERAL INDUSTRY —————— $17.30
CONSTRUCTION INDUSTRY ———— $13.30

WE DO NOT SHIP TO P.O. BOX-WE MUST HAVE A STREET ADDRESS
***Payment must be money order or check. Please, no cash.***

**NAME** Richard A. Milleville

**~~COMPANY NAME~~** 15417 W Echo Canyon Dr

**STREET ADDRESS** ↵

**CITY** Surprise **STATE** AZ **ZIP** 85374

**TELEPHONE NUMBER** 623-556-0393

I wish to purchase:

______ copies of the **GENERAL INDUSTRY STANDARDS**

1 copies of the **CONSTRUCTION INDUSTRY STANDARDS**

TOTAL PURCHASE PRICE 13.30 + SHIPPING/HANDLING 3.60

Make checks payable to: **INDUSTRIAL COMMISSION OF ARIZONA**

MAIL TO:

Division of Occupational Safety & Health
Industrial Commission of Arizona
P.O. Box 19070
Phoenix AZ 85005-9070

Phone: (602) 542-5795
Fax: (602) 542-1614
(See reverse side for shipping and handling rates)

03 JUN 16 AM 10:01
A - SAFETY

Date Received: June 16 03
Date Mailed: June 16 03
Receiver name: Richard
Check # 1853
employee sign: [signature] 6/13

**Authority:** Sec. 107, Contract Work Hours and Safety Standards Act (40 U.S.C. 333); secs. 4, 6, and 8, Occupational Safety and Health Act of 1970 (29 U.S.C. 653, 655, 657); Secretary of Labor's Order No. 12-71 (36 FR 8754), 8-76 (41 FR 25059), 9-83 (48 FR 35736), or 6-96 (62 FR 111), as applicable; and 29 CFR part 1911.

### §1926.20 General safety and health provisions.

1926.20(a) ***Contractor requirements.*** (1) Section 107 of the Act requires that it shall be a condition of each contract which is entered into under legislation subject to Reorganization Plan Number 14 of 1950 (64 Stat. 1267), as defined in §1926.12, and is for construction, alteration, and/or repair, including painting and decorating, that no contractor or subcontractor for any part of the contract work shall require any laborer or mechanic employed in the performance of the contract to work in surroundings or under working conditions which are unsanitary, hazardous, or dangerous to his health or safety.

1926.20(b) ***Accident prevention responsibilities.*** (1) It shall be the responsibility of the employer to initiate and maintain such programs as may be necessary to comply with this part.

1926.20(b)(2) Such programs shall provide for frequent and regular inspections of the job sites, materials, and equipment to be made by competent persons designated by the employers.

1926.20(b)(3) The use of any machinery, tool, material, or equipment which is not in compliance with any applicable requirement of this part is prohibited. Such machine, tool, material, or equipment shall either be identified as unsafe by tagging or locking the controls to render them inoperable or shall be physically removed from its place of operation.

1926.20(b)(4) The employer shall permit only those employees qualified by training or experience to operate equipment and machinery.

1926.20(c) The standards contained in this part shall apply with respect to employments performed in a workplace in a State, the District of Columbia, the Commonwealth of Puerto Rico, the Virgin Islands, American Samoa, Guam, Trust Territory of the Pacific Islands, Wake Island, Outer Continental Shelf lands defined in the Outer Continental Shelf Lands Act, Johnston Island, and the Canal Zone.

1926.20(d)(1) If a particular standard is specifically applicable to a condition, practice, means, method, operation, or process, it shall prevail over any different general standard which might otherwise be applicable to the same condition, practice, means, method, operation, or process.

1926.20(d)(2) On the other hand, any standard shall apply according to its terms to any employment and place of employment in any industry, even though particular standards are also prescribed for the industry to the extent that none of such particular standards applies.

1926.20(e) In the event a standard protects on its face a class of persons larger than employees, the standard shall be applicable under this part only to employees and their employment and places of employment.

[44 FR 8577, Feb. 9, 1979; 44 FR 20940, Apr. 6, 1979, as amended at 58 FR 35078, June 30, 1993]

### §1926.21 Safety training and education.

1926.21(a) ***General requirements.*** The Secretary shall, pursuant to section 107(f) of the Act, establish and supervise programs for the education and training of employers and employees in the recognition, avoidance and prevention of unsafe conditions in employments covered by the act.

1926.21(b) ***Employer responsibility.*** (1) The employer should avail himself of the safety and health training programs the Secretary provides.

1926.21(b)(2) The employer shall instruct each employee in the recognition and avoidance of unsafe conditions and the regulations applicable to his work environment to control or eliminate any hazards or other exposure to illness or injury.

1926.21(b)(3) Employees required to handle or use poisons, caustics, and other harmful substances shall be instructed regarding the safe handling and use, and be made aware of the potential hazards, personal hygiene, and personal protective measures required.

1926.21(b)(4) In job site areas where harmful plants or animals are present, employees who may be exposed shall be instructed regarding the potential hazards, and how to avoid injury, and the first aid procedures to be used in the event of injury.

1926.21(b)(5) Employees required to handle or use flammable liquids, gases, or toxic materials shall be instructed in the safe handling and use of these materials and made aware of the specific requirements contained in subparts D, F, and other applicable subparts of this part.

1926.21(b)(6)(i) All employees required to enter into confined or enclosed spaces shall be instructed as to the nature of the hazards involved, the necessary precautions to be taken, and in the use of protective and emergency equipment required. The employer shall comply with any specific regulations that apply to work in dangerous or potentially dangerous areas.

1926.21(b)(6)(ii) For purposes of paragraph (b)(6)(i) of this section, *confined or enclosed space* means any space having a limited means of egress, which is subject to the accumulation of toxic or flammable contaminants or has an oxygen deficient atmosphere. Confined or enclosed spaces include, but are not limited to, storage tanks, process vessels, bins, boilers, ventilation or exhaust ducts, sewers, underground utility vaults, tunnels, pipelines, and open top spaces more than 4 feet in depth such as pits, tubs, vaults, and vessels.

### §1926.22 Recording and reporting of injuries. [Reserved]

### §1926.23 First aid and medical attention.

First aid services and provisions for medical care shall be made available by the employer for every employee covered by these regulations. Regulations prescribing specific requirements for first aid, medical attention, and emergency facilities are contained in subpart D of this part.

### §1926.24 Fire protection and prevention.

The employer shall be responsible for the development and maintenance of an effective fire protection and prevention program at the job site throughout all phases of the construction, repair, alteration, or demolition work. The employer shall ensure the availability of the fire protection and suppression equipment required by subpart F of this part.

### §1926.25 Housekeeping.

1926.25(a) During the course of construction, alteration, or repairs, form and scrap lumber with protruding nails, and all other debris, shall be kept cleared from work areas, passageways, and stairs, in and around buildings or other structures.

1926.25(b) Combustible scrap and debris shall be removed at regular intervals during the course of construction. Safe means shall be provided to facilitate such removal.

1926.25(c) Containers shall be provided for the collection and separation of waste, trash, oily and used rags, and other refuse. Containers used for garbage and other oily, flammable, or hazardous wastes, such as caustics, acids, harmful dusts, etc. shall be equipped with covers. Garbage and other waste shall be disposed of at frequent and regular intervals.

### §1926.26 Illumination.

Construction areas, aisles, stairs, ramps, runways, corridors, offices, shops, and storage areas where work is in progress shall be lighted with either natural or artificial illumination. The minimum illumination requirements for work areas are contained in subpart D of this part.

### §1926.27 Sanitation.

Health and sanitation requirements for drinking water are contained in subpart D of this part.

### §1926.28 Personal protective equipment.

1926.28(a) The employer is responsible for requiring the wearing of appropriate personal protective equipment in all operations where there is an exposure to hazardous conditions or where this part indicates the need for using such equipment to reduce the hazards to the employees.

1926.28(b) Regulations governing the use, selection, and maintenance of personal protective and lifesaving equipment are described under subpart E of this part.

### §1926.29 Acceptable certifications.

1926.29(a) ***Pressure vessels.*** Current and valid certification by an insurance company or regulatory authority shall be deemed as acceptable evidence of safe installation, inspection, and testing of pressure vessels provided by the employer.

1926.29(b) ***Boilers.*** Boilers provided by the employer shall be deemed to be in compliance with the requirements of this part when evidence of current and valid certification by an insurance company or regulatory authority attesting to the safe installation, inspection, and testing is presented.

1926.29(c) ***Other requirements.*** Regulations prescribing specific requirements for other types of pressure vessels and similar equipment are contained in subparts F and O of this part.

### §1926.30 Shipbuilding and ship repairing.

1926.30(a) ***General.*** Shipbuilding, ship repairing, alterations, and maintenance performed on ships under Government contract, except naval ship construction, is work subject to the Act.

1926.30(b) ***Applicable safety and health standards.*** For the purpose of work carried out under this section, the safety and health regulations in part 1915 of this title, Shipyard Employment, shall apply.

[44 FR 8577, Feb. 9, 1979; 44 FR 20940, Apr. 6, 1979, as amended at 61 FR 9249, Mar. 7, 1996]

### §1926.31 Incorporation by reference.

1926.31(a) The standards of agencies of the U.S. Government, and organizations which are not agencies of the U.S. Government which are incorporated by reference in this part, have the same force and effect as other standards in this part. Only the mandatory provisions (i.e., provisions containing the word "shall" or other mandatory language) of standards incorporated by reference are adopted as standards under the Occupational Safety and Health Act. The locations where these standards may be examined are as follows:

1926.31(a)(1) Offices of the Occupational Safety and Health Administration, U.S. Department of Labor, Frances Perkins Building, Washington, DC 20210.

1926.31(a)(2) The Regional and Field Offices of the Occupational Safety and Health Administration, which are listed in the U.S. Government Manual.

1926.31(b) Any changes in the specifications, standards and codes incorporated by reference in this part and an official historic file of such changes are available at the offices referred to in paragraph (a) of this section. All questions as to the applicability of such changes should also be referred to these offices.

[44 FR 8577, Feb. 9, 1979; 44 FR 20940, Apr. 6, 1979, as amended at 61 FR 9249, Mar. 7, 1996; 63 FR 33468, June 18, 1998]

### §1926.32 Definitions.

The following definitions shall apply in the application of the regulations in this part:

1926.32(a) *Act* means section 107 of the Contract Work Hours and Safety Standards Act, commonly known as the Construction Safety Act (86 Stat. 96; 40 U.S.C. 333).

1926.32(b) *ANSI* means American National Standards Institute.

1926.32(c) *Approved* means sanctioned, endorsed, accredited, certified, or accepted as satisfactory by a duly constituted and nationally recognized authority or agency.

1926.32(d) *Authorized person* means a person approved or assigned by the employer to perform a specific type of duty or duties or to be at a specific location or locations at the jobsite.

1926.32(e) *Administration* means the Occupational Safety and Health Administration.

1926.32(f) *Competent person* means one who is capable of identifying existing and predictable hazards in the surroundings or working conditions which are unsanitary, hazardous, or dangerous to employees, and who has authorization to take prompt corrective measures to eliminate them.

1926.32(g) *Construction work.* For purposes of this section, *Construction work* means work for construction, alteration, and/or repair, including painting and decorating.

1926.32(h) *Defect* means any characteristic or condition which tends to weaken or reduce the strength of the tool, object, or structure of which it is a part.

1926.32(i) *Designated person* means "authorized person" as defined in paragraph (d) of this section.

1926.32(j) *Employee* means every laborer or mechanic under the Act regardless of the contractual relationship which may be alleged to exist between the laborer and mechanic and the contractor or subcontractor who engaged him. "Laborer and mechanic" are not defined in the Act, but the identical terms are used in the Davis-Bacon Act (40 U.S.C. 276a), which provides for minimum wage protection on Federal and federally assisted construction contracts. The use of the same term in a statute which often applies concurrently with section 107 of the Act has considerable precedential value in ascertaining the meaning of "laborer and mechanic" as used in the Act. *Laborer* generally means one who performs manual labor or who labors at an occupation requiring physical strength; *mechanic* generally means a worker skilled with tools. See 18 Comp. Gen. 341.

1926.32(k) *Employer* means contractor or subcontractor within the meaning of the Act and of this part.

1926.32(l) *Hazardous substance* means a substance which, by reason of being explosive, flammable, poisonous, corrosive, oxidizing, irritating, or otherwise harmful, is likely to cause death or injury.

1926.32(m) *Qualified* means one who, by possession of a recognized degree, certificate, or professional standing, or who by extensive knowledge, training, and experience, has successfully demonstrated his ability to solve or resolve problems relating to the subject matter, the work, or the project.

1926.32(n) *Safety factor* means the ratio of the ultimate breaking strength of a member or piece of material or equipment to the actual working stress or safe load when in use.

1926.32(o) *Secretary* means the Secretary of Labor.

1926.32(p) *SAE* means Society of Automotive Engineers.

1926.32(q) *Shall* means mandatory.

1926.32(r) *Should* means recommended.

1926.32(s) *Suitable* means that which fits, and has the qualities or qualifications to meet a given purpose, occasion, condition, function, or circumstance.

[44 FR 8577, Feb. 9, 1979; 44 FR 20940, Apr. 6, 1979, as amended at 58 FR 35078, June 30, 1993]

### §1926.33 Access to employee exposure and medical records.

**Note:** The requirements applicable to construction work under this section are identical to those set forth at §1910.1020 of this chapter.

[61 FR 31431, June 20, 1996]

[Section 1910.1020 and Appendices are printed below—CCH.]

### §1910.1020 Access to employee exposure and medical records.

1910.1020(a) ***Purpose.*** The purpose of this section is to provide employees and their designated representatives a right of access to relevant exposure and medical records; and to provide representatives of the Assistant Secretary a right of access to these records in order to fulfill responsibilities under the Occupational Safety and Health Act. Access by employees, their representatives, and the Assistant Secretary is necessary to yield both direct and indirect improvements in the detection, treatment, and prevention of occupational disease. Each employer is responsible for assuring compliance with this section, but the activities involved in complying with the access to medical records provisions can be carried out, on behalf of the employer, by the physician or other health care personnel in charge of employee medical records. Except as expressly provided, nothing in this section is intended to affect existing legal and ethical obligations concerning the maintenance and confidentiality of employee medical information, the duty to disclose information to a patient/employee or any other aspect of the medical-care relationship, or affect existing legal obligations concerning the protection of trade secret information.

1910.1020(b) ***Scope and application.*** (1) This section applies to each general industry, maritime, and construction employer who makes, maintains, contracts for, or has access to employee exposure or medical records, or analyses thereof, pertaining to employees exposed to toxic substances or harmful physical agents.

1910.1020(b)(2) This section applies to all employee exposure and medical records, and analyses thereof, of such employees, whether or not the records are mandated by specific occupational safety and health standards.

1910.1020(b)(3) This section applies to all employee exposure and medical records, and analyses thereof, made or maintained in any manner, including on an in-house of contractual (e.g., fee-for-service) basis. Each employer shall assure that the preservation and access requirements of this section are complied with regardless of the manner in which the records are made or maintained.

1910.1020(c) ***Definitions.*** (1) *Access* means the right and opportunity to examine and copy.

1910.1020(c)(2) *Analysis using exposure or medical records* means any compilation of data or any statistical study based at least in part on information collected from individual employee exposure or medical records or information collected from health insurance claims records, provided that either the analysis has been reported to the employer or no further work is currently being done by the person responsible for preparing the analysis.

1910.1020(c)(3) *Designated representative* means any individual or organization to whom an employee gives written authorization to exercise a right of access. For the purposes of access to employee exposure records and analyses using exposure or medical records, a recognized or certified collective bargaining agent shall be treated automatically as a designated representative without regard to written employee authorization.

1910.1020(c)(4) *Employee* means a current employee, a former employee, or an employee being assigned or transferred to work where there will be exposure to toxic substances or harmful physical agents. In the case of a deceased or legally incapacitated employee, the employee's legal representative may directly exercise all the employee's rights under this section.

1910.1020(c)(5) *Employee exposure record* means a record containing any of the following kinds of information:

1910.1020(c)(5)(i) Environmental (workplace) monitoring or measuring of a toxic substance or harmful physical agent, including personal, area, grab, wipe, or other form of sampling, as well as related collection and analytical methodologies, calculations, and other background data relevant to interpretation of the results obtained;

1910.1020(c)(5)(ii) Biological monitoring results which directly assess the absorption of a toxic substance or harmful physical agent by body systems (e.g., the level of a chemical in the blood, urine, breath, hair, fingernails, etc) but not including results which assess the biological effect of a substance or agent or which assess an employee's use of alcohol or drugs;

1910.1020(c)(5)(iii) Material safety data sheets indicating that the material may pose a hazard to human health; or

1910.1020(c)(5)(iv) In the absence of the above, a chemcial inventory or any other record which reveals where and when used and the identity (e.g., chemical, common, or trade name) of a toxic substance or harmful physical agent.

1910.1020(c)(6)(i) *Employee medical record* means a record concerning the health status of an employee which is made or maintained by a physician, nurse, or other health care personnel or technician, including:

1910.1020(c)(6)(i)(A) Medical and employment questionnaires or histories (including job description and occupational exposures),

1910.1020(c)(6)(i)(B) The results of medical examinations (pre-employment, pre-assignment, periodic, or episodic) and laboratory tests (including chest and other X-ray examinations taken for the purposes of establishing a base-line or detecting occupational illness, and all biological monitoring not defined as an "employee exposure record"),

1910.1020(c)(6)(i)(C) Medical opinions, diagnoses, progress notes, and recommendations,

1910.1020(c)(6)(i)(D) First aid records,

1910.1020(c)(6)(i)(E) Descriptions of treatments and prescriptions, and

1910.1020(c)(6)(i)(F) Employee medical complaints.

1910.1020(c)(6)(ii) "Employee medical record" does not include medical information in the form of:

1910.1020(c)(6)(ii)(A) Physical specimens (e.g., blood or urine samples) which are routinely discarded as a part of normal medical practice; or

1910.1020(c)(6)(ii)(B) Records concerning health insurance claims if maintained separately from the employer's medical program and its records, and not accessible to the employer by employee name or other direct personal identifier (e.g., social security number, payroll number, etc.); or

1910.1020(c)(6)(ii)(C) Records created solely in preparation for litigation which are privileged from discovery under the applicable rules of procedure or evidence; or

1910.1020(c)(6)(ii)(D) Records concerning voluntary employee assistance programs (alcohol, drug abuse, or personal counseling programs) if maintained separately from the employer's medical program and its records.

1910.1020(c)(7) *Employer* means a current employer, a former employer, or a successor employer.

1910.1020(c)(8) *Exposure* or *exposed* means that an employee is subjected to a toxic substance or harmful physical agent in the course of employment through any route of entry (inhalation, ingestion, skin contact or absorption, etc.), and includes past exposure and potential (e.g., accidental or possible) exposure, but does not include situations where the employer can demonstrate that the toxic substance or harmful physical agent is not used, handled, stored, generated, or present in the workplace in any manner different from typical non-occupational situations.

1910.1020(c)(9) *Health Professional* means a physician, occupational health nurse, industrial hygienist, toxicologist, or epidemiologist, providing medical or other occupational health services to exposed employees.

1910.1020(c)(10) *Record* means any item, collection, or grouping of information regardless of the form or process by which it is maintained (e.g., paper document, microfiche, microfilm, X-ray film, or automated data processing).

1910.1020(c)(11) *Specific chemical identity* means the chemical name, Chemical Abstracts Service (CAS) Registry Number, or any other information that reveals the precise chemical designation of the substance.

1910.1020(c)(12)(i) *Specific written consent* means a written authorization containing the following:

1910.1020(c)(12)(i)(A) The name and signature of the employee authorizing the release of medical information,

1910.1020(c)(12)(i)(B) The date of the written authorization,

1910.1020(c)(12)(i)(C) The name of the individual or organization that is authorized to release the medical information,

1910.1020(c)(12)(i)(D) The name of the designated representative (individual or organization) that is authorized to receive the released information,

1910.1020(c)(12)(i)(E) A general description of the medical information that is authorized to be released,

1910.1020(c)(12)(i)(F) A general description of the purpose for the release of the medical information, and

1910.1020(c)(12)(i)(G) A date or condition upon which the written authorization will expire (if less than one year).

1910.1020(c)(12)(ii) A written authorization does not operate to authorize the release of medical information not in existence on the date of written authorization, unless the release of future information is expressly authorized, and does not operate for more than one year from the date of written authorization.

1910.1020(c)(12)(iii) A written authorization may be revoked in writing prospectively at any time.

1910.1020(c)(13) *Toxic substance or harmful physical agent* means any chemical substance, biological agent (bacteria, virus, fungus, etc.), or physical stress (noise, heat, cold, vibration, repetitive motion, ionizing and non-ionizing radiation, hypo-or hyperbaric pressure, etc.) which:

1910.1020(c)(13)(i) Is listed in the latest printed edition of the National Institute for Occupational Safety and Health (NIOSH) Registry of Toxic Effects of Chemical Substances (RTECS), which is incorporated by reference as specified in §1910.6; or

1910.1020(c)(13)(ii) Has yielded positive evidence of an acute or chronic health hazard in testing conducted by, or known to, the employer; or

1910.1020(c)(13)(iii) Is the subject of a material safety data sheet kept by or known to the employer indicating that the material may pose a hazard to human health.

1910.1020(c)(14) *Trade secret* means any confidential formula, pattern, process, device, or information or compilation of information that is used in an employer's business and that gives the employer an opportunity to obtain an advantage over competitors who do not know or use it.

1910.1020(d) ***Preservation of records.*** (1) Unless a specific occupational safety and health standard provides a different period of time, each employer shall assure the preservation and retention of records as follows:

1910.1020(d)(1)(i) *Employee medical records.* The medical record for each employee shall be preserved and maintained for at least the duration of employment plus thirty (30) years, except that the following types of records need not be retained for any specified period:

1910.1020(d)(1)(i)(A) Health insurance claims records maintained separately from the employer's medical program and its records,

1910.1020(d)(1)(i)(B) First aid records (not including medical histories) of one-time treatment and subsequent observation of minor scratches, cuts, burns, splinters, and the like which do not involve medical treatment, loss of consciousness, restriction of work or motion, or transfer to another job, if made on-site by a non-physician and if maintained separately from the employer's medical program and its records, and

1910.1020(d)(1)(i)(C) The medical records of employees who have worked for less than (1) year for the employer need not be retained beyond the term of employment if they are provided to the employee upon the termination of employment.

1910.1020(d)(1)(ii) *Employee exposure records.* Each employee exposure record shall be preserved and maintained for at least thirty (30) years, except that:

1910.1020(d)(1)(ii)(A) Background data to environmental (workplace) monitoring or measuring, such as laboratory reports and worksheets, need only be retained for one (1) year as long as the sampling results, the collection methodology (sampling plan), a description of the analytical and mathematical methods used, and a summary of other background data relevant to interpretation of the results obtained, are retained for at least thirty (30) years; and

1910.1020(d)(1)(ii)(B) Material safety data sheets and paragraph (c)(5)(iv) records concerning the identity of a substance or agent need not be retained for any specified period as long as some record of the identity (chemical name if known) of the substance or agent, where it was used, and when it was used is retained for at least thirty (30) years;[1] and

1910.1020(d)(1)(ii)(C) Biological monitoring results designated as exposure records by specific occupational safety and health standards shall be preserved and maintained as required by the specific standard.

1910.1020(d)(1)(iii) *Analyses using exposure or medical records.* Each analysis using exposure or medial records shall be preserved and maintained for at least thirty (30) years.

1910.1020(d)(2) Nothing in this section is intended to mandate the form, manner, or process by which an employer preserves a record as long as the information contained in the record is preserved and retrievable, except that chest X-ray films shall be preserved in their original state.

1910.1020(e) ***Access to records***—(1) *General.* (i) Whenever an employee or designated representative requests access to a record, the employer shall assure that access is provided in a reasonable time, place, and manner. If the employer cannot reasonably provide access to the record within fifteen (15) working days, the employer shall within the fifteen (15) working days apprise the employee or designated representative requesting the record of the reason for the delay and the earliest date when the record can be made available.

1910.1020(e)(1)(ii) The employer may require of the requester only such information as should be readily known to the requester and which may be necessary to locate or identify the records being requested (e.g. dates and locations where the employee worked during the time period in question).

1910.1020(e)(1)(iii) Whenever an employee or designated representative requests a copy of a record, the employer shall assure that either:

1910.1020(e)(1)(iii)(A) A copy of the record is provided without cost to the employee or representative,

1910.1020(e)(1)(iii)(B) The necessary mechanical copying facilities (e.g., photocopying) are made available without cost to the employee or representative for copying the record, or

1910.1020(e)(1)(iii)(C) The record is loaned to the employee or representative for a reasonable time to enable a copy to be made.

1910.1020(e)(1)(iv) In the case of an original X-ray, the employer may restrict access to on-site examination or make other suitable arrangements for the temporary loan of the X-ray.

1910.1020(e)(1)(v) Whenever a record has been previously provided without cost to an employee or designated representative, the employer may charge reasonable, non-discriminatory administrative costs (i.e., search and copying expenses but not including overhead expenses) for a request by the employee or designated representative for additional copies of the record, except that

1910.1020(e)(1)(v)(A) An employer shall not charge for an initial request for a copy of new information that has been added to a record which was previously provided; and

1910.1020(e)(1)(v)(B) An employer shall not charge for an initial request by a recognized or certified collective bargaining agent for a copy of an employee exposure record or an analysis using exposure or medical records.

1910.1020(e)(1)(vi) Nothing in this section is intended to preclude employees and collective bargaining agents from collectively bargaining to obtain access to information in addition to that available under this section.

1910.1020(e)(2) *Employee and designated representative access*—(i) *Employee exposure records.* (A) Except as limited by paragraph (f) of this section, each employer shall, upon request, assure the access to each employee and designated representative to employee exposure records relevant to the employee. For the purpose of this section, an exposure record relevant to the employee consists of:

1910.1020(e)(2)(i)(A)(*1*) A record which measures or monitors the amount of a toxic substance or harmful physical agent to which the employee is or has been exposed;

1910.1020(e)(2)(i)(A)(2) In the absence of such directly relevant records, such records of other employees with past or present job duties or working conditions related to or similar to those of the employee to the extent necessary to reasonably indicate the amount and nature of the toxic substances or harmful physical agents to which the employee is or has been subjected, and

1910.1020(e)(2)(i)(A)(3) Exposure records to the extent necessary to reasonably indicate the amount and nature of the toxic substances or harmful physical agents at workplaces or under working conditions to which the employee is being assigned or transferred.

1910.1020(e)(2)(i)(B) Requests by designated representatives for unconsented access to employee exposure records shall be in writing and shall specify with reasonable particularity:

1910.1020(e)(2)(i)(B)(*1*) The records requested to be disclosed; and

1910.1020(e)(2)(i)(B)(2) The occupational health need for gaining access to these records.

---

1 Material safety data sheets must be kept for those chemicals currently in use that are effected by the Hazard Communication Standard in accordance with 29 CFR 1910.1200(g).

**1910.1020(e)(2)(ii)** *Employee medical records.* (A) Each employer shall, upon request, assure the access of each employee to employee medical records of which the employee is the subject, except as provided in paragraph (e)(2)(ii)(D) of this section.

**1910.1020(e)(2)(ii)(B)** Each employer shall, upon request, assure the access of each designated representative to the employee medical records of any employee who has given the designated representative specific written consent. Appendix A to this section contains a sample form which may be used to establish specific written consent for access to employee medical records.

**1910.1020(e)(2)(ii)(C)** Whenever access to employee medical records is requested, a physician representing the employer may recommend that the employee or designated representative:

**1910.1020(e)(2)(ii)(C)(*1*)** Consult with the physician for the purposes of reviewing and discussing the records requested,

**1910.1020(e)(2)(ii)(C)(2)** Accept a summary of material facts and opinions in lieu of the records requested, or

**1910.1020(e)(2)(ii)(C)(3)** Accept release of the requested records only to a physician or other designated representative.

**1910.1020(e)(2)(ii)(D)** Whenever an employee requests access to his or her employee medical records, and a physician representing the employer believes that direct employee access to information contained in the records regarding a specific diagnosis of a terminal illness or a psychiatric condition could be detrimental to the employee's health, the employer may inform the employee that access will only be provided to a designated representative of the employee having specific written consent, and deny the employee's request for direct access to this information only. Where a designated representative with specific written consent requests access to information so withheld, the employer shall assure the access of the designated representative to this information, even when it is known that the designated representative will give the information to the employee.

**1910.1020(e)(2)(ii)(E)** A physician, nurse, or other responsible health care personnel maintaining medical records may delete from requested medical records the identity of a family member, personal friend, or fellow employee who has provided confidential information concerning an employee's health status.

**1910.1020(e)(2)(iii)** *Analyses using exposure or medical records.* (A) Each employee shall, upon request, assure the access of each employee and designated representative to each analysis using exposure or medical records concerning the employee's working conditions or workplace.

**1910.1020(e)(2)(iii)(B)** Whenever access is requested to an analysis which reports the contents of employee medical records by either direct identifier (name, address, social security number, payroll number, etc.) or by information which could reasonably be used under the circumstances indirectly to identify specific employees (exact age, height, weight, race, sex, date of initial employment, job title, etc.), the employer shall assure that personal identifiers are removed before access is provided. If the employer can demonstrate that removal of personal identifiers from an analysis is not feasible, access to the personally identifiable portions of the analysis need not be provided.

**1910.1020(e)(3)** *OSHA access.* (i) Each employer shall, upon request, and without derogation of any rights under the Constitution or the Occupational Safety and Health Act of 1970, 29 U.S.C. 651 *et seq.*, that the employer chooses to exercise, assure the prompt access of representatives of the Assistant Secretary of Labor for Occupational Safety and Health to employee exposure and medical records and to analyses using exposure or medical records. Rules of agency practice and procedure governing OSHA access to employee medical records are contained in 29 CFR 1913.10.

**1910.1020(e)(3)(ii)** Whenever OSHA seeks access to personally identifiable employee medical information by presenting to the employer a written access order pursuant to 29 CFR 1913.10(d), the employer shall prominently post a copy of the written access order and its accompanying cover letter for at least fifteen (15) working days.

**1910.1020(f)** ***Trade secrets.*** (1) Except as provided in paragraph (f)(2) of this section, nothing in this section precludes an employer from deleting from records requested by a health professional, employee, or designated representative any trade secret data which discloses manufacturing processes, or discloses the percentage of a chemical substance in mixture, as long as the health professional, employee, or designated representative is notified that information has been deleted. Whenever deletion of trade secret information substantially impairs evaluation of the place where or the time when exposure to a toxic substance or harmful physical agent occurred, the employer shall provide alternative information which is sufficient to permit the requesting party to identify where and when exposure occurred.

**1910.1020(f)(2)** The employer may withhold the specific chemical identity, including the chemical name and other specific identification of a toxic substance from a disclosable record provided that:

**1910.1020(f)(2)(i)** The claim that the information withheld is a trade secret can be supported;

**1910.1020(f)(2)(ii)** All other available information on the properties and effects of the toxic substance is disclosed;

**1910.1020(f)(2)(iii)** The employer informs the requesting party that the specific chemical identity is being withheld as a trade secret; and

**1910.1020(f)(2)(iv)** The specific chemical identity is made available to health professionals, employees and designated representatives in accordance with the specific applicable provisions of this paragraph.

**1910.1020(f)(3)** Where a treating physician or nurse determines that a medical emergency exists and the specific chemical identity of a toxic substance is necessary for emergency or first-aid treatment, the employer shall immediately disclose the specific chemical identity of a trade secret chemical to the treating physician or nurse, regardless of the existence of a written statement of need or a confidentiality agreement. The employer may require a written statement of need and confidentiality agreement, in accordance with the provisions of paragraphs (f)(4) and (f)(5), as soon as circumstances permit.

**1910.1020(f)(4)** In non-emergency situations, an employer shall, upon request, disclose a specific chemical identity, otherwise permitted to be withheld under paragraph (f)(2) of this section, to a health professional, employee, or designated representative if:

**1910.1020(f)(4)(i)** The request is in writing;

**1910.1020(f)(4)(ii)** The request describes with reasonable detail one or more of the following occupational health needs for the information:

**1910.1020(f)(4)(ii)(A)** To assess the hazards of the chemicals to which employees will be exposed;

**1910.1020(f)(4)(ii)(B)** To conduct or assess sampling of the workplace atmosphere to determine employee exposure levels;

**1910.1020(f)(4)(ii)(C)** To conduct pre-assignment or periodic medical surveillance of exposed employees;

**1910.1020(f)(4)(ii)(D)** To provide medical treatment to exposed employees;

**1910.1020(f)(4)(ii)(E)** To select or assess appropriate personal protective equipment for exposed employees;

**1910.1020(f)(4)(ii)(F)** To design or assess engineering controls or other protective measures for exposed employees; and

**1910.1020(f)(4)(ii)(G)** To conduct studies to determine the health effects of exposure.

**1910.1020(f)(4)(iii)** The request explains in detail why the disclosure of the specific chemical identity is essential and that, in lieu thereof, the disclosure of the following information would not enable the health professional, employee or designated representative to provide the occupational health services described in paragraph (f)(4)(ii) of this section:

**1910.1020(f)(4)(iii)(A)** The properties and effects of the chemical;

**1910.1020(f)(4)(iii)(B)** Measures for controlling workers' exposure to the chemical;

**1910.1020(f)(4)(iii)(C)** Methods of monitoring and analyzing worker exposure to the chemical; and,

**1910.1020(f)(4)(iii)(D)** Methods of diagnosing and treating harmful exposures to the chemical;

**1910.1020(f)(4)(iv)** The request includes a description of the procedures to be used to maintain the confidentiality of the disclosed information; and,

**1910.1020(f)(4)(v)** The health professional, employee, or designated representative and the employer or contractor of the services of the health professional or designated representative agree in a written confidentiality agreement that the health professional, employee or designated representative will not use the trade secret information for any purpose other than the health need(s) asserted and agree not to release the information under any circumstances other than to OSHA, as provided in paragraph (f)(9) of this section, except as authorized by the terms of the agreement or by the employer.

**1910.1020(f)(5)** The confidentiality agreement authorized by paragraph (f)(4)(iv) of this section:

**1910.1020(f)(5)(i)** May restrict the use of the information to the health purposes indicated in the written statement of need;

**1910.1020(f)(5)(ii)** May provide for appropriate legal remedies in the event of a breach of the agreement, including stipulation of a reasonable pre-estimate of likely damages; and,

**1910.1020(f)(5)(iii)** May not include requirements for the posting of a penalty bond.

**1910.1020(f)(6)** Nothing in this section is meant to preclude the parties from pursuing non-contractual remedies to the extent permitted by law.

**1910.1020(f)(7)** If the health professional, employee or designated representative receiving the trade secret information decides that there is a need to disclose it to OSHA, the employer who provided the information shall be informed by the health professional prior to, or at the same time as, such disclosure.

**1910.1020(f)(8)** If the employer denies a written request for disclosure of a specific chemical identity, the denial must:

**1910.1020(f)(8)(i)** Be provided to the health professional, employee or designated representative within thirty days of the request;

**1910.1020(f)(8)(ii)** Be in writing;

**1910.1020(f)(8)(iii)** Include evidence to support the claim that the specific chemical identity is a trade secret;

**1910.1020(f)(8)(iv)** State the specific reasons why the request is being denied; and,

**1910.1020(f)(8)(v)** Explain in detail how alternative information may satisfy the specific medical or occupational health need without revealing the specific chemical identity.

**1910.1020(f)(9)** The health professional, employee, or designated representative whose request for information is denied under paragraph (f)(4) of this section may refer the request and the written denial of the request to OSHA for consideration.

**1910.1020(f)(10)** When a heath [*sic*] professional employee, or designated representative refers a denial to OSHA under paragraph (f)(9) of this section, OSHA shall consider the evidence to determine if:

**1910.1020(f)(10)(i)** The employer has supported the claim that the specific chemical identity is a trade secret;

**1910.1020(f)(10)(ii)** The health professional employee, or designated representative has supported the claim that there is a medical or occupational health need for the information; and

**1910.1020(f)(10)(iii)** The health professional, employee or designated representative has demonstrated adequate means to protect the confidentiality.

**1910.1020(f)(11)(i)** If OSHA determines that the specific chemical identity requested under paragraph (f)(4) of this section is not a *bona fide* trade secret, or that it is a trade secret but the requesting health professional, employee or designated representatives has a legitimate medical or occupational health need for the information, has executed a written confidentiality agreement, and has shown adequate means for complying with the terms of such agreement, the employer will be subject to citation by OSHA.

**1910.1020(f)(11)(ii)** If an employer demonstrates to OSHA that the execution of a confidentiality agreement would not provide sufficient protection against the potential harm from the unauthorized disclosure of a trade secret specific chemical identity, the Assistant Secretary may issue such orders or impose such additional limitations or conditions upon the disclosure of the requested chemical information as may be appropriate to assure that the occupational health needs are met without an undue risk of harm to the employer.

**1910.1020(f)(12)** Notwithstanding the existence of a trade secret claim, an employer shall, upon request, disclose to the Assistant Secretary any information which this section requires the employer to make available. Where there is a trade secret claim, such claim shall be made no later than at the time the information is provided to the Assistant Secretary so that suitable determinations of trade secret status can be made and the necessary protections can be implemented.

**1910.1020(f)(13)** Nothing in this paragraph shall be construed as requiring the disclosure under any circumstances of process or percentage of mixture information which is trade secret.

**1910.1020(g)** ***Employee information.*** (1) Upon an employee's first entering into employment, and at least annually thereafter, each employer shall inform current employees covered by this section of the following:

**1910.1020(g)(1)(i)** The existence, location, and availability of any records covered by this section;

**1910.1020(g)(1)(ii)** The person responsible for maintaining and providing access to records; and

**1910.1020(g)(1)(iii)** Each employee's rights of access to these records.

**1910.1020(g)(2)** Each employer shall keep a copy of this section and its appendices, and make copies readily available, upon request, to employees. The employer shall also distribute to current employees any informational materials concerning this section which are made available to the employer by the Assistant Secretary of Labor for Occupational Safety and Health.

**1910.1020(h)** ***Transfer of records.*** (1) Whenever an employer is ceasing to do business, the employer shall transfer all records subject to this section to the successor employer. The successor employer shall receive and maintain these records.

**1910.1020(h)(2)** Whenever an employer is ceasing to do business and there is no successor employer to receive and maintain the records subject to this standard, the employer shall notify affected current employees of their rights of access to records at least three (3) months prior to the cessation of the employer's business.

**1910.1020(h)(3)** Whenever an employer either is ceasing to do business and there is no successor employer to receive and maintain the records, or intends to dispose of any records required to be preserved for at least thirty (30) years, the employer shall:

**1910.1020(h)(3)(i)** Transfer the records to the Director of the National Institute for Occupational Safety and Health (NIOSH) if so required by a specific occupational safety and health standard; or

**1910.1020(h)(3)(ii)** Notify the Director of NIOSH in writing of the impending disposal of records at least three (3) months prior to the disposal of the records.

**1910.1020(h)(4)** Where an employer regularly disposes of records required to be preserved for at least thirty (30) years, the employer may, with at least (3) months notice, notify the Director of NIOSH on an annual basis of the records intended to be disposed of in the coming year.

**1910.1020(i)** ***Appendices.*** The information contained in appendices A and B to this section is not intended, by itself, to create any additional obligations not otherwise imposed by this section nor detract from any existing obligation.

### APPENDIX A TO §1910.20 [sic §1910.1020]—SAMPLE AUTHORIZATION LETTER FOR THE RELEASE OF EMPLOYEE MEDICAL RECORD INFORMATION TO A DESIGNATED REPRESENTATIVE (NON-MANDATORY)

I, ______________ (full name of worker/patient), hereby authorize ______________ (individual or organization holding the medical records) to release to ______________ (individual or organization authorized to receive the medical information), the following medical information from my personal medical records:

(Describe generally the information desired to be released)

______________________________________________

______________________________________________

______________________________________________

I give my permission for this medical information to be used for the following purpose:

________________________________________

________________________________________

but I do not give permission for any other use or re-disclosure of this information.

**NOTE:** Several extra lines are provided below so that you can place additional restrictions on this authorization letter if you want to. You may, however, leave these lines blank. On the other hand, you may want to (1) specify a particular expiration date for this letter (if less than one year); (2) describe medical information to be created in the future that you intend to be covered by this authorization letter; or (3) describe portions of the medical information in your records which you do not intend to be released as a result of this letter.)

________________________________________

________________________________________

________________________________________

________________________________________

________________________________________

Full name of Employee or Legal Representative

________________________________________

Signature of Employee or Legal Representative

________________________________________

Date of Signature ____________________

## APPENDIX B TO §1910.20 [sic §1910.1020]—AVAILABILITY OF NIOSH REGISTRY OF TOXIC EFFECTS OF CHEMICAL SUBSTANCES (RTECS) (NON-MANDATORY)

The final regulation, 29 CFR 1910.20 [sic 1910.1020], applies to all employee exposure and medical records, and analyses thereof, of employees exposed to toxic substances or harmful physical agents (paragraph (b)(2)). The term *toxic substance or harmful physical agent* is defined by paragraph (c)(13) to encompass chemical substances, biological agents, and physical stresses for which there is evidence of harmful health effects. The regulation uses the latest printed edition of the National Institute for Occupational Safety and Health (NIOSH) Registry of Toxic Effects of Chemical Substances (RTECS) as one of the chief sources of information as to whether evidence of harmful health effects exists. If a substance is listed in the latest printed RTECS, the regulation applies to exposure and medical records (and analyses of these records) relevant to employees exposed to the substance.

It is appropriate to note that the final regulation does not require that employers purchase a copy of RTECS, and many employers need not consult RTECS to ascertain whether their employee exposure or medical records are subject to the rule. Employers who do not currently have the latest printed edition of the NIOSH RTECS, however, may desire to obtain a copy. The RTECS is issued in an annual printed edition as mandated by section 20(a)(6) of the Occupational Safety and Health Act (29 U.S.C. 669(a)(6)).

The Introduction to the 1980 printed edition describes the RTECS as follows:

"The 1980 edition of the Registry of Toxic Effects of Chemical Substances, formerly known as the Toxic Substances list, is the ninth revision prepared in compliance with the requirements of Section 20(a)(6) of the Occupational Safety and Health Act of 1970 (Public Law 91-596). The original list was completed on June 28, 1971, and has been updated annually in book format. Beginning in October 1977, quarterly revisions have been provided in microfiche. This edition of the Registry contains 168,096 listings of chemical substances: 45,156 are names of different chemicals with their associated toxicity data and 122,940 are synonyms. This edition includes approximately 5,900 new chemical compounds that did not appear in the 1979 Registry. (p. xi)

"The Registry's purposes are many, and it serves a variety of users. It is a single source document for basic toxicity information and for other data, such as chemical identifiers ad information necessary for the preparation of safety directives and hazard evaluations for chemical substances. The various types of toxic effects linked to literature citations provide researchers and occupational health scientists with an introduction to the toxicological literature, making their own review of the toxic hazards of a given substance easier. By presenting data on the lowest reported doses that produce effects by several routes of entry in various species, the Registry furnishes valuable information to those responsible for preparing safety data sheets for chemical substances in the workplace. Chemical and production engineers can use the Registry to identify the hazards which may be associated with chemical intermediates in the development of final products, and thus can more readily select substitutes or alternative processes which may be less hazardous. Some organizations, including health agencies and chemical companies, have included the NIOSH Registry accession numbers with the listing of chemicals in their files to reference toxicity information associated with those chemicals. By including foreign language chemical names, a start has been made toward providing rapid identification of substances produced in other countries. (p. xi)

"In this edition of the Registry, the editors intend to identify "all known toxic substances" which may exist in the environment and to provide pertinent data on the toxic effects from known doses entering an organism by any route described. (p xi)

"It must be reemphasized that the entry of a substance in the Registry does not automatically mean that it must be avoided. A listing does mean, however, that the substance has the documented potential of being harmful if misused, and care must be exercised to prevent tragic consequences. Thus, the Registry lists many substances that are common in everyday life and are in nearly every household in the United States. One can name a variety of such dangerous substances: prescription and non-prescription drugs; food additives; pesticide concentrates, sprays, and dusts; fungicides; herbicides; paints; glazes, dyes; bleaches and other household cleaning agents; alkalies; and various solvents and diluents. The list is extensive because chemicals have become an integral part of our existence."

The RTECS printed edition may be purchased from the Superintendent of Documents, U.S. Government Printing Office (GPO), Washington, DC 20402 (202-783-3238).

Some employers may desire to subscribe to the quarterly update to the RTECS which is published in a microfiche edition. An annual subscription to the quarterly microfiche may be purchased from the GPO (Order the "Microfiche Edition, Registry of Toxic Effects of Chemical Substances"). Both the printed edition and the microfiche edition of RTECS are available for review at many university and public libraries throughout the country. The latest RTECS editions may also be examined at the OSHA Technical Data Center, Room N2439—Rear, United States Department of Labor, 200 Constitution Avenue, NW., Washington, DC 20210 (202-523-9700), or at any OSHA Regional or Area Office (*See,* major city telephone directories under United States Government Labor Department).

[53 FR 38163, Sept. 29, 1988; 53 FR 49981, Dec. 13, 1988, as amended at 54 FR 24333, June 7, 1989; 55 FR 26431, June 28, 1990; 61 FR 9235, Mar. 7, 1996. Redesignated at 61 FR 31430, June 20, 1996]

### §1926.34 Means of egress.

**1926.34(a)** ***General.*** In every building or structure exits shall be so arranged and maintained as to provide free and unobstructed egress from all parts of the building or structure at all times when it is occupied. No lock or fastening to prevent free escape from the inside of any building shall be installed except in mental, penal, or corrective institutions where supervisory personnel is continually on duty and effective provisions are made to remove occupants in case of fire or other emergency.

**1926.34(b)** ***Exit marking.*** Exits shall be marked by a readily visible sign. Access to exits shall be marked by readily visible signs in all cases where the exit or way to reach it is not immediately visible to the occupants.

**1926.34(c)** ***Maintenance and workmanship.*** Means of egress shall be continually maintained free of all obstructions or impediments to full instant use in the case of fire or other emergency.

[58 FR 35083, June 30, 1993]

### §1926.35 Employee emergency action plans.

**1926.35(a)** ***Scope and application.*** This section applies to all emergency action plans required by a particular OSHA standard. The emergency action plan shall be in writing (except as provided in the last sentence of paragraph (e)(3) of this section) and shall cover those designated actions employers and employees must take to ensure employee safety from fire and other emergencies.

**1926.35(b)** ***Elements.*** The following elements, at a minimum, shall be included in the plan:

**1926.35(b)(1)** Emergency escape procedures and emergency escape route assignments;

**1926.35(b)(2)** Procedures to be followed by employees who remain to operate critical plant operations before they evacuate;

**1926.35(b)(3)** Procedures to account for all employees after emergency evacuation has been completed;

**1926.35(b)(4)** Rescue and medical duties for those employees who are to perform them;

**1926.35(b)(5)** The preferred means of reporting fires and other emergencies; and

**1926.35(b)(6)** Names or regular job titles of persons or departments who can be contacted for further information or explanation of duties under the plan.

**1926.35(c)** ***Alarm system.*** (1) The employer shall establish an employee alarm system which complies with §1926.159.

**1926.35(c)(2)** If the employee alarm system is used for alerting fire brigade members, or for other purposes, a distinctive signal for each purpose shall be used.

**1926.35(d)** ***Evacuation.*** The employer shall establish in the emergency action plan the types of evacuation to be used in emergency circumstances.

**1926.35(e)** ***Training.*** (1) Before implementing the emergency action plan, the employer shall designate and train a sufficient number of persons to assist in the safe and orderly emergency evacuation of employees.

**1926.35(e)(2)** The employer shall review the plan with each employee covered by the plan at the following times:

**1926.35(e)(2)(i)** Initially when the plan is developed,

**1926.35(e)(2)(ii)** Whenever the employee's responsibilities or designated actions under the plan change, and

**1926.35(e)(2)(iii)** Whenever the plan is changed.

**1926.35(e)(3)** The employer shall review with each employee upon initial assignment those parts of the plan which the employee must know to protect the employee in the event of an emergency. The written plan shall be kept at the workplace and made available for employee review. For those employers with 10 or fewer employees the plan may be communicated orally to employees and the employer need not maintain a written plan.

[58 FR 35083, June 30, 1993]

**Authority:** Sec. 107, Contract Work Hours and Safety Standards Act (40 U.S.C. 333); secs. 4, 6, and 8, Occupational Safety and Health Act of 1970 (29 U.S.C. 653, 655, 657); Secretary of Labor's Order No. 12-71 (36 FR 8754), 8-76 (41 FR 25059), 9-83 (48 FR 35736), or 6-96 (62 FR 111), as applicable; and 29 CFR part 1911.

### §1926.50 Medical services and first aid.

**1926.50(a)** The employer shall insure the availability of medical personnel for advice and consultation on matters of occupational health.

**1926.50(b)** Provisions shall be made prior to commencement of the project for prompt medical attention in case of serious injury.

**1926.50(c)** In the absence of an infirmary, clinic, hospital, or physician, that is reasonably accessible in terms of time and distance to the worksite, which is available for the treatment of injured employees, a person who has a valid certificate in first-aid training from the U.S. Bureau of Mines, the American Red Cross, or equivalent training that can be verified by documentary evidence, shall be available at the worksite to render first aid.

**1926.50(d)(1)** First aid supplies shall be easily accessible when required.

**1926.50(d)(2)** The contents of the first aid kit shall be placed in a weatherproof container with individual sealed packages for each type of item, and shall be checked by the employer before being sent out on each job and at least weekly on each job to ensure that the expended items are replaced.

**1926.50(e)** Proper equipment for prompt transportation of the injured person to a physician or hospital, or a communication system for contacting necessary ambulance service, shall be provided.

**1926.50(f)** In areas where 911 is not available, the telephone numbers of the physicians, hospitals, or ambulances shall be conspicuously posted.

**1926.50(g)** Where the eyes or body of any person may be exposed to injurious corrosive materials, suitable facilities for quick drenching or flushing of the eyes and body shall be provided within the work area for immediate emergency use.

### Appendix A to §1926.50—First aid Kits (Non-Mandatory)

First aid supplies are required to be easily accessible under paragraph §1926.50(d)(1). An example of the minimal contents of a generic first aid kit is described in American National Standard (ANSI) Z308.1-1978 "Minimum Requirements for Industrial Unit-Type First-aid Kits". The contents of the kit listed in the ANSI standard should be adequate for small work sites. When larger operations or multiple operations are being conducted at the same location, employers should determine the need for additional first aid kits at the worksite, additional types of first aid equipment and supplies and additional quantities and types of supplies and equipment in the first aid kits.

In a similar fashion, employers who have unique or changing first-aid needs in their workplace, may need to enhance their first-aid kits. The employer can use the OSHA 200 log, OSHA 101's or other reports to identify these unique problems. Consultation from the local Fire/Rescue Department, appropriate medical professional, or local emergency room may be helpful to employers in these circumstances. By assessing the specific needs of their workplace, employers can ensure that reasonably anticipated supplies are available. Employers should assess the specific needs of their worksite periodically and augment the first aid kit appropriately.

If it is reasonably anticipated employees will be exposed to blood or other potentially infectious materials while using first-aid supplies, employers should provide personal protective equipment (PPE). Appropriate PPE includes gloves, gowns, face shields, masks and eye protection (see "Occupational Exposure to Blood borne Pathogens", 29 CFR 1910.1030(d)(3)) (56 FR 64175).

[44 FR 8577, Feb. 9, 1979; 44 FR 20940, Apr. 6, 1979, as amended at 49 FR 18295, Apr. 30, 1984; 58 FR 35084, June 30, 1993; 61 FR 5510, Feb. 13, 1996; 63 FR 33469, June 18, 1998]

### §1926.51 Sanitation.

**1926.51(a)** ***Potable water.*** (1) An adequate supply of potable water shall be provided in all places of employment.

**1926.51(a)(2)** Portable containers used to dispense drinking water shall be capable of being tightly closed, and equipped with a tap. Water shall not be dipped from containers.

**1926.51(a)(3)** Any container used to distribute drinking water shall be clearly marked as to the nature of its contents and not used for any other purpose.

**1926.51(a)(4)** The common drinking cup is prohibited.

**1926.51(a)(5)** Where single service cups (to be used but once) are supplied, both a sanitary container for the unused cups and a receptacle for disposing of the used cups shall be provided.

**1926.51(a)(6)** *Potable water* means water which meets the quality standards prescribed in the U.S. Public Health Service Drinking Water Standards, published in 42 CFR part 72, or water which is approved for drinking purposes by the State or local authority having jurisdiction.

**1926.51(b)** ***Nonpotable water.*** (1) Outlets for nonpotable water, such as water for industrial or firefighting purposes only, shall be identified by signs meeting the requirements of subpart G of this part, to indicate clearly that the water is unsafe and is not to be used for drinking, washing, or cooking purposes.

**1926.51(b)(2)** There shall be no cross-connection, open or potential, between a system furnishing potable water and a system furnishing nonpotable water.

**1926.51(c)** ***Toilets at construction jobsites.*** (1) Toilets shall be provided for employees according to the following table:

**TABLE D-1**

| *Number of employees* | *Minimum number of facilities* |
|---|---|
| 20 or less | 1. |
| 20 or more | 1 toilet seat and 1 urinal per 40 workers. |
| 200 or more | 1 toilet seat and 1 urinal per 50 workers. |

**1926.51(c)(2)** Under temporary field conditions, provisions shall be made to assure not less than one toilet facility is available.

**1926.51(c)(3)** Job sites, not provided with a sanitary sewer, shall be provided with one of the following toilet facilities unless prohibited by local codes:

**1926.51(c)(3)(i)** Privies (where their use will not contaminate ground or surface water);

**1926.51(c)(3)(ii)** Chemical toilets;

**1926.51(c)(3)(iii)** Recirculating toilets;

**1926.51(c)(3)(iv)** Combustion toilets.

**1926.51(c)(4)** The requirements of this paragraph (c) for sanitation facilities shall not apply to mobile crews having transportation readily available to nearby toilet facilities.

**1926.51(d)** ***Food handling.*** (1) All employees' food service facilities and operations shall meet the applicable laws, ordinances, and regulations of the jurisdictions in which they are located.

**1926.51(d)(2)** All employee food service facilities and operations shall be carried out in accordance with sound hygienic principles. In all places of employment where all or part of the food service is provided, the food dispensed shall be wholesome, free from spoilage, and shall be processed, prepared, handled, and stored in such a manner as to be protected against contamination.

**1926.51(e)** ***Temporary sleeping quarters.*** When temporary sleeping quarters are provided, they shall be heated, ventilated, and lighted.

**1926.51(f)** ***Washing facilities.*** (1) The employer shall provide adequate washing facilities for employees engaged in the application of paints, coating, herbicides, or insecticides, or in other operations where contaminants may be harmful to the employees. Such facilities shall be in near proximity to the worksite and shall be so equipped as to enable employees to remove such substances.

**1926.51(f)(2)** *General.* Washing facilities shall be maintained in a sanitary condition.

**1926.51(f)(3)** *Lavatories.* (i) Lavatories shall be made available in all places of employment. The requirements of this subdivision do not apply to mobile crews or to normally unattended work locations if employees working at these locations have transportation readily available to nearby washing facilities which meet the other requirements of this paragraph.

**1926.51(f)(3)(ii)** Each lavatory shall be provided with hot and cold running water, or tepid running water.

**1926.51(f)(3)(iii)** Hand soap or similar cleansing agents shall be provided.

**1926.51(f)(3)(iv)** Individual hand towels or sections thereof, of cloth or paper, warm air blowers or clean individual sections of continuous cloth toweling, convenient to the lavatories, shall be provided.

**1926.51(f)(4)** *Showers.* (i) Whenever showers are required by a particular standard, the showers shall be provided in accordance with paragraphs (f)(4) (ii) through (v) of this section.

**1926.51(f)(4)(ii)** One shower shall be provided for each 10 employees of each sex, or numerical fraction thereof, who are required to shower during the same shift.

**1926.51(f)(4)(iii)** Body soap or other appropriate cleansing agents convenient to the showers shall be provided as specified in paragraph (f)(3)(iii) of this section.

**1926.51(f)(4)(iv)** Showers shall be provided with hot and cold water feeding a common discharge line.

**1926.51(f)(4)(v)** Employees who use showers shall be provided with individual clean towels.

**1926.51(g)** ***Eating and drinking areas.*** No employee shall be allowed to consume food or beverages in a toilet room nor in any area exposed to a toxic material.

**1926.51(h)** ***Vermin control.*** Every enclosed workplace shall be so constructed, equipped, and maintained, so far as reasonably practicable, as to prevent the entrance or harborage of rodents, insects, and other vermin. A continuing and effective extermination program shall be instituted where their presence is detected.

**1926.51(i)** ***Change rooms.*** Whenever employees are required by a particular standard to wear protective clothing because of the possibility of contamination with toxic materials, change rooms equipped with storage facilities for street clothes and separate storage facilities for the protective clothing shall be provided.

[44 FR 8577, Feb. 9, 1979; 44 FR 20940, Apr. 6, 1979, as amended at 58 FR 35084, June 30, 1993]

## §1926.52 Occupational noise exposure.

**1926.52(a)** Protection against the effects of noise exposure shall be provided when the sound levels exceed those shown in Table D-2 of this section when measured on the A-scale of a standard sound level meter at slow response.

**1926.52(b)** When employees are subjected to sound levels exceeding those listed in Table D-2 of this section, feasible administrative or engineering controls shall be utilized. If such controls fail to reduce sound levels within the levels of the table, personal protective equipment as required in subpart E, shall be provided and used to reduce sound levels within the levels of the table.

**1926.52(c)** If the variations in noise level involve maxima at intervals of 1 second or less, it is to be considered continuous.

**1926.52(d)(1)** In all cases where the sound levels exceed the values shown herein, a continuing, effective hearing conservation program shall be administered.

### TABLE D-2—PERMISSIBLE NOISE EXPOSURES

| Duration per day, hours: | Sound level dBA slow response |
|---|---|
| 8 | 90 |
| 6 | 92 |
| 4 | 95 |
| 3 | 97 |
| 2 | 100 |
| 1 1/2 | 102 |
| 1 | 105 |
| 1/2 | 110 |
| 1/4 or less | 115 |

**1926.52(d)(2)(i)** When the daily noise exposure is composed of two or more periods of noise exposure of different levels, their combined effect should be considered, rather than the individual effect of each. Exposure to different levels for various periods of time shall be computed according to the formula set forth in paragraph (d)(2)(ii) of this section.

**1926.52(d)(2)(ii)**

$$F_e = (T_1 / L_1) + (T_2 / L_2) + \dots + (T_n / L_n)$$

Where:

$F_e$ = *The equivalent noise exposure factor.*

$T$ = The period of noise exposure at any essentially constant level.

$L$ = The duration of the permissible noise exposure at the constant level (from Table D-2).

If the value of $F_e$ exceeds unity (1) the exposure exceeds permissible levels.

**1926.52(d)(2)(iii)** A sample computation showing an application of the formula in paragraph (d)(2)(ii) of this section is as follows. An employee is exposed at these levels for these periods:

110 db A $^1/_4$ hour.

100 db A $^1/_2$ hour.

90 db A $1^1/_2$ hours.

$F_e = (^1/_4 / ^1/_2) + (^1/_2 / 2) + (1^1/_2 / 8)$

$F_e = 0.500+0.25+0.188$

$F_e = 0.938$

*Since the value of* $F_e$ does not exceed unity, the exposure is within permissible limits.

**1926.52(e)** Exposure to impulsive or impact noise should not exceed 140 dB peak sound pressure level.

## §1926.53 Ionizing radiation.

**1926.53(a)** In construction and related activities involving the use of sources of ionizing radiation, the pertinent provisions of the Nuclear Regulatory Commission's Standards for Protection Against Radiation (10 CFR part 20), relating to protection against occupational radiation exposure, shall apply.

**1926.53(b)** Any activity which involves the use of radioactive materials or X-rays, whether or not under license from the Nuclear Regulatory Commission, shall be performed by competent persons specially trained in the proper and safe operation of such equipment. In the case of materials used under Commission license, only persons actually licensed, or competent persons under direction and supervision of the licensee, shall perform such work.

**1926.53(c)—(r)** [Reserved]

**Note:** The requirements applicable to construction work under paragraphs (c) through (r) of this section are identical to those set forth at paragraphs (a) through (p) of §1910.1096 of this chapter.

[44 FR 8577, Feb. 9, 1979; 44 FR 20940, Apr. 6, 1979, as amended at 61 FR 5510, Feb. 13, 1996; 61 FR 31431, June 20, 1996]

[Section 1910.1096 (a) through (p) is printed below—CCH.]

## §1910.1096 Ionizing radiation.

**1910.1096(a)** ***Definitions applicable to this section.*** (1) *Radiation* includes alpha rays, beta rays, gamma rays, X-rays, neutrons, high-speed electrons, high-speed protons, and other atomic particles; but such term does not include sound or radio waves, or visible light, or infrared or ultraviolet light.

**1910.1096(a)(2)** *Radioactive material* means any material which emits, by spontaneous nuclear disintegration, corpuscular or electromagnetic emanations.

**1910.1096(a)(3)** *Restricted area* means any area access to which is controlled by the employer for purposes of protection of individuals from exposure to radiation or radioactive materials.

**1910.1096(a)(4)** *Unrestricted area* means any area access to which is not controlled by the employer for purposes of protection of individuals from exposure to radiation or radioactive materials.

**1910.1096(a)(5)** *Dose* means the quantity of ionizing radiation absorbed, per unit of mass, by the body or by any portion of the body. When the provisions in this section specify a dose during a period of time, the dose is the total quantity of radiation absorbed, per unit of mass, by the body or by any portion of the body during such period of time. Several different units of dose are in current use. Definitions of units used in this section are set forth in paragraphs (a) (6) and (7) of this section.

1910.1096(a)(6) *Rad* means a measure of the dose of any ionizing radiation to body tissues in terms of the energy absorbed per unit of mass of the tissue. One rad is the dose corresponding to the absorption of 100 ergs per gram of tissue (1 millirad (mrad)=0.001 rad).

1910.1096(a)(7) *Rem* means a measure of the dose of any ionizing radiation to body tissue in terms of its estimated biological effect relative to a dose of 1 roentgen (r) of X-rays (1 millirem (mrem)=0.001 rem). The relation of the rem to other dose units depends upon the biological effect under consideration and upon the conditions for irradiation. Each of the following is considered to be equivalent to a dose of 1 rem:

1910.1096(a)(7)(i) A dose of 1 roentgen due to X- or gamma radiation;

1910.1096(a)(7)(ii) A dose of 1 rad due to X-, gamma, or beta radiation;

1910.1096(a)(7)(iii) A dose of 0.1 rad due to neutrons or high energy protons;

1910.1096(a)(7)(iv) A dose of 0.05 rad due to particles heavier than protons and with sufficient energy to reach the lens of the eye;

1910.1096(a)(7)(v) If it is more convenient to measure the neutron flux, or equivalent, than to determine the neutron dose in rads, as provided in paragraph (a)(7)(iii) of this section, 1 rem of neutron radiation may, for purposes of the provisions in this section be assumed to be equivalent to 14 million neutrons per square centimeter incident upon the body; or, if there is sufficient information to estimate with reasonable accuracy the approximate distribution in energy of the neutrons, the incident number of neutrons per square centimeter equivalent to 1 rem may be estimated from Table G-17:

### TABLE G-17—NEUTRON FLUX DOSE EQUIVALENTS

| Neutron energy (million electron volts (Mev)) | Number of neutrons per square centimeter equivalent to a dose of 1 rem (neutrons/cm²) | Average flux to deliver 100 millirem in 40 hours (neutrons/ cm² per sec.) |
|---|---|---|
| Thermal | $970 \times 10^6$ | 670 |
| 0.0001 | $720 \times 10^6$ | 500 |
| 0.005 | $820 \times 10^6$ | 570 |
| 0.02 | $400 \times 10^6$ | 280 |
| 0.1 | $120 \times 10^6$ | 80 |
| 0.5 | $43 \times 10^6$ | 30 |
| 1.0 | $26 \times 10^6$ | 18 |
| 2.5 | $29 \times 10^6$ | 20 |
| 5.0 | $26 \times 10^6$ | 18 |
| 7.5 | $24 \times 10^6$ | 17 |
| 10 | $24 \times 10^6$ | 17 |
| 10 to 30 | $14 \times 10^6$ | 10 |

1910.1096(a)(8) For determining exposures to X- or gamma rays up to 3 Mev., the dose limits specified in this section may be assumed to be equivalent to the "air dose". For the purpose of this section *air dose* means that the dose is measured by a properly calibrated appropriate instrument in air at or near the body surface in the region of the highest dosage rate.

1910.1096(b) ***Exposure of individuals to radiation in restricted areas.*** (1) Except as provided in paragraph (b)(2) of this section, no employer shall possess, use, or transfer sources of ionizing radiation in such a manner as to cause any individual in a restricted area to receive in any period of one calendar quarter from sources in the employer's possession or control a dose in excess of the limits specified in Table G-18:

### TABLE G-18

| | Rems per calendar quarter |
|---|---|
| Whole body: | |
| Head and trunk; active blood-forming organs; lens of eyes; or gonads | 1 1/4 |
| Hands and forearms; feet and ankles | 18 3/4 |
| Skin of whole body | 7 1/2 |

1910.1096(b)(2) An employer may permit an individual in a restricted area to receive doses to the whole body greater than those permitted under subparagraph (1) of this paragraph, so long as:

1910.1096(b)(2)(i) During any calendar quarter the dose to the whole body shall not exceed 3 rems; and

1910.1096(b)(2)(ii) The dose to the whole body, when added to the accumulated occupational dose to the whole body, shall not exceed 5 (N-18) rems, where "N" equals the individual's age in years at his last birthday; and

1910.1096(b)(2)(iii) The employer maintains adequate past and current exposure records which show that the addition of such a dose will not cause the individual to exceed the amount authorized in this subparagraph. As used in this subparagraph *Dose to the whole body* shall be deemed to include any dose to the whole body, gonad, active bloodforming organs, head and trunk, or lens of the eye.

1910.1096(b)(3) No employer shall permit any employee who is under 18 years of age to receive in any period of one calendar quarter a dose in excess of 10 percent of the limits specified in Table G-18.

1910.1096(b)(4) *Calendar quarter* means any 3-month period determined as follows:

1910.1096(b)(4)(i) The first period of any year may begin on any date in January: *Provided,* That the second, third, and fourth periods accordingly begin on the same date in April, July, and October, respectively, and that the fourth period extends into January of the succeeding year, if necessary to complete a 3-month quarter. During the first year of use of this method of determination, the first period for that year shall also include any additional days in January preceding the starting date for the first period; or

1910.1096(b)(4)(ii) The first period in a calendar year of 13 complete, consecutive calendar weeks; the second period in a calendar year of 13 complete, consecutive weeks; the third period in a calendar year of 13 complete, consecutive calendar weeks; the fourth period in a calendar year of 13 complete, consecutive calendar weeks. If at the end of a calendar year there are any days not falling within a complete calendar week of that year, such days shall be included within the last complete calendar week of that year. If at the beginning of any calendar year there are days not falling within a complete calendar week of that year, such days shall be included within the last complete calendar week of the previous year; or

1910.1096(b)(4)(iii) The four periods in a calendar year may consist of the first 14 complete, consecutive calendar weeks; the next 12 complete, consecutive calendar weeks, the next 14 complete, consecutive calendar weeks, and the last 12 complete, consecutive calendar weeks. If at the end of a calendar year there are any days not falling within a complete calendar week of that year, such days shall be included (for purposes of this section) within the last complete calendar week of the year. If at the beginning of any calendar year there are days not falling within a complete calendar week of that year, such days shall be included (for purposes of this section) within the last complete week of the previous year.

1910.1096(c) ***Exposure to airborne radioactive material.*** (1) No employer shall possess, use or transport radioactive material in such a manner as to cause any employee, within a restricted area, to be exposed to airborne radioactive material in an average concentration in excess of the limits specified in Table 1 of appendix B to 10 CFR part 20. The limits given in Table 1 are for exposure to the concentrations specified for 40 hours in any work-week of 7 consecutive days. In any such period where the number of hours of exposure is less than 40, the limits specified in the table may be increased proportionately. In any such period where the number of hours of exposure is greater than 40, the limits specified in the table shall be decreased proportionately.

**1910.1096(c)(2)** No employer shall possess, use, or transfer radioactive material in such a manner as to cause any individual within a restricted area, who is under 18 years of age, to be exposed to airborne radioactive material in an average concentration in excess of the limits specified in Table II of appendix B to 10 CFR part 20. For purposes of this paragraph, concentrations may be averaged over periods not greater than 1 week.

**1910.1096(c)(3)** *Exposed* as used in this paragraph means that the individual is present in an airborne concentration. No allowance shall be made for the use of protective clothing or equipment, or particle size.

**1910.1096(d)** ***Precautionary procedures and personal monitoring.*** (1) Every employer shall make such surveys as may be necessary for him to comply with the provisions in this section. *Survey* means an evaluation of the radiation hazards incident to the production, use, release, disposal, or presence of radioactive materials or other sources of radiation under a specific set of conditions. When appropriate, such evaluation includes a physical survey of the location of materials and equipment, and measurements of levels of radiation or concentrations of radioactive material present.

**1910.1096(d)(2)** Every employer shall supply appropriate personnel monitoring equipment, such as film badges, pocket chambers, pocket dosimeters, or film rings, and shall require the use of such equipment by:

**1910.1096(d)(2)(i)** Each employee who enters a restricted area under such circumstances that he receives, or is likely to receive, a dose in any calendar quarter in excess of 25 percent of the applicable value specified in paragraph (b)(1) of this section; and

**1910.1096(d)(2)(ii)** Each employee under 18 years of age who enters a restricted area under such circumstances that he receives, or is likely to receive, a dose in any calendar quarter in excess of 5 percent of the applicable value specified in paragraph (b)(1) of this section; and

**1910.1096(d)(2)(iii)** Each employee who enters a high radiation area.

**1910.1096(d)(3)** As used in this section:

**1910.1096(d)(3)(i)** *Personnel monitoring equipment* means devices designed to be worn or carried by an individual for the purpose of measuring the dose received (e.g., film badges, pocket chambers, pocket dosimeters, film rings, etc.);

**1910.1096(d)(3)(ii)** *Radiation area* means any area, accessible to personnel, in which there exists radiation at such levels that a major portion of the body could receive in any 1 hour a dose in excess of 5 millirem, or in any 5 consecutive days a dose in excess of 100 millirem; and

**1910.1096(d)(3)(iii)** *High radiation area* means any area, accessible to personnel, in which there exists radiation at such levels that a major portion of the body could receive in any one hour a dose in excess of 100 millirem.

**1910.1096(e)** ***Caution signs, labels, and signals***—(1) *General.* (i) Symbols prescribed by this paragraph shall use the conventional radiation caution colors (magenta or purple on yellow background). The symbol prescribed by this paragraph is the conventional three-bladed design:

**FIGURE G-10**

**RADIATION SYMBOL**

1. Cross-hatched area is to be magenta or purple.
2. Background is to be yellow.

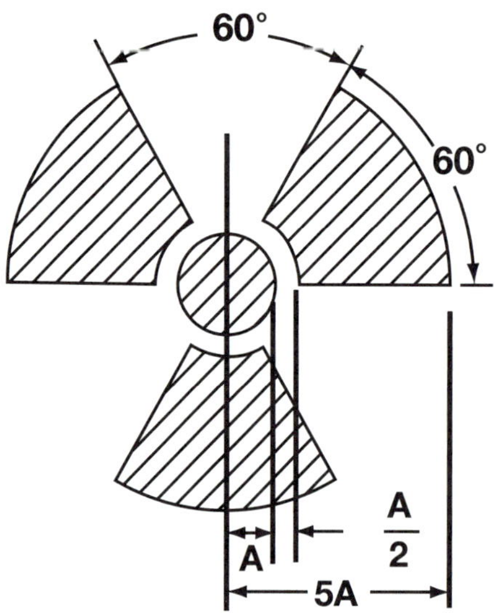


**1910.1096(e)(1)(ii)** [Reserved]

**1910.1096(e)(2)** *Radiation area.* Each radiation area shall be conspicuously posted with a sign or signs bearing the radiation caution symbol described in subparagraph (1) of this paragraph and the words:

**CAUTION**

**RADIATION AREA**

**1910.1096(e)(3)** *High radiation area.* (i) Each high radiation area shall be conspicuously posted with a sign or signs bearing the radiation caution symbol and the words:

**CAUTION**

**HIGH RADIATION AREA**

**1910.1096(e)(3)(ii)** Each high radiation area shall be equipped with a control device which shall either cause the level of radiation to be reduced below that at which an individual might receive a dose of 100 millirems in 1 hour upon entry into the area or shall energize a conspicuous visible or audible alarm signal in such a manner that the individual entering and the employer or a supervisor of the activity are made aware of the entry. In the case of a high radiation area established for a period of 30 days or less, such control device is not required.

**1910.1096(e)(4)** *Airborne radioactivity area.* (i) As used in the provisions of this section, *airborne radioactivity area* means:

**1910.1096(e)(4)(i)*(a)*** Any room, enclosure, or operating area in which airborne radioactive materials, composed wholly or partly of radioactive material, exist in concentrations in excess of the amounts specified in column 1 of Table 1 of appendix B to 10 CFR part 20 or

**1910.1096(e)(4)(i)*(b)*** Any room, enclosure, or operating area in which airborne radioactive materials exist in concentrations which, averaged over the number of hours in any week during which individuals are in the area, exceed 25 percent of the amounts specified in column 1 of Table 1 of appendix B to 10 CFR part 20.

**1910.1096(e)(4)(ii)** Each airborne radioactivity area shall be conspicuously posted with a sign or signs bearing the radiation caution symbol described in paragraph (e)(1) of this section and the words:

**CAUTION**

**AIRBORNE RADIOACTIVITY AREA**

**1910.1096(e)(5)** *Additional requirements.* (i) Each area or room in which radioactive material is used or stored and which contains any radioactive material (other than natural uranium or thorium) in any amount exceeding 10 times the quantity of such material specified in appendix C to 10 CFR part 20 shall be conspicuously posted with a sign or signs bearing the radiation caution symbol described in paragraph (e)(1) of this section and the words:

**CAUTION**

**RADIOACTIVE MATERIALS**

**1910.1096(e)(5)(ii)** Each area or room in which natural uranium or thorium is used or stored in an amount exceeding 100 times the quantity of such material specified in 10 CFR part 20 shall be conspicuously posted with a sign or signs bearing the radiation caution symbol described in paragraph (e)(1) of this section and the words:

**CAUTION**

**RADIOACTIVE MATERIALS**

**1910.1096(e)(6)** *Containers.* (i) Each container in which is transported, stored, or used a quantity of any radioactive material (other than natural uranium or thorium) greater than the quantity of such material specified in appendix C to 10 CFR part 20 shall bear a durable, clearly visible label bearing the radiation caution symbol described in paragraph (e)(1) of this section and the words:

**CAUTION**

**RADIOACTIVE MATERIALS**

**1910.1096(e)(6)(ii)** Each container in which natural uranium or thorium is transported, stored, or used in a quantity greater than 10 times the quantity specified in appendix C to 10 CFR part 20 shall bear a durable, clearly visible label bearing the radiation caution symbol described in paragraph (e)(1) of this section and the words:

**CAUTION**

**RADIOACTIVE MATERIALS**

**1910.1096(e)(6)(iii)** Notwithstanding the provisions of paragraphs (e)(6) (i) and (ii) of this section a label shall not be required:

**1910.1096(e)(6)(iii)*(a)*** If the concentration of the material in the container does not exceed that specified in column 2 of Table 1 of appendix B to 10 CFR part 20, or

**1910.1096(e)(6)(iii)*(b)*** For laboratory containers, such as beakers, flasks, and test tubes, used transiently in laboratory procedures, when the user is present.

**1910.1096(e)(6)(iv)** Where containers are used for storage, the labels required in this subparagraph shall state also the quantities and kinds of radioactive materials in the containers and the date of measurement of the quantities.

**1910.1096(f)** ***Immediate evacuation warning signal***—(1) *Signal characteristics.* (i) The signal shall be a midfrequency complex sound wave amplitude modulated at a subsonic frequency. The complex sound wave in free space shall have a fundamental frequency (f1) between 450 and 500 hertz (Hz) modulated at a subsonic rate between 4 and 5 hertz.

**1910.1096(f)(1)(ii)** The signal generator shall not be less than 75 decibels at every location where an individual may be present whose immediate, rapid, and complete evacuation is essential.

**1910.1096(f)(1)(iii)** A sufficient number of signal units shall be installed such that the requirements of paragraph (f)(1)(ii) of this section are met at every location where an individual may be present whose immediate, rapid, and complete evacuation is essential.

**1910.1096(f)(1)(iv)** The signal shall be unique in the plant or facility in which it is installed.

**1910.1096(f)(1)(v)** The minimum duration of the signal shall be sufficient to insure that all affected persons hear the signal.

**1910.1096(f)(1)(vi)** The signal-generating system shall respond automatically to an initiating event without requiring any human action to sound the signal.

**1910.1096(f)(2)** *Design objectives.* (i) The signal-generating system shall be designed to incorporate components which enable the system to produce the desired signal each time it is activated within one-half second of activation.

**1910.1096(f)(2)(ii)** The signal-generating system shall be provided with an automatically activated secondary power supply which is adequate to simultaneously power all emergency equipment to which it is connected, if operation during power failure is necessary, except in those systems using batteries as the primary source of power.

**1910.1096(f)(2)(iii)** All components of the signal-generating system shall be located to provide maximum practicable protection against damage in case of fire, explosion, corrosive atmosphere, or other environmental extremes consistent with adequate system performance.

**1910.1096(f)(2)(iv)** The signal-generating system shall be designed with the minimum number of components necessary to make it function as intended, and should utilize components which do not require frequent servicing such as lubrication or cleaning.

**1910.1096(f)(2)(v)** Where several activating devices feed activating information to a central signal generator, failure of any activating device shall not render the signal-generator system inoperable to activating information from the remaining devices.

**1910.1096(f)(2)(vi)** The signal-generating system shall be designed to enhance the probability that alarm occurs only when immediate evacuation is warranted. The number of false alarms shall not be so great that the signal will come to be disregarded and shall be low enough to minimize personal injuries or excessive property damage that might result from such evacuation.

**1910.1096(f)(3)** *Testing.* (i) Initial tests, inspections, and checks of the signal-generating system shall be made to verify that the fabrication and installation were made in accordance with design plans and specifications and to develop a thorough knowledge of the performance of the system and all components under normal and hostile conditions.

**1910.1096(f)(3)(ii)** Once the system has been placed in service, periodic tests, inspections, and checks shall be made to minimize the possibility of malfunction.

**1910.1096(f)(3)(iii)** Following significant alterations or revisions to the system, tests and checks similar to the initial installation tests shall be made.

**1910.1096(f)(3)(iv)** Tests shall be designed to minimize hazards while conducting the tests.

**1910.1096(f)(3)(v)** Prior to normal operation the signal-generating system shall be checked physically and functionally to assure reliability and to demonstrate accuracy and performance. Specific tests shall include:

**1910.1096(f)(3)(v)*(a)*** All power sources.

**1910.1096(f)(3)(v)*(b)*** Calibration and calibration stability.

**1910.1096(f)(3)(v)*(c)*** Trip levels and stability.

**1910.1096(f)(3)(v)*(d)*** Continuity of function with loss and return of required services such as AC or DC power, air pressure, etc.

**1910.1096(f)(3)(v)*(e)*** All indicators.

**1910.1096(f)(3)(v)*(f)*** Trouble indicator circuits and signals, where used.

**1910.1096(f)(3)(v)*(g)*** Air pressure (if used)

**1910.1096(f)(3)(v)*(h)*** Determine that sound level of the signal is within the limit of paragraph (f)(1)(ii) of this section at all points that require immediate evacuation.

**1910.1096(f)(3)(vi)** In addition to the initial startup and operating tests, periodic scheduled performance tests and status checks must be made to insure that the system is at all times operating within design limits and capable of the required response. Specific periodic tests or checks or both shall include:

**1910.1096(f)(3)(vi)*(a)*** Adequacy of signal activation device.

**1910.1096(f)(3)(vi)*(b)*** All power sources.

**1910.1096(f)(3)(vi)*(c)*** Function of all alarm circuits and trouble indicator circuits including trip levels.

**1910.1096(f)(3)(vi)*(d)*** Air pressure (if used).

**1910.1096(f)(3)(vi)*(e)*** Function of entire system including operation without power where required.

1910.1096(f)(3)(vi)*(f)* Complete operational tests including sounding of the signal and determination that sound levels are adequate.

1910.1096(f)(3)(vii) Periodic tests shall be scheduled on the basis of need, experience, difficulty, and disruption of operations. The entire system should be operationally tested at least quarterly.

1910.1096(f)(3)(viii) All employees whose work may necessitate their presence in an area covered by the signal shall be made familiar with the actual sound of the signal—preferably as it sounds at their work location. Before placing the system into operation, all employees normally working in the area shall be made acquainted with the signal by actual demonstration at their work locations.

1910.1096(g) ***Exceptions from posting requirements.*** Notwithstanding the provisions of paragraph (e) of this section:

1910.1096(g)(1) A room or area is not required to be posted with a caution sign because of the presence of a sealed source, provided the radiation level 12 inches from the surface of the source container or housing does not exceed 5 millirem per hour.

1910.1096(g)(2) Rooms or other areas in onsite medical facilities are not required to be posted with caution signs because of the presence of patients containing radioactive material, provided that there are personnel in attendance who shall take the precautions necessary to prevent the exposure of any individual to radiation or radioactive material in excess of the limits established in the provisions of this section.

1910.1096(g)(3) Caution signs are not required to be posted at areas or rooms containing radioactive materials for periods of less than 8 hours: *Provided*, That

1910.1096(g)(3)(i) The materials are constantly attended during such periods by an individual who shall take the precautions necessary to prevent the exposure of any individual to radiation or radioactive materials in excess of the limits established in the provisions of this section; and

1910.1096(g)(3)(ii) Such area or room is subject to the employer's control.

1910.1096(h) ***Exemptions for radioactive materials packaged for shipment.*** Radioactive materials packaged and labeled in accordance with regulations of the Department of Transportation published in 49 CFR Chapter I, are exempt from the labeling and posting requirements of this subpart during shipment, provided that the inside containers are labeled in accordance with the provisions of paragraph (e) of this section.

1910.1096(i) ***Instruction of personnel, posting.*** (1) Employers regulated by the Nuclear Regulatory Commission shall be governed by 10 CFR part 20 standards. Employers in a State named in paragraph (p)(3) of this section shall be governed by the requirements of the laws and regulations of that State. All other employers shall be regulated by the following:

1910.1096(i)(2) All individuals working in or frequenting any portion of a radiation area shall be informed of the occurrence of radioactive materials or of radiation in such portions of the radiation area; shall be instructed in the safety problems associated with exposure to such materials or radiation and in precautions or devices to minimize exposure; shall be instructed in the applicable provisions of this section for the protection of employees from exposure to radiation or radioactive materials; and shall be advised of reports of radiation exposure which employees may request pursuant to the regulations in this section.

1910.1096(i)(3) Each employer to whom this section applies shall post a current copy of its provisions and a copy of the operating procedures applicable to the work conspicuously in such locations as to insure that employees working in or frequenting radiation areas will observe these documents on the way to and from their place of employment, or shall keep such documents available for examination of employees upon request.

1910.1096(j) ***Storage of radioactive materials.*** Radioactive materials stored in a nonradiation area shall be secured against unauthorized removal from the place of storage.

1910.1096(k) ***Waste disposal.*** No employer shall dispose of radioactive material except by transfer to an authorized recipient, or in a manner approved by the Nuclear Regulatory Commission or a State named in paragraph (p)(3) of this section.

1910.1096(l) ***Notification of incidents***—(1) *Immediate notification.* Each employer shall immediately notify the Assistant Secretary of Labor or his duly authorized representative, for employees not protected by the Nuclear Regulatory Commission by means of 10 CFR part 20; paragraph (p)(2) of this section, or the requirements of the laws and regulations of States named in paragraph (p)(3) of this section, by telephone or telegraph of any incident involving radiation which may have caused or threatens to cause:

1910.1096(l)(1)(i) Exposure of the whole body of any individual to 25 rems or more of radiation; exposure of the skin of the whole body of any individual to 150 rems or more of radiation; or exposure of the feet, ankles, hands, or forearms of any individual to 375 rems or more of radiation; or

1910.1096(l)(1)(ii) The release of radioactive material in concentrations which, if averaged over a period of 24 hours, would exceed 5,000 times the limit specified for such materials in Table II of appendix B to 10 CFR part 20.

1910.1096(l)(2) *Twenty-four hour notification.* Each employer shall within 24 hours following its occurrence notify the Assistant Secretary of Labor or his duly authorized representative for employees not protected by the Nuclear Regulatory Commission by means of 10 CFR part 20; paragraph (p)(2) of this section, or the requirements of the laws and applicable regulations of States named in paragraph (p)(3) of this section, by telephone or telegraph of any incident involving radiation which may have caused or threatens to cause:

1910.1096(l)(2)(i) Exposure of the whole body of any individual to 5 rems or more of radiation; exposure of the skin of the whole body of any individual to 30 rems or more of radiation; or exposure of the feet, ankles, hands, or forearms to 75 rems or more of radiation; or

1910.1096(l)(2)(ii) [Reserved]

1910.1096(m) ***Reports of overexposure and excessive levels and concentrations.*** (1) In addition to any notification required by paragraph (l) of this section each employer shall make a report in writing within 30 days to the Assistant Secretary of Labor or his duly authorized representative, for employees not protected by the Nuclear Regulatory Commission by means of 10 CFR part 20; or under paragraph (p)(2) of this section, or the requirements of the laws and regulations of States named in paragraph (p)(3) of this section, of each exposure of an individual to radiation or concentrations of radioactive material in excess of any applicable limit in this section. Each report required under this paragraph shall describe the extent of exposure of persons to radiation or to radioactive material; levels of radiation and concentration of radioactive material involved, the cause of the exposure, levels of concentrations; and corrective steps taken or planned to assure against a recurrence.

1910.1096(m)(2) In any case where an employer is required pursuant to the provisions of this paragraph to report to the U.S. Department of Labor any exposure of an individual to radiation or to concentrations of radioactive material, the employer shall also notify such individual of the nature and extent of exposure. Such notice shall be in writing and shall contain the following statement: "You should preserve this report for future reference."

1910.1096(n) ***Records.*** (1) Every employer shall maintain records of the radiation exposure of all employees for whom personnel monitoring is required under paragraph (d) of this section and advise each of his employees of his individual exposure on at least an annual basis.

1910.1096(n)(2) Every employer shall maintain records in the same units used in tables in paragraph (b) of this section and appendix B to 10 CFR part 20.

1910.1096(o) ***Disclosure to former employee of individual employee's record.*** (1) At the request of a former employee an employer shall furnish to the employee a report of the employee's exposure to radiation as shown in records maintained by the employer pursuant to paragraph (n)(1) of this section. Such report shall be furnished within 30 days from the time the request is made, and shall cover each calendar quarter of the individual's employment involving exposure to radiation or such lesser period as may be requested by the employee. The report shall also include the results of any calculations and analysis of radioactive material deposited in the body of the employee. The report shall be in writing and contain the following statement: "You should preserve this report for future reference."

1910.1096(o)(2) [Reserved]

1910.1096(p) ***Nuclear Regulatory Commission licensees—NRC contractors operating NRC plants and facilities—NRC Agreement***

***State licensees or registrants.*** (1) Any employer who possesses or uses source material, byproduct material, or special nuclear material, as defined in the Atomic Energy Act of 1954, as amended, under a license issued by the Nuclear Regulatory Commission and in accordance with the requirements of 10 CFR part 20 shall be deemed to be in compliance with the requirements of this section with respect to such possession and use.

1910.1096(p)(2) NRC contractors operating NRC plants and facilities: Any employer who possesses or uses source material, byproduct material, special nuclear material, or other radiation sources under a contract with the Nuclear Regulatory Commission for the operation of NRC plants and facilities and in accordance with the standards, procedures, and other requirements for radiation protection established by the Commission for such contract pursuant to the Atomic Energy Act of 1954 as amended (42 U.S.C. 2011 et seq.), shall be deemed to be in compliance with the requirements of this section with respect to such possession and use.

1910.1096(p)(3) NRC-agreement State licensees or registrants:

1910.1096(p)(3)(i) *Atomic Energy Act sources.* Any employer who possesses or uses source material, byproduct material, or special nuclear material, as defined in the Atomic Energy Act of 1954, as amended (42 U.S.C. 2011 et seq.), and has either registered such sources with, or is operating under a license issued by, a State which has an agreement in effect with the Nuclear Regulatory Commission pursuant to section 274(b) (42 U.S.C. 2021(b)) of the Atomic Energy Act of 1954, as amended, and in accordance with the requirements of that State's laws and regulations shall be deemed to be in compliance with the radiation requirements of this section, insofar as his possession and use of such material is concerned, unless the Secretary of Labor, after conference with the Nuclear Regulatory Commission, shall determine that the State's program for control of these radiation sources is incompatible with the requirements of this section. Such agreements currently are in effect only in the States of Alabama, Arkansas, California, Kansas, Kentucky, Florida, Mississippi, New Hampshire, New York, North Carolina, Texas, Tennessee, Oregon, Idaho, Arizona, Colorado, Louisiana, Nebraska, Washington, Maryland, North Dakota, South Carolina, and Georgia.

1910.1096(p)(3)(ii) *Other sources.* Any employer who possesses or uses radiation sources other than source material, byproduct material, or special nuclear material, as defined in the Atomic Energy Act of 1954, as amended (42 U.S.C. 2011 et seq.), and has either registered such sources with, or is operating under a license issued by a State which has an agreement in effect with the Nuclear Regulatory Commission pursuant to section 274(b) (42 U.S.C. 2021(b)) of the Atomic Energy Act of 1954, as amended, and in accordance with the requirements of that State's laws and regulations shall be deemed to be in compliance with the radiation requirements of this section, insofar as his possession and use of such material is concerned, provided the State's program for control of these radiation sources is the subject of a currently effective determination by the Assistant Secretary of Labor that such program is compatible with the requirements of this section. Such determinations currently are in effect only in the States of Alabama, Arkansas, California, Kansas, Kentucky, Florida, Mississippi, New Hampshire, New York, North Carolina, Texas, Tennessee, Oregon, Idaho, Arizona, Colorado, Louisiana, Nebraska, Washington, Maryland, North Dakota, South Carolina, and Georgia.

[39 FR 23502, June 27, 1974, as amended at 43 FR 49746, Oct. 24, 1978; 43 FR 51759, Nov. 7, 1978; 49 FR 18295, Apr. 30, 1984; 58 FR 35309, June 30, 1993. Redesignated at 61 FR 31430, June 20, 1996]

## §1926.54 Nonionizing radiation.

1926.54(a) Only qualified and trained employees shall be assigned to install, adjust, and operate laser equipment.

1926.54(b) Proof of qualification of the laser equipment operator shall be available and in possession of the operator at all times.

1926.54(c) Employees, when working in areas in which a potential exposure to direct or reflected laser light greater than 0.005 watts (5 milliwatts) exists, shall be provided with antilaser eye protection devices as specified in subpart E of this part.

1926.54(d) Areas in which lasers are used shall be posted with standard laser warning placards.

1926.54(e) Beam shutters or caps shall be utilized, or the laser turned off, when laser transmission is not actually required. When the laser is left unattended for a substantial period of time, such as during lunch hour, overnight, or at change of shifts, the laser shall be turned off.

1926.54(f) Only mechanical or electronic means shall be used as a detector for guiding the internal alignment of the laser.

1926.54(g) The laser beam shall not be directed at employees.

1926.54(h) When it is raining or snowing, or when there is dust or fog in the air, the operation of laser systems shall be prohibited where practicable; in any event, employees shall be kept out of range of the area of source and target during such weather conditions.

1926.54(i) Laser equipment shall bear a label to indicate maximum output.

1926.54(j) Employees shall not be exposed to light intensities above:

1926.54(j)(1) Direct staring: 1 micro-watt per square centimeter;

1926.54(j)(2) Incidental observing: 1 milliwatt per square centimeter;

1926.54(j)(3) Diffused reflected light: 2 1/2 watts per square centimeter.

1926.54(k) Laser unit in operation should be set up above the heads of the employees, when possible.

1926.54(l) Employees shall not be exposed to microwave power densities in excess of 10 milliwatts per square centimeter.

## §1926.55 Gases, vapors, fumes, dusts, and mists.

1926.55(a) Exposure of employees to inhalation, ingestion, skin absorption, or contact with any material or substance at a concentration above those specified in the "Threshold Limit Values of Airborne Contaminants for 1970" of the American Conference of Governmental Industrial Hygienists, shall be avoided. See Appendix A to this section.

1926.55(b) To achieve compliance with paragraph (a) of this section, administrative or engineering controls must first be implemented whenever feasible. When such controls are not feasible to achieve full compliance, protective equipment or other protective measures shall be used to keep the exposure of employees to air contaminants within the limits prescribed in this section. Any equipment and technical measures used for this purpose must first be approved for each particular use by a competent industrial hygienist or other technically qualified person. Whenever respirators are used, their use shall comply with §1926.103.

1926.55(c) Paragraphs (a) and (b) of this section do not apply to the exposure of employees to airborne asbestos, tremolite, anthophyllite, or actinolite dust. Whenever any employee is exposed to airborne asbestos, tremolite, anthophyllite, or actinolite dust, the requirements of §1910.1101 or §1926.58 of this title shall apply.

1926.55(d) Paragraphs (a) and (b) of this section do not apply to the exposure of employees to formaldehyde. Whenever any employee is exposed to formaldehyde, the requirements of §1910.1048 of this title shall apply.

## Appendix A to §1926.55— 1970 American Conference of Governmental Industrial Hygienists' Threshold Limit Values of Airborne Contaminants

### THRESHOLD LIMIT VALUES OF AIRBORNE CONTAMINANTS FOR CONSTRUCTION

| Substance | CAS No.[d] | ppm[a] | mg/m[3] [b] | Skin Designation |
|---|---|---|---|---|
| Abate, see Temephos. | | | | |
| Acetaldehyde | 75-07-0 | 200 | 360 | |
| Acetic acid | 64-19-7 | 10 | 25 | |
| Acetic anhydride | 108-24-7 | 5 | 20 | |
| Acetone | 67-64-1 | 1000 | 2400 | |
| Acetonitrile | 75-05-8 | 40 | 70 | |
| 2-Acetylaminofluorine; see 1926.1114 | 53-96-3 | | | |
| Acetylene | 74-86-2 | E | | |
| Acetylene dichloride; see 1,2-Dichloroethylene. | | | | |
| Acetylene tetrabromide | 79-27-6 | 1 | 14 | |
| Acrolein | 107-02-8 | 0.1 | 0.25 | |
| Acrylamide | 79-06-1 | ........ | 0.3 | X |
| Acrylonitrile; see 1926.1145 | 107-13-1 | | | |
| Aldrin | 309-00-2 | ........ | 0.25 | X |
| Allyl alcohol | 107-18-6 | 2 | 5 | X |
| Allyl chloride | 107-05-1 | 1 | 3 | |
| Allyl glycidyl ether (AGE) | 106-92-3 | (C)10 | (C)45 | |
| Allyl propyl disulfide | 2179-59-1 | 2 | 12 | |
| alpha-Alumina | 1344-28-1 | | | |
| Total dust | | ........ | | |
| Respirable fraction | | ........ | | |
| Alundum; see alpha-Alumina | | | | |
| 4-Aminodiphenyl; see 1926.1111 | 92-67-1 | | | |
| 2-Aminoethanol; see Ethanolamine. | | | | |
| 2-Aminopyridine | 504-29-0 | 0.5 | 2 | |
| Ammonia | 7664-41-7 | 50 | 35 | |
| Ammonium sulfamate | 7773-06-0 | | | |
| Total dust | | ........ | 15 | |
| Respirable fraction | | ........ | 5 | |
| n-Amyl acetate | 628-63-7 | 100 | 525 | |
| sec-Amyl acetate | 626-38-0 | 125 | 650 | |
| Aniline and homologs | 62-53-3 | 5 | 19 | X |
| Anisidine (o-, p-isomers) | 29191-52-4 | ........ | 0.5 | X |
| Antimony and compounds (as Sb) | 7440-36-0 | ........ | 0.5 | |
| ANTU (alpha Naphthylthiourea) | 86-88-4 | ........ | 0.3 | |
| Argon | 7440-37-1 | E | | |
| Arsenic, inorganic compounds (as As); see 1926.1118 | 7440-38-2 | | | |
| Arsenic, organic compounds (as As) | 7440-38-2 | ........ | 0.5 | |
| Arsine | 7784-42-1 | 0.05 | 0.2 | |
| Asbestos; see 1926.58 [now 1926.1101.—CCH.] | | | | |
| Azinphos-methyl | 86-50-0 | ........ | 0.2 | X |
| Barium, soluble compounds (as Ba) | 7440-39-3 | ........ | 0.5 | |
| Benzene[g]; see 1926.1128 | 71-43-2 | | | |
| Benzidine; see 1926.1110 | 92-87-5 | | | |
| p-Benzoquinone; see Quinone. | | | | |
| Benzo(a)pyrene; see Coal tar pitch volatiles. | | | | |
| Benzoyl peroxide | 94-36-0 | ........ | 5 | |
| Benzyl chloride | 100-44-7 | 1 | 5 | |
| Beryllium and beryllium compounds (as Be) | 7440-41-7 | | 0.002 | |
| Biphenyl; see Diphenyl. | | | | |
| Bisphenol A; see Diglycidyl ether. | | | | |
| Boron oxide | 1303-86-2 | | | |
| Total dust | | ........ | 15 | |
| Boron tribromide | 10294-33-4 | 1 | 10 | |
| Boron trifluoride | 7637-07-2 | (C)1 | (C)3 | |
| Bromine | 7726-95-6 | 0.1 | 0.7 | |
| Bromine pentafluoride | 7789-30-2 | 0.1 | 0.7 | |
| Bromoform | 75-25-2 | 0.5 | 5 | X |
| Butadiene (1,3-Butadiene); See 29 CFR 1910.1051; 29 CFR 1910.19(*l*) | 106-99-0 | 1 ppm/5 ppm STEL | | |
| Butanethiol; see Butyl mercaptan. | | | | |
| 2-Butanone (Methyl ethyl ketone) | 78-93-3 | 200 | 590 | |
| 2-Butoxyethanol | 111-76-2 | 50 | 240 | X |
| n-Butyl-acetate | 123-86-4 | 150 | 710 | |
| sec-Butyl acetate | 105-46-4 | 200 | 950 | |
| tert-Butyl acetate | 540-88-5 | 200 | 950 | |
| n-Butyl alcohol | 71-36-3 | 100 | 300 | |
| sec-Butyl alcohol | 78-92-2 | 150 | 450 | |
| tert-Butyl alcohol | 75-65-0 | 100 | 300 | |
| Butylamine | 109-73-9 | (C)5 | (C)15 | X |
| tert-Butyl chromate (as $CrO_3$) | 1189-85-1 | ........ | (C)0.1 | X |
| n-Butyl glycidyl ether (BGE) | 2426-08-6 | 50 | 270 | |
| Butyl mercaptan | 109-79-5 | .05 | 1.5 | |
| p-tert-Butyltoluene | 98-51-1 | 10 | 60 | |
| Cadmium (as Cd); see 1926.1127 | 7440-43-9 | | | |

## THRESHOLD LIMIT VALUES OF AIRBORNE CONTAMINANTS FOR CONSTRUCTION

| Substance | CAS No.[d] | ppm[a] | mg/m$^3$ [b] | Skin Designation |
|---|---|---|---|---|
| Calcium carbonate | 1317-65-3 | | | |
| Total dust | | ........ | | |
| Respirable fraction | | ........ | | |
| Calcium oxide | 1305-78-8 | ........ | 5 | |
| Calcium sulfate | 7778-18-9 | | | |
| Total dust | | ........ | 15 | |
| Respirable fraction | | ........ | 5 | |
| Camphor, synthetic | 76-22-2 | ........ | 2 | |
| Carbaryl (Sevin) | 63-25-2 | ........ | 5 | |
| Carbon black | 1333-86-4 | ........ | 3.5 | |
| Carbon dioxide | 124-38-9 | 5000 | 9000 | |
| Carbon disulfide | 75-15-0 | 20 | 60 | X |
| Carbon monoxide | 630-08-0 | 50 | 55 | |
| Carbon tetrachloride | 56-23-5 | 10 | 65 | X |
| Cellulose | 9004-34-6 | | | |
| Total dust | | ........ | | |
| Respirable fraction | | ........ | | |
| Chlordane | 57-74-9 | ........ | 0.5 | X |
| Chlorinated camphene | 8001-35-2 | ........ | 0.5 | X |
| Chlorinated diphenyl oxide | 55720-99-5 | ........ | 0.5 | |
| Chlorine | 7782-50-5 | 1 | 3 | |
| Chlorine dioxide | 10049-04-4 | 0.1 | 0.3 | |
| Chlorine trifluoride | 7790-91-2 | (C)0.1 | (C)0.4 | |
| Chloroacetaldehyde | 107-20-0 | (C)1 | (C)3 | |
| a-Chloroacetophenone (Phenacyl chloride) | 532-27-4 | 0.05 | 0.3 | |
| Chlorobenzene | 108-90-7 | 75 | 350 | |
| o-Chlorobenzylidene malononitrile | 2698-41-1 | 0.05 | 0.4 | |
| Chlorobromomethane | 74-97-5 | 200 | 1050 | |
| 2-Chloro-1,3-butadiene; see beta-Chloroprene. | | | | |
| Chlorodiphenyl (42% Chlorine) (PCB) | 53469-21-9 | ........ | 1 | X |
| Chlorodiphenyl (54% Chlorine) (PCB) | 11097-69-1 | ........ | 0.5 | X |
| 1-Chloro-2,3-epoxypropane; see Epichlorohydrin. | | | | |
| 2-Chloroethanol; see Ethylene chlorohydrin. | | | | |
| Chloroethylene; see Vinyl chloride. | | | | |
| Chloroform (Trichloromethane) | 67-66-3 | (C) 50 | (C) 240 | |
| bis(Chloromethyl) ether; see 1926.1108 | 542-88-1 | | | |
| Chloromethyl methyl ether; see 1926.1106 | 107-30-2 | | | |
| 1-Chloro-1-nitropropane | 600-25-9 | 20 | 100 | |
| Chloropicrin | 76-06-2 | 0.1 | 0.7 | |
| beta-Chloroprene | 126-99-8 | 25 | 90 | X |
| Chromic acid and chromates (as $CrO_3$) | Varies with compound | | 0.1 | |
| Chromium (II) compounds. (as Cr) | 7440-47-3 | ........ | 0.5 | |
| Chromium (III) compounds. (as Cr) | 7440-47-3 | ........ | 0.5 | |
| Chromium metal and insol. salts (as Cr) | 7440-47-3 | ........ | 1 | |
| Chrysene; see Coal tar pitch volatiles. | | | | |
| Coal tar pitch volatiles (benzene soluble fraction), anthracene, BaP, phenanthrene, acridine, chrysene, pyrene. | 65996-93-2 | ........ | 0.2 | |
| Cobalt metal, dust, and fume (as Co) | 7440-48-4 | ........ | 0.1 | |
| Coke oven emissions; see 1926.1129. | | | | |
| Copper | 7440-50-8 | | | |
| Fume (as Cu) | | ........ | 0.1 | |
| Dusts and mists (as Cu) | | ........ | 1 | |
| Corundum; see Emery. | | | | |
| Cotton dust (raw) | | ........ | 1 | |
| Crag herbicide (Sesone) | 136-78-7 | | | |
| Total dust | | ........ | 15 | |
| Respirable fraction | | ........ | 5 | |
| Cresol, all isomers | 1319-77-3 | 5 | 22 | X |
| Crotonaldehyde | 123-73-9; 4170-30-3 | 2 | 6 | |
| Cumene | 98-82-8 | 50 | 245 | X |
| Cyanides (as CN) | Varies with compound | ........ | 5 | X |
| Cyanogen | 460-19-5 | 10 | | |
| Cyclohexane | 110-82-7 | 300 | 1050 | |
| Cyclohexanol | 108-93-0 | 50 | 200 | |
| Cyclohexanone | 108-94-1 | 50 | 200 | |
| Cyclohexene | 110-83-8 | 300 | 1015 | |
| Cyclonite | 121-82-4 | ........ | 1.5 | X |
| Cyclopentadiene | 542-92-7 | 75 | 200 | |
| 2,4-D (Dichlorophenoxyacetic acid) | 94-75-7 | ........ | 10 | |
| DDT, see Dichlorodiphenyltrichloroethane | ........ | ........ | ........ | |
| DDVP, see Dichlorvos | ........ | ........ | ........ | |
| Decaborane | 17702-41-9 | 0.05 | 0.3 | X |
| Demeton (Systox) | 8065-48-3 | ........ | 0.1 | X |
| Diacetone alcohol (4-Hydroxy-4-methyl-2-pentanone) | 123-42-2 | 50 | 240 | |
| 1,2-Diaminoethane; see Ethylenediamine. | | | | |

## THRESHOLD LIMIT VALUES OF AIRBORNE CONTAMINANTS FOR CONSTRUCTION

| Substance | CAS No.[d] | ppm[a] | mg/m³ [b] | Skin Designation |
|---|---|---|---|---|
| Diazomethane | 334-88-3 | 0.2 | 0.4 | |
| Diborane | 19287-45-7 | 0.1 | 0.1 | |
| 1,2-Dibromo-3-chloropropane (DBCP); see 1926.1144 | 96-12-8 | | | – |
| 1,2-Dibromoethane; see Ethylene dibromide. | | | | |
| Dibutyl phosphate | 107-66-4 | 1 | 5 | |
| Dibutyl phthalate | 84-74-2 | ........ | 5 | |
| Dichloroacetylene | 7572-29-4 | (C)0.1 | (C)0.4 | |
| o-Dichlorobenzene | 95-50-1 | (C)50 | (C)300 | |
| p-Dichlorobenzene | 106-46-7 | 75 | 450 | |
| 3,3′-Dichlorobenzidine; see 1926.1107 | 91-94-1 | | | |
| Dichlorodifluoromethane | 75-71-8 | 1000 | 4950 | |
| 1,3-Dichloro-5,5-dimethyl hydantoin | 118-52-5 | ........ | 0.2 | |
| Dichlorodiphenyltrichloroethane (DDT) | 50-29-3 | ........ | 1 | X |
| 1,1-Dichloroethane | 75-34-3 | 100 | 400 | |
| 1,2-Dichloroethane; see Ethylene dichloride. | | | | |
| 1,2-Dichloroethylene | 540-59-0 | 200 | 790 | |
| Dichloroethyl ether | 111-44-4 | (C)15 | (C)90 | X |
| Dichloromethane; see Methylene chloride. | | | | |
| Dichloromonofluoromethane | 75-43-4 | 1000 | 4200 | |
| 1,1-Dichloro-1-nitroethane | 594-72-9 | (C)10 | (C)60 | |
| 1,2-Dichloropropane; see Propylene dichloride. | | | | |
| Dichlorotetrafluoroethane | 76-14-2 | 1000 | 7000 | |
| Dichlorvos (DDVP) | 62-73-7 | ........ | 1 | X |
| Dieldrin | 60-57-1 | ........ | 0.25 | X |
| Diethylamine | 109-89-7 | 25 | 75 | |
| 2-Diethylaminoethanol | 100-37-8 | 10 | 50 | X |
| Diethylene triamine | 111-40-0 | (C)10) | C42 | X |
| Diethyl ether; see Ethyl ether. | | | | |
| Difluorodibromomethane | 75-61-6 | 100 | 860 | |
| Diglycidyl ether (DGE) | 2238-07-5 | (C)0.5 | (C)2.8 | |
| Dihydroxybenzene; see Hydroquinone. | | | | |
| Diisobutyl ketone | 108-83-8 | 50 | 290 | |
| Diisopropylamine | 108-18-9 | 5 | 20 | X |
| 4-Dimethylaminoazobenzene; see 1926.1115 | 60-11-7 | | | |
| Dimethoxymethane; see Methylal. | | | | |
| Dimethyl acetamide | 127-19-5 | 10 | 35 | X |
| Dimethylamine | 124-40-3 | 10 | 18 | |
| Dimethylaminobenzene; see Xylidine. | | | | |
| Dimethylaniline (N,N-Dimethylaniline) | 121-69-7 | 5 | 25 | X |
| Dimethylbenzene; see Xylene. | | | | |
| Dimethyl-1,2-dibromo-2,2-dichloroethyl phosphate | 300-76-5 | ........ | 3 | |
| Dimethylformamide | 68-12-2 | 10 | 30 | X |
| 2,6-Dimethyl-4-heptanone; see Diisobutyl ketone. | | | | |
| 1,1-Dimethylhydrazine | 57-14-7 | 0.5 | 1 | X |
| Dimethylphthalate | 131-11-3 | ........ | 5 | |
| Dimethyl sulfate | 77-78-1 | 1 | 5 | X |
| Dinitrobenzene (all isomers) | | | 1 | X |
| (ortho) | 528-29-0 | | | |
| (meta) | 99-65-0 | | | |
| (para) | 100-25-4 | | | |
| Dinitro-o-cresol | 534-52-1 | ........ | 0.2 | X |
| Dinitrotoluene | 25321-14-6 | ........ | 1.5 | X |
| Dioxane (Diethylene dioxide) | 123-91-1 | 100 | 360 | X |
| Diphenyl (Biphenyl) | 92-52-4 | 0.2 | 1 | |
| Diphenylamine | 122-39-4 | ........ | 10 | |
| Diphenylmethane diisocyanate; see Methylene bisphenyl isocyanate. | | | | |
| Dipropylene glycol methyl ether | 34590-94-8 | 100 | 600 | X |
| Di-sec octyl phthalate (Di-(2-ethylhexyl) phthalate) | 117-81-7 | ........ | 5 | |
| Emery | 12415-34-8 | | | |
| Total dust | | ........ | | |
| Respirable fraction | | ........ | | |
| Endosulfan | 115-29-7 | ........ | 0.1 | X |
| Endrin | 72-20-8 | ........ | 0.1 | X |
| Epichlorohydrin | 106-89-8 | 5 | 19 | X |
| EPN | 2104-64-5 | ........ | 0.5 | X |
| 1,2-Epoxypropane; see Propylene oxide. | | | | |
| 2,3-Epoxy-1-propanol; see Glycidol. | | | | |
| Ethane | 74-84-0 | | | |
| Ethanethiol; see Ethyl mercaptan. | | | | |
| Ethanolamine | 141-43-5 | 3 | 6 | |
| 2-Ethoxyethanol (Cellosolve) | 110-80-5 | 200 | 740 | X |
| 2-Ethoxyethyl acetate (Cellosolve acetate) | 111-15-9 | 100 | 540 | X |
| Ethyl acetate | 141-78-6 | 400 | 1400 | |
| Ethyl acrylate | 140-88-5 | 25 | 100 | X |
| Ethyl alcohol (Ethanol) | 64-17-5 | 1000 | 1900 | |
| Ethylamine | 75-04-7 | 10 | 18 | |
| Ethyl amyl ketone (5-Methyl-3-heptanone) | 541-85-5 | 25 | 130 | |
| Ethyl benzene | 100-41-4 | 100 | 435 | |
| Ethyl bromide | 74-96-4 | 200 | 890 | |
| Ethyl butyl ketone (3-Heptanone) | 106-35-4 | 50 | 230 | |

## THRESHOLD LIMIT VALUES OF AIRBORNE CONTAMINANTS FOR CONSTRUCTION

| Substance | CAS No.[d] | ppm[a] | mg/m$^3$ [b] | Skin Designation |
|---|---|---|---|---|
| Ethyl chloride | 75-00-3 | 1000 | 2600 | |
| Ethyl ether | 60-29-7 | 400 | 1200 | |
| Ethyl formate | 109-94-4 | 100 | 300 | |
| Ethyl mercaptan | 75-08-1 | 0.5 | 1 | |
| Ethyl silicate | 78-10-4 | 100 | 850 | |
| Ethylene | E | | | |
| Ethylene chlorohydrin | 107-07-3 | 5 | 16 | X |
| Ethylenediamine | 107-15-3 | 10 | 25 | |
| Ethylene dibromide | 106-93-4 | (C)25 | (C)90 | |
| Ethylene dichloride (1,2-Dichloroethane) | 107-06-2 | 50 | 200 | |
| Ethylene glycol dinitrate | 628-96-6 | (C)0.2 | (C)1 | X |
| Ethylene glycol methyl acetate; see Methyl cellosolve acetate. | | | | |
| Ethyleneimine; see 1926.1112 | 151-56-4 | | | |
| Ethylene oxide; see 1926.1147 | 75-21-8 | | | |
| Ethylidene chloride; see 1,1-Dichloroethane. | | | | |
| N-Ethylmorpholine | 100-74-3 | 20 | 94 | X |
| Ferbam | 14484-64-1 | | | |
| Total dust | | ......... | 15 | |
| Ferrovanadium dust | 12604-58-9 | ......... | 1 | |
| Fibrous glass | | | 15 | |
| Total dust | | | | |
| Respirable fraction | | | | |
| Fluorides (as F) | Varies with compound | ......... | 2.5 | |
| Fluorine | 7782-41-4 | 0.1 | 0.2 | |
| Fluorotrichloromethane (Trichlorofluoromethanes) | 75-69-4 | 1000 | 5600 | |
| Formaldehyde; see 1926.1148 | 50-00-0 | | | |
| Formic acid | 64-18-6 | 5 | 9 | |
| Furfural | 98-01-1 | 5 | 20 | X |
| Furfuryl alcohol | 98-00-0 | 50 | 200 | |
| Gasoline | 8006-61-9 | | A[3] | |
| Glycerin (mist) | 56-81-5 | | | |
| Total dust | | ......... | | |
| Respirable fraction | | ......... | | |
| Glycidol | 556-52-5 | 50 | 150 | |
| Glycol monoethyl ether; see 2-Ethoxyethanol. | | | | |
| Graphite, natural, respirable dust | 7782-42-5 | ([2]) | ([2]) | |
| Graphite, synthetic | | | | |
| Total dust | | | | |
| Respirable fraction | | ......... | | |
| Guthion; see Azinphos methyl. | | ......... | | |
| Gypsum | 13397-24-5 | | | |
| Total dust | | ......... | | |
| Respirable fraction | | ......... | | |
| Hafnium | 7440-58-6 | ......... | 0.5 | |
| Helium | 7440-59-7 | E | | |
| Heptachlor | 76-44-8 | ......... | 0.5 | X |
| Heptane (n-Heptane) | 142-82-5 | 500 | 2000 | |
| Hexachloroethane | 67-72-1 | 1 | 10 | X |
| Hexachloronaphthalene | 1335-87-1 | ......... | 0.2 | X |
| n-Hexane | 110-54-3 | 500 | 1800 | |
| 2-Hexanone (Methyl n-butyl ketone) | 591-78-6 | 100 | 410 | |
| Hexone (Methyl isobutyl ketone) | 108-10-1 | 100 | 410 | |
| sec-Hexyl acetate | 108-84-9 | 50 | 300 | |
| Hydrazine | 302-01-2 | 1 | 1.3 | X |
| Hydrogen | 1333-74-0 | E | | |
| Hydrogen bromide | 10035-10-6 | 3 | 10 | |
| Hydrogen chloride | 7647-01-0 | (C)5 | (C)7 | |
| Hydrogen cyanide | 74-90-8 | 10 | 11 | X |
| Hydrogen fluoride (as F) | 7664-39-3 | 3 | 2 | |
| Hydrogen peroxide | 7722-84-1 | 1 | 1.4 | |
| Hydrogen selenide (as Se) | 7783-07-5 | 0.05 | 0.2 | – |
| Hydrogen sulfide | 7783-06-4 | | ([2]) | |
| Hydroquinone | 123-31-9 | ......... | 2 | |
| Indene | 95-13-6 | 10 | 45 | |
| Indium and compounds (as In) | 7440-74-6 | | 0.1 | |
| Iodine | 7553-56-2 | (C)0.1 | (C)1 | |
| Iron oxide fume | 1309-37-1 | ......... | 10 | |
| Iron salts (soluble) (as Fe) | Varies with compound | | 1 | |
| Isoamyl acetate | 123-92-2 | 100 | 525 | |
| Isoamyl alcohol (primary and secondary) | 123-51-3 | 100 | 360 | |
| Isobutyl acetate | 110-19-0 | 150 | 700 | |
| Isobutyl alcohol | 78-83-1 | 100 | 300 | |
| Isophorone | 78-59-1 | 25 | 140 | |
| Isopropyl acetate | 108-21-4 | 250 | 950 | |
| Isopropyl alcohol | 67-63-0 | 400 | 980 | |
| Isopropylamine | 75-31-0 | 5 | 12 | |
| Isopropyl ether | 108-20-3 | 500 | 2100 | |
| Isopropyl glycidyl ether (IGE) | 4016-14-2 | 50 | 240 | |

## THRESHOLD LIMIT VALUES OF AIRBORNE CONTAMINANTS FOR CONSTRUCTION

| Substance | CAS No.[d] | ppm[a] | mg/m³ [b] | Skin Designation |
|---|---|---|---|---|
| Kaolin | 1332-58-7 | | | |
| Total dust | | ........ | | |
| Respirable fraction | | ........ | | |
| Ketene | 463-51-4 | 0.5 | 0.9 | |
| Lead, inorganic (as Pb); see 1926.62 | 7439-92-1 | | | |
| Limestone | 1317-65-3 | | | |
| Total dust | | ........ | | |
| Respirable fraction | | ........ | | |
| Lindane | 58-89-9 | ........ | 0.5 | X |
| Lithium hydride | 7580-67-8 | ........ | 0.025 | |
| L.P.G. (Liquefied petroleum gas) | 68476-85-7 | 1000 | 1800 | |
| Magnesite | 546-93-0 | | | |
| Total dust | | ........ | | |
| Respirable fraction | | ........ | | |
| Magnesium oxide fume | 1309-48-4 | | | |
| Total particulate | | 15 | – | |
| Malathion | 121-75-5 | | | |
| Total dust | | ........ | 15 | X |
| Maleic anhydride | 108-31-6 | 0.25 | | |
| Manganese compounds (as Mn) | 7439-96-5 | ........ | (C)5 | |
| Manganese fume (as Mn) | 7439-96-5 | ........ | (C)5 | |
| Marble | 1317-65-3 | | | |
| Total dust | | ........ | | |
| Respirable fraction | | ........ | | |
| Mercury (aryl and inorganic) (as Hg) | 7439-97-6 | | 0.1 | |
| Mercury (organo) alkyl compounds (as Hg) | 7439-97-6 | | 0.1 | |
| Mercury (vapor) (as Hg) | 7439-97-6 | | 0.1 | |
| Mesityl oxide | 141-79-7 | 25 | 100 | |
| Methane | 74-82-8 | E | | |
| Methanethiol; see Methyl mercaptan. | | | | |
| Methoxychlor | 72-43-5 | | | |
| Total dust | | ........ | 15 | |
| 2-Methoxyethanol (Methyl cellosolve) | 109-86-4 | 25 | 80 | X |
| 2-Methoxyethyl acetate (Methyl cellosolve acetate) | 110-49-6 | 25 | 120 | X |
| Methyl acetate | 79-20-9 | 200 | 610 | |
| Methyl acetylene (Propyne) | 74-99-7 | 1000 | 1650 | |
| Methyl acetylene-propadiene mixture (MAPP) | | 1000 | 1800 | |
| Methyl acrylate | 96-33-3 | 10 | 35 | X |
| Methylal (Dimethoxy-methane) | 109-87-5 | 1000 | 3100 | |
| Methyl alcohol | 67-56-1 | 200 | 260 | |
| Methylamine | 74-89-5 | 10 | 12 | |
| Methyl amyl alcohol; see Methyl isobutyl carbinol. | | | | |
| Methyl n-amyl ketone | 110-43-0 | 100 | 465 | |
| Methyl bromide | 74-83-9 | (C)20 | (C)80 | X |
| Methyl butyl ketone; see 2-Hexanone. | | | | |
| Methyl cellosolve; see 2-Methoxyethanol. | | | | |
| Methyl cellosolve acetate; see 2-Methoxyethyl acetate. | | | | |
| Methyl chloride | 74-87-3 | 100 | 210 | |
| Methyl chloroform (1,1,1-Trichloroethane) | 71-55-6 | 350 | 1900 | |
| Methylcyclohexane | 108-87-2 | 500 | 2000 | |
| Methylcyclohexanol | 25639-42-3 | 100 | 470 | |
| o-Methylcyclohexanone | 583-60-8 | 100 | 460 | X |
| Methylene chloride: see § 1910.1052 | | | | |
| Methylenedianiline (MDA) | 101-77-9 | ........ | ........ | |
| Methyl ethyl ketone (MEK); see 2-Butanone. | | | | |
| Methyl formate | 107-31-3 | 100 | 250 | |
| Methyl hydrazine (Monomethyl hydrazine) | 60-34-4 | (C)0.2 | (C)0.35 | X |
| Methyl iodide | 74-88-4 | 5 | 28 | X |
| Methyl isoamyl ketone | 110-12-3 | 100 | 475 | |
| Methyl isobutyl carbinol | 108-11-2 | 25 | 100 | X |
| Methyl isobutyl ketone; see Hexone. | | | | |
| Methyl isocyanate | 624-83-9 | 0.02 | 0.05 | X |
| Methyl mercaptan | 74-93-1 | 0.5 | 1 | |
| Methyl methacrylate | 80-62-6 | 100 | 410 | – |
| Methyl propyl ketone; see 2-Pentanone. | | | | |
| Methyl silicate | 681-84-5 | (C) 5 | (C) 30 | |
| alpha-Methyl styrene | 98-83-9 | (C)100 | (C)480 | |
| Methylene bisphenyl isocyanate (MDI) | 101-68-8 | (C)0.02 | (C)0.2 | |
| Mica; see Silicates. | | | | |
| Molybdenum (as Mo) | 7439-98-7 | | | |
| Soluble compounds | | ........ | 5 | |
| Insoluble compounds. | | | | |
| Total dust | | ........ | 15 | |
| Monomethyl aniline | 100-61-8 | 2 | 9 | X |
| Monomethyl hydrazine; see Methyl hydrazine. | | | | |
| Morpholine | 110-91-8 | 20 | 70 | X |
| Naphtha (Coal tar) | 8030-30-6 | 100 | 400 | |
| Naphthalene | 91-20-3 | 10 | 50 | |
| alpha-Naphthylamine; see 1926.1104 | 134-32-7 | | | |
| beta-Naphthylamine; see 1926.1109 | 91-59-8 | | | |

## THRESHOLD LIMIT VALUES OF AIRBORNE CONTAMINANTS FOR CONSTRUCTION

| Substance | CAS No.[d] | ppm[a] | mg/m$^3$ [b] | Skin Designation |
|---|---|---|---|---|
| Neon | 7440-01-9 | E | | |
| Nickel carbonyl (as Ni) | 13463-39-3 | 0.001 | 0.007 | |
| Nickel, metal and insoluble compounds (as Ni) | 7440-02-0 | ......... | 1 | |
| Nickel, soluble compounds (as Ni) | 7440-02-0 | ......... | 1 | |
| Nicotine | 54-11-5 | ......... | 0.5 | X |
| Nitric acid | 7697-37-2 | 2 | 5 | |
| Nitric oxide | 10102-43-9 | 25 | 30 | |
| p-Nitroaniline | 100-01-6 | 1 | 6 | X |
| Nitrobenzene | 98-95-3 | 1 | 5 | X |
| p-Nitrochlorobenzene | 100-00-5 | ......... | 1 | X |
| 4-Nitrodiphenyl; see 1926.1103 | 92-93-3 | | | |
| Nitroethane | 79-24-3 | 100 | 310 | |
| Nitrogen | 7727-37-9 | E | | |
| Nitrogen dioxide | 10102-44-0 | (C)5 | (C)9 | |
| Nitrogen trifluoride | 7783-54-2 | 10 | 29 | |
| Nitroglycerin | 55-63-0 | (C)0.2 | (C)2 | X |
| Nitromethane | 75-52-5 | 100 | 250 | |
| 1-Nitropropane | 108-03-2 | 25 | 90 | |
| 2-Nitropropane | 79-46-9 | 25 | 90 | |
| N-Nitrosodimethylamine; see 1926.1116. | | | | |
| Nitrotoluene (all isomers) | | 5 | 30 | X |
| o-isomer | 88-72-2 | | | |
| m-isomer | 99-08-1 | | | |
| p-isomer | 99-99-0 | | | |
| Nitrotrichloromethane; see Chloropicrin. | | | | |
| Nitrous oxide | 10024-97-2 | E | | |
| Octachloronaphthalene | 2234-13-1 | ......... | 0.1 | X |
| Octane | 111-65-9 | 400 | 1900 | |
| Oil mist, mineral | 8012-95-1 | ......... | 5 | |
| Osmium tetroxide (as Os) | 20816-12-0 | ......... | 0.002 | |
| Oxalic acid | 144-62-7 | ......... | 1 | |
| Oxygen difluoride | 7783-41-7 | 0.05 | 0.1 | |
| Ozone | 10028-15-6 | 0.1 | 0.2 | |
| Paraquat, respirable dust | 4685-14-7; 1910-42-5; 2074-50-2 | ......... | 0.5 | X |
| Parathion | 56-38-2 | ......... | 0.1 | X |
| Particulates not otherwise regulated. | | | | |
| Total dust organic and inorganic | | ......... | 15 | |
| PCB; see Chlorodiphenyl (42% and 54% chlorine). | | | | |
| Pentaborane | 19624-22-7 | 0.005 | 0.01 | |
| Pentachloronaphthalene | 1321-64-8 | ......... | 0.5 | X |
| Pentachlorophenol | 87-86-5 | ......... | 0.5 | X |
| Pentaerythritol | 115-77-5 | | | |
| Total dust | | ......... | | |
| Respirable fraction | | ......... | | |
| Pentane | 109-66-0 | 500 | 1500 | |
| 2-Pentanone (Methyl propyl ketone) | 107-87-9 | 200 | 700 | |
| Perchloroethylene (Tetrachloroethylene) | 127-18-4 | 100 | 670 | |
| Perchloromethyl mercaptan | 594-42-3 | 0.1 | 0.8 | |
| Perchloryl fluoride | 7616-94-6 | 3 | 13.5 | |
| Petroleum distillates (Naphtha) (Rubber Solvent) | | | A[3] | |
| Phenol | 108-95-2 | 5 | 19 | X |
| p-Phenylene diamine | 106-50-3 | ......... | 0.1 | X |
| Phenyl ether, vapor | 101-84-8 | 1 | 7 | |
| Phenyl ether-biphenyl mixture, vapor | | 1 | 7 | |
| Phenylethylene; see Styrene. | | | | |
| Phenyl glycidyl ether (PGE) | 122-60-1 | 10 | 60 | |
| Phenylhydrazine | 100-63-0 | 5 | 22 | X |
| Phosdrin (Mevinphos) | 7786-34-7 | ......... | 0.1 | X |
| Phosgene (Carbonyl chloride) | 75-44-5 | 0.1 | 0.4 | |
| Phosphine | 7803-51-2 | 0.3 | 0.4 | |
| Phosphoric acid | 7664-38-2 | ......... | 1 | |
| Phosphorus (yellow) | 7723-14-0 | ......... | 0.1 | |
| Phosphorus pentachloride | 10026-13-8 | ......... | 1 | |
| Phosphorus pentasulfide | 1314-80-3 | ......... | 1 | |
| Phosphorus trichloride | 7719-12-2 | 0.5 | 3 | |
| Phthalic anhydride | 85-44-9 | 2 | 12 | |
| Picric acid | 88-89-1 | ......... | 0.1 | X |
| Pindone (2-Pivalyl-1,3-indandione) | 83-26-1 | ......... | 0.1 | |
| Plaster of Paris | 26499-65-0 | | | |
| Total dust | | ......... | | |
| Respirable fraction | | ......... | | |
| Platinum (as Pt) | 7440-06-4 | | | |
| Metal | | ......... | ......... | |
| Soluble salts | | ......... | 0.002 | |
| Polytetrafluoroethylene decomposition products | | | A[2] | |
| Portland cement | 65997-15-1 | | | |
| Total dust | | – | 15 | – |
| Respirable fraction | | 5 | | |

## THRESHOLD LIMIT VALUES OF AIRBORNE CONTAMINANTS FOR CONSTRUCTION

| Substance | CAS No.[d] | ppm[a] | mg/m$^3$ [b] | Skin Designation |
|---|---|---|---|---|
| Propane | 74-98-6 | E | ........ | |
| Propergyl alcohol | 107-19-7 | 1 | | X |
| beta-Propriolactone; see 1926.1113 | 57-57-8 | | | |
| Propionic acid | 79-09-4 | ........ | ........ | |
| n-Propyl acetate | 109-60-4 | 200 | 840 | |
| n-Propyl alcohol | 71-23-8 | 200 | 500 | |
| n-Propyl nitrate | 627-13-4 | 25 | 110 | |
| Propylene dichloride | 78-87-5 | 75 | 350 | |
| Propylene imine | 75-55-8 | 2 | 5 | X |
| Propylene oxide | 75-56-9 | 100 | 240 | |
| Propyne; see Methyl acetylene. | | | | |
| Pyrethrum | 8003-34-7 | ........ | 5 | |
| Pyridine | 110-86-1 | 5 | 15 | |
| Quinone | 106-51-4 | 0.1 | 0.4 | |
| RDX; see Cyclonite. | | | | |
| Rhodium (as Rh), metal fume and insoluble compounds | 7440-16-6 | ........ | 0.1 | |
| Rhodium (as Rh), soluble compounds | 7440-16-6 | ........ | 0.001 | |
| Ronnel | 299-84-3 | ........ | 10 | |
| Rotenone | 83-79-4 | ........ | 5 | |
| Rouge | | | | |
| Total dust | | ........ | | |
| Respirable fraction | | ........ | | |
| Selenium compounds (as Se) | 7782-49-2 | ........ | 0.2 | |
| Selenium hexafluoride (as Se) | 7783-79-1 | 0.05 | 0.4 | |
| Silica, amorphous, precipitated and gel | 112926-00-8 | (2) | (2) | (2) |
| Silica, amorphous, diatomaceous earth, containing less than 1% crystalline silica. | 61790-53-2 | (2) | (2) | (2) |
| Silica, crystalline cristobalite, respirable dust | 14464-46-1 | (2) | (2) | (2) |
| Silica, crystalline quartz, respirable dust | 14808-60-7 | (2) | (2) | (2) |
| Silica, crystalline tripoli (as quartz), respirable dust | 1317-95-9 | (2) | (2) | (2) |
| Silica, crystalline tridymite, respirable dust | 15468-32-3 | (2) | (2) | (2) |
| Silica, fused, respirable dust | 60676-86-0 | (2) | (2) | (2) |
| Silicates (less than 1% crystalline silica) | | | | |
| Mica (respirable dust) | 12001-26-2 | (2) | (2) | (2) |
| Soapstone, total dust | | (2) | (2) | (2) |
| Soapstone, respirable dust | | (2) | (2) | (2) |
| Talc (containing asbestos); use asbestos limit; see 1926.58 | | | | |
| Talc (containing no asbestos), respirable dust | 14807-96-6 | (2) | (2) | (2) |
| Tremolite, asbestiform; see 1926.58 | | | | |
| Silicon carbide | 409-21-2 | | | |
| Total dust | | ........ | | |
| Respirable fraction | | ........ | | |
| Silver, metal and soluble compounds (as Ag) | 7440-22-4 | ........ | 0.01 | |
| Soapstone; see Silicates. | | | | |
| Sodium fluoroacetate | 62-74-8 | ........ | 0.05 | X |
| Sodium hydroxide | 1310-73-2 | ........ | 2 | |
| Starch | 9005-25-8 | | | |
| Total dust | | ........ | | |
| Respirable fraction | | ........ | | |
| Stibine | 7803-52-3 | 0.1 | 0.5 | |
| Stoddard solvent | 8052-41-3 | 200 | 1150 | |
| Strychnine | 57-24-9 | ........ | 0.15 | |
| Styrene | 100-42-5 | (C) 100 | (C) 420 | – |
| Sucrose | 57-50-1 | | | |
| Total dust | | ........ | | |
| Respirable fraction | | ........ | | |
| Sulfur dioxide | 7446-09-5 | 5 | 13 | |
| Sulfur hexafluoride | 2551-62-4 | 1000 | 6000 | |
| Sulfuric acid | 7664-93-9 | ........ | 1 | |
| Sulfur monochloride | 10025-67-9 | 1 | 6 | |
| Sulfur pentafluoride | 5714-22-7 | 0.025 | 0.25 | |
| Sulfuryl fluoride | 2699-79-8 | 5 | 20 | |
| Systox; see Demeton. | | | | |
| 2,4,5-T (2,4,5-trichlorophenoxyacetic acid) | 93-76-5 | ........ | 10 | |
| Talc; see Silicates. | | | | |
| Tantalum, metal and oxide dust | 7440-25-7 | ........ | 5 | |
| TEDP (Sulfotep) | 3689-24-5 | ........ | 0.2 | X |
| Teflon decomposition products | | | A2 | |
| Tellurium and compounds (as Te) | 13494-80-9 | ........ | 0.1 | |
| Tellurium hexafluoride (as Te) | 7783-80-4 | 0.02 | 0.2 | |
| Temephos | 3383-96-8 | | | |
| Total dust | | ........ | | |
| Respirable fraction | | ........ | | |
| TEPP (Tetraethyl pyrophosphate) | 107-49-3 | ........ | 0.05 | X |
| Terphenyls | 26140-60-3 | (C)1 | (C)9 | |
| 1,1,1,2-Tetrachloro-2,2-difluoroethane | 76-11-9 | 500 | 4170 | |
| 1,1,2,2-Tetrachloro-1,2-difluoroethane | 76-12-0 | 500 | 4170 | |
| 1,1,2,2-Tetrachloroethane | 79-34-5 | 5 | 35 | X |
| Tetrachloroethylene; see Perchloroethylene. | | | | |
| Tetrachloromethane; see Carbon tetrachloride. | | | | |
| Tetrachloronaphthalene | 1335-88-2 | ........ | 2 | X |

## THRESHOLD LIMIT VALUES OF AIRBORNE CONTAMINANTS FOR CONSTRUCTION

| Substance | CAS No.[d] | ppm[a] | mg/m³ [b] | Skin Designation |
|---|---|---|---|---|
| Tetraethyl lead (as Pb) | 78-00-2 | ......... | 0.1 | X |
| Tetrahydrofuran | 109-99-9 | 200 | 590 | |
| Tetramethyl lead (as Pb) | 75-74-1 | ......... | 0.15 | X |
| Tetramethyl succinonitrile | 3333-52-6 | 0.5 | 3 | X |
| Tetranitromethane | 509-14-8 | 1 | 8 | |
| Tetryl (2,4,6-Trinitrophenylmethylnitramine) | 479-45-8 | ......... | 1.5 | X |
| Thallium, soluble compounds (as Tl) | 7440-28-0 | ......... | 0.1 | X |
| Thiram | 137-26-8 | ......... | 5 | |
| Tin, inorganic compounds (except oxides) (as Sn) | 7440-31-5 | ......... | 2 | |
| Tin, organic compounds (as Sn) | 7440-31-5 | ......... | 0.1 | |
| Tin oxide (as Sn) | 21651-19-4 | | | |
| Total dust | | | | |
| Respirable fraction | | | | |
| Titanium dioxide | 13463-67-7 | | | |
| Total dust | | ......... | | |
| Toluene | 108-88-3 | 200 | 750 | – |
| Toluene-2,4-diisocyanate (TDI) | 584-84-9 | (C)0.02 | (C)0.14 | |
| o-Toluidine | 95-53-4 | 5 | 22 | X |
| Toxaphene; see Chlorinated camphene. | | | | |
| Tremolite; see Silicates. | | | | |
| Tributyl phosphate | 126-73-8 | ......... | 5 | |
| 1,1,1-Trichloroethane; see Methyl chloroform. | | | | |
| 1,1,2-Trichloroethane | 79-00-5 | 10 | 45 | X |
| Trichloroethylene | 79-01-6 | 100 | 535 | |
| Trichloromethane; see Chloroform. | | | | |
| Trichloronaphthalene | 1321-65-9 | ......... | 5 | X |
| 1,2,3-Trichloropropane | 96-18-4 | 50 | 300 | |
| 1,1,2-Trichloro-1,2,2-trifluoroethane | 76-13-1 | 1000 | 7600 | |
| Triethylamine | 121-44-8 | 25 | 100 | |
| Trifluorobromomethane | 75-63-8 | 1000 | 6100 | |
| Trimethyl benzene | 25551-13-7 | 25 | 120 | – |
| 2,4,6-Trinitrophenol; see Picric acid. | | | | |
| 2,4,6-Trinitrophenylmethylnitramine; see Tetryl. | | | | |
| 2,4,6-Trinitrotoluene (TNT) | 118-96-7 | ......... | 1.5 | X |
| Triorthocresyl phosphate | 78-30-8 | ......... | 0.1 | |
| Triphenyl phosphate | 115-86-6 | ......... | 3 | |
| Tungsten (asW) | 7440-33-7 | | | |
| Insoluble compounds | | | 5 | |
| Soluble compounds | | | 1 | |
| Turpentine | 8006-64-2 | 100 | 560 | |
| Uranium (as U) | 7440-61-1 | | | |
| Soluble compounds | | ......... | 0.2 | |
| Insoluble compounds | | ......... | 0.2 | |
| Vanadium | 1314-62-1 | | | |
| Respirable dust (as $V_2O_5$) | | ......... | (C)0.5 | |
| Fume (as $V_2O_5$) | | ......... | (C)0.1 | |
| Vegetable oil mist | | | | |
| Total dust | | ......... | | |
| Respirable fraction | | ......... | | |
| Vinyl benzene; see Styrene. | | | | |
| Vinyl chloride; see 1926.1117 | 75-01-4 | | | |
| Vinyl cyanide; see Acrylonitrile. | | | | |
| Vinyl toluene | 25013-15-4 | 100 | 480 | |
| Warfarin | 81 81 2 | ......... | 0.1 | |
| Xylenes (o-, m-, p-isomers) | 1330-20-7 | 100 | 435 | |
| Xylidine | 1300-73-8 | 5 | 25 | X |
| Yttrium | 7440-65-5 | ......... | 1 | |
| Zinc chloride fume | 7646-85-7 | ......... | 1 | |
| Zinc oxide fume | 1314-13-2 | ......... | 5 | |
| Zinc oxide | 1314-13-2 | | | |
| Total dust | | ......... | 15 | |
| Respirable fraction | | ......... | 5 | |
| Zirconium compounds (as Zr) | 7440-67-7 | ......... | 5 | |

MINERAL DUSTS

| Substance | mppcf [j] |
|---|---|
| SILICA: | |
| Crystalline | |
| Quartz. Threshold Limit calculated from the formula | $\frac{250^{(k)}}{\%SiO_2+5}$ |
| Cristobalite. | |
| Amorphous, including natural diatomaceous earth | 20 |
| SILICATES (less than 1% crystalline silica) | |
| Mica | 20 |
| Portland cement | 50 |
| Soapstone | 20 |
| Talc (non-asbestiform) | 20 |
| Talc (fibrous), use asbestos limit | -- |
| Graphite (natural) | 15 |
| Inert or Nuisance Particulates:[m] | |

[ Inert or Nuisance Dusts includes all mineral, inorganic, and organic dusts as indicated by examples in TLV's Appendix D] 50 (or 15 mg/m³ whichever is the smaller) of total dust <1% SiO²

Conversion factors.
mppcf × 35.3 = million particles per cubic meter = particles per c.c.

Footnotes
[1] [Reserved]
[2] See Mineral Dusts Table.
[3] Use Asbestos Limit § 1926.58.
[4] See 1926.58.

* The PELs are 8-hour TWAs unless otherwise noted; a (C) designation denotes a ceiling limit.
** As determined from breathing-zone air samples.
[a] Parts of vapor or gas per million parts of contaminated air by volume at 25° C and 760 torr.
[b] Milligrams of substance per cubic meter of air. When entry is in this column only, the value is exact; when listed with a ppm entry, it is approximate.
[c] [Reserved]
[d] The CAS number is for information only. Enforcement is based on the substance name. For an entry covering more than one metal compound, measured as the metal, the CAS number for the metal is given—not CAS numbers for the individual compounds.
[e-f] [Reserved]
[g] For sectors excluded from § 1926.1128 the limit is 10 ppm TWA.
[h] [Removed.]
[i] [Reserved]
[j] Millions of particles per cubic foot of air, based on impinger samples counted by lightfield techniques.
[k] The percentage of crystalline silica in the formula is the amount determined from airborne samples, except in those instances in which other methods have been shown to be applicable.
[l] [Reserved]
[m] Covers all organic and inorganic particulates not otherwise regulated. Same as Particulates Not Otherwise Regulated.
The 1970 TLV uses letter designations instead of a numerical value as follows:
A [1] [Reserved]
A [2] Polytetrafluoroethylene decomposition products. Because these products decompose in part by hydrolysis in alkaline solution, they can be quantitatively determined in air as fluoride to provide an index of exposure. No TLV is recommended pending determination of the toxicity of the products, but air concentrations should be minimal.
A [3] Gasoline and/or Petroleum Distillates. The composition of these materials varies greatly and thus a single TLV for all types of these materials is no longer applicable. The content of benzene, other aromatics and additives should be determined to arrive at the appropriate TLV.
E Simple asphyxiants. The limiting factor is the available oxygen which shall be at least 19.5% and be within the requirement addressing explosion in part 1926.

[39 FR 22801, June 24, 1974, as amended at 51 FR 37007, Oct. 17, 1986; 52 FR 46312, Dec. 4, 1987; 58 FR 35089, June 30, 1993; 61 FR 9249, 9250, Mar. 7, 1996; 61 FR 56856, Nov. 4, 1996; 62 FR 1619, Jan. 10, 1997]

## §1926.56 Illumination.

**1926.56(a)** ***General.*** Construction areas, ramps, runways, corridors, offices, shops, and storage areas shall be lighted to not less than the minimum illumination intensities listed in Table D-3 while any work is in progress:

### TABLE D-3—MINIMUM ILLUMINATION INTENSITIES IN FOOT-CANDLES

| Foot-candles | Area or Operation |
|---|---|
| 5 | General construction area lighting. |
| 3 | General construction areas, concrete placement, excavation and waste areas, accessways, active storage areas, loading platforms, refueling, and field maintenance areas. |
| 5 | Indoors: warehouses, corridors, hallways, and exitways. |
| 5 | Tunnels, shafts, and general underground work areas: (Exception: minimum of 10 foot-candles is required at tunnel and shaft heading during drilling, mucking, and scaling. Bureau of Mines approved cap lights shall be acceptable for use in the tunnel heading) |
| 10 | General construction plant and shops (e.g., batch plants, screening plants, mechanical and electrical equipment rooms, carpenter shops, rigging lofts and active store rooms, barracks or living quarters, locker or dressing rooms, mess halls, and indoor toilets and workrooms). |
| 30 | First aid stations, infirmaries, and offices. |

**1926.56(b)** ***Other areas.*** For areas or operations not covered above, refer to the American National Standard A11.1-1965, R1970, Practice for Industrial Lighting, for recommended values of illumination.

## §1926.57 Ventilation.

**1926.57(a)** ***General.*** Whenever hazardous substances such as dusts, fumes, mists, vapors, or gases exist or are produced in the course of construction work, their concentrations shall not exceed the limits specified in §1926.55(a). When ventilation is used as an engineering control method, the system shall be installed and operated according to the requirements of this section.

**1926.57(b)** ***Local exhaust ventilation.*** Local exhaust ventilation when used as described in (a) shall be designed to prevent dispersion into the air of dusts, fumes, mists, vapors, and gases in concentrations causing harmful exposure. Such exhaust systems shall be so designed that dusts, fumes, mists, vapors, or gases are not drawn through the work area of employees.

**1926.57(c)** ***Design and operation.*** Exhaust fans, jets, ducts, hoods, separators, and all necessary appurtenances, including refuse receptacles, shall be so designed, constructed, maintained and operated as to ensure the required protection by maintaining a volume and velocity of exhaust air sufficient to gather dusts, fumes, vapors, or gases from said equipment or process, and to convey them to suitable points of safe disposal, thereby preventing their dispersion in harmful quantities into the atmosphere where employees work.

**1926.57(d)** ***Duration of operations.*** (1) The exhaust system shall be in operation continually during all operations which it is designed to serve. If the employee remains in the contaminated zone, the system shall continue to operate after the cessation of said operations, the length of time to depend upon the individual circumstances and effectiveness of the general ventilation system.

**1926.57(d)(2)** Since dust capable of causing disability is, according to the best medical opinion, of microscopic size, tending to remain for hours in suspension in still air, it is essential that the exhaust system be continued in operation for a time after the work process or equipment served by the same shall have ceased, in order to ensure the removal of the harmful elements to the required extent. For the same reason, employees wearing respiratory equipment should not remove same immediately until the atmosphere seems clear.

**1926.57(e)** ***Disposal of exhaust materials.*** The air outlet from every dust separator, and the dusts, fumes, mists, vapors, or gases collected by an exhaust or ventilating system shall discharge to the outside atmosphere. Collecting systems which return air to work area may be used if concentrations which accumulate in the work area air do not result in harmful exposure to employees. Dust and refuse discharged from an exhaust system shall be disposed of in such a manner that it will not result in harmful exposure to employees.

**1926.57(f)** ***Abrasive blasting***— (1) *Definitions applicable to this paragraph— (i) Abrasive.* A solid substance used in an abrasive blasting operation.

**1926.57(f)(1)(ii)** *Abrasive-blasting respirator.* A respirator constructed so that it covers the wearer's head, neck, and shoulders to protect the wearer from rebounding abrasive.

**1926.57(f)(1)(iii)** *Blast cleaning barrel.* A complete enclosure which rotates on an axis, or which has an internal moving tread to tumble the parts, in order to expose various surfaces of the parts to the action of an automatic blast spray.

**1926.57(f)(1)(iv)** *Blast cleaning room.* A complete enclosure in which blasting operations are performed and where the operator works inside of the room to operate the blasting nozzle and direct the flow of the abrasive material.

**1926.57(f)(1)(v)** *Blasting cabinet.* An enclosure where the operator stands outside and operates the blasting nozzle through an opening or openings in the enclosure.

**1926.57(f)(1)(vi)** *Clean air.* Air of such purity that it will not cause harm or discomfort to an individual if it is inhaled for extended periods of time.

**1926.57(f)(1)(vii)** *Dust collector.* A device or combination of devices for separating dust from the air handled by an exhaust ventilation system.

**1926.57(f)(1)(viii)** *Exhaust ventilation system.* A system for removing contaminated air from a space, comprising two or more of the following elements (A) enclosure or hood, (B) duct work, (C) dust collecting equipment, (D) exhauster, and (E) discharge stack.

**1926.57(f)(1)(ix)** *Particulate-filter respirator.* An air purifying respirator, commonly referred to as a dust or a fume respirator, which removes most of the dust or fume from the air passing through the device.

**1926.57(f)(1)(x)** *Respirable dust.* Airborne dust in sizes capable of passing through the upper respiratory system to reach the lower lung passages.

**1926.57(f)(1)(xi)** *Rotary blast cleaning table.* An enclosure where the pieces to be cleaned are positioned on a rotating table and are passed automatically through a series of blast sprays.

**1926.57(f)(1)(xii)** *Abrasive blasting.* The forcible application of an abrasive to a surface by pneumatic pressure, hydraulic pressure, or centrifugal force.

**1926.57(f)(2)** *Dust hazards from abrasive blasting.* (i) Abrasives and the surface coatings on the materials blasted are shattered and pulverized during blasting operations and the dust formed will contain particles of respirable size. The composition and toxicity of the dust from these sources shall be considered in making an evaluation of the potential health hazards.

**1926.57(f)(2)(ii)** The concentration of respirable dust or fume in the breathing zone of the abrasive-blasting operator or any other worker shall be kept below the levels specified in §1926.55 or other pertinent sections of this part.

**1926.57(f)(2)(iii)** Organic abrasives which are combustible shall be used only in automatic systems. Where flammable or explosive dust mixtures may be present, the construction of the equipment, including the exhaust system and all electric wiring, shall conform to the requirements of American National Standard Installation of Blower and Exhaust Systems for Dust, Stock, and Vapor Removal or Conveying, Z33.1-1961 (NFPA 91-1961), and subpart S of this part. The blast nozzle shall be bonded and grounded to prevent the build up of static charges. Where flammable or explosive dust mixtures may be present, the abrasive blasting enclosure, the ducts, and the dust collector shall be constructed with loose panels or explosion venting areas, located on sides away from any occupied area, to provide for pressure relief in case of explosion, following the principles set forth in the National Fire Protection Association Explosion Venting Guide. NFPA 68-1954.

**1926.57(f)(3)** *Blast-cleaning enclosures.* (i) Blast-cleaning enclosures shall be exhaust ventilated in such a way that a continuous inward flow of air will be maintained at all openings in the enclosure during the blasting operation.

**1926.57(f)(3)(i)(A)** All air inlets and access openings shall be baffled or so arranged that by the combination of inward air flow and baffling the escape of abrasive or dust particles into an adjacent work area will be minimized and visible spurts of dust will not be observed.

**1926.57(f)(3)(i)(B)** The rate of exhaust shall be sufficient to provide prompt clearance of the dust-laden air within the enclosure after the cessation of blasting.

**1926.57(f)(3)(i)(C)** Before the enclosure is opened, the blast shall be turned off and the exhaust system shall be run for a sufficient period of time to remove the dusty air within the enclosure.

**1926.57(f)(3)(i)(D)** Safety glass protected by screening shall be used in observation windows, where hard deep-cutting abrasives are used.

**1926.57(f)(3)(i)(E)** Slit abrasive-resistant baffles shall be installed in multiple sets at all small access openings where dust might escape, and shall be inspected regularly and replaced when needed.

**1926.57(f)(3)(i)(E)*(1)*** Doors shall be flanged and tight when closed.

**1926.57(f)(3)(i)(E)*(2)*** Doors on blast-cleaning rooms shall be operable from both inside and outside, except that where there is a small operator access door, the large work access door may be closed or opened from the outside only.

**1926.57(f)(4)** *Exhaust ventilation systems.* (i) The construction, installation, inspection, and maintenance of exhaust systems shall conform to the principles and requirements set forth in American National Standard Fundamentals Governing the Design and Operation of Local Exhaust Systems, Z9.2-1960, and ANSI Z33.1-1961.

**1926.57(f)(4)(i)*(a)*** When dust leaks are noted, repairs shall be made as soon as possible.

**1926.57(f)(4)(i)*(b)*** The static pressure drop at the exhaust ducts leading from the equipment shall be checked when the installation is completed and periodically thereafter to assure continued satisfactory operation. Whenever an appreciable change in the pressure drop indicates a partial blockage, the system shall be cleaned and returned to normal operating condition.

**1926.57(f)(4)(ii)** In installations where the abrasive is recirculated, the exhaust ventilation system for the blasting enclosure shall not be relied upon for the removal of fines from the spent abrasive instead of an abrasive separator. An abrasive separator shall be provided for the purpose.

**1926.57(f)(4)(iii)** The air exhausted from blast-cleaning equipment shall be discharged through dust collecting equipment. Dust collectors shall be set up so that the accumulated dust can be emptied and removed without contaminating other working areas.

**1926.57(f)(5)** *Personal protective equipment.* (i) Employers must use only respirators approved by NIOSH under 42 CFR part 84 for protecting employees from dusts produced during abrasive-blasting operations.

**1926.57(f)(5)(ii)** Abrasive-blasting respirators shall be worn by all abrasive-blasting operators:

**1926.57(f)(5)(ii)(A)** When working inside of blast-cleaning rooms, or

**1926.57(f)(5)(ii)(B)** When using silica sand in manual blasting operations where the nozzle and blast are not physically separated from the operator in an exhaust ventilated enclosure, or

**1926.57(f)(5)(ii)(C)** Where concentrations of toxic dust dispersed by the abrasive blasting may exceed the limits set in §1926.55 or other pertinent sections of this part and the nozzle and blast are not physically separated from the operator in an exhaust-ventilated enclosure.

**1926.57(f)(5)(iii)** Properly fitted particulate-filter respirators, commonly referred to as dust-filter respirators, may be used for short, intermittent, or occasional dust exposures such as cleanup, dumping of dust collectors, or unloading shipments of sand at a receiving point when it is not feasible to control the dust by enclosure, exhaust ventilation, or other means. The respirators used must be approved by NIOSH under 42 CFR part 84 for protection against the specific type of dust encountered.

**1926.57(f)(5)(iv)** A respiratory protection program as defined and described in §1926.103, shall be established wherever it is necessary to use respiratory protective equipment.

**1926.57(f)(5)(v)** Operators shall be equipped with heavy canvas or leather gloves and aprons or equivalent protection to protect them from the impact of abrasives. Safety shoes shall be worn to protect against foot injury where heavy pieces of work are handled.

**1926.57(f)(5)(v)(A)** Safety shoes shall conform to the requirements of American National Standard for Men's Safety-Toe Footwear, Z41.1-1967.

**1926.57(f)(5)(v)(B)** Equipment for protection of the eyes and face shall be supplied to the operator when the respirator design does not provide such protection and to any other personnel working in the vicinity of abrasive blasting operations. This equipment shall conform to the requirements of §1926.102.

**1926.57(f)(6)** *Air supply and air compressors.* Air for abrasive-blasting respirators must be free of harmful quantities of dusts, mists, or noxious gases, and must meet the requirements for supplied-air quality and use specified in 29 CFR 1910.134(i).

**1926.57(f)(7)** *Operational procedures and general safety.* Dust shall not be permitted to accumulate on the floor or on ledges outside of an abrasive-blasting enclosure, and dust spills shall be cleaned up promptly. Aisles and walkways shall be kept clear of steel shot or similar abrasive which may create a slipping hazard.

**1926.57(f)(8)** *Scope.* This paragraph applies to all operations where an abrasive is forcibly applied to a surface by pneumatic or hydraulic pressure, or by centrifugal force. It does not apply to steam blasting, or steam cleaning, or hydraulic cleaning methods where work is done without the aid of abrasives.

**1926.57(g)** ***Grinding, polishing, and buffing operations***— (1) Definitions applicable to this paragraph—

**1926.57(g)(1)(i)** *Abrasive cutting-off wheels.* Organic-bonded wheels, the thickness of which is not more than one forty-eighth of their diameter for those up to, and including, 20 inches (50.8 cm) in diameter, and not more than one-sixtieth of their diameter for those larger than 20 inches (50.8 cm) in diameter, used for a multitude of operations variously known as cutting, cutting off, grooving, slotting, coping, and jointing, and the like. The wheels may be "solid" consisting of organic-bonded abrasive material throughout, "steel centered" consisting of a steel disc with a rim of organic-bonded material moulded around the periphery, or of the "inserted tooth" type consisting of a steel disc with organic-bonded abrasive teeth or inserts mechanically secured around the periphery.

**1926.57(g)(1)(ii)** *Belts.* All power-driven, flexible, coated bands used for grinding, polishing, or buffing purposes.

**1926.57(g)(1)(iii)** *Branch pipe.* The part of an exhaust system piping that is connected directly to the hood or enclosure.

**1926.57(g)(1)(iv)** *Cradle.* A movable fixture, upon which the part to be ground or polished is placed.

**1926.57(g)(1)(v)** *Disc wheels.* All power-driven rotatable discs faced with abrasive materials, artificial or natural, and used for grinding or polishing on the side of the assembled disc.

**1926.57(g)(1)(vi)** *Entry loss.* The loss in static pressure caused by air flowing into a duct or hood. It is usually expressed in inches of water gauge.

**1926.57(g)(1)(vii)** *Exhaust system.* A system consisting of branch pipes connected to hoods or enclosures, one or more header pipes, an exhaust fan, means for separating solid contaminants from the air flowing in the system, and a discharge stack to outside.

**1926.57(g)(1)(viii)** *Grinding wheels.* All power-driven rotatable grinding or abrasive wheels, except disc wheels as defined in this standard, consisting of abrasive particles held together by artificial or natural bonds and used for peripheral grinding.

**1926.57(g)(1)(ix)** *Header pipe (main pipe).* A pipe into which one or more branch pipes enter and which connects such branch pipes to the remainder of the exhaust system.

**1926.57(g)(1)(x)** *Hoods and enclosures.* The partial or complete enclosure around the wheel or disc through which air enters an exhaust system during operation.

**1926.57(g)(1)(xi)** *Horizontal double-spindle disc grinder.* A grinding machine carrying two power-driven, rotatable, coaxial, horizontal spindles upon the inside ends of which are mounted abrasive disc wheels used for grinding two surfaces simultaneously.

**1926.57(g)(1)(xii)** Horizontal single-spindle disc grinder. A grinding machine carrying an abrasive disc wheel upon one or both ends of a power-driven, rotatable single horizontal spindle.

**1926.57(g)(1)(xiii)** *Polishing and buffing wheels.* All power-driven rotatable wheels composed all or in part of textile fabrics, wood, felt, leather, paper, and may be coated with abrasives on the periphery of the wheel for purposes of polishing, buffing, and light grinding.

**1926.57(g)(1)(xiv)** *Portable grinder.* Any power-driven rotatable grinding, polishing, or buffing wheel mounted in such manner that it may be manually manipulated.

**1926.57(g)(1)(xv)** *Scratch brush wheels.* All power-driven rotatable wheels made from wire or bristles, and used for scratch cleaning and brushing purposes.

**1926.57(g)(1)(xvi)** *Swing-frame grinder.* Any power-driven rotatable grinding, polishing, or buffing wheel mounted in such a manner that the wheel with its supporting framework can be manipulated over stationary objects.

**1926.57(g)(1)(xvii)** *Velocity pressure (vp).* The kinetic pressure in the direction of flow necessary to cause a fluid at rest to flow at a given velocity. It is usually expressed in inches of water gauge.

**1926.57(g)(1)(xviii)** *Vertical spindle disc grinder.* A grinding machine having a vertical, rotatable power-driven spindle carrying a horizontal abrasive disc wheel.

**1926.57(g)(2)** *Application.* Wherever dry grinding, dry polishing or buffing is performed, and employee exposure, without regard to the use of respirators, exceeds the permissible exposure limits prescribed in §1926.55 or other pertinent sections of this part, a local exhaust ventilation system shall be provided and used to maintain employee exposures within the prescribed limits.

**1926.57(g)(3)** *Hood and branch pipe requirements.* (i) Hoods connected to exhaust systems shall be used, and such hoods shall be designed, located, and placed so that the dust or dirt particles shall fall or be projected into the hoods in the direction of the air flow. No wheels, discs, straps, or belts shall be operated in such manner and in such direction as to cause the dust and dirt particles to be thrown into the operator's breathing zone.

**1926.57(g)(3)(ii)** Grinding wheels on floor stands, pedestals, benches, and special-purpose grinding machines and abrasive cutting-off wheels shall have not less than the minimum exhaust volumes shown in Table D-57.1 with a recommended minimum duct velocity of 4,500 feet per minute in the branch and 3,500 feet per minute in the main. The entry losses from all hoods except the vertical-spindle disc grinder hood, shall equal 0.65 velocity pressure for a straight takeoff and 0.45 velocity pressure for a tapered takeoff. The entry loss for the vertical-spindle disc grinder hood is shown in figure D-57.1 (following paragraph (g) of this section).

### TABLE D-57.1— GRINDING AND ABRASIVE CUTTING-OFF WHEELS

| Wheel diameter, inches (cm) | Wheel width, inches (cm) | Minimum exhaust volume (feet$^3$/min.) |
|---|---|---|
| To 9 (22.86) | 1 1/2 (3.81) | 220 |
| Over 9 to 16 (22.86 to 40.64) | 2 (5.08) | 390 |
| Over 16 to 19 (40.64 to 48.26) | 3 (7.62) | 500 |
| Over 19 to 24 (48.26 to 60.96) | 4 (10.16) | 610 |
| Over 24 to 30 (60.96 to 76.2) | 5 (12.7) | 880 |
| Over 30 to 36 (76.2 to 91.44) | 6 (15.24) | 1,200 |

For any wheel wider than wheel diameters shown in Table D-57.1, increase the exhaust volume by the ratio of the new width to the width shown.

***Example:*** If wheel width = 4 1/2 inches (11.43 cm), then 4.5 ÷ 4 × 610 = 686 (rounded to 690).

**1926.57(g)(3)(iii)** Scratch-brush wheels and all buffing and polishing wheels mounted on floor stands, pedestals, benches, or special-purpose machines shall have not less than the minimum exhaust volume shown in Table D-57.2.

### TABLE D-57.2—BUFFING AND POLISHING WHEELS

| Wheel diameter, inches (cm) | Wheel width, inches (cm) | Minimum exhaust volume (feet$^3$/min.) |
|---|---|---|
| To 9 (22.86) | 2 (5.08) | 300 |
| Over 9 to 16 (22.86 to 40.64) | 3 (7.62) | 500 |
| Over 16 to 19 (40.64 to 48.26) | 4 (10.16) | 610 |
| Over 19 to 24 (48.26 to 60.96) | 5 (12.7) | 740 |
| Over 24 to 30 (60.96 to 76.2) | 6 (15.24) | 1,040 |
| Over 30 to 36 (76.2 to 91.44) | 6 (15.24) | 1,200 |

**1926.57(g)(3)(iv)** Grinding wheels or discs for horizontal single-spindle disc grinders shall be hooded to collect the dust or dirt generated by the grinding operation and the hoods shall be connected to branch pipes having exhaust volumes as shown in Table D-57.3.

### TABLE D-57.3— HORIZONTAL SINGLE-SPINDLE DISC GRINDER

| Disc diameter, inches (cm): | Exhaust volume (ft.$^3$/min.) |
|---|---|
| Up to 12 (30.48) | 220 |
| Over 12 to 19 (30.48 to 48.26) | 390 |
| Over 19 to 30 (48.26 to 76.2) | 610 |
| Over 30 to 36 (76.2 to 91.44) | 880 |

**1926.57(g)(3)(v)** Grinding wheels or discs for horizontal double-spindle disc grinders shall have a hood enclosing the grinding chamber and the hood shall be connected to one or more branch pipes having exhaust volumes as shown in Table D-57.4.

### TABLE D-57.4— HORIZONTAL DOUBLE-SPINDLE DISC GRINDER

| Disc diameter, inches (cm): | Exhaust volume (ft.$^3$/min.) |
|---|---|
| Up to 19 (48.26) | 610 |
| Over 19 to 25 (48.26 to 63.5) | 880 |
| Over 25 to 30 (63.5 to 76.2) | 1,200 |
| Over 30 to 53 (76.2 to 134.62) | 1,770 |
| Over 53 to 72 (134.62 to 182.88) | 6,280 |

**1926.57(g)(3)(vi)** Grinding wheels or discs for vertical single-spindle disc grinders shall be encircled with hoods to remove the dust generated in the operation. The hoods shall be connected to one or more branch pipes having exhaust volumes as shown in Table D-57.5.

## TABLE D-57.5—VERTICAL SPINDLE DISC GRINDER

| Disc diameter inches (cm) | One-half or more of disc covered | | Disc not covered | |
|---|---|---|---|---|
| | Number[1] | Exhaust foot $^3$/min. | Number[1] | Exhaust foot $^3$/min. |
| Up to 20 (50.8) | 1 | 500 | 2 | 780 |
| Over 20 to 30 (50.8 to 76.2) | 2 | 780 | 2 | 1,480 |
| Over 30 to 53 (76.2 to 134.62) | 2 | 1,770 | 4 | 3,530 |
| Over 53 to 72 (134.62 to 182.88) | 2 | 3,140 | 5 | 6,010 |

[1] Number of exhaust outlets around periphery of hood, or equal distribution provided by other means.

**1926.57(g)(3)(vii)** Grinding and polishing belts shall be provided with hoods to remove dust and dirt generated in the operations and the hoods shall be connected to branch pipes having exhaust volumes as shown in Table D-57.6.

## TABLE D-57.6—GRINDING AND POLISHING BELTS

| Belts width, inches (cm): | Exhaust volume (ft.$^3$/min.) |
|---|---|
| Up to 3 (7.62) | 220 |
| Over 3 to 5 (7.62 to 12.7) | 300 |
| Over 5 to 7 (12.7 to 17.78) | 390 |
| Over 7 to 9 (17.78 to 22.86) | 500 |
| Over 9 to 11 (22.86 to 27.94) | 610 |
| Over 11 to 13 (27.94 to 33.02) | 740 |

**1926.57(g)(3)(viii)** Cradles and swing-frame grinders. Where cradles are used for handling the parts to be ground, polished, or buffed, requiring large partial enclosures to house the complete operation, a minimum average air velocity of 150 feet per minute shall be maintained over the entire opening of the enclosure. Swing-frame grinders shall also be exhausted in the same manner as provided for cradles. (See fig. D-57.3)

**1926.57(g)(3)(ix)** Where the work is outside the hood, air volumes must be increased as shown in American Standard Fundamentals Governing the Design and Operation of Local Exhaust Systems, Z9.2-1960 (section 4, exhaust hoods).

**1926.57(g)(4)** *Exhaust systems.* (i) Exhaust systems for grinding, polishing, and buffing operations should be designed in accordance with American Standard Fundamentals Governing the Design and Operation of Local Exhaust Systems, Z9.2-1960.

**1926.57(g)(4)(ii)** Exhaust systems for grinding, polishing, and buffing operations shall be tested in the manner described in American Standard Fundamentals Governing the Design and Operation of Local Exhaust Systems, Z9.2-1960.

**1926.57(g)(4)(iii)** All exhaust systems shall be provided with suitable dust collectors.

**1926.57(g)(5)** *Hood and enclosure design.* (i) (A) It is the dual function of grinding and abrasive cutting-off wheel hoods to protect the operator from the hazards of bursting wheels as well as to provide a means for the removal of dust and dirt generated. All hoods shall be not less in structural strength than specified in the American National Standard Safety Code for the Use, Care, and Protection of Abrasive Wheels, B7.1-1970.

**1926.57(g)(5)(i)(B)** Due to the variety of work and types of grinding machines employed, it is necessary to develop hoods adaptable to the particular machine in question, and such hoods shall be located as close as possible to the operation.

**1926.57(g)(5)(ii)** Exhaust hoods for floor stands, pedestals, and bench grinders shall be designed in accordance with figure D-57.2. The adjustable tongue shown in the figure shall be kept in working order and shall be adjusted within one-fourth inch (0.635 cm) of the wheel periphery at all times.

**1926.57(g)(5)(iii)** Swing-frame grinders shall be provided with exhaust booths as indicated in figure D-57.3.

**1926.57(g)(5)(iv)** Portable grinding operations, whenever the nature of the work permits, shall be conducted within a partial enclosure. The opening in the enclosure shall be no larger than is actually required in the operation and an average face air velocity of not less than 200 feet per minute shall be maintained.

**1926.57(g)(5)(v)** Hoods for polishing and buffing and scratch-brush wheels shall be constructed to conform as closely to figure D-57.4 as the nature of the work will permit.

**1926.57(g)(5)(vi)** Cradle grinding and polishing operations shall be performed within a partial enclosure similar to figure D-57.5. The operator shall be positioned outside the working face of the opening of the enclosure. The face opening of the enclosure should not be any greater in area than that actually required for the performance of the operation and the average air velocity into the working face of the enclosure shall not be less than 150 feet per minute.

**1926.57(g)(5)(vii)** Hoods for horizontal single-spindle disc grinders shall be constructed to conform as closely as possible to the hood shown in figure D-57.6. It is essential that there be a space between the back of the wheel and the hood, and a space around the periphery of the wheel of at least 1 inch (2.54 cm) in order to permit the suction to act around the wheel periphery. The opening on the side of the disc shall be no larger than is required for the grinding operation, but must never be less than twice the area of the branch outlet.

**1926.57(g)(5)(viii)** Horizontal double-spindle disc grinders shall have a hood encircling the wheels and grinding chamber similar to that illustrated in figure D-57.7. The openings for passing the work into the grinding chamber should be kept as small as possible, but must never be less than twice the area of the branch outlets.

**1926.57(g)(5)(ix)** Vertical-spindle disc grinders shall be encircled with a hood so constructed that the heavy dust is drawn off a surface of the disc and the lighter dust exhausted through a continuous slot at the top of the hood as shown in figure D-57.1.

**1926.57(g)(5)(x)** Grinding and polishing belt hoods shall be constructed as close to the operation as possible. The hood should extend almost to the belt, and 1-inch (2.54 cm) wide openings should be provided on either side. Figure D-57.8 shows a typical hood for a belt operation.

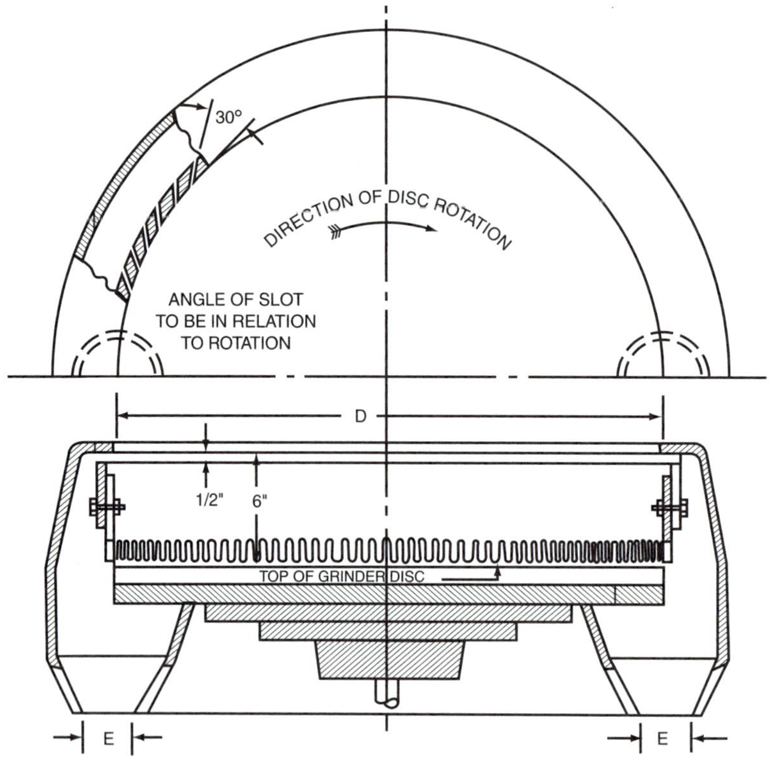


| Dia. *D*.inches (cm) | | Exhaust *E* | | Volume Exhausted at 4,500 ft/min | Note |
|---|---|---|---|---|---|
| Min. | Max. | No Pipes | Dia. | $ft^3$/min | |
| .............. | 20 (50.8) | 1 | 4 1/4 (10.795) | 500 | When one-half or more of the disc can be hooded, use exhaust ducts as shown at the left. |
| Over 20 (50.8) ... | 30 (76.2) | 2 | 4 (10.16) | 780 | |
| Over 30 (76.2) ... | 72 (182.88) | 2 | 6 (15.24) | 1,770 | |
| Over 53 (134.62) . | 72 (182.88) | 2 | 8 (20.32) | 3,140 | |
| .............. | 20 (50.8) | 2 | 4 (10.16) | 780 | When no hood can be used over disc, use exhaust ducts as shown at left. |
| Over 20 (50.8) ... | 20 (50.8) | 2 | 4 (10.16) | 780 | |
| Over 30 (76.2) ... | 30 (76.2) | 2 | 5 1/2 (13.97) | 1,480 | |
| Over 53 (134.62) . | 53 (134.62) | 4 | 6 (15.24) | 3,530 | |
| | 72 (182.88) | 5 | 7 (17.78) | 6,010 | |

Entry loss = 1.0 slot velocity pressure + 0.5 branch velocity pressure.
Minimum slot velocity = 2,000 ft/min—1/2-inch slot width.

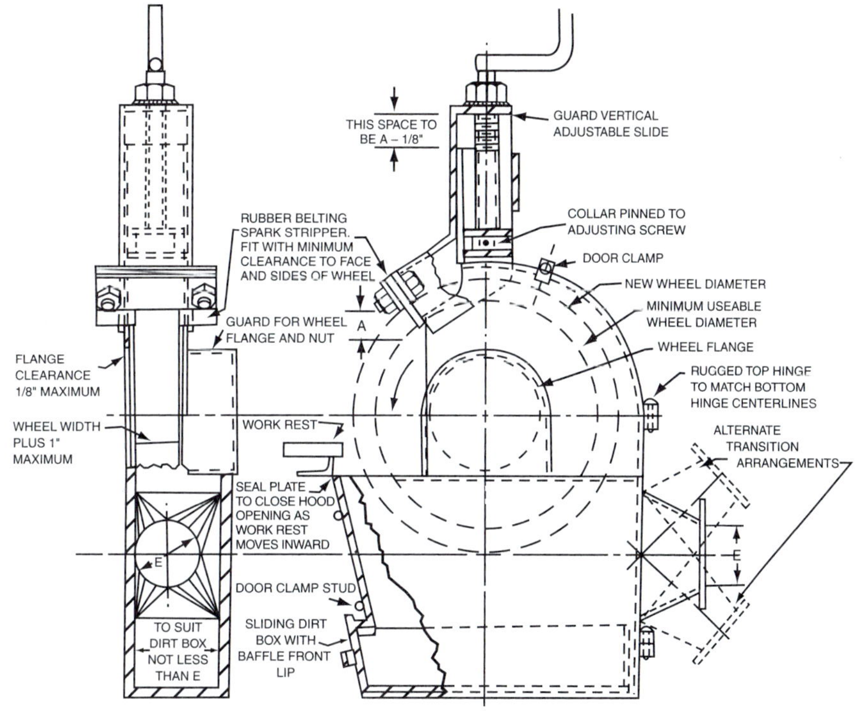


| Wheel dimension, inches (centimeters) | | | Exhaust outlet, inches (centimeters) *E* | Volume of air at 4,500 ft/min |
|---|---|---|---|---|
| Diameter | | Width, Max | | |
| Min = *d* | Max = *D* | | | |
| | 9 (22.86) | 1 1/2 (3.81) | 3 | 220 |
| Over 9 (22.86) | 16 (40.64) | 2 (5.08) | 4 | 390 |
| Over 16 (40.64) | 19 (48.26) | 3 (7.62) | 4 1/2 | 500 |
| Over 19 (48.26) | 24 (60.96) | 4 (10.16) | 5 | 610 |
| Over 24 (60.96) | 30 (76.2) | 5 (12.7) | 6 | 880 |
| Over 30 (76.2) | 36 (91.44) | 6 (15.24) | 7 | 1,200 |

Entry loss = 0.45 velocity pressure for tapered takeoff 0.65 velocity pressure for straight takeoff.

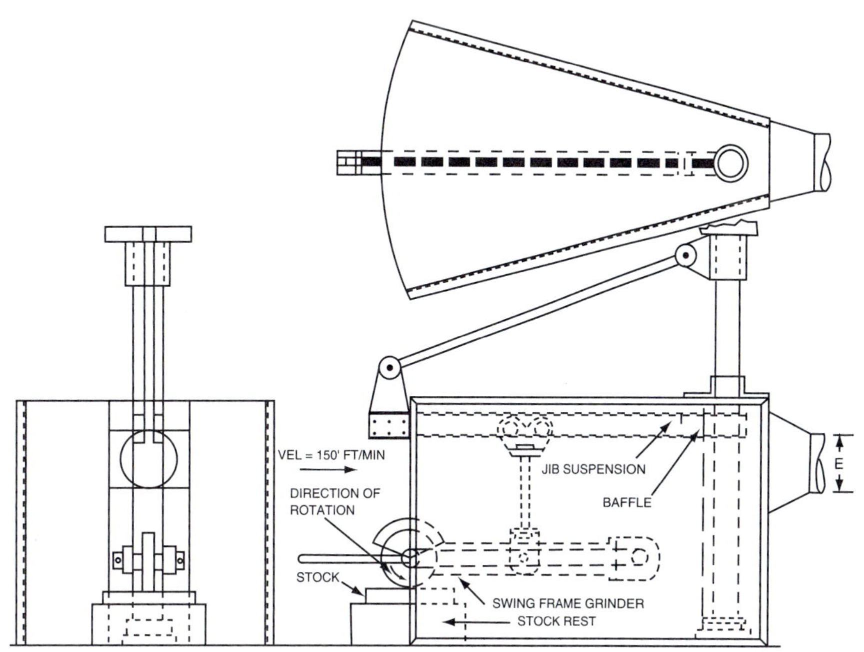


**FIGURE D-57.3—A METHOD OF APPLYING AN EXHAUST ENCLOSURE TO SWING-FRAME GRINDERS**

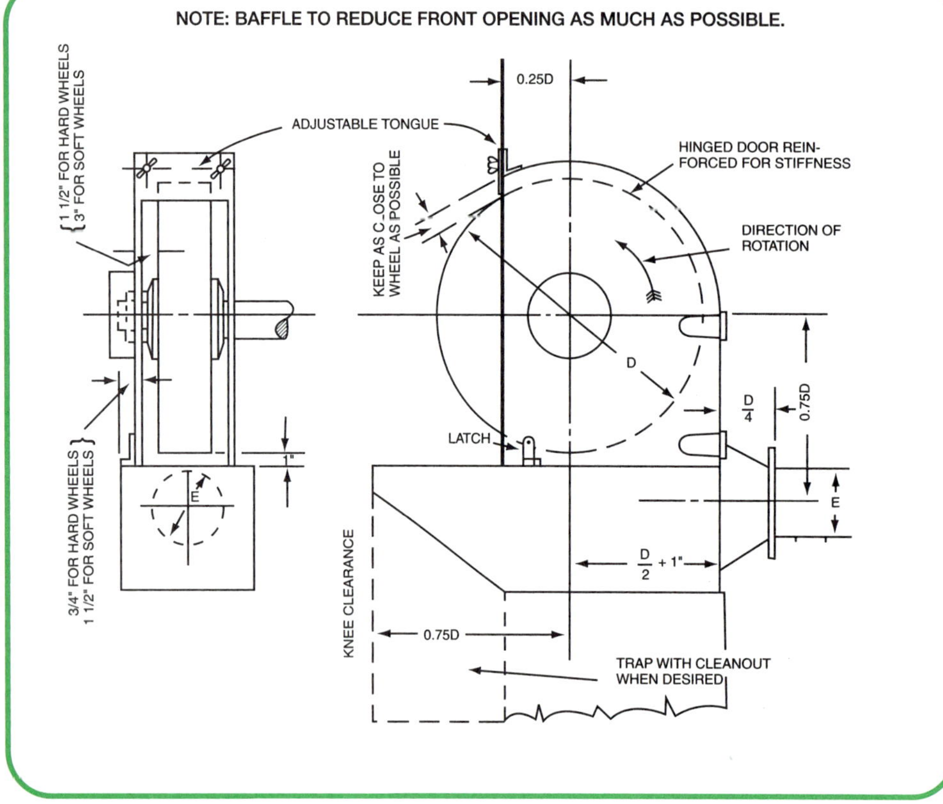


## STANDARD BUFFING AND POLISHING HOOD

| Wheel dimension, inches (centimeters) | | | Exhaust outlet, inches *E* | Volume of air at 4,500 ft/min |
|---|---|---|---|---|
| Diameter | | Width, Max | | |
| Min = *d* | Max = *D* | | | |
| | 9 (22.86) | 2 (5.08) | 3½ (3.81) | 300 |
| Over 9 (22.86) | 16 (40.64) | 3 (5.08) | 4 | 500 |
| Over 16 (40.64) | 19 (48.26) | 4 (11.43) | 5 | 610 |
| Over 19 (48.26) | 24 (60.96) | 5 (12.7) | 5½ | 740 |
| Over 24 (60.96) | 30 (76.2) | 6 (15.24) | 6½ | 1,040 |
| Over 30 (76.2) | 36 (91.44) | 6 (15.24) | 7 | 1,200 |

Entry loss = 0.15 velocity pressure for tapered takeoff 0.65 velocity pressure for straight takeoff.

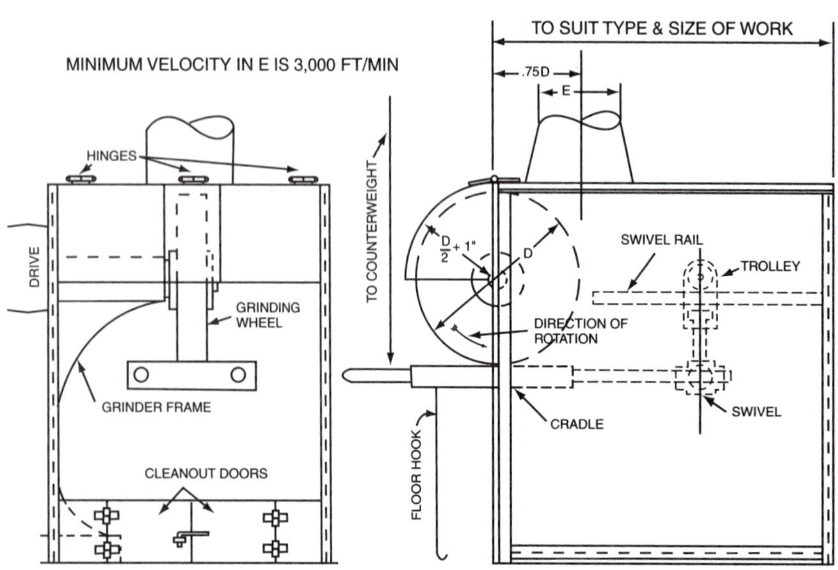


MINIMUM VELOCITY AT ENCLOSURE FACE IS 150 FT/MIN

**FIGURE D-57.5—CRADLE POLISHING OR GRINDING ENCLOSURE**

**ENTRY LOSS = 0.45 VELOCITY PRESSURE FOR TAPERED TAKEOFF**

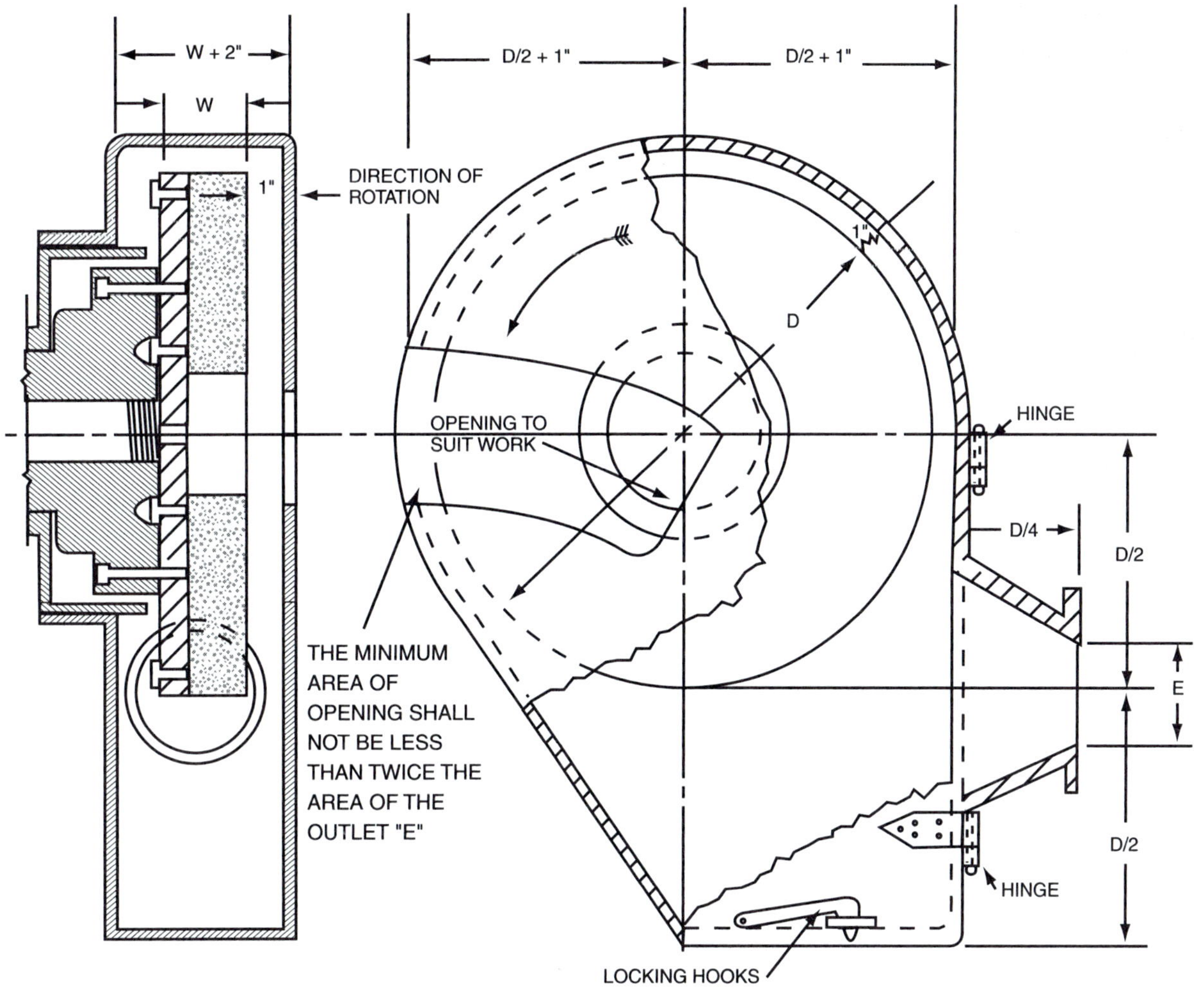


| Dia *D*, inches (centimeters) | | Exhaust *E*, dia. inches (cm) | Volume exhausted at 4,500 ft/min $ft^3$/min |
|---|---|---|---|
| Min. | Max. | | |
| | 12 (30.48) | 3 (7.6) | 220 |
| Over 12 (30.48) | 19 (48.26) | 4 (10.16) | 390 |
| Over 19 (48.26) | 30 (76.2) | 5 (12.7) | 610 |
| Over 30 (76.2) | 36 (91.44) | 6 (15.24) | 880 |

NOTE: If grinding wheels are used for disc grinding purposes, hoods must conform to structural strength and materials as described in 9.1.
Entry loss = 0.45 velocity pressure for tapered takeoff.

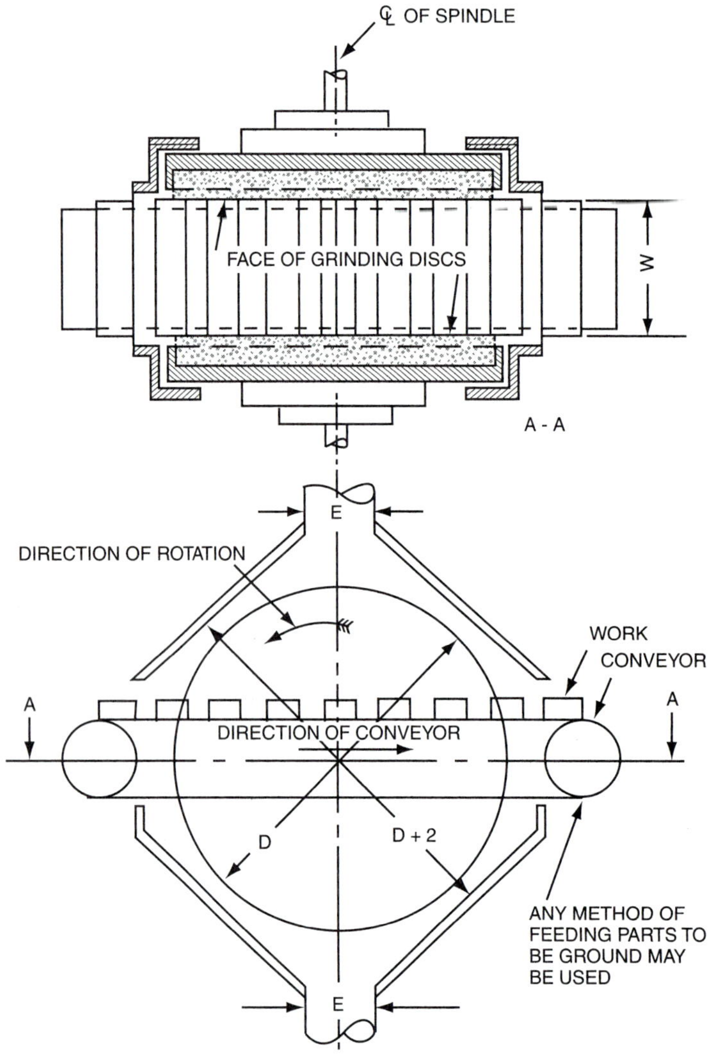


| Disc dia. inches (centimeters) | | Exhaust E | | Volume exhaust at 4,500 ft/min. | Note |
|---|---|---|---|---|---|
| Min. | Max. | No Pipes | Dia. | $ft^3$/min. | |
| | 19 (48.26) | 1 | 5 | 610 | |
| Over 19 (48.26) ......... | 25 (63.5) | 1 | 6 | 880 | When width "W" permits, exhaust ducts should be as near heaviest grinding as possible. |
| Over 25 (63.5) .......... | 30 (76.2) | 1 | 7 | 1,200 | |
| Over 30 (76.2) .......... | 53 (134.62) | 2 | 6 | 1,770 | |
| Over 53 (134.62) ........ | 72 (182.88) | 4 | 8 | 6,280 | |

Entry loss = 0.45 velocity pressure for tapered takeoff.

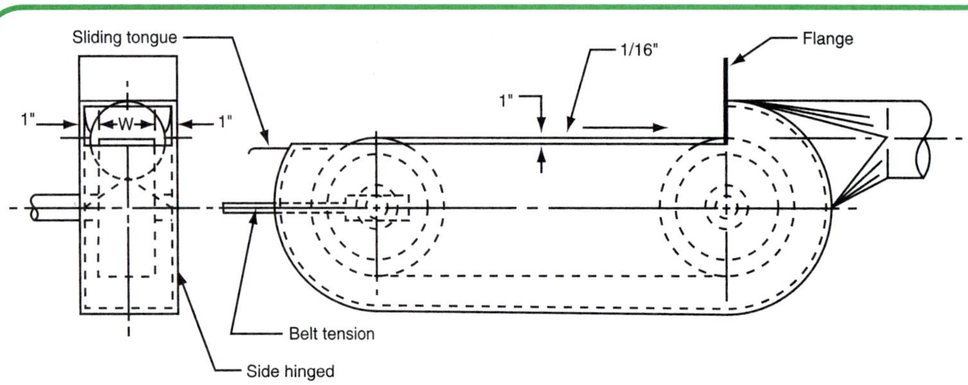


**FIGURE D–57.8. — A TYPICAL HOOD FOR A BELT OPERATION**

Entry loss = 0.45 velocity pressure for tapered takeoff

| Belt width W. inches (centimeters) | Exhaust volume. ft.[1]/min |
|---|---|
| Up to 3 (7.62) | 220 |
| 3 to 5 (7.62 to 12.7) | 300 |
| 5 to 7 (12.7 to 17.78) | 390 |
| 7 to 9 (17.78 to 22.86) | 500 |
| 9 to 11 (22.86 to 27.94) | 610 |
| 11 to 13 (27.94 to 33.02) | 740 |

Minimum duct velocity = 4,500 ft/min branch, 3,500 ft/min main.

Entry loss = 0.45 velocity pressure for tapered takeoff; 0.65 velocity pressure for straight takeoff.

**1926.57(g)(6)** *Scope.* This paragraph (g), prescribes the use of exhaust hood enclosures and systems in removing dust, dirt, fumes, and gases generated through the grinding, polishing, or buffing of ferrous and nonferrous metals.

**1926.57(h)** ***Spray finishing operations***— (1) *Definitions applicable to this paragraph*—(i) *Spray-finishing operations.* Spray-finishing operations are employment of methods wherein organic or inorganic materials are utilized in dispersed form for deposit on surfaces to be coated, treated, or cleaned. Such methods of deposit may involve either automatic, manual, or electrostatic deposition but do not include metal spraying or metallizing, dipping, flow coating, roller coating, tumbling, centrifuging, or spray washing and degreasing as conducted in self-contained washing and degreasing machines or systems.

**1926.57(h)(1)(ii)** *Spray booth.* Spray booths are defined and described in §1926.66(a). (See sections 103, 104, and 105 of the Standard for Spray Finishing Using Flammable and Combustible Materials, NFPA No. 33-1969).

**1926.57(h)(1)(iii)** *Spray room.* A spray room is a room in which spray-finishing operations not conducted in a spray booth are performed separately from other areas.

**1926.57(h)(1)(iv)** *Minimum maintained velocity.* Minimum maintained velocity is the velocity of air movement which must be maintained in order to meet minimum specified requirements for health and safety.

**1926.57(h)(2)** *Location and application.* Spray booths or spray rooms are to be used to enclose or confine all operations. Spray-finishing operations shall be located as provided in sections 201 through 206 of the Standard for Spray Finishing Using Flammable and Combustible Materials, NFPA No. 33-1969.

**1926.57(h)(3)** *Design and construction of spray booths.* (i) Spray booths shall be designed and constructed in accordance with §1926.66(b)(1) through (4) and (6) through (10) (see sections 301-304 and 306-310 of the Standard for Spray Finishing Using Flammable and Combustible Materials, NFPA No. 33-1969), for general construction specifications. For a more detailed discussion of fundamentals relating to this subject, see ANSI Z9.2-1960

**1926.57(h)(3)(i)(A)** Lights, motors, electrical equipment, and other sources of ignition shall conform to the requirements of §1926.66(b)(10) and (c). (See section 310 and chapter 4 of the Standard for Spray Finishing Using Flammable and Combustible Materials NFPA No. 33-1969.)

**1926.57(h)(3)(i)(B)** In no case shall combustible material be used in the construction of a spray booth and supply or exhaust duct connected to it.

**1926.57(h)(3)(ii)** Unobstructed walkways shall not be less than 6 1/2 feet (1.976 m) high and shall be maintained clear of obstruction from any work location in the booth to a booth exit or open booth front. In booths where the open front is the only exit, such exits shall be not less than 3 feet (0.912 m) wide. In booths having multiple exits, such exits shall not be less than 2 feet (0.608 m) wide, provided that the maximum distance from the work location to the exit is 25 feet (7.6 m) or less. Where booth exits are provided with doors, such doors shall open outward from the booth.

**1926.57(h)(3)(iii)** Baffles, distribution plates, and dry-type overspray collectors shall conform to the requirements of §1926.66(b)(4) and (5). (See sections 304 and 305 of the Standard for Spray Finishing Using Flammable and Combustible Materials, NFPA No. 33-1969.)

**1926.57(h)(3)(iii)(A)** Overspray filters shall be installed and maintained in accordance with the requirements of §1926.66(b)(5), (see section 305 of the Standard for Spray Finishing Using Flammable and Combustible Materials, NFPA No. 33-1969), and shall only be in a location easily accessible for inspection, cleaning, or replacement.

**1926.57(h)(3)(iii)(B)** Where effective means, independent of the overspray filters, are installed which will result in design air distribution across the booth cross section, it is permissible to operate the booth without the filters in place.

**1926.57(h)(3)(iv)(A)** For wet or water-wash spray booths, the water-chamber enclosure, within which intimate contact of contaminated air and cleaning water or other cleaning medium is maintained, if made of steel, shall be 18 gage or heavier and adequately protected against corrosion.

**1926.57(h)(3)(iv)(B)** Chambers may include scrubber spray nozzles, headers, troughs, or other devices. Chambers shall be provided with adequate means for creating and maintaining scrubbing action for removal of particulate matter from the exhaust air stream.

**1926.57(h)(3)(v)** Collecting tanks shall be of welded steel construction or other suitable non-combustible material. If pits are used as collecting tanks, they shall be concrete, masonry, or other material having similar properties.

**1926.57(h)(3)(v)(A)** Tanks shall be provided with weirs, skimmer plates, or screens to prevent sludge and floating paint from entering the pump suction box. Means for automatically maintaining the proper water level shall also be provided. Fresh water inlets shall not be submerged. They shall terminate at least one pipe diameter above the safety overflow level of the tank.

**1926.57(h)(3)(v)(B)** Tanks shall be so constructed as to discourage accumulation of hazardous deposits.

**1926.57(h)(3)(vi)** Pump manifolds, risers, and headers shall be adequately sized to insure sufficient water flow to provide efficient operation of the water chamber.

**1926.57(h)(4)** *Design and construction of spray rooms.* (i) Spray rooms, including floors, shall be constructed of masonry, concrete, or other noncombustible material.

**1926.57(h)(4)(ii)** Spray rooms shall have noncombustible fire doors and shutters.

**1926.57(h)(4)(iii)** Spray rooms shall be adequately ventilated so that the atmosphere in the breathing zone of the operator shall be maintained in accordance with the requirements of paragraph (h)(6)(ii) of this section.

**1926.57(h)(4)(iv)** Spray rooms used for production spray-finishing operations shall conform to the requirements for spray booths.

**1926.57(h)(5)** *Ventilation.* (i) Ventilation shall be provided in accordance with provisions of §1926.66(d) (see chapter 5 of the Standard for Spray Finishing Using Flammable or Combustible Materials, NFPA No. 33-1969), and in accordance with the following:

**1926.57(h)(5)(i)(A)** Where a fan plenum is used to equalize or control the distribution of exhaust air movement through the booth, it shall be of sufficient strength or rigidity to withstand the differential air pressure or other superficially imposed loads for which the equipment is designed and also to facilitate cleaning. Construction specifications shall be at least equivalent to those of paragraph (h)(5)(iii) of this section.

**1926.57(h)(5)(i)(B)** [Reserved]

**1926.57(h)(5)(ii)** Inlet or supply ductwork used to transport makeup air to spray booths or surrounding areas shall be constructed of noncombustible materials.

**1926.57(h)(5)(ii)(A)** If negative pressure exists within inlet ductwork, all seams and joints shall be sealed if there is a possibility of infiltration of harmful quantities of noxious gases, fumes, or mists from areas through which ductwork passes.

**1926.57(h)(5)(ii)(B)** Inlet ductwork shall be sized in accordance with volume flow requirements and provide design air requirements at the spray booth.

**1926.57(h)(5)(ii)(C)** Inlet ductwork shall be adequately supported throughout its length to sustain at least its own weight plus any negative pressure which is exerted upon it under normal operating conditions.

**1926.57(h)(5)(iii)** [Reserved]

**1926.57(h)(5)(iii)(A)** Exhaust ductwork shall be adequately supported throughout its length to sustain its weight plus any normal accumulation in interior during normal operating conditions and any negative pressure exerted upon it.

**1926.57(h)(5)(iii)(B)** Exhaust ductwork shall be sized in accordance with good design practice which shall include consideration of fan capacity, length of duct, number of turns and elbows, variation in size, volume, and character of materials being exhausted. See American National Standard Z9.2-1960 for further details and explanation concerning elements of design.

**1926.57(h)(5)(iii)(C)** Longitudinal joints in sheet steel ductwork shall be either lock-seamed, riveted, or welded. For other than steel construction, equivalent securing of joints shall be provided.

**1926.57(h)(5)(iii)(D)** Circumferential joints in ductwork shall be substantially fastened together and lapped in the direction of airflow. At least every fourth joint shall be provided with connecting flanges, bolted together, or of equivalent fastening security.

**1926.57(h)(5)(iii)(E)** Inspection or clean-out doors shall be provided for every 9 to 12 feet (2.736 to 3.648 m) of running length for ducts up to 12 inches (0.304 m) in diameter, but the distance between cleanout doors may be greater for larger pipes. (See 8.3.21 of American National Standard Z9.1-1951.) A clean-out door or doors shall be provided for servicing the fan, and where necessary, a drain shall be provided.

**1926.57(h)(5)(iii)(F)** Where ductwork passes through a combustible roof or wall, the roof or wall shall be protected at the point of penetration by open space or fire-resistive material between the duct and the roof or wall. When ducts pass through firewalls, they shall be provided with automatic fire dampers on both sides of the wall, except that three-eighth-inch steel plates may be used in lieu of automatic fire dampers for ducts not exceeding 18 inches (45.72 cm) in diameter.

**1926.57(h)(5)(iii)(G)** Ductwork used for ventilating any process covered in this standard shall not be connected to ducts ventilating any other process or any chimney or flue used for conveying any products of combustion.

**1926.57(h)(6)** *Velocity and air flow requirements.* (i) Except where a spray booth has an adequate air replacement system, the velocity of air into all openings of a spray booth shall be not less than that specified in Table D-57.7 for the operating conditions specified. An adequate air replacement system is one which introduces replacement air upstream or above the object being sprayed and is so designed that the velocity of air in the booth cross section is not less than that specified in Table D-57.7 when measured upstream or above the object being sprayed.

## TABLE D-57.7—MINIMUM MAINTAINED VELOCITIES INTO SPRAY BOOTHS

| Operating conditions for objects completely inside booth | Crossdraft, f.p.m. | Airflow velocities, f.p.m. Design | Airflow velocities, f.p.m. Range |
|---|---|---|---|
| Electrostatic and automatic airless operation contained in booth without operator. | Negligible | 50 large booth | 50- 75 |
| | | 100 small booth | 75-125 |
| Air-operated guns, manual or automatic | Up to 50 | 100 large booth | 75-125 |
| | | 150 small booth | 125-175 |
| Air-operated guns, manual or automatic | Up to 100 | 150 large booth | 125-175 |
| | | 200 small booth | 150-250 |

NOTES:

(1) Attention is invited to the fact that the effectiveness of the spray booth is dependent upon the relationship of the depth of the booth to its height and width.

(2) Crossdrafts can be eliminated through proper design and such design should be sought. Crossdrafts in excess of 100 fpm (feet per minute) should not be permitted.

(3) Excessive air pressures result in loss of both efficiency and material waste in addition to creating a backlash that may carry overspray and fumes into adjacent work areas.

(4) Booths should be designed with velocities shown in the column headed "Design." However, booths operating with velocities shown in the column headed "Range" are in compliance with this standard.

**1926.57(h)(6)(ii)** In addition to the requirements in paragraph (h)(6)(i) of this section the total air volume exhausted through a spray booth shall be such as to dilute solvent vapor to at least 25 percent of the lower explosive limit of the solvent being sprayed. An example of the method of calculating this volume is given below.

***Example:*** To determine the lower explosive limits of the most common solvents used in spray finishing, see Table D-57.8. Column 1 gives the number of cubic feet of vapor per gallon of solvent and column 2 gives the lower explosive limit (LEL) in percentage by volume of air. Note that the quantity of solvent will be diminished by the quantity of solids and nonflammables contained in the finish.

To determine the volume of air in cubic feet necessary to dilute the vapor from 1 gallon of solvent to 25 percent of the lower explosive limit, apply the following formula:

Dilution volume required per gallon of solvent = 4 (100-LEL) (cubic feet of vapor per gallon) ÷ LEL

Using toluene as the solvent.

(1) LEL of toluene from Table D-57.8, column 2, is 1.4 percent.

(2) Cubic feet of vapor per gallon from Table D-57.8, column 1, is 30.4 cubic feet per gallon.

(3) Dilution volume required =

4(100-1.4) 30.4 ÷ 1.4 = 8,564 cubic feet.

(4) To convert to cubic feet per minute of required ventilation, multiply the dilution volume required per gallon of solvent by the number of gallons of solvent evaporated per minute.

## TABLE D-57.8—LOWER EXPLOSIVE LIMIT OF SOME COMMONLY USED SOLVENTS

| Solvent | Cubic feet per gallon of vapor of liquid at 70°F. (21.11°C) | Lower explosive limit in percent by volume of air at 70°F. (21.11°C) |
|---|---|---|
| | *Column 1* | *Column 2* |
| Acetone | 44.0 | 2.6 |
| Amyl Acetate (iso) | 21.6 | [1] 1.0 |
| Amyl Alcohol (n) | 29.6 | 1.2 |
| Amyl Alcohol (iso). | 29.6 | 1.2 |
| Benzene | 36.8 | [1] 1.4 |
| Butyl Acetate (n) | 24.8 | 1.7 |
| Butyl Alcohol (n) | 35.2 | 1.4 |
| Butyl Cellosolve | 24.8 | 1.1 |
| Cellosolve | 33.6 | 1.8 |
| Cellosolve Acetate | 23.2 | 1.7 |
| Cyclohexanone | 31.2 | [1] 1.1 |
| 1,1 Dichloroethylene | 42.4 | 5.9 |
| 1,2 Dichloroethylene | 42.4 | 9.7 |
| Ethyl Acetate | 32.8 | 2.5 |
| Ethyl Alcohol | 55.2 | 4.3 |
| Ethyl Lactate | 28.0 | [1] 1.5 |
| Methyl Acetate | 40.0 | 3.1 |
| Methyl Alcohol | 80.8 | 7.3 |
| Methyl Cellosolve | 40.8 | 2.5 |
| Methyl Ethyl Ketone | 36.0 | 1.8 |
| Methyl n-Propyl Ketone | 30.4 | 1.5 |
| Naphtha (VM&P) (76° Naphtha) | 22.4 | 0.9 |
| Naphtha (100° Flash) Safety Solvent—Stoddard Solvent | 23.2 | 1.0 |
| Propyl Acetate (n) | 27.2 | 2.8 |
| Propyl Acetate (iso) | 28.0 | 1.1 |
| Propyl Alcohol (n) | 44.8 | 2.1 |
| Propyl Alcohol (iso) | 44.0 | 2.0 |
| Toluene | 30.4 | 1.4 |
| Turpentine | 20.8 | 0.8 |
| Xylene (o) | 26.4 | 1.0 |

[1] At 212°F (100°C).

**1926.57(h)(6)(iii)(A)** When an operator is in a booth downstream of the object being sprayed, an air-supplied respirator or other type of respirator approved by NIOSH under 42 CFR Part 84 for the material being sprayed should be used by the operator.

**1926.57(h)(6)(iii)(B)** Where downdraft booths are provided with doors, such doors shall be closed when spray painting.

**1926.57(h)(7)** *Make-up air.* (i) Clean fresh air, free of contamination from adjacent industrial exhaust systems, chimneys, stacks, or vents, shall be supplied to a spray booth or room in quantities equal to the volume of air exhausted through the spray booth.

**1926.57(h)(7)(ii)** Where a spray booth or room receives make-up air through self-closing doors, dampers, or louvers, they shall be fully open at all times when the booth or room is in use for spraying. The velocity of air through such doors, dampers, or louvers shall not exceed 200 feet per minute. If the fan characteristics are such that the required air flow through the booth will be provided, higher velocities through the doors, dampers, or louvers may be used.

**1926.57(h)(7)(iii)(A)** Where the air supply to a spray booth or room is filtered, the fan static pressure shall be calculated on the assumption that the filters are dirty to the extent that they require cleaning or replacement.

**1926.57(h)(7)(iii)(B)** The rating of filters shall be governed by test data supplied by the manufacturer of the filter. A pressure gage shall be installed to show the pressure drop across the filters. This gage shall be marked to show the pressure drop at which the filters require cleaning or replacement. Filters shall be replaced or cleaned whenever the pressure drop across them becomes excessive or whenever the air flow through the face of the booth falls below that specified in Table D-57.7.

**1926.57(h)(7)(iv)(A)** Means for heating make-up air to any spray booth or room, before or at the time spraying is normally performed, shall be provided in all places where the outdoor temperature may be expected to remain below 55°F. (12.77 °C.) for appreciable periods of time during the operation of the booth except where adequate and safe means of radiant heating for all operating personnel affected is provided. The replacement air during the heating seasons shall be maintained at not less than 65 °F. (18.33°C.) at the point of entry into the spray booth or spray room. When otherwise unheated make-up air would be at a temperature of more than 10 ° F. below room temperature, its temperature shall be regulated as provided in section 3.6.3 of ANSI Z9.2-1960.

**1926.57(h)(7)(iv)(B)** As an alternative to an air replacement system complying with the preceding section, general heating of the building in which the spray room or booth is located may be employed provided that all occupied parts of the building are maintained at not less than 65 °F. (18.33 °C.) when the exhaust system is in operation or the general heating system supplemented by other sources of heat may be employed to meet this requirement.

**1926.57(h)(7)(iv)(C)** No means of heating make-up air shall be located in a spray booth.

**1926.57(h)(7)(iv)(D)** Where make-up air is heated by coal or oil, the products of combustion shall not be allowed to mix with the make-up air, and the products of combustion shall be conducted outside the building through a flue terminating at a point remote from all points where make-up air enters the building.

**1926.57(h)(7)(iv)(E)** Where make-up air is heated by gas, and the products of combustion are not mixed with the make-up air but are conducted through an independent flue to a point outside the building remote from all points where make-up air enters the building, it is not necessary to comply with paragraph (h)(7)(iv)(F) of this section.

**1926.57(h)(7)(iv)(F)** Where make-up air to any manually operated spray booth or room is heated by gas and the products of combustion are allowed to mix with the supply air, the following precautions must be taken:

**1926.57(h)(7)(iv)(F)*(1)*** The gas must have a distinctive and strong enough odor to warn workmen in a spray booth or room of its presence if in an unburned state in the make-up air.

**1926.57(h)(7)(iv)(F)*(2)*** The maximum rate of gas supply to the make-up air heater burners must not exceed that which would yield in excess of 200 p.p.m. (parts per million) of carbon monoxide or 2,000 p.p.m. of total combustible gases in the mixture if the unburned gas upon the occurrence of flame failure were mixed with all of the make-up air supplied.

**1926.57(h)(7)(iv)(F)*(3)*** A fan must be provided to deliver the mixture of heated air and products of combustion from the plenum chamber housing the gas burners to the spray booth or room.

**1926.57(h)(8)** *Scope.* Spray booths or spray rooms are to be used to enclose or confine all spray finishing operations covered by this paragraph (h). This paragraph does not apply to the spraying of the exteriors of buildings, fixed tanks, or similar structures, nor to small portable spraying apparatus not used repeatedly in the same location.

**1926.57(i)** ***Open surface tanks***— (1) *General.* (i) This paragraph applies to all operations involving the immersion of materials in liquids, or in the vapors of such liquids, for the purpose of cleaning or altering the surface or adding to or imparting a finish thereto or changing the character of the materials, and their subsequent removal from the liquid or vapor, draining, and drying. These operations include washing, electroplating, anodizing, pickling, quenching, dying, dipping, tanning, dressing, bleaching, degreasing, alkaline cleaning, stripping, rinsing, digesting, and other similar operations.

**1926.57(i)(1)(ii)** Except where specific construction specifications are prescribed in this section, hoods, ducts, elbows, fans, blowers, and all other exhaust system parts, components, and supports thereof shall be so constructed as to meet conditions of service and to facilitate maintenance and shall conform in construction to the specifications contained in American National Standard Fundamentals Governing the Design and Operation of Local Exhaust Systems, Z9.2-1960.

**1926.57(i)(2)** *Classification of open-surface tank operations.* (i) Open-surface tank operations shall be classified into 16 classes, numbered A-1 to D-4, inclusive.

**1926.57(i)(2)(ii)** *Determination of class.* Class is determined by two factors, hazard potential designated by a letter from A to D, inclusive, and rate of gas, vapor, or mist evolution designated by a number from 1 to 4, inclusive (for example, B.3).

**1926.57(i)(2)(iii)** Hazard potential is an index, on a scale of from A to D, inclusive, of the severity of the hazard associated with the substance contained in the tank because of the toxic, flammable, or explosive nature of the vapor, gas, or mist produced therefrom. The toxic hazard is determined from the concentration, measured in parts by volume of a gas or vapor, per million parts by volume of contaminated air (p.p.m.), or in milligrams of mist per cubic meter of air (mg./m.$^3$), below which ill effects are unlikely to occur to the exposed worker. The concentrations shall be those in §1926.55 or other pertinent sections of this part.

**1926.57(i)(2)(iv)** The relative fire or explosion hazard is measured in degrees Fahrenheit in terms of the closed-cup flash point of the substance in the tank. Detailed information on the prevention of fire hazards in dip tanks may be found in Dip Tanks Containing Flammable or Combustible Liquids, NFPA No. 34-1966, National Fire Protection Association. Where the tank contains a mixture of liquids, other than organic solvents, whose effects are additive, the hygienic standard of the most toxic component (for example, the one having the lowest p.p.m. or mg./m.$^3$ shall be used, except where such substance constitutes an insignificantly small fraction of the mixture. For mixtures of organic solvents, their combined effect, rather than that of either individually, shall determine the hazard potential. In the absence of information to the contrary, the effects shall be considered as additive. If the sum of the ratios of the airborne concentration of each contaminant to the toxic concentration of that contaminant exceeds unity, the toxic concentration shall be considered to have been exceeded. (See Note A to paragraph (i)(2)(v) of this section.)

**1926.57(i)(2)(v)** Hazard potential shall be determined from Table D-57.9, with the value indicating greater hazard being used. When the hazardous material may be either a vapor with a threshhold limit value *(TLV)* in p.p.m. or a mist with a *TLV* in mg./m.$^3$, the *TLV* indicating the greater hazard shall be used (for example, A takes precedence over B or C; B over C; C over D).

**Note A:**

$$(c_1 \div TLV_1) + (c_2 \div TLV_2) + (c_3 \div TLV_3) + \ldots (c_N \div TLV_N)1$$

Where:

c = Concentration measured at the operation in p.p.m.

## TABLE D-57.9— DETERMINATION OF HAZARD POTENTIAL

| Hazard potential | Toxicity Group: Gas or vapor (p.p.m.) | Mist (mg./m[3]) | Flash point in degrees F. (C.) |
|---|---|---|---|
| A | 0-10 | 0-0.1 | |
| B | 11-100 | 0.11-1.0 | Under 100 (37.77) |
| C | 101-500 | 1.1-10 | 100-200 (37.77-93.33) |
| D | Over 500 | Over 10 | Over 200 (93.33) |

**1926.57(i)(2)(vi)** Rate of gas, vapor, or mist evolution is a numerical index, on a scale of from 1 to 4, inclusive, both of the relative capacity of the tank to produce gas, vapor, or mist and of the relative energy with which it is projected or carried upwards from the tank. Rate is evaluated in terms of

**1926.57(i)(2)(vi)(A)** The temperature of the liquid in the tank in degrees Fahrenheit;

**1926.57(i)(2)(vi)(B)** The number of degrees Fahrenheit that this temperature is below the boiling point of the liquid in degrees Fahrenheit;

**1926.57(i)(2)(vi)(C)** The relative evaporation of the liquid in still air at room temperature in an arbitrary scale—fast, medium, slow, or nil; and

**1926.57(i)(2)(vi)(D)** The extent that the tank gases or produces mist in an arbitrary scale—high, medium, low, and nil. (See Table D-57.10, Note 2.) Gassing depends upon electrochemical or mechanical processes, the effects of which have to be individually evaluated for each installation (see Table D-57.10, Note 3).

**1926.57(i)(2)(vii)** Rate of evolution shall be determined from Table D-57.10. When evaporation and gassing yield different rates, the lowest numerical value shall be used.

## TABLE D-57.10—DETERMINATION OF RATE OF GAS, VAPOR, OR MIST EVOLUTION[1]

| Rate | Liquid temperature, °F. (°C) | Degrees below boiling point | Relative evaporation[2] | Gassing[3] |
|---|---|---|---|---|
| 1 | Over 200 (93.33) | 0-20 | Fast | High. |
| 2 | 150-200 (65.55-93.33) | 21-50 | Medium | Medium. |
| 3 | 94-149 (34.44-65) | 51-100 | Slow | Low. |
| 4 | Under 94 (34.44) | Over 100 | Nil | Nil. |

1 In certain classes of equipment, specifically vapor degreasers, an internal condenser or vapor level thermostat is used to prevent the vapor from leaving the tank during normal operation. In such cases, rate of vapor evolution from the tank into the workroom is not dependent upon the factors listed in the table, but rather upon abnormalities of operating procedure, such as carryout of vapors from excessively fast action, dragout of liquid by entrainment in parts, contamination of solvent by water and other materials, or improper heat balance. When operating procedure is excellent, effective rate of evolution may be taken as 4. When operating procedure is average, the effective rate of evolution may be taken as 3. When operation is poor, a rate of 2 or 1 is indicated, depending upon observed conditions.

2 Relative evaporation rate is determined according to the methods described by A. K. Doolittle in Industrial and Engineering Chemistry, vol. 27, p. 1169, (3) where time for 100-percent evaporation is as follows: Fast: 0-3 hours; Medium: 3-12 hours; Slow: 12-50 hours; Nil: more than 50 hours.

3 Gassing means the formation by chemical or electrochemical action of minute bubbles of gas under the surface of the liquid in the tank and is generally limited to aqueous solutions.

**1926.57(i)(3)** *Ventilation.* Where ventilation is used to control potential exposures to workers as defined in paragraph (i)(2)(iii) of this section, it shall be adequate to reduce the concentration of the air contaminant to the degree that a hazard to the worker does not exist. Methods of ventilation are discussed in American National Standard Fundamentals Governing the Design and Operation of Local Exhaust Systems, Z9.2-1960.

**1926.57(i)(4)** *Control requirements.* (i) Control velocities shall conform to Table D-57.11 in all cases where the flow of air past the breathing or working zone of the operator and into the hoods is undisturbed by local environmental conditions, such as open windows, wall fans, unit heaters, or moving machinery.

**1926.57(i)(4)(ii)** All tanks exhausted by means of hoods which

**1926.57(i)(4)(ii)(A)** Project over the entire tank;

**1926.57(i)(4)(ii)(B)** Are fixed in position in such a location that the head of the workman, in all his normal operating positions while working at the tank, is in front of all hood openings; and

**1926.57(i)(4)(ii)(C)** Are completely enclosed on at least two sides, shall be considered to be exhausted through an enclosing hood.

**1926.57(i)(4)(ii)(D)** The quantity of air in cubic feet per minute necessary to be exhausted through an enclosing hood shall be not less than the product of the control velocity times the net area of all openings in the enclosure through which air can flow into the hood.

## TABLE D-57.11—CONTROL VELOCITIES IN FEET PER MINUTE (F.P.M.) FOR UNDISTURBED LOCATIONS

| Class | Enclosing hood: One open side | Enclosing hood: Two open sides | Lateral exhaust[1] | Canopy hood[2]: Three open sides | Canopy hood[2]: Four open sides |
|---|---|---|---|---|---|
| B-1 and A-2 | 100 | 150 | 150 | Do not use | Do not use |
| A-3[2], B-1, B-2, and C-1 | 75 | 100 | 100 | 125 | 175 |
| B-3, C-2, and D-1[3] | 65 | 90 | 75 | 100 | 150 |
| B4[2], C-3, and D-2[3] | 50 | 75 | 50 | 75 | 125 |
| A-4, C-4, D-3[3], and D-4 | | | | | |

1. See Table G-15 for computation of ventilation rate.
2. Do not use canopy hood for Hazard Potential A processes.
3. Where complete control of hot water is desired, design as next highest class.
4. General room ventilation required.

**1926.57(i)(4)(iii)** All tanks exhausted by means of hoods which do not project over the entire tank, and in which the direction of air movement into the hood or hoods is substantially horizontal, shall be considered to be laterally exhausted. The quantity of air in cubic feet per minute necessary to be laterally exhausted per square foot of tank area in order to maintain the required control velocity shall be determined from Table D-57.12 for all variations in ratio of tank width *(W)* to tank length *(L)*. The total quantity of air in cubic feet per minute required to be exhausted per tank shall be not less than the product of the area of tank surface times the cubic feet per minute per square foot of tank area, determined from Table D-57.12.

**1926.57(i)(4)(iii)(A)** For lateral exhaust hoods over 42 inches (1.06 m) wide, or where it is desirable to reduce the amount of air removed from the workroom, air supply slots or orifices shall be provided along the side or the center of the tank opposite from the exhaust slots. The design of such systems shall meet the following criteria:

**1926.57(i)(4)(iii)(A)*(1)*** The supply air volume plus the entrained air shall not exceed 50 percent of the exhaust volume.

**1926.57(i)(4)(iii)(A)*(2)*** The velocity of the supply airstream as it reaches the effective control area of the exhaust slot shall be less than the effective velocity over the exhaust slot area.

## TABLE D-57.12—MINIMUM VENTILATION RATE IN CUBIC FEET OF AIR PER MINUTE PER SQUARE FOOT OF TANK AREA FOR LATERAL EXHAUST

| Required minimum control velocity, f.p.m. (from Table D-57.11) | C.f.m. per sq. ft. to maintain required minimum velocities at following ratios (tank width (*W*)/tank length (*L*)).[1,2] | | | | |
|---|---|---|---|---|---|
| | 0.0-0.09 | 0.1-0.24 | 0.25-0.49 | 0.5-0.99 | 1.0-2.0 |
| Hood along one side or two parallel sides of tank when one hood is against a wall or baffle.[2] Also for a manifold along tank centerline.[3] | | | | | |
| 50 | 50 | 60 | 75 | 90 | 100 |
| 75 | 75 | 90 | 110 | 130 | 150 |
| 100 | 100 | 125 | 150 | 175 | 200 |
| 150 | 150 | 190 | 225 | 260 | 300 |
| Hood along one side or two parallel sides of free standing tank not against wall or baffle. | | | | | |
| 50 | 75 | 90 | 100 | 110 | 125 |
| 75 | 110 | 130 | 150 | 170 | 190 |
| 100 | 150 | 175 | 200 | 225 | 250 |
| 150 | 225 | 260 | 300 | 340 | 375 |

[1] It is not practicable to ventilate across the long dimension of a tank whose ratio $W/L$ exceeds 2.0.
It is undesirable to do so when $W/L$ exceeds 1.0. For circular tanks with lateral exhaust along up to ½ the circumference, use $W/L=1.0$; for over one-half the circumference use $W/L=0.5$.

[2] Baffle is a vertical plate the same length as the tank, and with the top of the plate as high as the tank is wide. If the exhaust hood is on the side of a tank against a building wall or close to it, it is perfectly baffled.

[3] Use $W/2$ as tank width in computing when manifold is along centerline, or when hoods are used on two parallel sides of a tank.
Tank Width (*W*) means the effective width over which the hood must pull air to operate (for example, where the hood face is set back from the edge of the tank, this set back must be added in measuring tank width). The surface area of tanks can frequently be reduced and better control obtained (particularly on conveyorized systems) by using covers extending from the upper edges of the slots toward the center of the tank.

**1926.57(i)(4)(iii)(A)*(3)*** The vertical height of the receiving exhaust hood, including any baffle, shall not be less than one-quarter the width of the tank.

**1926.57(i)(4)(iii)(A)*(4)*** The supply airstream shall not be allowed to impinge on obstructions between it and the exhaust slot in such a manner as to significantly interfere with the performance of the exhaust hood.

**1926.57(i)(4)(iii)(A)*(5)*** Since most failure of push-pull systems result from excessive supply air volumes and pressures, methods of measuring and adjusting the supply air shall be provided. When satisfactory control has been achieved, the adjustable features of the hood shall be fixed so that they will not be altered.

**1926.57(i)(4)(iv)** All tanks exhausted by means of hoods which project over the entire tank, and which do not conform to the definition of enclosing hoods, shall be considered to be overhead canopy hoods. The quantity of air in cubic feet per minute necessary to be exhausted through a canopy hood shall be not less than the product of the control velocity times the net area of all openings between the bottom edges of the hood and the top edges of the tank.

**1926.57(i)(4)(v)** The rate of vapor evolution (including steam or products of combustion) from the process shall be estimated. If the rate of vapor evolution is equal to or greater than 10 percent of the calculated exhaust volume required, the exhaust volume shall be increased in equal amount.

**1926.57(i)(5)** *Spray cleaning and degreasing.* Wherever spraying or other mechanical means are used to disperse a liquid above an open-surface tank, control must be provided for the airborne spray. Such operations shall be enclosed as completely as possible. The inward air velocity into the enclosure shall be sufficient to prevent the discharge of spray into the workroom. Mechanical baffles may be used to help prevent the discharge of spray. Spray painting operations are covered by paragraph (h) of this section.

**1926.57(i)(6)** *Control means other than ventilation.* Tank covers, foams, beads, chips, or other materials floating on the tank surface so as to confine gases, mists, or vapors to the area under the cover or to the foam, bead, or chip layer; or surface tension depressive agents added to the liquid in the tank to minimize mist formation, or any combination thereof, may all be used as gas, mist, or vapor control means for open-surface tank operations, provided that they effectively reduce the concentrations of hazardous materials in the vicinity of the worker below the limits set in accordance with paragraph (i)(2) of this section.

**1926.57(i)(7)** *System design.* (i) The equipment for exhausting air shall have sufficient capacity to produce the flow of air required in each of the hoods and openings of the system.

**1926.57(i)(7)(ii)** The capacity required in paragraph (i)(7)(i) of this section shall be obtained when the airflow producing equipment is operating against the following pressure losses, the sum of which is the static pressure:

**1926.57(i)(7)(ii)(A)** Entrance losses into the hood.

**1926.57(i)(7)(ii)(B)** Resistance to airflow in branch pipe including bends and transformations.

**1926.57(i)(7)(ii)(C)** Entrance loss into the main pipe.

**1926.57(i)(7)(ii)(D)** Resistance to airflow in main pipe including bends and transformations.

**1926.57(i)(7)(ii)(E)** Resistance of mechanical equipment; that is, filters, washers, condensers, absorbers, etc., plus their entrance and exit losses.

**1926.57(i)(7)(ii)(F)** Resistance in outlet duct and discharge stack.

**1926.57(i)(7)(iii)** Two or more operations shall not be connected to the same exhaust system where either one or the combination of the substances removed may constitute a fire, explosion, or chemical reaction hazard in the duct system. Traps or other devices shall be provided to insure that condensate in ducts does not drain back into any tank.

**1926.57(i)(7)(iv)** The exhaust system, consisting of hoods, ducts, air mover, and discharge outlet, shall be designed in accordance with American National Standard Fundamentals Governing the Design and Operation of Local Exhaust Systems, Z9.2-1960, or the manual, Industrial Ventilation, published by the American Conference of Governmental Industrial Hygienists 1970. Airflow and pressure loss data provided by the manufacturer of any air cleaning device shall be included in the design calculations.

**1926.57(i)(8)** *Operation.* (i) The required airflow shall be maintained at all times during which gas, mist, or vapor is emitted from the tank, and at all times the tank, the draining, or the drying area is in operation or use. When the system is first installed, the airflow from each hood shall be measured by means of a pitot traverse in the exhaust duct and corrective action taken if the flow is less than that required. When the proper flow is obtained, the hood static pressure shall be measured and recorded. At intervals of not more than 3 months operation, or after a prolonged shutdown period, the hoods and duct system shall be inspected for evidence of corrosion or damage. In any case where the airflow is found to be less than required, it shall be increased to the required value. (Information on airflow and static pressure measurement and calculations may be found in American National Standard Fundamental Governing the Design and Operation of Local Exhaust Systems, Z9.2-1960, or in the manual, Industrial Ventilation, published by the American Conference of Governmental Industrial Hygienists.)

**1926.57(i)(8)(ii)** The exhaust system shall discharge to the outer air in such a manner that the possibility of its effluent entering any building is at a minimum. Recirculation shall only be through a device for contaminant removal which will prevent the creation of a health hazard in the room or area to which the air is recirculated.

**1926.57(i)(8)(iii)** A volume of outside air in the range of 90 percent to 110 percent of the exhaust volume shall be provided to each room having exhaust hoods. The outside air supply shall enter the workroom in such a manner as not to be detrimental to any exhaust hood. The airflow of the makeup air system shall be measured on installation. Corrective action shall be taken when the airflow is below that required. The makeup air shall be uncontaminated.

**1926.57(i)(9)** *Personal protection.* (i) All employees working in and around open-surface tank operations must be instructed as to the hazards of their respective jobs, and in the personal protection and first aid procedures applicable to these hazards.

**1926.57(i)(9)(ii)** All persons required to work in such a manner that their feet may become wet shall be provided with rubber or other impervious boots or shoes, rubbers, or wooden-soled shoes sufficient to keep feet dry.

**1926.57(i)(9)(iii)** All persons required to handle work wet with a liquid other than water shall be provided with gloves impervious to such a liquid and of a length sufficient to prevent entrance of liquid into the tops of the gloves. The interior of gloves shall be kept free from corrosive or irritating contaminants.

**1926.57(i)(9)(iv)** All persons required to work in such a manner that their clothing may become wet shall be provided with such aprons, coats, jackets, sleeves, or other garments made of rubber, or of other materials impervious to liquids other than water, as are required to keep their clothing dry. Aprons shall extend well below the top of boots to prevent liquid splashing into the boots. Provision of dry, clean, cotton clothing along with rubber shoes or short boots and an apron impervious to liquids other than water shall be considered a satisfactory substitute where small parts are cleaned, plated, or acid dipped in open tanks and rapid work is required.

**1926.57(i)(9)(v)** Whenever there is a danger of splashing, for example, when additions are made manually to the tanks, or when acids and chemicals are removed from the tanks, the employees so engaged shall be required to wear either tight-fitting chemical goggles or an effective face shield. See §1926.102.

**1926.57(i)(9)(vi)** When, during the emergencies specified in paragraph (i)(11)(v) of this section, employees must be in areas where concentrations of air contaminants are greater than the limits set by paragraph (i)(2)(iii) of this section or oxygen concentrations are less than 19.5 percent, they must use respirators that reduce their exposure to a level below these limits or that provide adequate oxygen. Such respirators must also be provided in marked, quickly-accessible storage compartments built for this purpose when the possibility exists of accidental release of hazardous concentrations of air contaminants. Respirators must be approved by NIOSH under 42 CFR part 84, selected by a competent industrial hygienist or other technically-qualified source, and used in accordance with 29 CFR 1926.103.

**1926.57(i)(9)(vii)** Near each tank containing a liquid which may burn, irritate, or otherwise be harmful to the skin if splashed upon the worker's body, there shall be a supply of clean cold water. The water pipe (carrying a pressure not exceeding 25 pounds (11.325 kg)) shall be provided with a quick opening valve and at least 48 inches (1.216 m) of hose not smaller than three-fourths inch, so that no time may be lost in washing off liquids from the skin or clothing. Alternatively, deluge showers and eye flushes shall be provided in cases where harmful chemicals may be splashed on parts of the body.

**1926.57(i)(9)(viii)** Operators with sores, burns, or other skin lesions requiring medical treatment shall not be allowed to work at their regular operations until so authorized by a physician. Any small skin abrasions, cuts, rash, or open sores which are found or reported shall be treated by a properly designated person so that chances of exposures to the chemicals are removed. Workers exposed to chromic acids shall have a periodic examination made of the nostrils and other parts of the body, to detect incipient ulceration.

**1926.57(i)(9)(ix)** Sufficient washing facilities, including soap, individual towels, and hot water, shall be provided for all persons required to use or handle any liquids which may burn, irritate, or otherwise be harmful to the skin, on the basis of at least one basin (or its equivalent) with a hot water faucet for every 10 employees. See §1926.51(f).

**1926.57(i)(9)(x)** Locker space or equivalent clothing storage facilities shall be provided to prevent contamination of street clothing.

**1926.57(i)(9)(xi)** First aid facilities specific to the hazards of the operations conducted shall be readily available.

**1926.57(i)(10)** *Special precautions for cyanide.* Dikes or other arrangements shall be provided to prevent the possibility of intermixing of cyanide and acid in the event of tank rupture.

**1926.57(i)(11)** *Inspection, maintenance, and installation.* (i) Floors and platforms around tanks shall be prevented from becoming slippery both by original type of construction and by frequent flushing. They shall be firm, sound, and of the design and construction to minimize the possibility of tripping.

**1926.57(i)(11)(ii)** Before cleaning the interior of any tank, the contents shall be drained off, and the cleanout doors shall be opened where provided. All pockets in tanks or pits, where it is possible for hazardous vapors to collect, shall be ventilated and cleared of such vapors.

**1926.57(i)(11)(iii)** Tanks which have been drained to permit employees to enter for the purposes of cleaning, inspection, or maintenance may contain atmospheres which are hazardous to life or health, through the presence of flammable or toxic air contaminants, or through the absence of sufficient oxygen. Before employees shall be permitted to enter any such tank, appropriate tests of the atmosphere shall be made to determine if the limits set by paragraph (i)(2)(iii) of this section are exceeded, or if the oxygen concentration is less than 19.5 percent.

**1926.57(i)(11)(iv)** If the tests made in accordance with paragraph (i)(11)(iii) of this section indicate that the atmosphere in the tank is unsafe, before any employee is permitted to enter the tank, the tank shall be ventilated until the hazardous atmosphere is removed, and ventilation shall be continued so as to prevent the occurrence of a hazardous atmosphere as long as an employee is in the tank.

**1926.57(i)(11)(v)** If, in emergencies, such as rescue work, it is necessary to enter a tank which may contain a hazardous atmosphere, suitable respirators, such as self-contained breathing apparatus; hose mask with blower, if there is a possibility of oxygen deficiency; or a gas mask, selected and operated in accordance with paragraph (i)(9)(vi) of this section, shall be used. If a contaminant in the tank can cause dermatitis, or be absorbed through the skin, the employee entering the tank shall also wear protective clothing. At least one trained standby employee, with suitable respirator, shall be present in the nearest uncontaminated area. The standby employee must be able to communicate with the employee in the tank and be able to haul him out of the tank with a lifeline if necessary.

**1926.57(i)(11)(vi)** Maintenance work requiring welding or open flame, where toxic metal fumes such as cadmium, chromium, or lead may be evolved, shall be done only with sufficient local exhaust ventilation to prevent the creation of a health hazard, or be done with respirators selected and used in accordance with paragraph (i)(9)(vi) of this section. Welding, or the use of open flames near any solvent cleaning equipment shall be permitted only after such equipment has first been thoroughly cleared of solvents and vapors.

**1926.57(i)(12)** *Vapor degreasing tanks.* (i) In any vapor degreasing tank equipped with a condenser or vapor level thermostat, the condenser or thermostat shall keep the level of vapors below the top edge of the tank by a distance at least equal to one-half the tank width, or at least 36 inches (0.912 m), whichever is shorter.

**1926.57(i)(12)(ii)** Where gas is used as a fuel for heating vapor degreasing tanks, the combustion chamber shall be of tight construction, except for such openings as the exhaust flue, and those that are necessary for supplying air for combustion. Flues shall be of corrosion-resistant construction and shall extend to the outer air. If mechanical exhaust is used on this flue, a draft diverter shall be used. Special precautions must be taken to prevent solvent fumes from entering the combustion air of this or any other heater when chlorinated or fluorinated hydrocarbon solvents (for example, trichloroethylene, Freon) are used.

**1926.57(i)(12)(iii)** Heating elements shall be so designed and maintained that their surface temperature will not cause the solvent or mixture to decompose, break down, or be converted into an excessive quantity of vapor.

**1926.57(i)(12)(iv)** Tanks or machines of more than 4 square feet (0.368 $m^2$) of vapor area, used for solvent cleaning or vapor degreasing, shall be equipped with suitable cleanout or sludge doors located near the bottom of each tank or still. These doors shall be so designed and gasketed that there will be no leakage of solvent when they are closed.

**1926.57(i)(13)** *Scope.* (i) This paragraph (i) applies to all operations involving the immersion of materials in liquids, or in the vapors of such liquids, for the purpose of cleaning or altering their surfaces, or adding or imparting a finish thereto, or changing the character of the materials, and their subsequent removal from the liquids or vapors, draining, and drying. Such operations include washing, electroplating, anodizing, pickling, quenching, dyeing, dipping, tanning, dressing, bleaching, degreasing, alkaline cleaning, stripping, rinsing, digesting, and other similar operations, but do not include molten materials handling operations, or surface coating operations.

1926.57(i)(13)(ii) *Molten materials handling operations* means all operations, other than welding, burning, and soldering operations, involving the use, melting, smelting, or pouring of metals, alloys, salts, or other similar substances in the molten state. Such operations also include heat treating baths, descaling baths, die casting stereotyping, galvanizing, tinning, and similar operations.

1926.57(i)(13)(iii) *Surface coating operations* means all operations involving the application of protective, decorative, adhesive, or strengthening coating or impregnation to one or more surfaces, or into the interstices of any object or material, by means of spraying, spreading, flowing, brushing, roll coating, pouring, cementing, or similar means; and any subsequent draining or drying operations, excluding open-tank operations.

[44 FR 8577, Feb. 9, 1979; 44 FR 20940, Apr. 6, 1979, as amended at 58 FR 35099, June 30, 1993; 61 FR 9250, Mar. 3, 1996; 63 FR 1295, Jan. 8, 1998]

### §1926.58 [Reserved]

### §1926.59 Hazard communication.

**Note:** The requirements applicable to construction work under this section are identical to those set forth at §1910.1200 of this chapter.

[61 FR 31431, June 20, 1996]

[Section 1910.1200 is printed below—CCH.]

### §1910.1200 Hazard communication.

1910.1200(a) ***Purpose.*** (1) The purpose of this section is to ensure that the hazards of all chemicals produced or imported are evaluated, and that information concerning their hazards is transmitted to employers and employees. This transmittal of information is to be accomplished by means of comprehensive hazard communication programs, which are to include container labeling and other forms of warning, material safety data sheets and employee training.

1910.1200(a)(2) This occupational safety and health standard is intended to address comprehensively the issue of evaluating the potential hazards of chemicals, and communicating information concerning hazards and appropriate protective measures to employees, and to preempt any legal requirements of a state, or political subdivision of a state, pertaining to this subject. Evaluating the potential hazards of chemicals, and communicating information concerning hazards and appropriate protective measures to employees, may include, for example, but is not limited to, provisions for: developing and maintaining a written hazard communication program for the workplace, including lists of hazardous chemicals present; labeling of containers of chemicals in the workplace, as well as of containers of chemicals being shipped to other workplaces; preparation and distribution of material safety data sheets to employees and downstream employers; and development and implementation of employee training programs regarding hazards of chemicals and protective measures. Under section 18 of the Act, no state or political subdivision of a state may adopt or enforce, through any court or agency, any requirement relating to the issue addressed by this Federal standard, except pursuant to a Federally-approved state plan.

1910.1200(b) ***Scope and application.*** (1) This section requires chemical manufacturers or importers to assess the hazards of chemicals which they produce or import, and all employers to provide information to their employees about the hazardous chemicals to which they are exposed, by means of a hazard communication program, labels and other forms of warning, material safety data sheets, and information and training. In addition, this section requires distributors to transmit the required information to employers. (Employers who do not produce or import chemicals need only focus on those parts of this rule that deal with establishing a workplace program and communicating information to their workers. Appendix E of this section is a general guide for such employers to help them determine their compliance obligations under the rule.)

1910.1200(b)(2) This section applies to any chemical which is known to be present in the workplace in such a manner that employees may be exposed under normal conditions of use or in a foreseeable emergency.

1910.1200(b)(3) This section applies to laboratories only as follows:

1910.1200(b)(3)(i) Employers shall ensure that labels on incoming containers of hazardous chemicals are not removed or defaced;

1910.1200(b)(3)(ii) Employers shall maintain any material safety data sheets that are received with incoming shipments of hazardous chemicals, and ensure that they are readily accessible during each workshift to laboratory employees when they are in their work areas;

1910.1200(b)(3)(iii) Employers shall ensure that laboratory employees are provided information and training in accordance with paragraph (h) of this section, except for the location and availability of the written hazard communication program under paragraph (h)(2)(iii) of this section; and,

1910.1200(b)(3)(iv) Laboratory employers that ship hazardous chemicals are considered to be either a chemical manufacturer or a distributor under this rule, and thus must ensure that any containers of hazardous chemicals leaving the laboratory are labeled in accordance with paragraph (f)(1) of this section, and that a material safety data sheet is provided to distributors and other employers in accordance with paragraphs (g)(6) and (g)(7) of this section.

1910.1200(b)(4) In work operations where employees only handle chemicals in sealed containers which are not opened under normal conditions of use (such as are found in marine cargo handling, warehousing, or retail sales), this section applies to these operations only as follows:

1910.1200(b)(4)(i) Employers shall ensure that labels on incoming containers of hazardous chemicals are not removed or defaced;

1910.1200(b)(4)(ii) Employers shall maintain copies of any material safety data sheets that are received with incoming shipments of the sealed containers of hazardous chemicals, shall obtain a material safety data sheet as soon as possible for sealed containers of hazardous chemicals received without a material safety data sheet if an employee requests the material safety data sheet, and shall ensure that the material safety data sheets are readily accessible during each work shift to employees when they are in their work area(s); and,

1910.1200(b)(4)(iii) Employers shall ensure that employees are provided with information and training in accordance with paragraph (h) of this section (except for the location and availability of the written hazard communication program under paragraph (h)(2)(iii) of this section), to the extent necessary to protect them in the event of a spill or leak of a hazardous chemical from a sealed container.

1910.1200(b)(5) This section does not require labeling of the following chemicals:

1910.1200(b)(5)(i) Any pesticide as such term is defined in the Federal Insecticide, Fungicide, and Rodenticide Act (7 U.S.C. 136 *et seq.*), when subject to the labeling requirements of that Act and labeling regulations issued under that Act by the Environmental Protection Agency;

1910.1200(b)(5)(ii) Any chemical substance or mixture as such terms are defined in the Toxic Substances Control Act (15 U.S.C. 2601 *et seq.*), when subject to the labeling requirements of that Act and labeling regulations issued under that Act by the Environmental Protection Agency.

1910.1200(b)(5)(iii) Any food, food additive, color additive, drug, cosmetic, or medical or veterinary device or product, including materials intended for use as ingredients in such products (*e.g.* flavors and fragrances), as such terms are defined in the Federal Food, Drug, and Cosmetic Act (21 U.S.C. 301 *et seq.*) or the Virus-Serum-Toxin Act of 1913 (21 U.S.C. 151 *et seq.*), and regulations issued under those Acts, when they are subject to the labeling requirements under those Acts by either the Food and Drug Administration or the Department of Agriculture;

1910.1200(b)(5)(iv) Any distilled spirits (beverage alcohols), wine, or malt beverage intended for nonindustrial use, as such terms are defined in the Federal Alcohol Administration Act (27 U.S.C. 201 *et seq.*) and regulations issued under that Act, when subject to the labeling requirements of that Act and labeling regulations issued under that Act by the Bureau of Alcohol, Tobacco, and Firearms;

1910.1200(b)(5)(v) Any consumer product or hazardous substance as those terms are defined in the Consumer Product Safety Act

(15 U.S.C. 2051 *et seq.*) and Federal Hazardous Substances Act (15 U.S.C. 1261 *et seq.*) respectively, when subject to a consumer product safety standard or labeling requirement of those Acts, or regulations issued under those Acts by the Consumer Product Safety Commission; and,

**1910.1200(b)(5)(vi)** Agricultural or vegetable seed treated with pesticides and labeled in accordance with the Federal Seed Act (7 U.S.C. 1551 *et seq.*) and the labeling regulations issued under that Act by the Department of Agriculture.

**1910.1200(b)(6)** This section does not apply to: (i) Any hazardous waste as such term is defined by the Solid Waste Disposal Act, as amended by the Resource Conservation and Recovery Act of 1976, as amended (42 U.S.C. 6901 *et seq.*), when subject to regulations issued under that Act by the Environmental Protection Agency;

**1910.1200(b)(6)(ii)** Any hazardous substance as such term is defined by the Comprehensive Environmental Response, Compensation and Liability ACT (CERCLA) (42 U.S.C. 9601 *et seq.*) when the hazardous substance is the focus of remedial or removal action being conducted under CERCLA in accordance with Environmental Protection Agency regulations;

**1910.1200(b)(6)(iii)** Tobacco or tobacco products;

**1910.1200(b)(6)(iv)** Wood or wood products, including lumber which will not be processed, where the chemical manufacturer or importer can establish that the only hazard they pose to employees is the potential for flammability or combustibility (wood or wood products which have been treated with a hazardous chemical covered by this standard, and wood which may be subsequently sawed or cut, generating dust, are not exempted);

**1910.1200(b)(6)(v)** Articles (as that term is defined in paragraph (c) of this section);

**1910.1200(b)(6)(vi)** Food or alcoholic beverages which are sold, used, or prepared in a retail establishment (such as a grocery store, restaurant, or drinking place), and foods intended for personal consumption by employees while in the workplace;

**1910.1200(b)(6)(vii)** Any drug, as that term is defined in the Federal Food, Drug, and Cosmetic Act (21 U.S.C. 301 *et seq.*), when it is in solid, final form for direct administration to the patient (*e.g.*, tablets or pills); drugs which are packaged by the chemical manufacturer for sale to consumers in a retail establishment (*e.g.*, over-the-counter drugs); and drugs intended for personal consumption by employees while in the workplace (*e.g.*, first aid supplies);

**1910.1200(b)(6)(viii)** Cosmetics which are packaged for sale to consumers in a retail establishment, and cosmetics intended for personal consumption by employees while in the workplace;

**1910.1200(b)(6)(ix)** Any consumer product or hazardous substance, as those terms are defined in the Consumer Product Safety Act (15 U.S.C. 2051 *et seq.*) and Federal Hazardous Substances Act (15 U.S.C. 1261 *et seq.*) respectively, where the employer can show that it is used in the workplace for the purpose intended by the chemical manufacturer or importer of the product, and the use results in a duration and frequency of exposure which is not greater than the range of exposures that could reasonably be experienced by consumers when used for the purpose intended;

**1910.1200(b)(6)(x)** Nuisance particulates where the chemical manufacturer or importer can establish that they do not pose any physical or health hazard covered under this section;

**1910.1200(b)(6)(xi)** Ionizing and nonionizing radiation; and,

**1910.1200(b)(6)(xii)** Biological hazards.

**1910.1200(c)** ***Definitions.***

*Article* means a manufactured item other than a fluid or particle: (i) which is formed to a specific shape or design during manufacture; (ii) which has end use function(s) dependent in whole or in part upon its shape or design during end use; and (iii) which under normal conditions of use does not release more than very small quantities, *e.g.*, minute or trace amounts of a hazardous chemical (as determined under paragraph (d) of this section), and does not pose a physical hazard or health risk to employees.

*Assistant Secretary* means the Assistant Secretary of Labor for Occupational Safety and Health, U.S. Department of Labor, or designee.

*Chemical* means any element, chemical compound or mixture of elements and/or compounds.

*Chemical manufacturer* means an employer with a workplace where chemical(s) are produced for use or distribution.

*Chemical name* means the scientific designation of a chemical in accordance with the nomenclature system developed by the International Union of Pure and Applied Chemistry (IUPAC) or the Chemical Abstracts Service (CAS) rules of nomenclature, or a name which will clearly identify the chemical for the purpose of conducting a hazard evaluation.

*Combustible liquid* means any liquid having a flashpoint at or above 100 °F (37.8 °C), but below 200 °F (93.3 °C), except any mixture having components with flashpoints of 200 °F (93.3 °C), or higher, the total volume of which make up 99 percent or more of the total volume of the mixture.

*Commercial account* means an arrangement whereby a retail distributor sells hazardous chemicals to an employer, generally in large quantities over time and/or at costs that are below the regular retail price.

*Common name* means any designation or identification such as code name, code number, trade name, brand name or generic name used to identify a chemical other than by its chemical name.

*Compressed gas* means:

(i) A gas or mixture of gases having, in a container, an absolute pressure exceeding 40 psi at 70 °F (21.1 °C); or

(ii) A gas or mixture of gases having, in a container, an absolute pressure exceeding 104 psi at 130 °F (54.4 °C) regardless of the pressure at 70 °F (21.1 °C); or

(iii) A liquid having a vapor pressure exceeding 40 psi at 100 °F (37.8 °C) as determined by ASTM D-323-72.

*Container* means any bag, barrel, bottle, box, can, cylinder, drum, reaction vessel, storage tank, or the like that contains a hazardous chemical. For purposes of this section, pipes or piping systems, and engines, fuel tanks, or other operating systems in a vehicle, are not considered to be containers.

*Designated representative* means any individual or organization to whom an employee gives written authorization to exercise such employee's rights under this section. A recognized or certified collective bargaining agent shall be treated automatically as a designated representative without regard to written employee authorization.

*Director* means the Director, National Institute for Occupational Safety and Health, U.S. Department of Health and Human Services, or designee.

*Distributor* means a business, other than a chemical manufacturer or importer, which supplies hazardous chemicals to other distributors or to employers.

*Employee* means a worker who may be exposed to hazardous chemicals under normal operating conditions or in foreseeable emergencies. Workers such as office workers or bank tellers who encounter hazardous chemicals only in non-routine, isolated instances are not covered.

*Employer* means a person engaged in a business where chemicals are either used, distributed, or are produced for use or distribution, including a contractor or subcontractor.

*Explosive* means a chemical that causes a sudden, almost instantaneous release of pressure, gas, and heat when subjected to sudden shock, pressure, or high temperature.

*Exposure* or *exposed* means that an employee is subjected in the course of employment to a chemical that is a physical or health hazard, and includes potential (*e.g.* accidental or possible) exposure. "Subjected" in terms of health hazards includes any route of entry (*e.g.* inhalation, ingestion, skin contact or absorption.)

*Flammable* means a chemical that falls into one of the following categories:

(i) *Aerosol, flammable* means an aerosol that, when tested by the method described in 16 CFR 1500.45, yields a flame projection exceeding 18 inches at full valve opening, or a flashback (a flame extending back to the valve) at any degree of valve opening;

(ii) *Gas, flammable* means: (A) A gas that, at ambient temperature and pressure, forms a flammable mixture with air at a concentration of thirteen (13) percent by volume or less; or (ii)(B) A gas that, at ambient temperature and pressure, forms a range of flammable mixtures with air wider than twelve (12) percent by volume, regardless of the lower limit;

(iii) *Liquid, flammable* means any liquid having a flashpoint below 100 °F (37.8 °C), except any mixture having components with flashpoints of 100 °F (37.8 °C) or higher, the total of which make up 99 percent or more of the total volume of the mixture.

(iv) *Solid, flammable* means a solid, other than a blasting agent or explosive as defined in §1910.109(a), that is liable to cause fire through friction, absorption of moisture, spontaneous chemical change, or retained heat from manufacturing or processing, or which can be ignited readily and when ignited burns so vigorously and persistently as to create a serious hazard. A chemical shall be considered to be a flammable solid if, when tested by the method described in 16 CFR 1500.44, it ignites and burns with a self-sustained flame at a rate greater than one-tenth of an inch per second along its major axis.

*Flashpoint* means the minimum temperature at which a liquid gives off a vapor in sufficient concentration to ignite when tested as follows:

(i) Tagliabue Closed Tester (See American National Standard Method of Test for Flash Point by Tag Closed Tester, Z11.24-1979 (ASTM D 56-79)) for liquids with a viscosity of less than 45 Saybolt Universal Seconds (SUS) at 100 °F (37.8 °C), that do not contain suspended solids and do not have a tendency to form a surface film under test; or

(ii) Pensky-Martens Closed Tester (see American National Standard Method of Test for Flash Point by Pensky-Martens Closed Tester, Z11.7-1979 (ASTM D 93-79)) for liquids with a viscosity equal to or greater than 45 SUS at 100 °F (37.8 °C), or that contain suspended solids, or that have a tendency to form a surface film under test; or

(iii) Setaflash Closed Tester (see American National Standard Method of Test for Flash Point by Setaflash Closed Tester (ASTM D 3278-78)).

Organic peroxides, which undergo autoaccelerating thermal decomposition, are excluded from any of the flashpoint determination methods specified above.

*Foreseeable emergency* means any potential occurrence such as, but not limited to, equipment failure, rupture of containers, or failure of control equipment which could result in an uncontrolled release of a hazardous chemical into the workplace.

*Hazardous chemical* means any chemical which is a physical hazard or a health hazard.

*Hazard warning* means any words, pictures, symbols, or combination thereof appearing on a label or other appropriate form of warning which convey the specific physical and health hazard(s), including target organ effects, of the chemical(s) in the container(s). (See the definitions for "physical hazard" and "health hazard" to determine the hazards which must be covered.)

*Health hazard* means a chemical for which there is statistically significant evidence based on at least one study conducted in accordance with established scientific principles that acute or chronic health effects may occur in exposed employees. The term "health hazard" includes chemicals which are carcinogens, toxic or highly toxic agents, reproductive toxins, irritants, corrosives, sensitizers, hepatotoxins, nephrotoxins, neurotoxins, agents which act on the hematopoietic system, and agents which damage the lungs, skin, eyes, or mucous membranes. Appendix A provides further definitions and explanations of the scope of health hazards covered by this section, and Appendix B describes the criteria to be used to determine whether or not a chemical is to be considered hazardous for purposes of this standard.

*Identity* means any chemical or common name which is indicated on the material safety data sheet (MSDS) for the chemical. The identity used shall permit cross-references to be made among the required list of hazardous chemicals, the label and the MSDS.

*Immediate use* means that the hazardous chemical will be under the control of and used only by the person who transfers it from a labeled container and only within the work shift in which it is transferred.

*Importer* means the first business with employees within the Customs Territory of the United States which receives hazardous chemicals produced in other countries for the purpose of supplying them to distributors or employers within the United States.

*Label* means any written, printed, or graphic material displayed on or affixed to containers of hazardous chemicals.

*Material safety data sheet (MSDS)* means written or printed material concerning a hazardous chemical which is prepared in accordance with paragraph (g) of this section.

*Mixture* means any combination of two or more chemicals if the combination is not, in whole or in part, the result of a chemical reaction.

*Organic peroxide* means an organic compound that contains the bivalent -O-O-structure and which may be considered to be a structural derivative of hydrogen peroxide where one or both of the hydrogen atoms has been replaced by an organic radical.

*Oxidizer* means a chemical other than a blasting agent or explosive as defined in §1910.109(a), that initiates or promotes combustion in other materials, thereby causing fire either of itself or through the release of oxygen or other gases.

*Physical hazard* means a chemical for which there is scientifically valid evidence that it is a combustible liquid, a compressed gas, explosive, flammable, an organic peroxide, an oxidizer, pyrophoric, unstable (reactive) or water-reactive.

*Produce* means to manufacture, process, formulate, blend, extract, generate, emit, or repackage.

*Pyrophoric* means a chemical that will ignite spontaneously in air at a temperature of 130 °F (54.4 °C) or below.

*Responsible party* means someone who can provide additional information on the hazardous chemical and appropriate emergency procedures, if necessary.

*Specific chemical identity* means the chemical name, Chemical Abstracts Service (CAS) Registry Number, or any other information that reveals the precise chemical designation of the substance.

*Trade secret* means any confidential formula, pattern, process, device, information or compilation of information that is used in an employer's business, and that gives the employer an opportunity to obtain an advantage over competitors who do not know or use it. Appendix D sets out the criteria to be used in evaluating trade secrets.

*Unstable (reactive)* means a chemical which in the pure state, or as produced or transported, will vigorously polymerize, decompose, condense, or will become self-reactive under conditions of shocks, pressure or temperature.

*Use* means to package, handle, react, emit, extract, generate as a byproduct, or transfer.

*Water-reactive* means a chemical that reacts with water to release a gas that is either flammable or presents a health hazard.

*Work area* means a room or defined space in a workplace where hazardous chemicals are produced or used, and where employees are present.

*Workplace* means an establishment, job site, or project, at one geographical lSocation containing one or more work areas.

**1910.1200(d)** ***Hazard determination.*** (1) Chemical manufacturers and importers shall evaluate chemicals produced in their workplaces or imported by them to determine if they are hazardous. Employers are not required to evaluate chemicals unless they choose not to rely on the evaluation performed by the chemical manufacturer or importer for the chemical to satisfy this requirement.

**1910.1200(d)(2)** Chemical manufacturers, importers or employers evaluating chemicals shall identify and consider the available scientific evidence concerning such hazards. For health hazards, evidence which is statistically significant and which is based on at least one positive study conducted in accordance with established scientific principles is considered to be sufficient to establish a hazardous effect if the results of the study meet the definitions of health hazards in this section. Appendix A shall be consulted for the scope of health hazards covered, and Appendix B shall be consulted for the criteria to be followed with respect to the completeness of the evaluation, and the data to be reported.

**1910.1200(d)(3)** The chemical manufacturer, importer or employer evaluating chemicals shall treat the following sources as establishing that the chemicals listed in them are hazardous:

**1910.1200(d)(3)(i)** 29 CFR part 1910, subpart Z, Toxic and Hazardous Substances, Occupational Safety and Health Administration (OSHA); or,

**1910.1200(d)(3)(ii)** *Threshold Limit Values for Chemical Substances and Physical Agents in the Work Environment*, American Conference of Governmental Industrial Hygienists (ACGIH) (latest edition). The chemical manufacturer, importer, or employer is still responsible for evaluating the hazards associated with the chemicals in these source lists in accordance with the requirements of this standard.

**1910.1200(d)(4)** Chemical manufacturers, importers and employers evaluating chemicals shall treat the following sources as establishing that a chemical is a carcinogen or potential carcinogen for hazard communication purposes:

**1910.1200(d)(4)(i)** National Toxicology Program (NTP), *Annual Report on Carcinogens* (latest edition);

**1910.1200(d)(4)(ii)** International Agency for Research on Cancer (IARC) *Monographs* (latest editions); or

**1910.1200(d)(4)(iii)** 29 CFR part 1910, subpart Z, Toxic and Hazardous Substances, Occupational Safety and Health Administration.

**Note:** The *Registry of Toxic Effects of Chemical Substances* published by the National Institute for Occupational Safety and Health indicates whether a chemical has been found by NTP or IARC to be a potential carcinogen.

**1910.1200(d)(5)** The chemical manufacturer, importer or employer shall determine the hazards of mixtures of chemicals as follows:

**1910.1200(d)(5)(i)** If a mixture has been tested as a whole to determine its hazards, the results of such testing shall be used to determine whether the mixture is hazardous;

**1910.1200(d)(5)(ii)** If a mixture has not been tested as a whole to determine whether the mixture is a health hazard, the mixture shall be assumed to present the same health hazards as do the components which comprise one percent (by weight or volume) or greater of the mixture, except that the mixture shall be assumed to present a carcinogenic hazard if it contains a component in concentrations of 0.1 percent or greater which is considered to be a carcinogen under paragraph (d)(4) of this section;

**1910.1200(d)(5)(iii)** If a mixture has not been tested as a whole to determine whether the mixture is a physical hazard, the chemical manufacturer, importer, or employer may use whatever scientifically valid data is available to evaluate the physical hazard potential of the mixture; and,

**1910.1200(d)(5)(iv)** If the chemical manufacturer, importer, or employer has evidence to indicate that a component present in the mixture in concentrations of less than one percent (or in the case of carcinogens, less than 0.1 percent) could be released in concentrations which would exceed an established OSHA permissible exposure limit or ACGIH Threshold Limit Value, or could present a health risk to employees in those concentrations, the mixture shall be assumed to present the same hazard.

**1910.1200(d)(6)** Chemical manufacturers, importers, or employers evaluating chemicals shall describe in writing the procedures they use to determine the hazards of the chemical they evaluate. The written procedures are to be made available, upon request, to employees, their designated representatives, the Assistant Secretary and the Director. The written description may be incorporated into the written hazard communication program required under paragraph (e) of this section.

**1910.1200(e)** ***Written hazard communication program.*** (1) Employers shall develop, implement, and maintain at each workplace, a written hazard communication program which at least describes how the criteria specified in paragraphs (f), (g), and (h) of this section for labels and other forms of warning, material safety data sheets, and employee information and training will be met, and which also includes the following:

**1910.1200(e)(1)(i)** A list of the hazardous chemicals known to be present using an identity that is referenced on the appropriate material safety data sheet (the list may be compiled for the workplace as a whole or for individual work areas); and,

**1910.1200(e)(1)(ii)** The methods the employer will use to inform employees of the hazards of non-routine tasks (for example, the cleaning of reactor vessels), and the hazards associated with chemicals contained in unlabeled pipes in their work areas.

**1910.1200(e)(2)** *Multi-employer workplaces.* Employers who produce, use, or store hazardous chemicals at a workplace in such a way that the employees of other employer(s) may be exposed (for example, employees of a construction contractor working on-site) shall additionally ensure that the hazard communication programs developed and implemented under this paragraph (e) include the following:

**1910.1200(e)(2)(i)** The methods the employer will use to provide the other employer(s) on-site access to material safety data sheets for each hazardous chemical the other employer(s)' employees may be exposed to while working;

**1910.1200(e)(2)(ii)** The methods the employer will use to inform the other employer(s) of any precautionary measures that need to be taken to protect employees during the workplace's normal operating conditions and in foreseeable emergencies; and,

**1910.1200(e)(2)(iii)** The methods the employer will use to inform the other employer(s) of the labeling system used in the workplace.

**1910.1200(e)(3)** The employer may rely on an existing hazard communication program to comply with these requirements, provided that it meets the criteria established in this paragraph (e).

**1910.1200(e)(4)** The employer shall make the written hazard communication program available, upon request, to employees, their designated representatives, the Assistant Secretary and the Director, in accordance with the requirements of 29 CFR 1910.20(e).

**1910.1200(e)(5)** Where employees must travel between workplaces during a workshift, *i.e.*, their work is carried out at more than one geographical location, the written hazard communication program may be kept at the primary workplace facility.

**1910.1200(f)** ***Labels and other forms of warning.*** (1) The chemical manufacturer, importer, or distributor shall ensure that each container of hazardous chemicals leaving the workplace is labeled, tagged or marked with the following information:

**1910.1200(f)(1)(i)** Identity of the hazardous chemical(s);

**1910.1200(f)(1)(ii)** Appropriate hazard warnings; and

**1910.1200(f)(1)(iii)** Name and address of the chemical manufacturer, importer, or other responsible party.

**1910.1200(f)(2)(i)** For solid metal (such as a steel beam or a metal casting), solid wood, or plastic items that are not exempted as articles due to their downstream use, or shipments of whole grain, the required label may be transmitted to the customer at the time of the initial shipment, and need not be included with subsequent shipments to the same employer unless the information on the label changes;

**1910.1200(f)(2)(ii)** The label may be transmitted with the initial shipment itself, or with the material safety data sheet that is to be provided prior to or at the time of the first shipment; and,

**1910.1200(f)(2)(iii)** This exception to requiring labels on every container of hazardous chemicals is only for the solid material itself, and does not apply to hazardous chemicals used in conjunction with, or known to be present with, the material and to which employees handling the items in transit may be exposed (for example, cutting fluids or pesticides in grains).

**1910.1200(f)(3)** Chemical manufacturers, importers, or distributors shall ensure that each container of hazardous chemicals leaving the workplace is labeled, tagged, or marked in accordance with this section in a manner which does not conflict with the requirements of the Hazardous Materials Transportation Act (49 U.S.C. 1801 *et seq.*) and regulations issued under that Act by the Department of Transportation.

**1910.1200(f)(4)** If the hazardous chemical is regulated by OSHA in a substance-specific health standard, the chemical manufacturer, importer, distributor or employer shall ensure that the labels or other forms of warning used are in accordance with the requirements of that standard.

**1910.1200(f)(5)** Except as provided in paragraphs (f)(6) and (f)(7) of this section, the employer shall ensure that each container of hazardous chemicals in the workplace is labeled, tagged or marked with the following information:

**1910.1200(f)(5)(i)** Identity of the hazardous chemical(s) contained therein; and,

**1910.1200(f)(5)(ii)** Appropriate hazard warnings, or alternatively, words, pictures, symbols, or combination thereof, which provide at least general information regarding the hazards of the chemicals, and which, in conjunction with the other information immediately available to employees under the hazard communication program, will provide employees with the specific information regarding the physical and health hazards of the hazardous chemical.

**1910.1200(f)(6)** The employer may use signs, placards, process sheets, batch tickets, operating procedures, or other such written materials in lieu of affixing labels to individual stationary process containers, as long as the alternative method identifies the containers to which it is applicable and conveys the information required by paragraph (f)(5) of this section to be on a label. The written materials shall be readily accessible to the employees in their work area throughout each work shift.

**1910.1200(f)(7)** The employer is not required to label portable containers into which hazardous chemicals are transferred from labeled containers, and which are intended only for the immediate use of the employee who performs the transfer. For purposes of this section, drugs which are dispensed by a pharmacy to a health care provider for direct administration to a patient are exempted from labeling.

**1910.1200(f)(8)** The employer shall not remove or deface existing labels on incoming containers of hazardous chemicals, unless the container is immediately marked with the required information.

**1910.1200(f)(9)** The employer shall ensure that labels or other forms of warning are legible, in English, and prominently displayed on the container, or readily available in the work area throughout each work shift. Employers having employees who speak other languages may add the information in their language to the material presented, as long as the information is presented in English as well.

**1910.1200(f)(10)** The chemical manufacturer, importer, distributor or employer need not affix new labels to comply with this section if existing labels already convey the required information.

**1910.1200(f)(11)** Chemical manufacturers, importers, distributors, or employers who become newly aware of any significant information regarding the hazards of a chemical shall revise the labels for the chemical within three months of becoming aware of the new information. Labels on containers of hazardous chemicals shipped after that time shall contain the new information. If the chemical is not currently produced or imported, the chemical manufacturer, importers, distributor, or employer shall add the information to the label before the chemical is shipped or introduced into the workplace again.

**1910.1200(g)** ***Material safety data sheets.*** (1) Chemical manufacturers and importers shall obtain or develop a material safety data sheet for each hazardous chemical they produce or import. Employers shall have a material safety data sheet in the workplace for each hazardous chemical which they use.

**1910.1200(g)(2)** Each material safety data sheet shall be in English (although the employer may maintain copies in other languages as well), and shall contain at least the following information:

**1910.1200(g)(2)(i)** The identity used on the label, and, except as provided for in paragraph (i) of this section on trade secrets:

**1910.1200(g)(2)(i)(A)** If the hazardous chemical is a single substance, its chemical and common name(s);

**1910.1200(g)(2)(i)(B)** If the hazardous chemical is a mixture which has been tested as a whole to determine its hazards, the chemical and common name(s) of the ingredients which contribute to these known hazards, and the common name(s) of the mixture itself; or,

**1910.1200(g)(2)(i)(C)** If the hazardous chemical is a mixture which has not been tested as a whole:

**1910.1200(g)(2)(i)(C)*(1)*** The chemical and common name(s) of all ingredients which have been determined to be health hazards, and which comprise 1% or greater of the composition, except that chemicals identified as carcinogens under paragraph (d) of this section shall be listed if the concentrations are 0.1% or greater; and,

**1910.1200(g)(2)(i)(C)*(2)*** The chemical and common name(s) of all ingredients which have been determined to be health hazards, and which comprise less than 1% (0.1% for carcinogens) of the mixture, if there is evidence that the ingredient(s) could be released from the mixture in concentrations which would exceed an established OSHA permissible exposure limit or ACGIH Threshold Limit Value, or could present a health risk to employees; and,

**1910.1200(g)(2)(i)(C)*(3)*** The chemical and common name(s) of all ingredients which have been determined to present a physical hazard when present in the mixture;

**1910.1200(g)(2)(ii)** Physical and chemical characteristics of the hazardous chemical (such as vapor pressure, flash point);

**1910.1200(g)(2)(iii)** The physical hazards of the hazardous chemical, including the potential for fire, explosion, and reactivity;

**1910.1200(g)(2)(iv)** The health hazards of the hazardous chemical, including signs and symptoms of exposure, and any medical conditions which are generally recognized as being aggravated by exposure to the chemical;

**1910.1200(g)(2)(v)** The primary route(s) of entry;

**1910.1200(g)(2)(vi)** The OSHA permissible exposure limit, ACGIH Threshold Limit Value, and any other exposure limit used or recommended by the chemical manufacturer, importer, or employer preparing the material safety data sheet, where available;

**1910.1200(g)(2)(vii)** Whether the hazardous chemical is listed in the National Toxicology Program (NTP) Annual Report on Carcinogens (latest edition) or has been found to be a potential carcinogen in the International Agency for Research on Cancer (IARC) Monographs (latest editions), or by OSHA;

**1910.1200(g)(2)(viii)** Any generally applicable precautions for safe handling and use which are known to the chemical manufacturer, importer or employer preparing the material safety data sheet, including appropriate hygienic practices, protective measures during repair and maintenance of contaminated equipment, and procedures for clean-up of spills and leaks;

**1910.1200(g)(2)(ix)** Any generally applicable control measures which are known to the chemical manufacturer, importer or employer preparing the material safety data sheet, such as appropriate engineering controls, work practices, or personal protective equipment;

**1910.1200(g)(2)(x)** Emergency and first aid procedures;

**1910.1200(g)(2)(xi)** The date of preparation of the material safety data sheet or the last change to it; and,

**1910.1200(g)(2)(xii)** The name, address and telephone number of the chemical manufacturer, importer, employer or other responsible party preparing or distributing the material safety data sheet, who can provide additional information on the hazardous chemical and appropriate emergency procedures, if necessary.

**1910.1200(g)(3)** If no relevant information is found for any given category on the material safety data sheet, the chemical manufacturer, importer or employer preparing the material safety data sheet shall mark it to indicate that no applicable information was found.

**1910.1200(g)(4)** Where complex mixtures have similar hazards and contents (i.e. the chemical ingredients are essentially the same, but the specific composition varies from mixture to mixture), the chemical manufacturer, importer or employer may prepare one material safety data sheet to apply to all of these similar mixtures.

**1910.1200(g)(5)** The chemical manufacturer, importer or employer preparing the material safety data sheet shall ensure that the information recorded accurately reflects the scientific evidence used in making the hazard determination. If the chemical manufacturer, importer or employer preparing the material safety data sheet becomes newly aware of any significant information regarding the hazards of a chemical, or ways to protect against the hazards, this new information shall be added to the material safety data sheet within three months. If the chemical is not currently being produced or imported the chemical manufacturer or importer shall add the information to the material safety data sheet before the chemical is introduced into the workplace again.

**1910.1200(g)(6)(i)** Chemical manufacturers or importers shall ensure that distributors and employers are provided an appropriate material safety data sheet with their initial shipment, and with the first shipment after a material safety data sheet is updated;

**1910.1200(g)(6)(ii)** The chemical manufacturer or importer shall either provide material safety data sheets with the shipped containers or send them to the distributor or employer prior to or at the time of the shipment;

**1910.1200(g)(6)(iii)** If the material safety data sheet is not provided with a shipment that has been labeled as a hazardous chemical, the distributor or employer shall obtain one from the chemical manufacturer or importer as soon as possible; and,

**1910.1200(g)(6)(iv)** The chemical manufacturer or importer shall also provide distributors or employers with a material safety data sheet upon request.

**1910.1200(g)(7)(i)** Distributors shall ensure that material safety data sheets, and updated information, are provided to other distributors and employers with their initial shipment and with the first shipment after a material safety data sheet is updated;

**1910.1200(g)(7)(ii)** The distributor shall either provide material safety data sheets with the shipped containers, or send them to the other distributor or employer prior to or at the time of the shipment;

**1910.1200(g)(7)(iii)** Retail distributors selling hazardous chemicals to employers having a commercial account shall provide a material safety data sheet to such employers upon request, and shall post a sign or otherwise inform them that a material safety data sheet is available;

**1910.1200(g)(7)(iv)** Wholesale distributors selling hazardous chemicals to employers over-the-counter may also provide material safety data sheets upon the request of the employer at the time of the over-the-counter purchase, and shall post a sign or otherwise inform such employers that a material safety data sheet is available;

**1910.1200(g)(7)(v)** If an employer without a commercial account purchases a hazardous chemical from a retail distributor not required to have material safety data sheets on file (*i.e.*, the retail distributor does not have commercial accounts and does not use the materials), the retail distributor shall provide the employer, upon request, with the name, address, and telephone number of the chemical manufacturer, importer, or distributor from which a material safety data sheet can be obtained;

**1910.1200(g)(7)(vi)** Wholesale distributors shall also provide material safety data sheets to employers or other distributors upon request; and,

**1910.1200(g)(7)(vii)** Chemical manufacturers, importers, and distributors need not provide material safety data sheets to retail distributors that have informed them that the retail distributor does not sell the product to commercial accounts or open the sealed container to use it in their own workplaces.

**1910.1200(g)(8)** The employer shall maintain in the workplace copies of the required material safety data sheets for each hazardous chemical, and shall ensure that they are readily accessible during each work shift to employees when they are in their work area(s). (Electronic access, microfiche, and other alternatives to maintaining paper copies of the material safety data sheets are permitted as long as no barriers to immediate employee access in each workplace are created by such options.)

**1910.1200(g)(9)** Where employees must travel between workplaces during a workshift, *i.e.*, their work is carried out at more than one geographical location, the material safety data sheets may be kept at the primary workplace facility. In this situation, the employer shall ensure that employees can immediately obtain the required information in an emergency.

**1910.1200(g)(10)** Material safety data sheets may be kept in any form, including operating procedures, and may be designed to cover groups of hazardous chemicals in a work area where it may be more appropriate to address the hazards of a process rather than individual hazardous chemicals. However, the employer shall ensure that in all cases the required information is provided for each hazardous chemical, and is readily accessible during each work shift to employees when they are in in their work area(s).

**1910.1200(g)(11)** Material safety data sheets shall also be made readily available, upon request, to designated representatives and to the Assistant Secretary, in accordance with the requirements of 29 CFR 1910.20(e). The Director shall also be given access to material safety data sheets in the same manner.

**1910.1200(h)** ***Employee information and training.*** (1) Employers shall provide employees with effective information and training on hazardous chemicals in their work area at the time of their initial assignment, and whenever a new physical or health hazard the employees have not previously been trained about is introduced into their work area. Information and training may be designed to cover categories of hazards (*e.g.*, flammability, carcinogenicity) or specific chemicals. Chemical-specific information must always be available through labels and material safety data sheets.

**1910.1200(h)(2)** *Information.* Employees shall be informed of:

**1910.1200(h)(2)(i)** The requirements of this section;

**1910.1200(h)(2)(ii)** Any operations in their work area where hazardous chemicals are present; and,

**1910.1200(h)(2)(iii)** The location and availability of the written hazard communication program, including the required list(s) of hazardous chemicals, and material safety data sheets required by this section.

**1910.1200(h)(3)** *Training.* Employee training shall include at least:

**1910.1200(h)(3)(i)** Methods and observations that may be used to detect the presence or release of a hazardous chemical in the work area (such as monitoring conducted by the employer, continuous monitoring devices, visual appearance or odor of hazardous chemicals when being released, etc.);

**1910.1200(h)(3)(ii)** The physical and health hazards of the chemicals in the work area;

**1910.1200(h)(3)(iii)** The measures employees can take to protect themselves from these hazards, including specific procedures the employer has implemented to protect employees from exposure to hazardous chemicals, such as appropriate work practices, emergency procedures, and personal protective equipment to be used; and,

**1910.1200(h)(3)(iv)** The details of the hazard communication program developed by the employer, including an explanation of the labeling system and the material safety data sheet, and how employees can obtain and use the appropriate hazard information.

**1910.1200(i)** ***Trade secrets.*** (1) The chemical manufacturer, importer, or employer may withhold the specific chemical identity, including the chemical name and other specific identification of a hazardous chemical, from the material safety data sheet, provided that:

**1910.1200(i)(1)(i)** The claim that the information withheld is a trade secret can be supported;

**1910.1200(i)(1)(ii)** Information contained in the material safety data sheet concerning the properties and effects of the hazardous chemical is disclosed;

**1910.1200(i)(1)(iii)** The material safety data sheet indicates that the specific chemical identity is being withheld as a trade secret; and,

**1910.1200(i)(1)(iv)** The specific chemical identity is made available to health professionals, employees, and designated representatives in accordance with the applicable provisions of this paragraph.

**1910.1200(i)(2)** Where a treating physician or nurse determines that a medical emergency exists and the specific chemical identity of a hazardous chemical is necessary for emergency or first-aid treatment, the chemical manufacturer, importer, or employer shall immediately disclose the specific chemical identity of a trade secret chemical to that treating physician or nurse, regardless of the existence of a written statement of need or a confidentiality agreement. The chemical manufacturer, importer, or employer may require a written statement of need and confidentiality agreement, in accordance with the provisions of paragraphs (i) (3) and (4) of this section, as soon as circumstances permit.

**1910.1200(i)(3)** In non-emergency situations, a chemical manufacturer, importer, or employer shall, upon request, disclose a specific chemical identity, otherwise permitted to be withheld under paragraph (i)(1) of this section, to a health professional (i.e. physician, industrial hygienist, toxicologist, epidemiologist, or occupational health nurse) providing medical or other occupational health services to exposed employee(s), and to employees or designated representatives, if:

**1910.1200(i)(3)(i)** The request is in writing;

**1910.1200(i)(3)(ii)** The request describes with reasonable detail one or more of the following occupational health needs for the information:

**1910.1200(i)(3)(ii)(A)** To assess the hazards of the chemicals to which employees will be exposed;

**1910.1200(i)(3)(ii)(B)** To conduct or assess sampling of the workplace atmosphere to determine employee exposure levels;

**1910.1200(i)(3)(ii)(C)** To conduct pre-assignment or periodic medical surveillance of exposed employees;

**1910.1200(i)(3)(ii)(D)** To provide medical treatment to exposed employees;

**1910.1200(i)(3)(ii)(E)** To select or assess appropriate personal protective equipment for exposed employees;

**1910.1200(i)(3)(ii)(F)** To design or assess engineering controls or other protective measures for exposed employees; and,

**1910.1200(i)(3)(ii)(G)** To conduct studies to determine the health effects of exposure.

**1910.1200(i)(3)(iii)** The request explains in detail why the disclosure of the specific chemical identity is essential and that, in lieu thereof, the disclosure of the following information to the health professional, employee, or designated representative, would not satisfy the purposes described in paragraph (i)(3)(ii) of this section:

**1910.1200(i)(3)(iii)(A)** The properties and effects of the chemical;

**1910.1200(i)(3)(iii)(B)** Measures for controlling workers' exposure to the chemical;

**1910.1200(i)(3)(iii)(C)** Methods of monitoring and analyzing worker exposure to the chemical; and,

**1910.1200(i)(3)(iii)(D)** Methods of diagnosing and treating harmful exposures to the chemical;

**1910.1200(i)(3)(iv)** The request includes a description of the procedures to be used to maintain the confidentiality of the disclosed information; and,

**1910.1200(i)(3)(v)** The health professional, and the employer or contractor of the services of the health professional (i.e. downstream employer, labor organization, or individual employee), employee, or designated representative, agree in a written confidentiality agreement that the health professional, employee, or designated representative, will not use the trade secret information for any purpose other than the health need(s) asserted and agree not to release the information under any circumstances other than to OSHA, as provided in paragraph (i)(6) of this section, except as authorized by the terms of the agreement or by the chemical manufacturer, importer, or employer.

**1910.1200(i)(4)** The confidentiality agreement authorized by paragraph (i)(3)(iv) of this section:

**1910.1200(i)(4)(i)** May restrict the use of the information to the health purposes indicated in the written statement of need;

**1910.1200(i)(4)(ii)** May provide for appropriate legal remedies in the event of a breach of the agreement, including stipulation of a reasonable pre-estimate of likely damages; and,

**1910.1200(i)(4)(iii)** May not include requirements for the posting of a penalty bond.

**1910.1200(i)(5)** Nothing in this standard is meant to preclude the parties from pursuing non-contractual remedies to the extent permitted by law.

**1910.1200(i)(6)** If the health professional, employee, or designated representative receiving the trade secret information decides that there is a need to disclose it to OSHA, the chemical manufacturer, importer, or employer who provided the information shall be informed by the health professional, employee, or designated representative prior to, or at the same time as, such disclosure.

**1910.1200(i)(7)** If the chemical manufacturer, importer, or employer denies a written request for disclosure of a specific chemical identity, the denial must:

**1910.1200(i)(7)(i)** Be provided to the health professional, employee, or designated representative, within thirty days of the request;

**1910.1200(i)(7)(ii)** Be in writing;

**1910.1200(i)(7)(iii)** Include evidence to support the claim that the specific chemical identity is a trade secret;

**1910.1200(i)(7)(iv)** State the specific reasons why the request is being denied; and,

**1910.1200(i)(7)(v)** Explain in detail how alternative information may satisfy the specific medical or occupational health need without revealing the specific chemical identity.

**1910.1200(i)(8)** The health professional, employee, or designated representative whose request for information is denied under paragraph (i)(3) of this section may refer the request and the written denial of the request to OSHA for consideration.

**1910.1200(i)(9)** When a health professional, employee, or designated representative refers the denial to OSHA under paragraph (i)(8) of this section, OSHA shall consider the evidence to determine if:

**1910.1200(i)(9)(i)** The chemical manufacturer, importer, or employer has supported the claim that the specific chemical identity is a trade secret;

**1910.1200(i)(9)(ii)** The health professional, employee, or designated representative has supported the claim that there is a medical or occupational health need for the information; and,

**1910.1200(i)(9)(iii)** The health professional, employee or designated representative has demonstrated adequate means to protect the confidentiality.

**1910.1200(i)(10)(i)** If OSHA determines that the specific chemical identity requested under paragraph (i)(3) of this section is not a *bona fide* trade secret, or that it is a trade secret, but the requesting health professional, employee, or designated representative has a legitimate medical or occupational health need for the information, has executed a written confidentiality agreement, and has shown adequate means to protect the confidentiality of the information, the chemical manufacturer, importer, or employer will be subject to citation by OSHA.

**1910.1200(i)(10)(ii)** If a chemical manufacturer, importer, or employer demonstrates to OSHA that the execution of a confidentiality agreement would not provide sufficient protection against the potential harm from the unauthorized disclosure of a trade secret specific chemical identity, the Assistant Secretary may issue such orders or impose such additional limitations or conditions upon the disclosure of the requested chemical information as may be appropriate to assure that the occupational health services are provided without an undue risk of harm to the chemical manufacturer, importer, or employer.

**1910.1200(i)(11)** If a citation for a failure to release specific chemical identity information is contested by the chemical manufacturer, importer, or employer, the matter will be adjudicated before the Occupational Safety and Health Review Commission in accordance with the Act's enforcement scheme and the applicable Commission rules of procedure. In accordance with the Commission rules, when a chemical manufacturer, importer, or employer continues to withhold the information during the contest, the Administrative Law Judge may review the citation and supporting documentation *in camera* or issue appropriate orders to protect the confidentiality of such matters.

**1910.1200(i)(12)** Notwithstanding the existence of a trade secret claim, a chemical manufacturer, importer, or employer shall, upon request, disclose to the Assistant Secretary any information which this section requires the chemical manufacturer, importer, or employer to make available. Where there is a trade secret claim, such claim shall be made no later than at the time the information is provided to the Assistant Secretary so that suitable determinations of trade secret status can be made and the necessary protections can be implemented.

**1910.1200(i)(13)** Nothing in this paragraph shall be construed as requiring the disclosure under any circumstances of process or percentage of mixture information which is a trade secret.

**1910.1200(j)** ***Effective dates.*** Chemical manufacturers, importers, distributors, and employers shall be in compliance with all provisions of this section by March 11, 1994.

**Note:** The effective date of the clarification that the exemption of wood and wood products from the Hazard Communication standard in paragraph (b)(6)(iv) only applies to wood and wood products including lumber which will not be processed, where the manufacturer or importer can establish that the only hazard they pose to employees is the potential for flammability or combustibility, and that the exemption does not apply to wood or wood products which have been treated with a hazardous chemical covered by this standard, and wood which may be subsequently sawed or cut generating dust has been stayed from March 11, 1994 to August 11, 1994.

## APPENDIX A TO §1910.1200— HEALTH HAZARD DEFINITIONS (MANDATORY)

Although safety hazards related to the physical characteristics of a chemical can be objectively defined in terms of testing requirements (e.g. flammability), health hazard definitions are less precise and more subjective. Health hazards may cause measurable changes in the body—such as decreased pulmonary function. These changes are generally indicated by the occurrence of signs and symptoms in the exposed employees—such as shortness of breath, a non-measurable, subjective feeling. Employees exposed to such hazards must be apprised of both the change in body function and the signs and symptoms that may occur to signal that change.

The determination of occupational health hazards is complicated by the fact that many of the effects or signs and symptoms occur commonly in non-occupationally exposed populations, so that effects of exposure are difficult to separate from normally occurring illnesses. Occasionally, a substance causes an effect that is rarely seen in the population at large, such as angiosarcomas caused by vinyl chloride exposure, thus making it easier to ascertain that the occupational exposure was the primary causative factor. More often, however, the effects are common, such as lung cancer. The situation is further complicated by the fact that most chemicals have not been adequately tested to determine their health hazard potential, and data do not exist to substantiate these effects.

There have been many attempts to categorize effects and to define them in various ways. Generally, the terms "acute" and "chronic" are used to delineate between effects on the basis of severity or duration. "Acute" effects usually occur rapidly as a result of short-term exposures, and are of short duration. "Chronic" effects generally occur as a result of long-term exposure, and are of long duration.

The acute effects referred to most frequently are those defined by the American National Standards Institute (ANSI) standard for Precautionary Labeling of Hazardous Industrial Chemicals (Z129.1-1988)—irritation, corrosivity, sensitization and lethal dose. Although these are important health effects, they do not adequately cover the considerable range of acute effects which may occur as a result of occupational exposure, such as, for example, narcosis.

Similarly, the term chronic effect is often used to cover only carcinogenicity, teratogenicity, and mutagenicity. These effects are obviously a concern in the workplace, but again, do not adequately cover the area of chronic effects, excluding, for example, blood dyscrasias (such as anemia), chronic bronchitis and liver atrophy.

The goal of defining precisely, in measurable terms, every possible health effect that may occur in the workplace as a result of chemical exposures cannot realistically be accomplished. This does not negate the need for employees to be informed of such effects and protected from them. Appendix B, which is also mandatory, outlines the principles and procedures of hazard assessment.

For purposes of this section, any chemicals which meet any of the following definitions, as determined by the criteria set forth in Appendix B are health hazards. However, this is not intended to be an exclusive categorization scheme. If there are available scientific data that involve other animal species or test methods, they must also be evaluated to determine the applicability of the HCS.7

1. *Carcinogen:* A chemical is considered to be a carcinogen if:

(a) It has been evaluated by the International Agency for Research on Cancer (IARC), and found to be a carcinogen or potential carcinogen; or

(b) It is listed as a carcinogen or potential carcinogen in the Annual Report on Carcinogens published by the National Toxicology Program (NTP) (latest edition); or,

(c) It is regulated by OSHA as a carcinogen.

2. *Corrosive:* A chemical that causes visible destruction of, or irreversible alterations in, living tissue by chemical action at the site of contact. For example, a chemical is considered to be corrosive if, when tested on the intact skin of albino rabbits by the method described by the U.S. Department of Transportation in appendix A to 49 CFR part 173, it destroys or changes irreversibly the structure of the tissue at the site of contact following an exposure period of four hours. This term shall not refer to action on inanimate surfaces.

3. *Highly toxic:* A chemical falling within any of the following categories:

(a) A chemical that has a median lethal dose (LD50) of 50 milligrams or less per kilogram of body weight when administered orally to albino rats weighing between 200 and 300 grams each.

(b) A chemical that has a median lethal dose (LD50) of 200 milligrams or less per kilogram of body weight when administered by continuous contact for 24 hours (or less if death occurs within 24 hours) with the bare skin of albino rabbits weighing between two and three kilograms each.

(c) A chemical that has a median lethal concentration (LC50) in air of 200 parts per million by volume or less of gas or vapor, or 2 milligrams per liter or less of mist, fume, or dust, when administered by continuous inhalation for one hour (or less if death occurs within one hour) to albino rats weighing between 200 and 300 grams each.

4. *Irritant:* A chemical, which is not corrosive, but which causes a reversible inflammatory effect on living tissue by chemical action at the site of contact. A chemical is a skin irritant if, when tested on the intact skin of albino rabbits by the methods of 16 CFR 1500.41 for four hours exposure or by other appropriate techniques, it results in an empirical score of five or more. A chemical is an eye irritant if so determined under the procedure listed in 16 CFR 1500.42 or other appropriate techniques.

5. *Sensitizer:* A chemical that causes a substantial proportion of exposed people or animals to develop an allergic reaction in normal tissue after repeated exposure to the chemical.

6. *Toxic.* A chemical falling within any of the following categories:

(a) A chemical that has a median lethal dose (LD50) of more than 50 milligrams per kilogram but not more than 500 milligrams per kilogram of body weight when administered orally to albino rats weighing between 200 and 300 grams each.

(b) A chemical that has a median lethal dose (LD50) of more than 200 milligrams per kilogram but not more than 1,000 milligrams per kilogram of body weight when administered by continuous contact for 24 hours (or less if death occurs within 24 hours) with the bare skin of albino rabbits weighing between two and three kilograms each.

(c) A chemical that has a median lethal concentration (LC 50) in air of more than 200 parts per million but not more than 2,000 parts per million by volume of gas or vapor, or more than two milligrams per liter but not more than 20 milligrams per liter of mist, fume, or dust, when administered by continuous inhalation for one hour (or less if death occurs within one hour) to albino rats weighing between 200 and 300 grams each.

7. *Target organ effects.*

The following is a target organ categorization of effects which may occur, including examples of signs and symptoms and chemicals which have been found to cause such effects. These examples are presented to illustrate the range and diversity of effects and hazards found in the workplace, and the broad scope employers must consider in this area, but are not intended to be all-inclusive.

a. Hepatotoxins: Chemicals which produce liver damage Signs & Symptoms: Jaundice; liver enlargement

Chemicals: Carbon tetrachloride; nitrosamines

b. Nephrotoxins: Chemicals which produce kidney damage Signs & Symptoms: Edema; proteinuria

Chemicals: Halogenated hydrocarbons; uranium

c. Neurotoxins: Chemicals which produce their primary toxic effects on the nervous system

Signs & Symptoms: Narcosis; behavioral changes; decrease in motor functions

Chemicals: Mercury; carbon disulfide

d. Agents which act on the blood or hemato-poietic system: Decrease hemoglobin function; deprive the body tissues of oxygen

Signs & Symptoms: Cyanosis; loss of consciousness

Chemicals: Carbon monoxide; cyanides

e. Agents which damage the lung: Chemicals which irritate or damage pulmonary tissue

Signs & Symptoms: Cough; tightness in chest; shortness of breath

Chemicals: Silica; asbestos

f. Reproductive toxins: Chemicals which affect the reproductive capabilities including chromosomal damage (mutations) and effects on fetuses (teratogenesis)

Signs & Symptoms: Birth defects; sterility

Chemicals: Lead; DBCP

g. Cutaneous hazards: Chemicals which affect the dermal layer of the body

Signs & Symptoms: Defatting of the skin; rashes; irritation

Chemicals: Ketones; chlorinated compounds

h. Eye hazards: Chemicals which affect the eye or visual capacity

Signs & Symptoms: Conjunctivitis; corneal damage

Chemicals: Organic solvents; acids

## APPENDIX B TO §1910.1200—HAZARD DETERMINATION (MANDATORY)

The quality of a hazard communication program is largely dependent upon the adequacy and accuracy of the hazard determination. The hazard determination requirement of this standard is performance-oriented. Chemical manufacturers, importers, and employers evaluating chemicals are not required to follow any specific methods for determining hazards, but they must be able to demonstrate that they have adequately ascertained the hazards of the chemicals produced or imported in accordance with the criteria set forth in this Appendix.

Hazard evaluation is a process which relies heavily on the professional judgment of the evaluator, particularly in the area of chronic hazards. The performance-orientation of the hazard determination does not diminish the duty of the chemical manufacturer, importer or employer to conduct a thorough evaluation, examining all relevant data and producing a scientifically defensible evaluation. For purposes of this standard, the following criteria shall be used in making hazard determinations that meet the requirements of this standard.

1. *Carcinogenicity:* As described in paragraph (d)(4) of this section and Appendix A of this section, a determination by the National Toxicology Program, the International Agency for Research on Cancer, or OSHA that a chemical is a carcinogen or potential carcinogen will be considered conclusive evidence for purposes of this section. In addition, however, all available scientific data on carcinogenicity must be evaluated in accordance with the provisions of this Appendix and the requirements of the rule.

2. *Human data:* Where available, epidemiological studies and case reports of adverse health effects shall be considered in the evaluation.

3. *Animal data:* Human evidence of health effects in exposed populations is generally not available for the majority of chemicals produced or used in the workplace. Therefore, the available results of toxicological testing in animal populations shall be used to predict the health effects that may be experienced by exposed workers. In particular, the definitions of certain acute hazards refer to specific animal testing results (see Appendix A).

4. *Adequacy and reporting of data.* The results of any studies which are designed and conducted according to established scientific principles, and which report statistically significant conclusions regarding the health effects of a chemical, shall be a sufficient basis for a hazard determination and reported on any material safety data sheet. *In vitro* studies alone generally do not form the basis for a definitive finding of hazard under the HCS since they have a positive or negative result rather than a statistically significant finding.

The chemical manufacturer, importer, or employer may also report the results of other scientifically valid studies which tend to refute the findings of hazard.

## APPENDIX C TO §1910.1200—INFORMATION SOURCES (ADVISORY) [Reserved]

[Appendix C to §1910.1200 removed at 61 FR 9245, Mar. 7, 1996, effective May 6, 1996.]

## APPENDIX D TO §1910.1200—DEFINITION OF "TRADE SECRET" (MANDATORY)

The following is a reprint of the *Restatement of Torts* section 757, comment *b* (1939):

*b. Definition of trade secret.* A trade secret may consist of any formula, pattern, device or compilation of information which is used in one's business, and which gives him an opportunity to obtain an advantage over competitors who do not know or use it. It may be a formula for a chemical compound, a process of manufacturing, treating or preserving materials, a pattern for a machine or other device, or a list of customers. It differs from other secret information in a business (see s759 of the *Restatement of Torts* which is not included in this Appendix) in that it is not simply information as to single or ephemeral events in the conduct of the business, as, for example, the amount or other terms of a secret bid for a contract or the salary of certain employees, or the security investments made or contemplated, or the date fixed for the announcement of a new policy or for bringing out a new model or the like. A trade secret is a process or device for continuous use in the operations of the business. Generally it relates to the production of goods, as, for example, a machine or formula for the production of an article. It may, however, relate to the sale of goods or to other operations in the business, such as a code for determining discounts, rebates or other concessions in a price list or catalogue, or a list of specialized customers, or a method of bookkeeping or other office management.

*Secrecy.* The subject matter of a trade secret must be secret. Matters of public knowledge or of general knowledge in an industry cannot be appropriated by one as his secret. Matters which are completely disclosed by the goods which one markets cannot be his secret. Substantially, a trade secret is known only in the particular business in which it is used. It is not requisite that only the proprietor of the business know it. He may, without losing his protection, communicate it to employees involved in its use. He may likewise communicate it to others pledged to secrecy. Others may also know of it independently, as, for example, when they have discovered the process or formula by independent invention and are keeping it secret. Nevertheless, a substantial element of secrecy must exist, so that, except by the use of improper means, there would be difficulty in acquiring the information. An exact definition of a trade secret is not possible. Some factors to be considered in determining whether given information is one's trade secret are: (1) The extent to which the information is known outside of his business; (2) the extent to which it is known by employees and others involved in his business; (3) the extent of measures taken by him to guard the secrecy of the information; (4) the value of the information to him and his competitors; (5) the amount of effort or money expended by him in developing the information; (6) the ease or difficulty with which the information could be properly acquired or duplicated by others.

*Novelty and prior art.* A trade secret may be a device or process which is patentable; but it need not be that. It may be a device or process which is clearly anticipated in the prior art or one which is merely a mechanical improvement that a good mechanic can make. Novelty and invention are not requisite for a trade secret as they are for patentability. These requirements are essential to patentability because a patent protects against unlicensed use of the patented device or process even by one who discovers it properly through independent research. The patent monopoly is a reward to the inventor. But such is not the case with a trade secret. Its protection is not based on a policy of rewarding or otherwise encouraging the development of secret processes or devices. The protection is merely against breach of faith and reprehensible means of learning another's secret. For this limited protection it is not appropriate to require also the kind of novelty and invention which is a requisite of patentability. The nature of the secret is, however, an important factor in determining the kind of relief that is appropriate against one who is subject to liability under the rule stated in this Section. Thus, if the secret consists of a device or process which is a novel invention, one who acquires the secret wrongfully is ordinarily enjoined from further use of it and is required to account for the profits derived from his past use. If, on the other hand, the secret consists of mechanical improvements that a good mechanic can make without resort to the secret, the wrongdoer's liability may be limited to damages, and an injunction against future use of the improvements made with the aid of the secret may be inappropriate.

## APPENDIX E TO §1910.1200—(ADVISORY)—GUIDELINES FOR EMPLOYER COMPLIANCE

The Hazard Communication Standard (HCS) is based on a simple concept—that employees have both a need and a right to know the hazards and identities of the chemicals they are exposed to when working. They also need to know what protective measures are available to prevent adverse effects from occurring. The HCS is designed to provide employees with the information they need.

Knowledge acquired under the HCS will help employers provide safer workplaces for their employees. When employers have information about the chemicals being used, they can take steps to reduce exposures, substitute less hazardous materials, and establish proper work practices. These efforts will help prevent the occurrence of work-related illnesses and injuries caused by chemicals.

The HCS addresses the issues of evaluating and communicating hazards to workers. Evaluation of chemical hazards involves a number of technical concepts, and is a process that requires the professional judgment of experienced experts. That's why the HCS is designed so that employers who simply use chemicals, rather than produce or import them, are not required to evaluate the hazards of those chemicals. Hazard determination is the responsibility of the producers and importers of the materials. Producers and importers of chemicals are then required to provide the hazard information to employers that purchase their products.

Employers that don't produce or import chemicals need only focus on those parts of the rule that deal with establishing a workplace program and communicating information to their workers. This appendix is a general guide for such employers to help them determine what's required under the rule. It does not supplant or substitute for the regulatory provisions, but rather provides a simplified outline of the steps an average employer would follow to meet those requirements.

### *1. Becoming Familiar With The Rule.*

OSHA has provided a simple summary of the HCS in a pamphlet entitled "Chemical Hazard Communication," OSHA Publication Number 3084. Some employers prefer to begin to become familiar with the rule's requirements by reading this pamphlet. A copy may be obtained from your local OSHA Area Office, or by contacting the OSHA Publications Office at (202) 523-9667.

The standard is long, and some parts of it are technical, but the basic concepts are simple. In fact, the requirements reflect what many employers have been doing for years. You may find that you are already largely in compliance with many of the provisions, and will simply have to modify your existing programs somewhat. If you are operating in an OSHA-approved State Plan State, you must comply with the State's requirements, which may be different than those of the Federal rule. Many of the State Plan States had hazard communication or "right-to-know" laws prior to promulgation of the Federal rule. Employers in State Plan States should contact their State OSHA offices for more information regarding applicable requirements.

The HCS requires information to be prepared and transmitted regarding all hazardous chemicals. The HCS covers both physical hazards (such as flammability), and health hazards (such as irritation, lung damage, and cancer). Most chemicals used in the workplace have some hazard potential, and thus will be covered by the rule.

One difference between this rule and many others adopted by OSHA is that this one is performance-oriented. That means that you have the flexibility to adapt the rule to the needs of your workplace, rather than having to follow specific, rigid requirements. It also means that you have to exercise more judgment to implement an appropriate and effective program.

The standard's design is simple. Chemical manufacturers and importers must evaluate the hazards of the chemicals they produce or import. Using that information, they must then prepare labels for containers, and more detailed technical bulletins called material safety data sheets (MSDS).

Chemical manufacturers, importers, and distributors of hazardous chemicals are all required to provide the appropriate labels and material safety data sheets to the employers to which they ship the chemicals. The information is to be provided automatically. Every container of hazardous chemicals you receive must be labeled, tagged, or marked with the required information. Your suppliers must also send you a properly completed material safety data sheet (MSDS) at the time of the first shipment of the chemical, and with the next shipment after the MSDS is updated with new and significant information about the hazards.

You can rely on the information received from your suppliers. You have no independent duty to analyze the chemical or evaluate the hazards of it.

Employers that "use" hazardous chemicals must have a program to ensure the information is provided to exposed employees. "Use" means to package, handle, react, or transfer. This is an intentionally broad scope, and includes any situation where a chemical is present in such a way that employees may be exposed under normal conditions of use or in a foreseeable emergency.

The requirements of the rule that deal specifically with the hazard communication program are found in this section in paragraphs (e), written hazard communication program; (f), labels and other forms of warning; (g), material safety data sheets; and (h), employee information and training. The requirements of these paragraphs should be the focus of your attention. Concentrate on becoming familiar with them, using paragraphs (b), scope and application, and (c), definitions, as references when needed to help explain the provisions.

There are two types of work operations where the coverage of the rule is limited. These are laboratories and operations where chemicals are only handled in sealed containers (*e.g.*, a warehouse). The limited provisions for these workplaces can be found in paragraph (b) of this section, scope and application. Basically, employers having these types of work operations need only keep labels on containers as they are received; maintain material safety data sheets that are received, and give employees access to them; and provide information and training for employees. Employers do not have to have written hazard communication programs and lists of chemicals for these types of operations.

The limited coverage of laboratories and sealed container operations addresses the obligation of an employer to the workers in the operations involved, and does not affect the employer's duties as a distributor of chemicals. For example, a distributor may have warehouse operations where employees would be protected under the limited sealed container provisions. In this situation, requirements for obtaining and maintaining MSDSs are limited to providing access to those received with containers while the substance is in the workplace, and requesting MSDSs when employees request access for those not received with the containers. However, as a distributor of hazardous chemicals, that employer will still have responsibilities for providing MSDSs to downstream customers at the time of the first shipment and when the MSDS is updated. Therefore, although they may not be required for the employees in the work operation, the distributor may, nevertheless, have to have MSDSs to satisfy other requirements of the rule.

### *2. Identify Responsible Staff*

Hazard communication is going to be a continuing program in your facility. Compliance with the HCS is not a "one shot deal." In order to have a successful program, it will be necessary to assign responsibility for both the initial and ongoing activities that have to be undertaken to comply with the rule. In some cases, these activities may already be part of current job assignments. For example, site supervisors are frequently responsible for on-the-job training sessions. Early identification of the responsible employees, and involvement of them in the development of your plan of action, will result in a more effective program design. Evaluation of the effectiveness of your program will also be enhanced by involvement of affected employees.

For any safety and health program, success depends on commitment at every level of the organization. This is particularly true for hazard communication, where success requires a change in behavior. This will only occur if employers understand the program, and are committed to its success, and if employees are motivated by the people presenting the information to them.

### *3. Identify Hazardous Chemicals in the Workplace.*

The standard requires a list of hazardous chemicals in the workplace as part of the written hazard communication program. The list will eventually serve as an inventory of everything for which an MSDS must be maintained. At this point, however, preparing the list will help you complete the rest of the program since it will give you some idea of the scope of the program required for compliance in your facility.

The best way to prepare a comprehensive list is to survey the workplace. Purchasing records may also help, and certainly employers should establish procedures to ensure that in the future purchasing procedures result in MSDSs being received before a material is used in the workplace.

The broadest possible perspective should be taken when doing the survey. Sometimes people think of "chemicals" as being only liquids in containers. The HCS covers chemicals in all physical forms—liquids, solids, gases, vapors, fumes, and mists—whether they are "contained" or not. The hazardous nature of the chemical and the potential for exposure are the factors which determine whether a chemical is covered. If it's not hazardous, it's not covered. If there is no potential for exposure (e.g., the chemical is inextricably bound and cannot be released), the rule does not cover the chemical.

Look around. Identify chemicals in containers, including pipes, but also think about chemicals generated in the work operations. For example, welding fumes, dusts, and exhaust fumes are all sources of chemical exposures. Read labels provided by suppliers for hazard information. Make a list of all chemicals in the workplace that are potentially hazardous. For your own information and planning, you may also want to note on the list the location(s) of the products within the workplace, and an indication of the hazards as found on the label. This will help you as you prepare the rest of your program.

Paragraph (b) of this section, scope and application, includes exemptions for various chemicals or workplace situations. After compiling the complete list of chemicals, you should review paragraph (b) of this section to determine if any of the items can be eliminated from the list because they are exempted materials. For example, food, drugs, and cosmetics brought into the workplace for employee consumption are exempt. So rubbing alcohol in the first aid kit would not be covered.

Once you have compiled as complete a list as possible of the potentially hazardous chemicals in the workplace, the next step is to determine if you have received material safety data sheets for all of them. Check your files against the inventory you have just compiled. If any are missing, contact your supplier and request one. It is a good idea to document these requests, either by copy of a letter or a note regarding telephone conversations. If you have MSDSs for chemicals that are not on your list, figure out why. Maybe you don't use the chemical anymore. Or maybe you missed it in your survey. Some suppliers do provide MSDSs for products that are not hazardous. These do not have to be maintained by you.

You should not allow employees to use any chemicals for which you have not received an MSDS. The MSDS provides information you need to ensure proper protective measures are implemented prior to exposure.

### 4. Preparing and Implementing a Hazard Communication Program

All workplaces where employees are exposed to hazardous chemicals must have a written plan which describes how the standard will be implemented in that facility. Preparation of a plan is not just a paper exercise—all of the elements must be implemented in the workplace in order to be in compliance with the rule. See paragraph (e) of this section for the specific requirements regarding written hazard communication programs. The only work operations which do not have to comply with the written plan requirements are laboratories and work operations where employees only handle chemicals in sealed containers. See paragraph (b) of this section, scope and application, for the specific requirements for these two types of workplaces.

The plan does not have to be lengthy or complicated. It is intended to be a blueprint for implementation of your program—an assurance that all aspects of the requirements have been addressed.

Many trade associations and other professional groups have provided sample programs and other assistance materials to affected employers. These have been very helpful to many employers since they tend to be tailored to the particular industry involved. You may wish to investigate whether your industry trade groups have developed such materials.

Although such general guidance may be helpful, you must remember that the written program has to reflect what you are doing in your workplace. Therefore, if you use a generic program it must be adapted to address the facility it covers. For example, the written plan must list the chemicals present at the site, indicate who is to be responsible for the various aspects of the program in your facility, and indicate where written materials will be made available to employees.

If OSHA inspects your workplace for compliance with the HCS, the OSHA compliance officer will ask to see your written plan at the outset of the inspection. In general, the following items will be considered in evaluating your program.

The written program must describe how the requirements for labels and other forms of warning, material safety data sheets, and employee information and training, are going to be met in your facility. The following discussion provides the type of information compliance officers will be looking for to decide whether these elements of the hazard communication program have been properly addressed:

#### A. Labels and Other Forms of Warning

In-plant containers of hazardous chemicals must be labeled, tagged, or marked with the identity of the material and appropriate hazard warnings. Chemical manufacturers, importers, and distributors are required to ensure that every container of hazardous chemicals they ship is appropriately labeled with such information and with the name and address of the producer or other responsible party. Employers purchasing chemicals can rely on the labels provided by their suppliers. If the material is subsequently transferred by the employer from a labeled container to another container, the employer will have to label that container unless it is subject to the portable container exemption. See paragraph (f) of this section for specific labeling requirements.

The primary information to be obtained from an OSHA-required label is an identity for the material, and appropriate hazard warnings. The identity is any term which appears on the label, the MSDS, and the list of chemicals, and thus links these three sources of information. The identity used by the supplier may be a common or trade name ("Black Magic Formula"), or a chemical name (1,1,1,-trichloroethane). The hazard warning is a brief statement of the hazardous effects of the chemical ("flammable," "causes lung damage"). Labels frequently contain other information, such as precautionary measures ("do not use near open flame"), but this information is provided voluntarily and is not required by the rule. Labels must be legible, and prominently displayed. There are no specific requirements for size or color, or any specified text.

With these requirements in mind, the compliance officer will be looking for the following types of information to ensure that labeling will be properly implemented in your facility:

1. Designation of person(s) responsible for ensuring labeling of in-plant containers;

2. Designation of person(s) responsible for ensuring labeling of any shipped containers;

3. Description of labeling system(s) used;

4. Description of written alternatives to labeling of in-plant containers (if used); and,

5. Procedures to review and update label information when necessary.

Employers that are purchasing and using hazardous chemicals—rather than producing or distributing them—will primarily be concerned with ensuring that every purchased container is labeled. If materials are transferred into other containers, the employer must ensure that these are labeled as well, unless they fall under the portable container exemption (paragraph (f)(7) of this section). In terms of labeling systems, you can simply choose to use the labels provided by your suppliers on the containers. These will generally be verbal text labels, and do not usually include numerical rating systems or symbols that require special training. The most important thing to remember is that this is a continuing duty—all in-plant containers of hazardous chemicals must always be labeled. Therefore, it is important to designate someone to be responsible for ensuring that the labels are maintained as required on the containers in your facility, and that newly purchased materials are checked for labels prior to use.

#### B. Material Safety Data Sheets

Chemical manufacturers and importers are required to obtain or develop a material safety data sheet for each hazardous chemical they produce or import. Distributors are responsible for ensuring that their customers are provided a copy of these MSDSs. Employers must have an MSDS for each hazardous chemical which they use. Employers may rely on the information received from their suppliers. The specific requirements for material safety data sheets are in paragraph (g) of this section.

There is no specified format for the MSDS under the rule, although there are specific information requirements. OSHA has developed a non-mandatory format, OSHA Form 174, which may be used by chemical manufacturers and importers to comply with the rule. The MSDS must be in English. You are entitled to receive from your supplier a data sheet which includes all of the information required under the rule. If you do not receive one automatically, you should request one. If you receive one that is obviously inadequate, with, for example, blank spaces that are not completed, you should request an appropriately completed one. If your request for a data sheet or for a corrected data sheet does not produce the information needed, you should contact your local OSHA Area Office for assistance in obtaining the MSDS.

The role of MSDSs under the rule is to provide detailed information on each hazardous chemical, including its potential hazardous effects, its physical and chemical characteristics, and recommendations for appropriate protective measures. This information should be useful to you as the employer responsible for designing protective programs, as well as to the workers. If you are not familiar with material safety data sheets and with chemical terminology, you may need to learn to use them yourself. A glossary of MSDS terms may be helpful in this regard. Generally speaking, most employers using hazardous chemicals will primarily be concerned with MSDS information regarding hazardous effects and recommended protective measures. Focus on the sections of the MSDS that are applicable to your situation.

MSDSs must be readily accessible to employees when they are in their work areas during their workshifts. This may be accomplished in many different ways. You must decide what is appropriate for your particular workplace. Some employers keep the MSDSs in a binder in a central location (*e.g.*, in the pick-up truck on a construction site). Others, particularly in workplaces with large numbers of chemicals, computerize the information and provide access through terminals. As long as employees can get the information when they need it, any approach may be used. The employees must have access to the MSDSs themselves—simply having a system where the information can be read to them over the phone is only permitted under the mobile worksite provision, paragraph (g)(9) of this section, when employees must travel between workplaces during the shift. In this situation, they have access to the MSDSs prior to leaving the primary worksite, and when they return, so the telephone system is simply an emergency arrangement.

In order to ensure that you have a current MSDS for each chemical in the plant as required, and that employee access is provided, the compliance officers will be looking for the following types of information in your written program:

1. Designation of person(s) responsible for obtaining and maintaining the MSDSs;

2. How such sheets are to be maintained in the workplace (*e.g.*, in notebooks in the work area(s) or in a computer with terminal access), and how employees can obtain access to them when they are in their work area during the work shift;

3. Procedures to follow when the MSDS is not received at the time of the first shipment;

4. For producers, procedures to update the MSDS when new and significant health information is found; and,

5. Description of alternatives to actual data sheets in the workplace, if used.

For employers using hazardous chemicals, the most important aspect of the written program in terms of MSDSs is to ensure that someone is responsible for obtaining and maintaining the MSDSs for every hazardous chemical in the workplace. The list of hazardous chemicals required to be maintained as part of the written program will serve as an inventory. As new chemicals are purchased, the list should be updated. Many companies have found it convenient to include on their purchase orders the name and address of the person designated in their company to receive MSDSs.

### C. Employee Information and Training

Each employee who may be "exposed" to hazardous chemicals when working must be provided information and trained prior to initial assignment to work with a hazardous chemical, and whenever the hazard changes. "Exposure" or "exposed" under the rule means that "an employee is subjected to a hazardous chemical in the course of employment through any route of entry (inhalation, ingestion, skin contact or absorption, etc.) and includes potential (*e.g.*, accidental or possible) exposure." See paragraph (h) of this section for specific requirements. Information and training may be done either by individual chemical, or by categories of hazards (such as flammability or carcinogenicity). If there are only a few chemicals in the workplace, then you may want to discuss each one individually. Where there are large numbers of chemicals, or the chemicals change frequently, you will probably want to train generally based on the hazard categories (*e.g.*, flammable liquids, corrosive materials, carcinogens). Employees will have access to the substance-specific information on the labels and MSDSs.

Information and training is a critical part of the hazard communication program. Information regarding hazards and protective measures are provided to workers through written labels and material safety data sheets. However, through effective information and training, workers will learn to read and understand such information, determine how it can be obtained and used in their own workplaces, and understand the risks of exposure to the chemicals in their workplaces as well as the ways to protect themselves. A properly conducted training program will ensure comprehension and understanding. It is not sufficient to either just read material to the workers, or simply hand them material to read. You want to create a climate where workers feel free to ask questions. This will help you to ensure that the information is understood. You must always remember that the underlying purpose of the HCS is to reduce the incidence of chemical source illnesses and injuries. This will be accomplished by modifying behavior through the provision of hazard information and information about protective measures. If your program works, you and your workers will better understand the chemical hazards within the workplace. The procedures you establish regarding, for example, purchasing, storage, and handling of these chemicals will improve, and thereby reduce the risks posed to employees exposed to the chemical hazards involved. Furthermore, your workers' comprehension will also be increased, and proper work practices will be followed in your workplace.

If you are going to do the training yourself, you will have to understand the material and be prepared to motivate the workers to learn. This is not always an easy task, but the benefits are worth the effort. More information regarding appropriate training can be found in OSHA Publication No. 2254 which contains voluntary training guidelines prepared by OSHA's Training Institute. A copy of this document is available from OSHA's Publications Office at (202) 219-4667.

In reviewing your written program with regard to information and training, the following items need to be considered:

1. Designation of person(s) responsible for conducting training;

2. Format of the program to be used (audiovisuals, classroom instruction, etc.);

3. Elements of the training program (should be consistent with the elements in paragraph (h) of this section); and,

4. Procedure to train new employees at the time of their initial assignment to work with a hazardous chemical, and to train employees when a new hazard is introduced into the workplace.

The written program should provide enough details about the employer's plans in this area to assess whether or not a good faith effort is being made to train employees. OSHA does not expect that every worker will be able to recite all of the information about each chemical in the workplace. In general, the most important aspects of training under the HCS are to ensure that employees are aware that they are exposed to hazardous chemicals, that they know how to read and use labels and material safety data sheets, and that, as a consequence of learning this information, they are following the appropriate protective measures established by the employer. OSHA compliance officers will be talking to employees to determine if they have received training, if they know they are exposed to hazardous chemicals, and if they know where to obtain substance-specific information on labels and MSDSs.

The rule does not require employers to maintain records of employee training, but many employers choose to do so. This may help you monitor your own program to ensure that all employees are appropriately trained. If you already have a training program, you may simply have to supplement it with whatever additional information is required under the HCS. For example, construction employers that are already in compliance with the construction training standard (29 CFR 1926.21) will have little extra training to do.

An employer can provide employees information and training through whatever means are found appropriate and protective. Although there would always have to be some training on-site (such as informing employees of the location and availability of the written program and MSDSs), employee training may be satisfied in part by general training about the requirements of the HCS and about chemical hazards on the job which is provided by, for example, trade associations, unions, colleges, and professional schools. In addition, previous training, education and experience of a worker may relieve the employer of some of the burdens of informing and training that worker. Regardless of the method relied upon, however, the employer is always ultimately responsible for ensuring that employees are adequately trained. If the compliance officer finds that the training is deficient, the employer will be cited for the deficiency regardless of who actually provided the training on behalf of the employer.

### *D. Other Requirements*

In addition to these specific items, compliance officers will also be asking the following questions in assessing the adequacy of the program:

Does a list of the hazardous chemicals exist in each work area or at a central location?

Are methods the employer will use to inform employees of the hazards of non-routine tasks outlined?

Are employees informed of the hazards associated with chemicals contained in unlabeled pipes in their work areas?

On multi-employer worksites, has the employer provided other employers with information about labeling systems and precautionary measures where the other employers have employees exposed to the initial employer's chemicals?

Is the written program made available to employees and their designated representatives?

If your program adequately addresses the means of communicating information to employees in your workplace, and provides answers to the basic questions outlined above, it will be found to be in compliance with the rule.

### 5. Checklist for Compliance

The following checklist will help to ensure you are in compliance with the rule:

Obtained a copy of the rule. _______

Read and understood the requirements. _______

Assigned responsibility for tasks. _______

Prepared an inventory of chemicals. _______

Ensured containers are labeled. _______

Obtained MSDS for each chemical. _______

Prepared written program. _______

Made MSDSs available to workers. _______

Conducted training of workers. _______

Established procedures to maintain current program. _______

Established procedures to evaluate effectiveness. _______

### 6. Further Assistance

If you have a question regarding compliance with the HCS, you should contact your local OSHA Area Office for assistance. In addition, each OSHA Regional Office has a Hazard Communication Coordinator who can answer your questions. Free consultation services are also available to assist employers, and information regarding these services can be obtained through the Area and Regional offices as well.

The telephone number for the OSHA office closest to you should be listed in your local telephone directory. If you are not able to obtain this information, you may contact OSHA's Office of Information and Consumer Affairs at (202) 219-8151 for further assistance in identifying the appropriate contacts.

[59 FR 6170, Feb. 9, 1994, as amended at 59 FR 17479, Apr. 13, 1994; 59 FR 65948, Dec. 22, 1994; 61 FR 9245, Mar. 7. 1996]

## §1926.60 Methylenedianiline.

**1926.60(a)** ***Scope and application.*** (1) This section applies to all construction work as defined in 29 CFR 1910.12(b), in which there is exposure to MDA, including but not limited to the following:

**1926.60(a)(i)** Construction, alteration, repair, maintenance, or renovation of structures, substrates, or portions thereof, that contain MDA;

**1926.60(a)(ii)** Installation or the finishing of surfaces with products containing MDA;

**1926.60(a)(iii)** MDA spill/emergency cleanup at construction sites; and

**1926.60(a)(iv)** Transportation, disposal, storage, or containment of MDA or products containing MDA on the site or location at which construction activities are performed.

**1926.60(a)(2)** Except as provided in paragraphs (a)(7) and (f)(5) of this section, this section does not apply to the processing, use, and handling of products containing MDA where initial monitoring indicates that the product is not capable of releasing MDA in excess of the action level under the expected conditions of processing, use, and handling which will cause the greatest possible release; and where no "dermal exposure to MDA" can occur.

**1926.60(a)(3)** Except as provided in paragraph (a)(7) of this section, this section does not apply to the processing, use, and handling of products containing MDA where objective data are reasonably relied upon which demonstrate the product is not capable of releasing MDA under the expected conditions of processing, use, and handling which will cause the greatest possible release; and where no "dermal exposure to MDA" can occur.

**1926.60(a)(4)** Except as provided in paragraph (a)(7) of this section, this section does not apply to the storage, transportation, distribution or sale of MDA in intact containers sealed in such a manner as to contain the MDA dusts, vapors, or liquids, except for the provisions of 29 CFR 1910.1200 and paragraph (e) of this section.

**1926.60(a)(5)** Except as provided in paragraph (a)(7) of this section, this section does not apply to materials in any form which contain less than 0.1% MDA by weight or volume.

**1926.60(a)(6)** Except as provided in paragraph (a)(7) of this section, this section does not apply to "finished articles containing MDA."

**1926.60(a)(7)** Where products containing MDA are exempted under paragraphs (a)(2) through (a)(6) of this section, the employer shall maintain records of the initial monitoring results or objective data supporting that exemption and the basis for the employer's reliance on the data, as provided in the recordkeeping provision of paragraph (o) of this section.

**1926.60(b)** ***Definitions.*** For the purpose of this section, the following definitions shall apply:

*Action level* means a concentration of airborne MDA of 5 ppb as an eight (8)-hour time-weighted average.

*Assistant Secretary* means the Assistant Secretary of Labor for Occupational Safety and Health, U.S. Department of Labor, or designee.

*Authorized person* means any person specifically authorized by the employer whose duties require the person to enter a regulated area, or any person entering such an area as a designated representative of employees for the purpose of exercising the right to observe monitoring and measuring procedures under paragraph (p) of this section, or any other person authorized by the Act or regulations issued under the Act.

*Container* means any barrel, bottle, can, cylinder, drum, reaction vessel, storage tank, commercial packaging or the like, but does not include piping systems.

*Decontamination area* means an area outside of but as near as practical to the regulated area, consisting of an equipment storage area, wash area, and clean change area, which is used for the decontamination of workers, materials, and equipment contaminated with MDA.

*Dermal exposure to MDA* occurs where employees are engaged in the handling, application or use of mixtures or materials containing MDA, with any of the following non-airborne forms of MDA:

(i) Liquid, powdered, granular, or flaked mixtures containing MDA in concentrations greater than 0.1% by weight or volume; and

(ii) Materials other than "finished articles" containing MDA in concentrations greater than 0.1% by weight or volume.

*Director* means the Director of the National Institute for Occupational Safety and Health, U.S. Department of Health and Human Services, or designee.

*Emergency* means any occurrence such as, but not limited to, equipment failure, rupture of containers, or failure of control equipment which results in an unexpected and potentially hazardous release of MDA.

*Employee exposure* means exposure to MDA which would occur if the employee were not using respirators or protective work clothing and equipment.

*Finished article containing MDA* is defined as a manufactured item:

(i) Which is formed to a specific shape or design during manufacture;

(ii) Which has end use function(s) dependent in whole or part upon its shape or design during end use; and

(iii) Where applicable, is an item which is fully cured by virtue of having been subjected to the conditions (temperature, time) necessary to complete the desired chemical reaction.

*Historical monitoring data* means monitoring data for construction jobs that meet the following conditions:

(i) The data upon which judgments are based are scientifically sound and were collected using methods that are sufficiently accurate and precise;

(ii) The processes and work practices that were in use when the historical monitoring data were obtained are essentially the same as those to be used during the job for which initial monitoring will not be performed;

(iii) The characteristics of the MDA-containing material being handled when the historical monitoring data were obtained are the same as those on the job for which initial monitoring will not be performed;

(iv) Environmental conditions prevailing when the historical monitoring data were obtained are the same as those on the job for which initial monitoring will not be performed; and

(v) Other data relevant to the operations, materials, processing, or employee exposures covered by the exception are substantially similar. The data must be scientifically sound, the characteristics of the MDA containing material must be similar and the environmental conditions comparable.

*4,4'Methylenedianiline or MDA* means the chemical; 4,4'-diaminodiphenylmethane, Chemical Abstract Service Registry number 101-77-9, in the form of a vapor, liquid, or solid. The definition also includes the salts of MDA.

*Regulated Areas* means areas where airborne concentrations of MDA exceed or can reasonably be expected to exceed, the permissible exposure limits, or where "dermal exposure to MDA" can occur.

*STEL* means short term exposure limit as determined by any 15-minute sample period.

**1926.60(c)** ***Permissible exposure limits.*** The employer shall assure that no employee is exposed to an airborne concentration of MDA in excess of ten parts per billion (10 ppb) as an 8-hour time-weighted average and a STEL of one hundred parts per billion (100 ppb).

**1926.60(d)** ***Communication among employers.*** On multi-employer worksites, an employer performing work involving the application of MDA or materials containing MDA for which establishment of one or more regulated areas is required shall inform other employers on the site of the nature of the employer's work with MDA and of the existence of, and requirements pertaining to, regulated areas.

**1926.60(e)** ***Emergency situations***— (1) *Written plan.* (i) A written plan for emergency situations shall be developed for each construction operation where there is a possibility of an emergency. The plan shall include procedures where the employer identifies emergency escape routes for his employees at each construction site before the construction operation begins. Appropriate portions of the plan shall be implemented in the event of an emergency.

**1926.60(e)(1)(ii)** The plan shall specifically provide that employees engaged in correcting emergency conditions shall be equipped with the appropriate personal protective equipment and clothing as required in paragraphs (i) and (j) of this section until the emergency is abated.

**1926.60(e)(1)(iii)** The plan shall specifically include provisions for alerting and evacuating affected employees as well as the applicable elements prescribed in 29 CFR 1910.38, "Employee emergency plans and fire prevention plans."

**1926.60(e)(2)** *Alerting employees.* Where there is the possibility of employee exposure to MDA due to an emergency, means shall be developed to promptly alert employees who have the potential to be directly exposed. Affected employees not engaged in correcting emergency conditions shall be evacuated immediately in the event that an emergency occurs. Means shall also be developed for alerting other employees who may be exposed as a result of the emergency.

**1926.60(f)** ***Exposure monitoring***—(1) *General.* (i) Determinations of employee exposure shall be made from breathing zone air samples that are representative of each employee's exposure to airborne MDA over an eight (8) hour period. Determination of employee exposure to the STEL shall be made from breathing zone air samples collected over a 15 minute sampling period.

**1926.60(f)(1)(ii)** Representative employee exposure shall be determined on the basis of one or more samples representing full shift exposure for each shift for each job classification in each work area where exposure to MDA may occur.

**1926.60(f)(1)(iii)** Where the employer can document that exposure levels are equivalent for similar operations in different work shifts, the employer shall only be required to determine representative employee exposure for that operation during one shift.

**1926.60(f)(2)** *Initial monitoring.* Each employer who has a workplace or work operation covered by this standard shall perform initial monitoring to determine accurately the airborne concentrations of MDA to which employees may be exposed unless:

**1926.60(f)(2)(i)** The employer can demonstrate, on the basis of objective data, that the MDA-containing product or material being handled cannot cause exposures above the standard's action level, even under worst-case release conditions; or

**1926.60(f)(2)(ii)** The employer has historical monitoring or other data demonstrating that exposures on a particular job will be below the action level.

**1926.60(f)(3)** *Periodic monitoring and monitoring frequency.* (i) If the monitoring required by paragraph (f)(2) of this section reveals employee exposure at or above the action level, but at or below the PELs, the employer shall repeat such monitoring for each such employee at least every six (6) months.

**1926.60(f)(3)(ii)** If the monitoring required by paragraph (f)(2) of this section reveals employee exposure above the PELs, the employer shall repeat such monitoring for each such employee at least every three (3) months.

**1926.60(f)(3)(iii)** Employers who are conducting MDA operations within a regulated area can forego periodic monitoring if the employees are all wearing supplied-air respirators while working in the regulated area.

**1926.60(f)(3)(iv)** The employer may alter the monitoring schedule from every three months to every six months for any employee for whom two consecutive measurements taken at least 7 days apart indicate that the employee exposure has decreased to below the PELs but above the action level.

**1926.60(f)(4)** *Termination of monitoring.* (i) If the initial monitoring required by paragraph (f)(2) of this section reveals employee exposure to be below the action level, the employer may discontinue the monitoring for that employee, except as otherwise required by paragraph (f)(5) of this section.

**1926.60(f)(4)(ii)** If the periodic monitoring required by paragraph (f)(3) of this section reveals that employee exposures, as indicated by at least two consecutive measurements taken at least 7 days apart, are below the action level the employer may discontinue the monitoring for that employee, except as otherwise required by paragraph (f)(5) of this section.

**1926.60(f)(5)** *Additional monitoring.* The employer shall institute the exposure monitoring required under paragraphs (f)(2) and (f)(3) of this section when there has been a change in production process, chemicals present, control equipment, personnel, or work practices which may result in new or additional exposures to MDA, or when the employer has any reason to suspect a change which may result in new or additional exposures.

**1926.60(f)(6)** *Accuracy of monitoring.* Monitoring shall be accurate, to a confidence level of 95 percent, to within plus or minus 25 percent for airborne concentrations of MDA.

**1926.60(f)(7)** *Employee notification of monitoring results.* (i) The employer shall, within 15 working days after the receipt of the results of any monitoring performed under this standard, notify each employee of these results, in writing, either individually or by posting of results in an appropriate location that is accessible to affected employees.

**1926.60(f)(7)(ii)** The written notification required by paragraph (f)(7)(i) of this section shall contain the corrective action being taken by the employer or any other protective measures which have been implemented to reduce the employee exposure to or below the PELs, wherever the PELs are exceeded.

**1926.60(f)(8)** *Visual monitoring.* The employer shall make routine inspections of employee hands, face and forearms potentially exposed to MDA. Other potential dermal exposures reported by the employee must be referred to the appropriate medical personnel for observation. If the employer determines that the employee has been exposed to MDA the employer shall:

**1926.60(f)(8)(i)** Determine the source of exposure;

**1926.60(f)(8)(ii)** Implement protective measures to correct the hazard; and

**1926.60(f)(8)(iii)** Maintain records of the corrective actions in accordance with paragraph (*n*) of this section.

**1926.60(g)** ***Regulated areas***— (1) *Establishment*— (i) *Airborne exposures.* The employer shall establish regulated areas where airborne concentrations of MDA exceed or can reasonably be expected to exceed, the permissible exposure limits.

**1926.60(g)(1)(ii)** *Dermal exposures.* Where employees are subject to "dermal exposure to MDA" the employer shall establish those work areas as regulated areas.

**1926.60(g)(2)** *Demarcation.* Regulated areas shall be demarcated from the rest of the workplace in a manner that minimizes the number of persons potentially exposed.

**1926.60(g)(3)** *Access.* Access to regulated areas shall be limited to authorized persons.

**1926.60(g)(4)** *Personal protective equipment and clothing.* Each person entering a regulated area shall be supplied with, and required to use, the appropriate personal protective clothing and equipment in accordance with paragraphs (i) and (j) of this section.

**1926.60(g)(5)** *Prohibited activities.* The employer shall ensure that employees do not eat, drink, smoke, chew tobacco or gum, or apply cosmetics in regulated areas.

**1926.60(h)** ***Methods of compliance***— (1) *Engineering controls and work practices and respirators.* (i) The employer shall use one or any combination of the following control methods to achieve compliance with the permissible exposure limits prescribed by paragraph (c) of this section:

**1926.60(h)(1)(i)(A)** Local exhaust ventilation equipped with HEPA filter dust collection systems;

**1926.60(h)(1)(i)(B)** General ventilation systems;

**1926.60(h)(1)(i)(C)** Use of workpractices; or

**1926.60(h)(1)(i)(D)** Other engineering controls such as isolation and enclosure that the Assistant Secretary can show to be feasible.

**1926.60(h)(1)(ii)** Wherever the feasible engineering controls and work practices "which can be instituted are not sufficient to reduce employee exposure to or below the PELs, the employer shall use them to reduce employee exposure to the lowest levels achievable by these controls and shall supplement them by the use of respiratory protective devices which comply with the requirements of paragraph (i) of this section.

**1926.60(h)(2)** *Special Provisions.* For workers engaged in spray application methods, respiratory protection must be used in addition to feasible engineering controls and work practices to reduce employee exposure to or below the PELs.

**1926.60(h)(3)** *Prohibitions.* Compressed air shall not be used to remove MDA, unless the compressed air is used in conjunction with an enclosed ventilation system designed to capture the dust cloud created by the compressed air.

**1926.60(h)(4)** *Employee rotation.* The employer shall not use employee rotation as a means of compliance with the exposure limits prescribed in paragraph (c) of this section.

**1926.60(h)(5)** *Compliance program.* (i) The employer shall establish and implement a written program to reduce employee exposure to or below the PELs by means of engineering and work practice controls, as required by paragraph (h)(1) of this section, and by use of respiratory protection where permitted under this section.

**1926.60(h)(5)(ii)** Upon request this written program shall be furnished for examination and copying to the Assistant Secretary, the Director, affected employees and designated employee representatives. The employer shall review and, as necessary, update such plans at least once every 12 months to make certain they reflect the current status of the program.

**1926.60(i)** ***Respiratory protection.*** (1) *General.* For employees who use respirators required by this section, the employer must provide respirators that comply with the requirements of this paragraph. Respirators must be used during:

**1926.60(i)(1)(i)** Periods necessary to install or implement feasible engineering and work-practice controls.

**1926.60(i)(1)(ii)** Work operations, such as maintenance and repair activities and spray-application processes, for which engineering and work-practice controls are not feasible.

**1926.60(i)(1)(iii)** Work operations for which feasible engineering and work-practice controls are not yet sufficient to reduce employee exposure to or below the PELs.

**1926.60(i)(1)(iv)** Emergencies.

**1926.60(i)(2)** *Respirator program.* The employer must implement a respiratory protection program in accordance with 29 CFR 1910.134 (b) through (d) (except (d)(1)(iii), and (f) through (m).

**1926.60(i)(3)** *Respirator selection.* (i) The employer must select the appropriate respirator from Table 1 of this section.

### TABLE 1.—RESPIRATORY PROTECTION FOR MDA

| Airborne concentration of MDA or condition of use | Respirator type |
|---|---|
| a. Less than or equal to 10 × PEL | (1) Half-Mask Respirator with HEPA [1] Cartridge.[2] |
| b. Less than or equal to 50 × PEL | (1) Full facepiece respirator with HEPA [1] Cartridge or Canister.[2] |
| c. Less than or equal to 1000 × PEL | (1) Full facepiece powered air-purifying respirator with HEPA [1] cartridges.[2] |
| d. Greater than 1000 × PEL or unknown | (1) Self-contained breathing apparatus with full facepiece in positive pressure mode. |
| concentration. | (2) Full facepiece positive pressure demand supplied-air respirator with auxiliary self-contained air supply. |
| e. Escape | (1) Any full facepiece air-purifying respirator with HEPA [1] cartridges.[2] |
| | (2) Any positive pressure or continuous flow self-contained breathing apparatus with full facepiece or hood. |
| f. Firefighting | (1) Full facepiece self-contained breathing apparatus in positive pressure mode. |

**NOTE:** Respirators assigned for higher environmental concentrations may be used at lower concentrations.

1 High Efficiency Particulate in Air filter (HEPA) means a filter that is at least 99.97 percent efficient against mono-dispersed particles of 0.3 micrometers or larger.

2 Combination HEPA/Organic Vapor Cartridges shall be used whenever MDA in liquid form or a process requiring heat is used.

**1926.60(i)(3)(ii)** An employee who cannot use a negative-pressure respirator must be given the option of using a positive-pressure respirator, or a supplied-air respirator operated in the continuous-flow or pressure-demand mode.

**1926.60(j)** ***Protective work clothing and equipment***— (1) *Provision and use.* Where employees are subject to dermal exposure to MDA, where liquids containing MDA can be splashed into the eyes, or where airborne concentrations of MDA are in excess of the PEL, the employer shall provide, at no cost to the employee, and ensure that the employee uses, appropriate protective work clothing and equipment which prevent contact with MDA such as, but not limited to:

**1926.60(j)(1)(i)** Aprons, coveralls or other full-body work clothing;

**1926.60(j)(1)(ii)** Gloves, head coverings, and foot coverings; and

**1926.60(j)(1)(iii)** Face shields, chemical goggles; or

**1926.60(j)(1)(iv)** Other appropriate protective equipment which comply with 29 CFR 1910.133.

**1926.60(j)(2)** *Removal and storage.* (i) The employer shall ensure that, at the end of their work shift, employees remove MDA-contaminated protective work clothing and equipment that is not routinely removed throughout the day in change areas provided in accordance with the provisions in paragraph (k) of this section.

**1926.60(j)(2)(ii)** The employer shall ensure that, during their work shift, employees remove all other MDA-contaminated protective work clothing or equipment before leaving a regulated area.

**1926.60(j)(2)(iii)** The employer shall ensure that no employee takes MDA-contaminated work clothing or equipment out of the decontamination areas, except those employees authorized to do so for the purpose of laundering, maintenance, or disposal.

**1926.60(j)(2)(iv)** MDA-contaminated work clothing or equipment shall be placed and stored and transported in sealed, impermeable bags, or other closed impermeable containers.

**1926.60(j)(2)(v)** Containers of MDA-contaminated protective work clothing or equipment which are to be taken out of decontamination areas or the workplace for cleaning, maintenance, or disposal, shall bear labels warning of the hazards of MDA.

**1926.60(j)(3)** *Cleaning and replacement.* (i) The employer shall provide the employee with clean protective clothing and equipment. The employer shall ensure that protective work clothing or equipment required by this paragraph is cleaned, laundered, repaired, or replaced at intervals appropriate to maintain its effectiveness.

**1926.60(j)(3)(ii)** The employer shall prohibit the removal of MDA from protective work clothing or equipment by blowing, shaking, or any methods which allow MDA to re-enter the workplace.

**1926.60(j)(3)(iii)** The employer shall ensure that laundering of MDA-contaminated clothing shall be done so as to prevent the release of MDA in the workplace.

**1926.60(j)(3)(iv)** Any employer who gives MDA-contaminated clothing to another person for laundering shall inform such person of the requirement to prevent the release of MDA.

**1926.60(j)(3)(v)** The employer shall inform any person who launders or cleans protective clothing or equipment contaminated with MDA of the potentially harmful effects of exposure.

**1926.60(j)(4)** *Visual Examination.* (i) The employer shall ensure that employees' work clothing is examined periodically for rips or tears that may occur during performance of work.

**1926.60(j)(4)(ii)** When rips or tears are detected, the protective equipment or clothing shall be repaired and replaced immediately.

**1926.60(k)** ***Hygiene facilities and practices***— (1) *General.* (i) The employer shall provide decontamination areas for employees required to work in regulated areas or required by paragraph (j)(1) of this section to wear protective clothing. *Exception:* In lieu of the decontamination area requirement specified in paragraph (k)(1)(i) of this section, the employer may permit employees engaged in small scale, short duration operations, to clean their protective clothing or dispose of the protective clothing before such employees leave the area where the work was performed.

**1926.60(k)(1)(ii)** *Change areas.* The employer shall ensure that change areas are equipped with separate storage facilities for protective clothing and street clothing, in accordance with 29 CFR 1910.141(e).

**1926.60(k)(1)(iii)** *Equipment area.* The equipment area shall be supplied with impermeable, labeled bags and containers for the containment and disposal of contaminated protective clothing and equipment.

**1926.60(k)(2)** *Shower area.* (i) Where feasible, shower facilities shall be provided which comply with 29 CFR 1910.141(d)(3) wherever the possibility of employee exposure to airborne levels of MDA in excess of the permissible exposure limit exists.

**1926.60(k)(2)(ii)** Where dermal exposure to MDA occurs, the employer shall ensure that materials spilled or deposited on the skin are removed as soon as possible by methods which do not facilitate the dermal absorption of MDA.

**1926.60(k)(3)** *Lunch Areas.* (i) Whenever food or beverages are consumed at the worksite and employees are exposed to MDA the employer shall provide clean lunch areas were MDA levels are below the action level and where no dermal exposure to MDA can occur.

**1926.60(k)(3)(ii)** The employer shall ensure that employees wash their hands and faces with soap and water prior to eating, drinking, smoking, or applying cosmetics.

**1926.60(k)(3)(iii)** The employer shall ensure that employees do not enter lunch facilities with contaminated protective work clothing or equipment.

**1926.60(l)** ***Communication of hazards to employees***— (1) *Signs and labels.* (i) The employer shall post and maintain legible signs demarcating regulated areas and entrances or accessways to regulated areas that bear the following legend:

**DANGER**

**MDA**

**MAY CAUSE CANCER**

**LIVER TOXIN**

**AUTHORIZED PERSONNEL ONLY**

**RESPIRATORS AND PROTECTIVE CLOTHING MAY BE REQUIRED TO BE WORN IN THIS AREA**

**1926.60(l)(1)(ii)** The employer shall ensure that labels or other appropriate forms of warning are provided for containers of MDA within the workplace. The labels shall comply with the requirements of 29 CFR 1910.1200(f) and shall include one of the following legends:

**1926.60(l)(1)(ii)(A)** For pure MDA

**DANGER**

**CONTAINS MDA**

**MAY CAUSE CANCER**

**LIVER TOXIN**

**1926.60(l)(1)(ii)(B)** For mixtures containing MDA

**DANGER**

**CONTAINS MDA**

**CONTAINS MATERIALS WHICH MAY CAUSE CANCER**

**LIVER TOXIN**

**1926.60(l)(2)** *Material safety data sheets* (MSDS). Employers shall obtain or develop, and shall provide access to their employees, to a material safety data sheet (MSDS) for MDA.

**1926.60(l)(3)** *Information and training.* (i) The employer shall provide employees with information and training on MDA, in accordance with 29 CFR 1910.1200(h), at the time of initial assignment and at least annually thereafter.

**1926.60(l)(3)(ii)** In addition to the information required under 29 CFR 1910.1200, the employer shall:

**1926.60(l)(3)(ii)(A)** Provide an explanation of the contents of this section, including appendices A and B of this section, and indicate to employees where a copy of the standard is available;

**1926.60(l)(3)(ii)(B)** Describe the medical surveillance program required under paragraph (n) of this section, and explain the information contained in appendix C of this section; and

**1926.60(l)(3)(ii)(C)** Describe the medical removal provision required under paragraph (n) of this section.

**1926.60(l)(4)** *Access to training materials.* (i) The employer shall make readily available to all affected employees, without cost, all written materials relating to the employee training program, including a copy of this regulation.

**1926.60(l)(4)(ii)** The employer shall provide to the Assistant Secretary and the Director, upon request, all information and training materials relating to the employee information and training program.

**1926.60(m)** ***Housekeeping.*** (1) All surfaces shall be maintained as free as practicable of visible accumulations of MDA.

**1926.60(m)(2)** The employer shall institute a program for detecting MDA leaks, spills, and discharges, including regular visual inspections of operations involving liquid or solid MDA.

**1926.60(m)(3)** All leaks shall be repaired and liquid or dust spills cleaned up promptly.

**1926.60(m)(4)** Surfaces contaminated with MDA may not be cleaned by the use of compressed air.

**1926.60(m)(5)** Shoveling, dry sweeping, and other methods of dry clean-up of MDA may be used where HEPA filtered vacuuming and/or wet cleaning are not feasible or practical.

**1926.60(m)(6)** Waste, scrap, debris, bags, containers, equipment, and clothing contaminated with MDA shall be collected and disposed of in a manner to prevent the re-entry of MDA into the workplace.

**1926.60(n)** ***Medical surveillance—*** (1) *General.* (i) The employer shall make available a medical surveillance program for employees exposed to MDA under the following circumstances:

**1926.60(n)(1)(i)(A)** Employees exposed at or above the action level for 30 or more days per year;

**1926.60(n)(1)(i)(B)** Employees who are subject to dermal exposure to MDA for 15 or more days per year;

**1926.60(n)(1)(i)(C)** Employees who have been exposed in an emergency situation;

**1926.60(n)(1)(i)(D)** Employees whom the employer, based on results from compliance with paragraph (f)(8) of this section, has reason to believe are being dermally exposed; and

**1926.60(n)(1)(i)(E)** Employees who show signs or symptoms of MDA exposure.

**1926.60(n)(1)(ii)** The employer shall ensure that all medical examinations and procedures are performed by or under the supervision of a licensed physician at a reasonable time and place, and provided without cost to the employee.

**1926.60(n)(2)** *Initial examinations.* (i) Within 150 days of the effective date of this standard, or before the time of initial assignment, the employer shall provide each employee covered by paragraph (n)(1)(i) of this section with a medical examination including the following elements:

**1926.60(n)(2)(i)(A)** A detailed history which includes:

**1926.60(n)(2)(i)(A)*(1)*** Past work exposure to MDA or any other toxic substances;

**1926.60(n)(2)(i)(A)*(2)*** A history of drugs, alcohol, tobacco, and medication routinely taken (duration and quantity); and

**1926.60(n)(2)(i)(A)*(3)*** A history of dermatitis, chemical skin sensitization, or previous hepatic disease.

**1926.60(n)(2)(i)(B)** A physical examination which includes all routine physical examination parameters, skin examination, and examination for signs of liver disease.

**1926.60(n)(2)(i)(C)** Laboratory tests including:

**1926.60(n)(2)(i)(C)*(1)*** Liver function tests and (2) Urinalysis.

**1926.60(n)(2)(i)(D)** Additional tests as necessary in the opinion of the physician.

**1926.60(n)(2)(ii)** No initial medical examination is required if adequate records show that the employee has been examined in accordance with the requirements of this section within the previous six months prior to the effective date of this standard or prior to the date of initial assignment.

**1926.60(n)(3)** *Periodic examinations.* (i) The employer shall provide each employee covered by this section with a medical examination at least annually following the initial examination. These periodic examinations shall include at least the following elements:

**1926.60(n)(3)(i)(A)** A brief history regarding any new exposure to potential liver toxins, changes in drug, tobacco, and alcohol intake, and the appearance of physical signs relating to the liver, and the skin;

**1926.60(n)(3)(i)(B)** The appropriate tests and examinations including liver function tests and skin examinations; and

**1926.60(n)(3)(i)(C)** Appropriate additional tests or examinations as deemed necessary by the physician.

**1926.60(n)(3)(ii)** If in the physician's opinion the results of liver function tests indicate an abnormality, the employee shall be removed from further MDA exposure in accordance with paragraph (n)(9) of this section. Repeat liver function tests shall be conducted on advice of the physician.

**1926.60(n)(4)** *Emergency examinations.* If the employer determines that the employee has been exposed to a potentially hazardous amount of MDA in an emergency situation under paragraph (e) of this section, the employer shall provide medical examinations in accordance with paragraphs (n)(3) (i) and (ii) of this section. If the results of liver function testing indicate an abnormality, the employee shall be removed in accordance with paragraph (n)(9) of this section. Repeat liver function tests shall be conducted on the advice of the physician. If the results of the tests are normal, tests must be repeated two to three weeks from the initial testing. If the results of the second set of tests are normal and on the advice of the physician, no additional testing is required.

**1926.60(n)(5)** *Additional examinations.* Where the employee develops signs and symptoms associated with exposure to MDA, the employer shall provide the employee with an additional medical examination including liver function tests. Repeat liver function tests shall be conducted on the advice of the physician. If the results of the tests are normal, tests must be repeated two to three weeks from the initial testing. If the results of the second set of tests are normal and on the advice of the physician, no additional testing is required.

**1926.60(n)(6)** *Multiple physician review mechanism.* (i) If the employer selects the initial physician who conducts any medical examination or consultation provided to an employee under this section, and the employee has signs or symptoms of occupational exposure to MDA (which could include an abnormal liver function test), and the employee disagrees with the opinion of the examining physician, and this opinion could affect the employee's job status, the employee may designate an appropriate and mutually acceptable second physician:

**1926.60(n)(6)(i)(A)** To review any findings, determinations or recommendations of the initial physician; and

**1926.60(n)(6)(i)(B)** To conduct such examinations, consultations, and laboratory tests as the second physician deems necessary to facilitate this review.

**1926.60(n)(6)(ii)** The employer shall promptly notify an employee of the right to seek a second medical opinion after each occasion that an initial physician conducts a medical examination or consultation pursuant to this section. The employer may condition its participation in, and payment for, the multiple physician review mechanism upon the employee doing the following within fifteen (15) days after receipt of the foregoing notification, or receipt of the initial physician's written opinion, whichever is later:

**1926.60(n)(6)(ii)(A)** The employee informing the employer that he or she intends to seek a second medical opinion, and

**1926.60(n)(6)(ii)(B)** The employee initiating steps to make an appointment with a second physician.

**1926.60(n)(6)(iii)** If the findings, determinations, or recommendations of the second physician differ from those of the initial physician, then the employer and the employee shall assure that efforts are made for the two physicians to resolve any disagreement.

**1926.60(n)(6)(iv)** If the two physicians have been unable to quickly resolve their disagreement, then the employer and the employee through their respective physicians shall designate a third physician:

**1926.60(n)(6)(iv)(A)** To review any findings, determinations, or recommendations of the prior physicians; and

**1926.60(n)(6)(iv)(B)** To conduct such examinations, consultations, laboratory tests, and discussions with the prior physicians as the third physician deems necessary to resolve the disagreement of the prior physicians.

**1926.60(n)(6)(v)** The employer shall act consistent with the findings, determinations, and recommendations of the second physician, unless the employer and the employee reach a mutually acceptable agreement.

**1926.60(n)(7)** *Information provided to the examining physician.* (i) The employer shall provide the following information to the examining physician:

**1926.60(n)(7)(i)(A)** A copy of this regulation and its appendices;

**1926.60(n)(7)(i)(B)** A description of the affected employee's duties as they relate to the employee's potential exposure to MDA;

**1926.60(n)(7)(i)(C)** The employee's current actual or representative MDA exposure level;

**1926.60(n)(7)(i)(D)** A description of any personal protective equipment used or to be used; and

**1926.60(n)(7)(i)(E)** Information from previous employment related medical examinations of the affected employee.

**1926.60(n)(7)(ii)** The employer shall provide the foregoing information to a second physician under this section upon request either by the second physician, or by the employee.

**1926.60(n)(8)** *Physician's written opinion.* (i) For each examination under this section, the employer shall obtain, and provide the employee with a copy of, the examining physician's written opinion within 15 days of its receipt. The written opinion shall include the following:

**1926.60(n)(8)(i)(A)** The occupationally pertinent results of the medical examination and tests;

**1926.60(n)(8)(i)(B)** The physician's opinion concerning whether the employee has any detected medical conditions which would place the employee at increased risk of material impairment of health from exposure to MDA;

**1926.60(n)(8)(i)(C)** The physician's recommended limitations upon the employee's exposure to MDA or upon the employee's use of protective clothing or equipment and respirators; and

**1926.60(n)(8)(i)(D)** A statement that the employee has been informed by the physician of the results of the medical examination and any medical conditions resulting from MDA exposure which require further explanation or treatment.

**1926.60(n)(8)(ii)** The written opinion obtained by the employer shall not reveal specific findings or diagnoses unrelated to occupational exposures.

**1926.60(n)(9)** *Medical removal*—(i) *Temporary medical removal of an employee*—(A) *Temporary removal resulting from occupational exposure.* The employee shall be removed from work environments in which exposure to MDA is at or above the action level or where dermal exposure to MDA may occur, following an initial examination (paragraph (n)(2) of this section), periodic examinations (paragraph (n)(3) of this section), an emergency situation (paragraph (n)(4) of this section), or an additional examination (paragraph (n)(5) of this section) in the following circumstances:

**1926.60(n)(9)(i)(A)*(1)*** When the employee exhibits signs and/or symptoms indicative of acute exposure to MDA; or

**1926.60(n)(9)(i)(A)*(2)*** When the examining physician determines that an employee's abnormal liver function tests are not associated with MDA exposure but that the abnormalities may be exacerbated as a result of occupational exposure to MDA.

**1926.60(n)(9)(i)(B)** *Temporary removal due to a final medical determination.* (*1*) The employer shall remove an employee from work having an exposure to MDA at or above the action level or where the potential for dermal exposure exists on each occasion that a final medical determination results in a medical finding, determination, or opinion that the employee has a detected medical condition which places the employee at increased risk of material impairment to health from exposure to MDA.

**1926.60(n)(9)(i)(B)*(2)*** For the purposes of this section, the phrase "final medical determination" shall mean the outcome of the physician review mechanism used pursuant to the medical surveillance provisions of this section.

**1926.60(n)(9)(i)(B)*(3)*** Where a final medical determination results in any recommended special protective measures for an employee, or limitations on an employee's exposure to MDA, the employer shall implement and act consistent with the recommendation.

**1926.60(n)(9)(ii)** *Return of the employee to former job status.* (A) The employer shall return an employee to his or her former job status:

**1926.60(n)(9)(ii)(A)*(1)*** When the employee no longer shows signs or symptoms of exposure to MDA, or upon the advice of the physician.

**1926.60(n)(9)(ii)*(A)(2)*** When a subsequent final medical determination results in a medical finding, determination, or opinion that the employee no longer has a detected medical condition which places the employee at increased risk of material impairment to health from exposure to MDA.

**1926.60(n)(9)(ii)(B)** For the purposes of this section, the requirement that an employer return an employee to his or her former job status is not intended to expand upon or restrict any rights an employee has or would have had, absent temporary medical removal, to a specific job classification or position under the terms of a collective bargaining agreement.

**1926.60(n)(9)(iii)** *Removal of other employee special protective measure or limitations.* The employer shall remove any limitations placed on an employee or end any special protective measures provided to an employee pursuant to a final medical determination when a subsequent final medical determination indicates that the limitations or special protective measures are no longer necessary.

**1926.60(n)(9)(iv)** *Employer options pending a final medical determination.* Where the physician review mechanism used pursuant to the medical surveillance provisions of this section, has not yet resulted in a final medical determination with respect to an employee, the employer shall act as follows:

**1926.60(n)(9)(iv)(A)** *Removal.* The employer may remove the employee from exposure to MDA, provide special protective measures to the employee, or place limitations upon the employee, consistent with the medical findings, determinations, or recommendations of the physician who has reviewed the employee's health status.

**1926.60(n)(9)(iv)(B)** *Return.* The employer may return the employee to his or her former job status, and end any special protective measures provided to the employee, consistent with the medical findings, determinations, or recommendations of any of the physicians who have reviewed the employee's health status, with two exceptions:

**1926.60(n)(9)(iv)(B)*(1)*** If the initial removal, special protection, or limitation of the employee resulted from a final medical determination which differed from the findings, determinations, or recommendations of the initial physician; or

**1926.60(n)(9)(iv)(B)*(2)*** The employee has been on removal status for the preceding six months as a result of exposure to MDA, then the employer shall await a final medical determination.

**1926.60(n)(9)(v)** *Medical removal protection benefits*—(A) *Provisions of medical removal protection benefits.* The employer shall provide to an employee up to six (6) months of medical removal protection benefits on each occasion that an employee is removed from exposure to MDA or otherwise limited pursuant to this section.

**1926.60(n)(9)(v)(B)** *Definition of medical removal protection benefits.* For the purposes of this section, the requirement that an employer provide medical removal protection benefits means that the employer shall maintain the earnings, seniority, and other employment rights and benefits of an employee as though the employee had not been removed from normal exposure to MDA or otherwise limited.

**1926.60(n)(9)(v)(C)** *Follow-up medical surveillance during the period of employee removal or limitations.* During the period of time that an employee is removed from normal exposure to MDA or otherwise limited, the employer may condition the provision of medical removal protection benefits upon the employee's participation in follow-up medical surveillance made available pursuant to this section.

**1926.60(n)(9)(v)(D)** *Workers' compensation claims.* If a removed employee files a claim for workers' compensation payments for a MDA-related disability, then the employer shall continue to provide medical removal protection benefits pending disposition of the claim. To the extent that an award is made to the employee for earnings lost during the period of removal, the employer's medical removal protection obligation shall be reduced by such amount. The employer shall receive no credit for workers' compensation payments received by the employee for treatment-related expenses.

**1926.60(n)(9)(v)(E)** *Other credits.* The employer's obligation to provide medical removal protection benefits to a removed employee shall be reduced to the extent that the employee receives compensation for earnings lost during the period of removal either from a publicly or employer-funded compensation program, or receives income from employment with any employer made possible by virtue of the employee's removal.

**1926.60(n)(9)(v)(F)** *Employees who do not recover within the 6 months of removal.* The employer shall take the following measures with respect to any employee removed from exposure to MDA:

**1926.60(n)(9)(v)(F)*(1)*** The employer shall make available to the employee a medical examination pursuant to this section to obtain a final medical determination with respect to the employee;

**1926.60(n)(9)(v)(F)*(2)*** The employer shall assure that the final medical determination obtained indicates whether or not the employee may be returned to his or her former job status, and, if not, what steps should be taken to protect the employee's health;

**1926.60(n)(9)(v)(F)*(3)*** Where the final medical determination has not yet been obtained, or once obtained indicates that the employee may not yet be returned to his or her former job status, the employer shall continue to provide medical removal protection benefits to the employee until either the employee is returned to former job status, or a final medical determination is made that the employee is incapable of ever safely returning to his or her former job status; and

**1926.60(n)(9)(v)(F)*(4)*** Where the employer acts pursuant to a final medical determination which permits the return of the employee to his or her former job status despite what would otherwise be an unacceptable liver function test, later questions concerning removing the employee again shall be decided by a final medical determination. The employer need not automatically remove such an employee pursuant to the MDA removal criteria provided by this section.

**1926.60(n)(9)(vi)** *Voluntary removal or restriction of an employee.* Where an employer, although not required by this section to do so, removes an employee from exposure to MDA or otherwise places limitations on an employee due to the effects of MDA exposure on the employee's medical condition, the employer shall provide medical removal protection benefits to the employee equal to that required by paragraph (n)(9)(v) of this section.

**1926.60(o)** ***Recordkeeping***— (1) *Objective data for exempted operations.* (i) Where the employer has relied on objective data that demonstrate that products made from or containing MDA are not capable of releasing MDA or do not present a dermal exposure problem under the expected conditions of processing, use, or handling to exempt such operations from the initial monitoring requirements under paragraph (f)(2) of this section, the employer shall establish and maintain an accurate record of objective data reasonably relied upon in support of the exemption.

**1926.60(o)(1)(ii)** The record shall include at least the following information:

**1926.60(o)(1)(ii)(A)** The product qualifying for exemption;

**1926.60(o)(1)(ii)(B)** The source of the objective data;

**1926.60(o)(1)(ii)(C)** The testing protocol, results of testing, and/or analysis of the material for the release of MDA;

**1926.60(o)(1)(ii)(D)** A description of the operation exempted and how the data support the exemption; and

**1926.60(o)(1)(ii)(E)** Other data relevant to the operations, materials, processing, or employee exposures covered by the exemption.

**1926.60(o)(1)(iii)** The employer shall maintain this record for the duration of the employer's reliance upon such objective data.

**1926.60(o)(2)** *Historical monitoring data.* (i) Where the employer has relied on historical monitoring data that demonstrate that exposures on a particular job will be below the action level to exempt such operations from the initial monitoring requirements under paragraph (f)(2) of this section, the employer shall establish and maintain an accurate record of historical monitoring data reasonably relied upon in support of the exception.

**1926.60(o)(2)(ii)** The record shall include information that reflect the following conditions:

**1926.60(o)(2)(ii)(A)** The data upon which judgments are based are scientifically sound and were collected using methods that are sufficiently accurate and precise;

**1926.60(o)(2)(ii)(B)** The processes and work practices that were in use when the historical monitoring data were obtained are essentially the same as those to be used during the job for which initial monitoring will not be performed;

**1926.60(o)(2)(ii)(C)** The characteristics of the MDA-containing material being handled when the historical monitoring data were obtained are the same as those on the job for which initial monitoring will not be performed;

**1926.60(o)(2)(ii)(D)** Environmental conditions prevailing when the historical monitoring data were obtained are the same as those on the job for which initial monitoring will not be performed; and

**1926.60(o)(2)(ii)(E)** Other data relevant to the operations, materials, processing, or employee exposures covered by the exception.

**1926.60(o)(2)(iii)** The employer shall maintain this record for the duration of the employer's reliance upon such historical monitoring data.

**1926.60(o)(3)** The employer may utilize the services of competent organizations such as industry trade associations and employee associations to maintain the records required by this section.

**1926.60(o)(4)** *Exposure measurements.* (i) The employer shall keep an accurate record of all measurements taken to monitor employee exposure to MDA.

**1926.60(o)(4)(ii)** This record shall include at least the following information:

**1926.60(o)(4)(ii)(A)** The date of measurement;

**1926.60(o)(4)(ii)(B)** The operation involving exposure to MDA;

**1926.60(o)(4)(ii)(C)** Sampling and analytical methods used and evidence of their accuracy;

**1926.60(o)(4)(ii)(D)** Number, duration, and results of samples taken;

**1926.60(o)(4)(ii)(E)** Type of protective devices worn, if any; and

**1926.60(o)(4)(ii)(F)** Name, social security number, and exposure of the employees whose exposures are represented.

**1926.60(o)(4)(iii)** The employer shall maintain this record for at least thirty (30) years, in accordance with 29 CFR 1910.20.

**1926.60(o)(5)** *Medical surveillance.* (i) The employer shall establish and maintain an accurate record for each employee subject to medical surveillance by paragraph (n) of this section, in accordance with 29 CFR 1910.20.

**1926.60(o)(5)(ii)** The record shall include at least the following information:

**1926.60(o)(5)(ii)(A)** The name and social security number of the employee;

**1926.60(o)(5)(ii)(B)** A copy of the employee's medical examination results, including the medical history, questionnaire responses, results of any tests, and physician's recommendations.

**1926.60(o)(5)(ii)(C)** Physician's written opinions;

**1926.60(o)(5)(ii)(D)** Any employee medical complaints related to exposure to MDA; and

**1926.60(o)(5)(ii)(E)** A copy of the information provided to the physician as required by paragraph (n) of this section.

**1926.60(o)(5)(iii)** The employer shall ensure that this record is maintained for the duration of employment plus thirty (30) years, in accordance with 29 CFR 1910.20.

**1926.60(o)(5)(iv)** A copy of the employee's medical removal and return to work status.

**1926.60(o)(6)** *Training records.* The employer shall maintain all employee training records for one (1) year beyond the last date of employment.

**1926.60(o)(7)** *Availability.* (i) The employer, upon written request, shall make all records required to be maintained by this section available to the Assistant Secretary and the Director for examination and copying.

**1926.60(o)(7)(ii)** The employer, upon request, shall make any exposure records required by paragraphs (f) and (n) of this section available for examination and copying to affected employees, former employees, designated representatives, and the Assistant Secretary, in accordance with 29 CFR 1910.20(a)-(e) and (g)-(i).

**1926.60(o)(7)(iii)** The employer, upon request, shall make employee medical records required by paragraphs (n) and (o) of this section available for examination and copying to the subject employee, anyone having the specific written consent of the subject employee, and the Assistant Secretary, in accordance with 29 CFR 1910.20.

**1926.60(o)(8)** *Transfer of records.* (i) The employer shall comply with the requirements concerning transfer of records set forth in 29 CFR 1910.20(h).

**1926.60(o)(8)(ii)** Whenever the employer ceases to do business and there is no successor employer to receive and retain the records for the prescribed period, the employer shall notify the Director at least 90 days prior to disposal and, upon request, transmit them to the Director.

**1926.60(p)** ***Observation of monitoring***— (1) *Employee observation.* The employer shall provide affected employees, or their designated representatives, an opportunity to observe the measuring or monitoring of employee exposure to MDA conducted pursuant to paragraph (f) of this section.

**1926.60(p)(2)** *Observation procedures.* When observation of the measuring or monitoring of employee exposure to MDA requires entry into areas where the use of protective clothing and equipment or respirators is required, the employer shall provide the observer with personal protective clothing and equipment or respirators required to be worn by employees working in the area, assure the use of such clothing and equipment or respirators, and require the observer to comply with all other applicable safety and health procedures.

**1926.60(q)** ***Effective date.*** This standard shall become effective on September 9, 1992.

**1926.60(r)** ***Appendices.*** The information contained in appendices A, B, C and D of this section is not intended by itself, to create any additional obligations not otherwise imposed by this standard nor detract from any existing obligation. The protocols for respiratory fit testing in appendix E of this section are mandatory.

**1926.60(s)** ***Startup dates.*** Compliance with all obligations of this standard commence September 9, 1992, except as follows:

**1926.60(s)(1)** Initial monitoring under paragraph (f)(2) of this section shall be completed as soon as possible but no later than December 8, 1992.

**1926.60(s)(2)** Medical examinations under paragraph (n) of this section shall be completed as soon as possible but no later than February 8, 1993.

**1926.60(s)(3)** Emergency plans required by paragraph (e) of this section shall be provided and available for inspection and copying as soon as possible but no later than January 7, 1993.

**1926.60(s)(4)** Initial training and education shall be completed as soon as possible but no later than January 7, 1993.

**1926.60(s)(5)** Decontamination and lunch areas under paragraph (k) of this section shall be in operation as soon as possible but no later than September 9, 1993.

**1926.60(s)(6)** Respiratory Protection required by paragraph (i) of this section shall be provided as soon as possible but no later than January 7, 1993.

**1926.60(s)(7)** Written compliance plans required by paragraph (h)(5) of this section shall be completed and available for inspection and copying as soon as possible but no later than January 7, 1993.

**1926.60(s)(8)** OSHA shall enforce the permissible exposure limits in paragraph (c) of this section no earlier than January 7, 1993.

**1926.60(s)(9)** Engineering controls needed to achieve the PELs must be in place September 9, 1993.

**1926.60(s)(10)** Personal protective clothing required by paragraph (j) of this section shall be available January 7, 1993.

## Appendix A to §1926.60— Substance Data Sheet, for 4-4' Methylenedianiline

**Note:** The requirements applicable to construction work under this Appendix A are identical to those set forth in Appendix A to §1910.1050 of this chapter.

[Appendix A to §1910.1050 is printed below—CCH.]

## APPENDIX A TO §1910.1050—SUBSTANCE DATA SHEET, FOR 4,4'-METHYLENEDIANILINE

### *I. Substance Identification*

A. Substance: Methylenedianiline (MDA)

B. Permissible Exposure:

1. Airborne: Ten parts per billion parts of air (10 ppb), time-weighted average (TWA) for an 8-hour workday and an action level of five parts per billion parts of air (5 ppb).

2. Dermal: Eye contact and skin contact with MDA are not permitted.

C. Appearance and odor: White to tan solid; amine odor

### *II. Health Hazard Data*

A. *Ways in which MDA affects your health.* MDA can affect your health if you inhale it, or if it comes in contact with your skin or eyes. MDA is also harmful if you happen to swallow it. Do not get MDA in eyes, on skin, or on clothing.

B. *Effects of overexposure.* 1. Short-term (acute) overexposure: Overexposure to MDA may produce fever, chills, loss of appetite, vomiting, jaundice. Contact may irritate skin, eyes and mucous membranes. Sensitization may occur.

2. *Long-term (chronic) exposure.* Repeated or prolonged exposure to MDA, even at relatively low concentrations, may cause cancer. In addition, damage to the liver, kidneys, blood, and spleen may occur with long term exposure.

3. *Reporting signs and symptoms.* You should inform your employer if you develop any signs or symptoms which you suspect are caused by exposure to MDA including yellow staining of the skin.

### *III. Protective Clothing and Equipment*

A. Respirators. Respirators are required for those operations in which engineering controls or work-practice controls are not adequate or feasible to reduce exposure to the permissible limit. If respirators are worn, they must have a label issued by the National Institute for Occupational Safety and Health under the provisions of 42 CFR part 84 stating that the respirators have been approved for this purpose, and cartridges and canisters must be replaced in accordance with the requirements of 29 CFR 1910.134. If you experience difficulty breathing while wearing a respirator, you can request a positivepressure respirator from your employer. You must be thoroughly trained to use the assigned respirator, and the training must be provided by your employer.

MDA does not have a detectable odor except at levels well above the permissible exposure limits. Do not depend on odor to warn you when a respirator canister is exhausted. If you can smell MDA while wearing a respirator, proceed immediately to fresh air. If you experience difficulty breathing while wearing a respirator, tell your employer.

B. *Protective Clothing.* You may be required to wear coveralls, aprons, gloves, face shields, or other appropriate protective clothing to prevent skin contact with MDA. Where protective clothing is required, your employer is required to provide clean garments to you, as necessary, to assure that the clothing protects you adequately. Replace or repair impervious clothing that has developed leaks.

MDA should never be allowed to remain on the skin. Clothing and shoes which are not impervious to MDA should not be allowed to become contaminated with MDA, and if they do, the clothing and shoes should be promptly removed and decontaminated. The clothing should be laundered to remove MDA or discarded. Once MDA penetrates shoes or other leather articles, they should not be worn again.

C. *Eye protection.* You must wear splashproof safety goggles in areas where liquid MDA may contact your eyes. Contact lenses should not be worn in areas where eye contact with MDA can occur. In addition, you must wear a face shield if your face could be splashed with MDA liquid.

### *IV. Emergency and First Aid Procedures*

A. *Eye and face exposure.* If MDA is splashed into the eyes, wash the eyes for at least 15 minutes. See a doctor as soon as possible.

B. *Skin exposure.* If MDA is spilled on your clothing or skin, remove the contaminated clothing and wash the exposed skin with large amounts of soap and water immediately. Wash contaminated clothing before you wear it again.

C. *Breathing.* If you or any other person breathes in large amounts of MDA, get the exposed person to fresh air at once. Apply artificial respiration if breathing has stopped. Call for medical assistance or a doctor as soon as possible. Never enter any vessel or confined space where the MDA concentration might be high without proper safety equipment and at least one other person present who will stay outside. A life line should be used.

D. *Swallowing.* If MDA has been swallowed and the patient is conscious, do not induce vomiting. Call for medical assistance or a doctor immediately.

### *V. Medical Requirements*

If you are exposed to MDA at a concentration at or above the action level for more than 30 days per year, or exposed to liquid mixtures more than 15 days per year, your employer is required to provide a medical examination, including a medical history and laboratory tests, within 60 days of the effective date of this standard and annually thereafter. These tests shall be provided without cost to you. In addition, if you are accidentally exposed to MDA (either by ingestion, inhalation, or skin/eye contact) under conditions known or suspected to constitute toxic exposure to MDA, your employer is required to make special examinations and tests available to you.

### *VI. Observation of Monitoring*

Your employer is required to perform measurements that are representative of your exposure to MDA and you or your designated representative are entitled to observe the monitoring procedure. You are entitled to observe the steps taken in the measurement procedure and to record the results obtained. When the monitoring procedure is taking place in an area where respirators or personal protective clothing and equipment are required to be worn, you and your representative must also be provided with, and must wear, the protective clothing and equipment.

### *VII. Access to Records*

You or your representative are entitled to see the records of measurements of your exposure to MDA upon written request to your employer. Your medical examination records can be furnished to your physician or designated representative upon request by you to your employer.

### VIII. Precautions for Safe Use, Handling and Storage

A. *Material is combustible.* Avoid strong acids and their anhydrides. Avoid strong oxidants. Consult supervisor for disposal requirements.

B. *Emergency clean-up.* Wear self-contained breathing apparatus and fully clothe the body in the appropriate personal protective clothing and equipment.

## Appendix B to §1926.60—Substance Technical Guidelines, MDA

**Note:** The requirements applicable to construction work under this Appendix B are identical to those set forth in Appendix B to §1910.1050 of this chapter.

[Appendix B to §1910.1050 is printed below—CCH.]

## APPENDIX B TO §1910.1050—SUBSTANCE TECHNICAL GUIDELINES, MDA

### I. Identification

A. Substance identification.

1. Synonyms: CAS No. 101-77-9. 4,4'-methylenedianiline; 4,4'methylenebisaniline; methylenedianiline; dianilinomethane.
2. Formula: $C_{13}H_{14}N_2$

### II. Physical Data

1. Appearance and Odor: White to tan solid; amine odor
2. Molecular Weight: 198.26
3. Boiling Point: 398-399 degrees C at 760 mm Hg
4. Melting Point: 88-93 degrees C (190-100 degrees F)
5. Vapor Pressure: 9 mmHg at 232 degrees C
6. Evaporation Rate (n-butyl acetate = 1): Negligible
7. Vapor Density (Air=1): Not Applicable
8. Volatile Fraction by Weight: Negligible
9. Specific Gravity (Water=1): Slight
10. Heat of Combustion: -8.40 kcal/g
11. Solubility in Water: Slightly soluble in cold water, very soluble in alcohol, benzene, ether, and many organic solvents.

### III. Fire, Explosion, and Reactivity Hazard Data

1. Flash Point: 190 degrees C (374 degrees F) Setaflash closed cup
2. Flash Point: 226 degrees C (439 degrees F) Cleveland open cup
3. Extinguishing Media: Water spray; Dry Chemical; Carbon dioxide.
4. Special Fire Fighting Procedures: Wear self-contained breathing apparatus and protective clothing to prevent contact with skin and eyes.
5. Unusual Fire and Explosion Hazards: Fire or excessive heat may cause production of hazardous decomposition products.

IV. Reactivity Data

1. Stability: Stable
2. Incompatibility: Strong oxidizers
3. Hazardous Decomposition Products: As with any other organic material, combustion may produce carbon monoxide. Oxides of nitrogen may also be present.
4. Hazardous Polymerization: Will not occur.

### V. Spill and Leak Procedures

1. Sweep material onto paper and place in fiber carton.
2. Package appropriately for safe feed to an incinerator or dissolve in compatible waste solvents prior to incineration.
3. Dispose of in an approved incinerator equipped with afterburner and scrubber or contract with licensed chemical waste disposal service.
4. Discharge treatment or disposal may be subject to federal, state, or local laws.
5. Wear appropriate personal protective equipment.

### VI. Special Storage and Handling Precautions

A. High exposure to MDA can occur when transferring the substance from one container to another. Such operations should be well ventilated and good work practices must be established to avoid spills.

B. Pure MDA is a solid with a low vapor pressure. Grinding or heating operations increase the potential for exposure.

C. Store away from oxidizing materials.

D. Employers shall advise employees of all areas and operations where exposure to MDA could occur.

### VII. Housekeeping and Hygiene Facilities

A. The workplace should be kept clean, orderly, and in a sanitary condition.

The employer should institute a leak and spill detection program for operations involving MDA in order to detect sources of fugitive MDA emissions.

B. Adequate washing facilities with hot and cold water are to be provided and maintained in a sanitary condition. Suitable cleansing agents should also be provided to assure the effective removal of MDA from the skin.

### VIII. Common Operations

Common operations in which exposure to MDA is likely to occur include the following: Manufacture of MDA; Manufacture of Methylene diisocyanate; Curing agent for epoxy resin structures; Wire coating operations; and filament winding.

## Appendix C to §1926.60—Medical Surveillance Guidelines for MDA

**Note:** The requirements applicable to construction work under this Appendix C are identical to those set forth in Appendix C to §1910.1050 of this chapter.

[Appendix C to §1910.1050 is printed below—CCH.]

## APPENDIX C TO §1910.1050—MEDICAL SURVEILLANCE GUIDELINES FOR MDA

### I. Route of Entry

Inhalation; skin absorption; ingestion. MDA can be inhaled, absorbed through the skin, or ingested.

### II. Toxicology

MDA is a suspect carcinogen in humans. There are several reports of liver disease in humans and animals resulting from acute exposure to MDA. A well documented case of an acute cardiomyopathy secondary to exposure to MDA is on record. Numerous human cases of hepatitis secondary to MDA are known. Upon direct contact MDA may also cause damage to the eyes. Dermatitis and skin sensitization have been observed. Almost all forms of acute environmental hepatic injury in humans involve the hepatic parenchyma and produce hepatocellular jaundice. This agent produces intrahepatic cholestasis. The clinical picture consists of cholestatic jaundice, preceded or accompanied by abdominal pain, fever, and chills. Onset in about 60% of all observed cases is abrupt with severe abdominal pain. In about 30% of observed cases, the illness presented and evolved more slowly and less dramatically, with only slight abdominal pain. In about 10% of the cases only jaundice was evident. The cholestatic nature of the jaundice is evident in the prominence of itching, the histologic predominance of bile stasis, and portal inflammatory infiltration, accompanied by only slight parenchymal injury in most cases, and by the moderately elevated transaminase values. Acute, high doses, however, have been known to cause hepatocellular damage resulting in elevated SGPT, SGOT, alkaline phosphatase and bilirubin.

Absorption through the skin is rapid. MDA is metabolized and excreted over a 48-hour period. Direct contact may be irritating to the skin, causing dermatitis. Also MDA which is deposited on the skin is not thoroughly removed through washing.

MDA may cause bladder cancer in humans. Animal data supporting this assumption is not available nor is conclusive human data. However, human data collected on workers at a helicopter manufacturing facility where MDA is used suggests a higher incidence of bladder cancer among exposed workers.

### III. Signs and Symptoms

Skin may become yellow from contact with MDA.

Repeated or prolonged contact with MDA may result in recurring dermatitis (red-itchy, cracked skin) and eye irritation. Inhalation, ingestion or absorption through the skin at high concentrations may result in hepatitis, causing symptoms such as fever and chills, nausea and vomiting, dark urine, anorexia, rash, right upper quadrant pain and jaundice. Corneal burns may occur when MDA is splashed in the eyes.

### IV. Treatment of Acute Toxic Effects/ Emergency Situation

If MDA gets into the eyes, immediately wash eyes with large amounts of water. If MDA is splashed on the skin, immediately wash contaminated skin with mild soap or detergent. Employee should be removed from exposure and given proper medical treatment. Medical tests required under the emergency section of the medical surveillance section (M)(4) must be conducted.

If the chemical is swallowed do not induce vomiting but remove by gastric lavage.

## Appendix D to §1926.60—Sampling and Analytical Methods for MDA Monitoring and Measurement Procedures

**Note:** The requirements applicable to construction work under this Appendix D are identical to those set forth in Appendix D to §1910.1050 of this chapter.

[57 FR 35681, Aug. 10, 1992, as amended at 57 FR 49649, Nov. 3, 1992; 61 FR 5510, Feb. 13, 1996; 61 FR 31431, June 20, 1996; 63 FR 1296, Jan. 8, 1998]

[Appendix D to §1910.1050 is printed below—CCH.]

## APPENDIX D TO §1910.1050—SAMPLING AND ANALYTICAL METHODS FOR MDA MONITORING AND MEASUREMENT PROCEDURES

Measurements taken for the purpose of determining employee exposure to MDA are best taken so that the representative average 8-hour exposure may be determined from a single 8-hour sample or two (2) 4-hour samples. Short-time interval samples (or grab samples) may also be used to determine average exposure level if a minimum of five measurements are taken in a random manner over the 8-hour work shift. Random sampling means that any portion of the work shift has the same chance of being sampled as any other. The arithmetic average of all such random samples taken on one work shift is an estimate of an employee's average level of exposure for that work shift. Air samples should be taken in the employee's breathing zone (air that would most nearly represent that inhaled by the employee).

There are a number of methods available for monitoring employee exposures to MDA. The method OSHA currently uses is included below.

The employer, however, has the obligation of selecting any monitoring method which meets the accuracy and precision requirements of the standard under his unique field conditions. The standard requires that the method of monitoring must have an accuracy, to a 95 percent confidence level, of not less than plus or minus 25 percent for the select PEL.

### OSHA Methodology

### Sampling Procedure

#### Apparatus

Samples are collected by use of a personal sampling pump that can be calibrated within ±5% of the recommended flow rate with the sampling filter in line.

Samples are collected on 37 mm Gelman type A/E glass fiber filters treated with sulfuric acid. The filters are prepared by soaking each filter with 0.5 mL of 0.26N $H_2SO_4$. (0.26 N $H_2SO_4$ can be prepared by diluting 1.5 mL of 36N $H_2SO_4$ to 200 mL with deionized water.) The filters are dried in an oven at 100 degrees C for one hour and then assembled into two-piece 37 mm polystyrene cassettes with backup pads. The cassettes are sealed with shrink bands and the ends are plugged with plastic plugs.

After sampling, the filters are carefully removed from the cassettes and individually transferred to small vials containing approximately 2 mL deionized water. The vials must be tightly sealed. The water can be added before or after the filters are transferred. The vials must be sealable and capable of holding at least 7 mL of liquid. Small glass scintillation vials with caps containing Teflon liners are recommended.

#### Reagents

Deionized water is needed for addition to the vials.

#### Sampling Technique

Immediately before sampling, remove the plastic plugs from the filter cassettes.

Attach the cassette to the sampling pump with flexible tubing and place the cassette in the employee's breathing zone.

After sampling, seal the cassettes with plastic plugs until the filters are transferred to the vials containing deionized water.

At some convenient time within 10 hours of sampling, transfer the sample filters to vials.

Seal the small vials lengthwise.

Submit at least one blank filter with each sample set. Blanks should be handled in the same manner as samples, but no air is drawn through them.

Record sample volumes (in L of air) for each sample, along with any potential interferences.

#### Retention Efficiency

A retention efficiency study was performed by drawing 100 L of air (80% relative humidity) at 1 L/min through sample filters that had been spiked with 0.814 μg MDA. Instead of using backup pads, blank acid-treated filters were used as backups in each cassette. Upon analysis, the top filters were found to have an average of 91.8% of the spiked amount. There was no MDA found on the bottom filters, so the amount lost was probably due to the slight instability of the MDA salt.

#### Extraction Efficiency

The average extraction efficiency for six filters spiked at the target concentration is 99.6%.

The stability of extracted and derivatized samples was verified by reanalyzing the above six samples the next day using fresh standards. The average extraction efficiency for the reanalyzed samples is 98.7%.

#### Recommended Air Volume and Sampling Rate

The recommended air volume is 100 L.

The recommended sampling rate is 1 L/min.

#### Interferences (Sampling)

MDI appears to be a positive interference. It was found that when MDI was spiked onto an acid-treated filter, the MDI converted to MDA after air was drawn through it.

Suspected interferences should be reported to the laboratory with submitted samples.

#### Safety Precautions (Sampling)

Attach the sampling equipment to the employees so that it will not interfere with work performance or safety.

Follow all safety procedures that apply to the work area being sampled.

### Analytical Procedure

Apparatus: The following are required for analysis.

A GC equipped with an electron capture detector. For this evaluation a Tracor 222 Gas Chromatograph equipped with a Nickel 63 High Temperature Electron Capture Detector and a Linearizer was used.

A GC column capable of separating the MDA derivative from the solvent and interferences. A 6 ft X 2 mm ID glass column packed with 3% OV-101 coated on 100/120 Gas Chrom Q was used in this evaluation.

A electronic integrator or some other suitable means of measuring peak areas or heights.

Small resealable vials with Teflon-lined caps capable of holding 4 mL.

A dispenser or pipet for toluene capable of delivering 2.0 mL.

Pipets (or repipets with plastic or Teflon tips) capable of delivering 1 mL for the sodium hydroxide and buffer solutions.

A repipet capable of delivering 25 μL HFAA.

Syringes for preparation of standards and injection of standards and samples into a GC.

Volumetric flasks and pipets to dilute the pure MDA in preparation of standards.

Disposable pipets to transfer the toluene layers after the samples are extracted.

### Reagents

0.5 NaOH prepared from reagent grade NaOH.

Toluene, pesticide grade. Burdick and Jackson distilled in glass toluene was used.

Heptafluorobutyric acid anhydride (HFAA). HFAA from Pierce Chemical Company was used.

pH 7.0 phosphate buffer, prepared from 136 g potassium dihydrogen phosphate and 1 L deionized water. The pH is adjusted to 7.0 with saturated sodium hydroxide solution.

4,4' -Methylenedianiline (MDA), reagent grade.

### Standard Preparation

Concentrated stock standards are prepared by diluting pure MDA with toluene. Analytical standards are prepared by injecting uL amounts of diluted stock standards into vials that contain 2.0 mL toluene.

25 uL HFAA are added to each vial and the vials are capped and shaken for 10 seconds.

After 10 min, 1 mL of buffer is added to each vial.

The vials are recapped and shaken for 10 seconds.

After allowing the layers to separate, aliquots of the toluene (upper) layers are removed with a syringe and analyzed by GC.

Analytical standard concentrations should bracket sample concentrations. Thus, if samples fall out of the range of prepared standards, additional standards must be prepared to ascertain detector response.

### Sample Preparation

The sample filters are received in vials containing deionized water.

1 mL of 0.5N NaOH and 2.0 mL toluene are added to each vial.

The vials are recapped and shaken for 10 min.

After allowing the layers to separate, approximately 1 mL aliquots of the toluene (upper) layers are transferred to separate vials with clean disposable pipets.

The toluene layers are treated and analyzed.

### Analysis

### GC conditions

Zone temperatures:

Column—220 degrees C

Injector—235 degrees C

Detector—335 degrees C

Gas flows, Ar/CH4 Column—28 mL/min

(95/5) Purge—40 mL/min

Injection volume: 5.0 uL

Column: 6 ft × 1/8 in ID glass, 3% OV-101 on 100/120 Gas Chrom Q

Retention time of MDA derivative: 3.5 min

### Chromatogram

Peak areas or heights are measured by an integrator or other suitable means.

A calibration curve is constructed by plotting response (peak areas or heights) of standard injections versus ug of MDA per sample. Sample concentrations must be bracketed by standards.

### Interferences (Analytical)

Any compound that gives an electron capture detector response and has the same general retention time as the HFAA derivative of MDA is a potential interference. Suspected interferences reported to the laboratory with submitted samples by the industrial hygienist must be considered before samples are derivatized.

GC parameters may be changed to possibly circumvent interferences.

Retention time on a single column is not considered proof of chemical identity. Analyte identity should be confirmed by GC/MS if possible.

### Calculations

The analyte concentration for samples is obtained from the calibration curve in terms of ug MDA per sample. The extraction efficiency is 100%. If any MDA is found on the blank, that amount is subtracted from the sample amounts. The air concentrations are calculated using the following formulae.

**μg/m³=(μg MDA per sample) (1000)/(L of air sampled)**

**ppb=(μg/m³) (24.46)/(198.3)=(μg/m³)(0.1233)**

where 24.46 is the molar volume at 25 degrees C and 760 mm Hg

### Safety Precautions (Analytical)

Avoid skin contact and inhalation of all chemicals.

Restrict the use of all chemicals to a fume hood if possible.

Wear safety glasses and a lab coat at all times while in the lab area.

## Appendix E to §1926.60—Qualitative and Quantitative Fit Testing Procedures [Removed.]

[§1926.60, App. E revised at 61 FR 31431, June 20, 1996, eff. June 30, 1996.]

## §1926.61 Retention of DOT markings, placards and labels.

**Note:** The requirements applicable to construction work under this section are identical to those set forth at §1910.1201 of this chapter.

[61 FR 31432, June 20, 1996]

[Section 1910.1201 is printed below—CCH.]

## §1910.1201 Retention of DOT markings, placards and labels.

**1910.1201(a)** Any employer who receives a package of hazardous material which is required to be marked, labeled or placarded in accordance with the U. S. Department of Transportation's Hazardous Materials Regulations (49 CFR Parts 171 through 180) shall retain those markings, labels and placards on the package until the packaging is sufficiently cleaned of residue and purged of vapors to remove any potential hazards.

**1910.1201(b)** Any employer who receives a freight container, rail freight car, motor vehicle, or transport vehicle that is required to be marked or placarded in accordance with the Hazardous Materials Regulations shall retain those markings and placards on the freight container, rail freight car, motor vehicle or transport vehicle until the hazardous materials which require the marking or placarding are sufficiently removed to prevent any potential hazards.

**1910.1201(c)** Markings, placards and labels shall be maintained in a manner that ensures that they are readily visible.

**1910.1201(d)** For non-bulk packages which will not be reshipped, the provisions of this section are met if a label or other acceptable marking is affixed in accordance with the Hazard Communication Standard (29 CFR 1910.1200).

**1910.1201(e)** For the purposes of this section, the term "hazardous material" and any other terms not defined in this section have the same definition as in the Hazardous Materials Regulations (49 CFR Parts 171 through 180).

[59 FR 36700, July 19, 1994]

## §1926.62 Lead.

**1926.62(a)** ***Scope.*** This section applies to all construction work where an employee may be occupationally exposed to lead. All construction work excluded from coverage in the general industry standard for lead by 29 CFR 1910.1025(a)(2) is covered by this standard. Construction work is defined as work for construction, alteration and/or repair, including painting and decorating. It includes but is not limited to the following:

**1926.62(a)(1)** Demolition or salvage of structures where lead or materials containing lead are present;

**1926.62(a)(2)** Removal or encapsulation of materials containing lead;

**1926.62(a)(3)** New construction, alteration, repair, or renovation of structures, substrates, or portions thereof, that contain lead, or materials containing lead;

**1926.62(a)(4)** Installation of products containing lead;

**1926.62(a)(5)** Lead contamination/emergency cleanup;

**1926.62(a)(6)** Transportation, disposal, storage, or containment of lead or materials containing lead on the site or location at which construction activities are performed, and

**1926.62(a)(7)** Maintenance operations associated with the construction activities described in this paragraph.

**1926.62(b)** ***Definitions.***

*Action level* means employee exposure, without regard to the use of respirators, to an airborne concentration of lead of 30 micrograms per cubic meter of air (30 $\mu g/m_3$) calculated as an 8-hour time-weighted average (TWA).

*Assistant Secretary* means the Assistant Secretary of Labor for Occupational Safety and Health, U.S. Department of Labor, or designee.

*Competent person* means one who is capable of identifying existing and predictable lead hazards in the surroundings or working conditions and who has authorization to take prompt corrective measures to eliminate them.

*Director* means the Director, National Institute for Occupational Safety and Health (NIOSH), U.S. Department of Health and Human Services, or designee.

*Lead* means metallic lead, all inorganic lead compounds, and organic lead soaps. Excluded from this definition are all other organic lead compounds.

*This section* means this standard.

**1926.62(c)** ***Permissible exposure limit.*** (1) The employer shall assure that no employee is exposed to lead at concentrations greater than fifty micrograms per cubic meter of air (50 $\mu g/m^3$) averaged over an 8-hour period.

**1926.62(c)(2)** If an employee is exposed to lead for more than 8 hours in any work day the employees' allowable exposure, as a time weighted average (TWA) for that day, shall be reduced according to the following formula:

Allowable employee exposure (in $\mu g/m^3$) = 400 divided by hours worked in the day.

**1926.62(c)(3)** When respirators are used to limit employee exposure as required under paragraph (c) of this section and all the requirements of paragraphs (e)(1) and (f) of this section have been met, employee exposure may be considered to be at the level provided by the protection factor of the respirator for those periods the respirator is worn. Those periods may be averaged with exposure levels during periods when respirators are not worn to determine the employee's daily TWA exposure.

**1926.62(d)** ***Exposure assessment***— (1) *General.* (i) Each employer who has a workplace or operation covered by this standard shall initially determine if any employee may be exposed to lead at or above the action level.

**1926.62(d)(1)(ii)** For the purposes of paragraph (d) of this section, employee exposure is that exposure which would occur if the employee were not using a respirator.

**1926.62(d)(1)(iii)** With the exception of monitoring under paragraph (d)(3), where monitoring is required under this section, the employer shall collect personal samples representative of a full shift including at least one sample for each job classification in each work area either for each shift or for the shift with the highest exposure level.

**1926.62(d)(1)(iv)** Full shift personal samples shall be representative of the monitored employee's regular, daily exposure to lead.

**1926.62(d)(2)** *Protection of employees during assessment of exposure.* (i) With respect to the lead related tasks listed in paragraph (d)(2)(i) of this section, where lead is present, until the employer performs an employee exposure assessment as required in paragraph (d) of this section and documents that the employee performing any of the listed tasks is not exposed above the PEL, the employer shall treat the employee as if the employee were exposed above the PEL, and not in excess of ten (10) times the PEL, and shall implement employee protective measures prescribed in paragraph (d)(2)(v) of this section. The tasks covered by this requirement are:

**1926.62(d)(2)(i)(A)** Where lead containing coatings or paint are present: Manual demolition of structures (e.g. dry wall), manual scraping, manual sanding, heat gun applications, and power tool cleaning with dust collection systems;

**1926.62(d)(2)(i)(B)** Spray painting with lead paint.

**1926.62(d)(2)(ii)** In addition, with regard to tasks not listed in paragraph (d)(2)(i), where the employee has any reason to believe that an employee performing the task may be exposed to lead in excess of the PEL, until the employer performs an employee exposure assessment as required by paragraph (d) of this section and documents that the employee's lead exposure is not above the PEL the employer shall treat the employee as if the employee were exposed above the PEL and shall implement employee protective measures as prescribed in paragraph (d)(2)(v) of this section.

**1926.62(d)(2)(iii)** With respect to the tasks listed in paragraph (d)(2)(iii) of this section, where lead is present, until the employer performs an employee exposure assessment as required in paragraph (d) of this section, and documents that the employee performing any of the listed tasks is not exposed in excess of 500 $\mu g/m^3$, the employer shall treat the employee as if the employee were exposed to lead in excess of 500 $\mu g/m^3$ and shall implement employee protective measures as prescribed in paragraph (d)(2)(v) of this section. Where the employer does establish that the employee is exposed to levels of lead below 500 $\mu g/m^3$, the employer may provide the exposed employee with the appropriate respirator prescribed for such use at such lower exposures, in accordance with Table 1 of this section. The tasks covered by this requirement are:

**1926.62(d)(2)(iii)(A)** Using lead containing mortar; lead burning

**1926.62(d)(2)(iii)(B)** Where lead containing coatings or paint are present: rivet busting; power tool cleaning without dust collection systems; cleanup activities where dry expendable abrasives are used; and abrasive blasting enclosure movement and removal.

**1926.62(d)(2)(iv)** With respect to the tasks listed in paragraph (d)(2)(iv) of this section, where lead is present, until the employer performs an employee exposure assessment as required in paragraph (d) of this section and documents that the employee performing any of the listed tasks is not exposed to lead in excess of 2,500 $\mu g/m^3$ (50 ×PEL), the employer shall treat the employee as if the employee were exposed to lead in excess of 2,500 $\mu g/m^3$ and shall implement employee protective measures as prescribed in paragraph (d)(2)(v) of this section. Where the employer does establish that the employee is exposed to levels of lead below 2,500 $\mu g$/m3, the employer may provide the exposed employee with the appropriate respirator prescribed for use at such lower exposures, in accordance with Table I of this section. Interim protection as described in this paragraph is required where lead containing coatings or paint are present on structures when performing:

**1926.62(d)(2)(iv)(A)** Abrasive blasting,

**1926.62(d)(2)(iv)(B)** Welding,

**1926.62(d)(2)(iv)(C)** Cutting, and

**1926.62(d)(2)(iv)(D)** Torch burning.

**1926.62(d)(2)(v)** Until the employer performs an employee exposure assessment as required under paragraph (d) of this section and determines actual employee exposure, the employer shall provide to employees performing the tasks described in paragraphs (d)(2)(i), (d)(2)(ii), (d)(2)(iii), and (d)(2)(iv) of this section with interim protection as follows:

**1926.62(d)(2)(v)(A)** Appropriate respiratory protection in accordance with paragraph (f) of this section.

**1926.62(d)(2)(v)(B)** Appropriate personal protective clothing and equipment in accordance with paragraph (g) of this section.

**1926.62(d)(2)(v)(C)** Change areas in accordance with paragraph (i)(2) of this section.

**1926.62(d)(2)(v)(D)** Hand washing facilities in accordance with paragraph (i)(5) of this section.

**1926.62(d)(2)(v)(E)** Biological monitoring in accordance with paragraph (j)(1)(i) of this section, to consist of blood sampling and analysis for lead and zinc protoporphyrin levels, and

**1926.62(d)(2)(v)(F)** Training as required under paragraph (l)(1)(i) of this section regarding 29 CFR 1926.59, Hazard Communication; training as required under paragraph (l)(2)(ii)(C) of this section, regarding use of respirators; and training in accordance with 29 CFR 1926.21, Safety training and education.

**1926.62(d)(3)** *Basis of initial determination.* (i) Except as provided under paragraphs (d)(3)(iii) and (d)(3)(iv) of this section the employer shall monitor employee exposures and shall base initial determinations on the employee exposure monitoring results and any of the following, relevant considerations:

**1926.62(d)(3)(i)(A)** Any information, observations, or calculations which would indicate employee exposure to lead;

**1926.62(d)(3)(i)(B)** Any previous measurements of airborne lead; and

**1926.62(d)(3)(i)(C)** Any employee complaints of symptoms which may be attributable to exposure to lead.

**1926.62(d)(3)(ii)** Monitoring for the initial determination where performed may be limited to a representative sample of the exposed employees who the employer reasonably believes are exposed to the greatest airborne concentrations of lead in the workplace.

**1926.62(d)(3)(iii)** Where the employer has previously monitored for lead exposures, and the data were obtained within the past 12 months during work operations conducted under workplace conditions closely resembling the processes, type of material, control methods, work practices, and environmental conditions used and prevailing in the employer's current operations, the employer may rely on such earlier monitoring results to satisfy the requirements of paragraphs (d)(3)(i) and (d)(6) of this section if the sampling and analytical methods meet the accuracy and confidence levels of paragraph (d)(10) of this section.

**1926.62(d)(3)(iv)** Where the employer has objective data, demonstrating that a particular product or material containing lead or a specific process, operation or activity involving lead cannot result in employee exposure to lead at or above the action level during processing, use, or handling, the employer may rely upon such data instead of implementing initial monitoring.

**1926.62(d)(3)(iv)(A)** The employer shall establish and maintain an accurate record documenting the nature and relevancy of objective data as specified in paragraph (n)(4) of this section, where used in assessing employee exposure in lieu of exposure monitoring.

**1926.62(d)(3)(iv)(B)** Objective data, as described in paragraph (d)(3)(iv) of this section, is not permitted to be used for exposure assessment in connection with paragraph (d)(2) of this section.

**1926.62(d)(4)** *Positive initial determination and initial monitoring.* (i) Where a determination conducted under paragraphs (d) (1), (2) and (3) of this section shows the possibility of any employee exposure at or above the action level the employer shall conduct monitoring which is representative of the exposure for each employee in the workplace who is exposed to lead.

**1926.62(d)(4)(ii)** Where the employer has previously monitored for lead exposure, and the data were obtained within the past 12 months during work operations conducted under workplace conditions closely resembling the processes, type of material, control methods, work practices, and environmental conditions used and prevailing in the employer's current operations, the employer may rely on such earlier monitoring results to satisfy the requirements of paragraph (d)(4)(i) of this section if the sampling and analytical methods meet the accuracy and confidence levels of paragraph (d)(10) of this section.

**1926.62(d)(5)** *Negative initial determination.* Where a determination, conducted under paragraphs (d) (1), (2), and (3) of this section is made that no employee is exposed to airborne concentrations of lead at or above the action level the employer shall make a written record of such determination. The record shall include at least the information specified in paragraph (d)(3)(i) of this section and shall also include the date of determination, location within the worksite, and the name and social security number of each employee monitored.

**1926.62(d)(6)** *Frequency.* (i) If the initial determination reveals employee exposure to be below the action level further exposure determination need not be repeated except as otherwise provided in paragraph (d)(7) of this section.

**1926.62(d)(6)(ii)** If the initial determination or subsequent determination reveals employee exposure to be at or above the action level but at or below the PEL the employer shall perform monitoring in accordance with this paragraph at least every 6 months. The employer shall continue monitoring at the required frequency until at least two consecutive measurements, taken at least 7 days apart, are below the action level at which time the employer may discontinue monitoring for that employee except as otherwise provided in paragraph (d)(7) of this section.

**1926.62(d)(6)(iii)** If the initial determination reveals that employee exposure is above the PEL the employer shall perform monitoring quarterly. The employer shall continue monitoring at the required frequency until at least two consecutive measurements, taken at least 7 days apart, are at or below the PEL but at or above the action level at which time the employer shall repeat monitoring for that employee at the frequency specified in paragraph (d)(6)(ii) of this section, except as otherwise provided in paragraph (d)(7) of this section. The employer shall continue monitoring at the required frequency until at least two consecutive measurements, taken at least 7 days apart, are below the action level at which time the employer may discontinue monitoring for that employee except as otherwise provided in paragraph (d)(7) of this section.

**1926.62(d)(7)** *Additional exposure assessments.* Whenever there has been a change of equipment, process, control, personnel or a new task has been initiated that may result in additional employees being exposed to lead at or above the action level or may result in employees already exposed at or above the action level being exposed above the PEL, the employer shall conduct additional monitoring in accordance with this paragraph.

**1926.62(d)(8)** *Employee notification.* (i) Within 5 working days after completion of the exposure assessment the employer shall notify each employee in writing of the results which represent that employee's exposure.

**1926.62(d)(8)(ii)** Whenever the results indicate that the representative employee exposure, without regard to respirators, is at or above the PEL the employer shall include in the written notice a statement that the employees exposure was at or above that level and a description of the corrective action taken or to be taken to reduce exposure to below that level.

**1926.62(d)(9)** *Accuracy of measurement.* The employer shall use a method of monitoring and analysis which has an accuracy (to a confidence level of 95%) of not less than plus or minus 25 percent for airborne concentrations of lead equal to or greater than 30 $\mu g/m^3$.

**1926.62(e)** ***Methods of compliance.*** (1) *Engineering and work practice controls.* The employer shall implement engineering and work practice controls, including administrative controls, to reduce and maintain employee exposure to lead to or below the permissible exposure limit to the extent that such controls are feasible. Wherever all feasible engineering and work practices controls that can be instituted are not sufficient to reduce employee exposure to or below the permissible exposure limit prescribed in paragraph (c) of this section, the employer shall nonetheless use them to reduce employee exposure to the lowest feasible level and shall supplement them by the use of respiratory protection that complies with the requirements of paragraph (f) of this section.

**1926.62(e)(2)** *Compliance program.* (i) Prior to commencement of the job each employer shall establish and implement a written compliance program to achieve compliance with paragraph (c) of this section.

**1926.62(e)(2)(ii)** Written plans for these compliance programs shall include at least the following:

**1926.62(e)(2)(ii)(A)** A description of each activity in which lead is emitted; e.g. equipment used, material involved, controls in place, crew size, employee job responsibilities, operating procedures and maintenance practices;

**1926.62(e)(2)(ii)(B)** A description of the specific means that will be employed to achieve compliance and, where engineering controls are required engineering plans and studies used to determine methods selected for controlling exposure to lead;

**1926.62(e)(2)(ii)(C)** A report of the technology considered in meeting the PEL;

**1926.62(e)(2)(ii)(D)** Air monitoring data which documents the source of lead emissions;

**1926.62(e)(2)(ii)(E)** A detailed schedule for implementation of the program, including documentation such as copies of purchase orders for equipment, construction contracts, etc.;

**1926.62(e)(2)(ii)(F)** A work practice program which includes items required under paragraphs (g), (h) and (i) of this section and incorporates other relevant work practices such as those specified in paragraph (e)(5) of this section;

**1926.62(e)(2)(ii)(G)** An administrative control schedule required by paragraph (e)(4) of this section, if applicable;

**1926.62(e)(2)(ii)(H)** A description of arrangements made among contractors on multi-contractor sites with respect to informing affected employees of potential exposure to lead and with respect to responsibility for compliance with this section as set-forth in §1926.16.

**1926.62(e)(2)(ii)(I)** Other relevant information.

**1926.62(e)(2)(iii)** The compliance program shall provide for frequent and regular inspections of job sites, materials, and equipment to be made by a competent person.

**1926.62(e)(2)(iv)** Written programs shall be submitted upon request to any affected employee or authorized employee representatives, to the Assistant Secretary and the Director, and shall be available at the worksite for examination and copying by the Assistant Secretary and the Director.

**1926.62(e)(2)(v)** Written programs shall be revised and updated at least every 6 months to reflect the current status of the program.

**1926.62(e)(3)** *Mechanical ventilation.* When ventilation is used to control lead exposure, the employer shall evaluate the mechanical performance of the system in controlling exposure as necessary to maintain its effectiveness.

**1926.62(e)(4)** *Administrative controls.* If administrative controls are used as a means of reducing employees TWA exposure to lead, the employer shall establish and implement a job rotation schedule which includes:

**1926.62(e)(4)(i)** Name or identification number of each affected employee;

**1926.62(e)(4)(ii)** Duration and exposure levels at each job or work station where each affected employee is located; and

**1926.62(e)(4)(iii)** Any other information which may be useful in assessing the reliability of administrative controls to reduce exposure to lead.

**1926.62(e)(5)** The employer shall ensure that, to the extent relevant, employees follow good work practices such as described in Appendix B of this section.

**1926.62(f)** ***Respiratory protection.*** (1) *General.* For employees who use respirators required by this section, the employer must provide respirators that comply with the requirements of this paragraph. Respirators must be used during:

**1926.62(f)(1)(i)** Periods when an employee's exposure to lead exceeds the PEL.

**1926.62(f)(1)(ii)** Work operations for which engineering and work-practice controls are not sufficient to reduce employee exposures to or below the PEL.

**1926.62(f)(1)(iii)** Periods when an employee requests a respirator.

**1926.62(f)(1)(iv)** Periods when respirators are required to provide interim protection of employees while they perform the operations specified in paragraph (d)(2) of this section.

**1926.62(f)(2)** *Respirator program.* (i) The employer must implement a respiratory protection program in accordance with 29 CFR 1910.134 (b) through (d) (except (d)(1)(iii)), and (f) through (m).

**1926.62(f)(2)(ii)** If an employee has breathing difficulty during fit testing or respirator use, the employer must provide the employee with a medical examination in accordance with paragraph (j)(3)(i)(B) of this section to determine whether or not the employee can use a respirator while performing the required duty.

**1926.62(f)(3)** *Respirator selection.* (i) The employer must select the appropriate respirator or combination of respirators from Table I of this section.

**1926.62(f)(3)(ii)** The employer must provide a powered air-purifying respirator when an employee chooses to use such a respirator and it will provide adequate protection to the employee.

## TABLE I.—RESPIRATORY PROTECTION FOR LEAD AEROSOLS

| Airborne concentration of lead or condition of use | Required respirator[1] |
|---|---|
| Not in excess of 500 µg/m³ ....... | ½ mask air purifying respirator with high efficiency filters.[2, 3] |
| | ½ mask supplied air respirator operated in demand (negative pressure) mode. |
| Not in excess of 1,250 µg/m³ ...... | Loose fitting hood or helmet powered air purifying respirator with high efficiency filters.[3] |
| | Hood or helmet supplied air respirator operated in a continuous-flow mode—e.g., type CE abrasive blasting respirators operated in a continuous-flow mode. |
| Not in excess of 2,500 µg/m³ ...... | Full facepiece air purifying respirator with high efficiency filters.[3] |
| | Tight fitting powered air purifying respirator with high efficiency filters.[3] |
| | Full facepiece supplied air respirator operated in demand mode. |
| | ½ mask or full facepiece supplied air respirator operated in a continuous-flow mode. |
| | Full facepiece self-contained breathing apparatus (SCBA) operated in demand mode. |
| Not in excess of 50,000 µg/m³ ..... | ½ mask supplied air respirator operated in pressure demand or other positive-pressure mode. |
| Not in excess of 100,000 µg/m³ .... | Full facepiece supplied air respirator operated in pressure demand or other positive-pressure mode—e.g., type CE abrasive blasting respirators operated in a positive-pressure mode. |
| Greater than 100,000 µg/m³ unknown concentration, or fire fighting. | Full facepiece SCBA operated in pressure demand or other positive-pressure mode. |

[1] Respirators specified for higher concentrations can be used at lower concentrations of lead.
[2] Full facepiece is required if the lead aerosols cause eye or skin irritation at the use concentrations.
[3] A high efficiency particulate filter (HEPA) means a filter that is 99.97 percent efficient against particles of 0.3 micron size or larger.

**1926.62(g)** ***Protective work clothing and equipment***— (1) *Provision and use.* Where an employee is exposed to lead above the PEL without regard to the use of respirators, where employees are exposed to lead compounds which may cause skin or eye irritation (e.g. lead arsenate, lead azide), and as interim protection for employees performing tasks as specified in paragraph (d)(2) of this section, the employer shall provide at no cost to the employee and assure that the employee uses appropriate protective work clothing and equipment that prevents contamination of the employee and the employee's garments such as, but not limited to:

**1926.62(g)(1)(i)** Coveralls or similar full-body work clothing;

**1926.62(g)(1)(ii)** Gloves, hats, and shoes or disposable shoe coverlets; and

**1926.62(g)(1)(iii)** Face shields, vented goggles, or other appropriate protective equipment which complies with §1910.133 of this chapter.

**1926.62(g)(2)** *Cleaning and replacement.* (i) The employer shall provide the protective clothing required in paragraph (g)(1) of this section in a clean and dry condition at least weekly, and daily to employees whose exposure levels without regard to a respirator are over 200 µg/m³ of lead as an 8-hour TWA.

**1926.62(g)(2)(ii)** The employer shall provide for the cleaning, laundering, and disposal of protective clothing and equipment required by paragraph (g)(1) of this section.

**1926.62(g)(2)(iii)** The employer shall repair or replace required protective clothing and equipment as needed to maintain their effectiveness.

**1926.62(g)(2)(iv)** The employer shall assure that all protective clothing is removed at the completion of a work shift only in change areas provided for that purpose as prescribed in paragraph (i)(2) of this section.

**1926.62(g)(2)(v)** The employer shall assure that contaminated protective clothing which is to be cleaned, laundered, or disposed of, is placed in a closed container in the change area which prevents dispersion of lead outside the container.

**1926.62(g)(2)(vi)** The employer shall inform in writing any person who cleans or launders protective clothing or equipment of the potentially harmful effects of exposure to lead.

**1926.62(g)(2)(vii)** The employer shall assure that the containers of contaminated protective clothing and equipment required by paragraph (g)(2)(v) of this section are labeled as follows:

Caution: Clothing contaminated with lead. Do not remove dust by blowing or shaking. Dispose of lead contaminated wash water in accordance with applicable local, state, or federal regulations.

**1926.62(g)(2)(viii)** The employer shall prohibit the removal of lead from protective clothing or equipment by blowing, shaking, or any other means which disperses lead into the air.

**1926.62(h)** ***Housekeeping***— (1) All surfaces shall be maintained as free as practicable of accumulations of lead.

**1926.62(h)(2)** Clean-up of floors and other surfaces where lead accumulates shall wherever possible, be cleaned by vacuuming or other methods that minimize the likelihood of lead becoming airborne.

**1926.62(h)(3)** Shoveling, dry or wet sweeping, and brushing may be used only where vacuuming or other equally effective methods have been tried and found not to be effective.

**1926.62(h)(4)** Where vacuuming methods are selected, the vacuums shall be equipped with HEPA filters and used and emptied in a manner which minimizes the reentry of lead into the workplace.

**1926.62(h)(5)** Compressed air shall not be used to remove lead from any surface unless the compressed air is used in conjunction with a ventilation system designed to capture the airborne dust created by the compressed air.

**1926.62(i)** ***Hygiene facilities and practices***—(1) The employer shall assure that in areas where employees are exposed to lead above the PEL without regard to the use of respirators, food or beverage is not present or consumed, tobacco products are not present or used, and cosmetics are not applied.

**1926.62(i)(2)** *Change areas.* (i) The employer shall provide clean change areas for employees whose airborne exposure to lead is above the PEL, and as interim protection for employees performing tasks as specified in paragraph (d)(2) of this section, without regard to the use of respirators.

**1926.62(i)(2)(ii)** The employer shall assure that change areas are equipped with separate storage facilities for protective work clothing and equipment and for street clothes which prevent cross-contamination.

**1926.62(i)(2)(iii)** The employer shall assure that employees do not leave the workplace wearing any protective clothing or equipment that is required to be worn during the work shift.

**1926.62(i)(3)** *Showers.* (i) The employer shall provide shower facilities, where feasible, for use by employees whose airborne exposure to lead is above the PEL.

**1926.62(i)(3)(ii)** The employer shall assure, where shower facilities are available, that employees shower at the end of the work shift and shall provide an adequate supply of cleansing agents and towels for use by affected employees.

**1926.62(i)(4)** *Eating facilities.* (i) The employer shall provide lunchroom facilities or eating areas for employees whose airborne exposure to lead is above the PEL, without regard to the use of respirators.

**1926.62(i)(4)(ii)** The employer shall assure that lunchroom facilities or eating areas are as free as practicable from lead contamination and are readily accessible to employees.

**1926.62(i)(4)(iii)** The employer shall assure that employees whose airborne exposure to lead is above the PEL, without regard to the use of a respirator, wash their hands and face prior to eating, drinking, smoking or applying cosmetics.

**1926.62(i)(4)(iv)** The employer shall assure that employees do not enter lunchroom facilities or eating areas with protective work clothing or equipment unless surface lead dust has been removed by vacuuming, downdraft booth, or other cleaning method that limits dispersion of lead dust.

**1926.62(i)(5)** *Hand washing facilities.* (i) The employer shall provide adequate handwashing facilities for use by employees exposed to lead in accordance with 29 CFR 1926.51(f).

**1926.62(i)(5)(ii)** Where showers are not provided the employer shall assure that employees wash their hands and face at the end of the work-shift.

**1926.62(j)** ***Medical surveillance***— (1) *General.* (i) The employer shall make available initial medical surveillance to employees occupationally exposed on any day to lead at or above the action level. Initial medical surveillance consists of biological monitoring in the form of blood sampling and analysis for lead and zinc protoporphyrin levels.

**1926.62(j)(1)(ii)** The employer shall institute a medical surveillance program in accordance with paragraphs (j)(2) and (j)(3) of this section for all employees who are or may be exposed by the employer at or above the action level for more than 30 days in any consecutive 12 months;

**1926.62(j)(1)(iii)** The employer shall assure that all medical examinations and procedures are performed by or under the supervision of a licensed physician.

**1926.62(j)(1)(iv)** The employer shall make available the required medical surveillance including multiple physician review under paragraph (j)(3)(iii) without cost to employees and at a reasonable time and place.

**1926.62(j)(2)** *Biological monitoring*—(i) *Blood lead and ZPP level sampling and analysis.* The employer shall make available biological monitoring in the form of blood sampling and analysis for lead and zinc protoporphyrin levels to each employee covered under paragraphs (j)(1)(i) and (ii) of this section on the following schedule:

**1926.62(j)(2)(i)(A)** For each employee covered under paragraph (j)(1)(ii) of this section, at least every 2 months for the first 6 months and every 6 months thereafter;

**1926.62(j)(2)(i)(B)** For each employee covered under paragraphs (j)(1) (i) or (ii) of this section whose last blood sampling and analysis indicated a blood lead level at or above 40 μg/dl, at least every two months. This frequency shall continue until two consecutive blood samples and analyses indicate a blood lead level below 40 μg/dl; and

**1926.62(j)(2)(i)(C)** For each employee who is removed from exposure to lead due to an elevated blood lead level at least monthly during the removal period.

**1926.62(j)(2)(ii)** *Follow-up blood sampling tests.* Whenever the results of a blood lead level test indicate that an employee's blood lead level exceeds the numerical criterion for medical removal under paragraph (k)(1)(i) of this section, the employer shall provide a second (follow-up) blood sampling test within two weeks after the employer receives the results of the first blood sampling test.

**1926.62(j)(2)(iii)** *Accuracy of blood lead level sampling and analysis.* Blood lead level sampling and analysis provided pursuant to this section shall have an accuracy (to a confidence level of 95 percent) within plus or minus 15 percent or 6 μg/dl, whichever is greater, and shall be conducted by a laboratory approved by OSHA.

**1926.62(j)(2)(iv)** *Employee notification.* (A) Within five working days after the receipt of biological monitoring results, the employer shall notify each employee in writing of his or her blood lead level; and

**1926.62(j)(2)(iv)(B)** The employer shall notify each employee whose blood lead level exceeds 40 μg/dl that the standard requires temporary medical removal with Medical Removal Protection benefits when an employee's blood lead level exceeds the numerical criterion for medical removal under paragraph (k)(1)(i) of this section.

**1926.62(j)(3)** *Medical examinations and consultations*—(i) *Frequency.* The employer shall make available medical examinations and consultations to each employee covered under paragraph (j)(1)(ii) of this section on the following schedule:

**1926.62(j)(3)(i)(A)** At least annually for each employee for whom a blood sampling test conducted at any time during the preceding 12 months indicated a blood lead level at or above 40 μg/dl;

**1926.62(j)(3)(i)(B)** As soon as possible, upon notification by an employee either that the employee has developed signs or symptoms commonly associated with lead intoxication, that the employee desires medical advice concerning the effects of current or past exposure to lead on the employee's ability to procreate a healthy child, that the employee is pregnant, or that the employee has demonstrated difficulty in breathing during a respirator fitting test or during use; and

**1926.62(j)(3)(i)(C)** As medically appropriate for each employee either removed from exposure to lead due to a risk of sustaining material impairment to health, or otherwise limited pursuant to a final medical determination.

**1926.62(j)(3)(ii)** *Content.* The content of medical examinations made available pursuant to paragraph (j)(3)(i)(B)-(C) of this section shall be determined by an examining physician and, if requested by an employee, shall include pregnancy testing or laboratory evaluation of male fertility. Medical examinations made available pursuant to paragraph (j)(3)(i)(A) of this section shall include the following elements:

**1926.62(j)(3)(ii)(A)** A detailed work history and a medical history, with particular attention to past lead exposure (occupational and non-occupational), personal habits (smoking, hygiene), and past gastrointestinal, hematologic, renal, cardiovascular, reproductive and neurological problems;

**1926.62(j)(3)(ii)(B)** A thorough physical examination, with particular attention to teeth, gums, hematologic, gastrointestinal, renal, cardiovascular, and neurological systems. Pulmonary status should be evaluated if respiratory protection will be used;

**1926.62(j)(3)(ii)(C)** A blood pressure measurement;

**1926.62(j)(3)(ii)(D)** A blood sample and analysis which determines:

**1926.62(j)(3)(ii)(D)*(1)*** Blood lead level;

**1926.62(j)(3)(ii)(D)*(2)*** Hemoglobin and hematocrit determinations, red cell indices, and examination of peripheral smear morphology;

**1926.62(j)(3)(ii)(D)*(3)*** Zinc protoporphyrin;

**1926.62(j)(3)(ii)(D)*(4)*** Blood urea nitrogen; and,

**1926.62(j)(3)(ii)(D)*(5)*** Serum creatinine;

**1926.62(j)(3)(ii)(E)** A routine urinalysis with microscopic examination; and

**1926.62(j)(3)(ii)(F)** Any laboratory or other test relevant to lead exposure which the examining physician deems necessary by sound medical practice.

**1926.62(j)(3)(iii)** *Multiple physician review mechanism.* (A) If the employer selects the initial physician who conducts any medical examination or consultation provided to an employee under this section, the employee may designate a second physician:

**1926.62(j)(3)(iii)(A)*(1)*** To review any findings, determinations or recommendations of the initial physician; and

**1926.62(j)(3)(iii)(A)*(2)*** To conduct such examinations, consultations, and laboratory tests as the second physician deems necessary to facilitate this review.

**1926.62(j)(3)(iii)(B)** The employer shall promptly notify an employee of the right to seek a second medical opinion after each occasion that an initial physician conducts a medical examination or consultation pursuant to this section. The employer may condition its participation in, and payment for, the multiple physician review mechanism upon the employee doing the following within fifteen (15) days after receipt of the foregoing notification, or receipt of the initial physician's written opinion, whichever is later:

**1926.62(j)(3)(iii)(B)*(1)*** The employee informing the employer that he or she intends to seek a second medical opinion, and

**1926.62(j)(3)(iii)(B)*(2)*** The employee initiating steps to make an appointment with a second physician.

**1926.62(j)(3)(iii)(C)** If the findings, determinations or recommendations of the second physician differ from those of the initial physician, then the employer and the employee shall assure that efforts are made for the two physicians to resolve any disagreement.

**1926.62(j)(3)(iii)(D)** If the two physicians have been unable to quickly resolve their disagreement, then the employer and the employee through their respective physicians shall designate a third physician:

**1926.62(j)(3)(iii)(D)*(1)*** To review any findings, determinations or recommendations of the prior physicians; and

**1926.62(j)(3)(iii)(D)*(2)*** To conduct such examinations, consultations, laboratory tests and discussions with the prior physicians as the third physician deems necessary to resolve the disagreement of the prior physicians.

**1926.62(j)(3)(iii)(E)** The employer shall act consistent with the findings, determinations and recommendations of the third physician, unless the employer and the employee reach an agreement which is otherwise consistent with the recommendations of at least one of the three physicians.

**1926.62(j)(3)(iv)** *Information provided to examining and consulting physicians.* (A) The employer shall provide an initial physician conducting a medical examination or consultation under this section with the following information:

**1926.62(j)(3)(iv)(A)*(1)*** A copy of this regulation for lead including all Appendices;

**1926.62(j)(3)(iv)(A)*(2)*** A description of the affected employee's duties as they relate to the employee's exposure;

**1926.62(j)(3)(iv)(A)*(3)*** The employee's exposure level or anticipated exposure level to lead and to any other toxic substance (if applicable);

**1926.62(j)(3)(iv)(A)*(4)*** A description of any personal protective equipment used or to be used;

**1926.62(j)(3)(iv)(A)*(5)*** Prior blood lead determinations; and

**1926.62(j)(3)(iv)(A)*(6)*** All prior written medical opinions concerning the employee in the employer's possession or control.

**1926.62(j)(3)(iv)(B)** The employer shall provide the foregoing information to a second or third physician conducting a medical examination or consultation under this section upon request either by the second or third physician, or by the employee.

**1926.62(j)(3)(v)** *Written medical opinions.* (A) The employer shall obtain and furnish the employee with a copy of a written medical opinion from each examining or consulting physician which contains only the following information:

**1926.62(j)(3)(v)(A)*(1)*** The physician's opinion as to whether the employee has any detected medical condition which would place the employee at increased risk of material impairment of the employee's health from exposure to lead;

**1926.62(j)(3)(v)(A)*(2)*** Any recommended special protective measures to be provided to the employee, or limitations to be placed upon the employee's exposure to lead;

**1926.62(j)(3)(v)(A)*(3)*** Any recommended limitation upon the employee's use of respirators, including a determination of whether the employee can wear a powered air purifying respirator if a physician determines that the employee cannot wear a negative pressure respirator; and

**1926.62(j)(3)(v)(A)*(4)*** The results of the blood lead determinations.

**1926.62(j)(3)(v)(B)** The employer shall instruct each examining and consulting physician to:

**1926.62(j)(3)(v)(B)*(1)*** Not reveal either in the written opinion or orally, or in any other means of communication with the employer, findings, including laboratory results, or diagnoses unrelated to an employee's occupational exposure to lead; and

**1926.62(j)(3)(v)(B)*(2)*** Advise the employee of any medical condition, occupational or nonoccupational, which dictates further medical examination or treatment.

**1926.62(j)(3)(vi)** *Alternate physician determination mechanisms.* The employer and an employee or authorized employee representative may agree upon the use of any alternate physician determination mechanism in lieu of the multiple physician review mechanism provided by paragraph (j)(3)(iii) of this section so long as the alternate mechanism is as expeditious and protective as the requirements contained in this paragraph.

**1926.62(j)(4)** *Chelation.* (i) The employer shall assure that any person whom he retains, employs, supervises or controls does not engage in prophylactic chelation of any employee at any time.

**1926.62(j)(4)(ii)** If therapeutic or diagnostic chelation is to be performed by any person in paragraph (j)(4)(i) of this section, the employer shall assure that it be done under the supervision of a licensed physician in a clinical setting with thorough and appropriate medical monitoring and that the employee is notified in writing prior to its occurrence.

**1926.62(k)** ***Medical removal protection***— (1) *Temporary medical removal and return of an employee*—(i) *Temporary removal due to elevated blood lead level.* The employer shall remove an employee from work having an exposure to lead at or above the action level on each occasion that a periodic and a follow-up blood sampling test conducted pursuant to this section indicate that the employee's blood lead level is at or above 50 µg/dl; and,

**1926.62(k)(1)(ii)** *Temporary removal due to a final medical determination.* (A) The employer shall remove an employee from work having an exposure to lead at or above the action level on each occasion that a final medical determination results in a medical finding, determination, or opinion that the employee has a detected medical condition which places the employee at increased risk of material impairment to health from exposure to lead.

**1926.62(k)(1)(ii)(B)** For the purposes of this section, the phrase *final medical determination* means the written medical opinion on the employees' health status by the examining physician or, where relevant, the outcome of the multiple physician review mechanism or alternate medical determination mechanism used pursuant to the medical surveillance provisions of this section.

**1926.62(k)(1)(ii)(C)** Where a final medical determination results in any recommended special protective measures for an employee, or limitations on an employee's exposure to lead, the employer shall implement and act consistent with the recommendation.

**1926.62(k)(1)(iii)** *Return of the employee to former job status.* (A) The employer shall return an employee to his or her former job status:

**1926.62(k)(1)(iii)(A)*(1)*** For an employee removed due to a blood lead level at or above 50 μg/dl when two consecutive blood sampling tests indicate that the employee's blood lead level is at or below 40 μg/dl;

**1926.62(k)(1)(iii)(A)*(2)*** For an employee removed due to a final medical determination, when a subsequent final medical determination results in a medical finding, determination, or opinion that the employee no longer has a detected medical condition which places the employee at increased risk of material impairment to health from exposure to lead.

**1926.62(k)(1)(iii)(B)** For the purposes of this section, the requirement that an employer return an employee to his or her former job status is not intended to expand upon or restrict any rights an employee has or would have had, absent temporary medical removal, to a specific job classification or position under the terms of a collective bargaining agreement.

**1926.62(k)(1)(iv)** *Removal of other employee special protective measure or limitations.* The employer shall remove any limitations placed on an employee or end any special protective measures provided to an employee pursuant to a final medical determination when a subsequent final medical determination indicates that the limitations or special protective measures are no longer necessary.

**1926.62(k)(1)(v)** *Employer options pending a final medical determination.* Where the multiple physician review mechanism, or alternate medical determination mechanism used pursuant to the medical surveillance provisions of this section, has not yet resulted in a final medical determination with respect to an employee, the employer shall act as follows:

**1926.62(k)(1)(v)(A)** *Removal.* The employer may remove the employee from exposure to lead, provide special protective measures to the employee, or place limitations upon the employee, consistent with the medical findings, determinations, or recommendations of any of the physicians who have reviewed the employee's health status.

**1926.62(k)(1)(v)(B)** *Return.* The employer may return the employee to his or her former job status, end any special protective measures provided to the employee, and remove any limitations placed upon the employee, consistent with the medical findings, determinations, or recommendations of any of the physicians who have reviewed the employee's health status, with two exceptions.

**1926.62(k)(1)(v)(B)*(1)*** If the initial removal, special protection, or limitation of the employee resulted from a final medical determination which differed from the findings, determinations, or recommendations of the initial physician or;

**1926.62(k)(1)(v)(B)*(2)*** If the employee has been on removal status for the preceding eighteen months due to an elevated blood lead level, then the employer shall await a final medical determination.

**1926.62(k)(2)** *Medical removal protection benefits*—(i) *Provision of medical removal protection benefits.* The employer shall provide an employee up to eighteen (18) months of medical removal protection benefits on each occasion that an employee is removed from exposure to lead or otherwise limited pursuant to this section.

**1926.62(k)(2)(ii)** *Definition of medical removal protection benefits.* For the purposes of this section, the requirement that an employer provide medical removal protection benefits means that, as long as the job the employee was removed from continues, the employer shall maintain the total normal earnings, seniority and other employment rights and benefits of an employee, including the employee's right to his or her former job status as though the employee had not been medically removed from the employee's job or otherwise medically limited.

**1926.62(k)(2)(iii)** *Follow-up medical surveillance during the period of employee removal or limitation.* During the period of time that an employee is medically removed from his or her job or otherwise medically limited, the employer may condition the provision of medical removal protection benefits upon the employee's participation in follow-up medical surveillance made available pursuant to this section.

**1926.62(k)(2)(iv)** *Workers' compensation claims.* If a removed employee files a claim for workers' compensation payments for a lead-related disability, then the employer shall continue to provide medical removal protection benefits pending disposition of the claim. To the extent that an award is made to the employee for earnings lost during the period of removal, the employer's medical removal protection obligation shall be reduced by such amount. The employer shall receive no credit for workers' compensation payments received by the employee for treatment-related expenses.

**1926.62(k)(2)(v)** *Other credits.* The employer's obligation to provide medical removal protection benefits to a removed employee shall be reduced to the extent that the employee receives compensation for earnings lost during the period of removal either from a publicly or employer-funded compensation program, or receives income from employment with another employer made possible by virtue of the employee's removal.

**1926.62(k)(2)(vi)** *Voluntary removal or restriction of an employee.* Where an employer, although not required by this section to do so, removes an employee from exposure to lead or otherwise places limitations on an employee due to the effects of lead exposure on the employee's medical condition, the employer shall provide medical removal protection benefits to the employee equal to that required by paragraph (k)(2) (i) and (ii) of this section.

**1926.62(l)** ***Employee information and training***— (1) *General.* (i) The employer shall communicate information concerning lead hazards according to the requirements of OSHA's Hazard Communication Standard for the construction industry, 29 CFR 1926.59, including but not limited to the requirements concerning warning signs and labels, material safety data sheets (MSDS), and employee information and training. In addition, employers shall comply with the following requirements:

**1926.62(l)(1)(ii)** For all employees who are subject to exposure to lead at or above the action level on any day or who are subject to exposure to lead compounds which may cause skin or eye irritation (e.g. lead arsenate, lead azide), the employer shall provide a training program in accordance with paragraph (l)(2) of this section and assure employee participation.

**1926.62(l)(1)(iii)** The employer shall provide the training program as initial training prior to the time of job assignment or prior to the start up date for this requirement, whichever comes last.

**1926.62(l)(1)(iv)** The employer shall also provide the training program at least annually for each employee who is subject to lead exposure at or above the action level on any day.

**1926.62(l)(2)** *Training program.* The employer shall assure that each employee is trained in the following:

**1926.62(l)(2)(i)** The content of this standard and its appendices;

**1926.62(l)(2)(ii)** The specific nature of the operations which could result in exposure to lead above the action level;

**1926.62(l)(2)(iii)** The purpose, proper selection, fitting, use, and limitations of respirators;

**1926.62(l)(2)(iv)** The purpose and a description of the medical surveillance program, and the medical removal protection program including information concerning the adverse health effects associated with excessive exposure to lead (with particular attention to the adverse reproductive effects on both males and females and hazards to the fetus and additional precautions for employees who are pregnant);

**1926.62(l)(2)(v)** The engineering controls and work practices associated with the employee's job assignment including training of employees to follow relevant good work practices described in Appendix B of this section;

**1926.62(l)(2)(vi)** The contents of any compliance plan in effect;

**1926.62(l)(2)(vii)** Instructions to employees that chelating agents should not routinely be used to remove lead from their bodies and should not be used at all except under the direction of a licensed physician; and

**1926.62(l)(2)(viii)** The employee's right of access to records under 29 CFR 1910.20.

**1926.62(l)(3)** *Access to information and training materials.* (i) The employer shall make readily available to all affected employees a copy of this standard and its appendices.

**1926.62(l)(3)(ii)** The employer shall provide, upon request, all materials relating to the employee information and training program to affected employees and their designated representatives, and to the Assistant Secretary and the Director.

**1926.62(m)** ***Signs***— (1) *General.* (i) The employer may use signs required by other statutes, regulations or ordinances in addition to, or in combination with, signs required by this paragraph.

**1926.62(m)(1)(i)** The employer shall assure that no statement appears on or near any sign required by this paragraph which contradicts or detracts from the meaning of the required sign.

**1926.62(m)(1)(ii)** *Signs.* (i) The employer shall post the following warning signs in each work area where an employees exposure to lead is above the PEL.

**WARNING**

**LEAD WORK AREA**

**POISON**

**NO SMOKING OR EATING**

**1926.62(m)(2)(ii)** The employer shall assure that signs required by this paragraph are illuminated and cleaned as necessary so that the legend is readily visible.

**1926.62(n)** ***Recordkeeping***— (1) *Exposure assessment.* (i) The employer shall establish and maintain an accurate record of all monitoring and other data used in conducting employee exposure assessments as required in paragraph (d) of this section.

**1926.62(n)(1)(ii)** Exposure monitoring records shall include:

**1926.62(n)(1)(ii)(A)** The date(s), number, duration, location and results of each of the samples taken if any, including a description of the sampling procedure used to determine representative employee exposure where applicable;

**1926.62(n)(1)(ii)(B)** A description of the sampling and analytical methods used and evidence of their accuracy;

**1926.62(m)(1)(ii)(C)** The type of respiratory protective devices worn, if any;

**1926.62(m)(1)(ii)(D)** Name, social security number, and job classification of the employee monitored and of all other employees whose exposure the measurement is intended to represent; and

**1926.62(m)(1)(ii)(E)** The environmental variables that could affect the measurement of employee exposure.

**1926.62(m)(1)(iii)** The employer shall maintain monitoring and other exposure assessment records in accordance with the provisions of 29 CFR 1910.20.

**1926.62(n)(2)** *Medical surveillance.* (i) The employer shall establish and maintain an accurate record for each employee subject to medical surveillance as required by paragraph (j) of this section.

**1926.62(n)(2)(ii)** This record shall include:

**1926.62(n)(2)(ii)(A)** The name, social security number, and description of the duties of the employee;

**1926.62(n)(2)(ii)(B)** A copy of the physician's written opinions;

**1926.62(n)(2)(ii)(C)** Results of any airborne exposure monitoring done on or for that employee and provided to the physician; and

**1926.62(n)(2)(ii)(D)** Any employee medical complaints related to exposure to lead.

**1926.62(n)(2)(iii)** The employer shall keep, or assure that the examining physician keeps, the following medical records:

**1926.62(n)(2)(iii)(A)** A copy of the medical examination results including medical and work history required under paragraph (j) of this section;

**1926.62(n)(2)(iii)(B)** A description of the laboratory procedures and a copy of any standards or guidelines used to interpret the test results or references to that information;

**1926.62(n)(2)(iii)(C)** A copy of the results of biological monitoring.

**1926.62(n)(2)(iv)** The employer shall maintain or assure that the physician maintains medical records in accordance with the provisions of 29 CFR 1910.20.

**1926.62(n)(3)** *Medical removals.* (i) The employer shall establish and maintain an accurate record for each employee removed from current exposure to lead pursuant to paragraph (k) of this section.

**1926.62(n)(3)(ii)** Each record shall include:

**1926.62(n)(3)(ii)(A)** The name and social security number of the employee;

**1926.62(n)(3)(ii)(B)** The date of each occasion that the employee was removed from current exposure to lead as well as the corresponding date on which the employee was returned to his or her former job status;

**1926.62(n)(3)(ii)(C)** A brief explanation of how each removal was or is being accomplished; and

**1926.62(n)(3)(ii)(D)** A statement with respect to each removal indicating whether or not the reason for the removal was an elevated blood lead level.

**1926.62(n)(3)(iii)** The employer shall maintain each medical removal record for at least the duration of an employee's employment.

**1926.62(n)(4)** *Objective data for exemption from requirement for initial monitoring.* (i) For purposes of this section, objective data are information demonstrating that a particular product or material containing lead or a specific process, operation, or activity involving lead cannot release dust or fumes in concentrations at or above the action level under any expected conditions of use. Objective data can be obtained from an industry-wide study or from laboratory product test results from manufacturers of lead containing products or materials. The data the employer uses from an industry-wide survey must be obtained under workplace conditions closely resembling the processes, types of material, control methods, work practices and environmental conditions in the employer's current operations.

**1926.62(n)(4)(ii)** The employer shall maintain the record of the objective data relied upon for at least 30 years.

**1926.62(n)(5)** *Availability.* The employer shall make available upon request all records required to be maintained by paragraph (n) of this section to affected employees, former employees, and their designated representatives, and to the Assistant Secretary and the Director for examination and copying.

**1926.62(n)(6)** *Transfer of records.* (i) Whenever the employer ceases to do business, the successor employer shall receive and retain all records required to be maintained by paragraph (n) of this section.

**1926.62(n)(6)(ii)** Whenever the employer ceases to do business and there is no successor employer to receive and retain the records required to be maintained by this section for the prescribed period, these records shall be transmitted to the Director.

**1926.62(n)(6)(iii)** At the expiration of the retention period for the records required to be maintained by this section, the employer shall notify the Director at least 3 months prior to the disposal of such records and shall transmit those records to the Director if requested within the period.

**1926.62(n)(6)(iv)** The employer shall also comply with any additional requirements involving transfer of records set forth in 29 CFR 1910.20(h).

**1926.62(o)** ***Observation of monitoring.*** (1) *Employee observation.* The employer shall provide affected employees or their designated representatives an opportunity to observe any monitoring of employee exposure to lead conducted pursuant to paragraph (d) of this section.

**1926.62(o)(2)** *Observation procedures.* (i) Whenever observation of the monitoring of employee exposure to lead requires entry into an area where the use of respirators, protective clothing or equipment is required, the employer shall provide the observer with and assure the use of such respirators, clothing and equipment, and shall require the observer to comply with all other applicable safety and health procedures.

**1926.62(o)(2)(ii)** Without interfering with the monitoring, observers shall be entitled to:

**1926.62(o)(2)(ii)(A)** Receive an explanation of the measurement procedures;

**1926.62(o)(2)(ii)(B)** Observe all steps related to the monitoring of lead performed at the place of exposure; and

**1926.62(o)(2)(ii)(C)** Record the results obtained or receive copies of the results when returned by the laboratory.

**1926.62(p)** ***Effective date.*** This standard (§1926.62) shall become effective June 3, 1993.

**1926.62(q)** ***Appendices.*** The information contained in the appendices to this section is not intended by itself, to create any additional obligations not otherwise imposed by this standard nor detract from any existing obligation.

**1926.62(r)** ***Startup dates.*** (1) The requirements of paragraphs (c) through (o) of this section, including administrative controls and feasible work practice controls, but not including engineering controls specified in paragraph (e)(1) of this section, shall be complied with as soon as possible, but no later than 60 days from the effective date of this section.

**1926.62(r)(2)** Feasible engineering controls specified by paragraph (e)(1) of this section shall be implemented as soon as possible, but no later than 120 days from the effective date of this section

## Appendix A to §1926.62—Substance Data Sheet for Occupational Exposure to Lead

### I. Substance Identification

A. *Substance:* Pure lead (Pb) is a heavy metal at room temperature and pressure and is a basic chemical element. It can combine with various other substances to form numerous lead compounds.

B. *Compounds covered by the standard:* The word *lead* when used in this interim final standard means elemental lead, all inorganic lead compounds and a class of organic lead compounds called lead soaps. This standard does not apply to other organic lead compounds.

C. *Uses:* Exposure to lead occurs in several different occupations in the construction industry, including demolition or salvage of structures where lead or lead-containing materials are present; removal or encapsulation of lead-containing materials, new construction, alteration, repair, or renovation of structures that contain lead or materials containing lead; installation of products containing lead. In addition, there are construction related activities where exposure to lead may occur, including transportation, disposal, storage, or containment of lead or materials containing lead on construction sites, and maintenance operations associated with construction activities.

D. *Permissible exposure:* The permissible exposure limit (PEL) set by the standard is 50 micrograms of lead per cubic meter of air (50 $\mu g/m^3$), averaged over an 8-hour workday.

E. *Action level:* The interim final standard establishes an action level of 30 micrograms of lead per cubic meter of air (30 $\mu g/m^3$), averaged over an 8-hour workday. The action level triggers several ancillary provisions of the standard such as exposure monitoring, medical surveillance, and training.

### II. Health Hazard Data

A. *Ways in which lead enters your body.* When absorbed into your body in certain doses, lead is a toxic substance. The object of the lead standard is to prevent absorption of harmful quantities of lead. The standard is intended to protect you not only from the immediate toxic effects of lead, but also from the serious toxic effects that may not become apparent until years of exposure have passed. Lead can be absorbed into your body by inhalation (breathing) and ingestion (eating). Lead (except for certain organic lead compounds not covered by the standard, such as tetraethyl lead) is not absorbed through your skin. When lead is scattered in the air as a dust, fume respiratory tract. Inhalation of airborne lead is generally the most important source of occupational lead absorption. You can also absorb lead through your digestive system if lead gets into your mouth and is swallowed. If you handle food, cigarettes, chewing tobacco, or make-up which have lead on them or handle them with hands contaminated with lead, this will contribute to ingestion. A significant portion of the lead that you inhale or ingest gets into your blood stream. Once in your blood stream, lead is circulated throughout your body and stored in various organs and body tissues. Some of this lead is quickly filtered out of your body and excreted, but some remains in the blood and other tissues. As exposure to lead continues, the amount stored in your body will increase if you are absorbing more lead than your body is excreting. Even though you may not be aware of any immediate symptoms of disease, this lead stored in your tissues can be slowly causing irreversible damage, first to individual cells, then to your organs and whole body systems.

B. *Effects of overexposure to lead*—(1) *Short term (acute) overexposure.* Lead is a potent, systemic poison that serves no known useful function once absorbed by your body. Taken in large enough doses, lead can kill you in a matter of days. A condition affecting the brain called acute encephalopathy may arise which develops quickly to seizures, coma, and death from cardiorespiratory arrest. A short term dose of lead can lead to acute encephalopathy. Short term occupational exposures of this magnitude are highly unusual, but not impossible. Similar forms of encephalopathy may, however, arise from extended, chronic exposure to lower doses of lead. There is no sharp dividing line between rapidly developing acute effects of lead, and chronic effects which take longer to acquire. Lead adversely affects numerous body systems, and causes forms of health impairment and disease which arise after periods of exposure as short as days or as long as several years.

(2) *Long-term (chronic) overexposure.* Chronic overexposure to lead may result in severe damage to your blood-forming, nervous, urinary and reproductive systems. Some common symptoms of chronic overexposure include loss of appetite, metallic taste in the mouth, anxiety, constipation, nausea, pallor, excessive tiredness, weakness, insomnia, headache, nervous irritability, muscle and joint pain or soreness, fine tremors, numbness, dizziness, hyperactivity and colic. In lead colic there may be severe abdominal pain. Damage to the central nervous system in general and the brain (encephalopathy) in particular is one of the most severe forms of lead poisoning. The most severe, often fatal, form of encephalopathy may be preceded by vomiting, a feeling of dullness progressing to drowsiness and stupor, poor memory, restlessness, irritability, tremor, and convulsions. It may arise suddenly with the onset of seizures, followed by coma, and death. There is a tendency for muscular weakness to develop at the same time. This weakness may progress to paralysis often observed as a characteristic "wrist drop" or "foot drop" and is a manifestation of a disease to the nervous system called peripheral neuropathy. Chronic overexposure to lead also results in kidney disease with few, if any, symptoms appearing until extensive and most likely permanent kidney damage has occurred. Routine laboratory tests reveal the presence of this kidney disease only after about two-thirds of kidney function is lost. When overt symptoms of urinary dysfunction arise, it is often too late to correct or prevent worsening conditions, and progression to kidney dialysis or death is possible. Chronic overexposure to lead impairs the reproductive systems of both men and women. Overexposure to lead may result in decreased sex drive, impotence and sterility in men. Lead can alter the structure of sperm cells raising the risk of birth defects. There is evidence of miscarriage and stillbirth in women whose husbands were exposed to lead or who were exposed to lead themselves. Lead exposure also may result in decreased fertility, and abnormal menstrual cycles in women. The course of pregnancy may be adversely affected by exposure to lead since lead crosses the placental barrier and poses risks to developing fetuses. Children born of parents either one of whom were exposed to excess lead levels are more likely to have birth defects, mental retardation, behavioral disorders or die during the first year of childhood. Overexposure to lead also disrupts the blood-forming system resulting in decreased hemoglobin (the substance in the blood that carries oxygen to the cells) and ultimately anemia. Anemia is characterized by weakness, pallor and fatigability as a result of decreased oxygen carrying capacity in the blood.

(3) *Health protection goals of the standard.* Prevention of adverse health effects for most workers from exposure to lead throughout a working lifetime requires that a worker's blood lead level (BLL, also expressed as PbB) be maintained at or below forty micrograms per deciliter of whole blood (40 $\mu g/dl$). The blood lead levels of workers (both male and female workers) who intend to have children should be maintained below 30 $\mu g/dl$ to minimize adverse reproductive health effects to the parents and to the developing fetus. The measurement of your blood lead level (BLL) is the most useful indicator of the amount of lead being absorbed by your body. Blood lead levels are most often reported in units of milligrams (mg) or micrograms ($\mu g$) of lead (1 mg = 1000 $\mu g$) per 100 grams (100g), 100 milliliters (100 ml) or deciliter (dl) of blood. These three units are essentially the same. Sometime BLLs are expressed in the form of mg% or $\mu g$%. This is a shorthand notation for 100g, 100 ml, or dl. (References to BLL measurements in this standard are expressed in the form of $\mu g/dl$.)

BLL measurements show the amount of lead circulating in your blood stream, but do not give any information about the amount of lead stored in your various tissues. BLL measurements merely show current absorption of lead, not the effect that lead is having on your body or the effects that past lead exposure may have already caused. Past research into lead-related diseases, however, has focused heavily on associations between BLLs and various diseases. As a result, your BLL is an important indicator of the likelihood that you will gradually acquire a lead-related health impairment or disease.

Once your blood lead level climbs above 40 $\mu g/dl$, your risk of disease increases. There is a wide variability of individual response to lead, thus it is difficult to say that a particular BLL in a given person will cause a particular effect. Studies have associated fatal encephalopathy with BLLs as low as 150 $\mu g/dl$. Other studies have shown other forms of diseases in some workers with BLLs well below 80 $\mu g/dl$. Your BLL is a crucial indicator of the risks to your health, but one other factor is also extremely important. This factor is the length of time you have had elevated BLLs. The longer you have an elevated BLL, the greater the risk that large quantities of lead are being gradually stored in your organs and tissues (body burden). The greater your overall body burden, the greater the chances of substantial permanent damage. The best way to prevent all forms of lead-related impairments and diseases—both short term and long term—is to maintain your BLL below 40 $\mu g/dl$. The provisions of the standard are designed with this end in mind.

Your employer has prime responsibility to assure that the provisions of the standard are complied with both by the company and by individual workers. You, as a worker, however, also have a responsibility to assist your employer in complying with the standard. You can play a key role in protecting your own health by learning about the lead hazards and their control, learning what the standard requires, following the standard where it governs your own actions, and seeing that your employer complies with provisions governing his or her actions.

(4) *Reporting signs and symptoms of health problems.* You should immediately notify your employer if you develop signs or symptoms associated with lead poisoning or if you desire medical advice concerning the effects of current or past exposure to lead or your ability to have a healthy child. You should also notify your employer if you have difficulty breathing during a respirator fit test or while wearing a respirator. In each of these cases, your employer must make available to you appropriate medical examinations or consultations. These must be provided at no cost to you and at a reasonable time and place. The standard contains a procedure whereby you can obtain a second opinion by a physician of your choice if your employer selected the initial physician.

## Appendix B to §1926.62—Employee Standard Summary

This appendix summarizes key provisions of the interim final standard for lead in construction that you as a worker should become familiar with.

### I. Permissible Exposure Limit (PEL)—Paragraph (C)

The standard sets a permissible exposure limit (PEL) of 50 micrograms of lead per cubic meter of air (50 $\mu g/m^3$), averaged over an 8-hour workday which is referred to as a time-weighted average (TWA). This is the highest level of lead in air to which you may be permissibly exposed over an 8-hour workday. However, since this is an 8-hour average, short exposures above the PEL are permitted so long as for each 8-hour work day your average exposure does not exceed this level. This interim final standard, however, takes into account the fact that your daily exposure to lead can extend beyond a typical 8-hour workday as the result of overtime or other alterations in your work schedule. To deal with this situation, the standard contains a formula which reduces your permissible exposure when you are exposed more than 8 hours. For example, if you are exposed to lead for 10 hours a day, the maximum permitted average exposure would be 40 $\mu g/m^3$.

### II. Exposure Assessment—Paragraph (D)

If lead is present in your workplace in any quantity, your employer is required to make an initial determination of whether any employee's exposure to lead exceeds the action level (30 $\mu g/m^3$ averaged over an 8-hour day). Employee exposure is that exposure which would occur if the employee were not using a respirator. This initial determination requires your employer to monitor workers' exposures unless he or she has objective data which can demonstrate conclusively that no employee will be exposed to lead in excess of the action level. Where objective data is used in lieu of actual monitoring the employer must establish and maintain an accurate record, documenting its relevancy in assessing exposure levels for current job conditions. If such objective data is available, the employer need proceed no further on employee exposure assessment until such time that conditions have changed and the determination is no longer valid.

Objective data may be compiled from various sources, e.g., insurance companies and trade associations and information from suppliers or exposure data collected from similar operations. Objective data may also comprise previously-collected sampling data including area monitoring. If it cannot be determined through using objective data that worker exposure is less than the action level, your employer must conduct monitoring or must rely on relevant previous personal sampling, if available. Where monitoring is required for the initial determination, it may be limited to a representative number of employees who are reasonably expected to have the highest exposure levels. If your employer has conducted appropriate air sampling for lead in the past 12 months, he or she may use these results, provided they are applicable to the same employee tasks and exposure conditions and meet the requirements for accuracy as specified in the standard. As with objective data, if such results are relied upon for the initial determination, your employer must establish and maintain a record as to the relevancy of such data to current job conditions.

If there have been any employee complaints of symptoms which may be attributable to exposure to lead or if there is any other information or observations which would indicate employee exposure to lead, this must also be considered as part of the initial determination.

If this initial determination shows that a reasonable possibility exists that any employee may be exposed, without regard to respirators, over the action level, your employer must set up an air monitoring program to determine the exposure level representative of each employee exposed to lead at your workplace. In carrying out this air monitoring program, your employer is not required to monitor the exposure of every employee, but he or she must monitor a representative number of employees and job types. Enough sampling must be done to enable each employee's exposure level to be reasonably represent full shift exposure. In addition, these air samples must be taken under conditions which represent each employee's regular, daily exposure to lead. Sampling performed in the past 12 months may be used to determine exposures above the action level if such sampling was conducted during work activities essentially similar to present work conditions.

The standard lists certain tasks which may likely result in exposures to lead in excess of the PEL and, in some cases, exposures in excess of 50 times the PEL. If you are performing any of these tasks, your employer must provide you with appropriate respiratory protection, protective clothing and equipment, change areas, hand washing facilities, biological monitoring, and training until such time that an exposure assessment is conducted which demonstrates that your exposure level is below the PEL.

If you are exposed to lead and air sampling is performed, your employer is required to notify you in writing within 5 working days of the air monitoring results which represent your exposure. If the results indicate that your exposure exceeds the PEL (without regard to your use of a respirator), then your employer must also notify you of this in writing, and provide you with a description of the corrective action that has been taken or will be taken to reduce your exposure.

Your exposure must be rechecked by monitoring, at least every six months if your exposure is at or over the action level but below the PEL. Your employer may discontinue monitoring for you if 2 consecutive measurements, taken at least 7 days apart, are at or below the action level. Air monitoring must be repeated every 3 months if you are exposed over the PEL. Your employer must continue monitoring for you at this frequency until 2 consecutive measurements, taken at least 7 days apart, are below the PEL but above the action level, at which time your employer must repeat monitoring of your exposure every six months and may discontinue monitoring only after your exposure drops to or below the action level. However, whenever there is a change of equipment, process, control, or personnel or a new type of job is added at your workplace which may result in new or additional exposure to lead, your employer must perform additional monitoring.

### III. Methods of Compliance—Paragraph (E)

Your employer is required to assure that no employee is exposed to lead in excess of the PEL as an 8-hour TWA. The interim final standard for lead in construction requires employers to institute engineering and work practice controls including administrative controls to the extent feasible to reduce employee exposure to lead. Where such controls are feasible but not adequate to reduce exposures below the PEL they must be used nonetheless to reduce exposures to the lowest level that can be accomplished by these means and then supplemented with appropriate respiratory protection.

Your employer is required to develop and implement a written compliance program prior to the commencement of any job where employee exposures may reach the PEL as an 8-hour TWA. The interim final standard identifies the various elements that must be included in the plan. For example, employers are required to include a description of operations in which lead is emitted, detailing other relevant information about the operation such as the type of equipment used, the type of material involved, employee job responsibilities, operating procedures and maintenance practices. In addition, your employer's compliance plan must specify the means that will be used to achieve compliance and, where engineering controls are required, include any engineering plans or studies that have been used to select the control methods. If administrative controls involving job rotation are used to reduce employee exposure to lead, the job rotation schedule must be included in the compliance plan. The plan must also detail the type of protective clothing and equipment, including respirators, housekeeping and hygiene practices that will be used to protect you from the adverse effects of exposure to lead.

The written compliance program must be made available, upon request, to affected employees and their designated representatives, the Assistant Secretary and the Director.

Finally, the plan must be reviewed and updated at least every 6 months to assure it reflects the current status in exposure control.

### IV. Respiratory Protection—Paragraph (F)

Your employer is required to provide and assure your use of respirators when your exposure to lead is not controlled below the PEL by other means.

The employer must pay the cost of the respirator. Whenever you request one, your employer is also required to provide you a respirator even if your air exposure level is not above the PEL. You might desire a respirator when, for example, you have received medical advice that your lead absorption should be decreased. Or, you may intend to have children in the near future, and want to reduce the level of lead in your body to minimize adverse reproductive effects. While respirators are the least satisfactory means of controlling your exposure, they are capable of providing significant protection if properly chosen, fitted, worn, cleaned, maintained, and replaced when they stop providing adequate protection.

Your employer is required to select respirators from the types listed in Table I of the Respiratory Protection section of the standard (§1926.62(f)). Any respirator chosen must be approved by the National Institute for Occupational Safety and Health (NIOSH) under the provisions of 42 CFR part 84. This respirator selection table will enable your employer to choose a type of respirator that will give you a proper amount of protection based on your airborne lead exposure. Your employer may select a type of respirator that provides greater protection than that required by the standard; that is, one recommended for a higher concentration of lead than is present in your workplace. For example, a powered air-purifying respirator (PAPR) is much more protective than a typical negative pressure respirator, and may also be more comfortable to wear. A PAPR has a filter, cartridge, or canister to clean the air, and a power source that continuously blows filtered air into your breathing zone. Your employer might make a PAPR available to you to ease the burden of having to wear a respirator for long periods of time. The standard provides that you can obtain a PAPR upon request.

Your employer must also start a Respiratory Protection Program. This program must include written procedures for the proper selection, use, cleaning, storage, and maintenance of respirators.

Your employer must ensure that your respirator facepiece fits properly. Proper fit of a respirator facepiece is critical to your protection from airborne lead. Obtaining a proper fit on each employee may require your employer to make available several different types of respirator masks. To ensure that your respirator fits properly and that facepiece leakage is minimal, your employer must give you either a qualitative or quantitative fit test as specified in Appendix A of the Respiratory Protection standard located at 29 CFR 1910.134.

You must also receive from your employer proper training in the use of respirators. Your employer is required to teach you how to wear a respirator, to know why it is needed, and to understand its limitations.

The standard provides that if your respirator uses filter elements, you must be given an opportunity to change the filter elements whenever an increase in breathing resistance is detected. You also must be permitted to periodically leave your work area to wash your face and respirator facepiece whenever necessary to prevent skin irritation. If you ever have difficulty in breathing during a fit test or while using a respirator, your employer must make a medical examination available to you to determine whether you can safely wear a respirator. The result of this examination may be to give you a positive pressure respirator (which reduces breathing resistance) or to provide alternative means of protection.

## V. Protective Work Clothing and Equipment—Paragraph (G)

If you are exposed to lead above the PEL as an 8-hour TWA, without regard to your use of a respirator, or if you are exposed to lead compounds such as lead arsenate or lead azide which can cause skin and eye irritation, your employer must provide you with protective work clothing and equipment appropriate for the hazard. If work clothing is provided, it must be provided in a clean and dry condition at least weekly, and daily if your airborne exposure to lead is greater than 200 $\mu g/m^3$. Appropriate protective work clothing and equipment can include coveralls or similar full-body work clothing, gloves, hats, shoes or disposable shoe coverlets, and face shields or vented goggles. Your employer is required to provide all such equipment at no cost to you. In addition, your employer is responsible for providing repairs and replacement as necessary, and also is responsible for the cleaning, laundering or disposal of protective clothing and equipment.

The interim final standard requires that your employer assure that you follow good work practices when you are working in areas where your exposure to lead may exceed the PEL. With respect to protective clothing and equipment, where appropriate, the following procedures should be observed prior to beginning work:

1. Change into work clothing and shoe covers in the clean section of the designated changing areas;

2. Use work garments of appropriate protective gear, including respirators before entering the work area; and

3. Store any clothing not worn under protective clothing in the designated changing area.

Workers should follow these procedures upon leaving the work area:

1. HEPA vacuum heavily contaminated protective work clothing while it is still being worn. At no time may lead be removed from protective clothing by any means which result in uncontrolled dispersal of lead into the air;

2. Remove shoe covers and leave them in the work area;

3. Remove protective clothing and gear in the dirty area of the designated changing area. Remove protective coveralls by carefully rolling down the garment to reduce exposure to dust.

4. Remove respirators last; and

5. Wash hands and face.

Workers should follow these procedures upon finishing work for the day (in addition to procedures described above):

1. Where applicable, place disposal coveralls and shoe covers with the abatement waste;

2. Contaminated clothing which is to be cleaned, laundered or disposed of must be placed in closed containers in the change room.

3. Clean protective gear, including respirators, according to standard procedures;

4. Wash hands and face again. If showers are available, take a shower and wash hair. If shower facilities are not available at the work site, shower immediately at home and wash hair.

## VI. Housekeeping—Paragraph (H)

Your employer must establish a housekeeping program sufficient to maintain all surfaces as free as practicable of accumulations of lead dust. Vacuuming is the preferred method of meeting this requirement, and the use of compressed air to clean floors and other surfaces is generally prohibited unless removal with compressed air is done in conjunction with ventilation systems designed to contain dispersal of the lead dust. Dry or wet sweeping, shoveling, or brushing may not be used except where vacuuming or other equally effective methods have been tried and do not work. Vacuums must be used equipped with a special filter called a high-efficiency particulate air (HEPA) filter and emptied in a manner which minimizes the reentry of lead into the workplace.

## VII. Hygiene Facilities and Practices—Paragraph (I)

The standard requires that hand washing facilities be provided where occupational exposure to lead occurs. In addition, change areas, showers (where feasible), and lunchrooms or eating areas are to be made available to workers exposed to lead above the PEL. Your employer must assure that except in these facilities, food and beverage is not present or consumed, tobacco products are not present or used, and cosmetics are not applied, where airborne exposures are above the PEL. Change rooms provided by your employer must be equipped with separate storage facilities for your protective clothing and equipment and street clothes to avoid cross-contamination. After showering, no required protective clothing or equipment worn during the shift may be worn home. It is important that contaminated clothing or equipment be removed in change areas and not be worn home or you will extend your exposure and expose your family since lead from your clothing can accumulate in your house, car, etc.

Lunchrooms or eating areas may not be entered with protective clothing or equipment unless surface dust has been removed by vacuuming, downdraft booth, or other cleaning method. Finally, workers exposed above the PEL must wash both their hands and faces prior to eating, drinking, smoking or applying cosmetics.

All of the facilities and hygiene practices just discussed are essential to minimize additional sources of lead absorption from inhalation or ingestion of lead that may accumulate on you, your clothes, or your possessions. Strict compliance with these provisions can virtually eliminate several sources of lead exposure which significantly contribute to excessive lead absorption.

### VIII. Medical Surveillance—Paragraph (J)

The medical surveillance program is part of the standard's comprehensive approach to the prevention of lead-related disease. Its purpose is to supplement the main thrust of the standard which is aimed at minimizing airborne concentrations of lead and sources of ingestion. Only medical surveillance can determine if the other provisions of the standard have affectively protected you as an individual. Compliance with the standard's provision will protect most workers from the adverse effects of lead exposure, but may not be satisfactory to protect individual workers (1) who have high body burdens of lead acquired over past years, (2) who have additional uncontrolled sources of non-occupational lead exposure, (3) who exhibit unusual variations in lead absorption rates, or (4) who have specific non-work related medical conditions which could be aggravated by lead exposure (e.g., renal disease, anemia). In addition, control systems may fail, or hygiene and respirator programs may be inadequate. Periodic medical surveillance of individual workers will help detect those failures. Medical surveillance will also be important to protect your reproductive ability-regardless of whether you are a man or woman.

All medical surveillance required by the interim final standard must be performed by or under the supervision of a licensed physician. The employer must provide required medical surveillance without cost to employees and at a reasonable time and place. The standard's medical surveillance program has two parts—periodic biological monitoring and medical examinations. Your employer's obligation to offer you medical surveillance is triggered by the results of the air monitoring program. Full medical surveillance must be made available to all employees who are or may be exposed to lead in excess of the action level for more than 30 days a year and whose blood lead level exceeds 40 µg/dl. Initial medical surveillance consisting of blood sampling and analysis for lead and zinc protoporphyrin must be provided to all employees exposed at any time (1 day) above the action level.

Biological monitoring under the standard must be provided at least every 2 months for the first 6 months and every 6 months thereafter until your blood lead level is below 40 µg/dl. A zinc protoporphyrin (ZPP) test is a very useful blood test which measures an adverse metabolic effect of lead on your body and is therefore an indicator of lead toxicity.

If your BLL exceeds 40 µg/dl the monitoring frequency must be increased from every 6 months to at least every 2 months and not reduced until two consecutive BLLs indicate a blood lead level below 40 µg/dl. Each time your BLL is determined to be over 40 µg/dl, your employer must notify you of this in writing within five working days of his or her receipt of the test results. The employer must also inform you that the standard requires temporary medical removal with economic protection when your BLL exceeds 50 µg/dl. (See Discussion of Medical Removal Protection-Paragraph (k).) Anytime your BLL exceeds 50 µg/dl your employer must make available to you within two weeks of receipt of these test results a second follow-up BLL test to confirm your BLL. If the two tests both exceed 50 µg/dl, and you are temporarily removed, then your employer must make successive BLL tests available to you on a monthly basis during the period of your removal.

Medical examinations beyond the initial one must be made available on an annual basis if your blood lead level exceeds 40 µg/dl at any time during the preceding year and you are being exposed above the airborne action level of 30 $\mu g/m_3$ for 30 or more days per year. The initial examination will provide information to establish a baseline to which subsequent data can be compared.

An initial medical examination to consist of blood sampling and analysis for lead and zinc protoporphyrin must also be made available (prior to assignment) for each employee being assigned for the first time to an area where the airborne concentration of lead equals or exceeds the action level at any time. In addition, a medical examination or consultation must be made available as soon as possible if you notify your employer that you are experiencing signs or symptoms commonly associated with lead poisoning or that you have difficulty breathing while wearing a respirator or during a respirator fit test. You must also be provided a medical examination or consultation if you notify your employer that you desire medical advice concerning the effects of current or past exposure to lead on your ability to procreate a healthy child.

Finally, appropriate follow-up medical examinations or consultations may also be provided for employees who have been temporarily removed from exposure under the medical removal protection provisions of the standard. (See Part IX, below.)

The standard specifies the minimum content of pre-assignment and annual medical examinations. The content of other types of medical examinations and consultations is left up to the sound discretion of the examining physician. Pre-assignment and annual medical examinations must include (1) a detailed work history and medical history; (2) a thorough physical examination, including an evaluation of your pulmonary status if you will be required to use a respirator; (3) a blood pressure measurement; and (4) a series of laboratory tests designed to check your blood chemistry and your kidney function. In addition, at any time upon your request, a laboratory evaluation of male fertility will be made (microscopic examination of a sperm sample), or a pregnancy test will be given.

The standard does not require that you participate in any of the medical procedures, tests, etc. which your employer is required to make available to you. Medical surveillance can, however, play a very important role in protecting your health. You are strongly encouraged, therefore, to participate in a meaningful fashion. The standard contains a multiple physician review mechanism which will give you a chance to have a physician of your choice directly participate in the medical surveillance program. If you are dissatisfied with an examination by a physician chosen by your employer, you can select a second physician to conduct an independent analysis. The two doctors would attempt to resolve any differences of opinion, and select a third physician to resolve any firm dispute. Generally your employer will choose the physician who conducts medical surveillance under the lead standard-unless you and your employer can agree on the choice of a physician or physicians. Some companies and unions have agreed in advance, for example, to use certain independent medical laboratories or panels of physicians. Any of these arrangements are acceptable so long as required medical surveillance is made available to workers.

The standard requires your employer to provide certain information to a physician to aid in his or her examination of you. This information includes (1) the standard and its appendices, (2) a description of your duties as they relate to occupational lead exposure, (3) your exposure level or anticipated exposure level, (4) a description of any personal protective equipment you wear, (5) prior blood lead level results, and (6) prior written medical opinions concerning you that the employer has. After a medical examination or consultation the physician must prepare a written report which must contain (1) the physician's opinion as to whether you have any medical condition which places you at increased risk of material impairment to health from exposure to lead, (2) any recommended special protective measures to be provided to you, (3) any blood lead level determinations, and (4) any recommended limitation on your use of respirators. This last element must include a determination of whether you can wear a powered air purifying respirator (PAPR) if you are found unable to wear a negative pressure respirator.

The medical surveillance program of the interim lead standard may at some point in time serve to notify certain workers that they have acquired a disease or other adverse medical condition as a result of occupational lead exposure. If this is true, these workers might have legal rights to compensation from public agencies, their employers, firms that supply hazardous products to their employers, or other persons. Some states have laws, including worker compensation laws, that disallow a worker who learns of a job-related health impairment to sue, unless the worker sues within a short period of time after learning of the impairment. (This period of time may be a matter of months or years.) An attorney can be consulted about these possibilities. It should be stressed that OSHA is in no way trying to either encourage or discourage claims or lawsuits. However, since results of the standard's medical surveillance program can significantly affect the legal remedies of a worker who has acquired a job-related disease or impairment, it is proper for OSHA to make you aware of this.

The medical surveillance section of the standard also contains provisions dealing with chelation. Chelation is the use of certain drugs (administered in pill form or injected into the body) to reduce the amount of lead absorbed in body tissues. Experience accumulated by the medical and scientific communities has largely confirmed the effectiveness of this type of therapy for the treatment of very severe lead poisoning. On the other hand, it has also been established that there can be a long list of extremely harmful side effects associated with the use of chelating agents. The medical community has balanced the advantages and disadvantages resulting from the use of chelating agents in various circumstances and has established when the use of these agents is acceptable. The standard includes these accepted limitations due to a history of abuse of chelation therapy by some lead companies. The most widely used chelating agents are calcium disodium EDTA, (Ca Na2 EDTA), Calcium Disodium Versenate (Versenate), and d-penicillamine (pencillamine or Cupramine).

The standard prohibits "prophylactic chelation" of any employee by any person the employer retains, supervises or controls. *Prophylactic chelation* is the routine use of chelating or similarly acting drugs to prevent elevated blood levels in workers who are occupationally exposed to lead, or the use of these drugs to routinely lower blood lead levels to predesignated concentrations believed to be "safe". It should be emphasized that where an employer takes a worker who has no symptoms of lead poisoning and has chelation carried out by a physician (either inside or outside of a hospital) solely to reduce the worker's blood lead level, that will generally be considered prophylactic chelation. The use of a hospital and a physician does not mean that prophylactic chelation is not being performed. Routine chelation to prevent increased or reduce current blood lead levels is unacceptable whatever the setting.

The standard allows the use of "therapeutic" or "diagnostic" chelation if administered under the supervision of a licensed physician in a clinical setting with thorough and appropriate medical monitoring. Therapeutic chelation responds to severe lead poisoning where there are marked symptoms. Diagnostic chelation involved giving a patient a dose of the drug then collecting all urine excreted for some period of time as an aid to the diagnosis of lead poisoning.

In cases where the examining physician determines that chelation is appropriate, you must be notified in writing of this fact before such treatment. This will inform you of a potentially harmful treatment, and allow you to obtain a second opinion.

## IX. Medical Removal Protection—Paragraph (K)

Excessive lead absorption subjects you to increased risk of disease. Medical removal protection (MRP) is a means of protecting you when, for whatever reasons, other methods, such as engineering controls, work practices, and respirators, have failed to provide the protection you need. MRP involves the temporary removal of a worker from his or her regular job to a place of significantly lower exposure without any loss of earnings, seniority, or other employment rights or benefits. The purpose of this program is to cease further lead absorption and allow your body to naturally excrete lead which has previously been absorbed. Temporary medical removal can result from an elevated blood lead level, or a medical opinion. For up to 18 months, or for as long as the job the employee was removed from lasts, protection is provided as a result of either form of removal. The vast majority of removed workers, however, will return to their former jobs long before this eighteen month period expires.

You may also be removed from exposure even if your blood lead level is below 50 μg/dl if a final medical determination indicates that you temporarily need reduced lead exposure for medical reasons. If the physician who is implementing your employers medical program makes a final written opinion recommending your removal or other special protective measures, your employer must implement the physician's recommendation. If you are removed in this manner, you may only be returned when the doctor indicates that it is safe for you to do so.

The standard does not give specific instructions dealing with what an employer must do with a removed worker. Your job assignment upon removal is a matter for you, your employer and your union (if any) to work out consistent with existing procedures for job assignments. Each removal must be accomplished in a manner consistent with existing collective bargaining relationships. Your employer is given broad discretion to implement temporary removals so long as no attempt is made to override existing agreements. Similarly, a removed worker is provided no right to veto an employer's choice which satisfies the standard.

In most cases, employers will likely transfer removed employees to other jobs with sufficiently low lead exposure. Alternatively, a worker's hours may be reduced so that the time weighted average exposure is reduced, or he or she may be temporarily laid off if no other alternative is feasible.

In all of these situation, MRP benefits must be provided during the period of removal—i.e., you continue to receive the same earnings, seniority, and other rights and benefits you would have had if you had not been removed. Earnings includes more than just your base wage; it includes overtime, shift differentials, incentives, and other compensation you would have earned if you had not been removed. During the period of removal you must also be provided with appropriate follow-up medical surveillance. If you were removed because your blood lead level was too high, you must be provided with a monthly blood test. If a medical opinion caused your removal, you must be provided medical tests or examinations that the doctor believes to be appropriate. If you do not participate in this follow up medical surveillance, you may lose your eligibility for MRP benefits.

When you are medically eligible to return to your former job, your employer must return you to your "former job status." This means that you are entitled to the position, wages, benefits, etc., you would have had if you had not been removed. If you would still be in your old job if no removal had occurred that is where you go back. If not, you are returned consistent with whatever job assignment discretion your employer would have had if no removal had occurred. MRP only seeks to maintain your rights, not expand them or diminish them.

If you are removed under MRP and you are also eligible for worker compensation or other compensation for lost wages, your employer's MRP benefits obligation is reduced by the amount that you actually receive from these other sources. This is also true if you obtain other employment during the time you are laid off with MRP benefits.

The standard also covers situations where an employer voluntarily removes a worker from exposure to lead due to the effects of lead on the employee's medical condition, even though the standard does not require removal. In these situations MRP benefits must still be provided as though the standard required removal. Finally, it is important to note that in all cases where removal is required, respirators cannot be used as a substitute. Respirators may be used before removal becomes necessary, but not as an alternative to a transfer to a low exposure job, or to a lay-off with MRP benefits.

## X. Employee Information and Training—Paragraph (L)

Your employer is required to provide an information and training program for all employees exposed to lead above the action level or who may suffer skin or eye irritation from lead compounds such as lead arsenate or lead azide. The program must train these employees regarding the specific hazards associated with their work environment, protective measures which can be taken, including the contents of any compliance plan in effect, the danger of lead to their bodies (including their reproductive systems), and their rights under the standard. All employees must be trained prior to initial assignment to areas where there is a possibility of exposure over the action level.

This training program must also be provided at least annually thereafter unless further exposure above the action level will not occur.

## XI. Signs—Paragraph (M)

The standard requires that the following warning sign be posted in work areas where the exposure to lead exceeds the PEL:

**WARNING**

**LEAD WORK AREA**

**POISON**

**NO SMOKING OR EATING**

These signs are to be posted and maintained in a manner which assures that the legend is readily visible.

## XII. Recordkeeping—Paragraph (N)

Your employer is required to keep all records of exposure monitoring for airborne lead. These records must include the name and job classification of employees measured, details of the sampling and analytical techniques, the results of this sampling, and the type of respiratory protection being worn by the person sampled. Such records are to be retained for at least 30 years. Your employer is also required to keep all records of biological monitoring and medical examination results. These records must include the names of the employees, the physician's written opinion, and a copy of the results of the examination. Medical records must be preserved and maintained for the duration of employment plus 30 years. However, if the employee's duration of employment is less than one year, the employer need not retain that employee's medical records beyond the period of employment if they are provided to the employee upon termination of employment.

Recordkeeping is also required if you are temporarily removed from your job under the medical removal protection program. This record must include your name and social security number, the date of your removal and return, how the removal was or is being accomplished, and whether or not the reason for the removal was an elevated blood lead level. Your employer is required to keep each medical removal record only for as long as the duration of an employee's employment.

The standard requires that if you request to see or copy environmental monitoring, blood lead level monitoring, or medical removal records, they must be made available to you or to a representative that you authorize. Your union also has access to these records. Medical records other than BLL's must also be provided upon request to you, to your physician or to any other person whom you may specifically designate. Your union does not have access to your personal medical records unless you authorize their access.

### XIII. Observation of Monitoring—Paragraph (O)

When air monitoring for lead is performed at your workplace as required by this standard, your employer must allow you or someone you designate to act as an observer of the monitoring. Observers are entitled to an explanation of the measurement procedure, and to record the results obtained. Since results will not normally be available at the time of the monitoring, observers are entitled to record or receive the results of the monitoring when returned by the laboratory. Your employer is required to provide the observer with any personal protective devices required to be worn by employees working in the area that is being monitored. The employer must require the observer to wear all such equipment and to comply with all other applicable safety and health procedures.

### XIV. Effective Date—Paragraph (P)

The standard's effective date is June 3, 1993. Employer obligations under the standard begin as of that date with full implementation of engineering controls as soon as possible but no later than within 4 months, and all other provisions completed as soon as possible, but no later than within 2 months from the effective date.

### XV. For Additional Information

A. A copy of the interim standard for lead in construction can be obtained free of charge by calling or writing the OSHA Office of Publications, room N-3101, United States Department of Labor, Washington, DC 20210: Telephone (202) 219-4667.

B. Additional information about the standard, its enforcement, and your employer's compliance can be obtained from the nearest OSHA Area Office listed in your telephone directory under United States Government/Department of Labor.

## Appendix C to §1926.62— Medical Surveillance Guidelines

### Introduction

The primary purpose of the Occupational Safety and Health Act of 1970 is to assure, so far as possible, safe and healthful working conditions for every working man and woman. The interim final occupational health standard for lead in construction is designed to protect workers exposed to inorganic lead including metallic lead, all inorganic lead compounds and organic lead soaps.

Under this interim final standard occupational exposure to inorganic lead is to be limited to 50 $\mu g/m_3$ (micrograms per cubic meter) based on an 8 hour time-weighted average (TWA). This permissible exposure limit (PEL) must be achieved through a combination of engineering, work practice and administrative controls to the extent feasible. Where these controls are in place but are found not to reduce employee exposures to or below the PEL, they must be used nonetheless, and supplemented with respirators to meet the 50 $\mu g/m_3$ exposure limit.

The standard also provides for a program of biological monitoring for employees exposed to lead above the action level at any time, and additional medical surveillance for all employees exposed to levels of inorganic lead above 30 $\mu g/m_3$ (TWA) for more than 30 days per year and whose BLL exceeds 40 $\mu g/dl$.

The purpose of this document is to outline the medical surveillance provisions of the interim standard for inorganic lead in construction, and to provide further information to the physician regarding the examination and evaluation of workers exposed to inorganic lead.

Section 1 provides a detailed description of the monitoring procedure including the required frequency of blood testing for exposed workers, provisions for medical removal protection (MRP), the recommended right of the employee to a second medical opinion, and notification and recordkeeping requirements of the employer. A discussion of the requirements for respirator use and respirator monitoring and OSHA's position on prophylactic chelation therapy are also included in this section.

Section 2 discusses the toxic effects and clinical manifestations of lead poisoning and effects of lead intoxication on enzymatic pathways in heme synthesis. The adverse effects on both male and female reproductive capacity and on the fetus are also discussed.

Section 3 outlines the recommended medical evaluation of the worker exposed to inorganic lead, including details of the medical history, physical examination, and recommended laboratory tests, which are based on the toxic effects of lead as discussed in Section 2.

Section 4 provides detailed information concerning the laboratory tests available for the monitoring of exposed workers. Included also is a discussion of the relative value of each test and the limitations and precautions which are necessary in the interpretation of the laboratory results.

### I. Medical Surveillance and Monitoring Requirements for Workers Exposed to Inorganic Lead

Under the interim final standard for inorganic lead in the construction industry, initial medical surveillance consisting of biological monitoring to include blood lead and ZPP level determination shall be provided to employees exposed to lead at or above the action level on any one day. In addition, a program of biological monitoring is to be made available to all employees exposed above the action level at any time and additional medical surveillance is to be made available to all employees exposed to lead above 30 $\mu g/m^3$ TWA for more than 30 days each year and whose BLL exceeds 40 $\mu g/dl$. This program consists of periodic blood sampling and medical evaluation to be performed on a schedule which is defined by previous laboratory results, worker complaints or concerns, and the clinical assessment of the examining physician.

Under this program, the blood lead level (BLL) of all employees who are exposed to lead above 30 $\mu g/m^3$ for more than 30 days per year or whose blood lead is above 40 $\mu g/dl$ but exposed for no more than 30 days per year is to be determined at least every two months for the first six months of exposure and every six months thereafter. The frequency is increased to every two months for employees whose last blood lead level was 40 $\mu g/dl$ or above. For employees who are removed from exposure to lead due to an elevated blood lead, a new blood lead level must be measured monthly. A zinc protoporphyrin (ZPP) measurement is strongly recommended on each occasion that a blood lead level measurement is made.

An annual medical examination and consultation performed under the guidelines discussed in Section 3 is to be made available to each employee exposed above 30 $\mu g/m^3$ for more than 30 days per year for whom a blood test conducted at any time during the preceding 12 months indicated a blood lead level at or above 40 $\mu g/dl$. Also, an examination is to be given to all employees prior to their assignment to an area in which airborne lead concentrations reach or exceed the 30 $\mu g/m^3$ for more than 30 days per year. In addition, a medical examination must be provided as soon as possible after notification by an employee that the employee has developed signs or symptoms commonly associated with lead intoxication, that the employee desires medical advice regarding lead exposure and the ability to procreate a healthy child, or that the employee has demonstrated difficulty in breathing during a respirator fitting test or during respirator use. An examination is also to be made available to each employee removed from exposure to lead due to a risk of sustaining material impairment to health, or otherwise limited or specially protected pursuant to medical recommendations.

Results of biological monitoring or the recommendations of an examining physician may necessitate removal of an employee from further lead exposure pursuant to the standard's medical removal protection (MRP) program. The object of the MRP program is to provide temporary medical removal to workers either with substantially elevated blood lead levels or otherwise at risk of sustaining material health impairment from continued substantial exposure to lead.

Under the standard's ultimate worker removal criteria, a worker is to be removed from any work having an eight hour TWA exposure to lead of 30 $\mu g/m^3$ when his or her blood lead level reaches 50 $\mu g/dl$ and is confirmed by a second follow-up blood lead level performed within two weeks after the employer receives the results of the first blood sampling test. Return of the employee to his or her job status depends on a worker's blood lead level declining to 40 $\mu g/dl$.

As part of the interim standard, the employer is required to notify in writing each employee whose blood lead level exceeds 40 $\mu g/dl$. In addition each such employee is to be informed that the standard requires medical removal with MRP benefits, discussed below, when an employee's blood lead level exceeds the above defined limit.

In addition to the above blood lead level criterion, temporary worker removal may also take place as a result of medical determinations and recommendations. Written medical opinions must be prepared after each examination pursuant to the standard. If the examining physician includes a medical finding, determination or opinion that the employee has a medical condition which places the employee at increased risk of material health impairment from exposure to lead, then the employee must be removed from exposure to lead at or above 30 $\mu g/m^3$. Alternatively, if the examining physician recommends special protective measures for an employee (e.g., use of a powered air purifying respirator) or recommends limitations on an employee's exposure to lead, then the employer must implement these recommendations.

Recommendations may be more stringent than the specific provisions of the standard. The examining physician, therefore, is given broad flexibility to tailor special protective procedures to the needs of individual employees. This flexibility extends to the evaluation and management of pregnant workers and male and female workers who are planning to raise children. Based on the history, physical examination, and laboratory studies, the physician might recommend special protective measures or medical removal for an employee who is pregnant or who is planning to conceive a child when, in the physician's judgment, continued exposure to lead at the current job would pose a significant risk. The return of the employee to his or her former job status, or the removal of special protections or limitations, depends upon the examining physician determining that the employee is no longer at increased risk of material impairment or that special measures are no longer needed.

During the period of any form of special protection or removal, the employer must maintain the worker's earnings, seniority, and other employment rights and benefits (as though the worker had not been removed) for a period of up to 18 months or for as long as the job the employee was removed from lasts if less than 18 months. This economic protection will maximize meaningful worker participation in the medical surveillance program, and is appropriate as part of the employer's overall obligation to provide a safe and healthful workplace. The provisions of MRP benefits during the employee's removal period may, however, be conditioned upon participation in medical surveillance.

The lead standard provides for a multiple physician review in cases where the employee wishes a second opinion concerning potential lead poisoning or toxicity. If an employee wishes a second opinion, he or she can make an appointment with a physician of his or her choice. This second physician will review the findings, recommendations or determinations of the first physician and conduct any examinations, consultations or tests deemed necessary in an attempt to make a final medical determination. If the first and second physicians do not agree in their assessment they must try to resolve their differences. If they cannot reach an agreement then they must designate a third physician to resolve the dispute.

The employer must provide examining and consulting physicians with the following specific information: A copy of the lead regulations and all appendices, a description of the employee's duties as related to exposure, the exposure level or anticipated level to lead and any other toxic substances (if applicable), a description of personal protective equipment used, blood lead levels, and all prior written medical opinions regarding the employee in the employer's possession or control. The employer must also obtain from the physician and provide the employee with a written medical opinion containing blood lead levels, the physicians's opinion as to whether the employee is at risk of material impairment to health, any recommended protective measures for the employee if further exposure is permitted, as well as any recommended limitations upon an employee's use of respirators.

Employers must instruct each physician not to reveal to the employer in writing or in any other way his or her findings, laboratory results, or diagnoses which are felt to be unrelated to occupational lead exposure. They must also instruct each physician to advise the employee of any occupationally or non-occupationally related medical condition requiring further treatment or evaluation.

The standard provides for the use of respirators where engineering and other primary controls are not effective. However, the use of respirator protection shall not be used in lieu of temporary medical removal due to elevated blood lead levels or findings that an employee is at risk of material health impairment. This is based on the numerous inadequacies of respirators including skin rash where the facepiece makes contact with the skin, unacceptable stress to breathing in some workers with underlying cardiopulmonary impairment, difficulty in providing adequate fit, the tendency for respirators to create additional hazards by interfering with vision, hearing, and mobility, and the difficulties of assuring the maximum effectiveness of a complicated work practice program involving respirators. Respirators do, however, serve a useful function where engineering and work practice controls are inadequate by providing supplementary, interim, or short-term protection, provided they are properly selected for the environment in which the employee will be working, properly fitted to the employee, maintained and cleaned periodically, and worn by the employee when required.

In its interim final standard on occupational exposure to inorganic lead in the construction industry, OSHA has prohibited prophylactic chelation. Diagnostic and therapeutic chelation are permitted only under the supervision of a licensed physician with appropriate medical monitoring in an acceptable clinical setting. The decision to initiate chelation therapy must be made on an individual basis and take into account the severity of symptoms felt to be a result of lead toxicity along with blood lead levels, ZPP levels, and other laboratory tests as appropriate. EDTA and penicillamine which are the primary chelating agents used in the therapy of occupational lead poisoning have significant potential side effects and their use must be justified on the basis of expected benefits to the worker. Unless frank and severe symptoms are present, therapeutic chelation is not recommended, given the opportunity to remove a worker from exposure and allow the body to naturally excrete accumulated lead. As a diagnostic aid, the chelation mobilization test using CA-EDTA has limited applicability. According to some investigators, the test can differentiate between lead-induced and other nephropathies. The test may also provide an estimation of the mobile fraction of the total body lead burden.

Employers are required to assure that accurate records are maintained on exposure assessment, including environmental monitoring, medical surveillance, and medical removal for each employee. Exposure assessment records must be kept for at least 30 years. Medical surveillance records must be kept for the duration of employment plus 30 years except in cases where the employment was less than one year. If duration of employment is less than one year, the employer need not retain this record beyond the term of employment if the record is provided to the employee upon termination of employment. Medical removal records also must be maintained for the duration of employment. All records required under the standard must be made available upon request to the Assistant Secretary of Labor for Occupational Safety and Health and the Director of the National Institute for Occupational Safety and Health. Employers must also make environmental and biological monitoring and medical removal records available to affected employees and to former employees or their authorized employee representatives. Employees or their specifically designated representatives have access to their entire medical surveillance records.

In addition, the standard requires that the employer inform all workers exposed to lead at or above 30 $\mu g/m_3$ of the provisions of the standard and all its appendices, the purpose and description of medical surveillance and provisions for medical removal protection if temporary removal is required. An understanding of the potential health effects of lead exposure by all exposed employees along with full understanding of their rights under the lead standard is essential for an effective monitoring program.

## II. Adverse Health Effects of Inorganic Lead

Although the toxicity of lead has been known for 2,000 years, the knowledge of the complex relationship between lead exposure and human response is still being refined. Significant research into the toxic properties of lead continues throughout the world, and it should be anticipated that our understanding of thresholds of effects and margins of safety will be improved in future years. The provisions of the lead standard are founded on two prime medical judgments: First, the prevention of adverse health effects from exposure to lead throughout a working lifetime requires that worker blood lead levels be maintained at or below 40 $\mu g/dl$ and second, the blood lead levels of workers, male or female, who intend to parent in the near future should be maintained below 30 $\mu g/dl$ to minimize adverse reproductive health effects to the parents and developing fetus. The adverse effects of lead on reproduction are being actively researched and OSHA encourages the physician to remain abreast of recent developments in the area to best advise pregnant workers or workers planning to conceive children.

The spectrum of health effects caused by lead exposure can be subdivided into five developmental stages: Normal, physiological changes of uncertain significance, pathophysiological changes, overt symptoms (morbidity), and mortality. Within this process there are no sharp distinctions, but rather a continuum of effects. Boundaries between categories overlap due to the wide variation of individual responses and exposures in the working population. OSHA's development of the lead standard focused on pathophysiological changes as well as later stages of disease.

1. Heme Synthesis Inhibition. The earliest demonstrated effect of lead involves its ability to inhibit at least two enzymes of the heme synthesis pathway at very low blood levels. Inhibition of delta aminolevulinic acid dehydrase (ALA-D) which catalyzes the conversion of delta-aminolevulinic acid (ALA) to protoporphyrin is observed at a blood lead level below 20 $\mu g/dl$. At a blood lead level of 40 $\mu g/dl$, more than 20% of the population would have 70% inhibition of ALA-D. There is an exponential increase in ALA excretion at blood lead levels greater than 40 $\mu g/dl$.

Another enzyme, ferrochelatase, is also inhibited at low blood lead levels. Inhibition of ferrochelatase leads to increased free erythrocyte protoporphyrin (FEP) in the blood which can then bind to zinc to yield zinc protoporphyrin. At a blood lead level of 50 $\mu g/dl$ or greater, nearly 100% of the population will have an increase in FEP. There is also an exponential relationship between blood lead levels greater than 40 μg/dl and the associated ZPP level, which has led to the development of the ZPP screening test for lead exposure.

While the significance of these effects is subject to debate, it is OSHA's position that these enzyme disturbances are early stages of a disease process which may eventually result in the clinical symptoms of lead poisoning. Whether or not the effects do progress to the later stages of clinical disease, disruption of these enzyme processes over a working lifetime is considered to be a material impairment of health.

One of the eventual results of lead-induced inhibition of enzymes in the heme synthesis pathway is anemia which can be asymptomatic if mild but associated with a wide array of symptoms including dizziness, fatigue, and tachycardia when more severe. Studies have indicated that lead levels as low as 50 μg/dl can be associated with a definite decreased hemoglobin, although most cases of lead-induced anemia, as well as shortened red-cell survival times, occur at lead levels exceeding 80 μg/dl. Inhibited hemoglobin synthesis is more common in chronic cases whereas shortened erythrocyte life span is more common in acute cases.

In lead-induced anemias, there is usually a reticulocytosis along with the presence of basophilic stippling, and ringed sideroblasts, although none of the above are pathognomonic for lead-induced anemia.

2. Neurological Effects. Inorganic lead has been found to have toxic effects on both the central and peripheral nervous systems. The earliest stages of lead-induced central nervous system effects first manifest themselves in the form of behavioral disturbances and central nervous system symptoms including irritability, restlessness, insomnia and other sleep disturbances, fatigue, vertigo, headache, poor memory, tremor, depression, and apathy. With more severe exposure, symptoms can progress to drowsiness, stupor, hallucinations, delirium, convulsions and coma.

The most severe and acute form of lead poisoning which usually follows ingestion or inhalation of large amounts of lead is acute encephalopathy which may arise precipitously with the onset of intractable seizures, coma, cardiorespiratory arrest, and death within 48 hours.

While there is disagreement about what exposure levels are needed to produce the earliest symptoms, most experts agree that symptoms definitely can occur at blood lead levels of 60 μg/dl whole blood and therefore recommend a 40 μg/dl maximum. The central nervous system effects frequently are not reversible following discontinued exposure or chelation therapy and when improvement does occur, it is almost always only partial.

The peripheral neuropathy resulting from lead exposure characteristically involves only motor function with minimal sensory damage and has a marked predilection for the extensor muscles of the most active extremity. The peripheral neuropathy can occur with varying degrees of severity. The earliest and mildest form which can be detected in workers with blood lead levels as low as 50 μg/dl is manifested by slowing of motor nerve conduction velocity often without clinical symptoms. With progression of the neuropathy there is development of painless extensor muscle weakness usually involving the extensor muscles of the fingers and hand in the most active upper extremity, followed in severe cases by wrist drop or, much less commonly, foot drop.

In addition to slowing of nerve conduction, electromyographical studies in patients with blood lead levels greater than 50 μg/dl have demonstrated a decrease in the number of acting motor unit potentials, an increase in the duration of motor unit potentials, and spontaneous pathological activity including fibrillations and fasciculations. Whether these effects occur at levels of 40 μg/dl is undetermined.

While the peripheral neuropathies can occasionally be reversed with therapy, again such recovery is not assured particularly in the more severe neuropathies and often improvement is only partial. The lack of reversibility is felt to be due in part to segmental demyelination.

3. Gastrointestinal. Lead may also affect the gastrointestinal system producing abdominal colic or diffuse abdominal pain, constipation, obstipation, diarrhea, anorexia, nausea and vomiting. Lead colic rarely develops at blood lead levels below 80 μg/dl.

4. Renal. Renal toxicity represents one of the most serious health effects of lead poisoning. In the early stages of disease nuclear inclusion bodies can frequently be identified in proximal renal tubular cells. Renal function remains normal and the changes in this stage are probably reversible. With more advanced disease there is progressive interstitial fibrosis and impaired renal function. Eventually extensive interstitial fibrosis ensues with sclerotic glomeruli and dilated and atrophied proximal tubules; all represent end stage kidney disease. Azotemia can be progressive, eventually resulting in frank uremia necessitating dialysis. There is occasionally associated hypertension and hyperuricemia with or without gout.

Early kidney disease is difficult to detect. The urinalysis is normal in early lead nephropathy and the blood urea nitrogen and serum creatinine increase only when two-thirds of kidney function is lost. Measurement of creatinine clearance can often detect earlier disease as can other methods of measurement of glomerular filtration rate. An abnormal Ca-EDTA mobilization test has been used to differentiate between lead-induced and other nephropathies, but this procedure is not widely accepted. A form of Fanconi syndrome with aminoaciduria, glycosuria, and hyperphosphaturia indicating severe injury to the proximal renal tubules is occasionally seen in children.

5. Reproductive effects. Exposure to lead can have serious effects on reproductive function in both males and females. In male workers exposed to lead there can be a decrease in sexual drive, impotence, decreased ability to produce healthy sperm, and sterility. Malformed sperm (teratospermia), decreased number of sperm (hypospermia), and sperm with decreased motility (asthenospermia) can all occur. Teratospermia has been noted at mean blood lead levels of 53 μg/dl and hypospermia and asthenospermia at 41 μg/dl. Furthermore, there appears to be a dose-response relationship for teratospermia in lead exposed workers.

Women exposed to lead may experience menstrual disturbances including dysmenorrhea, menorrhagia and amenorrhea. Following exposure to lead, women have a higher frequency of sterility, premature births, spontaneous miscarriages, and stillbirths.

Germ cells can be affected by lead and cause genetic damage in the egg or sperm cells before conception and result in failure to implant, miscarriage, stillbirth, or birth defects.

Infants of mothers with lead poisoning have a higher mortality during the first year and suffer from lowered birth weights, slower growth, and nervous system disorders.

Lead can pass through the placental barrier and lead levels in the mother's blood are comparable to concentrations of lead in the umbilical cord at birth. Transplacental passage becomes detectable at 12-14 weeks of gestation and increases until birth.

There is little direct data on damage to the fetus from exposure to lead but it is generally assumed that the fetus and newborn would be at least as susceptible to neurological damage as young children. Blood lead levels of 50-60 μg/dl in children can cause significant neurobehavioral impairments and there is evidence of hyperactivity at blood levels as low as 25 μg/dl. Given the overall body of literature concerning the adverse health effects of lead in children, OSHA feels that the blood lead level in children should be maintained below 30 μg/dl with a population mean of 15 μg/dl. Blood lead levels in the fetus and newborn likewise should not exceed 30 μg/dl.

Because of lead's ability to pass through the placental barrier and also because of the demonstrated adverse effects of lead on reproductive function in both the male and female as well as the risk of genetic damage of lead on both the ovum and sperm, OSHA recommends a 30 μg/dl maximum permissible blood lead level in both males and females who wish to bear children.

6. Other toxic effects. Debate and research continue on the effects of lead on the human body. Hypertension has frequently been noted in occupationally exposed individuals although it is difficult to assess whether this is due to lead's adverse effects on the kidney or if some other mechanism is involved. Vascular and electrocardiographic changes have been detected but have not been well characterized. Lead is thought to impair thyroid function and interfere with the pituitary-adrenal axis, but again these effects have not been well defined.

## III. Medical Evaluation

The most important principle in evaluating a worker for any occupational disease including lead poisoning is a high index of suspicion on the part of the examining physician. As discussed in Section 2, lead can affect numerous organ systems and produce a wide array of signs and symptoms, most of which are non-specific and subtle in nature at least in the early stages of disease. Unless serious concern for lead toxicity is present, many of the early clues to diagnosis may easily be overlooked.

The crucial initial step in the medical evaluation is recognizing that a worker's employment can result in exposure to lead. The worker will frequently be able to define exposures to lead and lead containing materials but often will not volunteer this information unless specifically asked. In other situations the worker may not know of any exposures to lead but the suspicion might be raised on the part of the physician because of the industry or occupation of the worker. Potential occupational exposure to lead and its compounds occur in many occupations in the construction industry, including demolition and salvaging operations, removal or encapsulation of materials containing lead, construction, alteration, repair or renovation of structures containing lead, transportation, disposal, storage or containment of lead or lead-containing materials on construction sites, and maintenance operations associated with construction activities.

Once the possibility for lead exposure is raised, the focus can then be directed toward eliciting information from the medical history, physical exam, and finally from laboratory data to evaluate the worker for potential lead toxicity.

A complete and detailed work history is important in the initial evaluation. A listing of all previous employment with information on job description, exposure to fumes or dust, known exposures to lead or other toxic substances, a description of any personal protective equipment used, and previous medical surveillance should all be included in the worker's record. Where exposure to lead is suspected, information concerning on-the-job personal hygiene, smoking or eating habits in work areas, laundry procedures, and use of any protective clothing or respiratory protection equipment should be noted. A complete work history is essential in the medical evaluation of a worker with suspected lead toxicity, especially when long term effects such as neurotoxicity and nephrotoxicity are considered.

The medical history is also of fundamental importance and should include a listing of all past and current medical conditions, current medications including proprietary drug intake, previous surgeries and hospitalizations, allergies, smoking history, alcohol consumption, and also non-occupational lead exposures such as hobbies (hunting, riflery). Also known childhood exposures should be elicited. Any previous history of hematological, neurological, gastrointestinal, renal, psychological, gynecological, genetic, or reproductive problems should be specifically noted.

A careful and complete review of systems must be performed to assess both recognized complaints and subtle or slowly acquired symptoms which the worker might not appreciate as being significant. The review of symptoms should include the following:

1. General—weight loss, fatigue, decreased appetite.

2. Head, Eyes, Ears, Nose, Throat (HEENT)—headaches, visual disturbances or decreased visual acuity, hearing deficits or tinnitus, pigmentation of the oral mucosa, or metallic taste in mouth.

3. Cardio-pulmonary—shortness of breath, cough, chest pains, palpitations, or orthopnea.

4. Gastrointestinal—nausea, vomiting, heartburn, abdominal pain, constipation or diarrhea.

5. Neurologic—irritability, insomnia, weakness (fatigue), dizziness, loss of memory, confusion, hallucinations, incoordination, ataxia, decreased strength in hands or feet, disturbances in gait, difficulty in climbing stairs, or seizures.

6. Hematologic—pallor, easy fatigability, abnormal blood loss, melena.

7. Reproductive (male and female and spouse where relevant)—history of infertility, impotence, loss of libido, abnormal menstrual periods, history of miscarriages, stillbirths, or children with birth defects.

8. Musculo-skeletal—muscle and joint pains.

The physical examination should emphasize the neurological, gastrointestinal, and cardiovascular systems. The worker's weight and blood pressure should be recorded and the oral mucosa checked for pigmentation characteristic of a possible Burtonian or lead line on the gingiva. It should be noted, however, that the lead line may not be present even in severe lead poisoning if good oral hygiene is practiced.

The presence of pallor on skin examination may indicate an anemia which, if severe, might also be associated with a tachycardia. If an anemia is suspected, an active search for blood loss should be undertaken including potential blood loss through the gastrointestinal tract.

A complete neurological examination should include an adequate mental status evaluation including a search for behavioral and psychological disturbances, memory testing, evaluation for irritability, insomnia, hallucinations, and mental clouding. Gait and coordination should be examined along with close observation for tremor. A detailed evaluation of peripheral nerve function including careful sensory and motor function testing is warranted. Strength testing particularly of extensor muscle groups of all extremities is of fundamental importance.

Cranial nerve evaluation should also be included in the routine examination.

The abdominal examination should include auscultation for bowel sounds and abdominal bruits and palpation for organomegaly, masses, and diffuse abdominal tenderness.

Cardiovascular examination should evaluate possible early signs of congestive heart failure. Pulmonary status should be addressed particularly if respirator protection is contemplated.

As part of the medical evaluation, the interim lead standard requires the following laboratory studies:

1. Blood lead level

2. Hemoglobin and hematocrit determinations, red cell indices, and examination of the peripheral blood smear to evaluate red blood cell morphology

3. Blood urea nitrogen

4. Serum creatinine

5. Routine urinalysis with microscopic examination.

6. A zinc protoporphyrin level.

In addition to the above, the physician is authorized to order any further laboratory or other tests which he or she deems necessary in accordance with sound medical practice. The evaluation must also include pregnancy testing or laboratory evaluation of male fertility if requested by the employee. Additional tests which are probably not warranted on a routine basis but may be appropriate when blood lead and ZPP levels are equivocal include delta aminolevulinic acid and coproporphyrin concentrations in the urine, and dark-field illumination for detection of basophilic stippling in red blood cells.

If an anemia is detected further studies including a careful examination of the peripheral smear, reticulocyte count, stool for occult blood, serum iron, total iron binding capacity, bilirubin, and, if appropriate, vitamin B12 and folate may be of value in attempting to identify the cause of the anemia.

If a peripheral neuropathy is suspected, nerve conduction studies are warranted both for diagnosis and as a basis to monitor any therapy.

If renal disease is questioned, a 24 hour urine collection for creatinine clearance, protein, and electrolytes may be indicated. Elevated uric acid levels may result from lead-induced renal disease and a serum uric acid level might be performed.

An electrocardiogram and chest x-ray may be obtained as deemed appropriate.

Sophisticated and highly specialized testing should not be done routinely and where indicated should be under the direction of a specialist.

## IV. Laboratory Evaluation

The blood lead level at present remains the single most important test to monitor lead exposure and is the test used in the medical surveillance program under the lead standard to guide employee medical removal. The ZPP has several advantages over the blood lead level. Because of its relatively recent development and the lack of extensive data concerning its interpretation, the ZPP currently remains an ancillary test.

This section will discuss the blood lead level and ZPP in detail and will outline their relative advantages and disadvantages. Other blood tests currently available to evaluate lead exposure will also be reviewed.

The blood lead level is a good index of current or recent lead absorption when there is no anemia present and when the worker has not taken any chelating agents. However, blood lead levels along with urinary lead levels do not necessarily indicate the total body burden of lead and are not adequate measures of past exposure. One reason for this is that lead has a high affinity for bone and up to 90% of the body's total lead is deposited there. A very important component of the total lead body burden is lead in soft tissue (liver, kidney, and brain). This fraction of the lead body burden, the biologically active lead, is not entirely reflected by blood lead levels since it is a function of the dynamics of lead absorption, distribution, deposition in bone and excretion. Following discontinuation of exposure to lead, the excess body burden is only slowly mobilized from bone and other relatively stable body stores and excreted. Consequently, a high blood lead level may only represent recent heavy exposure to lead without a significant total body excess and likewise a low blood lead level does not exclude an elevated total body burden of lead.

Also due to its correlation with recent exposures, the blood lead level may vary considerably over short time intervals.

To minimize laboratory error and erroneous results due to contamination, blood specimens must be carefully collected after thorough cleaning of the skin with appropriate methods using lead-free blood containers and analyzed by a reliable laboratory. Under the standard, samples must be analyzed in laboratories which are approved by OSHA. Analysis is to be made using atomic absorption spectrophotometry, anodic stripping voltammetry or any method which meets the accuracy requirements set forth by the standard.

The determination of lead in urine is generally considered a less reliable monitoring technique than analysis of whole blood primarily due to individual variability in urinary excretion capacity as well as the technical difficulty of obtaining accurate 24 hour urine collections. In addition, workers with renal insufficiency, whether due to lead or some other cause, may have decreased lead clearance and consequently urine lead levels may underestimate the true lead burden. Therefore, urine lead levels should not be used as a routine test.

The zinc protoporphyrin test, unlike the blood lead determination, measures an adverse metabolic effect of lead and as such is a better indicator of lead toxicity than the level of blood lead itself. The level of ZPP reflects lead absorption over the preceding 3 to 4 months, and therefore is a better indicator of lead body burden. The ZPP requires more time than the blood lead to read significantly elevated levels; the return to normal after discontinuing lead exposure is also slower. Furthermore, the ZPP test is simpler, faster, and less expensive to perform and no contamination is possible. Many investigators believe it is the most reliable means of monitoring chronic lead absorption.

Zinc protoporphyrin results from the inhibition of the enzyme ferrochelatase which catalyzes the insertion of an iron molecule into the protoporphyrin molecule, which then becomes heme. If iron is not inserted into the molecule then zinc, having a greater affinity for protoporphyrin, takes the place of the iron, forming ZPP.

An elevation in the level of circulating ZPP may occur at blood lead levels as low as 20-30 μg/dl in some workers. Once the blood lead level has reached 40 μg/dl there is more marked rise in the ZPP value from its normal range of less than 100 μg/dl100 ml. Increases in blood lead levels beyond 40 μg/100 g are associated with exponential increases in ZPP.

Whereas blood lead levels fluctuate over short time spans, ZPP levels remain relatively stable. ZPP is measured directly in red blood cells and is present for the cell's entire 120 day life-span. Therefore, the ZPP level in blood reflects the average ZPP production over the previous 3-4 months and consequently the average lead exposure during that time interval.

It is recommended that a hematocrit be determined whenever a confirmed ZPP of 50 μg/100 ml whole blood is obtained to rule out a significant underlying anemia. If the ZPP is in excess of 100 μg/100 ml and not associated with abnormal elevations in blood lead levels, the laboratory should be checked to be sure that blood leads were determined using atomic absorption spectrophotometry anodic stripping voltammetry, or any method which meets the accuracy requirements set forth by the standard by an OSHA approved laboratory which is experienced in lead level determinations. Repeat periodic blood lead studies should be obtained in all individuals with elevated ZPP levels to be certain that an associated elevated blood lead level has not been missed due to transient fluctuations in blood leads.

ZPP has a characteristic fluorescence spectrum with a peak at 594 nm which is detectable with a hematofluorimeter. The hematofluorimeter is accurate and portable and can provide on-site, instantaneous results for workers who can be frequently tested via a finger prick.

However, careful attention must be given to calibration and quality control procedures. Limited data on blood lead-ZPP correlations and the ZPP levels which are associated with the adverse health effects discussed in Section 2 are the major limitations of the test. Also it is difficult to correlate ZPP levels with environmental exposure and there is some variation of response with age and sex. Nevertheless, the ZPP promises to be an important diagnostic test for the early detection of lead toxicity and its value will increase as more data is collected regarding its relationship to other manifestations of lead poisoning.

Levels of delta-aminolevulinic acid (ALA) in the urine are also used as a measure of lead exposure. Increasing concentrations of ALA are believed to result from the inhibition of the enzyme delta-aminolevulinic acid dehydrase (ALA-D). Although the test is relatively easy to perform, inexpensive, and rapid, the disadvantages include variability in results, the necessity to collect a complete 24 hour urine sample which has a specific gravity greater than 1.010, and also the fact that ALA decomposes in the presence of light.

The pattern of porphyrin excretion in the urine can also be helpful in identifying lead intoxication. With lead poisoning, the urine concentrations of coproporphyrins I and II, porphobilinogen and uroporphyrin I rise. The most important increase, however, is that of coproporphyrin III; levels may exceed 5,000 μg/l in the urine in lead poisoned individuals, but its correlation with blood lead levels and ZPP are not as good as those of ALA. Increases in urinary porphyrins are not diagnostic of lead toxicity and may be seen in porphyria, some liver diseases, and in patients with high reticulocyte counts.

Summary. The Occupational Safety and Health Administration's interim standard for inorganic lead in the construction industry places significant emphasis on the medical surveillance of all workers exposed to levels of inorganic lead above 30 $\mu g/m^3$ TWA. The physician has a fundamental role in this surveillance program, and in the operation of the medical removal protection program.

Even with adequate worker education on the adverse health effects of lead and appropriate training in work practices, personal hygiene and other control measures, the physician has a primary responsibility for evaluating potential lead toxicity in the worker. It is only through a careful and detailed medical and work history, a complete physical examination and appropriate laboratory testing that an accurate assessment can be made. Many of the adverse health effects of lead toxicity are either irreversible or only partially reversible and therefore early detection of disease is very important.

This document outlines the medical monitoring program as defined by the occupational safety and health standard for inorganic lead. It reviews the adverse health effects of lead poisoning and describes the important elements of the history and physical examinations as they relate to these adverse effects. Finally, the appropriate laboratory testing for evaluating lead exposure and toxicity is presented.

It is hoped that this review and discussion will give the physician a better understanding of the OSHA standard with the ultimate goal of protecting the health and well-being of the worker exposed to lead under his or her care.

[58 FR 26627, May 4, 1993, as amended at 58 FR 34218, June 24, 1993; 61 FR 5510, Feb. 13, 1996; 63 FR 1296, Jan. 8, 1998]

## §1926.64 Process safety management of highly hazardous chemicals.

*Purpose.* This section contains requirements for preventing or minimizing the consequences of catastrophic releases of toxic, reactive, flammable, or explosive chemicals. These releases may result in toxic, fire or explosion hazards.

**1926.64(a)** ***Application.*** (1) This section applies to the following:

**1926.64(a)(i)** A process which involves a chemical at or above the specified threshold quantities listed in Appendix A to this section;

**1926.64(a)(ii)** A process which involves a flammable liquid or gas (as defined in §1926.59(c) of this part) on site in one location, in a quantity of 10,000 pounds (4535.9 kg) or more except for:

**1926.64(a)(ii)(A)** Hydrocarbon fuels used solely for workplace consumption as a fuel (e.g., propane used for comfort heating, gasoline for vehicle refueling), if such fuels are not a part of a process containing another highly hazardous chemical covered by this standard;

**1926.64(a)(ii)(B)** Flammable liquids stored in atmospheric tanks or transferred which are kept below their normal boiling point without benefit of chilling or refrigeration.

**1926.64(a)(2)** This section does not apply to:

**1926.64(a)(2)(i)** Retail facilities;

**1926.64(a)(2)(ii)** Oil or gas well drilling or servicing operations; or,

**1926.64(a)(2)(iii)** Normally unoccupied remote facilities.

**1926.64(b)** ***Definitions.*** *Atmospheric tank* means a storage tank which has been designed to operate at pressures from atmospheric through 0.5 p.s.i.g. (pounds per square inch gauge, 3.45 Kpa).

*Boiling point* means the boiling point of a liquid at a pressure of 14.7 pounds per square inch absolute (p.s.i.a.) (760 mm.). For the purposes of this section, where an accurate boiling point is unavailable for the material in question, or for mixtures which do not have a constant boiling point, the 10 percent point of a distillation performed in accordance with the Standard Method of Test for Distillation of Petroleum Products, ASTM D-86-62, may be used as the boiling point of the liquid.

*Catastrophic release* means a major uncontrolled emission, fire, or explosion, involving one or more highly hazardous chemicals, that presents serious danger to employees in the workplace.

*Facility* means the buildings, containers or equipment which contain a process.

*Highly hazardous chemical* means a substance possessing toxic, reactive, flammable, or explosive properties and specified by paragraph (a)(1) of this section.

*Hot work* means work involving electric or gas welding, cutting, brazing, or similar flame or spark-producing operations.

*Normally unoccupied remote facility* means a facility which is operated, maintained or serviced by employees who visit the facility only periodically to check its operation and to perform necessary operating or maintenance tasks. No employees are permanently stationed at the facility. Facilities meeting this definition are not contiguous with, and must be geographically remote from all other buildings, processes or persons.

*Process* means any activity involving a highly hazardous chemical including any use, storage, manufacturing, handling, or the on-site movement of such chemicals, or combination of these activities. For purposes of this definition, any group of vessels which are interconnected and separate vessels which are located such that a highly hazardous chemical could be involved in a potential release shall be considered a single process.

*Replacement in kind* means a replacement which satisfies the design specification.

*Trade secret* means any confidential formula, pattern, process, device, information or compilation of information that is used in an employer's business, and that gives the employer an opportunity to obtain an advantage over competitors who do not know or use it. Appendix D contained in §1926.59 sets out the criteria to be used in evaluating trade secrets.

**1926.64(c)** ***Employee participation.*** (1) Employers shall develop a written plan of action regarding the implementation of the employee participation required by this paragraph.

**1926.64(c)(2)** Employers shall consult with employees and their representatives on the conduct and development of process hazards analyses and on the development of the other elements of process safety management in this standard.

**1926.64(c)(3)** Employers shall provide to employees and their representatives access to process hazard analyses and to all other information required to be developed under this standard.

**1926.64(d)** ***Process safety information.*** In accordance with the schedule set forth in paragraph (e)(1) of this section, the employer shall complete a compilation of written process safety information before conducting any process hazard analysis required by the standard. The compilation of written process safety information is to enable the employer and the employees involved in operating the process to identify and understand the hazards posed by those processes involving highly hazardous chemicals. This process safety information shall include information pertaining to the hazards of the highly hazardous chemicals used or produced by the process, information pertaining to the technology of the process, and information pertaining to the equipment in the process.

**1926.64(d)(1)** *Information pertaining to the hazards of the highly hazardous chemicals in the process.* This information shall consist of at least the following:

**1926.64(d)(1)(i)** Toxicity information;

**1926.64(d)(1)(ii)** Permissible exposure limits;

**1926.64(d)(1)(iii)** Physical data;

**1926.64(d)(1)(iv)** Reactivity data:

**1926.64(d)(1)(v)** Corrosivity data;

**1926.64(d)(1)(vi)** Thermal and chemical stability data; and

**1926.64(d)(1)(vii)** Hazardous effects of inadvertent mixing of different materials that could foreseeably occur.

**Note:** Material Safety Data Sheets meeting the requirements of 29 CFR 1926.59(g) may be used to comply with this requirement to the extent they contain the information required by this subparagraph.

**1926.64(d)(2)** *Information pertaining to the technology of the process.* (i) Information concerning the technology of the process shall include at least the following:

**1926.64(d)(2)(i)(A)** A block flow diagram or simplified process flow diagram (see Appendix B to this section);

**1926.64(d)(2)(i)(B)** Process chemistry;

**1926.64(d)(2)(i)(C)** Maximum intended inventory;

**1926.64(d)(2)(i)(D)** Safe upper and lower limits for such items as temperatures, pressures, flows or compositions; and,

**1926.64(d)(2)(i)(E)** An evaluation of the consequences of deviations, including those affecting the safety and health of employees.

**1926.64(d)(2)(ii)** Where the original technical information no longer exists, such information may be developed in conjunction with the process hazard analysis in sufficient detail to support the analysis.

**1926.64(d)(3)** *Information pertaining to the equipment in the process.* (i) Information pertaining to the equipment in the process shall include:

**1926.64(d)(3)(i)(A)** Materials of construction;

**1926.64(d)(3)(i)(B)** Piping and instrument diagrams (P&ID's);

**1926.64(d)(3)(i)(C)** Electrical classification;

**1926.64(d)(3)(i)(D)** Relief system design and design basis;

**1926.64(d)(3)(i)(E)** Ventilation system design;

**1926.64(d)(3)(i)(F)** Design codes and standards employed;

**1926.64(d)(3)(i)(G)** Material and energy balances for processes built after May 26, 1992; and,

**1926.64(d)(3)(i)(H)** Safety systems (e.g. interlocks, detection or suppression systems).

**1926.64(d)(3)(ii)** The employer shall document that equipment complies with recognized and generally accepted good engineering practices.

**1926.64(d)(3)(iii)** For existing equipment designed and constructed in accordance with codes, standards, or practices that are no longer in general use, the employer shall determine and document that the equipment is designed, maintained, inspected, tested, and operating in a safe manner.

**1926.64(e)** ***Process hazard analysis.*** (1) The employer shall perform an initial process hazard analysis (hazard evaluation) on processes covered by this standard. The process hazard analysis shall be appropriate to the complexity of the process and shall identify, evaluate, and control the hazards involved in the process. Employers shall determine and document the priority order for conducting process hazard analyses based on a rationale which includes such considerations as extent of the process hazards, number of potentially affected employees, age of the process, and operating history of the process. The process hazard analysis shall be conducted as soon as possible, but not later than the following schedule:

**1926.64(e)(1)(i)** No less than 25 percent of the initial process hazards analyses shall be completed by May 26, 1994;

**1926.64(e)(1)(ii)** No less than 50 percent of the initial process hazards analyses shall be completed by May 26, 1995;

**1926.64(e)(1)(iii)** No less than 75 percent of the initial process hazards analyses shall be completed by May 26, 1996;

**1926.64(e)(1)(iv)** All initial process hazards analyses shall be completed by May 26, 1997.

**1926.64(e)(1)(v)** Process hazards analyses completed after May 26, 1987 which meet the requirements of this paragraph are acceptable as initial process hazards analyses. These process hazard analyses shall be updated and revalidated, based on their completion date, in accordance with paragraph (e)(6) of this standard.

**1926.64(e)(2)** The employer shall use one or more of the following methodologies that are appropriate to determine and evaluate the hazards of the process being analyzed.

**1926.64(e)(2)(i)** What-If;

**1926.64(e)(2)(ii)** Checklist;

**1926.64(e)(2)(iii)** What-If/Checklist;

**1926.64(e)(2)(iv)** Hazard and Operability Study (HAZOP);

**1926.64(e)(2)(v)** Failure Mode and Effects Analysis (FMEA);

**1926.64(e)(2)(vi)** Fault-Tree Analysis; or

**1926.64(e)(2)(vii)** An appropriate equivalent methodology.

**1926.64(e)(3)** The process hazard analysis shall address:

**1926.64(e)(3)(i)** The hazards of the process;

**1926.64(e)(3)(ii)** The identification of any previous incident which had a likely potential for catastrophic consequences in the workplace;

**1926.64(e)(3)(iii)** Engineering and administrative controls applicable to the hazards and their interrelationships such as appropriate application of detection methodologies to provide early warning of releases. (Acceptable detection methods might include process monitoring and control instrumentation with alarms, and detection hardware such as hydrocarbon sensors.);

**1926.64(e)(3)(iv)** Consequences of failure of engineering and administrative controls;

**1926.64(e)(3)(v)** Facility siting;

**1926.64(e)(3)(vi)** Human factors; and

**1926.64(e)(3)(vii)** A qualitative evaluation of a range of the possible safety and health effects of failure of controls on employees in the workplace.

**1926.64(e)(4)** The process hazard analysis shall be performed by a team with expertise in engineering and process operations, and the team shall include at least one employee who has experience and knowledge specific to the process being evaluated. Also, one member of the team must be knowledgeable in the specific process hazard analysis methodology being used.

**1926.64(e)(5)** The employer shall establish a system to promptly address the team's findings and recommendations; assure that the recommendations are resolved in a timely manner and that the resolution is documented; document what actions are to be taken; complete actions as soon as possible; develop a written schedule of when these actions are to be completed; communicate the actions to operating, maintenance and other employees whose work assignments are in the process and who may be affected by the recommendations or actions.

**1926.64(e)(6)** At least every five (5) years after the completion of the initial process hazard analysis, the process hazard analysis shall be updated and revalidated by a team meeting the requirements in paragraph (e)(4) of this section, to assure that the process hazard analysis is consistent with the current process.

**1926.64(e)(7)** Employers shall retain process hazards analyses and updates or revalidations for each process covered by this section, as well as the documented resolution of recommendations described in paragraph (e)(5) of this section for the life of the process.

**1926.64(f)** ***Operating procedures.*** (1) The employer shall develop and implement written operating procedures that provide clear instructions for safely conducting activities involved in each covered process consistent with the process safety information and shall address at least the following elements.

**1926.64(f)(1)(i)** Steps for each operating phase:

**1926.64(f)(1)(i)(A)** Initial startup;

**1926.64(f)(1)(i)(B)** Normal operations;

**1926.64(f)(1)(i)(C)** Temporary operations;

**1926.64(f)(1)(i)(D)** Emergency shutdown including the conditions under which emergency shutdown is required, and the assignment of shutdown responsibility to qualified operators to ensure that emergency shutdown is executed in a safe and timely manner.

**1926.64(f)(1)(i)(E)** Emergency operations;

**1926.64(f)(1)(i)(F)** Normal shutdown; and,

**1926.64(f)(1)(i)(G)** Startup following a turnaround, or after an emergency shutdown.

**1926.64(f)(1)(ii)** Operating limits:

**1926.64(f)(1)(ii)(A)** Consequences of deviation; and

**1926.64(f)(1)(ii)(B)** Steps required to correct or avoid deviation.

**1926.64(f)(1)(iii)** Safety and health considerations:

**1926.64(f)(1)(iii)(A)** Properties of, and hazards presented by, the chemicals used in the process;

**1926.64(f)(1)(iii)(B)** Precautions necessary to prevent exposure, including engineering controls, administrative controls, and personal protective equipment;

**1926.64(f)(1)(iii)(C)** Control measures to be taken if physical contact or airborne exposure occurs;

**1926.64(f)(1)(iii)(D)** Quality control for raw materials and control of hazardous chemical inventory levels; and,

**1926.64(f)(1)(iii)(E)** Any special or unique hazards.

**1926.64(f)(1)(iv)** Safety systems and their functions.

**1926.64(f)(2)** Operating procedures shall be readily accessible to employees who work in or maintain a process.

**1926.64(f)(3)** The operating procedures shall be reviewed as often as necessary to assure that they reflect current operating practice, including changes that result from changes in process chemicals, technology, and equipment, and changes to facilities. The employer shall certify annually that these operating procedures are current and accurate.

**1926.64(f)(4)** The employer shall develop and implement safe work practices to provide for the control of hazards during operations such as lockout/tagout; confined space entry; opening process equipment or piping; and control over entrance into a facility by maintenance, contractor, laboratory, or other support personnel. These safe work practices shall apply to employees and contractor employees.

**1926.64(g)** ***Training***— (1) Initial training. (i) Each employee presently involved in operating a process, and each employee before being involved in operating a newly assigned process, shall be trained in an overview of the process and in the operating procedures as specified in paragraph (f) of this section. The training shall include emphasis on the specific safety and health hazards, emergency operations including shutdown, and safe work practices applicable to the employee's job tasks.

**1926.64(g)(1)(ii)** In lieu of initial training for those employees already involved in operating a process on May 26, 1992, an employer may certify in writing that the employee has the required knowledge, skills, and abilities to safely carry out the duties and responsibilities as specified in the operating procedures.

**1926.64(g)(2)** *Refresher training.* Refresher training shall be provided at least every three years, and more often if necessary, to each employee involved in operating a process to assure that the employee understands and adheres to the current operating procedures of the process. The employer, in consultation with the employees involved in operating the process, shall determine the appropriate frequency of refresher training.

**1926.64(g)(3)** *Training documentation.* The employer shall ascertain that each employee involved in operating a process has received and understood the training required by this paragraph. The employer shall prepare a record which contains the identity of the employee, the date of training, and the means used to verify that the employee understood the training.

**1926.64(h)** ***Contractors***— (1) Application. This paragraph applies to contractors performing maintenance or repair, turnaround, major renovation, or specialty work on or adjacent to a covered process. It does not apply to contractors providing incidental services which do not influence process safety, such as janitorial work, food and drink services, laundry, delivery or other supply services.

**1926.64(h)(2)** *Employer responsibilities.* (i) The employer, when selecting a contractor, shall obtain and evaluate information regarding the contract employer's safety performance and programs.

**1926.64(h)(2)(ii)** The employer shall inform contract employers of the known potential fire, explosion, or toxic release hazards related to the contractor's work and the process.

**1926.64(h)(2)(iii)** The employer shall explain to contract employers the applicable provisions of the emergency action plan required by paragraph (n) of this section.

**1926.64(h)(2)(iv)** The employer shall develop and implement safe work practices consistent with paragraph (f)(4) of this section, to control the entrance, presence and exit of contract employers and contract employees in covered process areas.

**1926.64(h)(2)(v)** The employer shall periodically evaluate the performance of contract employers in fulfilling their obligations as specified in paragraph (h)(3) of this section.

**1926.64(h)(2)(vi)** The employer shall maintain a contract employee injury and illness log related to the contractor's work in process areas.

**1926.64(h)(3)** *Contract employer responsibilities.* (i) The contract employer shall assure that each contract employee is trained in the work practices necessary to safely perform his/her job.

**1926.64(h)(3)(ii)** The contract employer shall assure that each contract employee is instructed in the known potential fire, explosion, or toxic release hazards related to his/her job and the process, and the applicable provisions of the emergency action plan.

**1926.64(h)(3)(iii)** The contract employer shall document that each contract employee has received and understood the training required by this paragraph. The contract employer shall prepare a record which contains the identity of the contract employee, the date of training, and the means used to verify that the employee understood the training.

**1926.64(h)(3)(iv)** The contract employer shall assure that each contract employee follows the safety rules of the facility including the safe work practices required by paragraph (f)(4) of this section.

**1926.64(h)(3)(v)** The contract employer shall advise the employer of any unique hazards presented by the contract employer's work, or of any hazards found by the contract employer's work.

**1926.64(i)** ***Pre-startup safety review.*** (1) The employer shall perform a pre-startup safety review for new facilities and for modified facilities when the modification is significant enough to require a change in the process safety information.

**1926.64(i)(2)** The pre-startup safety review shall confirm that prior to the introduction of highly hazardous chemicals to a process:

**1926.64(i)(2)(i)** Construction and equipment is in accordance with design specifications;

**1926.64(i)(2)(ii)** Safety, operating, maintenance, and emergency procedures are in place and are adequate;

**1926.64(i)(2)(iii)** For new facilities, a process hazard analysis has been performed and recommendations have been resolved or implemented before startup; and modified facilities meet the requirements contained in management of change, paragraph (l).

**1926.64(i)(2)(iv)** Training of each employee involved in operating a process has been completed.

**1926.64(j)** ***Mechanical integrity—*** (1) Application. Paragraphs (j)(2) through (j)(6) of this section apply to the following process equipment:

**1926.64(j)(1)(i)** Pressure vessels and storage tanks;

**1926.64(j)(1)(ii)** Piping systems (including piping components such as valves);

**1926.64(j)(1)(iii)** Relief and vent systems and devices;

**1926.64(j)(1)(iv)** Emergency shutdown systems;

**1926.64(j)(1)(v)** Controls (including monitoring devices and sensors, alarms, and interlocks) and,

**1926.64(j)(1)(vi)** Pumps.

**1926.64(j)(2)** *Written procedures.* The employer shall establish and implement written procedures to maintain the on-going integrity of process equipment.

**1926.64(j)(3)** *Training for process maintenance activities.* The employer shall train each employee involved in maintaining the on-going integrity of process equipment in an overview of that process and its hazards and in the procedures applicable to the employee's job tasks to assure that the employee can perform the job tasks in a safe manner.

**1926.64(j)(4)** *Inspection and testing.* (i) Inspections and tests shall be performed on process equipment.

**1926.64(j)(4)(ii)** Inspection and testing procedures shall follow recognized and generally accepted good engineering practices.

**1926.64(j)(4)(iii)** The frequency of inspections and tests of process equipment shall be consistent with applicable manufacturers' recommendations and good engineering practices, and more frequently if determined to be necessary by prior operating experience.

**1926.64(j)(4)(iv)** The employer shall document each inspection and test that has been performed on process equipment. The documentation shall identify the date of the inspection or test, the name of the person who performed the inspection or test, the serial number or other identifier of the equipment on which the inspection or test was performed, a description of the inspection or test performed, and the results of the inspection or test.

**1926.64(j)(5)** *Equipment deficiencies.* The employer shall correct deficiencies in equipment that are outside acceptable limits (defined by the process safety information in paragraph (d) of this section) before further use or in a safe and timely manner when necessary means are taken to assure safe operation.

**1926.64(j)(6)** *Quality assurance.* (i) In the construction of new plants and equipment, the employer shall assure that equipment as it is fabricated is suitable for the process application for which they will be used.

**1926.64(j)(6)(ii)** Appropriate checks and inspections shall be performed to assure that equipment is installed properly and consistent with design specifications and the manufacturer's instructions.

**1926.64(j)(6)(iii)** The employer shall assure that maintenance materials, spare parts and equipment are suitable for the process application for which they will be used.

**1926.64(k)** ***Hot work permit.*** (1) The employer shall issue a hot work permit for hot work operations conducted on or near a covered process.

**1926.64(k)(2)** The permit shall document that the fire prevention and protection requirements in 29 CFR 1926.352 have been implemented prior to beginning the hot work operations; it shall indicate the date(s) authorized for hot work; and identify the object on which hot work is to be performed. The permit shall be kept on file until completion of the hot work operations.

**1926.64(l)** ***Management of change.*** (1) The employer shall establish and implement written procedures to manage changes (except for "replacements in kind") to process chemicals, technology, equipment, and procedures; and, changes to facilities that affect a covered process.

**1926.64(l)(2)** The procedures shall assure that the following considerations are addressed prior to any change:

**1926.64(l)(2)(i)** The technical basis for the proposed change;

**1926.64(l)(2)(ii)** Impact of change on safety and health;

**1926.64(l)(2)(iii)** Modifications to operating procedures;

**1926.64(l)(2)(iv)** Necessary time period for the change; and,

**1926.64(l)(2)(v)** Authorization requirements for the proposed change.

**1926.64(l)(3)** Employees involved in operating a process and maintenance and contract employees whose job tasks will be affected by a change in the process shall be informed of, and trained in, the change prior to start-up of the process or affected part of the process.

**1926.64(l)(4)** If a change covered by this paragraph results in a change in the process safety information required by paragraph (d) of this section, such information shall be updated accordingly.

**1926.64(l)(5)** If a change covered by this paragraph results in a change in the operating procedures or practices required by paragraph (f) of this section, such procedures or practices shall be updated accordingly.

**1926.64(m)** ***Incident investigation.*** (1) The employer shall investigate each incident which resulted in, or could reasonably have resulted in a catastrophic release of highly hazardous chemical in the workplace.

**1926.64(m)(2)** An incident investigation shall be initiated as promptly as possible, but not later than 48 hours following the incident.

**1926.64(m)(3)** An incident investigation team shall be established and consist of at least one person knowledgeable in the process involved, including a contract employee if the incident involved work of the contractor, and other persons with appropriate knowledge and experience to thoroughly investigate and analyze the incident.

**1926.64(m)(4)** A report shall be prepared at the conclusion of the investigation which includes at a minimum:

**1926.64(m)(4)(i)** Date of incident;

**1926.64(m)(4)(ii)** Date investigation began;

**1926.64(m)(4)(iii)** A description of the incident;

**1926.64(m)(4)(iv)** The factors that contributed to the incident; and,

**1926.64(m)(4)(v)** Any recommendations resulting from the investigation.

**1926.64(m)(5)** The employer shall establish a system to promptly address and resolve the incident report findings and recommendations. Resolutions and corrective actions shall be documented.

**1926.64(m)(6)** The report shall be reviewed with all affected personnel whose job tasks are relevant to the incident findings including contract employees where applicable.

**1926.64(m)(7)** Incident investigation reports shall be retained for five years.

**1926.64(n)** ***Emergency planning and response.*** The employer shall establish and implement an emergency action plan for the entire plant in accordance with the provisions of 29 CFR 1926.35(a). In addition, the emergency action plan shall include procedures for handling small releases. Employers covered under this standard may also be subject to the hazardous waste and emergency response provisions contained in 29 CFR 1926.65(a), (p) and (q).

**1926.64(o)** ***Compliance audits.*** (1) Employers shall certify that they have evaluated compliance with the provisions of this section at least every three years to verify that the procedures and practices developed under the standard are adequate and are being followed.

**1926.64(o)(2)** The compliance audit shall be conducted by at least one person knowledgeable in the process.

**1926.64(o)(3)** A report of the findings of the audit shall be developed.

**1926.64(o)(4)** The employer shall promptly determine and document an appropriate response to each of the findings of the compliance audit, and document that deficiencies have been corrected.

**1926.64(o)(5)** Employers shall retain the two (2) most recent compliance audit reports.

**1926.64(p)** ***Trade secrets.*** (1) Employers shall make all information necessary to comply with the section available to those persons responsible for compiling the process safety information (required by paragraph (d) of this section), those assisting in the development of the process hazard analysis (required by paragraph (e) of this section), those responsible for developing the operating procedures (required by paragraph (f) of this section), and those involved in incident investigations (required by paragraph (m) of this section), emergency planning and response (paragraph (n) of this section) and compliance audits (paragraph (o) of this section) without regard to possible trade secret status of such information.

**1926.64(p)(2)** Nothing in this paragraph shall preclude the employer from requiring the persons to whom the information is made available under paragraph (p)(1) of this section to enter into confidentiality agreements not to disclose the information as set forth in 29 CFR 1926.59.

**1926.64(p)(3)** Subject to the rules and procedures set forth in 29 CFR 1926.59(i)(1) through (12), employees and their designated representatives shall have access to trade secret information contained within the process hazard analysis and other documents required to be developed by this standard.

## Appendix A to §1926.64—List of Highly Hazardous Chemicals, Toxics and Reactives (Mandatory)

This Appendix contains a listing of toxic and reactive highly hazardous chemicals which present a potential for a catastrophic event at or above the threshold quantity.

| CHEMICAL NAME | CAS * | TQ ** |
|---|---|---|
| Acetaldehyde | 75-07-0 | 2500 |
| Acrolein (2-Propenal) | 107-02-8 | 150 |
| Acrylyl Chloride | 814-68-6 | 250 |
| Allyl Chloride | 107-05-1 | 1000 |
| Allylamine | 107-11-9 | 1000 |
| Alkylaluminums | Varies | 5000 |
| Ammonia, Anhydrous | 7664-41-7 | 10000 |
| Ammonia solutions (.44% ammonia by weight) | 7664-41-7 | 15000 |
| Ammonium Perchlorate | 7790-98-9 | 500 |
| Ammonium Permanganate | 7787-36-2 | 7500 |
| Arsine (also called Arsenic Hydride) | 7784-42-1 | 100 |
| Bis(Chloromethyl) Ether | 542-88-1 | 100 |
| Boron Trichloride | 10294-34-5 | 2500 |
| Boron Trifluoride | 7637-07-2 | 250 |
| Bromine | 7726-95-6 | 1500 |
| Bromine Chloride | 13863-41-7 | 1500 |
| Bromine Pentafluoride | 7789-30-2 | 2500 |
| Bromine Trifluoride | 7787-71-5 | 15000 |
| 3-Bromopropyne (also called Propargyl Bromide) | 106-96-7 | 100 |
| Butyl Hydroperoxide (Tertiary) | 75-91-2 | 5000 |
| Butyl Perbenzoate (Tertiary) | 614-45-9 | 7500 |
| Carbonyl Chloride (see Phosgene) | 75-44-5 | 100 |
| Carbonyl Fluoride | 353-50-4 | 2500 |
| Cellulose Nitrate (concentration .12.6% nitrogen) | 9004-70-0 | 2500 |
| Chlorine | 7782-50-5 | 1500 |
| Chlorine Dioxide | 10049-04-4 | 1000 |
| Chlorine Pentrafluoride | 13637-63-3 | 1000 |
| Chlorine Trifluoride | 7790-91-2 | 1000 |
| Chlorodiethylaluminum (also called Diethylaluminum Chloride) | 96-10-6 | 5000 |
| 1-Chloro-2,4-Dinitrobenzene | 97-00-7 | 5000 |
| Chloromethyl Methyl Ether | 107-30-2 | 500 |
| Chloropicrin | 76-06-2 | 500 |
| Chloropicrin and Methyl Bromide mixture | None | 1500 |
| Chloropicrin and Methyl Chloride mixture | None | 1500 |
| Cumene Hydroperoxide | 80-15-9 | 5000 |
| Cyanogen | 460-19-5 | 2500 |
| Cyanogen Chloride | 506-77-4 | 500 |
| Cyanuric Fluoride | 675-14-9 | 100 |
| Diacetyl Peroxide (Concentration .70%) | 110-22-5 | 5000 |
| Diazomethane | 334-88-3 | 500 |
| Dibenzoyl Peroxide | 94-36-0 | 7500 |
| Diborane | 19287-45-7 | 100 |
| Dibutyl Peroxide (Tertiary) | 110-05-4 | 5000 |
| Dichloro Acetylene | 7572-29-4 | 250 |
| Dichlorosilane | 4109-96-0 | 2500 |
| Diethylzinc | 557-20-0 | 10000 |
| Diisopropyl Peroxydicarbonate | 105-64-6 | 7500 |
| Dilauroyl Peroxide | 105-74-8 | 7500 |
| Dimethyldichlorosilane | 75-78-5 | 1000 |
| Dimethylhydrazine, 1,1- | 57-14-7 | 1000 |
| Dimethylamine, Anhydrous | 124-40-3 | 2500 |
| 2,4-Dinitroaniline | 97-02-9 | 5000 |
| Ethyl Methyl Ketone Peroxide (also Methyl Ethyl Ketone Peroxide; concentration .60%) | 1338-23-4 | 5000 |
| Ethyl Nitrite | 109-95-5 | 5000 |
| Ethylamine | 75-04-7 | 7500 |
| Ethylene Fluorohydrin | 371-62-0 | 100 |
| Ethylene Oxide | 75-21-8 | 5000 |
| Ethyleneimine | 151-56-4 | 1000 |
| Fluorine | 7782-41-4 | 1000 |
| Formaldehyde (Formalin) | 50-00-0 | 1000 |
| Furan | 110-00-9 | 500 |
| Hexafluoroacetone | 684-16-2 | 5000 |
| Hydrochloric Acid, Anhydrous | 7647-01-0 | 5000 |
| Hydrofluoric Acid, Anhydrous | 7664-39-3 | 1000 |
| Hydrogen Bromide | 10035-10-6 | 5000 |
| Hydrogen Chloride | 7647-01-0 | 5000 |
| Hydrogen Cyanide, Anhydrous | 74-90-8 | 1000 |
| Hydrogen Fluoride | 7664-39-3 | 1000 |
| Hydrogen Peroxide (52% by weight or greater) | 7722-84-1 | 7500 |
| Hydrogen Selenide | 7783-07-5 | 150 |
| Hydrogen Sulfide | 7783-06-4 | 1500 |
| Hydroxylamine | 7803-49-8 | 2500 |
| Iron, Pentacarbonyl | 13463-40-6 | 250 |
| Isopropylamine | 75-31-0 | 5000 |
| Ketene | 463-51-4 | 100 |
| Methacrylaldehyde | 78-85-3 | 1000 |
| Methacryloyl Chloride | 920-46-7 | 150 |
| Methacryloyloxyethyl Isocyanate | 30674-80-7 | 100 |
| Methyl Acrylonitrile | 126-98-7 | 250 |
| Methylamine, Anhydrous | 74-89-5 | 1000 |
| Methyl Bromide | 74-83-9 | 2500 |
| Methyl Chloride | 74-87-3 | 15000 |

| CHEMICAL NAME | CAS * | TQ ** |
|---|---|---|
| Methyl Chloroformate | 79-22-1 | 500 |
| Methyl Ethyl Ketone Peroxide (concentration .60%) | 1338-23-4 | 5000 |
| Methyl Fluoroacetate | 453-18-9 | 100 |
| Methyl Fluorosulfate | 421-20-5 | 100 |
| Methyl Hydrazine | 60-34-4 | 100 |
| Methyl Iodide | 74-88-4 | 7500 |
| Methyl Isocyanate | 624-83-9 | 250 |
| Methyl Mercaptan | 74-93-1 | 5000 |
| Methyl Vinyl Ketone | 79-84-4 | 100 |
| Methyltrichlorosilane | 75-79-6 | 500 |
| Nickel Carbonyl (Nickel Tetracarbonyl) | 13463-39-3 | 150 |
| Nitric Acid (94.5% by weight or greater) | 7697-37-2 | 500 |
| Nitric Oxide | 10102-43-9 | 250 |
| Nitroaniline (para Nitroaniline) | 100-01-6 | 5000 |
| Nitromethane | 75-52-5 | 2500 |
| Nitrogen Dioxide | 10102-44-0 | 250 |
| Nitrogen Oxides (NO; $NO_2$; $N_2O_4$; $N_2O_3$) | 10102-44-0 | 250 |
| Nitrogen Tetroxide (also called Nitrogen Peroxide) | 10544-72-6 | 250 |
| Nitrogen Trifluoride | 7783-54-2 | 5000 |
| Nitrogen Trioxide | 10544-73-7 | 250 |
| Oleum (65% to 80% by weight; also called Fuming Sulfuric Acid) | 8014-94-7 | 1000 |
| Osmium Tetroxide | 20816-12-0 | 100 |
| Oxygen Difluoride (Fluorine Monoxide) | 7783-41-7 | 100 |
| Ozone | 10028-15-6 | 100 |
| Pentaborane | 19624-22-7 | 100 |
| Peracetic Acid (concentration .60% Acetic Acid; also called Peroxyacetic Acid) | 79-21-0 | 1000 |
| Perchloric Acid (concentration .60% by weight) | 7601-90-3 | 5000 |
| Perchloromethyl Mercaptan | 594-42-3 | 150 |
| Perchloryl Fluoride | 7616-94-6 | 5000 |
| Peroxyacetic Acid (concentration .60% Acetic Acid; also called Peracetic Acid) | 79-21-0 | 1000 |
| Phosgene (also called Carbonyl Chloride) | 75-44-5 | 100 |
| Phosphine (Hydrogen Phosphide) | 7803-51-2 | 100 |
| Phosphorus Oxychloride (also called Phosphoryl Chloride) | 10025-87-3 | 1000 |
| Phosphorus Trichloride | 7719-12-2 | 1000 |
| Phosphoryl Chloride (also called Phosphorus Oxychloride) | 10025-87-3 | 1000 |
| Propargyl Bromide | 106-96-7 | 100 |
| Propyl Nitrate | 627-3-4 | 2500 |
| Sarin | 107-44-8 | 100 |
| Selenium Hexafluoride | 7783-79-1 | 1000 |
| Stibine (Antimony Hydride) | 7803-52-3 | 500 |
| Sulfur Dioxide (liquid) | 7446-09-5 | 1000 |
| Sulfur Pentafluoride | 5714-22-7 | 250 |
| Sulfur Tetrafluoride | 7783-60-0 | 250 |
| Sulfur Trioxide (also called Sulfuric Anhydride) | 7446-11-9 | 1000 |
| Sulfuric Anhydride (also called Sulfur Trioxide) | 7446-11-9 | 1000 |
| Tellurium Hexafluoride | 7783-80-4 | 250 |
| Tetrafluoroethylene | 116-14-3 | 5000 |
| Tetrafluorohydrazine | 10036-47-2 | 5000 |
| Tetramethyl Lead | 75-74-1 | 1000 |
| Thionyl Chloride | 7719-09-7 | 250 |
| Trichloro (chloromethyl) Silane | 1558-25-4 | 100 |
| Trichloro (dichlorophenyl) Silane | 27137-85-5 | 2500 |
| Trichlorosilane | 10025-78-2 | 5000 |
| Trifluorochloroethylene | 79-38-9 | 10000 |
| Trimethyoxysilane | 2487-90-3 | 1500 |

* Chemical Abstract Service Number.

** Threshold Quantity in Pounds (Amount necessary to be covered by this standard).

## Appendix B to §1926.64—Block Flow Diagram and Simplified Process Flow Diagram (Nonmandatory)

### EXAMPLE OF A BLOCK FLOW DIAGRAM

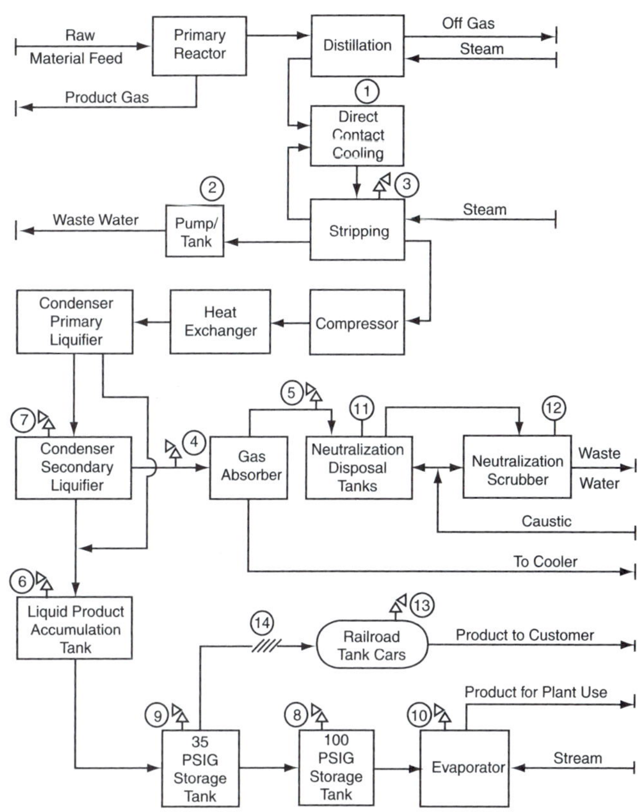


### EXAMPLE OF A PROCESS FLOW DIAGRAM

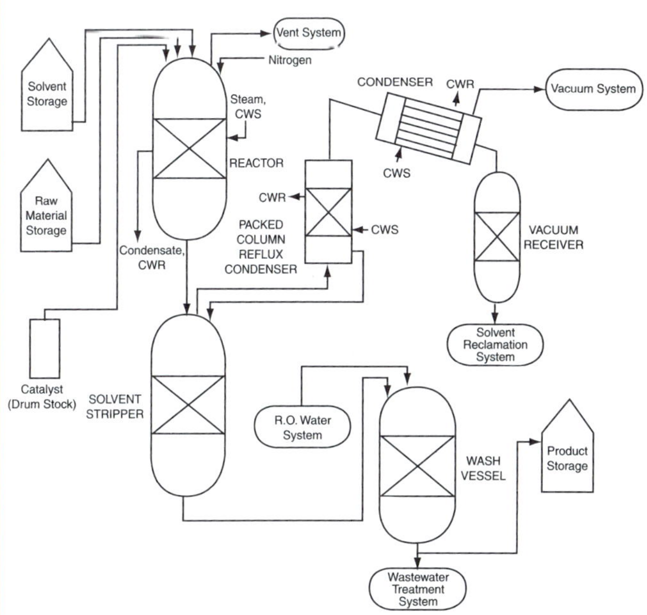

## Appendix C to §1926.64— Compliance Guidelines and Recommendations for Process Safety Management (Nonmandatory)

This appendix serves as a nonmandatory guideline to assist employers and employees in complying with the requirements of this section, as well as provides other helpful recommendations and information. Examples presented in this appendix are not the only means of achieving the performance goals in the standard. This appendix neither adds nor detracts from the requirements of the standard.

1. *Introduction to Process Safety Management.* The major objective of process safety management of highly hazardous chemicals is to prevent unwanted releases of hazardous chemicals especially into locations which could expose employees and others to serious hazards. An effective process safety management program requires a systematic approach to evaluating the whole process. Using this approach the process design, process technology, operational and maintenance activities and procedures, nonroutine activities and procedures, emergency preparedness plans and procedures, training programs, and other elements which impact the process are all considered in the evaluation. The various lines of defense that have been incorporated into the design and operation of the process to prevent or mitigate the release of hazardous chemicals need to be evaluated and strengthened to assure their effectiveness at each level. Process safety management is the proactive identification, evaluation and mitigation or prevention of chemical releases that could occur as a result of failures in process, procedures or equipment.

The process safety management standard targets highly hazardous chemicals that have the potential to cause a catastrophic incident. This standard as a whole is to aid employers in their efforts to prevent or mitigate episodic chemical releases that could lead to a catastrophe in the workplace and possibly to the surrounding community. To control these types of hazards, employers need to develop the necessary expertise, experiences, judgement and proactive initiative within their workforce to properly implement and maintain an effective process safety management program as envisioned in the OSHA standard. This OSHA standard is required by the Clean Air Act Amendments as is the Environmental Protection Agency's Risk Management Plan. Employers, who merge the two sets of requirements into their process safety management program, will better assure full compliance with each as well as enhancing their relationship with the local community.

While OSHA believes process safety management will have a positive effect on the safety of employees in workplaces and also offers other potential benefits to employers (increased productivity), smaller businesses which may have limited resources available to them at this time, might consider alternative avenues of decreasing the risks associated with highly hazardous chemicals at their workplaces. One method which might be considered is the reduction in the inventory of the highly hazardous chemical. This reduction in inventory will result in a reduction of the risk or potential for a catastrophic incident. Also, employers including small employers may be able to establish more efficient inventory control by reducing the quantities of highly hazardous chemicals on site below the established threshold quantities. This reduction can be accomplished by ordering smaller shipments and maintaining the minimum inventory necessary for efficient and safe operation. When reduced inventory is not feasible, then the employer might consider dispersing inventory to several locations on site. Dispersing storage into locations where a release in one location will not cause a release in another location is a practical method to also reduce the risk or potential for catastrophic incidents.

2. *Employee Involvement in Process Safety Management.* Section 304 of the Clean Air Act Amendments states that employers are to consult with their employees and their representatives regarding the employers efforts in the development and implementation of the process safety management program elements and hazard assessments. Section 304 also requires employers to train and educate their employees and to inform affected employees of the findings from incident investigations required by the process safety management program. Many employers, under their safety and health programs, have already established means and methods to keep employees and their representatives informed about relevant safety and health issues and employers may be able to adapt these practices and procedures to meet their obligations under this standard. Employers who have not implemented an occupational safety and health program may wish to form a safety and health committee of employees and management representatives to help the employer meet the obligations specified by this standard. These committees can become a significant ally in helping the employer to implement and maintain an effective process safety management program for all employees.

3. *Process Safety Information.* Complete and accurate written information concerning process chemicals, process technology, and process equipment is essential to an effective process safety management program and to a process hazards analysis. The compiled information will be a necessary resource to a variety of users including the team that will perform the process hazards analysis as required under paragraph (e); those developing the training programs and the operating procedures; contractors whose employees will be working with the process; those conducting the pre-startup reviews; local emergency preparedness planners; and insurance and enforcement officials.

The information to be compiled about the chemicals, including process intermediates, needs to be comprehensive enough for an accurate assessment of the fire and explosion characteristics, reactivity hazards, the safety and health hazards to workers, and the corrosion and erosion effects on the process equipment and monitoring tools. Current material safety data sheet (MSDS) information can be used to help meet this requirement which must be supplemented with process chemistry information including runaway reaction and over pressure hazards if applicable.

Process technology information will be a part of the process safety information package and it is expected that it will include diagrams of the type shown in Appendix B of this section as well as employer established criteria for maximum inventory levels for process chemicals; limits beyond which would be considered upset conditions; and a qualitative estimate of the consequences or results of deviation that could occur if operating beyond the established process limits. Employers are encouraged to use diagrams which will help users understand the process.

A block flow diagram is used to show the major process equipment and interconnecting process flow lines and show flow rates, stream composition, temperatures, and pressures when necessary for clarity.

The block flow diagram is a simplified diagram.

Process flow diagrams are more complex and will show all main flow streams including valves to enhance the understanding of the process, as well as pressures and temperatures on all feed and product lines within all major vessels, in and out of headers and heat exchangers, and points of pressure and temperature control. Also, materials of construction information, pump capacities and pressure heads, compressor horsepower and vessel design pressures and temperatures are shown when necessary for clarity. In addition, major components of control loops are usually shown along with key utilities on process flow diagrams.

Piping and instrument diagrams (P&Ids) may be the more appropriate type of diagrams to show some of the above details and to display the information for the piping designer and engineering staff. The P&IDs are to be used to describe the relationships between equipment and instrumentation as well as other relevant information that will enhance clarity. Computer software programs which do P&Ids or other diagrams useful to the information package, may be used to help meet this requirement.

The information pertaining to process equipment design must be documented. In other words, what were the codes and standards relied on to establish good engineering practice. These codes and standards are published by such organizations as the American Society of Mechanical Engineers, American Petroleum Institute, American National Standards Institute, National Fire Protection Association, American Society for Testing and Materials, National Board of Boiler and Pressure Vessel Inspectors, National Association of Corrosion Engineers, American Society of Exchange Manufacturers Association, and model building code groups.

In addition, various engineering societies issue technical reports which impact process design. For example, the American Institute of Chemical Engineers has published technical reports on topics such as two phase flow for venting devices. This type of technically recognized report would constitute good engineering practice.

For existing equipment designed and constructed many years ago in accordance with the codes and standards available at that time and no longer in general use today, the employer must document which codes and standards were used and that the design and construction along with the testing, inspection and operation are still suitable for the intended use. Where the process technology requires a design which departs from the applicable codes and standards, the employer must document that the design and construction is suitable for the intended purpose.

4. *Process Hazard Analysis.* A process hazard analysis (PHA), sometimes called a process hazard evaluation, is one of the most important elements of the process safety management program. A PHA is an organized and systematic effort to identify and analyze the significance of potential hazards

associated with the processing or handling of highly hazardous chemicals. A PHA provides information which will assist employers and employees in making decisions for improving safety and reducing the consequences of unwanted or unplanned releases of hazardous chemicals. A PHA is directed toward analyzing potential causes and consequences of fires, explosions, releases of toxic or flammable chemicals and major spills of hazardous chemicals. The PHA focuses on equipment, instrumentation, utilities, human actions (routine and nonroutine), and external factors that might impact the process. These considerations assist in determining the hazards and potential failure points or failure modes in a process.

The selection of a PHA methodology or technique will be influenced by many factors including the amount of existing knowledge about the process. Is it a process that has been operated for a long period of time with little or no innovation and extensive experience has been generated with its use? Or, is it a new process or one which has been changed frequently by the inclusion of innovative features? Also, the size and complexity of the process will influence the decision as to the appropriate PHA methodology to use. All PHA methodologies are subject to certain limitations. For example, the checklist methodology works well when the process is very stable and no changes are made, but it is not as effective when the process has undergone extensive change. The checklist may miss the most recent changes and consequently the changes would not be evaluated. Another limitation to be considered concerns the assumptions made by the team or analyst. The PHA is dependent on good judgement and the assumptions made during the study need to be documented and understood by the team and reviewer and kept for a future PHA.

The team conducting the PHA need to understand the methodology that is going to be used. A PHA team can vary in size from two people to a number of people with varied operational and technical backgrounds. Some team members may only be a part of the team for a limited time. The team leader needs to be fully knowledgeable in the proper implementation of the PHA methodology that is to be used and should be impartial in the evaluation. The other full or part time team members need to provide the team with expertise in areas such as process technology, process design, operating procedures and practices, including how the work is actually performed, alarms, emergency procedures, instrumentation, maintenance procedures, both routine and nonroutine tasks, including how the tasks are authorized, procurement of parts and supplies, safety and health, and any other relevant subject as the need dictates. At least one team member must be familiar with the process.

The ideal team will have an intimate knowledge of the standards, codes, specifications and regulations applicable to the process being studied. The selected team members need to be compatible and the team leader needs to be able to manage the team and the PHA study. The team needs to be able to work together while benefiting from the expertise of others on the team or outside the team, to resolve issues, and to forge a consensus on the findings of the study and the recommendations.

The application of a PHA to a process may involve the use of different methodologies for various parts of the process. For example, a process involving a series of unit operations of varying sizes, complexities, and ages may use different methodologies and team members for each operation. Then the conclusions can be integrated into one final study and evaluation. A more specific example is the use of a checklist PHA for a standard boiler or heat exchanger and the use of a Hazard and Operability PHA for the overall process. Also, for batch type processes like custom batch operations, a generic PHA of a representative batch may be used where there are only small changes of monomer or other ingredient ratios and the chemistry is documented for the full range and ratio of batch ingredients. Another process that might consider using a generic type of PHA is a gas plant. Often these plants are simply moved from site to site and therefore, a generic PHA may be used for these movable plants. Also, when an employer has several similar size gas plants and no sour gas is being processed at the site, then a generic PHA is feasible as long as the variations of the individual sites are accounted for in the PHA. Finally, when an employer has a large continuous process which has several control rooms for different portions of the process such as for a distillation tower and a blending operation, the employer may wish to do each segment separately and then integrate the final results.

Additionally, small businesses which are covered by this rule, will often have processes that have less storage volume, less capacity, and less complicated than processes at a large facility. Therefore, OSHA would anticipate that the less complex methodologies would be used to meet the process hazard analysis criteria in the standard. These process hazard analyses can be done in less time and with a few people being involved. A less complex process generally means that less data, P&IDs, and process information is needed to perform a process hazard analysis.

Many small businesses have processes that are not unique, such as cold storage lockers or water treatment facilities. Where employer associations have a number of members with such facilities, a generic PHA, evolved from a checklist or what-if questions, could be developed and used by each employer effectively to reflect his/her particular process; this would simplify compliance for them.

When the employer has a number of processes which require a PHA, the employer must set up a priority system of which PHAs to conduct first. A preliminary or gross hazard analysis may be useful in prioritizing the processes that the employer has determined are subject to coverage by the process safety management standard. Consideration should first be given to those processes with the potential of adversely affecting the largest number of employees. This prioritizing should consider the potential severity of a chemical release, the number of potentially affected employees, the operating history of the process such as the frequency of chemical releases, the age of the process and any other relevant factors. These factors would suggest a ranking order and would suggest either using a weighing factor system or a systematic ranking method. The use of a preliminary hazard analysis would assist an employer in determining which process should be of the highest priority and thereby the employer would obtain the greatest improvement in safety at the facility.

Detailed guidance on the content and application of process hazard analysis methodologies is available from the American Institute of Chemical Engineers' Center for Chemical Process Safety (see Appendix D).

5. *Operating Procedures and Practices.* Operating procedures describe tasks to be performed, data to be recorded, operating conditions to be maintained, samples to be collected, and safety and health precautions to be taken. The procedures need to be technically accurate, understandable to employees, and revised periodically to ensure that they reflect current operations. The process safety information package is to be used as a resource to better assure that the operating procedures and practices are consistent with the known hazards of the chemicals in the process and that the operating parameters are accurate. Operating procedures should be reviewed by engineering staff and operating personnel to ensure that they are accurate and provide practical instructions on how to actually carry out job duties safely.

Operating procedures will include specific instructions or details on what steps are to be taken or followed in carrying out the stated procedures. These operating instructions for each procedure should include the applicable safety precautions and should contain appropriate information on safety implications. For example, the operating procedures addressing operating parameters will contain operating instructions about pressure limits, temperature ranges, flow rates, what to do when an upset condition occurs, what alarms and instruments are pertinent if an upset condition occurs, and other subjects. Another example of using operating instructions to properly implement operating procedures is in starting up or shutting down the process. In these cases, different parameters will be required from those of normal operation. These operating instructions need to clearly indicate the distinctions between startup and normal operations such as the appropriate allowances for heating up a unit to reach the normal operating parameters. Also the operating instructions need to describe the proper method for increasing the temperature of the unit until the normal operating temperature parameters are achieved.

Computerized process control systems add complexity to operating instructions. These operating instructions need to describe the logic of the software as well as the relationship between the equipment and the control system; otherwise, it may not be apparent to the operator.

Operating procedures and instructions are important for training operating personnel. The operating procedures are often viewed as the standard operating practices (SOPs) for operations. Control room personnel and operating staff, in general, need to have a full understanding of operating procedures. If workers are not fluent in English then procedures and instructions need to be prepared in a second language understood by the workers. In addition, operating procedures need to be changed when there is a change in the process as a result of the management of change procedures. The consequences of operating procedure changes need to be fully evaluated and the information conveyed to the personnel. For example, mechanical changes to the process made by the maintenance department (like changing a valve from steel to brass or other subtle changes) need to be evaluated to determine if operating procedures and practices also need to be changed. All management of change actions must be coordinated and integrated with current operating procedures and operating personnel must be oriented to the changes in procedures before the change is made. When the process is shutdown in order to make a change, then the operating procedures must be updated before startup of the process.

Training in how to handle upset conditions must be accomplished as well as what operating personnel are to do in emergencies such as when a pump seal fails or a pipeline ruptures. Communication between operating personnel and workers performing work within the process area, such as nonroutine tasks, also must be maintained. The hazards of the tasks are to be conveyed to operating personnel in accordance with established procedures and to those performing the actual tasks. When the work is completed, operating personnel should be informed to provide closure on the job.

6. *Employee Training.* All employees, including maintenance and contractor employees, involved with highly hazardous chemicals need to fully understand the safety and health hazards of the chemicals and processes they work with for the protection of themselves, their fellow employees and the citizens of nearby communities. Training conducted in compliance with 1926.59, the Hazard Communication standard, will help employees to be more knowledgeable about the chemicals they work with as well as familiarize them with reading and understanding MSDS. However, additional training in subjects such as operating procedures and safety work practices, emergency evacuation and response, safety procedures, routine and nonroutine work authorization activities, and other areas pertinent to process safety and health will need to be covered by an employer's training program.

In establishing their training programs, employers must clearly define the employees to be trained and what subjects are to be covered in their training. Employers in setting up their training program will need to clearly establish the goals and objectives they wish to achieve with the training that they provide to their employees. The learning goals or objectives should be written in clear measurable terms before the training begins. These goals and objectives need to be tailored to each of the specific training modules or segments. Employers should describe the important actions and conditions under which the employee will demonstrate competence or knowledge as well as what is acceptable performance.

Hands-on-training where employees are able to use their senses beyond listening, will enhance learning. For example, operating personnel, who will work in a control room or at control panels, would benefit by being trained at a simulated control panel or panels. Upset conditions of various types could be displayed on the simulator, and then the employee could go through the proper operating procedures to bring the simulator panel back to the normal operating parameters. A training environment could be created to help the trainee feel the full reality of the situation but, of course, under controlled conditions. This realistic type of training can be very effective in teaching employees correct procedures while allowing them to also see the consequences of what might happens if they do not follow established operating procedures. Other training techniques using videos or on-the- job training can also be very effective for teaching other job tasks, duties, or other important information. An effective training program will allow the employee to fully participate in the training process and to practice their skill or knowledge.

Employers need to periodically evaluate their training programs to see if the necessary skills, knowledge, and routines are being properly understood and implemented by their trained employees. The means or methods for evaluating the training should be developed along with the training program goals and objectives. Training program evaluation will help employers to determine the amount of training their employees understood, and whether the desired results were obtained. If, after the evaluation, it appears that the trained employees are not at the level of knowledge and skill that was expected, the employer will need to revise the training program, provide retraining, or provide more frequent refresher training sessions until the deficiency is resolved. Those who conducted the training and those who received the training should also be consulted as to how best to improve the training process. If there is a language barrier, the language known to the trainees should be used to reinforce the training messages and information.

Careful consideration must be given to assure that employees including maintenance and contract employees receive current and updated training. For example, if changes are made to a process, impacted employees must be trained in the changes and understand the effects of the changes on their job tasks (e.g., any new operating procedures pertinent to their tasks). Additionally, as already discussed the evaluation of the employee's absorption of training will certainly influence the need for training.

7. *Contractors.* Employers who use contractors to perform work in and around processes that involve highly hazardous chemicals, will need to establish a screening process so that they hire and use contractors who accomplish the desired job tasks without compromising the safety and health of employees at a facility. For contractors, whose safety performance on the job is not known to the hiring employer, the employer will need to obtain information on injury and illness rates and experience and should obtain contractor references. Additionally, the employer must assure that the contractor has the appropriate job skills, knowledge and certifications (such as for pressure vessel welders). Contractor work methods and experiences should be evaluated. For example, does the contractor conducting demolition work swing loads over operating processes or does the contractor avoid such hazards?

Maintaining a site injury and illness log for contractors is another method employers must use to track and maintain current knowledge of work activities involving contract employees working on or adjacent to covered processes. Injury and illness logs of both the employer's employees and contract employees allow an employer to have full knowledge of process injury and illness experience. This log will also contain information which will be of use to those auditing process safety management compliance and those involved in incident investigations.

Contract employees must perform their work safely. Considering that contractors often perform very specialized and potentially hazardous tasks such as confined space entry activities and nonroutine repair activities it is quite important that their activities be controlled while they are working on or near a covered process. A permit system or work authorization system for these activities would also be helpful to all affected employers. The use of a work authorization system keeps an employer informed of contract employee activities, and as a benefit the employer will have better coordination and more management control over the work being performed in the process area. A well run and well maintained process where employee safety is fully recognized will benefit all of those who work in the facility whether they be contract employees or employees of the owner.

8. *Pre-Startup Safety.* For new processes, the employer will find a PHA helpful in improving the design and construction of the process from a reliability and quality point of view. The safe operation of the new process will be enhanced by making use of the PHA recommendations before final installations are completed. P&IDs are to be completed along with having the operating procedures in place and the operating staff trained to run the process before startup. The initial startup procedures and normal operating procedures need to be fully evaluated as part of the pre-startup review to assure a safe transfer into the normal operating mode for meeting the process parameters.

For existing processes that have been shutdown for turnaround, or modification, etc., the employer must assure that any changes other than "replacement in kind" made to the process during shutdown go through the management of change procedures. P&IDs will need to be updated as necessary, as well as operating procedures and instructions. If the changes made to the process during shutdown are significant and impact the training program, then operating personnel as well as employees engaged in routine and nonroutine work in the process area may need some refresher or additional training in light of the changes. Any incident investigation recommendations, compliance audits or PHA recommendations need to be reviewed as well to see what impacts they may have on the process before beginning the startup.

9. *Mechanical Integrity.* Employers will need to review their maintenance programs and schedules to see if there are areas where "breakdown" maintenance is used rather than an on-going mechanical integrity program. Equipment used to process, store, or handle highly hazardous chemicals needs to be designed, constructed, installed and maintained to minimize the risk of releases of such chemicals. This requires that a mechanical integrity program be in place to assure the continued integrity of process equipment. Elements of a mechanical integrity program include the identification and categorization of equipment and instrumentation, inspections and tests, testing and inspection frequencies, development of maintenance procedures, training of maintenance personnel, the establishment of criteria for acceptable test results, documentation of test and inspection results, and documentation of manufacturer recommendations as to meantime to failure for equipment and instrumentation.

The first line of defense an employer has available is to operate and maintain the process as designed, and to keep the chemicals contained. This line of defense is backed up by the next line of defense which is the controlled release of chemicals through venting to scrubbers or flares, or to surge or overflow tanks which are designed to receive such chemicals, etc. These lines of defense are the primary lines of defense or means to prevent unwanted releases. The secondary lines of defense would include fixed fire protection systems like sprinklers, water spray, or deluge systems, monitor guns, etc., dikes, designed drainage systems, and other systems which would control or mitigate hazardous chemicals once an unwanted release occurs. These primary and secondary lines of defense are what the mechanical integrity program needs to protect and strengthen these primary and secondary lines of defenses where appropriate.

The first step of an effective mechanical integrity program is to compile and categorize a list of process equipment and instrumentation for inclusion in the program. This list would include pressure vessels, storage tanks, process piping, relief and vent systems, fire protection system components, emergency shutdown systems and alarms and interlocks and pumps. For the categorization of instrumentation and the listed equipment the employer would prioritize which pieces of equipment require closer scrutiny than others. Meantime to failure of various instrumentation and equipment parts would be known from the manufacturers data or the employer's experience with the parts, which would then influence the inspection and testing frequency and associated procedures. Also, applicable codes and standards such as the National Board Inspection Code, or those from the American Society for Testing and Material, American Petroleum Institute, National Fire Protection Association, American National Standards Institute, American Society of Mechanical Engineers, and other groups, provide information to help establish an effective testing and inspection frequency, as well as appropriate methodologies.

The applicable codes and standards provide criteria for external inspections for such items as foundation and supports, anchor bolts, concrete or steel supports, guy wires, nozzles and sprinklers, pipe hangers, grounding connections, protective coatings and insulation, and external metal surfaces of piping and vessels, etc. These codes and standards also provide information on methodologies for internal inspection, and a frequency formula based on the corrosion rate of the materials of construction. Also, erosion both internal and external needs to be considered along with corrosion effects for piping and valves. Where the corrosion rate is not known, a maximum inspection frequency is recommended, and methods of developing the corrosion rate are available in the codes. Internal inspections need to cover items such as vessel shell, bottom and head; metallic linings; nonmetallic linings; thickness measurements for vessels and piping; inspection for erosion, corrosion, cracking and bulges; internal equipment like trays, baffles, sensors and screens for erosion, corrosion or cracking and other deficiencies. Some of these inspections may be performed by state or local government inspectors under state and local statutes. However, each employer needs to develop procedures to ensure that tests and inspections are conducted properly and that consistency is maintained even where different employees may be involved. Appropriate training is to be provided to maintenance personnel to ensure that they understand the preventive maintenance program procedures, safe practices, and the proper use and application of special equipment or unique tools that may be required. This training is part of the overall training program called for in the standard.

A quality assurance system is needed to help ensure that the proper materials of construction are used, that fabrication and inspection procedures are proper, and that installation procedures recognize field installation concerns. The quality assurance program is an essential part of the mechanical integrity program and will help to maintain the primary and secondary lines of defense that have been designed into the process to prevent unwanted chemical releases or those which control or mitigate a release. "As built" drawings, together with certifications of coded vessels and other equipment, and materials of construction need to be verified and retained in the quality assurance documentation. Equipment installation jobs need to be properly inspected in the field for use of proper materials and procedures and to assure that qualified craftsmen are used to do the job. The use of appropriate gaskets, packing, bolts, valves, lubricants and welding rods need to be verified in the field. Also, procedures for installation of safety devices need to be verified, such as the torque on the bolts on ruptured disc installations, uniform torque on flange bolts, proper installation of pump seals, etc. If the quality of parts is a problem, it may be appropriate to conduct audits of the equipment supplier's facilities to better assure proper purchases of required equipment which is suitable for its intended service. Any changes in equipment that may become necessary will need to go through the management of change procedures.

10. *Nonroutine Work Authorizations.* Nonroutine work which is conducted in process areas needs to be controlled by the employer in a consistent manner. The hazards identified involving the work that is to be accomplished must be communicated to those doing the work, but also to those operating personnel whose work could affect the safety of the process. A work authorization notice or permit must have a procedure that describes the steps the maintenance supervisor, contractor representative or other person needs to follow to obtain the necessary clearance to get the job started. The work authorization procedures need to reference and coordinate, as applicable, lockout/tagout procedures, line breaking procedures, confined space entry procedures and hot work authorizations. This procedure also needs to provide clear steps to follow once the job is completed in order to provide closure for those that need to know the job is now completed and equipment can be returned to normal.

11. *Managing Change.* To properly manage changes to process chemicals, technology, equipment and facilities, one must define what is meant by change. In this process safety management standard, change includes all modifications to equipment, procedures, raw materials and processing conditions other than "replacement in kind." These changes need to be properly managed by identifying and reviewing them prior to implementation of the change. For example, the operating procedures contain the operating parameters (pressure limits, temperature ranges, flow rates, etc.) and the importance of operating within these limits. While the operator must have the flexibility to maintain safe operation within the established parameters, any operation outside of these parameters requires review and approval by a written management of change procedure.

Management of change covers such as changes in process technology and changes to equipment and instrumentation. Changes in process technology can result from changes in production rates, raw materials, experimentation, equipment unavailability, new equipment, new product development, change in catalyst and changes in operating conditions to improve yield or quality. Equipment changes include among others change in materials of construction, equipment specifications, piping pre-arrangements, experimental equipment, computer program revisions and changes in alarms and interlocks. Employers need to establish means and methods to detect both technical changes and mechanical changes.

Temporary changes have caused a number of catastrophes over the years, and employers need to establish ways to detect temporary changes as well as those that are permanent. It is important that a time limit for temporary changes be established and monitored since, without control, these changes may tend to become permanent. Temporary changes are subject to the management of change provisions. In addition, the management of change procedures are used to insure that the equipment and procedures are returned to their original or designed conditions at the end of the temporary change. Proper documentation and review of these changes is invaluable in assuring that the safety and health considerations are being incorporated into the operating procedures and the process.

Employers may wish to develop a form or clearance sheet to facilitate the processing of changes through the management of change procedures. A typical change form may include a description and the purpose of the change, the technical basis for the change, safety and health considerations, documentation of changes for the operating procedures, maintenance procedures, inspection and testing, P&IDs, electrical classification, training and communications, pre-startup inspection, duration if a temporary change, approvals and authorization. Where the impact of the change is minor and well understood, a check list reviewed by an authorized person with proper communication to others who are affected may be sufficient. However, for a more complex or significant design change, a hazard evaluation procedure with approvals by operations, maintenance, and safety departments may be appropriate. Changes in documents such as P&IDs, raw materials, operating procedures, mechanical integrity programs, electrical classifications, etc., need to be noted so that these revisions can be made permanent when the drawings and procedure manuals are updated. Copies of process changes need to be kept in an accessible location to ensure that design changes are available to operating personnel as well as to PHA team members when a PHA is being done or one is being updated.

12. *Investigation of Incidents.* Incident investigation is the process of identifying the underlying causes of incidents and implementing steps to prevent similar events from occurring. The intent of an incident investigation is for employers to learn from past experiences and thus avoid repeating past mistakes. The incidents for which OSHA expects employers to become aware and to investigate are the types of events which result in or could reasonably have resulted in a catastrophic release. Some of the events are sometimes referred to as "near misses," meaning that a serious consequence did not occur, but could have.

Employers need to develop in-house capability to investigate incidents that occur in their facilities. A team needs to be assembled by the employer and trained in the techniques of investigation including how to conduct interviews of witnesses, needed documentation and report writing. A multi-disciplinary team is better able to gather the facts of the event and to analyze them and develop plausible scenarios as to what happened, and why. Team members should be selected on the basis of their training, knowledge and ability to contribute to a team effort to fully investigate the incident. Employees in the process area where the incident occurred should be consulted, interviewed or made a member of the team. Their knowledge of the events form a significant set of facts about the incident which occurred. The report, its findings and recommendations are to be shared with those who can benefit from the information. The cooperation of employees is essential to an effective incident investigation. The focus of the investigation should be to obtain facts, and not to place blame. The team and the investigation process should clearly deal with all involved individuals in a fair, open and consistent manner.

13. *Emergency Preparedness.* Each employer must address what actions employees are to take when there is an unwanted release of highly hazardous chemicals. Emergency preparedness or the employer's tertiary (third) lines of defense are those that will be relied on along with the secondary lines of defense when the primary lines of defense which are used to prevent an unwanted release fail to stop the release. Employers will need to decide if they want employees to handle and stop small or minor incidental releases. Whether they wish to mobilize the available resources at the plant and have them brought to bear on a more significant release. Or whether employers want their employees to evacuate the danger area and promptly escape to a preplanned safe zone area, and allow the local community emergency response organizations to handle the release. Or whether the employer wants to use some combination of these actions. Employers will need to select how many different emergency preparedness or tertiary lines of defense they plan to have and then develop the necessary plans and procedures, and appropriately train employees in their emergency duties and responsibilities and then implement these lines of defense.

Employers at a minimum must have an emergency action plan which will facilitate the prompt evacuation of employees when an unwanted release of highly hazardous chemical. This means that the employer will have a plan that will be activated by an alarm system to alert employees when to evacuate and, that employees who are physically impaired, will have the necessary support and assistance to get them to the safe zone as well. The intent of these requirements is to alert and move employees to a safe zone quickly. Delaying alarms or confusing alarms are to be avoided. The use of process control centers or similar process buildings in the process area as safe areas is discouraged. Recent catastrophes have shown that a large life loss has occurred in these structures because of where they have been sited and because they are not necessarily designed to withstand over-pressures from shockwaves resulting from explosions in the process area.

Unwanted incidental releases of highly hazardous chemicals in the process area must be addressed by the employer as to what actions employees are to take. If the employer wants employees to evacuate the area, then the emergency action plan will be activated. For outdoor processes where wind direction is important for selecting the safe route to a refuge area, the employer should place a wind direction indicator such as a wind sock or pennant at the highest point that can be seen throughout the process area. Employees can move in the direction of cross wind to upwind to gain safe access to the refuge area by knowing the wind direction.

If the employer wants specific employees in the release area to control or stop the minor emergency or incidental release, these actions must be planned for in advance and procedures developed and implemented. Preplanning for handling incidental releases for minor emergencies in the process area needs to be done, appropriate equipment for the hazards must be provided, and training conducted for those employees who will perform the emergency work before they respond to handle an actual release. The employer's training program, including the Hazard Communication standard training is to address the training needs for employees who are expected to handle incidental or minor releases.

Preplanning for releases that are more serious than incidental releases is another important line of defense to be used by the employer. When a serious release of a highly hazardous chemical occurs, the employer through preplanning will have determined in advance what actions employees are to take. The evacuation of the immediate release area and other areas as necessary would be accomplished under the emergency action plan. If the employer wishes to use plant personnel such as a fire brigade, spill control team, a hazardous materials team, or use employees to render aid to those in the immediate release area and control or mitigate the incident, these actions are covered by 1926.65, the Hazardous Waste Operations and Emergency Response (HAZWOPER) standard. If outside assistance is necessary, such as through mutual aid agreements between employers or local government emergency response organizations, these emergency responders are also covered by HAZWOPER. The safety and health protections required for emergency responders are the responsibility of their employers and of the on-scene incident commander.

Responders may be working under very hazardous conditions and therefore the objective is to have them competently led by an on-scene incident commander and the commander's staff, properly equipped to do their assigned work safely, and fully trained to carry out their duties safely before they respond to an emergency. Drills, training exercises, or simulations with the local community emergency response planners and responder organizations is one means to obtain better preparedness. This close cooperation and coordination between plant and local community emergency preparedness managers will also aid the employer in complying with the Environmental Protection Agency's Risk Management Plan criteria.

One effective way for medium to large facilities to enhance coordination and communication during emergencies for on plant operations and with local community organizations is for employers to establish and equip an emergency control center. The emergency control center would be sited in a safe zone area so that it could be occupied throughout the duration of an emergency. The center would serve as the major communication link between the on-scene incident commander and plant or corporate management as well as with the local community officials. The communication equipment in the emergency control center should include a network to receive and transmit information by telephone, radio or other means. It is important to have a backup communication network in case of power failure or one communication means fails. The center should also be equipped with the plant layout and community maps, utility drawings including fire water, emergency lighting, appropriate reference materials such as a government agency notification list, company personnel phone list, SARA Title III reports and material safety data sheets, emergency plans and procedures manual, a listing with the location of emergency response equipment, mutual aid information, and access to meteorological or weather condition data and any dispersion modeling data.

14. *Compliance Audits.* Employers need to select a trained individual or assemble a trained team of people to audit the process safety management system and program. A small process or plant may need only one knowledgeable person to conduct an audit. The audit is to include an evaluation of the design and effectiveness of the process safety management system and a field inspection of the safety and health conditions and practices to verify that the employer's systems are effectively implemented. The audit should be conducted or lead by a person knowledgeable in audit techniques and who is impartial towards the facility or area being audited. The essential elements of an audit program include planning, staffing, conducting the audit, evaluation and corrective action, follow-up and documentation.

Planning in advance is essential to the success of the auditing process. Each employer needs to establish the format, staffing, scheduling and verification methods prior to conducting the audit. The format should be designed to provide the lead auditor with a procedure or checklist which details the requirements of each section of the standard. The names of the audit team members should be listed as part of the format as well. The checklist, if properly designed, could serve as the verification sheet which provides the auditor with the necessary information to expedite the review and assure that no requirements of the standard are omitted. This verification sheet format could also identify those elements that will require evaluation or a response to correct deficiencies. This sheet could also be used for developing the follow-up and documentation requirements.

The selection of effective audit team members is critical to the success of the program. Team members should be chosen for their experience, knowledge, and training and should be familiar with the processes and with auditing techniques, practices and procedures. The size of the team will vary depending on the size and complexity of the process under consideration. For a large, complex, highly instrumented plant, it may be desirable to have team members with expertise in process engineering and design, process chemistry, instrumentation and computer controls, electrical hazards and classifications, safety and health disciplines, maintenance, emergency preparedness, warehousing or shipping, and process safety auditing. The team may use part-time members to provide for the depth of expertise required as well as for what is actually done or followed, compared to what is written.

An effective audit includes a review of the relevant documentation and process safety information, inspection of the physical facilities, and interviews with all levels of plant personnel. Utilizing the audit procedure and checklist developed in the preplanning stage, the audit team can systematically analyze compliance with the provisions of the standard and any other corporate policies that are relevant. For example, the audit team will review all aspects of the training program as part of the overall audit. The team will review the written training program for adequacy of content, frequency of training, effectiveness of training in terms of its goals and objectives as well as to how it fits into meeting the standard's requirements, documentation, etc. Through interviews, the team can determine the employee's knowledge and awareness of the safety procedures, duties, rules, emergency response assignments, etc. During the inspection, the team can observe actual practices such as safety and health policies, procedures, and work authorization practices. This approach enables the team to identify deficiencies and determine where corrective actions or improvements are necessary.

An audit is a technique used to gather sufficient facts and information, including statistical information, to verify compliance with standards. Auditors should select as part of their preplanning a sample size sufficient to give a degree of confidence that the audit reflects the level of compliance with the standard. The audit team, through this systematic analysis, should document areas which require corrective action as well as those areas where the process safety management system is effective and working in an effective manner. This provides a record of the audit procedures and findings, and serves as a baseline of operation data for future audits. It will assist future auditors in determining changes or trends from previous audits.

Corrective action is one of the most important parts of the audit. It includes not only addressing the identified deficiencies, but also planning, followup, and documentation. The corrective action process normally begins with a management review of the audit findings. The purpose of this review is to determine what actions are appropriate, and to establish priorities, timetables, resource allocations and requirements and responsibilities. In some cases, corrective action may involve a simple change in procedure or minor maintenance effort to remedy the concern. Management of change procedures need to be used, as appropriate, even for what may seem to be a minor change. Many of the deficiencies can be acted on promptly, while some may require engineering studies or indepth review of actual procedures and practices. There may be instances where no action is necessary and this is a valid response to an audit finding. All actions taken, including an explanation where no action is taken on a finding, needs to be documented as to what was done and why.

It is important to assure that each deficiency identified is addressed, the corrective action to be taken noted, and the audit person or team responsible be properly documented by the employer. To control the corrective action process, the employer should consider the use of a tracking system. This tracking system might include periodic status reports shared with affected levels of management, specific reports such as completion of an engineering study, and a final implementation report to provide closure for audit findings that have been through management of change, if appropriate, and then shared with affected employees and management. This type of tracking system provides the employer with the status of the corrective action. It also provides the documentation required to verify that appropriate corrective actions were taken on deficiencies identified in the audit.

### Appendix D to §1926.64— Sources of Further Information (Nonmandatory)


1. Center for Chemical Process Safety, American Institute of Chemical Engineers, 345 East 47th Street, New York, NY 10017, (212) 705-7319.

2. "Guidelines for Hazard Evaluation Procedures," American Institute of Chemical Engineers; 345 East 47th Street, New York, NY 10017.

3. "Guidelines for Technical Management of Chemical Process Safety," Center for Chemical Process Safety of the American Institute of Chemical Engineers; 345 East 47th Street, New York, NY 10017.

4. "Evaluating Process Safety in the Chemical Industry," Chemical Manufacturers Association; 2501 M Street NW, Washington, DC 20037.

5. "Safe Warehousing of Chemicals," Chemical Manufacturers Association; 2501 M Street NW, Washington, DC 20037.

6. "Management of Process Hazards," American Petroleum Institute (API Recommended Practice 750); 1220 L Street, N.W., Washington, DC 20005.

7. "Improving Owner and Contractor Safety Performance," American Petroleum Institute (API Recommended Practice 2220); API, 1220 L Street N.W., Washington, DC 20005.

8. Chemical Manufacturers Association (CMA's Manager Guide), First Edition, September 1991; CMA, 2501 M Street, N.W., Washington, DC 20037.

9. "Improving Construction Safety Performance," Report A- 3, The Business Roundtable; The Business Roundtable, 200 Park Avenue, New York, NY 10166. (Report includes criteria to evaluate contractor safety performance and criteria to enhance contractor safety performance).

10. "Recommended Guidelines for Contractor Safety and Health," Texas Chemical Council; Texas Chemical Council, 1402 Nueces Street, Austin, TX 78701-1534.

11. "Loss Prevention in the Process Industries," Volumes I and II; Frank P. Lees, Butterworth; London 1983.

12. "Safety and Health Program Management Guidelines," 1989; U.S. Department of Labor, Occupational Safety and Health Administration.

13. "Safety and Health Guide for the Chemical Industry," 1986, (OSHA 3091); U.S. Department of Labor, Occupational Safety and Health Administration; 200 Constitution Avenue, N.W., Washington, DC 20210.

14. "Review of Emergency Systems," June 1988; U.S. Environmental Protection Agency (EPA), Office of Solid Waste and Emergency Response, Washington, DC 20460.

15. "Technical Guidance for Hazards Analysis, Emergency Planning for Extremely Hazardous Substances," December 1987; U.S. Environmental Protection Agency (EPA), Federal Emergency Management Administration (FEMA) and U.S. Department of Transportation (DOT), Washington, DC 20460.

16. "Accident Investigation...A New Approach," 1983, National Safety Council; 444 North Michigan Avenue, Chicago, IL 60611-3991.

17. "Fire & Explosion Index Hazard Classification Guide," 6th Edition, May 1987, Dow Chemical Company; Midland, Michigan 48674.

18. "Chemical Exposure Index," May 1988, Dow Chemical Company; Midland, Michigan 48674.


[58 FR 35115, June 30, 1993]

### §1926.65 Hazardous waste operations and emergency response.

**1926.65(a)** ***Scope, application, and definitions***— (1) Scope. This section covers the following operations, unless the employer can demonstrate that the operation does not involve employee exposure or the reasonable possibility for employee exposure to safety or health hazards:

**1926.65(a)(1)(i)** Clean-up operations required by a governmental body, whether Federal, state, local or other involving hazardous substances that are conducted at uncontrolled hazardous waste sites (including, but not limited to, the EPA's National Priority Site List (NPL), state priority site lists, sites recommended for the EPA NPL, and initial investigations of government identified sites which are conducted before the presence or absence of hazardous substances has been ascertained);

**1926.65(a)(1)(ii)** Corrective actions involving clean-up operations at sites covered by the Resource Conservation and Recovery Act of 1976 (RCRA) as amended (42 U.S.C. 6901 *et seq.*);

**1926.65(a)(1)(iii)** Voluntary clean-up operations at sites recognized by Federal, state, local or other governmental bodies as uncontrolled hazardous waste sites;

**1926.65(a)(1)(iv)** Operations involving hazardous wastes that are conducted at treatment, storage, and disposal (TSD) facilities regulated by 40 CFR parts 264 and 265 pursuant to RCRA; or by agencies under agreement with U.S.E.P.A. to implement RCRA regulations; and

**1926.65(a)(1)(v)** Emergency response operations for releases of, or substantial threats of releases of, hazardous substances without regard to the location of the hazard.

**1926.65(a)(2)** *Application.* (i) All requirements of part 1910 and part 1926 of title 29 of the Code of Federal Regulations apply pursuant to their terms to hazardous waste and emergency response operations whether covered by this section or not. If there is a conflict or overlap, the provision more protective of employee safety and health shall apply without regard to 29 CFR 1926.20(e)(1).

**1926.65(a)(2)(ii)** Hazardous substance clean-up operations within the scope of paragraphs (a)(1)(i) through (a)(1)(iii) of this section must comply with all paragraphs of this section except paragraphs (p) and (q).

**1926.65(a)(2)(iii)** Operations within the scope of paragraph (a)(1)(iv) of this section must comply only with the requirements of paragraph (p) of this section.

**Notes and Exceptions:** (A) All provisions of paragraph (p) of this section cover any treatment, storage or disposal (TSD) operation regulated by 40

CFR parts 264 and 265 or by state law authorized under RCRA, and required to have a permit or interim status from EPA pursuant to 40 CFR 270.1 or from a state agency pursuant to RCRA.

(B) Employers who are not required to have a permit or interim status because they are conditionally exempt small quantity generators under 40 CFR 261.5 or are generators who qualify under 40 CFR 262.34 for exemptions from regulation under 40 CFR parts 264, 265 and 270 ("excepted employers") are not covered by paragraphs (p)(1) through (p)(7) of this section. Excepted employers who are required by the EPA or state agency to have their employees engage in emergency response or who direct their employees to engage in emergency response are covered by paragraph (p)(8) of this section, and cannot be exempted by (p)(8)(i) of this section. Excepted employers who are not required to have employees engage in emergency response, who direct their employees to evacuate in the case of such emergencies and who meet the requirements of paragraph (p)(8)(i) of this section are exempt from the balance of paragraph (p)(8) of this section.

(C) If an area is used primarily for treatment, storage or disposal, any emergency response operations in that area shall comply with paragraph (p)(8) of this section. In other areas not used primarily for treatment, storage, or disposal, any emergency response operations shall comply with paragraph (q) of this section. Compliance with the requirements of paragraph (q) of this section shall be deemed to be in compliance with the requirements of paragraph (p)(8) of this section.

**1926.65(a)(2)(iv)** Emergency response operations for releases of, or substantial threats of releases of, hazardous substances which are not covered by paragraphs (a)(1)(i) through (a)(1)(iv) of this section must only comply with the requirements of paragraph (q) of this section.

**1926.65(a)(3)** *Definitions—Buddy system* means a system of organizing employees into work groups in such a manner that each employee of the work group is designated to be observed by at least one other employee in the work group. The purpose of the buddy system is to provide rapid assistance to employees in the event of an emergency.

*Clean-up operation* means an operation where hazardous substances are removed, contained, incinerated, neutralized, stabilized, cleared-up, or in any other manner processed or handled with the ultimate goal of making the site safer for people or the environment.

*Decontamination* means the removal of hazardous substances from employees and their equipment to the extent necessary to preclude the occurrence of foreseeable adverse health affects.

*Emergency response* or *responding to emergencies* means a response effort by employees from outside the immediate release area or by other designated responders (i.e., mutual-aid groups, local fire departments, etc.) to an occurrence which results, or is likely to result, in an uncontrolled release of a hazardous substance. Responses to incidental releases of hazardous substances where the substance can be absorbed, neutralized, or otherwise controlled at the time of release by employees in the immediate release area, or by maintenance personnel are not considered to be emergency responses within the scope of this standard. Responses to releases of hazardous substances where there is no potential safety or health hazard (i.e., fire, explosion, or chemical exposure) are not considered to be emergency responses.

*Facility* means (A) any building, structure, installation, equipment, pipe or pipeline (including any pipe into a sewer or publicly owned treatment works), well, pit, pond, lagoon, impoundment, ditch, storage container, motor vehicle, rolling stock, or aircraft, or (B) any site or area where a hazardous substance has been deposited, stored, disposed of, or placed, or otherwise come to be located; but does not include any consumer product in consumer use or any water-borne vessel.

*Hazardous materials response (HAZMAT) team* means an organized group of employees, designated by the employer, who are expected to perform work to handle and control actual or potential leaks or spills of hazardous substances requiring possible close approach to the substance. The team members perform responses to releases or potential releases of hazardous substances for the purpose of control or stabilization of the incident. A HAZMAT team is not a fire brigade nor is a typical fire brigade a HAZMAT team. A HAZMAT team, however, may be a separate component of a fire brigade or fire department.

*Hazardous substance* means any substance designated or listed under paragraphs (A) through (D) of this definition, exposure to which results or may result in adverse affects on the health or safety of employees:

(A) Any substance defined under section 101(14) of CERCLA;

(B) Any biological agent and other disease-causing agent which after release into the environment and upon exposure, ingestion, inhalation, or assimilation into any person, either directly from the environment or indirectly by ingestion through food chains, will or may reasonably be anticipated to cause death, disease, behavioral abnormalities, cancer, genetic mutation, physiological malfunctions (including malfunctions in reproduction) or physical deformations in such persons or their offspring;

(C) Any substance listed by the U.S. Department of Transportation as hazardous materials under 49 CFR 172.101 and appendices; and

(D) Hazardous waste as herein defined.

*Hazardous waste* means—(A) A waste or combination of wastes as defined in 40 CFR 261.3, or

(B) Those substances defined as hazardous wastes in 49 CFR 171.8.

*Hazardous waste operation* means any operation conducted within the scope of this standard.

*Hazardous waste site* or *Site* means any facility or location within the scope of this standard at which hazardous waste operations take place.

*Health hazard* means a chemical, mixture of chemicals or a pathogen for which there is statistically significant evidence based on at least one study conducted in accordance with established scientific principles that acute or chronic health effects may occur in exposed employees. The term *health hazard* includes chemicals which are carcinogens, toxic or highly toxic agents, reproductive toxins, irritants, corrosives, sensitizers, heptaotoxins, nephrotoxins, neurotoxins, agents which act on the hematopoietic system, and agents which damage the lungs, skin, eyes, or mucous membranes. It also includes stress due to temperature extremes. Further definition of the terms used above can be found in appendix A to 29 CFR 1926.59.

*IDLH* or *Immediately dangerous to life or health* means an atmospheric concentration of any toxic, corrosive or asphyxiant substance that poses an immediate threat to life or would cause irreversible or delayed adverse health effects or would interfere with an individual's ability to escape from a dangerous atmosphere.

*Oxygen deficiency* means that concentration of oxygen by volume below which atmosphere supplying respiratory protection must be provided. It exists in atmospheres where the percentage of oxygen by volume is less than 19.5 percent oxygen.

*Permissible exposure limit* means the exposure, inhalation or dermal permissible exposure limit specified either in §1926.55, elsewhere in subpart D, or in other pertinent sections of this part.

*Published exposure level* means the exposure limits published in "NIOSH Recommendations for Occupational Health Standards" dated 1986 incorporated by reference, or if none is specified, the exposure limits published in the standards specified by the American Conference of Governmental Industrial Hygienists in their publication "Threshold Limit Values and Biological Exposure Indices for 1987-88" dated 1987 incorporated by reference.

*Post emergency response* means that portion of an emergency response performed after the immediate threat of a release has been stabilized or eliminated and clean-up of the site has begun. If post emergency response is performed by an employer's own employees who were part of the initial emergency response, it is considered to be part of the initial response and not post emergency response. However, if a group of an employer's own employees, separate from the group providing initial response, performs the clean-up operation, then the separate group of employees would be considered to be performing post-emergency response and subject to paragraph (q)(11) of this section.

*Qualified person* means a person with specific training, knowledge and experience in the area for which the person has the responsibility and the authority to control.

*Site safety and health supervisor (or official)* means the individual located on a hazardous waste site who is responsible to the employer and has the authority and knowledge necessary to implement the site safety and health plan and verify compliance with applicable safety and health requirements.

*Small quantity generator* means a generator of hazardous wastes who in any calendar month generates no more than 1,000 kilograms (2,205 pounds) of hazardous waste in that month.

*Uncontrolled hazardous waste site,* means an area identified as an uncontrolled hazardous waste site by a governmental body, whether Federal, state, local or other where an accumulation of hazardous substances creates a threat to the health and safety of individuals or the environment or both.

Some sites are found on public lands such as those created by former municipal, county or state landfills where illegal or poorly managed waste disposal has taken place. Other sites are found on private property, often belonging to generators or former generators of hazardous substance wastes. Examples of such sites include, but are not limited to, surface impoundments, landfills, dumps, and tank or drum farms. Normal operations at TSD sites are not covered by this definition.

**1926.65(b)** ***Safety and health program.***

**Note to (B):** Safety and health programs developed and implemented to meet other Federal, state, or local regulations are considered acceptable in meeting this requirement if they cover or are modified to cover the topics required in this paragraph. An additional or separate safety and health program is not required by this paragraph.

*General.* (i) Employers shall develop and implement a written safety and health program for their employees involved in hazardous waste operations. The program shall be designed to identify, evaluate, and control safety and health hazards, and provide for emergency response for hazardous waste operations.

**1926.65(b)(ii)** The written safety and health program shall incorporate the following:

**1926.65(b)(ii)(A)** An organizational structure;

**1926.65(b)(ii)(B)** A comprehensive workplan;

**1926.65(b)(ii)(C)** A site-specific safety and health plan which need not repeat the employer's standard operating procedures required in paragraph (b)(1)(ii)(F) of this section;

**1926.65(b)(ii)(D)** The safety and health training program;

**1926.65(b)(ii)(E)** The medical surveillance program;

**1926.65(b)(ii)(F)** The employer's standard operating procedures for safety and health; and

**1926.65(b)(ii)(G)** Any necessary interface between general program and site specific activities.

**1926.65(b)(iii)** *Site excavation.* Site excavations created during initial site preparation or during hazardous waste operations shall be shored or sloped as appropriate to prevent accidental collapse in accordance with subpart P of 29 CFR part 1926.

**1926.65(b)(iv)** *Contractors and sub-contractors.* An employer who retains contractor or sub-contractor services for work in hazardous waste operations shall inform those contractors, sub-contractors, or their representatives of the site emergency response procedures and any potential fire, explosion, health, safety or other hazards of the hazardous waste operation that have been identified by the employer, including those identified in the employer's information program.

**1926.65(b)(v)** *Program availability.* The written safety and health program shall be made available to any contractor or subcontractor or their representative who will be involved with the hazardous waste operation; to employees; to employee designated representatives; to OSHA personnel, and to personnel of other Federal, state, or local agencies with regulatory authority over the site.

**1926.65(b)(2)** *Organizational structure part of the site program*— (i) The organizational structure part of the program shall establish the specific chain of command and specify the overall responsibilities of supervisors and employees. It shall include, at a minimum, the following elements:

**1926.65(b)(2)(i)(A)** A general supervisor who has the responsibility and authority to direct all hazardous waste operations.

**1926.65(b)(2)(i)(B)** A site safety and health supervisor who has the responsibility and authority to develop and implement the site safety and health plan and verify compliance.

**1926.65(b)(2)(i)(C)** All other personnel needed for hazardous waste site operations and emergency response and their general functions and responsibilities.

**1926.65(b)(2)(i)(D)** The lines of authority, responsibility, and communication.

**1926.65(b)(2)(ii)** The organizational structure shall be reviewed and updated as necessary to reflect the current status of waste site operations.

**1926.65(b)(3)** *Comprehensive workplan part of the site program.* The comprehensive workplan part of the program shall address the tasks and objectives of the site operations and the logistics and resources required to reach those tasks and objectives.

**1926.65(b)(3)(i)** The comprehensive workplan shall address anticipated clean-up activities as well as normal operating procedures which need not repeat the employer's procedures available elsewhere.

**1926.65(b)(3)(ii)** The comprehensive workplan shall define work tasks and objectives and identify the methods for accomplishing those tasks and objectives.

**1926.65(b)(3)(iii)** The comprehensive workplan shall establish personnel requirements for implementing the plan.

**1926.65(b)(3)(iv)** The comprehensive workplan shall provide for the implementation of the training required in paragraph (c) of this section.

**1926.65(b)(3)(v)** The comprehensive workplan shall provide for the implementation of the required informational programs required in paragraph (i) of this section.

**1926.65(b)(3)(vi)** The comprehensive workplan shall provide for the implementation of the medical surveillance program described in paragraph (f) of this section.

**1926.65(b)(4)** *Site-specific safety and health plan part of the program*— (i) *General.* The site safety and health plan, which must be kept on site, shall address the safety and health hazards of each phase of site operation and include the requirements and procedures for employee protection.

**1926.65(b)(4)(ii)** *Elements.* The site safety and health plan, as a minimum, shall address the following:

**1926.65(b)(4)(ii)(A)** A safety and health risk or hazard analysis for each site task and operation found in the workplan.

**1926.65(b)(4)(ii)(B)** Employee training assignments to assure compliance with paragraph (e) of this section.

**1926.65(b)(4)(ii)(C)** Personal protective equipment to be used by employees for each of the site tasks and operations being conducted as required by the personal protective equipment program in paragraph (g)(5) of this section.

**1926.65(b)(4)(ii)(D)** Medical surveillance requirements in accordance with the program in paragraph (f) of this section.

**1926.65(b)(4)(ii)(E)** Frequency and types of air monitoring, personnel monitoring, and environmental sampling techniques and instrumentation to be used, including methods of maintenance and calibration of monitoring and sampling equipment to be used.

**1926.65(b)(4)(ii)(F)** Site control measures in accordance with the site control program required in paragraph (d) of this section.

**1926.65(b)(4)(ii)(G)** Decontamination procedures in accordance with paragraph (k) of this section.

**1926.65(b)(4)(ii)(H)** An emergency response plan meeting the requirements of paragraph (l) of this section for safe and effective responses to emergencies, including the necessary PPE and other equipment.

**1926.65(b)(4)(ii)(I)** Confined space entry procedures.

**1926.65(b)(4)(ii)(J)** A spill containment program meeting the requirements of paragraph (j) of this section.

**1926.65(b)(4)(iii)** *Pre-entry briefing.* The site specific safety and health plan shall provide for pre-entry briefings to be held prior to initiating any site activity, and at such other times as necessary to ensure that employees are apprised of the site safety and health plan and that this plan is being followed. The information and data obtained from site characterization and analysis work required in paragraph (c) of this section shall be used to prepare and update the site safety and health plan.

**1926.65(b)(4)(iv)** *Effectiveness of site safety and health plan.* Inspections shall be conducted by the site safety and health supervisor or, in the absence of that individual, another individual who is knowledgeable in occupational safety and health, acting on behalf of the employer as necessary to determine the effectiveness of the site safety and health plan. Any deficiencies in the effectiveness of the site safety and health plan shall be corrected by the employer.

**1926.65(c)** ***Site characterization and analysis***— (1) General. Hazardous waste sites shall be evaluated in accordance with this paragraph to identify specific site hazards and to determine the appropriate safety and health control procedures needed to protect employees from the identified hazards.

**1926.65(c)(2)** *Preliminary evaluation.* A preliminary evaluation of a site's characteristics shall be performed prior to site entry by a qualified person in order to aid in the selection of appropriate employee protection methods prior to site entry. Immediately after initial site entry, a more detailed evaluation of the site's specific characteristics shall be performed by a qualified person in order to further identify existing site hazards and to further aid in the selection of the appropriate engineering controls and personal protective equipment for the tasks to be performed.

**1926.65(c)(3)** *Hazard identification.* All suspected conditions that may pose inhalation or skin absorption hazards that are immediately dangerous to life or health (IDLH), or other conditions that may cause death or serious harm, shall be identified during the preliminary survey and evaluated during the detailed survey. Examples of such hazards include, but are not limited to, confined space entry, potentially explosive or flammable situations, visible vapor clouds, or areas where biological indicators such as dead animals or vegetation are located.

**1926.65(c)(4)** *Required information.* The following information to the extent available shall be obtained by the employer prior to allowing employees to enter a site:

**1926.65(c)(4)(i)** Location and approximate size of the site.

**1926.65(c)(4)(ii)** Description of the response activity and/or the job task to be performed.

**1926.65(c)(4)(iii)** Duration of the planned employee activity.

**1926.65(c)(4)(iv)** Site topography and accessibility by air and roads.

**1926.65(c)(4)(v)** Safety and health hazards expected at the site.

**1926.65(c)(4)(vi)** Pathways for hazardous substance dispersion.

**1926.65(c)(4)(vii)** Present status and capabilities of emergency response teams that would provide assistance to hazardous waste clean-up site employees at the time of an emergency.

**1926.65(c)(4)(viii)** Hazardous substances and health hazards involved or expected at the site, and their chemical and physical properties.

**1926.65(c)(5)** *Personal protective equipment.* Personal protective equipment (PPE) shall be provided and used during initial site entry in accordance with the following requirements:

**1926.65(c)(5)(i)** Based upon the results of the preliminary site evaluation, an ensemble of PPE shall be selected and used during initial site entry which will provide protection to a level of exposure below permissible exposure limits and published exposure levels for known or suspected hazardous substances and health hazards, and which will provide protection against other known and suspected hazards identified during the preliminary site evaluation. If there is no permissible exposure limit or published exposure level, the employer may use other published studies and information as a guide to appropriate personal protective equipment.

**1926.65(c)(5)(ii)** If positive-pressure self-contained breathing apparatus is not used as part of the entry ensemble, and if respiratory protection is warranted by the potential hazards identified during the preliminary site evaluation, an escape self-contained breathing apparatus of at least five minute's duration shall be carried by employees during initial site entry.

**1926.65(c)(5)(iii)** If the preliminary site evaluation does not produce sufficient information to identify the hazards or suspected hazards of the site, an ensemble providing protection equivalent to Level B PPE shall be provided as minimum protection, and direct reading instruments shall be used as appropriate for identifying IDLH conditions. (See appendix B for a description of Level B hazards and the recommendations for Level B protective equipment.)

**1926.65(c)(5)(iv)** Once the hazards of the site have been identified, the appropriate PPE shall be selected and used in accordance with paragraph (g) of this section.

**1926.65(c)(6)** *Monitoring.* The following monitoring shall be conducted during initial site entry when the site evaluation produces information that shows the potential for ionizing radiation or IDLH conditions, or when the site information is not sufficient reasonably to eliminate these possible conditions:

**1926.65(c)(6)(i)** Monitoring with direct reading instruments for hazardous levels of ionizing radiation.

**1926.65(c)(6)(ii)** Monitoring the air with appropriate direct reading test equipment (i.e., combustible gas meters, detector tubes) for IDLH and other conditions that may cause death or serious harm (combustible or explosive atmospheres, oxygen deficiency, toxic substances).

**1926.65(c)(6)(iii)** Visually observing for signs of actual or potential IDLH or other dangerous conditions.

**1926.65(c)(6)(iv)** An ongoing air monitoring program in accordance with paragraph (h) of this section shall be implemented after site characterization has determined the site is safe for the start-up of operations.

**1926.65(c)(7)** *Risk identification.* Once the presence and concentrations of specific hazardous substances and health hazards have been established, the risks associated with these substances shall be identified. Employees who will be working on the site shall be informed of any risks that have been identified. In situations covered by the Hazard Communication Standard, 29 CFR 1926.59, training required by that standard need not be duplicated.

**Note to (c)(7).** Risks to consider include, but are not limited to:

(a) Exposures exceeding the permissible exposure limits and published exposure levels.

(b) IDLH concentrations.

(c) Potential skin absorption and irritation sources.

(d) Potential eye irritation sources.

(e) Explosion sensitivity and flammability ranges.

(f) Oxygen deficiency.

**1926.65(c)(8)** *Employee notification.* Any information concerning the chemical, physical, and toxicologic properties of each substance known or expected to be present on site that is available to the employer and relevant to the duties an employee is expected to perform shall be made available to the affected employees prior to the commencement of their work activities. The employer may utilize information developed for the hazard communication standard for this purpose.

**1926.65(d)** ***Site control***— (1) General. Appropriate site control procedures shall be implemented to control employee exposure to hazardous substances before clean-up work begins.

**1926.65(d)(2)** *Site control program.* A site control program for protecting employees which is part of the employer's site safety and health program required in paragraph (b) of this section shall be developed during the planning stages of a hazardous waste clean-up operation and modified as necessary as new information becomes available.

**1926.65(d)(3)** *Elements of the site control program.* The site control program shall, as a minimum, include: A site map; site work zones; the use of a "buddy system"; site communications including alerting means for emergencies; the standard operating procedures or safe work practices; and, identification of the nearest medical assistance. Where these requirements are covered elsewhere they need not be repeated.

**1926.65(e)** ***Training***— (1) General. (i) All employees working on site (such as but not limited to equipment operators, general laborers and others) exposed to hazardous substances, health hazards, or safety hazards and their supervisors and management responsible for the site shall receive training meeting the requirements of this paragraph before they are permitted to engage in hazardous waste operations that could expose them to hazardous substances, safety, or health hazards, and they shall receive review training as specified in this paragraph.

**1926.65(e)(1)(ii)** Employees shall not be permitted to participate in or supervise field activities until they have been trained to a level required by their job function and responsibility.

**1926.65(e)(2)** *Elements to be covered.* The training shall thoroughly cover the following:

**1926.65(e)(2)(i)** Names of personnel and alternates responsible for site safety and health;

**1926.65(e)(2)(ii)** Safety, health and other hazards present on the site;

**1926.65(e)(2)(iii)** Use of personal protective equipment;

**1926.65(e)(2)(iv)** Work practices by which the employee can minimize risks from hazards;

**1926.65(e)(2)(v)** Safe use of engineering controls and equipment on the site;

**1926.65(e)(2)(vi)** Medical surveillance requirements, including recognition of symptoms and signs which might indicate overexposure to hazards; and

**1926.65(e)(2)(vii)** The contents of paragraphs (G) through (J) of the site safety and health plan set forth in paragraph (b)(4)(ii) of this section.

**1926.65(e)(3)** *Initial training.* (i) General site workers (such as equipment operators, general laborers and supervisory personnel) engaged in hazardous substance removal or other activities which expose or potentially expose workers to hazardous substances and health hazards shall receive a minimum of 40 hours of instruction off the site, and a minimum of three days actual field experience under the direct supervision of a trained, experienced supervisor.

**1926.65(e)(3)(ii)** Workers on site only occasionally for a specific limited task (such as, but not limited to, ground water monitoring, land surveying, or geo-physical surveying) and who are unlikely to be exposed over permissible exposure limits and published exposure limits shall receive a minimum of 24 hours of instruction off the site, and the minimum of one day actual field experience under the direct supervision of a trained, experienced supervisor.

**1926.65(e)(3)(iii)** Workers regularly on site who work in areas which have been monitored and fully characterized indicating that exposures are under permissible exposure limits and published exposure limits where respirators are not necessary, and the characterization indicates that there are no health hazards or the possibility of an emergency developing, shall receive a minimum of 24 hours of instruction off the site and the minimum of one day actual field experience under the direct supervision of a trained, experienced supervisor.

**1926.65(e)(3)(iv)** Workers with 24 hours of training who are covered by paragraphs (e)(3)(ii) and (e)(3)(iii) of this section, and who become general site workers or who are required to wear respirators, shall have the additional 16 hours and two days of training necessary to total the training specified in paragraph (e)(3)(i).

**1926.65(e)(4)** *Management and supervisor training.* On-site management and supervisors directly responsible for, or who supervise employees engaged in, hazardous waste operations shall receive 40 hours initial training, and three days of supervised field experience (the training may be reduced to 24 hours and one day if the only area of their responsibility is employees covered by paragraphs (e)(3)(ii) and (e)(3)(iii)) and at least eight additional hours of specialized training at the time of job assignment on such topics as, but not limited to, the employer's safety and health program and the associated employee training program, personal protective equipment program, spill containment program, and health hazard monitoring procedure and techniques.

**1926.65(e)(5)** *Qualifications for trainers.* Trainers shall be qualified to instruct employees about the subject matter that is being presented in training. Such trainers shall have satisfactorily completed a training program for teaching the subjects they are expected to teach, or they shall have the academic credentials and instructional experience necessary for teaching the subjects. Instructors shall demonstrate competent instructional skills and knowledge of the applicable subject matter.

**1926.65(e)(6)** *Training certification.* Employees and supervisors that have received and successfully completed the training and field experience specified in paragraphs (e)(1) through (e)(4) of this section shall be certified by their instructor or the head instructor and trained supervisor as having successfully completed the necessary training. A written certificate shall be given to each person so certified. Any person who has not been so certified or who does not meet the requirements of paragraph (e)(9) of this section shall be prohibited from engaging in hazardous waste operations.

**1926.65(e)(7)** *Emergency response.* Employees who are engaged in responding to hazardous emergency situations at hazardous waste clean-up sites that may expose them to hazardous substances shall be trained in how to respond to such expected emergencies.

**1926.65(e)(8)** *Refresher training.* Employees specified in paragraph (e)(1) of this section, and managers and supervisors specified in paragraph (e)(4) of this section, shall receive eight hours of refresher training annually on the items specified in paragraph (e)(2) and/or (e)(4) of this section, any critique of incidents that have occurred in the past year that can serve as training examples of related work, and other relevant topics.

**1926.65(e)(9)** *Equivalent training.* Employers who can show by documentation or certification that an employee's work experience and/or training has resulted in training equivalent to that training required in paragraphs (e)(1) through (e)(4) of this section shall not be required to provide the initial training requirements of those paragraphs to such employees and shall provide a copy of the certification or documentation to the employee upon request. However, certified employees or employees with equivalent training new to a site shall receive appropriate, site specific training before site entry and have appropriate supervised field experience at the new site. Equivalent training includes any academic training or the training that existing employees might have already received from actual hazardous waste site work experience.

**1926.65(f)** ***Medical surveillance***— (1) General. Employers engaged in operations specified in paragraphs (a)(1)(i) through (a)(1)(iv) of this section and not covered by (a)(2)(iii) exceptions and employers of employees specified in paragraph (q)(9) shall institute a medical surveillance program in accordance with this paragraph.

**1926.65(f)(2)** *Employees covered.* The medical surveillance program shall be instituted by the employer for the following employees:

**1926.65(f)(2)(i)** All employees who are or may be exposed to hazardous substances or health hazards at or above the permissible exposure limits or, if there is no permissible exposure limit, above the published exposure levels for these substances, without regard to the use of respirators, for 30 days or more a year;

**1926.65(f)(2)(ii)** All employees who wear a respirator for 30 days or more a year or as required by §1926.103;

**1926.65(f)(2)(iii)** All employees who are injured, become ill or develop signs or symptoms due to possible overexposure involving hazardous substances or health hazards from an emergency response or hazardous waste operation; and

**1926.65(f)(2)(iv)** Members of HAZMAT teams.

**1926.65(f)(3)** *Frequency of medical examinations and consultations.* Medical examinations and consultations shall be made available by the employer to each employee covered under paragraph (f)(2) of this section on the following schedules:

**1926.65(f)(3)(i)** For employees covered under paragraphs (f)(2)(i), (f)(2)(ii), and (f)(2)(iv):

**1926.65(f)(3)(i)(A)** Prior to assignment;

**1926.65(f)(3)(i)(B)** At least once every twelve months for each employee covered unless the attending physician believes a longer interval (not greater than biennially) is appropriate;

**1926.65(f)(3)(i)(C)** At termination of employment or reassignment to an area where the employee would not be covered if the employee has not had an examination within the last six months;

**1926.65(f)(3)(i)(D)** As soon as possible upon notification by an employee that the employee has developed signs or symptoms indicating possible overexposure to hazardous substances or health hazards, or that the employee has been injured or exposed above the permissible exposure limits or published exposure levels in an emergency situation;

**1926.65(f)(3)(i)(E)** At more frequent times, if the examining physician determines that an increased frequency of examination is medically necessary.

**1926.65(f)(3)(ii)** For employees covered under paragraph (f)(2)(iii) and for all employees including those of employers covered by paragraph (a)(1)(v) who may have been injured, received a health impairment, developed signs or symptoms which may have resulted from exposure to hazardous substances resulting from an emergency incident, or exposed during an emergency incident to hazardous substances at concentrations above the permissible exposure limits or the published exposure levels without the necessary personal protective equipment being used:

**1926.65(f)(3)(ii)(A)** As soon as possible following the emergency incident or development of signs or symptoms;

**1926.65(f)(3)(ii)(B)** At additional times, if the examining physician determines that follow-up examinations or consultations are medically necessary.

**1926.65(f)(4)** *Content of medical examinations and consultations.* (i) Medical examinations required by paragraph (f)(3) of this section shall include a medical and work history (or updated history if one is in the employee's file) with special emphasis on symptoms related to the handling of hazardous substances and health hazards, and to fitness for duty including the ability to wear any required PPE under conditions (i.e., temperature extremes) that may be expected at the work site.

**1926.65(f)(4)(ii)** The content of medical examinations or consultations made available to employees pursuant to paragraph (f) shall be determined by the attending physician. The guidelines in the *Occupational Safety and Health Guidance Manual for Hazardous Waste Site Activities* (See appendix D, Reference #10) should be consulted.

**1926.65(f)(5)** *Examination by a physician and costs.* All medical examinations and procedures shall be performed by or under the supervision of a licensed physician, preferably one knowledgeable in occupational medicine, and shall be provided without cost to the employee, without loss of pay, and at a reasonable time and place.

**1926.65(f)(6)** *Information provided to the physician.* The employer shall provide one copy of this standard and its appendices to the attending physician, and in addition the following for each employee:

**1926.65(f)(6)(i)** A description of the employee's duties as they relate to the employee's exposures.

**1926.65(f)(6)(ii)** The employee's exposure levels or anticipated exposure levels.

**1926.65(f)(6)(iii)** A description of any personal protective equipment used or to be used.

**1926.65(f)(6)(iv)** Information from previous medical examinations of the employee which is not readily available to the examining physician.

**1926.65(f)(6)(v)** Information required by §1926.103.

**1926.65(f)(7)** *Physician's written opinion.* (i) The employer shall obtain and furnish the employee with a copy of a written opinion from the attending physician containing the following:

**1926.65(f)(7)(i)(A)** The physician's opinion as to whether the employee has any detected medical conditions which would place the employee at increased risk of material impairment of the employee's health from work in hazardous waste operations or emergency response, or from respirator use.

**1926.65(f)(7)(i)(B)** The physician's recommended limitations upon the employee's assigned work.

**1926.65(f)(7)(i)(C)** The results of the medical examination and tests if requested by the employee.

**1926.65(f)(7)(i)(D)** A statement that the employee has been informed by the physician of the results of the medical examination and any medical conditions which require further examination or treatment.

**1926.65(f)(7)(ii)** The written opinion obtained by the employer shall not reveal specific findings or diagnoses unrelated to occupational exposures.

**1926.65(f)(8)** *Recordkeeping.* (i) An accurate record of the medical surveillance required by paragraph (f) of this section shall be retained. This record shall be retained for the period specified and meet the criteria of 29 CFR 1926.33.

**1926.65(f)(8)(ii)** The record required in paragraph (f)(8)(i) of this section shall include at least the following information:

**1926.65(f)(8)(ii)(A)** The name and social security number of the employee;

**1926.65(f)(8)(ii)(B)** Physician's written opinions, recommended limitations, and results of examinations and tests;

**1926.65(f)(8)(ii)(C)** Any employee medical complaints related to exposure to hazardous substances;

**1926.65(f)(8)(ii)(D)** A copy of the information provided to the examining physician by the employer, with the exception of the standard and its appendices.

**1926.65(g)** ***Engineering controls, work practices, and personal protective equipment for employee protection.*** Engineering controls, work practices, personal protective equipment, or a combination of these shall be implemented in accordance with this paragraph to protect employees from exposure to hazardous substances and safety and health hazards.

**1926.65(g)(1)** *Engineering controls, work practices and PPE for substances regulated either in §1926.55, elsewhere in subpart D, or in other pertinent sections of this part.* (i) Engineering controls and work practices shall be instituted to reduce and maintain employee exposure to or below the permissible exposure limits for substances regulated either in §1926.55 or other pertinent sections of this part, except to the extent that such controls and practices are not feasible.

**Note to (g)(1)(i):** Engineering controls which may be feasible include the use of pressurized cabs or control booths on equipment, and/or the use of remotely operated material handling equipment. Work practices which may be feasible are removing all non-essential employees from potential exposure during opening of drums, wetting down dusty operations and locating employees upwind of possible hazards.

**1926.65(g)(1)(ii)** Whenever engineering controls and work practices are not feasible or not required, any reasonable combination of engineering controls, work practices and PPE shall be used to reduce and maintain employee exposures to or below the permissible exposure limits or dose limits for substances regulated either in §1926.55 or other pertinent sections of this part.

**1926.65(g)(1)(iii)** The employer shall not implement a schedule of employee rotation as a means of compliance with permissible exposure limits or dose limits except when there is no other feasible way of complying with the airborne or dermal dose limits for ionizing radiation.

**1926.65(g)(1)(iv)** The provisions of subpart D shall be followed.

**1926.65(g)(2)** *Engineering controls, work practices, and PPE for substances not regulated either in §1926.55, elsewhere in subpart D, or in other pertinent sections of this part.* An appropriate combination of engineering controls, work practices and personal protective equipment shall be used to reduce and maintain employee exposure to or below published exposure levels for hazardous substances and health hazards not regulated either in §1926.55, elsewhere in subpart D, or in other pertinent sections of this part. The employer may use the published literature and MSDS as a guide in making the employer's determination as to what level of protection the employer believes is appropriate for hazardous substances and health hazards for which there is no permissible exposure limit or published exposure limit.

**1926.65(g)(3)** *Personal protective equipment selection.* (i) Personal protective equipment (PPE) shall be selected and used which will protect employees from the hazards and potential hazards they are likely to encounter as identified during the site characterization and analysis.

**1926.65(g)(3)(ii)** Personal protective equipment selection shall be based on an evaluation of the performance characteristics of the PPE relative to the requirements and limitations of the site, the task-specific conditions and duration, and the hazards and potential hazards identified at the site.

**1926.65(g)(3)(iii)** Positive pressure self-contained breathing apparatus, or positive pressure air-line respirators equipped with an escape air supply, shall be used when chemical exposure levels present will create a substantial possibility of immediate death, immediate serious illness or injury, or impair the ability to escape.

**1926.65(g)(3)(iv)** Totally-encapsulating chemical protective suits (protection equivalent to Level A protection as recommended in appendix B) shall be used in conditions where skin absorption of a hazardous substance may result in a substantial possibility of immediate death, immediate serious illness or injury, or impair the ability to escape.

**1926.65(g)(3)(v)** The level of protection provided by PPE selection shall be increased when additional information on site conditions indicates that increased protection is necessary to reduce employee exposures below permissible exposure limits and published exposure levels for hazardous substances and health hazards. (See appendix B for guidance on selecting PPE ensembles.)

**Note to (G)(3):** The level of employee protection provided may be decreased when additional information or site conditions show that decreased protection will not result in hazardous exposures to employees.

**1926.65(g)(3)(vi)** Personal protective equipment shall be selected and used to meet the requirements of subpart E of this part and additional requirements specified in this section.

**1926.65(g)(4)** *Totally-encapsulating chemical protective suits.* (i) Totally-encapsulating suits shall protect employees from the particular hazards which are identified during site characterization and analysis.

**1926.65(g)(4)(ii)** Totally-encapsulating suits shall be capable of maintaining positive air pressure. (See appendix A for a test method which may be used to evaluate this requirement.)

**1926.65(g)(4)(iii)** Totally-encapsulating suits shall be capable of preventing inward test gas leakage of more than 0.5 percent. (See appendix A for a test method which may be used to evaluate this requirement.)

**1926.65(g)(5)** *Personal protective equipment (PPE) program.* A written personal protective equipment program, which is part of the employer's safety and health program required in paragraph (b) of this section or required in paragraph (p)(1) of this section and which is also a part of the site-specific safety and health plan shall be established. The PPE program shall address the elements listed below. When elements, such as donning and doffing procedures, are provided by the manufacturer of a piece of equipment and are attached to the plan, they need not be rewritten into the plan as long as they adequately address the procedure or element.

**1926.65(g)(5)(i)** PPE selection based upon site hazards,

**1926.65(g)(5)(ii)** PPE use and limitations of the equipment,

**1926.65(g)(5)(iii)** Work mission duration,

**1926.65(g)(5)(iv)** PPE maintenance and storage,

**1926.65(g)(5)(v)** PPE decontamination and disposal,

**1926.65(g)(5)(vi)** PPE training and proper fitting,

**1926.65(g)(5)(vii)** PPE donning and doffing procedures,

**1926.65(g)(5)(viii)** PPE inspection procedures prior to, during, and after use,

**1926.65(g)(5)(ix)** Evaluation of the effectiveness of the PPE program, and

**1926.65(g)(5)(x)** Limitations during temperature extremes, heat stress, and other appropriate medical considerations.

**1926.65(h) *Monitoring*—** (1) General. (i) Monitoring shall be performed in accordance with this paragraph where there may be a question of employee exposure to hazardous concentrations of hazardous substances in order to assure proper selection of engineering controls, work practices and personal protective equipment so that employees are not exposed to levels which exceed permissible exposure limits, or published exposure levels if there are no permissible exposure limits, for hazardous substances.

**1926.65(h)(1)(ii)** Air monitoring shall be used to identify and quantify airborne levels of hazardous substances and safety and health hazards in order to determine the appropriate level of employee protection needed on site.

**1926.65(h)(2)** *Initial entry.* Upon initial entry, representative air monitoring shall be conducted to identify any IDLH condition, exposure over permissible exposure limits or published exposure levels, exposure over a radioactive material's dose limits or other dangerous condition such as the presence of flammable atmospheres or oxygen-deficient environments.

**1926.65(h)(3)** *Periodic monitoring.* Periodic monitoring shall be conducted when the possibility of an IDLH condition or flammable atmosphere has developed or when there is indication that exposures may have risen over permissible exposure limits or published exposure levels since prior monitoring. Situations where it shall be considered whether the possibility that exposures have risen are as follows:

**1926.65(h)(3)(i)** When work begins on a different portion of the site.

**1926.65(h)(3)(ii)** When contaminants other than those previously identified are being handled.

**1926.65(h)(3)(iii)** When a different type of operation is initiated (e.g., drum opening as opposed to exploratory well drilling).

**1926.65(h)(3)(iv)** When employees are handling leaking drums or containers or working in areas with obvious liquid contamination (e.g., a spill or lagoon).

**1926.65(h)(4)** *Monitoring of high risk employees.* After the actual clean-up phase of any hazardous waste operation commences; for example, when soil, surface water or containers are moved or disturbed; the employer shall monitor those employees likely to have the highest exposures to hazardous substances and health hazards likely to be present above permissible exposure limits or published exposure levels by using personal sampling frequently enough to characterize employee exposures. If the employees likely to have the highest exposure are over permissible exposure limits or published exposure limits, then monitoring shall continue to determine all employees likely to be above those limits. The employer may utilize a representative sampling approach by documenting that the employees and chemicals chosen for monitoring are based on the criteria stated above.

**Note to (h):** It is not required to monitor employees engaged in site characterization operations covered by paragraph (c) of this section.

**1926.65(i) *Informational programs.*** Employers shall develop and implement a program, which is part of the employer's safety and health program required in paragraph (b) of this section, to inform employees, contractors, and subcontractors (or their representative) actually engaged in hazardous waste operations of the nature, level and degree of exposure likely as a result of participation in such hazardous waste operations. Employees, contractors and subcontractors working outside of the operations part of a site are not covered by this standard.

**1926.65(j) *Handling drums and containers*—** (1) General. (i) Hazardous substances and contaminated soils, liquids, and other residues shall be handled, transported, labeled, and disposed of in accordance with this paragraph.

**1926.65(j)(1)(ii)** Drums and containers used during the clean-up shall meet the appropriate DOT, OSHA, and EPA regulations for the wastes that they contain.

**1926.65(j)(1)(iii)** When practical, drums and containers shall be inspected and their integrity shall be assured prior to being moved. Drums or containers that cannot be inspected before being moved because of storage conditions (i.e., buried beneath the earth, stacked behind other drums, stacked several tiers high in a pile, etc.) shall be moved to an accessible location and inspected prior to further handling.

**1926.65(j)(1)(iv)** Unlabelled drums and containers shall be considered to contain hazardous substances and handled accordingly until the contents are positively identified and labeled.

**1926.65(j)(1)(v)** Site operations shall be organized to minimize the amount of drum or container movement.

**1926.65(j)(1)(vi)** Prior to movement of drums or containers, all employees exposed to the transfer operation shall be warned of the potential hazards associated with the contents of the drums or containers.

**1926.65(j)(1)(vii)** U.S. Department of Transportation specified salvage drums or containers and suitable quantities of proper absorbent shall be kept available and used in areas where spills, leaks, or ruptures may occur.

**1926.65(j)(1)(viii)** Where major spills may occur, a spill containment program, which is part of the employer's safety and health program required in paragraph (b) of this section, shall be implemented to contain and isolate the entire volume of the hazardous substance being transferred.

**1926.65(j)(1)(ix)** Drums and containers that cannot be moved without rupture, leakage, or spillage shall be emptied into a sound container using a device classified for the material being transferred.

**1926.65(j)(1)(x)** A ground-penetrating system or other type of detection system or device shall be used to estimate the location and depth of buried drums or containers.

**1926.65(j)(1)(xi)** Soil or covering material shall be removed with caution to prevent drum or container rupture.

**1926.65(j)(1)(xii)** Fire extinguishing equipment meeting the requirements of subpart F of this part shall be on hand and ready for use to control incipient fires.

**1926.65(j)(2)** *Opening drums and containers.* The following procedures shall be followed in areas where drums or containers are being opened:

**1926.65(j)(2)(i)** Where an airline respirator system is used, connections to the source of air supply shall be protected from contamination and the entire system shall be protected from physical damage.

**1926.65(j)(2)(ii)** Employees not actually involved in opening drums or containers shall be kept a safe distance from the drums or containers being opened.

**1926.65(j)(2)(iii)** If employees must work near or adjacent to drums or containers being opened, a suitable shield that does not interfere with the work operation shall be placed between the employee and the drums or containers being opened to protect the employee in case of accidental explosion.

**1926.65(j)(2)(iv)** Controls for drum or container opening equipment, monitoring equipment, and fire suppression equipment shall be located behind the explosion-resistant barrier.

**1926.65(j)(2)(v)** When there is a reasonable possibility of flammable atmospheres being present, material handling equipment and hand tools shall be of the type to prevent sources of ignition.

**1926.65(j)(2)(vi)** Drums and containers shall be opened in such a manner that excess interior pressure will be safely relieved. If pressure can not be relieved from a remote location, appropriate shielding shall be placed between the employee and the drums or containers to reduce the risk of employee injury.

**1926.65(j)(2)(vii)** Employees shall not stand upon or work from drums or containers.

**1926.65(j)(3)** *Material handling equipment.* Material handling equipment used to transfer drums and containers shall be selected, positioned and operated to minimize sources of ignition related to the equipment from igniting vapors released from ruptured drums or containers.

**1926.65(j)(4)** *Radioactive wastes.* Drums and containers containing radioactive wastes shall not be handled until such time as their hazard to employees is properly assessed.

**1926.65(j)(5)** *Shock sensitive wastes.* As a minimum, the following special precautions shall be taken when drums and containers containing or suspected of containing shock-sensitive wastes are handled:

**1926.65(j)(5)(i)** All non-essential employees shall be evacuated from the area of transfer.

**1926.65(j)(5)(ii)** Material handling equipment shall be provided with explosive containment devices or protective shields to protect equipment operators from exploding containers.

**1926.65(j)(5)(iii)** An employee alarm system capable of being perceived above surrounding light and noise conditions shall be used to signal the commencement and completion of explosive waste handling activities.

**1926.65(j)(5)(iv)** Continuous communications (i.e., portable radios, hand signals, telephones, as appropriate) shall be maintained between the employee-in-charge of the immediate handling area and both the site safety and health supervisor and the command post until such time as the handling operation is completed. Communication equipment or methods that could cause shock sensitive materials to explode shall not be used.

**1926.65(j)(5)(v)** Drums and containers under pressure, as evidenced by bulging or swelling, shall not be moved until such time as the cause for excess pressure is determined and appropriate containment procedures have been implemented to protect employees from explosive relief of the drum.

**1926.65(j)(5)(vi)** Drums and containers containing packaged laboratory wastes shall be considered to contain shock-sensitive or explosive materials until they have been characterized.

CAUTION: Shipping of shock sensitive wastes may be prohibited under U.S. Department of Transportation regulations. Employers and their shippers should refer to 49 CFR 173.21 and 173.50.

**1926.65(j)(6)** *Laboratory waste packs.* In addition to the requirements of paragraph (j)(5) of this section, the following precautions shall be taken, as a minimum, in handling laboratory waste packs (lab packs):

**1926.65(j)(6)(i)** Lab packs shall be opened only when necessary and then only by an individual knowledgeable in the inspection, classification, and segregation of the containers within the pack according to the hazards of the wastes.

**1926.65(j)(6)(ii)** If crystalline material is noted on any container, the contents shall be handled as a shock-sensitive waste until the contents are identified.

**1926.65(j)(7)** *Sampling of drum and container contents.* Sampling of containers and drums shall be done in accordance with a sampling procedure which is part of the site safety and health plan developed for and available to employees and others at the specific worksite.

**1926.65(j)(8)** *Shipping and transport.* (i) Drums and containers shall be identified and classified prior to packaging for shipment.

**1926.65(j)(8)(ii)** Drum or container staging areas shall be kept to the minimum number necessary to identify and classify materials safely and prepare them for transport.

**1926.65(j)(8)(iii)** Staging areas shall be provided with adequate access and egress routes.

**1926.65(j)(8)(iv)** Bulking of hazardous wastes shall be permitted only after a thorough characterization of the materials has been completed.

**1926.65(j)(9)** *Tank and vault procedures.* (i) Tanks and vaults containing hazardous substances shall be handled in a manner similar to that for drums and containers, taking into consideration the size of the tank or vault.

**1926.65(j)(9)(ii)** Appropriate tank or vault entry procedures as described in the employer's safety and health plan shall be followed whenever employees must enter a tank or vault.

**1926.65(k)** ***Decontamination***— (1) General. Procedures for all phases of decontamination shall be developed and implemented in accordance with this paragraph.

**1926.65(k)(2)** *Decontamination procedures.* (i) A decontamination procedure shall be developed, communicated to employees and implemented before any employees or equipment may enter areas on site where potential for exposure to hazardous substances exists.

**1926.65(k)(2)(ii)** Standard operating procedures shall be developed to minimize employee contact with hazardous substances or with equipment that has contacted hazardous substances.

**1926.65(k)(2)(iii)** All employees leaving a contaminated area shall be appropriately decontaminated; all contaminated clothing and equipment leaving a contaminated area shall be appropriately disposed of or decontaminated.

**1926.65(k)(2)(iv)** Decontamination procedures shall be monitored by the site safety and health supervisor to determine their effectiveness. When such procedures are found to be ineffective, appropriate steps shall be taken to correct any deficiencies.

**1926.65(k)(3)** *Location.* Decontamination shall be performed in geographical areas that will minimize the exposure of uncontaminated employees or equipment to contaminated employees or equipment.

**1926.65(k)(4)** *Equipment and solvents.* All equipment and solvents used for decontamination shall be decontaminated or disposed of properly.

**1926.65(k)(5)** *Personal protective clothing and equipment.* (i) Protective clothing and equipment shall be decontaminated, cleaned, laundered, maintained or replaced as needed to maintain their effectiveness.

**1926.65(k)(5)(ii)** Employees whose non-impermeable clothing becomes wetted with hazardous substances shall immediately remove that clothing and proceed to shower. The clothing shall be disposed of or decontaminated before it is removed from the work zone.

**1926.65(k)(6)** *Unauthorized employees.* Unauthorized employees shall not remove protective clothing or equipment from change rooms.

**1926.65(k)(7)** *Commercial laundries or cleaning establishments.* Commercial laundries or cleaning establishments that decontaminate protective clothing or equipment shall be informed of the potentially harmful effects of exposures to hazardous substances.

**1926.65(k)(8)** *Showers and change rooms.* Where the decontamination procedure indicates a need for regular showers and change rooms outside of a contaminated area, they shall be provided and meet the requirements of 29 CFR 1910.141. If temperature conditions prevent the effective use of water, then other effective means for cleansing shall be provided and used.

**1926.65(l)** ***Emergency response by employees at uncontrolled hazardous waste sites***— (1) Emergency response plan. (i) An emergency response plan shall be developed and implemented by all employers within the scope of paragraphs (a)(1) (i)-(ii) of this section to handle anticipated emergencies prior to the commencement of hazardous waste operations. The plan shall be in writing and available for inspection and copying by employees, their representatives, OSHA personnel and other governmental agencies with relevant responsibilities.

**1926.65(l)(1)(ii)** Employers who will evacuate their employees from the danger area when an emergency occurs, and who do not permit any of their employees to assist in handling the emergency, are exempt from the requirements of this paragraph if they provide an emergency action plan complying with §1926.35 of this part.

**1926.65(l)(2)** *Elements of an emergency response plan.* The employer shall develop an emergency response plan for emergencies which shall address, as a minimum, the following:

**1926.65(l)(2)(i)** Pre-emergency planning.

**1926.65(l)(2)(ii)** Personnel roles, lines of authority, and communication.

**1926.65(l)(2)(iii)** Emergency recognition and prevention.

**1926.65(l)(2)(iv)** Safe distances and places of refuge.

**1926.65(l)(2)(v)** Site security and control.

**1926.65(l)(2)(vi)** Evacuation routes and procedures.

**1926.65(l)(2)(vii)** Decontamination procedures which are not covered by the site safety and health plan.

**1926.65(l)(2)(viii)** Emergency medical treatment and first aid.

**1926.65(l)(2)(ix)** Emergency alerting and response procedures.

**1926.65(l)(2)(x)** Critique of response and follow-up.

**1926.65(l)(2)(xi)** PPE and emergency equipment.

**1926.65(l)(3)** *Procedures for handling emergency incidents.* (i) In addition to the elements for the emergency response plan required in paragraph (1)(2) of this section, the following elements shall be included for emergency response plans:

**1926.65(l)(3)(i)(A)** Site topography, layout, and prevailing weather conditions.

**1926.65(l)(3)(i)(B)** Procedures for reporting incidents to local, state, and federal governmental agencies.

**1926.65(l)(3)(ii)** The emergency response plan shall be a separate section of the Site Safety and Health Plan.

**1926.65(l)(3)(iii)** The emergency response plan shall be compatible and integrated with the disaster, fire and/or emergency response plans of local, state, and federal agencies.

**1926.65(l)(3)(iv)** The emergency response plan shall be rehearsed regularly as part of the overall training program for site operations.

**1926.65(l)(3)(v)** The site emergency response plan shall be reviewed periodically and, as necessary, be amended to keep it current with new or changing site conditions or information.

**1926.65(l)(3)(vi)** An employee alarm system shall be installed in accordance with 29 CFR 1926.159 to notify employees of an emergency situation; to stop work activities if necessary; to lower background noise in order to speed communication; and to begin emergency procedures.

**1926.65(l)(3)(vii)** Based upon the information available at time of the emergency, the employer shall evaluate the incident and the site response capabilities and proceed with the appropriate steps to implement the site emergency response plan.

**1926.65(m)** ***Illumination.*** Areas accessible to employees shall be lighted to not less than the minimum illumination intensities listed in the following Table D-65.1 while any work is in progress:

### TABLE D-65.1—MINIMUM ILLUMINATION INTENSITIES IN FOOT-CANDLES

| Foot-candles | Area or operations |
|---|---|
| 5 | General site areas. |
| 3 | Excavation and waste areas, accessways, active storage areas, loading platforms, refueling, and field maintenance areas. |
| 5 | Indoors: Warehouses, corridors, hallways, and exitways. |
| 5 | Tunnels, shafts, and general underground work areas. (Exception: minimum of 10 foot-candles is required at tunnel and shaft heading during drilling, mucking, and scaling. Mine Safety and Health Administration approved cap lights shall be acceptable for use in the tunnel heading. |
| 10 | General shops (e.g., mechanical and electrical equipment rooms, active storerooms, barracks or living quarters, locker or dressing rooms, dining areas, and indoor toilets and workrooms. |
| 30 | First aid stations, infirmaries, and offices. |

**1926.65(n)** ***Sanitation at temporary workplaces***— (1) Potable water. (i) An adequate supply of potable water shall be provided on the site.

**1926.65(n)(1)(ii)** Portable containers used to dispense drinking water shall be capable of being tightly closed, and equipped with a tap. Water shall not be dipped from containers.

**1926.65(n)(1)(iii)** Any container used to distribute drinking water shall be clearly marked as to the nature of its contents and not used for any other purpose.

**1926.65(n)(1)(iv)** Where single service cups (to be used but once) are supplied, both a sanitary container for the unused cups and a receptacle for disposing of the used cups shall be provided.

**1926.65(n)(2)** *Nonpotable water.* (i) Outlets for nonpotable water, such as water for firefighting purposes, shall be identified to indicate clearly that the water is unsafe and is not to be used for drinking, washing, or cooking purposes.

**1926.65(n)(2)(ii)** There shall be no cross-connection, open or potential, between a system furnishing potable water and a system furnishing nonpotable water.

**1926.65(n)(3)** *Toilet facilities.* (i) Toilets shall be provided for employees according to the following Table D-65.2.

### TABLE D-65.2—TOILET FACILITIES

| Number of employees | Minimum number of facilities |
|---|---|
| 20 or fewer | One. |
| More than 20, fewer than 200. | One toilet seat and one urinal per 40 employees. |
| More than 200 | One toilet seat and one urinal per 50 employees. |

**1926.65(n)(3)(ii)** Under temporary field conditions, provisions shall be made to assure that at least one toilet facility is available.

**1926.65(n)(3)(iii)** Hazardous waste sites not provided with a sanitary sewer shall be provided with the following toilet facilities unless prohibited by local codes:

**1926.65(n)(3)(iii)(A)** Chemical toilets;

**1926.65(n)(3)(iii)(B)** Recirculating toilets;

**1926.65(n)(3)(iii)(C)** Combustion toilets; or

**1926.65(n)(3)(iii)(D)** Flush toilets.

**1926.65(n)(3)(iv)** The requirements of this paragraph for sanitation facilities shall not apply to mobile crews having transportation readily available to nearby toilet facilities.

**1926.65(n)(3)(v)** Doors entering toilet facilities shall be provided with entrance locks controlled from inside the facility.

**1926.65(n)(4)** *Food handling.* All food service facilities and operations for employees shall meet the applicable laws, ordinances, and regulations of the jurisdictions in which they are located.

**1926.65(n)(5)** *Temporary sleeping quarters.* When temporary sleeping quarters are provided, they shall be heated, ventilated, and lighted.

**1926.65(n)(6)** *Washing facilities.* The employer shall provide adequate washing facilities for employees engaged in operations where hazardous substances may be harmful to employees. Such facilities shall be in near proximity to the worksite; in areas where exposures are below permissible exposure limits and published exposure levels and which are under the controls of the employer; and shall be so equipped as to enable employees to remove hazardous substances from themselves.

**1926.65(n)(7)** *Showers and change rooms.* When hazardous waste clean-up or removal operations commence on a site and the duration of the work will require six months or greater time to complete, the employer shall provide showers and change rooms for all employees exposed to hazardous substances and health hazards involved in hazardous waste clean-up or removal operations.

**1926.65(n)(7)(i)** Showers shall be provided and shall meet the requirements of 29 CFR 1926.51(f)(4).

**1926.65(n)(7)(ii)** Change rooms shall be provided and shall meet the requirements of 29 CFR 1926.51(i). Change rooms shall consist of two separate change areas separated by the shower area required in paragraph (n)(7)(i) of this section. One change area, with an exit leading off the worksite, shall provide employees with a clean area where they can remove, store, and put on street clothing. The second area, with an exit to the worksite, shall provide employees with an area where they can put on, remove and store work clothing and personal protective equipment.

**1926.65(n)(7)(iii)** Showers and change rooms shall be located in areas where exposures are below the permissible exposure limits and published exposure levels. If this cannot be accomplished, then a ventilation system shall be provided that will supply air that is below the permissible exposure limits and published exposure levels.

**1926.65(n)(7)(iv)** Employers shall assure that employees shower at the end of their work shift and when leaving the hazardous waste site.

**1926.65(o)** ***New technology programs.*** (1) The employer shall develop and implement procedures for the introduction of effective new technologies and equipment developed for the improved protection of employees working with hazardous waste clean-up operations, and the same shall be implemented as part of the site safety and health program to assure that employee protection is being maintained.

**1926.65(o)(2)** New technologies, equipment or control measures available to the industry, such as the use of foams, absorbents, adsorbents, neutralizers, or other means to suppress the level of air contaminates while excavating the site or for spill control, shall be evaluated by employers or their representatives. Such an evaluation shall be done to determine the effectiveness of the new methods, materials, or equipment before implementing their use on a large scale for enhancing employee protection. Information and data from manufacturers or suppliers may be used as part of the employer's evaluation effort. Such evaluations shall be made available to OSHA upon request.

**1926.65(p)** ***Certain operations conducted under the Resource Conservation and Recovery Act of 1976 (RCRA).*** Employers conducting operations at treatment, storage and disposal (TSD) facilities specified in paragraph (a)(1)(iv) of this section shall provide and implement the programs specified in this paragraph. See the "Notes and Exceptions" to paragraph (a)(2)(iii) of this section for employers not covered.)".

**1926.65(p)(1)** *Safety and health program.* The employer shall develop and implement a written safety and health program for employees involved in hazardous waste operations that shall be available for inspection by employees, their representatives and OSHA personnel. The program shall be designed to identify, evaluate and control safety and health hazards in their facilities for the purpose of employee protection, to provide for emergency response meeting the requirements of paragraph (p)(8) of this section and to address as appropriate site analysis, engineering controls, maximum exposure limits, hazardous waste handling procedures and uses of new technologies.

**1926.65(p)(2)** *Hazard communication program.* The employer shall implement a hazard communication program meeting the requirements of 29 CFR 1926.59 as part of the employer's safety and program.

**Note to 1926.65:** The exemption for hazardous waste provided in §1926.59 is applicable to this section.

**1926.65(p)(3)** *Medical surveillance program.* The employer shall develop and implement a medical surveillance program meeting the requirements of paragraph (f) of this section.

**1926.65(p)(4)** *Decontamination program.* The employer shall develop and implement a decontamination procedure meeting the requirements of paragraph (k) of this section.

**1926.65(p)(5)** *New technology program.* The employer shall develop and implement procedures meeting the requirements of paragraph (o) of this section for introducing new and innovative equipment into the workplace.

**1926.65(p)(6)** *Material handling program.* Where employees will be handling drums or containers, the employer shall develop and implement procedures meeting the requirements of paragraphs (j)(1) (ii) through (viii) and (xi) of this section, as well as (j)(3) and (j)(8) of this section prior to starting such work.

**1926.65(p)(7)** *Training program*—(i) *New employees.* The employer shall develop and implement a training program, which is part of the employer's safety and health program, for employees exposed to health hazards or hazardous substances at TSD operations to enable the employees to perform their assigned duties and functions in a safe and healthful manner so as not endanger themselves or other employees. The initial training shall be for 24 hours and refresher training shall be for eight hours annually. Employees who have received the initial training required by this paragraph shall be given a written certificate attesting that they have successfully completed the necessary training.

**1926.65(p)(7)(ii)** *Current employees.* Employers who can show by an employee's previous work experience and/or training that the employee has had training equivalent to the initial training required by this paragraph, shall be considered as meeting the initial training requirements of this paragraph as to that employee. Equivalent training includes the training that existing employees might have already received from actual site work experience. Current employees shall receive eight hours of refresher training annually.

**1926.65(p)(7)(iii)** *Trainers.* Trainers who teach initial training shall have satisfactorily completed a training course for teaching the subjects they are expected to teach or they shall have the academic credentials and instruction experience necessary to demonstrate a good command of the subject matter of the courses and competent instructional skills.

**1926.65(p)(8)** *Emergency response program*—(i) *Emergency response plan.* An emergency response plan shall be developed and implemented by all employers. Such plans need not duplicate any of the subjects fully addressed in the employer's contingency planning required by permits, such as those issued by the U.S. Environmental Protection Agency, provided that the contingency plan is made part of the emergency response plan. The emergency response plan shall be a written portion of the employers safety and health program required in paragraph (p)(1) of this section. Employers who will evacuate their employees from the worksite location when an emergency occurs and who do not permit any of their employees to assist in handling the emergency are exempt from the requirements of paragraph (p)(8) if they provide an emergency action plan complying with §1926.35 of this part.

**1926.65(p)(8)(ii)** *Elements of an emergency response plan.* The employer shall develop an emergency response plan for emergencies which shall address, as a minimum, the following areas to the extent that they are not addressed in any specific program required in this paragraph:

**1926.65(p)(8)(ii)(A)** Pre-emergency planning and coordination with outside parties.

**1926.65(p)(8)(ii)(B)** Personnel roles, lines of authority, and communication.

**1926.65(p)(8)(ii)(C)** Emergency recognition and prevention.

**1926.65(p)(8)(ii)(D)** Safe distances and places of refuge.

**1926.65(p)(8)(ii)(E)** Site security and control.

**1926.65(p)(8)(ii)(F)** Evacuation routes and procedures.

**1926.65(p)(8)(ii)(G)** Decontamination procedures.

**1926.65(p)(8)(ii)(H)** Emergency medical treatment and first aid.

**1926.65(p)(8)(ii)(I)** Emergency alerting and response procedures.

**1926.65(p)(8)(ii)(J)** Critique of response and follow-up.

**1926.65(p)(8)(ii)(K)** PPE and emergency equipment.

**1926.65(p)(8)(iii)** *Training.* (A) Training for emergency response employees shall be completed before they are called upon to perform in real emergencies. Such training shall include the elements of the emergency response plan, standard operating procedures the employer has established for the job, the personal protective equipment to be worn and procedures for handling emergency incidents.

*Exception #1:* An employer need not train all employees to the degree specified if the employer divides the work force in a manner such that a sufficient number of employees who have responsibility to control emergencies have the training specified, and all other employees, who may first respond to an emergency incident, have sufficient awareness training to recognize that an emergency response situation exists and that they are instructed in that case to summon the fully trained employees and not attempt control activities for which they are not trained.

*Exception #2:* An employer need not train all employees to the degree specified if arrangements have been made in advance for an outside fully-trained emergency response team to respond in a reasonable period and all employees, who may come to the incident first, have sufficient awareness training to recognize that an emergency response situation exists and they have been instructed to call the designated outside fully-trained emergency response team for assistance.

**1926.65(p)(8)(iii)(B)** Employee members of TSD facility emergency response organizations shall be trained to a level of competence in the recognition of health and safety hazards to protect themselves and other employees. This would include training in the methods used to minimize the risk from safety and health hazards; in the safe use of control equipment; in the selection and use of appropriate personal protective equipment; in the safe operating procedures to be used at the incident scene; in the techniques of coordination with other employees to minimize risks; in the appropriate response to over exposure from health hazards or injury to themselves and other employees; and in the recognition of subsequent symptoms which may result from over exposures.

**1926.65(p)(8)(iii)(C)** The employer shall certify that each covered employee has attended and successfully completed the training required in paragraph (p)(8)(iii) of this section, or shall certify the employee's competency at least yearly. The method used to demonstrate competency for certification of training shall be recorded and maintained by the employer.

**1926.65(p)(8)(iv)** *Procedures for handling emergency incidents.* (A) In addition to the elements for the emergency response plan required in paragraph (p)(8)(ii) of this section, the following elements shall be included for emergency response plans to the extent that they do not repeat any information already contained in the emergency response plan:

**1926.65(p)(8)(iv)(A)*(1)*** Site topography, layout, and prevailing weather conditions.

**1926.65(p)(8)(iv)(A)(2)** Procedures for reporting incidents to local, state, and federal governmental agencies.

**1926.65(p)(8)(iv)(B)** The emergency response plan shall be compatible and integrated with the disaster, fire and/or emergency response plans of local, state, and federal agencies.

**1926.65(p)(8)(iv)(C)** The emergency response plan shall be rehearsed regularly as part of the overall training program for site operations.

**1926.65(p)(8)(iv)(D)** The site emergency response plan shall be reviewed periodically and, as necessary, be amended to keep it current with new or changing site conditions or information.

**1926.65(p)(8)(iv)(E)** An employee alarm system shall be installed in accordance with 29 CFR 1926.159 to notify employees of an emergency situation; to stop work activities if necessary; to lower background noise in order to speed communication; and to begin emergency procedures.

**1926.65(p)(8)(iv)(F)** Based upon the information available at time of the emergency, the employer shall evaluate the incident and the site response capabilities and proceed with the appropriate steps to implement the site emergency response plan.

**1926.65(q)** ***Emergency response to hazardous substance releases.*** This paragraph covers employers whose employees are engaged in emergency response no matter where it occurs except that it does not cover employees engaged in operations specified in paragraphs (a)(1)(i) through (a)(1)(iv) of this section. Those emergency response organizations who have developed and implemented programs equivalent to this paragraph for handling releases of hazardous substances pursuant to section 303 of the Superfund Amendments and Reauthorization Act of 1986 (Emergency Planning and Community Right-to-Know Act of 1986, 42 U.S.C. 11003) shall be deemed to have met the requirements of this paragraph.

**1926.65(q)(1)** *Emergency response plan.* An emergency response plan shall be developed and implemented to handle anticipated emergencies prior to the commencement of emergency response operations. The plan shall be in writing and available for inspection and copying by employees, their representatives and OSHA personnel. Employers who will evacuate their employees from the danger area when an emergency occurs, and who do not permit any of their employees to assist in handling the emergency, are exempt from the requirements of this paragraph if they provide an emergency action plan in accordance with §1926.35 of this part.

**1926.65(q)(2)** *Elements of an emergency response plan.* The employer shall develop an emergency response plan for emergencies which shall address, as a minimum, the following to the extent that they are not addressed elsewhere:

**1926.65(q)(2)(i)** Pre-emergency planning and coordination with outside parties.

**1926.65(q)(2)(ii)** Personnel roles, lines of authority, training, and communication.

**1926.65(q)(2)(iii)** Emergency recognition and prevention.

**1926.65(q)(2)(iv)** Safe distances and places of refuge.

**1926.65(q)(2)(v)** Site security and control.

**1926.65(q)(2)(vi)** Evacuation routes and procedures.

**1926.65(q)(2)(vii)** Decontamination.

**1926.65(q)(2)(viii)** Emergency medical treatment and first aid.

**1926.65(q)(2)(ix)** Emergency alerting and response procedures.

**1926.65(q)(2)(x)** Critique of response and follow-up.

**1926.65(q)(2)(xi)** PPE and emergency equipment.

**1926.65(q)(2)(xii)** Emergency response organizations may use the local emergency response plan or the state emergency response plan or both, as part of their emergency response plan to avoid duplication. Those items of the emergency response plan that are being properly addressed by the SARA Title III plans may be substituted into their emergency plan or otherwise kept together for the employer and employee's use.

**1926.65(q)(3)** *Procedures for handling emergency response.* (i) The senior emergency response official responding to an emergency shall become the individual in charge of a site-specific Incident Command System (ICS). All emergency responders and their communications shall be coordinated and controlled through the individual in charge of the ICS assisted by the senior official present for each employer.

**Note to (g)(3)(i): [sic (q)(3)(1)]** The *senior official* at an emergency response is the most senior official on the site who has the responsibility for controlling the operations at the site. Initially it is the senior officer on the first-due piece of responding emergency apparatus to arrive on the incident scene. As more senior officers arrive (i.e., battalion chief, fire chief, state law enforcement official, site coordinator, etc.) the position is passed up the line of authority which has been previously established.

**1926.65(q)(3)(ii)** The individual in charge of the ICS shall identify, to the extent possible, all hazardous substances or conditions present and shall address as appropriate site analysis, use of engineering controls, maximum exposure limits, hazardous substance handling procedures, and use of any new technologies.

**1926.65(q)(3)(iii)** Based on the hazardous substances and/or conditions present, the individual in charge of the ICS shall implement appropriate emergency operations, and assure that the personal protective equipment worn is appropriate for the hazards to be encountered. However, personal protective equipment shall meet, at a minimum, the criteria contained in 29 CFR 1926.97 when worn while performing fire fighting operations beyond the incipient stage for any incident.

**1926.65(q)(3)(iv)** Employees engaged in emergency response and exposed to hazardous substances presenting an inhalation hazard or potential inhalation hazard shall wear positive pressure self-contained breathing apparatus while engaged in emergency response, until such time that the individual in charge of the ICS determines through the use of air monitoring that a decreased level of respiratory protection will not result in hazardous exposures to employees.

**1926.65(q)(3)(v)** The individual in charge of the ICS shall limit the number of emergency response personnel at the emergency site, in those areas of potential or actual exposure to incident or site hazards, to those who are actively performing emergency operations. However, operations in hazardous areas shall be performed using the buddy system in groups of two or more.

**1926.65(q)(3)(vi)** Back-up personnel shall stand by with equipment ready to provide assistance or rescue. Advance first aid support personnel, as a minimum, shall also stand by with medical equipment and transportation capability.

**1926.65(q)(3)(vii)** The individual in charge of the ICS shall designate a safety official, who is knowledgeable in the operations being implemented at the emergency response site, with specific responsibility to identify and evaluate hazards and to provide direction with respect to the safety of operations for the emergency at hand.

**1926.65(q)(3)(viii)** When activities are judged by the safety official to be an IDLH condition and/or to involve an imminent danger condition, the safety official shall have the authority to alter, suspend, or terminate those activities. The safety official shall immediately inform the individual in charge of the ICS of any actions needed to be taken to correct these hazards at the emergency scene.

**1926.65(q)(3)(ix)** After emergency operations have terminated, the individual in charge of the ICS shall implement appropriate decontamination procedures.

**1926.65(q)(3)(x)** When deemed necessary for meeting the tasks at hand, approved self-contained compressed air breathing apparatus may be used with approved cylinders from other approved self-contained compressed air breathing apparatus provided that such cylinders are of the same capacity and pressure rating. All compressed air cylinders used with self-contained breathing apparatus shall meet U.S. Department of Transportation and National Institute for Occupational Safety and Health criteria.

**1926.65(q)(4)** *Skilled support personnel.* Personnel, not necessarily an employer's own employees, who are skilled in the operation of certain equipment, such as mechanized earth moving or digging equipment or crane and hoisting equipment, and who are needed temporarily to perform immediate emergency support work that cannot reasonably be performed in a timely fashion by an employer's own employees, and who will be or may be exposed to the hazards at an emergency response scene, are not required to meet the training required in this paragraph for the employer's regular employees. However, these personnel shall be given an initial briefing at the site prior to their participation in any emergency response. The initial briefing shall include instruction in the wearing of appropriate personal protective equipment, what chemical hazards are involved, and what duties are to be performed. All other appropriate safety and health precautions provided to the employer's own employees shall be used to assure the safety and health of these personnel.

**1926.65(q)(5)** *Specialist employees.* Employees who, in the course of their regular job duties, work with and are trained in the hazards of specific hazardous substances, and who will be called upon to provide technical advice or assistance at a hazardous substance release incident to the individual in charge, shall receive training or demonstrate competency in the area of their specialization annually.

**1926.65(q)(6)** *Training.* Training shall be based on the duties and function to be performed by each responder of an emergency response organization. The skill and knowledge levels required for all new responders, those hired after the effective date of this standard, shall be conveyed to them through training before they are permitted to take part in actual emergency operations on an incident. Employees who participate, or are expected to participate, in emergency response, shall be given training in accordance with the following paragraphs:

**1926.65(q)(6)(i)** *First responder awareness level.* First responders at the awareness level are individuals who are likely to witness or discover a hazardous substance release and who have been trained to initiate an emergency response sequence by notifying the proper authorities of the release. They would take no further action beyond notifying the authorities of the release. First responders at the awareness level shall have sufficient training or have had sufficient experience to objectively demonstrate competency in the following areas:

**1926.65(q)(6)(i)(A)** An understanding of what hazardous substances are, and the risks associated with them in an incident.

**1926.65(q)(6)(i)(B)** An understanding of the potential outcomes associated with an emergency created when hazardous substances are present.

**1926.65(q)(6)(i)(C)** The ability to recognize the presence of hazardous substances in an emergency.

**1926.65(q)(6)(i)(D)** The ability to identify the hazardous substances, if possible.

**1926.65(q)(6)(i)(E)** An understanding of the role of the first responder awareness individual in the employer's emergency response plan including site security and control and the U.S. Department of Transportation's Emergency Response Guidebook.

**1926.65(q)(6)(i)(F)** The ability to realize the need for additional resources, and to make appropriate notifications to the communication center.

**1926.65(q)(6)(ii)** *First responder operations level.* First responders at the operations level are individuals who respond to releases or potential releases of hazardous substances as part of the initial response to the site for the purpose of protecting nearby persons, property, or the environment from the effects of the release. They are trained to respond in a defensive fashion without actually trying to stop the release. Their function is to contain the release from a safe distance, keep it from spreading, and prevent exposures. First responders at the operational level shall have received at least eight hours of training or have had sufficient experience to objectively demonstrate competency in the following areas in addition to those listed for the awareness level and the employer shall so certify:

**1926.65(q)(6)(ii)(A)** Knowledge of the basic hazard and risk assessment techniques.

**1926.65(q)(6)(ii)(B)** Know how to select and use proper personal protective equipment provided to the first responder operational level.

**1926.65(q)(6)(ii)(C)** An understanding of basic hazardous materials terms.

**1926.65(q)(6)(ii)(D)** Know how to perform basic control, containment and/or confinement operations within the capabilities of the resources and personal protective equipment available with their unit.

**1926.65(q)(6)(ii)(E)** Know how to implement basic decontamination procedures.

**1926.65(q)(6)(ii)(F)** An understanding of the relevant standard operating procedures and termination procedures.

**1926.65(q)(6)(iii)** *Hazardous materials technician.* Hazardous materials technicians are individuals who respond to releases or potential releases for the purpose of stopping the release. They assume a more aggressive role than a first responder at the operations level in that they will approach the point of release in order to plug, patch or otherwise stop the release of a hazardous substance. Hazardous materials technicians shall have received at least 24 hours of training equal to the first responder operations level and in addition have competency in the following areas and the employer shall so certify:

**1926.65(q)(6)(iii)(A)** Know how to implement the employer's emergency response plan.

**1926.65(q)(6)(iii)(B)** Know the classification, identification and verification of known and unknown materials by using field survey instruments and equipment.

**1926.65(q)(6)(iii)(C)** Be able to function within an assigned role in the Incident Command System.

**1926.65(q)(6)(iii)(D)** Know how to select and use proper specialized chemical personal protective equipment provided to the hazardous materials technician.

**1926.65(q)(6)(iii)(E)** Understand hazard and risk assessment techniques.

**1926.65(q)(6)(iii)(F)** Be able to perform advance control, containment, and/or confinement operations within the capabilities of the resources and personal protective equipment available with the unit.

**1926.65(q)(6)(iii)(G)** Understand and implement decontamination procedures.

**1926.65(q)(6)(iii)(H)** Understand termination procedures.

**1926.65(q)(6)(iii)(I)** Understand basic chemical and toxicological terminology and behavior.

**1926.65(q)(6)(iv)** *Hazardous materials specialist.* Hazardous materials specialists are individuals who respond with and provide support to hazardous materials technicians. Their duties parallel those of the hazardous materials technician, however, those duties require a more directed or specific knowledge of the various substances they may be called upon to contain. The hazardous materials specialist would also act as the site liaison with Federal, state, local and other government authorities in regards to site activities. Hazardous materials specialists shall have received at least 24 hours of training equal to the technician level and in addition have competency in the following areas and the employer shall so certify:

**1926.65(q)(6)(iv)(A)** Know how to implement the local emergency response plan.

**1926.65(q)(6)(iv)(B)** Understand classification, identification and verification of known and unknown materials by using advanced survey instruments and equipment.

**1926.65(q)(6)(iv)(C)** Know of the state emergency response plan.

**1926.65(q)(6)(iv)(D)** Be able to select and use proper specialized chemical personal protective equipment provided to the hazardous materials specialist.

**1926.65(q)(6)(iv)(E)** Understand in-depth hazard and risk techniques.

**1926.65(q)(6)(iv)(F)** Be able to perform specialized control, containment, and/or confinement operations within the capabilities of the resources and personal protective equipment available.

**1926.65(q)(6)(iv)(G)** Be able to determine and implement decontamination procedures.

**1926.65(q)(6)(iv)(H)** Have the ability to develop a site safety and control plan.

**1926.65(q)(6)(iv)(I)** Understand chemical, radiological and toxicological terminology and behavior.

**1926.65(q)(6)(v)** *On scene incident commander.* Incident commanders, who will assume control of the incident scene beyond the first responder awareness level, shall receive at least 24 hours of training equal to the first responder operations level and in addition have competency in the following areas and the employer shall so certify:

**1926.65(q)(6)(v)(A)** Know and be able to implement the employer's incident command system.

**1926.65(q)(6)(v)(B)** Know how to implement the employer's emergency response plan.

**1926.65(q)(6)(v)(C)** Know and understand the hazards and risks associated with employees working in chemical protective clothing.

**1926.65(q)(6)(v)(D)** Know how to implement the local emergency response plan.

**1926.65(q)(6)(v)(E)** Know of the state emergency response plan and of the Federal Regional Response Team.

**1926.65(q)(6)(v)(F)** Know and understand the importance of decontamination procedures.

**1926.65(q)(7)** *Trainers.* Trainers who teach any of the above training subjects shall have satisfactorily completed a training course for teaching the subjects they are expected to teach, such as the courses offered by the U.S. National Fire Academy, or they shall have the training and/or academic credentials and instructional experience necessary to demonstrate competent instructional skills and a good command of the subject matter of the courses they are to teach.

**1926.65(q)(8)** *Refresher training.* (i) Those employees who are trained in accordance with paragraph (q)(6) of this section shall receive annual refresher training of sufficient content and duration to maintain their competencies, or shall demonstrate competency in those areas at least yearly.

**1926.65(q)(8)(ii)** A statement shall be made of the training or competency, and if a statement of competency is made, the employer shall keep a record of the methodology used to demonstrate competency.

**1926.65(q)(9)** *Medical surveillance and consultation.* (i) Members of an organized and designated HAZMAT team and hazardous materials specialists shall receive a baseline physical examination and be provided with medical surveillance as required in paragraph (f) of this section.

**1926.65(q)(9)(ii)** Any emergency response employees who exhibits signs or symptoms which may have resulted from exposure to hazardous substances during the course of an emergency incident, either immediately or subsequently, shall be provided with medical consultation as required in paragraph (f)(3)(ii) of this section.

**1926.65(q)(10)** *Chemical protective clothing.* Chemical protective clothing and equipment to be used by organized and designated HAZMAT team members, or to be used by hazardous materials specialists, shall meet the requirements of paragraphs (g) (3) through (5) of this section.

**1926.65(q)(11)** *Post-emergency response operations.* Upon completion of the emergency response, if it is determined that it is necessary to remove hazardous substances, health hazards, and materials contaminated with them (such as contaminated soil or other elements of the natural environment) from the site of the incident, the employer conducting the clean-up shall comply with one of the following:

**1926.65(q)(11)(i)** Meet all of the requirements of paragraphs (b) through (o) of this section; or

**1926.65(q)(11)(ii)** Where the clean-up is done on plant property using plant or workplace employees, such employees shall have completed the training requirements of the following: 29 CFR 1926.35, 1926.59, and 1926.103, and other appropriate safety and health training made necessary by the tasks that they are expected to be performed such as personal protective equipment and decontamination procedures. All equipment to be used in the performance of the clean-up work shall be in serviceable condition and shall have been inspected prior to use.

## APPENDICES TO §1926.65—HAZARDOUS WASTE OPERATIONS AND EMERGENCY RESPONSE

**Note:** The following appendices serve as non-mandatory guidelines to assist employees and employers in complying with the appropriate requirements of this section. However §1926.65(g) makes mandatory in certain circumstances the use of Level A and Level B PPE protection.

## Appendix A to §1926.65— Personal Protective Equipment Test Methods

This appendix sets forth the non-mandatory examples of tests which may be used to evaluate compliance with §1926.65(g)(4) (ii) and (iii). Other tests and other challenge agents may be used to evaluate compliance.

### A. Totally-encapsulating chemical protective suit pressure test

1.0—Scope

1.1 This practice measures the ability of a gas tight totally-encapsulating chemical protective suit material, seams, and closures to maintain a fixed positive pressure. The results of this practice allow the gas tight integrity of a totally-encapsulating chemical protective suit to be evaluated.

1.2 Resistance of the suit materials to permeation, penetration, and degradation by specific hazardous substances is not determined by this test method.

2.0—Definition of terms

2.1 *Totally-encapsulated chemical protective suit (TECP suit)* means a full body garment which is constructed of protective clothing materials; covers the wearer's torso, head, arms, legs and respirator; may cover the wearer's hands and feet with tightly attached gloves and boots; completely encloses the wearer and respirator by itself or in combination with the wearer's gloves and boots.

2.2 *Protective clothing material* means any material or combination of materials used in an item of clothing for the purpose of isolating parts of the body from direct contact with a potentially hazardous liquid or gaseous chemicals.

2.3 *Gas tight* means, for the purpose of this test method, the limited flow of a gas under pressure from the inside of a TECP suit to atmosphere at a prescribed pressure and time interval.

3.0—Summary of test method

3.1 The TECP suit is visually inspected and modified for the test. The test apparatus is attached to the suit to permit inflation to the pre-test suit expansion pressure for removal of suit wrinkles and creases. The pressure is lowered to the test pressure and monitored for three minutes. If the pressure drop is excessive, the TECP suit fails the test and is removed from service. The test is repeated after leak location and repair.

4.0—Required Supplies

4.1 Source of compressed air.

4.2 Test apparatus for suit testing, including a pressure measurement device with a sensitivity of at least 1/4 inch water gauge.

4.3 Vent valve closure plugs or sealing tape.

4.4 Soapy water solution and soft brush.

4.5 Stop watch or appropriate timing device.

5.0—Safety Precautions

5.1 Care shall be taken to provide the correct pressure safety devices required for the source of compressed air used.

6.0—Test Procedure

6.1 Prior to each test, the tester shall perform a visual inspection of the suit. Check the suit for seam integrity by visually examining the seams and gently pulling on the seams. Ensure that all air supply lines, fittings, visor, zippers, and valves are secure and show no signs of deterioration.

6.1.1 Seal off the vent valves along with any other normal inlet or exhaust points (such as umbilical air line fittings or face piece opening) with tape or other appropriate means (caps, plugs, fixture, etc.). Care should be exercised in the sealing process not to damage any of the suit components.

6.1.2 Close all closure assemblies.

6.1.3 Prepare the suit for inflation by providing an improvised connection point on the suit for connecting an airline. Attach the pressure test apparatus to the suit to permit suit inflation from a compressed air source equipped with a pressure indicating regulator. The leak tightness of the pressure test apparatus should be tested before and after each test by closing off the end of the tubing attached to the suit and assuring a pressure of three inches water gauge for three minutes can be maintained. If a component is removed for the test, that component shall be replaced and a second test conducted with another component removed to permit a complete test of the ensemble.

6.1.4 The pre-test expansion pressure (A) and the suit test pressure (B) shall be supplied by the suit manufacturer, but in no case shall they be less than: (A)=three inches water gauge; and (B)=two inches water gauge. The ending suit pressure (C) shall be no less than 80 percent of the test pressure (B); i.e., the pressure drop shall not exceed 20 percent of the test pressure (B).

6.1.5 Inflate the suit until the pressure inside is equal to pressure (A), the pre-test expansion suit pressure. Allow at least one minute to fill out the wrinkles in the suit. Release sufficient air to reduce the suit pressure to pressure (B), the suit test pressure. Begin timing. At the end of three minutes, record the suit pressure as pressure (C), the ending suit pressure. The difference between the suit test pressure and the ending suit test pressure (B-C) shall be defined as the suit pressure drop.

6.1.6 If the suit pressure drop is more than 20 percent of the suit test pressure (B) during the three-minute test period, the suit fails the test and shall be removed from service.

7.0—Retest Procedure

7.1 If the suit fails the test check for leaks by inflating the suit to pressure (A) and brushing or wiping the entire suit (including seams, closures, lens gaskets, glove-to-sleeve joints, etc.) with a mild soap and water solution. Observe the suit for the formation of soap bubbles, which is an indication of a leak. Repair all identified leaks.

7.2 Retest the TECP suit as outlined in Test procedure 6.0.

8.0—Report

8.1 Each TECP suit tested by this practice shall have the following information recorded:

8.1.1 Unique identification number, identifying brand name, date of purchase, material of construction, and unique fit features, e.g., special breathing apparatus.

8.1.2 The actual values for test pressures (A), (B), and (C) shall be recorded along with the specific observation times. If the ending pressure (C) is less than 80 percent of the test pressure (B), the suit shall be identified as failing the test. When possible, the specific leak location shall be identified in the test records. Retest pressure data shall be recorded as an additional test.

8.1.3 The source of the test apparatus used shall be identified and the sensitivity of the pressure gauge shall be recorded.

8.1.4 Records shall be kept for each pressure test even if repairs are being made at the test location.

**CAUTION**

Visually inspect all parts of the suit to be sure they are positioned correctly and secured tightly before putting the suit back into service. Special care should be taken to examine each exhaust valve to make sure it is not blocked.

Care should also be exercised to assure that the inside and outside of the suit is completely dry before it is put into storage.

## B. Totally-encapsulating chemical protective suit qualitative leak test

1.0—Scope

1.1 This practice semi-qualitatively tests gas tight totally-encapsulating chemical protective suit integrity by detecting inward leakage of ammonia vapor. Since no modifications are made to the suit to carry out this test, the results from this practice provide a realistic test for the integrity of the entire suit.

1.2 Resistance of the suit materials to permeation, penetration, and degradation is not determined by this test method. ASTM test methods are available to test suit materials for these characteristics and the tests are usually conducted by the manufacturers of the suits.

2.0—Definition of terms

2.1 *Totally-encapsulated chemical protective suit (TECP suit)* means a full body garment which is constructed of protective clothing materials; covers the wearer's torso, head, arms, legs and respirator; may cover the wearer's hands and feet with tightly attached gloves and boots; completely encloses the wearer and respirator by itself or in combination with the wearer's gloves, and boots.

2.2 *Protective clothing material* means any material or combination of materials used in an item of clothing for the purpose of isolating parts of the body from direct contact with a potentially hazardous liquid or gaseous chemicals.

2.3 *Gas tight* means, for the purpose of this test method, the limited flow of a gas under pressure from the inside of a TECP suit to atmosphere at a prescribed pressure and time interval.

2.4 *Intrusion Coefficient* means a number expressing the level of protection provided by a gas tight totally-encapsulating chemical protective suit. The intrusion coefficient is calculated by dividing the test room challenge agent concentration by the concentration of challenge agent found inside the suit. The accuracy of the intrusion coefficient is dependent on the challenge agent monitoring methods. The larger the intrusion coefficient the greater the protection provided by the TECP suit.

3.0—Summary of recommended practice

3.1 The volume of concentrated aqueous ammonia solution (ammonia hydroxide $NH_4OH$) required to generate the test atmosphere is determined using the directions outlined in 6.1. The suit is donned by a person wearing the appropriate respiratory equipment (either a positive pressure self-contained breathing apparatus or a positive pressure supplied air respirator) and worn inside the enclosed test room. The concentrated aqueous ammonia solution is taken by the suited individual into the test room and poured into an open plastic pan. A two-minute evaporation period is observed before the test room concentration is measured, using a high range ammonia length of stain detector tube. When the ammonia vapor reaches a concentration of between 1000 and 1200 ppm, the suited individual starts a standardized exercise protocol to stress and flex the suit. After this protocol is completed, the test room concentration is measured again. The suited individual exits the test room and his stand-by person measures the ammonia concentration inside the suit using a low range ammonia length of stain detector tube or other more sensitive ammonia detector. A stand-by person is required to observe the test individual during the test procedure; aid the person in donning and doffing the TECP suit; and monitor the suit interior. The intrusion coefficient of the suit can be calculated by dividing the average test area concentration by the interior suit concentration. A colorimetric ammonia indicator strip of bromophenol blue or equivalent is placed on the inside of the suit face piece lens so that the suited individual is able to detect a color change and know if the suit has a significant leak. If a color change is observed the individual shall leave the test room immediately.

4.0—Required supplies

4.1 A supply of concentrated aqueous ammonium hydroxide (58% by weight).

4.2 A supply of bromophenol/blue indicating paper or equivalent, sensitive to 5-10 ppm ammonia or greater over a two-minute period of exposure. [pH 3.0 (yellow) to pH 4.6 (blue)]

4.3 A supply of high range (0.5-10 volume percent) and low range (5-700 ppm) detector tubes for ammonia and the corresponding sampling pump. More sensitive ammonia detectors can be substituted for the low range detector tubes to improve the sensitivity of this practice.

4.4 A shallow plastic pan (PVC) at least 12":14":1" and a half pint plastic container (PVC) with tightly closing lid.

4.5 A graduated cylinder or other volumetric measuring device of at least 50 milliliters in volume with an accuracy of at least ±1 milliliters.

### 5.0—SAFETY PRECAUTIONS

5.1 Concentrated aqueous ammonium hydroxide, $NH_4$ OH, is a corrosive volatile liquid requiring eye, skin, and respiratory protection. The person conducting the test shall review the MSDS for aqueous ammonia.

5.2 Since the established permissible exposure limit for ammonia is 35 ppm as a 15 minute STEL, only persons wearing a positive pressure self-contained breathing apparatus or a positive pressure supplied air respirator shall be in the chamber. Normally only the person wearing the totally-encapsulating suit will be inside the chamber. A stand-by person shall have a positive pressure self-contained breathing apparatus, or a positive pressure supplied air respirator available to enter the test area should the suited individual need assistance.

5.3 A method to monitor the suited individual must be used during this test. Visual contact is the simplest but other methods using communication devices are acceptable.

5.4 The test room shall be large enough to allow the exercise protocol to be carried out and then to be ventilated to allow for easy exhaust of the ammonia test atmosphere after the test(s) are completed.

5.5 Individuals shall be medically screened for the use of respiratory protection and checked for allergies to ammonia before participating in this test procedure.

### 6.0—TEST PROCEDURE

6.1.1 Measure the test area to the nearest foot and calculate its volume in cubic feet. Multiply the test area volume by 0.2 milliliters of concentrated aqueous ammonia solution per cubic foot of test area volume to determine the approximate volume of concentrated aqueous ammonia required to generate 1000 ppm in the test area.

6.1.2 Measure this volume from the supply of concentrated aqueous ammonia and place it into a closed plastic container.

6.1.3 Place the container, several high range ammonia detector tubes, and the pump in the clean test pan and locate it near the test area entry door so that the suited individual has easy access to these supplies.

6.2.1 In a non-contaminated atmosphere, open a pre-sealed ammonia indicator strip and fasten one end of the strip to the inside of the suit face shield lens where it can be seen by the wearer. Moisten the indicator strip with distilled water. Care shall be taken not to contaminate the detector part of the indicator paper by touching it. A small piece of masking tape or equivalent should be used to attach the indicator strip to the interior of the suit face shield.

6.2.2 If problems are encountered with this method of attachment, the indicator strip can be attached to the outside of the respirator face piece lens being used during the test.

6.3 Don the respiratory protective device normally used with the suit, and then don the TECP suit to be tested. Check to be sure all openings which are intended to be sealed (zippers, gloves, etc.) are completely sealed. DO NOT, however, plug off any venting valves.

6.4 Step into the enclosed test room such as a closet, bathroom, or test booth, equipped with an exhaust fan. No air should be exhausted from the chamber during the test because this will dilute the ammonia challenge concentrations.

6.5 Open the container with the pre-measured volume of concentrated aqueous ammonia within the enclosed test room, and pour the liquid into the empty plastic test pan. Wait two minutes to allow for adequate volatilization of the concentrated aqueous ammonia. A small mixing fan can be used near the evaporation pan to increase the evaporation rate of the ammonia solution.

6.6 After two minutes a determination of the ammonia concentration within the chamber should be made using the high range colorimetric detector tube. A concentration of 1000 ppm ammonia or greater shall be generated before the exercises are started.

6.7 To test the integrity of the suit the following four minute exercise protocol should be followed:

6.7.1 Raising the arms above the head with at least 15 raising motions completed in one minute.

6.7.2 Walking in place for one minute with at least 15 raising motions of each leg in a one-minute period.

6.7.3 Touching the toes with a least 10 complete motions of the arms from above the head to touching of the toes in a one-minute period.

6.7.4 Knee bends with at least 10 complete standing and squatting motions in a one-minute period.

6.8 If at any time during the test the colorimetric indicating paper should change colors, the test should be stopped and section 6.10 and 6.12 initiated (See ¶4.2).

6.9 After completion of the test exercise, the test area concentration should be measured again using the high range colorimetric detector tube.

6.10 Exit the test area.

6.11 The opening created by the suit zipper or other appropriate suit penetration should be used to determine the ammonia concentration in the suit with the low range length of stain detector tube or other ammonia monitor. The internal TECP suit air should be sampled far enough from the enclosed test area to prevent a false ammonia reading.

6.12 After completion of the measurement of the suit interior ammonia concentration the test is concluded and the suit is doffed and the respirator removed.

6.13 The ventilating fan for the test room should be turned on and allowed to run for enough time to remove the ammonia gas. The fan shall be vented to the outside of the building.

6.14 Any detectable ammonia in the suit interior (five ppm ammonia ($NH_3$) or more for the length of stain detector tube) indicates that the suit has failed the test. When other ammonia detectors are used a lower level of detection is possible, and it should be specified as the pass/fail criteria.

6.15 By following this test method, an intrusion coefficient of approximately 200 or more can be measured with the suit in a completely operational condition. If the intrusion coefficient is 200 or more, then the suit is suitable for emergency response and field use.

### 7.0—Retest procedures

7.1 If the suit fails this test, check for leaks by following the pressure test in test A above.

7.2 Retest the TECP suit as outlined in the test procedure 6.0.

### 8.0—Report

8.1 Each gas tight totally-encapsulating chemical protective suit tested by this practice shall have the following information recorded.

8.1.1 Unique identification number, identifying brand name, date of purchase, material of construction, and unique suit features; e.g., special breathing apparatus.

8.1.2 General description of test room used for test.

8.1.3 Brand name and purchase date of ammonia detector strips and color change data.

8.1.4 Brand name, sampling range, and expiration date of the length of stain ammonia detector tubes. The brand name and model of the sampling pump should also be recorded. If another type of ammonia detector is used, it should be identified along with its minimum detection limit for ammonia.

8.1.5 Actual test results shall list the two test area concentrations, their average, the interior suit concentration, and the calculated intrusion coefficient. Retest data shall be recorded as an additional test.

8.2 The evaluation of the data shall be specified as "suit passed" or "suit failed," and the date of the test. Any detectable ammonia (five ppm or greater for the length of stain detector tube) in the suit interior indicates the suit has failed this test. When other ammonia detectors are used, a lower level of detection is possible and it should be specified as the pass fail criteria.

**CAUTION**

Visually inspect all parts of the suit to be sure they are positioned correctly and secured tightly before putting the suit back into service. Special care should be taken to examine each exhaust valve to make sure it is not blocked.

Care should also be exercised to assure that the inside and outside of the suit is completely dry before it is put into storage.

## Appendix B to §1926.65—General Description and Discussion of the Levels of Protection and Protective Gear

This appendix sets forth information about personal protective equipment (PPE) protection levels which may be used to assist employers in complying with the PPE requirements of this section.

As required by the standard, PPE must be selected which will protect employees from the specific hazards which they are likely to encounter during their work on-site.

Selection of the appropriate PPE is a complex process which should take into consideration a variety of factors. Key factors involved in this process are identification of the hazards, or suspected hazards; their routes of potential hazard to employees (inhalation, skin absorption, ingestion, and eye or skin contact); and the performance of the PPE *materials* (and seams) in providing a barrier to these hazards. The amount of protection provided by PPE is material-hazard specific. That is, protective equipment materials will protect well against some hazardous substances and poorly, or not at all, against others. In many instances, protective equipment materials cannot be found which will provide continuous protection from the particular hazardous substance. In these cases the breakthrough time of the protective material should exceed the work durations.

Other factors in this selection process to be considered are matching the PPE to the employee's work requirements and task-specific conditions. The durability of PPE materials, such as tear strength and seam strength, should be considered in relation to the employee's tasks. The effects of PPE in relation to heat stress and task duration are a factor in selecting and using PPE. In some cases layers of PPE may be necessary to provide sufficient protection, or to protect expensive PPE inner garments, suits or equipment.

The more that is known about the hazards at the site, the easier the job of PPE selection becomes. As more information about the hazards and conditions at the site becomes available, the site supervisor can make decisions to up-grade or down-grade the level of PPE protection to match the tasks at hand.

The following are guidelines which an employer can use to begin the selection of the appropriate PPE. As noted above, the site information may suggest the use of combinations of PPE selected from the different protection levels (i.e., A, B, C, or D) as being more suitable to the hazards of the work. It should be cautioned that the listing below does not fully address the performance of the specific PPE material in relation to the specific hazards at the job site, and that PPE selection, evaluation and re-selection is an ongoing process until sufficient information about the hazards and PPE performance is obtained.

*Part A.* Personal protective equipment is divided into four categories based on the degree of protection afforded. (See Part B of this appendix for further explanation of Levels A, B, C, and D hazards.)

I. *Level A*— To be selected when the greatest level of skin, respiratory, and eye protection is required.

The following constitute Level A equipment; it may be used as appropriate;

1. Positive pressure, full face-piece self-contained breathing apparatus (SCBA), or positive pressure supplied air respirator with escape

SCBA, approved by the National Institute for Occupational Safety and Health (NIOSH).

2. Totally-encapsulating chemical-protective suit.

3. Coveralls.[1]

4. Long underwear.[1]

5. Gloves, outer, chemical-resistant.

6. Gloves, inner, chemical-resistant.

7. Boots, chemical-resistant, steel toe and shank.

8. Hard hat (under suit).[1]

9. Disposable protective suit, gloves and boots (depending on suit construction, may be worn over totally-encapsulating suit).

II. *Level B*—The highest level of respiratory protection is necessary but a lesser level of skin protection is needed.

The following constitute Level B equipment; it may be used as appropriate.

1. Positive pressure, full-facepiece self-contained breathing apparatus (SCBA), or positive pressure supplied air respirator with escape SCBA (NIOSH approved).

2. Hooded chemical-resistant clothing (overalls and long-sleeved jacket; coveralls; one or two-piece chemical-splash suit; disposable chemical-resistant overalls).

3. Coveralls.[1]

4. Gloves, outer, chemical-resistant.

5. Gloves, inner, chemical-resistant.

6. Boots, outer, chemical-resistant steel toe and shank.

7. Boot-covers, outer, chemical-resistant (disposable).[1]

8. Hard hat.[1]

9. [Reserved]

10. Face shield.[1]

III. *Level C*—The concentration(s) and type(s) of airborne substance(s) is known and the criteria for using air purifying respirators are met.

The following constitute Level C equipment; it may be used as appropriate.

1. Full-face or half-mask, air purifying respirators (NIOSH approved).

2. Hooded chemical-resistant clothing (overalls; two-piece chemical-splash suit; disposable chemical-resistant overalls).

3. Coveralls.[1]

4. Gloves, outer, chemical-resistant.

5. Gloves, inner, chemical-resistant.

6. Boots (outer), chemical-resistant steel toe and shank.[1]

7. Boot-covers, outer, chemical-resistant (disposable).[1]

8. Hard hat.[1]

9. Escape mask.[1]

10. Face shield.[1]

IV. *Level D*—A work uniform affording minimal protection, used for nuisance contamination only.

The following constitute Level D equipment; it may be used as appropriate:

1. Coveralls.

2. Gloves.[1]

3. Boots/shoes, chemical-resistant steel toe and shank.

4. Boots, outer, chemical-resistant (disposable).[1]

5. Safety glasses or chemical splash goggles*.

6. Hard hat.[1]

7. Escape mask.[1]

8. Face shield.[1]

*Part B.* The types of hazards for which levels A, B, C, and D protection are appropriate are described below:

I. *Level A*—Level A protection should be used when:

1. The hazardous substance has been identified and requires the highest level of protection for skin, eyes, and the respiratory system based on either the measured (or potential for) high concentration of atmospheric vapors, gases, or particulates; or the site operations and work functions involve a high potential for splash, immersion, or exposure to unexpected vapors, gases, or particulates of materials that are harmful to skin or capable of being absorbed through the skin;

2. Substances with a high degree of hazard to the skin are known or suspected to be present, and skin contact is possible; or

3. Operations are being conducted in confined, poorly ventilated areas, and the absence of conditions requiring Level A have not yet been determined.

II. *Level B*—Level B protection should be used when:

1. The type and atmospheric concentration of substances have been identified and require a high level of respiratory protection, but less skin protection;

2. The atmosphere contains less than 19.5 percent oxygen; or

3. The presence of incompletely identified vapors or gases is indicated by a direct-reading organic vapor detection instrument, but vapors and gases are not suspected of containing high levels of chemicals harmful to skin or capable of being absorbed through the skin

**Note:** This involves atmospheres with IDLH concentrations of specific substances that present severe inhalation hazards and that do not represent a severe skin hazard; or that do not meet the criteria for use of air-purifying respirators.

III. *Level C*—Level C protection should be used when:

1. The atmospheric contaminants, liquid splashes, or other direct contact will not adversely affect or be absorbed through any exposed skin;

2. The types of air contaminants have been identified, concentrations measured, and an air-purifying respirator is available that can remove the contaminants; and

3. All criteria for the use of air-purifying respirators are met.

IV. *Level D*—Level D protection should be used when:

1. The atmosphere contains no known hazard; and

2. Work functions preclude splashes, immersion, or the potential for unexpected inhalation of or contact with hazardous levels of any chemicals.

**Note:** As stated before, combinations of personal protective equipment other than those described for Levels A, B, C, and D protection may be more appropriate and may be used to provide the proper level of protection.

As an aid in selecting suitable chemical protective clothing, it should be noted that the National Fire Protection Association (NFPA) has developed standards on chemical protective clothing. The standards that have been adopted by include:

NFPA 1991—Standard on Vapor-Protective Suits for Hazardous Chemical Emergencies (EPA Level A Protective Clothing).

NFPA 1992—Standard on Liquid Splash-Protective Suits for Hazardous Chemical Emergencies (EPA Level B Protective Clothing).

NFPA 1993—Standard on Liquid Splash-Protective Suits for Non-emergency, Non-flammable Hazardous Chemical Situations (EPA Level B Protective Clothing).

These standards apply documentation and performance requirements to the manufacture of chemical protective suits. Chemical protective suits meeting these requirements are labelled as compliant with the appropriate standard. It is recommended that chemical protective suits that meet these standards be used.

## Appendix C to §1926.65—Compliance Guidelines

1. *Occupational Safety and Health Program.* Each hazardous waste site clean-up effort will require an occupational safety and health program headed by the site coordinator or the employer's representative. The purpose of the program will be the protection of employees at the site and will be an extension of the employer's overall safety and health program. The program will need to be developed before work begins on the site and implemented as work proceeds as stated in paragraph (b). The program is to facilitate coordination and communication of safety and health issues among personnel responsible for the

[1] Optional, as applicable

various activities which will take place at the site. It will provide the overall means for planning and implementing the needed safety and health training and job orientation of employees who will be working at the site. The program will provide the means for identifying and controlling worksite hazards and the means for monitoring program effectiveness. The program will need to cover the responsibilities and authority of the site coordinator or the employer's manager on the site for the safety and health of employees at the site, and the relationships with contractors or support services as to what each employer's safety and health responsibilities are for their employees on the site. Each contractor on the site needs to have its own safety and health program so structured that it will smoothly interface with the program of the site coordinator or principal contractor.

Also those employers involved with treating, storing or disposal of hazardous waste as covered in paragraph (p) must have implemented a safety and health program for their employees. This program is to include the hazard communication program required in paragraph (p)(1) and the training required in paragraphs (p)(7) and (p)(8) as parts of the employers comprehensive overall safety and health program. This program is to be in writing.

Each site or workplace safety and health program will need to include the following: (1) Policy statements of the line of authority and accountability for implementing the program, the objectives of the program and the role of the site safety and health supervisor or manager and staff; (2) means or methods for the development of procedures for identifying and controlling workplace hazards at the site; (3) means or methods for the development and communication to employees of the various plans, work rules, standard operating procedures and practices that pertain to individual employees and supervisors; (4) means for the training of supervisors and employees to develop the needed skills and knowledge to perform their work in a safe and healthful manner; (5) means to anticipate and prepare for emergency situations; and (6) means for obtaining information feedback to aid in evaluating the program and for improving the effectiveness of the program. The management and employees should be trying continually to improve the effectiveness of the program thereby enhancing the protection being afforded those working on the site.

Accidents on the site or workplace should be investigated to provide information on how such occurrences can be avoided in the future. When injuries or illnesses occur on the site or workplace, they will need to be investigated to determine what needs to be done to prevent this incident from occurring again. Such information will need to be used as feedback on the effectiveness of the program and the information turned into positive steps to prevent any reoccurrence. Receipt of employee suggestions or complaints relating to safety and health issues involved with site or workplace activities is also a feedback mechanism that can be used effectively to improve the program and may serve in part as an evaluative tool(s).

For the development and implementation of the program to be the most effective, professional safety and health personnel should be used. Certified Safety Professionals, Board Certified Industrial Hygienists or Registered Professional Safety Engineers are good examples of professional stature for safety and health managers who will administer the employer's program.

2. *Training.* The training programs for employees subject to the requirements of paragraph (e) of this standard should address: the safety and health hazards employees should expect to find on hazardous waste clean-up sites; what control measures or techniques are effective for those hazards; what monitoring procedures are effective in characterizing exposure levels; what makes an effective employer's safety and health program; what a site safety and health plan should include; hands on training with personal protective equipment and clothing they may be expected to use; the contents of the OSHA standard relevant to the employee's duties and function; and, employee's responsibilities under OSHA and other regulations. Supervisors will need training in their responsibilities under the safety and health program and its subject areas such as the spill containment program, the personal protective equipment program, the medical surveillance program, the emergency response plan and other areas.

The training programs for employees subject to the requirements of paragraph (p) of this standard should address: the employers safety and health program elements impacting employees; the hazard communication program; the medical surveillance program; the hazards and the controls for such hazards that employees need to know for their job duties and functions. All require annual refresher training.

The training programs for employees covered by the requirements of paragraph (q) of this standard should address those competencies required for the various levels of response such as: the hazards associated with hazardous substances; hazard identification and awareness; notification of appropriate persons; the need for and use of personal protective equipment including respirators; the decontamination procedures to be used; preplanning activities for hazardous substance incidents including the emergency reponse plan; company standard operating procedures for hazardous substance emergency responses; the use of the incident command system and other subjects. Hands-on training should be stressed whenever possible. Critiques done after an incident which include an evaluation of what worked and what did not and how could the incident be better handled the next time may be counted as training time.

For hazardous materials specialists (usually members of hazardous materials teams), the training should address the care, use and/or testing of chemical protective clothing including totally encapsulating suits, the medical surveillance program, the standard operating procedures for the hazardous materials team including the use of plugging and patching equipment and other subject areas.

Officers and leaders who may be expected to be in charge at an incident should be fully knowledgeable of their company's incident command system. They should know where and how to obtain additional assistance and be familiar with the local district's emergency response plan and the state emergency response plan.

Specialist employees such as technical experts, medical experts or environmental experts that work with hazardous materials in their regular jobs, who may be sent to the incident scene by the shipper, manufacturer or governmental agency to advise and assist the person in charge of the incident should have training on an annual basis. Their training should include the care and use of personal protective equipment including respirators; knowledge of the incident command system and how they are to relate to it; and those areas needed to keep them current in their respective field as it relates to safety and health involving specific hazardous substances.

Those skilled support personnel, such as employees who work for public works departments or equipment operators who operate bulldozers, sand trucks, backhoes, etc., who may be called to the incident scene to provide emergency support assistance, should have at least a safety and health briefing before entering the area of potential or actual exposure. These skilled support personnel, who have not been a part of the emergency response plan and do not meet the training requirements, should be made aware of the hazards they face and should be provided all necessary protective clothing and equipment required for their tasks.

There are two National Fire Protection Association standards, NFPA 472—"Standard for Professional Competence of Responders to Hazardous Material Incidents" and NFPA 471—"Recommended Practice for Responding to Hazardous Material Incidents", which are excellent resource documents to aid fire departments and other emergency response organizations in developing their training program materials. NFPA 472 provides guidance on the skills and knowledge needed for first responder awareness level, first responder operations level, hazmat technicians, and hazmat specialist. It also offers guidance for the officer corp who will be in charge of hazardous substance incidents.

3. *Decontamination.* Decontamination procedures should be tailored to the specific hazards of the site, and may vary in complexity and number of steps, depending on the level of hazard and the employee's exposure to the hazard. Decontamination procedures and PPE decontamination methods will vary depending upon the specific substance, since one procedure or method may not work for all substances. Evaluation of decontamination methods and procedures should be performed, as necessary, to assure that employees are not exposed to hazards by re-using PPE. References in appendix D may be used for guidance in establishing an effective decontamination program. In addition, the U.S. Coast Guard's Manual, "Policy Guidance for Response to Hazardous Chemical Releases," U.S. Department of Transportation, Washington, DC (COMDTINST M16465.30) is a good reference for establishing an effective decontamination program.

4. *Emergency response plans.* States, along with designated districts within the states, will be developing or have developed local emergency response plans. These state and district plans should be utilized in the emergency response plans called for in the standard. Each employer should assure that its emergency response plan is compatible with the local plan. The major reference being used to aid in developing the state and local district plans is the *Hazardous Materials Emergency Planning Guide*, NRT-1. The current Emergency Response Guidebook from the U.S. Department of Transportation, CMA's CHEMTREC and the Fire Service Emergency Management Handbook may also be used as resources.

Employers involved with treatment, storage, and disposal facilities for hazardous waste, which have the required contingency plan called for by their permit, would not need to duplicate the same planning elements. Those items of the emergency response plan that are properly addressed in the contingency plan may be substituted into the emergency response plan required in 1926.65 or otherwise kept together for employer and employee use.

5. *Personal protective equipment programs.* The purpose of personal protective clothing and equipment (PPE) is to shield or isolate individuals from the chemical, physical, and biologic hazards that may be encountered at a hazardous substance site.

As discussed in appendix B, no single combination of protective equipment and clothing is capable of protecting against all hazards. Thus PPE should be used in conjunction with other protective methods and its effectiveness evaluated periodically.

The use of PPE can itself create significant worker hazards, such as heat stress, physical and psychological stress, and impaired vision, mobility, and communication. For any given situation, equipment and clothing should be selected that provide an adequate level of protection. However, over-protection, as well as under-protection, can be hazardous and should be avoided where possible.

Two basic objectives of any PPE program should be to protect the wearer from safety and health hazards, and to prevent injury to the wearer from incorrect use and/or malfunction of the PPE. To accomplish these goals, a comprehensive PPE program should include hazard identification, medical monitoring, environmental surveillance, selection, use, maintenance, and decontamination of PPE and its associated training.

The written PPE program should include policy statements, procedures, and guidelines. Copies should be made available to all employees, and a reference copy should be made available at the worksite. Technical data on equipment, maintenance manuals, relevant regulations, and other essential information should also be collected and maintained.

6. *Incident command system (ICS).* Paragraph 1926.65(q)(3)(ii) requires the implementation of an ICS. The ICS is an organized approach to effectively control and *manage* operations at an emergency incident. The individual in charge of the ICS is the senior official responding to the incident. The ICS is not much different than the "command post" approach used for many years by the fire service. During large complex fires involving several companies and many pieces of apparatus, a command post would be established. This enabled *one* individual to be in charge of managing the incident, rather than having several officers from different companies making separate, and sometimes conflicting, decisions. The individual in charge of the command post would delegate responsibility for performing various tasks to subordinate officers. Additionally, all communications were routed through the command post to reduce the number of radio transmissions and eliminate confusion. However, strategy, tactics, and all decisions were made by one individual.

The ICS is a very similar system, except it is implemented for emergency response to all incidents, both large and small, that involve hazardous substances.

For a small incident, the individual in charge of the ICS may perform many tasks of the ICS. There may not be any, or little, delegation of tasks to subordinates. For example, in response to a small incident, the individual in charge of the ICS, in addition to normal command activities, may become the safety officer and may designate only one employee (with proper equipment) as a back-up to provide assistance if needed. OSHA does recommend, however, that at least two employees be designated as back-up personnel since the assistance needed may include rescue.

To illustrate the operation of the ICS, the following scenario might develop during a small incident, such as an overturned tank truck with a small leak of flammable liquid.

The first responding senior officer would implement and take command of the ICS. That person would size-up the incident and determine if additional personnel and apparatus were necessary; would determine what actions to take to control the leak; and, determine the proper level of personal protective equipment. If additional assistance is not needed, the individual in charge of the ICS would implement actions to stop and control the leak using the fewest number of personnel that can effectively accomplish the tasks. The individual in charge of the ICS then would designate himself as the safety officer and two other employees as a back-up in case rescue may become necessary. In this scenario, decontamination procedures would not be necessary.

A large complex incident may require many employees and difficult, time-consuming efforts to control. In these situations, the individual in charge of the ICS will want to delegate different tasks to subordinates in order to maintain a span of control that will keep the number of subordinates, that are reporting, to a manageable level.

Delegation of task at large incidents may be by location, where the incident scene is divided into sectors, and subordinate officers coordinate activities within the sector that they have been assigned.

Delegation of tasks can also be by function. Some of the functions that the individual in charge of the ICS may want to delegate at a large incident are: medical services; evacuation; water supply; resources (equipment, apparatus); media relations; safety; and, site control (integrate activities with police for crowd and traffic control). Also for a large incident, the individual in charge of the ICS will designate several employees as back-up personnel; and a number of safety officers to monitor conditions and recommend safety precautions.

Therefore, no matter what size or complexity an incident may be, by implementing an ICS there will be *one individual in charge* who makes the decisions and gives directions; and, all actions, and communications are coordinated through one central point of command. Such a system should reduce confusion, improve safety, organize and coordinate actions, and should facilitate effective management of the incident.

7. *Site Safety and Control Plans.* The safety and security of response personnel and others in the area of an emergeny response incident site should be of primary concern to the incident commander. The use of a site safety and control plan could greatly assist those in charge of assuring the safety and health of employees on the site.

A comprehensive site safety and control plan should include the following: summary analysis of hazards on the site and a risk analysis of those hazards; site map or sketch; site work zones (clean zone, transition or decontamination zone, work or hot zone); use of the buddy system; site communications; command post or command center; standard operating procedures and safe work practices; medical assistance and triage area; hazard monitoring plan (air contaminate monitoring, etc.); decontamination procedures and area; and other relevant areas. This plan should be a part of the employer's emergency response plan or an extension of it to the specific site.

8. *Medical surveillance programs.* Workers handling hazardous substances may be exposed to toxic chemicals, safety hazards, biologic hazards, and radiation. Therefore, a medical surveillance program is essential to assess and monitor workers' health and fitness for employment in hazardous waste operations and during the course of work; to provide emergency and other treatment as needed; and to keep accurate records for future reference.

The *Occupational Safety and Health Guidance Manual for Hazardous Waste Site Activities* developed by the National Institute for Occupational Safety and Health (NIOSH), the Occupational Safety and Health Administration (OSHA), the U.S. Coast Guard (USCG), and the Environmental Protection Agency (EPA); October 1985 provides an excellent example of the types of medical testing that should be done as part of a medical surveillance program.

9. *New Technology and Spill Containment Programs.* Where hazardous substances may be released by spilling from a container that will expose employees to the hazards of the materials, the employer will need to implement a program to contain and control the spilled material. Diking and ditching, as well as use of absorbents like diatomaceous earth, are traditional techniques which have proven to be effective over the years. However, in recent years new products have come into the marketplace, the use of which complement and increase the effectiveness of these traditional methods. These new products also provide emergency responders and others with additional tools or agents to use to reduce the hazards of spilled materials.

These agents can be rapidly applied over a large area and can be uniformly applied or otherwise can be used to build a small dam, thus improving the workers' ability to control spilled material. These application techniques enhance the intimate contact between the agent and the spilled material allowing for the quickest effect by the agent or quickest control of the spilled material. Agents are available to solidify liquid spilled materials, to suppress vapor generation from spilled materials, and to do both. Some special agents, which when applied as recommended by the manufacturer, will react in a controlled manner with the spilled material to neutralize acids or caustics, or greatly reduce the level of hazard of the spilled material.

There are several modern methods and devices for use by emergency response personnel or others involved with spill control efforts to safely apply spill control agents to control spilled material hazards. These include portable pressurized applicators similar to hand-held portable fire extinguishing devices, and nozzle and hose systems similar to portable fire fighting foam systems which allow the operator to apply the agent without having to come into contact with the spilled material. The operator is able to apply the agent to the spilled material from a remote position.

The solidification of liquids provides for rapid containment and isolation of hazardous substance spills. By directing the agent at run-off points or at the edges of the spill, the reactant solid will automatically create a barrier to slow or stop the spread of the material. Clean-up of hazardous substances is greatly improved when solidifying agents, acid or caustic neutralizers, or activated carbon adsorbents are used. Properly applied, these agents can totally solidify liquid hazardous substances or neutralize or absorb them, which results in materials which are less hazardous and easier to handle, transport, and dispose of. The concept of spill treatment, to create less hazardous substances, will improve the safety and level of protection of employees working at spill clean-up operations or emergency response operations to spills of hazardous substances.

The use of vapor suppression agents for volatile hazardous substances, such as flammable liquids and those substances which present an inhalation hazard, is important for protecting workers. The rapid and uniform distribution of the agent over the surface of the spilled material can provide quick vapor knockdown. There are temporary and long-term foam-type agents which are effective on vapors and dusts, and activated carbon adsorption agents which are effective for vapor control and soaking-up of the liquid. The proper use of hose lines or hand-held portable pressurized applicators provides good mobility and permits the worker to deliver the agent from a safe distance without having to step into the untreated spilled material. Some of these systems can be recharged in the field to provide coverage of larger spill areas than the design limits of a single charged applicator unit. Some of the more effective agents can solidify the liquid flammable hazardous substances and at the same time elevate the flashpoint above 140°F so the resulting substance may be handled as a nonhazardous waste material if it meets the U.S. Environmental Protection Agency's 40 CFR part 261 requirements (See particularly §261.21).

All workers performing hazardous substance spill control work are expected to wear the proper protective clothing and equipment for the materials present and to follow the employer's established standard operating procedures for spill control. All involved workers need to be trained in the established operating procedures; in the use and care of spill control equipment; and in the associated hazards and control of such hazards of spill containment work.

These new tools and agents are the things that employers will want to evaluate as part of their new technology program. The treatment of spills of hazardous substances or wastes at an emergency incident as part of the immediate spill containment and control efforts is sometimes acceptable to EPA and a permit exception is described in 40 CFR 264.1(g)(8) and 265.1(c)(11).

## Appendix D to §1926.65—References

The following references may be consulted for further information on the subject of this standard:


1. OSHA Instruction DFO CPL 2.70—January 29, 1986, *Special Emphasis Program: Hazardous Waste Sites.*

2. OSHA Instruction DFO CPL 2-2.37A—January 29, 1986, *Technical Assistance and Guidelines for Superfund and Other Hazardous Waste Site Activities.*

3. OSHA Instruction DTS CPL 2.74—January 29, 1986, *Hazardous Waste Activity Form, OSHA 175.*

4. *Hazardous Waste Inspections Reference Manual,* U.S. Department of Labor, Occupational Safety and Health Administration, 1986.

5. Memorandum of Understanding Among the National Institute for Occupational Safety and Health, the Occupational Safety and Health Administration, the United States Coast Guard, and the United States Environmental Protection Agency, *Guidance for Worker Protection During Hazardous Waste Site Investigations and Clean-up and Hazardous Substance Emergencies.* December 18, 1980.

6. *National Priorities List,* 1st Edition, October 1984; U.S. Environmental Protection Agency, Revised periodically.

7. *The Decontamination of Response Personnel,* Field Standard Operating Procedures (F.S.O.P.) 7; U.S. Environmental Protection Agency, Office of Emergency and Remedial Response, Hazardous Response Support Division, December 1984.

8. *Preparation of a Site Safety Plan,* Field Standard Operating Procedures (F.S.O.P.) 9; U.S. Environmental Protection Agency, Office of Emergency and Remedial Response, Hazardous Response Support Division, April 1985.

9. *Standard Operating Safety Guidelines;* U.S. Environmental Protection Agency, Office of Emergency and Remedial Response, Hazardous Response Support Division, Environmental Response Team; November 1984.

10. *Occupational Safety and Health Guidance Manual for Hazardous Waste Site Activities,* National Institute for Occupational Safety and Health (NIOSH), Occupational Safety and Health Administration (OSHA), U.S. Coast Guard (USCG), and Environmental Protection Agency (EPA); October 1985.

11. *Protecting Health and Safety at Hazardous Waste Sites: An Overview,* U.S. Environmental Protection Agency, EPA/625/9-85/006; September 1985.

12. *Hazardous Waste Sites and Hazardous Substance Emergencies,* NIOSH Worker Bulletin, U.S. Department of Health and Human Services, Public Health Service, Centers for Disease Control, National Institute for Occupational Safety and Health; December 1982.

13. *Personal Protective Equipment for Hazardous Materials Incidents: A Selection Guide;* U.S. Department of Health and Human Services, Public Health Service, Centers for Disease Control, National Institute for Occupational Safety and Health; October 1984.

14. *Fire Service Emergency Management Handbook,* International Association of Fire Chiefs Foundation, 101 East Holly Avenue, Unit 10B, Sterling, VA 22170, January 1985.

15. *Emergency Response Guidebook,* U.S Department of Transportation, Washington, DC, 1987.

16. *Report to the Congress on Hazardous Materials Training, Planning and Preparedness,* Federal Emergency Management Agency, Washington, DC, July 1986.

17. *Workbook for Fire Command,* Alan V. Brunacini and J. David Beageron, National Fire Protection Association, Batterymarch Park, Quincy, MA 02269, 1985.

18. *Fire Command,* Alan V. Brunacini, National Fire Protection Association, Batterymarch Park,, Quincy, MA 02269, 1985.

19. *Incident Command System,* Fire Protection Publications, Oklahoma State University, Stillwater, OK 74078, 1983.

20. *Site Emergency Response Planning,* Chemical Manufacturers Association, Washington, DC 20037, 1986.

21. *Hazardous Materials Emergency Planning Guide,* NRT-1, Environmental Protection Agency, Washington, DC, March 1987.

22. *Community Teamwork: Working Together to Promote Hazardous Materials Transportation Safety.* U.S. Department of Transportation, Washington, DC, May 1983.

23. *Disaster Planning Guide for Business and Industry,* Federal Emergency Management Agency, Publication No. FEMA 141, August 1987.


## Appendix to §1926.65—Training Curriculum Guidelines

The following non-mandatory general criteria may be used for assistance in developing site-specific training curriculum used to meet the training requirements of 29 CFR 1926.65(e); 29 CFR 1926.65(p)(7), (p)(8)(iii); and 29 CFR 1926.65(q)(6), (q)(7), and (q)(8). These are generic guidelines and they are not presented as a complete training curriculum for any specific employer. Site-specific training programs must be developed on the basis of a needs assessment of the hazardous waste site, RCRA/TSDF, or emergency response operation in accordance with 29 CFR 1926.65.

*It is noted that* the legal requirements are set forth in the regulatory text of §1926.65. The guidance set forth here presents a highly effective program that in the areas covered would meet or exceed the regulatory requirements. In addition, other approaches could meet the regulatory requirements.

*Suggested General Criteria*

*Definitions:*

*Competent* means possessing the skills, knowledge, experience, and judgment to perform assigned tasks or activities satisfactorily as determined by the employer.

*Demonstration* means the showing by actual use of equipment or procedures.

*Hands-on training* means training in a simulated work environment that permits each student to have experience performing tasks, making decisions, or using equipment appropriate to the job assignment for which the training is being conducted.

*Initial training* means training required prior to beginning work.

*Lecture* means an interactive discourse with a class lead by an instructor.

*Proficient* means meeting a stated level of achievement.

*Site-specific* means individual training directed to the operations of a specific job site.

*Training hours* means the number of hours devoted to lecture, learning activities, small group work sessions, demonstration, evaluations, or hands-on experience.

*Suggested Core Criteria:*

1. *Training facility.* The training facility should have available sufficient resources, equipment, and site locations to perform didactic and hands-on training when appropriate. Training facilities should have sufficient organization, support staff, and services to conduct training in each of the courses offered.

2. *Training Director.* Each training program should be under the direction of a training director who is responsible for the program. The Training Director should have a minimum of two years of employee education experience.

3. *Instructors.* Instructors should be deem competent on the basis of previous documented experience in their area of instruction, successful completion of a "train-the-trainer" program specific to the topics they will teach, and an evaluation of instructional competence by the Training Director.

Instructors should be required to maintain professional competency by participating in continuing education or professional development programs or by completing successfully an annual refresher course and having an annual review by the Training Director.

The annual review by the Training Director should include observation of an instructor's delivery, a review of those observations with the trainer, and an analysis of any instructor or class evaluations completed by the students during the previous year.

4. *Course materials.* The Training Director should approve all course materials to be used by the training provider. Course materials should be reviewed and updated at least annually. Materials and equipment should be in good working order and maintained properly.

All written and audio-visual materials in training curricula should be peer reviewed by technically competent outside reviewers or by a standing advisory committee.

Reviews should possess expertise in the following disciplines were applicable: occupational health, industrial hygiene and safety, chemical/environmental engineering, employee education, or emergency response. One or more of the peer reviewers should be a employee experienced in the work activities to which the training is directed.

5. *Students.* The program for accepting students should include:

a. Assurance that the student is or will be involved in work where chemical exposures are likely and that the student possesses the skills necessary to perform the work.

b. A policy on the necessary medical clearance.

6. *Ratios.* Student-instructor ratios should not exceed 30 students per instructor. Hands-on activity requiring the use of personal protective equipment should have the following student-instructor ratios. For Level C or Level D personal protective equipment the ratio should be 10 students per instructor. For Level A or Level B personal protective equipment the ratio should be 5 students per instructor.

7. *Proficiency assessment.* Proficiency should be evaluated and documented by the use of a written assessment and a skill demonstration selected and developed by the Training Director and training staff. The assessment and demonstration should evaluate the knowledge and individual skills developed in the course of training. The level of minimum achievement necessary for proficiency shall be specified in writing by the Training Director.

If a written test is used, there should be a minimum of 50 questions. If a written test is used in combination with a skills demonstration, a minimum of 25 questions should be used. If a skills demonstration is used, the tasks chosen and the means to rate successful completion should be fully documented by the Training Director.

The content of the written test or of the skill demonstration shall be relevant to the objectives of the course. The written test and skill demonstration should be updated as necessary to reflect changes in the curriculum and any update should be approved by the Training Director.

The proficiency assessment methods, regardless of the approach or combination of approaches used, should be justified, document and approved by the Training Director.

The proficiency of those taking the additional courses for supervisors should be evaluated and document by using proficiency assessment methods acceptable to the Training Director. These proficiency assessment methods must reflect the additional responsibilities borne by supervisory personnel in hazardous waste operations or emergency response.

8. *Course certificate.* Written documentation should be provided to each student who satisfactorily completes the training course. The documentation should include:

a. Student's name.

b. Course title.

c. Course date.

d. Statement that the student has successfully completed the course.

e. Name and address of the training provider.

f. An individual identification number for the certificate.

g. List of the levels of personal protective equipment used by the student to complete the course.

This documentation may include a certificate and an appropriate wallet-sized laminated card with a photograph of the student and the above information. When such course certificate cards are used, the individual identification number for the training certificate should be shown on the card.

9. *Recordkeeping.* Training providers should maintain records listing the dates courses were presented, the names of the individual course attenders, the names of those students successfully completing each course, and the number of training certificates issued to each successful student. These records should be maintained for a minimum of five years after the date an individual participated in a training program offered by the training provider. These records should be available and provided upon the student's request or as mandated by law.

10. *Program quality control.* The Training Director should conduct or direct an annual written audit of the training program. Program modifications to address deficiencies, if any, should be documented, approved, and implemented by the training provider. The audit and the program modification documents should be maintained at the training facility.

### Suggested Program Quality Control Criteria

Factors listed here are suggested criteria for determining the quality and appropriateness of employee health and safety training for hazardous waste operations and emergency response.

#### A. Training Plan.

Adequacy and appropriateness of the training program's curriculum development, instructor training, distribution of course materials, and direct student training should be considered, including

1. The duration of training, course content, and course schedules/agendas;

2. The different training requirements of the various target populations, as specified in the appropriate generic training curriculum;

3. The process for the development of curriculum, which includes appropriate technical input, outside review, evaluation, program pretesting.

4. The adequate and appropriate inclusion of hands-on, demonstration, and instruction methods;

5. Adequate monitoring of student safety, progress, and performance during the training.

#### B. Program management, Training Director, staff, and consultants.

Adequacy and appropriateness of staff performance and delivering an effective training program should be considered, including

1. Demonstration of the training director's leadership in assuring quality of health and safety training.

2. Demonstration of the competency of the staff to meet the demands of delivering high quality hazardous waste employee health and safety training.

3. Organization charts establishing clear lines of authority.

4. Clearly defined staff duties including the relationship of the training staff to the overall program.

5. Evidence that the training organizational structure suits the needs of the training program.

6. Appropriateness and adequacy of the training methods used by the instructors.

7. Sufficiency of the time committed by the training director and staff to the training program.

8. Adequacy of the ratio of training staff to students.

9. Availability and commitment of the training program of adequate human and equipment resources in the areas of

a. Health effects,

b. Safety,

c. Personal protective equipment (PPE),

d. Operational procedures,

e. Employee protection practices/procedures.

10. Appropriateness of management controls.

11. Adequacy of the organization and appropriate resources assigned to assure appropriate training.

12. In the case of multiple-site training programs, adequacy of satellite centers management.

### C. Training facilities and resources.

Adequacy and appropriateness of the facilities and resources for supporting the training program should be considered, including,

1. Space and equipment to conduct the training.

2. Facilities for representative hands-on training.

3. In the case of multiple-site programs, equipment and facilities at the satellite centers.

4. Adequacy and appropriateness of the quality control and evaluations program to account for instructor performance.

5. Adequacy and appropriateness of the quality control and evaluation program to ensure appropriate course evaluation, feedback, updating, and corrective action.

6. Adequacy and appropriateness of disciplines and expertise being used within the quality control and evaluation program.

7. Adequacy and appropriateness of the role of student evaluations to provide feedback for training program improvement.

### D. Quality control and evaluation.

Adequacy and appropriateness of quality control and evaluation plans for training programs should be considered, including:

1. A balanced advisory committee and/or competent outside reviewers to give overall policy guidance;

2. Clear and adequate definition of the composition and active programmatic role of the advisory committee or outside reviewers.

3. Adequacy of the minutes or reports of the advisory committee or outside reviewers' meetings or written communication.

4. Adequacy and appropriateness of the quality control and evaluations program to account for instructor performance.

5. Adequacy and appropriateness of the quality control and evaluation program to ensure appropriate course evaluation, feedback, updating, and corrective action.

6. Adequacy and appropriateness of disciplines and expertise being used within the quality control and evaluation program.

7. Adequacy and appropriateness of the role of student evaluations to provide feedback for training program improvement.

### E. Students

Adequacy and appropriateness of the program for accepting students should be considered, including

1. Assurance that the student already possess the necessary skills for their job, including necessary documentation.

2. Appropriateness of methods the program uses to ensure that recruits are capable of satisfactorily completing training.

3. Review and compliance with any medical clearance policy.

### F. Institutional Environment and Administrative Support

The adequacy and appropriateness of the institutional environment and administrative support system for the training program should be considered, including

1. Adequacy of the institutional commitment to the employee training program.

2. Adequacy and appropriateness of the administrative structure and administrative support

### G. Summary of Evaluation Questions

Key questions for evaluating the quality and appropriateness of an overall training program should include the following:

1. Are the program objectives clearly stated?

2. Is the program accomplishing its objectives?

3. Are appropriate facilities and staff available?

4. Is there an appropriate mix of classroom, demonstration, and hands-on training?

5. Is the program providing quality employee health and safety training that fully meets the intent of regulatory requirements?

6. What are the program's main strengths?

7. What are the program's main weaknesses?

8. What is recommended to improve the program?

9. Are instructors instructing according to their training outlines?

10. Is the evaluation tool current and appropriate for the program content?

11. Is the course material current and relevant to the target group?

### Suggested Training Curriculum Guidelines

The following training curriculum guidelines are for those operations specifically identified in 29 CFR 1926.65 as requiring training. Issues such as qualifications of instructors, training certification, and similar criteria appropriate to all categories of operations addressed in 1926.65 have been covered in the preceding section and are not re-addressed in each of the generic guidelines. Basic core requirements for training programs that are addressed include

1. General Hazardous Waste Operations

2. RCRA operations—Treatment, storage, and disposal facilities.

3. Emergency Response.

### A. General Hazardous Waste Operations and Site-specific Training

1. *Off-site training.*

Minimum training course content for hazardous waste operations, required by 29 CFR 1926.65(e), should include the following topics or procedures:

a. *Regulatory knowledge.*

(1) A review of 29 CFR 1926.65 and the core elements of an occupational safety and health program.

(2) The content of a medical surveillance program as outlined in 29 CFR 1926.65(f).

(3) The content of an effective site safety and health plan consistent with the requirements of 29 CFR 1926.65(b)(4)(ii).

(4) Emergency response plan and procedures as outlined in 29 CFR 1910.38 and 29 CFR 1926.65(l).

(5) Adequate illumination.

(6) Sanitation recommendation and equipment.

(7) Review and explanation of OSHA's hazard-communication standard (29 CFR 1910.1200) and lock-out-tag-out standard (29 CFR 1910.147).

(8) Review of other applicable standards including but not limited to those in the construction standards (29 CFR Part 1926).

(9) Rights and responsibilities of employers and employees under applicable OSHA and EPA laws.

b. *Technical knowledge.*

(1) Type of potential exposures to chemical, biological, and radiological hazards; types of human responses to these hazards and recognition of those responses; principles of toxicology and information about acute and chronic hazards; health and safety considerations of new technology.

(2) Fundamentals of chemical hazards including but not limited to vapor pressure, boiling points, flash points, ph, other physical and chemical properties.

(3) Fire and explosion hazards of chemicals.

(4) General safety hazards such as but not limited to electrical hazards, powered equipment hazards, motor vehicle hazards, walking-working surface hazards, excavation hazards, and hazards associated with working in hot and cold temperature extremes.

(5) Review and knowledge of confined space entry procedures in 29 CFR 1910.146.

(6) Work practices to minimize employee risk from site hazards.

(7) Safe use of engineering controls, equipment, and any new relevant safety technology or safety procedures.

(8) Review and demonstration of competency with air sampling and monitoring equipment that may be used in a site monitoring program.

(9) Container sampling procedures and safeguarding; general drum and container handling procedures including special requirement for laboratory waste packs, shock-sensitive wastes, and radioactive wastes.

(10) The elements of a spill control program.

(11) Proper use and limitations of material handling equipment.

(12) Procedures for safe and healthful preparation of containers for shipping and transport.

(13) Methods of communication including those used while wearing respiratory protection.

c. *Technical skills.*

(1) Selection, use maintenance, and limitations of personal protective equipment including the components and procedures for carrying out a respirator program to comply with 29 CFR 1910.134.

(2) Instruction in decontamination programs including personnel, equipment, and hardware; hands-on training including level A, B, and C ensembles and appropriate decontamination lines; field activities including the donning and doffing of protective equipment to a level commensurate with the employee's anticipated job function and responsibility and to the degree required by potential hazards.

(3) Sources for additional hazard information; exercises using relevant manuals and hazard coding systems.

d. *Additional suggested items.*

(1) A laminated, dated card or certificate with photo, denoting limitations and level of protection for which the employee is trained should be issued to those students successfully completing a course.

(2) Attendance should be required at all training modules, with successful completion of exercises and a final written or oral examination with at least 50 questions.

(3) A minimum of one-third of the program should be devoted to hands-on exercises.

(4) A curriculum should be established for the 8-hour refresher training required by 29 CFR 1926.65(e)(8), with delivery of such courses directed toward those areas of previous training that need improvement or reemphasis.

(5) A curriculum should be established for the required 8-hour training for supervisors. Demonstrated competency in the skills and knowledge provided in a 40-hour course should be a prerequisite for supervisor training.

2. *Refresher training.*

The 8-hour annual refresher training required in 29 CFR 1926.65(e)(8) should be conducted by qualified training providers. Refresher training should include at a minimum the following topics and procedures:

(a) Review of and retraining on relevant topics covered in the 40-hour program, as appropriate, using reports by the students on their work experiences.

(b) Update on developments with respect to material covered in the 40-hour course.

(c) Review of changes to pertinent provisions of EPA or OSHA standards or laws.

(d) Introduction of additional subject areas as appropriate.

(e) Hands-on review of new or altered PPE or decontamination equipment or procedures. Review of new developments in personal protective equipment.

(f) Review of newly developed air and contaminant monitoring equipment.

3. *On-site training.*

a. The employer should provide employees engaged in hazardous waste site activities with information and training prior to initial assignment into their work area, as follows:

(1) The requirements of the hazard communication program including the location and availability of the written program, required lists of hazardous chemicals, and material safety data sheets.

(2) Activities and locations in their work area where hazardous substance may be present.

(3) Methods and observations that may be used to detect the present or release of a hazardous chemical in the work area (such as monitoring conducted by the employer, continuous monitoring devices, visual appearances, or other evidence (sight, sound or smell) of hazardous chemicals being released, and applicable alarms from monitoring devices that record chemical releases.

(4) The physical and health hazards of substances known or potentially present in the work area.

(5) The measures employees can take to help protect themselves from work-site hazards, including specific procedures the employer has implemented.

(6) An explanation of the labeling system and material safety data sheets and how employees can obtain and use appropriate hazard information.

(7) The elements of the confined space program including special PPE, permits, monitoring requirements, communication procedures, emergency response, and applicable lock-out procedures.

b. The employer should provide hazardous waste employees information and training and should provide a review and access to the site safety and plan as follows:

(1) Names of personnel and alternate responsible for site safety and health.

(2) Safety and health hazards present on the site.

(3) Selection, use, maintenance, and limitations of personal protective equipment specific to the site.

(4) Work practices by which the employee can minimize risks from hazards.

(5) Safe use of engineering controls and equipment available on site.

(6) Safe decontamination procedures established to minimize employee contact with hazardous substances, including:

(A) Employee decontamination,

(B) Clothing decontamination, and

(C) Equipment decontamination.

(7) Elements of the site emergency response plan, including:

(A) Pre-emergency planning.

(B) Personnel roles and lines of authority and communication.

(C) Emergency recognition and prevention.

(D) Safe distances and places of refuge.

(E) Site security and control.

(F) Evacuation routes and procedures.

(G) Decontamination procedures not covered by the site safety and health plan.

(H) Emergency medical treatment and first aid.

(I) Emergency equipment and procedures for handling emergency incidents.

c. The employer should provide hazardous waste employees information and training on personal protective equipment used at the site, such as the following:

(1) PPE to be used based upon known or anticipated site hazards.

(2) PPE limitations of materials and construction; limitations during temperature extremes, heat stress, and other appropriate medical considerations; use and limitations of respirator equipment as well as documentation procedures as outlined in 29 CFR 1910.134.

(3) PPE inspection procedures prior to, during, and after use.

(4) PPE donning and doffing procedures.

(5) PPE decontamination and disposal procedures.

(6) PPE maintenance and storage.

(7) Task duration as related to PPE limitations.

d. The employer should instruct the employee about the site medical surveillance program relative to the particular site, including

(1) Specific medical surveillance programs that have been adapted for the site.

(2) Specific signs and symptoms related to exposure to hazardous materials on the site.

(3) The frequency and extent of periodic medical examinations that will be used on the site.

(4) Maintenance and availability of records.

(5) Personnel to be contacted and procedures to be followed when signs and symptoms of exposures are recognized.

e. The employees will review and discuss the site safety plan as part of the training program. The location of the site safety plan and all written programs should be discussed with employees including a discussion of the mechanisms for access, review, and references described

### B. RCRA Operations Training for Treatment, Storage and Disposal Facilities.

1. As a minimum, the training course required in 29 CFR 1926.65 (p) should include the following topics:

(a) Review of the applicable paragraphs of 29 CFR 1926.65 and the elements of the employer's occupational safety and health plan.

(b) Review of relevant hazards such as, but not limited to, chemical, biological, and radiological exposures; fire and explosion hazards; thermal extremes; and physical hazards.

(c) General safety hazards including those associated with electrical hazards, powered equipment hazards, lock-out-tag-out procedures, motor vehicle hazards and walking-working surface hazards.

(d) Confined-space hazards and procedures.

(e) Work practices to minimize employee risk from workplace hazards.

(f) Emergency response plan and procedures including first aid meeting the requirements of paragraph (p)(8).

(g) A review of procedures to minimize exposure to hazardous waste and various type of waste streams, including the materials handling program and spill containment program.

(h) A review of hazard communication programs meeting the requirements of 29 CFR 1910.1200.

(i) A review of medical surveillance programs meeting the requirements of 29 CFR 1926.65(p)(3) including the recognition of signs and symptoms of overexposure to hazardous substance including known synergistic interactions.

(j) A review of decontamination programs and procedures meeting the requirements of 29 CFR 1926.65(p)(4).

(k) A review of an employer's requirements to implement a training program and its elements.

(l) A review of the criteria and programs for proper selection and use of personal protective equipment, including respirators.

(m) A review of the applicable appendices to 29 CFR 1926.65.

(n) Principles of toxicology and biological monitoring as they pertain to occupational health.

(o) Rights and responsibilities of employees and employers under applicable OSHA and EPA laws.

(p) Hands-on exercises and demonstrations of competency with equipment to illustrate the basic equipment principles that may be used during the performance of work duties, including the donning and doffing of PPE.

(q) Sources of reference, efficient use of relevant manuals, and knowledge of hazard coding systems to include information contained in hazardous waste manifests.

(r) At least 8 hours of hands-on training.

(s) Training in the job skills required for an employee's job function and responsibility before they are permitted to participate in or supervise field activities.

2. The individual employer should provide hazardous waste employees with information and training prior to an employee's initial assignment into a work area. The training and information should cover the following topics:

(a) The Emergency response plan and procedures including first aid.

(b) A review of the employer's hazardous waste handling procedures including the materials handling program and elements of the spill containment program, location of spill response kits or equipment, and the names of those trained to respond to releases.

(c) The hazardous communication program meeting the requirements of 29 CFR 1910.1200.

(d) A review of the employer's medical surveillance program including the recognition of signs and symptoms of exposure to relevant hazardous substance including known synergistic interactions.

(e) A review of the employer's decontamination program and procedures.

(f) An review of the employer's training program and the parties responsible for that program.

(g) A review of the employer's personal protective equipment program including the proper selection and use of PPE based upon specific site hazards.

(h) All relevant site-specific procedures addressing potential safety and health hazards. This may include, as appropriate, biological and radiological exposures, fire and explosion hazards, thermal hazards, and physical hazards such as electrical hazards, powered equipment hazards, lock-out-tag-out hazards, motor vehicle hazards, and walking-working surface hazards.

(i) Safe use engineering controls and equipment on site.

(j) Names of personnel and alternates responsible for safety and health.

### C. Emergency response training.

Federal OSHA standards in 29 CFR 1926.65(q) are directed toward private sector emergency responders. Therefore, the guidelines provided in this portion of the appendix are directed toward that employee population. However, they also impact indirectly through State OSHA or USEPA regulations some public sector emergency responders. Therefore, the guidelines provided in this portion of the appendix may be applied to both employee populations.

States with OSHA state plans must cover their employees with regulations at least as effective as the Federal OSHA standards. Public employees in states without approved state OSHA programs covering hazardous waste operations and emergency response are covered by the U.S. EPA under 40 CFR 311, a regulation virtually identical to §1926.65.

Since this is a non-mandatory appendix and therefore not an enforceable standard, OSHA recommends that those employers, employees or volunteers in public sector emergency response organizations outside Federal OSHA jurisdiction consider the following criteria in developing their own training programs. A unified approach to training at the community level between emergency response organizations covered by Federal OSHA and those not covered directly by Federal OSHA can help ensure an effective community response to the release or potential release of hazardous substances in the community.

a. *General considerations.*

Emergency response organizations are required to consider the topics listed in §1926.65(q)(6). Emergency response organizations may use some or all of the following topics to supplement those mandatory topics when developing their response training programs. Many of the topics would require an interaction between the response provider and the individuals responsible for the site where the response would be expected.

(1) Hazard recognition, including:

(A) Nature of hazardous substances present,

(B) Practical applications of hazard recognition, including presentations on biology, chemistry, and physics.

(2) Principles of toxicology, biological monitoring, and risk assessment.

(3) Safe work practices and general site safety.

(4) Engineering controls and hazardous waste operations.

(5) Site safety plans and standard operating procedures.

(6) Decontamination procedures and practices.

(7) Emergency procedures, first aid, and self-rescue.

(8) Safe use of field equipment.

(9) Storage, handling, use and transportation of hazardous substances.

(10) Use, care, and limitations of personal protective equipment.

(11) Safe sampling techniques.

(12) Rights and responsibilities of employees under OSHA and other related laws concerning right-to-know, safety and health, compensations and liability.

(13) Medical monitoring requirements.

(14) Community relations.

b. *Suggested criteria for specific courses.*

(1) *First responder awareness level.*

(A) Review of and demonstration of competency in performing the applicable skills of 29 CFR 1926.65(q).

(B) Hands-on experience with the U.S. Department of Transportation's *Emergency Response Guidebook* (ERG) and familiarization with OSHA standard 29 CFR 1926.60.

(C) Review of the principles and practices for analyzing an incident to determine both the hazardous substances present and the basic hazard and response information for each hazardous substance present.

(D) Review of procedures for implementing actions consistent with the local emergency response plan, the organization's standard operating procedures, and the current edition of DOT's ERG including emergency notification procedures and follow-up communications.

(E) Review of the expected hazards including fire and explosions hazards, confined space hazards, electrical hazards, powered equipment hazards, motor vehicle hazards, and walking-working surface hazards.

(F) Awareness and knowledge of the competencies for the First Responder at the Awareness Level covered in the National Fire Protection Association's Standard No. 472, *Professional Competence of Responders to Hazardous Materials Incidents.*

(2) *First responder operations level.*

(A) Review of and demonstration of competency in performing the applicable skills of 29 CFR 1926.65(q).

(B) Hands-on experience with the U.S. Department of Transportation's *Emergency Response Guidebook* (ERG), manufacturer material safety data sheets, CHEMTREC/CANUTEC, shipper or manufacturer contacts and other relevant sources of information addressing hazardous substance releases. Familiarization with OSHA standard 29 CFR 1926.60.

(C) Review of the principles and practices for analyzing an incident to determine the hazardous substances present, the likely behavior of the hazardous substance and its container, the types of hazardous substance transportation containers and vehicles, the types and selection of the appropriate defensive strategy for containing the release.

(D) Review of procedures for implementing continuing response actions consistent with the local emergency response plan, the organization's standard operating procedures, and the current edition of DOT's ERG including extended emergency notification procedures and follow-up communications.

(E) Review of the principles and practice for proper selection and use of personal protective equipment.

(F) Review of the principles and practice of personnel and equipment decontamination.

(G) Review of the expected hazards including fire and explosions hazards, confined space hazards, electrical hazards, powered equipment hazards, motor vehicle hazards, and walking-working surface hazards.

(H) Awareness and knowledge of the competencies for the First Responder at the Operations Level covered in the National Fire Protection Association's Standard No. 472, *Professional Competence of Responders to Hazardous Materials Incidents.*

(3) *Hazardous materials technician.*

(A) Review of and demonstration of competency in performing the applicable skills of 29 CFR 1926.65(q).

(B) Hands-on experience with written and electronic information relative to response decision making including but not limited to the U.S. Department of Transportation's *Emergency Response Guidebook* (ERG), manufacturer material safety data sheets, CHEMTREC/CANUTEC, shipper or manufacturer contacts, computer data bases and response models, and other relevant sources of information addressing hazardous substance releases. Familiarization with 29 CFR 1926.60.

(C) Review of the principles and practices for analyzing an incident to determine the hazardous substances present, their physical and chemical properties, the likely behavior of the hazardous substance and its container, the types of hazardous substance transportation containers and vehicles involved in the release, the appropriate strategy for approaching release sites and containing the release.

(D) Review of procedures for implementing continuing response actions consistent with the local emergency response plan, the organization's standard operating procedures, and the current edition of DOT's ERG including extended emergency notification procedures and follow-up communications.

(E) Review of the principles and practice for proper selection and use of personal protective equipment.

(F) Review of the principles and practices of establishing exposure zones, proper decontamination and medical surveillance stations and procedures.

(G) Review of the expected hazards including fire and explosions hazards, confined space hazards, electrical hazards, powered equipment hazards, motor vehicle hazards, and walking-working surface hazards.

(H) Awareness and knowledge of the competencies for the Hazardous Materials Technician covered in the National Fire Protection Association's Standard No. 472, *Professional Competence of Responders to Hazardous Materials Incidents.*

(4) *Hazardous materials specialist.*

(A) Review of and demonstration of competency in performing the applicable skills of 29 CFR 1926.65(q).

(B) Hands-on experience with retrieval and use of written and electronic information relative to response decision making including but not limited to the U.S. Department of Transportation's *Emergency Response Guidebook* (ERG), manufacturer material safety data sheets, CHEMTREC/CANUTEC, shipper or manufacturer contacts, computer data bases and response models, and other relevant sources of information addressing hazardous substance releases. Familiarization with 29 CFR 1926.60.

(C) Review of the principles and practices for analyzing an incident to determine the hazardous substances present, their physical and chemical properties, and the likely behavior of the hazardous substance and its container, vessel, or vehicle.

(D) Review of the principles and practices for identification of the types of hazardous substance transportation containers, vessels and vehicles involved in the release; selecting and using the various types of equipment available for plugging or patching transportation containers, vessels or vehicles; organizing and directing the use of multiple teams of hazardous material technicians and selecting the appropriate strategy for approaching release sites and containing or stopping the release.

(E) Review of procedures for implementing continuing response actions consistent with the local emergency response plan, the organization's standard operating procedures, including knowledge of the available public and private response resources, establishment of an incident command post, direction of hazardous material technician teams, and extended emergency notification procedures and follow-up communications.

(F) Review of the principles and practice for proper selection and use of personal protective equipment.

(G) Review of the principles and practices of establishing exposure zones and proper decontamination, monitoring and medical surveillance stations and procedures.

(H) Review of the expected hazards including fire and explosions hazards, confined space hazards, electrical hazards, powered equipment hazards, motor vehicle hazards, and walking-working surface hazards.

(I) Awareness and knowledge of the competencies for the Off-site Specialist Employee covered in the National Fire Protection Association's Standard No. 472, *Professional Competence of Responders to Hazardous Materials Incidents*.

(5) *Incident commander.*

The incident commander is the individual who, at any one time, is responsible for and in control of the response effort. This individual is the person responsible for the direction and coordination of the response effort. An incident commander's position should be occupied by the most senior, appropriately trained individual present at the response site. Yet, as necessary and appropriate by the level of response provided, the position may be occupied by many individuals during a particular response as the need for greater authority, responsibility, or training increases. It is possible for the first responder at the awareness level to assume the duties of incident commander until a more senior and appropriately trained individual arrives at the response site.

Therefore, any emergency responder expected to perform as an incident commander should be trained to fulfill the obligations of the position at the level of response they will be providing including the following:

(A) Ability to analyze a hazardous substance incident to determine the magnitude of the response problem.

(B) Ability to plan and implement an appropriate response plan within the capabilities of available personnel and equipment.

(C) Ability to implement a response to favorably change the outcome of the incident in a manner consistent with the local emergency response plan and the organization's standard operating procedures.

(D) Ability to evaluate the progress of the emergency response to ensure that the response objectives are being met safely, effectively, and efficiently.

(E) Ability to adjust the response plan to the conditions of the response and to notify higher levels of response when required by the changes to the response plan.

[58 FR 35129, June 30, 1993, as amended at 59 FR 43275, Aug. 22, 1994: 61 FR 5510, Feb. 13, 1996]

## §1926.66 Criteria for design and construction of spray booths.

**1926.66(a)** ***Definitions applicable to this section***— (1) Aerated solid powders. Aerated powders shall mean any powdered material used as a coating material which shall be fluidized within a container by passing air uniformly from below. It is common practice to fluidize such materials to form a fluidized powder bed and then dip the part to be coated into the bed in a manner similar to that used in liquid dipping. Such beds are also used as sources for powder spray operations.

**1926.66(a)(2)** *Spraying area.* Any area in which dangerous quantities of flammable vapors or mists, or combustible residues, dusts, or deposits are present due to the operation of spraying processes.

**1926.66(a)(3)** *Spray booth.* A power-ventilated structure provided to enclose or accommodate a spraying operation to confine and limit the escape of spray, vapor, and residue, and to safely conduct or direct them to an exhaust system.

**1926.66(a)(4)** *Waterwash spray booth.* A spray booth equipped with a water washing system designed to minimize dusts or residues entering exhaust ducts and to permit the recovery of overspray finishing material.

**1926.66(a)(5)** *Dry spray booth.* A spray booth not equipped with a water washing system as described in paragraph (a)(4) of this section. A dry spray booth may be equipped with

**1926.66(a)(5)(i)** Distribution or baffle plates to promote an even flow of air through the booth or cause the deposit of overspray before it enters the exhaust duct; or

**1926.66(a)(5)(ii)** Overspray dry filters to minimize dusts; or

**1926.66(a)(5)(iii)** Overspray dry filters to minimize dusts or residues entering exhaust ducts; or

**1926.66(a)(5)(iv)** Overspray dry filter rolls designed to minimize dusts or residues entering exhaust ducts; or

**1926.66(a)(5)(v)** Where dry powders are being sprayed, with powder collection systems so arranged in the exhaust to capture oversprayed material.

**1926.66(a)(6)** *Fluidized bed.* A container holding powder coating material which is aerated from below so as to form an air-supported expanded cloud of such material through which the preheated object to be coated is immersed and transported.

**1926.66(a)(7)** *Electrostatic fluidized bed.* A container holding powder coating material which is aerated from below so as to form an air-supported expanded cloud of such material which is electrically charged with a charge opposite to the charge of the object to be coated; such object is transported, through the container immediately above the charged and aerated materials in order to be coated.

**1926.66(a)(8)** *Approved.* Shall mean approved and listed by a nationally recognized testing laboratory.

**1926.66(a)(9)** *Listed.* See "approved" in paragraph (a)(8) of this section.

**1926.66(b)** ***Spray booths***— (1) Construction. Spray booths shall be substantially constructed of steel, securely and rigidly supported, or of concrete or masonry except that aluminum or other substantial noncombustible material may be used for intermittent or low volume spraying. Spray booths shall be designed to sweep air currents toward the exhaust outlet.

**1926.66(b)(2)** *Interiors.* The interior surfaces of spray booths shall be smooth and continuous without edges and otherwise designed to prevent pocketing of residues and facilitate cleaning and washing without injury.

**1926.66(b)(3)** *Floors.* The floor surface of a spray booth and operator's working area, if combustible, shall be covered with noncombustible material of such character as to facilitate the safe cleaning and removal of residues.

**1926.66(b)(4)** *Distribution or baffle plates.* Distribution or baffle plates, if installed to promote an even flow of air through the booth or cause the deposit of overspray before it enters the exhaust duct, shall be of noncombustible material and readily removable or accessible on both sides for cleaning. Such plates shall not be located in exhaust ducts.

**1926.66(b)(5)** *Dry type overspray collectors—(exhaust air filters).* In conventional dry type spray booths, overspray dry filters or filter rolls, if installed, shall conform to the following:

**1926.66(b)(5)(i)** The spraying operations except electrostatic spraying operations shall be so designed, installed and maintained that the average air velocity over the open face of the booth (or booth cross section during spraying operations) shall be not less than 100 linear feet per minute. Electrostatic spraying operations may be conducted with an air velocity over the open face of the booth of not less than 60 linear feet per minute, or more, depending on the volume of the finishing material being applied and its flammability and explosion characteristics. Visible gauges or audible alarm or pressure activated devices shall be installed to indicate or insure that the required air velocity is maintained. Filter rolls shall be inspected to insure proper replacement of filter media.

**1926.66(b)(5)(ii)** All discarded filter pads and filter rolls shall be immediately removed to a safe, well-detached location or placed in a water-filled metal container and disposed of at the close of the day's operation unless maintained completely in water.

**1926.66(b)(5)(iii)** The location of filters in a spray booth shall be so as to not reduce the effective booth enclosure of the articles being sprayed.

**1926.66(b)(5)(iv)** Space within the spray booth on the downstream and upstream sides of filters shall be protected with approved automatic sprinklers.

**1926.66(b)(5)(v)** Filters or filter rolls shall not be used when applying a spray material known to be highly susceptible to spontaneous heating and ignition.

**1926.66(b)(5)(vi)** Clean filters or filter rolls shall be noncombustible or of a type having a combustibility not in excess of class 2 filters as listed by Underwriters' Laboratories, Inc. Filters and filter rolls shall not be alternately used for different types of coating materials, where the combination of materials may be conducive to spontaneous ignition.

**1926.66(b)(6)** *Frontal area.* Each spray booth having a frontal area larger than 9 square feet shall have a metal deflector or curtain not less than 2

1/2 inches (5.35 cm) deep installed at the upper outer edge of the booth over the opening.

**1926.66(b)(7)** *Conveyors.* Where conveyors are arranged to carry work into or out of spray booths, the openings therefor shall be as small as practical.

**1926.66(b)(8)** *Separation of operations.* Each spray booth shall be separated from other operations by not less than 3 feet (0.912 m), or by a greater distance, or by such partition or wall as to reduce the danger from juxtaposition of hazardous operations. See also paragraph (c)(1) of this section.

**1926.66(b)(9)** *Cleaning.* Spray booths shall be so installed that all portions are readily accessible for cleaning. A clear space of not less than 3 feet (0.912 m) on all sides shall be kept free from storage or combustible construction.

**1926.66(b)(10)** *Illumination.* When spraying areas are illuminated through glass panels or other transparent materials, only fixed lighting units shall be used as a source of illumination. Panels shall effectively isolate the spraying area from the area in which the lighting unit is located, and shall be of a noncombustible material of such a nature or so protected that breakage will be unlikely. Panels shall be so arranged that normal accumulations of residue on the exposed surface of the panel will not be raised to a dangerous temperature by radiation or conduction from the source of illumination.

**1926.66(c)** ***Electrical and other sources of ignition***— (1) Conformance. All electrical equipment, open flames and other sources of ignition shall conform to the requirements of this paragraph, except as follows:

**1926.66(c)(1)(i)** Electrostatic apparatus shall conform to the requirements of paragraphs (e) and (f) of this section;

**1926.66(c)(1)(ii)** Drying, curing, and fusion apparatus shall conform to the requirements of paragraph (g) of this section;

**1926.66(c)(1)(iii)** [Reserved]

**1926.66(c)(1)(iv)** Powder coating equipment shall conform to the requirements of paragraph (c)(1) of this section.

**1926.66(c)(2)** *Minimum separation.* There shall be no open flame or spark producing equipment in any spraying area nor within 20 feet (6.08 m) thereof, unless separated by a partition.

**1926.66(c)(3)** *Hot surfaces.* Space-heating appliances, steampipes, or hot surfaces shall not be located in a spraying area where deposits of combustible residues may readily accumulate.

**1926.66(c)(4)** *Wiring conformance.* Electrical wiring and equipment shall conform to the provisions of this paragraph and shall otherwise be in accordance with subpart S of this part.

**1926.66(c)(5)** *Combustible residues, areas.* Unless specifically approved for locations containing both deposits of readily ignitable residue and explosive vapors, there shall be no electrical equipment in any spraying area, whereon deposits of combustible residues may readily accumulate, except wiring in rigid conduit or in boxes or fittings containing no taps, splices, or terminal connections.

**1926.66(c)(6)** *Wiring type approved.* Electrical wiring and equipment not subject to deposits of combustible residues but located in a spraying area as herein defined shall be of explosion-proof type approved for Class I, group D locations and shall otherwise conform to the provisions of subpart S of this part, for Class I, Division 1, Hazardous Locations. Electrical wiring, motors, and other equipment outside of but within 20 feet (6.08 m) of any spraying area, and not separated therefrom by partitions, shall not produce sparks under normal operating conditions and shall otherwise conform to the provisions of subpart S of this part for Class I, Division 2 Hazardous Locations.

**1926.66(c)(7)** *Lamps.* Electric lamps outside of, but within 20 feet (6.08 m) of any spraying area, and not separated therefrom by a partition, shall be totally enclosed to prevent the falling of hot particles and shall be protected from mechanical injury by suitable guards or by location.

**1926.66(c)(8)** *Portable lamps.* Portable electric lamps shall not be used in any spraying area during spraying operations. Portable electric lamps, if used during cleaning or repairing operations, shall be of the type approved for hazardous Class I locations.

**1926.66(c)(9)** *Grounding.* (i) All metal parts of spray booths, exhaust ducts, and piping systems conveying flammable or combustible liquids or aerated solids shall be properly electrically grounded in an effective and permanent manner.

**1926.66(d)** ***Ventilation***— (1) Conformance. Ventilating and exhaust systems shall be in accordance with the Standard for Blower and Exhaust Systems for Vapor Removal, NFPA No. 91-1961, where applicable and shall also conform to the provisions of this section.

**1926.66(d)(2)** *General.* All spraying areas shall be provided with mechanical ventilation adequate to remove flammable vapors, mists, or powders to a safe location and to confine and control combustible residues so that life is not endangered. Mechanical ventilation shall be kept in operation at all times while spraying operations are being conducted and for a sufficient time thereafter to allow vapors from drying coated articles and drying finishing material residue to be exhausted.

**1926.66(d)(3)** *Independent exhaust.* Each spray booth shall have an independent exhaust duct system discharging to the exterior of the building, except that multiple cabinet spray booths in which identical spray finishing material is used with a combined frontal area of not more than 18 square feet may have a common exhaust. If more than one fan serves one booth, all fans shall be so interconnected that one fan cannot operate without all fans being operated.

**1926.66(d)(4)** *Fan-rotating element.* The fan-rotating element shall be nonferrous or nonsparking or the casing shall consist of or be lined with such material. There shall be ample clearance between the fan-rotating element and the fan casing to avoid a fire by friction, necessary allowance being made for ordinary expansion and loading to prevent contact between moving parts and the duct or fan housing. Fan blades shall be mounted on a shaft sufficiently heavy to maintain perfect alignment even when the blades of the fan are heavily loaded, the shaft preferably to have bearings outside the duct and booth. All bearings shall be of the self-lubricating type, or lubricated from the outside duct.

**1926.66(d)(5)** *Electric motors.* Electric motors driving exhaust fans shall not be placed inside booths or ducts. See also paragraph (c) of this section.

**1926.66(d)(6)** *Belts.* Belts shall not enter the duct or booth unless the belt and pulley within the duct or booth are thoroughly enclosed.

**1926.66(d)(7)** *Exhaust ducts.* Exhaust ducts shall be constructed of steel and shall be substantially supported. Exhaust ducts without dampers are preferred; however, if dampers are installed, they shall be maintained so that they will be in a full open position at all times the ventilating system is in operation.

**1926.66(d)(7)(i)** Exhaust ducts shall be protected against mechanical damage and have a clearance from unprotected combustible construction or other combustible material of not less than 18 inches (45.72 cm).

**1926.66(d)(7)(ii)** If combustible construction is provided with the following protection applied to all surfaces within 18 inches (45.72 cm), clearances may be reduced to the distances indicated:

| | |
|---|---|
| (*a*) 28-gage sheet metal on 1/4-inch asbestos mill board. | 12 inches (30.48 cm). |
| (*b*) 28-gage sheet metal on 1/8-inch asbestos mill board spaced out 1 inch (2.54 cm) on noncombustible spacers. | 9 inches (22.86 cm). |
| (*c*) 22-gage sheet metal on 1-inch rockwool batts reinforced with wire mesh or the equivalent. | 3 inches (7.62 cm). |
| (*d*) Where ducts are protected with an approved automatic sprinkler system, properly maintained, the clearance required in paragraph (d)(7)(i) of this section may be reduced to 6 inches (15.24 cm). | |

**1926.66(d)(8)** *Discharge clearance.* Unless the spray booth exhaust duct terminal is from a water-wash spray booth, the terminal discharge point shall be not less than 6 feet from any combustible exterior wall or roof nor discharge in the direction of any combustible construction or unprotected opening in any noncombustible exterior wall within 25 feet (7.6 m).

**1926.66(d)(9)** *Air exhaust.* Air exhaust from spray operations shall not be directed so that it will contaminate makeup air being introduced into the spraying area or other ventilating intakes, nor directed so as to create a nuisance. Air exhausted from spray operations shall not be recirculated.

**1926.66(d)(10)** *Access doors.* When necessary to facilitate cleaning, exhaust ducts shall be provided with an ample number of access doors.

**1926.66(d)(11)** *Room intakes.* Air intake openings to rooms containing spray finishing operations shall be adequate for the efficient operation of exhaust fans and shall be so located as to minimize the creation of dead air pockets.

**1926.66(d)(12)** *Drying spaces.* Freshly sprayed articles shall be dried only in spaces provided with adequate ventilation to prevent the formation of explosive vapors. In the event adequate and reliable ventilation is not provided such drying spaces shall be considered a spraying area.

**1926.66(e)** ***Fixed electrostatic apparatus***— (1) Conformance. Where installation and use of electrostatic spraying equipment is used, such installation and use shall conform to all other paragraphs of this section, and shall also conform to the requirements of this paragraph.

**1926.66(e)(2)** *Type approval.* Electrostatic apparatus and devices used in connection with coating operations shall be of approved types.

**1926.66(e)(3)** *Location.* Transformers, power packs, control apparatus, and all other electrical portions of the equipment, with the exception of high-voltage grids, electrodes, and electrostatic atomizing heads and their connections, shall be located outside of the spraying area, or shall otherwise conform to the requirements of paragraph (c) of this section.

**1926.66(e)(4)** *Support.* Electrodes and electrostatic atomizing heads shall be adequately supported in permanent locations and shall be effectively insulated from the ground. Electrodes and electrostatic atomizing heads which are permanently attached to their bases, supports, or reciprocators, shall be deemed to comply with this section. Insulators shall be nonporous and noncombustible.

**1926.66(e)(5)** *Insulators, grounding.* High-voltage leads to electrodes shall be properly insulated and protected from mechanical injury or exposure to destructive chemicals. Electrostatic atomizing heads shall be effectively and permanently supported on suitable insulators and shall be effectively guarded against accidental contact or grounding. An automatic means shall be provided for grounding the electrode system when it is electrically deenergized for any reason. All insulators shall be kept clean and dry.

**1926.66(e)(6)** *Safe distance.* A safe distance shall be maintained between goods being painted and electrodes or electrostatic atomizing heads or conductors of at least twice the sparking distance. A suitable sign indicating this safe distance shall be conspicuously posted near the assembly.

**1926.66(e)(7)** *Conveyors required.* Goods being painted using this process are to be supported on conveyors. The conveyors shall be so arranged as to maintain safe distances between the goods and the electrodes or electrostatic atomizing heads at all times. Any irregularly shaped or other goods subject to possible swinging or movement shall be rigidly supported to prevent such swinging or movement which would reduce the clearance to less than that specified in paragraph (e)(6) of this section.

**1926.66(e)(8)** *Prohibition.* This process is not acceptable where goods being coated are manipulated by hand. When finishing materials are applied by electrostatic equipment which is manipulated by hand, see paragraph (f) of this section for applicable requirements.

**1926.66(e)(9)** *Fail-safe controls.* Electrostatic apparatus shall be equipped with automatic controls which will operate without time delay to disconnect the power supply to the high voltage transformer and to signal the operator under any of the following conditions:

**1926.66(e)(9)(i)** Stoppage of ventilating fans or failure of ventilating equipment from any cause.

**1926.66(e)(9)(ii)** Stoppage of the conveyor carrying goods through the high voltage field.

**1926.66(e)(9)(iii)** Occurrence of a ground or of an imminent ground at any point on the high voltage system.

**1926.66(e)(9)(iv)** Reduction of clearance below that specified in paragraph (e)(6) of this section.

**1926.66(e)(10)** *Guarding.* Adequate booths, fencing, railings, or guards shall be so placed about the equipment that they, either by their location or character or both, assure that a safe isolation of the process is maintained from plant storage or personnel. Such railings, fencing, and guards shall be of conducting material, adequately grounded.

**1926.66(e)(11)** *Ventilation.* Where electrostatic atomization is used the spraying area shall be so ventilated as to insure safe conditions from a fire and health standpoint.

**1926.66(e)(12)** *Fire protection.* All areas used for spraying, including the interior of the booth, shall be protected by automatic sprinklers where this protection is available. Where this protection is not available, other approved automatic extinguishing equipment shall be provided.

**1926.66(f)** ***Electrostatic hand spraying equipment***— (1) Application. This paragraph shall apply to any equipment using electrostatically charged elements for the atomization and/or, precipitation of materials for coatings on articles, or for other similar purposes in which the atomizing device is hand held and manipulated during the spraying operation.

**1926.66(f)(2)** *Conformance.* Electrostatic hand spraying equipment shall conform with the other provisions of this section.

**1926.66(f)(3)** *Equipment approval and specifications.* Electrostatic hand spray apparatus and devices used in connection with coating operations shall be of approved types. The high voltage circuits shall be designed so as to not produce a spark of sufficient intensity to ignite any vapor-air mixtures nor result in appreciable shock hazard upon coming in contact with a grounded object under all normal operating conditions. The electrostatically charged exposed elements of the handgun shall be capable of being energized only by a switch which also controls the coating material supply.

**1926.66(f)(4)** *Electrical support equipment.* Transformers, powerpacks, control apparatus, and all other electrical portions of the equipment, with the exception of the handgun itself and its connections to the power supply shall be located outside of the spraying area or shall otherwise conform to the requirements of paragraph (c) of this section.

**1926.66(f)(5)** *Spray gun ground.* The handle of the spraying gun shall be electrically connected to ground by a metallic connection and to be so constructed that the operator in normal operating position is in intimate electrical contact with the grounded handle.

**1926.66(f)(6)** *Grounding-general.* All electrically conductive objects in the spraying area shall be adequately grounded. This requirement shall apply to paint containers, wash cans, and any other objects or devices in the area. The equipment shall carry a prominent permanently installed warning regarding the necessity for this grounding feature.

**1926.66(f)(7)** *Maintenance of grounds.* Objects being painted or coated shall be maintained in metallic contact with the conveyor or other grounded support. Hooks shall be regularly cleaned to insure this contact and areas of contact shall be sharp points or knife edges where possible. Points of support of the object shall be concealed from random spray where feasible and where the objects being sprayed are supported from a conveyor, the point of attachment to the conveyor shall be so located as to not collect spray material during normal operation.

**1926.66(f)(8)** *Interlocks.* The electrical equipment shall be so interlocked with the ventilation of the spraying area that the equipment cannot be operated unless the ventilation fans are in operation.

**1926.66(f)(9)** *Ventilation.* The spraying operation shall take place within a spray area which is adequately ventilated to remove solvent vapors released from the operation.

**1926.66(g)** ***Drying, curing, or fusion apparatus***— (1) Conformance. Drying, curing, or fusion apparatus in connection with spray application of flammable and combustible finishes shall conform to the Standard for Ovens and Furnaces, NFPA 86A-1969, where applicable and shall also conform with the following requirements of this paragraph.

**1926.66(g)(2)** *Alternate use prohibited.* Spray booths, rooms, or other enclosures used for spraying operations shall not alternately be used for the purpose of drying by any arrangement which will cause a material increase in the surface temperature of the spray booth, room, or enclosure.

**1926.66(g)(3)** *Adjacent system interlocked.* Except as specifically provided in paragraph (g)(4) of this section, drying, curing, or fusion units utilizing a heating system having open flames or which may produce sparks shall not be installed in a spraying area, but may be installed adjacent thereto when equipped with an interlocked ventilating system arranged to:

**1926.66(g)(3)(i)** Thoroughly ventilate the drying space before the heating system can be started;

**1926.66(g)(3)(ii)** Maintain a safe atmosphere at any source of ignition;

**1926.66(g)(3)(iii)** Automatically shut down the heating system in the event of failure of the ventilating system.

**1926.66(g)(4)** *Alternate use permitted.* Automobile refinishing spray booths or enclosures, otherwise installed and maintained in full conformity with this section, may alternately be used for drying with portable electrical infrared drying apparatus when conforming with the following:

**1926.66(g)(4)(i)** Interior (especially floors) of spray enclosures shall be kept free of overspray deposits.

**1926.66(g)(4)(ii)** During spray operations, the drying apparatus and electrical connections and wiring thereto shall not be located within spray enclosure nor in any other location where spray residues may be deposited thereon.

**1926.66(g)(4)(iii)** The spraying apparatus, the drying apparatus, and the ventilating system of the spray enclosure shall be equipped with suitable interlocks so arranged that:

**1926.66(g)(4)(iii)*(a)*** The spraying apparatus cannot be operated while the drying apparatus is inside the spray enclosure.

**1926.66(g)(4)(iii)*(b)*** The spray enclosure will be purged of spray vapors for a period of not less than 3 minutes before the drying apparatus can be energized.

**1926.66(g)(4)(iii)*(c)*** The ventilating system will maintain a safe atmosphere within the enclosure during the drying process and the drying apparatus will automatically shut off in the event of failure of the ventilating system.

**1926.66(g)(4)(iv)** All electrical wiring and equipment of the drying apparatus shall conform with the applicable sections of subpart S of this part. Only equipment of a type approved for Class I, Division 2 hazardous locations shall be located within 18 inches (45.72 cm) of floor level. All metallic parts of the drying apparatus shall be properly electrically bonded and grounded.

**1926.66(g)(4)(v)** The drying apparatus shall contain a prominently located, permanently attached warning sign indicating that ventilation should be maintained during the drying period and that spraying should not be conducted in the vicinity that spray will deposit on apparatus.

[58 FR 35149, June 30, 1993]

**Authority:** Sec. 107, Contract Work Hours and Safety Standards Act (Construction Safety Act) (40 U.S.C. 333); secs. 4, 6, and 8 of the Occupational Safety and Health Act of 1970 (29 U.S.C. 653, 655, 657); Secretary of Labor's Orders 12-71 (36 FR 8754), 8-76 (41 FR 25059), 9-83 (48 FR 35736), 1-90 (55 FR 9033), or 6-96 (62 FR 111), as applicable; and 29 CFR part 11.

### §1926.95 Criteria for personal protective equipment.

**1926.95(a)** ***Application.*** Protective equipment, including personal protective equipment for eyes, face, head, and extremities, protective clothing, respiratory devices, and protective shields and barriers, shall be provided, used, and maintained in a sanitary and reliable condition wherever it is necessary by reason of hazards of processes or environment, chemical hazards, radiological hazards, or mechanical irritants encountered in a manner capable of causing injury or impairment in the function of any part of the body through absorption, inhalation or physical contact.

**1926.95(b)** ***Employee-owned equipment.*** Where employees provide their own protective equipment, the employer shall be responsible to assure its adequacy, including proper maintenance, and sanitation of such equipment.

**1926.95(c)** ***Design.*** All personal protective equipment shall be of safe design and construction for the work to be performed.

[58 FR 35152, June 30, 1993]

### §1926.96 Occupational foot protection.

Safety-toe footwear for employees shall meet the requirements and specifications in American National Standard for Men's Safety-Toe Footwear, Z41.1-1967.

[58 FR 35152, June 30, 1993]

### §§1926.97—1926.98 [Reserved]

### §1926.100 Head protection.

**1926.100(a)** Employees working in areas where there is a possible danger of head injury from impact, or from falling or flying objects, or from electrical shock and burns, shall be protected by protective helmets.

**1926.100(b)** Helmets for the protection of employees against impact and penetration of falling and flying objects shall meet the specifications contained in American National Standards Institute, Z89.1-1969, Safety Requirements for Industrial Head Protection.

**1926.100(c)** Helmets for the head protection of employees exposed to high voltage electrical shock and burns shall meet the specifications contained in American National Standards Institute, Z89.2-1971.

### §1926.101 Hearing protection.

**1926.101(a)** Wherever it is not feasible to reduce the noise levels or duration of exposures to those specified in Table D-2, Permissible Noise Exposures, in §1926.52, ear protective devices shall be provided and used.

**1926.101(b)** Ear protective devices inserted in the ear shall be fitted or determined individually by competent persons.

**1926.101(c)** Plain cotton is not an acceptable protective device.

### §1926.102 Eye and face protection.

**1926.102(a)** ***General.*** (1) Employees shall be provided with eye and face protection equipment when machines or operations present potential eye or face injury from physical, chemical, or radiation agents.

**1926.102(a)(2)** Eye and face protection equipment required by this Part shall meet the requirements specified in American National Standards Institute, Z87.1-1968, Practice for Occupational and Educational Eye and Face Protection.

**1926.102(a)(3)** Employees whose vision requires the use of corrective lenses in spectacles, when required by this regulation to wear eye protection, shall be protected by goggles or spectacles of one of the following types:

**1926.102(a)(3)(i)** Spectacles whose protective lenses provide optical correction;

**1926.102(a)(3)(ii)** Goggles that can be worn over corrective spectacles without disturbing the adjustment of the spectacles;

**1926.102(a)(3)(iii)** Goggles that incorporate corrective lenses mounted behind the protective lenses.

**1926.102(a)(4)** Face and eye protection equipment shall be kept clean and in good repair. The use of this type equipment with structural or optical defects shall be prohibited.

**1926.102(a)(5)** Table E-1 shall be used as a guide in the selection of face and eye protection for the hazards and operations noted.

**TABLE E-1—EYE AND FACE PROTECTOR SELECTION GUIDE**

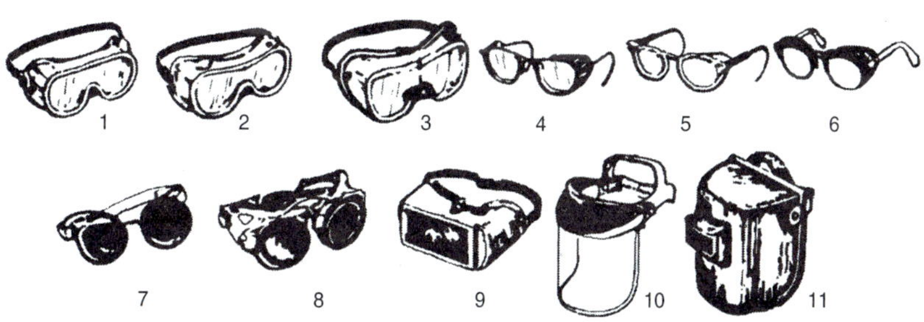


1. GOGGLES, Flexible Fitting, Regular Ventilation
2. GOGGLES, Flexible Fitting, Hooded Ventilation
3. GOGGLES, Cushioned Fitting, Rigid Body
*4. SPECTACLES, Metal Frame, with Sideshields
*5. SPECTACLES, Plastic Frame, with Sideshields
*6. SPECTACLES, Metal-Plastic Frame, with Sideshields

**7. WELDING GOGGLES, Eyecup Type, Tinted Lenses (Illustrated)
7A. CHIPPING GOGGLES, Eyecup Type, Clear Safety Lenses (Not Illustrated)
** 8. WELDING GOGGLES, Coverspec Type, Tinted Lenses (Illustrated)
8A. CHIPPING GOGGLES, Coverspec Type, Clear Safety Lenses (Not Illustrated)
** 9. WELDING GOGGLES, Coverspec Type, Tinted Plate Lens
10. FACE SHIELD (Available with Plastic or Mesh Window)
**11. WELDING HELMETS

* Non-side shield spectacles are available for limited hazard use requiring only frontal protection.

**See Table E-2, in paragraph (b) of this section, Filter Lens Shade Numbers for Protection Against Radiant Energy.

| Applications | | |
|---|---|---|
| Operation | Hazards | Recommended protectors: bold type numbers signify preferred protection |
| Acetylene—Burning, Acetylene—Cutting, Acetylene—Welding. | Sparks, harmful rays, molten metal, flying particles | 7, 8, 9. |
| Chemical Handling . . | Splash, acid burns, fumes | 2, 10 (For severe exposure add 10 over 2). |
| Chipping . . . . . . . . . . | Flying particles | 1, 3, 4, 5, 6, 7A, 8A. |
| Electric (arc) welding . | Sparks, intense rays, molten metal | 9, 11 (11 in combination with 4, 5, 6, in tinted lenses, advisable). |
| Furnace operations . . . | Glare, heat, molten metal | 7, 8, 9 (For severe exposure add 10). |
| Grinding—Light . . . . | Flying particles | 1, 3, 4, 5, 6, 10. |
| Grinding—Heavy . . . | Flying particles | 1, 3, 7A, 8A (For severe exposure add 10). |
| Laboratory . . . . . . . . . | Chemical splash, glass breakage | 2 (10 when in combination with 4, 5, 6). |
| Machining . . . . . . . . . | Flying particles | 1, 3, 4, 5, 6, 10. |
| Molten metals . . . . . . | Heat, glare, sparks, splash | 7, 8 (10 in combination with 4, 5, 6, in tinted lenses). |
| Spot welding . . . . . . . | Flying particles, sparks | 1, 3, 4, 5, 6, 10. |

**1926.102(a)(6)** Protectors shall meet the following minimum requirements:

**1926.102(a)(6)(i)** They shall provide adequate protection against the particular hazards for which they are designed.

**1926.102(a)(6)(ii)** They shall be reasonably comfortable when worn under the designated conditions.

**1926.102(a)(6)(iii)** They shall fit snugly and shall not unduly interfere with the movements of the wearer.

**1926.102(a)(6)(iv)** They shall be durable.

**1926.102(a)(6)(v)** They shall be capable of being disinfected.

**1926.102(a)(6)(vi)** They shall be easily cleanable.

**1926.102(a)(7)** Every protector shall be distinctly marked to facilitate identification only of the manufacturer.

**1926.102(a)(8)** When limitations or precautions are indicated by the manufacturer, they shall be transmitted to the user and care taken to see that such limitations and precautions are strictly observed.

**1926.102(b)** ***Protection against radiant energy***—(1) *Selection of shade numbers for welding filter.* Table E-2 shall be used as a guide for the selection of the proper shade numbers of filter lenses or plates used in welding. Shades more dense than those listed may be used to suit the individual's needs.

### TABLE E-2—FILTER LENS SHADE NUMBERS FOR PROTECTION AGAINST RADIANT ENERGY

| Welding operation | Shade number |
|---|---|
| Shielded metal-arc welding 1/16-, 3/32-, 1/8-, 5/32-inch diameter electrodes | 10 |
| Gas-shielded arc welding (nonferrous) 1/16-, 3/32-, 1/8-, 5/32-inch diameter electrodes | 11 |
| Gas-shielded arc welding (ferrous) 1/16-, 3/32-, 1/8-, 5/32-inch diameter electrodes | 12 |
| Shielded metal-arc welding 3/16-, 7/32-, 1/4-inch diameter electrodes | 12 |
| 5/16-, 3/8-inch diameter electrodes | 14 |
| Atomic hydrogen welding | 10-14 |
| Carbon-arc welding | 14 |
| Soldering | 2 |
| Torch brazing | 3 or 4 |
| Light cutting, up to 1 inch | 3 or 4 |
| Medium cutting, 1 inch to 6 inches | 4 or 5 |
| Heavy cutting, over 6 inches | 5 or 6 |
| Gas welding (light), up to 1/8 inch | 4 or 5 |
| Gas welding (medium), 1/8-inch to 1/2 inch | 5 or 6 |
| Gas welding (heavy), over 1/2 inch | 6 or 8 |

**1926.102(b)(2)** *Laser protection.* (i) Employees whose occupation or assignment requires exposure to laser beams shall be furnished suitable laser safety goggles which will protect for the specific wavelength of the laser and be of optical density (O.D.) adequate for the energy involved. Table E-3 lists the maximum power or energy density for which adequate protection is afforded by glasses of optical densities from 5 through 8.

### TABLE E-3—SELECTING LASER SAFETY GLASS

| Intensity, CW maximum power density (watts/cm$^2$) | Attenuation | |
|---|---|---|
| | Optical density (O.D.) | Attenuation factor |
| $10^{-2}$ | 5 | $10^5$ |
| $10^{-1}$ | 6 | $10^6$ |
| 1.0 | 7 | $10^7$ |
| 10.0 | 8 | $10^8$ |

Output levels falling between lines in this table shall require the higher optical density.

**1926.102(b)(2)(ii)** All protective goggles shall bear a label identifying the following data:

**1926.102(b)(2)(ii)*(a)*** The laser wavelengths for which use is intended;

**1926.102(b)(2)(ii)*(b)*** The optical density of those wavelengths;

**1926.102(b)(2)(ii)*(c)*** The visible light transmission.

[44 FR 8577, Feb. 9, 1979; 44 FR 20940, Apr. 6, 1979, as amended at 58 FR 35160, June 30, 1993]

## §1926.103 Respiratory protection.

**Note:** The requirements applicable to construction work under this section are identical to those set forth at 29 CFR 1910.134 of this chapter.

[63 FR 1297; Jan. 8, 1998]

[Section 1910.134 is printed below—CCH.]

## §1910.134 Respiratory protection.

This section applies to General Industry (part 1910), Shipyards (part 1915), Marine Terminals (part 1917), Longshoring (part 1918), and Construction (part 1926).

**1910.134(a)** ***Permissible practice.*** (1) In the control of those occupational diseases caused by breathing air contaminated with harmful dusts, fogs, fumes, mists, gases, smokes, sprays, or vapors, the primary objective shall be to prevent atmospheric contamination. This shall be accomplished as far as feasible by accepted engineering control measures (for example, enclosure or confinement of the operation, general and local ventilation, and substitution of less toxic materials). When effective engineering controls are not feasible, or while they are being instituted, appropriate respirators shall be used pursuant to this section.

**1910.134(a)(2)** Respirators shall be provided by the employer when such equipment is necessary to protect the health of the employee. The employer shall provide the respirators which are applicable and suitable for the purpose intended. The employer shall be responsible for the establishment and maintenance of a respiratory protection program which shall include the requirements outlined in paragraph (c) of this section.

**1910.134(b)** ***Definitions.*** The following definitions are important terms used in the respiratory protection standard in this section.

*Air-purifying respirator* means a respirator with an air-purifying filter, cartridge, or canister that removes specific air contaminants by passing ambient air through the air-purifying element.

*Assigned protection factor (APF)* [Reserved]

*Atmosphere-supplying respirator* means a respirator that supplies the respirator user with breathing air from a source independent of the ambient atmosphere, and includes supplied-air respirators (SARs) and self-contained breathing apparatus (SCBA) units.

*Canister or cartridge* means a container with a filter, sorbent, or catalyst, or combination of these items, which removes specific contaminants from the air passed through the container.

*Demand respirator* means an atmosphere-supplying respirator that admits breathing air to the facepiece only when a negative pressure is created inside the facepiece by inhalation.

*Emergency situation* means any occurrence such as, but not limited to, equipment failure, rupture of containers, or failure of control equipment that may or does result in an uncontrolled significant release of an airborne contaminant.

*Employee exposure* means exposure to a concentration of an airborne contaminant that would occur if the employee were not using respiratory protection.

*End-of-service-life indicator (ESLI)* means a system that warns the respirator user of the approach of the end of adequate respiratory protection, for example, that the sorbent is approaching saturation or is no longer effective.

*Escape-only respirator* means a respirator intended to be used only for emergency exit.

*Filter or air purifying element* means a component used in respirators to remove solid or liquid aerosols from the inspired air.

*Filtering facepiece* (dust mask) means a negative pressure particulate respirator with a filter as an integral part of the facepiece or with the entire facepiece composed of the filtering medium.

*Fit factor* means a quantitative estimate of the fit of a particular respirator to a specific individual, and typically estimates the ratio of the concentration of a substance in ambient air to its concentration inside the respirator when worn.

*Fit test* means the use of a protocol to qualitatively or quantitatively evaluate the fit of a respirator on an individual. (See also Qualitative fit test QLFT and Quantitative fit test QNFT.)

*Helmet* means a rigid respiratory inlet covering that also provides head protection against impact and penetration.

*High efficiency particulate air (HEPA) filter* means a filter that is at least 99.97% efficient in removing monodisperse particles of 0.3 micrometers in diameter. The equivalent NIOSH 42 CFR 84 particulate filters are the N100, R100, and P100 filters.

*Hood* means a respiratory inlet covering that completely covers the head and neck and may also cover portions of the shoulders and torso.

*Immediately dangerous to life or health (IDLH)* means an atmosphere that poses an immediate threat to life, would cause irreversible adverse health effects, or would impair an individual's ability to escape from a dangerous atmosphere.

*Interior structural firefighting* means the physical activity of fire suppression, rescue or both, inside of buildings or enclosed structures which are involved in a fire situation beyond the incipient stage. (See 29 CFR 1910.155)

*Loose-fitting facepiece* means a respiratory inlet covering that is designed to form a partial seal with the face.

*Maximum use concentration (MUC)* [Reserved].

*Negative pressure respirator (tight fitting)* means a respirator in which the air pressure inside the facepiece is negative during inhalation with respect to the ambient air pressure outside the respirator.

*Oxygen deficient atmosphere* means an atmosphere with an oxygen content below 19.5% by volume.

*Physician or other licensed health care professional (PLHCP)* means an individual whose legally permitted scope of practice (*i.e.*, license, registration, or certification) allows him or her to independently provide, or be delegated the responsibility to provide, some or all of the health care services required by paragraph (e) of this section.

*Positive pressure respirator* means a respirator in which the pressure inside the respiratory inlet covering exceeds the ambient air pressure outside the respirator.

*Powered air-purifying respirator (PAPR)* means an air-purifying respirator that uses a blower to force the ambient air through air-purifying elements to the inlet covering.

*Pressure demand respirator* means a positive pressure atmosphere-supplying respirator that admits breathing air to the facepiece when the positive pressure is reduced inside the facepiece by inhalation.

*Qualitative fit test (QLFT)* means a pass/fail fit test to assess the adequacy of respirator fit that relies on the individual's response to the test agent.

*Quantitative fit test (QNFT)* means an assessment of the adequacy of respirator fit by numerically measuring the amount of leakage into the respirator.

*Respiratory inlet covering* means that portion of a respirator that forms the protective barrier between the user's respiratory tract and an air-purifying device or breathing air source, or both. It may be a facepiece, helmet, hood, suit, or a mouthpiece respirator with nose clamp.

*Self-contained breathing apparatus (SCBA)* means an atmosphere-supplying respirator for which the breathing air source is designed to be carried by the user.

*Service life* means the period of time that a respirator, filter or sorbent, or other respiratory equipment provides adequate protection to the wearer.

*Supplied-air respirator (SAR) or airline respirator* means an atmosphere-supplying respirator for which the source of breathing air is not designed to be carried by the user.

*This section* means this respiratory protection standard.

*Tight-fitting facepiece* means a respiratory inlet covering that forms a complete seal with the face.

*User seal check* means an action conducted by the respirator user to determine if the respirator is properly seated to the face.

**1910.134(c)** ***Respiratory protection program.*** This paragraph requires the employer to develop and implement a written respiratory protection program with required worksite-specific procedures and elements for required respirator use. The program must be administered by a suitably trained program administrator. In addition, certain program elements may be required for voluntary use to prevent potential hazards associated with the use of the respirator. The Small Entity Compliance Guide contains criteria for the selection of a program administrator and a sample program that meets the requirements of this paragraph. Copies of the Small Entity Compliance Guide will be available on or about April 8, 1998 from the Occupational Safety and Health Administration's Office of Publications, Room N 3101, 200 Constitution Avenue, NW, Washington, DC, 20210 (202-219-4667).

**1910.134(c)(1)** In any workplace where respirators are necessary to protect the health of the employee or whenever respirators are required by the employer, the employer shall establish and implement a written respiratory protection program with worksite-specific procedures. The program shall be updated as necessary to reflect those changes in workplace conditions that affect respirator use. The employer shall include in the program the following provisions of this section, as applicable:

**1910.134(c)(1)(i)** Procedures for selecting respirators for use in the workplace;

**1910.134(c)(1)(ii)** Medical evaluations of employees required to use respirators;

**1910.134(c)(1)(iii)** Fit testing procedures for tight-fitting respirators;

**1910.134(c)(1)(iv)** Procedures for proper use of respirators in routine and reasonably foreseeable emergency situations;

**1910.134(c)(1)(v)** Procedures and schedules for cleaning, disinfecting, storing, inspecting, repairing, discarding, and otherwise maintaining respirators;

**1010.134(c)(1)(vi)** Procedures to ensure adequate air quality, quantity, and flow of breathing air for atmosphere-supplying respirators;

**1910.134(c)(1)(vii)** Training of employees in the respiratory hazards to which they are potentially exposed during routine and emergency situations;

**1910.134(c)(1)(viii)** Training of employees in the proper use of respirators, including putting on and removing them, any limitations on their use, and their maintenance; and

**1910.134(c)(1)(ix)** Procedures for regularly evaluating the effectiveness of the program.

**1910.134(c)(2)** Where respirator use is not required:

**1910.134(c)(2)(i)** An employer may provide respirators at the request of employees or permit employees to use their own respirators, if the employer determines that such respirator use will not in itself create a hazard. If the employer determines that any voluntary respirator use is permissible, the employer shall provide the respirator users with the information contained in Appendix D to this section ("Information for Employees Using Respirators When Not Required Under the Standard"); and

**1910.134(c)(2)(ii)** In addition, the employer must establish and implement those elements of a written respiratory protection program necessary to ensure that any employee using a respirator voluntarily is medically able to use that respirator, and that the respirator is cleaned, stored, and maintained so that its use does not present a health hazard to the user. Exception: Employers are not required to include in a written respiratory protection program those employees whose only use of respirators involves the voluntary use of filtering facepieces (dust masks).

**1910.134(c)(3)** The employer shall designate a program administrator who is qualified by appropriate training or experience that is commensurate with the complexity of the program to administer or oversee the respiratory protection program and conduct the required evaluations of program effectiveness.

**1910.134(c)(4)** The employer shall provide respirators, training, and medical evaluations at no cost to the employee.

**1910.134(d)** ***Selection of respirators.*** This paragraph requires the employer to evaluate respiratory hazard(s) in the workplace, identify relevant workplace and user factors, and base respirator selection on these factors. The paragraph also specifies appropriately protective respirators for use in IDLH atmospheres, and limits the selection and use of air-purifying respirators.

**1910.134(d)(1)** *General requirements.* (i) The employer shall select and provide an appropriate respirator based on the respiratory hazard(s) to which the worker is exposed and workplace and user factors that affect respirator performance and reliability.

**1910.134(d)(1)(ii)** The employer shall select a NIOSH-certified respirator. The respirator shall be used in compliance with the conditions of its certification.

**1910.134(d)(1)(iii)** The employer shall identify and evaluate the respiratory hazard(s) in the workplace; this evaluation shall include a reasonable estimate of employee exposures to respiratory hazard(s) and an identification of the contaminant's chemical state and physical form. Where the employer cannot identify or reasonably estimate the employee exposure, the employer shall consider the atmosphere to be IDLH.

**1910.134(d)(1)(iv)** The employer shall select respirators from a sufficient number of respirator models and sizes so that the respirator is acceptable to, and correctly fits, the user.

**1910.134(d)(2)** *Respirators for IDLH atmospheres.* (i) The employer shall provide the following respirators for employee use in IDLH atmospheres:

**1910.134(d)(2)(i)(A)** A full facepiece pressure demand SCBA certified by NIOSH for a minimum service life of thirty minutes, or

**1910.134(d)(2)(i)(B)** A combination full facepiece pressure demand supplied-air respirator (SAR) with auxiliary self-contained air supply.

**1910.134(d)(2)(ii)** Respirators provided only for escape from IDLH atmospheres shall be NIOSH-certified for escape from the atmosphere in which they will be used.

**1910.134(d)(2)(iii)** All oxygen deficient atmospheres shall be considered IDLH. Exception: If the employer demonstrates that, under all foreseeable conditions, the oxygen concentration can be maintained within the ranges specified in Table II of this section (i.e., for the altitudes set out in the table), then any atmosphere-supplying respirator may be used.

**1910.134(d)(3)** *Respirators for atmospheres that are not IDLH.* (i) The employer shall provide a respirator that is adequate to protect the health of the employee and ensure compliance with all other OSHA statutory and regulatory requirements, under routine and reasonably foreseeable emergency situations.

**1910.134(d)(3)(i)(A)** *Assigned Protection Factors (APFs)* [Reserved]

**1910.134(d)(3)(i)(B)** *Maximum Use Concentration (MUC)* [Reserved]

**1910.134(d)(3)(ii)** The respirator selected shall be appropriate for the chemical state and physical form of the contaminant.

**1910.134(d)(3)(iii)** For protection against gases and vapors, the employer shall provide:

**1910.134(d)(3)(iii)(A)** An atmosphere-supplying respirator, or

**1910.134(d)(3)(iii)(B)** An air-purifying respirator, provided that:

**1910.134(d)(3)(iii)(B)*(1)*** The respirator is equipped with an end-of-service-life indicator (ESLI) certified by NIOSH for the contaminant; or

**1910.134(d)(3)(iii)(B)*(2)*** If there is no ESLI appropriate for conditions in the employer's workplace, the employer implements a change schedule for canisters and cartridges that is based on objective information or data that will ensure that canisters and cartridges are changed before the end of their service life. The employer shall describe in the respirator program the information and data relied upon and the basis for the canister and cartridge change schedule and the basis for reliance on the data.

**1910.134(d)(3)(iv)** For protection against particulates, the employer shall provide:

**1910.134(d)(3)(iv)(A)** An atmosphere-supplying respirator; or

**1910.134(d)(3)(iv)(B)** An air-purifying respirator equipped with a filter certified by NIOSH under 30 CFR part 11 as a high efficiency particulate air (HEPA) filter, or an air-purifying respirator equipped with a filter certified for particulates by NIOSH under 42 CFR part 84; or

**1910.134(d)(3)(iv)(C)** For contaminants consisting primarily of particles with mass median aerodynamic diameters (MMAD) of at least 2 micrometers, an air-purifying respirator equipped with any filter certified for particulates by NIOSH.

## TABLE I—ASSIGNED PROTECTION FACTORS [RESERVED]

## TABLE II

| Altitude (ft.) | Oxygen deficient Atmospheres (% $0_2$) for which the employer may rely on atmosphere-supplying respirators |
|---|---|
| Less than 3,001 | 16.0-19.5 |
| 3,001-4,000 | 16.4-19.5 |
| 4,001-5,000 | 17.1-19.5 |
| 5,001-6,000 | 17.8-19.5 |
| 6,001-7,000 | 18.5-19.5 |
| 7,001-8,000[1] | 19.3-19.5. |

[1] Above 8,000 feet the exception does not apply. Oxygen-enriched breathing air must be supplied above 14,000 feet.

**1910.134(e)** ***Medical evaluation.*** Using a respirator may place a physiological burden on employees that varies with the type of respirator worn, the job and workplace conditions in which the respirator is used, and the medical status of the employee. Accordingly, this paragraph specifies the minimum requirements for medical evaluation that employers must implement to determine the employee's ability to use a respirator.

**1910.134(e)(1)** *General.* The employer shall provide a medical evaluation to determine the employee's ability to use a respirator, before the employee is fit tested or required to use the respirator in the workplace. The employer may discontinue an employee's medical evaluations when the employee is no longer required to use a respirator.

**1910.134(e)(2)** *Medical evaluation procedures.* (i) The employer shall identify a physician or other licensed health care professional (PLHCP) to perform medical evaluations using a medical questionnaire or an initial medical examination that obtains the same information as the medical questionnaire.

**1910.134(e)(2)(ii)** The medical evaluation shall obtain the information requested by the questionnaire in Sections 1 and 2, Part A of Appendix C of this section.

**1910.134(e)(3)** *Follow-up medical examination.* (i) The employer shall ensure that a follow-up medical examination is provided for an employee who gives a positive response to any question among questions 1 through 8 in Section 2, Part A of Appendix C or whose initial medical examination demonstrates the need for a follow-up medical examination.

**1910.134(e)(3)(ii)** The follow-up medical examination shall include any medical tests, consultations, or diagnostic procedures that the PLHCP deems necessary to make a final determination.

**1910.134(e)(4)** *Administration of the medical questionnaire and examinations.* (i) The medical questionnaire and examinations shall be administered confidentially during the employee's normal working hours or at a time and place convenient to the employee. The medical questionnaire shall be administered in a manner that ensures that the employee understands its content.

**1910.134(e)(4)(ii)** The employer shall provide the employee with an opportunity to discuss the questionnaire and examination results with the PLHCP.

**1910.134(e)(5)** *Supplemental information for the PLHCP.* (i) The following information must be provided to the PLHCP before the PLHCP makes a recommendation concerning an employee's ability to use a respirator:

**1910.134(e)(5)(i)(A)** The type and weight of the respirator to be used by the employee;

**1910.134(e)(5)(i)(B)** The duration and frequency of respirator use (including use for rescue and escape);

**1910.134(e)(5)(i)(C)** The expected physical work effort;

**1910.134(e)(5)(i)(D)** Additional protective clothing and equipment to be worn; and

**1910.134(e)(5)(i)(E)** Temperature and humidity extremes that may be encountered.

**1910.134(e)(5)(ii)** Any supplemental information provided previously to the PLHCP regarding an employee need not be provided for a subsequent medical evaluation if the information and the PLHCP remain the same.

**1910.134(e)(5)(iii)** The employer shall provide the PLHCP with a copy of the written respiratory protection program and a copy of this section.

**Note to paragraph** (e)(5)(iii): When the employer replaces a PLHCP, the employer must ensure that the new PLHCP obtains this information, either by providing the documents directly to the PLHCP or having the documents transferred from the former PLHCP to the new PLHCP. However, OSHA does not expect employers to have employees medically reevaluated solely because a new PLHCP has been selected.

**1910.134(e)(6)** *Medical determination.* In determining the employee's ability to use a respirator, the employer shall:

**1910.134(e)(6)(i)** Obtain a written recommendation regarding the employee's ability to use the respirator from the PLHCP. The recommendation shall provide only the following information:

**1910.134(e)(6)(i)(A)** Any limitations on respirator use related to the medical condition of the employee, or relating to the workplace conditions in which the respirator will be used, including whether or not the employee is medically able to use the respirator;

**1910.134(e)(6)(i)(B)** The need, if any, for follow-up medical evaluations; and

**1910.134(e)(6)(i)(C)** A statement that the PLHCP has provided the employee with a copy of the PLHCP's written recommendation.

**1910.134(e)(6)(ii)** If the respirator is a negative pressure respirator and the PLHCP finds a medical condition that may place the employee's health at increased risk if the respirator is used, the employer shall provide a PAPR if the PLHCP's medical evaluation finds that the employee can use such a respirator; if a subsequent medical evaluation finds that the employee is medically able to use a negative pressure respirator, then the employer is no longer required to provide a PAPR.

**1910.134(e)(7)** *Additional medical evaluations.* At a minimum, the employer shall provide additional medical evaluations that comply with the requirements of this section if:

**1910.134(e)(7)(i)** An employee reports medical signs or symptoms that are related to ability to use a respirator;

**1910.134(e)(7)(ii)** A PLHCP, supervisor, or the respirator program administrator informs the employer that an employee needs to be reevaluated;

**1910.134(e)(7)(iii)** Information from the respiratory protection program, including observations made during fit testing and program evaluation, indicates a need for employee reevaluation; or

**1910.134(e)(7)(iv)** A change occurs in workplace conditions (e.g., physical work effort, protective clothing, temperature) that may result in a substantial increase in the physiological burden placed on an employee.

**1910.134(f)** ***Fit testing.*** This paragraph requires that, before an employee may be required to use any respirator with a negative or positive pressure tight-fitting facepiece, the employee must be fit tested with the same make, model, style, and size of respirator that will be used. This paragraph specifies the kinds of fit tests allowed, the procedures for conducting them, and how the results of the fit tests must be used.

**1910.134(f)(1)** The employer shall ensure that employees using a tight-fitting facepiece respirator pass an appropriate qualitative fit test (QLFT) or quantitative fit test (QNFT) as stated in this paragraph.

**1910.134(f)(2)** The employer shall ensure that an employee using a tight-fitting facepiece respirator is fit tested prior to initial use of the respirator, whenever a different respirator facepiece (size, style, model or make) is used, and at least annually thereafter.

**1910.134(f)(3)** The employer shall conduct an additional fit test whenever the employee reports, or the employer, PLHCP, supervisor, or program administrator makes visual observations of, changes in the employee's physical condition that could affect respirator fit. Such conditions include, but are not limited to, facial scarring, dental changes, cosmetic surgery, or an obvious change in body weight.

**1910.134(f)(4)** If after passing a QLFT or QNFT, the employee subsequently notifies the employer, program administrator, supervisor, or PLHCP that the fit of the respirator is unacceptable, the employee shall be given a reasonable opportunity to select a different respirator facepiece and to be retested.

**1910.134(f)(5)** The fit test shall be administered using an OSHA-accepted QLFT or QNFT protocol. The OSHA-accepted QLFT and QNFT protocols and procedures are contained in Appendix A of this section.

**1910.134(f)(6)** QLFT may only be used to fit test negative pressure air-purifying respirators that must achieve a fit factor of 100 or less.

**1910.134(f)(7)** If the fit factor, as determined through an OSHA-accepted QNFT protocol, is equal to or greater than 100 for tight-fitting half facepieces, or equal to or greater than 500 for tight-fitting full facepieces, the QNFT has been passed with that respirator.

**1910.134(f)(8)** Fit testing of tight-fitting atmosphere-supplying respirators and tight-fitting powered air-purifying respirators shall be accomplished by performing quantitative or qualitative fit testing in the negative pressure mode, regardless of the mode of operation (negative or positive pressure) that is used for respiratory protection.

**1910.134(f)(8)(i)** Qualitative fit testing of these respirators shall be accomplished by temporarily converting the respirator user's actual facepiece into a negative pressure respirator with appropriate filters, or by using an identical negative pressure air-purifying respirator facepiece with the same sealing surfaces as a surrogate for the atmosphere-supplying or powered air-purifying respirator facepiece.

**1910.134(f)(8)(ii)** Quantitative fit testing of these respirators shall be accomplished by modifying the facepiece to allow sampling inside the facepiece in the breathing zone of the user, midway between the nose and mouth. This requirement shall be accomplished by installing a permanent sampling probe onto a surrogate facepiece, or by using a sampling adapter designed to temporarily provide a means of sampling air from inside the facepiece.

**1910.134(f)(8)(iii)** Any modifications to the respirator facepiece for fit testing shall be completely removed, and the facepiece restored to NIOSH-approved configuration, before that facepiece can be used in the workplace.

**1910.134(g)** ***Use of respirators.*** This paragraph requires employers to establish and implement procedures for the proper use of respirators. These requirements include prohibiting conditions that may result in facepiece seal leakage, preventing employees from removing respirators in hazardous environments, taking actions to ensure continued effective respirator operation throughout the work shift, and establishing procedures for the use of respirators in IDLH atmospheres or in interior structural firefighting situations.

**1910.134(g)(1)** *Facepiece seal protection.* (i) The employer shall not permit respirators with tight-fitting facepieces to be worn by employees who have:

**1910.134(g)(1)(i)(A)** Facial hair that comes between the sealing surface of the facepiece and the face or that interferes with valve function; or

**1910.134(g)(1)(i)(B)** Any condition that interferes with the face-to-facepiece seal or valve function.

**1910.134(g)(1)(ii)** If an employee wears corrective glasses or goggles or other personal protective equipment, the employer shall ensure that such equipment is worn in a manner that does not interfere with the seal of the facepiece to the face of the user.

**1910.134(g)(1)(iii)** For all tight-fitting respirators, the employer shall ensure that employees perform a user seal check each time they put on the respirator using the procedures in Appendix B-1 or procedures recommended by the respirator manufacturer that the employer demonstrates are as effective as those in Appendix B-1 of this section.

**1910.134(g)(2)** *Continuing respirator effectiveness.* (i) Appropriate surveillance shall be maintained of work area conditions and degree of employee exposure or stress. When there is a change in work area conditions or degree of employee exposure or stress that may affect respirator effectiveness, the employer shall reevaluate the continued effectiveness of the respirator.

**1910.134(g)(2)(ii)** The employer shall ensure that employees leave the respirator use area:

**1910.134(g)(2)(ii)(A)** To wash their faces and respirator facepieces as necessary to prevent eye or skin irritation associated with respirator use; or

**1910.134(g)(2)(ii)(B)** If they detect vapor or gas breakthrough, changes in breathing resistance, or leakage of the facepiece; or

**1910.134(g)(2)(ii)(C)** To replace the respirator or the filter, cartridge, or canister elements.

**1910.134(g)(2)(iii)** If the employee detects vapor or gas breakthrough, changes in breathing resistance, or leakage of the facepiece, the employer must replace or repair the respirator before allowing the employee to return to the work area.

**1910.134(g)(3)** *Procedures for IDLH atmospheres.* For all IDLH atmospheres, the employer shall ensure that:

**1910.134(g)(3)(i)** One employee or, when needed, more than one employee is located outside the IDLH atmosphere;

**1910.134(g)(3)(ii)** Visual, voice, or signal line communication is maintained between the employee(s) in the IDLH atmosphere and the employee(s) located outside the IDLH atmosphere;

**1910.134(g)(3)(iii)** The employee(s) located outside the IDLH atmosphere are trained and equipped to provide effective emergency rescue;

**1910.134(g)(3)(iv)** The employer or designee is notified before the employee(s) located outside the IDLH atmosphere enter the IDLH atmosphere to provide emergency rescue;

**1910.134(g)(3)(v)** The employer or designee authorized to do so by the employer, once notified, provides necessary assistance appropriate to the situation;

**1910.134(g)(3)(vi)** Employee(s) located outside the IDLH atmospheres are equipped with:

**1910.134(g)(3)(vi)(A)** Pressure demand or other positive pressure SCBAs, or a pressure demand or other positive pressure supplied-air respirator with auxiliary SCBA; and either

**1910.134(g)(3)(vi)(B)** Appropriate retrieval equipment for removing the employee(s) who enter(s) these hazardous atmospheres where retrieval equipment would contribute to the rescue of the employee(s) and would not increase the overall risk resulting from entry; or

**1910.134(g)(3)(vi)(C)** Equivalent means for rescue where retrieval equipment is not required under paragraph (g)(3)(vi)(B).

**1910.134(g)(4)** *Procedures for interior structural firefighting.* In addition to the requirements set forth under paragraph (g)(3), in interior structural fires, the employer shall ensure that:

**1910.134(g)(4)(i)** At least two employees enter the IDLH atmosphere and remain in visual or voice contact with one another at all times;

**1910.134(g)(4)(ii)** At least two employees are located outside the IDLH atmosphere; and

**1910.134(g)(4)(iii)** All employees engaged in interior structural firefighting use SCBAs.

**Note 1 to paragraph** (g): One of the two individuals located outside the IDLH atmosphere may be assigned to an additional role, such as incident commander in charge of the emergency or safety officer, so long as this individual is able to perform assistance or rescue activities without jeopardizing the safety or health of any firefighter working at the incident.

**Note 2 to paragraph** (g): Nothing in this section is meant to preclude firefighters from performing emergency rescue activities before an entire team has assembled.

**1910.134(h)** ***Maintenance and care of respirators.*** This paragraph requires the employer to provide for the cleaning and disinfecting, storage, inspection, and repair of respirators used by employees.

**1910.134(h)(1)** *Cleaning and disinfecting.* The employer shall provide each respirator user with a respirator that is clean, sanitary, and in good working order. The employer shall ensure that respirators are cleaned and disinfected using the procedures in Appendix B-2 of this section, or procedures recommended by the respirator manufacturer, provided that such procedures are of equivalent effectiveness. The respirators shall be cleaned and disinfected at the following intervals:

**1910.134(h)(1)(i)** Respirators issued for the exclusive use of an employee shall be cleaned and disinfected as often as necessary to be maintained in a sanitary condition;

**1910.134(h)(1)(ii)** Respirators issued to more than one employee shall be cleaned and disinfected before being worn by different individuals;

**1910.134(h)(1)(iii)** Respirators maintained for emergency use shall be cleaned and disinfected after each use; and

**1910.134(h)(1)(iv)** Respirators used in fit testing and training shall be cleaned and disinfected after each use.

**1910.134(h)(2)** *Storage.* The employer shall ensure that respirators are stored as follows:

**1910.134(h)(2)(i)** All respirators shall be stored to protect them from damage, contamination, dust, sunlight, extreme temperatures, excessive moisture, and damaging chemicals, and they shall be packed or stored to prevent deformation of the facepiece and exhalation valve.

**1910.134(h)(2)(ii)** In addition to the requirements of paragraph (h)(2)(i) of this section, emergency respirators shall be:

**1910.134(h)(2)(ii)(A)** Kept accessible to the work area;

**1910.134(h)(2)(ii)(B)** Stored in compartments or in covers that are clearly marked as containing emergency respirators; and

**1910.134(h)(2)(ii)(C)** Stored in accordance with any applicable manufacturer instructions.

**1910.134(h)(3)** *Inspection.* (i) The employer shall ensure that respirators are inspected as follows:

**1910.134(h)(3)(i)(A)** All respirators used in routine situations shall be inspected before each use and during cleaning;

**1910.134(h)(3)(i)(B)** All respirators maintained for use in emergency situations shall be inspected at least monthly and in accordance with the manufacturer's recommendations, and shall be checked for proper function before and after each use; and

**1910.134(h)(3)(i)(C)** Emergency escape-only respirators shall be inspected before being carried into the workplace for use.

**1910.134(h)(3)(ii)** The employer shall ensure that respirator inspections include the following:

**1910.134(h)(3)(ii)(A)** A check of respirator function, tightness of connections, and the condition of the various parts including, but not limited to, the facepiece, head straps, valves, connecting tube, and cartridges, canisters or filters; and

**1910.134(h)(3)(ii)(B)** A check of elastomeric parts for pliability and signs of deterioration.

**1910.134(h)(3)(iii)** In addition to the requirements of paragraphs (h)(3)(i) and (ii) of this section, self-contained breathing apparatus shall be inspected monthly. Air and oxygen cylinders shall be maintained in a fully charged state and shall be recharged when the pressure falls to 90% of the manufacturer's recommended pressure level. The employer shall determine that the regulator and warning devices function properly.

**1910.134(h)(3)(iv)** For respirators maintained for emergency use, the employer shall:

**1910.134(h)(3)(iv)(A)** Certify the respirator by documenting the date the inspection was performed, the name (or signature) of the person who made the inspection, the findings, required remedial action, and a serial number or other means of identifying the inspected respirator; and

**1910.134(h)(3)(iv)(B)** Provide this information on a tag or label that is attached to the storage compartment for the respirator, is kept with the respirator, or is included in inspection reports stored as paper or electronic files. This information shall be maintained until replaced following a subsequent certification.

**1910.134(h)(4)** *Repairs.* The employer shall ensure that respirators that fail an inspection or are otherwise found to be defective are removed from service, and are discarded or repaired or adjusted in accordance with the following procedures:

**1910.134(h)(4)(i)** Repairs or adjustments to respirators are to be made only by persons appropriately trained to perform such operations and shall use only the respirator manufacturer's NIOSH-approved parts designed for the respirator;

**1910.134(h)(4)(ii)** Repairs shall be made according to the manufacturer's recommendations and specifications for the type and extent of repairs to be performed; and

**1910.134(h)(4)(iii)** Reducing and admission valves, regulators, and alarms shall be adjusted or repaired only by the manufacturer or a technician trained by the manufacturer.

**1910.134(i)** ***Breathing air quality and use.*** This paragraph requires the employer to provide employees using atmosphere-supplying respirators (supplied-air and SCBA) with breathing gases of high purity.

**1910.134(i)(1)** The employer shall ensure that compressed air, compressed oxygen, liquid air, and liquid oxygen used for respiration accords with the following specifications:

**1910.134(i)(1)(i)** Compressed and liquid oxygen shall meet the United States Pharmacopoeia requirements for medical or breathing oxygen; and

**1910.134(i)(1)(ii)** Compressed breathing air shall meet at least the requirements for Grade D breathing air described in ANSI/Compressed Gas Association Commodity Specification for Air, G-7.1-1989, to include:

**1910.134(i)(1)(ii)(A)** Oxygen content (v/v) of 19.5-23.5%;

**1910.134(i)(1)(ii)(B)** Hydrocarbon (condensed) content of 5 milligrams per cubic meter of air or less;

**1910.134(i)(1)(ii)(C)** Carbon monoxide (CO) content of 10 ppm or less;

**1910.134(i)(1)(ii)(D)** Carbon dioxide content of 1,000 ppm or less; and

**1910.134(i)(1)(ii)(E)** Lack of noticeable odor.

**1910.134(i)(2)** The employer shall ensure that compressed oxygen is not used in atmosphere-supplying respirators that have previously used compressed air.

**1910.134(i)(3)** The employer shall ensure that oxygen concentrations greater than 23.5% are used only in equipment designed for oxygen service or distribution.

**1910.134(i)(4)** The employer shall ensure that cylinders used to supply breathing air to respirators meet the following requirements:

**1910.134(i)(4)(i)** Cylinders are tested and maintained as prescribed in the Shipping Container Specification Regulations of the Department of Transportation (49 CFR part 173 and part 178);

**1910.134(i)(4)(ii)** Cylinders of purchased breathing air have a certificate of analysis from the supplier that the breathing air meets the requirements for Grade D breathing air; and

**1910.134(i)(4)(iii)** The moisture content in the cylinder does not exceed a dew point of -50 °F (-45.6 °C) at 1 atmosphere pressure.

**1910.134(i)(5)** The employer shall ensure that compressors used to supply breathing air to respirators are constructed and situated so as to:

**1910.134(i)(5)(i)** Prevent entry of contaminated air into the air-supply system;

**1910.134(i)(5)(ii)** Minimize moisture content so that the dew point at 1 atmosphere pressure is 10 degrees F (5.56 °C) below the ambient temperature;

**1910.134(i)(5)(iii)** Have suitable in-line air-purifying sorbent beds and filters to further ensure breathing air quality. Sorbent beds and filters shall be maintained and replaced or refurbished periodically following the manufacturer's instructions.

**1910.134(i)(5)(iv)** Have a tag containing the most recent change date and the signature of the person authorized by the employer to perform the change. The tag shall be maintained at the compressor.

**1910.134(i)(6)** For compressors that are not oil-lubricated, the employer shall ensure that carbon monoxide levels in the breathing air do not exceed 10 ppm.

**1910.134(i)(7)** For oil-lubricated compressors, the employer shall use a high-temperature or carbon monoxide alarm, or both, to monitor carbon monoxide levels. If only high-temperature alarms are used, the air supply shall be monitored at intervals sufficient to prevent carbon monoxide in the breathing air from exceeding 10 ppm.

**1910.134(i)(8)** The employer shall ensure that breathing air couplings are incompatible with outlets for nonrespirable worksite air or other gas systems. No asphyxiating substance shall be introduced into breathing air lines.

**1910.134(i)(9)** The employer shall use breathing gas containers marked in accordance with the NIOSH respirator certification standard, 42 CFR part 84.

**1910.134(j)** ***Identification of filters, cartridges, and canisters.*** The employer shall ensure that all filters, cartridges and canisters used in the workplace are labeled and color coded with the NIOSH approval label and that the label is not removed and remains legible.

**1910.134(k)** ***Training and information.*** This paragraph requires the employer to provide effective training to employees who are required to use respirators. The training must be comprehensive, understandable, and recur annually, and more often if necessary. This paragraph also requires the employer to provide the basic information on respirators in Appendix D of this section to employees who wear respirators when not required by this section or by the employer to do so.

**1910.134(k)(1)** The employer shall ensure that each employee can demonstrate knowledge of at least the following:

**1910.134(k)(1)(i)** Why the respirator is necessary and how improper fit, usage, or maintenance can compromise the protective effect of the respirator;

**1910.134(k)(1)(ii)** What the limitations and capabilities of the respirator are;

**1910.134(k)(1)(iii)** How to use the respirator effectively in emergency situations, including situations in which the respirator malfunctions;

**1910.134(k)(1)(iv)** How to inspect, put on and remove, use, and check the seals of the respirator;

**1910.134(k)(1)(v)** What the procedures are for maintenance and storage of the respirator;

**1910.134(k)(1)(vi)** How to recognize medical signs and symptoms that may limit or prevent the effective use of respirators; and

**1910.134(k)(1)(vii)** The general requirements of this section.

**1910.134(k)(2)** The training shall be conducted in a manner that is understandable to the employee.

**1910.134(k)(3)** The employer shall provide the training prior to requiring the employee to use a respirator in the workplace.

**1910.134(k)(4)** An employer who is able to demonstrate that a new employee has received training within the last 12 months that addresses the elements specified in paragraph (k)(1)(i) through (vii) is not required to repeat such training provided that, as required by paragraph (k)(1), the employee can demonstrate knowledge of those element(s). Previous training not repeated initially by the employer must be provided no later than 12 months from the date of the previous training.

**1910.134(k)(5)** Retraining shall be administered annually, and when the following situations occur:

**1910.134(k)(5)(i)** Changes in the workplace or the type of respirator render previous training obsolete;

**1910.134(k)(5)(ii)** Inadequacies in the employee's knowledge or use of the respirator indicate that the employee has not retained the requisite understanding or skill; or

**1910.134(k)(5)(iii)** Any other situation arises in which retraining appears necessary to ensure safe respirator use.

**1910.134(k)(6)** The basic advisory information on respirators, as presented in Appendix D of this section, shall be provided by the employer in any written or oral format, to employees who wear respirators when such use is not required by this section or by the employer.

**1910.134(l)** ***Program evaluation.*** This section requires the employer to conduct evaluations of the workplace to ensure that the written respiratory protection program is being properly implemented, and to consult employees to ensure that they are using the respirators properly.

**1910.134(l)(1)** The employer shall conduct evaluations of the workplace as necessary to ensure that the provisions of the current written program are being effectively implemented and that it continues to be effective.

**1910.134(l)(2)** The employer shall regularly consult employees required to use respirators to assess the employees' views on program effectiveness and to identify any problems. Any problems that are identified during this assessment shall be corrected. Factors to be assessed include, but are not limited to:

**1910.134(l)(2)(i)** Respirator fit (including the ability to use the respirator without interfering with effective workplace performance);

**1910.134(l)(2)(ii)** Appropriate respirator selection for the hazards to which the employee is exposed;

**1910.134(l)(2)(iii)** Proper respirator use under the workplace conditions the employee encounters; and

**1910.134(l)(2)(iv)** Proper respirator maintenance.

**1910.134(m)** ***Recordkeeping.*** This section requires the employer to establish and retain written information regarding medical evaluations, fit testing, and the respirator program. This information will facilitate employee involvement in the respirator program, assist the employer in auditing the adequacy of the program, and provide a record for compliance determinations by OSHA.

**1910.134(m)(1)** *Medical evaluation.* Records of medical evaluations required by this section must be retained and made available in accordance with 29 CFR 1910.1020.

**1910.134(m)(2)** *Fit testing.* (i) The employer shall establish a record of the qualitative and quantitative fit tests administered to an employee including:

**1910.134(m)(2)(i)(A)** The name or identification of the employee tested;

**1910.134(m)(2)(i)(B)** Type of fit test performed;

**1910.134(m)(2)(i)(C)** Specific make, model, style, and size of respirator tested;

**1910.134(m)(2)(i)(D)** Date of test; and

**1910.134(m)(2)(i)(E)** The pass/fail results for QLFTs or the fit factor and strip chart recording or other recording of the test results for QNFTs.

**1910.134(m)(2)(ii)** Fit test records shall be retained for respirator users until the next fit test is administered.

**1910.134(m)(3)** A written copy of the current respirator program shall be retained by the employer.

**1910.134(m)(4)** Written materials required to be retained under this paragraph shall be made available upon request to affected employees and to the Assistant Secretary or designee for examination and copying.

**1910.134(n)** ***Dates***— (1) *Effective date.* This section is effective April 8, 1998. The obligations imposed by this section commence on the effective date unless otherwise noted in this paragraph.

Compliance with obligations that do not commence on the effective date shall occur no later than the applicable start-up date.

**1910.134(n)(2)** *Compliance dates.* All obligations of this section commence on the effective date except as follows:

**1910.134(n)(2)(i)** The determination that respirator use is required (paragraph (a)) shall be completed no later than September 8, 1998.

**1910.134(n)(2)(ii)** Compliance with provisions of this section for all other provisions shall be completed no later than October 5, 1998.

**1910.134(n)(3)** The provisions of 29 CFR 1910.134 and 29 CFR 1926.103, contained in the 29 CFR parts 1900 to 1910.99 and the 29 CFR part 1926 editions, revised as of July 1, 1997, are in effect and enforceable until October 5, 1998, or during any administrative or judicial stay of the provisions of this section.

**1910.134(n)(4)** *Existing respiratory protection programs.* If, in the 12 month period preceding April 8, 1998, the employer has conducted annual respirator training, fit testing, respirator program evaluation, or medical evaluations, the employer may use the results of those activities to comply with the corresponding provisions of this section, providing that these activities were conducted in a manner that meets the requirements of this section.

**1910.134(o)** ***Appendices.*** (1) Compliance with Appendix A, Appendix B-1, Appendix B-2, and Appendix C of this section is mandatory.

**1910.134(o)(2)** Appendix D of this section is non-mandatory and is not intended to create any additional obligations not otherwise imposed or to detract from any existing obligations.

## APPENDIX A TO §1910.134— FIT TESTING PROCEDURES (MANDATORY)

### *Part I. OSHA-Accepted Fit Test Protocols*

### A. Fit Testing Procedures—General Requirements

The employer shall conduct fit testing using the following procedures. The requirements in this appendix apply to all OSHA-accepted fit test methods, both QLFT and QNFT.

1. The test subject shall be allowed to pick the most acceptable respirator from a sufficient number of respirator models and sizes so that the respirator is acceptable to, and correctly fits, the user.

2. Prior to the selection process, the test subject shall be shown how to put on a respirator, how it should be positioned on the face, how to set strap tension and how to determine an acceptable fit. A mirror shall be available to assist the subject in evaluating the fit and positioning of the respirator. This instruction may not constitute the subject's formal training on respirator use, because it is only a review.

3. The test subject shall be informed that he/she is being asked to select the respirator that provides the most acceptable fit. Each respirator represents a different size and shape, and if fitted and used properly, will provide adequate protection.

4. The test subject shall be instructed to hold each chosen facepiece up to the face and eliminate those that obviously do not give an acceptable fit.

5. The more acceptable facepieces are noted in case the one selected proves unacceptable; the most comfortable mask is donned and worn at least five minutes to assess comfort. Assistance in assessing comfort can be given by discussing the points in the following item A.6. If the test subject is not familiar with using a particular respirator, the test subject shall be directed to don the mask several times and to adjust the straps each time to become adept at setting proper tension on the straps.

6. Assessment of comfort shall include a review of the following points with the test subject and allowing the test subject adequate time to determine the comfort of the respirator:

(a) Position of the mask on the nose

(b) Room for eye protection

(c) Room to talk

(d) Position of mask on face and cheeks

7. The following criteria shall be used to help determine the adequacy of the respirator fit:

(a) Chin properly placed;

(b) Adequate strap tension, not overly tightened;

(c) Fit across nose bridge;

(d) Respirator of proper size to span distance from nose to chin;

(e) Tendency of respirator to slip;

(f) Self-observation in mirror to evaluate fit and respirator position.

8. The test subject shall conduct a user seal check, either the negative and positive pressure seal checks described in Appendix B-1 of this section or those recommended by the respirator manufacturer which provide equivalent protection to the procedures in Appendix B-1. Before conducting the negative and positive pressure checks, the subject shall be told to seat the mask on the face by moving the head from side-to-side and up and down slowly while taking in a few slow deep breaths. Another facepiece shall be selected and retested if the test subject fails the user seal check tests.

9. The test shall not be conducted if there is any hair growth between the skin and the facepiece sealing surface, such as stubble beard growth, beard, mustache or sideburns which cross the respirator sealing surface. Any type of apparel which interferes with a satisfactory fit shall be altered or removed.

10. If a test subject exhibits difficulty in breathing during the tests, she or he shall be referred to a physician or other licensed health care professional, as appropriate, to determine whether the test subject can wear a respirator while performing her or his duties.

11. If the employee finds the fit of the respirator unacceptable, the test subject shall be given the opportunity to select a different respirator and to be retested.

12. Exercise regimen. Prior to the commencement of the fit test, the test subject shall be given a description of the fit test and the test subject's responsibilities during the test procedure. The description of the process shall include a description of the test exercises that the subject will be performing. The respirator to be tested shall be worn for at least 5 minutes before the start of the fit test.

13. The fit test shall be performed while the test subject is wearing any applicable safety equipment that may be worn during actual respirator use which could interfere with respirator fit.

14. Test Exercises. (a) The following test exercises are to be performed for all fit testing methods prescribed in this appendix, except for the CNP method. A separate fit testing exercise regimen is contained in the CNP protocol. The test subject shall perform exercises, in the test environment, in the following manner:

(1) Normal breathing. In a normal standing position, without talking, the subject shall breathe normally.

(2) Deep breathing. In a normal standing position, the subject shall breathe slowly and deeply, taking caution so as not to hyperventilate.

(3) Turning head side to side. Standing in place, the subject shall slowly turn his/her head from side to side between the extreme positions on each side. The head shall be held at each extreme momentarily so the subject can inhale at each side.

(4) Moving head up and down. Standing in place, the subject shall slowly move his/her head up and down. The subject shall be instructed to inhale in the up position (i.e., when looking toward the ceiling).

(5) Talking. The subject shall talk out loud slowly and loud enough so as to be heard clearly by the test conductor. The subject can read from a prepared text such as the Rainbow Passage, count backward from 100, or recite a memorized poem or song.

### *Rainbow Passage*

When the sunlight strikes raindrops in the air, they act like a prism and form a rainbow. The rainbow is a division of white light into many beautiful colors. These take the shape of a long round arch, with its path high above, and its two ends apparently beyond the horizon. There is, according to legend, a boiling pot of gold at one end. People look, but no one ever finds it. When a man looks for something beyond reach, his friends say he is looking for the pot of gold at the end of the rainbow.

(6) Grimace. The test subject shall grimace by smiling or frowning. (This applies only to QNFT testing; it is not performed for QLFT)

(7) Bending over. The test subject shall bend at the waist as if he/she were to touch his/her toes. Jogging in place shall be substituted for this exercise in those test environments such as shroud type QNFT or QLFT units that do not permit bending over at the waist.

(8) Normal breathing. Same as exercise (1).

(b) Each test exercise shall be performed for one minute except for the grimace exercise which shall be performed for 15 seconds. The test subject shall be questioned by the test conductor regarding the comfort of the respirator upon completion of the protocol. If it has become unacceptable, another model of respirator shall be tried. The respirator shall not be adjusted once the fit test exercises begin. Any adjustment voids the test, and the fit test must be repeated.

## *B. Qualitative Fit Test (QLFT) Protocols*

### 1. General

(a) The employer shall ensure that persons administering QLFT are able to prepare test solutions, calibrate equipment and perform tests properly, recognize invalid tests, and ensure that test equipment is in proper working order.

(b) The employer shall ensure that QLFT equipment is kept clean and well maintained so as to operate within the parameters for which it was designed.

### 2. Isoamyl Acetate Protocol

**Note:** This protocol is not appropriate to use for the fit testing of particulate respirators. If used to fit test particulate respirators, the respirator must be equipped with an organic vapor filter.

#### (a) Odor Threshold Screening

Odor threshold screening, performed without wearing a respirator, is intended to determine if the individual tested can detect the odor of isoamyl acetate at low levels.

(1) Three 1 liter glass jars with metal lids are required.

(2) Odor-free water (e.g., distilled or spring water) at approximately 25 °C (77 °F) shall be used for the solutions.

(3) The isoamyl acetate (IAA) (also known at isopentyl acetate) stock solution is prepared by adding 1 ml of pure IAA to 800 ml of odor-free water in a 1 liter jar, closing the lid and shaking for 30 seconds. A new solution shall be prepared at least weekly.

(4) The screening test shall be conducted in a room separate from the room used for actual fit testing. The two rooms shall be well-ventilated to prevent the odor of IAA from becoming evident in the general room air where testing takes place.

(5) The odor test solution is prepared in a second jar by placing 0.4 ml of the stock solution into 500 ml of odor-free water using a clean dropper or pipette. The solution shall be shaken for 30 seconds and allowed to stand for two to three minutes so that the IAA concentration above the liquid may reach equilibrium. This solution shall be used for only one day.

(6) A test blank shall be prepared in a third jar by adding 500 cc of odor-free water.

(7) The odor test and test blank jar lids shall be labeled (e.g., 1 and 2) for jar identification. Labels shall be placed on the lids so that they can be peeled off periodically and switched to maintain the integrity of the test.

(8) The following instruction shall be typed on a card and placed on the table in front of the two test jars (i.e., 1 and 2): "The purpose of this test is to determine if you can smell banana oil at a low concentration. The two bottles in front of you contain water. One of these bottles also contains a small amount of banana oil. Be sure the covers are on tight, then shake each bottle for two seconds. Unscrew the lid of each bottle, one at a time, and sniff at the mouth of the bottle. Indicate to the test conductor which bottle contains banana oil."

(9) The mixtures used in the IAA odor detection test shall be prepared in an area separate from where the test is performed, in order to prevent olfactory fatigue in the subject.

(10) If the test subject is unable to correctly identify the jar containing the odor test solution, the IAA qualitative fit test shall not be performed.

(11) If the test subject correctly identifies the jar containing the odor test solution, the test subject may proceed to respirator selection and fit testing.

### (b) Isoamyl Acetate Fit Test

(1) The fit test chamber shall be a clear 55-gallon drum liner suspended inverted over a 2-foot diameter frame so that the top of the chamber is about 6 inches above the test subject's head. If no drum liner is available, a similar chamber shall be constructed using plastic sheeting. The inside top center of the chamber shall have a small hook attached.

(2) Each respirator used for the fitting and fit testing shall be equipped with organic vapor cartridges or offer protection against organic vapors.

(3) After selecting, donning, and properly adjusting a respirator, the test subject shall wear it to the fit testing room. This room shall be separate from the room used for odor threshold screening and respirator selection, and shall be well-ventilated, as by an exhaust fan or lab hood, to prevent general room contamination.

(4) A copy of the test exercises and any prepared text from which the subject is to read shall be taped to the inside of the test chamber.

(5) Upon entering the test chamber, the test subject shall be given a 6-inch by 5-inch piece of paper towel, or other porous, absorbent, single-ply material, folded in half and wetted with 0.75 ml of pure IAA. The test subject shall hang the wet towel on the hook at the top of the chamber. An IAA test swab or ampule may be substituted for the IAA wetted paper towel provided it has been demonstrated that the alternative IAA source will generate an IAA test atmosphere with a concentration equivalent to that generated by the paper towel method.

(6) Allow two minutes for the IAA test concentration to stabilize before starting the fit test exercises. This would be an appropriate time to talk with the test subject; to explain the fit test, the importance of his/her cooperation, and the purpose for the test exercises; or to demonstrate some of the exercises.

(7) If at any time during the test, the subject detects the banana-like odor of IAA, the test is failed. The subject shall quickly exit from the test chamber and leave the test area to avoid olfactory fatigue.

(8) If the test is failed, the subject shall return to the selection room and remove the respirator. The test subject shall repeat the odor sensitivity test, select and put on another respirator, return to the test area and again begin the fit test procedure described in (b) (1) through (7) above. The process continues until a respirator that fits well has been found. Should the odor sensitivity test be failed, the subject shall wait at least 5 minutes before retesting. Odor sensitivity will usually have returned by this time.

(9) If the subject passes the test, the efficiency of the test procedure shall be demonstrated by having the subject break the respirator face seal and take a breath before exiting the chamber.

(10) When the test subject leaves the chamber, the subject shall remove the saturated towel and return it to the person conducting the test, so that there is no significant IAA concentration buildup in the chamber during subsequent tests. The used towels shall be kept in a self-sealing plastic bag to keep the test area from being contaminated.

## 3. Saccharin Solution Aerosol Protocol

The entire screening and testing procedure shall be explained to the test subject prior to the conduct of the screening test.

(a) Taste threshold screening. The saccharin taste threshold screening, performed without wearing a respirator, is intended to determine whether the individual being tested can detect the taste of saccharin.

(1) During threshold screening as well as during fit testing, subjects shall wear an enclosure about the head and shoulders that is approximately 12 inches in diameter by 14 inches tall with at least the front portion clear and that allows free movements of the head when a respirator is worn. An enclosure substantially similar to the 3M hood assembly, parts # FT 14 and # FT 15 combined, is adequate.

(2) The test enclosure shall have a 3/4-inch (1.9 cm) hole in front of the test subject's nose and mouth area to accommodate the nebulizer nozzle.

(3) The test subject shall don the test enclosure. Throughout the threshold screening test, the test subject shall breathe through his/her slightly open mouth with tongue extended. The subject is instructed to report when he/she detects a sweet taste.

(4) Using a DeVilbiss Model 40 Inhalation Medication Nebulizer or equivalent, the test conductor shall spray the threshold check solution into the enclosure. The nozzle is directed away from the nose and mouth of the person. This nebulizer shall be clearly marked to distinguish it from the fit test solution nebulizer.

(5) The threshold check solution is prepared by dissolving 0.83 gram of sodium saccharin USP in 100 ml of warm water. It can be prepared by putting 1 ml of the fit test solution (see (b)(5) below) in 100 ml of distilled water.

(6) To produce the aerosol, the nebulizer bulb is firmly squeezed so that it collapses completely, then released and allowed to fully expand.

(7) Ten squeezes are repeated rapidly and then the test subject is asked whether the saccharin can be tasted. If the test subject reports tasting the sweet taste during the ten squeezes, the screening test is completed. The taste threshold is noted as ten regardless of the number of squeezes actually completed.

(8) If the first response is negative, ten more squeezes are repeated rapidly and the test subject is again asked whether the saccharin is tasted. If the test subject reports tasting the sweet taste during the second ten squeezes, the screening test is completed. The taste threshold is noted as twenty regardless of the number of squeezes actually completed.

(9) If the second response is negative, ten more squeezes are repeated rapidly and the test subject is again asked whether the saccharin is tasted. If the test subject reports tasting the sweet taste during the third set of ten squeezes, the screening test is completed. The taste threshold is noted as thirty regardless of the number of squeezes actually completed.

(10) The test conductor will take note of the number of squeezes required to solicit a taste response.

(11) If the saccharin is not tasted after 30 squeezes (step 10), the test subject is unable to taste saccharin and may not perform the saccharin fit test.

**Note to paragraph 3.** (a): If the test subject eats or drinks something sweet before the screening test, he/she may be unable to taste the weak saccharin solution.

(12) If a taste response is elicited, the test subject shall be asked to take note of the taste for reference in the fit test.

(13) Correct use of the nebulizer means that approximately 1 ml of liquid is used at a time in the nebulizer body.

(14) The nebulizer shall be thoroughly rinsed in water, shaken dry, and refilled at least each morning and afternoon or at least every four hours.

(b) Saccharin solution aerosol fit test procedure.

(1) The test subject may not eat, drink (except plain water), smoke, or chew gum for 15 minutes before the test.

(2) The fit test uses the same enclosure described in 3. (a) above.

(3) The test subject shall don the enclosure while wearing the respirator selected in section I. A. of this appendix. The respirator shall be properly adjusted and equipped with a particulate filter(s).

(4) A second DeVilbiss Model 40 Inhalation Medication Nebulizer or equivalent is used to spray the fit test solution into the enclosure. This nebulizer shall be clearly marked to distinguish it from the screening test solution nebulizer.

(5) The fit test solution is prepared by adding 83 grams of sodium saccharin to 100 ml of warm water.

(6) As before, the test subject shall breathe through the slightly open mouth with tongue extended, and report if he/she tastes the sweet taste of saccharin.

(7) The nebulizer is inserted into the hole in the front of the enclosure and an initial concentration of saccharin fit test solution is sprayed into the enclosure using the same number of squeezes (either 10, 20 or 30 squeezes) based on the number of squeezes required to elicit a taste response as noted during the screening test. A minimum of 10 squeezes is required.

(8) After generating the aerosol, the test subject shall be instructed to perform the exercises in section I. A. 14. of this appendix.

(9) Every 30 seconds the aerosol concentration shall be replenished using one half the original number of squeezes used initially (e.g., 5, 10 or 15).

(10) The test subject shall indicate to the test conductor if at any time during the fit test the taste of saccharin is detected. If the test subject does not report tasting the saccharin, the test is passed.

(11) If the taste of saccharin is detected, the fit is deemed unsatisfactory and the test is failed. A different respirator shall be tried and the entire test procedure is repeated (taste threshold screening and fit testing).

(12) Since the nebulizer has a tendency to clog during use, the test operator must make periodic checks of the nebulizer to ensure that it is not clogged. If clogging is found at the end of the test session, the test is invalid.

### 4. Bitrex™ (Denatonium Benzoate) Solution Aerosol Qualitative Fit Test Protocol

The Bitrex™ (Denatonium benzoate) solution aerosol QLFT protocol uses the published saccharin test protocol because that protocol is widely accepted. Bitrex is routinely used as a taste aversion agent in household liquids which children should not be drinking and is endorsed by the American Medical Association, the National Safety Council, and the American Association of Poison Control Centers. The entire screening and testing procedure shall be explained to the test subject prior to the conduct of the screening test.

(a) Taste Threshold Screening.

The Bitrex taste threshold screening, performed without wearing a respirator, is intended to determine whether the individual being tested can detect the taste of Bitrex.

(1) During threshold screening as well as during fit testing, subjects shall wear an enclosure about the head and shoulders that is approximately 12 inches (30.5 cm) in diameter by 14 inches (35.6 cm) tall. The front portion of the enclosure shall be clear from the respirator and allow free movement of the head when a respirator is worn. An enclosure substantially similar to the 3M hood assembly, parts # FT 14 and # FT 15 combined, is adequate.

(2) The test enclosure shall have a 3/4-inch (1.9 cm) hole in front of the test subject's nose and mouth area to accommodate the nebulizer nozzle.

(3) The test subject shall don the test enclosure. Throughout the threshold screening test, the test subject shall breathe through his or her slightly open mouth with tongue extended. The subject is instructed to report when he/she detects a bitter taste.

(4) Using a DeVilbiss Model 40 Inhalation Medication Nebulizer or equivalent, the test conductor shall spray the Threshold Check Solution into the enclosure. This Nebulizer shall be clearly marked to distinguish it from the fit test solution nebulizer.

(5) The Threshold Check Solution is prepared by adding 13.5 milligrams of Bitrex to 100 ml of 5% salt (NaCl) solution in distilled water.

(6) To produce the aerosol, the nebulizer bulb is firmly squeezed so that the bulb collapses completely, and is then released and allowed to fully expand.

(7) An initial ten squeezes are repeated rapidly and then the test subject is asked whether the Bitrex can be tasted. If the test subject reports tasting the bitter taste during the ten squeezes, the screening test is completed. The taste threshold is noted as ten regardless of the number of squeezes actually completed.

(8) If the first response is negative, ten more squeezes are repeated rapidly and the test subject is again asked whether the Bitrex is tasted. If the test subject reports tasting the bitter taste during the second ten squeezes, the screening test is completed. The taste threshold is noted as twenty regardless of the number of squeezes actually completed.

(9) If the second response is negative, ten more squeezes are repeated rapidly and the test subject is again asked whether the Bitrex is tasted. If the test subject reports tasting the bitter taste during the third set of ten squeezes, the screening test is completed. The taste threshold is noted as thirty regardless of the number of squeezes actually completed.

(10) The test conductor will take note of the number of squeezes required to solicit a taste response.

(11) If the Bitrex is not tasted after 30 squeezes (step 10), the test subject is unable to taste Bitrex and may not perform the Bitrex fit test.

(12) If a taste response is elicited, the test subject shall be asked to take note of the taste for reference in the fit test.

(13) Correct use of the nebulizer means that approximately 1 ml of liquid is used at a time in the nebulizer body.

(14) The nebulizer shall be thoroughly rinsed in water, shaken to dry, and refilled at least each morning and afternoon or at least every four hours.

(b) Bitrex Solution Aerosol Fit Test Procedure.

(1) The test subject may not eat, drink (except plain water), smoke, or chew gum for 15 minutes before the test.

(2) The fit test uses the same enclosure as that described in 4. (a) above.

(3) The test subject shall don the enclosure while wearing the respirator selected according to section I. A. of this appendix. The respirator shall be properly adjusted and equipped with any type particulate filter(s).

(4) A second DeVilbiss Model 40 Inhalation Medication Nebulizer or equivalent is used to spray the fit test solution into the enclosure. This nebulizer shall be clearly marked to distinguish it from the screening test solution nebulizer.

(5) The fit test solution is prepared by adding 337.5 mg of Bitrex to 200 ml of a 5% salt (NaCl) solution in warm water.

(6) As before, the test subject shall breathe through his or her slightly open mouth with tongue extended, and be instructed to report if he/she tastes the bitter taste of Bitrex..

(7) The nebulizer is inserted into the hole in the front of the enclosure and an initial concentration of the fit test solution is sprayed into the enclosure using the same number of squeezes (either 10, 20 or 30 squeezes) based on the number of squeezes required to elicit a taste response as noted during the screening test.

(8) After generating the aerosol, the test subject shall be instructed to perform the exercises in section I. A. 14. of this appendix.

(9) Every 30 seconds the aerosol concentration shall be replenished using one half the number of squeezes used initially (e.g., 5, 10 or 15).

(10) The test subject shall indicate to the test conductor if at any time during the fit test the taste of Bitrex is detected. If the test subject does not report tasting the Bitrex, the test is passed.

(11) If the taste of Bitrex is detected, the fit is deemed unsatisfactory and the test is failed. A different respirator shall be tried and

the entire test procedure is repeated (taste threshold screening and fit testing).

### 5. Irritant Smoke (Stannic Chloride) Protocol

This qualitative fit test uses a person's response to the irritating chemicals released in the "smoke" produced by a stannic chloride ventilation smoke tube to detect leakage into the respirator.

#### (a) General Requirements and Precautions

(1) The respirator to be tested shall be equipped with high efficiency particulate air (HEPA) or P100 series filter(s).

(2) Only stannic chloride smoke tubes shall be used for this protocol.

(3) No form of test enclosure or hood for the test subject shall be used.

(4) The smoke can be irritating to the eyes, lungs, and nasal passages. The test conductor shall take precautions to minimize the test subject's exposure to irritant smoke. Sensitivity varies, and certain individuals may respond to a greater degree to irritant smoke. Care shall be taken when performing the sensitivity screening checks that determine whether the test subject can detect irritant smoke to use only the minimum amount of smoke necessary to elicit a response from the test subject.

(5) The fit test shall be performed in an area with adequate ventilation to prevent exposure of the person conducting the fit test or the build-up of irritant smoke in the general atmosphere.

#### (b) Sensitivity Screening Check

The person to be tested must demonstrate his or her ability to detect a weak concentration of the irritant smoke.

(1) The test operator shall break both ends of a ventilation smoke tube containing stannic chloride, and attach one end of the smoke tube to a low flow air pump set to deliver 200 milliliters per minute, or an aspirator squeeze bulb. The test operator shall cover the other end of the smoke tube with a short piece of tubing to prevent potential injury from the jagged end of the smoke tube.

(2) The test operator shall advise the test subject that the smoke can be irritating to the eyes, lungs, and nasal passages and instruct the subject to keep his/her eyes closed while the test is performed.

(3) The test subject shall be allowed to smell a weak concentration of the irritant smoke before the respirator is donned to become familiar with its irritating properties and to determine if he/she can detect the irritating properties of the smoke. The test operator shall carefully direct a small amount of the irritant smoke in the test subject's direction to determine that he/she can detect it.

#### (c) Irritant Smoke Fit Test Procedure

(1) The person being fit tested shall don the respirator without assistance, and perform the required user seal check(s).

(2) The test subject shall be instructed to keep his/her eyes closed.

(3) The test operator shall direct the stream of irritant smoke from the smoke tube toward the faceseal area of the test subject, using the low flow pump or the squeeze bulb. The test operator shall begin at least 12 inches from the facepiece and move the smoke stream around the whole perimeter of the mask. The operator shall gradually make two more passes around the perimeter of the mask, moving to within six inches of the respirator.

(4) If the person being tested has not had an involuntary response and/or detected the irritant smoke, proceed with the test exercises.

(5) The exercises identified in section I.A. 14. of this appendix shall be performed by the test subject while the respirator seal is being continually challenged by the smoke, directed around the perimeter of the respirator at a distance of six inches.

(6) If the person being fit tested reports detecting the irritant smoke at any time, the test is failed. The person being retested must repeat the entire sensitivity check and fit test procedure.

(7) Each test subject passing the irritant smoke test without evidence of a response (involuntary cough, irritation) shall be given a second sensitivity screening check, with the smoke from the same smoke tube used during the fit test, once the respirator has been removed, to determine whether he/she still reacts to the smoke. Failure to evoke a response shall void the fit test.

(8) If a response is produced during this second sensitivity check, then the fit test is passed.

### C. Quantitative Fit Test (QNFT) Protocols

The following quantitative fit testing procedures have been demonstrated to be acceptable: Quantitative fit testing using a non-hazardous test aerosol (such as corn oil, polyethylene glycol 400 [PEG 400], di-2-ethyl hexyl sebacate [DEHS], or sodium chloride) generated in a test chamber, and employing instrumentation to quantify the fit of the respirator; Quantitative fit testing using ambient aerosol as the test agent and appropriate instrumentation (condensation nuclei counter) to quantify the respirator fit; Quantitative fit testing using controlled negative pressure and appropriate instrumentation to measure the volumetric leak rate of a facepiece to quantify the respirator fit.

### 1. General

(a) The employer shall ensure that persons administering QNFT are able to calibrate equipment and perform tests properly, recognize invalid tests, calculate fit factors properly and ensure that test equipment is in proper working order.

(b) The employer shall ensure that QNFT equipment is kept clean, and is maintained and calibrated according to the manufacturer's instructions so as to operate at the parameters for which it was designed.

### 2. Generated Aerosol Quantitative Fit Testing Protocol

(a) Apparatus.

(1) Instrumentation. Aerosol generation, dilution, and measurement systems using particulates (corn oil, polyethylene glycol 400 [PEG 400], di-2-ethyl hexyl sebacate [DEHS] or sodium chloride) as test aerosols shall be used for quantitative fit testing.

(2) Test chamber. The test chamber shall be large enough to permit all test subjects to perform freely all required exercises without disturbing the test agent concentration or the measurement apparatus. The test chamber shall be equipped and constructed so that the test agent is effectively isolated from the ambient air, yet uniform in concentration throughout the chamber.

(3) When testing air-purifying respirators, the normal filter or cartridge element shall be replaced with a high efficiency particulate air (HEPA) or P100 series filter supplied by the same manufacturer.

(4) The sampling instrument shall be selected so that a computer record or strip chart record may be made of the test showing the rise and fall of the test agent concentration with each inspiration and expiration at fit factors of at least 2,000. Integrators or computers that integrate the amount of test agent penetration leakage into the respirator for each exercise may be used provided a record of the readings is made.

(5) The combination of substitute air-purifying elements, test agent and test agent concentration shall be such that the test subject is not exposed in excess of an established exposure limit for the test agent at any time during the testing process, based upon the length of the exposure and the exposure limit duration.

(6) The sampling port on the test specimen respirator shall be placed and constructed so that no leakage occurs around the port (e.g., where the respirator is probed), a free air flow is allowed into the sampling line at all times, and there is no interference with the fit or performance of the respirator. The in-mask sampling device (probe) shall be designed and used so that the air sample is drawn from the breathing zone of the test subject, midway between the nose and mouth and with the probe extending into the facepiece cavity at least 1/4 inch.

(7) The test setup shall permit the person administering the test to observe the test subject inside the chamber during the test.

(8) The equipment generating the test atmosphere shall maintain the concentration of test agent constant to within a 10 percent variation for the duration of the test.

(9) The time lag (interval between an event and the recording of the event on the strip chart or computer or integrator) shall be kept to a minimum. There shall be a clear association between the occurrence of an event and its being recorded.

(10) The sampling line tubing for the test chamber atmosphere and for the respirator sampling port shall be of equal diameter and of the same material. The length of the two lines shall be equal.

(11) The exhaust flow from the test chamber shall pass through an appropriate filter (i.e., high efficiency particulate filter) before release.

(12) When sodium chloride aerosol is used, the relative humidity inside the test chamber shall not exceed 50 percent.

(13) The limitations of instrument detection shall be taken into account when determining the fit factor.

(14) Test respirators shall be maintained in proper working order and be inspected regularly for deficiencies such as cracks or missing valves and gaskets.

(b) Procedural Requirements.

(1) When performing the initial user seal check using a positive or negative pressure check, the sampling line shall be crimped closed in order to avoid air pressure leakage during either of these pressure checks.

(2) The use of an abbreviated screening QLFT test is optional. Such a test may be utilized in order to quickly identify poor fitting respirators that passed the positive and/or negative pressure test and reduce the amount of QNFT time. The use of the CNC QNFT instrument in the count mode is another optional method to obtain a quick estimate of fit and eliminate poor fitting respirators before going on to perform a full QNFT.

(3) A reasonably stable test agent concentration shall be measured in the test chamber prior to testing. For canopy or shower curtain types of test units, the determination of the test agent's stability may be established after the test subject has entered the test environment.

(4) Immediately after the subject enters the test chamber, the test agent concentration inside the respirator shall be measured to ensure that the peak penetration does not exceed 5 percent for a half mask or 1 percent for a full facepiece respirator.

(5) A stable test agent concentration shall be obtained prior to the actual start of testing.

(6) Respirator restraining straps shall not be over-tightened for testing. The straps shall be adjusted by the wearer without assistance from other persons to give a reasonably comfortable fit typical of normal use. The respirator shall not be adjusted once the fit test exercises begin.

(7) The test shall be terminated whenever any single peak penetration exceeds 5 percent for half masks and 1 percent for full facepiece respirators. The test subject shall be refitted and retested.

(8) Calculation of fit factors.

(i) The fit factor shall be determined for the quantitative fit test by taking the ratio of the average chamber concentration to the concentration measured inside the respirator for each test exercise except the grimace exercise.

(ii) The average test chamber concentration shall be calculated as the arithmetic average of the concentration measured before and after each test (i.e., 7 exercises) or the arithmetic average of the concentration measured before and after each exercise or the true average measured continuously during the respirator sample.

(iii) The concentration of the challenge agent inside the respirator shall be determined by one of the following methods:

(*A*) Average peak penetration method means the method of determining test agent penetration into the respirator utilizing a strip chart recorder, integrator, or computer. The agent penetration is determined by an average of the peak heights on the graph or by computer integration, for each exercise except the grimace exercise. Integrators or computers that calculate the actual test agent penetration into the respirator for each exercise will also be considered to meet the requirements of the average peak penetration method.

(*B*) Maximum peak penetration method means the method of determining test agent penetration in the respirator as determined by strip chart recordings of the test. The highest peak penetration for a given exercise is taken to be representative of average penetration into the respirator for that exercise.

(*C*) Integration by calculation of the area under the individual peak for each exercise except the grimace exercise. This includes computerized integration.

(*D*) The calculation of the overall fit factor using individual exercise fit factors involves first converting the exercise fit factors to penetration values, determining the average, and then converting that result back to a fit factor. This procedure is described in the following equation:

$$\text{Overall Fit Factor} = \frac{\text{Number of exercises}}{1/ff_1 + 1/ff_2 + 1/ff_3 + 1/ff_4 + 1/ff_5 + 1/ff_6 + 1/ff_7 + 1/ff_8}$$

Where $ff_1$, $ff_2$, $ff_3$, etc. are the fit factors for exercises 1, 2, 3, etc.

(9) The test subject shall not be permitted to wear a half mask or quarter facepiece respirator unless a minimum fit factor of 100 is obtained, or a full facepiece respirator unless a minimum fit factor of 500 is obtained.

(10) Filters used for quantitative fit testing shall be replaced whenever increased breathing resistance is encountered, or when the test agent has altered the integrity of the filter media.

### 3. Ambient aerosol condensation nuclei counter (CNC) quantitative fit testing protocol.

The ambient aerosol condensation nuclei counter (CNC) quantitative fit testing (Portacount™) protocol quantitatively fit tests respirators with the use of a probe. The probed respirator is only used for quantitative fit tests. A probed respirator has a special sampling device, installed on the respirator, that allows the probe to sample the air from inside the mask. A probed respirator is required for each make, style, model, and size that the employer uses and can be obtained from the respirator manufacturer or distributor. The CNC instrument manufacturer, TSI Inc., also provides probe attachments (TSI sampling adapters) that permit fit testing in an employee's own respirator. A minimum fit factor pass level of at least 100 is necessary for a half-mask respirator and a minimum fit factor pass level of at least 500 is required for a full facepiece negative pressure respirator. The entire screening and testing procedure shall be explained to the test subject prior to the conduct of the screening test.

(a) Portacount Fit Test Requirements. (1) Check the respirator to make sure the sampling probe and line are properly attached to the facepiece and that the respirator is fitted with a particulate filter capable of preventing significant penetration by the ambient particles used for the fit test (e.g., NIOSH 42 CFR 84 series 100, series 99, or series 95 particulate filter) per manufacturer's instruction.

(2) Instruct the person to be tested to don the respirator for five minutes before the fit test starts. This purges the ambient particles trapped inside the respirator and permits the wearer to make certain the respirator is comfortable. This individual shall already have been trained on how to wear the respirator properly.

(3) Check the following conditions for the adequacy of the respirator fit: Chin properly placed; Adequate strap tension, not overly tightened; Fit across nose bridge; Respirator of proper size to span distance from nose to chin; Tendency of the respirator to slip; Self-observation in a mirror to evaluate fit and respirator position.

(4) Have the person wearing the respirator do a user seal check. If leakage is detected, determine the cause. If leakage is from a poorly fitting facepiece, try another size of the same model respirator, or another model of respirator.

(5) Follow the manufacturer's instructions for operating the Portacount and proceed with the test.

(6) The test subject shall be instructed to perform the exercises in section I. A. 14. of this appendix.

(7) After the test exercises, the test subject shall be questioned by the test conductor regarding the comfort of the respirator upon completion of the protocol. If it has become unacceptable, another model of respirator shall be tried.

(b) Portacount Test Instrument.

(1) The Portacount will automatically stop and calculate the overall fit factor for the entire set of exercises. The overall fit factor is what counts. The Pass or Fail message will indicate whether or not the test was successful. If the test was a Pass, the fit test is over.

(2) Since the pass or fail criterion of the Portacount is user programmable, the test operator shall ensure that the pass or fail criterion meet the requirements for minimum respirator performance in this Appendix.

(3) A record of the test needs to be kept on file, assuming the fit test was successful. The record must contain the test subject's name; overall fit factor; make, model, style, and size of respirator used; and date tested.

### 4. Controlled negative pressure (CNP) quantitative fit testing protocol.

The CNP protocol provides an alternative to aerosol fit test methods. The CNP fit test method technology is based on exhausting air from a temporarily sealed respirator facepiece to generate and then maintain a constant negative pressure inside the facepiece. The rate of air exhaust is controlled so that a constant negative pressure is maintained in the respirator during the fit test. The level of pressure is selected to replicate the mean inspiratory pressure that causes leakage into the respirator under normal use conditions. With pressure held constant, air flow out of the respirator is equal to air flow into the respirator. Therefore, measurement of the exhaust stream that is required to hold the pressure in the temporarily sealed respirator constant yields a direct measure of leakage air flow into the respirator. The CNP fit test method measures leak rates through the facepiece as a method for determining the facepiece fit for negative pressure respirators. The CNP instrument manufacturer Dynatech Nevada also provides attachments (sampling manifolds) that replace the filter cartridges to permit fit testing in an employee's own respirator. To perform the test, the test subject closes his or her mouth and holds his/her breath, after which an air pump removes air from the respirator facepiece at a pre-selected constant pressure. The facepiece fit is expressed as the leak rate through the facepiece, expressed as milliliters per minute. The quality and validity of the CNP fit tests are determined by the degree to which the in-mask pressure tracks the test pressure during the system measurement time of approximately five seconds. Instantaneous feedback in the form of a real-time pressure trace of the in-mask pressure is provided and used to determine test validity and quality. A minimum fit factor pass level of 100 is necessary for a half-mask respirator and a minimum fit factor of at least 500 is required for a full facepiece respirator. The entire screening and testing procedure shall be explained to the test subject prior to the conduct of the screening test.

(a) CNP Fit Test Requirements.

(1) The instrument shall have a non-adjustable test pressure of 15.0 mm water pressure.

(2) The CNP system defaults selected for test pressure shall be set at - 15 mm of water (-0.58 inches of water) and the modeled inspiratory flow rate shall be 53.8 liters per minute for performing fit tests.

**Note:** CNP systems have built-in capability to conduct fit testing that is specific to unique work rate, mask, and gender situations that might apply in a specific workplace. Use of system default values, which were selected to represent respirator wear with medium cartridge resistance at a low-moderate work rate, will allow inter-test comparison of the respirator fit.)

(3) The individual who conducts the CNP fit testing shall be thoroughly trained to perform the test.

(4) The respirator filter or cartridge needs to be replaced with the CNP test manifold. The inhalation valve downstream from the manifold either needs to be temporarily removed or propped open.

(5) The test subject shall be trained to hold his or her breath for at least 20 seconds.

(6) The test subject shall don the test respirator without any assistance from the individual who conducts the CNP fit test.

(7) The QNFT protocol shall be followed according to section I. C. 1. of this appendix with an exception for the CNP test exercises.

(b) CNP Test Exercises.

(1) Normal breathing. In a normal standing position, without talking, the subject shall breathe normally for 1 minute. After the normal breathing exercise, the subject needs to hold head straight ahead and hold his or her breath for 10 seconds during the test measurement.

(2) Deep breathing. In a normal standing position, the subject shall breathe slowly and deeply for 1 minute, being careful not to hyperventilate. After the deep breathing exercise, the subject shall hold his or her head straight ahead and hold his or her breath for 10 seconds during test measurement.

(3) Turning head side to side. Standing in place, the subject shall slowly turn his or her head from side to side between the extreme positions on each side for 1 minute. The head shall be held at each extreme momentarily so the subject can inhale at each side. After the turning head side to side exercise, the subject needs to hold head full left and hold his or her breath for 10 seconds during test measurement. Next, the subject needs to hold head full right and hold his or her breath for 10 seconds during test measurement.

(4) Moving head up and down. Standing in place, the subject shall slowly move his or her head up and down for 1 minute. The subject shall be instructed to inhale in the up position (i.e., when looking toward the ceiling). After the moving head up and down exercise, the subject shall hold his or her head full up and hold his or her breath for 10 seconds during test measurement. Next, the subject shall hold his or her head full down and hold his or her breath for 10 seconds during test measurement.

(5) Talking. The subject shall talk out loud slowly and loud enough so as to be heard clearly by the test conductor. The subject can read from a prepared text such as the Rainbow Passage, count backward from 100, or recite a memorized poem or song for 1 minute. After the talking exercise, the subject shall hold his or her head straight ahead and hold his or her breath for 10 seconds during the test measurement.

(6) Grimace. The test subject shall grimace by smiling or frowning for 15 seconds.

(7) Bending Over. The test subject shall bend at the waist as if he or she were to touch his or her toes for 1 minute. Jogging in place shall be substituted for this exercise in those test environments such as shroud-type QNFT units that prohibit bending at the waist. After the bending over exercise, the subject shall hold his or her head straight ahead and hold his or her breath for 10 seconds during the test measurement.

(8) Normal Breathing. The test subject shall remove and re-don the respirator within a one-minute period. Then, in a normal standing position, without talking, the subject shall breathe normally for 1 minute. After the normal breathing exercise, the subject shall hold his or her head straight ahead and hold his or her breath for 10 seconds during the test measurement. After the test exercises, the test subject shall be questioned by the test conductor regarding the comfort of the respirator upon completion of the protocol. If it has become unacceptable, another model of a respirator shall be tried.

(c) CNP Test Instrument.

(1) The test instrument shall have an effective audio warning device when the test subject fails to hold his or her breath during the test. The test shall be terminated whenever the test subject failed to hold his or her breath. The test subject may be refitted and retested.

(2) A record of the test shall be kept on file, assuming the fit test was successful. The record must contain the test subject's name; overall fit factor; make, model, style and size of respirator used; and date tested.

### *Part II. New Fit Test Protocols*

A. Any person may submit to OSHA an application for approval of a new fit test protocol. If the application meets the following criteria, OSHA will initiate a rulemaking proceeding under section 6(b)(7) of the OSH Act to determine whether to list the new protocol as an approved protocol in this Appendix A.

B. The application must include a detailed description of the proposed new fit test protocol. This application must be supported by either:

1. A test report prepared by an independent government research laboratory (e.g., Lawrence Livermore National Laboratory, Los Alamos National Laboratory, the National Institute for Standards and Technology) stating that the laboratory has tested the protocol and had found it to be accurate and reliable; or

2. An article that has been published in a peer-reviewed industrial hygiene journal describing the protocol and explaining how test data support the protocol's accuracy and reliability.

C. If OSHA determines that additional information is required before the Agency commences a rulemaking proceeding under this section, OSHA will so notify the applicant and afford the applicant the opportunity to submit the supplemental information. Initiation of a rulemaking proceeding will be deferred until OSHA has received and evaluated the supplemental information.

## APPENDIX B-1 TO §1910.134: USER SEAL CHECK PROCEDURES (MANDATORY)

The individual who uses a tight-fitting respirator is to perform a user seal check to ensure that an adequate seal is achieved each time the respirator is put on. Either the positive and negative pressure checks listed in this appendix, or the respirator manufacturer's recommended user seal check method shall be used. User seal checks are not substitutes for qualitative or quantitative fit tests.

### I. Facepiece Positive and/or Negative Pressure Checks

A. *Positive pressure check.* Close off the exhalation valve and exhale gently into the facepiece. The face fit is considered satisfactory if a slight positive pressure can be built up inside the facepiece without any evidence of outward leakage of air at the seal. For most respirators this method of leak testing requires the wearer to first remove the exhalation valve cover before closing off the exhalation valve and then carefully replacing it after the test.

B. *Negative pressure check.* Close off the inlet opening of the canister or cartridge(s) by covering with the palm of the hand(s) or by replacing the filter seal(s), inhale gently so that the facepiece collapses slightly, and hold the breath for ten seconds. The design of the inlet opening of some cartridges cannot be effectively covered with the palm of the hand. The test can be performed by covering the inlet opening of the cartridge with a thin latex or nitrile glove. If the facepiece remains in its slightly collapsed condition and no inward leakage of air is detected, the tightness of the respirator is considered satisfactory.

### II. Manufacturer's Recommended User Seal Check Procedures

The respirator manufacturer's recommended procedures for performing a user seal check may be used instead of the positive and/or negative pressure check procedures provided that the employer demonstrates that the manufacturer's procedures are equally effective.

## APPENDIX B-2 TO §1910.134: RESPIRATOR CLEANING PROCEDURES (MANDATORY)

These procedures are provided for employer use when cleaning respirators. They are general in nature, and the employer as an alternative may use the cleaning recommendations provided by the manufacturer of the respirators used by their employees, provided such procedures are as effective as those listed here in Appendix B-2. Equivalent effectiveness simply means that the procedures used must accomplish the objectives set forth in Appendix B-2, i.e., must ensure that the respirator is properly cleaned and disinfected in a manner that prevents damage to the respirator and does not cause harm to the user.

### I. Procedures for Cleaning Respirators

A. Remove filters, cartridges, or canisters. Disassemble facepieces by removing speaking diaphragms, demand and pressure-demand valve assemblies, hoses, or any components recommended by the manufacturer. Discard or repair any defective parts.

B. Wash components in warm (43 °C [110 °F] maximum) water with a mild detergent or with a cleaner recommended by the manufacturer. A stiff bristle (not wire) brush may be used to facilitate the removal of dirt.

C. Rinse components thoroughly in clean, warm (43 °C [110 °F] maximum), preferably running water. Drain.

D. When the cleaner used does not contain a disinfecting agent, respirator components should be immersed for two minutes in one of the following:

1. Hypochlorite solution (50 ppm of chlorine) made by adding approximately one milliliter of laundry bleach to one liter of water at 43 °C (110 °F); or,

2. Aqueous solution of iodine (50 ppm iodine) made by adding approximately 0.8 milliliters of tincture of iodine (6-8 grams ammonium and/or potassium iodide/100 cc of 45% alcohol) to one liter of water at 43 °C (110 °F); or,

3. Other commercially available cleansers of equivalent disinfectant quality when used as directed, if their use is recommended or approved by the respirator manufacturer.

E. Rinse components thoroughly in clean, warm (43 °C [110 °F] maximum), preferably running water. Drain. The importance of thorough rinsing cannot be overemphasized. Detergents or disinfectants that dry on facepieces may result in dermatitis. In addition, some disinfectants may cause deterioration of rubber or corrosion of metal parts if not completely removed.

F. Components should be hand-dried with a clean lint-free cloth or air-dried.

G. Reassemble facepiece, replacing filters, cartridges, and canisters where necessary.

H. Test the respirator to ensure that all components work properly.

# APPENDIX C TO §1910.134: OSHA RESPIRATOR MEDICAL EVALUATION QUESTIONNAIRE (MANDATORY)

## OSHA RESPIRATOR MEDICAL EVALUATION QUESTIONNAIRE (MANDATORY)

To the employer: Answers to questions in Section 1, and to question 9 in Section 2 of Part A, do not require a medical examination.

To the employee:

Can you read (circle one): Yes/No

Your employer must allow you to answer this questionnaire during normal working hours, or at a time and place that is convenient to you. To maintain your confidentiality, your employer or supervisor must not look at or review your answers, and your employer must tell you how to deliver or send this questionnaire to the health care professional who will review it.

Part A. Section 1. (Mandatory) The following information must be provided by every employee who has been selected to use any type of respirator (please print).

1. Today's date: ______________________________
2. Your name: ______________________________
3. Your age (to nearest year): _______
4. Sex (circle one): Male/Female
5. Your height: _____ ft. _____ in.
6. Your weight: _____ lbs.
7. Your job title: ______________________________
8. A phone number where you can be reached by the health care professional who reviews this questionnaire (include the Area Code): ______________________________
9. The best time to phone you at this number: _____
10. Has your employer told you how to contact the health care professional who will review this questionnaire (circle one): Yes/No
11. Check the type of respirator you will use (you can check more than one category):
    - a. _____ N, R, or P disposable respirator (filter-mask, non-cartridge type only).
    - b. _____ Other type (for example, half- or full-facepiece type, powered-air purifying, supplied-air, self-contained breathing apparatus).
12. Have you worn a respirator (circle one): Yes/No

    7 If "yes," what type(s): ______________________________

Part A. Section 2. (Mandatory) Questions 1 through 9 below must be answered by every employee who has been selected to use any type of respirator (please circle "yes" or "no").

1. Do you *currently* smoke tobacco, or have you smoked tobacco in the last month: Yes/No
2. Have you *ever had* any of the following conditions?
    - a. Seizures (fits): Yes/No
    - b. Diabetes (sugar disease): Yes/No
    - c. Allergic reactions that interfere with your breathing: Yes/No
    - d. Claustrophobia (fear of closed-in places): Yes/No
    - e. Trouble smelling odors: Yes/No
3. Have you *ever had* any of the following pulmonary or lung problems?
    - a. Asbestosis: Yes/No
    - b. Asthma: Yes/No
    - c. Chronic bronchitis: Yes/No
    - d. Emphysema: Yes/No
    - e. Pneumonia: Yes/No
    - f. Tuberculosis: Yes/No
    - g. Silicosis: Yes/No
    - h. Pneumothorax (collapsed lung): Yes/No
    - i. Lung cancer: Yes/No
    - j. Broken ribs: Yes/No
    - k. Any chest injuries or surgeries: Yes/No
    - l. Any other lung problem that you've been told about: Yes/No
4. Do you *currently* have any of the following symptoms of pulmonary or lung illness?
    - a. Shortness of breath: Yes/No
    - b. Shortness of breath when walking fast on level ground or walking up a slight hill or incline: Yes/No
    - c. Shortness of breath when walking with other people at an ordinary pace on level ground: Yes/No
    - d. Have to stop for breath when walking at your own pace on level ground: Yes/No
    - e. Shortness of breath when washing or dressing yourself: Yes/No
    - f. Shortness of breath that interferes with your job: Yes/No
    - g. Coughing that produces phlegm (thick sputum): Yes/No
    - h. Coughing that wakes you early in the morning: Yes/No
    - i. Coughing that occurs mostly when you are lying down: Yes/No
    - j. Coughing up blood in the last month: Yes/No
    - k. Wheezing: Yes/No
    - l. Wheezing that interferes with your job: Yes/No
    - m. Chest pain when you breathe deeply: Yes/No
    - n. Any other symptoms that you think may be related to lung problems: Yes/No
5. Have you *ever had* any of the following cardiovascular or heart problems?
    - a. Heart attack: Yes/No
    - b. Stroke: Yes/No
    - c. Angina: Yes/No
    - d. Heart failure: Yes/No
    - e. Swelling in your legs or feet (not caused by walking): Yes/No
    - f. Heart arrhythmia (heart beating irregularly): Yes/No
    - g. High blood pressure: Yes/No
    - h. Any other heart problem that you've been told about: Yes/No
6. Have you *ever had* any of the following cardiovascular or heart symptoms?
    - a. Frequent pain or tightness in your chest: Yes/No
    - b. Pain or tightness in your chest during physical activity: Yes/No
    - c. Pain or tightness in your chest that interferes with your job: Yes/No
    - d. In the past two years, have you noticed your heart skipping or missing a beat: Yes/No
    - e. Heartburn or indigestion that is not related to eating: Yes/No
    - f. Any other symptoms that you think may be related to heart or circulation problems: Yes/No
7. Do you *currently* take medication for any of the following problems?
    - a. Breathing or lung problems: Yes/No
    - b. Heart trouble: Yes/No
    - c. Blood pressure: Yes/No
    - d. Seizures (fits): Yes/No
8. If you've used a respirator, have you *ever had* any of the following problems?
   (If you've never used a respirator, check the following space and go to question 9:)
    - a. Eye irritation: Yes/No
    - b. Skin allergies or rashes: Yes/No
    - c. Anxiety: Yes/No
    - d. General weakness or fatigue: Yes/No
    - e. Any other problem that interferes with your use of a respirator: Yes/No
9. Would you like to talk to the health care professional who will review this questionnaire about your answers to this questionnaire: Yes/No

Questions 10 to 15 below must be answered by every employee who has been selected to use either a full-facepiece respirator or a self-contained breathing apparatus (SCBA). For employees who have been selected to use other types of respirators, answering these questions is voluntary.

10. Have you *ever lost* vision in either eye (temporarily or permanently): Yes/No
11. Do you *currently* have any of the following vision problems?
    - a. Wear contact lenses: Yes/No
    - b. Wear glasses: Yes/No
    - c. Color blind: Yes/No
    - d. Any other eye or vision problem: Yes/No
12. Have you *ever had* an injury to your ears, including a broken ear drum: Yes/No
13. Do you *currently* have any of the following hearing problems?
    - a. Difficulty hearing: Yes/No
    - b. Wear a hearing aid: Yes/No
    - c. Any other hearing or ear problem: Yes/No
14. Have you *ever had* a back injury: Yes/No
15. Do you *currently* have any of the following musculoskeletal problems?
    - a. Weakness in any of your arms, hands, legs, or feet: Yes/No
    - b. Back pain: Yes/No
    - c. Difficulty fully moving your arms and legs: Yes/No
    - d. Pain or stiffness when you lean forward or backward at the waist: Yes/No
    - e. Difficulty fully moving your head up or down: Yes/No
    - f. Difficulty fully moving your head side to side: Yes/No
    - g. Difficulty bending at your knees: Yes/No
    - h. Difficulty squatting to the ground: Yes/No
    - i. Climbing a flight of stairs or a ladder carrying more than 25 lbs: Yes/No
    - j. Any other muscle or skeletal problem that interferes with using a respirator: Yes/No

Part B Any of the following questions, and other questions not listed, may be added to the questionnaire at the discretion of the health care professional who will review the questionnaire.

1. In your present job, are you working at high altitudes (over 5,000 feet) or in a place that has lower than normal amounts of oxygen: Yes/No

   If "yes," do you have feelings of dizziness, shortness of breath, pounding in your chest, or other symptoms when you're working under these conditions: Yes/No
2. At work or at home, have you ever been exposed to hazardous solvents, hazardous airborne chemicals (*e.g.*, gases, fumes, or dust), or have you come into skin contact with hazardous chemicals: Yes/No

If "yes," name the chemicals if you know them:

3. Have you ever worked with any of the materials, or under any of the conditions, listed below:
   a. Asbestos: Yes/No
   b. Silica ( *e.g.*, in sandblasting): Yes/No
   c. Tungsten/cobalt ( *e.g.*, grinding or welding this material): Yes/No
   d. Beryllium: Yes/No
   e. Aluminum: Yes/No
   f. Coal (for example, mining): Yes/No
   g. Iron: Yes/No
   h. Tin: Yes/No
   i. Dusty environments: Yes/No
   j. Any other hazardous exposures: Yes/No
   If "yes," describe these exposures:

4. List any second jobs or side businesses you have:

5. List your previous occupations:

6. List your current and previous hobbies:

7. Have you been in the military services? Yes/No
   If "yes," were you exposed to biological or chemical agents (either in training or combat): Yes/No
8. Have you ever worked on a HAZMAT team? Yes/No
9. Other than medications for breathing and lung problems, heart trouble, blood pressure, and seizures mentioned earlier in this questionnaire, are you taking any other medications for any reason (including over-the-counter medications): Yes/No
   If "yes," name the medications if you know them:

10. Will you be using any of the following items with your respirator(s)?
    a. HEPA Filters: Yes/No
    b. Canisters (for example, gas masks): Yes/No
    c. Cartridges: Yes/No
11. How often are you expected to use the respirator(s) (circle "yes" or "no" for all answers that apply to you)?:
    a. Escape only (no rescue): Yes/No
    b. Emergency rescue only: Yes/No
    c. Less than 5 hours *per week:* Yes/No
    d. Less than 2 hours *per day:* Yes/No
    e. 2 to 4 hours *per day:* Yes/No
    f. Over 4 hours *per day:* Yes/No
12. During the period you are using the respirator(s), is your work effort:
    a. *Light* (less than 200 kcal per hour): Yes/No
    If "yes," how long does this period last during the average shift: ____ hrs. ____ mins.
    Examples of a light work effort are *sitting* while writing, typing, drafting, or performing light assembly work; or *standing* while operating a drill press (1-3 lbs.) or controlling machines.
    b. *Moderate* (200 to 350 kcal per hour): Yes/No
    If "yes," how long does this period last during the average shift: ____ hrs. ____ mins.
    Examples of moderate work effort are *sitting* while nailing or filing; *driving* a truck or bus in urban traffic; *standing* while drilling, nailing, performing assembly work, or transferring a moderate load (about 35 lbs.) at trunk level; *walking* on a level surface about 2 mph or down a 5-degree grade about 3 mph; or *pushing* a wheelbarrow with a heavy load (about 100 lbs.) on a level surface.
    c. *Heavy* (above 350 kcal per hour): Yes/No
    If "yes," how long does this period last during the average shift: ____ hrs. ____ mins.
    Examples of heavy work are *lifting* a heavy load (about 50 lbs.) from the floor to your waist or shoulder; *working* on a loading dock; *shoveling; standing* while bricklaying or chipping castings; *walking* up an 8-degree grade about 2 mph; *climbing* stairs with a heavy load (about 50 lbs.).
13. Will you be wearing protective clothing and/or equipment (other than the respirator) when you're using your respirator: Yes/No
    If "yes," describe this protective clothing and/or equipment:

14. Will you be working under hot conditions (temperature exceeding 77°F): Yes/No
15. Will you be working under humid conditions: Yes/No
16. Describe the work you'll be doing while you're using your respirator(s):

17. Describe any special or hazardous conditions you might encounter when you're using your respirator(s) (for example, confined spaces, life-threatening gases):

18. Provide the following information, if you know it, for each toxic substance that you'll be exposed to when you're using your respirator(s):

    Name of the first toxic substance:
    Estimated maximum exposure level per shift: ____
    Duration of exposure per shift ____
    Name of the second toxic substance:

    Estimated maximum exposure level per shift: ____
    Duration of exposure per shift: ____
    Name of the third toxic substance:

    Estimated maximum exposure level per shift: ____
    Duration of exposure per shift: ____
    The name of any other toxic substances that you'll be exposed to while using your respirator:

19. Describe any special responsibilities you'll have while using your respirator(s) that may affect the safety and well-being of others (for example, rescue, security):

## APPENDIX D TO §1910.134 (MANDATORY) INFORMATION FOR EMPLOYEES USING RESPIRATORS WHEN NOT REQUIRED UNDER THE STANDARD

Respirators are an effective method of protection against designated hazards when properly selected and worn. Respirator use is encouraged, even when exposures are below the exposure limit, to provide an additional level of comfort and protection for workers. However, if a respirator is used improperly or not kept clean, the respirator itself can become a hazard to the worker. Sometimes, workers may wear respirators to avoid exposures to hazards, even if the amount of hazardous substance does not exceed the limits set by OSHA standards. If your employer provides respirators for your voluntary use, of if you provide your own respirator, you need to take certain precautions to be sure that the respirator itself does not present a hazard.

You should do the following:

1. Read and heed all instructions provided by the manufacturer on use, maintenance, cleaning and care, and warnings regarding the respirators limitations.

2. Choose respirators certified for use to protect against the contaminant of concern. NIOSH, the National Institute for Occupational Safety and Health of the U.S. Department of Health and Human Services, certifies respirators. A label or statement of certification should appear on the respirator or respirator packaging. It will tell you what the respirator is designed for and how much it will protect you.

3. Do not wear your respirator into atmospheres containing contaminants for which your respirator is not designed to protect against. For example, a respirator designed to filter dust particles will not protect you against gases, vapors, or very small solid particles of fumes or smoke.

4. Keep track of your respirator so that you do not mistakenly use someone else's respirator.

[63 FR 1270, Jan. 8, 1998; 63 FR 20098, 20099, Apr. 23, 1998]

## §1926.104 Safety belts, lifelines, and lanyards.

**1926.104(a)** Lifelines, safety belts, and lanyards shall be used only for employee safeguarding. Any lifeline, safety belt, or lanyard actually subjected to in-service loading, as distinguished from static load testing, shall be immediately removed from service and shall not be used again for employee safeguarding.

**1926.104(b)** Lifelines shall be secured above the point of operation to an anchorage or structural member capable of supporting a minimum dead weight of 5,400 pounds.

**1926.104(c)** Lifelines used on rock-scaling operations, or in areas where the lifeline may be subjected to cutting or abrasion, shall be a minimum of 7/8-inch wire core manila rope. For all other lifeline applications, a minimum of 3/4-inch manila or equivalent, with a minimum breaking strength of 5,400 pounds, shall be used.

**1926.104(d)** Safety belt lanyard shall be a minimum of 1/2-inch nylon, or equivalent, with a maximum length to provide for a fall of no greater than 6 feet. The rope shall have a nominal breaking strength of 5,400 pounds.

**1926.104(e)** All safety belt and lanyard hardware shall be drop forged or pressed steel, cadmium plated in accordance with type 1, Class B plating specified in Federal Specification QQ-P-416. Surface shall be smooth and free of sharp edges.

**1926.104(f)** All safety belt and lanyard hardware, except rivets, shall be capable of withstanding a tensile loading of 4,000 pounds without cracking, breaking, or taking a permanent deformation.

## §1926.105 Safety nets.

**1926.105(a)** Safety nets shall be provided when workplaces are more than 25 feet above the ground or water surface, or other surfaces where the use of ladders, scaffolds, catch platforms, temporary floors, safety lines, or safety belts is impractical.

**1926.105(b)** Where safety net protection is required by this part, operations shall not be undertaken until the net is in place and has been tested.

**1926.105(c)(1)** Nets shall extend 8 feet beyond the edge of the work surface where employees are exposed and shall be installed as close under the work surface as practical but in no case more than 25 feet below such work surface. Nets shall be hung with sufficient clearance to prevent user's contact with the surfaces or structures below. Such clearances shall be determined by impact load testing.

**1926.105(c)(2)** It is intended that only one level of nets be required for bridge construction.

**1926.105(d)** The mesh size of nets shall not exceed 6 inches by 6 inches. All new nets shall meet accepted performance standards of 17,500 foot-pounds minimum impact resistance as determined and certified by the manufacturers, and shall bear a label of proof test. Edge ropes shall provide a minimum breaking strength of 5,000 pounds.

**1926.105(e)** Forged steel safety hooks or shackles shall be used to fasten the net to its supports.

**1926.105(f)** Connections between net panels shall develop the full strength of the net.

## §1926.106 Working over or near water.

**1926.106(a)** Employees working over or near water, where the danger of drowning exists, shall be provided with U.S. Coast Guard-approved life jacket or buoyant work vests.

**1926.106(b)** Prior to and after each use, the buoyant work vests or life preservers shall be inspected for defects which would alter their strength or buoyancy. Defective units shall not be used.

**1926.106(c)** Ring buoys with at least 90 feet of line shall be provided and readily available for emergency rescue operations. Distance between ring buoys shall not exceed 200 feet.

**1926.106(d)** At least one lifesaving skiff shall be immediately available at locations where employees are working over or adjacent to water.

## §1926.107 Definitions applicable to this subpart.

**1926.107(a)** *Contaminant* means any material which by reason of its action upon, within, or to a person is likely to cause physical harm.

**1926.107(b)** *Lanyard* means a rope, suitable for supporting one person. One end is fastened to a safety belt or harness and the other end is secured to a substantial object or a safety line.

**1926.107(c)** *Lifeline* means a rope, suitable for supporting one person, to which a lanyard or safety belt (or harness) is attached.

**1926.107(d)** *O.D.* means optical density and refers to the light refractive characteristics of a lens.

**1926.107(e)** *Radiant energy* means energy that travels outward in all directions from its sources.

**1926.107(f)** *Safety belt* means a device, usually worn around the waist which, by reason of its attachment to a lanyard and lifeline or a structure, will prevent a worker from falling.

[44 FR 8577, Feb. 9, 1979]

**Authority:** Sec. 107, Contract Work Hours and Safety Standards Act (40 U.S.C. 333); secs. 4, 6, and 8, Occupational Safety and Health Act of 1970 (29 U.S.C. 653, 655, 657); Secretary of Labor's Order No. 12-71 (36 FR 8754), 8-76 (41 FR 25059), 9-83 (48 FR 35736), or 6-96 (62 FR 111) as applicable; and 29 CFR part 1911.

## §1926.150 Fire protection.

1926.150(a) ***General requirements.*** (1) The employer shall be responsible for the development of a fire protection program to be followed throughout all phases of the construction and demolition work, and he shall provide for the firefighting equipment as specified in this subpart. As fire hazards occur, there shall be no delay in providing the necessary equipment.

1926.150(a)(2) Access to all available firefighting equipment shall be maintained at all times.

1926.150(a)(3) All firefighting equipment, provided by the employer, shall be conspicuously located.

1926.150(a)(4) All firefighting equipment shall be periodically inspected and maintained in operating condition. Defective equipment shall be immediately replaced.

1926.150(a)(5) As warranted by the project, the employer shall provide a trained and equipped firefighting organization (Fire Brigade) to assure adequate protection to life.

1926.150(b) ***Water supply.*** (1) A temporary or permanent water supply, of sufficient volume, duration, and pressure, required to properly operate the firefighting equipment shall be made available as soon as combustible materials accumulate.

1926.150(b)(2) Where underground water mains are to be provided, they shall be installed, completed, and made available for use as soon as practicable.

1926.150(c) ***Portable firefighting equipment***—(1) *Fire extinguishers and small hose lines.* (i) A fire extinguisher, rated not less than 2A, shall be provided for each 3,000 square feet of the protected building area, or major fraction thereof. Travel distance from any point of the protected area to the nearest fire extinguisher shall not exceed 100 feet.

1926.150(c)(1)(ii) One 55-gallon open drum of water with two fire pails may be substituted for a fire extinguisher having a 2A rating.

1926.150(c)(1)(iii) A 1/2-inch diameter garden-type hose line, not to exceed 100 feet in length and equipped with a nozzle, may be substituted for a 2A-rated fire extinguisher, providing it is capable of discharging a minimum of 5 gallons per minute with a minimum hose stream range of 30 feet horizontally. The garden-type hose lines shall be mounted on conventional racks or reels. The number and location of hose racks or reels shall be such that at least one hose stream can be applied to all points in the area.

1926.150(c)(1)(iv) One or more fire extinguishers, rated not less than 2A, shall be provided on each floor. In multistory buildings, at least one fire extinguisher shall be located adjacent to stairway.

1926.150(c)(1)(v) Extinguishers and water drums, subject to freezing, shall be protected from freezing.

1926.150(c)(1)(vi) A fire extinguisher, rated not less than 10B, shall be provided within 50 feet of wherever more than 5 gallons of flammable or combustible liquids or 5 pounds of flammable gas are being used on the jobsite. This requirement does not apply to the integral fuel tanks of motor vehicles.

1926.150(c)(1)(vii) Carbon tetrachloride and other toxic vaporizing liquid fire extinguishers are prohibited.

1926.150(c)(1)(viii) Portable fire extinguishers shall be inspected periodically and maintained in accordance with Maintenance and Use of Portable Fire Extinguishers, NFPA No. 10A-1970.

1926.150(c)(1)(ix) Fire extinguishers which have been listed or approved by a nationally recognized testing laboratory, shall be used to meet the requirements of this subpart.

### TABLE F-1—FIRE EXTINGUISHERS DATA

| | WATER TYPE | | | | FOAM | CARBON DIOXIDE | DRY CHEMICAL | | | |
|---|---|---|---|---|---|---|---|---|---|---|
| | | | | | | | SODIUM OR POTASSIUM BICARBONATE | | MULTI-PURPOSE ABC | |
| | STORED PRESSURE | CARTRIDGE OPERATED | WATER PUMP TANK | SODA ACID | FOAM | CO2 | CARTRIDGE OPERATED | STORED PRESSURE | STORED PRESSURE | CARTRIDGE OPERATED |
| CLASS A FIRES WOOD, PAPER, TRASH HAVING GLOWING EMBERS (ORDINARY A COMBUSTIBLES) | YES | YES | YES | YES | YES | NO (But will control small surface fires) | NO (But will control small surface fires) | NO (But will control small surface fires) | YES | YES |
| CLASS B FIRES FLAMMABLE LIQUIDS, GASOLINE, OIL, PAINTS GREASE, ETC. (FLAMMABLE B LIQUIDS) | NO | NO | NO | NO | YES | YES | YES | YES | YES | YES |
| CLASS C FIRES ELECTRICAL EQUIPMENT (ELECTRICAL C EQUIPMENT) | NO | NO | NO | NO | NO | YES | YES | YES | YES | YES |
| CLASS D FIRES COMBUSTIBLE METALS (COMBUSTIBLE D METALS) | SPECIAL EXTINGUISHING AGENTS APPROVED BY RECOGNIZED TESTING LABORATORIES | | | | | | | | | |
| METHOD OF OPERATION | PULL PIN-SQUEEZE HANDLE | TURN UPSIDE DOWN AND BUMP | PUMP HANDLE | TURN UPSIDE DOWN | TURN UPSIDE DOWN | PULL PIN-SQUEEZE LEVER | RUPTURE CARTRIDGE SQUEEZE LEVER | PULL PIN-SQUEEZE HANDLE | PULL PIN-SQUEEZE HANDLE | RUPTURE CARTRIDGE SQUEEZE LEVER |
| RANGE | 30'-40' | 30'-40' | 30'-40' | 30'-40' | 30'-40' | 3'-8' | 5'-20' | 5'-20' | 5'-20' | 5'-20' |
| MAINTENANCE | CHECK AIR PRESSURE GAUGE MONTHLY | WEIGH GAS CARTRIDGE ADD WATER IF REQURED ANNUALLY | DISCHARGE AND FILL WITH WATER ANNUALLY | DISCHARGE ANNUALLY—RECHARGE | DISCHARGE ANNUALLY—RECHARGE | WEIGH SEMI-ANNUALLY | WEIGH GAS CARTRIDGE—CHECK CONDITION OF DRY CHEMICAL ANNNUALLY | CHECK PRESSURE GAUGE AND CONDITION OF DRY CHEMICAL ANNUALLY | CHECK PRESSURE GAUGE AND CONDITION OF DRY CHEMICAL ANNUALLY | WEIGH GAS CARTRIDGE—CHECK CONDITION OF DRY CHEMICAL ANNNUALLY |

1926.150(c)(1)(x) Table F-1 may be used as a guide for selecting the appropriate portable fire extinguishers.

1926.150(c)(2) *Fire hose and connections.* (i) One hundred feet, or less, of 1 1/2-inch hose, with a nozzle capable of discharging water at 25 gallons or more per minute, may be substituted for a fire extinguisher rated not more than 2A in the designated area provided that the hose line can reach all points in the area.

1926.150(c)(2)(ii) If fire hose connections are not compatible with local firefighting equipment, the contractor shall provide adapters, or equivalent, to permit connections.

**1926.150(c)(2)(iii)** During demolition involving combustible materials, charged hose lines, supplied by hydrants, water tank trucks with pumps, or equivalent, shall be made available.

**1926.150(d)** ***Fixed firefighting equipment***—(1) *Sprinkler protection.* (i) If the facility being constructed includes the installation of automatic sprinkler protection, the installation shall closely follow the construction and be placed in service as soon as applicable laws permit following completion of each story.

**1926.150(d)(1)(ii)** During demolition or alterations, existing automatic sprinkler installations shall be retained in service as long as reasonable. The operation of sprinkler control valves shall be permitted only by properly authorized persons. Modification of sprinkler systems to permit alterations or additional demolition should be expedited so that the automatic protection may be returned to service as quickly as possible. Sprinkler control valves shall be checked daily at close of work to ascertain that the protection is in service.

**1926.150(d)(2)** *Standpipes.* In all structures in which standpipes are required, or where standpipes exist in structures being altered, they shall be brought up as soon as applicable laws permit, and shall be maintained as construction progresses in such a manner that they are always ready for fire protection use. The standpipes shall be provided with Siamese fire department connections on the outside of the structure, at the street level, which shall be conspicuously marked. There shall be at least one standard hose outlet at each floor.

**1926.150(e)** ***Fire alarm devices.*** (1) An alarm system, e.g., telephone system, siren, etc., shall be established by the employer whereby employees on the site and the local fire department can be alerted for an emergency.

**1926.150(e)(2)** The alarm code and reporting instructions shall be conspicuously posted at phones and at employee entrances.

**1926.150(f)** ***Fire cutoffs.*** (1) Fire walls and exit stairways, required for the completed buildings, shall be given construction priority. Fire doors, with automatic closing devices, shall be hung on openings as soon as practicable.

**1926.150(f)(2)** Fire cutoffs shall be retained in buildings undergoing alterations or demolition until operations necessitate their removal.

[44 FR 8577, Feb. 9, 1979; 44 FR 20940, Apr. 6, 1979, as amended at 58 FR 35162, June 30, 1993; 61 FR 31432, June 20, 1996]

### §1926.151 Fire prevention.

**1926.151(a)** ***Ignition hazards.*** (1) Electrical wiring and equipment for light, heat, or power purposes shall be installed in compliance with the requirements of subpart K of this part.

**1926.151(a)(2)** Internal combustion engine powered equipment shall be so located that the exhausts are well away from combustible materials. When the exhausts are piped to outside the building under construction, a clearance of at least 6 inches shall be maintained between such piping and combustible material.

**1926.151(a)(3)** Smoking shall be prohibited at or in the vicinity of operations which constitute a fire hazard, and shall be conspicuously posted: "No Smoking or Open Flame."

**1926.151(a)(4)** Portable battery powered lighting equipment, used in connection with the storage, handling, or use of flammable gases or liquids, shall be of the type approved for the hazardous locations.

**1926.151(a)(5)** The nozzle of air, inert gas, and steam lines or hoses, when used in the cleaning or ventilation of tanks and vessels that contain hazardous concentrations of flammable gases or vapors, shall be bonded to the tank or vessel shell. Bonding devices shall not be attached or detached in hazardous concentrations of flammable gases or vapors.

**1926.151(b)** ***Temporary buildings.*** (1) No temporary building shall be erected where it will adversely affect any means of exit.

**1926.151(b)(2)** Temporary buildings, when located within another building or structure, shall be of either noncombustible construction or of combustible construction having a fire resistance of not less than 1 hour.

**1926.151(b)(3)** Temporary buildings, located other than inside another building and not used for the storage, handling, or use of flammable or combustible liquids, flammable gases, explosives, or blasting agents, or similar hazardous occupancies, shall be located at a distance of not less than 10 feet from another building or structure. Groups of temporary buildings, not exceeding 2,000 square feet in aggregate, shall, for the purposes of this part, be considered a single temporary building.

**1926.151(c)** ***Open yard storage.*** (1) Combustible materials shall be piled with due regard to the stability of piles and in no case higher than 20 feet.

**1926.151(c)(2)** Driveways between and around combustible storage piles shall be at least 15 feet wide and maintained free from accumulation of rubbish, equipment, or other articles or materials. Driveways shall be so spaced that a maximum grid system unit of 50 feet by 150 feet is produced.

**1926.151(c)(3)** The entire storage site shall be kept free from accumulation of unnecessary combustible materials. Weeds and grass shall be kept down and a regular procedure provided for the periodic cleanup of the entire area.

**1926.151(c)(4)** When there is a danger of an underground fire, that land shall not be used for combustible or flammable storage.

**1926.151(c)(5)** Method of piling shall be solid wherever possible and in orderly and regular piles. No combustible material shall be stored outdoors within 10 feet of a building or structure.

**1926.151(c)(6)** Portable fire extinguishing equipment, suitable for the fire hazard involved, shall be provided at convenient, conspicuously accessible locations in the yard area. Portable fire extinguishers, rated not less than 2A, shall be placed so that maximum travel distance to the nearest unit shall not exceed 100 feet.

**1926.151(d)** ***Indoor storage.*** (1) Storage shall not obstruct, or adversely affect, means of exit.

**1926.151(d)(2)** All materials shall be stored, handled, and piled with due regard to their fire characteristics.

**1926.151(d)(3)** Noncompatible materials, which may create a fire hazard, shall be segregated by a barrier having a fire resistance of at least 1 hour.

**1926.151(d)(4)** Material shall be piled to minimize the spread of fire internally and to permit convenient access for firefighting. Stable piling shall be maintained at all times. Aisle space shall be maintained to safely accommodate the widest vehicle that may be used within the building for firefighting purposes.

**1926.151(d)(5)** Clearance of at least 36 inches shall be maintained between the top level of the stored material and the sprinkler deflectors.

**1926.151(d)(6)** Clearance shall be maintained around lights and heating units to prevent ignition of combustible materials.

**1926.151(d)(7)** A clearance of 24 inches shall be maintained around the path of travel of fire doors unless a barricade is provided, in which case no clearance is needed. Material shall not be stored within 36 inches of a fire door opening.

[44 FR 8577, Feb. 9, 1979; 44 FR 20940, Apr. 6, 1979, as amended at 51 FR 25318, July 11, 1986]

### §1926.152 Flammable and combustible liquids.

**1926.152(a)** ***General requirements.*** (1) Only approved containers and portable tanks shall be used for storage and handling of flammable and combustible liquids. Approved safety cans or Department of Transportation approved containers shall be used for the handling and use of flammable liquids in quantities of 5 gallons or less, except that this shall not apply to those flammable liquid materials which are highly viscid (extremely hard to pour), which may be used and handled in original shipping containers. For quantities of one gallon or less, the original container may be used, for storage, use and handling of flammable liquids.

**1926.152(a)(2)** Flammable or combustible liquids shall not be stored in areas used for exits, stairways, or normally used for the safe passage of people.

**1926.152(b)** ***Indoor storage of flammable and combustible liquids.*** (1) No more than 25 gallons of flammable or combustible liquids shall be stored in a room outside of an approved storage cabinet. For storage of liquefied petroleum gas, see §1926.153.

**1926.152(b)(2)** Quantities of flammable and combustible liquid in excess of 25 gallons shall be stored in an acceptable or approved cabinet meeting the following requirements:

**1926.152(b)(2)(i)** Acceptable wooden storage cabinets shall be constructed in the following manner, or equivalent: The bottom, sides, and top shall be constructed of an exterior grade of plywood at least 1 inch in thickness, which shall not break down or delaminate under standard fire test conditions. All joints shall be rabbeted and shall be fastened in two directions with flathead wood screws. When more than one door is used, there shall be a rabbeted overlap of not less than 1 inch. Steel hinges shall be mounted in such a manner as to not lose their holding capacity due to loosening or burning out of the screws when subjected to fire. Such cabinets shall be painted inside and out with fire retardant paint.

**1926.152(b)(2)(ii)** Approved metal storage cabinets will be acceptable.

**1926.152(b)(2)(iii)** Cabinets shall be labeled in conspicuous lettering, "Flammable—Keep Fire Away."

**1926.152(b)(3)** Not more than 60 gallons of flammable or 120 gallons of combustible liquids shall be stored in any one storage cabinet. Not more than three such cabinets may be located in a single storage area. Quantities in excess of this shall be stored in an inside storage room.

**1926.152(b)(4)(i)** Inside storage rooms shall be constructed to meet the required fire-resistive rating for their use. Such construction shall comply with the test specifications set forth in Standard Methods of Fire Test of Building Construction and Material, NFPA 251-1969.

**1926.152(b)(4)(ii)** Where an automatic extinguishing system is provided, the system shall be designed and installed in an approved manner. Openings to other rooms or buildings shall be provided with noncombustible liquid-tight raised sills or ramps at least 4 inches in height, or the floor in the storage area shall be at least 4 inches below the surrounding floor. Openings shall be provided with approved self-closing fire doors. The room shall be liquid-tight where the walls join the floor. A permissible alternate to the sill or ramp is an open-grated trench, inside of the room, which drains to a safe location. Where other portions of the building or other buildings are exposed, windows shall be protected as set forth in the Standard for Fire Doors and Windows, NFPA No. 80-1970, for Class E or F openings. Wood of at least 1-inch nominal thickness may be used for shelving, racks, dunnage, scuffboards, floor overlay, and similar installations.

**1926.152(b)(4)(iii)** Materials which will react with water and create a fire hazard shall not be stored in the same room with flammable or combustible liquids.

**1926.152(b)(4)(iv)** Storage in inside storage rooms shall comply with Table F-2 following:

### TABLE F-2

| Fire protection provided | Fire resistance | Maximum size | Total allowable quantities gals./ sq. ft./floor area |
|---|---|---|---|
| Yes | 2 hrs | 500 sq. ft | 10 |
| No | 2 hrs | 500 sq. ft | 4 |
| Yes | 1 hr | 150 sq. ft | 5 |
| No | 1 hr | 150 sq. ft | 2 |

**NOTE:** Fire protection system shall be sprinkler, water spray, carbon dioxide or other system approved by a nationally recognized testing laboratory for this purpose.

**1926.152(b)(4)(v)** Electrical wiring and equipment located in inside storage rooms shall be approved for Class I, Division 1, Hazardous Locations. For definition of Class I, Division 1, Hazardous Locations, see §1926.449.

**1926.152(b)(4)(vi)** Every inside storage room shall be provided with either a gravity or a mechanical exhausting system. Such system shall commence not more than 12 inches above the floor and be designed to provide for a complete change of air within the room at least 6 times per hour. If a mechanical exhausting system is used, it shall be controlled by a switch located outside of the door. The ventilating equipment and any lighting fixtures shall be operated by the same switch. An electric pilot light shall be installed adjacent to the switch if flammable liquids are dispensed within the room. Where gravity ventilation is provided, the fresh air intake, as well as the exhausting outlet from the room, shall be on the exterior of the building in which the room is located.

**1926.152(b)(4)(vii)** In every inside storage room there shall be maintained one clear aisle at least 3 feet wide. Containers over 30 gallons capacity shall not be stacked one upon the other.

**1926.152(b)(4)(viii)** Flammable and combustible liquids in excess of that permitted in inside storage rooms shall be stored outside of buildings in accordance with paragraph (c) of this section.

**1926.152(b)(5)** *Quantity.* The quantity of flammable or combustible liquids kept in the vicinity of spraying operations shall be the minimum required for operations and should ordinarily not exceed a supply for 1 day or one shift. Bulk storage of portable containers of flammable or combustible liquids shall be in a separate, constructed building detached from other important buildings or cut off in a standard manner.

**1926.152(c)** ***Storage outside buildings.*** (1) Storage of containers (not more than 60 gallons each) shall not exceed 1,100 gallons in any one pile or area. Piles or groups of containers shall be separated by a 5-foot clearance. Piles or groups of containers shall not be nearer than 20 feet to a building.

**1926.152(c)(2)** Within 200 feet of each pile of containers, there shall be a 12-foot-wide access way to permit approach of fire control apparatus.

**1926.152(c)(3)** The storage area shall be graded in a manner to divert possible spills away from buildings or other exposures, or shall be surrounded by a curb or earth dike at least 12 inches high. When curbs or dikes are used, provisions shall be made for draining off accumulations of ground or rain water, or spills of flammable or combustible liquids. Drains shall terminate at a safe location and shall be accessible to operation under fire conditions.

**1926.152(c)(4)** Outdoor portable tank storage: (i) Portable tanks shall not be nearer than 20 feet from any building. Two or more portable tanks, grouped together, having a combined capacity in excess of 2,200 gallons, shall be separated by a 5-foot-clear area. Individual portable tanks exceeding 1,100 gallons shall be separated by a 5-foot-clear area.

**1926.152(c)(4)(ii)** Within 200 feet of each portable tank, there shall be a 12-foot-wide access way to permit approach of fire control apparatus.

**1926.152(c)(5)** Storage areas shall be kept free of weeds, debris, and other combustible material not necessary to the storage.

**1926.152(c)(6)** Portable tanks, not exceeding 660 gallons, shall be provided with emergency venting and other devices, as required by chapters III and IV of NFPA 30-1969, The Flammable and Combustible Liquids Code.

**1926.152(c)(7)** Portable tanks, in excess of 660 gallons, shall have emergency venting and other devices, as required by chapters II and III of The Flammable and Combustible Liquids Code, NFPA 30-1969.

**1926.152(d)** ***Fire control for flammable or combustible liquid storage.*** (1) At least one portable fire extinguisher, having a rating of not less than 20-B units, shall be located outside of, but not more than 10 feet from, the door opening into any room used for storage of more than 60 gallons of flammable or combustible liquids.

**1926.152(d)(2)** At least one portable fire extinguisher having a rating of not less than 20-B units shall be located not less than 25 feet, nor more than 75 feet, from any flammable liquid storage area located outside.

**1926.152(d)(3)** When sprinklers are provided, they shall be installed in accordance with the Standard for the Installation of Sprinkler Systems, NFPA 13-1969.

**1926.152(d)(4)** At least one portable fire extinguisher having a rating of not less than 20-B:C units shall be provided on all tank trucks or other vehicles used for transporting and/or dispensing flammable or combustible liquids.

**1926.152(e)** ***Dispensing liquids.*** (1) Areas in which flammable or combustible liquids are transferred at one time, in quantities greater than 5 gallons from one tank or container to another tank or container, shall be separated from other operations by 25 feet distance or by construction having a fire resistance of at least 1 hour. Drainage or other means shall be provided to control spills. Adequate natural or mechanical ventilation shall be provided to maintain the concentration of flammable vapor at or below 10 percent of the lower flammable limit.

**1926.152(e)(2)** Transfer of flammable liquids from one container to another shall be done only when containers are electrically interconnected (bonded).

**1926.152(e)(3)** Flammable or combustible liquids shall be drawn from or transferred into vessels, containers, or tanks within a building or outside only through a closed piping system, from safety cans, by means of a device drawing through the top, or from a container, or portable tanks, by gravity or pump, through an approved self-closing valve. Transferring by means of air pressure on the container or portable tanks is prohibited.

**1926.152(e)(4)** The dispensing units shall be protected against collision damage.

**1926.152(e)(5)** Dispensing devices and nozzles for flammable liquids shall be of an approved type.

**1926.152(f)** ***Handling liquids at point of final use.*** (1) Flammable liquids shall be kept in closed containers when not actually in use.

**1926.152(f)(2)** Leakage or spillage of flammable or combustible liquids shall be disposed of promptly and safely.

**1926.152(f)(3)** Flammable liquids may be used only where there are no open flames or other sources of ignition within 50 feet of the operation, unless conditions warrant greater clearance.

**1926.152(g)** ***Service and refueling areas.*** (1) Flammable or combustible liquids shall be stored in approved closed containers, in tanks located underground, or in aboveground portable tanks.

**1926.152(g)(2)** The tank trucks shall comply with the requirements covered in the Standard for Tank Vehicles for Flammable and Combustible Liquids, NFPA No. 385-1966.

**1926.152(g)(3)** The dispensing hose shall be an approved type.

**1926.152(g)(4)** The dispensing nozzle shall be an approved automatic-closing type without a latch-open device.

**1926.152(g)(5)** Underground tanks shall not be abandoned.

**1926.152(g)(6)** Clearly identified and easily accessible switch(es) shall be provided at a location remote from dispensing devices to shut off the power to all dispensing devices in the event of an emergency.

**1926.152(g)(7)(i)** Heating equipment of an approved type may be installed in the lubrication or service area where there is no dispensing or transferring of flammable liquids, provided the bottom of the heating unit is at least 18 inches above the floor and is protected from physical damage.

**1926.152(g)(7)(ii)** Heating equipment installed in lubrication or service areas, where flammable liquids are dispensed, shall be of an approved type for garages, and shall be installed at least 8 feet above the floor.

**1926.152(g)(8)** There shall be no smoking or open flames in the areas used for fueling, servicing fuel systems for internal combustion engines, receiving or dispensing of flammable or combustible liquids.

**1926.152(g)(9)** Conspicuous and legible signs prohibiting smoking shall be posted.

**1926.152(g)(10)** The motors of all equipment being fueled shall be shut off during the fueling operation.

**1926.152(g)(11)** Each service or fueling area shall be provided with at least one fire extinguisher having a rating of not less than 20-B:C located so that an extinguisher will be within 75 feet of each pump, dispenser, underground fill pipe opening, and lubrication or service area.

**1926.152(h)** ***Scope.*** This section applies to the handling, storage, and use of flammable and combustible liquids with a flashpoint below 200° (93.33°C). This section does not apply to:

**1926.152(h)(1)** Bulk transportation of flammable and combustible liquids; and

**1926.152(h)(2)** Storage, handling, and use of fuel oil tanks and containers connected with oil burning equipment.

**1926.152(i)** ***Tank storage***—(1) *Design and construction of tanks*—(i) *Materials.* (A) Tanks shall be built of steel except as provided in paragraphs (i)(1)(i) (B) through (E) of this section.

**1926.152(i)(1)(i)(B)** Tanks may be built of materials other than steel for installation underground or if required by the properties of the liquid stored. Tanks located above ground or inside buildings shall be of noncombustible construction.

**1926.152(i)(1)(i)(C)** Tanks built of materials other than steel shall be designed to specifications embodying principles recognized as good engineering design for the material used.

**1926.152(i)(1)(i)(D)** Unlined concrete tanks may be used for storing flammable or combustible liquids having a gravity of 40°API or heavier. Concrete tanks with special lining may be used for other services provided the design is in accordance with sound engineering practice.

**1926.152(i)(1)(i)(E)** [Reserved]

**1926.152(i)(1)(i)(F)** Special engineering consideration shall be required if the specific gravity of the liquid to be stored exceeds that of water or if the tanks are designed to contain flammable or combustible liquids at a liquid temperature below 0°F.

**1926.152(i)(1)(ii)** *Fabrication.* (A) [Reserved]

**1926.152(i)(1)(ii)(B)** Metal tanks shall be welded, riveted, and caulked, brazed, or bolted, or constructed by use of a combination of these methods. Filler metal used in brazing shall be nonferrous metal or an alloy having a melting point above 1000°F. and below that of the metal joined.

**1926.152(i)(1)(iii)** *Atmospheric tanks.* (A) Atmospheric tanks shall be built in accordance with acceptable good standards of design. Atmospheric tanks may be built in accordance with:

**1926.152(i)(1)(iii)(A)(*1*)** Underwriters' Laboratories, Inc., Subjects No. 142, Standard for Steel Aboveground Tanks for Flammable and Combustible Liquids, 1968; No. 58, Standard for Steel Underground Tanks for Flammable and Combustible Liquids, Fifth Edition, December 1961; or No. 80, Standard for Steel Inside Tanks for Oil-Burner Fuel, September 1963.

**1926.152(i)(1)(iii)(A)(2)** American Petroleum Institute Standards No. 12A, Specification for Oil Storage Tanks with Riveted Shells, Seventh Edition, September 1951, or No. 650, Welded Steel Tanks for Oil Storage, Third Edition, 1966.

**1926.152(i)(1)(iii)(A)(3)** American Petroleum Institute Standards No. 12B, Specification for Bolted Production Tanks, Eleventh Edition, May 1958, and Supplement 1, March 1962; No. 12D, Specification for Large Welded Production Tanks, Seventh Edition, August 1957; or No. 12F, Specification for Small Welded Production Tanks, Fifth Edition, March 1961. Tanks built in accordance with these standards shall be used only as production tanks for storage of crude petroleum in oil-producing areas.

**1926.152(i)(1)(iii)(B)** Tanks designed for underground service not exceeding 2,500 gallons (9,462.5 L) capacity may be used aboveground.

**1926.152(i)(1)(iii)(C)** Low-pressure tanks and pressure vessels may be used as atmospheric tanks.

**1926.152(i)(1)(iii)(D)** Atmospheric tanks shall not be used for the storage of a flammable or combustible liquid at a temperature at or above its boiling point.

**1926.152(i)(1)(iv)** *Low pressure tanks.* (A) The normal operating pressure of the tank shall not exceed the design pressure of the tank.

**1926.152(i)(1)(iv)(B)** Low-pressure tanks shall be built in accordance with acceptable standards of design. Low-pressure tanks may be built in accordance with:

**1926.152(i)(1)(iv)(B)(*1*)** American Petroleum Institute Standard No. 620. Recommended Rules for the Design and Construction of Large, Welded, Low-Pressure Storage Tanks, Third Edition, 1966.

**1926.152(i)(1)(iv)(B)(2)** The principles of the Code for Unfired Pressure Vessels, Section VIII of the ASME Boiler and Pressure Vessels Code, 1968.

**1926.152(i)(1)(iv)(C)** Atmospheric tanks built according to Underwriters' Laboratories, Inc., requirements in paragraph (i)(1)(iii)(A) of this section and shall be limited to 2.5 p.s.i.g. under emergency venting conditions.

This paragraph may be used for operating pressures not exceeding 1 p.s.i.g.

**1926.152(i)(1)(iv)(D)** Pressure vessels may be used as low-pressure tanks.

**1926.152(i)(1)(v)** *Pressure vessels.* (A) The normal operating pressure of the vessel shall not exceed the design pressure of the vessel.

**1926.152(i)(1)(v)(B)** Pressure vessels shall be built in accordance with the Code for Unfired Pressure Vessels, Section VIII of the ASME Boiler and Pressure Vessel Code 1968.

**1926.152(i)(1)(vi)** *Provisions for internal corrosion.* When tanks are not designed in accordance with the American Petroleum Institute, American Society of Mechanical Engineers, or the Underwriters' Laboratories, Inc.'s, standards, or if corrosion is anticipated beyond that provided for in the design formulas used, additional metal thickness or suitable protective coatings or linings shall be provided to compensate for the corrosion loss expected during the design life of the tank.

**1926.152(i)(2)** *Installation of outside aboveground tanks.* (i) [Reserved]

**1926.152(i)(2)(ii)** *Spacing (shell-to-shell) between aboveground tanks.* (A) The distance between any two flammable or combustible liquid storage tanks shall not be less than 3 feet (0.912 m).

**1926.152(i)(2)(ii)(B)** Except as provided in paragraph (i)(2)(ii)(C) of this section, the distance between any two adjacent tanks shall not be less than one-sixth the sum of their diameters. When the diameter of one tank is less than one-half the diameter of the adjacent tank, the distance between the two tanks shall not be less than one-half the diameter of the smaller tank.

**1926.152(i)(2)(ii)(C)** Where crude petroleum in conjunction with production facilities are located in noncongested areas and have capacities not exceeding 126,000 gallons (3,000 barrels), the distance between such tanks shall not be less than 3 feet (0.912 m).

**1926.152(i)(2)(ii)(D)** Where unstable flammable or combustible liquids are stored, the distance between such tanks shall not be less than one-half the sum of their diameters.

**1926.152(i)(2)(ii)(E)** When tanks are compacted in three or more rows or in an irregular pattern, greater spacing or other means shall be provided so that inside tanks are accessible for firefighting purposes.

**1926.152(i)(2)(ii)(F)** The minimum separation between a liquefied petroleum gas container and a flammable or combustible liquid storage tank shall be

20 feet (6.08 m), except in the case of flammable or combustible liquid tanks operating at pressures exceeding 2.5 p.s.i.g. or equipped with emergency venting which will permit pressures to exceed 2.5 p.s.i.g. in which case the provisions of paragraphs (i)(2)(ii) (A) and (B) of this section shall apply. Suitable means shall be taken to prevent the accumulation of flammable or combustible liquids under adjacent liquefied petroleum gas containers such as by diversion curbs or grading. When flammable or combustible liquid storage tanks are within a diked area, the liquefied petroleum gas containers shall be outside the diked area and at least 10 feet (3.04 m) away from the centerline of the wall of the diked area. The foregoing provisions shall not apply when liquefied petroleum gas containers of 125 gallons (473.125 L) or less capacity are installed adjacent to fuel oil supply tanks of 550 gallons (2,081.75 L) or less capacity.

**1926.152(i)(2)(iii)** [Reserved]

**1926.152(i)(2)(iv)** *Normal venting for aboveground tanks.* (A) Atmospheric storage tanks shall be adequately vented to prevent the development of vacuum or pressure sufficient to distort the roof of a cone roof tank or exceeding the design pressure in the case of other atmospheric tanks, as a result of filling or emptying, and atmospheric temperature changes.

**1926.152(i)(2)(iv)(B)** Normal vents shall be sized either in accordance with: *(1)* The American Petroleum Institute Standard 2000 (1968), Venting Atmospheric and Low-Pressure Storage Tanks; or *(2)* other accepted standard; or *(3)* shall be at least as large as the filling or withdrawal connection, whichever is larger but in no case less than 1 1/4 inch (3.175 cm) nominal inside diameter.

**1926.152(i)(2)(iv)(C)** Low-pressure tanks and pressure vessels shall be adequately vented to prevent development of pressure or vacuum, as a result of filling or emptying and atmospheric temperature changes, from exceeding the design pressure of the tank or vessel. Protection shall also be provided to prevent overpressure from any pump discharging into the tank or vessel when the pump discharge pressure can exceed the design pressure of the tank or vessel.

**1926.152(i)(2)(iv)(D)** If any tank or pressure vessel has more than one fill or withdrawal connection and simultaneous filling or withdrawal can be made, the vent size shall be based on the maximum anticipated simultaneous flow.

**1926.152(i)(2)(iv)(E)** Unless the vent is designed to limit the internal pressure 2.5 p.s.i. or less, the outlet of vents and vent drains shall be arranged to discharge in such a manner as to prevent localized overheating of any part of the tank in the event vapors from such vents are ignited.

**1926.152(i)(2)(iv)(F)** Tanks and pressure vessels storing Class IA liquids shall be equipped with venting devices which shall be normally closed except when venting to pressure or vacuum conditions. Tanks and pressure vessels storing Class IB and IC liquids shall be equipped with venting devices which shall be normally closed except when venting under pressure or vacuum conditions, or with approved flame arresters.

*Exemption:* Tanks of 3,000 bbls. (84 $m^3$) capacity or less containing crude petroleum in crude-producing areas; and, outside aboveground atmospheric tanks under 1,000 gallons (3,785 L) capacity containing other than Class IA flammable liquids may have open vents. (See paragraph (i)(2)(vi)(B) of this section.)

**1926.152(i)(2)(iv)(G)** Flame arresters or venting devices required in paragraph (i)(2)(iv)(F) of this section may be omitted for Class IB and IC liquids where conditions are such that their use may, in case of obstruction, result in tank damage.

**1926.152(i)(2)(v)** *Emergency relief venting for fire exposure for aboveground tanks.* (A) Every aboveground storage tank shall have some form of construction or device that will relieve excessive internal pressure caused by exposure fires.

**1926.152(i)(2)(v)(B)** In a vertical tank the construction referred to in paragraph (i)(2)(v)(A) of this section may take the form of a floating roof, lifter roof, a weak roof-to-shell seam, or other approved pressure relieving construction. The weak roof-to-shell seam shall be constructed to fail preferential to any other seam.

**1926.152(i)(2)(v)(C)** Where entire dependence for emergency relief is placed upon pressure relieving devices, the total venting capacity of both normal and emergency vents shall be enough to prevent rupture of the shell or bottom of the tank if vertical, or of the shell or heads if horizontal. If unstable liquids are stored, the effects of heat or gas resulting from polymerization, decomposition, condensation, or self-reactivity shall be taken into account. The total capacity of both normal and emergency venting devices shall be not less than that derived from Table F-10 except as provided in paragraph (i)(2)(v) (E) or (F) of this section. Such device may be a self-closing manhole cover, or one using long bolts that permit the cover to lift under internal pressure, or an additional or larger relief valve or valves. The wetted area of the tank shall be calculated on the basis of 55 percent of the total exposed area of a sphere or spheroid, 75 percent of the total exposed area of a horizontal tank and the first 30 feet (9.12 m) above grade of the exposed shell area of a vertical tank.

### TABLE F-10—WETTED AREA VERSUS CUBIC FEET (METERS) FREE AIR PER HOUR

[14.7 psia and 60 ° F. (15.55 ° C)]

| Square feet | ($m^2$) | CFH | ($m^3$H) |
|---|---|---|---|
| 20 | (1.84) | 21,100 | (590.8) |
| 30 | (2.76) | 31,600 | (884.8) |
| 40 | (3.68) | 42,100 | (1,178.8) |
| 50 | (4.6) | 52,700 | (1,475.6) |
| 60 | (5.52) | 63,200 | (1,769.6) |
| 70 | (6.44) | 73,700 | (2,063.6) |
| 80 | (7.36) | 84,200 | (2,357.6) |
| 90 | (8.28) | 94,800 | (2,654.4) |
| 100 | (9.2) | 105,000 | (2,940) |
| 120 | (11.04) | 126,000 | (3,528) |
| 140 | (12.88) | 147,000 | (4,116) |
| 160 | (14.72) | 168,000 | (4,704) |
| 180 | (16.56) | 190,000 | (5,320) |
| 200 | (18.4) | 211,000 | (5,908) |
| 250 | (23) | 239,000 | (6,692) |
| 300 | (27.6) | 265,000 | (7,420) |
| 350 | (32.2) | 288,000 | (8,064) |
| 400 | (36.8) | 312,000 | (8,736) |
| 500 | (46) | 354,000 | (9,912) |
| 600 | (55.2) | 392,000 | (10,976) |
| 700 | (64.4) | 428,000 | (11,984) |
| 800 | (73.6) | 462,000 | (12,936) |
| 900 | (82.8) | 493,000 | (13,804) |
| 1,000 | (90.2) | 524,000 | (14,672) |
| 1,200 | (110.4) | 557,000 | (15,596) |
| 1,400 | (128.8) | 587,000 | (16,436) |
| 1,600 | (147.2) | 614,000 | (17,192) |
| 1,800 | (165.6) | 639,000 | (17,892) |
| 2,000 | (180.4) | 662,000 | (18,536) |
| 2,400 | (220.8) | 704,000 | (19,712) |
| 2,800 | (257.6) | 742,000 | (20,776) |
| | and over | | |

**1926.152(i)(2)(v)(D)** For tanks and storage vessels designed for pressure over 1 p.s.i.g., the total rate of venting shall be determined in accordance with Table F-10, except that when the exposed wetted area of the surface is greater than 2,800 square feet (257.6 $m^2$), the total rate of venting shall be calculated by the following formula:

$$CFH = 1{,}107A^{0.82}$$

Where:

*CFH* = Venting requirement, in cubic feet (meters) of free air per hour.

A = Exposed wetted surface, in square feet ($m^2$).

**Note:** The foregoing formula is based on $Q=21{,}000A^{0.82}$.

**1926.152(i)(2)(v)(E)** The total emergency relief venting capacity for any specific stable liquid may be determined by the following formula:

$$V = 1337 \div L\sqrt{M}$$

*V* = Cubic feet (meters) of free air per hour from Table F-10.

*L* = Latent heat of vaporization of specific liquid in B.t.u. per pound.

*M* = Molecular weight of specific liquids.

**1926.152(i)(2)(v)(F)** The required airflow rate of paragraph (i)(2)(v) (C) or (E) of this section may be multiplied by the appropriate factor listed in the following schedule when protection is provided as indicated. Only one factor may be used for any one tank.

0.5 for drainage in accordance with paragraph (i)(2)(vii)(B) of this section for tanks over 200 square feet (18.4 $m^2$) of wetted area.

0.3 for approved water spray.

0.3 for approved insulation.

0.15 for approved water spray with approved insulation.

**1926.152(i)(2)(v)(G)** The outlet of all vents and vent drains on tanks equipped with emergency venting to permit pressures exceeding 2.5 p.s.i.g. shall be arranged to discharge in such a way as to prevent localized overheating of any part of the tank, in the event vapors from such vents are ignited.

**1926.152(i)(2)(v)(H)** Each commercial tank venting device shall have stamped on it the opening pressure, the pressure at which the valve reaches the full open position, and the flow capacity at the latter pressure, expressed in cubic feet (meters) per hour of air at 60°F. (15.55°C) and at a pressure of 14.7 p.s.i.a.

**1926.152(i)(2)(v)(I)** The flow capacity of tank venting devices 12 inches (30.48 cm) and smaller in nominal pipe size shall be determined by actual test of each type and size of vent. These flow tests may be conducted by the manufacturer if certified by a qualified impartial observer, or may be conducted by an outside agency. The flow capacity of tank venting devices larger than 12 inches (30.48 cm) nominal pipe size, including manhole covers with long bolts or equivalent, may be calculated provided that the opening pressure is actually measured, the rating pressure and corresponding free orifice area are stated, the word "calculated" appears on the nameplate, and the computation is based on a flow coefficient of 0.5 applied to the rated orifice area.

**1926.152(i)(2)(vi)** *Vent piping for aboveground tanks.* (A) Vent piping shall be constructed in accordance with paragraph (c) of this section.

**1926.152(i)(2)(vi)(B)** Where vent pipe outlets for tanks storing Class I liquids are adjacent to buildings or public ways, they shall be located so that the vapors are released at a safe point outside of buildings and not less than 12 feet (3.648 m) above the adjacent ground level. In order to aid their dispersion, vapors shall be discharged upward or horizontally away from closely adjacent walls. Vent outlets shall be located so that flammable vapors will not be trapped by eaves or other obstructions and shall be at least 5 feet (1.52 m) from building openings.

**1926.152(i)(2)(vi)(C)** When tank vent piping is manifolded, pipe sizes shall be such as to discharge, within the pressure limitations of the system, the vapors they may be required to handle when manifolded tanks are subject to the same fire exposure.

**1926.152(i)(2)(vii)** *Drainage, dikes, and walls for aboveground tanks*—(A) *Drainage and diked areas.* The area surrounding a tank or a group of tanks shall be provided with drainage as in paragraph (i)(2)(vii)(B) of this section, or shall be diked as provided in (i)(2)(vii)(C) of this section, to prevent accidental discharge of liquid from endangering adjoining property or reaching waterways.

**1926.152(i)(2)(vii)(B)** *Drainage.* Where protection of adjoining property or waterways is by means of a natural or manmade drainage system, such systems shall comply with the following:

**1926.152(i)(2)(vii)(B)(*1*)** [Reserved]

**1926.152(i)(2)(vii)(B)(*2*)** The drainage system shall terminate in vacant land or other area or in an impounding basin having a capacity not smaller than that of the largest tank served. This termination area and the route of the drainage system shall be so located that, if the flammable or combustible liquids in the drainage system are ignited, the fire will not seriously expose tanks or adjoining property.

**1926.152(i)(2)(vii)(C)** *Diked areas.* Where protection of adjoining property or waterways is accomplished by retaining the liquid around the tank by means of a dike, the volume of the diked area shall comply with the following requirements:

**1926.152(i)(2)(vii)(C)(*1*)** Except as provided in paragraph (i)(2)(vii)(C)(2) of this section, the volumetric capacity of the diked area shall not be less than the greatest amount of liquid that can be released from the largest tank within the diked area, assuming a full tank. The capacity of the diked area enclosing more than one tank shall be calculated by deducting the volume of the tanks other than the largest tank below the height of the dike.

**1926.152(i)(2)(vii)(C)(*2*)** For a tank or group of tanks with fixed roofs containing crude petroleum with boilover characteristics, the volumetric capacity of the diked area shall be not less than the capacity of the largest tank served by the enclosure, assuming a full tank. The capacity of the diked enclosure shall be calculated by deducting the volume below the height of the dike of all tanks within the enclosure.

**1926.152(i)(2)(vii)(C)(*3*)** Walls of the diked area shall be of earth, steel, concrete or solid masonry designed to be liquidtight and to withstand a full hydrostatic head. Earthen walls 3 feet (0.912 m) or more in height shall have a flat section at the top not less than 2 feet (0.608 m) wide. The slope of an earthen wall shall be consistent with the angle of repose of the material of which the wall is constructed.

**1926.152(i)(2)(vii)(C)(*4*)** The walls of the diked area shall be restricted to an average height of 6 feet (1.824 m) above interior grade.

**1926.152(i)(2)(vii)(C)(*5*)** [Reserved]

**1926.152(i)(2)(vii)(C)(*6*)** No loose combustible material, empty or full drum or barrel, shall be permitted within the diked area.

**1926.152(i)(2)(viii)** *Tank openings other than vents for aboveground tanks.*

**1926.152(i)(2)(viii)(A)-(C)** [Reserved]

**1926.152(i)(2)(viii)(D)** Openings for gaging shall be provided with a vaportight cap or cover.

**1926.152(i)(2)(viii)(E)** For Class IB and Class IC liquids other than crude oils, gasolines, and asphalts, the fill pipe shall be so designed and installed as to minimize the possibility of generating static electricity. A fill pipe entering the top of a tank shall terminate within 6 inches (15.24 cm) of the bottom of the tank and shall be installed to avoid excessive vibration.

**1926.152(i)(2)(viii)(F)** Filling and emptying connections which are made and broken shall be located outside of buildings at a location free from any source of ignition and not less than 5 feet (1.52 m) away from any building opening. Such connection shall be closed and liquidtight when not in use. The connection shall be properly identified.

**1926.152(i)(3)** *Installation of underground tanks*—(i) *Location.* Excavation for underground storage tanks shall be made with due care to avoid undermining of foundations of existing structures. Underground tanks or tanks under buildings shall be so located with respect to existing building foundations and supports that the loads carried by the latter cannot be transmitted to the tank. The distance from any part of a tank storing Class I liquids to the nearest wall of any basement or pit shall be not less than 1 foot (0.304 m), and to any property line that may be built upon, not less than 3 feet (0.912 m). The distance from any part of a tank storing Class II or Class III liquids to the nearest wall of any basement, pit or property line shall be not less than 1 foot (0.304 m).

**1926.152(i)(3)(ii)** *Depth and cover.* Underground tanks shall be set on firm foundations and surrounded with at least 6 inches (15.24 cm) of noncorrosive, inert materials such as clean sand, earth, or gravel well tamped in place. The tank shall be placed in the hole with care since dropping or rolling the tank into the hole can break a weld, puncture or damage the tank, or scrape off the protective coating of coated tanks. Tanks shall be covered with a minimum of 2 feet (0.608 m) of earth, or shall be covered with not less than 1 foot (0.304 m) of earth, on top of which shall be placed a slab of reinforced concrete not less than 4 inches (10.16 cm) thick. When underground tanks are, or are likely to be, subject to traffic, they shall be protected against damage from vehicles passing over them by at least 3 feet (0.912 m) of earth cover, or 18 inches (45.72 cm) of well-tamped earth, plus 6 inches (15.24 cm) of reinforced concrete or 8 inches (20.32 cm) of asphaltic concrete. When asphaltic or reinforced concrete paving is used as part of the protection, it shall extend at least 1 foot (0.304 m) horizontally beyond the outline of the tank in all directions.

**1926.152(i)(3)(iii)** *Corrosion protection.* Corrosion protection for the tank and its piping shall be provided by one or more of the following methods:

**1926.152(i)(3)(iii)(A)** Use of protective coatings or wrappings;

**1926.152(i)(3)(iii)(B)** Cathodic protection; or,

**1926.152(i)(3)(iii)(C)** Corrosion resistant materials of construction.

**1926.152(i)(3)(iv)** *Vents.* (A) Location and arrangement of vents for Class I liquids. Vent pipes from tanks storing Class I liquids shall be so located that the discharge point is outside of buildings, higher than the fill pipe opening, and not less than 12 feet (3.648 m) above the adjacent ground level. Vent pipes shall discharge only upward in order to disperse vapors. Vent pipes 2 inches (5.08 cm) or less in nominal inside diameter shall not be obstructed by devices that will cause excessive back pressure. Vent pipe outlets shall be so located that flammable vapors will not enter building openings, or be trapped under eaves or other obstructions. If the vent pipe is less than 10 feet (3.04 m) in length, or greater than 2 inches (5.08 cm) in nominal inside diameter, the outlet shall be provided with a vacuum and pressure relief device or there shall be an approved flame arrester located in the vent line at the outlet or within the approved distance from the outlet.

**1926.152(i)(3)(iv)(B)** Size of vents. Each tank shall be vented through piping adequate in size to prevent blow-back of vapor or liquid at the fill opening while the tank is being filled. Vent pipes shall be not less than 1 1/4 inch (3.175 cm) nominal inside diameter.

## TABLE F-11—VENT LINE DIAMETERS

| *Maximum flow GPM (L)* | *Pipe length* [1] | | | | | |
|---|---|---|---|---|---|---|
| | *50 feet* | *(15.2 m)* | *100 feet* | *(30.4 m)* | *200 feet* | *(60.8 m)* |
| | Inches | (cm) | Inches | (cm) | Inches | (cm) |
| 100 (378.5) | 1 1/4 | (3.175) | 1 1/4 | (3.175) | 1 1/4 | (3.175) |
| 200 (757) | 1 1/4 | (3.175) | 1 1/4 | (3.175) | 1 1/4 | (3.175) |
| 300 (1,135.5) | 1 1/4 | (3.175) | 1 1/4 | (3.175) | 1 1/2 | (3.81) |
| 400 (1,514) | 1 1/4 | (3.175) | 1 1/2 | (3.81) | 2 | (5.08) |
| 500 (1,892.5) | 1 1/2 | (3.81) | 1 1/2 | (3.81) | 2 | (5.08) |
| 600 (2,271) | 1 1/2 | (3.81) | 2 | (5.08) | 2 | (5.08) |
| 700 (2,649.5) | 2 | (5.08) | 2 | (5.08) | 2 | (5.08) |
| 800 (3,028) | 2 | (5.08) | 2 | (5.08) | 3 | (7.62) |
| 900 (3,406.5) | 2 | (5.08) | 2 | (5.08) | 3 | (7.62) |
| 1.000 (3,785) | 2 | (5.08) | 2 | (5.08) | 3 | (7.62) |

[1] Vent lines of 50 ft. (15.2 m), 100 ft. (30.4 m), and 200 ft. (60.8 m) of pipe plus 7 ells.

**1926.152(i)(3)(iv)(C)** Location and arrangement of vents for Class II or Class III liquids. Vent pipes from tanks storing Class II or Class III flammable liquids shall terminate outside of the building and higher than the fill pipe opening. Vent outlets shall be above normal snow level. They may be fitted with return bends, coarse screens or other devices to minimize ingress of foreign material.

**1926.152(i)(3)(iv)(D)** Vent piping shall be constructed in accordance with paragraph (3)(iv)(C) of this section. Vent pipes shall be so laid as to drain toward the tank without sags or traps in which liquid can collect. They shall be located so that they will not be subjected to physical damage. The tank end of the vent pipe shall enter the tank through the top.

**1926.152(i)(3)(iv)(E)** When tank vent piping is manifolded, pipe sizes shall be such as to discharge, within the pressure limitations of the system, the vapors they may be required to handle when manifolded tanks are filled simultaneously.

**1926.152(i)(3)(v)** *Tank openings other than vents.* (A) Connections for all tank openings shall be vapor or liquid tight.

**1926.152(i)(3)(v)(B)** Openings for manual gaging, if independent of the fill pipe, shall be provided with a liquid-tight cap or cover. If inside a building, each such opening shall be protected against liquid overflow and possible vapor release by means of a spring loaded check valve or other approved device.

**1926.152(i)(3)(v)(C)** Fill and discharge lines shall enter tanks only through the top. Fill lines shall be sloped toward the tank.

**1926.152(i)(3)(v)(D)** For Class IB and Class IC liquids other than crude oils, gasolines, and asphalts, the fill pipe shall be so designed and installed as to minimize the possibility of generating static electricity by terminating within 6 inches (15.24 cm) of the bottom of the tank.

**1926.152(i)(3)(v)(E)** Filling and emptying connections which are made and broken shall be located outside of buildings at a location free from any source of ignition and not less than 5 feet (1.52 m) away from any building opening. Such connection shall be closed and liquidtight when not in use. The connection shall be properly identified.

**1926.152(i)(4)** *Installation of tanks inside of buildings—*(i) *Location.* Tanks shall not be permitted inside of buildings except as provided in paragraphs (e), (g), (h), or (i) of this section.

**1926.152(i)(4)(ii)** *Vents.* Vents for tanks inside of buildings shall be as provided in paragraphs (i)(2) (iv), (v), (vi)(B), and (3)(iv) of this section, except that emergency venting by the use of weak roof seams on tanks shall not be permitted. Vents shall discharge vapors outside the buildings.

**1926.152(i)(4)(iii)** *Vent piping.* Vent piping shall be constructed in accordance with paragraph (c) of this section.

**1926.152(i)(4)(iv)** *Tank openings other than vents.* (A) Connections for all tank openings shall be vapor or liquidtight. Vents are covered in paragraph (i)(4)(ii) of this section.

**1926.152(i)(4)(iv)(B)** Each connection to a tank inside of buildings through which liquid can normally flow shall be provided with an internal or an external valve located as close as practical to the shell of the tank. Such valves, when external, and their connections to the tank shall be of steel except when the chemical characteristics of the liquid stored are incompatible with steel. When materials other than steel are necessary, they shall be suitable for the pressures, structural stresses, and temperatures involved, including fire exposures.

**1926.152(i)(4)(iv)(C)** Flammable or combustible liquid tanks located inside of buildings, except in one-story buildings designed and protected for flammable or combustible liquid storage, shall be provided with an automatic-closing heat-actuated valve on each withdrawal connection below the liquid level, except for connections used for emergency disposal, to prevent continued flow in the event of fire in the vicinity of the tank. This function may be incorporated in the valve required in paragraph (i)(4)(iv)(B) of this section, and if a separate valve, shall be located adjacent to the valve required in paragraph (i)(4)(iv)(B) of this section.

**1926.152(i)(4)(iv)(D)** Openings for manual gaging, if independent of the fill pipe (see paragraph (i)(4)(iv)(F) of this section), shall be provided with a vaportight cap or cover. Each such opening shall be protected against liquid overflow and possible vapor release by means of a spring loaded check valve or other approved device.

**1926.152(i)(4)(iv)(E)** For Class IB and Class IC liquids other than crude oils, gasolines, and asphalts, the fill pipe shall be so designed and installed as to minimize the possibility of generating static electricity by terminating within 6 inches (15.24 cm) of the bottom of the tank.

**1926.152(i)(4)(iv)(F)** The fill pipe inside of the tank shall be installed to avoid excessive vibration of the pipe.

**1926.152(i)(4)(iv)(G)** The inlet of the fill pipe shall be located outside of buildings at a location free from any source of ignition and not less than 5 feet (1.52 m) away from any building opening. The inlet of the fill pipe shall be closed and liquidtight when not in use. The fill connection shall be properly identified.

**1926.152(i)(4)(iv)(H)** Tanks inside buildings shall be equipped with a device, or other means shall be provided, to prevent overflow into the building.

**1926.152(i)(5)** *Supports, foundations, and anchorage for all tank locations—*(i) *General.* Tank supports shall be installed on firm foundations. Tank supports shall be of concrete, masonry, or protected steel. Single wood timber supports (not cribbing) laid horizontally may be used for outside aboveground tanks if not more than 12 inches (30.48 cm) high at their lowest point.

**1926.152(i)(5)(ii)** *Fire resistance.* Steel supports or exposed piling shall be protected by materials having a fire resistance rating of not less than 2 hours, except that steel saddles need not be protected if less than 12 inches (30.48 cm) high at their lowest point. Water spray protection or its equivalent may be used in lieu of fire-resistive materials to protect supports.

**1926.152(i)(5)(iii)** *Spheres.* The design of the supporting structure for tanks such as spheres shall receive special engineering consideration.

**1926.152(i)(5)(iv)** *Load distribution.* Every tank shall be so supported as to prevent the excessive concentration of loads on the supporting portion of the shell.

**1926.152(i)(5)(v)** *Foundations.* Tanks shall rest on the ground or on foundations made of concrete, masonry, piling, or steel. Tank foundations shall be designed to minimize the possibility of uneven settling of the tank and to minimize corrosion in any part of the tank resting on the foundation.

**1926.152(i)(5)(vi)** *Flood areas.* Where a tank is located in an area that may be subjected to flooding, the applicable precautions outlined in this subdivision shall be observed.

**1926.152(i)(5)(vi)(A)** No aboveground vertical storage tank containing a flammable or combustible liquid shall be located so that the allowable liquid level within the tank is below the established maximum flood stage, unless the tank is provided with a guiding structure such as described in paragraphs (i)(5)(vi) (M), (N), and (O) of this section.

**1926.152(i)(5)(vi)(B)** Independent water supply facilities shall be provided at locations where there is no ample and dependable public water supply available for loading partially empty tanks with water.

**1926.152(i)(5)(vi)(C)** In addition to the preceding requirements, each tank so located that more than 70 percent, but less than 100 percent, of its allowable liquid storage capacity will be submerged at the established maximum flood stage, shall be safeguarded by one of the following methods: Tank shall be raised, or its height shall be increased, until its top extends above the maximum flood stage a distance equivalent to 30 percent or more of its allowable liquid storage capacity: *Provided, however,* That the submerged part of the tank shall not exceed two and one-half times the diameter. Or, as an alternative to the foregoing, adequate noncombustible structural guides, designed to permit the tank to float vertically without loss of product, shall be provided.

**1926.152(i)(5)(vi)(D)** Each horizontal tank so located that more than 70 percent of its storage capacity will be submerged at the established flood stage, shall be anchored, attached to a foundation of concrete or of steel and concrete, of sufficient weight to provide adequate load for the tank when filled with flammable or combustible liquid and submerged by flood waters to the established flood stage, or adequately secured by other means.

**1926.152(i)(5)(vi)(E)** [Reserved]

**1926.152(i)(5)(vi)(F)** At locations where there is no ample and dependable water supply, or where filling of underground tanks with liquids is impracticable because of the character of their contents, their use, or for other reasons, each tank shall be safeguarded against movement when empty and submerged by high ground water or flood waters by anchoring, weighting with concrete or other approved solid loading material, or securing by other means. Each such tank shall be so constructed and installed that it will safely resist external pressures due to high ground water or flood waters.

**1926.152(i)(5)(vi)(G)** At locations where there is an ample and dependable water supply available, underground tanks containing flammable or combustible liquids, so installed that more than 70 percent of their storage capacity will be submerged at the maximum flood stage, shall be so anchored, weighted, or secured by other means, as to prevent movement of such tanks when filled with flammable or combustible liquids, and submerged by flood waters to the established flood stage.

**1926.152(i)(5)(vi)(H)** Pipe connections below the allowable liquid level in a tank shall be provided with valves or cocks located as closely as practicable to the tank shell. Such valves and their connections to tanks shall be of steel or other material suitable for use with the liquid being stored. Cast iron shall not be permitted.

**1926.152(i)(5)(vi)(I)** At locations where an independent water supply is required, it shall be entirely independent of public power and water supply. Independent source of water shall be available when flood waters reach a level not less than 10 feet (3.04 m) below the bottom of the lowest tank on a property.

**1926.152(i)(5)(vi)(J)** The self-contained power and pumping unit shall be so located or so designed that pumping into tanks may be carried on continuously throughout the rise in flood waters from a level 10 feet (3.04 m) below the lowest tank to the level of the potential flood stage.

**1926.152(i)(5)(vi)(K)** Capacity of the pumping unit shall be such that the rate of rise of water in all tanks shall be equivalent to the established potential average rate of rise of flood waters at any stage.

**1926.152(i)(5)(vi)(L)** Each independent pumping unit shall be tested periodically to insure that it is in satisfactory operating condition.

**1926.152(i)(5)(vi)(M)** Structural guides for holding floating tanks above their foundations shall be so designed that there will be no resistance to the free rise of a tank, and shall be constructed of noncombustible material.

**1926.152(i)(5)(vi)(N)** The strength of the structure shall be adequate to resist lateral movement of a tank subject to a horizontal force in any direction equivalent to not less than 25 pounds per square foot (1.05 kg $m^2$) acting on the projected vertical cross-sectional area of the tank.

**1926.152(i)(5)(vi)(O)** Where tanks are situated on exposed points or bends in a shoreline where swift currents in flood waters will be present, the structures shall be designed to withstand a unit force of not less than 50 pounds per square foot (2.1 kg $m^2$).

**1926.152(i)(5)(vi)(P)** The filling of a tank to be protected by water loading shall be started as soon as flood waters reach a dangerous flood stage. The rate of filling shall be at least equal to the rate of rise of the floodwaters (or the established average potential rate of rise).

**1926.152(i)(5)(vi)(Q)** Sufficient fuel to operate the water pumps shall be available at all times to insure adequate power to fill all tankage with water.

**1926.152(i)(5)(vi)(R)** All valves on connecting pipelines shall be closed and locked in closed position when water loading has been completed.

**1926.152(i)(5)(vi)(S)** Where structural guides are provided for the protection of floating tanks, all rigid connections between tanks and pipelines shall be disconnected and blanked off or blinded before the floodwaters reach the bottom of the tank, unless control valves and their connections to the tank are of a type designed to prevent breakage between the valve and the tank shell.

**1926.152(i)(5)(vi)(T)** All valves attached to tanks other than those used in connection with water loading operations shall be closed and locked.

**1926.152(i)(5)(vi)(U)** If a tank is equipped with a swing line, the swing pipe shall be raised to and secured at its highest position.

**1926.152(i)(5)(vi)(V)** Inspections. The Assistant Secretary or his designated representative shall make periodic inspections of all plants where the storage of flammable or combustible liquids is such as to require compliance with the foregoing requirements, in order to assure the following:

**1926.152(i)(5)(vi)(V)(*1*)** That all flammable or combustible liquid storage tanks are in compliance with these requirements and so maintained.

**1926.152(i)(5)(vi)(V)(*2*)** That detailed printed instructions of what to do in flood emergencies are properly posted.

**1926.152(i)(5)(vi)(V)(*3*)** That station operators and other employees depended upon to carry out such instructions are thoroughly informed as to the location and operation of such valves and other equipment necessary to effect these requirements.

**1926.152(i)(5)(vii)** *Earthquake areas.* In areas subject to earthquakes, the tank supports and connections shall be designed to resist damage as a result of such shocks.

**1926.152(i)(6)** *Sources of ignition.* In locations where flammable vapors may be present, precautions shall be taken to prevent ignition by eliminating or controlling sources of ignition. Sources of ignition may include open flames, lightning, smoking, cutting and welding, hot surfaces, frictional heat, sparks (static, electrical, and mechanical), spontaneous ignition, chemical and physical-chemical reactions, and radiant heat.

**1926.152(i)(7)** *Testing*—(i) *General.* All tanks, whether shop built or field erected, shall be strength tested before they are placed in service in accordance with the applicable paragraphs of the code under which they were built. The American Society of Mechanical Engineers (ASME) code stamp, American Petroleum Institute (API) monogram, or the label of the Underwriters' Laboratories, Inc., on a tank shall be evidence of compliance with this strength test. Tanks not marked in accordance with the above codes shall be strength tested before they are placed in service in accordance with good engineering principles and reference shall be made to the sections on testing in the codes listed in paragraphs (i)(1) (iii)(A), (iv)(B), or (v)(B) of this section.

**1926.152(i)(7)(ii)** *Strength.* When the vertical length of the fill and vent pipes is such that when filled with liquid the static head imposed upon the bottom of the tank exceeds 10 pounds per square inch (68.94 kPa), the tank and related piping shall be tested hydrostatically to a pressure equal to the static head thus imposed.

**1926.152(i)(7)(iii)** *Tightness.* In addition to the strength test called for in paragraphs (i)(7) (i) and (ii) of this section, all tanks and connections shall be tested for tightness. Except for underground tanks, this tightness test shall be made at operating pressure with air, inert gas, or water prior to placing the tank in service. In the case of field-erected tanks the strength test may be considered to be the test for tank tightness. Underground tanks and piping, before being covered, enclosed, or placed in use, shall be tested for tightness hydrostatically, or with air pressure at not less than 3 pounds per square inch (20.68 kPa) and not more than 5 pounds per square inch (34.47 kPa).

**1926.152(i)(7)(iv)** *Repairs.* All leaks or deformations shall be corrected in an acceptable manner before the tank is placed in service. Mechanical caulking is not permitted for correcting leaks in welded tanks except pinhole leaks in the roof.

**1926.152(i)(7)(v)** *Derated operations.* Tanks to be operated at pressures below their design pressure may be tested by the applicable provisions of paragraphs (i)(7) (i) or (ii) of this section, based upon the pressure developed under full emergency venting of the tank.

**1926.152(j)** ***Piping, valves, and fittings***—(1) *General*—(i) *Design.* The design (including selection of materials) fabrication, assembly, test, and inspection of piping systems containing flammable or combustible liquids shall be suitable for the expected working pressures and structural stresses. Conformity with the applicable provisions of Pressure Piping, ANSI B31 series and the provisions of this paragraph, shall be considered prima facie evidence of compliance with the foregoing provisions.

**1926.152(j)(1)(ii)** *Exceptions.* This paragraph does not apply to any of the following:

**1926.152(j)(1)(ii)(A)** Tubing or casing on any oil or gas wells and any piping connected directly thereto.

**1926.152(j)(1)(ii)(B)** Motor vehicle, aircraft, boat, or portable or stationary engines.

**1926.152(j)(1)(ii)(C)** Piping within the scope of any applicable boiler and pressures vessel code.

**1926.152(j)(1)(iii)** *Definitions.* As used in this paragraph, piping systems consist of pipe, tubing, flanges, bolting, gaskets, valves, fittings, the pressure containing parts of other components such as expansion joints and strainers, and devices which serve such purposes as mixing, separating, snubbing, distributing, metering, or controlling flow.

**1926.152(j)(2)** *Materials for piping, valves, and fittings*—(i) *Required materials.* Materials for piping, valves, or fittings shall be steel, nodular iron, or malleable iron, except as provided in paragraphs (j)(2) (ii), (iii) and (iv) of this section.

**1926.152(j)(2)(ii)** *Exceptions.* Materials other than steel, nodular iron, or malleable iron may be used underground, or if required by the properties of the flammable or combustible liquid handled. Material other than steel, nodular iron, or malleable iron shall be designed to specifications embodying principles recognized as good engineering practices for the material used.

**1926.152(j)(2)(iii)** *Linings.* Piping, valves, and fittings may have combustible or noncombustible linings.

**1926.152(j)(2)(iv)** *Low-melting materials.* When low-melting point materials such as aluminum and brass or materials that soften on fire exposure such as plastics, or non-ductile materials such as cast iron, are necessary, special consideration shall be given to their behavior on fire exposure. If such materials are used in above ground piping systems or inside buildings, they shall be suitably protected against fire exposure or so located that any spill resulting from the failure of these materials could not unduly expose persons, important buildings or structures or can be readily controlled by remote valves.

**1926.152(j)(3)** *Pipe joints.* Joints shall be made liquid tight. Welded or screwed joints or approved connectors shall be used. Threaded joints and connections shall be made up tight with a suitable lubricant or piping compound. Pipe joints dependent upon the friction characteristics of combustible materials for mechanical continuity of piping shall not be used inside buildings. They may be used outside of buildings above or below ground. If used above ground, the piping shall either be secured to prevent disengagement at the fitting or the piping system shall be so designed that any spill resulting from such disengagement could not unduly expose persons, important buildings or structures, and could be readily controlled by remote valves.

**1926.152(j)(4)** *Supports.* Piping systems shall be substantially supported and protected against physical damage and excessive stresses arising from settlement, vibration, expansion, or contraction.

**1926.152(j)(5)** *Protection against corrosion.* All piping for flammable or combustible liquids, both aboveground and underground, where subject to external corrosion, shall be painted or otherwise protected.

**1926.152(j)(6)** *Valves.* Piping systems shall contain a sufficient number of valves to operate the system properly and to protect the plant. Piping systems in connection with pumps shall contain a sufficient number of valves to control properly the flow of liquid in normal operation and in the event of physical damage. Each connection to pipelines, by which equipments such as tankcars or tank vehicles discharge liquids by means of pumps into storage tanks, shall be provided with a check valve for automatic protection against backflow if the piping arrangement is such that backflow from the system is possible.

**1926.152(j)(7)** *Testing.* All piping before being covered, enclosed, or placed in use shall be hydrostatically tested to 150 percent of the maximum anticipated pressure of the system, or pneumatically tested to 110 percent of the maximum anticipated pressure of the system, but not less than 5 pounds per square inch gage at the highest point of the system. This test shall be maintained for a sufficient time to complete visual inspection of all joints and connections, but for at least 10 minutes.

**1926.152(k)** ***Marine service stations***—(1) *Dispensing.* (i) The dispensing area shall be located away from other structures so as to provide room for safe ingress and egress of craft to be fueled. Dispensing units shall in all cases be at least 20 feet (6.08 m) from any activity involving fixed sources of ignition.

**1926.152(k)(1)(ii)** Dispensing shall be by approved dispensing units with or without integral pumps and may be located on open piers, wharves, or floating docks or on shore or on piers of the solid fill type.

**1926.152(k)(1)(iii)** Dispensing nozzles shall be automatic-closing without a hold-open latch.

**1926.152(k)(2)** *Tanks and pumps.* (i) Tanks, and pumps not integral with the dispensing unit, shall be on shore or on a pier of the solid fill type, except as provided in paragraphs (k)(2) (ii) and (iii) of this section.

**1926.152(k)(2)(ii)** Where shore location would require excessively long supply lines to dispensers, tanks may be installed on a pier provided that applicable portions of paragraph (b) of this section relative to spacing, diking, and piping are complied with and the quantity so stored does not exceed 1,100 gallons (4,163.5 L) aggregate capacity.

**1926.152(k)(2)(iii)** Shore tanks supplying marine service stations may be located above ground, where rock ledges or high water table make underground tanks impractical.

**1926.152(k)(2)(iv)** Where tanks are at an elevation which would produce gravity head on the dispensing unit, the tank outlet shall be equipped with a pressure control valve positioned adjacent to and outside the tank block valve specified in §1926.152(c)(8) of this section, so adjusted that liquid cannot flow by gravity from the tank in case of piping or hose failure.

**1926.152(k)(3)** *Piping.* (i) Piping between shore tanks and dispensing units shall be as described in paragraph (k)(2)(iii) of this section, except that, where dispensing is from a floating structure, suitable lengths of oil-resistant flexible hose may be employed between the shore piping and the piping on the floating structure as made necessary by change in water level or shoreline.

**TABLE F-19—ELECTRICAL EQUIPMENT HAZARDOUS AREAS—SERVICE STATIONS**

| Location | Class I Group D division | Extent of classified area |
|---|---|---|
| Underground tank: | | |
| Fill opening | 1 | Any pit, box or space below grade level, any part of which is within the Division 1 or 2 classified area. |
| | 2 | Up to 18 inches (45.72 cm) above grade level within a horizontal radius of 10 feet (3.04m) from a loose fill connection and within a horizontal radius of 5 feet (1.52 m) from a tight fill connection. |
| Vent—Discharging upward | 1 | Within 3 feet (0.912 m) of open end of vent, extending in all directions. |
| | 2 | Area between 3 feet (0.912 m) and 5 feet (1.52 m) of open end of vent, extending in all directions. |
| Dispenser: | | |
| Pits | 1 | Any pit, box or space below grade level, any part of which is within the Division 1 or 2 classified area. |
| Dispenser enclosure | 1 | The area 4 feet (1.216 m) vertically above base within the enclosure and 18 inches (45.72 cm) horizontally in all directions. |
| Outdoor | 2 | Up to 18 inches (45.72 cm) above grade level within 20 feet (6.08 m) horizontally of any edge of enclosure. |
| Indoor: | | |
| With mechanical ventilation | 2 | Up to 18 inches (45.72 cm) above grade or floor level within 20 feet (6.08 m) horizontally of any edge of enclosure. |
| With gravity ventilation | 2 | Up to 18 inches (45.72 cm) above grade or floor level within 25 feet (7.6 m) horizontally of any edge of enclosure. |
| Remote pump—Outdoor | 1 | Any pit, box or space below grade level if any part is within a horizontal distance of 10 feet (3.04m) from any edge of pump. |
| | 2 | Within 3 feet (0.912 m) of any edge of pump, extending in all directions. Also up to 18 inches (45.72 cm) above grade level within 10 feet (3.04m) horizontally from any edge of pump. |
| Remote pump—Indoor | 1 | Entire area within any pit. |
| | 2 | Within 5 feet (1.52 m) of any edge of pump, extending in all directions. Also up to 3 feet (3.04 m [sic 0.912 m]) above floor or grade level within 25 feet (6.08 m [sic 7.6 m]) horizontally from any edge of pump. |
| Lubrication or service room | 1 | Entire area within any pit. |
| | 2 | Area up to 18 inches (45.72 cm) above floor or grade level within entire lubrication room. |
| Dispenser for Class I Liquids | 2 | Within 3 feet (0.912 m) of any fill or dispensing point, extending in all directions. |
| Special enclosure inside building per § 1910.106(f)(1)(ii). | 1 | Entire enclosure. |
| Sales, storage and rest rooms | (1) | If there is any opening to these rooms within the extent of a Division 1 area, the entire room shall be classified as Division 1. |

[1] Ordinary.

**1926.152(k)(3)(ii)** A readily accessible valve to shut off the supply from shore shall be provided in each pipeline at or near the approach to the pier and at the shore end of each pipeline adjacent to the point where flexible hose is attached.

**1926.152(k)(3)(iii)** Piping shall be located so as to be protected from physical damage.

**1926.152(k)(3)(iv)** Piping handling Class I liquids shall be grounded to control stray currents.

**1926.152(k)(4)** *Definition; as used in this section:* Marine service station shall mean that portion of a property where flammable or combustible liquids used as fuels are stored and dispensed from fixed equipment on shore, piers, wharves, or floating docks into the fuel tanks of self-propelled craft, and shall include all facilities used in connection therewith.

[44 FR 8577, Feb. 9, 1979; 44 FR 20940, Apr. 6, 1979, as amended at 51 FR 25318, July 11, 1986; 58 FR 35162, June 30, 1993; 63 FR 33469, June 18, 1998]

## §1926.153 Liquefied petroleum gas (LP-Gas).

**1926.153(a)** ***Approval of equipment and systems.*** (1) Each system shall have containers, valves, connectors, manifold valve assemblies, and regulators of an approved type.

**1926.153(a)(2)** All cylinders shall meet the Department of Transportation specification identification requirements published in 49 CFR Part 178, Shipping Container Specifications.

**1926.153(a)(3)** *Definition.* As used in this section, *Containers*—All vessels, such as tanks, cylinders, or drums, used for transportation or storing liquefied petroleum gases.

**1926.153(b)** ***Welding on LP-Gas containers.*** Welding is prohibited on containers.

**1926.153(c)** ***Container valves and container accessories.*** (1) Valves, fittings, and accessories connected directly to the container, including primary shut off valves, shall have a rated working pressure of at least 250 p.s.i.g. and shall be of material and design suitable for LP-Gas service.

**1926.153(c)(2)** Connections to containers, except safety relief connections, liquid level gauging devices, and plugged openings, shall have shutoff valves located as close to the container as practicable.

**1926.153(d)** ***Safety devices.*** (1) Every container and every vaporizer shall be provided with one or more approved safety relief valves or devices. These valves shall be arranged to afford free vent to the outer air with discharge not less than 5 feet horizontally away from any opening into a building which is below such discharge.

**1926.153(d)(2)** Shutoff valves shall not be installed between the safety relief device and the container, or the equipment or piping to which the safety relief device is connected, except that a shutoff valve may be used where the arrangement of this valve is such that full required capacity flow through the safety relief device is always afforded.

**1926.153(d)(3)** Container safety relief devices and regulator relief vents shall be located not less than 5 feet in any direction from air openings into sealed combustion system appliances or mechanical ventilation air intakes.

**1926.153(e)** ***Dispensing.*** (1) Filling of fuel containers for trucks or motor vehicles from bulk storage containers shall be performed not less than 10 feet from the nearest masonry-walled building, or not less than 25 feet from the nearest building or other construction and, in any event, not less than 25 feet from any building opening.

**1926.153(e)(2)** Filling of portable containers or containers mounted on skids from storage containers shall be performed not less than 50 feet from the nearest building.

**1926.153(f)** ***Requirements for appliances.*** (1) LP-Gas consuming appliances shall be approved types.

**1926.153(f)(2)** Any appliance that was originally manufactured for operation with a gaseous fuel other than LP-Gas, and is in good condition, may be used with LP-Gas only after it is properly converted, adapted, and tested for performance with LP-Gas before the appliance is placed in use.

**1926.153(g)** ***Containers and regulating equipment installed outside of buildings or structures.*** Containers shall be upright upon firm foundations or otherwise firmly secured. The possible effect on the outlet piping of settling shall be guarded against by a flexible connection or special fitting.

**1926.153(h)** ***Containers and equipment used inside of buildings or structures.*** (1) When operational requirements make portable use of containers necessary, and their location outside of buildings or structures is impracticable, containers and equipment shall be permitted to be used inside of buildings or structures in accordance with paragraphs (h)(2) through (11) of this section.

**1926.153(h)(2)** *Containers in use* means connected for use.

**1926.153(h)(3)** Systems utilizing containers having a water capacity greater than 2 1/2 pounds (nominal 1 pound LP-Gas capacity) shall be equipped with excess flow valves. Such excess flow valves shall be either integral with the container valves or in the connections to the container valve outlets.

**1926.153(h)(4)** Regulators shall be either directly connected to the container valves or to manifolds connected to the container valves. The regulator shall be suitable for use with LP-Gas. Manifolds and fittings connecting containers to pressure regulator inlets shall be designed for at least 250 p.s.i.g. service pressure.

**1926.153(h)(5)** Valves on containers having water capacity greater than 50 pounds (nominal 20 pounds LP-Gas capacity) shall be protected from damage while in use or storage.

**1926.153(h)(6)** Aluminum piping or tubing shall not be used.

**1926.153(h)(7)** Hose shall be designed for a working pressure of at least 250 p.s.i.g. Design, construction, and performance of hose, and hose connections shall have their suitability determined by listing by a nationally recognized testing agency. The hose length shall be as short as practicable. Hoses shall be long enough to permit compliance with spacing provisions of paragraphs (h)(1) through (13) of this section, without kinking or straining, or causing hose to be so close to a burner as to be damaged by heat.

**1926.153(h)(8)** Portable heaters, including salamanders, shall be equipped with an approved automatic device to shut off the flow of gas to the main burner, and pilot if used, in the event of flame failure. Such heaters, having inputs above 50,000 B.t.u. per hour, shall be equipped with either a pilot, which must be lighted and proved before the main burner can be turned on, or an electrical ignition system.

**Note:** The provisions of this subparagraph do not apply to portable heaters under 7,500 B.t.u. per hour input when used with containers having a maximum water capacity of 2 1/2 pounds.

**1926.153(h)(9)** Container valves, connectors, regulators, manifolds, piping, and tubing shall not be used as structural supports for heaters.

**1926.153(h)(10)** Containers, regulating equipment, manifolds, pipe, tubing, and hose shall be located to minimize exposure to high temperatures or physical damage.

**1926.153(h)(11)** Containers having a water capacity greater than 2 1/2 pounds (nominal 1 pound LP-Gas capacity) connected for use shall stand on a firm and substantially level surface and, when necessary, shall be secured in an upright position.

**1926.153(h)(12)** The maximum water capacity of individual containers shall be 245 pounds (nominal 100 pounds LP-Gas capacity).

**1926.153(h)(13)** For temporary heating, heaters (other than integral heater-container units) shall be located at least 6 feet from any LP-Gas container. This shall not prohibit the use of heaters specifically designed for attachment to the container or to a supporting standard, provided they are designed and installed so as to prevent direct or radiant heat application from the heater onto the containers. Blower and radiant type heaters shall not be directed toward any LP-Gas container within 20 feet.

**1926.153(h)(14)** If two or more heater-container units, of either the integral or nonintegral type, are located in an unpartitioned area on the same floor, the container or containers of each unit shall be separated from the container or containers of any other unit by at least 20 feet.

**1926.153(h)(15)** When heaters are connected to containers for use in an unpartitioned area on the same floor, the total water capacity of containers, manifolded together for connection to a heater or heaters, shall not be greater than 735 pounds (nominal 300 pounds LP-Gas capacity). Such manifolds shall be separated by at least 20 feet.

**1926.153(h)(16)** Storage of containers awaiting use shall be in accordance with paragraphs (j) and (k) of this section.

**1926.153(i)** ***Multiple container systems.*** (1) Valves in the assembly of multiple container systems shall be arranged so that replacement of containers can be made without shutting off the flow of gas in the system. This provision is not to be construed as requiring an automatic changeover device.

**1926.153(i)(2)** Heaters shall be equipped with an approved regulator in the supply line between the fuel cylinder and the heater unit. Cylinder connectors shall be provided with an excess flow valve to minimize the flow of gas in the event the fuel line becomes ruptured.

**1926.153(i)(3)** Regulators and low-pressure relief devices shall be rigidly attached to the cylinder valves, cylinders, supporting standards, the building walls, or otherwise rigidly secured, and shall be so installed or protected from the elements.

**1926.153(j)** ***Storage of LPG containers.*** Storage of LPG within buildings is prohibited.

**1926.153(k)** ***Storage outside of buildings.*** (1) Storage outside of buildings, for containers awaiting use, shall be located from the nearest building or group of buildings, in accordance with the following:

**TABLE F-3**

| Quantity of LP-Gas stored: | Distance (feet) |
|---|---|
| 500 lbs. or less | 0 |
| 501 to 6,000 lbs | 10 |
| 6,001 to 10,000 lbs | 20 |
| Over 10,000 lbs | 25 |

**1926.153(k)(2)** Containers shall be in a suitable ventilated enclosure or otherwise protected against tampering.

**1926.153(l)** ***Fire protection.*** Storage locations shall be provided with at least one approved portable fire extinguisher having a rating of not less than 20-B:C.

**1926.153(m)** ***Systems utilizing containers other than DOT containers***—(1) Application. This paragraph applies specifically to systems utilizing storage containers other than those constructed in accordance with DOT specifications. Paragraph (b) of this section applies to this paragraph unless otherwise noted in paragraph (b) of this section.

**1926.153(m)(2)** *Design pressure and classification of storage containers.* Storage containers shall be designed and classified in accordance with Table F-31.

**TABLE F-31**

| Container type | Minimum design pressure of container, lb. per sq. in. gage | | |
|---|---|---|---|
| | For gases with vapor press. Not to exceed lb. per sq. in. gage at 100° F. (37.8° C.) | 1949 and earlier editions of ASME Code (Par. U-68, U-69) | 1949 edition of ASME Code (Par. U-200, U-201); 1950, 1952, 1956, 1959, 1962, 1965, and 1968 (Division 1) editions of ASME Code; All editions of API-ASME Code[3] |
| [1] 80 | [1] 80 | [1] 80 | [1] 100 |
| 100 | 100 | 100 | 125 |
| 125 | 125 | 125 | 156 |
| 150 | 150 | 150 | 187 |
| 175 | 175 | 175 | 219 |
| [2] 200 | 215 | 200 | 250 |

[1] New storage containers of the 80 type have not been authorized since Dec. 31, 1947.

[2] Container type may be increased by increments of 25. The minimum design pressure of containers shall be 100% of the container type designation when constructed under 1949 or earlier editions of the ASME Code (Par. U-68 and U-69). The minimum design pressure of containers shall be 125% of the container type designation when constructed under: (1) the 1949 ASME Code (Par. U-200 and U-201), (2) 1950, 1952, 1956, 1959, 1962, 1965, and 1968 (Division 1) editions of the ASME Code, and (3) all editions of the API-ASME Code.

[3] Construction of containers under the API-ASME Code is not authorized after July 1, 1961.

**1926.153(m)(3)** Containers with foundations attached (portable or semiportable b containers with suitable steel "runners" or "skids" and popularly known in the industry as "skid tanks") shall be designed, installed, and used in accordance with these rules subject to the following provisions:

**1926.153(m)(3)(i)** If they are to be used at a given general location for a temporary period not to exceed 6 months they need not have fire-resisting foundations or saddles but shall have adequate ferrous metal supports.

**1926.153(m)(3)(ii)** They shall not be located with the outside bottom of the container shell more than 5 feet (1.52 m) above the surface of the ground unless fire-resisting supports are provided.

**1926.153(m)(3)(iii)** The bottom of the skids shall not be less than 2 inches (5.08 cm) or more than 12 inches (30.48 cm) below the outside bottom of the container shell.

**1926.153(m)(3)(iv)** Flanges, nozzles, valves, fittings, and the like, having communication with the interior of the container, shall be protected against physical damage.

**1926.153(m)(3)(v)** When not permanently located on fire-resisting foundations, piping connections shall be sufficiently flexible to minimize the possibility of breakage or leakage of connections if the container settles, moves, or is otherwise displaced.

**1926.153(m)(3)(vi)** Skids, or lugs for attachment of skids, shall be secured to the container in accordance with the code or rules under which the container is designed and built (with a minimum factor of safety of four) to withstand loading in any direction equal to four times the weight of the container and attachments when filled to the maximum permissible loaded weight.

**1926.153(m)(4)** Field welding where necessary shall be made only on saddle plates or brackets which were applied by the manufacturer of the tank.

**1926.153(n)** When LP-Gas and one or more other gases are stored or used in the same area, the containers shall be marked to identify their content. Marking shall be in compliance with American National Standard Z48.1-1954, "Method of Marking Portable Compressed Gas Containers To Identify the Material Contained."

**1926.153(o)** ***Damage from vehicles.*** When damage to LP-Gas systems from vehicular traffic is a possibility, precautions against such damage shall be taken.

[44 FR 8577, Feb. 9, 1979; 44 FR 20940, Apr. 6, 1979, as amended at 58 FR 35170, June 30, 1993]

## §1926.154 Temporary heating devices.

**1926.154(a)** ***Ventilation.*** (1) Fresh air shall be supplied in sufficient quantities to maintain the health and safety of workmen. Where natural means of fresh air supply is inadequate, mechanical ventilation shall be provided.

**1926.154(a)(2)** When heaters are used in confined spaces, special care shall be taken to provide sufficient ventilation in order to ensure proper combustion, maintain the health and safety of workmen, and limit temperature rise in the area.

**1926.154(b)** ***Clearance and mounting.*** (1) Temporary heating devices shall be installed to provide clearance to combustible material not less than the amount shown in Table F-4.

**1926.154(b)(2)** Temporary heating devices, which are listed for installation with lesser clearances than specified in Table F-4, may be installed in accordance with their approval.

**TABLE F-4**

| Heating appliances | Minimum clearance, (inches) | | |
|---|---|---|---|
| | Sides | Rear | Chimney Connector |
| Room heater, circulating type ... | 12 | 12 | 18 |
| Room heater, radiant type ........ | 36 | 36 | 18 |

**1926.154(b)(3)** Heaters not suitable for use on wood floors shall not be set directly upon them or other combustible materials. When such heaters are used, they shall rest on suitable heat insulating material or at least 1-inch concrete, or equivalent. The insulating material shall extend beyond the heater 2 feet or more in all directions.

**1926.154(b)(4)** Heaters used in the vicinity of combustible tarpaulins, canvas, or similar coverings shall be located at least 10 feet from the coverings. The coverings shall be securely fastened to prevent ignition or upsetting of the heater due to wind action on the covering or other material.

**1926.154(c)** ***Stability.*** Heaters, when in use, shall be set horizontally level, unless otherwise permitted by the manufacturer's markings.

**1926.154(d)** ***Solid fuel salamanders.*** Solid fuel salamanders are prohibited in buildings and on scaffolds.

**1926.154(e)** ***Oil-fired heaters.*** (1) Flammable liquid-fired heaters shall be equipped with a primary safety control to stop the flow of fuel in the event of flame failure. Barometric or gravity oil feed shall not be considered a primary safety control.

**1926.154(e)(2)** Heaters designed for barometric or gravity oil feed shall be used only with the integral tanks.

**1926.154(e)(3)** [Reserved]

**1926.154(e)(4)** Heaters specifically designed and approved for use with separate supply tanks may be directly connected for gravity feed, or an automatic pump, from a supply tank.

## §1926.155 Definitions applicable to this subpart.

**1926.155(a)** *Approved*, for the purpose of this subpart, means equipment that has been listed or approved by a nationally recognized testing laboratory such as Factory Mutual Engineering Corp., or Underwriters' Laboratories, Inc., or Federal agencies such as Bureau of Mines, or U.S. Coast Guard, which issue approvals for such equipment.

**1926.155(b)** *Closed container* means a container so sealed by means of a lid or other device that neither liquid nor vapor will escape from it at ordinary temperatures.

**1926.155(c)** *Combustible liquids* mean any liquid having a flash point at or above 140 °F. (60 °C.), and below 200 °F. (93.4 °C.).

**1926.155(d)** *Combustion* means any chemical process that involves oxidation sufficient to produce light or heat.

**1926.155(e)** *Fire brigade* means an organized group of employees that are knowledgeable, trained, and skilled in the safe evacuation of employees during emergency situations and in assisting in fire fighting operations.

**1926.155(f)** *Fire resistance* means so resistant to fire that, for specified time and under conditions of a standard heat intensity, it will not fail structurally and will not permit the side away from the fire to become hotter than a specified temperature. For purposes of this part, fire resistance shall be determined by the Standard Methods of Fire Tests of Building Construction and Materials, NFPA 251-1969.

**1926.155(g)** *Flammable* means capable of being easily ignited, burning intensely, or having a rapid rate of flame spread.

**1926.155(h)** *Flammable liquids* means any liquid having a flash point below 140 °F. and having a vapor pressure not exceeding 40 pounds per square inch (absolute) at 100 °F.

**1926.155(i)** *Flash point* of the liquid means the temperature at which it gives off vapor sufficient to form an ignitable mixture with the air near the surface of the liquid or within the vessel used as determined by appropriate test procedure and apparatus as specified below.

**1926.155(i)(1)** The flash point of liquids having a viscosity less than 45 Saybolt Universal Second(s) at 100 °F. (37.8 °C.) and a flash point below 175 °F. (79.4 °C.) shall be determined in accordance with the Standard Method of Test for Flash Point by the Tag Closed Tester, ASTM D-56-69.

**1926.155(i)(2)** The flash point of liquids having a viscosity of 45 Saybolt Universal Second(s) or more at 175 °F. (79.4 °C.) or higher shall be determined in accordance with the Standard Method of Test for Flash Point by the Pensky Martens Closed Tester, ASTM D-93-69.

**1926.155(j)** *Liquefied petroleum gases, LPG* and *LP Gas* mean and include any material which is composed predominantly of any of the following hydrocarbons, or mixtures of them, such as propane, propylene, butane (normal butane or iso-butane), and butylenes.

**1926.155(k)** *Portable tank* means a closed container having a liquid capacity more than 60 U.S. gallons, and not intended for fixed installation.

**1926.155(l)** *Safety can* means an approved closed container, of not more than 5 gallons capacity, having a flash-arresting screen, spring-closing lid and spout cover and so designed that it will safely relieve internal pressure when subjected to fire exposure.

**1926.155(m)** *Vapor pressure* means the pressure, measured in pounds per square inch (absolute), exerted by a volatile liquid as determined by the "Standard Method of Test for Vapor Pressure of Petroleum Products (Reid Method)." (ASTM D-323-58).

**Authority:** Sec. 107, Contract Work Hours and Safety Standards Act (Construction Safety Act) (40 U.S.C. 333); secs. 4, 6, 8, Occupational Safety and Health Act of 1970 (29 U.S.C. 653, 655, 657); Secretary of Labor's Order No. 1271 (36 FR 8754), 8-76 (41 FR 25059), or 9-83 (48 FR 35736), as applicable.

### §1926.200 Accident prevention signs and tags.

**1926.200(a)** ***General.*** Signs and symbols required by this subpart shall be visible at all times when work is being performed, and shall be removed or covered promptly when the hazards no longer exist.

**1926.200(b)** ***Danger signs.*** (1) Danger signs (see Figure G-1) shall be used only where an immediate hazard exists.

**1926.200(b)(2)** Danger signs shall have red as the predominating color for the upper panel; black outline on the borders; and a white lower panel for additional sign wording.

**1926.200(c)** ***Caution signs.*** (1) Caution signs (see Figure G-2) shall be used only to warn against potential hazards or to caution against unsafe practices.

**1926.200(c)(2)** Caution signs shall have yellow as the predominating color; black upper panel and borders: yellow lettering of "caution" on the black panel; and the lower yellow panel for additional sign wording. Black lettering shall be used for additional wording.

FIGURE G-1

FIGURE G-2

**1926.200(c)(3)** Standard color of the background shall be yellow; and the panel, black with yellow letters. Any letters used against the yellow background shall be black. The colors shall be those of opaque glossy samples as specified in Table 1 of American National Standard Z53.1-1967.

**1926.200(d)** ***Exit signs.*** Exit signs, when required, shall be lettered in legible red letters, not less than 6 inches high, on a white field and the principal stroke of the letters shall be at least three-fourths inch in width.

**1926.200(e)** ***Safety instruction signs.*** Safety instruction signs, when used, shall be white with green upper panel with white letters to convey the principal message. Any additional wording on the sign shall be black letters on the white background.

**1926.200(f)** ***Directional signs.*** Directional signs, other than automotive traffic signs specified in paragraph (g) of this section, shall be white with a black panel and a white directional symbol. Any additional wording on the sign shall be black letters on the white background.

**1926.200(g)** ***Traffic signs.*** (1) Construction areas shall be posted with legible traffic signs at points of hazard.

**1926.200(g)(2)** All traffic control signs or devices used for protection of construction workmen shall conform to American National Standards Institute D6.1-1971, Manual on Uniform Traffic Control Devices for Streets and Highways.

**1926.200(h)** ***Accident prevention tags.*** (1) Accident prevention tags shall be used as a temporary means of warning employees of an existing hazard, such as defective tools, equipment, etc. They shall not be used in place of, or as a substitute for, accident prevention signs.

**1926.200(h)(2)** Specifications for accident prevention tags similar to those in Table G-1 shall apply.

**TABLE G-1**

| White tag - White letters on red square | White tag - White letters on red oval with a black square | Yellow tag - Yellow letters on a black background | White tag - White letters on black background |
|---|---|---|---|

| Basic Stock (Background) | Safety Colors (Ink) | Copy Specification (Letters) |
|---|---|---|
| White | Red | Do Not Operate |
| White | Black and Red | Danger |
| Yellow | Black | Caution |
| White | Black | Out of Order Do Not Use |

**1926.200(i)** ***Additional rules.*** American National Standards Institute (ANSI) Z35.1-1968, Specifications for Accident Prevention Signs, and Z35.2-1968, Specifications for Accident Prevention Tags, contain rules which are additional to the rules prescribed in this section. The employer shall comply with ANSI Z35.1-1968 and Z35.2-1968 with respect to rules not specifically prescribed in this subpart.

[44 FR 8577, Feb. 9, 1979; 44 FR 20940, Apr. 6, 1979, as amended at 58 FR 35173, June 30, 1993]

## §1926.201 Signaling.

**1926.201(a)** ***Flagmen.*** (1) When operations are such that signs, signals, and barricades do not provide the necessary protection on or adjacent to a highway or street, flagmen or other appropriate traffic controls shall be provided.

**1926.201(a)(2)** Signaling directions by flagmen shall conform to American National Standards Institute D6.1-1971, Manual on Uniform Traffic Control Devices for Streets and Highways.

**1926.201(a)(3)** Hand signaling by flagmen shall be by use of red flags at least 18 inches square or sign paddles, and in periods of darkness, red lights.

**1926.201(a)(4)** Flagmen shall be provided with and shall wear a red or orange warning garment while flagging. Warning garments worn at night shall be of reflectorized material.

**1926.201(b)** ***Crane and hoist signals.*** Regulations for crane and hoist signaling will be found in applicable American National Standards Institute standards.

## §1926.202 Barricades.

Barricades for protection of employees shall conform to the portions of the American National Standards Institute D6.1-1971, Manual on Uniform Traffic Control Devices for Streets and Highways, relating to barricades.

## §1926.203 Definitions applicable to this subpart.

**1926.203(a)** *Barricade* means an obstruction to deter the passage of persons or vehicles.

**1926.203(b)** *Signs* are the warnings of hazard, temporarily or permanently affixed or placed, at locations where hazards exist.

**1926.203(c)** *Signals* are moving signs, provided by workers, such as flagmen, or by devices, such as flashing lights, to warn of possible or existing hazards.

**1926.203(d)** *Tags* are temporary signs, usually attached to a piece of equipment or part of a structure, to warn of existing or immediate hazards.

**Authority:** Sec. 107, Contract Work Hours and Safety Standards Act (Construction Safety Act) (40 U.S.C. 333); Secs. 4, 6, 8, Occupational Safety and Health Act of 1970 (29 U.S.C. 653, 655, 657); Secretary of Labor's Order No. 12-71 (36 FR 8754), 8-76 (41 FR 25059), 9-83 (48 FR 35736), or 1-90 (50 FR 9033), as applicable. Section 1926.250 also issued under 29 CFR Part 1911.

### §1926.250 General requirements for storage.

**1926.250(a)** ***General.*** (1) All materials stored in tiers shall be stacked, racked, blocked, interlocked, or otherwise secured to prevent sliding, falling or collapse.

**1926.250(a)(2)** Maximum safe load limits of floors within buildings and structures, in pounds per square foot, shall be conspicuously posted in all storage areas, except for floor or slab on grade. Maximum safe loads shall not be exceeded.

**1926.250(a)(3)** Aisles and passageways shall be kept clear to provide for the free and safe movement of material handling equipment or employees. Such areas shall be kept in good repair.

**1926.250(a)(4)** When a difference in road or working levels exist, means such as ramps, blocking, or grading shall be used to ensure the safe movement of vehicles between the two levels.

**1926.250(b)** ***Material storage.*** (1) Material stored inside buildings under construction shall not be placed within 6 feet of any hoistway or inside floor openings, nor within 10 feet of an exterior wall which does not extend above the top of the material stored.

**1926.250(b)(2)** Each employee required to work on stored material in silos, hoppers, tanks, and similar storage areas shall be equipped with personal fall arrest equipment meeting the requirements of subpart M of this part.

**1926.250(b)(3)** Noncompatible materials shall be segregated in storage.

**1926.250(b)(4)** Bagged materials shall be stacked by stepping back the layers and cross-keying the bags at least every 10 bags high.

**1926.250(b)(5)** Materials shall not be stored on scaffolds or runways in excess of supplies needed for immediate operations.

**1926.250(b)(6)** Brick stacks shall not be more than 7 feet in height. When a loose brick stack reaches a height of 4 feet, it shall be tapered back 2 inches in every foot of height above the 4-foot level.

**1926.250(b)(7)** When masonry blocks are stacked higher than 6 feet, the stack shall be tapered back one-half block per tier above the 6-foot level.

**1926.250(b)(8)** Lumber:

**1926.250(b)(8)(i)** Used lumber shall have all nails withdrawn before stacking.

**1926.250(b)(8)(ii)** Lumber shall be stacked on level and solidly supported sills.

**1926.250(b)(8)(iii)** Lumber shall be so stacked as to be stable and self-supporting.

**1926.250(b)(8)(iv)** Lumber piles shall not exceed 20 feet in height provided that lumber to be handled manually shall not be stacked more than 16 feet high.

**1926.250(b)(9)** Structural steel, poles, pipe, bar stock, and other cylindrical materials, unless racked, shall be stacked and blocked so as to prevent spreading or tilting.

**1926.250(c)** ***Housekeeping.*** Storage areas shall be kept free from accumulation of materials that constitute hazards from tripping, fire, explosion, or pest harborage. Vegetation control will be exercised when necessary.

**1926.250(d)** ***Dockboards (bridge plates).*** (1) Portable and powered dockboards shall be strong enough to carry the load imposed on them.

**1926.250(d)(2)** Portable dockboards shall be secured in position, either by being anchored or equipped with devices which will prevent their slipping.

**1926.250(d)(3)** Handholds, or other effective means, shall be provided on portable dockboards to permit safe handling.

**1926.250(d)(4)** Positive protection shall be provided to prevent railroad cars from being moved while dockboards or bridge plates are in position.

[44 FR 8577, Feb. 9, 1979; 44 FR 20940, Apr. 6, 1979, as amended at 49 FR 18295, Apr. 30, 1984; 54 FR 24334, June 7, 1989; 58 FR 35173, June 30, 1993; 59 FR 40729, Aug. 9, 1994; 61 FR 5510, Feb. 13, 1996]

### §1926.251 Rigging equipment for material handling.

**1926.251(a)** ***General.*** (1) Rigging equipment for material handling shall be inspected prior to use on each shift and as necessary during its use to ensure that it is safe. Defective rigging equipment shall be removed from service.

**1926.251(a)(2)** Rigging equipment shall not be loaded in excess of its recommended safe working load, as prescribed in Tables H-1 through H-20 in this subpart, following §1926.252(e) for the specific equipment.

**1926.251(a)(3)** Rigging equipment, when not in use, shall be removed from the immediate work area so as not to present a hazard to employees.

**1926.251(a)(4)** Special custom design grabs, hooks, clamps, or other lifting accessories, for such units as modular panels, prefabricated structures and similar materials, shall be marked to indicate the safe working loads and shall be proof-tested prior to use to 125 percent of their rated load.

**1926.251(a)(5)** *Scope.* This section applies to slings used in conjunction with other material handling equipment for the movement of material by hoisting, in employments covered by this part. The types of slings covered are those made from alloy steel chain, wire rope, metal mesh, natural or synthetic fiber rope (conventional three strand construction), and synthetic web (nylon, polyester, and polypropylene).

**1926.251(a)(6)** *Inspections.* Each day before being used, the sling and all fastenings and attachments shall be inspected for damage or defects by a competent person designated by the employer. Additional inspections shall be performed during sling use, where service conditions warrant. Damaged or defective slings shall be immediately removed from service.

**1926.251(b)** ***Alloy steel chains.*** (1) Welded alloy steel chain slings shall have permanently affixed durable identification stating size, grade, rated capacity, and sling manufacturer.

**1926.251(b)(2)** Hooks, rings, oblong links, pear-shaped links, welded or mechanical coupling links, or other attachments, when used with alloy steel chains, shall have a rated capacity at least equal to that of the chain.

**1926.251(b)(3)** Job or shop hooks and links, or makeshift fasteners, formed from bolts, rods, etc., or other such attachments, shall not be used.

**1926.251(b)(4)** Rated capacity (working load limit) for alloy steel chain slings shall conform to the values shown in Table H-1.

**1926.251(b)(5)** Whenever wear at any point of any chain link exceeds that shown in Table H-2, the assembly shall be removed from service.

**1926.251(b)(6)** *Inspections.* (i) In addition to the inspection required by other paragraphs of this section, a thorough periodic inspection of alloy steel chain slings in use shall be made on a regular basis, to be determined on the basis of (A) frequency of sling use; (B) severity of service conditions; (C) nature of lifts being made; and (D) experience gained on the service life of slings used in similar circumstances. Such inspections shall in no event be at intervals greater than once every 12 months.

**1926.251(b)(6)(ii)** The employer shall make and maintain a record of the most recent month in which each alloy steel chain sling was thoroughly inspected, and shall make such record available for examination.

**1926.251(c)** ***Wire rope.*** (1) Tables H-3 through H-14 shall be used to determine the safe working loads of various sizes and classifications of improved plow steel wire rope and wire rope slings with various types of terminals. For sizes, classifications, and grades not included in these tables, the safe working load recommended by the manufacturer for specific, identifiable products shall be followed, provided that a safety factor of not less than 5 is maintained.

**1926.251(c)(2)** Protruding ends of strands in splices on slings and bridles shall be covered or blunted.

**1926.251(c)(3)** Wire rope shall not be secured by knots, except on haul back lines on scrapers.

**1926.251(c)(4)** The following limitations shall apply to the use of wire rope:

**1926.251(c)(4)(i)** An eye splice made in any wire rope shall have not less than three full tucks. However, this requirement shall not operate to preclude the use of another form of splice or connection which can be shown to be as efficient and which is not otherwise prohibited.

**1926.251(c)(4)(ii)** Except for eye splices in the ends of wires and for endless rope slings, each wire rope used in hoisting or lowering, or in pulling loads, shall consist of one continuous piece without knot or splice.

**1926.251(c)(4)(iii)** Eyes in wire rope bridles, slings, or bull wires shall not be formed by wire rope clips or knots.

**1926.251(c)(4)(iv)** Wire rope shall not be used if, in any length of eight diameters, the total number of visible broken wires exceeds 10 percent of the total number of wires, or if the rope shows other signs of excessive wear, corrosion, or defect.

**1926.251(c)(5)** When U-bolt wire rope clips are used to form eyes, Table H-20 shall be used to determine the number and spacing of clips.

**1926.251(c)(5)(i)** When used for eye splices, the U-bolt shall be applied so that the "U" section is in contact with the dead end of the rope.

**1926.251(c)(6)** Slings shall not be shortened with knots or bolts or other makeshift devices.

**1926.251(c)(7)** Sling legs shall not be kinked.

**1926.251(c)(8)** Slings used in a basket hitch shall have the loads balanced to prevent slippage.

**1926.251(c)(9)** Slings shall be padded or protected from the sharp edges of their loads.

**1926.251(c)(10)** Hands or fingers shall not be placed between the sling and its load while the sling is being tightened around the load.

**1926.251(c)(11)** Shock loading is prohibited.

**1926.251(c)(12)** A sling shall not be pulled from under a load when the load is resting on the sling.

**1926.251(c)(13)** *Minimum sling lengths.* (i) Cable laid and 6 × 19 and 6 × 37 slings shall have a minimum clear length of wire rope 10 times the component rope diameter between splices, sleeves or end fittings.

**1926.251(c)(13)(ii)** Braided slings shall have a minimum clear length of wire rope 40 times the component rope diameter between the loops or end fittings.

**1926.251(c)(13)(iii)** Cable laid grommets, strand laid grommets and endless slings shall have a minimum circumferential length of 96 times their body diameter.

**1926.251(c)(14)** *Safe operating temperatures.* Fiber core wire rope slings of all grades shall be permanently removed from service if they are exposed to temperatures in excess of 200°F (93.33°C). When nonfiber core wire rope slings of any grade are used at temperatures above 400°F (204.44°C) or below minus 60°F (15.55°C), recommendations of the sling manufacturer regarding use at that temperature shall be followed.

**1926.251(c)(15)** *End attachments.* (i) Welding of end attachments, except covers to thimbles, shall be performed prior to the assembly of the sling.

**1926.251(c)(15)(ii)** All welded end attachments shall not be used unless proof tested by the manufacturer or equivalent entity at twice their rated capacity prior to initial use. The employer shall retain a certificate of the proof test, and make it available for examination.

**1926.251(d)** ***Natural rope, and synthetic fiber***—(1) *General.* When using natural or synthetic fiber rope slings, Tables H-15, 16, 17, and 18 shall apply.

**1926.251(d)(2)** All splices in rope slings provided by the employer shall be made in accordance with fiber rope manufacturers recommendations.

**1926.251(d)(2)(i)** In manila rope, eye splices shall contain at least three full tucks, and short splices shall contain at least six full tucks (three on each side of the centerline of the splice).

**1926.251(d)(2)(ii)** In layed synthetic fiber rope, eye splices shall contain at least four full tucks, and short splices shall contain at least eight full tucks (four on each side of the centerline of the splice).

**1926.251(d)(2)(iii)** Strand end tails shall not be trimmed short (flush with the surface of the rope) immediately adjacent to the full tucks. This precaution applies to both eye and short splices and all types of fiber rope. For fiber ropes under 1-inch diameter, the tails shall project at least six rope diameters beyond the last full tuck. For fiber ropes 1-inch diameter and larger, the tails shall project at least 6 inches beyond the last full tuck. In applications where the projecting tails may be objectionable, the tails shall be tapered and spliced into the body of the rope using at least two additional tucks (which will require a tail length of approximately six rope diameters beyond the last full tuck).

**1926.251(d)(2)(iv)** For all eye splices, the eye shall be sufficiently large to provide an included angle of not greater than 60° at the splice when the eye is placed over the load or support.

**1926.251(d)(2)(v)** Knots shall not be used in lieu of splices.

**1926.251(d)(3)** *Safe operating temperatures.* Natural and synthetic fiber rope slings, except for wet frozen slings, may be used in a temperature range from minus 20°F (-28.88°C) to plus 180°F (82.2°C) without decreasing the working load limit. For operations outside this temperature range and for wet frozen slings, the sling manufacturer's recommendations shall be followed.

**1926.251(d)(4)** *Splicing.* Spliced fiber rope slings shall not be used unless they have been spliced in accordance with the following minimum requirements and in accordance with any additional recommendations of the manufacturer:

**1926.251(d)(4)(i)** In manila rope, eye splices shall consist of at least three full tucks, and short splices shall consist of at least six full tucks, three on each side of the splice center line.

**1926.251(d)(4)(ii)** In synthetic fiber rope, eye splices shall consist of at least four full tucks, and short splices shall consist of at least eight full tucks, four on each side of the center line.

**1926.251(d)(4)(iii)** Strand end tails shall not be trimmed flush with the surface of the rope immediately adjacent to the full tucks. This applies to all types of fiber rope and both eye and short splices. For fiber rope under 1 inch (2.54 cm) in diameter, the tail shall project at least six rope diameters beyond the last full tuck. For fiber rope 1 inch (2.54 cm) in diameter and larger, the tail shall project at least 6 inches (15.24 cm) beyond the last full tuck. Where a projecting tail interferes with the use of the sling, the tail shall be tapered and spliced into the body of the rope using at least two additional tucks (which will require a tail length of approximately six rope diameters beyond the last full tuck).

**1926.251(d)(4)(iv)** Fiber rope slings shall have a minimum clear length of rope between eye splices equal to 10 times the rope diameter.

**1926.251(d)(4)(v)** Knots shall not be used in lieu of splices.

**1926.251(d)(4)(vi)** Clamps not designed specifically for fiber ropes shall not be used for splicing.

**1926.251(d)(4)(vii)** For all eye splices, the eye shall be of such size to provide an included angle of not greater than 60 degrees at the splice when the eye is placed over the load or support.

**1926.251(d)(5)** *End attachments.* Fiber rope slings shall not be used if end attachments in contact with the rope have sharp edges or projections.

**1926.251(d)(6)** *Removal from service.* Natural and synthetic fiber rope slings shall be immediately removed from service if any of the following conditions are present:

**1926.251(d)(6)(i)** Abnormal wear.

**1926.251(d)(6)(ii)** Powdered fiber between strands.

**1926.251(d)(6)(iii)** Broken or cut fibers.

**1926.251(d)(6)(iv)** Variations in the size or roundness of strands.

**1926.251(d)(6)(v)** Discoloration or rotting.

**1926.251(d)(6)(vi)** Distortion of hardware in the sling.

**1926.251(e)** ***Synthetic webbing (nylon, polyester, and polypropylene).*** (1) The employer shall have each synthetic web sling marked or coded to show:

**1926.251(e)(1)(i)** Name or trademark of manufacturer.

**1926.251(e)(1)(ii)** Rated capacities for the type of hitch.

**1926.251(e)(1)(iii)** Type of material.

**1926.251(e)(2)** Rated capacity shall not be exceeded.

**1926.251(e)(3)** *Webbing.* Synthetic webbing shall be of uniform thickness and width and selvage edges shall not be split from the webbing's width.

**1926.251(e)(4)** *Fittings.* Fittings shall be:

**1926.251(e)(4)(i)** Of a minimum breaking strength equal to that of the sling; and

**1926.251(e)(4)(ii)** Free of all sharp edges that could in any way damage the webbing.

**1926.251(e)(5)** *Attachment of end fittings to webbing and formation of eyes.* Stitching shall be the only method used to attach end fittings to webbing and to form eyes. The thread shall be in an even pattern and contain a sufficient number of stitches to develop the full breaking strength of the sling.

**1926.251(e)(6)** *Environmental conditions.* When synthetic web slings are used, the following precautions shall be taken:

**1926.251(e)(6)(i)** Nylon web slings shall not be used where fumes, vapors, sprays, mists or liquids of acids or phenolics are present.

**1926.251(e)(6)(ii)** Polyester and polypropylene web slings shall not be used where fumes, vapors, sprays, mists or liquids of caustics are present.

**1926.251(e)(6)(iii)** Web slings with aluminum fittings shall not be used where fumes, vapors, sprays, mists or liquids of caustics are present.

**1926.251(e)(7)** *Safe operating temperatures.* Synthetic web slings of polyester and nylon shall not be used at temperatures in excess of 180°F (82.2°C). Polypropylene web slings shall not be used at temperatures in excess of 200°F (93.33°C).

**1926.251(e)(8)** *Removal from service.* Synthetic web slings shall be immediately removed from service if any of the following conditions are present:

**1926.251(e)(8)(i)** Acid or caustic burns;

**1926.251(e)(8)(ii)** Melting or charring of any part of the sling surface;

**1926.251(e)(8)(iii)** Snags, punctures, tears or cuts;

**1926.251(e)(8)(iv)** Broken or worn stitches; or

**1926.251(e)(8)(v)** Distortion of fittings.

**1926.251(f)** ***Shackles and hooks.*** (1) Table H-19 shall be used to determine the safe working loads of various sizes of shackles, except that higher safe working loads are permissible when recommended by the manufacturer for specific, identifiable products, provided that a safety factor of not less than 5 is maintained.

**1926.251(f)(2)** The manufacturer's recommendations shall be followed in determining the safe working loads of the various sizes and types of specific and identifiable hooks. All hooks for which no applicable manufacturer's recommendations are available shall be tested to twice the intended safe working load before they are initially put into use. The employer shall maintain a record of the dates and results of such tests.

## TABLE H-1—RATED CAPACITY (WORKING LOAD LIMIT), FOR ALLOY STEEL CHAIN SLINGS[1]

Rated Capacity (Working Load Limit), Pounds

[Horizontal angles shown in parentheses] (2)

| Chain size (inches) | Single branch sling—90° loading | Double sling vertical angle (1) | | | Triple and quadruple sling vertical angle (1) | | |
|---|---|---|---|---|---|---|---|
| | | 30° (60°) | 45° (45°) | 60° (30°) | 30° (60°) | 45° (45°) | 60° (30°) |
| 1/4 | 3,250 | 5,560 | 4,550 | 3,250 | 8,400 | 6,800 | 4,900 |
| 3/8 | 6,600 | 11,400 | 9,300 | 6,600 | 17,000 | 14,000 | 9,900 |
| 1/2 | 11,250 | 19,500 | 15,900 | 11,250 | 29,000 | 24,000 | 17,000 |
| 5/8 | 16,500 | 28,500 | 23,300 | 16,500 | 43,000 | 35,000 | 24,500 |
| 3/4 | 23,000 | 39,800 | 32,500 | 23,000 | 59,500 | 48,500 | 34,500 |
| 7/8 | 28,750 | 49,800 | 40,600 | 28,750 | 74,500 | 61,000 | 43,000 |
| 1 | 38,750 | 67,100 | 54,800 | 38,750 | 101,000 | 82,000 | 58,000 |
| 1 1/8 | 44,500 | 77,000 | 63,000 | 44,500 | 115,500 | 94,500 | 66,500 |
| 1 1/4 | 57,500 | 99,500 | 81,000 | 57,500 | 149,000 | 121,500 | 86,000 |
| 1 3/8 | 67,000 | 116,000 | 94,000 | 67,000 | 174,000 | 141,000 | 100,500 |
| 1 1/2 | 80,000 | 138,000 | 112,500 | 80,000 | 207,000 | 169,000 | 119,500 |
| 1 3/4 | 100,000 | 172,000 | 140,000 | 100,000 | 258,000 | 210,000 | 150,000 |

[1] Other grades of proof tested steel chain include Proof Coil, BBB Coil and Hi-Test Chain. These grades are not recommended for overhead lifting and therefore are not covered by this code.

(1) Rating of multileg slings adjusted for angle of loading measured as the included angle between the inclined leg and the vertical.

(2) Rating of multileg slings adjusted for angle of loading between the inclined leg and the horizontal plane of the load.

## TABLE H-2—MAXIMUM ALLOWABLE WEAR AT ANY POINT OF LINK

| Chain size (inches) | Maximum allowable wear (inch) | Chain size (inches) | Maximum allowable wear (inch) |
|---|---|---|---|
| 1/4 | 3/64 | 1 | 3/16 |
| 3/8 | 5/64 | 1 1/8 | 7/32 |
| 1/2 | 7/64 | 1 1/4 | 1/4 |
| 5/8 | 9/64 | 1 3/8 | 9/32 |
| 3/4 | 5/32 | 1 1/2 | 5/16 |
| 7/8 | 11/64 | 1 3/4 | 11/32 |

## TABLE H-3—RATED CAPACITIES FOR SINGLE LEG SLINGS

6 × 19 and 6 × 37 Classification Improved Plow Steel Grade Rope with Fiber Core (FC)

| *Rope* | | *Rated capacities, tons (2,000 lb.)* | | | | | | | | |
|---|---|---|---|---|---|---|---|---|---|---|
| | | *Vertical* | | | *Choker* | | | *Vertical basket* [1] | | |
| *Dia. (inches)* | *Constr.* | *HT* | *MS* | *S* | *HT* | *MS* | *S* | *HT* | *MS* | *S* |
| 1/4 | 6 × 19 | 0.49 | 0.51 | 0.55 | 0.37 | 0.38 | 0.41 | 0.99 | 1.0 | 1.1 |
| 5/16 | 6 × 19 | 0.76 | 0.79 | 0.85 | 0.57 | 0.59 | 0.64 | 1.5 | 1.6 | 1.7 |
| 3/8 | 6 × 19 | 1.1 | 1.1 | 1.2 | 0.80 | 0.85 | 0.91 | 2.1 | 2.2 | 2.4 |
| 7/16 | 6 × 19 | 1.4 | 1.5 | 1.6 | 1.1 | 1.1 | 1.2 | 2.9 | 3.0 | 3.3 |
| 1/2 | 6 × 19 | 1.8 | 2.0 | 2.1 | 1.4 | 1.5 | 1.6 | 3.7 | 3.9 | 4.3 |
| 9/16 | 6 × 19 | 2.3 | 2.5 | 2.7 | 1.7 | 1.9 | 2.0 | 4.6 | 5.0 | 5.4 |
| 5/8 | 6 × 19 | 2.8 | 3.1 | 3.3 | 2.1 | 2.3 | 2.5 | 5.6 | 6.2 | 6.7 |
| 3/4 | 6 × 19 | 3.9 | 4.4 | 4.8 | 2.9 | 3.3 | 3.6 | 7.8 | 8.8 | 9.5 |
| 7/8 | 6 × 19 | 5.1 | 5.9 | 6.4 | 3.9 | 4.5 | 4.8 | 10.0 | 12.0 | 13.0 |
| 1 | 6 × 19 | 6.7 | 7.7 | 8.4 | 5.0 | 5.8 | 6.3 | 13.0 | 15.0 | 17.0 |
| 1 1/8 | 6 × 19 | 8.4 | 9.5 | 10.0 | 6.3 | 7.1 | 7.9 | 17.0 | 19.0 | 21.0 |
| 1 1/4 | 6 × 37 | 9.8 | 11.0 | 12.0 | 7.4 | 8.3 | 9.2 | 20.0 | 22.0 | 25.0 |
| 1 3/8 | 6 × 37 | 12.0 | 13.0 | 15.0 | 8.9 | 10.0 | 11.0 | 24.0 | 27.0 | 30.0 |
| 1 1/2 | 6 × 37 | 14.0 | 16.0 | 17.0 | 10.0 | 12.0 | 13.0 | 28.0 | 32.0 | 35.0 |
| 1 5/8 | 6 × 37 | 16.0 | 18.0 | 21.0 | 12.0 | 14.0 | 15.0 | 33.0 | 37.0 | 41.0 |
| 1 3/4 | 6 × 37 | 19.0 | 21.0 | 24.0 | 14.0 | 16.0 | 18.0 | 38.0 | 43.0 | 48.0 |
| 2 | 6 × 37 | 25.0 | 28.0 | 31.0 | 18.0 | 21.0 | 23.0 | 49.0 | 55.0 | 62.0 |

[1] These values only apply when the D/d ratio for HT slings is 10 or greater, and for MS and S Slings is 20 or greater where:
D = Diameter of curvature around which the body of the sling is bent. d = Diameter of rope.

HT = Hand Tucked Splice and Hidden Tuck Splice. For hidden tuck splice (IWRC) use values in HT columns.
MS = Mechanical Splice
S = Swaged or Zinc Poured Socket

## TABLE H-4—RATED CAPACITIES FOR SINGLE LEG SLINGS

6 × 19 and 6 × 37 Classification Improved Plow Steel Grade Rope with Independent Wire Rope Core (IWRC)

| *Rope* | | *Rated capacities, tons (2,000 lb.)* | | | | | | | | |
|---|---|---|---|---|---|---|---|---|---|---|
| | | *Vertical* | | | *Choker* | | | *Vertical basket* [1] | | |
| *Dia. (inches)* | *Constr.* | *HT* | *MS* | *S* | *HT* | *MS* | *S* | *HT* | *MS* | *S* |
| 1/4 | 6 × 19 | 0.53 | 0.56 | 0.59 | 0.40 | 0.42 | 0.44 | 1.0 | 1.1 | 1.2 |
| 5/16 | 6 × 19 | 0.81 | 0.87 | 0.92 | 0.61 | 0.65 | 0.69 | 1.6 | 1.7 | 1.8 |
| 3/8 | 6 × 19 | 1.1 | 1.2 | 1.3 | 0.86 | 0.93 | 0.98 | 2.3 | 2.5 | 2.6 |
| 7/16 | 6 × 19 | 1.5 | 1.7 | 1.8 | 1.2 | 1.3 | 1.3 | 3.1 | 3.4 | 3.5 |
| 1/2 | 6 × 19 | 2.0 | 2.2 | 2.3 | 1.5 | 1.6 | 1.7 | 3.9 | 4.4 | 4.6 |
| 9/16 | 6 × 19 | 2.5 | 2.7 | 2.9 | 1.8 | 2.1 | 2.2 | 4.9 | 5.5 | 5.8 |
| 5/8 | 6 × 19 | 3.0 | 3.4 | 3.6 | 2.2 | 2.5 | 2.7 | 6.0 | 6.8 | 7.2 |
| 3/4 | 6 × 19 | 4.2 | 4.9 | 5.1 | 3.1 | 3.6 | 3.8 | 8.4 | 9.7 | 10.0 |
| 7/8 | 6 × 19 | 5.5 | 6.6 | 6.9 | 4.1 | 4.9 | 5.2 | 11.0 | 13.0 | 14.0 |
| 1 | 6 × 19 | 7.2 | 8.5 | 9.0 | 5.4 | 6.4 | 6.7 | 14.0 | 17.0 | 18.0 |
| 1 1/8 | 6 × 19 | 9.0 | 10.0 | 11.0 | 6.8 | 7.8 | 8.5 | 18.0 | 21.0 | 23.0 |
| 1 1/4 | 6 × 37 | 10.0 | 12.0 | 13.0 | 7.9 | 9.2 | 9.9 | 21.0 | 24.0 | 26.0 |
| 1 3/8 | 6 × 37 | 13.0 | 15.0 | 16.0 | 9.6 | 11.0 | 12.0 | 25.0 | 29.0 | 32.0 |
| 1 1/2 | 6 × 37 | 15.0 | 17.0 | 19.0 | 11.0 | 13.0 | 14.0 | 30.0 | 35.0 | 38.0 |
| 1 5/8 | 6 × 37 | 18.0 | 20.0 | 22.0 | 13.0 | 15.0 | 17.0 | 35.0 | 41.0 | 44.0 |
| 1 3/4 | 6 × 37 | 20.0 | 24.0 | 26.0 | 15.0 | 18.0 | 19.0 | 41.0 | 47.0 | 51.0 |
| 2 | 6 × 37 | 26.0 | 30.0 | 33.0 | 20.0 | 23.0 | 25.0 | 53.0 | 61.0 | 66.0 |

[1] These values only apply when the D/d ratio for HT slings is 10 or greater, and for MS and S Slings is 20 or greater where:
D = Diameter of curvature around which the body of the sling is bent. d = Diameter of rope.

HT = Hand Tucked Splice: For hidden tuck splice (IWRC) use Table H-3 values in HT column.
MS = Mechanical Splice.
S = Swaged or Zinc Poured Socket.

## TABLE H-5—RATED CAPACITIES FOR SINGLE LEG SLINGS

Cable Laid Rope—Mechanical Splice Only
7 × 7 × 7 and 7 × 7 × 19 Construction Galvanized Aircraft Grade Rope
7 × 6 × 19 IWRC Construction Improved Plow Steel Grade Rope

| *Rope* | | *Rated capacities, tons (2,000 lb.)* | | |
|---|---|---|---|---|
| *Dia. (inches)* | *Constr.* | *Vertical* | *Choker* | *Vertical basket* [1] |
| 1/4 | 7 × 7 × 7 | 0.50 | 0.38 | 1.0 |
| 3/8 | 7 × 7 × 7 | 1.1 | 0.81 | 2.2 |
| 1/2 | 7 × 7 × 7 | 1.8 | 1.4 | 3.7 |
| 5/8 | 7 × 7 × 7 | 2.8 | 2.1 | 5.5 |
| 3/4 | 7 × 7 × 7 | 3.8 | 2.9 | 7.6 |
| 5/8 | 7 × 7 × 19 | 2.9 | 2.2 | 5.8 |
| 3/4 | 7 × 7 × 19 | 4.1 | 3.0 | 8.1 |
| 7/8 | 7 × 7 × 19 | 5.4 | 4.0 | 11.0 |
| 1 | 7 × 7 × 19 | 6.9 | 5.1 | 14.0 |
| 1 1/8 | 7 × 7 × 19 | 8.2 | 6.2 | 16.0 |
| 1 1/4 | 7 × 7 × 19 | 9.9 | 7.4 | 20.0 |
| 3/4 | [2] 7 × 6 × 19 | 3.8 | 2.8 | 7.6 |
| 7/8 | [2] 7 × 6 × 19 | 5.0 | 3.8 | 10.0 |
| 1 | [2] 7 × 6 × 19 | 6.4 | 4.8 | 13.0 |
| 1 1/8 | [2] 7 × 6 × 19 | 7.7 | 5.8 | 15.0 |
| 1 1/4 | [2] 7 × 6 × 19 | 9.2 | 6.9 | 18.0 |
| 1 5/16 | [2] 7 × 6 × 19 | 10.0 | 7.5 | 20.0 |
| 1 3/8 | [2] 7 × 6 × 19 | 11.0 | 8.2 | 22.0 |
| 1 1/2 | [2] 7 × 6 × 19 | 13.0 | 9.6 | 26.0 |

[1] These values only apply when the D/d ratio in 10 or greater where: D=Diameter of curvature around which the body of the sling is bent. d=Diameter of rope.
[2] IWRC.

## TABLE H-6—RATED CAPACITIES FOR SINGLE LEG SLINGS

8-Part and 6-Part Braided Rope
6 × 7 and 6 × 19 Construction Improved Plow Steel Grade Rope
7 × 7 Construction Galvanized Aircraft Grade Rope

| *Component ropes* | | *Rated capacities, tons (2,000 lb.)* | | | | | |
|---|---|---|---|---|---|---|---|
| | | *Vertical* | | *Choker* | | *Basket vertical to 30°* [1] | |
| *Diameter (inches)* | *Constr.* | *8-Part* | *6-Part* | *8-Part* | *6-Part* | *8-Part* | *6-Part* |
| 3/32 | 6 × 7 | 0.42 | 0.32 | 0.32 | 0.24 | 0.74 | 0.55 |
| 1/8 | 6 × 7 | 0.76 | 0.57 | 0.57 | 0.42 | 1.3 | 0.98 |
| 3/16 | 6 × 7 | 1.7 | 1.3 | 1.3 | 0.94 | 2.9 | 2.2 |
| 3/32 | 7 × 7 | 0.51 | 0.39 | 0.38 | 0.29 | 0.89 | 0.67 |
| 1/8 | 7 × 7 | 0.95 | 0.71 | 0.71 | 0.53 | 1.6 | 1.2 |
| 3/16 | 7 × 7 | 2.1 | 1.5 | 1.5 | 1.2 | 3.6 | 2.7 |
| 3/16 | 6 × 19 | 1.7 | 1.3 | 1.3 | 0.98 | 3.0 | 2.2 |
| 1/4 | 6 × 19 | 3.1 | 2.3 | 2.3 | 1.7 | 5.3 | 4.0 |
| 5/16 | 6 × 19 | 4.8 | 3.6 | 3.6 | 2.7 | 8.3 | 6.2 |
| 3/8 | 6 × 19 | 6.8 | 5.1 | 5.1 | 3.8 | 12.0 | 8.9 |
| 7/16 | 6 × 19 | 9.3 | 6.9 | 6.9 | 5.2 | 16.0 | 12.0 |
| 1/2 | 6 × 19 | 12.0 | 9.0 | 9.0 | 6.7 | 21.0 | 15.0 |
| 9/16 | 6 × 19 | 15.0 | 11.0 | 11.0 | 8.5 | 26.0 | 20.0 |
| 5/8 | 6 × 19 | 19.0 | 14.0 | 14.0 | 10.0 | 32.0 | 24.0 |
| 3/4 | 6 × 19 | 27.0 | 20.0 | 20.0 | 15.0 | 46.0 | 35.0 |
| 7/8 | 6 × 19 | 36.0 | 27.0 | 27.0 | 20.0 | 62.0 | 47.0 |
| 1 | 6 × 19 | 47.0 | 35.0 | 35.0 | 26.0 | 81.0 | 61.0 |

[1] These values only apply when the D/d ratio is 20 or greater where: D=Diameter of curvature around which the body of the sling is bent. d=Diameter of component rope.

## TABLE H-7—RATED CAPACITIES FOR 2-LEG AND 3-LEG BRIDLE SLINGS

6 × 19 and 6 × 37 Classification Improved Plow Steel Grade Rope with Fiber Core (FC)

| *Rope* | | *Rated capacities, tons (2,000 lb.)* | | | | | | | | | | | |
|---|---|---|---|---|---|---|---|---|---|---|---|---|---|
| | | *2-leg bridle slings* | | | | | | *3-leg bridle slings* | | | | | |
| *Dia. (inches)* | *Constr.* | *30° [1] (60°) [2]* | | *45° angle* | | *60° [1] (30°) [2]* | | *30° [1] (60°) [2]* | | *45° angle* | | *60° [1] (30°) [2]* | |
| | | *HT* | *MS* | *HT* | *MS* | *HT* | *MS* | *HT* | *MS* | *HT* | *MS* | *HT* | *MS* |
| 1/4 | 6 × 19 | 0.85 | 0.88 | 0.70 | 0.72 | 0.49 | 0.51 | 1.3 | 1.3 | 1.0 | 1.1 | 0.74 | 0.7 |
| 5/16 | 6 × 19 | 1.3 | 1.4 | 1.1 | 1.1 | 0.76 | 0.79 | 2.0 | 2.0 | 1.6 | 1.7 | 1.1 | 1.2 |
| 3/8 | 6 × 19 | 1.8 | 1.9 | 1.5 | 1.6 | 1.1 | 1.1 | 2.8 | 2.9 | 2.3 | 2.4 | 1.6 | 1.7 |
| 7/16 | 6 × 19 | 2.5 | 2.6 | 2.0 | 2.2 | 1.4 | 1.5 | 3.7 | 4.0 | 3.0 | 3.2 | 2.1 | 2.3 |
| 1/2 | 6 × 19 | 3.2 | 3.4 | 2.6 | 2.8 | 1.8 | 2.0 | 4.8 | 5.1 | 3.9 | 4.2 | 2.8 | 3.0 |
| 9/16 | 6 × 19 | 4.0 | 4.3 | 3.2 | 3.5 | 2.3 | 2.5 | 6.0 | 6.5 | 4.9 | 5.3 | 3.4 | 3.7 |
| 5/8 | 6 × 19 | 4.8 | 5.3 | 4.0 | 4.4 | 2.8 | 3.1 | 7.3 | 8.0 | 5.9 | 6.5 | 4.2 | 4.6 |
| 3/4 | 6 × 19 | 6.8 | 7.6 | 5.5 | 6.2 | 3.9 | 4.4 | 10.0 | 11.0 | 8.3 | 9.3 | 5.8 | 6.6 |
| 7/8 | 6 × 19 | 8.9 | 10.0 | 7.3 | 8.4 | 5.1 | 5.9 | 13.0 | 15.0 | 11.0 | 13.0 | 7.7 | 8.9 |
| 1 | 6 × 19 | 11.0 | 13.0 | 9.4 | 11.0 | 6.7 | 7.7 | 17.0 | 20.0 | 14.0 | 16.0 | 10.0 | 11.0 |
| 1 1/8 | 6 × 19 | 14.0 | 16.0 | 12.0 | 13.0 | 8.4 | 9.5 | 22.0 | 24.0 | 18.0 | 20.0 | 13.0 | 14.0 |
| 1 1/4 | 6 × 37 | 17.0 | 19.0 | 14.0 | 16.0 | 9.8 | 11.0 | 25.0 | 29.0 | 21.0 | 23.0 | 15.0 | 17.0 |
| 1 3/8 | 6 × 37 | 20.0 | 23.0 | 17.0 | 19.0 | 12.0 | 13.0 | 31.0 | 35.0 | 25.0 | 28.0 | 18.0 | 20.0 |
| 1 1/2 | 6 × 37 | 24.0 | 27.0 | 20.0 | 22.0 | 14.0 | 16.0 | 36.0 | 41.0 | 30.0 | 33.0 | 21.0 | 24.0 |
| 1 5/8 | 6 × 37 | 28.0 | 32.0 | 23.0 | 26.0 | 16.0 | 18.0 | 43.0 | 48.0 | 35.0 | 39.0 | 25.0 | 28.0 |
| 1 3/4 | 6 × 37 | 33.0 | 37.0 | 27.0 | 30.0 | 19.0 | 21.0 | 49.0 | 56.0 | 40.0 | 45.0 | 28.0 | 32.0 |
| 2 | 6 × 37 | 43.0 | 48.0 | 35.0 | 39.0 | 25.0 | 28.0 | 64.0 | 72.0 | 52.0 | 59.0 | 37.0 | 41.0 |

HT = Hand Tucked Splice.
MS = Mechanical Splice.
[1] Vertical angles.
[2] Horizontal angles.

## TABLE H-8—RATED CAPACITIES FOR 2-LEG AND 3-LEG BRIDLE SLINGS

6 × 19 and 6 × 37 Classification Improved Plow Steel Grade Rope with Independent Wire Rope Core (IWRC)

| *Rope* | | *Rated capacities, tons (2,000 lb.)* | | | | | | | | | | | |
|---|---|---|---|---|---|---|---|---|---|---|---|---|---|
| | | *2-leg bridle slings* | | | | | | *3-leg bridle slings* | | | | | |
| *Dia. (inches)* | *Constr.* | *30° [1] (60°) [2]* | | *45° angle* | | *60° [1] (30°) [2]* | | *30° [1] (60°) [2]* | | *45° angle* | | *60° [1] (30°) [2]* | |
| | | *HT* | *MS* | *HT* | *MS* | *HT* | *MS* | *HT* | *MS* | *HT* | *MS* | *HT* | *MS* |
| 1/4 | 6 × 19 | 0.92 | 0.97 | 0.75 | 0.79 | 0.53 | 0.56 | 1.4 | 1.4 | 1.1 | 1.2 | 0.79 | 0.84 |
| 5/16 | 6 × 19 | 1.4 | 1.5 | 1.1 | 1.2 | 1.81 | 0.87 | 2.1 | 2.3 | 1.7 | 1.8 | 1.2 | 1.3 |
| 3/8 | 6 × 19 | 2.0 | 2.1 | 1.6 | 1.8 | 1.1 | 1.2 | 3.0 | 3.2 | 2.4 | 2.6 | 1.7 | 1.9 |
| 7/16 | 6 × 19 | 2.7 | 2.9 | 2.2 | 2.4 | 1.5 | 1.7 | 4.0 | 4.4 | 3.3 | 3.6 | 2.3 | 2.5 |
| 1/2 | 6 × 19 | 3.4 | 3.8 | 2.8 | 3.1 | 2.0 | 2.2 | 5.1 | 5.7 | 4.2 | 4.6 | 3.0 | 3.3 |
| 9/16 | 6 × 19 | 4.3 | 4.8 | 3.5 | 3.9 | 2.5 | 2.7 | 6.4 | 7.1 | 5.2 | 5.8 | 3.7 | 4.1 |
| 5/8 | 6 × 19 | 5.2 | 5.9 | 4.2 | 4.8 | 3.0 | 3.4 | 7.8 | 8.8 | 6.4 | 7.2 | 4.5 | 5.1 |
| 3/4 | 6 × 19 | 7.3 | 8.4 | 5.9 | 6.9 | 4.2 | 4.9 | 11.0 | 13.0 | 8.9 | 10.0 | 6.3 | 7.3 |
| 7/8 | 6 × 19 | 9.6 | 11.0 | 7.8 | 9.3 | 5.5 | 6.6 | 14.0 | 17.0 | 12.0 | 14.0 | 8.3 | 9.9 |
| 1 | 6 × 19 | 12.0 | 15.0 | 10.0 | 12.0 | 7.2 | 8.5 | 19.0 | 22.0 | 15.0 | 18.0 | 11.0 | 13.0 |
| 1 1/8 | 6 × 19 | 16.0 | 18.0 | 13.0 | 15.0 | 9.0 | 10.0 | 23.0 | 27.0 | 19.0 | 22.0 | 13.0 | 16.0 |
| 1 1/4 | 6 × 37 | 18.0 | 21.0 | 15.0 | 17.0 | 10.0 | 12.0 | 27.0 | 32.0 | 22.0 | 26.0 | 16.0 | 18.0 |
| 1 3/8 | 6 × 37 | 22.0 | 25.0 | 18.0 | 21.0 | 13.0 | 15.0 | 33.0 | 38.0 | 27.0 | 31.0 | 19.0 | 22.0 |
| 1 1/2 | 6 × 37 | 26.0 | 30.0 | 21.0 | 25.0 | 15.0 | 17.0 | 39.0 | 45.0 | 32.0 | 37.0 | 23.0 | 26.0 |
| 1 5/8 | 6 × 37 | 31.0 | 35.0 | 25.0 | 29.0 | 18.0 | 20.0 | 46.0 | 53.0 | 38.0 | 43.0 | 27.0 | 31.0 |
| 1 3/4 | 6 × 37 | 35.0 | 41.0 | 29.0 | 33.0 | 20.0 | 24.0 | 53.0 | 61.0 | 43.0 | 50.0 | 31.0 | 35.0 |
| 2 | 6 × 37 | 46.0 | 53.0 | 37.0 | 43.0 | 26.0 | 30.0 | 68.0 | 79.0 | 56.0 | 65.0 | 40.0 | 46.0 |

HT = Hand Tucked Splice.
MS = Mechanical Splice.
[1] Vertical angles.
[2] Horizontal angles.

## TABLE H-9—RATED CAPACITIES FOR 2-LEG AND 3-LEG BRIDLE SLINGS

Cable Laid Rope—Mechanical Splice Only

7×7×7 and 7×7×19 Construction Galvanized Aircraft Grade Rope
7×6×19 IWRC Construction Improved Plow Steel Grade Rope

| Rope | | Rated capacities, tons (2,000 lb.) | | | | | |
|---|---|---|---|---|---|---|---|
| | | 2-leg bridle sling | | | 3-leg bridle sling | | |
| Dia. (inches) | Constr. | 30° [1] (60°) [2] | 45° angle | 60° [1] (30°) [2] | 30° [1] (60°) [2] | 45° angle | 60° [1] (30°) [2] |
| 1/4 | 7×7×7 | 0.87 | 0.71 | 0.50 | 1.3 | 1.1 | 0.75 |
| 3/8 | 7×7×7 | 1.9 | 1.5 | 1.1 | 2.8 | 2.3 | 1.6 |
| 1/2 | 7×7×7 | 3.2 | 2.6 | 1.8 | 4.8 | 3.9 | 2.8 |
| 5/8 | 7×7×7 | 4.8 | 3.9 | 2.8 | 7.2 | 5.9 | 4.2 |
| 3/4 | 7×7×7 | 6.6 | 5.4 | 3.8 | 9.9 | 8.1 | 5.7 |
| 5/8 | 7×7×19 | 5.0 | 4.1 | 2.9 | 7.5 | 6.1 | 4.3 |
| 3/4 | 7×7×19 | 7.0 | 5.7 | 4.1 | 10.0 | 8.6 | 6.1 |
| 7/8 | 7×7×19 | 9.3 | 7.6 | 5.4 | 14.0 | 11.0 | 8.1 |
| 1 | 7×7×19 | 12.0 | 9.7 | 6.9 | 18.0 | 14.0 | 10.0 |
| 1 1/8 | 7×7×19 | 14.0 | 12.0 | 8.2 | 21.0 | 17.0 | 12.0 |
| 1 1/4 | 7×7×19 | 17.0 | 14.0 | 9.9 | 26.0 | 21.0 | 15.0 |
| 3/4 | 7×6×19 IWRC | 6.6 | 5.4 | 3.8 | 9.9 | 8.0 | 5.7 |
| 7/8 | 7×6×19 IWRC | 8.7 | 7.1 | 5.0 | 13.0 | 11.0 | 7.5 |
| 1 | 7×6×19 IWRC | 11.0 | 9.0 | 6.4 | 17.0 | 13.0 | 9.6 |
| 1 1/8 | 7×6×19 IWRC | 13.0 | 11.0 | 7.7 | 20.0 | 16.0 | 11.0 |
| 1 1/4 | 7×6×19 IWRC | 16.0 | 13.0 | 9.2 | 24.0 | 20.0 | 14.0 |
| 1 5/16 | 7×6×19 IWRC | 17.0 | 14.0 | 10.0 | 26.0 | 21.0 | 15.0 |
| 1 3/8 | 7×6×19 IWRC | 19.0 | 15.0 | 11.0 | 28.0 | 23.0 | 16.0 |
| 1 1/2 | 7×6×19 IWRC | 22.0 | 18.0 | 13.0 | 33.0 | 27.0 | 19.0 |

[1] Vertical angles.
[2] Horizontal angles.

## TABLE H-10—RATED CAPACITIES FOR 2-LEG AND 3-LEG BRIDLE SLINGS

8-Part and 6-Part Braided Rope

6×7 and 6×19 Construction Improved Plow Steel Grade Rope
7×7 Construction Galvanized Aircraft Grade Rope

| Rope | | Rated capacities, tons (2,000 lb.) | | | | | | | | | | | |
|---|---|---|---|---|---|---|---|---|---|---|---|---|---|
| | | 2-leg bridle slings | | | | | | 3-leg bridle slings | | | | | |
| Dia. (inches) | Constr. | 30° [1] (60°) [2] | | 45° Angle | | 60° [1] (30°) [2] | | 30° [1] (60°) [2] | | 45° Angle | | 60° [1] (30°) [2] | |
| | | 8-Part | 6-Part | 8-Part | 6-Part | 8-Part | 6-Part | 8-Part | 6-Part | 8-Part | 6-Part | 8-Part | 6-Part |
| 3/32 | 6×7 | 0.74 | 0.55 | 0.60 | 0.45 | 0.42 | 0.32 | 1.1 | 0.83 | 0.90 | 0.68 | 0.64 | 0.48 |
| 1/8 | 6×7 | 1.3 | 0.98 | 1.1 | 0.80 | 0.76 | 0.57 | 2.0 | 1.5 | 1.6 | 1.2 | 1.1 | 0.85 |
| 3/16 | 6×7 | 2.9 | 2.2 | 2.4 | 1.8 | 1.7 | 1.3 | 4.4 | 3.3 | 3.6 | 2.7 | 2.5 | 1.9 |
| 3/32 | 7×7 | 0.89 | 0.67 | 0.72 | 0.55 | 0.51 | 0.39 | 1.3 | 1.0 | 1.1 | 0.82 | 0.77 | 0.58 |
| 1/8 | 7×7 | 1.6 | 1.2 | 1.3 | 1.0 | 0.95 | 0.71 | 2.5 | 1.8 | 2.0 | 1.5 | 1.4 | 1.1 |
| 3/16 | 7×7 | 3.6 | 2.7 | 2.9 | 2.2 | 2.1 | 1.5 | 5.4 | 4.0 | 4.4 | 3.3 | 3.1 | 2.3 |
| 3/16 | 6×19 | 3.0 | 2.2 | 2.4 | 1.8 | 1.7 | 1.3 | 4.5 | 3.4 | 3.7 | 2.8 | 2.6 | 1.9 |
| 1/4 | 6×19 | 5.3 | 4.0 | 4.3 | 3.2 | 3.1 | 2.3 | 8.0 | 6.0 | 6.5 | 4.9 | 4.6 | 3.4 |
| 5/16 | 6×19 | 8.3 | 6.2 | 6.7 | 5.0 | 4.8 | 3.6 | 12.0 | 9.3 | 10.0 | 7.6 | 7.1 | 5.4 |
| 3/8 | 6×19 | 12.0 | 8.9 | 9.7 | 7.2 | 6.8 | 5.1 | 18.0 | 13.0 | 14.0 | 11.0 | 10.0 | 7.7 |
| 7/16 | 6×19 | 16.0 | 12.0 | 13.0 | 9.8 | 9.3 | 6.9 | 24.0 | 18.0 | 20.0 | 15.0 | 14.0 | 10.0 |
| 1/2 | 6×19 | 21.0 | 15.0 | 17.0 | 13.0 | 12.0 | 9.0 | 31.0 | 23.0 | 25.0 | 19.0 | 18.0 | 13.0 |
| 9/16 | 6×19 | 26.0 | 20.0 | 21.0 | 16.0 | 15.0 | 11.0 | 39.0 | 29.0 | 32.0 | 24.0 | 23.0 | 17.0 |
| 5/8 | 6×19 | 32.0 | 24.0 | 26.0 | 20.0 | 19.0 | 14.0 | 48.0 | 36.0 | 40.0 | 30.0 | 28.0 | 21.0 |
| 3/4 | 6×19 | 46.0 | 35.0 | 38.0 | 28.0 | 27.0 | 20.0 | 69.0 | 52.0 | 56.0 | 42.0 | 40.0 | 30.0 |
| 7/8 | 6×19 | 62.0 | 47.0 | 51.0 | 38.0 | 36.0 | 27.0 | 94.0 | 70.0 | 76.0 | 57.0 | 54.0 | 40.0 |
| 1 | 6×19 | 81.0 | 61.0 | 66.0 | 50.0 | 47.0 | 35.0 | 22.0 | 91.0 | 99.0 | 74.0 | 70.0 | 53.0 |

[1] Vertical angles.
[2] Horizontal angles.

## TABLE H-11—RATED CAPACITIES FOR STRAND LAID GROMMET—HAND TUCKED

Improved Plow Steel Grade Rope

| Rope body | | Rated capacities, tons (2,000 lb.) | | |
|---|---|---|---|---|
| Dia. (inches) | Constr. | Vertical | Choker | Vertical basket [1] |
| 1/4 | 7 × 19 | 0.85 | 0.64 | 1.7 |
| 5/16 | 7 × 19 | 1.3 | 1.0 | 2.6 |
| 3/8 | 7 × 19 | 1.9 | 1.4 | 3.8 |
| 7/16 | 7 × 19 | 2.6 | 1.9 | 5.2 |
| 1/2 | 7 × 19 | 3.3 | 2.5 | 6.7 |
| 9/16 | 7 × 19 | 4.2 | 3.1 | 8.4 |
| 5/8 | 7 × 19 | 5.2 | 3.9 | 10.00 |
| 3/4 | 7 × 19 | 7.4 | 5.6 | 15.0 |
| 7/8 | 7 × 19 | 10.0 | 7.5 | 20.0 |
| 1 | 7 × 19 | 13.0 | 9.7 | 26.0 |
| 1 1/8 | 7 × 19 | 16.0 | 12.0 | 32.0 |
| 1 1/4 | 7 × 37 | 18.0 | 14.0 | 37.0 |
| 1 3/8 | 7 × 37 | 22.0 | 16.0 | 44.0 |
| 1 1/2 | 7 × 37 | 26.0 | 19.0 | 52.0 |

[1] These values only apply when the D/d ratio is 5 or greater where: D = Diameter of curvature around which rope is bent. d = Diameter of rope body.

## TABLE H-12—RATED CAPACITIES FOR CABLE LAID GROMMET—HAND TUCKED

7 × 6 × 7 and 7 × 6 × 19 Construction Improved Plow Steel Grade Rope

7 × 7 × 7 Construction Galvanized Aircraft Grade Rope

| Cable body | | Rated capacities, tons (2,000 lb.) | | |
|---|---|---|---|---|
| Dia. (inches) | Constr. | Vertical | Choker | Vertical basket [1] |
| 3/8 | 7 × 6 × 7 | 1.3 | 0.95 | 2.5 |
| 9/16 | 7 × 6 × 7 | 2.8 | 2.1 | 5.6 |
| 5/8 | 7 × 6 × 7 | 3.8 | 2.8 | 7.6 |
| 3/8 | 7 × 7 × 7 | 1.6 | 1.2 | 3.2 |
| 9/16 | 7 × 7 × 7 | 3.5 | 2.6 | 6.9 |
| 5/8 | 7 × 7 × 7 | 4.5 | 3.4 | 9.0 |
| 5/8 | 7 × 6 × 19 | 3.9 | 3.0 | 7.9 |
| 3/4 | 7 × 6 × 19 | 5.1 | 3.8 | 10.0 |
| 15/16 | 7 × 6 × 19 | 7.9 | 5.9 | 16.0 |
| 1 1/8 | 7 × 6 × 19 | 11.0 | 8.4 | 22.0 |
| 1 5/16 | 7 × 6 × 19 | 15.0 | 11.0 | 30.0 |
| 1 1/2 | 7 × 6 × 19 | 19.0 | 14.0 | 39.0 |
| 1 11/16 | 7 × 6 × 19 | 24.0 | 18.0 | 49.0 |
| 1 7/8 | 7 × 6 × 19 | 30.0 | 22.0 | 60.0 |
| 2 1/4 | 7 × 6 × 19 | 42.0 | 31.0 | 84.0 |
| 2 5/8 | 7 × 6 × 19 | 56.0 | 42.0 | 112.0 |

[1] These values only apply when the D/d ratio is 5 or greater where: D = Diameter of curvature around which cable body is bent. d = Diameter of cable body.

## TABLE H-13—RATED CAPACITIES FOR STRAND LAID ENDLESS SLINGS—MECHANICAL JOINT

Improved Plow Steel Grade Rope

| Rope body | | Rated capacities, tons (2,000 lb.) | | |
|---|---|---|---|---|
| Dia. (inches) | Constr. | Vertical | Choker | Vertical basket [1] |
| 1/4 | [2] 6 × 19 | 0.92 | 0.69 | 1.8 |
| 3/8 | [2] 6 × 19 | 2.0 | 1.5 | 4.1 |
| 1/2 | [2] 6 × 19 | 3.6 | 2.7 | 7.2 |
| 5/8 | [2] 6 × 19 | 5.6 | 4.2 | 11.0 |
| 3/4 | [2] 6 × 19 | 8.0 | 6.0 | 16.0 |
| 7/8 | [2] 6 × 19 | 11.0 | 8.1 | 21.0 |
| 1 | [2] 6 × 19 | 14.0 | 10.0 | 28.0 |
| 1 1/8 | [2] 6 × 19 | 18.0 | 13.0 | 35.0 |
| 1 1/4 | [2] 6 × 37 | 21.0 | 15.0 | 41.0 |
| 1 3/8 | [2] 6 × 37 | 25.0 | 19.0 | 50.0 |
| 1 1/2 | [2] 6 × 37 | 29.0 | 22.0 | 59.0 |

[1] These values only apply when the D/d ratio is 5 or greater where: D = Diameter of curvature around which rope is bent. d = Diameter of rope body.
[2] IWRC.

## TABLE H-14—RATED CAPACITIES FOR CABLE LAID ENDLESS SLINGS—MECHANICAL JOINT

7 × 7 × 7 and 7 × 7 × 19 Construction Galvanized Aircraft Grade Rope

7 × 6 × 19 IWRC Construction Improved Plow Steel Grade Rope

| Cable body | | Rated capacities, tons (2,000 lb.) | | |
|---|---|---|---|---|
| Dia. (inches) | Constr. | Vertical | Choker | Vertical basket [1] |
| 1/4 | 7 × 7 × 7 | 0.83 | 0.62 | 1.6 |
| 3/8 | 7 × 7 × 7 | 1.8 | 1.3 | 3.5 |
| 1/2 | 7 × 7 × 7 | 3.0 | 2.3 | 6.1 |
| 5/8 | 7 × 7 × 7 | 4.5 | 3.4 | 9.1 |
| 3/4 | 7 × 7 × 7 | 6.3 | 4.7 | 12.0 |
| 5/8 | 7 × 7 × 19 | 4.7 | 3.5 | 9.5 |
| 3/4 | 7 × 7 × 19 | 6.7 | 5.0 | 13.0 |
| 7/8 | 7 × 7 × 19 | 8.9 | 6.6 | 18.0 |
| 1 | 7 × 7 × 19 | 11.0 | 8.5 | 22.0 |
| 1 1/8 | 7 × 7 × 19 | 14.0 | 10.0 | 28.0 |
| 1 1/4 | 7 × 7 × 19 | 17.0 | 12.0 | 33.0 |
| 3/4 | [2] 7 × 6 × 19 | 6.2 | 4.7 | 12.0 |
| 7/8 | [2] 7 × 6 × 19 | 8.3 | 6.2 | 16.0 |
| 1 | [2] 7 × 6 × 19 | 10.0 | 7.9 | 21.0 |
| 1 1/8 | [2] 7 × 6 × 19 | 13.0 | 9.7 | 26.0 |
| 1 1/4 | [2] 7 × 6 × 19 | 16.0 | 12.0 | 31.0 |
| 1 3/8 | [2] 7 × 6 × 19 | 18.0 | 14.0 | 37.0 |
| 1 1/2 | [2] 7 × 6 × 19 | 22.0 | 16.0 | 43.0 |

[1] These values only apply when the D/d ratio is 5 or greater where: D = Diameter of curvature around which cable body is bent. d = Diameter of cable body.

[2] IWRC.

## TABLE H-15—MANILA ROPE SLINGS

[Angle of rope to vertical shown in parentheses]

| Rope diameter nominal in inches | Nominal weight per 100 ft. in pounds | Minimum breaking strength in pounds | Rated capacity in pounds (safety factor = 5) | | | | | | | | | | | |
|---|---|---|---|---|---|---|---|---|---|---|---|---|---|---|
| | | | Eye and eye sling | | | | | | Endless sling | | | | | |
| | | | Vertical hitch | Choker hitch | Basket hitch; angle of rope to horizontal | | | | Vertical hitch | Choker hitch | Basket hitch; angle of rope to horizontal | | | |
| | | | | | 90° (0°) | 60° (30°) | 45° (45°) | 30° (60°) | | | 90° (0°) | 60° (30°) | 45° (45°) | 30° (60°) |
| 1/2 | 7.5 | 2,650 | 550 | 250 | 1,100 | 900 | 750 | 550 | 950 | 500 | 1,900 | 1,700 | 1,400 | 950 |
| 9/16 | 10.4 | 3,450 | 700 | 350 | 1,400 | 1,200 | 1,000 | 700 | 1,200 | 600 | 2,500 | 2,200 | 1,800 | 1,200 |
| 5/8 | 13.3 | 4,400 | 900 | 450 | 1,800 | 1,500 | 1,200 | 900 | 1,600 | 800 | 3,200 | 2,700 | 2,200 | 1,600 |
| 3/4 | 16.7 | 5,400 | 1,100 | 550 | 2,200 | 1,900 | 1,500 | 1,100 | 2,000 | 950 | 3,900 | 3,400 | 2,800 | 2,000 |
| 13/16 | 19.5 | 6,500 | 1,300 | 650 | 2,600 | 2,300 | 1,800 | 1,300 | 2,300 | 1,200 | 4,700 | 4,100 | 3,300 | 2,300 |
| 7/8 | 22.5 | 7,700 | 1,500 | 750 | 3,100 | 2,700 | 2,200 | 1,500 | 2,800 | 1,400 | 5,600 | 4,800 | 3,900 | 2,800 |
| 1 | 27.0 | 9,000 | 1,800 | 900 | 3,600 | 3,100 | 2,600 | 1,800 | 3,200 | 1,600 | 6,500 | 5,600 | 4,600 | 3,200 |
| 1 1/16 | 31.3 | 10,500 | 2,100 | 1,100 | 4,200 | 3,600 | 3,000 | 2,100 | 3,800 | 1,900 | 7,600 | 6,600 | 5,400 | 3,800 |
| 1 1/8 | 36.0 | 12,000 | 2,400 | 1,200 | 4,800 | 4,200 | 3,400 | 2,400 | 4,300 | 2,200 | 8,600 | 7,500 | 6,100 | 4,300 |
| 1 1/4 | 41.7 | 13,500 | 2,700 | 1,400 | 5,400 | 4,700 | 3,800 | 2,700 | 4,900 | 2,400 | 9,700 | 8,400 | 6,900 | 4,900 |
| 1 5/16 | 47.9 | 15,000 | 3,000 | 1,500 | 6,000 | 5,200 | 4,300 | 3,000 | 5,400 | 2,700 | 11,000 | 9,400 | 7,700 | 5,400 |
| 1 1/2 | 59.9 | 18,500 | 3,700 | 1,850 | 7,400 | 6,400 | 5,200 | 3,700 | 6,700 | 3,300 | 13,500 | 11,500 | 9,400 | 6,700 |
| 1 5/8 | 74.6 | 22,500 | 4,500 | 2,300 | 9,000 | 7,800 | 6,400 | 4,500 | 8,100 | 4,100 | 16,000 | 14,000 | 11,500 | 8,000 |
| 1 3/4 | 89.3 | 26,500 | 5,300 | 2,700 | 10,500 | 9,200 | 7,500 | 5,300 | 9,500 | 4,800 | 19,000 | 16,500 | 13,500 | 9,500 |
| 2 | 107.5 | 31,000 | 6,200 | 3,100 | 12,500 | 10,500 | 8,800 | 6,200 | 11,000 | 5,600 | 22,500 | 19,500 | 16,000 | 11,000 |
| 2 1/3 | 125.0 | 36,000 | 7,200 | 3,600 | 14,500 | 12,500 | 10,000 | 7,200 | 13,000 | 6,500 | 26,000 | 22,500 | 18,500 | 13,000 |
| 2 1/4 | 146.0 | 41,000 | 8,200 | 4,100 | 16,500 | 14,000 | 11,500 | 8,200 | 15,000 | 7,400 | 29,500 | 25,500 | 21,000 | 15,000 |
| 2 1/2 | 166.7 | 46,500 | 9,300 | 4,700 | 18,500 | 16,000 | 13,000 | 9,300 | 16,500 | 8,400 | 33,500 | 29,000 | 23,500 | 16,500 |
| 2 5/8 | 190.8 | 52,000 | 10,500 | 5,200 | 21,000 | 18,000 | 14,500 | 10,500 | 18,500 | 9,500 | 37,500 | 32,500 | 26,500 | 18,500 |

## TABLE H-16—NYLON ROPE SLINGS

[Angle of rope to vertical shown in parentheses]

| Rope diameter nominal in inches | Nominal weight per 100 ft. in pounds | Minimum breaking strength in pounds | Rated capacity in pounds (safety factor = 9) | | | | | | | | | | | |
|---|---|---|---|---|---|---|---|---|---|---|---|---|---|---|
| | | | Eye and eye sling | | | | | | Endless sling | | | | | |
| | | | | | Basket hitch; angle of rope to horizontal | | | | | | Basket hitch; angle of rope to horizontal | | | |
| | | | Vertical hitch | Choker hitch | 90° (0°) | 60° (30°) | 45° (45°) | 30° (60°) | Vertical hitch | Choker hitch | 90° (0°) | 60° (30°) | 45° (45°) | 30° (60°) |
| 1/2 | 6.5 | 6,080 | 700 | 350 | 1,400 | 1,200 | 950 | 700 | 1,200 | 600 | 2,400 | 2,100 | 1,700 | 1,200 |
| 9/16 | 8.3 | 7,600 | 850 | 400 | 1,700 | 1,500 | 1,200 | 850 | 1,500 | 750 | 3,000 | 2,600 | 2,200 | 1,500 |
| 5/8 | 10.5 | 9,880 | 1,100 | 550 | 2,200 | 1,900 | 1,600 | 1,100 | 2,000 | 1,000 | 4,000 | 3,400 | 2,800 | 2,000 |
| 3/4 | 14.5 | 13,490 | 1,500 | 750 | 3,000 | 2,600 | 2,100 | 1,500 | 2,700 | 1,400 | 5,400 | 4,700 | 3,800 | 2,700 |
| 13/16 | 17.0 | 16,150 | 1,800 | 900 | 3,600 | 3,100 | 2,600 | 1,800 | 3,200 | 1,600 | 6,400 | 5,600 | 4,600 | 3,200 |
| 7/8 | 20.0 | 19,000 | 2,100 | 1,100 | 4,200 | 3,700 | 3,000 | 2,100 | 3,800 | 1,900 | 7,600 | 6,600 | 5,400 | 3,800 |
| 1 | 26.0 | 23,750 | 2,600 | 1,300 | 5,300 | 4,600 | 3,700 | 2,600 | 4,800 | 2,400 | 9,500 | 8,200 | 6,700 | 4,800 |
| 1 1/16 | 29.0 | 27,360 | 3,000 | 1,500 | 6,100 | 5,300 | 4,300 | 3,000 | 5,500 | 2,700 | 11,000 | 9,500 | 7,700 | 5,500 |
| 1 1/8 | 34.0 | 31,350 | 3,500 | 1,700 | 7,000 | 6,000 | 5,000 | 3,500 | 6,300 | 3,100 | 12,500 | 11,000 | 8,900 | 6,300 |
| 1 1/4 | 40.0 | 35,625 | 4,000 | 2,000 | 7,900 | 6,900 | 5,600 | 4,000 | 7,100 | 3,600 | 14,500 | 12,500 | 10,000 | 7,100 |
| 1 5/16 | 45.0 | 40,850 | 4,500 | 2,300 | 9,100 | 7,900 | 6,400 | 4,500 | 8,200 | 4,100 | 16,500 | 14,000 | 12,000 | 8,200 |
| 1 1/2 | 55.0 | 50,350 | 5,600 | 2,800 | 11,000 | 9,700 | 7,900 | 5,600 | 10,000 | 5,000 | 20,000 | 17,500 | 14,000 | 10,000 |
| 1 5/8 | 68.0 | 61,750 | 6,900 | 3,400 | 13,500 | 12,000 | 9,700 | 6,900 | 12,500 | 6,200 | 24,500 | 21,500 | 17,500 | 12,500 |
| 1 3/4 | 83.0 | 74,100 | 8,200 | 4,100 | 16,500 | 14,500 | 11,500 | 8,200 | 15,000 | 7,400 | 29,500 | 27,500 | 21,000 | 15,000 |
| 2 | 95.0 | 87,400 | 9,700 | 4,900 | 19,500 | 17,000 | 13,500 | 9,700 | 17,500 | 8,700 | 35,000 | 30,500 | 24,500 | 17,500 |
| 2 1/8 | 109.0 | 100,700 | 11,000 | 5,600 | 22,500 | 19,500 | 16,000 | 11,000 | 20,000 | 10,000 | 40,500 | 35,000 | 28,500 | 20,000 |
| 2 1/4 | 129.0 | 118,750 | 13,000 | 6,600 | 26,500 | 23,000 | 18,500 | 13,000 | 24,000 | 12,000 | 47,500 | 41,000 | 33,500 | 24,000 |
| 2 1/2 | 149.0 | 133,000 | 15,000 | 7,400 | 29,500 | 25,500 | 21,000 | 15,000 | 26,500 | 13,500 | 53,000 | 46,000 | 37,500 | 26,500 |
| 2 5/8 | 168.0 | 153,900 | 17,100 | 8,600 | 34,000 | 29,500 | 24,000 | 17,000 | 31,000 | 15,500 | 61,500 | 53,500 | 43,500 | 31,000 |

## TABLE H-17—POLYESTER ROPE SLINGS

[Angle of rope to vertical shown in parentheses]

| Rope diameter nominal in inches | Nominal weight per 100 ft. in pounds | Minimum breaking strength in pounds | Rated capacity in pounds (safety factor = 9) | | | | | | | | | | | |
|---|---|---|---|---|---|---|---|---|---|---|---|---|---|---|
| | | | Eye and eye sling | | | | | | Endless sling | | | | | |
| | | | | | Basket hitch; angle of rope to horizontal | | | | | | Basket hitch; angle of rope to horizontal | | | |
| | | | Vertical hitch | Choker hitch | 90° (0°) | 60° (30°) | 45° (45°) | 30° (60°) | Vertical hitch | Choker hitch | 90° (0°) | 60° (30°) | 45° (45°) | 30° (60°) |
| 1/2 | 8.0 | 6,080 | 700 | 350 | 1,400 | 1,200 | 950 | 700 | 1,200 | 600 | 2,400 | 2,100 | 1,700 | 1,200 |
| 9/16 | 10.2 | 7,600 | 850 | 400 | 1,700 | 1,500 | 1,200 | 850 | 1,500 | 750 | 3,000 | 2,600 | 2,200 | 1,500 |
| 5/8 | 13.0 | 9,500 | 1,100 | 550 | 2,100 | 1,800 | 1,500 | 1,100 | 1,900 | 950 | 3,800 | 3,300 | 2,700 | 1,900 |
| 3/4 | 17.5 | 11,875 | 1,300 | 650 | 2,600 | 2,300 | 1,900 | 1,300 | 2,400 | 1,200 | 4,800 | 4,100 | 3,400 | 2,400 |
| 13/16 | 21.0 | 14,725 | 1,600 | 800 | 3,300 | 2,800 | 2,300 | 1,600 | 2,900 | 1,500 | 5,900 | 5,100 | 4,200 | 2,900 |
| 7/8 | 25.0 | 17,100 | 1,900 | 950 | 3,800 | 3,300 | 2,700 | 1,900 | 3,400 | 1,700 | 6,800 | 5,900 | 4,800 | 3,400 |
| 1 | 30.5 | 20,900 | 2,300 | 1,200 | 4,600 | 4,000 | 3,300 | 2,300 | 4,200 | 2,100 | 8,400 | 7,200 | 5,900 | 4,200 |
| 1 1/16 | 34.5 | 24,225 | 2,700 | 1,300 | 5,400 | 4,700 | 3,800 | 2,700 | 4,800 | 2,400 | 9,700 | 8,400 | 6,900 | 4,800 |
| 1 1/8 | 40.0 | 28,025 | 3,100 | 1,600 | 6,200 | 5,400 | 4,400 | 3,100 | 5,600 | 2,800 | 11,000 | 9,700 | 7,900 | 5,600 |
| 1 1/4 | 46.3 | 31,540 | 3,500 | 1,800 | 7,000 | 6,100 | 5,000 | 3,500 | 6,300 | 3,200 | 12,500 | 11,000 | 8,900 | 6,300 |
| 1 5/16 | 52.5 | 35,625 | 4,000 | 2,000 | 7,900 | 6,900 | 5,600 | 4,000 | 7,100 | 3,600 | 14,500 | 12,500 | 10,000 | 7,100 |
| 1 1/2 | 66.8 | 44,460 | 4,900 | 2,500 | 9,900 | 8,600 | 7,000 | 4,900 | 8,900 | 4,400 | 18,000 | 15,500 | 12,500 | 8,900 |
| 1 5/8 | 82.0 | 54,150 | 6,000 | 3,000 | 12,000 | 10,400 | 8,500 | 6,000 | 11,000 | 5,400 | 21,500 | 19,000 | 15,500 | 11,000 |
| 1 3/4 | 98.0 | 64,410 | 7,200 | 3,600 | 14,500 | 12,500 | 10,000 | 7,200 | 13,000 | 6,400 | 26,000 | 22,500 | 18,000 | 13,000 |
| 2 | 118.0 | 76,000 | 8,400 | 4,200 | 17,000 | 14,500 | 12,000 | 8,400 | 15,000 | 7,600 | 30,500 | 26,500 | 21,500 | 15,000 |
| 2 1/8 | 135.0 | 87,400 | 9,700 | 4,900 | 19,500 | 17,000 | 13,500 | 9,700 | 17,500 | 8,700 | 35,000 | 30,500 | 24,500 | 17,500 |
| 2 1/4 | 157.0 | 101,650 | 11,500 | 5,700 | 22,500 | 19,500 | 16,000 | 11,500 | 20,500 | 10,000 | 40,500 | 35,000 | 29,000 | 20,500 |
| 2 1/2 | 181.0 | 115,900 | 13,000 | 6,400 | 26,000 | 22,500 | 18,000 | 13,000 | 23,000 | 11,500 | 46,500 | 40,000 | 33,000 | 23,000 |
| 2 5/8 | 205.0 | 130,150 | 14,500 | 7,200 | 29,000 | 25,000 | 20,500 | 14,500 | 26,000 | 13,000 | 52,000 | 45,000 | 37,000 | 26,000 |

### TABLE H-18—POLYPROPYLENE ROPE SLINGS

[Angle of rope to vertical shown in parentheses]

| | | | Rated capacity in pounds (safety factor = 6) | | | | | | | | | | | |
|---|---|---|---|---|---|---|---|---|---|---|---|---|---|---|
| | | | Eye and eye sling | | | | | | Endless sling | | | | | |
| | | | | | Basket hitch; angle of rope to horizontal | | | | | | Basket hitch; angle of rope to horizontal | | | |
| Rope diameter nominal in inches | Nominal weight per 100 ft. in pounds | Minimum breaking strength in pounds | Vertical hitch | Choker hitch | 90 ° (0 °) | 60 ° (30 °) | 45 ° (45 °) | 30 ° (60 °) | Vertical hitch | Choker hitch | 90 ° (0 °) | 60 ° (30 °) | 45 ° (45 °) | 30 ° (60 °) |
| 1/2 | 4.7 | 3,900 | 650 | 350 | 1,300 | 1,200 | 950 | 650 | 1,200 | 600 | 2,400 | 2,100 | 1,700 | 1,200 |
| 9/16 | 6.1 | 4,845 | 800 | 400 | 1,600 | 1,400 | 1,100 | 800 | 1,500 | 750 | 2,900 | 2,500 | 2,100 | 1,500 |
| 5/8 | 7.5 | 5,890 | 1,000 | 500 | 2,000 | 1,700 | 1,400 | 1,000 | 1,800 | 900 | 3,500 | 3,100 | 2,500 | 1,800 |
| 3/4 | 10.7 | 8,075 | 1,300 | 700 | 2,700 | 2,300 | 1,900 | 1,300 | 2,400 | 1,200 | 4,900 | 4,200 | 3,400 | 2,400 |
| 13/16 | 12.7 | 9,405 | 1,600 | 800 | 3,100 | 2,700 | 2,200 | 1,600 | 2,800 | 1,400 | 5,600 | 4,900 | 4,000 | 2,800 |
| 7/8 | 15.0 | 10,925 | 1,800 | 900 | 3,600 | 3,200 | 2,600 | 1,800 | 3,300 | 1,600 | 6,600 | 5,700 | 4,600 | 3,300 |
| 1 | 18.0 | 13,300 | 2,200 | 1,100 | 4,400 | 3,800 | 3,100 | 2,200 | 4,000 | 2,000 | 8,000 | 6,900 | 5,600 | 4,000 |
| 1 1/16 | 20.4 | 15,200 | 2,500 | 1,300 | 5,100 | 4,400 | 3,600 | 2,500 | 4,600 | 2,300 | 9,100 | 7,900 | 6,500 | 4,600 |
| 1 1/8 | 23.7 | 17,385 | 2,900 | 1,500 | 5,800 | 5,000 | 4,100 | 2,900 | 5,200 | 2,600 | 10,500 | 9,000 | 7,400 | 5,200 |
| 1 1/4 | 27.0 | 19,950 | 3,300 | 1,700 | 6,700 | 5,800 | 4,700 | 3,300 | 6,000 | 3,000 | 12,000 | 10,500 | 8,500 | 6,000 |
| 1 5/16 | 30.5 | 22,325 | 3,700 | 1,900 | 7,400 | 6,400 | 5,300 | 3,700 | 6,700 | 3,400 | 13,500 | 11,500 | 9,500 | 6,700 |
| 1 1/2 | 38.5 | 28,215 | 4,700 | 2,400 | 9,400 | 8,100 | 6,700 | 4,700 | 8,500 | 4,200 | 17,000 | 14,500 | 12,000 | 8,500 |
| 1 5/8 | 47.5 | 34,200 | 5,700 | 2,900 | 11,500 | 9,900 | 8,100 | 5,700 | 10,500 | 5,100 | 20,500 | 18,000 | 14,500 | 10,500 |
| 1 3/4 | 57.0 | 40,850 | 6,800 | 3,400 | 13,500 | 12,000 | 9,600 | 6,800 | 12,500 | 6,100 | 24,500 | 21,000 | 17,500 | 12,500 |
| 2 | 69.0 | 49,400 | 8,200 | 4,100 | 16,500 | 14,500 | 11,500 | 8,200 | 15,000 | 7,400 | 29,500 | 25,500 | 21,000 | 15,000 |
| 2 1/8 | 80.0 | 57,950 | 9,700 | 4,800 | 19,500 | 16,500 | 13,500 | 9,700 | 17,500 | 8,700 | 35,000 | 30,100 | 24,500 | 17,500 |
| 2 1/4 | 92.0 | 65,500 | 11,000 | 5,500 | 22,000 | 19,000 | 15,500 | 11,000 | 19,500 | 9,900 | 39,500 | 34,000 | 28,000 | 19,500 |
| 2 1/2 | 107.0 | 76,000 | 12,500 | 6,300 | 25,500 | 22,000 | 18,000 | 12,500 | 23,000 | 11,500 | 45,500 | 39,500 | 32,500 | 23,000 |
| 2 5/8 | 120.0 | 85,500 | 14,500 | 7,100 | 28,500 | 24,500 | 20,000 | 14,500 | 25,500 | 13,000 | 51,500 | 44,500 | 36,500 | 25,500 |

### TABLE H-19—SAFE WORKING LOADS FOR SHACKLES

[In tons of 2,000 pounds]

| Material size (inches) | Pin diameter (inches) | Safe working load |
|---|---|---|
| 1/2 | 5/8 | 1.4 |
| 5/8 | 3/4 | 2.2 |
| 3/4 | 7/8 | 3.2 |
| 7/8 | 1 | 4.3 |
| 1 | 1 1/8 | 5.6 |
| 1 1/8 | 1 1/4 | 6.7 |
| 1 1/4 | 1 3/8 | 8.2 |
| 1 3/8 | 1 1/2 | 10.0 |
| 1 1/2 | 1 5/8 | 11.9 |
| 1 3/4 | 2 | 16.2 |
| 2 | 2 1/4 | 21.2 |

### TABLE H-20—NUMBER AND SPACING OF U-BOLT WIRE ROPE CLIPS

| Improved plow steel, rope diameter (inches) | Number of clips | | Minimum spacing (inches) |
|---|---|---|---|
| | Drop forged | Other material | |
| 1/2 | 3 | 4 | 3 |
| 5/8 | 3 | 4 | 3 3/4 |
| 3/4 | 4 | 5 | 4 1/2 |
| 7/8 | 4 | 5 | 5 1/4 |
| 1 | 5 | 6 | 6 |
| 1 1/8 | 6 | 6 | 6 3/4 |
| 1 1/4 | 6 | 7 | 7 1/2 |
| 1 3/8 | 7 | 7 | 8 1/4 |
| 1 1/2 | 7 | 8 | 9 |

[44 FR 8577, Feb. 9, 1979; 44 FR 20940, Apr. 6, 1979, as amended at 58 FR 35173, June 30, 1993]

## §1926.252 Disposal of waste materials.

**1926.252(a)** Whenever materials are dropped more than 20 feet to any point lying outside the exterior walls of the building, an enclosed chute of wood, or equivalent material, shall be used. For the purpose of this paragraph, an enclosed chute is a slide, closed in on all sides, through which material is moved from a high place to a lower one.

**1926.252(b)** When debris is dropped through holes in the floor without the use of chutes, the area onto which the material is dropped shall be completely enclosed with barricades not less than 42 inches high and not less than 6 feet back from the projected edge of the opening above. Signs warning of the hazard of falling materials shall be posted at each level. Removal shall not be permitted in this lower area until debris handling ceases above.

**1926.252(c)** All scrap lumber, waste material, and rubbish shall be removed from the immediate work area as the work progresses.

**1926.252(d)** Disposal of waste material or debris by burning shall comply with local fire regulations.

**1926.252(e)** All solvent waste, oily rags, and flammable liquids shall be kept in fire resistant covered containers until removed from worksite.

**Authority:** Sec. 107, Contract Work Hours and Safety Standards Act (Construction Safety Act) (40 U.S.C. 333); Secs. 4, 6 and 8 Occupational Safety and Health Act of 1970 (29 U.S.C. 653, 655, and 657); Secretary of Labor's Order No. 12-71 (36 FR 8754), 8-76 (41 FR 25059), 9-83 (48 FR 35736), or 1-90 (55 FR 9033), as applicable; and 29 CFR part 1911.

### §1926.300 General requirements.

**1926.300(a)** ***Condition of tools.*** All hand and power tools and similar equipment, whether furnished by the employer or the employee, shall be maintained in a safe condition.

**1926.300(b)** ***Guarding.*** (1) When power operated tools are designed to accommodate guards, they shall be equipped with such guards when in use.

**1926.300(b)(2)** Belts, gears, shafts, pulleys, sprockets, spindles, drums, fly wheels, chains, or other reciprocating, rotating or moving parts of equipment shall be guarded if such parts are exposed to contact by employees or otherwise create a hazard. Guarding shall meet the requirements as set forth in American National Standards Institute, B15.1-1953 (R1958), Safety Code for Mechanical Power-Transmission Apparatus.

**1926.300(b)(3)** *Types of guarding.* One or more methods of machine guarding shall be provided to protect the operator and other employees in the machine area from hazards such as those created by point of operation, ingoing nip points, rotating parts, flying chips and sparks. Examples of guarding methods are—barrier guards, two-hand tripping devices, electronic safety devices, etc.

**1926.300(b)(4)** *Point of operation guarding.* (i) Point of operation is the area on a machine where work is actually performed upon the material being processed.

**1926.300(b)(4)(ii)** The point of operation of machines whose operation exposes an employee to injury, shall be guarded. The guarding device shall be in conformity with any appropriate standards therefor, or, in the absence of applicable specific standards, shall be so designed and constructed as to prevent the operator from having any part of his body in the danger zone during the operating cycle.

**1926.300(b)(4)(iii)** Special handtools for placing and removing material shall be such as to permit easy handling of material without the operator placing a hand in the danger zone. Such tools shall not be in lieu of other guarding required by this section, but can only be used to supplement protection provided.

**1926.300(b)(4)(iv)** The following are some of the machines which usually require point of operation guarding:

**1926.300(b)(4)(iv)(*a*)** Guillotine cutters.

**1926.300(b)(4)(iv)(*b*)** Shears.

**1926.300(b)(4)(iv)(*c*)** Alligator shears.

**1926.300(b)(4)(iv)(*d*)** Power presses.

**1926.300(b)(4)(iv)(*e*)** Milling machines.

**1926.300(b)(4)(iv)(*f*)** Power saws.

**1926.300(b)(4)(iv)(*g*)** Jointers.

**1926.300(b)(4)(iv)(*h*)** Portable power tools.

**1926.300(b)(4)(iv)(*i*)** Forming rolls and calenders.

**1926.300(b)(5)** *Exposure of blades.* When the periphery of the blades of a fan is less than 7 feet (2.128 m) above the floor or working level, the blades shall be guarded. The guard shall have openings no larger than 1/2 inch (1.27 cm).

**1926.300(b)(6)** *Anchoring fixed machinery.* Machines designed for a fixed location shall be securely anchored to prevent walking or moving.

**1926.300(b)(7)** *Guarding of abrasive wheel machinery—exposure adjustment.* Safety guards of the types described in paragraphs (b) (8) and (9) of this section, where the operator stands in front of the opening, shall be constructed so that the peripheral protecting member can be adjusted to the constantly decreasing diameter of the wheel. The maximum angular exposure above the horizontal plane of the wheel spindle as specified in paragraphs (b) (8) and (9) of this section shall never be exceeded, and the distance between the wheel periphery and the adjustable tongue or the end of the peripheral member at the top shall never exceed 1/4 inch (0.635 cm). (See Figures I-1 through I-6.)

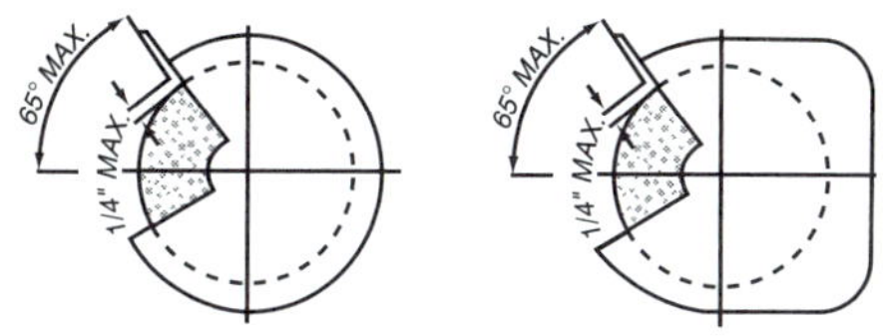


**FIGURE I-1** **FIGURE I-2**

***Correct***

Showing adjustable tongue giving required angle protection for all sizes of wheel used.

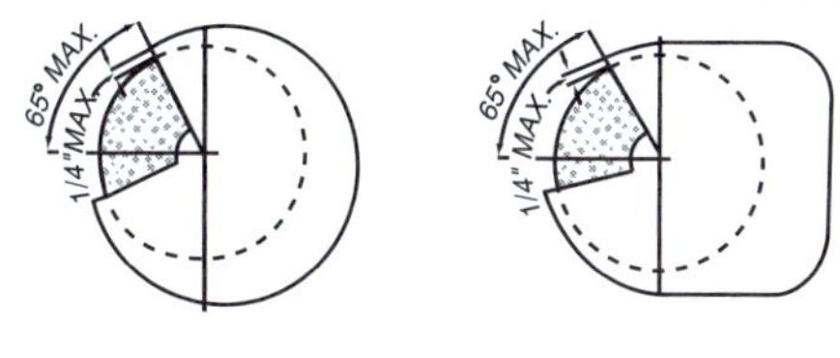


**FIGURE I-3** **FIGURE I-4**

***Correct***

Showing movable guard with opening small enough to give required protection for the smallest size wheel used.

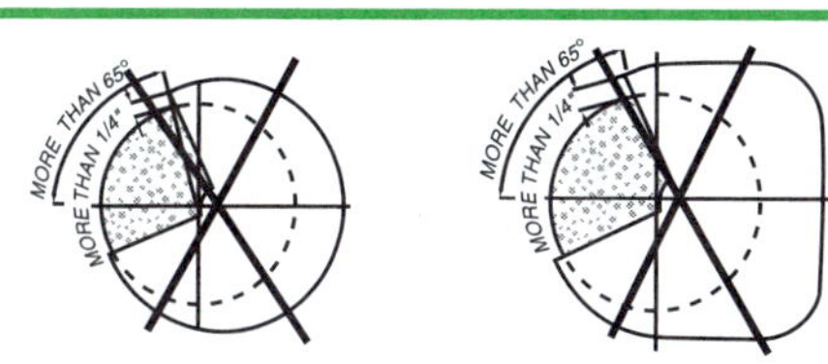


**FIGURE I-5** **FIGURE I-6**

***Incorrect***

Showing movable guard with size of opening correct for full size wheel but too large for smaller wheel.

**1926.300(b)(8)** *Bench and floor stands.* The angular exposure of the grinding wheel periphery and sides for safety guards used on machines known as bench and floor stands should not exceed 90° or one-fourth of the periphery. This exposure shall begin at a point not more than 65° above the horizontal plane of the wheel spindle. (See Figures I-7 and I-8 and paragraph (b)(7) of this section.)

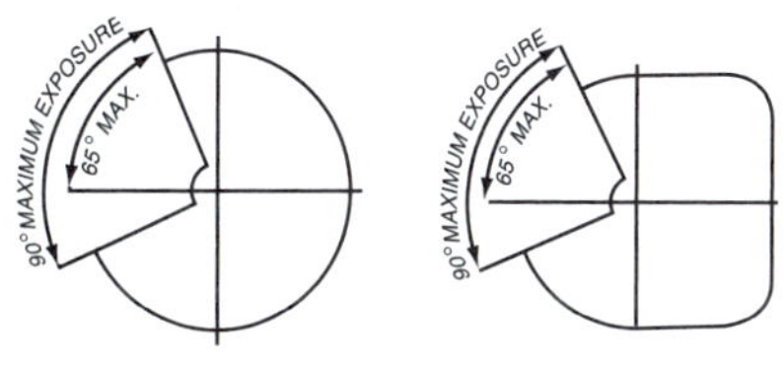


**FIGURE I-7** **FIGURE I-8**

Wherever the nature of the work requires contact with the wheel below the horizontal plane of the spindle, the exposure shall not exceed 125° (See Figures I-9 and I-10.)

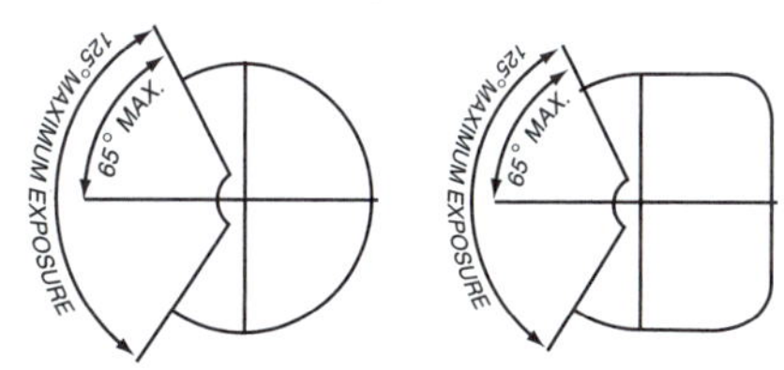


**FIGURE I-9** **FIGURE I-10**

**1926.300(b)(9)** *Cylindrical grinders.* The maximum angular exposure of the grinding wheel periphery and sides for safety guards used on cylindrical grinding machines shall not exceed 180 °. This exposure shall begin at a point not more than 65° above the horizontal plane of the wheel spindle. (See Figures I-11 and I-12 and paragraph (b)(7) of this section.)

FIGURE I-11

FIGURE I-12

**1926.300(c)** ***Personal protective equipment.*** Employees using hand and power tools and exposed to the hazard of falling, flying, abrasive, and splashing objects, or exposed to harmful dusts, fumes, mists, vapors, or gases shall be provided with the particular personal protective equipment necessary to protect them from the hazard. All personal protective equipment shall meet the requirements and be maintained according to subparts D and E of this part.

**1926.300(d)** ***Switches.*** (1) All hand-held powered platen sanders, grinders with wheels 2-inch diameter or less, routers, planers, laminate trimmers, nibblers, shears, scroll saws, and jigsaws with blade shanks one-fourth of an inch wide or less may be equipped with only a positive "on-off" control.

**1926.300(d)(2)** All hand-held powered drills, tappers, fastener drivers, horizontal, vertical, and angle grinders with wheels greater than 2 inches in diameter, disc sanders, belt sanders, reciprocating saws, saber saws, and other similar operating powered tools shall be equipped with a momentary contact "on-off" control and may have a lock-on control provided that turnoff can be accomplished by a single motion of the same finger or fingers that turn it on.

**1926.300(d)(3)** All other hand-held powered tools, such as circular saws, chain saws, and percussion tools without positive accessory holding means, shall be equipped with a constant pressure switch that will shut off the power when the pressure is released.

**1926.300(d)(4)** The requirements of this paragraph shall become effective on July 15, 1972.

**1926.300(d)(5)** Exception: This paragraph does not apply to concrete vibrators, concrete breakers, powered tampers, jack hammers, rock drills, and similar hand operated power tools.

[44 FR 8577, Feb. 9, 1979; 44 FR 20940, Apr. 6, 1979, as amended at 58 FR 35175, June 30, 1993; 61 FR 9250, Mar. 7, 1996]

## §1926.301 Hand tools.

**1926.301(a)** Employers shall not issue or permit the use of unsafe hand tools.

**1926.301(b)** Wrenches, including adjustable, pipe, end, and socket wrenches shall not be used when jaws are sprung to the point that slippage occurs.

**1926.301(c)** Impact tools, such as drift pins, wedges, and chisels, shall be kept free of mushroomed heads.

**1926.301(d)** The wooden handles of tools shall be kept free of splinters or cracks and shall be kept tight in the tool.

## §1926.302 Power-operated hand tools.

**1926.302(a)** ***Electric power-operated tools.*** (1) Electric power operated tools shall either be of the approved double-insulated type or grounded in accordance with subpart K of this part.

**1926.302(a)(2)** The use of electric cords for hoisting or lowering tools shall not be permitted.

**1926.302(b)** ***Pneumatic power tools.*** (1) Pneumatic power tools shall be secured to the hose or whip by some positive means to prevent the tool from becoming accidentally disconnected.

**1926.302(b)(2)** Safety clips or retainers shall be securely installed and maintained on pneumatic impact (percussion) tools to prevent attachments from being accidentally expelled.

**1926.302(b)(3)** All pneumatically driven nailers, staplers, and other similar equipment provided with automatic fastener feed, which operate at more than 100 p.s.i. pressure at the tool shall have a safety device on the muzzle to prevent the tool from ejecting fasteners, unless the muzzle is in contact with the work surface.

**1926.302(b)(4)** Compressed air shall not be used for cleaning purposes except where reduced to less than 30 p.s.i. and then only with effective chip guarding and personal protective equipment which meets the requirements of subpart E of this part. The 30 p.s.i. requirement does not apply for concrete form, mill scale and similar cleaning purposes.

**1926.302(b)(5)** The manufacturer's safe operating pressure for hoses, pipes, valves, filters, and other fittings shall not be exceeded.

**1926.302(b)(6)** The use of hoses for hoisting or lowering tools shall not be permitted.

**1926.302(b)(7)** All hoses exceeding 1/2-inch inside diameter shall have a safety device at the source of supply or branch line to reduce pressure in case of hose failure.

**1926.302(b)(8)** Airless spray guns of the type which atomize paints and fluids at high pressures (1,000 pounds or more per square inch) shall be equipped with automatic or visible manual safety devices which will prevent pulling of the trigger to prevent release of the paint or fluid until the safety device is manually released.

**1926.302(b)(9)** In lieu of the above, a diffuser nut which will prevent high pressure, high velocity release, while the nozzle tip is removed, plus a nozzle tip guard which will prevent the tip from coming into contact with the operator, or other equivalent protection, shall be provided.

**1926.302(b)(10)** *Abrasive blast cleaning nozzles.* The blast cleaning nozzles shall be equipped with an operating valve which must be held open manually. A support shall be provided on which the nozzle may be mounted when it is not in use.

**1926.302(c)** ***Fuel powered tools.*** (1) All fuel powered tools shall be stopped while being refueled, serviced, or maintained, and fuel shall be transported, handled, and stored in accordance with subpart F of this part.

**1926.302(c)(2)** When fuel powered tools are used in enclosed spaces, the applicable requirements for concentrations of toxic gases and use of personal protective equipment, as outlined in subparts D and E of this part, shall apply.

**1926.302(d)** ***Hydraulic power tools.*** (1) The fluid used in hydraulic powered tools shall be fire-resistant fluids approved under Schedule 30 of the U.S. Bureau of Mines, Department of the Interior, and shall retain its operating characteristics at the most extreme temperatures to which it will be exposed.

**1926.302(d)(2)** The manufacturer's safe operating pressures for hoses, valves, pipes, filters, and other fittings shall not be exceeded.

**1926.302(e)** ***Powder-actuated tools.*** (1) Only employees who have been trained in the operation of the particular tool in use shall be allowed to operate a powder-actuated tool.

**1926.302(e)(2)** The tool shall be tested each day before loading to see that safety devices are in proper working condition. The method of testing shall be in accordance with the manufacturer's recommended procedure.

**1926.302(e)(3)** Any tool found not in proper working order, or that develops a defect during use, shall be immediately removed from service and not used until properly repaired.

**1926.302(e)(4)** Personal protective equipment shall be in accordance with subpart E of this part.

**1926.302(e)(5)** Tools shall not be loaded until just prior to the intended firing time. Neither loaded nor empty tools are to be pointed at any employees. Hands shall be kept clear of the open barrel end.

**1926.302(e)(6)** Loaded tools shall not be left unattended.

**1926.302(e)(7)** Fasteners shall not be driven into very hard or brittle materials including, but not limited to, cast iron, glazed tile, surface-hardened steel, glass block, live rock, face brick, or hollow tile.

**1926.302(e)(8)** Driving into materials easily penetrated shall be avoided unless such materials are backed by a substance that will prevent the pin or fastener from passing completely through and creating a flying missile hazard on the other side.

**1926.302(e)(9)** No fastener shall be driven into a spalled area caused by an unsatisfactory fastening.

**1926.302(e)(10)** Tools shall not be used in an explosive or flammable atmosphere.

**1926.302(e)(11)** All tools shall be used with the correct shield, guard, or attachment recommended by the manufacturer.

**1926.302(e)(12)** Powder-actuated tools used by employees shall meet all other applicable requirements of American National Standards Institute, A10.3-1970, Safety Requirements for Explosive-Actuated Fastening Tools.

[44 FR 8577, Feb. 9, 1979; 44 FR 20940, Apr. 6, 1979, as amended at 58 FR 35175, June 30, 1993]

## §1926.303 Abrasive wheels and tools.

**1926.303(a)** ***Power.*** All grinding machines shall be supplied with sufficient power to maintain the spindle speed at safe levels under all conditions of normal operation.

**1926.303(b)** ***Guarding.*** (1) Grinding machines shall be equipped with safety guards in conformance with the requirements of American National Standards Institute, B7.1-1970, Safety Code for the Use, Care and Protection of Abrasive Wheels, and paragraph (d) of this section.

**1926.303(b)(2)** *Guard design.* The safety guard shall cover the spindle end, nut, and flange projections. The safety guard shall be mounted so as to maintain proper alignment with the wheel, and the strength of the fastenings shall exceed the strength of the guard, except:

**1926.303(b)(2)(i)** Safety guards on all operations where the work provides a suitable measure of protection to the operator, may be so constructed that the spindle end, nut, and outer flange are exposed; and where the nature of the work is such as to entirely cover the side of the wheel, the side covers of the guard may be omitted; and

**1926.303(b)(2)(ii)** The spindle end, nut, and outer flange may be exposed on machines designed as portable saws.

**1926.303(c)** ***Use of abrasive wheels.*** (1) Floor stand and bench mounted abrasive wheels, used for external grinding, shall be provided with safety guards (protection hoods). The maximum angular exposure of the grinding wheel periphery and sides shall be not more than 90°, except that when work requires contact with the wheel below the horizontal plane of the spindle, the angular exposure shall not exceed 125°. In either case, the exposure shall begin not more than 65° above the horizontal plane of the spindle. Safety guards shall be strong enough to withstand the effect of a bursting wheel.

**1926.303(c)(2)** Floor and bench-mounted grinders shall be provided with work rests which are rigidly supported and readily adjustable. Such work rests shall be kept at a distance not to exceed one-eighth inch from the surface of the wheel.

**1926.303(c)(3)** Cup type wheels used for external grinding shall be protected by either a revolving cup guard or a band type guard in accordance with the provisions of the American National Standards Institute, B7.1-1970 Safety Code for the Use, Care, and Protection of Abrasive Wheels. All other portable abrasive wheels used for external grinding, shall be provided with safety guards (protection hoods) meeting the requirements of paragraph (c)(5) of this section, except as follows:

**1926.303(c)(3)(i)** When the work location makes it impossible, a wheel equipped with safety flanges, as described in paragraph (c)(6) of this section, shall be used;

**1926.303(c)(3)(ii)** When wheels 2 inches or less in diameter which are securely mounted on the end of a steel mandrel are used.

**1926.303(c)(4)** Portable abrasive wheels used for internal grinding shall be provided with safety flanges (protection flanges) meeting the requirements of paragraph (c)(6) of this section, except as follows:

**1926.303(c)(4)(i)** When wheels 2 inches or less in diameter which are securely mounted on the end of a steel mandrel are used;

**1926.303(c)(4)(ii)** If the wheel is entirely within the work being ground while in use.

**1926.303(c)(5)** When safety guards are required, they shall be so mounted as to maintain proper alignment with the wheel, and the guard and its fastenings shall be of sufficient strength to retain fragments of the wheel in case of accidental breakage. The maximum angular exposure of the grinding wheel periphery and sides shall not exceed 180°.

**1926.303(c)(6)** When safety flanges are required, they shall be used only with wheels designed to fit the flanges. Only safety flanges, of a type and design and properly assembled so as to ensure that the pieces of the wheel will be retained in case of accidental breakage, shall be used.

**1926.303(c)(7)** All abrasive wheels shall be closely inspected and ring-tested before mounting to ensure that they are free from cracks or defects.

**1926.303(c)(8)** Grinding wheels shall fit freely on the spindle and shall not be forced on. The spindle nut shall be tightened only enough to hold the wheel in place.

**1926.303(c)(9)** All employees using abrasive wheels shall be protected by eye protection equipment in accordance with the requirements of subpart E of this part, except when adequate eye protection is afforded by eye shields which are permanently attached to the bench or floor stand.

**1926.303(d)** ***Other requirements.*** All abrasive wheels and tools used by employees shall meet other applicable requirements of American National Standards Institute, B7.1-1970, Safety Code for the Use, Care and Protection of Abrasive Wheels.

**1926.303(e)** ***Work rests.*** On offhand grinding machines, work rests shall be used to support the work. They shall be of rigid construction and designed to be adjustable to compensate for wheel wear. Work rests shall be kept adjusted closely to the wheel with a maximum opening of 1/8 inch (0.3175 cm) to prevent the work from being jammed between the wheel and the rest, which may cause wheel breakage. The work rest shall be securely clamped after each adjustment. The adjustment shall not be made with the wheel in motion.

[44 FR 8577, Feb. 9, 1979; 44 FR 20940, Apr. 6, 1979, as amended at 58 FR 35175, June 30, 1993]

## §1926.304 Woodworking tools.

**1926.304(a)** ***Disconnect switches.*** All fixed power driven woodworking tools shall be provided with a disconnect switch that can either be locked or tagged in the off position.

**1926.304(b)** ***Speeds.*** The operating speed shall be etched or otherwise permanently marked on all circular saws over 20 inches in diameter or operating at over 10,000 peripheral feet per minute. Any saw so marked shall not be operated at a speed other than that marked on the blade. When a marked saw is retensioned for a different speed, the marking shall be corrected to show the new speed.

**1926.304(c)** ***Self-feed.*** Automatic feeding devices shall be installed on machines whenever the nature of the work will permit. Feeder attachments shall have the feed rolls or other moving parts covered or guarded so as to protect the operator from hazardous points.

**1926.304(d)** ***Guarding.*** All portable, power-driven circular saws shall be equipped with guards above and below the base plate or shoe. The upper guard shall cover the saw to the depth of the teeth, except for the minimum arc required to permit the base to be tilted for bevel cuts. The lower guard shall cover the saw to the depth of the teeth, except for the minimum arc required to allow proper retraction and contact with the work. When the tool is withdrawn from the work, the lower guard shall automatically and instantly return to the covering position.

**1926.304(e)** ***Personal protective equipment.*** All personal protective equipment provided for use shall conform to subpart E of this part.

**1926.304(f)** ***Other requirements.*** All woodworking tools and machinery shall meet other applicable requirements of American National Standards Institute, 01.1-1961, Safety Code for Woodworking Machinery.

**1926.304(g)** ***Radial saws.*** (1) The upper hood shall completely enclose the upper portion of the blade down to a point that will include the end of the saw arbor. The upper hood shall be constructed in such a manner and of such material that it will protect the operator from flying splinters, broken saw teeth, etc., and will deflect sawdust away from the operator. The sides of the lower exposed portion of the blade shall be guarded to the full diameter of the blade by a device that will automatically adjust itself to the thickness of the stock and remain in contact with stock being cut to give maximum protection possible for the operation being performed.

**1926.304(h)** ***Hand-fed crosscut table saws.*** (1) Each circular crosscut table saw shall be guarded by a hood which shall meet all the requirements of paragraph (i)(1) of this section for hoods for circular ripsaws.

**1926.304(i)** ***Hand-fed ripsaws.*** (1) Each circular hand-fed ripsaw shall be guarded by a hood which shall completely enclose that portion of the saw above the table and that portion of the saw above the material being cut. The hood and mounting shall be arranged so that the hood will automatically adjust itself to the thickness of and remain in contact with the material being cut but it shall not offer any considerable resistance to insertion of material to saw or to passage of the material being sawed. The hood shall be made of adequate strength to resist blows and strains incidental to reasonable operation, adjusting, and handling, and shall be so designed as to protect the operator from flying splinters and broken saw teeth. It shall be made of material that is soft enough so that it will be unlikely to cause tooth breakage. The hood shall be so mounted as to insure that its operation will be positive, reliable, and in true alignment with the saw; and the mounting shall be adequate in strength to resist any reasonable side thrust or other force tending to throw it out of line.

[44 FR 8577, Feb. 9, 1979; 44 FR 20940, Apr. 6, 1979, as amended at 58 FR 35175, June 30, 1993; 61 FR 9251, Mar. 7, 1996]

## §1926.305 Jacks—lever and ratchet, screw, and hydraulic.

**1926.305(a)** ***General requirements.*** (1) The manufacturer's rated capacity shall be legibly marked on all jacks and shall not be exceeded.

**1926.305(a)(2)** All jacks shall have a positive stop to prevent overtravel.

**1926.305(b)** [Reserved]

**1926.305(c)** ***Blocking.*** When it is necessary to provide a firm foundation, the base of the jack shall be blocked or cribbed. Where there is a possibility of slippage of the metal cap of the jack, a wood block shall be placed between the cap and the load.

**1926.305(d)(1)** ***Operation and maintenance.*** (i) After the load has been raised, it shall be cribbed, blocked, or otherwise secured at once.

**1926.305(d)(1)(ii)** Hydraulic jacks exposed to freezing temperatures shall be supplied with an adequate antifreeze liquid.

**1926.305(d)(1)(iii)** All jacks shall be properly lubricated at regular intervals.

**1926.305(d)(1)(iv)** Each jack shall be thoroughly inspected at times which depend upon the service conditions. Inspections shall be not less frequent than the following:

**1926.305(d)(1)(iv)(*a*)** For constant or intermittent use at one locality, once every 6 months,

**1926.305(d)(1)(iv)(*b*)** For jacks sent out of shop for special work, when sent out and when returned,

**1926.305(d)(1)(iv)(c)** For a jack subjected to abnormal load or shock, immediately before and immediately thereafter.

**1926.305(d)(1)(v)** Repair or replacement parts shall be examined for possible defects.

**1926.305(d)(1)(vi)** Jacks which are out of order shall be tagged accordingly, and shall not be used until repairs are made.

[44 FR 8577, Feb. 9, 1979; 44 FR 20940, Apr. 6, 1979, as amended at 55 FR 42328, Oct. 18, 1990; 58 FR 35176, June 30, 1993]

## §1926.306 Air receivers.

**1926.306(a)** ***General requirements***—(1) *Application.* This section applies to compressed air receivers, and other equipment used in providing and utilizing compressed air for performing operations such as cleaning, drilling, hoisting, and chipping. On the other hand, however, this section does not deal with the special problems created by using compressed air to convey materials nor the problems created when men work in compressed air as in tunnels and caissons. This section is not intended to apply to compressed air machinery and equipment used on transportation vehicles such as steam railroad cars, electric railway cars, and automotive equipment.

**1926.306(a)(2)** *New and existing equipment.* (i) All new air receivers installed after the effective date of these regulations shall be constructed in accordance with the 1968 edition of the A.S.M.E. Boiler and Pressure Vessel Code Section VIII.

**1926.306(a)(2)(ii)** All safety valves used shall be constructed, installed, and maintained in accordance with the A.S.M.E. Boiler and Pressure Vessel Code, Section VIII Edition 1968.

**1926.306(b)** ***Installation and equipment requirements***—(1) *Installation.* Air receivers shall be so installed that all drains, handholes, and manholes therein are easily accessible. Under no circumstances shall an air receiver be buried underground or located in an inaccessible place.

**1926.306(b)(2)** *Drains and traps.* A drain pipe and valve shall be installed at the lowest point of every air receiver to provide for the removal of accumulated oil and water. Adequate automatic traps may be installed in addition to drain valves. The drain valve on the air receiver shall be opened and the receiver completely drained frequently and at such intervals as to prevent the accumulation of excessive amounts of liquid in the receiver.

**1926.306(b)(3)** *Gages and valves.* (i) Every air receiver shall be equipped with an indicating pressure gage (so located as to be readily visible) and with one or more spring-loaded safety valves. The total relieving capacity of such safety valves shall be such as to prevent pressure in the receiver from exceeding the maximum allowable working pressure of the receiver by more than 10 percent.

**1926.306(b)(3)(ii)** No valve of any type shall be placed between the air receiver and its safety valve or valves.

**1926.306(b)(3)(iii)** Safety appliances, such as safety valves, indicating devices and controlling devices, shall be constructed, located, and installed so that they cannot be readily rendered inoperative by any means, including the elements.

**1926.306(b)(3)(iv)** All safety valves shall be tested frequently and at regular intervals to determine whether they are in good operating condition.

[58 FR 35176, June 30, 1993]

## §1926.307 Mechanical power-transmission apparatus.

**1926.307(a)** ***General requirements.*** (1) This section covers all types and shapes of power-transmission belts, except the following when operating at two hundred and fifty (250) feet per minute or less: (i) Flat belts 1 inch (2.54 cm) or less in width, (ii) flat belts 2 inches (5.08 cm) or less in width which are free from metal lacings or fasteners, (iii) round belts 1/2 inch (1.27 cm) or less in diameter; and (iv) single strand V-belts, the width of which is thirteen thirty-seconds (13/32) inch or less.

**1926.307(a)(2)** Vertical and inclined belts (paragraphs (e) (3) and (4) of this section) if not more than 2 1/2 inches (6.35 cm) wide and running at a speed of less than one thousand (1,000) feet per minute, and if free from metal lacings or fastenings may be guarded with a nip-point belt and pulley guard.

**1926.307(a)(3)** For the Textile Industry, because of the presence of excessive deposits of lint, which constitute a serious fire hazard, the sides and face sections only of nip-point belt and pulley guards are required, provided the guard shall extend at least 6 inches (15.24 cm) beyond the rim of the pulley on the in-running and off-running sides of the belt and at least 2 inches (5.08 cm) away from the rim and face of the pulley in all other directions.

**1926.307(a)(4)** This section covers the principal features with which power transmission safeguards shall comply.

**1926.307(b)** ***Prime-mover guards***—(1) *Flywheels.* Flywheels located so that any part is 7 feet (2.128 m) or less above floor or platform shall be guarded in accordance with the requirements of this subparagraph:

**1926.307(b)(1)(i)** With an enclosure of sheet, perforated, or expanded metal, or woven wire;

**1926.307(b)(1)(ii)** With guard rails placed not less than 15 inches (38.1 cm) nor more than 20 inches (50.8 cm) from rim. When flywheel extends into pit or is within 12 inches (30.48 cm) of floor, a standard toeboard shall also be provided;

**1926.307(b)(1)(iii)** When the upper rim of flywheel protrudes through a working floor, it shall be entirely enclosed or surrounded by a guardrail and toeboard.

**1926.307(b)(1)(iv)** For flywheels with smooth rims 5 feet (1.52 m) or less in diameter, where the preceding methods cannot be applied, the following may be used: A disk attached to the flywheel in such manner as to cover the spokes of the wheel on the exposed side and present a smooth surface and edge, at the same time providing means for periodic inspection. An open space, not exceeding 4 inches (10.16 cm) in width, may be left between the outside edge of the disk and the rim of the wheel if desired, to facilitate turning the wheel over. Where a disk is used, the keys or other dangerous projections not covered by disk shall be cut off or covered. This subdivision does not apply to flywheels with solid web centers.

**1926.307(b)(1)(v)** Adjustable guard to be used for starting engine or for running adjustment may be provided at the flywheel of gas or oil engines. A slot opening for jack bar will be permitted.

**1926.307(b)(1)(vi)** Wherever flywheels are above working areas, guards shall be installed having sufficient strength to hold the weight of the flywheel in the event of a shaft or wheel mounting failure.

**1926.307(b)(2)** *Cranks and connecting rods.* Cranks and connecting rods, when exposed to contact, shall be guarded in accordance with paragraphs (m) and (n) of this section, or by a guardrail as described in paragraph (o)(5) of this section.

**1926.307(b)(3)** *Tail rods or extension piston rods.* Tail rods or extension piston rods shall be guarded in accordance with paragraphs (m) and (o) of this section, or by a guardrail on sides and end, with a clearance of not less than 15 (38.1 cm) nor more than 20 inches (50.8 cm) when rod is fully extended.

**1926.307(c)** ***Shafting***—(1) *Installation.* (i) Each continuous line of shafting shall be secured in position against excessive endwise movement.

**1926.307(c)(1)(ii)** Inclined and vertical shafts, particularly inclined idler shafts, shall be securely held in position against endwise thrust.

**1926.307(c)(2)** *Guarding horizontal shafting.* (i) All exposed parts of horizontal shafting 7 feet (2.128 m) or less from floor or working platform, excepting runways used exclusively for oiling, or running adjustments, shall be protected by a stationary casing enclosing shafting completely or by a trough enclosing sides and top or sides and bottom of shafting as location requires.

**1926.307(c)(2)(ii)** Shafting under bench machines shall be enclosed by a stationary casing, or by a trough at sides and top or sides and bottom, as location requires. The sides of the trough shall come within at least 6 inches (15.24 cm) of the underside of table, or if shafting is located near floor within 6 inches (15.24 cm) of floor. In every case the sides of trough shall extend at least 2 inches (5.08 cm) beyond the shafting or protuberance.

**1926.307(c)(3)** *Guarding vertical and inclined shafting.* Vertical and inclined shafting 7 feet (2.128 m) or less from floor or working platform, excepting maintenance runways, shall be enclosed with a stationary casing in accordance with requirements of paragraphs (m) and (o) of this section.

**1926.307(c)(4)** *Projecting shaft ends.* (i) Projecting shaft ends shall present a smooth edge and end and shall not project more than one-half the diameter of the shaft unless guarded by nonrotating caps or safety sleeves.

**1926.307(c)(4)(ii)** Unused keyways shall be filled up or covered.

**1926.307(c)(5)** *Power-transmission apparatus located in basements.* All mechanical power transmission apparatus located in basements, towers, and rooms used exclusively for power transmission equipment shall be guarded in accordance with this section, except that the requirements for safeguarding belts, pulleys, and shafting need not be complied with when the following requirements are met:

**1926.307(c)(5)(i)** The basement, tower, or room occupied by transmission equipment is locked against unauthorized entrance.

**1926.307(c)(5)(ii)** The vertical clearance in passageways between the floor and power transmission beams, ceiling, or any other objects, is not less than 5 ft. 6 in. (1.672 m).

**1926.307(c)(5)(iii)** The intensity of illumination conforms to the requirements of ANSI A11.1-1965 (R-1970).

**1926.307(c)(5)(iv)** [Reserved]

**1926.307(c)(5)(v)** The route followed by the oiler is protected in such manner as to prevent accident.

**1926.307(d)** ***Pulleys***—(1) *Guarding.* Pulleys, any parts of which are 7 feet (2.128 m) or less from the floor or working platform, shall be guarded in accordance with the standards specified in paragraphs (m) and (o) of this section. Pulleys serving as balance wheels (*e.g.*, punch presses) on which the point of contact between belt and pulley is more than 6 ft. 6 in. (1.976 m) from the floor or platform may be guarded with a disk covering the spokes.

**1926.307(d)(2)** *Location of pulleys.* (i) Unless the distance to the nearest fixed pulley, clutch, or hanger exceeds the width of the belt used, a guide shall be provided to prevent the belt from leaving the pulley on the side where insufficient clearance exists.

**1926.307(d)(2)(ii)** [Reserved]

**1926.307(d)(3)** *Broken pulleys.* Pulleys with cracks, or pieces broken out of rims, shall not be used.

**1926.307(d)(4)** *Pulley speeds.* Pulleys intended to operate at rim speed in excess of manufacturers normal recommendations shall be specially designed and carefully balanced for the speed at which they are to operate.

**1926.307(e)** ***Belt, rope, and chain drives***—(1) *Horizontal belts and ropes.* (i) Where both runs of horizontal belts are 7 feet (2.128 m) or less from the floor level, the guard shall extend to at least 15 inches (38.1 cm) above the belt or to a standard height (see Table O-12), except that where both runs of a horizontal belt are 42 inches (106.68 cm) or less from the floor, the belt shall be fully enclosed.

**1926.307(e)(1)(ii)** In powerplants or power-development rooms, a guardrail may be used in lieu of the guard required by paragraph (e)(1)(i) of this section.

**1926.307(e)(2)** *Overhead horizontal belts.* (i) Overhead horizontal belts, with lower parts 7 feet (2.128 m) or less from the floor or platform, shall be guarded on sides and bottom in accordance with paragraph (o)(3) of this section.

**1926.307(e)(2)(ii)** Horizontal overhead belts more than 7 feet (2.128 m) above floor or platform shall be guarded for their entire length under the following conditions:

**1926.307(e)(2)(ii)(*a*)** If located over passageways or work places and traveling 1,800 feet or more per minute.

**1926.307(e)(2)(ii)(*b*)** If center to center distance between pulleys is 10 feet (3.04 m) or more.

**1926.307(e)(2)(ii)(c)** If belt is 8 inches (20.32 cm) or more in width.

**1926.307(e)(2)(iii)** Where the upper and lower runs of horizontal belts are so located that passage of persons between them would be possible, the passage shall be either:

**1926.307(e)(2)(iii)(*a*)** Completely barred by a guardrail or other barrier in accordance with paragraphs (m) and (o) of this section; or

**1926.307(e)(2)(iii)(*b*)** Where passage is regarded as necessary, there shall be a platform over the lower run guarded on either side by a railing completely filled in with wire mesh or other filler, or by a solid barrier. The upper run shall be so guarded as to prevent contact therewith either by the worker or by objects carried by him. In powerplants only the lower run of the belt need be guarded.

**1926.307(e)(2)(iv)** Overhead chain and link belt drives are governed by the same rules as overhead horizontal belts and shall be guarded in the same manner as belts.

**1926.307(e)(3)** *Vertical and inclined belts.* (i) Vertical and inclined belts shall be enclosed by a guard conforming to standards in paragraphs (m) and (o) of this section.

**1926.307(e)(3)(ii)** All guards for inclined belts shall be arranged in such a manner that a minimum clearance of 7 feet (2.128 m) is maintained between belt and floor at any point outside of guard.

**1926.307(e)(4)** *Vertical belts.* Vertical belts running over a lower pulley more than 7 feet (2.128 m) above floor or platform shall be guarded at the bottom in the same manner as horizontal overhead belts, if conditions are as stated in paragraphs (e)(2)(ii) *(a)* and *(c)* of this section.

**1926.307(e)(5)** *Cone-pulley belts.* (i) The cone belt and pulley shall be equipped with a belt shifter so constructed as to adequately guard the nip point of the belt and pulley. If the frame of the belt shifter does not adequately guard the nip point of the belt and pulley, the nip point shall be further protected by means of a vertical guard placed in front of the pulley and extending at least to the top of the largest step of the cone.

**1926.307(e)(5)(ii)** If the belt is of the endless type or laced with rawhide laces, and a belt shifter is not desired, the belt will be considered guarded if the nip point of the belt and pulley is protected by a nip point guard located in front of the cone extending at least to the top of the largest step of the cone, and formed to show the contour of the cone in order to give the nip point of the belt and pulley the maximum protection.

**1926.307(e)(5)(iii)** If the cone is located less than 3 feet (0.912 m) from the floor or working platform, the cone pulley and belt shall be guarded to a height of 3 feet (0.912 m) regardless of whether the belt is endless or laced with rawhide.

**1926.307(e)(6)** *Belt tighteners.* (i) Suspended counterbalanced tighteners and all parts thereof shall be of substantial construction and securely fastened; the bearings shall be securely capped. Means must be provided to prevent tightener from falling, in case the belt breaks.

**1926.307(e)(6)(ii)** Where suspended counterweights are used and not guarded by location, they shall be so encased as to prevent accident.

**1926.307(f)** ***Gears, sprockets, and chains***—(1) *Gears.* Gears shall be guarded in accordance with one of the following methods:

**1926.307(f)(1)(i)** By a complete enclosure; or

**1926.307(f)(1)(ii)** By a standard guard as described in paragraph (o) of this section, at least 7 feet (2.128 m) high extending 6 inches (15.24 cm) above the mesh point of the gears; or

**1926.307(f)(1)(iii)** By a band guard covering the face of gear and having flanges extended inward beyond the root of the teeth on the exposed side or sides. Where any portion of the train of gears guarded by a band guard is less than 6 feet (1.824 m) from the floor a disk guard or a complete enclosure to the height of 6 feet (1.824 m) shall be required.

**1926.307(f)(2)** *Hand-operated gears.* Paragraph (f)(1) of this section does not apply to hand-operated gears used only to adjust machine parts and which do not continue to move after hand power is removed. However, the guarding of these gears is highly recommended.

**1926.307(f)(3)** *Sprockets and chains.* All sprocket wheels and chains shall be enclosed unless they are more than 7 feet (2.128 m) above the floor or platform. Where the drive extends over other machine or working areas, protection against falling shall be provided. This subparagraph does not apply to manually operated sprockets.

**1926.307(f)(4)** *Openings for oiling.* When frequent oiling must be done, openings with hinged or sliding self-closing covers shall be provided. All points not readily accessible shall have oil feed tubes if lubricant is to be added while machinery is in motion.

**1926.307(g)** ***Guarding friction drives.*** The driving point of all friction drives when exposed to contact shall be guarded, all arm or spoke friction drives and all web friction drives with holes in the web shall be entirely enclosed, and all projecting belts on friction drives where exposed to contact shall be guarded.

**1926.307(h)** ***Keys, setscrews, and other projections.*** (1) All projecting keys, setscrews, and other projections in revolving parts shall be removed or made flush or guarded by metal cover. This subparagraph does not apply to keys or setscrews within gear or sprocket casings or other enclosures, nor to keys, setscrews, or oilcups in hubs of pulleys less than 20 inches (50.8 cm) in diameter where they are within the plane of the rim of the pulley.

**1926.307(h)(2)** It is recommended, however, that no projecting setscrews or oilcups be used in any revolving pulley or part of machinery.

**1926.307(i)** ***Collars and couplings***—(1) *Collars.* All revolving collars, including split collars, shall be cylindrical, and screws or bolts used in collars shall not project beyond the largest periphery of the collar.

**1926.307(i)(2)** *Couplings.* Shaft couplings shall be so constructed as to present no hazard from bolts, nuts, setscrews, or revolving surfaces. Bolts, nuts, and setscrews will, however, be permitted where they are covered with safety sleeves or where they are used parallel with the shafting and are countersunk or else do not extend beyond the flange of the coupling.

**1926.307(j)** ***Bearings and facilities for oiling.*** All drip cups and pans shall be securely fastened.

**1926.307(k)** ***Guarding of clutches, cutoff couplings, and clutch pulleys***—(1) *Guards.* Clutches, cutoff couplings, or clutch pulleys having projecting parts, where such clutches are located 7 feet (2.128 m) or less above the floor or working platform, shall be enclosed by a stationary guard constructed in accordance with this section. A "U" type guard is permissible.

**1926.307(k)(2)** *Engine rooms.* In engine rooms a guardrail, preferably with toeboard, may be used instead of the guard required by paragraph (k)(1) of this section, provided such a room is occupied only by engine room attendants.

**1926.307(l)** ***Belt shifters, clutches, shippers, poles, perches, and fasteners***—(1) *Belt shifters.* (i) Tight and loose pulleys on all new installations made on or after August 31, 1971, shall be equipped with a permanent belt shifter provided with mechanical means to prevent belt from creeping from loose to tight pulley. It is recommended that old installations be changed to conform to this rule.

**1926.307(l)(1)(ii)** Belt shifter and clutch handles shall be rounded and be located as far as possible from danger of accidental contact, but within easy reach of the operator. Where belt shifters are not directly located over a machine or bench, the handles shall be cut off 6 ft. 6 in. (1.976 m) above floor level.

**1926.307(l)(2)** *Belt shippers and shipper poles.* The use of belt poles as substitutes for mechanical shifters is not recommended.

**1926.307(l)(3)** *Belt perches.* Where loose pulleys or idlers are not practicable, belt perches in form of brackets, rollers, etc., shall be used to keep idle belts away from the shafts.

**1926.307(l)(4)** *Belt fasteners.* Belts which of necessity must be shifted by hand and belts within 7 feet (2.128 m) of the floor or working platform which are not guarded in accordance with this section shall not be fastened with metal in any case, nor with any other fastening which by construction or wear will constitute an accident hazard.

**1926.307(m)** ***Standard guards—general requirements***—(1) *Materials.* (i) Standard conditions shall be secured by the use of the following materials. Expanded metal, perforated or solid sheet metal, wire mesh on a frame of angle iron, or iron pipe securely fastened to floor or to frame of machine.

**1926.307(m)(1)(ii)** All metal should be free from burrs and sharp edges.

**1926.307(m)(2)** *Methods of manufacture.* (i) Expanded metal, sheet or perforated metal, and wire mesh shall be securely fastened to frame.

**1926.307(n)** [Reserved]

**1926.307(o)** ***Approved materials***—(1) *Minimum requirements.* The materials and dimensions specified in this paragraph shall apply to all guards, except horizontal overhead belts, rope, cable, or chain guards more than 7 feet (2.128 m) above floor, or platform.

**1926.307(o)(1)(i)** [Reserved]

**1926.307(o)(1)(i)(a)** All guards shall be rigidly braced every 3 feet (0.912 m) or fractional part of their height to some fixed part of machinery or building structure. Where guard is exposed to contact with moving equipment additional strength may be necessary.

**1926.307(o)(2)** *Wood guards.* (i) Wood guards may be used in the woodworking and chemical industries, in industries where the presence of fumes or where manufacturing conditions would cause the rapid deterioration of metal guards; also in construction work and in locations outdoors where extreme cold or extreme heat make metal guards and railings undesirable. In all other industries, wood guards shall not be used.

**1926.307(o)(3)** *Guards for horizontal overhead belts.* (i) Guards for horizontal overhead belts shall run the entire length of the belt and follow the line of the pulley to the ceiling or be carried to the nearest wall, thus enclosing the belt effectively. Where belts are so located as to make it impracticable to carry the guard to wall or ceiling, construction of guard shall be such as to enclose completely the top and bottom runs of belt and the face of pulleys.

**1926.307(o)(3)(ii)** [Reserved]

**1926.307(o)(3)(iii)** Suitable reinforcement shall be provided for the ceiling rafters or overhead floor beams, where such is necessary, to sustain safely the weight and stress likely to be imposed by the guard. The interior surface of all guards, by which is meant the surface of the guard with which a belt will come in contact, shall be smooth and free from all projections of any character, except where construction demands it; protruding shallow roundhead rivets may be used. Overhead belt guards shall be at least one-quarter wider than belt which they protect, except that this clearance need not in any case exceed 6 inches (15.24 cm) on each side. Overhead rope drive and block and roller-chain-drive guards shall be not less than 6 inches (15.24 cm) wider than the drive on each side. In overhead silent chain-drive guards where the chain is held from lateral displacement on the sprockets, the side clearances required on drives of 20 inch (50.8 cm) centers or under shall be not less than 1/4 inch (0.635 cm) from the nearest moving chain part, and on drives of over 20 inch (50.8 cm) centers a minimum of 1/2 inch (1.27 cm) from the nearest moving chain part.

**1926.307(o)(4)** *Guards for horizontal overhead rope and chain drives.* Overhead-rope and chain-drive guard construction shall conform to the rules for overhead-belt guard.

**1926.307(o)(5)** *Guardrails and toeboards.* (i) Guardrail shall be 42 inches (106.68 cm) in height, with midrail between top rail and floor.

**1926.307(o)(5)(ii)** Posts shall be not more than 8 feet (2.432 m) apart; they are to be permanent and substantial, smooth, and free from protruding nails, bolts, and splinters. If made of pipe, the post shall be 1 1/4 inches (3.175 cm) inside diameter, or larger. If made of metal shapes or bars, their section shall be equal in strength to that of 1 1/2 (3.81 cm) by 1 1/2 (3.81 cm) by 3/16 inch angle iron. If made of wood, the posts shall be two by four (2 × 4) inches or larger. The upper rail shall be two by four (2 × 4) inches, or two one by four (1 × 4) strips, one at the top and one at the side of posts. The midrail may be one by four (1 × 4) inches or more. Where panels are fitted with expanded metal or wire mesh as noted in Table O-12 the middle rails may be omitted. Where guard is exposed to contact with moving equipment, additional strength may be necessary.

**1926.307(o)(5)(iii)** Toeboards shall be 4 inches (10.16 cm) or more in height, of wood, metal, or of metal grill not exceeding 1 inch (2.54 cm) mesh.

**1926.307(p)** ***Care of equipment***—(1) *General.* All power-transmission equipment shall be inspected at intervals not exceeding 60 days and be kept in good working condition at all times.

**1926.307(p)(2)** *Shafting.* (i) Shafting shall be kept in alignment, free from rust and excess oil or grease.

**1926.307(p)(2)(ii)** Where explosives, explosive dusts, flammable vapors or flammable liquids exist, the hazard of static sparks from shafting shall be carefully considered.

**1926.307(p)(3)** *Bearings.* Bearings shall be kept in alignment and properly adjusted.

**1926.307(p)(4)** *Hangers.* Hangers shall be inspected to make certain that all supporting bolts and screws are tight and that supports of hanger boxes are adjusted properly.

**1926.307(p)(5)** *Pulleys.* (i) Pulleys shall be kept in proper alignment to prevent belts from running off.

**1926.307(p)(6)** *Care of belts.*

**1926.307(p)(6)(i)** [Reserved]

**1926.307(p)(6)(ii)** Inspection shall be made of belts, lacings, and fasteners and such equipment kept in good repair.

**1926.307(p)(7)** *Lubrication.* The regular oilers shall wear tight-fitting clothing. Machinery shall be oiled when not in motion, wherever possible.

[58 FR 35176, June 30, 1993]

**Authority:** Sec. 107, Contract Work Hours and Safety Standards Act (Construction Safety Act) (40 U.S.C. 333); secs. 4, 6, 8, Occupational Safety and Health Act of 1970 (29 U.S.C. 653, 655, 657); Secretary of Labor's Order No. 12-71 (36 FR 8754), 8-76 (41 FR 25059), or 9-83 (48 FR 35736), as applicable.

### §1926.350 Gas welding and cutting.

**1926.350(a)** ***Transporting, moving, and storing compressed gas cylinders.*** (1) Valve protection caps shall be in place and secured.

**1926.350(a)(2)** When cylinders arc hoisted, they shall be secured on a cradle, slingboard, or pallet. They shall not be hoisted or transported by means of magnets or choker slings.

**1926.350(a)(3)** Cylinders shall be moved by tilting and rolling them on their bottom edges. They shall not be intentionally dropped, struck, or permitted to strike each other violently.

**1926.350(a)(4)** When cylinders are transported by powered vehicles, they shall be secured in a vertical position.

**1926.350(a)(5)** Valve protection caps shall not be used for lifting cylinders from one vertical position to another. Bars shall not be used under valves or valve protection caps to pry cylinders loose when frozen. Warm, not boiling, water shall be used to thaw cylinders loose.

**1926.350(a)(6)** Unless cylinders are firmly secured on a special carrier intended for this purpose, regulators shall be removed and valve protection caps put in place before cylinders are moved.

**1926.350(a)(7)** A suitable cylinder truck, chain, or other steadying device shall be used to keep cylinders from being knocked over while in use.

**1926.350(a)(8)** When work is finished, when cylinders are empty, or when cylinders are moved at any time, the cylinder valve shall be closed.

**1926.350(a)(9)** Compressed gas cylinders shall be secured in an upright position at all times except, if necessary, for short periods of time while cylinders are actually being hoisted or carried.

**1926.350(a)(10)** Oxygen cylinders in storage shall be separated from fuel-gas cylinders or combustible materials (especially oil or grease), a minimum distance of 20 feet (6.1 m) or by a noncombustible barrier at least 5 feet (1.5 m) high having a fire-resistance rating of at least one-half hour.

**1926.350(a)(11)** Inside of buildings, cylinders shall be stored in a well-protected, well-ventilated, dry location, at least 20 feet (6.1 m) from highly combustible materials such as oil or excelsior. Cylinders should be stored in definitely assigned places away from elevators, stairs, or gangways. Assigned storage places shall be located where cylinders will not be knocked over or damaged by passing or falling objects, or subject to tampering by unauthorized persons. Cylinders shall not be kept in unventilated enclosures such as lockers and cupboards.

**1926.350(a)(12)** The in-plant handling, storage, and utilization of all compressed gases in cylinders, portable tanks, rail tankcars, or motor vehicle cargo tanks shall be in accordance with Compressed Gas Association Pamphlet P-1-1965.

**1926.350(b)** ***Placing cylinders.*** (1) Cylinders shall be kept far enough away from the actual welding or cutting operation so that sparks, hot slag, or flame will not reach them. When this is impractical, fire resistant shields shall be provided.

**1926.350(b)(2)** Cylinders shall be placed where they cannot become part of an electrical circuit. Electrodes shall not be struck against a cylinder to strike an arc.

**1926.350(b)(3)** Fuel gas cylinders shall be placed with valve end up whenever they are in use. They shall not be placed in a location where they would be subject to open flame, hot metal, or other sources of artificial heat.

**1926.350(b)(4)** Cylinders containing oxygen or acetylene or other fuel gas shall not be taken into confined spaces.

**1926.350(c)** ***Treatment of cylinders.*** (1) Cylinders, whether full or empty, shall not be used as rollers or supports.

**1926.350(c)(2)** No person other than the gas supplier shall attempt to mix gases in a cylinder. No one except the owner of the cylinder or person authorized by him, shall refill a cylinder. No one shall use a cylinder's contents for purposes other than those intended by the supplier. All cylinders used shall meet the Department of Transportation requirements published in 49 CFR part 178, subpart C, Specification for Cylinders.

**1926.350(c)(3)** No damaged or defective cylinder shall be used.

**1926.350(d)** ***Use of fuel gas.*** The employer shall thoroughly instruct employees in the safe use of fuel gas, as follows:

**1926.350(d)(1)** Before a regulator to a cylinder valve is connected, the valve shall be opened slightly and closed immediately. (This action is generally termed "cracking" and is intended to clear the valve of dust or dirt that might otherwise enter the regulator.) The person cracking the valve shall stand to one side of the outlet, not in front of it. The valve of a fuel gas cylinder shall not be cracked where the gas would reach welding work, sparks, flame, or other possible sources of ignition.

**1926.350(d)(2)** The cylinder valve shall always be opened slowly to prevent damage to the regulator. For quick closing, valves on fuel gas cylinders shall not be opened more than 1 1/2 turns. When a special wrench is required, it shall be left in position on the stem of the valve while the cylinder is in use so that the fuel gas flow can be shut off quickly in case of an emergency. In the case of manifolded or coupled cylinders, at least one such wrench shall always be available for immediate use. Nothing shall be placed on top of a fuel gas cylinder, when in use, which may damage the safety device or interfere with the quick closing of the valve.

**1926.350(d)(3)** Fuel gas shall not be used from cylinders through torches or other devices which are equipped with shutoff valves without reducing the pressure through a suitable regulator attached to the cylinder valve or manifold.

**1926.350(d)(4)** Before a regulator is removed from a cylinder valve, the cylinder valve shall always be closed and the gas released from the regulator.

**1926.350(d)(5)** If, when the valve on a fuel gas cylinder is opened, there is found to be a leak around the valve stem, the valve shall be closed and the gland nut tightened. If this action does not stop the leak, the use of the cylinder shall be discontinued, and it shall be properly tagged and removed from the work area. In the event that fuel gas should leak from the cylinder valve, rather than from the valve stem, and the gas cannot be shut off, the cylinder shall be properly tagged and removed from the work area. If a regulator attached to a cylinder valve will effectively stop a leak through the valve seat, the cylinder need not be removed from the work area.

**1926.350(d)(6)** If a leak should develop at a fuse plug or other safety device, the cylinder shall be removed from the work area.

**1926.350(e)** ***Fuel gas and oxygen manifolds.*** (1) Fuel gas and oxygen manifolds shall bear the name of the substance they contain in letters at least 1-inch high which shall be either painted on the manifold or on a sign permanently attached to it.

**1926.350(e)(2)** Fuel gas and oxygen manifolds shall be placed in safe, well ventilated, and accessible locations. They shall not be located within enclosed spaces.

**1926.350(e)(3)** Manifold hose connections, including both ends of the supply hose that lead to the manifold, shall be such that the hose cannot be interchanged between fuel gas and oxygen manifolds and supply header connections. Adapters shall not be used to permit the interchange of hose. Hose connections shall be kept free of grease and oil.

**1926.350(e)(4)** When not in use, manifold and header hose connections shall be capped.

**1926.350(e)(5)** Nothing shall be placed on top of a manifold, when in use, which will damage the manifold or interfere with the quick closing of the valves.

**1926.350(f)** ***Hose.*** (1) Fuel gas hose and oxygen hose shall be easily distinguishable from each other. The contrast may be made by different colors or by surface characteristics readily distinguishable by the sense of touch. Oxygen and fuel gas hoses shall not be interchangeable. A single hose having more than one gas passage shall not be used.

**1926.350(f)(2)** When parallel sections of oxygen and fuel gas hose are taped together, not more than 4 inches out of 12 inches shall be covered by tape.

**1926.350(f)(3)** All hose in use, carrying acetylene, oxygen, natural or manufactured fuel gas, or any gas or substance which may ignite or enter into combustion, or be in any way harmful to employees, shall be inspected at the beginning of each working shift. Defective hose shall be removed from service.

**1926.350(f)(4)** Hose which has been subject to flashback, or which shows evidence of severe wear or damage, shall be tested to twice the normal pressure to which it is subject, but in no case less than 300 p.s.i. Defective hose, or hose in doubtful condition, shall not be used.

**1926.350(f)(5)** Hose couplings shall be of the type that cannot be unlocked or disconnected by means of a straight pull without rotary motion.

**1926.350(f)(6)** Boxes used for the storage of gas hose shall be ventilated.

**1926.350(f)(7)** Hoses, cables, and other equipment shall be kept clear of passageways, ladders and stairs.

**1926.350(g)** ***Torches.*** (1) Clogged torch tip openings shall be cleaned with suitable cleaning wires, drills, or other devices designed for such purpose.

**1926.350(g)(2)** Torches in use shall be inspected at the beginning of each working shift for leaking shutoff valves, hose couplings, and tip connections. Defective torches shall not be used.

**1926.350(g)(3)** Torches shall be lighted by friction lighters or other approved devices, and not by matches or from hot work.

**1926.350(h)** ***Regulators and gauges.*** Oxygen and fuel gas pressure regulators, including their related gauges, shall be in proper working order while in use.

**1926.350(i)** ***Oil and grease hazards.*** Oxygen cylinders and fittings shall be kept away from oil or grease. Cylinders, cylinder caps and valves, couplings, regulators, hose, and apparatus shall be kept free from oil or greasy substances and shall not be handled with oily hands or gloves. Oxygen shall not be directed at oily surfaces, greasy clothes, or within a fuel oil or other storage tank or vessel.

**1926.350(j)** ***Additional rules.*** For additional details not covered in this subpart, applicable technical portions of American National Standards Institute, Z49.1-1967, Safety in Welding and Cutting, shall apply.

[44 FR 8577, Feb. 9, 1979; 44 FR 20940, Apr. 6, 1979, as amended at 55 FR 42328, Oct. 18, 1990; 58 FR 35179, June 30, 1993]

## §1926.351 Arc welding and cutting.

**1926.351(a)** ***Manual electrode holders.*** (1) Only manual electrode holders which are specifically designed for arc welding and cutting, and are of a capacity capable of safely handling the maximum rated current required by the electrodes, shall be used.

**1926.351(a)(2)** Any current-carrying parts passing through the portion of the holder which the arc welder or cutter grips in his hand, and the outer surfaces of the jaws of the holder, shall be fully insulated against the maximum voltage encountered to ground.

**1926.351(b)** ***Welding cables and connectors.*** (1) All arc welding and cutting cables shall be of the completely insulated, flexible type, capable of handling the maximum current requirements of the work in progress, taking into account the duty cycle under which the arc welder or cutter is working.

**1926.351(b)(2)** Only cable free from repair or splices for a minimum distance of 10 feet from the cable end to which the electrode holder is connected shall be used, except that cables with standard insulated connectors or with splices whose insulating quality is equal to that of the cable are permitted.

**1926.351(b)(3)** When it becomes necessary to connect or splice lengths of cable one to another, substantial insulated connectors of a capacity at least equivalent to that of the cable shall be used. If connections are effected by means of cable lugs, they shall be securely fastened together to give good electrical contact, and the exposed metal parts of the lugs shall be completely insulated.

**1926.351(b)(4)** Cables in need of repair shall not be used. When a cable, other than the cable lead referred to in paragraph (b)(2) of this section, becomes worn to the extent of exposing bare conductors, the portion thus exposed shall be protected by means of rubber and friction tape or other equivalent insulation.

**1926.351(c)** ***Ground returns and machine grounding.*** (1) A ground return cable shall have a safe current carrying capacity equal to or exceeding the specified maximum output capacity of the arc welding or cutting unit which it services. When a single ground return cable services more than one unit, its safe current-carrying capacity shall equal or exceed the total specified maximum output capacities of all the units which it services.

**1926.351(c)(2)** Pipelines containing gases or flammable liquids, or conduits containing electrical circuits, shall not be used as a ground return. For welding on natural gas pipelines, the technical portions of regulations issued by the Department of Transportation, Office of Pipeline Safety, 49 CFR Part 192, Minimum Federal Safety Standards for Gas Pipelines, shall apply.

**1926.351(c)(3)** When a structure or pipeline is employed as a ground return circuit, it shall be determined that the required electrical contact exists at all joints. The generation of an arc, sparks, or heat at any point shall cause rejection of the structures as a ground circuit.

**1926.351(c)(4)** When a structure or pipeline is continuously employed as a ground return circuit, all joints shall be bonded, and periodic inspections shall be conducted to ensure that no condition of electrolysis or fire hazard exists by virtue of such use.

**1926.351(c)(5)** The frames of all arc welding and cutting machines shall be grounded either through a third wire in the cable containing the circuit conductor or through a separate wire which is grounded at the source of the current. Grounding circuits, other than by means of the structure, shall be checked to ensure that the circuit between the ground and the grounded power conductor has resistance low enough to permit sufficient current to flow to cause the fuse or circuit breaker to interrupt the current.

**1926.351(c)(6)** All ground connections shall be inspected to ensure that they are mechanically strong and electrically adequate for the required current.

**1926.351(d)** ***Operating instructions.*** Employers shall instruct employees in the safe means of arc welding and cutting as follows:

**1926.351(d)(1)** When electrode holders are to be left unattended, the electrodes shall be removed and the holders shall be so placed or protected that they cannot make electrical contact with employees or conducting objects.

**1926.351(d)(2)** Hot electrode holders shall not be dipped in water; to do so may expose the arc welder or cutter to electric shock.

**1926.351(d)(3)** When the arc welder or cutter has occasion to leave his work or to stop work for any appreciable length of time, or when the arc welding or cutting machine is to be moved, the power supply switch to the equipment shall be opened.

**1926.351(d)(4)** Any faulty or defective equipment shall be reported to the supervisor.

**1926.351(d)(5)** See §1926.406(c) for additional requirements.

**1926.351(e)** ***Shielding.*** Whenever practicable, all arc welding and cutting operations shall be shielded by noncombustible or flameproof screens which will protect employees and other persons working in the vicinity from the direct rays of the arc.

[44 FR 8577, Feb. 9, 1979; 44 FR 20940, Apr. 6, 1979, as amended at 51 FR 25318, July 11, 1986]

## §1926.352 Fire prevention.

**1926.352(a)** When practical, objects to be welded, cut, or heated shall be moved to a designated safe location or, if the objects to be welded, cut, or heated cannot be readily moved, all movable fire hazards in the vicinity shall be taken to a safe place, or otherwise protected.

**1926.352(b)** If the object to be welded, cut, or heated cannot be moved and if all the fire hazards cannot be removed, positive means shall be taken to confine the heat, sparks, and slag, and to protect the immovable fire hazards from them.

**1926.352(c)** No welding, cutting, or heating shall be done where the application of flammable paints, or the presence of other flammable compounds, or heavy dust concentrations creates a hazard.

**1926.352(d)** Suitable fire extinguishing equipment shall be immediately available in the work area and shall be maintained in a state of readiness for instant use.

**1926.352(e)** When the welding, cutting, or heating operation is such that normal fire prevention precautions are not sufficient, additional personnel shall be assigned to guard against fire while the actual welding, cutting, or heating operation is being performed, and for a sufficient period of time after completion of the work to ensure that no possibility of fire exists. Such personnel shall be instructed as to the specific anticipated fire hazards and how the firefighting equipment provided is to be used.

**1926.352(f)** When welding, cutting, or heating is performed on walls, floors, and ceilings, since direct penetration of sparks or heat transfer may introduce a fire hazard to an adjacent area, the same precautions shall be taken on the opposite side as are taken on the side on which the welding is being performed.

**1926.352(g)** For the elimination of possible fire in enclosed spaces as a result of gas escaping through leaking or improperly closed torch valves, the gas supply to the torch shall be positively shut off at some point outside the enclosed space whenever the torch is not to be used or whenever the torch is left unattended for a substantial period of time, such as during the lunch period. Overnight and at the change of shifts, the torch and hose shall be removed from the confined space. Open end fuel gas and oxygen hoses shall be immediately removed from enclosed spaces when they are disconnected from the torch or other gas-consuming device.

**1926.352(h)** Except when the contents are being removed or transferred, drums, pails, and other containers which contain or have contained flammable liquids shall be kept closed. Empty containers shall be removed to a safe area apart from hot work operations or open flames.

**1926.352(i)** Drums containers, or hollow structures which have contained toxic or flammable substances shall, before welding, cutting, or heating is undertaken on them, either be filled with water or thoroughly cleaned of such substances and ventilated and tested. For welding, cutting and heating on steel pipelines containing natural gas, the pertinent portions of regulations issued by the Department of Transportation, Office of Pipeline Safety, 49 CFR Part 192, Minimum Federal Safety Standards for Gas Pipelines, shall apply.

**1926.352(j)** Before heat is applied to a drum, container, or hollow structure, a vent or opening shall be provided for the release of any built-up pressure during the application of heat.

## §1926.353 Ventilation and protection in welding, cutting, and heating.

**1926.353(a)** ***Mechanical ventilation.*** For purposes of this section, mechanical ventilation shall meet the following requirements:

**1926.353(a)(1)** Mechanical ventilation shall consist of either general mechanical ventilation systems or local exhaust systems.

**1926.353(a)(2)** General mechanical ventilation shall be of sufficient capacity and so arranged as to produce the number of air changes necessary to maintain welding fumes and smoke within safe limits, as defined in subpart D of this part.

**1926.353(a)(3)** Local exhaust ventilation shall consist of freely movable hoods intended to be placed by the welder or burner as close as practicable to the work. This system shall be of sufficient capacity and so arranged as to remove fumes and smoke at the source and keep the concentration of them in the breathing zone within safe limits as defined in subpart D of this part.

**1926.353(a)(4)** Contaminated air exhausted from a working space shall be discharged into the open air or otherwise clear of the source of intake air.

**1926.353(a)(5)** All air replacing that withdrawn shall be clean and respirable.

**1926.353(a)(6)** Oxygen shall not be used for ventilation purposes, comfort cooling, blowing dust from clothing, or for cleaning the work area.

**1926.353(b)** ***Welding, cutting, and heating in confined spaces.*** (1) Except as provided in paragraph (b)(2) of this section, and paragraph (c)(2) of this section, either general mechanical or local exhaust ventilation meeting the requirements of paragraph (a) of this section shall be provided whenever welding, cutting, or heating is performed in a confined space.

**1926.353(b)(2)** When sufficient ventilation cannot be obtained without blocking the means of access, employees in the confined space shall be protected by air line respirators in accordance with the requirements of subpart E of this part, and an employee on the outside of such a confined space shall be assigned to maintain communication with those working within it and to aid them in an emergency.

**1926.353(b)(3)** *Lifelines.* Where a welder must enter a confined space through a manhole or other small opening, means shall be provided for quickly removing him in case of emergency. When safety belts and lifelines are used for this purpose they shall be so attached to the welder's body that his body cannot be jammed in a small exit opening. An attendant with a pre-planned rescue procedure shall be stationed outside to observe the welder at all times and be capable of putting rescue operations into effect.

**1926.353(c)** ***Welding, cutting, or heating of metals of toxic significance.*** (1) Welding, cutting, or heating in any enclosed spaces involving the metals specified in this subparagraph shall be performed with either general mechanical or local exhaust ventilation meeting the requirements of paragraph (a) of this section:

**1926.353(c)(1)(i)** Zinc-bearing base or filler metals or metals coated with zinc-bearing materials;

**1926.353(c)(1)(ii)** Lead base metals;

**1926.353(c)(1)(iii)** Cadmium-bearing filler materials;

**1926.353(c)(1)(iv)** Chromium-bearing metals or metals coated with chromium-bearing materials.

**1926.353(c)(2)** Welding, cutting, or heating in any enclosed spaces involving the metals specified in this subparagraph shall be performed with local exhaust ventilation in accordance with the requirements of paragraph (a) of this section, or employees shall be protected by air line respirators in accordance with the requirements of subpart E of this part:

**1926.353(c)(2)(i)** Metals containing lead, other than as an impurity, or metals coated with lead-bearing materials;

**1926.353(c)(2)(ii)** Cadmium-bearing or cadmium-coated base metals;

**1926.353(c)(2)(iii)** Metals coated with mercury-bearing metals;

**1926.353(c)(2)(iv)** Beryllium-containing base or filler metals. Because of its high toxicity, work involving beryllium shall be done with both local exhaust ventilation and air line respirators.

**1926.353(c)(3)** Employees performing such operations in the open air shall be protected by filter-type respirators in accordance with the requirements of subpart E of this part, except that employees performing such operations on beryllium-containing base or filler metals shall be protected by air line respirators in accordance with the requirements of subpart E of this part.

**1926.353(c)(4)** Other employees exposed to the same atmosphere as the welders or burners shall be protected in the same manner as the welder or burner.

**1926.353(d)** ***Inert-gas metal-arc welding.*** (1) Since the inert-gas metal-arc welding process involves the production of ultra-violet radiation of intensities of 5 to 30 times that produced during shielded metal-arc welding, the decomposition of chlorinated solvents by ultraviolet rays, and the liberation of toxic fumes and gases, employees shall not be permitted to engage in, or be exposed to the process until the following special precautions have been taken:

**1926.353(d)(1)(i)** The use of chlorinated solvents shall be kept at least 200 feet, unless shielded, from the exposed arc, and surfaces prepared with chlorinated solvents shall be thoroughly dry before welding is permitted on such surfaces.

**1926.353(d)(1)(ii)** Employees in the area not protected from the arc by screening shall be protected by filter lenses meeting the requirements of subpart E of this part. When two or more welders are exposed to each other's arc, filter lens goggles of a suitable type, meeting the requirements of subpart E of this part, shall be worn under welding helmets. Hand shields to protect the welder against flashes and radiant energy shall be used when either the helmet is lifted or the shield is removed.

**1926.353(d)(1)(iii)** Welders and other employees who are exposed to radiation shall be suitably protected so that the skin is covered completely to prevent burns and other damage by ultraviolet rays. Welding helmets and hand shields shall be free of leaks and openings, and free of highly reflective surfaces.

**1926.353(d)(1)(iv)** When inert-gas metal-arc welding is being performed on stainless steel, the requirements of paragraph (c)(2) of this section shall be met to protect against dangerous concentrations of nitrogen dioxide.

**1926.353(e)** ***General welding, cutting, and heating.*** (1) Welding, cutting, and heating, not involving conditions or materials described in paragraph (b), (c), or (d) of this section, may normally be done without mechanical ventilation or respiratory protective equipment, but where, because of unusual physical or atmospheric conditions, an unsafe accumulation of contaminants exists, suitable mechanical ventilation or respiratory protective equipment shall be provided.

**1926.353(e)(2)** Employees performing any type of welding, cutting, or heating shall be protected by suitable eye protective equipment in accordance with the requirements of subpart E of this part.

[44 FR 8577, Feb. 9, 1979; 44 FR 20940, Apr. 6, 1979, as amended at 55 FR 42328, Oct. 18, 1990; 58 FR 35179, June 30, 1993]

### §1926.354 Welding, cutting, and heating in way of preservative coatings.

**1926.354(a)** Before welding, cutting, or heating is commenced on any surface covered by a preservative coating whose flammability is not known, a test shall be made by a competent person to determine its flammability. Preservative coatings shall be considered to be highly flammable when scrapings burn with extreme rapidity.

**1926.354(b)** Precautions shall be taken to prevent ignition of highly flammable hardened preservative coatings. When coatings are determined to be highly flammable, they shall be stripped from the area to be heated to prevent ignition.

**1926.354(c)** Protection against toxic preservative coatings: (1) In enclosed spaces, all surfaces covered with toxic preservatives shall be stripped of all toxic coatings for a distance of at least 4 inches from the area of heat application, or the employees shall be protected by air line respirators, meeting the requirements of subpart E of this part.

**1926.354(c)(2)** In the open air, employees shall be protected by a respirator, in accordance with requirements of subpart E of this part.

**1926.354(d)** The preservative coatings shall be removed a sufficient distance from the area to be heated to ensure that the temperature of the unstripped metal will not be appreciably raised. Artificial cooling of the metal surrounding the heating area may be used to limit the size of the area required to be cleaned.

**Authority:** Sections 6 and 8 of the Occupational Safety and Health Act of 1970 (29 U.S.C. 655 and 657); sec. 107, Contract Work Hours and Safety Standards Act (40 U.S.C. 333); Secretary of Labor's Order No. 9-83 (48 FR 35736) or 1-90 (55 FR 9033), as applicable; 29 CFR part 1911.

**Source:** 51 FR 25318, July 11, 1986, unless otherwise noted.

## General

### §1926.400 Introduction.

This subpart addresses electrical safety requirements that are necessary for the practical safeguarding of employees involved in construction work and is divided into four major divisions and applicable definitions as follows:

**1926.400(a)** ***Installation safety requirements.*** Installation safety requirements are contained in §§1926.402 through 1926.408. Included in this category are electric equipment and installations used to provide electric power and light on jobsites.

**1926.400(b)** ***Safety-related work practices.*** Safety-related work practices are contained in §§1926.416 and 1926.417. In addition to covering the hazards arising from the use of electricity at jobsites, these regulations also cover the hazards arising from the accidental contact, direct or indirect, by employees with all energized lines, above or below ground, passing through or near the jobsite.

**1926.400(c)** ***Safety-related maintenance and environmental considerations.*** Safety-related maintenance and environmental considerations are contained in §§1926.431 and 1926.432.

**1926.400(d)** ***Safety requirements for special equipment.*** Safety requirements for special equipment are contained in §1926.441.

**1926.400(e)** ***Definitions.*** Definitions applicable to this subpart are contained in §1926.449.

### §1926.401 [Reserved]

## Installation Safety Requirements

### §1926.402 Applicability.

**1926.402(a)** ***Covered.*** Sections 1926.402 through 1926.408 contain installation safety requirements for electrical equipment and installations used to provide electric power and light at the jobsite. These sections apply to installations, both temporary and permanent, used on the jobsite; but these sections do not apply to existing permanent installations that were in place before the construction activity commenced.

**Note:** If the electrical installation is made in accordance with the National Electrical Code ANSI/NFPA 70-1984, exclusive of Formal Interpretations and Tentative Interim Amendments, it will be deemed to be in compliance with §§1926.403 through 1926.408, except for §§1926.404(b)(1) and 1926.405(a)(2)(ii) (E), (F), (G), and (J).

**1926.402(b)** ***Not covered.*** Sections 1926.402 through 1926.408 do not cover installations used for the generation, transmission, and distribution of electric energy, including related communication, metering, control, and transformation installations. (However, these regulations do cover portable and vehicle-mounted generators used to provide power for equipment used at the jobsite.) See subpart V of this part for the construction of power distribution and transmission lines.

### §1926.403 General requirements.

**1926.403(a)** ***Approval.*** All electrical conductors and equipment shall be approved.

**1926.403(b)** ***Examination, installation, and use of equipment***—(1) *Examination.* The employer shall ensure that electrical equipment is free from recognized hazards that are likely to cause death or serious physical harm to employees. Safety of equipment shall be determined on the basis of the following considerations:

**1926.403(b)(1)(i)** Suitability for installation and use in conformity with the provisions of this subpart. Suitability of equipment for an identified purpose may be evidenced by listing, labeling, or certification for that identified purpose.

**1926.403(b)(1)(ii)** Mechanical strength and durability, including, for parts designed to enclose and protect other equipment, the adequacy of the protection thus provided.

**1926.403(b)(1)(iii)** Electrical insulation.

**1926.403(b)(1)(iv)** Heating effects under conditions of use.

**1926.403(b)(1)(v)** Arcing effects.

**1926.403(b)(1)(vi)** Classification by type, size, voltage, current capacity, specific use.

**1926.403(b)(1)(vii)** Other factors which contribute to the practical safeguarding of employees using or likely to come in contact with the equipment.

**1926.403(b)(2)** *Installation and use.* Listed, labeled, or certified equipment shall be installed and used in accordance with instructions included in the listing, labeling, or certification.

**1926.403(c)** ***Interrupting rating.*** Equipment intended to break current shall have an interrupting rating at system voltage sufficient for the current that must be interrupted.

**1926.403(d)** ***Mounting and cooling of equipment***—(1) *Mounting.* Electric equipment shall be firmly secured to the surface on which it is mounted. Wooden plugs driven into holes in masonry, concrete, plaster, or similar materials shall not be used.

**1926.403(d)(2)** *Cooling.* Electrical equipment which depends upon the natural circulation of air and convection principles for cooling of exposed surfaces shall be installed so that room air flow over such surfaces is not prevented by walls or by adjacent installed equipment. For equipment designed for floor mounting, clearance between top surfaces and adjacent surfaces shall be provided to dissipate rising warm air. Electrical equipment provided with ventilating openings shall be installed so that walls or other obstructions do not prevent the free circulation of air through the equipment.

**1926.403(e)** ***Splices.*** Conductors shall be spliced or joined with splicing devices designed for the use or by brazing, welding, or soldering with a fusible metal or alloy. Soldered splices shall first be so spliced or joined as to be mechanically and electrically secure without solder and then soldered. All splices and joints and the free ends of conductors shall be covered with an insulation equivalent to that of the conductors or with an insulating device designed for the purpose.

**1926.403(f)** ***Arcing parts.*** Parts of electric equipment which in ordinary operation produce arcs, sparks, flames, or molten metal shall be enclosed or separated and isolated from all combustible material.

**1926.403(g)** ***Marking.*** Electrical equipment shall not be used unless the manufacturer's name, trademark, or other descriptive marking by which the organization responsible for the product may be identified is placed on the equipment and unless other markings are provided giving voltage, current, wattage, or other ratings as necessary. The marking shall be of sufficient durability to withstand the environment involved.

**1926.403(h)** ***Identification of disconnecting means and circuits.*** Each disconnecting means required by this subpart for motors and appliances shall be legibly marked to indicate its purpose, unless located and arranged so the purpose is evident. Each service, feeder, and branch circuit, at its disconnecting means or overcurrent device, shall be legibly marked to indicate its purpose, unless located and arranged so the purpose is evident. These markings shall be of sufficient durability to withstand the environment involved.

**1926.403(i)** ***600 Volts, nominal, or less.*** This paragraph applies to equipment operating at 600 volts, nominal, or less.

**1926.403(i)(1)** *Working space about electric equipment.* Sufficient access and working space shall be provided and maintained about all electric equipment to permit ready and safe operation and maintenance of such equipment.

**1926.403(i)(1)(i)** *Working clearances.* Except as required or permitted elsewhere in this subpart, the dimension of the working space in the direction of access to live parts operating at 600 volts or less and likely to require examination, adjustment, servicing, or maintenance while alive shall not be less than indicated in Table K-1. In addition to the dimensions shown in Table K-1, workspace shall not be less than 30 inches (762 mm) wide in front of the electric equipment. Distances shall be measured from the live parts if they are exposed, or from the enclosure front or opening if the live parts are enclosed. Walls constructed of concrete, brick, or tile are considered to be grounded. Working space is not required in back of assemblies such as dead-front switchboards or motor control centers where there are no renewable or adjustable parts such as fuses or switches on the back and where all connections are accessible from locations other than the back.

### TABLE K-1—WORKING CLEARANCES

| *Nominal voltage to ground* | *Minimum clear distance for conditions* [1] | | |
|---|---|---|---|
| | *(a)* | *(b)* | *(c)* |
| | Feet [2] | Feet [2] | Feet [2] |
| 0-150 | 3 | 3 | 3 |
| 151-600 | 3 | $3\frac{1}{2}$ | 4 |

[1] Conditions (a), (b), and (c) are as follows: (a) Exposed live parts on one side and no live or grounded parts on the other side of the working space, or exposed live parts on both sides effectively guarded by insulating material. Insulated wire or insulated busbars operating at not over 300 volts are not considered live parts. (b) Exposed live parts on one side and grounded parts on the other side. (c) Exposed live parts on both sides of the workspace [not guarded as provided in Condition (a)] with the operator between.

[2] Note.—For International System of Units (SI): one foot=0.3048m.

**1926.403(i)(1)(ii)** *Clear spaces.* Working space required by this subpart shall not be used for storage. When normally enclosed live parts are exposed for inspection or servicing, the working space, if in a passageway or general open space, shall be guarded.

**1926.403(i)(1)(iii)** *Access and entrance to working space.* At least one entrance shall be provided to give access to the working space about electric equipment.

**1926.403(i)(1)(iv)** *Front working space.* Where there are live parts normally exposed on the front of switchboards or motor control centers, the working space in front of such equipment shall not be less than 3 feet (914 mm).

**1926.403(i)(1)(v)** *Headroom.* The minimum headroom of working spaces about service equipment, switchboards, panelboards, or motor control centers shall be 6 feet 3 inches (1.91 m).

**1926.403(i)(2)** *Guarding of live parts.* (i) Except as required or permitted elsewhere in this subpart, live parts of electric equipment operating at 50 volts or more shall be guarded against accidental contact by cabinets or other forms of enclosures, or by any of the following means:

**1926.403(i)(2)(i)(A)** By location in a room, vault, or similar enclosure that is accessible only to qualified persons.

**1926.403(i)(2)(i)(B)** By partitions or screens so arranged that only qualified persons will have access to the space within reach of the live parts. Any openings in such partitions or screens shall be so sized and located that persons are not likely to come into accidental contact with the live parts or to bring conducting objects into contact with them.

**1926.403(i)(2)(i)(C)** By location on a balcony, gallery, or platform so elevated and arranged as to exclude unqualified persons.

**1926.403(i)(2)(i)(D)** By elevation of 8 feet (2.44 m) or more above the floor or other working surface and so installed as to exclude unqualified persons.

**1926.403(i)(2)(ii)** In locations where electric equipment would be exposed to physical damage, enclosures or guards shall be so arranged and of such strength as to prevent such damage.

**1926.403(i)(2)(iii)** Entrances to rooms and other guarded locations containing exposed live parts shall be marked with conspicuous warning signs forbidding unqualified persons to enter.

**1926.403(j)** ***Over 600 volts, nominal.*** (1) *General.* Conductors and equipment used on circuits exceeding 600 volts, nominal, shall comply with all applicable provisions of paragraphs (a) through (g) of this section and with the following provisions which supplement or modify those requirements. The provisions of paragraphs (j)(2), (j)(3), and (j)(4) of this section do not apply to equipment on the supply side of the service conductors.

**1926.403(j)(2)** *Enclosure for electrical installations.* Electrical installations in a vault, room, closet or in an area surrounded by a wall, screen, or fence, access to which is controlled by lock and key or other equivalent means, are considered to be accessible to qualified persons only. A wall, screen, or fence less than 8 feet (2.44 m) in height is not considered adequate to prevent access unless it has other features that provide a degree of isolation equivalent to an 8-foot (2.44-m) fence. The entrances to all buildings, rooms or enclosures containing exposed live parts or exposed conductors operating at over 600 volts, nominal, shall be kept locked or shall be under the observation of a qualified person at all times.

**1926.403(j)(2)(i)** *Installations accessible to qualified persons only.* Electrical installations having exposed live parts shall be accessible to qualified persons only and shall comply with the applicable provisions of paragraph (j)(3) of this section.

**1926.403(j)(2)(ii)** *Installations accessible to unqualified persons.* Electrical installations that are open to unqualified persons shall be made with metal-enclosed equipment or shall be enclosed in a vault or in an area, access to which is controlled by a lock. Metal-enclosed switchgear, unit substations, transformers, pull boxes, connection boxes, and other similar associated equipment shall be marked with appropriate caution signs. If equipment is exposed to physical damage from vehicular traffic, guards shall be provided to prevent such damage. Ventilating or similar openings in metal-enclosed equipment shall be designed so that foreign objects inserted through these openings will be deflected from energized parts.

**1926.403(j)(3)** *Workspace about equipment.* Sufficient space shall be provided and maintained about electric equipment to permit ready and safe operation and maintenance of such equipment. Where energized parts are exposed, the minimum clear workspace shall not be less than 6 feet 6 inches (1.98 m) high (measured vertically from the floor or platform), or less than 3 feet (914 mm) wide (measured parallel to the equipment). The depth shall be as required in Table K-2. The workspace shall be adequate to permit at least a 90-degree opening of doors or hinged panels.

**1926.403(j)(3)(i)** *Working space.* The minimum clear working space in front of electric equipment such as switchboards, control panels, switches, circuit breakers, motor controllers, relays, and similar equipment shall not be less than specified in Table K-2 unless otherwise specified in this subpart. Distances shall be measured from the live parts if they are exposed, or from the enclosure front or opening if the live parts are enclosed. However, working space is not required in back of equipment such as deadfront switchboards or control assemblies where there are no renewable or adjustable parts (such as fuses or switches) on the back and where all connections are accessible from locations other than the back. Where rear access is required to work on de-energized parts on the back of enclosed equipment, a minimum working space of 30 inches (762 mm) horizontally shall be provided.

### TABLE K-2—MINIMUM DEPTH OF CLEAR WORKING SPACE IN FRONT OF ELECTRIC EQUIPMENT

| Nominal voltage to ground | Conditions [1] | | |
|---|---|---|---|
| | (a) | (b) | (c) |
| | Feet [2] | Feet [2] | Feet [2] |
| 601-2,500 | 3 | 4 | 5 |
| 2,501 to 9,000 | 4 | 5 | 6 |
| 9,001 to 25,000 | 5 | 6 | 9 |
| 25,001 to 75kV | 6 | 8 | 10 |
| Above 75kV | 8 | 10 | 12 |

[1] Conditions (a), (b), and (c) are as follows: (a) Exposed live parts on one side and no live or grounded parts on the other side of the working space, or exposed live parts on both sides effectively guarded by insulating material. Insulated wire or insulated busbars operating at not over 300 volts are not considered live parts. (b) Exposed live parts on one side and grounded parts on the other side. Walls constructed of concrete, brick, or tile are considered to be grounded surfaces. (c) Exposed live parts on both sides of the workspace [not guarded as provided in Condition (a)] with the operator between.

[2] NOTE.—For SI units: one foot=0.3048m.

**1926.403(j)(3)(ii)** *Lighting outlets and points of control.* The lighting outlets shall be so arranged that persons changing lamps or making repairs on the lighting system will not be endangered by live parts or other equipment. The points of control shall be so located that persons are not likely to come in contact with any live part or moving part of the equipment while turning on the lights.

**1926.403(j)(3)(iii)** *Elevation of unguarded live parts.* Unguarded live parts above working space shall be maintained at elevations not less than specified in Table K-3.

### TABLE K-3—ELEVATIONS OF UNGUARDED ENERGIZED PARTS ABOVE WORKING SPACE

| Nominal voltage between phases | Minimum elevation |
|---|---|
| 601-7,500 | 8 feet 6 inches.[1] |
| 7,501-35,000 | 9 feet. |
| Over 35kV | 9 feet + 0.37 inches per kV above 35kV. |

[1] NOTE.—For SI units: one inch=25.4 mm; one foot=0.3048m.

**1926.403(j)(4)** *Entrance and access to workspace.* At least one entrance not less than 24 inches (610 mm) wide and 6 feet 6 inches (1.98 m) high shall be provided to give access to the working space about electric equipment. On switchboard and control panels exceeding 48 inches (1.22 m) in width, there shall be one entrance at each end of such board where practicable. Where bare energized parts at any voltage or insulated energized parts above 600 volts are located adjacent to such entrance, they shall be guarded.

[51 FR 25318, July 11, 1986, as amended at 61 FR 5510, Feb. 13, 1996]

## §1926.404 Wiring design and protection.

**1926.404(a)** ***Use and identification of grounded and grounding conductors***—(1) *Identification of conductors.* A conductor used as a grounded conductor shall be identifiable and distinguishable from all other conductors. A conductor used as an equipment grounding conductor shall be identifiable and distinguishable from all other conductors.

**1926.404(a)(2)** *Polarity of connections.* No grounded conductor shall be attached to any terminal or lead so as to reverse designated polarity.

**1926.404(a)(3)** *Use of grounding terminals and devices.* A grounding terminal or grounding-type device on a receptacle, cord connector, or attachment plug shall not be used for purposes other than grounding.

**1926.404(b)** ***Branch circuits***—(1) *Ground-fault protection*—(i) *General.* The employer shall use either ground fault circuit interrupters as specified in paragraph (b)(1)(ii) of this section or an assured equipment grounding conductor program as specified in paragraph (b)(1)(iii) of this section to protect employees on construction sites. These requirements are in addition to any other requirements for equipment grounding conductors.

**1926.404(b)(1)(ii)** *Ground-fault circuit interrupters.* All 120-volt, single-phase, 15- and 20-ampere receptacle outlets on construction sites, which are not a part of the permanent wiring of the building or structure and which are in use by employees, shall have approved ground-fault circuit interrupters for personnel protection. Receptacles on a two-wire, single-phase portable or vehicle-mounted generator rated not more than 5kW, where the circuit conductors of the generator are insulated from the generator frame and all other grounded surfaces, need not be protected with ground-fault circuit interrupters.

**1926.404(b)(1)(iii)** *Assured equipment grounding conductor program.* The employer shall establish and implement an assured equipment grounding conductor program on construction sites covering all cord sets, receptacles which are not a part of the building or structure, and equipment connected by cord and plug which are available for use or used by employees. This program shall comply with the following minimum requirements:

**1926.404(b)(1)(iii)(A)** A written description of the program, including the specific procedures adopted by the employer, shall be available at the jobsite for inspection and copying by the Assistant Secretary and any affected employee.

**1926.404(b)(1)(iii)(B)** The employer shall designate one or more competent persons (as defined in §1926.32(f)) to implement the program.

**1926.404(b)(1)(iii)(C)** Each cord set, attachment cap, plug and receptacle of cord sets, and any equipment connected by cord and plug, except cord sets and receptacles which are fixed and not exposed to damage, shall be visually inspected before each day's use for external defects, such as deformed or missing pins or insulation damage, and for indications of possible internal damage. Equipment found damaged or defective shall not be used until repaired.

**1926.404(b)(1)(iii)(D)** The following tests shall be performed on all cord sets, receptacles which are not a part of the permanent wiring of the building or structure, and cord- and plug-connected equipment required to be grounded:

**1926.404(b)(1)(iii)(D)*(1)*** All equipment grounding conductors shall be tested for continuity and shall be electrically continuous.

**1926.404(b)(1)(iii)(D)*(2)*** Each receptacle and attachment cap or plug shall be tested for correct attachment of the equipment grounding conductor. The equipment grounding conductor shall be connected to its proper terminal.

**1926.404(b)(1)(iii)(E)** All required tests shall be performed:

**1926.404(b)(1)(iii)(E)*(1)*** Before first use;

**1926.404(b)(1)(iii)(E)*(2)*** Before equipment is returned to service following any repairs;

**1926.404(b)(1)(iii)(E)*(3)*** Before equipment is used after any incident which can be reasonably suspected to have caused damage (for example, when a cord set is run over); and

**1926.404(b)(1)(iii)(E)*(4)*** At intervals not to exceed 3 months, except that cord sets and receptacles which are fixed and not exposed to damage shall be tested at intervals not exceeding 6 months.

**1926.404(b)(1)(iii)(F)** The employer shall not make available or permit the use by employees of any equipment which has not met the requirements of this paragraph (b)(1)(iii) of this section.

**1926.404(b)(1)(iii)(G)** Tests performed as required in this paragraph shall be recorded. This test record shall identify each receptacle, cord set, and cord- and plug-connected equipment that passed the test and shall indicate the last date it was tested or the interval for which it was tested. This record shall be kept by means of logs, color coding, or other effective means and shall be maintained until replaced by a more current record. The record shall be made available on the jobsite for inspection by the Assistant Secretary and any affected employee.

**1926.404(b)(2)** *Outlet devices.* Outlet devices shall have an ampere rating not less than the load to be served and shall comply with the following:

**1926.404(b)(2)(i)** *Single receptacles.* A single receptacle installed on an individual branch circuit shall have an ampere rating of not less than that of the branch circuit.

**1926.404(b)(2)(ii)** *Two or more receptacles.* Where connected to a branch circuit supplying two or more receptacles or outlets, receptacle ratings shall conform to the values listed in Table K-4.

**1926.404(b)(2)(iii)** *Receptacles used for the connection of motors.* The rating of an attachment plug or receptacle used for cord- and plug-connection of a motor to a branch circuit shall not exceed 15 amperes at 125 volts or 10 amperes at 250 volts if individual overload protection is omitted.

**TABLE K-4—RECEPTACLE RATINGS FOR VARIOUS SIZE CIRCUITS**

| *Circuit rating amperes* | *Receptacle rating amperes* |
|---|---|
| 15 | Not over 15. |
| 20 | 15 or 20. |
| 30 | 30. |
| 40 | 40 or 50. |
| 50 | 50. |

**1926.404(c)** ***Outside conductors and lamps***—(1) *600 volts, nominal, or less.* Paragraphs (c)(1)(i) through (c)(1)(iv) of this section apply to branch circuit, feeder, and service conductors rated 600 volts, nominal, or less and run outdoors as open conductors.

**1926.404(c)(1)(i)** *Conductors on poles.* Conductors supported on poles shall provide a horizontal climbing space not less than the following:

**1926.404(c)(1)(i)(A)** Power conductors below communication conductors—30 inches (762 mm) .

**1926.404(c)(1)(i)(B)** Power conductors alone or above communication conductors: 300 volts or less—24 inches (610 mm); more than 300 volts—30 inches (762 mm).

**1926.404(c)(1)(i)(C)** Communication conductors below power conductors: with power conductors 300 volts or less—24 inches (610 mm); more than 300 volts—30 inches (762 mm).

**1926.404(c)(1)(ii)** *Clearance from ground.* Open conductors shall conform to the following minimum clearances:

**1926.404(c)(1)(ii)(A)** 10 feet (3.05 m)—above finished grade, sidewalks, or from any platform or projection from which they might be reached.

**1926.404(c)(1)(ii)(B)** 12 feet (3.66 m)—over areas subject to vehicular traffic other than truck traffic.

**1926.404(c)(1)(ii)(C)** 15 feet (4.57 m)—over areas other than those specified in paragraph (c)(1)(ii)(D) of this section that are subject to truck traffic.

**1926.404(c)(1)(ii)(D)** 18 feet (5.49 m)—over public streets, alleys, roads, and driveways.

**1926.404(c)(1)(iii)** *Clearance from building openings.* Conductors shall have a clearance of at least 3 feet (914 mm) from windows, doors, fire escapes, or similar locations. Conductors run above the top level of a window are considered to be out of reach from that window and, therefore, do not have to be 3 feet (914 mm) away.

**1926.404(c)(1)(iv)** *Clearance over roofs.* Conductors above roof space accessible to employees on foot shall have a clearance from the highest point of the roof surface of not less than 8 feet (2.44 m) vertical clearance for insulated conductors, not less than 10 feet (3.05 m) vertical or diagonal clearance for covered conductors, and not less than 15 feet (4.57 m) for bare conductors, except that:

**1926.404(c)(1)(iv)(A)** Where the roof space is also accessible to vehicular traffic, the vertical clearance shall not be less than 18 feet (5.49 m), or

**1926.404(c)(1)(iv)(B)** Where the roof space is not normally accessible to employees on foot, fully insulated conductors shall have a vertical or diagonal clearance of not less than 3 feet (914 mm), or

**1926.404(c)(1)(iv)(C)** Where the voltage between conductors is 300 volts or less and the roof has a slope of not less than 4 inches (102 mm) in 12 inches (305 mm), the clearance from roofs shall be at least 3 feet (914 mm), or

**1926.404(c)(1)(iv)(D)** Where the voltage between conductors is 300 volts or less and the conductors do not pass over more than 4 feet (1.22 m) of the overhang portion of the roof and they are terminated at a through-the-roof raceway or support, the clearance from roofs shall be at least 18 inches (457 mm).

**1926.404(c)(2)** *Location of outdoor lamps.* Lamps for outdoor lighting shall be located below all live conductors, transformers, or other electric equipment, unless such equipment is controlled by a disconnecting means that can be locked in the open position or unless adequate clearances or other safeguards are provided for relamping operations.

**1926.404(d)** ***Services***—(1) *Disconnecting means*—(i) *General.* Means shall be provided to disconnect all conductors in a building or other structure from the service-entrance conductors. The disconnecting means shall plainly indicate whether it is in the open or closed position and shall be installed at a readily accessible location nearest the point of entrance of the service-entrance conductors.

**1926.404(d)(1)(ii)** *Simultaneous opening of poles.* Each service disconnecting means shall simultaneously disconnect all ungrounded conductors.

**1926.404(d)(2)** *Services over 600 volts, nominal.* The following additional requirements apply to services over 600 volts, nominal.

**1926.404(d)(2)(i)** *Guarding.* Service-entrance conductors installed as open wires shall be guarded to make them accessible only to qualified persons.

**1926.404(d)(2)(ii)** *Warning signs.* Signs warning of high voltage shall be posted where unauthorized employees might come in contact with live parts.

**1926.404(e)** ***Overcurrent protection***—(1) *600 volts, nominal, or less.* The following requirements apply to overcurrent protection of circuits rated 600 volts, nominal, or less.

**1926.404(e)(1)(i)** *Protection of conductors and equipment.* Conductors and equipment shall be protected from overcurrent in accordance with their ability to safely conduct current. Conductors shall have sufficient ampacity to carry the load.

**1926.404(e)(1)(ii)** *Grounded conductors.* Except for motor-running overload protection, overcurrent devices shall not interrupt the continuity of the grounded conductor unless all conductors of the circuit are opened simultaneously.

**1926.404(e)(1)(iii)** *Disconnection of fuses and thermal cutouts.* Except for devices provided for current-limiting on the supply side of the service disconnecting means, all cartridge fuses which are accessible to other than qualified persons and all fuses and thermal cutouts on circuits over 150 volts to ground shall be provided with disconnecting means. This disconnecting means shall be installed so that the fuse or thermal cutout can be disconnected from its supply without disrupting service to equipment and circuits unrelated to those protected by the overcurrent device.

**1926.404(e)(1)(iv)** *Location in or on premises.* Overcurrent devices shall be readily accessible. Overcurrent devices shall not be located where they could create an employee safety hazard by being exposed to physical damage or located in the vicinity of easily ignitible material.

**1926.404(e)(1)(v)** *Arcing or suddenly moving parts.* Fuses and circuit breakers shall be so located or shielded that employees will not be burned or otherwise injured by their operation.

**1926.404(e)(1)(vi)** *Circuit breakers*—(A) Circuit breakers shall clearly indicate whether they are in the open (off) or closed (on) position.

**1926.404(e)(1)(vi)(B)** Where circuit breaker handles on switchboards are operated vertically rather than horizontally or rotationally, the up position of the handle shall be the closed (on) position.

**1926.404(e)(1)(vi)(C)** If used as switches in 120-volt, fluorescent lighting circuits, circuit breakers shall be marked "SWD."

**1926.404(e)(2)** *Over 600 volts, nominal.* Feeders and branch circuits over 600 volts, nominal, shall have short-circuit protection.

**1926.404(f)** ***Grounding.*** Paragraphs (f)(1) through (f)(11) of this section contain grounding requirements for systems, circuits, and equipment.

**1926.404(f)(1)** *Systems to be grounded.* The following systems which supply premises wiring shall be grounded:

**1926.404(f)(1)(i)** *Three-wire DC systems.* All 3-wire DC systems shall have their neutral conductor grounded.

**1926.404(f)(1)(ii)** *Two-wire DC systems.* Two-wire DC systems operating at over 50 volts through 300 volts between conductors shall be grounded unless they are rectifier-derived from an AC system complying with paragraphs (f)(1)(iii), (f)(1)(iv), and (f)(1)(v) of this section.

**1926.404(f)(1)(iii)** *AC circuits, less than 50 volts.* AC circuits of less than 50 volts shall be grounded if they are installed as overhead conductors outside of buildings or if they are supplied by transformers and the transformer primary supply system is ungrounded or exceeds 150 volts to ground.

**1926.404(f)(1)(iv)** *AC systems, 50 volts to 1000 volts.* AC systems of 50 volts to 1000 volts shall be grounded under any of the following conditions, unless exempted by paragraph (f)(1)(v) of this section:

**1926.404(f)(1)(iv)(A)** If the system can be so grounded that the maximum voltage to ground on the ungrounded conductors does not exceed 150 volts;

**1926.404(f)(1)(iv)(B)** If the system is nominally rated 480Y/277 volt, 3-phase, 4-wire in which the neutral is used as a circuit conductor;

**1926.404(f)(1)(iv)(C)** If the system is nominally rated 240/120 volt, 3-phase, 4-wire in which the midpoint of one phase is used as a circuit conductor; or

**1926.404(f)(1)(iv)(D)** If a service conductor is uninsulated.

**1926.404(f)(1)(v)** *Exceptions.* AC systems of 50 volts to 1000 volts are not required to be grounded if the system is separately derived and is supplied by a transformer that has a primary voltage rating less than 1000 volts, provided all of the following conditions are met:

**1926.404(f)(1)(v)(A)** The system is used exclusively for control circuits,

**1926.404(f)(1)(v)(B)** The conditions of maintenance and supervision assure that only qualified persons will service the installation,

**1926.404(f)(1)(v)(C)** Continuity of control power is required, and

**1926.404(f)(1)(v)(D)** Ground detectors are installed on the control system.

**1926.404(f)(2)** *Separately derived systems.* Where paragraph (f)(1) of this section requires grounding of wiring systems whose power is derived from generator, transformer, or converter windings and has no direct electrical connection, including a solidly connected grounded circuit conductor, to supply conductors originating in another system, paragraph (f)(5) of this section shall also apply.

**1926.404(f)(3)** *Portable and vehicle-mounted generators*—(i) *Portable generators.* Under the following conditions, the frame of a portable generator need not be grounded and may serve as the grounding electrode for a system supplied by the generator:

**1926.404(f)(3)(i)(A)** The generator supplies only equipment mounted on the generator and/or cord- and plug-connected equipment through receptacles mounted on the generator, and

**1926.404(f)(3)(i)(B)** The noncurrent-carrying metal parts of equipment and the equipment grounding conductor terminals of the receptacles are bonded to the generator frame.

**1926.404(f)(3)(ii)** *Vehicle-mounted generators.* Under the following conditions the frame of a vehicle may serve as the grounding electrode for a system supplied by a generator located on the vehicle:

**1926.404(f)(3)(ii)(A)** The frame of the generator is bonded to the vehicle frame, and

**1926.404(f)(3)(ii)(B)** The generator supplies only equipment located on the vehicle and/or cord- and plug-connected equipment through receptacles mounted on the vehicle or on the generator, and

**1926.404(f)(3)(ii)(C)** The noncurrent-carrying metal parts of equipment and the equipment grounding conductor terminals of the receptacles are bonded to the generator frame, and

**1926.404(f)(3)(ii)(D)** The system complies with all other provisions of this section.

**1926.404(f)(3)(iii)** *Neutral conductor bonding.* A neutral conductor shall be bonded to the generator frame if the generator is a component of a separately derived system. No other conductor need be bonded to the generator frame.

**1926.404(f)(4)** *Conductors to be grounded.* For AC premises wiring systems the identified conductor shall be grounded.

**1926.404(f)(5)** *Grounding connections*—(i) *Grounded system.* For a grounded system, a grounding electrode conductor shall be used to connect both the equipment grounding conductor and the grounded circuit conductor to the grounding electrode. Both the equipment grounding conductor and the grounding electrode conductor shall be connected to the grounded circuit conductor on the supply side of the service disconnecting means, or on the supply side of the system disconnecting means or overcurrent devices if the system is separately derived.

**1926.404(f)(5)(ii)** *Ungrounded systems.* For an ungrounded service-supplied system, the equipment grounding conductor shall be connected to the grounding electrode conductor at the service equipment. For an ungrounded separately derived system, the equipment grounding conductor shall be connected to the grounding electrode conductor at, or ahead of, the system disconnecting means or overcurrent devices.

**1926.404(f)(6)** *Grounding path.* The path to ground from circuits, equipment, and enclosures shall be permanent and continuous.

**1926.404(f)(7)** *Supports, enclosures, and equipment to be grounded*—(i) *Supports and enclosures for conductors.* Metal cable trays, metal raceways, and metal enclosures for conductors shall be grounded, except that:

**1926.404(f)(7)(i)(A)** Metal enclosures such as sleeves that are used to protect cable assemblies from physical damage need not be grounded; and

**1926.404(f)(7)(i)(B)** Metal enclosures for conductors added to existing installations of open wire, knob-and-tube wiring, and nonmetallic-sheathed cable need not be grounded if all of the following conditions are met:

**1926.404(f)(7)(i)(B)*(1)*** Runs are less than 25 feet (7.62 m);

**1926.404(f)(7)(i)(B)*(2)*** Enclosures are free from probable contact with ground, grounded metal, metal laths, or other conductive materials; and

**1926.404(f)(7)(i)(B)*(3)*** Enclosures are guarded against employee contact.

**1926.404(f)(7)(ii)** *Service equipment enclosures.* Metal enclosures for service equipment shall be grounded.

**1926.404(f)(7)(iii)** *Fixed equipment.* Exposed noncurrent-carrying metal parts of fixed equipment which may become energized shall be grounded under any of the following conditions:

**1926.404(f)(7)(iii)(A)** If within 8 feet (2.44 m) vertically or 5 feet (1.52 m) horizontally of ground or grounded metal objects and subject to employee contact.

**1926.404(f)(7)(iii)(B)** If located in a wet or damp location and subject to employee contact.

**1926.404(f)(7)(iii)(C)** If in electrical contact with metal.

**1926.404(f)(7)(iii)(D)** If in a hazardous (classified) location.

**1926.404(f)(7)(iii)(E)** If supplied by a metal-clad, metal-sheathed, or grounded metal raceway wiring method.

**1926.404(f)(7)(iii)(F)** If equipment operates with any terminal at over 150 volts to ground; however, the following need not be grounded:

**1926.404(f)(7)(iii)(F)*(1)*** Enclosures for switches or circuit breakers used for other than service equipment and accessible to qualified persons only;

**1926.404(f)(7)(iii)(F)*(2)*** Metal frames of electrically heated appliances which are permanently and effectively insulated from ground; and

**1926.404(f)(7)(iii)(F)*(3)*** The cases of distribution apparatus such as transformers and capacitors mounted on wooden poles at a height exceeding 8 feet (2.44 m) above ground or grade level.

**1926.404(f)(7)(iv)** *Equipment connected by cord and plug.* Under any of the conditions described in paragraphs (f)(7)(iv)(A) through (f)(7)(iv)(C) of this section, exposed noncurrent-carrying metal parts of cord- and plug-connected equipment which may become energized shall be grounded:

**1926.404(f)(7)(iv)(A)** If in a hazardous (classified) location (see §1926.407).

**1926.404(f)(7)(iv)(B)** If operated at over 150 volts to ground, except for guarded motors and metal frames of electrically heated appliances if the appliance frames are permanently and effectively insulated from ground.

**1926.404(f)(7)(iv)(C)** If the equipment is one of the types listed in paragraphs (f)(7)(iv)(C)(*1*) through (f)(7)(iv)(C)(*5*) of this section. However, even though the equipment may be one of these types, it need not be grounded if it is exempted by paragraph (f)(7)(iv)(C)(*6*).

**1926.404(f)(7)(iv)(C)*(1)*** Hand held motor-operated tools;

**1926.404(f)(7)(iv)(C)*(2)*** Cord- and plug-connected equipment used in damp or wet locations or by employees standing on the ground or on metal floors or working inside of metal tanks or boilers;

**1926.404(f)(7)(iv)(C)*(3)*** Portable and mobile X-ray and associated equipment;

**1926.404(f)(7)(iv)(C)*(4)*** Tools likely to be used in wet and/or conductive locations; and

**1926.404(f)(7)(iv)(C)*(5)*** Portable hand lamps.

**1926.404(f)(7)(iv)(C)*(6)*** Tools likely to be used in wet and/or conductive locations need not be grounded if supplied through an isolating transformer with an ungrounded secondary of not over 50 volts. Listed or labeled portable tools and appliances protected by a system of double insulation, or its equivalent, need not be grounded. If such a system is employed, the equipment shall be distinctively marked to indicate that the tool or appliance utilizes a system of double insulation.

**1926.404(f)(7)(v)** *Nonelectrical equipment.* The metal parts of the following nonelectrical equipment shall be grounded: Frames and tracks of electrically operated cranes; frames of nonelectrically driven elevator cars to which electric conductors are attached; hand-operated metal shifting ropes or cables of electric elevators, and metal partitions, grill work, and similar metal enclosures around equipment of over IkV between conductors.

**1926.404(f)(8)** *Methods of grounding equipment*—(i) *With circuit conductors.* Noncurrent-carrying metal parts of fixed equipment, if required to be grounded by this subpart, shall be grounded by an equipment grounding conductor which is contained within the same raceway, cable, or cord, or runs with or encloses the circuit conductors. For DC circuits only, the equipment grounding conductor may be run separately from the circuit conductors.

**1926.404(f)(8)(ii)** *Grounding conductor.* A conductor used for grounding fixed or movable equipment shall have capacity to conduct safely any fault current which may be imposed on it.

**1926.404(f)(8)(iii)** *Equipment considered effectively grounded.* Electric equipment is considered to be effectively grounded if it is secured to, and in electrical contact with, a metal rack or structure that is provided for its support and the metal rack or structure is grounded by the method specified for the noncurrent-carrying metal parts of fixed equipment in paragraph (f)(8)(i) of this section. Metal car frames supported by metal hoisting cables attached to or running over metal sheaves or drums of grounded elevator machines are also considered to be effectively grounded.

**1926.404(f)(9)** *Bonding.* If bonding conductors are used to assure electrical continuity, they shall have the capacity to conduct any fault current which may be imposed.

**1926.404(f)(10)** *Made electrodes.* If made electrodes are used, they shall be free from nonconductive coatings, such as paint or enamel; and, if practicable, they shall be embedded below permanent moisture level. A single electrode consisting of a rod, pipe or plate which has a resistance to ground greater than 25 ohms shall be augmented by one additional electrode installed no closer than 6 feet (1.83 m) to the first electrode.

**1926.404(f)(11)** *Grounding of systems and circuits of 1000 volts and over (high voltage)*—(i) *General.* If high voltage systems are grounded, they shall comply with all applicable provisions of paragraphs (f)(1) through (f)(10) of this section as supplemented and modified by this paragraph (f)(11).

**1926.404(f)(11)(ii)** *Grounding of systems supplying portable or mobile equipment.* Systems supplying portable or mobile high voltage equipment, other than substations installed on a temporary basis, shall comply with the following:

**1926.404(f)(11)(ii)(A)** Portable and mobile high voltage equipment shall be supplied from a system having its neutral grounded through an impedance. If a delta-connected high voltage system is used to supply the equipment, a system neutral shall be derived.

**1926.404(f)(11)(ii)(B)** Exposed noncurrent-carrying metal parts of portable and mobile equipment shall be connected by an equipment grounding conductor to the point at which the system neutral impedance is grounded.

**1926.404(f)(11)(ii)(C)** Ground-fault detection and relaying shall be provided to automatically de-energize any high voltage system component which has developed a ground fault. The continuity of the equipment grounding conductor shall be continuously monitored so as to de-energize automatically the high voltage feeder to the portable equipment upon loss of continuity of the equipment grounding conductor.

**1926.404(f)(11)(ii)(D)** The grounding electrode to which the portable or mobile equipment system neutral impedance is connected shall be isolated from and separated in the ground by at least 20 feet (6.1 m) from any other system or equipment grounding electrode, and there shall be no direct connection between the grounding electrodes, such as buried pipe, fence or like objects.

**1926.404(f)(11)(iii)** *Grounding of equipment.* All noncurrent-carrying metal parts of portable equipment and fixed equipment including their associated fences, housings, enclosures, and supporting structures shall be grounded. However, equipment which is guarded by location and isolated from ground need not be grounded. Additionally, pole-mounted distribution apparatus at a height exceeding 8 feet (2.44 m) above ground or grade level need not be grounded.

[51 FR 25318, July 11, 1986, as amended at 54 FR 24334, June 7, 1989; 61 FR 5510, Feb. 13, 1996]

### §1926.405 Wiring methods, components, and equipment for general use.

**1926.405(a)** ***Wiring methods.*** The provisions of this paragraph do not apply to conductors which form an integral part of equipment such as motors, controllers, motor control centers and like equipment.

**1926.405(a)(1)** *General requirements*—(i) *Electrical continuity of metal raceways and enclosures.* Metal raceways, cable armor, and other metal enclosures for conductors shall be metallically joined together into a continuous electric conductor and shall be so connected to all boxes, fittings, and cabinets as to provide effective electrical continuity.

**1926.405(a)(1)(ii)** *Wiring in ducts.* No wiring systems of any type shall be installed in ducts used to transport dust, loose stock or flammable vapors. No wiring system of any type shall be installed in any duct used for vapor removal or in any shaft containing only such ducts.

**1926.405(a)(2)** *Temporary wiring*—(i) *Scope.* The provisions of paragraph (a)(2) of this section apply to temporary electrical power and lighting wiring methods which may be of a class less than would be required for a permanent installation. Except as specifically modified in paragraph (a)(2) of this section, all other requirements of this subpart for permanent wiring shall apply to temporary wiring installations. Temporary wiring shall be removed immediately upon completion of construction or the purpose for which the wiring was installed.

**1926.405(a)(2)(ii)** *General requirements for temporary wiring*—(A) Feeders shall originate in a distribution center. The conductors shall be run as multiconductor cord or cable assemblies or within raceways; or, where not subject to physical damage, they may be run as open conductors on insulators not more than 10 feet (3.05 m) apart.

**1926.405(a)(2)(ii)(B)** Branch circuits shall originate in a power outlet or panelboard. Conductors shall be run as multiconductor cord or cable assemblies or open conductors, or shall be run in raceways. All conductors shall be protected by overcurrent devices at their ampacity. Runs of open conductors shall be located where the conductors will not be subject to physical damage, and the conductors shall be fastened at intervals not exceeding 10 feet (3.05 m). No branch-circuit conductors shall be laid on the floor. Each branch circuit that supplies receptacles or fixed equipment shall contain a separate equipment grounding conductor if the branch circuit is run as open conductors.

**1926.405(a)(2)(ii)(C)** Receptacles shall be of the grounding type. Unless installed in a complete metallic raceway, each branch circuit shall contain a separate equipment grounding conductor, and all receptacles shall be electrically connected to the grounding conductor. Receptacles for uses other than temporary lighting shall not be installed on branch circuits which supply temporary lighting. Receptacles shall not be connected to the same ungrounded conductor of multiwire circuits which supply temporary lighting.

**1926.405(a)(2)(ii)(D)** Disconnecting switches or plug connectors shall be installed to permit the disconnection of all ungrounded conductors of each temporary circuit.

**1926.405(a)(2)(ii)(E)** All lamps for general illumination shall be protected from accidental contact or breakage. Metal-case sockets shall be grounded.

**1926.405(a)(2)(ii)(F)** Temporary lights shall not be suspended by their electric cords unless cords and lights are designed for this means of suspension.

**1926.405(a)(2)(ii)(G)** Portable electric lighting used in wet and/or other conductive locations, as for example, drums, tanks, and vessels, shall be operated at 12 volts or less. However, 120-volt lights may be used if protected by a ground-fault circuit interrupter.

**1926.405(a)(2)(ii)(H)** A box shall be used wherever a change is made to a raceway system or a cable system which is metal clad or metal sheathed.

**1926.405(a)(2)(ii)(I)** Flexible cords and cables shall be protected from damage. Sharp corners and projections shall be avoided. Flexible cords and cables may pass through doorways or other pinch points, if protection is provided to avoid damage.

**1926.405(a)(2)(ii)(J)** Extension cord sets used with portable electric tools and appliances shall be of three-wire type and shall be designed for hard or extra-hard usage. Flexible cords used with temporary and portable lights shall be designed for hard or extra-hard usage.

**Note:** The National Electrical Code, ANSI/NFPA 70, in Article 400, Table 400-4, lists various types of flexible cords, some of which are noted as being designed for hard or extra-hard usage. Examples of these types of flexible cords include hard service cord (types S, ST, SO, STO) and junior hard service cord (types SJ, SJO, SJT, SJTO).

**1926.405(a)(2)(iii)** *Guarding.* For temporary wiring over 600 volts, nominal, fencing, barriers, or other effective means shall be provided to prevent access of other than authorized and qualified personnel.

**1926.405(b)** ***Cabinets, boxes, and fittings.*** (1) *Conductors entering boxes, cabinets, or fittings.* Conductors entering boxes, cabinets, or fittings shall be protected from abrasion, and openings through which conductors enter shall be effectively closed. Unused openings in cabinets, boxes, and fittings shall also be effectively closed.

**1926.405(b)(2)** *Covers and canopies.* All pull boxes, junction boxes, and fittings shall be provided with covers. If metal covers are used, they shall be grounded. In energized installations each outlet box shall have a cover, faceplate, or fixture canopy. Covers of outlet boxes having holes through which flexible cord pendants pass shall be provided with bushings designed for the purpose or shall have smooth, well-rounded surfaces on which the cords may bear.

**1926.405(b)(3)** *Pull and junction boxes for systems over 600 volts, nominal.* In addition to other requirements in this section for pull and junction boxes, the following shall apply to these boxes for systems over 600 volts, nominal:

**1926.405(b)(3)(i)** *Complete enclosure.* Boxes shall provide a complete enclosure for the contained conductors or cables.

**1926.405(b)(3)(ii)** *Covers.* Boxes shall be closed by covers securely fastened in place. Underground box covers that weigh over 100 pounds (43.6 kg) meet this requirement. Covers for boxes shall be permanently marked "HIGH VOLTAGE." The marking shall be on the outside of the box cover and shall be readily visible and legible.

**1926.405(c)** ***Knife switches.*** Single-throw knife switches shall be so connected that the blades are dead when the switch is in the open position. Single-throw knife switches shall be so placed that gravity will not tend to close them. Single-throw knife switches approved for use in the inverted position shall be provided with a locking device that will ensure that the blades remain in the open position when so set. Double-throw knife switches may be mounted so that the throw will be either vertical or horizontal. However, if the throw is vertical, a locking device shall be provided to ensure that the blades remain in the open position when so set.

**1926.405(d)** ***Switchboards and panelboards.*** Switchboards that have any exposed live parts shall be located in permanently dry locations and accessible only to qualified persons. Panelboards shall be mounted in cabinets, cutout boxes, or enclosures designed for the purpose and shall be dead front. However, panelboards other than the dead front externally-operable type are permitted where accessible only to qualified persons. Exposed blades of knife switches shall be dead when open.

**1926.405(e)** ***Enclosures for damp or wet locations*** (1) *Cabinets, fittings, and boxes.* Cabinets, cutout boxes, fittings, boxes, and panelboard enclosures in damp or wet locations shall be installed so as to prevent moisture or water from entering and accumulating within the enclosures. In wet locations the enclosures shall be weatherproof.

**1926.405(e)(2)** *Switches and circuit breakers.* Switches, circuit breakers, and switchboards installed in wet locations shall be enclosed in weatherproof enclosures.

**1926.405(f)** ***Conductors for general wiring.*** All conductors used for general wiring shall be insulated unless otherwise permitted in this subpart. The conductor insulation shall be of a type that is suitable for the voltage, operating temperature, and location of use. Insulated conductors shall be distinguishable by appropriate color or other means as being grounded conductors, ungrounded conductors, or equipment grounding conductors.

**1926.405(g)** ***Flexible cords and cables***—(1) *Use of flexible cords and cables*—(i) *Permitted uses.* Flexible cords and cables shall be suitable for conditions of use and location. Flexible cords and cables shall be used only for:

**1926.405(g)(1)(i)(A)** Pendants;

**1926.405(g)(1)(i)(B)** Wiring of fixtures;

**1926.405(g)(1)(i)(C)** Connection of portable lamps or appliances;

**1926.405(g)(1)(i)(D)** Elevator cables;

**1926.405(g)(1)(i)(E)** Wiring of cranes and hoists;

**1926.405(g)(1)(i)(F)** Connection of stationary equipment to facilitate their frequent interchange;

**1926.405(g)(1)(i)(G)** Prevention of the transmission of noise or vibration; or

**1926.405(g)(1)(i)(H)** Appliances where the fastening means and mechanical connections are designed to permit removal for maintenance and repair.

**1926.405(g)(1)(ii)** *Attachment plugs for cords.* If used as permitted in paragraphs (g)(1)(i)(C), (g)(1)(i)(F), or (g)(1)(i)(H) of this section, the flexible cord shall be equipped with an attachment plug and shall be energized from a receptacle outlet.

**1926.405(g)(1)(iii)** *Prohibited uses.* Unless necessary for a use permitted in paragraph (g)(1)(i) of this section, flexible cords and cables shall not be used:

**1926.405(g)(1)(iii)(A)** As a substitute for the fixed wiring of a structure;

**1926.405(g)(1)(iii)(B)** Where run through holes in walls, ceilings, or floors;

**1926.405(g)(1)(iii)(C)** Where run through doorways, windows, or similar openings, except as permitted in paragraph (a)(2)(ii)(1) of this section;

**1926.405(g)(1)(iii)(D)** Where attached to building surfaces; or

**1926.405(g)(1)(iii)(E)** Where concealed behind building walls, ceilings, or floors.

**1926.405(g)(2)** *Identification, splices, and terminations*—(i) *Identification.* A conductor of a flexible cord or cable that is used as a grounded conductor or an equipment grounding conductor shall be distinguishable from other conductors.

**1926.405(g)(2)(ii)** *Marking.* Type SJ, SJO, SJT, SJTO, S, SO, ST, and STO cords shall not be used unless durably marked on the surface with the type designation, size, and number of conductors.

**1926.405(g)(2)(iii)** *Splices.* Flexible cords shall be used only in continuous lengths without splice or tap. Hard service flexible cords No. 12 or larger may be repaired if spliced so that the splice retains the insulation, outer sheath properties, and usage characteristics of the cord being spliced.

**1926.405(g)(2)(iv)** *Strain relief.* Flexible cords shall be connected to devices and fittings so that strain relief is provided which will prevent pull from being directly transmitted to joints or terminal screws.

**1926.405(g)(2)(v)** *Cords passing through holes.* Flexible cords and cables shall be protected by bushings or fittings where passing through holes in covers, outlet boxes, or similar enclosures.

**1926.405(h)** ***Portable cables over 600 volts, nominal.*** Multiconductor portable cable for use in supplying power to portable or mobile equipment at over 600 volts, nominal, shall consist of No. 8 or larger conductors employing flexible stranding. Cables operated at over 2000 volts shall be shielded for the purpose of confining the voltage stresses to the insulation. Grounding conductors shall be provided. Connectors for these cables shall be of a locking type with provisions to prevent their opening or closing while energized. Strain relief shall be provided at connections and terminations. Portable cables shall not be operated with splices unless the splices are of the permanent molded, vulcanized, or other equivalent type. Termination enclosures shall be marked with a high voltage hazard warning, and terminations shall be accessible only to authorized and qualified personnel.

**1926.405(i)** ***Fixture wires***—(1) *General.* Fixture wires shall be suitable for the voltage, temperature, and location of use. A fixture wire which is used as a grounded conductor shall be identified.

**1926.405(i)(2)** *Uses permitted.* Fixture wires may be used:

**1926.405(i)(2)(i)** For installation in lighting, fixtures and in similar equipment where enclosed or protected and not subject to bending or twisting in use; or

**1926.405(i)(2)(ii)** For connecting lighting fixtures to the branch-circuit conductors supplying the fixtures.

**1926.405(i)(3)** *Uses not permitted.* Fixture wires shall not be used as branch-circuit conductors except as permitted for Class 1 power-limited circuits.

**1926.405(j)** ***Equipment for general use***—(1) *Lighting fixtures, lampholders, lamps, and receptacles*—(i) *Live parts.* Fixtures, lampholders, lamps, rosettes, and receptacles shall have no live parts normally exposed to employee contact. However, rosettes and cleat-type lampholders and receptacles located at least 8 feet (2.44 m) above the floor may have exposed parts.

**1926.405(j)(1)(ii)** *Support.* Fixtures, lampholders, rosettes, and receptacles shall be securely supported. A fixture that weighs more than 6 pounds (2.72 kg) or exceeds 16 inches (406 mm) in any dimension shall not be supported by the screw shell of a lampholder.

**1926.405(j)(1)(iii)** *Portable lamps.* Portable lamps shall be wired with flexible cord and an attachment plug of the polarized or grounding type. If the portable lamp uses an Edison-based lampholder, the grounded conductor shall be identified and attached to the screw shell and the identified blade of the attachment plug. In addition, portable handlamps shall comply with the following:

**1926.405(j)(1)(iii)(A)** Metal shell, paperlined lampholders shall not be used;

**1926.405(j)(1)(iii)(B)** Handlamps shall be equipped with a handle of molded composition or other insulating material;

**1926.405(j)(1)(iii)(C)** Handlamps shall be equipped with a substantial guard attached to the lampholder or handle;

**1926.405(j)(1)(iii)(D)** Metallic guards shall be grounded by the means of an equipment grounding conductor run within the power supply cord.

**1926.405(j)(1)(iv)** *Lampholders.* Lampholders of the screw-shell type shall be installed for use as lampholders only. Lampholders installed in wet or damp locations shall be of the weatherproof type.

**1926.405(j)(1)(v)** *Fixtures.* Fixtures installed in wet or damp locations shall be identified for the purpose and shall be installed so that water cannot enter or accumulate in wireways, lampholders, or other electrical parts.

**1926.405(j)(2)** *Receptacles, cord connectors, and attachment plugs (caps)*—(i) *Configuration.* Receptacles, cord connectors, and attachment plugs shall be constructed so that no receptacle or cord connector will accept an attachment plug with a different voltage or current rating than that for which the device is intended. However, a 20-ampere T-slot receptacle or cord connector may accept a 15-ampere attachment plug of the same voltage rating. Receptacles connected to circuits having different voltages, frequencies, or types of current (ac or dc) on the same premises shall be of such design that the attachment plugs used on these circuits are not interchangeable.

**1926.405(j)(2)(ii)** *Damp and wet locations.* A receptacle installed in a wet or damp location shall be designed for the location.

**1926.405(j)(3)** *Appliances*—(i) *Live parts.* Appliances, other than those in which the current-carrying parts at high temperatures are necessarily exposed, shall have no live parts normally exposed to employee contact.

**1926.405(j)(3)(ii)** *Disconnecting means.* A means shall be provided to disconnect each appliance.

**1926.405(j)(3)(iii)** *Rating.* Each appliance shall be marked with its rating in volts and amperes or volts and watts.

**1926.405(j)(4)** *Motors.* This paragraph applies to motors, motor circuits, and controllers.

**1926.405(j)(4)(i)** *In sight from.* If specified that one piece of equipment shall be "in sight from" another piece of equipment, one shall be visible and not more than 50 feet (15.2 m) from the other.

**1926.405(j)(4)(ii)** *Disconnecting means*—(A) A disconnecting means shall be located in sight from the controller location. The controller disconnecting means for motor branch circuits over 600 volts, nominal, may be out of sight of the controller, if the controller is marked with a warning label giving the location and identification of the disconnecting means which is to be locked in the open position.

**1926.405(j)(4)(ii)(B)** The disconnecting means shall disconnect the motor and the controller from all ungrounded supply conductors and shall be so designed that no pole can be operated independently.

**1926.405(j)(4)(ii)(C)** If a motor and the driven machinery are not in sight from the controller location, the installation shall comply with one of the following conditions:

**1926.405(j)(4)(ii)(C)*(1)*** The controller disconnecting means shall be capable of being locked in the open position.

**1926.405(j)(4)(ii)(C)*(2)*** A manually operable switch that will disconnect the motor from its source of supply shall be placed in sight from the motor location.

**1926.405(j)(4)(ii)(D)** The disconnecting means shall plainly indicate whether it is in the open (off) or closed (on) position.

**1926.405(j)(4)(ii)(E)** The disconnecting means shall be readily accessible. If more than one disconnect is provided for the same equipment, only one need be readily accessible.

**1926.405(j)(4)(ii)(F)** An individual disconnecting means shall be provided for each motor, but a single disconnecting means may be used for a group of motors under any one of the following conditions:

**1926.405(j)(4)(ii)(F)*(1)*** If a number of motors drive special parts of a single machine or piece of apparatus, such as a metal or woodworking machine, crane, or hoist;

**1926.405(j)(4)(ii)(F)*(2)*** If a group of motors is under the protection of one set of branch-circuit protective devices; or

**1926.405(j)(4)(ii)(F)*(3)*** If a group of motors is in a single room in sight from the location of the disconnecting means.

**1926.405(j)(4)(iii)** *Motor overload, short-circuit, and ground-fault protection.* Motors, motor-control apparatus, and motor branch-circuit conductors shall be protected against overheating due to motor overloads or failure to start, and against short-circuits or ground faults. These provisions do not require overload protection that will stop a motor where a shutdown is likely to introduce additional or increased hazards, as in the case of fire pumps, or where continued operation of a motor is necessary for a safe shutdown of equipment or process and motor overload sensing devices are connected to a supervised alarm.

**1926.405(j)(4)(iv)** *Protection of live parts—all voltages*—(A) Stationary motors having commutators, collectors, and brush rigging located inside of motor end brackets and not conductively connected to supply circuits operating at more than 150 volts to ground need not have such parts guarded. Exposed live parts of motors and controllers operating at 50 volts or more between terminals shall be guarded against accidental contact by any of the following:

**1926.405(j)(4)(iv)(A)*(1)*** By installation in a room or enclosure that is accessible only to qualified persons;

**1926.405(j)(4)(iv)(A)*(2)*** By installation on a balcony, gallery, or platform, so elevated and arranged as to exclude unqualified persons; or

**1926.405(j)(4)(iv)(A)*(3)*** By elevation 8 feet (2.44 m) or more above the floor.

**1926.405(j)(4)(iv)(B)** Where live parts of motors or controllers operating at over 150 volts to ground are guarded against accidental contact only by location, and where adjustment or other attendance may be necessary during the operation of the apparatus, insulating mats or platforms shall be provided so that the attendant cannot readily touch live parts unless standing on the mats or platforms.

**1926.405(j)(5)** *Transformers*—(i) *Application.* The following paragraphs cover the installation of all transformers, except:

**1926.405(j)(5)(i)(A)** Current transformers;

**1926.405(j)(5)(i)(B)** Dry-type transformers installed as a component part of other apparatus;

**1926.405(j)(5)(i)(C)** Transformers which are an integral part of an X-ray, high frequency, or electrostatic-coating apparatus;

**1926.405(j)(5)(i)(D)** Transformers used with Class 2 and Class 3 circuits, sign and outline lighting, electric discharge lighting, and power-limited fire-protective signaling circuits.

**1926.405(j)(5)(ii)** *Operating voltage.* The operating voltage of exposed live parts of transformer installations shall be indicated by warning signs or visible markings on the equipment or structure.

**1926.405(j)(5)(iii)** *Transformers over 35 kV.* Dry-type, high fire point liquid-insulated, and askarel-insulated transformers installed indoors and rated over 35 kV shall be in a vault.

**1926.405(j)(5)(iv)** *Oil-insulated transformers.* If they present a fire hazard to employees, oil-insulated transformers installed indoors shall be in a vault.

**1926.405(j)(5)(v)** *Fire protection.* Combustible material, combustible buildings and parts of buildings, fire escapes, and door and window openings shall be safeguarded from fires which may originate in oil-insulated transformers attached to or adjacent to a building or combustible material.

**1926.405(j)(5)(vi)** *Transformer vaults.* Transformer vaults shall be constructed so as to contain fire and combustible liquids within the vault and to prevent unauthorized access. Locks and latches shall be so arranged that a vault door can be readily opened from the inside.

**1926.405(j)(5)(vii)** *Pipes and ducts.* Any pipe or duct system foreign to the vault installation shall not enter or pass through a transformer vault.

**1926.405(j)(5)(viii)** *Material storage.* Materials shall not be stored in transformer vaults.

**1926.405(j)(6)** *Capacitors*—(i) *Drainage of stored charge.* All capacitors, except surge capacitors or capacitors included as a component part of other apparatus, shall be provided with an automatic means of draining the stored charge and maintaining the discharged state after the capacitor is disconnected from its source of supply.

**1926.405(j)(6)(ii)** *Over 600 volts.* Capacitors rated over 600 volts, nominal, shall comply with the following additional requirements:

**1926.405(j)(6)(ii)(A)** Isolating or disconnecting switches (with no interrupting rating) shall be interlocked with the load interrupting device or shall be provided with prominently displayed caution signs to prevent switching load current.

**1926.405(j)(6)(ii)(B)** For series capacitors the proper switching shall be assured by use of at least one of the following:

**1926.405(j)(6)(ii)(B)*(1)*** Mechanically sequenced isolating and bypass switches,

**1926.405(j)(6)(ii)(B)*(2)*** Interlocks, or

**1926.405(j)(6)(ii)(B)*(3)*** Switching procedure prominently displayed at the switching location.

[51 FR 25318, July 11, 1986, as amended at 61 FR 5510, Feb. 13, 1996]

## §1926.406 Specific purpose equipment and installations.

**1926.406(a)** ***Cranes and hoists.*** This paragraph applies to the installation of electric equipment and wiring used in connection with cranes, monorail hoists, hoists, and all runways.

**1926.406(a)(1)** *Disconnecting means*—(i) *Runway conductor disconnecting means.* A readily accessible disconnecting means shall be provided between the runway contact conductors and the power supply.

**1926.406(a)(1)(ii)** *Disconnecting means for cranes and monorail hoists.* A disconnecting means, capable of being locked in the open position, shall be provided in the leads from the runway contact conductors or other power supply on any crane or monorail hoist.

**1926.406(a)(1)(ii)(A)** If this additional disconnecting means is not readily accessible from the crane or monorail hoist operating station, means shall be provided at the operating station to open the power circuit to all motors of the crane or monorail hoist.

**1926.406(a)(1)(ii)(B)** The additional disconnect may be omitted if a monorail hoist or hand-propelled crane bridge installation meets all of the following:

**1926.406(a)(1)(ii)(B)*(1)*** The unit is floor controlled;

**1926.406(a)(1)(ii)(B)*(2)*** The unit is within view of the power supply disconnecting means; and

**1926.406(a)(1)(ii)(B)*(3)*** No fixed work platform has been provided for servicing the unit.

**1926.406(a)(2)** *Control.* A limit switch or other device shall be provided to prevent the load block from passing the safe upper limit of travel of any hoisting mechanism.

**1926.406(a)(3)** *Clearance.* The dimension of the working space in the direction of access to live parts which may require examination, adjustment, servicing, or maintenance while alive shall be a minimum of 2 feet 6 inches (762 mm). Where controls are enclosed in cabinets, the door(s) shall open at least 90 degrees or be removable, or the installation shall provide equivalent access.

**1926.406(a)(4)** *Grounding.* All exposed metal parts of cranes, monorail hoists, hoists and accessories including pendant controls shall be metallically joined together into a continuous electrical conductor so that the entire crane or hoist will be grounded in accordance with §1926.404(f). Moving parts, other than removable accessories or attachments, having metal-to-metal bearing surfaces shall be considered to be electrically connected to each other through the bearing surfaces for grounding purposes. The trolley frame and bridge frame shall be considered as electrically grounded through the bridge and trolley wheels and its respective tracks unless conditions such as paint or other insulating materials prevent reliable metal-to-metal contact. In this case a separate bonding conductor shall be provided.

**1926.406(b)** ***Elevators, escalators, and moving walks***—(1) *Disconnecting means.* Elevators, escalators, and moving walks shall have a single means for disconnecting all ungrounded main power supply conductors for each unit.

**1926.406(b)(2)** *Control panels.* If control panels are not located in the same space as the drive machine, they shall be located in cabinets with doors or panels capable of being locked closed.

**1926.406(c)** ***Electric welders—disconnecting means***—(1) *Motor-generator, AC transformer, and DC rectifier arc welders.* A disconnecting means shall be provided in the supply circuit for each motor-generator arc welder, and for each AC transformer and DC rectifier arc welder which is not equipped with a disconnect mounted as an integral part of the welder.

**1926.406(c)(2)** *Resistance welders.* A switch or circuit breaker shall be provided by which each resistance welder and its control equipment can be isolated from the supply circuit. The ampere rating of this disconnecting means shall not be less than the supply conductor ampacity.

**1926.406(d)** ***X-Ray equipment***—(1) *Disconnecting means*—(i) *General.* A disconnecting means shall be provided in the supply circuit. The disconnecting means shall be operable from a location readily accessible from the X-ray control. For equipment connected to a 120-volt branch circuit of 30 amperes or less, a grounding-type attachment plug cap and receptacle of proper rating may serve as a disconnecting means.

**1926.406(d)(1)(ii)** *More than one piece of equipment.* If more than one piece of equipment is operated from the same high-voltage circuit, each piece or each group of equipment as a unit shall be provided with a high-voltage switch or equivalent disconnecting means. This disconnecting means shall be constructed, enclosed, or located so as to avoid contact by employees with its live parts.

**1926.406(d)(2)** *Control—Radiographic and fluoroscopic types.* Radiographic and fluoroscopic-type equipment shall be effectively enclosed or shall have interlocks that deenergize the equipment automatically to prevent ready access to live current-carrying parts.

## §1926.407 Hazardous (classified) locations.

**1926.407(a)** ***Scope.*** This section sets forth requirements for electric equipment and wiring in locations which are classified depending on the properties of the flammable vapors, liquids or gases, or combustible dusts or fibers which may be present therein and the likelihood that a flammable or combustible concentration or quantity is present. Each room, section or area shall be considered individually in determining its classification. These hazardous (classified) locations are assigned six designations as follows:

Class I, Division 1

Class I, Division 2

Class II, Division 1

Class II, Division 2

Class III, Division 1

Class III, Division 2

For definitions of these locations see §1926.449. All applicable requirements in this subpart apply to all hazardous (classified) locations, unless modified by provisions of this section.

**1926.407(b)** ***Electrical installations.*** Equipment, wiring methods, and installations of equipment in hazardous (classified) locations shall be approved as intrinsically safe or approved for the hazardous (classified) location or safe for the hazardous (classified) location. Requirements for each of these options are as follows:

**1926.407(b)(1)** *Intrinsically safe.* Equipment and associated wiring approved as intrinsically safe is permitted in any hazardous (classified) location included in its listing or labeling.

**1926.407(b)(2)** *Approved for the hazardous (classified) location*—(i) *General.* Equipment shall be approved not only for the class of location but also for the ignitible or combustible properties of the specific gas, vapor, dust, or fiber that will be present.

**Note:** NFPA 70, the National Electrical Code, lists or defines hazardous gases, vapors, and dusts by "Groups" characterized by their ignitible or combustible properties.

**1926.407(b)(2)(ii)** *Marking.* Equipment shall not be used unless it is marked to show the class, group, and operating temperature or temperature range, based on operation in a 40-degree C ambient, for which it is approved. The temperature marking shall not exceed the ignition temperature of the specific gas, vapor, or dust to be encountered. However, the following provisions modify this marking requirement for specific equipment:

**1926.407(b)(2)(ii)(A)** Equipment of the non-heat-producing type (such as junction boxes, conduit, and fitting) and equipment of the heat-producing type having a maximum temperature of not more than 100 degrees C (212 degrees F) need not have a marked operating temperature or temperature range.

**1926.407(b)(2)(ii)(B)** Fixed lighting fixtures marked for use only in Class I, Division 2 locations need not be marked to indicate the group.

**1926.407(b)(2)(ii)(C)** Fixed general-purpose equipment in Class I locations, other than lighting fixtures, which is acceptable for use in Class I, Division 2 locations need not be marked with the class, group, division, or operating temperature.

**1926.407(b)(2)(ii)(D)** Fixed dust-tight equipment, other than lighting fixtures, which is acceptable for use in Class II, Division 2 and Class III locations need not be marked with the class, group, division, or operating temperature.

**1926.407(b)(3)** *Safe for the hazardous (classified) location.* Equipment which is safe for the location shall be of a type and design which the employer demonstrates will provide protection from the hazards arising from the combustibility and flammability of vapors, liquids, gases, dusts, or fibers.

**Note:** The National Electrical Code, NFPA 70, contains guidelines for determining the type and design of equipment and installations which will meet this requirement. The guidelines of this document address electric wiring, equipment, and systems installed in hazardous (classified) locations and contain specific provisions for the following: wiring methods, wiring connections, conductor insulation, flexible cords, sealing and drainage, transformers, capacitors, switches, circuit breakers, fuses, motor controllers, receptacles, attachment plugs, meters, relays, instruments, resistors, generators, motors, lighting fixtures, storage battery charging equipment, electric cranes, electric hoists and similar equipment, utilization equipment, signaling systems, alarm systems, remote control systems, local loud speaker and communication systems, ventilation piping, live parts, lightning surge protection, and grounding. Compliance with these guidelines will constitute one means, but not the only means, of compliance with this paragraph.

**1926.407(c)** ***Conduits.*** All conduits shall be threaded and shall be made wrench-tight. Where it is impractical to make a threaded joint tight, a bonding jumper shall be utilized.

[51 FR 25318, July 11, 1986, as amended at 61 FR 5510, Feb. 13, 1996]

## §1926.408 Special systems.

**1926.408(a)** ***Systems over 600 volts, nominal.*** Paragraphs (a)(1) through (a)(4) of this section contain general requirements for all circuits and equipment operated at over 600 volts.

**1926.408(a)(1)** *Wiring methods for fixed installations*—(i) *Above ground.* Aboveground conductors shall be installed in rigid metal conduit, in intermediate metal conduit, in cable trays, in cablebus, in other suitable raceways, or as open runs of metal-clad cable designed for the use and purpose. However, open runs of non-metallic-sheathed cable or of bare conductors or busbars may be installed in locations which are accessible only to qualified persons. Metallic shielding components, such as tapes, wires, or braids for conductors, shall be grounded. Open runs of insulated wires and cables having a bare lead sheath or a braided outer covering shall be supported in a manner designed to prevent physical damage to the braid or sheath.

**1926.408(a)(1)(ii)** *Installations emerging from the ground.* Conductors emerging from the ground shall be enclosed in raceways. Raceways installed on poles shall be of rigid metal conduit, intermediate metal conduit, PVC schedule 80 or equivalent extending from the ground line up to a point 8 feet (2.44 m) above finished grade. Conductors entering a building shall be protected by an enclosure from the ground line to the point of entrance. Metallic enclosures shall be grounded.

**1926.408(a)(2)** *Interrupting and isolating devices*—(i) *Circuit breakers.* Circuit breakers located indoors shall consist of metal-enclosed or fire-resistant, cell-mounted units. In locations accessible only to qualified personnel, open mounting of circuit breakers is permitted. A means of indicating the open and closed position of circuit breakers shall be provided.

**1926.408(a)(2)(ii)** *Fused cutouts.* Fused cutouts installed in buildings or transformer vaults shall be of a type identified for the purpose. They shall be readily accessible for fuse replacement.

**1926.408(a)(2)(iii)** *Equipment isolating means.* A means shall be provided to completely isolate equipment for inspection and repairs. Isolating means which are not designed to interrupt the load current of the circuit shall be either interlocked with a circuit interrupter or provided with a sign warning against opening them under load.

**1926.408(a)(3)** *Mobile and portable equipment*—(i) *Power cable connections to mobile machines.* A metallic enclosure shall be provided on the mobile machine for enclosing the terminals of the power cable. The enclosure shall include provisions for a solid connection for the ground wire(s) terminal to ground effectively the machine frame. The method of cable termination used shall prevent any strain or pull on the cable from stressing the electrical connections. The enclosure shall have provision for locking so only authorized qualified persons may open it and shall be marked with a sign warning of the presence of energized parts.

**1926.408(a)(3)(ii)** *Guarding live parts.* All energized switching and control parts shall be enclosed in effectively grounded metal cabinets or enclosures. Circuit breakers and protective equipment shall have the operating means projecting through the metal cabinet or enclosure so these units can be reset without locked doors being opened. Enclosures and metal cabinets shall be locked so that only authorized qualified persons have access and shall be marked with a sign warning of the presence of energized parts. Collector ring assemblies on revolving-type machines (shovels, draglines, etc.) shall be guarded.

**1926.408(a)(4)** *Tunnel installations*—(i) *Application.* The provisions of this paragraph apply to installation and use of high-voltage power distribution and utilization equipment which is associated with tunnels and which is portable and/or mobile, such as substations, trailers, cars, mobile shovels, draglines, hoists, drills, dredges, compressors, pumps, conveyors, and underground excavators.

**1926.408(a)(4)(ii)** *Conductors.* Conductors in tunnels shall be installed in one or more of the following:

**1926.408(a)(4)(ii)(A)** Metal conduit or other metal raceway,

**1926.408(a)(4)(ii)(B)** Type MC cable, or

**1926.408(a)(4)(ii)(C)** Other suitable multiconductor cable.

Conductors shall also be so located or guarded as to protect them from physical damage. Multiconductor portable cable may supply mobile equipment. An equipment grounding conductor shall be run with circuit conductors inside the metal raceway or inside the multiconductor cable jacket. The equipment grounding conductor may be insulated or bare.

**1926.408(a)(4)(iii)** *Guarding live parts.* Bare terminals of transformers, switches, motor controllers, and other equipment shall be enclosed to prevent accidental contact with energized parts. Enclosures for use in tunnels shall be drip-proof, weatherproof, or submersible as required by the environmental conditions.

**1926.408(a)(4)(iv)** *Disconnecting means.* A disconnecting means that simultaneously opens all ungrounded conductors shall be installed at each transformer or motor location.

**1926.408(a)(4)(v)** *Grounding and bonding.* All nonenergized metal parts of electric equipment and metal raceways and cable sheaths shall be grounded and bonded to all metal pipes and rails at the portal and at intervals not exceeding 1000 feet (305 m) throughout the tunnel.

**1926.408(b)** ***Class 1, Class 2, and Class 3 remote control, signaling, and power-limited circuits***—(1) *Classification.* Class 1, Class 2, or Class 3 remote control, signaling, or power-limited circuits are characterized by their usage and electrical power limitation which differentiates them from light and power circuits. These circuits are classified in accordance with their respective voltage and power limitations as summarized in paragraphs (b)(1)(i) through (b)(1)(iii) of this section.

**1926.408(b)(1)(i)** *Class 1 circuits*—(A) A Class 1 power-limited circuit is supplied from a source having a rated output of not more than 30 volts and 1000 volt-amperes.

**1926.408(b)(1)(i)(B)** A Class 1 remote control circuit or a Class 1 signaling circuit has a voltage which does not exceed 600 volts; however, the power output of the source need not be limited.

**1926.408(b)(1)(ii)** *Class 2 and Class 3 circuits*—(A) Power for Class 2 and Class 3 circuits is limited either inherently (in which no overcurrent protection is required) or by a combination of a power source and overcurrent protection.

**1926.408(b)(1)(ii)(B)** The maximum circuit voltage is 150 volts AC or DC for a Class 2 inherently limited power source, and 100 volts AC or DC for a Class 3 inherently limited power source.

**1926.408(b)(1)(ii)(C)** The maximum circuit voltage is 30 volts AC and 60 volts DC for a Class 2 power source limited by overcurrent protection, and 150 volts AC or DC for a Class 3 power source limited by overcurrent protection.

**1926.408(b)(1)(iii)** *Application.* The maximum circuit voltages in paragraphs (b)(1)(i) and (b)(1)(ii) of this section apply to sinusoidal AC or continuous DC power sources, and where wet contact occurrence is not likely.

**1926.408(b)(2)** *Marking.* A Class 2 or Class 3 power supply unit shall not be used unless it is durably marked where plainly visible to indicate the class of supply and its electrical rating.

**1926.408(c)** ***Communications systems***—(1) *Scope.* These provisions for communication systems apply to such systems as central-station-connected and non-central-station-connected telephone circuits, radio receiving and transmitting equipment, and outside wiring for fire and burglar alarm, and similar central station systems. These installations need not comply with the provisions of §§1926.403 through 1926.408(b), except §1926.404(c)(1)(ii) and §1926.407.

**1926.408(c)(2)** *Protective devices*—(i) *Circuits exposed to power conductors.* Communication circuits so located as to be exposed to accidental contact with light or power conductors operating at over 300 volts shall have each circuit so exposed provided with an approved protector.

**1926.408(c)(2)(ii)** *Antenna lead-ins.* Each conductor of a lead-in from an outdoor antenna shall be provided with an antenna discharge unit or other means that will drain static charges from the antenna system.

**1926.408(c)(3)** *Conductor location*—(i) *Outside of buildings*—(A) Receiving distribution lead-in or aerial-drop cables attached to buildings and lead-in conductors to radio transmitters shall be so installed as to avoid the possibility of accidental contact with electric light or power conductors.

**1926.408(c)(3)(i)(B)** The clearance between lead-in conductors and any lightning protection conductors shall not be less than 6 feet (1.83 m).

**1926.408(c)(3)(ii)** *On poles.* Where practicable, communication conductors on poles shall be located below the light or power conductors. Communications conductors shall not be attached to a crossarm that carries light or power conductors.

**1926.408(c)(3)(iii)** *Inside of buildings.* Indoor antennas, lead-ins, and other communication conductors attached as open conductors to the inside of buildings shall be located at least 2 inches (50.8 mm) from conductors of any light or power or Class 1 circuits unless a special and equally protective method of conductor separation is employed.

**1926.408(c)(4)** *Equipment location.* Outdoor metal structures supporting antennas, as well as self-supporting antennas such as vertical rods or dipole structures, shall be located as far away from overhead conductors of electric light and power circuits of over 150 volts to ground as necessary to avoid the possibility of the antenna or structure falling into or making accidental contact with such circuits.

**1926.408(c)(5)** *Grounding*—(i) *Lead-in conductors.* If exposed to contact with electric light or power conductors, the metal sheath of aerial cables entering buildings shall be grounded or shall be interrupted close to the entrance to the building by an insulating joint or equivalent device. Where protective devices are used, they shall be grounded.

**1926.408(c)(5)(ii)** *Antenna structures.* Masts and metal structures supporting antennas shall be permanently and effectively grounded without splice or connection in the grounding conductor.

**1926.408(c)(5)(iii)** *Equipment enclosures.* Transmitters shall be enclosed in a metal frame or grill or separated from the operating space by a barrier, all metallic parts of which are effectively connected to ground. All external metal handles and controls accessible to the operating personnel shall be effectively grounded. Unpowered equipment and enclosures shall be considered grounded where connected to an attached coaxial cable with an effectively grounded metallic shield.

[51 FR 25318, July 11, 1986, as amended at 61 FR 5510, Feb. 13, 1996]

### §§1926.409—1926.415 [Reserved]

## Safety-Related Work Practices

### §1926.416 General requirements.

**1926.416(a)** ***Protection of employees***—(1) No employer shall permit an employee to work in such proximity to any part of an electric power circuit that the employee could contact the electric power circuit in the course of work, unless the employee is protected against electric shock by deenergizing the circuit and grounding it or by guarding it effectively by insulation or other means.

**1926.416(a)(2)** In work areas where the exact location of underground electric powerlines is unknown, employees using jack-hammers, bars, or other hand tools which may contact a line shall be provided with insulated protective gloves.

**1926.416(a)(3)** Before work is begun the employer shall ascertain by inquiry or direct observation, or by instruments, whether any part of an energized electric power circuit, exposed or concealed, is so located that the performance of the work may bring any person, tool, or machine into physical or electrical contact with the electric power circuit. The employer shall post and maintain proper warning signs where such a circuit exists. The employer shall advise employees of the location of such lines, the hazards involved, and the protective measures to be taken.

**1926.416(b)** ***Passageways and open spaces***—(1) Barriers or other means of guarding shall be provided to ensure that workspace for electrical equipment will not be used as a passageway during periods when energized parts of electrical equipment are exposed.

**1926.416(b)(2)** Working spaces, walkways, and similar locations shall be kept clear of cords so as not to create a hazard to employees.

**1926.416(c)** ***Load ratings.*** In existing installations, no changes in circuit protection shall be made to increase the load in excess of the load rating of the circuit wiring.

**1926.416(d)** ***Fuses.*** When fuses are installed or removed with one or both terminals energized, special tools insulated for the voltage shall be used.

**1926.416(e)** ***Cords and cables.*** (1) Worn or frayed electric cords or cables shall not be used.

**1926.416(e)(2)** Extension cords shall not be fastened with staples, hung from nails, or suspended by wire.

[44 FR 8577, Feb. 9, 1979; 44 FR 20940, Apr. 6, 1979, as amended at 55 FR 42328, Oct. 18, 1990; 58 FR 35179, June 30, 1993; 61 FR 9251, Mar. 7, 1996; 61 FR 41738, Aug. 12, 1996]

### §1926.417 Lockout and tagging of circuits.

**1926.417(a)** ***Controls.*** Controls that are to be deactivated during the course of work on energized or deenergized equipment or circuits shall be tagged.

**1926.417(b)** ***Equipment and circuits.*** Equipment or circuits that are deenergized shall be rendered inoperative and shall have tags attached at all points where such equipment or circuits can be energized.

**1926.417(c)** ***Tags.*** Tags shall be placed to identify plainly the equipment or circuits being worked on.

[44 FR 8577, Feb. 9, 1979; 44 FR 20940, Apr. 6, 1979, as amended at 55 FR 42328, Oct. 18, 1990; 58 FR 35181, June 30, 1993; 61 FR 9251, Mar. 7, 1996; 61 FR 41739, Aug. 12, 1996]

### §§1926.418—1926.430 [Reserved]

## Safety-Related Maintenance and Environmental Considerations

### §1926.431 Maintenance of equipment.

The employer shall ensure that all wiring components and utilization equipment in hazardous locations are maintained in a dust-tight, dust-ignition-proof, or explosion-proof condition, as appropriate. There shall be no loose or missing screws, gaskets, threaded connections, seals, or other impairments to a tight condition.

### §1926.432 Environmental deterioration of equipment.

**1926.432(a)** ***Deteriorating agents***—(1) Unless identified for use in the operating environment, no conductors or equipment shall be located:

**1926.432(a)(1)(i)** In damp or wet locations;

**1926.432(a)(1)(ii)** Where exposed to gases, fumes, vapors, liquids, or other agents having a deteriorating effect on the conductors or equipment; or

**1926.432(a)(1)(iii)** Where exposed to excessive temperatures.

**1926.432(a)(2)** Control equipment, utilization equipment, and busways approved for use in dry locations only shall be protected against damage from the weather during building construction.

**1926.432(b)** ***Protection against corrosion.*** Metal raceways, cable armor, boxes, cable sheathing, cabinets, elbows, couplings, fittings, supports, and support hardware shall be of materials appropriate for the environment in which they are to be installed.

### §§1926.433—1926.440 [Reserved]

## Safety Requirements for Special Equipment

### §1926.441 Batteries and battery charging.

**1926.441(a)** ***General requirements***—(1) Batteries of the unsealed type shall be located in enclosures with outside vents or in well ventilated rooms and shall be arranged so as to prevent the escape of fumes, gases, or electrolyte spray into other areas.

**1926.441(a)(2)** Ventilation shall be provided to ensure diffusion of the gases from the battery and to prevent the accumulation of an explosive mixture.

**1926.441(a)(3)** Racks and trays shall be substantial and shall be treated to make them resistant to the electrolyte.

**1926.441(a)(4)** Floors shall be of acid resistant construction unless protected from acid accumulations.

**1926.441(a)(5)** Face shields, aprons, and rubber gloves shall be provided for workers handling acids or batteries.

**1926.441(a)(6)** Facilities for quick drenching of the eyes and body shall be provided within 25 feet (7.62 m) of battery handling areas.

**1926.441(a)(7)** Facilities shall be provided for flushing and neutralizing spilled electrolyte and for fire protection.

**1926.441(b)** ***Charging***—(1) Battery charging installations shall be located in areas designated for that purpose.

**1926.441(b)(2)** Charging apparatus shall be protected from damage by trucks.

**1926.441(b)(3)** When batteries are being charged, the vent caps shall be kept in place to avoid electrolyte spray. Vent caps shall be maintained in functioning condition.

### §§1926.442—1926.448 [Reserved]

## Definitions

### §1926.449 Definitions applicable to this subpart.

The definitions given in this section apply to the terms used in subpart K. The definitions given here for "approved" and "qualified person" apply, instead of the definitions given in §1926.32, to the use of these terms in subpart K.

*Acceptable.* An installation or equipment is acceptable to the Assistant Secretary of Labor, and approved within the meaning of this subpart K:

(a) If it is accepted, or certified, or listed, or labeled, or otherwise determined to be safe by a qualified testing laboratory capable of determining the suitability of materials and equipment for installation and use in accordance with this standard; or

(b) With respect to an installation or equipment of a kind which no qualified testing laboratory accepts, certifies, lists, labels, or determines to be safe, if it is inspected or tested by another Federal agency, or by a State, municipal, or other local authority responsible for enforcing occupational safety provisions of the National Electrical Code, and found in compliance with those provisions; or

(c) With respect to custom-made equipment or related installations which are designed, fabricated for, and intended for use by a particular customer, if it is determined to be safe for its intended use by its manufacturer on the basis of test data which the employer keeps and makes available for inspection to the Assistant Secretary and his authorized representatives.

*Accepted.* An installation is "accepted" if it has been inspected and found to be safe by a qualified testing laboratory.

*Accessible.* (As applied to wiring methods.) Capable of being removed or exposed without damaging the building structure or finish, or not permanently closed in by the structure or finish of the building. (See *"concealed"* and *"exposed."*)

*Accessible.* (As applied to equipment.) Admitting close approach; not guarded by locked doors, elevation, or other effective means. (See *"Readily accessible."*)

*Ampacity.* The current in amperes a conductor can carry continuously under the conditions of use without exceeding its temperature rating.

*Appliances.* Utilization equipment, generally other than industrial, normally built in standardized sizes or types, which is installed or connecetcd as a unit to perform one or more functions.

*Approved.* Acceptable to the authority enforcing this subpart. The authority enforcing this subpart is the Assistant Secretary of Labor for Occupational Safety and Health. The definition of "acceptable" indicates what is acceptable to the Assistant Secretary of Labor, and therefore approved within the meaning of this subpart.

*Askarel.* A generic term for a group of nonflammable synthetic chlorinated hydrocarbons used as electrical insulating media. Askarels of various compositional types are used. Under arcing conditions the gases produced, while consisting predominantly of noncombustible hydrogen chloride, can include varying amounts of combustible gases depending upon the askarel type.

*Attachment plug (Plug cap)(Cap).* A device which, by insertion in a receptacle, establishes connection between the conductors of the attached flexible cord and the conductors connected permanently to the receptacle.

*Automatic.* Self-acting, operating by its own mechanism when actuated by some impersonal influence, as for example, a change in current strength, pressure, temperature, or mechanical configuration.

*Bare conductor.* See *"Conductor."*

*Bonding.* The permanent joining of metallic parts to form an electrically conductive path which will assure electrical continuity and the capacity to conduct safely any current likely to be imposed.

*Bonding jumper.* A reliable conductor to assure the required electrical conductivity between metal parts required to be electrically connected.

*Branch circuit.* The circuit conductors between the final overcurrent device protecting the circuit and the outlet(s).

*Building.* A structure which stands alone or which is cut off from adjoining structures by fire walls with all openings therein protected by approved fire doors.

*Cabinet.* An enclosure designed either for surface or flush mounting, and provided with a frame, mat, or trim in which a swinging door or doors are or may be hung.

*Certified.* Equipment is "certified" if it:

(a) Has been tested and found by a qualified testing laboratory to meet applicable test standards or to be safe for use in a specified manner, and

(b) Is of a kind whose production is periodically inspected by a qualified testing laboratory. Certified equipment must bear a label, tag, or other record of certification.

*Circuit breaker*—(a) (600 volts nominal, or less.) A device designed to open and close a circuit by nonautomatic means and to open the circuit automatically on a predetermined overcurrent without injury to itself when properly applied within its rating.

(b) (Over 600 volts, nominal.) A switching device capable of making, carrying, and breaking currents under normal circuit conditions, and also making, carrying for a specified time, and breaking currents under specified abnormal circuit conditions, such as those of short circuit.

*Class I locations.* Class I locations are those in which flammable gases or vapors are or may be present in the air in quantities sufficient to produce explosive or ignitible mixtures. Class I locations include the following:

(a) *Class I, Division 1.* A Class I, Division 1 location is a location:

(1) In which ignitible concentrations of flammable gases or vapors may exist under normal operating conditions; or

(2) In which ignitible concentrations of such gases or vapors may exist frequently because of repair or maintenance operations or because of leakage; or

(3) In which breakdown or faulty operation of equipment or processes might release ignitible concentrations of flammable gases or vapors, and might also cause simultaneous failure of electric equipment.

**Note:** This classification usually includes locations where volatile flammable liquids or liquefied flammable gases are transferred from one container to another; interiors of spray booths and areas in the vicinity of spraying and painting operations where volatile flammable solvents are used; locations containing open tanks or vats of volatile flammable liquids; drying rooms or compartments for the evaporation of flammable solvents; inadequately ventilated pump rooms for flammable gas or for volatile flammable liquids; and all other locations where ignitible concentrations of flammable vapors or gases are likely to occur in the course of normal operations.

(b) *Class I, Division 2.* A Class I, Division 2 location is a location:

(1) In which volatile flammable liquids or flammable gases are handled, processed, or used, but in which the hazardous liquids, vapors, or gases will normally be confined within closed containers or closed systems from which they can escape only in case of accidental rupture or breakdown of such containers or systems, or in case of abnormal operation of equipment; or

(2) In which ignitible concentrations of gases or vapors are normally prevented by positive mechanical ventilation, and which might become hazardous through failure or abnormal operations of the ventilating equipment; or

(3) That is adjacent to a Class I, Division 1 location, and to which ignitible concentrations of gases or vapors might occasionally be communicated unless such communication is prevented by adequate positive-pressure ventilation from a source of clean air, and effective safeguards against ventilation failure are provided.

**Note:** This classification usually includes locations where volatile flammable liquids or flammable gases or vapors are used, but which would become hazardous only in case of an accident or of some unusual operating condition. The quantity of flammable material that might escape in case of accident, the adequacy of ventilating equipment, the total area involved, and the record of the industry or business with respect to explosions or fires are all factors that merit consideration in determining the classification and extent of each location.

Piping without valves, checks, meters, and similar devices would not ordinarily introduce a hazardous condition even though used for flammable liquids or gases. Locations used for the storage of flammable liquids or of liquefied or compressed gases in sealed containers would not normally be considered hazardous unless also subject to other hazardous conditions.

Electrical conduits and their associated enclosures separated from process fluids by a single seal or barrier are classed as a Division 2 location if the outside of the conduit and enclosures is a nonhazardous location.

*Class II locations.* Class II locations are those that are hazardous because of the presence of combustible dust. Class II locations include the following:

(a) *Class II, Division 1.* A Class II, Division 1 location is a location:

(1) In which combustible dust is or may be in suspension in the air under normal operating conditions, in quantities sufficient to produce explosive or ignitible mixtures; or

(2) Where mechanical failure or abnormal operation of machinery or equipment might cause such explosive or ignitible mixtures to be produced, and might also provide a source of ignition through simultaneous failure of electric equipment, operation of protection devices, or from other causes, or

(3) In which combustible dusts of an electrically conductive nature may be present.

**Note:** Combustible dusts which are electrically nonconductive include dusts produced in the handling and processing of grain and grain products, pulverized sugar and cocoa, dried egg and milk powders, pulverized spices, starch and pastes, potato and woodflour, oil meal from beans and seed, dried hay, and other organic materials which may produce combustible dusts when processed or handled. Dusts containing magnesium or aluminum are particularly hazardous and the use of extreme caution is necessary to avoid ignition and explosion.

(b) *Class II, Division 2.* A Class II, Division 2 location is a location in which:

(1) Combustible dust will not normally be in suspension in the air in quantities sufficient to produce explosive or ignitible mixtures, and dust accumulations are normally insufficient to interfere with the normal operation of electrical equipment or other apparatus; or

(2) Dust may be in suspension in the air as a result of infrequent malfunctioning of handling or processing equipment, and dust accumulations resulting therefrom may be ignitible by abnormal operation or failure of electrical equipment or other apparatus.

**Note:** This classification includes locations where dangerous concentrations of suspended dust would not be likely but where dust accumulations might form on or in the vicinity of electric equipment. These areas may contain equipment from which appreciable quantities of dust would escape under abnormal operating conditions or be adjacent to a Class II Division 1 location, as described above, into which an explosive or ignitible concentration of dust may be put into suspension under abnormal operating conditions.

*Class III locations.* Class III locations are those that are hazardous because of the presence of easily ignitible fibers or flyings but in which such fibers or flyings are not likely to be in suspension in the air in quantities sufficient to produce ignitible mixtures. Class 111 locations include the following:

(a) *Class III, Division 1.* A Class III, Division 1 location is a location in which easily ignitible fibers or materials producing combustible flyings are handled, manufactured, or used.

**Note:** Easily ignitible fibers and flyings include rayon, cotton (including cotton linters and cotton waste), sisal or henequen, istle, jute, hemp, tow, cocoa fiber, oakum, baled waste kapok, Spanish moss, excelsior, sawdust, woodchips, and other material of similar nature.

(b) *Class III, Division 2.* A Class III, Division 2 location is a location in which easily ignitible fibers are stored or handled, except in process of manufacture.

*Collector ring.* A collector ring is an assembly of slip rings for transferring electrical energy from a stationary to a rotating member.

*Concealed.* Rendered inaccessible by the structure or finish of the building. Wires in concealed raceways are considered concealed, even though they may become accessible by withdrawing them. [See *"Accessible.* (As applied to wiring methods.)"]

*Conductor*—(a) *Bare.* A conductor having no covering or electrical insulation whatsoever.

(b) *Covered.* A conductor encased within material of composition or thickness that is not recognized as electrical insulation.

(c) *Insulated.* A conductor encased within material of composition and thickness that is recognized as electrical insulation.

*Controller.* A device or group of devices that serves to govern, in some predetermined manner, the electric power delivered to the apparatus to which it is connected.

*Covered conductor.* See *"Conductor."*

*Cutout.* (Over 600 volts, nominal.) An assembly of a fuse support with either a fuseholder, fuse carrier, or disconnecting blade. The fuseholder or fuse carrier may include a conducting element (fuse link), or may act as the disconnecting blade by the inclusion of a nonfusible member.

*Cutout box.* An enclosure designed for surface mounting and having swinging doors or covers secured directly to and telescoping with the walls of the box proper. (See *"Cabinet."*)

*Damp location.* See *"Location."*

*Dead front.* Without live parts exposed to a person on the operating side of the equipment.

*Device.* A unit of an electrical system which is intended to carry but not utilize electric energy.

*Disconnecting means.* A device, or group of devices, or other means by which the conductors of a circuit can be disconnected from their source of supply.

*Disconnecting (or Isolating) switch.* (Over 600 volts, nominal.) A mechanical switching device used for isolating a circuit or equipment from a source of power.

*Dry location.* See *"Location."*

*Enclosed.* Surrounded by a case, housing, fence or walls which will prevent persons from accidentally contacting energized parts.

*Enclosure.* The case or housing of apparatus, or the fence or walls surrounding an installation to prevent personnel from accidentally contacting energized parts, or to protect the equipment from physical damage.

*Equipment.* A general term including material, fittings, devices, appliances, fixtures, apparatus, and the like, used as a part of, or in connection with, an electrical installation.

*Equipment grounding conductor.* See *"Grounding conductor, equipment."*

*Explosion-proof apparatus.* Apparatus enclosed in a case that is capable of withstanding an explosion of a specified gas or vapor which may occur within it and of preventing the ignition of a specified gas or vapor surrounding the enclosure by sparks, flashes, or explosion of the gas or vapor within, and which operates at such an external temperature that it will not ignite a surrounding flammable atmosphere.

*Exposed.* (As applied to live parts.) Capable of being inadvertently touched or approached nearer than a safe distance by a person. It is applied to parts not suitably guarded, isolated, or insulated. (See *"Accessible* and *"Concealed."*)

*Exposed.* (As applied to wiring methods.) On or attached to the surface or behind panels designed to allow access. [See *"Accessible.* (As applied to wiring methods.)"]

*Exposed.* (For the purposes of §1926.408(d), Communications systems.) Where the circuit is in such a position that in case of failure of supports or insulation, contact with another circuit may result.

*Externally operable.* Capable of being operated without exposing the operator to contact with live parts.

*Feeder.* All circuit conductors between the service equipment, or the generator switchboard of an isolated plant, and the final branch-circuit overcurrent device.

*Festoon lighting.* A string of outdoor lights suspended between two points more than 15 feet (4.57 m) apart.

*Fitting.* An accessory such as a locknut, bushing, or other part of a wiring system that is intended primarily to perform a mechanical rather than an electrical function.

*Fuse.* (Over 600 volts, nominal.) An overcurrent protective device with a circuit opening fusible part that is heated and severed by the passage of overcurrent through it. A fuse comprises all the parts that form a unit capable of performing the prescribed functions. It may or may not be the complete device necessary to connect it into an electrical circuit.

*Ground.* A conducting connection, whether intentional or accidental, between an electrical circuit or equipment and the earth, or to some conducting body that serves in place of the earth.

*Grounded.* Connected to earth or to some conducting body that serves in place of the earth.

*Grounded, effectively* (Over 600 volts, nominal.) Permanently connected to earth through a ground connection of sufficiently low impedance and having sufficient ampacity that ground fault current which may occur cannot build up to voltages dangerous to personnel.

*Grounded conductor.* A system or circuit conductor that is intentionally grounded.

*Grounding conductor.* A conductor used to connect equipment or the grounded circuit of a wiring system to a grounding electrode or electrodes.

*Grounding conductor, equipment.* The conductor used to connect the noncurrent-carrying metal parts of equipment, raceways, and other enclosures to the system grounded conductor and/or the grounding electrode conductor at the service equipment or at the source of a separately derived system.

*Grounding electrode conductor.* The conductor used to connect the grounding electrode to the equipment grounding conductor and/or to the grounded conductor of the circuit at the service equipment or at the source of a separately derived system.

*Ground-fault circuit interrupter.* A device for the protection of personnel that functions to deenergize a circuit or portion thereof within an established period of time when a current to ground exceeds some predetermined value that is less than that required to operate the overcurrent protective device of the supply circuit.

*Guarded.* Covered, shielded, fenced, enclosed, or otherwise protected by means of suitable covers, casings, barriers, rails, screens, mats, or platforms to remove the likelihood of approach to a point of danger or contact by persons or objects.

*Hoistway.* Any shaftway, hatchway, well hole, or other vertical opening or space in which an elevator or dumbwaiter is designed to operate.

*Identified (conductors or terminals).* Identified, as used in reference to a conductor or its terminal, means that such conductor or terminal can be recognized as grounded.

*Identified (for the use).* Recognized as suitable for the specific purpose, function, use, environment, application, etc. where described as a requirement in this standard. Suitability of equipment for a specific purpose, environment, or application is determined by a qualified testing laboratory where such identification includes labeling or listing.

*Insulated conductor.* See "*Conductor.*"

*Interrupter switch.* (Over 600 volts, nominal.) A switch capable of making, carrying, and interrupting specified currents.

*Intrinsically safe equipment and associated wiring.* Equipment and associated wiring in which any spark or thermal effect, produced either normally or in specified fault conditions, is incapable, under certain prescribed test conditions, of causing ignition of a mixture of flammable or combustible material in air in its most easily ignitible concentration.

*Isolated.* Not readily accessible to persons unless special means for access are used.

*Isolated power system.* A system comprising an isolating transformer or its equivalent, a line isolation monitor, and its ungrounded circuit conductors.

*Labeled.* Equipment or materials to which has been attached a label, symbol or other identifying mark of a qualified testing laboratory which indicates compliance with appropriate standards or performance in a specified manner.

*Lighting outlet.* An outlet intended for the direct connection of a lampholder, a lighting fixture, or a pendant cord terminating in a lampholder.

*Listed.* Equipment or materials included in a list published by a qualified testing laboratory whose listing states either that the equipment or material meets appropriate standards or has been tested and found suitable for use in a specified manner.

*Location*—(a) *Damp location.* Partially protected locations under canopies, marquees, roofed open porches, and like locations, and interior locations subject to moderate degrees of moisture, such as some basements.

(b) *Dry location.* A location not normally subject to dampness or wetness. A location classified as dry may be temporarily subject to dampness or wetness, as in the case of a building under construction.

(c) *Wet location.* Installations underground or in concrete slabs or masonry in direct contact with the earth, and locations subject to saturation with water or other liquids, such as locations exposed to weather and unprotected.

*Mobile X-ray.* X-ray equipment mounted on a permanent base with wheels and/or casters for moving while completely assembled.

*Motor control center.* An assembly of one or more enclosed sections having a common power bus and principally containing motor control units.

*Outlet.* A point on the wiring system at which current is taken to supply utilization equipment.

*Overcurrent.* Any current in excess of the rated current of equipment or the ampacity of a conductor. It may result from overload (see definition), short circuit, or ground fault. A current in excess of rating may be accommodated by certain equipment and conductors for a given set of conditions. Hence the rules for overcurrent protection are specific for particular situations.

*Overload.* Operation of equipment in excess of normal, full load rating, or of a conductor in excess of rated ampacity which, when it persists for a sufficient length of time, would cause damage or dangerous overheating. A fault, such as a short circuit or ground fault, is not an overload. (See *"Overcurrent."*)

*Panelboard.* A single panel or group of panel units designed for assembly in the form of a single panel; including buses, automatic overcurrent devices, and with or without switches for the control of light, heat, or power circuits; designed to be placed in a cabinet or cutout box placed in or against a wall or partition and accessible only from the front. (See *"Switchboard."*)

*Portable X-ray.* X-ray equipment designed to be hand-carried.

*Power fuse.* (Over 600 volts, nominal.) See *"Fuse."*

*Power outlet.* An enclosed assembly which may include receptacles, circuit breakers, fuseholders, fused switches, buses and watt-hour meter mounting means; intended to serve as a means for distributing power required to operate mobile or temporarily installed equipment.

*Premises wiring system.* That interior and exterior wiring, including power, lighting, control, and signal circuit wiring together with all of its associated hardware, fittings, and wiring devices, both permanently and temporarily installed, which extends from the load end of the service drop, or load end of the service lateral conductors to the outlet(s). Such wiring does not include wiring internal to appliances, fixtures, motors, controllers, motor control centers, and similar equipment.

*Qualified person.* One familiar with the construction and operation of the equipment and the hazards involved.

*Qualified testing laboratory.* A properly equipped and staffed testing laboratory which has capabilities for and which provides the following services:

(a) Experimental testing for safety of specified items of equipment and materials referred to in this standard to determine compliance with appropriate test standards or performance in a specified manner;

(b) Inspecting the run of such items of equipment and materials at factories for product evaluation to assure compliance with the test standards;

(c) Service-value determinations through field inspections to monitor the proper use of labels on products and with authority for recall of the label in the event a hazardous product is installed;

(d) Employing a controlled procedure for identifying the listed and/or labeled equipment or materials tested; and

(e) Rendering creditable reports or findings that are objective and without bias of the tests and test methods employed.

*Raceway.* A channel designed expressly for holding wires, cables, or busbars, with additional functions as permitted in this subpart. Raceways may be of metal or insulating material, and the term includes rigid metal conduit, rigid nonmetallic conduit, intermediate metal conduit, liquidtight flexible metal conduit, flexible metallic tubing, flexible metal conduit, electrical metallic tubing, underfloor raceways, cellular concrete floor raceways, cellular metal floor raceways, surface raceways, wireways, and busways.

*Readily accessible.* Capable of being reached quickly for operation, renewal, or inspections, without requiring those to whom ready access is requisite to climb over or remove obstacles or to resort to portable ladders, chairs, etc. (See *"Accessible."*)

*Receptacle.* A receptacle is a contact device installed at the outlet for the connection of a single attachment plug. A single receptacle is a single contact device with no other contact device on the same yoke. A multiple receptacle is a single device containing two or more receptacles.

*Receptacle outlet.* An outlet where one or more receptacles are installed.

*Remote-control circuit.* Any electric circuit that controls any other circuit through a relay or an equivalent device.

*Sealable equipment.* Equipment enclosed in a case or cabinet that is provided with a means of sealing or locking so that live parts cannot be made accessible without opening the enclosure. The equipment may or may not be operable without opening the enclosure.

*Separately derived system.* A premises wiring system whose power is derived from generator, transformer, or converter windings and has no direct electrical connection, including a solidly connected grounded circuit conductor, to supply conductors originating in another system.

*Service.* The conductors and equipment for delivering energy from the electricity supply system to the wiring system of the premises served.

*Service conductors.* The supply conductors that extend from the street main or from transformers to the service equipment of the premises supplied.

*Service drop.* The overhead service conductors from the last pole or other aerial support to and including the splices, if any, connecting to the service-entrance conductors at the building or other structure.

*Service-entrance conductors, overhead system.* The service conductors between the terminals of the service equipment and a point usually outside the building, clear of building walls, where joined by tap or splice to the service drop.

*Service-entrance conductors, underground system.* The service conductors between the terminals of the service equipment and the point of connection to the service lateral. Where service equipment is located outside the building walls, there may be no service-entrance conductors, or they may be entirely outside the building.

*Service equipment.* The necessary equipment, usually consisting of a circuit breaker or switch and fuses, and their accessories, located near the point of entrance of supply conductors to a building or other structure, or an otherwise defined area, and intended to constitute the main control and means of cutoff of the supply.

*Service raceway.* The raceway that encloses the service-entrance conductors.

*Signaling circuit.* Any electric circuit that energizes signaling equipment.

*Switchboard.* A large single panel, frame, or assembly of panels which have switches, buses, instruments, overcurrent and other protective devices mounted on the face or back or both. Switchboards are generally accessible from the rear as well as from the front and are not intended to be installed in cabinets. (See *"Panelboard."*)

*Switches*—(a) *General-use switch.* A switch intended for use in general distribution and branch circuits. It is rated in amperes, and it is capable of interrupting its rated current at its rated voltage.

(b) *General-use snap switch.* A form of general-use switch so constructed that it can be installed in flush device boxes or on outlet box covers, or otherwise used in conjunction with wiring systems recognized by this subpart.

(c) *Isolating switch.* A switch intended for isolating an electric circuit from the source of power. It has no interrupting rating, and it is intended to be operated only after the circuit has been opened by some other means.

(d) *Motor-circuit switch.* A switch, rated in horsepower, capable of interrupting the maximum operating overload current of a motor of the same horsepower rating as the switch at the rated voltage.

*Switching devices.* (Over 600 volts, nominal.) Devices designed to close and/or open one or more electric circuits. Included in this category are circuit breakers, cutouts, disconnecting (or isolating) switches, disconnecting means, and interrupter switches.

*Transportable X-ray.* X-ray equipment installed in a vehicle or that may readily be disassembled for transport in a vehicle.

*Utilization equipment.* Utilization equipment means equipment which utilizes electric energy for mechanical, chemical, heating, lighting, or similar useful purpose.

*Utilization system.* A utilization system is a system which provides electric power and light for employee workplaces, and includes the premises wiring system and utilization equipment.

*Ventilated.* Provided with a means to permit circulation of air sufficient to remove an excess of heat, fumes, or vapors.

*Volatile flammable liquid.* A flammable liquid having a flash point below 38 degrees C (100 degrees F) or whose temperature is above its flash point, or a Class II combustible liquid having a vapor pressure not exceeding 40 psia (276 kPa) at 38° C (100° F) whose temperature is above its flash point.

*Voltage.* (Of a circuit.) The greatest root-mean-square (effective) difference of potential between any two conductors of the circuit concerned.

*Voltage, nominal.* A nominal value assigned to a circuit or system for the purpose of conveniently designating its voltage class (as 120/240, 480Y/277, 600, etc.). The actual voltage at which a circuit operates can vary from the nominal within a range that permits satisfactory operation of equipment.

*Voltage to ground.* For grounded circuits, the voltage between the given conductor and that point or conductor of the circuit that is grounded; for ungrounded circuits, the greatest voltage between the given conductor and any other conductor of the circuit.

*Watertight.* So constructed that moisture will not enter the enclosure.

*Weatherproof.* So constructed or protected that exposure to the weather will not interfere with successful operation. Rainproof, raintight, or watertight equipment can fulfill the requirements for weatherproof where varying weather conditions other than wetness, such as snow, ice, dust, or temperature extremes, are not a factor.

*Wet location.* See *"Location."*

**Authority:** Section 107, Contract Work Hours and Safety Standards Act (Construction Safety Act) (40 U.S.C. 333); Secs. 4, 6, 8, Occupational Safety and Health Act of 1970 (29 U.S.C. 653, 655, 657); Secretary of Labor's Order No. 1-90 (55 FR 9033); and 29 CFR Part 1911.

**Source:** 61 FR 46104, Aug. 30, 1996, unless otherwise noted.

### §1926.450 Scope, application and definitions applicable to this subpart.

**1926.450(a)** ***Scope and application.*** This subpart applies to all scaffolds used in workplaces covered by this part. It does not apply to crane or derrick suspended personnel platforms, which are covered by §1926.550(g). The criteria for aerial lifts are set out exclusively in §1926.453.

**1926.450(b)** ***Definitions.*** *Adjustable suspension scaffold* means a suspension scaffold equipped with a hoist(s) that can be operated by an employee(s) on the scaffold.

*Bearer (putlog)* means a horizontal transverse scaffold member (which may be supported by ledgers or runners) upon which the scaffold platform rests and which joins scaffold uprights, posts, poles, and similar members.

*Boatswains' chair* means a single-point adjustable suspension scaffold consisting of a seat or sling designed to support one employee in a sitting position.

*Body belt (safety belt)* means a strap with means both for securing it about the waist and for attaching it to a lanyard, lifeline, or deceleration device.

*Body harness* means a design of straps which may be secured about the employee in a manner to distribute the fall arrest forces over at least the thighs, pelvis, waist, chest and shoulders, with means for attaching it to other components of a personal fall arrest system.

*Brace* means a rigid connection that holds one scaffold member in a fixed position with respect to another member, or to a building or structure.

*Bricklayers' square scaffold* means a supported scaffold composed of framed squares which support a platform.

*Carpenters' bracket scaffold* means a supported scaffold consisting of a platform supported by brackets attached to building or structural walls.

*Catenary scaffold* means a suspension scaffold consisting of a platform supported by two essentially horizontal and parallel ropes attached to structural members of a building or other structure. Additional support may be provided by vertical pickups.

*Chimney hoist* means a multi-point adjustable suspension scaffold used to provide access to work inside chimneys. (See "Multi-point adjustable suspension scaffold".)

*Cleat* means a structural block used at the end of a platform to prevent the platform from slipping off its supports. Cleats are also used to provide footing on sloped surfaces such as crawling boards.

*Competent person* means one who is capable of identifying existing and predictable hazards in the surroundings or working conditions which are unsanitary, hazardous, or dangerous to employees, and who has authorization to take prompt corrective measures to eliminate them.

*Continuous run scaffold (Run scaffold)* means a two- point or multi-point adjustable suspension scaffold constructed using a series of interconnected braced scaffold members or supporting structures erected to form a continuous scaffold.

*Coupler* means a device for locking together the tubes of a tube and coupler scaffold.

*Crawling board (chicken ladder)* means a supported scaffold consisting of a plank with cleats spaced and secured to provide footing, for use on sloped surfaces such as roofs.

*Deceleration device* means any mechanism, such as a rope grab, rip-stitch lanyard, specially-woven lanyard, tearing or deforming lanyard, or automatic self-retracting lifeline lanyard, which dissipates a substantial amount of energy during a fall arrest or limits the energy imposed on an employee during fall arrest.

*Double pole (independent pole) scaffold* means a supported scaffold consisting of a platform(s) resting on cross beams (bearers) supported by ledgers and a double row of uprights independent of support (except ties, guys, braces) from any structure.

*Equivalent* means alternative designs, materials or methods to protect against a hazard which the employer can demonstrate will provide an equal or greater degree of safety for employees than the methods, materials or designs specified in the standard.

*Exposed power lines* means electrical power lines which are accessible to employees and which are not shielded from contact. Such lines do not include extension cords or power tool cords.

*Eye* or *Eye splice* means a loop with or without a thimble at the end of a wire rope.

*Fabricated decking and planking* means manufactured platforms made of wood (including laminated wood, and solid sawn wood planks), metal or other materials.

*Fabricated frame scaffold (tubular welded frame scaffold)* means a scaffold consisting of a platform(s) supported on fabricated end frames with integral posts, horizontal bearers, and intermediate members.

*Failure* means load refusal, breakage, or separation of component parts. Load refusal is the point where the ultimate strength is exceeded.

*Float (ship) scaffold* means a suspension scaffold consisting of a braced platform resting on two parallel bearers and hung from overhead supports by ropes of fixed length.

*Form scaffold* means a supported scaffold consisting of a platform supported by brackets attached to formwork.

*Guardrail system* means a vertical barrier, consisting of, but not limited to, toprails, midrails, and posts, erected to prevent employees from falling off a scaffold platform or walkway to lower levels.

*Hoist* means a manual or power-operated mechanical device to raise or lower a suspended scaffold.

*Horse scaffold* means a supported scaffold consisting of a platform supported by construction horses (saw horses). Horse scaffolds constructed of metal are sometimes known as trestle scaffolds.

*Independent pole scaffold* (see "Double pole scaffold").

*Interior hung scaffold* means a suspension scaffold consisting of a platform suspended from the ceiling or roof structure by fixed length supports.

*Ladder jack scaffold* means a supported scaffold consisting of a platform resting on brackets attached to ladders.

*Ladder stand* means a mobile, fixed-size, self-supporting ladder consisting of a wide flat tread ladder in the form of stairs.

*Landing* means a platform at the end of a flight of stairs.

*Large area scaffold* means a pole scaffold, tube and coupler scaffold, systems scaffold, or fabricated frame scaffold erected over substantially the entire work area. For example: a scaffold erected over the entire floor area of a room.

*Lean-to scaffold* means a supported scaffold which is kept erect by tilting it toward and resting it against a building or structure.

*Lifeline* means a component consisting of a flexible line that connects to an anchorage at one end to hang vertically (vertical lifeline), or that connects to anchorages at both ends to stretch horizontally (horizontal lifeline), and which serves as a means for connecting other components of a personal fall arrest system to the anchorage.

*Lower levels* means areas below the level where the employee is located and to which an employee can fall. Such areas include, but are not limited to, ground levels, floors, roofs, ramps, runways, excavations, pits, tanks, materials, water, and equipment.

*Masons' adjustable supported scaffold* (see "Self-contained adjustable scaffold").

*Masons' multi-point adjustable suspension scaffold* means a continuous run suspension scaffold designed and used for masonry operations.

*Maximum intended load* means the total load of all persons, equipment, tools, materials, transmitted loads, and other loads reasonably anticipated to be applied to a scaffold or scaffold component at any one time.

*Mobile scaffold* means a powered or unpowered, portable, caster or wheel-mounted supported scaffold.

*Multi-level suspended scaffold* means a two-point or multi-point adjustable suspension scaffold with a series of platforms at various levels resting on common stirrups.

*Multi-point adjustable suspension scaffold* means a suspension scaffold consisting of a platform(s) which is suspended by more than two ropes from overhead supports and equipped with means to raise and lower the platform to desired work levels. Such scaffolds include chimney hoists.

*Needle beam scaffold* means a platform suspended from needle beams.

*Open sides and ends* means the edges of a platform that are more than 14 inches (36 cm) away horizontally from a sturdy, continuous, vertical surface (such as a building wall) or a sturdy, continuous horizontal surface (such as a floor), or a point of access. Exception: For plastering and lathing operations the horizontal threshold distance is 18 inches (46 cm).

*Outrigger* means the structural member of a supported scaffold used to increase the base width of a scaffold in order to provide support for and increased stability of the scaffold.

*Outrigger beam (Thrustout)* means the structural member of a suspension scaffold or outrigger scaffold which provides support for the scaffold by extending the scaffold point of attachment to a point out and away from the structure or building.

*Outrigger scaffold* means a supported scaffold consisting of a platform resting on outrigger beams (thrustouts) projecting beyond the wall or face of the building or structure, the inboard ends of which are secured inside the building or structure.

*Overhand bricklaying* means the process of laying bricks and masonry units such that the surface of the wall to be jointed is on the opposite side of the wall from the mason, requiring the mason to lean over the wall to complete the work. It includes mason tending and electrical installation incorporated into the brick wall during the overhand bricklaying process.

*Personal fall arrest system* means a system used to arrest an employee's fall. It consists of an anchorage, connectors, a body belt or body harness and may include a lanyard, deceleration device, lifeline, or combinations of these.

*Platform* means a work surface elevated above lower levels. Platforms can be constructed using individual wood planks, fabricated planks, fabricated decks, and fabricated platforms.

*Pole scaffold* (see definitions for "Single-pole scaffold" and "Double (independent) pole scaffold").

*Power operated hoist* means a hoist which is powered by other than human energy.

*Pump jack scaffold* means a supported scaffold consisting of a platform supported by vertical poles and movable support brackets.

*Qualified* means one who, by possession of a recognized degree, certificate, or professional standing, or who by extensive knowledge, training, and experience, has successfully demonstrated his/her ability to solve or resolve problems related to the subject matter, the work, or the project.

*Rated load* means the manufacturer's specified maximum load to be lifted by a hoist or to be applied to a scaffold or scaffold component.

*Repair bracket scaffold* means a supported scaffold consisting of a platform supported by brackets which are secured in place around the circumference or perimeter of a chimney, stack, tank or other supporting structure by one or more wire ropes placed around the supporting structure.

*Roof bracket scaffold* means a rooftop supported scaffold consisting of a platform resting on angular-shaped supports.

*Runner (ledger or ribbon)* means the lengthwise horizontal spacing or bracing member which may support the bearers.

*Scaffold* means any temporary elevated platform (supported or suspended) and its supporting structure (including points of anchorage), used for supporting employees or materials or both.

*Self-contained adjustable scaffold* means a combination supported and suspension scaffold consisting of an adjustable platform(s) mounted on an independent supporting frame(s) not a part of the object being worked on, and which is equipped with a means to permit the raising and lowering of the platform(s). Such systems include rolling roof rigs, rolling outrigger systems, and some masons' adjustable supported scaffolds.

*Shore scaffold* means a supported scaffold which is placed against a building or structure and held in place with props.

*Single-point adjustable suspension scaffold* means a suspension scaffold consisting of a platform suspended by one rope from an overhead support and equipped with means to permit the movement of the platform to desired work levels.

*Single-pole scaffold* means a supported scaffold consisting of a platform(s) resting on bearers, the outside ends of which are supported on runners secured to a single row of posts or uprights, and the inner ends of which are supported on or in a structure or building wall.

*Stair tower (Scaffold stairway/tower)* means a tower comprised of scaffold components and which contains internal stairway units and rest platforms. These towers are used to provide access to scaffold platforms and other elevated points such as floors and roofs.

*Stall load* means the load at which the prime-mover of a power-operated hoist stalls or the power to the prime-mover is automatically disconnected.

*Step, platform, and trestle ladder scaffold* means a platform resting directly on the rungs of step ladders or trestle ladders.

*Stilts* means a pair of poles or similar supports with raised footrests, used to permit walking above the ground or working surface.

*Stonesetters' multi-point adjustable suspension scaffold* means a continuous run suspension scaffold designed and used for stonesetters' operations.

*Supported scaffold* means one or more platforms supported by outrigger beams, brackets, poles, legs, uprights, posts, frames, or similar rigid support.

*Suspension scaffold* means one or more platforms suspended by ropes or other non-rigid means from an overhead structure(s).

*System scaffold* means a scaffold consisting of posts with fixed connection points that accept runners, bearers, and diagonals that can be interconnected at predetermined levels.

*Tank builders' scaffold* means a supported scaffold consisting of a platform resting on brackets that are either directly attached to a cylindrical tank or attached to devices that are attached to such a tank.

*Top plate bracket scaffold* means a scaffold supported by brackets that hook over or are attached to the top of a wall. This type of scaffold is similar to carpenters' bracket scaffolds and form scaffolds and is used in residential construction for setting trusses.

*Tube and coupler scaffold* means a supported or suspended scaffold consisting of a platform(s) supported by tubing, erected with coupling devices connecting uprights, braces, bearers, and runners.

*Tubular welded frame scaffold* (see "Fabricated frame scaffold").

*Two-point suspension scaffold (swing stage)* means a suspension scaffold consisting of a platform supported by hangers (stirrups) suspended by two ropes from overhead supports and equipped with means to permit the raising and lowering of the platform to desired work levels.

*Unstable objects* means items whose strength, configuration, or lack of stability may allow them to become dislocated and shift and therefore may not properly support the loads imposed on them. Unstable objects do not constitute a safe base support for scaffolds, platforms, or employees. Examples include, but are not limited to, barrels, boxes, loose brick, and concrete blocks.

*Vertical pickup* means a rope used to support the horizontal rope in catenary scaffolds.

*Walkway* means a portion of a scaffold platform used only for access and not as a work level.

*Window jack scaffold* means a platform resting on a bracket or jack which projects through a window opening.

## §1926.451 General requirements.

This section does not apply to aerial lifts, the criteria for which are set out exclusively in §1926.453.

**1926.451(a)** ***Capacity.*** (1) Except as provided in paragraphs (a)(2), (a)(3), (a)(4), (a)(5) and (g) of this section, each scaffold and scaffold component shall be capable of supporting, without failure, its own weight and at least 4 times the maximum intended load applied or transmitted to it.

**1926.451(a)(2)** Direct connections to roofs and floors, and counterweights used to balance adjustable suspension scaffolds, shall be capable of resisting at least 4 times the tipping moment imposed by the scaffold operating at the rated load of the hoist, or 1.5 (minimum) times the tipping moment imposed by the scaffold operating at the stall load of the hoist, whichever is greater.

**1926.451(a)(3)** Each suspension rope, including connecting hardware, used on non-adjustable suspension scaffolds shall be capable of supporting, without failure, at least 6 times the maximum intended load applied or transmitted to that rope.

**1926.451(a)(4)** Each suspension rope, including connecting hardware, used on adjustable suspension scaffolds shall be capable of supporting, without failure, at least 6 times the maximum intended load applied or transmitted to that rope with the scaffold operating at either the rated load of the hoist, or 2 (minimum) times the stall load of the hoist, whichever is greater.

**1926.451(a)(5)** The stall load of any scaffold hoist shall not exceed 3 times its rated load.

**1926.451(a)(6)** Scaffolds shall be designed by a qualified person and shall be constructed and loaded in accordance with that design. Non-mandatory Appendix A to this subpart contains examples of criteria that will enable an employer to comply with paragraph (a) of this section.

**1926.451(b)** ***Scaffold platform construction.*** (1) Each platform on all working levels of scaffolds shall be fully planked or decked between the front uprights and the guardrail supports as follows:

**1926.451(b)(1)(i)** Each platform unit (e.g., scaffold plank, fabricated plank, fabricated deck, or fabricated platform) shall be installed so that the space between adjacent units and the space between the platform and the uprights is no more than 1 inch (2.5 cm) wide, except where the employer can demonstrate that a wider space is necessary (for example, to fit around uprights when side brackets are used to extend the width of the platform).

**1926.451(b)(1)(ii)** Where the employer makes the demonstration provided for in paragraph (b)(1)(i) of this section, the platform shall be planked or decked as fully as possible and the remaining open space between the platform and the uprights shall not exceed 9 1/2 inches (24.1 cm).

Exception to paragraph (b)(1): The requirement in paragraph (b)(1) to provide full planking or decking does not apply to platforms used solely as walkways or solely by employees performing scaffold erection or dismantling. In these situations, only the planking that the employer establishes is necessary to provide safe working conditions is required.

**1926.451(b)(2)** Except as provided in paragraphs (b)(2)(i) and (b)(2)(ii) of this section, each scaffold platform and walkway shall be at least 18 inches (46 cm) wide.

**1926.451(b)(2)(i)** Each ladder jack scaffold, top plate bracket scaffold, roof bracket scaffold, and pump jack scaffold shall be at least 12 inches (30 cm) wide. There is no minimum width requirement for boatswains' chairs.

**Note to paragraph (b)(2)(i):** Pursuant to an administrative stay effective November 29, 1996 and published in the FEDERAL REGISTER on November 25, 1996, the requirement in paragraph (b)(2)(i) that roof bracket scaffolds be at least 12 inches wide is stayed until November 25, 1997 or until rulemaking regarding the minimum width of roof bracket scaffolds has been completed, whichever is later.

**1926.451(b)(2)(ii)** Where scaffolds must be used in areas that the employer can demonstrate are so narrow that platforms and walkways cannot be at least 18 inches (46 cm) wide, such platforms and walkways shall be as wide as feasible, and employees on those platforms and walkways shall be protected from fall hazards by the use of guardrails and/or personal fall arrest systems.

**1926.451(b)(3)** Except as provided in paragraphs (b)(3) (i) and (ii) of this section, the front edge of all platforms shall not be more than 14 inches (36 cm) from the face of the work, unless guardrail systems are erected along the front edge and/or personal fall arrest systems are used in accordance with paragraph (g) of this section to protect employees from falling.

**1926.451(b)(3)(i)** The maximum distance from the face for outrigger scaffolds shall be 3 inches (8 cm);

**1926.451(b)(3)(ii)** The maximum distance from the face for plastering and lathing operations shall be 18 inches (46 cm).

**1926.451(b)(4)** Each end of a platform, unless cleated or otherwise restrained by hooks or equivalent means, shall extend over the centerline of its support at least 6 inches (15 cm).

**1926.451(b)(5)(i)** Each end of a platform 10 feet or less in length shall not extend over its support more than 12 inches (30 cm) unless the platform is designed and installed so that the cantilevered portion of the platform is able to support employees and/or materials without tipping, or has guardrails which block employee access to the cantilevered end.

**1926.451(b)(5)(ii)** Each platform greater than 10 feet in length shall not extend over its support more than 18 inches (46 cm), unless it is designed and installed so that the cantilevered portion of the platform is able to support employees without tipping, or has guardrails which block employee access to the cantilevered end.

**1926.451(b)(6)** On scaffolds where scaffold planks are abutted to create a long platform, each abutted end shall rest on a separate support surface. This provision does not preclude the use of common support members, such as "T" sections, to support abutting planks, or hook on platforms designed to rest on common supports.

**1926.451(b)(7)** On scaffolds where platforms are overlapped to create a long platform, the overlap shall occur only over supports, and shall not be less than 12 inches (30 cm) unless the platforms are nailed together or otherwise restrained to prevent movement.

**1926.451(b)(8)** At all points of a scaffold where the platform changes direction, such as turning a corner, any platform that rests on a bearer at an angle other than a right angle shall be laid first, and platforms which rest at right angles over the same bearer shall be laid second, on top of the first platform.

**1926.451(b)(9)** Wood platforms shall not be covered with opaque finishes, except that platform edges may be covered or marked for identification. Platforms may be coated periodically with wood preservatives, fire-retardant finishes, and slip-resistant finishes; however, the coating may not obscure the top or bottom wood surfaces.

**1926.451(b)(10)** Scaffold components manufactured by different manufacturers shall not be intermixed unless the components fit together without force and the scaffold's structural integrity is maintained by the user. Scaffold components manufactured by different manufacturers shall not be modified in order to intermix them unless a competent person determines the resulting scaffold is structurally sound.

**1926.451(b)(11)** Scaffold components made of dissimilar metals shall not be used together unless a competent person has determined that galvanic action will not reduce the strength of any component to a level below that required by paragraph (a)(1) of this section.

**1926.451(c)** ***Criteria for supported scaffolds.*** (1) Supported scaffolds with a height to base width (including outrigger supports, if used) ratio of more than four to one (4:1) shall be restrained from tipping by guying, tying, bracing, or equivalent means, as follows:

**1926.451(c)(1)(i)** Guys, ties, and braces shall be installed at locations where horizontal members support both inner and outer legs.

**1926.451(c)(1)(ii)** Guys, ties, and braces shall be installed according to the scaffold manufacturer's recommendations or at the closest horizontal member to the 4:1 height and be repeated vertically at locations of horizontal members every 20 feet (6.1 m) or less thereafter for scaffolds 3 feet (0.91 m) wide or less, and every 26 feet (7.9 m) or less thereafter for scaffolds greater than 3 feet (0.91 m) wide. The top guy, tie or brace of completed scaffolds shall be placed no further than the 4:1 height from the top. Such guys, ties and braces shall be installed at each end of the scaffold and at horizontal intervals not to exceed 30 feet (9.1 m) (measured from one end [not both] towards the other).

**1926.451(c)(1)(iii)** Ties, guys, braces, or outriggers shall be used to prevent the tipping of supported scaffolds in all circumstances where an eccentric load, such as a cantilevered work platform, is applied or is transmitted to the scaffold.

**1926.451(c)(2)** Supported scaffold poles, legs, posts, frames, and uprights shall bear on base plates and mud sills or other adequate firm foundation.

**1926.451(c)(2)(i)** Footings shall be level, sound, rigid, and capable of supporting the loaded scaffold without settling or displacement.

**1926.451(c)(2)(ii)** Unstable objects shall not be used to support scaffolds or platform units.

**1926.451(c)(2)(iii)** Unstable objects shall not be used as working platforms.

**1926.451(c)(2)(iv)** Front-end loaders and similar pieces of equipment shall not be used to support scaffold platforms unless they have been specifically designed by the manufacturer for such use.

**1926.451(c)(2)(v)** Fork-lifts shall not be used to support scaffold platforms unless the entire platform is attached to the fork and the fork-lift is not moved horizontally while the platform is occupied.

**1926.451(c)(3)** Supported scaffold poles, legs, posts, frames, and uprights shall be plumb and braced to prevent swaying and displacement.

**1926.451(d)** ***Criteria for suspension scaffolds.*** (1) All suspension scaffold support devices, such as outrigger beams, cornice hooks, parapet clamps, and similar devices, shall rest on surfaces capable of supporting at least 4 times the load imposed on them by the scaffold operating at the rated load of the hoist (or at least 1.5 times the load imposed on them by the scaffold at the stall capacity of the hoist, whichever is greater).

**1926.451(d)(2)** Suspension scaffold outrigger beams, when used, shall be made of structural metal or equivalent strength material, and shall be restrained to prevent movement.

**1926.451(d)(3)** The inboard ends of suspension scaffold outrigger beams shall be stabilized by bolts or other direct connections to the floor or roof deck, or they shall have their inboard ends stabilized by counterweights, except masons' multi-point adjustable suspension scaffold outrigger beams shall not be stabilized by counterweights.

**1926.451(d)(3)(i)** Before the scaffold is used, direct connections shall be evaluated by a competent person who shall confirm, based on the evaluation, that the supporting surfaces are capable of supporting the loads to be imposed. In addition, masons' multi-point adjustable suspension scaffold connections shall be designed by an engineer experienced in such scaffold design.

**1926.451(d)(3)(ii)** Counterweights shall be made of non-flowable material. Sand, gravel and similar materials that can be easily dislocated shall not be used as counterweights.

**1926.451(d)(3)(iii)** Only those items specifically designed as counterweights shall be used to counterweight scaffold systems. Construction materials such as, but not limited to, masonry units and rolls of roofing felt, shall not be used as counterweights.

**1926.451(d)(3)(iv)** Counterweights shall be secured by mechanical means to the outrigger beams to prevent accidental displacement.

**1926.451(d)(3)(v)** Counterweights shall not be removed from an outrigger beam until the scaffold is disassembled.

**1926.451(d)(3)(vi)** Outrigger beams which are not stabilized by bolts or other direct connections to the floor or roof deck shall be secured by tiebacks.

**1926.451(d)(3)(vii)** Tiebacks shall be equivalent in strength to the suspension ropes.

**1926.451(d)(3)(viii)** Outrigger beams shall be placed perpendicular to its bearing support (usually the face of the building or structure). However, where the employer can demonstrate that it is not possible to place an outrigger beam perpendicular to the face of the building or structure because of obstructions that cannot be moved, the outrigger beam may be placed at some other angle, provided opposing angle tiebacks are used.

**1926.451(d)(3)(ix)** Tiebacks shall be secured to a structurally sound anchorage on the building or structure. Sound anchorages include structural members, but do not include standpipes, vents, other piping systems, or electrical conduit.

**1926.451(d)(3)(x)** Tiebacks shall be installed perpendicular to the face of the building or structure, or opposing angle tiebacks shall be installed. Single tiebacks installed at an angle are prohibited.

**1926.451(d)(4)** Suspension scaffold outrigger beams shall be:

**1926.451(d)(4)(i)** Provided with stop bolts or shackles at both ends;

**1926.451(d)(4)(ii)** Securely fastened together with the flanges turned out when channel iron beams are used in place of I-beams;

**1926.451(d)(4)(iii)** Installed with all bearing supports perpendicular to the beam center line;

**1926.451(d)(4)(iv)** Set and maintained with the web in a vertical position; and

**1926.451(d)(4)(v)** When an outrigger beam is used, the shackle or clevis with which the rope is attached to the outrigger beam shall be placed directly over the center line of the stirrup.

**1926.451(d)(5)** Suspension scaffold support devices such as cornice hooks, roof hooks, roof irons, parapet clamps, or similar devices shall be:

**1926.451(d)(5)(i)** Made of steel, wrought iron, or materials of equivalent strength;

**1926.451(d)(5)(ii)** Supported by bearing blocks; and

**1926.451(d)(5)(iii)** Secured against movement by tiebacks installed at right angles to the face of the building or structure, or opposing angle tiebacks shall be installed and secured to a structurally sound point of anchorage on the building or structure. Sound points of anchorage include structural members, but do not include standpipes, vents, other piping systems, or electrical conduit.

**1926.451(d)(5)(iv)** Tiebacks shall be equivalent in strength to the hoisting rope.

**1926.451(d)(6)** When winding drum hoists are used on a suspension scaffold, they shall contain not less than four wraps of the suspension rope at the lowest point of scaffold travel. When other types of hoists are used, the suspension ropes shall be long enough to allow the scaffold to be lowered to the level below without the rope end passing through the hoist, or the rope end shall be configured or provided with means to prevent the end from passing through the hoist.

**1926.451(d)(7)** The use of repaired wire rope as suspension rope is prohibited.

**1926.451(d)(8)** Wire suspension ropes shall not be joined together except through the use of eye splice thimbles connected with shackles or coverplates and bolts.

**1926.451(d)(9)** The load end of wire suspension ropes shall be equipped with proper size thimbles and secured by eyesplicing or equivalent means.

**1926.451(d)(10)** Ropes shall be inspected for defects by a competent person prior to each workshift and after every occurrence which could affect a rope's integrity. Ropes shall be replaced if any of the following conditions exist:

**1926.451(d)(10)(i)** Any physical damage which impairs the function and strength of the rope.

**1926.451(d)(10)(ii)** Kinks that might impair the tracking or wrapping of rope around the drum(s) or sheave(s).

**1926.451(d)(10)(iii)** Six randomly distributed broken wires in one rope lay or three broken wires in one strand in one rope lay.

**1926.451(d)(10)(iv)** Abrasion, corrosion, scrubbing, flattening or peening causing loss of more than one-third of the original diameter of the outside wires.

**1926.451(d)(10)(v)** Heat damage caused by a torch or any damage caused by contact with electrical wires.

**1926.451(d)(10)(vi)** Evidence that the secondary brake has been activated during an overspeed condition and has engaged the suspension rope.

**1926.451(d)(11)** Swaged attachments or spliced eyes on wire suspension ropes shall not be used unless they are made by the wire rope manufacturer or a qualified person.

**1926.451(d)(12)** When wire rope clips are used on suspension scaffolds:

**1926.451(d)(12)(i)** There shall be a minimum of 3 wire rope clips installed, with the clips a minimum of 6 rope diameters apart;

**1926.451(d)(12)(ii)** Clips shall be installed according to the manufacturer's recommendations;

**1926.451(d)(12)(iii)** Clips shall be retightened to the manufacturer's recommendations after the initial loading;

**1926.451(d)(12)(iv)** Clips shall be inspected and retightened to the manufacturer's recommendations at the start of each workshift thereafter;

**1926.451(d)(12)(v)** U-bolt clips shall not be used at the point of suspension for any scaffold hoist;

**1926.451(d)(12)(vi)** When U-bolt clips are used, the U-bolt shall be placed over the dead end of the rope, and the saddle shall be placed over the live end of the rope.

**1926.451(d)(13)** Suspension scaffold power-operated hoists and manual hoists shall be tested by a qualified testing laboratory.

**1926.451(d)(14)** Gasoline-powered equipment and hoists shall not be used on suspension scaffolds.

**1926.451(d)(15)** Gears and brakes of power-operated hoists used on suspension scaffolds shall be enclosed.

**1926.451(d)(16)** In addition to the normal operating brake, suspension scaffold power-operated hoists and manually operated hoists shall have a braking device or locking pawl which engages automatically when a hoist makes either of the following uncontrolled movements: an instantaneous change in momentum or an accelerated overspeed.

**1926.451(d)(17)** Manually operated hoists shall require a positive crank force to descend.

**1926.451(d)(18)** Two-point and multi-point suspension scaffolds shall be tied or otherwise secured to prevent them from swaying, as determined to be necessary based on an evaluation by a competent person. Window cleaners' anchors shall not be used for this purpose.

**1926.451(d)(19)** Devices whose sole function is to provide emergency escape and rescue shall not be used as working platforms. This provision does not preclude the use of systems which are designed to function both as suspension scaffolds and emergency systems.

**1926.451(e)** ***Access.*** This paragraph applies to scaffold access for all employees. Access requirements for employees erecting or dismantling supported scaffolds are specifically addressed in paragraph (e)(9) of this section.

**1926.451(e)(1)** When scaffold platforms are more than 2 feet (0.6 m) above or below a point of access, portable ladders, hook-on ladders, attachable ladders, stair towers (scaffold stairways/towers), stairway type ladders (such as ladder stands), ramps, walkways, integral prefabricated scaffold access, or direct access from another scaffold, structure, personnel hoist, or similar surface shall be used. Crossbraces shall not be used as a means of access.

**1926.451(e)(2)** Portable, hook-on, and attachable ladders (Additional requirements for the proper construction and use of portable ladders are contained in subpart X of this part —Stairways and Ladders):

**1926.451(e)(2)(i)** Portable, hook-on, and attachable ladders shall be positioned so as not to tip the scaffold;

**1926.451(e)(2)(ii)** Hook-on and attachable ladders shall be positioned so that their bottom rung is not more than 24 inches (61 cm) above the scaffold supporting level;

**1926.451(e)(2)(iii)** When hook-on and attachable ladders are used on a supported scaffold more than 35 feet (10.7 m) high, they shall have rest platforms at 35-foot (10.7 m) maximum vertical intervals.

**1926.451(e)(2)(iv)** Hook-on and attachable ladders shall be specifically designed for use with the type of scaffold used;

**1926.451(e)(2)(v)** Hook-on and attachable ladders shall have a minimum rung length of 11 1/2 inches (29 cm); and

**1926.451(e)(2)(vi)** Hook-on and attachable ladders shall have uniformly spaced rungs with a maximum spacing between rungs of 16 3/4 inches.

**1926.451(e)(3)** Stairway-type ladders shall:

**1926.451(e)(3)(i)** Be positioned such that their bottom step is not more than 24 inches (61 cm) above the scaffold supporting level;

**1926.451(e)(3)(ii)** Be provided with rest platforms at 12 foot (3.7 m) maximum vertical intervals;

**1926.451(e)(3)(iii)** Have a minimum step width of 16 inches (41 cm), except that mobile scaffold stairway-type ladders shall have a minimum step width of 11 1/2 inches (30 cm); and

**1926.451(e)(3)(iv)** Have slip-resistant treads on all steps and landings.

**1926.451(e)(4)** Stairtowers (scaffold stairway/towers) shall be positioned such that their bottom step is not more than 24 inches (61 cm.) above the scaffold supporting level.

**1926.451(e)(4)(i)** A stairrail consisting of a toprail and a midrail shall be provided on each side of each scaffold stairway.

**1926.451(e)(4)(ii)** The toprail of each stairrail system shall also be capable of serving as a handrail, unless a separate handrail is provided.

**1926.451(e)(4)(iii)** Handrails, and toprails that serve as handrails, shall provide an adequate handhold for employees grasping them to avoid falling.

**1926.451(e)(4)(iv)** Stairrail systems and handrails shall be surfaced to prevent injury to employees from punctures or lacerations, and to prevent snagging of clothing.

**1926.451(e)(4)(v)** The ends of stairrail systems and handrails shall be constructed so that they do not constitute a projection hazard.

**1926.451(e)(4)(vi)** Handrails, and toprails that are used as handrails, shall be at least 3 inches (7.6 cm) from other objects.

**1926.451(e)(4)(vii)** Stairrails shall be not less than 28 inches (71 cm) nor more than 37 inches (94 cm) from the upper surface of the stairrail to the surface of the tread, in line with the face of the riser at the forward edge of the tread.

**1926.451(e)(4)(viii)** A landing platform at least 18 inches (45.7 cm) wide by at least 18 inches (45.7 cm) long shall be provided at each level.

**1926.451(e)(4)(ix)** Each scaffold stairway shall be at least 18 inches (45.7 cm) wide between stairrails.

**1926.451(e)(4)(x)** Treads and landings shall have slip-resistant surfaces.

**1926.451(e)(4)(xi)** Stairways shall be installed between 40 degrees and 60 degrees from the horizontal.

**1926.451(e)(4)(xii)** Guardrails meeting the requirements of paragraph (g)(4) of this section shall be provided on the open sides and ends of each landing.

**1926.451(e)(4)(xiii)** Riser height shall be uniform, within 1/4 inch, (0.6 cm) for each flight of stairs. Greater variations in riser height are allowed for the top and bottom steps of the entire system, not for each flight of stairs.

**1926.451(e)(4)(xiv)** Tread depth shall be uniform, within 1/4 inch, for each flight of stairs.

**1926.451(e)(5)** Ramps and walkways. (i) Ramps and walkways 6 feet (1.8 m) or more above lower levels shall have guardrail systems which comply with subpart M of this part—Fall Protection;

**1926.451(e)(5)(ii)** No ramp or walkway shall be inclined more than a slope of one (1) vertical to three (3) horizontal (20 degrees above the horizontal).

**1926.451(e)(5)(iii)** If the slope of a ramp or a walkway is steeper than one (1) vertical in eight (8) horizontal, the ramp or walkway shall have cleats not more than fourteen (14) inches (35 cm) apart which are securely fastened to the planks to provide footing.

**1926.451(e)(6)** Integral prefabricated scaffold access frames shall:

**1926.451(e)(6)(i)** Be specifically designed and constructed for use as ladder rungs;

**1926.451(e)(6)(ii)** Have a rung length of at least 8 inches (20 cm);

**1926.451(e)(6)(iii)** Not be used as work platforms when rungs are less than 11 1/2 inches in length, unless each affected employee uses fall protection, or a positioning device, which complies with §1926.502;

**1926.451(e)(6)(iv)** Be uniformly spaced within each frame section;

**1926.451(e)(6)(v)** Be provided with rest platforms at 35-foot (10.7 m) maximum vertical intervals on all supported scaffolds more than 35 feet (10.7 m) high; and

**1926.451(e)(6)(vi)** Have a maximum spacing between rungs of 16 3/4 inches (43 cm). Non-uniform rung spacing caused by joining end frames together is allowed, provided the resulting spacing does not exceed 16 3/4 inches (43 cm).

**1926.451(e)(7)** Steps and rungs of ladder and stairway type access shall line up vertically with each other between rest platforms.

**1926.451(e)(8)** Direct access to or from another surface shall be used only when the scaffold is not more than 14 inches (36 cm) horizontally and not more than 24 inches (61 cm) vertically from the other surface.

**1926.451(e)(9)** Effective September 2, 1997, access for employees erecting or dismantling supported scaffolds shall be in accordance with the following:

**1926.451(e)(9)(i)** The employer shall provide safe means of access for each employee erecting or dismantling a scaffold where the provision of safe access is feasible and does not create a greater hazard. The employer shall have a competent person determine whether it is feasible or would pose a greater hazard to provide, and have employees use a safe means of access. This determination shall be based on site conditions and the type of scaffold being erected or dismantled.

**1926.451(e)(9)(ii)** Hook-on or attachable ladders shall be installed as soon as scaffold erection has progressed to a point that permits safe installation and use.

**1926.451(e)(9)(iii)** When erecting or dismantling tubular welded frame scaffolds, (end) frames, with horizontal members that are parallel, level and are not more than 22 inches apart vertically may be used as climbing devices for access, provided they are erected in a manner that creates a usable ladder and provides good hand hold and foot space.

**1926.451(e)(9)(iv)** Cross braces on tubular welded frame scaffolds shall not be used as a means of access or egress.

**1926.451(f)** ***Use.*** (1) Scaffolds and scaffold components shall not be loaded in excess of their maximum intended loads or rated capacities, whichever is less.

**1926.451(f)(2)** The use of shore or lean-to scaffolds is prohibited.

**1926.451(f)(3)** Scaffolds and scaffold components shall be inspected for visible defects by a competent person before each work shift, and after any occurrence which could affect a scaffold's structural integrity.

**1926.451(f)(4)** Any part of a scaffold damaged or weakened such that its strength is less than that required by paragraph (a) of this section shall be immediately repaired or replaced, braced to meet those provisions, or removed from service until repaired.

**1926.451(f)(5)** Scaffolds shall not be moved horizontally while employees are on them, unless they have been designed by a registered professional engineer specifically for such movement or, for mobile scaffolds, where the provisions of §1926.452(w) are followed.

**1926.451(f)(6)** The clearance between scaffolds and power lines shall be as follows: Scaffolds shall not be erected, used, dismantled, altered, or moved such that they or any conductive material handled on them might come closer to exposed and energized power lines than as follows:

| Insulated lines voltage | Minimum distance | Alternatives |
|---|---|---|
| Less than 300 volts | 3 feet (0.9 m). | |
| 300 volts to 50 kv | 10 feet (3.1 m) | |
| More than 50 kv | 10 feet (3.1 m) plus 0.4 inches (1.0 cm) for each 1 kv over 50 kv. | 2 times the length of the line insulator, but never less than 10 feet (3.1 m). |

| Uninsulated lines voltage | Minimum distance | Alternatives |
|---|---|---|
| Less than 50 kv | 10 feet (3.1 m). | |
| More than 50 kv | 10 feet (3.1 m) plus 0.4 inches (1.0 cm) for each 1 kv over 50 kv. | 2 times the length of the line insulator, but never less than 10 feet (3.1 m). |

EXCEPTION TO PARAGRAPH (f)(6): Scaffolds and materials may be closer to power lines than specified above where such clearance is necessary for performance of work, and only after the utility company, or electrical system operator, has been notified of the need to work closer and the utility company, or electrical system operator, has deenergized the lines, relocated the lines, or installed protective coverings to prevent accidental contact with the lines.

**1926.451(f)(7)** Scaffolds shall be erected, moved, dismantled, or altered only under the supervision and direction of a competent person qualified in scaffold erection, moving, dismantling or alteration. Such activities shall be performed only by experienced and trained employees selected for such work by the competent person.

**1926.451(f)(8)** Employees shall be prohibited from working on scaffolds covered with snow, ice, or other slippery material except as necessary for removal of such materials.

**1926.451(f)(9)** Where swinging loads are being hoisted onto or near scaffolds such that the loads might contact the scaffold, tag lines or equivalent measures to control the loads shall be used.

**1926.451(f)(10)** Suspension ropes supporting adjustable suspension scaffolds shall be of a diameter large enough to provide sufficient surface area for the functioning of brake and hoist mechanisms.

**1926.451(f)(11)** Suspension ropes shall be shielded from heat-producing processes. When acids or other corrosive substances are used on a scaffold, the ropes shall be shielded, treated to protect against the corrosive substances, or shall be of a material that will not be damaged by the substance being used.

**1926.451(f)(12)** Work on or from scaffolds is prohibited during storms or high winds unless a competent person has determined that it is safe for employees to be on the scaffold and those employees are protected by a personal fall arrest system or wind screens. Wind screens shall not be used unless the scaffold is secured against the anticipated wind forces imposed.

**1926.451(f)(13)** Debris shall not be allowed to accumulate on platforms.

**1926.451(f)(14)** Makeshift devices, such as but not limited to boxes and barrels, shall not be used on top of scaffold platforms to increase the working level height of employees.

**1926.451(f)(15)** Ladders shall not be used on scaffolds to increase the working level height of employees, except on large area scaffolds where employers have satisfied the following criteria:

**1926.451(f)(15)(i)** When the ladder is placed against a structure which is not a part of the scaffold, the scaffold shall be secured against the sideways thrust exerted by the ladder;

**1926.451(f)(15)(ii)** The platform units shall be secured to the scaffold to prevent their movement;

**1926.451(f)(15)(iii)** The ladder legs shall be on the same platform or other means shall be provided to stabilize the ladder against unequal platform deflection, and

**1926.451(f)(15)(iv)** The ladder legs shall be secured to prevent them from slipping or being pushed off the platform.

**1926.451(f)(16)** Platforms shall not deflect more than 1/60 of the span when loaded.

**1926.451(f)(17)** To reduce the possibility of welding current arcing through the suspension wire rope when performing welding from suspended scaffolds, the following precautions shall be taken, as applicable:

**1926.451(f)(17)(i)** An insulated thimble shall be used to attach each suspension wire rope to its hanging support (such as cornice hook or outrigger). Excess suspension wire rope and any additional independent lines from grounding shall be insulated;

**1926.451(f)(17)(ii)** The suspension wire rope shall be covered with insulating material extending at least 4 feet (1.2 m) above the hoist. If there is a tail line below the hoist, it shall be insulated to prevent contact with the platform. The portion of the tail line that hangs free below the scaffold shall be guided or retained, or both, so that it does not become grounded;

**1926.451(f)(17)(iii)** Each hoist shall be covered with insulated protective covers;

**1926.451(f)(17)(iv)** In addition to a work lead attachment required by the welding process, a grounding conductor shall be connected from the scaffold to the structure. The size of this conductor shall be at least the size of the welding process work lead, and this conductor shall not be in series with the welding process or the work piece;

**1926.451(f)(17)(v)** If the scaffold grounding lead is disconnected at any time, the welding machine shall be shut off; and

**1926.451(f)(17)(vi)** An active welding rod or uninsulated welding lead shall not be allowed to contact the scaffold or its suspension system.

**1926.451(g)** ***Fall protection.*** (1) Each employee on a scaffold more than 10 feet (3.1 m) above a lower level shall be protected from falling to that lower level. Paragraphs (g)(1) (i) through (vii) of this section establish the types of fall protection to be provided to the employees on each type of scaffold. Paragraph (g)(2) of this section addresses fall protection for scaffold erectors and dismantlers.

**Note to paragraph (g)(1):** The fall protection requirements for employees installing suspension scaffold support systems on floors, roofs, and other elevated surfaces are set forth in subpart M of this part.

**1926.451(g)(1)(i)** Each employee on a boatswains' chair, catenary scaffold, float scaffold, needle beam scaffold, or ladder jack scaffold shall be protected by a personal fall arrest system;

**1926.451(g)(1)(ii)** Each employee on a single-point or two-point adjustable suspension scaffold shall be protected by both a personal fall arrest system and guardrail system;

**1926.451(g)(1)(iii)** Each employee on a crawling board (chicken ladder) shall be protected by a personal fall arrest system, a guardrail system (with minimum 200 pound toprail capacity), or by a three-fourth inch

(1.9 cm) diameter grabline or equivalent handhold securely fastened beside each crawling board;

**1926.451(g)(1)(iv)** Each employee on a self-contained adjustable scaffold shall be protected by a guardrail system (with minimum 200 pound toprail capacity) when the platform is supported by the frame structure, and by both a personal fall arrest system and a guardrail system (with minimum 200 pound toprail capacity) when the platform is supported by ropes;

**1926.451(g)(1)(v)** Each employee on a walkway located within a scaffold shall be protected by a guardrail system (with minimum 200 pound toprail capacity) installed within 9 1/2 inches (24.1 cm) of and along at least one side of the walkway.

**1926.451(g)(1)(vi)** Each employee performing overhand bricklaying operations from a supported scaffold shall be protected from falling from all open sides and ends of the scaffold (except at the side next to the wall being laid) by the use of a personal fall arrest system or guardrail system (with minimum 200 pound toprail capacity).

**1926.451(g)(1)(vii)** For all scaffolds not otherwise specified in paragraphs (g)(1)(i) through (g)(1)(vi) of this section, each employee shall be protected by the use of personal fall arrest systems or guardrail systems meeting the requirements of paragraph (g)(4) of this section.

**1926.451(g)(2)** Effective September 2, 1997, the employer shall have a competent person determine the feasibility and safety of providing fall protection for employees erecting or dismantling supported scaffolds. Employers are required to provide fall protection for employees erecting or dismantling supported scaffolds where the installation and use of such protection is feasible and does not create a greater hazard.

**1926.451(g)(3)** In addition to meeting the requirements of §1926.502(d), personal fall arrest systems used on scaffolds shall be attached by lanyard to a vertical lifeline, horizontal lifeline, or scaffold structural member. Vertical lifelines shall not be used when overhead components, such as overhead protection or additional platform levels, are part of a single-point or two-point adjustable suspension scaffold.

**1926.451(g)(3)(i)** When vertical lifelines are used, they shall be fastened to a fixed safe point of anchorage, shall be independent of the scaffold, and shall be protected from sharp edges and abrasion. Safe points of anchorage include structural members of buildings, but do not include standpipes, vents, other piping systems, electrical conduit, outrigger beams, or counterweights.

**1926.451(g)(3)(ii)** When horizontal lifelines are used, they shall be secured to two or more structural members of the scaffold, or they may be looped around both suspension and independent suspension lines (on scaffolds so equipped) above the hoist and brake attached to the end of the scaffold. Horizontal lifelines shall not be attached only to the suspension ropes.

**1926.451(g)(3)(iii)** When lanyards are connected to horizontal lifelines or structural members on a single-point or two-point adjustable suspension scaffold, the scaffold shall be equipped with additional independent support lines and automatic locking devices capable of stopping the fall of the scaffold in the event one or both of the suspension ropes fail. The independent support lines shall be equal in number and strength to the suspension ropes.

**1926.451(g)(3)(iv)** Vertical lifelines, independent support lines, and suspension ropes shall not be attached to each other, nor shall they be attached to or use the same point of anchorage, nor shall they be attached to the same point on the scaffold or personal fall arrest system.

**1926.451(g)(4)** Guardrail systems installed to meet the requirements of this section shall comply with the following provisions (guardrail systems built in accordance with Appendix A to this subpart will be deemed to meet the requirements of paragraphs (g)(4) (vii), (viii), and (ix) of this section):

**1926.451(g)(4)(i)** Guardrail systems shall be installed along all open sides and ends of platforms. Guardrail systems shall be installed before the scaffold is released for use by employees other than erection/dismantling crews.

**1926.451(g)(4)(ii)** The top edge height of toprails or equivalent member on supported scaffolds manufactured or placed in service after January 1, 2000 shall be installed between 38 inches (0.97 m) and 45 inches (1.2 m) above the platform surface. The top edge height on supported scaffolds manufactured and placed in service before January 1, 2000, and on all suspended scaffolds where both a guardrail and a personal fall arrest system are required shall be between 36 inches (0.9 m) and 45 inches (1.2 m). When conditions warrant, the height of the top edge may exceed the 45-inch height, provided the guardrail system meets all other criteria of paragraph (g)(4).

**1926.451(g)(4)(iii)** When midrails, screens, mesh, intermediate vertical members, solid panels, or equivalent structural members are used, they shall be installed between the top edge of the guardrail system and the scaffold platform.

**1926.451(g)(4)(iv)** When midrails are used, they shall be installed at a height approximately midway between the top edge of the guardrail system and the platform surface.

**1926.451(g)(4)(v)** When screens and mesh are used, they shall extend from the top edge of the guardrail system to the scaffold platform, and along the entire opening between the supports.

**1926.451(g)(4)(vi)** When intermediate members (such as balusters or additional rails) are used, they shall not be more than 19 inches (48 cm) apart.

**1926.451(g)(4)(vii)** Each toprail or equivalent member of a guardrail system shall be capable of withstanding, without failure, a force applied in any downward or horizontal direction at any point along its top edge of at least 100 pounds (445 n) for guardrail systems installed on single-point adjustable suspension scaffolds or two-point adjustable suspension scaffolds, and at least 200 pounds (890 n) for guardrail systems installed on all other scaffolds.

**1926.451(g)(4)(viii)** When the loads specified in paragraph (g)(4)(vii) of this section are applied in a downward direction, the top edge shall not drop below the height above the platform surface that is prescribed in paragraph (g)(4)(ii) of this section.

**1926.451(g)(4)(ix)** Midrails, screens, mesh, intermediate vertical members, solid panels, and equivalent structural members of a guardrail system shall be capable of withstanding, without failure, a force applied in any downward or horizontal direction at any point along the midrail or other member of at least 75 pounds (333 n) for guardrail systems with a minimum 100 pound toprail capacity, and at least 150 pounds (666 n) for guardrail systems with a minimum 200 pound toprail capacity.

**1926.451(g)(4)(x)** Suspension scaffold hoists and non-walk-through stirrups may be used as end guardrails, if the space between the hoist or stirrup and the side guardrail or structure does not allow passage of an employee to the end of the scaffold.

**1926.451(g)(4)(xi)** Guardrails shall be surfaced to prevent injury to an employee from punctures or lacerations, and to prevent snagging of clothing.

**1926.451(g)(4)(xii)** The ends of all rails shall not overhang the terminal posts except when such overhang does not constitute a projection hazard to employees.

**1926.451(g)(4)(xiii)** Steel or plastic banding shall not be used as a toprail or midrail.

**1926.451(g)(4)(xiv)** Manila or plastic (or other synthetic) rope being used for toprails or midrails shall be inspected by a competent person as frequently as necessary to ensure that it continues to meet the strength requirements of paragraph (g) of this section.

**1926.451(g)(4)(xv)** Crossbracing is acceptable in place of a midrail when the crossing point of two braces is between 20 inches (0.5 m) and 30 inches (0.8 m) above the work platform or as a toprail when the crossing point of two braces is between 38 inches (0.97 m) and 48 inches (1.3 m) above the work platform. The end points at each upright shall be no more than 48 inches (1.3 m) apart.

**1926.451(h)** ***Falling object protection.*** (1) In addition to wearing hardhats each employee on a scaffold shall be provided with additional protection from falling hand tools, debris, and other small objects through the installation of toeboards, screens, or guardrail systems, or through the erection of debris nets, catch platforms, or canopy structures that contain or deflect the falling objects. When the falling objects are too large, heavy or massive to be contained or deflected by any of the above-listed measures, the employer shall place such potential falling objects away from the edge of the surface from which they could fall and shall secure those materials as necessary to prevent their falling.

**1926.451(h)(2)** Where there is a danger of tools, materials, or equipment falling from a scaffold and striking employees below, the following provisions apply:

**1926.451(h)(2)(i)** The area below the scaffold to which objects can fall shall be barricaded, and employees shall not be permitted to enter the hazard area; or

**1926.451(h)(2)(ii)** A toeboard shall be erected along the edge of platforms more than 10 feet (3.1 m) above lower levels for a distance sufficient to protect employees below, except on float (ship) scaffolds where an edging of 3/4 × 1 1/2 inch (2 × 4 cm) wood or equivalent may be used in lieu of toeboards;

**1926.451(h)(2)(iii)** Where tools, materials, or equipment are piled to a height higher than the top edge of the toeboard, paneling or screening extending from the toeboard or platform to the top of the guardrail shall be erected for a distance sufficient to protect employees below; or

**1926.451(h)(2)(iv)** A guardrail system shall be installed with openings small enough to prevent passage of potential falling objects; or

**1926.451(h)(2)(v)** A canopy structure, debris net, or catch platform strong enough to withstand the impact forces of the potential falling objects shall be erected over the employees below.

**1926.451(h)(3)** Canopies, when used for falling object protection, shall comply with the following criteria:

**1926.451(h)(3)(i)** Canopies shall be installed between the falling object hazard and the employees.

**1926.451(h)(3)(ii)** When canopies are used on suspension scaffolds for falling object protection, the scaffold shall be equipped with additional independent support lines equal in number to the number of points supported, and equivalent in strength to the strength of the suspension ropes.

**1926.451(h)(3)(iii)** Independent support lines and suspension ropes shall not be attached to the same points of anchorage.

**1926.451(h)(4)** Where used, toeboards shall be:

**1926.451(h)(4)(i)** Capable of withstanding, without failure, a force of at least 50 pounds (222 n) applied in any downward or horizontal direction at any point along the toeboard (toeboards built in accordance with Appendix A to this subpart will be deemed to meet this requirement); and

**1926.451(h)(4)(ii)** At least three and one-half inches (9 cm) high from the top edge of the toeboard to the level of the walking/working surface. Toeboards shall be securely fastened in place at the outermost edge of the platform and have not more than 1/4 inch (0.7 cm) clearance above the walking/working surface. Toeboards shall be solid or with openings not over one inch (2.5 cm) in the greatest dimension.

[61 FR 46107, Aug. 30, 1996, as corrected and amended at 61 FR 59831, 59832, Nov. 25, 1996]

**Effective Date Note:** At 61 FR 59832, Nov. 25, 1996, §1926.451(b)(2)(i) was amended and certain requirements stayed until Nov. 25, 1997, or until further rulemaking has been completed, whichever is later.

### §1926.452 Additional requirements applicable to specific types of scaffolds.

In addition to the applicable requirements of §1926.451, the following requirements apply to the specific types of scaffolds indicated. Scaffolds not specifically addressed by §1926.452, such as but not limited to systems scaffolds, must meet the requirements of §1926.451.

**1926.452(a)** ***Pole scaffolds.*** (1) When platforms are being moved to the next level, the existing platform shall be left undisturbed until the new bearers have been set in place and braced, prior to receiving the new platforms.

**1926.452(a)(2)** Crossbracing shall be installed between the inner and outer sets of poles on double pole scaffolds.

**1926.452(a)(3)** Diagonal bracing in both directions shall be installed across the entire inside face of double-pole scaffolds used to support loads equivalent to a uniformly distributed load of 50 pounds (222 kg) or more per square foot (929 square cm).

**1926.452(a)(4)** Diagonal bracing in both directions shall be installed across the entire outside face of all double- and single-pole scaffolds.

**1926.452(a)(5)** Runners and bearers shall be installed on edge.

**1926.452(a)(6)** Bearers shall extend a minimum of 3 inches (7.6 cm) over the outside edges of runners.

**1926.452(a)(7)** Runners shall extend over a minimum of two poles, and shall be supported by bearing blocks securely attached to the poles.

**1926.452(a)(8)** Braces, bearers, and runners shall not be spliced between poles.

**1926.452(a)(9)** Where wooden poles are spliced, the ends shall be squared and the upper section shall rest squarely on the lower section. Wood splice plates shall be provided on at least two adjacent sides, and shall extend at least 2 feet (0.6 m) on either side of the splice, overlap the abutted ends equally, and have at least the same cross-sectional areas as the pole. Splice plates of other materials of equivalent strength may be used.

**1926.452(a)(10)** Pole scaffolds over 60 feet in height shall be designed by a registered professional engineer, and shall be constructed and loaded in accordance with that design. Non-mandatory Appendix A to this subpart contains examples of criteria that will enable an employer to comply with design and loading requirements for pole scaffolds under 60 feet in height.

**1926.452(b)** ***Tube and coupler scaffolds.*** (1) When platforms are being moved to the next level, the existing platform shall be left undisturbed until the new bearers have been set in place and braced prior to receiving the new platforms.

**1926.452(b)(2)** Transverse bracing forming an "X" across the width of the scaffold shall be installed at the scaffold ends and at least at every third set of posts horizontally (measured from only one end) and every fourth runner vertically. Bracing shall extend diagonally from the inner or outer posts or runners upward to the next outer or inner posts or runners. Building ties shall be installed at the bearer levels between the transverse bracing and shall conform to the requirements of §1926.451(c)(1).

**1926.452(b)(3)** On straight run scaffolds, longitudinal bracing across the inner and outer rows of posts shall be installed diagonally in both directions, and shall extend from the base of the end posts upward to the top of the scaffold at approximately a 45 degree angle. On scaffolds whose length is greater than their height, such bracing shall be repeated beginning at least at every fifth post. On scaffolds whose length is less than their height, such bracing shall be installed from the base of the end posts upward to the opposite end posts, and then in alternating directions until reaching the top of the scaffold. Bracing shall be installed as close as possible to the intersection of the bearer and post or runner and post.

**1926.452(b)(4)** Where conditions preclude the attachment of bracing to posts, bracing shall be attached to the runners as close to the post as possible.

**1926.452(b)(5)** Bearers shall be installed transversely between posts, and when coupled to the posts, shall have the inboard coupler bear directly on the runner coupler. When the bearers are coupled to the runners, the couplers shall be as close to the posts as possible.

**1926.452(b)(6)** Bearers shall extend beyond the posts and runners, and shall provide full contact with the coupler.

**1926.452(b)(7)** Runners shall be installed along the length of the scaffold, located on both the inside and outside posts at level heights (when tube and coupler guardrails and midrails are used on outside posts, they may be used in lieu of outside runners).

**1926.452(b)(8)** Runners shall be interlocked on straight runs to form continuous lengths, and shall be coupled to each post. The bottom runners and bearers shall be located as close to the base as possible.

**1926.452(b)(9)** Couplers shall be of a structural metal, such as drop-forged steel, malleable iron, or structural grade aluminum. The use of gray cast iron is prohibited.

**1926.452(b)(10)** Tube and coupler scaffolds over 125 feet in height shall be designed by a registered professional engineer, and shall be constructed and loaded in accordance with such design. Non-mandatory Appendix A to this subpart contains examples of criteria that will enable an employer to comply with design and loading requirements for tube and coupler scaffolds under 125 feet in height.

**1926.452(c)** ***Fabricated frame scaffolds*** (tubular welded frame scaffolds). (1) When moving platforms to the next level, the existing platform shall be left undisturbed until the new end frames have been set in place and braced prior to receiving the new platforms.

**1926.452(c)(2)** Frames and panels shall be braced by cross, horizontal, or diagonal braces, or combination thereof, which secure vertical members together laterally. The cross braces shall be of such length as will automatically square and align vertical members so that the erected scaffold is always plumb, level, and square. All brace connections shall be secured.

**1926.452(c)(3)** Frames and panels shall be joined together vertically by coupling or stacking pins or equivalent means.

**1926.452(c)(4)** Where uplift can occur which would displace scaffold end frames or panels, the frames or panels shall be locked together vertically by pins or equivalent means.

**1926.452(c)(5)** Brackets used to support cantilevered loads shall:

**1926.452(c)(5)(i)** Be seated with side-brackets parallel to the frames and end-brackets at 90 degrees to the frames;

**1926.452(c)(5)(ii)** Not be bent or twisted from these positions; and

**1926.452(c)(5)(iii)** Be used only to support personnel, unless the scaffold has been designed for other loads by a qualified engineer and built to withstand the tipping forces caused by those other loads being placed on the bracket-supported section of the scaffold.

**1926.452(c)(6)** Scaffolds over 125 feet (38.0 m) in height above their base plates shall be designed by a registered professional engineer, and shall be constructed and loaded in accordance with such design.

**1926.452(d)** ***Plasterers', decorators', and large area scaffolds.*** Scaffolds shall be constructed in accordance with paragraphs (a), (b), or (c) of this section, as appropriate.

**1926.452(e)** ***Bricklayers' square scaffolds (squares).*** (1) Scaffolds made of wood shall be reinforced with gussets on both sides of each corner.

**1926.452(e)(2)** Diagonal braces shall be installed on all sides of each square.

**1926.452(e)(3)** Diagonal braces shall be installed between squares on the rear and front sides of the scaffold, and shall extend from the bottom of each square to the top of the next square.

**1926.452(e)(4)** Scaffolds shall not exceed three tiers in height, and shall be so constructed and arranged that one square rests directly above the other. The upper tiers shall stand on a continuous row of planks laid across the next lower tier, and shall be nailed down or otherwise secured to prevent displacement.

**1926.452(f)** ***Horse scaffolds.*** (1) Scaffolds shall not be constructed or arranged more than two tiers or 10 feet (3.0 m) in height, whichever is less.

**1926.452(f)(2)** When horses are arranged in tiers, each horse shall be placed directly over the horse in the tier below.

**1926.452(f)(3)** When horses are arranged in tiers, the legs of each horse shall be nailed down or otherwise secured to prevent displacement.

**1926.452(f)(4)** When horses are arranged in tiers, each tier shall be crossbraced.

**1926.452(g)** ***Form scaffolds and carpenters' bracket scaffolds.*** (1) Each bracket, except those for wooden bracket-form scaffolds, shall be attached to the supporting formwork or structure by means of one or more of the following: nails; a metal stud attachment device; welding; hooking over a secured structural supporting member, with the form wales either bolted to the form or secured by snap ties or tie bolts extending through the form and securely anchored; or, for carpenters' bracket scaffolds only, by a bolt extending through to the opposite side of the structure's wall.

**1926.452(g)(2)** Wooden bracket-form scaffolds shall be an integral part of the form panel.

**1926.452(g)(3)** Folding type metal brackets, when extended for use, shall be either bolted or secured with a locking-type pin.

**1926.452(h)** ***Roof bracket scaffolds.*** (1) Scaffold brackets shall be constructed to fit the pitch of the roof and shall provide a level support for the platform.

**1926.452(h)(2)** Brackets (including those provided with pointed metal projections) shall be anchored in place by nails unless it is impractical to use nails. When nails are not used, brackets shall be secured in place with first-grade manila rope of at least three-fourth inch (1.9 cm) diameter, or equivalent.

**1926.452(i)** ***Outrigger scaffolds.*** (1) The inboard end of outrigger beams, measured from the fulcrum point to the extreme point of anchorage, shall be not less than one and one-half times the outboard end in length.

**1926.452(i)(2)** Outrigger beams fabricated in the shape of an I-beam or channel shall be placed so that the web section is vertical.

**1926.452(i)(3)** The fulcrum point of outrigger beams shall rest on secure bearings at least 6 inches (15.2 cm) in each horizontal dimension.

**1926.452(i)(4)** Outrigger beams shall be secured in place against movement, and shall be securely braced at the fulcrum point against tipping.

**1926.452(i)(5)** The inboard ends of outrigger beams shall be securely anchored either by means of braced struts bearing against sills in contact with the overhead beams or ceiling, or by means of tension members secured to the floor joists underfoot, or by both.

**1926.452(i)(6)** The entire supporting structure shall be securely braced to prevent any horizontal movement.

**1926.452(i)(7)** To prevent their displacement, platform units shall be nailed, bolted, or otherwise secured to outriggers.

**1926.452(i)(8)** Scaffolds and scaffold components shall be designed by a registered professional engineer and shall be constructed and loaded in accordance with such design.

**1926.452(j)** ***Pump jack scaffolds.*** (1) Pump jack brackets, braces, and accessories shall be fabricated from metal plates and angles. Each pump jack bracket shall have two positive gripping mechanisms to prevent any failure or slippage.

**1926.452(j)(2)** Poles shall be secured to the structure by rigid triangular bracing or equivalent at the bottom, top, and other points as necessary. When the pump jack has to pass bracing already installed, an additional brace shall be installed approximately 4 feet (1.2 m) above the brace to be passed, and shall be left in place until the pump jack has been moved and the original brace reinstalled.

**1926.452(j)(3)** When guardrails are used for fall protection, a workbench may be used as the toprail only if it meets all the requirements in paragraphs (g)(4) (ii), (vii), (viii), and (xiii) of §1926.451.

**1926.452(j)(4)** Work benches shall not be used as scaffold platforms.

**1926.452(j)(5)** When poles are made of wood, the pole lumber shall be straight-grained, free of shakes, large loose or dead knots, and other defects which might impair strength.

**1926.452(j)(6)** When wood poles are constructed of two continuous lengths, they shall be joined together with the seam parallel to the bracket.

**1926.452(j)(7)** When two by fours are spliced to make a pole, mending plates shall be installed at all splices to develop the full strength of the member.

**1926.452(k)** ***Ladder jack scaffolds.*** (1) Platforms shall not exceed a height of 20 feet (6.1 m).

**1926.452(k)(2)** All ladders used to support ladder jack scaffolds shall meet the requirements of subpart X of this part—Stairways and Ladders, except that job-made ladders shall not be used to support ladder jack scaffolds.

**1926.452(k)(3)** The ladder jack shall be so designed and constructed that it will bear on the side rails and ladder rungs or on the ladder rungs alone. If bearing on rungs only, the bearing area shall include a length of at least 10 inches (25.4 cm) on each rung.

**1926.452(k)(4)** Ladders used to support ladder jacks shall be placed, fastened, or equipped with devices to prevent slipping.

**1926.452(k)(5)** Scaffold platforms shall not be bridged one to another.

**1926.452(l)** ***Window jack scaffolds.*** (1) Scaffolds shall be securely attached to the window opening.

**1926.452(l)(2)** Scaffolds shall be used only for the purpose of working at the window opening through which the jack is placed.

**1926.452(l)(3)** Window jacks shall not be used to support planks placed between one window jack and another, or for other elements of scaffolding.

**1926.452(m)** ***Crawling boards (chicken ladders).*** (1) Crawling boards shall extend from the roof peak to the eaves when used in connection with roof construction, repair, or maintenance.

**1926.452(m)(2)** Crawling boards shall be secured to the roof by ridge hooks or by means that meet equivalent criteria (e.g., strength and durability).

**1926.452(n)** ***Step, platform, and trestle ladder scaffolds.*** (1) Scaffold platforms shall not be placed any higher than the second highest rung or step of the ladder supporting the platform.

**1926.452(n)(2)** All ladders used in conjunction with step, platform and trestle ladder scaffolds shall meet the pertinent requirements of subpart X of this part—Stairways and Ladders, except that job-made ladders shall not be used to support such scaffolds.

**1926.452(n)(3)** Ladders used to support step, platform, and trestle ladder scaffolds shall be placed, fastened, or equipped with devices to prevent slipping.

**1926.452(n)(4)** Scaffolds shall not be bridged one to another.

**1926.452(o)** ***Single-point adjustable suspension scaffolds.*** (1) When two single-point adjustable suspension scaffolds are combined to form a two-point adjustable suspension scaffold, the resulting two-point scaffold shall comply with the requirements for two-point adjustable suspension scaffolds in paragraph (p) of this section.

**1926.452(o)(2)** The supporting rope between the scaffold and the suspension device shall be kept vertical unless all of the following conditions are met:

**1926.452(o)(2)(i)** The rigging has been designed by a qualified person, and

**1926.452(o)(2)(ii)** The scaffold is accessible to rescuers, and

**1926.452(o)(2)(iii)** The supporting rope is protected to ensure that it will not chafe at any point where a change in direction occurs, and

**1926.452(o)(2)(iv)** The scaffold is positioned so that swinging cannot bring the scaffold into contact with another surface.

**1926.452(o)(3)** Boatswains' chair tackle shall consist of correct size ball bearings or bushed blocks containing safety hooks and properly "eye-spliced" minimum five-eighth (5/8) inch (1.6 cm) diameter first-grade manila rope, or other rope which will satisfy the criteria (e.g., strength and durability) of manila rope.

**1926.452(o)(4)** Boatswains' chair seat slings shall be reeved through four corner holes in the seat; shall cross each other on the underside of the seat; and shall be rigged so as to prevent slippage which could cause an out-of-level condition.

**1926.452(o)(5)** Boatswains' chair seat slings shall be a minimum of five-eight (5/8) inch (1.6 cm) diameter fiber, synthetic, or other rope which will satisfy the criteria (e.g., strength, slip resistance, durability, etc.) of first grade manila rope.

**1926.452(o)(6)** When a heat-producing process such as gas or arc welding is being conducted, boatswains' chair seat slings shall be a minimum of three-eight (3/8) inch (1.0 cm) wire rope.

**1926.452(o)(7)** Non-cross-laminated wood boatswains' chairs shall be reinforced on their underside by cleats securely fastened to prevent the board from splitting.

**1926.452(p)** ***Two-point adjustable suspension scaffolds (swing stages).*** The following requirements do not apply to two-point adjustable suspension scaffolds used as masons' or stonesetters' scaffolds. Such scaffolds are covered by paragraph (q) of this section.

**1926.452(p)(1)** Platforms shall not be more than 36 inches (0.9 m) wide unless designed by a qualified person to prevent unstable conditions.

**1926.452(p)(2)** The platform shall be securely fastened to hangers (stirrups) by U-bolts or by other means which satisfy the requirements of §1926.451(a).

**1926.452(p)(3)** The blocks for fiber or synthetic ropes shall consist of at least one double and one single block. The sheaves of all blocks shall fit the size of the rope used.

**1926.452(p)(4)** Platforms shall be of the ladder-type, plank-type, beam-type, or light-metal type. Light metal-type platforms having a rated capacity of 750 pounds or less and platforms 40 feet (12.2 m) or less in length shall be tested and listed by a nationally recognized testing laboratory.

**1926.452(p)(5)** Two-point scaffolds shall not be bridged or otherwise connected one to another during raising and lowering operations unless the bridge connections are articulated (attached), and the hoists properly sized.

**1926.452(p)(6)** Passage may be made from one platform to another only when the platforms are at the same height, are abutting, and walk-through stirrups specifically designed for this purpose are used.

**1926.452(q)** ***Multi-point adjustable suspension scaffolds, stonesetters' multi-point adjustable suspension scaffolds, and masons' multi-point adjustable suspension scaffolds.*** (1) When two or more scaffolds are used they shall not be bridged one to another unless they are designed to be bridged, the bridge connections are articulated, and the hoists are properly sized.

**1926.452(q)(2)** If bridges are not used, passage may be made from one platform to another only when the platforms are at the same height and are abutting.

**1926.452(q)(3)** Scaffolds shall be suspended from metal outriggers, brackets, wire rope slings, hooks, or means that meet equivalent criteria (e.g., strength, durability).

**1926.452(r)** ***Catenary scaffolds.*** (1) No more than one platform shall be placed between consecutive vertical pickups, and no more than two platforms shall be used on a catenary scaffold.

**1926.452(r)(2)** Platforms supported by wire ropes shall have hook-shaped stops on each end of the platforms to prevent them from slipping off the wire ropes. These hooks shall be so placed that they will prevent the platform from falling if one of the horizontal wire ropes breaks.

**1926.452(r)(3)** Wire ropes shall not be tightened to the extent that the application of a scaffold load will overstress them.

**1926.452(r)(4)** Wire ropes shall be continuous and without splices between anchors.

**1926.452(s)** ***Float (ship) scaffolds.*** (1) The platform shall be supported by a minimum of two bearers, each of which shall project a minimum of 6 inches (15.2 cm) beyond the platform on both sides. Each bearer shall be securely fastened to the platform.

**1926.452(s)(2)** Rope connections shall be such that the platform cannot shift or slip.

**1926.452(s)(3)** When only two ropes are used with each float:

**1926.452(s)(3)(i)** They shall be arranged so as to provide four ends which are securely fastened to overhead supports.

**1926.452(s)(3)(ii)** Each supporting rope shall be hitched around one end of the bearer and pass under the platform to the other end of the bearer where it is hitched again, leaving sufficient rope at each end for the supporting ties.

**1926.452(t)** ***Interior hung scaffolds.*** (1) Scaffolds shall be suspended only from the roof structure or other structural member such as ceiling beams.

**1926.452(t)(2)** Overhead supporting members (roof structure, ceiling beams, or other structural members) shall be inspected and checked for strength before the scaffold is erected.

**1926.452(t)(3)** Suspension ropes and cables shall be connected to the overhead supporting members by shackles, clips, thimbles, or other means that meet equivalent criteria (e.g., strength, durability).

**1926.452(u)** ***Needle beam scaffolds.*** (1) Scaffold support beams shall be installed on edge.

**1926.452(u)(2)** Ropes or hangers shall be used for supports, except that one end of a needle beam scaffold may be supported by a permanent structural member.

**1926.452(u)(3)** The ropes shall be securely attached to the needle beams.

**1926.452(u)(4)** The support connection shall be arranged so as to prevent the needle beam from rolling or becoming displaced.

**1926.452(u)(5)** Platform units shall be securely attached to the needle beams by bolts or equivalent means. Cleats and overhang are not considered to be adequate means of attachment.

**1926.452(v)** ***Multi-level suspended scaffolds.*** (1) Scaffolds shall be equipped with additional independent support lines, equal in number to the number of points supported, and of equivalent strength to the suspension ropes, and rigged to support the scaffold in the event the suspension rope(s) fail.

**1926.452(v)(2)** Independent support lines and suspension ropes shall not be attached to the same points of anchorage.

**1926.452(v)(3)** Supports for platforms shall be attached directly to the support stirrup and not to any other platform.

**1926.452(w)** ***Mobile scaffolds.*** (1) Scaffolds shall be braced by cross, horizontal, or diagonal braces, or combination thereof, to prevent racking or collapse of the scaffold and to secure vertical members together laterally so as to automatically square and align the vertical members. Scaffolds shall be plumb, level, and squared. All brace connections shall be secured.

**1926.452(w)(1)(i)** Scaffolds constructed of tube and coupler components shall also comply with the requirements of paragraph (b) of this section;

**1926.452(w)(1)(ii)** Scaffolds constructed of fabricated frame components shall also comply with the requirements of paragraph (c) of this section.

**1926.452(w)(2)** Scaffold casters and wheels shall be locked with positive wheel and/or wheel and swivel locks, or equivalent means, to prevent movement of the scaffold while the scaffold is used in a stationary manner.

**1926.452(w)(3)** Manual force used to move the scaffold shall be applied as close to the base as practicable, but not more than 5 feet (1.5 m) above the supporting surface.

**1926.452(w)(4)** Power systems used to propel mobile scaffolds shall be designed for such use. Forklifts, trucks, similar motor vehicles or add-on motors shall not be used to propel scaffolds unless the scaffold is designed for such propulsion systems.

**1926.452(w)(5)** Scaffolds shall be stabilized to prevent tipping during movement.

**1926.452(w)(6)** Employees shall not be allowed to ride on scaffolds unless the following conditions exist:

**1926.452(w)(6)(i)** The surface on which the scaffold is being moved is within 3 degrees of level, and free of pits, holes, and obstructions;

**1926.452(w)(6)(ii)** The height to base width ratio of the scaffold during movement is two to one or less, unless the scaffold is designed and constructed to meet or exceed nationally recognized stability test requirements such as those listed in paragraph (x) of Appendix A to this subpart (ANSI/SIA A92.5 and A92.6);

**1926.452(w)(6)(iii)** Outrigger frames, when used, are installed on both sides of the scaffold;

**1926.452(w)(6)(iv)** When power systems are used, the propelling force is applied directly to the wheels, and does not produce a speed in excess of 1 foot per second (.3 mps); and

**1926.452(w)(6)(v)** No employee is on any part of the scaffold which extends outward beyond the wheels, casters, or other supports.

**1926.452(w)(7)** Platforms shall not extend outward beyond the base supports of the scaffold unless outrigger frames or equivalent devices are used to ensure stability.

**1926.452(w)(8)** Where leveling of the scaffold is necessary, screw jacks or equivalent means shall be used.

**1926.452(w)(9)** Caster stems and wheel stems shall be pinned or otherwise secured in scaffold legs or adjustment screws.

**1926.452(w)(10)** Before a scaffold is moved, each employee on the scaffold shall be made aware of the move.

**1926.452(x)** ***Repair bracket scaffolds.*** (1) Brackets shall be secured in place by at least one wire rope at least 1/2 inch (1.27 cm) in diameter.

**1926.452(x)(2)** Each bracket shall be attached to the securing wire rope (or ropes) by a positive locking device capable of preventing the unintentional detachment of the bracket from the rope, or by equivalent means.

**1926.452(x)(3)** Each bracket, at the contact point between the supporting structure and the bottom of the bracket, shall be provided with a shoe (heel block or foot) capable of preventing the lateral movement of the bracket.

**1926.452(x)(4)** Platforms shall be secured to the brackets in a manner that will prevent the separation of the platforms from the brackets and the movement of the platforms or the brackets on a completed scaffold.

**1926.452(x)(5)** When a wire rope is placed around the structure in order to provide a safe anchorage for personal fall arrest systems used by employees erecting or dismantling scaffolds, the wire rope shall meet the requirements of subpart M of this part, but shall be at least 5/16 inch (0.8 cm) in diameter.

**1926.452(x)(6)** Each wire rope used for securing brackets in place or as an anchorage for personal fall arrest systems shall be protected from damage due to contact with edges, corners, protrusions, or other discontinuities of the supporting structure or scaffold components.

**1926.452(x)(7)** Tensioning of each wire rope used for securing brackets in place or as an anchorage for personal fall arrest systems shall be by means of a turnbuckle at least 1 inch (2.54 cm) in diameter, or by equivalent means.

**1926.452(x)(8)** Each turnbuckle shall be connected to the other end of its rope by use of an eyesplice thimble of a size appropriate to the turnbuckle to which it is attached.

**1926.452(x)(9)** U-bolt wire rope clips shall not be used on any wire rope used to secure brackets or to serve as an anchor for personal fall arrest systems.

**1926.452(x)(10)** The employer shall ensure that materials shall not be dropped to the outside of the supporting structure.

**1926.452(x)(11)** Scaffold erection shall progress in only one direction around any structure.

**1926.452(y)** ***Stilts.*** Stilts, when used, shall be used in accordance with the following requirements:

**1926.452(y)(1)** An employee may wear stilts on a scaffold only if it is a large area scaffold.

**1926.452(y)(2)** When an employee is using stilts on a large area scaffold where a guardrail system is used to provide fall protection, the guardrail system shall be increased in height by an amount equal to the height of the stilts being used by the employee.

**1926.452(y)(3)** Surfaces on which stilts are used shall be flat and free of pits, holes and obstructions, such as debris, as well as other tripping and falling hazards.

**1926.452(y)(4)** Stilts shall be properly maintained. Any alteration of the original equipment shall be approved by the manufacturer.

## §1926.453 Aerial lifts.

**1926.453(a)** ***General requirements.*** (1) Unless otherwise provided in this section, aerial lifts acquired for use on or after January 22, 1973 shall be designed and constructed in conformance with the applicable requirements of the American National Standards for "Vehicle Mounted Elevating and Rotating Work Platforms," ANSI A92.2-1969, including appendix. Aerial lifts acquired before January 22, 1973 which do not meet the requirements of ANSI A92.2-1969, may not be used after January 1, 1976, unless they shall have been modified so as to conform with the applicable design and construction requirements of ANSI A92.2-1969. Aerial lifts include the following types of vehicle-mounted aerial devices used to elevate personnel to job-sites above ground:

**1926.453(a)(1)(i)** Extensible boom platforms;

**1926.453(a)(1)(ii)** Aerial ladders;

**1926.453(a)(1)(iii)** Articulating boom platforms;

**1926.453(a)(1)(iv)** Vertical towers; and

**1926.453(a)(1)(v)** A combination of any such devices. Aerial equipment may be made of metal, wood, fiberglass reinforced plastic (FRP), or other material; may be powered or manually operated; and are deemed to be aerial lifts whether or not they are capable of rotating about a substantially vertical axis.

**1926.453(a)(2)** Aerial lifts may be "field modified" for uses other than those intended by the manufacturer provided the modification has been certified in writing by the manufacturer or by any other equivalent entity, such as a nationally recognized testing laboratory, to be in conformity with all applicable provisions of ANSI A92.2-1969 and this section and to be at least as safe as the equipment was before modification.

**1926.453(b)** ***Specific requirements.*** (1) *Ladder trucks and tower trucks.* Aerial ladders shall be secured in the lower traveling position by the locking device on top of the truck cab, and the manually operated device at the base of the ladder before the truck is moved for highway travel.

**1926.453(b)(2)** *Extensible and articulating boom platforms.* (i) Lift controls shall be tested each day prior to use to determine that such controls are in safe working condition.

**1926.453(b)(2)(ii)** Only authorized persons shall operate an aerial lift.

**1926.453(b)(2)(iii)** Belting off to an adjacent pole, structure, or equipment while working from an aerial lift shall not be permitted.

**1926.453(b)(2)(iv)** Employees shall always stand firmly on the floor of the basket, and shall not sit or climb on the edge of the basket or use planks, ladders, or other devices for a work position.

**1926.453(b)(2)(v)** A body belt shall be worn and a lanyard attached to the boom or basket when working from an aerial lift.

**Note to paragraph (b)(2)(v):** As of January 1, 1998, subpart M of this part (§1926.502(d)) provides that body belts are not acceptable as part of a personal fall arrest system. The use of a body belt in a tethering system or in a restraint system is acceptable and is regulated under §1926.502(e).

**1926.453(b)(2)(vi)** Boom and basket load limits specified by the manufacturer shall not be exceeded.

**1926.453(b)(2)(vii)** The brakes shall be set and when outriggers are used, they shall be positioned on pads or a solid surface. Wheel chocks shall be installed before using an aerial lift on an incline, provided they can be safely installed.

**1926.453(b)(2)(viii)** An aerial lift truck shall not be moved when the boom is elevated in a working position with men in the basket, except for equipment which is specifically designed for this type of operation in accordance with the provisions of paragraphs (a) (1) and (2) of this section.

**1926.453(b)(2)(ix)** Articulating boom and extensible boom platforms, primarily designed as personnel carriers, shall have both platform (upper) and lower controls. Upper controls shall be in or beside the platform within easy reach of the operator. Lower controls shall provide for overriding the upper controls. Controls shall be plainly marked as to their function. Lower level controls shall not be operated unless permission has been obtained from the employee in the lift, except in case of emergency.

**1926.453(b)(2)(x)** Climbers shall not be worn while performing work from an aerial lift.

**1926.453(b)(2)(xi)** The insulated portion of an aerial lift shall not be altered in any manner that might reduce its insulating value.

**1926.453(b)(2)(xii)** Before moving an aerial lift for travel, the boom(s) shall be inspected to see that it is properly cradled and outriggers are in stowed position except as provided in paragraph (b)(2)(viii) of this section.

**1926.453(b)(3)** *Electrical tests.* All electrical tests shall conform to the requirements of ANSI A92.2-1969 section 5. However equivalent d.c.; voltage tests may be used in lieu of the a.c. voltage specified in A92.2-1969; d.c. voltage tests which are approved by the equipment manufacturer or equivalent entity shall be considered an equivalent test for the purpose of this paragraph (b)(3).

**1926.453(b)(4)** *Bursting safety factor.* The provisions of the American National Standards Institute standard ANSI A92.2-1969, section 4.9 Bursting Safety Factor shall apply to all critical hydraulic and pneumatic components. Critical components are those in which a failure would result in a free fall or free rotation of the boom. All noncritical components shall have a bursting safety factor of at least 2 to 1.

**1926.453(b)(5)** *Welding standards.* All welding shall conform to the following standards as applicable:

**1926.453(b)(5)(i)** Standard Qualification Procedure, AWS B3.0-41.

**1926.453(b)(5)(ii)** Recommended Practices for Automotive Welding Design, AWS D8.4-61.

**1926.453(b)(5)(iii)** Standard Qualification of Welding Procedures and Welders for Piping and Tubing, AWS D10.9-69.

**1926.453(b)(5)(iv)** Specifications for Welding Highway and Railway Bridges, AWS D2.0-69.

**Note to §1926.453:** Non-mandatory Appendix C to this subpart lists examples of national consensus standards that are considered to provide employee protection equivalent to that provided through the application of ANSI A92.2-1969, where appropriate. This incorporation by reference was approved by the Director of the Federal Register in accordance with 5 U.S.C. 552(a) and 1 CFR part 51. Copies may be obtained from the American National Standards Institute. Copies may be inspected at the Docket Office, Occupational Safety and Health Administration, U.S. Department of Labor, 200 Constitution Avenue, NW., room N2634, Washington, DC or at the Office of the Federal Register, 800 North Capitol Street, NW., suite 700, Washington, DC.

[61 FR 46116, Aug. 30, 1996; 61 FR 59832, Nov. 25, 1996]

## §1926.454 Training requirements.

This section supplements and clarifies the requirements of §1926.21(b)(2) as these relate to the hazards of work on scaffolds.

**1926.454(a)** The employer shall have each employee who performs work while on a scaffold trained by a person qualified in the subject matter to recognize the hazards associated with the type of scaffold being used and to understand the procedures to control or minimize those hazards. The training shall include the following areas, as applicable:

**1926.454(a)(1)** The nature of any electrical hazards, fall hazards and falling object hazards in the work area;

**1926.454(a)(2)** The correct procedures for dealing with electrical hazards and for erecting, maintaining, and disassembling the fall protection systems and falling object protection systems being used;

**1926.454(a)(3)** The proper use of the scaffold, and the proper handling of materials on the scaffold;

**1926.454(a)(4)** The maximum intended load and the load-carrying capacities of the scaffolds used; and

**1926.454(a)(5)** Any other pertinent requirements of this subpart.

**1926.454(b)** The employer shall have each employee who is involved in erecting, disassembling, moving, operating, repairing, maintaining, or inspecting a scaffold trained by a competent person to recognize any hazards associated with the work in question. The training shall include the following topics, as applicable:

**1926.454(b)(1)** The nature of scaffold hazards;

**1926.454(b)(2)** The correct procedures for erecting, disassembling, moving, operating, repairing, inspecting, and maintaining the type of scaffold in question;

**1926.454(b)(3)** The design criteria, maximum intended load-carrying capacity and intended use of the scaffold;

**1926.454(b)(4)** Any other pertinent requirements of this subpart.

**1926.454(c)** When the employer has reason to believe that an employee lacks the skill or understanding needed for safe work involving the erection, use or dismantling of scaffolds, the employer shall retrain each such employee so that the requisite proficiency is regained. Retraining is required in at least the following situations:

**1926.454(c)(1)** Where changes at the worksite present a hazard about which an employee has not been previously trained; or

**1926.454(c)(2)** Where changes in the types of scaffolds, fall protection, falling object protection, or other equipment present a hazard about which an employee has not been previously trained; or

**1926.454(c)(3)** Where inadequacies in an affected employee's work involving scaffolds indicate that the employee has not retained the requisite proficiency.

## Non-Mandatory Appendices

### (Non-mandatory) Appendix A to Subpart L—Scaffold Specifications

This Appendix provides non-mandatory guidelines to assist employers in complying with the requirements of subpart L of this part. An employer may use these guidelines and tables as a starting point for designing scaffold systems. However, the guidelines do not provide all the information necessary to build a complete system, and the employer is still responsible for designing and assembling these components in such a way that the completed system will meet the requirements of §1926.451(a). Scaffold components which are not selected and loaded in accordance with this Appendix, and components for which no specific guidelines or tables are given in this Appendix (e.g., joints, ties, components for wood pole scaffolds more than 60 feet in height, components for heavy-duty horse scaffolds, components made with other materials, and components with other dimensions, etc.) must be designed and constructed in accordance with the capacity requirements of §1926.451(a), and loaded in accordance with §1926.451(d)(1).

### Index to Appendix A for Subpart L

1. General guidelines and tables.

2. Specific guidelines and tables.

(a) Pole scaffolds:

Single-pole wood pole scaffolds.

Independent wood pole scaffolds.

(b) Tube and coupler scaffolds.

(c) Fabricated frame scaffolds.

(d) Plasterers', decorators' and large area scaffolds.

(e) Bricklayers' square scaffolds.

(f) Horse scaffolds.

(g) Form scaffolds and carpenters' bracket scaffolds.

(h) Roof bracket scaffolds.

(i) Outrigger scaffolds (one level).

(j) Pump jack scaffolds.

(k) Ladder jack scaffolds.

(l) Window jack scaffolds.

(m) Crawling boards (chicken ladders).

(n) Step, platform and trestle ladder scaffolds.

(o) Single-point adjustable suspension scaffolds.

(p) Two-point adjustable suspension scaffolds.

(q)(1) Stonesetters' multi-point adjustable suspension scaffolds.

(2) Masons' multi-point adjustable suspension scaffolds.

(r) Catenary scaffolds.

(s) Float (ship) scaffolds.

(t) Interior hung scaffolds.

(u) Needle beam scaffolds.

(v) Multi-level suspension scaffolds.

(w) Mobile scaffolds.

(x) Repair bracket scaffolds.

(y) Stilts.

(z) Tank builders' scaffolds.

## 1. GENERAL GUIDELINES AND TABLES

(a) The following tables, and the tables in Part 2—Specific guidelines and tables, assume that all load-carrying timber members (except planks) of the scaffold are a minimum of 1,500 lb-f/in$^2$ (stress grade) construction grade lumber. All dimensions are nominal sizes as provided in the American Softwood Lumber Standards, dated January 1970, except that, where rough sizes are noted, only rough or undressed lumber of the size specified will satisfy minimum requirements.

(b) Solid sawn wood used as scaffold planks shall be selected for such use following the grading rules established by a recognized lumber grading association or by an independent lumber grading inspection agency. Such planks shall be identified by the grade stamp of such association or agency. The association or agency and the grading rules under which the wood is graded shall be certified by the Board of Review, American Lumber Standard Committee, as set forth in the American Softwood Lumber Standard of the U.S. Department of Commerce.

(i) Allowable spans shall be determined in compliance with the National Design Specification for Wood Construction published by the National Forest Products Association; paragraph 5 of ANSI A10.8-1988 Scaffolding-Safety Requirements published by the American National Standards Institute; or for 2 x 10 inch (nominal) or 2 x 9 inch (rough) solid sawn wood planks, as shown in the following table:

| Maximum intended nominal load (lb/ft$^2$) | Maximum permissible span using full thickness undressed lumber (ft) | Maximum permissible span using nominal thickness lumber (ft) |
|---|---|---|
| 25 | 10 | 8 |
| 50 | 8 | 6 |
| 75 | 6 | |

(ii) The maximum permissible span for 1 1/4 x 9-inch or wider wood plank of full thickness with a maximum intended load of 50 lb/ft.$^2$ shall be 4 feet.

(c) Fabricated planks and platforms may be used in lieu of solid sawn wood planks. Maximum spans for such units shall be as recommended by the manufacturer based on the maximum intended load being calculated as follows:

| Rated load capacity | Intended load |
|---|---|
| Light-duty | • 25 pounds per square foot applied uniformly over the entire span area. |
| Medium-duty | • 50 pounds per square foot applied uniformly over the entire span area. |
| Heavy-duty | • 75 pounds per square foot applied uniformly over the entire span area. |
| One-person | • 250 pounds placed at the center of the span (total 250 pounds). |
| Two-person | • 250 pounds placed 18 inches to the left and right of the center of the span (total 500 pounds). |
| Three-person | • 250 pounds placed at the center of the span and 250 pounds placed 18 inches to the left and right of the center of the span (total 750 pounds). |

**Note:** Platform units used to make scaffold platforms intended for light-duty use shall be capable of supporting at least 25 pounds per square foot applied uniformly over the entire unit-span area, or a 250-pound point load placed on the unit at the center of the span, whichever load produces the greater shear force.

(d) Guardrails shall be as follows:

(i) Toprails shall be equivalent in strength to 2 inch by 4 inch lumber; or

1 1/4 inch × 1/8 inch structural angle iron; or

1 inch × .070 inch wall steel tubing; or 1.990 inch × .058 inch wall aluminum tubing.

(ii) Midrails shall be equivalent in strength to 1 inch by 6 inch lumber; or

1 1/4 inch × 1 1/4 inch × 1/8 inch structural angle iron; or

1 inch × .070 inch wall steel tubing; or

1.990 inch × .058 inch wall aluminum tubing.

(iii) Toeboards shall be equivalent in strength to 1 inch by 4 inch lumber; or

1 1/4 inch × 1 1/4 inch structural angle iron; or

1 inch × .070 inch wall steel tubing; or

1.990 inch × .058 inch wall aluminum tubing.

(iv) Posts shall be equivalent in strength to 2 inch by 4 inch lumber; or

1 1/4 inch × 1 1/4 inch × 1/8 structural angle iron; or

1 inch × .070 inch wall steel tubing; or

1.990 inch × .058 inch wall aluminum tubing.

(v) Distance between posts shall not exceed 8 feet.

(e) Overhead protection shall consist of 2 inch nominal planking laid tight, or 3/4-inch plywood.

(f) Screen installed between toeboards and midrails or toprails shall consist of No. 18 gauge U.S. Standard wire one inch mesh.

## 2. SPECIFIC GUIDELINES AND TABLES.

(a) Pole Scaffolds.

## SINGLE WOOD POLE SCAFFOLDS

| | Light duty up to 20 feet high | Light duty up to 60 feet high | Medium duty up to 60 feet high | Heavy duty up to 60 feet high |
|---|---|---|---|---|
| Maximum intended load (lbs/ft$^2$) | 25 | 25 | 50 | 75 |
| Poles or uprights | 2 × 4 in | 4 × 4 in | 4 × 4 in | 4 × 6 in. |
| Maximum pole spacing (longitudinal) | 6 feet | 10 feet | 8 feet | 6 feet |
| Maximum pole spacing (transverse) | 5 feet | 5 feet | 5 feet | 5 feet |
| Runners | 1 × 4 in | 1 ¼ × 9 in | 2 × 10 in | 2 × 10 in. |
| Bearers and maximum spacing of bearers: | | | | |
| 3 feet | 2 × 4 in | 2 × 4 in | 2 × 10 in. or 3 × 4 in | 2 × 10 in. or 3 × 5 in. |
| 5 feet | 2 × 6 in. or 3 × 4 in | 2 × 6 in. or 3 × 4 in. (rough). | 2 × 10 in. or 3 × 4 in | 2 × 10 in. or 3 × 5 in. |
| 6 feet | | | 2 × 10 in. or 3 × 4 in | 2 × 10 in. or 3 × 5 in. |
| 8 feet | | | 2 × 10 in. or 3 × 4 in | |
| Planking | 1 ¼ × 9 in | 2 × 10 in | 2 × 10 in | 2 × 10 in. |
| Maximum vertical spacing of horizontal members. | 7 feet | 9 feet | 7 feet | 6 ft. 6 in. |
| Bracing horizontal | 1 × 4 in | 1 × 4 in | 1 × 6 in. or 1 ¼ × 4 in | 2 × 4 in. |
| Bracing diagonal | 1 × 4 in | 1 × 4 in | 1 × 4 in | 2 × 4 in. |
| Tie-ins | 1 × 4 in | 1 × 4 in | 1 × 4 in | 1 × 4 in. |

**Note:** All members except planking are used on edge. All wood bearers shall be reinforced with 3/16 × 2 inch steel strip, or the equivalent, secured to the lower edges for the entire length of the bearer.

## INDEPENDENT WOOD POLE SCAFFOLDS

| | Light duty up to 20 feet high | Light duty up to 60 feet high | Medium duty up to 60 feet high | Heavy duty up to 60 feet high |
|---|---|---|---|---|
| Maximum intended load | 25 lbs/ft$^2$ | 25 lbs/ft$^2$ | 50 lbs/ft$^2$ | 75 lbs/ft$^2$. |
| Poles or uprights | 2 × 4 in | 4 × 4 in | 4 × 4 in | 4 × 4 in. |
| Maximum pole spacing (longitudinal) | 6 feet | 10 feet | 8 feet | 6 feet. |
| Maximum (transverse) | 6 feet | 10 feet | 8 feet | 8 feet. |
| Runners | 1 ¼ × 4 in | 1 ¼ × 9 in | 2 × 10 in | 2 × 10 in. |
| Bearers and maximum spacing of bearers: | | | | |
| 3 feet | 2 × 4 in | 2 × 4 in | 2 × 10 in | 2 × 10 in. (rough). |
| 6 feet | 2 × 6 in. or 3 × 4 in | 2 × 10 in. (rough) or 3 × 8 in. | 2 × 10 in | 2 × 10 in. (rough). |
| 8 feet | 2 × 6 in. or 3 × 4 in | 2 × 10 in. (rough) or 3 × 8 in. | 2 × 10 in | |
| 10 feet | 2 × 6 in. or 3 × 4 in | 2 × 10 in.—(rough) or 3 × 3 in. | | |
| Planking | 1 ¼ × 9 in | 2 × 10 in | 2 × 10 in | 2 × 10 in. |
| Maximum vertical spacing of horizontal members. | 7 feet | 7 feet | 6 feet | 6 feet. |
| Bracing horizontal | 1 × 4 in | 1 × 4 in | 1 × 6 in. or 1 ¼ × 4 in | 2 × 4 in. |
| Bracing diagonal | 1 × 4 in | 1 × 4 in | 1 × 4 in | 2 × 4 in. |
| Tie-ins | 1 × 4 in | 1 × 4 in | 1 × 4 in | 1 × 4 in. |

**Note:** All members except planking are used on edge. All wood bearers shall be reinforced with 3/16 × 2 inch steel strip, or the equivalent, secured to the lower edges for the entire length of the bearer.

(b) ***Tube and coupler scaffolds.***

## MINIMUM SIZE OF MEMBERS

| | Light duty | Medium duty | Heavy duty |
|---|---|---|---|
| Maximum intended load | 25 lbs/ft$^2$ | 50 lbs/ft$^2$ | 75 lbs/ft$^2$. |
| Posts, runners and braces | Nominal 2 in. (1.90 inches) OD steel tube or pipe. | Nominal 2 in. (1.90 inches) OD steel tube or pipe. | Nominal 2 in. (1.90 inches) OD steel tube or pipe. |
| Bearers | Nominal 2 in. (1.90 inches OD steel tube or pipe and a maximum post spacing of 4 ft. × 10 ft. | Nominal 2. in. (.190 inches) OD steel tube or pipe and a maximum post spacing of 4 ft. × 7 ft. or Nominal 2 ½ in. (2.375 in.). OD steel tube or pipe and a maximum post spacing of 6 ft. × 8 ft.*. | Nominal 2 ½ in. (2.375 in.). OD steel tube or pipe and a maximum post spacing of 6 ft. × 6 ft. |
| Maximum runner spacing vertically | 6 ft. 6 in | 6 ft. 6 in | 6 ft. 6 in. |

* Bearers shall be installed in the direction of the shorter dimension.

**Note:** Longitudinal diagonal bracing shall be installed at an angle of 45° (± 5 °).

## MAXIMUM NUMBER OF PLANKED LEVELS

| | Maximum number of additional planked levels | | | Maximum height of scaffold (in feet) |
|---|---|---|---|---|
| | Light duty | Medium duty | Heavy duty | |
| Number of Working Levels: | | | | |
| 1 .................. | 16 | 11 | 6 | 125 |
| 2 .................. | 11 | 1 | 0 | 125 |
| 3 .................. | 6 | 0 | 0 | 125 |
| 4 .................. | 1 | 0 | 0 | 125 |

(c) *Fabricated frame scaffolds.* Because of their prefabricated nature, no additional guidelines or tables for these scaffolds are being adopted in this appendix.

(d) *Plasterers', decorators', and large area scaffolds.* The guidelines for pole scaffolds or tube and coupler scaffolds (Appendix A (a) and (b)) may be applied.

(e) *Bricklayers' square scaffolds.*

Maximum intended load: 50 lb/ft.$^2$*

Maximum width: 5 ft.

Maximum height: 5 ft.

Gussets: 1 × 6 in.

Braces: 1 × 8 in.

Legs: 2 × 6 in.

Bearers (horizontal members): 2 × 6 in.

*(f) Horse scaffolds.*

Maximum intended load (light duty): 25 lb/ft.$^2$**

Maximum intended load (medium duty): 50 lb/ft.$^2$**

Horizontal members or bearers:

Light duty: 2 × 4 in.

Medium duty: 3 × 4 in.

Legs: 2 × 4 in.

Longitudinal brace between legs: 1 × 6 in.

Gusset brace at top of legs: 1 × 8 in.

Half diagonal braces: 2 × 4 in.

(g) *Form scaffolds and carpenters' bracket scaffolds.*

(1) Brackets shall consist of a triangular shaped frame made of wood with a cross-section not less than 2 inches by 3 inches, or of 1 1/4 inch × 1 1/4 inch × 1/8 inch structural angle iron.

(2) Bolts used to attach brackets to structures shall not be less than 5/8 inches in diameter.

(3) Maximum bracket spacing shall be 8 feet on centers.

(4) No more than two employees shall occupy any given 8 feet of a bracket or form scaffold at any one time. Tools and materials shall not exceed 75 pounds in addition to the occupancy.

(5) Wooden figure-four scaffolds:

Maximum intended load: 25 lb/ft.$^2$

Uprights: 2 × 4 in. or 2 × 6 in.

Bearers (two): 1 × 6 in.

Braces: 1 × 6 in.

Maximum length of bearers (unsupported): 3 ft. 6 in.

(i) Outrigger bearers shall consist of two pieces of 1 × 6 inch lumber nailed on opposite sides of the vertical support.

(ii) Bearers for wood figure-four brackets shall project not more than 3 feet 6 inches from the outside of the form support, and shall be braced and secured to prevent tipping or turning. The knee or angle brace shall intersect the bearer at least 3 feet from the form at an angle of approximately 45 degrees, and the lower end shall be nailed to a vertical support.

(6) Metal bracket scaffolds:

Maximum intended load: 25 lb/ft.$^2$

Uprights: 2 × 4 inch

Bearers: As designed.

Braces: As designed.

(7) Wood bracket scaffolds:

Maximum intended load: 25 lb/ft.$^2$

Uprights: 2 × 4 in or 2 × 6 in

Bearers: 2 × 6 in

Maximum scaffold width: 3 ft 6 in

Braces: 1 × 6 in

(h) *Roof bracket scaffolds.* No specific guidelines or tables are given.

(i) *Outrigger scaffolds (single level).* No specific guidelines or tables are given.

(j) *Pump jack scaffolds.* Wood poles shall not exceed 30 feet in height. Maximum intended load—500 lbs between poles; applied at the center of the span. Not more than two employees shall be on a pump jack scaffold at one time between any two supports. When 2 × 4's are spliced together to make a 4 × 4 inch wood pole, they shall be spliced with "10 penny" common nails no more than 12 inches center to center, staggered uniformly from the opposite outside edges.

(k) *Ladder jack scaffolds.* Maximum intended load—25 lb/ft$^2$. However, not more than two employees shall occupy any platform at any one time. Maximum span between supports shall be 8 feet.

(l) *Window jack scaffolds.* Not more than one employee shall occupy a window jack scaffold at any one time.

(m) *Crawling boards (chicken ladders).* Crawling boards shall be not less than 10 inches wide and 1 inch thick, with cleats having a minimum 1 × 1 1/2 inch cross-sectional area. The cleats shall be equal in length to the width of the board and spaced at equal intervals not to exceed 24 inches.

(n) *Step, platform, and trestle ladder scaffolds.* No additional guidelines or tables are given.

(o) *Single-point adjustable suspension scaffolds.* Maximum intended load—250 lbs. Wood seats for boatswains' chairs shall be not less than 1 inch thick if made of non-laminated wood, or 5/8 inches thick if made of marine quality plywood.

(p) *Two-point adjustable suspension scaffolds.* (1) In addition to direct connections to buildings (except window cleaners' anchors) acceptable ways to prevent scaffold sway include angulated roping and static lines. Angulated roping is a system of platform suspension in which the upper wire rope sheaves or suspension points are closer to the plane of the building face than the corresponding attachment points on the platform, thus causing the platform to press against the face of the building. Static lines are separate ropes secured at their top and bottom ends closer to the plane of the building face than the outermost edge of the platform. By drawing the static line taut, the platform is drawn against the face of the building.

(2) On suspension scaffolds designed for a working load of 500 pounds, no more than two employees shall be permitted on the scaffold at one time. On suspension scaffolds with a working load of 750 pounds, no more than three employees shall be permitted on the scaffold at one time.

(3) Ladder-type platforms. The side stringer shall be of clear straight-grained spruce. The rungs shall be of straight-grained oak, ash, or hickory, at least 1 1/8 inches in diameter, with 7/8 inch tenons mortised into the side stringers at least 7/8 inch. The stringers shall be tied together with tie rods not less than 1/4 inch in diameter, passing through the stringers and riveted up tight against washers on both ends. The flooring strips shall be spaced not more than 5/8 inch apart, except at the side rails where the space may be 1 inch. Ladder-type platforms shall be constructed in accordance with the following table:

* The squares shall be set not more than 8 feet apart for light duty scaffolds and not more than 5 feet apart for medium duty scaffolds.

** Horses shall be spaced not more than 8 feet apart for light duty loads, and not more than 5 feet apart for medium duty loads.

## SCHEDULE FOR LADDER-TYPE PLATFORMS

| Length of Platform | 12 feet | 14 & 16 feet | 18 & 20 feet. |
|---|---|---|---|
| Side stringers, minimum cross section (finished sizes): | | | |
| At ends | 1 3/4 × 2 3/4 in | 1 3/4 × 2 3/4 in | 1 3/4 × 3 in. |
| At middle | 1 3/4 × 3 3/4 in | 1 3/4 × 3 3/4 in | 1 3/4 × 4 in. |
| Reinforcing strip (minimum) | A 1/8 × 7/8 inch steel reinforcing strip shall be attached to the side or underside, full length. | | |
| Rungs | Rungs shall be 1 1/8 inch minimum diameter with at least 7/8 inch in diameter tenons, and the maximum spacing shall be 12 inches to center. | | |
| Tie rods: | | | |
| Number (minimum) | 3 | 4 | 4 |
| Diameter (minimum) | 1/4 inch | 1/4 inch | 1/4 inch |
| Flooring, minimum finished size | 1/2 × 2 3/4 in | 1/2 × 2 3/4 in | 1/2 × 2 3/4 in. |

## SCHEDULE FOR LADDER-TYPE PLATFORMS

| Length of Platform | 22 & 24 ft | 28 & 30 ft. |
|---|---|---|
| Side stringers, minimum cross section (finished sizes): | | |
| At ends | 1 3/4 × 3 in | 1 3/4 × 3 1/2 in. |
| At middle | 1 3/4 × 4 1/4 in | 1 3/4 × 5 in. |
| Reinforcing strip (minimum). | A 1/8 × 7/8 -inch steel reinforcing strip shall be attached to the side or underside, full length. | |
| Rungs | Rungs shall be 1 1/8 inch minimum diameter with at least 7/8 inch in diameter tenons, and the maximum spacing shall be 12 inches to center. Tie rods. | |
| Number (minimum) | 5 | 6. |
| Diameter (minimum) | 1/4 in | 1/4 in. |
| Flooring, minimum finished size. | 1/2 × 2 3/4 in. | 1/2 × 2 3/4 in. |

(4) Plank-Type Platforms. Plank-type platforms shall be composed of not less than nominal 2 × 8 inch unspliced planks, connected together on the underside with cleats at intervals not exceeding 4 feet, starting 6 inches from each end. A bar or other effective means shall be securely fastened to the platform at each end to prevent the platform from slipping off the hanger. The span between hangers for plank-type platforms shall not exceed 10 feet.

(5) Beam-Type Platforms. Beam platforms shall have side stringers of lumber not less than 2 × 6 inches set on edge. The span between hangers shall not exceed 12 feet when beam platforms are used. The flooring shall be supported on 2 × 6 inch cross beams, laid flat and set into the upper edge of the stringers with a snug fit, at intervals of not more than 4 feet, securely nailed to the cross beams. Floor-boards shall not be spaced more than 1/2 inch apart.

(1) *Multi-point adjustable suspension scaffolds and stonesetters' multi-point adjustable suspension scaffolds.* No specific guidelines or tables are given for these scaffolds.

(2) *Masons' multi-point adjustable suspension scaffolds.* Maximum intended load—50 lb/ft$^2$. Each outrigger beam shall be at least a standard 7 inch, 15.3 pound steel I-beam, at least 15 feet long. Such beams shall not project more than 6 feet 6 inches beyond the bearing point. Where the overhang exceeds 6 feet 6 inches, outrigger beams shall be composed of stronger beams or multiple beams.

(r) *Catenary scaffolds.* (1) Maximum intended load—500 lbs.

(2) Not more than two employees shall be permitted on the scaffold at one time.

(3) Maximum capacity of come-along shall be 2,000 lbs.

(4) Vertical pickups shall be spaced not more than 50 feet apart.

(5) Ropes shall be equivalent in strength to at least 1/2 inch (1.3 cm) diameter improved plow steel wire rope.

(s) *Float (ship) scaffolds.* (1) Maximum intended load—750 lbs.

(2) Platforms shall be made of 3/4 inch plywood, equivalent in rating to American Plywood Association Grade B-B, Group I, Exterior.

(3) Bearers shall be made from 2×4 inch, or 1×10 inch rough lumber. They shall be free of knots and other flaws.

(4) Ropes shall be equivalent in strength to at least 1 inch (2.5 cm) diameter first grade manila rope.

(t) *Interior hung scaffolds.*

Bearers (use on edge): 2×10 in.

Maximum intended load: Maximum span

25 lb/ft.$^2$: 10 ft.

50 lb/ft.$^2$: 10 ft.

75 lb/ft.$^2$: 7 ft.

(u) *Needle beam scaffolds.*

Maximum intended load: 25 lb/ft.$^2$

Beams: 4×6 in.

Maximum platform span: 8 ft.

Maximum beam span: 10 ft.

(1) Ropes shall be attached to the needle beams by a scaffold hitch or an eye splice. The loose end of the rope shall be tied by a bowline knot or by a round turn and a half hitch.

(2) Ropes shall be equivalent in strength to at least 1 inch (2.5 cm) diameter first grade manila rope.

(v) *Multi-level suspension scaffolds.* No additional guidelines or tables are being given for these scaffolds.

(w) *Mobile Scaffolds.* Stability test as described in the ANSI A92 series documents, as appropriate for the type of scaffold, can be used to establish stability for the purpose of §1926.452(w)(6).

(x) *Repair bracket scaffolds.* No additional guidelines or tables are being given for these scaffolds.

(y) *Stilts.* No specific guidelines or tables are given.

(z) *Tank builder's scaffold.*

(1) The maximum distance between brackets to which scaffolding and guardrail supports are attached shall be no more than 10 feet 6 inches.

(2) Not more than three employees shall occupy a 10 feet 6 inch span of scaffold planking at any time.

(3) A taut wire or synthetic rope supported on the scaffold brackets shall be installed at the scaffold plank level between the innermost edge of the scaffold platform and the curved plate structure of the tank shell to serve as a safety line in lieu of an inner guardrail assembly where the space between the scaffold platform and the tank exceeds 12 inches (30.48 cm). In the event the open space on either side of the rope exceeds 12 inches (30.48 cm), a second wire or synthetic rope appropriately placed, or guardrails in accordance with §1926.451(e)(4), shall be installed in order to reduce that open space to less than 12 inches (30.48 cm).

(4) Scaffold planks of rough full-dimensioned 2-inch (5.1 cm)×12-inch (30.5 cm) Douglas Fir or Southern Yellow Pine of Select Structural Grade shall be used. Douglas Fir planks shall have a fiber stress of at least 1900 lb/in$^2$ (130,929 n/cm$^2$) and a modulus of elasticity of at least 1,900,000 lb/in$^2$ (130,929,000 n/cm$^2$), while Yellow Pine planks shall have a fiber stress of at least 2500 lb/in$^2$ (172,275 n/cm$^2$) and a modulus of elasticity of at least 2,000,000 lb/in$^2$ (137,820,000 n/cm $^2$).

(5) Guardrails shall be constructed of a taut wire or synthetic rope, and shall be supported by angle irons attached to brackets welded to the steel plates. These guardrails shall comply with §1926.451(e)(4). Guardrail supports shall be located at no greater than 10 feet 6 inch intervals.

### Non-Mandatory Appendix B to Subpart L—Criteria for Determining the Feasibility of Providing Safe Access and Fall Protection for Scaffold Erectors and Dismantlers [Reserved]

### Non-Mandatory Appendix C to Subpart L—List of National Consensus Standards

ANSI/SIA A92.2-1990 Vehicle-Mounted Elevating and Rotating Aerial Devices

ANSI/SIA A92.3-1990 Manually Propelled Elevating Aerial Platforms

ANSI/SIA A92.5-1990 Boom Supported Elevating Work Platforms

ANSI/SIA A92.6-1990 Self-Propelled Elevating Work Platforms

ANSI/SIA A92.7-1990 Airline Ground Support Vehicle-Mounted Vertical Lift Devices

ANSI/SIA A92.8-1993 Vehicle-Mounted Bridge Inspection and Maintenance Devices

ANSI/SIA A92.9-1993 Mast-Climbing Work Platforms

### Non-Mandatory Appendix D to Subpart L—List of Training Topics for Scaffold Erectors and Dismantlers

This Appendix D is provided to serve as a guide to assist employers when evaluating the training needs of employees erecting or dismantling supported scaffolds.

The Agency believes that employees erecting or dismantling scaffolds should be trained in the following topics:

- *General Overview of Scaffolding*
  - regulations and standards
  - erection/dismantling planning
  - PPE and proper procedures
  - fall protection
  - materials handling
  - access
  - working platforms
  - foundations
  - guys, ties and braces
- *Tubular Welded Frame Scaffolds*
  - specific regulations and standards
  - components
  - parts inspection
  - erection/dismantling planning
  - guys, ties and braces
  - fall protection
  - general safety
  - access and platforms
  - erection/dismantling procedures
  - rolling scaffold assembly
  - putlogs
- *Tube and Clamp Scaffolds*
  - specific regulations and standards
  - components
  - parts inspection
  - erection/dismantling planning
  - guys, ties and braces
  - fall protection
  - general safety
  - access and platforms
  - erection/dismantling procedures
  - buttresses, cantilevers, & bridges
- *System Scaffolds*
  - specific regulations and standards
  - components
  - parts inspection
  - erection/dismantling planning
  - guys, ties and braces
  - fall protection
  - general safety
  - access and platforms
  - erection/dismantling procedures
  - buttresses, cantilevers, & bridges

Scaffold erectors and dismantlers should all receive the general overview, and, in addition, specific training for the type of supported scaffold being erected or dismantled.

### (Non-Mandatory) Appendix E to Subpart L—Drawings and Illustrations

This Appendix provides drawings of particular types of scaffolds and scaffold components, and graphic illustrations of bracing patterns and tie spacing patterns.

This Appendix is intended to provide visual guidance to assist the user in complying with the requirements of subpart L, part 1926.

## BRACING—TUBE & COUPLER SCAFFOLDS

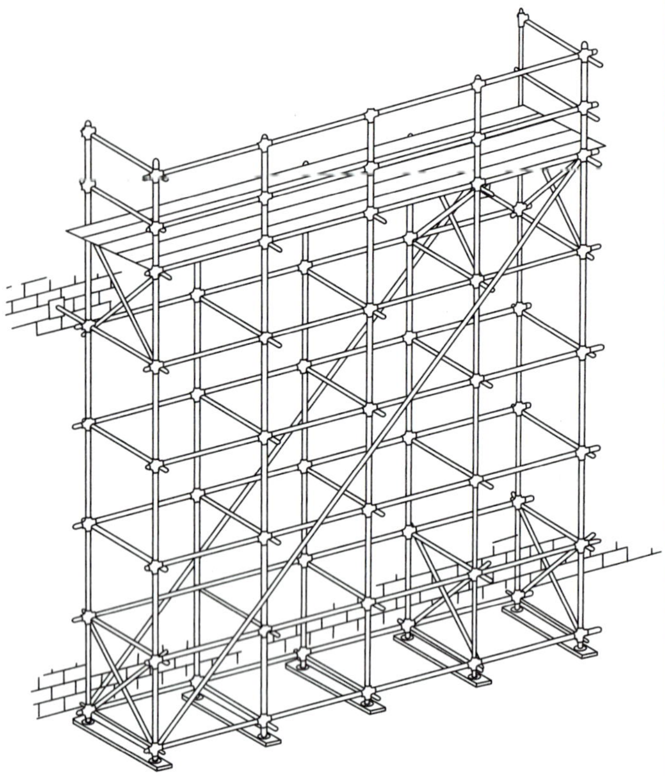

## MAXIMUM VERTICAL TIE SPACING WIDER THAN 3'–0" BASES

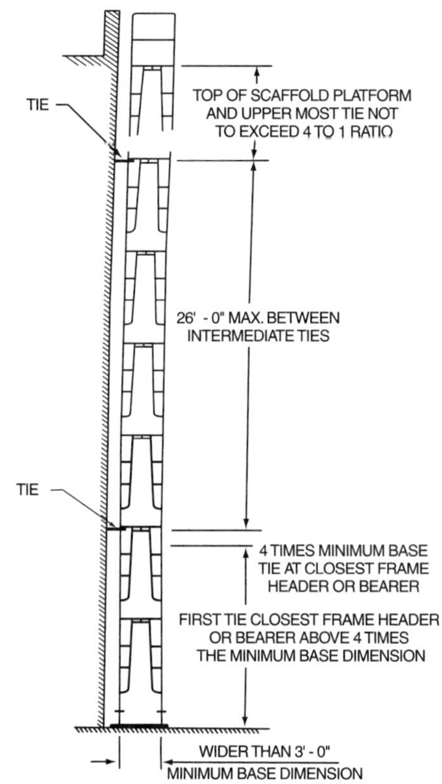


## SUSPENDED SCAFFOLD PLATFORM WELDING PRECAUTIONS

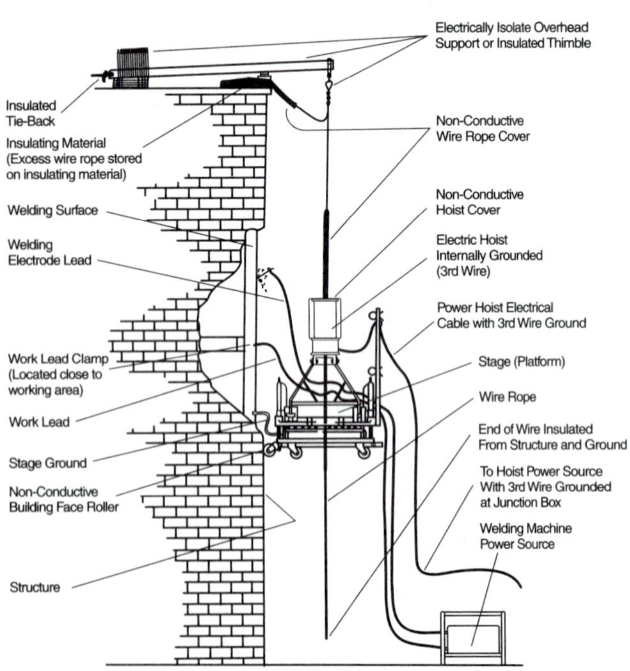


**HOISTS MUST BE ELECTRONICALLY ISOLATED FROM SCAFFOLD**

## MAXIMUM VERTICAL TIE SPACING 3'–0" AND NARROWER BASES

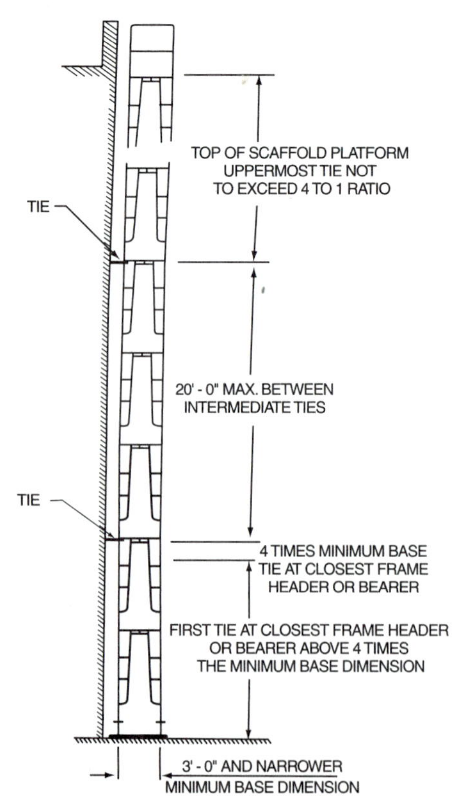

**SYSTEM SCAFFOLD**

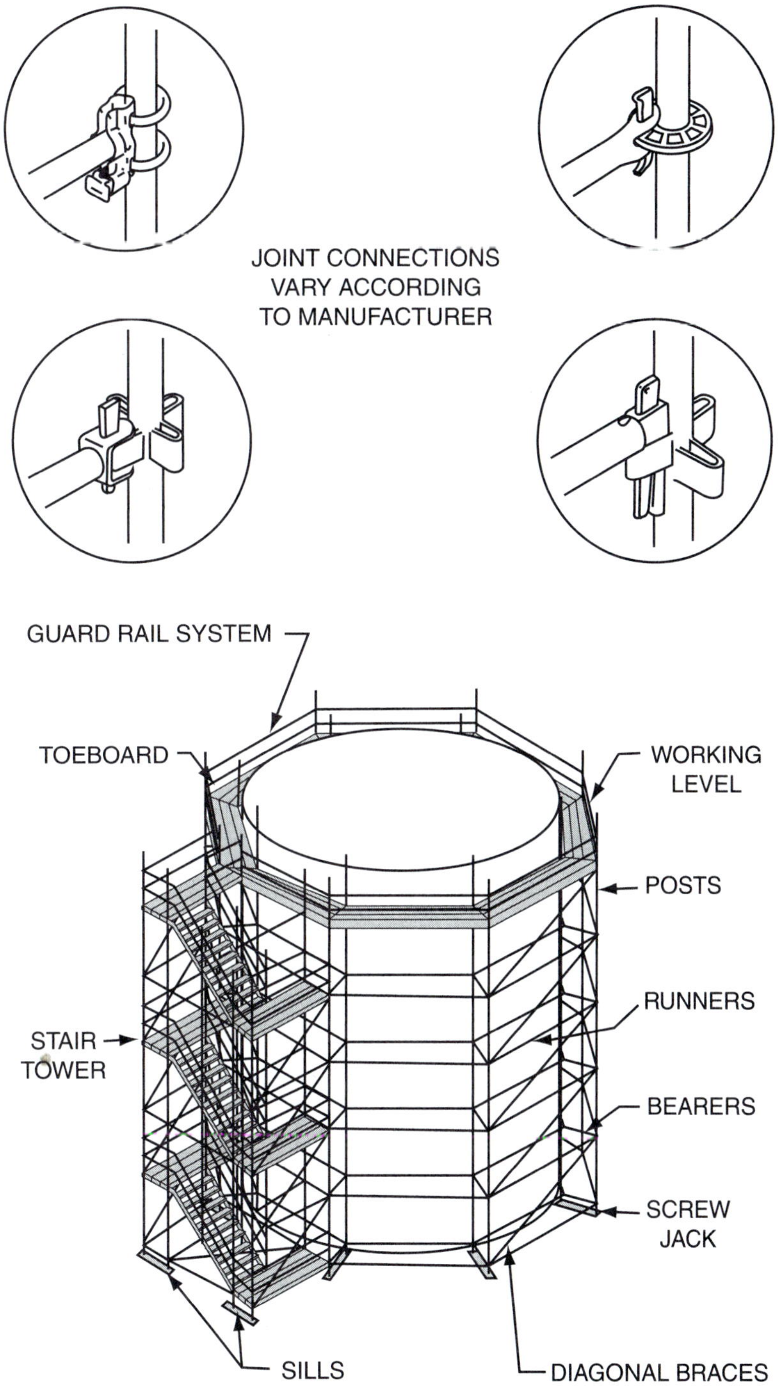

Grade stamp courtesy of Southern Pine Inspection Bureau

Grade stamp courtesy of West Coast Lumber Inspection Bureau

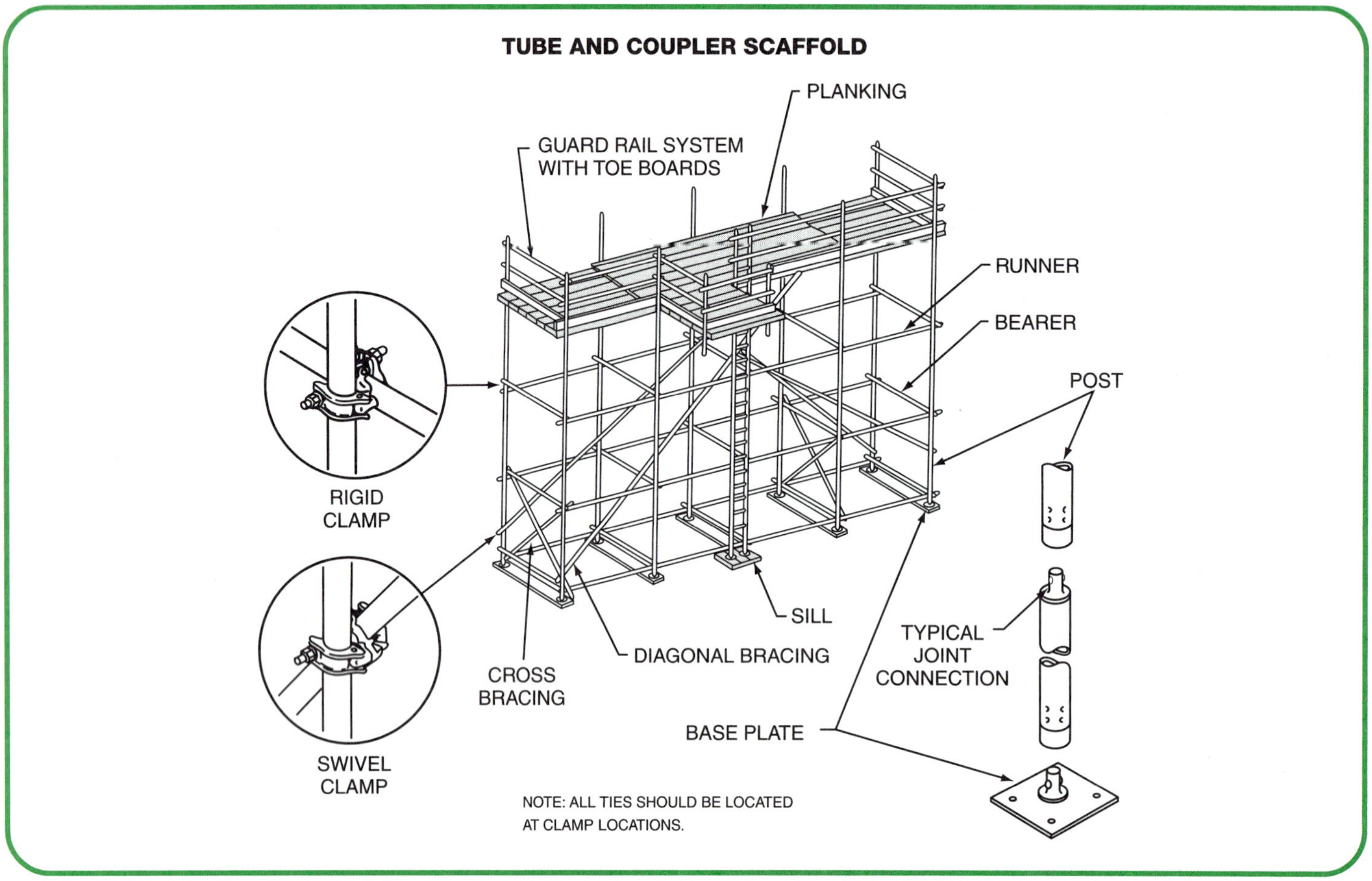


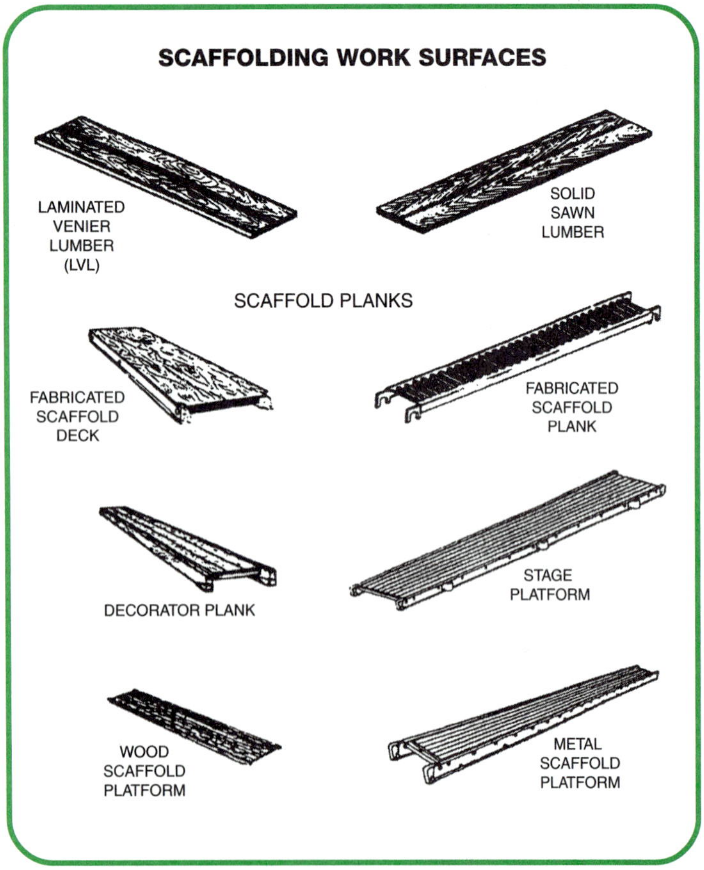


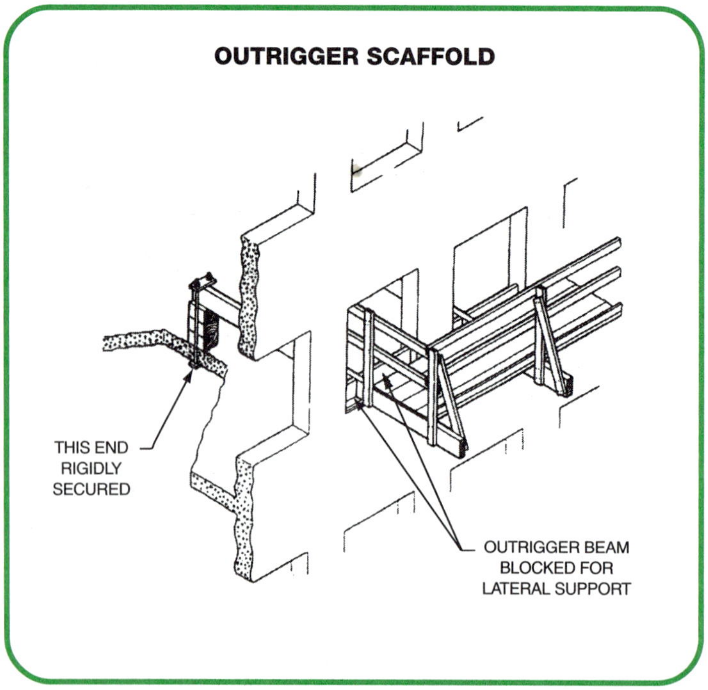


[61 FR 46122, Aug. 30, 1996; 61 FR 59832, Nov. 25, 1996]

**Authority:** Sec. 107, Contract Work Hours and Safety Standards Act (Construction Safety Act) (40 U.S.C. 333); Sec. 4, 6, 8, Occupational Safety and Health Act of 1970 (29 U.S.C. 653, 655, 657); Secretary of Labor's Orders Nos. 1-90 (55 FR 9033), 6-96 (62 FR 111); and 3-2000 (65 FR 50017), as applicable, and 29 CFR Part 1911.

**Source:** 59 FR 40730, Aug. 9, 1994, unless otherwise noted. Amended 66 FR 5195, Jan. 18, 2001.

### §1926.500 Scope, application, and definitions applicable to this subpart.

**1926.500(a)** ***Scope and application.*** (1) This subpart sets forth requirements and criteria for fall protection in construction workplaces covered under 29 CFR part 1926. Exception: The provisions of this subpart do not apply when employees are making an inspection, investigation, or assessment of workplace conditions prior to the actual start of construction work or after all construction work has been completed.

**1926.500(a)(2)** Section 1926.501 sets forth those workplaces, conditions, operations, and circumstances for which fall protection shall be provided except as follows:

**1926.500(a)(2)(i)** Requirements relating to fall protection for employees working on scaffolds are provided in subpart L of this part.

**1926.500(a)(2)(ii)** Requirements relating to fall protection for employees working on certain cranes and derricks are provided in subpart N of this part.

**1926.500(a)(2)(iii)** Fall protection requirements for employees performing steel erection work (except for towers and tanks) are provided in subpart R of this part.

**1926.500(a)(2)(iv)** Requirements relating to fall protection for employees working on certain types of equipment used in tunneling operations are provided in subpart S of this part.

**1926.500(a)(2)(v)** Requirements relating to fall protection for employees engaged in the erection of tanks and communication and broadcast towers are provided in §1926.105.

**1926.500(a)(2)(vi)** Requirements relating to fall protection for employees engaged in the construction of electric transmission and distribution lines and equipment are provided in subpart V of this part.

**1926.500(a)(2)(vii)** Requirements relating to fall protection for employees working on stairways and ladders are provided in subpart X of this part.

**1926.500(a)(3)** Section 1926.502 sets forth the requirements for the installation, construction, and proper use of fall protection required by part 1926, except as follows:

**1926.500(a)(3)(i)** Performance requirements for guardrail systems used on scaffolds and performance requirements for falling object protection used on scaffolds are provided in subpart L of this part.

**1926.500(a)(3)(ii)** Performance requirements for stairways, stairrail systems, and handrails are provided in subpart X of this part.

**1926.500(a)(3)(iii)** Additional performance requirements for personal climbing equipment, lineman's body belts, safety straps, and lanyards are provided in subpart V of this part.

**1926.500(a)(3)(iv)** Section 1926.502 does not apply to the erection of tanks and communication and broadcast towers. (Note: Section 1926.104 sets the criteria for body belts, lanyards and lifelines used for fall protection during tank and communication and broadcast tower erection. Paragraphs (b),(c) and (f) of §1926.107 provide definitions for the pertinent terms.)

**1926.500(a)(4)** Section 1926.503 sets forth requirements for training in the installation and use of fall protection systems, except in relation to steel erection activities.

**1926.500(b)** ***Definitions.***

*Anchorage* means a secure point of attachment for lifelines, lanyards or deceleration devices.

*Body belt* (safety belt) means a strap with means both for securing it about the waist and for attaching it to a lanyard, lifeline, or deceleration device.

*Body harness* means straps which may be secured about the employee in a manner that will distribute the fall arrest forces over at least the thighs, pelvis, waist, chest and shoulders with means for attaching it to other components of a personal fall arrest system.

*Buckle* means any device for holding the body belt or body harness closed around the employee's body.

*Connector* means a device which is used to couple (connect) parts of the personal fall arrest system and positioning device systems together. It may be an independent component of the system, such as a carabiner, or it may be an integral component of part of the system (such as a buckle or deering sewn into a body belt or body harness, or a snap-hook spliced or sewn to a lanyard or self-retracting lanyard).

*Controlled access zone* (CAZ) means an area in which certain work (e.g., overhand bricklaying) may take place without the use of guardrail systems, personal fall arrest systems, or safety net systems and access to the zone is controlled.

*Dangerous equipment* means equipment (such as pickling or galvanizing tanks, degreasing units, machinery, electrical equipment, and other units) which, as a result of form or function, may be hazardous to employees who fall onto or into such equipment.

*Deceleration device* means any mechanism, such as a rope grab, rip-stitch lanyard, specially-woven lanyard, tearing or deforming lanyards, automatic self-retracting lifelines/lanyards, etc., which serves to dissipate a substantial amount of energy during a fall arrest, or otherwise limit the energy imposed on an employee during fall arrest.

*Deceleration distance* means the additional vertical distance a falling employee travels, excluding lifeline elongation and free fall distance, before stopping, from the point at which the deceleration device begins to operate. It is measured as the distance between the location of an employee's body belt or body harness attachment point at the moment of activation (at the onset of fall arrest forces) of the deceleration device during a fall, and the location of that attachment point after the employee comes to a full stop.

*Equivalent* means alternative designs, materials, or methods to protect against a hazard which the employer can demonstrate will provide an equal or greater degree of safety for employees than the methods, materials or designs specified in the standard.

*Failure* means load refusal, breakage, or separation of component parts. Load refusal is the point where the ultimate strength is exceeded.

*Free fall* means the act of falling before a personal fall arrest system begins to apply force to arrest the fall.

*Free fall distance* means the vertical displacement of the fall arrest attachment point on the employee's body belt or body harness between onset of the fall and just before the system begins to apply force to arrest the fall. This distance excludes deceleration distance, and lifeline/lanyard elongation, but includes any deceleration device slide distance or self-retracting lifeline/lanyard extension before they operate and fall arrest forces occur.

*Guardrail system* means a barrier erected to prevent employees from falling to lower levels.

*Hole* means a gap or void 2 inches (5.1 cm) or more in its least dimension, in a floor, roof, or other walking/working surface.

*Infeasible* means that it is impossible to perform the construction work using a conventional fall protection system (i.e., guardrail system, safety net system, or personal fall arrest system) or that it is technologically impossible to use any one of these systems to provide fall protection.

*Lanyard* means a flexible line of rope, wire rope, or strap which generally has a connector at each end for connecting the body belt or body harness to a deceleration device, lifeline, or anchorage.

*Leading edge* means the edge of a floor, roof, or formwork for a floor or other walking/working surface (such as the deck) which changes location as additional floor, roof, decking, or formwork sections are placed, formed, or constructed. A leading edge is considered to be an "unprotected side and edge" during periods when it is not actively and continuously under construction.

*Lifeline* means a component consisting of a flexible line for connection to an anchorage at one end to hang vertically (vertical lifeline), or for connection to anchorages at both ends to stretch horizontally (horizontal lifeline), and which serves as a means for connecting other components of a personal fall arrest system to the anchorage.

*Low-slope roof* means a roof having a slope less than or equal to 4 in 12 (vertical to horizontal).

*Lower levels* means those areas or surfaces to which an employee can fall. Such areas or surfaces include, but are not limited to, ground levels, floors, platforms, ramps, runways, excavations, pits, tanks, material, water, equipment, structures, or portions thereof.

*Mechanical equipment* means all motor or human propelled wheeled equipment used for roofing work, except wheelbarrows and mopcarts.

*Opening* means a gap or void 30 inches (76 cm) or more high and 18 inches (48 cm) or more wide, in a wall or partition, through which employees can fall to a lower level.

*Overhand bricklaying and related work* means the process of laying bricks and masonry units such that the surface of the wall to be jointed is on the opposite side of the wall from the mason, requiring the mason to lean over the wall to complete the work. Related work includes mason tending and electrical installation incorporated into the brick wall during the overhand bricklaying process.

*Personal fall arrest system* means a system used to arrest an employee in a fall from a working level. It consists of an anchorage, connectors, a body belt or body harness and may include a lanyard, deceleration device, lifeline, or suitable combinations of these. As of January 1, 1998, the use of a body belt for fall arrest is prohibited.

*Positioning device system* means a body belt or body harness system rigged to allow an employee to be supported on an elevated vertical surface, such as a wall, and work with both hands free while leaning.

*Rope grab* means a deceleration device which travels on a lifeline and automatically, by friction, engages the lifeline and locks so as to arrest the fall of an employee. A rope grab usually employs the principle of inertial locking, cam/level locking, or both.

*Roof* means the exterior surface on the top of a building. This does not include floors or formwork which, because a building has not been completed, temporarily become the top surface of a building.

*Roofing work* means the hoisting, storage, application, and removal of roofing materials and equipment, including related insulation, sheet metal, and vapor barrier work, but not including the construction of the roof deck.

*Safety-monitoring system* means a safety system in which a competent person is responsible for recognizing and warning employees of fall hazards.

*Self-retracting lifeline/lanyard* means a deceleration device containing a drum-wound line which can be slowly extracted from, or retracted onto, the drum under slight tension during normal employee movement, and which, after onset of a fall, automatically locks the drum and arrests the fall.

*Snaphook* means a connector comprised of a hook-shaped member with a normally closed keeper, or similar arrangement, which may be opened to permit the hook to receive an object and, when released, automatically closes to retain the object. Snaphooks are generally one of two types:

(1) The locking type with a self-closing, self-locking keeper which remains closed and locked until unlocked and pressed open for connection or disconnection; or

(2) The non-locking type with a self-closing keeper which remains closed until pressed open for connection or disconnection. As of January 1, 1998, the use of a non-locking snaphook as part of personal fall arrest systems and positioning device systems is prohibited.

*Steep roof* means a roof having a slope greater than 4 in 12 (vertical to horizontal).

*Toeboard* means a low protective barrier that will prevent the fall of materials and equipment to lower levels and provide protection from falls for personnel.

*Unprotected sides and edges* means any side or edge (except at entrances to points of access) of a walking/working surface, e.g., floor, roof, ramp, or runway where there is no wall or guardrail system at least 39 inches (1.0 m) high.

*Walking/working surface* means any surface, whether horizontal or vertical on which an employee walks or works, including, but not limited to, floors, roofs, ramps, bridges, runways, formwork and concrete reinforcing steel but not including ladders, vehicles, or trailers, on which employees must be located in order to perform their job duties.

*Warning line system* means a barrier erected on a roof to warn employees that they are approaching an unprotected roof side or edge, and which designates an area in which roofing work may take place without the use of guardrail, body belt, or safety net systems to protect employees in the area.

*Work area* means that portion of a walking/working surface where job duties are being performed.

[59 FR 40730, Aug. 9, 1994, as amended at 60 FR 39255, Aug. 2, 1995, and at 66 FR 5195 (Jan. 18, 2001)]

### §1926.501 Duty to have fall protection.

**1926.501(a)** ***General.*** (1) This section sets forth requirements for employers to provide fall protection systems. All fall protection required by this section shall conform to the criteria set forth in §1926.502 of this subpart.

**1926.501(a)(2)** The employer shall determine if the walking/working surfaces on which its employees are to work have the strength and structural integrity to support employees safely. Employees shall be allowed to work on those surfaces only when the surfaces have the requisite strength and structural integrity.

**1926.501(b)(1)** ***Unprotected sides and edges.*** Each employee on a walking/working surface (horizontal and vertical surface) with an unprotected side or edge which is 6 feet (1.8 m) or more above a lower level shall be protected from falling by the use of guardrail systems, safety net systems, or personal fall arrest systems.

**1926.501(b)(2)** *Leading edges.* (i) Each employee who is constructing a leading edge 6 feet (1.8 m) or more above lower levels shall be protected from falling by guardrail systems, safety net systems, or personal fall arrest systems. Exception: When the employer can demonstrate that it is infeasible or creates a greater hazard to use these systems, the employer shall develop and implement a fall protection plan which meets the requirements of paragraph (k) of §1926.502.

**Note:** There is a presumption that it is feasible and will not create a greater hazard to implement at least one of the above-listed fall protection systems. Accordingly, the employer has the burden of establishing that it is appropriate to implement a fall protection plan which complies with §1926.502(k) for a particular workplace situation, in lieu of implementing any of those systems.

**1926.501(b)(2)(ii)** Each employee on a walking/working surface 6 feet (1.8 m) or more above a lower level where leading edges are under construction, but who is not engaged in the leading edge work, shall be protected from falling by a guardrail system, safety net system, or personal fall arrest system. If a guardrail system is chosen to provide the fall protection, and a controlled access zone has already been established for leading edge work, the control line may be used in lieu of a guardrail along the edge that parallels the leading edge.

**1926.501(b)(3)** *Hoist areas.* Each employee in a hoist area shall be protected from falling 6 feet (1.8 m) or more to lower levels by guardrail systems or personal fall arrest systems. If guardrail systems, [or chain, gate, or guardrail] or portions thereof, are removed to facilitate the hoisting operation (e.g., during landing of materials), and an employee must lean through the access opening or out over the edge of the access opening (to receive or guide equipment and materials, for example), that employee shall be protected from fall hazards by a personal fall arrest system.

**1926.501(b)(4)** *Holes.* (i) Each employee on walking/working surfaces shall be protected from falling through holes (including skylights) more than 6 feet (1.8 m) above lower levels, by personal fall arrest systems, covers, or guardrail systems erected around such holes.

**1926.501(b)(4)(ii)** Each employee on a walking/working surface shall be protected from tripping in or stepping into or through holes (including skylights) by covers.

**1926.501(b)(4)(iii)** Each employee on a walking/working surface shall be protected from objects falling through holes (including skylights) by covers.

**1926.501(b)(5)** *Formwork and reinforcing steel.* Each employee on the face of formwork or reinforcing steel shall be protected from falling 6 feet (1.8 m) or more to lower levels by personal fall arrest systems, safety net systems, or positioning device systems.

**1926.501(b)(6)** *Ramps, runways, and other walkways.* Each employee on ramps, runways, and other walkways shall be protected from falling 6 feet (1.8 m) or more to lower levels by guardrail systems.

**1926.501(b)(7)** *Excavations.* (i) Each employee at the edge of an excavation 6 feet (1.8 m) or more in depth shall be protected from falling by guardrail systems, fences, or barricades when the excavations are not readily seen because of plant growth or other visual barrier;

**1926.501(b)(7)(ii)** Each employee at the edge of a well, pit, shaft, and similar excavation 6 feet (1.8 m) or more in depth shall be protected from falling by guardrail systems, fences, barricades, or covers.

**1926.501(b)(8)** *Dangerous equipment.* (i) Each employee less than 6 feet (1.8 m) above dangerous equipment shall be protected from falling into or onto the dangerous equipment by guardrail systems or by equipment guards.

**1926.501(b)(8)(ii)** Each employee 6 feet (1.8 m) or more above dangerous equipment shall be protected from fall hazards by guardrail systems, personal fall arrest systems, or safety net systems.

**1926.501(b)(9)** *Overhand bricklaying and related work.* (i) Except as otherwise provided in paragraph (b) of this section, each employee performing overhand bricklaying and related work 6 feet (1.8 m) or more above lower levels, shall be protected from falling by guardrail systems, safety net systems, personal fall arrest systems, or shall work in a controlled access zone.

**1926.501(b)(9)(ii)** Each employee reaching more than 10 inches (25 cm) below the level of the walking/working surface on which they are working, shall be protected from falling by a guardrail system, safety net system, or personal fall arrest system.

**Note:** Bricklaying operations performed on scaffolds are regulated by subpart L—Scaffolds of this part.

**1926.501(b)(10)** *Roofing work on Low-slope roofs.* Except as otherwise provided in paragraph (b) of this section, each employee engaged in roofing activities on low-slope roofs, with unprotected sides and edges 6 feet (1.8 m) or more above lower levels shall be protected from falling by guardrail systems, safety net systems, personal fall arrest systems, or a combination of warning line system and guardrail system, warning line system and safety net system, or warning line system and personal fall arrest system, or warning line system and safety monitoring system. Or, on roofs 50-feet (15.25 m) or less in width (see Appendix A to subpart M of this part), the use of a safety monitoring system alone [i.e. without the warning line system] is permitted.

**1926.501(b)(11)** *Steep roofs.* Each employee on a steep roof with unprotected sides and edges 6 feet (1.8 m) or more above lower levels shall be protected from falling by guardrail systems with toeboards, safety net systems, or personal fall arrest systems.

**1926.501(b)(12)** *Precast concrete erection.* Each employee engaged in the erection of precast concrete members (including, but not limited to the erection of wall panels, columns, beams, and floor and roof "tees") and related operations such as grouting of precast concrete members, who is 6 feet (1.8 m) or more above lower levels shall be protected from falling by guardrail systems, safety net systems, or personal fall arrest systems, unless another provision in paragraph (b) of this section provides for an alternative fall protection measure. Exception: When the employer can demonstrate that it is infeasible or creates a greater hazard to use these systems, the employer shall develop and implement a fall protection plan which meets the requirements of paragraph (k) of §1926.502.

**Note:** There is a presumption that it is feasible and will not create a greater hazard to implement at least one of the above-listed fall protection systems. Accordingly, the employer has the burden of establishing that it is appropriate to implement a fall protection plan which complies with §1926.502(k) for a particular workplace situation, in lieu of implementing any of those systems.

**1926.501(b)(13)** *Residential construction.* Each employee engaged in residential construction activities 6 feet (1.8 m) or more above lower levels shall be protected by guardrail systems, safety net system, or personal fall arrest system unless another provision in paragraph (b) of this section provides for an alternative fall protection measure. Exception: When the employer can demonstrate that it is infeasible or creates a greater hazard to use these systems, the employer shall develop and implement a fall protection plan which meets the requirements of paragraph (k) of §1926.502.

**Note:** There is a presumption that it is feasible and will not create a greater hazard to implement at least one of the above-listed fall protection systems. Accordingly, the employer has the burden of establishing that it is appropriate to implement a fall protection plan which complies with §1926.502(k) for a particular workplace situation, in lieu of implementing any of those systems.

**1926.501(b)(14)** *Wall openings.* Each employee working on, at, above, or near wall openings (including those with chutes attached) where the outside bottom edge of the wall opening is 6 feet (1.8 m) or more above lower levels and the inside bottom edge of the wall opening is less than 39 inches (1.0 m) above the walking/working surface, shall be protected from falling by the use of a guardrail system, a safety net system, or a personal fall arrest system.

**1926.501(b)(15)** *Walking/working surfaces not otherwise addressed.* Except as provided in §1926.500(a)(2) or in §1926.501(b)(1) through (b)(14), each employee on a walking/working surface 6 feet (1.8 m) or more above lower levels shall be protected from falling by a guardrail system, safety net system, or personal fall arrest system.

**1926.501(c)** ***Protection from falling objects.*** When an employee is exposed to falling objects, the employer shall have each employee wear a hard hat and shall implement one of the following measures:

**1926.501(c)(1)** Erect toeboards, screens, or guardrail systems to prevent objects from falling from higher levels; or,

**1926.501(c)(2)** Erect a canopy structure and keep potential fall objects far enough from the edge of the higher level so that those objects would not go over the edge if they were accidentally displaced; or,

**1926.501(c)(3)** Barricade the area to which objects could fall, prohibit employees from entering the barricaded area, and keep objects that may fall far enough away from the edge of a higher level so that those objects would not go over the edge if they were accidentally displaced.

## §1926.502 Fall protection systems criteria and practices.

**1926.502(a)** ***General.*** (1) Fall protection systems required by this part shall comply with the applicable provisions of this section.

**1926.502(a)(2)** Employers shall provide and install all fall protection systems required by this subpart for an employee, and shall comply with all other pertinent requirements of this subpart before that employee begins the work that necessitates the fall protection.

**1926.502(b)** ***Guardrail systems.*** Guardrail systems and their use shall comply with the following provisions:

**1926.502(b)(1)** Top edge height of top rails, or equivalent guardrail system members, shall be 42 inches (1.1 m) plus or minus 3 inches (8 cm) above the walking/working level. When conditions warrant, the height of the top edge may exceed the 45-inch height, provided the guardrail system meets all other criteria of this paragraph.

**Note:** When employees are using stilts, the top edge height of the top rail, or equivalent member, shall be increased an amount equal to the height of the stilts.

**1926.502(b)(2)** Midrails, screens, mesh, intermediate vertical members, or equivalent intermediate structural members shall be installed between the top edge of the guardrail system and the walking/working surface when there is no wall or parapet wall at least 21 inches (53 cm) high.

**1926.502(b)(2)(i)** Midrails, when used, shall be installed at a height midway between the top edge of the guardrail system and the walking/working level.

**1926.502(b)(2)(ii)** Screens and mesh, when used, shall extend from the top rail to the walking/working level and along the entire opening between top rail supports.

**1926.502(b)(2)(iii)** Intermediate members (such as balusters), when used between posts, shall be not more than 19 inches (48 cm) apart.

**1926.502(b)(2)(iv)** Other structural members (such as additional midrails and architectural panels) shall be installed such that there are no openings in the guardrail system that are more than 19 inches (.5 m) wide.

**1926.502(b)(3)** Guardrail systems shall be capable of withstanding, without failure, a force of at least 200 pounds (890 N) applied within 2 inches (5.1 cm) of the top edge, in any outward or downward direction, at any point along the top edge.

**1926.502(b)(4)** When the 200 pound (890 N) test load specified in paragraph (b)(3) of this section is applied in a downward direction, the top edge of the guardrail shall not deflect to a height less than 39 inches (1.0 m) above the walking/working level. Guardrail system components selected and constructed in accordance with the Appendix B to subpart M of this part will be deemed to meet this requirement.

**1926.502(b)(5)** Midrails, screens, mesh, intermediate vertical members, solid panels, and equivalent structural members shall be capable of withstanding, without failure, a force of at least 150 pounds

(666 N) applied in any downward or outward direction at any point along the midrail or other member.

**1926.502(b)(6)** Guardrail systems shall be so surfaced as to prevent injury to an employee from punctures or lacerations, and to prevent snagging of clothing.

**1926.502(b)(7)** The ends of all top rails and midrails shall not overhang the terminal posts, except where such overhang does not constitute a projection hazard.

**1926.502(b)(8)** Steel banding and plastic banding shall not be used as top rails or midrails.

**1926.502(b)(9)** Top rails and midrails shall be at least one-quarter inch (0.6 cm) nominal diameter or thickness to prevent cuts and lacerations. If wire rope is used for top rails, it shall be flagged at not more than 6-foot (1.8 m) intervals with high-visibility material.

**1926.502(b)(10)** When guardrail systems are used at hoisting areas, a chain, gate or removable guardrail section shall be placed across the access opening between guardrail sections when hoisting operations are not taking place.

**1926.502(b)(11)** When guardrail systems are used at holes, they shall be erected on all unprotected sides or edges of the hole.

**1926.502(b)(12)** When guardrail systems are used around holes used for the passage of materials, the hole shall have not more than two sides provided with removable guardrail sections to allow the passage of materials. When the hole is not in use, it shall be closed over with a cover, or a guardrail system shall be provided along all unprotected sides or edges.

**1926.502(b)(13)** When guardrail systems are used around holes which are used as points of access (such as ladderways), they shall be provided with a gate, or be so offset that a person cannot walk directly into the hole.

**1926.502(b)(14)** Guardrail systems used on ramps and runways shall be erected along each unprotected side or edge.

**1926.502(b)(15)** Manila, plastic or synthetic rope being used for top rails or midrails shall be inspected as frequently as necessary to ensure that it continues to meet the strength requirements of paragraph (b)(3) of this section.

**1926.502(c)** ***Safety net systems.*** Safety net systems and their use shall comply with the following provisions:

**1926.502(c)(1)** Safety nets shall be installed as close as practicable under the walking/working surface on which employees are working, but in no case more than 30 feet (9.1 m) below such level. When nets are used on bridges, the potential fall area from the walking/working surface to the net shall be unobstructed.

**1926.502(c)(2)** Safety nets shall extend outward from the outermost projection of the work surface as follows:

| Vertical distance from working level to horizontal plane of net | Minimum required horizontal distance of outer edge of net from the edge of the working surface. |
|---|---|
| Up to 5 feet | 8 feet. |
| More than 5 feet up to 10 feet | 10 feet. |
| More than 10 feet | 13 feet. |

**1926.502(c)(3)** Safety nets shall be installed with sufficient clearance under them to prevent contact with the surface or structures below when subjected to an impact force equal to the drop test specified in paragraph (c)(4) of this section.

**1926.502(c)(4)** Safety nets and their installations shall be capable of absorbing an impact force equal to that produced by the drop test specified in paragraph (c)(4)(i) of this section.

**1926.502(c)(4)(i)** Except as provided in paragraph (c)(4)(ii) of this section, safety nets and safety net installations shall be drop-tested at the jobsite after initial installation and before being used as a fall protection system, whenever relocated, after major repair, and at 6-month intervals if left in one place. The drop-test shall consist of a 400 pound (180 kg) bag of sand 30 ± 2 inches (76 ± 5 cm) in diameter dropped into the net from the highest walking/working surface at which employees are exposed to fall hazards, but not from less than 42 inches (1.1 m) above that level.

**1926.502(c)(4)(ii)** When the employer can demonstrate that it is unreasonable to perform the drop-test required by paragraph (c)(4)(i) of this section, the employer (or a designated competent person) shall certify that the net and net installation is in compliance with the provisions of paragraphs (c)(3) and (c)(4)(i) of this section by preparing a certification record prior to the net being used as a fall protection system. The certification record must include an identification of the net and net installation for which the certification record is being prepared; the date that it was determined that the identified net and net installation were in compliance with paragraph (c)(3) of this section and the signature of the person making the determination and certification. The most recent certification record for each net and net installation shall be available at the jobsite for inspection.

**1926.502(c)(5)** Defective nets shall not be used. Safety nets shall be inspected at least once a week for wear, damage, and other deterioration. Defective components shall be removed from service. Safety nets shall also be inspected after any occurrence which could affect the integrity of the safety net system.

**1926.502(c)(6)** Materials, scrap pieces, equipment, and tools which have fallen into the safety net shall be removed as soon as possible from the net and at least before the next work shift.

**1926.502(c)(7)** The maximum size of each safety net mesh opening shall not exceed 36 square inches (230 cm$^2$) nor be longer than 6 inches (15 cm) on any side, and the opening, measured center-to-center of mesh ropes or webbing, shall not be longer than 6 inches (15 cm). All mesh crossings shall be secured to prevent enlargement of the mesh opening.

**1926.502(c)(8)** Each safety net (or section of it) shall have a border rope for webbing with a minimum breaking strength of 5,000 pounds (22.2 kN).

**1926.502(c)(9)** Connections between safety net panels shall be as strong as integral net components and shall be spaced not more than 6 inches (15 cm) apart.

**1926.502(d)** ***Personal fall arrest systems.*** Personal fall arrest systems and their use shall comply with the provisions set forth below. Effective January 1, 1998, body belts are not acceptable as part of a personal fall arrest system. Note: The use of a body belt in a positioning device system is acceptable and is regulated under paragraph (e) of this section.

**1926.502(d)(1)** Connectors shall be drop forged, pressed or formed steel, or made of equivalent materials.

**1926.502(d)(2)** Connectors shall have a corrosion-resistant finish, and all surfaces and edges shall be smooth to prevent damage to interfacing parts of the system.

**1926.502(d)(3)** Dee-rings and snaphooks shall have a minimum tensile strength of 5,000 pounds (22.2 kN).

**1926.502(d)(4)** Dee-rings and snaphooks shall be proof-tested to a minimum tensile load of 3,600 pounds (16 kN) without cracking, breaking, or taking permanent deformation.

**1926.502(d)(5)** Snaphooks shall be sized to be compatible with the member to which they are connected to prevent unintentional disengagement of the snaphook by depression of the snaphook keeper by the connected member, or shall be a locking type snaphook designed and used to prevent disengagement of the snaphook by the contact of the snaphook keeper by the connected member. Effective January 1, 1998, only locking type snaphooks shall be used.

**1926.502(d)(6)** Unless the snaphook is a locking type and designed for the following connections, snaphooks shall not be engaged:

**1926.502(d)(6)(i)** directly to webbing, rope or wire rope;

**1926.502(d)(6)(ii)** to each other;

**1926.502(d)(6)(iii)** to a Dee-ring to which another snaphook or other connector is attached;

**1926.502(d)(6)(iv)** to a horizontal lifeline; or

**1926.502(d)(6)(v)** to any object which is incompatibly shaped or dimensioned in relation to the snaphook such that unintentional disengagement could occur by the connected object being able to depress the snaphook keeper and release itself.

**1926.502(d)(7)** On suspended scaffolds or similar work platforms with horizontal lifelines which may become vertical lifelines, the devices used to connect to a horizontal lifeline shall be capable of locking in both directions on the lifeline.

**1926.502(d)(8)** Horizontal lifelines shall be designed, installed, and used, under the supervision of a qualified person, as part of a complete personal fall arrest system, which maintains a safety factor of at least two.

**1926.502(d)(9)** Lanyards and vertical lifelines shall have a minimum breaking strength of 5,000 pounds (22.2 kN).

**1926.502(d)(10)(i)** Except as provided in paragraph (d)(10)(ii) of this section, when vertical lifelines are used, each employee shall be attached to a separate lifeline.

**1926.502(d)(10)(ii)** During the construction of elevator shafts, two employees may be attached to the same lifeline in the hoistway, provided both employees are working atop a false car that is equipped with guardrails; the strength of the lifeline is 10,000 pounds [5,000 pounds per employee attached] (44.4 kN); and all other criteria specified in this paragraph for lifelines have been met.

**1926.502(d)(11)** Lifelines shall be protected against being cut or abraded.

**1926.502(d)(12)** Self-retracting lifelines and lanyards which automatically limit free fall distance to 2 feet (0.61 m) or less shall be capable of sustaining a minimum tensile load of 3,000 pounds (13.3 kN) applied to the device with the lifeline or lanyard in the fully extended position.

**1926.502(d)(13)** Self-retracting lifelines and lanyards which do not limit free fall distance to 2 feet (0.61 m) or less, ripstitch lanyards, and tearing and deforming lanyards shall be capable of sustaining a minimum tensile load of 5,000 pounds (22.2 kN) applied to the device with the lifeline or lanyard in the fully extended position.

**1926.502(d)(14)** Ropes and straps (webbing) used in lanyards, lifelines, and strength components of body belts and body harnesses shall be made from synthetic fibers.

**1926.502(d)(15)** Anchorages used for attachment of personal fall arrest equipment shall be independent of any anchorage being used to support or suspend platforms and capable of supporting at least 5,000 pounds (22.2 kN) per employee attached, or shall be designed, installed, and used as follows:

**1926.502(d)(15)(i)** as part of a complete personal fall arrest system which maintains a safety factor of at least two; and

**1926.502(d)(15)(ii)** under the supervision of a qualified person.

**1926.502(d)(16)** Personal fall arrest systems, when stopping a fall, shall:

**1926.502(d)(16)(i)** limit maximum arresting force on an employee to 900 pounds (4 kN) when used with a body belt;

**1926.502(d)(16)(ii)** limit maximum arresting force on an employee to 1,800 pounds (8 kN) when used with a body harness;

**1926.502(d)(16)(iii)** be rigged such that an employee can neither free fall more than 6 feet (1.8 m), nor contact any lower level;

**1926.502(d)(16)(iv)** bring an employee to a complete stop and limit maximum deceleration distance an employee travels to 3.5 feet (1.07 m); and,

**1926.502(d)(16)(v)** have sufficient strength to withstand twice the potential impact energy of an employee free falling a distance of 6 feet (1.8 m), or the free fall distance permitted by the system, whichever is less.

**Note:** If the personal fall arrest system meets the criteria and protocols contained in Appendix C to subpart M, and if the system is being used by an employee having a combined person and tool weight of less than 310 pounds (140 kg), the system will be considered to be in compliance with the provisions of paragraph (d)(16) of this section. If the system is used by an employee having a combined tool and body weight of 310 pounds (140 kg) or more, then the employer must appropriately modify the criteria and protocols of the Appendix to provide proper protection for such heavier weights, or the system will not be deemed to be in compliance with the requirements of paragraph (d)(16) of this section.

**1926.502(d)(17)** The attachment point of the body belt shall be located in the center of the wearer's back. The attachment point of the body harness shall be located in the center of the wearer's back near shoulder level, or above the wearer's head.

**1926.502(d)(18)** Body belts, harnesses, and components shall be used only for employee protection (as part of a personal fall arrest system or positioning device system) and not to hoist materials.

**1926.502(d)(19)** Personal fall arrest systems and components subjected to impact loading shall be immediately removed from service and shall not be used again for employee protection until inspected and determined by a competent person to be undamaged and suitable for reuse.

**1926.502(d)(20)** The employer shall provide for prompt rescue of employees in the event of a fall or shall assure that employees are able to rescue themselves.

**1926.502(d)(21)** Personal fall arrest systems shall be inspected prior to each use for wear, damage and other deterioration, and defective components shall be removed from service.

**1926.502(d)(22)** Body belts shall be at least one and five-eighths (1 5/8) inches (4.1 cm) wide.

**1926.502(d)(23)** Personal fall arrest systems shall not be attached to guardrail systems, nor shall they be attached to hoists except as specified in other subparts of this part.

**1926.502(d)(24)** When a personal fall arrest system is used at hoist areas, it shall be rigged to allow the movement of the employee only as far as the edge of the walking/working surface.

**1926.502(e)** ***Positioning device systems.*** Positioning device systems and their use shall conform to the following provisions:

**1926.502(e)(1)** Positioning devices shall be rigged such that an employee cannot free fall more than 2 feet (.6 m).

**1926.502(e)(2)** Positioning devices shall be secured to an anchorage capable of supporting at least twice the potential impact load of an employee's fall or 3,000 pounds (13.3 kN), whichever is greater.

**1926.502(e)(3)** Connectors shall be drop forged, pressed or formed steel, or made of equivalent materials.

**1926.502(e)(4)** Connectors shall have a corrosion-resistant finish, and all surfaces and edges shall be smooth to prevent damage to interfacing parts of this system.

**1926.502(e)(5)** Connecting assemblies shall have a minimum tensile strength of 5,000 pounds (22.2 kN)

**1926.502(e)(6)** Dee-rings and snaphooks shall be proof-tested to a minimum tensile load of 3,600 pounds (16 kN) without cracking, breaking, or taking permanent deformation.

**1926.502(e)(7)** Snaphooks shall be sized to be compatible with the member to which they are connected to prevent unintentional disengagement of the snaphook by depression of the snaphook keeper by the connected member, or shall be a locking type snaphook designed and used to prevent disengagement of the snaphook by the contact of the snaphook keeper by the connected member. As of January 1, 1998, only locking type snaphooks shall be used.

**1926.502(e)(8)** Unless the snaphook is a locking type and designed for the following connections, snaphooks shall not be engaged:

**1926.502(e)(8)(i)** directly to webbing, rope or wire rope;

**1926.502(e)(8)(ii)** to each other;

**1926.502(e)(8)(iii)** to a Dee-ring to which another snaphook or other connector is attached;

**1926.502(e)(8)(iv)** to a horizontal lifeline; or

**1926.502(e)(8)(v)** to any object which is incompatibly shaped or dimensioned in relation to the snaphook such that unintentional disengagement could occur by the connected object being able to depress the snaphook keeper and release itself.

**1926.502(e)(9)** Positioning device systems shall be inspected prior to each use for wear, damage, and other deterioration, and defective components shall be removed from service.

**1926.502(e)(10)** Body belts, harnesses, and components shall be used only for employee protection (as part of a personal fall arrest system or positioning device system) and not to hoist materials.

**1926.502(f)** ***Warning line systems.*** Warning line systems [See §1926.501(b)(10)] and their use shall comply with the following provisions:

**1926.502(f)(1)** The warning line shall be erected around all sides of the roof work area.

**1926.502(f)(1)(i)** When mechanical equipment is not being used, the warning line shall be erected not less than 6 feet (1.8 m) from the roof edge.

**1926.502(f)(1)(ii)** When mechanical equipment is being used, the warning line shall be erected not less than 6 feet (1.8 m) from the roof edge which is parallel to the direction of mechanical equipment operation, and not less than 10 feet (3.1 m) from the roof edge which is perpendicular to the direction of mechanical equipment operation.

**1926.502(f)(1)(iii)** Points of access, materials handling areas, storage areas, and hoisting areas shall be connected to the work area by an access path formed by two warning lines.

**1926.502(f)(1)(iv)** When the path to a point of access is not in use, a rope, wire, chain, or other barricade, equivalent in strength and height to the warning line, shall be placed across the path at the point where the path intersects the warning line erected around the work area, or the path shall be offset such that a person cannot walk directly into the work area.

**1926.502(f)(2)** Warning lines shall consist of ropes, wires, or chains, and supporting stanchions erected as follows:

**1926.502(f)(2)(i)** The rope, wire, or chain shall be flagged at not more than 6-foot (1.8 m) intervals with high-visibility material;

**1926.502(f)(2)(ii)** The rope, wire, or chain shall be rigged and supported in such a way that its lowest point (including sag) is no less than 34 inches (.9 m) from the walking/working surface and its highest point is no more than 39 inches (1.0 m) from the walking/working surface;

**1926.502(f)(2)(iii)** After being erected, with the rope, wire, or chain attached, stanchions shall be capable of resisting, without tipping over, a force of at least 16 pounds (71 N) applied horizontally against the stanchion, 30 inches (.8 m) above the walking/working surface, perpendicular to the warning line, and in the direction of the floor, roof, or platform edge;

**1926.502(f)(2)(iv)** The rope, wire, or chain shall have a minimum tensile strength of 500 pounds (2.22 kN), and after being attached to the stanchions, shall be capable of supporting, without breaking, the loads applied to the stanchions as prescribed in paragraph (f)(2)(iii) of this section; and

**1926.502(f)(2)(v)** The line shall be attached at each stanchion in such a way that pulling on one section of the line between stanchions will not result in slack being taken up in adjacent sections before the stanchion tips over.

**1926.502(f)(3)** No employee shall be allowed in the area between a roof edge and a warning line unless the employee is performing roofing work in that area.

**1926.502(f)(4)** Mechanical equipment on roofs shall be used or stored only in areas where employees are protected by a warning line system, guardrail system, or personal fall arrest system.

**1926.502(g)** ***Controlled access zones.*** Controlled access zones [See §1926.501(b)(9) and §1926.502(k)] and their use shall conform to the following provisions.

**1926.502(g)(1)** When used to control access to areas where leading edge and other operations are taking place the controlled access zone shall be defined by a control line or by any other means that restricts access.

**1926.502(g)(1)(i)** When control lines are used, they shall be erected not less than 6 feet (1.8 m) nor more than 25 feet (7.7 m) from the unprotected or leading edge, except when erecting precast concrete members.

**1926.502(g)(1)(ii)** When erecting precast concrete members, the control line shall be erected not less than 6 feet (1.8 m) nor more than 60 feet (18 m) or half the length of the member being erected, whichever is less, from the leading edge.

**1926.502(g)(1)(iii)** The control line shall extend along the entire length of the unprotected or leading edge and shall be approximately parallel to the unprotected or leading edge.

**1926.502(g)(1)(iv)** The control line shall be connected on each side to a guardrail system or wall.

**1926.502(g)(2)** When used to control access to areas where overhand bricklaying and related work are taking place:

**1926.502(g)(2)(i)** The controlled access zone shall be defined by a control line erected not less than 10 feet (3.1 m) nor more than 15 feet (4.5 m) from the working edge.

**1926.502(g)(2)(ii)** The control line shall extend for a distance sufficient for the controlled access zone to enclose all employees performing overhand bricklaying and related work at the working edge and shall be approximately parallel to the working edge.

**1926.502(g)(2)(iii)** Additional control lines shall be erected at each end to enclose the controlled access zone.

**1926.502(g)(2)(iv)** Only employees engaged in overhand bricklaying or related work shall be permitted in the controlled access zone.

**1926.502(g)(3)** Control lines shall consist of ropes, wires, tapes, or equivalent materials, and supporting stanchions as follows:

**1926.502(g)(3)(i)** Each line shall be flagged or otherwise clearly marked at not more than 6-foot (1.8 m) intervals with high-visibility material.

**1926.502(g)(3)(ii)** Each line shall be rigged and supported in such a way that its lowest point (including sag) is not less than 39 inches (1 m) from the walking/working surface and its highest point is not more than 45 inches (1.3 m) [50 inches (1.3 m) when overhand bricklaying operations are being performed] from the walking/working surface.

**1926.502(g)(3)(iii)** Each line shall have a minimum breaking strength of 200 pounds (.88 kN).

**1926.502(g)(4)** On floors and roofs where guardrail systems are not in place prior to the beginning of overhand bricklaying operations, controlled access zones shall be enlarged, as necessary, to enclose all points of access, material handling areas, and storage areas.

**1926.502(g)(5)** On floors and roofs where guardrail systems are in place, but need to be removed to allow overhand bricklaying work or leading edge work to take place, only that portion of the guardrail necessary to accomplish that day's work shall be removed.

**1926.502(h)** ***Safety monitoring systems.*** Safety monitoring systems [See §§1926.501(b)(10) and 1926.502(k)] and their use shall comply with the following provisions:

**1926.502(h)(1)** The employer shall designate a competent person to monitor the safety of other employees and the employer shall ensure that the safety monitor complies with the following requirements:

**1926.502(h)(1)(i)** The safety monitor shall be competent to recognize fall hazards;

**1926.502(h)(1)(ii)** The safety monitor shall warn the employee when it appears that the employee is unaware of a fall hazard or is acting in an unsafe manner;

**1926.502(h)(1)(iii)** The safety monitor shall be on the same walking/working surface and within visual sighting distance of the employee being monitored;

**1926.502(h)(1)(iv)** The safety monitor shall be close enough to communicate orally with the employee; and

**1926.502(h)(1)(v)** The safety monitor shall not have other responsibilities which could take the monitor's attention from the monitoring function.

**1926.502(h)(2)** Mechanical equipment shall not be used or stored in areas where safety monitoring systems are being used to monitor employees engaged in roofing operations on low-slope roofs.

**1926.502(h)(3)** No employee, other than an employee engaged in roofing work [on low-sloped roofs] or an employee covered by a fall protection plan, shall be allowed in an area where an employee is being protected by a safety monitoring system.

**1926.502(h)(4)** Each employee working in a controlled access zone shall be directed to comply promptly with fall hazard warnings from safety monitors.

**1926.502(i)** ***Covers.*** Covers for holes in floors, roofs, and other walking/working surfaces shall meet the following requirements:

**1926.502(i)(1)** Covers located in roadways and vehicular aisles shall be capable of supporting, without failure, at least twice the maximum axle load of the largest vehicle expected to cross over the cover.

**1926.502(i)(2)** All other covers shall be capable of supporting, without failure, at least twice the weight of employees, equipment, and materials that may be imposed on the cover at any one time.

**1926.502(i)(3)** All covers shall be secured when installed so as to prevent accidental displacement by the wind, equipment, or employees.

**1926.502(i)(4)** All covers shall be color coded or they shall be marked with the word "HOLE" or "COVER" to provide warning of the hazard.

**Note:** This provision does not apply to cast iron manhole covers or steel grates used on streets or roadways.

**1926.502(j)** ***Protection from falling objects.*** Falling object protection shall comply with the following provisions:

**1926.502(j)(1)** Toeboards, when used as falling object protection, shall be erected along the edge of the overhead walking/working surface for a distance sufficient to protect employees below.

**1926.502(j)(2)** Toeboards shall be capable of withstanding, without failure, a force of at least 50 pounds (222 N) applied in any downward or outward direction at any point along the toeboard.

**1926.502(j)(3)** Toeboards shall be a minimum of 3 1/2 inches (9 cm) in vertical height from their top edge to the level of the walking/working surface. They shall have not more than 1/4 inch (0.6 cm) clearance above the walking/working surface. They shall be solid or have openings not over 1 inch (2.5 cm) in greatest dimension.

**1926.502(j)(4)** Where tools, equipment, or materials are piled higher than the top edge of a toeboard, paneling or screening shall be erected from the walking/working surface or toeboard to the top of a guardrail system's top rail or midrail, for a distance sufficient to protect employees below.

**1926.502(j)(5)** Guardrail systems, when used as falling object protection, shall have all openings small enough to prevent passage of potential falling objects.

**1926.502(j)(6)** During the performance of overhand bricklaying and related work:

**1926.502(j)(6)(i)** No materials or equipment except masonry and mortar shall be stored within 4 feet (1.2 m) of the working edge.

**1926.502(j)(6)(ii)** Excess mortar, broken or scattered masonry units, and all other materials and debris shall be kept clear from the work area by removal at regular intervals.

**1926.502(j)(7)** During the performance of roofing work:

**1926.502(j)(7)(i)** Materials and equipment shall not be stored within 6 feet (1.8 m) of a roof edge unless guardrails are erected at the edge.

**1926.502(j)(7)(ii)** Materials which are piled, grouped, or stacked near a roof edge shall be stable and self-supporting.

**1926.502(j)(8)** Canopies, when used as falling object protection, shall be strong enough to prevent collapse and to prevent penetration by any objects which may fall onto the canopy.

**1926.502(k)** ***Fall protection plan.*** This option is available only to employees engaged in leading edge work, precast concrete erection work, or residential construction work (See §1926.501(b)(2), (b)(12), and (b)(13)) who can demonstrate that it is infeasible or it creates a greater hazard to use conventional fall protection equipment. The fall protection plan must conform to the following provisions.

**1926.502(k)(1)** The fall protection plan shall be prepared by a qualified person and developed specifically for the site where the leading edge work, precast concrete work, or residential construction work is being performed and the plan must be maintained up to date.

**1926.502(k)(2)** Any changes to the fall protection plan shall be approved by a qualified person.

**1926.502(k)(3)** A copy of the fall protection plan with all approved changes shall be maintained at the job site.

**1926.502(k)(4)** The implementation of the fall protection plan shall be under the supervision of a competent person.

**1926.502(k)(5)** The fall protection plan shall document the reasons why the use of conventional fall protection systems (guardrail systems, personal fall arrest systems, or safety nets systems) are infeasible or why their use would create a greater hazard.

**1926.502(k)(6)** The fall protection plan shall include a written discussion of other measures that will be taken to reduce or eliminate the fall hazard for workers who cannot be provided with protection from the conventional fall protection systems. For example, the employer shall discuss the extent to which scaffolds, ladders, or vehicle mounted work platforms can be used to provide a safer working surface and thereby reduce the hazard of falling.

**1926.502(k)(7)** The fall protection plan shall identify each location where conventional fall protection methods cannot be used. These locations shall then be classified as controlled access zones and the employer must comply with the criteria in paragraph (g) of this section.

**1926.502(k)(8)** Where no other alternative measure has been implemented, the employer shall implement a safety monitoring system in conformance with §1926.502(h).

**1926.502(k)(9)** The fall protection plan must include a statement which provides the name or other method of identification for each employee who is designated to work in controlled access zones. No other employees may enter controlled access zones.

**1926.502(k)(10)** In the event an employee falls, or some other related, serious incident occurs, (e.g., a near miss) the employer shall investigate the circumstances of the fall or other incident to determine if the fall protection plan needs to be changed (e.g. new practices, procedures, or training) and shall implement those changes to prevent similar types of falls or incidents.

### §1926.503 Training requirements.

The following training provisions supplement and clarify the requirements of §1926.21 regarding the hazards addressed in subpart M of this part.

**1926.503(a)** ***Training Program.*** (1) The employer shall provide a training program for each employee who might be exposed to fall hazards. The program shall enable each employee to recognize the hazards of falling and shall train each employee in the procedures to be followed in order to minimize these hazards.

**1926.503(a)(2)** The employer shall assure that each employee has been trained, as necessary, by a competent person qualified in the following areas:

**1926.503(a)(2)(i)** The nature of fall hazards in the work area;

**1926.503(a)(2)(ii)** The correct procedures for erecting, maintaining, disassembling, and inspecting the fall protection systems to be used;

**1926.503(a)(2)(iii)** The use and operation of guardrail systems, personal fall arrest systems, safety net systems, warning line systems, safety monitoring systems, controlled access zones, and other protection to be used;

**1926.503(a)(2)(iv)** The role of each employee in the safety monitoring system when this system is used;

**1926.503(a)(2)(v)** The limitations on the use of mechanical equipment during the performance of roofing work on low-sloped roofs;

**1926.503(a)(2)(vi)** The correct procedures for the handling and storage of equipment and materials and the erection of overhead protection; and

**1926.503(a)(2)(vii)** The role of employees in fall protection plans;

**1926.503(a)(2)(viii)** The standards contained in this subpart.

**1926.503(b)** ***Certification of training.*** (1) The employer shall verify compliance with paragraph (a) of this section by preparing a written certification record. The written certification record shall contain the name or other identity of the employee trained, the date(s) of the training, and the signature of the person who conducted the training or the signature of the employer. If the employer relies on training conducted by another employer or completed prior to the effective date of this section, the certification record shall indicate the date the employer determined the prior training was adequate rather than the date of actual training.

**1926.503(b)(2)** The latest training certification shall be maintained.

**1926.503(c)** ***Retraining.*** When the employer has reason to believe that any affected employee who has already been trained does not have the understanding and skill required by paragraph (a) of this section, the employer shall retrain each such employee. Circumstances where retraining is required include, but are not limited to, situations where:

**1926.503(c)(1)** Changes in the workplace render previous training obsolete; or

**1926.503(c)(2)** Changes in the types of fall protection systems or equipment to be used render previous training obsolete; or

**1926.503(c)(3)** Inadequacies in an affected employee's knowledge or use of fall protection systems or equipment indicate that the employee has not retained the requisite understanding or skill.

**Note:** The following appendices to subpart M of this part serve as nonmandatory guidelines to assist employers in complying with the appropriate requirements of subpart M of this part.

### Appendix A to Subpart M—Determining Roof Widths Non-mandatory Guidelines for Complying With §1926.501(b)(10)

(1) This Appendix serves as a guideline to assist employers complying with the requirements of §1926.501(b)(10). Section 1910.501(b)(10) [sic 1926.501(b)(10)] allows the use of a safety monitoring system alone as a means of providing fall protection during the performance of roofing operations on low-sloped roofs 50 feet (15.25 m) or less in width. Each example in the appendix shows a roof plan or plans and indicates where each roof or roof area is to be measured to determine its width. Section views or elevation views are shown where appropriate. Some examples

show "correct" and "incorrect" subdivisions of irregularly shaped roofs divided into smaller, regularly shaped areas. In all examples, the dimension selected to be the width of an area is the lesser of the two primary dimensions of the area, as viewed from above. Example A shows that on a simple rectangular roof, width is the lesser of the two primary overall dimensions. This is also the case with roofs which are sloped toward or away from the roof center, as shown in Example B.

(2) Many roofs are not simple rectangles. Such roofs may be broken down into subareas as shown in Example C. The process of dividing a roof area can produce many different configurations. Example C gives the general rule of using dividing lines of minimum length to minimize the size and number of the areas which are potentially less than 50 feet (15.25 m) wide. The intent is to minimize the number of roof areas where safety monitoring systems alone are sufficient protection.

(3) Roofs which are comprised of several separate, non-contiguous roof areas, as in Example D, may be considered as a series of individual roofs. Some roofs have penthouses, additional floors, courtyard openings, or similar architectural features; Example E shows how the rule for dividing roofs into subareas is applied to such configurations. Irregular, non-rectangular roofs must be considered on an individual basis, as shown in Example F.

**EXAMPLE A**

Rectangular Shaped Roofs

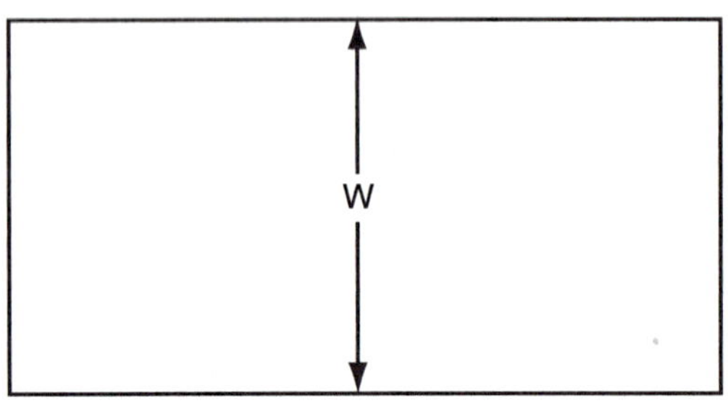


PLAN VIEW

**EXAMPLE B**

Sloped Rectangular Shaped Roofs

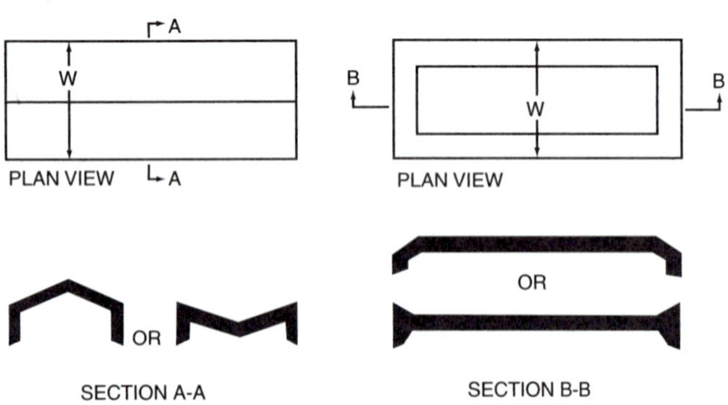


**EXAMPLE C**

Irregularly Shaped Roofs With Rectangular Shaped Sections

Such roofs are to be divided into sub-areas by using dividing lines of minimum length to minimize the size and number of the areas which are potentially less than or equal to 50 feet (15.25 meters) in width, in order to limit the size of roof areas where the safety monitoring system alone can be used [1926.502(b)(10)]. Dotted lines are used in the examples to show the location of dividing lines. Ⓦ denotes incorrect measurements of width.

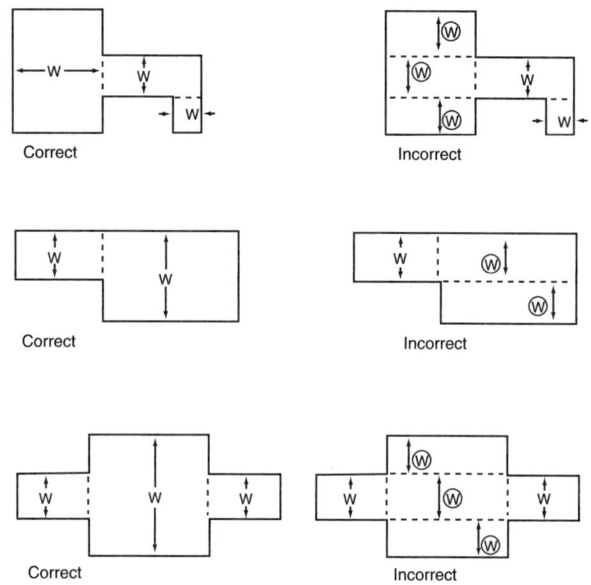


**EXAMPLE D**

Separate, Non-Contiguous Roof Areas

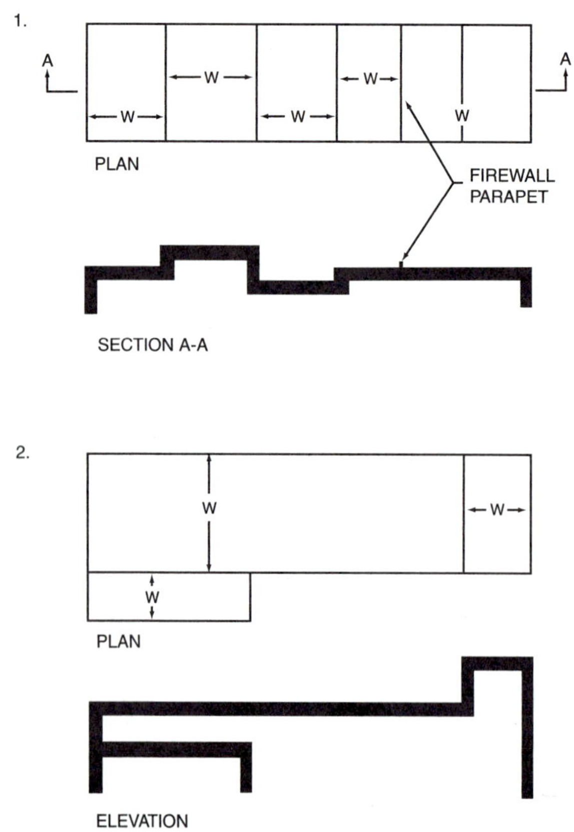

## EXAMPLE E

### Roofs With Penthouses, Open Courtyards, Additional Floors, Etc.

Such roofs are to be divided into sub-areas by using dividing lines of minimum length to minimize the size and number of the areas which are potentially less than or equal to 50 feet (15.25 meters) in width, in order to limit the size of roof areas where the safety monitoring system alone can be used [1926.502(b)(10)]. Dotted lines are used in the examples to show the location of dividing lines. W denotes incorrect measurements of width.

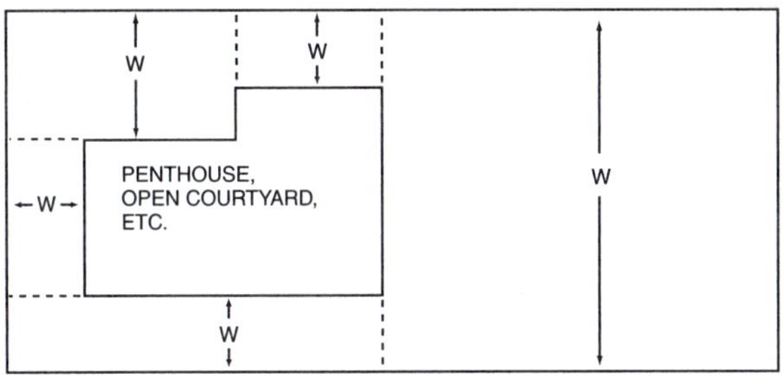


Correct

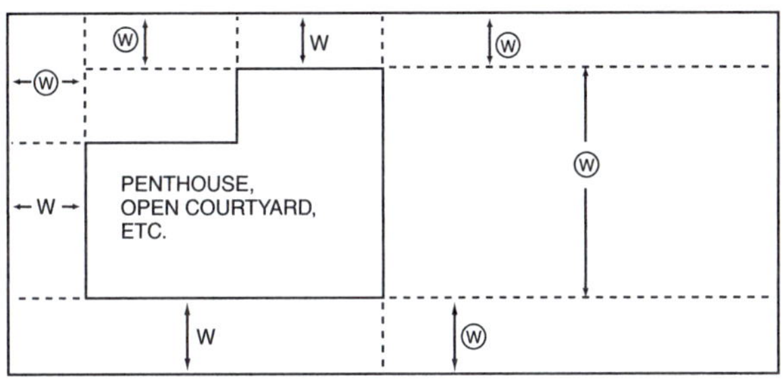


Incorrect

## EXAMPLE F

### Irregular, Non-Rectangular Shaped Roofs

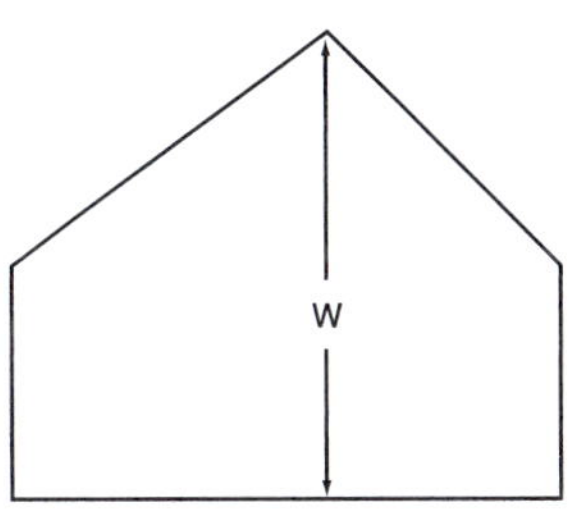


PLAN

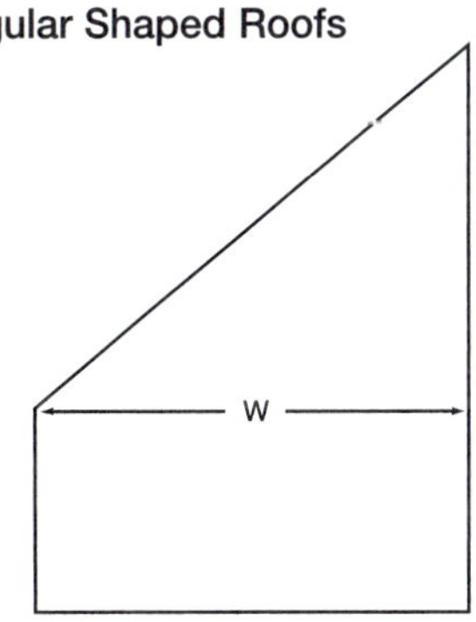


PLAN

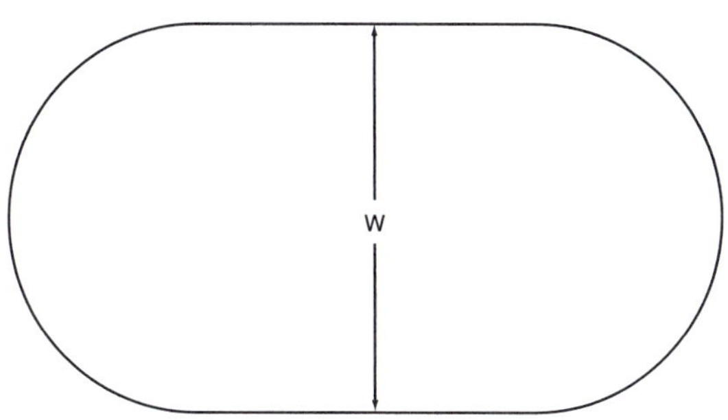


PLAN

## Appendix B to Subpart M—Guardrail Systems

### Non-Mandatory Guidelines for Complying with §1926.502(b)

The standard requires guardrail systems and components to be designed and built to meet the requirements of §1926.502(b)(3), (4), and (5). This Appendix serves as a non-mandatory guideline to assist employers in complying with these requirements. An employer may use these guidelines as a starting point for designing guardrail systems. However, the guidelines do not provide all the information necessary to build a complete system, and the employer is still responsible for designing and assembling these components in such a way that the completed system will meet the requirements of §1926.502(b)(3), (4), and (5). Components for which no specific guidelines are given in this Appendix (e.g., joints, base connections, components made with other materials, and components with other dimensions) must also be designed and constructed in such a way that the completed system meets the requirements of §1926.502.

(1) For wood railings: Wood components shall be minimum 1500 lb-ft/in $^2$ fiber (stress grade) construction grade lumber; the posts shall be at least 2-inch by 4-inch (5 cm x 10 cm) lumber spaced not more than 8 feet (2.4 m) apart on centers; the top rail shall be at least 2-inch by 4-inch (5 cm x 10 cm) lumber, the intermediate rail shall be at least 1-inch by 6-inch (2.5 cm x 15 cm) lumber. All lumber dimensions are nominal sizes as provided by the American Softwood Lumber Standards, dated January 1970.

(2) For pipe railings: posts, top rails, and intermediate railings shall be at least one and one-half inches nominal diameter (schedule 40 pipe) with posts spaced not more than 8 feet (2.4 m) apart on centers.

(3) For structural steel railings: posts, top rails, and intermediate rails shall be at least 2-inch by 2-inch (5 cm x 10 cm) by 3/8-inch (1.1 cm) angles, with posts spaced not more than 8 feet (2.4 m) apart on centers.

## Appendix C to Subpart M—Personal Fall Arrest Systems

### Non-Mandatory Guidelines for Complying With §1926.502(d)

I. *Test methods for personal fall arrest systems and positioning device systems*—(a) *General.* This appendix serves as a non-mandatory guideline to assist employers comply with the requirements in §1926.502(d). Paragraphs (b), (c), (d) and (e) of this Appendix describe test procedures which may be used to determine compliance with the requirements in §1926.502(d)(16). As noted in Appendix D of this subpart, the test methods listed here in Appendix C can also be used to assist employers comply with the requirements in §1926.502(e)(3) and (4) for positioning device systems.

(b) *General conditions for all tests in the Appendix to §1926.502(d).* (1) Lifelines, lanyards and deceleration devices should be attached to an anchorage and connected to the body-belt or body harness in the same manner as they would be when used to protect employees.

(2) The anchorage should be rigid, and should not have a deflection greater than 0.04 inches (1 mm) when a force of 2,250 pounds (10 kN) is applied.

(3) The frequency response of the load measuring instrumentation should be 500 Hz.

(4) The test weight used in the strength and force tests should be a rigid, metal, cylindrical or torso-shaped object with a girth of 38 inches plus or minus 4 inches (96 cm plus or minus 10 cm).

(5) The lanyard or lifeline used to create the free fall distance should be supplied with the system, or in its absence, the least elastic lanyard or lifeline available to be used with the system.

(6) The test weight for each test should be hoisted to the required level and should be quickly released without having any appreciable motion imparted to it.

(7) The system's performance should be evaluated taking into account the range of environmental conditions for which it is designed to be used.

(8) Following the test, the system need not be capable of further operation.

(c) *Strength test.* (1) During the testing of all systems, a test weight of 300 pounds plus or minus 5 pounds (135 kg plus or minus 2.5 kg) should be used. (See paragraph (b)(4) of this section.)

(2) The test consists of dropping the test weight once. A new unused system should be used for each test.

(3) For lanyard systems, the lanyard length should be 6 feet plus or minus 2 inches (1.83 m plus or minus 5 cm) as measured from the fixed anchorage to the attachment on the body belt or body harness.

(4) For rope-grab-type deceleration systems, the length of the lifeline above the centerline of the grabbing mechanism to the lifeline's anchorage point should not exceed 2 feet (0.61 m).

(5) For lanyard systems, for systems with deceleration devices which do not automatically limit free fall distance to 2 feet (0.61 m) or less, and for systems with deceleration devices which have a connection distance in excess of 1 foot (0.3 m) (measured between the centerline of the lifeline and the attachment point to the body belt or harness), the test weight should be rigged to free fall a distance of 7.5 feet (2.3 m) from a point that is 1.5 feet (.46 m) above the anchorage point, to its hanging location (6 feet below the anchorage). The test weight should fall without interference, obstruction, or hitting the floor or ground during the test. In some cases a non-elastic wire lanyard of sufficient length may need to be added to the system (for test purposes) to create the necessary free fall distance.

(6) For deceleration device systems with integral lifelines or lanyards which automatically limit free fall distance to 2 feet (0.61 m) or less, the test weight should be rigged to free fall a distance of 4 feet (1.22 m).

(7) Any weight which detaches from the belt or harness has failed the strength test.

(d) *Force test*—(1) *General.* The test consists of dropping the respective test weight once as specified in paragraph (d)(2)(i) or (d)(3)(i) of this section. A new, unused system should be used for each test.

(2) *For lanyard systems.* (i) A test weight of 220 pounds plus or minus 3 pounds (100 kg plus or minus 1.6 kg) should be used. (See paragraph (b)(4) of this appendix).

(ii) Lanyard length should be 6 feet plus or minus two inches (1.83 m plus or minus 5 cm) as measured from the fixed anchorage to the attachment on the body belt or body harness.

(iii) The test weight should fall free from the anchorage level to its hanging location (a total of 6 feet (1.83 m) free fall distance) without interference, obstruction, or hitting the floor or ground during the test.

(3) *For all other systems.* (i) A test weight of 220 pounds plus or minus 3 pounds (100 kg plus or minus 1.6 kg) should be used. (See paragraph (b)(4) of this appendix.)

(ii) The free fall distance to be used in the test should be the maximum fall distance physically permitted by the system during normal use conditions, up to a maximum free fall distance for the test weight of 6 feet (1.83 m), except as follows:

(A) For deceleration systems which have a connection link or lanyard, the test weight should free fall a distance equal to the connection distance (measured between the centerline of the lifeline and the attachment point to the body belt or harness).

(B) For deceleration device systems with integral lifelines or lanyards which automatically limit free fall distance to 2 feet (0.61 m) or less, the test weight should free fall a distance equal to that permitted by the system in normal use. (For example, to test a system with a self-retracting lifeline or lanyard, the test weight should be supported and the system allowed to retract the lifeline or lanyard as it would in normal use. The test weight would then be released and the force and deceleration distance measured).

(4) A system fails the force test if the recorded maximum arresting force exceeds 1,260 pounds (5.6 kN) when using a body belt, and/or exceeds 2,520 pounds (11.2 kN) when using a body harness.

(5) The maximum elongation and deceleration distance should be recorded during the force test.

(e) *Deceleration device tests.* (1) *General.* The device should be evaluated or tested under the environmental conditions, (such as rain, ice, grease, dirt, type of lifeline, etc.), for which the device is designed.

(2) *Rope-grab-type deceleration devices.* (i) Devices should be moved on a lifeline 1,000 times over the same length of line a distance of not less than 1 foot (30.5 cm), and the mechanism should lock each time.

(ii) Unless the device is permanently marked to indicate the type(s) of lifeline which must be used, several types (different diameters and different materials), of lifelines should be used to test the device.

(3) *Other self-activating-type deceleration devices.* The locking mechanisms of other self-activating-type deceleration devices designed for more than one arrest should lock each of 1,000 times as they would in normal service.

II. *Additional non-mandatory guidelines for personal fall arrest systems.* The following information constitutes additional guidelines for use in complying with requirements for a personal fall arrest system.

(a) *Selection and use considerations.* (1) The kind of personal fall arrest system selected should match the particular work situation, and any possible free fall distance should be kept to a minimum. Consideration should be given to the particular work environment. For example, the presence of acids, dirt, moisture, oil, grease, etc., and their effect on the system, should be evaluated. Hot or cold environments may also have an adverse effect on the system. Wire rope should not be used where an electrical hazard is anticipated. As required by the standard, the employer must plan to have means available to promptly rescue an employee should a fall occur, since the suspended employee may not be able to reach a work level independently.

(2) Where lanyards, connectors, and lifelines are subject to damage by work operations such as welding, chemical cleaning, and sandblasting, the component should be protected, or other securing systems should be used. The employer should fully evaluate the work conditions and environment (including seasonal weather changes) before selecting the appropriate personal fall protection system. Once in use, the system's effectiveness should be monitored. In some cases, a program for cleaning and maintenance of the system may be necessary.

(b) *Testing considerations.* Before purchasing or putting into use a personal fall arrest system, an employer should obtain from the supplier information about the system based on its performance during testing so that the employer can know if the system meets this standard. Testing should be done using recognized test methods. This Appendix contains test methods recognized for evaluating the performance of fall arrest systems. Not all systems may need to be individually tested; the performance of some systems may be based on data and calculations derived from testing of similar systems, provided that enough information is available to demonstrate similarity of function and design.

(c) *Component compatibility considerations.* Ideally, a personal fall arrest system is designed, tested, and supplied as a complete system. However, it is common practice for lanyards, connectors, lifelines, deceleration devices, body belts and body harnesses to be interchanged since some components wear out before others. The employer and employee should realize that not all components are interchangeable. For instance, a lanyard should not be connected between a body belt (or harness) and a deceleration device of the self-retracting type since this can result in additional free fall for which the system was not designed. Any substitution or change to a personal fall arrest system should be fully evaluated or tested by a competent person to determine that it meets the standard, before the modified system is put in use.

(d) *Employee training considerations.* Thorough employee training in the selection and use of personal fall arrest systems is imperative. Employees must be trained in the safe use of the system. This should include the following: application limits; proper anchoring and tie-off techniques; estimation of free fall distance, including determination of deceleration distance, and total fall distance to prevent striking a lower level; methods of use; and inspection and storage of the system. Careless or improper use of the equipment can result in serious injury or death. Employers and employees should become familiar with the material in this Appendix, as well as manufacturer's recommendations, before a system is used. Of uppermost importance is the reduction in strength caused by certain tie-offs (such as using knots, tying around sharp edges, etc.) and maximum permitted free fall distance. Also, to be stressed are the importance of inspections prior to use, the limitations of the equipment, and unique conditions at the worksite which may be important in determining the type of system to use.

(e) *Instruction considerations.* Employers should obtain comprehensive instructions from the supplier as to the system's proper use and application, including, where applicable:

(1) The force measured during the sample force test;

(2) The maximum elongation measured for lanyards during the force test;

(3) The deceleration distance measured for deceleration devices during the force test;

(4) Caution statements on critical use limitations;

(5) Application limits;

(6) Proper hook-up, anchoring and tie-off techniques, including the proper dee-ring or other attachment point to use on the body belt and harness for fall arrest;

(7) Proper climbing techniques;

(8) Methods of inspection, use, cleaning, and storage; and

(9) Specific lifelines which may be used. This information should be provided to employees during training.

(f) *Rescue considerations.* As required by §1926.502(d)(20), when personal fall arrest systems are used, the employer must assure that employees can be promptly rescued or can rescue themselves should a fall occur. The availability of rescue personnel, ladders or other rescue equipment should be evaluated. In some situations, equipment which allows employees to rescue themselves after the fall has been arrested may be desirable, such as devices which have descent capability.

(g) *Inspection considerations.* As required by §1926.502(d)(21), personal fall arrest systems must be regularly inspected. Any component with any significant defect, such as cuts, tears, abrasions, mold, or undue stretching; alterations or additions which might affect its efficiency; damage due to deterioration; contact with fire, acids, or other corrosives; distorted hooks or faulty hook springs; tongues unfitted to the shoulder of buckles; loose or damaged mountings; non-functioning parts; or wearing or internal deterioration in the ropes must be withdrawn from service immediately, and should be tagged or marked as unusable, or destroyed.

(h) *Tie-off considerations.* (1) One of the most important aspects of personal fall protection systems is fully planning the system *before* it is put into use. Probably the most overlooked component is planning for suitable anchorage points. Such planning should ideally be done before the structure or building is constructed so that anchorage points can be incorporated during construction for use later for window cleaning or other building maintenance. If properly planned, these anchorage points may be used *during* construction, as well as afterwards.

(i) Properly planned anchorages should be used if they are available. In some cases, anchorages must be installed immediately prior to use. In such cases, a registered professional engineer with experience in designing fall protection systems, or another qualified person with appropriate education and experience should design an anchor point to be installed.

(ii) In other cases, the Agency recognizes that there will be a need to devise an anchor point from existing structures. Examples of what might be appropriate anchor points are steel members or I-beams if an acceptable strap is available for the connection (do not use a lanyard with a snaphook clipped onto itself); large eye-bolts made of an appropriate grade steel; guardrails or railings if they have been designed for use as an anchor point; or masonry or wood members only if the attachment point is substantial and precautions have been taken to assure that bolts or other connectors will not pull through. A qualified person should be used to evaluate the suitable of these "make shift" anchorages with a focus on proper strength.

(2) Employers and employees should at all times be aware that the strength of a personal fall arrest system is based on its being attached to an anchoring system which does not reduce the strength of the system (such as a properly dimensioned eye-bolt/snap-hook anchorage). Therefore, if a means of attachment is used that will reduce the strength of the system, that component should be replaced by a stronger one, but one that will also maintain the appropriate maximum arrest force characteristics.

(3) Tie-off using a knot in a rope lanyard or lifeline (at any location) can reduce the lifeline or lanyard strength by 50 percent or more. Therefore, a stronger lanyard or lifeline should be used to compensate for the weakening effect of the knot, or the lanyard length should be reduced (or the tie-off location raised) to minimize free fall distance, or the lanyard or lifeline should be replaced by one which has an appropriately incorporated connector to eliminate the need for a knot.

(4) Tie-off of a rope lanyard or lifeline around an "H" or "I" beam or similar support can reduce its strength as much as 70 percent due to the cutting action of the beam edges. Therefore, use should be made of a webbing lanyard or wire core lifeline around the beam; or the lanyard or lifeline should be protected from the edge; or free fall distance should be greatly minimized.

(5) Tie-off where the line passes over or around rough or sharp surfaces reduces strength drastically. Such a tie-off should be avoided or an alternative tie-off rigging should be used. Such alternatives may include use of a snap-hook/dee ring connection, wire rope tie-off, an effective padding of the surfaces, or an abrasion-resistance strap around or over the problem surface.

(6) Horizontal lifelines may, depending on their geometry and angle of sag, be subjected to greater loads than the impact load imposed by an attached component. When the angle of horizontal lifeline sag is less than 30 degrees, the impact force imparted to the lifeline by an attached lanyard is greatly amplified. For example, with a sag angle of 15 degrees, the force amplification is about 2:1 and at 5 degrees sag, it is about 6:1. Depending on the angle of sag, and the line's elasticity, the strength of the horizontal lifeline and the anchorages to which it is attached should be increased a number of times over that of the lanyard. Extreme care should be taken in considering a horizontal lifeline for multiple tie-offs. The reason for this is that in multiple tie-offs to a horizontal lifeline, if one employee falls, the movement of the falling employee and the horizontal lifeline during arrest of the fall may cause other employees to fall also. Horizontal lifeline and anchorage strength should be increased for each additional employee to be tied off. For these and other reasons, the design of systems using horizontal lifelines must only be done by qualified persons. Testing of installed lifelines and anchors prior to use is recommended.

(7) The strength of an eye-bolt is rated along the axis of the bolt and its strength is greatly reduced if the force is applied at an angle to this axis (in the direction of shear). Also, care should be exercised in selecting the proper diameter of the eye to avoid accidental disengagement of snap-hooks not designed to be compatible for the connection.

(8) Due to the significant reduction in the strength of the lifeline/lanyard (in some cases, as much as a 70 percent reduction), the sliding hitch knot (prusik) should not be used for lifeline/lanyard connections except in emergency situations where no other available system is practical. The "one-and-one" sliding hitch knot should never be used because it is unreliable in stopping a fall. The "two-and-two," or "three-and-three" knot (preferable) may be used in emergency situations; however, care should be taken to limit free fall distance to a minimum because of reduced lifeline/lanyard strength.

(i) *Vertical lifeline considerations.* As required by the standard, each employee must have a separate lifeline [except employees engaged in constructing elevator shafts who are permitted to have two employees on one lifeline] when the lifeline is vertical. The reason for this is that in multiple tie-offs to a single lifeline, if one employee falls, the movement of the lifeline during the arrest of the fall may pull other employees' lanyards, causing them to fall as well.

(j) *Snap-hook considerations.* (1) Although not required by this standard for all connections until January 1, 1998, locking snaphooks designed for connection to suitable objects (of sufficient strength) are highly recommended in lieu of the nonlocking type. Locking snaphooks incorporate a positive locking mechanism in addition to the spring loaded keeper, which will not allow the keeper to open under moderate pressure without someone first releasing the mechanism. Such a feature, properly designed, effectively prevents roll-out from occurring.

(2) As required by §1926.502(d)(6), the following connections must be avoided (unless properly designed locking snaphooks are used) because they are conditions which can result in roll-out when a nonlocking snaphook is used:

(i) Direct connection of a snaphook to a horizontal lifeline.

(ii) Two (or more) snaphooks connected to one dee-ring.

(iii) Two snaphooks connected to each other.

(iv) A snaphook connected back on its integral lanyard.

(v) A snaphook connected to a webbing loop or webbing lanyard.

(vi) Improper dimensions of the dee-ring, rebar, or other connection point in relation to the snaphook dimensions which would allow the snaphook keeper to be depressed by a turning motion of the snaphook.

(k) *Free fall considerations.* The employer and employee should at all times be aware that a system's maximum arresting force is evaluated under normal use conditions established by the manufacturer, and in no case using a free fall distance in excess of 6 feet (1.8 m). A few extra feet of free fall can significantly increase the arresting force on the employee, possibly to the point of causing injury. Because of this, the free fall distance should be kept at a minimum, and, as required by the standard, in no case greater than 6 feet (1.8 m). To help assure this, the tie-off attachment point to the lifeline or anchor should be located at or above the connection point of the fall arrest equipment to belt or harness. (Since otherwise additional free fall distance is added to the length of the connecting means (i.e. lanyard)). Attaching to the working surface will often result in a free fall greater than 6 feet (1.8 m). For instance, if a 6 foot (1.8 m) lanyard is used, the total free fall distance will be the distance from the working level to the body belt (or harness) attachment point plus the 6 feet (1.8 m) of lanyard length. Another important consideration is that the arresting force which the fall system must withstand also goes up with greater distances of free fall, possibly exceeding the strength of the system.

(l) *Elongation and deceleration distance considerations.* Other factors involved in a proper tie-off are elongation and deceleration distance. During the arresting of a fall, a lanyard will experience a length of stretching or elongation, whereas activation of a deceleration device will result in a certain stopping distance. These distances should be available with the lanyard or device's instructions and must be added to the free fall distance to arrive at the total fall distance before an employee is fully stopped. The additional stopping distance may be very significant if the lanyard or deceleration device is attached near or at the end of a long lifeline, which may itself add considerable distance due to its own elongation. As required by the standard, sufficient distance to allow for all of these factors must also be maintained between the employee and obstructions below, to prevent an injury due to impact before the system fully arrests the fall. In addition, a minimum of 12 feet (3.7 m) of lifeline should be allowed below the securing point of a rope grab type deceleration device, and the end terminated to prevent the device from sliding off the lifeline. Alternatively, the lifeline should extend to the ground or the next working level below. These measures are suggested to prevent the worker from inadvertently moving past the end of the lifeline and having the rope grab become disengaged from the lifeline.

(m) *Obstruction considerations.* The location of the tie-off should also consider the hazard of obstructions in the potential fall path of the employee. Tie-offs which minimize the possibilities of exaggerated swinging should be considered. In addition, when a body belt is used, the employee's body will go through a horizontal position to a jack-knifed position during the arrest of all falls. Thus, obstructions which might interfere with this motion should be avoided or a severe injury could occur.

(n) *Other considerations.* Because of the design of some personal fall arrest systems, additional considerations may be required for proper tie-off. For example, heavy deceleration devices of the self-retracting type should be secured overhead in order to avoid the weight of the device having to be supported by the employee. Also, if self-retracting equipment is connected to a horizontal lifeline, the sag in the lifeline should be minimized to prevent the device from sliding down the lifeline to a position which creates a swing hazard during fall arrest. In all cases, manufacturer's instructions should be followed.

## Appendix D to Subpart M—Positioning Device Systems

### Non-Mandatory Guidelines for Complying With §1926.502(e)

I. *Testing Methods For Positioning Device Systems.* This appendix serves as a non-mandatory guideline to assist employers comply with the requirements for positioning device systems in §1926.502(e). Paragraphs (b), (c), (d) and (e) of Appendix C of subpart M relating to §1926.502(d)—Personal Fall Arrest Systems—set forth test procedures which may be used, along with the procedures listed below, to determine compliance with the requirements for positioning device systems in §1926.502(e)(3) and (4) of subpart M.

(a) *General.* (1) Single strap positioning devices shall have one end attached to a fixed anchorage and the other end connected to a body belt or harness in the same manner as they would be used to protect employees. Double strap positioning devices, similar to window cleaner's belts, shall have one end of the strap attached to a fixed anchorage and the other end shall hang free. The body belt or harness shall be attached to the strap in the same manner as it would be used to protect employees. The two strap ends shall be adjusted to their maximum span.

(2) The fixed anchorage shall be rigid, and shall not have a deflection greater than .04 inches (1 mm) when a force of 2,250 pounds (10 kN) is applied.

(3) During the testing of all systems, a test weight of 250 pounds plus or minus 3 pounds (113 kg plus or minus 1.6 kg) shall be used. The weight shall be a rigid object with a girth of 38 inches plus or minus 4 inches (96 cm plus or minus 10 cm).

(4) Each test shall consist of dropping the specified weight one time without failure of the system being tested. A new system shall be used for each test.

(5) The test weight for each test shall be hoisted exactly 4 feet (1.2 m above its "at rest" position), and shall be dropped so as to permit a vertical free fall of 4 feet (1.2 m).

(6) The test is failed whenever any breakage or slippage occurs which permits the weight to fall free of the system.

(7) Following the test, the system need not be capable of further operation; however, all such incapacities shall be readily apparent.

II. *Inspection Considerations.* As required in §1926.502(e)(5), positioning device systems must be regularly inspected. Any component with any significant defect, such as cuts, tears, abrasions, mold, or undue stretching; alterations or additions which might affect its efficiency; damage due to deterioration; contact with fire, acids, or other corrosives; distorted hooks or faulty hook springs; tongues unfitted to the shoulder of buckles; loose or damaged mountings; non-functioning parts; or wearing or internal deterioration in the ropes must be withdrawn from service immediately, and should be tagged or marked as unusable, or destroyed.

## Appendix E to Subpart M—Sample Fall Protection Plan

### Non-Mandatory Guidelines for Complying With §1926.502(k)

Employers engaged in leading edge work, precast concrete construction work and residential construction work who can demonstrate that it is infeasible or creates a greater hazard to use conventional fall protection systems must develop and follow a fall protection plan. Below are sample fall protection plans developed for precast concrete construction and residential work that could be tailored to be site specific for other precast concrete or residential jobsite. This sample plan can be modified to be used for other work involving leading edge work. The sample plan outlines the elements that must be addressed in any fall protection plan. The reasons outlined in this sample fall protection plan are for illustrative purposes only and are not necessarily a valid, acceptable rationale (unless the conditions at the job site are the same as those covered by these sample plans) for not using conventional fall protection systems for a particular precast concrete or residential construction worksite. However, the sample plans provide guidance to employers on the type of information that is required to be discussed in fall protection plans.

### SAMPLE FALL PROTECTION PLANS

### Fall Protection Plan For Precast/Prestress Concrete Structures

This Fall Protection Plan is specific for the following project:

Location of Job ____________________

Erecting Company ____________________

Date Plan Prepared or Modified ____________________

Plan Prepared By ____________________

Plan Approved By ____________________

Plan Supervised By ____________________

The following Fall Protection Plan is a sample program prepared for the prevention of injuries associated with falls. A Fall Protection Plan must be developed and evaluated on a site by site basis. It is recommended that erectors discuss the written Fall Protection Plan with their OSHA Area Office prior to going on a jobsite.

### I. STATEMENT OF COMPANY POLICY

(Company Name) is dedicated to the protection of its employees from on-the-job injuries. All employees of (Company Name) have the responsibility to work safely on the job. The purpose of this plan is: (a) To supplement our standard safety policy by providing safety standards specifically designed to cover fall protection on this job and; (b) to ensure that each employee is trained and made aware of the safety provisions which are to be implemented by this plan prior to the start of erection.

This Fall Protection Plan addresses the use of other than conventional fall protection at a number of areas on the project, as well as identifying specific activities that require non-conventional means of fall protection. These areas include:

a. Connecting activity (point of erection).

b. Leading edge work.

c. Unprotected sides or edge.

d. Grouting.

This plan is designed to enable employers and employees to recognize the fall hazards on this job and to establish the procedures that are to be followed in order to prevent falls to lower levels or through holes and openings in walking/working surfaces. Each employee will be trained in these procedures and strictly adhere to them except when doing so would expose the employee to a greater hazard. If, in the employees opinion, this is the case, the employee is to notify the foreman of the concern and the concern addressed before proceeding.

Safety policy and procedure on any one project cannot be administered, implemented, monitored and enforced by any one individual. The total objective of a safe, accident free work environment can only be accomplished by a dedicated, concerted effort by every individual involved with the project from management down to the last employee. Each employee must understand their value to the company; the costs of accidents, both monetary, physical, and emotional; the objective of the safety policy and procedures; the safety rules that apply to the safety policy and procedures; and what their individual role is in administering, implementing, monitoring, and compliance of their safety policy and procedures. This allows for a more personal approach to compliance through planning, training, understanding and cooperative effort, rather than by strict enforcement. If for any reason an unsafe act persists, strict enforcement will be implemented.

It is the responsibility of (name of competent person) to implement this Fall Protection Plan. (Name of Competent Person) is responsible for continual observational safety checks of their work operations and to enforce the safety policy and procedures. The foreman also is responsible to correct any unsafe acts or conditions immediately. It is the responsibility of the employee to understand and adhere to the procedures of this plan and to follow the instructions of the foreman. It is also the responsibility of the employee to bring to managements attention any unsafe or hazardous conditions or acts that may cause injury to either themselves or any other employees. Any changes to this Fall Protection Plan must be approved by (name of Qualified Person).

## II. FALL PROTECTION SYSTEMS TO BE USED ON THIS PROJECT

Where conventional fall protection is infeasible or creates a greater hazard at the leading edge and during initial connecting activity, we plan to do this work using a safety monitoring system and expose only a minimum number of employees for the time necessary to actually accomplish the job. The maximum number of workers to be monitored by one safety monitor is six (6). We are designating the following trained employees as designated erectors and they are permitted to enter the controlled access zones and work without the use of conventional fall protection.

Safety monitor:

Designated erector:

Designated erector:

Designated erector:

Designated erector:

Designated erector:

Designated erector:

The safety monitor shall be identified by wearing an orange hard hat. The designated erectors will be identified by one of the following methods:

1. They will wear a blue colored arm band, or

2. They will wear a blue colored hard hat, or

3. They will wear a blue colored vest.

Only individuals with the appropriate experience, skills, and training will be authorized as designated erectors. All employees that will be working as designated erectors under the safety monitoring system shall have been trained and instructed in the following areas:

1. Recognition of the fall hazards in the work area (at the leading edge and when making initial connections—point of erection).

2. Avoidance of fall hazards using established work practices which have been made known to the employees.

3. Recognition of unsafe practices or working conditions that could lead to a fall, such as windy conditions.

4. The function, use, and operation of safety monitoring systems, guardrail systems, body belt/harness systems, control zones and other protection to be used.

5. The correct procedure for erecting, maintaining, disassembling and inspecting the system(s) to be used.

6. Knowledge of construction sequence or the erection plan.

A conference will take place prior to starting work involving all members of the erection crew, crane crew and supervisors of any other concerned contractors. This conference will be conducted by the precast concrete erection supervisor in charge of the project. During the pre-work conference, erection procedures and sequences pertinent to this job will be thoroughly discussed and safety practices to be used throughout the project will be specified. Further, all personnel will be informed that the controlled access zones are off limits to all personnel other than those designated erectors specifically trained to work in that area.

### Safety Monitoring System

A safety monitoring system means a fall protection system in which a competent person is responsible for recognizing and warning employees of fall hazards. The duties of the safety monitor are to:

1. Warn by voice when approaching the open edge in an unsafe manner.

2. Warn by voice if there is a dangerous situation developing which cannot be seen by another person involved with product placement, such as a member getting out of control.

3. Make the designated erectors aware they are in a dangerous area.

4. Be competent in recognizing fall hazards.

5. Warn employees when they appear to be unaware of a fall hazard or are acting in an unsafe manner.

6. Be on the same walking/working surface as the monitored employees and within visual sighting distance of the monitored employees.

7. Be close enough to communicate orally with the employees.

8. Not allow other responsibilities to encumber monitoring.

If the safety monitor becomes too encumbered with other responsibilities, the monitor shall (1) stop the erection process; and (2) turn over other responsibilities to a designated erector; or (3) turn over the safety monitoring function to another designated, competent person. The safety monitoring system shall not be used when the wind is strong enough to cause loads with large surface areas to swing out of radius, or result in loss of control of the load, or when weather conditions cause the walking-working surfaces to become icy or slippery.

### Control Zone System

A controlled access zone means an area designated and clearly marked, in which leading edge work may take place without the use of guardrail, safety net or personal fall arrest systems to protect the employees in the area. Control zone systems shall comply with the following provisions:

1. When used to control access to areas where leading edge and other operations are taking place the controlled access zone shall be defined by a control line or by any other means that restricts access.

When control lines are used, they shall be erected not less than 6 feet (1.8 m) nor more than 60 feet (18 m) or half the length of the member being erected, whichever is less, from the leading edge.

2. The control line shall extend along the entire length of the unprotected or leading edge and shall be approximately parallel to the unprotected or leading edge.

3. The control line shall be connected on each side to a guardrail system or wall.

4. Control lines shall consist of ropes, wires, tapes, or equivalent materials, and supporting stanchions as follows:

5. Each line shall be flagged or otherwise clearly marked at not more than 6-foot (1.8 m) intervals with high- visibility material.

6. Each line shall be rigged and supported in such a way that its lowest point (including sag) is not less than 39 inches (1 m) from the walking/working surface and its highest point is not more than 45 inches (1.3 m) from the walking/working surface.

7. Each line shall have a minimum breaking strength of 200 pounds (.88 kN).

### Holes

All openings greater than 12 in. x 12 in. will have perimeter guarding or covering. All predetermined holes will have the plywood covers made in the precasters' yard and shipped with the member to the jobsite. Prior to cutting holes on the job, proper protection for the hole must be provided to protect the workers. Perimeter guarding or covers will not be removed without the approval of the erection foreman.

Precast concrete column erection through the existing deck requires that many holes be provided through this deck. These are to be covered and protected. Except for the opening being currently used to erect a column, all opening protection is to be left undisturbed. The opening being uncovered to erect a column will become part of the point of erection and will be addressed as part of this Fall Protection Plan. This uncovering is to be done at the erection foreman's direction and will only occur immediately prior to "feeding" the column through the opening. Once the end of the column is through the slab opening, there will no longer exist a fall hazard at this location.

## III. IMPLEMENTATION OF FALL PROTECTION PLAN

The structure being erected is a multistory total precast concrete building consisting of columns, beams, wall panels and hollow core slabs and double tee floor and roof members.

The following is a list of the products and erection situations on this job:

### *Columns*

For columns 10 ft to 36 ft long, employees disconnecting crane hooks from columns will work from a ladder and wear a body belt/harness with lanyard and be tied off when both hands are needed to disconnect. For tying off, a vertical lifeline will be connected to the lifting eye at the top of the column, prior to lifting, to be used with a manually operated or mobile rope grab. For columns too high for the use of a ladder, 36 ft and higher, an added cable will be used to reduce the height of the disconnecting point so that a ladder can be used. This cable will be left in place until a point in erection that it can be removed safely. In some cases, columns will be unhooked from the crane by using an erection tube or shackle with a pull pin which is released from the ground after the column is stabilized.

The column will be adequately connected and/or braced to safely support the weight of a ladder with an employee on it.

### *Inverted Tee Beams*

Employees erecting inverted tee beams, at a height of 6 to 40 ft, will erect the beam, make initial connections, and final alignment from a ladder. If the employee needs to reach over the side of the beam to bar or make an adjustment to the alignment of the beam, they will mount the beam and be tied off to the lifting device in the beam after ensuring the load has been stabilized on its bearing. To disconnect the crane from the beam an employee will stand a ladder against the beam. Because the use of ladders is not practical at heights above 40 ft, beams will be initially placed with the use of tag lines and their final alignment made by a person on a manlift or similar employee positioning systems.

### *Spandrel Beams*

Spandrel beams at the exterior of the building will be aligned as closely as possible with the use of tag lines with the final placement of the spandrel beam made from a ladder at the open end of the structure. A ladder will be used to make the initial connections and a ladder will be used to disconnect the crane. The other end of the beam will be placed by the designated erector from the double tee deck under the observation of the safety monitor.

The beams will be adequately connected and/or braced to safely support the weight of a ladder with an employee on it.

### *Floor and Roof Members*

During installation of the precast concrete floor and/or roof members, the work deck continuously increases in area as more and more units are being erected and positioned. Thus, the unprotected floor/roof perimeter is constantly modified with the leading edge changing location as each member is installed. The fall protection for workers at the leading edge shall be assured by properly constructed and maintained control zone lines not more than 60 ft away from the leading edge supplemented by a safety monitoring system to ensure the safety of all designated erectors working within the area defined by the control zone lines.

The hollow core slabs erected on the masonry portion of the building will be erected and grouted using the safety monitoring system. Grout will be placed in the space between the end of the slab and face shell of the concrete masonry by dumping from a wheelbarrow. The grout in the keyways between the slabs will be dumped from a wheelbarrow and then spread with long handled tools, allowing the worker to stand erect facing toward the unprotected edge and back from any work deck edge.

Whenever possible, the designated erectors will approach the incoming member at the leading edge only after it is below waist height so that the member itself provides protection against falls.

Except for the situations described below, when the arriving floor or roof member is within 2 to 3 inches of its final position, the designated erectors can then proceed to their position of erection at each end of the member under the control of the safety monitor. Crane hooks will be unhooked from double tee members by designated erectors under the direction and supervision of the safety monitor.

Designated erectors, while waiting for the next floor or roof member, will be constantly under the control of the safety monitor for fall protection and are directed to stay a minimum of six (6) ft from the edge. In the event a designated erector must move from one end of a member, which has just been placed at the leading edge, they must first move away from the leading edge a minimum of six (6) ft and then progress to the other end while maintaining the minimum distance of six (6) ft at all times.

Erection of double tees, where conditions require bearing of one end into a closed pocket and the other end on a beam ledge, restricting the tee legs from going directly into the pockets, require special considerations. The tee legs that are to bear in the closed pocket must hang lower than those at the beam bearing. The double tee will be "two-lined" in order to elevate one end higher than the other to allow for the low end to be ducked into the closed pocket using the following procedure.

The double tee will be rigged with a standard four-way spreader off of the main load line. An additional choker will be attached to the married point of the two-legged spreader at the end of the tee that is to be elevated. The double tee will be hoisted with the main load line and swung into a position as close as possible to the tee's final bearing elevation. When the tee is in this position and stabilized, the whip line load block will be lowered to just above the tee deck. At this time, two erectors will walk out on the suspended tee deck at midspan of the tee member and pull the load block to the end of the tee to be elevated and attach the additional choker to the load block. The possibility of entanglement with the crane lines and other obstacles during this two lining process while raising and lowering the crane block on that second line could be hazardous to an encumbered employee. Therefore, the designated erectors will not tie off during any part of this process. While the designated erectors are on the double tee, the safety monitoring system will be used. After attaching the choker, the two erectors then step back on the previously erected tee deck and signal the crane operator to hoist the load with the whip line to the elevation that will allow for enough clearance to let the low end tee legs slide into the pockets when the main load line is lowered. The erector, who is handling the lowered end of the tee at the closed pocket bearing, will step out on the suspended tee. An erection bar will then be placed between the end of the tee leg and the inside face of the pocketed spandrel member. The tee is barred away from the pocketed member to reduce the friction and lateral force against the pocketed member. As the tee is being lowered, the other erector remains on the tee which was previously erected to handle the other end. At this point the tee is slowly lowered by the crane to a point where the tee legs can freely slide into the pockets. The erector working the lowered end of the tee must keep pressure on the bar between the tee and the face of the pocketed spandrel member to very gradually let the tee legs slide into the pocket to its proper bearing dimension. The tee is then slowly lowered into its final erected position.

The designated erector should be allowed onto the suspended double tee, otherwise there is no control over the horizontal movement of the double tee and this movement could knock the spandrel off of its bearing or the column out of plumb. The control necessary to prevent hitting the spandrel can only be done safely from the top of the double tee being erected.

Loadbearing Wall Panels: The erection of the loadbearing wall panels on the elevated decks requires the use of a safety monitor and a controlled access zone that is a minimum of 25 ft and a maximum of 1/2 the length of the wall panels away from the unprotected edge, so that designated erectors can move freely and unencumbered when receiving the panels. Bracing, if required for stability, will be installed by ladder. After the braces are secured, the crane will be disconnected from the wall by using a ladder. The wall to wall connections will also be performed from a ladder.

Non-Loadbearing Panels (Cladding): The locating of survey lines, panel layout and other installation prerequisites (prewelding, etc.) for non-loadbearing panels (cladding) will not commence until floor perimeter and floor openings have been protected. In some areas, it is necessary because of panel configuration to remove the perimeter protection as the cladding is being installed. Removal of perimeter protection will be performed on a bay to bay basis, just ahead of cladding erection to minimize temporarily unprotected floor edges. Those workers within 6 ft of the edge, receiving and positioning the cladding when the perimeter protection is removed shall be tied off.

### Detailing

Employees exposed to falls of six (6) feet or more to lower levels, who are not actively engaged in leading edge work or connecting activity, such as welding, bolting, cutting, bracing, guying, patching, painting or other operations, and who are working less than six (6) ft from an unprotected edge will be tied off at all times or guardrails will be installed. Employees engaged in these activities but who are more than six (6) ft from an unprotected edge as defined by the control zone lines, do not require fall protection but a warning line or control lines must be erected to remind employees they are approaching an area where fall protection is required.

## IV. CONVENTIONAL FALL PROTECTION CONSIDERED FOR THE POINT OF ERECTION OR LEADING EDGE ERECTION OPERATIONS

### A. Personal Fall Arrest Systems

In this particular erection sequence and procedure, personal fall arrest systems requiring body belt/harness systems, lifelines and lanyards will not reduce possible hazards to workers and will create offsetting hazards during their usage at the leading edge of precast/prestressed concrete construction.

Leading edge erection and initial connections are conducted by employees who are specifically trained to do this type of work and are trained to recognize the fall hazards. The nature of such work normally exposes the employee to the fall hazard for a short period of time and installation of fall protection systems for a short duration is not feasible because it exposes the installers of the system to the same fall hazard, but for a longer period of time.

1. It is necessary that the employee be able to move freely without encumbrance in order to guide the sections of precast concrete into their final position without having lifelines attached which will restrict the employees ability to move about at the point of erection.

2. A typical procedure requires 2 or more workers to maneuver around each other as a concrete member is positioned to fit into the structure. If they are each attached to a lifeline, part of their attention must be diverted from their main task of positioning a member weighing several tons to the task of avoiding entanglements of their lifelines or avoiding tripping over lanyards. Therefore, if these workers are attached to lanyards, more fall potential would result than from not using such a device.

In this specific erection sequence and procedure, retractable lifelines do not solve the problem of two workers becoming tangled. In fact, such a tangle could prevent the lifeline from retracting as the worker moved, thus potentially exposing the worker to a fall greater than 6 ft. Also, a worker crossing over the lifeline of another worker can create a hazard because the movement of one person can unbalance the other. In the event of a fall by one person there is a likelihood that the other person will be caused to fall as well. In addition, if contamination such as grout (during hollow core grouting) enters the retractable housing it can cause excessive wear and damage to the device and could clog the retracting mechanism as the lanyard is dragged across the deck. Obstructing the cable orifice can defeat the devices shock absorbing function, produce cable slack and damage, and adversely affect cable extraction and retraction.

3. Employees tied to a lifeline can be trapped and crushed by moving structural members if the employee becomes restrained by the lanyard or retractable lifeline and cannot get out of the path of the moving load.

The sudden movement of a precast concrete member being raised by a crane can be caused by a number of factors. When this happens, a connector may immediately have to move a considerable distance to avoid injury. If a tied off body belt/harness is being used, the connector could be trapped. Therefore, there is a greater risk of injury if the connector is tied to the structure for this specific erection sequence and procedure.

When necessary to move away from a retractable device, the worker cannot move at a rate greater than the device locking speed typically 3.5 to 4.5 ft/sec. When moving toward the device it is necessary to move at a rate which does not permit cable slack to build up. This slack may cause cable retraction acceleration and cause a worker to lose their balance by applying a higher than normal jerking force on the body when the cable suddenly becomes taut after building up momentum. This slack can also cause damage to the internal spring-loaded drum, uneven coiling of cable on the drum, and possible cable damage.

The factors causing sudden movements for this location include:

#### (a) Cranes

(1) Operator error.

(2) Site conditions (soft or unstable ground).

(3) Mechanical failure.

(4) Structural failure.

(5) Rigging failure.

(6) Crane signal/radio communication failure.

#### (b) Weather Conditions

(1) Wind (strong wind/sudden gusting)—particularly a problem with the large surface areas of precast concrete members.

(2) Snow/rain (visibility).

(3) Fog (visibility).

(4) Cold—causing slowed reactions or mechanical problems.

(c) Structure/Product Conditions.

(1) Lifting Eye failure.

(2) Bearing failure or slippage.

(3) Structure shifting.

(4) Bracing failure.

(5) Product failure.

(d) Human Error.

(1) Incorrect tag line procedure.

(2) Tag line hang-up.

(3) Incorrect or misunderstood crane signals.

(4) Misjudged elevation of member.

(5) Misjudged speed of member.

(6) Misjudged angle of member.

4. Anchorages or special attachment points could be cast into the precast concrete members if sufficient preplanning and consideration of erectors position is done before the members are cast. Any hole or other attachment must be approved by the engineer who designed the member. It is possible that some design restrictions will not allow a member to be weakened by an additional hole; however, it is anticipated that such situations would be the exception, not the rule. Attachment points, other than on the deck surface, will require removal and/or patching. In order to remove and/or patch these points, requires the employee to be exposed to an additional fall hazard at an unprotected perimeter. The fact that attachment points could be available anywhere on the structure does not eliminate the hazards of using these points for tying off as discussed above. A logical point for tying off on double tees would be using the lifting loops, except that they must be cut off to eliminate a tripping hazard at an appropriate time.

5. Providing attachment at a point above the walking/working surface would also create fall exposures for employees installing their devices. Final positioning of a precast concrete member requires it to be moved in such a way that it must pass through the area that would be occupied by the lifeline and the lanyards attached to the point above. Resulting entanglements of lifelines and lanyards on a moving member could pull employees from the work surface. Also, the structure is being created and, in most cases, there is no structure above the members being placed.

(a) Temporary structural supports, installed to provide attaching points for lifelines limit the space which is essential for orderly positioning, alignment and placement of the precast concrete members. To keep the lanyards a reasonable and manageable length, lifeline supports would necessarily need to be in proximity to the positioning process. A sudden shift of the precast concrete member being positioned because of wind pressure or crane movement could make it strike the temporary supporting structure, moving it suddenly and causing tied off employees to fall.

(b) The time in manhours which would be expended in placing and maintaining temporary structural supports for lifeline attaching points could exceed the expended manhours involved in placing the precast concrete members. No protection could be provided for the employees erecting the temporary structural supports and these supports would have to be moved for each successive step in the construction process, thus greatly increasing the employees exposure to the fall hazard.

(c) The use of a cable strung horizontally between two columns to provide tie off lines for erecting or walking a beam for connecting work is not feasible and creates a greater hazard on this multi-story building for the following reasons:

(1) If a connector is to use such a line, it must be installed between the two columns. To perform this installation requires an erector to have more fall exposure time attaching the cable to the columns than would be spent to make the beam to column connection itself.

(2) If such a line is to be installed so that an erector can walk along a beam, it must be overhead or below him. For example, if a connector must walk along a 24 in. wide beam, the presence of a line next to the connector at waist level, attached directly to the columns, would prevent the connector from centering their weight over the beam and balancing themselves. Installing the line above the connector might be possible on the first level of a two-story column; however, the column may extend only a few feet above the floor level at the second level or be flush with the floor level. Attaching the line to the side of the beam could be a solution; however, it would require the connector to attach the lanyard below foot level which would most likely extend a fall farther than 6 ft.

(3) When lines are strung over every beam, it becomes more and more difficult for the crane operator to lower a precast concrete member into position without the member becoming fouled. Should the member become entangled, it could easily dislodge the line from a column. If a worker is tied to it at the time, a fall could be caused.

6. The ANSI A10.14-1991 American National Standard for Construction and Demolition Operations—Requirements for Safety Belts, Harnesses, Lanyards and Lifelines for Construction and Demolition Use, states that the anchor point of a lanyard or deceleration device should, if possible, be located above the wearer's belt or harness attachment. ANSI A10.14 also states that a suitable anchorage point is one which is located as high as possible to prevent contact with an obstruction below should the worker fall. Most manufacturers also warn in the user's handbook that the safety block/retractable lifeline must be positioned above the D-ring (above the work space of the intended user) and OSHA recommends that fall arrest and restraint equipment be used in accordance with the manufacturer's instructions.

Attachment of a retractable device to a horizontal cable near floor level or using the inserts in the floor or roof members may result in increased free fall due to the dorsal D-ring of the full-body harness riding higher than the attachment point of the snaphook to the cable or insert (e.g., 6 foot tall worker with a dorsal D-ring at 5 feet above the floor or surface, reduces the working length to only one foot, by placing the anchorage five feet away from the fall hazard). In addition, impact loads may exceed maximum fall arrest forces (MAF) because the fall arrest D-ring would be 4 to 5 feet higher than the safety block/retractable lifeline anchored to the walking-working surface; and the potential for swing hazards is increased.

Manufacturers also require that workers not work at a level where the point of snaphook attachment to the body harness is above the device because this will increase the free fall distance and the deceleration distance and will cause higher forces on the body in the event of an accidental fall.

Manufacturers recommend an anchorage for the retractable lifeline which is immovably fixed in space and is independent of the user's support systems. A moveable anchorage is one which can be moved around (such as equipment or wheeled vehicles) or which can deflect substantially under shock loading (such as a horizontal cable or very flexible beam). In the case of a very flexible anchorage, a shock load applied to the anchorage during fall arrest can cause oscillation of the flexible anchorage such that the retractable brake mechanism may undergo one or more cycles of locking/unlocking/locking (ratchet effect) until the anchorage deflection is dampened. Therefore, use of a moveable anchorage involves critical engineering and safety factors and should only be considered after fixed anchorage has been determined to be not feasible.

Horizontal cables used as an anchorage present an additional hazard due to amplification of the horizontal component of maximum arrest force (of a fall) transmitted to the points where the horizontal cable is attached to the structure. This amplification is due to the angle of sag of a horizontal cable and is most severe for small angles of sag. For a cable sag angle of 2 degrees the horizontal force on the points of cable attachment can be amplified by a factor of 15.

It is also necessary to install the retractable device vertically overhead to minimize swing falls. If an object is in the worker's swing path (or that of the cable) hazardous situations exist: (1) due to the swing, horizontal speed of the user may be high enough to cause injury when an obstacle in the swing fall path is struck by either the user or the cable; (2) the total vertical fall distance of the user may be much greater than if the user had fallen only vertically without a swing fall path.

With retractable lines, overconfidence may cause the worker to engage in inappropriate behavior, such as approaching the perimeter of a floor or roof at a distance appreciably greater than the shortest distance between the anchorage point and the leading edge. Though the retractable lifeline may arrest a worker's fall before he or she has fallen a few feet, the lifeline may drag along the edge of the floor or beam and swing the worker like a pendulum until the line has moved to a position where the distance between the anchorage point and floor edge is the shortest distance between those two points. Accompanying this pendulum swing is a lowering of the worker, with the attendant danger that he or she may violently impact the floor or some obstruction below.

The risk of a cable breaking is increased if a lifeline is dragged sideways across the rough surface or edge of a concrete member at the same moment that the lifeline is being subjected to a maximum impact loading during a fall. The typical 3/16 in. cable in a retractable lifeline has a breaking strength of from 3000 to 3700 lbs.

7. The competent person, who can take into account the specialized operations being performed on this project, should determine when and where a designated erector cannot use a personal fall arrest system.

### B. Safety Net Systems

The nature of this particular precast concrete erection worksite precludes the safe use of safety nets where point of erection or leading edge work must take place.

1. To install safety nets in the interior high bay of the single story portion of the building poses rigging attachment problems. Structural members do not exist to which supporting devices for nets can be attached in the area where protection is required. As the erection operation advances, the location of point of erection or leading edge work changes constantly as each member is attached to the structure. Due to this constant change it is not feasible to set net sections and build separate structures to support the nets.

2. The nature of the erection process for the precast concrete members is such that an installed net would protect workers as they position and secure only one structural member. After each member is stabilized the net would have to be moved to a new location (this could mean a move of 8 to 10 ft or the possibility of a move to a different level or area of the structure) to protect workers placing the next piece in the construction sequence. The result would be the installation and dismantling of safety nets repeatedly throughout the normal work day. As the time necessary to install a net, test, and remove it is significantly greater than the time necessary to position and secure a precast concrete member, the exposure time for the worker installing the safety net would be far longer than for the workers whom the net is intended to protect. The time exposure repeats itself each time the nets and supporting hardware must be moved laterally or upward to provide protection at the point of erection or leading edge.

3. Strict interpretation of §1926.502(c) requires that operations shall not be undertaken until the net is in place and has been tested. With the point of erection constantly changing, the time necessary to install and test a safety net significantly exceeds the time necessary to position and secure the concrete member.

4. Use of safety nets on exposed perimeter wall openings and opensided floors, causes attachment points to be left in architectural concrete which must be patched and filled with matching material after the net supporting hardware is removed. In order to patch these openings, additional numbers of employees must be suspended by swing stages, boatswain chairs or other devices, thereby increasing the amount of fall exposure time to employees.

5. Installed safety nets pose an additional hazard at the perimeter of the erected structure where limited space is available in which members can be turned after being lifted from the ground by the crane. There would be a high probability that the member being lifted could become entangled in net hardware, cables, etc.

6. The use of safety nets where structural wall panels are being erected would prevent movement of panels to point of installation. To be effective, nets would necessarily have to provide protection across the area where structural supporting wall panels would be set and plumbed before roof units could be placed.

7. Use of a tower crane for the erection of the high rise portion of the structure poses a particular hazard in that the crane operator cannot see or judge the proximity of the load in relation to the structure or nets. If the signaler is looking through nets and supporting structural devices while giving instructions to the crane operator, it is not possible to judge precise relationships between the load and the structure itself or to nets and supporting structural devices. This could cause the load to become entangled in the net or hit the structure causing potential damage.

### C. Guardrail Systems

On this particular worksite, guardrails, barricades, ropes, cables or other perimeter guarding devices or methods on the erection floor will pose problems to safe erection procedures. Typically, a floor or roof is erected by placing 4 to 10 ft wide structural members next to one another and welding or grouting them together. The perimeter of a floor and roof changes each time a new member is placed into position. It is unreasonable and virtually impossible to erect guardrails and toe boards at the ever changing leading edge of a floor or roof.

1. To position a member safely it is necessary to remove all obstructions extending above the floor level near the point of erection. Such a procedure allows workers to swing a new member across the erected surface as necessary to position it properly without worrying about knocking material off of this surface.

Hollow core slab erection on the masonry wall requires installation of the perimeter protection where the masonry wall has to be constructed. This means the guardrail is installed then subsequently removed to continue the masonry construction. The erector will be exposed to a fall hazard for a longer period of time while installing and removing perimeter protection than while erecting the slabs.

In hollow core work, as in other precast concrete erection, others are not typically on the work deck until the precast concrete erection is complete. The deck is not complete until the leveling, aligning, and grouting of the joints is done. It is normal practice to keep others off the deck until at least the next day after the installation is complete to allow the grout to harden.

2. There is no permanent boundary until all structural members have been placed in the floor or roof. At the leading edge, workers are operating at the temporary edge of the structure as they work to position the next member in the sequence. Compliance with the standard would require a guardrail and toe board be installed along this edge. However, the presence of such a device would prevent a new member from being swung over the erected surface low enough to allow workers to control it safely during the positioning process. Further, these employees would have to work through the guardrail to align the new member and connect it to the structure. The guardrail would not protect an employee who must lean through it to do the necessary work, rather it would hinder the employee to such a degree that a greater hazard is created than if the guardrail were absent.

3. Guardrail requirements pose a hazard at the leading edge of installed floor or roof sections by creating the possibility of employees being caught between guardrails and suspended loads. The lack of a clear work area in which to guide the suspended load into position for placement and welding of members into the existing structure creates still further hazards.

4. Where erection processes require precast concrete stairways or openings to be installed as an integral part of the overall erection process, it must also be recognized that guardrails or handrails must not project above the surface of the erection floor. Such guardrails should be terminated at the level of the erection floor to avoid placing hazardous obstacles in the path of a member being positioned.

## V. OTHER FALL PROTECTION MEASURES CONSIDERED FOR THIS JOB

The following is a list and explanation of other fall protection measures available and an explanation of limitations for use on this particular jobsite. If during the course of erecting the building the employee sees an area that could be erected more safely by the use of these fall protection measures, the foreman should be notified.

A. Scaffolds are not used because:

1. The leading edge of the building is constantly changing and the scaffolding would have to be moved at very frequent intervals. Employees erecting and dismantling the scaffolding would be exposed to fall hazards for a greater length of time than they would by merely erecting the precast concrete member.

2. A scaffold tower could interfere with the safe swinging of a load by the crane.

3. Power lines, terrain and site do not allow for the safe use of scaffolding.

B. Vehicle mounted platforms are not used because:

1. A vehicle mounted platform will not reach areas on the deck that are erected over other levels.

2. The leading edge of the building is usually over a lower level of the building and this lower level will not support the weight of a vehicle mounted platform.

3. A vehicle mounted platform could interfere with the safe swinging of a load by the crane, either by the crane swinging the load over or into the equipment.

4. Power lines and surrounding site work do not allow for the safe use of a vehicle mounted platform.

C. Crane suspended personnel platforms are not used because:

1. A second crane close enough to suspend any employee in the working and erecting area could interfere with the safe swinging of a load by the crane hoisting the product to be erected.

2. Power lines and surrounding site work do not allow for the safe use of a second crane on the job.

## VI. ENFORCEMENT

Constant awareness of and respect for fall hazards, and compliance with all safety rules are considered conditions of employment. The jobsite Superintendent, as well as individuals in the Safety and Personnel Department, reserve the right to issue disciplinary warnings to employees, up to and including termination, for failure to follow the guidelines of this program.

## VII. ACCIDENT INVESTIGATIONS

All accidents that result in injury to workers, regardless of their nature, shall be investigated and reported. It is an integral part of any safety program that documentation take place as soon as possible so that the cause and means of prevention can be identified to prevent a reoccurrence.

In the event that an employee falls or there is some other related, serious incident occurring, this plan shall be reviewed to determine if additional practices, procedures, or training need to be implemented to prevent similar types of falls or incidents from occurring.

## VIII. CHANGES TO PLAN

Any changes to the plan will be approved by (name of the qualified person). This plan shall be reviewed by a qualified person as the job progresses to determine if additional practices, procedures or training needs to be implemented by the competent person to improve or provide additional fall protection. Workers shall be notified and trained, if necessary, in the new procedures. A copy of this plan and all approved changes shall be maintained at the jobsite.

### Sample Fall Protection Plan for Residential Construction

**(INSERT COMPANY NAME)**

This Fall Protection Plan Is Specific For The Following Project:

Location of Job ______________________

Date Plan Prepared or Modified ______________________

Plan Prepared By ______________________

Plan Approved By ______________________

Plan Supervised By ______________________

The following Fall Protection Plan is a sample program prepared for the prevention of injuries associated with falls. A Fall Protection Plan must be developed and evaluated on a site by site basis. It is recommended that builders discuss the written Fall Protection Plan with their OSHA Area Office prior to going on a jobsite.

## I. STATEMENT OF COMPANY POLICY

(Your company name here) is dedicated to the protection of its employees from on-the-job injuries. All employees of (Your company name here) have the responsibility to work safely on the job. The purpose of the plan is to supplement our existing safety and health program and to ensure that every employee who works for (Your company name here) recognizes workplace fall hazards and takes the appropriate measures to address those hazards.

This Fall Protection Plan addresses the use of conventional fall protection at a number of areas on the project, as well as identifies specific activities that require non-conventional means of fall protection. During the construction of residential buildings under 48 feet in height, it is sometimes infeasible or it creates a greater hazard to use conventional fall protection systems at specific areas or for specific tasks. The areas or tasks may include, but are not limited to:

a. Setting and bracing of roof trusses and rafters;

b. Installation of floor sheathing and joists;

c. Roof sheathing operations; and

d. Erecting exterior walls.

In these cases, conventional fall protection systems may not be the safest choice for builders. This plan is designed to enable employers and employees to recognize the fall hazards associated with this job and to establish the safest procedures that are to be followed in order to prevent falls to lower levels or through holes and openings in walking/working surfaces.

Each employee will be trained in these procedures and will strictly adhere to them except when doing so would expose the employee to a greater hazard. If, in the employee's opinion, this is the case, the employee is to notify the competent person of their concern and have the concern addressed before proceeding.

It is the responsibility of (name of competent person) to implement this Fall Protection Plan. Continual observational safety checks of work operations and the enforcement of the safety policy and procedures shall be regularly enforced. The crew supervisor or foreman (insert name) is responsible for correcting any unsafe practices or conditions immediately.

It is the responsibility of the employer to ensure that all employees understand and adhere to the procedures of this plan and to follow the instructions of the crew supervisor. It is also the responsibility of the employee to bring to management's attention any unsafe or hazardous conditions or practices that may cause injury to either themselves or any other employees.

Any changes to the Fall Protection Plan must be approved by (name of qualified person).

## II. FALL PROTECTION SYSTEMS TO BE USED ON THIS JOB

Installation of roof trusses/rafters, exterior wall erection, roof sheathing, floor sheathing and joist/truss activities will be conducted by employees who are specifically trained to do this type of work and are trained to recognize the fall hazards. The nature of such work normally exposes the employee to the fall hazard for a short period of time. This Plan details how (Your company name here) will minimize these hazards.

### *Controlled Access Zones*

When using the Plan to implement the fall protection options available, workers must be protected through limited access to high hazard locations. Before any non-conventional fall protection systems are used as part of the work plan, a controlled access zone (CAZ) shall be clearly defined by the competent person as an area where a recognized hazard exists. The demarcation of the CAZ shall be communicated by the competent person in a recognized manner, either through signs, wires, tapes, ropes or chains.

(Your company name here) shall take the following steps to ensure that the CAZ is clearly marked or controlled by the competent person:

- All access to the CAZ must be restricted to authorized entrants;
- All workers who are permitted in the CAZ shall be listed in the appropriate sections of the Plan (or be visibly identifiable by the competent person) prior to implementation;
- The competent person shall ensure that all protective elements of the CAZ be implemented prior to the beginning of work.

### *Installation Procedures for Roof Truss and Rafter Erection*

During the erection and bracing of roof trusses/rafters, conventional fall protection may present a greater hazard to workers. On this job, safety nets, guardrails and personal fall arrest systems will not provide adequate fall protection because the nets will cause the walls to collapse, while there are no suitable attachment or anchorage points for guardrails or personal fall arrest systems.

On this job, requiring workers to use a ladder for the entire installation process will cause a greater hazard because the worker must stand on the ladder with his back or side to the front of the ladder. While erecting the truss or rafter the worker will need both hands to maneuver the truss and therefore cannot hold onto the ladder. In addition, ladders cannot be adequately protected from movement while trusses are being maneuvered into place. Many workers may experience additional fatigue because of the increase in overhead work with heavy materials, which can also lead to a greater hazard.

Exterior scaffolds cannot be utilized on this job because the ground, after recent backfilling, cannot support the scaffolding. In most cases, the erection and dismantling of the scaffold would expose workers to a greater fall hazard than erection of the trusses/rafters.

On all walls eight feet or less, workers will install interior scaffolds along the interior wall below the location where the trusses/rafters will be erected. "Sawhorse" scaffolds constructed of 46 inch sawhorses and 2x10 planks will often allow workers to be elevated high enough to allow for the erection of trusses and rafters without working on the top plate of the wall.

In structures that have walls higher than eight feet and where the use of scaffolds and ladders would create a greater hazard, safe working procedures will be utilized when working on the top plate and will be monitored by the crew supervisor. During all stages of truss/rafter erection the stability of the trusses/rafters will be ensured at all times.

(Your company name here) shall take the following steps to protect workers who are exposed to fall hazards while working from the top plate installing trusses/rafters:

- Only the following trained workers will be allowed to work on the top plate during roof truss or rafter installation:

________________________________________

________________________________________

________________________________________

- Workers shall have no other duties to perform during truss/rafter erection procedures;
- All trusses/rafters will be adequately braced before any worker can use the truss/rafter as a support;
- Workers will remain on the top plate using the previously stabilized truss/rafter as a support while other trusses/rafters are being erected;
- Workers will leave the area of the secured trusses only when it is necessary to secure another truss/rafter;
- The first two trusses/rafters will be set from ladders leaning on side walls at points where the walls can support the weight of the ladder; and
- A worker will climb onto the interior top plate via a ladder to secure the peaks of the first two trusses/rafters being set.

The workers responsible for detaching trusses from cranes and/or securing trusses at the peaks traditionally are positioned at the peak of the trusses/rafters. There are also situations where workers securing rafters to ridge beams will be positioned on top of the ridge beam.

(Your company name here) shall take the following steps to protect workers who are exposed to fall hazards while securing trusses/rafters at the peak of the trusses/ridge beam:

- Only the following trained workers will be allowed to work at the peak during roof truss or rafter installation:

________________________________________

________________________________________

________________________________________

- Once truss or rafter installation begins, workers not involved in that activity shall not stand or walk below or adjacent to the roof opening or exterior walls in any area where they could be struck by falling objects;
- Workers shall have no other duties than securing/bracing the trusses/ridge beam;

- Workers positioned at the peaks or in the webs of trusses or on top of the ridge beam shall work from a stable position, either by sitting on a "ridge seat" or other equivalent surface that provides additional stability or by positioning themselves in previously stabilized trusses/rafters and leaning into and reaching through the trusses/rafters;
- Workers shall not remain on or in the peak/ridge any longer than necessary to safely complete the task.

### Roof Sheathing Operations

Workers typically install roof sheathing after all trusses/rafters and any permanent truss bracing is in place. Roof structures are unstable until some sheathing is installed, so workers installing roof sheathing cannot be protected from fall hazards by conventional fall protection systems until it is determined that the roofing system can be used as an anchorage point. At that point, employees shall be protected by a personal fall arrest system.

Trusses/rafters are subject to collapse if a worker falls while attached to a single truss with a belt/harness. Nets could also cause collapse, and there is no place to attach guardrails.

All workers will ensure that they have secure footing before they attempt to walk on the sheathing, including cleaning shoes/boots of mud or other slip hazards.

To minimize the time workers must be exposed to a fall hazard, materials will be staged to allow for the quickest installation of sheathing.

(Your company name here) shall take the following steps to protect workers who are exposed to fall hazards while installing roof sheathing:

- Once roof sheathing installation begins, workers not involved in that activity shall not stand or walk below or adjacent to the roof opening or exterior walls in any area where they could be struck by falling objects;
- The competent person shall determine the limits of this area, which shall be clearly communicated to workers prior to placement of the first piece of roof sheathing;
- The competent person may order work on the roof to be suspended for brief periods as necessary to allow other workers to pass through such areas when this would not create a greater hazard;
- Only qualified workers shall install roof sheathing;
- The bottom row of roof sheathing may be installed by workers standing in truss webs;
- After the bottom row of roof sheathing is installed, a slide guard extending the width of the roof shall be securely attached to the roof. Slide guards are to be constructed of no less than nominal 4" height capable of limiting the uncontrolled slide of workers. Workers should install the slide guard while standing in truss webs and leaning over the sheathing;
- Additional rows of roof sheathing may be installed by workers positioned on previously installed rows of sheathing. A slide guard can be used to assist workers in retaining their footing during successive sheathing operations; and
- Additional slide guards shall be securely attached to the roof at intervals not to exceed 13 feet as successive rows of sheathing are installed. For roofs with pitches in excess of 9-in-12, slide guards will be installed at four-foot intervals.
- When wet weather (rain, snow, or sleet) are present, roof sheathing operations shall be suspended unless safe footing can be assured for those workers installing sheathing.
- When strong winds (above 40 miles per hour) are present, roof sheathing operations are to be suspended unless wind breakers are erected.

### Installation of Floor Joists and Sheathing

During the installation of floor sheathing/joists (leading edge construction), the following steps shall be taken to protect workers:

- Only the following trained workers will be allowed to install floor joists or sheathing:

________________________________________

________________________________________

________________________________________

- Materials for the operations shall be conveniently staged to allow for easy access to workers;
- The first floor joists or trusses will be rolled into position and secured either from the ground, ladders or sawhorse scaffolds;
- Each successive floor joist or truss will be rolled into place and secured from a platform created from a sheet of plywood laid over the previously secured floor joists or trusses;
- Except for the first row of sheathing which will be installed from ladders or the ground, workers shall work from the established deck; and
- Any workers not assisting in the leading edge construction while leading edges still exist (e.g. cutting the decking for the installers) shall not be permitted within six feet of the leading edge under construction.

### Erection of Exterior Walls

During the construction and erection of exterior walls, employers shall take the following steps to protect workers:

- Only the following trained workers will be allowed to erect exterior walls:

________________________________________

________________________________________

________________________________________

- A painted line six feet from the perimeter will be clearly marked prior to any wall erection activities to warn of the approaching unprotected edge;
- Materials for operations shall be conveniently staged to minimize fall hazards; and
- Workers constructing exterior walls shall complete as much cutting of materials and other preparation as possible away from the edge of the deck.

## III. ENFORCEMENT

Constant awareness of and respect for fall hazards, and compliance with all safety rules are considered conditions of employment. The crew supervisor or foreman, as well as individuals in the Safety and Personnel Department, reserve the right to issue disciplinary warnings to employees, up to and including termination, for failure to follow the guidelines of this program.

## IV. ACCIDENT INVESTIGATIONS

All accidents that result in injury to workers, regardless of their nature, shall be investigated and reported. It is an integral part of any safety program that documentation take place as soon as possible so that the cause and means of prevention can be identified to prevent a reoccurrence.

In the event that an employee falls or there is some other related, serious incident occurring, this plan shall be reviewed to determine if additional practices, procedures, or training need to be implemented to prevent similar types of falls or incidents from occurring.

## V. CHANGES TO PLAN

Any changes to the plan will be approved by (name of the qualified person). This plan shall be reviewed by a qualified person as the job progresses to determine if additional practices, procedures or training needs to be implemented by the competent person to improve or provide additional fall protection. Workers shall be notified and trained, if necessary, in the new procedures. A copy of this plan and all approved changes shall be maintained at the jobsite.

[59 FR 40730, Aug. 9, 1994]

# Subpart N— Cranes, Derricks, Hoists, Elevators, and Conveyors

**Authority:** Sec. 107, Contract Work Hours and Safety Standards Act (Construction Safety Act) (40 U.S.C. 333); secs. 4, 6, 8, Occupational Safety and Health Act of 1970 (29 U.S.C. 653, 655, 657); Secretary of Labor's Order No. 12-71 (36 FR 8754), 8-76 (41 FR 25059), or 9-83 (49 FR 35736), as applicable. Section 1926.550 also issued under 29 CFR Part 1911.

## §1926.550 Cranes and derricks.

**1926.550(a)** ***General requirements.*** (1) The employer shall comply with the manufacturer's specifications and limitations applicable to the operation of any and all cranes and derricks. Where manufacturer's specifications are not available, the limitations assigned to the equipment shall be based on the determinations of a qualified engineer competent in this field and such determinations will be appropriately documented and recorded. Attachments used with cranes shall not exceed the capacity, rating, or scope recommended by the manufacturer.

**1926.550(a)(2)** Rated load capacities, and recommended operating speeds, special hazard warnings, or instruction, shall be conspicuously posted on all equipment. Instructions or warnings shall be visible to the operator while he is at his control station.

**1926.550(a)(3)** [Reserved]

**1926.550(a)(4)** Hand signals to crane and derrick operators shall be those prescribed by the applicable ANSI standard for the type of crane in use. An illustration of the signals shall be posted at the job site.

**1926.550(a)(5)** The employer shall designate a competent person who shall inspect all machinery and equipment prior to each use, and during use, to make sure it is in safe operating condition. Any deficiencies shall be repaired, or defective parts replaced, before continued use.

**1926.550(a)(6)** A thorough, annual inspection of the hoisting machinery shall be made by a competent person, or by a government or private agency recognized by the U.S. Department of Labor. The employer shall maintain a record of the dates and results of inspections for each hoisting machine and piece of equipment.

**1926.550(a)(7)** Wire rope shall be taken out of service when any of the following conditions exist:

**1926.550(a)(7)(i)** In running ropes, six randomly distributed broken wires in one lay or three broken wires in one strand in one lay;

**1926.550(a)(7)(ii)** Wear of one-third the original diameter of outside individual wires. Kinking, crushing, bird caging, or any other damage resulting in distortion of the rope structure;

**1926.550(a)(7)(iii)** Evidence of any heat damage from any cause;

**1926.550(a)(7)(iv)** Reductions from nominal diameter of more than one-sixty-fourth inch for diameters up to and including five-sixteenths inch, one-thirty-second inch for diameters three-eighths inch to and including one-half inch, three-sixty-fourths inch for diameters nine-sixteenths inch to and including three- fourths inch, one-sixteenth inch for diameters seven-eighths inch to 1 1/8 inches inclusive, three-thirty-seconds inch for diameters 1 1/4 to 1 1/2 inches inclusive;

**1926.550(a)(7)(v)** In standing ropes, more than two broken wires in one lay in sections beyond end connections or more than one broken wire at an end connection.

**1926.550(a)(7)(vi)** Wire rope safety factors shall be in accordance with American National Standards Institute B30.5-1968 or SAE J959-1966.

**1926.550(a)(8)** Belts, gears, shafts, pulleys, sprockets, spindles, drums, fly wheels, chains, or other reciprocating, rotating, or other moving parts or equipment shall be guarded if such parts are exposed to contact by employees, or otherwise create a hazard. Guarding shall meet the requirements of the American National Standards Institute B15.1-1958 Rev., Safety Code for Mechanical Power Transmission Apparatus.

**1926.550(a)(9)** Accessible areas within the swing radius of the rear of the rotating superstructure of the crane, either permanently or temporarily mounted, shall be barricaded in such a manner as to prevent an employee from being struck or crushed by the crane.

**1926.550(a)(10)** All exhaust pipes shall be guarded or insulated in areas where contact by employees is possible in the performance of normal duties.

**1926.550(a)(11)** Whenever internal combustion engine powered equipment exhausts in enclosed spaces, tests shall be made and recorded to see that employees are not exposed to unsafe concentrations of toxic gases or oxygen deficient atmospheres.

**1926.550(a)(12)** All windows in cabs shall be of safety glass, or equivalent, that introduces no visible distortion that will interfere with the safe operation of the machine.

**1926.550(a)(13)(i)** Where necessary for rigging or service requirements, a ladder, or steps, shall be provided to give access to a cab roof.

**1926.550(a)(13)(ii)** Guardrails, handholds, and steps shall be provided on cranes for easy access to the car and cab, conforming to American National Standards Institute B30.5.

**1926.550(a)(13)(iii)** Platforms and walkways shall have anti-skid surfaces.

**1926.550(a)(14)** Fuel tank filler pipe shall be located in such a position, or protected in such manner, as to not allow spill or overflow to run onto the engine, exhaust, or electrical equipment of any machine being fueled.

**1926.550(a)(14)(i)** An accessible fire extinguisher of 5BC rating, or higher, shall be available at all operator stations or cabs of equipment.

**1926.550(a)(14)(ii)** All fuels shall be transported, stored, and handled to meet the rules of subpart F of this part. When fuel is transported by vehicles on public highways, Department of Transportation rules contained in 49 CFR Parts 177 and 393 concerning such vehicular transportation are considered applicable.

**1926.550(a)(15)** Except where electrical distribution and transmission lines have been deenergized and visibly grounded at point of work or where insulating barriers, not a part of or an attachment to the equipment or machinery, have been erected to prevent physical contact with the lines, equipment or machines shall be operated proximate to power lines only in accordance with the following:

**1926.550(a)(15)(i)** For lines rated 50 kV. or below, minimum clearance between the lines and any part of the crane or load shall be 10 feet;

**1926.550(a)(15)(ii)** For lines rated over 50 kV., minimum clearance between the lines and any part of the crane or load shall be 10 feet plus 0.4 inch for each 1 kV. over 50 kV., or twice the length of the line insulator, but never less than 10 feet;

**1926.550(a)(15)(iii)** In transit with no load and boom lowered, the equipment clearance shall be a minimum of 4 feet for voltages less than 50 kV., and 10 feet for voltages over 50 kV., up to and including 345 kV., and 16 feet for voltages up to and including 750 kV.

**1926.550(a)(15)(iv)** A person shall be designated to observe clearance of the equipment and give timely warning for all operations where it is difficult for the operator to maintain the desired clearance by visual means;

**1926.550(a)(15)(v)** Cage-type boom guards, insulating links, or proximity warning devices may be used on cranes, but the use of such devices shall not alter the requirements of any other regulation of this part even if such device is required by law or regulation;

**1926.550(a)(15)(vi)** Any overhead wire shall be considered to be an energized line unless and until the person owning such line or the electrical utility authorities indicate that it is not an energized line and it has been visibly grounded;

**1926.550(a)(15)(vii)** Prior to work near transmitter towers where an electrical charge can be induced in the equipment or materials being handled, the transmitter shall be de-energized or tests shall be made to determine if electrical charge is induced on the crane. The following precautions shall be taken when necessary to dissipate induced voltages:

**1926.550(a)(15)(vii)*(a)*** The equipment shall be provided with an electrical ground directly to the upper rotating structure supporting the boom; and

**1926.550(a)(15)(vii)*(b)*** Ground jumper cables shall be attached to materials being handled by boom equipment when electrical charge is induced while working near energized transmitters. Crews shall be provided with nonconductive poles having large alligator clips or other similar protection to attach the ground cable to the load.

**1926.550(a)(15)(vii)*(c)*** Combustible and flammable materials shall be removed from the immediate area prior to operations.

**1926.550(a)(16)** No modifications or additions which affect the capacity or safe operation of the equipment shall be made by the employer without the manufacturer's written approval. If such modifications or changes are made, the capacity, operation, and maintenance instruction plates, tags, or decals, shall be changed accordingly. In no case shall the original safety factor of the equipment be reduced.

**1926.550(a)(17)** The employer shall comply with Power Crane and Shovel Association Mobile Hydraulic Crane Standard No. 2.

**1926.550(a)(18)** Sideboom cranes mounted on wheel or crawler tractors shall meet the requirements of SAE J743a-1964.

**1926.550(a)(19)** All employees shall be kept clear of loads about to be lifted and of suspended loads.

**1926.550(b)** ***Crawler, locomotive, and truck cranes.*** (1) All jibs shall have positive stops to prevent their movement of more than 5° above the straight line of the jib and boom on conventional type crane booms. The use of cable type belly slings does not constitute compliance with this rule.

**1926.550(b)(2)** All crawler, truck, or locomotive cranes in use shall meet the applicable requirements for design, inspection, construction, testing, maintenance and operation as prescribed in the ANSI B30.5-1968, Safety Code for Crawler, Locomotive and Truck Cranes. However, the written, dated, and signed inspection reports and records of the monthly inspection of critical items prescribed in section 5-2.1.5 of the ANSI B30.5-1968 standard are not required. Instead, the employer shall prepare a certification record which includes the date the crane items were inspected; the signature of the person who inspected the crane items; and a serial number, or other identifier, for the crane inspected. The most recent certification record shall be maintained on file until a new one is prepared.

**1926.550(c)** ***Hammerhead tower cranes.*** (1) Adequate clearance shall be maintained between moving and rotating structures of the crane and fixed objects to allow the passage of employees without harm.

**1926.550(c)(2)** Each employee required to perform duties on the horizontal boom of hammerhead tower cranes shall be protected against falling by guardrails or by a personal fall arrest system in conformance with subpart M of this part.

**1926.550(c)(3)** Buffers shall be provided at both ends of travel of the trolley.

**1926.550(c)(4)** Cranes mounted on rail tracks shall be equipped with limit switches limiting the travel of the crane on the track and stops or buffers at each end of the tracks.

**1926.550(c)(5)** All hammerhead tower cranes in use shall meet the applicable requirements for design, construction, installation, testing, maintenance, inspection, and operation as prescribed by the manufacturer.

**1926.550(d)** ***Overhead and gantry cranes.*** (1) The rated load of the crane shall be plainly marked on each side of the crane, and if the crane has more than one hoisting unit, each hoist shall have its rated load marked on it or its load block, and this marking shall be clearly legible from the ground or floor.

**1926.550(d)(2)** Bridge trucks shall be equipped with sweeps which extend below the top of the rail and project in front of the truck wheels.

**1926.550(d)(3)** Except for floor-operated cranes, a gong or other effective audible warning signal shall be provided for each crane equipped with a power traveling mechanism.

**1926.550(d)(4)** All overhead and gantry cranes in use shall meet the applicable requirements for design, construction, installation, testing, maintenance, inspection, and operation as prescribed in the ANSI B30.2.0-1967, Safety Code for Overhead and Gantry Cranes.

**1926.550(e)** ***Derricks.*** All derricks in use shall meet the applicable requirements for design, construction, installation, inspection, testing, maintenance, and operation as prescribed in American National Standards Institute B30.6-1969, Safety Code for Derricks.

**1926.550(f)** ***Floating cranes and derricks***—(1) *Mobile cranes mounted on barges.* (i) When a mobile crane is mounted on a barge, the rated load of the crane shall not exceed the original capacity specified by the manufacturer.

**1926.550(f)(1)(ii)** A load rating chart, with clearly legible letters and figures, shall be provided with each crane, and securely fixed at a location easily visible to the operator.

**1926.550(f)(1)(iii)**When load ratings are reduced to stay within the limits for list of the barge with a crane mounted on it, a new load rating chart shall be provided.

**1926.550(f)(1)(iv)** Mobile cranes on barges shall be positively secured.

**1926.550(f)(2)** *Permanently mounted floating cranes and derricks.* (i) When cranes and derricks are permanently installed on a barge, the capacity and limitations of use shall be based on competent design criteria.

**1926.550(f)(2)(ii)** A load rating chart with clearly legible letters and figures shall be provided and securely fixed at a location easily visible to the operator.

**1926.550(f)(2)(iii)** Floating cranes and floating derricks in use shall meet the applicable requirements for design, construction, installation, testing, maintenance, and operation as prescribed by the manufacturer.

**1926.550(f)(3)** *Protection of employees working on barges.* The employer shall comply with the applicable requirements for protection of employees working onboard marine vessels specified in §1926.605.

**1926.550(g)** ***Crane or derrick suspended personnel platforms***—(1) *Scope, application and definitions*—(i) *Scope and application.* This standard applies to the design, construction, testing, use and maintenance of personnel platforms, and the hoisting of personnel platforms on the load lines of cranes or derricks.

**1926.550(g)(1)(ii)** *Definitions.* For the purposes of this paragraph (g), the following definitions apply:

**1926.550(g)(1)(ii)(A)** *Failure* means load refusal, breakage, or separation of components.

**1926.550(g)(1)(ii)(B)** *Hoist* (or hoisting) means all crane or derrick functions such as lowering, lifting, swinging, booming in and out or up and down, or suspending a personnel platform.

**1926.550(g)(1)(ii)(C)** *Load refusal* means the point where the ultimate strength is exceeded.

**1926.550(g)(1)(ii)(D)** *Maximum intended load* means the total load of all employees, tools, materials, and other loads reasonably anticipated to be applied to a personnel platform or personnel platform component at any one time.

**1926.550(g)(1)(ii)(E)** *Runway* means a firm, level surface designed, prepared and designated as a path of travel for the weight and configuration of the crane being used to lift and travel with the crane suspended platform. An existing surface may be used as long as it meets these criteria.

**1926.550(g)(2)** *General requirements.* The use of a crane or derrick to hoist employees on a personnel platform is prohibited, except when the erection, use, and dismantling of conventional means of reaching the worksite, such as a personnel hoist, ladder, stairway, aerial lift, elevating work platform or scaffold, would be more hazardous, or is not possible because of structural design or worksite conditions.

**1926.550(g)(3)** *Cranes and derricks*—(i) *Operational criteria.* (A) Hoisting of the personnel platform shall be performed in a slow, controlled, cautious manner with no sudden movements of the crane or derrick, or the platform.

**1926.550(g)(3)(i)(B)** Load lines shall be capable of supporting, without failure, at least seven times the maximum intended load, except that where rotation resistant rope is used, the lines shall be capable of supporting without failure, at least ten times the maximum intended load. The required design factor is achieved by taking the current safety factor of 3.5 (required under paragraph (b)(2)) of this section and applying the 50 per cent derating of the crane capacity which is required by paragraph (g)(3)(i)(F) of this section.

**1926.550(g)(3)(i)(C)** Load and boom hoist drum brakes, swing brakes, and locking devices such as pawls or dogs shall be engaged when the occupied personnel platform is in a stationary working position.

**1926.550(g)(3)(i)(D)** The crane shall be uniformly level within one percent of level grade and located on firm footing. Cranes equipped with outriggers shall have them all fully deployed following manufacturer's specifications, insofar as applicable, when hoisting employees.

**1926.550(g)(3)(i)(E)** The total weight of the loaded personnel platform and related rigging shall not exceed 50 percent of the rated capaciy for the radius and configuration of the crane or derrick.

**1926.550(g)(3)(i)(F)** The use of machines having live booms (booms in which lowering is controlled by a brake without aid from other devices which slow the lowering speeds) is prohibited.

**1926.550(g)(3)(ii)** *Instruments and components.* (A) Cranes and derricks with variable angle booms shall be equipped with a boom angle indicator, readily visible to the operator.

**1926.550(g)(3)(ii)(B)** Cranes with telescoping booms shall be equipped with a device to indicate clearly to the operator, at all times, the boom's extended length, or an accurate determination of the load radius to be used during the lift shall be made prior to hoisting personnel.

**1926.550(g)(3)(ii)(C)** A positive acting device shall be used which prevents contact between the load block or overhaul ball and the boom tip (anti-two-blocking device), or a system shall be used which deactivates the hoisting action before damage occurs in the event of a two-blocking situation (two block damage prevention feature).

**1926.550(g)(3)(ii)(D)** The load line hoist drum shall have a system or device on the power train, other than the load hoist brake, which regulates the lowering rate of speed of the hoist mechanism (controlled load lowering.) Free fall is prohibited.

**1926.550(g)(4)** *Personnel platforms*—(i) *Design criteria.* (A) The personnel platform and suspension system shall be designed by a qualified engineer or a qualified person competent in structural design

**1926.550(g)(4)(i)(B)** The suspension system shall be designed to minimize tipping of the platform due to movement of employees occupying the platform.

**1926.550(g)(4)(i)(C)** The personnel platform itself, except the guardrail system and personal fall arrest system anchorages, shall be capable of supporting, without failure, its own weight and at least five times the maximum intended load. Criteria for guardrail systems and personal fall arrest system anchorages are contained in subpart M of this Part.

**1926.550(g)(4)(ii)** *Platform specifications.* (A) Each personnel platform shall be equipped with a guardrail system which meets the requirements of subpart M, and, shall be enclosed at least from the toeboard to mid-rail with either solid construction or expanded metal having openings no greater than 1/2 inch (1.27 cm).

**1926.550(g)(4)(ii)(B)** A grab rail shall be installed inside the entire perimeter of the personnel platform.

**1926.550(g)(4)(ii)(C)** Access gates, if installed, shall not swing outward during hoisting.

**1926.550(g)(4)(ii)(D)** Access gates, including sliding or folding gates, shall be equipped with a restraining device to prevent accidental opening.

**1926.550(g)(4)(ii)(E)** Headroom shall be provided which allows employees to stand upright in the platform.

**1926.550(g)(4)(ii)(F)** In addition to the use of hard hats, employees shall be protected by overhead protection on the personnel platform when employees are exposed to falling objects.

**1926.550(g)(4)(ii)(G)** All rough edges exposed to contact by employees shall be surfaced or smoothed in order to prevent injury to employees from punctures or lacerations.

**1926.550(g)(4)(ii)(H)** All welding of the personnel platform and its components shall be performed by a qualified welder familiar with the weld grades, types and material specified in the platform design.

**1926.550(g)(4)(ii)(I)** The personnel platform shall be conspicuously posted with a plate or other permanent marking which indicates the weight of the platform and its rated load capacity or maximum intended load.

**1926.550(g)(4)(iii)** *Personnel platform loading.* (A) The personnel platform shall not be loaded in excess of its rated load capacity. When a personnel platform does not have a rated load capacity then the personnel platform shall not be loaded in excess of its maximum intended load.

**1926.550(g)(4)(iii)(B)** The number of employees occupying the personnel platform shall not exceed the number required for the work being performed.

**1926.550(g)(4)(iii)(C)** Personnel platforms shall be used only for employees, their tools, and the materials necessary to do their work, and shall not be used to hoist only materials or tools when not hoisting personnel.

**1926.550(g)(4)(iii)(D)** Materials and tools for use during a personnel lift shall be secured to prevent displacement.

**1926.550(g)(4)(iii)(E)** Materials and tools for use during a personnel lift shall be evenly distributed within the confines of the platform while the platform is suspended.

**1926.550(g)(4)(iv)** *Rigging.* (A) When a wire rope bridle is used to connect the personnel platform to the load line, each bridle leg shall be connected to a master link or shackle in such a manner to ensure that the load is evenly divided among the bridle legs.

**1926.550(g)(4)(iv)(B)** Hooks on overhaul ball assemblies, lower load blocks, or other attachment assemblies shall be of a type that can be closed and locked, eliminating the hook throat opening. Alternatively, an alloy anchor type shackle with a bolt, nut and retaining pin may be used.

**1926.550(g)(4)(iv)(C)** Wire rope, shackles, rings, master links, and other rigging hardware must be capable of supporting, without failure, at least five times the maximum intended load applied or transmitted to that component. Where rotation resistant rope is used, the slings shall be capable of supporting without failure at least ten times the maximum intended load.

**1926.550(g)(4)(iv)(D)** All eyes in wire rope slings shall be fabricated with thimbles.

**1926.550(g)(4)(iv)(E)** Bridles and associated rigging for attaching the personnel platform to the hoist line shall be used only for the platform and the necessary employees, their tools and the materials necessary to do their work, and shall not be used for any other purpose when not hoisting personnel.

**1926.550(g)(5)** *Trial lift, inspection, and proof testing.* (i) A trial lift with the unoccupied personnel platform loaded at least to the anticipated liftweight shall be made from ground level, or any other location where employees will enter the platform, to each location at which the personnel platform is to be hoisted and positioned. This trial lift shall be performed immediately prior to placing personnel on the platform. The operator shall determine that all systems, controls and safety devices are activated and functioning properly; that no interferences exist; and that all configurations necessary to reach those work locations will allow the operator to remain under the 50 percent limit of the hoist's rated capacity. Materials and tools to be used during the actual lift can be loaded in the platform, as provided in paragraphs (g)(4)(iii) (D), and (E) of this section for the trial lift. A single trial lift may be performed at one time for all locations that are to be reached from a single set up position.

**1926.550(g)(5)(ii)** The trial lift shall be repeated prior to hoisting employees whenever the crane or derrick is moved and set up in a new location or returned to a previously used location. Additionally, the trial lift shall be repeated when the lift route is changed unless the operator determines that the route change is not significant (i.e. the route change would not affect the safety of hoisted employees.)

**1926.550(g)(5)(iii)** After the trial lift, and just prior to hoisting personnel, the platform shall be hoisted a few inches and inspected to ensure that it is secure and properly balanced. Employees shall not be hoisted unless the following conditions are determined to exist:

**1926.550(g)(5)(iii)(A)** Hoist ropes shall be free of kinks;

**1926.550(g)(5)(iii)(B)** Multiple part lines shall not be twisted around each other;

**1926.550(g)(5)(iii)(C)** The primary attachment shall be centered over the platform; and

**1926.550(g)(5)(iii)(D)** The hoisting system shall be inspected if the load rope is slack to ensure all ropes are properly stated on drums and in sheaves.

**1926.550(g)(5)(iv)** A visual inspection of the crane or derrick, rigging, personnel platform, and the crane or derrick base support or ground shall be conducted by a competent person immediately after the trial lift to determine whether the testing has exposed any defect or produced any adverse effect upon any component or structure.

**1926.550(g)(5)(v)** Any defects found during inspections which create a safety hazard shall be corrected before hoisting personnel.

**1926.550(g)(5)(vi)** At each job site, prior to hoisting employees on the personnel platform, and after any repair or modification, the platform and rigging shall be proof tested to 125 percent of the platform's rated capacity by holding it in a suspended position for five minutes with the test load evenly distributed on the platform (this may be done concurrently with the trial lift). After prooftesting, a competent person shall inspect the platform and rigging. Any deficiencies found shall be corrected and another proof test shall be conducted. Personnel hoisting shall not be conducted until the proof testing requirements are satisfied.

**1926.550(g)(6)** *Work practices.* (i) Employees shall keep all parts of the body inside the platform during raising, lowering, and positioning. This provision does not apply to an occupant of the platform performing the duties of a signal person.

**1926.550(g)(6)(ii)** Before employees exit or enter a hoisted personnel platform that is not landed, the platform shall be secured to the structure where the work is to be performed, unless securing to the structure creates an unsafe situation.

**1926.550(g)(6)(iii)** Tag lines shall be used unless their use creates an unsafe condition.

**1926.550(g)(6)(iv)** The crane or derrick operator shall remain at the controls at all times when the crane engine is running and the platform is occupied.

**1926.550(g)(6)(v)** Hoisting of employees shall be promptly discontinued upon indication of any dangerous weather conditions or other impending danger.

**1926.550(g)(6)(vi)** Employees being hoisted shall remain in continuous sight of and in direct communication with the operator or signal person. In those situations where direct visual contact with the operator is not possible, and the use of a signal person would create a greater hazard for that person, direct communication alone such as by radio may be used.

**1926.550(g)(6)(vii)** Except over water, employees occupying the personnel platform shall use a body belt/harness system with lanyard appropriately attached to the lower load block or overhaul ball, or to a structural member within the personnel platform capable of supporting a fall impact for employees using the anchorage. When working over water, the requirements of §1926.106 shall apply.

**1926.550(g)(6)(viii)** No lifts shall be made on another of the crane's or derrick's loadlines while personnel are suspended on a platform.

**1926.550(g)(7)** *Traveling.* (i) Hoisting of employees while the crane is traveling is prohibited, except for portal, tower and locomotive cranes, or where the employer demonstrates that there is no less hazardous way to perform the work.

**1926.550(g)(7)(ii)** Under any circumstances where a crane would travel while hoisting personnel, the employer shall implement the following procedures to safeguard employees:

**1926.550(g)(7)(ii)(A)** Crane travel shall be restricted to a fixed track or runway;

**1926.550(g)(7)(ii)(B)** Travel shall be limited to the load radius of the boom used during the lift; and

**1926.550(g)(7)(ii)(C)** The boom must be parallel to the direction of travel.

**1926.550(g)(7)(ii)(D)** A complete trial run shall be performed to test the route of travel before employees are allowed to occupy the platform. This trial run can be performed at the same time as the trial lift required by paragraph (g)(5)(i) of this section which tests the route of the lift.

**1926.550(g)(7)(ii)(E)** If travel is done with a rubber tired-carrier, the condition and air pressure of the tires shall be checked. The chart capacity for lifts on rubber shall be used for application of the 50 percent reduction of rated capacity. Notwithstanding paragraph (g)(3)(i)(E) of this section, outriggers may be partially retracted as necessary for travel.

**1926.550(g)(8)** *Pre-lift meeting.* (i) A meeting attended by the crane or derrick operator, signal person(s) (if necessary for the lift), employee(s) to be lifted, and the person responsible for the task to be performed shall be held to review the appropriate requirements of paragraph (g) of this section and the procedures to be followed.

**1926.550(g)(8)(ii)** This meeting shall be held prior to the trial lift at each new work location, and shall be repeated for any employees newly assigned to the operation.

[44 FR 8577, Feb. 9, 1979; 44 FR 20940, Apr. 6, 1979, as amended at 52 FR 36382, Sept. 28, 1987; 53 FR 29139, Aug. 2, 1988; 54 FR 15406, Apr. 18, 1989; 54 FR 24334, June 7, 1989; 58 FR 35183, June 30, 1993; 59 FR 40730, Aug. 9, 1994; 61 FR 5510, Feb. 13, 1996]

### §1926.551 Helicopters.

**1926.551(a)** ***Helicopter regulations.*** Helicopter cranes shall be expected to comply with any applicable regulations of the Federal Aviation Administration.

**1926.551(b)** ***Briefing.*** Prior to each day's operation a briefing shall be conducted. This briefing shall set forth the plan of operation for the pilot and ground personnel.

**1926.551(c)** ***Slings and tag lines.*** Load shall be properly slung. Tag lines shall be of a length that will not permit their being drawn up into rotors. Pressed sleeve, swedged eyes, or equivalent means shall be used for all freely suspended loads to prevent hand splices from spinning open or cable clamps from loosening.

**1926.551(d)** ***Cargo hooks.*** All electrically operated cargo hooks shall have the electrical activating device so designed and installed as to prevent inadvertent operation. In addition, these cargo hooks shall be equipped with an emergency mechanical control for releasing the load. The hooks shall be tested prior to each day's operation to determine that the release functions properly, both electrically and mechanically.

**1926.551(e)** ***Personal protective equipment.*** (1) Personal protective equipment for employees receiving the load shall consist of complete eye protection and hard hats secured by chinstraps.

**1926.551(e)(2)** Loose-fitting clothing likely to flap in the downwash, and thus be snagged on hoist line, shall not be worn.

**1926.551(f)** ***Loose gear and objects.*** Every practical precaution shall be taken to provide for the protection of the employees from flying objects in the rotor downwash. All loose gear within 100 feet of the place of lifting the load, depositing the load, and all other areas susceptible to rotor downwash shall be secured or removed.

**1926.551(g)** ***Housekeeping.*** Good housekeeping shall be maintained in all helicopter loading and unloading areas.

**1926.551(h)** ***Operator responsibility.*** The helicopter operator shall be responsible for size, weight, and manner in which loads are connected to the helicopter. If, for any reason, the helicopter operator believes the lift cannot be made safely, the lift shall not be made.

**1926.551(i)** ***Hooking and unhooking loads.*** When employees are required to perform work under hovering craft, a safe means of access shall be provided for employees to reach the hoist line hook and engage or disengage cargo slings. Employees shall not perform work under hovering craft except when necessary to hook or unhook loads.

**1926.551(j)** ***Static charge.*** Static charge on the suspended load shall be dissipated with a grounding device before ground personnel touch the suspended load, or protective rubber gloves shall be worn by all ground personnel touching the suspended load.

**1926.551(k)** ***Weight limitation.*** The weight of an external load shall not exceed the manufacturer's rating.

**1926.551(l)** ***Ground lines.*** Hoist wires or other gear, except for pulling lines or conductors that are allowed to "pay out" from a container or roll off a reel, shall not be attached to any fixed ground structure, or allowed to foul on any fixed structure.

**1926.551(m)** ***Visibility.*** When visibility is reduced by dust or other conditions, ground personnel shall exercise special caution to keep clear of main and stabilizing rotors. Precautions shall also be taken by the employer to eliminate as far as practical reduced visibility.

**1926.551(n)** ***Signal systems.*** Signal systems between aircrew and ground personnel shall be understood and checked in advance of hoisting the load. This applies to either radio or hand signal systems. Hand signals shall be as shown in Figure N-1.

**FIGURE N-1 HELICOPTER HAND SIGNAL**

**1926.551(o)** ***Approach distance.*** No unauthorized person shall be allowed to approach within 50 feet of the helicopter when the rotor blades are turning.

**1926.551(p)** ***Approaching helicopter.*** Whenever approaching or leaving a helicopter with blades rotating, all employees shall remain in full view of the pilot and keep in a crouched position. Employees shall avoid the area from the cockpit or cabin rearward unless authorized by the helicopter operator to work there.

**1926.551(q)** ***Personnel.*** Sufficient ground personnel shall be provided when required for safe helicopter loading and unloading operations.

**1926.551(r)** ***Communications.*** There shall be constant reliable communication between the pilot, and a designated employee of the ground crew who acts as a signalman during the period of loading and unloading. This signalman shall be distinctly recognizable from other ground personnel.

**1926.551(s)** ***Fires.*** Open fires shall not be permitted in an area that could result in such fires being spread by the rotor downwash.

## §1926.552 Material hoists, personnel hoists, and elevators.

**1926.552(a)** ***General requirements.*** (1) The employer shall comply with the manufacturer's specifications and limitations applicable to the operation of all hoists and elevators. Where manufacturer's specifications are not available, the limitations assigned to the equipment shall be based on the determinations of a professional engineer competent in the field.

**1926.552(a)(2)** Rated load capacities, recommended operating speeds, and special hazard warnings or instructions shall be posted on cars and platforms.

**1926.552(a)(3)** Wire rope shall be removed from service when any of the following conditions exists:

**1926.552(a)(3)(i)** In hoisting ropes, six randomly distributed broken wires in one rope lay or three broken wires in one strand in one rope lay;

**1926.552(a)(3)(ii)** Abrasion, scrubbing, flattening, or peening, causing loss of more than one-third of the original diameter of the outside wires;

**1926.552(a)(3)(iii)** Evidence of any heat damage resulting from a torch or any damage caused by contact with electrical wires;

**1926.552(a)(3)(iv)** Reduction from nominal diameter of more than three sixty-fourths inch for diameters up to and including three-fourths inch; one-sixteenth inch for diameters seven-eights to 1 1/8 inches; and three thirty-seconds inch for diameters 1 1/4 to 1 1/2 inches.

**1926.552(a)(4)** Hoisting ropes shall be installed in accordance with the wire rope manufacturers' recommendations.

**1926.552(a)(5)** The installation of live booms on hoists is prohibited.

**1926.552(a)(6)** The use of endless belt-type manlifts on construction shall be prohibited.

**1926.552(b)** ***Material hoists.*** (1)(i) Operating rules shall be established and posted at the operator's station of the hoist. Such rules shall include signal system and allowable line speed for various loads. Rules and notices shall be posted on the car frame or crosshead in a conspicuous location, including the statement "No Riders Allowed."

**1926.552(b)(1)(ii)** No person shall be allowed to ride on material hoists except for the purposes of inspection and maintenance.

**1926.552(b)(2)** All entrances of the hoistways shall be protected by substantial gates or bars which shall guard the full width of the landing entrance. All hoistway entrance bars and gates shall be painted with diagonal contrasting colors, such as black and yellow stripes.

**1926.552(b)(2)(i)** Bars shall be not less than 2- by 4-inch wooden bars or the equivalent, located 2 feet from the hoistway line. Bars shall be located not less than 36 inches nor more than 42 inches above the floor.

**1926.552(b)(2)(ii)** Gates or bars protecting the entrances to hoistways shall be equipped with a latching device.

**1926.552(b)(3)** Overhead protective covering of 2-inch planking, 3/4-inch plywood, or other solid material of equivalent strength, shall be provided on the top of every material hoist cage or platform.

**1926.552(b)(4)** The operator's station of a hoisting machine shall be provided with overhead protection equivalent to tight planking not less than 2 inches thick. The support for the overhead protection shall be of equal strength.

**1926.552(b)(5)** Hoist towers may be used with or without an enclosure on all sides. However, whichever alternative is chosen, the following applicable conditions shall be met:

**1926.552(b)(5)(i)** When a hoist tower is enclosed, it shall be enclosed on all sides for its entire height with a screen enclosure of 1/2-inch mesh, No. 18 U.S. gauge wire or equivalent, except for landing access.

**1926.552(b)(5)(ii)** When a hoist tower is not enclosed, the hoist platform or car shall be totally enclosed (caged) on all sides for the full height between the floor and the overhead protective covering with 1/2-inch mesh of No. 14 U.S. gauge wire or equivalent. The hoist platform enclosure shall include the required gates for loading and unloading. A 6-foot high enclosure shall be provided on the unused sides of the hoist tower at ground level.

**1926.552(b)(6)** Car arresting devices shall be installed to function in case of rope failure.

**1926.552(b)(7)** All material hoist towers shall be designed by a licensed professional engineer.

**1926.552(b)(8)** All material hoists shall conform to the requirements of ANSI A10.5-1969, Safety Requirements for Material Hoists.

**1926.552(c)** ***Personnel hoists.*** (1) Hoist towers outside the structure shall be enclosed for the full height on the side or sides used for entrance and exit to the structure. At the lowest landing, the enclosure on the sides not used for exit or entrance to the structure shall be enclosed to a height of at least 10 feet. Other sides of the tower adjacent to floors or scaffold platforms shall be enclosed to a height of 10 feet above the level of such floors or scaffolds.

**1926.552(c)(2)** Towers inside of structures shall be enclosed on all four sides throughout the full height.

**1926.552(c)(3)** Towers shall be anchored to the structure at intervals not exceeding 25 feet. In addition to tie-ins, a series of guys shall be installed. Where tie-ins are not practical the tower shall be anchored by means of guys made of wire rope at least one-half inch in diameter, securely fastened to anchorage to ensure stability.

**1926.552(c)(4)** Hoistway doors or gates shall be not less than 6 feet 6 inches high and shall be provided with mechanical locks which cannot be operated from the landing side, and shall be accessible only to persons on the car.

**1926.552(c)(5)** Cars shall be permanently enclosed on all sides and the top, except sides used for entrance and exit which have car gates or doors.

**1926.552(c)(6)** A door or gate shall be provided at each entrance to the car which shall protect the full width and height of the car entrance opening.

**1926.552(c)(7)** Overhead protective covering of 2-inch planking, 3/4-inch plywood or other solid material or equivalent strength shall be provided on the top of every personnel hoist.

**1926.552(c)(8)** Doors or gates shall be provided with electric contacts which do not allow movement of the hoist when door or gate is open.

**1926.552(c)(9)** Safeties shall be capable of stopping and holding the car and rated load when traveling at governor tripping speed.

**1926.552(c)(10)** Cars shall be provided with a capacity and data plate secured in a conspicuous place on the car or crosshead.

**1926.552(c)(11)** Internal combustion engines shall not be permitted for direct drive.

**1926.552(c)(12)** Normal and final terminal stopping devices shall be provided.

**1926.552(c)(13)** An emergency stop switch shall be provided in the car and marked "Stop."

**1926.552(c)(14)** Ropes: (i) The minimum number of hoisting ropes used shall be three for traction hoists and two for drum-type hoists.

**1926.552(c)(14)(ii)** The minimum diameter of hoisting and counterweight wire ropes shall be 1/2-inch.

**1926.552(c)(14)(iii)** Safety factors:

### MINIMUM FACTORS OF SAFETY FOR SUSPENSION WIRE ROPES

| Rope speed in feet per minute | Minimum factor of safety |
|---|---|
| 50 | 7.60 |
| 75 | 7.75 |
| 100 | 7.95 |
| 125 | 8.10 |
| 150 | 8.25 |
| 175 | 8.40 |
| 200 | 8.60 |
| 225 | 8.75 |
| 250 | 8.90 |
| 300 | 9.20 |
| 350 | 9.50 |
| 400 | 9.75 |
| 450 | 10.00 |
| 500 | 10.25 |
| 550 | 10.45 |
| 600 | 10.70 |

**1926.552(c)(15)** Following assembly and erection of hoists, and before being put in service, an inspection and test of all functions and safety devices shall be made under the supervision of a competent person. A similar inspection and test is required following major alteration of an existing installation. All hoists shall be inspected and tested at not more than 3-month intervals. The employer shall prepare a certification record which includes the date the inspection and test of all functions and safety devices was performed; the signature of the person who performed the inspection and test; and a serial number, or other identifier, for the hoist that was inspected and tested. The most recent certification record shall be maintained on file.

**1926.552(c)(16)** All personnel hoists used by employees shall be constructed of materials and components which meet the specifications for materials, construction, safety devices, assembly, and structural integrity as stated in the American National Standard A10.4-1963, Safety Requirements for Workmen's Hoists. The requirements of this paragraph (c)(16) do not apply to cantilever type personnel hoists.

**1926.552(c)(17)(i)** Personnel hoists used in bridge tower construction shall be approved by a registered professional engineer and erected under the supervision of a qualified engineer competent in this field.

**1926.552(c)(17)(ii)** When a hoist tower is not enclosed, the hoist platform or car shall be totally enclosed (caged) on all sides for the full height between the floor and the overhead protective covering with 3/4-inch mesh of No. 14 U.S. gauge wire or equivalent. The hoist platform enclosure shall include the required gates for loading and unloading.

**1926.552(c)(17)(iii)** These hoists shall be inspected and maintained on a weekly basis. Whenever the hoisting equipment is exposed to winds exceeding 35 miles per hour it shall be inspected and put in operable condition before reuse.

**1926.552(c)(17)(iv)** Wire rope shall be taken out of service when any of the following conditions exist:

**1926.552(c)(17)(iv)*(a)*** In running ropes, six randomly distributed broken wires in one lay or three broken wires in one strand in one lay;

**1926.552(c)(17)(iv)*(b)*** Wear of one-third the original diameter of outside individual wires. Kinking, crushing, bird caging, or any other damage resulting in distortion of the rope structure;

**1926.552(c)(17)(iv)*(c)*** Evidence of any heat damage from any cause;

**1926.552(c)(17)(iv)*(d)*** Reductions from nominal diameter of more than three-sixty-fourths inch for diameters to and including three-fourths inch, one-sixteenth inch for diameters seven-eights inch to 1 1/8 inches inclusive, three-thirty-seconds inch for diameters 1 1/4 to 1 1/2 inches inclusive;

**1926.552(c)(17)(iv)*(e)*** In standing ropes, more than two broken wires in one lay in sections beyond end connections or more than one broken wire at an end connection.

**1926.552(d)** Permanent elevators under the care and custody of the employer and used by employees for work covered by this Act shall comply with the requirements of American National Standards Institute A17.1-1965 with addenda A17.1a-1967, A17.1b-1968, A17.1c-1969, A17.1d-1970, and inspected in accordance with A17.2-1960 with addenda A17.2a-1965, A17.2b-1967.

[44 FR 8577, Feb. 9, 1979; 44 FR 20940, Apr. 6, 1979, as amended at 52 FR 36382, Sept. 28, 1987]

## §1926.553 Base-mounted drum hoists.

**1926.553(a)** ***General requirements.*** (1) Exposed moving parts such as gears, projecting screws, setscrews, chain, cables, chain sprockets, and reciprocating or rotating parts, which constitute a hazard, shall be guarded.

**1926.553(a)(2)** All controls used during the normal operation cycle shall be located within easy reach of the operator's station.

**1926.553(a)(3)** Electric motor operated hoists shall be provided with:

**1926.553(a)(3)(i)** A device to disconnect all motors from the line upon power failure and not permit any motor to be restarted until the controller handle is brought to the "off" position;

**1926.553(a)(3)(ii)** Where applicable, an overspeed preventive device;

**1926.553(a)(3)(iii)** A means whereby remotely operated hoists stop when any control is ineffective.

**1926.553(a)(4)** All base-mounted drum hoists in use shall meet the applicable requirements for design, construction, installation, testing, inspection, maintenance, and operations, as prescribed by the manufacturer.

**1926.553(b)** ***Specific requirements.*** [Reserved]

## §1926.554 Overhead hoists.

**1926.554(a)** ***General requirements.*** (1) The safe working load of the overhead hoist, as determined by the manufacturer, shall be indicated on the hoist, and this safe working load shall not be exceeded.

**1926.554(a)(2)** The supporting structure to which the hoist is attached shall have a safe working load equal to that of the hoist.

**1926.554(a)(3)** The support shall be arranged so as to provide for free movement of the hoist and shall not restrict the hoist from lining itself up with the load.

**1926.554(a)(4)** The hoist shall be installed only in locations that will permit the operator to stand clear of the load at all times.

**1926.554(a)(5)** Air hoists shall be connected to an air supply of sufficient capacity and pressure to safely operate the hoist. All air hoses supplying air shall be positively connected to prevent their becoming disconnected during use.

**1926.554(a)(6)** All overhead hoists in use shall meet the applicable requirements for construction, design, installation, testing, inspection, maintenance, and operation, as prescribed by the manufacturer.

**1926.554(b)** ***Specific requirements.*** [Reserved]

## §1926.555 Conveyors.

**1926.555(a)** ***General requirements.*** (1) Means for stopping the motor or engine shall be provided at the operator's station. Conveyor systems shall be equipped with an audible warning signal to be sounded immediately before starting up the conveyor.

**1926.555(a)(2)** If the operator's station is at a remote point, similar provisions for stopping the motor or engine shall be provided at the motor or engine location.

**1926.555(a)(3)** Emergency stop switches shall be arranged so that the conveyor cannot be started again until the actuating stop switch has been reset to running or "on" position.

**1926.555(a)(4)** Screw conveyors shall be guarded to prevent employee contact with turning flights.

**1926.555(a)(5)** Where a conveyor passes over work areas, aisles, or thoroughfares, suitable guards shall be provided to protect employees required to work below the conveyors.

**1926.555(a)(6)** All crossovers, aisles, and passageways shall be conspicuously marked by suitable signs, as required by subpart G of this part.

**1926.555(a)(7)** Conveyors shall be locked out or otherwise rendered inoperable, and tagged out with a "Do Not Operate" tag during repairs and when operation is hazardous to employees performing maintenance work.

**1926.555(a)(8)** All conveyors in use shall meet the applicable requirements for design, construction, inspection, testing, maintenance, and operation, as prescribed in the ANSI B20.1-1957, Safety Code for Conveyors, Cableways, and Related Equipment.

## §1926.556 Aerial lifts. [Removed.]

[§1926.556 removed at 61 FR 46131, Aug. 30, 1996, eff. Nov. 29, 1996.]

**Authority:** Section 107, Construction Work Hours and Safety Standards Act (Construction Safety Act) (40 U.S.C. 333); Secs. 4, 6, 8, Occupational Safety and Health Act of 1970 (29 U.S.C. 653, 655, 657); Secretary of Labor's Order No. 12-71 (36 FR 8754), 8-76 (41 FR 25059), 9-83 (48 FR 35736), 1-90 (55 FR 9033), or 6-96 (62 FR 111), as applicable. Section 1926.602 also issued under 29 CFR part 1911.

## §1926.600 Equipment.

**1926.600(a)** ***General requirements.*** (1) All equipment left unattended at night, adjacent to a highway in normal use, or adjacent to construction areas where work is in progress, shall have appropriate lights or reflectors, or barricades equipped with appropriate lights or reflectors, to identify the location of the equipment.

**1926.600(a)(2)** A safety tire rack, cage, or equivalent protection shall be provided and used when inflating, mounting, or dismounting tires installed on split rims, or rims equipped with locking rings or similar devices.

**1926.600(a)(3)(i)** Heavy machinery, equipment, or parts thereof, which are suspended or held aloft by use of slings, hoists, or jacks shall be substantially blocked or cribbed to prevent falling or shifting before employees are permitted to work under or between them. Bulldozer and scraper blades, end-loader buckets, dump bodies, and similar equipment, shall be either fully lowered or blocked when being repaired or when not in use. All controls shall be in a neutral position, with the motors stopped and brakes set, unless work being performed requires otherwise.

**1926.600(a)(3)(ii)** Whenever the equipment is parked, the parking brake shall be set. Equipment parked on inclines shall have the wheels chocked and the parking brake set.

**1926.600(a)(4)** The use, care and charging of all batteries shall conform to the requirements of subpart K of this part.

**1926.600(a)(5)** All cab glass shall be safety glass, or equivalent, that introduces no visible distortion affecting the safe operation of any machine covered by this subpart.

**1926.600(a)(6)** All equipment covered by this subpart shall comply with the requirements of §1926.550(a)(15) when working or being moved in the vicinity of power lines or energized transmitters.

**1926.600(a)(7)** *Rolling railroad cars.* Derail and/or bumper blocks shall be provided on spur railroad tracks where a rolling car could contact other cars being worked, enter a building, work or traffic area.

**1926.600(b)** ***Specific requirements.*** [Reserved]

[44 FR 8577, Feb. 9, 1979; 44 FR 20940, Apr. 6, 1979, as amended at 58 FR 35183, June 30, 1993]

## §1926.601 Motor vehicles.

**1926.601(a)** ***Coverage.*** Motor vehicles as covered by this part are those vehicles that operate within an off-highway jobsite, not open to public traffic. The requirements of this section do not apply to equipment for which rules are prescribed in §1926.602.

**1926.601(b)** ***General requirements.*** (1) All vehicles shall have a service brake system, an emergency brake system, and a parking brake system. These systems may use common components, and shall be maintained in operable condition.

**1926.601(b)(2)(i)** Whenever visibility conditions warrant additional light, all vehicles, or combinations of vehicles, in use shall be equipped with at least two headlights and two taillights in operable condition.

**1926.601(b)(2)(ii)** All vehicles, or combination of vehicles, shall have brake lights in operable condition regardless of light conditions.

**1926.601(b)(3)** All vehicles shall be equipped with an adequate audible warning device at the operator's station and in an operable condition.

**1926.601(b)(4)** No employer shall use any motor vehicle equipment having an obstructed view to the rear unless:

**1926.601(b)(4)(i)** The vehicle has a reverse signal alarm audible above the surrounding noise level or:

**1926.601(b)(4)(ii)** The vehicle is backed up only when an observer signals that it is safe to do so.

**1926.601(b)(5)** All vehicles with cabs shall be equipped with windshields and powered wipers. Cracked and broken glass shall be replaced. Vehicles operating in areas or under conditions that cause fogging or frosting of the windshields shall be equipped with operable defogging or defrosting devices.

**1926.601(b)(6)** All haulage vehicles, whose pay load is loaded by means of cranes, power shovels, loaders, or similar equipment, shall have a cab shield and/or canopy adequate to protect the operator from shifting or falling materials.

**1926.601(b)(7)** Tools and material shall be secured to prevent movement when transported in the same compartment with employees.

**1926.601(b)(8)** Vehicles used to transport employees shall have seats firmly secured and adequate for the number of employees to be carried.

**1926.601(b)(9)** Seat belts and anchorages meeting the requirements of 49 CFR part 571 (Department of Transportation, Federal Motor Vehicle Safety Standards) shall be installed in all motor vehicles.

**1926.601(b)(10)** Trucks with dump bodies shall be equipped with positive means of support, permanently attached, and capable of being locked in position to prevent accidental lowering of the body while maintenance or inspection work is being done.

**1926.601(b)(11)** Operating levers controlling hoisting or dumping devices on haulage bodies shall be equipped with a latch or other device which will prevent accidental starting or tripping of the mechanism.

**1926.601(b)(12)** Trip handles for tailgates of dump trucks shall be so arranged that, in dumping, the operator will be in the clear.

**1926.601(b)(13)(i)** All rubber-tired motor vehicle equipment manufactured on or after May 1, 1972, shall be equipped with fenders. All rubber-tired motor vehicle equipment manufactured before May 1, 1972, shall be equipped with fenders not later than May 1, 1973.

**1926.601(b)(13)(ii)** Mud flaps may be used in lieu of fenders whenever motor vehicle equipment is not designed for fenders.

**1926.601(b)(14)** All vehicles in use shall be checked at the beginning of each shift to assure that the following parts, equipment, and accessories are in safe operating condition and free of apparent damage that could cause failure while in use: service brakes, including trailer brake connections; parking system (hand brake); emergency stopping system (brakes); tires; horn; steering mechanism; coupling devices; seat belts; operating controls; and safety devices. All defects shall be corrected before the vehicle is placed in service. These requirements also apply to equipment such as lights, reflectors, windshield wipers, defrosters, fire extinguishers, etc., where such equipment is necessary.

## §1926.602 Material handling equipment.

**1926.602(a)** ***Earthmoving equipment; General.*** (1) These rules apply to the following types of earthmoving equipment: scrapers, loaders, crawler or wheel tractors, bulldozers, off-highway trucks, graders, agricultural and industrial tractors, and similar equipment. The promulgation of specific rules for compactors and rubber-tired "skid-steer" equipment is reserved pending consideration of standards currently being developed.

**1926.602(a)(2)** *Seat belts.* (i) Seat belts shall be provided on all equipment covered by this section and shall meet the requirements of the Society of Automotive Engineers, J386-1969, Seat Belts for Construction Equipment. Seat belts for agricultural and light industrial tractors shall meet the seat belt requirements of Society of Automotive Engineers J333a-1970, Operator Protection for Agricultural and Light Industrial Tractors.

**1926.602(a)(2)(ii)** Seat belts need not be provided for equipment which is designed only for standup operation.

**1926.602(a)(2)(iii)** Seat belts need not be provided for equipment which does not have roll-over protective structure (ROPS) or adequate canopy protection.

**1926.602(a)(3)** *Access roadways and grades.* (i) No employer shall move or cause to be moved construction equipment or vehicles upon any access roadway or grade unless the access roadway or grade is constructed and maintained to accommodate safely the movement of the equipment and vehicles involved.

**1926.602(a)(3)(ii)** Every emergency access ramp and berm used by an employer shall be constructed to restrain and control runaway vehicles.

**1926.602(a)(4)** *Brakes.* All earthmoving equipment mentioned in this §1926.602(a) shall have a service braking system capable of stopping and holding the equipment fully loaded, as specified in Society of Automotive Engineers SAE-J237, Loader Dozer-1971, J236, Graders-1971, and J319b, Scrapers-1971. Brake systems for self-propelled rubber-tired off-highway equipment manufactured after January 1, 1972 shall meet the applicable minimum performance criteria set forth in the following Society of Automotive Engineers Recommended Practices:

| | |
|---|---|
| Self-Propelled Scrapers | SAE J319b-1971. |
| Self-Propelled Graders | SAE J236-1971. |
| Trucks and Wagons | SAE J166-1971. |
| Front End Loaders and Dozers | SAE J237-1971. |

**1926.602(a)(5)** *Fenders.* Pneumatic-tired earth-moving haulage equipment (trucks, scrapers, tractors, and trailing units) whose maximum speed exceeds 15 miles per hour, shall be equipped with fenders on all wheels to meet the requirements of Society of Automotive Engineers SAE J321a-1970, Fenders for Pneumatic-Tired Earthmoving Haulage Equipment. An employer may, of course, at any time seek to show under §1926.2, that the uncovered wheels present no hazard to personnel from flying materials.

**1926.602(a)(6)** *Rollover protective structures (ROPS).* See subpart W of this part for requirements for rollover protective structures and overhead protection.

**1926.602(a)(7)** *Rollover protective structures for off-highway trucks.* The promulgation of standards for rollover protective structures for off-highway trucks is reserved pending further study and development.

**1926.602(a)(8)** *Specific effective dates—brakes and fenders.* (i) Equipment mentioned in paragraph (a)(4) and (5) of this section, and manufactured after January 1, 1972, which is used by any employer after that date, shall comply with the applicable rules prescribed therein concerning brakes and fenders. Equipment mentioned in paragraphs (a) (4) and (5) of this section, and manufactured before January 1, 1972, which is used by any employer after that date, shall meet the applicable rules prescribed herein not later than June 30, 1973. It should be noted that, as permitted under §1926.2, employers may request variations from the applicable brakes and fender standards required by this subpart. Employers wishing to seek variations from the applicable brakes and fenders rules may submit any requests for variations after the publication of this document in the FEDERAL REGISTER. Any statements intending to meet the requirements of §1926.2(b)(4), should specify how the variation would protect the safety of the employees by providing for any compensating restrictions on the operation of equipment.

**1926.602(a)(8)(ii)** Notwithstanding the provisions of paragraphs (a)(5) and (a)(8)(i) of this section, the requirement that fenders be installed on pneumatic-tired earthmoving haulage equipment, is suspended pending reconsideration of the requirement.

**1926.602(a)(9)** *Audible alarms.* (i) All bidirectional machines, such as rollers, compacters, front-end loaders, bulldozers, and similar equipment, shall be equipped with a horn, distinguishable from the surrounding noise level, which shall be operated as needed when the machine is moving in either direction. The horn shall be maintained in an operative condition.

**1926.602(a)(9)(ii)** No employer shall permit earthmoving or compacting equipment which has an obstructed view to the rear to be used in reverse gear unless the equipment has in operation a reverse signal alarm distinguishable from the surrounding noise level or an employee signals that it is safe to do so.

**1926.602(a)(10)** *Scissor points.* Scissor points on all front-end loaders, which constitute a hazard to the operator during normal operation, shall be guarded.

**1926.602(b)** ***Excavating and other equipment.*** (1) Tractors covered in paragraph (a) of this section shall have seat belts as required for the operators when seated in the normal seating arrangement for tractor operation, even though back-hoes, breakers, or other similar attachments are used on these machines for excavating or other work.

**1926.602(b)(2)** For the purposes of this subpart and of subpart N of this part, the nomenclatures and descriptions for measurement of dimensions of machinery and attachments shall be as described in Society of Automotive Engineers 1970 Handbook, pages 1088 through 1103.

**1926.602(b)(3)** The safety requirements, ratios, or limitations applicable to machines or attachment usage covered in Power Crane and Shovel Associations Standards No. 1 and No. 2 of 1968, and No. 3 of 1969, shall be complied with, and shall apply to cranes, machines, and attachments under this part.

**1926.602(c)** ***Lifting and hauling equipment (other than equipment covered under subpart N of this part).*** (1) Industrial trucks shall meet the requirements of §1926.600 and the following:

**1926.602(c)(1)(i)** Lift trucks, stackers, etc., shall have the rated capacity clearly posted on the vehicle so as to be clearly visible to the operator. When auxiliary removable counterweights are provided by the manufacturer, corresponding alternate rated capacities also shall be clearly shown on the vehicle. These ratings shall not be exceeded.

**1926.602(c)(1)(ii)** No modifications or additions which affect the capacity or safe operation of the equipment shall be made without the manufacturer's written approval. If such modifications or changes are made, the capacity, operation, and maintenance instruction plates, tags, or decals shall be changed accordingly. In no case shall the original safety factor of the equipment be reduced.

**1926.602(c)(1)(iii)** If a load is lifted by two or more trucks working in unison, the proportion of the total load carried by any one truck shall not exceed its capacity.

**1926.602(c)(1)(iv)** Steering or spinner knobs shall not be attached to the steering wheel unless the steering mechanism is of a type that prevents road reactions from causing the steering handwheel to spin. The steering knob shall be mounted within the periphery of the wheel.

**1926.602(c)(1)(v)** All high lift rider industrial trucks shall be equipped with overhead guards which meet the configuration and structural requirements as defined in paragraph 421 of American National Standards Institute B56.1-1969, Safety Standards for Powered Industrial Trucks.

**1926.602(c)(1)(vi)** All industrial trucks in use shall meet the applicable requirements of design, construction, stability, inspection, testing, maintenance, and operation, as defined in American National Standards Institute B56.1-1969, Safety Standards for Powered Industrial Trucks.

**1926.602(c)(1)(vii)** Unauthorized personnel shall not be permitted to ride on powered industrial trucks. A safe place to ride shall be provided where riding of trucks is authorized.

**1926.602(c)(1)(viii)** Whenever a truck is equipped with vertical only, or vertical and horizontal controls elevatable with the lifting carriage or forks for lifting personnel, the following additional precautions shall be taken for the protection of personnel being elevated.

**1926.602(c)(1)(viii)(A)** Use of a safety platform firmly secured to the lifting carriage and/or forks.

**1926.602(c)(1)(viii)(B)** Means shall be provided whereby personnel on the platform can shut off power to the truck.

**1926.602(c)(1)(viii)(C)** Such protection from falling objects as indicated necessary by the operating conditions shall be provided.

**1926.602(d)** ***Powered industrial truck operator training.***

**Note:** The requirements applicable to construction work under this paragraph are identical to those set forth at §1910.178(l) of this chapter.

[44 FR 8577, Feb. 9, 1979; 44 FR 20940, Apr. 6, 1979, as amended at 58 FR 35183, June 30, 1993; 63 FR 66274, Dec. 1, 1998]

[Section 1910.178(l) is printed below—CCH.]

## §1910.178 Powered industrial trucks.

**1910.178(l)** ***Operator training.*** (1) *Safe operation.* (i) The employer shall ensure that each powered industrial truck operator is competent to operate a powered industrial truck safely, as demonstrated by the successful completion of the training and evaluation specified in this paragraph (l).

**1910.178(l)(1)(ii)** Prior to permitting an employee to operate a powered industrial truck (except for training purposes), the employer shall ensure that each operator has successfully completed the training required by this paragraph (l), except as permitted by paragraph (l)(5).

**1910.178(l)(2)** *Training program implementation.* (i) Trainees may operate a powered industrial truck only:

**1910.178(l)(2)(i)(A)** Under the direct supervision of persons who have the knowledge, training, and experience to train operators and evaluate their competence; and

**1910.178(l)(2)(i)(B)** Where such operation does not endanger the trainee or other employees.

**1910.178(l)(2)(ii)** Training shall consist of a combination of formal instruction (e.g., lecture, discussion, interactive computer learning, video tape, written material), practical training (demonstrations performed by the trainer and practical exercises performed by the trainee), and evaluation of the operator's performance in the workplace.

**1910.178(l)(2)(iii)** All operator training and evaluation shall be conducted by persons who have the knowledge, training, and experience to train powered industrial truck operators and evaluate their competence.

**1910.178(l)(3)** *Training program content.* Powered industrial truck operators shall receive initial training in the following topics, except in topics which the employer can demonstrate are not applicable to safe operation of the truck in the employer's workplace.

**1910.178(l)(3)(i)** Truck-related topics:

**1910.178(l)(3)(i)(A)** Operating instructions, warnings, and precautions for the types of truck the operator will be authorized to operate;

**1910.178(l)(3)(i)(B)** Differences between the truck and the automobile;

**1910.178(l)(3)(i)(C)** Truck controls and instrumentation: where they are located, what they do, and how they work;

**1910.178(l)(3)(i)(D)** Engine or motor operation;

**1910.178(l)(3)(i)(E)** Steering and maneuvering;

**1910.178(l)(3)(i)(F)** Visibility (including restrictions due to loading);

**1910.178(l)(3)(i)(G)** Fork and attachment adaptation, operation, and use limitations;

**1910.178(l)(3)(i)(H)** Vehicle capacity;

**1910.178(l)(3)(i)(I)** Vehicle stability;

**1910.178(l)(3)(i)(J)** Any vehicle inspection and maintenance that the operator will be required to perform;

**1910.178(l)(3)(i)(K)** Refueling and/or charging and recharging of batteries;

**1910.178(l)(3)(i)(L)** Operating limitations;

**1910.178(l)(3)(i)(M)** Any other operating instructions, warnings, or precautions listed in the operator's manual for the types of vehicle that the employee is being trained to operate.

**1910.178(l)(3)(ii)** Workplace-related topics:

**1910.178(l)(3)(ii)(A)** Surface conditions where the vehicle will be operated;

**1910.178(l)(3)(ii)(B)** Composition of loads to be carried and load stability;

**1910.178(l)(3)(ii)(C)** Load manipulation, stacking, and unstacking;

**1910.178(l)(3)(ii)(D)** Pedestrian traffic in areas where the vehicle will be operated;

**1910.178(l)(3)(ii)(E)** Narrow aisles and other restricted places where the vehicle will be operated;

**1910.178(l)(3)(ii)(F)** Hazardous (classified) locations where the vehicle will be operated;

**1910.178(l)(3)(ii)(G)** Ramps and other sloped surfaces that could affect the vehicle's stability;

**1910.178(l)(3)(ii)(H)** Closed environments and other areas where insufficient ventilation or poor vehicle maintenance could cause a buildup of carbon monoxide or diesel exhaust;

**1910.178(l)(3)(ii)(I)** Other unique or potentially hazardous environmental conditions in the workplace that could affect safe operation.

**1910.178(l)(3)(iii)** The requirements of this section.

**1910.178(l)(4)** *Refresher training and evaluation.* (i) Refresher training, including an evaluation of the effectiveness of that training, shall be conducted as required by paragraph (l)(4)(ii) to ensure that the operator has the knowledge and skills needed to operate the powered industrial truck safely.

**1910.178(l)(4)(ii)** Refresher training in relevant topics shall be provided to the operator when:

**1910.178(l)(4)(ii)(A)** The operator has been observed to operate the vehicle in an unsafe manner;

**1910.178(l)(4)(ii)(B)** The operator has been involved in an accident or near-miss incident;

**1910.178(l)(4)(ii)(C)** The operator has received an evaluation that reveals that the operator is not operating the truck safely;

**1910.178(l)(4)(ii)(D)** The operator is assigned to drive a different type of truck; or

**1910.178(l)(4)(ii)(E)** A condition in the workplace changes in a manner that could affect safe operation of the truck.

**1910.178(l)(4)(iii)** An evaluation of each powered industrial truck operator's performance shall be conducted at least once every three years.

**1910.178(l)(5)** *Avoidance of duplicative training.* If an operator has previously received training in a topic specified in paragraph (l)(3) of this section, and such training is appropriate to the truck and working conditions encountered, additional training in that topic is not required if the operator has been evaluated and found competent to operate the truck safely.

**1910.178(l)(6)** *Certification.* The employer shall certify that each operator has been trained and evaluated as required by this paragraph (l). The certification shall include the name of the operator, the date of the training, the date of the evaluation, and the identity of the person(s) performing the training or evaluation.

**1910.178(l)(7)** *Dates.* The employer shall ensure that operators of powered industrial trucks are trained, as appropriate, by the dates shown in the following table.

| If the employee was hired: | The initial training and evaluation of that employee must be completed: |
|---|---|
| Before December 1, 1999 ..... | By December 1, 1999. |
| After December 1, 1999 ........ | Before the employee is assigned to operate a powered industrial truck. |

**1910.178(l)(8)** Appendix A to this section provides non-mandatory guidance to assist employers in implementing this paragraph (l). This appendix does not add to, alter, or reduce the requirements of this section.

## §1926.603 Pile driving equipment.

**1926.603(a)** ***General requirements.*** (1) Boilers and piping systems which are a part of, or used with, pile driving equipment shall meet the applicable requirements of the American Society of Mechanical Engineers, Power Boilers (section I).

**1926.603(a)(2)** All pressure vessels which are a part of, or used with, pile driving equipment shall meet the applicable requirements of the American Society of Mechanical Engineers, Pressure Vessels (section VIII).

**1926.603(a)(3)** Overhead protection, which will not obscure the vision of the operator and which meets the requirements of subpart N of this part, shall be provided. Protection shall be the equivalent of 2-inch planking or other solid material of equivalent strength.

**1926.603(a)(4)** Stop blocks shall be provided for the leads to prevent the hammer from being raised against the head block.

**1926.603(a)(5)** A blocking device, capable of safely supporting the weight of the hammer, shall be provided for placement in the leads under the hammer at all times while employees are working under the hammer.

**1926.603(a)(6)** Guards shall be provided across the top of the head block to prevent the cable from jumping out of the sheaves.

**1926.603(a)(7)** When the leads must be inclined in the driving of batter piles, provisions shall be made to stabilize the leads.

**1926.603(a)(8)** Fixed leads shall be provided with ladder, and adequate rings, or similar attachment points, so that the loft worker may engage his safety belt lanyard to the leads. If the leads are provided with loft platforms(s), such platform(s) shall be protected by standard guardrails.

**1926.603(a)(9)** Steam hose leading to a steam hammer or jet pipe shall be securely attached to the hammer with an adequate length of at least 1/4-inch diameter chain or cable to prevent whipping in the event the joint at the hammer is broken. Air hammer hoses shall be provided with the same protection as required for steam lines.

**1926.603(a)(10)** Safety chains, or equivalent means, shall be provided for each hose connection to prevent the line from thrashing around in case the coupling becomes disconnected.

**1926.603(a)(11)** Steam line controls shall consist of two shutoff valves, one of which shall be a quick-acting lever type within easy reach of the hammer operator.

**1926.603(a)(12)** Guys, outriggers, thrustouts, or counterbalances shall be provided as necessary to maintain stability of pile driver rigs.

**1926.603(b)** ***Pile driving from barges and floats.*** Barges or floats supporting pile driving operations shall meet the applicable requirements of §1926.605.

**1926.603(c)** ***Pile driving equipment.*** (1) Engineers and winchmen shall accept signals only from the designated signalmen.

**1926.603(c)(2)** All employees shall be kept clear when piling is being hoisted into the leads.

**1926.603(c)(3)** When piles are being driven in an excavated pit, the walls of the pit shall be sloped to the angle of repose or sheet-piled and braced.

**1926.603(c)(4)** When steel tube piles are being "blown out", employees shall be kept well beyond the range of falling materials.

**1926.603(c)(5)** When it is necessary to cut off the tops of driven piles, pile driving operations shall be suspended except where the cutting operations are located at least twice the length of the longest pile from the driver.

**1926.603(c)(6)** When driving jacked piles, all access pits shall be provided with ladders and bulkheaded curbs to prevent material from falling into the pit.

## §1926.604 Site clearing.

**1926.604(a)** ***General requirements.*** (1) Employees engaged in site clearing shall be protected from hazards of irritant and toxic plants and suitably instructed in the first aid treatment available.

**1926.604(a)(2)** All equipment used in site clearing operations shall be equipped with rollover guards meeting the requirements of this subpart. In addition, rider-operated equipment shall be equipped with an overhead and rear canopy guard meeting the following requirements:

**1926.604(a)(2)(i)** The overhead covering on this canopy structure shall be of not less than 1/8-inch steel plate or 1/4-inch woven wire mesh with openings no greater than 1 inch, or equivalent.

**1926.604(a)(2)(ii)** The opening in the rear of the canopy structure shall be covered with not less than 1/4-inch woven wire mesh with openings no greater than 1 inch.

**1926.604(b)** ***Specific requirements.*** [Reserved]

## §1926.605 Marine operations and equipment.

**1926.605(a)** ***Material handling operations.*** (1) Operations fitting the definition of "material handling" shall be performed in conformance with applicable requirements of part 1918, "Safety and Health Regulations for Longshoring" of this chapter. The term "longshoring operations" means the loading, unloading, moving, or handling of construction materials, equipment and supplies, etc. into, in, on, or out of any vessel from a fixed structure or shore-to-vessel, vessel-to-shore or fixed structure or vessel-to-vessel.

**1926.605(b)** ***Access to barges.*** (1) Ramps for access of vehicles to or between barges shall be of adequate strength, provided with side boards, well maintained, and properly secured.

**1926.605(b)(2)** Unless employees can step safely to or from the wharf, float, barge, or river towboat, either a ramp, meeting the requirements of paragraph (b)(1) of this section, or a safe walkway, shall be provided.

**1926.605(b)(3)** Jacob's ladders shall be of the double rung or flat tread type. They shall be well maintained and properly secured.

**1926.605(b)(4)** A Jacob's ladder shall either hang without slack from its lashings or be pulled up entirely.

**1926.605(b)(5)** When the upper end of the means of access rests on or is flush with the top of the bulwark, substantial steps properly secured and equipped with at least one substantial hand rail approximately 33 inches in height, shall be provided between the top of the bulwark and the deck.

**1926.605(b)(6)** Obstructions shall not be laid on or across the gangway.

**1926.605(b)(7)** The means of access shall be adequately illuminated for its full length.

**1926.605(b)(8)** Unless the structure makes it impossible, the means of access shall be so located that the load will not pass over employees.

**1926.605(c)** ***Working surfaces of barges.*** (1) Employees shall not be permitted to walk along the sides of covered lighters or barges with coamings more than 5 feet high, unless there is a 3-foot clear walkway, or a grab rail, or a taut handline is provided.

**1926.605(c)(2)** Decks and other working surfaces shall be maintained in a safe condition.

**1926.605(c)(3)** Employees shall not be permitted to pass fore and aft, over, or around deckloads, unless there is a safe passage.

**1926.605(c)(4)** Employees shall not be permitted to walk over deckloads from rail to coaming unless there is a safe passage. If it is necessary to stand at the outboard or inboard edge of the deckload where less than 24 inches of bulwark, rail, coaming, or other protection exists, all employees shall be provided with a suitable means of protection against falling from the deckload.

**1926.605(d)** ***First-aid and lifesaving equipment.*** (1) Provisions for rendering first aid and medical assistance shall be in accordance with subpart D of this part.

**1926.605(d)(2)** The employer shall ensure that there is in the vicinity of each barge in use at least one U.S. Coast Guard-approved 30-inch lifering with not less than 90 feet of line attached, and at least one portable or permanent ladder which will reach the top of the apron to the surface of the water. If the above equipment is not available at the pier, the employer shall furnish it during the time that he is working the barge.

**1926.605(d)(3)** Employees walking or working on the unguarded decks of barges shall be protected with U.S. Coast Guard-approved work vests or buoyant vests.

**1926.605(e)** ***Commercial diving operations.*** Commercial diving operations shall be subject to subpart T of part 1910, §§1910.401-1910.441, of this chapter.

[39 FR 22801, June 24, 1974, as amended at 42 FR 37674, July 22, 1977]

[Sections 1910.401-1910.441 are printed below—CCH.]

## §1910.401 Scope and application.

**1910.401(a)** ***Scope.*** (1) This subpart (standard) applies to every place of employment within the waters of the United States, or within any State, the District of Columbia, the Commonwealth of Puerto Rico, the Virgin Islands, American Samoa, Guam, the Trust Territory of the Pacific Islands, Wake Island, Johnston Island, the Canal Zone, or within the Outer Continental Shelf lands as defined in the Outer Continental Shelf Lands Act (67 Stat. 462, 43 U.S.C. 1331), where diving and related support operations are performed.

**1910.401(a)(2)** This standard applies to diving and related support operations conducted in connection with all types of work and employments, including general industry, construction, ship repairing, shipbuilding, shipbreaking and longshoring. However, this standard does not apply to any diving operation:

**1910.401(a)(2)(i)** Performed solely for instructional purposes, using open-circuit, compressed-air SCUBA and conducted within the no-decompression limits;

**1910.401(a)(2)(ii)** Performed solely for search, rescue, or related public safety purposes by or under the control of a governmental agency; or

**1910.401(a)(2)(iii)** Governed by 45 CFR part 46 (Protection of Human Subjects, U.S. Department of Health and Human Services) or equivalent rules or regulations established by another federal agency, which regulate research, development, or related purposes involving human subjects.

**1910.401(a)(2)(iv)** Defined as scientific diving and which is under the direction and control of a diving program containing at least the following elements:

**1910.401(a)(2)(iv)(A)** Diving safety manual which includes at a minimum: Procedures covering all diving operations specific to the program; procedures for emergency care, including recompression and evacuation; and criteria for diver training and certification.

**1910.401(a)(2)(iv)(B)** Diving control (safety) board, with the majority of its members being active divers, which shall at a minimum have the authority to: Approve and monitor diving projects; review and revise the diving safety manual; assure compliance with the manual; certify the depths to which a diver has been trained; take disciplinary action for unsafe practices; and, assure adherence to the buddy system (a diver is accompanied by and is in continuous contact with another diver in the water) for SCUBA diving.

**1910.401(b)** ***Application in emergencies.*** An employer may deviate from the requirements of this standard to the extent necessary to prevent or minimize a situation which is likely to cause death, serious physical harm, or major environmental damage, provided that the employer:

**1910.401(b)(1)** Notifies the Area Director, Occupational Safety and Health Administration within 48 hours of the onset of the emergency situation indicating the nature of the emergency and extent of the deviation from the prescribed regulations; and

**1910.401(b)(2)** Upon request from the Area Director, submits such information in writing.

**1910.401(c)** ***Employer obligation.*** The employer shall be responsible for compliance with:

**1910.401(c)(1)** All provisions of this standard of general applicability; and

**1910.401(c)(2)** All requirements pertaining to specific diving modes to the extent diving operations in such modes are conducted.

[42 FR 37668, July 22, 1977, as amended at 47 FR 53365, Nov. 26, 1982; 58 FR 35310, June 30, 1993]

### §1910.402 Definitions.

As used in this standard, the listed terms are defined as follows:

*Acfm:* Actual cubic feet per minute.

*ASME Code or equivalent:* ASME (American Society of Mechanical Engineers) Boiler and Pressure Vessel Code, Section VIII, or an equivalent code which the employer can demonstrate to be equally effective.

*ATA:* Atmosphere absolute.

*Bell:* An enclosed compartment, pressurized (closed bell) or unpressurized (open bell), which allows the diver to be transported to and from the underwater work area and which may be used as a temporary refuge during diving operations.

*Bottom time:* The total elasped time measured in minutes from the time when the diver leaves the surface in descent to the time that the diver begins ascent.

*Bursting pressure:* The pressure at which a pressure containment device would fail structurally.

*Cylinder:* A pressure vessel for the storage of gases.

*Decompression chamber:* A pressure vessel for human occupancy such as a surface decompression chamber, closed bell, or deep diving system used to decompress divers and to treat decompression sickness.

*Decompression sickness:* A condition with a variety of symptoms which may result from gas or bubbles in the tissues of divers after pressure reduction.

*Decompression table:* A profile or set of profiles of depth-time relationships for ascent rates and breathing mixtures to be followed after a specific depth-time exposure or exposures.

*Dive location:* A surface or vessel from which a diving operation is conducted.

*Dive location reserve breathing gas:* A supply system of air or mixed-gas (as appropriate) at the dive location which is independent of the primary supply system and sufficient to support divers during the planned decompression.

*Dive team:* Divers and support employees involved in a diving operation, including the designated person-in-charge.

*Diver:* An employee working in water using underwater apparatus which supplies compressed breathing gas at the ambient pressure.

*Diver-carried reserve breathing gas:* A diver-carried supply of air or mixed gas (as appropriate) sufficient under standard operating conditions to allow the diver to reach the surface, or another source of breathing gas, or to be reached by a standby diver.

*Diving mode:* A type of diving requiring specific equipment, procedures and techniques (SCUBA, surface-supplied air, or mixed gas).

*Fsw:* Feet of seawater (or equivalent static pressure head).

*Heavy gear:* Diver-worn deep-sea dress including helmet, breastplate, dry suit, and weighted shoes.

*Hyperbaric conditions:* Pressure conditions in excess of surface pressure.

*Inwater stage:* A suspended underwater platform which supports a diver in the water.

*Liveboating:* The practice of supporting a surfaced-supplied air or mixed gas diver from a vessel which is underway.

*Mixed-gas diving:* A diving mode in which the diver is supplied in the water with a breathing gas other than air.

*No-decompression limits:* The depth-time limits of the "no-decompression limits and repetitive dive group designation table for no-decompression air dives", U.S. Navy Diving Manual or equivalent limits which the employer can demonstrate to be equally effective.

*Psi(g):* Pounds per square inch (gauge).

*Scientific diving* means diving performed solely as a necessary part of a scientific, research, or educational activity by employees whose sole purpose for diving is to perform scientific research tasks. Scientific diving does not include performing any tasks usually associated with commercial diving such as: Placing or removing heavy objects underwater; inspection of pipelines and similar objects; construction; demolition; cutting or welding; or the use of explosives.

*SCUBA diving:* A diving mode independent of surface supply in which the diver uses open circuit self-contained underwater breathing apparatus.

*Standby diver:* A diver at the dive location available to assist a diver in the water.

*Surface-supplied air diving:* A diving mode in which the diver in the water is supplied from the dive location with compressed air for breathing.

*Treatment table:* A depth-time and breathing gas profile designed to treat decompression sickness.

*Umbilical:* The composite hose bundle between a dive location and a diver or bell, or between a diver and a bell, which supplies the diver or bell with breathing gas, communications, power, or heat as appropriate to the diving mode or conditions, and includes a safety line between the diver and the dive location.

*Volume tank:* A pressure vessel connected to the outlet of a compressor and used as an air reservoir.

*Working pressure:* The maximum pressure to which a pressure containment device may be exposed under standard operating conditions.

[42 FR 37668, July 22, 1977, as amended at 47 FR 53365, Nov. 26, 1982]

## Personnel Requirements

### §1910.410 Qualifications of dive team.

**1910.410(a)** ***General.*** (1) Each dive team member shall have the experience or training necessary to perform assigned tasks in a safe and healthful manner.

**1910.410(a)(2)** Each dive team member shall have experience or training in the following:

**1910.410(a)(2)(i)** The use of tools, equipment and systems relevant to assigned tasks;

**1910.410(a)(2)(ii)** Techniques of the assigned diving mode: and

**1910.410(a)(2)(iii)** Diving operations and emergency procedures.

**1910.410(a)(3)** All dive team members shall be trained in cardiopulmonary resuscitation and first aid (American Red Cross standard course or equivalent).

**1910.410(a)(4)** Dive team members who are exposed to or control the exposure of others to hyperbaric conditions shall be trained in diving-related physics and physiology.

**1910.410(b)** ***Assignments.*** (1) Each dive team member shall be assigned tasks in accordance with the employee's experience or training, except that limited additional tasks may be assigned to an employee undergoing training provided that these tasks are performed under the direct supervision of an experienced dive team member.

**1910.410(b)(2)** The employer shall not require a dive team member to be exposed to hyperbaric conditions against the employee's will, except when necessary to complete decompression or treatment procedures.

**1910.410(b)(3)** The employer shall not permit a dive team member to dive or be otherwise exposed to hyperbaric conditions for the duration of any temporary physical impairment or condition which is known to the employer and is likely to affect adversely the safety or health of a dive team member.

**1910.410(c)** ***Designated person-in-charge.*** (1) The employer or an employee designated by the employer shall be at the dive location in charge of all aspects of the diving operation affecting the safety and health of dive team members.

**1910.410(c)(2)** The designated person-in-charge shall have experience and training in the conduct of the assigned diving operation.

## General Operations Procedures

### §1910.420 Safe practices manual.

**1910.420(a)** ***General.*** The employer shall develop and maintain a safe practices manual which shall be made available at the dive location to each dive team member.

**1910.420(b)** ***Contents.*** (1) The safe practices manual shall contain a copy of this standard and the employer's policies for implementing the requirements of this standard.

**1910.420(b)(2)** For each diving mode engaged in, the safe practices manual shall include:

**1910.420(b)(2)(i)** Safety procedures and checklists for diving operations;

**1910.420(b)(2)(ii)** Assignments and responsibilities of the dive team members;

**1910.420(b)(2)(iii)** Equipment procedures and checklists; and

**1910.420(b)(2)(iv)** Emergency procedures for fire, equipment failure, adverse environmental conditions, and medical illness and injury.

[42 FR 37668, July 22, 1977, as amended at 49 FR 18295, Apr. 30, 1984]

### §1910.421 Pre-dive procedures.

**1910.421(a)** ***General.*** The employer shall comply with the following requirements prior to each diving operation, unless otherwise specified.

**1910.421(b)** ***Emergency aid.*** A list shall be kept at the dive location of the telephone or call numbers of the following:

**1910.421(b)(1)** An operational decompression chamber (if not at the dive location);

**1910.421(b)(2)** Accessible hospitals;

**1910.421(b)(3)** Available physicians;

**1910.421(b)(4)** Available means of transportation; and

**1910.421(b)(5)** The nearest U.S. Coast Guard Rescue Coordination Center.

**1910.421(c)** ***First aid supplies.*** (1) A first aid kit appropriate for the diving operation and approved by a physician shall be available at the dive location.

**1910.421(c)(2)** When used in a decompression chamber or bell, the first aid kitshall be suitable for use under hyperbaric conditions.

**1910.421(c)(3)** In addition to any other first aid supplies, an American Red Cross standard first aid handbook or equivalent, and a bag-type manual resuscitator with transparent mask and tubing shall be available at the dive location.

**1910.421(d)** ***Planning and assessment.*** Planning of a diving operation shall include an assessment of the safety and health aspects of the following:

**1910.421(d)(1)** Diving mode;

**1910.421(d)(2)** Surface and underwater conditions and hazards;

**1910.421(d)(3)** Breathing gas supply (including reserves);

**1910.421(d)(4)** Thermal protection;

**1910.421(d)(5)** Diving equipment and systems;

**1910.421(d)(6)** Dive team assignments and physical fitness of dive team members (including any impairment known to the employer);

**1910.421(d)(7)** Repetitive dive designation or residual inert gas status of dive team members;

**1910.421(d)(8)** Decompression and treatment procedures (including altitude corrections); and

**1910.421(d)(9)** Emergency procedures.

**1910.421(e)** ***Hazardous activities.*** To minimize hazards to the dive team, diving operations shall be coordinated with other activities in the vicinity which are likely to interfere with the diving operation.

**1910.421(f)** ***Employee briefing.*** (1) Dive team members shall be briefed on:

**1910.421(f)(1)(i)** The tasks to be undertaken;

**1910.421(f)(1)(ii)** Safety procedures for the diving mode;

**1910.421(f)(1)(iii)** Any unusual hazards or environmental conditions likely to affect the safety of the diving operation; and

**1910.421(f)(1)(iv)** Any modifications to operating procedures necessitated by the specific diving operation.

**1910.421(f)(2)** Prior to making individual dive team member assignments, the employer shall inquire into the dive team member's current state of physical fitness, and indicate to the dive team member the procedure for reporting physical problems or adverse physiological effects during and after the dive.

**1910.421(g)** ***Equipment inspection.*** The breathing gas supply system including reserve breathing gas supplies, masks, helmets, thermal protection, and bell handling mechanism (when appropriate) shall be inspected prior to each dive.

**1910.421(h)** ***Warning signal.*** When diving from surfaces other than vessels in areas capable of supporting marine traffic, a rigid replica of the international code flag "A" at least one meter in height shall be displayed at the dive location in a manner which allows all-round visibility, and shall be illuminated during night diving operations.

[42 FR 37668, July 22, 1977, as amended at 47 FR 14706, Apr. 6, 1982; 54 FR 24334, June 7, 1989]

### §1910.422 Procedures during dive.

**1910.422(a)** ***General.*** The employer shall comply with the following requirements which are applicable to each diving operation unless otherwise specified.

**1910.422(b)** ***Water entry and exit.*** (1) A means capable of supporting the diver shall be provided for entering and exiting the water.

**1910.422(b)(2)** The means provided for exiting the water shall extend below the water surface.

**1910.422(b)(3)** A means shall be provided to assist an injured diver from the water or into a bell.

**1910.422(c)** ***Communications.*** (1) An operational two-way voice communication system shall be used between:

**1910.422(c)(1)(i)** Each surface-supplied air or mixed-gas diver and a dive team member at the dive location or bell (when provided or required); and

**1910.422(c)(1)(ii)** The bell and the dive location.

**1910.422(c)(2)** An operational, two-way communication system shall be available at the dive location to obtain emergency assistance.

**1910.422(d)** ***Decompression tables.*** Decompression, repetitive, and no-decompression tables (as appropriate) shall be at the dive location.

**1910.422(e)** ***Dive profiles.*** A depth-time profile, including when appropriate any breathing gas changes, shall be maintained for each diver during the dive including decompression.

**1910.422(f)** ***Hand-held power tools and equipment.*** (1) Hand-held electrical tools and equipment shall be de-energized before being placed into or retrieved from the water.

**1910.422(f)(2)** Hand-held power tools shall not be supplied with power from the dive location until requested by the diver.

**1910.422(g)** ***Welding and burning.*** (1) A current supply switch to interrupt the current flow to the welding or burning electrode shall be:

**1910.422(g)(1)(i)** Tended by a dive team member in voice communication with the diver performing the welding or burning; and

**1910.422(g)(1)(ii)** Kept in the open position except when the diver is welding or burning.

**1910.422(g)(2)** The welding machine frame shall be grounded.

**1910.422(g)(3)** Welding and burning cables, electrode holders, and connections shall be capable of carrying the maximum current required by the work, and shall be properly insulated.

**1910.422(g)(4)** Insulated gloves shall be provided to divers performing welding and burning operations.

**1910.422(g)(5)** Prior to welding or burning on closed compartments, structures or pipes, which contain a flammable vapor or in which a flammable vapor may be generated by the work, they shall be vented, flooded, or purged with a mixture of gases which will not support combustion.

**1910.422(h)** ***Explosives.*** (1) Employers shall transport, store, and use explosives in accordance with this section and the applicable provisions of §1910.109 and §1926.912 of Title 29 of the Code of Federal Regulations.

**1910.422(h)(2)** Electrical continuity of explosive circuits shall not be tested until the diver is out of the water.

**1910.422(h)(3)** Explosives shall not be detonated while the diver is in the water.

**1910.422(i)** ***Termination of dive.*** The working interval of a dive shall be terminated when:

**1910.422(i)(1)** A diver requests termination;

**1910.422(i)(2)** A diver fails to respond correctly to communications or signals from a dive team member;

**1910.422(i)(3)** Communications are lost and can not be quickly re-established between the diver and a dive team member at the dive location, and between the designated person-in-charge and the person controlling the vessel in liveboating operations; or

**1910.422(i)(4)** A diver begins to use diver-carried reserve breathing gas or the dive-location reserve breathing gas.

## §1910.423 Post-dive procedures.

**1910.423(a)** ***General.*** The employer shall comply with the following requirements which are applicable after each diving operation, unless otherwise specified.

**1910.423(b)** ***Precautions.*** (1) After the completion of any dive, the employer shall:

**1910.423(b)(1)(i)** Check the physical condition of the diver;

**1910.423(b)(1)(ii)** Instruct the diver to report any physical problems or adverse physiological effects including symptoms of decompression sickness;

**1910.423(b)(1)(iii)** Advise the diver of the location of a decompression chamber which is ready for use; and

**1910.423(b)(1)(iv)** Alert the diver to the potential hazards of flying after diving.

**1910.423(b)(2)** For any dive outside the no-decompression limits, deeper than 100 fsw or using mixed gas as a breathing mixture, the employer shall instruct the diver to remain awake and in the vicinity of the decompression chamber which is at the dive location for at least one hour after the dive (including decompression or treatment as appropriate).

**1910.423(c)** ***Recompression capability.*** (1) A decompression chamber capable of recompressing the diver at the surface to a minimum of 165 fsw (6 ATA) shall be available at the dive location for:

**1910.423(c)(1)(i)** Surface-supplied air diving to depths deeper than 100 fsw and shallower than 220 fsw;

**1910.423(c)(1)(ii)** Mixed gas diving shallower than 300 fsw; or

**1910.423(c)(1)(iii)** Diving outside the no-decompression limits shallower than 300 fsw.

**1910.423(c)(2)** A decompression chamber capable of recompressing the diver at the surface to the maximum depth of the dive shall be available at the dive location for dives deeper than 300 fsw.

**1910.423(c)(3)** The decompression chamber shall be:

**1910.423(c)(3)(i)** Dual-lock;

**1910.423(c)(3)(ii)** Multiplace; and

**1910.423(c)(3)(iii)** Located within 5 minutes of the dive location.

**1910.423(c)(4)** The decompression chamber shall be equipped with:

**1910.423(c)(4)(i)** A pressure gauge for each pressurized compartment designed for human occupancy;

**1910.423(c)(4)(ii)** A built-in-breathing-system with a minimum of one mask per occupant;

**1910.423(c)(4)(iii)** A two-way voice communication system between occupants and a dive team member at the dive location;

**1910.423(c)(4)(iv)** A viewport; and

**1910.423(c)(4)(v)** Illumination capability to light the interior.

**1910.423(c)(5)** Treatment tables, treatment gas appropriate to the diving mode, and sufficient gas to conduct treatment shall be available at the dive location.

**1910.423(c)(6)** A dive team member shall be available at the dive location during and for at least one hour after the dive to operate the decompression chamber (when required or provided).

**1910.423(d)** ***Record of dive.*** (1) The following information shall be recorded and maintained for each diving operation:

**1910.423(d)(1)(i)** Names of dive team members including designated person-in-charge;

**1910.423(d)(1)(ii)** Date, time, and location;

**1910.423(d)(1)(iii)** Diving modes used;

**1910.423(d)(1)(iv)** General nature of work performed;

**1910.423(d)(1)(v)** Approximate underwater and surface conditions (visibility, water temperature and current); and

**1910.423(d)(1)(vi)** Maximum depth and bottom time for each diver.

**1910.423(d)(2)** For each dive outside the no-decompression limits, deeper than 100 fsw or using mixed gas, the following additional information shall be recorded and maintained:

**1910.423(d)(2)(i)** Depth-time and breathing gas profiles;

**1910.423(d)(2)(ii)** Decompression table designation (including modification); and

**1910.423(d)(2)(iii)** Elapsed time since last pressure exposure if less than 24 hours or repetitive dive designation for each diver.

**1910.423(d)(3)** For each dive in which decompression sickness is suspected or symptoms are evident, the following additional information shall be recorded and maintained:

**1910.423(d)(3)(i)** Description of decompression sickness symptoms (including depth and time of onset); and

**1910.423(d)(3)(ii)** Description and results of treatment.

**1910.423(e)** ***Decompression procedure assessment.*** The employer shall:

**1910.423(e)(1)** Investigate and evaluate each incident of decompression sickness based on the recorded information, consideration of the past performance of decompression table used, and individual susceptibility;

**1910.423(e)(2)** Take appropriate corrective action to reduce the probability of recurrence of decompression sickness; and

1910.423(e)(3) Prepare a written evaluation of the decompression procedure assessment, including any corrective action taken, within 45 days of the incident of decompression sickness.

[42 FR 37668, July 22, 1977, as amended at 49 FR 18295, Apr. 30, 1984]

## Specific Operations Procedures

### §1910.424 SCUBA diving.

1910.424(a) ***General.*** Employers engaged in SCUBA diving shall comply with the following requirements, unless otherwise specified.

1910.424(b) ***Limits.*** SCUBA diving shall not be conducted:

1910.424(b)(1) At depths deeper than 130 fsw;

1910.424(b)(2) At depths deeper than 100 fsw or outside the no-decompression limits unless a decompression chamber is ready for use;

1910.424(b)(3) Against currents exceeding one (1) knot unless line-tended; or

1910.424(b)(4) In enclosed or physically confining spaces unless line-tended.

1910.424(c) ***Procedures.*** (1) A standby diver shall be available while a diver is in the water.

1910.424(c)(2) A diver shall be line-tended from the surface, or accompanied by another diver in the water in continuous visual contact during the diving operations.

1910.424(c)(3) A diver shall be stationed at the underwater point of entry when diving is conducted in enclosed or physically confining spaces.

1910.424(c)(4) A diver-carried reserve breathing gas supply shall be provided for each diver consisting of:

1910.424(c)(4)(i) A manual reserve (J valve); or

1910.424(c)(4)(ii) An independent reserve cylinder with a separate regulator or connected to the underwater breathing apparatus.

1910.424(c)(5) The valve of the reserve breathing gas supply shall be in the closed position prior to the dive.

### §1910.425 Surface-supplied air diving.

1910.425(a) ***General.*** Employers engaged in surface-supplied air diving shall comply with the following requirements, unless otherwise specified.

1910.425(b) ***Limits.*** (1) Surface-supplied air diving shall not be conducted at depths deeper than 190 fsw, except that dives with bottom times of 30 minutes or less may be conducted to depths of 220 fsw.

1910.425(b)(2) A decompression chamber shall be ready for use at the dive location for any dive outside the no-decompression limits or deeper than 100 fsw.

1910.425(b)(3) A bell shall be used for dives with an inwater decompression time greater than 120 minutes, except when heavy gear is worn or diving is conducted in physically confining spaces.

1910.425(c) ***Procedures.*** (1) Each diver shall be continuously tended while in the water.

1910.425(c)(2) A diver shall be stationed at the underwater point of entry when diving is conducted in enclosed or physically confining spaces.

1910.425(c)(3) Each diving operation shall have a primary breathing gas supply sufficient to support divers for the duration of the planned dive including decompression.

1910.425(c)(4) For dives deeper than 100 fsw or outside the no-decompression limits:

1910.425(c)(4)(i) A separate dive team member shall tend each diver in the water;

1910.425(c)(4)(ii) A standby diver shall be available while a diver is in the water;

1910.425(c)(4)(iii) A diver-carried reserve breathing gas supply shall be provided for each diver except when heavy gear is worn; and

1910.425(c)(4)(iv) A dive-location reserve breathing gas supply shall be provided.

1910.425(c)(5) For heavy-gear diving deeper than 100 fsw or outside the no-decompression limits:

1910.425(c)(5)(i) An extra breathing gas hose capable of supplying breathing gas to the diver in the water shall be available to the standby diver.

1910.425(c)(5)(ii) An inwater stage shall be provided to divers in the water.

1910.425(c)(6) Except when heavy gear is worn or where physical space does not permit, a diver-carried reserve breathing gas supply shall be provided whenever the diver is prevented by the configuration of the dive area from ascending directly to the surface.

### §1910.426 Mixed-gas diving.

1910.426(a) ***General.*** Employers engaged in mixed-gas diving shall comply with the following requirements, unless otherwise specified.

1910.426(b) ***Limits.*** Mixed-gas diving shall be conducted only when:

1910.426(b)(1) A decompression chamber is ready for use at the dive location; and

1910.426(b)(1)(i) A bell is used at depths greater than 220 fsw or when the dive involves inwater decompression time of greater than 120 minutes, except when heavy gear is worn or when diving in physically confining spaces; or

1910.426(b)(1)(ii) A closed bell is used at depths greater than 300 fsw, except when diving is conducted in physically confining spaces.

1910.426(c) ***Procedures.*** (1) A separate dive team member shall tend each diver in the water.

1910.426(c)(2) A standby diver shall be available while a diver is in the water.

1910.426(c)(3) A diver shall be stationed at the underwater point of entry when diving is conducted in enclosed or physically confining spaces.

1910.426(c)(4) Each diving operation shall have a primary breathing gas supply sufficient to support divers for the duration of the planned dive including decompression.

1910.426(c)(5) Each diving operation shall have a dive-location reserve breathing gas supply.

1910.426(c)(6) When heavy gear is worn:

1910.426(c)(6)(i) An extra breathing gas hose capable of supplying breathing gas to the diver in the water shall be available to the standby diver; and

1910.426(c)(6)(ii) An inwater stage shall be provided to divers in the water.

1910.426(c)(7) An inwater stage shall be provided for divers without access to a bell for dives deeper than 100 fsw or outside the no-decompression limits.

1910.426(c)(8) When a closed bell is used, one dive team member in the bell shall be available and tend the diver in the water.

1910.426(c)(9) Except when heavy gear is worn or where physical space does not permit, a diver-carried reserve breathing gas supply shall be provided for each diver:

1910.426(c)(9)(i) Diving deeper than 100 fsw or outside the no-decompression limits; or

1910.426(c)(9)(ii) Prevented by the configuration of the dive area from directly ascending to the surface.

### §1910.427 Liveboating.

1910.427(a) ***General.*** Employers engaged in diving operations involving liveboating shall comply with the following requirements.

1910.427(b) ***Limits.*** Diving operations involving liveboating shall not be conducted:

1910.427(b)(1) With an inwater decompression time of greater than 120 minutes;

1910.427(b)(2) Using surface-supplied air at depths deeper than 190 fsw, except that dives with bottom times of 30 minutes or less may be conducted to depths of 220 fsw;

1910.427(b)(3) Using mixed gas at depths greater than 220 fsw;

1910.427(b)(4) In rough seas which significantly inpede diver mobility or work function; or

**1910.427(b)(5)** In other than daylight hours.

**1910.427(c)** ***Procedures.*** (1) The propeller of the vessel shall be stopped before the diver enters or exits the water.

**1910.427(c)(2)** A device shall be used which minimizes the possibility of entanglement of the diver's hose in the propeller of the vessel.

**1910.427(c)(3)** Two-way voice communication between the designated person-in-charge and the person controlling the vessel shall be available while the diver is in the water.

**1910.427(c)(4)** A standby diver shall be available while a diver is in the water.

**1910.427(c)(5)** A diver-carried reserve breathing gas supply shall be carried by each diver engaged in liveboating operations.

## Equipment Procedures and Requirements

### §1910.430 Equipment.

**1910.430(a)** ***General.*** (1) All employers shall comply with the following requirements, unless otherwise specified.

**1910.430(a)(2)** Each equipment modification, repair, test, calibration or maintenance service shall be recorded by means of a tagging or logging system, and include the date and nature of work performed, and the name or initials of the person performing the work.

**1910.430(b)** ***Air compressor system.*** (1) Compressors used to supply air to the diver shall be equipped with a volume tank with a check valve on the inlet side, a pressure gauge, a relief valve, and a drain valve.

**1910.430(b)(2)** Air compressor intakes shall be located away from areas containing exhaust or other contaminants.

**1910.430(b)(3)** Respirable air supplied to a diver shall not contain:

**1910.430(b)(3)(i)** A level of carbon monoxide (CO) greater than 20 p/m;

**1910.430(b)(3)(ii)** A level of carbon dioxide ($CO_2$) greater than 1,000 p/m;

**1910.430(b)(3)(iii)** A level of oil mist greater than 5 milligrams per cubic meter; or

**1910.430(b)(3)(iv)** A noxious or pronounced odor.

**1910.430(b)(4)** The output of air compressor systems shall be tested for air purity every 6 months by means of samples taken at the connection to the distribution system, except that non-oil lubricated compressors need not be tested for oil mist.

**1910.430(c)** ***Breathing gas supply hoses.*** (1) Breathing gas supply hoses shall:

**1910.430(c)(1)(i)** Have a working pressure at least equal to the working pressure of the total breathing gas system;

**1910.430(c)(1)(ii)** Have a rated bursting pressure at least equal to 4 times the working pressure;

**1910.430(c)(1)(iii)** Be tested at least annually to 1.5 times their working pressure; and

**1910.430(c)(1)(iv)** Have their open ends taped, capped or plugged when not in use.

**1910.430(c)(2)** Breathing gas supply hose connectors shall:

**1910.430(c)(2)(i)** Be made of corrosion-resistant materials;

**1910.430(c)(2)(ii)** Have a working pressure at least equal to the working pressure of the hose to which they are attached; and

**1910.430(c)(2)(iii)** Be resistant to accidental disengagement.

**1910.430(c)(3)** Umbilicals shall:

**1910.430(c)(3)(i)** Be marked in 10-ft. increments to 100 feet beginning at the diver's end, and in 50 ft. increments thereafter;

**1910.430(c)(3)(ii)** Be made of kink-resistant materials; and

**1910.430(c)(3)(iii)** Have a working pressure greater than the pressure equivalent to the maximum depth of the dive (relative to the supply source) plus 100 psi.

**1910.430(d)** ***Buoyancy control.*** (1) Helmets or masks connected directly to the dry suit or other buoyancy-changing equipment shall be equipped with an exhaust valve.

**1910.430(d)(2)** A dry suit or other buoyancy-changing equipment not directly connected to the helmet or mask shall be equipped with an exhaust valve.

**1910.430(d)(3)** When used for SCUBA diving, a buoyancy compensator shall have an inflation source separate from the breathing gas supply.

**1910.430(d)(4)** An inflatable flotation device capable of maintaining the diver at the surface in a face-up position, having a manually activated inflation source independent of the breathing supply, an oral inflation device, and an exhaust valve shall be used for SCUBA diving.

**1910.430(e)** ***Compressed gas cylinders.*** Compressed gas cylinders shall:

**1910.430(e)(1)** Be designed, constructed and maintained in accordance with the applicable provisions of 29 CFR 1910.101 and 1910.169 through 1910.171.

**1910.430(e)(2)** Be stored in a ventilated area and protected from excessive heat;

**1910.430(e)(3)** Be secured from falling; and

**1910.430(e)(4)** Have shut-off valves recessed into the cylinder or protected by a cap, except when in use or manifolded, or when used for SCUBA diving.

**1910.430(f)** ***Decompression chambers.*** (1) Each decompression chamber manufactured after the effective date of this standard, shall be built and maintained in accordance with the ASME Code or equivalent.

**1910.430(f)(2)** Each decompression chamber manufactured prior to the effective date of this standard shall be maintained in conformity with the code requirements to which it was built, or equivalent.

**1910.430(f)(3)** Each decompression chamber shall be equipped with:

**1910.430(f)(3)(i)** Means to maintain the atmosphere below a level of 25 percent oxygen by volume;

**1910.430(f)(3)(ii)** Mufflers on intake and exhaust lines, which shall be regularly inspected and maintained;

**1910.430(f)(3)(iii)** Suction guards on exhaust line openings; and

**1910.430(f)(3)(iv)** A means for extinguishing fire, and shall be maintained to minimize sources of ignition and combustible material.

**1910.430(g)** ***Gauges and timekeeping devices.*** (1) Gauges indicating diver depth which can be read at the dive location shall be used for all dives except SCUBA.

**1910.430(g)(2)** Each depth gauge shall be deadweight tested or calibrated against a master reference gauge every 6 months, and when there is a discrepancy greater than two percent (2 percent) of full scale between any two equivalent gauges.

**1910.430(g)(3)** A cylinder pressure gauge capable of being monitored by the diver during the dive shall be worn by each SCUBA diver.

**1910.430(g)(4)** A timekeeping device shall be available at each dive location.

**1910.430(h)** ***Masks and helmets.*** (1) Surface-supplied air and mixed-gas masks and helmets shall have:

**1910.430(h)(1)(i)** A non-return valve at the attachment point between helmet or mask and hose which shall close readily and positively; and

**1910.430(h)(1)(ii)** An exhaust valve.

**1910.430(h)(2)** Surface-supplied air masks and helmets shall have a minimum ventilation rate capability of 4.5 acfm at any depth at which they are operated or the capability of maintaining the diver's inspired carbon dioxide partial pressure below 0.02 ATA when the diver is producing carbon dioxide at the rate of 1.6 standard liters per minute.

**1910.430(i)** ***Oxygen safety.*** (1) Equipment used with oxygen or mixtures containing over forty percent (40%) by volume oxygen shall be designed for oxygen service.

**1910.430(i)(2)** Components (except umbilicals) exposed to oxygen or mixtures containing over forty percent (40%) by volume oxygen shall be cleaned of flammable materials before use.

**1910.430(i)(3)** Oxygen systems over 125 psig and compressed air systems over 500 psig shall have slow-opening shut-off valves.

**1910.430(j) *Weights and harnesses.*** (1) Except when heavy gear is worn, divers shall be equipped with a weight belt or assembly capable of quick release.

**1910.430(j)(2)** Except when heavy gear is worn or in SCUBA diving, each diver shall wear a safety harness with:

**1910.430(j)(2)(i)** A positive buckling device;

**1910.430(j)(2)(ii)** An attachment point for the umbilical to prevent strain on the mask or helmet; and

**1910.430(j)(2)(iii)** A lifting point to distribute the pull force of the line over the diver's body.

[39 FR 23502, June 27, 1974, as amended at 49 FR 18295, Apr. 30, 1984; 51 FR 33033, Sept. 18, 1986]

## Recordkeeping

### §1910.440 Recordkeeping requirements.

**1910.440(a)(1)** [Reserved]

**1910.440(a)(2)** The employer shall record the occurrence of any diving-related injury or illness which requires any dive team member to be hospitalized for 24 hours or more, specifying the circumstances of the incident and the extent of any injuries or illnesses.

**1910.440(b) *Availability of records.*** (1) Upon the request of the Assistant Secretary of Labor for Occupational Safety and Health, or the Director, National Institute for Occupational Safety and Health, Department of Health and Human Services of their designees, the employer shall make available for inspection and copying any record or document required by this standard.

**1910.440(b)(2)** Records and documents required by this standard shall be provided upon request to employees, designated representatives, and the Assistant Secretary in accordance with 29 CFR 1910.20(a)-(e) and (g)-(i). Safe practices manuals (§1910.420), depth-time profiles (§1910.422), recordings of dives (§1910.423), decompression procedure assessment evaluations (§1910.423), and records of hospitalizations (§1910.440) shall be provided in the same manner as employee exposure records or analyses using exposure or medical records. Equipment inspections and testing records which pertain to employees (§1910.430) shall also be provided upon request to employees and their designated representatives.

**1910.440(b)(3)** Records and documents required by this standard shall be retained by the employer for the following period:

**1910.440(b)(3)(i)** Dive team member medical records (physician's reports) (§1910.411)—5 years;

**1910.440(b)(3)(ii)** Safe practices manual (§1910.420)—current document only;

**1910.440(b)(3)(iii)** Depth-time profile (§1910.422)—until completion of the recording of dive, or until completion of decompression procedure assessment where there has been an incident of decompression sickness;

**1910.440(b)(3)(iv)** Recording of dive (§1910.423)—1 year, except 5 years where there has been an incident of decompression sickness;

**1910.440(b)(3)(v)** Decompression procedure assessment evaluations (§1910.423)—5 years;

**1910.440(b)(3)(vi)** Equipment inspections and testing records (§1910.430)—current entry or tag, or until equipment is withdrawn from service;

**1910.440(b)(3)(vii)** Records of hospitalizations (§1910.440)—5 years.

**1910.440(b)(4)** After the expiration of the retention period of any record required to be kept for five (5) years, the employer shall forward such records to the National Institute for Occupational Safety and Health, Department of Health and Human Services. The employer shall also comply with any additional requirements set forth at 29 CFR 1910.20(h).

**1910.440(b)(5)** In the event the employer ceases to do business:

**1910.440(b)(5)(i)** The successor employer shall receive and retain all dive and employee medical records required by this standard; or

**1910.440(b)(5)(ii)** If there is no successor employer, dive and employee medical records shall be forwarded to the National Institute for Occupational Safety and Health, Department of Health and Human Services.

[42 FR 37668, July 22, 1977, as amended at 45 FR 35281, May 23, 1980; 47 FR 14706, Apr. 6, 1982; 51 FR 34562, Sept. 29, 1986; 61 FR 9242, Mar. 7, 1996]

### §1910.441 Effective date.

This standard shall be effective on October 20, 1977, except that for provisions where decompression chambers or bells are required and such equipment is not yet available, employers shall comply as soon as possible thereafter but in no case later than 6 months after the effective date of the standard.

### §1926.606 Definitions applicable to this subpart.

**1926.606(a)** *Apron*—The area along the waterfront edge of the pier or wharf.

**1926.606(b)** *Bulwark*—The side of a ship above the upper deck.

**1926.606(c)** *Coaming*—The raised frame, as around a hatchway in the deck, to keep out water.

**1926.606(d)** *Jacob's ladder*—A marine ladder of rope or chain with wooden or metal rungs.

**1926.606(e)** *Rail*, for the purpose of §1926.605, means a light structure serving as a guard at the outer edge of a ship's deck.

**Authority:** Sec. 107, Contract Worker Hours and Safety Standards Act (Construction Safety Act) (40 U.S.C. 333); Secs. 4, 6, 8, Occupational Safety and Health Act of 1970 (29 U.S.C. 653, 655, 657); Secretary of Labor's Order No. 12-71 (36 FR 8754), 8-76 (41 FR 25059), or 9-83 (48 FR 35736), as applicable, and 29 CFR part 1911.

**Source:** 54 FR 45959, Oct. 31, 1989, unless otherwise noted.

## §1926.650 Scope, application, and definitions applicable to this subpart.

1926.650(a) ***Scope and application.*** This subpart applies to all open excavations made in the earth's surface. Excavations are defined to include trenches.

1926.650(b) ***Definitions applicable to this subpart.***

*Accepted engineering practices* means those requirements which are compatible with standards of practice required by a registered professional engineer.

*Aluminum Hydraulic Shoring* means a pre-engineered shoring system comprised of aluminum hydraulic cylinders (crossbraces) used in conjunction with vertical rails (uprights) or horizontal rails (walers). Such system is designed, specifically to support the sidewalls of an excavation and prevent cave-ins.

*Bell-bottom pier hole* means a type of shaft or footing excavation, the bottom of which is made larger than the cross section above to form a belled shape.

*Benching* (Benching system) means a method of protecting employees from cave-ins by excavating the sides of an excavation to form one or a series of horizontal levels or steps, usually with vertical or near-vertical surfaces between levels.

*Cave-in* means the separation of a mass of soil or rock material from the side of an excavation, or the loss of soil from under a trench shield or support system, and its sudden movement into the excavation, either by falling or sliding, in sufficient quantity so that it could entrap, bury, or otherwise injure and immobilize a person.

*Competent person* means one who is capable of identifying existing and predictable hazards in the surroundings, or working conditions which are unsanitary, hazardous, or dangerous to employees, and who has authorization to take prompt corrective measures to eliminate them.

*Cross braces* mean the horizontal members of a shoring system installed perpendicular to the sides of the excavation, the ends of which bear against either uprights or wales.

*Excavation* means any man-made cut, cavity, trench, or depression in an earth surface, formed by earth removal.

*Faces* or *sides* means the vertical or inclined earth surfaces formed as a result of excavation work.

*Failure* means the breakage, displacement, or permanent deformation of a structural member or connection so as to reduce its structural integrity and its supportive capabilities.

*Hazardous atmosphere* means an atmosphere which by reason of being explosive, flammable, poisonous, corrosive, oxidizing, irritating, oxygen deficient, toxic, or otherwise harmful, may cause death, illness, or injury.

*Kickout* means the accidental release or failure of a cross brace.

*Protective system* means a method of protecting employees from cave-ins, from material that could fall or roll from an excavation face or into an excavation, or from the collapse of adjacent structures. Protective systems include support systems, sloping and benching systems, shield systems, and other systems that provide the necessary protection.

*Ramp* means an inclined walking or working surface that is used to gain access to one point from another, and is constructed from earth or from structural materials such as steel or wood.

*Registered Professional Engineer* means a person who is registered as a professional engineer in the state where the work is to be performed. However, a professional engineer, registered in any state is deemed to be a "registered professional engineer" within the meaning of this standard when approving designs for "manufactured protective systems" or "tabulated data" to be used in interstate commerce.

*Sheeting* means the members of a shoring system that retain the earth in position and in turn are supported by other members of the shoring system.

*Shield* (Shield system) means a structure that is able to withstand the forces imposed on it by a cave-in and thereby protect employees within the structure. Shields can be permanent structures or can be designed to be portable and moved along as work progresses. Additionally, shields can be either premanufactured or job-built in accordance with §1926.652(c)(3) or (c)(4). Shields used in trenches are usually referred to as "trench boxes" or "trench shields."

*Shoring* (Shoring system) means a structure such as a metal hydraulic, mechanical or timber shoring system that supports the sides of an excavation and which is designed to prevent cave-ins.

*Sides.* See "Faces."

*Sloping* (Sloping system) means a method of protecting employees from cave-ins by excavating to form sides of an excavation that are inclined away from the excavation so as to prevent cave-ins. The angle of incline required to prevent a cave-in varies with differences in such factors as the soil type, environmental conditions of exposure, and application of surcharge loads.

*Stable rock* means natural solid mineral material that can be excavated with vertical sides and will remain intact while exposed. Unstable rock is considered to be stable when the rock material on the side or sides of the excavation is secured against caving-in or movement by rock bolts or by another protective system that has been designed by a registered professional engineer.

*Structural ramp* means a ramp built of steel or wood, usually used for vehicle access. Ramps made of soil or rock are not considered structural ramps.

*Support system* means a structure such as underpinning, bracing, or shoring, which provides support to an adjacent structure, underground installation, or the sides of an excavation.

*Tabulated data* means tables and charts approved by a registered professional engineer and used to design and construct a protective system.

*Trench* (Trench excavation) means a narrow excavation (in relation to its length) made below the surface of the ground. In general, the depth is greater than the width, but the width of a trench (measured at the bottom) is not greater than 15 feet (4.6 m). If forms or other structures are installed or constructed in an excavation so as to reduce the dimension measured from the forms or structure to the side of the excavation to 15 feet (4.6 m) or less (measured at the bottom of the excavation), the excavation is also considered to be a trench.

*Trench box.* See "Shield."

*Trench shield.* See "Shield."

*Uprights* means the vertical members of a trench shoring system placed in contact with the earth and usually positioned so that individual members do not contact each other. Uprights placed so that individual members are closely spaced, in contact with or interconnected to each other, are often called "sheeting."

*Wales* means horizontal members of a shoring system placed parallel to the excavation face whose sides bear against the vertical members of the shoring system or earth.

## §1926.651 Specific excavation requirements.

1926.651(a) ***Surface encumbrances.*** All surface encumbrances that are located so as to create a hazard to employees shall be removed or supported, as necessary, to safeguard employees.

1926.651(b) ***Underground installations.*** (1) The estimated location of utility installations, such as sewer, telephone, fuel, electric, water lines, or any other underground installations that reasonably may be expected to be encountered during excavation work, shall be determined prior to opening an excavation.

1926.651(b)(2) Utility companies or owners shall be contacted within established or customary local response times, advised of the proposed work, and asked to establish the location of the utility underground installations prior to the start of actual excavation. When utility companies or owners cannot respond to a request to locate underground utility installations within 24 hours (unless a longer period is required by state or local law), or cannot establish the exact location of these installations, the employer may proceed, provided the employer does so with caution, and provided detection equipment or other acceptable means to locate utility installations are used.

**1926.651(b)(3)** When excavation operations approach the estimated location of underground installations, the exact location of the installations shall be determined by safe and acceptable means.

**1926.651(b)(4)** While the excavation is open, underground installations shall be protected, supported or removed as necessary to safeguard employees.

**1926.651(c)** ***Access and egress***— (1) *Structural ramps.* (i) Structural ramps that are used solely by employees as a means of access or egress from excavations shall be designed by a competent person. Structural ramps used for access or egress of equipment shall be designed by a competent person qualified in structural design, and shall be constructed in accordance with the design.

**1926.651(c)(1)(ii)** Ramps and runways constructed of two or more structural members shall have the structural members connected together to prevent displacement.

**1926.651(c)(1)(iii)** Structural members used for ramps and runways shall be of uniform thickness.

**1926.651(c)(1)(iv)** Cleats or other appropriate means used to connect runway structural members shall be attached to the bottom of the runway or shall be attached in a manner to prevent tripping.

**1926.651(c)(1)(v)** Structural ramps used in lieu of steps shall be provided with cleats or other surface treatments on the top surface to prevent slipping.

**1926.651(c)(2)** *Means of egress from trench excavations.* A stairway, ladder, ramp or other safe means of egress shall be located in trench excavations that are 4 feet (1.22 m) or more in depth so as to require no more than 25 feet (7.62 m) of lateral travel for employees.

**1926.651(d)** ***Exposure to vehicular traffic.*** Employees exposed to public vehicular traffic shall be provided with, and shall wear, warning vests or other suitable garments marked with or made of reflectorized or high-visibility material.

**1926.651(e)** ***Exposure to falling loads.*** No employee shall be permitted underneath loads handled by lifting or digging equipment. Employees shall be required to stand away from any vehicle being loaded or unloaded to avoid being struck by any spillage or falling materials. Operators may remain in the cabs of vehicles being loaded or unloaded when the vehicles are equipped, in accordance with §1926.601(b)(6), to provide adequate protection for the operator during loading and unloading operations.

**1926.651(f)** ***Warning system for mobile equipment.*** When mobile equipment is operated adjacent to an excavation, or when such equipment is required to approach the edge of an excavation, and the operator does not have a clear and direct view of the edge of the excavation, a warning system shall be utilized such as barricades, hand or mechanical signals, or stop logs. If possible, the grade should be away from the excavation.

**1926.651(g)** ***Hazardous atmospheres***—(1) *Testing and controls.* In addition to the requirements set forth in subparts D and E of this part (29 CFR 1926.50-1926.107) to prevent exposure to harmful levels of atmospheric contaminants and to assure acceptable atmospheric conditions, the following requirements shall apply:

**1926.651(g)(1)(i)** Where oxygen deficiency (atmospheres containing less than 19.5 percent oxygen) or a hazardous atmosphere exists or could reasonably be expected to exist, such as in excavations in landfill areas or excavations in areas where hazardous substances are stored nearby, the atmospheres in the excavation shall be tested before employees enter excavations greater than 4 feet (1.22 m) in depth.

**1926.651(g)(1)(ii)** Adequate precautions shall be taken to prevent employee exposure to atmospheres containing less than 19.5 percent oxygen and other hazardous atmospheres. These precautions include providing proper respiratory protection or ventilation in accordance with subparts D and E of this part respectively.

**1926.651(g)(1)(iii)** Adequate precaution shall be taken such as providing ventilation, to prevent employee exposure to an atmosphere containing a concentration of a flammable gas in excess of 20 percent of the lower flammable limit of the gas.

**1926.651(g)(1)(iv)** When controls are used that are intended to reduce the level of atmospheric contaminants to acceptable levels, testing shall be conducted as often as necessary to ensure that the atmosphere remains safe.

**1926.651(g)(2)** *Emergency rescue equipment.* (i) Emergency rescue equipment, such as breathing apparatus, a safety harness and line, or a basket stretcher, shall be readily available where hazardous atmospheric conditions exist or may reasonably be expected to develop during work in an excavation. This equipment shall be attended when in use.

**1926.651(g)(2)(ii)** Employees entering bell-bottom pier holes, or other similar deep and confined footing excavations, shall wear a harness with a lifeline securely attached to it. The lifeline shall be separate from any line used to handle materials, and shall be individually attended at all times while the employee wearing the lifeline is in the excavation.

**1926.651(h)** ***Protection from hazards associated with water accumulation.*** (1) Employees shall not work in excavations in which there is accumulated water, or in excavations in which water is accumulating, unless adequate precautions have been taken to protect employees against the hazards posed by water accumulation. The precautions necessary to protect employees adequately vary with each situation, but could include special support or shield systems to protect from cave-ins, water removal to control the level of accumulating water, or use of a safety harness and lifeline.

**1926.651(h)(2)** If water is controlled or prevented from accumulating by the use of water removal equipment, the water removal equipment and operations shall be monitored by a competent person to ensure proper operation.

**1926.651(h)(3)** If excavation work interrupts the natural drainage of surface water (such as streams), diversion ditches, dikes, or other suitable means shall be used to prevent surface water from entering the excavation and to provide adequate drainage of the area adjacent to the excavation. Excavations subject to runoff from heavy rains will require an inspection by a competent person and compliance with paragraphs (h)(1) and (h)(2) of this section.

**1926.651(i)** ***Stability of adjacent structures.*** (1) Where the stability of adjoining buildings, walls, or other structures is endangered by excavation operations, support systems such as shoring, bracing, or underpinning shall be provided to ensure the stability of such structures for the protection of employees.

**1926.651(i)(2)** Excavation below the level of the base or footing of any foundation or retaining wall that could be reasonably expected to pose a hazard to employees shall not be permitted except when:

**1926.651(i)(2)(i)** A support system, such as underpinning, is provided to ensure the safety of employees and the stability of the structure; or

**1926.651(i)(2)(ii)** The excavation is in stable rock; or

**1926.651(i)(2)(iii)** A registered professional engineer has approved the determination that the structure is sufficently removed from the excavation so as to be unaffected by the excavation activity; or

**1926.651(i)(2)(iv)** A registered professional engineer has approved the determination that such excavation work will not pose a hazard to employees.

**1926.651(i)(3)** Sidewalks, pavements, and appurtenant structure shall not be undermined unless a support system or another method of protection is provided to protect employees from the possible collapse of such structures.

**1926.651(j)** ***Protection of employees from loose rock or soil.*** (1) Adequate protection shall be provided to protect employees from loose rock or soil that could pose a hazard by falling or rolling from an excavation face. Such protection shall consist of scaling to remove loose material; installation of protective barricades at intervals as necessary on the face to stop and contain falling material; or other means that provide equivalent protection.

**1926.651(j)(2)** Employees shall be protected from excavated or other materials or equipment that could pose a hazard by falling or rolling into excavations. Protection shall be provided by placing and keeping such materials or equipment at least 2 feet (.61 m) from the edge of excavations, or by the use of retaining devices that are sufficient to prevent materials or equipment from falling or rolling into excavations, or by a combination of both if necessary.

**1926.651(k)** ***Inspections.*** (1) Daily inspections of excavations, the adjacent areas, and protective systems shall be made by a competent person for evidence of a situation that could result in possible cave-ins, indications of failure of protective systems, hazardous atmospheres, or other hazardous conditions. An inspection shall be conducted by the competent person prior to the start of work and as needed throughout the shift. Inspections shall also be made after every rainstorm or other hazard increasing occurrence. These inspections are only required when employee exposure can be reasonably anticipated.

**1926.651(k)(2)** Where the competent person finds evidence of a situation that could result in a possible cave-in, indications of failure of protective systems, hazardous atmospheres, or other hazardous conditions, exposed employees shall be removed from the hazardous area until the necessary precautions have been taken to ensure their safety.

**1926.651(l)** Walkways shall be provided where employees or equipment are required or permitted to cross over excavations. Guardrails which comply with §1926.502(b) shall be provided where walkways are 6 feet (1.8 m) or more above lower levels.

[54 FR 45959, Oct. 31, 1989, as amended by 59 FR 40730, Aug. 9, 1994]

## §1926.652 Requirements for protective systems.

**1926.652(a)** ***Protection of employees in excavations.*** (1) Each employee in an excavation shall be protected from cave-ins by an adequate protective system designed in accordance with paragraph (b) or (c) of this section except when:

**1926.652(a)(1)(i)** Excavations are made entirely in stable rock; or

**1926.652(a)(1)(ii)** Excavations are less than 5 feet (1.52m) in depth and examination of the ground by a competent person provides no indication of a potential cave-in.

**1926.652(a)(2)** Protective systems shall have the capacity to resist without failure all loads that are intended or could reasonably be expected to be applied or transmitted to the system.

**1926.652(b)** ***Design of sloping and benching systems.*** The slopes and configurations of sloping and benching systems shall be selected and constructed by the employer or his designee and shall be in accordance with the requirements of paragraph (b)(1); or, in the alternative, paragraph (b)(2); or, in the alternative, paragraph (b)(3), or, in the alternative, paragraph (b)(4), as follows:

**1926.652(b)(1)** *Option (1)—Allowable configurations and slopes.* (i) Excavations shall be sloped at an angle not steeper than one and one-half horizontal to one vertical (34 degrees measured from the horizontal), unless the employer uses one of the other options listed below.

**1926.652(b)(1)(ii)** Slopes specified in paragraph (b)(1)(i) of this section, shall be excavated to form configurations that are in accordance with the slopes shown for Type C soil in Appendix B to this subpart.

**1926.652(b)(2)** *Option (2)—Determination of slopes and configurations using Appendices A and B.* Maximum allowable slopes, and allowable configurations for sloping and benching systems, shall be determined in accordance with the conditions and requirements set forth in appendices A and B to this subpart.

**1926.652(b)(3)** *Option (3)—Designs using other tabulated data.* (i) Designs of sloping or benching systems shall be selected from and be in accordance with tabulated data, such as tables and charts.

**1926.652(b)(3)(ii)** The tabulated data shall be in written form and shall include all of the following:

**1926.652(b)(3)(ii)(A)** Identification of the parameters that affect the selection of a sloping or benching system drawn from such data;

**1926.652(b)(3)(ii)(B)** Identification of the limits of use of the data, to include the magnitude and configuration of slopes determined to be safe;

**1926.652(b)(3)(ii)(C)** Explanatory information as may be necessary to aid the user in making a correct selection of a protective system from the data.

**1926.652(b)(3)(iii)** At least one copy of the tabulated data which identifies the registered professional engineer who approved the data, shall be maintained at the jobsite during construction of the protective system. After that time the data may be stored off the jobsite, but a copy of the data shall be made available to the Secretary upon request.

**1926.652(b)(4)** *Option (4)—Design by a registered professional engineer.* (i) Sloping and benching systems not utilizing Option (1) or Option (2) or Option (3) under paragraph (b) of this section shall be approved by a registered professional engineer.

**1926.652(b)(4)(ii)** Designs shall be in written form and shall include at least the following:

**1926.652(b)(4)(ii)(A)** The magnitude of the slopes that were determined to be safe for the particular project;

**1926.652(b)(4)(ii)(B)** The configurations that were determined to be safe for the particular project; and

**1926.652(b)(4)(ii)(C)** The identity of the registered professional engineer approving the design.

**1926.652(b)(4)(iii)** At least one copy of the design shall be maintained at the jobsite while the slope is being constructed. After that time the design need not be at the jobsite, but a copy shall be made available to the Secretary upon request.

**1926.652(c)** ***Design of support systems, shield systems, and other protective systems.*** Designs of support systems shield systems, and other protective systems shall be selected and constructed by the employer or his designee and shall be in accordance with the requirements of paragraph (c)(1); or, in the alternative, paragraph (c)(2); or, in the alternative, paragraph (c)(3); or, in the alternative, paragraph (c)(4) as follows:

**1926.652(c)(1)** *Option (1)—Designs using Appendices A, C and D.* Designs for timber shoring in trenches shall be determined in accordance with the conditions and requirements set forth in Appendices A and C to this subpart. Designs for aluminum hydraulic shoring shall be in accordance with paragraph (c)(2) of this section, but if manufacturer's tabulated data cannot be utilized, designs shall be in accordance with Appendix D.

**1926.652(c)(2)** *Option (2)—Designs Using Manufacturer's Tabulated Data.* (i) Design of support systems, shield systems, or other protective systems that are drawn from manufacturer's tabulated data shall be in accordance with all specifications, recommendations, and limitations issued or made by the manufacturer.

**1926.652(c)(2)(ii)** Deviation from the specifications, recommendations, and limitations issued or made by the manufacturer shall only be allowed after the manufacturer issues specific written approval.

**1926.652(c)(2)(iii)** Manufacturer's specifications, recommendations, and limitations, and manufacturer's approval to deviate from the specifications, recommendations, and limitations shall be in written form at the jobsite during construction of the protective system. After that time this data may be stored off the jobsite, but a copy shall be made available to the Secretary upon request.

**1926.652(c)(3)** *Option (3)—Designs using other tabulated data.* (i) Designs of support systems, shield systems, or other protective systems shall be selected from and be in accordance with tabulated data, such as tables and charts.

**1926.652(c)(3)(ii)** The tabulated data shall be in written form and include all of the following:

**1926.652(c)(3)(ii)(A)** Identification of the parameters that affect the selection of a protective system drawn from such data;

**1926.652(c)(3)(ii)(B)** Identification of the limits of use of the data;

**1926.652(c)(3)(ii)(C)** Explanatory information as may be necessary to aid the user in making a correct selection of a protective system from the data.

**1926.652(c)(3)(iii)** At least one copy of the tabulated data, which identifies the registered professional engineer who approved the data, shall be maintained at the jobsite during construction of the protective system. After that time the data may be stored off the jobsite, but a copy of the data shall be made available to the Secretary upon request.

**1926.652(c)(4)** *Option (4)—Design by a registered professional engineer.* (i) Support systems, shield systems, and other protective systems not utilizing Option 1, Option 2 or Option 3, above, shall be approved by a registered professional engineer.

**1926.652(c)(4)(ii)** Designs shall be in written form and shall include the following:

**1926.652(c)(4)(ii)(A)** A plan indicating the sizes, types, and configurations of the materials to be used in the protective system; and

**1926.652(c)(4)(ii)(B)** The identity of the registered professional engineer approving the design.

**1926.652(c)(4)(iii)** At least one copy of the design shall be maintained at the jobsite during construction of the protective system. After that time, the design may be stored off the jobsite, but a copy of the design shall be made available to the Secretary upon request.

**1926.652(d)** ***Materials and equipment.*** (1) Materials and equipment used for protective systems shall be free from damage or defects that might impair their proper function.

**1926.652(d)(2)** Manufactured materials and equipment used for protective systems shall be used and maintained in a manner that is consistent with the recommendations of the manufacturer, and in a manner that will prevent employee exposure to hazards.

**1926.652(d)(3)** When material or equipment that is used for protective systems is damaged, a competent person shall examine the material or equipment and evaluate its suitability for continued use. If the competent person cannot assure the material or equipment is able to support the intended loads or is otherwise suitable for safe use, then such material or equipment shall be removed from service, and shall be evaluated and approved by a registered professional engineer before being returned to service.

**1926.652(e)** ***Installation and removal of support***— (1) *General.* (i) Members of support systems shall be securely connected together to prevent sliding, falling, kickouts, or other predictable failure.

**1926.652(e)(1)(ii)** Support systems shall be installed and removed in a manner that protects employees from cave-ins, structural collapses, or from being struck by members of the support system.

**1926.652(e)(1)(iii)** Individual members of support systems shall not be subjected to loads exceeding those which those members were designed to withstand.

**1926.652(e)(1)(iv)** Before temporary removal of individual members begins, additional precautions shall be taken to ensure the safety of employees, such as installing other structural members to carry the loads imposed on the support system.

**1926.652(e)(1)(v)** Removal shall begin at, and progress from, the bottom of the excavation. Members shall be released slowly so as to note any indication of possible failure of the remaining members of the structure or possible cave-in of the sides of the excavation.

**1926.652(e)(1)(vi)** Backfilling shall progress together with the removal of support systems from excavations.

**1926.652(e)(2)** *Additional requirements for support systems for trench excavations.* (i) Excavation of material to a level no greater than 2 feet (.61 m) below the bottom of the members of a support system shall be permitted, but only if the system is designed to resist the forces calculated for the full depth of the trench, and there are no indications while the trench is open of a possible loss of soil from behind or below the bottom of the support system.

**1926.652(e)(2)(ii)** Installation of a support system shall be closely coordinated with the excavation of trenches.

**1926.652(f)** ***Sloping and benching systems.*** Employees shall not be permitted to work on the faces of sloped or benched excavations at levels above other employees except when employees at the lower levels are adequately protected from the hazard of falling, rolling, or sliding material or equipment.

**1926.652(g)** ***Shield systems***— (*1*) *General.* (i) Shield systems shall not be subjected to loads exceeding those which the system was designed to withstand.

**1926.652(g)(1)(ii)** Shields shall be installed in a manner to restrict lateral or other hazardous movement of the shield in the event of the application of sudden lateral loads.

**1926.652(g)(1)(iii)** Employees shall be protected from the hazard of cave-ins when entering or exiting the areas protected by shields.

**1926.652(g)(1)(iv)** Employees shall not be allowed in shields when shields are being installed, removed, or moved vertically.

**1926.652(g)(2)** *Additional requirement for shield systems used in trench excavations.* Excavations of earth material to a level not greater than 2 feet (.61 m) below the bottom of a shield shall be permitted, but only if the shield is designed to resist the forces calculated for the full depth of the trench, and there are no indications while the trench is open of a possible loss of soil from behind or below the bottom of the shield.

## Appendix A to Subpart P—Soil Classification

(a) *Scope and application*—(1) *Scope.* This appendix describes a method of classifying soil and rock deposits based on site and environmental conditions, and on the structure and composition of the earth deposits. The appendix contains definitions, sets forth requirements, and describes acceptable visual and manual tests for use in classifying soils.

(2) *Application.* This appendix applies when a sloping or benching system is designed in accordance with the requirements set forth in §1926.652(b)(2) as a method of protection for employees from cave-ins. This appendix also applies when timber shoring for excavations is designed as a method of protection from cave-ins in accordance with appendix C to subpart P of part 1926, and when aluminum hydraulic shoring is designed in accordance with appendix D. This Appendix also applies if other protective systems are designed and selected for use from data prepared in accordance with the requirements set forth in §1926.652(c), and the use of the data is predicated on the use of the soil classification system set forth in this appendix.

(b) *Definitions.* The definitions and examples given below are based on, in whole or in part, the following: American Society for Testing Materials (ASTM) Standards D653-85 and D2488; The Unified Soils Classification System, The U.S. Department of Agriculture (USDA) Textural Classification Scheme; and The National Bureau of Standards Report BSS-121.

*Cemented soil* means a soil in which the particles are held together by a chemical agent, such as calcium carbonate, such that a hand-size sample cannot be crushed into powder or individual soil particles by finger pressure.

*Cohesive soil* means clay (fine grained soil), or soil with a high clay content, which has cohesive strength. Cohesive soil does not crumble, can be excavated with vertical sideslopes, and is plastic when moist. Cohesive soil is hard to break up when dry, and exhibits significant cohesion when submerged. Cohesive soils include clayey silt, sandy clay, silty clay, clay and organic clay.

*Dry soil* means soil that does not exhibit visible signs of moisture content.

*Fissured* means a soil material that has a tendency to break along definite planes of fracture with little resistance, or a material that exhibits open cracks, such as tension cracks, in an exposed surface.

*Granular soil* means gravel, sand, or silt, (coarse grained soil) with little or no clay content. Granular soil has no cohesive strength. Some moist granular soils exhibit apparent cohesion. Granular soil cannot be molded when moist and crumbles easily when dry.

*Layered system* means two or more distinctly different soil or rock types arranged in layers. Micaceous seams or weakened planes in rock or shale are considered layered.

*Moist soil* means a condition in which a soil looks and feels damp. Moist cohesive soil can easily be shaped into a ball and rolled into small diameter threads before crumbling. Moist granular soil that contains some cohesive material will exhibit signs of cohesion between particles.

*Plastic* means a property of a soil which allows the soil to be deformed or molded without cracking, or appreciable volume change.

*Saturated soil* means a soil in which the voids are filled with water. Saturation does not require flow. Saturation, or near saturation, is necessary for the proper use of instruments such as a pocket penetrometer or sheer vane.

*Soil classification system* means, for the purpose of this subpart, a method of categorizing soil and rock deposits in a hierarchy of Stable Rock, Type A, Type B, and Type C, in decreasing order of stability. The categories are determined based on an analysis of the properties and performance characteristics of the deposits and the environmental conditions of exposure.

*Stable rock* means natural solid mineral matter that can be excavated with vertical sides and remain intact while exposed.

*Submerged soil* means soil which is underwater or is free seeping.

*Type A* means cohesive soils with an unconfined compressive strength of 1.5 ton per square foot (tsf) (144 kPa) or greater. Examples of cohesive soils are: clay, silty clay, sandy clay, clay loam and, in some cases, silty clay loam and sandy clay loam. Cemented soils such as caliche and hardpan are also considered Type A. However, no soil is Type A if:

(i) The soil is fissured; or

(ii) The soil is subject to vibration from heavy traffic, pile driving, or similar effects; or

(iii) The soil has been previously disturbed; or

(iv) The soil is part of a sloped, layered system where the layers dip into the excavation on a slope of four horizontal to one vertical (4H:1V) or greater; or

(v) The material is subject to other factors that would require it to be classified as a less stable material.

*Type B* means:

(i) Cohesive soil with an unconfined compressive strength greater than 0.5 tsf (48 kPa) but less than 1.5 tsf (144 kPa); or

(ii) Granular cohesionless soils including: angular gravel (similar to crushed rock), silt, silt loam, sandy loam and, in some cases, silty clay loam and sandy clay loam.

(iii) Previously disturbed soils except those which would otherwise be classed as Type C soil.

(iv) Soil that meets the unconfined compressive strength or cementation requirements for Type A, but is fissured or subject to vibration; or

(v) Dry rock that is not stable; or

(vi) Material that is part of a sloped, layered system where the layers dip into the excavation on a slope less steep than four horizontal to one vertical (4H:1V), but only if the material would otherwise be classified as Type B.

*Type C* means:

(i) Cohesive soil with an unconfined compressive strength of 0.5 tsf (48 kPa) or less; or

(ii) Granular soils including gravel, sand, and loamy sand; or

(iii) Submerged soil or soil from which water is freely seeping; or

(iv) Submerged rock that is not stable, or

(v) Material in a sloped, layered system where the layers dip into the excavation or a slope of four horizontal to one vertical (4H:1V) or steeper.

*Unconfined compressive strength* means the load per unit area at which a soil will fail in compression. It can be determined by laboratory testing, or estimated in the field using a pocket penetrometer, by thumb penetration tests, and other methods.

*Wet soil* means soil that contains significantly more moisture than moist soil, but in such a range of values that cohesive material will slump or begin to flow when vibrated. Granular material that would exhibit cohesive properties when moist will lose those cohesive properties when wet.

(c) *Requirements*—(1) *Classification of soil and rock deposits.* Each soil and rock deposit shall be classified by a competent person as Stable Rock, Type A, Type B, or Type C in accordance with the definitions set forth in paragraph (b) of this appendix.

(2) *Basis of classification.* The classification of the deposits shall be made based on the results of at least one visual and at least one manual analysis. Such analyses shall be conducted by a competent person using tests described in paragraph (d) below, or in other recognized methods of soil classification and testing such as those adopted by the America Society for Testing Materials, or the U.S. Department of Agriculture textural classification system.

(3) *Visual and manual analyses.* The visual and manual analyses, such as those noted as being acceptable in paragraph (d) of this appendix, shall be designed and conducted to provide sufficient quantitative and qualitative information as may be necessary to identify properly the properties, factors, and conditions affecting the classification of the deposits.

(4) *Layered systems.* In a layered system, the system shall be classified in accordance with its weakest layer. However, each layer may be classified individually where a more stable layer lies under a less stable layer.

(5) *Reclassification.* If, after classifying a deposit, the properties, factors, or conditions affecting its classification change in any way, the changes shall be evaluated by a competent person. The deposit shall be reclassified as necessary to reflect the changed circumstances.

(d) *Acceptable visual and manual tests.*—(1) *Visual tests.* Visual analysis is conducted to determine qualitative information regarding the excavation site in general, the soil adjacent to the excavation, the soil forming the sides of the open excavation, and the soil taken as samples from excavated material.

(i) Observe samples of soil that are excavated and soil in the sides of the excavation. Estimate the range of particle sizes and the relative amounts of the particle sizes. Soil that is primarily composed of fine-grained material is cohesive material. Soil composed primarily of coarse-grained sand or gravel is granular material.

(ii) Observe soil as it is excavated. Soil that remains in clumps when excavated is cohesive. Soil that breaks up easily and does not stay in clumps is granular.

(iii) Observe the side of the opened excavation and the surface area adjacent to the excavation. Crack-like openings such as tension cracks could indicate fissured material. If chunks of soil spall off a vertical side, the soil could be fissured. Small spalls are evidence of moving ground and are indications of potentially hazardous situations.

(iv) Observe the area adjacent to the excavation and the excavation itself for evidence of existing utility and other underground structures, and to identify previously disturbed soil.

(v) Observe the opened side of the excavation to identify layered systems. Examine layered systems to identify if the layers slope toward the excavation. Estimate the degree of slope of the layers.

(vi) Observe the area adjacent to the excavation and the sides of the opened excavation for evidence of surface water, water seeping from the sides of the excavation, or the location of the level of the water table.

(vii) Observe the area adjacent to the excavation and the area within the excavation for sources of vibration that may affect the stability of the excavation face.

(2) *Manual tests.* Manual analysis of soil samples is conducted to determine quantitative as well as qualitative properties of soil and to provide more information in order to classify soil properly.

(i) *Plasticity.* Mold a moist or wet sample of soil into a ball and attempt to roll it into threads as thin as 1/8-inch in diameter. Cohesive material can be successfully rolled into threads without crumbling. For example, if at least a two inch (50 mm) length of 1/8-inch thread can be held on one end without tearing, the soil is cohesive.

(ii) *Dry strength.* If the soil is dry and crumbles on its own or with moderate pressure into individual grains or fine powder, it is granular (any combination of gravel, sand, or silt). If the soil is dry and falls into clumps which break up into smaller clumps, but the smaller clumps can only be broken up with difficulty, it may be clay in any combination with gravel, sand or silt. If the dry soil breaks into clumps which do not break up into small clumps and which can only be broken with difficulty, and there is no visual indication the soil is fissured, the soil may be considered unfissured.

(iii) *Thumb penetration.* The thumb penetration test can be used to estimate the unconfined compressive strength of cohesive soils. (This test is based on the thumb penetration test described in American Society for Testing and Materials (ASTM) Standard designation D2488—"Standard Recommended Practice for Description of Soils (Visual—Manual Procedure).") Type A soils with an unconfined compressive strength of 1.5 tsf can be readily indented by the thumb; however, they can be penetrated by the thumb only with very great effort. Type C soils with an unconfined compressive strength of 0.5 tsf can be easily penetrated several inches by the thumb, and can be molded by light finger pressure. This test should be conducted on an undisturbed soil sample, such as a large clump of spoil, as soon as practicable after excavation to keep to a miminum the effects of exposure to drying influences. If the excavation is later exposed to wetting influences (rain, flooding), the classification of the soil must be changed accordingly.

(iv) *Other strength tests.* Estimates of unconfined compressive strength of soils can also be obtained by use of a pocket penetrometer or by using a hand-operated shearvane.

(v) *Drying test.* The basic purpose of the drying test is to differentiate between cohesive material with fissures, unfissured cohesive material, and granular material. The procedure for the drying test involves drying a sample of soil that is approximately one inch thick (2.54 cm) and six inches (15.24 cm) in diameter until it is thoroughly dry:

(A) If the sample develops cracks as it dries, significant fissures are indicated.

(B) Samples that dry without cracking are to be broken by hand. If considerable force is necessary to break a sample, the soil has significant cohesive material content. The soil can be classified as a unfissured cohesive material and the unconfined compressive strength should be determined.

(C) If a sample breaks easily by hand, it is either a fissured cohesive material or a granular material. To distinguish between the two, pulverize the dried clumps of the sample by hand or by stepping on them. If the clumps do not pulverize easily, the material is cohesive with fissures. If they pulverize easily into very small fragments, the material is granular.

## Appendix B to Subpart P—Sloping and Benching

(a) *Scope and application.* This appendix contains specifications for sloping and benching when used as methods of protecting employees working in excavations from cave-ins. The requirements of this appendix apply when the design of sloping and benching protective systems is to be performed in accordance with the requirements set forth in §1926.652(b)(2).

(b) *Definitions.*

*Actual slope* means the slope to which an excavation face is excavated.

*Distress* means that the soil is in a condition where a cave-in is imminent or is likely to occur. Distress is evidenced by such phenomena as the development of fissures in the face of or adjacent to an open excavation; the subsidence of the edge of an excavation; the slumping of material from the face or the bulging or heaving of material from the bottom of an excavation; the spalling of material from the face of an excavation; and ravelling, i.e., small amounts of material such as pebbles or little clumps of material suddenly separating from the face of an excavation and trickling or rolling down into the excavation.

*Maximum allowable slope* means the steepest incline of an excavation face that is acceptable for the most favorable site conditions as protection against cave-ins, and is expressed as the ratio of horizontal distance to vertical rise (H:V).

*Short term exposure* means a period of time less than or equal to 24 hours that an excavation is open.

(c) *Requirements*—(1) *Soil classification.* Soil and rock deposits shall be classified in accordance with appendix A to subpart P of part 1926.

(2) *Maximum allowable slope.* The maximum allowable slope for a soil or rock deposit shall be determined from Table B-1 of this appendix.

(3) *Actual slope.* (i) The actual slope shall not be steeper than the maximum allowable slope.

(ii) The actual slope shall be less steep than the maximum allowable slope, when there are signs of distress. If that situation occurs, the slope shall be cut back to an actual slope which is at least 1/2 horizontal to one vertical ( 1/2H:1V) less steep than the maximum allowable slope.

(iii) When surcharge loads from stored material or equipment, operating equipment, or traffic are present, a competent person shall determine the degree to which the actual slope must be reduced below the maximum allowable slope, and shall assure that such reduction is achieved. Surcharge loads from adjacent structures shall be evaluated in accordance with §1926.651(i).

(4) *Configurations.* Configurations of sloping and benching systems shall be in accordance with Figure B-1.

## TABLE B-1—MAXIMUM ALLOWABLE SLOPES

| Soil or Rock Type | Maximum Allowable Slopes (H:V)[1] For Excavations Less Than 20 Feet Deep[3] |
|---|---|
| Stable Rock | Vertical (90°) |
| Type A[2] | 3/4:1 (53°) |
| Type B | 1:1 (45°) |
| Type C | 1 1/2:1 (34°) |

**NOTES:**

1 Numbers shown in parentheses next to maximum allowable slopes are angles expressed in degrees from the horizontal. Angles have been rounded off.

2 A short-term maximum allowable slope of 1/2 H:1V (63°) is allowed in excavations in Type A soil that are 12 feet (3.67 m) or less in depth. Short-term maximum allowable slopes for excavations greater than 12 feet (3.67 m) in depth shall be 3/4 H:1V (53°).

3 Sloping or benching for excavations greater than 20 feet deep shall be designed by a registered professional engineer.

## FIGURE B-1—SLOPE CONFIGURATIONS

(All slopes stated below are in the horizontal to vertical ratio)

**B—1.1 Excavations made in Type A soil.**

1. All simple slope excavation 20 feet or less in depth shall have a maximum allowable slope of ³/₄:1.

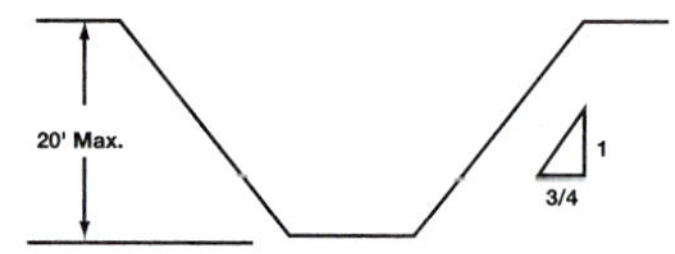


SIMPLE SLOPE—GENERAL

Exception: Simple slope excavations which are open 24 hours or less (short term) and which are 12 feet or less in depth shall have a maximum allowable slope of ¹/₂:1.

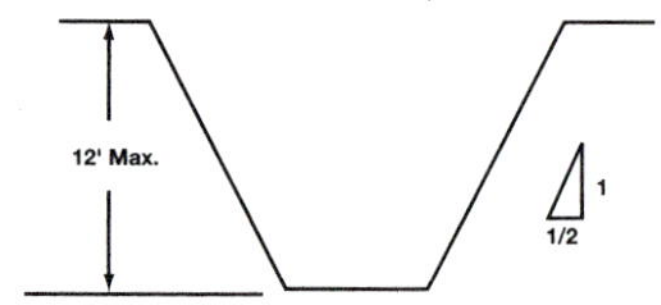


SIMPLE SLOPE—SHORT TERM

2. All benched excavations 20 feet or less in depth shall have a maximum allowable slope of ³/₄ to 1 and maximum bench dimensions as follows:

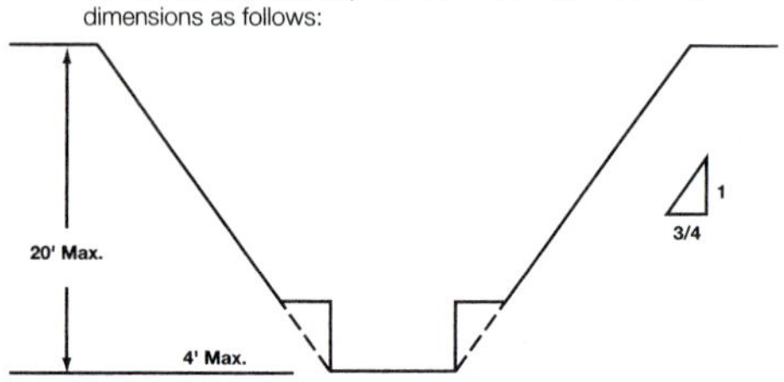


SIMPLE BENCH

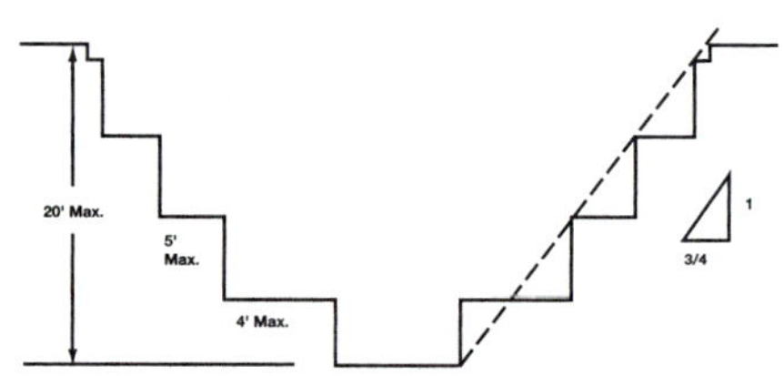


MULTIPLE BENCH

3. All excavations 8 feet or less in depth which have unsupported vertically sided lower portions shall have a maximum vertical side of 3¹/₂ feet.

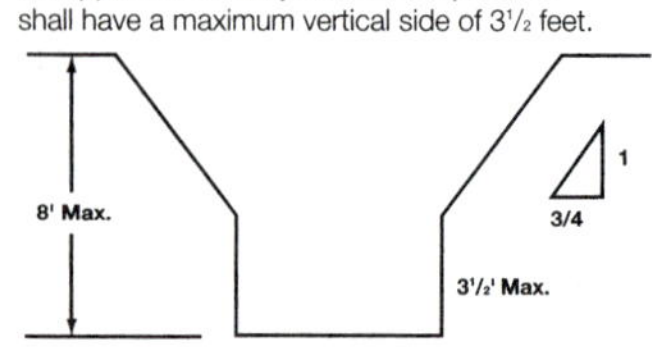


UNSUPPORTED VERTICALLY SIDED LOWER PORTION—MAXIMUM 8 FEET IN DEPTH

All excavations more than 8 feet but not more than 12 feet in depth which unsupported vertically sided lower portions shall have a maximum allowable slope of 1:1 and a maximum vertical side of 3¹/₂ feet.

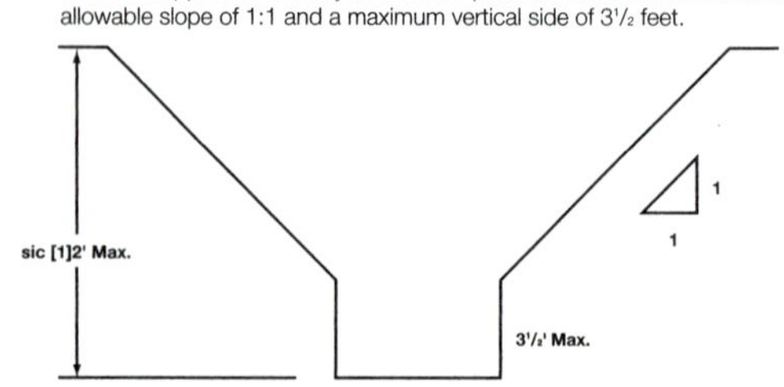


UNSUPPORTED VERTICALLY SIDED LOWER PORTION—MAXIMUM 12 FEET IN DEPTH

All excavations 20 feet or less in depth which have vertically sided lower portions that are supported or shielded shall have a maximum allowable slope of $^3/_4$:1. The support or shield system must extend at least 18 inches above the top of the vertical side.

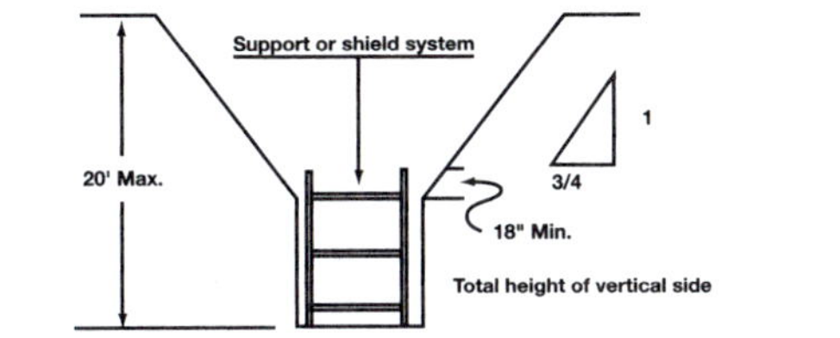


SUPPORTED OR SHIELDED VERTICALLY SIDED LOWER PORTION

4. All other simple slope, compound slope, and vertically sided lower portion excavations shall be in accordance with the other options permitted under §1926.652(b).

### B-1.2 Excavations Made in Type B Soil

1. All simple slope excavations 20 feet or less in depth shall have a maximum allowable slope of 1:1.

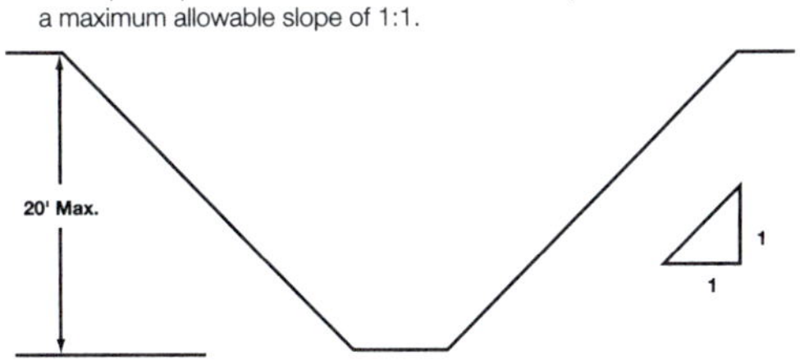


SIMPLE SLOPE

2. All benched excavations 20 feet or less in depth shall have a maximum allowable slope of 1:1 and maximum bench dimensions as follows:

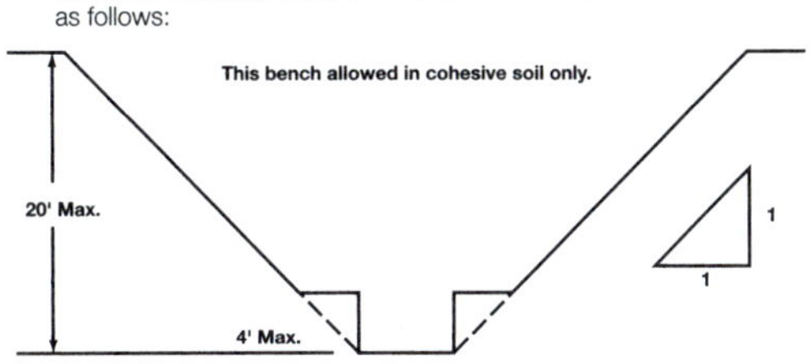


SINGLE BENCH

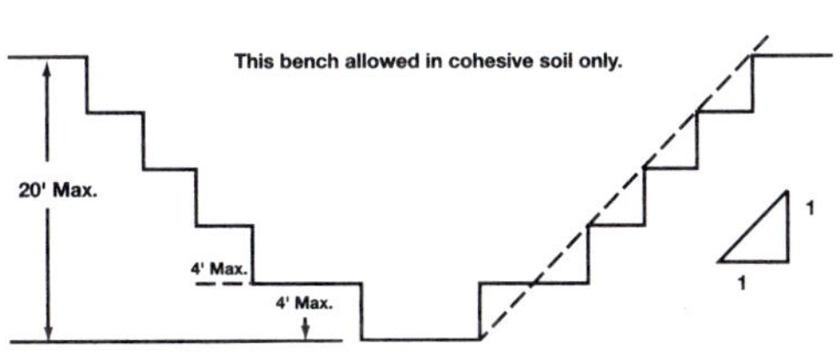


MULTIPLE BENCH

3. All excavations 20 feet or less in depth which have vertically sided lower portions shall be shielded or supported to a height at least 18 inches above the top of the vertical side. All such excavations shall have a maximum allowable slope of 1:1.

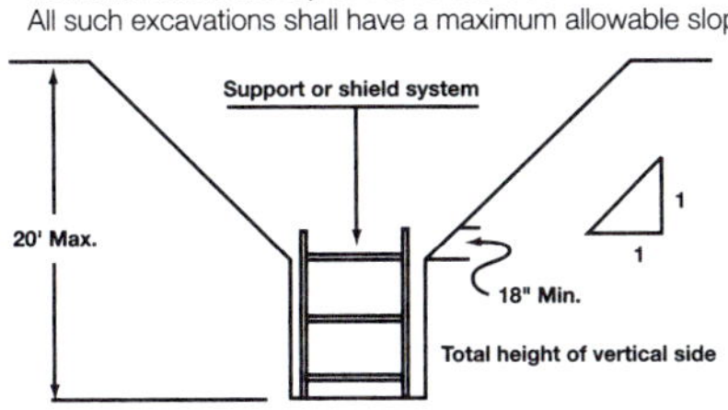


VERTICALLY SIDED LOWER PORTION

4. All other sloped excavations shall be in accordance with the other options permitted in §1926.652(b).

### B—1.3 Excavations Made in Type C Soil.

1. All simple slope excavations 20 feet or less in depth shall have a maximum allowable slope of $1^1/_2$:1.

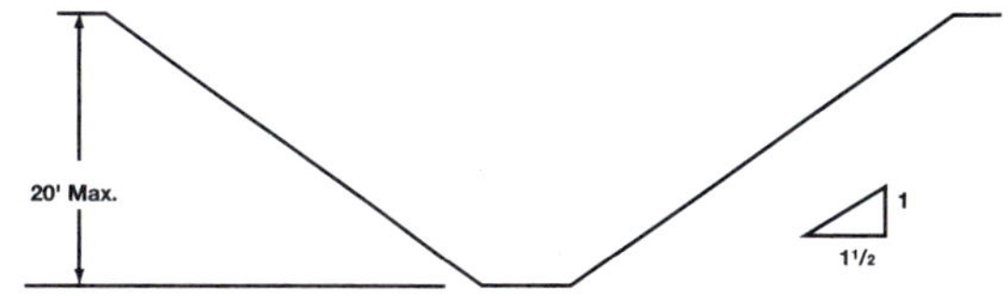


SIMPLE SLOPE

2. All excavations 20 feet or less in depth which have vertically sided lower portions shall be shielded or supported to a height at least 18 inches above the top of the vertical side. All such excavations shall have a maximum allowable slope of $1^1/_2$:1.

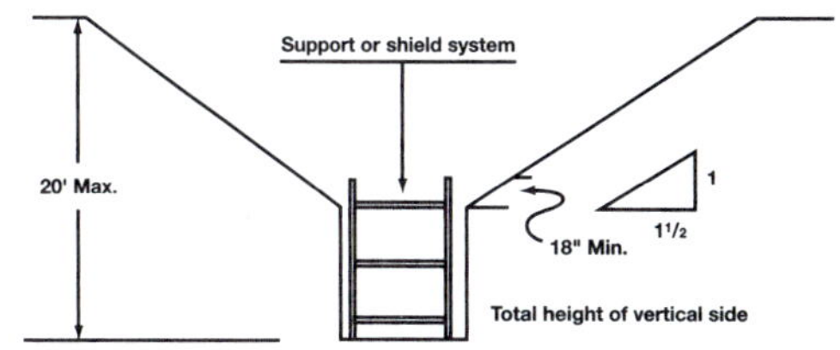


VERTICALLY SIDED LOWER PORTION

3. All other sloped excavations shall be in accordance with the other options permitted in §1926.652(b).

### B—1.4 Excavations Made in Layered Soils

1. All excavations 20 feet or less in depth made in layered soils shall have a maximum allowable slope for each layer as set forth below.

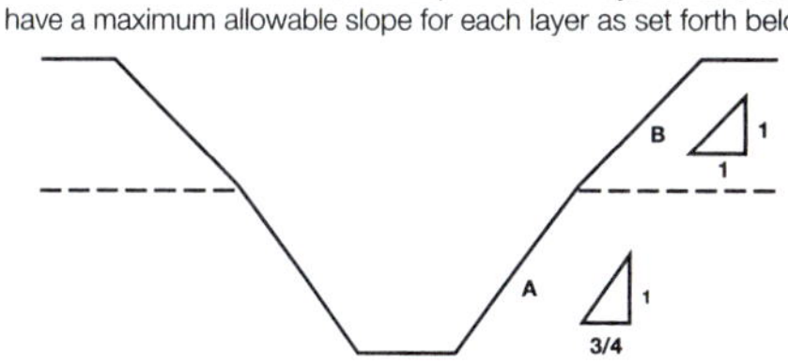


B OVER A

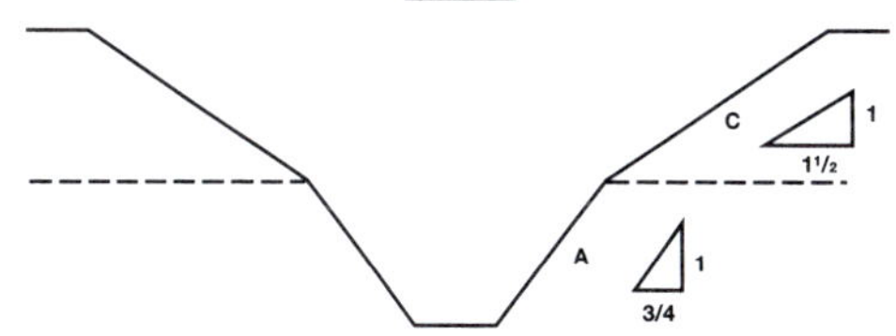


C OVER A

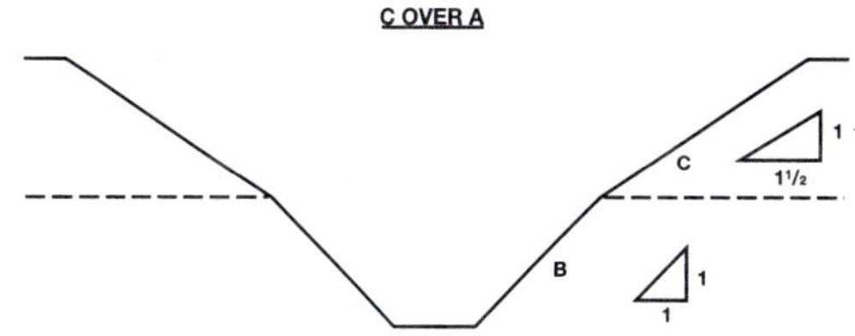


C OVER B

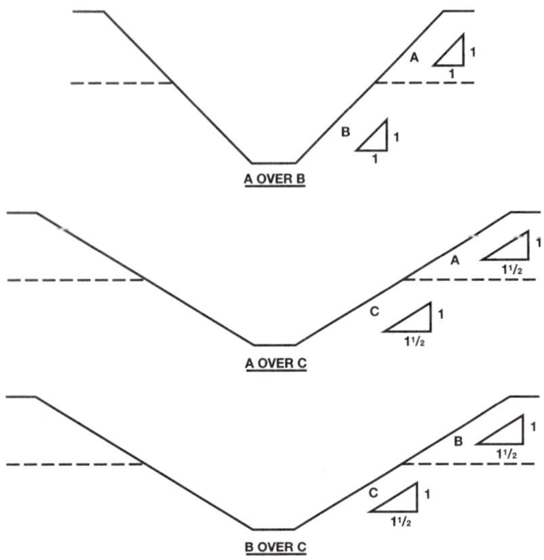


2. All other sloped excavations shall be in accordance with the other options permitted in §1926.652(b).

## Appendix C to Subpart P—Timber Shoring for Trenches

(a) *Scope.* This appendix contains information that can be used timber shoring is provided as a method of protection from cave-ins in trenches that do not exceed 20 feet (6.1 m) in depth. This appendix must be used when design of timber shoring protective systems is to be performed in accordance with §1926.652(c)(1). Other timber shoring configurations; other systems of support such as hydraulic and pneumatic systems; and other protective systems such as sloping, benching, shielding, and freezing systems must be designed in accordance with the requirements set forth in §1926.652(b) and §1926.652(c).

(b) *Soil Classification.* In order to use the data presented in this appendix, the soil type or types in which the excavation is made must first be determined using the soil classification method set forth in appendix A of subpart P of this part.

(c) *Presentation of Information.* Information is presented in several forms as follows:

(1) Information is presented in tabular form in Tables C-1.1, C-1.2, and C-1.3, and Tables C-2.1, C-2.2 and C-2.3 following paragraph (g) of the appendix. Each table presents the minimum sizes of timber members to use in a shoring system, and each table contains data only for the particular soil type in which the excavation or portion of the excavation is made. The data are arranged to allow the user the flexibility to select from among several acceptable configurations of members based on varying the horizontal spacing of the crossbraces. Stable rock is exempt from shoring requirements and therefore, no data are presented for this condition.

(2) Information concerning the basis of the tabular data and the limitations of the data is presented in paragraph (d) of this appendix, and on the tables themselves.

(3) Information explaining the use of the tabular data is presented in paragraph (e) of this appendix.

(4) Information illustrating the use of the tabular data is presented in paragraph (f) of this appendix.

(5) Miscellaneous notations regarding Tables C-1.1 through C-1.3 and Tables C-2.1 through C-2.3 are presented in paragraph (g) of this appendix.

(d) *Basis and limitations of the data.*—(1) *Dimensions of timber members.* (i) The sizes of the timber members listed in Tables C-1.1 through C-1.3 are taken from the National Bureau of Standards (NBS) report, "Recommended Technical Provisions for Construction Practice in Shoring and Sloping of Trenches and Excavations." In addition, where NBS did not recommend specific sizes of members, member sizes are based on an analysis of the sizes required for use by existing codes and on empirical practice.

(ii) The required dimensions of the members listed in Tables C-1.1 through C-1.3 refer to actual dimensions and not nominal dimensions of the timber. Employers wanting to use nominal size shoring are directed to Tables C-2.1 through C-2.3, or have this choice under §1926.652(c)(3), and are referred to The Corps of Engineers, The Bureau of Reclamation or data from other acceptable sources.

(2) *Limitation of application.* (i) It is not intended that the timber shoring specification apply to every situation that may be experienced in the field. These data were developed to apply to the situations that are most commonly experienced in current trenching practice. Shoring systems for use in situations that are not covered by the data in this appendix must be designed as specified in §1926.652(c).

(ii) When any of the following conditions are present, the members specified in the tables are not considered adequate. Either an alternate timber shoring system must be designed or another type of protective system designed in accordance with §1926.652.

(A) When loads imposed by structures or by stored material adjacent to the trench weigh in excess of the load imposed by a two-foot soil surcharge. The term "adjacent" as used here means the area within a horizontal distance from the edge of the trench equal to the depth of the trench.

(B) When vertical loads imposed on cross braces exceed a 240-pound gravity load distributed on a one-foot section of the center of the crossbrace.

(C) When surcharge loads are present from equipment weighing in excess of 20,000 pounds.

(D) When only the lower portion of a trench is shored and the remaining portion of the trench is sloped or benched unless: The sloped portion is sloped at an angle less steep than three horizontal to one vertical; or the members are selected from the tables for use at a depth which is determined from the top of the overall trench, and not from the toe of the sloped portion.

(e) *Use of Tables.* The members of the shoring system that are to be selected using this information are the cross braces, the uprights, and the wales, where wales are required. Minimum sizes of members are specified for use in different types of soil. There are six tables of information, two for each soil type. The soil type must first be determined in accordance with the soil classification system described in appendix A to subpart P of part 1926. Using the appropriate table, the selection of the size and spacing of the members is then made. The selection is based on the depth and width of the trench where the members are to be installed and, in most instances, the selection is also based on the horizontal spacing of the crossbraces. Instances where a choice of horizontal spacing of crossbracing is available, the horizontal spacing of the crossbraces must be chosen by the user before the size of any member can be determined. When the soil type, the width and depth of the trench, and the horizontal spacing of the crossbraces are known, the size and vertical spacing of the crossbraces, the size and vertical spacing of the wales, and the size and horizontal spacing of the uprights can be read from the appropriate table.

(f) *Examples to Illustrate the Use of Tables C-1.1 through C-1.3.*

(1) *Example 1.*

A trench dug in Type A soil is 13 feet deep and five feet wide.

From *Table C-1.1,* for acceptable arrangements of timber can be used.

### Arrangement #B1

Space 4×4 crossbraces at six feet horizontally and four feet vertically.

Wales are not required.

Space 3×8 uprights at six feet horizontally. This arrangement is commonly called "skip shoring."

### Arrangement #B2

Space 4×6 crossbraces at eight feet horizontally and four feet vertically.

Space 8×8 wales at four feet vertically.

Space 2×6 uprights at four feet horizontally.

### Arrangement #B3

Space 6×6 crossbraces at 10 feet horizontally and four feet vertically.

Space 8×10 wales at four feet vertically.

Space 2×6 uprights at five feet horizontally.

### Arrangement #B4

Space 6×6 crossbraces at 12 feet horizontally and four feet vertically.

Space 10×10 wales at four feet vertically.

Spaces 3×8 uprights at six feet horizontally.

(2) *Example 2.*

A trench dug in Type B soil in 13 feet deep and five feet wide. From Table C-1.2 three acceptable arrangements of members are listed.

### Arrangement #B1

Space 6x6 crossbraces at six feet horizontally and five feet vertically.

Space 8x8 wales at five feet vertically.

Space 2x6 uprights at two feet horizontally.

### Arrangement #B2

Space 6x8 crossbraces at eight feet horizontally and five feet vertically.

Space 10x10 wales at five feet vertically.

Space 2x6 uprights at two feet horizontally.

### Arrangement #B3

Space 8x8 crossbraces at 10 feet horizontally and five feet vertically.

Space 10x12 wales at five feet vertically.

Space 2x6 uprights at two feet vertically.

(3) *Example 3.*

A trench dug in Type C soil is 13 feet deep and five feet wide.

From Table C-1.3 two acceptable arrangements of members can be used.

### Arrangement #B1

Space 8x8 crossbraces at six feet horizontally and five feet vertically.

Space 10x12 wales at five feet vertically.

Position 2x6 uprights as closely together as possible.

If water must be retained use special tongue and groove uprights to form tight sheeting.

### Arrangement #B2

Space 8x10 crossbraces at eight feet horizontally and five feet vertically.

Space 12x12 wales at five feet vertically.

Position 2x6 uprights in a close sheeting configuration unless water pressure must be resisted. Tight sheeting must be used where water must be retained.

(4) *Example 4.*

A trench dug in Type C soil is 20 feet deep and 11 feet wide. The size and spacing of members for the section of trench that is over 15 feet in depth is determined using Table C-1.3. Only one arrangement of members is provided.

Space 8x10 crossbraces at six feet horizontally and five feet vertically.

Space 12x12 wales at five feet vertically.

Use 3x6 tight sheeting.

Use of Tables C-2.1 through C-2.3 would follow the same procedures.

(g) *Notes for all Tables.*

1. Member sizes at spacings other than indicated are to be determined as specified in §1926.652(c), "Design of Protective Systems."

2. When conditions are saturated or submerged use Tight Sheeting. Tight Sheeting refers to the use of specially-edged timber planks (e.g., tongue and groove) at least three inches thick, steel sheet piling, or similar construction that when driven or placed in position provide a tight wall to resist the lateral pressure of water and to prevent the loss of backfill material. Close Sheeting refers to the placement of planks side-by-side allowing as little space as possible between them.

3. All spacing indicated is measured center to center.

4. Wales to be installed with greater dimension horizontal.

5. If the vertical distance from the center of the lowest crossbrace to the bottom of the trench exceeds two and one-half feet, uprights shall be firmly embedded or a mudsill shall be used. Where uprights are embedded, the vertical distance from the center of the lowest crossbrace to the bottom of the trench shall not exceed 36 inches. When mudsills are used, the vertical distance shall not exceed 42 inches. Mudsills are wales that are installed at the toe of the trench side.

6. Trench jacks may be used in lieu of or in combination with timber crossbraces.

7. Placement of crossbraces. When the vertical spacing of crossbraces is four feet, place the top crossbrace no more than two feet below the top of the trench. When the vertical spacing of crossbraces is five feet, place the top crossbrace no more than 2.5 feet below the top of the trench.

## TABLE C-1.1—TIMBER TRENCH SHORING—MINIMUM TIMBER REQUIREMENTS*

Soil Type A $P_a = 25 \times H + 72$ psf (2 ft Surcharge)

| Depth of Trench (feet) | Size (Actual) and Spacing of Members ** | | | | | | | | | | | | | |
|---|---|---|---|---|---|---|---|---|---|---|---|---|---|---|
| | Cross Braces | | | | | | | Wales | | Uprights | | | | |
| | Horiz. Spacing (feet) | Width of Trench (feet) | | | | | Vert. Spacing (feet) | Size (in) | Vert. Spacing (feet) | Maximum Allowable Horizontal Spacing (feet) | | | | |
| | | Up to 4 | Up to 6 | Up to 9 | Up to 12 | Up to 15 | | | | Close | 4 | 5 | 6 | 8 |
| 5 | Up to 6 | 4x4 | 4x4 | 4x6 | 6x6 | 6x6 | 4 | Not Req'd | — | | | | 2x6 | |
| To | Up to 8 | 4x4 | 4x4 | 4x6 | 6x6 | 6x6 | 4 | Not Req'd | — | | | | | 2x8 |
| 10 | Up to 10 | 4x6 | 4x6 | 4x6 | 6x6 | 6x6 | 4 | 8x8 | 4 | | | 2x6 | | |
| | Up to 12 | 4x6 | 4x6 | 6x6 | 6x6 | 6x6 | 4 | 8x8 | 4 | | | | 2x6 | |
| 10 | Up to 6 | 4x4 | 4x4 | 4x6 | 6x6 | 6x6 | 4 | Not Req'd | — | | | | 3x8 | |
| To | Up to 8 | 4x6 | 4x6 | 6x6 | 6x6 | 6x6 | 4 | 8x8 | 4 | | 2x6 | | | |
| 15 | Up to 10 | 6x6 | 6x5 | 6x6 | 6x8 | 6x8 | 4 | 8x10 | 4 | | | 2x6 | | |
| | Up to 12 | 6x6 | 6x6 | 6x6 | 6x8 | 6x8 | 4 | 10x10 | 4 | | | | 3x8 | |
| 15 | Up to 6 | 6x6 | 6x6 | 6x6 | 6x8 | 6x8 | 4 | 6x8 | 4 | 3x6 | | | | |
| To | Up to 8 | 6x6 | 6x6 | 6x6 | 6x8 | 6x8 | 4 | 8x8 | 4 | 3x6 | | | | |
| 20 | Up to 10 | 8x8 | 8x8 | 8x8 | 8x8 | 8x10 | 4 | 8x10 | 4 | 3x6 | | | | |
| | Up to 12 | 8x8 | 8x8 | 8x8 | 8x8 | 8x10 | 4 | 10x10 | 4 | 3x6 | | | | |
| Over 20 | See Note 1 | | | | | | | | | | | | | |

* Mixed oak or equivalent with a bending strength not less than 850 psi.

** Manufactured members of equivalent strength may be substituted for wood.

## TABLE C-1.2—TIMBER TRENCH SHORING—MINIMUM TIMBER REQUIREMENTS*

Soil Type B $P_a = 45 \times H + 72$ psf (2 ft Surcharge)

| Depth of Trench (feet) | Size (Actual) and Spacing of Members ** | | | | | | | | | | | | | |
|---|---|---|---|---|---|---|---|---|---|---|---|---|---|---|
| | Cross Braces | | | | | | | Wales | | Uprights | | | | |
| | Horiz. Spacing (feet) | Width of Trench (feet) Up to 4 | Up to 6 | Up to 9 | Up to 12 | Up to 15 | Vert. Spacing (feet) | Size (in) | Vert. Spacing (feet) | Maximum Allowable Horizontal Spacing (feet) Close | 2 | 3 | | |
| 5 | Up to 6 | 4×6 | 4×6 | 6×6 | 6×6 | 6×6 | 5 | 6×8 | 5 | | | 2×6 | | |
| To | Up to 8 | 6×6 | 6×6 | 6×6 | 6×8 | 6×8 | 5 | 8×10 | 5 | | | 2×6 | | |
| 10 | Up to 10 | 6×6 | 6×6 | 6×6 | 6×8 | 6×8 | 5 | 10×10 | 5 | | | 2×6 | | |
| | See Note 1 | | | | | | | | | | | | | |
| 10 | Up to 6 | 6×6 | 6×6 | 6×6 | 6×8 | 6×8 | 5 | 8×8 | 5 | | 2×6 | | | |
| To | Up to 8 | 6×8 | 6×8 | 6×8 | 8×8 | 8×8 | 5 | 10×10 | 5 | | 2×6 | | | |
| 15 | Up to 10 | 8×8 | 8×8 | 8×8 | 8×8 | 8×10 | 5 | 10×12 | 5 | | 2×6 | | | |
| | See Note 1 | | | | | | | | | | | | | |
| 15 | Up to 6 | 6×8 | 6×8 | 6×8 | 8×8 | 8×8 | 5 | 8×10 | 5 | 3×6 | | | | |
| To | Up to 8 | 8×8 | 8×8 | 8×8 | 8×8 | 8×10 | 5 | 10×12 | 5 | 3×6 | | | | |
| 20 | Up to 10 | 8×10 | 8×10 | 8×10 | 8×10 | 10×10 | 5 | 12×12 | 5 | 3×6 | | | | |
| | See Note 1 | | | | | | | | | | | | | |
| Over 20 | See Note 1 | | | | | | | | | | | | | |

* Mixed oak or equivalent with a bending strength not less than 850 psi.
** Manufactured members of equivalent strength may be substituted for wood.

## TABLE C-1.3—TIMBER TRENCH SHORING—MINIMUM TIMBER REQUIREMENTS*

Soil Type C $P_a = 80 \times H + 72$ psf (2 ft Surcharge)

| Depth of Trench (feet) | Size (Actual) and Spacing of Members ** | | | | | | | | | | | | | |
|---|---|---|---|---|---|---|---|---|---|---|---|---|---|---|
| | Cross Braces | | | | | | | Wales | | Uprights | | | | |
| | Horiz. Spacing (feet) | Width of Trench (feet) Up to 4 | Up to 6 | Up to 9 | Up to 12 | Up to 15 | Vert. Spacing (feet) | Size (in) | Vert. Spacing (feet) | Maximum Allowable Horizontal Spacing (feet) (See Note 2) Close | | | | |
| 5 | Up to 6 | 6×8 | 6×8 | 6×8 | 8×8 | 8×8 | 5 | 8×10 | 5 | 2×6 | | | | |
| To | Up to 8 | 8×8 | 8×8 | 8×8 | 8×8 | 8×10 | 5 | 10×12 | 5 | 2×6 | | | | |
| 10 | Up to 10 | 8×10 | 8×10 | 8×10 | 8×10 | 10×10 | 5 | 12×12 | 5 | 2×6 | | | | |
| | See Note 1 | | | | | | | | | | | | | |
| 10 | Up to 6 | 8×8 | 8×8 | 8×8 | 8×8 | 8×10 | 5 | 10×12 | 5 | 2×6 | | | | |
| To | Up to 8 | 8×10 | 8×10 | 8×10 | 8×10 | 10×10 | 5 | 12×12 | 5 | 2×6 | | | | |
| 15 | See Note 1 | | | | | | | | | | | | | |
| | See Note 1 | | | | | | | | | | | | | |
| 15 | Up to 6 | 8×10 | 8×10 | 8×10 | 8×10 | 10×10 | 5 | 12×12 | 5 | 3×6 | | | | |
| To | See Note 1 | | | | | | | | | | | | | |
| 20 | See Note 1 | | | | | | | | | | | | | |
| | See Note 1 | | | | | | | | | | | | | |
| Over 20 | See Note 1 | | | | | | | | | | | | | |

* Mixed oak or equivalent with a bending strength not less than 850 psi.
** Manufactured members of equivalent strength may be substituted for wood.

## TABLE C-2.1—TIMBER TRENCH SHORING—MINIMUM TIMBER REQUIREMENTS*

Soil Type A $P_a = 25 \times H + 72$ psf (2 ft Surcharge)

| Depth of Trench (feet) | Size (S4S) and Spacing of Members ** | | | | | | | | | | | | | |
|---|---|---|---|---|---|---|---|---|---|---|---|---|---|---|
| | Cross Braces | | | | | | | Wales | | Uprights | | | | |
| | Horiz. Spacing (feet) | Width of Trench (feet) Up to 4 | Up to 6 | Up to 9 | Up to 12 | Up to 15 | Vert. Spacing (feet) | Size (in) | Vert. Spacing (feet) | Maximum Allowable Horizontal Spacing (feet) Close | 4 | 5 | 6 | 8 |
| 5 | Up to 6 | 4×4 | 4×4 | 4×4 | 4×4 | 4×6 | 4 | Not Req'd | Not Req'd | | | | 4×6 | |
| To | Up to 8 | 4×4 | 4×4 | 4×4 | 4×6 | 4×6 | 4 | Not Req'd | Not Req'd | | | | | 4×8 |
| 10 | Up to 10 | 4×6 | 4×6 | 4×6 | 6×6 | 6×6 | 4 | 8×8 | 4 | | | 4×6 | | |
| | Up to 12 | 4×6 | 4×6 | 4×6 | 6×6 | 6×6 | 4 | 8×8 | 4 | | | | 4×6 | |
| 10 | Up to 6 | 4×4 | 4×4 | 4×4 | 6×6 | 6×6 | 4 | Not Req'd | Not Req'd | | | | 4×10 | |
| To | Up to 8 | 4×6 | 4×6 | 4×6 | 6×6 | 6×6 | 4 | 6×8 | 4 | | 4×6 | | | |
| 15 | Up to 10 | 6×6 | 6×6 | 6×6 | 6×6 | 6×6 | 4 | 8×8 | 4 | | | 4×8 | | |
| | Up to 12 | 6×6 | 6×6 | 6×6 | 6×6 | 6×6 | 4 | 8×10 | 4 | | 4×6 | | 4×10 | |
| 15 | Up to 6 | 6×6 | 6×6 | 6×6 | 6×6 | 6×6 | 4 | 6×8 | 4 | 3×6 | | | | |
| To | Up to 8 | 6×6 | 6×6 | 6×6 | 6×6 | 6×6 | 4 | 8×8 | 4 | 3×6 | 4×12 | | | |
| 20 | Up to 10 | 6×6 | 6×6 | 6×6 | 6×6 | 6×8 | 4 | 8×10 | 4 | 3×6 | | | | |
| | Up to 12 | 6×6 | 6×6 | 6×6 | 6×8 | 6×8 | 4 | 8×12 | 4 | 3×6 | 4×12 | | | |
| Over 20 | See Note 1 | | | | | | | | | | | | | |

* Douglas fir or equivalent with a bending strength not less than 1500 psi.
** Manufactured members of equivalent strength may be substituted for wood.

## TABLE C-2.2—TIMBER TRENCH SHORING—MINIMUM TIMBER REQUIREMENTS*

Soil Type B $P_a = 45 \times H + 72$ psf (2 ft Surcharge)

| Depth of Trench (feet) | Size (S4S) and Spacing of Members ** | | | | | | | | | | | | | |
|---|---|---|---|---|---|---|---|---|---|---|---|---|---|---|
| | Cross Braces | | | | | | | Wales | | Uprights | | | | |
| | Horiz. Spacing (feet) | Width of Trench (feet) Up to 4 | Up to 6 | Up to 9 | Up to 12 | Up to 15 | Vert. Spacing (feet) | Size (in) | Vert. Spacing (feet) | Maximum Allowable Horizontal Spacing (feet) Close | 2 | 3 | 4 | 6 |
| 5 | Up to 6 | 4×6 | 4×6 | 4×6 | 6×6 | 6×6 | 5 | 6×8 | 5 | | | 3×12<br>4×8 | | 4×12 |
| To | Up to 8 | 4×6 | 4×6 | 6×6 | 6×6 | 6×6 | 5 | 8×8 | 5 | | 3×8 | | 4×8 | |
| 10 | Up to 10 | 4×6 | 4×6 | 6×6 | 6×6 | 6×8 | 5 | 8×10 | 5 | | | 4×8 | | |
| | See Note 1 | | | | | | | | | | | | | |
| 10 | Up to 6 | 6×6 | 6×6 | 6×6 | 6×8 | 6×8 | 5 | 8×8 | 5 | 3×6 | 4×10 | | | |
| To | Up to 8 | 6×8 | 6×8 | 6×8 | 8×8 | 8×8 | 5 | 10×10 | 5 | 3×6 | 4×10 | | | |
| 15 | Up to 10 | 6×8 | 6×8 | 8×8 | 8×8 | 8×8 | 5 | 10×12 | 5 | 3×6 | 4×10 | | | |
| | See Note 1 | | | | | | | | | | | | | |
| 15 | Up to 6 | 6×8 | 6×8 | 6×8 | 6×8 | 8×8 | 5 | 8×10 | 5 | 4×6 | | | | |
| To | Up to 8 | 6×8 | 6×8 | 6×8 | 8×8 | 8×8 | 5 | 10×12 | 5 | 4×6 | | | | |
| 20 | Up to 10 | 8×8 | 8×8 | 8×8 | 8×8 | 8×8 | 5 | 12×12 | 5 | 4×6 | | | | |
| | See Note 1 | | | | | | | | | | | | | |
| Over 20 | See Note 1 | | | | | | | | | | | | | |

* Douglas fir or equivalent with a bending strength not less than 1500 psi.
** Manufactured members of equivalent strength may be substituted for wood.

## TABLE C-2.3
## TIMBER TRENCH SHORING—MINIMUM TIMBER REQUIREMENTS*

Soil Type C $P_a = 80 \times H + 72$ psf (2 ft Surcharge)

| Depth of Trench (feet) | Size (S4S) and Spacing of Members ** | | | | | | | | | | | | | | |
|---|---|---|---|---|---|---|---|---|---|---|---|---|---|---|---|
| | Cross Braces | | | | | | | Wales | | Uprights | | | | | |
| | Horiz. Spacing (feet) | Width of Trench (feet) | | | | | Vert. Spacing (feet) | Size (in) | Vert. Spacing (feet) | Maximum Allowable Horizontal Spacing (feet) | | | | | |
| | | Up to 4 | Up to 6 | Up to 9 | Up to 12 | Up to 15 | | | | Close | | | | | |
| 5 | Up to 6 | 6X6 | 6X6 | 6X6 | 6X6 | 8X8 | 5 | 8X8 | 5 | 3X6 | | | | | |
| To | Up to 8 | 6X6 | 6X6 | 6X6 | 8X8 | 8X8 | 5 | 10X10 | 5 | 3X6 | | | | | |
| 10 | Up to 10 | 6X6 | 6X6 | 8X8 | 8X8 | 8X8 | 5 | 10X12 | 5 | 3X6 | | | | | |
| | See Note 1 | | | | | | | | | | | | | | |
| 10 | Up to 6 | 6X8 | 6X8 | 6X8 | 8X8 | 8X8 | 5 | 10X10 | 5 | 4X6 | | | | | |
| To | Up to 8 | 8X8 | 8X8 | 8X8 | 8X8 | 8X8 | 5 | 12X12 | 5 | 4X6 | | | | | |
| 15 | See Note 1 | | | | | | | | | | | | | | |
| | See Note 1 | | | | | | | | | | | | | | |
| 15 | Up to 6 | 8X8 | 8X8 | 8X8 | 8X10 | 8X10 | 5 | 10X12 | 5 | 4X6 | | | | | |
| To | See Note 1 | | | | | | | | | | | | | | |
| 20 | See Note 1 | | | | | | | | | | | | | | |
| | See Note 1 | | | | | | | | | | | | | | |
| Over 20 | See Note 1 | | | | | | | | | | | | | | |

* Douglas fir or equivalent with a bending strength not less than 1500 psi.
** Manufactured members of equivalent strength may be substituted for wood.

## Appendix D to Subpart P— Aluminum Hydraulic Shoring for Trenches

(a) *Scope.* This appendix contains information that can be used when aluminum hydraulic shoring is provided as a method of protection against cave-ins in trenches that do not exceed 20 feet (6.1m) in depth. This appendix must be used when design of the aluminum hydraulic protective system cannot be performed in accordance with §1926.652(c)(2).

(b) *Soil Classification.* In order to use data presented in this appendix, the soil type or types in which the excavation is made must first be determined using the soil classification method set forth in appendix A of subpart P of part 1926.

(c) *Presentation of Information.* Information is presented in several forms as follows:

(1) Information is presented in tabular form in Tables D-1.1, D-1.2, D-1.3 and E-1.4. Each table presents the maximum vertical and horizontal spacings that may be used with various aluminum member sizes and various hydraulic cylinder sizes. Each table contains data only for the particular soil type in which the excavation or portion of the excavation is made. Tables D-1.1 and D-1.2 are for vertical shores in Types A and B soil. Tables D-1.3 and D1.4 are for horizontal waler systems in Types B and C soil.

(2) Information concerning the basis of the tabular data and the limitations of the data is presented in paragraph (d) of this appendix.

(3) Information explaining the use of the tabular data is presented in paragraph (e) of this appendix.

(4) Information illustrating the use of the tabular data is presented in paragraph (f) of this appendix.

(5) Miscellaneous notations (footnotes) regarding Table D-1.1 through D-1.4 are presented in paragraph (g) of this appendix.

(6) Figures, illustrating typical installations of hydraulic shoring, are included just prior to the Tables. The illustrations page is entitled "Aluminum Hydraulic Shoring; Typical Installations."

(d) *Basis and limitations of the data.*

(1) Vertical shore rails and horizontal wales are those that meet the Section Modulus requirements in the D-1 Tables. Aluminum material is 6061-T6 or material of equivalent strength and properties.

(2) Hydraulic cylinders specifications. (i) 2-inch cylinders shall be a minimum 2-inch inside diameter with a minimum safe working capacity of no less than 18,000 pounds axial compressive load at maximum extension. Maximum extension is to include full range of cylinder extensions as recommended by product manufacturer.

(ii) 3-inch cylinders shall be a minimum 3-inch inside diameter with a safe working capacity of not less than 30,000 pounds axial compressive load at extensions as recommended by product manufacturer.

(3) Limitation of application.

(i) It is not intended that the aluminum hydraulic specification apply to every situation that may be experienced in the field. These data were developed to apply to the situations that are most commonly experienced in current trenching practice. Shoring systems for use in situations that are not covered by the data in this appendix must be otherwise designed as specified in §1926.652(c).

(ii) When any of the following conditions are present, the members specified in the Tables are not considered adequate. In this case, an alternative aluminum hydraulic shoring system or other type of protective system must be designed in accordance with §1926.652.

(A) When vertical loads imposed on cross braces exceed a 100 Pound gravity load distributed on a one foot section of the center of the hydraulic cylinder.

(B) When surcharge loads are present from equipment weighing in excess of 20,000 pounds.

(C) When only the lower portion or a trench is shored and the remaining portion of the trench is sloped or benched unless: The sloped portion is sloped at an angle less steep than three horizontal to one vertical; or the members are selected from the tables for use at a depth which is determined from the top of the overall trench, and not from the toe of the sloped portion.

(e) *Use of Tables D-1.1, D-1.2, D-1.3 and D-1.4.* The members of the shoring system that are to be selected using this information are the hydraulic cylinders, and either the vertical shores or the horizontal wales. When a waler system is used the vertical timber sheeting to be used is also selected from these tables. The Tables D-1.1 and D-1.2 for vertical shores are used in Type A and B soils that do not require sheeting. Type B soils that may require sheeting, and Type C soils that always require sheeting are found in the horizontal wale Tables D-1.3 and D-1.4. The soil type must first be determined in accordance with the soil classification system described in appendix A to subpart P of part 1926. Using the appropriate table, the selection of the size and spacing of the members is made. The selection is based on the depth and width of the trench where the members are to be installed. In these tables the vertical spacing is held constant at four feet on center. The tables show the maximum horizontal spacing of cylinders allowed for each size of wale in the waler system tables, and in the vertical shore tables, the hydraulic cylinder horizontal spacing is the same as the vertical shore spacing.

(f) *Example to Illustrate the Use of the Tables:*

(1) Example 1:

A trench dug in Type A soil is 6 feet deep and 3 feet wide. From Table D-1.1: Find vertical shores and 2 inch diameter cylinders spaced 8 feet on center (o.c.) horizontally and 4 feet on center (o.c.) vertically. (See Figures 1 & 3 for typical installations.)

(2) Example 2:

A trench is dug in Type B soil that does not require sheeting, 13 feet deep and 5 feet wide. From Table D-1.2: Find vertical shores and 2 inch diameter cylinders spaced 6.5 feet o.c. horizontally and 4 feet o.c. vertically. (See Figures 1 & 3 for typical installations.)

(3) A trench is dug in Type B soil that does not require sheeting, but does experience some minor raveling of the trench face. The trench is 16 feet deep and 9 feet wide. From Table D-1.2: Find vertical shores and 2 inch diameter cylinder (with special oversleeves as designated by footnote #B2) spaced 5.5 feet o.c. horizontally and 4 feet o.c. vertically, plywood (per footnote (g)(7) to the D-1 Table) should be used behind the shores. (See Figures 2 & 3 for typical installations.)

(4) Example 4: A trench is dug in previously disturbed Type B soil, with characteristics of a Type C soil, and will require sheeting. The trench is 18 feet deep and 12 feet wide. 8 foot horizontal spacing between cylinders is desired for working space. From Table D-1.3: Find horizontal wale with a section modulus of 14.0 spaced at 4 feet o.c. vertically and 3 inch diameter cylinder spaced at 9 feet maximum o.c. horizontally. 3×12 timber sheeting is required at close spacing vertically. (See Figure 4 for typical installation.)

(5) Example 5: A trench is dug in Type C soil, 9 feet deep and 4 feet wide. Horizontal cylinder spacing in excess of 6 feet is desired for working space. From Table D-1.4: Find horizontal wale with a section modulus of 7.0 and 2 inch diameter cylinders spaced at 6.5 feet o.c. horizontally. Or, find horizontal wale with a 14.0 section modulus and 3 inch diameter cylinder spaced at 10 feet o.c. horizontally. Both wales are spaced 4 feet o.c. vertically. 3×12 timber sheeting is required at close spacing vertically. (See Figure 4 for typical installation.)

(g) *Footnotes, and general notes, for Tables D-1.1, D-1.2, D-1.3, and D-1.4.*

(1) For applications other than those listed in the tables, refer to §1926.652(c)(2) for use of manufacturer's tabulated data. For trench depths in excess of 20 feet, refer to §1926.652(c)(2) and §1926.652(c)(3).

(2) 2 inch diameter cylinders, at this width, shall have structural steel tube (3.5×3.5×0.1875) oversleeves, or structural oversleeves of manufacturer's specification, extending the full, collapsed length.

(3) Hydraulic cylinders capacities. (i) 2 inch cylinders shall be a minimum 2-inch inside diameter with a safe working capacity of not less than 18,000 pounds axial compressive load at maximum extension. Maximum extension is to include full range of cylinder extensions as recommended by product manufacturer.

(ii) 3-inch cylinders shall be a minimum 3-inch inside diameter with a safe work capacity of not less than 30,000 pounds axial compressive load at maximum extension. Maximum extension is to include full range of cylinder extensions as recommended by product manufacturer.

(4) All spacing indicated is measured center to center.

(5) Vertical shoring rails shall have a minimum section modulus of 0.40 inch.

(6) When vertical shores are used, there must be a minimum of three shores spaced equally, horizontally, in a group.

(7) Plywood shall be 1.125 in. thick softwood or 0.75 inch. thick, 14 ply, arctic white birch (Finland form). Please note that plywood is not intended as a structural member, but only for prevention of local raveling (sloughing of the trench face) between shores.

(8) See Appendix C for timber specifications.

(9) Wales are calculated for simple span conditions.

(10) See Appendix D, item (d), for basis and limitations of the data.

## ALUMINUM HYDRAULIC SHORING, TYPICAL INSTALLAIONS

**FIGURE NO. 1**
VERTICAL ALUMINUM HYDRAULIC SHORING (SPOT BRACING)

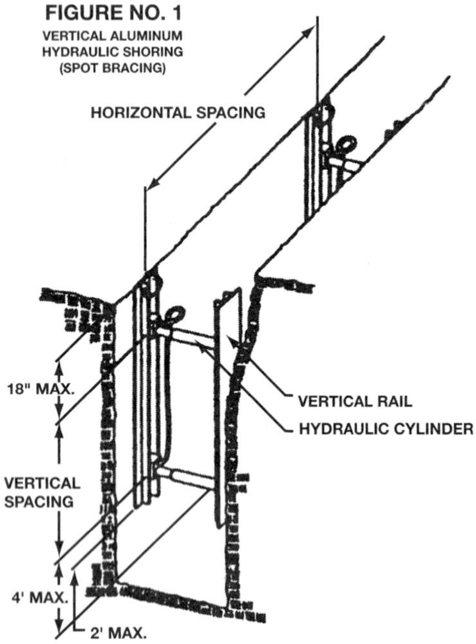


**FIGURE NO. 2**
VERTICAL ALUMINUM HYDRAULIC SHORING (WITH PLYWOOD)

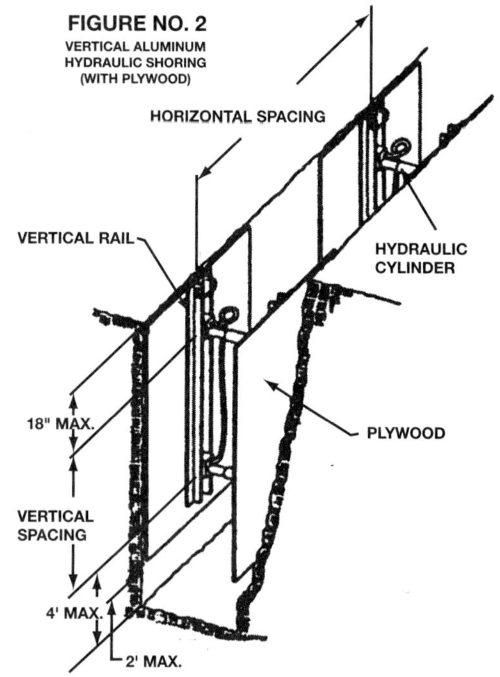


**FIGURE NO. 3**
VERTICAL ALUMINUM HYDRAULIC SHORING (STACKED)

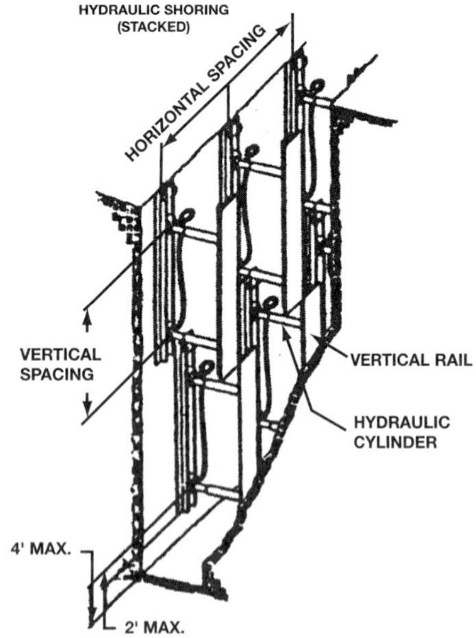


**FIGURE NO. 4**
ALUMINUM HYDRAULIC SHORING WALER SYSTEM (TYPICAL)

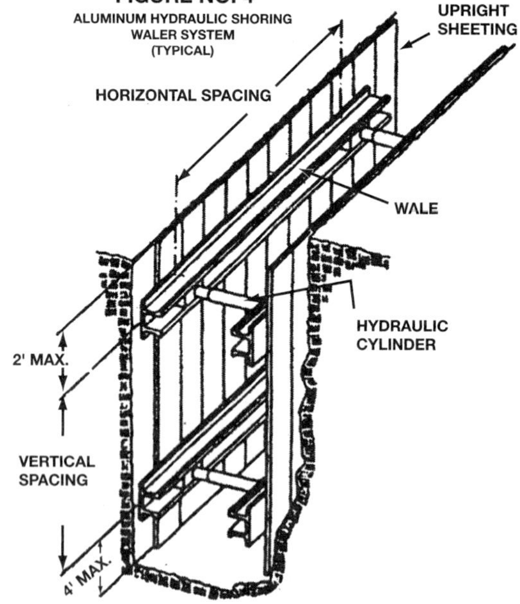

## TABLE D-1.1—ALUMINUM HYDRAULIC SHORING

Vertical Shores
For Soil Type A

| Depth of Trench (feet) | Maximum Horizontal Spacing (feet) | Maximum Vertical Spacing (feet) | Hydraulic Cylinders — Width of Trench (feet): Up to 8 | Over 8 Up to 12 | Over 12 Up to 15 |
|---|---|---|---|---|---|
| Over 5 Up to 10 | 8 | | | | |
| Over 10 Up to 15 | 8 | 4 | 2 inch Diameter | 2 inch Diameter Note (2) | 3 inch Diameter |
| Over 15 Up to 20 | 7 | | | | |
| Over 20 | Note (1) | | | | |

Footnotes to tables, and general notes on hydraulic shoring, are found in Appendix D, Item (g)
Note (1): See Appendix D, Item (g)(1)
Note (2): See Appendix D, Item (g)(2)

## TABLE D-1.2—ALUMINUM HYDRAULIC SHORING

Vertical Shores
For Soil Type B

| Depth of Trench (feet) | Maximum Horizontal Spacing (feet) | Maximum Vertical Spacing (feet) | Hydraulic Cylinders — Width of Trench (feet): Up to 8 | Over 8 Up to 12 | Over 12 Up to 15 |
|---|---|---|---|---|---|
| Over 5 Up to 10 | 8 | | | | |
| Over 10 Up to 15 | 6.5 | 4 | 2 inch Diameter | 2 inch Diameter Note (2) | 3 inch Diameter |
| Over 15 Up to 20 | 5.5 | | | | |
| Over 20 | Note (1) | | | | |

Footnotes to tables, and general notes on hydraulic shoring, are found in Appendix D, Item (g)
Note (1): See Appendix D, Item (g)(1)
Note (2): See Appendix D, Item (g)(2)

## TABLE D-1.3—ALUMINUM HYDRAULIC SHORING

Waler Systems
For Soil Type B

| Depth of Trench (feet) | Wales: Vertical Spacing (feet) | Wales: Section Modulus* ($in^3$) | Hydraulic Cylinders — Width of Trench (feet): Up to 8 — Horiz. Spacing | Up to 8 — Cylinder Diameter | Over 8 Up to 12 — Horiz. Spacing | Over 8 Up to 12 — Cylinder Diameter | Over 12 Up to 15 — Horiz. Spacing | Over 12 Up to 15 — Cylinder Diameter | Timber Uprights — Max. Horiz. Spacing (on Center): Solid Sheet | 2 ft. | 3 ft. |
|---|---|---|---|---|---|---|---|---|---|---|---|
| | | 3.5 | 8.0 | 2 in | 8.0 | 2 in Note (2) | 8.0 | 3 in | | | |
| Over 5 Up to 10 | 4 | 7.0 | 9.0 | 2 in | 9.0 | 2 in Note (2) | 9.0 | 3 in | — | — | 3×12 |
| | | 14.0 | 12.0 | 3 in | 12.0 | 3 in | 12.0 | 3 in | | | |
| | | 3.5 | 6.0 | 2 in | 6.0 | 2 in Note (2) | 6.0 | 3 in | | | |
| Over 10 Up to 15 | 4 | 7.0 | 8.0 | 3 in | 8.0 | 3 in | 8.0 | 3 in | — | 3×12 | — |
| | | 14.0 | 10.0 | 3 in | 10.0 | 3 in | 10.0 | 3 in | | | |
| | | 3.5 | 5.5 | 2 in | 5.5 | 2 in Note (2) | 5.5 | 3 in | | | |
| Over 15 Up to 20 | 4 | 7.0 | 6.0 | 3 in | 6.0 | 3 in | 6.0 | 3 in | 3×12 | — | — |
| | | 14.0 | 9.0 | 3 in | 9.0 | 3 in | 9.0 | 3 in | | | |
| Over 20 | Note (1) | | | | | | | | | | |

Footnotes to tables, and general notes on hydraulic shoring, are found in Appendix D, Item (g)
Notes (1): See Appendix D, Item (g)(1)
Notes (2): See Appendix D, Item (g)(2)
* Consult product manufacturer and/or qualified engineer for Section Modulus of available wales.

## TABLE D-1.4—ALUMINUM HYDRAULIC SHORING

Waler Systems
For Soil Type C

| Depth of Trench (feet) | Wales: Vertical Spacing (feet) | Wales: Section* Modulus (in³) | Hydraulic Cylinders, Width of Trench (feet): Up to 8, Horiz. Spacing | Up to 8, Cylinder Diameter | Over 8 Up to 12, Horiz. Spacing | Over 8 Up to 12, Cylinder Diameter | Over 12 Up to 15, Horiz. Spacing | Over 12 Up to 15, Cylinder Diameter | Timber Uprights, Max. Horiz. Spacing (on Center): Solid Sheet | 2 ft. | 3 ft. |
|---|---|---|---|---|---|---|---|---|---|---|---|
| Over 5 Up to 10 | 4 | 3.5 | 6.0 | 2 in | 6.0 | 2 in Note (2) | 6.0 | 3 in | 3×12 | — | — |
| | | 7.0 | 6.5 | 2 in | 6.5 | 2 in Note (2) | 6.5 | 3 in | | | |
| | | 14.0 | 10.0 | 3 in | 10.0 | 3 in | 10.0 | 3 in | | | |
| Over 10 Up to 15 | 4 | 3.5 | 4.0 | 2 in | 4.0 | 2 in Note (2) | 4.0 | 3 in | 3×12 | — | — |
| | | 7.0 | 5.5 | 3 in | 5.5 | 3 in | 5.5 | 3 in | | | |
| | | 14.0 | 8.0 | 3 in | 8.0 | 3 in | 8.0 | 3 in | | | |
| Over 15 Up to 20 | 4 | 3.5 | 3.5 | 2 in | 3.5 | 2 in Note (2) | 3.5 | 3 in | 3×12 | — | — |
| | | 7.0 | 5.0 | 3 in | 5.0 | 3 in | 5.0 | 3 in | | | |
| | | 14.0 | 6.0 | 3 in | 6.0 | 3 in | 6.0 | 3 in | | | |
| Over 20 | Note (1) | | | | | | | | | | |

Footnotes to tables, and general notes on hydraulic shoring, are found in Appendix D, Item (g)
Notes (1): See Appendix D, Item (g)(1)
Notes (2): See Appendix D, Item (g)(2)
* Consult product manufacturer and/or qualified engineer for Section Modulus of available wales.

### Appendix E to Subpart P—Alternatives to Timber Shoring

**FIGURE 1—ALUMINUM HYDRAULIC SHORING**

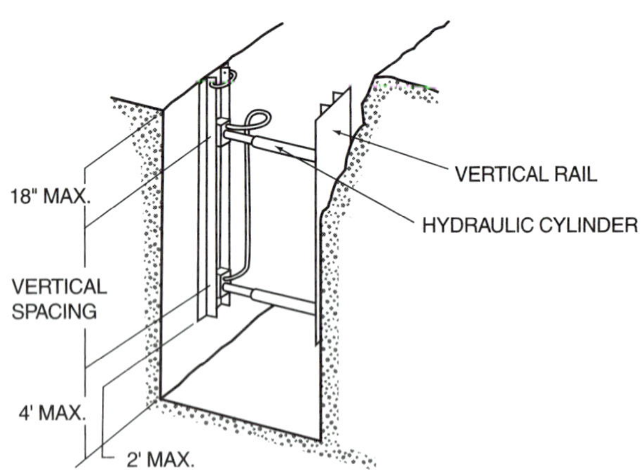


**FIGURE 2—PNEUMATIC/HYDRAULIC SHORING**

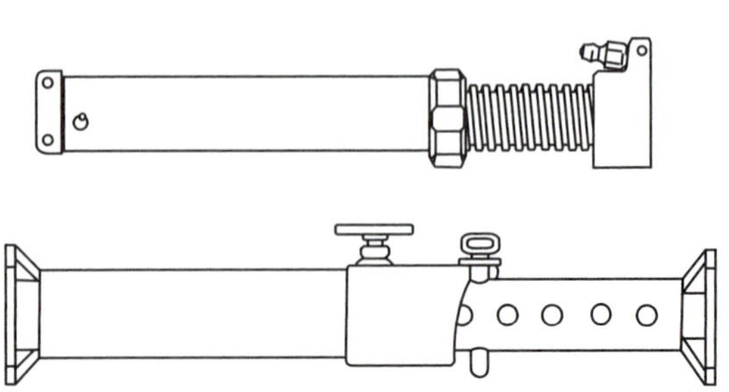

**FIGURE 3—TRENCH JACKS (SCREW JACKS)**

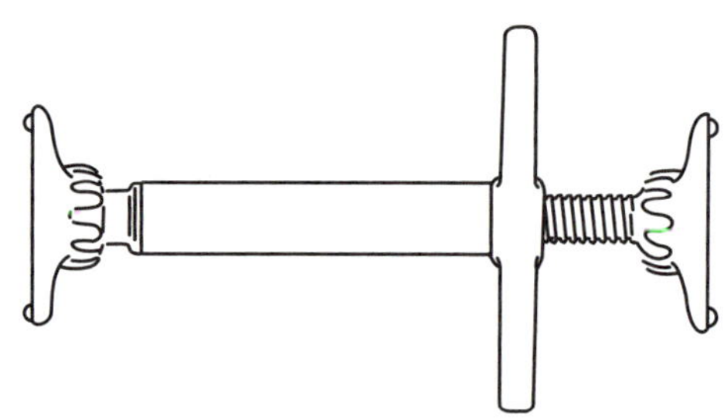

**FIGURE 4—TRENCH SHIELDS**

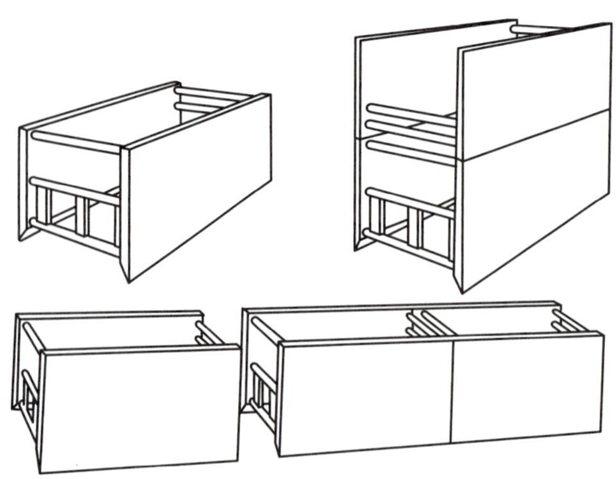

## Appendix F to Subpart P— Selection of Protective Systems

The following figures are a graphic summary of the requirements contained in subpart P for excavations 20 feet or less in depth. Protective systems for use in excavations more than 20 feet in depth must be designed by a registered professional engineer in accordance with §1926.652 (b) and (c).

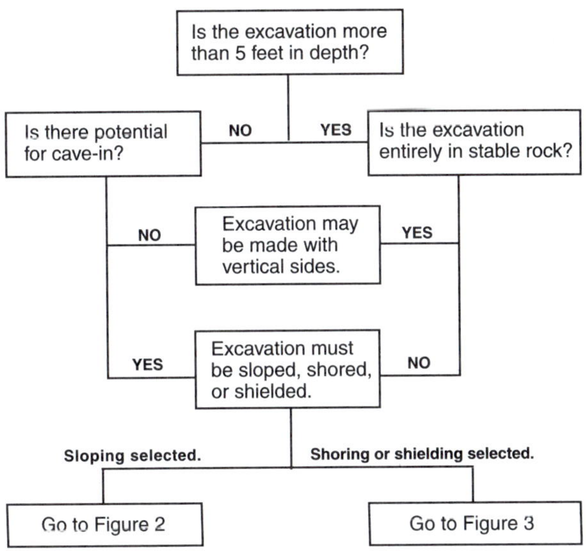


**FIGURE 1—PRELIMINARY DECISIONS**

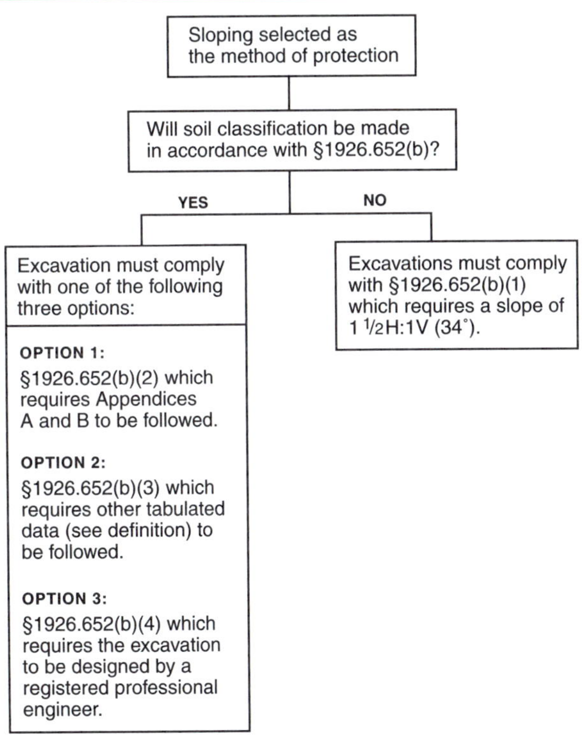


**FIGURE 2—SLOPING OPTIONS**

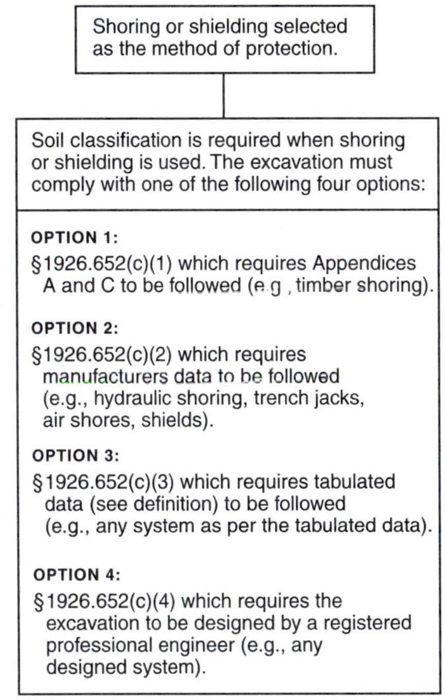


**FIGURE 3—SHORING AND SHIELDING OPTIONS**

[Subpart P revised and added at 54 FR 45959, extended Oct. 31, 1989; effective date extended to March 5, 1990 at 54 FR 53055, Dec. 27, 1989.]

**Authority:** Sec. 107, Contract Work Hours and Safety Standards Act (Construction Safety Act) (40 U.S.C. 333); Secs. 4, 6 and 8 Occupational Safety and Health Act of 1970 (29 U.S.C. 653, 655, and 657); Secretary of Labor's Order No. 12-71 (36 FR 8754), 8-76 (41 FR 25059), 9-83 (48 FR 35736), or 1-90 (55 FR 9033), as applicable; and 29 CFR part 1911.

**Source:** 53 FR 22643, June 16, 1988, unless otherwise noted.

### §1926.700 Scope, application, and definitions applicable to this subpart.

**1926.700(a)** ***Scope and application.*** This subpart sets forth requirements to protect all construction employees from the hazards associated with concrete and masonry construction operations performed in workplaces covered under 29 CFR part 1926. In addition to the requirements in subpart Q, other relevant provisions in parts 1910 and 1926 apply to concrete and masonry construction operations.

**1926.700(b)** ***Definitions applicable to this subpart.*** In addition to the definitions set forth in §1926.32, the following definitions apply to this subpart.

**1926.700(b)(1)** *Bull float* means a tool used to spread out and smooth concrete.

**1926.700(b)(2)** *Formwork* means the total system of support for freshly placed or partially cured concrete, including the mold or sheeting (form) that is in contact with the concrete as well as all supporting members including shores, reshores, hardware, braces, and related hardware.

**1926.700(b)(3)** *Lift slab* means a method of concrete construction in which floor, and roof slabs are cast on or at ground level and, using jacks, lifted into position.

**1926.700(b)(4)** *Limited access zone* means an area alongside a masonry wall, which is under construction, and which is clearly demarcated to limit access by employees.

**1926.700(b)(5)** *Precast concrete* means concrete members (such as walls, panels, slabs, columns, and beams) which have been formed, cast, and cured prior to final placement in a structure.

**1926.700(b)(6)** *Reshoring* means the construction operation in which shoring equipment (also called reshores or reshoring equipment) is placed, as the original forms and shores are removed, in order to support partially cured concrete and construction loads.

**1926.700(b)(7)** *Shore* means a supporting member that resists a compressive force imposed by a load.

**1926.700(b)(8)** *Vertical slip forms* means forms which are jacked vertically during the placement of concrete.

**1926.700(b)(9)** *Jacking operation* means the task of lifting a slab (or group of slabs) vertically from one location to another (e.g., from the casting location to a temporary (parked) location, or from a temporary location to another temporary location, or to its final location in the structure), during the construction of a building/structure where the lift-slab process is being used.

[53 FR 22643, June 16, 1988, as amended at 55 FR 42328, Oct. 18, 1990]

### §1926.701 General requirements

**1926.701(a)** ***Construction loads.*** No construction loads shall be placed on a concrete structure or portion of a concrete structure unless the employer determines, based on information received from a person who is qualified in structural design, that the structure or portion of the structure is capable of supporting the loads.

**1926.701(b)** ***Reinforcing steel.*** All protruding reinforcing steel, onto and into which employees could fall, shall be guarded to eliminate the hazard of impalement.

**1926.701(c)** ***Post-tensioning operations.*** (1) No employee (except those essential to the post-tensioning operations) shall be permitted to be behind the jack during tensioning operations.

**1926.701(c)(2)** Signs and barriers shall be erected to limit employee access to the post-tensioning area during tensioning operations.

**1926.701(d)** ***Riding concrete buckets.*** No employee shall be permitted to ride concrete buckets.

**1926.701(e)** ***Working under loads.*** (1) No employee shall be permitted to work under concrete buckets while buckets are being elevated or lowered into position.

**1926.701(e)(2)** To the extent practical, elevated concrete buckets shall be routed so that no employee, or the fewest number of employees, are exposed to the hazards associated with falling concrete buckets.

**1926.701(f)** ***Personal protective equipment.*** No employee shall be permitted to apply a cement, sand, and water mixture through a pneumatic hose unless the employee is wearing protective head and face equipment.

[53 FR 22643, June 16, 1988, as amended at 59 FR 40730, Aug. 9, 1994]

### §1926.702 Requirements for equipment and tools.

**1926.702(a)** ***Bulk cement storage.*** (1) Bulk storage bins, containers, and silos shall be equipped with the following:

**1926.702(a)(1)(i)** Conical or tapered bottoms; and

**1926.702(a)(1)(ii)** Mechanical or pneumatic means of starting the flow of material.

**1926.702(a)(2)** No employee shall be permitted to enter storage facilities unless the ejection system has been shut down, locked out, and tagged to indicate that the ejection system is not to be operated.

**1926.702(b)** ***Concrete mixers.*** Concrete mixers with one cubic yard (.8 $m^3$) or larger loading skips shall be equipped with the following:

**1926.702(b)(1)** A mechanical device to clear the skip of materials; and

**1926.702(b)(2)** Guardrails installed on each side of the skip.

**1926.702(c)** ***Power concrete trowels.*** Powered and rotating type concrete troweling machines that are manually guided shall be equipped with a control switch that will automatically shut off the power whenever the hands of the operator are removed from the equipment handles.

**1926.702(d)** ***Concrete buggies.*** Concrete buggy handles shall not extend beyond the wheels on either side of the buggy.

**1926.702(e)** ***Concrete pumping systems.*** (1) Concrete pumping systems using discharge pipes shall be provided with pipe supports designed for 100 percent overload.

**1926.702(e)(2)** Compressed air hoses used on concrete pumping system shall be provided with positive fail-safe joint connectors to prevent separation of sections when pressurized.

**1926.702(f)** ***Concrete buckets.*** (1) Concrete buckets equipped with hydraulic or pneumatic gates shall have positive safety latches or similar safety devices installed to prevent premature or accidental dumping.

**1926.702(f)(2)** Concrete buckets shall be designed to prevent concrete from hanging up on top and the sides.

**1926.702(g)** ***Tremies.*** Sections of tremies and similar concrete conveyances shall be secured with wire rope (or equivalent materials) in addition to the regular couplings or connections.

**1926.702(h)** ***Bull floats.*** Bull float handles, used where they might contact energized electrical conductors, shall be constructed of nonconductive material or insulated with a nonconductive sheath whose electrical and mechanical characteristics provide the equivalent protection of a handle constructed of nonconductive material.

**1926.702(i)** ***Masonry saws.*** (1) Masonry saws shall be guarded with a semicircular enclosure over the blade.

**1926.702(i)(2)** A method for retaining blade fragments shall be incorporated in the design of the semicircular enclosure.

**1926.702(j)** ***Lockout/Tagout Procedures.*** (1) No employee shall be permitted to perform maintenance or repair activity on equipment (such as compressors, mixers, screens or pumps used for concrete and masonry construction activities) where the inadvertent operation of the equipment could occur and cause injury, unless all potentially hazardous energy sources have been locked out and tagged.

**1926.702(j)(2)** Tags shall read *Do Not Start* or similar language to indicate that the equipment is not to be operated.

## §1926.703 Requirements for cast-in-place concrete.

**1926.703(a)** ***General requirements for formwork.*** (1) Formwork shall be designed, fabricated, erected, supported, braced and maintained so that it will be capable of supporting without failure all vertical and lateral loads that may reasonably be anticipated to be applied to the formwork. Formwork which is designed, fabricated, erected, supported, braced and maintained in conformance with the Appendix to this section will be deemed to meet the requirements of this paragraph.

**1926.703(a)(2)** Drawings or plans, including all revisions, for the jack layout, formwork (including shoring equipment), working decks, and scaffolds, shall be available at the jobsite.

**1926.703(b)** ***Shoring and reshoring.*** (1) All shoring equipment (including equipment used in reshoring operations) shall be inspected prior to erection to determine that the equipment meets the requirements specified in the formwork drawings.

**1926.703(b)(2)** Shoring equipment found to be damaged such that its strength is reduced to less than that required by §1926.703(a)(1) shall not be used for shoring.

**1926.703(b)(3)** Erected shoring equipment shall be inspected immediately prior to, during, and immediately after concrete placement.

**1926.703(b)(4)** Shoring equipment that is found to be damaged or weakened after erection, such that its strength is reduced to less than that required by §1926.703(a)(1), shall be immediately reinforced.

**1926.703(b)(5)** The sills for shoring shall be sound, rigid, and capable of carrying the maximum intended load.

**1926.703(b)(6)** All base plates, shore heads, extension devices, and adjustment screws shall be in firm contact, and secured when necessary, with the foundation and the form.

**1926.703(b)(7)** Eccentric loads on shore heads and similar members shall be prohibited unless these members have been designed for such loading.

**1926.703(b)(8)** Whenever single post shores are used one on top of another (tiered), the employer shall comply with the following specific requirements in addition to the general requirements for formwork:

**1926.703(b)(8)(i)** The design of the shoring shall be prepared by a qualified designer and the erected shoring shall be inspected by an engineer qualified in structural design.

**1926.703(b)(8)(ii)** The single post shores shall be vertically aligned.

**1926.703(b)(8)(iii)** The single post shores shall be spliced to prevent misalignment.

**1926.703(b)(8)(iv)** The single post shores shall be adequately braced in two mutually perpendicular directions at the splice level. Each tier shall also be diagonally braced in the same two directions.

**1926.703(b)(9)** Adjustment of single post shores to raise formwork shall not be made after the placement of concrete.

**1926.703(b)(10)** Reshoring shall be erected, as the original forms and shores are removed, whenever the concrete is required to support loads in excess of its capacity.

**1926.703(c)** ***Vertical slip forms.*** (1) The steel rods or pipes on which jacks climb or by which the forms are lifted shall be—

**1926.703(c)(1)(i)** Specifically designed for that purpose; and

**1926.703(c)(1)(ii)** Adequately braced where not encased in concrete.

**1926.703(c)(2)** Forms shall be designed to prevent excessive distortion of the structure during the jacking operation.

**1926.703(c)(3)** All vertical slip forms shall be provided with scaffolds or work platforms where employees are required to work or pass.

**1926.703(c)(4)** Jacks and vertical supports shall be positioned in such a manner that the loads do not exceed the rated capacity of the jacks.

**1926.703(c)(5)** The jacks or other lifting devices shall be provided with mechanical dogs or other automatic holding devices to support the slip forms whenever failure of the power supply or lifting mechanism occurs.

**1926.703(c)(6)** The form structure shall be maintained within all design tolerances specified for plumbness during the jacking operation.

**1926.703(c)(7)** The predetermined safe rate of lift shall not be exceeded.

**1926.703(d)** ***Reinforcing steel.*** (1) Reinforcing steel for walls, piers, columns, and similar vertical structures shall be adequately supported to prevent overturning and to prevent collapse.

**1926.703(d)(2)** Employers shall take measures to prevent unrolled wire mesh from recoiling. Such measures may include, but are not limited to, securing each end of the roll or turning over the roll.

**1926.703(e)** ***Removal of formwork.*** (1) Forms and shores (except those used for slabs on grade and slip forms) shall not be removed until the employer determines that the concrete has gained sufficient strength to support its weight and superimposed loads. Such determination shall be based on compliance with one of the following:

**1926.703(e)(1)(i)** The plans and specifications stipulate conditions for removal of forms and shores, and such conditions have been followed, or

**1926.703(e)(1)(ii)** The concrete has been properly tested with an appropriate ASTM standard test method designed to indicate the concrete compressive strength, and the test results indicate that the concrete has gained sufficient strength to support its weight and superimposed loads.

**1926.703(e)(2)** Reshoring shall not be removed until the concrete being supported has attained adequate strength to support its weight and all loads in place upon it.

### Appendix to §1926.703(a)(1)

### GENERAL REQUIREMENTS FOR FORMWORK

(This Appendix is non-mandatory.)

This appendix serves as a non-mandatory guideline to assist employers in complying with the formwork requirements in §1926.703(a)(1). Formwork which has been designed, fabricated, erected, braced, supported and maintained in accordance with Sections 6 and 7 of the American National Standard for Construction and Demolition Operations—Concrete and Masonry Work, ANSI A10.9-1983, shall be deemed to be in compliance with the provision of §1926.703(a)(1).

[53 FR 22643, June 16, 1988, as amended at 61 FR 5510, Feb. 13, 1996]

## §1926.704 Requirements for precast concrete.

**1926.704(a)** Precast concrete wall units, structural framing, and tilt-up wall panels shall be adequately supported to prevent overturning and to prevent collapse until permanent connections are completed.

**1926.704(b)** Lifting inserts which are embedded or otherwise attached to tilt-up precast concrete members shall be capable of supporting at least two times the maximum intended load applied or transmitted to them.

**1926.704(c)** Lifting inserts which are embedded or otherwise attached to precast concrete members, other than the tilt-up members, shall be capable of supporting at least four times the maximum intended load applied or transmitted to them.

**1926.704(d)** Lifting hardware shall be capable of supporting at least five times the maximum intended load applied or transmitted to the lifting hardware.

**1926.704(e)** No employee shall be permitted under precast concrete members being lifted or tilted into position except those employees required for the erection of those members.

[53 FR 22643, June 16, 1988, as amended at 54 FR 41088, Oct. 5, 1989]

## §1926.705 Requirements for lift-slab construction operations.

**1926.705(a)** Lift-slab operations shall be designed and planned by a registered professional engineer who has experience in lift-slab construction. Such plans and designs shall be implemented by the employer and shall include detailed instructions and sketches indicating the prescribed method of erection. These plans and designs shall also include provisions for ensuring lateral stability of the building/structure during construction.

**1926.705(b)** Jacks/lifting units shall be marked to indicate their rated capacity as established by the manufacturer.

**1926.705(c)** Jacks/lifting units shall not be loaded beyond their rated capacity as established by the manufacturer.

**1926.705(d)** Jacking equipment shall be capable of supporting at least two and one-half times the load being lifted during jacking operations and the equipment shall not be overloaded. For the purpose of this provision, jacking equipment includes any load bearing component which is used to carry out the lifting operation(s). Such equipment includes, but is not limited, to the following: threaded rods, lifting attachments, lifting nuts, hook-up collars, T-caps, shearheads, columns, and footings.

**1926.705(e)** Jacks/lifting units shall be designed and installed so that they will neither lift nor continue to lift when they are loaded in excess of their rated capacity.

**1926.705(f)** Jacks/lifting units shall have a safety device installed which will cause the jacks/lifting units to support the load in any position in the event any jack/lifting unit malfunctions or loses its lifting ability.

**1926.705(g)** Jacking operations shall be synchronized in such a manner to ensure even and uniform lifting of the slab. During lifting, all points at which the slab is supported shall be kept within 1/2 inch of that needed to maintain the slab in a level position.

**1926.705(h)** If leveling is automatically controlled, a device shall be installed that will stop the operation when the 1/2 inch tolerance set forth in paragraph (g) of this section is exceeded or where there is a malfunction in the jacking (lifting) system.

**1926.705(i)** If leveling is maintained by manual controls, such controls shall be located in a central location and attended by a competent person while lifting is in progress. In addition to meeting the definition in §1926.32(f), the competent person must be experienced in the lifting operation and with the lifting equipment being used.

**1926.705(j)** The maximum number of manually controlled jacks/lifting units on one slab shall be limited to a number that will permit the operator to maintain the slab level within specified tolerances of paragraph (g) of this section, but in no case shall that number exceed 14.

**1926.705(k)(1)** No employee, except those essential to the jacking operation, shall be permitted in the building/structure while any jacking operation is taking place unless the building/structure has been reinforced sufficiently to ensure its integrity during erection. The phrase "reinforced sufficiently to ensure its integrity" used in this paragraph means that a registered professional engineer, independent of the engineer who designed and planned the lifting operation, has determined from the plans that if there is a loss of support at any jack location, that loss will be confined to that location and the structure as a whole will remain stable.

**1926.705(k)(2)** Under no circumstances, shall any employee who is not essential to the jacking operation be permitted immediately beneath a slab while it is being lifted.

**1926.705(k)(3)** For the purpose of paragraph (k) of this section, a jacking operation begins when a slab or group of slabs is lifted and ends when such slabs are secured (with either temporary connections or permanent connections).

**1926.705(k)(4)** Employers who comply with appendix A to §1926.705 shall be considered to be in compliance with the provisions of paragraphs (k)(1) through (k)(3) of this section.

**1926.705(l)** When making temporary connections to support slabs, wedges shall be secured by tack welding, or an equivalent method of securing the wedges to prevent them from falling out of position. Lifting rods may not be released until the wedges at that column have been secured.

**1926.705(m)** All welding on temporary and permanent connections shall be performed by a certified welder, familiar with the welding requirements specified in the plans and specifications for the lift-slab operation.

**1926.705(n)** Load transfer from jacks/lifting units to building columns shall not be executed until the welds on the column shear plates (weld blocks) are cooled to air temperature.

**1926.705(o)** Jacks/lifting units shall be positively secured to building columns so that they do not become dislodged or dislocated.

**1926.705(p)** Equipment shall be designed and installed so that the lifting rods cannot slip out of position or the employer shall institute other measures, such as the use of locking or blocking devices, which will provide positive connection between the lifting rods and attachments and will prevent components from disengaging during lifting operations.

## Appendix to §1926.705—Lift-Slab Operations

(This Appendix is non-mandatory.)

In paragraph 1926.705(k), OSHA requires employees to be removed from the building/structure during jacking operations unless an independent registered professional engineer, other than the engineer who designed and planned the lifting operation, has determined that the building/structure has been sufficiently reinforced to insure the integrity of the building/structure. One method to comply with this provision is for the employer to ensure that continuous bottom steel is provided in every slab and in both directions through every wall or column head area. (Column head area means the distance between lines that are one and one half times the thickness of the slab or drop panel. These lines are located outside opposite faces of the outer edges of the shearhead sections—See Figure 1). The amount of bottom steel shall be established by assuming loss of support at a given lifting jack and then determining the steel necessary to carry, by catenary action over the span between surrounding supports, the slab service dead load *plus* any service dead and live loads likely to be acting on the slab during jacking. In addition, the surrounding supports must be capable of resisting any additional load transferred to them as a result of the loss of support at the lifting jack considered.

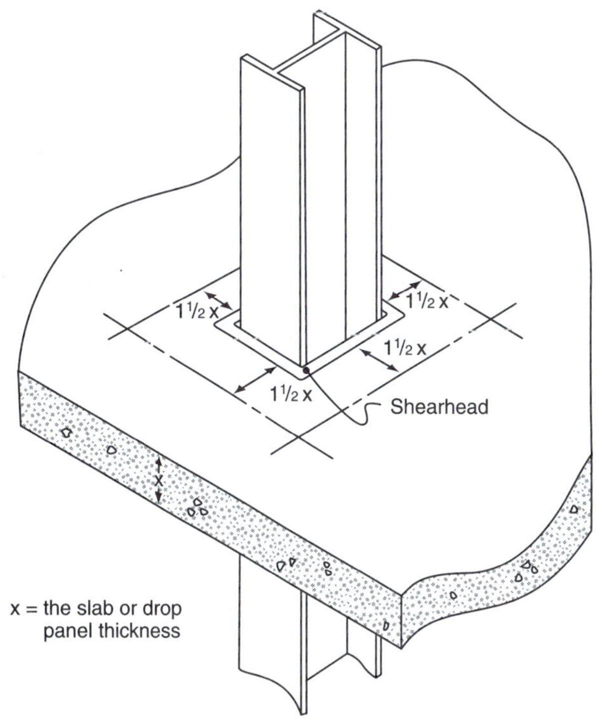


**FIGURE 1—COLUMN HEAD AREA**

[55 FR 42328, Oct. 18, 1990]

### §1926.706 Requirements for masonry construction.

**1926.706(a)** A limited access zone shall be established whenever a masonry wall is being constructed. The limited access zone shall conform to the following.

**1926.706(a)(1)** The limited access zone shall be established prior to the start of construction of the wall.

**1926.706(a)(2)** The limited access zone shall be equal to the height of the wall to be constructed plus four feet, and shall run the entire length of the wall.

**1926.706(a)(3)** The limited access zone shall be established on the side of the wall which will be unscaffolded.

**1926.706(a)(4)** The limited access zone shall be restricted to entry by employees actively engaged in constructing the wall. No other employees shall be permitted to enter the zone.

**1926.706(a)(5)** The limited access zone shall remain in place until the wall is adequately supported to prevent overturning and to prevent collapse unless the height of wall is over eight feet, in which case, the limited access zone shall remain in place until the requirements of paragraph (b) of this section have been met.

**1926.706(b)** All masonry walls over eight feet in height shall be adequately braced to prevent overturning and to prevent collapse unless the wall is adequately supported so that it will not overturn or collapse. The bracing shall remain in place until permanent supporting elements of the structure are in place.

## Appendix A to Subpart Q— References to Subpart Q of Part 1926

(This Appendix is non-mandatory.)

The following non-mandatory references provide information which can be helpful in understanding and complying with the requirements contained in subpart Q.

- Accident Prevention Manual for Industrial Operations; Eighth Edition; National Safety Council.
- Building Code Requirements for Reinforced Concrete (ACI 318-83).
- Formwork for Concrete (ACI SP-4).
- Recommended Practice for Concrete Formwork (ACI 347-78).
- Safety Requirements for Concrete and Masonry Work (ANSI A10.9-1983).
- Standard Test Method for Compressive Strength of Cylindrical Concrete Specimens (ASTM C39-86).
- Standard Test Method for Making and Curing Concrete Test Specimens in the Field (ASTM C31-85).
- Standard Test Method for Penetration Resistance of Hardened Concrete (ASTM C803-82).
- Standard Test Method for Compressive Strength of Concrete Cylinders Cast In-Place in Cylindrical Molds (ASTM C873-85).
- Standard Method for Developing Early Age Compressive Test Values and Projecting Later Age Strengths (ASTM C918-80).
- Recommended Practice for Inspection and Testing Agencies for Concrete, Steel and Bituminous Materials as Used in Construction (ASTM E329-77).
- Method of Making and Curing Concrete Test Specimens in the Laboratory (ASTM C192-88).
- Methods of Obtaining and Testing Drilled Cores and Sawed Beams of Concrete (ASTM C42-87).
- Methods of Securing, Preparing and Testing Specimens from Hardened Lightweight Insulating Concrete for Compressive Strength (ASTM C513-86).
- Test Method for Comprehensive Strength of Lightweight Insulating Concrete (ASTM C495-86).
- Method of Making, Accelerating Curing, and Testing of Concrete Compression Test Specimens (ASTM C684-81).
- Test Method for Compressive Strength of Concrete Using Portions of Beams Broken in Flexure (ASTM C116-68 (1980)).

**Authority:** Sec. 107, Contract Work Hours and Safety Standards Act (Construction Safety Act) (40 U.S.C. 333); Sec. 4, 6, and 8, Occupational Safety and Health Act of 1970 (29 U.S.C. 653, 655, 657); Secretary of Labor's Order No. 3-2000 (65 FR 50017), and 29 CFR part 1911.

## §1926.750 Scope.

**1926.750(a)** This subpart sets forth requirements to protect employees from the hazards associated with steel erection activities involved in the construction, alteration, and/or repair of single and multi-story buildings, bridges, and other structures where steel erection occurs. The requirements of this subpart apply to employers engaged in steel erection unless otherwise specified. This subpart does not cover electrical transmission towers, communication and broadcast towers, or tanks.

**Note to paragraph (a):** Examples of structures where steel erection may occur include but are not limited to the following: Single and multi-story buildings; systems-engineered metal buildings; lift slab/tilt-up structures; energy exploration structures; energy production, transfer and storage structures and facilities; auditoriums; malls; amphitheaters; stadiums; power plants; mills; chemical process structures; bridges; trestles; overpasses; underpasses; viaducts; aqueducts; aerospace facilities and structures; radar and communication structures; light towers; signage; billboards; scoreboards; conveyor systems; conveyor supports and related framing; stairways; stair towers; fire escapes; draft curtains; fire containment structures; monorails; aerialways; catwalks; curtain walls; window walls; store fronts; elevator fronts; entrances; skylights; metal roofs; industrial structures; hi-bay structures; rail, marine and other transportation structures; sound barriers; water process and water containment structures; air and cable supported structures; space frames; geodesic domes; canopies; racks and rack support structures and frames; platforms; walkways; balconies; atriums; penthouses; car dumpers; stackers/reclaimers; cranes and craneways; bins; hoppers; ovens; furnaces; stacks; amusement park structures and rides; and artistic and monumental structures.

**1926.750(b)(1)** Steel erection activities include hoisting, laying out, placing, connecting, welding, burning, guying, bracing, bolting, plumbing and rigging structural steel, steel joists and metal buildings; installing metal decking, curtain walls, window walls, siding systems, miscellaneous metals, ornamental iron and similar materials; and moving point-to-point while performing these activities.

**1926.750(b)(2)** The following activities are covered by this subpart when they occur during and are a part of steel erection activities: rigging, hoisting, laying out, placing, connecting, guying, bracing, dismantling, burning, welding, bolting, grinding, sealing, caulking, and all related activities for construction, alteration and/or repair of materials and assemblies such as structural steel; ferrous metals and alloys; non-ferrous metals and alloys; glass; plastics and synthetic composite materials; structural metal framing and related bracing and assemblies; anchoring devices; structural cabling; cable stays; permanent and temporary bents and towers; falsework for temporary supports of permanent steel members; stone and other non-precast concrete architectural materials mounted on steel frames; safety systems for steel erection; steel and metal joists; metal decking and raceway systems and accessories; metal roofing and accessories; metal siding; bridge flooring; cold formed steel framing; elevator beams; grillage; shelf racks; multi-purpose supports; crane rails and accessories; miscellaneous, architectural and ornamental metals and metal work; ladders; railings; handrails; fences and gates; gratings; trench covers; floor plates; castings; sheet metal fabrications; metal panels and panel wall systems; louvers; column covers; enclosures and pockets; stairs; perforated metals; ornamental iron work, expansion control including bridge expansion joint assemblies; slide bearings; hydraulic structures; fascias; soffit panels; penthouse enclosures; skylights; joint fillers; gaskets; sealants and seals; doors; windows; hardware; detention/security equipment and doors, windows and hardware; conveying systems; building specialties; building equipment; machinery and plant equipment, furnishings and special construction.

**1926.750(c)** The duties of controlling contractors under this subpart include, but are not limited to, the duties specified in §§1926.752 (a) and (c), 1926.755(b)(2), 1926.759(b), and 1926.760(e).

[66 FR 5195, Jan. 18, 2001]

## §1926.751 Definitions.

*Anchored bridging* means that the steel joist bridging is connected to a bridging terminus point.

*Bolted diagonal bridging* means diagonal bridging that is bolted to a steel joist or joists.

*Bridging clip* means a device that is attached to the steel joist to allow the bolting of the bridging to the steel joist.

*Bridging terminus point* means a wall, a beam, tandem joists (with all bridging installed and a horizontal truss in the plane of the top chord) or other element at an end or intermediate point(s) of a line of bridging that provides an anchor point for the steel joist bridging.

*Choker* means a wire rope or synthetic fiber rigging assembly that is used to attach a load to a hoisting device.

*Cold forming* means the process of using press brakes, rolls, or other methods to shape steel into desired cross sections at room temperature.

*Column* means a load-carrying vertical member that is part of the primary skeletal framing system. Columns do not include posts.

*Competent person* (also defined in §1926.32) means one who is capable of identifying existing and predictable hazards in the surroundings or working conditions which are unsanitary, hazardous, or dangerous to employees, and who has authorization to take prompt corrective measures to eliminate them.

*Connector* means an employee who, working with hoisting equipment, is placing and connecting structural members and/or components.

*Constructibility* means the ability to erect structural steel members in accordance with subpart R without having to alter the over-all structural design.

*Construction load (for joist erection)* means any load other than the weight of the employee(s), the joists and the bridging bundle.

*Controlled Decking Zone* (CDZ) means an area in which certain work (for example, initial installation and placement of metal decking) may take place without the use of guardrail systems, personal fall arrest systems, fall restraint systems, or safety net systems and where access to the zone is controlled.

*Controlled load lowering* means lowering a load by means of a mechanical hoist drum device that allows a hoisted load to be lowered with maximum control using the gear train or hydraulic components of the hoist mechanism. Controlled load lowering requires the use of the hoist drive motor, rather than the load hoist brake, to lower the load.

*Controlling contractor* means a prime contractor, general contractor, construction manager or any other legal entity which has the overall responsibility for the construction of the project—its planning, quality and completion.

*Critical lift* means a lift that (1) exceeds 75 percent of the rated capacity of the crane or derrick, or (2) requires the use of more than one crane or derrick.

*Decking hole* means a gap or void more than 2 inches (5.1 cm) in its least dimension and less than 12 inches (30.5 cm) in its greatest dimension in a floor, roof or other walking/working surface. Pre-engineered holes in cellular decking (for wires, cables, etc.) are not included in this definition.

*Derrick floor* means an elevated floor of a building or structure that has been designated to receive hoisted pieces of steel prior to final placement.

*Double connection* means an attachment method where the connection point is intended for two pieces of steel which share common bolts on either side of a central piece.

*Double connection seat* means a structural attachment that, during the installation of a double connection, supports the first member while the second member is connected.

*Erection bridging* means the bolted diagonal bridging that is required to be installed prior to releasing the hoisting cables from the steel joists.

*Fall restraint system* means a fall protection system that prevents the user from falling any distance. The system is comprised of either a body belt or body harness, along with an anchorage, connectors and other necessary equipment. The other components typically include a lanyard, and may also include a lifeline and other devices.

*Final interior perimeter* means the perimeter of a large permanent open space within a building such as an atrium or courtyard. This does not include openings for stairways, elevator shafts, etc.

*Girt (in systems-engineered metal buildings)* means a "Z" or "C" shaped member formed from sheet steel spanning between primary framing and supporting wall material.

*Headache ball* means a weighted hook that is used to attach loads to the hoist load line of the crane.

*Hoisting equipment* means commercially manufactured lifting equipment designed to lift and position a load of known weight to a location at some known elevation and horizontal distance from the equipment's center of rotation. "Hoisting equipment" includes but is not limited to cranes, derricks, tower cranes, barge-mounted derricks or cranes, gin poles and gantry hoist systems. A "come-a-long" (a mechanical device, usually consisting of a chain or cable attached at each end, that is used to facilitate movement of materials through leverage) is not considered "hoisting equipment."

*Leading edge* means the unprotected side and edge of a floor, roof, or formwork for a floor or other walking/working surface (such as deck) which changes location as additional floor, roof, decking or formwork sections are placed, formed or constructed.

*Metal decking* means a commercially manufactured, structural grade, cold rolled metal panel formed into a series of parallel ribs; for this subpart, this includes metal floor and roof decks, standing seam metal roofs, other metal roof systems and other products such as bar gratings, checker plate, expanded metal panels, and similar products. After installation and proper fastening, these decking materials serve a combination of functions including, but not limited to: a structural element designed in combination with the structure to resist, distribute and transfer loads, stiffen the structure and provide a diaphragm action; a walking/working surface; a form for concrete slabs; a support for roofing systems; and a finished floor or roof.

*Multiple lift rigging* means a rigging assembly manufactured by wire rope rigging suppliers that facilitates the attachment of up to five independent loads to the hoist rigging of a crane.

*Opening* means a gap or void 12 inches (30.5 cm) or more in its least dimension in a floor, roof or other walking/working surface. For the purposes of this subpart, skylights and smoke domes that do not meet the strength requirements of §1926.754(e)(3) shall be regarded as openings.

*Permanent floor* means a structurally completed floor at any level or elevation (including slab on grade).

*Personal fall arrest system* means a system used to arrest an employee in a fall from a working level. A personal fall arrest system consists of an anchorage, connectors, a body harness and may include a lanyard, deceleration device, lifeline, or suitable combination of these. The use of a body belt for fall arrest is prohibited.

*Positioning device system* means a body belt or body harness rigged to allow an employee to be supported on an elevated, vertical surface, such as a wall or column and work with both hands free while leaning.

*Post* means a structural member with a longitudinal axis that is essentially vertical, that: (1) weighs 300 pounds or less and is axially loaded (a load presses down on the top end), or (2) is not axially loaded, but is laterally restrained by the above member. Posts typically support stair landings, wall framing, mezzanines and other substructures.

*Project structural engineer of record* means the registered, licensed professional responsible for the design of structural steel framing and whose seal appears on the structural contract documents.

*Purlin (in systems-engineered metal buildings)* means a "Z" or "C" shaped member formed from sheet steel spanning between primary framing and supporting roof material.

*Qualified person* (also defined in §1926.32) means one who, by possession of a recognized degree, certificate, or professional standing, or who by extensive knowledge, training, and experience, has successfully demonstrated the ability to solve or resolve problems relating to the subject matter, the work, or the project.

*Safety deck attachment* means an initial attachment that is used to secure an initially placed sheet of decking to keep proper alignment and bearing with structural support members.

*Shear connector* means headed steel studs, steel bars, steel lugs, and similar devices which are attached to a structural member for the purpose of achieving composite action with concrete.

*Steel erection* means the construction, alteration or repair of steel buildings, bridges and other structures, including the installation of metal decking and all planking used during the process of erection.

*Steel joist* means an open web, secondary load-carrying member of 144 feet (43.9 m) or less, designed by the manufacturer, used for the support of floors and roofs. This does not include structural steel trusses or cold-formed joists.

*Steel joist girder* means an open web, primary load-carrying member, designed by the manufacturer, used for the support of floors and roofs. This does not include structural steel trusses.

*Steel truss* means an open web member designed of structural steel components by the project structural engineer of record. For the purposes of this subpart, a steel truss is considered equivalent to a solid web structural member.

*Structural steel* means a steel member, or a member made of a substitute material (such as, but not limited to, fiberglass, aluminum or composite members). These members include, but are not limited to, steel joists, joist girders, purlins, columns, beams, trusses, splices, seats, metal decking, girts, and all bridging, and cold formed metal framing which is integrated with the structural steel framing of a building.

*Systems-engineered metal building* means a metal, field-assembled building system consisting of framing, roof and wall coverings. Typically, many of these components are cold-formed shapes. These individual parts are fabricated in one or more manufacturing facilities and shipped to the job site for assembly into the final structure. The engineering design of the system is normally the responsibility of the systems-engineered metal building manufacturer.

*Tank* means a container for holding gases, liquids or solids.

*Unprotected sides and edges* means any side or edge (except at entrances to points of access) of a walking/working surface, for example a, floor, roof, ramp or runway, where there is no wall or guardrail system at least 39 inches (1.0 m) high.

[66 FR 5195, Jan. 18, 2001]

### §1926.752 Site layout, site-specific erection plan and construction sequence.

**1926.752(a)** ***Approval to begin steel erection.*** Before authorizing the commencement of steel erection, the controlling contractor shall ensure that the steel erector is provided with the following written notifications:

**1926.752(a)(1)** The concrete in the footings, piers and walls and the mortar in the masonry piers and walls has attained, on the basis of an appropriate ASTM standard test method of field-cured samples, either 75 percent of the intended minimum compressive design strength or sufficient strength to support the loads imposed during steel erection.

**1926.752(a)(2)** Any repairs, replacements and modifications to the anchor bolts were conducted in accordance with §1926.755(b).

**1926.752(b)** ***Commencement of steel erection.*** A steel erection contractor shall not erect steel unless it has received written notification that the concrete in the footings, piers and walls or the mortar in the masonry piers and walls has attained, on the basis of an appropriate ASTM standard test method of field-cured samples, either 75 percent of the intended minimum compressive design strength or sufficient strength to support the loads imposed during steel erection.

**1926.752(c)** ***Site layout.*** The controlling contractor shall ensure that the following is provided and maintained:

**1926.752(c)(1)** Adequate access roads into and through the site for the safe delivery and movement of derricks, cranes, trucks, other necessary equipment, and the material to be erected and means and methods for pedestrian and vehicular control. Exception: this requirement does not apply to roads outside of the construction site.

**1926.752(c)(2)** A firm, properly graded, drained area, readily accessible to the work with adequate space for the safe storage of materials and the safe operation of the erector's equipment.

**1926.752(d)** ***Pre-planning of overhead hoisting operations.*** All hoisting operations in steel erection shall be pre-planned to ensure that the requirements of §1926.753(d) are met.

**1926.752(e)** ***Site-specific erection plan.*** Where employers elect, due to conditions specific to the site, to develop alternate means and methods that provide employee protection in accordance with §1926.753(c)(5), §1926.757(a)(4) or §1926.757(e)(4), a site-specific erection plan shall be developed by a qualified person and be available at the work site. Guidelines for establishing a site-specific erection plan are contained in Appendix A to this subpart.

[66 FR 5195, Jan. 18, 2001]

## §1926.753 Hoisting and rigging.

**1926.753(a)** All the provisions of §1926.550 apply to hoisting and rigging with the exception of §1926.550(g)(2).

**1926.753(b)** In addition, paragraphs (c) through (e) of this section apply regarding the hazards associated with hoisting and rigging.

**1926.753(c)** ***General.*** (1) Pre-shift visual inspection of cranes.

**1926.753(c)(1)(i)** Cranes being used in steel erection activities shall be visually inspected prior to each shift by a competent person; the inspection shall include observation for deficiencies during operation. At a minimum this inspection shall include the following:

**1926.753(c)(1)(i)(A)** All control mechanisms for maladjustments;

**1926.753(c)(1)(i)(B)** Control and drive mechanism for excessive wear of components and contamination by lubricants, water or other foreign matter;

**1926.753(c)(1)(i)(C)** Safety devices, including but not limited to boom angle indicators, boom stops, boom kick out devices, anti-two block devices, and load moment indicators where required;

**1926.753(c)(1)(i)(D)** Air, hydraulic, and other pressurized lines for deterioration or leakage, particularly those which flex in normal operation;

**1926.753(c)(1)(i)(E)** Hooks and latches for deformation, chemical damage, cracks, or wear;

**1926.753(c)(1)(i)(F)** Wire rope reeving for compliance with hoisting equipment manufacturer's specifications;

**1926.753(c)(1)(i)(G)** Electrical apparatus for malfunctioning, signs of excessive deterioration, dirt, or moisture accumulation;

**1926.753(c)(1)(i)(H)** Hydraulic system for proper fluid level;

**1926.753(c)(1)(i)(I)** Tires for proper inflation and condition;

**1926.753(c)(1)(i)(J)** Ground conditions around the hoisting equipment for proper support, including ground settling under and around outriggers, ground water accumulation, or similar conditions;

**1926.753(c)(1)(i)(K)** The hoisting equipment for level position; and

**1926.753(c)(1)(i)(L)** The hoisting equipment for level position after each move and setup.

**1926.753(c)(1)(ii)** If any deficiency is identified, an immediate determination shall be made by the competent person as to whether the deficiency constitutes a hazard.

**1926.753(c)(1)(iii)** If the deficiency is determined to constitute a hazard, the hoisting equipment shall be removed from service until the deficiency has been corrected.

**1926.753(c)(1)(iv)** The operator shall be responsible for those operations under the operator's direct control. Whenever there is any doubt as to safety, the operator shall have the authority to stop and refuse to handle loads until safety has been assured.

**1926.753(c)(2)** A qualified rigger (a rigger who is also a qualified person) shall inspect the rigging prior to each shift in accordance with §1926.251.

**1926.753(c)(3)** The headache ball, hook or load shall not be used to transport personnel except as provided in paragraph (c)(4) of this section.

**1926.753(c)(4)** Cranes or derricks may be used to hoist employees on a personnel platform when work under this subpart is being conducted, provided that all provisions of §1926.550 (except for §1926.550(g)(2)) are met.

**1926.753(c)(5)** Safety latches on hooks shall not be deactivated or made inoperable except:

**1926.753(c)(5)(i)** When a qualified rigger has determined that the hoisting and placing of purlins and single joists can be performed more safely by doing so; or

**1926.753(c)(5)(ii)** When equivalent protection is provided in a site-specific erection plan.

**1926.753(d)** ***Working under loads.***

**1926.753(d)(1)** Routes for suspended loads shall be pre-planned to ensure that no employee is required to work directly below a suspended load except for:

**1926.753(d)(1)(i)** Employees engaged in the initial connection of the steel; or

**1926.753(d)(1)(ii)** Employees necessary for the hooking or unhooking of the load.

**1926.753(d)(2)** When working under suspended loads, the following criteria shall be met:

**1926.753(d)(2)(i)** Materials being hoisted shall be rigged to prevent unintentional displacement;

**1926.753(d)(2)(ii)** Hooks with self-closing safety latches or their equivalent shall be used to prevent components from slipping out of the hook; and

**1926.753(d)(2)(iii)** All loads shall be rigged by a qualified rigger

**1926.753(e)** ***Multiple lift rigging procedure.***

**1926.753(e)(1)** A multiple lift shall only be performed if the following criteria are met:

**1926.753(e)(1)(i)** A multiple lift rigging assembly is used;

**1926.753(e)(1)(ii)** A maximum of five members are hoisted per lift;

**1926.753(e)(1)(iii)** Only beams and similar structural members are lifted; and

**1926.753(e)(1)(iv)** All employees engaged in the multiple lift have been trained in these procedures in accordance with §1926.761(c)(1).

**1926.753(e)(1)(v)** No crane is permitted to be used for a multiple lift where such use is contrary to the manufacturer's specifications and limitations.

**1926.753(e)(2)** Components of the multiple lift rigging assembly shall be specifically designed and assembled with a maximum capacity for total assembly and for each individual attachment point. This capacity, certified by the manufacturer or a qualified rigger, shall be based on the manufacturer's specifications with a 5 to 1 safety factor for all components.

**1926.753(e)(3)** The total load shall not exceed:

**1926.753(e)(3)(i)** The rated capacity of the hoisting equipment specified in the hoisting equipment load charts;

**1926.753(e)(3)(ii)** The rigging capacity specified in the rigging rating chart.

**1926.753(e)(4)** The multiple lift rigging assembly shall be rigged with members:

**1926.753(e)(4)(i)** Attached at their center of gravity and maintained reasonably level;

**1926.753(e)(4)(ii)** Rigged from top down; and

**1926.753(e)(4)(iii)** Rigged at least 7 feet (2.1 m) apart.

**1926.753(e)(5)** The members on the multiple lift rigging assembly shall be set from the bottom up.

**1926.753(e)(6)** Controlled load lowering shall be used whenever the load is over the connectors.

[66 FR 5195, Jan. 18, 2001]

## §1926.754 Structural steel assembly.

**1926.754(a)** Structural stability shall be maintained at all times during the erection process.

**1926.754(b)** The following additional requirements shall apply for multistory structures:

**1926.754(b)(1)** The permanent floors shall be installed as the erection of structural members progresses, and there shall be not more than eight stories between the erection floor and the upper-most permanent floor, except where the structural integrity is maintained as a result of the design.

**1926.754(b)(2)** At no time shall there be more than four floors or 48 feet (14.6 m), whichever is less, of unfinished bolting or welding above the foundation or uppermost permanently secured floor, except where the structural integrity is maintained as a result of the design.

**1926.754(b)(3)** A fully planked or decked floor or nets shall be maintained within two stories or 30 feet (9.1 m), whichever is less, directly under any erection work being performed.

**1926.754(c)** ***Walking/working surfaces.***

**1926.754(c)(1)** *Shear connectors and other similar devices.*

**1926.754(c)(1)(i)** *Tripping hazards.* Shear connectors (such as headed steel studs, steel bars or steel lugs), reinforcing bars, deformed anchors or threaded studs shall not be attached to the top flanges of beams, joists or beam attachments so that they project vertically from or horizontally across the top flange of the member until after the metal decking, or other walking/working surface, has been installed.

**1926.754(c)(1)(ii)** *Installation of shear connectors on composite floors, roofs and bridge decks.* When shear connectors are used in construction of composite floors, roofs and bridge decks, employees shall lay out and install the shear connectors after the metal decking has been installed, using the metal decking as a working platform. Shear connectors shall not be installed from within a controlled decking zone (CDZ), as specified in §1926.760(c)(8).

**1926.754(c)(2)** *Slip resistance of metal decking.* [Reserved]

**1926.754(c)(3)** *Slip resistance of skeletal structural steel.* Workers shall not be permitted to walk the top surface of any structural steel member installed after July 18, 2006 that has been coated with paint or similar material unless documentation or certification that the coating has achieved a minimum average slip resistance of .50 when measured with an English XL tribometer or equivalent tester on a wetted surface at a testing laboratory is provided. Such documentation or certification shall be based on the appropriate ASTM standard test method conducted by a laboratory capable of performing the test. The results shall be available at the site and to the steel erector. (Appendix B to this subpart references appropriate ASTM standard test methods that may be used to comply with this paragraph (c)(3)).

**1926.754(d)** ***Plumbing-up.***

**1926.754(d)(1)** When deemed necessary by a competent person, plumbing-up equipment shall be installed in conjunction with the steel erection process to ensure the stability of the structure.

**1926.754(d)(2)** When used, plumbing-up equipment shall be in place and properly installed before the structure is loaded with construction material such as loads of joists, bundles of decking or bundles of bridging.

**1926.754(d)(3)** Plumbing-up equipment shall be removed only with the approval of a competent person.

**1926.754(e)** ***Metal decking.***—(1) *Hoisting, landing and placing of metal decking bundles.*

**1926.754(e)(1)(i)** Bundle packaging and strapping shall not be used for hoisting unless specifically designed for that purpose.

**1926.754(e)(1)(ii)** If loose items such as dunnage, flashing, or other materials are placed on the top of metal decking bundles to be hoisted, such items shall be secured to the bundles.

**1926.754(e)(1)(iii)** Bundles of metal decking on joists shall be landed in accordance with §1926.757(e)(4).

**1926.754(e)(1)(iv)** Metal decking bundles shall be landed on framing members so that enough support is provided to allow the bundles to be unbanded without dislodging the bundles from the supports.

**1926.754(e)(1)(v)** At the end of the shift or when environmental or jobsite conditions require, metal decking shall be secured against displacement.

**1926.754(e)(2)** *Roof and floor holes and openings.* Metal decking at roof and floor holes and openings shall be installed as follows:

**1926.754(e)(2)(i)** Framed metal deck openings shall have structural members turned down to allow continuous deck installation except where not allowed by structural design constraints or constructibility.

**1926.754(e)(2)(ii)** Roof and floor holes and openings shall be decked over. Where large size, configuration or other structural design does not allow openings to be decked over (such as elevator shafts, stair wells, etc.) employees shall be protected in accordance with §1926.760(a)(1).

**1926.754(e)(2)(iii)** Metal decking holes and openings shall not be cut until immediately prior to being permanently filled with the equipment or structure needed or intended to fulfill its specific use and which meets the strength requirements of paragraph (e)(3) of this section, or shall be immediately covered.

**1926.754(e)(3)** *Covering roof and floor openings.*

**1926.754(e)(3)(i)** Covers for roof and floor openings shall be capable of supporting, without failure, twice the weight of the employees, equipment and materials that may be imposed on the cover at any one time.

**1926.754(e)(3)(ii)** All covers shall be secured when installed to prevent accidental displacement by the wind, equipment or employees.

**1926.754(e)(3)(iii)** All covers shall be painted with high-visibility paint or shall be marked with the word "HOLE" or "COVER" to provide warning of the hazard.

**1926.754(e)(3)(iv)** Smoke dome or skylight fixtures that have been installed, are not considered covers for the purpose of this section unless they meet the strength requirements of paragraph (e)(3)(i) of this section.

**1926.754(e)(4)** *Decking gaps around columns.* Wire mesh, exterior plywood, or equivalent, shall be installed around columns where planks or metal decking do not fit tightly. The materials used must be of sufficient strength to provide fall protection for personnel and prevent objects from falling through.

**1926.754(e)(5)** *Installation of metal decking.* (i) Except as provided in §1926.760(c), metal decking shall be laid tightly and immediately secured upon placement to prevent accidental movement or displacement.

**1926.754(e)(5)(ii)** During initial placement, metal decking panels shall be placed to ensure full support by structural members.

**1926.754(e)(6)** Derrick floors. (i) A derrick floor shall be fully decked and/or planked and the steel member connections completed to support the intended floor loading.

**1926.754(e)(6)(ii)** Temporary loads placed on a derrick floor shall be distributed over the underlying support members so as to prevent local overloading of the deck material.

[66 FR 5195, Jan. 18, 2001]

## §1926.755 Column anchorage.

**1926.755(a)** ***General requirements for erection stability.*** (1) All columns shall be anchored by a minimum of 4 anchor rods (anchor bolts).

**1926.755(a)(2)** Each column anchor rod (anchor bolt) assembly, including the column-to-base plate weld and the column foundation, shall be designed to resist a minimum eccentric gravity load of 300 pounds (136.2 kg) located 18 inches (.46m) from the extreme outer face of the column in each direction at the top of the column shaft.

**1926.755(a)(3)** Columns shall be set on level finished floors, pre-grouted leveling plates, leveling nuts, or shim packs which are adequate to transfer the construction loads.

**1926.755(a)(4)** All columns shall be evaluated by a competent person to determine whether guying or bracing is needed; if guying or bracing is needed, it shall be installed.

**1926.755(b)** ***Repair, replacement or field modification of anchor rods (anchor bolts).***

**1926.755(b)(1)** Anchor rods (anchor bolts) shall not be repaired, replaced or field-modified without the approval of the project structural engineer of record.

**1926.755(b)(2)** Prior to the erection of a column, the controlling contractor shall provide written notification to the steel erector if there has been any repair, replacement or modification of the anchor rods (anchor bolts) of that column.

[66 FR 5195, Jan. 18, 2001]

## §1926.756 Beams and columns.

**1926.756(a)** ***General.*** (1) During the final placing of solid web structural members, the load shall not be released from the hoisting line until the members are secured with at least two bolts per connection, of the same size and strength as shown in the erection drawings, drawn up wrench-tight or the equivalent as specified by the project structural engineer of record, except as specified in paragraph (b) of this section.

**1926.756(a)(2)** A competent person shall determine if more than two bolts are necessary to ensure the stability of cantilevered members; if additional bolts are needed, they shall be installed.

**1926.756(b)** ***Diagonal bracing.*** Solid web structural members used as diagonal bracing shall be secured by at least one bolt per connection drawn up wrench-tight or the equivalent as specified by the project structural engineer of record.

**1926.756(c)(1)** ***Double connections at columns and/or at beam webs over a column.*** When two structural members on opposite sides of a column web, or a beam web over a column, are connected sharing common connection holes, at least one bolt with its wrench-tight nut shall remain connected to the first member unless a shop-attached or field-attached seat or equivalent connection device is supplied with the member to secure the first member and prevent the column from being displaced (See Appendix H to this subpart for examples of equivalent connection devices).

**1926.756(c)(2)** If a seat or equivalent device is used, the seat (or device) shall be designed to support the load during the double connection process. It shall be adequately bolted or welded to both a supporting member and the first member before the nuts on the shared bolts are removed to make the double connection.

**1926.756(d)** ***Column splices.*** Each column splice shall be designed to resist a minimum eccentric gravity load of 300 pounds (136.2 kg) located 18 inches (.46 m) from the extreme outer face of the column in each direction at the top of the column shaft.

**1926.756(e)** ***Perimeter columns.*** Perimeter columns shall not be erected unless:

**1926.756(e)(1)** The perimeter columns extend a minimum of 48 inches (1.2 m) above the finished floor to permit installation of perimeter safety cables prior to erection of the next tier, except where constructibility does not allow (see Appendix F to this subpart);

**1926.756(e)(2)** The perimeter columns have holes or other devices in or attached to perimeter columns at 42-45 inches (107-114 cm) above the finished floor and the midpoint between the finished floor and the top cable to permit installation of perimeter safety cables required by §1926.760(a)(2), except where constructibility does not allow. (See Appendix F to this subpart).

[66 FR 5195, Jan. 18, 2001]

## §1926.757 Open web steel joists.

**1926.757(a)** ***General.*** (1) Except as provided in paragraph (a)(2) of this section, where steel joists are used and columns are not framed in at least two directions with solid web structural steel members, a steel joist shall be field-bolted at the column to provide lateral stability to the column during erection. For the installation of this joist:

**1926.757(a)(1)(i)** A vertical stabilizer plate shall be provided on each column for steel joists. The plate shall be a minimum of 6 inch by 6 inch (152 mm by 152 mm) and shall extend at least 3 inches (76 mm) below the bottom chord of the joist with a 13/16 inch (21 mm) hole to provide an attachment point for guying or plumbing cables.

**1926.757(a)(1)(ii)** The bottom chords of steel joists at columns shall be stabilized to prevent rotation during erection.

**1926.757(a)(1)(iii)** Hoisting cables shall not be released until the seat at each end of the steel joist is field-bolted, and each end of the bottom chord is restrained by the column stabilizer plate.

**1926.757(a)(2)** Where constructibility does not allow a steel joist to be installed at the column:

**1926.757(a)(2)(i)** an alternate means of stabilizing joists shall be installed on both sides near the column and shall:

**1926.757(a)(2)(i)(A)** provide stability equivalent to paragraph (a)(1) of this section;

**1926.757(a)(2)(i)(B)** be designed by a qualified person;

**1926.757(a)(2)(i)(C)** be shop installed; and

**1926.757(a)(2)(i)(D)** be included in the erection drawings.

**1926.757(a)(2)(ii)** hoisting cables shall not be released until the seat at each end of the steel joist is field-bolted and the joist is stabilized.

**1926.757(a)(3)** Where steel joists at or near columns span 60 feet (18.3 m) or less, the joist shall be designed with sufficient strength to allow one employee to release the hoisting cable without the need for erection bridging.

**1926.757(a)(4)** Where steel joists at or near columns span more than 60 feet (18.3 m), the joists shall be set in tandem with all bridging installed unless an alternative method of erection, which provides equivalent stability to the steel joist, is designed by a qualified person and is included in the site-specific erection plan.

**1926.757(a)(5)** A steel joist or steel joist girder shall not be placed on any support structure unless such structure is stabilized.

**1926.757(a)(6)** When steel joist(s) are landed on a structure, they shall be secured to prevent unintentional displacement prior to installation.

**1926.757(a)(7)** No modification that affects the strength of a steel joist or steel joist girder shall be made without the approval of the project structural engineer of record.

**1926.757(a)(8)** *Field-bolted joists.* (i) Except for steel joists that have been pre-assembled into panels, connections of individual steel joists to steel structures in bays of 40 feet (12.2 m) or more shall be fabricated to allow for field bolting during erection.

**1926.757(a)(8)(ii)** These connections shall be field-bolted unless constructibility does not allow.

**1926.757(a)(9)** Steel joists and steel joist girders shall not be used as anchorage points for a fall arrest system unless written approval to do so is obtained from a qualified person.

**1926.757(a)(10)** A bridging terminus point shall be established before bridging is installed. (See Appendix C to this subpart.)

**1926.757(b)** ***Attachment of steel joists and steel joist girders.*** (1) Each end of "K" series steel joists shall be attached to the support structure with a minimum of two 1/8-inch (3 mm) fillet welds 1 inch (25 mm) long or with two 1/2-inch (13 mm) bolts, or the equivalent.

**1926.757(b)(2)** Each end of "LH" and "DLH" series steel joists and steel joist girders shall be attached to the support structure with a minimum of two 1/4-inch (6 mm) fillet welds 2 inches (51 mm) long, or with two 3/4-inch (19 mm) bolts, or the equivalent.

**1926.757(b)(3)** Except as provided in paragraph (b)(4) of this section, each steel joist shall be attached to the support structure, at least at one end on both sides of the seat, immediately upon placement in the final erection position and before additional joists are placed.

**1926.757(b)(4)** Panels that have been pre-assembled from steel joists with bridging shall be attached to the structure at each corner before the hoisting cables are released.

**1926.757(c)** ***Erection of steel joists.*** (1) Both sides of the seat of one end of each steel joist that requires bridging under Tables A and B shall be attached to the support structure before hoisting cables are released.

**1926.757(c)(2)** For joists over 60 feet, both ends of the joist shall be attached as specified in paragraph (b) of this section and the provisions of paragraph (d) of this section met before the hoisting cables are released.

**1926.757(c)(3)** On steel joists that do not require erection bridging under Tables A and B, only one employee shall be allowed on the joist until all bridging is installed and anchored.

### TABLE A.—ERECTION BRIDGING FOR SHORT SPAN JOISTS

| Joist | Span |
|---|---|
| 8L1 | NM |
| 10K1 | NM |
| 12K1 | 23-0 |
| 12K3 | NM |
| 12K5 | NM |
| 14K1 | 27-0 |
| 14K3 | NM |
| 14K4 | NM |
| 14K6 | NM |
| 16K2 | 29-0 |
| 16K3 | 30-0 |
| 16K4 | 32-0 |
| 16K5 | 32-0 |
| 16K6 | NM |
| 16K7 | NM |
| 16K9 | NM |
| 18K3 | 31-0 |
| 18K4 | 32-0 |
| 18K5 | 33-0 |
| 18K6 | 35-0 |
| 18K7 | NM |
| 18K9 | NM |
| 18K10 | NM |
| 20K3 | 32-0 |
| 20K4 | 34-0 |
| 20K5 | 34-0 |

| Joist | Span |
|---|---|
| 20K6 | 36-0 |
| 20K7 | 39-0 |
| 20K9 | 39-0 |
| 20K10 | NM |
| 22K4 | 34-0 |
| 22K5 | 35-0 |
| 22K6 | 36-0 |
| 22K7 | 40-0 |
| 22K9 | 40-0 |
| 22K10 | 40-0 |
| 22K11 | 40-0 |
| 24K4 | 36-0 |
| 24K5 | 38-0 |
| 24K6 | 39-0 |
| 24K7 | 43-0 |
| 24K8 | 43-0 |
| 24K9 | 44-0 |
| 24K10 | NM |
| 24K12 | NM |
| 26K5 | 38-0 |
| 26K6 | 39-0 |
| 26K7 | 43-0 |
| 26K8 | 44-0 |
| 26K9 | 45-0 |
| 26K10 | 49-0 |
| 26K12 | NM |
| 28K6 | 40-0 |
| 28K7 | 43-0 |
| 28K8 | 44-0 |
| 28K9 | 45-0 |
| 28K10 | 49-0 |
| 28K12 | 53-0 |
| 30K7 | 44-0 |
| 30K8 | 45-0 |
| 30K9 | 45-0 |
| 30K10 | 50-0 |
| 30K11 | 52-0 |
| 30K12 | 54-0 |
| 10KCS1 | NM |
| 10KCS2 | NM |
| 10KCS3 | NM |
| 12KCS1 | NM |
| 12KCS2 | NM |
| 12KCS3 | NM |
| 14KCS1 | NM |
| 14KCS2 | NM |
| 14KCS3 | NM |
| 16KCS2 | NM |
| 16KCS3 | NM |
| 16KCS4 | NM |
| 16KCS5 | NM |
| 18KCS2 | 35-0 |
| 18KCS3 | NM |
| 18KCS4 | NM |
| 18KCS5 | NM |
| 20KCS2 | 36-0 |
| 20KCS3 | 39-0 |
| 20KCS4 | NM |
| 20KCS5 | NM |
| 22KCS2 | 36-0 |
| 22KCS3 | 40-0 |
| 22KCS4 | NM |
| 22KCS5 | NM |
| 24KCS2 | 39-0 |
| 24KCS3 | 44-0 |
| 24KCS4 | NM |
| 24KCS5 | NM |
| 26KCS2 | 39-0 |
| 26KCS3 | 44-0 |
| 26KCS4 | NM |
| 26KCS5 | NM |
| 28KCS2 | 40-0 |
| 28KCS3 | 45-0 |
| 28KCS4 | 53-0 |
| 28KCS5 | 53-0 |
| 30KC53 | 45-0 |
| 30KCS4 | 54-0 |
| 30KCS5 | 54-0 |

NM = diagonal bolted bridging not mandatory for joists under 40 feet.

## TABLE B.—ERECTION BRIDGING FOR LONG SPAN JOISTS

| Joist | Span |
|---|---|
| 18LH02 | 33-0. |
| 18LH03 | NM. |
| 18LH04 | NM. |
| 18LH05 | NM. |
| 18LH06 | NM. |
| 18LH07 | NM. |
| 18LH08 | NM. |
| 18LH09 | NM. |
| 20LH02 | 33-0. |
| 20LH03 | 38-0. |
| 20LH04 | NM. |
| 20LH05 | NM. |
| 20LH06 | NM. |
| 20LH07 | NM. |
| 20LH08 | NM. |
| 20LH09 | NM. |
| 20LH10 | NM. |
| 24LH03 | 35-0. |
| 24LH04 | 39-0. |
| 24LH05 | 40-0. |
| 24LH06 | 45-0. |
| 24LH07 | NM. |
| 24LH08 | NM. |
| 24LH09 | NM. |
| 24LH10 | NM. |
| 24LH11 | NM. |
| 28LH05 | 42-0. |
| 28LH06 | 42-0. |
| 28LH07 | NM. |
| 28LH08 | NM. |
| 28LH09 | NM. |
| 28LH10 | NM. |
| 28LH11 | NM. |
| 28LH12 | NM. |
| 28LH13 | NM. |
| 32LH06 | 47-0 through 60-0. |
| 32LH07 | 47-0 through 60-0. |
| 32LH08 | 55-0 through 60-0. |
| 32LH09 | NM through 60-0. |
| 32LH10 | NM through 60-0. |
| 32LH11 | NM through 60-0. |
| 32LH12 | NM through 60-0. |
| 32LH13 | NM through 60-0. |
| 32LH14 | NM through 60-0. |
| 32LH15 | NM through 60-0. |
| 36LH07 | 47-0 through 60-0. |
| 36LH08 | 47-0 through 60-0. |
| 36LH09 | 57-0 through 60-0. |
| 36LH10 | NM through 60-0. |
| 36LH11 | NM through 60-0. |

| Joist | Span |
|---|---|
| 36LH12 | NM through 60-0. |
| 36LH13 | NM through 60-0. |
| 36LH14 | NM through 60-0. |
| 36LH15 | NM through 60-0. |

NM = diagonal bolted bridging not mandatory for joists under 40 feet.

**1926.757(c)(4)** Employees shall not be allowed on steel joists where the span of the steel joist is equal to or greater than the span shown in Tables A and B except in accordance with §1926.757(d).

**1926.757(c)(5)** When permanent bridging terminus points cannot be used during erection, additional temporary bridging terminus points are required to provide stability. (See appendix C of this subpart.)

**1926.757(d)** ***Erection bridging.*** (1) Where the span of the steel joist is equal to or greater than the span shown in Tables A and B, the following shall apply:

**1926.757(d)(1)(i)** A row of bolted diagonal erection bridging shall be installed near the midspan of the steel joist;

**1926.757(d)(1)(ii)** Hoisting cables shall not be released until this bolted diagonal erection bridging is installed and anchored; and

**1926.757(d)(1)(iii)** No more than one employee shall be allowed on these spans until all other bridging is installed and anchored.

**1926.757(d)(2)** Where the span of the steel joist is over 60 feet (18.3 m) through 100 feet (30.5 m), the following shall apply:

**1926.757(d)(2)(i)** All rows of bridging shall be bolted diagonal bridging;

**1926.757(d)(2)(ii)** Two rows of bolted diagonal erection bridging shall be installed near the third points of the steel joist;

**1926.757(d)(2)(iii)** Hoisting cables shall not be released until this bolted diagonal erection bridging is installed and anchored; and

**1926.757(d)(2)(iv)** No more than two employees shall be allowed on these spans until all other bridging is installed and anchored.

**1926.757(d)(3)** Where the span of the steel joist is over 100 feet (30.5 m) through 144 feet (43.9 m), the following shall apply:

**1926.757(d)(3)(i)** All rows of bridging shall be bolted diagonal bridging;

**1926.757(d)(3)(ii)** Hoisting cables shall not be released until all bridging is installed and anchored; and

**1926.757(d)(3)(iii)** No more than two employees shall be allowed on these spans until all bridging is installed and anchored.

**1926.757(d)(4)** For steel members spanning over 144 feet (43.9 m), the erection methods used shall be in accordance with §1926.756.

**1926.757(d)(5)** Where any steel joist specified in paragraphs (c)(2) and (d)(1), (d)(2), and (d)(3) of this section is a bottom chord bearing joist, a row of bolted diagonal bridging shall be provided near the support(s). This bridging shall be installed and anchored before the hoisting cable(s) is released.

**1926.757(d)(6)** When bolted diagonal erection bridging is required by this section, the following shall apply:

**1926.757(d)(6)(i)** The bridging shall be indicated on the erection drawing;

**1926.757(d)(6)(ii)** The erection drawing shall be the exclusive indicator of the proper placement of this bridging;

**1926.757(d)(6)(iii)** Shop-installed bridging clips, or functional equivalents, shall be used where the bridging bolts to the steel joists;

**1926.757(d)(6)(iv)** When two pieces of bridging are attached to the steel joist by a common bolt, the nut that secures the first piece of bridging shall not be removed from the bolt for the attachment of the second; and

**1926.757(d)(6)(v)** Bridging attachments shall not protrude above the top chord of the steel joist.

**1926.757(e)** ***Landing and placing loads.*** (1) During the construction period, the employer placing a load on steel joists shall ensure that the load is distributed so as not to exceed the carrying capacity of any steel joist.

**1926.757(e)(2)** Except for paragraph (e)(4) of this section, no construction loads are allowed on the steel joists until all bridging is installed and anchored and all joist-bearing ends are attached.

**1926.757(e)(3)** The weight of a bundle of joist bridging shall not exceed a total of 1,000 pounds (454 kg). A bundle of joist bridging shall be placed on a minimum of three steel joists that are secured at one end. The edge of the bridging bundle shall be positioned within 1 foot (.30 m) of the secured end.

**1926.757(e)(4)** No bundle of decking may be placed on steel joists until all bridging has been installed and anchored and all joist bearing ends attached, unless all of the following conditions are met:

**1926.757(e)(4)(i)** The employer has first determined from a qualified person and documented in a site-specific erection plan that the structure or portion of the structure is capable of supporting the load;

**1926.757(e)(4)(ii)** The bundle of decking is placed on a minimum of three steel joists;

**1926.757(e)(4)(iii)** The joists supporting the bundle of decking are attached at both ends;

**1926.757(e)(4)(iv)** At least one row of bridging is installed and anchored;

**1926.757(e)(4)(v)** The total weight of the bundle of decking does not exceed 4,000 pounds (1816 kg); and

**1926.757(e)(4)(vi)** Placement of the bundle of decking shall be in accordance with paragraph (e)(5) of this section.

**1926.757(e)(5)** The edge of the construction load shall be placed within 1 foot (.30 m) of the bearing surface of the joist end.

[66 FR 5195, Jan. 18, 2001]

## §1926.758 Systems-engineered metal buildings.

**1926.758(a)** All of the requirements of this subpart apply to the erection of systems-engineered metal buildings except §§1926.755 (column anchorage) and 1926.757 (open web steel joists).

**1926.758(b)** Each structural column shall be anchored by a minimum of four anchor rods (anchor bolts).

**1926.758(c)** Rigid frames shall have 50 percent of their bolts or the number of bolts specified by the manufacturer (whichever is greater) installed and tightened on both sides of the web adjacent to each flange before the hoisting equipment is released.

**1926.758(d)** Construction loads shall not be placed on any structural steel framework unless such framework is safely bolted, welded or otherwise adequately secured.

**1926.758(e)** In girt and eave strut-to-frame connections, when girts or eave struts share common connection holes, at least one bolt with its wrench-tight nut shall remain connected to the first member unless a manufacturer-supplied, field-attached seat or similar connection device is present to secure the first member so that the girt or eave strut is always secured against displacement.

**1926.758(f)** Both ends of all steel joists or cold-formed joists shall be fully bolted and/or welded to the support structure before:

**1926.758(f)(1)** Releasing the hoisting cables;

**1926.758(f)(2)** Allowing an employee on the joists; or

**1926.758(f)(3)** Allowing any construction loads on the joists.

**1926.758(g)** Purlins and girts shall not be used as an anchorage point for a fall arrest system unless written approval is obtained from a qualified person.

**1926.758(h)** Purlins may only be used as a walking/working surface when installing safety systems, after all permanent bridging has been installed and fall protection is provided.

**1926.758(i)** Construction loads may be placed only within a zone that is within 8 feet (2.5 m) of the center-line of the primary support member.

[66 FR 5195, Jan. 18, 2001]

## §1926.759 Falling object protection.

**1926.759(a)** ***Securing loose items aloft.*** All materials, equipment, and tools, which are not in use while aloft, shall be secured against accidental displacement.

**1926.759(b)** ***Protection from falling objects other than materials being hoisted.*** The controlling contractor shall bar other construction processes below steel erection unless overhead protection for the employees below is provided.

[66 FR 5195, Jan. 18, 2001]

## §1926.760 Fall protection.

**1926.760(a)** ***General requirements.*** (1) Except as provided by paragraph (a)(3) of this section, each employee engaged in a steel erection activity who is on a walking/working surface with an unprotected side or edge more than 15 feet (4.6 m) above a lower level shall be protected from fall hazards by guardrail systems, safety net systems, personal fall arrest systems, positioning device systems or fall restraint systems.

**1926.760(a)(2)** *Perimeter safety cables.* On multi-story structures, perimeter safety cables shall be installed at the final interior and exterior perimeters of the floors as soon as the metal decking has been installed.

**1926.760(a)(3)** Connectors and employees working in controlled decking zones shall be protected from fall hazards as provided in paragraphs (b) and (c) of this section, respectively.

**1926.760(b)** ***Connectors.*** Each connector shall:

**1926.760(b)(1)** Be protected in accordance with paragraph (a)(1) of this section from fall hazards of more than two stories or 30 feet (9.1 m) above a lower level, whichever is less;

**1926.760(b)(2)** Have completed connector training in accordance with §1926.761; and

**1926.760(b)(3)** Be provided, at heights over 15 and up to 30 feet above a lower level, with a personal fall arrest system, positioning device system or fall restraint system and wear the equipment necessary to be able to be tied off; or be provided with other means of protection from fall hazards in accordance with paragraph (a)(1) of this section.

**1926.760(c)** ***Controlled Decking Zone (CDZ).*** A controlled decking zone may be established in that area of the structure over 15 and up to 30 feet above a lower level where metal decking is initially being installed and forms the leading edge of a work area. In each CDZ, the following shall apply:

**1926.760(c)(1)** Each employee working at the leading edge in a CDZ shall be protected from fall hazards of more than two stories or 30 feet (9.1 m), whichever is less.

**1926.760(c)(2)** Access to a CDZ shall be limited to only those employees engaged in leading edge work.

**1926.760(c)(3)** The boundaries of a CDZ shall be designated and clearly marked. The CDZ shall not be more than 90 feet (27.4 m) wide and 90 (27.4 m) feet deep from any leading edge. The CDZ shall be marked by the use of control lines or the equivalent. Examples of acceptable procedures for demarcating CDZ's can be found in Appendix D to this subpart.

**1926.760(c)(4)** Each employee working in a CDZ shall have completed CDZ training in accordance with §1926.761.

**1926.760(c)(5)** Unsecured decking in a CDZ shall not exceed 3,000 square feet (914.4 m$^2$).

**1926.760(c)(6)** Safety deck attachments shall be performed in the CDZ from the leading edge back to the control line and shall have at least two attachments for each metal decking panel.

**1926.760(c)(7)** Final deck attachments and installation of shear connectors shall not be performed in the CDZ.

**1926.760(d)** ***Criteria for fall protection equipment.*** (1) Guardrail systems, safety net systems, personal fall arrest systems, positioning device systems and their components shall conform to the criteria in §1926.502 (see Appendix G to this subpart).

**1926.760(d)(2)** Fall arrest system components shall be used in fall restraint systems and shall conform to the criteria in §1926.502 (see Appendix G). Either body belts or body harnesses shall be used in fall restraint systems.

**1926.760(d)(3)** Perimeter safety cables shall meet the criteria for guardrail systems in §1926.502 (see Appendix G).

**1926.760(e)** ***Custody of fall protection.*** Fall protection provided by the steel erector shall remain in the area where steel erection activity has been completed, to be used by other trades, only if the controlling contractor or its authorized representative:

**1926.760(e)(1)** Has directed the steel erector to leave the fall protection in place; and

**1926.760(e)(2)** Has inspected and accepted control and responsibility of the fall protection prior to authorizing persons other than steel erectors to work in the area.

[66 FR 5195, Jan. 18, 2001]

## §1926.761 Training.

The following provisions supplement the requirements of §1926.21 regarding the hazards addressed in this subpart.

**1926.761(a)** ***Training personnel.*** Training required by this section shall be provided by a qualified person(s).

**1926.761(b)** ***Fall hazard training.*** The employer shall provide a training program for all employees exposed to fall hazards. The program shall include training and instruction in the following areas:

**1926.761(b)(1)** The recognition and identification of fall hazards in the work area;

**1926.761(b)(2)** The use and operation of guardrail systems (including perimeter safety cable systems), personal fall arrest systems, positioning device systems, fall restraint systems, safety net systems, and other protection to be used;

**1926.761(b)(3)** The correct procedures for erecting, maintaining, disassembling, and inspecting the fall protection systems to be used;

**1926.761(b)(4)** The procedures to be followed to prevent falls to lower levels and through or into holes and openings in walking/working surfaces and walls; and

**1926.761(b)(5)** The fall protection requirements of this subpart.

**1926.761(c)** ***Special training programs.*** In addition to the training required in paragraphs (a) and (b) of this section, the employer shall provide special training to employees engaged in the following activities.

**1926.761(c)(1)** *Multiple lift rigging procedure.* The employer shall ensure that each employee who performs multiple lift rigging has been provided training in the following areas:

**1926.761(c)(1)(i)** The nature of the hazards associated with multiple lifts; and

**1926.761(c)(1)(ii)** The proper procedures and equipment to perform multiple lifts required by §1926.753(e).

**1926.761(c)(2)** *Connector procedures.* The employer shall ensure that each connector has been provided training in the following areas:

**1926.761(c)(2)(i)** The nature of the hazards associated with connecting; and

**1926.761(c)(2)(ii)** The establishment, access, proper connecting techniques and work practices required by §1926.756(c) and §1926.760(b).

**1926.761(c)(3)** *Controlled Decking Zone Procedures.* Where CDZs are being used, the employer shall assure that each employee has been provided training in the following areas:

**1926.761(c)(3)(i)** The nature of the hazards associated with work within a controlled decking zone; and

**1926.761(c)(3)(ii)** The establishment, access, proper installation techniques and work practices required by §1926.760(c) and §1926.754(e).

[66 FR 5195, Jan. 18, 2001]

## Appendix A to Subpart R— Guidelines for Establishing the Components of a Site-specific Erection Plan: Non-mandatory Guidelines for Complying with §1926.752(e).

(a) ***General.*** This appendix serves as a guideline to assist employers who elect to develop a site-specific erection plan in accordance with §1926.752(e) with alternate means and methods to provide employee protection in accordance with §1926.752(e), §1926.753(c)(5), §1926.757(a)(4) and §1926.757(e)(4).

(b) ***Development of a site-specific erection plan.*** Pre-construction conference(s) and site inspection(s) are held between the erector and the controlling contractor, and others such as the project engineer and fabricator before the start of steel erection. The purpose of such conference(s) is to develop and review the site-specific erection plan that will meet the requirements of this section.

(c) ***Components of a site-specific erection plan.*** In developing a site-specific erection plan, a steel erector considers the following elements:

(1) The sequence of erection activity, developed in coordination with the controlling contractor, that includes the following:

(i) Material deliveries;

(ii) Material staging and storage; and

(iii) Coordination with other trades and construction activities.

(2) A description of the crane and derrick selection and placement procedures, including the following:

(i) Site preparation;

(ii) Path for overhead loads; and

(iii) Critical lifts, including rigging supplies and equipment.

(3) A description of steel erection activities and procedures, including the following:

(i) Stability considerations requiring temporary bracing and guying;

(ii) Erection bridging terminus point;

(iii) Anchor rod (anchor bolt) notifications regarding repair, replacement and modifications;

(iv) Columns and beams (including joists and purlins);

(v) Connections;

(vi) Decking; and

(vii) Ornamental and miscellaneous iron.

(4) A description of the fall protection procedures that will be used to comply with §1926.760.

(5) A description of the procedures that will be used to comply with §1926.759.

(6) A description of the special procedures required for hazardous non-routine tasks.

(7) A certification for each employee who has received training for performing steel erection operations as required by §1926.761.

(8) A list of the qualified and competent persons.

(9) A description of the procedures that will be utilized in the event of rescue or emergency response.

(d) ***Other plan information.*** The plan:

(1) Includes the identification of the site and project; and

(2) Is signed and dated by the qualified person(s) responsible for its preparation and modification.

[66 FR 5195, Jan. 18, 2001]

### Appendix B to Subpart R—Acceptable Test Methods for Testing Slip-Resistance of Walking/Working Surfaces (§1926.754(c)(3)). Non-Mandatory Guidelines for Complying With §1926.754(c)(3).

The following references provide acceptable test methods for complying with the requirements of §1926.754(c)(3).

- Standard Test Method for Using a Portable Inclineable Articulated Strut Slip Tester (PIAST)(ASTM F1677-96)
- Standard Test Method for Using a Variable Incidence Tribometer (VIT)(ASTM F1679-96)

[66 FR 5195, Jan. 18, 2001]

### Appendix C to Subpart R—Illustrations of Bridging Terminus Points: Non-mandatory Guidelines for Complying with §§1926.757(a)(10) and 1926.757(c)(5).

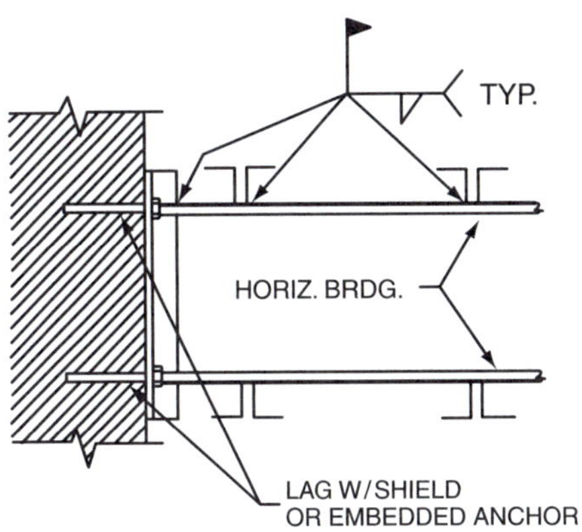


**HORIZONTAL BRIDGING TERMINUS AT WALL**

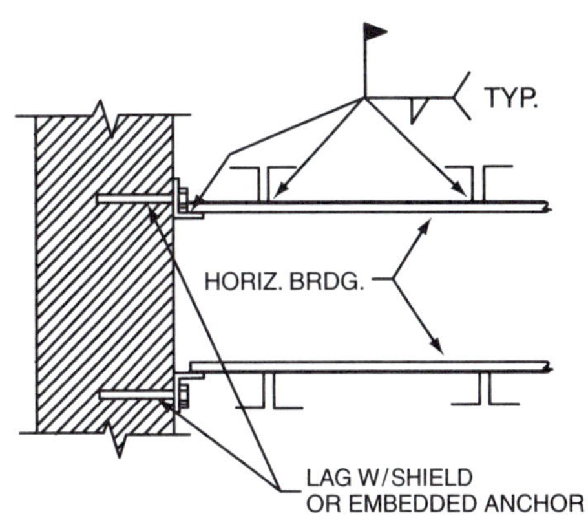


**HORIZONTAL BRIDGING TERMINUS AT WALL**

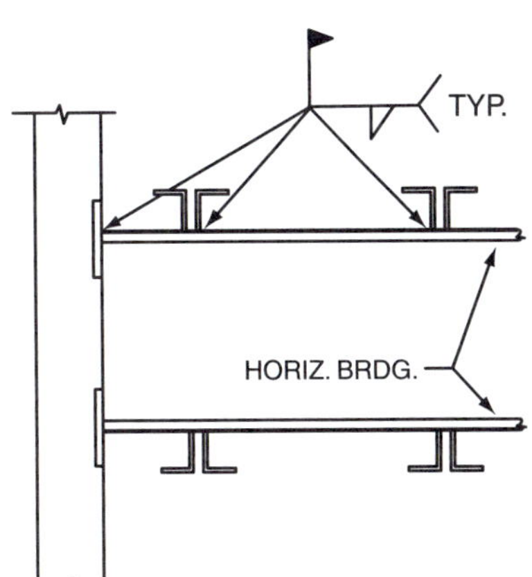


**HORIZONTAL BRIDGING TERMINUS AT PANEL WALL**

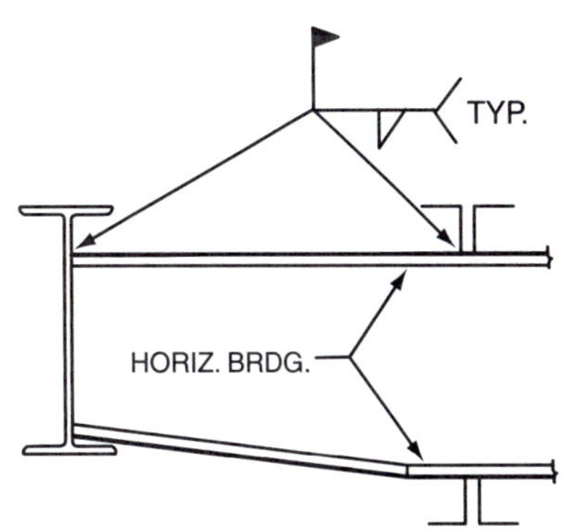


**HORIZONTAL BRIDGING TERMINUS AT STRUCTURAL SHAPE**

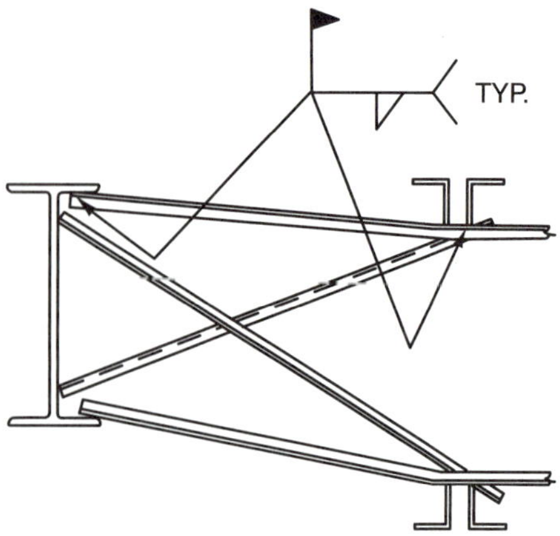


HORIZONTAL BRIDGING TERMINUS AT STRUCTURAL SHAPE WITH OPTIONAL "X-BRIDGING"

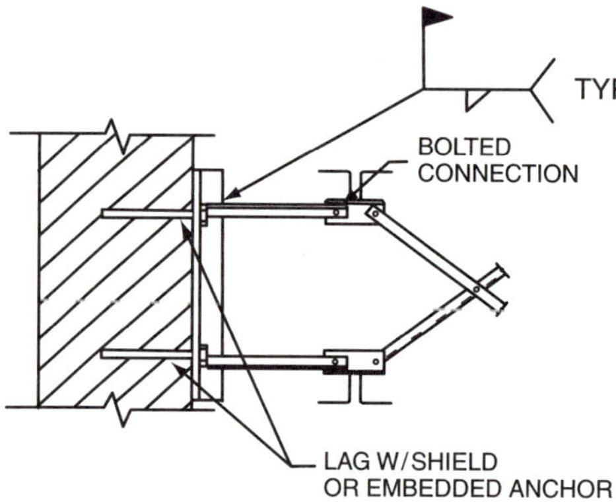


BOLTED DIAGONAL BRIDGING TERMINUS AT WALL

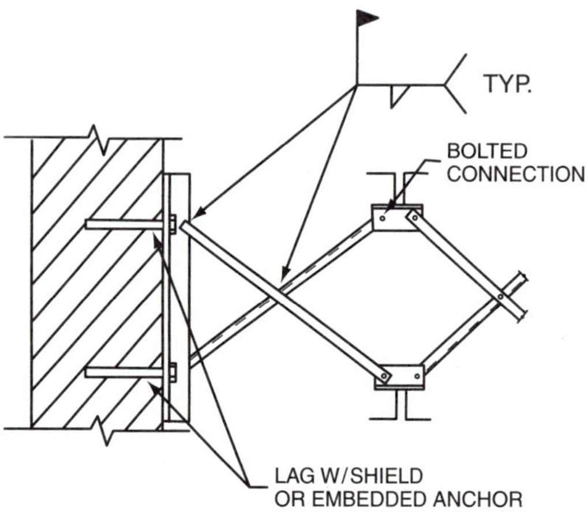


BOLTED DIAGONAL BRIDGING TERMINUS AT WALL

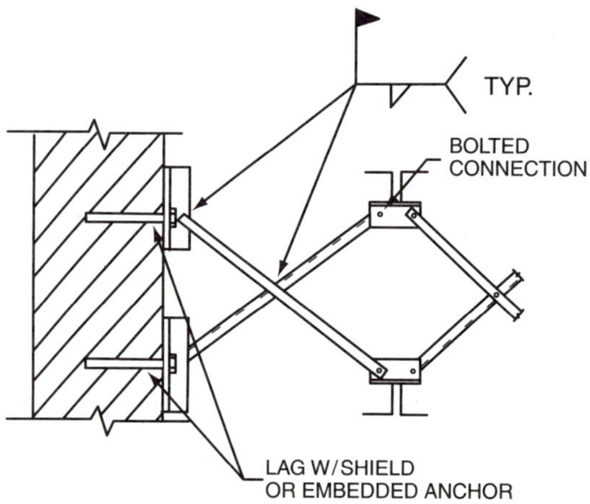


BOLTED DIAGONAL BRIDGING TERMINUS AT WALL

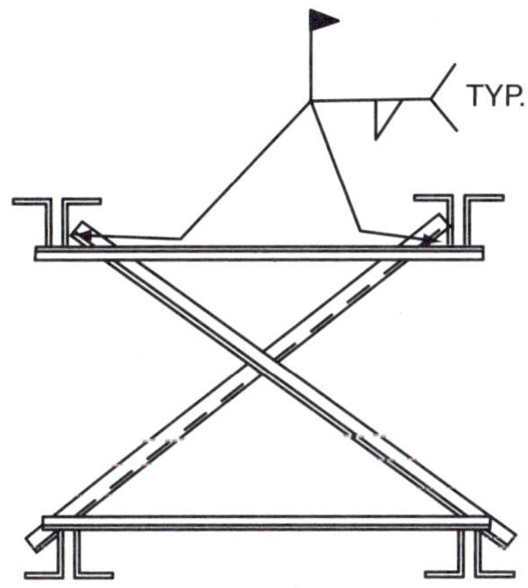


JOISTS PAIR BRIDGING
TERMINUS POINT

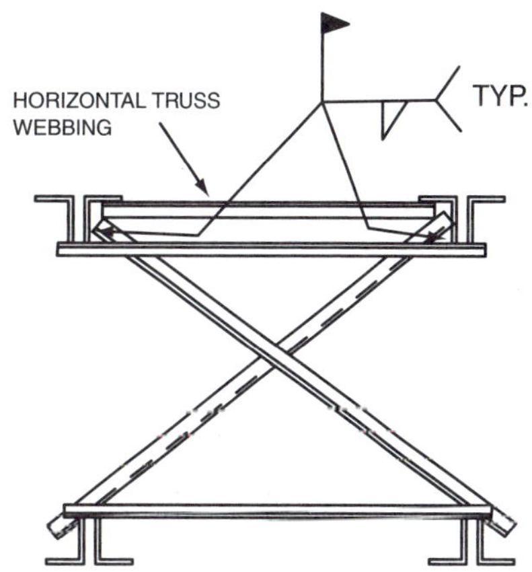


JOISTS PAIR BRIDGING
TERMINUS POINT
W / HORIZ. TRUSS

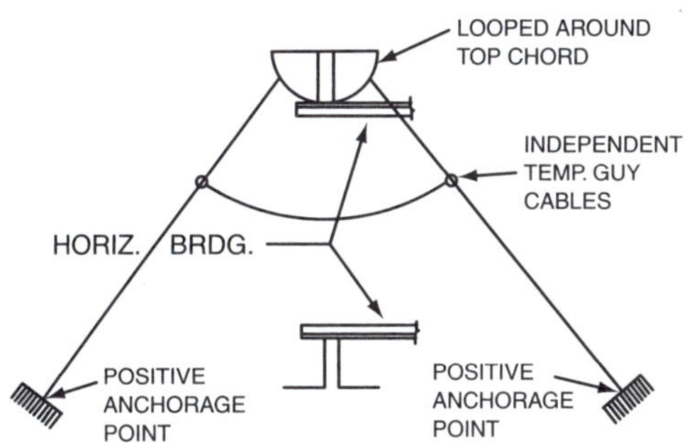


HORIZONTAL BRIDGING
TERMINUS POINT
SECURED BY TEMP.
GUY CABLES

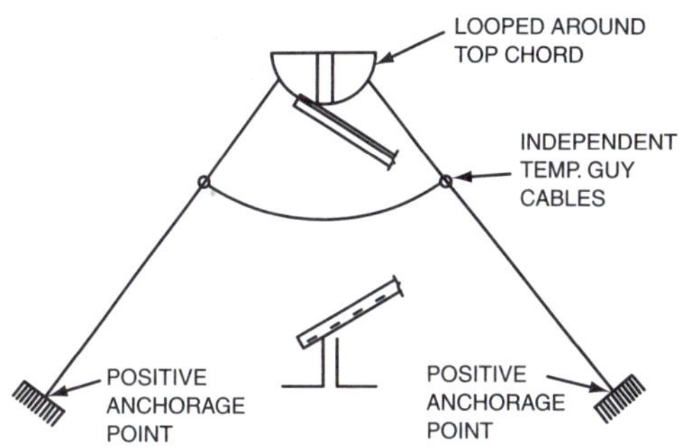


DIAGONAL BRIDGING
TERMINUS POINT
SECURED BY TEMP.
GUY CABLES

[66 FR 5195, Jan. 18, 2001]

### Appendix D to Subpart R—Illustration of the Use of Control Lines to Demarcate Controlled Decking Zones (CDZs): Non-mandatory Guidelines for Complying with §1926.760(c)(3)

(1) When used to control access to areas where leading edge and initial securement of metal deck and other operations connected with leading edge work are taking place, the controlled decking zone (CDZ) is defined by a control line or by any other means that restricts access.

(i) A control line for a CDZ is erected not less than 6 feet (1.8 m) nor more than 90 feet (27.4 m) from the leading edge.

(ii) Control lines extend along the entire length of the unprotected or leading edge and are approximately parallel to the unprotected or leading edge.

(iii) Control lines are connected on each side to a guardrail system, wall, stanchion or other suitable anchorage.

(2) Control lines consist of ropes, wires, tapes, or equivalent materials, and supporting stanchions as follows:

(i) Each line is rigged and supported in such a way that its lowest point (including sag) is not less than 39 inches (1.0 m) from the walking/working surface and its highest point is not more than 45 inches (1.3 m) from the walking/working surface.

(ii) Each line has a minimum breaking strength of 200 pounds (90.8 kg).

[66 FR 5195, Jan. 18, 2001]

### Appendix E to Subpart R—Training: Non-mandatory Guidelines for Complying with §1926.761

The training requirements of §1926.761 will be deemed to have been met if employees have completed a training course on steel erection, including instruction in the provisions of this standard, that has been approved by the U.S. Department of Labor Bureau of Apprenticeship.

[66 FR 5195, Jan. 18, 2001]

### Appendix F to Subpart R—Perimeter Columns: Non-Mandatory Guidelines for Complying with §1926.756(e) To Protect the Unprotected Side or Edge of a Walking/Working Surface

In multi-story structures, when holes in the column web are used for perimeter safety cables, the column splice must be placed sufficiently high so as not to interfere with any attachments to the column necessary for the column splice. Column splices are recommended to be placed at every other or fourth levels as design allows. Column splices at third levels are detrimental to the erection process and should be avoided if possible.

[66 FR 5195, Jan. 18, 2001]

## Appendix G to Subpart R— §1926.502 (b)-(e) Fall Protection Systems Criteria and Practices

(b) ***"Guardrail systems."*** Guardrail systems and their use shall comply with the following provisions:

(1) Top edge height of top rails, or equivalent guardrail system members, shall be 42 inches (1.1 m) plus or minus 3 inches (8 cm) above the walking/working level. When conditions warrant, the height of the top edge may exceed the 45-inch height, provided the guardrail system meets all other criteria of this paragraph (§1926.502(b)).

**Note:** When employees are using stilts, the top edge height of the top rail, or equivalent member, shall be increased an amount equal to the height of the stilts.

(2) Midrails, screens, mesh, intermediate vertical members, or equivalent intermediate structural members shall be installed between the top edge of the guardrail system and the walking/working surface when there is no wall or parapet wall at least 21 inches (53 cm) high.

(i) Midrails, when used, shall be installed at a height midway between the top edge of the guardrail system and the walking/working level.

(ii) Screens and mesh, when used, shall extend from the top rail to the walking/working level and along the entire opening between top rail supports.

(iii) Intermediate members (such as balusters), when used between posts, shall be not more than 19 inches (48 cm) apart.

(iv) Other structural members (such as additional midrails and architectural panels) shall be installed such that there are no openings in the guardrail system that are more than 19 inches (.5 m) wide.

(3) Guardrail systems shall be capable of withstanding, without failure, a force of at least 200 pounds (890 N) applied within 2 inches (5.1 cm) of the top edge, in any outward or downward direction, at any point along the top edge.

(4) When the 200 pound (890 N) test load specified in paragraph (b)(3) of this section (§1926.502) is applied in a downward direction, the top edge of the guardrail shall not deflect to a height less than 39 inches (1.0 m) above the walking/working level. Guardrail system components selected and constructed in accordance with the appendix B to subpart M of this part will be deemed to meet this requirement.

(5) Midrails, screens, mesh, intermediate vertical members, solid panels, and equivalent structural members shall be capable of withstanding, without failure, a force of at least 150 pounds (666 N) applied in any downward or outward direction at any point along the midrail or other member.

(6) Guardrail systems shall be so surfaced as to prevent injury to an employee from punctures or lacerations, and to prevent snagging of clothing.

(7) The ends of all top rails and midrails shall not overhang the terminal posts, except where such overhang does not constitute a projection hazard.

(8) Steel banding and plastic banding shall not be used as top rails or midrails.

(9) Top rails and midrails shall be at least one-quarter inch (0.6 cm) nominal diameter or thickness to prevent cuts and lacerations. If wire rope is used for top rails, it shall be flagged at not more than 6-foot intervals with high-visibility material.

(10) When guardrail systems are used at hoisting areas, a chain, gate or removable guardrail section shall be placed across the access opening between guardrail sections when hoisting operations are not taking place.

(11) When guardrail systems are used at holes, they shall be erected on all unprotected sides or edges of the hole.

(12) When guardrail systems are used around holes used for the passage of materials, the hole shall have not more than two sides provided with removable guardrail sections to allow the passage of materials. When the hole is not in use, it shall be closed over with a cover, or a guardrail system shall be provided along all unprotected sides or edges.

(13) When guardrail systems are used around holes which are used as points of access (such as ladderways), they shall be provided with a gate, or be so offset that a person cannot walk directly into the hole.

(14) Guardrail systems used on ramps and runways shall be erected along each unprotected side or edge.

(15) Manila, plastic or synthetic rope being used for top rails or midrails shall be inspected as frequently as necessary to ensure that it continues to meet the strength requirements of paragraph (b)(3) of this section (§1926.502).

(c) ***Safety net systems.*** Safety net systems and their use shall comply with the following provisions:

(1) Safety nets shall be installed as close as practicable under the walking/working surface on which employees are working, but in no case more than 30 feet (9.1 m) below such level. When nets are used on bridges, the potential fall area from the walking/working surface to the net shall be unobstructed.

(2) Safety nets shall extend outward from the outermost projection of the work surface as follows:

| Vertical distance from working level to horizontal plane of net | Minimum required horizontal distance of outer edge of net from the edge of the working surface |
|---|---|
| Up to 5 feet | 8 feet |
| More than 5 feet up to 10 feet | 10 feet |
| More than 10 feet | 13 feet |

(3) Safety nets shall be installed with sufficient clearance under them to prevent contact with the surface or structures below when subjected to an impact force equal to the drop test specified in paragraph (4) of this section [§1926.502].

(4) Safety nets and their installations shall be capable of absorbing an impact force equal to that produced by the drop test specified in paragraph (c)(4)(i) of this section [§1926.502].

(i) Except as provided in paragraph (c)(4)(ii) of this section (§1926.502), safety nets and safety net installations shall be drop-tested at the jobsite after initial installation and before being used as a fall protection system, whenever relocated, after major repair, and at 6-month intervals if left in one place. The drop-test shall consist of a 400 pound (180 kg) bag of sand 30 + or - 2 inches (76 + or - 5 cm) in diameter dropped into the net from the highest walking/working surface at which employees are exposed to fall hazards, but not from less than 42 inches (1.1 m) above that level.

(ii) When the employer can demonstrate that it is unreasonable to perform the drop-test required by paragraph (c)(4)(i) of this section (§1926.502), the employer (or a designated competent person) shall certify that the net and net installation is in compliance with the provisions of paragraphs (c)(3) and (c)(4)(i) of this section (§1926.502) by preparing a certification record prior to the net being used as a fall protection system. The certification record must include an identification of the net and net installation for which the certification record is being prepared; the date that it was determined that the identified net and net installation were in compliance with paragraph (c)(3) of this section (§1926.502) and the signature of the person making the determination and certification. The most recent certification record for each net and net installation shall be available at the jobsite for inspection.

(5) Defective nets shall not be used. Safety nets shall be inspected at least once a week for wear, damage, and other deterioration. Defective components shall be removed from service. Safety nets shall also be inspected after any occurrence which could affect the integrity of the safety net system.

(6) Materials, scrap pieces, equipment, and tools which have fallen into the safety net shall be removed as soon as possible from the net and at least before the next work shift.

(7) The maximum size of each safety net mesh opening shall not exceed 36 square inches (230 cm) nor be longer than 6 inches (15 cm) on any side, and the opening, measured center-to-center of mesh ropes or webbing, shall not be longer than 6 inches (15 cm). All mesh crossings shall be secured to prevent enlargement of the mesh opening.

(8) Each safety net (or section of it) shall have a border rope for webbing with a minimum breaking strength of 5,000 pounds (22.2 kN).

(9) Connections between safety net panels shall be as strong as integral net components and shall be spaced not more than 6 inches (15 cm) apart.

(d) ***"Personal fall arrest systems."*** Personal fall arrest systems and their use shall comply with the provisions set forth below. Effective January 1, 1998, body belts are not acceptable as part of a personal fall arrest system.

**Note:** The use of a body belt in a positioning device system is acceptable and is regulated under paragraph (e) of this section (§1926.502).

(1) Connectors shall be drop forged, pressed or formed steel, or made of equivalent materials.

(2) Connectors shall have a corrosion-resistant finish, and all surfaces and edges shall be smooth to prevent damage to interfacing parts of the system.

(3) Dee-rings and snaphooks shall have a minimum tensile strength of 5,000 pounds (22.2 kN).

(4) Dee-rings and snaphooks shall be proof-tested to a minimum tensile load of 3,600 pounds (16 kN) without cracking, breaking, or taking permanent deformation.

(5) Snaphooks shall be sized to be compatible with the member to which they are connected to prevent unintentional disengagement of the snaphook by depression of the snaphook keeper by the connected member, or shall be a locking type snaphook designed and used to prevent disengagement of the snaphook by the contact of the snaphook keeper by the connected member. Effective January 1, 1998, only locking type snaphooks shall be used.

(6) Unless the snaphook is a locking type and designed for the following connections, snaphooks shall not be engaged:

(i) directly to webbing, rope or wire rope;

(ii) to each other;

(iii) to a dee-ring to which another snaphook or other connector is attached;

(iv) to a horizontal lifeline; or

(v) to any object which is incompatibly shaped or dimensioned in relation to the snaphook such that unintentional disengagement could occur by the connected object being able to depress the snaphook keeper and release itself.

(7) On suspended scaffolds or similar work platforms with horizontal lifelines which may become vertical lifelines, the devices used to connect to a horizontal lifeline shall be capable of locking in both directions on the lifeline.

(8) Horizontal lifelines shall be designed, installed, and used, under the supervision of a qualified person, as part of a complete personal fall arrest system, which maintains a safety factor of at least two.

(9) Lanyards and vertical lifelines shall have a minimum breaking strength of 5,000 pounds (22.2 kN).

(10)(i) Except as provided in paragraph (d)(10)(ii) of this section [§1926.502], when vertical lifelines are used, each employee shall be attached to a separate lifeline.

(ii) During the construction of elevator shafts, two employees may be attached to the same lifeline in the hoistway, provided both employees are working atop a false car that is equipped with guardrails; the strength of the lifeline is 10,000 pounds [5,000 pounds per employee attached] (44.4 kN); and all other criteria specified in this paragraph for lifelines have been met.

(11) Lifelines shall be protected against being cut or abraded.

(12) Self-retracting lifelines and lanyards which automatically limit free fall distance to 2 feet (0.61 m) or less shall be capable of sustaining a minimum tensile load of 3,000 pounds (13.3 kN) applied to the device with the lifeline or lanyard in the fully extended position.

(13) Self-retracting lifelines and lanyards which do not limit free fall distance to 2 feet (0.61 m) or less, ripstitch lanyards, and tearing and deforming lanyards shall be capable of sustaining a minimum tensile load of 5,000 pounds (22.2 kN) applied to the device with the lifeline or lanyard in the fully extended position.

(14) Ropes and straps (webbing) used in lanyards, lifelines, and strength components of body belts and body harnesses shall be made from synthetic fibers.

(15) Anchorages used for attachment of personal fall arrest equipment shall be independent of any anchorage being used to support or suspend platforms and capable of supporting at least 5,000 pounds (22.2 kN) per employee attached, or shall be designed, installed, and used as follows:

(i) as part of a complete personal fall arrest system which maintains a safety factor of at least two; and

(ii) under the supervision of a qualified person.

(16) Personal fall arrest systems, when stopping a fall, shall:

(i) limit maximum arresting force on an employee to 900 pounds (4 kN) when used with a body belt;

(ii) limit maximum arresting force on an employee to 1,800 pounds (8 kN) when used with a body harness;

(iii) be rigged such that an employee can neither free fall more than 6 feet (1.8 m), nor contact any lower level;

(iv) bring an employee to a complete stop and limit maximum deceleration distance an employee travels to 3.5 feet (1.07 m); and,

(v) have sufficient strength to withstand twice the potential impact energy of an employee free falling a distance of 6 feet (1.8 m), or the free fall distance permitted by the system, whichever is less.

**Note:** If the personal fall arrest system meets the criteria and protocols contained in Appendix C to subpart M, and if the system is being used by an employee having a combined person and tool weight of less than 310 pounds (140 kg), the system will be considered to be in compliance with the provisions of paragraph (d)(16) of this section [§1926.502]. If the system is used by an employee having a combined tool and body weight of 310 pounds (140 kg) or more, then the employer must appropriately modify the criteria and protocols of the Appendix to provide proper protection for such heavier weights, or the system will not be deemed to be in compliance with the requirements of paragraph (d)(16) of this section (§1926.502).

(17) The attachment point of the body belt shall be located in the center of the wearer's back. The attachment point of the body harness shall be located in the center of the wearer's back near shoulder level, or above the wearer's head.

(18) Body belts, harnesses, and components shall be used only for employee protection (as part of a personal fall arrest system or positioning device system) and not to hoist materials.

(19) Personal fall arrest systems and components subjected to impact loading shall be immediately removed from service and shall not be used again for employee protection until inspected and determined by a competent person to be undamaged and suitable for reuse.

(20) The employer shall provide for prompt rescue of employees in the event of a fall or shall assure that employees are able to rescue themselves.

(21) Personal fall arrest systems shall be inspected prior to each use for wear, damage and other deterioration, and defective components shall be removed from service.

(22) Body belts shall be at least one and five-eighths (1 5/8) inches (4.1 cm) wide.

(23) Personal fall arrest systems shall not be attached to guardrail systems, nor shall they be attached to hoists except as specified in other subparts of this Part.

(24) When a personal fall arrest system is used at hoist areas, it shall be rigged to allow the movement of the employee only as far as the edge of the walking/working surface.

(e) ***Positioning device systems.*** Positioning device systems and their use shall conform to the following provisions:

(1) Positioning devices shall be rigged such that an employee cannot free fall more than 2 feet (.9 m).

(2) Positioning devices shall be secured to an anchorage capable of supporting at least twice the potential impact load of an employee's fall or 3,000 pounds (13.3 kN), whichever is greater.

(3) Connectors shall be drop forged, pressed or formed steel, or made of equivalent materials.

(4) Connectors shall have a corrosion-resistant finish, and all surfaces and edges shall be smooth to prevent damage to interfacing parts of this system.

(5) Connecting assemblies shall have a minimum tensile strength of 5,000 pounds (22.2 kN)

(6) Dee-rings and snaphooks shall be proof-tested to a minimum tensile load of 3,600 pounds (16 kN) without cracking, breaking, or taking permanent deformation.

(7) Snaphooks shall be sized to be compatible with the member to which they are connected to prevent unintentional disengagement of the snaphook by depression of the snaphook keeper by the connected member, or shall be a locking type snaphook designed and used to prevent disengagement of the snaphook by the contact of the snaphook keeper by the connected member. As of January 1, 1998, only locking type snaphooks shall be used.

(8) Unless the snaphook is a locking type and designed for the following connections, snaphooks shall not be engaged:

(i) directly to webbing, rope or wire rope;

(ii) to each other;

(iii) to a dee-ring to which another snaphook or other connector is attached;

(iv) to a horizontal lifeline; or to depress the snaphook keeper and release itself.

(v) to any object which is incompatibly shaped or dimensioned in relation to the snaphook such that unintentional disengagement could occur by the connected object being able to depress the snaphook keeper and release itself.

(9) Positioning device systems shall be inspected prior to each use for wear, damage, and other deterioration, and defective components shall be removed from service.

(10) Body belts, harnesses, and components shall be used only for employee protection (as part of a personal fall arrest system or positioning device system) and not to hoist materials.

[66 FR 5195, Jan. 18, 2001]

### Appendix H to Subpart R—Double Connections: Illustration of a Clipped End Connection and a Staggered Connection: Non-Mandatory Guidelines for Complying with §1926.756(c)(1).

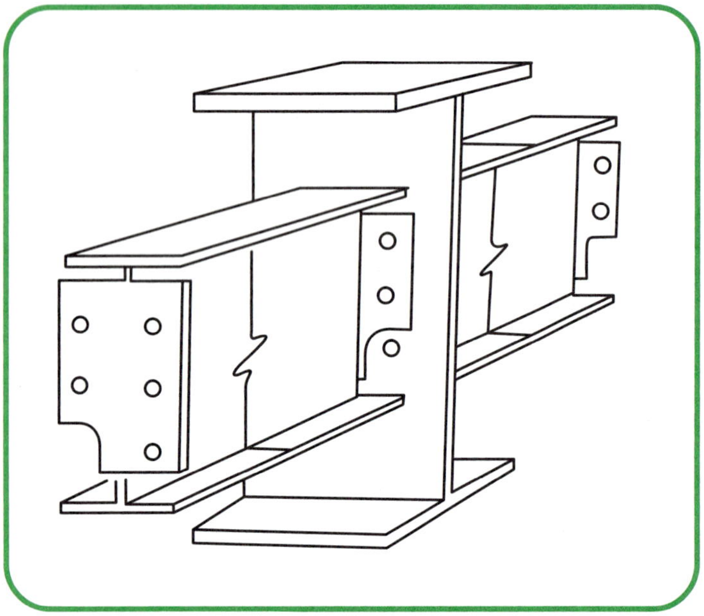

Clipped end connections are connection material on the end of a structural member which has a notch at the bottom and/or top to allow the bolt(s) of the first member placed on the opposite side of the central member to remain in place. The notch(es) fits around the nut or bolt head of the opposing member to allow the second member to be bolted up without removing the bolt(s) holding the first member.

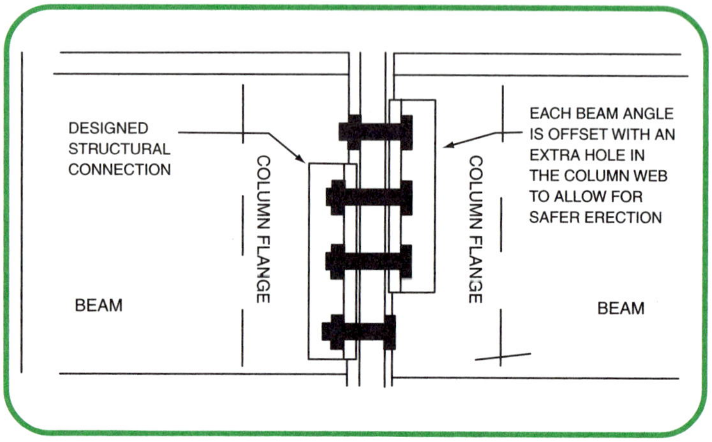


Staggered connections are connection material on a structural member in which all of the bolt holes in the common member web are not shared by the two incoming members in the final connection. The extra hole in the column web allows the erector to maintain at least a one bolt connection at all times while making the double connection.

[66 FR 5195, Jan. 18, 2001]

**Authority:** Sec. 107, Contract Work Hours and Safety Standards Act (40 U.S.C. 333); secs. 4, 6, and 8 of the Occupational Safety and Health Act of 1970 (29 U.S.C. 653, 655, 657); Secretary of Labor's Orders 12-71 (36 FR 8754), 8-76 (41 FR 25059), 9-83 (48 FR 35736), 1-90 (55 FR 9033), or 6-96 (62 FR 111), as applicable.

## §1926.800 Underground construction.

**1926.800(a)** ***Scope and application.*** (1) This section applies to the construction of underground tunnels, shafts, chambers, and passageways. This section also applies to cut-and-cover excavations which are both physically connected to ongoing underground construction operations within the scope of this section, and covered in such a manner as to create conditions characteristic of underground construction.

**1926.800(a)(2)** This section does not apply to the following:

**1926.800(a)(2)(i)** Excavation and trenching operations covered by subpart P of this part, such as foundation operations for above-ground structures that are not physically connected to underground construction operations, and surface excavation; nor

**1926.800(a)(2)(ii)** Underground electrical transmission and distribution lines, as addressed in subpart V of this part.

**1926.800(b)** ***Access and egress.*** (1) The employer shall provide and maintain safe means of access and egress to all work stations.

**1926.800(b)(2)** The employer shall provide access and egress in such a manner that employees are protected from being struck by excavators, haulage machines, trains and other mobile equipment.

**1926.800(b)(3)** The employer shall control access to all openings to prevent unauthorized entry underground. Unused chutes, manways, or other openings shall be tightly covered, bulkheaded, or fenced off, and shall be posted with warning signs indicating "Keep Out" or similar language. Completed or unused sections of the underground facility shall be barricaded.

**1926.800(c)** ***Check-in/check-out.*** The employer shall maintain a check-in/check-out procedure that will ensure that above-ground personnel can determine an accurate count of the number of persons underground in the event of an emergency. However, this procedure is not required when the construction of underground facilities designed for human occupancy has been sufficiently completed so that the permanent environmental controls are effective, and when the remaining construction activity will not cause any environmental hazard or structural failure within the facilities.

**1926.800(d)** ***Safety instruction.*** All employees shall be instructed in the recognition and avoidance of hazards associated with underground construction activities including, where appropriate, the following subjects:

**1926.800(d)(1)** Air monitoring;

**1926.800(d)(2)** Ventilation;

**1926.800(d)(3)** Illumination;

**1926.800(d)(4)** Communications;

**1926.800(d)(5)** Flood control;

**1926.800(d)(6)** Mechanical equipment;

**1926.800(d)(7)** Personal protective equipment;

**1926.800(d)(8)** Explosives;

**1926.800(d)(9)** Fire prevention and protection; and

**1926.800(d)(10)** Emergency procedures, including evacuation plans and check-in/check-out systems.

**1926.800(e)** ***Notification.*** (1) Oncoming shifts shall be informed of any hazardous occurrences or conditions that have affected or might affect employee safety, including liberation of gas, equipment failures, earth or rock slides, cave-ins, floodings, fires or explosions.

**1926.800(e)(2)** The employer shall establish and maintain direct communications for coordination of activities with other employers whose operations at the jobsite affect or may affect the safety of employees underground.

**1926.800(f)** ***Communications.*** (1) When natural unassisted voice communication is ineffective, a power-assisted means of voice communication shall be used to provide communication between the work face, the bottom of the shaft, and the surface.

**1926.800(f)(2)** Two effective means of communication, at least one of which shall be voice communication, shall be provided in all shafts which are being developed or used either for personnel access or for hoisting. Additional requirements for hoist operator communication are contained in paragraph (t)(3)(xiv) of this section.

**1926.800(f)(3)** Powered communication systems shall operate on an independent power supply, and shall be installed so that the use of or disruption of any one phone or signal location will not disrupt the operation of the system from any other location.

**1926.800(f)(4)** Communication systems shall be tested upon initial entry of each shift to the underground, and as often as necessary at later times, to ensure that they are in working order.

**1926.800(f)(5)** Any employee working alone underground in a hazardous location, who is both out of the range of natural unassisted voice communication and not under observation by other persons, shall be provided with an effective means of obtaining assistance in an emergency.

**1926.800(g)** ***Emergency provisions***—(1) *Hoisting capability.* When a shaft is used as a means of egress, the employer shall make advance arrangements for power-assisted hoisting capability to be readily available in an emergency, unless the regular hoisting means can continue to function in the event of an electrical power failure at the jobsite. Such hoisting means shall be designed so that the load hoist drum is powered in both directions of rotation and so that the brake is automatically applied upon power release or failure.

**1926.800(g)(2)** *Self-rescuers.* The employer must provide self-rescuers approved by the National Institute for Occupational Safety and Health under 42 CFR part 84. The respirators must be immediately available to all employees at work stations in underground areas where employees might be trapped by smoke or gas. The selection, issuance, use, and care of respirators must be in accordance with 29 CFR 1926.103.

**1926.800(g)(3)** *Designated person.* At least one designated person shall be on duty above ground whenever any employee is working underground. This designated person shall be responsible for securing immediate aid and keeping an accurate count of employees underground in case of emergency. The designated person must not be so busy with other responsibilities that the counting function is encumbered.

**1926.800(g)(4)** *Emergency lighting.* Each employee underground shall have an acceptable portable hand lamp or cap lamp in his or her work area for emergency use, unless natural light or an emergency lighting system provides adequate illumination for escape.

**1926.800(g)(5)** *Rescue teams.* (i) On jobsites where 25 or more employees work underground at one time, the employer shall provide (or make arrangements in advance with locally available rescue services to provide) at least two 5-person rescue teams, one on the jobsite or within one-half hour travel time from the entry point, and the other within 2 hours travel time.

**1926.800(g)(5)(ii)** On jobsites where less than 25 employees work underground at one time, the employer shall provide (or make arrangements in advance with locally available rescue services to provide) at least one 5-person rescue team to be either on the jobsite or within one-half hour travel time from the entry point.

**1926.800(g)(5)(iii)** Rescue team members shall be qualified in rescue procedures, the use and limitations of breathing apparatus, and the use of firefighting equipment. Qualifications shall be reviewed not less than annually.

**1926.800(g)(5)(iv)** On jobsites where flammable or noxious gases are encountered or anticipated in hazardous quantities, rescue team members shall practice donning and using self-contained breathing apparatus monthly.

**1926.800(g)(5)(v)** The employer shall ensure that rescue teams are familiar with conditions at the jobsite.

**1926.800(h)** ***Hazardous classifications***—(1) *Potentially gassy operations.* Underground construction operations shall be classified as potentially gassy if either:

**1926.800(h)(1)(i)** Air monitoring discloses 10 percent or more of the lower explosive limit for methane or other flammable gases measured at 12 inches (304.8 mm) ±0.25 inch (6.35 mm) from the roof, face, floor or walls in any underground work area for more than a 24-hour period; or

**1926.800(h)(1)(ii)** The history of the geographical area or geological formation indicates that 10 percent or more of the lower explosive limit for methane or other flammable gases is likely to be encountered in such underground operations.

**1926.800(h)(2)** *Gassy operations.* Underground construction operations shall be classified as gassy if:

**1926.800(h)(2)(i)** Air monitoring discloses 10 percent or more of the lower explosive limit for methane or other flammable gases measured at 12 inches (304.8 mm) ±0.25 inch (6.35 mm) from the roof, face, floor or walls in any underground work area for three consecutive days; or

**1926.800(h)(2)(ii)** There has been an ignition of methane or of other flammable gases emanating from the strata that indicates the presence of such gases; or

**1926.800(h)(2)(iii)** The underground construction operation is both connected to an underground work area which is currently classified as gassy and is also subject to a continuous course of air containing the flammable gas concentration.

**1926.800(h)(3)** *Declassification to potentially gassy operations.* Underground construction gassy operations may be declassified to Potentially Gassy when air monitoring results remain under 10 percent of the lower explosive limit for methane or other flammable gases for three consecutive days.

**1926.800(i)** ***Gassy operations-additional requirements.*** (1) Only acceptable equipment, maintained in suitable condition, shall be used in gassy operations.

**1926.800(i)(2)** Mobile diesel-powered equipment used in gassy operations shall be either approved in accordance with the requirements of 30 CFR part 36 (formerly Schedule 31) by MSHA, or shall be demonstrated by the employer to be fully equivalent to such MSHA-approved equipment, and shall be operated in accordance with that part.

**1926.800(i)(3)** Each entrance to a gassy operation shall be prominently posted with signs notifying all entrants of the gassy classification.

**1926.800(i)(4)** Smoking shall be prohibited in all gassy operations and the employer shall be responsible for collecting all personal sources of ignition, such as matches and lighters, from all persons entering a gassy operation.

**1926.800(i)(5)** A fire watch as described in §1926.352(e) shall be maintained when hot work is performed.

**1926.800(i)(6)** Once an operation has met the criteria in paragraph (h)(2) warranting classification as gassy, all operations in the affected area, except the following, shall be discontinued until the operation either is in compliance with all of the gassy operation requirements or has been declassified in accordance with paragraph (h)(3) of this section:

**1926.800(i)(6)(i)** Operations related to the control of the gas concentration;

**1926.800(i)(6)(ii)** Installation of new equipment, or conversion of existing equipment, to comply with this paragraph (i); and

**1926.800(i)(6)(iii)** Installation of above-ground controls for reversing the air flow.

**1926.800(j)** ***Air quality and monitoring***—(1) *General.* Air quality limits and control requirements for construction are found in §1926.55, except as modified by this section.

**1926.800(j)(1)(i)(A)** The employer shall assign a competent person who shall perform all air monitoring required by this section.

**1926.800(j)(1)(i)(B)** Where this paragraph requires monitoring of airborne contaminants "as often as necessary," the competent person shall make a reasonable determination as to which substances to monitor and how frequently to monitor, considering at least the following factors:

**1926.800(j)(1)(i)(B)(*1*)** Location of jobsite: Proximity to fuel tanks, sewers, gas lines, old landfills, coal deposits, and swamps;

**1926.800(j)(1)(i)(B)(*2*)** Geology: Geological studies of the jobsite, particularly involving the soil type and its permeability;

**1926.800(j)(1)(i)(B)(*3*)** History: Presence of air contaminants in nearby jobsites, changes in levels of substances monitored on the prior shift; and

**1926.800(j)(1)(i)(B)(*4*)** Work practices and jobsite conditions: The use of diesel engines, use of explosives, use of fuel gas, volume and flow of ventilation, visible atmospheric conditions, decompression of the atmosphere, welding, cutting and hot work, and employees' physical reactions to working underground.

**1926.800(j)(1)(ii)(A)** The atmosphere in all underground work areas shall be tested as often as necessary to assure that the atmosphere at normal atmospheric pressure contains at least 19.5 percent oxygen and no more than 22 percent oxygen.

**1926.800(j)(1)(ii)(B)** Tests for oxygen content shall be made before tests for air contaminants.

**1926.800(j)(1)(iii)(A)** The atmosphere in all underground work areas shall be tested quantitatively for carbon monoxide, nitrogen dioxide, hydrogen sulfide, and other toxic gases, dusts, vapors, mists, and fumes as often as necessary to ensure that the permissible exposure limits prescribed in §1926.55 are not exceeded.

**1926.800(j)(1)(iii)(B)** The atmosphere in all underground work areas shall be tested quantitatively for methane and other flammable gases as often as necessary to determine:

**1926.800(j)(1)(iii)(B)(*1*)** Whether action is to be taken under paragraphs (j)(1)(vii), (viii), and (ix), of this section; and

**1926.800(j)(1)(iii)(B)(*2*)** Whether an operation is to be classified potentially gassy or gassy under paragraph (h) of this section.

**1926.800(j)(1)(iii)(C)** If diesel-engine or gasoline-engine driven ventilating fans or compressors are used, an initial test shall be made of the inlet air of the fan or compressor, with the engines operating, to ensure that the air supply is not contaminated by engine exhaust.

**1926.800(j)(1)(iii)(D)** Testing shall be performed as often as necessary to ensure that the ventilation requirements of paragraph (k) of this section are met.

**1926.800(j)(1)(iv)** When rapid excavation machines are used, a continuous flammable gas monitor shall be operated at the face with the sensor(s) placed as high and close to the front of the machine's cutter head as practicable.

**1926.800(j)(1)(v)(A)** Whenever air monitoring indicates the presence of 5 ppm or more of hydrogen sulfide, a test shall be conducted in the affected underground work area(s), at least at the beginning and midpoint of each shift, until the concentration of hydrogen sulfide has been less than 5 ppm for 3 consecutive days.

**1926.800(j)(1)(v)(B)** Whenever hydrogen sulfide is detected in an amount exceeding 10 ppm, a continuous sampling and indicating hydrogen sulfide monitor shall be used to monitor the affected work area.

**1926.800(j)(1)(v)(C)** Employees shall be informed when a concentration of 10 ppm hydrogen sulfide is exceeded.

**1926.800(j)(1)(v)(D)** The continuous sampling and indicating hydrogen sulfide monitor shall be designed, installed, and maintained to provide a visual and aural alarm when the hydrogen sulfide concentration reaches 20 ppm to signal that additional measures, such as respirator use, increased ventilation, or evacuation, might be necessary to maintain hydrogen sulfide exposure below the permissible exposure limit.

**1926.800(j)(1)(vi)** When the competent person determines, on the basis of air monitoring results or other information, that air contaminants may be present in sufficient quantity to be dangerous to life, the employer shall:

**1926.800(j)(1)(vi)(A)** Prominently post a notice at all entrances to the underground jobsite to inform all entrants of the hazardous condition; and

**1926.800(j)(1)(vi)(B)** Ensure that the necessary precautions are taken.

**1926.800(j)(1)(vii)** Whenever five percent or more of the lower explosive limit for methane or other flammable gases is detected in any underground work area(s) or in the air return, steps shall be taken to increase ventilation air volume or otherwise control the gas concentration, unless the employer is operating in accordance with the potentially gassy or gassy operation requirements. Such additional ventilation controls may be discontinued when gas concentrations are reduced below five percent of the lower explosive limit, but shall be reinstituted whenever the five percent level is exceeded.

**1926.800(j)(1)(viii)** Whenever 10 percent or more of the lower explosive limit for methane or other flammable gases is detected in the vicinity of welding, cutting, or other hot work, such work shall be suspended until the concentration of such flammable gas is reduced to less than 10 percent of the lower explosive limit.

**1926.800(j)(1)(ix)** Whenever 20 percent or more of the lower explosive limit for methane or other flammable gases is detected in any underground work area(s) or in the air return:

**1926.800(j)(1)(ix)(A)** All employees, except those necessary to eliminate the hazard, shall be immediately withdrawn to a safe location above ground; and

**1926.800(j)(1)(ix)(B)** Electrical power, except for acceptable pumping and ventilation equipment, shall be cut off to the area endangered by the flammable gas until the concentration of such gas is reduced to less than 20 percent of the lower explosive limit.

**1926.800(j)(2)** *Additional monitoring for potentially gassy and gassy operations.* Operations which meet the criteria for potentially gassy and gassy operations set forth in paragraph (h) of this section shall be subject to the additional monitoring requirements of this paragraph.

**1926.800(j)(2)(i)** A test for oxygen content shall be conducted in the affected underground work areas and work areas immediately adjacent to such areas at least at the beginning and midpoint of each shift.

**1926.800(j)(2)(ii)** When using rapid excavation machines, continuous automatic flammable gas monitoring equipment shall be used to monitor the air at the heading, on the rib, and in the return air duct. The continuous monitor shall signal the heading, and shut down electric power in the affected underground work area, except for acceptable pumping and ventilation equipment, when 20 percent or more of the lower explosive limit for methane or other flammable gases is encountered.

**1926.800(j)(2)(iii)** A manual flammable gas monitor shall be used as needed, but at least at the beginning and midpoint of each shift, to ensure that the limits prescribed in paragraphs (h) and (j) are not exceeded. In addition, a manual electrical shut down control shall be provided near the heading.

**1926.800(j)(2)(iv)** Local gas tests shall be made prior to and continuously during any welding, cutting, or other hot work.

**1926.800(j)(2)(v)** In underground operations driven by drill-and-blast methods, the air in the affected area shall be tested for flammable gas prior to re-entry after blasting, and continuously when employees are working underground.

**1926.800(j)(3)** *Recordkeeping.* A record of all air quality tests shall be maintained above ground at the worksite and be made available to the Secretary of Labor upon request. The record shall include the location, date, time, substance and amount monitored. Records of exposures to toxic substances shall be retained in accordance with §1910.20 of this chapter. All other air quality test records shall be retained until completion of the project.

**1926.800(k)** ***Ventilation.*** (1)(i) Fresh air shall be supplied to all underground work areas in sufficient quantities to prevent dangerous or harmful accumulation of dusts, fumes, mists, vapors or gases.

**1926.800(k)(1)(ii)** Mechanical ventilation shall be provided in all underground work areas except when the employer can demonstrate that natural ventilation provides the necessary air quality through sufficient air volume and air flow.

**1926.800(k)(2)** A minimum of 200 cubic feet (5.7 $m^3$) of fresh air per minute shall be supplied for each employee underground.

**1926.800(k)(3)** The linear velocity of air flow in the tunnel bore, in shafts, and in all other underground work areas shall be at least 30 feet (9.15 m) per minute where blasting or rock drilling is conducted, or where other conditions likely to produce dust, fumes, mists, vapors, or gases in harmful or explosive quantities are present.

**1926.800(k)(4)** The direction of mechanical air flow shall be reversible.

**1926.800(k)(5)** Following blasting, ventilation systems shall exhaust smoke and fumes to the outside atmosphere before work is resumed in affected areas.

**1926.800(k)(6)** Ventilation doors shall be designed and installed so that they remain closed when in use, regardless of the direction of the air flow.

**1926.800(k)(7)** When ventilation has been reduced to the extent that hazardous levels of methane or flammable gas may have accumulated, a competent person shall test all affected areas after ventilation has been restored and shall determine whether the atmosphere is within flammable limits before any power, other than for acceptable equipment, is restored or work is resumed.

**1926.800(k)(8)** Whenever the ventilation system has been shut down with all employees out of the underground area, only competent persons authorized to test for air contaminants shall be allowed underground until the ventilation has been restored and all affected areas have been tested for air contaminants and declared safe.

**1926.800(k)(9)** When drilling rock or concrete, appropriate dust control measures shall be taken to maintain dust levels within limits set in §1926.55. Such measures may include, but are not limited to, wet drilling, the use of vacuum collectors, and water mix spray systems.

**1926.800(k)(10)(i)** Internal combustion engines, except diesel-powered engines on mobile equipment, are prohibited underground.

**1926.800(k)(10)(ii)** Mobile diesel-powered equipment used underground in atmospheres other than gassy operations shall be either approved by MSHA in accordance with the provisions of 30 CFR part 32 (formerly Schedule 24), or shall be demonstrated by the employer to be fully equivalent to such MSHA-approved equipment, and shall be operated in accordance with that part. (Each brake horsepower of a diesel engine requires at least 100 cubic feet (28.32 $m^3$) of air per minute for suitable operation in addition to the air requirements for personnel. Some engines may require a greater amount of air to ensure that the allowable levels of carbon monoxide, nitric oxide, and nitrogen dioxide are not exceeded.)

**1926.800(k)(11)** Potentially gassy or gassy operations shall have ventilation systems installed which shall:

**1926.800(k)(11)(i)** Be constructed of fire-resistant materials; and

**1926.800(k)(11)(ii)** Have acceptable electrical systems, including fan motors.

**1926.800(k)(12)** Gassy operations shall be provided with controls located above ground for reversing the air flow of ventilation systems.

**1926.800(k)(13)** In potentially gassy or gassy operations, wherever mine-type ventilation systems using an offset main fan installed on the surface are used, they shall be equipped with explosion-doors or a weak-wall having an area at least equivalent to the cross-sectional area of the airway.

**1926.800(l)** ***Illumination.*** (1) Illumination requirements applicable to underground construction operations are found in *Table D-3* of §1926.56 of this part.

**1926.800(l)(2)** Only acceptable portable lighting equipment shall be used within 50 feet (15.24 m) of any underground heading during explosives handling.

**1926.800(m)** ***Fire prevention and control.*** Fire prevention and protection requirements applicable to underground construction operations are found in subpart F of this part, except as modified by the following additional standards.

**1926.800(m)(1)** Open flames and fires are prohibited in all underground construction operations except as permitted for welding, cutting and other hot work operations in paragraph (n) of this section.

**1926.800(m)(2)(i)** Smoking may be allowed only in areas free of fire and explosion hazards.

**1926.800(m)(2)(ii)** Readily visible signs prohibiting smoking and open flames shall be posted in areas having fire or explosion hazards.

**1926.800(m)(3)** The employer may store underground no more than a 24-hour supply of diesel fuel for the underground equipment used at the worksite.

**1926.800(m)(4)** The piping of diesel fuel from the surface to an underground location is permitted only if:

**1926.800(m)(4)(i)** Diesel fuel is contained at the surface in a tank whose maximum capacity is no more than the amount of fuel required to supply for a 24-hour period the equipment serviced by the underground fueling station; and

**1926.800(m)(4)(ii)** The surface tank is connected to the underground fueling station by an acceptable pipe or hose system that is controlled at the surface by a valve, and at the shaft bottom by a hose nozzle; and

**1926.800(m)(4)(iii)** The pipe is empty at all times except when transferring diesel fuel from the surface tank to a piece of equipment in use underground; and

**1926.800(m)(4)(iv)** Hoisting operations in the shaft are suspended during refueling operations if the supply piping in the shaft is not protected from damage.

**1926.800(m)(5)(i)** Gasoline shall not be carried, stored, or used underground.

**1926.800(m)(5)(ii)** Acetylene, liquefied petroleum gas, and Methylacetylene Propadiene Stabilized gas may be used underground only for welding, cutting and other hot work, and only in accordance with subpart J of this part, and paragraphs (j), (k), (m), and (n) of this section.

**1926.800(m)(6)** Oil, grease, and diesel fuel stored underground shall be kept in tightly sealed containers in fire-resistant areas at least 300 feet (91.44m) from underground explosive magazines, and at least 100 feet (30.48m) from shaft stations and steeply inclined passageways. Storage areas shall be positioned or diked so that the contents of ruptured or overturned containers will not flow from the storage area.

**1926.800(m)(7)** Flammable or combustible materials shall not be stored above ground within 100 feet (30.48m) of any access opening to any underground operation. Where this is not feasible because of space limitations at the jobsite, such materials may be located within the 100-foot limit, provided that:

**1926.800(m)(7)(i)** They are located as far as practicable from the opening; and

**1926.800(m)(7)(ii)** Either a fire-resistant barrier of not less than one-hour rating is placed between the stored material and the opening, or additional precautions are taken which will protect the materials from ignition sources.

**1926.800(m)(8)** Fire-resistant hydraulic fluids shall be used in hydraulically-actuated underground machinery and equipment unless such equipment is protected by a fire suppression system or by multi-purpose fire extinguisher(s) rated at of sufficient capacity for the type and size of hydraulic equipment involved, but rated at least 4A:40B:C.

**1926.800(m)(9)(i)** Electrical installations in underground areas where oil, grease, or diesel fuel are stored shall be used only for lighting fixtures.

**1926.800(m)(9)(ii)** Lighting fixtures in storage areas, or within 25 feet (7.62m) of underground areas where oil, grease, or diesel fuel are stored, shall be approved for Class I, Division 2 locations, in accordance with subpart K of this part.

**1926.800(m)(10)** Leaks and spills of flammable or combustible fluids shall be cleaned up immediately.

**1926.800(m)(11)** A fire extinguisher of at least 4A:40B:C rating or other equivalent extinguishing means shall be provided at the head pulley and at the tail pulley of underground belt conveyors.

**1926.800(m)(12)** Any structure located underground or within 100 feet (30.48m) of an opening to the underground shall be constructed of material having a fire-resistance rating of at least one hour.

**1926.800(n)** ***Welding, cutting, and other hot work.*** In addition to the requirements of subpart J of this part, the following requirements shall apply to underground welding, cutting, and other hot work.

**1926.800(n)(1)** No more than the amount of fuel gas and oxygen cylinders necessary to perform welding, cutting, or other hot work during the next 24-hour period shall be permitted underground.

**1926.800(n)(2)** Noncombustible barriers shall be installed below welding, cutting, or other hot work being done in or over a shaft or raise.

**1926.800(o)** ***Ground support***—(1) *Portal areas.* Portal openings and access areas shall be guarded by shoring, fencing, head walls, shotcreting or other equivalent protection to ensure safe access of employees and equipment. Adjacent areas shall be scaled or otherwise secured to prevent loose soil, rock, or fractured materials from endangering the portal and access area.

**1926.800(o)(2)** *Subsidence areas.* The employer shall ensure ground stability in hazardous subsidence areas by shoring, by filling in, or by erecting barricades and posting warning signs to prevent entry.

**1926.800(o)(3)** *Underground areas.* (i)(A) A competent person shall inspect the roof, face, and walls of the work area at the start of each shift and as often as necessary to determine ground stability.

**1926.800(o)(3)(i)(B)** Competent persons conducting such inspections shall be protected from loose ground by location, ground support or equivalent means.

**1926.800(o)(3)(ii)** Ground conditions along haulageways and travelways shall be inspected as frequently as necessary to ensure safe passage.

**1926.800(o)(3)(iii)** Loose ground that might be hazardous to employees shall be taken down, scaled or supported.

**1926.800(o)(3)(iv)(A)** Torque wrenches shall be used wherever bolts that depend on torsionally applied force are used for ground support.

**1926.800(o)(3)(iv)(B)** A competent person shall determine whether rock bolts meet the necessary torque, and shall determine the testing frequency in light of the bolt system, ground conditions and the distance from vibration sources.

**1926.800(o)(3)(v)** Suitable protection shall be provided for employees exposed to the hazard of loose ground while installing ground support systems.

**1926.800(o)(3)(vi)** Support sets shall be installed so that the bottoms have sufficient anchorage to prevent ground pressures from dislodging the support base of the sets. Lateral bracing (collar bracing, tie rods, or spreaders) shall be provided between immediately adjacent sets to ensure added stability.

**1926.800(o)(3)(vii)** Damaged or dislodged ground supports that create a hazardous condition shall be promptly repaired or replaced. When replacing supports, the new supports shall be installed before the damaged supports are removed.

**1926.800(o)(3)(viii)** A shield or other type of support shall be used to maintain a safe travelway for employees working in dead-end areas ahead of any support replacement operation.

**1926.800(o)(4)** *Shafts.* (i) Shafts and wells over 5 feet (1.53 m) in depth that employees must enter shall be supported by a steel casing, concrete pipe, timber, solid rock or other suitable material.

**1926.800(o)(4)(ii)(A)** The full depth of the shaft shall be supported by casing or bracing except where the shaft penetrates into solid rock having characteristics that will not change as a result of exposure. Where the shaft passes through earth into solid rock, or through solid rock into earth, and where there is potential for shear, the casing or bracing shall extend at least 5 feet (1.53 m) into the solid rock. When the shaft terminates in solid rock, the casing or bracing shall extend to the end of the shaft or 5 feet (1.53 m) into the solid rock, whichever is less.

**1926.800(o)(4)(ii)(B)** The casing or bracing shall extend 42 inches (1.07 m) plus or minus 3 inches (8 cm) above ground level, except that the minimum casing height may be reduced to 12 inches (0.3 m), provided that a standard railing is installed; that the ground adjacent to the top of the shaft is sloped away from the shaft collar to prevent entry of liquids; and that effective barriers are used to prevent mobile equipment operating near the shaft from jumping over the 12 inch (0.3 m) barrier.

**1926.800(o)(4)(iii)** After blasting operations in shafts, a competent person shall determine if the walls, ladders, timbers, blocking, or wedges have loosened. If so, necessary repairs shall be made before employees other than those assigned to make the repairs are allowed in or below the affected areas.

**1926.800(p)** ***Blasting.*** This paragraph applies in addition to the requirements for blasting and explosives operations, including handling of misfires, which are found in subpart U of this part.

**1926.800(p)(1)** Blasting wires shall be kept clear of electrical lines, pipes, rails, and other conductive material, excluding earth, to prevent explosives initiation or employee exposure to electric current.

**1926.800(p)(2)** Following blasting, an employee shall not enter a work area until the air quality meets the requirements of paragraph (j) of this section.

**1926.800(q)** ***Drilling.*** (1) A competent person shall inspect all drilling and associated equipment prior to each use. Equipment defects affecting safety shall be corrected before the equipment is used.

**1926.800(q)(2)** The drilling area shall be inspected for hazards before the drilling operation is started.

**1926.800(q)(3)** Employees shall not be allowed on a drill mast while the drill bit is in operation or the drill machine is being moved.

**1926.800(q)(4)** When a drill machine is being moved from one drilling area to another, drill steel, tools, and other equipment shall be secured and the mast shall be placed in a safe position.

**1926.800(q)(5)** Receptacles or racks shall be provided for storing drill steel located on jumbos.

**1926.800(q)(6)** Employees working below jumbo decks shall be warned whenever drilling is about to begin.

**1926.800(q)(7)** Drills on columns shall be anchored firmly before starting drilling, and shall be retightened as necessary thereafter.

**1926.800(q)(8)(i)** The employer shall provide mechanical means on the top deck of a jumbo for lifing unwieldy or heavy material.

**1926.800(q)(8)(ii)** When jumbo decks are over 10 feet (3.05 m) in height, the employer shall install stairs wide enough for two persons.

**1926.800(q)(8)(iii)** Jumbo decks more than 10 feet (3.05 m) in height shall be equipped with guardrails on all open sides, excluding access openings of platforms, unless an adjacent surface provides equivalent fall protection.

**1926.800(q)(8)(iv)(A)** Only employees assisting the operator shall be allowed to ride on jumbos, unless the jumbo meets the requirements of paragraph (r)(6)(ii) of this section.

**1926.800(q)(8)(iv)(B)** Jumbos shall be chocked to prevent movement while employees are working on them.

**1926.800(q)(8)(v)(A)** Walking and working surfaces of jumbos shall be maintained to prevent the hazards of slipping, tripping and falling.

**1926.800(q)(8)(v)(B)** Jumbo decks and stair treads shall be designed to be slip-resistent and secured to prevent accidental displacement.

**1926.800(q)(9)** Scaling bars shall be available at scaling operations and shall be maintained in good condition at all times. Blunted or severely worn bars shall not be used.

**1926.800(q)(10)(i)** Blasting holes shall not be drilled through blasted rock (muck) or water.

**1926.800(q)(10)(ii)** Employees in a shaft shall be protected either by location or by suitable barrier(s) if powered mechanical loading equipment is used to remove muck containing unfired explosives.

**1926.800(q)(11)** A caution sign reading "Buried Line," or similar wording shall be posted where air lines are buried or otherwise hidden by water or debris.

**1926.800(r)** ***Haulage.*** (1)(i) A competent person shall inspect haulage equipment before each shift.

**1926.800(r)(1)(ii)** Equipment defects affecting safety and health shall be corrected before the equipment is used.

**1926.800(r)(2)** Powered mobile haulage equipment shall have suitable means of stopping.

**1926.800(r)(3)(i)** Power mobile haulage equipment, including trains, shall have audible warning devices to warn employees to stay clear. The operator shall sound the warning device before moving the equipment and whenever necessary during travel.

**1926.800(r)(3)(ii)** The operator shall assure that lights which are visible to employees at both ends of any mobile equipment, including a train, are turned on whenever the equipment is operating.

**1926.800(r)(4)** In those cabs where glazing is used, the glass shall be safety glass, or its equivalent, and shall be maintained and cleaned so that vision is not obstructed.

**1926.800(r)(5)** Anti-roll back devices or brakes shall be installed on inclined conveyor drive units to prevent conveyors from inadvertently running in reverse.

**1926.800(r)(6)(i)(A)** Employees shall not be permitted to ride a power-driven chain, belt, or bucket conveyor unless the conveyor is specifically designed for the transportation of persons.

**1926.800(r)(6)(i)(B)** Endless belt-type manlifts are prohibited in underground construction.

**1926.800(r)(6)(i)(C)** General requirements also applicable to underground construction for use of conveyors in construction are found in §1926.555 of this part.

**1926.800(r)(6)(ii)** No employee shall ride haulage equipment unless it is equipped with seating for each passenger and protects passengers from being struck, crushed, or caught between other equipment or surfaces. Members of train crews may ride on a locomotive if it is equipped with handholds and nonslip steps or footboards. Requirements applicable to Underground Construction for motor vehicle transportation of employees are found in §1926.601 of this part.

**1926.800(r)(7)** Powered mobile haulage equipment, including trains, shall not be left unattended unless the master switch or motor is turned off; operating controls are in neutral or park position; and the brakes are set, or equivalent precautions are taken to prevent rolling.

**1926.800(r)(8)** Whenever rails serve as a return for a trolley circuit, both rails shall be bonded at every joint and crossbonded every 200 feet (60.96 m).

**1926.800(r)(9)** When dumping cars by hand, the car dumps shall have tiedown chains, bumper blocks, or other locking or holding devices to prevent the cars from overturning.

**1926.800(r)(10)** Rocker-bottom or bottom-dump cars shall be equipped with positive locking devices to prevent unintended dumping.

**1926.800(r)(11)** Equipment to be hauled shall be loaded and secured to prevent sliding or dislodgement.

**1926.800(r)(12)(i)** Mobile equipment, including rail-mounted equipment, shall be stopped for manual connecting or service work.

**1926.800(r)(12)(ii)** Employees shall not reach between moving cars during coupling operations.

**1926.800(r)(12)(iii)** Couplings shall not be aligned, shifted or cleaned on moving cars or locomotives.

**1926.800(r)(13)(i)** Safety chains or other connections shall be used in addition to couplers to connect man cars or powder cars whenever the locomotive is uphill of the cars.

**1926.800(r)(13)(ii)** When the grade exceeds one percent and there is a potential for runaway cars, safety chains or other connections shall be used in addition to couplers to connect haulage cars or, as an alternative, the locomotive must be downhill of the train.

**1926.800(r)(13)(iii)** Such safety chains or other connections shall be capable of maintaining connection between cars in the event of either coupler disconnect, failure or breakage.

**1926.800(r)(14)** Parked rail equipment shall be chocked, blocked, or have brakes set to prevent inadvertent movement.

**1926.800(r)(15)** Berms, bumper blocks, safety hooks, or equivalent means shall be provided to prevent overtravel and overturning of haulage equipment at dumping locations.

**1926.800(r)(16)** Bumper blocks or equivalent stopping devices shall be provided at all track dead ends.

**1926.800(r)(17)(i)** Only small handtools, lunch pails or similar small items may be transported with employees in man-cars, or on top of a locomotive.

**1926.800(r)(17)(ii)** When small hand tools or other small items are carried on top of a locomotive, the top shall be designed or modified to retain them while traveling.

**1926.800(r)(18)(i)** Where switching facilities are available, occupied personnel-cars shall be pulled, not pushed. If personnel-cars must be pushed and visibility of the track ahead is hampered, then a qualified person shall be stationed in the lead car to give signals to the locomotive operator.

**1926.800(r)(18)(ii)** Crew trips shall consist of personnel-loads only.

**1926.800(s)** ***Electrical safety.*** This paragraph applies in addition to the general requirements for electrical safety which are found in subpart K of this part.

**1926.800(s)(1)** Electric power lines shall be insulated or located away from water lines, telephone lines, air lines, or other conductive materials so that a damaged circuit will not energize the other systems.

**1926.800(s)(2)** Lighting circuits shall be located so that movement of personnel or equipment will not damage the circuits or disrupt service.

**1926.800(s)(3)** Oil-filled transformers shall not be used underground unless they are located in a fire-resistant enclosure suitably vented to the outside and surrounded by a dike to retain the contents of the transformers in the event of rupture.

**1926.800(t)** ***Hoisting unique to underground construction.*** Except as modified by this paragraph (t), the following provisions of subpart N of this part apply: Requirements for cranes are found in §1926.550 of this part. Paragraph (g) of §1926.550 applies to crane-hoisting of personnel, except that the limitation in paragraph (g)(2) does not apply to the routine access of employees to the underground via a shaft. Requirements for material hoists are found in §1926.552(a) and (b) of this part. Requirements for personnel hoists are found in the personnel hoist requirements of §1926.552(a) and (c) of this part and in the elevator requirement of §1926.552(a) and (d) of this part.

**1926.800(t)(1)** *General requirements for cranes and hoists.* (i) Materials, tools, and supplies being raised or lowered, whether within a cage or otherwise, shall be secured or stacked in a manner to prevent the load from shifting, snagging or falling into the shaft.

**1926.800(t)(1)(ii)** A warning light suitably located to warn employees at the shaft bottom and subsurface shaft entrances shall flash whenever a load is above the shaft bottom or subsurface entrances, or the load is being moved in the shaft. This paragraph does not apply to fully enclosed hoistways.

**1926.800(t)(1)(iii)** Whenever a hoistway is not fully enclosed and employees are at the shaft bottom, conveyances or equipment shall be stopped at least 15 feet (4.57 m) above the bottom of the shaft and held there until the signalman at the bottom of the shaft directs the operator to continue lowering the load, except that the load may be lowered without stopping if the load or conveyance is within full view of a bottom signalman who is in constant voice communication with the operator.

**1926.800(t)(1)(iv)(A)** Before maintenance, repairs, or other work is commenced in the shaft served by a cage, skip, or bucket, the operator and other employees in the area shall be informed and given suitable instructions.

**1926.800(t)(1)(iv)(B)** A sign warning that work is being done in the shaft shall be installed at the shaft collar, at the operator's station, and at each underground landing.

**1926.800(t)(1)(v)** Any connection between the hoisting rope and the cage or skip shall be compatible with the type of wire rope used for hoisting.

**1926.800(t)(1)(vi)** Spin-type connections, where used, shall be maintained in a clean condition and protected from foreign matter that could affect their operation.

**1926.800(t)(1)(vii)** Cage, skip, and load connections to the hoist rope shall be made so that the force of the hoist pull, vibration, misalignment, release of lift force, or impact will not disengage the connection. Moused or latched open-throat hooks do not meet this requirement.

**1926.800(t)(1)(viii)** When using wire rope wedge sockets, means shall be provided to prevent wedge escapement and to ensure that the wedge is properly seated.

**1926.800(t)(2)** *Additional requirements for cranes.* Cranes shall be equipped with a limit switch to prevent overtravel at the boom tip. Limit switches are to be used only to limit travel of loads when operational controls malfunction and shall not be used as a substitute for other operational controls.

**1926.800(t)(3)** *Additional requirements for hoists.* (i) Hoists shall be designed so that the load hoist drum is powered in both directions of rotation, and so that brakes are automatically applied upon power release or failure.

**1926.800(t)(3)(ii)** Control levers shall be of the "deadman type" which return automatically to their center (neutral) position upon release.

**1926.800(t)(3)(iii)** When a hoist is used for both personnel hoisting and material hoisting, load and speed ratings for personnel and for materials shall be assigned to the equipment.

**1926.800(t)(3)(iv)** Material hoisting may be performed at speeds higher than the rated speed for personnel hoisting if the hoist and components have been designed for such higher speeds and if shaft conditions permit.

**1926.800(t)(3)(v)** Employees shall not ride on top of any cage, skip or bucket except when necessary to perform inspection or maintenance of the hoisting system, in which case they shall be protected by a body belt/harness system to prevent falling.

**1926.800(t)(3)(vi)** Personnel and materials (other than small tools and supplies secured in a manner that will not create a hazards to employees) shall not be hoisted together in the same conveyance. However, if the operator is protected from the shifting of materials, then the operator may ride with materials in cages or skips which are designed to be controlled by an operator within the cage or skip.

**1926.800(t)(3)(vii)** Line speed shall not exceed the design limitations of the systems.

**1926.800(t)(3)(viii)** Hoists shall be equipped with landing level indicators at the operator's station. Marking of the hoist rope does not satisfy this requirement.

**1926.800(t)(3)(ix)** Whenever glazing is used in the hoist house, it shall be safety glass, or its equivalent, and be free of distortions and obstructions.

**1926.800(t)(3)(x)** A fire extinguisher that is rated at least 2A:10B:C (multi-purpose, dry chemical) shall be mounted in each hoist house.

**1926.800(t)(3)(xi)** Hoist controls shall be arranged so that the operator can perform all operating cycle functions and reach the emergency power cutoff without having to reach beyond the operator's normal operating position.

**1926.800(t)(3)(xii)** Hoists shall be equipped with limit switches to prevent overtravel at the top and bottom of the hoistway.

**1926.800(t)(3)(xiii)** Limit switches are to be used only to limit travel of loads when operational controls malfunction and shall not be used as a substitute for other operational controls.

**1926.800(t)(3)(xiv)** Hoist operators shall be provided with a closed-circuit voice communication system to each landing station, with speaker-microphones so located that the operator can communicate with individual landing stations during hoist use.

**1926.800(t)(3)(xv)** When sinking shafts 75 feet (22.86 m) or less in depth, cages, skips, and buckets that may swing, bump, or snag against shaft sides or other structural protrusions shall be guided by fenders, rails, ropes, or a combination of those means.

**1926.800(t)(3)(xvi)** When sinking shafts more than 75 feet (22.86 m) in depth, all cages, skips, and buckets shall be rope or rail-guided to within a rail length from the sinking operation.

**1926.800(t)(3)(xvii)** Cages, skips, and buckets in all completed shafts, or in all shafts being used as completed shafts, shall be rope or rail-guided for the full length of their travel.

**1926.800(t)(3)(xviii)** Wire rope used in load lines of material hoists shall be capable of supporting, without failure, at least five times the maximum intended load or the factor recommended by the rope manufacturer, whichever is greater. Refer to §1926.552(c)(14)(iii) of this part for design factors for wire rope used in personnel hoists. The design factor shall be calculated by dividing the breaking strength of wire rope, as reported in the manufacturer's rating tables, by the total static load, including the weight of the wire rope in the shaft when fully extended.

**1926.800(t)(3)(xix)** A competent person shall visually check all hoisting machinery, equipment, anchorages, and hoisting rope at the beginning of each shift and during hoist use, as necessary.

**1926.800(t)(3)(xx)** Each safety device shall be checked by a competent person at least weekly during hoist use to ensure suitable operation and safe condition.

**1926.800(t)(3)(xxi)** In order to ensure suitable operation and safe condition of all functions and safety devices, each hoist assembly shall be inspected and load-tested to 100 percent of its rated capacity: at the time of installation; after any repairs or alterations affecting its structural integrity; after the operation of any safety device; and annually when in use. The employer shall prepare a certification record which includes the date each inspection and load-test was performed; the signature of the person who performed the inspection and test; and a serial number or other identifier for the hoist that was inspected and tested. The most recent certification record shall be maintained on file until completion of the project.

**1926.800(t)(3)(xxii)** Before hoisting personnel or material, the operator shall perform a test run of any cage or skip whenever it has been out of service for one complete shift, and whenever the assembly or components have been repaired or adjusted.

**1926.800(t)(3)(xxiii)** Unsafe conditions shall be corrected before using the equipment.

**1926.800(t)(4)** *Additional requirements for personnel hoists.* (i) Hoist drum systems shall be equipped with at least two means of stopping the load, each of which shall be capable of stopping and holding 150 percent of the hoist's rated line pull. A broken-rope safety, safety catch, or arrestment device is not a permissible means of stopping under this paragraph.

**1926.800(t)(4)(ii)** The operator shall remain within sight and sound of the signals at the operator's station.

**1926.800(t)(4)(iii)** All sides of personnel cages shall be enclosed by one-half inch (12.70 mm) wire mesh (not less than No. 14 gauge or equivalent) to a height of not less than 6 feet (1.83 m). However, when the cage or skip is being used as a work platform, its sides may be reduced in height to 42 inches 1.07 m) when the conveyance is not in motion.

**1926.800(t)(4)(iv)** All personnel cages shall be provided with positive locking door that does not open outward.

**1926.800(t)(4)(v)** All personnel cages shall be provided with a protective canopy. The canopy shall be made of steel plate, at least 3/16-inch (4.763 mm) in thickness, or material of equivalent strength and impact resistance. The canopy shall be sloped to the outside, and so designed that a section may be readily pushed upward to afford emergency egress. The canopy shall cover the top in such a manner as to protect those inside from objects falling in the shaft.

**1926.800(t)(4)(vi)** Personnel platforms operating on guide rails or guide ropes shall be equipped with broken-rope safety devices, safety catches or arrestment devices that will stop and hold 150 percent of the weight of the personnel platform and its maximum rated load.

**1926.800(t)(4)(vii)** During sinking operations in shafts where guides and safeties are not yet used, the travel speed of the personnel platform shall not exceed 200 feet (60.96 m) per minute. Governor controls set for 200 feet (60.96 m) per minute shall be installed in the control system and shall be used during personnel hoisting.

**1926.800(t)(4)(viii)** The personnel platform may travel over the controlled length of the hoistway at rated speeds up to 600 feet (182.88 m) per minute during sinking operations in shafts where guides and safeties are used.

**1926.800(t)(4)(ix)** The personnel platform may travel at rated speeds greater than 600 feet (182.88 m) per minute in completed shafts.

**1926.800(u)** ***Definitions.*** "Accept"—Any device, equipment, or appliance that is either approved by MSHA and maintained in permissible condition, or is listed or labeled for the class and location under subpart K of this part.

"Rapid Excavation Machine"—Tunnel boring machines, shields, roadheaders, or any other similar excavation machine.

[54 FR 23850, June 2, 1989; 58 FR 35311, June 30, 1993, as amended at 61 FR 5510, Feb. 13, 1996; 63 FR 1297, Jan. 8, 1998]

### §1926.801 Caissons.

**1926.801(a)** Wherever, in caisson work in which compressed air is used, and the working chamber is less than 11 feet in length, and when such caissons are at any time suspended or hung while work is in progress so that the bottom of the excavation is more than 9 feet below the deck of the working chamber, a shield shall be erected therein for the protection of the employees.

**1926.801(b)** Shafts shall be subjected to a hydrostatic or air-pressure test, at which pressure they shall be tight. The shaft shall be stamped on the outside shell about 12 inches from each flange to show the pressure to which they have been subjected.

**1926.801(c)** Whenever a shaft is used, it shall be provided, where space permits, with a safe, proper, and suitable staircase for its entire length, including landing platforms, not more than 20 feet apart. Where this is impracticable, suitable ladders shall be installed with landing platforms located about 20 feet apart to break the climb.

**1926.801(d)** All caissons having a diameter or side greater than 10 feet shall be provided with a man lock and shaft for the exclusive use of employees.

**1926.801(e)** In addition to the gauge in the locks, an accurate gauge shall be maintained on the outer and inner side of each bulkhead. These gauges shall be accessible at all times and kept in accurate working order.

**1926.801(f)** In caisson operations where employees are exposed to compressed air working environments, the requirements contained in §1926.803 shall be complied with.

### §1926.802 Cofferdams.

**1926.802(a)** If overtopping of the cofferdam by high waters is possible, means shall be provided for controlled flooding of the work area.

**1926.802(b)** Warning signals for evacuation of employees in case of emergency shall be developed and posted.

**1926.802(c)** Cofferdam walkways, bridges, or ramps with at least two means of rapid exit shall be provided with guardrails as specified in subpart M of this part.

**1926.802(d)** Cofferdams located close to navigable shipping channels shall be protected from vessels in transit, where possible.

### §1926.803 Compressed air.

**1926.803(a)** ***General provisions.*** (1) There shall be present, at all times, at least one competent person designated by and representing the employer, who shall be familiar with this subpart in all respects, and responsible for full compliance with these and other applicable subparts.

**1926.803(a)(2)** Every employee shall be instructed in the rules and regulations which concern his safety or the safety of others.

**1926.803(b)** ***Medical attendance, examination, and regulations.*** (1) There shall be retained one or more licensed physicians familiar with and experienced in the physical requirements and the medical aspects of compressed air work and the treatment of decompression illness. He shall be available at all times while work is in progress in order to provide medical supervision of employees employed in compressed air work. He shall himself be physically qualified and be willing to enter a pressurized environment.

**1926.803(b)(2)** No employee shall be permitted to enter a compressed air environment until he has been examined by the physician and reported by him to be physically qualified to engage in such work.

**1926.803(b)(3)** In the event an employee is absent from work for 10 days, or is absent due to sickness or injury, he shall not resume work until he is reexamined by the physician, and his physical condition reported, as provided in this paragraph, to be such as to permit him to work in compressed air.

**1926.803(b)(4)** After an employee has been employed continuously in compressed air for a period designated by the physician, but not to exceed 1 year, he shall be reexamined by the physician to determine if he is still physically qualified to engage in compressed air work.

**1926.803(b)(5)** Such physician shall at all times keep a complete and full record of examinations made by him. The physician shall also keep an accurate record of any decompression illness or other illness or injury incapacitating any employee for work, and of all loss of life that occurs in the operation of a tunnel, caisson, or other compartment in which compressed air is used.

**1926.803(b)(6)** Records shall be available for the inspection of the Secretary or his representatives, and a copy thereof shall be forwarded to OSHA within 48 hours following the occurrence of the accident, death, injury, or decompression illness. It shall state as fully as possible the cause of said death or decompression illness, and the place where the injured or sick employee was taken, and such other relative information as may be required by the Secretary.

**1926.803(b)(7)** A fully equipped first aid station shall be provided at each tunnel project regardless of the number of persons employed. An ambulance or transportation suitable for a litter case shall be at each project.

**1926.803(b)(8)** Where tunnels are being excavated from portals more than 5 road miles apart, a first aid station and transportation facilities shall be provided at each portal.

**1926.803(b)(9)** A medical lock shall be established and maintained in immediate working order whenever air pressure in the working chamber is increased above the normal atmosphere.

**1926.803(b)(10)** The medical lock shall:

**1926.803(b)(10)(i)** Have at least 6 feet of clear headroom at the center, and be subdivided into not less than two compartments;

**1926.803(b)(10)(ii)** Be readily accessible to employees working under compressed air;

**1926.803(b)(10)(iii)** Be kept ready for immediate use for at least 5 hours subsequent to the emergence of any employee from the working chamber;

**1926.803(b)(10)(iv)** Be properly heated, lighted and ventilated;

**1926.803(b)(10)(v)** Be maintained in a sanitary condition;

**1926.803(b)(10)(vi)** Have a nonshatterable port through which the occupant(s) may be kept under constant observation;

**1926.803(b)(10)(vii)** Be designed for a working pressure of 75 p.s.i.g.

**1926.803(b)(10)(viii)** Be equipped with internal controls which may be overridden by external controls;

**1926.803(b)(10)(ix)** Be provided with air pressure gauges to show the air pressure within each compartment to observers inside and outside the medical lock.

**1926.803(b)(10)(x)** Be equipped with a manual type sprinkler system that can be activated inside the lock or by the outside lock tender.

**1926.803(b)(10)(xi)** Be provided with oxygen lines and fittings leading into external tanks. The lines shall be fitted with check valves to prevent reverse flow. The oxygen system inside the chamber shall be of a closed circuit design and be so designed as to automatically shut off the oxygen supply whenever the fire system is activated.

**1926.803(b)(10)(xii)** Be in constant charge of an attendant under the direct control of the retained physician. The attendant shall be trained in the use of the lock and suitably instructed regarding steps to be taken in the treatment of employee exhibiting symptoms compatible with a diagnosis of decompression illness;

**1926.803(b)(10)(xiii)** Be adjacent to an adequate emergency medical facility;

**1926.803(b)(10)(xiv)** The medical facility shall be equipped with demand-type oxygen inhalation equipment approved by the U.S. Bureau of Mines;

**1926.803(b)(10)(xv)** Be capable of being maintained at a temperature, in use, not to exceed 90°F. nor be less than 70°F.; and

**1926.803(b)(10)(xvi)** Be provided with sources of air, free of oil and carbon monoxide, for normal and emergency use, which are capable of raising the air pressure in the lock from 0 to 75 p.s.i.g. in 5 minutes.

**1926.803(b)(11)** Identification badges shall be furnished to all employees, indicating that the wearer is a compressed air worker. A permanent record shall be kept of all identification badges issued. The badge shall give the employee's name, address of the medical lock, the telephone number of the licensed physician for the compressed air project, and contain instructions that in case of emergency of unknown or doubtful cause or illness, the wearer shall be rushed to the medical lock. The badge shall be worn at all times—off the job, as well as on the job.

**1926.803(c)** ***Telephone and signal communication.*** (1) Effective and reliable means of communication, such as bells, whistles, or telephones, shall be maintained, at all times between all the following locations:

**1926.803(c)(1)(i)** The working chamber face;

**1926.803(c)(1)(ii)** The working chamber side of the man lock near the door;

**1926.803(c)(1)(iii)** The interior of the man lock;

**1926.803(c)(1)(iv)** Lock attendant's station;

**1926.803(c)(1)(v)** The compressor plant;

**1926.803(c)(1)(vi)** The first-aid station;

**1926.803(c)(1)(vii)** The emergency lock (if one is required); and

**1926.803(c)(1)(viii)** The special decompression chamber (if one is required).

**1926.803(d)** ***Signs and records.*** (1) The time of decompression shall be posted in each man lock as follows:

**TIME OF DECOMPRESSION FOR THIS LOCK**

_____ pounds to _____ pounds in _____ minutes.

_____ pounds to _____ pounds in _____ minutes.

(Signed by) ______________________________ (Superintendent)

This form shall be posted in the Man Lock at all times.

**1926.803(d)(2)** Any code of signals used shall be conspicuously posted near workplace entrances and such other locations as may be necessary to bring them to the attention of all employees concerned.

**1926.803(d)(3)** For each 8-hour shift, a record of employees employed under air pressure shall be kept by an employee who shall remain outside the lock near the entrance. This record shall show the period each employee spends in the air chamber and the time taken from decompression. A copy shall be submitted to the appointed physician after each shift.

**1926.803(e)** ***Compression.*** (1) Every employee going under air pressure for the first time shall be instructed on how to avoid excessive discomfort.

**1926.803(e)(2)** During the compression of employees, the pressure shall not be increased to more than 3 p.s.i.g. within the first minute. The pressure shall be held at 3 p.s.i.g. and again at 7 p.s.i.g. sufficiently long to determine if any employees are experiencing discomfort.

**1926.803(e)(3)** After the first minute the pressure shall be raised uniformly and at a rate not to exceed 10 p.s.i. per minute.

**1926.803(e)(4)** If any employee complains of discomfort, the pressure shall be held to determine if the symptoms are relieved. If, after 5 minutes the discomfort does not disappear, the lock attendant shall gradually reduce the pressure until the employee signals that the discomfort has ceased. If he does not indicate that the discomfort has disappeared, the lock attendant shall reduce the pressure to atmospheric and the employee shall be released from the lock.

**1926.803(e)(5)** No employee shall be subjected to pressure exceeding 50 pounds per square inch except in emergency.

**1926.803(f)** ***Decompression.*** (1) Decompression to normal condition shall be in accordance with the Decompression Tables in Appendix A of this subpart.

**1926.803(f)(2)** In the event it is necessary for an employee to be in compressed air more than once in a 24-hour period, the appointed physician shall be responsible for the establishment of methods and procedures of decompression applicable to repetitive exposures.

**1926.803(f)(3)** If decanting is necessary, the appointed physician shall establish procedures before any employee is permitted to be decompressed by decanting methods. The period of time that the employees spend at atmospheric pressure between the decompression following the shift and recompression shall not exceed 5 minutes.

**1926.803(g)** ***Man locks and special decompression chambers***—(1) *Man locks.* (i) Except in emergency, no employees employed in compressed air shall be permitted to pass from the working chamber to atmospheric pressure until after decompression, in accordance with the procedures in this subpart.

**1926.803(g)(1)(ii)** The lock attendant in charge of a man lock shall be under the direct supervision of the appointed physician. He shall be stationed at the lock controls on the free air side during the period of compression and decompression and shall remain at the lock control station whenever there are men in the working chamber or in the man lock.

**1926.803(g)(1)(iii)** Except where air pressure in the working chamber is below 12 p.s.i.g., each man lock shall be equipped with automatic controls which, through taped programs, cams, or similar apparatus, shall automatically regulate decompressions. It shall also be equipped with manual controls to permit the lock attendant to override the automatic mechanism in the event of an emergency, as provided in paragraph (g)(1)(viii) of this section.

**1926.803(g)(1)(iv)** A manual control, which can be used in the event of an emergency, shall be placed inside the man lock.

**1926.803(g)(1)(v)** A clock, thermometer, and continuous recording pressure gauge with a 4-hour graph shall be installed outside of each man lock and shall be changed prior to each shift's decompression. The chart shall be of sufficient size to register a legible record of variations in pressure within the man lock and shall be visible to the lock attendant. A copy of each graph shall be submitted to the appointed physician after each shift. In addition, a pressure gauge, clock, and thermometer shall also be installed in each man lock. Additional fittings shall be provided so that test gauges may be attached whenever necessary.

**1926.803(g)(1)(vi)** Except where air pressure is below 12 p.s.i.g. and there is no danger of rapid flooding, all caissons having a working area greater than 150 square feet, and each bulkhead in tunnels of 14 feet or more in diameter, or equivalent area, shall have at least two locks in perfect working condition, one of which shall be used exclusively as a man lock, the other, as a materials lock.

**1926.803(g)(1)(vii)** Where only a combination man-and-materials lock is required, this single lock shall be of sufficient capacity to hold the employees constituting two successive shifts.

**1926.803(g)(1)(viii)** Emergency locks shall be large enough to hold an entire heading shift and a limit maintained of 12 p.s.i.g. There shall be a chamber available for oxygen decompression therapy to 28 p.s.i.g.

**1926.803(g)(1)(ix)** The man lock shall be large enough so that those using it are not compelled to be in a cramped position, and shall not have less than 5 feet clear head room at the center and a minimum of 30 cubic feet of air space per occupant.

**1926.803(g)(1)(x)** Locks on caissons shall be so located that the bottom door shall be not less than 3 feet above the water level surrounding the caisson on the outside. (The water level, where it is affected by tides, is construed to mean high tide.)

**1926.803(g)(1)(xi)** In addition to the pressure gauge in the locks, an accurate pressure gauge shall be maintained on the outer and inner side of each bulkhead. These gauges shall be accessible at all times and shall be kept in accurate working order.

**1926.803(g)(1)(xii)** Man locks shall have an observation port at least 4 inches in diameter located in such a position that all occupants of the man lock may be observed from the working chamber and from the free air side of the lock.

**1926.803(g)(1)(xiii)** Adequate ventilation in the lock shall be provided.

**1926.803(g)(1)(xiv)** Man locks shall be maintained at a minimum temperature of 70 °F.

**1926.803(g)(1)(xv)** When locks are not in use and employees are in the working chamber, lock doors shall be kept open to the working chamber, where practicable.

**1926.803(g)(1)(xvi)** Provision shall be made to allow for rescue parties to enter the tunnel if the working force is disabled.

**1926.803(g)(1)(xvii)** A special decompression chamber of sufficient size to accommodate the entire force of employees being decompressed at the end of a shift shall be provided whenever the regularly established working period requires a total time of decompression exceeding 75 minutes.

**1926.803(g)(2)** *Special decompression chamber.* (i) The headroom in the special decompression chamber shall be not less than a minimum 7 feet and the cubical content shall provide at least 50 cubic feet of airspace for each employee. For each occupant, there shall be provided 4 square feet of free walking area and 3 square feet of seating space, exclusive of area required for lavatory and toilet facilities. The rated capacity shall be based on the stated minimum space per employee and shall be posted at the chamber entrance. The posted capacity shall not be exceeded, except in case of emergency.

**1926.803(g)(2)(ii)** Each special decompression chamber shall be equipped with the following:

**1926.803(g)(2)(ii)(*a*)** A clock or clocks suitably placed so that the attendant and the chamber occupants can readily ascertain the time;

**1926.803(g)(2)(ii)(*b*)** Pressure gauges which will indicate to the attendants and to the chamber occupants the pressure in the chamber;

**1926.803(g)(2)(ii)(*c*)** Valves to enable the attendant to control the supply and discharge of compressed air into and from the chamber;

**1926.803(g)(2)(ii)(*d*)** Valves and pipes, in connection with the air supply and exhaust, arranged so that the chamber pressure can be controlled from within and without;

**1926.803(g)(2)(ii)(*e*)** Effective means of oral intercommunication between the attendant, occupants of the chamber, and the air compressor plant; and

**1926.803(g)(2)(ii)(*f*)** An observation port at the entrance to permit observation of the chamber occupants.

**1926.803(g)(2)(iii)** Seating facilities in special decompression chambers shall be so arranged as to permit a normal sitting posture without cramping. Seating space, not less than 18 inches by 24 inches wide, shall be provided per occupant.

**1926.803(g)(2)(iv)** Adequate toilet and washing facilities, in a screened or enclosed recess, shall be provided. Toilet bowls shall have a built-in protector on the rim so that an air space is created when the seat lid is closed.

**1926.803(g)(2)(v)** Fresh and pure drinking water shall be available. This may be accomplished by either piping water into the special decompression chamber and providing drinking fountains, or by providing individual canteens, or by some other sanitary means. Community drinking vessels are prohibited.

**1926.803(g)(2)(vi)** No refuse or discarded material of any kind shall be permitted to accumulate, and the chamber shall be kept clean.

**1926.803(g)(2)(vii)** Unless the special decompression chamber is serving as the man lock to atmospheric pressure, the special decompression chamber shall be situated, where practicable, adjacent to the man lock on the atmospheric pressure side of the bulkhead. A passageway shall be provided, connecting the special chamber with the man lock, to permit employees in the process of decompression to move from the man lock to the special chamber without a reduction in the ambient pressure from that designated for the next stage of decompression. The passageway shall be so arranged as to not interfere with the normal operation of the man lock, nor with the release of the occupants of the special chamber to atmospheric pressure upon the completion of the decompression procedure.

**1926.803(h)** ***Compressor plant and air supply.*** (1) At all times there shall be a thoroughly experienced, competent, and reliable person on duty at the air control valves as a gauge tender who shall regulate the pressure in the working areas. During tunneling operations, one gauge tender may regulate the pressure in not more than two headings: *Provided*, That the gauge and controls are all in one location. In caisson work, there shall be a gauge tender for each caisson.

**1926.803(h)(2)** The low air compressor plant shall be of sufficient capacity to not only permit the work to be done safely, but shall also provide a margin to meet emergencies and repairs.

**1926.803(h)(3)** Low air compressor units shall have at least two independent and separate sources of power supply and each shall be capable of operating the entire low air plant and its accessory systems.

**1926.803(h)(4)** The capacity, arrangement, and number of compressors shall be sufficient to maintain the necessary pressure without overloading the equipment and to assure maintenance of such pressure in the working chamber during periods of breakdown, repair, or emergency.

**1926.803(h)(5)** Switching from one independent source of power supply to the other shall be done periodically to ensure the workability of the apparatus in an emergency.

**1926.803(h)(6)** Duplicate low-pressure air feedlines and regulating valves shall be provided between the source of air supply and a point beyond the locks with one of the lines extending to within 100 feet of the working face.

**1926.803(h)(7)** All high- and low-pressure air supply lines shall be equipped with check valves.

**1926.803(h)(8)** Low-pressure air shall be regulated automatically. In addition, manually operated valves shall be provided for emergency conditions.

**1926.803(h)(9)** The air intakes for all air compressors shall be located at a place where fumes, exhaust, gases, and other air contaminants will be at a minimum.

**1926.803(h)(10)** Gauges indicating the pressure in the working chamber shall be installed in the compressor building, the lock attendant's station, and at the employer's field office.

**1926.803(i)** ***Ventilation and air quality.*** (1) Exhaust valves and exhaust pipes shall be provided and operated so that the working chamber shall be well ventilated, and there shall be no pockets of dead air. Outlets may be required at intermediate points along the main low-pressure air supply line to the heading to eliminate such pockets of dead air. Ventilating air shall be not less than 30 cubic feet per minute.

**1926.803(i)(2)** The air in the workplace shall be analyzed by the employer not less than once each shift, and records of such tests shall be kept on file at the place where the work is in progress. The test results shall be within the threshold limit values specified in subpart D of this part, for hazardous gases, and within 10 percent of the lower explosive limit of flammable gases. If these limits are not met, immediate action to correct the situation shall be taken by the employer.

**1926.803(i)(3)** The temperature of all working chambers which are subjected to air pressure shall, by means of after-coolers or other suitable devices, be maintained at a temperature not to exceed 85 °F.

**1926.803(i)(4)** Forced ventilation shall be provided during decompression. During the entire decompression period, forced ventilation through chemical or mechanical air purifying devices that will ensure a source of fresh air shall be provided.

**1926.803(i)(5)** Whenever heat-producing machines (moles, shields) are used in compressed air tunnel operations, a positive means of removing the heat build-up at the heading shall be provided.

**1926.803(j)** ***Electricity.*** (1) All lighting in compressed-air chambers shall be by electricity exclusively, and two independent electric-lighting systems with independent sources of supply shall be used. The emergency source shall be arranged to become automatically operative in the event of failure of the regularly used source.

**1926.803(j)(2)** The minimum intensity of light on any walkway, ladder, stairway, or working level shall be not less than 10 foot-candles, and in all workplaces the lighting shall at all times be such as to enable employees to see clearly.

**1926.803(j)(3)** All electrical equipment and wiring for light and power circuits shall comply with the requirements of subpart K of this part for use in damp, hazardous, high temperature, and compressed air environments.

**1926.803(j)(4)** External parts of lighting fixtures and all other electrical equipment, when within 8 feet of the floor, shall be constructed of noncombustible, nonabsorptive, insulating materials, except that metal may be used if it is effectively grounded.

**1926.803(j)(5)** Portable lamps shall be equipped with noncombustible, nonabsorptive, insulating sockets, approved handles, basket guards, and approved cords.

**1926.803(j)(6)** The use of worn or defective portable and pendant conductors is prohibited.

**1926.803(k)** ***Sanitation.*** (1) Sanitary, heated, lighted, and ventilated dressing rooms and drying rooms shall be provided for all employees engaged in compressed air work. Such rooms shall contain suitable benches and lockers. Bathing accommodations (showers at the ratio of one to 10 employees per shift), equipped with running hot and cold water, and suitable and adequate toilet accommodations, shall be provided. One toilet for each 15 employees, or fractional part thereof, shall be provided.

**1926.803(k)(2)** When the toilet bowl is shut by a cover, there should be an air space so that the bowl or bucket does not implode when pressure is increased.

**1926.803(k)(3)** All parts of caissons and other working compartments shall be kept in a sanitary condition.

**1926.803(l)** ***Fire prevention and protection.*** (1) Firefighting equipment shall be available at all times and shall be maintained in working condition.

**1926.803(l)(2)** While welding or flame-cutting is being done in compressed air, a firewatch with a fire hose or approved extinguisher shall stand by until such operation is completed.

**1926.803(l)(3)** Shafts and caissons containing flammable material of any kind, either above or below ground, shall be provided with a waterline and a fire hose connected thereto, so arranged that all points of the shaft or caisson are within reach of the hose stream.

**1926.803(l)(4)** Fire hose shall be at least 1 1/2 inches in nominal diameter; the water pressure shall at all times be adequate for efficient operation of the type of nozzle used; and the water supply shall be such as to ensure an uninterrupted flow. Fire hose, when not in use, shall be located or guarded to prevent injury thereto.

**1926.803(l)(5)** The power house, compressor house, and all buildings housing ventilating equipment, shall be provided with at least one hose connection in the water line, with a fire hose connected thereto. A fire hose shall be maintained within reach of structures of wood over or near shafts.

**1926.803(l)(6)** Tunnels shall be provided with a 2-inch minimum diameter water line extending into the working chamber and to within 100 feet of the working face. Such line shall have hose outlets with 100 feet of fire hose attached and maintained as follows: One at the working face; one immediately inside of the bulkhead of the working chamber; and one immediately outside such bulkhead. In addition, hose outlets shall be provided at 200-foot intervals throughout the length of the tunnel, and 100 feet of fire hose shall be attached to the outlet nearest to any location where flammable material is being kept or stored or where any flame is being used.

**1926.803(l)(7)** In addition to fire hose protection required by this subpart, on every floor of every building not under compressed air, but used in connection with the compressed air work, there shall be provided at least one approved fire extinguisher of the proper type for the hazard involved. At least two approved fire extinguishers shall be provided in the working chamber as follows: One at the working face and one immediately inside the bulkhead (pressure side). Extinguishers in the working chamber shall use water as the primary extinguishing agent and shall not use any extinguishing agent which could be harmful to the employees in the working chamber. The fire extinguisher shall be protected from damage.

**1926.803(l)(8)** Highly combustible materials shall not be used or stored in the working chamber. Wood, paper, and similar combustible material shall not be used in the working chamber in quantities which could cause a fire hazard. The compressor building shall be constructed of non-combustible material.

**1926.803(l)(9)** Man locks shall be equipped with a manual type fire extinguisher system that can be activated inside the man lock and also by the outside lock attendant. In addition, a fire hose and portable fire extinguisher shall be provided inside and outside the man lock. The portable fire extinguisher shall be the dry chemical type.

**1926.803(l)(10)** Equipment, fixtures, and furniture in man locks and special decompression chambers shall be constructed of noncombustible materials. Bedding, etc., shall be chemically treated so as to be fire resistant.

**1926.803(l)(11)** Head frames shall be constructed of structural steel or open frame-work fireproofed timber. Head houses and other temporary surface buildings or structures within 100 feet of the shaft, caisson, or tunnel opening shall be built of fire-resistant materials.

**1926.803(l)(12)** No oil, gasoline, or other combustible material shall be stored within 100 feet of any shaft, caisson, or tunnel opening, except that oils may be stored in suitable tanks in isolated fireproof buildings, provided such buildings are not less than 50 feet from any shaft, caisson, or tunnel opening, or any building directly connected thereto.

**1926.803(l)(13)** Positive means shall be taken to prevent leaking flammable liquids from flowing into the areas specifically mentioned in the preceding paragraph.

**1926.803(l)(14)** All explosives used in connection with compressed air work shall be selected, stored, transported, and used as specified in subpart U of this part.

**1926.803(m)** ***Bulkheads and safety screens.*** (1) Intermediate bulkheads with locks, or intermediate safety screens or both, are required where there is the danger of rapid flooding.

**1926.803(m)(2)** In tunnels 16 feet or more in diameter, hanging walkways shall be provided from the face to the man lock as high in the tunnel as practicable, with at least 6 feet of head room. Walkways shall be constructed of noncombustible material. Standard railings shall be securely installed throughout the length of all walkways on open sides in accordance with subpart M of this part. Where walkways are ramped under safety screens, the walkway surface shall be skidproofed by cleats or by equivalent means.

**1926.803(m)(3)** Bulkheads used to contain compressed air shall be tested, where practicable, to prove their ability to resist the highest air pressure which may be expected to be used.

[44 FR 8577, Feb. 9, 1979; 44 FR 20940, Apr. 6, 1979, as amended at 47 FR 14696, 14706, Apr. 6, 1982; 51 FR 25318, July 11, 1986; 61 FR 5510, Feb. 13, 1996]

### §1926.804 Definitions applicable to this subpart.

**1926.804(a)** *Bulkhead*—An airtight structure separating the working chamber from free air or from another chamber under a lesser pressure than the working pressure.

**1926.804(b)** *Caisson*—A wood, steel, concrete or reinforced concrete, air- and water-tight chamber in which it is possible for men to work under air pressure greater than atmospheric pressure to excavate material below water level.

**1926.804(c)** *Decanting*—A method used for decompressing under emergency circumstances. In this procedure, the employees are brought to atmospheric pressure with a very high gas tension in the tissues and then immediately recompressed in a second and separate chamber or lock.

**1926.804(d)** *Emergency locks*—A lock designed to hold and permit the quick passage of an entire shift of employees.

**1926.804(e)** *High air*—Air pressure used to supply power to pneumatic tools and devices.

**1926.804(f)** *Low air*—Air supplied to pressurize working chambers and locks.

**1926.804(g)** *Man lock*—A chamber through which men pass from one air pressure environment into another.

**1926.804(h)** *Materials lock*—A chamber through which materials and equipment pass from one air pressure environment into another.

**1926.804(i)** *Medical lock*—A special chamber in which employees are treated for decompression illness. It may also be used in preemployment physical examinations to determine the adaptability of the prospective employee to changes in pressure.

**1926.804(j)** *Normal condition*—One during which exposure to compressed air is limited to a single continuous working period followed by a single decompression in any given 24-hour period; the total time of exposure to compressed air during the single continuous working period is not interrupted by exposure to normal atmospheric pressure, and a second exposure to compressed air does not occur until at least 12 consecutive hours of exposure to normal atmospheric pressure has elapsed since the employee has been under pressure.

**1926.804(k)** *Pressure*—A force acting on a unit area. Usually shown as pounds per square inch. (p.s.i.)

**1926.804(l)** *Absolute pressure* (p.s.i.a.)—The sum of the atmospheric pressure and gauge pressure (p.s.i.g.).

**1926.804(m)** *Atmospheric pressure*—The pressure of air at sea level, usually 14.7 p.s.i.a. (1 atmosphere), or 0 p.s.i.g.

**1926.804(n)** *Gauge pressure* (p.s.i.g.)—Pressure measured by a gauge and indicating the pressure exceeding atmospheric.

**1926.804(o)** *Safety screen*—An air- and water-tight diaphragm placed across the upper part of a compressed air tunnel between the face and bulkhead, in order to prevent flooding the crown of the tunnel between the safety screen and the bulkhead, thus providing a safe means of refuge and exit from a flooding or flooded tunnel.

**1926.804(p)** *Special decompression chamber*—A chamber to provide greater comfort of employees when the total decompression time exceeds 75 minutes.

**1926.804(q)** *Working chamber*—The space or compartment under air pressure in which the work is being done.

### Appendix A to Subpart S—Decompression Tables

1. *Explanation.* The decompression tables are computed for working chamber pressures from 0 to 14 pounds, and from 14 to 50 pounds per square inch gauge inclusive by 2-pound increments and for exposure times for each pressure extending from one-half to over 8 hours inclusive. Decompressions will be conducted by two or more stages with a maximum of four stages, the latter for a working chamber pressure of 40 pounds per square inch gauge or over.

Stage 1 consists of a reduction in ambient pressure ranging from 10 to a maximum of 16 pounds per square inch, but in no instance will the pressure be reduced below 4 pounds at the end of stage 1. This reduction in pressure in stage 1 will always take place at a rate not greater than 5 pounds per minute.

Further reduction in pressure will take place during stage 2 and subsequent stages as required at a slower rate, but in no event at a rate greater than 1 pound per minute.

Decompression Table No. 1 indicates in the body of the table the total decompression time in minutes for various combinations of working chamber pressure and exposure time.

Decompression Table No. 2 indicates for the same various combinations of working chamber pressure and exposure time the following:

a. The number of stages required;

b. The reduction in pressure and the terminal pressure for each required stage;

c. The time in minutes through which the reduction in pressure is accomplished for each required stage;

d. The pressure reduction rate in minutes per pound for each required stage;

IMPORTANT NOTE: The Pressure Reduction in Each Stage is Accomplished at a Uniform Rate, Do Not Interpolate Between Values Shown on the Tables. Use the Next Higher Value of Working Chamber Pressure or Exposure Time Should the Actual Working Chamber Pressure or the Actual Exposure Time, Respectively, Fall Between Those for Which Calculated Values Are Shown in the Body of the Tables.

| Examples: | Minutes |
|---|---|
| Example No. 1: 4 hours working period at 20 pounds gauge. | |
| Decompression Table No. 1: 20 pounds for 4 hours, total decompression time | 43 |
| Decompression Table No. 2: | |
| Stage 1: Reduce pressure from 20 pounds to 4 pounds at the uniform rate of 5 pounds per minute. | |
| Elapsed time stage 1: 16/5 | 3 |
| Stage 2 (final stage): Reduce pressure at a uniform rate from 4 pounds to 0-pounds gage over a period of 40 minutes. | |
| Rate—0.10 pounds per minute or 10 minutes per pound. | |
| Stage 2 (final) elapsed time | 40 |
| Total time | 43 |
| Example No. 2: 5-hour working period at 24 pounds gage. | |
| Decompression Table No. 1: 24 pounds for 5 hours, total decompression time | 117 |
| Decompression Table No. 2: | |
| Stage 1: Reduce pressure from 24 pounds to 8 pounds at the uniform rate of 5 pounds per minute. | |
| Elapsed time stage 1: 16/5 | 3 |
| Stage 2: Reduce pressure at a uniform rate from 8 pounds to 4 pounds over a period of 4 minutes. Rate, 1 pound per minute elapsed time, stage 2 | 4 |
| Transfer men to special decompression chamber maintaining the 4-pound pressure during the transfer operation. | |
| Stage 3 (final stage): In the special decompression chamber, reduce the pressure at a uniform rate from 4 pounds to 0-pound gage over a period of 110 minutes. Rate, 0.037 pound per minute or 27.5 minutes per pound. Stage 3 (final) elapsed time | 110 |
| Total time | 117 |

## DECOMPRESSION TABLE NO. 1—TOTAL DECOMPRESSION TIME

| WORK PRESSURE p.s.i.g. | WORKING PERIOD HOURS | | | | | | | | | | |
|---|---|---|---|---|---|---|---|---|---|---|---|
| | 1/2 | 1 | 1 1/2 | 2 | 3 | 4 | 5 | 6 | 7 | 8 | Over 8 |
| 9-12 | 3 | 3 | 3 | 3 | 3 | 3 | 3 | 3 | 3 | 3 | 3 |
| 14 | 6 | 6 | 6 | 6 | 6 | 6 | 6 | 6 | 16 | 16 | 33 |
| 16 | 7 | 7 | 7 | 7 | 7 | 7 | 17 | 33 | 48 | 48 | 62 |
| 18 | 7 | 7 | 7 | 8 | 11 | 17 | 48 | 63 | 63 | 73 | 87 |
| 20 | 7 | 7 | 8 | 15 | 15 | 43 | 63 | 73 | 83 | 103 | 113 |
| 22 | 9 | 9 | 16 | 24 | 38 | 68 | 93 | 103 | 113 | 128 | 133 |
| 24 | 11 | 12 | 23 | 27 | 52 | 92 | 117 | 122 | 127 | 137 | 151 |
| 26 | 13 | 14 | 29 | 34 | 69 | 104 | 126 | 141 | 142 | 142 | 163 |
| 28 | 15 | 23 | 31 | 41 | 98 | 127 | 143 | 153 | 153 | 165 | 183 |
| 30 | 17 | 28 | 38 | 62 | 105 | 143 | 165 | 168 | 178 | 188 | 204 |
| 32 | 19 | 35 | 43 | 85 | 126 | 163 | 178 | 193 | 203 | 213 | 226 |
| 34 | 21 | 39 | 58 | 98 | 151 | 178 | 195 | 218 | 223 | 233 | 248 |
| 36 | 24 | 44 | 63 | 113 | 170 | 198 | 223 | 233 | 243 | 253 | 273 |
| 38 | 28 | 49 | 73 | 128 | 178 | 203 | 223 | 238 | 253 | 263 | 278 |
| 40 | 31 | 49 | 84 | 143 | 183 | 213 | 233 | 248 | 258 | 278 | 288 |
| 42 | 37 | 56 | 102 | 144 | 189 | 215 | 245 | 260 | 263 | 268 | 293 |
| 44 | 43 | 64 | 118 | 154 | 199 | 234 | 254 | 264 | 269 | 269 | 293 |
| 46 | 44 | 74 | 139 | 171 | 214 | 244 | 269 | 274 | 289 | 299 | 318 |
| 48 | 51 | 89 | 144 | 189 | 229 | 269 | 299 | 309 | 319 | 319 | .. |
| 50 | 58 | 94 | 164 | 209 | 249 | 279 | 309 | 329 | .... | ..... | ..... |

## DECOMPRESSION TABLE NO. 2

(Do not interpolate, use next higher value for conditions not computed.)

| Working chamber pressure p.s.i.g. | Working period Hours | Decompression data: Stage No. | Pressure reduction p.s.i.g. From | Pressure reduction p.s.i.g. To | Time in stage Minutes | Pressure reduction rate Min/Pound | Total time decompress Minutes |
|---|---|---|---|---|---|---|---|
| 14 | ½ | 1 | 14 | 4 | 2 | 0.20 | 6 |
| | | 2 | 4 | 0 | 4 | 1.00 | 6 |
| | 1 | 1 | 14 | 4 | 2 | 0.20 | 6 |
| | | 2 | 4 | 0 | 4 | 1.00 | 6 |
| | 1½ | 1 | 14 | 4 | 2 | 0.20 | 6 |
| | | 2 | 4 | 0 | 4 | 1.00 | 6 |
| | 2 | 1 | 14 | 4 | 2 | 0.20 | 6 |
| | | 2 | 4 | 0 | 4 | 1.00 | 6 |
| | 3 | 1 | 14 | 4 | 2 | 0.20 | 6 |
| | | 2 | 4 | 0 | 4 | 1.00 | 6 |
| | 4 | 1 | 14 | 0 | 2 | 0.20 | 6 |
| | | 2 | 4 | 0 | 4 | 1.00 | 6 |
| | 5 | 1 | 14 | 4 | 2 | 0.20 | 6 |
| | | 2 | 4 | 0 | 4 | 1.00 | 6 |
| | 6 | 1 | 14 | 4 | 2 | 0.20 | |
| | | 2 | 4 | 0 | 4 | 1.00 | 6 |
| | 7 | 1 | 14 | 4 | 2 | 0.20 | |
| | | 2 | 4 | 0 | 14 | 3.50 | 16 |
| | 8 | 1 | 14 | 4 | 2 | 0.20 | |
| | | 2 | 4 | 0 | 14 | 3.50 | 16 |
| | Over 8 | 1 | 14 | 4 | 2 | 0.20 | |
| | | 2 | 4 | 0 | 30 | 7.50 | 32 |
| 16 | ½ | 1 | 16 | 4 | 3 | 0.20 | |
| | | 2 | 4 | 0 | 4 | 1.00 | 7 |
| | 1 | 1 | 16 | 4 | 3 | 0.20 | 7 |
| | | 2 | 4 | 0 | 4 | 1.00 | 7 |
| | 1½ | 1 | 16 | 4 | 3 | 0.20 | |
| | | 2 | 4 | 0 | 4 | 1.00 | 7 |
| | 2 | 1 | 16 | 4 | 3 | 0.20 | |
| | | 2 | 4 | 0 | 4 | 1.00 | 7 |
| | 3 | 1 | 16 | 4 | 3 | 0.20 | |
| | | 2 | 4 | 0 | 4 | 1.00 | 7 |
| | 4 | 1 | 14 | 4 | 3 | 0.20 | |
| | | 2 | 4 | 0 | 4 | 1.00 | 7 |
| | 5 | 1 | 14 | 4 | 3 | 0.20 | 7 |
| | | 2 | 4 | 0 | 4 | 3.50 | 17 |
| | 6 | 1 | 14 | 4 | 3 | 0.20 | |
| | | 2 | 4 | 0 | 30 | 7.50 | 33 |
| | 7 | 1 | 14 | 4 | 3 | 0.20 | |
| | | 2 | 4 | 0 | 45 | 11.25 | 48 |
| | 8 | 1 | 14 | 4 | 3 | 0.20 | |
| | | 2 | 4 | 0 | 45 | 11.25 | 48 |
| | Over 8 | 1 | 14 | 4 | 3 | 0.20 | |
| | | 2 | 4 | 0 | 60 | 15.00 | 63 |
| 18 | ½ | 1 | 18 | 4 | 3 | 0.20 | |
| | | 2 | 4 | 0 | 4 | 1.00 | 7 |
| | 1 | 1 | 18 | 4 | 3 | 0.20 | |
| | | 2 | 4 | 0 | 4 | 1.00 | 7 |
| | 1½ | 1 | 18 | 4 | 3 | 0.20 | |
| | | 2 | 4 | 0 | 4 | 1.00 | 7 |
| | 2 | 1 | 18 | 4 | 3 | 0.20 | |
| | | 2 | 4 | 0 | 5 | 1.25 | 8 |
| | 3 | 1 | 18 | 4 | 3 | 0.20 | |
| | | 2 | 4 | 0 | 8 | 2.00 | 11 |
| | 4 | 1 | 18 | 4 | 3 | 0.20 | |
| | | 2 | 4 | 0 | 14 | 3.50 | 17 |
| | 5 | 1 | 18 | 4 | 3 | 0.20 | |
| | | 2 | 4 | 0 | 45 | 11.25 | 48 |
| | 6 | 1 | 18 | 4 | 3 | 0.20 | |
| | | 2 | 4 | 0 | 60 | 15.00 | 63 |
| | 7 | 1 | 18 | 4 | 3 | 0.20 | |
| | | 2 | 4 | 0 | 60 | 15.00 | 63 |
| | 8 | 1 | 18 | 4 | 3 | 0.20 | |
| | | 2 | 4 | 0 | 70 | 17.50 | 73 |
| | Over 8 | 1 | 18 | 4 | 3 | 0.20 | |
| | | 2 | 4 | 0 | 84 | 21.00 | 87 |
| 20 | ½ | 1 | 20 | 4 | 3 | 0.20 | |
| | | 2 | 4 | 0 | 4 | 1.00 | 7 |
| | 1 | 1 | 20 | 4 | 3 | 0.20 | |
| | | 2 | 4 | 0 | 4 | 1.00 | 7 |
| | 1½ | 1 | 20 | 4 | 3 | 0.20 | |
| | | 2 | 4 | 0 | 5 | 1.25 | 8 |
| | 2 | 1 | 20 | 4 | 3 | 0.20 | |
| | | 2 | 4 | 0 | 12 | 3.00 | 15 |

## DECOMPRESSION TABLE NO. 2

(Do not interpolate, use next higher value for conditions not computed.)

| Working chamber pressure p.s.i.g. | Working period Hours | Decompression data | | | | | |
|---|---|---|---|---|---|---|---|
| | | Stage No. | Pressure reduction p.s.i.g. | | Time in stage Minutes | Pressure reduction rate Min/Pound | Total time decompress Minutes |
| | | | From | To | | | |
| | 3 | 1 | 20 | 4 | 3 | 0.20 | |
| | | 2 | 4 | 0 | 12 | 3.00 | 15 |
| | 4 | 1 | 20 | 4 | 3 | 0.20 | |
| | | 2 | 4 | 0 | 40 | 10.00 | 43 |
| | 5 | 1 | 20 | 4 | 3 | 0.20 | |
| | | 2 | 4 | 0 | 60 | 15.00 | 63 |
| | 6 | 1 | 20 | 4 | 3 | 0.20 | |
| | | 2 | 4 | 0 | 70 | 17.50 | 73 |
| | 7 | 1 | 20 | 4 | 3 | 0.20 | |
| | | 2 | 4 | 0 | 80 | 20.00 | 83 |
| | 8 | 1 | 20 | 4 | 3 | 0.20 | |
| | | 2 | 4 | 0 | 100 | 25.00 | 103 |
| | Over 8 | 1 | 20 | 4 | 3 | 0.20 | |
| | | 2 | 4 | 0 | 110 | 27.50 | 113 |
| 22 | 1/2 | 1 | 22 | 6 | 3 | 0.20 | |
| | | 2 | 6 | 0 | 6 | 1.00 | 9 |
| | 1 | 1 | 22 | 6 | 3 | 0.20 | |
| | | 2 | 6 | 0 | 6 | 1.00 | 9 |
| | 1 1/2 | 1 | 22 | 6 | 3 | 0.20 | |
| | | 2 | 6 | 0 | 13 | 2.20 | 16 |
| | 2 | 1 | 22 | 6 | 3 | 0.20 | |
| | | 2 | 6 | 0 | 21 | 3.50 | 24 |
| | 3 | 1 | 22 | 6 | 3 | 0.20 | |
| | | 2 | 6 | 0 | 35 | 5.85 | 38 |
| | 4 | 1 | 22 | 6 | 3 | 0.20 | |
| | | 2 | 6 | 0 | 65 | 10.83 | 68 |
| | 5 | 1 | 22 | 6 | 3 | 0.20 | |
| | | 2 | 6 | 0 | 90 | 15.00 | 93 |
| | 6 | 1 | 22 | 6 | 3 | 0.20 | |
| | | 2 | 6 | 0 | 100 | 16.67 | 103 |
| | 7 | 1 | 22 | 6 | 3 | 0.20 | |
| | | 2 | 6 | 0 | 110 | 18.35 | 113 |
| | 8 | 1 | 22 | 6 | 3 | 0.20 | |
| | | 2 | 6 | 0 | 125 | 20.80 | 128 |
| | Over 8 | 1 | 22 | 6 | 3 | 0.20 | |
| | | 2 | 6 | 0 | 130 | 21.70 | 133 |
| 24 | 1/2 | 1 | 24 | 8 | 3 | 0.20 | |
| | | 2 | 8 | 4 | 4 | 1.00 | |
| | | 3 | 4 | 0 | 4 | 1.00 | 11 |
| | 1 | 1 | 24 | 8 | 3 | 0.20 | |
| | | 2 | 8 | 4 | 4 | 1.00 | |
| | | 3 | 4 | 0 | 5 | 1.25 | 12 |
| | 1 1/2 | 1 | 24 | 8 | 3 | 0.20 | |
| | | 2 | 8 | 4 | 4 | 1.00 | |
| | | 3 | 4 | 0 | 16 | 4.00 | 23 |
| | 2 | 1 | 24 | 8 | 3 | 0.20 | |
| | | 2 | 8 | 4 | 4 | 1.00 | |
| | | 3 | 4 | 0 | 20 | 5.00 | 27 |
| | 3 | 1 | 24 | 8 | 3 | 0.20 | |
| | | 2 | 8 | 4 | 4 | 1.00 | |
| | | 3 | 4 | 0 | 45 | 11.25 | 52 |
| | 4 | 1 | 24 | 8 | 3 | 0.20 | |
| | | 2 | 8 | 4 | 4 | 1.00 | |
| | | 3 | 4 | 0 | 85 | 21.25 | 92 |
| | 5 | 1 | 24 | 8 | 3 | 0.20 | |
| | | 2 | 8 | 4 | 4 | 1.00 | |
| | | 3 | 4 | 0 | 110 | 27.50 | 117 |
| | 6 | 1 | 24 | 8 | 3 | 0.20 | |
| | | 2 | 8 | 4 | 4 | 1.00 | |
| | | 3 | 4 | 0 | 115 | 28.80 | 122 |
| | 7 | 1 | 24 | 8 | 3 | 0.20 | |
| | | 2 | 8 | 4 | 4 | 1.00 | |
| | | 3 | 4 | 0 | 120 | 30.00 | 127 |
| | 8 | 1 | 24 | 8 | 3 | 0.20 | |
| | | 2 | 8 | 4 | 4 | 1.00 | |
| | | 3 | 4 | 0 | 130 | 32.50 | 137 |
| | Over 8 | 1 | 24 | 8 | 3 | 0.20 | |
| | | 2 | 8 | 4 | 8 | 2.00 | |
| | | 3 | 4 | 0 | 140 | 35.00 | 151 |
| 26 | 1/2 | 1 | 26 | 10 | 3 | 0.20 | |
| | | 2 | 10 | 4 | 6 | 1.00 | |
| | | 3 | 4 | 0 | 4 | 1.00 | 13 |
| | 1 | 1 | 26 | 10 | 3 | 0.20 | |
| | | 2 | 10 | 4 | 6 | 1.00 | |

## DECOMPRESSION TABLE NO. 2

(Do not interpolate, use next higher value for conditions not computed.)

| Working chamber pressure p.s.i.g. | Working period Hours | Decompression data | | | | | |
|---|---|---|---|---|---|---|---|
| | | Stage No. | Pressure reduction p.s.i.g. From | Pressure reduction p.s.i.g. To | Time in stage Minutes | Pressure reduction rate Min/Pound | Total time decompress Minutes |
| | | 3 | 4 | 0 | 5 | 1.25 | 14 |
| | 1½ | 1 | 26 | 10 | 3 | 0.20 | |
| | | 2 | 10 | 4 | 6 | 1.00 | |
| | | 3 | 4 | 0 | 20 | 5.00 | 29 |
| | 2 | 1 | 26 | 10 | 3 | 0.20 | |
| | | 2 | 10 | 4 | 6 | 1.00 | |
| | | 3 | 4 | 0 | 25 | 6.25 | 34 |
| | 3 | 1 | 26 | 10 | 3 | 0.20 | |
| | | 2 | 10 | 4 | 6 | 1.00 | |
| | | 3 | 4 | 0 | 60 | 15.00 | 69 |
| | 4 | 1 | 26 | 10 | 3 | 0.20 | |
| | | 2 | 10 | 4 | 6 | 1.00 | |
| | | 3 | 4 | 0 | 95 | 23.75 | 104 |
| | 5 | 1 | 26 | 10 | 3 | 0.20 | |
| | | 2 | 10 | 4 | 8 | 1.33 | |
| | | 3 | 4 | 0 | 115 | 28.80 | 126 |
| | 6 | 1 | 26 | 10 | 3 | 0.20 | |
| | | 2 | 10 | 4 | 8 | 1.33 | |
| | | 3 | 4 | 0 | 130 | 32.50 | 141 |
| | 7 | 1 | 26 | 10 | 3 | 2.20 | |
| | | 2 | 10 | 4 | 9 | 1.50 | |
| | | 3 | 4 | 0 | 130 | 32.50 | 142 |
| | 8 | 1 | 26 | 10 | 3 | 0.20 | |
| | | 2 | 10 | 4 | 9 | 1.50 | |
| | | 3 | 4 | 0 | 130 | 32.50 | 142 |
| | Over 8 | 1 | 26 | 10 | 3 | 0.20 | |
| | | 2 | 10 | 4 | 30 | 5.00 | |
| | | 3 | 4 | 0 | 130 | 32.50 | 163 |
| 28 | ½ | 1 | 28 | 12 | 3 | 0.20 | |
| | | 2 | 12 | 4 | 8 | 1.00 | |
| | | 3 | 4 | 0 | 4 | 1.00 | 15 |
| | 1 | 1 | 28 | 12 | 3 | 0.20 | |
| | | 2 | 12 | 4 | 8 | 1.00 | |
| | | 3 | 4 | 0 | 12 | 3.00 | 23 |
| | 1½ | 1 | 28 | 12 | 3 | 0.20 | |
| | | 2 | 12 | 4 | 8 | 1.00 | |
| | | 3 | 4 | 0 | 20 | 5.00 | 31 |
| | 2 | 1 | 28 | 12 | 3 | 0.20 | |
| | | 2 | 12 | 4 | 8 | 1.00 | |
| | | 3 | 4 | 0 | 30 | 7.50 | 41 |
| | 3 | 1 | 28 | 12 | 3 | 0.20 | |
| | | 2 | 12 | 4 | 10 | 1.25 | |
| | | 3 | 4 | 0 | 85 | 21.20 | 98 |
| | 4 | 1 | 28 | 12 | 3 | 0.20 | |
| | | 2 | 12 | 4 | 14 | 1.75 | |
| | | 3 | 4 | 0 | 110 | 27.50 | 127 |
| | 5 | 1 | 28 | 12 | 3 | 0.20 | |
| | | 2 | 12 | 4 | 20 | 2.50 | |
| | | 3 | 4 | 0 | 120 | 30.00 | 143 |
| | 6 | 1 | 28 | 12 | 3 | 0.20 | |
| | | 2 | 12 | 4 | 20 | 2.50 | |
| | | 3 | 4 | 0 | 130 | 32.50 | 153 |
| | 7 | 1 | 28 | 12 | 3 | 0.20 | |
| | | 2 | 12 | 4 | 20 | 2.50 | |
| | | 3 | 4 | 0 | 120 | 32.50 | 153 |
| | 8 | 1 | 28 | 12 | 3 | 0.20 | |
| | | 2 | 12 | 4 | 32 | 4.00 | |
| | | 3 | 4 | 0 | 130 | 32.50 | 165 |
| | Over 8 | 1 | 28 | 12 | 3 | 0.20 | |
| | | 2 | 12 | 4 | 50 | 6.25 | |
| | | 3 | 4 | 0 | 130 | 32.50 | 183 |
| 30 | ½ | 1 | 30 | 14 | 3 | 0.20 | |
| | | 2 | 14 | 4 | 10 | 1.00 | |
| | | 3 | 4 | 0 | 4 | 1.00 | 17 |
| | 1 | 1 | 30 | 14 | 3 | 0.20 | |
| | | 2 | 14 | 4 | 10 | 1.00 | |
| | | 3 | 4 | 0 | 15 | 3.75 | 28 |
| | 1½ | 1 | 30 | 14 | 3 | 0.20 | |
| | | 2 | 14 | 4 | 10 | 1.00 | |
| | | 3 | 4 | 0 | 25 | 6.25 | 38 |
| | 2 | 1 | 30 | 14 | 3 | 0.20 | |
| | | 2 | 14 | 4 | 14 | 1.40 | |
| | | 3 | 4 | 0 | 45 | 11.25 | 62 |
| | 3 | 1 | 30 | 14 | 3 | 0.20 | |

## DECOMPRESSION TABLE NO. 2

(Do not interpolate, use next higher value for conditions not computed.)

| Working chamber pressure p.s.i.g. | Working period Hours | Decompression data: Stage No. | Pressure reduction p.s.i.g. From | Pressure reduction p.s.i.g. To | Time in stage Minutes | Pressure reduction rate Min/Pound | Total time decompress Minutes |
|---|---|---|---|---|---|---|---|
| | | 2 | 14 | 4 | 17 | 1.70 | |
| | | 3 | 4 | 0 | 85 | 21.20 | 105 |
| | 4 | 1 | 30 | 14 | 3 | 0.20 | |
| | | 2 | 14 | 4 | 30 | 3.00 | |
| | | 3 | 4 | 0 | 110 | 27.50 | 143 |
| | 5 | 1 | 30 | 14 | 3 | 0.20 | |
| | | 2 | 14 | 4 | 35 | 3.50 | |
| | | 3 | 4 | 0 | 130 | 32.50 | 165 |
| | 6 | 1 | 30 | 14 | 3 | 0.20 | |
| | | 2 | 14 | 4 | 35 | 3.50 | |
| | | 3 | 4 | 0 | 130 | 32.50 | 168 |
| | 7 | 1 | 30 | 14 | 3 | 0.20 | |
| | | 2 | 14 | 4 | 45 | 4.50 | |
| | | 3 | 4 | 0 | 130 | 32.50 | 178 |
| | 8 | 1 | 30 | 14 | 3 | 0.20 | |
| | | 2 | 14 | 4 | 55 | 5.50 | |
| | | 3 | 4 | 0 | 130 | 32.50 | 188 |
| | Over 8 | 1 | 30 | 14 | 3 | 0.20 | |
| | | 2 | 14 | 4 | 71 | 7.10 | |
| | | 3 | 4 | 0 | 130 | 32.50 | 204 |
| 32 | ½ | 1 | 32 | 16 | 3 | 0.20 | |
| | | 2 | 16 | 4 | 12 | 1.00 | |
| | | 3 | 4 | 0 | 4 | 1.00 | 19 |
| | 1 | 1 | 32 | 16 | 3 | 0.20 | |
| | | 2 | 16 | 4 | 12 | 1.00 | |
| | | 3 | 4 | 0 | 20 | 5.00 | 35 |
| | 1½ | 1 | 32 | 16 | 3 | 0.20 | |
| | | 2 | 16 | 4 | 15 | 1.25 | |
| | | 3 | 4 | 0 | 25 | 6.25 | 43 |
| | 2 | 1 | 32 | 16 | 3 | 0.20 | |
| | | 2 | 16 | 4 | 22 | 1.83 | |
| | | 3 | 4 | 0 | 60 | 15.00 | 85 |
| | 3 | 1 | 32 | 16 | 3 | 0.20 | |
| | | 2 | 16 | 4 | 28 | 2.33 | |
| | | 3 | 4 | 0 | 95 | 23.75 | 126 |
| | 4 | 1 | 32 | 16 | 3 | 0.20 | |
| | | 2 | 16 | 4 | 40 | 3.33 | |
| | | 3 | 4 | 0 | 120 | 30.00 | 163 |
| | 5 | 1 | 32 | 16 | 3 | 0.20 | |
| | | 2 | 16 | 4 | 45 | 3.75 | |
| | | 3 | 4 | 0 | 130 | 32.50 | 178 |
| | 6 | 1 | 32 | 16 | 3 | 0.20 | |
| | | 2 | 16 | 4 | 60 | 5.00 | |
| | | 3 | 4 | 0 | 130 | 32.50 | 193 |
| | 7 | 1 | 32 | 16 | 3 | 0.20 | |
| | | 2 | 16 | 4 | 70 | 5.83 | |
| | | 3 | 4 | 0 | 130 | 32.50 | 203 |
| | 8 | 1 | 32 | 16 | 3 | 0.20 | |
| | | 2 | 16 | 4 | 80 | 6.67 | |
| | | 3 | 4 | 0 | 130 | 32.50 | 213 |
| | Over 8 | 1 | 32 | 16 | 3 | 0.20 | |
| | | 2 | 16 | 4 | 93 | 7.75 | |
| | | 3 | 4 | 0 | 130 | 32.50 | 226 |
| 34 | ½ | 1 | 34 | 18 | 3 | 0.20 | |
| | | 2 | 18 | 4 | 14 | 1.00 | |
| | | 3 | 4 | 0 | 4 | 1.00 | 21 |
| | 1 | 1 | 34 | 18 | 3 | 0.20 | |
| | | 2 | 18 | 4 | 14 | 1.00 | |
| | | 3 | 4 | 0 | 22 | 5.50 | 39 |
| | 1½ | 1 | 34 | 18 | 3 | 0.20 | |
| | | 2 | 18 | 4 | 25 | 1.80 | |
| | | 3 | 4 | 0 | 30 | 7.50 | 58 |
| | 2 | 1 | 34 | 18 | 3 | 0.20 | |
| | | 2 | 18 | 4 | 35 | 2.50 | |
| | | 3 | 4 | 0 | 60 | 15.00 | 98 |
| | 3 | 1 | 34 | 18 | 3 | 0.20 | |
| | | 2 | 18 | 4 | 43 | 3.10 | |
| | | 3 | 4 | 0 | 105 | 26.25 | 151 |
| | 4 | 1 | 34 | 18 | 3 | 0.20 | |
| | | 2 | 18 | 4 | 55 | 3.93 | |
| | | 3 | 4 | 0 | 120 | 30.00 | 178 |
| | 5 | 1 | 34 | 18 | 3 | 0.20 | |
| | | 2 | 18 | 4 | 62 | 4.43 | |
| | | 3 | 4 | 0 | 130 | 32.50 | 195 |

## DECOMPRESSION TABLE NO. 2

(Do not interpolate, use next higher value for conditions not computed.)

| Working chamber pressure p.s.i.g. | Working period Hours | Decompression data: Stage No. | Pressure reduction p.s.i.g. From | Pressure reduction p.s.i.g. To | Time in stage Minutes | Pressure reduction rate Min/Pound | Total time decompress Minutes |
|---|---|---|---|---|---|---|---|
| | 6 | 1 | 34 | 18 | 3 | 0.20 | |
| | | 2 | 18 | 4 | 85 | 6.07 | |
| | | 3 | 4 | 0 | 130 | 32.50 | 218 |
| | 7 | 1 | 34 | 18 | 3 | 0.20 | |
| | | 2 | 18 | 4 | 90 | 6.43 | |
| | | 3 | 4 | 0 | 130 | 32.50 | 223 |
| | 8 | 1 | 34 | 18 | 3 | 0.20 | |
| | | 2 | 18 | 4 | 100 | 7.15 | |
| | | 3 | 4 | 0 | 130 | 32.50 | 233 |
| | Over 8 | 1 | 34 | 18 | 3 | 0.20 | |
| | | 2 | 18 | 4 | 115 | 8.23 | |
| | | 3 | 4 | 0 | 130 | 32.50 | 248 |
| 36 | 1/2 | 1 | 36 | 20 | 3 | 0.20 | |
| | | 2 | 20 | 4 | 16 | 1.00 | |
| | | 3 | 4 | 0 | 5 | 1.25 | 24 |
| | 1 | 1 | 36 | 20 | 3 | 0.20 | |
| | | 2 | 20 | 4 | 16 | 1.00 | |
| | | 3 | 4 | 0 | 25 | 6.25 | 44 |
| | 1 1/2 | 1 | 36 | 20 | 3 | 0.20 | |
| | | 2 | 20 | 4 | 30 | 1.88 | |
| | | 3 | 4 | 0 | 30 | 7.50 | 63 |
| | 2 | 1 | 36 | 20 | 3 | 0.20 | |
| | | 2 | 20 | 4 | 40 | 2.50 | |
| | | 3 | 4 | 0 | 70 | 17.50 | 113 |
| | 3 | 1 | 36 | 20 | 3 | 0.20 | |
| | | 2 | 20 | 4 | 52 | 3.25 | |
| | | 3 | 4 | 0 | 115 | 28.75 | 170 |
| | 4 | 1 | 36 | 20 | 3 | 0.20 | |
| | | 2 | 20 | 4 | 65 | 4.06 | |
| | | 3 | 4 | 0 | 130 | 32.50 | 198 |
| | 5 | 1 | 36 | 20 | 3 | 0.20 | |
| | | 2 | 20 | 4 | 90 | 5.63 | |
| | | 3 | 4 | 0 | 130 | 32.50 | 223 |
| | 6 | 1 | 36 | 20 | 3 | 0.20 | |
| | | 2 | 20 | 4 | 100 | 6.25 | |
| | | 3 | 4 | 0 | 130 | 32.50 | 233 |
| | 7 | 1 | 36 | 20 | 3 | 0.20 | |
| | | 2 | 20 | 4 | 110 | 6.88 | |
| | | 3 | 4 | 0 | 130 | 32.50 | 243 |
| | 8 | 1 | 36 | 20 | 3 | 0.20 | |
| | | 2 | 20 | 4 | 120 | 7.50 | |
| | | 3 | 4 | 0 | 130 | 32.50 | 253 |
| | Over 8 | 1 | 36 | 20 | 3 | 0.20 | |
| | | 2 | 20 | 4 | 140 | 8.75 | |
| | | 3 | 4 | 0 | 130 | 32.50 | 273 |
| 38 | 1/2 | 1 | 38 | 22 | 3 | 0.20 | |
| | | 2 | 22 | 6 | 16 | 1.00 | |
| | | 3 | 6 | 0 | 9 | 1.50 | 28 |
| | 1 | 1 | 38 | 22 | 3 | 0.20 | |
| | | 2 | 22 | 6 | 16 | 1.00 | |
| | | 3 | 6 | 0 | 30 | 5.00 | 49 |
| | 1 1/2 | 1 | 38 | 22 | 3 | 0.20 | |
| | | 2 | 22 | 6 | 20 | 1.25 | |
| | | 3 | 6 | 0 | 50 | 8.34 | 73 |
| | 2 | 1 | 38 | 22 | 3 | 0.20 | |
| | | 2 | 22 | 6 | 30 | 1.88 | |
| | | 3 | 6 | 0 | 95 | 15.83 | 128 |
| | 3 | 1 | 38 | 22 | 3 | 0.20 | |
| | | 2 | 22 | 6 | 35 | 2.19 | |
| | | 3 | 6 | 0 | 140 | 23.35 | 178 |
| | 4 | 1 | 38 | 22 | 3 | 0.20 | |
| | | 2 | 22 | 6 | 50 | 3.12 | |
| | | 3 | 6 | 0 | 150 | 25.00 | 203 |
| | 5 | 1 | 38 | 22 | 3 | 0.20 | |
| | | 2 | 22 | 6 | 55 | 3.44 | |
| | | 3 | 6 | 0 | 165 | 27.50 | 223 |
| | 6 | 1 | 38 | 22 | 3 | 0.20 | |
| | | 2 | 22 | 6 | 70 | 4.38 | |
| | | 3 | 6 | 0 | 165 | 27.50 | 238 |
| | 7 | 1 | 38 | 22 | 3 | 0.20 | |
| | | 2 | 22 | 6 | 85 | 5.32 | |
| | | 3 | 6 | 0 | 165 | 27.50 | 253 |
| | 8 | 1 | 38 | 22 | 3 | 0.20 | |
| | | 2 | 22 | 6 | 95 | 5.93 | |

## DECOMPRESSION TABLE NO. 2

(Do not interpolate, use next higher value for conditions not computed.)

| Working chamber pressure p.s.i.g. | Working period Hours | Decompression data | | | | | |
|---|---|---|---|---|---|---|---|
| | | Stage No. | Pressure reduction p.s.i.g. From | Pressure reduction p.s.i.g. To | Time in stage Minutes | Pressure reduction rate Min/Pound | Total time decompress Minutes |
| | | 3 | 6 | 0 | 165 | 27.50 | 263 |
| | Over 8 | 1 | 38 | 22 | 3 | 0.20 | |
| | | 2 | 22 | 6 | 110 | 6.88 | |
| | | 3 | 6 | 0 | 165 | 27.50 | 278 |
| 40 | ½ | 1 | 40 | 24 | 3 | 0.20 | |
| | | 2 | 24 | 8 | 16 | 1.00 | |
| | | 3 | 8 | 4 | 4 | 1.00 | |
| | | 4 | 4 | 0 | 8 | 2.00 | 31 |
| | 1 | 1 | 40 | 24 | 3 | 0.20 | |
| | | 2 | 24 | 8 | 16 | 1.00 | |
| | | 3 | 8 | 4 | 5 | 1.25 | |
| | | 4 | 4 | 0 | 25 | 6.25 | 49 |
| | 1½ | 1 | 40 | 24 | 3 | 0.20 | |
| | | 2 | 24 | 8 | 16 | 1.00 | |
| | | 3 | 8 | 4 | 20 | 5.00 | |
| | | 4 | 4 | 0 | 45 | 11.25 | 84 |
| | 2 | 2 | 40 | 24 | 3 | 0.20 | |
| | | 1 | 24 | 8 | 25 | 1.56 | |
| | | 3 | 8 | 4 | 20 | 5.00 | |
| | | 4 | 4 | 0 | 95 | 23.75 | 143 |
| | 3 | 1 | 40 | 24 | 3 | 0.20 | |
| | | 2 | 24 | 8 | 30 | 1.88 | |
| | | 3 | 8 | 4 | 30 | 7.50 | |
| | | 4 | 4 | 0 | 120 | 30.00 | 183 |
| | 4 | 1 | 40 | 24 | 3 | 0.20 | |
| | | 2 | 24 | 8 | 45 | 2.81 | |
| | | 3 | 8 | 4 | 35 | 8.75 | |
| | | 4 | 4 | 0 | 130 | 32.50 | 213 |
| | 5 | 1 | 40 | 24 | 3 | 0.20 | |
| | | 2 | 24 | 8 | 47 | 2.94 | |
| | | 3 | 8 | 4 | 53 | 13.25 | |
| | | 4 | 4 | 0 | 130 | 32.50 | 233 |
| | 6 | 1 | 40 | 24 | 3 | 0.20 | |
| | | 2 | 24 | 8 | 55 | 3.44 | |
| | | 3 | 8 | 4 | 60 | 15.00 | |
| | | 4 | 4 | 0 | 130 | 32.50 | 248 |
| | 7 | 1 | 40 | 24 | 3 | 0.20 | |
| | | 2 | 24 | 8 | 65 | 4.06 | |
| | | 3 | 8 | 4 | 60 | 15.00 | |
| | | 4 | 4 | 0 | 130 | 32.50 | 258 |
| | 8 | 1 | 40 | 24 | 3 | 0.20 | |
| | | 2 | 24 | 8 | 75 | 4.70 | |
| | | 3 | 8 | 4 | 60 | 15.00 | |
| | | 4 | 4 | 0 | 130 | 32.50 | 268 |
| | Over 8 | 1 | 40 | 24 | 3 | 0.20 | |
| | | 2 | 24 | 8 | 95 | 5.93 | |
| | | 3 | 8 | 4 | 60 | 15.00 | |
| | | 4 | 4 | 0 | 130 | 32.50 | 288 |
| 42 | ½ | 1 | 42 | 26 | 3 | 0.20 | |
| | | 2 | 26 | 10 | 16 | 1.00 | |
| | | 3 | 10 | 4 | 6 | 1.00 | |
| | | 4 | 4 | 0 | 12 | 3.00 | 37 |
| | 1 | 1 | 42 | 26 | 3 | 0.20 | |
| | | 2 | 26 | 10 | 16 | 1.00 | |
| | | 3 | 10 | 4 | 12 | 2.00 | |
| | | 4 | 4 | 0 | 25 | 6.25 | 56 |
| | 1½ | 1 | 42 | 26 | 3 | 0.20 | |
| | | 2 | 26 | 10 | 16 | 1.00 | |
| | | 3 | 10 | 4 | 23 | 3.83 | |
| | | 4 | 4 | 0 | 60 | 15.00 | 102 |
| | 2 | 1 | 42 | 26 | 3 | 0.20 | |
| | | 2 | 26 | 10 | 16 | 1.00 | |
| | | 3 | 10 | 4 | 30 | 5.00 | |
| | | 4 | 4 | 0 | 95 | 23.75 | 144 |
| | 3 | 1 | 42 | 26 | 3 | 0.20 | |
| | | 2 | 26 | 10 | 16 | 1.00 | |
| | | 3 | 10 | 4 | 50 | 8.34 | |
| | | 4 | 4 | 0 | 120 | 30.00 | 189 |
| | 4 | 1 | 42 | 26 | 3 | 0.20 | |
| | | 2 | 26 | 10 | 17 | 1.06 | |
| | | 3 | 10 | 4 | 65 | 10.83 | |
| | | 4 | 4 | 0 | 130 | 32.50 | 215 |
| | 5 | 1 | 42 | 26 | 3 | 0.20 | |
| | | 2 | 26 | 10 | 27 | 1.69 | |

## DECOMPRESSION TABLE NO. 2

(Do not interpolate, use next higher value for conditions not computed.)

| Working chamber pressure p.s.i.g. | Working period Hours | Decompression data | | | | | |
|---|---|---|---|---|---|---|---|
| | | Stage No. | Pressure reduction p.s.i.g. | | Time in stage Minutes | Pressure reduction rate Min/Pound | Total time decompress Minutes |
| | | | From | To | | | |
| | | 3 | 10 | 4 | 85 | 14.18 | |
| | | 4 | 4 | 0 | 130 | 32.50 | 245 |
| | 6 | 1 | 42 | 26 | 3 | 0.20 | |
| | | 2 | 26 | 10 | 27 | 1.69 | |
| | | 3 | 10 | 4 | 100 | 16.67 | |
| | | 4 | 4 | 0 | 130 | 32.50 | 260 |
| | 7 | 1 | 42 | 26 | 3 | 0.20 | |
| | | 2 | 26 | 10 | 30 | 1.88 | |
| | | 3 | 10 | 4 | 100 | 16.67 | |
| | | 4 | 4 | 0 | 130 | 32.50 | 263 |
| | 8 | 1 | 42 | 26 | 3 | 0.20 | |
| | | 2 | 26 | 10 | 35 | 2.19 | |
| | | 3 | 10 | 4 | 100 | 16.67 | |
| | | 4 | 4 | 0 | 130 | 32.50 | 268 |
| | Over 8 | 1 | 42 | 26 | 3 | 0.20 | |
| | | 2 | 26 | 10 | 60 | 3.75 | |
| | | 3 | 10 | 4 | 100 | 16.67 | |
| | | 4 | 4 | 0 | 130 | 32.50 | 293 |
| 44 | 1/2 | 1 | 44 | 28 | 3 | 0.20 | |
| | | 2 | 28 | 12 | 16 | 1.00 | |
| | | 3 | 12 | 4 | 8 | 1.00 | |
| | | 4 | 4 | 0 | 16 | 4.00 | 43 |
| | 1 | 1 | 44 | 28 | 3 | 0.20 | |
| | | 2 | 28 | 12 | 16 | 1.00 | |
| | | 3 | 12 | 4 | 20 | 2.50 | |
| | | 4 | 4 | 0 | 25 | 6.25 | 64 |
| | 1 1/2 | 1 | 44 | 28 | 3 | 0.20 | |
| | | 2 | 28 | 12 | 16 | 1.00 | |
| | | 3 | 12 | 4 | 27 | 3.38 | |
| | | 4 | 4 | 0 | 72 | 18.00 | 118 |
| | 2 | 1 | 44 | 28 | 3 | 0.20 | |
| | | 2 | 28 | 12 | 16 | 1.00 | |
| | | 3 | 12 | 4 | 40 | 5.00 | |
| | | 4 | 4 | 0 | 95 | 23.75 | 154 |
| | 3 | 1 | 44 | 28 | 3 | 0.20 | |
| | | 2 | 28 | 12 | 16 | 1.00 | |
| | | 3 | 12 | 4 | 60 | 7.50 | |
| | | 4 | 4 | 0 | 120 | 30.00 | 199 |
| | 4 | 1 | 44 | 28 | 3 | 0.20 | |
| | | 2 | 28 | 12 | 16 | 1.00 | |
| | | 3 | 12 | 4 | 85 | 10.62 | |
| | | 4 | 4 | 0 | 130 | 32.50 | 234 |
| | 5 | 1 | 44 | 28 | 3 | 0.20 | |
| | | 2 | 28 | 12 | 16 | 1.00 | |
| | | 3 | 12 | 4 | 105 | 13.13 | |
| | | 4 | 4 | 0 | 130 | 32.50 | 254 |
| | 6 | 1 | 44 | 28 | 3 | 0.20 | |
| | | 2 | 28 | 12 | 16 | 1.00 | |
| | | 3 | 12 | 4 | 115 | 14.38 | |
| | | 4 | 4 | 0 | 130 | 32.50 | 264 |
| | 7 | 1 | 44 | 28 | 3 | 0.20 | |
| | | 2 | 28 | 12 | 16 | 1.00 | |
| | | 3 | 12 | 4 | 120 | 15.00 | |
| | | 4 | 4 | 0 | 130 | 32.50 | 269 |
| | 8 | 1 | 44 | 28 | 3 | 0.20 | |
| | | 2 | 28 | 12 | 16 | 1.00 | |
| | | 3 | 12 | 4 | 120 | 15.00 | |
| | | 4 | 4 | 0 | 130 | 32.50 | 269 |
| | Over 8 | 1 | 44 | 28 | 3 | 0.20 | |
| | | 2 | 28 | 12 | 40 | 2.50 | |
| | | 3 | 12 | 4 | 120 | 15.00 | |
| | | 4 | 4 | 0 | 130 | 32.50 | 293 |
| 46 | 1/2 | 1 | 46 | 30 | 3 | 0.20 | |
| | | 2 | 30 | 14 | 16 | 1.00 | |
| | | 3 | 14 | 4 | 10 | 1.00 | |
| | | 4 | 4 | 0 | 15 | 3.75 | 44 |
| | 1 | 1 | 46 | 30 | 3 | 0.20 | |
| | | 2 | 30 | 14 | 16 | 1.00 | |
| | | 3 | 14 | 4 | 25 | 2.50 | |
| | | 4 | 4 | 0 | 30 | 7.50 | 74 |
| | 1 1/2 | 1 | 46 | 30 | 3 | 0.20 | |
| | | 2 | 30 | 14 | 16 | 1.00 | |
| | | 3 | 14 | 4 | 35 | 3.50 | |
| | | 4 | 4 | 0 | 85 | 21.20 | 139 |

## DECOMPRESSION TABLE NO. 2

(Do not interpolate, use next higher value for conditions not computed.)

| Working chamber pressure p.s.i.g. | Working period Hours | Decompression data | | | | | |
|---|---|---|---|---|---|---|---|
| | | Stage No. | Pressure reduction p.s.i.g. | | Time in stage Minutes | Pressure reduction rate Min/Pound | Total time decompress Minutes |
| | | | From | To | | | |
| | 2 | 1 | 46 | 30 | 3 | 0.20 | |
| | | 2 | 30 | 14 | 16 | 1.00 | |
| | | 3 | 14 | 4 | 47 | 4.70 | |
| | | 4 | 4 | 0 | 105 | 25.25 | 171 |
| | 3 | 1 | 46 | 30 | 3 | 0.20 | |
| | | 2 | 30 | 14 | 16 | 1.00 | |
| | | 3 | 14 | 4 | 65 | 6.50 | |
| | | 4 | 4 | 0 | 130 | 32.50 | 214 |
| | 4 | 1 | 46 | 30 | 3 | 0.20 | |
| | | 2 | 30 | 14 | 16 | 1.00 | |
| | | 3 | 14 | 4 | 95 | 9.50 | |
| | | 4 | 4 | 0 | 130 | 32.50 | 244 |
| | 5 | 1 | 46 | 30 | 3 | 0.20 | |
| | | 2 | 30 | 14 | 16 | 1.00 | |
| | | 3 | 14 | 4 | 120 | 12.00 | |
| | | 4 | 4 | 0 | 130 | 32.50 | 269 |
| | 6 | 1 | 46 | 30 | 3 | 0.20 | |
| | | 2 | 30 | 14 | 16 | 1.00 | |
| | | 3 | 14 | 4 | 125 | 12.50 | |
| | | 4 | 4 | 0 | 130 | 32.50 | 274 |
| | 7 | 1 | 46 | 30 | 3 | 0.20 | |
| | | 2 | 30 | 14 | 16 | 1.00 | |
| | | 3 | 14 | 4 | 140 | 14.00 | |
| | | 4 | 4 | 0 | 130 | 32.50 | 289 |
| | 8 | 1 | 46 | 30 | 3 | 0.20 | |
| | | 2 | 30 | 14 | 16 | 1.00 | |
| | | 3 | 14 | 4 | 150 | 15.00 | |
| | | 4 | 4 | 0 | 130 | 32.50 | 299 |
| | Over 8 | 1 | 46 | 30 | 3 | 0.20 | |
| | | 2 | 30 | 14 | 25 | 1.56 | |
| | | 3 | 14 | 4 | 160 | 16.00 | |
| | | 4 | 4 | 0 | 130 | 32.50 | 318 |
| 48 | ½ | 1 | 48 | 32 | 3 | 0.20 | |
| | | 2 | 32 | 16 | 16 | 1.00 | |
| | | 3 | 16 | 4 | 12 | 1.00 | |
| | | 4 | 4 | 0 | 20 | 5.00 | 51 |
| | 1 | 1 | 48 | 32 | 3 | 0.20 | |
| | | 2 | 32 | 16 | 16 | 1.00 | |
| | | 3 | 16 | 4 | 35 | 2.92 | |
| | | 4 | 4 | 0 | 35 | 8.75 | 89 |
| | 1½ | 1 | 48 | 32 | 3 | 0.20 | |
| | | 2 | 32 | 16 | 16 | 1.00 | |
| | | 3 | 16 | 4 | 45 | 3.75 | |
| | | 4 | 4 | 0 | 80 | 20.00 | 144 |
| | 2 | 1 | 48 | 32 | 3 | 0.20 | |
| | | 2 | 32 | 16 | 16 | 1.00 | |
| | | 3 | 16 | 4 | 60 | 5.00 | |
| | | 4 | 4 | 0 | 110 | 27.50 | 189 |
| | 3 | 1 | 48 | 32 | 3 | 0.20 | |
| | | 2 | 32 | 16 | 16 | 1.00 | |
| | | 3 | 16 | 4 | 90 | 7.50 | |
| | | 4 | 4 | 0 | 120 | 30.00 | 299 |
| | 4 | 1 | 48 | 32 | 3 | 0.20 | |
| | | 2 | 32 | 16 | 16 | 1.00 | |
| | | 3 | 16 | 4 | 120 | 10.00 | |
| | | 4 | 4 | 0 | 130 | 32.50 | 269 |
| | 5 | 1 | 48 | 32 | 3 | 0.20 | |
| | | 2 | 32 | 16 | 16 | 1.00 | |
| | | 3 | 16 | 4 | 140 | 11.67 | |
| | | 4 | 4 | 0 | 130 | 32.50 | 299 |
| | 6 | 1 | 48 | 32 | 3 | 0.20 | |
| | | 2 | 32 | 16 | 16 | 1.00 | |
| | | 3 | 16 | 4 | 160 | 13.33 | |
| | | 4 | 4 | 0 | 130 | 32.50 | 309 |
| | 7 | 1 | 48 | 32 | 3 | 0.20 | |
| | | 2 | 32 | 16 | 16 | 1.00 | |
| | | 3 | 16 | 4 | 170 | 14.17 | |
| | | 4 | 4 | 0 | 130 | 32.50 | 319 |
| | 8 | 1 | 48 | 32 | 3 | 0.20 | |
| | | 2 | 32 | 16 | 16 | 1.00 | |
| | | 3 | 16 | 4 | 170 | 14.17 | |
| | | 4 | 4 | 0 | 130 | 32.50 | 319 |
| 50 | ½ | 1 | 50 | 34 | 3 | 0.20 | |
| | | 2 | 34 | 18 | 16 | 1.00 | |

## DECOMPRESSION TABLE NO. 2

(Do not interpolate, use next higher value for conditions not computed.)

| Working chamber pressure p.s.i.g. | Working period Hours | Decompression data | | | | | |
|---|---|---|---|---|---|---|---|
| | | Stage No. | Pressure reduction p.s.i.g. | | Time in stage Minutes | Pressure reduction rate Min/Pound | Total time decompress Minutes |
| | | | From | To | | | |
| | | 3 | 18 | 4 | 14 | 1.00 | |
| | | 4 | 4 | 0 | 25 | 6.25 | 58 |
| | 1 | 1 | 50 | 34 | 3 | 0.20 | |
| | | 2 | 34 | 18 | 16 | 1.00 | |
| | | 3 | 18 | 4 | 40 | 2.86 | |
| | | 4 | 4 | 0 | 35 | 8.75 | 94 |
| | 1½ | 1 | 50 | 34 | 3 | 0.20 | |
| | | 2 | 34 | 18 | 16 | 1.00 | |
| | | 3 | 18 | 4 | 55 | 3.93 | |
| | | 4 | 4 | 0 | 90 | 22.50 | 164 |
| | 2 | 1 | 50 | 34 | 3 | 0.20 | |
| | | 2 | 34 | 18 | 16 | 1.00 | |
| | | 3 | 18 | 4 | 70 | 5.00 | |
| | | 4 | 4 | 0 | 120 | 30.00 | 209 |
| | 3 | 1 | 50 | 34 | 3 | 0.20 | |
| | | 2 | 34 | 18 | 16 | 1.00 | |
| | | 3 | 18 | 4 | 100 | 7.15 | |
| | | 4 | 4 | 0 | 130 | 32.50 | 249 |
| | 4 | 1 | 50 | 34 | 3 | 0.20 | |
| | | 2 | 34 | 18 | 16 | 1.00 | |
| | | 3 | 18 | 4 | 130 | 8.58 | |
| | | 4 | 4 | 0 | 130 | 32.50 | 279 |
| | 5 | 1 | 50 | 34 | 3 | 0.20 | |
| | | 2 | 34 | 18 | 16 | 1.00 | |
| | | 3 | 18 | 4 | 160 | 11.42 | |
| | | 4 | 4 | 0 | 130 | 32.50 | 309 |
| | 6 | 1 | 50 | 34 | 3 | 0.20 | |
| | | 2 | 34 | 18 | 16 | 1.00 | |
| | | 3 | 18 | 4 | 180 | 12.85 | |
| | | 4 | 4 | 0 | 130 | 32.50 | 329 |

[44 FR 8577, Feb. 9, 1979; 44 FR 20940, Apr. 6. 1979; as amended at 58 FR 35311, June 30, 1993]

**Authority:** Sec. 107, Contract Work Hours and Safety Standards Act (Construction Safety Act) (40 U.S.C. 333); secs. 4, 6, 8, Occupational Safety and Health Act of 1970 (29 U.S.C. 653, 655, 657); Secretary of Labor's Order No. 12-71 (36 FR 8754), 8-76 (41 FR 25059), or 9-83 (48 FR 35736), as applicable.

### §1926.850 Preparatory operations.

**1926.850(a)** Prior to permitting employees to start demolition operations, an engineering survey shall be made, by a competent person, of the structure to determine the condition of the framing, floors, and walls, and possibility of unplanned collapse of any portion of the structure. Any adjacent structure where employees may be exposed shall also be similarly checked. The employer shall have in writing evidence that such a survey has been performed.

**1926.850(b)** When employees are required to work within a structure to be demolished which has been damaged by fire, flood, explosion, or other cause, the walls or floor shall be shored or braced.

**1926.850(c)** All electric, gas, water, steam, sewer, and other service lines shall be shut off, capped, or otherwise controlled, outside the building line before demolition work is started. In each case, any utility company which is involved shall be notified in advance.

**1926.850(d)** If it is necessary to maintain any power, water or other utilities during demolition, such lines shall be temporarily relocated, as necessary, and protected.

**1926.850(e)** It shall also be determined if any type of hazardous chemicals, gases, explosives, flammable materials, or similarly dangerous substances have been used in any pipes, tanks, or other equipment on the property. When the presence of any such substances is apparent or suspected, testing and purging shall be performed and the hazard eliminated before demolition is started.

**1926.850(f)** Where a hazard exists from fragmentation of glass, such hazards shall be removed.

**1926.850(g)** Where a hazard exists to employees falling through wall openings, the opening shall be protected to a height of approximately 42 inches.

**1926.850(h)** When debris is dropped through holes in the floor without the use of chutes, the area onto which the material is dropped shall be completely enclosed with barricades not less than 42 inches high and not less than 6 feet back from the projected edge of the opening above. Signs, warning of the hazard of falling materials, shall be posted at each level. Removal shall not be permitted in this lower area until debris handling ceases above.

**1926.850(i)** All floor openings, not used as material drops, shall be covered over with material substantial enough to support the weight of any load which may be imposed. Such material shall be properly secured to prevent its accidental movement.

**1926.850(j)** Except for the cutting of holes in floors for chutes, holes through which to drop materials, preparation of storage space, and similar necessary preparatory work, the demolition of exterior walls and floor construction shall begin at the top of the structure and proceed downward. Each story of exterior wall and floor construction shall be removed and dropped into the storage space before commencing the removal of exterior walls and floors in the story next below.

**1926.850(k)** Employee entrances to multi-story structures being demolished shall be completely protected by sidewalk sheds or canopies, or both, providing protection from the face of the building for a minimum of 8 feet. All such canopies shall be at least 2 feet wider than the building entrances or openings (1 foot wider on each side thereof), and shall be capable of sustaining a load of 150 pounds per square foot.

### §1926.851 Stairs, passageways, and ladders.

**1926.851(a)** Only those stairways, passageways, and ladders, designated as means of access to the structure of a building, shall be used. Other access ways shall be entirely closed at all times.

**1926.851(b)** All stairs, passageways, ladders and incidental equipment thereto, which are covered by this section, shall be periodically inspected and maintained in a clean safe condition.

**1926.851(c)** In a multistory building, when a stairwell is being used, it shall be properly illuminated by either natural or artificial means, and completely and substantially covered over at a point not less than two floors below the floor on which work is being performed, and access to the floor where the work is in progress shall be through a properly lighted, protected, and separate passageway.

### §1926.852 Chutes.

**1926.852(a)** No material shall be dropped to any point lying outside the exterior walls of the structure unless the area is effectively protected.

**1926.852(b)** All materials chutes, or sections thereof, at an angle of more than 45° from the horizontal, shall be entirely enclosed, except for openings equipped with closures at or about floor level for the insertion of materials. The openings shall not exceed 48 inches in height measured along the wall of the chute. At all stories below the top floor, such openings shall be kept closed when not in use.

**1926.852(c)** A substantial gate shall be installed in each chute at or near the discharge end. A competent employee shall be assigned to control the operation of the gate, and the backing and loading of trucks.

**1926.852(d)** When operations are not in progress, the area surrounding the discharge end of a chute shall be securely closed off.

**1926.852(e)** Any chute opening, into which workmen dump debris, shall be protected by a substantial guardrail approximately 42 inches above the floor or other surface on which the men stand to dump the material. Any space between the chute and the edge of openings in the floors through which it passes shall be solidly covered over.

**1926.852(f)** Where the material is dumped from mechanical equipment or wheelbarrows, a securely attached toeboard or bumper, not less than 4 inches thick and 6 inches high, shall be provided at each chute opening.

**1926.852(g)** Chutes shall be designed and constructed of such strength as to eliminate failure due to impact of materials or debris loaded therein.

### §1926.853 Removal of materials through floor openings.

Any openings cut in a floor for the disposal of materials shall be no larger in size than 25 percent of the aggregate of the total floor area, unless the lateral supports of the removed flooring remain in place. Floors weakened or otherwise made unsafe by demolition operations shall be shored to carry safely the intended imposed load from demolition operations.

### §1926.854 Removal of walls, masonry sections, and chimneys.

**1926.854(a)** Masonry walls, or other sections of masonry, shall not be permitted to fall upon the floors of the building in such masses as to exceed the safe carrying capacities of the floors.

**1926.854(b)** No wall section, which is more than one story in height, shall be permitted to stand alone without lateral bracing, unless such wall was originally designed and constructed to stand without such lateral support, and is in a condition safe enough to be self-supporting. All walls shall be left in a stable condition at the end of each shift.

**1926.854(c)** Employees shall not be permitted to work on the top of a wall when weather conditions constitute a hazard.

**1926.854(d)** Structural or load-supporting members on any floor shall not be cut or removed until all stories above such a floor have been demolished and removed. This provision shall not prohibit the cutting of floor beams for the disposal of materials or for the installation of equipment, provided that the requirements of §§1926.853 and 1926.855 are met.

**1926.854(e)** Floor openings within 10 feet of any wall being demolished shall be planked solid, except when employees are kept out of the area below.

**1926.854(f)** In buildings of "skeleton-steel" construction, the steel framing may be left in place during the demolition of masonry. Where this is done, all steel beams, girders, and similar structural supports shall be cleared of all loose material as the masonry demolition progresses downward.

**1926.854(g)** Walkways or ladders shall be provided to enable employees to safely reach or leave any scaffold or wall.

**1926.854(h)** Walls, which serve as retaining walls to support earth or adjoining structures, shall not be demolished until such earth has been properly braced or adjoining structures have been properly underpinned.

**1926.854(i)** Walls, which are to serve as retaining walls against which debris will be piled, shall not be so used unless capable of safely supporting the imposed load.

## §1926.855 Manual removal of floors.

**1926.855(a)** Openings cut in a floor shall extend the full span of the arch between supports.

**1926.855(b)** Before demolishing any floor arch, debris and other material shall be removed from such arch and other adjacent floor area. Planks not less than 2 inches by 10 inches in cross section, full size undressed, shall be provided for, and shall be used by employees to stand on while breaking down floor arches between beams. Such planks shall be so located as to provide a safe support for the workmen should the arch between the beams collapse. The open space between planks shall not exceed 16 inches.

**1926.855(c)** Safe walkways, not less than 18 inches wide, formed of planks not less than 2 inches thick if wood, or of equivalent strength if metal, shall be provided and used by workmen when necessary to enable them to reach any point without walking upon exposed beams.

**1926.855(d)** Stringers of ample strength shall be installed to support the flooring planks, and the ends of such stringers shall be supported by floor beams or girders, and not by floor arches alone.

**1926.855(e)** Planks shall be laid together over solid bearings with the ends overlapping at least 1 foot.

**1926.855(f)** When floor arches are being removed, employees shall not be allowed in the area directly underneath, and such an area shall be barricaded to prevent access to it.

**1926.855(g)** Demolition of floor arches shall not be started until they, and the surrounding floor area for a distance of 20 feet, have been cleared of debris and any other unnecessary materials.

## §1926.856 Removal of walls, floors, and material with equipment.

**1926.856(a)** Mechanical equipment shall not be used on floors or working surfaces unless such floors or surfaces are of sufficient strength to support the imposed load.

**1926.856(b)** Floor openings shall have curbs or stop-logs to prevent equipment from running over the edge.

**1926.856(c)** Mechanical equipment used shall meet the requirements specified in subparts N and O of this part.

## §1926.857 Storage.

**1926.857(a)** The storage of waste material and debris on any floor shall not exceed the allowable floor loads.

**1926.857(b)** In buildings having wooden floor construction, the flooring boards may be removed from not more than one floor above grade to provide storage space for debris, provided falling material is not permitted to endanger the stability of the structure.

**1926.857(c)** When wood floor beams serve to brace interior walls or free-standing exterior walls, such beams shall be left in place until other equivalent support can be installed to replace them.

**1926.857(d)** Floor arches, to an elevation of not more than 25 feet above grade, may be removed to provide storage area for debris: *Provided*, That such removal does not endanger the stability of the structure.

**1926.857(e)** Storage space into which material is dumped shall be blocked off, except for openings necessary for the removal of material. Such openings shall be kept closed at all times when material is not being removed.

## §1926.858 Removal of steel construction.

**1926.858(a)** When floor arches have been removed, planking in accordance with §1926.855(b) shall be provided for the workers engaged in razing the steel framing.

**1926.858(b)** Cranes, derricks, and other hoisting equipment used shall meet the requirements specified in subpart N of this part.

**1926.858(c)** Steel construction shall be dismantled column length by column length, and tier by tier (columns may be in two-story lengths).

**1926.858(d)** Any structural member being dismembered shall not be overstressed.

## §1926.859 Mechanical demolition.

**1926.859(a)** No workers shall be permitted in any area, which can be adversely affected by demolition operations, when balling or clamming is being performed. Only those workers necessary for the performance of the operations shall be permitted in this area at any other time.

**1926.859(b)** The weight of the demolition ball shall not exceed 50 percent of the crane's rated load, based on the length of the boom and the maximum angle of operation at which the demolition ball will be used, or it shall not exceed 25 percent of the nominal breaking strength of the line by which it is suspended, whichever results in a lesser value.

**1926.859(c)** The crane boom and loadline shall be as short as possible.

**1926.859(d)** The ball shall be attached to the loadline with a swivel-type connection to prevent twisting of the loadline, and shall be attached by positive means in such manner that the weight cannot become accidentally disconnected.

**1926.859(e)** When pulling over walls or portions thereof, all steel members affected shall have been previously cut free.

**1926.859(f)** All roof cornices or other such ornamental stonework shall be removed prior to pulling walls over.

**1926.859(g)** During demoliton, continuing inspections by a competent person shall be made as the work progresses to detect hazards resulting from weakened or deteriorated floors, or walls, or loosened material. No employee shall be permitted to work where such hazards exist until they are corrected by shoring, bracing, or other effective means.

## §1926.860 Selective demolition by explosives.

Selective demolition by explosives shall be conducted in accordance with the applicable sections of subpart U of this part.

**Authority:** Sec. 107, Contract Work Hours and Safety Standards Act (40 U.S.C. 333); secs. 4, 6, 8, Occupational Safety and Health Act of 1970 (29 U.S.C. 653, 655, 657); Secretary of Labor's Order No. 12-71 (36 FR 8754), 8-76 (41 FR 25059), 9-83 (48 FR 35736), or 6-96 (62 FR 111), as applicable; and 29 CFR part 1911.

### §1926.900 General provisions.

**1926.900(a)** The employer shall permit only authorized and qualified persons to handle and use explosives.

**1926.900(b)** Smoking, firearms, matches, open flame lamps, and other fires, flame or heat producing devices and sparks shall be prohibited in or near explosive magazines or while explosives are being handled, transported or used.

**1926.900(c)** No person shall be allowed to handle or use explosives while under the influence of intoxicating liquors, narcotics, or other dangerous drugs.

**1926.900(d)** All explosives shall be accounted for at all times. Explosives not being used shall be kept in a locked magazine, unavailable to persons not authorized to handle them. The employer shall maintain an inventory and use record of all explosives. Appropriate authorities shall be notified of any loss, theft, or unauthorized entry into a magazine.

**1926.900(e)** No explosives or blasting agents shall be abandoned.

**1926.900(f)** No fire shall be fought where the fire is in imminent danger of contact with explosives. All employees shall be removed to a safe area and the fire area guarded against intruders.

**1926.900(g)** Original containers, or Class II magazines, shall be used for taking detonators and other explosives from storage magazines to the blasting area.

**1926.900(h)** When blasting is done in congested areas or in proximity to a structure, railway, or highway, or any other installation that may be damaged, the blaster shall take special precautions in the loading, delaying, initiation, and confinement of each blast with mats or other methods so as to control the throw of fragments, and thus prevent bodily injury to employees.

**1926.900(i)** Employees authorized to prepare explosive charges or conduct blasting operations shall use every reasonable precaution including, but not limited to, visual and audible warning signals, flags, or barricades, to ensure employee safety.

**1926.900(j)** Insofar as possible, blasting operations above ground shall be conducted between sunup and sundown.

**1926.900(k)** Due precautions shall be taken to prevent accidental discharge of electric blasting caps from current induced by radar, radio transmitters, lightning, adjacent powerlines, dust storms, or other sources of extraneous electricity. These precautions shall include:

**1926.900(k)(1)** Detonators shall be short-circuited in holes which have been primed and shunted until wired into the blasting circuit.

**1926.900(k)(2)** The suspension of all blasting operations and removal of persons from the blasting area during the approach and progress of an electric storm;

**1926.900(k)(3)(i)** The prominent display of adequate signs, warning against the use of mobile radio transmitters, on all roads within 1,000 feet of blasting operations. Whenever adherence to the 1,000-foot distance would create an operational handicap, a competent person shall be consulted to evaluate the particular situation, and alternative provisions may be made which are adequately designed to prevent any premature firing of electric blasting caps. A description of any such alternatives shall be reduced to writing and shall be certified as meeting the purposes of this subdivision by the competent person consulted. The description shall be maintained at the construction site during the duration of the work, and shall be available for inspection by representatives of the Secretary of Labor.

**1926.900(k)(3)(ii)** Specimens of signs which would meet the requirements of paragraph (k)(3) of this section are the following:

About 48" x 48"

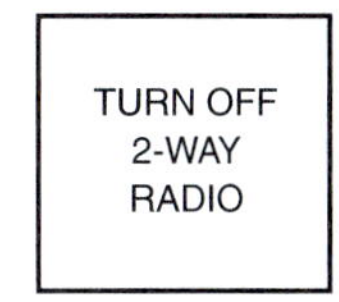


About 42" x 36"

**1926.900(k)(4)** Ensuring that mobile radio transmitters which are less than 100 feet away from electric blasting caps, in other than original containers, shall be deenergized and effectively locked;

**1926.900(k)(5)** Compliance with the recommendations of The Institute of the Makers of Explosives with regard to blasting in the vicinity of radio transmitters as stipulated in Radio Frequency Energy—A Potential Hazard in the Use of Electric Blasting Caps, IME Publication No. 20, March 1971.

**1926.900(l)** Empty boxes and paper and fiber packing materials, which have previously contained high explosives, shall not be used again for any purpose, but shall be destroyed by burning at an approved location.

**1926.900(m)** Explosives, blasting agents, and blasting supplies that are obviously deteriorated or damaged shall not be used.

**1926.900(n)** Delivery and issue of explosives shall only be made by and to authorized persons and into authorized magazines or approved temporary storage or handling areas.

**1926.900(o)** Blasting operations in the proximity of overhead power lines, communication lines, utility services, or other services and structures shall not be carried on until the operators and/or owners have been notified and measures for safe control have been taken.

**1926.900(p)** The use of black powder shall be prohibited.

**1926.900(q)** All loading and firing shall be directed and supervised by competent persons thoroughly experienced in this field.

**1926.900(r)** All blasts shall be fired electrically with an electric blasting machine or properly designed electric power source, except as provided in §1926.906(a) and (r).

**1926.900(s)** Buildings used for the mixing of blasting agents shall conform to the requirements of this section.

**1926.900(s)(1)** Buildings shall be of noncombustible construction or sheet metal on wood studs.

**1926.900(s)(2)** Floors in a mixing plant shall be of concrete or of other nonabsorbent materials.

**1926.900(s)(3)** All fuel oil storage facilities shall be separated from the mixing plant and located in such a manner that in case of tank rupture, the oil will drain away from the mixing plant building.

**1926.900(s)(4)** The building shall be well ventilated.

**1926.900(s)(5)** Heating units which do not depend on combustion processes, when properly designed and located, may be used in the building. All direct sources of heat shall be provided exclusively from units located outside the mixing building.

**1926.900(s)(6)** All internal-combustion engines used for electric power generation shall be located outside the mixing plant building, or shall be properly ventilated and isolated by a firewall. The exhaust systems on all such engines shall be located so any spark emission cannot be a hazard to any materials in or adjacent to the plant.

**1926.900(t)** Buildings used for the mixing of water gels shall conform to the requirements of this subdivision.

**1926.900(t)(1)** Buildings shall be of noncombustible construction or sheet metal on wood studs.

**1926.900(t)(2)** Floors in a mixing plant shall be of concrete or of other nonabsorbent materials.

**1926.900(t)(3)** Where fuel oil is used all fuel oil storage facilities shall be separated from the mixing plant and located in such a manner that in case of tank rupture, the oil will drain away from the mixing plant building.

**1926.900(t)(4)** The building shall be well ventilated.

**1926.900(t)(5)** Heating units that do not depend on combustion processes, when properly designed and located, may be used in the building. All direct sources of heat shall be provided exclusively from units located outside of the mixing building.

**1926.900(t)(6)** All internal-combustion engines used for electric power generation shall be located outside the mixing plant building, or shall be properly ventilated and isolated by a firewall. The exhaust systems on all such engines shall be located so any spark emission cannot be a hazard to any materials in or adjacent to the plant.

[44 FR 8577, Feb. 9, 1979; 44 FR 20940, Apr. 6, 1979, as amended at 58 FR 35183, June 30, 1993]

### §1926.901 Blaster qualifications.

**1926.901(a)** A blaster shall be able to understand and give written and oral orders.

**1926.901(b)** A blaster shall be in good physical condition and not be addicted to narcotics, intoxicants, or similar types of drugs.

**1926.901(c)** A blaster shall be qualified, by reason of training, knowledge, or experience, in the field of transporting, storing, handling, and use of explosives, and have a working knowledge of State and local laws and regulations which pertain to explosives.

**1926.901(d)** Blasters shall be required to furnish satisfactory evidence of competency in handling explosives and performing in a safe manner the type of blasting that will be required.

**1926.901(e)** The blaster shall be knowledgeable and competent in the use of each type of blasting method used.

### §1926.902 Surface transportation of explosives.

**1926.902(a)** Transportation of explosives shall meet the provisions of Department of Transportation regulations contained in 46 CFR parts 146-149, Water Carriers; 49 CFR parts 171-179, Highways and Railways; 49 CFR part 195, Pipelines; and 49 CFR parts 390-397, Motor Carriers.

**1926.902(b)** Motor vehicles or conveyances transporting explosives shall only be driven by, and be in the charge of, a licensed driver who is physically fit. He shall be familiar with the local, State, and Federal regulation governing the transportation of explosives.

**1926.902(c)** No person shall smoke, or carry matches or any other flame-producing device, nor shall firearms or loaded cartridges be carried while in or near a motor vehicle or conveyance transporting explosives.

**1926.902(d)** Explosives, blasting agents, and blasting supplies shall not be transported with other materials or cargoes. Blasting caps (including electric) shall not be transported in the same vehicle with other explosives.

**1926.902(e)** Vehicles used for transporting explosives shall be strong enough to carry the load without difficulty, and shall be in good mechanical condition.

**1926.902(f)** When explosives are transported by a vehicle with an open body, a Class II magazine or original manufacturer's container shall be securely mounted on the bed to contain the cargo.

**1926.902(g)** All vehicles used for the transportation of explosives shall have tight floors and any exposed spark-producing metal on the inside of the body shall be covered with wood, or other nonsparking material, to prevent contact with containers of explosives.

**1926.902(h)** Every motor vehicle or conveyance used for transporting explosives shall be marked or placarded on both sides, the front, and the rear with the word "Explosives" in red letters, not less than 4 inches in height, on white background. In addition to such marking or placarding, the motor vehicle or conveyance may display, in such a manner that it will be readily visible from all directions, a red flag 18 inches by 30 inches, with the word "Explosives" painted, stamped, or sewed thereon, in white letters, at least 6 inches in height.

**1926.902(i)** Each vehicle used for transportation of explosives shall be equipped with a fully charged fire extinguisher, in good condition. An Underwriters Laboratory-approved extinguisher of not less than 10-ABC rating will meet the minimum requirement. The driver shall be trained in the use of the extinguisher on his vehicle.

**1926.902(j)** Motor vehicles or conveyances carrying explosives, blasting agents, or blasting supplies, shall not be taken inside a garage or shop for repairs or servicing.

**1926.902(k)** No motor vehicle transporting explosives shall be left unattended.

[44 FR 8577, Feb. 9, 1979; 44 FR 20940, Apr. 6, 1979, as amended at 58 FR 35311, June 30, 1993]

### §1926.903 Underground transportation of explosives.

**1926.903(a)** All explosives or blasting agents in transit underground shall be taken to the place of use or storage without delay.

**1926.903(b)** The quantity of explosives or blasting agents taken to an underground loading area shall not exceed the amount estimated to be necessary for the blast.

**1926.903(c)** Explosives in transit shall not be left unattended.

**1926.903(d)** The hoist operator shall be notified before explosives or blasting agents are transported in a shaft conveyance.

**1926.903(e)** Trucks used for the transportation of explosives underground shall have the electrical system checked weekly to detect any failures which may constitute an electrical hazard. A certification record which includes the date of the inspection; the signature of the person who performed the inspection; and a serial number, or other identifier, of the truck inspected shall be prepared and the most recent certification record shall be maintained on file.

**1926.903(f)** The installation of auxiliary lights on truck beds, which are powered by the truck's electrical system, shall be prohibited.

**1926.903(g)** Explosives and blasting agents shall be hoisted, lowered, or conveyed in a powder car. No other materials, supplies, or equipment shall be transported in the same conveyance at the same time.

**1926.903(h)** No one, except the operator, his helper, and the powderman, shall be permitted to ride on a conveyance transporting explosives and blasting agents.

**1926.903(i)** No person shall ride in any shaft conveyance transporting explosives and blasting agents.

**1926.903(j)** No explosives or blasting agents shall be transported on any locomotive. At least two car lengths shall separate the locomotive from the powder car.

**1926.903(k)** No explosives or blasting agents shall be transported on a man haul trip.

**1926.903(l)** The car or conveyance containing explosives or blasting agents shall be pulled, not pushed, whenever possible.

**1926.903(m)** The powder car or conveyance especially built for the purpose of transporting explosives or blasting agents shall bear a reflectorized sign on each side with the word "Explosives" in letters, not less than 4 inches in height; upon a background of sharply contrasting color.

**1926.903(n)** Compartments for transporting detonators and explosives in the same car or conveyance shall be physically separated by a distance of 24 inches or by a solid partition at least 6 inches thick.

**1926.903(o)** Detonators and other explosives shall not be transported at the same time in any shaft conveyance.

**1926.903(p)** Explosives, blasting agents, or blasting supplies shall not be transported with other materials.

**1926.903(q)** Explosives or blasting agents, not in original containers, shall be placed in a suitable container when transported manually.

**1926.903(r)** Detonators, primers, and other explosives shall be carried in separate containers when transported manually.

[44 FR 8577, Feb. 9, 1979; 44 FR 20940, Apr. 6, 1979, as amended at 52 FR 36382, Sept. 28, 1987]

### §1926.904 Storage of explosives and blasting agents.

**1926.904(a)** Explosives and related materials shall be stored in approved facilities required under the applicable provisions of the Bureau of Alcohol, Tobacco and Firearms regulations contained in 27 CFR part 55, Commerce in Explosives.

**1926.904(b)** Blasting caps, electric blasting caps, detonating primers, and primed cartridges shall not be stored in the same magazine with other explosives or blasting agents.

**1926.904(c)** Smoking and open flames shall not be permitted within 50 feet of explosives and detonator storage magazine.

**1926.904(d)** No explosives or blasting agents shall be permanently stored in any underground operation until the operation has been developed to the point where at least two modes of exit have been provided.

**1926.904(e)** Permanent underground storage magazines shall be at least 300 feet from any shaft, adit, or active underground working area.

**1926.904(f)** Permanent underground magazines containing detonators shall not be located closer than 50 feet to any magazine containing other explosives or blasting agents.

[44 FR 8577, Feb. 9, 1979; 44 FR 20940, Apr. 6, 1979, as amended at 58 FR 35311, June 30, 1993]

## §1926.905 Loading of explosives or blasting agents.

**1926.905(a)** Procedures that permit safe and efficient loading shall be established before loading is started.

**1926.905(b)** All drill holes shall be sufficiently large to admit freely the insertion of the cartridges of explosives.

**1926.905(c)** Tamping shall be done only with wood rods or plastic tamping poles without exposed metal parts, but nonsparking metal connectors may be used for jointed poles. Violent tamping shall be avoided. The primer shall never be tamped.

**1926.905(d)** No holes shall be loaded except those to be fired in the next round of blasting. After loading, all remaining explosives and detonators shall be immediately returned to an authorized magazine.

**1926.905(e)** Drilling shall not be started until all remaining butts of old holes are examined for unexploded charges, and if any are found, they shall be refired before work proceeds.

**1926.905(f)** No person shall be allowed to deepen drill holes which have contained explosives or blasting agents.

**1926.905(g)** No explosives or blasting agents shall be left unattended at the blast site.

**1926.905(h)** Machines and all tools not used for loading explosives into bore holes shall be removed from the immediate location of holes before explosives are delivered. Equipment shall not be operated within 50 feet of loaded holes.

**1926.905(i)** No activity of any nature other than that which is required for loading holes with explosives shall be permitted in a blast area.

**1926.905(j)** Powerlines and portable electric cables for equipment being used shall be kept a safe distance from explosives or blasting agents being loaded into drill holes. Cables in the proximity of the blast area shall be deenergized and locked out by the blaster.

**1926.905(k)** Holes shall be checked prior to loading to determine depth and conditions. Where a hole has been loaded with explosives but the explosives have failed to detonate, there shall be no drilling within 50 feet of the hole.

**1926.905(l)** When loading a long line of holes with more than one loading crew, the crews shall be separated by practical distance consistent with efficient operation and supervision of crews.

**1926.905(m)** No explosive shall be loaded or used underground in the presence of combustible gases or combustible dusts.

**1926.905(n)** No explosives other than those in Fume Class 1, as set forth by the Institute of Makers of Explosives, shall be used; however, explosives complying with the requirements of Fume Class 2 and Fume Class 3 may be used if adequate ventilation has been provided.

**1926.905(o)** All blast holes in open work shall be stemmed to the collar or to a point which will confine the charge.

**1926.905(p)** Warning signs, indicating a blast area, shall be maintained at all approaches to the blast area. The warning sign lettering shall not be less than 4 inches in height on a contrasting background.

**1926.905(q)** A bore hole shall never be sprung when it is adjacent to or near a hole that is loaded. Flashlight batteries shall not be used for springing holes.

**1926.905(r)** Drill holes which have been sprung or chambered, and which are not water-filled, shall be allowed to cool before explosives are loaded.

**1926.905(s)** No loaded holes shall be left unattended or unprotected.

**1926.905(t)** The blaster shall keep an accurate, up-to-date record of explosives, blasting agents, and blasting supplies used in a blast and shall keep an accurate running inventory of all explosives and blasting agents stored on the operation.

**1926.905(u)** When loading blasting agents pneumatically over electric blasting caps, semiconductive delivery hose shall be used and the equipment shall be bonded and grounded.

[44 FR 8577, Feb. 9, 1979; 44 FR 20940, Apr. 6, 1979, as amended at 58 FR 35184, June 30, 1993]

## §1926.906 Initiation of explosive charges—electric blasting.

**1926.906(a)** Electric blasting caps shall not be used where sources of extraneous electricity make the use of electric blasting caps dangerous. Blasting cap leg wires shall be kept short-circuited (shunted) until they are connected into the circuit for firing.

**1926.906(b)** Before adopting any system of electrical firing, the blaster shall conduct a thorough survey for extraneous currents, and all dangerous currents shall be eliminated before any holes are loaded.

**1926.906(c)** In any single blast using electric blasting caps, all caps shall be of the same style or function, and of the same manufacture.

**1926.906(d)** Electric blasting shall be carried out by using blasting circuits or power circuits in accordance with the electric blasting cap manufacturer's recommendations, or an approved contractor or his designated representative.

**1926.906(e)** When firing a circuit of electric blasting caps, care must be exercised to ensure that an adequate quantity of delivered current is available, in accordance with the manufacturer's recommendations.

**1926.906(f)** Connecting wires and lead wires shall be insulated single solid wires of sufficient current-carrying capacity.

**1926.906(g)** Bus wires shall be solid single wires of sufficient current-carrying capacity.

**1926.906(h)** When firing electrically, the insulation on all firing lines shall be adequate and in good condition.

**1926.906(i)** A power circuit used for firing electric blasting caps shall not be grounded.

**1926.906(j)** In underground operations when firing from a power circuit, a safety switch shall be placed in the permanent firing line at intervals. This switch shall be made so it can be locked only in the "Off" position and shall be provided with a short-circuiting arrangement of the firing lines to the cap circuit.

**1926.906(k)** In underground operations there shall be a "lightning" gap of at least 5 feet in the firing system ahead of the main firing switch; that is, between this switch and the source of power. This gap shall be bridged by a flexible jumper cord just before firing the blast.

**1926.906(l)** When firing from a power circuit, the firing switch shall be locked in the open or "Off" position at all times, except when firing. It shall be so designed that the firing lines to the cap circuit are automatically short-circuited when the switch is in the "Off" position. Keys to this switch shall be entrusted only to the blaster.

**1926.906(m)** Blasting machines shall be in good condition and the efficiency of the machine shall be tested periodically to make certain that it can deliver power at its rated capacity.

**1926.906(n)** When firing with blasting machines, the connections shall be made as recommended by the manufacturer of the electric blasting caps used.

**1926.906(o)** The number of electric blasting caps connected to a blasting machine shall not be in excess of its rated capacity. Furthermore, in primary blasting, a series circuit shall contain no more caps than the limits recommended by the manufacturer of the electric blasting caps in use.

**1926.906(p)** The blaster shall be in charge of the blasting machines, and no other person shall connect the leading wires to the machine.

**1926.906(q)** Blasters, when testing circuits to charged holes, shall use only blasting galvanometers or other instruments that are specifically designed for this purpose.

**1926.906(r)** Whenever the possibility exists that a leading line or blasting wire might be thrown over a live powerline by the force of an explosion, care shall be taken to see that the total length of wires are kept too short to hit the lines, or that the wires are securely anchored to the ground. If neither of these requirements can be satisfied, a nonelectric system shall be used.

**1926.906(s)** In electrical firing, only the man making leading wire connections shall fire the shot. All connections shall be made from the bore hole back to the source of firing current, and the leading wires shall remain shorted and not be connected to the blasting machine or other source of current until the charge is to be fired.

**1926.906(t)** After firing an electric blast from a blasting machine, the leading wires shall be immediately disconnected from the machine and short-circuited.

[44 FR 8577, Feb. 9, 1979; 44 FR 20940, Apr. 6, 1979, as amended at 63 FR 33469, June 18, 1998]

### §1926.907 Use of safety fuse.

**1926.907(a)** Safety fuse shall only be used where sources of extraneous electricity make the use of electric blasting caps dangerous. The use of a fuse that has been hammered or injured in any way shall be forbidden.

**1926.907(b)** The hanging of a fuse on nails or other projections which will cause a sharp bend to be formed in the fuse is prohibited.

**1926.907(c)** Before capping safety fuse, a short length shall be cut from the end of the supply reel so as to assure a fresh cut end in each blasting cap.

**1926.907(d)** Only a cap crimper of approved design shall be used for attaching blasting caps to safety fuse. Crimpers shall be kept in good repair and accessible for use.

**1926.907(e)** No unused cap or short capped fuse shall be placed in any hole to be blasted; such unused detonators shall be removed from the working place and destroyed.

**1926.907(f)** No fuse shall be capped, or primers made up, in any magazine or near any possible source of ignition.

**1926.907(g)** No one shall be permitted to carry detonators or primers of any kind on his person.

**1926.907(h)** The minimum length of safety fuse to be used in blasting shall be as required by State law, but shall not be less than 30 inches.

**1926.907(i)** At least two men shall be present when multiple cap and fuse blasting is done by hand lighting methods.

**1926.907(j)** Not more than 12 fuses shall be lighted by each blaster when hand lighting devices are used. However, when two or more safety fuses in a group are lighted as one by means of igniter cord, or other similar fuse-lighting devices, they may be considered as one fuse.

**1926.907(k)** The so-called "drop fuse" method of dropping or pushing a primer or any explosive with a lighted fuse attached is forbidden.

**1926.907(l)** Cap and fuse shall not be used for firing mudcap charges unless charges are separated sufficiently to prevent one charge from dislodging other shots in the blast.

**1926.907(m)** When blasting with safety fuses, consideration shall be given to the length and burning rate of the fuse. Sufficient time, with a margin of safety, shall always be provided for the blaster to reach a place of safety.

### §1926.908 Use of detonating cord.

**1926.908(a)** Care shall be taken to select a detonating cord consistent with the type and physical condition of the bore hole and stemming and the type of explosives used.

**1926.908(b)** Detonating cord shall be handled and used with the same respect and care given other explosives.

**1926.908(c)** The line of detonating cord extending out of a bore hole or from a charge shall be cut from the supply spool before loading the remainder of the bore hole or placing additional charges.

**1926.908(d)** Detonating cord shall be handled and used with care to avoid damaging or severing the cord during and after loading and hooking-up.

**1926.908(e)** Detonating cord connections shall be competent and positive in accordance with approved and recommended methods. Knot-type or other cord-to-cord connections shall be made only with detonating cord in which the explosive core is dry.

**1926.908(f)** All detonating cord trunklines and branchlines shall be free of loops, sharp kinks, or angles that direct the cord back toward the oncoming line of detonation.

**1926.908(g)** All detonating cord connections shall be inspected before firing the blast.

**1926.908(h)** When detonating cord millisecond-delay connectors or short-interval-delay electric blasting caps are used with detonating cord, the practice shall conform strictly to the manufacturer's recommendations.

**1926.908(i)** When connecting a blasting cap or an electric blasting cap to detonating cord, the cap shall be taped or otherwise attached securely along the side or the end of the detonating cord, with the end of the cap containing the explosive charge pointed in the direction in which the detonation is to proceed.

**1926.908(j)** Detonators for firing the trunkline shall not be brought to the loading area nor attached to the detonating cord until everything else is in readiness for the blast.

### §1926.909 Firing the blast.

**1926.909(a)** A code of blasting signals equivalent to Table U-1, shall be posted on one or more conspicuous places at the operation, and all employees shall be required to familiarize themselves with the code and conform to it. Danger signs shall be placed at suitable locations.

**1926.909(b)** Before a blast is fired, a loud warning signal shall be given by the blaster in charge, who has made certain that all surplus explosives are in a safe place and all employees, vehicles, and equipment are at a safe distance, or under sufficient cover.

**1926.909(c)** Flagmen shall be safely stationed on highways which pass through the danger zone so as to stop traffic during blasting operations.

**1926.909(d)** It shall be the duty of the blaster to fix the time of blasting.

**1926.909(e)** Before firing an underground blast, warning shall be given, and all possible entries into the blasting area, and any entrances to any working place where a drift, raise, or other opening is about to hole through, shall be carefully guarded. The blaster shall make sure that all employees are out of the blast area before firing a blast.

**TABLE U-1**

| |
|---|
| WARNING SIGNAL—A 1-minute series of long blasts 5 minutes prior to blast signal. |
| BLAST SIGNAL—A series of short blasts 1 minute prior to the shot. |
| ALL CLEAR SIGNAL—A prolonged blast following the inspection of blast area. |

### §1926.910 Inspection after blasting.

**1926.910(a)** Immediately after the blast has been fired, the firing line shall be disconnected from the blasting machine, or where power switches are used, they shall be locked open or in the off position.

**1926.910(b)** Sufficient time shall be allowed, not less than 15 minutes in tunnels, for the smoke and fumes to leave the blasted area before returning to the shot. An inspection of the area and the surrounding rubble shall be made by the blaster to determine if all charges have been exploded before employees are allowed to return to the operation, and in tunnels, after the muck pile has been wetted down.

### §1926.911 Misfires.

**1926.911(a)** If a misfire is found, the blaster shall provide proper safeguards for excluding all employees from the danger zone.

**1926.911(b)** No other work shall be done except that necessary to remove the hazard of the misfire and only those employees necessary to do the work shall remain in the danger zone.

**1926.911(c)** No attempt shall be made to extract explosives from any charged or misfired hole; a new primer shall be put in and the hole reblasted. If refiring of the misfired hole presents a hazard, the explosives may be removed by washing out with water or, where the misfire is under water, blown out with air.

**1926.911(d)** If there are any misfires while using cap and fuse, all employees shall remain away from the charge for at least 1 hour. Misfires shall be handled under the direction of the person in charge of the blasting. All wires shall be carefully traced and a search made for unexploded charges.

**1926.911(e)** No drilling, digging, or picking shall be permitted until all missed holes have been detonated or the authorized representative has approved that work can proceed.

## §1926.912 Underwater blasting.

**1926.912(a)** A blaster shall conduct all blasting operations, and no shot shall be fired without his approval.

**1926.912(b)** Loading tubes and casings of dissimilar metals shall not be used because of possible electric transient currents from galvanic action of the metals and water.

**1926.912(c)** Only water-resistant blasting caps and detonating cords shall be used for all marine blasting. Loading shall be done through a nonsparking metal loading tube when tube is necessary.

**1926.912(d)** No blast shall be fired while any vessel under way is closer than 1,500 feet to the blasting area. Those on board vessels or craft moored or anchored within 1,500 feet shall be notified before a blast is fired.

**1926.912(e)** No blast shall be fired while any swimming or diving operations are in progress in the vicinity of the blasting area. If such operations are in progress, signals and arrangements shall be agreed upon to assure that no blast shall be fired while any person is in the water.

**1926.912(f)** Blasting flags shall be displayed.

**1926.912(g)** The storage and handling of explosives aboard vessels used in underwater blasting operations shall be according to provisions outlined herein on handling and storing explosives.

**1926.912(h)** When more than one charge is placed under water, a float device shall be attached to an element of each charge in such manner that it will be released by the firing. Misfires shall be handled in accordance with the requirements of §1926.911.

## §1926.913 Blasting in excavation work under compressed air.

**1926.913(a)** Detonators and explosives shall not be stored or kept in tunnels, shafts, or caissons. Detonators and explosives for each round shall be taken directly from the magazines to the blasting zone and immediately loaded. Detonators and explosives left over after loading a round shall be removed from the working chamber before the connecting wires are connected up.

**1926.913(b)** When detonators or explosives are brought into an air lock, no employee except the powderman, blaster, lock tender and the employees necessary for carrying, shall be permitted to enter the air lock. No other material, supplies, or equipment shall be locked through with the explosives.

**1926.913(c)** Detonators and explosives shall be taken separately into pressure working chambers.

**1926.913(d)** The blaster or powderman shall be responsible for the receipt, unloading, storage, and on-site transportation of explosives and detonators.

**1926.913(e)** All metal pipes, rails, air locks, and steel tunnel lining shall be electrically bonded together and grounded at or near the portal or shaft, and such pipes and rails shall be cross-bonded together at not less than 1,000-foot intervals throughout the length of the tunnel. In addition, each low air supply pipe shall be grounded at its delivery end.

**1926.913(f)** The explosives suitable for use in wet holes shall be water-resistant and shall be Fume Class 1.

**1926.913(g)** When tunnel excavation in rock face is approaching mixed face, and when tunnel excavation is in mixed face, blasting shall be performed with light charges and with light burden on each hole. Advance drilling shall be performed as tunnel excavation in rock face approaches mixed face, to determine the general nature and extent of rock cover and the remaining distance ahead to soft ground as excavation advances.

## §1926.914 Definitions applicable to this subpart.

**1926.914(a)** *American Table of Distances* (also known as Quantity Distance Tables) means American Table of Distances for Storage of Explosives as revised and approved by the Institute of the Makers of Explosives, June 5, 1964.

**1926.914(b)** *Approved storage facility*—A facility for the storage of explosive materials conforming to the requirements of this part and covered by a license or permit issued under authority of the Bureau of Alcohol, Tobacco and Firearms. (See 27 CFR Part 55)

**1926.914(c)** *Blast area*—The area in which explosives loading and blasting operations are being conducted.

**1926.914(d)** *Blaster*—The person or persons authorized to use explosives for blasting purposes and meeting the qualifications contained in §1926.901.

**1926.914(e)** *Blasting agent*—A blasting agent is any material or mixture consisting of a fuel and oxidizer used for blasting, but not classified an explosive and in which none of the ingredients is classified as an explosive provided the furnished (mixed) product cannot be detonated with a No. 8 test blasting cap when confined. A common blasting agent presently in use is a mixture of ammonium nitrate ($NH_4$ $NO_3$) and carbonaceous combustibles, such as fuel oil or coal, and may either be procured, premixed and packaged from explosives companies or mixed in the field.

**1926.914(f)** *Blasting cap*—A metallic tube closed at one end, containing a charge of one or more detonating compounds, and designed for and capable of detonation from the sparks or flame from a safety fuse inserted and crimped into the open end.

**1926.914(g)** *Block holing*—The breaking of boulders by firing a charge of explosives that has been loaded in a drill hole.

**1926.914(h)** *Conveyance*—Any unit for transporting explosives or blasting agents, including but not limited to trucks, trailers, rail cars, barges, and vessels.

**1926.914(i)** *Detonating cord*—A flexible cord containing a center core of high explosives which when detonated, will have sufficient strength to detonate other cap-sensitive explosives with which it is in contact.

**1926.914(j)** *Detonator*—Blasting caps, electric blasting caps, delay electric blasting caps, and nonelectric delay blasting caps.

**1926.914(k)** *Electric blasting cap*—A blasting cap designed for and capable of detonation by means of an electric current.

**1926.914(l)** *Electric blasting circuitry*—

**1926.914(l)(1)** Bus wire. An expendable wire, used in parallel or series, in parallel circuits, to which are connected the leg wires of electric blasting caps.

**1926.914(l)(2)** Connecting wire. An insulated expendable wire used between electric blasting caps and the leading wires or between the bus wire and the leading wires.

**1926.914(l)(3)** Leading wire. An insulated wire used between the electric power source and the electric blasting cap circuit.

**1926.914(l)(4)** Permanent blasting wire. A permanently mounted insulated wire used between the electric power source and the electric blasting cap circuit.

**1926.914(m)** *Electric delay blasting caps*—Caps designed to detonate at a predetermined period of time after energy is applied to the ignition system.

**1926.914(n)** *Explosives*—(1) Any chemical compound, mixture, or device, the primary or common purpose of which is to function by explosion; that is, with substantially instantaneous release of gas and heat, unless such compound, mixture or device is otherwise specifically classified by the U.S. Department of Transportation.

**1926.914(n)(2)** All material which is classified as Class A, Class B, and Class C Explosives by the U.S. Department of Transportation.

**1926.914(n)(3)** Classification of explosives by the U.S. Department of Transportation is as follows:

*Class A Explosives.* Possessing detonating hazard, such as dynamite, nitroglycerin, picric acid, lead azide, fulminate of mercury, black powder, blasting caps, and detonating primers.

*Class B Explosives.* Possessing flammable hazard, such as propellant explosives, including some smokeless propellants.

*Class C Explosives.* Include certain types of manufactured articles which contain Class A or Class B explosives, or both, as components, but in restricted quantities.

**1926.914(o)** *Fuse lighters*—Special devices for the purpose of igniting safety fuse.

**1926.914(p)** *Magazine*—Any building or structure, other than an explosives manufacturing building, used for the storage of explosives.

**1926.914(q)** *Misfire*—An explosive charge which failed to detonate.

**1926.914(r)** *Mud-capping* (sometimes known as bulldozing, adobe blasting, or dobying). The blasting of boulders by placing a quantity of explosives against a rock, boulder, or other object without confining the explosives in a drill hole.

**1926.914(s)** *Nonelectric delay blasting cap*—A blasting cap with an integral delay element in conjunction with and capable of being detonated by a detonation impulse or signal from miniaturized detonating cord.

**1926.914(t)** *Primary blasting*—The blasting operation by which the original rock formation is dislodged from its natural location.

**1926.914(u)** *Primer*—A cartridge or container of explosives into which a detonator or detonating cord is inserted or attached.

**1926.914(v)** *Safety fuse*—A flexible cord containing an internal burning medium by which fire is conveyed at a continuous and uniform rate for the purpose of firing blasting caps.

**1926.914(w)** *Secondary blasting*—The reduction of oversize material by the use of explosives to the dimension required for handling, including mudcapping and blockholing.

**1926.914(x)** *Stemming*—A suitable inert incombustible material or device used to confine or separate explosives in a drill hole, or to cover explosives in mud-capping.

**1926.914(y)** *Springing*—The creation of a pocket in the bottom of a drill hole by the use of a moderate quantity of explosives in order that larger quantities or explosives may be inserted therein.

**1926.914(z)** *Water gels, or slurry explosives*—A wide variety of materials used for blasting. They all contain substantial proportions of water and high proportions of ammonium nitrate, some of which is in solution in the water. Two broad classes of water gels are: (1) Those which are sensitized by a material classed as an explosive, such as TNT or smokeless powder, and (2) those which contain no ingredient classified as an explosive; these are sensitized with metals such as aluminum or with other fuels. Water gels may be premixed at an explosives plant or mixed at the site immediately before delivery into the bore hole.

**1926.914(aa)** *Semiconductive hose.* Semiconductive hose—a hose with an electrical resistance high enough to limit flow of stray electric currents to safe levels, yet not so high as to prevent drainage of static electric charges to ground; hose of not more than 2 megohms resistance over its entire length and of not less than 5,000 ohms per foot meets the requirement.

[44 FR 8577, Feb. 9, 1979; 44 FR 20940, Apr. 6, 1979, as amended at 58 FR 35184 and 35311, June 30, 1993]

**Authority:** Sec. 107, Contract Work Hours and Safety Standards Act (Construction Safety Act) (40 U.S.C. 333); Secs. 4, 6, 8, Occupational Safety and Health Act of 1970 (29 U.S.C. 653, 655, 657); Secretary of Labor's Order No. 12-71 (36 FR 8754), 8-76 (41 FR 25059), 9-83 (48 FR 35736), or 1-90 (55 FR 9033), as applicable. Section 1926.951 also issued under 29 CFR Part 1911.

### §1926.950 General requirements.

1926.950(a) ***Application.*** The occupational safety and health standards contained in this subpart V shall apply to the construction of electric transmission and distribution lines and equipment.

1926.950(a)(1) As used in this subpart V the term "construction" includes the erection of new electric transmission and distribution lines and equipment, and the alteration, conversion, and improvement of existing electric transmission and distribution lines and equipment.

1926.950(a)(2) Existing electric transmission and distribution lines and electrical equipment need not be modified to conform to the requirements of applicable standards in this subpart V, until such work as described in paragraph (a)(1) of this section is to be performed on such lines or equipment.

1926.950(a)(3) The standards set forth in this subpart V provide minimum requirements for safety and health. Employers may require adherence to additional standards which are not in conflict with the standards contained in this subpart V.

1926.950(b) ***Initial inspections, tests, or determinations.*** (1) Existing conditions shall be determined before starting work, by an inspection or a test. Such conditions shall include, but not be limited to, energized lines and equipment, conditions of poles, and the location of circuits and equipment, including power and communication lines, CATV and fire alarm circuits.

1926.950(b)(2) Electric equipment and lines shall be considered energized until determined to be deenergized by tests or other appropriate methods or means.

1926.950(b)(3) Operating voltage of equipment and lines shall be determined before working on or near energized parts.

1926.950(c) ***Clearances.*** The provisions of paragraph (c) (1) or (2) of this section shall be observed.

1926.950(c)(1) No employee shall be permitted to approach or take any conductive object without an approved insulating handle closer to exposed energized parts than shown in Table V-1, unless:

1926.950(c)(1)(i) The employee is insulated or guarded from the energized part (gloves or gloves with sleeves rated for the voltage involved shall be considered insulation of the employee from the energized part), or

1926.950(c)(1)(ii) The energized part is insulated or guarded from him and any other conductive object at a different potential, or

1926.950(c)(1)(iii) The employee is isolated, insulated, or guarded from any other conductive object(s), as during live-line bare-hand work.

1926.950(c)(2)(i) The minimum working distance and minimum clear hot stick distances stated in Table V-1 shall not be violated. The minimum clear hot stick distance is that for the use of live-line tools held by linemen when performing live-line work.

1926.950(c)(2)(ii) Conductor support tools, such as link sticks, strain carriers, and insulator cradles, may be used: *Provided,* That the clear insulation is at least as long as the insulator string or the minimum distance specified in Table V-1 for the operating voltage.

**TABLE V-1**

**ALTERNATING CURRENT—MINIMUM DISTANCES**

| Voltage range (phase to phase) kilovolt | Minimum working and clear hot stick distance |
|---|---|
| 2.1 to 15 | 2 ft. 0 in. |
| 15.1 to 35 | 2 ft. 4 in. |
| 35.1 to 46 | 2 ft. 6 in. |
| 46.1 to 72.5 | 3 ft. 0 in. |
| 72.6 to 121 | 3 ft. 4 in. |
| 138 to 145 | 3 ft. 6 in. |
| 161 to 169 | 3 ft. 8 in. |
| 230 to 242 | 5 ft. 0 in. |
| 345 to 362 | [1]7 ft. 0 in. |
| 500 to 552 | [1]11 ft. 0 in. |
| 700 to 765 | [1]15 ft. 0 in. |

1 **NOTE:** For 345—362 kv., 500—552 kv., and 700—765 kv., the minimum working distance and the minimum clear hot stick distance may be reduced provided that such distances are not less than the shortest distance between the energized part and a grounded surface.

1926.950(d) ***Deenergizing lines and equipment.*** (1) When deenergizing lines and equipment operated in excess of 600 volts, and the means of disconnecting from electric energy is not visibly open or visibly locked out, the provisions of paragraphs (d)(1) (i) through (vii) of this section shall be complied with:

1926.950(d)(1)(i) The particular section of line or equipment to be deenergized shall be clearly identified, and it shall be isolated from all sources of voltage.

1926.950(d)(1)(ii) Notification and assurance from the designated employee shall be obtained that:

1926.950(d)(1)(ii)*(a)* All switches and disconnectors through which electric energy may be supplied to the particular section of line or equipment to be worked have been deenergized;

1926.950(d)(1)(ii)*(b)* All switches and disconnectors are plainly tagged indicating that men are at work;

1926.950(d)(1)(ii)*(c)* And that where design of such switches and disconnectors permits, they have been rendered inoperable.

1926.950(d)(1)(iii) After all designated switches and disconnectors have been opened, rendered inoperable, and tagged, visual inspection or tests shall be conducted to insure that equipment or lines have been deenergized.

1926.950(d)(1)(iv) Protective grounds shall be applied on the disconnected lines or equipment to be worked on.

1926.950(d)(1)(v) Guards or barriers shall be erected as necessary to adjacent energized lines.

1926.950(d)(1)(vi) When more than one independent crew requires the same line or equipment to be deenergized, a prominent tag for each such independent crew shall be placed on the line or equipment by the designated employee in charge.

1926.950(d)(1)(vii) Upon completion of work on deenergized lines or equipment, each designated employee in charge shall determine that all employees in his crew are clear, that protective grounds installed by his crew have been removed, and he shall report to the designated authority that all tags protecting his crew may be removed.

1926.950(d)(2) When a crew working on a line or equipment can clearly see that the means of disconnecting from electric energy are visibly open or visibly locked-out, the provisions of paragraphs (d)(i), and (ii) of this section shall apply:

1926.950(d)(2)(i) Guards or barriers shall be erected as necessary to adjacent energized lines.

1926.950(d)(2)(ii) Upon completion of work on deenergized lines or equipment, each designated employee in charge shall determine that all employees in his crew are clear, that protective grounds installed by his crew have been removed, and he shall report to the designated authority that all tags protecting his crew may be removed.

**1926.950(e)** ***Emergency procedures and first aid.*** (1) The employer shall provide training or require that his employees are knowledgeable and proficient in:

**1926.950(e)(1)(i)** Procedures involving emergency situations, and

**1926.950(e)(1)(ii)** First-aid fundamentals including resuscitation.

**1926.950(e)(2)** In lieu of paragraph (e)(1) of this section the employer may comply with the provisions of §1926.50(c) regarding first-aid requirements.

**1926.950(f)** ***Night work.*** When working at night, spotlights or portable lights for emergency lighting shall be provided as needed to perform the work safely.

**1926.950(g)** ***Work near and over water.*** When crews are engaged in work over or near water and when danger of drowning exists, suitable protection shall be provided as stated in §1926.104, or §1926.105, or §1926.106.

**1926.950(h)** ***Sanitation facilities.*** The requirements of §1926.51 of subpart D of this part shall be complied with for sanitation facilities.

**1926.950(i)** ***Hydraulic fluids.*** All hydraulic fluids used for the insulated sections of derrick trucks, aerial lifts, and hydraulic tools which are used on or around energized lines and equipment shall be of the insulating type. The requirements for fire resistant fluids of §1926.302(d)(1) do not apply to hydraulic tools covered by this paragraph.

## §1926.951 Tools and protective equipment.

**1926.951(a)** ***Protective equipment.*** (1)(i) Rubber protective equipment shall be in accordance with the provisions of the American National Standards Institute (ANSI), ANSI J6 series, as follows:

| *Item* | *Standard* |
|---|---|
| Rubber insulating gloves. | J6.6-1971. |
| Rubber matting for use around electric apparatus. | J6.7-1935 (R1971). |
| Rubber insulating blankets | J6.4-1971. |
| Rubber insulating hoods | J6.2-1950 (R1971). |
| Rubber insulating line hose | J6.1-1950 (R1971). |
| Rubber insulating sleeves | J6.5-1971. |

**1926.951(a)(1)(ii)** Rubber protective equipment shall be visually inspected prior to use.

**1926.951(a)(1)(iii)** In addition, an "air" test shall be performed for rubber gloves prior to use.

**1926.951(a)(1)(iv)** Protective equipment of material other than rubber shall provide equal or better electrical and mechanical protection.

**1926.951(a)(2)** Protective hats shall be in accordance with the provisions of ANSI Z89.2-1971 Industrial Protective Helmets for Electrical Workers, Class B, and shall be worn at the jobsite by employees who are exposed to the hazards of falling objects, electric shock, or burns.

**1926.951(b)** ***Personal climbing equipment.*** (1) Body belts with straps or lanyards shall be worn to protect employees working at elevated locations on poles, towers, or other structures except where such use creates a greater hazard to the safety of the employees, in which case other safeguards shall be employed.

**1926.951(b)(2)** Body belts and safety straps shall meet the requirements of §1926.959. In addition to being used as an employee safeguarding item, body belts with approved tool loops may be used for the purpose of holding tools. Body belts shall be free from additional metal hooks and tool loops other than those permitted in §1926.959.

**1926.951(b)(3)** Body belts and straps shall be inspected before use each day to determine that they are in safe working condition.

**1926.951(b)(4)(i)** Lifelines and lanyards shall comply with the provisions of §1926.502.

**1926.951(b)(4)(i)(ii)** Safety lines are not intended to be subjected to shock loading and are used for emergency rescue such as lowering a man to the ground. Such safety lines shall be a minimum of one-half-inch diameter and three or four strand first-grade manila or its equivalent in strength (2,650 lb.) and durability.

**1926.951(b)(5)** Defective ropes shall be replaced.

**1926.951(c)** ***Ladders.*** (1) Portable metal or conductive ladders shall not be used near energized lines or equipment except as may be necessary in specialized work such as in high voltage substations where nonconductive ladders might present a greater hazard than conductive ladders. Conductive or metal ladders shall be prominently marked as conductive and all necessary precautions shall be taken when used in specialized work.

**1926.951(c)(2)** Hook or other type ladders used in structures shall be positively secured to prevent the ladder from being accidentally displaced.

**1926.951(d)** ***Live-line tools.*** (1) Only live-line tool poles having a manufacturer's certification to withstand the following minimum tests shall be used:

**1926.951(d)(1)(i)** 100,000 volts per foot of length for 5 minutes when the tool is made of fiberglass; or

**1926.951(d)(1)(ii)** 75,000 volts per foot of length for 3 minutes when the tool is made of wood; or

**1926.951(d)(1)(iii)** Other tests equivalent to paragraph (d) (i) or (ii) of this section as appropriate.

**1926.951(d)(2)** All live-line tools shall be visually inspected before use each day. Tools to be used shall be wiped clean and if any hazardous defects are indicated such tools shall be removed from service.

**1926.951(e)** ***Measuring tapes or measuring ropes.*** Measuring tapes or measuring ropes which are metal or contain conductive strands shall not be used when working on or near energized parts.

**1926.951(f)** ***Handtools.*** (1) Switches for all powered hand tools shall comply with §1926.300(d).

**1926.951(f)(2)** All portable electric handtools shall:

**1926.951(f)(2)(i)** Be equipped with three-wire cord having the ground wire permanently connected to the tool frame and means for grounding the other end; or

**1926.951(f)(2)(ii)** Be of the double insulated type and permanently labeled as "Double Insulated"; or

**1926.951(f)(2)(iii)** Be connected to the power supply by means of an isolating transformer, or other isolated power supply.

**1926.951(f)(3)** All hydraulic tools which are used on or around energized lines or equipment shall use nonconducting hoses having adequate strength for the normal operating pressures. It should be noted that the provisions of §1926.302(d)(2) shall also apply.

**1926.951(f)(4)** All pneumatic tools which are used on or around energized lines or equipment shall:

**1926.951(f)(4)(i)** Have nonconducting hoses having adequate strength for the normal operating pressures, and

**1926.951(f)(4)(ii)** Have an accumulator on the compressor to collect moisture.

[44 FR 8577, Feb. 9, 1979; 44 FR 20940, Apr. 6, 1979, as amended at 59 FR 40730, Aug. 9, 1994]

## §1926.952 Mechanical equipment.

**1926.952(a)** ***General.*** (1) Visual inspections shall be made of the equipment to determine that it is in good condition each day the equipment is to be used.

**1926.952(a)(2)** Tests shall be made at the beginning of each shift during which the equipment is to be used to determine that the brakes and operating systems are in proper working condition.

**1926.952(a)(3)** No employer shall use any motor vehicle equipment having an obstructed view to the rear unless:

**1926.952(a)(3)(i)** The vehicle has a reverse signal alarm audible above the surrounding noise level or:

**1926.952(a)(3)(ii)** The vehicle is backed up only when an observer signals that it is safe to do so.

**1926.952(b)** ***Aerial lifts.*** (1) The provisions of §1926.556, subpart N of this part, shall apply to the utilization of aerial lifts.

**1926.952(b)(2)** When working near energized lines or equipment, aerial lift trucks shall be grounded or barricaded and considered as energized equipment, or the aerial lift truck shall be insulated for the work being performed.

**1926.952(b)(3)** Equipment or material shall not be passed between a pole or structure and an aerial lift while an employee working from the basket is within reaching distance of energized conductors or equipment that are not covered with insulating protective equipment.

**1926.952(c)** ***Derrick trucks, cranes and other lifting equipment.*** (1) All derrick trucks, cranes and other lifting equipment shall comply with subpart N and O of this part except:

**1926.952(c)(1)(i)** As stated in §1926.550(a)(15) (i) and (ii) relating to clearance (for clearances in this subpart see Table V-1) and

**1926.952(c)(1)(ii)** Derrick truck (electric line trucks) shall not be required to comply with §1926.550(a)(7)(vi), (a)(17), (b)(2), and (e).

**1926.952(c)(2)** With the exception of equipment certified for work on the proper voltage, mechanical equipment shall not be operated closer to any energized line or equipment than the clearances set forth in §1926.950(c) unless:

**1926.952(c)(2)(i)** An insulated barrier is installed between the energized part and the mechanical equipment, or

**1926.952(c)(2)(ii)** The mechanical equipment is grounded, or

**1926.952(c)(2)(iii)** The mechanical equipment is insulated, or

**1926.952(c)(2)(iv)** The mechanical equipment is considered as energized.

## §1926.953 Material handling.

**1926.953(a)** ***Unloading.*** Prior to unloading steel, poles, cross arms and similar material, the load shall be thoroughly examined to ascertain if the load has shifted, binders or stakes have broken or the load is otherwise hazardous to employees.

**1926.953(b)** ***Pole hauling.*** (1) During pole hauling operations, all loads shall be secured to prevent displacement and a red flag shall be displayed at the trailing end of the longest pole.

**1926.953(b)(2)** Precautions shall be exercised to prevent blocking of roadways or endangering other traffic.

**1926.953(b)(3)** When hauling poles during the hours of darkness, illuminated warning devices shall be attached to the trailing end of the longest pole.

**1926.953(c)** ***Storage.*** (1) No materials or equipment shall be stored under energized bus, energized lines, or near energized equipment, if it is practical to store them elsewhere.

**1926.953(c)(2)** When materials or equipment are stored under energized lines or near energized equipment, applicable clearances shall be maintained as stated in Table V-1; and extraordinary caution shall be exercised when moving materials near such energized equipment.

**1926.953(d)** ***Tag line.*** Where hazards to employees exist tag lines or other suitable devices shall be used to control loads being handled by hoisting equipment.

**1926.953(e)** ***Oil filled equipment.*** During construction or repair of oil filled equipment the oil may be stored in temporary containers other than those required in §1926.152, such as pillow tanks.

**1926.953(f)** ***Framing.*** During framing operations, employees shall not work under a pole or a structure suspended by a crane, A-frame or similar equipment unless the pole or structure is adequately supported.

**1926.953(g)** ***Attaching the load.*** The hoist rope shall not be wrapped around the load. This provision shall not apply to electric construction crews when setting or removing poles.

## §1926.954 Grounding for protection of employees.

**1926.954(a)** ***General.*** All conductors and equipment shall be treated as energized until tested or otherwise determined to be deenergized or until grounded.

**1926.954(b)** ***New construction.*** New lines or equipment may be considered deenergized and worked as such where:

**1926.954(b)(1)** The lines or equipment are grounded, or

**1926.954(b)(2)** The hazard of induced voltages is not present, and adequate clearances or other means are implemented to prevent contact with energized lines or equipment and the new lines or equipment.

**1926.954(c)** ***Communication conductors.*** Bare wire communication conductors on power poles or structures shall be treated as energized lines unless protected by insulating materials.

**1926.954(d)** ***Voltage testing.*** Deenergized conductors and equipment which are to be grounded shall be tested for voltage. Results of this voltage test shall determine the subsequent procedures as required in §1926.950(d).

**1926.954(e)** ***Attaching grounds.*** (1) When attaching grounds, the ground end shall be attached first, and the other end shall be attached and removed by means of insulated tools or other suitable devices.

**1926.954(e)(2)** When removing grounds, the grounding device shall first be removed fromt the line or equipment using insulating tools or other suitable devices.

**1926.954(f)** Grounds shall be placed between work location and all sources of energy and as close as practicable to the work location, or grounds shall be placed at the work location. If work is to be performed at more than one location in a line section, the line section must be grounded and short circuited at one location in the line section and the conductor to be worked on shall be grounded at each work location. The minimum distance shown in Table V-1 shall be maintained from ungrounded conductors at the work location. Where the making of a ground is impracticable, or the conditions resulting therefrom would be more hazardous than working on the lines or equipment without grounding, the grounds may be omitted and the line or equipment worked as energized.

**1926.954(g)** ***Testing without grounds.*** Grounds may be temporarily removed only when necessary for test purposes and extreme caution shall be exercised during the test procedures.

**1926.954(h)** ***Grounding electrode.*** When grounding electrodes are utilized, such electrodes shall have a resistance to ground low enough to remove the danger of harm to personnel or permit prompt operation of protective devices.

**1926.954(i)** ***Grounding to tower.*** Grounding to tower shall be made with a tower clamp capable of conducting the anticipated fault current.

**1926.954(j)** ***Ground lead.*** A ground lead, to be attached to either a tower ground or driven ground, shall be capable of conducting the anticipated fault current and shall have a minimum conductance of No. 2 AWG copper.

## §1926.955 Overhead lines.

**1926.955(a)** ***Overhead lines.*** (1) When working on or with overhead lines the provisions of paragraphs (a) (2) through (8) of this section shall be complied with in addition to other applicable provisions of this subpart.

**1926.955(a)(2)** Prior to climbing poles, ladders, scaffolds, or other elevated structures, an inspection shall be made to determine that the structures are capable of sustaining the additional or unbalanced stresses to which they will be subjected.

**1926.955(a)(3)** Where poles or structures may be unsafe for climbing, they shall not be climbed until made safe by guying, bracing, or other adequate means.

**1926.955(a)(4)** Before installing or removing wire or cable, strains to which poles and structures will be subjected shall be considered and necessary action taken to prevent failure of supporting structures.

**1926.955(a)(5)(i)** When setting, moving, or removing poles using cranes, derricks, gin poles, A-frames, or other mechanized equipment near energized lines or equipment, precautions shall be taken to avoid contact with energized lines or equipment, except in bare-hand live-line work, or where barriers or protective devices are used.

**1926.955(a)(5)(ii)** Equipment and machinery operating adjacent to energized lines or equipment shall comply with §1926.952(c)(2).

**1926.955(a)(6)(i)** Unless using suitable protective equipment for the voltage involved, employees standing on the ground shall avoid contacting equipment or machinery working adjacent to energized lines or equipment.

**1926.955(a)(6)(ii)** Lifting equipment shall be bonded to an effective ground or it shall be considered energized and barricaded when utilized near energized equipment or lines.

**1926.955(a)(7)** Pole holes shall not be left unattended or unguarded in areas where employees are currently working.

**1926.955(a)(8)** Tag lines shall be of a nonconductive type when used near energized lines.

**1926.955(b)** ***Metal tower construction.*** (1) When working in unstable material the excavation for pad- or pile-type footings in excess of 5 feet deep shall be either sloped to the angle of repose as required in §1926.652 or shored if entry is required. Ladders shall be provided for access to pad- or pile-type footing excavations in excess of 4 feet.

**1926.955(b)(2)** When working in unstable material provision shall be made for cleaning out auger-type footings without requiring an employee to enter the footing unless shoring is used to protect the employee.

**1926.955(b)(3)(i)** A designated employee shall be used in directing mobile equipment adjacent to footing excavations.

**1926.955(b)(3)(ii)** No one shall be permitted to remain in the footing while equipment is being spotted for placement.

**1926.955(b)(3)(iii)** Where necessary to assure the stability of mobile equipment the location of use for such equipment shall be graded and leveled.

**1926.955(b)(4)(i)** Tower assembly shall be carried out with a minimum exposure of employees to falling objects when working at two or more levels on a tower.

**1926.955(b)(4)(ii)** Guy lines shall be used as necessary to maintain sections or parts of sections in position and to reduce the possibility of tipping.

**1926.955(b)(4)(iii)** Members and sections being assembled shall be adequately supported.

**1926.955(b)(5)** When assembling and erecting towers the provisions of paragraphs (b)(5) (i), (ii) and (iii) of this section shall be complied with:

**1926.955(b)(5)(i)** The construction of transmission towers and the erecting of poles, hoisting machinery, site preparation machinery, and other types of construction machinery shall conform to the applicable requirements of this part.

**1926.955(b)(5)(ii)** No one shall be permitted under a tower which is in the process of erection or assembly, except as may be required to guide and secure the section being set.

**1926.955(b)(5)(iii)** When erecting towers using hoisting equipment adjacent to energized transmission lines, the lines shall be deenergized when practical. If the lines are not deenergized, extraordinary caution shall be exercised to maintain the minimum clearance distances required by §1926.950(c), including Table V-1.

**1926.955(b)(6)(i)** Erection cranes shall be set on firm level foundations and when the cranes are so equipped outriggers shall be used.

**1926.955(b)(6)(ii)** Tag lines shall be utilized to maintain control of tower sections being raised and positioned, except where the use of such lines would create a greater hazard.

**1926.955(b)(6)(iii)** The loadline shall not be detached from a tower section until the section is adequately secured.

**1926.955(b)(6)(iv)** Except during emergency restoration procedures erection shall be discontinued in the event of high wind or other adverse weather conditions which would make the work hazardous.

**1926.955(b)(6)(v)** Equipment and rigging shall be regularly inspected and maintained in safe operating condition.

**1926.955(b)(7)** Adequate traffic control shall be maintained when crossing highways and railways with equipment as required by the provisions of §1926.200(g)(1) and (2).

**1926.955(b)(8)** A designated employee shall be utilized to determine that required clearance is maintained in moving equipment under or near energized lines.

**1926.955(c)** ***Stringing or removing deenergized conductors.*** (1) When stringing or removing deenergized conductors, the provisions of paragraphs (c) (2) through (12) of this section shall be complied with.

**1926.955(c)(2)** Prior to stringing operations a briefing shall be held setting forth the plan of operation and specifying the type of equipment to be used, grounding devices and procedures to be followed, crossover methods to be employed, and the clearance authorization required.

**1926.955(c)(3)** Where there is a possibility of the conductor accidentally contacting an energized circuit or receiving a dangerous induced voltage buildup, to further protect the employee from the hazards of the conductor, the conductor being installed or removed shall be grounded or provisions made to insulate or isolate the employee.

**1926.955(c)(4)(i)** If the existing line is deenergized, proper clearance authorization shall be secured and the line grounded on both sides of the crossover or, the line being strung or removed shall be considered and worked as energized.

**1926.955(c)(4)(ii)** When crossing over energized conductors in excess of 600 volts, rope nets or guard structures shall be installed unless provision is made to isolate or insulate the workman or the energized conductor. Where practical the automatic reclosing feature of the circuit interrupting device shall be made inoperative. In addition, the line being strung shall be grounded on either side of the crossover or considered and worked as energized.

**1926.955(c)(5)** Conductors being strung in or removed shall be kept under positive control by the use of adequate tension reels, guard structures, tielines, or other means to prevent accidental contact with energized circuits.

**1926.955(c)(6)** Guard structure members shall be sound and of adequate dimension and strength, and adequately supported.

**1926.955(c)(7)(i)** Catch-off anchors, rigging, and hoists shall be of ample capacity to prevent loss of the lines.

**1926.955(c)(7)(ii)** The manufacturer's load rating shall not be exceeded for stringing lines, pulling lines, sock connections, and all load-bearing hardware and accessories.

**1926.955(c)(7)(iii)** Pulling lines and accessories shall be inspected regularly and replaced or repaired when damaged or when dependability is doubtful. The provisions of §1926.251(c)(4)(ii) (concerning splices) shall not apply.

**1926.955(c)(8)** Conductor grips shall not be used on wire rope unless designed for this application.

**1926.955(c)(9)** While the conductor or pulling line is being pulled (in motion) employees shall not be permitted directly under overhead operations, nor shall any employee be permitted on the crossarm.

**1926.955(c)(10)** A transmission clipping crew shall have a minimum of two structures clipped in between the crew and the conductor being sagged. When working on bare conductors, clipping and tying crews shall work between grounds at all times. The grounds shall remain intact until the conductors are clipped in, except on dead end structures.

**1926.955(c)(11)(i)** Except during emergency restoration procedures, work from structures shall be discontinued when adverse weather (such as high wind or ice on structures) makes the work hazardous.

**1926.955(c)(11)(ii)** Stringing and clipping operations shall be discontinued during the progress of an electrical storm in the immediate vicinity.

**1926.955(c)(12)(i)** Reel handling equipment, including pulling and braking machines, shall have ample capacity, operate smoothly, and be leveled and aligned in accordance with the manufacturer's operating instructions.

**1926.955(c)(12)(ii)** Reliable communications between the reel tender and pulling rig operator shall be provided.

**1926.955(c)(12)(iii)** Each pull shall be snubbed or dead ended at both ends before subsequent pulls.

**1926.955(d)** ***Stringing adjacent to energized lines.*** (1) Prior to stringing parallel to an existing energized transmission line a competent determination shall be made to ascertain whether dangerous induced voltage buildups will occur, particularly during switching and ground fault conditions. When there is a possibility that such dangerous induced voltage may exist the employer shall comply with the provisions of paragraphs (d) (2) through (9) of this section in addition to the provisions of paragraph (c) of this §1926.955, unless the line is worked as energized.

**1926.955(d)(2)** When stringing adjacent to energized lines the tension stringing method or other methods which preclude unintentional contact between the lines being pulled and any employee shall be used.

**1926.955(d)(3)** All pulling and tensioning equipment shall be isolated, insulated, or effectively grounded.

**1926.955(d)(4)** A ground shall be installed between the tensioning reel setup and the first structure in order to ground each bare conductor, subconductor, and overhead ground conductor during stringing operations.

**1926.955(d)(5)** During stringing operations, each bare conductor, subconductor, and overhead ground conductor shall be grounded at the first tower adjacent to both the tensioning and pulling setup and in increments so that no point is more than 2 miles from a ground.

**1926.955(d)(5)(i)** The grounds shall be left in place until conductor installation is completed.

**1926.955(d)(5)(ii)** Such grounds shall be removed as the last phase of aerial cleanup.

**1926.955(d)(5)(iii)** Except for moving type grounds, the grounds shall be placed and removed with a hot stick.

**1926.955(d)(6)** Conductors, subconductors, and overhead ground conductors shall be grounded at all dead-end or catch-off points.

**1926.955(d)(7)** A ground shall be located at each side and within 10 feet of working areas where conductors, subconductors, or overhead ground conductors are being spliced at ground level. The two ends to be spliced shall be bonded to each other. It is recommended that splicing be carried out on either an insulated platform or on a conductive metallic grounding mat bonded to both grounds. When a grounding mat is used, it is recommended that the grounding mat be roped off and an insulated walkway provided for access to the mat.

**1926.955(d)(8)(i)** All conductors, subconductors, and overhead ground conductors shall be bonded to the tower at any isolated tower where it may be necessary to complete work on the transmission line.

**1926.955(d)(8)(ii)** Work on dead-end towers shall require grounding on all deenergized lines.

**1926.955(d)(8)(iii)** Grounds may be removed as soon as the work is completed: *Provided*, That the line is not left open circuited at the isolated tower at which work is being completed.

**1926.955(d)(9)** When performing work from the structures, clipping crews and all others working on conductors, subconductors, or overhead ground conductors shall be protected by individual grounds installed at every work location.

**1926.955(e)** ***Live-line bare-hand work.*** In addition to any other applicable standards contained elsewhere in this subpart all live-line bare-hand work shall be performed in accordance with the following requirements:

**1926.955(e)(1)** Employees shall be instructed and trained in the live-line bare-hand technique and the safety requirements pertinent thereto before being permitted to use the technique on energized circuits.

**1926.955(e)(2)** Before using the live-line bare-hand technique on energized high-voltage conductors or parts, a check shall be made of:

**1926.955(e)(2)(i)** The voltage rating of the circuit on which the work is to be performed;

**1926.955(e)(2)(ii)** The clearances to ground of lines and other energized parts on which work is to be performed; and

**1926.955(e)(2)(iii)** The voltage limitations of the aerial-lift equipment intended to be used.

**1926.955(e)(3)** Only equipment designed, tested, and intended for live-line bare-hand work shall be used.

**1926.955(e)(4)** All work shall be personally supervised by a person trained and qualified to perform live-line bare-hand work.

**1926.955(e)(5)** The automatic reclosing feature of circuit interrupting devices shall be made inoperative where practical before working on any energized line or equipment.

**1926.955(e)(6)** Work shall not be performed during the progress of an electrical storm in the immediate vicinity.

**1926.955(e)(7)** A conductive bucket liner or other suitable conductive device shall be provided for bonding the insulated aerial device to the energized line or equipment.

**1926.955(e)(7)(i)** The employee shall be connected to the bucket liner by use of conductive shoes, leg clips, or other suitable means.

**1926.955(e)(7)(ii)** Where necessary, adequate electrostatic shielding for the voltage being worked or conductive clothing shall be provided.

**1926.955(e)(8)** Only tools and equipment intended for live-line bare-hand work shall be used, and such tools and equipment shall be kept clean and dry.

**1926.955(e)(9)** Before the boom is elevated, the outriggers on the aerial truck shall be extended and adjusted to stabilize the truck and the body of the truck shall be bonded to an effective ground, or barricaded and considered as energized equipment.

**1926.955(e)(10)** Before moving the aerial lift into the work position, all controls (ground level and bucket) shall be checked and tested to determine that they are in proper working condition.

**1926.955(e)(11)** Arm current tests shall be made before starting work each day, each time during the day when higher voltage is going to be worked and when changed conditions indicate a need for additional tests. Aerial buckets used for bare-hand live-line work shall be subjected to an arm current test. This test shall consist of placing the bucket in contact with an energized source equal to the voltage to be worked upon for a minimum time of three (3) minutes. the leakage current shall not exceed 1 microampere per kilo-volt of nominal line-to-line voltage. Work operations shall be suspended immediately upon any indication of a malfunction in the equipment.

**1926.955(e)(12)** All aerial lifts to be used for live-line bare-hand work shall have dual controls (lower and upper) as required by paragraph (e)(12) (i) and (ii) of this section.

**1926.955(e)(12)(i)** The upper controls shall be within easy reach of the employee in the basket. If a two basket type lift is used access to the controls shall be within easy reach from either basket.

**1926.955(e)(12)(ii)** The lower set of controls shall be located near base of the boom that will permit over-ride operation of equipment at any time.

**1926.955(e)(13)** Ground level lift control shall not be operated unless permission has been obtained from the employee in lift, except in case of emergency.

**1926.955(e)(14)** Before the employee contacts the energized part to be worked on, the conductive bucket liner shall be bonded to the energized conductor by means of a positive connection which shall remain attached to the energized conductor until the work on the energized circuit is completed.

**1926.955(e)(15)** The minimum clearance distances for live-line bare-hand work shall be as specified in Table V-2. These minimum clearance distances shall be maintained from all grounded objects and from lines and equipment at a different potential than that to which the insulated aerial device is bonded unless such grounded objects or other lines and equipment are covered by insulated guards. These distances shall be maintained when approaching, leaving, and when bonded to the energized circuit.



## TABLE V-2
## MINIMUM CLEARANCE DISTANCES FOR LIVE-LINE BARE-HAND WORK (ALTERNATING CURRENT)

| Voltage range (phase-to-phase) kilovolts | Distance in feet and inches for maximum voltage | |
|---|---|---|
| | Phase to ground | Phase to phase |
| 2.1 to 15 | 2′0″ | 2′0″ |
| 15.1 to 35 | 2′4″ | 2′4″ |
| 35.1 to 46 | 2′6″ | 2′6″ |
| 46.1 to 72.5 | 3′0″ | 3′0″ |
| 72.6 to 121 | 3′4″ | 4′6″ |
| 138 to 145 | 3′6″ | 5′0″ |
| 161 to 169 | 3′8″ | 5′6″ |
| 230 to 242 | 5′0″ | 8′4″ |
| 345 to 362 | [1] 7′0″ | [1] 13′4″ |
| 500 to 552 | [1] 11′0″ | [1] 20′0″ |
| 700 to 765 | [1] 15′0″ | [1] 31′0″ |

[1] For 345-362 kv., 500-552 kv., and 700-765 kv., the minimum clearance distance may be reduced provided the distances are not made less than the shortest distance between the energized part and a grounded surface.

**1926.955(e)(16)** When approaching, leaving, or bonding to an energized circuit the minimum distances in Table V-2 shall be maintained between all parts of the insulated boom assembly and any grounded parts (including the lower arm or portions of the truck).

**1926.955(e)(17)** When positioning the bucket alongside an energized bushing or insulator string, the minimum line-to-ground clearances of Table V-2 must be maintained between all parts of the bucket and the grounded end of the bushing or insulator string.

**1926.955(e)(18)(i)** The use of handlines between buckets, booms, and the ground is prohibited.

**1926.955(e)(18)(ii)** No conductive materials over 36 inches long shall be placed in the bucket, except for appropriate length jumpers, armor rods, and tools.

**1926.955(e)(18)(iii)** Nonconductive-type handlines may be used from line to ground when not supported from the bucket.

**1926.955(e)(19)** The bucket and upper insulated boom shall not be overstressed by attempting to lift or support weights in excess of the manufacturer's rating.

**1926.955(e)(20)(i)** A minimum clearance table (as shown in table V-2) shall be printed on a plate of durable nonconductive material, and mounted in the buckets or its vicinity so as to be visible to the operator of the boom.

**1926.955(e)(20)(ii)** It is recommended that insulated measuring sticks be used to verify clearance distances.

## §1926.956 Underground lines.

**1926.956(a)** ***Guarding and ventilating street opening used for access to underground lines or equipment.*** (1) Appropriate warning signs shall be promptly placed when covers of manholes, handholes, or vaults are removed. What is an appropriate warning sign is dependent upon the nature and location of the hazards involved.

**1926.956(a)(2)** Before an employee enters a street opening, such as a manhole or an unvented vault, it shall be promptly protected with a barrier, temporary cover, or other suitable guard.

**1926.956(a)(3)** When work is to be performed in a manhole or unvented vault:

**1926.956(a)(3)(i)** No entry shall be permitted unless forced ventilation is provided or the atmosphere is found to be safe by testing for oxygen deficiency and the presence of explosive gases or fumes;

**1926.956(a)(3)(ii)** Where unsafe conditions are detected, by testing or other means, the work area shall be ventilated and otherwise made safe before entry;

**1926.956(a)(3)(iii)** Provisions shall be made for an adequate continuous supply of air.

**1926.956(b)** ***Work in manholes.*** (1) While work is being performed in manholes, an employee shall be available in the immediate vicinity to render emergency assistance as may be required. This shall not preclude the employee in the immediate vicinity from occasionally entering a manhole to provide assistance, other than emergency. This requirement does not preclude a qualified employee, working alone, from entering for brief periods of time, a manhole where energized cables or equipment are in service, for the purpose of inspection, housekeeping, taking readings, or similar work if such work can be performed safely.

**1926.956(b)(2)** When open flames must be used or smoking is permitted in manholes, extra precautions shall be taken to provide adequate ventilation.

**1926.956(b)(3)** Before using open flames in a manhole or excavation in an area where combustible gases or liquids may be present, such as near a gasoline service station, the atmosphere of the manhole or excavation shall be tested and found safe or cleared of the combustible gases or liquids.

**1926.956(c)** ***Trenching and excavating.*** (1) During excavation or trenching, in order to prevent the exposure of employees to the hazards created by damage to dangerous underground facilities, efforts shall be made to determine the location of such facilities and work conducted in a manner designed to avoid damage.

**1926.956(c)(2)** Trenching and excavation operations shall comply with §§1926.651 and 1926.652.

**1926.956(c)(3)** When underground facilities are exposed (electric, gas, water, telephone, etc.) they shall be protected as necessary to avoid damage.

**1926.956(c)(4)** Where multiple cables exist in an excavation, cables other than the one being worked on shall be protected as necessary.

**1926.956(c)(5)** When multiple cables exist in an excavation, the cable to be worked on shall be identified by electrical means unless its identity is obvious by reason of distinctive appearance.

**1926.956(c)(6)** Before cutting into a cable or opening a splice, the cable shall be identified and verified to be the proper cable.

**1926.956(c)(7)** When working on buried cable or on cable in manholes, metallic sheath continuity shall be maintained by bonding across the opening or by equivalent means.

## §1926.957 Construction in energized substations.

**1926.957(a)** ***Work near energized equipment facilities.*** (1) When construction work is performed in an energized substation, authorization shall be obtained from the designated, authorized person before work is started.

**1926.957(a)(2)** When work is to be done in an energized substation, the following shall be determined:

**1926.957(a)(2)(i)** What facilities are energized, and

**1926.957(a)(2)(ii)** What protective equipment and precautions are necessary for the safety of personnel.

**1926.957(a)(3)** Extraordinary caution shall be exercised in the handling of busbars, tower steel, materials, and equipment in the vicinity of energized facilities. The requirements set forth in §1926.950(c), shall be complied with.

**1926.957(b)** ***Deenergized equipment or lines.*** When it is necessary to deenergize equipment or lines for protection of employees, the requirements of §1926.950(d) shall be complied with.

**1926.957(c)** ***Barricades and barriers.*** (1) Barricades or barriers shall be installed to prevent accidental contact with energized lines or equipment.

**1926.957(c)(2)** Where appropriate, signs indicating the hazard shall be posted near the barricade or barrier. These signs shall comply with §1926.200.

**1926.957(d)** ***Control panels.*** (1) Work on or adjacent to energized control panels shall be performed by designated employees.

**1926.957(d)(2)** Precaution shall be taken to prevent accidental operation of relays or other protective devices due to jarring, vibration, or improper wiring.

**1926.957(e)** ***Mechanized equipment.*** (1) Use of vehicles, gin poles, cranes, and other equipment in restricted or hazardous areas shall at all times be controlled by designated employees.

**1926.957(e)(2)** All mobile cranes and derricks shall be effectively grounded when being moved or operated in close proximity to energized lines or equipment, or the equipment shall be considered energized.

**1926.957(e)(3)** Fenders shall not be required for lowboys used for transporting large electrical equipment, transformers, or breakers.

**1926.957(f)** ***Storage.*** The storage requirements of §1926.953(c) shall be complied with.

**1926.957(g)** ***Substation fences.*** (1) When a substation fence must be expanded or removed for construction purposes, a temporary fence affording similar protection when the site is unattended, shall be provided. Adequate interconnection with ground shall be maintained between temporary fence and permanent fence.

**1926.957(g)(2)** All gates to all unattended substations shall be locked, except when work is in progress.

**1926.957(h)** ***Footing excavation.*** (1) Excavation for auger, pad and piling type footings for structures and towers shall require the same precautions as for metal tower construction (see §1926.955(b)(1)).

**1926.957(h)(2)** No employee shall be permitted to enter an unsupported auger-type excavation in unstable material for any purpose. Necessary cleanout in such cases shall be accomplished without entry.

## §1926.958 External load helicopters.

In all operations performed using a rotorcraft for moving or placing external loads, the provisions of §1926.551 of subpart N of this part shall be complied with.

## §1926.959 Lineman's body belts, safety straps, and lanyards.

**1926.959(a)** ***General requirements.*** The requirements of paragraphs (a) and (b) of this section shall be complied with for all lineman's body belts, safety straps and lanyards acquired for use after the effective date of this subpart.

**1926.959(a)(1)** Hardware for lineman's body belts, safety straps, and lanyards shall be drop forged or pressed steel and have a corrosive resistive finish tested to American Society for Testing and Materials B117-64 (50-hour test). Surfaces shall be smooth and free of sharp edges.

**1926.959(a)(2)** All buckles shall withstand a 2,000-pound tensile test with a maximum permanent deformation no greater than one sixty-fourth inch.

**1926.959(a)(3)** D rings shall withstand a 5,000-pound tensile test without failure. Failure of a D ring shall be considered cracking or breaking.

**1926.959(a)(4)** Snaphooks shall withstand a 5,000-pound tensile test without failure. Failure of a snaphook shall be distortion sufficient to release the keeper.

**1926.959(b)** ***Specific requirements.*** (1)(i) All fabric used for safety straps shall withstand an A.C. dielectric test of not less than 25,000 volts per foot "dry" for 3 minutes, without visible deterioration.

**1926.959(b)(1)(ii)** All fabric and leather used shall be tested for leakage current and shall not exceed 1 milliampere when a potention of 3,000 volts is applied to the electrodes positioned 12 inches apart.

**1926.959(b)(1)(iii)** Direct current tests may be permitted in lieu of alternating current tests.

**1926.959(b)(2)** The cushion part of the body belt shall:

**1926.959(b)(2)(i)** Contain no exposed rivets on the inside;

**1926.959(b)(2)(ii)** Be at least three (3) inches in width;

**1926.959(b)(2)(iii)** Be at least five thirty-seconds (5/32) inch thick, if made of leather; and

**1926.959(b)(2)(iv)** Have pocket tabs that extended at least 1 1/2 inches down and three (3) inches back of the inside of circle of each D ring for riveting on plier or tool pockets. On shifting D belts, this measurement for pocket tabs shall be taken when the D ring section is centered.

**1926.959(b)(3)** A maximum of four (4) tool loops shall be so situated on the body belt that four (4) inches of the body belt in the center of the back, measuring from D ring to D ring, shall be free of tool loops, and any other attachments.

**1926.959(b)(4)** Suitable copper, steel, or equivalent liners shall be used around bar of D rings to prevent wear between these members and the leather or fabric enclosing them.

**1926.959(b)(5)** All stitching shall be of a minimum 42-pound weight nylon or equivalent thread and shall be lock stitched. Stitching parallel to an edge shall not be less than three-sixteenths (3/16) inch from edge of narrowest member caught by the thread. The use of cross stitching on leather is prohibited.

**1926.959(b)(6)** The keeper of snaphooks shall have a spring tension that will not allow the keeper to begin to open with a weight of 2 1/2 pounds or less, but the keeper of snaphooks shall begin to open with a weight of four (4) pounds, when the weight is supported on the keeper against the end of the nose.

**1926.959(b)(7)** Testing of lineman's safety straps, body belts and lanyards shall be in accordance with the following procedure:

**1926.959(b)(7)(i)** Attach one end of the safety strap or lanyard to a rigid support, the other end shall be attached to a 250-pound canvas bag of sand:

**1926.959(b)(7)(ii)** Allow the 250-pound canvas bag of sand to free fall 4 feet for (safety strap test) and 6 feet for (lanyard test); in each case stopping the fall of the 250-pound bag:

**1926.959(b)(7)(iii)** Failure of the strap or lanyard shall be indicated by any breakage, or slippage sufficient to permit the bag to fall free of the strap or lanyard. The entire "body belt assembly" shall be tested using one D ring. A safety strap or lanyard shall be used that is capable of passing the "impact loading test" and attached as required in paragraph (b)(7)(i) of this section. The body belt shall be secured to the 250-pound bag of sand at a point to simulate the waist of a man and allowed to drop as stated in paragraph (b)(7)(ii) of this section. Failure of the body belt shall be indicated by any breakage, or slippage sufficient to permit the bag to fall free of the body belt.

## §1926.960 Definitions applicable to this subpart.

**1926.960(a)** *Alive or live (energized).* The term means electrically connected to a source of potential difference, or electrically charged so as to have a potential significantly different from that of the earth in the vicinity. The term "live" is sometimes used in place of the term "current-carrying," where the intent is clear, to avoid repetition of the longer term.

**1926.960(b)** *Automatic circuit recloser.* The term means a self-controlled device for automatically interrupting and reclosing an alternating current circuit with a predetermined sequence of opening and reclosing followed by resetting, hold closed, or lockout operation.

**1926.960(c)** *Barrier.* The term means a physical obstruction which is intended to prevent contact with energized lines or equipment.

**1926.960(d)** *Barricade.* The term means a physical obstruction such as tapes, screens, or cones intended to warn and limit access to a hazardous area.

**1926.960(e)** *Bond.* The term means an electrical connection from one conductive element to another for the purpose of minimizing potential differences or providing suitable conductivity for fault current or for mitigation of leak-age current and electrolytic action.

**1926.960(f)** *Bushing.* The term means an insulating structure including a through conductor, or providing a passageway for such a conductor, with provision for mounting on a barrier, conducting or otherwise, for the purpose of insulating the conductor from the barrier and conducting current from one side of the barrier to the other.

**1926.960(g)** *Cable.* The term means a conductor with insulation, or a stranded conductor with or without insulation and other coverings (single-conductor cable) or a combination of conductors insulated from one another (multiple-conductor cable).

**1926.960(h)** *Cable sheath.* The term means a protective covering applied to cables.

**Note:** A cable sheath may consist of multiple layers of which one or more is conductive.

**1926.960(i)** *Circuit.* The term means a conductor or system of conductors through which an electric current is intended to flow.

**1926.960(j)** *Communication lines.* The term means the conductors and their supporting or containing structures which are used for public or private signal or communication service, and which operate at potentials not exceeding 400 volts to ground or 750 volts between any two points of the circuit, and the transmitted power of which does not exceed 150 watts. When operating at less than 150 volts no limit is placed on the capacity of the system.

**Note:** Telephone, telegraph, railroad signal, data, clock, fire, police-alarm, community television antenna, and other systems conforming with the above are included. Lines used for signaling purposes, but not included under the above definition, are considered as supply lines of the same voltage and are to be so run.

**1926.960(k)** *Conductor.* The term means a material, usually in the form of a wire, cable, or bus bar suitable for carrying an electric current.

**1926.960(l)** *Conductor shielding.* The term means an envelope which encloses the conductor of a cable and provides an equipotential surface in contact with the cable insulation.

**1926.960(m)** *Current-carrying part.* The term means a conducting part intended to be connected in an electric circuit to a source of voltage. Non-current-carrying parts are those not intended to be so connected.

**1926.960(n)** *Dead (deenergized).* The term means free from any electrical connection to a source of potential difference and from electrical charges: Not having a potential difference from that of earth.

**Note:** The term is used only with reference to current-carrying parts which are sometimes alive (energized).

**1926.960(o)** *Designated employee.* The term means a qualified person delegated to perform specific duties under the conditions existing.

**1926.960(p)** *Effectively grounded.* The term means intentionally connected to earth through a ground connection or connections of sufficiently low impedance and having sufficient current-carrying capacity to prevent the buildup of voltages which may result in undue hazard to connected equipment or to persons.

**1926.960(q)** *Electric line trucks.* The term means a truck used to transport men, tools, and material, and to serve as a traveling workshop for electric power line construction and maintenance work. It is sometimes equipped with a boom and auxiliary equipment for setting poles, digging holes, and elevating material or men.

**1926.960(r)** *Enclosed.* The term means surrounded by a case, cage, or fence, which will protect the contained equipment and prevent accidental contact of a person with live parts.

**1926.960(s)** *Equipment.* This is a general term which includes fittings, devices, appliances, fixtures, apparatus, and the like, used as part of, or in connection with, an electrical power transmission and distribution system, or communication systems.

**1926.960(t)** *Exposed.* The term means not isolated or guarded.

**1926.960(u)** *Electric supply lines.* The term means those conductors used to transmit electric energy and their necessary supporting or containing structures. Signal lines of more than 400 volts to ground are always supply lines within the meaning of the rules, and those of less than 400 volts to ground may be considered as supply lines, if so run and operated throughout.

**1926.960(v)** *Guarded.* The term means protected by personnel, covered, fenced, or enclosed by means of suitable casings, barrier rails, screens, mats, platforms, or other suitable devices in accordance with standard barricading techniques designed to prevent dangerous approach or contact by persons or objects.

**Note:** Wires, which are insulated but not otherwise protected, are not considered as guarded.

**1926.960(w)** *Ground.* (*Reference*). The term means that conductive body, usually earth, to which an electric potential is referenced.

**1926.960(x)** *Ground* (*as a noun*). The term means a conductive connection whether intentional or accidental, by which an electric circuit or equipment is connected to reference ground.

**1926.960(y)** *Ground* (*as a verb*). The term means the connecting or establishment of a connection, whether by intention or accident of an electric circuit or equipment to reference ground.

**1926.960(z)** *Grounding electrode* (*ground electrode*). The term grounding electrode means a conductor embedded in the earth, used for maintaining ground potential on conductors connected to it, and for dissipating into the earth current conducted to it.

**1926.960(aa)** *Grounding electrode resistance.* The term means the resistance of the grounding electrode to earth.

**1926.960(bb)** *Grounding electrode conductor* (*grounding conductor*). The term means a conductor used to connect equipment or the grounded circuit of a wiring system to a grounding electrode.

**1926.960(cc)** *Grounded conductor.* The term means a system or circuit conductor which is intentionally grounded.

**1926.960(dd)** *Grounded system.* The term means a system of conductors in which at least one conductor or point (usually the middle wire, or neutral point of transformer or generator windings) is intentionally grounded, either solidly or through a current-limiting device (not a current-interrupting device).

**1926.960(ee)** *Hotline tools and ropes.* The term means those tools and ropes which are especially designed for work on energized high voltage lines and equipment. Insulated aerial equipment especially designed for work on energized high voltage lines and equipment shall be considered hot line.

**1926.960(ff)** *Insulated.* The term means separated from other conducting surfaces by a dielectric substance (including air space) offering a high resistance to the passage of current.

**Note:** When any object is said to be insulated, it is understood to be insulated in suitable manner for the conditions to which it is subjected. Otherwise, it is within the purpose of this subpart, uninsulated. Insulating covering of conductors is one means of making the conductor insulated.

**1926.960(gg)** *Insulation* (*as applied to cable*). The term means that which is relied upon to insulate the conductor from other conductors or conducting parts or from ground.

**1926.960(hh)** *Insulation shielding.* The term means an envelope which encloses the insulation of a cable and provides an equipotential surface in contact with cable insulation.

**1926.960(ii)** *Isolated.* The term means an object that is not readily accessible to persons unless special means of access are used.

**1926.960(jj)** *Manhole.* The term means a subsurface enclosure which personnel may enter and which is used for the purpose of installing, operating, and maintaining equipment and/or cable.

**1926.960(kk)** *Pulling tension.* The term means the longitudinal force exerted on a cable during installation.

**1926.960(ll)** *Qualified person.* The term means a person who by reason of experience or training is familiar with the operation to be performed and the hazards involved.

**1926.960(mm)** *Switch.* The term means a device for opening and closing or changing the connection of a circuit. In these rules, a switch is understood to be manually operable, unless otherwise stated.

**1926.960(nn)** *Tag.* The term means a system or method of identifying circuits, systems or equipment for the purpose of alerting persons that the circuit, system or equipment is being worked on.

**1926.960(oo)** *Unstable material.* The term means earth material, other than running, that because of its nature or the influence of related conditions, cannot be depended upon to remain in place without extra support, such as would be furnished by a system of shoring.

**1926.960(pp)** *Vault.* The term means an enclosure above or below ground which personnel may enter and is used for the purpose of installing, operating, and/or maintaining equipment and/or cable.

**1926.960(qq)** *Voltage.* The term means the effective (rms) potential difference between any two conductors or between a conductor and ground. Voltages are expressed in nominal values. The nominal voltage of a system or circuit is the value assigned to a system or circuit of a given voltage class for the purpose of convenient designation. The operating voltage of the system may vary above or below this value.

**1926.960(rr)** *Voltage of an effectively grounded circuit.* The term means the voltage between any conductor and ground unless otherwise indicated.

**1926.960(ss)** *Voltage of a circuit not effectively grounded.* The term means the voltage between any two conductors. If one circuit is directly connected to and supplied from another circuit of higher voltage (as in the case of an autotransformer), both are considered as of the higher voltage, unless the circuit of lower voltage is effectively grounded, in which case its voltage is not determined by the circuit of higher voltage. Direct connection implies electric connection as distinguished from connection merely through electromagnetic or electrostatic induction.

# Subpart W—Rollover Protective Structures; Overhead Protection



**Authority:** Sec. 107, Contract Work Hours and Safety Standards Act (Construction Safety Act), 40 U.S.C. 333; secs. 4, 6, and 8, Occupational Safety and Health Act of 1970 (29 U.S.C. 653, 655, 657); Secretary of Labor's Order No. 12-71 (36 FR 8754), 8-76 (41 FR 25059), 9-83 (48 FR 35736), or 1-90 (55 FR 9033), as applicable.

## §1926.1000 Rollover protective structures (ROPS) for material handling equipment.

**1926.1000(a)** ***Coverage.*** (1) This section applies to the following types of material handling equipment: To all rubber-tired, self-propelled scrapers, rubber-tired front-end loaders, rubber-tired dozers, wheel-type agricultural and industrial tractors, crawler tractors, crawler-type loaders, and motor graders, with or without attachments, that are used in construction work. This requirement does not apply to sideboom pipelaying tractors.

**1926.1000(a)(2)** The promulgation of specific standards for rollover protective structures for compactors and rubber-tired skid-steer equipment is reserved pending consideraton of standards currently being developed.

**1926.1000(b)** ***Equipment manufactured on or after September 1, 1972.*** Material handling machinery described in paragraph (a) of this section and manufactured on or after September 1, 1972, shall be equipped with rollover protective structures which meet the minimum performance standards prescribed in §§1926.1001 and 1926.1002, as applicable.

**1926.1000(c)** ***Equipment manufactured before September 1, 1972.*** (1) All material handling equipment described in paragraph (a) of this section and manufactured or placed in service (owned or operated by the employer) prior to September 1, 1972, shall be fitted with rollover protective structures no later than the dates listed below:

**1926.1000(c)(1)(i)** Machines manufactured on or after January 1, 1972, shall be fitted no later than April 1, 1973.

**1926.1000(c)(1)(ii)** Machines manufactured between July 1, 1971, and December 31, 1971, shall be fitted no later than July 1, 1973.

**1926.1000(c)(1)(iii)** Machines manufactured between July 1, 1970, and June 30, 1971, shall be fitted no later than January 1, 1974.

**1926.1000(c)(1)(iv)** Machines manufactured between July 1, 1969, and June 30, 1970, shall be fitted no later than July 1, 1974.

**1926.1000(c)(1)(v)** Machines manufactured before July 1, 1969: Reserved pending further study, development, and review.

**1926.1000(c)(2)** Rollover protective structures and supporting attachment shall meet the minimum performance criteria detailed in §§1926.1001 and 1926.1002, as applicable or shall be designed, fabricated, and installed in a manner which will support, based on the ultimate strength of the metal, at least two times the weight of the prime mover applied at the point of impact.

**1926.1000(c)(2)(i)** The design objective shall be to minimize the likelihood of a complete overturn and thereby minimize the possibility of the operator being crushed as a result of a rollover or upset.

**1926.1000(c)(2)(ii)** The design shall provide a vertical clearance of at least 52 inches from the work deck to the ROPS at the point of ingress or egress.

**1926.1000(d)** ***Remounting.*** ROPS removed for any reason, shall be remounted with equal quality, or better, bolts or welding as required for the original mounting.

**1926.1000(e)** ***Labeling.*** Each ROPS shall have the following information permanently affixed to the structure:

**1926.1000(e)(1)** Manufacturer or fabricator's name and address;

**1926.1000(e)(2)** ROPS model number, if any;

**1926.1000(e)(3)** Machine make, model, or series number that the structure is designed to fit.

**1926.1000(f)** ***Machines meeting certain existing governmental requirements.*** Any machine in use, equipped with rollover protective structures, shall be deemed in compliance with this section if it meets the rollover protective structure requirements of the State of California, the U.S. Army Corps of Engineers, or the Bureau of Reclamation of the U.S. Department of the Interior in effect on April 5, 1972. The requirements in effect are:

**1926.1000(f)(1)** State of California: Construction Safety Orders, issued by the Department of Industrial Relations pursuant to Division 5, Labor Code, §6312, State of California.

**1926.1000(f)(2)** U.S. Army Corps of Engineers: General Safety Requirements, EM-385-1-1 (March 1967).

**1926.1000(f)(3)** Bureau of Reclamation, U.S. Department of the Interior: Safety and Health Regulations for Construction. Part II (September 1971).

## §1926.1001 Minimum performance criteria for rollover protective structures for designated scrapers, loaders, dozers, graders, and crawler tractors.

**1926.1001(a)** ***General.*** This section prescribes minimum performance criteria for rollover protective structures (ROPS) for rubber-tired self-propelled scrapers; rubber-tired front-end loaders and rubber-tired dozers; crawler tractors, and crawler-type loaders, and motor graders. The vehicle and ROPS as a system shall have the structural characteristics prescribed in paragraph (f) of this section for each type of machine described in this paragraph.

**1926.1001(b)** The static laboratory test prescribed herein will determine the adequacy of the structures used to protect the operator under the following conditions:

**1926.1001(b)(1)** For rubber-tired self-propelled scrapers, rubber-tired front-end loaders, and rubber-tired dozers: Operating between 0 and 10 miles per hour over hard clay where rollover would be limited to a maximum roll angle of 360° down a slope of 30° maximum.

**1926.1001(b)(2)** For motor graders: Operating between 0 and 10 miles per hour over hard clay where rollover would be limited to 360° down a slope of 30° maximum.

**1926.1001(b)(3)** For crawler tractors and crawler-type loaders: Operating between 0 and 10 miles per hour over hard clay where rollover would be limited to a maximum roll angle of 360° down a slope of 45°.

**1926.1001(c)** ***Facilities and apparatus.*** (1) The following material is necessary:

**1926.1001(c)(1)(i)** Material, equipment, and tiedown means adequate to insure that the ROPS and its vehicle frame absorb the applied energy.

**1926.1001(c)(1)(ii)** Equipment necessary to measure and apply loads to the ROPS. Adequate means to measure deflections and lengths should also be provided.

**1926.1001(c)(1)(iii)** Recommended, but not mandatory, types of test setups are illustrated in Figure W-1 for all types of equipment to which this section applies; and in Figure W-2 for rubber-tired self-propelled scrapers; Figure W-3 for rubber-tired front-end loaders, rubber-tired dozers, and motor graders; and Figure W-4 for crawler tractors and crawler-type loaders.

**1926.1001(c)(2)** Table W-1 contains a listing of the required apparatus for all types of equipment described in paragraph (a) of this section.

**TABLE W-1**

| Means to measure | Accuracy |
|---|---|
| Deflection of ROPS, inches | ±5% of deflection measured. |
| Vehicle weight, pounds | ±5% of the weight measured. |
| Force applied to frame, pounds | ±5% of force measured. |
| Dimensions of critical zone, inches. | ±0.5 in. |

**1926.1001(d)** ***Vehicle condition.*** The ROPS to be tested must be attached to the vehicle structure in the same manner as it will be attached during vehicle use. A totally assembled vehicle is not required. However, the vehicle structure and frame which support the ROPS must represent the actual vehicle installation. All normally detachable windows, panels, or nonstructural fittings shall be removed so that they do not contribute to the strength of the ROPS.

**1926.1001(e)** ***Test procedure.*** The test procedure shall include the following, in the sequence indicated:

**1926.1001(e)(1)** Energy absorbing capabilities of ROPS shall be verified when loaded laterally by incrementally applying a distributed load to the longitudinal outside top member of the ROPS, as shown in Figure W-1, W-

2, or W-3, as applicable. The distributed load must be applied so as to result in approximately uniform deflection of the ROPS. The load increments should correspond with approximately 0.5 in. ROPS deflection increment in the direction of the load application, measured at the ROPS top edge. Should the operator's seat be offcenter, the load shall be applied on the offcenter side. For each applied load increment, the total load (lb.) versus corresponding deflection (in.) shall be plotted, and the area under the load -deflection curve shall be calculated. This area is equal to the energy (in.-lb.) absorbed by the ROPS. For a typical load-deflection curve and calculation method, see Figure W-5.

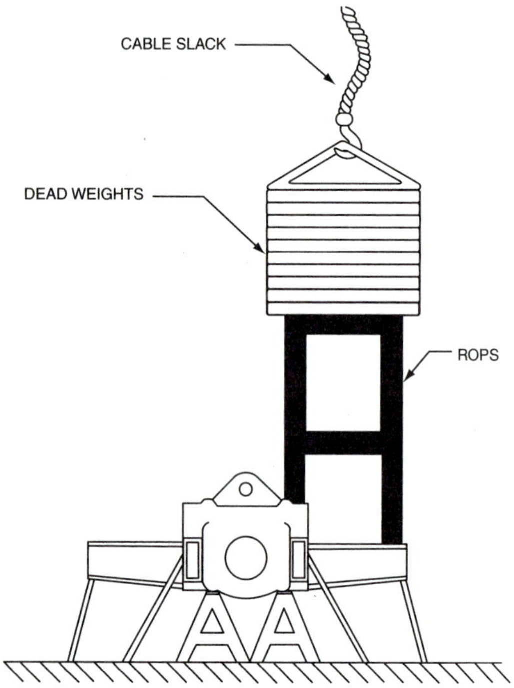


**FIGURE W-1— VERTICAL LOADING SETUP FOR ALL TYPES OF EQUIPMENT DESCRIBED IN §1518.1001(a)**

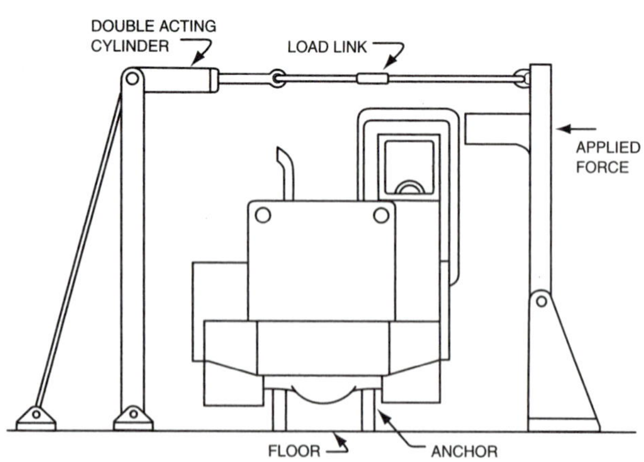


**FIGURE W-2—TEST SETUP FOR RUBBER-TIRED SELF-PROPELLED SCRAPERS.**

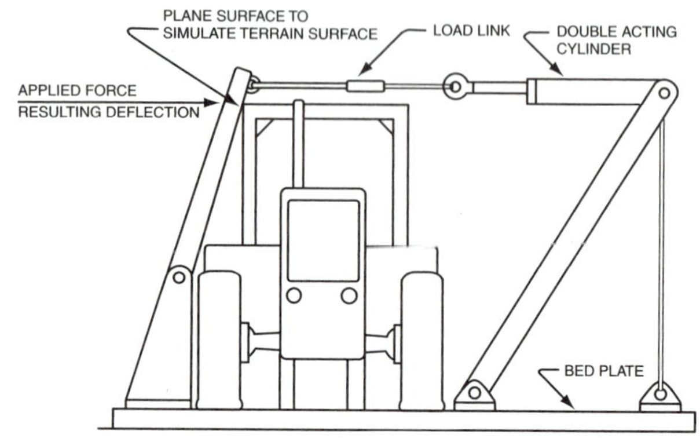


**FIGURE W-3—TEST SETUP FOR RUBBER-TIRED FRONT-END LOADERS, RUBBER-TIRED DOZERS, AND MOTOR GRADERS**

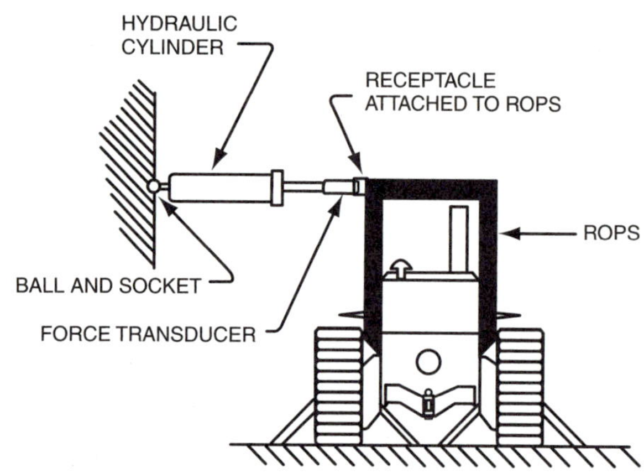


**FIGURE W-4—SIDE-LOADING SETUP FOR CRAWLER TRACTORS AND CRAWLER LOADERS.**

Incremental loading shall be continued until the ROPS has absorbed the amount of energy and the minimum applied load specified under paragraph (f) of this section has been reached or surpassed.

**1926.1001(e)(2)** To cover the possibility of the vehicle coming to rest on its top, the support capability shall be verified by applying a distributed vertical load to the top of the ROPS so as to result in approximately uniform deflection (see Figure W-1). The load magnitude is specified in paragraph (f)(2)(iii) of this section.

**1926.1001(e)(3)** The low temperature impact strength of the material used in the ROPS shall be verified by suitable material tests or material certification (see paragraph (f)(2)(iv) of this section).

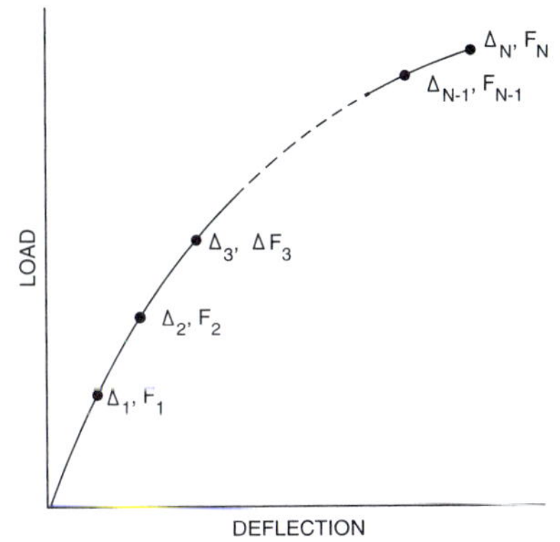


Δ - TOTAL DEFLECTION
F - FORCE APPLIED

$$\text{AREA} = \frac{\Delta_1 F_1}{2} + (\Delta_2 - \Delta_1)\frac{F_1 + F_2}{2} + (\Delta_3 - \Delta_2)\frac{F_2 + F_3}{2} + \ldots + (\Delta_N - \Delta_{N-1})\frac{F_{N-1} + F_N}{2}$$

**FIGURE W-5—DETERMINATION OF ENERGY AREA UNDER FORCE DEFLECTION CURVE FOR ALL TYPES OF ROPS EQUIPMENT DEFINED IN §1926.1001.**

**1926.1001(f)** ***Performance requirements***—(1) *General performance requirements.* (i) No repairs or straightening of any member shall be carried out between each prescribed test.

**1926.1001(f)(1)(ii)** During each test, no part of the ROPS shall enter the critical zone as detailed in SAE J397 (1969). Deformation of the ROPS shall not allow the plane of the ground to enter this zone.

**1926.1001(f)(2)** *Specific performance requirements.* (i) The energy requirement for purposes of meeting the requirements of paragraph (e)(1) of this section is to be determined by referring to the plot of the energy versus weight of vehicle (see Figure W-6 for rubber-tired self-propelled scrapers; Figure W-7 for rubber-tired front-end loaders and rubber-tired dozers; Figure W-8 for crawler tractors and crawler-type loaders; and Figure W-9 for motor graders). For purposes of this section, force and weight are measured as pounds (lb.); energy (U) is measured as inch-pounds.

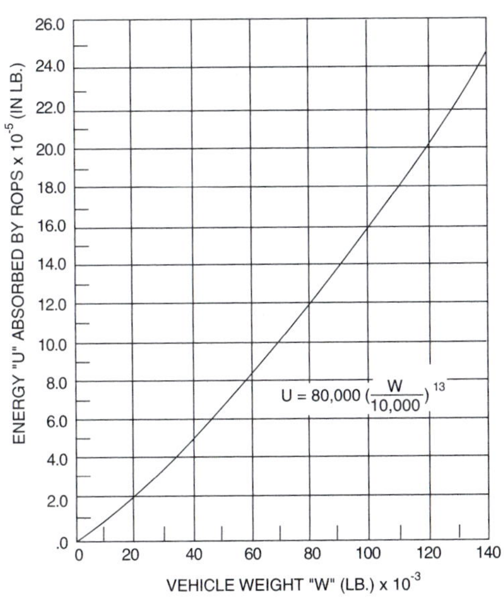


**FIGURE W-6—ENERGY ABSORBED VERSUS VEHICLE WEIGHT.**

**1926.1001(f)(2)(ii)** The applied load must attain at least a value which is determined by multiplying the vehicle weight by the corresponding factor shown in Figure W-10 for rubber-tired self-propelled scrapers; in Figure W-11 for rubber-tired front-end loaders and rubber-tired dozers; in Figure W-12 for crawler tractors and crawler-type loaders; and in Figure W-13 for motor graders.

**1926.1001(f)(2)(iii)** The load magnitude for purposes of compliance with paragraph (e)(2) of this section is equal to the vehicle weight. The test of load magnitude shall only be made after the requirements of paragraph (f)(2)(i) of this section are met.

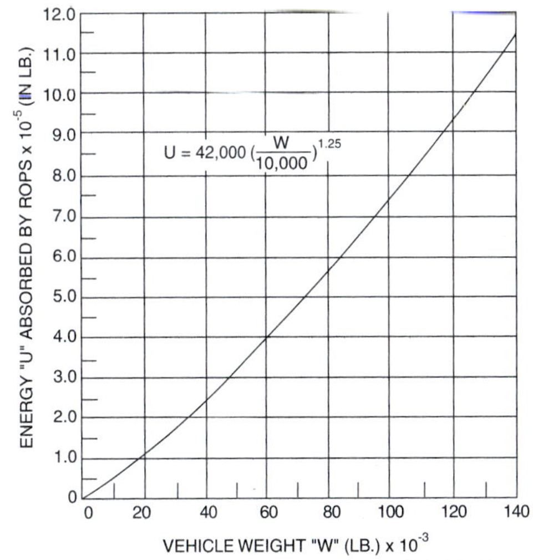


**FIGURE W-7—ENERGY ABSORBED VERSUS VEHICLE WEIGHT**

**1926.1001(f)(2)(iv)** Material used in the ROPS must have the capability of performing at zero degrees Fahrenheit, or exhibit Charpy V notch impact strength of 8 foot-pounds at minus 20°Fahrenheit.

This is a standard Charpy specimen as described in American Society of Testing and Materials A 370, Methods and Definitions for Mechanical Testing of Steel Products (available at each Regional Office of the Occupational Safety and Health Administration). The purpose of this requirement is to reduce the tendency of brittle fracture associated with dynamic loading, low temperature operation, and stress raisers which cannot be entirely avoided on welded structures.

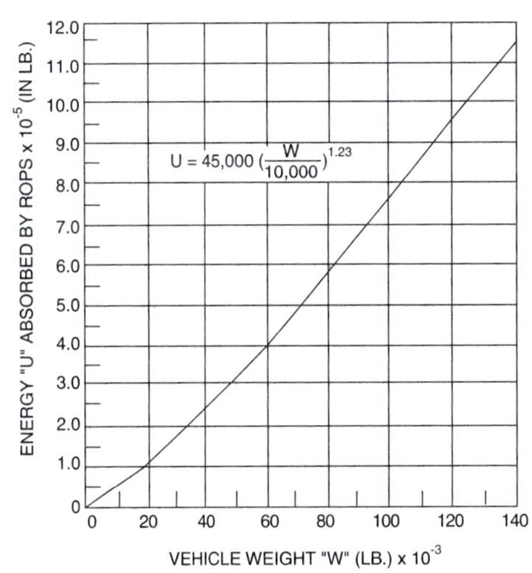


**FIGURE W-8—ENERGY ABSORBED VERSUS VEHICLE WEIGHT**

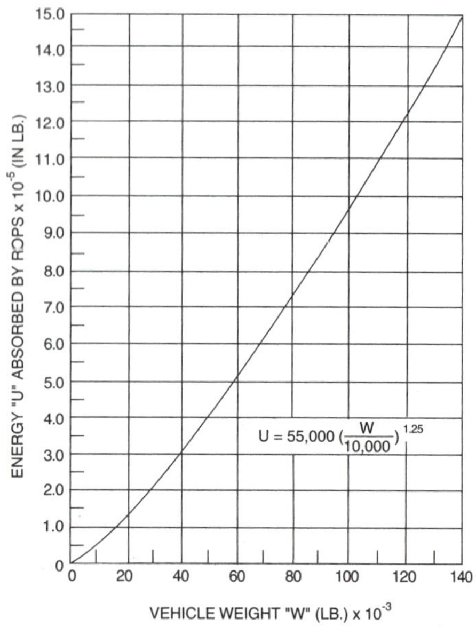


**FIGURE W-9—ENERGY ABSORBED VERSUS VEHICLE WEIGHT**

**1926.1001(g)** ***Definitions.*** For purposes of this section, "vehicle weight" means the manufacturer's maximum weight of the prime mover for rubber-tired self-propelled scrapers. For other types of equipment to which this section applies, "vehicle weight" means the manufacturer's maximum recommended weight of the vehicle plus the heaviest attachment.

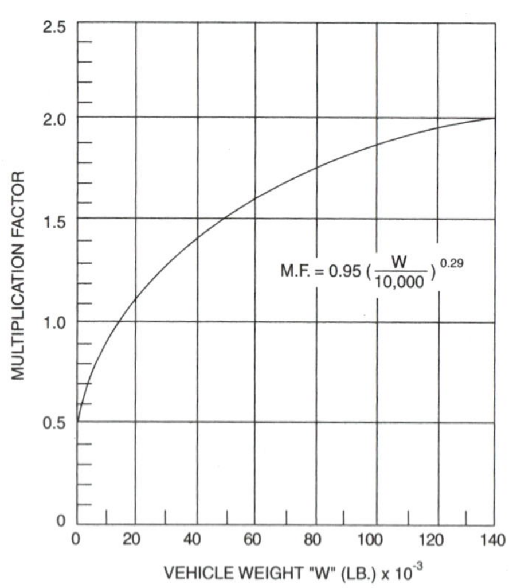


**FIGURE W-10—MINIMUM HORIZONTAL LOAD FACTOR FOR SELF-PROPELLED SCRAPERS**

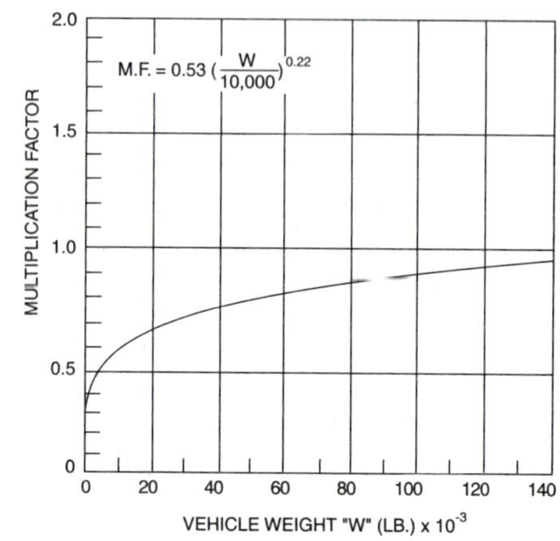


**FIGURE W-11—MINIMUM HORIZONTAL LOAD FACTOR FOR RUBBER-TIRED LOADERS AND DOZERS.**

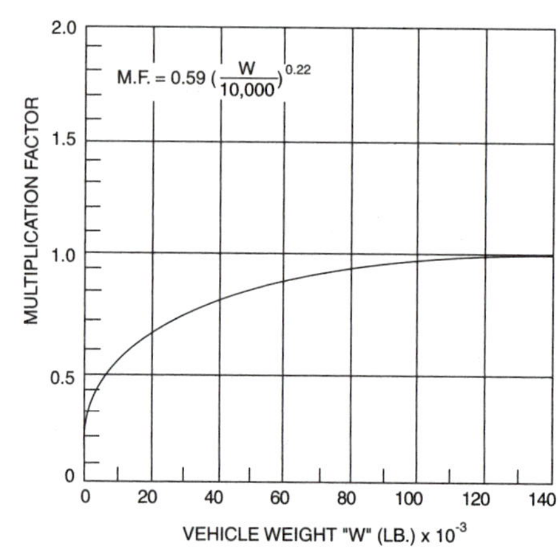


**FIGURE W-12—MINIMUM HORIZONTAL LOAD FACTOR FOR CRAWLER TRACTORS AND CRAWLER-TYPE LOADERS**

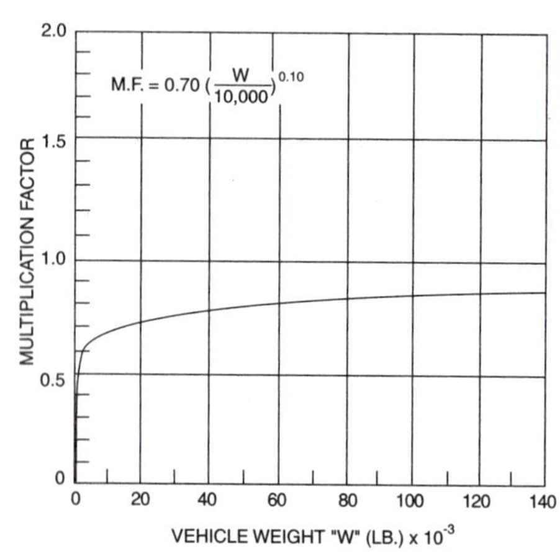


**FIGURE W-13—MINIMUM HORIZONTAL LOAD FACTOR FOR MOTOR GRADERS**

**1926.1001(h)** ***Source of standard.*** This standard is derived from, and restates, the following Society of Automotive Engineers Recommended Practices: SAE J320a, Minimum Performance Criteria for Roll-Over Protective Structure for Rubber-Tired, Self-Propelled Scrapers; SAE J394, Minimum Performance Criteria for Roll-Over Protective Structure for Rubber-Tired Front End Loaders and Rubber-Tired Dozers; SAE J395, Minimum Performance Criteria for Roll-Over Protective Structure for Crawler Tractors and Crawler-Type Loaders; and SAE J396, Minimum Performance Criteria for Roll-Over Protective Structure for Motor Graders. These recommended practices shall be resorted to in the event that questions of interpretation arise. The recommended practices appear in the 1971 SAE Handbook, which may be examined in each of the Regional Offices of the Occupational Safety and Health Administration.

## §1926.1002 Protective frames (roll-over protective structures, known as ROPS) for wheel-type agricultural and industrial tractors used in construction.

**1926.1002(a)** ***General.*** (1) The purpose of this section is to set forth requirements for frames for the protection of operators of wheel type agricultural and industrial tractors to minimize the possibility of operator injury resulting from accidental upsets during normal operation.

These frames shall meet the test and performance requirements of the Society of Automotive Engineers Standard J334a-1970, Protective Frame Test Procedures and Performance Requirements, which is incorporated by reference. The incorporation by reference was approved by the Director of the Federal Register in accordance with 5 U.S.C. 552(a) and 1 CFR part 51. Copies may be obtained from the Society of Automotive Engineers, 485 Lexington Avenue, New York, NY 10017. Copies may be inspected at the OSHA Docket Office, U.S. Department of Labor, 200 Constitution Ave., NW., Room N2634, or at the Office of the FEDERAL REGISTER, 800 North Capitol St., NW., Suite 700, Washington, D.C. The standard also appears in the 1971 SAE Handbook, which may be examined in each of OSHA's Regional Offices. With respect to agricultural and industrial tractors, the provisions of §§1926.1001 and 1926.1003 for rubber-tired dozers and rubber-tired loaders may be utilized in lieu of the requirements of this section.

**1926.1002(a)(2)** The protective frame which is the subject of this standard is a structure mounted to the tractor that extends above the operator's seat and conforms generally to Figure W-14.

**1926.1002(a)(3)** If an overhead weather shield is attached to the protective frame, it may be in place during tests: *Provided,* That it does not contribute to the strength of the protective frame. If such an overhead weather shield is attached, it must meet the requirements of paragraph (i) of this section.

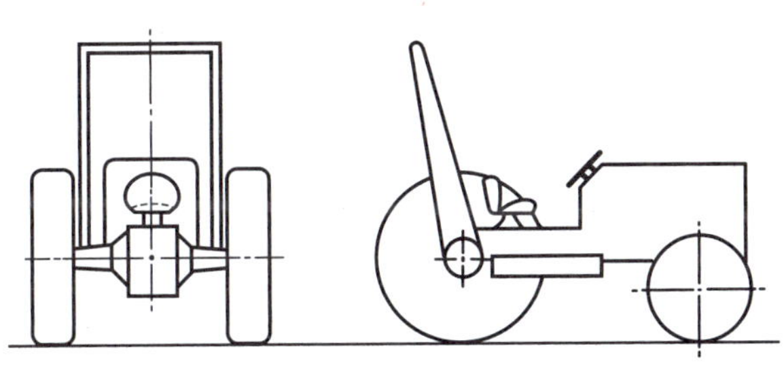

**FIGURE W-14. TYPICAL FRAME CONFIGURATION.**

**1926.1002(a)(4)** For overhead protection requirements, see §1926.1003.

**1926.1002(a)(5)** If protective enclosures are used on wheel-type agricultural and industrial tractors, they shall meet the requirements of Society of Automotive Engineers Standard J168 (July 1970), Protective Enclosures, Test Procedures, and Performance Requirements. This standard appears in the 1971 SAE Handbook and may be examined in each Regional Office of the Occupational Safety and Health Administration.

**1926.1002(b)** ***Applicability.*** The requirements of this section apply to wheel-type agricultural tractors used in construction work and to wheel-type industrial tractors used in construction work. See paragraph (j) of this section for definitions of agricultural tractors and industrial tractors.

**1926.1002(c)—(i)** [Reserved]

**1926.1002(j)** ***Definitions applicable to this section.*** (1) SAE J333a, Operator Protection for Wheel-Type Agricultural and Industrial Tractors (July 1970) defines *agricultural tractor* as a "wheel-type vehicle of more than 20 engine horsepower designed to furnish the power to pull, carry, propel, or drive implements that are designed for agricultural usage." Since this Part 1926 applies only to construction work, the following definition of "agricultural tractor" is adopted for purposes of this subpart: "Agricultural tractor" means a wheel-type vehicle of more than 20 engine horsepower, used in construction work, which is designed to furnish the power to pull, propel, or drive implements.

**1926.1002(j)(2)** *Industrial tractor* means that class of wheeled type tractor of more than 20 engine horsepower (other than rubber-tired loaders and dozers described in §1926.1001), used in operations such as landscaping, construction services, loading, digging, grounds keeping, and highway maintenance.

[44 FR 8577, Feb. 9, 1979; 44 FR 20940, Apr. 6, 1979, as amended at 61 FR 9251, Mar. 7, 1996]

## §1926.1003 Overhead protection for operators of agricultural and industrial tractors.

**1926.1003(a)** ***General***—(1) *Purpose.* When overhead protection is provided on wheel-type agricultural and industrial tractors, the overhead protection shall be designed and installed according to the requirements contained in the test and performance requirements of Society of Automotive Engineers Standard J167-1970, Protective Frame with Overhead Protection-Test Procedures and Performance Requirements, which pertains to overhead protection requirements and is incorporated by reference. The incorporation by reference was approved by the Director of the Federal Register in accordance with 5 U.S.C. 552(a) and 1 CFR part 51. Copies may be obtained from the Society of Automotive Engineers, 485 Lexington Avenue, New York, NY 10017. Copies may be inspected at the OSHA Docket Office, U.S. Department of Labor, 200 Constitution Ave., NW., Room N2634, or at the Office of the Federal Register, 800 North Capitol St., NW., Suite 700, Washington, D.C. The standard also appears in the 1971 SAE Handbook, which may be examined in each of OSHA's Regional Offices. The provisions of §1926.1001 for rubber-tired dozers and rubber-tired loaders may be used in lieu of the standards contained in this section. The purpose of the standard is to minimize the possibility of operator injury resulting from overhead hazards such as flying and falling objects, and at the same time to minimize the possibility of operator injury from the cover itself in the event of accidental upset.

**1926.1003(a)(2)** *Applicability.* This standard applies to wheel-type agricultural tractors used in construction work and to wheel-type industrial tractors used in construction work. See §1926.1002(b) and (j). In the case of machines to which §1926.604 (relating to site clearing) also applies, the overhead protection may be either the type of protection provided in §1926.604 or the type of protection provided by this section.

**1926.1003(b)** ***Overhead protection.*** When overhead protection is installed on wheel-type agricultural or industrial tractors used in construction work, it shall meet the requirements of this paragraph. The overhead protection may be constructed of a solid material. If grid or mesh is used, the largest permissible opening shall be such that the maximum circle which can be inscribed between the elements of the grid or mesh is 1.5 in. (38 mm.) in diameter. The overhead protection shall not be installed in such a way as to become a hazard in the case of upset.

[44 FR 8577, Feb. 9, 1979; 44 FR 20940, Apr. 6, 1979, as amended at 61 FR 9251, Mar. 7, 1996]

# Subpart X—Stairways and Ladders

**Authority:** Section 107, Contract Work Hours and Safety Standards Act (Construction Safety Act) (40 U.S.C. 333); Secs. 4, 6, 8, Occupational Safety and Health Act of 1970 (29 U.S.C. 653, 655, 657); Secretary of Labor's Order No. 1-90 (55 FR 9033); and 29 CFR part 1911.

**Source:** 55 FR 47687, Nov. 14, 1990, unless otherwise noted.

## §1926.1050 Scope, application, and definitions applicable to this subpart.

**1926.1050(a)** ***Scope and application.*** This subpart applies to all stairways and ladders used in construction, alteration, repair (including painting and decorating), and demolition workplaces covered under 29 CFR part 1926, and also sets forth, in specified circumstances, when ladders and stairways are required to be provided. Additional requirements for ladders used on or with scaffolds are contained in subpart L—Scaffolds

**1926.1050(b)** ***Definitions.*** *Cleat* means a ladder crosspiece of rectangular cross section placed on edge upon which a person may step while ascending or descending a ladder.

*Double-cleat ladder* means a ladder similar in construction to a single-cleat ladder, but with a center rail to allow simultaneous two-way traffic for employees ascending or descending.

*Equivalent* means alternative designs, materials, or methods that the employer can demonstrate will provide an equal or greater degree of safety for employees than the method or item specified in the standard.

*Extension trestle ladder* means a self-supporting portable ladder, adjustable in length, consisting of a trestle ladder base and a vertically adjustable extension section, with a suitable means for locking the ladders together.

*Failure* means load refusal, breakage, or separation of component parts. Load refusal is the point where the structural members lose their ability to carry the loads.

*Fixed ladder* means a ladder that cannot be readily moved or carried because it is an integral part of a building or structure. A *side-step fixed ladder* is a fixed ladder that requires a person getting off at the top to step to the side of the ladder side rails to reach the landing. A *through fixed ladder* is a fixed ladder that requires a person getting off at the top to step between the side rails of the ladder to reach the landing.

*Handrail* means a rail used to provide employees with a handhold for support.

*Individual-rung/step ladders* means ladders without a side rail or center rail support. Such ladders are made by mounting individual steps or rungs directly to the side or wall of the structure.

*Job-made ladder* means a ladder that is fabricated by employees, typically at the construction site, and is not commercially manufactured. This definition does not apply to any individual-rung/step ladders.

*Ladder stand.* A mobile fixed size self-supporting ladder consisting of a wide flat tread ladder in the form of stairs. The assembly may include handrails.

*Lower levels* means those areas to which an employee can fall from a stairway or ladder. Such areas include ground levels, floors, roofs, ramps, runways, excavations, pits, tanks, material, water, equipment, and similar surfaces. It does not include the surface from which the employee falls.

*Maximum intended load* means the total load of all employees, equipment, tools, materials, transmitted loads, and other loads anticipated to be applied to a ladder component at any one time.

*Nosing* means that portion of a tread projecting beyond the face of the riser immediately below.

*Point of access* means all areas used by employees for work-related passage from one area or level to another. Such open areas include doorways, 1passageways, stairway openings, studded walls, and various other permanent or temporary openings used for such travel.

*Portable ladder* means a ladder that can be readily moved or carried.

*Riser height* means the vertical distance from the top of a tread to the top of the next higher tread or platform/landing or the distance from the top of a platform/landing to the top of the next higher tread or platform/landing.

*Side-step fixed ladder.* See "Fixed ladder."

*Single-cleat ladder* means a ladder consisting of a pair of side rails, connected together by cleats, rungs, or steps.

*Single-rail ladder* means a portable ladder with rungs, cleats, or steps mounted on a single rail instead of the normal two rails used on most other ladders.

*Spiral stairway* means a series of steps attached to a vertical pole and progressing upward in a winding fashion within a cylindrical space.

*Stairrail system* means a vertical barrier erected along the unprotected sides and edges of a stairway to prevent employees from falling to lower levels. The top surface of a stairrail system may also be a "handrail."

*Step stool (ladder type)* means a self-supporting, foldable, portable ladder, nonadjustable in length, 32 inches or less in overall size, with flat steps and without a pail shelf, designed to be climbed on the ladder top cap as well as all steps. The side rails may continue above the top cap.

*Through fixed ladder.* See "Fixed ladder."

*Tread depth* means the horizontal distance from front to back of a tread (excluding nosing, if any).

*Unprotected sides and edges* means any side or edge (except at entrances to points of access) of a stairway where there is no stairrail system or wall 36 inches (.9 m) or more in height, and any side or edge (except at entrances to points of access) of a stairway landing, or ladder platform where there is no wall or guardrail system 39 inches (1 m) or more in height.

[55 FR 47687, Nov. 14, 1990; 56 FR 2585, Jan. 23, 1991, as amended at 58 FR 35184, June 30, 1993]

## §1926.1051 General requirements.

**1926.1051(a)** A stairway or ladder shall be provided at all personnel points of access where there is a break in elevation of 19 inches (48 cm) or more, and no ramp, runway, sloped embankment, or personnel hoist is provided.

**1926.1051(a)(1)** Employees shall not use any spiral stairways that will not be a permanent part of the structure on which construction work is being performed.

**1926.1051(a)(2)** A double-cleated ladder or two or more separate ladders shall be provided when ladders are the only mean of access or exit from a working area for 25 or more employees, or when a ladder is to serve simultaneous two-way traffic.

**1926.1051(a)(3)** When a building or structure has only one point of access between levels, that point of access shall be kept clear to permit free passage of employees. When work must be performed or equipment must be used such that free passage at that point of access is restricted, a second point of access shall be provided and used.

**1926.1051(a)(4)** When a building or structure has two or more points of access between levels, at least one point of access shall be kept clear to permit free passage of employees.

**1926.1051(b)** Employers shall provide and install all stairway and ladder fall protection systems required by this subpart and shall comply with all other pertinent requirements of this subpart before employees begin the work that necessitates the installation and use of stairways, ladders, and their respective fall protection systems.

## §1926.1052 Stairways.

**1926.1052(a)** ***General.*** The following requirements apply to all stairways as indicated:

**1926.1052(a)(1)** Stairways that will not be a permanent part of the structure on which construction work is being performed shall have landings of not less than 30 inches (76 cm) in the direction of travel and extend at least 22 inches (56 cm) in width at every 12 feet (3.7 m) or less of vertical rise.

**1926.1052(a)(2)** Stairs shall be installed between 30° and 50° from horizontal.

**1926.1052(a)(3)** Riser height and tread depth shall be uniform within each flight of stairs, including any foundation structure used as one or more treads of the stairs. Variations in riser height or tread depth shall not be over 1/4-inch (0.6 cm) in any stairway system.

**1926.1052(a)(4)** Where doors or gates open directly on a stairway, a platform shall be provided, and the swing of the door shall not reduce the effective width of the platform to less than 20 inches (51 cm).

**1926.1052(a)(5)** Metal pan landings and metal pan treads, when used, shall be secured in place before filling with concrete or other material.

**1926.1052(a)(6)** All parts of stairways shall be free of hazardous projections, such as protruding nails.

**1926.1052(a)(7)** Slippery conditions on stairways shall be eliminated before the stairways are used to reach other levels.

**1926.1052(b)** ***Temporary service.*** The following requirements apply to all stairways as indicated:

**1926.1052(b)(1)** Except during stairway construction, foot traffic is prohibited on stairways with pan stairs where the treads and/or landings are to be filled in with concrete or other material at a later date, unless the stairs are temporarily fitted with wood or other solid material at least to the top edge of each pan. Such temporary treads and landings shall be replaced when worn below the level of the top edge of the pan.

**1926.1052(b)(2)** Except during stairway construction, foot traffic is prohibited on skeleton metal stairs where permanent treads and/or landings are to be installed at a later date, unless the stairs are fitted with secured temporary treads and landings long enough to cover the entire tread and/or landing area.

**1926.1052(b)(3)** Treads for temporary service shall be made of wood or other solid material, and shall be installed the full width and depth of the stair.

**1926.1052(c)** ***Stairrails and handrails.*** The following requirements apply to all stairways as indicated:

**1926.1052(c)(1)** Stairways having four or more risers or rising more than 30 inches (76 cm), whichever is less, shall be equipped with:

**1926.1052(c)(1)(i)** At least one handrail; and

**1926.1052(c)(1)(ii)** One stairrail system along each unprotected side or edge.

**Note:** When the top edge of a stairrail system also serves as a handrail, paragraph (c)(7) of this section applies.

**1926.1052(c)(2)** Winding and spiral stairways shall be equipped with a handrail offset sufficiently to prevent walking on those portions of the stairways where the tread width is less than 6 inches (15 cm).

**1926.1052(c)(3)** The height of stairrails shall be as follows:

**1926.1052(c)(3)(i)** Stairrails installed after March 15, 1991, shall be not less than 36 inches (91.5 cm) from the upper surface of the stairrail system to the surface of the tread, in line with the face of the riser at the forward edge of the tread.

**1926.1052(c)(3)(ii)** Stairrails installed before March 15, 1991, shall be not less than 30 inches (76 cm) nor more than 34 inches (86 cm) from the upper surface of the stairrail system to the surface of the tread, in line with the face of the riser at the forward edge of the tread.

**1926.1052(c)(4)** Midrails, screens, mesh, intermediate vertical members, or equivalent intermediate structural members, shall be provided between the top rail of the stairrail system and the stairway steps.

**1926.1052(c)(4)(i)** Midrails, when used, shall be located at a height midway between the top edge of the stairrail system and the stairway steps.

**1926.1052(c)(4)(ii)** Screens or mesh, when used, shall extend from the top rail to the stairway step, and along the entire opening between top rail supports.

**1926.1052(c)(4)(iii)** When intermediate vertical members, such as balusters, are used between posts, they shall be not more than 19 inches (48 cm) apart.

**1926.1052(c)(4)(iv)** Other structural members, when used, shall be installed such that there are no openings in the stairrail system that are more than 19 inches (48 cm) wide.

**1926.1052(c)(5)** Handrails and the top rails of stairrail systems shall be capable of withstanding, without failure, a force of at least 200 pounds (890 n) applied within 2 inches (5 cm) of the top edge, in any downward or outward direction, at any point along the top edge.

**1926.1052(c)(6)** The height of handrails shall be not more than 37 inches (94 cm) nor less than 30 inches (76 cm) from the upper surface of the handrail to the surface of the tread, in line with the face of the riser at the forward edge of the tread.

**1926.1052(c)(7)** When the top edge of a stairrail system also serves as a handrail, the height of the top edge shall be not more than 37 inches (94 cm) nor less than 36 inches (91.5 cm) from the upper surface of the stairrail system to the surface of the tread, in line with the face of the riser at the forward edge of the tread.

**1926.1052(c)(8)** Stairrail systems and handrails shall be so surfaced as to prevent injury to employees from punctures or lacerations, and to prevent snagging of clothing.

**1926.1052(c)(9)** Handrails shall provide an adequate handhold for employees grasping them to avoid falling.

**1926.1052(c)(10)** The ends of stairrail systems and handrails shall be constructed so as not to constitute a projection hazard.

**1926.1052(c)(11)** Handrails that will not be a permanent part of the structure being built shall have a minimum clearance of 3 inches (8 cm) between the handrail and walls, stairrail systems, and other objects.

**1926.1052(c)(12)** Unprotected sides and edges of stairway landings shall be provided with guardrail systems. Guardrail system criteria are contained in subpart M of this part.

[55 FR 47687, Nov. 14, 1990; 56 FR 2585, Jan. 23, 1991; 56 FR 5061, Feb. 7, 1991; 56 FR 41794, Aug. 23, 1991]

## §1926.1053 Ladders.

**1926.1053(a)** ***General.*** The following requirements apply to all ladders as indicated, including job-made ladders.

**1926.1053(a)(1)** Ladders shall be capable of supporting the following loads without failure:

**1926.1053(a)(1)(i)** Each self-supporting portable ladder: At least four times the maximum intended load, except that each extra-heavy-duty type 1A metal or plastic ladder shall sustain at least 3.3 times the maximum intended load. The ability of a ladder to sustain the loads indicated in this paragraph shall be determined by applying or transmitting the requisite load to the ladder in a downward vertical direction. Ladders built and tested in conformance with the applicable provisions of appendix A of this subpart will be deemed to meet this requirement.

**1926.1053(a)(1)(ii)** Each portable ladder that is not self-supporting: At least four times the maximum intended load, except that each extra-heavy-duty type 1A metal or plastic ladders shall sustain at least 3.3 times the maximum intended load. The ability of a ladder to sustain the loads indicated in this paragraph shall be determined by applying or transmitting the requisite load to the ladder in a downward vertical direction when the ladder is placed at an angle of 75 1/2 degrees from the horizontal. Ladders built and tested in conformance with the applicable provisions of appendix A will be deemed to meet this requirement.

**1926.1053(a)(1)(iii)** Each fixed ladder: At least two loads of 250 pounds (114 kg) each, concentrated between any two consecutive attachments (the number and position of additional concentrated loads of 250 pounds (114 kg) each, determined from anticipated usage of the ladder, shall also be included), plus anticipated loads caused by ice buildup, winds, rigging, and impact loads resulting from the use of ladder safety devices. Each step or rung shall be capable of supporting a single concentrated load of at least 250 pounds (114 kg) applied in the middle of the step or rung. Ladders built in conformance with the applicable provisions of appendix A will be deemed to meet this requirement.

**1926.1053(a)(2)** Ladder rungs, cleats, and steps shall be parallel, level, and uniformly spaced when the ladder is in position for use.

**1926.1053(a)(3)(i)** Rungs, cleats, and steps of portable ladders (except as provided below) and fixed ladders (including individual-rung/step ladders) shall be spaced not less than 10 inches (25 cm) apart, nor more than 14 inches (36 cm) apart, as measured between center lines of the rungs, cleats, and steps.

**1926.1053(a)(3)(ii)** Rungs, cleats, and steps of step stools shall be not less than 8 inches (20 cm) apart, nor more than 12 inches (31 cm) apart, as measured between center lines of the rungs, cleats, and steps.

**1926.1053(a)(3)(iii)** Rungs, cleats, and steps of the base section of extension trestle ladders shall not be less than 8 inches (20 cm) nor more than 18 inches (46 cm) apart, as measured between center lines of the rungs, cleats, and steps. The rung spacing on the extension section of the extension trestle ladder shall be not less than 6 inches (15 cm) nor more than 12 inches (31 cm), as measured between center lines of the rungs, cleats, and steps.

**1926.1053(a)(4)(i)** The minimum clear distance between the sides of individual-rung/step ladders and the minimum clear distance between the side rails of other fixed ladders shall be 16 inches (41 cm).

**1926.1053(a)(4)(ii)** The minimum clear distance between side rails for all portable ladders shall be 11 1/2 inches (29 cm).

**1926.1053(a)(5)** The rungs of individual-rung/step ladders shall be shaped such that employees' feet cannot slide off the end of the rungs.

**1926.1053(a)(6)(i)** The rungs and steps of fixed metal ladders manufactured after March 15, 1991, shall be corrugated, knurled, dimpled, coated with skid-resistant material, or otherwise treated to minimize slipping.

**1926.1053(a)(6)(ii)** The rungs and steps of portable metal ladders shall be corrugated, knurled, dimpled, coated with skid-resistant material, or otherwise treated to minimize slipping.

**1926.1053(a)(7)** Ladders shall not be tied or fastened together to provide longer sections unless they are specifically designed for such use.

**1926.1053(a)(8)** A metal spreader or locking device shall be provided on each stepladder to hold the front and back sections in an open position when the ladder is being used.

**1926.1053(a)(9)** When splicing is required to obtain a given length of side rail, the resulting side rail must be at least equivalent in strength to a one-piece side rail made of the same material.

**1926.1053(a)(10)** Except when portable ladders are used to gain access to fixed ladders (such as those on utility towers, billboards, and other structures where the bottom of the fixed ladder is elevated to limit access), when two or more separate ladders are used to reach an elevated work area, the ladders shall be offset with a platform or landing between the ladders. (The requirements to have guardrail systems with toeboards for falling object and overhead protection on platforms and landings are set forth in subpart M of this part.)

**1926.1053(a)(11)** Ladder components shall be surfaced so as to prevent injury to an employee from punctures or lacerations, and to prevent snagging of clothing.

**1926.1053(a)(12)** Wood ladders shall not be coated with any opaque covering, except for identification or warning labels which may be placed on one face only of a side rail.

**1926.1053(a)(13)** The minimum perpendicular clearance between fixed ladder rungs, cleats, and steps, and any obstruction behind the ladder shall be 7 inches (18 cm), except in the case of an elevator pit ladder, for which a minimum perpendicular clearance of 4 1/2 inches (11 cm) is required.

**1926.1053(a)(14)** The minimum perpendicular clearance between the center line of fixed ladder rungs, cleats, and steps, and any obstruction on the climbing side of the ladder shall be 30 inches (76 cm), except as provided in paragraph (a)(15) of this section.

**1926.1053(a)(15)** When unavoidable obstructions are encountered, the minimum perpendicular clearance between the centerline of fixed ladder rungs, cleats, and steps, and the obstruction on the climbing side of the ladder may be reduced to 24 inches (61 cm), provided that a deflection device is installed to guide employees around the obstruction.

**1926.1053(a)(16)** Through fixed ladders at their point of access/egress shall have a step-across distance of not less than 7 inches (18 cm) nor more than 12 inches (30 cm) as measured from the centerline of the steps or rungs to the nearest edge of the landing area. If the normal step-across distance exceeds 12 inches (30 cm), a landing platform shall be provided to reduce the distance to the specified limit.

**1926.1053(a)(17)** Fixed ladders without cages or wells shall have a clear width to the nearest permanent object of at least 15 inches (38 cm) on each side of the centerline of the ladder.

**1926.1053(a)(18)** Fixed ladders shall be provided with cages, wells, ladder safety devices, or self-retracting lifelines where the length of climb is less than 24 feet (7.3 m) but the top of the ladder is at a distance greater than 24 feet (7.3 m) above lower levels.

**1926.1053(a)(19)** Where the total length of a climb equals or exceeds 24 feet (7.3 m), fixed ladders shall be equipped with one of the following:

**1926.1053(a)(19)(i)** Ladder safety devices; or

**1926.1053(a)(19)(ii)** Self-retracting lifelines, and rest platforms at intervals not to exceed 150 feet (45.7 m); or

**1926.1053(a)(19)(iii)** A cage or well, and multiple ladder sections, each ladder section not to exceed 50 feet (15.2 m) in length. Ladder sections shall be offset from adjacent sections, and landing platforms shall be provided at maximum intervals of 50 feet (15.2 m).

**1926.1053(a)(20)** Cages for fixed ladders shall conform to all of the following:

**1926.1053(a)(20)(i)** Horizontal bands shall be fastened to the side rails of rail ladders, or directly to the structure, building, or equipment for individual-rung ladders;

**1926.1053(a)(20)(ii)** Vertical bars shall be on the inside of the horizontal bands and shall be fastened to them;

**1926.1053(a)(20)(iii)** Cages shall extend not less than 27 inches (68 cm), or more than 30 inches (76 cm) from the centerline of the step or rung (excluding the flare at the bottom of the cage), and shall not be less than 27 inches (68 cm) in width;

**1926.1053(a)(20)(iv)** The inside of the cage shall be clear of projections;

**1926.1053(a)(20)(v)** Horizontal bands shall be spaced not more than 4 feet (1.2 m) on center vertically;

**1926.1053(a)(20)(vi)** Vertical bars shall be spaced at intervals not more than 9 1/2 inches (24 cm) on center horizontally;

**1926.1053(a)(20)(vii)** The bottom of the cage shall be at a level not less than 7 feet (2.1 m) nor more than 8 feet (2.4 m) above the point of access to the bottom of the ladder. The bottom of the cage shall be flared not less than 4 inches (10 cm) all around within the distance between the bottom horizontal band and the next higher band;

**1926.1053(a)(20)(viii)** The top of the cage shall be a minimum of 42 inches (1.1 m) above the top of the platform, or the point of access at the top of the ladder, with provision for access to the platform or other point of access.

**1926.1053(a)(21)** Wells for fixed ladders shall conform to all of the following:

**1926.1053(a)(21)(i)** They shall completely encircle the ladder;

**1926.1053(a)(21)(ii)** They shall be free of projections;

**1926.1053(a)(21)(iii)** Their inside face on the climbing side of the ladder shall extend not less than 27 inches (68 cm) nor more than 30 inches (76 cm) from the centerline of the step or rung;

**1926.1053(a)(21)(iv)** The inside clear width shall be at least 30 inches (76 cm);

**1926.1053(a)(21)(v)** The bottom of the wall on the access side shall start at a level not less than 7 feet (2.1 m) nor more than 8 feet (2.4 m) above the point of access to the bottom of the ladder.

**1926.1053(a)(22)** Ladder safety devices, and related support systems, for fixed ladders shall conform to all of the following:

**1926.1053(a)(22)(i)** They shall be capable of withstanding without failure a drop test consisting of an 18-inch (41 cm) drop of a 500-pound (226 kg) weight;

**1926.1053(a)(22)(ii)** They shall permit the employee using the device to ascend or descend without continually having to hold, push or pull any part of the device, leaving both hands free for climbing;

**1926.1053(a)(22)(iii)** They shall be activated within 2 feet (.61 m) after a fall occurs, and limit the descending velocity of an employee to 7 feet/sec. (2.1 m/sec.) or less;

**1926.1053(a)(22)(iv)** The connection between the carrier or lifeline and the point of attachment to the body belt or harness shall not exceed 9 inches (23 cm) in length.

**1926.1053(a)(23)** The mounting of ladder safety devices for fixed ladders shall conform to the following:

**1926.1053(a)(23)(i)** Mountings for rigid carriers shall be attached at each end of the carrier, with intermediate mountings, as necessary, spaced along the entire length of the carrier, to provide the strength necessary to stop employees' falls.

**1926.1053(a)(23)(ii)** Mountings for flexible carriers shall be attached at each end of the carrier. When the system is exposed to wind, cable guides for flexible carriers shall be installed at a minimum spacing of 25 feet (7.6 m) and maximum spacing of 40 feet (12.2 m) along the entire length of the carrier, to prevent wind damage to the system.

**1926.1053(a)(23)(iii)** The design and installation of mountings and cable guides shall not reduce the design strength of the ladder.

**1926.1053(a)(24)** The side rails of through or side-step fixed ladders shall extend 42 inches (1.1 m) above the top of the access level or landing platform served by the ladder. For a parapet ladder, the access level shall be the roof if the parapet is cut to permit passage through the parapet; if the parapet is continuous, the access level shall be the top of the parapet.

**1926.1053(a)(25)** For through-fixed-ladder extensions, the steps or rungs shall be omitted from the extension and the extension of the side rails shall be flared to provide not less than 24 inches (61 cm) nor more than 30 inches (76 cm) clearance between side rails. Where ladder safety devices are provided, the maximum clearance between side rails of the extensions shall not exceed 36 inches (91 cm).

**1926.1053(a)(26)** For side-step fixed ladders, the side rails and the steps or rungs shall be continuous in the extension.

**1926.1053(a)(27)** Individual-rung/step ladders, except those used where their access openings are covered with manhole covers or hatches, shall extend at least 42 inches (1.1 m) above an access level or landing platform either by the continuation of the rung spacings as horizontal grab bars or by providing vertical grab bars that shall have the same lateral spacing as the vertical legs of the rungs.

**1926.1053(b)** ***Use.*** The following requirements apply to the use of all ladders, including job-made ladders, except as otherwise indicated:

**1926.1053(b)(1)** When portable ladders are used for access to an upper landing surface, the ladder side rails shall extend at least 3 feet (.9 m) above the upper landing surface to which the ladder is used to gain access; or, when such an extension is not possible because of the ladder's length, then the ladder shall be secured at its top to a rigid support that will not deflect, and a grasping device, such as a grabrail, shall be provided to assist employees in mounting and dismounting the ladder. In no case shall the extension be such that ladder deflection under a load would, by itself, cause the ladder to slip off its support.

**1926.1053(b)(2)** Ladders shall be maintained free of oil, grease, and other slipping hazards.

**1926.1053(b)(3)** Ladders shall not be loaded beyond the maximum intended load for which they were built, nor beyond their manufacturer's rated capacity.

**1926.1053(b)(4)** Ladders shall be used only for the purpose for which they were designed.

**1926.1053(b)(5)(i)** Non-self-supporting ladders shall be used at an angle such that the horizontal distance from the top support to the foot of the ladder is approximately one-quarter of the working length of the ladder (the distance along the ladder between the foot and the top support).

**1926.1053(b)(5)(ii)** Wood job-made ladders with spliced side rails shall be used at an angle such that the horizontal distance is one-eighth the working length of the ladder.

**1926.1053(b)(5)(iii)** Fixed ladders shall be used at a pitch no greater than 90 degrees from the horizontal, as measured to the back side of the ladder.

**1926.1053(b)(6)** Ladders shall be used only on stable and level surfaces unless secured to prevent accidental displacement.

**1926.1053(b)(7)** Ladders shall not be used on slippery surfaces unless secured or provided with slip-resistant feet to prevent accidental displacement. Slip-resistant feet shall not be used as a substitute for care in placing, lashing, or holding a ladder that is used upon slippery surfaces including, but not limited to, flat metal or concrete surfaces that are constructed so they cannot be prevented from becoming slippery.

**1926.1053(b)(8)** Ladders placed in any location where they can be displaced by workplace activities or traffic, such as in passageways, doorways, or driveways, shall be secured to prevent accidental displacement, or a barricade shall be used to keep the activities or traffic away from the ladder.

**1926.1053(b)(9)** The area around the top and bottom of ladders shall be kept clear.

**1926.1053(b)(10)** The top of a non-self-supporting ladder shall be placed with the two rails supported equally unless it is equipped with a single support attachment.

**1926.1053(b)(11)** Ladders shall not be moved, shifted, or extended while occupied.

**1926.1053(b)(12)** Ladders shall have nonconductive siderails if they are used where the employee or the ladder could contact exposed energized electrical equipment, except as provided in §1926.951(c)(1) of this part.

**1926.1053(b)(13)** The top or top step of a stepladder shall not be used as a step.

**1926.1053(b)(14)** Cross-bracing on the rear section of stepladders shall not be used for climbing unless the ladders are designed and provided with steps for climbing on both front and rear sections.

**1926.1053(b)(15)** Ladders shall be inspected by a competent person for visible defects on a periodic basis and after any occurrence that could affect their safe use.

**1926.1053(b)(16)** Portable ladders with structural defects, such as, but not limited to, broken or missing rungs, cleats, or steps, broken or split rails, corroded components, or other faulty or defective components, shall either be immediately marked in a manner that readily identifies them as defective, or be tagged with "Do Not Use" or similar language, and shall be withdrawn from service until repaired.

**1926.1053(b)(17)** Fixed ladders with structural defects, such as, but not limited to, broken or missing rungs, cleats, or steps, broken or split rails, or corroded components, shall be withdrawn from service until repaired. The requirement to withdraw a defective ladder from service is satisfied if the ladder is either:

**1926.1053(b)(17)(i)** Immediately tagged with "Do Not Use" or similar language,

**1926.1053(b)(17)(ii)** Marked in a manner that readily identifies it as defective;

**1926.1053(b)(17)(iii)** Or blocked (such as with a plywood attachment that spans several rungs).

**1926.1053(b)(18)** Ladder repairs shall restore the ladder to a condition meeting its original design criteria, before the ladder is returned to use.

**1926.1053(b)(19)** Single-rail ladders shall not be used.

**1926.1053(b)(20)** When ascending or descending a ladder, the user shall face the ladder.

**1926.1053(b)(21)** Each employee shall use at least one hand to grasp the ladder when progressing up and/or down the ladder.

**1926.1053(b)(22)** An employee shall not carry any object or load that could cause the employee to lose balance and fall.

[55 FR 47687, Nov. 14, 1990; 56 FR 2585, Jan. 23, 1991, as amended at 56 FR 41794, Aug. 23, 1991]

### §§1926.1054—1926.1059 [Reserved]

### §1926.1060 Training requirements.

The following training provisions clarify the requirements of §1926.21(b)(2), regarding the hazards addressed in subpart X.

**1926.1060(a)** The employer shall provide a training program for each employee using ladders and stairways, as necessary. The program shall enable each employee to recognize hazards related to ladders and stairways, and shall train each employee in the procedures to be followed to minimize these hazards.

**1926.1060(a)(1)** The employer shall ensure that each employee has been trained by a competent person in the following areas, as applicable:

**1926.1060(a)(1)(i)** The nature of fall hazards in the work area;

**1926.1060(a)(1)(ii)** The correct procedures for erecting, maintaining, and disassembling the fall protection systems to be used;

**1926.1060(a)(1)(iii)** The proper construction, use, placement, and care in handling of all stairways and ladders;

**1926.1060(a)(1)(iv)** The maximum intended load-carrying capacities of ladders used; and

**1926.1060(a)(1)(v)** The standards contained in this subpart.

**1926.1060(b)** Retraining shall be provided for each employee as necessary so that the employee maintains the understanding and knowledge acquired through compliance with this section.

### APPENDIX A TO SUBPART X—LADDERS

This appendix serves as a non-mandatory guideline to assist employers in complying with the ladder loading and strength requirements of §1926.1053(a)(1). A ladder designed and built in accordance with the applicable national consensus standards, as set forth below, will be considered to meet the requirements of §1926.1053(a)(1):

- Manufactured portable wood ladders: American National Standards Institute (ANSI) A14.1-1982—American National Standard for Ladders-Portable Wood-Safety Requirements.
- Manufactured portable metal ladders: ANSI A14.2-1982—American National Standard for Ladders—Portable Metal-Safety Requirements.
- Manufactured fixed ladders: ANSI A14.3-1984—American National Standard for Ladders-Fixed-Safety Requirements.
- Job-made ladders: ANSI A14.4-1979—Safety Requirements for Job-Made Ladders.
- Plastic ladders: ANSI A14.5-1982—American National Standard for Ladders-Portable Reinforced Plastic-Safety Requirements.

**Authority:** Sections 4, 6, and 8, Occupational Safety and Health Act of 1970 (29 U.S.C. 653, 655, 657); sec. 107, Contract Work Hours and Safety Standards Act (the Construction Safety Act) (40 U.S.C. 333); sec. 41, Longshore and Harbor Workers' Compensation Act (33 U.S.C. 941); Secretary of Labor's Order No. 12-71 (36 FR 8754), 8-76 (41 FR 25059), 9-83 (48 FR 35736), or 1-90 (55 FR 9033), as applicable; 29 CFR part 1911.

**Source:** 58 FR 35184, June 30, 1993, unless otherwise noted.

## GENERAL

### §1926.1071 Scope and application.

**Note:** The requirements applicable to construction work under this section are identical to those set forth at §1910.401 of this chapter.

[61 FR 31432, June 20, 1996]

[Section 1910.401 is printed below—CCH.]

### §1910.401 Scope and application.

**1910.401(a)** ***Scope.*** (1) This subpart (standard) applies to every place of employment within the waters of the United States, or within any State, the District of Columbia, the Commonwealth of Puerto Rico, the Virgin Islands, American Samoa, Guam, the Trust Territory of the Pacific Islands, Wake Island, Johnston Island, the Canal Zone, or within the Outer Continental Shelf lands as defined in the Outer Continental Shelf Lands Act (67 Stat. 462, 43 U.S.C. 1331), where diving and related support operations are performed.

**1910.401(a)(2)** This standard applies to diving and related support operations conducted in connection with all types of work and employments, including general industry, construction, ship repairing, shipbuilding, shipbreaking and longshoring. However, this standard does not apply to any diving operation:

**1910.401(a)(2)(i)** Performed solely for instructional purposes, using open-circuit, compressed-air SCUBA and conducted within the no-decompression limits;

**1910.401(a)(2)(ii)** Performed solely for search, rescue, or related public safety purposes by or under the control of a governmental agency; or

**1910.401(a)(2)(iii)** Governed by 45 CFR part 46 (Protection of Human Subjects, U.S. Department of Health and Human Services) or equivalent rules or regulations established by another federal agency, which regulate research, development, or related purposes involving human subjects.

**1910.401(a)(2)(iv)** Defined as scientific diving and which is under the direction and control of a diving program containing at least the following elements:

**1910.401(a)(2)(iv)(A)** Diving safety manual which includes at a minimum: Procedures covering all diving operations specific to the program; procedures for emergency care, including recompression and evacuation; and criteria for diver training and certification.

**1910.401(a)(2)(iv)(B)** Diving control (safety) board, with the majority of its members being active divers, which shall at a minimum have the authority to: Approve and monitor diving projects; review and revise the diving safety manual; assure compliance with the manual; certify the depths to which a diver has been trained; take disciplinary action for unsafe practices; and, assure adherence to the buddy system (a diver is accompanied by and is in continuous contact with another diver in the water) for SCUBA diving.

**1910.401(b)** ***Application in emergencies.*** An employer may deviate from the requirements of this standard to the extent necessary to prevent or minimize a situation which is likely to cause death, serious physical harm, or major environmental damage, provided that the employer:

**1910.401(b)(1)** Notifies the Area Director, Occupational Safety and Health Administration within 48 hours of the onset of the emergency situation indicating the nature of the emergency and extent of the deviation from the prescribed regulations; and

**1910.401(b)(2)** Upon request from the Area Director, submits such information in writing.

**1910.401(c)** ***Employer obligation.*** The employer shall be responsible for compliance with:

**1910.401(c)(1)** All provisions of this standard of general applicability; and

**1910.401(c)(2)** All requirements pertaining to specific diving modes to the extent diving operations in such modes are conducted.

[42 FR 37668, July 22, 1977, as amended at 47 FR 53365, Nov. 26, 1982; 58 FR 35310, June 30, 1993]

### §1926.1072 Definitions.

**Note:** The provisions applicable to construction work under this section are identical to those set forth at §1910.402 of this chapter.

[61 FR 31432, June 20, 1996]

[Section 1910.402 is printed below—CCH.]

### §1910.402 Definitions.

As used in this standard, the listed terms are defined as follows:

*Acfm:* Actual cubic feet per minute.

*ASME Code or equivalent:* ASME (American Society of Mechanical Engineers) Boiler and Pressure Vessel Code, Section VIII, or an equivalent code which the employer can demonstrate to be equally effective.

*ATA:* Atmosphere absolute.

*Bell:* An enclosed compartment, pressurized (closed bell) or unpressurized (open bell), which allows the diver to be transported to and from the underwater work area and which may be used as a temporary refuge during diving operations.

*Bottom time:* The total elasped time measured in minutes from the time when the diver leaves the surface in descent to the time that the diver begins ascent.

*Bursting pressure:* The pressure at which a pressure containment device would fail structurally.

*Cylinder:* A pressure vessel for the storage of gases.

*Decompression chamber:* A pressure vessel for human occupancy such as a surface decompression chamber, closed bell, or deep diving system used to decompress divers and to treat decompression sickness.

*Decompression sickness:* A condition with a variety of symptoms which may result from gas or bubbles in the tissues of divers after pressure reduction.

*Decompression table:* A profile or set of profiles of depth-time relationships for ascent rates and breathing mixtures to be followed after a specific depth-time exposure or exposures.

*Dive location:* A surface or vessel from which a diving operation is conducted.

*Dive-location reserve breathing gas:* A supply system of air or mixed-gas (as appropriate) at the dive location which is independent of the primary supply system and sufficient to support divers during the planned decompression.

*Dive team:* Divers and support employees involved in a diving operation, including the designated person-in-charge.

*Diver:* An employee working in water using underwater apparatus which supplies compressed breathing gas at the ambient pressure.

*Diver-carried reserve breathing gas:* A diver-carried supply of air or mixed gas (as appropriate) sufficient under standard operating conditions to allow the diver to reach the surface, or another source of breathing gas, or to be reached by a standby diver.

*Diving mode:* A type of diving requiring specific equipment, procedures and techniques (SCUBA, surface-supplied air, or mixed gas).

*Fsw:* Feet of seawater (or equivalent static pressure head).

*Heavy gear:* Diver-worn deep-sea dress including helmet, breastplate, dry suit, and weighted shoes.

*Hyperbaric conditions:* Pressure conditions in excess of surface pressure.

*Inwater stage:* A suspended underwater platform which supports a diver in the water.

*Liveboating:* The practice of supporting a surfaced-supplied air or mixed gas diver from a vessel which is underway.

*Mixed-gas diving:* A diving mode in which the diver is supplied in the water with a breathing gas other than air.

*No-decompression limits:* The depth-time limits of the "no-decompression limits and repetitive dive group designation table for no-decompression air dives", U.S. Navy Diving Manual or equivalent limits which the employer can demonstrate to be equally effective.

*Psi(g):* Pounds per square inch (gauge).

*Scientific diving* means diving performed solely as a necessary part of a scientific, research, or educational activity by employees whose sole purpose for diving is to perform scientific research tasks. Scientific diving does not include performing any tasks usually associated with commercial diving such as: Placing or removing heavy objects underwater; inspection of pipelines and similar objects; construction; demolition; cutting or welding; or the use of explosives.

*SCUBA diving:* A diving mode independent of surface supply in which the diver uses open circuit self-contained underwater breathing apparatus.

*Standby diver:* A diver at the dive location available to assist a diver in the water.

*Surface-supplied air diving:* A diving mode in which the diver in the water is supplied from the dive location with compressed air for breathing.

*Treatment table:* A depth-time and breathing gas profile designed to treat decompression sickness.

*Umbilical:* The composite hose bundle between a dive location and a diver or bell, or between a diver and a bell, which supplies the diver or bell with breathing gas, communications, power, or heat as appropriate to the diving mode or conditions, and includes a safety line between the diver and the dive location.

*Volume tank:* A pressure vessel connected to the outlet of a compressor and used as an air reservoir.

*Working pressure:* The maximum pressure to which a pressure containment device may be exposed under standard operating conditions.

[42 FR 37668, July 22, 1977, as amended at 47 FR 53365, Nov. 26, 1982]

## PERSONNEL REQUIREMENTS

### §1926.1076 Qualifications of dive team.

**Note:** The requirements applicable to construction work under this section are identical to those set forth at §1910.410 of this chapter.

[61 FR 31432, June 20, 1996]

[Section 1910.410 is printed below—CCH.]

### §1910.410 Qualifications of dive team.

**1910.410(a)** ***General.*** (1) Each dive team member shall have the experience or training necessary to perform assigned tasks in a safe and healthful manner.

**1910.410(a)(2)** Each dive team member shall have experience or training in the following:

**1910.410(a)(2)(i)** The use of tools, equipment and systems relevant to assigned tasks;

**1910.410(a)(2)(ii)** Techniques of the assigned diving mode: and

**1910.410(a)(2)(iii)** Diving operations and emergency procedures.

**1910.410(a)(3)** All dive team members shall be trained in cardiopulmonary resuscitation and first aid (American Red Cross standard course or equivalent).

**1910.410(a)(4)** Dive team members who are exposed to or control the exposure of others to hyperbaric conditions shall be trained in diving-related physics and physiology.

**1910.410(b)** ***Assignments.*** (1) Each dive team member shall be assigned tasks in accordance with the employee's experience or training, except that limited additional tasks may be assigned to an employee undergoing training provided that these tasks are performed under the direct supervision of an experienced dive team member.

**1910.410(b)(2)** The employer shall not require a dive team member to be exposed to hyperbaric conditions against the employee's will, except when necessary to complete decompression or treatment procedures.

**1910.410(b)(3)** The employer shall not permit a dive team member to dive or be otherwise exposed to hyperbaric conditions for the duration of any temporary physical impairment or condition which is known to the employer and is likely to affect adversely the safety or health of a dive team member.

**1910.410(c)** ***Designated person-in-charge.*** (1) The employer or an employee designated by the employer shall be at the dive location in charge of all aspects of the diving operation affecting the safety and health of dive team members.

**1910.410(c)(2)** The designated person-in-charge shall have experience and training in the conduct of the assigned diving operation.

## GENERAL OPERATIONS PROCEDURES

### §1926.1080 Safe practices manual.

**Note:** The requirements applicable to construction work under this section are identical to those set forth at §1910.420 of this chapter.

[61 FR 31432, June 20, 1996]

[Section 1910.420 is printed below—CCH.]

### §1910.420 Safe practices manual.

**1910.420(a)** ***General.*** The employer shall develop and maintain a safe practices manual which shall be made available at the dive location to each dive team member.

**1910.420(b)** ***Contents.*** (1) The safe practices manual shall contain a copy of this standard and the employer's policies for implementing the requirements of this standard.

**1910.420(b)(2)** For each diving mode engaged in, the safe practices manual shall include:

**1910.420(b)(2)(i)** Safety procedures and checklists for diving operations;

**1910.420(b)(2)(ii)** Assignments and responsibilities of the dive team members;

**1910.420(b)(2)(iii)** Equipment procedures and checklists; and

**1910.420(b)(2)(iv)** Emergency procedures for fire, equipment failure, adverse environmental conditions, and medical illness and injury.

[42 FR 37668, July 22, 1977, as amended at 49 FR 18295, Apr. 30, 1984]

### §1926.1081 Pre-dive procedures.

**Note:** The requirements applicable to construction work under this section are identical to those set forth at §1910.421 of this chapter.

[61 FR 31432, June 20, 1996]

[Section 1910.421 is printed below—CCH.]

### §1910.421 Pre-dive procedures.

**1910.421(a)** ***General.*** The employer shall comply with the following requirements prior to each diving operation, unless otherwise specified.

**1910.421(b)** ***Emergency aid.*** A list shall be kept at the dive location of the telephone or call numbers of the following:

**1910.421(b)(1)** An operational decompression chamber (if not at the dive location);

**1910.421(b)(2)** Accessible hospitals;

**1910.421(b)(3)** Available physicians;

**1910.421(b)(4)** Available means of transportation; and

**1910.421(b)(5)** The nearest U.S. Coast Guard Rescue Coordination Center.

**1910.421(c)** ***First aid supplies.*** (1) A first aid kit appropriate for the diving operation and approved by a physician shall be available at the dive location.

**1910.421(c)(2)** When used in a decompression chamber or bell, the first aid kitshall be suitable for use under hyperbaric conditions.

**1910.421(c)(3)** In addition to any other first aid supplies, an American Red Cross standard first aid handbook or equivalent, and a bag-type manual resuscitator with transparent mask and tubing shall be available at the dive location.

**1910.421(d)** ***Planning and assessment.*** Planning of a diving operation shall include an assessment of the safety and health aspects of the following:

**1910.421(d)(1)** Diving mode;

**1910.421(d)(2)** Surface and underwater conditions and hazards;

**1910.421(d)(3)** Breathing gas supply (including reserves);

**1910.421(d)(4)** Thermal protection;

**1910.421(d)(5)** Diving equipment and systems;

**1910.421(d)(6)** Dive team assignments and physical fitness of dive team members (including any impairment known to the employer);

**1910.421(d)(7)** Repetitive dive designation or residual inert gas status of dive team members;

**1910.421(d)(8)** Decompression and treatment procedures (including altitude corrections); and

**1910.421(d)(9)** Emergency procedures.

**1910.421(e)** ***Hazardous activities.*** To minimize hazards to the dive team, diving operations shall be coordinated with other activities in the vicinity which are likely to interfere with the diving operation.

**1910.421(f)** ***Employee briefing.*** (1) Dive team members shall be briefed on:

**1910.421(f)(1)(i)** The tasks to be undertaken;

**1910.421(f)(1)(ii)** Safety procedures for the diving mode;

**1910.421(f)(1)(iii)** Any unusual hazards or environmental conditions likely to affect the safety of the diving operation; and

**1910.421(f)(1)(iv)** Any modifications to operating procedures necessitated by the specific diving operation.

**1910.421(f)(2)** Prior to making individual dive team member assignments, the employer shall inquire into the dive team member's current state of physical fitness, and indicate to the dive team member the procedure for reporting physical problems or adverse physiological effects during and after the dive.

**1910.421(g)** ***Equipment inspection.*** The breathing gas supply system including reserve breathing gas supplies, masks, helmets, thermal protection, and bell handling mechanism (when appropriate) shall be inspected prior to each dive.

**1910.421(h)** ***Warning signal.*** When diving from surfaces other than vessels in areas capable of supporting marine traffic, a rigid replica of the international code flag "A" at least one meter in height shall be displayed at the dive location in a manner which allows all-round visibility, and shall be illuminated during night diving operations.

[42 FR 37668, July 22, 1977, as amended at 47 FR 14706, Apr. 6, 1982; 54 FR 24334, June 7, 1989]

## §1926.1082 Procedures during dive.

**Note:** The requirements applicable to construction work under this section are identical to those set forth at §1910.422 of this chapter.

[61 FR 31432, June 20, 1996]

[Section 1910.422 is printed below—CCH.]

## §1910.422 Procedures during dive.

**1910.422(a)** ***General.*** The employer shall comply with the following requirements which are applicable to each diving operation unless otherwise specified.

**1910.422(b)** ***Water entry and exit.*** (1) A means capable of supporting the diver shall be provided for entering and exiting the water.

**1910.422(b)(2)** The means provided for exiting the water shall extend below the water surface.

**1910.422(b)(3)** A means shall be provided to assist an injured diver from the water or into a bell.

**1910.422(c)** ***Communications.*** (1) An operational two-way voice communication system shall be used between:

**1910.422(c)(1)(i)** Each surface-supplied air or mixed-gas diver and a dive team member at the dive location or bell (when provided or required); and

**1910.422(c)(1)(ii)** The bell and the dive location.

**1910.422(c)(2)** An operational, two-way communication system shall be available at the dive location to obtain emergency assistance.

**1910.422(d)** ***Decompression tables.*** Decompression, repetitive, and no-decompression tables (as appropriate) shall be at the dive location.

**1910.422(e)** ***Dive profiles.*** A depth-time profile, including when appropriate any breathing gas changes, shall be maintained for each diver during the dive including decompression.

**1910.422(f)** ***Hand-held power tools and equipment.*** (1) Hand-held electrical tools and equipment shall be de-energized before being placed into or retrieved from the water.

**1910.422(f)(2)** Hand-held power tools shall not be supplied with power from the dive location until requested by the diver.

**1910.422(g)** ***Welding and burning.*** (1) A current supply switch to interrupt the current flow to the welding or burning electrode shall be:

**1910.422(g)(1)(i)** Tended by a dive team member in voice communication with the diver performing the welding or burning; and

**1910.422(g)(1)(ii)** Kept in the open position except when the diver is welding or burning.

**1910.422(g)(2)** The welding machine frame shall be grounded.

**1910.422(g)(3)** Welding and burning cables, electrode holders, and connections shall be capable of carrying the maximum current required by the work, and shall be properly insulated.

**1910.422(g)(4)** Insulated gloves shall be provided to divers performing welding and burning operations.

**1910.422(g)(5)** Prior to welding or burning on closed compartments, structures or pipes, which contain a flammable vapor or in which a flammable vapor may be generated by the work, they shall be vented, flooded, or purged with a mixture of gases which will not support combustion.

**1910.422(h)** ***Explosives.*** (1) Employers shall transport, store, and use explosives in accordance with this section and the applicable provisions of §1910.109 and §1926.912 of Title 29 of the Code of Federal Regulations.

**1910.422(h)(2)** Electrical continuity of explosive circuits shall not be tested until the diver is out of the water.

**1910.422(h)(3)** Explosives shall not be detonated while the diver is in the water.

**1910.422(i)** ***Termination of dive.*** The working interval of a dive shall be terminated when:

**1910.422(i)(1)** A diver requests termination;

**1910.422(i)(2)** A diver fails to respond correctly to communications or signals from a dive team member;

**1910.422(i)(3)** Communications are lost and can not be quickly re-established between the diver and a dive team member at the dive location, and between the designated person-in-charge and the person controlling the vessel in liveboating operations; or

**1910.422(i)(4)** A diver begins to use diver-carried reserve breathing gas or the dive-location reserve breathing gas.

## §1926.1083 Post-dive procedures.

**Note:** The requirements applicable to construction work under this section are identical to those set forth at §1910.423 of this chapter.

[61 FR 31432, June 20, 1996]

[Section 1910.423 is printed below—CCH.]

## §1910.423 Post-dive procedures.

**1910.423(a)** ***General.*** The employer shall comply with the following requirements which are applicable after each diving operation, unless otherwise specified.

**1910.423(b)** ***Precautions.*** (1) After the completion of any dive, the employer shall:

**1910.423(b)(1)(i)** Check the physical condition of the diver;

**1910.423(b)(1)(ii)** Instruct the diver to report any physical problems or adverse physiological effects including symptoms of decompression sickness;

**1910.423(b)(1)(iii)** Advise the diver of the location of a decompression chamber which is ready for use; and

**1910.423(b)(1)(iv)** Alert the diver to the potential hazards of flying after diving.

**1910.423(b)(2)** For any dive outside the no-decompression limits, deeper than 100 fsw or using mixed gas as a breathing mixture, the employer shall instruct the diver to remain awake and in the vicinity of the decompression chamber which is at the dive location for at least one hour after the dive (including decompression or treatment as appropriate).

**1910.423(c)** ***Recompression capability.*** (1) A decompression chamber capable of recompressing the diver at the surface to a minimum of 165 fsw (6 ATA) shall be available at the dive location for:

**1910.423(c)(1)(i)** Surface-supplied air diving to depths deeper than 100 fsw and shallower than 220 fsw;

**1910.423(c)(1)(ii)** Mixed gas diving shallower than 300 fsw; or

**1910.423(c)(1)(iii)** Diving outside the no-decompression limits shallower than 300 fsw.

**1910.423(c)(2)** A decompression chamber capable of recompressing the diver at the surface to the maximum depth of the dive shall be available at the dive location for dives deeper than 300 fsw.

**1910.423(c)(3)** The decompression chamber shall be:

**1910.423(c)(3)(i)** Dual-lock;

**1910.423(c)(3)(ii)** Multiplace; and

**1910.423(c)(3)(iii)** Located within 5 minutes of the dive location.

**1910.423(c)(4)** The decompression chamber shall be equipped with:

**1910.423(c)(4)(i)** A pressure gauge for each pressurized compartment designed for human occupancy;

**1910.423(c)(4)(ii)** A built-in-breathing-system with a minimum of one mask per occupant;

**1910.423(c)(4)(iii)** A two-way voice communication system between occupants and a dive team member at the dive location;

**1910.423(c)(4)(iv)** A viewport; and

**1910.423(c)(4)(v)** Illumination capability to light the interior.

**1910.423(c)(5)** Treatment tables, treatment gas appropriate to the diving mode, and sufficient gas to conduct treatment shall be available at the dive location.

**1910.423(c)(6)** A dive team member shall be available at the dive location during and for at least one hour after the dive to operate the decompression chamber (when required or provided).

**1910.423(d)** ***Record of dive.*** (1) The following information shall be recorded and maintained for each diving operation:

**1910.423(d)(1)(i)** Names of dive team members including designated person-in-charge;

**1910.423(d)(1)(ii)** Date, time, and location;

**1910.423(d)(1)(iii)** Diving modes used;

**1910.423(d)(1)(iv)** General nature of work performed;

**1910.423(d)(1)(v)** Approximate underwater and surface conditions (visibility, water temperature and current); and

**1910.423(d)(1)(vi)** Maximum depth and bottom time for each diver.

**1910.423(d)(2)** For each dive outside the no-decompression limits, deeper than 100 fsw or using mixed gas, the following additional information shall be recorded and maintained:

**1910.423(d)(2)(i)** Depth-time and breathing gas profiles;

**1910.423(d)(2)(ii)** Decompression table designation (including modification); and

**1910.423(d)(2)(iii)** Elapsed time since last pressure exposure if less than 24 hours or repetitive dive designation for each diver.

**1910.423(d)(3)** For each dive in which decompression sickness is suspected or symptoms are evident, the following additional information shall be recorded and maintained:

**1910.423(d)(3)(i)** Description of decompression sickness symptoms (including depth and time of onset); and

**1910.423(d)(3)(ii)** Description and results of treatment.

**1910.423(e)** ***Decompression procedure assessment.*** The employer shall:

**1910.423(e)(1)** Investigate and evaluate each incident of decompression sickness based on the recorded information, consideration of the past performance of decompression table used, and individual susceptibility;

**1910.423(e)(2)** Take appropriate corrective action to reduce the probability of recurrence of decompression sickness; and

**1910.423(e)(3)** Prepare a written evaluation of the decompression procedure assessment, including any corrective action taken, within 45 days of the incident of decompression sickness.

[42 FR 37668, July 22, 1977, as amended at 49 FR 18295, Apr. 30, 1984]

## SPECIFIC OPERATIONS PROCEDURES

### §1926.1084 SCUBA diving.

**Note:** The requirements applicable to construction work under this section are identical to those set forth at §1910.424 of this chapter.

[61 FR 31432, June 20, 1996]

[Section 1910.424 is printed below—CCH.]

### §1910.424 SCUBA diving.

**1910.424(a)** ***General.*** Employers engaged in SCUBA diving shall comply with the following requirements, unless otherwise specified.

**1910.424(b)** ***Limits.*** SCUBA diving shall not be conducted:

**1910.424(b)(1)** At depths deeper than 130 fsw;

**1910.424(b)(2)** At depths deeper than 100 fsw or outside the no-decompression limits unless a decompression chamber is ready for use;

**1910.424(b)(3)** Against currents exceeding one (1) knot unless line-tended; or

**1910.424(b)(4)** In enclosed or physically confining spaces unless line-tended.

**1910.424(c)** ***Procedures.*** (1) A standby diver shall be available while a diver is in the water.

**1910.424(c)(2)** A diver shall be line-tended from the surface, or accompanied by another diver in the water in continuous visual contact during the diving operations.

**1910.424(c)(3)** A diver shall be stationed at the underwater point of entry when diving is conducted in enclosed or physically confining spaces.

**1910.424(c)(4)** A diver-carried reserve breathing gas supply shall be provided for each diver consisting of:

**1910.424(c)(4)(i)** A manual reserve (J valve); or

**1910.424(c)(4)(ii)** An independent reserve cylinder with a separate regulator or connected to the underwater breathing apparatus.

**1910.424(c)(5)** The valve of the reserve breathing gas supply shall be in the closed position prior to the dive.

### §1926.1085 Surface-supplied air diving.

**Note:** The requirements applicable to construction work under this section are identical to those set forth at §1910.425 of this chapter.

[61 FR 31432, June 20, 1996]

[Section 1910.425 is printed below—CCH.]

### §1910.425 Surface-supplied air diving.

**1910.425(a)** ***General.*** Employers engaged in surface-supplied air diving shall comply with the following requirements, unless otherwise specified.

**1910.425(b)** ***Limits.*** (1) Surface-supplied air diving shall not be conducted at depths deeper than 190 fsw, except that dives with bottom times of 30 minutes or less may be conducted to depths of 220 fsw.

1910.425(b)(2) A decompression chamber shall be ready for use at the dive location for any dive outside the no-decompression limits or deeper than 100 fsw.

1910.425(b)(3) A bell shall be used for dives with an inwater decompression time greater than 120 minutes, except when heavy gear is worn or diving is conducted in physically confining spaces.

1910.425(c) ***Procedures.*** (1) Each diver shall be continuously tended while in the water.

1910.425(c)(2) A diver shall be stationed at the underwater point of entry when diving is conducted in enclosed or physically confining spaces.

1910.425(c)(3) Each diving operation shall have a primary breathing gas supply sufficient to support divers for the duration of the planned dive including decompression.

1910.425(c)(4) For dives deeper than 100 fsw or outside the no-decompression limits:

1910.425(c)(4)(i) A separate dive team member shall tend each diver in the water;

1910.425(c)(4)(ii) A standby diver shall be available while a diver is in the water;

1910.425(c)(4)(iii) A diver-carried reserve breathing gas supply shall be provided for each diver except when heavy gear is worn; and

1910.425(c)(4)(iv) A dive-location reserve breathing gas supply shall be provided.

1910.425(c)(5) For heavy-gear diving deeper than 100 fsw or outside the no-decompression limits:

1910.425(c)(5)(i) An extra breathing gas hose capable of supplying breathing gas to the diver in the water shall be available to the standby diver.

1910.425(c)(5)(ii) An inwater stage shall be provided to divers in the water.

1910.425(c)(6) Except when heavy gear is worn or where physical space does not permit, a diver-carried reserve breathing gas supply shall be provided whenever the diver is prevented by the configuration of the dive area from ascending directly to the surface.

### §1926.1086 Mixed-gas diving.

**Note:** The requirements applicable to construction work under this section are identical to those set forth at §1910.426 of this chapter.

[61 FR 31432, June 20, 1996]

[Section 1910.426 is printed below—CCH.]

### §1910.426 Mixed-gas diving.

1910.426(a) ***General.*** Employers engaged in mixed-gas diving shall comply with the following requirements, unless otherwise specified.

1910.426(b) ***Limits.*** Mixed-gas diving shall be conducted only when:

1910.426(b)(1) A decompression chamber is ready for use at the dive location; and

1910.426(b)(1)(i) A bell is used at depths greater than 220 fsw or when the dive involves inwater decompression time of greater than 120 minutes, except when heavy gear is worn or when diving in physically confining spaces; or

1910.426(b)(1)(ii) A closed bell is used at depths greater than 300 fsw, except when diving is conducted in physically confining spaces.

1910.426(c) ***Procedures.*** (1) A separate dive team member shall tend each diver in the water.

1910.426(c)(2) A standby diver shall be available while a diver is in the water.

1910.426(c)(3) A diver shall be stationed at the underwater point of entry when diving is conducted in enclosed or physically confining spaces.

1910.426(c)(4) Each diving operation shall have a primary breathing gas supply sufficient to support divers for the duration of the planned dive including decompression.

1910.426(c)(5) Each diving operation shall have a dive-location reserve breathing gas supply.

1910.426(c)(6) When heavy gear is worn:

1910.426(c)(6)(i) An extra breathing gas hose capable of supplying breathing gas to the diver in the water shall be available to the standby diver; and

1910.426(c)(6)(ii) An inwater stage shall be provided to divers in the water.

1910.426(c)(7) An inwater stage shall be provided for divers without access to a bell for dives deeper than 100 fsw or outside the no-decompression limits.

1910.426(c)(8) When a closed bell is used, one dive team member in the bell shall be available and tend the diver in the water.

1910.426(c)(9) Except when heavy gear is worn or where physical space does not permit, a diver-carried reserve breathing gas supply shall be provided for each diver:

1910.426(c)(9)(i) Diving deeper than 100 fsw or outside the no-decompression limits; or

1910.426(c)(9)(ii) Prevented by the configuration of the dive area from directly ascending to the surface.

### §1926.1087 Liveboating.

**Note:** The requirements applicable to construction work under this section are identical to those set forth at §1910.427 of this chapter.

[61 FR 31432, June 20, 1996]

[Section 1910.427 is printed below—CCH.]

### §1910.427 Liveboating.

1910.427(a) ***General.*** Employers engaged in diving operations involving liveboating shall comply with the following requirements.

1910.427(b) ***Limits.*** Diving operations involving liveboating shall not be conducted:

1910.427(b)(1) With an inwater decompression time of greater than 120 minutes;

1910.427(b)(2) Using surface-supplied air at depths deeper than 190 fsw, except that dives with bottom times of 30 minutes or less may be conducted to depths of 220 fsw;

1910.427(b)(3) Using mixed gas at depths greater than 220 fsw;

1910.427(b)(4) In rough seas which significantly inpede diver mobility or work function; or

1910.427(b)(5) In other than daylight hours.

1910.427(c) ***Procedures.*** (1) The propeller of the vessel shall be stopped before the diver enters or exits the water.

1910.427(c)(2) A device shall be used which minimizes the possibility of entanglement of the diver's hose in the propeller of the vessel.

1910.427(c)(3) Two-way voice communication between the designated person-in-charge and the person controlling the vessel shall be available while the diver is in the water.

1910.427(c)(4) A standby diver shall be available while a diver is in the water.

1910.427(c)(5) A diver-carried reserve breathing gas supply shall be carried by each diver engaged in liveboating operations.

## EQUIPMENT PROCEDURES AND REQUIREMENTS

### §1926.1090 Equipment.

**Note:** The requirements applicable to construction work under this section are identical to those set forth at §1910.430 of this chapter.

[61 FR 31432, June 20, 1996]

[Section 1910.430 is printed below—CCH.]

### §1910.430 Equipment.

1910.430(a) ***General.*** (1) All employers shall comply with the following requirements, unless otherwise specified.

**1910.430(a)(2)** Each equipment modification, repair, test, calibration or maintenance service shall be recorded by means of a tagging or logging system, and include the date and nature of work performed, and the name or initials of the person performing the work.

**1910.430(b)** ***Air compressor system.*** (1) Compressors used to supply air to the diver shall be equipped with a volume tank with a check valve on the inlet side, a pressure gauge, a relief valve, and a drain valve.

**1910.430(b)(2)** Air compressor intakes shall be located away from areas containing exhaust or other contaminants.

**1910.430(b)(3)** Respirable air supplied to a diver shall not contain:

**1910.430(b)(3)(i)** A level of carbon monoxide (CO) greater than 20 p/m;

**1910.430(b)(3)(ii)** A level of carbon dioxide ($CO_2$) greater than 1,000 p/m;

**1910.430(b)(3)(iii)** A level of oil mist greater than 5 milligrams per cubic meter; or

**1910.430(b)(3)(iv)** A noxious or pronounced odor.

**1910.430(b)(4)** The output of air compressor systems shall be tested for air purity every 6 months by means of samples taken at the connection to the distribution system, except that non-oil lubricated compressors need not be tested for oil mist.

**1910.430(c)** ***Breathing gas supply hoses.*** (1) Breathing gas supply hoses shall:

**1910.430(c)(1)(i)** Have a working pressure at least equal to the working pressure of the total breathing gas system;

**1910.430(c)(1)(ii)** Have a rated bursting pressure at least equal to 4 times the working pressure;

**1910.430(c)(1)(iii)** Be tested at least annually to 1.5 times their working pressure; and

**1910.430(c)(1)(iv)** Have their open ends taped, capped or plugged when not in use.

**1910.430(c)(2)** Breathing gas supply hose connectors shall:

**1910.430(c)(2)(i)** Be made of corrosion-resistant materials;

**1910.430(c)(2)(ii)** Have a working pressure at least equal to the working pressure of the hose to which they are attached; and

**1910.430(c)(2)(iii)** Be resistant to accidental disengagement.

**1910.430(c)(3)** Umbilicals shall:

**1910.430(c)(3)(i)** Be marked in 10-ft. increments to 100 feet beginning at the diver's end, and in 50 ft. increments thereafter;

**1910.430(c)(3)(ii)** Be made of kink-resistant materials; and

**1910.430(c)(3)(iii)** Have a working pressure greater than the pressure equivalent to the maximum depth of the dive (relative to the supply source) plus 100 psi.

**1910.430(d)** ***Buoyancy control.*** (1) Helmets or masks connected directly to the dry suit or other buoyancy-changing equipment shall be equipped with an exhaust valve.

**1910.430(d)(2)** A dry suit or other buoyancy-changing equipment not directly connected to the helmet or mask shall be equipped with an exhaust valve.

**1910.430(d)(3)** When used for SCUBA diving, a buoyancy compensator shall have an inflation source separate from the breathing gas supply.

**1910.430(d)(4)** An inflatable flotation device capable of maintaining the diver at the surface in a face-up position, having a manually activated inflation source independent of the breathing supply, an oral inflation device, and an exhaust valve shall be used for SCUBA diving.

**1910.430(e)** ***Compressed gas cylinders.*** Compressed gas cylinders shall:

**1910.430(e)(1)** Be designed, constructed and maintained in accordance with the applicable provisions of 29 CFR 1910.101 and 1910.169 through 1910.171.

**1910.430(e)(2)** Be stored in a ventilated area and protected from excessive heat;

**1910.430(e)(3)** Be secured from falling; and

**1910.430(e)(4)** Have shut-off valves recessed into the cylinder or protected by a cap, except when in use or manifolded, or when used for SCUBA diving.

**1910.430(f)** ***Decompression chambers.*** (1) Each decompression chamber manufactured after the effective date of this standard, shall be built and maintained in accordance with the ASME Code or equivalent.

**1910.430(f)(2)** Each decompression chamber manufactured prior to the effective date of this standard shall be maintained in conformity with the code requirements to which it was built, or equivalent.

**1910.430(f)(3)** Each decompression chamber shall be equipped with:

**1910.430(f)(3)(i)** Means to maintain the atmosphere below a level of 25 percent oxygen by volume;

**1910.430(f)(3)(ii)** Mufflers on intake and exhaust lines, which shall be regularly inspected and maintained;

**1910.430(f)(3)(iii)** Suction guards on exhaust line openings; and

**1910.430(f)(3)(iv)** A means for extinguishing fire, and shall be maintained to minimize sources of ignition and combustible material.

**1910.430(g)** ***Gauges and timekeeping devices.*** (1) Gauges indicating diver depth which can be read at the dive location shall be used for all dives except SCUBA.

**1910.430(g)(2)** Each depth gauge shall be deadweight tested or calibrated against a master reference gauge every 6 months, and when there is a discrepancy greater than two percent (2 percent) of full scale between any two equivalent gauges.

**1910.430(g)(3)** A cylinder pressure gauge capable of being monitored by the diver during the dive shall be worn by each SCUBA diver.

**1910.430(g)(4)** A timekeeping device shall be available at each dive location.

**1910.430(h)** ***Masks and helmets.*** (1) Surface-supplied air and mixed-gas masks and helmets shall have:

**1910.430(h)(1)(i)** A non-return valve at the attachment point between helmet or mask and hose which shall close readily and positively; and

**1910.430(h)(1)(ii)** An exhaust valve.

**1910.430(h)(2)** Surface-supplied air masks and helmets shall have a minimum ventilation rate capability of 4.5 acfm at any depth at which they are operated or the capability of maintaining the diver's inspired carbon dioxide partial pressure below 0.02 ATA when the diver is producing carbon dioxide at the rate of 1.6 standard liters per minute.

**1910.430(i)** ***Oxygen safety.*** (1) Equipment used with oxygen or mixtures containing over forty percent (40%) by volume oxygen shall be designed for oxygen service.

**1910.430(i)(2)** Components (except umbilicals) exposed to oxygen or mixtures containing over forty percent (40%) by volume oxygen shall be cleaned of flammable materials before use.

**1910.430(i)(3)** Oxygen systems over 125 psig and compressed air systems over 500 psig shall have slow-opening shut-off valves.

**1910.430(j)** ***Weights and harnesses.*** (1) Except when heavy gear is worn, divers shall be equipped with a weight belt or assembly capable of quick release.

**1910.430(j)(2)** Except when heavy gear is worn or in SCUBA diving, each diver shall wear a safety harness with:

**1910.430(j)(2)(i)** A positive buckling device;

**1910.430(j)(2)(ii)** An attachment point for the umbilical to prevent strain on the mask or helmet; and

**1910.430(j)(2)(iii)** A lifting point to distribute the pull force of the line over the diver's body.

[39 FR 23502, June 27, 1974, as amended at 49 FR 18295, Apr. 30, 1984; 51 FR 33033, Sept. 18, 1986]

## RECORDKEEPING

### §1926.1091 Recordkeeping requirements.

**Note:** The requirements applicable to construction work under this section are identical to those set forth at §1910.440 of this chapter.

[61 FR 31432, June 20, 1996]

[Section 1910.440 is printed below—CCH.]

### §1910.440 Recordkeeping requirements.

**1910.440(a)(1)** [Reserved]

**1910.440(a)(2)** The employer shall record the occurrence of any diving-related injury or illness which requires any dive team member to be hospitalized for 24 hours or more, specifying the circumstances of the incident and the extent of any injuries or illnesses.

**1910.440(b)** ***Availability of records.*** (1) Upon the request of the Assistant Secretary of Labor for Occupational Safety and Health, or the Director, National Institute for Occupational Safety and Health, Department of Health and Human Services of their designees, the employer shall make available for inspection and copying any record or document required by this standard.

**1910.440(b)(2)** Records and documents required by this standard shall be provided upon request to employees, designated representatives, and the Assistant Secretary in accordance with 29 CFR 1910.20(a)-(e) and (g)-(i). Safe practices manuals (§1910.420), depth-time profiles (§1910.422), recordings of dives (§1910.423), decompression procedure assessment evaluations (§1910.423), and records of hospitalizations (§1910.440) shall be provided in the same manner as employee exposure records or analyses using exposure or medical records. Equipment inspections and testing records which pertain to employees (§1910.430) shall also be provided upon request to employees and their designated representatives.

**1910.440(b)(3)** Records and documents required by this standard shall be retained by the employer for the following period:

**1910.440(b)(3)(i)** Dive team member medical records (physician's reports) (§1910.411)—5 years;

**1910.440(b)(3)(ii)** Safe practices manual (§1910.420)—current document only;

**1910.440(b)(3)(iii)** Depth-time profile (§1910.422)—until completion of the recording of dive, or until completion of decompression procedure assessment where there has been an incident of decompression sickness;

**1910.440(b)(3)(iv)** Recording of dive (§1910.423)—1 year, except 5 years where there has been an incident of decompression sickness;

**1910.440(b)(3)(v)** Decompression procedure assessment evaluations (§1910.423)—5 years;

**1910.440(b)(3)(vi)** Equipment inspections and testing records (§1910.430)—current entry or tag, or until equipment is withdrawn from service;

**1910.440(b)(3)(vii)** Records of hospitalizations (§1910.440)—5 years.

**1910.440(b)(4)** After the expiration of the retention period of any record required to be kept for five (5) years, the employer shall forward such records to the National Institute for Occupational Safety and Health, Department of Health and Human Services. The employer shall also comply with any additional requirements set forth at 29 CFR 1910.20(h).

**1910.440(b)(5)** In the event the employer ceases to do business:

**1910.440(b)(5)(i)** The successor employer shall receive and retain all dive and employee medical records required by this standard; or

**1910.440(b)(5)(ii)** If there is no successor employer, dive and employee medical records shall be forwarded to the National Institute for Occupational Safety and Health, Department of Health and Human Services.

[42 FR 37668, July 22, 1977, as amended at 45 FR 35281, May 23, 1980; 47 FR 14706, Apr. 6, 1982; 51 FR 34562, Sept. 29, 1986; 61 FR 9242, Mar. 7, 1996]

### §1926.1092 Effective date.

**Note:** The requirements applicable to construction work under this section are identical to those set forth at §1910.441 of this chapter.

[61 FR 31432, June 20, 1996]

[Section 1910.441 is printed below—CCH.]

### §1910.441 Effective date.

This standard shall be effective on October 20, 1977, except that for provisions where decompression chambers or bells are required and such equipment is not yet available, employers shall comply as soon as possible thereafter but in no case later than 6 months after the effective date of the standard.

### Appendix A to Subpart Y— Examples of Conditions Which May Restrict or Limit Exposure to Hyperbaric Conditions

**Note:** The requirements applicable to construction work under this appendix A are identical to those set forth at Appendix A to Subpart T of part 1910 of this chapter.

[61 FR 31432, June 20, 1996]

[Appendix A to Subpart T from Part 1910 is printed below—CCH.]

### Appendix A to Subpart T to Part 1910— Examples of Conditions Which May Restrict or Limit Exposure to Hyperbaric Conditions

The following disorders may restrict or limit occupational exposure to hyperbaric conditions depending on severity, presence of residual effects, response to therapy, number of occurrences, diving mode, or degree and duration of isolation.

History of seizure disorder other than early febrile convulsions.

Malignancies (active) unless treated and without recurrence for 5 yrs.

Chronic inability to equalize sinus and/or middle ear pressure.

Cystic or cavitary disease of the lungs.

Impaired organ function caused by alcohol or drug use.

Conditions requiring continuous medication for control (e.g., antihistamines, steroids, barbiturates, moodaltering drugs, or insulin).

Meniere's disease.

Hemoglobinopathies.

Obstructive or restrictive lung disease.

Vestibular end organ destruction.

Pneumothorax.

Cardiac abnormalities (e.g., pathological heart block, valvular disease, intraventricular conduction defects other than isolated right bundle branch block, angina pectoris, arrhythmia, coronary artery disease).

Juxta-articular osteonecrosis.

### Appendix B to Subpart Y— Guidelines for Scientific Diving

**Note:** The requirements applicable to construction work under this appendix B are identical to those set forth at Appendix B to Subpart T of part 1910 of this chapter.

[61 FR 31433, June 20, 1996]

[Appendix B to Subpart T from Part 1910 is printed below—CCH.]

### Appendix B to Subpart T to Part 1910— Guidelines for Scientific Diving

This appendix contains guidelines that will be used in conjunction with §1910.401(a)(2)(iv) to determine those scientific diving programs which are exempt from the requirements for commercial diving. The guidelines are as follows:

1. The Diving Control Board consists of a majority of active scientific divers and has autonomous and absolute authority over the scientific diving program's operations.

2. The purpose of the project using scientific diving is the advancement of science; therefore, information and data resulting from the project are non-proprietary.

3. The tasks of a scientific diver are those of an observer and data gatherer. Construction and trouble-shooting tasks traditionally associated with commercial diving are not included within scientific diving.

4. Scientific divers, based on the nature of their activities, must use scientific expertise in studying the underwater environment and, therefore, are scientists or scientists in training.

[50 FR 1050, Jan. 9, 1985]

**Authority:** Sec. 107, Contract Work Hours and Safety Standards Act (40 U.S.C. 333); Secs. 4, 6, 8, Occupational Safety and Health Act of 1970 (29 U.S.C. 653, 655, 657); Secretary of Labor's Order Nos. 12-71 (36 FR 8754), 8-76 (41 FR 25059), 9-83 (48 FR 35736), 1-90 (55 FR 9033), or 6-96 (62 FR 111), as applicable; 29 CFR part 1911.

Section 1926.1102 not issued under 29 U.S.C. 655 or 29 CFR part 1911; also issued under 5 U.S.C. 553.

**Source:** 58 FR 35190, June 30, 1993, unless otherwise noted.

### §1926.1100 [Reserved]

### §1926.1101 Asbestos.

**1926.1101(a)** ***Scope and application.*** This section regulates asbestos exposure in all work as defined in 29 CFR 1910.12(b), including but not limited to the following:

**1926.1101(a)(1)** Demolition or salvage of structures where asbestos is present;

**1926.1101(a)(2)** Removal or encapsulation of materials containing asbestos;

**1926.1101(a)(3)** Construction, alteration, repair, maintenance, or renovation of structures, substrates, or portions thereof, that contain asbestos;

**1926.1101(a)(4)** Installation of products containing asbestos;

**1926.1101(a)(5)** Asbestos spill/emergency cleanup; and

**1926.1101(a)(6)** Transportation, disposal, storage, containment of and housekeeping activities involving asbestos or products containing asbestos, on the site or location at which construction activities are performed.

**1926.1101(a)(7)** Coverage under this standard shall be based on the nature of the work operation involving asbestos exposure.

**1926.1101(a)(8)** This section does not apply to asbestos-containing asphalt roof coatings, cements and mastics.

**1926.1101(b)** ***Definitions.***

*Aggressive method* means removal or disturbance of building material by sanding, abrading, grinding or other method that breaks, crumbles, or disintegrates intact ACM.

*Amended water* means water to which surfactant (wetting agent) has been added to increase the ability of the liquid to penetrate ACM.

*Asbestos* includes chrysotile, amosite, crocidolite, tremolite asbestos, anthophyllite asbestos, actinolite asbestos, and any of these minerals that has been chemically treated and/or altered. For purposes of this standard, "asbestos" includes PACM, as defined below.

*Asbestos-containing material (ACM)*, means any material containing more than one percent asbestos.

*Assistant Secretary* means the Assistant Secretary of Labor for Occupational Safety and Health, U.S. Department of Labor, or designee.

*Authorized person* means any person authorized by the employer and required by work duties to be present in regulated areas.

*Building/facility owner* is the legal entity, including a lessee, which exercises control over management and record keeping functions relating to a building and/or facility in which activities covered by this standard take place.

*Certified Industrial Hygienist* (CIH) means one certified in the practice of industrial hygiene by the American Board of Industrial Hygiene.

*Class I asbestos work* means activities involving the removal of TSI and surfacing ACM and PACM.

*Class II asbestos work* means activities involving the removal of ACM which is not thermal system insulation or surfacing material. This includes, but is not limited to, the removal of asbestos-containing wallboard, floor tile and sheeting, roofing and siding shingles, and construction mastics.

*Class III asbestos work* means repair and maintenance operations, where "ACM", including TSI and surfacing ACM and PACM, is likely to be disturbed.

*Class IV asbestos work* means maintenance and custodial activities during which employees contact but do not disturb ACM or PACM and activities to clean up dust, waste and debris resulting from Class I, II, and III activities.

*Clean room* means an uncontaminated room having facilities for the storage of employees' street clothing and uncontaminated materials and equipment.

*Closely resemble* means that the major workplace conditions which have contributed to the levels of historic asbestos exposure, are no more protective than conditions of the current workplace.

*Competent person* means, in addition to the definition in 29 CFR 1926.32(f), one who is capable of identifying existing asbestos hazards in the workplace and selecting the appropriate control strategy for asbestos exposure, who has the authority to take prompt corrective measures to eliminate them, as specified in 29 CFR 1926.32(f): in addition, for Class I and Class II work who is specially trained in a training course which meets the criteria of EPA's Model Accreditation Plan (40 CFR part 763) for supervisor, or its equivalent and, for Class III and Class IV work, who is trained in a manner consistent with EPA requirements for training of local education agency maintenance and custodial staff as set forth at 40 CFR 763.92(a)(2).

*Critical barrier* means one or more layers of plastic sealed over all openings into a work area or any other similarly placed physical barrier sufficient to prevent airborne asbestos in a work area from migrating to an adjacent area.

*Decontamination area* means an enclosed area adjacent and connected to the regulated area and consisting of an equipment room, shower area, and clean room, which is used for the decontamination of workers, materials, and equipment that are contaminated with asbestos.

*Demolition* means the wrecking or taking out of any load-supporting structural member and any related razing, removing, or stripping of asbestos products.

*Director* means the Director, National Institute for Occupational Safety and Health, U.S. Department of Health and Human Services, or designee.

*Disturbance* means activities that disrupt the matrix of ACM or PACM, crumble or pulverize ACM or PACM, or generate visible debris from ACM or PACM. In no event shall the amount of ACM or PACM so disturbed exceed that which can be contained in one glove bag or waste bag which shall not exceed 60 inches in length and width.

*Employee exposure* means that exposure to airborne asbestos that would occur if the employee were not using respiratory protective equipment.

*Equipment room (change room)* means a contaminated room located within the decontamination area that is supplied with impermeable bags or containers for the disposal of contaminated protective clothing and equipment.

*Fiber* means a particulate form of asbestos, 5 micrometers or longer, with a length-to-diameter ratio of at least 3 to 1.

*Glovebag* means not more than a 60 x 60 inch impervious plastic bag-like enclosure affixed around an asbestos-containing material, with glove-like appendages through which material and tools may be handled.

*High-efficiency particulate air (HEPA) filter* means a filter capable of trapping and retaining at least 99.97 percent of all mono-dispersed particles of 0.3 micrometers in diameter.

*Homogeneous area* means an area of surfacing material or thermal system insulation that is uniform in color and texture.

*Industrial hygienist* means a professional qualified by education, training, and experience to anticipate, recognize, evaluate and develop controls for occupational health hazards.

*Intact* means that the ACM has not crumbled, been pulverized, or otherwise deteriorated so that the asbestos is no longer likely to be bound with its matrix.

*Modification* for purposes of paragraph (g)(6)(ii), means a changed or altered procedure, material or component of a control system, which replaces a procedure, material or component of a required system. Omitting a procedure or component, or reducing or diminishing the stringency or strength of a material or component of the control system is not a "modification" for purposes of paragraph (g)(6) of this section.

*Negative Initial Exposure Assessment* means a demonstration by the employer, which complies with the criteria in paragraph (f)(2)(iii) of this section, that employee exposure during an operation is expected to be consistently below the PELs.

*PACM* means "presumed asbestos containing material".

*Presumed Asbestos Containing Material* means thermal system insulation and surfacing material found in buildings constructed no later than 1980. The designation of a material as "PACM" may be rebutted pursuant to paragraph (k)(5) of this section.

*Project Designer* means a person who has successfully completed the training requirements for an abatement project designer established by 40 U.S.C. 763.90(g).

*Regulated area* means: an area established by the employer to demarcate areas where Class I, II, and III asbestos work is conducted, and any adjoining area where debris and waste from such asbestos work accumulate; and a work area within which airborne concentrations of asbestos, exceed or there is a reasonable possibility they may exceed the permissible exposure limit. Requirements for regulated areas are set out in paragraph (e) of this section.

*Removal* means all operations where ACM and/or PACM is taken out or stripped from structures or substrates, and includes demolition operations.

*Renovation* means the modifying of any existing structure, or portion thereof.

*Repair* means overhauling, rebuilding, reconstructing, or reconditioning of structures or substrates, including encapsulation or other repair of ACM or PACM attached to structures or substrates.

*Surfacing material* means material that is sprayed, troweled-on or otherwise applied to surfaces (such as acoustical plaster on ceilings and fireproofing materials on structural members, or other materials on surfaces for acoustical, fireproofing, and other purposes).

*Surfacing ACM* means surfacing material which contains more than 1% asbestos.

*Thermal system insulation (TSI)* means ACM applied to pipes, fittings, boilers, breeching, tanks, ducts or other structural components to prevent heat loss or gain.

*Thermal system insulation ACM* is thermal system insulation which contains more than 1% asbestos.

**1926.1101(c)** ***Permissible exposure limits (PELS).*** (1) *Time-weighted average limit (TWA).* The employer shall ensure that no employee is exposed to an airborne concentration of asbestos in excess of 0.1 fiber per cubic centimeter of air as an eight (8) hour time-weighted average (TWA), as determined by the method prescribed in Appendix A to this section, or by an equivalent method.

**1926.1101(c)(2)** *Excursion limit.* The employer shall ensure that no employee is exposed to an airborne concentration of asbestos in excess of 1.0 fiber per cubic centimeter of air (1 f/cc) as averaged over a sampling period of thirty (30) minutes, as determined by the method prescribed in Appendix A to this section, or by an equivalent method.

**1926.1101(d)** ***Multi-employer worksites.*** (1) On multi-employer worksites, an employer performing work requiring the establishment of a regulated area shall inform other employers on the site of the nature of the employer's work with asbestos and/or PACM, of the existence of and requirements pertaining to regulated areas, and the measures taken to ensure that employees of such other employers are not exposed to asbestos.

**1926.1101(d)(2)** Asbestos hazards at a multi-employer work site shall be abated by the contractor who created or controls the source of asbestos contamination. For example, if there is a significant breach of an enclosure containing Class I work, the employer responsible for erecting the enclosure shall repair the breach immediately.

**1926.1101(d)(3)** In addition, all employers of employees exposed to asbestos hazards shall comply with applicable protective provisions to protect their employees. For example, if employees working immediately adjacent to a Class I asbestos job are exposed to asbestos due to the inadequate containment of such job, their employer shall either remove the employees from the area until the enclosure breach is repaired; or perform an initial exposure assessment pursuant to (f) of this section.

**1926.1101(d)(4)** All employers of employees working adjacent to regulated areas established by another employer on a multi-employer work-site, shall take steps on a daily basis to ascertain the integrity of the enclosure and/or the effectiveness of the control method relied on by the primary asbestos contractor to assure that asbestos fibers do not migrate to such adjacent areas.

**1926.1101(d)(5)** All general contractors on a construction project which includes work covered by this standard shall be deemed to exercise general supervisory authority over the work covered by this standard, even though the general contractor is not qualified to serve as the asbestos "competent person" as defined by paragraph (b) of this section. As supervisor of the entire project, the general contractor shall ascertain whether the asbestos contractor is in compliance with this standard, and shall require such contractor to come into compliance with this standard when necessary.

**1926.1101(e)** ***Regulated areas.*** (1) All Class I, II and III asbestos work shall be conducted within regulated areas. All other operations covered by this standard shall be conducted within a regulated area where airborne concentrations of asbestos exceed, or there is a reasonable possibility they may exceed a PEL. Regulated areas shall comply with the requirements of paragraphs (2), (3),(4) and (5) of this section.

**1926.1101(e)(2)** *Demarcation.* The regulated area shall be demarcated in any manner that minimizes the number of persons within the area and protects persons outside the area from exposure to airborne asbestos. Where critical barriers or negative pressure enclosures are used, they may demarcate the regulated area. Signs shall be provided and displayed pursuant to the requirements of paragraph (k)(7) of this section.

**1926.1101(e)(3)** *Access.* Access to regulated areas shall be limited to authorized persons and to persons authorized by the Act or regulations issued pursuant thereto.

**1926.1101(e)(4)** *Respirators.* All persons entering a regulated area where employees are required pursuant to paragraph (h)(1) of this section to wear respirators shall be supplied with a respirator selected in accordance with paragraph (h)(2) of this section.

**1926.1101(e)(5)** *Prohibited activities.* The employer shall ensure that employees do not eat, drink, smoke, chew tobacco or gum, or apply cosmetics in the regulated area.

**1926.1101(e)(6)** *Competent Persons.* The employer shall ensure that all asbestos work performed within regulated areas is supervised by a competent person, as defined in paragraph (b) of this section. The duties of the competent person are set out in paragraph (o) of this section.

**1926.1101(f)** ***Exposure assessments and monitoring***— (1) *General monitoring criteria.* (i) Each employer who has a workplace or work operation where exposure monitoring is required under this section shall perform monitoring to determine accurately the airborne concentrations of asbestos to which employees may be exposed.

**1926.1101(f)(1)(ii)** Determinations of employee exposure shall be made from breathing zone air samples that are representative of the 8-hour TWA and 30-minute short-term exposures of each employee.

**1926.1101(f)(1)(iii)** Representative 8-hour TWA employee exposure shall be determined on the basis of one or more samples representing full-shift exposure for employees in each work area. Representative 30-minute short-term employee exposures shall be determined on the basis of one or more samples representing 30 minute exposures associated with operations that are most likely to produce exposures above the excursion limit for employees in each work area.

**1926.1101(f)(2)** *Initial Exposure Assessment.* (i) Each employer who has a workplace or work operation covered by this standard shall ensure that a "competent person" conducts an exposure assessment immediately before or at the initiation of the operation to ascertain expected exposures during that operation or workplace. The assessment must be completed in time to comply with requirements which are triggered by exposure data or the lack of a "negative exposure assessment," and to provide information necessary to assure that all control systems planned are appropriate for that operation and will work properly.

**1926.1101(f)(2)(ii)** Basis of Initial Exposure Assessment: Unless a negative exposure assessment has been made pursuant to paragraph (f)(2)(iii) of this section, the initial exposure assessment shall, if feasible, be based on monitoring conducted pursuant to paragraph (f)(1)(iii) of this section. The assessment shall take into consideration both the monitoring results and all observations, information or calculations which indicate employee exposure to asbestos, including any previous monitoring conducted in the workplace, or of the operations of the employer which indicate the levels of airborne asbestos likely to be encountered on the job. For Class I asbestos work, until the employer conducts exposure monitoring and documents that employees on that job will not be exposed in excess of the PELs, or otherwise makes a negative exposure assessment pursuant to paragraph (f)(2)(iii) of this section, the employer shall presume that employees are exposed in excess of the TWA and excursion limit.

**1926.1101(f)(2)(iii)** Negative Exposure Assessment: For any one specific asbestos job which will be performed by employees who have been trained in compliance with the standard, the employer may demonstrate that employee exposures will be below the PELs by data which conform to the following criteria;

**1926.1101(f)(2)(iii)(A)** Objective data demonstrating that the product or material containing asbestos minerals or the activity involving such product or material cannot release airborne fibers in concentrations exceeding the TWA and excursion limit under those work conditions having the greatest potential for releasing asbestos; or

**1926.1101(f)(2)(iii)(B)** Where the employer has monitored prior asbestos jobs for the PEL and the excursion limit within 12 months of the current or projected job, the monitoring and analysis were performed in compliance with the asbestos standard in effect; and the data were obtained during work operations conducted under workplace conditions "closely resembling" the processes, type of material, control methods, work practices, and environmental conditions used and prevailing in the employer's current operations, the operations were conducted by employees whose training and experience are no more extensive than that of employees performing the current job, and these data show that under the conditions prevailing and which will prevail in the current workplace there is a high degree of certainty that employee exposures will not exceed the TWA and excursion limit; or

**1926.1101(f)(2)(iii)(C)** The results of initial exposure monitoring of the current job made from breathing zone air samples that are representative of the 8-hour TWA and 30-minute short-term exposures of each employee covering operations which are most likely during the performance of the entire asbestos job to result in exposures over the PELs.

**1926.1101(f)(3)** *Periodic monitoring*—(i) *Class I and II operations.* The employer shall conduct daily monitoring that is representative of the exposure of each employee who is assigned to work within a regulated area who is performing Class I or II work, unless the employer pursuant to (f)(2)(iii) of this section, has made a negative exposure assessment for the entire operation.

**1926.1101(f)(3)(ii)** All operations under the standard other than Class I and II operations. The employer shall conduct periodic monitoring of all work where exposures are expected to exceed a PEL, at intervals sufficient to document the validity of the exposure prediction.

**1926.1101(f)(3)(iii)** Exception: When all employees required to be monitored daily are equipped with supplied-air respirators operated in the pressure demand mode, or other positive pressure mode respirator, the employer may dispense with the daily monitoring required by this paragraph. However, employees performing Class I work using a control method which is not listed in paragraph (g)(4) (i), (ii), or (iii) of this section or using a modification of a listed control method, shall continue to be monitored daily even if they are equipped with supplied-air respirators.

**1926.1101(f)(4)** *Termination of monitoring.* (i) If the periodic monitoring required by paragraph (f)(3) of this section reveals that employee exposures, as indicated by statistically reliable measurements, are below the permissible exposure limit and excursion limit the employer may discontinue monitoring for those employees whose exposures are represented by such monitoring.

**1926.1101(f)(4)(ii)** Additional monitoring. Notwithstanding the provisions of paragraph (f) (2) and (3), and (f)(4) of this section, the employer shall institute the exposure monitoring required under paragraph (f)(3) of this section whenever there has been a change in process, control equipment, personnel or work practices that may result in new or additional exposures above the permissible exposure limit and/or excursion limit or when the employer has any reason to suspect that a change may result in new or additional exposures above the permissible exposure limit and/or excursion limit. Such additional monitoring is required regardless of whether a "negative exposure assessment" was previously produced for a specific job.

**1926.1101(f)(5)** *Employee notification of monitoring results.* (i) The employer shall notify affected employees of the monitoring results that represent that employee's exposure as soon as possible following receipt of monitoring results.

**1926.1101(f)(5)(ii)** The employer shall notify affected employees of the results of monitoring representing the employee's exposure in writing either individually or by posting at a centrally located place that is accessible to affected employees.

**1926.1101(f)(6)** *Observation of monitoring.* (i) The employer shall provide affected employees and their designated representatives an opportunity to observe any monitoring of employee exposure to asbestos conducted in accordance with this section.

**1926.1101(f)(6)(ii)** When observation of the monitoring of employee exposure to asbestos requires entry into an area where the use of protective clothing or equipment is required, the observer shall be provided with and be required to use such clothing and equipment and shall comply with all other applicable safety and health procedures.

**1926.1101(g)** ***Methods of compliance.***(1) Engineering controls and work practices for all operations covered by this section. The employer shall use the following engineering controls and work practices in all operations covered by this section, regardless of the levels of exposure:

**1926.1101(g)(1)(i)** Vacuum cleaners equipped with HEPA filters to collect all debris and dust containing ACM and PACM, except as provided in paragraph (g)(8)(ii) of this section in the case of roofing material.

**1926.1101(g)(1)(ii)** Wet methods, or wetting agents, to control employee exposures during asbestos handling, mixing, removal, cutting, application, and cleanup, except where employers demonstrate that the use of wet methods is infeasible due to for example, the creation of electrical hazards, equipment malfunction, and, in roofing, except as provided in paragraph (g)(8)(ii) of this section; and

**1926.1101(g)(1)(iii)** Prompt clean-up and disposal of wastes and debris contaminated with asbestos in leak-tight containers except in roofing operations, where the procedures specified in paragraph (g)(8)(ii) of this section apply.

**1926.1101(g)(2)** In addition to the requirements of paragraph (g)(1) of this section, the employer shall use the following control methods to achieve compliance with the TWA permissible exposure limit and excursion limit prescribed by paragraph (c) of this section;

**1926.1101(g)(2)(i)** Local exhaust ventilation equipped with HEPA filter dust collection systems;

**1926.1101(g)(2)(ii)** Enclosure or isolation of processes producing asbestos dust;

**1926.1101(g)(2)(iii)** Ventilation of the regulated area to move contaminated air away from the breathing zone of employees and toward a filtration or collection device equipped with a HEPA filter;

**1926.1101(g)(2)(iv)** Use of other work practices and engineering controls that the Assistant Secretary can show to be feasible.

**1926.1101(g)(2)(v)** Wherever the feasible engineering and work practice controls described above are not sufficient to reduce employee exposure to or below the permissible exposure limit and/or excursion limit prescribed in paragraph (c) of this section, the employer shall use them to reduce employee exposure to the lowest levels attainable by these controls and shall supplement them by the use of respiratory protection that complies with the requirements of paragraph (h) of this section.

**1926.1101(g)(3)** *Prohibitions.* The following work practices and engineering controls shall not be used for work related to asbestos or for work which disturbs ACM or PACM, regardless of measured levels of asbestos exposure or the results of initial exposure assessments:

**1926.1101(g)(3)(i)** High-speed abrasive disc saws that are not equipped with point of cut ventilator or enclosures with HEPA filtered exhaust air.

**1926.1101(g)(3)(ii)** Compressed air used to remove asbestos, or materials containing asbestos, unless the compressed air is used in conjunction with an enclosed ventilation system designed to capture the dust cloud created by the compressed air.

**1926.1101(g)(3)(iii)** Dry sweeping, shoveling or other dry clean-up of dust and debris containing ACM and PACM.

**1926.1101(g)(3)(iv)** Employee rotation as a means of reducing employee exposure to asbestos.

**1926.1101(g)(4)** *Class I Requirements.* In addition to the provisions of paragraphs (g) (1) and (2) of this section, the following engineering controls and work practices and procedures shall be used.

**1926.1101(g)(4)(i)** All Class I work, including the installation and operation of the control system shall be supervised by a competent person as defined in paragraph (b) of this section;

**1926.1101(g)(4)(ii)** For all Class I jobs involving the removal of more than 25 linear or 10 square feet of thermal system insulation or surfacing material; for all other Class I jobs, where the employer cannot produce a negative exposure assessment pursuant to paragraph (f)(2)(iii) of this section, or where employees are working in areas adjacent to the regulated area, while the Class I work is being performed, the employer shall use one of the following methods to ensure that airborne asbestos does not migrate from the regulated area:

**1926.1101(g)(4)(ii)(A)** Critical barriers shall be placed over all the openings to the regulated area, except where activities are performed outdoors; or

**1926.1101(g)(4)(ii)(B)** The employer shall use another barrier or isolation method which prevents the migration of airborne asbestos from the regulated area, as verified by perimeter area surveillance during each work shift at each boundary of the regulated area, showing no visible asbestos dust; and perimeter area monitoring showing that clearance levels contained in 40 CFR part 763, subpt. E, of the EPA Asbestos in Schools Rule are met, or that perimeter area levels, measured by Phase Contrast Microscopy (PCM) are no more than background levels representing the same area before the asbestos work began. The results of such monitoring shall be made known to the employer no later than 24 hours from the end of the work shift represented by such monitoring. Exception: For work completed outdoors where employees are not working in areas adjacent to the regulated areas, this paragraph (g)(4)(ii) is satisfied when the specific control methods in paragraph (g)(5) of this section are used.

**1926.1101(g)(4)(iii)** For all Class I jobs, HVAC systems shall be isolated in the regulated area by sealing with a double layer of 6 mil plastic or the equivalent;

**1926.1101(g)(4)(iv)** For all Class I jobs, impermeable dropcloths shall be placed on surfaces beneath all removal activity;

**1926.1101(g)(4)(v)** For all Class I jobs, all objects within the regulated area shall be covered with impermeable dropcloths or plastic sheeting which is secured by duct tape or an equivalent.

**1926.1101(g)(4)(vi)** For all Class I jobs where the employer cannot produce a negative exposure assessment, or where exposure monitoring shows that a PEL is exceeded, the employer shall ventilate the regulated area to move contaminated air away from the breathing zone of employees toward a HEPA filtration or collection device.

**1926.1101(g)(5)** *Specific control methods for Class I work.* In addition, Class I asbestos work shall be performed using one or more of the following control methods pursuant to the limitations stated below:

**1926.1101(g)(5)(i)** Negative Pressure Enclosure (NPE) systems: NPE systems may be used where the configuration of the work area does not make the erection of the enclosure infeasible, with the following specifications and work practices.

**1926.1101(g)(5)(i)(A)** *Specifications:*

**1926.1101(g)(5)(i)(A)*(1)*** The negative pressure enclosure (NPE) may be of any configuration,

**1926.1101(g)(5)(i)(A)*(2)*** At least 4 air changes per hour shall be maintained in the NPE,

**1926.1101(g)(5)(i)(A)*(3)*** A minimum of -0.02 column inches of water pressure differential, relative to outside pressure, shall be maintained within the NPE as evidenced by manometric measurements,

**1926.1101(g)(5)(i)(A)*(4)*** The NPE shall be kept under negative pressure throughout the period of its use, and

**1926.1101(g)(5)(i)(A)*(5)*** Air movement shall be directed away from employees performing asbestos work within the enclosure, and toward a HEPA filtration or a collection device.

**1926.1101(g)(5)(i)(B)** *Work Practices:*

**1926.1101(g)(5)(i)(B)*(1)*** Before beginning work within the enclosure and at the beginning of each shift, the NPE shall be inspected for breaches and smoke-tested for leaks, and any leaks sealed.

**1926.1101(g)(5)(i)(B)*(2)*** Electrical circuits in the enclosure shall be deactivated, unless equipped with ground-fault circuit interrupters.

**1926.1101(g)(5)(ii)** Glove bag systems may be used to remove PACM and/or ACM from straight runs of piping and elbows and other connections with the following specifications and work practices:

**1926.1101(g)(5)(ii)(A)** *Specifications:*

**1926.1101(g)(5)(ii)(A)*(1)*** Glovebags shall be made of 6 mil thick plastic and shall be seamless at the bottom.

**1926.1101(g)(5)(ii)(A)*(2)*** Glovebags used on elbows and other connections must be designed for that purpose and used without modifications.

**1926.1101(g)(5)(ii)(B)** *Work Practices:*

**1926.1101(g)(5)(ii)(B)*(1)*** Each glovebag shall be installed so that it completely covers the circumference of pipe or other structure where the work is to be done.

**1926.1101(g)(5)(ii)(B)*(2)*** Glovebags shall be smoke-tested for leaks and any leaks sealed prior to use.

**1926.1101(g)(5)(ii)(B)*(3)*** Glovebags may be used only once and may not be moved.

**1926.1101(g)(5)(ii)(B)*(4)*** Glovebags shall not be used on surfaces whose temperature exceeds 150 °F.

**1926.1101(g)(5)(ii)(B)*(5)*** Prior to disposal, glovebags shall be collapsed by removing air within them using a HEPA vacuum.

**1926.1101(g)(5)(ii)(B)*(6)*** Before beginning the operation, loose and friable material adjacent to the glovebag/box operation shall be wrapped and sealed in two layers of six mil plastic or otherwise rendered intact,

**1926.1101(g)(5)(ii)(B)*(7)*** Where system uses attached waste bag, such bag shall be connected to collection bag using hose or other material which shall withstand pressure of ACM waste and water without losing its integrity:

**1926.1101(g)(5)(ii)(B)*(8)*** Sliding valve or other device shall separate waste bag from hose to ensure no exposure when waste bag is disconnected:

**1926.1101(g)(5)(ii)(B)*(9)*** At least two persons shall perform Class I glovebag removal operations.

**1926.1101(g)(5)(iii)** *Negative Pressure Glove Bag Systems.* Negative pressure glove bag systems may be used to remove ACM or PACM from piping.

**1926.1101(g)(5)(iii)(A)** *Specifications:* In addition to specifications for glove bag systems above, negative pressure glove bag systems shall attach HEPA vacuum systems or other devices to bag to prevent collapse during removal.

**1926.1101(g)(5)(iii)(B)** *Work Practices:* (*1*) The employer shall comply with the work practices for glove bag systems in paragraph (g)(5)(ii)(B)(*4*) of this section.

**1926.1101(g)(5)(iii)(B)*(2)*** The HEPA vacuum cleaner or other device used to prevent collapse of bag during removal shall run continually during the operation until it is completed at which time the bag shall be collapsed prior to removal of the bag from the pipe.

**1926.1101(g)(5)(iii)(B)*(3)*** Where a separate waste bag is used along with a collection bag and discarded after one use, the collection bag may be reused if rinsed clean with amended water before reuse.

**1926.1101(g)(5)(iv)** Negative Pressure Glove Box Systems: Negative pressure glove boxes may be used to remove ACM or PACM from pipe runs with the following specifications and work practices.

**1926.1101(g)(5)(iv)(A)** *Specifications:*

**1926.1101(g)(5)(iv)(A)*(1)*** Glove boxes shall be constructed with rigid sides and made from metal or other material which can withstand the weight of the ACM and PACM and water used during removal:

**1926.1101(g)(5)(iv)(A)*(2)*** A negative pressure generator shall be used to create negative pressure in the system:

**1926.1101(g)(5)(iv)(A)*(3)*** An air filtration unit shall be attached to the box:

**1926.1101(g)(5)(iv)(A)*(4)*** The box shall be fitted with gloved apertures:

**1926.1101(g)(5)(iv)(A)*(5)*** An aperture at the base of the box shall serve as a bagging outlet for waste ACM and water:

**1926.1101(g)(5)(iv)(A)*(6)*** A back-up generator shall be present on site:

**1926.1101(g)(5)(iv)(A)*(7)*** Waste bags shall consist of 6 mil thick plastic double-bagged before they are filled or plastic thicker than 6 mil.

**1926.1101(g)(5)(iv)(B)** *Work practices:*

**1926.1101(g)(5)(iv)(B)*(1)*** At least two persons shall perform the removal:

**1926.1101(g)(5)(iv)(B)*(2)*** The box shall be smoke-tested for leaks and any leaks sealed prior to each use.

**1926.1101(g)(5)(iv)(B)*(3)*** Loose or damaged ACM adjacent to the box shall be wrapped and sealed in two layers of 6 mil plastic prior to the job, or otherwise made intact prior to the job.

**1926.1101(g)(5)(iv)(B)*(4)*** A HEPA filtration system shall be used to maintain pressure barrier in box.

**1926.1101(g)(5)(v)** *Water Spray Process System.* A water spray process system may be used for removal of ACM and PACM from cold line piping if, employees carrying out such process have completed a 40-hour separate training course in its use, in addition to training required for employees performing Class I work. The system shall meet the following specifications and shall be performed by employees using the following work practices.

**1926.1101(g)(5)(v)(A)** *Specifications:*

**1926.1101(g)(5)(v)(A)*(1)*** Piping shall be surrounded on 3 sides by rigid framing,

**1926.1101(g)(5)(v)(A)*(2)*** A 360 degree water spray, delivered through nozzles supplied by a high pressure separate water line, shall be formed around the piping.

**1926.1101(g)(5)(v)(A)*(3)*** The spray shall collide to form a fine aerosol which provides a liquid barrier between workers and the ACM and PACM.

**1926.1101(g)(5)(v)(B)** *Work Practices:*

**1926.1101(g)(5)(v)(B)*(1)*** The system shall be run for at least 10 minutes before removal begins.

**1926.1101(g)(5)(v)(B)*(2)*** All removal shall take place within the water barrier.

**1926.1101(g)(5)(v)(B)*(3)*** The system shall be operated by at least three persons, one of whom shall not perform removal, but shall check equipment, and ensure proper operation of the system.

**1926.1101(g)(5)(v)(B)*(4)*** After removal, the ACM and PACM shall be bagged while still inside the water barrier.

**1926.1101(g)(5)(vi)** A small walk-in enclosure which accommodates no more than two persons (mini-enclosure) may be used if the disturbance or removal can be completely contained by the enclosure with the following specifications and work practices.

**1926.1101(g)(5)(vi)(A)** *Specifications:*

**1926.1101(g)(5)(vi)(A)*(1)*** The fabricated or job-made enclosure shall be constructed of 6 mil plastic or equivalent:

**1926.1101(g)(5)(vi)(A)*(2)*** The enclosure shall be placed under negative pressure by means of a HEPA filtered vacuum or similar ventilation unit:

**1926.1101(g)(5)(vi)(B)** *Work practices:*

**1926.1101(g)(5)(vi)(B)*(1)*** Before use, the mini-enclosure shall be inspected for leaks and smoke-tested to detect breaches, and any breaches sealed.

**1926.1101(g)(5)(vi)(B)*(2)*** Before reuse, the interior shall be completely washed with amended water and HEPA-vacuumed..

**1926.1101(g)(5)(vi)(B)*(3)*** During use, air movement shall be directed away from the employee's breathing zone within the mini-enclosure.

**1926.1101(g)(6)** *Alternative control methods for Class I work.* Class I work may be performed using a control method which is not referenced in paragraph (g)(5) of this section, or which modifies a control method referenced in paragraph (g)(5)of this section, if the following provisions are complied with:

**1926.1101(g)(6)(i)** The control method shall enclose, contain or isolate the processes or source of airborne asbestos dust, or otherwise capture or redirect such dust before it enters the breathing zone of employees.

**1926.1101(g)(6)(ii)** A certified industrial hygienist or licensed professional engineer who is also qualified as a project designer as defined in paragraph (b) of this section, shall evaluate the work area, the projected work practices and the engineering controls and shall certify in writing that the planned control method is adequate to reduce direct and indirect employee exposure to below the PELs under worst-case conditions of use, and that the planned control method will prevent asbestos contamination outside the regulated area, as measured by clearance sampling which meets the requirements of EPA's Asbestos in Schools rule issued under AHERA, or perimeter monitoring which meets the criteria in paragraph (g)(4)(ii)(B) of this section.

**1926.1101(g)(6)(ii)(A)** Where the TSI or surfacing material to be removed is 25 linear or 10 square feet or less, the evaluation required in paragraph (g)(6) of this section may be performed by a "competent person", and may omit consideration of perimeter or clearance monitoring otherwise required.

**1926.1101(g)(6)(ii)(B)** The evaluation of employee exposure required in paragraph (g)(6) of this section, shall include and be based on sampling and analytical data representing employee exposure during the use of such method under worst-case conditions and by employees whose training and experience are equivalent to employees who are to perform the current job.

**1926.1101(g)(6)(iii)** Before work which involves the removal of more than 25 linear or 10 square feet of thermal system insulation or surfacing material is begun using an alternative method which has been the subject of a paragraph (g)(6) of this section required evaluation and certification, the employer shall send a copy of such evaluation and certification to the national office of OSHA, Office of Technical Support, Room N3653, 200 Constitution Avenue, NW, Washington, DC 20210. The submission shall not constitute approval by OSHA.

**1926.1101(g)(7)** Work Practices and Engineering Controls for Class II work.

**1926.1101(g)(7)(i)** All Class II work shall be supervised by a competent person as defined in paragraph (b) of this section.

**1926.1101(g)(7)(ii)** For all indoor Class II jobs, where the employer has not produced a negative exposure assessment pursuant to paragraph (f)(2)(iii) of this section, or where during the job, changed conditions indicate there may be exposure above the PEL or where the employer does not remove the ACM in a substantially intact state, the employer shall use one of the following methods to ensure that airborne asbestos does not migrate from the regulated area;

**1926.1101(g)(7)(ii)(A)** Critical barriers shall be placed over all openings to the regulated area; or,

**1926.1101(g)(7)(ii)(B)** The employer shall use another barrier or isolation method which prevents the migration of airborne asbestos from the regulated area, as verified by perimeter area monitoring or clearance monitoring which meets the criteria set out in paragraph (g)(4)(ii)(B) of this section.

**1926.1101(g)(7)(ii)(C)** Impermeable dropcloths shall be placed on surfaces beneath all removal activity;

**1926.1101(g)(7)(iii)** [Reserved]

**1926.1101(g)(7)(iv)** All Class II asbestos work shall be performed using the work practices and requirements set out above in paragraph (g)(1) (i) through (g)(1)(iii) of this section.

**1926.1101(g)(8)** *Additional Controls for Class II work.* Class II asbestos work shall also be performed by complying with the work practices and controls designated for each type of asbestos work to be performed, set out in this paragraph. Where more than one control method may be used for a type of asbestos work, the employer may choose one or a combination of designated control methods. Class II work also may be performed using a method allowed for Class I work, except that glove bags and glove boxes are allowed if they fully enclose the Class II material to be removed.

**1926.1101(g)(8)(i)** For removing vinyl and asphalt flooring materials which contain ACM or for which in buildings constructed no later than 1980, the employer has not verified the absence of ACM pursuant to paragraph (g)(8)(i)(I) of this section. The employer shall ensure that employees comply with the following work practices and that employees are trained in these practices pursuant to paragraph (k)(9):

**1926.1101(g)(8)(i)(A)** Flooring or its backing shall not be sanded.

**1926.1101(g)(8)(i)(B)** Vacuums equipped with HEPA filter, disposable dust bag, and metal floor tool (no brush) shall be used to clean floors.

**1926.1101(g)(8)(i)(C)** Resilient sheeting shall be removed by cutting with wetting of the snip point and wetting during delamination. Rip-up of resilient sheet floor material is prohibited.

**1926.1101(g)(8)(i)(D)** All scraping of residual adhesive and/or backing shall be performed using wet methods.

**1926.1101(g)(8)(i)(E)** Dry sweeping is prohibited.

**1926.1101(g)(8)(i)(F)** Mechanical chipping is prohibited unless performed in a negative pressure enclosure which meets the requirements of paragraph (g)(5)(i) of this section.

**1926.1101(g)(8)(i)(G)** Tiles shall be removed intact, unless the employer demonstrates that intact removal is not possible.

**1926.1101(g)(8)(i)(H)** When tiles are heated and can be removed intact, wetting may be omitted.

**1926.1101(g)(8)(i)(I)** Resilient flooring material including associated mastic and backing shall be assumed to be asbestos-containing unless an industrial hygienist determines that it is asbestos-free using recognized analytical techniques.

**1926.1101(g)(8)(ii)** For removing roofing material which contains ACM the employer shall ensure that the following work practices are followed:

**1926.1101(g)(8)(ii)(A)** Roofing material shall be removed in an intact state to the extent feasible.

**1926.1101(g)(8)(ii)(B)** Wet methods shall be used to remove roofing materials that are not intact, or that will be rendered not intact during removal, unless such wet methods are not feasible or will create safety hazards.

**1926.1101(g)(8)(ii)(C)** Cutting machines shall be continuously misted during use, unless a competent person determines that misting substantially decreases worker safety.

**1926.1101(g)(8)(ii)(D)** When removing built-up roofs with asbestos-containing roofing felts and an aggregate surface using a power roof cutter, all dust resulting from the cutting operation shall be collected by a HEPA dust collector, or shall be HEPA vacuumed by vacuuming along the cut line. When removing built-up roofs with asbestos-containing roofing felts and a smooth surface using a power roof cutter, the dust resulting from the cutting operation shall be collected either by a HEPA dust collector or HEPA vacuuming along the cut line, or by gently sweeping and then carefully and completely wiping up the still-wet dust and debris left along the cut line. The dust and debris shall be immediately bagged or placed in covered containers.

**1926.1101(g)(8)(ii)(E)** Asbestos-containing material that has been removed from a roof shall not be dropped or thrown to the ground. Unless the material is carried or passed to the ground by hand, it shall be lowered to the ground via covered, dust-tight chute, crane or hoist:

**1926.1101(g)(8)(ii)(E)*(1)*** Any ACM that is not intact shall be lowered to the ground as soon as is practicable, but in any event no later than the end of the work shift. While the material remains on the roof it shall either be kept wet, placed in an impermeable waste bag, or wrapped in plastic sheeting.

**1926.1101(g)(8)(ii)(E)*(2)*** Intact ACM shall be lowered to the ground as soon as is practicable, but in any event no later than the end of the work shift.

**1926.1101(g)(8)(ii)(F)** Upon being lowered, unwrapped material shall be transferred to a closed receptacle in such manner so as to preclude the dispersion of dust.

**1926.1101(g)(8)(ii)(G)** Roof level heating and ventilation air intake sources shall be isolated or the ventilation system shall be shut down.

**1926.1101(g)(8)(ii)(H)** Notwithstanding any other provision of this section, removal or repair of sections of intact roofing less than 25 square feet in area does not require use of wet methods or HEPA vacuuming as long as manual methods which do not render the material non-intact are used to remove the material and no visible dust is created by the removal method used. In determining whether a job involves less than 25 square feet, the employer shall include all removal and repair work performed on the same roof on the same day.

**1926.1101(g)(8)(iii)** When removing cementitious asbestos-containing siding and shingles or transite panels containing ACM on building exteriors (other than roofs, where paragraph (g)(8)(ii) of this section applies) the employer shall ensure that the following work practices are followed:

**1926.1101(g)(8)(iii)(A)** Cutting, abrading or breaking siding, shingles, or transite panels, shall be prohibited unless the employer can demonstrate that methods less likely to result in asbestos fiber release cannot be used.

**1926.1101(g)(8)(iii)(B)** Each panel or shingle shall be sprayed with amended water prior to removal.

**1926.1101(g)(8)(iii)(C)** Unwrapped or unbagged panels or shingles shall be immediately lowered to the ground via covered dust-tight chute, crane or hoist, or placed in an impervious waste bag or wrapped in plastic sheeting and lowered to the ground no later than the end of the work shift.

**1926.1101(g)(8)(iii)(D)** Nails shall be cut with flat, sharp instruments.

**1926.1101(g)(8)(iv)** When removing gaskets containing ACM, the employer shall ensure that the following work practices are followed:

**1926.1101(g)(8)(iv)(A)** If a gasket is visibly deteriorated and unlikely to be removed intact, removal shall be undertaken within a glovebag as described in paragraph (g)(5)(ii) of this section.

**1926.1101(g)(8)(iv)(B)** [Reserved]

**1926.1101(g)(8)(iv)(C)** The gasket shall be immediately placed in a disposal container.

**1926.1101(g)(8)(iv)(D)** Any scraping to remove residue must be performed wet.

**1926.1101(g)(8)(v)** When performing any other Class II removal of asbestos containing material for which specific controls have not been listed in paragraph (g)(8)(iv) (A) through (D) of this section, the employer shall ensure that the following work practices are complied with.

**1926.1101(g)(8)(v)(A)** The material shall be thoroughly wetted with amended water prior to and during its removal.

**1926.1101(g)(8)(v)(B)** The material shall be removed in an intact state unless the employer demonstrates that intact removal is not possible.

**1926.1101(g)(8)(v)(C)** Cutting, abrading or breaking the material shall be prohibited unless the employer can demonstrate that methods less likely to result in asbestos fiber release are not feasible.

**1926.1101(g)(8)(v)(D)** Asbestos-containing material removed, shall be immediately bagged or wrapped, or kept wetted until transferred to a closed receptacle, no later than the end of the work shift.

**1926.1101(g)(8)(vi)** *Alternative Work Practices and Controls.* Instead of the work practices and controls listed in paragraph (g)(8) (i) through (v) of this section, the employer may use different or modified engineering and work practice controls if the following provisions are complied with.

**1926.1101(g)(8)(vi)(A)** The employer shall demonstrate by data representing employee exposure during the use of such method under conditions which closely resemble the conditions under which the method is to be used, that employee exposure will not exceed the PELs under any anticipated circumstances.

**1926.1101(g)(8)(vi)(B)** A competent person shall evaluate the work area, the projected work practices and the engineering controls, and shall certify in writing, that the different or modified controls are adequate to reduce direct and indirect employee exposure to below the PELs under all expected conditions of use and that the method meets the requirements of this standard. The evaluation shall include and be based on data representing employee exposure during the use of such method under conditions which closely resemble the conditions under which the method is to be used for the current job, and by employees whose training and experience are equivalent to employees who are to perform the current job.

**1926.1101(g)(9)** *Work Practices and Engineering Controls for Class III asbestos work.* Class III asbestos work shall be conducted using engineering and work practice controls which minimize the exposure to employees performing the asbestos work and to bystander employees.

**1926.1101(g)(9)(i)** The work shall be performed using wet methods.

**1926.1101(g)(9)(ii)** To the extent feasible, the work shall be performed using local exhaust ventilation.

**1926.1101(g)(9)(iii)** Where the disturbance involves drilling, cutting, abrading, sanding, chipping, breaking, or sawing of thermal system insulation or surfacing material, the employer shall use impermeable dropcloths, and shall isolate the operation using mini-enclosures or glove bag systems pursuant to paragraph (g)(5) of this section or another isolation method.

**1926.1101(g)(9)(iv)** Where the employer does not produce a "negative exposure assessment" for a job, or where monitoring results show the PEL has been exceeded, the employer shall contain the area using impermeable dropcloths and plastic barriers or their equivalent, or shall isolate the operation using a control system listed in and in compliance with paragraph (g)(5) of this section.

**1926.1101(g)(9)(v)** Employees performing Class III jobs, which involve the disturbance of thermal system insulation or surfacing material, or where the employer does not produce a "negative exposure assessment" or where monitoring results show a PEL has been exceeded, shall wear respirators which are selected, used and fitted pursuant to provisions of paragraph (h) of this section.

**1926.1101(g)(10)** *Class IV asbestos work.* Class IV asbestos jobs shall be conducted by employees trained pursuant to the asbestos awareness training program set out in paragraph (k)(9) of this section. In addition, all Class IV jobs shall be conducted in conformity with the requirements set out in paragraph (g)(1) of this section, mandating wet methods, HEPA vacuums, and prompt clean up of debris containing ACM or PACM.

**1926.1101(g)(10)(i)** Employees cleaning up debris and waste in a regulated area where respirators are required shall wear respirators which are selected, used and fitted pursuant to provisions of paragraph (h) of this section.

**1926.1101(g)(10)(ii)** Employers of employees who clean up waste and debris in, and employers in control of, areas where friable thermal system insulation or surfacing material is accessible, shall assume that such waste and debris contain asbestos.

**1926.1101(g)(11)** *Alternative methods of compliance for installation, removal, repair, and maintenance of certain roofing and pipeline coating materials.* Notwithstanding any other provision of this section, an employer who complies with all provisions of this paragraph (g)(11) when installing, removing, repairing, or maintaining intact pipeline asphaltic wrap, or roof flashings which contain asbestos fibers encapsulated or coated by bituminous or resinous compounds shall be deemed to be in compliance with this section. If an employer does not comply with all provisions of this paragraph (g)(11) or if during the course of the job the material does not remain intact, the provisions of paragraph (g)(8) of this section apply instead of this paragraph (g)(11).

**1926.1101(g)(11)(i)** Before work begins and as needed during the job, a competent person who is capable of identifying asbestos hazards in the workplace and selecting the appropriate control strategy for asbestos exposure, and who has the authority to take prompt corrective measures to eliminate such hazards, shall conduct an inspection of the worksite and determine that the roofing material is intact and will likely remain intact.

**1926.1101(g)(11)(ii)** All employees performing work covered by this paragraph (g)(11) shall be trained in a training program that meets the requirements of paragraph (k)(9)(viii) of this section.

**1926.1101(g)(11)(iii)** The material shall not be sanded, abraded, or ground. Manual methods which do not render the material non-intact shall be used.

**1926.1101(g)(11)(iv)** Material that has been removed from a roof shall not be dropped or thrown to the ground. Unless the material is carried or passed to the ground by hand, it shall be lowered to the ground via covered, dust-tight chute, crane or hoist. All such material shall be removed from the roof as soon as is practicable, but in any event no later than the end of the work shift.

**1926.1101(g)(11)(v)** Where roofing products which have been labeled as containing asbestos pursuant to paragraph (k)(8) of this section are installed on non-residential roofs during operations covered by this paragraph (g)(11), the employer shall notify the building owner of the presence and location of such materials no later than the end of the job.

**1926.1101(g)(11)(vi)** All removal or disturbance of pipeline asphaltic wrap shall be performed using wet methods.

**1926.1101(h)** ***Respiratory protection.*** (1) *General.* For employees who use respirators required by this section, the employer must provide respirators that comply with the requirements of this paragraph. Respirators must be used during:

**1926.1101(h)(1)(i)** Class I asbestos work.

**1926.1101(h)(1)(ii)** Class II asbestos work when ACM is not removed in a substantially intact state.

**1926.1101(h)(1)(iii)** Class II and III asbestos work that is not performed using wet methods, except for removal of ACM from sloped roofs when a negative-exposure assessment has been conducted and ACM is removed in an intact state.

**1926.1101(h)(1)(iv)** Class II and III asbestos work for which a negative-exposure assessment has not been conducted.

**1926.1101(h)(1)(v)** Class III asbestos work when TSI or surfacing ACM or PACM is being disturbed.

**1926.1101(h)(1)(vi)** Class IV asbestos work performed within regulated areas where employees who are performing other work are required to use respirators.

**1926.1101(h)(1)(vii)** Work operations covered by this section for which employees are exposed above the TWA or excursion limit.

**1926.1101(h)(1)(viii)** Emergencies.

**1926.1101(h)(2)** *Respirator program.* (i) The employer must implement a respiratory protection program in accordance with 29 CFR 1910.134(b) through (d) (except (d)(1)(iii)), and (f) through (m).

**1926.1101(h)(2)(ii)** No employee shall be assigned to asbestos work that requires respirator use if, based on their most recent medical examination, the examining physician determines that the employee will be unable to function normally while using a respirator, or that the safety or health of the employee or other employees will be impaired by the employee's respirator use. Such employees must be assigned to another job or given the opportunity to transfer to a different position that they can perform. If such a transfer position is available, it must be with the same employer, in the same geographical area, and with the same seniority, status, rate of pay, and other job benefits the employee had just prior to such transfer.

**1926.1101(h)(3)** *Respirator selection.* (i) The employer must select the appropriate respirator from Table 1 of this section.

## TABLE 1—RESPIRATORY PROTECTION FOR ASBESTOS FIBERS

| Airborne concentrations of asbestos or conditions of use | Required respirator |
|---|---|
| Not in excess of 1 f/cc (10 × PEL), or otherwise as required independent of exposure pursuant to paragraph (h)(2)(iv) of this section. | Half-mask air purifying respirator other than a disposable respirator, equipped with high efficiency filters. |
| Not in excess of 5 f/cc (50 × PEL) | Full facepiece air-purifying respirator equipped with high efficiency filters. |
| Not in excess of 10 f/cc (100 × PEL) | Any powered air-purifying respirator equipped with high efficiency filters or any supplied air respirator operated in continuous flow mode. |
| Not in excess of 100 f/cc (1,000 × PEL) or unknown concentration. | Full facepiece supplied air respirator operated in pressure demand mode. |
| Greater than 100 f/cc (1,000 × PEL) or unknown concentration. | Full facepiece supplied air respirator operated in pressure demand mode, equipped with an auxiliary positive pressure self-contained breathing apparatus. |

Note:

a. Respirators assigned for high environmental concentrations may be used at lower concentrations, or when required respirator use is independent of concentration.

b. A high efficiency filter means a filter that is at least 99.97 percent efficient against mono-dispersed particles of 0.3 micrometers in diameter or larger.

**1926.1101(h)(3)(ii)** The employer must provide an employee with a tight-fitting, powered air-purifying respirator instead of a negative-pressure respirator from Table 1 when the employee chooses to use this type of respirator and such a respirator will provide adequate protection to the employee.

**1926.1101(h)(3)(iii)** The employer must provide a half-mask air-purifying respirator, other than a disposable respirator, that is equipped with high-efficiency filters when the employee performs:

**1926.1101(h)(3)(iii)(A)** Class II and III asbestos work and a negative-exposure assessment has not been conducted by the employer.

**1926.1101(h)(3)(iii)(B)** Class III asbestos work when TSI or surfacing ACM or PACM is being disturbed.

**1926.1101(h)(3)(iv)** In addition to the above selection criteria, when employees are in a regulated area where Class I work is being performed, a negative exposure assessment of the area has not been produced, and the exposure assessment of the area indicates the exposure level will not exceed 1 f/cc as an 8-hour time weighted average, employers must provide the employees with one of the following respirators:

**1926.1101(h)(3)(iv)(A)** A tight-fitting powered air-purifying respirator equipped with high efficiency filters;

**1926.1101(h)(3)(iv)(B)** A full facepiece supplied-air respirator operated in the pressure-demand mode equipped with HEPA egress cartridges; or

**1926.1101(h)(3)(iv)(C)** A full facepiece supplied-air respirator operated in the pressure-demand mode equipped with an auxiliary positive pressure self-contained breathing apparatus. A full facepiece supplied-air respirator operated in the pressure-demand mode equipped with an auxiliary positive pressure self-contained breathing apparatus must be provided under such conditions when the exposure assessment indicates exposure levels above 1 f/cc as an 8-hour time weighted average.

**1926.1101(i)** ***Protective clothing***— (1) *General.* The employer shall provide or require the use of protective clothing, such as coveralls or similar whole-body clothing, head coverings, gloves, and foot coverings for any employee exposed to airborne concentrations of asbestos that exceed the TWA and/or excursion limit prescribed in paragraph (c) of this section, or for which a required negative exposure assessment is not produced, or for any employee performing Class I operations which involve the removal of over 25 linear or 10 square feet of TSI or surfacing ACM and PACM.

**1926.1101(i)(2)** *Laundering.* (i) The employer shall ensure that laundering of contaminated clothing is done so as to prevent the release of airborne asbestos in excess of the TWA or excursion limit prescribed in paragraph (c) of this section.

**1926.1101(i)(2)(ii)** Any employer who gives contaminated clothing to another person for laundering shall inform such person of the requirement in paragraph (i)(2)(i) of this section to effectively prevent the release of airborne asbestos in excess of the TWA and excursion limit prescribed in paragraph (c) of this section.

**1926.1101(i)(3)** *Contaminated clothing.* Contaminated clothing shall be transported in sealed impermeable bags, or other closed, impermeable containers, and be labeled in accordance with paragraph (k) of this section.

**1926.1101(i)(4)** *Inspection of protective clothing.* (i) The competent person shall examine worksuits worn by employees at least once per workshift for rips or tears that may occur during performance of work.

**1926.1101(i)(4)(ii)** When rips or tears are detected while an employee is working, rips and tears shall be immediately mended, or the worksuit shall be immediately replaced.

**1926.1101(j)** ***Hygiene facilities and practices for employees.*** (1) Requirements for employees performing Class I asbestos jobs involving over 25 linear or 10 square feet of TSI or surfacing ACM and PACM.

**1926.1101(j)(1)(i)** *Decontamination areas.* The employer shall establish a decontamination area that is adjacent and connected to the regulated area for the decontamination of such employees. The decontamination area shall consist of an equipment room, shower area, and clean room in series. The employer shall ensure that employees enter and exit the regulated area through the decontamination area.

**1926.1101(j)(1)(i)(A)** *Equipment room.* The equipment room shall be supplied with impermeable, labeled bags and containers for the containment and disposal of contaminated protective equipment.

**1926.1101(j)(1)(i)(B)** *Shower area.* Shower facilities shall be provided which comply with 29 CFR 1910.141(d)(3), unless the employer can demonstrate that they are not feasible. The showers shall be adjacent both to the equipment room and the clean room, unless the employer can demonstrate that this location is not feasible. Where the employer can demonstrate that it is not feasible to locate the shower between the equipment room and the clean room, or where the work is performed outdoors, the employers shall ensure that employees:

**1926.1101(j)(1)(i)(B)*(1)*** Remove asbestos contamination from their worksuits in the equipment room using a HEPA vacuum before proceeding to a shower that is not adjacent to the work area; or

**1926.1101(j)(1)(i)(B)*(2)*** Remove their contaminated worksuits in the equipment room, then don clean worksuits, and proceed to a shower that is not adjacent to the work area.

**1926.1101(j)(1)(i)(C)** *Clean change room.* The clean room shall be equipped with a locker or appropriate storage container for each employee's use. When the employer can demonstrate that it is not feasible to provide a clean change area adjacent to the work area or where the work is performed outdoors, the employer may permit employees engaged in Class I asbestos jobs to clean their protective clothing with a portable HEPA-equipped vacuum before such employees leave the regulated area. Following showering, such employees however must then change into street clothing in clean change areas provided by the employer which otherwise meet the requirements of this section.

**1926.1101(j)(1)(ii)** *Decontamination area entry procedures.* The employer shall ensure that employees:

**1926.1101(j)(1)(ii)(A)** Enter the decontamination area through the clean room;

**1926.1101(j)(1)(ii)(B)** Remove and deposit street clothing within a locker provided for their use; and

**1926.1101(j)(1)(ii)(C)** Put on protective clothing and respiratory protection before leaving the clean room.

**1926.1101(j)(1)(ii)(D)** Before entering the regulated area, the employer shall ensure that employees pass through the equipment room.

**1926.1101(j)(1)(iii)** *Decontamination area exit procedures.* The employer shall ensure that:

**1926.1101(j)(1)(iii)(A)** Before leaving the regulated area, employees shall remove all gross contamination and debris from their protective clothing.

**1926.1101(j)(1)(iii)(B)** Employees shall remove their protective clothing in the equipment room and deposit the clothing in labeled impermeable bags or containers.

**1926.1101(j)(1)(iii)(C)** Employees shall not remove their respirators in the equipment room.

**1926.1101(j)(1)(iii)(D)** Employees shall shower prior to entering the clean room.

**1926.1101(j)(1)(iii)(E)** After showering, employees shall enter the clean room before changing into street clothes.

**1926.1101(j)(1)(iv)** *Lunch Areas.* Whenever food or beverages are consumed at the worksite where employees are performing Class I asbestos work, the employer shall provide lunch areas in which the airborne concentrations of asbestos are below the permissible exposure limit and/or excursion limit.

**1926.1101(j)(2)** Requirements for Class I work involving less than 25 linear or 10 square feet of TSI or surfacing ACM and PACM, and for Class II and Class III asbestos work operations where exposures exceed a PEL or where there is no negative exposure assessment produced before the operation.

**1926.1101(j)(2)(i)** The employer shall establish an equipment room or area that is adjacent to the regulated area for the decontamination of employees and their equipment which is contaminated with asbestos which shall consist of an area covered by a impermeable drop cloth on the floor or horizontal working surface.

**1926.1101(j)(2)(ii)** The area must be of sufficient size as to accommodate cleaning of equipment and removing personal protective equipment without spreading contamination beyond the area (as determined by visible accumulations).

**1926.1101(j)(2)(iii)** Work clothing must be cleaned with a HEPA vacuum before it is removed.

**1926.1101(j)(2)(iv)** All equipment and surfaces of containers filled with ACM must be cleaned prior to removing them from the equipment room or area.

**1926.1101(j)(2)(v)** The employer shall ensure that employees enter and exit the regulated area through the equipment room or area.

**1926.1101(j)(3)** *Requirements for Class IV work.* Employers shall ensure that employees performing Class IV work within a regulated area comply with the hygiene practice required of employees performing work which has a higher classification within that regulated area. Otherwise employers of employees cleaning up debris and material which is TSI or surfacing ACM or identified as PACM shall provide decontamination facilities for such employees which are required by paragraph (j)(2) of this section.

**1926.1101(j)(4)** *Smoking in work areas.* The employer shall ensure that employees do not smoke in work areas where they are occupationally exposed to asbestos because of activities in that work area.

**1926.1101(k)** ***Communication of hazards.*** (1) This section applies to the communication of information concerning asbestos hazards in construction activities to facilitate compliance with this standard. Most asbestos-related construction activities involve previously installed building materials. Building owners often are the only and/or best sources of information concerning them. Therefore, they, along with employers of potentially exposed employees, are assigned specific information conveying and retention duties under this section. Installed Asbestos Containing Building Material. Employers and building owners shall identify TSI and sprayed or troweled on surfacing materials in buildings as asbestos-containing, unless they determine in compliance with paragraph (k)(5) of this section that the material is not asbestos-containing. Asphalt and vinyl flooring material installed no later than 1980 must also be considered as asbestos containing unless the employer, pursuant to paragraph (g)(8)(i)(I) of this section determines that it is not asbestos-containing. If the employer/building owner has actual knowledge, or should have known through the exercise of due diligence, that other materials are asbestos-containing, they too must be treated as such. When communicating information to employees pursuant to this standard, owners and employers shall identify "PACM" as ACM. Additional requirements relating to communication of asbestos work on multi-employer worksites are set out in paragraph (d) of this section.

**1926.1101(k)(2)** *Duties of building and facility owners.* (i) Before work subject to this standard is begun, building and facility owners shall determine the presence, location, and quantity of ACM and/or PACM at the work site pursuant to paragraph (k)(1) of this section. (ii) Building and/or facility owners shall notify the following persons of the presence, location and quantity of ACM or PACM, at the work sites in their buildings and facilities. Notification either shall be in writing, or shall consist of a personal communication between the owner and the person to whom notification must be given or their authorized representatives:

**1926.1101(k)(2)(ii)(A)** Prospective employers applying or bidding for work whose employees reasonably can be expected to work in or adjacent to areas containing such material;

**1926.1101(k)(2)(ii)(B)** Employees of the owner who will work in or adjacent to areas containing such material:

**1926.1101(k)(2)(ii)(C)** On multi-employer worksites, all employers of employees who will be performing work within or adjacent to areas containing such materials;

**1926.1101(k)(2)(ii)(D)** Tenants who will occupy areas containing such material.

**1926.1101(k)(3)** Duties of employers whose employees perform work subject to this standard in or adjacent to areas containing ACM and PACM. Building/facility owners whose employees perform such work shall comply with these provisions to the extent applicable.

**1926.1101(k)(3)(i)** Before work in areas containing ACM and PACM is begun; employers shall identify the presence, location, and quantity of ACM, and/or PACM therein pursuant to paragraph (k)(1) of this section.

**1926.1101(k)(3)(ii)** Before work under this standard is performed employers of employees who will perform such work shall inform the following persons of the location and quantity of ACM and/or PACM present in the area and the precautions to be taken to insure that airborne asbestos is confined to the area.

**1926.1101(k)(3)(ii)(A)** Owners of the building/facility;

**1926.1101(k)(3)(ii)(B)** Employees who will perform such work and employers of employees who work and/or will be working in adjacent areas.

**1926.1101(k)(3)(iii)** Within 10 days of the completion of such work, the employer whose employees have performed work subject to this standard, shall inform the building/facility owner and employers of employees who will be working in the area of the current location and quantity of PACM and/or ACM remaining in the area and final monitoring results, if any.

**1926.1101(k)(4)** In addition to the above requirements, all employers who discover ACM and/or PACM on a worksite shall convey information concerning the presence, location and quantity of such newly discovered ACM and/or PACM to the owner and to other employers of employees working at the work site, within 24 hours of the discovery.

**1926.1101(k)(5)** Criteria to rebut the designation of installed material as PACM. (i) At any time, an employer and/or building owner may demonstrate, for purposes of this standard, that PACM does not contain asbestos. Building owners and/or employers are not required to communicate information about the presence of building material for which such a demonstration pursuant to the requirements of paragraph (k)(5)(ii) of this section has been made. However, in all such cases, the information, data and analysis supporting the determination that PACM does not contain asbestos, shall be retained pursuant to paragraph (n) of this section.

**1926.1101(k)(5)(ii)** An employer or owner may demonstrate that PACM does not contain more than 1% asbestos by the following: (A) Having a completed inspection conducted pursuant to the requirements of AHERA (40 CFR Part 763, subpart E) which demonstrates that the material is not ACM; or

**1926.1101(k)(5)(ii)(B)** Performing tests of the material containing PACM which demonstrate that no ACM is present in the material. Such tests shall include analysis of bulk samples collected in the manner described in 40 CFR 763.86. The tests, evaluation and sample collection shall be conducted by an accredited inspector or by a CIH. Analysis of samples shall be performed by persons or laboratories with proficiency demonstrated by current successful participation in a nationally recognized testing program such as the National Voluntary Laboratory Accreditation Program (NVLAP) or the National Institute for Standards and Technology (NIST) or the Round Robin for bulk samples administered by the American Industrial Hygiene Association (AIHA) or an equivalent nationally-recognized round robin testing program.

**1926.1101(k)(5)(iii)** The employer and/or building owner may demonstrate that flooring material including associated mastic and backing does not contain asbestos, by a determination of an industrial hygienist based upon recognized analytical techniques showing that the material is not ACM.

**1926.1101(k)(6)** At the entrance to mechanical rooms/areas in which employees reasonably can be expected to enter and which contain ACM and/or PACM, the building owner shall post signs which identify the material which is present, its location, and appropriate work practices which, if followed, will ensure that ACM and/or PACM will not be disturbed. The employer shall ensure, to the extent feasible, that employees who come in contact with these signs can comprehend them. Means to ensure employee comprehension may include the use of foreign languages, pictographs, graphics, and awareness training.

**1926.1101(k)(7)** *Signs.* (i) Warning signs that demarcate the regulated area shall be provided and displayed at each location where a regulated area is required to be established by paragraph (e) of this section. Signs shall be posted at such a distance from such a location that an employee may read the signs and take necessary protective steps before entering the area marked by the signs.

**1926.1101(k)(7)(ii)(A)** The warning signs required by paragraph (k)(7) of this section shall bear the following information:

**DANGER**

**ASBESTOS**

**CANCER AND LUNG DISEASE HAZARD**

**AUTHORIZED PERSONNEL ONLY**

**1926.1101(k)(7)(ii)(B)** In addition, where the use of respirators and protective clothing is required in the regulated area under this section, the warning signs shall include the following:

**RESPIRATORS AND PROTECTIVE CLOTHING ARE REQUIRED IN THIS AREA**

**1926.1101(k)(7)(iii)** The employer shall ensure that employees working in and contiguous to regulated areas comprehend the warning signs required to be posted by paragraph (k)(7)(i) of this section. Means to ensure employee comprehension may include the use of foreign languages, pictographs and graphics.

**1926.1101(k)(8)** *Labels.* (i) Labels shall be affixed to all products containing asbestos and to all containers containing such products, including waste containers. Where feasible, installed asbestos products shall contain a visible label.

**1926.1101(k)(8)(ii)** Labels shall be printed in large, bold letters on a contrasting background.

**1926.1101(k)(8)(iii)** Labels shall be used in accordance with the requirements of 29 CFR 1910.1200(f) of OSHA's Hazard Communication standard, and shall contain the following information:

**DANGER**

**CONTAINS ASBESTOS**

**FIBERS**

**AVOID CREATING DUST**

**CANCER AND LUNG**

**DISEASE HAZARD**

**1926.1101(k)(8)(iv)** [Reserved]

**1926.1101(k)(8)(v)** Labels shall contain a warning statement against breathing asbestos fibers.

**1926.1101(k)(8)(vi)** The provisions for labels required by paragraphs (k)(8)(i) through (k)(8)(iii) of this section do not apply where:

**1926.1101(k)(8)(vi)(A)** Asbestos fibers have been modified by a bonding agent, coating, binder, or other material, provided that the manufacturer can demonstrate that, during any reasonably foreseeable use, handling, storage, disposal, processing, or transportation, no airborne concentrations of asbestos fibers in excess of the permissible exposure limit and/or excursion limit will be released, or

**1926.1101(k)(8)(vi)(B)** Asbestos is present in a product in concentrations less than 1.0 percent.

**1926.1101(k)(8)(vii)** When a building owner or employer identifies previously installed PACM and/or ACM, labels or signs shall be affixed or posted so that employees will be notified of what materials contain PACM and/or ACM. The employer shall attach such labels in areas where they will clearly be noticed by employees who are likely to be exposed, such as at the entrance to mechanical room/areas. Signs required by paragraph (k)(6) of this section may be posted in lieu of labels so long as they contain information required for labelling. The employer shall ensure, to the extent feasible, that employees who come in contact with these signs or labels can comprehend them. Means to ensure employee comprehension may include the use of foreign languages, pictographs, graphics, and awareness training.

**1926.1101(k)(9)** *Employee Information and Training.* (i) The employer shall, at no cost to the employee, institute a training program for all employees who are likely to be exposed in excess of a PEL and for all employees who perform Class I through IV asbestos operations, and shall ensure their participation in the program.

**1926.1101(k)(9)(ii)** Training shall be provided prior to or at the time of initial assignment and at least annually thereafter.

**1926.1101(k)(9)(iii)** Training for Class I operations and for Class II operations that require the use of critical barriers (or equivalent isolation methods) and/or negative pressure enclosures under this section shall be the equivalent in curriculum, training method and length to the EPA Model Accreditation Plan (MAP) asbestos abatement workers training (40 CFR Part 763, subpart E, Appendix C).

**1926.1101(k)(9)(iv)** Training for other Class II work.

**1926.1101(k)(9)(iv)(A)** For work with asbestos containing roofing materials, flooring materials, siding materials, ceiling tiles, or transite panels, training shall include at a minimum all the elements included in paragraph (k)(9)(viii) of this section and in addition, the specific work practices and engineering controls set forth in paragraph (g) of this section which specifically relate to that category. Such course shall include "hands-on" training and shall take at least 8 hours.

**1926.1101(k)(9)(iv)(B)** An employee who works with more than one of the categories of material specified in paragraph (k)(9)(iv)(A) of this section shall receive training in the work practices applicable to each category of material that the employee removes and each removal method that the employee uses.

**1926.1101(k)(9)(iv)(C)** For Class II operations not involving the categories of material specified in paragraph (k)(9)(iv)(A) of this section, training shall be provided which shall include at a minimum all the elements included in paragraph (k)(9)(viii) of this section and in addition, the specific work practices and engineering controls set forth in paragraph (g) of this section which specifically relate to the category of material being removed, and shall include "hands-on" training in the work practices applicable to each category of material that the employee removes and each removal method that the employee uses.

**1926.1101(k)(9)(v)** Training for Class III employees shall be consistent with EPA requirements for training of local education agency maintenance and custodial staff as set forth at 40 CFR 763.92(a)(2). Such a course shall also include "hands-on" training and shall take at least 16 hours. Exception: For Class III operations for which the competent person determines that the EPA curriculum does not adequately cover the training needed to perform that activity, training shall include as a minimum all the elements included in paragraph (k)(9)(viii) of this section and in addition, the specific work practices and engineering controls set forth in paragraph (g) of this section which specifically relate to that activity, and shall include "hands-on" training in the work practices applicable to each category of material that the employee disturbs.

**1926.1101(k)(9)(vi)** Training for employees performing Class IV operations shall be consistent with EPA requirements for training of local education agency maintenance and custodial staff as set forth at 40 CFR 763.92(a)(1). Such a course shall include available information concerning the locations of thermal system insulation and surfacing ACM/PACM, and asbestos-containing flooring material, or flooring material where the absence of asbestos has not yet been certified; and instruction in recognition of damage, deterioration, and delamination of asbestos containing building materials. Such course shall take at least 2 hours.

**1926.1101(k)(9)(vii)** Training for employees who are likely to be exposed in excess of the PEL and who are not otherwise required to be trained under paragraph (k)(9)(iii) through (vi) of this section, shall meet the requirements of paragraph (k)(9)(viii) of this section.

**1926.1101(k)(9)(viii)** The training program shall be conducted in a manner that the employee is able to understand. In addition to the content required by provisions in paragraphs (k)(9)(iii) through (vi) of this section, the employer shall ensure that each such employee is informed of the following:

**1926.1101(k)(9)(viii)(A)** Methods of recognizing asbestos, including the requirement in paragraph (k)(1) of this section to presume that certain building materials contain asbestos;

**1926.1101(k)(9)(viii)(B)** The health effects associated with asbestos exposure;

**1926.1101(k)(9)(viii)(C)** The relationship between smoking and asbestos in producing lung cancer;

**1926.1101(k)(9)(viii)(D)** The nature of operations that could result in exposure to asbestos, the importance of necessary protective controls to minimize exposure including, as applicable, engineering controls, work practices, respirators, housekeeping procedures, hygiene facilities, protective clothing, decontamination procedures, emergency procedures, and waste disposal procedures, and any necessary instruction in the use of these controls and procedures; where Class III and IV work will be or is performed, the contents of EPA 20T-2003, "Managing Asbestos In-Place" July 1990 or its equivalent in content;

**1926.1101(k)(9)(viii)(E)** The purpose, proper use, fitting instructions, and limitations of respirators as required by 29 CFR 1910.134;

**1926.1101(k)(9)(viii)(F)** The appropriate work practices for performing the asbestos job;

**1926.1101(k)(9)(viii)(G)** Medical surveillance program requirements;

**1926.1101(k)(9)(viii)(H)** The content of this standard including appendices;

**1926.1101(k)(9)(viii)(I)** The names, addresses and phone numbers of public health organizations which provide information, materials and/or conduct programs concerning smoking cessation. The employer may distribute the list of such organizations contained in Appendix J to this section, to comply with this requirement; and

**1926.1101(k)(9)(viii)(J)** The requirements for posting signs and affixing labels and the meaning of the required legends for such signs and labels.

**1926.1101(k)(10)** Access to training materials. (i) The employer shall make readily available to affected employees without cost, written materials relating to the employee training program, including a copy of this regulation.

**1926.1101(k)(10)(ii)** The employer shall provide to the Assistant Secretary and the Director, upon request, all information and training materials relating to the employee information and training program.

**1926.1101(k)(10)(iii)** The employer shall inform all employees concerning the availability of self-help smoking cessation program material. Upon employee request, the employer shall distribute such material, consisting of NIH Publication No. 89-1647, or equivalent self-help material, which is approved or published by a public health organization listed in Appendix J to this section.

**1926.1101(l)** ***Housekeeping***— (1) *Vacuuming.* Where vacuuming methods are selected, HEPA filtered vacuuming equipment must be used. The equipment shall be used and emptied in a manner that minimizes the reentry of asbestos into the workplace.

**1926.1101(l)(2)** *Waste disposal.* Asbestos waste, scrap, debris, bags, containers, equipment, and contaminated clothing consigned for disposal shall be collected and disposed of in sealed, labeled, impermeable bags or other closed, labeled, impermeable containers except in roofing operations, where the procedures specified in paragraph (g)(8)(ii) of this section apply.

**1926.1101(l)(3)** *Care of asbestos-containing flooring material.* (i) All vinyl and asphalt flooring material shall be maintained in accordance with this paragraph unless the building/facility owner demonstrates, pursuant to paragraph (g)(8)(i)(I) of this section that the flooring does not contain asbestos.

**1926.1101(l)(3)(ii)** Sanding of flooring material is prohibited.

**1926.1101(l)(3)(iii)** Stripping of finishes shall be conducted using low abrasion pads at speeds lower than 300 rpm and wet methods.

**1926.1101(l)(3)(iv)** Burnishing or dry buffing may be performed only on flooring which has sufficient finish so that the pad cannot contact the flooring material.

**1926.1101(l)(4)** Waste and debris and accompanying dust in an area containing accessible thermal system insulation or surfacing ACM/PACM or visibly deteriorated ACM:

**1926.1101(l)(4)(i)** Shall not be dusted or swept dry, or vacuumed without using a HEPA filter;

**1926.1101(l)(4)(ii)** Shall be promptly cleaned up and disposed of in leak tight containers.

**1926.1101(m)** ***Medical surveillance.*** (1) General—(i) *Employees covered.* (A) The employer shall institute a medical surveillance program for all employees who for a combined total of 30 or more days per year are engaged in Class I, II and III work or are exposed at or above a permissible exposure limit. For purposes of this paragraph, any day in which a worker engages in Class II or Class III operations or a combination thereof on intact material for one hour or less (taking into account the entire time spent on the removal operation, including cleanup) and, while doing so, adheres fully to the work practices specified in this standard, shall not be counted.

**1926.1101(m)(1)(i)(B)** For employees otherwise required by this standard to wear a negative pressure respirator, employers shall ensure employees are physically able to perform the work and use the equipment. This determination shall be made under the supervision of a physician.

**1926.1101(m)(1)(ii)** *Examination.* (A) The employer shall ensure that all medical examinations and procedures are performed by or under the supervision of a licensed physician, and are provided at no cost to the employee and at a reasonable time and place.

**1926.1101(m)(1)(ii)(B)** Persons other than such licensed physicians who administer the pulmonary function testing required by this section shall complete a training course in spirometry sponsored by an appropriate academic or professional institution.

**1926.1101(m)(2)** *Medical examinations and consultations.* (i) Frequency. The employer shall make available medical examinations and consultations to each employee covered under paragraph (m)(1)(i) of this section on the following schedules:

**1926.1101(m)(2)(i)(A)** Prior to assignment of the employee to an area where negative-pressure respirators are worn;

**1926.1101(m)(2)(i)(B)** When the employee is assigned to an area where exposure to asbestos may be at or above the permissible exposure limit for 30 or more days per year, or engage in Class I, II, or III work for a combined total of 30 or more days per year, a medical examination must be given within 10 working days following the thirtieth day of exposure;

**1926.1101(m)(2)(i)(C)** And at least annually thereafter.

**1926.1101(m)(2)(i)(D)** If the examining physician determines that any of the examinations should be provided more frequently than specified, the employer shall provide such examinations to affected employees at the frequencies specified by the physician.

**1926.1101(m)(2)(i)(E)** Exception: No medical examination is required of any employee if adequate records show that the employee has been examined in accordance with this paragraph within the past 1-year period.

**1926.1101(m)(2)(ii)** *Content.* Medical examinations made available pursuant to paragraphs (m)(2)(i)(A) through (m)(2)(i)(C) of this section shall include:

**1926.1101(m)(2)(ii)(A)** A medical and work history with special emphasis directed to the pulmonary, cardiovascular, and gastrointestinal systems.

**1926.1101(m)(2)(ii)(B)** On initial examination, the standardized questionnaire contained in Part 1 of Appendix D to this section, and, on annual examination, the abbreviated standardized questionnaire contained in Part 2 of Appendix D to this section.

**1926.1101(m)(2)(ii)(C)** A physical examination directed to the pulmonary and gastrointestinal systems, including a chest roentgenogram to be administered at the discretion of the physician, and pulmonary function tests of forced vital capacity (FVC) and forced expiratory volume at one second (FEV(1)). Interpretation and classification of chest shall be conducted in accordance with Appendix E to this section.

**1926.1101(m)(2)(ii)(D)** Any other examinations or tests deemed necessary by the examining physician.

**1926.1101(m)(3)** *Information provided to the physician.* The employer shall provide the following information to the examining physician:

**1926.1101(m)(3)(i)** A copy of this standard and Appendices D, E, and I to this section;

**1926.1101(m)(3)(ii)** A description of the affected employee's duties as they relate to the employee's exposure;

**1926.1101(m)(3)(iii)** The employee's representative exposure level or anticipated exposure level;

**1926.1101(m)(3)(iv)** A description of any personal protective and respiratory equipment used or to be used; and

**1926.1101(m)(3)(v)** Information from previous medical examinations of the affected employee that is not otherwise available to the examining physician.

**1926.1101(m)(4)** *Physician's written opinion.* (i) The employer shall obtain a written opinion from the examining physician. This written opinion shall contain the results of the medical examination and shall include:

**1926.1101(m)(4)(i)(A)** The physician's opinion as to whether the employee has any detected medical conditions that would place the employee at an increased risk of material health impairment from exposure to asbestos;

**1926.1101(m)(4)(i)(B)** Any recommended limitations on the employee or on the use of personal protective equipment such as respirators; and

**1926.1101(m)(4)(i)(C)** A statement that the employee has been informed by the physician of the results of the medical examination and of any medical conditions that may result from asbestos exposure.

**1926.1101(m)(4)(i)(D)** A statement that the employee has been informed by the physician of the increased risk of lung cancer attributable to the combined effect of smoking and asbestos exposure.

**1926.1101(m)(4)(ii)** The employer shall instruct the physician not to reveal in the written opinion given to the employer specific findings or diagnoses unrelated to occupational exposure to asbestos.

**1926.1101(m)(4)(iii)** The employer shall provide a copy of the physician's written opinion to the affected employee within 30 days from its receipt.

**1926.1101(n)** ***Recordkeeping.*** (1) Objective data relied on pursuant to paragraph (f) to this section.

**1926.1101(n)(1)(i)** Where the employer has relied on objective data that demonstrates that products made from or containing asbestos or the activity involving such products or material are not capable of releasing fibers of asbestos in concentrations at or above the permissible exposure limit and/or excursion limit under the expected conditions of processing, use, or handling to satisfy the requirements of paragraph (f), the employer shall establish and maintain an accurate record of objective data reasonably relied upon in support of the exemption.

**1926.1101(n)(1)(ii)** The record shall include at least the following information:

**1926.1101(n)(1)(ii)(A)** The product qualifying for exemption;

**1926.1101(n)(1)(ii)(B)** The source of the objective data;

**1926.1101(n)(1)(ii)(C)** The testing protocol, results of testing, and/or analysis of the material for the release of asbestos;

**1926.1101(n)(1)(ii)(D)** A description of the operation exempted and how the data support the exemption; and

**1926.1101(n)(1)(ii)(E)** Other data relevant to the operations, materials, processing, or employee exposures covered by the exemption.

**1926.1101(n)(1)(iii)** The employer shall maintain this record for the duration of the employer's reliance upon such objective data.

**1926.1101(n)(2)** *Exposure measurements.* (i) The employer shall keep an accurate record of all measurements taken to monitor employee exposure to asbestos as prescribed in paragraph (f) of this section.

NOTE: The employer may utilize the services of competent organizations such as industry trade associations and employee associations to maintain the records required by this section.

**1926.1101(n)(2)(ii)** This record shall include at least the following information:

**1926.1101(n)(2)(ii)(A)** The date of measurement;

**1926.1101(n)(2)(ii)(B)** The operation involving exposure to asbestos that is being monitored;

**1926.1101(n)(2)(ii)(C)** Sampling and analytical methods used and evidence of their accuracy;

**1926.1101(n)(2)(ii)(D)** Number, duration, and results of samples taken;

**1926.1101(n)(2)(ii)(E)** Type of protective devices worn, if any; and

**1926.1101(n)(2)(ii)(F)** Name, social security number, and exposure of the employees whose exposures are represented.

**1926.1101(n)(2)(iii)** The employer shall maintain this record for at least thirty (30) years, in accordance with 29 CFR 1910.20.

**1926.1101(n)(3)** *Medical surveillance.* (i) The employer shall establish and maintain an accurate record for each employee subject to medical surveillance by paragraph (m) of this section, in accordance with 29 CFR 1910.20.

**1926.1101(n)(3)(ii)** The record shall include at least the following information:

**1926.1101(n)(3)(ii)(A)** The name and social security number of the employee;

**1926.1101(n)(3)(ii)(B)** A copy of the employee's medical examination results, including the medical history, questionnaire responses, results of any tests, and physician's recommendations.

**1926.1101(n)(3)(ii)(C)** Physician's written opinions;

**1926.1101(n)(3)(ii)(D)** Any employee medical complaints related to exposure to asbestos; and

**1926.1101(n)(3)(ii)(E)** A copy of the information provided to the physician as required by paragraph (m) of this section.

**1926.1101(n)(3)(iii)** The employer shall ensure that this record is maintained for the duration of employment plus thirty (30) years, in accordance with 29 CFR 1910.20.

**1926.1101(n)(4)** *Training records.* The employer shall maintain all employee training records for one (1) year beyond the last date of employment by that employer.

**1926.1101(n)(5)** *Data to Rebut PACM.* Where the building owner and employer have relied on data to demonstrate that PACM is not asbestos-containing, such data shall be maintained for as long as they are relied upon to rebut the presumption.

**1926.1101(n)(6)** *Records of required notifications.* Where the building owner has communicated and received information concerning the identification, location and quantity of ACM and PACM, written records of such notifications and their content shall be maintained by the building owner for the duration of ownership and shall be transferred to successive owners of such buildings/facilities.

**1926.1101(n)(7)** *Availability.* (i) The employer, upon written request, shall make all records required to be maintained by this section available to the Assistant Secretary and the Director for examination and copying.

**1926.1101(n)(7)(ii)** The employer, upon request, shall make any exposure records required by paragraphs (f) and (n) of this section available for examination and copying to affected employees, former employees, designated representatives, and the Assistant Secretary, in accordance with 29 CFR 1910.20(a) through (e) and (g) through (i).

**1926.1101(n)(7)(iii)** The employer, upon request, shall make employee medical records required by paragraphs (m) and (n) of this section available for examination and copying to the subject employee, anyone having the specific written consent of the subject employee, and the Assistant Secretary, in accordance with 29 CFR 1910.20.

**1926.1101(n)(8)** *Transfer of records.* (i) The employer shall comply with the requirements concerning transfer of records set forth in 29 CFR 1910.20(h).

**1926.1101(n)(8)(ii)** Whenever the employer ceases to do business and there is no successor employer to receive and retain the records for the prescribed period, the employer shall notify the Director at least 90 days prior to disposal and, upon request, transmit them to the Director.

**1926.1101(o)** ***Competent person.*** (1) General. On all construction worksites covered by this standard, the employer shall designate a competent person, having the qualifications and authorities for ensuring worker safety and health required by subpart C, General Safety and Health Provisions for Construction (29 CFR 1926.20 through 1926.32).

**1926.1101(o)(2)** *Required inspections by the competent person.* Section 1926.20(b)(2) which requires health and safety prevention programs to provide for frequent and regular inspections of the job sites, materials, and equipment to be made by competent persons, is incorporated.

**1926.1101(o)(3)** *Additional inspections.* In addition, the competent person shall make frequent and regular inspections of the job sites, in order to perform the duties set out below in paragraph (o)(3)(i) of this section. For Class I jobs, on-site inspections shall be made at least once during each work shift, and at any time at employee request. For Class II, III, and IV jobs, on-site inspections shall be made at intervals sufficient to assess whether conditions have changed, and at any reasonable time at employee request.

**1926.1101(o)(3)(i)** On all worksites where employees are engaged in Class I or II asbestos work, the competent person designated in accordance with paragraph (e)(6) of this section shall perform or supervise the following duties, as applicable:

**1926.1101(o)(3)(i)(A)** Set up the regulated area, enclosure, or other containment;

**1926.1101(o)(3)(i)(B)** Ensure (by on-site inspection) the integrity of the enclosure or containment;

**1926.1101(o)(3)(i)(C)** Set up procedures to control entry to and exit from the enclosure and/or area;

**1926.1101(o)(3)(i)(D)** Supervise all employee exposure monitoring required by this section and ensure that it is conducted as required by paragraph (f) of this section;

**1926.1101(o)(3)(i)(E)** Ensure that employees working within the enclosure and/or using glove bags wear respirators and protective clothing as required by paragraphs (h) and (i) of this section;

**1926.1101(o)(3)(i)(F)** Ensure through on-site supervision, that employees set up, use, and remove engineering controls, use work practices and personal protective equipment in compliance with all requirements;

**1926.1101(o)(3)(i)(G)** Ensure that employees use the hygiene facilities and observe the decontamination procedures specified in paragraph (j) of this section;

**1926.1101(o)(3)(i)(H)** Ensure that through on-site inspection, engineering controls are functioning properly and employees are using proper work practices; and,

**1926.1101(o)(3)(i)(I)** Ensure that notification requirement in paragraph (k) of this section are met.

**1926.1101(o)(3)(ii)** [Reserved]

**1926.1101(o)(4)** *Training for the competent person.* (i) For Class I and II asbestos work the competent person shall be trained in all aspects of asbestos removal and handling, including: abatement, installation, removal and handling; the contents of this standard; the identification of asbestos; removal procedures, where appropriate; and other practices for reducing the hazard. Such training shall be obtained in a comprehensive course for supervisors that meets the criteria of EPA's Model Accreditation Plan (40 CFR part 763, subpart E, Appendix C), such as a course conducted by an EPA-approved or state-approved training provider, certified by EPA or a state, or a course equivalent in stringency, content, and length.

**1926.1101(o)(4)(ii)** For Class III and IV asbestos work, the competent person shall be trained in aspects of asbestos handling appropriate for the nature of the work, to include procedures for setting up glove bags and mini-enclosures, practices for reducing asbestos exposures, use of wet methods, the contents of this standard, and the identification of asbestos. Such training shall include successful completion of a course that is consistent with EPA requirements for training of local education agency maintenance and custodial staff as set forth at 40 CFR 763.92(a)(2), or its equivalent in stringency, content and length. Competent persons for Class III and IV work, may also be trained pursuant to the requirements of paragraph (o)(4)(i) of this section.

**1926.1101(p)** ***Appendices.*** (1) Appendices A, C, D, and E to this section are incorporated as part of this section and the contents of these appendices are mandatory.

**1926.1101(p)(2)** Appendices B, F, H, I, J, and K to this section are informational and are not intended to create any additional obligations not otherwise imposed or to detract from any existing obligations.

**1926.1101(q)** ***Dates.*** (1) This standard shall become effective October 11, 1994.

**1926.1101(q)(2)** The provisions of 29 CFR 1926.58 remain in effect until the start-up dates of the equivalent provisions of this standard.

**1926.1101(q)(3)** *Start-up dates.* All obligations of this standard commence on the effective date except as follows:

**1926.1101(q)(3)(i)** *Methods of compliance.* The engineering and work practice controls required by paragraph (g) of this section shall be implemented by October 1, 1995.

**1926.1101(q)(3)(ii)** *Respiratory protection.* Respiratory protection required by paragraph (h) of this section shall be provided by October 1, 1995.

**1926.1101(q)(3)(iii)** *Hygiene facilities and practices for employees.* Hygience facilities and practices required by paragraph (j) of this section shall be provided by October 1, 1995.

**1926.1101(q)(3)(iv)** *Communication of hazards.* Identification, notification, labeling and sign posting, and training required by paragraph (k) of this section shall be provided by October 1, 1995.

**1926.1101(q)(3)(v)** *Housekeeping.* Housekeeping practices and controls required by paragraph (l) of this section shall be provided by October 1, 1995.

**1926.1101(q)(3)(vi)** Medical surveillance required by paragraph (m) of this section shall be provided by October 1, 1995.

**1926.1101(q)(3)(vii)** The designation and training of competent persons required by paragraph (o) of this section shall be completed by October 1, 1995.

## Appendix A to §1926.1101—OSHA Reference Method—Mandatory

This mandatory appendix specifies the procedure for analyzing air samples for asbestos and specifies quality control procedures that must be implemented by laboratories performing the analysis. The sampling and analytical methods described below represent the elements of the available monitoring methods (such as Appendix B of this regulation, the most current version of the OSHA method ID-160, or the most current version of the NIOSH Method 7400). All employers who are required to conduct air monitoring under paragraph (f) of the standard are required to utilize analytical laboratories that use this procedure, or an equivalent method, for collecting and analyzing samples.

### Sampling and Analytical Procedure

1. The sampling medium for air samples shall be mixed cellulose ester filter membranes. These shall be designated by the manufacturer as suitable for asbestos counting. See below for rejection of blanks.

2. The preferred collection device shall be the 25-mm diameter cassette with an open-faced 50-mm electrically conductive extension cowl. The 37-mm cassette may be used if necessary but only if written justification for the need to use the 37-mm filter cassette accompanies the sample results in the employee's exposure monitoring record. Do not reuse or reload cassettes for asbestos sample collection.

3. An air flow rate between 0.5 liter/min and 2.5 liters/min shall be selected for the 25/mm cassette. If the 37-mm cassette is used, an air flow rate between 1 liter/min and 2.5 liters/min shall be selected.

4. Where possible, a sufficient air volume for each air sample shall be collected to yield between 100 and 1,300 fibers per square millimeter on the membrane filter. If a filter darkens in appearance or if loose dust is seen on the filter, a second sample shall be started.

5. Ship the samples in a rigid container with sufficient packing material to prevent dislodging the collected fibers. Packing material that has a high electrostatic charge on its surface (e.g., expanded polystyrene) cannot be used because such material can cause loss of fibers to the sides of the cassette.

6. Calibrate each personal sampling pump before and after use with a representative filter cassette installed between the pump and the calibration devices.

7. Personal samples shall be taken in the "breathing zone" of the employee (i.e., attached to or near the collar or lapel near the worker's face).

8. Fiber counts shall be made by positive phase contrast using a microscope with an 8 to 10 X eyepiece and a 40 to 45 X objective for a total magnification of approximately 400 X and a numerical aperture of 0.65 to 0.75. The microscope shall also be fitted with a green or blue filter.

9. The microscope shall be fitted with a Walton-Beckett eyepiece graticule calibrated for a field diameter of 100 micrometers (+/- 2 micrometers).

10. The phase-shift detection limit of the microscope shall be about 3 degrees measured using the HSE phase shift test slide as outlined below.

a. Place the test slide on the microscope stage and center it under the phase objective.

b. Bring the blocks of grooved lines into focus.

**Note:** The slide consists of seven sets of grooved lines (ca. 20 grooves to each block) in descending order of visibility from sets 1 to 7, seven being the least visible. The requirements for asbestos counting are that the microscope optics must resolve the groooved lines in set 3 completely, although they may appear somewhat faint, and that the grooved lines in sets 6 and 7 must be invisible. Sets 4 and 5 must be at least partially visible but may vary slightly in visibility between microscopes. A microscope that fails to meet these requirements has either too low or too high a resolution to be used for asbestos counting.

c. If the image deteriorates, clean and adjust the microscope optics. If the problem persists, cosult the microscope manufacturer.

11. Each set of samples taken will include 10% field blanks or a minimum of 2 field blanks. These blanks must come from the same lot as the filters used for sample collection. The field blank results shall be averaged and subtracted from the analytical results before reporting. A set consists of any sample or group of samples for which an evaluation for this standard must be made. Any samples represented by a field blank having a fiber count in excess of the detection limit of the method being used shall be rejected.

12. The samples shall be mounted by the acetone/triacetin method or a method with an equivalent index of refraction and similar clarity.

13. Observe the following counting rules.

a. Count only fibers equal to or longer than 5 micrometers. Measure the length of curved fibers along the curve.

b. In the absence of other information, count all particles as asbestos, that have a length-to-width ratio (aspect ratio) of 3:1 or greater.

c. Fibers lying entirely within the boundary of the Walton-Beckett graticule field shall receive a count of 1. Fibers crossing the boundary once, having one end within the circle, shall receive the count of one half (1/2). Do not count any fiber that crosses the graticule boundary more than once. Reject and do not count any other fibers even though they may be visible outside the graticule area.

d. Count bundles of fibers as one fiber unless individual fibers can be identified by observing both ends of an individual fiber.

e. Count enough graticule fields to yield 100 fibers. Count a minimum of 20 fields; stop counting at 100 fields regardless of fiber count.

14. Blind recounts shall be conducted at the rate of 10 percent.

### Quality Control Procedures

1. Intralaboratory program. Each laboratory and/or each company with more than one microscopist counting slides shall establish a statistically designed quality assurance program involving blind recounts and comparisons between microscopists to monitor the variability of counting by each microscopist and between microscopists. In a company with more than one laboratory, the program shall include all laboratories, and shall also evaluate the laboratory-to-laboratory variability.

2a. Interlaboratory program. Each laboratory analyzing asbestos samples for compliance determination shall implement an interlaboratory quality assurance program that, as a minimum, includes participation of at least two other independent laboratories. Each laboratory shall participate in round robin testing at least once every 6 months with at least all the other laboratories in its interlaboratory quality assurance group. Each laboratory shall submit slides typical of its own workload for use in this program. The round robin shall be designed and results analyzed using appropriate statistical methodology.

b. All laboratories should also participate in a national sample testing scheme such as the Proficiency Analytical Testing Program (PAT), or the Asbestos Registry sponsored by the American Industrial Hygiene Association (AIHA).

3. All individuals performing asbestos analysis must have taken the NIOSH course for sampling and evaluating airborne asbestos dust or an equivalent course.

4. When the use of different microscopes contributes to differences between counters and laboratories, the effect of the different microscope shall be evaluated and the microscope shall be replaced, as necessary.

5. Current results of these quality assurance programs shall be posted in each laboratory to keep the microscopists informed.

## Appendix B to §1926.1101—Sampling and Analysis (Non-Mandatory)

Matrix Air:

OSHA Permissible Exposure Limits:

| | |
|---|---|
| Time Weighted Average | 0.1 fiber/cc |
| Excursion Level (30 minutes) | 1.0 fiber cc |

Collection Procedure:

A known volume of air is drawn through a 25-mm diameter cassette containing a mixed-cellulose ester filter. The cassette must be equipped with an electrically conductive 50-mm extension cowl. The sampling time and rate are chosen to give a fiber density of between 100 to 1,300 fibers/$mm^2$ on the filter.

| | |
|---|---|
| Recommended Sampling Rate liters/minute (L/min) | 0.5 to 5.0 |

Recommended Air Volumes:

| | |
|---|---|
| Minimum | 25 L |
| Maximum | 2,400 L |

Analytical Procedure:

A portion of the sample filter is cleared and prepared for asbestos fiber counting by Phase Contrast Microscopy (PCM) at 400X.

Commercial manufacturers and products mentioned in this method are for descriptive use only and do not constitute endorsements by USDOL-OSHA. Similar products from other sources can be substituted.

### 1. Introduction

This method describes the collection of airborne asbestos fibers using calibrated sampling pumps with mixed-cellulose ester (MCE) filters and analysis by phase contrast microscopy (PCM). Some terms used are unique to this method and are defined below: *Asbestos:* A term for naturally occurring fibrous minerals. Asbestos includes chrysotile, crocidolite, amosite (cummingtonite-grunerite asbestos), tremolite asbestos, actinolite asbestos, anthophyllite asbestos, and any of these minerals that have been chemically treated and/or altered. The precise chemical formulation of each species will vary with the location from which it was mined. Nominal compositions are listed:

| | |
|---|---|
| Chrysotile | $Mg_3\ Si_2\ O_5(OH)_4$ |
| Crocidolite | $Na_2\ Fe_3^{2} + Fe_2^{3} + Si_8\ O_{22}(OH)_2$ |
| Amosite | $(Mg,Fe)_7\ Si_8\ O_{22}(OH)_2$ |
| Tremolite-actinolite | $Ca_2(Mg,Fe)_5\ Si_8\ O_{22}(OH)_2$ |
| Anthophyllite | $(Mg,Fe)_7\ Si_8\ O_{22}(OH)_2$ |

*Asbestos Fiber:* A fiber of asbestos which meets the criteria specified below for a fiber.

*Aspect Ratio:* The ratio of the length of a fiber to it's diameter (e.g. 3:1, 5:1 aspect ratios).

*Cleavage Fragments:* Mineral particles formed by comminution of minerals, especially those characterized by parallel sides and a moderate aspect ratio (usually less than 20:1).

*Detection Limit:* The number of fibers necessary to be 95% certain that the result is greater than zero.

*Differential Counting:* The term applied to the practice of excluding certain kinds of fibers from the fiber count because they do not appear to be asbestos.

*Fiber:* A particle that is 5 μ m or longer, with a length-to-width ratio of 3 to 1 or longer.

*Field:* The area within the graticule circle that is superimposed on the microscope image.

*Set:* The samples which are taken, submitted to the laboratory, analyzed, and for which, interim or final result reports are generated.

*Tremolite, Anthophyllite, and Actinolite:* The non-asbestos form of these minerals which meet the definition of a fiber. It includes any of these minerals that have been chemically treated and/or altered.

*Walton-Beckett Graticule:* An eyepiece graticule specifically designed for asbestos fiber counting. It consists of a circle with a projected diameter of 100±2 μ m (area of about 0.00785 $mm^2$) with a crosshair having tic-marks at 3-μ m intervals in one direction and 5-μ m in the orthogonal direction. There are marks around the periphery of the circle to demonstrate the proper sizes and shapes of fibers. This design is reproduced in Figure 1. The disk is placed in one of the microscope eyepieces so that the design is superimposed on the field of view.

### 1.1. History

Early surveys to determine asbestos exposures were conducted using impinger counts of total dust with the counts expressed as million particles per cubic foot. The British Asbestos Research Council recommended filter membrane counting in 1969. In July 1969, the Bureau of Occupational Safety and Health published a filter membrane method for counting asbestos fibers in the United States. This method was refined by NIOSH and published as P & CAM 239. On May 29, 1971, OSHA specified filter membrane sampling with phase contrast counting for evaluation of asbestos exposures at work sites in the United States. The use of this technique was again required by OSHA in 1986. Phase contrast microscopy has continued to be the method of choice for the measurement of occupational exposure to asbestos.

### 1.2. Principle

Air is drawn through a MCE filter to capture airborne asbestos fibers. A wedge shaped portion of the filter is removed, placed on a glass microscope slide and made transparent. A measured area (field) is viewed by PCM. All the fibers meeting defined criteria for asbestos are counted and considered a measure of the airborne asbestos concentration.

### 1.3. Advantages and Disadvantages

There are four main advantages of PCM over other methods:

(1) The technique is specific for fibers. Phase contrast is a fiber counting technique which excludes non-fibrous particles from the analysis.

(2) The technique is inexpensive and does not require specialized knowledge to carry out the analysis for total fiber counts.

(3) The analysis is quick and can be performed on-site for rapid determination of air concentrations of asbestos fibers.

(4) The technique has continuity with historical epidemiological studies so that estimates of expected disease can be inferred from long-term determinations of asbestos exposures.

The main disadvantage of PCM is that it does not positively identify asbestos fibers. Other fibers which are not asbestos may be included in the count unless differential counting is performed. This requires a great deal of experience to adequately differentiate asbestos from non-asbestos fibers. Positive identification of asbestos must be performed by polarized light or electron microscopy techniques. A further disadvantage of PCM is that the smallest visible fibers are about 0.2 μ m in diameter while the finest asbestos fibers may be as small as 0.02 μ m in diameter. For some exposures, substantially more fibers may be present than are actually counted.

### 1.4. Workplace Exposure

Asbestos is used by the construction industry in such products as shingles, floor tiles, asbestos cement, roofing felts, insulation and acoustical products. Non-construction uses include brakes, clutch facings, paper, paints, plastics, and fabrics. One of the most significant exposures in the workplace is the removal and encapsulation of asbestos in schools, public buildings, and homes. Many workers have the potential to be exposed to asbestos during these operations.

About 95% of the asbestos in commercial use in the United States is chrysotile. Crocidolite and amosite make up most of the remainder. Anthophyllite and tremolite or actinolite are likely to be encountered as contaminants in various industrial products.

### 1.5. Physical Properties

Asbestos fiber possesses a high tensile strength along its axis, is chemically inert, non-combustible, and heat resistant. It has a high electrical resistance and good sound absorbing properties. It can be weaved into cables, fabrics or other textiles, and also matted into asbestos papers, felts, or mats.

## *2. Range and Detection Limit*

2.1. The ideal counting range on the filter is 100 to 1,300 fibers/$mm^2$. With a Walton-Beckett graticule this range is equivalent to 0.8 to 10 fibers/field. Using NIOSH counting statistics, a count of 0.8 fibers/field would give an approximate coefficient of variation (CV) of 0.13.

2.2. The detection limit for this method is 4.0 fibers per 100 fields or 5.5 fibers/$mm^2$. This was determined using an equation to estimate the maximum CV possible at a specific concentration (95% confidence) and a Lower Control Limit of zero. The CV value was then used to determine a corresponding concentration from historical CV vs fiber relationships. As an example:

**Lower Control Limit (95% Confidence) = AC—1.645(CV)(AC)**

Where:

AC = Estimate of the airborne fiber concentration (fibers/cc) Setting the Lower Control Limit=0 and solving for CV:

0 = AC—1.645(CV)(AC)

CV = 0.61

This value was compared with CV vs. count curves. The count at which CV = 0.61 for Leidel-Busch counting statistics or for an OSHA Salt Lake Technical Center (OSHA-SLTC) CV curve (see Appendix A for further information) was 4.4 fibers or 3.9 fibers per 100 fields, respectively. Although a lower detection limit of 4 fibers per 100 fields is supported by the OSHA-SLTC data, both data sets support the 4.5 fibers per 100 fields value.

## *3. Method Performance—Precision and Accuracy*

Precision is dependent upon the total number of fibers counted and the uniformity of the fiber distribution on the filter. A general rule is to count at least 20 and not more than 100 fields. The count is discontinued when 100 fibers are counted, provided that 20 fields have already been counted. Counting more than 100 fibers results in only a small gain in precision. As the total count drops below 10 fibers, an accelerated loss of precision is noted.

At this time, there is no known method to determine the absolute accuracy of the asbestos analysis. Results of samples prepared through the Proficiency Analytical Testing (PAT) Program and analyzed by the OSHA-SLTC showed no significant bias when compared to PAT reference values. The PAT samples were analyzed from 1987 to 1989 (N=36) and the concentration range was from 120 to 1,300 fibers/$mm^2$.

## *4. Interferences*

Fibrous substances, if present, may interfere with asbestos analysis.

Some common fibers are:

fiberglass

anhydrite

plant fibers

perlite veins

gypsum

some synthetic fibers

membrane structures

sponge spicules

diatoms

microorganisms

wollastonite

The use of electron microscopy or optical tests such as polarized light, and dispersion staining may be used to differentiate these materials from asbestos when necessary.

## *5. Sampling*

### 5.1. Equipment

*5.1.1.* Sample assembly (The assembly is shown in Figure 3). Conductive filter holder consisting of a 25-mm diameter, 3-piece cassette having a 50-mm long electrically conductive extension cowl. Backup pad, 25-mm, cellulose. Membrane filter, mixed-cellulose ester (MCE), 25-mm, plain, white, 0.4 to 1.2-μ m pore size.

NOTES:

(a) DO NOT RE-USE CASSETTES.

(b) Fully conductive cassettes are required to reduce fiber loss to the sides of the cassette due to electrostatic attraction.

(c) Purchase filters which have been selected by the manufacturer for asbestos counting or analyze representative filters for fiber background before use. Discard the filter lot if more than 4 fibers/100 fields are found.

(d) To decrease the possibility of contamination, the sampling system (filter-backup pad-cassette) for asbestos is usually preassembled by the manufacturer.

(e) Other cassettes, such as the Bell-mouth, may be used within the limits of their validation.

*5.1.2.* Gel bands for sealing cassettes.

*5.1.3.* Sampling pump.

Each pump must be a battery operated, self-contained unit small enough to be placed on the monitored employee and not interfere with the work being performed. The pump must be capable of sampling at the collection rate for the required sampling time.

*5.1.4.* Flexible tubing, 6-mm bore.

*5.1.5.* Pump calibration.

Stopwatch and bubble tube/burette or electronic meter.

5.2. Sampling Procedure

*5.2.1.* Seal the point where the base and cowl of each cassette meet with a gel band or tape.

*5.2.2.* Charge the pumps completely before beginning.

*5.2.3.* Connect each pump to a calibration cassette with an appropriate length of 6-mm bore plastic tubing. Do not use luer connectors—the type of cassette specified above has built-in adapters.

*5.2.4.* Select an appropriate flow rate for the situation being monitored. The sampling flow rate must be between 0.5 and 5.0 L/min for personal sampling and is commonly set between 1 and 2 L/min. Always choose a flow rate that will not produce overloaded filters.

*5.2.5.* Calibrate each sampling pump before and after sampling with a calibration cassette in-line (Note: This calibration cassette should be from the same lot of cassettes used for sampling). Use a primary standard (e.g. bubble burette) to calibrate each pump. If possible, calibrate at the sampling site.

NOTE: If sampling site calibration is not possible, environmental influences may affect the flow rate. The extent is dependent on the type of pump used. Consult with the pump manufacturer to determine dependence on environmental influences. If the pump is affected by temperature and pressure changes, correct the flow rate using the formula shown in the section "Sampling Pump Flow Rate Corrections" at the end of this appendix.

*5.2.6.* Connect each pump to the base of each sampling cassette with flexible tubing. Remove the end cap of each cassette and take each air sample open face. Assure that each sample cassette is held open side down in the employee's breathing zone during sampling. The distance from the nose/mouth of the employee to the cassette should be about 10 cm. Secure the cassette on the collar or lapel of the employee using spring clips or other similar devices.

*5.2.7.* A suggested minimum air volume when sampling to determine TWA compliance is 25 L. For Excursion Limit (30 min sampling time) evaluations, a minimum air volume of 48 L is recommended.

*5.2.8.* The most significant problem when sampling for asbestos is overloading the filter with non-asbestos dust. Suggested maximum air sample volumes for specific environments are:

| Environment | Air Vol. (L) |
|---|---|
| Asbestos removal operations (visible dust) | 100. |
| Asbestos removal operations (little dust) | 240. |
| Office environments | 400 to 2,400. |

*CAUTION:* Do not overload the filter with dust. High levels of non-fibrous dust particles may obscure fibers on the filter and lower the count or make counting impossible. If more than about 25 to 30% of the field area is obscured with dust, the result may be biased low. Smaller air volumes may be necessary when there is excessive non-asbestos dust in the air.

While sampling, observe the filter with a small flashlight. If there is a visible layer of dust on the filter, stop sampling, remove and seal the cassette, and replace with a new sampling assembly. The total dust loading should not exceed 1 mg.

*5.2.9.* Blank samples are used to determine if any contamination has occurred during sample handling. Prepare two blanks for the first 1 to 20 samples. For sets containing greater than 20 samples, prepare blanks as 10% of the samples. Handle blank samples in the same manner as air samples with one exception: Do not draw any air through the blank samples. Open the blank cassette in the place where the sample cassettes are mounted on the employee. Hold it open for about 30 seconds. Close and seal the cassette appropriately. Store blanks for shipment with the sample cassettes.

*5.2.10.* Immediately after sampling, close and seal each cassette with the base and plastic plugs. Do not touch or puncture the filter membrane as this will invalidate the analysis.

*5.2.11* Attach and secure a sample seal around each sample cassette in such a way as to assure that the end cap and base plugs cannot be removed without destroying the seal. Tape the ends of the seal together since the seal is not long enough to be wrapped end-to-end. Also wrap tape around the cassette at each joint to keep the seal secure.

*5.3.1.* Send the samples to the laboratory with paperwork requesting asbestos analysis. List any known fibrous interferences present during sampling on the paperwork. Also, note the workplace operation(s) sampled.

*5.3.2.* Secure and handle the samples in such that they will not rattle during shipment nor be exposed to static electricity. Do not ship samples in expanded polystyrene peanuts, vermiculite, paper shreds, or excelsior. Tape sample cassettes to sheet bubbles and place in a container that will cushion the samples in such a manner that they will not rattle.

*5.3.3.* To avoid the possibility of sample contamination, always ship bulk samples in separate mailing containers.

## *6. Analysis*

### 6.1. Safety Precautions

*6.1.1.* Acetone is extremely flammable and precautions must be taken not to ignite it. Avoid using large containers or quantities of acetone. Transfer the solvent in a ventilated laboratory hood. Do not use acetone near any open flame. For generation of acetone vapor, use a spark free heat source.

*6.1.2.* Any asbestos spills should be cleaned up immediately to prevent dispersal of fibers. Prudence should be exercised to avoid contamination of laboratory facilities or exposure of personnel to asbestos. Asbestos spills should be cleaned up with wet methods and/or a High Efficiency Particulate-Air (HEPA) filtered vacuum.

*CAUTION:* Do not use a vacuum without a HEPA filter—It will disperse fine asbestos fibers in the air.

### 6.2. Equipment

*6.2.1.* Phase contrast microscope with binocular or trinocular head.

*6.2.2.* Widefield or Huygenian 10X eyepieces (NOTE: The eyepiece containing the graticule must be a focusing eyepiece. Use a 40X phase objective with a numerical aperture of 0.65 to 0.75).

*6.2.3.* Kohler illumination (if possible) with green or blue filter.

*6.2.4.* Walton-Beckett Graticule, type G-22 with 100 ±2 μ m projected diameter.

*6.2.5.* Mechanical stage. A rotating mechanical stage is convenient for use with polarized light.

*6.2.6.* Phase telescope.

*6.2.7.* Stage micrometer with 0.01-mm subdivisions.

*6.2.8.* Phase-shift test slide, mark II (Available from PTR optics Ltd., and also McCrone).

*6.2.9.* Precleaned glass slides, 25 mm X 75 mm. One end can be frosted for convenience in writing sample numbers, etc., or paste-on labels can be used.

*6.2.10.* Cover glass 1 1/2.

*6.2.11.* Scalpel (10, curved blade).

*6.2.12.* Fine tipped forceps.

*6.2.13.* Aluminum block for clearing filter (see Appendix D and Figure 4).

*6.2.14.* Automatic adjustable pipette, 100- to 500-μ L.

*6.2.15.* Micropipette, 5 μ L.

### 6.3. Reagents

*6.3.1.* Acetone (HPLC grade).

*6.3.2.* Triacetin (glycerol triacetate).

*6.3.3.* Lacquer or nail polish.

### 6.4. Standard Preparation

A way to prepare standard asbestos samples of known concentration has not been developed. It is possible to prepare replicate samples of nearly equal concentration. This has been performed through the PAT program. These asbestos samples are distributed by the AIHA to participating laboratories.

Since only about one-fourth of a 25-mm sample membrane is required for an asbestos count, any PAT sample can serve as a "standard" for replicate counting.

### 6.5. Sample Mounting

NOTE: See Safety Precautions in Section 6.1. before proceeding. The objective is to produce samples with a smooth (non-grainy) background in a medium with a refractive index of approximately 1.46. The technique below collapses the filter for easier focusing and produces permanent mounts which are useful for quality control and interlaboratory comparison.

An aluminum block or similar device is required for sample preparation.

*6.5.1.* Heat the aluminum block to about 70 °C. The hot block should not be used on any surface that can be damaged by either the heat or from exposure to acetone.

*6.5.2.* Ensure that the glass slides and cover glasses are free of dust and fibers.

*6.5.3.* Remove the top plug to prevent a vacuum when the cassette is opened. Clean the outside of the cassette if necessary. Cut the seal and/or tape on the cassette with a razor blade. Very carefully separate the base from the extension cowl, leaving the filter and backup pad in the base.

*6.5.4.* With a rocking motion cut a triangular wedge from the filter using the scalpel. This wedge should be one-sixth to one-fourth of the filter. Grasp the filter wedge with the forceps on the perimeter of the filter which was clamped between the cassette pieces. DO NOT TOUCH the filter with your finger. Place the filter on the glass slide sample side up. Static electricity will usually keep the filter on the slide until it is cleared.

*6.5.5.* Place the tip of the micropipette containing about 200 μ L acetone into the aluminum block. Insert the glass slide into the receiving slot in the aluminum block. Inject the acetone into the block with slow, steady pressure on the plunger while holding the pipette firmly in place. Wait 3 to 5 seconds for the filter to clear, then remove the pipette and slide from the aluminum block.

*6.5.6.* Immediately (less than 30 seconds) place 2.5 to 3.5 μ L of triacetin on the filter (NOTE: Waiting longer than 30 seconds will result in increased index of refraction and decreased contrast between the fibers and the preparation. This may also lead to separation of the cover slip from the slide).

*6.5.7.* Lower a cover slip gently onto the filter at a slight angle to reduce the possibility of forming air bubbles. If more than 30 seconds have elapsed between acetone exposure and triacetin application, glue the edges of the cover slip to the slide with lacquer or nail polish.

*6.5.8.* If clearing is slow, warm the slide for 15 min on a hot plate having a surface temperature of about 50 °C to hasten clearing. The top of the hot block can be used if the slide is not heated too long.

*6.5.9.* Counting may proceed immediately after clearing and mounting are completed.

### 6.6. Sample Analysis

Completely align the microscope according to the manufacturer's instructions. Then, align the microscope using the following general alignment routine at the beginning of every counting session and more often if necessary.

#### 6.6.1. Alignment

(1) Clean all optical surfaces. Even a small amount of dirt can significantly degrade the image.

(2) Rough focus the objective on a sample.

(3) Close down the field iris so that it is visible in the field of view. Focus the image of the iris with the condenser focus. Center the image of the iris in the field of view.

(4) Install the phase telescope and focus on the phase rings. Critically center the rings. Misalignment of the rings results in astigmatism which will degrade the image.

(5) Place the phase-shift test slide on the microscope stage and focus on the lines. The analyst must see line set 3 and should see at least parts of 4 and 5 but, not see line set 6 or 6. A microscope/microscopist combination which does not pass this test may not be used.

#### 6.6.2. Counting Fibers

(1) Place the prepared sample slide on the mechanical stage of the microscope. Position the center of the wedge under the objective lens and focus upon the sample.

(2) Start counting from one end of the wedge and progress along a radial line to the other end (count in either direction from perimeter to wedge tip). Select fields randomly, without looking into the eyepieces, by slightly advancing the slide in one direction with the mechanical stage control.

(3) Continually scan over a range of focal planes (generally the upper 10 to 15 μ m of the filter surface) with the fine focus control during each field count. Spend at least 5 to 15 seconds per field.

(4) Most samples will contain asbestos fibers with fiber diameters less than 1 μ m. Look carefully for faint fiber images. The small diameter fibers will be very hard to see. However, they are an important contribution to the total count.

(5) Count only fibers equal to or longer than 5 μ m. Measure the length of curved fibers along the curve.

(6) Count fibers which have a length to width ratio of 3:1 or greater.

(7) Count all the fibers in at least 20 fields. Continue counting until either 100 fibers are counted or 100 fields have been viewed; whichever occurs first. Count all the fibers in the final field.

(8) Fibers lying entirely within the boundary of the Walton-Beckett graticule field shall receive a count of 1. Fibers crossing the boundary once, having one end within the circle shall receive a count of 1/2. Do not count any fiber that crosses the graticule boundary more than once. Reject and do not count any other fibers even though they may be visible outside the graticule area. If a fiber touches the circle, it is considered to cross the line.

(9) Count bundles of fibers as one fiber unless individual fibers can be clearly identified and each individual fiber is clearly not connected to another counted fiber. See Figure 1 for counting conventions.

(10) Record the number of fibers in each field in a consistent way such that filter non-uniformity can be assessed.

(11) Regularly check phase ring alignment.

(12) When an agglomerate (mass of material) covers more than 25% of the field of view, reject the field and select another. Do not include it in the number of fields counted.

(13) Perform a "blind recount" of 1 in every 10 filter wedges (slides). Relabel the slides using a person other than the original counter.

### 6.7. Fiber Identification

As previously mentioned in Section 1.3., PCM does not provide positive confirmation of asbestos fibers. Alternate differential counting techniques should be used if discrimination is desirable. Differential counting may include primary discrimination based on morphology, polarized light analysis of fibers, or modification of PCM data by Scanning Electron or Transmission Electron Microscopy.

A great deal of experience is required to routinely and correctly perform differential counting. It is discouraged unless it is legally necessary. Then, only if a fiber is obviously not asbestos should it be excluded from the count. Further discussion of this technique can be found in reference 8.10.

If there is a question whether a fiber is asbestos or not, follow the rule:

"WHEN IN DOUBT, COUNT."

### 6.8. Analytical Recommendations—Quality Control System

*6.8.1.* All individuals performing asbestos analysis must have taken the NIOSH course for sampling and evaluating airborne asbestos or an equivalent course.

*6.8.2.* Each laboratory engaged in asbestos counting shall set up a slide trading arrangement with at least two other laboratories in order to compare performance and eliminate inbreeding of error. The slide exchange occurs at least semiannually. The round robin results shall be posted where all analysts can view individual analyst's results.

*6.8.3.* Each laboratory engaged in asbestos counting shall participate in the Proficiency Analytical Testing Program, the Asbestos Analyst Registry or equivalent.

*6.8.4.* Each analyst shall select and count prepared slides from a "slide bank". These are quality assurance counts. The slide bank shall be prepared using uniformly distributed samples taken from the workload. Fiber densities should cover the entire range routinely analyzed by the laboratory. These slides are counted blind by all counters to establish an original standard deviation. This historical distribution is compared with the quality assurance counts. A counter must have 95% of all quality control samples counted within three standard deviations of the historical mean. This count is then integrated into a new historical mean and standard deviation for the slide.

The analyses done by the counters to establish the slide bank may be used for an interim quality control program if the data are treated in a proper statistical fashion.

## *7. Calculations*

7.1. Calculate the estimated airborne asbestos fiber concentration on the filter sample using the following formula:

$$AC = \frac{\left[\left(\frac{FB}{FL}\right) - \left(\frac{BFB}{BFL}\right)\right] \times ECA}{1000 \times FR \times T \times MFA}$$

where:

AC = Airborne fiber concentration

FB = Total number of fibers greater than 5 µ m counted

FL = Total number of fields counted on the filter

BFB = Total number of fibers greater than 5 mm counted in the blank

BFL = Total number of fields counted on the blank

ECA = Effective collecting area of filter (385 $mm^2$ nominal for a 25-mm filter.)

FR = Pump flow rate (L/min)

MFA = Microscope count field area ($mm^2$). This is 0.00785 $mm^2$ for a Walton-Beckett Graticule.

T = Sample collection time (min)

1,000 = Conversion of L to cc

**Note:** The collection area of a filter is seldom equal to 385 $mm^2$. It is appropriate for laboratories to routinely monitor the exact diameter using an inside micrometer. The collection area is calculated according to the formula:

**Area = † (d/2)²**

### 7.2. Short-Cut Calculation

Since a given analyst always has the same interpupillary distance, the number of fields per filter for a particular analyst will remain constant for a given size filter. The field size for that analyst is constant (i.e. the analyst is using an assigned microscope and is not changing the reticle).

For example, if the exposed area of the filter is always 385 $mm^2$ and the size of the field is always 0.00785 $mm^2$ the number of fields per filter will always be 49,000. In addition it is necessary to convert liters of air to cc. These three constants can then be combined such that ECA/(1,000 × MFA)=49. The previous equation simplifies to:

$$AC = \frac{\left(\frac{FB}{FL}\right) - \left(\frac{BFB}{BFL}\right) \times 49}{FR \times T}$$

### 7.3. Recount Calculations

As mentioned in step 13 of Section 6.6.2., a "blind recount" of 10% of the slides is performed. In all cases, differences will be observed between the first and second counts of the same filter wedge. Most of these differences will be due to chance alone, that is, due to the random variability (precision) of the count method. Statistical recount criteria enables one to decide whether observed differences can be explained due to chance alone or are probably due to systematic differences between analysts, microscopes, or other biasing factors.

The following recount criterion is for a pair of counts that estimate AC in fibers/cc. The criterion is given at the type-I error level. That is, there is 5% maximum risk that we will reject a pair of counts for the reason that one might be biased, when the large observed difference is really due to chance.

Reject a pair of counts if:

$$\left|\sqrt{AC_2} - \sqrt{AC_1}\right| > 2.78 \times \left(\sqrt{AC_{AVG}}\right) \times CV_{FB}$$

Where:

$AC_1$ = lower estimated airborne fiber concentration

$AC_2$ = higher estimated airborne fiber concentration

$AC_{avg}$ = average of the two concentration estimates

$CV_{FB}$ = CV for the average of the two concentration estimates

If a pair of counts are rejected by this criterion then, recount the rest of the filters in the submitted set. Apply the test and reject any other pairs failing the test. Rejection shall include a memo to the industrial hygienist stating that the sample failed a statistical test for homogeneity and the true air concentration may be significantly different than the reported value.

### 7.4. Reporting Results

Report results to the industrial hygienist as fibers/cc. Use two significant figures. If multiple analyses are performed on a sample, an average of the results is to be reported unless any of the results can be rejected for cause.

## *8. References*

8.1. Dreesen, W.C., et al., *U.S. Public Health Service: A Study of Asbestosis in the Asbestos Textile Industry* (Public Health Bulletin No. 241), U.S. Treasury Dept., Washington, DC, 1938.

8.2. *Asbestos Research Council: The Measurement of Airborne Asbestos Dust by the Membrane Filter Method* (Technical Note), Asbestos Research Council, Rockdale, Lancashire, Great Britain, 1969.

8.3. Bayer, S.G., Zumwalde, R.D., Brown, T.A., *Equipment and Procedure for Mounting Millipore Filters and Counting Asbestos Fibers by Phase Contrast Microscopy*, Bureau of Occupational Health, U.S. Dept. of Health, Education and Welfare, Cincinnati, OH, 1969.

8.4. *NIOSH Manual of Analytical Methods*, 2nd ed., Vol. 1 (DHEW/NIOSH Pub. No. 77-157-A). National Institute for Occupational Safety and Health, Cincinnati, OH, 1977. pp. 239-1—239-21.

8.5. *Asbestos*, Code of Federal Regulations 29 CFR 1910.1001. 1971.

8.6. *Occupational Exposure to Asbestos, Tremolite, Anthophyllite, and Actinolite. Final Rule*, Federal Register 51:119 (20 June 1986). pp. 22612-22790.

8.7. *Asbestos, Tremolite, Anthophyllite, and Actinolite,* Code of Federal Regulations 1910.1001. 1988. pp. 711-752.

8.8. *Criteria for a Recommended Standard—Occupational Exposure to Asbestos* (DHEW/NIOSH Pub. No. HSM 72-10267), National Institute for Occupational Safety and Health, NIOSH, Cincinnati, OH, 1972. pp. III-1—III-24.

8.9. Leidel, N.A., Bayer, S.G., Zumwalde, R.D., Busch, K.A., *USPHS/NIOSH Membrane Filter Method for Evaluating Airborne Asbestos Fibers* (DHEW/NIOSH Pub. No. 79-127). National Institute for Occupational Safety and Health, Cincinnati, OH, 1979.

8.10. Dixon, W.C., *Applications of Optical Microscopy in Analysis of Asbestos and Quartz,* Analytical Techniques in Occupational Health Chemistry, edited by D.D. Dollberg and A.W. Verstuyft. Wash. DC: American Chemical Society, (ACS Symposium Series 120) 1980. pp. 13-41.

## Quality Control

The OSHA asbestos regulations require each laboratory to establish a quality control program. The following is presented as an example of how the OSHA-SLTC constructed its internal CV curve as part of meeting this requirement. Data is from 395 samples collected during OSHA compliance inspections and analyzed from October 1980 through April 1986.

Each sample was counted by 2 to 5 different counters independently of one another. The standard deviation and the CV statistic was calculated for each sample. This data was then plotted on a graph of CV vs. fibers/mm². A least squares regression was performed using the following equation:

$$\mathbf{CV = antilog_{10}[A(log_{10}(x))^2 + B(log_{10}(x)) + C]}$$

where:

x = the number of fibers/mm²

Application of least squares gave:

A = 0.182205

B = 0.973343

C = 0.327499

Using these values, the equation becomes:

$$CV = antilog_{10}[0.182205(log_{10}(x))^2 - 0.973343(log_{10}(x)) + 0.327499]$$

## Sampling Pump Flow Rate Corrections

This correction is used if a difference greater than 5% in ambient temperature and/or pressure is noted between calibration and sampling sites and the pump does not compensate for the differences.

$$Q_{act} = Q_{cal} \times \sqrt{\left(\frac{P_{cal}}{P_{act}}\right) \times \left(\frac{T_{act}}{T_{cal}}\right)}$$

Where:

$Q_{act}$ = actual flow rate

$Q_{cal}$ = calibrated flow rate (if a rotameter was used, the rotameter value)

$P_{cal}$ = uncorrected air pressure at calibration

$P_{act}$ = uncorrected air pressure at sampling site

$T_{act}$ = temperature at sampling site (K)

$T_{cal}$ = temperature at calibration (K)

## Walton-Beckett Graticule

When ordering the Graticule for asbestos counting, specify the exact disc diameter needed to fit the ocular of the microscope and the diameter (mm) of the circular counting area. Instructions for measuring the dimensions necessary are listed:

(1) Insert any available graticule into the focusing eyepiece and focus so that the graticule lines are sharp and clear.

(2) Align the microscope.

(3) Place a stage micrometer on the microscope object stage and focus the microscope on the graduated lines.

(4) Measure the magnified grid length, PL (μ m), using the stage micrometer.

(5) Remove the graticule from the microscope and measure its actual grid length, AL (mm). This can be accomplished by using a mechanical stage fitted with verniers, or a jeweler's loupe with a direct reading scale.

(6) Let D = 100 μ m. Calculate the circle diameter, $d_c$ (mm), for the Walton-Beckett graticule and specify the diameter when making a purchase:

$$d_c = \frac{AL \times D}{PL}$$

***Example:*** If PL =108 μ m, AL = 2.93 mm and D = 100 μ m, then,

$$d_c = \frac{2.93 \times 100}{108} = 2.71\text{mm}$$

(7) Each eyepiece-objective-reticle combination on the microscope must be calibrated. Should any of the three be changed (by zoom adjustment, disassembly, replacement, etc.), the combination must be recalibrated. Calibration may change if interpupillary distance is changed.

Measure the field diameter, D (acceptable range: 100±2 μ m) with a stage micrometer upon receipt of the graticule from the manufacturer. Determine the field area (mm²).

Field Area = † $(D/2)^2$

If D = 100 m m = 0.1 mm, then

Field Area = † $(0.1\ \mu m/2)^2 = 0.00785\ mm^2$

The Graticule is available from: Graticules Ltd., Morley Road, Tonbridge TN9 IRN, Kent, England (Telephone 011-44-732-359061). Also available from PTR Optics Ltd., 145 Newton Street, Waltham, MA 02154 [telephone (617) 891-6000] or McCrone Accessories and Components, 2506 S. Michigan Ave., Chicago, IL 60616 [phone (312)-842-7100]. The graticule is custom made for each microscope.

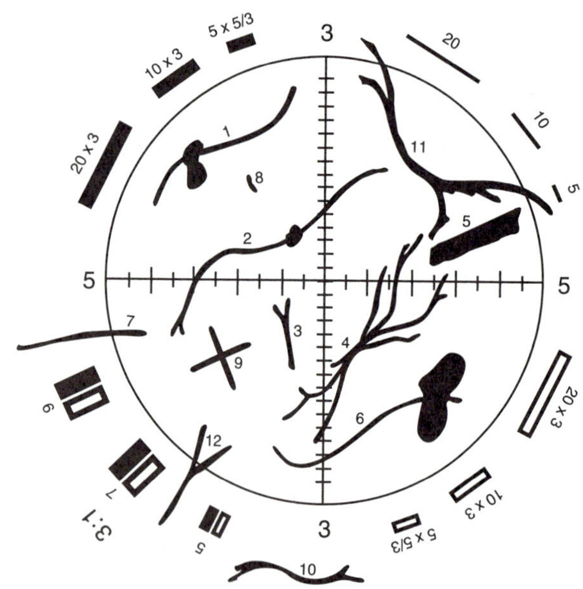


**FIGURE 1: WALTON-BECKETT GRATICULE WITH SOME EXPLANATORY FIBERS.**

**COUNTS FOR THE FIBERS IN THE FIGURE**

| Structure No. | Count | Explanation |
|---|---|---|
| 1 to 6 | 1 | Single fibers all contained within the Circle. |
| 7 | 1/2 | Fiber crosses circle once. |
| 8 | 0 | Fiber too short. |
| 9 | 2 | Two crossing fibers. |
| 10 | 0 | Fiber outside graticule. |
| 11 | 0 | Fiber crosses graticule twice. |
| 12 | 1/2 | Although split, fiber only crosses once. |

## APPENDIX C TO §1926.1101 [RESERVED]

## APPENDIX D TO §1926.1101—Medical Questionnaires; Mandatory

This mandatory appendix contains the medical questionnaires that must be administered to all employees who are exposed to asbestos above the permissible exposure limit, and who will therefore be included in their employer's medical surveillance program. Part 1 of the appendix contains the Initial Medical Questionnaire, which must be obtained for all new hires who will be covered by the medical surveillance requirements. Part 2 includes the abbreviated Periodical Medical Questionnaire, which must be administered to all employees who are provided periodic medical examinations under the medical surveillance provisions of the standard.

### Part 1
### INITIAL MEDICAL QUESTIONNAIRE

1. NAME ____________________
2. SOCIAL SECURITY # ______________ 3. CLOCK NUMBER ______________
4. PRESENT OCCUPATION ______________ 5. PLANT ______________
6. ADDRESS ______________ 7. ______________ (Zip Code)
8. TELEPHONE NUMBER ______________ 9. INTERVIEWER ______________
10. Date ______________ 11. Date of Birth ______________ Month Day Year

12. Place of Birth ____________________
13. Sex 1. Male _____ 2. Female _____
14. What is your marital status? 1. Single _____ 2. Married _____ 3. Widowed _____ 4. Separated/Divorced _____
15. Race 1. White _____ 2. Black _____ 3. Asian _____ 4. Hispanic _____ 5. Indian _____ 6. Other _____
16. What is the highest grade completed in school? ______________ (For example 12 years is completion of high school)

**OCCUPATIONAL HISTORY**

17 A. Have you ever worked full time (30 hours per week or more) for 6 months or more? 1. Yes _____ 2. No _____

IF YES TO 17A:

B. Have you ever worked for a year or more in any dusty job?
1. Yes _____ 2. No _____ 3. Does Not Apply _____
Specify job/industry ______________ Total Years Worked ______________
Was dust exposure: 1. Mild _____ 2. Moderate _____ 3. Severe _____

C. Have you ever been exposed to gas or chemical fumes in your work? 1. Yes _____ 2. No _____
Specify job/industry ______________ Total Years Worked ______________
Was dust exposure: 1. Mild _____ 2. Moderate _____ 3. Severe _____

D. What has been your usual occupation or job—the one you have worked at the longest?
1. Job occupation ______________
2. Number of years employed in this occupation ______________
3. Position/job title ______________
4. Business, field or industry ______________

**(Record on lines the years in which you have worked in any of these industries, e.g. 1960-1969)**

| Have you ever worked: | YES | NO |
|---|---|---|
| E. In a mine? | ❑ | ❑ |
| F. In a quarry? | ❑ | ❑ |
| G. In a foundry? | ❑ | ❑ |
| H. In a pottery? | ❑ | ❑ |
| I. In a cotton, flax or hemp mill? | ❑ | ❑ |
| J. With asbestos? | ❑ | ❑ |

| 18. **PAST MEDICAL HISTORY** | YES | NO |
|---|---|---|
| A. Do you consider yourself to be in good health? | ❑ | ❑ |
| If "NO" state reason ______________ | | |
| B. Have you any defect of vision? | ❑ | ❑ |
| If "YES" state nature of defect ______________ | | |
| C. Have you any hearing defect? | ❑ | ❑ |
| If "YES" state nature of defect ______________ | | |
| D. Are you suffering from or have you ever suffered from: | | |
| a. Epilepsy (or fits, seizures, convulsions)? | ❑ | ❑ |
| b. Rheumatic fever? | ❑ | ❑ |
| c. Kidney disease? | ❑ | ❑ |
| d. Bladder disease? | ❑ | ❑ |
| e. Diabetes? | ❑ | ❑ |
| f. Jaundice? | ❑ | ❑ |

19. **CHEST COLDS AND CHEST ILLNESSES**

19 A. If you get a cold, does it *usually* go to your chest? 1. Yes ______ 2. No ______ 3. Don't get colds ______
(Usually means more than 1/2 the time)

20 A. During the past 3 years, have you had any chest illnesses that have kept you off work, indoors at home, or in bed?
1. Yes ______ 2. No ______
IF YES TO 20A:
B. Did you produce phlegm with any of these chest illnesses?
1. Yes ______ 2. No ______ 3. Does Not Apply ______
C. In the last 3 years, how many such illnesses with (increased) phlegm did you have which lasted a week or more?
Number of illnesses ______ No such illnesses ______

21. Did you have any lung trouble before the age of 16? 1. Yes ______ 2. No ______

22. Have you ever had any of the following?
1 A. Attacks of bronchitis? 1. Yes ______ 2. No ______
IF YES TO 1A:
B. Was it confirmed by a doctor? 1. Yes ______ 2. No ______ 3. Does Not Apply ______
C. At what age was your first attack? Age in Years ______ Does Not Apply ______
2 A. Pneumonia (include bronchopneumonia)? 1. Yes ______ 2. No ______
IF YES TO 2A:
B. Was it confirmed by a doctor? 1. Yes ______ 2. No ______ 3. Does Not Apply ______
C. At what age did you first have it? Age in Years ______ Does Not Apply ______
3 A. Hay Fever? 1. Yes ______ 2. No ______
IF YES TO 3A:
B. Was it confirmed by a doctor? 1. Yes ______ 2. No ______ 3. Does Not Apply ______
C. At what age did it start? Age in Years ______ Does Not Apply ______

23 A. Have you ever had chronic bronchitis? 1. Yes ______ 2. No ______
IF YES TO 23A:
B. Do you still have it? 1. Yes ______ 2. No ______ 3. Does Not Apply ______
C. Was it confirmed by a doctor? 1. Yes ______ 2. No ______ 3. Does Not Apply ______
D. At what age did it start? Age in Years ______ Does Not Apply ______

24 A. Have you ever had emphysema? 1. Yes ______ 2. No ______
IF YES TO 24A:
B. Do you still have it? 1. Yes ______ 2. No ______ 3. Does Not Apply ______
C. Was it confirmed by a doctor? 1. Yes ______ 2. No ______ 3. Does Not Apply ______
D. At what age did it start? Age in Years ______ Does Not Apply ______

25 A. Have you ever had asthma? 1. Yes ______ 2. No ______
IF YES TO 25A:
B. Do you still have it? 1. Yes ______ 2. No ______ 3. Does Not Apply ______
C. Was it confirmed by a doctor? 1. Yes ______ 2. No ______ 3. Does Not Apply ______
D. At what age did it start? Age in Years ______ Does Not Apply ______
E. If you no longer have it, at what age did it stop? Age Stopped ______ Does Not Apply ______

26. Have you ever had:
A. Any other chest illness? 1. Yes ______ 2. No ______
If yes, please specify ______
B. Any chest operations? 1. Yes ______ 2. No ______
If yes, please specify ______
C. Any chest injuries? 1. Yes ______ 2. No ______
If yes, please specify ______

27 A. Has a doctor ever told you that you had heart trouble? 1. Yes ______ 2. No ______
IF YES TO 27A:
B. Have you ever had treatment for heart trouble in the past 10 years?
1. Yes ______ 2. No ______ 3. Does Not Apply ______

28 A. Has a doctor ever told you that you had high blood pressure? 1. Yes ______ 2. No ______
IF YES TO 28A:
B. Have you had any treatment for high blood pressure (hypertension) in the past 10 years?
1. Yes ______ 2. No ______ 3. Does Not Apply ______

29. When did you last have your chest X-rayed? (Year) ______ ___ ___ ___ ___
25 26 27 28

30. Where did you last have your chest X-rayed (if known)? ______
What was the outcome? ______

**FAMILY HISTORY**

31. Were either of your natural parents ever told by a doctor that they had a chronic lung condition such as:

| | FATHER 1. Yes | 2. No | 3. Don't Know | MOTHER 1. Yes | 2. No | 3. Don't Know |
|---|---|---|---|---|---|---|
| A. Chronic Bronchitis? | ___ | ___ | ___ | ___ | ___ | ___ |
| B. Emphysema? | ___ | ___ | ___ | ___ | ___ | ___ |
| C. Asthma? | ___ | ___ | ___ | ___ | ___ | ___ |
| D. Lung cancer? | ___ | ___ | ___ | ___ | ___ | ___ |
| E. Other chest conditions? | ___ | ___ | ___ | ___ | ___ | ___ |
| F. Is parent currently alive? | ___ | ___ | ___ | ___ | ___ | ___ |
| G. Please Specify | ___ | Age if Living | | ___ | Age if Living | |
| | ___ | Age at Death | | ___ | Age at Death | |
| | ___ | Don't Know | | ___ | Don't Know | |
| H. Please specify cause of death | ______ | | | ______ | | |

**COUGH**

32 A. Do you usually have a cough? (Count a cough with first smoke or on first going out of doors. Exclude clearing of throat.) [If no, skip to question 32C.] 1. Yes ______ 2. No ______
B. Do you usually cough as much as 4 to 6 times a day 4 or more days out of the week? 1. Yes ______ 2. No ______
C. Do you usually cough at all on getting up or first thing in the morning? 1. Yes ______ 2. No ______
D. Do you usually cough at all during the rest of the day or at night? 1. Yes ______ 2. No ______
IF YES TO ANY OF ABOVE (32A, B, C, or D), ANSWER THE FOLLOWING.
IF NO TO ALL, CHECK **DOES NOT APPLY** AND SKIP TO NEXT PAGE.
E. Do you usually cough like this on most days for 3 consecutive months or more during the year?
1. Yes ______ 2. No ______ 3. Does Not Apply ______
F. For how many years have you had the cough? Number of Years ______ Does Not Apply ______

33 A. Do you usually bring up phlegm from your chest? 1. Yes ______ 2. No ______
(Count phlegm with the first smoke or on first going out of doors. Exclude phlegm from the nose. Count swallowed phlegm.) (If no, skip to 33C)
B. Do you usually bring up phlegm like this as much as twice a day 4 or more days out of the week?
1. Yes ______ 2. No ______
C. Do you usually bring up phlegm at all on getting up or first thing in the morning?
1. Yes ______ 2. No ______
D. Do you usually bring up phlegm at all during the rest of the day or at night?
1. Yes ______ 2. No ______
IF YES TO ANY OF ABOVE (33A, B, C, or D), ANSWER THE FOLLOWING.
IF NO TO ALL, CHECK **DOES NOT APPLY** AND SKIP TO 34A.
E. Do you bring up phlegm like this on most days for 3 consecutive months or more during the year?
1. Yes ______ 2. No ______ 3. Does Not Apply ______
F. For how many years have you had trouble with phlegm? Number of years ______ Does not apply ______

**EPISODES OF COUGH AND PHLEGM**

34 A. Have you had periods or episodes of (increased *) cough and phlegm lasting for 3 weeks or more each year?
* (For persons who usually have cough and/or phlegm) 1. Yes ______ 2. No ______
IF YES TO 34A:
B. For how long have you had at least 1 such episode per year? Number of years ______ Does not apply ______

**WHEEZING**

35 A. Does your chest ever sound wheezy or whistling
1. When you have a cold? 1. Yes ______ 2. No ______
2. Occasionally apart from colds? 1. Yes ______ 2. No ______
3. Most days or nights? 1. Yes ______ 2. No ______
IF YES TO 1, 2, or 3 in 35A:
B. For how many years has this been present? Number of years ______ Does not apply ______

36 A. Have you ever had an attack of wheezing that has made you feel short of breath?
1. Yes ______ 2. No ______
IF YES TO 36A
B. How old were you when you had your first such attack? Age in years ______ Does not apply ______
C. Have you had 2 or more such episodes? 1. Yes ______ 2. No ______ 3. Does not apply ______
D. Have you ever required medicine or treatment for the(se) attack(s)?
1. Yes ______ 2. No ______ 3. Does not apply ______

**BREATHLESSNESS**

37. If disabled from walking by any condition other than heart or lung disease, please describe and proceed to question 39A.
    Nature of condition(s) ______

38 A. Are you troubled by shortness of breath when hurrying on the level or walking up a slight hill?
    1. Yes ______ 2. No ______
    IF YES TO 38A
    B. Do you have to walk slower than people of your age on the level because of breathlessness?
    1. Yes ______ 2. No ______ 3. Does not apply ______
    C. Do you ever have to stop for breath when walking at your own pace on the level?
    1. Yes ______ 2. No ______ 3. Does not apply ______
    D. Do you ever have to stop for breath after walking about 100 yards (or after a few minutes) on the level?
    1. Yes ______ 2. No ______ 3. Does not apply ______
    E. Are you too breathless to leave the house or breathless on dressing or climbing one flight of stairs?
    1. Yes ______ 2. No ______ 3. Does not apply ______

**TOBACCO SMOKING**

39 A. Have you ever smoked cigarettes? (No means less than 20 packs of cigarettes or 12 oz. of tobacco in a lifetime or less than 1 cigarette a day for 1 year.) 1. Yes ______ 2. No ______
    IF YES TO 39A
    B. Do you now smoke cigarettes (as of one month ago) 1. Yes ______ 2. No ______ 3. Does not apply ______
    C. How old were you when you first started regular cigarette smoking? Age in Years ______ Does not apply ______
    D. If you have stopped smoking cigarettes completely, how old were you when you stopped?
    Age stopped ______ Check if still smoking ______ Does not apply ______
    E. How many cigarettes do you smoke per day now? Cigarettes per day ______ Does not apply ______
    F. On the average of the entire time you smoked, how many cigarettes did you smoke per day?
    Cigarettes per day ______ Does not apply ______
    G. Do or did you inhale the cigarette smoke?
    1. Does not apply ______ 2. Not at all ______ 3. Slightly ______ 4. Moderately ______ 5. Deeply ______

40 A. Have you ever smoked a pipe regularly? 1. Yes ______ 2. No ______
    (Yes means more than 12 oz. of tobacco in a lifetime.)

    IF YES TO 40A:
    **FOR PERSONS WHO HAVE EVER SMOKED A PIPE**
    B. 1. How old were you when you started to smoke a pipe regularly? Age ______
    2. If you have stopped smoking a pipe completely, how old were you when you stopped.
    Age stopped ______ Check if still smoking pipe ______ Does not apply ______
    C. On the average over the entire time you smoked a pipe, how much pipe tobacco did you smoke per week?
    ______ oz. per week (a standard pouch of tobacco contains 1 1/2 oz.) ______ Does not apply
    D. How much pipe tobacco are you smoking now? ______ oz. per week ______ Not currently smoking a pipe
    E. Do you or did you inhale the pipe smoke?
    1. Never smoked ______ 2. Not at all ______ 3. Slightly ______ 4. Moderately ______ 5. Deeply ______

41 A. Have you ever smoked cigars regularly? 1. Yes ______ 2. No ______
    (Yes means more than 1 cigar a week for a year)
    IF YES TO 41A:
    **FOR PERSONS WHO HAVE EVER SMOKED CIGARS**
    B. 1. How old were you when you started smoking cigars regularly? Age ______
    2. If you have stopped smoking cigars completely, how old were you when you stopped?
    Age stopped ______ Check if still smoking cigars ______ Does not apply ______
    C. On the average over the entire time you smoked cigars, how many cigars did you smoke per week?
    Cigars per week ______ Does not apply ______
    D. How many cigars are you smoking per week now?
    Cigars per week ______ Check if not smoking cigars currently ______
    E. Do or did you inhale the cigar smoke?
    1. Never smoked ______ 2. Not at all ______ 3. Slightly ______ 4. Moderately ______ 5. Deeply ______

Signature ______ Date ______

## Part 2
## PERIODIC MEDICAL QUESTIONNAIRE

1. NAME ______
2. SOCIAL SECURITY # ______ 3. CLOCK NUMBER ______
4. PRESENT OCCUPATION ______ 5. PLANT ______
6. ADDRESS ______ 7. ______ (Zip Code)
8. TELEPHONE NUMBER ______ 9. INTERVIEWER ______
10. Date ______
11. What is your marital status? 1. Single ______ 2. Married ______ 3. Widowed ______ 4. Separated/Divorced ______

12 **OCCUPATIONAL HISTORY**
12 A. In the past year, did you work full time (30 hours per week or more) for 6 months or more?
    1. Yes ______ 2. No ______
IF YES TO 12A:
12 B. In the past year, did you work in a dusty job?
    1. Yes ______ 2. No ______ 3. Does Not Apply ______
12 C. Was dust exposure: 1. Mild ______ 2. Moderate ______ 3. Severe ______
12 D. In the past year, were you exposed to gas or chemical fumes in your work? 1. Yes ______ 2. No ______
12 E. Was exposure: 1. Mild ______ 2. Moderate ______ 3. Severe ______
12 F. In the past year, what was your: 1. Job/occupation? ______ 2. Position/job title? ______

13. **RECENT MEDICAL HISTORY**
13 A. Do you consider yourself to be in good health? Yes ______ No ______
    If NO, state reason ______
13 B. In the past year, have you developed:

| | Yes | No | | Yes | No |
|---|---|---|---|---|---|
| Epilepsy? | ______ | ______ | Diabetes? | ______ | ______ |
| Rheumatic fever? | ______ | ______ | Jaundice? | ______ | ______ |
| Kidney disease? | ______ | ______ | Cancer? | ______ | ______ |
| Bladder disease? | ______ | ______ | | | |

14. **CHEST COLDS AND CHEST ILLNESSES**
14 A. If you get a cold, does it *usually* go to your chest? (Usually means more than 1/2 the time)
    1. Yes ______ 2. No ______ 3. Don't get colds ______
15 A. During the past year, have you had any chest illnesses that have kept you off work, indoors at home, or in bed?
    1. Yes ______ 2. No ______ 3. Does not apply ______
    IF YES TO 15A:
15 B. Did you produce phlegm with any of these chest illnesses?
    1. Yes ______ 2. No ______ 3. Does not apply ______
15 C. In the past year, how many such illnesses with (increased) phlegm did you have which lasted a week or more?
    Number of illnesses ______ No such illnesses ______

16. **RESPIRATORY SYSTEM**
    In the past year have you had:

| | Yes or No | Further Comment on Positive Answers |
|---|---|---|
| Asthma | ____ | ______ |
| Bronchitis | ____ | ______ |
| Hay Fever | ____ | ______ |
| Other Allergies | ____ | ______ |
| Pneumonia | ____ | ______ |
| Tuberculosis | ____ | ______ |
| Chest Surgery | ____ | ______ |
| Other Lung Problems | ____ | ______ |
| Heart Disease | ____ | ______ |

| Do you have: | Yes or No | Further Comment on Positive Answers |
|---|---|---|
| Frequent colds | ____ | ______ |
| Chronic cough | ____ | ______ |
| Shortness of breath when walking or climbing one flight of stairs | ____ | ______ |
| Do you: Wheeze | ____ | ______ |
| Cough up phlegm | ____ | ______ |
| Smoke cigarettes | ____ | Packs per day ______ How many years ______ |

Date ______ Signature ______

### Appendix E to §1926.1101—Interpretation and Classification of Chest Roentgenograms—Mandatory

(a) Chest roentgenograms shall be interpreted and classified in accordance with a professionally accepted classification system and recorded on an interpretation form following the format of the CDC/NIOSH (M) 2.8 form. As a minimum, the content within the bold lines of this form (items 1 through 4) shall be included. This form is not to be submitted to NIOSH..

(b) Roentgenograms shall be interpreted and classified only by a B-reader, a board eligible/certified radiologist, or an experienced physician with known s.

(c) All interpreters, whenever interpreting chest roentgenograms made under this section, shall have immediately available for reference a complete set of the ILO-U/C International Classification of Radiographs for Pneumoconioses, 1980.

### Appendix F to §1926.1101—Work Practices And Engineering Controls For Class I Asbestos Operations (Non-Mandatory)

This is a non-mandatory appendix to the asbestos standards for construction and for shipyards. It describes criteria and procedures for erecting and using negative pressure enclosures for Class I Asbestos Work, when NPEs are used as an allowable control method to comply with paragraph (g)(5)(i) of this section. Many small and variable details are involved in the erection of a negative pressure enclosure. OSHA and most participants in the rulemaking agreed that only the major, more performance oriented criteria should be made mandatory. These criteria are set out in paragraph (g) of this section. In addition, this appendix includes these mandatory specifications and procedures in its guidelines in order to make this appendix coherent and helpful. The mandatory nature of the criteria which appear in the regulatory text is not changed because they are included in this "non-mandatory" appendix. Similarly, the additional criteria and procedures included as guidelines in the appendix, do not become mandatory because mandatory criteria are also included in these comprehensive guidelines.

In addition, none of the criteria, both mandatory and recommended, are meant to specify or imply the need for use of patented or licensed methods or equipment. Recommended specifications included in this attachment should not discourage the use of creative alternatives which can be shown to reliably achieve the objectives of negative-pressure enclosures.

Requirements included in this appendix, cover general provisions to be followed in all asbestos jobs, provisions which must be followed for all Class I asbestos jobs, and provisions governing the construction and testing of negative pressure enclosures. The first category includes the requirement for use of wet methods, HEPA vacuums, and immediate bagging of waste; Class I work must conform to the following provisions:

- oversight by competent person
- use of critical barriers over all openings to work area
- isolation of HVAC systems
- use of impermeable dropcloths and coverage of all objects within regulated areas

In addition, more specific requirements for NPEs include:

- maintenance of 0.02 inches water gauge within enclosure
- manometric measurements
- air movement away from employees performing removal work
- smoke testing or equivalent for detection of leaks and air direction
- deactivation of electrical circuits, if not provided with ground-fault circuit interrupters.

#### Planning the Project

The standard requires that an exposure assessment be conducted before the asbestos job is begun [§1926.1101(f)(1)]. Information needed for that assessment, includes data relating to prior similar jobs, as applied to the specific variables of the current job. The information needed to conduct the assessment will be useful in planning the project, and in complying with any reporting requirements under this standard, when significant changes are being made to a control system listed in the standard, [see also those of USEPA (40 CFR 61, subpart M). Thus, although the standard does not explicitly require the preparation of a written asbestos removal plan, the usual constituents of such a plan, i.e., a description of the enclosure, the equipment, and the procedures to be used throughout the project, must be determined before the enclosure can be erected. The following information should be included in the planning of the system:

A physical description of the work area;

A description of the approximate amount of material to be removed;

A schedule for turning off and sealing existing ventilation systems;

Personnel hygiene procedures;

A description of personal protective equipment and clothing to be worn by employees;

A description of the local exhaust ventilation systems to be used and how they are to be tested;

A description of work practices to be observed by employees;

An air monitoring plan;

A description of the method to be used to transport waste material; and

The location of the dump site.

#### Materials and Equipment Necessary for Asbestos Removal

Although individual asbestos removal projects vary in terms of the equipment required to accomplish the removal of the materials, some equipment and materials are common to most asbestos removal operations.

Plastic sheeting used to protect horizontal surfaces, seal HVAC openings or to seal vertical openings and ceilings should have a minimum thickness of 6 mils. Tape or other adhesive used to attach plastic sheeting should be of sufficient adhesive strength to support the weight of the material plus all stresses encountered during the entire duration of the project without becoming detached from the surface.

Other equipment and materials which should be available at the beginning of each project are:

—HEPA Filtered Vacuum is essential for cleaning the work area after the asbestos has been removed. It should have a long hose capable of reaching out-of-the-way places, such as areas above ceiling tiles, behind pipes, etc.

—Portable air ventilation systems installed to provide the negative air pressure and air removal from the enclosure must be equipped with a HEPA filter. The number and capacity of units required to ventilate an enclosure depend on the size of the area to be ventilated. The filters for these systems should be designed in such a manner that they can be replaced when the air flow volume is reduced by the build-up of dust in the filtration material. Pressure monitoring devices with alarms and strip chart recorders attached to each system to indicate the pressure differential and the loss due to dust buildup on the filter are recommended.

—Water sprayers should be used to keep the asbestos material as saturated as possible during removal; the sprayers will provide a fine mist that minimizes the impact of the spray on the material.

—Water used to saturate the asbestos containing material can be amended by adding at least 15 milliliters (1/4 ounce) of wetting agent in 1 liter (1 pint) of water. An example of a wetting agent is a 50/50 mixture of polyoxyethylene ether and polyoxyethylene polyglycol ester.

—Backup power supplies are recommended, especially for ventilation systems.

—Shower and bath water should be with mixed hot and cold water faucets. Water that has been used to clean personnel or equipment should either be filtered or be collected and discarded as asbestos waste. Soap and shampoo should be provided to aid in removing dust from the workers' skin and hair.

—See paragraphs (h) and (i) of this section for appropriate respiratory protection and protective clothing.

—See paragraph (k) of this section for required signs and labels.

#### Preparing the Work Area

Disabling HVAC Systems: The power to the heating, ventilation, and air conditioning systems that service the restricted area must be deactivated and locked off. All ducts, grills, access ports, windows and vents must be sealed off with two layers of plastic to prevent entrainment of contaminated air.

Operating HVAC Systems in the Restricted Area: If components of a HVAC system located in the restricted area are connected to a system that will service another zone during the project, the portion of the duct in the restricted area must be sealed and pressurized. Necessary precautions include caulking the duct joints, covering all cracks and openings with two layers of sheeting, and pressurizing the duct throughout the duration of the project by restricting the return air flow. The power to the fan supplying the positive pressure should be locked "on" to prevent pressure loss.

Sealing Elevators: If an elevator shaft is located in the restricted area, it should be either shut down or isolated by sealing with two layers of plastic sheeting. The sheeting should provide enough slack to accommodate the pressure changes in the shaft without breaking the air-tight seal.

Removing Mobile Objects: All movable objects should be cleaned and removed from the work area before an enclosure is constructed unless moving the objects creates a hazard. Mobile objects will be assumed to be contaminated and should be either cleaned with amended water and a HEPA vacuum and then removed from the area or wrapped and then disposed of as hazardous waste.

Cleaning and Sealing Surfaces: After cleaning with water and a HEPA vacuum, surfaces of stationary objects should be covered with two layers of plastic sheeting. The sheeting should be secured with duct tape or an equivalent method to provide a tight seal around the object.

Bagging Waste: In addition to the requirement for immediate bagging of waste for disposal, it is further recommended that the waste material be double-bagged and sealed in plastic bags designed for asbestos disposal. The bags should be stored in a waste storage area that can be controlled by the workers conducting the removal. Filters removed from air handling units and rubbish removed from the area are to be bagged and handled as hazardous waste.

### Constructing the Enclosure

The enclosure should be constructed to provide an air-tight seal around ducts and openings into existing ventilation systems and around penetrations for electrical conduits, telephone wires, water lines, drain pipes, etc. Enclosures should be both airtight and watertight except for those openings designed to provide entry and/or air flow control.

Size: An enclosure should be the minimum volume to encompass all of the working surfaces yet allow unencumbered movement by the worker(s), provide unrestricted air flow past the worker(s), and ensure walking surfaces can be kept free of tripping hazards.

Shape: The enclosure may be any shape that optimizes the flow of ventilation air past the worker(s).

Structural Integrity: The walls, ceilings and floors must be supported in such a manner that portions of the enclosure will not fall down during normal use.

Openings: It is not necessary that the structure be airtight; openings may be designed to direct air flow. Such openings should be located at a distance from active removal operations. They should be designed to draw air into the enclosure under all anticipated circumstances. In the event that negative pressure is lost, they should be fitted with either HEPA filters to trap dust or automatic trap doors that prevent dust from escaping the enclosure. Openings for exits should be controlled by an airlock or a vestibule.

Barrier Supports: Frames should be constructed to support all unsupported spans of sheeting.

Sheeting: Walls, barriers, ceilings, and floors should be lined with two layers of plastic sheeting having a thickness of at least 6 mil.

Seams: Seams in the sheeting material should be minimized to reduce the possibilities of accidental rips and tears in the adhesive or connections. All seams in the sheeting should overlap, be staggered and not be located at corners or wall-to-floor joints.

Areas Within an Enclosure: Each enclosure consists of a work area, a decontamination area, and waste storage area. The work area where the asbestos removal operations occur should be separated from both the waste storage area and the contamination control area by physical curtains, doors, and/or airflow patterns that force any airborne contamination back into the work area.

See paragraph (j) of this section for requirements for hygiene facilities.

During egress from the work area, each worker should step into the equipment room, clean tools and equipment, and remove gross contamination from clothing by wet cleaning and HEPA vacuuming. Before entering the shower area, foot coverings, head coverings, hand coverings, and coveralls are removed and placed in impervious bags for disposal or cleaning. Airline connections from airline respirators with HEPA disconnects and power cables from powered air-purifying respirators (PAPRs) will be disconnected just prior to entering the shower room.

### Establishing Negative Pressure Within the Enclosure

Negative Pressure: Air is to be drawn into the enclosure under all anticipated conditions and exhausted through a HEPA filter for 24 hours a day during the entire duration of the project.

Air Flow Tests: Air flow patterns will be checked before removal operations begin, at least once per operating shift and any time there is a question regarding the integrity of the enclosure. The primary test for air flow is to trace air currents with smoke tubes or other visual methods. Flow checks are made at each opening and at each doorway to demonstrate that air is being drawn into the enclosure and at each worker's position to show that air is being drawn away from the breathing zone.

Monitoring Pressure Within the Enclosure: After the initial air flow patterns have been checked, the static pressure must be monitored within the enclosure. Monitoring may be made using manometers, pressure gauges, or combinations of these devices. It is recommended that they be attached to alarms and strip chart recorders at points identified by the design engineer.

Corrective Actions: If the manometers or pressure gauges demonstrate a reduction in pressure differential below the required level, work should cease and the reason for the change investigated and appropriate changes made. The air flow patterns should be retested before work begins again.

Pressure Differential: The design parameters for static pressure differentials between the inside and outside of enclosures typically range from 0.02 to 0.10 inches of water gauge, depending on conditions. All zones inside the enclosure must have less pressure than the ambient pressure outside of the enclosure (-0.02 inches water gauge differential). Design specifications for the differential vary according to the size, configuration, and shape of the enclosure as well as ambient and mechanical air pressure conditions around the enclosure.

Air Flow Patterns: The flow of air past each worker shall be enhanced by positioning the intakes and exhaust ports to remove contaminated air from the worker's breathing zone, by positioning HEPA vacuum cleaners to draw air from the worker's breathing zone, by forcing relatively uncontaminated air past the worker toward an exhaust port, or by using a combination of methods to reduce the worker's exposure.

Air Handling Unit Exhaust: The exhaust plume from air handling units should be located away from adjacent personnel and intakes for HVAC systems.

Air Flow Volume: The air flow volume (cubic meters per minute) exhausted (removed) from the workplace must exceed the amount of makeup air supplied to the enclosure. The rate of air exhausted from the enclosure should be designed to maintain a negative pressure in the enclosure and air movement past each worker. The volume of air flow removed from the enclosure should replace the volume of the container at every 5 to 15 minutes. Air flow volume will need to be relatively high for large enclosures, enclosures with awkward shapes, enclosures with multiple openings, and operations employing several workers in the enclosure.

Air Flow Velocity: At each opening, the air flow velocity must visibly "drag" air into the enclosure. The velocity of air flow within the enclosure must be adequate to remove airborne contamination from each worker's breathing zone without disturbing the asbestos-containing material on surfaces.

Airlocks: Airlocks are mechanisms on doors and curtains that control the air flow patterns in the doorways. If air flow occurs, the patterns through doorways must be such that the air flows toward the inside of the enclosure. Sometimes vestibules, double doors, or double curtains are used to prevent air movement through the doorways. To use a vestibule, a worker enters a chamber by opening the door or curtain and then closing the entry before opening the exit door or curtain.

Airlocks should be located between the equipment room and shower room, between the shower room and the clean room, and between the waste storage area and the outside of the enclosure. The air flow between adjacent rooms must be checked using smoke tubes or other visual tests to ensure the flow patterns draw air toward the work area without producing eddies.

### Monitoring for Airborne Concentrations

In addition to the breathing zone samples taken as outlined in paragraph (f) of this section, samples of air should be taken to demonstrate the integrity of the enclosure, the cleanliness of the clean room and shower area, and the effectiveness of the HEPA filter. If the clean room is shown to be contaminated, the room must be relocated to an uncontaminated area.

Samples taken near the exhaust of portable ventilation systems must be done with care.

### General Work Practices

Preventing dust dispersion is the primary means of controlling the spread of asbestos within the enclosure. Whenever practical, the point of removal should be isolated, enclosed, covered, or shielded from the workers in the area. Waste asbestos containing materials must be bagged during or immediately after removal; the material must remain saturated until the waste container is sealed.

Waste material with sharp points or corners must be placed in hard airtight containers rather than bags.

Whenever possible, large components should be sealed in plastic sheeting and removed intact.

Bags or containers of waste will be moved to the waste holding area, washed, and wrapped in a bag with the appropriate labels.

### Cleaning the Work Area

Surfaces within the work area should be kept free of visible dust and debris to the extent feasible. Whenever visible dust appears on surfaces, the surfaces within the enclosure must be cleaned by wiping with a wet sponge, brush, or cloth and then vacuumed with a HEPA vacuum.

All surfaces within the enclosure should be cleaned before the exhaust ventilation system is deactivated and the enclosure is disassembled. An approved encapsulant may be sprayed onto areas after the visible dust has been removed.

## Appendix G to §1926.1101 [Reserved]

## Appendix H to §1926.1101—Substance Technical Information for Asbestos. Non-Mandatory

### I. Substance Identification

A. Substance: "Asbestos" is the name of a class of magnesium-silicate minerals that occur in fibrous form. Minerals that are included in this group are chrysotile, crocidolite, amosite, anthophyllite asbestos, tremolite asbestos, and actinolite asbestos.

B. Asbestos is and was used in the manufacture of heat-resistant clothing, automotive brake and clutch linings, and a variety of building materials including floor tiles, roofing felts, ceiling tiles, asbestos-cement pipe and sheet, and fire-resistant drywall. Asbestos is also present in pipe and boiler insulation materials and in sprayed-on materials located on beams, in crawlspaces, and between walls.

C. The potential for an asbestos-containing product to release breathable fibers depends largely on its degree of friability. Friable means that the material can be crumbled with hand pressure and is therefore likely to emit fibers. The fibrous fluffy sprayed-on materials used for fireproofing, insulation, or sound proofing are considered to be friable, and they readily release airborne fibers if disturbed. Materials such as vinyl-asbestos floor tile or roofing felt are considered non-friable if intact and generally do not emit airborne fibers unless subjected to sanding, sawing and other aggressive operations. Asbestos-cement pipe or sheet can emit airborne fibers if the materials are cut or sawed, or if they are broken.

D. Permissible exposure: Exposure to airborne asbestos fibers may not exceed 0.1 fibers per cubic centimeter of air (0.1 f/cc) averaged over the 8-hour workday, and 1 fiber per cubic centimeter of air (1.0 f/cc) averaged over a 30 minute work period.

### II. Health Hazard Data

A. Asbestos can cause disabling respiratory disease and various types of cancers if the fibers are inhaled. Inhaling or ingesting fibers from contaminated clothing or skin can also result in these diseases. The symptoms of these diseases generally do not appear for 20 or more years after initial exposure.

B. Exposure to asbestos has been shown to cause lung cancer, mesothelioma, and cancer of the stomach and colon. Mesothelioma is a rare cancer of the thin membrane lining of the chest and abdomen. Symptoms of mesothelioma include shortness of breath, pain in the walls of the chest, and/or abdominal pain.

### III. Respirators and Protective Clothing

A. Respirators: You are required to wear a respirator when performing tasks that result in asbestos exposure that exceeds the permissible exposure limit (PEL) of 0.1 f/cc and when performing certain designated operations. Air-purifying respirators equipped with a high-efficiency particulate air (HEPA) filter can be used where airborne asbestos fiber concentrations do not exceed 1.0 f/cc; otherwise, more protective respirators such as air-supplied, positive-pressure, full facepiece respirators must be used. Disposable respirators or dust masks are not permitted to be used for asbestos work. For effective protection, respirators must fit your face and head snugly. Your employer is required to conduct a fit test when you are first assigned a respirator and every 6 months thereafter. Respirators should not be loosened or removed in work situations where their use is required.

B. Protective Clothing: You are required to wear protective clothing in work areas where asbestos fiber concentrations exceed the permissible exposure limit (PEL) of 0.1 f/cc.

### IV. Disposal Procedures and Clean-up

A. Wastes that are generated by processes where asbestos is present include:

1. Empty asbestos shipping containers.
2. Process wastes such as cuttings, trimmings, or reject materials.
3. Housekeeping waste from wet-sweeping or HEPA-vacuuming.
4. Asbestos fireproofing or insulating material that is removed from buildings.
5. Asbestos-containing building products removed during building renovation or demolition.
6. Contaminated disposable protective clothing.

B. Empty shipping bags can be flattened under exhaust hoods and packed into airtight containers for disposal. Empty shipping drums are difficult to clean and should be sealed.

C. Vacuum bags or disposable paper filters should not be cleaned, but should be sprayed with a fine water mist and placed into a labeled waste container.

D. Process waste and housekeeping waste should be wetted with water or a mixture of water and surfactant prior to packaging in disposable containers.

E. Asbestos-containing material that is removed from buildings must be disposed of in leak-tight 6-mil plastic bags, plastic-lined cardboard containers, or plastic-lined metal containers. These wastes, which are removed while wet, should be sealed in containers before they dry out to minimize the release of asbestos fibers during handling.

### V. Access to Information

A. Each year, your employer is required to inform you of the information contained in this standard and appendices for asbestos. In addition, your employer must instruct you in the proper work practices for handling asbestos-containing materials, and the correct use of protective equipment.

B. Your employer is required to determine whether you are being exposed to asbestos. Your employer must treat exposure to thermal system insulation and sprayed-on and troweled-on surfacing material as asbestos exposure, unless results of laboratory analysis show that the material does not contain asbestos. You or your representative has the right to observe employee measurements and to record the results obtained. Your employer is required to inform you of your exposure, and, if you are exposed above the permissible exposure limit, he or she is required to inform you of the actions that are being taken to reduce your exposure to within the permissible limit.

C. Your employer is required to keep records of your exposures and medical examinations. These exposure records must be kept for at least thirty (30) years. Medical records must be kept for the period of your employment plus thirty (30) years.

D. Your employer is required to release your exposure and medical records to your physician or designated representative upon your written request.

## Appendix I to §1926.1101—Medical Surveillance Guidelines for Asbestos, Non-Mandatory

### I. Route of Entry

Inhalation, ingestion.

### II. Toxicology

Clinical evidence of the adverse effects associated with exposure to asbestos is present in the form of several well-conducted epidemiological studies of occupationally exposed workers, family contacts of workers, and persons living near asbestos mines. These studies have shown a definite association between exposure to asbestos and an increased incidence of lung cancer, pleural and peritoneal mesothelioma, gastrointestinal cancer, and asbestosis. The latter is a disabling fibrotic lung disease that is caused only by exposure to asbestos. Exposure to asbestos has also been associated with an increased incidence of esophageal, kidney, laryngeal, pharyngeal, and buccal cavity cancers. As with other known chronic occupational diseases, disease associated with asbestos generally appears about 20 years following the first occurrence of exposure: There are no known acute effects associated with exposure to asbestos.

Epidemiological studies indicate that the risk of lung cancer among exposed workers who smoke cigarettes is greatly increased over the risk of lung cancer among non-exposed smokers or exposed nonsmokers. These studies suggest that cessation of smoking will reduce the risk of lung cancer for a person exposed to asbestos but will not reduce it to the same level of risk as that existing for an exposed worker who has never smoked.

### III. Signs and Symptoms of Exposure-Related Disease

The signs and symptoms of lung cancer or gastrointestinal cancer induced by exposure to asbestos are not unique, except that a chest X-ray of an exposed patient with lung cancer may show pleural plaques, pleural calcification, or pleural fibrosis. Symptoms characteristic of mesothelioma include shortness of breath, pain in the walls of the chest, or abdominal pain. Mesothelioma has a much longer latency period compared with lung cancer (40 years versus 15-20 years), and mesothelioma is therefore more likely to be found among workers who were first exposed to asbestos at an early age. Mesothelioma is always fatal.

Asbestosis is pulmonary tibrosis caused by the accumulation of asbestos fibers in the lungs. Symptoms include shortness of breath, coughing, fatigue, and vague feelings of sickness. When the fibrosis worsens, shortness of breath occurs even at rest. The diagnosis of asbestosis is based on a history of exposure to asbestos, the presence of characteristics radiologic changes, end-inspiratory crackles (rales), and other clinical features of fibrosing lung disease. Pleural plaques and thickening are observed on X-rays taken during the early stages of the disease. Asbestosis is often a progressive disease even in the absence of continued exposure, although this appears to be a highly individualized characteristic. In severe cases, death may be caused by respiratory or cardiac failure.

### IV. Surveillance and Preventive Considerations

As noted above, exposure to asbestos has been linked to an increased risk of lung cancer, mesothelioma, gastrointestinal cancer, and asbestosis among occupationally exposed workers. Adequate screening tests to determine an employee's potential for developing serious chronic diseases, such as a cancer, from exposure to asbestos do not presently exist. However, some tests, particularly chest X-rays and pulmonary function tests, may indicate that an employee has been overexposed to asbestos increasing his or her risk of developing exposure related chronic diseases. It is important for the physician to become familiar with the operating conditions in which occupational exposure to asbestos is likely to occur. This is particularly important in evaluating medical and work histories and in conducting physical examinations. When an active employee has been identified as having been overexposed to asbestos measures taken by the employer to eliminate or mitigate further exposure should also lower the risk of serious long-term consequences.

The employer is required to institute a medical surveillance program for all employees who are or will be exposed to asbestos at or above the permissible exposure limit (0.1 fiber per cubic centimeter of air). All examinations and procedures must be performed by or under the supervision of a licensed physician, at a reasonable time and place, and at no cost to the employee.

Although broad latitude is given to the physician in prescribing specific tests to be included in the medical surveillance program, OSHA requires inclusion of the following elements in the routine examination:

(i) Medical and work histories with special emphasis directed to symptoms of the respiratory system, cardiovascular system, and digestive tract.

(ii) Completion of the respiratory disease questionnaire contained in Appendix D.

(iii) A physical examination including a chest roentgenogram and pulmonary function test that includes measurement of the employee's forced vital capacity (FVC) and forced expiratory volume at one second ($FEV_1$).

(iv) Any laboratory or other test that the examining physician deems by sound medical practice to be necessary.

The employer is required to make the prescribed tests available at least annually to those employees covered; more often than specified if recommended by the examining physician; and upon termination of employment.

The employer is required to provide the physician with the following information: A copy of this standard and appendices; a description of the employee's duties as they relate to asbestos exposure; the employee's representative level of exposure to asbestos; a description of any personal protective and respiratory equipment used; and information from previous medical examinations of the affected employee that is not otherwise available to the physician. Making this information available to the physician will aid in the evaluation of the employee's health in relation to assigned duties and fitness to wear personal protective equipment, if required.

The employer is required to obtain a written opinion from the examining physician containing the results of the medical examination; the physician's opinion as to whether the employee has any detected medical conditions that would place the employee at an increased risk of exposure-related disease; any recommended limitations on the employee or on the use of personal protective equipment; and a statement that the employee has been informed by the physician of the results of the medical examination and of any medical conditions related to asbestos exposure that require further explanation or treatment. This written opinion must not reveal specific findings or diagnoses unrelated to exposure to asbestos, and a copy of the opinion must be provided to the affected employee.

## Appendix J to §1926.1101—Smoking Cessation Program Information for Asbestos—Non-Mandatory

The following organizations provide smoking cessation information.

1. The National Cancer Institute operates a toll-free Cancer Information Service (CIS) with trained personnel to help you. Call 1-800-4-CANCER* to reach the CIS office serving your area, or write: Office of Cancer Communications, National Cancer Institute, National Institutes of Health, Building 31 Room 10A24, Bethesda, Maryland 20892.

2. American Cancer Society, 3340 Peachtree Road, N.E., Atlanta, Georgia 30026, (404) 320-3333

The American Cancer Society (ACS) is a voluntary organization composed of 58 divisions and 3,100 local units. Through "The Great American Smokeout" in November, the annual Cancer Crusade in April, and numerous educational materials, ACS helps people learn about the health hazards of smoking and become successful ex-smokers.

3. American Heart Association, 7320 Greenville Avenue, Dallas, Texas 75231, (214) 750-5300

The American Heart Association (AHA) is a voluntary organization with 130,000 members (physicians, scientists, and laypersons) in 55 state and regional groups. AHA produces a variety of publications and audiovisual materials about the effects of smoking on the heart. AHA also has developed a guidebook for incorporating a weight-control component into smoking cessation programs.

4. American Lung Association, 1740 Broadway, New York, New York 10019, (212) 245-8000

A voluntary organization of 7,500 members (physicians, nurses, and laypersons), the American Lung Association (ALA) conducts numerous public information programs about the health effects of smoking. ALA has 59 state and 85 local units. The organization actively supports legislation and information campaigns for non-smokers' rights and provides help for smokers who want to quit, for example, through "Freedom From Smoking," a self-help smoking cessation program.

5. Office on Smoking and Health, U.S. Department of Health and Human Services, 5600 Fishers Lane, Park Building, Room 110, Rockville, Maryland 20857

The Office on Smoking and Health (OSH) is the Department of Health and Human Services' lead agency in smoking control. OSH has sponsored distribution of publications on smoking-related topics, such as free flyers on relapse after initial quitting, helping a friend or family member quit smoking, the health hazards of smoking, and the effects of parental smoking on teenagers.

*In Hawaii, on Oahu call 524-1234 (call collect from neighboring islands),

Spanish-speaking staff members are available during daytime hours to callers from the following areas: California, Florida, Georgia, Illinois, New Jersey (area code 201), New York, and Texas. Consult your local telephone directory for listings of local chapters.

## Appendix K to §1926.1101—Polarized Light Microscopy of Asbestos (Non-Mandatory)

Method number:

ID-191

Matrix: Bulk

Collection Procedure:

Collect approximately 1 to 2 grams of each type of material and place into separate 20 mL scintillation vials.

Analytical Procedure:

A portion of each separate phase is analyzed by gross examination, phase-polar examination, and central stop dispersion microscopy.

Commercial manufacturers and products mentioned in this method are for descriptive use only and do not constitute endorsements by USDOL-OSHA. Similar products from other sources may be substituted.

### 1. Introduction

This method describes the collection and analysis of asbestos bulk materials by light microscopy techniques including phase- polar illumination and central-stop dispersion microscopy. Some terms unique to asbestos analysis are defined below:

*Amphibole:* A family of minerals whose crystals are formed by long, thin units which have two thin ribbons of double chain silicate with a brucite ribbon in between. The shape of each unit is similar to an "I beam". Minerals important in asbestos analysis include cummingtonite-grunerite, crocidolite, tremolite-actinolite and anthophyllite.

*Asbestos:* A term for naturally occurring fibrous minerals. Asbestos includes chrysotile, cummingtonite-grunerite asbestos (amosite), anthophyllite asbestos, tremolite asbestos, crocidolite, actinolite asbestos and any of these minerals which have been chemically treated or altered. The precise chemical formulation of each species varies with the location from which it was mined. Nominal compositions are listed:

| | |
|---|---|
| Chrysotile | $Mg_3Si_2O_5(OH)_4$ |
| Crocidolite (Riebeckite asbestos) | $Na_2Fe_3{}^2{+}Fe_2{}^3{+}Si_8O_{22}(OH)_2$ |
| Cummingtonite-Grunerite asbestos (Amosite) | $(Mg,Fe)_7\ Si_8O_{22}(OH)_2$ |
| Tremolite-Actinolite asbestos | $Ca_2(Mg,Fe)_5\ Si_8O_{22}(OH)_2$ |
| Anthophyllite asbestos | $(Mg,Fe)_7\ Si_8O_{22}\ (OH)_2$ |

*Asbestos Fiber:* A fiber of asbestos meeting the criteria for a fiber. (See section 3.5. of this Appendix).

*Aspect Ratio:* The ratio of the length of a fiber to its diameter usually defined as "length : width", e.g. 3:1.

*Brucite:* A sheet mineral with the composition $Mg(OH)_2$.

*Central Stop Dispersion Staining (microscope):* This is a dark field microscope technique that images particles using only light refracted by the particle, excluding light that travels through the particle unrefracted. This is usually accomplished with a McCrone objective or other arrangement which places a circular stop with apparent aperture equal to the objective aperture in the back focal plane of the microscope.

*Cleavage Fragments:* Mineral particles formed by the comminution of minerals, especially those characterized by relatively parallel sides and moderate aspect ratio.

*Differential Counting:* The term applied to the practice of excluding certain kinds of fibers from a phase contrast asbestos count because they are not asbestos.

*Fiber:* A particle longer than or equal to 5 µm with a length to width ratio greater than or equal to 3:1. This may include cleavage fragments. (see section 3.5 of this appendix).

*Phase Contrast:* Contrast obtained in the microscope by causing light scattered by small particles to destructively interfere with unscattered light, thereby enhancing the visibility of very small particles and particles with very low intrinsic contrast.

*Phase Contrast Microscope:* A microscope configured with a phase mask pair to create phase contrast. The technique which uses this is called Phase Contrast Microscopy (PCM).

*Phase-Polar Analysis:* This is the use of polarized light in a phase contrast microscope. It is used to see the same size fibers that are visible in air filter analysis. Although fibers finer than 1 µ m are visible, analysis of these is inferred from analysis of larger bundles that are usually present.

*Phase-Polar Microscope:* The phase-polar microscope is a phase contrast microscope which has an analyzer, a polarizer, a first order red plate and a rotating phase condenser all in place so that the polarized light image is enhanced by phase contrast.

*Sealing Encapsulant:* This is a product which can be applied, preferably by spraying, onto an asbestos surface which will seal the surface so that fibers cannot be released.

*Serpentine:* A mineral family consisting of minerals with the general composition $Mg_3(Si_2\ O_5(OH)_4$ having the magnesium in brucite layer over a silicate layer. Minerals important in asbestos analysis included in this family are chrysotile, lizardite, antigorite.

### 1.1. History

Light microscopy has been used for well over 100 years for the determination of mineral species. This analysis is carried out using specialized polarizing microscopes as well as bright field microscopes. The identification of minerals is an on-going process with many new minerals described each year. The first recorded use of asbestos was in Finland about 2500 B.C. where the material was used in the mud wattle for the wooden huts the people lived in as well as strengthening for pottery. Adverse health aspects of the mineral were noted nearly 2000 years ago when Pliny the Younger wrote about the poor health of slaves in the asbestos mines. Although known to be injurious for centuries, the first modern references to its toxicity were by the British Labor Inspectorate when it banned asbestos dust from the workplace in 1898. Asbestosis cases were described in the literature after the turn of the century. Cancer was first suspected in the mid 1930's and a causal link to mesothelioma was made in 1965. Because of the public concern for worker and public safety with the use of this material, several different types of analysis were applied to the determination of asbestos content. Light microscopy requires a great deal of experience and craft. Attempts were made to apply less subjective methods to the analysis. X-ray diffraction was partially successful in determining the mineral types but was unable to separate out the fibrous portions from the non-fibrous portions. Also, the minimum detection limit for asbestos analysis by X-ray diffraction (XRD) is about 1%. Differential Thermal Analysis (DTA) was no more successful. These provide useful corroborating information when the presence of asbestos has been shown by microscopy; however, neither can determine the difference between fibrous and non-fibrous minerals when both habits are present. The same is true of Infrared Absorption (IR).

When electron microscopy was applied to asbestos analysis, hundreds of fibers were discovered present too small to be visible in any light microscope. There are two different types of electron microscope used for asbestos analysis: Scanning Electron Microscope (SEM) and Transmission Electron Microscope (TEM). Scanning Electron Microscopy is useful in identifying minerals. The SEM can provide two of the three pieces of information required to identify fibers by electron microscopy: morphology and chemistry. The third is structure as determined by Selected Area Electron Diffraction—SAED which is performed in the TEM. Although the resolution of the SEM is sufficient for very fine fibers to be seen, accuracy of chemical analysis that can be performed on the fibers varies with fiber diameter in fibers of less than 0.2 µ m diameter. The TEM is a powerful tool to identify fibers too small to be resolved by light microscopy and should be used in conjunction with this method when necessary. The TEM can provide all three pieces of information required for fiber identification. Most fibers thicker than 1 µ m can adequately be defined in the light microscope. The light microscope remains as the best instrument for the determination of mineral type. This is because the minerals under investigation were first described analytically with the light microscope. It is inexpensive and gives positive identification for most samples analyzed. Further, when optical techniques are inadequate, there is ample indication that alternative techniques should be used for complete identification of the sample.

### 1.2. Principle

Minerals consist of atoms that may be arranged in random order or in a regular arrangement. Amorphous materials have atoms in random order while crystalline materials have long range order. Many materials are transparent to light, at least for small particles or for thin sections. The properties of these materials can be investigated by the effect that the material has on light passing through it. The six asbestos minerals are all crystalline with particular properties that have been identified and cataloged. These six minerals are anisotropic. They have a regular array of atoms, but the arrangement is not the same in all directions. Each major direction of the crystal presents a different regularity. Light photons travelling in each of these main directions will encounter different electrical neighborhoods, affecting the path and time of travel. The techniques outlined in this method use the fact that light traveling through fibers or crystals in different directions will behave differently, but predictably. The behavior of the light as it travels through a crystal can be measured and compared with known or determined values to identify the mineral species. Usually, Polarized Light Microscopy (PLM) is performed with strain-free objectives on a bright-field microscope platform. This would limit the resolution of the microscope to about 0.4 μm. Because OSHA requires the counting and identification of fibers visible in phase contrast, the phase contrast platform is used to visualize the fibers with the polarizing elements added into the light path. Polarized light methods cannot identify fibers finer than about 1μ m in diameter even though they are visible. The finest fibers are usually identified by inference from the presence of larger, identifiable fiber bundles. When fibers are present, but not identifiable by light microscopy, use either SEM or TEM to determine the fiber identity.

### 1.3. Advantages and Disadvantages

The advantages of light microcopy are:

(a) Basic identification of the materials was first performed by light microscopy and gross analysis. This provides a large base of published information against which to check analysis and analytical technique.

(b) The analysis is specific to fibers. The minerals present can exist in asbestiform, fibrous, prismatic, or massive varieties all at the same time. Therefore, bulk methods of analysis such as X-ray diffraction, IR analysis, DTA, etc. are inappropriate where the material is not known to be fibrous.

(c) The analysis is quick, requires little preparation time, and can be performed on-site if a suitably equipped microscope is available.

The disadvantages are:

(a) Even using phase-polar illumination, not all the fibers present may be seen. This is a problem for very low asbestos concentrations where agglomerations or large bundles of fibers may not be present to allow identification by inference.

(b) The method requires a great degree of sophistication on the part of the microscopist. An analyst is only as useful as his mental catalog of images. Therefore, a microscopist's accuracy is enhanced by experience. The mineralogical training of the analyst is very important. It is the basis on which subjective decisions are made.

(c) The method uses only a tiny amount of material for analysis. This may lead to sampling bias and false results (high or low). This is especially true if the sample is severely inhomogeneous.

(d) Fibers may be bound in a matrix and not distinguishable as fibers so identification cannot be made.

### 1.4. Method Performance

*1.4.1.* This method can be used for determination of asbestos content from 0 to 100% asbestos. The detection limit has not been adequately determined, although for selected samples, the limit is very low, depending on the number of particles examined. For mostly homogeneous, finely divided samples, with no difficult fibrous interferences, the detection limit is below 1%. For inhomogeneous samples (most samples), the detection limit remains undefined. NIST has conducted proficiency testing of laboratories on a national scale. Although each round is reported statistically with an average, control limits, etc., the results indicate a difficulty in establishing precision especially in the low concentration range. It is suspected that there is significant bias in the low range especially near 1%. EPA tried to remedy this by requiring a mandatory point counting scheme for samples less than 10%. The point counting procedure is tedious, and may introduce significant biases of its own. It has not been incorporated into this method.

*1.4.2.* The precision and accuracy of the quantitation tests performed in this method are unknown. Concentrations are easier to determine in commercial products where asbestos was deliberately added because the amount is usually more than a few percent. An analyst's results can be "calibrated" against the known amounts added by the manufacturer. For geological samples, the degree of homogeneity affects the precision.

*1.4.3.* The performance of the method is analyst dependent. The analyst must choose carefully and not necessarily randomly the portions for analysis to assure that detection of asbestos occurs when it is present. For this reason, the analyst must have adequate training in sample preparation, and experience in the location and identification of asbestos in samples. This is usually accomplished through substantial on-the-job training as well as formal education in mineralogy and microscopy.

### 1.5. Interferences

Any material which is long, thin, and small enough to be viewed under the microscope can be considered an interference for asbestos. There are literally hundreds of interferences in workplaces. The techniques described in this method are normally sufficient to eliminate the interferences. An analyst's success in eliminating the interferences depends on proper training.

Asbestos minerals belong to two mineral families: the serpentines and the amphiboles. In the serpentine family, the only common fibrous mineral is chrysotile. Occasionally, the mineral antigorite occurs in a fibril habit with morphology similar to the amphiboles. The amphibole minerals consist of a score of different minerals of which only five are regulated by federal standard: amosite, crocidolite, anthophyllite asbestos, tremolite asbestos and actinolite asbestos. These are the only amphibole minerals that have been commercially exploited for their fibrous properties; however, the rest can and do occur occasionally in asbestiform habit.

In addition to the related mineral interferences, other minerals common in building material may present a problem for some microscopists: gypsum, anhydrite, brucite, quartz fibers, talc fibers or ribbons, wollastonite, perlite, attapulgite, etc. Other fibrous materials commonly present in workplaces are: fiberglass, mineral wool, ceramic wool, refractory ceramic fibers, kevlar, nomex, synthetic fibers, graphite or carbon fibers, cellulose (paper or wood) fibers, metal fibers, etc.

Matrix embedding material can sometimes be a negative interference. The analyst may not be able to easily extract the fibers from the matrix in order to use the method. Where possible, remove the matrix before the analysis, taking careful note of the loss of weight. Some common matrix materials are: vinyl, rubber, tar, paint, plant fiber, cement, and epoxy. A further negative interference is that the asbestos fibers themselves may be either too small to be seen in Phase contrast Microscopy (PCM) or of a very low fibrous quality, having the appearance of plant fibers. The analyst's ability to deal with these materials increases with experience.

### 1.6. Uses and Occupational Exposure

Asbestos is ubiquitous in the environment. More than 40% of the land area of the United States is composed of minerals which may contain asbestos. Fortunately, the actual formation of great amounts of asbestos is relatively rare. Nonetheless, there are locations in which environmental exposure can be severe such as in the Serpentine Hills of California.

There are thousands of uses for asbestos in industry and the home. Asbestos abatement workers are the most current segment of the population to have occupational exposure to great amounts of asbestos. If the material is undisturbed, there is no exposure. Exposure occurs when the asbestos-containing material is abraded or otherwise disturbed during maintenance operations or some other activity. Approximately 95% of the asbestos in place in the United States is chrysotile.

Amosite and crocidolite make up nearly all the difference. Tremolite and anthophyllite make up a very small percentage. Tremolite is found in extremely small amounts in certain chrysotile deposits. Actinolite exposure is probably greatest from environmental sources, but has been identified in vermiculite containing, sprayed-on insulating materials which may have been certified as asbestos-free.

### 1.7. Physical and Chemical Properties

The nominal chemical compositions for the asbestos minerals were given in Section 1. Compared to cleavage fragments of the same minerals, asbestiform fibers possess a high tensile strength along the fiber axis. They are chemically inert, non-combustible, and heat resistant. Except for chrysotile, they are insoluble in Hydrochloric acid (HCl). Chrysotile is slightly soluble in HCl. Asbestos has high electrical resistance and good sound absorbing characteristics. It can be woven into cables, fabrics or other textiles, or matted into papers, felts, and mats.

1.8. Toxicology (This Section is for Information Only and Should Not Be Taken as OSHA Policy)

Possible physiologic results of respiratory exposure to asbestos are mesothelioma of the pleura or peritoneum, interstitial fibrosis, asbestosis, pneumoconiosis, or respiratory cancer. The possible consequences of asbestos exposure are detailed in the NIOSH Criteria Document or in the OSHA Asbestos Standards 29 CFR 1910.1001 and 29 CFR 1926.1101 and 29 CFR 1915.1001.

## 2. Sampling Procedure

### 2.1. Equipment for sampling

(a) Tube or cork borer sampling device

(b) Knife

(c) 20 mL scintillation vial or similar vial

(d) Sealing encapsulant

### 2.2. Safety Precautions

Asbestos is a known carcinogen. Take care when sampling. While in an asbestos-containing atmosphere, a properly selected and fit-tested respirator should be worn. Take samples in a manner to cause the least amount of dust. Follow these general guidelines:

(a) Do not make unnecessary dust.

(b) Take only a small amount (1 to 2 g).

(c) Tightly close the sample container.

(d) Use encapsulant to seal the spot where the sample was taken, if necessary.

### 2.3. Sampling Procedure

Samples of any suspect material should be taken from an inconspicuous place. Where the material is to remain, seal the sampling wound with an encapsulant to eliminate the potential for exposure from the sample site. Microscopy requires only a few milligrams of material. The amount that will fill a 20 mL scintillation vial is more than adequate. Be sure to collect samples from all layers and phases of material. If possible, make separate samples of each different phase of the material. This will aid in determining the actual hazard. *DO NOT USE ENVELOPES, PLASTIC OR PAPER BAGS OF ANY KIND TO COLLECT SAMPLES.* The use of plastic bags presents a contamination hazard to laboratory personnel and to other samples. When these containers are opened, a bellows effect blows fibers out of the container onto everything, including the person opening the container.

If a cork-borer type sampler is available, push the tube through the material all the way, so that all layers of material are sampled. Some samplers are intended to be disposable. These should be capped and sent to the laboratory. If a non-disposable cork borer is used, empty the contents into a scintillation vial and send to the laboratory. Vigorously and completely clean the cork borer between samples.

### 2.4 Shipment

Samples packed in glass vials must not touch or they might break in shipment.

(a) Seal the samples with a sample seal over the end to guard against tampering and to identify the sample.

(b) Package the bulk samples in separate packages from the air samples. They may cross-contaminate each other and will invalidate the results of the air samples.

(c) Include identifying paperwork *with* the samples, but not in contact with the suspected asbestos.

(d) To maintain sample accountability, ship the samples by certified mail, overnight express, or hand carry them to the laboratory.

## 3. Analysis

The analysis of asbestos samples can be divided into two major parts: sample preparation and microscopy. Because of the different asbestos uses that may be encountered by the analyst, each sample may need different preparation steps. The choices are outlined below. There are several different tests that are performed to identify the asbestos species and determine the percentage. They will be explained below.

### 3.1. Safety

(a) Do not create unnecessary dust. Handle the samples in HEPA-filter equipped hoods. If samples are received in bags, envelopes or other inappropriate container, open them only in a hood having a face velocity at or greater than 100 fpm. Transfer a small amount to a scintillation vial and only handle the smaller amount.

(b) Open samples in a hood, never in the open lab area.

(c) Index of refraction oils can be toxic. Take care not to get this material on the skin. Wash immediately with soap and water if this happens.

(d) Samples that have been heated in the muffle furnace or the drying oven may be hot. Handle them with tongs until they are cool enough to handle.

(e) Some of the solvents used, such as THF (tetrahydrofuran), are toxic and should only be handled in an appropriate fume hood and according to instructions given in the Material Safety Data Sheet (MSDS).

### 3.2. Equipment

(a) Phase contrast microscope with 10x, 16x and 40x objectives, 10x wide-field eyepieces, G-22 Walton-Beckett graticule, Whipple disk, polarizer, analyzer and first order red or gypsum plate, 100 Watt illuminator, rotating position condenser with oversize phase rings, central stop dispersion objective, Kohler illumination and a rotating mechanical stage.

(b) Stereo microscope with reflected light illumination, transmitted light illumination, polarizer, analyzer and first order red or gypsum plate, and rotating stage.

(c) Negative pressure hood for the stereo microscope

(d) Muffle furnace capable of 600 °C

(e) Drying oven capable of 50—150 °C

(f) Aluminum specimen pans

(g) Tongs for handling samples in the furnace

(h) High dispersion index of refraction oils (Special for dispersion staining.)

n = 1.550

n = 1.585

n = 1.590

n = 1.605

n = 1.620

n = 1.670

n = 1.680

n = 1.690

(i) A set of index of refraction oils from about n=1.350 to n=2.000 in n=0.005 increments. (Standard for Becke line analysis.)

(j) Glass slides with painted or frosted ends 1x3 inches 1mm (thick, precleaned.

(k) Cover Slips 22x22 mm, 1 1/2

(l) Paper clips or dissection needles

(m) Hand grinder

(n) Scalpel with both 10 and 11 blades

(o) 0.1 molar HCl

(p) Decalcifying solution (Baxter Scientific Products) Ethylenediaminetetraacetic Acid

| | |
|---|---|
| Tetrasodium | 0.7 g/l |
| Sodium Potassium Tartrate | 8.0 mg/liter |
| Hydrochloric Acid | 99.2 g/liter |
| Sodium Tartrate | 0.14 g/liter |

(q) Tetrahydrofuran (THF)

(r) Hotplate capable of 60 °C

(s) Balance

(t) Hacksaw blade

(u) Ruby mortar and pestle

### 3.3. Sample Pre-Preparation

Sample preparation begins with pre-preparation which may include chemical reduction of the matrix, heating the sample to dryness or heating in the muffle furnace. The end result is a sample which has been reduced to a powder that is sufficiently fine to fit under the cover slip. Analyze different phases of samples separately, e.g., tile and the tile mastic should be analyzed separately as the mastic may contain asbestos while the tile may not.

#### (a) *Wet Samples*

Samples with a high water content will not give the proper dispersion colors and must be dried prior to sample mounting. Remove the lid of the scintillation vial, place the bottle in the drying oven and heat at 100 °C to dryness (usually about 2 h). Samples which are not submitted to the lab in glass must be removed and placed in glass vials or aluminum weighing pans before placing them in the drying oven.

### (b) *Samples With Organic Interference—Muffle Furnace*

These may include samples with tar as a matrix, vinyl asbestos tile, or any other organic that can be reduced by heating. Remove the sample from the vial and weigh in a balance to determine the weight of the submitted portion. Place the sample in a muffle furnace at 500 °C for 1 to 2 h or until all obvious organic material has been removed. Retrieve, cool and weigh again to determine the weight loss on ignition. This is necessary to determine the asbestos content of the submitted sample, because the analyst will be looking at a reduced sample.

NOTE: Heating above 600 °C will cause the sample to undergo a structural change which, given sufficient time, will convert the chrysotile to forsterite. Heating even at lower temperatures for 1 to 2 h may have a measurable effect on the optical properties of the minerals. If the analyst is unsure of what to expect, a sample of standard asbestos should be heated to the same temperature for the same length of time so that it can be examined for the proper interpretation.

### (c) *Samples With Organic Interference—THF*

Vinyl asbestos tile is the most common material treated with this solvent, although, substances containing tar will sometimes yield to this treatment. Select a portion of the material and then grind it up if possible. Weigh the sample and place it in a test tube. Add sufficient THF to dissolve the organic matrix. This is usually about 4 to 5 mL. *Remember, THF is highly flammable.* Filter the remaining material through a tared silver membrane, dry and weigh to determine how much is left after the solvent extraction. Further process the sample to remove carbonate or mount directly.

### (d) *Samples With Carbonate Interference*

Carbonate material is often found on fibers and sometimes must be removed in order to perform dispersion microscopy. Weigh out a portion of the material and place it in a test tube. Add a sufficient amount of 0.1 M HCl or decalcifying solution in the tube to react all the carbonate as evidenced by gas formation; i.e., when the gas bubbles stop, add a little more solution. If no more gas forms, the reaction is complete. Filter the material out through a tared silver membrane, dry and weigh to determine the weight lost.

## 3.4. Sample Preparation

Samples must be prepared so that accurate determination can be made of the asbestos type and amount present. The following steps are carried out in the low-flow hood (a low-flow hood has less than 50 fpm flow):

(1) If the sample has large lumps, is hard, or cannot be made to lie under a cover slip, the grain size must be reduced. Place a small amount between two slides and grind the material between them or grind a small amount in a clean mortar and pestle. The choice of whether to use an alumina, ruby, or diamond mortar depends on the hardness of the material. Impact damage can alter the asbestos mineral if too much mechanical shock occurs. (Freezer mills can completely destroy the observable crystallinity of asbestos and should not be used). For some samples, a portion of material can be shaved off with a scalpel, ground off with a hand grinder or hack saw blade.

The preparation tools should either be disposable or cleaned thoroughly. Use vigorous scrubbing to loosen the fibers during the washing. Rinse the implements with copious amounts of water and air-dry in a dust-free environment.

(2) If the sample is powder or has been reduced as in (1) above, it is ready to mount. Place a glass slide on a piece of optical tissue and write the identification on the painted or frosted end. Place two drops of index of refraction medium n=1.550 on the slide. (The medium n=1.550 is chosen because it is the matching index for chrysotile. Dip the end of a clean paper-clip or dissecting needle into the droplet of refraction medium *on the slide* to moisten it. Then dip the probe into the powder sample. Transfer what sticks on the probe to the slide. The material on the end of the probe should have a diameter of about 3 mm for a good mount. If the material is very fine, less sample may be appropriate. For non-powder samples such as fiber mats, forceps should be used to transfer a small amount of material to the slide. Stir the material in the medium on the slide, spreading it out and making the preparation as uniform as possible. Place a cover-slip on the preparation by gently lowering onto the slide and allowing it to fall "trapdoor" fashion on the preparation to push out any bubbles. Press gently on the cover slip to even out the distribution of particulate on the slide. If there is insufficient mounting oil on the slide, one or two drops may be placed near the edge of the coverslip on the slide. Capillary action will draw the necessary amount of liquid into the preparation. Remove excess oil with the point of a laboratory wiper.

Treat at least two different areas of each phase in this fashion. Choose representative areas of the sample. It may be useful to select particular areas or fibers for analysis. This is useful to identify asbestos in severely inhomogeneous samples.

When it is determined that amphiboles may be present, repeat the above process using the appropriate high-dispersion oils until an identification is made or all six asbestos minerals have been ruled out. Note that percent determination must be done in the index medium 1.550 because amphiboles tend to disappear in their matching mediums.

## 3.5. Analytical procedure

NOTE: This method presumes some knowledge of mineralogy and optical petrography.

The analysis consists of three parts: The determination of whether there is asbestos present, what type is present and the determination of how much is present. The general flow of the analysis is:

(1) Gross examination.

(2) Examination under polarized light on the stereo microscope.

(3) Examination by phase-polar illumination on the compound phase microscope.

(4) Determination of species by dispersion stain. Examination by Becke line analysis may also be used; however, this is usually more cumbersome for asbestos determination.

(5) Difficult samples may need to be analyzed by SEM or TEM, or the results from those techniques combined with light microscopy for a definitive identification.

Identification of a particle as asbestos requires that it be asbestiform. Description of particles should follow the suggestion of Campbell. (Figure 1)

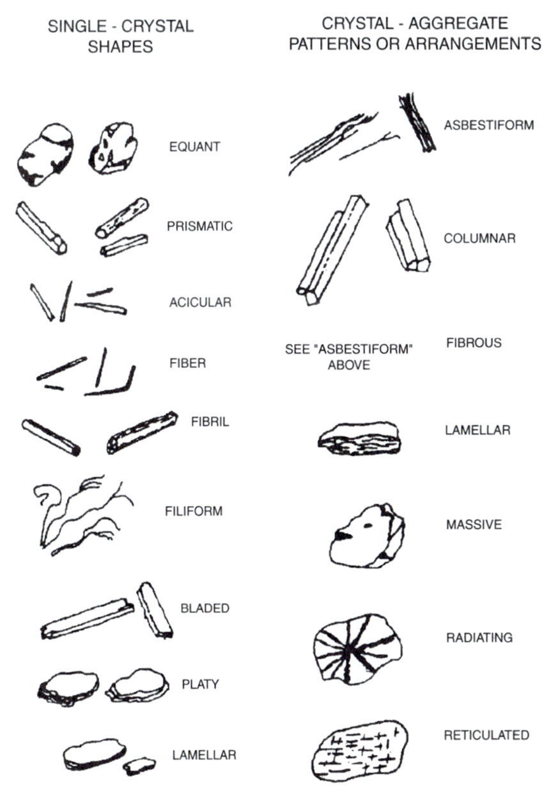


**FIGURE 1—PARTICLE DEFINITIONS SHOWING MINERAL GROWTH HABITS. FROM THE U.S. BUREAU OF MINES**

For the purpose of regulation, the mineral must be one of the six minerals covered and must be in the asbestos growth habit. Large specimen samples of asbestos generally have the gross appearance of wood. Fibers are easily parted from it. Asbestos fibers are very long compared with their widths. The fibers have a very high tensile strength as demonstrated by bending without breaking. Asbestos fibers exist in bundles that are easily parted, show longitudinal fine structure and may be tufted at the ends showing "bundle of sticks" morphology. In the microscope some of these properties may not be observable. Amphiboles do not always show striations along their length even when they are asbestos. Neither will they always show tufting. They generally do not show a curved nature except for very long fibers. Asbestos and asbestiform minerals are usually characterized in groups by extremely high aspect ratios (greater than 100:1). While aspect ratio analysis is useful for characterizing populations of fibers, it cannot be used to identify individual fibers of intermediate to short aspect ratio. Observation of many fibers is often necessary to determine whether a sample consists of "cleavage fragments" or of asbestos fibers.

Most cleavage fragments of the asbestos minerals are easily distinguishable from true asbestos fibers. This is because true cleavage fragments usually have larger diameters than 1 μm. Internal structure of particles larger than this usually shows them to have no internal fibrillar structure. In addition, cleavage fragments of the monoclinic amphiboles show inclined extinction under crossed polars with no compensator. Asbestos fibers usually show extinction at zero degrees or ambiguous extinction if any at all. Morphologically, the larger cleavage fragments are obvious by their blunt or stepped ends showing prismatic habit. Also, they tend to be acicular rather than filiform.

Where the particles are less than 1 μm in diameter and have an aspect ratio greater than or equal to 3:1, it is recommended that the sample be analyzed by SEM or TEM if there is any question whether the fibers are cleavage fragments or asbestiform particles.

Care must be taken when analyzing by electron microscopy because the interferences are different from those in light microscopy and may structurally be very similar to asbestos. The classic interference is between anthophyllite and biopyribole or intermediate fiber. Use the same morphological clues for electron microscopy as are used for light microscopy, e.g. fibril splitting, internal longitudinal striation, fraying, curvature, etc.

(1) Gross examination:

Examine the sample, preferably in the glass vial. Determine the presence of any obvious fibrous component. Estimate a percentage based on previous experience and current observation. Determine whether any pre-preparation is necessary. Determine the number of phases present. This step may be carried out or augmented by observation at 6 to 40× under a stereo microscope.

(2) After performing any necessary pre-preparation, prepare slides of each phase as described above. Two preparations of the same phase in the same index medium can be made side-by-side on the same glass for convenience. Examine with the polarizing stereo microscope. Estimate the percentage of asbestos based on the amount of birefringent fiber present.

(3) Examine the slides on the phase-polar microscopes at magnifications of 160 and 400×. Note the morphology of the fibers. Long, thin, very straight fibers with little curvature are indicative of fibers from the amphibole family. Curved, wavy fibers are usually indicative of chrysotile. Estimate the percentage of asbestos on the phase-polar microscope under conditions of crossed polars and a gypsum plate. Fibers smaller than 1.0 μm in thickness must be identified by inference to the presence of larger, identifiable fibers and morphology. If no larger fibers are visible, electron microscopy should be performed. At this point, only a tentative identification can be made. Full identification must be made with dispersion microscopy. Details of the tests are included in the appendices.

(4) Once fibers have been determined to be present, they must be identified. Adjust the microscope for dispersion mode and observe the fibers. The microscope has a rotating stage, one polarizing element, and a system for generating dark-field dispersion microscopy (see Section 4.6. of this appendix). Align a fiber with its length parallel to the polarizer and note the color of the Becke lines. Rotate the stage to bring the fiber length perpendicular to the polarizer and note the color. Repeat this process for every fiber or fiber bundle examined. The colors must be consistent with the colors generated by standard asbestos reference materials for a positive identification. In n=1.550, amphiboles will generally show a yellow to straw-yellow color indicating that the fiber indices of refraction are higher than the liquid. If long, thin fibers are noted and the colors are yellow, prepare further slides as above in the suggested matching liquids listed below:

| Type of asbestos | Index of refraction |
|---|---|
| Chrysotile | n=1.550. |
| Amosite | n=1.670 r 1.680. |
| Crocidolite | n=1.690. |
| Anthophyllite | n=1.605 nd 1.620. |
| Tremolite | n=1.605 and 1.620. |
| Actinolite | n=1.620. |

Where more than one liquid is suggested, the first is preferred; however, in some cases this liquid will not give good dispersion color. Take care to avoid interferences in the other liquid; e.g., wollastonite in n=1.620 will give the same colors as tremolite. In n=1.605 wollastonite will appear yellow in all directions. Wollastonite may be determined under crossed polars as it will change from blue to yellow as it is rotated along its fiber axis by tapping on the cover slip. Asbestos minerals will not change in this way.

Determination of the angle of extinction may, when present, aid in the determination of anthophyllite from tremolite. True asbestos fibers usually have 0° extinction or ambiguous extinction, while cleavage fragments have more definite extinction.

Continue analysis until both preparations have been examined and all present species of asbestos are identified. If there are no fibers present, or there is less than 0.1% present, end the analysis with the minimum number of slides (2).

(5) Some fibers have a coating on them which makes dispersion microscopy very difficult or impossible. Becke line analysis or electron microscopy may be performed in those cases. Determine the percentage by light microscopy. TEM analysis tends to overestimate the actual percentage present.

(6) Percentage determination is an estimate of occluded area, tempered by gross observation. Gross observation information is used to make sure that the high magnification microscopy does not greatly over- or underestimate the amount of fiber present. This part of the analysis requires a great deal of experience. Satisfactory models for asbestos content analysis have not yet been developed, although some models based on metallurgical grain-size determination have found some utility. Estimation is more easily handled in situations where the grain sizes visible at about 160× are about the same and the sample is relatively homogeneous.

View all of the area under the cover slip to make the percentage determination. View the fields while moving the stage, paying attention to the clumps of material. These are not usually the best areas to perform dispersion microscopy because of the interference from other materials. But, they are the areas most likely to represent the accurate percentage in the sample. Small amounts of asbestos require slower scanning and more frequent analysis of individual fields.

Report the area occluded by asbestos as the concentration. This estimate does not generally take into consideration the difference in density of the different species present in the sample. For most samples this is adequate. Simulation studies with similar materials must be carried out to apply microvisual estimation for that purpose and is beyond the scope of this procedure.

(7) Where successive concentrations have been made by chemical or physical means, the amount reported is the percentage of the material in the "as submitted" or original state. The percentage determined by microscopy is multiplied by the fractions remaining after pre-preparation steps to give the percentage in the original sample. For example:

Step 1. 60% remains after heating at 550 °C for 1 h.

Step 2. 30% of the residue of step 1 remains after dissolution of carbonate in 0.1 m HCl.

Step 3. Microvisual estimation determines that 5% of the sample is chrysotile asbestos.

The reported result is:

R = (Microvisual result in percent) × (Fraction remaining after step 2) × (Fraction remaining of original sample after step 1)

R = (5) × (.30) × (.60) = 0.9%

(8) Report the percent and type of asbestos present. For samples where asbestos was identified, but is less than 1.0%, report "Asbestos present, less than 1.0%." There must have been at least two observed fibers or fiber bundles in the two preparations to be reported as present. For samples where asbestos was not seen, report as "None Detected."

## Auxiliary Information

Because of the subjective nature of asbestos analysis, certain concepts and procedures need to be discussed in more depth. This information will help the analyst understand why some of the procedures are carried out the way they are.

### 4.1. Light

Light is electromagnetic energy. It travels from its source in packets called quanta. It is instructive to consider light as a plane wave. The light has a direction of travel. Perpendicular to this and mutually perpendicular to each other, are two vector components. One is the magnetic vector and the other is the electric vector. We shall only be concerned with the electric vector. In this description, the interaction of the vector and the mineral will describe all the observable phenomena. From a light source such a microscope illuminator, light travels in all different direction from the filament.

In any given direction away from the filament, the electric vector is perpendicular to the direction of travel of a light ray. While perpendicular, its orientation is random about the travel axis. If the electric vectors from all the light rays were lined up by passing the light through a filter that would only let light rays with electric vectors oriented in one direction pass, the light would then be *POLARIZED.*

Polarized light interacts with matter in the direction of the electric vector. This is the polarization direction. Using this property it is possible to use polarized light to probe different materials and identify them by how they interact with light.

The speed of light in a vacuum is a constant at about $2.99\times10^8$ m/s. When light travels in different materials such as air, water, minerals or oil, it does not travel at this speed. It travels slower. This slowing is a function of both the material through which the light is traveling and the wavelength or frequency of the light. In general, the more dense the material, the slower the light travels. Also, generally, the higher the frequency, the slower the light will travel. The ratio of the speed of light in a vacuum to that in a material is called the index of refraction (n). It is usually measured at 589 nm (the sodium D line). If white light (light containing all the visible wavelengths) travels through a material, rays of longer wavelengths will travel faster than those of shorter wavelengths, this separation is called dispersion. Dispersion is used as an identifier of materials as described in Section 4.6.

### 4.2. Material Properties

Materials are either amorphous or crystalline. The difference between these two descriptions depends on the positions of the atoms in them. The atoms in amorphous materials are randomly arranged with no long range order. An example of an amorphous material is glass. The atoms in crystalline materials, on the other hand, are in regular arrays and have long range order. Most of the atoms can be found in highly predictable locations. Examples of crystalline material are salt, gold, and the asbestos minerals.

It is beyond the scope of this method to describe the different types of crystalline materials that can be found, or the full description of the classes into which they can fall. However, some general crystallography is provided below to give a foundation to the procedures described.

With the exception of anthophyllite, all the asbestos minerals belong to the monoclinic crystal type. The unit cell is the basic repeating unit of the crystal and for monoclinic crystals can be described as having three unequal sides, two 90° angles and one angle not equal to 90°. The orthorhombic group, of which anthophyllite is a member has three unequal sides and three 90° angles. The unequal sides are a consequence of the complexity of fitting the different atoms into the unit cell. Although the atoms are in a regular array, that array is not symmetrical in all directions. There is long range order in the three major directions of the crystal. However, the order is different in each of the three directions. This has the effect that the index of refraction is different in each of the three directions. Using polarized light, we can investigate the index of refraction in each of the directions and identify the mineral or material under investigation. The indices $\alpha$, $\beta$, and $\gamma$ are used to identify the lowest, middle, and highest index of refraction respectively. The x direction, associated with a is called the fast axis. Conversely, the z direction is associated with $\gamma$ and is the slow direction. Crocidolite has a along the fiber length making it "length-fast". The remainder of the asbestos minerals have the $\gamma$ axis along the fiber length. They are called "length-slow". This orientation to fiber length is used to aid in the identification of asbestos.

### 4.3. Polarized Light Technique

Polarized light microscopy as described in this section uses the phase-polar microscope described in Section 3.2. A phase contrast microscope is fitted with two polarizing elements, one below and one above the sample. The polarizers have their polarization directions at right angles to each other. Depending on the tests performed, there may be a compensator between these two polarizing elements. A compensator is a piece of mineral with known properties that "compensates" for some deficiency in the optical train. Light emerging from a polarizing element has its electric vector pointing in the polarization direction of the element. The light will not be subsequently transmitted through a second element set at a right angle to the first element. Unless the light is altered as it passes from one element to the other, there is no transmission of light.

### 4.4. Angle of Extinction

Crystals which have different crystal regularity in two or three main directions are said to be anisotropic. They have a different index of refraction in each of the main directions. When such a crystal is inserted between the crossed polars, the field of view is no longer dark but shows the crystal in color. The color depends on the properties of the crystal. The light acts as if it travels through the crystal along the optical axes. If a crystal optical axis were lined up along one of the polarizing directions (either the polarizer or the analyzer) the light would appear to travel only in that direction, and it would blink out or go dark. The difference in degrees between the fiber direction and the angle at which it blinks out is called the angle of extinction. When this angle can be measured, it is useful in identifying the mineral. The procedure for measuring the angle of extinction is to first identify the polarization direction in the microscope. A commercial alignment slide can be used to establish the polarization directions or use anthophyllite or another suitable mineral. This mineral has a zero degree angle of extinction and will go dark to extinction as it aligns with the polarization directions. When a fiber of anthophyllite has gone to extinction, align the eyepiece reticle or graticule with the fiber so that there is a visual cue as to the direction of polarization in the field of view. Tape or otherwise secure the eyepiece in this position so it will not shift.

After the polarization direction has been identified in the field of view, move the particle of interest to the center of the field of view and align it with the polarization direction. For fibers, align the fiber along this direction. Note the angular reading of the rotating stage. Looking at the particle, rotate the stage until the fiber goes dark or "blinks out". Again note the reading of the stage. The difference in the first reading and the second is an angle of extinction.

The angle measured may vary as the orientation of the fiber changes about its long axis. Tables of mineralogical data usually report the maximum angle of extinction. Asbestos forming minerals, when they exhibit an angle of extinction, usually do show an angle of extinction close to the reported maximum, or as appropriate depending on the substitution chemistry.

### 4.5. Crossed Polars with Compensator

When the optical axes of a crystal are not lined up along one of the polarizing directions (either the polarizer or the analyzer) part of the light travels along one axis and part travels along the other visible axis. This is characteristic of birefringent materials.

The color depends on the difference of the two visible indices of refraction and the thickness of the crystal. The maximum difference available is the difference between the a and the $\gamma$ axes. This maximum difference is usually tabulated as the birefringence of the crystal.

For this test, align the fiber at 45° to the polarization directions in order to maximize the contribution to each of the optical axes. The colors seen are called retardation colors. They arise from the recombination of light which has traveled through the two separate directions of the crystal. One of the rays is retarded behind the other since the light in that direction travels slower. On recombination, some of the colors which make up white light are enhanced by constructive interference and some are suppressed by destructive interference. The result is a color dependent on the difference between the indices and the thickness of the crystal. The proper colors, thicknesses, and retardations are shown on a Michel-Levy chart. The three items, retardation, thickness and birefringence are related by the following relationship:

$$\mathbf{R = t(n_\gamma - n_\alpha)}$$

R = retardation, t = crystal thickness in $\mu$ m,

and

$n_{\alpha,\gamma}$ = indices of refraction.

Examination of the equation for asbestos minerals reveals that the visible colors for almost all common asbestos minerals and fiber sizes are shades of gray and black. The eye is relatively poor at discriminating different shades of gray. It is very good at discriminating different colors. In order to compensate for the low retardation, a compensator is added to the light train between the polarization elements. The compensator used for this test is a gypsum plate of known thickness and birefringence. Such a compensator when oriented at 45° to the polarizer direction, provides a retardation of 530 nm of the 530 nm wavelength color. This enhances the red color and gives the background a characteristic red to red-magenta color. If this "full-wave" compensator is in place when the asbestos preparation is inserted into the light train, the colors seen on the fibers are quite different. Gypsum, like asbestos has a fast axis and a slow axis. When a fiber is aligned with its fast axis in the same direction as the fast axis of the gypsum plate, the ray vibrating in the slow direction is retarded by both the asbestos and the gypsum. This results in a higher retardation than would be present for either of the two minerals. The color seen is a second order blue. When the fiber is rotated 90° using the rotating stage, the slow direction of the fiber is now aligned with the fast direction of the gypsum and the fast direction of the fiber is aligned with the slow direction of the gypsum. Thus, one ray vibrates faster in the fast direction of the gypsum, and slower in the slow direction of the fiber; the other ray will vibrate slower in the slow direction of the gypsum and faster in the fast direction of the fiber. In this case, the effect is subtractive and the color seen is a first order yellow. As long as the fiber thickness does not add appreciably to the color, the same basic colors will be seen for all asbestos types except crocidolite. In crocidolite the colors will be weaker, may be in the opposite directions, and will be altered by the blue absorption color natural to crocidolite. Hundreds of other materials will give the same colors as asbestos, and therefore, this test is not definitive for asbestos. The test is useful in discriminating against fiberglass or other amorphous fibers such as some synthetic fibers. Certain synthetic fibers will show retardation colors different than asbestos; however, there are some forms of polyethylene and aramid which will show morphology and retardation colors similar to asbestos minerals. This test must be supplemented with a positive identification test when birefringent fibers are present which can not be excluded by morphology. This test is relatively ineffective for use on fibers less than 1 $\mu$ m in diameter. For positive confirmation TEM or SEM should be used if no larger bundles or fibers are visible.

## 4.6. Dispersion Staining

Dispersion microscopy or dispersion staining is the method of choice for the identification of asbestos in bulk materials. Becke line analysis is used by some laboratories and yields the same results as does dispersion staining for asbestos and can be used in lieu of dispersion staining. Dispersion staining is performed on the same platform as the phase-polar analysis with the analyzer and compensator removed. One polarizing element remains to define the direction of the light so that the different indices of refraction of the fibers may be separately determined. Dispersion microscopy is a dark-field technique when used for asbestos. Particles are imaged with scattered light. Light which is unscattered is blocked from reaching the eye either by the back field image mask in a McCrone objective or a back field image mask in the phase condenser. The most convenient method is to use the rotating phase condenser to move an oversized phase ring into place. The ideal size for this ring is for the central disk to be just larger than the objective entry aperture as viewed in the back focal plane. The larger the disk, the less scattered light reaches the eye. This will have the effect of diminishing the intensity of dispersion color and will shift the actual color seen. The colors seen vary even on microscopes from the same manufacturer. This is due to the different bands of wavelength exclusion by different mask sizes. The mask may either reside in the condenser or in the objective back focal plane. It is imperative that the analyst determine by experimentation with asbestos standards what the appropriate colors should be for each asbestos type. The colors depend also on the temperature of the preparation and the exact chemistry of the asbestos. Therefore, some slight differences from the standards should be allowed. This is not a serious problem for commercial asbestos uses. This technique is used for identification of the indices of refraction for fibers by recognition of color. There is no direct numerical readout of the index of refraction. Correlation of color to actual index of refraction is possible by referral to published conversion tables. This is not necessary for the analysis of asbestos. Recognition of appropriate colors along with the proper morphology are deemed sufficient to identify the commercial asbestos minerals. Other techniques including SEM, TEM, and XRD may be required to provide additional information in order to identify other types of asbestos.

Make a preparation in the suspected matching high dispersion oil, e.g., n=1.550 for chrysotile. Perform the preliminary tests to determine whether the fibers are birefringent or not. Take note of the morphological character. Wavy fibers are indicative of chrysotile while long, straight, thin, frayed fibers are indicative of amphibole asbestos. This can aid in the selection of the appropriate matching oil. The microscope is set up and the polarization direction is noted as in Section 4.4. Align a fiber with the polarization direction. Note the color. This is the color parallel to the polarizer. Then rotate the fiber rotating the stage 90° so that the polarization direction is across the fiber. This is the perpendicular position. Again note the color. Both colors must be consistent with standard asbestos minerals in the correct direction for a positive identification of asbestos. If only one of the colors is correct while the other is not, the identification is not positive. If the colors in both directions are bluish-white, the analyst has chosen a matching index oil which is higher than the correct matching oil, e.g. the analyst has used n=1.620 where chrysotile is present. The next lower oil (Section 3.5.) should be used to prepare another specimen. If the color in both directions is yellow-white to straw-yellow-white, this indicates that the index of the oil is lower than the index of the fiber, e.g. the preparation is in n=1.550 while anthophyllite is present. Select the next higher oil (Section 3.5.) and prepare another slide. Continue in this fashion until a positive identification of all asbestos species present has been made or all possible asbestos species have been ruled out by negative results in this test. Certain plant fibers can have similar dispersion colors as asbestos. Take care to note and evaluate the morphology of the fibers or remove the plant fibers in pre-preparation. Coating material on the fibers such as carbonate or vinyl may destroy the dispersion color. Usually, there will be some outcropping of fiber which will show the colors sufficient for identification. When this is not the case, treat the sample as described in Section 3.3. and then perform dispersion staining. Some samples will yield to Becke line analysis if they are coated or electron microscopy can be used for identification.

## 5. References

5.1. Crane, D.T., *Asbestos in Air,* OSHA method ID160, Revised November 1992.

5.2. Ford, W.E., *Dana's Textbook of Mineralogy;* Fourth Ed.; John Wiley and Son, New York, 1950, p. vii.

5.3. Selikoff, I.J., Lee, D.H.K., *Asbestos and Disease,* Academic Press, New York, 1978, pp. 3,20.

5.4. *Women Inspectors of Factories.* Annual Report for 1898, H.M. Statistical Office, London, p. 170 (1898).

5.5. Selikoff,.I.J., Lee, D.H.K., *Asbestos and Disease,* Academic Press, New York, 1978, pp. 26,30.

5.6. Campbell, W.J., et al, *Selected Silicate Minerals and Their Asbestiform Varieties,* United States Department of the Interior, Bureau of Mines, Information Circular 8751, 1977.

5.7. *Asbestos,* Code of Federal Regulations, 29 CFR 1910.1001 and 29 CFR 1926.58.

5.8. *National Emission Standards for Hazardous Air Pollutants; Asbestos NESHAP Revision, Federal Register, Vol. 55, No. 224, 20 November 1990, p. 48410.*

5.9. Ross, M. *The Asbestos Minerals: Definitions, Description, Modes of Formation, Physical and Chemical Properties and Health Risk to the Mining Community,* Nation Bureau of Standards Special Publication, Washington, DC, 1977.

5.10. Lilis, R., Fibrous Zeolites and Endemic Mesothelioma in Cappadocia, Turkey, *J. Occ Medicine,* 1981, 23,(8),548-550.

5.11. *Occupational Exposure to Asbestos—1972,* U.S. Department of Health Education and Welfare, Public Health Service, Center for Disease Control, National Institute for Occupational Safety and Health, HSM-72-10267.

5.12. Campbell,W.J., et al, Relationship of Mineral Habit to Size *Characteristics for Tremolite Fragments and Fibers,* United States Department of the Interior, Bureau of Mines, Information Circular 8367, 1979.

5.13. Mefford, D., DCM Laboratory, Denver, private communication, July 1987.

5.14. Deer, W.A., Howie, R.A., Zussman, J., *Rock Forming Minerals,* Longman, Thetford, UK, 1974.

5.15. Kerr, P.F., *Optical Mineralogy;* Third Ed. McGraw-Hill, New York, 1959.

5.16. Veblen, D.R. (Ed.), *Amphiboles and Other Hydrous Pyriboles—Mineralogy, Reviews in Mineralogy,* Vol 9A, Michigan, 1982, pp 1-102.

5.17. Dixon, W.C., *Applications of Optical Microscopy in the Analysis of Asbestos* and Quartz, ACS Symposium Series, No. 120, Analytical Techniques in Occupational Health Chemistry, 1979.

5.18. Polarized Light Microscopy, McCrone Research Institute, Chicago, 1976.

5.19. Asbestos Identification, McCrone Research Institute, G & G printers, Chicago, 1987.

5.20. McCrone, W.C., Calculation of Refractive Indices from Dispertion Staining Data, The Microscope, No 37, Chicago, 1989.

5.21. Levadie, B. (Ed.), *Asbestos and Other Health Related Silicates*, ASTM Technical Publication 834, ASTM, Philadelphia 1982.

5.22. Steel, E. and Wylie, A., Riordan, P.H. (Ed.), Mineralogical Characteristics of Asbestos, *Geology of Asbestos Deposits*, pp. 93-101, SME-AIME, 1981.

5.23. Zussman, J., The Mineralogy of Asbestos, *Asbestos: Properties, Applications and Hazards*, pp. 45-67 Wiley, 1979.

[51 FR 22756, June 20, 1986, as amended at 52 FR 17755, May 12, 1987; 53 FR 27346, July 20, 1988; 55 FR 3732, Feb. 5, 1990; 57 FR 24331, June 8, 1992. Redesignated at 59 FR 41131, Aug. 10, 1994 and further amended at 59 FR 41144, 41150, 41152, and 41153, Aug. 10, 1994; 60 FR 33345, June 28, 1995; 60 FR 33995, June 29, 1995; 60 FR 36044, July 13, 1995; 60 FR 50412, 50413, Sept. 29, 1995; 61 FR 5510, Feb. 13, 1996; 61 FR 43458, Aug. 23, 1996; 63 FR 1298, Jan. 8, 1998; 63 FR 20099, Apr. 23, 1998; 63 FR 35138, June 29, 1998]

## §1926.1102 Coal tar pitch volatiles; interpretation of term.

**Note:** The requirements applicable to construction work under this section are identical to those set forth at §1910.1002 of this chapter.

[61 FR 31433, June 20, 1996]

[Section 1910.1002 is printed below—CCH.]

## §1910.1002 Coal tar pitch volatiles; interpretation of term.

As used in §1910.1000 (Table Z-1), coal tar pitch volatiles include the fused polycyclic hydrocarbons which volatilize from the distillation residues of coal, petroleum (excluding asphalt), wood, and other organic matter. Asphalt (CAS 8052-42-4, and CAS 64742-93-4) is not covered under the "coal tar pitch volatiles" standard.

[48 FR 2768, Jan. 21, 1983]

## §1926.1103 13 carcinogens (4-Nitrobiphenyl, etc.).

**Note:** The requirements applicable to construction work under this section are identical to those set forth at §1910.1003 of this chapter.

[61 FR 31433, June 20, 1996]

[Section 1910.1003 is printed below—CCH.]

## §1910.1003 13 Carcinogens (4-Nitrobiphenyl, etc.).

1910.1003(a) ***Scope and application.*** (1) This section applies to any area in which the 13 carcinogens addressed by this section are manufactured, processed, repackaged, released, handled, or stored, but shall not apply to transshipment in sealed containers, except for the labeling requirements under paragraphs (e)(2), (3) and (4) of this section. The 13 carcinogens are the following:

4-Nitrobiphenyl, Chemical Abstracts Service Register Number (CAS No.) 92933;

alpha-Naphthylamine, CAS No. 134327;

methyl chloromethyl ether, CAS No. 107302;

3,'-Dichlorobenzidine (and its salts) CAS No. 91941;

bis-Chloromethyl ether, CAS No. 542881;

beta-Naphthylamine, CAS No. 91598;

Benzidine, CAS No. 92875;

4-Aminodiphenyl, CAS No. 92671;

Ethyleneimine, CAS No. 151564;

beta-Propiolactone, CAS No. 57578;

2-Acetylaminofluorene, CAS No. 53963;

4-Dimethylaminoazo-benezene, CAS No. 60117; and N-Nitrosodimethylamine, CAS No. 62759.

1910.1003(a)(2) This section shall not apply to the following:

1910.1003(a)(2)(i) Solid or liquid mixtures containing less than 0.1 percent by weight or volume of 4-Nitrobiphenyl; methyl chloromethyl ether; bis-chloromethyl ether; beta-Naphthylamine; benzidine or 4-Aminodiphenyl; and

1910.1003(a)(2)(ii) Solid or liquid mixtures containing less than 1.0 percent by weight or volume of alpha-Naphthylamine; 3,'- Dichlorobenzidine (and its salts); Ethyleneimine; beta-Propiolactone; 2-Acetylaminofluorene; 4-Dimethylaminoazobenzene, or N-Nitrosodimethylamine.

1910.1003(b) ***Definitions.*** For the purposes of this section:

*Absolute filter* is one capable of retaining 99.97 percent of a mono disperse aerosol of 0.3 μm particles.

*Authorized employee* means an employee whose duties require him to be in the regulated area and who has been specifically assigned by the employer.

*Clean change room* means a room where employees put on clean clothing and/or protective equipment in an environment free of the 13 carcinogens addressed by this section. The clean change room shall be contiguous to and have an entry from a shower room, when the shower room facilities are otherwise required in this section.

*Closed system* means an operation involving a carcinogen addressed by this section where containment prevents the release of the material into regulated areas, non-regulated areas, or the external environment.

*Decontamination* means the inactivation of a carcinogen addressed by this section or its safe disposal.

*Director* means the Director, National Institute for Occupational Safety and Health, or any person directed by him or the Secretary of Health and Human Services to act for the Director.

*Disposal* means the safe removal of the carcinogens addressed by this section from the work environment.

*Emergency* means an unforeseen circumstance or set of circumstances resulting in the release of a carcinogen addressed by this section that may result in exposure to or contact with the material.

*External environment* means any environment external to regulated and nonregulated areas.

*Isolated system* means a fully enclosed structure other than the vessel of containment of a carcinogen addressed by this section that is impervious to the passage of the material and would prevent the entry of the carcinogen addressed by this section into regulated areas, nonregulated areas, or the external environment, should leakage or spillage from the vessel of containment occur.

*Laboratory-type hood* is a device enclosed on the three sides and the top and bottom, designed and maintained so as to draw air inward at an average linear face velocity of 150 feet per minute with a minimum of 125 feet per minute; designed, constructed, and maintained in such a way that an operation involving a carcinogen addressed by this section within the hood does not require the insertion of any portion of any employee's body other than his hands and arms.

*Nonregulated area* means any area under the control of the employer where entry and exit is neither restricted nor controlled.

*Open-vessel system* means an operation involving a carcinogen addressed by this section in an open vessel that is not in an isolated system, a laboratory-type hood, nor in any other system affording equivalent protection against the entry of the material into regulated areas, non-regulated areas, or the external environment.

*Protective clothing* means clothing designed to protect an employee against contact with or exposure to a carcinogen addressed by this section.

*Regulated area* means an area where entry and exit is restricted and controlled.

1910.1003(c) ***Requirements for areas containing a carcinogen addressed by this section.*** A regulated area shall be established by an employer where a carcinogen addressed by this section is manufactured, processed, used, repackaged, released, handled or stored. All such areas shall be controlled in accordance with the requirements for the following category or categories describing the operation involved:

1910.1003(c)(1) *Isolated systems.* Employees working with a carcinogen addressed by this section within an isolated system such as a "glove box" shall wash their hands and arms upon completion of the assigned task and before engaging in other activities not associated with the isolated system.

**1910.1003(c)(2)** *Closed system operation.* (i) Within regulated areas where the carcinogens addressed by this section are stored in sealed containers, or contained in a closed system, including piping systems, with any sample ports or openings closed while the carcinogens addressed by this section are contained within, access shall be restricted to authorized employees only.

**1910.1003(c)(2)(ii)** Employees exposed to 4-Nitrobiphenyl; alpha-Naphthylamine; 3,'-Dichlorobenzidine (and its salts); beta-Naphthylamine; benzidine; 4-Aminodiphenyl; 2-Acetylaminofluorene; 4-Dimethylaminoazo-benzene; and N-Nitrosodimethylamine shall be required to wash hands, forearms, face, and neck upon each exit from the regulated areas, close to the point of exit, and before engaging in other activities.

**1910.1003(c)(3)** *Open-vessel system operations.* Open-vessel system operations as defined in paragraph (b)(13) of this section are prohibited.

**1910.1003(c)(4)** *Transfer from a closed system, charging or discharging point operations, or otherwise opening a closed system.* In operations involving "laboratory-type hoods," or in locations where the carcinogens addressed by this section are contained in an otherwise "closed system," but is transferred, charged, or discharged into other normally closed containers, the provisions of this paragraph shall apply.

**1910.1003(c)(4)(i)** Access shall be restricted to authorized employees only.

**1910.1003(c)(4)(ii)** Each operation shall be provided with continuous local exhaust ventilation so that air movement is always from ordinary work areas to the operation. Exhaust air shall not be discharged to regulated areas, nonregulated areas or the external environment unless decontaminated. Clean makeup air shall be introduced in sufficient volume to maintain the correct operation of the local exhaust system.

**1910.1003(c)(4)(iii)** Employees shall be provided with, and required to wear, clean, full body protective clothing (smocks, coveralls, or long-sleeved shirt and pants), shoe covers and gloves prior to entering the regulated area.

**1910.1003(c)(4)(iv)** Employees engaged in handling operations involving the carcinogens addressed by this section must be provided with, and required to wear and use a half-face filter-type respirator with filters for dusts, mists, and fumes, or air-purifying canisters or cartridges. A respirator affording higher levels of protection than this respirator may be substituted.

**1910.1003(c)(4)(v)** Prior to each exit from a regulated area, employees shall be required to remove and leave protective clothing and equipment at the point of exit and at the last exit of the day, to place used clothing and equipment in impervious containers at the point of exit for purposes of decontamination or disposal. The contents of such impervious containers shall be identified, as required under paragraphs (e), (2), (3), and (4) of this section.

**1910.1003(c)(4)(vi)** Drinking fountains are prohibited in the regulated area.

**1910.1003(c)(4)(vii)** Employees shall be required to wash hands, forearms, face, and neck on each exit from the regulated area, close to the point of exit, and before engaging in other activities and employees exposed to 4-Nitrobiphenyl; alpha-Naphthylamine; 3,'-Dichlorobenzidine (and its salts); beta-Naphthylamine; Benzidine; 4-Aminodiphenyl; 2-Acetylaminofluorene; 4-Dimethylaminoazo-benzene; and N-Nitrosodimethylamine shall be required to shower after the last exit of the day.

**1910.1003(c)(5)** *Maintenance and decontamination activities.* In cleanup of leaks of spills, maintenance, or repair operations on contaminated systems or equipment, or any operations involving work in an area where direct contact with a carcinogen addressed by this section could result, each authorized employee entering that area shall:

**1910.1003(c)(5)(i)** Be provided with and required to wear clean, impervious garments, including gloves, boots, and continuous-air supplied hood in accordance with §1910.134;

**1910.1003(c)(5)(ii)** Be decontaminated before removing the protective garments and hood;

**1910.1003(c)(5)(iii)** Be required to shower upon removing the protective garments and hood.

**1910.1003(d)** ***General regulated area requirements***—(1) *Respirator program.* The employer must implement a respiratory protection program in accordance with 29 CFR 1910.134 (b), (c), (d) (except (d)(1)(iii) and (iv), and (d)(3)), and (e) through (m).

**1910.1003(d)(2)** *Emergencies.* In an emergency, immediate measures including, but not limited to, the requirements of paragraphs (d)(2)(i) through (v) of this section shall be implemented.

**1910.1003(d)(2)(i)** The potentially affected area shall be evacuated as soon as the emergency has been determined.

**1910.1003(d)(2)(ii)** Hazardous conditions created by the emergency shall be eliminated and the potentially affected area shall be decontaminated prior to the resumption of normal operations.

**1910.1003(d)(2)(iii)** Special medical surveillance by a physician shall be instituted within 24 hours for employees present in the potentially affected area at the time of the emergency. A report of the medical surveillance and any treatment shall be included in the incident report, in accordance with paragraph (f)(2) of this section.

**1910.1003(d)(2)(iv)** Where an employee has a known contact with a carcinogen addressed by this section, such employee shall be required to shower as soon as possible, unless contraindicated by physical injuries.

**1910.1003(d)(2)(v)** An incident report on the emergency shall be reported as provided in paragraph (f)(2) of this section.

**1910.1003(d)(2)(vi)** Emergency deluge showers and eyewash fountains supplied with running potable water shall be located near, within sight of, and on the same level with locations where a direct exposure to Ethyleneimine or beta-Propiolactone only would be most likely as a result of equipment failure or improper work practice.

**1910.1003(d)(3)** *Hygiene facilities and practices.* (i) Storage or consumption of food, storage or use of containers of beverages, storage or application of cosmetics, smoking, storage of smoking materials, tobacco products or other products for chewing, or the chewing of such products are prohibited in regulated areas.

**1910.1003(d)(3)(ii)** Where employees are required by this section to wash, washing facilities shall be provided in accordance with §1910.141(d)(1) and (2) (ii) through (vii).

**1910.1003(d)(3)(iii)** Where employees are required by this section to shower, shower facilities shall be provided in accordance with §1910.141(d)(3).

**1910.1003(d)(3)(iv)** Where employees wear protective clothing and equipment, clean change rooms shall be provided for the number of such employees required to change clothes, in accordance with §1910.141(e).

**1910.1003(d)(3)(v)** Where toilets are in regulated areas, such toilets shall be in a separate room.

**1910.1003(d)(4)** *Contamination control.* (i) Except for outdoor systems, regulated areas shall be maintained under pressure negative with respect to nonregulated areas. Local exhaust ventilation may be used to satisfy this requirement. Clean makeup air in equal volume shall replace air removed.

**1910.1003(d)(4)(ii)** Any equipment, material, or other item taken into or removed from a regulated area shall be done so in a manner that does not cause contamination in nonregulated areas or the external environment.

**1910.1003(d)(4)(iii)** Decontamination procedures shall be established and implemented to remove carcinogens addressed by this section from the surfaces of materials, equipment, and the decontamination facility.

**1910.1003(d)(4)(iv)** Dry sweeping and dry mopping are prohibited for 4-Nitrobiphenyl; alpha-Naphthylamine; 3,'-Dichlorobenzidine (and its salts); beta-Naphthylamine; Benzidine; 4-Aminodiphenyl; 2-Acetylaminofluorene; 4-Dimethylaminoazo-benzene and N-Nitrosodimethylamine.

**1910.1003(e)** ***Signs, information and training***—(1) *Signs*—(i) Entrances to regulated areas shall be posted with signs bearing the legend:

**CANCER-SUSPECT AGENT**

**AUTHORIZED PERSONNEL ONLY**

**1910.1003(e)(1)(ii)** Entrances to regulated areas containing operations covered in paragraph (c)(5) of this section shall be posted with signs bearing the legend:

**CANCER-SUSPECT AGENT EXPOSED IN THIS AREA**

**IMPERVIOUS SUIT INCLUDING GLOVES, BOOTS, AND AIR-SUPPLIED HOOD REQUIRED AT ALL TIMES**

**AUTHORIZED PERSONNEL ONLY**

**1910.1003(e)(1)(iii)** Appropriate signs and instructions shall be posted at the entrance to, and exit from, regulated areas, informing employees of the procedures that must be followed in entering and leaving a regulated area.

**1910.1003(e)(2)** *Container contents identification.* (i) Containers of a carcinogen addressed by this section and containers required under paragraphs (c)(4)(v) and (c)(6) (vii)(B) and (viii)(B) of this section that are accessible only to and handled only by authorized employees, or by other employees trained in accordance with paragraph (e)(5) of this section, may have contents identification limited to a generic or proprietary name or other proprietary identification of the carcinogen and percent.

**1910.1003(e)(2)(ii)** Containers of a carcinogen addressed by this section and containers required under paragraphs (c)(4)(v) and (c)(6) (vii)(B) and (viii)(B) of this section that are accessible to or handled by employees other than authorized employees or employees trained in accordance with paragraph (e)(5) of this section shall have contents identification that includes the full chemical name and Chemical Abstracts Service Registry number as listed in paragraph (a)(1) of this section.

**1910.1003(e)(2)(iii)** Containers shall have the warning words "CANCER-SUSPECT AGENT" displayed immediately under or adjacent to the contents identification.

**1910.1003(e)(2)(iv)** Containers whose contents are carcinogens addressed by this section with corrosive or irritating properties shall have label statements warning of such hazards noting, if appropriate, particularly sensitive or affected portions of the body.

**1910.1003(e)(3)** *Lettering.* Lettering on signs and instructions required by paragraph (e)(1) shall be a minimum letter height of 2 inches (5 cm). Labels on containers required under this section shall not be less than one-half the size of the largest lettering on the package, and not less than 8-point type in any instance. *Provided,* That no such required lettering need be more than 1 inch (2.5 cm) in height.

**1910.1003(e)(4)** *Prohibited statements.* No statement shall appear on or near any required sign, label, or instruction that contradicts or detracts from the effect of any required warning, information, or instruction.

**1910.1003(e)(5)** *Training and indoctrination.* (i) Each employee prior to being authorized to enter a regulated area, shall receive a training and indoctrination program including, but not necessarily limited to:

**1910.1003(e)(5)(i)(A)** The nature of the carcinogenic hazards of a carcinogen addressed by this section, including local and systemic toxicity;

**1910.1003(e)(5)(i)(B)** The specific nature of the operation involving a carcinogen addressed by this section that could result in exposure;

**1910.1003(e)(5)(i)(C)** The purpose for and application of the medical surveillance program, including, as appropriate, methods of self-examination;

**1910.1003(e)(5)(i)(D)** The purpose for and application of decontamination practices and purposes;

**1910.1003(e)(5)(i)(E)** The purpose for and significance of emergency practices and procedures;

**1910.1003(e)(5)(i)(F)** The employee's specific role in emergency procedures;

**1910.1003(e)(5)(i)(G)** Specific information to aid the employee in recognition and evaluation of conditions and situations which may result in the release of a carcinogen addressed by this section;

**1910.1003(e)(5)(i)(H)** The purpose for and application of specific first aid procedures and practices;

**1910.1003(e)(5)(i)(I)** A review of this section at the employee's first training and indoctrination program and annually thereafter.

**1910.1003(e)(5)(ii)** Specific emergency procedures shall be prescribed, and posted, and employees shall be familiarized with their terms, and rehearsed in their application.

**1910.1003(e)(5)(iii)** All materials relating to the program shall be provided upon request to authorized representatives of the Assistant Secretary and the Director.

**1910.1003(f)** ***Reports***—(1) *Operations.* The information required in paragraphs (f)(1) (i) through (iv) of this section shall be reported in writing to the nearest OSHA Area Director. Any changes in such information shall be similarly reported in writing within 15 calendar days of such change:

**1910.1003(f)(1)(i)** A brief description and in-plant location of the area(s) regulated and the address of each regulated area;

**1910.1003(f)(1)(ii)** The name(s) and other identifying information as to the presence of a carcinogen addressed by this section in each regulated area;

**1910.1003(f)(1)(iii)** The number of employees in each regulated area, during normal operations including maintenance activities; and

**1910.1003(f)(1)(iv)** The manner in which carcinogens addressed by this section are present in each regulated area; for example, whether it is manufactured, processed, used, repackaged, released, stored, or otherwise handled.

**1910.1003(f)(2)** *Incidents.* Incidents that result in the release of a carcinogen addressed by this section into any area where employees may be potentially exposed shall be reported in accordance with this paragraph.

**1910.1003(f)(2)(i)** A report of the occurrence of the incident and the facts obtainable at that time including a report on any medical treatment of affected employees shall be made within 24 hours to the nearest OSHA Area Director.

**1910.1003(f)(2)(ii)** A written report shall be filed with the nearest OSHA Area Director within 15 calendar days thereafter and shall include:

**1910.1003(f)(2)(ii)(A)** A specification of the amount of material released, the amount of time involved, and an explanation of the procedure used in determining this figure;

**1910.1003(f)(2)(ii)(B)** A description of the area involved, and the extent of known and possible employee exposure and area contamination;

**1910.1003(f)(2)(ii)(C)** A report of any medical treatment of affected employees, and any medical surveillance program implemented; and

**1910.1003(f)(2)(ii)(D)** An analysis of the circumstances of the incident and measures taken or to be taken, with specific completion dates, to avoid further similar releases.

**1910.1003(g)** ***Medical surveillance.*** At no cost to the employee, a program of medical surveillance shall be established and implemented for employees considered for assignment to enter regulated areas, and for authorized employees.

**1910.1003(g)(1)** *Examinations.* (i) Before an employee is assigned to enter a regulated area, a preassignment physical examination by a physician shall be provided. The examination shall include the personal history of the employee, family and occupational background, including genetic and environmental factors.

**1910.1003(g)(1)(ii)** Authorized employees shall be provided periodic physical examinations, not less often than annually, following the preassignment examination.

**1910.1003(g)(1)(iii)** In all physical examinations, the examining physician shall consider whether there exist conditions of increased risk, including reduced immunological competence, those undergoing treatment with steroids or cytotoxic agents, pregnancy, and cigarette smoking.

**1910.1003(g)(2)** *Records.* (i) Employers of employees examined pursuant to this paragraph shall cause to be maintained complete and accurate records of all such medical examinations. Records shall be maintained for the duration of the employee's employment. Upon termination of the employee's employment, including retirement or death, or in the event that the employer ceases business without a successor, records, or notarized true copies thereof, shall be forwarded by registered mail to the Director.

**1910.1003(g)(2)(ii)** Records required by this paragraph shall be provided upon request to employees, designated representatives, and the Assistant Secretary in accordance with 29 CFR 1910.20(a) through (e) and (g) through (i). These records shall also be provided upon request to the Director.

**1910.1003(g)(2)(iii)** Any physician who conducts a medical examination required by this paragraph shall furnish to the employer a statement of the employee's suitability for employment in the specific exposure.

[61 FR 9242, Mar. 7, 1996, as amended at 63 FR 1286, Jan. 8, 1998; 63 FR 20099, Apr. 23, 1998]

### §1926.1104 alpha-Naphthylamine.

**Note:** The requirements applicable to construction work under this section are identical to those set forth at §1910.1003 of this chapter.

[61 FR 31433, June 20, 1996]

[Section 1910.1003 is printed above at §1926.1103—CCH.]

### §1926.1105 [Reserved]

### §1926.1106 Methyl chloromethyl ether.

**Note:** The requirements applicable to construction work under this section are identical to those set forth at §1910.1003 of this chapter.

[61 FR 31433, June 20, 1996]

[Section 1910.1003 is printed following §1926.1103—CCH.]

### §1926.1107 3,3'-Dichlorobenzidiene (and its salts).

**Note:** The requirements applicable to construction work under this section are identical to those set forth at §1910.1003 of this chapter.

[61 FR 31433, June 20, 1996]

[Section 1910.1003 is printed following §1926.1103—CCH.]

### §1926.1108 bis-Chloromethyl ether.

**Note:** The requirements applicable to construction work under this section are identical to those set forth at §1910.1003 of this chapter.

[61 FR 31433, June 20, 1996]

[Section 1910.1003 is printed following §1926.1103—CCH.]

### §1926.1109 beta-Naphthylamine.

**Note:** The requirements applicable to construction work under this section are identical to those set forth at §1910.1003 of this chapter.

[61 FR 31433, June 20, 1996]

[Section 1910.1003 is printed following §1926.1103—CCH.]

### §1926.1110 Benzidine.

**Note:** The requirements applicable to construction work under this section are identical to those set forth at §1910.1003 of this chapter.

[61 FR 31433, June 20, 1996]

[Section 1910.1003 is printed following §1926.1103—CCH.]

### §1926.1111 4-Aminodiphenyl.

**Note:** The requirements applicable to construction work under this section are identical to those set forth at §1910.1003 of this chapter.

[61 FR 31433, June 20, 1996]

[Section 1910.1003 is printed following §1926.1103—CCH.]

### §1926.1112 Ethyleneimine.

**Note:** The requirements applicable to construction work under this section are identical to those set forth at §1910.1003 of this chapter.

[61 FR 31433, June 20, 1996]

[Section 1910.1003 is printed following §1926.1103—CCH.]

### §1926.1113 beta-Propiolactone.

**Note:** The requirements applicable to construction work under this section are identical to those set forth at §1910.1003 of this chapter.

[61 FR 31433, June 20, 1996]

[Section 1910.1003 is printed following §1926.1103—CCH.]

### §1926.1114 2-Acetylaminofluorene.

**Note:** The requirements applicable to construction work under this section are identical to those set forth at §1910.1003 of this chapter.

[61 FR 31433, June 20, 1996]

[Section 1910.1003 is printed following §1926.1103—CCH.]

### §1926.1115 4-Dimethylaminoazobenzene.

**Note:** The requirements applicable to construction work under this section are identical to those set forth at §1910.1003 of this chapter.

[61 FR 31433, June 20, 1996]

[Section 1910.1003 is printed following §1926.1103—CCH.]

### §1926.1116 N-Nitrosodimethylamine.

**Note:** The requirements applicable to construction work under this section are identical to those set forth at §1910.1003 of this chapter.

[61 FR 31433, June 20, 1996]

[Section 1910.1003 is printed following §1926.1103—CCH.]

### §1926.1117 Vinyl chloride.

**Note:** The requirements applicable to construction work under this section are identical to those set forth at §1910.1017 of this chapter.

[61 FR 31433, June 20, 1996]

[Section 1910.1017 is printed below—CCH.]

### §1910.1017 Vinyl chloride.

**1910.1017(a)** ***Scope and application.*** (1) This section includes requirements for the control of employee exposure to vinyl chloride (chloroethene), Chemical Abstracts Service Registry No. 75014.

**1910.1017(a)(2)** This section applies to the manufacture, reaction, packaging, repackaging, storage, handling or use of vinyl chloride or polyvinyl chloride, but does not apply to the handling or use of fabricated products made of polyvinyl chloride.

**1910.1017(a)(3)** This section applies to the transportation of vinyl chloride or polyvinyl chloride except to the extent that the Department of Transportation may regulate the hazards covered by this section.

**1910.1017(b)** ***Definitions.*** (1) *Action level* means a concentration of vinyl chloride of 0.5 ppm averaged over an 8-hour work day.

**1910.1017(b)(2)** *Assistant Secretary* means the Assistant Secretary of Labor for Occupational Safety and Health, U.S. Department of Labor, or his designee.

**1910.1017(b)(3)** *Authorized person* means any person specifically authorized by the employer whose duties require him to enter a regulated area or any person entering such an area as a designated representative of employees for the purpose of exercising an opportunity to observe monitoring and measuring procedures.

**1910.1017(b)(4)** *Director* means the Director, National Institute for Occupational Safety and Health, U.S. Department of Health and Human Services, or his designee.

**1910.1017(b)(5)** *Emergency* means any occurrence such as, but not limited to, equipment failure, or operation of a relief device which is likely to, or does, result in massive release of vinyl chloride.

**1910.1017(b)(6)** *Fabricated product* means a product made wholly or partly from polyvinyl chloride, and which does not require further processing at temperatures, and for times, sufficient to cause mass melting of the polyvinyl chloride resulting in the release of vinyl chloride.

**1910.1017(b)(7)** *Hazardous operation* means any operation, procedure, or activity where a release of either vinyl chloride liquid or gas might be expected as a consequence of the operation or because of an accident in the operation, which would result in an employee exposure in excess of the permissible exposure limit.

**1910.1017(b)(8)** *OSHA Area Director* means the Director for the Occupational Safety and Health Administration Area Office having jurisdiction over the geographic area in which the employer's establishment is located.

**1910.1017(b)(9)** *Polyvinyl chloride* means polyvinyl chloride homopolymer or copolymer before such is converted to a fabricated product.

**1910.1017(b)(10)** *Vinyl chloride* means vinyl chloride monomer.

**1910.1017(c)** ***Permissible exposure limit.*** (1) No employee may be exposed to vinyl chloride at concentrations greater than 1 ppm averaged over any 8-hour period, and

**1910.1017(c)(2)** No employee may be exposed to vinyl chloride at concentrations greater than 5 ppm averaged over any period not exceeding 15 minutes.

**1910.1017(c)(3)** No employee may be exposed to vinyl chloride by direct contact with liquid vinyl chloride.

**1910.1017(d)** ***Monitoring.*** (1) A program of initial monitoring and measurement shall be undertaken in each establishment to determine if there is any employee exposed, without regard to the use of respirators, in excess of the action level.

**1910.1017(d)(2)** Where a determination conducted under paragraph (d)(1) of this section shows any employee exposures, without regard to the use of respirators, in excess of the action level, a program for determining exposures for each such employee shall be established. Such a program:

**1910.1017(d)(2)(i)** Shall be repeated at least monthly where any employee is exposed, without regard to the use of respirators, in excess of the permissible exposure limit.

**1910.1017(d)(2)(ii)** Shall be repeated not less than quarterly where any employee is exposed, without regard to the use of respirators, in excess of the action level.

**1910.1017(d)(2)(iii)** May be discontinued for any employee only when at least two consecutive monitoring determinations, made not less than 5 working days apart, show exposures for that employee at or below the action level.

**1910.1017(d)(3)** Whenever there has been a production, process or control change which may result in an increase in the release of vinyl chloride, or the employer has any other reason to suspect that any employee may be exposed in excess of the action level, a determination of employee exposure under paragraph (d)(1) of this section shall be performed.

**1910.1017(d)(4)** The method of monitoring and measurement shall have an accuracy (with a confidence level of 95 percent) of not less than plus or minus 50 percent from 0.25 through 0.5 ppm, plus or minus 35 percent from over 0.5 ppm through 1.0 ppm, and plus or minus 25 percent over 1.0 ppm. (Methods meeting these accuracy requirements are available in the "NIOSH Manual of Analytical Methods").

**1910.1017(d)(5)** Employees or their designated representatives shall be afforded reasonable opportunity to observe the monitoring and measuring required by this paragraph.

**1910.1017(e)** ***Regulated area.*** (1) A regulated area shall be established where:

**1910.1017(e)(1)(i)** Vinyl chloride or polyvinyl chloride is manufactured, reacted, repackaged, stored, handled or used; and

**1910.1017(e)(1)(ii)** Vinyl chloride concentrations are in excess of the permissible exposure limit.

**1910.1017(e)(2)** Access to regulated areas shall be limited to authorized persons.

**1910.1017(f)** ***Methods of compliance.*** Employee exposures to vinyl chloride shall be controlled to at or below the permissible exposure limit provided in paragraph (c) of this section by engineering, work practice, and personal protective controls as follows:

**1910.1017(f)(1)** Feasible engineering and work practice controls shall immediately be used to reduce exposures to at or below the permissible exposure limit.

**1910.1017(f)(2)** Wherever feasible engineering and work practice controls which can be instituted immediately are not sufficient to reduce exposures to at or below the permissible exposure limit, they shall nonetheless be used to reduce exposures to the lowest practicable level, and shall be supplemented by respiratory protection in accordance with paragraph (g) of this section. A program shall be established and implemented to reduce exposures to at or below the permissible exposure limit, or to the greatest extent feasible, solely by means of engineering and work practice controls, as soon as feasible.

**1910.1017(f)(3)** Written plans for such a program shall be developed and furnished upon request for examination and copying to authorized representatives of the Assistant Secretary and the Director. Such plans shall be updated at least every six months.

**1910.1017(g)** ***Respiratory protection***—(1) *General.* For employees who use respirators required by this section, the employer must provide respirators that comply with the requirements of this paragraph.

**1910.1017(g)(2)** *Respirator program.* The employer must implement a respiratory protection program in accordance with 29 CFR 1910.134 (b) through (d) (except (d)(1)(iii), and (d)(3)(iii)(B)(*1*) and (*2*)), and (f) through (m).

**1910.1017(g)(3)** *Respirator selection.* (i) Respirators must be selected from the following table:

| Atmospheric concentration of vinyl chloride | Required apparatus |
|---|---|
| (i) Unknown, or above 3,600 p/m. .... | Open-circuit, self-contained breathing apparatus, pressure demand type, with full facepiece. |
| (ii) Not over 3,600 p/m .................... | (A) Combination type C supplied air respirator, pressure demand type, with full or half facepiece, and auxiliary self-contained air supply; or |
| (iii) Not over 1,000 p/m .................... | (B) Combination type, supplied air respirator continuous flow type, with full or half facepiece, and auxiliary self-contained air supply. Type C, supplied air respirator, continuous flow type, with full or half facepiece, helmet or hood. |
| (iv) Not over 100 p/m ........................ | (A) Combination type C supplied air respirator demand type, with full facepiece, and auxiliary self-contained air supply; or |
| | (B) Open-circuit self-contained breathing apparatus with full facepiece, in demand mode; or |
| | Type (C) supplied air respirator, demand type, with full facepiece. |
| (v) Not over 25 p/m .......................... | (A) A powered air-purifying respirator with hood, helmet, full or half facepiece, and a canister which provides a service life of at least 4 hours for concentrations of vinyl chloride up to 25 p/m, or |
| | (B) Gas mask, front- or back-mounted canister which provides a service life of at least 4 hours for concentrations of vinyl chloride up to 25 p/m. |
| (vi) Not over 10 p/m .......................... | (A) Combination type C supplied-air respirator, demand type, with half facepiece, and auxiliary self-contained air supply; or |
| | (B) Type C supplied-air respirator, demand type, with half facepiece; or |
| | (C) Any chemical cartridge respirator with an organic vapor cartridge which provides a service life of at least 1 hour for concentrations of vinyl chloride up to 10 p/m. |

**1910.1017(g)(3)(ii)** When air-purifying respirators are used:

**1910.1017(g)(3)(ii)(A)** Air-purifying canisters or cartridges must be replaced prior to the expiration of their service life or the end of the shift in which they are first used, whichever occurs first.

**1910.1017(g)(3)(ii)(B)** A continuous-monitoring and alarm system must be provided when concentrations of vinyl chloride could reasonably exceed the allowable concentrations for the devices in use. Such a system must be used to alert employees when vinyl chloride concentrations exceed the allowable concentrations for the devices in use.

**1910.1017(g)(3)(iii)** Respirators specified for higher concentrations may be used for lower concentrations.

**1910.1017(h)** ***Hazardous operations.*** (1) Employees engaged in hazardous operations, including entry of vessels to clean polyvinyl chloride residue from vessel walls, shall be provided and required to wear and use;

**1910.1017(h)(1)(i)** Respiratory protection in accordance with paragraphs (c) and (g) of this section; and

**1910.1017(h)(1)(ii)** Protective garments to prevent skin contact with liquid vinyl chloride or with polyvinyl chloride residue from vessel walls. The protective garments shall be selected for the operation and its possible exposure conditions.

**1910.1017(h)(2)** Protective garments shall be provided clean and dry for each use.

**1910.1017(i)** ***Emergency situations.*** A written operational plan for emergency situations shall be developed for each facility storing, handling, or otherwise using vinyl chloride as a liquid or compressed gas. Appropriate portions of the plan shall be implemented in the event of an emergency. The plan shall specifically provide that:

**1910.1017(i)(1)** Employees engaged in hazardous operations or correcting situations of existing hazardous releases shall be equipped as required in paragraph (h) of this section;

**1910.1017(i)(2)** Other employees not so equipped shall evacuate the area and not return until conditions are controlled by the methods required in paragraph (f) of this section and the emergency is abated.

**1910.1017(j)** ***Training.*** Each employee engaged in vinyl chloride or polyvinyl chloride operations shall be provided training in a program relating to the hazards of vinyl chloride and precautions for its safe use.

**1910.1017(j)(1)** The program shall include:

**1910.1017(j)(1)(i)** The nature of the health hazard from chronic exposure to vinyl chloride including specifically the carcinogenic hazard;

**1910.1017(j)(1)(ii)** The specific nature of operations which could result in exposure to vinyl chloride in excess of the permissible limit and necessary protective steps;

**1910.1017(j)(1)(iii)** The purpose for, proper use, and limitations of respiratory protective devices;

**1910.1017(j)(1)(iv)** The fire hazard and acute toxicity of vinyl chloride, and the necessary protective steps;

**1910.1017(j)(1)(v)** The purpose for and a description of the monitoring program;

**1910.1017(j)(1)(vi)** The purpose for, and a description of, the medical surveillance program;

**1910.1017(j)(1)(vii)** Emergency procedures;

**1910.1017(j)(1)(viii)** Specific information to aid the employee in recognition of conditions which may result in the release of vinyl chloride; and

**1910.1017(j)(1)(ix)** A review of this standard at the employee's first training and indoctrination program, and annually thereafter.

**1910.1017(j)(2)** All materials relating to the program shall be provided upon request to the Assistant Secretary and the Director.

**1910.1017(k)** ***Medical surveillance.*** A program of medical surveillance shall be instituted for each employee exposed, without regard to the use of respirators, to vinyl chloride in excess of the action level. The program shall provide each such employee with an opportunity for examinations and tests in accordance with this paragraph. All medical examinations and procedures shall be performed by or under the supervision of a licensed physician, and shall be provided without cost to the employee.

**1910.1017(k)(1)** At the time of initial assignment, or upon institution of medical surveillance;

**1910.1017(k)(1)(i)** A general physical examination shall be performed, with specific attention to detecting enlargement of liver, spleen or kidneys, or dysfunction in these organs, and for abnormalities in skin, connective tissues and the pulmonary system (See Appendix A).

**1910.1017(k)(1)(ii)** A medical history shall be taken, including the following topics:

**1910.1017(k)(1)(ii)(A)** Alcohol intake;

**1910.1017(k)(1)(ii)(B)** Past history of hepatitis;

**1910.1017(k)(1)(ii)(C)** Work history and past exposure to potential hepatotoxic agents, including drugs and chemicals;

**1910.1017(k)(1)(ii)(D)** Past history of blood transfusions; and

**1910.1017(k)(1)(II)(E)** Past history of hospitalizations.

**1910.1017(k)(1)(iii)** A serum specimen shall be obtained and determinations made of:

**1910.1017(k)(1)(iii)(A)** Total bilirubin;

**1910.1017(k)(1)(iii)(B)** Alkaline phosphatase;

**1910.1017(k)(1)(iii)(C)** Serum glutamic oxalacetic transaminase (SGOT);

**1910.1017(k)(1)(iii)(D)** Serum glutamic pyruvic transaminase (SGPT); and

**1910.1017(k)(1)(iii)(E)** Gamma glustamyl transpeptidase.

**1910.1017(k)(2)** Examinations provided in accordance with this paragraph shall be performed at least:

**1910.1017(k)(2)(i)** Every 6 months for each employee who has been employed in vinyl chloride or polyvinyl chloride manufacturing for 10 years or longer; and

**1910.1017(k)(2)(ii)** Annually for all other employees.

**1910.1017(k)(3)** Each employee exposed to an emergency shall be afforded appropriate medical surveillance.

**1910.1017(k)(4)** A statement of each employee's suitability for continued exposure to vinyl chloride including use of protective equipment and respirators, shall be obtained from the examining physician promptly after any examination. A copy of the physician's statement shall be provided each employee.

**1910.1017(k)(5)** If any employee's health would be materially impaired by continued exposure, such employee shall be withdrawn from possible contact with vinyl chloride.

**1910.1017(k)(6)** Laboratory analyses for all biological specimens included in medical examinations shall be performed in laboratories licensed under 42 CFR Part 74.

**1910.1017(k)(7)** If the examining physician determines that alternative medical examinations to those required by paragraph (k)(1) of this section will provide at least equal assurance of detecting medical conditions pertinent to the exposure to vinyl chloride, the employer may accept such alternative examinations as meeting the requirements of paragraph (k)(1) of this section, if the employer obtains a statement from the examining physician setting forth the alternative examinations and the rationale for substitution. This statement shall be available upon request for examination and copying to authorized representatives of the Assistant Secretary and the Director.

**1910.1017(l)** ***Signs and labels.*** (1) Entrances to regulated areas shall be posted with legible signs bearing the legend:

**CANCER-SUSPECT AGENT AREA**
**AUTHORIZED PERSONNEL ONLY**

**1910.1017(l)(2)** Areas containing hazardous operations or where an emergency currently exists shall be posted with legible signs bearing the legend:

**CANCER-SUSPECT AGENT IN THIS AREA**
**PROTECTIVE EQUIPMENT REQUIRED**
**AUTHORIZED PERSONNEL ONLY**

**1910.1017(l)(3)** Containers of polyvinyl chloride resin waste from reactors or other waste contaminated with vinyl chloride shall be legibly labeled:

**CONTAMINATED WITH VINYL CHLORIDE**

**CANCER-SUSPECT AGENT**

**1910.1017(l)(4)** Containers of polyvinyl chloride shall be legibly labeled:

**POLYVINYL CHLORIDE (OR TRADE NAME)**

**Contains**

**VINYL CHLORIDE**

**VINYL CHLORIDE IS A CANCER-SUSPECT AGENT**

**1910.1017(l)(5)** Containers of vinyl chloride shall be legibly labeled either:

**1910.1017(l)(5)(i)**

**VINYL CHLORIDE**

**EXTREMELY FLAMMABLE GAS UNDER PRESSURE**

**CANCER SUSPECT AGENT**

or (ii) In accordance with 49 CFR Parts 170 through 189, with the additional legend:

**CANCER-SUSPECT AGENT**

applied near the label or placard.

**1910.1017(l)(6)** No statement shall appear on or near any required sign, label or instruction which contradicts or detracts from the effect of, any required warning, information or instruction.

**1910.1017(m)** ***Records.*** (1) All records maintained in accordance with this section shall include the name and social security number of each employee where relevant.

**1910.1017(m)(2)** Records of required monitoring and measuring and medical records shall be provided upon request to employees, designated representatives, and the Assistant Secretary in accordance with 29 CFR 1910.20(a) through (e) and (g) through (i). These records shall be provided upon request to the Director. Authorized personnel rosters shall also be provided upon request to the Assistant Secretary and the Director.

**1910.1017(m)(2)(i)** Monitoring and measuring records shall:

**1910.1017(m)(2)(i)(A)** State the date of such monitoring and measuring and the concentrations determined and identify the instruments and methods used;

**1910.1017(m)(2)(i)(B)** Include any additional information necessary to determine individual employee exposures where such exposures are determined by means other than individual monitoring of employees; and

**1910.1017(m)(2)(i)(C)** Be maintained for not less than 30 years.

**1910.1017(m)(2)(ii)** [Reserved]

**1910.1017(m)(2)(iii)** Medical records shall be maintained for the duration of the employment of each employee plus 20 years, or 30 years, whichever is longer.

**1910.1017(m)(3)** In the event that the employer ceases to do business and there is no successor to receive and retain his records for the prescribed period, these records shall be transmitted by registered mail to the Director, and each employee individually notified in writing of this transfer. The employer shall also comply with any additional requirements set forth in 29 CFR 1910.20(h).

**1910.1017(n)** ***Reports.*** (1) Not later than 1 month after the establishment of a regulated area, the following information shall be reported to the OSHA Area Director. Any changes to such information shall be reported within 15 days.

**1910.1017(n)(1)(i)** The address and location of each establishment which has one or more regulated areas; and

**1910.1017(n)(1)(ii)** The number of employees in each regulated area during normal operations, including maintenance.

**1910.1017(n)(2)** Emergencies, and the facts obtainable at that time, shall be reported within 24 hours to the OSHA Area Director. Upon request of the Area Director, the employer shall submit additional information in writing relevant to the nature and extent of employee exposures and measures taken to prevent future emergencies of similar nature.

**1910.1017(n)(3)** Within 10 working days following any monitoring and measuring which discloses that any employee has been exposed, without regard to the use of respirators, in excess of the permissible exposure limit, each such employee shall be notified in writing of the results of the exposure measurement and the steps being taken to reduce the exposure to within the permissible exposure limit.

**1910.1017(o)** ***Effective dates.*** (1) Until April 1, 1975, the provisions currently set forth in §1910.93q of this part shall apply.

**1910.1017(o)(2)** Effective April 1, 1975, the provisions set forth in §1910.93q of this part shall apply.

## APPENDIX A TO §1910.1017—SUPPLEMENTARY MEDICAL INFORMATION

When required tests under paragraph (k)(1) of this section show abnormalities, the tests should be repeated as soon as practicable, preferably within 3 to 4 weeks. If tests remain abnormal, consideration should be given to withdrawal of the employee from contact with vinyl chloride, while a more comprehensive examination is made.

Additional tests which may be useful:

A. For kidney dysfunction: urine examination for albumin, red blood cells, and exfoliative abnormal cells.

B. Pulmonary system: Forced vital capacity, Forced expiratory volume at 1 second, and chest roentgenogram (posterior anterior, 14 × 17 inches).

C. Additional serum tests: Lactic acid dehydrogenase, lactic acid dehydrogenase isoenzyme, protein determination, and protein electrophoresis.

D. For a more comprehensive examination on repeated abnormal serum tests: Hepatitis B antigen, and liver scanning.

[39 FR 35896, Oct. 4, 1974; 39 FR 41848, Dec. 3, 1974, as amended at 40 FR 13211, Mar. 25, 1975. Redesignated at 40 FR 23072, May 28, 1975 and amended at 43 FR 49751, Oct. 24, 1978; 45 FR 35282, May 23, 1980; 54 FR 24334, June 7, 1989; 58 FR 35310, June 30, 1993; 61 FR 5508, Feb. 13, 1996; 63 FR 1286, Jan. 8, 1998]

## §1926.1118 Inorganic arsenic.

**Note:** The requirements applicable to construction work under this section are identical to those set forth at §1910.1018 of this chapter.

[61 FR 31433, June 20, 1996]

[Section 1910.1018 is printed below—CCH.]

## §1910.1018 Inorganic arsenic.

**1910.1018(a)** ***Scope and application.*** This section applies to all occupational exposures to inorganic arsenic except that this section does not apply to employee exposures in agriculture or resulting from pesticide application, the treatment of wood with preservatives or the utilization of arsenically preserved wood.

**1910.1018(b)** ***Definitions.*** *Action level* means a concentration of inorganic arsenic of 5 micrograms per cubic meter of air (5 $\mu g/m^3$) averaged over any eight (8) hour period.

*Assistant Secretary* means the Assistant Secretary of Labor for Occupational Safety and Health, U.S. Department of Labor, or designee.

*Authorized person* means any person specifically authorized by the employer whose duties require the person to enter a regulated area, or any person entering such an area as a designated representative of employees for the purpose of exercising the right to observe monitoring and measuring procedures under paragraph (e) of this section.

*Director* means the Director, National Institute for Occupational Safety and Health, U.S. Department of Health and Human Services, or designee.

*Inorganic arsenic* means copper aceto- arsenite and all inorganic compounds containing arsenic except arsine, measured as arsenic (As).

**1910.1018(c)** ***Permissible exposure limit.*** The employer shall assure that no employee is exposed to inorganic arsenic at concentrations greater than 10 micrograms per cubic meter of air (10 $\mu g/m^3$), averaged over any 8-hour period.

**1910.1018(d)** ***Notification of use.*** (1) By October 1, 1978 or within 60 days after the introduction of inorganic arsenic into the workplace, every employer who is required to establish a regulated area in his workplaces shall report in writing to the OSHA area office for each such workplace:

**1910.1018(d)(1)(i)** The address of each such workplace;

**1910.1018(d)(1)(ii)** The approximate number of employees who will be working in regulated areas; and

**1910.1018(d)(1)(iii)** A brief summary of the operations creating the exposure and the actions which the employer intends to take to reduce exposures.

**1910.1018(d)(2)** Whenever there has been a significant change in the information required by paragraph (d)(1) of this section the employer shall report the changes in writing within 60 days to the OSHA area office.

**1910.1018(e)** ***Exposure monitoring***—(1) *General.* (i) Determinations of airborne exposure levels shall be made from air samples that are representative of each employee's exposure to inorganic arsenic over an eight (8) hour period.

**1910.1018(e)(1)(ii)** For the purposes of this section, employee exposure is that exposure which would occur if the employee were not using a respirator.

**1910.1018(e)(1)(iii)** The employer shall collect full shift (for at least 7 continuous hours) personal samples including at least one sample for each shift for each job classification in each work area.

**1910.1018(e)(2)** *Initial monitoring.* Each employer who has a workplace or work operation covered by this standard shall monitor each such workplace and work operation to accurately determine the airborne concentration of inorganic arsenic to which employees may be exposed.

**1910.1018(e)(3)** *Frequency.* (i) If the initial monitoring reveals employee exposure to be below the action level the measurements need not be repeated except as otherwise provided in paragraph (e)(4) of this section.

**1910.1018(e)(3)(ii)** If the initial monitoring, required by this section, or subsequent monitoring reveals employee exposure to be above the permissible exposure limit, the employer shall repeat monitoring at least quarterly.

**1910.1018(e)(3)(iii)** If the initial monitoring, required by this section, or subsequent monitoring reveals employee exposure to be above the action level and below the permissible exposure limit the employer shall repeat monitoring at least every six months.

**1910.1018(e)(3)(iv)** The employer shall continue monitoring at the required frequency until at least two consecutive measurements, taken at least seven (7) days apart, are below the action level at which time the employer may discontinue monitoring for that employee until such time as any of the events in paragraph (e)(4) of this section occur.

**1910.1018(e)(4)** *Additional monitoring.* Whenever there has been a production, process, control or personal change which may result in new or additional exposure to inorganic arsenic, or whenever the employer has any other reason to suspect a change which may result in new or additional exposures to inorganic arsenic, additional monitoring which complies with paragraph (e) of this section shall be conducted.

**1910.1018(e)(5)** *Employee notification.* (i) Within five (5) working days after the receipt of monitoring results, the employer shall notify each employee in writing of the results which represent that employee's exposures.

**1910.1018(e)(5)(ii)** Whenever the results indicate that the representative employee exposure exceeds the permissible exposure limit, the employer shall include in the written notice a statement that the permissible exposure limit was exceeded and a description of the corrective action taken to reduce exposure to or below the permissible exposure limit.

**1910.1018(e)(6)** *Accuracy of measurement.* (i) The employer shall use a method of monitoring and measurement which has an accuracy (with a confidence level of 95 percent) of not less than plus or minus 25 percent for concentrations of inorganic arsenic greater than or equal to 10 $\mu g/m^3$.

**1910.1018(e)(6)(ii)** The employer shall use a method of monitoring and measurement which has an accuracy (with confidence level of 95 percent) of not less than plus or minus 35 percent for concentrations of inorganic arsenic greater than 5 $\mu g/m^3$ but less than 10 $\mu g/m^3$.

**1910.1018(f)** ***Regulated area***—(1) *Establishment.* The employer shall establish regulated areas where worker exposures to inorganic arsenic, without regard to the use of respirators, are in excess of the permissible limit.

**1910.1018(f)(2)** *Demarcation.* Regulated areas shall be demarcated and segregated from the rest of the workplace in any manner that minimizes the number of persons who will be exposed to inorganic arsenic.

**1910.1018(f)(3)** *Access.* Access to regulated areas shall be limited to authorized persons or to persons otherwise authorized by the Act or regulations issued pursuant thereto to enter such areas.

**1910.1018(f)(4)** *Provision of respirators.* All persons entering a regulated area shall be supplied with a respirator, selected in accordance with paragraph (h)(2) of this section.

**1910.1018(f)(5)** *Prohibited activities.* The employer shall assure that in regulated areas, food or beverages are not consumed, smoking products, chewing tobacco and gum are not used and cosmetics are not applied, except that these activities may be conducted in the lunchrooms, change rooms and showers required under paragraph (m) of this section. Drinking water may be consumed in the regulated area.

**1910.1018(g)** ***Methods of compliance***—(1) *Controls.* (i) The employer shall institute at the earliest possible time but not later than December 31, 1979, engineering and work practice controls to reduce exposures to or below the permissible exposure limit, except to the extent that the employer can establish that such controls are not feasible.

**1910.1018(g)(1)(ii)** Where engineering and work practice controls are not sufficient to reduce exposures to or below the permissible exposure limit, they shall nonetheless be used to reduce exposures to the lowest levels achievable by these controls and shall be supplemented by the use of respirators in accordance with paragraph (h) of this section and other necessary personal protective equipment. Employee rotation is not required as a control strategy before respiratory protection is instituted.

**1910.1018(g)(2)** *Compliance Program.* (i) The employer shall establish and implement a written program to reduce exposures to or below the permissible exposure limit by means of engineering and work practice controls.

**1910.1018(g)(2)(ii)** Written plans for these compliance programs shall include at least the following:

**1910.1018(g)(2)(ii)(A)** A description of each operation in which inorganic arsenic is emitted; e.g. machinery used, material processed, controls in place, crew size, operating procedures and maintenance practices;

**1910.1018(g)(2)(ii)(B)** Engineering plans and studies used to determine methods selected for controlling exposure to inorganic arsenic;

**1910.1018(g)(2)(ii)(C)** A report of the technology considered in meeting the permissible exposure limit;

**1910.1018(g)(2)(ii)(D)** Monitoring data;

**1910.1018(g)(2)(ii)(E)** A detailed schedule for implementation of the engineering controls and work practices that cannot be implemented immediately and for the adaption and implementation of any additional engineering and work practices necessary to meet the permissible exposure limit;

**1910.1018(g)(2)(ii)(F)** Whenever the employer will not achieve the permissible exposure limit with engineering controls and work practices by December 31, 1979, the employer shall include in the compliance plan an analysis of the effectiveness of the various controls, shall install engineering controls and institute work practices on the quickest schedule feasible, and shall include in the compliance plan and implement a program to minimize the discomfort and maximize the effectiveness of respirator use; and

**1910.1018(g)(2)(ii)(G)** Other relevant information.

**1910.1018(g)(2)(iii)** Written plans for such a program shall be submitted upon request to the Assistant Secretary and the Director, and shall be available at the worksite for examination and copying by the Assistant Secretary, Director, any affected employee or authorized employee representatives.

**1910.1018(g)(2)(iv)** The plans required by this paragraph shall be revised and updated at least every 6 months to reflect the current status of the program.

**1910.1018(h)** ***Respiratory protection***—(1) *General.* For employees who use respirators required by this section, the employer must provide respirators that comply with the requirements of this paragraph. Respirators must be used during:

**1910.1018(h)(1)(i)** Periods necessary to install or implement feasible engineering or work-practice controls.

**1910.1018(h)(1)(ii)** Work operations, such as maintenance and repair activities, for which the employer establishes that engineering and work-practice controls are not feasible.

**1910.1018(h)(1)(iii)** Work operations for which engineering and work-practice controls are not yet sufficient to reduce employee exposures to or below the permissible exposure limit.

**1910.1018(h)(1)(iv)** Emergencies.

**1910.1018(h)(2)** *Respirator program.* (i) The employer must implement a respiratory protection program in accordance with 29 CFR 1910.134(b) through (d) (except (d)(1)(iii)), and (f) through (m).

**1910.1018(h)(2)(ii)** If an employee exhibits breathing difficulty during fit testing or respirator use, they must be examined by a physician trained in pulmonary medicine to determine whether they can use a respirator while performing the required duty.

**1910.1018(h)(3)** *Respirator selection.* (i) The employer must use Table I of this section to select the appropriate respirator or combination of respirators for inorganic arsenic compounds without significant vapor pressure, and Table II of this section to select the appropriate respirator or combination of respirators for inorganic arsenic compounds that have significant vapor pressure.

**1910.1018(h)(3)(ii)** When employee exposures exceed the permissible exposure limit for inorganic arsenic and also exceed the relevant limit for other gases (for example, sulfur dioxide), an air-purifying respirator provided to the employee as specified by this section must have a combination high-efficiency filter with an appropriate gas sorbent. (See footnote in Table 1 of this section.)

**1910.1018(h)(3)(iii)** Employees required to use respirators may choose, and the employer must provide, a powered air-purifying respirator if it will provide proper protection. In addition, the employer must provide a combination dust and acid-gas respirator to employees who are exposed to gases over the relevant exposure limits.

## TABLE I—RESPIRATORY PROTECTION FOR INORGANIC ARSENIC PARTICULATE EXCEPT FOR THOSE WITH SIGNIFICANT VAPOR PRESSURE

| *Concentration of inorganic arsenic (as As) or condition of use* | *Required respirator* |
|---|---|
| (i) Unknown or greater or lesser than 20,000 μg/m(3) (20 mg/m(3)) or firefighting. | (A) Any full facepiece self-contained breathing apparatus operated in positive pressure mode. |
| (ii) Not greater than 20,000 μg/m(3) (20 mg/m(3)). | (A) Supplied air respirator with full facepiece, hood, or helmet or suit and operated in positive pressure mode. |
| (iii) Not greater than 10,000 μg/m(3) (10 mg/m(3)). | (A) Powered air-purifying respirators in all inlet face coverings with high efficiency filters.[1] |
| | (B) Half-mask supplied air respirators operated in positive pressure mode. |
| (iv) Not greater than 500 μg/m(3). | (A) Full facepiece air-purifying respirator equipped with high-efficiency filter.[1] |
| | (B) Any full facepiece supplied air respirator. |
| | (C) Any full facepiece self-contained breathing apparatus. |
| (v) Not greater than 100 μg/m(3). | (A) Half-mask air-purifying respirator equipped with high-efficiency filter.[1] |
| | (B) Any half-mask supplied air respirator. |

[1] High-efficiency filter-99.97 pct efficiency against 0.3 micrometer monodisperse diethyl-hexyl phthalate (DOP) particles.

## TABLE II—RESPIRATORY PROTECTION FOR INORGANIC ARSENICALS (SUCH AS ARSENIC TRICHLORIDE[2] AND ARSENIC PHOSPHIDE) WITH SIGNIFICANT VAPOR PRESSURE

| *Concentration of inorganic arsenic (as As) or condition of use* | *Required respirator* |
|---|---|
| (i) Unknown or greater or lesser than 20,000 μg/m(3) (20 mg/m(3)) or firefighting. | (A) Any full facepiece self-contained breathing apparatus operated in positive pressure mode. |
| (ii) Not greater than 20,000 μg/m(3) (20 mg/m(3)). | (A) Supplied air respirator with full facepiece, hood, or helmet or suit and operated in positive pressure mode. |
| (iii) Not greater than 10,000 μg/m(3) (10 mg/m(3)). | (A) Half-mask[2] supplied air respirator operated in positive pressure mode. |
| (iv) Not greater than 500 μg/m(3). | (A) Front or back mounted gas mask equipped with high-efficiency filter[1] and acid gas canister. |
| | (B) Any full facepiece supplied air respirator. |
| | (C) Any full facepiece self-contained breathing apparatus. |
| (v) Not greater than 100 μg/m(3). | (A) Half-mask air-purifying respirator equipped with high efficiency filter[1] and acid gas cartridge. |
| | (B) Any half-mask supplied air respirator. |

[1] High-efficiency filter-99.97 pct efficiency against 0.3 micrometer monodisperse diethyl-hexyl phthalate (DOP) particles.
[2] Half-mask respirators shall not be used for protection against arsenic trichloride, as it is rapidly absorbed through the skin.

**1910.1018(i)** [Reserved]

**1910.1018(j)** ***Protective work clothing and equipment***—(1) *Provision and use.* Where the possibility of skin or eye irritation from inorganic arsenic exists, and for all workers working in regulated areas, the employer shall provide at no cost to the employee and assure that employees use appropriate and clean protective work clothing and equipment such as, but not limited to:

**1910.1018(j)(1)(i)** Coveralls or similar full-body work clothing;

**1910.1018(j)(1)(ii)** Gloves, and shoes or coverlets;

**1910.1018(j)(1)(iii)** Face shields or vented goggles when necessary to prevent eye irritation, which comply with the requirements of §1910.133(a)(2)-(6); and

**1910.1018(j)(1)(iv)** Impervious clothing for employees subject to exposure to arsenic trichloride.

**1910.1018(j)(2)** *Cleaning and replacement.* (i) The employer shall provide the protective clothing required in paragraph (j) (1) of this section in a freshly laundered and dry condition at least weekly, and daily if the employee works in areas where exposures are over 100 $\mu g/m^3$ of inorganic arsenic or in areas where more frequent washing is needed to prevent skin irritation.

**1910.1018(j)(2)(ii)** The employer shall clean, launder, or dispose of protective clothing required by paragraph (j) (1) of this section.

**1910.1018(j)(2)(iii)** The employer shall repair or replace the protective clothing and equipment as needed to maintain their effectiveness.

**1910.1018(j)(2)(iv)** The employer shall assure that all protective clothing is removed at the completion of a work shift only in change rooms prescribed in paragraph (m) (1) of this section.

**1910.1018(j)(2)(v)** The employer shall assure that contaminated protective clothing which is to be cleaned, laundered, or disposed of, is placed in a closed container in the change-room which prevents dispersion of inorganic arsenic outside the container.

**1910.1018(j)(2)(vi)** The employer shall inform in writing any person who cleans or launders clothing required by this section, of the potentially harmful effects including the carcinogenic effects of exposure to inorganic arsenic.

**1910.1018(j)(2)(vii)** The employer shall assure that the containers of contaminated protective clothing and equipment in the workplace or which are to be removed from the workplace are labelled as follows:

CAUTION: Clothing contaminated with inorganic arsenic; do not remove dust by blowing or shaking. Dispose of inorganic arsenic contaminated wash water in accordance with applicable local, State or Federal regulations.

**1910.1018(j)(2)(viii)** The employer shall prohibit the removal of inorganic arsenic from protective clothing or equipment by blowing or shaking.

**1910.1018(k)** ***Housekeeping***—(1) *Surfaces.* All surfaces shall be maintained as free as practicable of accumulations of inorganic arsenic.

**1910.1018(k)(2)** *Cleaning floors.* Floors and other accessible surfaces contaminated with inorganic arsenic may not be cleaned by the use of compressed air, and shoveling and brushing may be used only where vacuuming or other relevant methods have been tried and found not to be effective.

**1910.1018(k)(3)** *Vacuuming.* Where vacuuming methods are selected, the vacuums shall be used and emptied in a manner to minimize the reentry of inorganic arsenic into the workplace.

**1910.1018(k)(4)** *Housekeeping plan.* A written housekeeping and maintenance plan shall be kept which shall list appropriate frequencies for carrying out housekeeping operations, and for cleaning and maintaining dust collection equipment. The plan shall be available for inspection by the Assistant Secretary.

**1910.1018(k)(5)** *Maintenance of equipment.* Periodic cleaning of dust collection and ventilation equipment and checks of their effectiveness shall be carried out to maintain the effectiveness of the system and a notation kept of the last check of effectiveness and cleaning or maintenance.

**1910.1018(l)** [Reserved]

**1910.1018(m)** ***Hygiene facilities and practices***—(1) *Change rooms.* The employer shall provide for employees working in regulated areas or subject to the possibility of skin or eye irritation from inorganic arsenic, clean change rooms equipped with storage facilities for street clothes and separate storage facilities for protective clothing and equipment in accordance with 29 CFR 1910.141(e).

**1910.1018(m)(2)** *Showers.* (i) The employer shall assure that employees working in regulated areas or subject to the possibility of skin or eye irritation from inorganic arsenic shower at the end of the work shift.

**1910.1018(m)(2)(ii)** The employer shall provide shower facilities in accordance with §1910.141(d)(3).

**1910.1018(m)(3)** *Lunchrooms.* (i) The employer shall provide for employees working in regulated areas, lunchroom facilities which have a temperature controlled, positive pressure, filtered air supply, and which are readily accessible to employees working in regulated areas.

**1910.1018(m)(3)(ii)** The employer shall assure that employees working in the regulated area or subject to the possibility of skin or eye irritation from exposure to inorganic arsenic wash their hands and face prior to eating.

**1910.1018(m)(4)** *Lavatories.* The employer shall provide lavatory facilities which comply with §1910.141(d)(1) and (2).

**1910.1018(m)(5)** *Vacuuming clothes.* The employer shall provide facilities for employees working in areas where exposure, without regard to the use of respirators, exceeds 100 $\mu g/m^3$ to vacuum their protective clothing and clean or change shoes worn in such areas before entering change rooms, lunchrooms or shower rooms required by paragraph (j) of this section and shall assure that such employees use such facilities.

**1910.1018(m)(6)** *Avoidance of skin irritation.* The employer shall assure that no employee is exposed to skin or eye contact with arsenic trichloride, or to skin or eye contact with liquid or particulate inorganic arsenic which is likely to cause skin or eye irritation.

**1910.1018(n)** ***Medical surveillance***—(1) *General*—(i) *Employees covered.* The employer shall institute a medical surveillance program for the following employees:

**1910.1018(n)(1)(i)(A)** All employees who are or will be exposed above the action level, without regard to the use of respirators, at least 30 days per year; and

**1910.1018(n)(1)(i)(B)** All employees who have been exposed above the action level, without regard to respirator use, for 30 days or more per year for a total of 10 years or more of combined employment with the employer or predecessor employers prior to or after the effective date of this standard. The determination of exposures prior to the effective date of this standard shall be based upon prior exposure records, comparison with the first measurements taken after the effective date of this standard, or comparison with records of exposures in areas with similar processes, extent of engineering controls utilized and materials used by that employer.

**1910.1018(n)(1)(ii)** *Examination by physician.* The employer shall assure that all medical examinations and procedures are performed by or under the supervision of a licensed physician, and shall be provided without cost to the employee, without loss of pay and at a reasonable time and place.

**1910.1018(n)(2)** *Initial examinations.* By December 1, 1978, for employees initially covered by the medical provisions of this section, or thereafter at the time of initial assignment to an area where the employee is likely to be exposed over the action level at least 30 days per year, the employer shall provide each affected employee an opportunity for a medical examination, including at least the following elements:

**1910.1018(n)(2)(i)** A work history and a medical history which shall include a smoking history and the presence and degree of respiratory symptoms such as breathlessness, cough, sputum production and wheezing.

**1910.1018(n)(2)(ii)** A medical examination which shall include at least the following:

**1910.1018(n)(2)(ii)(A)** A 14" by 17" posterior-anterior chest X-ray and International Labor Office UICC/Cincinnati (ILO U/C) rating;

**1910.1018(n)(2)(ii)(B)** A nasal and skin examination; and

**1910.1018(n)(2)(ii)(C)** Other examinations which the physician believes appropriate because of the employees exposure to inorganic arsenic or because of required respirator use.

**1910.1018(n)(3)** *Periodic examinations.* (i) The employer shall provide the examinations specified in paragraphs (n)(2)(i) and (n)(2)(ii) at least annually for covered employees who are under 45 years of age with fewer than 10 years of exposure over the action level without regard to respirator use.

**1910.1018(n)(3)(ii)** The employer shall provide the examinations specified in paragraphs (n)(2)(i) and (n)(2)(ii)(B) and (C) of this section at least semiannually, and the x-ray requirement specified in paragraph (n)(2)(ii)(A) of this section at least annually, for other covered employees.

**1910.1018(n)(3)(iii)** Whenever a covered employee has not taken the examinations specified in paragraphs (n)(2)(i) and (n)(2)(ii) of this section within six (6) months preceding the termination of employment, the employer shall provide such examinations to the employee upon termination of employment.

**1910.1018(n)(4)** *Additional examinations.* If the employee for any reason develops signs or symptoms commonly associated with exposure to inorganic arsenic the employer shall provide an appropriate examination and emergency medical treatment.

**1910.1018(n)(5)** *Information provided to the physician.* The employer shall provide the following information to the examining physician:

**1910.1018(n)(5)(i)** A copy of this standard and its appendices;

**1910.1018(n)(5)(ii)** A description of the affected employee's duties as they relate to the employee's exposure;

**1910.1018(n)(5)(iii)** The employee's representative exposure level or anticipated exposure level;

**1910.1018(n)(5)(iv)** A description of any personal protective equipment used or to be used; and

**1910.1018(n)(5)(v)** Information from previous medical examinations of the affected employee which is not readily available to the examining physician.

**1910.1018(n)(6)** *Physician's written opinion.* (i) The employer shall obtain a written opinion from the examining physician which shall include:

**1910.1018(n)(6)(i)(A)** The results of the medical examination and tests performed;

**1910.1018(n)(6)(i)(B)** The physician's opinion as to whether the employee has any detected medical conditions which would place the employee at increased risk of material impairment of the employee's health from exposure to inorganic arsenic;

**1910.1018(n)(6)(i)(C)** Any recommended limitations upon the employee's exposure to inorganic arsenic or upon the use of protective clothing or equipment such as respirators; and

**1910.1018(n)(6)(i)(D)** A statement that the employee has been informed by the physician of the results of the medical examination and any medical conditions which require further explanation or treatment.

**1910.1018(n)(6)(ii)** The employer shall instruct the physician not to reveal in the written opinion specific findings or diagnoses unrelated to occupational exposure.

**1910.1018(n)(6)(iii)** The employer shall provide a copy of the written opinion to the affected employee.

**1910.1018(o)** ***Employee information and training***—(1) *Training program.* (i) The employer shall institute a training program for all employees who are subject to exposure to inorganic arsenic above the action level without regard to respirator use, or for whom there is the possibility of skin or eye irritation from inorganic arsenic. The employer shall assure that those employees participate in the training program.

**1910.1018(o)(1)(ii)** The training program shall be provided by October 1, 1978, for employees covered by this provision, at the time of initial assignment for those subsequently covered by this provision, and at least annually for other covered employees thereafter; and the employer shall assure that each employee is informed of the following:

**1910.1018(o)(1)(ii)(A)** The information contained in Appendix A;

**1910.1018(o)(1)(ii)(B)** The quantity, location, manner of use, storage, sources of exposure, and the specific nature of operations which could result in exposure to inorganic arsenic as well as any necessary protective steps;

**1910.1018(o)(1)(ii)(C)** The purpose, proper use, and limitation of respirators;

**1910.1018(o)(1)(ii)(D)** The purpose and a description of the medical surveillance program as required by paragraph (n) of this section;

**1910.1018(o)(1)(ii)(E)** The engineering controls and work practices associated with the employee's job assignment; and

**1910.1018(o)(1)(ii)(F)** A review of this standard.

**1910.1018(o)(2)** *Access to training materials.* (i) The employer shall make readily available to all affected employees a copy of this standard and its appendices.

**1910.1018(o)(2)(ii)** The employer shall provide; upon request, all materials relating to the employee information and training program to the Assistant Secretary and the Director.

**1910.1018(p)** ***Signs and labels***—(1) *General.* (i) The employer may use labels or signs required by other statutes, regulations, or ordinances in addition to, or in combination with, signs and labels required by this paragraph.

**1910.1018(p)(1)(ii)** The employer shall assure that no statement appears on or near any sign or label required by this paragraph which contradicts or detracts from the meaning of the required sign or label.

**1910.1018(p)(2)** *Signs.* (i) The employer shall post signs demarcating regulated areas bearing the legend;

**DANGER**

**INORGANIC ARSENIC**

**CANCER HAZARD**

**AUTHORIZED PERSONNEL ONLY**

**NO SMOKING OR EATING**

**RESPIRATOR REQUIRED**

**1910.1018(p)(2)(ii)** The employer shall assure that signs required by this paragraph are illuminated and cleaned as necessary so that the legend is readily visible.

**1910.1018(p)(3)** *Labels.* The employer shall apply precautionary labels to all shipping and storage containers of inorganic arsenic, and to all products containing inorganic arsenic except when the inorganic arsenic in the product is bound in such a manner so as to make unlikely the possibility of airborne exposure to inorganic arsenic. (Possible examples of products not requiring labels are semiconductors, light emitting diodes and glass). The label shall bear the following legend:

**DANGER**

**CONTAINS INORGANIC ARSENIC**

**CANCER HAZARD**

**HARMFUL IF INHALED OR SWALLOWED**

**USE ONLY WITH ADEQUATE VENITLATION**

**OR RESPIRATORY PROTECTION**

**1910.1018(q)** ***Recordkeeping***—(1) *Exposure monitoring.* (i) The employer shall establish and maintain an accurate record of all monitoring required by paragraph (e) of this section.

**1910.1018(q)(1)(ii)** This record shall include:

**1910.1018(q)(1)(ii)(A)** The date(s), number, duration location, and results of each of the samples taken, including a description of the sampling procedure used to determine representative employee exposure where applicable;

**1910.1018(q)(1)(ii)(B)** A description of the sampling and analytical methods used and evidence of their accuracy;

**1910.1018(q)(1)(ii)(C)** The type of respiratory protective devices worn, if any;

**1910.1018(q)(1)(ii)(D)** Name, social security number, and job classification of the employees monitored and of all other employees whose exposure the measurement is intended to represent; and

**1910.1018(q)(1)(ii)(E)** The environmental variables that could affect the measurement of the employee's exposure.

**1910.1018(q)(1)(iii)** The employer shall maintain these monitoring records for at least 40 years or for the duration of employment plus 20 years, whichever, is longer.

**1910.1018(q)(2)** *Medical surveillance.* (i) The employer shall establish and maintain an accurate record for each employee subject to medical surveillance as required by paragraph (n) of this section.

**1910.1018(q)(2)(ii)** This record shall include:

**1910.1018(q)(2)(ii)(A)** The name, social security number, and description of duties of the employee;

**1910.1018(q)(2)(ii)(B)** A copy of the physician's written opinions;

**1910.1018(q)(2)(ii)(C)** Results of any exposure monitoring done for that employee and the representative exposure levels supplied to the physician; and

**1910.1018(q)(2)(ii)(D)** Any employee medical complaints related to exposure to inorganic arsenic.

**1910.1018(q)(2)(iii)** The employer shall in addition keep, or assure that the examining physician keeps, the following medical records;

**1910.1018(q)(2)(iii)(A)** A copy of the medical examination results including medical and work history required under paragraph (n) of this section;

**1910.1018(q)(2)(iii)(B)** A description of the laboratory procedures and a copy of any standards or guidelines used to interpret the test results or references to that information;

**1910.1018(q)(2)(iii)(C)** The initial X-ray;

**1910.1018(q)(2)(iii)(D)** The X-rays for the most recent 5 years; and

**1910.1018(q)(2)(iii)(E)** Any X-rays with a demonstrated abnormality and all subsequent X-rays;

**1910.1018(q)(2)(iv)** The employer shall maintain or assure that the physician maintains those medical records for at least 40 years, or for the duration of employment plus 20 years whichever is longer.

**1910.1018(q)(3)** *Availability.* (i) The employer shall make available upon request all records required to be maintained by paragraph (q) of this section to the Assistant Secretary and the Director for examination and copying.

**1910.1018(q)(3)(ii)** Records required by this paragraph shall be provided upon request to employees, designated representatives, and the Assistant Secretary in accordance with 29 CFR 1910.20(a) through (e) and (g) through (i).

**1910.1018(q)(4)** *Transfer of records.* (i) Whenever the employer ceases to do business, the successor employer shall receive and retain all records required to be maintained by this section.

**1910.1018(q)(4)(ii)** Whenever the employer ceases to do business and there is no successor employer to receive and retain the records required to be maintained by this section for the prescribed period, these records shall be transmitted to the Director.

**1910.1018(q)(4)(iii)** At the expiration of the retention period for the records required to be maintained by this section, the employer shall notify the Director at least 3 months prior to the disposal of such records and shall transmit those records to the Director if he requests them within that period.

**1910.1018(q)(4)(iv)** The employer shall also comply with any additional requirements involving the transfer of records set in 29 CFR 1910.20(h).

**1910.1018(r)** ***Observation of monitoring***—(1) *Employee observation.* The employer shall provide affected employees or their designated representatives an opportunity to observe any monitoring of employee exposure to inorganic arsenic conducted pursuant to paragraph (e) of this section.

**1910.1018(r)(2)** *Observation procedures.* (i) Whenever observation of the monitoring of employee exposure to inorganic arsenic requires entry into an area where the use of respirators, protective clothing, or equipment is required, the employer shall provide the observer with and assure the use of such respirators, clothing, and such equipment, and shall require the observer to comply with all other applicable safety and health procedures.

**1910.1018(r)(2)(ii)** Without interfering with the monitoring, observers shall be entitled to;

**1910.1018(r)(2)(ii)(A)** Receive an explanation of the measurement procedures;

**1910.1018(r)(2)(ii)(B)** Observe all steps related to the monitoring of inorganic arsenic performed at the place of exposure; and

**1910.1018(r)(2)(ii)(C)** Record the results obtained or receive copies of the results when returned by the laboratory.

**1910.1018(s)** ***Effective date.*** This standard shall become effective August 1, 1978.

**1910.1018(t)** ***Appendices.*** The information contained in the appendices to this section is not intended by itself, to create any additional obligations not otherwise imposed by this standard nor detract from any existing obligation.

**1910.1018(u)** ***Startup dates***—(1) *General.* The startup dates of requirements of this standard shall be the effective date of this standard unless another startup date is provided for either in other paragraphs of this section or in this paragraph.

**1910.1018(u)(2)** *Monitoring.* Initial monitoring shall be commenced on August 1, 1978, and shall be completed by September 15, 1978.

**1910.1018(u)(3)** *Regulated areas.* Regulated areas required to be established as a result of initial monitoring shall be set up as soon as possible after the results of that monitoring is known and no later than October 1, 1978.

**1910.1018(u)(4)** *Compliance program.* The written program required by paragraph (g)(2) as a result of initial monitoring shall be made available for inspection and copying as soon as possible and no later than December 1, 1978.

**1910.1018(u)(5)** *Hygiene and lunchroom facilities.* Construction plans for change- rooms, showers, lavatories, and lunchroom facilities shall be completed no later than December 1, 1978, and these facilities shall be constructed and in use no later than July 1, 1979. However, if as part of the compliance plan it is predicted by an independent engineering firm that engineering controls and work practices will reduce exposures below the permissible exposure limit by December 31, 1979, for affected employees, then such facilities need not be completed until 1 year after the engineering controls are completed or December 31, 1980, whichever is earlier, if such controls have not in fact succeeded in reducing exposure to below the permissible exposure limit.

**1910.1018(u)(6)** *Summary of startup dates set forth elsewhere in this standard.*

**STARTUP DATES**

August 1, 1978—Respirator use over 500 $\mu g/m^3$.

**AS SOON AS POSSIBLE BUT NO LATER THAN**

September 15, 1978—Completion of initial monitoring.

October 1, 1978—Complete establishment of regulated areas. Respirator use for employees exposed above 50 $\mu g/m^3$. Completion of initial training. Notification of use.

December 1, 1978—Respirator use over 10 $\mu g/m^3$. Completion of initial medical. Completion of compliance plan. Optional use of powered air-purifying respirators.

July 1, 1979—Completion of lunch rooms and hygiene facilities.

December 31, 1979—Completion of engineering controls.

All other requirements of the standard have as their startup date August 1, 1978.

## APPENDIX A TO §1910.1018—INORGANIC ARSENIC SUBSTANCE INFORMATION SHEET

### I. SUBSTANCE IDENTIFICATION

A. *Substance.* Inorganic Arsenic.

B. *Definition.* Copper acetoarsenite, arsenic and all inorganic compounds containing arsenic except arsine, measured as arsenic (As).

C. *Permissible Exposure Limit.* 10 micrograms per cubic meter of air as determined as an average over an 8-hour period. No employee may be exposed to any skin or eye contact with arsenic trichloride or to skin or eye contact likely to cause skin or eye irritation.

D. *Regulated Areas.* Only employees authorized by your employer should enter a regulated area.

### II. HEALTH HAZARD DATA

A. *Comments.* The health hazard of inorganic arsenic is high.

B. *Ways in which the chemical affects your body.* Exposure to airborne concentrations of inorganic arsenic may cause lung cancer, and can be a skin irritant. Inorganic arsenic may also affect your body if swallowed. One compound in particular, arsenic trichloride, is especially dangerous because it can be absorbed readily through the skin. Because inorganic arsenic is a poison, you should wash your hands thoroughly prior to eating or smoking.

### III. PROTECTIVE CLOTHING AND EQUIPMENT

A. *Respirators.* Respirators will be provided by your employer at no cost to you for routine use if your employer is in the process of implementing engineering and work practice controls or where engineering and work practice controls are not feasible or insufficient. You must wear respirators for non-routine activities or in emergency situations where you are likely to be exposed to levels of inorganic arsenic in excess of the permissible exposure limit. Since how well your respirator fits your face is very important, your employer is required to conduct fit tests to make sure the respirator seals properly when you wear it. These tests are simple and rapid and will be explained to you during training sessions.

B. *Protective clothing.* If you work in a regulated area, your employer is required to provide at no cost to you, and you must wear, appropriate, clean, protective clothing and equipment. The purpose of this equipment is to prevent you from bringing to your home arsenic-contaminated dust and to protect your body from repeated skin contact with inorganic arsenic likely to cause skin irritation. This clothing should include such items as coveralls or similar full-body clothing, gloves, shoes or coverlets, and aprons. Protective equipment should include face shields or vented goggles, where eye irritation may occur.

### IV. HYGIENE FACILITIES AND PRACTICES

You must not eat, drink, smoke, chew gum or tobacco, or apply cosmetics in the regulated area, except that drinking water is permitted. If you work in a regulated area your employer is required to provide lunchrooms and other areas for these purposes.

If you work in a regulated area, your employer is required to provide showers, washing facilities, and change rooms. You must wash your face, and hands before eating and must shower at the end of the work shift. Do not take used protective clothing out of change rooms without your employer's permission. Your employer is required to provide for laundering or cleaning of your protective clothing.

### V. SIGNS AND LABELS

Your employer is required to post warning signs and labels for your protection. Signs must be posted in regulated areas. The signs must warn that a cancer hazard is present, that only authorized employees may enter the area, and that no smoking or eating is allowed, and that respirators must be worn.

### VI. MEDICAL EXAMINATIONS

If your exposure to arsenic is over the Action Level (5 mg/m$^3$)—(including all persons working in regulated areas) at least 30 days per year, or you have been exposed to arsenic for more than 10 years over the Action Level, your employer is required to provide you with a medical examination. The examination shall be every 6 months for employees over 45 years old or with more than 10 years exposure over the Action Level and annually for other covered employees. The medical examination must include a medical history; a chest x-ray; a skin examination and a nasal examination. The examining physician will provide a written opinion to your employer containing the results of your medical exams. You should also receive a copy of this opinion. The physician must not tell your employer any conditions he detects unrelated to occupational exposure to arsenic but must tell you those conditions.

### VII. OBSERVATION OF MONITORING

Your employer is required to monitor your exposure to arsenic and you or your representatives are entitled to observe the monitoring procedure. You are entitled to receive an explanation of the measurement procedure, and to record the results obtained. When the monitoring procedure is taking place in an area where respirators or personal protective clothing and equipment are required to be worn, you must also be provided with and must wear the protective clothing and equipment.

### VIII. ACCESS TO RECORDS

You or your representative are entitled to records of your exposure to inorganic arsenic and your medical examination records if you request your employer to provide them.

### IX. TRAINING AND NOTIFICATION

Additional information on all of these items plus training as to hazards of exposure to inorganic arsenic and the engineering and work practice controls associated with your job will also be provided by your employer. If you are exposed over the permissible exposure limit, your employer must inform you of that fact and the actions he is taking to reduce your exposures.

## APPENDIX B TO §1910.1018—SUBSTANCE TECHNICAL GUIDELINES

### ARSENIC, ARSENIC TRIOXIDE, ARSENIC TRICHLORIDE (THREE EXAMPLES)

#### *I. Physical and chemical properties*

A. Arsenic (metal).

1. Formula: As.

2. Appearance: Gray metal.

3. Melting point: Sublimes without melting at 613C.

4. Specific Gravity. (H20 – 1):5.73.

5. Solubility in water: Insoluble.

B. Arsenic Trioxide.

1. Formula: As203, (As406).

2. Appearance: White powder.

3. Melting point: 315C.

4. Specific Gravity (H20 = 1):3.74.

5. Solubility in water: 3.7 grams in 100cc of water at 20c.

C. Arsenic Trichloride (liquid).

1. Formula: AsC13.

2. Appearance: Colorless or pale yellow liquid.

3. Melting point: -8.5C.

4. Boiling point: 130.2C.

5. Specific Gravity (H20 = 1):2.16 at 20C.

6. Vapor Pressure: 10mm Hg at 23.5C.

7. Solubility in Water: Decomposes in water.

#### *II. Fire, explosion and reactivity data.*

A. Fire: Arsenic, arsenic Trioxide and Arsenic Trichloride are nonflammable.

B. Reactivity:

1. Conditions Contributing to instability: Heat.

2. Incompatibility: Hydrogen gas can react with inorganic arsenic to form the highly toxic gas arsine.

#### *III. Monitoring and Measurement Procedures*

Samples collected should be full shift (at least 7-hour) samples. Sampling should be done using a personal sampling pump at a flow rate of 2 liters per minute. Samples should be collected on 0.8 micrometer pore size membrane filter (37mm diameter). Volatile arsenicals such as arsenic trichloride can be most easily collected in a midget bubbler filled with 15 ml. of 0.1 N NaOH.

The method of sampling and analysis should have an accuracy of not less ±25 percent (with a confidence limit of 95 percent) for 10 micrograms per cubic meter of air (10 $\mu$g/m$^3$) and ±35 percent (with a confidence limit of 95 percent) for concentrations of inorganic arsenic between 5 and 10 $\mu$g/m$^3$.

## APPENDIX C TO §1910.1018—MEDICAL SURVEILLANCE GUIDELINES

### I. GENERAL

Medical examinations are to be provided for all employees exposed to levels of inorganic arsenic above the action level (5 $\mu$g/m$^3$) for at least 30 days per year (which would include among others, all employees, who work in regulated areas). Examinations are also to be provided to all employees who have had 10 years or more exposure above the action level for more than 30 days per year while working for the present or predecessor employer though they may no longer be exposed above the level.

An initial medical examination is to be provided to all such employees by December 1, 1978. In addition, an initial medical examination is to be provided to all employees who are first assigned to areas in which worker exposure will probably exceed 5 $\mu$g/m$^3$ (after the effective date of this standard) at the time of initial assignment.

In addition to its immediate diagnostic usefulness, the initial examination will provide a baseline for comparing future test results. The initial examination must include as a minimum the following elements:

(1) A work and medical history, including a smoking history, and presence and degree of respiratory symptoms such as breathlessness, cough, sputum production, and wheezing;

(2) A 14" by 17" posterior-anterior chest X-ray and an International Labor Office UICC/Cincinnati (ILO U/C) rating;

(3) A nasal and skin examination; and

(4) Other examinations which the physician believes appropriate because of the employee's exposure to inorganic arsenic or because of required respirator use.

Periodic examinations are also to be provided to the employees listed above. The periodic examinations shall be given annually for those covered employees 45 years of age or less with fewer than 10 years employment in areas where employee exposure exceeds the action level (5 $\mu g/m^3$). Periodic examinations need not include sputum cytology and only an updated medical history is required.

Periodic examinations for other covered employees, shall be provided every six (6) months. These examinations shall include all tests required in the initial examination, except that the medical history need only be updated.

The examination contents are minimum requirements. Additional tests such as lateral and oblique X-rays or pulmonary function tests may be useful. For workers exposed to three arsenicals which are associated with lymphatic cancer, copper acetoarsenite, potassium arsenite, or sodium arsenite the examination should also include palpation of superficial lymph nodes and complete blood count.

## II. NONCARCINOGENIC EFFECTS

The OSHA standard is based on minimizing risk of exposed workers dying of lung cancer from exposure to inorganic arsenic. It will also minimize skin cancer from such exposures.

The following three sections quoted from "Occupational Diseases: A Guide to Their Recognition", Revised Edition, June 1977, National Institute for Occupational Safety and Health is included to provide information on the nonneoplastic effects of exposure to inorganic arsenic. Such effects should not occur if the OSHA standards are followed.

A. *Local*—Trivalent arsenic compounds are corrosive to the skin. Brief contact has no effect but prolonged contact results in a local hyperemia and later vesicular or pustular eruption. The moist mucous membranes are most sensitive to the irritant action. Conjunctiva, moist and macerated areas of skin, the eyelids, the angles of the ears, nose, mouth, and respiratory mucosa are also vulnerable to the irritant effects. The wrists are common sites of dermatitis, as are the genitalia if personal hygiene is poor. Perforations of the nasal septum may occur. Arsenic trioxide and pentoxide are capable of producing skin sensitization and contact dermatitis. Arsenic is also capable of producing keratoses, especially of the palms and soles.

B. *Systemic*—The acute toxic effects of arsenic are generally seen following ingestion of inorganic arsenical compounds. This rarely occurs in an industrial setting. Symptoms develop within 1/2 to 4 hours following ingestion and are usually characterized by constriction of the throat followed by dysphagia, epigastric pain, vomiting, and watery diarrhea. Blood may appear in vomitus and stools. If the amount ingested is sufficiently high, shock may develop due to severe fluid loss, and death may ensue in 24 hours. If the acute effects are survived, exfoliative dermatitis and peripheral neuritis may develop.

Cases of acute arsenical poisoning due to inhalation are exceedingly rare in industry. When it does occur, respiratory tract symptoms—cough, chest pain, dyspnea—giddiness, headache, and extreme general weakness precede gastrointestinal symptoms. The acute toxic symptoms of trivalent arsenical poisoning are due to severe inflammation of the mucous membranes and greatly increased permeability of the blood capillaries.

Chronic arsenical poisoning due to ingestion is rare and generally confined to patients taking prescribed medications. However, it can be a concomitant of inhaled inorganic arsenic from swallowed sputum and improper eating habits. Symptoms are weight loss, nausea and diarrhea alternating with constipation, pigmentation and eruption of the skin, loss of hair, and peripheral neuritis. Chronic hepatitis and cirrhosis have been described. Polyneuritis may be the salient feature, but more frequently there are numbness and parasthenias of "glove and stocking" distribution. The skin lesions are usually melanotic and keratotic and may occasionally take the form of an intradermal cancer of the squamous cell type, but without infiltrative properties. Horizontal white lines (striations) on the fingernails and toenails are commonly seen in chronic arsenical poisoning and are considered to be a diagnostic accompaniment of arsenical polyneuritis.

Inhalation of inorganic arsenic compounds is the most common cause of chronic poisoning in the industrial situation. This condition is divided into three phases based on signs and symptoms.

First Phase: The worker complains of weakness, loss of appetite, some nausea, occasional vomiting, a sense of heaviness in the stomach, and some diarrhea.

Second Phase: The worker complains of conjunctivitis, a catarrhal state of the mucous membranes of the nose, larynx, and respiratory passage. Coryza, hoarseness, and mild tracheobronchitis may occur. Perforation of the nasal septum is common, and is probably the most typical lesion of the upper respiratory tract in occupational exposure to arsenical dust. Skin lesions, eczematoid and allergic in type, are common.

Third Phase: The worker complains of symptoms of peripheral neuritis, initially of hands and feet, which is essentially sensory. In more severe cases, motor paralyses occur; the first muscles affected are usually the toe extensors and the peronei. In only the most severe cases will paralysis of flexor muscles of the feet or of the extensor muscles of hands occur.

Liver damage from chronic arsenical poisoning is still debated, and as yet the question is unanswered. In cases of chronic and acute arsenical poisoning, toxic effects to the myocardium have been reported based on EKG changes. These findings, however, are now largely discounted and the EKG changes are ascribed to electrolyte disturbances concomitant with arsenicalism. Inhalation of arsenic trioxide and other inorganic arsenical dusts does not give rise to radiological evidence or pneumoconiosis. Arsenic does have a depressant effect upon the bone marrow, with disturbances of both erythropoiesis and myelopoiesis.

## BIBLIOGRAPHY

Dinman, B. D. 1960. Arsenic; chronic human intoxication. J. Occup. Med. 2:137.

Elkins, H. B. 1959. The Chemistry of Industrial Toxicology, 2nd ed. John Wiley and Sons, New York.

Holmquist, L. 1951. Occupational arsenical dermatitis; a study among employees at a copper-ore smelting works including investigations of skin reactions to contact with arsenic compounds. Acta. Derm. Venereol. (Supp. 26) 31:1.

Pinto, S. S., and C. M. McGill. 1953. Arsenic trioxide exposure in industry. Ind. Med. Surg. 22:281.

Pinto, S. S., and K. W. Nelson. 1976. Arsenic toxicology and industrial exposure. Annu. Rev. Pharmacol. Toxicol. 16:95.

Vallee, B. L., D. D. Ulmer, and W. E. C. Wacker. 1960. Arsenic toxicology and biochemistry. AMA Arch. Indust. Health 21:132.

[39 FR 23502, June 27, 1974, as amended at 43 FR 19624, May 5, 1978; 43 FR 28472, June 30, 1978; 45 FR 35282, May 23, 1980; 54 FR 24334, June 7, 1989; 58 FR 35310, June 30, 1993; 61 FR 5508, Feb. 13, 1996; 61 FR 9245, Mar. 7, 1996; 63 FR 1286, Jan. 8, 1998; 63 FR 33468, June 18, 1998]

### §1926.1127 Cadmium.

**1926.1127(a)** ***Scope.*** This standard applies to all occupational exposures to cadmium and cadmium compounds, in all forms, in all construction work where an employee may potentially be exposed to cadmium. Construction work is defined as work involving construction, alteration and/or repair, including but not limited to the following:

**1926.1127(a)(1)** Wrecking, demolition or salvage of structures where cadmium or materials containing cadmium are present;

**1926.1127(a)(2)** Use of cadmium containing-paints and cutting, brazing, burning, grinding or welding on surfaces that were painted with cadmium-containing paints;

**1926.1127(a)(3)** Construction, alteration, repair, maintenance, or renovation of structures, substrates, or portions thereof, that contain cadmium, or materials containing cadmium;

**1926.1127(a)(4)** Cadmium welding; cutting and welding cadmium-plated steel; brazing or welding with cadmium alloys;

**1926.1127(a)(5)** Installation of products containing cadmium;

**1926.1127(a)(6)** Electrical grounding with cadmium welding, or electrical work using cadmium-coated conduit;

**1926.1127(a)(7)** Maintaining or retrofitting cadmium-coated equipment;

**1926.1127(a)(8)** Cadmium contamination/emergency cleanup; and

**1926.1127(a)(9)** Transportation, disposal, storage, or containment of cadmium or materials containing cadmium on the site or location at which construction activities are performed.

**1926.1127(b)** ***Definitions.***

*Action level* (AL) is defined as an airborne concentration of cadmium of 2.5 micrograms per cubic meter of air (2.5 $\mu$ g/m$^3$), calculated as an 8-hour time-weighted average (TWA).

*Assistant Secretary* means the Assistant Secretary of Labor for Occupational Safety and Health, U.S. Department of Labor, or designee.

*Authorized person* means any person authorized by the employer and required by work duties to be present in regulated areas or any person authorized by the OSH Act or regulations issued under it to be in regulated areas.

*Competent person*, in accordance with 29 CFR 1926.32(f), means a person designated by the employer to act on the employer's behalf who is capable of identifying existing and potential cadmium hazards in the workplace and the proper methods to control them in order to protect workers, and has the authority necessary to take prompt corrective measures to eliminate or control such hazards. The duties of a competent person include at least the following: Determining prior to the performance of work whether cadmium is present in the workplace; establishing, where necessary, regulated areas and assuring that access to and from those areas is limited to authorized employees; assuring the adequacy of any employee exposure monitoring required by this standard; assuring that all employees exposed to air cadmium levels above the PEL wear appropriate personal protective equipment and are trained in the use of appropriate methods of exposure control; assuring that proper hygiene facilities are provided and that workers are trained to use those facilities; and assuring that the engineering controls required by this standard are implemented, maintained in proper operating condition, and functioning properly.

*Director* means the Director of the National Institute for Occupational Safety and Health (NIOSH), U.S. Department of Health and Human Services, or designee.

*Employee exposure* and similar language referring to the air cadmium level to which an employee is exposed means the exposure to airborne cadmium that would occur if the employee were not using respiratory protective equipment.

*Final medical determination* is the written medical opinion of the employee's health status by the examining physician under paragraphs (l)(3)-(12) of this section or, if multiple physician review under paragraph (l)(13) of this section or the alternative physician determination under paragraph (l)(14) of this section is invoked, it is the final, written medical finding, recommendation or determination that emerges from that process.

*High-efficiency Particulate Air [HEPA] filter* means a filter capable of trapping and retaining at least 99.97 percent of mono-dispersed particles of 0.3 micrometers in diameter.

*Regulated area* means an area demarcated by the employer where an employee's exposure to airborne concentrations of cadmium exceeds, or can reasonably be expected to exceed the permissible exposure limit (PEL).

*This section* means this cadmium standard.

**1926.1127(c)** ***Permissible Exposure Limit (PEL).*** The employer shall assure that no employee is exposed to an airborne concentration of cadmium in excess of five micrograms per cubic meter of air (5 $\mu$ g/m$^3$), calculated as an eight-hour time-weighted average exposure (TWA).

**1926.1127(d)** ***Exposure Monitoring***— (1) *General.* (i) Prior to the performance of any construction work where employees may be potentially exposed to cadmium, the employer shall establish the applicability of this standard by determining whether cadmium is present in the workplace and whether there is the possibility that employee exposures will be at or above the action level. The employer shall designate a competent person who shall make this determination. Investigation and material testing techniques shall be used, as appropriate, in the determination. Investigation shall include a review of relevant plans, past reports, material safety data sheets, and other available records, and consultations with the property owner and discussions with appropriate individuals and agencies.

**1926.1127(d)(1)(ii)** Where cadmium has been determined to be present in the workplace, and it has been determined that there is a possibility the employee's exposure will be at or above the action level, the competent person shall identify employees potentially exposed to cadmium at or above the action level.

**1926.1127(d)(1)(iii)** Determinations of employee exposure shall be made from breathing-zone air samples that reflect the monitored employee's regular, daily 8-hour TWA exposure to cadmium.

**1926.1127(d)(1)(iv)** Eight-hour TWA exposures shall be determined for each employee on the basis of one or more personal breathing-zone air samples reflecting full shift exposure on each shift, for each job classification, in each work area. Where several employees perform the same job tasks, in the same job classification, on the same shift, in the same work area, and the length, duration, and level of cadmium exposures are similar, an employer may sample a representative fraction of the employees instead of all employees in order to meet this requirement. In representative sampling, the employer shall sample the employee(s) expected to have the highest cadmium exposures.

**1926.1127(d)(2)** *Specific.* (i) Initial monitoring. Except as provided for in paragraph (d)(2)(iii) of this section, where a determination conducted under paragraph (d)(1)(i) of this section shows the possibility of employee exposure to cadmium at or above the action level, the employer shall conduct exposure monitoring as soon as practicable that is representative of the exposure for each employee in the workplace who is or may be exposed to cadmium at or above the action level.

**1926.1127(d)(2)(ii)** In addition, if the employee periodically performs tasks that may expose the employee to a higher concentration of airborne cadmium, the employee shall be monitored while performing those tasks.

**1926.1127(d)(2)(iii)** Where the employer has objective data, as defined in paragraph (n)(2) of this section, demonstrating that employee exposure to cadmium will not exceed airborne concentrations at or above the action level under the expected conditions of processing, use, or handling, the employer may rely upon such data instead of implementing initial monitoring.

**1926.1127(d)(2)(iv)** Where a determination conducted under paragraphs (d)(1) or (d)(2) of this section is made that a potentially exposed employee is not exposed to airborne concentrations of cadmium at or above the action level, the employer shall make a written record of such determination. The record shall include at least the monitoring data developed under paragraphs (d)(2)(i)-(iii) of this section, where applicable, and shall also include the date of determination, and the name and social security number of each employee.

**1926.1127(d)(3)** *Monitoring frequency (periodic monitoring).* (i) If the initial monitoring or periodic monitoring reveals employee exposures to be at or above the action level, the employer shall monitor at a frequency and pattern needed to assure that the monitoring results reflect with reasonable accuracy the employee's typical exposure levels, given the variability in the tasks performed, work practices, and environmental conditions on the job site, and to assure the adequacy of respiratory selection and the effectiveness of engineering and work practice controls.

**1926.1127(d)(3)(ii)** If the initial monitoring or the periodic monitoring indicates that employee exposures are below the action level and that result is confirmed by the results of another monitoring taken at least seven days later, the employer may discontinue the monitoring for those employees whose exposures are represented by such monitoring.

**1926.1127(d)(4)** *Additional monitoring.* The employer also shall institute the exposure monitoring required under paragraphs (d)(2)(i) and (d)(3) of this section whenever there has been a change in the raw materials, equipment, personnel, work practices, or finished products that may result in additional employees being exposed to cadmium at or above the action level or in employees already exposed to cadmium at or above the action level being exposed above the PEL, or whenever the employer or competent person has any reason to suspect that any other change might result in such further exposure.

**1926.1127(d)(5)** *Employee notification of monitoring results.* (i) No later than five working days after the receipt of the results of any monitoring performed under this section, the employer shall notify each affected employee individually in writing of the results. In addition, within the same time period, the employer shall post the results of the exposure monitoring in an appropriate location that is accessible to all affected employees.

**1926.1127(d)(5)(ii)** Wherever monitoring results indicate that employee exposure exceeds the PEL, the employer shall include in the written notice a statement that the PEL has been exceeded and a description of the corrective action being taken by the employer to reduce employee exposure to or below the PEL.

**1926.1127(d)(6)** *Accuracy of measurement.* The employer shall use a method of monitoring and analysis that has an accuracy of not less than plus or minus 25 percent (+ 25%), with a confidence level of 95 percent, for airborne concentrations of cadmium at or above the action level and the permissible exposure limit.

**1926.1127(e)** ***Regulated areas***— (1) *Establishment.* The employer shall establish a regulated area wherever an employee's exposure to airborne concentrations of cadmium is, or can reasonably be expected to be in excess of the permissible exposure limit (PEL).

**1926.1127(e)(2)** *Demarcation.* Regulated areas shall be demarcated from the rest of the workplace in any manner that adequately establishes and alerts employees of the boundaries of the regulated area, including employees who are or may be incidentally in the regulated areas, and that protects persons outside the area from exposure to airborne concentrations of cadmium in excess of the PEL.

**1926.1127(e)(3)** *Access.* Access to regulated areas shall be limited to authorized persons.

**1926.1127(e)(4)** *Provision of respirators.* Each person entering a regulated area shall be supplied with and required to use a respirator, selected in accordance with paragraph (g)(2) of this section.

**1926.1127(e)(5)** *Prohibited activities.* The employer shall assure that employees do not eat, drink, smoke, chew tobacco or gum, or apply cosmetics in regulated areas, or carry the products associated with any of these activities into regulated areas or store such products in those areas.

**1926.1127(f)** ***Methods of compliance***— (1) *Compliance hierarchy.* (i) Except as specified in paragraph (f)(1)(ii) of this section, the employer shall implement engineering and work practice controls to reduce and maintain employee exposure to cadmium at or below the PEL, except to the extent that the employer can demonstrate that such controls are not feasible.

**1926.1127(f)(1)(ii)** The requirement to implement engineering controls to achieve the PEL does not apply where the employer demonstrates the following:

**1926.1127(f)(1)(ii)(A)** The employee is only intermittently exposed; and

**1926.1127(f)(1)(ii)(B)** The employee is not exposed above the PEL on 30 or more days per year (12 consecutive months).

**1926.1127(f)(1)(iii)** Wherever engineering and work practice controls are not sufficient to reduce employee exposure to or below the PEL, the employer nonetheless shall implement such controls to reduce exposures to the lowest levels achievable. The employer shall supplement such controls with respiratory protection that complies with the requirements of paragraph (g) of this section and the PEL.

**1926.1127(f)(1)(iv)** The employer shall not use employee rotation as a method of compliance.

**1926.1127(f)(2)** *Specific operations*—(i) *Abrasive blasting.* Abrasive blasting on cadmium or cadmium-containing materials shall be conducted in a manner that will provide adequate protection.

**1926.1127(f)(2)(ii)** *Heating cadmium and cadmium-containing materials.* Welding, cutting, and other forms of heating of cadmium or cadmium-containing materials shall be conducted in accordance with the requirements of 29 CFR 1926.353 and 29 CFR 1926.354, where applicable.

**1926.1127(f)(3)** *Prohibitions.* (i) High speed abrasive disc saws and similar abrasive power equipment shall not be used for work on cadmium or cadmium-containing materials unless they are equipped with appropriate engineering controls to minimize emissions, if the exposure levels are above the PEL.

**1926.1127(f)(3)(ii)** Materials containing cadmium shall not be applied by spray methods, if exposures are above the PEL, unless employees are protected with supplied-air respirators with full facepiece, hood, helmet, suit, operated in positive pressure mode and measures are instituted to limit overspray and prevent contamination of adjacent areas.

**1926.1127(f)(4)** *Mechanical ventilation.* (i) When ventilation is used to control exposure, measurements that demonstrate the effectiveness of the system in controlling exposure, such as capture velocity, duct velocity, or static pressure shall be made as necessary to maintain its effectiveness.

**1926.1127(f)(4)(ii)** Measurements of the system's effectiveness in controlling exposure shall be made as necessary within five working days of any change in production, process, or control that might result in a significant increase in employee exposure to cadmium.

**1926.1127(f)(4)(iii)** Recirculation of air. If air from exhaust ventilation is recirculated into the workplace, the system shall have a high efficiency filter and be monitored to assure effectiveness.

**1926.1127(f)(4)(iv)** Procedures shall be developed and implemented to minimize employee exposure to cadmium when maintenance of ventilation systems and changing of filters is being conducted.

**1926.1127(f)(5)** *Compliance program.* (i) Where employee exposure to cadmium exceeds the PEL and the employer is required under paragraph (f)(1) of this section to implement controls to comply with the PEL, prior to the commencement of the job the employer shall establish and implement a written compliance program to reduce employee exposure to or below the PEL. To the extent that engineering and work practice controls cannot reduce exposures to or below the PEL, the employer shall include in the written compliance program the use of appropriate respiratory protection to achieve compliance with the PEL.

**1926.1127(f)(5)(ii)** Written compliance programs shall be reviewed and updated as often and as promptly as necessary to reflect significant changes in the employer's compliance status or significant changes in the lowest air cadmium level that is technologically feasible.

**1926.1127(f)(5)(iii)** A competent person shall review the comprehensive compliance program initially and after each change.

**1926.1127(f)(5)(iv)** Written compliance programs shall be provided upon request for examination and copying to the Assistant Secretary, the Director, affected employees, and designated employee representatives.

**1926.1127(g)** ***Respirator protection.*** (1) *General.* For employees who use respirators required by this section, the employer must provide respirators that comply with the requirements of this paragraph. Respirators must be used during:

**1926.1127(g)(1)(i)** Periods necessary to install or implement feasible engineering and work-practice controls when employee exposures exceed the PEL.

**1926.1127(g)(1)(ii)** Maintenance and repair activities, and brief or intermittent work operations, for which employee exposures exceed the PEL and engineering and work-practice controls are not feasible or are not required.

**1926.1127(g)(1)(iii)** Work operations in the regulated areas specified in paragraph (e) of this section.

**1926.1127(g)(1)(iv)** Work operations for which the employer has implemented all feasible engineering and work-practice controls, and such controls are not sufficient to reduce employee exposures to or below the PEL.

**1926.1127(g)(1)(v)** Work operations for which an employee, who is exposed to cadmium at or above the action level, requests a respirator.

**1926.1127(g)(1)(vi)** Work operations for which engineering controls are not required by paragraph (f)(1)(ii) of this section to reduce employee exposures that exceed the PEL.

**1926.1127(g)(1)(vii)** Emergencies.

**1926.1127(g)(2)** *Respirator program.* (i) The employer must implement a respiratory protection program in accordance with 29 CFR 1910.134(b) through (d) (except (d)(1)(iii)), and (f) through (m).

**1926.1127(g)(2)(ii)** If an employee exhibits breathing difficulty during fit testing or respirator use, the employer must provide the employee with a medical examination in accordance with paragraph (l)(6)(ii) of this section to determine if the employee can use a respirator while performing the required duties.

**1926.1127(g)(2)(iii)** No employee must use a respirator when, based on their most recent medical examination, the examining physician determines that the employee will be unable to continue to function normally while using a respirator. If the physician determines the employee must be limited in, or removed from, their current job because of the employee's inability to use a respirator, the job limitation or removal must be conducted in accordance with paragraphs (l) (11) and (12) of this section.

**1926.1127(g)(3)** *Respirator selection.* (i) The employer must select the appropriate respirator from Table 1 of this section.

## TABLE 1.—RESPIRATORY PROTECTION FOR CADMIUM

| Airborne concentration or condition of use[a] | Required respirator type[b] |
|---|---|
| 10 × or less | A half mask, air-purifying respirator equipped with a HEPA [c] filter.[d] |
| 25 × or less | A powered air-purifying respirator ("PAPR") with a loose-fitting hood or helmet equipped with a HEPA filter, or a supplied-air respirator with a loose-fitting hood or helmet facepiece operated in the continuous flow mode. |
| 50 × or less | A full facepiece air-purifying respirator equipped with a HEPA filter, or a powered air-purifying respirator with a tight-fitting half mask equipped with a HEPA filter, or a supplied air respirator with a tight-fitting half mask operated in the continuous flow mode. |
| 250 × or less | A powered air-purifying respirator with a tight-fitting full facepiece equipped with a HEPA filter, or a supplied-air respirator with a tight-fitting full facepiece operated in the continuous flow mode. |
| 1000 × or less | A supplied-air respirator with half mask or full facepiece operated in the pressure demand or other positive pressure mode. |
| >1000 × or unknown concentrations | A self-contained breathing apparatus with a full facepiece operated in the pressure demand or other positive pressure mode, or a supplied-air respirator with a full facepiece operated in the pressure demand or other positive pressure mode and equipped with an auxiliary escape type self contained breathing apparatus operated in the pressure demand mode. |
| Fire fighting | A self-contained breathing apparatus with full facepiece operated in the pressure demand or other positive pressure mode. |

[a] Concentrations expressed as multiple of the PEL.

[b] Respirators assigned for higher environmental concentrations may be used at lower exposure levels. Quantitative fit testing is required for all tight-fitting air purifying respirators where airborne concentration of cadmium exceeds 10 times the TWA PEL (10 × 5 ug/m(3)=50 ug/m(3)). A full facepiece respirator is required when eye irritation is experienced.

[c] HEPA means High-efficiency Particulate Air.

[d] Fit testing, qualitative or quantitative, is required.

**SOURCE:** *Respiratory Decision Logic,* NIOSH, 1987.

**1926.1127(g)(3)(ii)** The employer must provide a powered air-purifying respirator instead of a negative-pressure respirator when an employee entitled to a respirator chooses to use this type of respirator and such a respirator will provide adequate protection to the employee.

**1926.1127(h)** ***Emergency situations.*** The employer shall develop and implement a written plan for dealing with emergency situations involving substantial releases of airborne cadmium. The plan shall include provisions for the use of appropriate respirators and personal protective equipment. In addition, employees not essential to correcting the emergency situation shall be restricted from the area and normal operations halted in that area until the emergency is abated.

**1926.1127(i)** ***Protective work clothing and equipment***—(1) *Provision and use.* If an employee is exposed to airborne cadmium above the PEL or where skin or eye irritation is associated with cadmium exposure at any level, the employer shall provide at no cost to the employee, and assure that the employee uses, appropriate protective work clothing and equipment that prevents contamination of the employee and the employee's garments. Protective work clothing and equipment includes, but is not limited to:

**1926.1127(i)(1)(i)** Coveralls or similar full-body work clothing;

**1926.1127(i)(1)(ii)** Gloves, head coverings, and boots or foot coverings; and

**1926.1127(i)(1)(iii)** Face shields, vented goggles, or other appropriate protective equipment that complies with 29 CFR 1910.133.

**1926.1127(i)(2)** *Removal and storage.* (i) The employer shall assure that employees remove all protective clothing and equipment contaminated with cadmium at the completion of the work shift and do so only in change rooms provided in accordance with paragraph (j)(1) of this section.

**1926.1127(i)(2)(ii)** The employer shall assure that no employee takes cadmium-contaminated protective clothing or equipment from the workplace, except for employees authorized to do so for purposes of laundering, cleaning, maintaining, or disposing of cadmium-contaminated protective clothing and equipment at an appropriate location or facility away from the workplace.

**1926.1127(i)(2)(iii)** The employer shall assure that contaminated protective clothing and equipment, when removed for laundering, cleaning, maintenance, or disposal, is placed and stored in sealed, impermeable bags or other closed, impermeable containers that are designed to prevent dispersion of cadmium dust.

**1926.1127(i)(2)(iv)** The employer shall assure that containers of contaminated protective clothing and equipment that are to be taken out of the change rooms or the workplace for laundering, cleaning, maintenance or disposal shall bear labels in accordance with paragraph (m)(2) of this section.

**1926.1127(i)(3)** *Cleaning, replacement, and disposal.* (i) The employer shall provide the protective clothing and equipment required by paragraph (i)(1) of this section in a clean and dry condition as often as necessary to maintain its effectiveness, but in any event at least weekly. The employer is responsible for cleaning and laundering the protective clothing and equipment required by this paragraph to maintain its effectiveness and is also responsible for disposing of such clothing and equipment.

**1926.1127(i)(3)(ii)** The employer also is responsible for repairing or replacing required protective clothing and equipment as needed to maintain its effectiveness. When rips or tears are detected while an employee is working they shall be immediately mended, or the worksuit shall be immediately replaced.

**1926.1127(i)(3)(iii)** The employer shall prohibit the removal of cadmium from protective clothing and equipment by blowing, shaking, or any other means that disperses cadmium into the air.

**1926.1127(i)(3)(iv)** The employer shall assure that any laundering of contaminated clothing or cleaning of contaminated equipment in the workplace is done in a manner that prevents the release of airborne cadmium in excess of the permissible exposure limit prescribed in paragraph (c) of this section.

**1926.1127(i)(3)(v)** The employer shall inform any person who launders or cleans protective clothing or equipment contaminated with cadmium of the potentially harmful effects of exposure to cadmium, and that the clothing and equipment should be laundered or cleaned in a manner to effectively prevent the release of airborne cadmium in excess of the PEL.

**1926.1127(j)** ***Hygiene areas and practices.*** (1) *General.* For employees whose airborne exposure to cadmium is above the PEL, the employer shall provide clean change rooms, handwashing facilities, showers, and lunchroom facilities that comply with 29 CFR 1926.51.

**1926.1127(j)(2)** *Change rooms.* The employer shall assure that change rooms are equipped with separate storage facilities for street clothes and for protective clothing and equipment, which are designed to prevent dispersion of cadmium and contamination of the employee's street clothes.

**1926.1127(j)(3)** *Showers and handwashing facilities.* (i) The employer shall assure that employees whose airborne exposure to cadmium is above the PEL shower during the end of the work shift.

**1926.1127(j)(3)(ii)** The employer shall assure that employees who are exposed to cadmium above the PEL wash their hands and faces prior to eating, drinking, smoking, chewing tobacco or gum, or applying cosmetics.

**1926.1127(j)(4)** *Lunchroom facilities.* (i) The employer shall assure that the lunchroom facilities are readily accessible to employees, that tables for eating are maintained free of cadmium, and that no employee in a lunchroom facility is exposed at any time to cadmium at or above a concentration of 2.5 $\mu g/m^3$.

**1926.1127(j)(4)(ii)** The employer shall assure that employees do not enter lunchroom facilities with protective work clothing or equipment unless surface cadmium has been removed from the clothing and equipment by HEPA vacuuming or some other method that removes cadmium dust without dispersing it.

**1926.1127(k)** ***Housekeeping.*** (1) All surfaces shall be maintained as free as practicable of accumulations of cadmium.

**1926.1127(k)(2)** All spills and sudden releases of material containing cadmium shall be cleaned up as soon as possible.

**1926.1127(k)(3)** Surfaces contaminated with cadmium shall, wherever possible, be cleaned by vacuuming or other methods that minimize the likelihood of cadmium becoming airborne.

**1926.1127(k)(4)** HEPA-filtered vacuuming equipment or equally effective filtration methods shall be used for vacuuming. The equipment shall be used and emptied in a manner that minimizes the reentry of cadmium into the workplace.

**1926.1127(k)(5)** Shoveling, dry or wet sweeping, and brushing may be used only where vacuuming or other methods that minimize the likelihood of cadmium becoming airborne have been tried and found not to be effective.

**1926.1127(k)(6)** Compressed air shall not be used to remove cadmium from any surface unless the compressed air is used in conjunction with a ventilation system designed to capture the dust cloud created by the compressed air.

**1926.1127(k)(7)** Waste, scrap, debris, bags, containers, personal protective equipment, and clothing contaminated with cadmium and consigned for disposal shall be collected and disposed of in sealed impermeable bags or other closed, impermeable containers. These bags and containers shall be labeled in accordance with paragraph (m)(2) of this section.

**1926.1127(l)** ***Medical Surveillance***— (1) *General*—(i) *Scope*—(A) *Currently exposed*—The employer shall institute a medical surveillance program for all employees who are or may be exposed at or above the action level and all employees who perform the following tasks, operations or jobs: Electrical grounding with cadmium welding; cutting, brazing, burning, grinding or welding on surfaces that were painted with cadmium-containing paints; electrical work using cadmium-coated conduit; use of cadmium containing paints; cutting and welding cadmium-plated steel; brazing or welding with cadmium alloys; fusing of reinforced steel by cadmium welding; maintaining or retrofitting cadmium-coated equipment; and, wrecking and demolition where cadmium is present. A medical surveillance program will not be required if the employer demonstrates that the employee:

**1926.1127(l)(1)(i)(A)*(1)*** Is not currently exposed by the employer to airborne concentrations of cadmium at or above the action level on 30 or more days per year (twelve consecutive months); and,

**1926.1127(l)(1)(i)(A)*(2)*** Is not currently exposed by the employer in those tasks on 30 or more days per year (twelve consecutive months).

**1926.1127(l)(1)(i)(B)** *Previously exposed*—The employer shall also institute a medical surveillance program for all employees who might previously have been exposed to cadmium by the employer prior to the effective date of this standard in tasks specified under paragraph (l)(1)(i)(A) of this section, unless the employer demonstrates that the employee did not in the years prior to the effective date of this section work in those tasks for the employer with exposure to cadmium for an aggregated total of more than 12 months.

**1926.1127(l)(1)(ii)** To determine an employee's fitness for using a respirator, the employer shall provide the limited medical examination specified in paragraph (l)(6) of this section.

**1926.1127(l)(1)(iii)** The employer shall assure that all medical examinations and procedures required by this section are performed by or under the supervision of a licensed physician, who has read and is familiar with the health effects section of appendix A to this section, the regulatory text of this section, the protocol for sample handling and lab selection in appendix F to this section, and the questionnaire of appendix D to this section.

**1926.1127(l)(1)(iv)** The employer shall provide the medical surveillance required by this section, including multiple physician review under paragraph (l)(13) of this section without cost to employees, and at a time and place that is reasonable and convenient to employees.

**1926.1127(l)(1)(v)** The employer shall assure that the collecting and handling of biological samples of cadmium in urine (CdU), cadmium in blood (CdB), and beta-2 microglobulin in urine ($\beta_2$-M) taken from employees under this section is done in a manner that assures their reliability and that analysis of biological samples of cadmium in urine (CdU), cadmium in blood (CdB), and beta-2 microglobulin in urine ($\beta_2$-M) taken from employees under this section is performed in laboratories withdemonstrated proficiency to perform the particular analysis. (See appendix F to this section.)

**1926.1127(l)(2)** *Initial Examination.* (i) For employees covered by medical surveillance under paragraph (l)(1)(i) of this section, the employer shall provide an initial medical examination. The examination shall be provided to those employees within 30 days after initial assignment to a job with exposure to cadmium or no later than 90 days after the effective date of this section, whichever date is later.

**1926.1127(l)(2)(ii)** The initial medical examination shall include:

**1926.1127(l)(2)(ii)(A)** A detailed medical and work history, with emphasis on: Past, present, and anticipated future exposure to cadmium; any history of renal, cardiovascular, respiratory, hematopoietic, reproductive, and/or musculo-skeletal system dysfunction; current usage of medication with potential nephrotoxic side-effects; and smoking history and current status; and

**1926.1127(l)(2)(ii)(B)** Biological monitoring that includes the following tests:

**1926.1127(l)(2)(ii)(B)*(1)*** Cadmium in urine (CdU), standardized to grams of creatinine (g/Cr);

**1926.1127(l)(2)(ii)(B)*(2)*** Beta-2 microglobulin in urine (β2-M), standardized to grams of creatinine (g/Cr), with pH specified, as described in Appendix F to this section; and

**1926.1127(l)(2)(ii)(B)*(3)*** Cadmium in blood (CdB), standardized to liters of whole blood (lwb).

**1926.1127(l)(2)(iii)** Recent Examination: An initial examination is not required to be provided if adequate records show that the employee has been examined in accordance with the requirements of paragraph (l)(2)(ii) of this section within the past 12 months. In that case, such records shall be maintained as part of the employee's medical record and the prior exam shall be treated as if it were an initial examination for the purposes of paragraphs (l)(3) and (4) of this section.

**1926.1127(l)(3)** *Actions triggered by initial biological monitoring.* (i) If the results of the biological monitoring tests in the initial examination show the employee's CdU level to be at or below 3μg/g Cr, $\beta_2$-M level to be at or below 300μg/g Cr and CdB level to be at or below 5μg/lwb, then:

**1926.1127(l)(3)(i)(A)** For employees who are subject to medical surveillance under paragraphs (l)(1)(i)(A) of this section because of current or anticipated exposure to cadmium, the employer shall provide the minimum level of periodic medical surveillance in accordance with the requirements in paragraph (l)(4)(i) of this section; and

**1926.1127(l)(3)(i)(B)** For employees who are subject to medical surveillance under paragraph (l)(1)(i)(B) of this section because of prior but not current exposure, the employer shall provide biological monitoring for CdU, $B_2$-M, and CdB one year after the initial biological monitoring and then the employer shall comply with the requirements of paragraph (l)(4)(vi) of this section.

**1926.1127(l)(3)(ii)** For all employees who are subject to medical surveillance under paragraph (l)(1)(i) of this section, if the results of the initial biological monitoring tests show the level of CdU to exceed 3 μ g/g Cr, the level of $\beta_2$-M to be in excess of 300 μ g/g Cr, or the level of CdB to be in excess of 5 μ g/lwb, the employer shall:

**1926.1127(l)(3)(ii)(A)** Within two weeks after receipt of biological monitoring results, reassess the employee's occupational exposure to cadmium as follows:

**1926.1127(l)(3)(ii)(A)*(1)*** Reassess the employee's work practices and personal hygiene;

**1926.1127(l)(3)(ii)(A)*(2)*** Reevaluate the employee's respirator use, if any, and the respirator program;

**1926.1127(l)(3)(ii)(A)*(3)*** Review the hygiene facilities;

**1926.1127(l)(3)(ii)(A)*(4)*** Reevaluate the maintenance and effectiveness of the relevant engineering controls;

**1926.1127(l)(3)(ii)(A)*(5)*** Assess the employee's smoking history and status;

**1926.1127(l)(3)(ii)(B)** Within 30 days after the exposure reassessment, specified in paragraph (l)(3)(ii)(A) of this section, take reasonable steps to correct any deficiencies found in the reassessment that may be responsible for the employee's excess exposure to cadmium; and,

**1926.1127(l)(3)(ii)(C)** Within 90 days after receipt of biological monitoring results, provide a full medical examination to the employee in accordance with the requirements of paragraph (l)(4)(ii) of this section. After completing the medical examination, the examining physician shall determine in a written medical opinion whether to medically remove the employee. If the physician determines that medical removal is not necessary, then until the employee's CdU level falls to or below 3 μ g/g Cr, $β_2$-M level falls to or below 300 μ g/g Cr and CdB level falls to or below 5 μ g/lwb, the employer shall:

**1926.1127(l)(3)(ii)(C)*(1)*** Provide biological monitoring in accordance with paragraph (l)(2)(ii)(B) of this section on a semiannual basis; and

**1926.1127(l)(3)(ii)(C)*(2)*** Provide annual medical examinations in accordance with paragraph (l)(4)(ii) of this section.

**1926.1127(l)(3)(iii)** For all employees who are subject to medical surveillance under paragraph (l)(1)(i) of this section, if the results of the initial biological monitoring tests show the level of CdU to be in excess of 15 μ g/g Cr, or the level of CdB to be in excess of 15 μ g/lwb, or the level of $β_2$ M to be in excess of 1,500 μ g/g Cr, the employer shall comply with the requirements of paragraphs (l)(3)(ii)(A)-(B) of this section. Within 90 days after receipt of biological monitoring results, the employer shall provide a full medical examination to the employee in accordance with the requirements of paragraph (l)(4)(ii) of this section. After completing the medical examination, the examining physician shall determine in a written medical opinion whether to medically remove the employee. However, if the initial biological monitoring results and the biological monitoring results obtained during the medical examination both show that: CdU exceeds 15 μ g/g Cr; or CdB exceeds 15 μ g/lwb; or $β_2$-M exceeds 1500 μ g/g Cr, and in addition CdU exceeds 3 μ g/g Cr or CdB exceeds 5 μ g/liter of whole blood, then the physician shall medically remove the employee from exposure to cadmium at or above the action level. If the second set of biological monitoring results obtained during the medical examination does not show that a mandatory removal trigger level has been exceeded, then the employee is not required to be removed by the mandatory provisions of this paragraph. If the employee is not required to be removed by the mandatory provisions of this paragraph or by the physician's determination, then until the employee's CdU level falls to or below 3 μ g/g Cr, $β_2$-M level falls to or below 300 μ g/g Cr and CdB level falls to or below 5 μ g/lwb, the employer shall:

**1926.1127(l)(3)(iii)(A)** Periodically reassess the employee's occupational exposure to cadmium;

**1926.1127(l)(3)(iii)(B)** Provide biological monitoring in accordance with paragraph (l)(2)(ii)(B) of this section on a quarterly basis; and

**1926.1127(l)(3)(iii)(C)** Provide semiannual medical examinations in accordance with paragraph (l)(4)(ii) of this section.

**1926.1127(l)(3)(iv)** For all employees to whom medical surveillance is provided, beginning on January 1, 1999, and in lieu of paragraph (l)(3)(iii) of this section, whenever the results of initial biological monitoring tests show the employee's CdU level to be in excess of 7 μ g/g Cr, or $β_2$-M level to be in excess of 750 μ g/g Cr, or CdB level to be in excess of 10 μ g/lwb, the employer shall comply with the requirements of paragraphs (l)(3)(ii)(A)-(B) of this section. Within 90 days after receipt of biological monitoring results, the employer shall provide a full medical examination to the employee in accordance with the requirements of paragraph (l)(4)(ii) of this section. After completing the medical examination, the examining physician shall determine in a written medical opinion whether to medically remove the employee. However, if the initial biological monitoring results and the biological monitoring results obtained during the medical examination both show that: CdU exceeds 7 μ g/g Cr; or CdB exceeds 10 μ g/lwb; or $β_2$-M exceeds 750 μ g/g Cr, and in addition CdU exceeds 3 μ g/g Cr or CdB exceeds 5 μ g/liter of whole blood, then the physician shall medically remove the employee from exposure to cadmium at or above the action level. If the second set of biological monitoring results obtained during the medical examination does not show that a mandatory removal trigger level has been exceeded, then the employee is not required to be removed by the mandatory provisions of this paragraph. If the employee is not required to be removed by the mandatory provisions of this paragraph or by the physician's determination, then until the employee's CdU level falls to or below 3 μ g/g Cr, $β_2$-M level falls to or below 300 μ g/g Cr and CdB level falls to or below 5 μ g/lwb, the employer shall:

**1926.1127(l)(3)(iv)(A)** Periodically reassess the employee's occupational exposure to cadmium;

**1926.1127(l)(3)(iv)(B)** Provide biological monitoring in accordance with paragraph (l)(2)(ii)(B) of this section on a quarterly basis; and

**1926.1127(l)(3)(iv)(C)** Provide semiannual medical examinations in accordance with paragraph (l)(4)(ii) of this section.

**1926.1127(l)(4)** *Periodic medical surveillance.* (i) For each employee who is covered by medical surveillance under paragraph (l)(1)(i)(A) of this section because of current or anticipated exposure to cadmium, the employer shall provide at least the minimum level of periodic medical surveillance, which consists of periodic medical examinations and periodic biological monitoring. A periodic medical examination shall be provided within one year after the initial examination required by paragraph (l)(2) of this section and thereafter at least biennially. Biological sampling shall be provided at least annually either as part of a periodic medical examination or separately as periodic biological monitoring.

**1926.1127(l)(4)(ii)** The periodic medical examination shall include:

**1926.1127(l)(4)(ii)(A)** A detailed medical and work history, or update thereof, with emphasis on: Past, present and anticipated future exposure to cadmium; smoking history and current status; reproductive history; current use of medications with potential nephrotoxic side-effects; any history of renal, cardiovascular, respiratory, hematopoietic, and/or musculo-skeletal system dysfunction; and as part of the medical and work history, for employees who wear respirators, questions 3-11 and 25-32 in appendix D to this section;

**1926.1127(l)(4)(ii)(B)** A complete physical examination with emphasis on: blood pressure, the respiratory system, and the urinary system;

**1926.1127(l)(4)(ii)(C)** A 14 inch by 17 inch, or a reasonably standard sized posterior-anterior chest X-ray (after the initial X-ray, the frequency of chest X-rays is to be determined by the examining physician);

**1926.1127(l)(4)(ii)(D)** Pulmonary function tests, including forced vital capacity (FVC) and forced expiratory volume at 1 second (FEV1);

**1926.1127(l)(4)(ii)(E)** Biological monitoring, as required in paragraph (l)(2)(ii)(B) of this section;

**1926.1127(l)(4)(ii)(F)** Blood analysis, in addition to the analysis required under paragraph (l)(2)(ii)(B) of this section, including blood urea nitrogen, complete blood count, and serum creatinine;

**1926.1127(l)(4)(ii)(G)** Urinalysis, in addition to the analysis required under paragraph (l)(2)(ii)(B) of this section, including the determination of albumin, glucose, and total and low molecular weight proteins;

**1926.1127(l)(4)(ii)(H)** For males over 40 years old, prostate palpation, or other at least as effective diagnostic test(s), and;

**1926.1127(l)(4)(ii)(I)** Any additional tests or procedures deemed appropriate by the examining physician.

**1926.1127(l)(4)(iii)** Periodic biological monitoring shall be provided in accordance with paragraph (l)(2)(ii)(B) of this section.

**1926.1127(l)(4)(iv)** If the results of periodic biological monitoring or the results of biological monitoring performed as part of the periodic medical examination show the level of the employee's CdU, $β_2$-M, or CdB to be in excess of the levels specified in paragraphs (l)(3)(ii) or (iii) of this section; or beginning on January 1, 1999, in excess of the levels specified in paragraphs (l)(3)(ii) or (iv), the employer shall take the appropriate actions specified in paragraphs (l)(3)(ii)-(iv) of this section, respectively.

**1926.1127(l)(4)(v)** For previously exposed employees under paragraph (l)(1)(i)(B) of this section:

**1926.1127(l)(4)(v)(A)** If the employee's levels of CdU did not exceed 3 μg/g Cr, CdB did not exceed 5 μg/lwb, and $β_2$-M did not exceed 300 μg/g Cr in the initial biological monitoring tests, and if the results of the followup biological monitoring required by paragraph (l)(3)(i)(B) of this section one year after the initial examination confirm the previous results, the employer may discontinue all periodic medical surveillance for that employee.

**1926.1127(l)(4)(v)(B)** If the initial biological monitoring results for CdU, CdB, or $β_2$-M were in excess of the levels specified in paragraph (l)(3)(i) of this section, but subsequent biological monitoring results required by paragraph (l)(3)(ii)-(iv) of this section show that the employee's CdU levels no longer exceed 3 μg/g Cr, CdB levels no longer exceed 5 μg/lwb, and $β_2$-M levels no longer exceed 300 μg/g Cr, the employer shall provide biological monitoring for CdU, CdB, and $β_2$-M one year after these most recent biological monitoring results. If the results of the followup biological monitoring specified in this paragraph, confirm the previous results, the employer may discontinue all periodic medical surveillance for that employee.

**1926.1127(l)(4)(v)(C)** However, if the results of the follow-up tests specified in paragraph (l)(4)(v)(A) or (B) of this section indicate that the level of the employee's CdU, $β_2$-M, or CdB exceeds these same levels, the employer is required to provide annual medical examinations in accordance with the provisions of paragraph (l)(4)(ii) of this section until the results of biological monitoring are consistently below these levels or the examining physician determines in a written medical opinion that further medical surveillance is not required to protect the employee's health.

**1926.1127(l)(4)(vi)** A routine, biennial medical examination is not required to be provided in accordance with paragraphs (l)(3)(i) and (l)(4) of this section if adequate medical records show that the employee has been examined in accordance with the requirements of paragraph (l)(4)(ii) of this section within the past 12 months. In that case, such records shall be maintained by the employer as part of the employee's medical record, and the next routine, periodic medical examination shall be made available to the employee within two years of the previous examination.

**1926.1127(l)(5)** *Actions triggered by medical examinations.* (i) If the results of a medical examination carried out in accordance with this section indicate any laboratory or clinical finding consistent with cadmium toxicity that does not require employer action under paragraphs (l)(2), (3) or (4) of this section, the employer shall take the following steps and continue to take them until the physician determines that they are no longer necessary.

**1926.1127(l)(5)(i)(A)** Periodically reassess: The employee's work practices and personal hygiene; the employee's respirator use, if any; the employee's smoking history and status; the respiratory protection program; the hygiene facilities; the maintenance and effectiveness of the relevant engineering controls; and take all reasonable steps to correct the deficiencies found in the reassessment that may be responsible for the employee's excess exposure to cadmium.

**1926.1127(l)(5)(i)(B)** Provide semi-annual medical reexaminations to evaluate the abnormal clinical sign(s) of cadmium toxicity until the results are normal or the employee is medically removed; and

**1926.1127(l)(5)(i)(C)** Where the results of tests for total proteins in urine are abnormal, provide a more detailed medical evaluation of the toxic effects of cadmium on the employee's renal system.

**1926.1127(l)(6)** *Examination for respirator use.* (i) To determine an employee's fitness for respirator use, the employer shall provide a medical examination that includes the elements specified in paragraph (l)(6)(i)(A)-(D) of this section. This examination shall be provided prior to the employee's being assigned to a job that requires the use of a respirator or no later than 90 days after this section goes into effect, whichever date is later, to any employee without a medical examination within the preceding 12 months that satisfies the requirements of this paragraph.

**1926.1127(l)(6)(i)(A)** A detailed medical and work history, or update thereof, with emphasis on: past exposure to cadmium; smoking history and current status; any history of renal, cardiovascular, respiratory, hematopoietic, and/or musculo-skeletal system dysfunction; a description of the job for which the respirator is required; and questions 3-11 and 25-32 in appendix D;

**1926.1127(l)(6)(i)(B)** A blood pressure test;

**1926.1127(l)(6)(i)(C)** Biological monitoring of the employee's levels of CdU, CdB and $\beta_2$-M in accordance with the requirements of paragraph (l)(2)(ii)(B) of this section, unless such results already have been obtained within the twelve months; and

**1926.1127(l)(6)(i)(D)** Any other test or precedure that the examining physician deems appropriate.

**1926.1127(l)(6)(ii)** After reviewing all the information obtained from the medical examination required in paragraph (l)(6)(i) of this section, the physician shall determine whether the employee is fit to wear a respirator.

**1926.1127(l)(6)(iii)** Whenever an employee has exhibited difficulty in breathing during a respirator fit test or during use of a respirator, the employer, as soon as possible, shall provide the employee with a periodic medical examination in accordance with paragraph (l)(4)(ii) of this section to determine the employee's fitness to wear a respirator.

**1926.1127(l)(6)(iv)** Where the results of the examination required under paragraphs (l)(6)(i), (ii), or (iii) of this section are abnormal, medical limitation or prohibition of respirator use shall be considered. If the employee is allowed to wear a respirator, the employee's ability to continue to do so shall be periodically evaluated by a physician.

**1926.1127(l)(7)** *Emergency Examinations.* (i) In addition to the medical surveillance required in paragraphs (l)(2)-(6) of this section, the employer shall provide a medical examination as soon as possible to any employee who may have been acutely exposed to cadmium because of an emergency.

**1926.1127(l)(7)(ii)** The examination shall include the requirements of paragraph (l)(4)(ii), of this section, with emphasis on the respiratory system, other organ systems considered appropriate by the examining physician, and symptoms of acute overexposure, as identified in paragraphs II(B)(1)-(2) and IV of appendix A of this section.

**1926.1127(l)(8)** *Termination of employment examination.* (i) At termination of employment, the employer shall provide a medical examination in accordance with paragraph (l)(4)(ii) of this section, including a chest X-ray where necessary, to any employee to whom at any prior time the employer was required to provide medical surveillance under paragraph (l)(1)(i) or (l)(7) of this section. However, if the last examination satisfied the requirements of paragraph (l)(4)(ii) of this section and was less than six months prior to the date of termination, no further examination is required unless otherwise specified in paragraph (l)(3) or (l)(5) of this section;

**1926.1127(l)(8)(ii)** In addition, if the employer has discontinued all periodic medical surveillance under paragraph (l)(4)(v) of this section, no termination of employment medical examination is required.

**1926.1127(l)(9)** *Information provided to the physician.* The employer shall provide the following information to the examining physician:

**1926.1127(l)(9)(i)** A copy of this standard and appendices;

**1926.1127(l)(9)(ii)** A description of the affected employee's former, current, and anticipated duties as they relate to the employee's occupational exposure to cadmium;

**1926.1127(l)(9)(iii)** The employee's former, current, and anticipated future levels of occupational exposure to cadmium;

**1926.1127(l)(9)(iv)** A description of any personal protective equipment, including respirators, used or to be used by the employee, including when and for how long the employee has used that equipment; and

**1926.1127(l)(9)(v)** Relevant results of previous biological monitoring and medical examinations.

**1926.1127(l)(10)** *Physician's written medical opinion.* (i) The employer shall promptly obtain a written, signed medical opinion from the examining physician for each medical examination performed on each employee. This written opinion shall contain:

**1926.1127(l)(10)(i)(A)** The physician's diagnosis for the employee;

**1926.1127(l)(10)(i)(B)** The physician's opinion as to whether the employee has any detected medical condition(s) that would place the employee at increased risk of material impairment to health from further exposure to cadmium, including any indications of potential cadmium toxicity;

**1926.1127(l)(10)(i)(C)** The results of any biological or other testing or related evaluations that directly assess the employee's absorption of cadmium;

**1926.1127(l)(10)(i)(D)** Any recommended removal from, or limitation on the activities or duties of the employee or on the employee's use of personal protective equipment, such as respirators;

**1926.1127(l)(10)(i)(E)** A statement that the physician has clearly and carefully explained to the employee the results of the medical examination, including all biological monitoring results and any medical conditions related to cadmium exposure that require further evaluation or treatment, and any limitation on the employee's diet or use of medications.

**1926.1127(l)(10)(ii)** The employer shall promptly obtain a copy of the results of any biological monitoring provided by an employer to an employee independently of a medical examination under paragraphs (l)(2) and (l)(4) of this section, and, in lieu of a written medical opinion, an explanation sheet explaining those results.

**1926.1127(l)(10)(iii)** The employer shall instruct the physician not to reveal orally or in the written medical opinion given to the employer specific findings or diagnoses unrelated to occupational exposure to cadmium.

**1926.1127(l)(11)** *Medical Removal Protection (MRP)*—(i) *General.* (A) The employer shall temporarily remove an employee from work where there is excess exposure to cadmium on each occasion that medical removal is required under paragraphs (l)(3), (l)(4), or (l)(6) of this section and on each occasion that a physician determines in a written medical opinion that the employee should be removed from such exposure. The physician's determination may be based on biological monitoring results, inability to wear a respirator, evidence of illness, other signs or symptoms of cadmium-related dysfunction or disease, or any other reason deemed medically sufficient by the physician.

**1926.1127(l)(11)(i)(B)** The employer shall medically remove an employee in accordance with paragraph (l)(11) of this section regardless of whether at the time of removal a job is available into which the removed employee may be transferred.

**1926.1127(l)(11)(i)(C)** Whenever an employee is medically removed under paragraph (l)(11) of this section, the employer shall transfer the removed employee to a job where the exposure to cadmium is within the permissible levels specified in that paragraph as soon as one becomes available.

**1926.1127(l)(11)(i)(D)** For any employee who is medically removed under the provisions of paragraph (l)(11)(i) of this section, the employer shall provide follow-up medical examinations semi-annually until, in a written medical opinion, the examining physician determines that either the employee may be returned to his/her former job status or the employee must be permanently removed from excess cadmium exposure.

**1926.1127(l)(11)(i)(E)** The employer may not return an employee who has been medically removed for any reason to his/her former job status until a physician determines in a written medical opinion that continued medical removal is no longer necessary to protect the employee's health.

**1926.1127(l)(11)(ii)** Where an employee is found unfit to wear a respirator under paragraph (l)(6)(ii) of this section, the employer shall remove the employee from work where exposure to cadmium is above the PEL.

**1926.1127(l)(11)(iii)** Where removal is based upon any reason other than the employee's inability to wear a respirator, the employer shall remove the employee from work where exposure to cadmium is at or above the action level.

**1926.1127(l)(11)(iv)** Except as specified in paragraph (l)(11)(v) of this section, no employee who was removed because his/her level of CdU, CdB and/or $\beta_2$-M exceeded the trigger levels in paragraph (l)(3) or (l)(4) of this section may be returned to work with exposure to cadmium at or above the action level until the employee's levels of CdU fall to or below 3 $\mu$ g/g Cr, CdB fall to or below 5 $\mu$ g/lwb, and $\beta_2$-M fall to or below 300 $\mu$ g/g Cr.

**1926.1127(l)(11)(v)** However, when in the examining physician's opinion continued exposure to cadmium will not pose an increased risk to the employee's health and there are special circumstances that make continued medical removal an inappropriate remedy, the physician shall fully discuss these matters with the employee, and then in a written determination may return a worker to his/her former job status despite what would otherwise be unacceptably high biological monitoring results. Thereafter and until such time as the employee's biological monitoring results have decreased to levels where he/she could have been returned to his/her former job status, the returned employee shall continue medical surveillance as if he/she were still on medical removal. Until such time, the employee is no longer subject to mandatory medical removal. Subsequent questions regarding the employee's medical removal shall be decided solely by a final medical determination.

**1926.1127(l)(11)(vi)** Where an employer, although not required by this section to do so, removes an employee from exposure to cadmium or otherwise places limitations on an employee due to the effects of cadmium exposure on the employee's medical condition, the employer shall provide the same medical removal protection benefits to that employee under paragraph (l)(12) of this section as would have been provided had the removal been required under paragraph (l)(11) of this section.

**1926.1127(l)(12)** *Medical removal protection benefits.* (i) The employer shall provide medical removal protection benefits to an employee for up to a maximum of 18 months each time, and while the employee is temporarily medically removed under paragraph (l)(11) of this section.

**1926.1127(l)(12)(ii)** For purposes of this section, the requirement that the employer provide medical removal protection benefits means that the employer shall maintain the total normal earnings, seniority, and all other employee rights and benefits of the removed employee, including the employee's right to his/her former job status, as if the employee had not been removed from the employee's job or otherwise medically limited.

**1926.1127(l)(12)(iii)** Where, after 18 months on medical removal because of elevated biological monitoring results, the employee's monitoring results have not declined to a low enough level to permit the employee to be returned to his/her former job status:

**1926.1127(l)(12)(iii)(A)** The employer shall make available to the employee a medical examination pursuant to this section in order to obtain a final medical determination as to whether the employee may be returned to his/her former job status or must be permanently removed from excess cadmium exposure; and

**1926.1127(l)(12)(iii)(B)** The employer shall assure that the final medical determination indicates whether the employee may be returned to his/her former job status and what steps, if any, should be taken to protect the employee's health;

**1926.1127(l)(12)(iv)** The employer may condition the provision of medical removal protection benefits upon the employee's participation in medical surveillance provided in accordance with this section.

**1926.1127(l)(13)** *Multiple physician review.* (i) If the employer selects the initial physician to conduct any medical examination or consultation provided to an employee under this section, the employee may designate a second physician to:

**1926.1127(l)(13)(i)(A)** Review any findings, determinations, or recommendations of the initial physician; and

**1926.1127(l)(13)(i)(B)** Conduct such examinations, consultations, and laboratory tests as the second physician deems necessary to facilitate this review.

**1926.1127(l)(13)(ii)** The employer shall promptly notify an employee of the right to seek a second medical opinion after each occasion that an initial physician provided by the employer conducts a medical examination or consultation pursuant to this section. The employer may condition its participation in, and payment for, multiple physician review upon the employee doing the following within fifteen (15) days after receipt of this notice, or receipt of the initial physician's written opinion, whichever is later:

**1926.1127(l)(13)(ii)(A)** Informing the employer that he or she intends to seek a medical opinion; and

**1926.1127(l)(13)(ii)(B)** Initiating steps to make an appointment with a second physician.

**1926.1127(l)(13)(iii)** If the findings, determinations, or recommendations of the second physician differ from those of the initial physician, then the employer and the employee shall assure that efforts are made for the two physicians to resolve any disagreement.

**1926.1127(l)(13)(iv)** If the two physicians have been unable to quickly resolve their disagreement, then the employer and the employee, through their respective physicians, shall designate a third physician to:

**1926.1127(l)(13)(iv)(A)** Review any findings, determinations, or recommendations of the other two physicians; and

**1926.1127(l)(13)(iv)(B)** Conduct such examinations, consultations, laboratory tests, and discussions with the other two physicians as the third physician deems necessary to resolve the disagreement among them.

**1926.1127(l)(13)(v)** The employer shall act consistently with the findings, determinations, and recommendations of the third physician, unless the employer and the employee reach an agreement that is consistent with the recommendations of at least one of the other two physicians.

**1926.1127(l)(14)** *Alternate physician determination.* The employer and an employee or designated employee representative may agree upon the use of any alternate form of physician determination in lieu of the multiple physician review provided by paragraph (l)(13) of this section, so long as the alternative is expeditious and at least as protective of the employee.

**1926.1127(l)(15)** *Information the employer must provide the employee.* (i) The employer shall provide a copy of the physician's written medical opinion to the examined employee within five working days after receipt thereof.

**1926.1127(l)(15)(ii)** The employer shall provide the employee with a copy of the employee's biological monitoring results and an explanation sheet explaining the results within five working days after receipt thereof.

**1926.1127(l)(15)(iii)** Within 30 days after a request by an employee, the employer shall provide the employee with the information the employer is required to provide the examining physician under paragraph (l)(9) of this section.

**1926.1127(l)(16)** *Reporting.* In addition to other medical events that are required to be reported on the OSHA Form No. 200, the employer shall report any abnormal condition or disorder caused by occupational exposure to cadmium associated with employment as specified in Chapter (V)(E) of the *Reporting Guidelines for Occupational Injuries and Illnesses.*

**1926.1127(m)** ***Communication of cadmium hazards to employees—*** (1) *General.* In communications concerning cadmium hazards, employers shall comply with the requirements of OSHA's Hazard Communication Standard for the construction industry, 29 CFR 1926.59, including but not limited to the requirements concerning warning signs and labels, material safety data sheets (MSDS), and employee information and training. In addition, employers shall comply with the following requirements:

**1926.1127(m)(2)** *Warning signs.* (i) Warning signs shall be provided and displayed in regulated areas. In addition, warning signs shall be posted at all approaches to regulated areas so that an employee may read the signs and take necessary protective steps before entering the area.

**1926.1127(m)(2)(ii)** Warning signs required by paragraph (m)(2)(i) of this section shall bear the following information:

**DANGER, CADMIUM, CANCER HAZARD,<br>CAN CAUSE LUNG AND KIDNEY DISEASE,<br>AUTHORIZED PERSONNEL ONLY,<br>RESPIRATORS REQUIRED IN THIS AREA**

**1926.1127(m)(2)(iii)** The employer shall assure that signs required by this paragraph are illuminated, cleaned, and maintained as necessary so that the legend is readily visible.

**1926.1127(m)(3)** *Warning labels.* (i) Shipping and storage containers containing cadmium, cadmium compounds, or cadmium contaminated clothing, equipment, waste, scrap, or debris shall bear appropriate warning labels, as specified in paragraph (m)(3)(ii) of this section.

**1926.1127(m)(3)(ii)** The warning labels shall include at least the following information:

**DANGER, CONTAINS CADMIUM, CANCER HAZARD,<br>AVOID CREATING DUST,<br>CAN CAUSE LUNG AND KIDNEY DISEASE**

**1926.1127(m)(3)(iii)** Where feasible, installed cadmium products shall have a visible label or other indication that cadmium is present.

**1926.1127(m)(4)** *Employee information and training.* (i) The employer shall institute a training program for all employees who are potentially exposed to cadmium, assure employee participation in the program, and maintain a record of the contents of such program.

**1926.1127(m)(4)(ii)** Training shall be provided prior to or at the time of initial assignment to a job involving potential exposure to cadmium and at least annually thereafter.

**1926.1127(m)(4)(iii)** The employer shall make the training program understandable to the employee and shall assure that each employee is informed of the following:

**1926.1127(m)(4)(iii)(A)** The health hazards associated with cadmium exposure, with special attention to the information incorporated in appendix A to this section;

**1926.1127(m)(4)(iii)(B)** The quantity, location, manner of use, release, and storage of cadmium in the workplace and the specific nature of operations that could result in exposure to cadmium, especially exposures above the PEL;

**1926.1127(m)(4)(iii)(C)** The engineering controls and work practices associated with the employee's job assignment;

**1926.1127(m)(4)(iii)(D)** The measures employees can take to protect themselves from exposure to cadmium, including modification of such habits as smoking and personal hygiene, and specific procedures the employer has implemented to protect employees from exposure to cadmium such as appropriate work practices, emergency procedures, and the provision of personal protective equipment;

**1926.1127(m)(4)(iii)(E)** The purpose, proper selection, fitting, proper use, and limitations of respirators and protective clothing;

**1926.1127(m)(4)(iii)(F)** The purpose and a description of the medical surveillance program required by paragraph (l) of this section;

**1926.1127(m)(4)(iii)(G)** The contents of this section and its appendices, and,

**1926.1127(m)(4)(iii)(H)** The employee's rights of access to records under §1926.33(g)(1) and (2).

**1926.1127(m)(4)(iv)** Additional access to information and training program and materials.

**1926.1127(m)(4)(iv)(A)** The employer shall make a copy of this section and its appendices readily available to all affected employees and shall provide a copy without cost if requested.

**1926.1127(m)(4)(iv)(B)** Upon request, the employer shall provide to the Assistant Secretary or the Director all materials relating to the employee information and the training program.

**1926.1127(m)(5)** *Multi-employer workplace.* In a multi-employer workplace, an employer who produces, uses, or stores cadmium in a manner that may expose employees of other employers to cadmium shall notify those employers of the potential hazard in accordance with paragraph (e) of the hazard communication standard for construction, 29 CFR 1926.59.

**1926.1127(n)** ***Recordkeeping***—(1) *Exposure monitoring.* (i) The employer shall establish and keep an accurate record of all air monitoring for cadmium in the workplace.

**1926.1127(n)(1)(ii)** This record shall include at least the following information:

**1926.1127(n)(1)(ii)(A)** The monitoring date, shift, duration, air volume, and results in terms of an 8 hour TWA of each sample taken, and if cadmium is not detected, the detection level;

**1926.1127(n)(1)(ii)(B)** The name, social security number, and job classification of all employees monitored and of all other employees whose exposures the monitoring result is intended to represent, including, where applicable, a description of how it was determined that the employee's monitoring result could be taken to represent other employee's exposures;

**1926.1127(n)(1)(ii)(C)** A description of the sampling and analytical methods used and evidence of their accuracy;

**1926.1127(n)(1)(ii)(D)** The type of respiratory protective device, if any, worn by the monitored employee and by any other employee whose exposure the monitoring result is intended to represent;

**1926.1127(n)(1)(ii)(E)** A notation of any other conditions that might have affected the monitoring results.

**1926.1127(n)(1)(ii)(F)** Any exposure monitoring or objective data that were used and the levels.

**1926.1127(n)(1)(iii)** The employer shall maintain this record for at least thirty (30) years, in accordance with §1926.33 of this part.

**1926.1127(n)(1)(iv)** The employer shall also provide a copy of the results of an employee's air monitoring prescribed in paragraph (d) of this section to an industry trade association and to the employee's union, if any, or, if either of such associations or unions do not exist, to another comparable organization that is competent to maintain such records and is reasonably accessible to employers and employees in the industry.

**1926.1127(n)(2)** *Objective data for exemption from requirement for initial monitoring.* (i) For purposes of this section, objective data are information demonstrating that a particular product or material containing cadmium or a specific process, operation, or activity involving cadmium cannot release dust or fumes in concentrations at or above the action level even under the worst-case release conditions. Objective data can be obtained from an industry-wide study or from laboratory product test results from manufacturers of cadmium-containing products or materials. The data the employer uses from an industry-wide survey must be obtained under workplace conditions closely resembling the processes, types of material, control methods, work practices and environmental conditions in the employer's current operations.

**1926.1127(n)(2)(ii)** The employer shall maintain the record for at least 30 years of the objective data relied upon.

**1926.1127(n)(3)** *Medical surveillance.* (i) The employer shall establish and maintain an accurate record for each employee covered by medical surveillance under paragraph (l)(1)(i) of this section.

**1926.1127(n)(3)(ii)** The record shall include at least the following information about the employee:

**1926.1127(n)(3)(ii)(A)** Name, social security number, and description of duties;

**1926.1127(n)(3)(ii)(B)** A copy of the physician's written opinions and of the explanation sheets for biological monitoring results;

**1926.1127(n)(3)(ii)(C)** A copy of the medical history, and the results of any physical examination and all test results that are required to be provided by this section, including biological tests, X-rays, pulmonary function tests, etc., or that have been obtained to further evaluate any condition that might be related to cadmium exposure;

**1926.1127(n)(3)(ii)(D)** The employee's medical symptoms that might be related to exposure to cadmium; and

**1926.1127(n)(3)(ii)(E)** A copy of the information provided to the physician as required by paragraph (l)(9) of this section.

**1926.1127(n)(3)(iii)** The employer shall assure that this record is maintained for the duration of employment plus thirty (30) years, in accordance with §1926.33 of this part.

**1926.1127(n)(3)(iv)** At the employee's request, the employer shall promptly provide a copy of the employee's medical record, or update as appropriate, to a medical doctor or a union specified by the employee.

**1926.1127(n)(4)** *Training.* The employer shall certify that employees have been trained by preparing a certification record which includes the identity of the person trained, the signature of the employer or the person who conducted the training, and the date the training was completed. The certification records shall be prepared at the completion of training and shall be maintained on file for one (1) year beyond the date of training of that employee.

**1926.1127(n)(5)** *Availability.* (i) Except as otherwise provided for in this section, access to all records required to be maintained by paragraphs (n)(1)-(4) of this section shall be in accordance with the provisions of §1926.33 of this part.

**1926.1127(n)(5)(ii)** Within 15 days after a request, the employer shall make an employee's medical records required to be kept by paragraph (n)(3) of this section available for examination and copying to the subject employee, to designated representatives, to anyone having the specific written consent of the subject employee, and after the employee's death or incapacitation, to the employee's family members.

**1926.1127(n)(6)** *Transfer of records.* Whenever an employer ceases to do business and there is no successor employer or designated organization to receive and retain records for the prescribed period, the employer shall comply with the requirements concerning transfer of records set forth in §1926.33(h) of this part.

**1926.1127(o)** ***Observation of monitoring***— (1) *Employee observation.* The employer shall provide affected employees or their designated representatives an opportunity to observe any monitoring of employee exposure to cadmium.

**1926.1127(o)(2)** *Observation procedures.* When observation of monitoring requires entry into an area where the use of protective clothing or equipment is required, the employer shall provide the observer with that clothing and equipment and shall assure that the observer uses such clothing and equipment and complies with all other applicable safety and health procedures.

**1926.1127(p)** ***Dates***— (1) *Effective date.* This section shall become effective on December 14, 1992.

**1926.1127(p)(2)** *Start-up dates.* All obligations of this section commence on the effective date except as follows:

**1926.1127(p)(2)(i)** *Exposure monitoring.* Except for small businesses (nineteen (19) or fewer employees), initial monitoring required by paragraph (d)(2) of this section shall be completed as soon as possible and in any event no later than 60 days after the effective date of this standard. For small businesses, initial monitoring required by paragraph (d)(2) of this section shall be completed as soon as possible and in any event no later than 120 days after the effective date of this standard.

**1926.1127(p)(2)(ii)** *The permissible exposure limit (PEL).* Except for small businesses, as defined under paragraph (p)(2)(i) of this section, the employer shall comply with the PEL established by paragraph (c) of this section as soon as possible and in any event no later than 90 days after the effective date of this section. For small businesses, the employer shall comply with the PEL established by paragraph (c) of this section as soon as possible and in any event no later than 150 days after the effective date of this section.

**1926.1127(p)(2)(iii)** *Regulated areas.* Except for small businesses, as defined under paragraph (p)(2)(i) of this section, regulated areas required to be established by paragraph (e) of this section shall be set up as soon as possible after the results of exposure monitoring are known and in any event no later than 90 days after the effective date of this section. For small businesses, regulated areas required to be established by paragraph (e) of this section shall be set up as soon as possible after the results of exposure monitoring are known and in any event no later than 150 days after the effective date of this section.

**1926.1127(p)(2)(iv)** *Respiratory protection.* Except for small businesses, as defined under paragraph (p)(2)(i) of this section, respiratory protection required by paragraph (g) of this section shall be provided as soon as possible and in any event no later than 90 days after the effective date of this section. For small businesses, respiratory protection required by paragraph (g) of this section shall be provided as soon as possible and in any event no later than 150 days after the effective date of this section.

**1926.1127(p)(2)(v)** *Compliance program.* Except for small businesses, as defined under paragraph (p)(2)(i) of this section, written compliance programs required by paragraph (f)(2) of this section shall be completed and available as soon as possible and in any event no later than 90 days after the effective date of this section. For small businesses, written compliance programs required by paragraph (f)(2) of this section shall be completed and available as soon as possible and in any event no later than 180 days after the effective date of this section.

**1926.1127(p)(2)(vi)** *Methods of compliance.* Except for small businesses, as defined under paragraph (p)(2)(i) of this section, the engineering controls required by paragraph (f)(1) of this section shall be implemented as soon as possible and in any event no later than 120 days after the effective date of this section. For small businesses, the engineering controls required by paragraph (f)(1) of this section shall be implemented as soon as possible and in any event no later than 240 days after the effective date of this section. Work practice controls shall be implemented as soon as possible. Work practice controls that are directly related to engineering controls to be implemented shall be implemented as soon as possible after such engineering controls are implemented.

**1926.1127(p)(2)(vii)** *Hygiene and lunchroom facilities.* Except for small businesses, as defined under paragraph (p)(2)(i) of this section, handwashing facilities, showers, change rooms and eating facilities required by paragraph (j) of this section, whether permanent or temporary, shall be provided as soon as possible and in any event no later than 60 days after the effective date of this section. For small businesses, handwashing facilities, showers, change rooms and eating facilities required by paragraph (j) of this section, whether permanent or temporary, shall be provided as soon as possible and in any event no later than 120 days after the effective date of this section.

**1926.1127(p)(2)(viii)** *Employee information and training.* Except for small businesses, as defined under paragraph (p)(2)(i) of this section, employee information and training required by paragraph (m)(4) of this section shall be aprovided as soon as possible and in any event no later than 90 days after the effective date of this standard. For small businesses, employee information and training required by paragraph (m)(4) of this section shall be provided as soon as possible and in any event no later than 180 days after the effective date of this standard.

**1926.1127(p)(2)(ix)** *Medical surveillance.* Except for small businesses, as defined under paragraph (p)(2)(i) of this section, initial medical examinations required by paragraph (l) of this section shall be provided as soon as possible and in any event no later than 90 days after the effective date of this standard. For small businesses, initial medical examinations required by paragraph (l) of this section shall be provided as soon as possible and in any event no later than 180 days after the effective date of this standard.

**1926.1127(q)** ***Appendices.*** (1) Appendix C to this section is incorporated as part of this section, and compliance with its contents is mandatory.

**1926.1127(q)(2)** Except where portions of appendices A, B, D, E, and F to this section are expressly incorporated in requirements of this section, these appendices are purely informational and are not intended to create any additional obligations not otherwise imposed or to detract from any existing obligations.

## Appendix A To §1926.1127— Substance Safety Data Sheet

**Note:** The requirements applicable to construction work under this Appendix A are identical to those set forth in Appendix A to §1910.1027 of this chapter.

[Appendix A to Section 1910.1027 is printed below—CCH.]

## APPENDIX A TO §1910.1027— SUBSTANCE SAFETY DATA SHEET

### CADMIUM

### *I. Substance Identification*

A. Substance: Cadmium.

B. 8-Hour, Time-weighted-average, Permissible Exposure Limit (TWA PEL): 1. TWA PEL: Five micrograms of cadmium per cubic meter of air 5 $\mu g/m^3$, time-weighted average (TWA) for an 8-hour workday.

C. Appearance: Cadmium metal—soft, blue-white, malleable, lustrous metal or grayish-white powder. Some cadmium compounds may also appear as a brown, yellow, or red powdery substance.

## *II. Health Hazard Data*

A. Routes of Exposure. Cadmium can cause local skin or eye irritation. Cadmium can affect your health if you inhale it or if you swallow it.

B. Effects of Overexposure.

1. Short-term (acute) exposure: Cadmium is much more dangerous by inhalation than by ingestion. High exposures to cadmium that may be immediately dangerous to life or health occur in jobs where workers handle large quantities of cadmium dust or fume; heat cadmium-containing compounds or cadmium-coated surfaces; weld with cadmium solders or cut cadmium-containing materials such as bolts.

2. Severe exposure may occur before symptoms appear. Early symptoms may include mild irritation of the upper respiratory tract, a sensation of constriction of the throat, a metallic taste and/or a cough. A period of 1-10 hours may precede the onset of rapidly progressing shortness of breath, chest pain, and flu-like symptoms with weakness, fever, headache, chills, sweating and muscular pain. Acute pulmonary edema usually develops within 24 hours and reaches a maximum by three days. If death from asphyxia does not occur, symptoms may resolve within a week.

3. Long-term (chronic) exposure. Repeated or long-term exposure to cadmium, even at relatively low concentrations, may result in kidney damage and an increased risk of cancer of the lung and of the prostate.

C. Emergency First Aid Procedures.

1. Eye exposure: Direct contact may cause redness or pain. Wash eyes immediately with large amounts of water, lifting the upper and lower eyelids. Get medical attention immediately.

2. Skin exposure: Direct contact may result in irritation. Remove contaminated clothing and shoes immediately. Wash affected area with soap or mild detergent and large amounts of water. Get medical attention immediately.

3. Ingestion: Ingestion may result in vomiting, abdominal pain, nausea, diarrhea, headache and sore throat. Treatment for symptoms must be administered by medical personnel. Under no circumstances should the employer allow any person whom he retains, employs, supervises or controls to engage in therapeutic chelation. Such treatment is likely to translocate cadmium from pulmonary or other tissue to renal tissue. Get medical attention immediately.

4. Inhalation: If large amounts of cadmium are inhaled, the exposed person must be moved to fresh air at once. If breathing has stopped, perform cardiopulmonary resuscitation. Administer oxygen if available. Keep the affected person warm and at rest. Get medical attention immediately.

5. Rescue: Move the affected person from the hazardous exposure. If the exposed person has been overcome, attempt rescue only after notifying at least one other person of the emergency and putting into effect established emergency procedures. Do not become a casualty yourself. Understand your emergency rescue procedures and know the location of the emergency equipment before the need arises.

## *III. Employee Information*

A. Protective Clothing and Equipment.

1. Respirators: You may be required to wear a respirator for non-routine activities; in emergencies; while your employer is in the process of reducing cadmium exposures through engineering controls; and where engineering controls are not feasible. If respirators are worn in the future, they must have a joint Mine Safety and Health Administration (MSHA) and National Institute for Occupational Safety and Health (NIOSH) label of approval. Cadmium does not have a detectable odor except at levels well above the permissible exposure limits. If you can smell cadmium while wearing a respirator, proceed immediately to fresh air. If you experience difficulty breathing while wearing a respirator, tell your employer.

2. Protective Clothing: You may be required to wear impermeable clothing, gloves, foot gear, a face shield, or other appropriate protective clothing to prevent skin contact with cadmium. Where protective clothing is required, your employer must provide clean garments to you as necessary to assure that the clothing protects you adequately. The employer must replace or repair protective clothing that has become torn or otherwise damaged.

3. Eye Protection: You may be required to wear splash-proof or dust resistant goggles to prevent eye contact with cadmium.

B. Employer Requirements.

1. Medical: If you are exposed to cadmium at or above the action level, your employer is required to provide a medical examination, laboratory tests and a medical history according to the medical surveillance provisions under paragraph (1) of this standard. (See summary chart and tables in this appendix A.) These tests shall be provided without cost to you. In addition, if you are accidentally exposed to cadmium under conditions known or suspected to constitute toxic exposure to cadmium, your employer is required to make special tests available to you.

2. Access to Records: All medical records are kept strictly confidential. You or your representative are entitled to see the records of measurements of your exposure to cadmium. Your medical examination records can be furnished to your personal physician or designated representative upon request by you to your employer.

3. Observation of Monitoring: Your employer is required to perform measurements that are representative of your exposure to cadmium and you or your designated representative are entitled to observe the monitoring procedure. You are entitled to observe the steps taken in the measurement procedure, and to record the results obtained. When the monitoring procedure is taking place in an area where respirators or personal protective clothing and equipment are required to be worn, you or your representative must also be provided with, and must wear the protective clothing and equipment.

C. Employee Requirements—You will not be able to smoke, eat, drink, chew gum or tobacco, or apply cosmetics while working with cadmium in regulated areas. You will also not be able to carry or store tobacco products, gum, food, drinks or cosmetics in regulated areas because these products easily become contaminated with cadmium from the workplace and can therefore create another source of unnecessary cadmium exposure.

Some workers will have to change out of work clothes and shower at the end of the day, as part of their workday, in order to wash cadmium from skin and hair. Handwashing and cadmium-free eating facilities shall be provided by the employer and proper hygiene should always be performed before eating. It is also recommended that you do not smoke or use tobacco products, because among other things, they naturally contain cadmium. For further information, read the labeling on such products.

## *IV. Physician Information*

A. Introduction.—The medical surveillance provisions of paragraph (1) generally are aimed at accomplishing three main interrelated purposes: First, identifying employees at higher risk of adverse health effects from excess, chronic exposure to cadmium; second, preventing cadmium-induced disease; and third, detecting and minimizing existing cadmium-induced disease. The core of medical surveillance in this standard is the early and periodic monitoring of the employee's biological indicators of: (a) Recent exposure to cadmium; (b) cadmium body burden; and (c) potential and actual kidney damage associated with exposure to cadmium.

The main adverse health effects associated with cadmium overexposure are lung cancer and kidney dysfunction. It is not yet known how to adequately biologically monitor human beings to specifically prevent cadmium-induced lung cancer. By contrast, the kidney can be monitored to provide prevention and early detection of cadmium-induced kidney damage. Since, for non-carcinogenic effects, the kidney is considered the primary target organ of chronic exposure to cadmium, the medical surveillance provisions of this standard effectively focus on cadmium-induced kidney disease. Within that focus, the aim, where possible, is to prevent the onset of such disease and, where necessary, to minimize such disease as may already exist. The by-products of successful prevention of kidney disease are anticipated to be the reduction and prevention of other cadmium-induced diseases.

B. Health Effects.—The major health effects associated with cadmium overexposure are described below.

1. Kidney: The most prevalent non-malignant disease observed among workers chronically exposed to cadmium is kidney dysfunction. Initially, such dysfunction is manifested as proteinuria. The proteinuria associated with cadmium exposure is most commonly characterized by excretion of low-molecular weight proteins (15,000 to 40,000 MW) accompanied by loss of electrolytes, uric acid, calcium, amino acids, and phosphate. The compounds commonly excreted include: beta-2-microglobulin ($\beta_2$-M), retinol binding protein (RBP), immunoglobulin light chains, and lysozyme. Excretion of low molecular weight pro-

teins are characteristic of damage to the proximal tubules of the kidney (Iwao *et al.*, 1980).

It has also been observed that exposure to cadmium may lead to urinary excretion of high-molecular weight proteins such as albumin, immunoglobulin G, and glycoproteins (Ex. 29). Excretion of high-molecular weight proteins is typically indicative of damage to the glomeruli of the kidney. Bernard *et al.*, (1979) suggest that damage to the glomeruli and damage to the proximal tubules of the kidney may both be linked to cadmium exposure but they may occur independently of each other.

Several studies indicate that the onset of low-molecular weight proteinuria is a sign of irreversible kidney damage (Friberg *et al.*, 1974; Roels *et al.*, 1982; Piscator 1984; Elinder *et al.*, 1985; Smith *et al.*, 1986). Above specific levels of $\beta_2$-M associated with cadmium exposure it is unlikely that $\beta_2$-M levels return to normal even when cadmium exposure is eliminated by removal of the individual from the cadmium work environment (Friberg, Ex. 29, 1990).

Some studies indicate that such proteinuria may be progressive; levels of $\beta_2$-M observed in the urine increase with time even after cadmium exposure has ceased. See, for example, Elinder *et al.*, 1985. Such observations, however, are not universal, and it has been suggested that studies in which proteinuria has not been observed to progress may not have tracked patients for a sufficiently long time interval (Jarup, Ex. 8-661).

When cadmium exposure continues after the onset of proteinuria, chronic nephrotoxicity may occur (Friberg, Ex. 29). Uremia results from the inability of the glomerulus to adequately filter blood. This leads to severe disturbance of electrolyte concentrations and may lead to various clinical complications including kidney stones (L-140-50).

After prolonged exposure to cadmium, glomerular proteinuria, glucosuria, aminoaciduria, phosphaturia, and hypercalciuria may develop (Exs. 8-86, 4-28, 14-18). Phosphate, calcium, glucose, and amino acids are essential to life, and under normal conditions, their excretion should be regulated by the kidney. Once low molecular weight proteinuria has developed, these elements dissipate from the human body. Loss of glomerular function may also occur, manifested by decreased glomerular filtration rate and increased serum creatinine. Severe cadmium-induced renal damage may eventually develop into chronic renal failure and uremia (Ex. 55).

Studies in which animals are chronically exposed to cadmium confirm the renal effects observed in humans (Friberg *et al.*, 1986). Animal studies also confirm problems with calcium metabolism and related skeletal effects which have been observed among humans exposed to cadmium in addition to the renal effects. Other effects commonly reported in chronic animal studies include anemia, changes in liver morphology, immunosup pression and hypertension. Some of these effects may be associated with co-factors. Hypertension, for example, appears to be associated with diet as well as cadmium exposure. Animals injected with cadmium have also shown testicular necrosis (Ex. 8-86B).

## 2. Biological Markers

It is universally recognized that the best measures of cadmium exposures and its effects are measurements of cadmium in biological fluids, especially urine and blood. Of the two, CdU is conventionally used to determine body burden of cadmium in workers without kidney disease. CdB is conventionally used to monitor for recent exposure to cadmium. In addition, levels of CdU and CdB historically have been used to predict the percent of the population likely to develop kidney disease (Thun *et al.*, Ex. L-140-50; WHO, Ex. 8-674; ACGIH, Exs. 8-667, 140-50).

The third biological parameter upon which OSHA relies for medical surveillance is Beta-2-microglobulin in urine ($\beta_2$-M), a low molecular weight protein. Excess $\beta_2$-M has been widely accepted by physicians and scientists as a reliable indicator of functional damage to the proximal tubule of the kidney (Exs. 8-447, 144-3-C, 4-47, L-140-45, 19-43-A).

Excess $\beta_2$-M is found when the proximal tubules can no longer reabsorb this protein in a normal manner. This failure of the proximal tubules is an early stage of a kind of kidney disease that commonly occurs among workers with excessive cadmium exposure. Used in conjunction with biological test results indicating abnormal levels of CdU and CdB, the finding of excess $\beta_2$-M can establish for an examining physician that any existing kidney disease is probably cadmium-related (Trs. 6/6/90, pp. 82-86, 122, 134). The upper limits of normal levels for cadmium in urine and cadmium in blood are 3 µg Cd/gram creatinine in urine and 5 µgCd/liter whole blood, respectively. These levels were derived from broad-based population studies.

Three issues confront the physicians in the use of $\beta_2$-M as a marker of kidney dysfunction and material impairment. First, there are a few other causes of elevated levels of $\beta_2$-M not related to cadmium exposures, some of which may be rather common diseases and some of which are serious diseases (e.g., myeloma or transient flu, Exs. 29 and 8-086). These can be medically evaluated as alternative causes (Friberg, Ex. 29). Also, there are other factors that can cause $\beta_2$-M to degrade so that low levels would result in workers with tubular dysfunction. For example, regarding the degradation of $\beta_2$-M, workers with acidic urine ($pH<6$) might have $\beta_2$-M levels that are within the "normal" range when in fact kidney dysfunction has occurred (Ex. L-140-1) and the low molecular weight proteins are degraded in acid urine. Thus, it is very important that the pH of urine be measured, that urine samples be buffered as necessary (See appendix F.), and that urine samples be handled correctly, i.e., measure the pH of freshly voided urine samples, then if necessary, buffer to $pH>6$ (or above for shipping purposes), measure pH again and then, perhaps, freeze the sample for storage and shipping. (See also appendix F.) Second, there is debate over the pathological significance of proteinuria, however, most world experts believe that $\beta_2$-M levels greater than 300 µg/g Cr are abnormal (Elinder, Ex. 55, Friberg, Ex. 29). Such levels signify kidney dysfunction that constitutes material impairment of health. Finally, detection of $\beta_2$-M at low levels has often been considered difficult, however, many laboratories have the capability of detecting excess $\beta_2$-M using simple kits, such as the Phadebas Delphia test, that are accurate to levels of 100 µg $\beta_2$-M/g Cr U (Ex. L-140-1).

Specific recommendations for ways to measure $\beta_2$-M and proper handling of urine samples to prevent degradation of $\beta_2$-M have been addressed by OSHA in appendix F, in the section on laboratory standardization. All biological samples must be analyzed in a laboratory that is proficient in the analysis of that particular analyte, under paragraph (l)(1)(iv). (See appendix F). Specifically, under paragraph (l)(1)(iv), the employer is to assure that the collecting and handling of biological samples of cadmium in urine (CdU), cadmium in blood (CdB), and beta-2 microglobulin in urine ($\beta_2$-M) taken from employees is collected in a manner that assures reliability. The employer must also assure that analysis of biological samples of cadmium in urine (CdU), cadmium in blood (CdB), and beta-2 microglobulin in urine ($\beta_2$-M) taken from employees is performed in laboratories with demonstrated proficiency for that particular analyte. (See appendix F.)

## 3. Lung and Prostate Cancer

The primary sites for cadmium-associated cancer appear to be the lung and the prostate (L-140-50). Evidence for an association between cancer and cadmium exposure derives from both epidemiological studies and animal experiments. Mortality from prostate cancer associated with cadmium is slightly elevated in several industrial cohorts, but the number of cases is small and there is not clear dose-response relationship. More substantive evidence exists for lung cancer.

The major epidemiological study of lung cancer was conducted by Thun *et al.*, (Ex. 4-68). Adequate data on cadmium exposures were available to allow evaluation of dose-response relationships between cadmium exposure and lung cancer. A statistically significant excess of lung cancer attributed to cadmium exposure was observed in this study even when confounding variables such as co-exposure to arsenic and smoking habits were taken into consideration (Ex. L-140-50).

The primary evidence for quantifying a link between lung cancer and cadmium exposure from animal studies derives from two rat bioassay studies; one by Takenaka *et al.*, (1983), which is a study of cadmium chloride and a second study by Oldiges and Glaser (1990) of four cadmium compounds.

Based on the above cited studies, the U.S. Environmental Protection Agency (EPA) classified cadmium as "B1", a probable human carcinogen, in 1985 (Ex. 4-4). The International Agency for Research on Cancer (IARC) in 1987 also recommended that cadmium be listed as "2A", a probable human carcinogen (Ex. 4-15). The

American Conference of Governmental Industrial Hygienists (ACGIH) has recently recommended that cadmium be labeled as a carcinogen. Since 1984, NIOSH has concluded that cadmium is possibly a human carcinogen and has recommended that exposures be controlled to the lowest level feasible.

### 4. Non-carcinogenic Effects

Acute pneumonitis occurs 10 to 24 hours after initial acute inhalation of high levels of cadmium fumes with symptoms such as fever and chest pain (Exs. 30, 8-86B). In extreme exposure cases pulmonary edema may develop and cause death several days after exposure. Little actual exposure measurement data is available on the level of airborne cadmium exposure that causes such immediate adverse lung effects, nonetheless, it is reasonable to believe a cadmium concentration of approximately 1 $mg/m^3$ over an eight hour period is "immediately dangerous" (55 FR 4052, ANSI; Ex. 8-86B).

In addition to acute lung effects and chronic renal effects, long term exposure to cadmium may cause other severe effects on the respiratory system. Reduced pulmonary function and chronic lung disease indicative of emphysema have been observed in workers who have had prolonged exposure to cadmium dust or fumes (Exs. 4-29, 4-22, 4-42, 4-50, 4-63). In a study of workers conducted by Kazantzis *et al.*, a statistically significant excess of worker deaths due to chronic bronchitis was found, which in his opinion was directly related to high cadmium exposures of 1 $mg/m^3$ or more (Tr. 6/8/90, pp. 156-157).

Cadmium need not be respirable to constitute a hazard. Inspirable cadmium particles that are too large to be respirable but small enough to enter the tracheobronchial region of the lung can lead to bronchoconstriction, chronic pulmonary disease, and cancer of that portion of the lung. All of these diseases have been associated with occupational exposure to cadmium (Ex. 8-86B). Particles that are constrained by their size to the extra-thoracic regions of the respiratory system such as the nose and maxillary sinuses can be swallowed through mucocillary clearance and be absorbed into the body (ACGIH, Ex. 8-692). The impaction of these particles in the upper airways can lead to anosmia, or loss of sense of smell, which is an early indication of overexposure among workers exposed to heavy metals. This condition is commonly reported among cadmium-exposed workers (Ex. 8-86-B).

## C. Medical Surveillance

In general, the main provisions of the medical surveillance section of the standard, under paragraphs (l)(1)-(17) of the regulatory text, are as follows:

1. Workers exposed above the action level are covered;

2. Workers with intermittent exposures are not covered;

3. Past workers who are covered receive biological monitoring for at least one year;

4. Initial examinations include a medical questionnaire and biological monitoring of cadmium in blood (CdB), cadmium in urine (CdU), and Beta-2-microglobulin in urine ($\beta_2$-M);

5. Biological monitoring of these three analytes is performed at least annually; full medical examinations are performed biennially;

6. Until five years from the effective date of the standard, medical removal is required when CdU is greater than 1 5 $\mu$g/gram creatinine (g Cr), or CdB is greater than 15 $\mu$g/liter whole blood (lwb), or $\beta_2$-M is greater than 1500 $\mu$g/g Cr, and CdB is greater than 5 $\mu$g/lwb or CdU is greater than 3 $\mu$g/g Cr;

7. Beginning five years after the standard is in effect, medical removal triggers will be reduced;

8. Medical removal protection benefits are to be provided for up to 18 months;

9. Limited initial medical examinations are required for respirator usage;

10. Major provisions are fully described under section (l) of the regulatory text; they are outlined here as follows:

A. Eligibility

B. Biological monitoring

C. Actions triggered by levels of CdU, CdB, and $\beta_2$-M (See Summary Charts and Tables in Attachment-1.)

D. Periodic medical surveillance

E. Actions triggered by periodic medical surveillance (See appendix A Summary Chart and Tables in Attachment-1.)

F. Respirator usage

G. Emergency medical examinations

H. Termination examination

I. Information to physician

J. Physician's medical opinion

K. Medical removal protection

L. Medical removal protection benefits

M. Multiple physician review

N. Alternate physician review

O. Information employer gives to employee

P. Recordkeeping

Q. Reporting on OSHA form 200

11. The above mentioned summary of the medical surveillance provisions, the summary chart, and tables for the actions triggered at different levels of CdU, CdB and $\beta_2$-M (in appendix A Attachment-1) are included only for the purpose of facilitating understanding of the provisions of paragraphs (l)(3) of the final cadmium standard. The summary of the provisions, the summary chart, and the tables do not add to or reduce the requirements in paragraph (l)(3).

## D. Recommendations to Physicians

1. It is strongly recommended that patients with tubular proteinuria are counseled on: The hazards of smoking; avoidance of nephrotoxins and certain prescriptions and over-the-counter medications that may exacerbate kidney symptoms; how to control diabetes and/or blood pressure; proper hydration, diet, and exercise (Ex. 19-2). A list of prominent or common nephrotoxins is attached. (See appendix A Attachment-2.)

2. DO NOT CHELATE; KNOW WHICH DRUGS ARE NEPHROTOXINS OR ARE ASSOCIATED WITH NEPHRITIS.

3. The gravity of cadmium-induced renal damage is compounded by the fact there is no medical treatment to prevent or reduce the accumulation of cadmium in the kidney (Ex. 8-619). Dr. Friberg, a leading world expert on cadmium toxicity, indicated in 1992, that there is no form of chelating agent that could be used without substantial risk. He stated that tubular proteinuria has to be treated in the same way as other kidney disorders (Ex. 29).

4. After the results of a workers' biological monitoring or medical examination are received the employer is required to provide an information sheet to the patient, briefly explaining the significance of the results. (See Attachment 3 of this appendix A.)

5. For additional information the physician is referred to the following additional resources:

a. The physician can always obtain a copy of the preamble, with its full discussion of the health effects, from OSHA's Computerized Information System (OCIS).

b. The Docket Officer maintains a record of the rulemaking. The Cadmium Docket (H-057A), is located at 200 Constitution Ave. NW., room N-2625, Washington, DC 20210; telephone: 202-219-7894.

c. The following articles and exhibits in particular from that docket (H-057A):

| Exhibit number | Author and paper title |
|---|---|
| 8-447 .......... | Lauwerys et. al., Guide for physicians, "Health Maintenance of Workers Exposed to Cadmium," published by the Cadmium Council. |
| 4-67 ............ | Takenaka, S., H. Oldiges, H. Konig, D. Hochrainer, G. Oberdorster. "Carcinogenicity of Cadmium Chloride Aerosols in Wistar Rats". JNCI 70:367-373, 1983. (32) |
| 4-68 ............ | Thun, M.J., T.M. Schnoor, A.B. Smith, W.E. Halperin, R.A. Lemen. "Mortality Among a Cohort of U.S. Cadmium Production Workers—An Update." JNCI 74(2):325-33, 1985. (8) |
| 4-25 ............ | Elinder, C.G., Kjellstrom, T., Hogstedt, C., et al., "Cancer Mortality of Cadmium Workers." Brit. J. Ind. Med. 42:651-655, 1985. (14) |
| 4-26 ............ | Ellis, K.J. et al., "Critical Concentrations of Cadmium in Human Renal Cortex: Dose Effect Studies to Cadmium Smelter Workers." J. Toxicol. Environ. Health 7:691-703, 1981. (76) |
| 4-27 ............ | Ellis, K.J., S.H. Cohn and T.J. Smith. "Cadmium Inhalation Exposure Estimates: Their Significance with Respect to Kidney and Liver Cadmium Burden." J. Toxicol. Environ. Health 15:173-187, 1985. |
| 4-28 ............ | Falck, F.Y., Jr., Fine, L.J., Smith, R.G., McClatchey, K.D., Annesley, T., England, B., and Schork, A.M. "Occupational Cadmium Exposure and Renal Status." Am. J. Ind. Med. 4:541, 1983. (64) |
| 8-86A.......... | Friberg, L., C.G. Elinder, et al., "Cadmium and Health a Toxicological and Epidemiological Appraisal, Volume I, Exposure, Dose, and Metabolism." CRC Press, Inc., Boca Raton, FL, 1986. (Available from the OSHA Technical Data Center) |
| 8-86B .......... | Friberg, L., C.G. Elinder, et al., "Cadmium and Health: A Toxicological and Epidemiological Appraisal, Volume II, Effects and Response." CRC Press, Inc., Boca Raton, FL, 1986. (Available from the OSHA Technical Data Center) |
| L-140-45...... | Elinder, C.G., "Cancer Mortality of Cadmium Workers", Brit. J. Ind. Med., 42, 651-655, 1985. |
| L-140-50...... | Thun, M., Elinder, C.G., Friberg, L, "Scientific Basis for an Occupational Standard for Cadmium, Am. J. Ind. Med., 20; 629-642, 1991. |


### *V. Information Sheet*

The information sheet (appendix A Attachment-3.) or an equally explanatory one should be provided to you after any biological monitoring results are reviewed by the physician, or where applicable, after any medical examination.

## ATTACHMENT 1—APPENDIX A SUMMARY CHART AND TABLES A AND B OF ACTIONS TRIGGERED BY BIOLOGICAL MONITORING

## APPENDIX A SUMMARY CHART: SECTION (1)(3) MEDICAL SURVEILLANCE

Categorizing Biological Monitoring Results

(A) Biological monitoring results categories are set forth in Appendix A Table A for the periods ending December 31, 1998 and for the period beginning January 1, 1999.

(B) The results of the biological monitoring for the initial medical exam and the subsequent exams shall determine an employee's biological monitoring result category.

Actions Triggered by Biological Monitoring

(A)

(i) The actions triggered by biological monitoring for an employee are set forth in Appendix A Table B.

(ii) The biological monitoring results for each employee under section (1)(3) shall determine the actions required for that employee. That is, for any employee in biological monitoring category C, the employer will perform all of the actions for which there is an X in column C of Appendix A Table B.

(iii) An employee is assigned the alphabetical category ("A" being the lowest) depending upon the test results of the three biological markers.

(iv) An employee is assigned category A if monitoring results for all three biological markers fall at or below the levels indicated in the table listed for category A.

(v) An employee is assigned category B if any monitoring result for any of the three biological markers fall within the range of levels indicated in the table listed for category B, providing no result exceeds the levels listed for category B.

(vi) An employee is assigned category C if any monitoring result for any of the three biological markers are above the levels listed for category C.

(B) The user of Appendix A Tables A and B should know that these tables are provided only to facilitate understanding of the relevant provisions of paragraph (l)(3) of this section. Appendix A Tables A and B are not meant to add to or subtract from the requirements of those provisions.

## APPENDIX A TABLE A—CATEGORIZATION OF BIOLOGICAL MONITORING RESULTS

### APPLICABLE THROUGH 1998 ONLY

| Biological marker | Monitoring result categories | | |
|---|---|---|---|
| | A | B | C |
| Cadmium in urine (CdU) ($\mu$g/g creatinine) | ≤3 | >3 and ≤15 | >15 |
| $\beta_2$-microglobulin ($\beta_2$-M) ($\mu$g/g creatinine) | ≤300 | >300 and ≤1500 | >1500* |
| Cadmium in blood (CdB) ($\mu$g/liter whole blood) | ≤5 | >5 and ≤15 | >15 |

* If an employee's $\beta_2$-M levels are above 1,500 $\mu$g/g creatinine, in order for mandatory medical removal to be required (See Appendix A Table B.), either the employee's CdU level must also be >3 $\mu$g/g creatinine or CdB level must also be >5 $\mu$g/liter whole blood.

### APPLICABLE BEGINNING JANUARY 1, 1999

| Biological marker | Monitoring result categories | | |
|---|---|---|---|
| | A | B | C |
| Cadmium in urine (CdU) ($\mu$g/g creatinine) | ≤3 | >3 and ≤7 | >7 |
| $\beta_2$-microglobulin ($\beta_2$-M) ($\mu$g/g creatinine) | ≤300 | >300 and ≤750 | >750* |
| Cadmium in blood (CdB) ($\mu$g/liter whole blood) | ≤5 | >5 and ≤10 | >10 |

* If an employee's $\beta_2$-M levels are above 750 $\mu$g/g creatinine, in order for mandatory medical removal to be required (See Appendix A Table B.), either the employee's CdU level must also be >3 $\mu$g/g creatinine or CdB level must also be >5 $\mu$g/liter whole blood.

## APPENDIX A TABLE B—ACTIONS DETERMINED BY BIOLOGICAL MONITORING

This table presents the actions required based on the monitoring result in Appendix A Table A. Each item is a separate requirement in citing non-compliance. For example, a medical examination within 90 days for an employee in category B is separate from the requirement to administer a periodic medical examination for category B employees on an annual basis.

| *Required actions* | *Monitoring result category* A[1] | B[1] | C[1] |
|---|---|---|---|
| (1) Biological monitoring: | | | |
| (a) Annual | X | | |
| (b) Semiannual | | X | |
| (c) Quarterly | | | X |
| (2) Medical examination: | | | |
| (a) Biennial | X | | |
| (b) Annual | | X | |
| (c) Semiannual | | | X |
| (d) Within 90 days | | X | X |
| (3) Assess within two weeks: | | | |
| (a) Excess cadmium exposure | | X | X |
| (b) Work practices | | X | X |
| (c) Personal hygiene | | X | X |
| (d) Respirator usage | | X | X |
| (e) Smoking history | | X | X |
| (f) Hygiene facilities | | X | X |
| (g) Engineering controls | | X | X |
| (h) Correct within 30 days | | X | X |
| (i) Periodically assess exposures | | | X |
| (4) Discretionary medical removal | | X | X |
| (5) Mandatory medical removal | | | X[2] |

[1] For all employees covered by medical surveillance exclusively because of exposures prior to the effective date of this standard, if they are in Category A, the employer shall follow the requirements of paragraphs (1)(3)(i)(B) and (1)(4)(v)(A). If they are in Category B or C, the employer shall follow the requirements of paragraphs (1)(4)(v)(B) - (C).

[2] See footnote Appendix A Table A.

## APPENDIX A—ATTACHMENT 2—LIST OF MEDICATIONS

A list of the more common medications that a physician, and the employee, may wish to review is likely to include some of the following: (1) Anticonvulsants: Paramethadione, phenytoin, trimethadone; (2) antihypertensive drugs: Captopril, methyldopa; (3) antimicrobials: Aminoglycosides, amphotericin B, cephalosporins, ethambutol; (4) antineoplastic agents: Cisplatin, methotrexate, mitomycin-C, nitrosoureas, radiation; (4) sulfonamide diuretics: Acetazolamide, chlorthalidone, furosemide, thiazides; (5) halogenated alkanes, hydrocarbons, and solvents that may occur in some settings: Carbon tetrachloride, ethylene glycol, toluene; iodinated radiographic contrast media; nonsteroidal anti-inflammatory drugs; and, (7) other miscellaneous compounds: Acetominophen, allopurinol, amphetamines, azathioprine, cimetidine, cyclosporine, lithium, methoxyflurane, methysergide, D-penicillamine, phenacetin, phenendione. A list of drugs associated with acute interstitial nephritis includes: (1) Antimicrobial drugs: Cephalosporins, chloramphenicol, colistin, erythromycin, ethambutol, isoniazid, para-aminosalicylic acid, penicillins, polymyxin B, rifampin, sulfonamides, tetracyclines, and vancomycin; (2) other miscellaneous drugs: Allopurinol, antipyrene, azathioprine, captopril, cimetidine, clofibrate, methyldopa, phenindione, phenylpropanolamine, phenytoin, probenecid, sulfinpyrazone, sulfonamid diuretics, triamterene; and, (3) metals: Bismuth, gold.

This list have been derived from commonly available medical textbooks (e.g., Ex. 14-18). The list has been included merely to facilitate the physician's, employer's, and employee's understanding. The list does not represent an official OSHA opinion or policy regarding the use of these medications for particular employees. The use of such medications should be under physician discretion.

## ATTACHMENT 3—BIOLOGICAL MONITORING AND MEDICAL EXAMINATION RESULTS

Employee ________________ Testing Date ________________

Cadmium in Urine ________________ μg/g Cr—Normal Levels: ≤3 μg/g Cr.

Cadmium in Blood ________________ μg/lwb—Normal Levels: ≤5 μg/lwb.

Beta-2-microglobulin in Urine ________ μg/g Cr—Normal Levels: ≤300 μg/g Cr.

Physical Examination Results: N/A ________________ Satisfactory __________ Unsatisfactory __________ (see physician again).

Physician's Review of Pulmonary Function Test: N/A ________

Normal ________ Abnormal ________.

Next biological monitoring or medical examination scheduled for ____________

The biological monitoring program has been designed for three main purposes: 1) to identify employees at risk of adverse health effects from excess, chronic exposure to cadmium; 2) to prevent cadmium-induced disease(s); and 3) to detect and minimize existing cadmium-induced disease(s).

The levels of cadmium in the urine and blood provide an estimate of the total amount of cadmium in the body. The amount of a specific protein in the urine (beta-2-microglobulin) indicates changes in kidney function. All three tests must be evaluated together. A single mildly elevated result may not be important if testing at a later time indicates that the results are normal and the workplace has been evaluated to decrease possible sources of cadmium exposure. The levels of cadmium or beta-2-microglobulin may change over a period of days to months and the time needed for those changes to occur is different for each worker.

If the results for biological monitoring are above specific "high levels" [cadmium urine greater than 10 micrograms per gram of creatinine (μg/g Cr), cadmium blood greater than 10 micrograms per liter of whole blood (μg/lwb), or beta-2-microglobulin greater than 1000 micrograms per gram of creatinine (μg/g Cr)], the worker has a much greater chance of developing other kidney diseases.

One way to measure for kidney function is by measuring beta-2-microglobulin in the urine. Beta-2-microglobulin is a protein which is normally found in the blood as it is being filtered in the kidney, and the kidney reabsorbs or returns almost all of the beta-2-microglobulin to the blood. A very small amount (less than 300 μg/g Cr in the urine) of beta-2-microglobulin is not reabsorbed into the blood, but is released in the urine. If cadmium damages the kidney, the amount of beta-2-microglobulin in the urine increases because the kidney cells are unable to reabsorb the beta-2-microglobulin normally. An increase in the amount of beta-2-microglobulin in the urine is a very early sign of kidney dysfunction. A small increase in beta-2-microglobulin in the urine will serve as an early warning sign that the worker may be absorbing cadmium from the air, cigarettes contaminated in the workplace, or eating in areas that are cadmium contaminated.

Even if cadmium causes permanent changes in the kidney's ability to reabsorb beta-2-microglobulin, and the beta-2-microglobulin is above the "high levels", the loss of kidney function may not lead to any serious health problems. Also, renal function naturally declines as people age. The risk for changes in kidney function for workers who have biological monitoring results between the "normal values" and the "high levels" is not well known. Some people are more cadmium-tolerant, while others are more cadmium-susceptible.

For anyone with even a slight increase of beta-2-microglobulin, cadmium in the urine, or cadmium in the blood, it is very important to protect the kidney from further damage. Kidney damage can come from other sources than excess cadmium-exposure so it is also recommended that if a worker's levels are "high" he/she should receive counseling about drinking more water; avoiding cadmium-tainted tobacco and certain medications (nephrotoxins, acetaminophen); controlling diet, vitamin intake, blood pressure and diabetes; etc.

## Appendix B to §1926.1127—Substance Technical Guidelines for Cadmium

**Note:** The requirements applicable to construction work under this Appendix B are identical to those set forth in Appendix B to §1910.1027 of this chapter.

[Appendix B to Section 1910.1027 is printed below—CCH.]

## APPENDIX B TO §1910.1027—SUBSTANCE TECHNICAL GUIDELINES FOR CADMIUM

I. Cadmium Metal

A. *Physical and Chemical Data.*

1. *Substance Identification.*

*Chemical name:* Cadmium.

*Formula:* Cd.

*Molecular Weight:* 112.4.

*Chemical Abstracts Service (CAS) Registry No.:* 7740-43-9.

*Other Identifiers:* RETCS EU9800000; EPA D006; DOT 2570 53.

*Synonyms:* Colloidal Cadmium: Kadmium (German): CI 77180.

2. *Physical data.*

*Boiling point:* (760 mm Hg): 765 degrees C.

*Melting point:* 321 degrees C.

*Specific Gravity:* ($H_2O$=;@ 20°C): 8.64.

*Solubility:* Insoluble in water; soluble in dilute nitric acid and in sulfuric acid.

*Appearance:* Soft, blue-white, malleable, lustrous metal or grayish-white powder.

B. *Fire, Explosion and Reactivity Data.*

1. *Fire.*

*Fire and Explosion Hazards:* The finely divided metal is pyrophoric, that is the dust is a severe fire hazard and moderate explosion hazard when exposed to heat or flame. Burning material reacts violently with extinguishing agents such as water, foam, carbon dioxide, and halons.

*Flash point:* Flammable (dust).

*Extinguishing media:* Dry sand, dry dolomite, dry graphite, or sodimum chloride.

2. *Reactivity.*

*Conditions contributing to instability:* Stable when kept in sealed containers under normal temperatures and pressure, but dust may ignite upon contact with air. Metal tarnishes in moist air.

*Incompatibilities:* Ammonium nitrate, fused: Reacts violently or explosively with cadmium dust below 20 °C. Hydrozoic acid: Violent explosion occurs after 30 minutes. Acids: Reacts violently, forms hydrogen gas. Oxidizing agents or metals: Strong reaction with cadmium dust. Nitryl fluoride at slightly elevated temperature: Glowing or white incandescence occurs. Selenium: Reacts exothermically. Ammonia: Corrosive reaction. Sulfur dioxide: Corrosive reaction. Fire extinguishing agents (water, foam, carbon dioxide, and halons): Reacts violently. Tellurium: Incandescent reaction in hydrogen atmosphere.

*Hazardous decomposition products:* The heated metal rapidly forms highly toxic, brownish fumes of oxides of cadmium.

C. *Spill, Leak and Disposal Procedures.*

1. *Steps to be taken if the materials is released or spilled.* Do not touch spilled material. Stop leak if you can do it without risk. Do not get water inside container. For large spills, dike spill for later disposal. Keep unnecessary people away. Isolate hazard area and deny entry. The Superfund Amendments and Reauthorization Act of 1986 Section 304 requires that a release equal to or greater than the reportable quantity for this substance (1 pound) must be immediately reported to the local emergency planning committee, the state emergency response commission, and the National Response Center (800) 424-8802; in Washington, DC metropolitan area (202) 426-2675.

II. Cadmium Oxide

A. *Physical and Chemical Date.*

1. *Substance identification.*

*Chemical name:* Cadmium Oxide.

*Formula:* CdO.

*Molecular Weight:* 128.4.

*CAS No.:* 1306-19-0.

*Other Identifiers:* RTECS EV1929500.

*Synonyms:* Kadmu tlenek (Polish).

2. *Physical data.*

*Boiling point (760 mm Hg):* 950 degrees C decomposes.

*Melting point:* 1500 °C.

*Specific Gravity:* ($H_2O$=1@20°C): 7.0.

*Solubility:* Insoluble in water; soluble in acids and alkalines.

*Appearance:* Red or brown crystals.

B. *Fire, Explosion and Reactivity Data.*

1. *Fire.*

*Fire and Explosion Hazards:* Negligible fire hazard when exposed to heat or flame.

*Flash point:* Nonflammable.

*Extinguishing media:* Dry chemical, carbon dioxide, water spray or foam.

2. *Reactivity.*

*Conditions contributing to instability:* Stable under normal temperatures and pressures.

*Incompatibilities:* Magnesium may reduce $CdO_2$ explosively on heating.

*Hazardous decomposition products:* Toxic fumes of cadmium.

C. *Spill Leak and Disposal Procedures.*

1. *Steps to be taken if the material is released or spilled.* Do not touch spilled material. Stop leak if you can do it without risk. For small spills, take up with sand or other absorbent material and place into containers for later disposal. For small dry spills, use a clean shovel to place material into clean, dry container and then cover. Move containers from spill area. For larger spills, dike far ahead of spill for later disposal. Keep unnecessary people away. Isolate hazard area and deny entry. The Superfund Amendments and Reauthorization Act of 1986 Section 304 requires that a release equal to or greater than the reportable quantity for this substance (1 pound) must be immediately reported to the local emergency planning com-

mittee, the state emergency response commission, and the National Response Center (800) 424-8802; in Washington, DC metropolitan area (202) 426-2675.

III. Cadmium Sulfide.

A. *Physical and Chemical Data.*

1. *Substance Identification.*

*Chemical name:* Cadmium sulfide.

*Formula:* CdS.

*Molecular weight:* 144.5.

*CAS No.* 1306-23-6.

*Other Identifiers:* RTECS EV3150000.

*Synonyms:* Aurora yellow; Cadmium Golden 366; Cadmium Lemon Yellow 527; Cadmium Orange; Cadmium Primrose 819; Cadmium Sulphide; Cadmium Yellow; Cadmium Yellow 000; Cadmium Yellow Conc. Deep; Cadmium Yellow Conc. Golden; Cadmium Yellow Conc. Lemon; Cadmium Yellow Conc. Primrose; Cadmium Yellow Oz. Dark; Cadmium Yellow Primrose 47-1400; Cadmium Yellow 10G Conc.; Cadmium Yellow 892; Cadmopur Golden Yellow N; Cadmopur Yellow: Capsebon; C.I. 77199; C.I. Pigment Orange 20; CI Pigment Yellow 37; Ferro Lemon Yellow; Ferro Orange Yellow; Ferro Yellow; Greenockite; NCI-C02711.

2. *Physical data.*

*Boiling point (760 mm. Hg):* sublines in $N_2$ at 980 °C.

*Melting point:* 1750 degrees C (100 atm).

*Specific Gravity:* ($H_2O$=1@ 20°C): 4.82.

*Solubility:* Slightly soluble in water; soluble in acid.

*Appearance:* Light yellow or yellow-orange crystals.

B. *Fire, Explosion and Reactivity Data.*

1. *Fire.*

*Fire and Explosion Hazards:* Neglible fire hazard when exposed to heat or flame.

*Flash point:* Nonflammable.

*Extinguishing media:* Dry chemical, carbon dioxide, water spray or foam.

2. *Reactivity.*

*Conditions contributing to instability:* Generally non-reactive under normal conditions. Reacts with acids to form toxic hydrogen sulfide gas.

*Incompatibilities:* Reacts vigorously with iodinemonochloride.

*Hazardous decomposition products:* Toxic fumes of cadmium and sulfur oxides.

C. *Spill Leak and Disposal Procedures.*

1. *Steps to be taken if the material is released or spilled.* Do not touch spilled material. Stop leak if you can do it without risk. For small, dry spills, with a clean shovel place material into clean, dry container and cover. Move containers from spill area. For larger spills, dike far ahead of spill for later disposal. Keep unnecessary people away. Isolate hazard and deny entry.

IV. *Cadmium Chloride.*

A. *Physical and Chemical Data.*

1. *Substance Identification.*

*Chemcail name:* Cadmium chloride.

*Formula:* $CdCl_2$.

*Molecular weight:* 183.3.

*CAS No.* 10108-64-2.

*Other Identifiers:* RTECS EY0175000.

*Synonyms:* Caddy; Cadmium dichloride; NA 2570 (DOT); UI-CAD; dichlorocadmium.

2. *Physical data.*

*Boiling point (760 mm Hg):* 960 degrees C.

*Melting point:* 568 degrees C.

*Specific Gravity:* ($H_2O$=1 @ 20 °C): 4.05.

*Solubility:* Soluble in water (140 g/100 cc); soluble in acetone.

*Appearance:* Small, white crystals.

B. *Fire, Explosion and Reactivity Data.*

1. *Fire.*

*Fire and Explosion Hazards:* Negligible fire and negligible explosion hazard in dust form when exposed to heat or flame.

*Flash point:* Nonflamable.

*Extinguishing media:* Dry chemical, carbon dioxide, water spray or foam.

2. *Reactivity.*

*Conditions contributing to instability:* Generally stable under normal temperatures and pressures.

*Incompatibilities:* Bromine triflouride rapidly attacks cadmium chloride. A mixture of potassium and cadmium chloride may produce a strong explosion on impact.

*Hazardous decomposition products:* Thermal ecompostion may release toxic fumes of hydrogen chloride, chloride, chlorine or oxides of cadmium.

C. *Spill Leak and Disposal Procedures.*

1. *Steps to be taken if the materials is released or spilled.* Do not touch spilled material. Stop leak if you can do it without risk. For small, dry spills, with a clean shovel place material into clean, dry container and cover. Move containers from spill area. For larger spills, dike far ahead of spill for later disposal. Keep unnecessary people away. Isolate hazard and deny entry. The Superfund Amendments and Reauthorization Act of 1986 Section 304 requires that a release equal to or greater than the reportable quantity for this substance (100 pounds) must be immediately reported to the local emergency planning committee, the state emergency response commission, and the National Response Center (800) 424-8802; in Washington, DC Metropolitan area (202) 426-2675.

## Appendix C to §1926.1127 [Reserved]

## Appendix C to §1910.1027—Qualitative and Quantitative Fit Testing Procedures [Reserved]

[§1910.1027, App. C, removed and reserved at 63 FR 1288, Jan. 8, 1998, eff. Apr. 8, 1998.]

## Appendix D to §1926.1127—Occupational Health History Interview With Reference to Cadmium Exposure

**Note:** The requirements applicable to construction work under this Appendix D are identical to those set forth in Appendix D to §1910.1027 of this chapter.

[Appendix D to Section 1910.1027 is printed below—CCH.]

# APPENDIX D TO §1910.1027— OCCUPATIONAL HEALTH HISTORY INTERVIEW WITH REFERENCE TO CADMIUM EXPOSURE

## OCCUPATIONAL HEALTH HISTORY INTERVIEW WITH REFERENCE TO CADMIUM EXPOSURE

*Directions*

(To be read by employee and signed prior to the interview)

Please answer the questions you will be asked as completely and carefully as you can. These questions are asked of everyone who works with cadmium. You will also be asked to give blood and urine samples. The doctor will give your employer a written opinion on whether you are physically capable of working with cadmium. Legally, the doctor cannot share personal information you may tell him/her with your employer. The following information is considered strictly confidential. The results of the tests will go to you, your doctor and your employer. You will also receive an information sheet explaining the results of any biological monitoring or physical examinations performed.

If you are just being hired, the results of this interview and examination will be used to:

(1) Establish your health status and see if working with cadmium might be expected to cause unusual problems,

(2) Determine your health status today and see if there are changes over time,

(3) See if you can wear a respirator safely.

If you are not a new hire:

OSHA says that everyone who works with cadmium can have periodic medical examinations performed by a doctor. The reasons for this are:

(a) If there are changes in your health, either because of cadmium or some other reason, to find them early,

(b) to prevent kidney damage.

Please sign below.

I have read these directions and understand them:

______________________________

Employee signature

______________________________

Date

Thank you for answering these questions.

(Suggested Format)

Name ______________________ Age ______________________

Social Security # ______________________ Company ______________________

Job ______________________

Type of Preplacement Exam: ❑ Periodic ❑ Termination ❑ Initial ❑ Other

Blood Pressure ______________________ Pulse Rate ______________________

1. How long have you worked at the job listed above? ❑ Not yet hired ❑ Number of months ❑ Number of years
2. Job Duties etc.

______________________________

______________________________

______________________________

3. Have you *ever* been told by a doctor that you had bronchitis? ❑ Yes ❑ No
   If yes, how long ago? ❑ Number of months ❑ Number of years
4. Have you *ever* been told by a doctor that you had emphysema? ❑ Yes ❑ No
   If yes, how long ago? ❑ Number of years ❑ Number of months
5. Have you ever been told by a doctor that you had other lung problems? ❑ Yes ❑ No
   If yes, please describe type of lung problems and when you had these problems

______________________________

6. In the past year, have you had a cough? ❑ Yes ❑ No
   If yes, did you cough up sputum? ❑ Yes ❑ No
   If yes, how long did the cough with sputum production last? ❑ Less than 3 months ❑ 3 months or longer
   If yes, for how many years have you had episodes of cough with sputum production lasting this long?
   ❑ Less than one ❑ 1 ❑ 2 ❑ Longer than 2
7. Have you ever smoked cigarettes? ❑ Yes ❑ No
8. Do you now smoke cigarettes? ❑ Yes ❑ No
9. If you smoke or have smoked cigarettes, for how many years have you smoked, or did you smoke?
   ❑ Less than 1 year Number of years ________
   What is or was the greatest number of packs per day that you have smoked? Number of packs ________
   If you quit smoking cigarettes, how many years ago did you quit? ❑ Less than 1 year Number of years ________
   How many packs a day do you now smoke? Number of packs per day ________
10. Have you ever been told by a doctor that you had a kidney or urinary tract disease or disorder? ❑ Yes ❑ No
11. Have you ever had any of these disorders?
    Kidney stones ❑ Yes ❑ No Protein in urine ❑ Yes ❑ No Blood in urine ❑ Yes ❑ No
    Difficulty urinating ❑ Yes ❑ No Other kidney/Urinary disorders ❑ Yes ❑ No
    Please describe problems, age, treatment, and follow up for any kidney or urinary problems you have had:

______________________________

______________________________

______________________________

12. Have you ever been told by a doctor or other health care provider who took your blood pressure that your blood pressure was high? ❑ Yes ❑ No
13. Have you ever been advised to take any blood pressure medication? ❑ Yes ❑ No
14. Are you presently taking any blood pressure medication? ❑ Yes ❑ No
15. Are you presently taking any other medication? ❑ Yes ❑ No
16. Please list any blood pressure or other medications and describe how long you have been taking each one:
    Medicine:

______________________________

______________________________

______________________________

______________________________

How Long Taken

______________________________

______________________________

______________________________

17. Have you ever been told by a doctor that you have diabetes? (sugar in your blood or urine) ❑ Yes ❑ No
    If yes, do you presently see a doctor about your diabetes? ❑ Yes ❑ No
    If yes, how do you control your blood sugar? ❑ Diet alone ❑ Diet plus oral medicine ❑ Diet plus insulin (injection)
18. Have you ever been told by a doctor that you had:
    Anemia ❑ Yes ❑ No A low blood count? ❑ Yes ❑ No
19. Do you presently feel that you tire or run out of energy sooner than normal or sooner than other people your age?
    ❑ Yes ❑ No
    If yes, for how long have you felt that you tire easily? ❑ Less than 1 year ❑ Number of years ________
20. Have you given blood within the last year? ❑ Yes ❑ No If yes, how many times? ❑ Number of times ________
    How long ago was the last time you gave blood? ❑ Less than 1 month ❑ Number of months ________
21. Within the last year have you had any injuries with heavy bleeding? ❑ Yes ❑ No
    If yes, how long ago? ❑ Less than 1 month ❑ Number of months ________
    Describe:

______________________________

______________________________

______________________________

22. Have you recently had any surgery? ❑ Yes ❑ No
    If yes, please describe:

______________________________

______________________________

______________________________

23. Have you seen any blood lately in your stool or after a bowel movement? ❑ Yes ❑ No
24. Have you ever had a test for blood in your stool? ❑ Yes ❑ No
    If yes, did the test show any blood in the stool? ❑ Yes ❑ No
    What further evaluation and treatment were done?

______________________________

______________________________

The following questions pertain to the ability to wear a respirator. Additional information for the physician can be found in The Respiratory Protective Devices Manual.

25. Have you ever been told by a doctor that you have asthma? ❑ Yes ❑ No
If yes, are you presently taking any medication for asthma? Mark all that apply. ❑ Shots ❑ Pills ❑ Inhaler
26. Have you ever had a heart attack? ❑ Yes ❑ No
If yes, how long ago? Number of years ______ Number of months ______
27. Have you ever had pains in your chest? ❑ Yes ❑ No
If yes, when did it usually happen? ❑ While resting ❑ While working ❑ While exercising ❑ Activity didn't matter
28. Have you ever had a thyroid problem? ❑ Yes ❑ No
29. Have you ever had a seizure or fits? ❑ Yes ❑ No
30. Have you ever had a stroke (cerebrovascular accident)? ❑ Yes ❑ No
31. Have you ever had a ruptured eardrum or a serious hearing problem? ❑ Yes ❑ No
32. Do you now have a claustrophobia, meaning fear of crowded or closed in spaces or any psychological problems that would make it hard for you to wear a respirator? ❑ Yes ❑ No

The following questions pertain to reproductive history.

33. Have you or your partner had a problem conceiving a child? ❑ Yes ❑ No
If yes, specify: ❑ Self ❑ Present mate ❑ Previous mate
34. Have you or your partner consulted a physician for a fertility or other reproductive problem? ❑ Yes ❑ No
If yes, specify who consulted the physician: ❑ Self ❑ Spouse/partner ❑ Self and partner
If yes, specify diagnosis made:
______
______
35. Have you or your partner ever conceived a child resulting in a miscarriage, still birth or deformed offspring?
❑ Yes ❑ No
If yes, specify: ❑ Miscarriage ❑ Still birth ❑ Deformed offspring
If outcome was a deformed offspring, please specify type:
______
36. Was this outcome a result of a pregnancy of: ❑ Yours with present partner ❑ Yours with a previous partner
37. Did the timing of any abnormal pregnancy outcome coincide with present employment? ❑ Yes ❑ No
List dates of occurrences:
______
______
38. What is the occupation of your spouse or partner?
______

**For Women Only**

39. Do you have menstrual periods? ❑ Yes ❑ No Have you had menstrual irregularities? ❑ Yes ❑ No
If yes, specify type:
______
If yes, what was the approximated date this problem began? ______
Approximate date problem stopped? ______

**For Men Only**

40. Have you ever been diagnosed by a physician as having prostate gland problem(s)? ❑ Yes ❑ No
If yes, please describe type of problem(s) and what was done to evaluate and treat the problem(s):
______
______
______
______
______

## Appendix E to §1926.1127—Cadmium in Workplace Atmospheres

**Note:** The requirements applicable to construction work under this Appendix E are identical to those set forth in Appendix E to §1910.1027 of this chapter.

[Appendix E to Section 1910.1027 is printed below—CCH.]

## APPENDIX E TO §1910.1027—CADMIUM IN WORKPLACE ATMOSPHERES

Method Number: ID-189

Matrix: Air

OSHA Permissible Exposure Limits: 5 μg/m$^3$ (TWA), 2.5 μg/m$^3$ (Action Level TWA)

Collection Procedure: A known volume of air is drawn through a 37-mm diameter filter cassette containing a 0.8-m mixed cellulose ester membrane filter (MCEF).

Recommended Air Volume: 960 L

Recommended Sampling Rate: 2.0 L/min

Analytical Procedure: Air filter samples are digested with nitric acid. After digestion, a small amount of hydrochloric acid is added. The samples are then diluted to volume with deionized water and analyzed by either flame atomic absorption spectroscopy (AAS) or flameless atomic absorption spectroscopy using a heated graphite furnace atomizer (AAS-HGA).

Detection Limits:

Qualitative: 0.2 μg/m$^3$ for a 200 L sample by Flame AAS, 0.007 μg/m$^3$ for a 60 L sample by AAS-HGA

Quantitative: 0.70 μg/m$^3$ for a 200 L sample by Flame AAS, 0.025 μg/m$^3$ for a 60 L sample by AAS-HGA

Precision and Accuracy: (Flame AAS Analysis and AAS-HGA Analysis):

Validation Level: 2.5 to 10 μg/m$^3$ for a 400 L air vol, 1.25 to 5.0 μg/m$^3$ for a 60 L air vol

$CV_1$ (pooled): 0.010, 0.043

Analytical Bias: +4.0%, -5.8%

Overall Analytical Error: ±6.0%, ±14.2%

Method Classification: Validated

Date: June, 1992

Inorganic Service Branch II, OSHA Salt Lake Technical Center, Salt Lake City, Utah

Commercial manufacturers and products mentioned in this method are for descriptive use only and do not constitute endorsements by USDOL-OSHA. Similar products from other sources can be substituted.

### 1. INTRODUCTION

#### 1.1. Scope

This method describes the collection of airborne elemental cadmium and cadmium compounds on 0.8-μm mixed cellulose ester membrane filters and their subsequent analysis by either flame atomic absorption spectroscopy (AAS) or flameless atomic absorption spectroscopy using a heated graphite furnace atomizer (AAS-HGA). It is applicable for both TWA and Action Level TWA Permissible Exposure Level (PEL) measurements. The two atomic absorption analytical techniques included in the method do not differentiate between cadmium fume and cadmium dust samples. They also do not differentiate between elemental cadmium and its compounds.

#### 1.2. Principle

Airborne elemental cadmium and cadmium compounds are collected on a 0.8-μm mixed cellulose ester membrane filter (MCEF). The air filter samples are digested with concentrated nitric acid to destroy the organic matrix and dissolve the cadmium analytes. After digestion, a small amount of concentrated hydrochloric acid is added to help dissolve other metals which may be present. The samples are diluted to volume with deionized water and then aspirated into the oxidizing air/acetylene flame of an atomic absorption spectrophotometer for analysis of elemental cadmium.

If the concentration of cadmium in a sample solution is too low for quantitation by this flame AAS analytical technique, and the sample is to be averaged with other samples for TWA calculations, aliquots of the sample and a matrix modifier are later injected onto a L'vov platform in a pyrolytically-coated graphite tube of a Zeeman atomic absorption spectrophotometer/graphite furnace assembly for analysis of elemental cadmium. The matrix modifier is added to stabilize the cadmium metal and minimize sodium chloride as an interference during the high temperature charring step of the analysis (5.1., 5.2.).

### 1.3. History

Previously, two OSHA sampling and analytical methods for cadmium were used concurrently (5.3., 5.4.). Both of these methods also required 0.8-μm mixed cellulose ester membrane filters for the collection of air samples. These cadmium air filter samples were analyzed by either flame atomic absorption spectroscopy (5.3.) or inductively coupled plasma/atomic emission spectroscopy (ICP-AES) (5.4.). Neither of these two analytical methods have adequate sensitivity for measuring workplace exposure to airborne cadmium at the new lower TWA and Action Level TWA PEL levels when consecutive samples are taken on one employee and the sample results need to be averaged with other samples to determine a single TWA.

The inclusion of two atomic absorption analytical techniques in the new sampling and analysis method for airborne cadmium permits quantitation of sample results over a broad range of exposure levels and sampling periods. The flame AAS analytical technique included in this method is similar to the previous procedure given in the General Metals Method ID-121 (5.3.) with some modifications. The sensitivity of the AAS-HGA analytical technique included in this method is adequate to measure exposure levels at 1/10 the Action Level TWA, or lower, when less than full-shift samples need to be averaged together.

### 1.4. Properties (5.5.)

Elemental cadmium is a silver-white, blue-tinged, lustrous metal which is easily cut with a knife. It is slowly oxidized by moist air to form cadmium oxide. It is insoluble in water, but reacts readily with dilute nitric acid. Some of the physical properties and other descriptive information of elemental cadmium are given below:

|  |  |
|---|---|
| CAS No. | 7440-43-9 |
| Atomic Number | 48 |
| Atomic Symbol | Cd |
| Atomic Weight | 112.41 |
| Melting Point | 321° C |
| Boiling Point | 765° C |
| Density | 8.65 g/mL (25° C) |

The properties of specific cadmium compounds are described in reference 5.5.

### 1.5. Method Performance

A synopsis of method performance is presented below. Further information can be found in Section 4.

1.5.1. The qualitative and quantitative detection limits for the flame AAS analytical technique are 0.04 μg (0.004 μg/mL) and 0.14 μg (0.014 μg/mL) cadmium, respectively, for a 10 mL solution volume. These correspond, respectively, to 0.2 $\mu g/m^3$ and 0.70 $\mu g/m^3$ for a 200 L air volume.

1.5.2. The qualitative and quantitative detection limits for the AAS-HGA analytical technique are 0.44 ng (0.044 ng/mL) and 1.5ng (0.15 ng/mL) cadmium, respectively, for a 10 mL solution volume. These correspond, respectively, to 0.007 $\mu g/m^3$ and 0.025 $\mu g/m^3$ for a 60 L air volume.

1.5.3. The average recovery by the flame AAS analytical technique of 17 spiked MCEF samples containing cadmium in the range of 0.5 to 2.0 times the TWA target concentration of 5 $\mu g/m^3$ (assuming a 400 L air volume) was 104.0% with a pooled coefficient of variation (CV1) of 0.010. The flame analytical technique exhibited a positive bias of +4.0% for the validated concentration range. The overall analytical error (OAE) for the flame AAS analytical technique was ±6.0%.

1.5.4. The average recovery by the AAS-HGA analytical technique of 18 spiked MCEF samples containing cadmium in the range of 0.5 to 2.0 times the Action Level TWA target concentration of 2.5 $\mu g/m^3$ (assuming a 60 L air volume) was 94.2% with a pooled coefficient of variation (CV1) of 0.043. The AAS-HGA analytical technique exhibited a negative bias of -5.8% for the validated concentration range. The overall analytical error (OAE) for the AAS-HGA analytical technique was ±14.2%.

1.5.5. Sensitivity in flame atomic absorption is defined as the characteristic concentration of an element required to produce a signal of 1% absorbance (0.0044 absorbance units). Sensitivity values are listed for each element by the atomic absorption spectrophotometer manufacturer and have proved to be a very valuable diagnostic tool to determine if instrumental parameters are optimized and if the instrument is performing up to specification. The sensitivity of the spectrophotometer used in the validation of the flame AAS analytical technique agreed with the manufacturer specifications (5.6.); the 2 μg/mL cadmium standard gave an absorbance reading of 0.350 abs. units.

1.5.6. Sensitivity in graphite furnace atomic absorption is defined in terms of the characteristic mass, the number of picograms required to give an integrated absorbance value of 0.0044 absorbance-second (5.7.). Data suggests that under Stabilized Temperature Platform Furnace (STPF) conditions (see Section 1.6.2.), characteristic mass values are transferable between properly functioning instruments to an accuracy of about 20% (5.2.). The characteristic mass for STPF analysis of cadmium with Zeeman background correction listed by the manufacturer of the instrument used in the validation of the AAS-HGA analytical technique was 0.35 pg. The experimental characteristic mass value observed during the determination of the working range and detection limits of the AAS-HGA analytical technique was 0.41 pg.

### 1.6. Interferences

1.6.1. High concentrations of silicate interfere in determining cadmium by flame AAS (5.6.). However, silicates are not significantly soluble in the acid matrix used to prepare the samples.

1.6.2. Interferences, such as background absorption, are reduced to a minimum in the AAS-HGA analytical technique by taking full advantage of the Stabilized Temperature Platform Furnace (STPF) concept. STPF includes all of the following parameters (5.2.):

a. Integrated Absorbance,

b. Fast Instrument Electronics and Sampling Frequency,

c. Background Correction,

d. Maximum Power Heating,

e. Atomization off the L'vov platform in a pyrolytically coated graphite tube,

f. Gas Stop during Atomization,

g. Use of Matrix Modifiers.

### 1.7. Toxicology (5.14.)

Information listed within this section is synopsis of current knowledge of the physiological effects of cadmium and is not intended to be used as the basis for OSHA policy. IARC classifies cadmium and certain of its compounds as Group 2A carcinogens (probably carcinogenic to humans). Cadmium fume is intensely irritating to the respiratory tract. Workplace exposure to cadmium can cause both chronic and acute effects. Acute effects include tracheobronchitis, pneumonitis, and pulmonary edema. Chronic effects include anemia, rhinitis/anosmia, pulmonary emphysema, proteinuria and lung cancer. The primary target organs for chronic disease are the kidneys (non-carcinogenic) and the lungs (carcinogenic).

## 2. SAMPLING

### 2.1. Apparatus

2.1.1. Filter cassette unit for air sampling: A 37-mm diameter mixed cellulose ester membrane filter with a pore size of 0.8-μm contained in a 37-mm polystyrene two- or three-piece cassette filter holder (part no. MAWP 037 A0, Millipore Corp., Bedford, MA). The filter is supported with a cellulose backup pad. The cassette is sealed prior to use with a shrinkable gel band.

2.1.2. A calibrated personal sampling pump whose flow is determined to an accuracy of ±5% at the recommended flow rate with the filter cassette unit in line.

### 2.2. Procedure

2.2.1. Attach the prepared cassette to the calibrated sampling pump (the backup pad should face the pump) using flexible tubing. Place the sampling device on the employee such that air is sampled from the breathing zone.

2.2.2. Collect air samples at a flow rate of 2.0 L/min. If the filter does not become overloaded, a full-shift (at least seven hours) sample is strongly recommended for TWA and Action Level TWA measurements with a maximum air volume of 960 L. If overloading occurs, collect consecutive air samples for shorter sampling periods to cover the full workshift.

2.2.3. Replace the end plugs into the filter cassettes immediately after sampling. Record the sampling conditions.

2.2.4. Securely wrap each sample filter cassette end-to-end with an OSHA Form 21 sample seal.

2.2.5. Submit at least one blank sample with each set of air samples. The blank sample should be handled the same as the other samples except that no air is drawn through it.

2.2.6. Ship the samples to the laboratory for analysis as soon as possible in a suitable container designed to prevent damage in transit.

## 3. ANALYSIS

### 3.1. Safety Precautions

3.1.1. Wear safety glasses, protective clothing and gloves at all times.

3.1.2. Handle acid solutions with care. Handle all cadmium samples and solutions with extra care (see Sect. 1.7.). Avoid their direct contact with work area surfaces, eyes, skin and clothes. Flush acid solutions which contact the skin or eyes with copious amounts of water.

3.1.3. Perform all acid digestions and acid dilutions in an exhaust hood while wearing a face shield. To avoid exposure to acid vapors, do not remove beakers containing concentrated acid solutions from the exhaust hood until they have returned to room temperature and have been diluted or emptied.

3.1.4. Exercise care when using laboratory glassware. Do not use chipped pipets, volumetric flasks, beakers or any glassware with sharp edges exposed in order to avoid the possibility of cuts or abrasions.

3.1.5. Never pipet by mouth.

3.1.6. Refer to the instrument instruction manuals and SOPs (5.8., 5.9.) for proper and safe operation of the atomic absorption spectrophotometer, graphite furnace atomizer and associated equipment.

3.1.7. Because metallic elements and other toxic substances are vaporized during AAS flame or graphite furnace atomizer operation, it is imperative that an exhaust vent be used. Always ensure that the exhaust system is operating properly during instrument use.

### 3.2. Apparatus for Sample and Standard Preparation

3.2.1. Hot plate, capable of reaching 150 °C, installed in an exhaust hood.

3.2.2. Phillips beakers, 125 mL.

3.2.3. Bottles, narrow-mouth, polyethylene or glass with leakproof caps: used for storage of standards and matrix modifier.

3.2.4. Volumetric flasks, volumetric pipets, beakers and other associated general laboratory glassware.

3.2.5. Forceps and other associated general laboratory equipment.

### 3.3. Apparatus for Flame AAS Analysis

3.3.1. Atomic absorption spectrophotometer consisting of a(an):

Nebulizer and burner head

Pressure regulating devices capable of maintaining constant oxidant and fuel pressures

Optical system capable of isolating the desired wavelength of radiation (228.8 nm)

Adjustable slit

Light measuring and amplifying device

Display, strip chart, or computer interface for indicating the amount of absorbed radiation

Cadmium hollow cathode lamp or electrodeless discharge lamp (EDL) and power supply

3.3.2. Oxidant: compressed air, filtered to remove water, oil and other foreign substances.

3.3.3. Fuel: standard commercially available tanks of acetylene dissolved in acetone; tanks should be equipped with flash arresters.

CAUTION: Do not use grades of acetylene containing solvents other than acetone because they may damage the PVC tubing used in some instruments.

3.3.4. Pressure-reducing valves: two gauge, two-stage pressure regulators to maintain fuel and oxidant pressures somewhat higher than the controlled operating pressures of the instrument.

3.3.5. Exhaust vent installed directly above the spectrophotometer burner head.

### 3.4. Apparatus for AAS-HGA Analysis

3.4.1. Atomic absorption spectrophotometer consisting of a(an):

Heated graphite furnace atomizer (HGA) with argon purge system

Pressure-regulating devices capable of maintaining constant argon purge pressure

Optical system capable of isolating the desired wavelength of radiation (228.8 nm)

Adjustable slit

Light measuring and amplifying device

Display, strip chart, or computer interface for indicating the amount of absorbed radiation (as integrated absorbance, peak area)

Background corrector: Zeeman or deuterium arc. The Zeeman background corrector is recommended

Cadmium hollow cathode lamp or electrodeless discharge lamp (EDL) and power supply

Autosampler capable of accurately injecting 5 to 20 µL sample aliquots onto the L'vov Platform in a graphite tube

3.4.2. Pyrolytically coated graphite tubes containing solid, pyrolytic L'vov platforms.

3.4.3. Polyethylene sample cups, 2.0 to 2.5 mL, for use with the autosampler.

3.4.4. Inert purge gas for graphite furnace atomizer: compressed gas cylinder of purified argon.

3.4.5. Two gauge, two-stage pressure regulator for the argon gas cylinder.

3.4.6. Cooling water supply for graphite furnace atomizer.

3.4.7. Exhaust vent installed directly above the graphite furnace atomizer.

### 3.5. Reagents

All reagents should be ACS analytical reagent grade or better.

3.5.1. Deionized water with a specific conductance of less than 10 µS.

3.5.2. Concentrated nitric acid, $HNO_3$.

3.5.3. Concentrated hydrochloric acid, HCl.

3.5.4. Ammonium phosphate, monobasic, $NH_4\ H_2\ PO_4$.

3.5.5. Magnesium nitrate, $Mg(NO_3)(NO_3)_2 \bullet 6H_2O$.

3.5.6. Diluting solution (4% $HNO_3$, 0.4% HCl): Add 40 mL $HNO_3$ and 4 mL HCl carefully to approximately 500 mL deionized water and dilute to 1 L with deionized water.

3.5.7. Cadmium standard stock solution, 1,000 µg/mL: Use a commercially available certified 1,000 µg/mL cadmium standard or, alternatively, dissolve 1.0000 g of cadmium metal in a minimum volume of 1:1 HCl and dilute to 1 L with 4% $HNO_3$. Observe expiration dates of commercial standards. Properly dispose of commercial standards with no expiration dates or prepared standards one year after their receipt or preparation date.

3.5.8. Matrix modifier for AAS-HGA analysis: Dissolve 1.0 g $NH_4H_2PO_4$ and 0.15 g $Mg(NO_3)_2 \bullet 6H_2O$ in approximately 200 mL deionized water. Add 1 mL $HNO_3$ and dilute to 500 mL with deionized water.

3.5.9. Nitric Acid, 1:1 $HNO_3$/DI $H_2O$ mixture: Carefully add a measured volume of concentrated $HNO_3$ to an equal volume of DI $H_2O$.

3.5.10. Nitric acid, 10% v/v: Carefully add 100 mL of concentrated $HNO_3$ to 500 mL of DI $H_2O$ and dilute to 1 L.

### 3.6. Glassware Preparation

3.6.1. Clean Phillips beakers by refluxing with 1:1 nitric acid on a hot plate in a fume hood. Thoroughly rinse with deionized water and invert the beakers to allow them to drain dry.

3.6.2. Rinse volumetric flasks and all other glassware with 10% nitric acid and deionized water prior to use.

### 3.7. Standard Preparation for Flame AAS Analysis

3.7.1. Dilute stock solutions: Prepare 1, 5, 10 and 100 μg/mL cadmium standard stock solutions by making appropriate serial dilutions of 1,000 μg/mL cadmium standard stock solution with the diluting solution described in Section 3.5.6.

3.7.2. Working standards: Prepare cadmium working standards in the range of 0.02 to 2.0 μg/mL by making appropriate serial dilutions of the dilute stock solutions with the same diluting solution. A suggested method of preparation of the working standards is given below.

| Working standard (μg/mL) | Std solution (μg/mL) | Aliquot (mL) | Final vol. (mL) |
|---|---|---|---|
| 0.02 | 1 | 10 | 500 |
| 0.05 | 5 | 5 | 500 |
| 0.1 | 10 | 5 | 500 |
| 0.2 | 10 | 10 | 500 |
| 0.5 | 10 | 25 | 500 |
| 1 | 100 | 5 | 500 |
| 2 | 100 | 10 | 500 |

Store the working standards in 500-mL, narrow-mouth polyethylene or glass bottles with leak proof caps. Prepare every twelve months.

### 3.8. Standard Preparation for AAS-HGA Analysis

3.8.1. Dilute stock solutions: Prepare 10, 100 and 1,000 ng/mL cadmium standard stock solutions by making appropriate ten-fold serial dilutions of the 1,000 μg/mL cadmium standard stock solution with the diluting solution described in Section 3.5.6.

3.8.2. Working standards: Prepare cadmium working standards in the range of 0.2 to 20 ng/mL by making appropriate serial dilutions of the dilute stock solutions with the same diluting solution. A suggested method of preparation of the working standards is given below.

| Working standard (ng/mL) | Std solution (ng/mL) | Aliquot (mL) | Final vol. (mL) |
|---|---|---|---|
| 0.2 | 10 | 2 | 100 |
| 0.5 | 10 | 5 | 100 |
| 1 | 10 | 10 | 100 |
| 2 | 100 | 2 | 100 |
| 5 | 100 | 5 | 100 |
| 10 | 100 | 10 | 100 |
| 20 | 1,000 | 2 | 100 |

Store the working standards in narrow-mouth polyethylene or glass bottles with leakproof caps. Prepare monthly.

### 3.9. Sample Preparation

3.9.1. Carefully transfer each sample filter with forceps from its filter cassette unit to a clean, separate 125-mL Phillips beaker along with any loose dust found in the cassette. Label each Phillips beaker with the appropriate sample number.

3.9.2. Digest the sample by adding 5 mL of concentrated nitric acid ($HNO_3$) to each Phillips beaker containing an air filter sample. Place the Phillips beakers on a hot plate in an exhaust hood and heat the samples until approximately 0.5 mL remains. The sample solution in each Phillips beaker should become clear. If it is not clear, digest the sample with another portion of concentrated nitric acid.

3.9.3. After completing the $HNO_3$ digestion and cooling the samples, add 40 μL (2 drops) of concentrated HCl to each air sample solution and then swirl the contents. Carefully add about 5 mL of deionized water by pouring it down the inside of each beaker.

3.9.4. Quantitatively transfer each cooled air sample solution from each Phillips beaker to a clean 10-mL volumetric flask. Dilute each flask to volume with deionized water and mix well.

### 3.10. Flame AAS Analysis

Analyze all of the air samples for their cadmium content by flame atomic absorption spectroscopy (AAS) according to the instructions given below.

3.10.1. Set up the atomic absorption spectrophotometer for the air/acetylene flame analysis of cadmium according to the SOP (5.8.) or the manufacturer's operational instructions. For the source lamp, use the cadmium hollow cathode or electrodeless discharge lamp operated at the manufacturer's recommended rating for continuous operation. Allow the lamp to warm up 10 to 20 min or until the energy output stabilizes. Optimize conditions such as lamp position, burner head alignment, fuel and oxidant flow rates, etc. See the SOP or specific instrument manuals for details. Instrumental parameters for the Perkin-Elmer Model 603 used in the validation of this method are given in Attachment 1.

3.10.2. Aspirate and measure the absorbance of a standard solution of cadmium. The standard concentration should be within the linear range. For the instrumentation used in the validation of this method a 2 μg/mL cadmium standard gives a net absorbance reading of about 0.350 abs. units (see Section 1.5.5.) when the instrument and the source lamp are performing to manufacturer specifications.

3.10.3. To increase instrument response, scale expand the absorbance reading of the aspirated 2 μg/mL working standard approximately four times. Increase the integration time to at least 3 seconds to reduce signal noise.

3.10.4. Autozero the instrument while aspirating a deionized water blank. Monitor the variation in the baseline absorbance reading (baseline noise) for a few minutes to insure that the instrument, source lamp and associated equipment are in good operating condition.

3.10.5. Aspirate the working standards and samples directly into the flame and record their absorbance readings. Aspirate the deionized water blank immediately after every standard or sample to correct for and monitor any baseline drift and noise. Record the baseline absorbance reading of each deionized water blank. Label each standard and sample reading and its accompanying baseline reading.

3.10.6. It is recommended that the entire series of working standards be analyzed at the beginning and end of the analysis of a set of samples to establish a concentration-response curve, ensure that the standard readings agree with each other and are reproducible. Also, analyze a working standard after every five or six samples to monitor the performance of the spectrophotometer. Standard readings should agree within ±10 to 15% of the readings obtained at the beginning of the analysis.

3.10.7. Bracket the sample readings with standards during the analysis. If the absorbance reading of a sample is above the absorbance reading of the highest working standard, dilute the sample with diluting solution and reanalyze. Use the appropriate dilution factor in the calculations.

3.10.8. Repeat the analysis of approximately 10% of the samples for a check of precision.

3.10.9. If possible, analyze quality control samples from an independent source as a check on analytical recovery and precision.

3.10.10. Record the final instrument settings at the end of the analysis. Date and label the output.

### 3.11. AAS-HGA Analysis

Initially analyze all of the air samples for their cadmium content by flame atomic absorption spectroscopy (AAS) according to the instructions given in Section 3.10. If the concentration of cadmium in a sample solution is less than three times the quantitative detection limit [0.04 μg/mL (40 ng/mL) for the instrumentation used in the validation] and the sample results are to be averaged with other samples for TWA calculations, proceed with the AAS-HGA analysis of the sample as described below.

3.11.1. Set up the atomic absorption spectrophotometer and HGA for flameless atomic absorption analysis of cadmium according to the SOP (5.9.) or the manufacturer's operational instructions and allow the instrument to stabilize. The graphite furnace atomizer is equipped with a pyrolytically coated graphite tube containing a pyrolytic platform. For the source lamp, use a cadmium hollow cathode or electrodeless discharge lamp operated at the manufacturer's recommended setting for graphite furnace operation. The Zeeman background corrector and EDL are recommended for use with the L'vov platform. Instrumental parameters for the Perkin-Elmer Model 5100 spectrophotometer and Zeeman HGA-600 graphite furnace used in the validation of this method are given in Attachment 2.

3.11.2. Optimize the energy reading of the spectrophotometer at 228.8 nm by adjusting the lamp position and the wavelength according to the manufacturer's instructions.

3.11.3. Set up the autosampler to inject a 5-μ L aliquot of the working standard, sample or reagent blank solution onto the L'vov platform along with a 10-μL overlay of the matrix modifier.

3.11.4. Analyze the reagent blank (diluting solution, Section 3.5.6.) and then autozero the instrument before starting the analysis of a set of samples. It is recommended that the reagent blank be analyzed several times during the analysis to assure the integrated absorbance (peak area) reading remains at or near zero.

3.11.5. Analyze a working standard approximately midway in the linear portion of the working standard range two or three times to check for reproducibility and sensitivity (see sections 1.5.5. and 1.5.6.) before starting the analysis of samples. Calculate the experimental characteristic mass value from the average integrated absorbance reading and injection volume of the analyzed working standard. Compare this value to the manufacturer's suggested value as a check of proper instrument operation.

3.11.6. Analyze the reagent blank, working standard, and sample solutions. Record and label the peak area (abs-sec) readings and the peak and background peak profiles on the printer/plotter.

3.11.7. It is recommended the entire series of working standards be analyzed at the beginning and end of the analysis of a set of samples. Establish a concentration-response curve and ensure standard readings agree with each other and are reproducible. Also, analyze a working standard after every five or six samples to monitor the performance of the system. Standard readings should agree within ±15% of the readings obtained at the beginning of the analysis.

3.11.8. Bracket the sample readings with standards during the analysis. If the peak area reading of a sample is above the peak area reading of the highest working standard, dilute the sample with the diluting solution and reanalyze. Use the appropriate dilution factor in the calculations.

3.11.9. Repeat the analysis of approximately 10% of the samples for a check of precision.

3.11.10. If possible, analyze quality control samples from an independent source as a check of analytical recovery and precision.

3.11.11. Record the final instrument settings at the end of the analysis. Date and label the output.

### 3.12. Calculations

**NOTE:** Standards used for HGA analysis are in ng/mL. Total amounts of cadmium from calculations will be in ng (not g) unless a prior conversion is made.

3.12.1. Correct for baseline drift and noise in flame AAS analysis by subtracting each baseline absorbance reading from its corresponding working standard or sample absorbance reading to obtain the net absorbance reading for each standard and sample.

3.12.2. Use a least squares regression program to plot a concentration-response curve of net absorbance reading (or peak area for HGA analysis) versus concentration (μg/mL or ng/mL) of cadmium in each working standard.

3.12.3. Determine the concentration (μg/mL or ng/mL) of cadmium in each sample from the resulting concentration-response curve. If the concentration of cadmium in a sample solution is less than three times the quantitative detection limit [0.04 μg/mL (40 ng/mL) for the instrumentation used in the validation of the method] and if consecutive samples were taken on one employee and the sample results are to be averaged with other samples to determine a single TWA, reanalyze the sample by AAS-HGA as described in Section 3.11. and report the AAS-HGA analytical results.

3.12.4. Calculate the total amount (μg or ng) of cadmium in each sample from the sample solution volume (mL):

$$\mathbf{W = (C)(sample\ vol,\ mL)(DF)}$$

Where:

W = Total cadmium in sample

C = Calculated concentration of cadmium

DF = Dilution Factor (if applicable)

3.12.5. Make a blank correction for each air sample by subtracting the total amount of cadmium in the corresponding blank sample from the total amount of cadmium in the sample.

3.12.6. Calculate the concentration of cadmium in an air sample (mg/m³ or μg/m³) by using one of the following equations:

$$\mathbf{mg/m^3 = W_{bc}/(Air\ vol\ sampled,\ L)}$$

or

$$\mathbf{\mu g/m^3 = (W_{bc})(1{,}000\ ng/\mu g)/(Air\ vol\ sampled,\ L)}$$

Where:

$W_{bc}$ = blank corrected total μg cadmium in the sample. (1μg = 1,000 ng)

## 4. BACKUP DATA

### 4.1. Introduction

4.1.1. The purpose of this evaluation is to determine the analytical method recovery, working standard range, and qualitative and quantitative detection limits of the two atomic absorption analytical techniques included in this method. The evaluation consisted of the following experiments:

1. An analysis of 24 samples (six samples each at 0.1, 0.5, 1 and 2 times the TWA-PEL) for the analytical method recovery study of the flame AAS analytical technique.

2. An analysis of 18 samples (six samples each at 0.5, 1 and 2 times the Action Level TWA-PEL) for the analytical method recovery study of the AAS-HGA analytical technique.

3. Multiple analyses of the reagent blank and a series of standard solutions to determine the working standard range and the qualitative and quantitative detection limits for both atomic absorption analytical techniques.

4.1.2. The analytical method recovery results at all test levels were calculated from concentration-response curves and statistically examined for outliers at the 99% confidence level. Possible outliers were determined using the Treatment of Outliers test (5.10.). In addition, the sample results of the two analytical techniques, at 0.5, 1.0 and 2.0 times their target concentrations, were tested for homogeneity of variances also at the 99% confidence level. Homogeneity of the coefficients of variation was determined using the Bartlett's test (5.11.). The overall analytical error (OAE) at the 95% confidence level was calculated using the equation (5.12.):

$$\mathbf{OAE = \pm[|Bias| + (1.96)(CV_1(pooled))(100\%)]}$$

4.1.3. A derivation of the International Union of Pure and Applied Chemistry (IUPAC) detection limit equation (5.13.) was used to determine the qualitative and quantitative detection limits for both atomic absorption analytical techniques:

$$\mathbf{C_{ld} = k(sd)/m\ (Equation\ 1)}$$

Where:

$C_{ld}$ = the smallest reliable detectable concentration an analytical instrument can determine at a given confidence level.

k = 3 for the Qualitative Detection Limit at the 99.86% Confidence Level

= 10 for the Quantitative Detection Limit at the 99.99% Confidence Level.

sd = standard deviation of the reagent blank (Rbl) readings.

m = analytical sensitivity or slope as calculated by linear regression.

4.1.4. Collection efficiencies of metallic fume and dust atmospheres on 0.8-μ m mixed cellulose ester membrane filters are well documented and have been shown to be excellent (5.11.). Since elemental cadmium and the cadmium component of cadmium compounds are nonvolatile, stability studies of cadmium spiked MCEF samples were not performed.

## 4.2. Equipment

4.2.1. A Perkin-Elmer (PE) Model 603 spectrophotometer equipped with a manual gas control system, a stainless steel nebulizer, a burner mixing chamber, a flow spoiler and a 10 cm. (one-slot) burner head was used in the experimental validation of the flame AAS analytical technique. A PE cadmium hollow cathode lamp, operated at the manufacturer's recommended current setting for continuous operation (4 mA), was used as the source lamp. Instrument parameters are listed in Attachment 1.

4.2.2. A PE Model 5100 spectrophotometer, Zeeman HGA-600 graphite furnace atomizer and AS-60 HGA autosampler were used in the experimental validation of the AAS-HGA analytical technique. The spectrophotometer was equipped with a PE Series 7700 professional computer and Model PR-310 printer. A PE System 2 cadmium electrodeless discharge lamp, operated at the manufacturer's recommended current setting for modulated operation (170 mA), was used as the source lamp. Instrument parameters are listed in Attachment 2.

## 4.3. Reagents

4.3.1. J.T. Baker Chem. Co. (Analyzed grade) concentrated nitric acid, 69.0-71.0%, and concentrated hydrochloric acid, 36.5-38.0%, were used to prepare the samples and standards.

4.3.2. Ammonium phosphate, monobasic, $NH_4 H_2 PO_4$ and magnesium nitrate, $Mg(NO_3)_2 6H_2O$, both manufactured by the Mallinckrodt Chem. Co., were used to prepare the matrix modifier for AAS-HGA analysis.

## 4.4. Standard Preparation for Flame AAS Analysis

4.4.1. Dilute stock solutions: Prepared 0.01, 0.1, 1, 10 and 100 μg/mL cadmium standard stock solutions by making appropriate serial dilutions of a commercially available 1,000 μg/mL cadmium standard stock solution (RICCA Chemical Co., Lot# A102) with the diluting solution (4% $HNO_3$, 0.4% HCl).

4.4.2. Analyzed Standards: Prepared cadmium standards in the range of 0.001 to 2.0 μg/mL by pipetting 2 to 10 mL of the appropriate dilute cadmium stock solution into a 100-mL volumetric flask and diluting to volume with the diluting solution. (See Section 3.7.2.)

## 4.5. Standard Preparation for AAS-HGA Analysis

4.5.1. Dilute stock solutions: Prepared 1, 10, 100 and 1,000 ng/mL cadmium standard stock solutions by making appropriate serial dilutions of a commercially available 1,000 μg/mL cadmium standard stock solution (J.T. Baker Chemical Co., Instra-analyzed, Lot# D22642) with the diluting solution (4% $HNO_3$, 0.4% HCl).

4.5.2. Analyzed Standards: Prepared cadmium standards in the range of 0.1 to 40 ng/mL by pipetting 2 to 10 mL of the appropriate dilute cadmium stock solution into a 100-mL volumetric flask and diluting to volume with the diluting solution. (See Section 3.8.2.)

## 4.6. Detection Limits and Standard Working Range for Flame AAS Analysis

4.6.1. Analyzed the reagent blank solution and the entire series of cadmium standards in the range of 0.001 to 2.0 μg/mL three to six times according to the instructions given in Section 3.10. The diluting solution (4% $HNO_3$, 0.4% HCl) was used as the reagent blank. The integration time on the PE 603 spectrophotometer was set to 3.0 seconds and a four-fold expansion of the absorbance reading of the 2.0 μg/mL cadmium standard was made prior to analysis. The 2.0 μg/mL standard gave a net absorbance reading of 0.350 abs. units prior to expansion in agreement with the manufacturer's specifications (5.6.).

4.6.2. The net absorbance readings of the reagent blank and the low concentration Cd standards from 0.001 to 0.1 μg/mL and the statistical analysis of the results are shown in Table I. The standard deviation, sd, of the six net absorbance readings of the reagent blank is 1.05 abs. units. The slope, m, as calculated by a linear regression plot of the net absorbance readings (shown in Table II) of the 0.02 to 1.0 μg/mL cadmium standards versus their concentration is 772.7 abs. units/(μg/mL).

4.6.3. If these values for sd and the slope, m, are used in Eqn. 1 (Sect. 4.1.3.), the qualitative and quantitative detection limits as determined by the IUPAC Method are:

$$C_{ld} = (3)(1.05 \text{ abs. units})/(772.7 \text{ abs. units}/(\mu g/mL))$$

$$= 0.0041\ \mu g/mL \text{ for the qualitative detection limit.}$$

$$C_{ld} = (10)(1.05 \text{ abs. units})/(772.7 \text{ abs. units}/\mu g/mL))$$

$$= 0.014\ \mu g/mL \text{ for the quantitative detection limit.}$$

The qualitative and quantitative detection limits for the flame AAS analytical technique are 0.041 μg and 0.14 μg cadmium, respectively, for a 10 mL solution volume. These correspond, respectively, to 0.2 μg/m$^3$ and 0.70 μg/m$^3$ for a 200 L air volume.

4.6.4. The recommended Cd standard working range for flame AAS analysis is 0.02 to 2.0 μg/mL. The net absorbance readings of the reagent blank and the recommended working range standards and the statistical analysis of the results are shown in Table II. The standard of lowest concentration in the working range, 0.02 μg/mL, is slightly greater than the calculated quantitative detection limit, 0.014 μg/mL. The standard of highest concentration in the working range, 2.0 μg/mL, is at the upper end of the linear working range suggested by the manufacturer (5.6.). Although the standard net absorbance readings are not strictly linear at concentrations above 0.5 μg/mL, the deviation from linearity is only about 10% at the upper end of the recommended standard working range. The deviation from linearity is probably caused by the four-fold expansion of the signal suggested in the method. As shown in Table II, the precision of the standard net absorbance readings are excellent throughout the recommended working range; the relative standard deviations of the readings range from 0.009 to 0.064.

## 4.7. Detection Limits and Standard Working Range for AAS-HGA Analysis

4.7.1. Analyzed the reagent blank solution and the entire series of cadmium standards in the range of 0.1 to 40 ng/mL according to the instructions given in Section 3.11. The diluting solution (4% $HNO_3$, 0.4% HCl) was used as the reagent blank. A fresh aliquot of the reagent blank and of each standard was used for every analysis. The experimental characteristic mass value was 0.41 pg, calculated from the average peak area (abs-sec) reading of the 5 ng/mL standard which is approximately midway in the linear portion of the working standard range. This agreed within 20% with the characteristic mass value, 0.35 pg, listed by the manufacturer of the instrument (5.2.).

4.7.2. The peak area (abs-sec) readings of the reagent blank and the low concentration Cd standards from 0.1 to 2.0 ng/mL and statistical analysis of the results are shown in Table III. Five of the reagent blank peak area readings were zero and the sixth reading was 1 and was an outlier. The near lack of a blank signal does not satisfy a strict interpretation of the IUPAC method for determining the detection limits. Therefore, the standard deviation of the six peak area readings of the 0.2 ng/mL cadmium standard, 0.75 abs-sec, was used to calculate the detection limits by the IUPAC method. The slope, m, as calculated by a linear regression plot of the peak area (abs-sec) readings (shown in Table IV) of the 0.2 to 10 ng/mL cadmium standards versus their concentration is 51.5 abs-sec/(ng/mL).

4.7.3. If 0.75 abs-sec (sd) and 51.5 abs-sec/(ng/mL) (m) are used in Eqn. 1 (Sect. 4.1.3.), the qualitative and quantitative detection limits as determined by the IUPAC method are:

$$C_{ld} = (3)(0.75 \text{ abs-sec})/(51.5 \text{ abs-sec})/(ng/mL)$$

$$= 0.044 \text{ ng/mL for the qualitative detection limit.}$$

$$C_{ld} = (10)(0.75 \text{ abs-sec})/(51.5 \text{ abs-sec})/(ng/mL) = 0.15 \text{ ng/mL for the quantitative detection limit.}$$

The qualitative and quantitative detection limits for the AAS-HGA analytical technique are 0.44 ng and 1.5 ng cadmium, respectively, for a 10 mL solution volume. These correspond, respectively, to 0.007 μg/m$^3$ and 0.025 μg/m$^3$ for a 60 L air volume.

4.7.4. The peak area (abs-sec) readings of the Cd standards from 0.2 to 40 ng/mL and the statistical analysis of the results are given in Table IV. The recommended standard working range for AAS-HGA analysis is 0.2 to 20 ng/mL. The standard of lowest concentration in the recommended working range is slightly greater than the calculated quantitative detection limit, 0.15 ng/mL. The deviation from linearity of the peak area readings of the 20 ng/mL standard, the highest concentration standard in the recommended working range, is approximately 10%. The deviations from linearity of the peak area readings of the 30 and 40 ng/mL standards are significantly greater than 10%. As shown in Table IV, the precision of the peak area readings are satisfactory throughout the recommended working range; the relative standard deviations of the readings range from 0.025 to 0.083.

### 4.8. Analytical Method Recovery for Flame AAS Analysis

4.8.1. Four sets of spiked MCEF samples were prepared by injecting 20 μL of 10, 50, 100 and 200 μg/mL dilute cadmium stock solutions on 37 mm diameter filters (part no. AAWP 037 00, Millipore Corp., Bedford, MA) with a calibrated micropipet. The dilute stock solutions were prepared by making appropriate serial dilutions of a commercially available 1,000 μg/mL cadmium standard stock solution (RICCA Chemical Co., Lot# A102) with the diluting solution (4% $HNO_3$, 0.4% HCl). Each set contained six samples and a sample blank. The amount of cadmium in the prepared sets were equivalent to 0.1, 0.5, 1.0 and 2.0 times the TWA PEL target concentration of 5 $\mu g/m^3$ for a 400 L air volume.

4.8.2. The air-dried spiked filters were digested and analyzed for their cadmium content by flame atomic absorption spectroscopy (AAS) following the procedure described in Section 3. The 0.02 to 2.0μg/mL cadmium standards (the suggested working range) were used in the analysis of the spiked filters.

4.8.3. The results of the analysis are given in Table V. One result at 0.5 times the TWA PEL target concentration was an outlier and was excluded from statistical analysis. Experimental justification for rejecting it is that the outlier value was probably due to a spiking error. The coefficients of variation for the three test levels at 0.5 to 2.0 times the TWA PEL target concentration passed the Bartlett's test and were pooled.

4.8.4. The average recovery of the six spiked filter samples at 0.1 times the TWA PEL target concentration was 118.2% with a coefficient of variation ($CV_1$) of 0.128. The average recovery of the spiked filter samples in the range of 0.5 to 2.0 times the TWA target concentration was 104.0% with a pooled coefficient of variation ($CV_1$) of 0.010. Consequently, the analytical bias found in these spiked sample results over the tested concentration range was +4.0% and the OAE was ±6.0%.

### 4.9. Analytical Method Recovery for AAS-HGA Analysis

4.9.1. Three sets of spiked MCEF samples were prepared by injecting 15μ L of 5, 10 and 20 μg/mL dilute cadmium stock solutions on 37 mm diameter filters (part no. AAWP 037 00, Millipore Corp., Bedford, MA) with a calibrated micropipet. The dilute stock solutions were prepared by making appropriate serial dilutions of a commercially available certified 1,000 μg/mL cadmium standard stock solution (Fisher Chemical Co., Lot# 913438-24) with the diluting solution (4% $HNO_3$,0.4% HCl). Each set contained six samples and a sample blank. The amount of cadmium in the prepared sets were equivalent to 0.5, 1 and 2 times the Action Level TWA target concentration of 2.5 $\mu g/m^3$ for a 60 L air volume.

4.9.2. The air-dried spiked filters were digested and analyzed for their cadmium content by flameless atomic absorption spectroscopy using a heated graphite furnace atomizer following the procedure described in Section 3. A five-fold dilution of the spiked filter samples at 2 times the Action Level TWA was made prior to their analysis. The 0.05 to 20 ng/mL cadmium standards were used in the analysis of the spiked filters.

4.9.3. The results of the analysis are given in Table VI. There were no outliers. The coefficients of variation for the three test levels at 0.5 to 2.0 times the Action Level TWA PEL passed the Bartlett's test and were pooled. The average recovery of the spiked filter samples was 94.2% with a pooled coefficient of variation ($CV_1$) of 0.043. Consequently, the analytical bias was -5.8% and the OAE was ±14.2%.

### 4.10. Conclusions

The experiments performed in this evaluation show the two atomic absorption analytical techniques included in this method to be precise and accurate and have sufficient sensitivity to measure airborne cadmium over a broad range of exposure levels and sampling periods.

## 5. REFERENCES

5.1. Slavin, W. Graphite Furnace AAS—A Source Book; Perkin-Elmer Corp., Spectroscopy Div.: Ridgefield, CT, 1984; p. 18 and pp. 83-90.

5.2. Grosser, Z., Ed.; Techniques in Graphite Furnace Atomic Absorption Spectrophotometry; Perkin-Elmer Corp., Spectroscopy Div.: Ridgefield, CT, 1985.

5.3. Occupational Safety and Health Administration Salt Lake Technical Center: Metal and Metalloid Particulate in Workplace Atmospheres (Atomic Absorption) (USDOL/OSHA Method No. ID-121). In OSHA Analytical Methods Manual 2nd ed. Cincinnati, OH: American Conference of Governmental Industrial Hygienists, 1991.

5.4. Occupational Safety and Health Administration Salt Lake Technical Center: Metal and Metalloid Particulate in Workplace Atmospheres (ICP) (USDOL/OSHA Method No. ID-125G). In OSHA Analytical Methods Manual 2nd ed. Cincinnati, OH: American Conference of Governmental Industrial Hygienists, 1991.

5.5. Windholz, M., Ed.; The Merck Index, 10th ed.; Merck & Co.: Rahway, NJ, 1983.

5.6. Analytical Methods for Atomic Absorption Spectrophotometry, The Perkin-Elmer Corporation: Norwalk, CT, 1982.

5.7. Slavin, W., D.C. Manning, G. Carnrick, and E. Pruszkowska: Properties of the Cadmium Determination with the Platform Furnace and Zeeman Background Correction. Spectrochim. Acta 38B:1157-1170 (1983).

5.8. Occupational Safety and Health Administration Salt Lake Technical Center: Standard Operating Procedure for Atomic Absorption. Salt Lake City, UT: USDOL/OSHA-SLTC, in progress.

5.9. Occupational Safety and Health Administration Salt Lake Technical Center: AAS-HGA Standard Operating Procedure. Salt Lake City, UT: USDOL/OSHA-SLTC, in progress.

5.10. Mandel, J.: Accuracy and Precision, Evaluation and Interpretation of Analytical Results, The Treatment of Outliers. In Treatise On Analytical Chemistry, 2nd ed., Vol.1, edited by I. M. Kolthoff and P. J. Elving. New York: John Wiley and Sons, 1978. pp. 282-285.

5.11. National Institute for Occupational Safety and Health: Documentation of the NIOSH Validation Tests by D. Taylor, R. Kupel, and J. Bryant (DHEW/NIOSH Pub. No. 77-185). Cincinnati, OH: National Institute for Occupational Safety and Health, 1977.

5.12. Occupational Safety and Health Administration Analytical Laboratory: Precision and Accuracy Data Protocol for Laboratory Validations. In OSHA Analytical Methods Manual 1st ed. Cincinnati, OH: American Conference of Governmental Industrial Hygienists (Pub. No. ISBN: 0-936712-66-X), 1985.

5.13. Long, G.L. and J.D. Winefordner: Limit of Detection—A Closer Look at the IUPAC Definition. Anal.Chem. 55:712A-724A (1983).

5.14. American Conference of Governmental Industrial Hygienists: Documentation of Threshold Limit Values and Biological Exposure Indices. 5th ed. Cincinnati, OH: American Conference of Governmental Industrial Hygienists, 1986.

## TABLE I—CD DETECTION LIMIT STUDY

[Flame AAS Analysis]

| STD (µg/mL) | Absorbance reading at 228.8 nm | | Statistical analysis |
|---|---|---|---|
| Reagent blank ...... | 5 | 2 | n=6. |
| | 4 | 3 | mean=3.50. |
| | 4 | 3 | std dev=1.05. |
| | | | CV=0.30. |
| 0.001 .................... | 6 | 6 | n=6. |
| | 2 | 4 | mean=5.00. |
| | 6 | 6 | std dev=1.67. |
| | | | CV=0.335. |
| 0.002 .................... | 5 | 7 | n=6. |
| | 7 | 3 | mean=5.50. |
| | 7 | 4 | std dev=1.76. |
| | | | CV=0.320. |
| 0.005 .................... | 7 | 7 | n=6. |
| | 8 | 8 | mean=7.33. |
| | 8 | 6 | std dev=0.817. |
| | | | CV=0.111. |
| 0.010 .................... | 10 | 9 | n=6. |
| | 10 | 13 | mean=10.3. |
| | 10 | 10 | std dev=1.37. |
| | | | CV=0.133. |
| 0.020 .................... | 20 | 23 | n=6. |
| | 20 | 22 | mean=20.8. |
| | 20 | 20 | std dev=1.33. |
| | | | CV=0.064. |
| 0.050 .................... | 42 | 42 | n=6. |
| | 42 | 42 | mean=42.5. |
| | 42 | 45 | std dev=1.22. |
| | | | CV=0.029. |
| 0.10 .................... | | 84 | n=3. |
| | | 80 | mean=82.3. |
| | | 83 | std dev=2.08. |
| | | | CV=0.025. |

## TABLE II—CD STANDARD WORKING RANGE STUDY

[Flame AAS Analysis]

| STD (µg/mL) | Absorbance reading at 228.8 nm | | Statistical analysis |
|---|---|---|---|
| Reagent blank ...... | 5 | 2 | n=6. |
| | 4 | 3 | mean=3.50. |
| | 4 | 3 | std dev=1.05. |
| | | | CV=0.30. |
| 0.020 .................... | 20 | 23 | n=6. |
| | 20 | 22 | mean=20.8. |
| | 20 | 20 | std dev=1.33. |
| | | | CV=0.064. |
| 0.050 .................... | 42 | 42 | n=6. |
| | 42 | 42 | mean=42.5. |
| | 42 | 45 | std dev=1.22. |
| | | | CV=0.029. |
| 0.10 .................... | | 84 | n=3. |
| | | 80 | mean=82.3. |
| | | 83 | std dev=2.08. |
| | | | CV=0.025. |
| 0.20 .................... | | 161 | n=3. |
| | | 161 | mean=160.0. |
| | | 158 | std dev=1.73. |
| | | | CV=0.011. |
| 0.50 .................... | | 391 | n=3. |
| | | 389 | mean=391.0. |
| | | 393 | std dev=2.00. |
| | | | CV=0.005. |
| 1.00 .................... | | 760 | n=3. |
| | | 748 | mean=753.3. |
| | | 752 | std dev=6.11. |
| | | | CV=0.008. |
| 2.00 .................... | | 1416 | n=3. |
| | | 1426 | mean=1414.3. |
| | | 1401 | std dev=12.6. |
| | | | CV=0.009. |

## TABLE III—CD DETECTION LIMIT STUDY

[AAS-HGA ANALYSIS]

| *STD (ng/mL)* | *Peak area readings×10[3] at 228.8 nm* | | *Statistical analysis* |
|---|---|---|---|
| Reagent blank ............... | 0 | 0 | n=6. |
| | 0 | 1 | mean=0.167. |
| | 0 | 0 | std dev=0.41. |
| | | | CV=2.45. |
| 0.1 ........................ | 8 | 6 | n=6. |
| | 5 | 7 | mean=7.7. |
| | 13 | 7 | std dev=2.8. |
| | | | CV=0.366. |
| 0.2 ........................ | 11 | 13 | n=6. |
| | 11 | 12 | mean=11.8. |
| | 12 | 12 | std dev=0.75. |
| | | | CV=0.064. |
| 0.5 ........................ | 28 | 33 | n=6. |
| | 26 | 28 | mean=28.8. |
| | 28 | 30 | std dev=2.4. |
| | | | CV=0.083. |
| 1.0 ........................ | 52 | 55 | n=6. |
| | 56 | 58 | mean=54.8. |
| | 54 | 54 | std dev=2.0. |
| | | | CV=0.037. |
| 2.0 ........................ | 101 | 112 | n=6. |
| | 110 | 110 | mean=108.8. |
| | 110 | 110 | std dev=3.9. |
| | | | CV=0.036. |

## TABLE IV—CD STANDARD WORKING RANGE STUDY

[AAS-HGA ANALYSIS]

| *STD (ng/mL)* | *Peak area readings×10[3] at 228.8 nm* | | *Statistical analysis* |
|---|---|---|---|
| 0.2 ........................ | 11 | 13 | n=6. |
| | 11 | 12 | mean=11.8. |
| | 12 | 12 | std dev=0.75. |
| | | | CV=0.064. |
| 0.5 ........................ | 28 | 33 | n=6. |
| | 26 | 28 | mean=28.8. |
| | 28 | 30 | std dev=2.4. |
| | | | CV=0.083. |
| 1.0 ........................ | 52 | 55 | n=6. |
| | 56 | 58 | mean=54.8. |
| | 54 | 54 | std dev=2.0. |
| | | | CV=0.037. |
| 2.0 ........................ | 101 | 112 | n=6. |
| | 110 | 110 | mean=108.8. |
| | 110 | 110 | std dev=3.9. |
| | | | CV=0.036. |
| 5.0 ........................ | 247 | 265 | n=6. |
| | 268 | 275 | mean=265.5. |
| | 259 | 279 | std dev=11.5. |
| | | | CV=0.044. |
| 10.0 ........................ | 495 | 520 | n=6. |
| | 523 | 513 | mean=516.7. |
| | 516 | 533 | std dev=12.7. |
| | | | CV=0.025. |
| 20.0 ........................ | 950 | 953 | n=6. |
| | 951 | 958 | mean=941.8. |
| | 949 | 890 | std dev=25.6. |
| | | | CV=0.027. |
| 30.0 ........................ | 1269 | 1291 | n=6. |
| | 1303 | 1307 | mean=1293. |
| | 1295 | 1290 | std dev=13.3. |
| | | | CV=0.010. |
| 40.0 ........................ | 1505 | 1567 | n=6. |
| | 1535 | 1567 | mean=1552. |
| | 1566 | 1572 | std dev=26.6. |
| | | | CV=0.017. |

## TABLE V—ANALYTICAL METHOD RECOVERY

[Flame AAS Analysis]

| Test level µg taken | 0.5 × µg found | Percent rec. | µg taken | 1.0 × µg found | Percent rec. | µg taken | 2.0 × µg found | Percent rec. |
|---|---|---|---|---|---|---|---|---|
| 1.00 | 1.0715 | 107.2 | 2.00 | 2.0688 | 103.4 | 4.00 | 4.1504 | 103.8 |
| 1.00 | 1.0842 | 108.4 | 2.00 | 2.0174 | 100.9 | 4.00 | 4.1108 | 102.8 |
| 1.00 | 1.0842 | 108.4 | 2.00 | 2.0431 | 102.2 | 4.00 | 4.0581 | 101.5 |
| 1.00 | * 1.0081 | * 100.8 | 2.00 | 2.0431 | 102.2 | 4.00 | 4.0844 | 102.1 |
| 1.00 | 1.0715 | 107.2 | 2.00 | 2.0174 | 100.9 | 4.00 | 4.1504 | 103.8 |
| 1.00 | 1.0842 | 108.4 | 2.00 | 2.0045 | 100.2 | 4.00 | 4.1899 | 104.7 |
| n = | | 5 | | | 6 | | | 6 |
| mean = | | 107.9 | | | 101.6 | | | 103.1 |
| std dev = | | 0.657 | | | 1.174 | | | 1.199 |
| $CV_1$ = | | 0.006 | | | 0.011 | | | 0.012 |
| | | $CV_1$ (pooled) = 0.010 | | | | | | |

* Rejected as an outlier—this value did not pass the outlier T-test at the 99% confidence level.

| Test level µg taken | 0.1 × µg found | Percent rec. |
|---|---|---|
| 0.200 | 0.2509 | 125.5 |
| 0.200 | 0.2509 | 125.5 |
| 0.200 | 0.2761 | 138.1 |
| 0.200 | 0.2258 | 112.9 |
| 0.200 | 0.2258 | 112.9 |
| 0.200 | 0.1881 | 94.1 |
| n = | | 6 |
| mean = | | 118.2 |
| std dev = | | 15.1 |
| $CV_1$ = | | 0.128 |

## TABLE VI—ANALYTICAL METHOD RECOVERY

[AAS-HGA Analysis]

| Test level ng taken | 0.5 × ng found | Percent rec. | ng taken | 1.0 × ng found | Percent rec. | ng taken | 2.0 × ng found | Percent rec. |
|---|---|---|---|---|---|---|---|---|
| 75 | 71.23 | 95.0 | 150 | 138.00 | 92.0 | 300 | 258.43 | 86.1 |
| 75 | 71.47 | 95.3 | 150 | 138.29 | 92.2 | 300 | 258.46 | 86.2 |
| 75 | 70.02 | 93.4 | 150 | 136.30 | 90.9 | 300 | 280.55 | 93.5 |
| 75 | 77.34 | 103.1 | 150 | 146.62 | 97.7 | 300 | 288.34 | 96.1 |
| 75 | 78.32 | 104.4 | 150 | 145.17 | 96.8 | 300 | 261.74 | 87.2 |
| 75 | 71.96 | 95.9 | 150 | 144.88 | 96.6 | 300 | 277.22 | 92.4 |
| n = | | 6 | | | 6 | | | 6 |
| mean = | | 97.9 | | | 94.4 | | | 90.3 |
| std dev = | | 4.66 | | | 2.98 | | | 4.30 |
| $CV_1$ = | | 0.048 | | | 0.032 | | | 0.048 |
| | | $CV_1$ (pooled) = 0.043 | | | | | | |

### ATTACHMENT 1

#### *Instrumental Parameters for Flame AAS Analysis*

Atomic Absorption Spectrophotometer (Perkin-Elmer Model 603)

Flame: Air/Acetylene—lean, blue

Oxidant Flow: 55

Fuel Flow: 32

Wavelength: 228.8 nm

Slit: 4 (0.7 nm)

Range: UV

Signal: Concentration (4 exp)

Integration Time: 3 sec

### ATTACHMENT 2

#### *Instrumental Parameters for HGA Analysis*

Atomic Absorption Spectrophotometer (Perkin-Elmer Model 5100)

Signal Type: Zeeman AA

Slitwidth: 0.7 nm

Wavelength: 228.8 nm

Measurement: Peak Area

Integration Time: 6.0 sec

*BOC Time: 5 sec*

BOC = Background Offset Correction.

### ZEEMAN GRAPHITE FURNACE (PERKIN-ELMER MODEL HGA-600)

| Step | Ramp time (sec) | Hold time (sec) | Temp. (°C) | Argon flow (mL/min) | Read (sec) |
|---|---|---|---|---|---|
| 1) Predry | 5 | 10 | 90 | 300 | ........ |
| 2) Dry | 30 | 10 | 140 | 300 | ........ |
| 3) Char | 10 | 20 | 900 | 300 | ........ |
| 4) Cool Down | 1 | 8 | 30 | 300 | ........ |
| 5) Atomize | 0 | 5 | 1600 | 0 | - 1 |
| 6) Burnout | 1 | 8 | 2500 | 300 | ........ |

## Appendix F to §1926.1127—Nonmandatory Protocol for Biological Monitoring

**Note:** The requirements applicable to construction work under this Appendix F are identical to those set forth in Appendix F to §1910.1027 of this chapter.

[57 FR 42452, Sept. 14, 1992, as amended at 57 FR 49272, Oct. 30, 1992; 58 FR 21787, Apr. 23, 1993. Redesignated and amended at 59 FR 215, Jan. 3, 1994; 61 FR 5510, Feb. 13, 1996; 61 FR 31433, 31434, June 20, 1996; 63 FR 1298, Jan. 8, 1998]

[Appendix F to Section 1910.1027 is printed below—CCH.]

## APPENDIX F TO §1910.1027—NONMANDATORY PROTOCOL FOR BIOLOGICAL MONITORING

### 1.00 Introduction

Under the final OSHA cadmium rule (29 CFR part 1910), monitoring of biological specimens and several periodic medical examinations are required for eligible employees. These medical examinations are to be conducted regularly, and medical monitoring is to include the periodic analysis of cadmium in blood (CDB), cadmium in urine (CDU) and beta-2-microglobulin in urine (B2MU). As CDU and B2MU are to be normalized to the concentration of creatinine in urine (CRTU), then CRTU must be analyzed in conjunction with CDU and B2MU analyses.

The purpose of this protocol is to provide procedures for establishing and maintaining the quality of the results obtained from the analyses of CDB, CDU and B2MU by commercial laboratories. Laboratories conforming to the provisions of this nonmandatory protocol shall be known as "participating laboratories." The biological monitoring data from these laboratories will be evaluated by physicians responsible for biological monitoring to determine the conditions under which employees may continue to work in locations exhibiting airborne-cadmium concentrations at or above defined actions levels (see paragraphs (l)(3) and (l)(4) of the final rule). These results also may be used to support a decision to remove workers from such locations.

Under the medical monitoring program for cadmium, blood and urine samples must be collected at defined intervals from workers by physicians responsible for medical monitoring; these samples are sent to commerical laboratories that perform the required analyses and report results of these analyses to the responsible physicians. To ensure the accuracy and reliability of these laboratory analyses, the laboratories to which samples are submitted should participate in an ongoing and efficacious proficiency testing program. Availability of proficiency testing programs may vary with the analyses performed.

To test proficiency in the analysis of CDB, CDU and B2MU, a laboratory should participate either in the interlaboratory comparison program operated by the Centre de Toxicologie du Quebec (CTQ) or an equivalent program. (Currently, no laboratory in the U.S. performs proficiency testing on CDB, CDU or B2MU.) Under this program, CTQ sends participating laboratories 18 samples of each analyte (CDB, CDU and/or B2MU) annually for analysis. Participating laboratories must return the results of these analyses to CTQ within four to five weeks after receiving the samples.

The CTQ program pools analytical results from many participating laboratories to derive consensus mean values for each of the samples distributed. Results reported by each laboratory then are compared against these consensus means for the analyzed samples to determine the relative performance of each laboratory. The proficiency of a participating laboratory is a function of the extent of agreement between results submitted by the participating laboratory and the consensus values for the set of samples analyzed.

Proficiency testing for CRTU analysis (which should be performed with CDU and B2MU analyses to evaluate the results properly) also is recommended. In the U.S., only the College of American Pathologists (CAP) currently conducts CRTU proficiency testing; participating laboratories should be accredited for CRTU analysis by the CAP.

Results of the proficiency evaluations will be forwarded to the participating laboratory by the proficiency-testing laboratory, as well as to physicians designated by the participating laboratory to receive this information. In addition, the participating laboratory should, on request, submit the results of their internal Quality Assurance/Quality Control (QA/QC) program for each analytic procedure (i.e., CDB, CDU and/or B2MU) to physicians designated to receive the proficiency results. For participating laboratories offering CDU and/or B2MU analyses, QA/QC documentation also should be provided for CRTU analysis. (Laboratories should provide QA/QC information regarding CRTU analysis directly to the requesting physician if they perform the analysis in-house; if CRTU analysis is performed by another laboratory under contract, this information should be provided to the physician by the contract laboratory.)

QA/QC information, along with the actual biological specimen measurements, should be provided to the responsible physician using standard formats. These physicians then may collate the QA/QC information with proficiency test results to compare the relative performance of laboratories, as well as to facilitate evaluation of the worker monitoring data. This information supports decisions made by the physician with regard to the biological monitoring program, and for mandating medical removal.

This protocol describes procedures that may be used by the responsible physicians to identify laboratories most likely to be proficient in the analysis of samples used in the biological monitoring of cadmium; also provided are procedures for record keeping and reporting by laboratories participating in proficiency testing programs, and recommendations to assist these physicians in interpreting analytical results determined by participating laboratories. As the collection and handling of samples affects the quality of the data, recommendations are made for these tasks. Specifications for analytical methods to be used in the medical monitoring program are included in this protocol as well.

In conclusion, this document is intended as a supplement to characterize and maintain the quality of medical monitoring data collected under the final cadmium rule promulgated by OSHA (29 CFR part 1910). OSHA has been granted authority under the Occupational Safety and Health Act of 1970 to protect workers from the effects of exposure to hazardous substances in the work place and to mandate adequate monitoring of workers to determine when adverse health effects may be occurring. This nonmandatory protocol is intended to provide guidelines and recommendations to improve the accuracy and reliability of the procedures used to analyze the biological samples collected as part of the medical monitoring program for cadmium.

### 2.0 Definitions

When the terms below appear in this protocol, use the following definitions.

*Accuracy:* A measure of the bias of a data set. Bias is a systematic error that is either inherent in a method or caused by some artifact or idiosyncracy of the measurement system. Bias is characterized by a consistent deviation (positive or negative) in the results from an accepted reference value.

*Arithmetic Mean:* The sum of measurements in a set divided by the number of measurements in a set.

*Blind Samples:* A quality control procedure in which the concentration of analyte in the samples should be unknown to the analyst at the time that the analysis is performed.

*Coefficient of Variation:* The ratio of the standard deviation of a set of measurements to the mean (arithmetic or geometric) of the measurements.

*Compliance Samples:* Samples from exposed workers sent to a participating laboratory for analysis.

*Control Charts:* Graphic representations of the results for quality control samples being analyzed by a participating laboratory.

*Control Limits:* Statistical limits which define when an analytic procedure exceeds acceptable parameters; control limits provide a method of assessing the accuracy of analysts, laboratories, and discrete analytic runs.

*Control Samples:* Quality control samples.

*F/T:* The measured amount of an analyte divided by the theoretical value (defined below) for that analyte in the sample analyzed; this ratio is a measure of the recovery for a quality control sample.

*Geometric Mean:* The natural antilog of the mean of a set of natural log-transformed data.

*Geometric Standard Deviation:* The antilog of the standard deviation of a set of natural log-transformed data.

*Limit of Detection:* Using a predefined level of confidence, this is the lowest measured value at which some of the measured material is likely to have come from the sample.

*Mean:* A central tendency of a set of data; in this protocol, this mean is defined as the *arithmetic mean* (see definition of *arithmetic mean* above) unless stated otherwise.

*Performance:* A measure of the overall quality of data reported by a laboratory.

*Pools:* Groups of quality-control samples to be established for each target value (defined below) of an analyte. For the protocol provided in attachment 3, for example, the theoretical value of the quality control samples of the pool must be within a range defined as plus or minus (±) 50% of the target value. Within each analyte pool, there must be quality control samples of at least 4 theoretical values.

*Precision:* The extent of agreement between repeated, independent measurements of the same quantity of an analyte.

*Proficiency:* The ability to satisfy a specified level of analyte performance.

*Proficiency Samples:* Specimens, the values of which are unknown to anyone at a participating laboratory, and which are submitted by a participating laboratory for proficiency testing.

*Quality or Data Quality:* A measure of the confidence in the measurement value.

*Quality Control (QC) Samples:* Specimens, the value of which is unknown to the analyst, but is known to the appropriate QA/QC personnel of a participating laboratory; when used as part of a laboratory QA/QC program, the theoretical values of these samples should not be known to the analyst until the analyses are complete. QC samples are to be run in sets consisting of one QC sample from each pool (see definition of "pools" above).

*Sensitivity:* For the purposes of this protocol, the limit of detection.

*Standard Deviation:* A measure of the distribution or spread of a data set about the mean; the standard deviation is equal to the positive square root of the variance, and is expressed in the same units as the original measurements in the data set.

*Standards:* Samples with values known by the analyst and used to calibrate equipment and to check calibration throughout an analytic run. In a laboratory QA/QC program, the values of the standards must exceed the values obtained for compliance samples such that the lowest standard value is near the limit of detection and the highest standard is higher than the highest compliance sample or QC sample. Standards of at least three different values are to be used for calibration, and should be constructed from at least 2 different sources.

*Target Value:* Those values of CDB, CDU or B2MU which trigger some action as prescribed in the medical surveillance section of the regulatory text of the final cadmium rule. For CDB, the target values are 5, 10 and 15 μg/l. For CDU, the target values are 3, 7, and 15 μg/g CRTU. For $B_2$ MU, the target values are 300, 750 and 1500 μg/g CRTU. (Note that target values may vary as a function of time.)

*Theoretical Value (or Theoretical Amount):* The reported concentration of a quality-control sample (or calibration standard) derived from prior characterizations of the sample.

*Value or Measurement Value:* The numerical result of a measurement.

*Variance:* A measure of the distribution or spread of a data set about the mean; the variance is the sum of the squares of the differences between the mean and each discrete measurement divided by one less than the number of measurements in the data set.

## 3.0 Protocol

This protocol provides procedures for characterizing and maintaining the quality of analytic results derived for the medical monitoring program mandated for workers under the final cadmium rule.

## 3.1 Overview

The goal of this protocol is to assure that medical monitoring data are of sufficient quality to facilitate proper interpretation. The data quality objectives (DQOs) defined for the medical monitoring program are summarized in Table 1. Based on available information, the DQOs presented in Table 1 should be achievable by the majority of laboratories offering the required analyses commercially; OSHA recommends that only laboratories meeting these DQOs be used for the analysis of biological samples collected for monitoring cadmium exposure.

**TABLE 1—RECOMMENDED DATA QUALITY OBJECTIVES (DQOs) FOR THE CADMIUM MEDICAL MONITORING PROGRAM**

| Analyte/concentration pool | Limit of detection | Precision (CV) (%) | Accuracy |
|---|---|---|---|
| Cadmium in blood | 0.5 μg/l | | ±1 μg/l or 15% of the mean. |
| ≤2 μg/l | | 40 | |
| >2μg/l | | 20 | |
| Cadmium in urine | 0.5 μg/g creatinine | | ±1 μg/l or 15% of the mean. |
| ≤2 μg/l creatinine | | 40 | |
| >2μg/l creatinine | | 20 | |
| β-2-microglobulin in urine: 100 μg/g creatine. | 100 μg/g creatinine | 5 | ±15% of the mean. |

To satisfy the DQOs presented in Table 1, OSHA provides the following guidelines:

1. Procedures for the collection and handling of blood and urine are specified (Section 3.4.1 of this protocol);

2. Preferred analytic methods for the analysis of CDB, CDU and B2MU are defined (and a method for the determination of CRTU also is specified since CDU and B2MU results are to be normalized to the level of CRTU).

3. Procedures are described for identifying laboratories likely to provide the required analyses in an accurate and reliable manner;

4. These guidelines (Sections 3.2.1 to 3.2.3, and Section 3.3) include recommendations regarding internal QA/QC programs for participating laboratories, as well as levels of proficiency through participation in an interlaboratory proficiency program;

5. Procedures for QA/QC record keeping (Section 3.3.2), and for reporting QC/QA results are described (Section 3.3.3); and,

6. Procedures for interpreting medical monitoring results are specified (Section 3.4.3).

Methods recommended for the biological monitoring of eligible workers are:

1. The method of Stoeppler and Brandt (1980) for CDB determinations (limit of detection: 0.5 μg/l);

2. The method of Pruszkowska et al. (1983) for CDU determinations (limit of detection: 0.5 μg/l of urine); and,

3. The Pharmacia Delphia test kit (Pharmacia 1990) for the determination of B2MU (limit of detection: 100 μg/l urine).

Because both CDU and B2MU should be reported in μg/g CRTU, an independent determination of CRTU is recommended. Thus, both the OSHA Salt Lake City Technical Center (OSLTC) method (OSHA, no date) and the Jaffe method (Du Pont, no date) for the determination of CRTU are specified under this protocol (i.e., either of these 2 methods may be used). Note that although detection limits are not reported for either of these CRTU methods, the range of measurements expected for CRTU (0.9-1.7 g/l) are well above the likely limit of detection for either of these methods (Harrison, 1987).

Laboratories using alternate methods should submit sufficient data to the responsible physicians demonstrating that the alternate method is capable of satisfying the defined data quality objectives of the program. Such laboratories also should submit a QA/QC plan that documents the performance of the alternate method in a manner entirely equivalent to the QA/QC plans proposed in Section 3.3.1.

### 3.2 Duties of the Responsible Physician

The responsible physician will evaluate biological monitoring results provided by participating laboratories to determine whether such laboratories are proficient and have satisfied the QA/QC recommendations. In determining which laboratories to employ for this purpose, these physicians should review proficiency and QA/QC data submitted to them by the participating laboratories.

Participating laboratories should demonstrate proficiency for each analyte (CDU, CDB and B2MU) sampled under the biological monitoring program. Participating laboratories involved in analyzing CDU and B2MU also should demonstrate proficiency for CRTU analysis, or provide evidence of a contract with a laboratory proficient in CRTU analysis.

#### 3.2.1 Recommendations for Selecting Among Existing Laboratories

OSHA recommends that existing laboratories providing commercial analyses for CDB, CDU and/or B2MU for the medical monitoring program satisfy the following criteria:

1. Should have performed commercial analyses for the appropriate analyte (CDB, CDU and/or B2MU) on a regular basis over the last 2 years;

2. Should provide the responsible physician with an internal QA/QC plan;

3. If performing CDU or B2MU analyses, the participating laboratory should be accredited by the CAP for CRTU analysis, and should be enrolled in the corresponding CAP survey (note that alternate credentials may be acceptable, but acceptability is to be determined by the responsible physician); and,

4. Should have enrolled in the CTQ interlaboratory comparison program for the appropriate analyte (CDB, CDU and/or B2MU).

Participating laboratories should submit appropriate documentation demonstrating compliance with the above criteria to the responsible physician. To demonstrate compliance with the first of the above criteria, participating laboratories should submit the following documentation for each analyte they plan to analyze (note that each document should cover a period of at least 8 consecutive quarters, and that the period designated by the term "regular analyses" is at least once a quarter):

1. Copies of laboratory reports providing results from regular analyses of the appropriate analyte (CDB, CDU and/or B2MU);

2. Copies of 1 or more signed and executed contracts for the provision of regular analyses of the appropriate analyte (CDB, CDU and/or B2MU); or,

3. Copies of invoices sent to 1 or more clients requesting payment for the provision of regular analyses of the appropriate analyte (CDB, CDU and/or B2MU). Whatever the form of documentation submitted, the specific analytic procedures conducted should be identified directly. The forms that are copied for submission to the responsible physician also should identify the laboratory which provided these analyses.

To demonstrate compliance with the second of the above criteria, a laboratory should submit to the responsible physician an internal QA/QC plan detailing the standard operating procedures to be adopted for satisfying the recommended QA/QC procedures for the analysis of each specific analyte (CDB, CDU and/or B2MU). Procedures for internal QA/QC programs are detailed in Section 3.3.1 below.

To satisfy the third of the above criteria, laboratories analyzing for CDU or B2MU also should submit a QA/QC plan for creatinine analysis (CRTU); the QA/QC plan and characterization analyses for CRTU must come from the laboratory performing the CRTU analysis, even if the CRTU analysis is being performed by a contract laboratory.

Laboratories enrolling in the CTQ program (to satisfy the last of the above criteria) must remit, with the enrollment application, an initial fee of approximately $100 per analyte. (Note that this fee is only an estimate, and is subject to revision without notice.) Laboratories should indicate on the application that they agree to have proficiency test results sent by the CTQ directly to the physicians designated by participating laboratories.

Once a laboratory's application is processed by the CTQ, the laboratory will be assigned a code number which will be provided to the laboratory on the initial confirmation form, along with identification of the specific analytes for which the laboratory is participating. Confirmation of participation will be sent by the CTQ to physicians designated by the applicant laboratory.

#### 3.2.2 Recommended Review of Laboratories Selected to Perform Analyses

Six months after being selected initially to perform analyte determinations, the status of participating laboratories should be reviewed by the responsible physicians. Such reviews should then be repeated every 6 months or whenever additional proficiency or QA/QC documentation is received (whichever occurs first).

As soon as the responsible physician has received the CTQ results from the first 3 rounds of proficiency testing (i.e., 3 sets of 3 samples each for CDB, CDU and/or B2MU) for a participating laboratory, the status of the laboratory's continued participation should be reviewed. Over the same initial 6-month period, participating laboratories also should provide responsible physicians the results of their internal QA/QC monitoring program used to assess performance for each analyte (CDB, CDU and/or B2MU) for which the laboratory performs determinations. This information should be submitted using appropriate forms and documentation.

The status of each participating laboratory should be determined for each analyte (i.e., whether the laboratory satisfies minimum proficiency guidelines based on the proficiency samples sent by the CTQ and the results of the laboratory's internal QA/QC program). To maintain competency for analysis of CDB, CDU and/or B2MU during the first review, the laboratory should satisfy performance requirements for at least 2 of the 3 proficiency samples provided in each of the 3 rounds completed over the 6-month period. Proficiency should be maintained for the analyte(s) for which the laboratory conducts determinations.

To continue participation for CDU and/or B2MU analyse, laboratories also should either maintain accreditation for CRTU analysis in the CAP program and participate in the CAP surveys, or they should contract the CDU and B2MU analyses to a laboratory which satisfies these requirements (or which can provide documentation of accreditation/participation in an equivalent program).

The performance requirement for CDB analysis is defined as an analytical result within ±1 μg/l blood or 15% of the consensus mean (whichever is greater). For samples exhibiting a consensus mean less than 1 μg/l, the performance requirement is defined as a concentration between the detection limit of the analysis and a maximum of 2 μg/l. The purpose for redefining the acceptable interval for low CDB values is to encourage proper reporting of the actual values obtained during measurement; laboratories, therefore, will not be penalized (in terms of a narrow range of acceptability) for reporting measured concentrations smaller than 1 μg/l.

The performance requirement for CDU analysis is defined as an analytical result within ±1 μg/l urine or 15% of the consensus mean (whichever is greater). For samples exhibiting a consensus mean less than 1 μg/l urine, the performance requirement is defined as a concentration between the detection limit of the analysis and a maximum of 2 μg/l urine. Laboratories also should demonstrate proficiency in creatinine analysis as defined by the CAP. Note that reporting CDU results, other than for the CTQ proficiency samples (i.e., compliance samples), should be accompanied with results of analyses for CRTU, and these 2 sets of results should be combined to provide a measure of CDU in units of μg/g CRTU.

The performance requirement for B2MU is defined as analytical results within ± 15% of the consensus mean. Note that reporting B2MU results, other than for CTQ proficiency samples (i.e., compliance samples), should be accompanied with results of analyses for CRTU, and these 2 sets of results should be combined to provide a measure of B2MU in units of μg/g CRTU.

There are no recommended performance checks for CRTU analyses. As stated previously, laboratories performing CRTU analysis in support of CDU or B2MU analyses should be accredited by the CAP, and participating in the CAP's survey for CRTU.

Following the first review, the status of each participating laboratory should be reevaluated at regular intervals (i.e., corresponding to receipt of results from each succeeding round of proficiency testing and submission of reports from a participating laboratory's internal QA/QC program).

After a year of collecting proficiency test results, the following proficiency criterion should be added to the set of criteria used to determine the participating laboratory's status (for analyzing CDB, CDU and/or B2MU): A participating laboratory should not fail performance requirements for more than 4 samples from the 6 most recent consecutive rounds used to assess proficiency for CDB, CDU and/or B2MU separately (i.e., a total of 18 discrete proficiency samples for each analyte). Note that this requirement does not replace, but supplements, the recommendation that a laboratory should satisfy the performance criteria for at least 2 of the 3 samples tested for each round of the program.

### 3.2.3 Recommendations for Selecting Among Newly-Formed Laboratories (or Laboratories that Previously Failed to Meet the Protocol Guidelines)

OSHA recommends that laboratories that have not previously provided commercial analyses of CDB, CDU and/or B2MU (or have done so for a period less than 2 years), or which have provided these analyses for 2 or more years but have not conformed previously with these protocol guidelines, should satisfy the following provisions for each analyte for which determinations are to be made prior to being selected to analyze biological samples under the medical monitoring program:

1. Submit to the responsible physician an internal QA/QC plan detailing the standard operating procedures to be adopted for satisfying the QA/QC guidelines (guidelines for internal QA/QC programs are detailed in Section 3.3.1);

2. Submit to the responsible physician the results of the initial characterization analyses for each analyte for which determinations are to be made;

3. Submit to the responsible physician the results, for the initial 6-month period, of the internal QA/QC program for each analyte for which determinations are to be made (if no commercial analyses have been conducted previously, a minimum of 2 mock standardization trials for each analyte should be completed per month for a 6-month period);

4. Enroll in the CTQ program for the appropriate analyte for which determinations are to be made, and arrange to have the CTQ program submit the initial confirmation of participation and proficiency test results directly to the designated physicians. Note that the designated physician should receive results from 3 completed rounds from the CTQ program before approving a laboratory for participation in the biological monitoring program;

5. Laboratories seeking participation for CDU and/or B2MU analyses should submit to the responsible physician documentation of accreditation by the CAP for CRTU analyses performed in conjunction with CDU and/or B2MU determinations (if CRTU analyses are conducted by a contract laboratory, this laboratory should submit proof of CAP accreditation to the responsible physician); and,

6. Documentation should be submitted on an appropriate form.

To participate in CDB, CDU and/or B2MU analyses, the laboratory should satisfy the above criteria for a minimum of 2 of the 3 proficiency samples provided in each of the 3 rounds of the CTQ program over a 6-month period; this procedure should be completed for each appropriate analyte. Proficiency should be maintained for each analyte to continue participation. Note that laboratories seeking participation for CDU or B2MU also should address the performance requirements for CRTU, which involves providing evidence of accreditation by the CAP and participation in the CAP surveys (or an equivalent program).

The performance requirement for CDB analysis is defined as an analytical result within ±1 μg/l or 15% of the consensus mean (whichever is greater). For samples exhibiting a consensus mean less than 1 μg/l, the performance requirement is defined as a concentration between the detection limit of the analysis and a maximum of 2 μg/l. The purpose of redefining the acceptable interval for low CDB values is to encourage proper reporting of the actual values obtained during measurement; laboratories, therefore, will not be penalized (in terms of a narrow range of acceptability) for reporting measured concentrations less than 1 μg/l.

The performance requirement for CDU analysis is defined as an analytical result within ±1 μg/l urine or 15% of the consensus mean (whichever is greater). For samples exhibiting a consensus mean less than 1 μg/l urine, the performance requirement is defined as a concentration that falls between the detection limit of the analysis and a maximum of 2 μg/l urine. Performance requirements for the companion CRTU analysis (defined by the CAP) also should be met. Note that reporting CDU results, other than for CTQ proficiency testing should be accompanied with results of CRTU analyses, and these 2 sets of results should be combined to provide a measure of CDU in units of μg/g CRTU.

The performance requirement for B2MU is defined as an analytical result within ±15% of the consensus mean. Note that reporting B2MU results, other than for CTQ proficiency testing should be accompanied with results of CRTU analysis, these 2 sets of results should be combined to provide a measure of B2MU in units of μg/g CRTU.

Once a new laboratory has been approved by the responsible physician for conducting analyte determinations, the status of this approval should be reviewed periodically by the responsible physician as per the criteria presented under Section 3.2.2.

Laboratories which have failed previously to gain approval of the responsible physician for conducting determinations of 1 or more analytes due to lack of compliance with the criteria defined above for existing laboratories (Section 3.2.1), may obtain approval by satisfying the criteria for newly-formed laboratories defined under this section; for these laboratories, the second of the above criteria may be satisfied by submitting a new set of characterization analyses for each analyte for which determinations are to be made.

Reevaluation of these laboratories is discretionary on the part of the responsible physician. Reevaluation, which normally takes about 6 months, may be expedited if the laboratory can achieve 100% compliance with the proficiency test criteria using the 6 samples of each analyte submitted to the CTQ program during the first 2 rounds of proficiency testing.

For laboratories seeking reevaluation for CDU or B2MU analysis, the guidelines for CRTU analyses also should be satisfied, including accreditation for CRTU analysis by the CAP, and participation in the CAP survey program (or accreditation/participation in an equivalent program).

### 3.2.4 Future Modifications to the Protocol Guidelines

As participating laboratories gain experience with analyses for CDB, CDU and B2MU, it is anticipated that the performance achievable by the majority of laboratories should improve until it approaches that reported by the research groups which developed each method. OSHA, therefore, may choose to recommend stricter performance guidelines in the future as the overall performance of participating laboratories improves.

### 3.3 Guidelines for Record Keeping and Reporting

To comply with these guidelines, participating laboratories should satisfy the above-stated performance and proficiency recommendations, as well as the following internal QA/QC, record keeping, and reporting provisions.

If a participating laboratory fails to meet the provisions of these guidelines, it is recommended that the responsible physician disapprove further analyses of biological samples by that laboratory until it demonstrates compliance with these guidelines. On disapproval, biological samples should be sent to a laboratory that can demonstrate compliance with these guidelines, at least until the former laboratory is reevaluated by the responsible physician and found to be in compliance.

The following record keeping and reporting procedures should be practiced by participating laboratories.

### 3.3.1 Internal Quality Assurance/Quality Control Procedures

Laboratories participating in the cadmium monitoring program should develop and maintain an internal quality assurance/quality control (QA/QC) program that incorporates procedures for establishing and maintaining control for each of the analytic procedures (determinations of CDB, CDU and/or B2MU) for which the laboratory is seeking participation. For laboratories analyzing CDU and/or B2MU, a QA/QC program for CRTU also should be established.

Written documentation of QA/QC procedures should be described in a formal QA/QC plan; this plan should contain the following information: Sample acceptance and handling procedures (i.e., chain-of-custody); sample preparation procedures; instrument parameters; calibration procedures; and, calculations. Documentation of QA/QC procedures should be sufficient to identify analytical problems, define criteria under which analysis of compliance samples will be suspended, and describe procedures for corrective actions.

#### 3.3.1.1 QA/QC procedures for establishing control of CDB and CDU analyses

The QA/QC program for CDB and CDU should address, at a minimum, procedures involved in calibration, establishment of control limits, internal QC analyses and maintaining control, and corrective-action protocols. Participating laboratory should develop and maintain procedures to assure that analyses of compliance samples are within control limits, and that these procedures are documented thoroughly in a QA/QC plan.

A nonmandatory QA/QC protocol is presented in Attachment 1. This attachment is illustrative of the procedures that should be addressed in a proper QA/QC program.

*Calibration.* Before any analytic runs are conducted, the analytic instrument should be calibrated. Calibration should be performed at the beginning of each day on which QC and/or compliance samples are run. Once calibration is established, QC or compliance samples may be run. Regardless of the type of samples run, about every fifth sample should serve as a standard to assure that calibration is being maintained.

Calibration is being maintained if the standard is within ±15% of its theoretical value. If a standard is more than ±15% of its theoretical value, the run has exceeded control limits due to calibration error; the entire set of samples then should be reanalyzed after recalibrating or the results should be recalculated based on a statistical curve derived from that set of standards.

It is essential that the value of the highest standard analyzed be higher than the highest sample analyzed; it may be necessary, therefore, to run a high standard at the end of the run, which has been selected based on results obtained over the course of the run (i.e., higher than any standard analyzed to that point).

Standards should be kept fresh; as samples age, they should be compared with new standards and replaced if necessary.

*Internal Quality Control Analyses.* Internal QC samples should be determined interspersed with analyses of compliance samples. At a minimum, these samples should be run at a rate of 5% of the compliance samples or at least one set of QC samples per analysis of compliance samples, whichever is greater. If only 2 samples are run, they should contain different levels of cadmium.

Internal QC samples may be obtained as commercially-available reference materials and/or they may be internally prepared. Internally-prepared samples should be well characterized and traced, or compared to a reference material for which a consensus value is available.

Levels of cadmium contained in QC samples should not be known to the analyst prior to reporting the results of the analysis.

Internal QC results should be plotted or charted in a manner which describes sample recovery and laboratory control limits.

*Internal Control Limits.* The laboratory protocol for evaluating internal QC analyses per control limits should be clearly defined. Limits may be based on statistical methods (e.g., as $2\sigma$ from the laboratory mean recovery), or on proficiency testing limits (e.g., ±1 μg or 15% of the mean, whichever is greater). Statistical limits that exceed ±40% should be reevaluated to determine the source error in the analysis.

When laboratory limits are exceeded, analytic work should terminate until the source of error is determined and corrected; compliance samples affected by the error should be reanalyzed. In addition, the laboratory protocol should address any unusual trends that develop which may be biasing the results. Numerous, consecutive results above or below laboratory mean recoveries, or outside laboratory statistical limits, indicate that problems may have developed.

*Corrective Actions.* The QA/QC plan should document in detail specific actions taken if control limits are exceeded or unusual trends develop. Corrective actions should be noted on an appropriate form, accompanied by supporting documentation.

In addition to these actions, laboratories should include whatever additional actions are necessary to assure that accurate data are reported to the responsible physicians.

*Reference Materials.* The following reference materials may be available:

##### Cadmium in Blood (CDB)

1. Centre de Toxicologie du Quebec, Le Centre Hospitalier de l'Universite Laval, 2705 boul. Laurier, Quebec, Que., Canada G1V 4G2. (Prepared 6 times per year at 1-15 μg Cd/l.)

2. H. Marchandise, Community Bureau of Reference-BCR, Directorate General XII, Commission of the European Communities, 200, rue de la Loi, B-1049, Brussels, Belgium. (Prepared as Bl CBM-1 at 5.37 μg Cd/l, and Bl CBM-2 at 12.38 μg Cd/l.)

3. Kaulson Laboratories Inc., 691 Bloomfield Ave., Caldwell, NJ 07006; tel: (201) 226-9494, FAX (201) 226-3244. (Prepared as #0141 [As, Cd, Hg, Pb] at 2 levels.)

##### Cadmium in Urine (CDU)

1. Centre de Toxicologie du Quebec, Le Centre Hospitalier de l'Universite Laval, 2705 boul. Laurier, Quebec, Que., Canada G1V 4G2. (Prepared 6 times per year.)

2. National Institute of Standards and Technology (NIST), Dept. of Commerce, Gaithersburg, MD; tel: (301) 975-6776. (Prepared as SRM 2670 freeze-dried urine [metals]; set includes normal and elevated levels of metals; cadmium is certified for elevated level of 88.0 μg/l in reconstituted urine.)

3. Kaulson Laboratories Inc., 691 Bloomfield Ave., Caldwell, NJ 07006; tel: (201) 226-9494, FAX (201) 226-3244. (Prepared as #0140 [As, Cd, Hg, Pb] at 2 levels.)

#### 3.3.1.2 QA/QC procedures for establishing control of B2MU

A written, detailed QA/QC plan for B2MU analysis should be developed. The QA/QC plan should contain a protocol similar to those protocols developed for the CDB/CDU analyses. Differences in analyses may warrant some differences in the QA/QC protocol, but procedures to ensure analytical integrity should be developed and followed.

Examples of performance summaries that can be provided include measurements of accuracy (i.e., the means of measured values versus target values for the control samples) and precision (i.e., based on duplicate analyses). It is recommended that the accuracy and precision measurements be compared to those reported as achievable by the Pharmacia Delphia kit (Pharmacia 1990) to determine if and when unsatisfactory analyses have arisen. If the measurement error of 1 or more of the control samples is more than 15%, the run exceeds control limits. Similarly, this decision is warranted when the average CV for duplicate samples is greater than 5%.

### 3.3.2 Procedures for Record Keeping

To satisfy reporting requirements for commercial analyses of CDB, CDU and/or B2MU performed for the medical monitoring program mandated under the cadmium rule, participating laboratories should maintain the following documentation for each analyte:

1. For each analytic instrument on which analyte determinations are made, records relating to the most recent calibration and QC sample analyses;

2. For these instruments, a tabulated record for each analyte of those determinations found to be within and outside of control limits over the past 2 years;

3. Results for the previous 2 years of the QC sample analyses conducted under the internal QA/QC program (this information should be: Provided for each analyte for which determinations are made and for each analytic instrument used for this purpose, sufficient to demonstrate that internal QA/QC programs are being executed properly, and consistent with data sent to responsible physicians.

4. Duplicate copies of monitoring results for each analyte sent to clients during the previous 5 years, as well as associated information; supporting material such as chain-of-custody forms also should be retained; and,

5. Proficiency test results and related materials received while participating in the CTQ interlaboratory program over the past 2 years; results also should be tabulated to provide a serial record of relative error (derived per Section 3.3.3 below).

### 3.3.3 Reporting Procedures

Participating laboratories should maintain these documents: QA/QC program plans; QA/QC status reports; CTQ proficiency program reports; and, analytical data reports. The information that should be included in these reports is summarized in Table 2; a copy of each report should be sent to the responsible physician.

**TABLE 2—REPORTING PROCEDURES FOR LABORATORIES PARTICIPATING IN THE CADMIUM MEDICAL MONITORING PROGRAM**

| *Report* | *Frequency (time frame)* | *Contents* |
|---|---|---|
| 1 QA/QC Program Plan | Once (initially). | A detailed description of the QA/QC protocol to be established by the laboratory to maintain control of analyte determinations. |
| 2 QA/QC Status Report | Every 2 months | Results of the QC samples incorporated into regular runs for each instrument (over the period since the last report). |
| 3 Proficiency Report | Attached to every data report. | Results from the last full year of proficiency samples submitted to the CTQ program and Results of the 100 most recent QC samples incorporated into regular runs for each instrument. |
| 4 Analytical Data Report | For all reports of data results. | Date the sample was received; Date the sample was analyzed; Appropriate chain-of-custody information; Types of analyses performed; Results of the requested analyses and Copy of the most current proficiency report. |

As noted in Section 3.3.1, a QA/QC program plan should be developed that documents internal QA/QC procedures (defined under Section 3.3.1) to be implemented by the participating laboratory for each analyte; this plan should provide a list identifying each instrument used in making analyte determinations.

A QA/QC status report should be written bimonthly for each analyte. In this report, the results of the QC program during the reporting period should be reported for each analyte in the following manner: The number (N) of QC samples analyzed during the period; a table of the target levels defined for each sample and the corresponding measured values; the mean of F/T value (as defined below) for the set of QC samples run during the period; and, use of $\bar{X} \pm \hat{\sigma}$ (as defined below) for the set of QC samples run during the period as a measure of precision.

As noted in Section 2, an F/T value for a QC sample is the ratio of the measured concentration of analyte to the established (i.e., reference) concentration of analyte for that QC sample. The equation below describes the derivation of the mean for F/T values, X, (with N being the total number of samples analyzed):

$$\bar{X} = \frac{\Sigma\left(F/T\right)}{N}$$

The standard deviation, $\hat{\sigma}$, for these measurements is derived using the following equation (note that $\hat{\sigma}$ is twice this value):

$$\hat{\sigma} = \left[\frac{\Sigma\left(F/T - \bar{X}\right)^2}{N-1}\right]^{\frac{1}{2}}$$

The nonmandatory QA/QC protocol (see Attachment 1) indicates that QC samples should be divided into several discrete pools, and a separate estimate of precision for each pools then should be derived. Several precision estimates should be provided for concentrations which differ in average value. These precision measures may be used to document improvements in performance with regard to the combined pool.

Participating laboratories should use the CTQ proficiency program for each analyte. Results of the this program will be sent by CTQ directly to physicians designated by the participating laboratories.

Proficiency results from the CTQ program are used to establish the accuracy of results from each participating laboratory, and should be provided to responsible physicians for use in trend analysis. A proficiency report consisting of these proficiency results should accompany data reports as an attachment.

For each analyte, the proficiency report should include the results from the 6 previous proficiency rounds in the following format:

1. Number (N) of samples analyzed;

2. Mean of the target levels, $(1/N)\Sigma_i$, with $T_i$ being a consensus mean for the sample;

3. Mean of the measurements, $(1/N)\Sigma_i$, with $M_i$ being a sample measurement;

4. A measure of error defined by:

$$(1/N)\Sigma(T_i - M_i)^2$$

Analytical data reports should be submitted to responsible physicians directly. For each sample, report the following information: The date the sample was received; the date the sample was analyzed; appropriate chain-of-custody information; the type(s) of analyses performed; and, the results of the analyses. This information should be reported on a form similar to the form provided an appropriate form. The most recent proficiency program report should accompany the analytical data reports (as an attachment).

Confidence intervals for the analytical results should be reported as $X \pm \hat{\sigma}$, with X being the measured value and $\hat{\sigma}$ the standard deviation calculated as described above.

For CDU or B2MU results, which are combined with CRTU measurements for proper reporting, the 95% confidence limits are derived from the limits for CDU or B2MU, (p), and the limits for CRTU, (q), as follows:

$$\frac{X}{Y} \pm \left(\frac{1}{Y^2}\right)(Y^2 \times p^2 + X^2 \times q^2)^{1/2}$$

For these calculations, $X \pm p$ is the measurement and confidence limits for CDU or B2MU, and $Y \pm q$ is the measurement and confidence limit for CRTU.

Participating laboratories should notify responsible physicians as soon as they receive information indicating a change in their accreditation status with the CTQ or the CAP. These physicians should not be expected to wait until formal notice of a status change has been received from the CTQ or the CAP.

### 3.4 Instructions to Physicians

Physicians responsible for the medical monitoring of cadmium-exposed workers must collect the biological samples from workers; they then should select laboratories to perform the required analyses, and should interpret the analytic results.

#### 3.4.1 Sample Collection and Holding Procedures

*Blood Samples.* The following procedures are recommended for the collection, shipment and storage of blood samples for CDB analysis to reduce analytical variablility; these recommendations were obtained primarily through personal communications with J.P. Weber of the CTQ (1991), and from reports by the Centers for Disease Control (CDC, 1986) and Stoeppler and Brandt (1980).

To the extent possible, blood samples should be collected from workers at the same time of day. Workers should shower or thoroughly wash their hands and arms before blood samples are drawn. The following materials are needed for blood sample collection: Alcohol wipes; sterile gauze sponges; band-aids; 20-gauge, 1.5-in. stainless steel needles (sterile); preprinted labels; tourniquets; vacutainer holders; 3-ml "metal free" vacutainer tubes (i.e., dark-blue caps), with EDTA as an anti-coagulant; and, styrofoam vacutainer shipping containers.

Whole blood samples are taken by venipuncture. Each blue-capped tube should be labeled or coded for the worker and company before the sample is drawn. (Blue-capped tubes are recommended instead of red-capped tubes because the latter may consist of red coloring pigment containing cadmium, which could contaminate the samples.) Immediately after sampling, the vacutainer tubes must be thoroughly mixed by inverting the tubes at least 10 times manually or mechanically using a Vortex device (for 15 sec). Samples should be refrigerated immediately or stored on ice until they can be packed for shipment to the participating laboratory for analysis.

The CDC recommends that blood samples be shipped with a "cool pak" to keep the samples cold during shipment. However, the CTQ routinely ships and receives blood samples for cadmium analysis that have not been kept cool during shipment. The CTQ has found no deterioration of cadmium in biological fluids that were shipped via parcel post without a cooling agent, even though these deliveries often take 2 weeks to reach their destination.

*Urine Samples.* The following are recommended procedures for the collection, shipment and storage of urine for CDU and B2MU analyses, and were obtained primarily through personal communications with J.P. Weber of the CTQ (1991), and from reports by the CDC (1986) and Stoeppler and Brandt (1980).

Single "spot" samples are recommended. As B2M can degrade in the bladder, workers should first empty their bladder and then drink a large glass of water at the start of the visit. Urine samples then should be collected within 1 hour. Separate samples should be collected for CDU and B2MU using the following materials: Sterile urine collection cups (250 ml); small sealable plastic bags; preprinted labels; 15-ml polypropylene or polyethylene screw-cap tubes; lab gloves ("metal free"); and, preservatives (as indicated).

The sealed collection cup should be kept in the plastic bag until collection time. The workers should wash their hands with soap and water before receiving the collection cup.

The collection cup should not be opened until just before voiding and the cup should be sealed immediately after filling. It is important that the inside of the container and cap are not touched by, or come into contact with, the body, clothing or other surfaces.

For CDU analyzes, the cup is swirled gently to resuspend any solids, and the 15-ml tube is filled with 10-12 ml urine. The CDC recommends the addition of 100 µl concentrated $HNO_3$ as a preservative before sealing the tube and then freezing the sample. The CTQ recommends minimal handling and does not acidify their interlaboratory urine reference materials prior to shipment, nor do they freeze the sample for shipment. At the CTQ, if the urine sample has much sediment, the sample is acidified in the lab to free any cadmium in the precipitate.

For B2M, the urine sample should be collected directly into a polyethylene bottle previously washed with dilute nitric acid. The pH of the urine should be measured and adjusted to 8.0 with 0.1 N NaOH immediately following collection. Samples should be frozen and stored at -20 °C until testing is performed. The B2M in the samples should be stable for 2 days when stored at 2-8 °C, and for at least 2 months at -20 °C. Repeated freezing and thawing should be avoided to prevent denaturing the B2M (Pharmacia 1990).

#### 3.4.2 Recommendations for Evaluating Laboratories

Using standard error data and the results of proficiency testing obtained from CTQ, responsible physicians can make an informed choice of which laboratory to select to analyze biological samples. In general, laboratories with small standard errors and little disparity between target and measured values tend to make precise and accurate sample determinations. Estimates of precision provided to the physicians with each set of monitoring results can be compared to previously-reported proficiency and precision estimates. The latest precision estimates should be at least as small as the standard error reported previously by the laboratory. Moreover, there should be no indication that precision is deteriorating (i.e., increasing values for the precision estimates). If precision is deteriorating, physicians may decide to use another laboratory for these analyses. QA/QC information provided by the participating laboratories to physicians can, therefore, assist physicians in evaluating laboratory performance.

#### 3.4.3 Use and Interpretation of Results

When the responsible physician has received the CDB, CDU and/or B2MU results, these results must be compared to the action levels discussed in the final rule for cadmium. The comparison of the sample results to action levels is straightforward. The measured value reported from the laboratory can be compared directly to the action levels; if the reported value exceeds an action level, the required actions must be initiated.

### 4.0 Background

Cadmium is a naturally-occurring environmental contaminant to which humans are continually exposed in food, water, and air. The average daily intake of cadmium by the U.S. population is estimated to be 10-20 µg/day. Most of this intake is via ingestion, for which absorption is estimated at 4-7% (Kowal et al. 1979). An additional nonoccupational source of cadmium is smoking tobacco; smoking a pack of cigarettes a day adds an additional 2-4 µg cadmium to the daily intake, assuming absorption via inhalation of 25-35% (Nordberg and Nordberg 1988; Friberg and Elinder 1988; Travis and Haddock 1980).

Exposure to cadmium fumes and dusts in an occupational setting where air concentrations are 20-50 µg/m $^3$ results in an additional daily intake of several hundred micrograms (Friberg and Elinder 1988, p. 563). In such a setting, occupational exposure to cadmium occurs primarily via inhalation, although additional exposure may occur through the ingestion of material via contaminated hands if workers eat or smoke without first washing. Some of the particles that are inhaled initially may be ingested when the material is deposited in the upper respiratory tract, where it may be cleared by mucociliary transport and subsequently swallowed.

Cadmium introduced into the body through inhalation or ingestion is transported by the albumin fraction of the blood plasma to the liver, where it accumulates and is stored principally as a bound form complexed with the protein metallothionein. Metallothionein-bound cadmium is the main form of cadmium subsequently transported to the kidney; it is these 2 organs, the liver and kidney, in which the majority of the cadmium body burden accumulates. As much as one half of the total body burden of cadmium may be found in the kidneys (Nordberg and Nordberg 1988).

Once cadmium has entered the body, elimination is slow; about 0.02% of the body burden is excreted per day via urinary/fecal elimination. The whole-body half-life of cadmium is 10-35 years, decreasing slightly with increasing age (Travis and Haddock 1980).

The continual accumulation of cadmium is the basis for its chronic noncarcinogenic toxicity. This accumulation makes the kidney the target organ in which cadmium toxicity usually is first observed (Piscator 1964). Renal damage may occur when cadmium levels in the kidney cortex approach 200 µg/g wet tissue-weight (Travis and Haddock 1980).

The kinetics and internal distribution of cadmium in the body are complex, and depend on whether occupational exposure to cadmium is ongoing or has terminated. In general, cadmium in blood is related principally to recent cadmium exposure, while cadmium in urine reflects cumulative exposure (i.e., total body burden) (Lauwerys et al. 1976; Friberg and Elinder 1988).

### 4.1 Health Effects

Studies of workers in a variety of industries indicate that chronic exposure to cadmium may be linked to several adverse health effects including kidney dysfunction, reduced pulmonary function, chronic lung disease and cancer (Federal Register 1990). The primary sites for cadmium-associated cancer appear to be the lung and the prostate.

*Cancer.* Evidence for an association between cancer and cadmium exposure comes from both epidemiological studies and animal experiments. Pott (1965) found a statistically significant elevation in the incidence of prostate cancer among a cohort of cadmium workers. Other epidemiology studies also report an elevated incidence of prostate cancer; however, the increases observed in these other studies were not statistically significant (Meridian Research, Inc. 1989).

One study (Thun et al. 1985) contains sufficiently quantitative estimates of cadmium exposure to allow evaluation of dose-response relationships between cadmium exposure and lung cancer. A statistically significant excess of lung cancer attributed to cadmium exposure was found in this study, even after accounting for confounding variables such as coexposure to arsenic and smoking habits (Meridian Research, Inc. 1989).

Evidence for quantifying a link between lung cancer and cadmium exposure comes from a single study (Takenaka et al. 1983). In this study, dose-response relationships developed from animal data were extrapolated to humans using a variety of models. OSHA chose the multistage risk model for estimating the risk of cancer for humans using these animal data. Animal injection studies also suggest an association between cadmium exposure and cancer, particularly observations of an increased incidence of tumors at sites remote from the point of injection. The International Agency for Research on Cancer (IARC) (Supplement 7, 1987) indicates that this, and related, evidence is sufficient to classify cadmium as an animal carcinogen. However, the results of these injection studies cannot be used to quantify risks attendant to human occupational exposures due to differences in routes of exposure (Meridian Research, Inc. 1989).

Based on the above-cited studies, the U.S. Environmental Protection Agency (EPA) classifies cadmium as "B1," a probable human carcinogen (USEPA 1985). IARC in 1987 recommended that cadmium be listed as a probable human carcinogen.

*Kidney Dysfunction.* The most prevalent nonmalignant effect observed among workers chronically exposed to cadmium is kidney dysfunction. Initially, such dysfunction is manifested by proteinuria (Meridian Research, Inc. 1989; Roth Associates, Inc. 1989). Proteinuria associated with cadmium exposure is most commonly characterized by excretion of low-molecular weight proteins (15,000-40,000 MW), accompanied by loss of electrolytes, uric acid, calcium, amino acids, and phosphate. Proteins commonly excreted include β-2-microglobulin (B2M), retinol-binding protein (RBP), immunoglobulin light chains, and lysozyme. Excretion of low molecular weight proteins is characteristic of damage to the proximal tubules of the kidney (Iwao et al. 1980).

Exposure to cadmium also may lead to urinary excretion of high-molecular weight proteins such as albumin, immunoglobulin G, and glycoproteins (Meridian Research, Inc. 1989; Roth Associates, Inc. 1989). Excretion of high-molecular weight proteins is indicative of damage to the glomeruli of the kidney. Bernard et al. (1979) suggest that cadmium-associated damage to the glomeruli and damage to the proximal tubules of the kidney develop independently of each other, but may occur in the same individual.

Several studies indicate that the onset of low-molecular weight proteinuria is a sign of irreversible kidney damage (Friberg et al. 1974; Roels et al. 1982; Piscator 1984; Elinder et al. 1985; Smith et al. 1986). For many workers, once sufficiently elevated levels of B2M are observed in association with cadmium exposure, such levels do not appear to return to normal even when cadmium exposure is eliminated by removal of the worker from the cadmium-contaminated work environment (Friberg, exhibit 29, 1990).

Some studies indicate that cadmium-induced proteinuria may be progressive; levels of B2MU increase even after cadmium exposure has ceased (Elinder et al. 1985). Other researchers have reached similar conclusions (Frieburg testimony, OSHA docket exhibit 29, Elinder testimony, OSHA docket exhibit 55, and OSHA docket exhibits 8-86B). Such observations are not universal, however (Smith et al. 1986; Tsuchiya 1976). Studies in which proteinuria has not been observed, however, may have initiated the reassessment too early (Meridian Research, Inc.1989; Roth Associates, Inc. 1989; Roels 1989).

A quantitative assessment of the risks of developing kidney dysfunction as a result of cadmium exposure was performed using the data from Ellis et al. (1984) and Falck et al. (1983). Meridian Research, Inc. (1989) and Roth Associates, Inc. (1989) employed several mathematical models to evaluate the data from the 2 studies, and the results indicate that cumulative cadmium exposure levels between 5 and 100 $\mu$g-years/m$^3$ correspond with a one-in-a-thousand probability of developing kidney dysfunction.

When cadmium exposure continues past the onset of early kidney damage (manifested as proteinuria), chronic nephrotoxicity may occur (Meridian Research, Inc. 1989; Roth Associates, Inc. 1989). Uremia, which is the loss of the glomerulus' ability to adequately filter blood, may result. This condition leads to severe disturbance of electrolyte concentrations, which may result in various clinical complications including atherosclerosis, hypertension, pericarditis, anemia, hemorrhagic tendencies, deficient cellular immunity, bone changes, and other problems. Progression of the disease may require dialysis or a kidney transplant.

Studies in which animals are chronically exposed to cadmium confirm the renal effects observed in humans (Friberg et al. 1986). Animal studies also confirm cadmium-related problems with calcium metabolism and associated skeletal effects, which also have been observed among humans. Other effects commonly reported in chronic animal studies include anemia, changes in liver morphology, immunosuppression and hypertension. Some of these effects may be associated with cofactors; hypertension, for example, appears to be associated with diet, as well as with cadmium exposure. Animals injected with cadmium also have shown testicular necrosis.

### 4.2 Objectives for Medical Monitoring

In keeping with the observation that renal disease tends to be the earliest clinical manifestation of cadmium toxicity, the final cadmium standard mandates that eligible workers must be medically monitored to prevent this condition (as well as cadmimum-induced cancer). The objectives of medical-monitoring, therefore, are to: Identify workers at significant risk of adverse health effects from excess, chronic exposure to cadmium; prevent future cases of cadmium-induced disease; detect and minimize existing cadmium-induced disease; and, identify workers most in need of medical intervention.

The overall goal of the medical monitoring program is to protect workers who may be exposed continuously to cadmium over a 45-year occupational lifespan. Consistent with this goal, the medical monitoring program should assure that:

1. Current exposure levels remain sufficiently low to prevent the accumulation of cadmium body burdens sufficient to cause disease in the future by monitoring CDB as an indicator of recent cadmium exposure;

2. Cumulative body burdens, especially among workers with undefined historical exposures, remain below levels potentially capable of leading to damage and disease by assessing CDU as an indicator of cumulative exposure to cadmium; and,

3. Health effects are not occurring among exposed workers by determining B2MU as an early indicator of the onset of cadmium-induced kidney disease.

### 4.3 Indicators of Cadmium Exposure and Disease

Cadmium is present in whole blood bound to albumin, in erythrocytes, and as a metallothionein-cadmium complex. The metallothionein-cadmium complex that represents the primary transport mechanism for cadmium delivery to the kidney. CDB concentrations in the general, nonexposed population average 1 $\mu$g Cd/l whole blood, with smokers exhibiting higher levels (see Section 5.1.6). Data presented in Section 5.1.6 shows that 95% of the general population not occupationally exposed to cadmium have CDB levels less than 5 $\mu$g Cd/l.

If total body burdens of cadmium remain low, CDB concentrations indicate recent exposure (i.e., daily intake). This conclusion is based on data showing that cigarette smokers exhibit CDB concentrations of 2-7 $\mu$g/l depending on the number of cigarettes smoked per day (Nordberg and Nordberg 1988), while CDB levels for those who quit smoking return to general population values (approximately 1 $\mu$g/l) within several weeks (Lauwerys et al. 1976). Based on these observations, Lauwerys et al. (1976) concluded that CDB has a biological half-life of a few weeks to less than 3 months. As indicated in Section 3.1.6, the upper 95th percentile for CDB levels observed among those who are not occupationally exposed to cadmium is 5 μg/l, which suggests that the absolute upper limit to the range reported for smokers by Nordberg and Nordberg may have been affected by an extreme value (i.e., beyond 2$\hat{\sigma}$ above the mean).

Among occupationally-exposed workers, the occupational history of exposure to cadmium must be evaluated to interpret CDB levels. New workers, or workers with low exposures to cadmium, exhibit CDB levels that are representative of recent exposures, similar to the general population. However, for workers with a history of chronic exposure to cadmium, who have accumulated significant stores of cadmium in the kidneys/liver, part of the CDB concentrations appear to indicate body burden. If such workers are removed from cadmium exposure, their CDB levels remain elevated, possibly for years, reflecting prior long-term accumulation of cadmium in body tissues. This condition tends to occur, however, only beyond some threshold exposure value, and possibly indicates the capacity of body tissues to accumulate cadmium which cannot be excreted readily (Friberg and Elinder 1988; Nordberg and Nordberg 1988).

CDU is widely used as an indicator of cadmium body burdens (Nordberg and Nordberg 1988). CDU is the major route of elimination and, when CDU is measured, it is commonly expressed either as μg Cd/l urine (unadjusted), μg Cd/l urine (adjusted for specific gravity), or μg Cd/g CRTU (see Section 5.2.1). The metabolic model for CDU is less complicated than CDB, since CDU is dependentin large part on the body (i.e., kidney) burden of cadmium. However, a small proportion of CDU still be attributed to recent cadmium exposure, particularly if exposure to high airborne concentrations of cadmium occurred. Note that CDU is subject to larger interindividual and day-to-day variations than CDB, so repeated measurements are recommended for CDU evaluations.

CDU is bound principally to metallothionein, regardless of whether the cadmium originates from metallothionein in plasma or from the cadmium pool accumulated in the renal tubules. Therefore, measurement of metallothionein in urine may provide information similar to CDU, while avoiding the contamination problems that may occur during collection and handling urine for cadmium analysis (Nordberg and Nordberg 1988). However, a commercial method for the determination of metallothionein at the sensitivity levels required under the final cadmium rule is not currently available; therefore, analysis of CDU is recommended.

Among the general population not occupationally exposed to cadmium, CDU levels average less than 1 μg/l (see Section 5.2.7). Normalized for creatinine (CRTU), the average CDU concentration of the general population is less than 1 μg/g CRTU. As cadmium accumulates over the lifespan, CDU increases with age. Also, cigarette smokers may eventually accumulate twice the cadmium body burden of nonsmokers, CDU is slightly higher in smokers than in nonsmokers, even several years after smoking cessation (Nordberg and Nordberg 1988). Despite variations due to age and smoking habits, 95% of those not occupationally exposed to cadmium exhibit levels of CDU less than 3 μg/g CRTU (based on the data presented in Section 5.2.7).

About 0.02% of the cadmium body burden is excreted daily in urine. When the critical cadmium concentration (about 200 ppm) in the kidney is reached, or if there is sufficient cadmium-induced kidney dysfunction, dramatic increases in CDU are observed (Nordberg and Nordberg 1988). Above 200 ppm, therefore, CDU concentrations cease to be an indicator of cadmium body burden, and are instead an index of kidney failure.

Proteinuria is an index of kidney dysfunction, and is defined by OSHA to be a material impairment. Several small proteins may be monitored as markers for proteinuria. Below levels indicative of proteinuria, these small proteins may be early indicators of increased risk of cadmium-induced renal tubular disease. Analytes useful for monitoring cadmium-induced renal tubular damage include:

1. β-2-Microglobulin (B2M), currently the most widely used assay for detecting kidney dysfunction, is the best characterized analyte available (Iwao et al. 1980; Chia et al. 1989);

2. Retinol Binding Protein (RBP) is more stable than B2M in acidic urine (i.e., B2M breakdown occurs if urinary pH is less than 5.5; such breakdown may result in false [i.e., low] B2M values [Bernard and Lauwerys, 1990]);

3. N-Acetyl-B-Glucosaminidase (NAG) is the analyte of an assay that is simple, inexpensive, reliable, and correlates with cadmium levels under 10 μg/g CRTU, but the assay is less sensitive than RBP or B2M (Kawada et al. 1989);

4. Metallothionein (MT) correlates with cadmium and B2M levels, and may be a better predictor of cadmium exposure than CDU and B2M (Kawada et al. 1989);

5. Tamm-Horsfall Glycoprotein (THG) increases slightly with elevated cadmium levels, but this elevation is small compared to increases in urinary albumin, RBP, or B2M (Bernard and Lauwerys 1990);

6. Albumin (ALB), determined by the biuret method, is not sufficiently sensitive to serve as an early indicator of the onset of renal disease (Piscator 1962);

7. Albumin (ALB), determined by the Amido Black method, is sensitive and reproducible, but involves a time-consuming procedure (Piscator 1962);

8. Glycosaminoglycan (GAG) increases among cadmium workers, but the significance of this effect is unknown because no relationship has been found between elevated GAG and other indices of tubular damage (Bernard and Lauwerys 1990);

9. Trehalase seems to increase earlier than B2M during cadmium exposure, but the procedure for analysis is complicated and unreliable (Iwata et al. 1988); and,

10. Kallikrein is observed at lower concentrations among cadmium-exposed workers than among normal controls (Roels et al. 1990).

Of the above analytes, B2M appears to be the most widely used and best characterized analyte to evaluate the presence/absence, as well as the extent of, cadmium-induced renal tubular damage (Kawada, Koyama, and Suzuki 1989; Shaikh and Smith 1984; Nogawa 1984). However, it is important that samples be collected and handled so as to minimize B2M degradation under acidic urine conditions.

The threshold value of B2MU commonly used to indicate the presence of kidney damage 300 μg/g CRTU (Kjellstrom et al. 1977a; Buchet et al. 1980; and Kowal and Zirkes 1983). This value represents the upper 95th or 97.5th percentile level of urinary excretion observed among those without tubular dysfunction (Elinder, exbt L-140-45, OSHA docket H057A). In agreement with these conclusions, the data presented in Section 5.3.7 of this protocol generally indicate that the level of 300 μg/g CRTU appears to define the boundary for kidney dysfunction. It is not clear, however, that this level represents the upper 95th percentile of values observed among those who fail to demonstrate proteinuria effects.

Although elevated B2MU levels appear to be a fairly specific indicator of disease associated with cadmium exposure, other conditions that may lead to elevated B2MU levels include high fevers from influenza, extensive physical exercise, renal disease unrelated to cadmium exposure, lymphomas, and AIDS (Iwao et al. 1980; Schardun and van Epps 1987). Elevated B2M levels observed in association with high fevers from influenza or from extensive physical exercise are transient, and will return to normal levels once the fever has abated or metabolic rates return to baseline values following exercise. The other conditions linked to elevated B2M levels can be diagnosed as part of a properly-designed medical examination. Consequently, monitoring B2M, when accompanied by regular medical examinations and CDB and CDU determinations (as indicators of present and past cadmium exposure), may serve as a specific, early indicator of cadmium-induced kidney damage.

### 4.4 Criteria for Medical Monitoring of Cadmium Workers

Medical monitoring mandated by the final cadmium rule includes a combination of regular medical examinations and periodic monitoring of 3 analytes: CDB, CDU and B2MU. As indicated above, CDB is monitored as an indicator of current cadmium exposure, while CDU serves as an indicator of the cadmium body burden; B2MU is assessed as an early marker of irreversible kidney damage and disease.

The final cadmium rule defines a series of action levels that have been developed for each of the 3 analytes to be monitored. These action levels serve to guide the responsible physician through a decision-making process. For each action level that is exceeded, a specific response is mandated. The sequence of action levels, and the attendant actions, are described in detail in the final cadmium rule.

Other criteria used in the medical decision-making process relate to tests performed during the medical examination (including a determination of the ability of a worker to wear a respirator). These criteria, however, are not affected by the results of the analyte determinations addressed in the above paragraphs and, consequently, will not be considered further in these guidelines.

### 4.5 Defining to Quality and Proficiency of the Analyte Determinations

As noted above in Sections 2 and 3, the quality of a measurement should be defined along with its value to properly interpret the results. Generally, it is necessary to know the accuracy and the precision of a measurement before it can be properly evaluated. The precision of the data from a specific laboratory indicates the extent to which the repeated measurements of the same sample vary within that laboratory. The accuracy of the data provides an indication of the extent to which these results deviate from average results determined from many laboratories performing the same measurement (i.e., in the absence of an independent determination of the true value of a measurement). Note that terms are defined operationally relative to the manner in which they will be used in this protocol. Formal definitions for the terms in italics used in this section can be found in the list of definitions (Section 2).

Another data quality criterion required to properly evaluate measurement results is the limit of detection of that measurement. For measurements to be useful, the range of the measurement which is of interest for biological monitoring purposes must lie entirely above the limit of detection defined for that measurement.

The overall quality of a laboratory's results is termed the performance of that laboratory. The degree to which a laboratory satisfies a minimum performance level is referred to as the proficiency of the laboratory. A successful medical monitoring program, therefore, should include procedures developed for monitoring and recording laboratory performance; these procedures can be used to identify the most proficient laboratories.

### 5.0 Overview of Medical Monitoring Tests for CDB, CDU, B2MU and CRTU

To evaluate whether available methods for assessing CDB, CDU, B2MU and CRTU are adequate for determining the parameters defined by the proposed action levels, it is necessary to review procedures available for sample collection, preparation and analysis. A variety of techniques for these purposes have been used historically for the determination of cadmium in biological matrices (including CDB and CDU), and for the determination of specific proteins in biological matrices (including B2MU). However, only the most recent techniques are capable of satisfying the required accuracy, precision and sensitivity (i.e., limit of detection) for monitoring at the levels mandated in the final cadmium rule, while still facilitating automated analysis and rapid processing.

### 5.1 Measuring Cadmium in Blood (CDB)

Analysis of biological samples for cadmium requires strict analytical discipline regarding collection and handling of samples. In addition to occupational settings, where cadmium contamination would be apparent, cadmium is a ubiquitous environmental contaminant, and much care should be exercised to ensure that samples are not contaminated during collection, preparation or analysis. Many common chemical reagents are contaminated with cadmium at concentrations that will interfere with cadmium analysis; because of the widespread use of cadmium compounds as colored pigments in plastics and coatings, the analyst should continually monitor each manufacturer's chemical reagents and collection containers to prevent contamination of samples.

Guarding against cadmium contamination of biological samples is particularly important when analyzing blood samples because cadmium concentrations in blood samples from nonexposed populations are generally less than 2 μg/l (2 ng/ml), while occupationally-exposed workers can be at medical risk to cadmium toxicity if blood concentrations exceed 5 μg/l (ACGIH 1991 and 1992). This narrow margin between exposed and unexposed samples requires that exceptional care be used in performing analytic determinations for biological monitoring for occupational cadmium exposure.

Methods for quantifying cadmium in blood have improved over the last 40 years primarily because of improvements in analytical instrumentation. Also, due to improvements in analytical techniques, there is less need to perform extensive multi-step sample preparations prior to analysis. Complex sample preparation was previously required to enhance method sensitivity (for cadmium), and to reduce interference by other metals or components of the sample.

#### 5.1.1 Analytical Techniques Used to Monitor Cadmium in Biological Matrices

**TABLE 3—COMPARISON OF ANALYTICAL PROCEDURES/INSTRUMENTATION FOR DETERMINATION OF CADMIUM IN BIOLOGICAL SAMPLES**

| *Analytical procedure* | *Limit of detection [ng/(g or ml)]* | *Specified biological matrix* | *Reference* | *Comments* |
|---|---|---|---|---|
| Flame Atomic Absorption Spectroscopy (FAAS). | ≥1.0 | Any matrix .... | Perkin-Elmer (1982). | Not sensitive enough for biomonitoring without extensive sample digestion, metal chelation and organic solvent extraction. |
| Graphite Furnace Atomic Absorption Spectroscopy (GFAAS). | 0.04 | Urine ........ | Pruszkowska et al. (1983). | Methods of choice for routine cadmium analysis. |
| | ≥0.20 | Blood ........ | Stoeppler and Brandt (1980). | |
| Inductively-Coupled Argon-Plasma Atomic Emission Spectroscopy (ICAP AES). | 2.0 | Any matrix .... | NIOSH (1984A) | Requires extensive sample preparation and concentration of metal with chelating resin. Advantage is simultaneous analyses for as many as 10 metals from 1 sample. |
| Neutron Activation Gamma Spectroscopy (NA). | 1.5 | In vivo (liver) .. | Ellis et al. (1983) | Only available *in vivo* method for direct determination of cadmium body tissue burdens; expensive; absolute determination of cadmium in reference materials. |
| Isotope Dilution Mass Spectroscopy (IDMS). | <1.0 | Any matrix .... | Michiels and DeBievre (1986). | Suitable for absolute determination of cadmium in reference materials; expensive. |
| Differential Pulse Anodic Stripping Voltammetry (DPASV). | <1.0 | Any matrix .... | Stoeppler and Brandt (1980). | Suitable for absolute determination of cadmium in reference materials; efficient method to check accuracy of analytical method. |

A number of analytical techniques have been used for determining cadmium concentrations in biological materials. A summary of the characteristics of the most widely employed techniques is presented in Table 3. The technique most suitable for medical monitoring for cadmium is atomic absorption spectroscopy (AAS).

To obtain a measurement using AAS, a light source (i.e., hollow cathode or lectrode-free discharge lamp) containing the element of interest as the cathode, is energized and the lamp emits a spectrum that is unique for that element. This light source is focused through a sample cell, and a selected wavelength is monitored by a monochrometer and photodetector cell. Any ground state atoms in the sample that match those of the lamp element and are in the path of the emitted light may absorb some of the light and decrease the amount of light that reaches the photodetector cell. The amount of light absorbed at each characteristic wavelength is proportional to the number of ground state atoms of the corresponding element that are in the pathway of the light between the source and detector.

To determine the amount of a specific metallic element in a sample using AAS, the sample is dissolved in a solvent and aspirated into a high-temperature flame as an aerosol. At high temperatures, the solvent is rapidly evaporated or decomposed and the solute is initially solidified; the majority of the sample elements then are transformed into an atomic vapor. Next, a light beam is focused above the flame and the amount of metal in the sample can be determined by measuring the degree of absorbance of the atoms of the target element released by the flame at a characteristic wavelength.

A more refined atomic absorption technique, flameless AAS, substitutes an electrothermal, graphite furnace for the flame. An aliquot (10-100 µl) of the sample is pipetted into the cold furnace, which is then heated rapidly to generate an atomic vapor of the element.

AAS is a sensitive and specific method for the elemental analysis of metals; its main drawback is nonspecific background absorbtion and scattering of the light beam by particles of the sample as it decomposes at high temperatures; nonspecific absorbance reduces the sensitivity of the analytical method. The problem of nonspecific absorbance and scattering can be reduced by extensive sample pretreatment, such as ashing and/or acid digestion of the sample to reduce its organic content.

Current AAS instruments employ background correction devices to adjust electronically for background absorbtion and scattering. A common method to correct for background effects is to use a deuterium arc lamp as a second light source. A continuum light source, such as the deuterium lamp, emits a broad spectrum of wavelengths instead of specific wavelengths characteristic of a particular element, as with the hollow cathode tube. With this system, light from the primary source and the continuum source are passed alternately through the sample cell. The target element effectively absorbs light only from the primary source (which is much brighter than the continuum source at the characteristic wavelengths), while the background matrix absorbs and scatters light from both sources equally. Therefore, when the ratio of the two beams is measured electronically, the effect of nonspecific background absorption and scattering is eliminated. A less common, but more sophisticated, backgrond correction system is based on the Zeeman effect, which uses a magnetically-activated light polarizer to compensate electronically for nonspecific absorbtion and scattering.

Atomic emission spectroscopy with inductively-coupled argon plasma (AES-ICAP) is widely used to analyze for metals. With this instrument, the sample is aspirated into an extremely hot argon plasma flame, which excites the metal atoms; emission spectra specific for the sample element then are generated. The quanta of emitted light passing through a monochrometer are amplified by photomultiplier tubes and measured by a photodetector to determine the amount of metal in the sample. An advantage of AES-ICAP over AAS is that multi-elemental analyses of a sample can be performed by simultaneously measuring specific elemental emission energies. However, AES-ICAP lacks the sensitivity of AAS, exhibiting a limit of detection which is higher than the limit of detection for graphite-furnace AAS (Table 3).

Neutron activation (NA) analysis and isotope dilution mass spectrometry (IDMS) are 2 additional, but highly specialized, methods that have been used for cadmium determinations. These methods are expensive because they require elaborate and sophisticated instrumentation.

NA analysis has the distinct advantage over other analytical methods of being able to determine cadmium body burdens in specific organs (e.g., liver, kidney) in vivo (Ellis et al. 1983). Neutron bombardment of the target transforms cadmium-113 to cadmium-114, which promptly decays ($<10^{-14}$ sec) to its ground state, emitting gamma rays that are measured using large gamma detectors; appropriate shielding and instrumentation are required when using this method.

IDMS analysis, a definitive but laborious method, is based on the change in the ratio of 2 isotopes of cadmium (cadmium 111 and 112) that occurs when a known amount of the element (with an artificially altered ratio of the same isotopes [i.e., a cadmium 111 "spike"] is added to a weighed aliquot of the sample (Michiels and De Bievre 1986).

### 5.1.2 Methods Developed for CDB Determinations

A variety of methods have been used for preparing and analyzing CDB samples; most of these methods rely on one of the analytical techniques described above. Among the earliest reports, Princi (1947) and Smith et al. (1955) employed a colorimetric procedure to analyze for CDB and CDU. Samples were dried and digested through several cycles with concentrated mineral acids ($HNO_3$ and $H_2SO_4$) and hydrogen peroxide ($H_2O_2$). The digest was neutralized, and the cadmium was complexed with diphenylthiocarbazone and extracted with chloroform. The dithizone-cadmium complex then was quantified using a spectrometer.

Colorimetric procedures for cadmium analyses were replaced by methods based on atomic absorption spectroscopy (AAS) in the early 1960s, but many of the complex sample preparation procedures were retained. Kjellstrom (1979) reports that in Japanese, American and Swedish laboratories during the early 1970s, blood samples were wet ashed with mineral acids or ashed at high temperature and wetted with nitric acid. The cadmium in the digest was complexed with metal chelators including diethyl dithiocarbamate (DDTC), ammonium pyrrolidine dithiocarbamate (APDC) or diphenylthiocarbazone (dithizone) in ammonia-citrate buffer and extracted with methyl isobutyl ketone (MIBK). The resulting solution then was analyzed by flame AAS or graphite-furnace AAS forcadmium determinations using deuterium-lamp background correction.

In the late 1970s, researchers began developing simpler preparation procedures. Roels et al. (1978) and Roberts and Clark (1986) developed simplified digestion procedures. Using the Roberts and Clark method, a 0.5 ml aliquot of blood is collected and transferred to a digestion tube containing 1 ml concentrated $HNO_3$. The blood is then digested at 110 °C for 4 hours. The sample is reduced in volume by continued heating, and 0.5 ml 30% $H_2O_2$ is added as the sample dries. The residue is dissolved in 5 ml dilute (1%) $HNO_3$, and 20 µl of sample is then analyzed by graphite-furnace AAS with deuterium-background correction.

The current trend in the preparation of blood samples is to dilute the sample and add matrix modifiers to reduce background interference, rather than digesting the sample to reduce organic content. The method of Stoeppler and Brandt (1980), and the abbreviated procedure published in the American Public Health Association's (APHA) *Methods for Biological Monitoring* (1988), are straightforward and are nearly identical. For the APHA method, a small aliquot (50-300 µl) of whole blood that has been stabilized with ethylenediaminetetraacetate (EDTA) is added to 1.0 ml 1 $MHNO_3$, vigorously shaken and centrifuged. Aliquots (10-25 µl) of the supernatant then are then analyzed by graphite-furnace AAS with appropriate background correction.

Using the method of Stoeppler and Brandt (1980), aliquots (50-200 µl) of whole blood that have been stabilized with EDTA are pipetted into clean polystyrene tubes and mixed with 150-600 µ l of 1 M $HNO_3$. After vigorous shaking, the solution is centrifuged and a 10-25 µl aliquot of the supernatant then is analyzed by graphite-furnace AAS with appropriate background correction.

Claeys-Thoreau (1982) and DeBenzo et al. (1990) diluted blood samples at a ratio of 1:10 with a matrix modifier (0.2% Triton X-100, a wetting agent) for direct determinations of CDB. DeBenzo et al. also demonstrated that aqueous standards of cadmium, instead of spiked, whole-blood samples, could be used to establish calibration curves if standards and samples are treated with additional small volumes of matrix modifiers (i.e., 1% $HNO_3$, 0.2% ammonium hydrogenphosphate and 1 mg/ml magnesium salts).

These direct dilution procedures for CDB analysis are simple and rapid. Laboratories can process more than 100 samples a day using a dedicated graphite-furnace AAS, an auto-sampler, and either a Zeeman- or a deuterium-background correction system. Several authors emphasize using optimum settings for graphite-furnace temperatures during the drying, charring, and atomization processes associated with the flameless AAS method, and the need to run frequent QC samples when performing automated analysis.

### 5.1.3 Sample Collection and Handling

Sample collection procedures are addressed primarily to identify ways to minimize the degree of variability that may be introduced by sample collection during medical monitoring. It is unclear at this point the extent to which collection procedures contribute to variability among CDB samples. Sources of variation that may result from sampling procedures include time-of-day effects and introduction of external contamination during the collection process. To minimize these sources, strict adherence to a sample collection protocol is recommended. Such a protocol must include provisions for thorough cleaning of the site from which blood will be extracted; also, every effort should be made to collect samples near the same time of day. It is also important to recognize that under the recent OSHA blood-borne pathogens standard (29 CFR 1910.1030), blood samples and certain body fluids must be handled and treated as if they are infectious.

### 5.1.4 Best Achievable Performance

The best achievable performance using a particular method for CDB determinations is assumed to be equivalent to the performance reported by research laboratories in which the method was developed.

For their method, Roberts and Clark (1986) demonstrated a limit of detection of 0.4 μg Cd/l in whole blood, with a linear response curve from 0.4 to 16.0 μg Cd/l. They report a coefficient of variation (CV) of 6.7% at 8.0 μg/l.

The APHA (1988) reports a range of 1.0-25 μg/l, with a CV of 7.3% (concentration not stated). Insufficient documentation was available to critique this method.

Stoeppler and Brandt (1980) achieved a detection limit of 0.2 μg Cd/l whole blood, with a linear range of 0.4-12.0 μg Cd/l, and a CV of 15-30%, for samples at <1.0 μg/l. Improved precision (CV of 3.8%) was reported for CDB concentrations at 9.3 μg/l.

### 5.1.5 General Method Performance

For any particular method, the performance expected from commercial laboratories may be somewhat lower than that reported by the research laboratory in which the method was developed. With participation in appropriate proficiency programs and use of a proper in-house QA/QC program incorporating provisions for regular corrective actions, the performance of commercial laboratories is expected to approach that reported by research laboratories. Also, the results reported for existing proficiency programs serve as a gauge of the likely level of performance that currently can be expected from commercial laboratories offering these analyses.

Weber (1988) reports on the results of the proficiency program run by the Centre de Toxicologie du Quebec (CTQ). As indicated previously, participants in that program receive 18 blood samples per year having cadmium concentrations ranging from 0.2-20 μg/l. Currently, 76 laboratories are participating in this program. The program is established for several analytes in addition to cadmium, and not all of these laboratories participate in the cadmium proficiency-testing program.

Under the CTQ program, cadmium results from individual laboratories are compared against the consensus mean derived for each sample. Results indicate that after receiving 60 samples (i.e., after participation for approximately three years), 60% of the laboratories in the program are able to report results that fall within ± 1 μg/l or 15% of the mean, whichever is greater. (For this procedure, the 15% criterion was applied to concentrations exceeding 7 μg/l.) On any single sample of the last 20 samples, the percentage of laboratories falling within the specified range is between 55 and 80%.

The CTQ also evaluates the performance of participating laboratories against a less severe standard: ±2 μg/l or 15% of the mean, whichever is greater (Weber 1988); 90% of participating laboratories are able to satisfy this standard after approximately 3 years in the program. (The 15% criterion is used for concentrations in excess of 13 μg/l.) On any single sample of the last 15 samples, the percentage of laboratories falling within the specified range is between 80 and 95% (except for a single test for which only 60% of the laboratories achieved the desired performance).

Based on the data presented in Weber (1988), the CV for analysis of CDB is nearly constant at 20% for cadmium concentrations exceeding 5 μg/l, and increases for cadmium concentrations below 5 μg/l. At 2 μg/l, the reported CV rises to approximately 40%. At 1 μg/l, the reported CV is approximately 60%.

Participating laboratories also tend to overestimate concentrations for samples exhibiting concentrations less than 2 μg/l (see Figure 11 of Weber 1988). This problem is due in part to the proficiency evaluation criterion that allows reporting a minimum ±2.0 μg/l for evaluated CDB samples. There is currently little economic or regulatory incentive for laboratories participating in the CTQ program to achieve greater accuracy for CDB samples containing cadmium at concentrations less than 2.0 μg/l, even if the laboratory has the experience and competency to distinguish among lower concentrations in the samples obtained from the CTQ.

The collective experience of international agencies and investigators demonstrate the need for a vigorous QC program to ensure that CDB values reported by participating laboratories are indeed reasonably accurate. As Friberg (1988) stated:

"Information about the quality of published data has often been lacking. This is of concern as assessment of metals in trace concentrations in biological media are fraught with difficulties from the collection, handling, and storage of samples to the chemical analyses. This has been proven over and over again from the results of interlaboratory testing and quality control exercises. Large variations in results were reported even from 'experienced' laboratories."

The UNEP/WHO global study of cadmium biological monitoring set a limit for CDB accuracy using the maximum allowable deviation method at Y=X ± (0.1 X +1) for a targeted concentration of 10 μg Cd/l (Friberg and Vahter 1983). The performance of participating laboratories over a concentration range of 1.5-12 μg/l was reported by Lind et al. (1987). Of the 3 QC runs conducted during 1982 and 1983, 1 or 2 of the 6 laboratories failed each run. For the years 1983 and 1985, between zero and 2 laboratories failed each of the consecutive QC runs.

In another study (Vahter and Friberg 1988), QC samples consisting of both external (unknown) and internal (stated) concentrations were distributed to laboratories participating in the epidemiology research. In this study, the maximum acceptable deviation between the regression analysis of reported results and reference values was set at Y=X±(0.05X+0.2) for a concentration range of 0.3-5.0 μg Cd/l. It is reported that only 2 of 5 laboratories had acceptable data after the first QC set, and only 1 of 5 laboratories had acceptable data after the second QC set. By the fourth QC set, however, all 5 laboratories were judged proficient.

The need for high quality CDB monitoring is apparent when the toxicological and biological characteristics of this metal are considered; an increase in CDB from 2 to 4 μg/l could cause a doubling of the cadmium accumulation in the kidney, a critical target tissue for selective cadmium accumulation (Nordberg and Nordberg 1988).

Historically, the CDC's internal QC program for CDB cadmium monitoring program has found achievable accuracy to be ±10% of the true value at CDB concentrations ≥5.0 μg/l (Paschal 1990). Data on the performance of laboratories participating in this program currently are not available.

### 5.1.6 Observed CDB Concentrations

As stated in Section 4.3, CDB concentrations are representative of ongoing levels of exposure to cadmium. Among those who have been exposed chronically to cadmium for extended periods, however, CDB may contain a component attributable to the general cadmium body burden.

#### 5.1.6.1 *CDB Concentrations Among Unexposed Samples*

Numerous studies have been conducted examining CDB concentrations in the general population, and in control groups used for comparison with cadmium-exposed workers. A number of reports have been published that present erroneously high values of CDB (Nordberg and Nordberg 1988). This problem was due to contamination of samples during sampling and analysis, and to errors in analysis. Early AAS methods were not sufficiently sensitive to accurately estimate CDB concentrations.

Table 4 presents results of recent studies reporting CDB levels for the general U.S. population not exposed occupationally to cadmium. Other surveys of tissue cadmium using U.S. samples and conducted as part of a cooperative effort among Japan, Sweden and the U.S., did not collect CDB data because standard analytical methodologies were unavailable, and because of analytic problems (Kjellstrom 1979; SWRI 1978).

## TABLE 4—BLOOD CADMIUM CONCENTRATIONS OF U.S. POPULATION NOT OCCUPATIONALLY EXPOSED TO CADMIUM [a]

| Study No. | No. in study (n) | Sex | Age | Smoking habits [b] | Arithmetic mean (± S.D.) [c] | Absolute range or (95% CI) [d] | Geometric mean (± GSD) [e] | Lower 95th percentile of distribution [f] | Upper 95th percentile of distribution [f] | Reference |
|---|---|---|---|---|---|---|---|---|---|---|
| 1 ....... | 80 | M | 4 to 69 ...... | NS,S | 1.13 | 0.35-3.3 | 0.98±1.71 | 0.4 | 2.4 | Kowal et al. (1979). |
| | 88 | F | 4 to 69 ...... | NS,S | 1.03 | 0.21-3.3 | 0.91±1.63 | 0.4 | 2.0 | |
| | 115 | M/F | 4 to 69 ...... | NS | 0.95 | 0.21-3.3 | 0.85±1.59 | 0.4 | 1.8 | |
| | 31 | M/F | 4 to 69 ...... | S | 1.54 | 0.4-3.3 | 1.37±1.65 | 0.6 | 3.2 | |
| 2 ....... | 10 | M | Adults ...... | (?) | 2.0±2.1 | (0.5-5.0) | | [g](0) | [g](5.8) | Ellis et al. (1983). |
| 3 ....... | 24 | M | Adults ...... | NS | | | 0.6±1/87 | 0.2 | 1.8 | Frieberg and Vahter (1983). |
| | 20 | M | Adults ...... | S | | | 1.2±2.13 | 0.3 | 4.4 | |
| | 64 | F | Adults ...... | NS | | | 0.5±1.85 | 0.2 | 1.4 | |
| | 39 | F | Adults ...... | S | | | 0.8±2.22 | 0.2 | 3.1 | |
| 4 ....... | 32 | M | Adults ...... | S,NS | | | 1.2±2.0 | 0.4 | 3.9 | Thun et al. (1989). |
| 5 ....... | 35 | M | Adults ...... | (?) | 2.1±2.1 | (0.5-7.3) | | [g](0) | [g](5.6) | Mueller et al. (1989). |

[a] Concentrations reported in μg Cd/I blood unless otherwise stated.
[b] NS—never smoked; S—current cigarette smoker.
[c] S.D.—Arithmetic Standard Deviation.
[d] C.I.—Confidence interval.
[e] GSD—Geometric Standard Deviation.
[f] Based on an assumed lognormal distribution.
[g] Based on an assumed normal distribution.

Arithmetic and/or geometric means and standard deviations are provided in Table 4 for measurements among the populations defined in each study listed. The range of reported measurements and/or the 95% upper and lower confidence intervals for the means are presented when this information was reported in a study. For studies reporting either an arithmetic or geometric standard deviation along with a mean, the lower and upper 95th percentile for the distribution also were derived and reported in the table.

The data provided in table 4 from Kowal et al. (1979) are from studies conducted between 1974 and 1976 evaluating CDB levels for the general population in Chicago, and are considered to be representative of the U.S. population. These studies indicate that the average CDB concentration among those not occupationally exposed to cadmium is approximately 1 μg/l.

In several other studies presented in Table 4, measurements are reported separately for males and females, and for smokers and nonsmokers. The data in this table indicate that similar CDB levels are observed among males and females in the general population, but that smokers tend to exhibit higher CDB levels than nonsmokers. Based on the Kowal et al. (1979) study, smokers not occupationally exposed to cadmium exhibit an average CDB level of 1.4 μg/l.

In general, nonsmokers tend to exhibit levels ranging to 2 μg/l, while levels observed among smokers range to 5 μg/l. Based on the data presented in Table 4, 95% of those not occupationally exposed to cadmium exhibit CDB levels less than 5 μg/l.

### 5.1.6.2 CDB Concentrations Among Exposed Workers

Table 5 is a summary of results from studies reporting CDB levels among workers exposed to cadmium in the work place. As in Table 4, arithmetic and/or geometric means and standard deviations are provided if reported in the listed studies. The absolute range, or the 95% confidence interval around the mean, of the data in each study are provided when reported. In addition, the lower and upper 95th percentile of the distribution are presented for each study in which a mean and corresponding standard deviation were reported. Table 5 also provides estimates of the duration, and level, of exposure to cadmium in the work place if these data were reported in the listed studies. The data presented in table 5 suggest that CDB levels are dose related. Sukuri et al. (1983) show that higher CDB levels are observed among workers experiencing higher work place exposure. This trend appears to be true of the studies listed in the table.

CDB levels reported in table 5 are higher among those showing signs of cadmium-related kidney damage than those showing no such damage. Lauwerys et al. (1976) report CDB levels among workers with kidney lesions that generally are above the levels reported for workers without kidney lesions. Ellis et al. (1983) report a similar observation comparing workers with and without renal dysfunction, although they found more overlap between the 2 groups than Lauwerys et al.

## TABLE 5—BLOOD CADMIUM IN WORKERS EXPOSED TO CADMIUM IN THE WORKPLACE

| Study number | Work environment (worker population monitored) | Number in study | Employment in years (mean) | Mean concentration of cadmium in air (µg/m³) | Concentrations of Cadmium in blood[a] | | | | | Reference |
|---|---|---|---|---|---|---|---|---|---|---|
| | | | | | Arithmetic mean (± S.D.)[b] | Absolute range or (95% C.I.)[c] | Geometric mean (GSD)[d] | Lower 95th percentile of range[e] ( )[f] | Upper 95th percentile of range[e] ( )[f] | |
| 1 ............ | Ni-Cd battery plant and Cd production plant: | | 3–40 | ≤90 | | | | | | Lauwerys et al. 1976. |
| | (Workers without kidney lesions) .. | 96 | ........ | ........ | 21.4±1.9 | ........ | ........ | (18) | (25) | |
| | (Workers with kidney lesions) ....... | 25 | ........ | ........ | 38.8±3.8 | ........ | ........ | (32) | (45) | |
| 2 ............ | Ni-Cd battery plant: | | | | | | | | | Adamsson et al. (1979). |
| | (Smokers) ........ | 7 | (5) | 10.1 | 22.7 | 7.3–67.2 | | | | |
| | (Nonsmokers) ........ | 8 | (9) | 7.0 | 7.0 | 4.9–10.5 | | | | |
| 3 ............ | Cadmium alloy plant: | | | | | | | | | Sukuri et al. 1982. |
| | (High exposure group) ........ | 7 | (10.6) | [1,000–5 yrs; | 20.8±7.1 | ........ | ........ | (7.3) | (34) | |
| | (Low exposure group) ........ | 9 | (7.3) | 40–5 yrs] | 7.1±1.1 | ........ | ........ | (5.1) | (9.1) | |
| 4 ............ | Retrospective study of workers with renal problems: | 19 | 15–41 | | | | | | ........ | Roels et al. 1982. |
| | (Before removal) ........ | ........ | (27.2) | ........ | 39.9±3.7 | 11–179 | ........ | (34) | (46) | |
| | (After removal) ........ | ........ | [g](4.2) | ........ | 14.1±5.6 | 5.7–27.4 | ........ | (4.4) | (24) | |
| 5 ............ | Cadmium production plant: | | | | | | | | | Ellis et al. 1983. |
| | (Workers without renal dysfunction). | 33 | 1–34 | ........ | 15±5.7 | 7–31 | ........ | (5.4) | (25) | |
| | (Workers with renal dysfunction) ... | 18 | 10–34 | ........ | 24±8.5 | 10–34 | ........ | (9.3) | (39) | |
| 6 ............ | Cd-Cu alloy plant ........ | 75 | Up to 39 | ........ | ........ | ........ | 8.8±1.1 | 7.5 | 10 | Mason et al. 1988. |
| 7 ............ | Cadmium recovery operation—Current (19) and former (26) workers. | 45 | (19.0) | ........ | ........ | ........ | 7.9±2.0 | 2.5 | 25 | Thun et al. 1989. |
| 8 ............ | Cadmium recovery operation | 40 | ........ | ........ | 10.2±5.3 | 2.2–18.8 | ........ | (1.3) | (19) | Mueller et al. 1989. |

[a] Concentrations reported in µg Cd/l blood unless otherwise stated.
[b] S.D.—Standard Deviation.
[c] C.I.—Confidence Interval.
[d] GSD—Geometric Standard Deviation.
[e] Based on an assumed lognormal distribution.
[f] Based on an assumed normal distribution.
[g] Years following removal.

The data in table 5 also indicate that CDB levels are higher among those experiencing current occupational exposure than those who have been removed from such exposure. Roels et al. (1982) indicate that CDB levels observed among workers experiencing ongoing exposure in the work place are almost entirely above levels observed among workers removed from such exposure. This finding suggests that CDB levels decrease once cadmium exposure has ceased.

A comparison of the data presented in tables 4 and 5 indicates that CDB levels observed among cadmium-exposed workers is significantly higher than levels observed among the unexposed groups. With the exception of 2 studies presented in table 5 (1 of which includes former workers in the sample group tested), the lower 95th percentile for CDB levels among exposed workers are greater than 5 µg/l, which is the value of the upper 95th percentile for CDB levels observed among those who are not occupationally exposed. Therefore, a CDB level of 5 µg/l represents a threshold above which significant work place exposure to cadmium may be occurring.

### 5.1.7 Conclusions and Recommendations for CDB

Based on the above evaluation, the following recommendations are made for a CDB proficiency program.

#### 5.1.7.1 Recommended method

The method of Stoeppler and Brandt (1980) should be adopted for analyzing CDB. This method was selected over other methods for its straightforward sample-preparation procedures, and because limitations of the method were described adequately. It also is the method used by a plurality of laboratories currently participating in the CTQ proficiency program. In a recent CTQ interlaboratory comparison report (CTQ 1991), analysis of the methods used by laboratories to measure CDB indicates that 46% (11 of 24) of the participating laboratories used the Stoeppler and Brandt methodology ($HNO_3$ deproteinization of blood followed by analysis of the supernatant by GF-AAS). Other CDB methods employed by participating laboratories identified in the CTQ report include dilution of blood (29%), acid digestion (12%) and miscellaneous methods (12%).

Laboratories may adopt alternate methods, but it is the responsibility of the laboratory to demonstrate that the alternate methods meet the data quality objectives defined for the Stoeppler and Brandt method (see Section 5.1.7.2 below).

#### 5.1.7.2 Data quality objectives

Based on the above evaluation, the following data quality objectives (DQOs) should facilitate interpretation of analytical results.

*Limit of Detection.* 0.5 µg/l should be achievable using the Stoeppler and Brandt method. Stoeppler and Brandt (1980) report a limit of detection equivalent to ≤0.2 µg/l in whole blood using 25 µ l aliquots of deproteinized, diluted blood samples.

*Accuracy.* Initially, some of the laboratories performing CDB measurements may be expected to satisfy criteria similar to the less severe criteria specified by the CTQ program, i.e., measurements within 2 g/l or 15% (whichever is greater) of the target value. About 60% of the laboratories enrolled in the CTQ program could meet this criterion on the first proficiency test (Weber 1988).

Currently, approximately 12 laboratories in the CTQ program are achieving an accuracy for CDB analysis within the more severe constraints of ±1 µg/l or 15% (whichever is greater). Later, as laboratories gain experience, they should achieve the level of accuracy exhibited by these 12 laboratories. The experience in the CTQ program has shown that, even without incentives, laboratories benefit from the feedback of the program; after they have analyzed 40-50 control samples from the program, performance improves to the point where about 60% of the laboratories can meet the stricter criterion of ±1 µg/l or 15% (Weber 1988). Thus, this stricter target accuracy is a reasonable DQO.

*Precision.* Although Stoeppler and Brandt (1980) suggest that a coefficient of variation (CV) near 1.3% (for a 10 µg/l concentration) is achievable for within-run reproducibility, it is recognized that other factors affecting within- and between-run comparability will increase the achievable CV. Stoeppler and Brandt (1980) observed CVs that were as high as 30% for low concentrations (0.4 µg/l), and CVs of less than 5% for higher concentrations.

For internal QC samples (see Section 3.3.1), laboratories should attain an overall precision near 25%. For CDB samples with concentrations less than 2 µg/l, a target precision of 40% is reasonable, while precisions of 20% should be achievable for concentrations greater than 2 µg/l. Although these values are more strict than values observed in the CTQ interlaboratory program reported by Webber (1988), they are within the achievable limits reported by Stoeppler and Brandt (1980).

### 5.1.7.3 Quality assurance/quality control

Commercial laboratories providing measurement of CDB should adopt an internal QA/QC program that incorporates the following components: Strict adherence to the selected method, including all calibration requirements; regular incorporation of QC samples during actual runs; a protocol for corrective actions, and documentation of these actions; and, participation in an interlaboratory proficiency program. Note that the nonmandatory QA/QC program presented in Attachment 1 is based on the Stoeppler and Brandt method for CDB analysis. Should an alternate method be adopted, the laboratory should develop a QA/QC program satisfying the provisions of Section 3.3.1.

## 5.2 Measuring Cadmium in Urine (CDU)

As in the case of CDB measurement, proper determination of CDU requires strict analytical discipline regarding collection and handling of samples. Because cadmium is both ubiquitous in the environment and employed widely in coloring agents for industrial products that may be used during sample collection, preparation and analysis, care should be exercised to ensure that samples are not contaminated during the sampling procedure.

Methods for CDU determination share many of the same features as those employed for the determination of CDB. Thus, changes and improvements to methods for measuring CDU over the past 40 years parallel those used to monitor CDB. The direction of development has largely been toward the simplification of sample preparation techniques made possible because of improvements in analytic techniques.

### 5.2.1 Units of CDU Measurement

Procedures adopted for reporting CDU concentrations are not uniform. In fact, the situation for reporting CDU is more complicated than for CDB, where concentrations are normalized against a unit volume of whole blood.

Concentrations of solutes in urine vary with several biological factors (including the time since last voiding and the volume of liquid consumed over the last few hours); as a result, solute concentrations should be normalized against another characteristic of urine that represents changes in solute concentrations. The 2 most common techniques are either to standardize solute concentrations against the concentration of creatinine, or to standardize solute concentrations against the specific gravity of the urine. Thus, CDU concentrations have been reported in the literature as "uncorrected" concentrations of cadmium per volume of urine (i.e., µg Cd/l urine), "corrected" concentrations of cadmium per volume of urine at a standard specific gravity (i.e., µg Cd/l urine at a specific gravity of 1.020), or "corrected" mass concentration per unit mass of creatinine (i.e., µg Cd/g creatinine). (CDU concentrations [whether uncorrected or corrected for specific gravity, or normalized to creatinine] occasionally are reported in nanomoles [i.e., nmoles] of cadmium per unit mass or volume. In this protocol, these values are converted to µg of cadmium per unit mass or volume using 89 nmoles of cadmium=10 µg.)

While it is agreed generally that urine values of analytes should be normalized for reporting purposes, some debate exists over what correction method should be used. The medical community has long favored normalization based on creatinine concentration, a common urinary constituent. Creatinine is a normal product of tissue catabolism, is excreted at a uniform rate, and the total amount excreted per day is constant on a day-to-day basis (NIOSH 1984b). While this correction method is accepted widely in Europe, and within some occupational health circles, Kowals (1983) argues that the use of specific gravity (i.e., total solids per unit volume) is more straightforward and practical (than creatinine) in adjusting CDU values for populations that vary by age or gender.

Kowals (1983) found that urinary creatinine (CRTU) is lower in females than males, and also varies with age. Creatinine excretion is highest in younger males (20-30 years old), decreases at middle age (50-60 years), and may rise slightly in later years. Thus, cadmium concentrations may be underestimated for some workers with high CRTU levels.

Within a single void urine collection, urine concentration of any analyte will be affected by recent consumption of large volumes of liquids, and by heavy physical labor in hot environments. The absolute amount of analyte excreted may be identical, but concentrations will vary widely so that urine must be corrected for specific gravity (i.e., to normalize concentrations to the quantity of total solute) using a fixed value (e.g., 1.020 or 1.024). However, since heavy-metal exposure may increase urinary protein excretion, there is a tendency to underestimate cadmium concentrations in samples with high specific gravities when specific-gravity corrections are applied.

Despite some shortcomings, reporting solute concentrations as a function of creatinine concentration is accepted generally; OSHA therefore recommends that CDU levels be reported as the mass of cadmium per unit mass of creatinine (µg/g CTRU).

Reporting CDU as µg/g CRTU requires an additional analytical process beyond the analysis of cadmium: Samples must be analyzed independently for creatinine so that results may be reported as the ratio of cadmium to creatinine concentrations found in the urine sample. Consequently, the overall quality of the analysis depends on the combined performance by a laboratory on these 2 determinations. The analysis used for CDU determinations is addressed below in terms of µg Cd/l, with analysis of creatinine addressed separately. Techniques for assessing creatinine are discussed in Section 5.4.

Techniques for deriving cadmium as a ratio of CRTU, and the confidence limits for independent measurements of cadmium and CRTU, are provided in Section 3.3.3.

### 5.2.2 Analytical Techniques Used to Monitor CDU

Analytical techniques used for CDU determinations are similar to those employed for CDB determinations; these techniques are summarized in Table 3. As with CDB monitoring, the technique most suitable for CDU determinations is atomic absorption spectroscopy (AAS). AAS methods used for CDU determinations typically employ a graphite furnace, with background correction made using either the deuterium-lamp or Zeeman techniques; Section 5.1.1 provides a detailed description of AAS methods.

### 5.2.3 Methods Developed for CDU Determinations

Princi (1947), Smith et al. (1955), Smith and Kench (1957), and Tsuchiya (1967) used colorimetric procedures similar to those described in the CDB section above to estimate CDU concentrations. In these methods, urine (50 ml) is reduced to dryness by heating in a sand bath and digested (wet ashed) with mineral acids. Cadmium then is complexed with dithiazone, extracted with chloroform and quantified by spectrophotometry. These early studies typically report reagent blank values equivalent to 0.3 µg Cd/l, and CDU concentrations among nonexposed control groups at maximum levels of 10 µg Cd/l—erroneously high values when compared to more recent surveys of cadmium concentrations in the general population.

By the mid-1970s, most analytical procedures for CDU analysis used either wet ashing (mineral acid) or high temperatures (>400 °C) to digest the organic matrix of urine, followed by cadmium chelation with APDC or DDTC solutions and extraction with MIBK. The resulting aliquots were analyzed by flame or graphite-furnace AAS (Kjellstrom 1979).

Improvements in control over temperature parameters with electrothermal heating devices used in conjunction with flameless AAS techniques, and optimization of temperature programs for controlling the drying, charring, and atomization processes in sample analyses, led to improved analytical detection of diluted urine samples without the need for sample digestion or ashing. Roels et al. (1978) successfully used a simple sample preparation, dilution of 1.0 ml aliquots of urine with 0.1 N $HNO_3$, to achieve accurate low-level determinations of CDU.

In the method described by Pruszkowska et al. (1983), which has become the preferred method for CDU analysis, urine samples were diluted at a ratio of 1:5 with water; diammonium hydrogenphosphate in dilute $HNO_3$ was used as a matrix modifier. The matrix modifier allows for a higher charring temperature without loss of cadmium through volatilization during preatomization. This procedure also employs a stabilized temperature platform in a graphite furnace, while nonspecific background absorbtion is corrected using the Zeeman technique. This method allows for an absolute detection limit of approximately 0.04 µg Cd/l urine.

### 5.2.4 Sample Collection and Handling

Sample collection procedures for CDU may contribute to variability observed among CDU measurements. Sources of variation attendant to sampling include time-of-day, the interval since ingestion of liquids, and the introduction of external contamination during the collection process. Therefore, to minimize contributions from these variables, strict adherence to a sample-collection protocol is recommended. This protocol should include provisions for normalizing the conditions under which urine is collected. Every effort also should be made to collect samples during the same time of day.

Collection of urine samples from an industrial work force for biological monitoring purposes usually is performed using "spot" (i.e., single-void) urine with the pH of the sample determined immediately. Logistic and sample-integrity problems arise when efforts are made to collect urine over long periods (e.g., 24 hrs). Unless single-void urines are used, there are numerous opportunities for measurement error because of poor control over sample collection, storage and environmental contamination.

To minimize the interval during which sample urine resides in the bladder, the following adaption to the "spot" collection procedure is recommended: The bladder should first be emptied, and then a large glass of water should be consumed; the sample may be collected within an hour after the water is consumed.

#### 5.2.5 Best Achievable Performance

Performance using a particular method for CDU determinations is assumed to be equivalent to the performance reported by the research laboratories in which the method was developed. Pruszkowska et al. (1983) report a detection limit of 0.04 μg/l CDU, with a CV of <4% between 0-5 μg/l. The CDC reports a minimum CDU detection limit of 0.07 μg/l using a modified method based on Pruszkowska et al. (1983). No CV is stated in this protocol; the protocol contains only rejection criteria for internal QC parameters used during accuracy determinations with known standards (Attachment 8 of exhibit 106 of OSHA docket H057A). Stoeppler and Brandt (1980) report a CDU detection limit of 0.2 μ/l for their methodology.

#### 5.2.6 General Method Performance

For any particular method, the expected initial performance from commercial laboratories may be somewhat lower than that reported by the research laboratory in which the method was developed. With participation in appropriate proficiency programs, and use of a proper in-house QA/QC program incorporating provisions for regular corrective actions, the performance of commercial laboratories may be expected to improve and approach that reported by a research laboratories. The results reported for existing proficiency programs serve to specify the initial level of performance that likely can be expected from commercial laboratories offering analysis using a particular method.

Weber (1988) reports on the results of the CTQ proficiency program, which includes CDU results for laboratories participating in the program. Results indicate that after receiving 60 samples (i.e., after participating in the program for approximately 3 years), approximately 80% of the participating laboratories report CDU results ranging between ±2 μg/l or 15% of the consensus mean, whichever is greater. On any single sample of the last 15 samples, the proportion of laboratories falling within the specified range is between 75 and 95%, except for a single test for which only 60% of the laboratories reported acceptable results. For each of the last 15 samples, approximately 60% of the laboratories reported results within ±1 μg or 15% of the mean, whichever is greater. The range of concentrations included in this set of samples was not reported.

Another report from the CTQ (1991) summarizes preliminary CDU results from their 1991 interlaboratory program. According to the report, for 3 CDU samples with values of 9.0, 16.8, 31.5 μg/l, acceptable results (target of ±2 μg/l or 15 % of the consensus mean, whichever is greater) were achieved by only 44-52% of the 34 laboratories participating in the CDU program. The overall CVs for these 3 CDU samples among the 34 participating laboratories were 31%, 25%, and 49%, respectively. The reason for this poor performance has not been determined.

A more recent report from the CTQ (Weber, private communication) indicates that 36% of the laboratories in the program have been able to achieve the target of ± 1 μg/l or 15% for more than 75% of the samples analyzed over the last 5 years, while 45% of participating laboratories achieved a target of ±2 μg/l or 15% for more than 75% of the samples analyzed over the same period.

Note that results reported in the interlaboratory programs are in terms of μg Cd/l of urine, unadjusted for creatinine. The performance indicated, therefore, is a measure of the performance of the cadmium portion of the analyses, and does not include variation that may be introduced during the analysis of CRTU.

#### 5.2.7 Observed CDU Concentrations

Prior to the onset of renal dysfunction, CDU concentrations provide a general indication of the exposure history (i.e., body burden) (see Section 4.3). Once renal dysfunction occurs, CDU levels appear to increase and are no longer indicative solely of cadmium body burden (Friberg and Elinder 1988).

##### *5.2.7.1 Range of CDU concentrations observed among unexposed samples*

Surveys of CDU concentrations in the general population were first reported from cooperative studies among industrial countries (i.e., Japan, U.S. and Sweden) conducted in the mid-1970s. In summarizing these data, Kjellstrom (1979) reported that CDU concentrations among Dallas, Texas men (age range: <9-59 years; smokers and nonsmokers) varied from 0.11-1.12 μg/l (uncorrected for creatinine or specific gravity). These CDU concentrations are intermediate between population values found in Sweden (range: 0.11-0.80 μg/l) and Japan (range: 0.14-2.32 μg/l).

Kowal and Zirkes (1983) reported CDU concentrations for almost 1,000 samples collected during 1978-79 from the general U.S. adult population (i.e., nine states; both genders; ages 20-74 years). They report that CDU concentrations are lognormally distributed; low levels predominated, but a small proportion of the population exhibited high levels. These investigators transformed the CDU concentrations values, and reported the same data 3 different ways: μg/l urine (unadjusted), μg/l (specific gravity adjusted to 1.020), and μg/g CRTU. These data are summarized in Tables 6 and 7.

Based on further statistical examination of these data, including the lifestyle characteristics of this group, Kowal (1988) suggested increased cadmium absorption (i.e., body burden) was correlated with low dietary intakes of calcium and iron, as well as cigarette smoking.

CDU levels presented in Table 6 are adjusted for age and gender. Results suggest that CDU levels may be slightly different among men and women (i.e., higher among men when values are unadjusted, but lower among men when the values are adjusted, for specific gravity or CRTU). Mean differences among men and women are small compared to the standard deviations, and therefore may not be significant. Levels of CDU also appear to increase with age. The data in Table 6 suggest as well that reporting CDU levels adjusted for specific gravity or as a function of CRTU results in reduced variability.

**TABLE 6—URINE CADMIUM CONCENTRATIONS IN THE U.S. ADULT POPULATION: NORMAL AND CONCENTRATION-ADJUSTED VALUES BY AGE AND SEX[1]**

| | Geometric means (and geometric standard deviations) | | |
|---|---|---|---|
| | Unadjusted (μg/l) | SG-adjusted[2] μg/l at 1.020) | Creatine-adjusted (μg/g) |
| Sex: | | | |
| Male (n=484) | 0.55 (2.9) | 0.73 (2.6) | 0.55 (2.7) |
| Female (n=498) | 0.49 (3.0) | 0.86 (2.7) | 0.78 (2.7) |
| Age: | | | |
| 20–29 (n=222) | 0.32 (3.0) | 0.43 (2.7) | 0.32 (2.7) |
| 30–39 (n=141) | 0.46 (3.2) | 0.70 (2.8) | 0.54 (2.7) |
| 40–49 (n=142) | 0.50 (3.0) | 0.81 (2.6) | 0.70 (2.7) |
| 50–59 (n=117) | 0.61 (2.9) | 0.99 (2.4) | 0.90 (2.3) |
| 60–69 (n=272) | 0.76 (2.6) | 1.16 (2.3) | 1.03 (2.3) |

[1] From Kowal and Zirkes 1983.
[2] SC-adjusted is adjusted for specific gravity.

## TABLE 7—URINE CADMIUM CONCENTRATIONS IN THE U.S. ADULT POPULATION: CUMULATIVE FREQUENCY DISTRIBUTION OF URINARY CADMIUM (N=982)[1]

| Range of concentrations | Unadjusted (µg/l) percent | SG-adjusted (µg/l at 1.020) percent | Creatine-adjusted (µg/g) percent |
|---|---|---|---|
| <0.5 | 43.9 | 28.0 | 35.8 |
| 0.6–1.0 | 71.7 | 56.4 | 65.6 |
| 1.1–1.5 | 84.4 | 74.9 | 81.4 |
| 1.6–2.0 | 91.3 | 84.7 | 88.9 |
| 2.1–3.0 | 97.3 | 94.4 | 95.8 |
| 3.1–4.0 | 98.8 | 97.4 | 97.2 |
| 4.1–5.0 | 99.4 | 98.2 | 97.9 |
| 5.1–10.0 | 99.6 | 99.4 | 99.3 |
| 10.0–20.0 | 99.8 | 99.6 | 99.6 |

[1] Source: Kowal and Zirkes (1983).

The data in the Table 6 indicate the geometric mean of CDU levels observed among the general population is 0.52 µ/g Cd/l urine (unadjusted), with a geometric standard deviation of 3.0. Normalized for creatinine, the geometric mean for the population is 0.66 µ/g CRTU, with a geometric standard deviation of 2.7. Table 7 provides the distributions of CDU concentrations for the general population studied by Kowal and Zirkes. The data in this table indicate that 95% of the CDU levels observed among those not occupationally exposed to cadmium are below 3 µ/g CRTU.

### *5.2.7.2 Range of CDU concentrations observed among exposed workers*

Table 8 is a summary of results from available studies of CDU concentrations observed among cadmium-exposed workers. In this table, arithmetic and/or geometric means and standard deviations are provided if reported in these studies. The absolute range for the data in each study, or the 95% confidence interval around the mean of each study, also are provided when reported. The lower and upper 95th percentile of the distribution are presented for each study in which a mean and corresponding standard deviation were reported. Table 8 also provides estimates of the years of exposure, and the levels of exposure, to cadmium in the work place if reported in these studies. Concentrations reported in this table are in µ/g CRTU, unless otherwise stated.

## TABLE 8—URINE CADMIUM CONCENTRATIONS IN WORKERS EXPOSED TO CADMIUM IN THE WORKPLACE

| Study number | Work environment (worker population monitored) | Number in Study (n) | Employment in years (mean) | Mean Concentration of cadmium in air (µg/m³) | Concentration of cadmium in Urine[a] Arithmetic mean (± S.D.)[b] | Absolute range or (95% C.I.)[c] | Geometric mean (GSD)[d] | Lower 95th percentile of range[e] ( )[f] | Upper 95th percentile of range[e] ( )[f] | Reference |
|---|---|---|---|---|---|---|---|---|---|---|
| 1 | Ni-Cd battery plant and Cd production plant. | | 3–40 | ≤ 90 | | | | | | Lauwerys et al. 1976. |
| | (Workers without kidney lesions). | 96 | | | 16.3±16.7 | | | (0) | (44) | |
| | (Workers with kidney lesions). | 25 | | | 48.2±42.6 | | | (0) | (120) | |
| 2 | Ni-Cd battery plant | | | | | | | | | Adamsson et al. (1979). |
| | (Smokers) | 7 | (5) | 10.1 | 5.5 | 1.0–14.7 | | | | |
| | (Nonsmokers) | 8 | (9) | 7.0 | 3.6 | 0.5–9.3 | | | | |
| 3 | Cadmium salts production facility. | 148 | (15.4) | | 15.8 | 2–150 | | | | Butchet et al. 1980. |
| 4 | Retrospective study of workers with renal problems. | 19 | 15–41 | | | | | | | Roels et al. 1982. |
| | (Before removal) | | (27.2) | | 39.4±28.1 | 10.8–117 | | (0) | (88) | |
| | (After removal) | | (4.2)[g] | | 16.4±9.0 | 80–42.3 | | (1.0) | (32) | |
| 5 | Cadmium production plant. | | | | | | | | | Ellis et al. 1983. |
| | (Workers without renal dysfunction). | 33 | 1–34 | | 9.4±6.9 | 2-27 | | (0) | (21) | |
| | (Workers with renal dysfunction). | 18 | 10–34 | | 22.8±12.7 | 8–55 | | (1) | (45) | |
| 6 | Cd-Cu alloy plant | 75 | Up to 39 | Note h | 6.9±9.4 | | | (0) | (23) | Mason et al. 1988. |
| 7 | Cadmium recovery operation. | 45 | (19) | 87 | 9.3±6.9 | | | (0) | (21) | Thun et al. 1989. |
| 8 | Pigment manufacturing plant. | 29 | (12.8) | 0.18–3.0 | | 0.2–9.5 | 1.1 | | | Mueller et al. 1989. |
| 9 | Pigment manufacturing plant. | 26 | (12.1) | ≤3.0 | | | 1.25±2.45 | 0.3 | 6 | Kawada et al. 1990. |

[a] Concentrations reported in µg/g Cr.
[b] S.D.—Standard Deviation.
[c] C.I.—Confidence Interval.
[d] GSD—Geometric Standard Deviation.
[e] Based on an assumed lognormal distribution.
[f] Based on an assumed normal distribution.
[g] Years following removal.
[h] Equivalent to 50 for 20–22 yrs

Data in Table 8 from Lauwerys et al. (1976) and Ellis et al. (1983) indicate that CDU concentrations are higher among those exhibiting kidney lesions or dysfunction than among those lacking these symptoms. Data from the study by Roels et al. (1982) indicate that CDU levels decrease among workers removed from occupational exposure to cadmium in comparison to workers experiencing ongoing exposure. In both cases, however, the distinction between the 2 groups is not as clear as with CDB; there is more overlap in CDU levels observed among each of the paired populations than is true for corresponding CDB levels. As with CDB levels, the data in Table 8 suggest increased CDU concentrations among workers who experienced increased overall exposure.

Although a few occupationally-exposed workers in the studies presented in Table 8 exhibit CDU levels below 3 µg/g CRTU, most of those workers exposed to cadmium levels in excess of the PEL defined in the final cadmium rule exhibit CDU levels above 3 µg/g CRTU; this level represents the upper 95th percentile of the CDU distribution observed among those who are not occupationally exposed to cadmium (Table 7).

The mean CDU levels reported in Table 8 among occupationallyexposed groups studied (except 2) exceed 3 μg/g CRTU. Correspondingly, the level of exposure reported in these studies (with 1 exception) are significantly higher than what workers will experience under the final cadmium rule. The 2 exceptions are from the studies by Mueller et al. (1989) and Kawada et al. (1990); these studies indicate that workers exposed to cadmium during pigment manufacture do not exhibit CDU levels as high as those levels observed among workers exposed to cadmium in other occupations. Exposure levels, however, were lower in the pigment manufacturing plants studied. Significantly, workers removed from occupational cadmium exposure for an average of 4 years still exhibited CDU levels in excess of 3 μg/g CRTU (Roels et al. 1982). In the single-exception study with a reported level of cadmium exposure lower than levels proposed in the final rule (i.e., the study of a pigment manufacturing plant by Kawada et al. 1990), most of the workers exhibited CDU levels less than 3 μg/g CRTU (i.e., the mean value was only 1.3 μg/g CRTU). CDU levels among workers with such limited cadmium exposure are expected to be significantly lower than levels of other studies reported in Table 8.

Based on the above data, a CDU level of 3 μg/g CRTU appear to represent a threshold above which significant work place exposure to cadmium occurs over the work span of those being monitored. Note that this threshold is not as distinct as the corresponding threshold described for CDB. In general, the variability associated with CDU measurements among exposed workers appears to be higher than the variability associated with CDB measurements among similar workers.

### 5.2.8 Conclusions and Recommendations for CDU

The above evaluation supports the following recommendations for a CDU proficiency program. These recommendations address only sampling and analysis procedures for CDU determinations specifically, which are to be reported as an unadjusted μg Cd/l urine. Normalizing this result to creatinine requires a second analysis for CRTU so that the ratio of the 2 measurements can be obtained. Creatinine analysis is addressed in Section 5.4. Formal procedures for combining the 2 measurements to derive a value and a confidence limit for CDU in μg/g CRTU are provided in Section 3.3.3.

#### 5.2.8.1 *Recommended method*

The method of Pruszkowska et al. (1983) should be adopted for CDU analysis. This method is recommended because it is simple, straightforward and reliable (i.e., small variations in experimental conditions do not affect the analytical results).

A synopsis of the methods used by laboratories to determine CDU under the interlaboratory program administered by the CTQ (1991) indicates that more than 78% (24 of 31) of the participating laboratories use a dilution method to prepare urine samples for CDU analysis. Laboratories may adopt alternate methods, but it is the responsibility of the laboratory to demonstrate that the alternate methods provide results of comparable quality to the Pruszkowska method.

#### 5.2.8.2 *Data quality objectives*

The following data quality objectives should facilitate interpretation of analytical results, and are achievable based on the above evaluation.

*Limit of Detection.* A level of 0.5 μg/l (i.e., corresponding to a detection limit of 0.5 μg/g CRTU, assuming 1 g CRT/l urine) should be achievable. Pruszkowska et al. (1983) achieved a limit of detection of 0.04 g/l for CDU based on the slope of the curve for their working standards (0.35 pg Cd/0.0044, A signal=1% absorbance using GF-AAS).

The CDC reports a minimum detection limit for CDU of 0.07 μg/l using a modified Pruszkowska method. This limit of detection was defined as 3 times the standard deviation calculated from 10 repeated measurements of a "low level" CDU test sample (Attachment 8 of exhibit 106 of OSHA docket H057A).

Stoeppler and Brandt (1980) report a limit of detection for CDU of 0.2 μg/l using an aqueous dilution (1:2) of the urine samples.

*Accuracy.* A recent report from the CTQ (Weber, private communication) indicates that 36% of the laboratories in the program achieve the target of ±1 μg/l or 15% for more than 75% of the samples analyzed over the last 5 years, while 45% of participating laboratories achieve a target of ±2 μg/l or 15% for more than 75% of the samples analyzed over the same period. With time and a strong incentive for improvement, it is expected that the proportion of laboratories successfully achieving the stricter level of accuracy should increase. It should be noted, however, these indices of performance do not include variations resulting from the ancillary measurement of CRTU (which is recommended for the proper recording of results). The low cadmium levels expected to be measured indicate that the analysis of creatinine will contribute relatively little to the overall variability observed among creatinine-normalized CDU levels (see Section 5.4). The initial target value for reporting CDU under this program, therefore, is set at ±1 μg/g CRTU or 15% (whichever is greater).

*Precision.* For internal QC samples (which are recommended as part of an internal QA/QC program, Section 3.3.1), laboratories should attain an overall precision of 25%. For CDB samples with concentrations less than 2 μg/l, a target precision of 40% is acceptable, while precisions of 20% should be achievable for CDU concentrations greater than 2 μg/l. Although these values are more stringent than those observed in the CTQ interlaboratory program reported by Webber (1988), they are well within limits expected to be achievable for the method as reported by Stoeppler and Brandt (1980).

#### 5.2.8.3 *Quality assurance/quality control*

Commercial laboratories providing CDU determinations should adopt an internal QA/QC program that incorporates the following components: Strict adherence to the selected method, including calibration requirements; regular incorporation of QC samples during actual runs; a protocol for corrective actions, and documentation of such actions; and, participation in an interlaboratory proficiency program. Note that the nonmandatory program presented in Attachment 1 as an example of an acceptable QA/QC program, is based on using the Pruszkowska method for CDU analysis. Should an alternate method be adopted by a laboratory, the laboratory should develop a QA/QC program equivalent to the nonmandatory program, and which satisfies the provisions of Section 3.3.1.

### 5.3 Monitoring β-2-Microglobulin in Urine (B2MU)

As indicated in Section 4.3, B2MU appears to be the best of several small proteins that may be monitored as early indicators of cadmium-induced renal damage. Several analytic techniques are available for measuring B2M.

#### 5.3.1 Units of B2MU Measurement

Procedures adopted for reporting B2MU levels are not uniform. In these guidelines, OSHA recommends that B2MU levels be reported as μg/g CRTU, similar to reporting CDU concentrations. Reporting B2MU normalized to the concentration of CRTU requires an additional analytical process beyond the analysis of B2M: Independent analysis for creatinine so that results may be reported as a ratio of the B2M and creatinine concentrations found in the urine sample. Consequently, the overall quality of the analysis depends on the combined performance on these 2 analyses. The analysis used for B2MU determinations is described in terms of μg B2M/l urine, with analysis of creatinine addressed separately. Techniques used to measure creatinine are provided in Section 5.4. Note that Section 3.3.3 provides techniques for deriving the value of B2M as function of CRTU, and the confidence limits for independent measurements of B2M and CRTU.

#### 5.3.2 Analytical Techniques Used to Monitor B2MU

One of the earliest tests used to measure B2MU was the radial immunodiffusion technique. This technique is a simple and specific method for identification and quantitation of a number of proteins found in human serum and other body fluids when the protein is not readily differentiated by standard electrophoretic procedures. A quantitative relationship exists between the concentration of a protein deposited in a well that is cut into a thin agarose layer containing the corresponding monospecific antiserum, and the distance that the resultant complex diffuses. The wells are filled with an unknown serum and the standard (or control), and incubated in a moist environment at room temperature. After the optimal point of diffusion has been reached, the diameters of the resulting precipition rings are measured. The diameter of a ring is related to the concentration of the constituent substance. For B2MU determinations required in the medical monitoring program, this method requires a process that may be insufficient to concentrate the protein to levels that are required for detection.

Radioimmunoassay (RIA) techniques are used widely in immunologic assays to measure the concentration of antigen or antibody in body-fluid samples. RIA procedures are based on competitive-binding techniques. If antigen concentration is being measured, the principle underlying the procedure is that radioactive-labeled antigen competes with the sample's unlabeled antigen for binding sites on a known amount of immobile antibody. When these 3 components are present in the system, an equilibrium exists. This equilibrium is followed by a separation of the free and bound forms of the antigen. Either free or bound radioactivelabeled antigen can be assessed to determine the amount of antigen in the sample. The analysis is performed by measuring the level of radiation emitted either by the bound complex following removal of the solution containing the free antigen, or by the isolated solution containing the residual-free antigen. The main advantage of the RIA method is the extreme sensitivity of detection for emitted radiation and the corresponding ability to detect trace amounts of antigen. Additionally, large numbers of tests can be performed rapidly.

The enzyme-linked immunosorbent assay (ELISA) techniques are similar to RIA techniques except that nonradioactive labels are employed. This technique is safe, specific and rapid, and is nearly as sensitive as RIA techniques. An enzyme-labeled antigen is used in the immunologic assay; the labeled antigen detects the presence and quantity of unlabeled antigen in the sample. In a representative ELISA test, a plastic plate is coated with antibody (e.g., antibody to B2M). The antibody reacts with antigen (B2M) in the urine and forms an antigen-antibody complex on the plate. A second anti-B2M antibody (i.e., labeled with an enzyme) is added to the mixture and forms an antibody-antigen-antibody complex. Enzyme activity is measured spectrophotometrically after the addition of a specific chromogenic substrate which is activated by the bound enzyme. The results of a typical test are calculated by comparing the spectrophotometric reading of a serum sample to that of a control or reference serum. In general, these procedures are faster and require less laboratory work than other methods.

In a fluorescent ELISA technique (such as the one employed in the Pharmacia Delphia test for B2M), the labeled enzyme is bound to a strong fluorescent dye. In the Pharmacia Delphia test, an antigen bound to a fluorescent dye competes with unlabeled antigen in the sample for a predetermined amount of specific, immobile antibody. Once equilibrium is reached, the immobile phase is removed from the labeled antigen in the sample solution and washed; an enhancement solution then is added that liberates the fluorescent dye from the bound antigen-antibody complex. The enhancement solution also contains a chelate that complexes with the fluorescent dye in solution; this complex increases the fluorescent properties of the dye so that it is easier to detect.

To determine the quantity of B2M in a sample using the Pharmacia Delphia test, the intensity of the fluorescence of the enhancement solution is measured. This intensity is proportional to the concentration of labeled antigen that bound to the immobile antibody phase during the initial competition with unlabeled antigen from the sample. Consequently, the intensity of the fluorescence is an inverse function of the concentration of antigen (B2M) in the original sample. The relationship between the fluorescence level and the B2M concentration in the sample is determined using a series of graded standards, and extrapolating these standards to find the concentration of the unknown sample.

### 5.3.3 Methods Developed for B2MU Determinations

B2MU usually is measured by radioimmunoassay (RIA) or enzymelinked immunosorbent assay (ELISA); however, other methods (including gel electrophoresis, radial immunodiffusion, and nephelometric assays) also have been described (Schardun and van Epps 1987). RIA and ELISA methods are preferred because they are sensitive at concentrations as low as micrograms per liter, require no concentration processes, are highly reliable and use only a small sample volume.

Based on a survey of the literature, the ELISA technique is recommended for monitoring B2MU. While RIAs provide greater sensitivity (typically about 1 μg/l, Evrin et al. 1971), they depend on the use of radioisotopes; use of radioisotopes requires adherence to rules and regulations established by the Atomic Energy Commission, and necessitates an expensive radioactivity counter for testing. Radioisotopes also have a relatively short half-life, which corresponds to a reduced shelf life, thereby increasing the cost and complexity of testing. In contrast, ELISA testing can be performed on routine laboratory spectrophotometers, do *not* necessitate adherence to additional rules and regulations governing the handling of radioactive substances, and the test kits have long shelf lives. Further, the range of sensitivity commonly achieved by the recommended ELISA test (i.e., the Pharmacia Delphia test) is approximately 100 μg/l (Pharmacia 1990), which is sufficient for monitoring B2MU levels resulting from cadmium exposure. Based on the studies listed in Table 9 (Section 5.3.7), the average range of B2M concentrations among the general, nonexposed population falls between 60 and 300 μg/g CRTU. The upper 95th percentile of distributions, derived from studies in Table 9 which reported standard deviations, range between 180 and 1,140 μg/g CRTU. Also, the Pharmacia Delphia test currently is the most widely used test for assessing B2MU.

### 5.3.4 Sample Collection and Handling

As with CDB or CDU, sample collection procedures are addressed primarily to identify ways to minimize the degree of variability introduced by sample collection during medical monitoring. It is unclear the extent to which sample collection contributes to B2MU variability. Sources of variation include time-of-day effects, the interval since consuming liquids and the quantity of liquids consumed, and the introduction of external contamination during the collection process. A special problem unique to B2M sampling is the sensitivity of this protein to degradation under acid conditions commonly found in the bladder. To minimize this problem, strict adherence to a sampling protocol is recommended. The protocol should include provisions for normalizing the conditions under which the urine is collected. Clearly, it is important to minimize the interval urine spends in the bladder. It also is recommended that every effort be made to collect samples during the same time of day.

Collection of urine samples for biological monitoring usually is performed using "spot" (i.e., single-void) urine. Logistics and sample integrity become problems when efforts are made to collect urine over extended periods (e.g., 24 hrs). Unless single-void urines are used, numerous opportunities exist for measurement error because of poor control over sample collection, storage and environmental contamination.

To minimize the interval that sample urine resides in the bladder, the following adaption to the "spot" collection procedure is recommended: The bladder should be emptied and then a large glass of water should be consumed; the sample then should be collected within an hour after the water is consumed.

### 5.3.5 Best Achievable Performance

The best achievable performance is assumed to be equivalent to the performance reported by the manufacturers of the Pharmacia Delphia test kits (Pharmacia 1990). According to the insert that comes with these kits, QC results should be within ±2 SDs of the mean for each control sample tested; a CV of less than or equal to 5.2% should be maintained. The total CV reported for test kits is less than or equal to 7.2%.

### 5.3.6 General Method Performance

Unlike analyses for CDB and CDU, the Pharmacia Delphia test is standardized in a commercial kit that controls for many sources of variation. In the absence of data to the contrary, it is assumed that the achievable performance reported by the manufacturer of this test kit will serve as an achievable performance objective. The CTQ proficiency testing program for B2MU analysis is expected to use the performance parameters defined by the test kit manufacturer as the basis of the B2MU proficiency testing program.

Note that results reported for the test kit are expressed in terms of μg B2M/l of urine, and have not been adjusted for creatinine. The indicated performance, therefore, is a measure of the performance of the B2M portion of the analyses only, and does not include variation that may have been introduced during the analysis of creatinine.

### 5.3.7 Observed B2MU Concentrations

As indicated in Section 4.3, the concentration of B2MU may serve as an early indicator of the onset of kidney damage associated with cadmium exposure.

### 5.3.7.1 Range of B2MU Concentrations Among Unexposed Samples

Most of the studies listed in Table 9 report B2MU levels for those who were not occupationally exposed to cadmium. Studies noted in the second column of this table (which contain the footnote "d") reported B2MU concentrations among cadmium-exposed workers who, nonetheless, showed *no* signs of proteinuria. These latter studies are included in this table because, as indicated in Section 4.3, monitoring B2MU is intended to provide advanced warning of the onset of kidney dysfunction associated with cadmium exposure, rather than to distinguish relative exposure. This table, therefore, indicates the range of B2MU levels observed among those who had no symptoms of renal dysfunction (including cadmium-exposed workers with none of these symptoms).

**TABLE 9—B-2—MICROGLOBULIN CONCENTRATIONS OBSERVED IN URINE AMONG THOSE NOT OCCUPATIONALLY EXPOSED TO CADMIUM**

| Study No. | No. in study | Geometric mean | Geometric standard deviation | Lower 95th percentile of distribution[a] | Upper 95th percentile of distribution[a] | Reference |
|---|---|---|---|---|---|---|
| 1 | 133 m[b] | 115 μg/g[c]. | 4.03 | 12 | 1,140 μg/g[c]. | Ishizaki et al. 1989. |
| 2 | 161 f[b] | 146 μg/g[c]. | 3.11 | 23 | 940 μg/g[c]. | Ishizaki et al. 1989. |
| 3 | 10 | 84 μg/g | | | | Ellis et al. 1983. |
| 4 | 203 | 76 μg/l | | | | Stewart and Hughes 1981. |
| 5 | 9 | 103 μg/g | | | | Chia et al. 1989. |
| 6 | 47[d] | 86 μg/L | 1.9 | 30 μg/1 | 250 μg/L | Kjellstrom et al. 1977. |
| 7 | 1,000[e] | 68.1 μg/gr Cr[f]. | 3.1 m & f | < 10 μg/gr Cr[h]. | 320 μg/gr Cr[h]. | Kowal 1983. |
| 8 | 87 | 71 μg/g[i] | | 7[h] | 200[h] | Buchet et al. 1980. |
| 9 | 10 | 0.073 mg/24h. | | | | Evrin et al. 1971. |
| 10 | 59 | 156 μg/g | 1.1[j] | 130 | 180 | Mason et al. 1988. |
| 11 | 8 | 118 μg/g | | | | Iwao et al. 1980. |
| 12 | 34 | 79 μg/g | | | | Wibowo et al. 1982. |
| 13 | 41 m | | | | 400 μg/gr Cr[k]. | Falck et al. 1983. |
| 14 | 35[n] | 67 | | | | Roels et al. 1991. |
| 15 | 31[d] | 63 | | | | Roels et al. 1991. |
| 16 | 36[d] | 77[i] | | | | Miksche et al. 1981. |
| 17 | 18[n] | 130 | | | | Kawada et al. 1989. |
| 18 | 32[p] | 122 | | | | Kawada et al. 1989. |
| 19 | 18[d] | 295 | 1.4 | 170 | 510 | Thun et al. 1989. |

a—Based on an assumed lognormal distribution.
b—m = males, f = females.
c—Aged general population from non-polluted area; 47.9% population aged 50–69; 52.1% ≥ 70 years of age; values reported in study.
d—Exposed workers without proteinuria.
e—492 females, 484 male.
f—Creatinine adjusted; males = 68.1 μg/g Cr, females = 64.3 μg/g Cr.
h—Reported in the study.
i—Arithmetic mean.
j—Geometric standard error.
k—Upper 95% tolerance limits: for Falck this is based on the 24 hour urine sample.
n—Controls.
p—Exposed synthetic resin and pigment workers without proteinuria; Cadmium in urine levels up to 10 μg/g Cr.

To the extent possible, the studies listed in Table 9 provide geometric means and geometric standard deviations for measurements among the groups defined in each study. For studies reporting a geometric standard deviation along with a mean, the lower and upper 95th percentile for these distributions were derived and reported in the table.

The data provided from 15 of the 19 studies listed in Table 9 indicate that the geometric mean concentration of B2M observed among those who were not occupationally exposed to cadmium is 70-170 μg/g CRTU. Data from the 4 remaining studies indicate that exposed workers who exhibit no signs of proteinuria show mean B2MU levels of 60-300 μg/g CRTU. B2MU values in the study by Thun et al. (1989), however, appear high in comparison to the other 3 studies. If this study is removed, B2MU levels for those who are not occupationally exposed to cadmium are similar to B2MU levels found among cadmium-exposed workers who exhibit no signs of kidney dysfunction. Although the mean is high in the study by Thun et al., the range of measurements reported in this study is within the ranges reported for the other studies.

Determining a reasonable upper limit from the range of B2M concentrations observed among those who do not exhibit signs of proteinuria is problematic. Elevated B2MU levels are among the signs used to define the onset of kidney dysfunction. Without access to the raw data from the studies listed in Table 9, it is necessary to rely on reported standard deviations to estimate an upper limit for normal B2MU concentrations (i.e., the upper 95th percentile for the distributions measured). For the 8 studies reporting a geometric standard deviation, the upper 95th percentiles for the distributions are 180-1140 μg/g CRTU. These values are in general agreement with the upper 95th percentile for the distribution (i.e., 631 μg/g CRTU) reported by Buchet et al. (1980). These upper limits also appear to be in general agreement with B2MU values (i.e., 100-690 μg/g CRTU) reported as the normal upper limit by Iwao et al. (1980), Kawada et al. (1989), Wibowo et al. (1982), and Schardun and van Epps (1987). These values must be compared to levels reported among those exhibiting kidney dysfunction to define a threshold level for kidney dysfunction related to cadmium exposure.

### 5.3.7.2 Range of B2MU Concentrations Among Exposed Workers

Table 10 presents results from studies reporting B2MU determinations among those occupationally exposed to cadmium in the work place; in some of these studies, kidney dysfunction was observed among exposed workers, while other studies did not make an effort to distinguish among exposed workers based on kidney dysfunction. As with Table 9, this table provides geometric means and geometric standard deviations for the groups defined in each study if available. For studies reporting a geometric standard deviation along with a mean, the lower and upper 95th percentiles for the distributions are derived and reported in the table.

Perkin-Elmer Corporation. (1977). *Analytical Methods Using the HGA Graphite Furnace.*

Pharmacia Diagnostics. (1990). Pharmacia DELFIA system B-2microglobulin kit insert. Uppsala, Sweden: Pharmacia Diagnostics.

Piscator M. (1962). Proteinuria in chronic cadmium poisoning. *Archives of Environmental Health,5*, 55-62.

Potts, C.L. (1965). Cadmium Proteinuria—The Health Battery Workers Exposed to Cadmium Oxide dust. Ann Occup Hyg, 3:55-61, 1965.

Princi F. (1947). A study of industrial exposures to cadmium. *Journal of Industrial Hygiene and Toxicology, 29*, 315-320.

Pruszkowska E, Carnick G, and Slavin W. (1983). Direct determination of cadmium in urine with use of a stabilized temperature platform furnace and Zeeman background correction. *Clinical Chemistry, 29*, 477-480.

Roberts C and Clark J. (1986). Improved determination of cadmium in blood and plasma by flameless atomic absorption spectroscopy. *Bulletin of Environmental Contamination and Toxicology, 36*, 496-499.

Roelandts I. (1989). Biological reference materials. *Soectrochimica Acta, 44B*, 281-290.

Roels H, Buchet R, Lauwerys R, Bruaux P, Clays-Thoreau F, Laafontaine A, Overschelde J, and Verduyn J. (1978). Lead and cadmium absorption among children near a nonferrous metal plant. *Environmental Research, 15*, 290-308.

Roels H, Djubgang J, Buchet J, Bernard A, and Lauwerys R. (1982). Evolution of cadmium-induced renal dysfunction in workers removed from exposure. *Scandanavian Journal of Work and Environmental Health, 8*, 191-200.

Roels H, Lauwerys R, and Buchet J. (1989). Health significance of cadmium induced renal dysfunction: A five year follow-up. *British Journal of Industrial Medicine, 46*, 755-764.

Roels J, Lauwerys R, Buchet J, Bernard A, Chettle D, Harvey T, and Al-Haddad I. (1981). In vivo measurements of liver and kidney cadmium in workers exposed to this metal: Its significance with respect to cadmium in blood and urine. *Environmental Research, 26*, 217-240.

Roels H, Lauwerys R, Buchet J, Bernard A, Lijnen P, and Houte G. (1990). Urinary kallikrein activity in workers exposed to cadmium, lead, r mercury vapor. *British Journal of Industrial Medicine, 47*, 331-337.

akurai H, Omae K, Toyama T, Higashi T, and Nakadate T. (1982). ross-sectional study of pulmonary function in cadmium alloy works. *Scandanavian Journal of Work and Environmental Health, 8*, 22-130.

hardun G and van Epps L. (1987). B-2-microglobulin: Its significe in the evaluation of renal function. *Kidney International, 32*, 5-641.

aikh Z, and Smith L. (1984). Biological indicators of cadmium exure and toxicity. *Experentia, 40*, 36-43.

ith J and Kench J. (1957). Observations on urinary cadmium protein excretion in men exposed to cadmium oxide dust and e. *British Journal of Industrial Medicine, 14*, 240-245.

h J, Kench J, and Lane R. (1955). Determination of Cadmium in e and observations on urinary cadmium and protein excretion in exposed to cadmium oxide dust. *British Journal of Industrial Medi- 12*, 698-701.

(Southwest Research Institute). (1978). The distribution of cad- and other metals in human tissues. Health Effects Research Research Triangle Park, NC, Population Studies Division. NTIS 3-285-200.

rt M and Hughes E. (1981). Urinary B-2-microglobulin in the cal monitoring of cadmium workers. *British Journal of Indus- edicine, 38*, 170-174.

ler K and Brandt M. (1980). Contributions to automated trace s. Part V. Determination of cadmium in whole blood and urine trothermal atomic absorption spectrophotometry. *Fresenius' ift fur Analytical Chemistry, 300*, 372-380.

ka et al. (1983). *Carcinogencity of Cd Chloride Aerosols in White* NCI 70: 367-373, 1983.

Thun M, Osorio A, Schober S, Hannon W, Lewis B, and Halperin W. (1989). Nephropathy in cadmium workers: Assessment of risk from airborne occupational exposure to cadmium. *British Journal of Industrial Medicine, 46*, 689-697.

Thun M, Schnorr T, Smith A, Halperin W, and Lemen R. (1985). Mortality among a cohort of US cadmium production workers—an update. *Journal of the National Cancer Institute, 74*, 325-333.

Travis D and Haddock A. (1980). Interpretation of the observed age-dependency of cadmium body burdens in man. *Environmental Research, 22*, 46-60.

Tsuchiya K. (1967). Proteinuria of workers exposed to cadmium fume. *Archives of Environmental Health, 14*, 875-880.

Tsuchiya K. (1976). Proteinuria of cadmium workers. *Journal of Occupational Medicine, 18*, 463-470.

Tsuchiya K, Iwao S, Sugita M, Sakurai H. (1979). Increased urinary B-2-microglobulin in cadmium exposure: Dose-effect relationship and biological significance of B-2-microglobulin. *Environmental Health Perspectives, 28*, 147-153.

USEPA. (1985). Updated Mutagenicity and Carcinogenicity Assessments of Cd: Addendum to the Health Assessment Document for Cd (May 1981). Final Report. June 1985.

Vahter M and Friberg L. (1988). Quality control in integrated human exposure monitoring of lead and cadmium. *Fresenius' Zeitschrift fur Analytical Chemistry, 332*, 726-731.

Weber J. (1988). An interlaboratory comparison programme for several toxic substances in blood and urine. *The Science of the Total Environment, 71*, 111-123.

Weber J. (1991a). Accuracy and precision of trace metal determinations in biological fluids. In K. Subramanian, G. Iyengar, and K. Okamot (Eds.), *Biological Trace Element Research*-Multidisciplinary Perspectives, ACS Symposium Series 445. Washington, DC: American Chemical Society.

Weber J. (1991b). Personal communication about interlaboratory program and shipping biological media samples for cadmium analyses.

Wibowo A, Herber R, van Deyck W, and Zielhuis R. (1982). Biological assessment of exposure in factories with second degree usage of cadmium compounds. *International Archives of Occupational Environmental Health, 49*, 265-273.

## ATTACHMENT 1: NONMANDATORY PROTOCOL FOR AN INTERNAL QUALITY ASSURANCE/ QUALITY CONTROL PROGRAM

The following is an example of the type of internal quality assurance/ quality control program that assures adequate control to satisfy OSHA requirements under this protocol. However, other approaches may also be acceptable.

As indicated in Section 3.3.1 of the protocol, the QA/QC program for CDB and CDU should address, at a minimum, the following:

- calibration;
- establishment of control limits;
- internal QC analyses and maintaining control; and
- corrective action protocols.

This illustrative program includes both initial characterization runs to establish the performance of the method and ongoing analysis of quality control samples intermixed with compliance samples to maintain control.

### Calibration

Before any analytical runs are conducted, the analytic instrument must be calibrated. This is to be done at the beginning of each day on which quality control samples and/or compliance samples are run. Once calibration is established, quality control samples or compliance samples may be run. Regardless of the type of samples run, every fifth sample must be a standard to assure that the calibration is holding.

Calibration is defined as holding if every standard is within plus or minus (±) 15% of its theoretical value. If a standard is more than plus or minus 15% of its theoretical value, then the run is out of control due to calibration error and the entire set of samples must either be reana-



**TABLE 10—B-2—MICROGLOBULIN CONCENTRATIONS OBSERVED IN URINE AMONG OCCUPATIONALLY-EXPOSED WORKERS**

| Study No. | N | Concentration of B-2-Microglobulin in urine: Geometric mean (µg/g) [a] | Geom std dev | L 95% of range [b] | U 95% of range [b] | Reference |
|---|---|---|---|---|---|---|
| 1 | 1,424 | 160 | 6.19 | 8.1 | 3,300 | Ishizaki et al., 1989. |
| 2 | 1,754 | 260 | 6.50 | 12 | 5,600 | Ishizaki et al., 1989. |
| 3 | 33 | 210 | … | … | … | Ellis et al., 1983. |
| 4 | 65 | 210 | … | … | … | Chia et al., 1989. |
| 5 | [c]44 | 5,700 | 6.49 | [d]300 | [d]98,000 | Kjellstrom et al., 1977. |
| 6 | 148 | [e]180 | … | [f]110 | [f]280 | Buchet et al., 1980. |
| 7 | 37 | 160 | 3.90 | 17 | 1,500 | Kenzaburo et al., 1979. |
| 8 | [c]45 | 3,300 | 8.7 | [d]310 | [d]89,000 | Mason et al., 1988. |
| 9 | [c]10 | 6,100 | 5.99 | [f]650 | [f]57,000 | Falck et al., 1983. |
| 10 | [c]11 | 3,900 | 2.96 | [d]710 | [d]15,000 | Elinder et al., 1985. |
| 11 | [c]12 | 300 | … | … | … | Roels et al., 1991. |
| 12 | [g]8 | 7,400 | … | … | … | Roels et al., 1991. |
| 13 | [c]23 | [h]1,800 | … | … | … | Roels et al., 1989. |
| 14 | 10 | 690 | … | … | … | Iwao et al., 1980. |
| 15 | 34 | 71 | … | … | … | Wibowo et al., 1982. |
| 16 | [c]15 | 4,700 | 6.49 | [d]590 | [d]93,000 | Thun et al., 1989. |

[a] Unless otherwise stated.

[b] Based on an assumed lognormal distribution.

[c] Among workers diagnosed as having renal dysfunction; for Elinder this means β 2 levels greater than 300 micrograms per gram creatinine (µg/gr Cr); for Roels, 1991, range = 31 – 35, 170 µg$\beta_2$/gr Cr and geometric mean = 63 among healthy workers; for Mason $\beta_2$ > 300 µg/gr Cr.

[d] Based on a detailed review of the data by OSHA.

[e] Arthmetic mean.

[f] Reported in the study.

[g] Retired workers.

[h] 1,800 µg$\beta_2$/gr Cr for first survey; second survey = 1,600; third survey = 2,600; fourth survey = 2,600; fifth survey = 2,600.

The data provided in Table 10 indicate that the mean B2MU concentration observed among workers experiencing occupational exposure to cadmium (but with undefined levels of proteinuria) is 160-7400 µg/g CRTU. One of these studies reports geometric means lower than this range (i.e., as low as 71 µg/g CRTU); an explanation for this wide spread in average concentrations is not available.

Seven of the studies listed in Table 10 report a range of B2MU levels among those diagnosed as having renal dysfunction. As indicated in this table, renal dysfunction (proteinuria) is defined in several of these studies by B2MU levels in excess of 300 µg/g CRTU (see footnote "c" of Table 10); therefore, the range of B2MU levels observed in these studies is a function of the operational definition used to identify those with renal dysfunction. Nevertheless, a B2MU level of 300 µg/g CRTU appears to be a meaningful threshold for identifying those having early signs of kidney damage. While levels much higher than 300 µg/g CRTU have been observed among those with renal dysfunction, the vast majority of those not occupationally exposed to cadmium exhibit much lower B2MU concentrations (see Table 9). Similarly, the vast majority of workers *not* exhibiting renal dysfunction are found to have levels below 300 µg/g CRTU (Table 9).

The 300 µg/g CRTU level for B2MU proposed in the above paragraph has support among researchers as the threshold level that distinguishes between cadmium-exposed workers with and without kidney dysfunction. For example, in the guide for physicians who must evaluate cadmium-exposed workers written for the Cadmium Council by Dr. Lauwerys, levels of B2M greater than 200-300 µg/g CRTU are considered to require additional medical evaluation for kidney dysfunction (exhibit 8-447, OSHA docket H057A). The most widely used test for measuring B2M (i.e., the Pharmacia Delphia test) defines B2MU levels above 300 µg/l as abnormal (exhibit L-140-1, OSHA docket H057A).

Dr. Elinder, chairman of the Department of Nephrology at the Karolinska Institute, testified at the hearings on the proposed cadmium rule. According to Dr. Elinder (exhibit L-140-45, OSHA docket H057A), the normal concentration of B2MU has been well documented (Evrin and Wibell 1972; Kjellstrom et al. 1977a; Elinder et al. 1978, 1983; Buchet et al. 1980; Jawaid et al. 1983; Kowal and Zirkes, 1983). Elinder stated that the upper 95 or 97.5 percentiles for B2MU among those without tubular dysfunction is below 300 µg/g CRTU (Kjellstrom et al. 1977a; Buchet et al. 1980; Kowal and Zirkes, 1983). Elinder defined levels of B2M above 300 µg/g CRTU as "slight" proteinuria.

### 5.3.8 Conclusions and Recommendations for B2MU

Based on the above evaluation, the following recommendations are made for a B2MU proficiency testing program. Note that the following discussion addresses only sampling and analysis for B2MU determinations (i.e., to be reported as an unadjusted µg B2M/l urine). Normalizing this result to creatinine requires a second analysis for CRTU (see Section 5.4) so that the ratio of the 2 measurements can be obtained.

#### 5.3.8.1 *Recommended method*

The Pharmacia Delphia method (Pharmacia 1990) should be adopted as the standard method for B2MU determinations. Laboratories may adopt alternate methods, but it is the responsibility of the laboratory to demonstrate that alternate methods provide results of comparable quality to the Pharmacia Delphia method.

#### 5.3.8.2 *Data quality objectives*

The following data quality objectives should facilitate interpretation of analytical results, and should be achievable based on the above evaluation.

*Limit of Detection.* A limit of 100 µg/l urine should be achievable, although the insert to the test kit (Pharmacia 1990) cites a detection limit of 150 µg/l; private conversations with representatives of Pharmacia, however, indicate that the lower limit of 100 µg/l should be achievable provided an additional standard of 100 µg/l B2M is run with the other standards to derive the calibration curve (Section 3.3.1.1). The lower detection limit is desirable due to the proximity of this detection limit to B2MU values defined for the cadmium medical monitoring program.

*Accuracy.* Because results from an interlaboratory proficiency testing program are not available currently, it is difficult to define an achievable level of accuracy. Given the general performance parameters defined by the insert to the test kits, however, an accuracy of ±15% of the target value appears achievable.

Due to the low levels of B2MU to be measured generally, it is anticipated that the analysis of creatinine will contribute relatively little to the overall variability observed among creatinine-normalized B2MU levels (see Section 5.4). The initial level of accuracy for reporting B2MU levels under this program should be set at ±15%.

*Precision.* Based on precision data reported by Pharmacia (1990), a precision value (i.e., CV) of 5% should be achievable over the defined range of the analyte. For internal QC samples (i.e., recommended as part of an internal QA/QC program, Section 3.3.1), laboratories should attain precision near 5% over the range of concentrations measured.

#### 5.3.8.3 *Quality assurance/quality control*

Commercial laboratories providing measurement of B2MU should adopt an internal QA/QC program that incorporates the following components: Strict adherence to the Pharmacia Delphia method, including calibration requirements; regular use of QC samples during routine runs; a protocol for corrective actions, and documentation of these actions; and, participation in an interlaboratory proficiency program. Procedures that may be used to address internal QC requirements are presented in Attachment 1. Due to differences between analyses for B2MU and CDB/CDU, specific values presented in Attachment 1 may have to be modified. Other components of the program (including characterization runs), however, can be adapted to a program for B2MU.

### 5.4 Monitoring Creatinine in Urine (CRTU)

Because CDU and B2MU should be reported relative to concentrations of CRTU, these concentrations should be determined in addition CDU and B2MU determinations.

#### 5.4.1 Units of CRTU Measurement

CDU should be reported as µg Cd/g CRTU, while B2MU should be reported as µg B2M/g CRTU. To derive the ratio of cadmium or B2M to creatinine, CRTU should be reported in units of g crtn/l of urine. Depending on the analytical method, it may be necessary to convert results of creatinine determinations accordingly.

#### 5.4.2 Analytical Techniques Used to Monitor CRTU

Of the techniques available for CRTU determinations, an absorbance spectrophotometric technique and a high-performance liquid chromatography (HPLC) technique are identified as acceptable in this protocol.

#### 5.4.3 Methods Developed for CRTU Determinations

CRTU analysise performed in support of either CDU or B2MU determinations should be performed using either of the following 2 methods:

1. The Du Pont method (i.e., Jaffe method), in which creatinine in a sample reacts with picrate under alkaline conditions, and the resulting red chromophore is monitored (at 510 nm) for a fixed interval to determine the rate of the reaction; this reaction rate is proportional to the concentration of creatinine present in the sample (a copy of this method is provided in Attachment 2 of this protocol); or,

2. The OSHA SLC Technical Center (OSLTC) method, in which creatinine in an aliquot of sample is separated using an HPLC column equipped with a UV detector; the resulting peak is quantified using an electrical integrator (a copy of this method is provided in Attachment 3 of this protocol).

#### 5.4.4 Sample Collection and Handling

CRTU samples should be segregated from samples collected for CDU or B2MU analysis. Sample-collection techniques have been described under Section 5.2.4. Samples should be preserved either to stabilize CDU (with $HNO_3$) or B2MU (with NaOH). Neither of these procedures should adversely affect CRTU analysis (see Attachment 3).

#### 5.4.5 General Method Performance

Data from the OSLTC indicate that a CV of 5% should be achievable using the OSLTC method (Septon, L private communication). The achievable accuracy of this method has not been determined.

Results reported in surveys conducted by the CAP (CAP 1991a, 1991b and 1992) indicate that a CV of 5% is achievable. The accuracy achievable for CRTU determinations has not been reported.

Laboratories performing creatinine analysis under this protocol should be CAP accredited and should be active participants in the CAP surveys.

#### 5.4.6 Observed CRTU Concentrations

Published data suggest the range of CRTU concentrations is 1.0-1.6 g in 24-hour urine samples (Harrison 1987). These values are equivalent to about 1 g/l urine.

#### 5.4.7 Conclusions and Recommendations for CRTU

##### 5.4.7.1 *Recommended method*

Use either the Jaffe method (Attachment 2) or the OSLTC method (Attachment 3). Alternate methods may be acceptable provided adequate performance is demonstrated in the CAP program.

##### 5.4.7.2 *Data quality objectives*

*Limit of Detection.* This value has not been formally defined; however, a value of 0.1 g/l urine should be readily achievable.

*Accuracy.* This value has not been defined formally; accuracy should be sufficient to retain accreditation from the CAP

*Precision.* A CV of 5% should be achievable using the recommended methods.

### 6.0 References


Adamsson E, Piscator M, and Nogawa K. (1979). Pulmonary and gastrointestinal exposure to cadmium oxide dust in a battery factory. *Environmental Health Perspectives, 28,* 219-222.

American Conference of Governmental Industrial Hygienists (ACGIH). (1986). *Documentation of the Threshold Limit Values and Biological Exposure Indices.* 5th edition. p. BEI-55.

Bernard A, Buchet J, Roels H, Masson P, and Lauwerys R. (1979). Renal excretion of proteins and enzymes in workers exposed to cadmium. *European Journal of Clinical Investigation, 9,* 11-22.

Bernard A and Lauwerys R. (1990). Early markers of cadmium nephrotoxicity: Biological significance and predictive value. *Toxocological and Environmental Chemistry, 27,* 65-72.

Braunwald E, Isselbacher K, Petersdorf R, Wilson J, Martin J, and Fauci A (Eds.). (1987). *Harrison's Principles of Internal Medicine.* New York: McGraw-Hill Book Company.

Buchet J, Roels H, Bernard I, and Lauwerys R. (1980). Assessment of renal funcion of workers exposed to inorganic lead, cadmium, or mercury vapor. *Journal of Occupational Medicine,* 22, 741-750.

CAP. (1991). Urine Chemistry, Series 1: Survey (Set U-B). College of American Pathologists.

CAP. (1991). Urine Chemistry, Series 1: Survey (Set U-C). College of American Pathologists.

CAP. (1992). Urine Chemistry, Series 1: Survey (Set U-A). College of American Pathologists.

CDC. (1986). Centers for Disease Control, Division of Environmental Health Laboratory Sciences, Center for Environmental Health, Atlanta, Georgia. Docket No. 106A. Lake Couer d'Alene, Idaho cadmium and lead study: 86-0030, Specimen collection and shipping protocol.

CDC. (1990). Centers for Disease Control, Nutritional Biochemistry Branch. 4/27/90 Draft SOP for Method 0360A "Determination of cadmium in urine by graphite furnace atomic absorption spectrometry with Zeeman background correction.

Centre de Toxicologie du Quebec. (1991). Interlaboratory comparison program report for run #2. Shipping date 3/11/91. Addition BLR 9/19.

Chia K, Ong C, Ong H, and Endo G. (1989). Renal tubular function of workers exposed to low levels of cadmium. *British Journal of Industrial Medicine, 46,* 165-170.

Claeys-Thoreau F. (1982). Determination of low levels of cadmium and lead in biological fluids with simple dilution by atomic absorption spectrophotometry using Zeeman effect background absorption and the L'Vov platform. *Atomic Spectroscopy, 3,* 188-191.

DeBenzo Z, Fraile R, and Carrion N. (1990). Electrothermal atomization atomic absorption spectrometry with stabilized aqueous standards for the determination of cadmium in whole blood. *Analytica Chimica Acta, 231,* 283-288.

Elinder C, Edling C, Lindberg E, Kagedal B, and Vesterberg O. (1985). Assessment of renal function in workers previously exposed to cadmium. *British Journal of Internal Medicine, 42,* 754.

Ellis K, Cohn S, and Smith T. (1985). Cadmium inhalation exposure estimates: Their significance with respect to kidney and liver cadmium burden. *Journal of Toxicology and Environmental Health, 15,* 173-187.





Ellis K, Yasumura S, Vartsky D, and Cohn S. (1983). Evaluation of biological indicators of body burden of cadmium in humans. *Fundamentals and Applied Toxicology, 3,* 169-174.

Ellis K, Yeun K, Yasumura S, and Cohn S. (1984). Doseresponse analysis of cadmium in man: Body burden vs kidney function. *Environmental Research, 33,* 216-226.

Evrin P, Peterson A, Wide I, and Berggard I. (1971). Radioimmunoassay of B-2-microglobulin in human biological fluids. *Scandanavian Journal of Clinical Laboratory Investigation, 28,* 439-443.

Falck F, Fine L, Smith R, Garvey J, Schork A, England B, McClatchey K, and Linton J. (1983). Metallothionein and occupational exposure to cadmium. *British Journal of Industrial Medicine, 40,* 305-313.

FEDERAL REGISTER. (1990). Occupational exposure to cadmium: Proposed rule. 55/22/4052-4147, February 6.

Friberg, Exhibit 29, (1990). Exhibit No. 29 of the OSHA Federal Docket H057A. Washington, DC.

Friberg L. (1988). Quality assurance. In T. Clarkson (Ed.), *Biological Monitoring of Toxic Metals* (pp. 103-105). New York: Plenum Press.

Friberg L, and Elinder C. (1988). Cadmium toxicity in humans. In *Essential and Trace Elements in Human Health and Disease* (pp. 559-587). Docket Number 8-660.

Friberg L, Elinder F, et al. (1986). *Cadmium and Health: A Toxicological and Epidemiological Appraisal. Volume II, Effects and Response.* Boca Raton, FL: CRC Press.

Friberg L, Piscator M, Nordberg G, and Kjellstrom T. (1974). *Cadmium in the Environment* (2nd ed.). Cleveland:CRC.

Friberg L and Vahter M. (1983). Assessment of exposure to lead and cadmium through biological monitoring: Results of a UNEP/WHO global study. *Environmental Research, 30,* 95-128.

Gunter E, and Miller D. (1986). Laboratory procedures used by the division of environmental health laboratory sciences center for environmental health, Centers for Disease Control for the hispanic health and nutrition examination survey (HHANES). Atlanta, GA: Centers for Disease Control.

Harrison. (1987). Harrison's Principles of Internal Medicine. Braunwald, E; Isselbacher, KJ; Petersdorf, RG; Wilson, JD; Martin, JB; and Fauci, AS Eds. Eleventh Ed. McGraw Hill Book Company. San Francisco.

Henry J. (1991). *Clinical Diagnosis and Management by Laboratory Methods* (18th edition). Philadelphia: WB Saunders Company.

IARC (1987). *IRAC Monographs on the Evaluation of Carcinogenic Risks to Humans. Overall Evaluation of Carcinogenicity: Update of Volume 1-42.* Supplemental 7, 1987.

Ishizaki M, Kido T, Honda R, Tsuritani I, Yamada Y, Nakagawa H, and Nogawa K. (1989). Dose-response relationship between urinary cadmium and B-2-microglobulin in a Japanese environmentally cadmium exposed population. *Toxicology, 58,* 121-131.

Iwao S, Tsuchiya K, and Sakurai H. (1980). Serum and urinary B-2-microglobulin among cadmium-exposed workers. *Journal of Occupational Medicine, 22,* 399-402.

Iwata K, Katoh T, Morikawa Y, Aoshima K, Nishijo M, Teranishi H, and Kasuya M. (1988). Urinary trehalase activity as an indicator of kidney injury due to environmental cadmium exposure. *Archives of Toxicology, 62,* 435-439.

Kawada T, Koyama H, and Suzuki S. (1989). Cadmium, NAG activity, and B-2-microglobulin in the urine of cadmium pigment workers. *British Journal of Industrial Medicine, 46,* 52-55.

Kawada T, Tohyama C, and Suzuki S. (1990). Significance of the excretion of urinary indicator proteins for a low level of occupational exposure to cadmium. *International Archives of Occupational Environmental Health, 62,* 95-100.

Kjellstrom T. (1979). Exposure and accumulation of cadmium in populations from Japan, the United States, and Sweden. *Environmental Health Perspectives, 28,* 169-197.

Kjellstrom T, Evrin P, and Rahnster B. (1977). Dose-response analysis of cadmium-induced tubular proteinuria. *Environmental Research, 13,* 303-317.

Kjellstrom T, Shiroishi K, and Evrin P. (1977). Urinary B-2microglobulin excretion among people exposed to cadmium in the general environment. *Environmental Research, 13,* 318-344.

Kneip T, & Crable J (Eds.). (1988). Method 107. Cadmium in blood. *Methods for biological monitoring* (pp.161-164). Washington, DC: American Public Health Association.

Kowal N. (1988). Urinary cadmium and B-2-microglobulin: Correlation with nutrition and smoking history. *Journal of Toxicology and Environmental Health, 25,* 179-183.

Kowal N, Johnson D, Kraemer D, and Pahren H. (1979). Normal levels of cadmium in diet, urine, blood, and tissues of inhabitants of the United States. *Journal of Toxicology and Environmental Health, 5,* 995-1014.

Kowal N and Zirkes M. (1983). Urinary cadmium and B-2microglobulin: Normal values and concentration adjustment. *Journal of Toxicology and Environmental Health, 11,* 607-624.

Lauwerys R, Buchet J, and Roels H. (1976). The relationship between cadmium exposure or body burden and the concentration of cadmium in blood and urine in man. *International Archives of Occupational and Environmental Health, 36,* 275-285

Lauwerys R, Roels H, Regniers, Buchet J, and Bernard A. (1979). Significance of cadmium concentration in blood and in urine in workers exposed to cadmium. *Environmental Research, 20,* 375-391.

Lind B, Elinder C, Friberg L, Nilsson B, Svartengren M, and Vahter M. (1987). Quality control in the analysis of lead and cadmium in blood. *Fresenius' Zeitschrift fur Analytical Chemistry, 326,* 647-655.

Mason H, Davison A, Wright A, Guthrie C, Fayers P, Venables K, Smith N, Chettle D, Franklin D, Scott M, Holden H, Gompertz D, and Newman-Taylor A. (1988). Relations between liver cadmium, cumulative exposure, and renal function in cadmium alloy workers *British Journal of Industrial Medicine, 45,* 793-802.

Meridian Research, Inc. (1989). *Quantitative Assessment of Canc Risks Associated with Occupational Exposure to Cd.* Prepared by M ridian Research, Inc. and Roth Associates, Inc. for the Occupatio Safety & Health Administration. June 12, 1989.

Meridian Research, Inc and Roth Associates, Inc. (1989). *Quan tive Assessment of the Risk of Kidney Dysfunction Associated Occupational Exposure to Cd.* Prepared by Meridian Research and Roth Associates, Inc. for the Occupational Safety & Healt ministration. July 31 1989.

Micheils E and DeBicvre P. (1986). Method 25-Determination mium in whole blood by isotope dilution mass spectrometry. I, Schuller P, and Fishbein L (Eds.), *Environmental Carcinog lected Methods of Analysis* (Vol. 8). Lyon, France: Interr Agency for Research on Cancer.

Mueller P, Smith S, Steinberg K, and Thun M. (1989). renal tubular effects in relation to urine cadmium leve *ron, 52,* 45-54.

NIOSH. (1984a). Elements in blood or tissues. Methoc sued 5/15/85 and Metals in urine. Method 8310 issue In P. Eller (Ed.), *NIOSH Manual of Analytical Methods* (Vo Cincinnati, Ohio: US-DHHS.

NIOSH. (1984b). Lowry L. Section F: Special considerat logical samples in *NIOSH Manual of Analytical Methods* ed). P. Eller (Ed.). Cincinnati, Ohio: US-DHHS.

Nordberg G and Nordberg M. (1988). Biological moni mium. In T. Clarkson, L. Friberg, G. Nordberg, and P *Biological Monitoring of Toxic Metals,* New York: Plen

Nogawa K. (1984). Biologic indicators of cadmium n persons with low-level cadmium exposure. *Environm spectives, 54,* 163-169.

OSLTC (no date). Analysis of Creatinine for the Cadmium and Beta-2-Microglobulin Concentratio Salt Lake Technical Center. Salt Lake City, UT. P tachment 8 of exhibit 106 of the OSHA docket H

Perkin-Elmer Corporation. (1982). *Analytical Me sorption Spectroscopy.*

lyzed after recalibrating or results should be recalculated based on a statistical curve derived from the measurement of all standards.

It is essential that the highest standard run is higher than the highest sample run. To assure that this is the case, it may be necessary to run a high standard at the end of the run, which is selected based on the results obtained over the course of the run.

All standards should be kept fresh, and as they get old, they should be compared with new standards and replaced if they exceed the new standards by ± 15%.

## INITIAL CHARACTERIZATION RUNS AND ESTABLISHING CONTROL

A participating laboratory should establish four pools of quality control samples for each of the analytes for which determinations will be made. The concentrations of quality control samples within each pool are to be centered around each of the four target levels for the particular analyte identified in Section 4.4 of the protocol.

Within each pool, at least 4 quality control samples need to be established with varying concentrations ranging between plus or minus 50% of the target value of that pool. Thus for the medium-high cadmium in blood pool, the theoretical values of the quality control samples may range from 5 to 15 μg/l, (the target value is 10 μg/l). At least 4 unique theoretical values must be represented in this pool.

The range of theoretical values of plus or minus 50% of the target value of a pool means that there will be overlap of the pools. For example, the range of values for the medium-low pool for cadmium in blood is 3.5 to 10.5 μg/l while the range of values for the medium-high pool is 5 to 15 μg/l. Therefore, it is possible for a quality control sample from the medium-low pool to have a higher concentration of cadmium than a quality control sample from the medium-high pool.

Quality control samples may be obtained as commercially available reference materials, internally prepared, or both. Internally prepared samples should be well characterized and traced or compared to a reference material for which a consensus value for concentration is available. Levels of analyte in the quality control samples must be concealed from the analyst prior to the reporting of analytical results. Potential sources of materials that may be used to construct quality control samples are listed in Section 3.3.1 of the protocol.

Before any compliance samples are analyzed, control limits must be established. Control limits should be calculated for every pool of each analyte for which determinations will be made and control charts should be kept for each pool of each analyte. A separate set of control charts and control limits should be established for each analytical instrument in a laboratory that will be used for analysis of compliance samples.

At the beginning of this QA/QC program, control limits should be based on the results of the analysis of 20 quality control samples from each pool of each analyte. For any given pool, the 20 quality control samples should be run on 20 different days. Although no more than one sample should be run from any single pool on a particular day, a laboratory may run quality control samples from different pools on the same day. This constitutes a set of initial characterization runs.

For each quality control sample analyzed, the value F/T (defined in the glossary) should be calculated. To calculate the control limits for a pool of an analyte, it is first necessary to calculate the mean, $\bar{X}$, of the F/T values for each quality control sample in a pool and then to calculate its standard deviation σ. Thus, for the control limit for a pool, $\bar{X}$ is calculated as:

$$\frac{\left(\sum \frac{F}{T}\right)}{N}$$

**and σ is calculated as**

$$\left[\frac{\sum\left(\frac{F}{T}-\bar{X}\right)^2}{N-1}\right]^{\frac{1}{2}}$$

Where N is the number of quality control samples run for a pool.

The control limit for a particular pool is then given by the mean plus or minus 2 standard deviations (X ± 3σ).

The control limits may be no greater than 40% of the mean F/T value. If three standard deviations are greater than 40% of the mean F/T value, then analysis of compliance samples may not begin.[1] Instead, an investigation into the causes of the large standard deviation should begin, and the inadequacies must be remedied. Then, control limits must be reestablished which will mean repeating the running 20 quality control samples from each pool over 20 days.

## Internal Quality Control Analyses and Maintaining Control

Once control limits have been established for each pool of an analyte, analysis of compliance samples may begin. During any run of compliance samples, quality control samples are to be interspersed at a rate of no less than 5% of the compliance sample workload. When quality control samples are run, however, they should be run in sets consisting of one quality control sample from each pool. Therefore, it may be necessary, at times, to intersperse quality control samples at a rate greater than 5%.

There should be at least one set of quality control samples run with any analysis of compliance samples. At a minimum, for example, 4 quality control samples should be run even if only 1 compliance sample is run. Generally, the number of quality control samples that should be run are a multiple of four with the minimum equal to the smallest multiple of four that is greater than 5% of the total number of samples to be run. For example, if 300 compliance samples of an analyte are run, then at least 16 quality control samples should be run (16 is the smallest multiple of four that is greater than 15, which is 5% of 300).

Control charts for each pool of an analyte (and for each instrument in the laboratory to be used for analysis of compliance samples) should be established by plotting F/T versus date as the quality control sample results are reported. On the graph there should be lines representing the control limits for the pool, the mean F/T limits for the pool, and the theoretical F/T of 1.000. Lines representing plus or minus (±) $\hat{\sigma}$ should also be represented on the charts. A theoretical example of a control chart is presented in Figure 1.

### FIGURE 1—THEORETICAL EXAMPLE OF A CONTROL CHART FOR A POOL OF AN ANALYTE

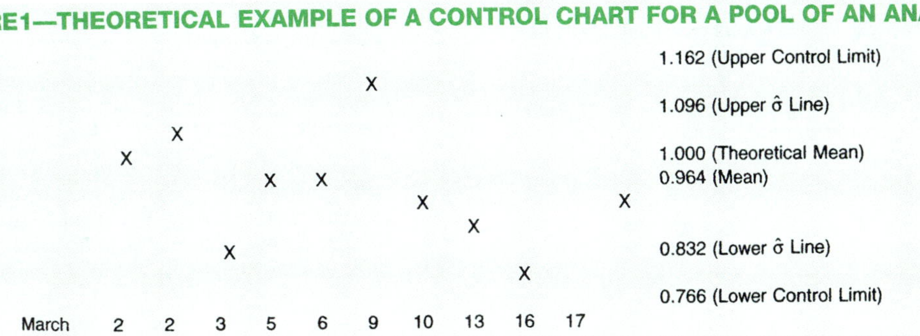


[1] Note that the value, "40%" may change over time as experience is gained with the program.

All quality control samples should be plotted on the chart, and the charts should be checked for visual trends. If a quality control sample falls above or below the control limits for its pool, then corrective steps must be taken (see the section on corrective actions below). Once a laboratory's program has been established, control limits should be updated every 2 months.

The updated control limits should be calculated from the results of the last 100 quality control samples run for each pool. If 100 quality control samples from a pool have not been run at the time of the update, then the limits should be based on as many as have been run provided at least 20 quality control samples from each pool have been run over 20 different days.

The trends that should be looked for on the control charts are:

1. 10 consecutive quality control samples falling above or below the mean;

2. 3 consecutive quality control samples falling more than $2\sigma$ from the mean (above or below the $2\sigma$ lines of the chart); or

3. the mean calculated to update the control limits falls more than 10% above or below the theoretical mean of 1.000.

If any of these trends is observed, then all analysis must be stopped, and an investigation into the causes of the errors must begin. Before the analysis of compliance samples may resume, the inadequacies must be remedied and the control limits must be reestablished for that pool of an analyte. Reestablishment of control limits will entail running 20 sets of quality control samples over 20 days.

Note that alternative procedures for defining internal quality control limits may also be acceptable. Limits may be based, for example, on proficiency testing, such as ± 1 μg or 15% of the mean (whichever is greater). These should be clearly defined.

## Corrective actions

Corrective action is the term used to describe the identification and remediation of errors occurring within an analysis. Corrective action is necessary whenever the result of the analysis of any quality control sample falls outside of the established control limits. The steps involved may include simple things like checking calculations of basic instrument maintenance, or it may involve more complicated actions like major instrument repair. Whatever the source of error, it must be identified and corrected (and a Corrective Action Report (CAR) must be completed. CARs should be kept on file by the laboratory.

## ATTACHMENT 2—CREATININE IN URINE (JAFFE PROCEDURE)

Intended use: The CREA pack is used in the Du Pont ACA ® discrete clinical analyzer to quantitatively measure creatinine in serum and urine.

Summary: The CREA method employs a modification of the kinetic Jaffe reaction reported by Larsen. This method has been reported to be less susceptible than conventional methods to interference from non-creatinine, Jaffe-positive compounds.[1]

A split sample comparison between the CREA method and a conventional Jaffe procedure on Autoanalyzer® showed a good correlation. (See Specific Performance Characteristics).

***Note:** Numbered subscripts refer to the bibliography and lettered subscripts refer to footnotes.

Autoanalyzer,® is a registered trademark of Technicon Corp., Tarrytown, NY.

Principles of Procedure: In the presence of a strong base such as NaOH, picrate reacts with creatinine to form a red chromophore. The rate of increasing absorbance at 510 nm due to the formation of this chromophore during a 17.07-second measurement period is directly proportional to the creatinine concentration in the sample.

$$\text{Creatinine + Picrate} \xrightarrow{\text{NaOH}} \text{Red chromophore (absorbs at 510 nm)}$$

Reagents:

| Compartment [a] | Form | Ingredient | Quantity[b] |
|---|---|---|---|
| No. 2, 3, & 4 | Liquid | Picrate | 0.11 mmol. |
| No. 6 | Liquid | NaOH (for pH adjustment)[c]. | |

[a] Compartments are numbered 1-7, with compartment #7 located closest to pack fill position #2.

[b] Nominal value at manufacture.

[c] See Precautions.

Precautions: Compartment #6 contains 75 μL of 10 N NaOH; avoid contact; skin irritant; rinse contacted area with water. Comply with OSHA'S Bloodborne Pathogens Standard while handling biological samples (29 CFR 1910.1039).

Used packs contain human body fluids; handle with appropriate care.

## FOR IN VITRO DIAGNOSTIC USE

## Mixing and Diluting:

Mixing and diluting are automatically performed by the ACA® discrete clinical analyzer. The sample cup must contain sufficient quantity to accommodate the sample volume plus the "dead volume"; precise cup filling is not required.

## SAMPLE CUP VOLUMES (μL)

| Analyzer | Standard | | Microsystem | |
|---|---|---|---|---|
| | Dead | Total | Dead | Total |
| II, III | 120 | 3000 | 10 | 500 |
| IV, SX | 120 | 3000 | 30 | 500 |
| V | 90 | 3000 | 10 | 500 |

Storage of Unprocessed Packs: Store at 2-8 °C. Do not freeze. Do not expose to temperatures above 35 °C or to direct sunlight.

Expiration: Refer to EXPIRATION DATE on the tray label.

Specimen Collection: Serum or urine can be collected and stored by normal procedures.[2]

## Known Interfering Substances[3]

- Serum Protein Influence—Serum protein levels exert a direct influence on the CREA assay. The following should be taken into account when this method is used for urine samples and when it is calibrated:

  Aqueous creatinine standards or urine specimens will give CREA results depressed by approximately 0.7 mg/dL [62 μmol/L][d] and will be less precise than samples containing more than 3 g/dL [30 g/L] protein.

  All urine specimens should be diluted with an albumin solution to give a final protein concentration of at least 3 g/dL [30 g/L]. Du Pont Enzyme Diluent (Cat. #790035-901) may be used for this purpose.

- High concentration of endrogenous bilirubin (>20 mg/dL [>342 μmol/L]) will give depressed CREA results (average depression 0.8 mg/dL [71 μmol/L]).[4]

- Grossly hemolyzed (hemoglobin >100 mg/dL [>62 μmol/L]) or visibly lipemic specimens may cause falsely elevated CREA results.[5, 6]

- The following cephalosporin antibiotics do not interfere with the CREA method when present at the concentrations indicated. Systematic inaccuracies (bias) due to these substances are less than or equal to 0.1 mg/dL [8.84 μmol/L] at CREA concentrations of approximately 1 mg/dL [88 μmol/L].

| *Antibiotic* | *Peak serum level* [7,8,9] | | *Drug concentration* | |
|---|---|---|---|---|
| | *mg/dL* | *[mmol/L]* | *mg/dL* | *[mmol/L]* |
| Cephaloridine | 1.4 | 0.3 | 25 | 6.0 |
| Cephalexin | 0.6-2.0 | 0.2-0.6 | 25 | 7.2 |
| Cephamandole | 1.3-2.5 | 0.3-0.5 | 25 | 4.9 |
| Cephapirin | 2.0 | D0.4 | 25 | 5.6 |
| Cephradine | 1.5-2.0 | 0.4-0.6 | 25 | 7.1 |
| Cefazolin | 2.5-5.0 | 0.55-1.1 | 50 | 11.0 |

- The following cephalosporin antibiotics have been shown to affect CREA results when present at the indicated concentrations.

System inaccuracies (bias) due to these substances are greater that 0.1 mg/dL [8.84 μmol/L] at CREA concentrations of:

| *Antibiotic* | *Peak serum level* [8,10] | | *Drug concentration* | | |
|---|---|---|---|---|---|
| | *md/dL* | *[mmol/L]* | *mg/dL* | *[mmol/L]* | *Effect* |
| Cephalothin | 1-6 | 0.2-1.5 | 100 | 25.2 | ↓20-25% |
| Cephoxitin | 2.0 | 0.5 | 5.0 | 1.2 | ↑35-40% |

- The single wavelength measurement used in this method eliminates interference from chromophores whose 510 nm absorbance is constant throughout the measurement period.
- Each laboratory should determine the acceptability of its own blood collection tubes and serum separation products. Variations in these products may exist between manufacturers and, at times, from lot to lot.

d. Systeme International d'unites (S.I. Units) are in brackets.

Procedure:

## TEST MATERIALS

| Item | II, III Du Pont Cat. No. | IV, SX Du Pont Cat. No. | V Du Pont Cat. No. |
|---|---|---|---|
| ACA® CREA Analytical Test Pack | 701976901 | 701976901 | 701976901 |
| Sample System Kit or | 710642901 | 710642901 | 713697901 |
| Micro Sample System Kit and | 702694901 | 710356901 | NA |
| Micro Sample System Holders | 702785000 | NA | NA |
| DYLUX® Photosensitive | ................ | ................ | ................ |
| Printer Paper | 700036000 | NA | NA |
| Thermal Printer Paper | NA | 710639901 | 713645901 |
| Du Pont Purified Water | 704209901 | 710615901 | 710815901 |
| Cell Wash Solution | 701864901 | 710664901 | 710864901 |

Test Steps: The operator need only load the sample kit and appropriate test pack(s) into a properly prepared ACA® discrete clinical analyzer. It automatically advances the pack(s) through the test steps and prints a result(s). See the Instrument Manual of the ACA® analyzer for details of mechanical travel of the test pack(s).

### Preset Creatinine (CREA)—Test Conditions

- Sample Volume: 200 μ L
- Diluent: Purified Water
- Temperature: 37.0 ± 0.1 °C
- Reaction Period: 29 seconds
- Type of Measurement: Rate
- Measurement Period: 17.07 seconds
- Wavelength: 510 nm
- Units: mg/dL [μmol/L]

CALIBRATION: The general calibration procedure is described in the Calibration/Verification chapter of the Manuals.

The following information should be considered when calibrating the CREA method.

- Assay Range: 0-20 mg/mL [0-1768 μmol/L][e].
- Reference Material: Protein containing primary standards[f] or secondary calibrators such as Du Pont Elevated Chemistry Control (Cat. #790035903) and Normal Chemistry Control (Cat. ●#790035905)[g].
- Suggested Calibration Levels: 1,5,20, mg/mL [88, 442, 1768 μmol/L].
- Calibration Scheme: 3 levels, 3 packs per level.
- Frequency: Each new pack lot. Every 3 months for any one pack lot.

[e] For the results in S.I. units [mmol/L] the conversion factory is 88.4.

[f] Refer to the Creatinine Standard Preparation and Calibration Procedure available on request from a Du Pont Representative.

[g] If the Du Pont Chemistry Controls are being used, prepare them according to the instructions on the product insert sheets.

## PRESET CREATININE (CREA) TEST CONDITIONS

| Item | ACA®, II, analyzer | ACA®, III, IV, SX, V, analyzer |
|---|---|---|
| Count by | One (1) [Five (5)] | NA. |
| Decimal point location. | 0.0 mg/dL [000. µmo/L] | 000.0 mg/dL [000 µmol/L]. |
| Assigned starting point or offset $C_0$. | 999.8 [9823.] | − 1.000 E1 [-8.840 E2]. |
| Scale factor or assigned. | 0.2000 mg/dL/count [h] | 2.004 E-1 [h]. |
| Linear term $C_1$[h] | [0.3536 µmol/L/count]. | [1.772E1]. |

[h] The preset scale factor (linear term) was derived from the molar absorptivity of the indicator and is based on an absorbance to activity relationship (sensitivity) of 0.596 (mA/min)/(U/L). Due to small differences in filters and electronic components between instruments, the actual scale factor (linear term) may differ slightly from that given above.

Quality Control: Two types of quality control procedures are recommended:

- General Instrument Check. Refer to the Filter Balance Procedure and the Absorbance Test Method described in the ACA Analyzer Instrument Manual. Refer also to the ABS Test Methodology literature.
- Creatinine Method Check. At least once daily run a CREA test on a solution of known creatinine activity such as an assayed control or calibration standard other than that used to calibrate the CREA method. For further details review the Quality Assurance Section of the Chemistry Manual. The result obtained should fall within acceptable limits defined by the day-to-day variability of the system as measured in the user's laboratory. (See SPECIFIC PERFORMANCE CHARACTERISTICS for guidance.) If the result falls outside the laboratory's acceptable limits, follow the procedure outlined in the Chemistry Troubleshooting Section of the Chemistry Manual.

A possible system malfunction is indicated when analysis of a sample with five consecutive test packs gives the following results:

| Level | SD |
|---|---|
| 1 mg/dL | >0.15 mg/dL. |
| [88 µmol/L] | [>13 µmol/L]. |
| 20 mg/dL | >0.68 mg/dL. |
| [1768 µmol/L] | [>60 µmol/L]. |

Refer to the procedure outlined in the Trouble Shooting Section of the Manual.

Results: The ACA® analyzer automatically calculates and prints the CREA result in mg/dL [µmol/L].

### *Limitation of Procedure: Results >20 mg/dL [1768 µmol/L]:*

- Dilute with suitable protein base diluent. Reassay. Correct for diluting before reporting.

The reporting system contains error messages to warn the operator of specific malfunctions. Any report slip containing a letter code or word immediately following the numerical value should not be reported. Refer to the Manual for the definition of error codes.

### *Reference Interval*

| Serum: [11,i] | |
|---|---|
| Males | 0.8-1.3 md/dL [71-115 mmol/L]. |
| Females | 0.6-1.0 md/dL [53-88 mmol/L] |

| Urine: [12] | |
|---|---|
| Males | 0.6-2.5 g/24 hr [53-221 mmol/24 hr] |
| Females | 0.6-1.5 g/24 hr [53-133 mmol/24 hr] |

[i] Reference interval data obtained from 200 apparently healthy individuals (71 males, 129 females) between the ages of 19 and 72.

Each laboratory should establish its own reference intervals for CREA as performed on the analyzer.

### *Specific Performance Characteristics*[j]

#### REPRODUCIBILITY[k]

| Material | Mean | Standard deviation (% CV) Within-run | Standard deviation (% CV) Between-day |
|---|---|---|---|
| Lyophilized | 1.3 | 0.05 (3.7) | 0.05 (3.7) |
| Control | [115] | [4.4] | [4.4] |
| Lyophilized | 20.6 | 0.12 (0.6) | 0.37 (1.8) |
| Control | [1821] | [10.6] | [32.7] |

[j] All specific performance characteristics tests were run after normal recommended equipment quality control checks were performed (see Instrument Manual).

[k] Specimens at each level were analyzed in duplicate for twenty days. The within-run and between-day standard deviations were calculated by the analysis of variance method.

#### CORRELATION—REGRESSION STATISTICS[l]

| Comparative method | Slope | Intercept | Correlation coefficient | n |
|---|---|---|---|---|
| Autoanalyzer® | 1.03 | 0.03[2.7] | 0.997 | 260 |

[l] Model equation for regression statistics is:

**Result of ACA® Analyzer = Slope (Comparative method result) + intercept**

### *Assay Range*[m]

0.0-20.0 mg/dl

[0-1768 µmol]

m. See REPRODUCIBILITY for method performance within the assay range.

### *Analytical Specificity*

See KNOWN INTERFERING SUBSTANCES section for details.

### BIBLIOGRAPHY

[1] Larsen, K, Clin Chem Acta 41, 209 (1972).

[2] Tietz, NW, Fundamentals of Clinical Chemistry, W. B. Saunders Co., Philadelphia, PA, 1976, pp 47-52, 1211.

[3] Supplementary information pertaining to the effects of various drugs and patient conditions on in vivo or in vitro diagnostic levels can be found in "Drug Interferences with Clinical Laboratory Tests," Clin. Chem 21(5) (1975), and "Effects of Disease on Clinical Laboratory Tests," Clin Chem, 26 (4) 1D-476D (1980).

[4] Watkins, R. Fieldkamp, SC, Thibert, RJ, and Zak, B, Clin Chem, 21, 1002 (1975).

[5] Kawas, EE, Richards, AH, and Bigger, R, An Evaluation of a Kinetic Creatinine Test for the Du Pont ACA, Du Pont Company, Wilmington, DE (February 1973). (Reprints available from DuPont Company, Diagnostic Systems)

[6] Westgard, JO, Effects of Hemolysis and Lipemia on ACA Creatinine Method, 0.200 L, Sample Size, Du Pont Company, Wilmington, DE (October 1972).

[7] Physicians' Desk Reference, Medical Economics Company, 33 Edition, 1979.

[8] Henry, JB, Clinical Diagnosis and Management by Laboratory Methods, W.B. Saunders Co., Philadelphia, PA 1979, Vol. III.

[9] Krupp, MA, Tierney, LM Jr., Jawetz, E, Roe, RI, Camargo, CA, Physicians Handbook, Lange Medical Publications, Los Altos, CA, 1982 pp 635-636.

[10] Sarah, AJ, Koch, TR, Drusano, GL, Celoxitin Falsely Elevates Creatinine Levels, JAMA 247, 205-206 (1982).

[11] Gadsden, RH, and Phelps, CA, A Normal Range Study of Amylase in Urine and Serum on the Du Pont ACA, Du Pont Company, Wilmington, DE (March 1978). (Reprints available from DuPont Company, Diagnostic Systems)

12 Dicht, JJ, Reference Intervals for Serum Amylase and Urinary Creatinine on the Du Pont ACA® Discrete Clinical Analyzer, Du Pont Company, Wilmington, DE (November 1984).

## ATTACHMENT 3— ANALYSIS OF CREATININE FOR THE NORMALIZATION OF CADMIUM AND BETA-2-MICROGLOBULIN CONCENTRATIONS IN URINE (OSLTC PROCEDURE).

Matrix: Urine.

Target concentration: 1.1 g/L (this amount is representative of creatinine concentrations found in urine).

Procedure: A 1.0 mL aliquot of urine is passed through a C18 SEP-PAK® (Waters Associates). Approximately 30 mL of HPLC (high performance liquid chromatography) grade water is then run through the SEP PAK. The resulting solution is diluted to volume in a 100-mL volumetric flask and analyzed by HPLC using an ultraviolet (UV) detector.

Special requirements: After collection, samples should be appropriately stabilized for cadmium (Cd) analysis by using 10% high purity (with low Cd background levels) nitric acid (exactly 1.0 mL of 10% nitric acid per 10 mL of urine) or stabilized for Beta-2-Microglobulin (B2M) by taking to pH 7 with dilute NaOH (exactly 1.0 mL of 0.11 N NaOH per 10 mL of urine). If not immediately analyzed, the samples should be frozen and shipped by overnight mail in an insulated container.

Dated: January 1992.

David B. Armitage,

Duane Lee,

*Chemists.*

Organic Service Branch II, OSHA Technical Center, Salt Lake City, Utah

### 1. General Discussion

#### 1.1 Background

1.1.1. History of procedure

Creatinine has been analyzed by several methods in the past. The earliest methods were of the wet chemical type. As an example, creatinine reacts with sodium picrate in basic solution to form a red complex, which is then analyzed colorimetrically (Refs. 5.1. and 5.2.).

Since industrial hygiene laboratories will be analyzing for Cd and B2M in urine, they will be normalizing those concentrations to the concentration of creatinine in urine. A literature search revealed several HPLC methods (Refs. 5.3., 5.4., 5.5. and 5.6.) for creatinine in urine and because many industrial hygiene laboratories have HPLC equipment, it was desirable to develop an industrial hygiene HPLC method for creatinine in urine. The method of Hausen, Fuchs, and Wachter was chosen as the starting point for method development. SEP-PAKs were used for sample clarification and cleanup in this method to protect the analytical column. The urine aliquot which has been passed through the SEP-PAK is then analyzed by reverse-phase HPLC using ion-pair techniques.

This method is very similar to that of Ogata and Taguchi (Ref. 5.6.), except they used centrifugation for sample clean-up. It is also of note that they did a comparison of their HPLC results to those of the Jaffe method (a picric acid method commonly used in the health care industry) and found a linear relationship of close to 1:1. This indicates that either HPLC or colorimetric methods may be used to measure creatinine concentrations in urine.

1.1.2. Physical properties (Ref. 5.7.)

Molecular weight: 113.12

Molecular formula: $C_4$-$H_7$-$N_3$-0

Chemical name: 2-amino-1,5-dihydro-1-methyl-4H-imidazol-4-one

CAS No.: 60-27-5

Melting point: 300 °C (decomposes)

Appearance: white powder

Solubility: soluble in water; slightly soluble in alcohol; practically insoluble in acetone, ether, and chloroform

Synonyms: 1-methylglycocyamidine, 1-methylhydantoin-2-imide

Structure: see Figure #1

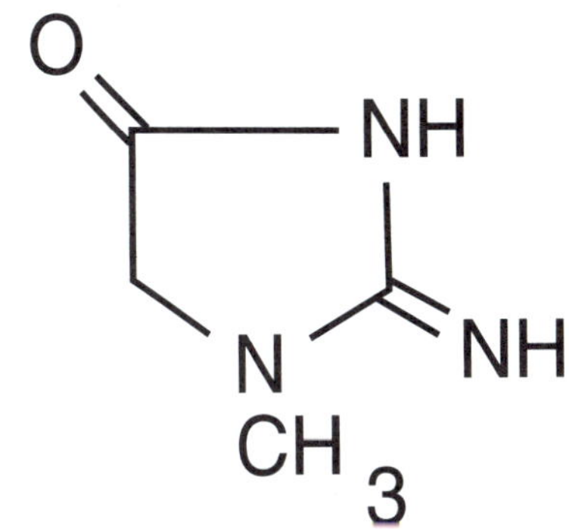


**FIGURE #1**

#### 1.2. Advantages

1.2.1. This method offers a simple, straightforward, and specific alternative method to the Jaffe method.

1.2.2. HPLC instrumentation is commonly found in many industrial hygiene laboratories.

### 2. Sample stabilization procedure

#### 2.1. Apparatus

Metal-free plastic container for urine sample.

#### 2.2. Reagents

2.2.1. Stabilizing Solution—

(1) Nitric acid (10%, high purity with low Cd background levels) for stabilizing urine for Cd analysis or

(2) NaOH, 0.11 N, for stabilizing urine for B2M analysis.

2.2.2. HPLC grade water

#### 2.3. Technique

2.3.1. Stabilizing solution is added to the urine sample (see section 2.2.1.). The stabilizing solution should be such that for each 10 mL of urine, add exactly 1.0 mL of stabilizer solution. (Never add water or urine to acid or base. Always add acid or base to water or urine.) Exactly 1.0 mL of 0.11 N NaOH added to 10 mL of urine should result in a pH of 7. Or add 1.0 mL of 10% nitric acid to 10 mL of urine.

2.3.2. After sample collection seal the plastic bottle securely and wrap it with an appropriate seal. Urine samples should be frozen and then shipped by overnight mail (if shipping is necessary) in an insulated container. (Do not fill plastic bottle too full. This will allow for expansion of contents during the freezing process.)

#### 2.4. The Effect of Preparation and Stabilization Techniques on Creatinine Concentrations

Three urine samples were prepared by making one sample acidic, not treating a second sample, and adjusting a third sample to pH 7. The samples were analyzed in duplicate by two different procedures. For the first procedure a 1.0 mL aliquot of urine was put in a 100-mL volumetric flask, diluted to volume with HPLC grade water, and then analyzed directly on an HPLC. The other procedure used SEP-PAKs. The SEP-PAK was rinsed with approximately 5 mL of methanol followed by approximately 10 mL of HPLC grade water and both rinses were discarded. Then, 1.0 mL of the urine sample was put through the SEP-PAK, followed by 30 mL of HPLC grade water. The urine and water were transferred to a 100-mL volumetric flask, diluted to volume with HPLC grade water, and analyzed by HPLC. These three urine samples were analyzed on the day they were obtained and then frozen. The results show that whether the urine is acidic, untreated or adjusted to pH 7, the resulting answer for creatinine is essentially unchanged. The purpose of stabilizing the urine by making it acidic or neutral is for the analysis of Cd or B2M respectively.

### COMPARISON OF PREPARATION AND STABILIZATION TECHNIQUES

| Sample | w/o SEP-PAK g/L creatinine | with SEP-PAK g/L creatinine |
|---|---|---|
| Acid | 1.10 | 1.10 |
| Acid | 1.11 | 1.10 |
| Untreated | 1.12 | 1.11 |
| Untreated | 1.11 | 1.12 |
| pH 7 | 1.08 | 1.02 |
| pH 7 | 1.11 | 1.08 |

### 2.5. Storage

After 4 days and 54 days of storage in a freezer, the samples were thawed, brought to room temperature and analyzed using the same procedures as in section 2.4. The results of several days of storage show that the resulting answer of creatinine is essentially unchanged.

### STORAGE DATA

| Sample | 4 days | | 54 days | |
|---|---|---|---|---|
| | w/o SEP-PAK g/L creatinine | with SEP-PAK g/L creatinine | w/o SEP-PAK g/L creatinine | with SEP-PAK g/L creatinine |
| Acid | 1.09 | 1.09 | 1.08 | 1.09 |
| Acid | 1.10 | 1.10 | 1.09 | 1.10 |
| Acid | ........ | ........ | 1.09 | 1.09 |
| Untreated | 1.13 | 1.14 | 1.09 | 1.11 |
| Untreated | 1.15 | 1.14 | 1.10 | 1.10 |
| Untreated | ........ | ........ | 1.09 | 1.10 |
| pH 7 | 1.14 | 1.13 | 1.12 | 1.12 |
| pH 7 | 1.14 | 1.13 | 1.12 | 1.12 |
| pH 7 | ........ | ........ | 1.12 | 1.12 |

### 2.6. Interferences

None.

### 2.7. Safety precautions

2.7.1. Make sure samples are properly sealed and frozen before shipment to avoid leakage.

2.7.2. Follow the appropriate shipping procedures.

The following modified special safety precautions are based on those recommended by the Centers for Disease Control (CDC) (Ref. 5.8.). and OSHA's Bloodborne Pathogens standard (29 CFR 1910.1039).

2.7.3. Wear gloves, lab coat, and safety glasses while handling all human urine products. Disposable plastic, glass, and paper (pipet tips, gloves, etc.) that contact urine should be placed in a biohazard autoclave bag. These bags should be kept in appropriate containers until sealed and autoclaved. Wipe down all work surfaces with 10% sodium hypochlorite solution when work is finished.

2.7.4. Dispose of all biological samples and diluted specimens in a biohazard autoclave bag at the end of the analytical run.

2.7.5. Special care should be taken when handling and dispensing nitric acid. Always remember to add acid to water (or urine). Nitric acid is a corrosive chemical capable of severe eye and skin damage. Wear metal-free gloves, a lab coat, and safety glasses. If the nitric acid comes in contact with any part of the body, quickly wash with copious quantities of water for at least 15 minutes.

2.7.6. Special care should be taken when handling and dispensing NaOH. Always remember to add base to water (or urine). NaOH can cause severe eye and skin damage. Always wear the appropriate gloves, a lab coat, and safety glasses. If the NaOH comes in contact with any part of the body, quickly wash with copious quantities of water for at least 15 minutes.

## 3. Analytical procedure

### 3.1. Apparatus

3.1.1. A high performance liquid chromatograph equipped with pump, sample injector and UV detector.

3.1.2. A C18 HPLC column; 25 cm x 4.6 mm I.D.

3.1.3. An electronic integrator, or some other suitable means of determining analyte response.

3.1.4. Stripchart recorder.

3.1.5. C18 SEP-PAKs (Waters Associates) or equivalent.

3.1.6. Luer-lock syringe for sample preparation (5 mL or 10 mL).

3.1.7. Volumetric pipettes and flasks for standard and sample preparation.

3.1.8. Vacuum system to aid sample preparation (optional).

### 3.2. Reagents

3.2.1. Water, HPLC grade.

3.2.2. Methanol, HPLC grade.

3.2.3. PIC B-7® (Waters Associates) in small vials.

3.2.4. Creatinine, anhydrous, Sigma hemical Corp., purity not listed.

3.2.5. 1-Heptanesulfonic acid, sodium salt monohydrate.

3.2.6. Phosphoric acid.

3.2.7. Mobile phase. It can be prepared by mixing one vial of PIC B-7 into a 1 L solution of 50% methanol and 50% water. The mobile phase can also be made by preparing a solution that is 50% methanol and 50% water with 0.005M heptanesulfonic acid and adjusting the pH of the solution to 3.5 with phosphoric acid.

### 3.3. Standard preparation

3.3.1. Stock standards are prepared by weighing 10 to 15 mg of creatinine. This is transferred to a 25-mL volumetric flask and diluted to volume with HPLC grade water.

3.3.2. Dilutions to a working range of 3 to 35 μg/mL are made in either HPLC grade water or HPLC mobile phase (standards give the same detector response in either solution).

### 3.4. Sample preparation

3.4.1. The C18 SEP-PAK is connected to a Luer-lock syringe. It is rinsed with 5 mL HPLC grade methanol and then 10 mL of HPLC grade water. These rinses are discarded.

3.4.2. Exactly 1.0 mL of urine is pipetted into the syringe. The urine is put through the SEP-PAK into a suitable container using a vacuum system.

3.4.3. The walls of the syringe are rinsed in several stages with a total of approximately 30 mL of HPLC grade water. These rinses are put through the SEP-PAK into the same container. The resulting solution is transferred to a 100-mL volumetric flask and then brought to volume with HPLC grade water.

### 3.5. Analysis (conditions and hardware are those used in this evaluation.)

3.5.1. Instrument conditions

Column: Zorbax® ODS, 5-6 μm particle size; 25 cm × 4.6 mm I.D.

Mobile phase: See Section 3.2.7.

Detector: Dual wavelength UV; 229 nm (primary) 254 nm (secondary)

Flow rate: 0.7 mL/ minute

Retention time: 7.2 minutes

Sensitivity: 0.05 AUFS

Injection volume: 20μl

3.5.2. Chromatogram (see Figure #2)

### CHROMATOGRAM OF A CREATININE STANDARD FIGURE #2

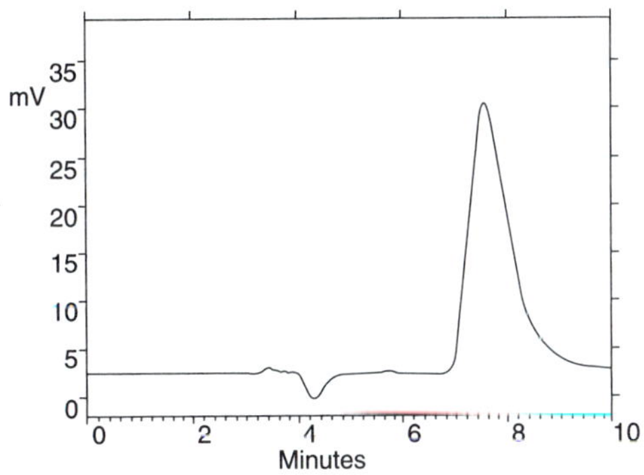


### 3.6. Interferences

3.6.1. Any compound that has the same retention time as creatinine and absorbs at 229 nm is an interference.

3.6.2. HPLC conditions may be varied to circumvent interferences. In addition, analysis at another UV wavelength (i.e. 254 nm) would allow a comparison of the ratio of response of a standard to that of a sample. Any deviations would indicate an interference.

### 3.7. Calculations

3.7.1. A calibration curve is constructed by plotting detector response versus standard concentration (See Figure #3).

3.7.2. The concentration of creatinine in a sample is determined by finding the concentration corresponding to its detector response. (See Figure #3).

### CALIBRATION CURVE FOR CREATININE FIGURE #3

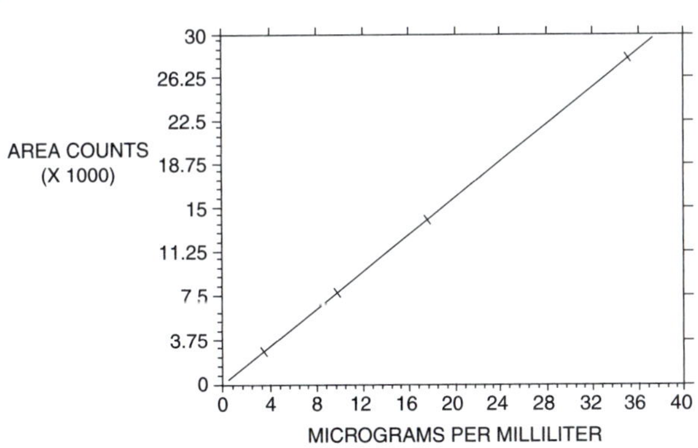


3.7.3. The µg/mL creatinine from section 3.7.2. is then multiplied by 100 (the dilution factor). This value is equivalent to the micrograms of creatinine in the 1.0 mL stabilized urine aliquot or the milligrams of creatinine per liter of urine. The desired units, g/L, is determined by the following relationship:

$$\mathrm{g/L} = \frac{\mu\mathrm{g/mL}}{1000} = \frac{\mathrm{mg/L}}{1000}$$

3.7.4. The resulting value for creatinine is used to normalize the urinary concentration of the desired analyte (A) (Cd or B2M) by using the following formula.

$$\frac{\mu\mathrm{g\ A/g}}{\mathrm{creatinine}} = \frac{\mu\mathrm{g\ A/L\ (experimental)}}{\mathrm{g/L\ creatinine}}$$

Where A is the desired analyte. The protocol of reporting such normalized results is µg A/g creatinine.

### 3.8. Safety precautions See section 2.7.

### 4. Conclusions

The determination of creatinine in urine by HPLC is a good alternative to the Jaffe method for industrial hygiene laboratories. Sample clarification with SEP-PAKs did not change the amount of creatinine found in urine samples. However, it does protect the analytical column. The results of this creatinine in urine procedure are unaffected by the pH of the urine sample under the conditions tested by this procedure. Therefore, no special measures are required for creatinine analysis whether the urine sample has been stabilized with 10% nitric acid for the Cd analysis or brought to a pH of 7 with 0.11 N NaOH for the B2M analysis.

### 5. References

5.1. Clark, L.C.; Thompson, H.L.; *Anal. Chem.* 1949, 21, 1218.

5.2. Peters, J.H.; *J. Biol. Chem.* 1942, 146, 176.

5.3. Hausen, V.A.; Fuchs, D.; Wachter, H.; *J. Clin. Chem. Clin. Biochem.* 1981, 19, 373-378.

5.4. Clark, P.M.S.; Kricka L.J.; Patel, A.; *J. Liq. Chrom.* 1980, 3(7), 1031-1046.

5.5. Ballerini, R.; Chinol, M.; Cambi, A.; *J. Chrom.* 1979, 179, 365-369.

5.6. Ogata, M.; Taguchi, T.; *Industrial Health* 1987, 25, 225-228.

5.7. "Merck Index", 11th ed.; Windholz, Martha Ed.; Merck: Rahway, N.J., 1989; p 403.

5.8. Kimberly, M.; *"Determination of Cadmium in Urine by Graphite Furnace Atomic Absorption Spectrometry with Zeeman Background Correction."*, Centers for Disease Control, Atlanta, Georgia, unpublished, update 1990.

[57 FR 42389, Sept. 14, 1992, as amended at 57 FR 49272, Oct. 30, 1992; 58 FR 21781, Apr. 23, 1993; 61 FR 5508, Feb. 13, 1996; 63 FR 1288, Jan. 8, 1998]

## §1926.1128 Benzene.

**Note:** The requirements applicable to construction work under this section are identical to those set forth at §1910.1028 of this chapter.

[61 FR 31434, June 20, 1996]

[Section 1910.1028 is printed below—CCH.]

## §1910.1028 Benzene.

**1910.1028(a)** ***Scope and application.*** (1) This section applies to all occupational exposures to benzene. Chemical Abstracts Service Registry No. 71-43-2, except as provided in paragraphs (a)(2) and (a)(3) of this section.

**1910.1028(a)(2)** This section does not apply to:

**1910.1028(a)(2)(i)** The storage, transportation, distribution, dispensing, sale or use of gasoline, motor fuels, or other fuels containing benzene subsequent to its final discharge from bulk wholesale storage facilities, except that operations where gasoline or motor fuels are dispensed for more than 4 hours per day in an indoor location are covered by this section.

**1910.1028(a)(2)(ii)** Loading and unloading operations at bulk wholesale storage facilities which use vapor control systems for all loading and unloading operations, except for the provisions of 29 CFR 1910.1200 as incorporated into this section and the emergency provisions of paragraphs (g) and (i)(4) of this section.

**1910.1028(a)(2)(iii)** The storage, transportation, distribution or sale of benzene or liquid mixtures containing more than 0.1 percent benzene in intact containers or in transportation pipelines while sealed in such a manner as to contain benzene vapors or liquid, except for the provisions of 29 CFR 1910.1200 as incorporated into this section and the emergency provisions of paragraphs (g) and (i)(4) of this section.

**1910.1028(a)(2)(iv)** Containers and pipelines carrying mixtures with less than 0.1 percent benzene and natural gas processing plants processing gas with less than 0.1 percent benzene.

**1910.1028(a)(2)(v)** Work operations where the only exposure to benzene is from liquid mixtures containing 0.5 percent or less of benzene by volume, or the vapors released from such liquids until September 12, 1988; work operations where the only exposure to benzene is from liquid mixtures containing 0.3 percent or less of benzene by volume or the vapors released from such liquids from September 12, 1988, to September 12, 1989; and work operations where the only exposure to benzene is from liquid mixtures containing 0.1 percent or less of benzene by volume or the vapors released from such liquids after September 12, 1989; except that tire building machine operators using solvents with more than 0.1 percent benzene are covered by paragraph (i) of this section.

**1910.1028(a)(2)(vi)** Oil and gas drilling, production and servicing operations.

**1910.1028(a)(2)(vii)** Coke oven batteries.

**1910.1028(a)(3)** The cleaning and repair of barges and tankers which have contained benzene are excluded from paragraph (f) methods of compliance, paragraph (e)(1) exposure monitoring-general, and paragraph (e)(6) accuracy of monitoring. Engineering and work practice controls shall be used to keep exposures below 10 ppm unless it is proven to be not feasible.

**1910.1028(b)** ***Definitions.***

*Action level* means an airborne concentration of benzene of 0.5 ppm calculated as an 8-hour time-weighted average.

*Assistant Secretary* means the Assistant Secretary of Labor for Occupational Safety and Health, U.S. Department of Labor, or designee.

*Authorized person* means any person specifically authorized by the employer whose duties require the person to enter a regulated area, or any person entering such an area as a designated representative of employees for the purpose of exercising the right to observe monitoring and measuring procedures under paragraph (l) of this section, or any other person authorized by the Act or regulations issued under the Act.

*Benzene* ($C_6$ $H_6$) (CAS Registry No. 71-43-2) means liquefied or gaseous benzene. It includes benzene contained in liquid mixtures and the benzene vapors released by these liquids. It does not include trace amounts of unreacted benzene contained in solid materials.

*Bulk wholesale storage facility* means a bulk terminal or bulk plant where fuel is stored prior to its delivery to wholesale customers.

*Container* means any barrel, bottle, can, cylinder, drum, reaction vessel, storage tank, or the like, but does not include piping systems.

*Day* means any part of a calendar day.

*Director* means the Director of the National Institute for Occupational Safety and Health, U.S. Department of Health and Human Services, or designee.

*Emergency* means any occurrence such as, but not limited to, equipment failure, rupture of containers, or failure of control equipment which may or does result in an unexpected significant release of benzene.

*Employee exposure* means exposure to airborne benzene which would occur if the employee were not using respiratory protective equipment.

*Regulated area* means any area where airborne concentrations of benzene exceed or can reasonably be expected to exceed, the permissible exposure limits, either the 8-hour time weighted average exposure of 1 ppm or the short-term exposure limit of 5 ppm for 15 minutes.

*Vapor control system* means any equipment used for containing the total vapors displaced during the loading of gasoline, motor fuel or other fuel tank trucks and the displacing of these vapors through a vapor processing system or balancing the vapor with the storage tank. This equipment also includes systems containing the vapors displaced from the storage tank during the unloading of the tank truck which balance the vapors back to the tank truck.

**1910.1028(c)** ***Permissible exposure limits (PELs)***— (1) *Time-weighted average limit (TWA).* The employer shall assure that no employee is exposed to an airborne concentration of benzene in excess of one part of benzene per million parts of air (1 ppm) as an 8-hour time-weighted average.

**1910.1028(c)(2)** *Short-term exposure limit (STEL).* The employer shall assure that no employee is exposed to an airborne concentration of benzene in excess of five (5) ppm as averaged over any 15 minute period.

**1910.1028(d)** ***Regulated areas.*** (1) The employer shall establish a regulated area wherever the airborne concentration of benzene exceeds or can reasonably be expected to exceed the permissible exposure limits, either the 8-hour time weighted average exposure of 1 ppm or the short-term exposure limit of 5 ppm for 15 minutes.

**1910.1028(d)(2)** Access to regulated areas shall be limited to authorized persons.

**1910.1028(d)(3)** Regulated areas shall be determined from the rest of the workplace in any manner that minimizes the number of employees exposed to benzene within the regulated area.

**1910.1028(e)** ***Exposure monitoring***— (1) *General.* (i) Determinations of employee exposure shall be made from breathing zone air samples that are representative of each employee's average exposure to airborne benzene.

**1910.1028(e)(1)(ii)** Representative 8-hour TWA employee exposures shall be determined on the basis of one sample or samples representing the full shift exposure for each job classification in each work area.

**1910.1028(e)(1)(iii)** Determinations of compliance with the STEL shall be made from 15 minute employee breathing zone samples measured at operations where there is reason to believe exposures are high, such as where tanks are opened, filled, unloaded or gauged; where containers or process equipment are opened and where benzene is used for cleaning or as a solvent in an uncontrolled situation. The employer may use objective data, such as measurements from brief period measuring devices, to determine where STEL monitoring is needed.

**1910.1028(e)(1)(iv)** Except for initial monitoring as required under paragraph (e)(2) of this section, where the employer can document that one shift will consistently have higher employee exposures for an operation, the employer shall only be required to determine representative employee exposure for that operation during the shift on which the highest exposure is expected.

**1910.1028(e)(2)** *Initial monitoring.* (i) Each employer who has a place of employment covered under paragraph (a)(1) of this section shall monitor each of these workplaces and work operations to determine accurately the airborne concentrations of benzene to which employees may be exposed.

**1910.1028(e)(2)(ii)** The initial monitoring required under paragraph (e)(2)(i) of this section shall be completed by 60 days after the effective date of this standard or within 30 days of the introduction of benzene into the workplace. Where the employer has monitored within one year prior to the effective date of this standard and the monitoring satisfies all other requirements of this section, the employer may rely on such earlier monitoring results to satisfy the requirements of paragraph (e)(2)(i) of this section.

**1910.1028(e)(3)** *Periodic monitoring and monitoring frequency.* (i) If the monitoring required by paragraph (e)(2)(i) of this section reveals employee exposure at or above the action level but at or below the TWA, the employer shall repeat such monitoring for each such employee at least every year.

**1910.1028(e)(3)(ii)** If the monitoring required by paragraph (e)(2)(i) of this section reveals employee exposure above the TWA, the employer shall repeat such monitoring for each such employee at least every six (6) months.

**1910.1028(e)(3)(iii)** The employer may alter the monitoring schedule from every six months to annually for any employee for whom two consecutive measurements taken at least 7 days apart indicate that the employee exposure has decreased to the TWA or below, but is at or above the action level.

**1910.1028(e)(3)(iv)** Monitoring for the STEL shall be repeated as necessary to evaluate exposures of employees subject to short term exposures.

**1910.1028(e)(4)** *Termination of monitoring.* (i) If the initial monitoring required by paragraph (e)(2)(i) of this section reveals employee exposure to be below the action level the employer may discontinue the monitoring for that employee, except as otherwise required by paragraph (e)(5) of this section.

**1910.1028(e)(4)(ii)** If the periodic monitoring required by paragraph (e)(3) of this section reveals that employee exposures, as indicated by at least two consecutive measurements taken at least 7 days apart, are below the action level the employer may discontinue the monitoring for that employee, except as otherwise required by paragraph (e)(5).

**1910.1028(e)(5)** *Additional monitoring.* (i) The employer shall institute the exposure monitoring required under paragraphs (e)(2) and (e)(3) of this section when there has been a change in the production, process, control equipment, personnel or work practices which may result in new or additional exposures to benzene, or when the employer has any reason to suspect a change which may result in new or additional exposures.

**1910.1028(e)(5)(ii)** Whenever spills, leaks, ruptures or other breakdowns occur that may lead to employee exposure, the employer shall monitor (using area or personal sampling) after the cleanup of the spill or repair of the leak, rupture or other breakdown to ensure that exposures have returned to the level that existed prior to the incident.

**1910.1028(e)(6)** *Accuracy of monitoring.* Monitoring shall be accurate, to a confidence level of 95 percent, to within plus or minus 25 percent for airborne concentrations of benzene.

**1910.1028(e)(7)** *Employee notification of monitoring results.* (i) The employer shall, within 15 working days after the receipt of the results of any monitoring performed under this standard, notify each employee of these results in writing either individually or by posting of results in an appropriate location that is accessible to affected employees.

**1910.1028(e)(7)(ii)** Whenever the PELs are exceeded, the written notification required by paragraph (e)(7)(i) of this section shall contain the corrective action being taken by the employer to reduce the employee exposure to or below the PEL, or shall refer to a document available to the employee which states the corrective actions to be taken.

**1910.1028(f)** ***Methods of compliance***— (1) *Engineering controls and work practices.* (i) The employer shall institute engineering controls and work practices to reduce and maintain employee exposure to benzene at or below the permissible exposure limits, except to the extent that the employer can establish that these controls are not feasible or where the provisions of paragraph (f)(1)(iii) or (g)(1) of this section apply.

**1910.1028(f)(1)(ii)** Wherever the feasible engineering controls and work practices which can be instituted are not sufficient to reduce employee exposure to or below the PELs, the employer shall use them to reduce employee exposure to the lowest levels achievable by these controls and shall supplement them by the use of respiratory protection which complies with the requirements of paragraph (g) of this section.

**1910.1028(f)(1)(iii)** Where the employer can document that benzene is used in a workplace less than a total of 30 days per year, the employer shall use engineering controls, work practice controls or respiratory protection or any combination of these controls to reduce employee exposure to benzene to or below the PELs, except that employers shall use engineering and work practice controls, if feasible, to reduce exposure to or below 10 ppm as an 8-hour TWA.

**1910.1028(f)(2)** *Compliance program.* (i) When any exposures are over the PEL, the employer shall establish and implement a written program to reduce employee exposure to or below the PEL primarily by means of engineering and work practice controls, as required by paragraph (f)(1) of this section.

**1910.1028(f)(2)(ii)** The written program shall include a schedule for development and implementation of the engineering and work practice controls. These plans shall be reviewed and revised as appropriate based on the most recent exposure monitoring data, to reflect the current status of the program.

**1910.1028(f)(2)(iii)** Written compliance programs shall be furnished upon request for examination and copying to the Assistant Secretary, the Director, affected employees and designated employee representatives.

**1910.1028(g)** ***Respiratory protection***— (1) *General.* For employees who use respirators required by this section, the employer must provide respirators that comply with the requirements of this paragraph. Respirators must be used during:

**1910.1028(g)(1)(i)** Periods necessary to install or implement feasible engineering and work-practice controls.

**1910.1028(g)(1)(ii)** Work operations for which the employer establishes that compliance with either the TWA or STEL through the use of engineering and work-practice controls is not feasible; for example, some maintenance and repair activities, vessel cleaning, or other operations for which engineering and work-practice con trols are infeasible because exposures are intermittent and limited in duration.

**1910.1028(g)(1)(iii)** Work operations for which feasible engineering and work-practice controls are not yet sufficient, or are not required under paragraph (f)(1)(iii) of this section, to reduce employee exposure to or below the PELs.

**1910.1028(g)(1)(iv)** Emergencies.

**1910.1028(g)(2)** *Respirator program.* (i) The employer must implement a respiratory protection program in accordance with 29 CFR 1910.134(b) through (d) (except (d)(1)(iii), (d)(3)(iii)(B)(1), and (2)), and (f) through (m).

**1910.1028(g)(2)(ii)** For air-purifying respirators, the employer must replace the air-purifying element at the expiration of its service life or at the beginning of each shift in which such elements are used, whichever comes first.

**1910.1028(g)(2)(iii)** If NIOSH approves an air-purifying element with an endof-service-life indicator for benzene, such an element may be used until the indicator shows no further useful life.

**1910.1028(g)(3)** *Respirator selection.* (i) The employer must select the appropriate respirator from Table 1 of this section.

**1910.1028(g)(3)(ii)** Any employee who cannot use a negative-pressure respirator must be allowed to use a respirator with less breathing resistance, such as a powered air-purifying respirator or supplied-air respirator.

## TABLE 1—RESPIRATORY PROTECTION FOR BENZENE

| Airborne concentration of benzene or condition of use | Respirator type |
|---|---|
| (a) Less than or equal to 10 ppm. | (1) Half-mask air-purifying respirator with organic vapor cartridge. |
| (b) Less than or equal to 50 ppm. | (1) Full facepiece respirator with organic vapor cartridges. |
| | (1) Full facepiece gas mask with chin style canister.[1] |
| (c) Less than or equal to 100 ppm. | (1) Full facepiece powered air-purifying respirator with organic vapor canister.[1] |
| (d) Less than or equal to 1,000 ppm. | (1) Supplied air respirator with full facepiece in positive-pressure mode. |
| (e) Greater than 1,000 ppm or unknown concentration. | (1) Self-contained breathing apparatus with full facepiece in positive pressure mode. |
| | (2) Full facepiece positive-pressure supplied-air respirator with auxiliary self-contained air supply. |
| (f) Escape | (1) Any organic vapor gas mask; or |
| | (2) Any self-contained breathing apparatus with full facepiece. |
| (g) Firefighting | (1) Full facepiece self-contained breathing apparatus in positive pressure mode. |

[1] Canisters must have a minimum service life of four (4) hours when tested at 150 ppm benzene, at a flow rate of 64 LPM, 25 deg. C, and 85% relative humidity for non-powered air purifying respirators. The flow rate shall be 115 LPM and 170 LPM respectively for tight fitting and loose fitting powered air-purifying respirators.

**1910.1028(h)** ***Protective clothing and equipment.*** Personal protective clothing and equipment shall be worn where appropriate to prevent eye contact and limit dermal exposure to liquid benzene. Protective clothing and equipment shall be provided by the employer at no cost to the employee and the employer shall assure its use where appropriate. Eye and face protection shall meet the requirements of 29 CFR 1910.133.

**1910.1028(i)** ***Medical surveillance***— (1) *General.* (i) The employer shall make available a medical surveillance program for employees who are or may be exposed to benzene at or above the action level 30 or more days per year; for employees who are or may be exposed to benzene at or above the PELs 10 or more days per year; for employees who have been exposed to more than 10 ppm of benzene for 30 or more days in a year prior to the effective date of the standard when employed by their current employer; and for employees involved in the tire building operations called tire building machine operators, who use solvents containing greater than 0.1 percent benzene.

**1910.1028(i)(1)(ii)** The employer shall assure that all medical examinations and procedures are performed by or under the supervision of a licensed physician and that all laboratory tests are conducted by an accredited laboratory.

**1910.1028(i)(1)(iii)** The employer shall assure that persons other than licensed physicians who administer the pulmonary function testing required by this section shall complete a training course in spirometry sponsored by an appropriate governmental, academic or professional institution.

**1910.1028(i)(1)(iv)** The employer shall assure that all examinations and procedures are provided without cost to the employee and at a reasonable time and place.

**1910.1028(i)(2)** *Initial examination.* (i) Within 60 days of the effective date of this standard, or before the time of initial assignment, the employer shall provide each employee covered by paragraph (i)(1)(i) of this section with a medical examination including the following elements:

**1910.1028(i)(2)(i)(A)** A detailed occupational history which includes:

**1910.1028(i)(2)(i)(A)*(1)*** Past work exposure to benzene or any other hematological toxins,

**1910.1028(i)(2)(i)(A)*(2)*** A family history of blood dyscrasias including hematological neoplasms;

**1910.1028(i)(2)(i)(A)*(3)*** A history of blood dyscrasias including genetic hemoglobin abnormalities, bleeding abnormalities, abnormal function of formed blood elements;

**1910.1028(i)(2)(i)(A)*(4)*** A history of renal or liver dysfunction;

**1910.1028(i)(2)(i)(A)*(5)*** A history of medicinal drugs routinely taken;

**1910.1028(i)(2)(i)(A)*(6)*** A history of previous exposure to ionizing radiation and

**1910.1028(i)(2)(i)(A)*(7)*** Exposure to marrow toxins outside of the current work situation.

**1910.1028(i)(2)(i)(B)** A complete physical examination.

**1910.1028(i)(2)(i)(C)** *Laboratory tests.* A complete blood count including a leukocyte count with differential, a quantitative thrombocyte count, hematocrit, hemoglobin, erythrocyte count and erythrocyte indices (MCV, MCH, MCHC). The results of these tests shall be reviewed by the examining physician.

**1910.1028(i)(2)(i)(D)** Additional tests as necessary in the opinion of the examining physician, based on alterations to the components of the blood or other signs which may be related to benzene exposure; and

**1910.1028(i)(2)(i)(E)** For all workers required to wear respirators for at least 30 days a year, the physical examination shall pay special attention to the cardiopulmonary system and shall include a pulmonary function test.

**1910.1028(i)(2)(ii)** No initial medical examination is required to satisfy the requirements of paragraph (i)(2)(i) of this section if adequate records show that the employee has been examined in accordance with the procedures of paragraph (i)(2)(i) of this section within the twelve months prior to the effective date of this standard.

**1910.1028(i)(3)** *Periodic examinations.* (i) The employer shall provide each employee covered under paragraph (i)(1)(i) of this section with a medical examination annually following the previous examination. These periodic examinations shall incude at least the following elements:

**1910.1028(i)(3)(i)(A)** A brief history regarding any new exposure to potential marrow toxins, changes in medicinal drug use, and the appearance of physical signs relating to blood disorders:

**1910.1028(i)(3)(i)(B)** A complete blood count including a leukocyte count with differential, quantitative thrombocyte count, hemoglobin, hematocrit, erythrocyte count and erythrocyte indices (MCV, MCH, MCHC); and

**1910.1028(i)(3)(i)(C)** Appropriate additional tests as necessary, in the opinion of the examining physician, in consequence of alterations in the components of the blood or other signs which may be related to benzene exposure.

**1910.1028(i)(3)(ii)** Where the employee develops signs and symptoms commonly associated with toxic exposure to benzene, the employer shall provide the employee with an additional medical examination which shall include those elements considered appropriate by the examining physician.

**1910.1028(i)(3)(iii)** For persons required to use respirators for at least 30 days a year, a pulmonary function test shall be performed every three (3) years. A specific evaluation of the cardiopulmonary system shall be made at the time of the pulmonary function test.

**1910.1028(i)(4)** *Emergency examinations.* (i) In addition to the surveillance required by (i)(1)(i), if an employee is exposed to benzene in an emergency situation, the employer shall have the employee provide a urine sample at the end of the employee's shift and have a urinary phenol test performed on the sample within 72 hours. The urine specific gravity shall be corrected to 1.024.

**1910.1028(i)(4)(ii)** If the result of the urinary phenol test is below 75 mg phenol/L of urine, no further testing is required.

**1910.1028(i)(4)(iii)** If the result of the urinary phenol test is equal to or greater than 75 mg phenol/L of urine, the employer shall provide the employee with a complete blood count including an erythrocyte count, leukocyte count with differential and thrombocyte count at monthly intervals for a duration of three (3) months following the emergency exposure.

**1910.1028(i)(4)(iv)** If any of the conditions specified in paragraph (i)(5)(i) of this section exists, then the further requirements of paragraph (i)(5) of this section shall be met and the employer shall, in addition, provide the employees with periodic examinations if directed by the physician.

**1910.1028(i)(5)** *Additional examinations and referrals.* (i) Where the results of the complete blood count required for the initial and periodic examinations indicate any of the following abnormal conditions exist, then the blood count shall be repeated within 2 weeks.

**1910.1028(i)(5)(i)(A)** The hemoglobin level or the hematocrit falls below the normal limit [outside the 95% confidence interval (C.I.)] as determined by the laboratory for the particular geographic area and/or these indices show a persistent downward trend from the individual's pre-exposure norms; provided these findings cannot be explained by other medical reasons.

**1910.1028(i)(5)(i)(B)** The thrombocyte (platelet) count varies more than 20 percent below the employee's most recent values or falls outside the normal limit (95% C.I.) as determined by the laboratory.

**1910.1028(i)(5)(i)(C)** The leukocyte count is below 4,000 per $mm^3$ or there is an abnormal differential count.

**1910.1028(i)(5)(ii)** If the abnormality persists, the examining physician shall refer the employee to a hematologist or an internist for further evaluation unless the physician has good reason to believe such referral is unnecessary. (See Appendix C for examples of conditions where a referral may be unnecessary.)

**1910.1028(i)(5)(iii)** The employer shall provide the hematologist or internist with the information required to be provided to the physician under paragraph (i)(6) of this section and the medical record required to be maintained by paragraph (k)(2)(ii) of this section.

**1910.1028(i)(5)(iv)** The hematologist's or internist's evaluation shall include a determination as to the need for additional tests, and the employer shall assure that these tests are provided.

**1910.1028(i)(6)** *Information provided to the physician.* The employer shall provide the following information to the examining physician:

**1910.1028(i)(6)(i)** A copy of this regulation and its appendices;

**1910.1028(i)(6)(ii)** A description of the affected employee's duties as they relate to the employee's exposure;

**1910.1028(i)(6)(iii)** The employee's actual or representative exposure level:

**1910.1028(i)(6)(iv)** A description of any personal protective equipment used or to be used; and

**1910.1028(i)(6)(v)** Information from previous employment-related medical examinations of the affected employee which is not otherwise available to the examining physician.

**1910.1028(i)(7)** *Physician's written opinions.* (i) For each examination under this section, the employer shall obtain and provide the employee with a copy of the examining physician's written opinion within 15 days of the examination. The written opinion shall be limited to the following information:

**1910.1028(i)(7)(i)(A)** The occupationally pertinent results of the medical examination and tests;

**1910.1028(i)(7)(i)(B)** The physician's opinion concerning whether the employee has any detected medical conditions which would place the employee's health at greater than normal risk of material impairment from exposure to benzene;

**1910.1028(i)(7)(i)(C)** The physician's recommended limitations upon the employee's exposure to benzene or upon the employee's use of protective clothing or equipment and respirators.

**1910.1028(i)(7)(i)(D)** A statement that the employee has been informed by the physician of the results of the medical examination and any medical conditions resulting from benzene exposure which require further explanation or treatment.

**1910.1028(i)(7)(ii)** The written opinion obtained by the employer shall not reveal specific records, findings and diagnoses that have no bearing on the employee's ability to work in a benzene-exposed workplace.

**1910.1028(i)(8)** *Medical removal plan.* (i) When a physician makes a referral to a hematologist/internist as required under paragraph (i)(5)(ii) of this section, the employee shall be removed from areas where exposures may exceed the action level until such time as the physician makes a determination under paragraph (i)(8)(ii) of this section.

**1910.1028(i)(8)(ii)** Following the examination and evaluation by the hematologist/internist, a decision to remove an employee from areas where benzene exposure is above the action level or to allow the employee to return to areas where benzene exposure is above the action level shall be made by the physician in consultation with the hematologist/internist. This decision shall be communicated in writing to the employer and employee. In the case of removal, the physician shall state the required probable duration of removal from occupational exposure to benzene above the action level and the requirements for future medical examinations to review the decision.

**1910.1028(i)(8)(iii)** For any employee who is removed pursuant to paragraph (i)(8)(ii) of this section, the employer shall provide a followup examination. The physician, in consultation with the hematologist/internist, shall make a decision within 6 months of the date the employee was removed as to whether the employee shall be returned to the usual job or whether the employee should be removed permanently.

**1910.1028(i)(8)(iv)** Whenever an employee is temporarily removed from benzene exposure pursuant to paragraph (i)(8)(i) or (i)(8)(ii) of this section, the employer shall transfer the employee to a comparable job for which the employee is qualified (or can be trained for in a short period) and where benzene exposures are as low as possible, but in no event higher than the action level. The employer shall maintain the employee's current wage rate, seniority and other benefits. If there is no such job available, the employer shall provide medical removal protection benefits until such a job becomes available or for 6 months, whichever comes first.

**1910.1028(i)(8)(v)** Whenever an employee is removed permanently from benzene exposure based on a physician's recommendation pursuant to paragraph (i)(8)(iii) of this section, the employee shall be given the opportunity to transfer to another position which is available or later becomes available for which the employee is qualified (or can be trained for in a short period) and where benzene exposures are as low as possible but in no event higher than the action level. The employer shall assure that such employee suffers no reduction in current wage rate, seniority or other benefits as a result of the transfer.

**1910.1028(i)(9)** *Medical removal protection benefits.* (i) The employer shall provide to an employee 6 months of medical removal protection benefits immediately following each occasion an employee is removed from exposure to benzene because of hematological findings pursuant to paragraphs (i)(8) (i) and (ii) of this section, unless the employee has been transferred to a comparable job where benzene exposures are below the action level.

**1910.1028(i)(9)(ii)** For the purposes of this section, the requirement that an employer provide medical removal protection benefits means that the employer shall maintain the current wage rate, seniority and other benefits of an employee as though the employee had not been removed.

**1910.1028(i)(9)(iii)** The employer's obligation to provide medical removal protection benefits to a removed employee shall be reduced to the extent that the employee receives compensation for earnings lost during the period of removal either from a publicly or employerfunded compensation program, or from employment with another employer made possible by virtue of the employee's removal.

**1910.1028(j)** ***Communication of benzene hazards to employees***—(1) *Signs and labels.* (i) The employer shall post signs at entrances to regulated areas. The signs shall bear the following legend:

**DANGER**

**BENZENE**

**CANCER HAZARD**

**FLAMMABLE—NO SMOKING**

**AUTHORIZED PERSONNEL ONLY**

**RESPIRATOR REQUIRED**

**1910.1028(j)(1)(ii)** The employer shall ensure that lables or other appropriate forms of warning are provided for containers of benzene within the workplace. There is no requirement to label pipes. The labels shall comply with the requirements of 29 CFR 1910.1200(f) and in addition shall include the following legend:

**DANGER**

**CONTAINS BENZENE**

**CANCER HAZARD**

**1910.1028(j)(2)** *Material safety data sheets.* (i) Employers shall obtain or develop, and shall provide access to their employees, to a material safety data sheet (MSDS) which addresses benzene and complies with 29 CFR 1910.1200.

**1910.1028(j)(2)(ii)** Employers who are manufacturers or importers shall:

**1910.1028(j)(2)(ii)(A)** Comply with paragraph (a) of this section, and

**1910.1028(j)(2)(ii)(B)** Comply with the requirement in OSHA's Hazard Communication Standard, 29 CFR 1910.1200, that they deliver to downstream employers an MSDS which addresses benzene.

**1910.1028(j)(3)** *Information and training.* (i) The employer shall provide employees with information and training at the time of their initial assignment to a work area where benzene is present. If exposures are above the action level, employees shall be provided with information and training at least annually thereafter.

**1910.1028(j)(3)(ii)** The training program shall be in accordance with the requirements of 29 CFR 1910.1200(h)(1) and (2), and shall include specific information on benzene for each category of information included in that section.

**1910.1028(j)(3)(iii)** In addition to the information required under 29 CFR 1910.1200, the employer shall:

**1910.1028(j)(3)(iii)(A)** Provide employees with an explanation of the contents of this section, including Appendices A and B, and indicate to them where the standard is available; and

**1910.1028(j)(3)(iii)(B)** Describe the medical surveillance program required under paragraph (i) of this section, and explain the information contained in Appendix C.

**1910.1028(k)** ***Recordkeeping***—(1) *Exposure measurements.* (i) The employer shall establish and maintain an accurate record of all measurements required by paragraph (e) of this section, in accordance with 29 CFR 1910.20.

**1910.1028(k)(1)(ii)** This record shall include:

**1910.1028(k)(1)(ii)(A)** The dates, number, duration, and results of each of the samples taken, including a description of the procedure used to determine representative employee exposures;

**1910.1028(k)(1)(ii)(B)** A description of the sampling and analytical methods used;

**1910.1028(k)(1)(ii)(C)** A description of the type of respiratory protective devices worn, if any; and

**1910.1028(k)(1)(ii)(D)** The name, social security number, job classification and exposure levels of the employee monitored and all other employees whose exposure the measurement is intended to represent.

**1910.1028(k)(1)(iii)** The employer shall maintain this record for at least 30 years, in accordance with 29 CFR 1910.20.

**1910.1028(k)(2)** *Medical surveillance.* (i) The employer shall establish and maintain an accurate record for each employee subject to medical surveillance required by paragraph (i) of this section, in accordance with 29 CFR 1910.20.

**1910.1028(k)(2)(ii)** This record shall include:

**1910.1028(k)(2)(ii)(A)** The name and social security number of the employee;

**1910.1028(k)(2)(ii)(B)** The employer's copy of the physician's written opinion on the initial, periodic and special examinations, including results of medical examinations and all tests, opinions and recommendations;

**1910.1028(k)(2)(ii)(C)** Any employee medical complaints related to exposure to benzene;

**1910.1028(k)(2)(ii)(D)** A copy of the information provided to the physician as required by paragraphs (i)(6) (ii) through (v) of this section; and

**1910.1028(k)(2)(ii)(E)** A copy of the employee's medical and work history related to exposure to benzene or any other hematologic toxins.

**1910.1028(k)(2)(iii)** The employer shall maintain this record for at least the duration of employment plus 30 years, in accordance with 29 CFR 1910.20.

**1910.1028(k)(3)** *Availability.* (i) The employer shall assure that all records required to be maintained by this section shall be made available upon request to the Assistant Secretary and the Director for examination and copying.

**1910.1028(k)(3)(ii)** Employee exposure monitoring records required by this paragraph shall be provided upon request for examination and copying to employees, employee representatives, and the Assistant Secretary in accordance with 29 CFR 1910.20(a) through (e) and (g) through (i).

**1910.1028(k)(3)(iii)** Employee medical records required by this paragraph shall be provided upon request for examination and copying, to the subject employee, to anyone having the specific written consent of the subject employee, and to the Assistant Secretary in accordance with 29 CFR 1910.20.

**1910.1028(k)(4)** *Transfer of records.* (i) The employer shall comply with the requirements involving transfer of records set forth in 29 CFR 1019.20(h).

**1910.1028(k)(4)(ii)** If the employer ceases to do business and there is no successor employer to receive and retain the records for the prescribed period, the employer shall notify the Director, at least three (3) months prior to disposal, and transmit them to the Director if required by the Director within that period.

**1910.1028(l)** ***Observation of monitoring***— (1) *Employee observation.* The employer shall provide affected employees, or their designated representatives, an opportunity to observe the measuring or monitoring of employee exposure to benzene conducted pursuant to paragraph (e) of this section.

**1910.1028(l)(2)** *Observation procedures.* When observation of the measuring or monitoring of employee exposure to benzene requires entry into areas where the use of protective clothing and equipment or respirators is required, the employer shall provide the observer with personal protective clothing and equipment or respirators required to be worn by employees working in the area, assure the use of such clothing and equipment or respirators, and require the observer to comply with all other applicable safety and health procedures.

**1910.1028(m)** ***Dates***— (1) *Effective date.* The standard shall become effective December 10, 1987.

**1910.1028(m)(2)** *Start-up dates.* (i) The requirements of paragraph (a) through (m) of this section, except the engineering control requirements of paragraph (f)(1) of this section shall be completed within sixty (60) days after the effective date of the standard.

**1910.1028(m)(2)(ii)** Engineering and work practice controls required by paragraph (f)(1) of this section shall be implemented no later than 2 years after the effective date of the standard.

**1910.1028(m)(2)(iii)** Coke and coal chemical operations may comply with paragraph (m)(2)(ii) of this section or alternately include within the compliance program required by paragraph (f)(2) of this section, a requirement to phase in engineering controls as equipment is repaired and replaced. For coke and coal chemical operations choosing the latter alternative, compliance with the engineering controls requirements of paragraph (f)(1) of this section shall be achieved no later than 5 years after the effective date of this standard and substantial compliance with the engineering control requirements shall be achieved within 3 years of the effective date of this standard.

**1910.1028(n)** ***Appendices.*** The information contained in Appendices A, B, C, and D is not intended, by itself, to create any additional obligations not otherwise imposed or to detract from any existing obligations. The protocols on respiratory fit testing in Appendix E are mandatory.

## APPENDIX A TO §1910.1028— SUBSTANCE SAFETY DATA SHEET, BENZENE

### I. SUBSTANCE IDENTIFICATION

A. Substance: Benzene.

B. Permissible Exposure: Except as to the use of gasoline, motor fuels and other fuels subsequent to discharge from bulk terminals and other exemptions specified in §1910.1028(a)(2):

1. Airborne: The maximum time-weighted average (TWA) exposure limit is 1 part of benzene vapor per million parts of air (1 ppm) for an 8-hour workday and the maximum short-term exposure limit (STEL) is 5 ppm for any 15-minute period.

2. Dermal: Eye contact shall be prevented and skin contact with liquid benzene shall be limited.

C. Appearance and odor: Benzene is a clear, colorless liquid with a pleasant, sweet odor. The odor of benzene does not provide adequate warning of its hazard.

### II. HEALTH HAZARD DATA

A. *Ways in which benzene affects your health.* Benzene can affect your health if you inhale it, or if it comes in contact with your skin or eyes. Benzene is also harmful if you happen to swallow it.

B. *Effects of overexposure.* 1. Short-term (acute) overexposure: If you are overexposed to high concentrations of benzene, well above the levels where its odor is first recognizable, you may feel breathless, irritable, euphoric, or giddy; you may experience irritation in eyes, nose, and respiratory tract. You may develop a headache, feel dizzy, nauseated, or intoxicated. Severe exposures may lead to convulsions and loss of consciousness.

2. Long-term (chronic) exposure. Repeated or prolonged exposure to benzene, even at relatively low concentrations, may result in various blood disorders, ranging from anemia to leukemia, an irreversible, fatal disease. Many blood disorders associated with benzene exposure may occur without symptoms.

### III. PROTECTIVE CLOTHING AND EQUIPMENT

A. *Respirators.* Respirators are required for those operations in which engineering controls or work practice controls are not feasible to reduce exposure to the permissible level. However, where employers can document that benzene is present in the workplace less than 30 days a year, respirators may be used in lieu of engineering controls. If respirators are worn, they must have joint Mine Safety and Health Administration and the National Institute for Occupational Safety and Health (NIOSH) seal of approval, and cartridge or canisters must be replaced before the end of their service life, or the end of the shift, whichever occurs first. If you experience difficulty breathing while wearing a respirator, you may request a positive pressure respirator from your employer. You must be thoroughly trained to use the assigned respirator, and the training will be provided by your employer.

B. *Protective Clothing.* You must wear appropriate protective clothing (such as boots, gloves, sleeves, aprons, etc.) over any parts of your body that could be exposed to liquid benzene.

C. *Eye and Face Protection.* You must wear splashproof safety goggles if it is possible that benzene may get into your eyes. In addition, you must wear a face shield if your face could be splashed with benzene liquid.

### IV. EMERGENCY AND FIRST AID PROCEDURES

A. *Eye and face exposure.* If benzene is splashed in your eyes, wash it out immediately with large amounts of water. If irritation persists or vision appears to be affected see a doctor as soon as possible.

B. *Skin exposure.* If benzene is spilled on your clothing or skin, remove the contaminated clothing and wash the exposed skin with large amounts of water and soap immediately. Wash contaminated clothing before you wear it again.

C. *Breathing.* If you or any other person breathes in large amounts of benzene, get the exposed person to fresh air at once. Apply artificial respiration if breathing has stopped. Call for medical assistance or a doctor as soon as possible. Never enter any vessel or confined space where the benzene concentration might be high without proper safety equipment and at least one other person present who will stay outside. A life line should be used.

D. *Swallowing.* If benzene has been swallowed and the patient is conscious, do not induce vomiting. Call for medical assistance or a doctor immediately.

## V. MEDICAL REQUIREMENTS

If you are exposed to benzene at a concentration at or above 0.5 ppm as an 8-hour time-weighted average, or have been exposed at or above 10 ppm in the past while employed by your current employer, your employer is required to provide a medical examination and history and laboratory tests within 60 days of the effective date of this standard and annually thereafter. These tests shall be provided without cost to you. In addition, if you are accidentally exposed to benzene (either by ingestion, inhalation, or skin/eye contact) under emergency conditions known or suspected to constitute toxic exposure to benzene, your employer is required to make special laboratory tests available to you.

## VI. OBSERVATION OF MONITORING

Your employer is required to perform measurements that are representative of your exposure to benzene and you or your designated representative are entitled to observe the monitoring procedure. You are entitled to observe the steps taken in the measurement procedure, and to record the results obtained. When the monitoring procedure is taking place in an area where respirators or personal protective clothing and equipment are required to be worn, you or your representative must also be provided with, and must wear the protective clothing and equipment.

## VII. ACCESS TO RECORDS

You or your representative are entitled to see the records of measurements of your exposure to benzene upon written request to your employer. Your medical examination records can be furnished to yourself, your physician or designated representative upon request by you to your employer.

## VIII. PRECAUTIONS FOR SAFE USE, HANDLING AND STORAGE

Benzene liquid is highly flammable. It should be stored in tightly closed containers in a cool, well ventilated area. Benzene vapor may form explosive mixtures in air. All sources of ignition must be controlled. Use nonsparking tools when opening or closing benzene containers. Fire extinguishers, where provided, must be readily available. Know where they are located and how to operate them. Smoking is prohibited in areas where benzene is used or stored. Ask your supervisor where benzene is used in your area and for additional plant safety rules.

# APPENDIX B TO §1910.1028—SUBSTANCE TECHNICAL GUIDLINES[sic], BENZENE

## I. PHYSICAL AND CHEMICAL DATA

A. Substance identification.

1. *Synonyms:* Benzol, benzole, coal naphtha, cyclohexatriene, phene, phenyl hydride, pyrobenzol. (Benzin, petroleum benzin and Benzine do not contain benzene).

2. *Formula:* $C_6 H_6$ (CAS Registry Number: 71-43-2)

B. Physical data.

1. Boiling Point (760 mm Hg); 80.1 ° C (176 °F)

2. Specific Gravity (water = 1): 0.879

3. Vapor Density (air = 1): 2.7

4. Melting Point: 5.5 °C (42 ° F)

5. Vapor Pressure at 20 ° C (68 ° F): 75 mm Hg

6. Solubility in Water: .06%

7. Evaporation Rate (ether = 1): 2.8

8. Appearance and Odor: Clear, colorless liquid with a distinctive sweet odor.

## II. FIRE, EXPLOSION, AND REACTIVITY HAZARD DATA

A. Fire.

1. Flash Point (closed cup): -11 ° C (12 ° F)

2. Autoignition Temperature: 580 ° C (1076 °F)

3. Flammable limits in Air. % by Volume: Lower: 1.3%, Upper: 7.5%

4. Extinguishing Media: Carbon dioxide, dry chemical, or foam.

5. Special Fire-Fighting procedures: Do not use solid stream of water, since stream will scatter and spread fire. Fine water spray can be used to keep fire-exposed containers cool.

6. Unusual fire and explosion hazards: Benzene is a flammable liquid. Its vapors can form explosive mixtures. All ignition sources must be controlled when benzene is used, handled, or stored. Where liquid or vapor may be released, such areas shall be considered as hazardous locations. Benzene vapors are heavier than air; thus the vapors may travel along the ground and be ignited by open flames or sparks at locations remote from the site at which benzene is handled.

7. Benzene is classified as a 1 B flammable liquid for the purpose of conforming to the requirements of 29 CFR 1910.106. A concentration exceeding 3,250 ppm is considered a potential fire explosion hazard. Locations where benzene may be present in quantities sufficient to produce explosive or ignitable mixtures are considered Class I Group D for the purposes of conforming to the requirements of 29 CFR 1910.309.

B. Reactivity.

1. Conditions contributing to instability: Heat.

2. Incompatibility: Heat and oxidizing materials.

3. Hazardous decomposition products: Toxic gases and vapors (such as carbon monoxide).

## III. SPILL AND LEAK PROCEDURES

A. Steps to be taken if the material is released or spilled. As much benzene as possible should be absorbed with suitable materials, such as dry sand or earth. That remaining must be flushed with large amounts of water. Do not flush benzene into a confined space, such as a sewer, because of explosion danger. Remove all ignition sources. Ventilate enclosed places.

B. Waste disposal method. Disposal methods must conform to other jurisdictional regulations. If allowed, benzene may be disposed of: (a) By absorbing it in dry sand or earth and disposing in a sanitary landfill; (b) if small quantities, by removing it to a safe location from buildings or other combustible sources, pouring it in dry sand or earth and cautiously igniting it; and (c) if large quantities, by atomizing it in a suitable combustion chamber.

## IV. MISCELLANEOUS PRECAUTIONS

A. High exposure to benzene can occur when transferring the liquid from one container to another. Such operations should be well ventilated and good work practices must be established to avoid spills.

B. Use non-sparking tools to open benzene containers which are effectively grounded and bonded prior to opening and pouring.

C. Employers must advise employees of all plant areas and operations where exposure to benzene could occur. Common operations in which high exposures to benzene may be encountered are: the primary production and utilization of benzene, and transfer of benzene.

# APPENDIX C TO §1910.1028—MEDICAL SURVEILLANCE GUIDELINES FOR BENZENE

## I. ROUTE OF ENTRY

Inhalation; skin absorption.

## II. TOXICOLOGY

Benzene is primarily an inhalation hazard. Systemic absorption may cause depression of the hematopoietic system, pancytopenia, aplastic anemia, and leukemia. Inhalation of high concentrations can affect central nervous system function. Aspiration of small amounts of liquid benzene immediately causes pulmonary edema and hemorrhage of pulmonary tissue. There is some absorption through the skin. Absorption may be more rapid in the case of abraded skin, and benzene may be more readily absorbed if it is present in a mixture or as a contaminant in solvents which are readily absorbed. The defatting action of benzene may produce primary irritation due to repeated or prolonged contact with the skin. High concentration are irritating to the eyes and the mucuous membranes of the nose, and respiratory tract.

## III. SIGNS AND SYMPTOMS

Direct skin contact with benzene may cause erythema. Repeated or prolonged contact may result in drying, scaling dermatitis, or development of secondary skin infections. In addition, there is benzene absorption through the skin. Local effects of benzene vapor or liquid on the eye are slight. Only at very high concentrations is there any smarting sensation in the eye. Inhalation of high concentrations of benzene may have an initial stimulatory effect on the central nervous system characterized by exhilaration, nervous excitation, and/or giddiness, followed by a period of depression, drowsiness, or fatigue. A sensation of tightness in the chest accompanied by breathlessness may occur and ultimately the victim may lose consciousness. Tremors, convulsions and death may follow from respiratory paralysis or circulatory collapse in a few minutes to several hours following severe exposures.

The detrimental effect on the blood-forming system of prolonged exposure to small quantities of benzene vapor is of extreme importance. The hematopoietic system is the chief target for benzene's toxic effects which are manifested by alterations in the levels of formed elements in the peripheral blood. These effects have occurred at concentrations of benzene which may not cause irritation of mucous membranes, or any unpleasant sensory effects. Early signs and symptoms of benzene morbidity are varied, often not readily noticed and non-specific. Subjective complaints of headache, dizziness, and loss of appetite may precede or follow clinical signs. Rapid pulse and low blood pressure, in addition to a physical appearance of anemia, may accompany a subjective complaint of shortness of breath and excessive tiredness. Bleeding from the nose, gums, or mucous membranes, and the development of purpuric spots (small bruises) may occur as the condition progresses. Clinical evidence of leukopenia, anemia, and thrombocytopenia, singly or in combination, has been frequently reported among the first signs.

Bone marrow may appear normal, aplastic, or hyperplastic, and may not, in all situations, correlate with peripheral blood forming tissues. Because of variations in the susceptibility to benzene morbidity, there is no "typical" blood picture. The onset of effects of prolonged benzene exposure may be delayed for many months or years after the actual exposure has ceased and identification or correlation with benzene exposure must be sought out in the occupational history.

## IV. TREATMENT OF ACUTE TOXIC EFFECTS

Remove from exposure immediately. Make sure you are adequately protected and do not risk being overcome by fumes. Give oxygen or artificial resuscitation if indicated. Flush eyes, wash skin if contaminated and remove all contaminated clothing. Symptoms of intoxication may persist following severe exposures. Recovery from mild exposures is usually rapid and complete.

## V. SURVEILLANCE AND PREVENTIVE CONSIDERATIONS

### A. General

The principal effects of benzene exposure which form the basis for this regulation are pathological changes in the hematopoietic system, reflected by changes in the peripheral blood and manifesting clinically as pancytopenia, aplastic anemia, and leukemia. Consequently, the medical surveillance program is designed to observe, on a regular basis, blood indices for early signs of these effects, and although early signs of leukemia are not usually available, emerging diagnostic technology and innovative regimes make consistent surveillance for leukemia, as well as other hematopoietic effects, essential.

Initial examinations are to be provided within 60 days of the effective date of this standard, or at the time of initial assignment, and periodic examinations annually thereafter. There are special provisions for medical tests in the event of hematologic abnormalities or for emergency situations.

The blood values which require referral to a hematologist or internist are noted in the standard in paragraph (i)(5). The standard specifies that blood abnormalities that persist must be referred "unless the physician has good reason to believe such referral is unnecessary" (paragraph (i)(5)). Examples of conditions that could make a referral unnecessary despite abnormal blood limits are iron or folate deficiency, menorrhagia, or blood loss due to some unrelated medical abnormality.

Symptoms and signs of benzene toxicity can be non-specific. Only a detailed history and appropriate investigative procedures will enable a physician to rule out or confirm conditions that place the employee at increased risk. To assist the examining physician with regard to which laboratory tests are necessary and when to refer an employee to the specialist, OSHA has established the following guidelines.

### B. Hematology Guidelines

A minimum battery of tests is to be performed by strictly standardized methods.

1. Red cell, white cell, platelet counts, white blood cell differential, hematacrit and red cell indices must be performed by an accredited laboratory. The normal ranges for the red cell and white cell counts are influenced by altitude, race, and sex, and therefore should be determined by the accredited laboratory in the specific area where the tests are performed.

Either a decline from an absolute normal or an individual's base line to a subnormal value or a rise to a supra-normal value, are indicative of potential toxicity, particularly if all blood parameters decline. The normal total white blood count is approximately 7,200/mm$^3$ plus or minus 3,000. For cigarette smokers the white count may be higher and the upper range may be 2,000 cells higher than normal for the laboratory. In addition, infection, allergies and some drugs may raise the white cell count. The normal platelet count is approximately 250,000 with a range of 140,000 to 400,000. Counts outside this range should be regarded as possible evidence of benzene toxicity.

Certain abnormalities found through routine screening are of greater significance in the benzene-exposed worker and require prompt consultation with a specialist, namely:

a. Thrombocytopenia.

b. A trend of decreasing white cell, red cell, or platelet indices in an individual over time is more worrisome than an isolated abnormal finding at one test time. The importance of trend highlights the need to compare an individual's test results to baseline and/or previous periodic tests.

c. A constellation or pattern of abnormalities in the different blood indices is of more significance than a single abnormality. A low white count not associated with any abnormalities in other cell indices may be a normal statistical variation, whereas if the low white count is accompanied by decreases in the platelet and/or red cell indices, such a pattern is more likely to be associated with benzene toxicity and merits thorough investigation.

Anemia, leukopenia, macrocytosis or an abnormal differential white blood cell count should alert the physician to further investigate and/or refer the patient if repeat tests confirm the abnormalities. If routine screening detects an abnormality, follow-up tests which may be helpful in establishing the etiology of the abnormality are the peripheral blood smear and the reticulocyte count.

The extreme range of normal for reticulocytes is 0.4 to 2.5 percent of the red cells, the usual range being 0.5 to 1.2 percent of the red cells, but the typical value is in the range of 0.8 to 1.0 percent. A decline in reticulocytes to levels of less than 0.4 percent is to be regarded as possible evidence (unless another specific cause is found) of benzene toxicity requiring accelerated surveillance. An increase in reticulocyte levels to about 2.5 percent may also be consistent with (but is not as characteristic of) benzene toxicity.

2. An important diagnostic test is a careful examination of the peripheral blood smear. As with reticulocyte count the smear should be with fresh uncoagulated blood obtained from a needle tip following venipuncture or from a drop of earlobe blood (capillary blood). If necessary, the smear may, under certain limited conditions, be made from a blood sample anticoagulated with EDTA (but never with oxalate or heparin). When the smear is to be prepared from a specimen of venous blood which has been collected by a commercial Vacutainer# type tube containing neutral EDTA, the smear should be made as soon as possible after the venesection. A delay of up to 12 hours is permissible between the drawing of the blood specimen into EDTA and the preparation of the smear if the blood is stored at refrigerator (not freezing) temperature.

3. The minimum mandatory observations to be made from the smear are:

a. The differential white blood cell count.

b. Description of abnormalities in the appearance of red cells.

c. Description of any abnormalities in the platelets.

d. A careful search must be made throughout of every blood smear for immature white cells such as band forms (in more than normal proportion, i.e., over 10 percent of the total differential count), any number of metamyelocytes, myelocytes or myeloblasts. Any nucleate or multinucleated red blood cells should be reported. Large "giant" platelets or fragments of megakaryocytes must be recognized.

An increase in the proportion of band forms among the neutrophilic granulocytes is an abnormality deserving special mention, for it may represent a change which should be considered as an early warning of benzene toxicity in the absence of other causative factors (most commonly infection). Likewise, the appearance of metamyelocytes, in the absence of another probable cause, is to be considered a possible indication of benzeneinduced toxicity.

An upward trend in the number of basophils, which normally do not exceed about 2.0 percent of the total white cells, is to be regarded as possible evidence of benzene toxicity. A rise in the eosinophil count is less specific but also may be suspicious of toxicity if the rises above 6.0 percent of the total white count.

The normal range of monocytes is from 2.0 to 8.0 percent of the total white count with an average of about 5.0 percent. About 20 percent of individuals reported to have mild but persisting abnormalities caused by exposure to benzene show a persistent monocytosis. The findings of a monocyte count which persists at more than 10 to 12 percent of the normal white cell count (when the total count is normal) or persistence of an absolute monocyte count in excess of 800/mm$^3$ should be regarded as a possible sign of benzene-induced toxicity.

A less frequent but more serious indication of benzene toxicity is the finding in the peripheral blood of the so-called "pseudo" (or acquired) Pelger-Huet anomaly. In this anomaly many, or sometimes the majority, of the neutrophilic granulocytes possess two round nuclear segements—less often one or three round segments—rather than three normally elongated segments. When this anomaly is not hereditary, it is often but not invariably predictive of subsequent leukemia. However, only about two percent of patients who ultimately develop acute myelogenous leukemia show the acquired Pelger-Huet anomaly. Other tests that can be administered to investigate blood abnormalities are discussed below; however, such procedures should be undertaken by the hematologist.

An uncommon sign, which cannot be detected from the smear, but can be elicited by a "sucrose water test" of peripheral blood, is transient paroxysmal nocturnal hemoglobinuria (PNH), which may first occur insidiously during a period of established aplastic anemia, and may be followed within one to a few years by the appearance of rapidly fatal acute myelogenous leukemia. Clinical detection of PNH, which occurs in only one or two percent of those destined to have acute myelogenous leukemia, may be difficult; if the "sucrose water test" is positive, the somewhat more definitive Ham test, also known as the acid-serum hemolysis test, may provide confirmation.

e. Individuals documented to have developed acute myelogenous leukemia years after initial exposure to benzene may have progressed through a preliminary phase of hematologic abnormality. In some instances pancytopenia (i.e., a lowering in the counts of all circulating blood cells of bone marrow origin, but not to the extent implied by the term "aplastic anemia") preceded leukemia for many years. Depression of a single blood cell type or platelets may represent a harbinger of aplasia or leukemia. The finding of two or more cytopenias, or pancytopenia in a benzene-exposed individual, must be regarded as highly suspicious of more advanced although still reversible, toxicity. "Pancytopenia" coupled with the appearance of immature cells (myelocytes, myeloblasts, erythroblasts, etc.), with abnormal cells (pseudo Pelger-Huet anomaly, atypical nuclear heterochromatin, etc.), or unexplained elevations of white blood cells must be regarded as evidence of benzene overexposure unless proved otherwise. Many severely aplastic patients manifested the ominous finding of 5-10 percent myeloblasts in the marrow, occasional myeloblasts and myelocytes in the blood and 20-30% monocytes. It is evident that isolated cytopenias, pancytopenias, and even aplastic anemias induced by benzene may be reversible and complete recovery has been reported on cessation of exposure. However, since any of these abnormalities is serious, the employee must immediately be removed from any possible exposure to benzene vapor. Certain tests may substantiate the employee's prospects for progression or regression. One such test would be an examination of the bone marrow, but the decision to perform a bone marrow aspiration or needle biopsy is made by the hematologist.

The findings of basophilic stippling in circulating red blood cells (usually found in 1 to 5% of red cells following marrow injury), and detection in the bone marrow of what are termed "ringed sideroblasts" must be taken seriously, as they have been noted in recent years to be premonitory signs of subsequent leukemia.

Recently peroxidase-staining of circulating or marrow neutrophil granulocytes, employing benzidine dihydrochloride, have revealed the disappearance of, or diminution in, peroxidase in a sizable proportion of the granulocytes, and this has been reported as an early sign of leukemia. However, relatively few patients have been studied to date. Granulocyte granules are normally strongly peroxidase positive. A steady decline in leukocyte alkaline phosphatase has also been reported as suggestive of early acute leukemia. Exposure to benzene may cause an early rise in serum iron, often but not always associated with a fall in the reticulocyte count. Thus, serial measurements of serum iron levels may provide a means of determining whether or not there is a trend representing sustained suppression of erythropoiesis.

Measurement of serum iron, determination of peroxidase and of alkaline phosphatase activity in peripheral granulocytes can be performed in most pathology laboratories. Peroxidase and alkaline phosphatase staining are usually undertaken when the index of suspecion for leukemia is high.

## APPENDIX D TO §1910.1028—SAMPLING AND ANALYTICAL METHODS FOR BENZENE MONITORING AND MEASUREMENT PROCEDURES

Measurements taken for the purpose of determining employee exposure to benzene are best taken so that the representative average 8-hour exposure may be determined from a single 8-hour sample or two (2) 4-hour samples. Short-time interval samples (or grab samples) may also be used to determine average exposure level if a minimum of five measurements are taken in a random manner over the 8-hour work shift. Random sampling means that any portion of the work shift has the same change of being sampled as any other. The arithmetic average of all such random samples taken on one work shift is an estimate of an employee's average level of exposure for that work shift. Air samples should be taken in the employee's breathing zone (air that would most nearly represent that inhaled by the employee). Sampling and analysis must be performed with procedures meeting the requirements of the standard.

There are a number of methods available for monitoring employee exposures to benzene. The sampling and analysis may be performed by collection of the benzene vaptor or charcoal absorption tubes, with subsequent chemical analysis by gas chromatography. Sampling and analysis may also be performed by portable direct reading instruments, real-time continuous monitoring systems, passive dosimeters or other suitable methods. The employer has the obligation of selecting a monitoring method which meets the accuracy and precision requirements of the standard under his unique field conditions. The standard requires that the method of monitoring must have an accuracy, to a 95 percent confidence level, of not less than plus or minus 25 percent for concentrations of benzene greater than or equal to 0.5 ppm.

The OSHA Laboratory modified NIOSH Method S311 and evaluated it at a benzene air concentration of 1 ppm. A procedure for determining the benzene concentration in bulk material samples was also evalauted. This work, reported in OSHA Laboratory Method No. 12, includes the following two analytical procedures:

### *I. OSHA Method 12 for Air Samples*

Analyte: Benzene

Matrix: Air

Procedure: Adsorption on charcoal, desorption with carbon disulfide, analysis by GC.

Detection limit: 0.04 ppm

Recommended air volume and sampling rate: 10L to 0.2 L/min.

1. Principle of the Method.

1.1 A known volume of air is drawn through a charcoal tube to trap the organic vapors present.

1.2. The charcoal in the tube is transferred to a small, stoppered vial, and the anlyte is desorbed with carbon disulfide.

1.3. An aliquot of the desorbed sample is injected into a gas chromatograph.

1.4 The area of the resulting peak is determined and compared with areas obtained from standards.

2. Advantages and disadvantages of the method.

2.1 The sampling device is small, portable, and involved no liquids. Interferences are minimal, and most of those which do occur can be eliminated by altering chromatographic conditions. The samples are analyzed by means of a quick, instrumental method.

2.2 The amount of sample which can be taken is limited by the number of milligrams that the tube will hold before overloading. When the sample value obtained for the backup section of the charcoal tube exceeds 25 percent of that found on the front section, the possibility of sample loss exists.

3. Apparatus.

3.1 A calibrated personal sampling pump whose flow can be determined within ±5 percent at the recommended flow rate.

3.2. Charcoal tubes: Glass with both ends flame sealed, 7 cm long with a 6-mm O.D. and a 4-mm I.D., containing 2 sections of 20/40 mesh activated charcoal separated by a 2-mm portion of urethane foam. The activated charcoal is prepared from coconut shells and is fired at 600 °C prior to packing. The adsorbing section contains 100 mg of charcoal, the back-up section 50 mg. A 3-mm portion of urethane foam is placed between the outlet end of the tube and the back-up section. A plug of silanized glass wool is placed in front of the adsorbing section. The pressure drop across the tube must be less than one inch of mercury at a flow rate of 1 liter per minute.

3.3. Gas chromatograph equipped with a flame ionization detector.

3.4. Column (10-ft × 1/8-in stainless steel) packed with 80/100 Supelcoport coated with 20 percent SP 2100, 0.1 percent CW 1500.

3.5. An electronic integrator or some other suitable method for measuring peak area.

3.6. Two-milliliter sample vials with Teflon-lined caps.

3.7. Microliter syringes: 10-microliter (10-μL syringe, and other convenient sizes for making standards, 1-μL syringe for sample injections.

3.8. Pipets: 1.0 mL delivery pipets

3.9. Volumetric flasks: convenient sizes for making standard solutions.

4. Reagents.

4.1. Chromatographic quality carbon disulfide ($CS_2$). Most commercially available carbon disulfide contains a trace of benzene which must be removed. It can be removed with the following procedure:

Heat under reflux for 2 to 3 hours, 500 mL of carbon disulfide, 10 mL concentrated sulfuric acid, and 5 drops of concentrated nitric acid. The benzene is converted to nitrobenzene. The carbon disulfide layer is removed, dried with anhydrous sodium sulfate, and distilled. The recovered carbon disulfide should be benzene free. (It has recently been determined that benzene can also be removed by passing the carbon disulfide through 13x molecular sieve).

4.2. Benzene, reagent grade.

4.3. p-Cymene, reagent grade, (internal standard).

4.4. Desorbing reagent. The desorbing reagent is prepared by adding 0.05 mL of p-cymene per milliliter of carbon disulfide. (The internal standard offers a convenient means correcting analytical response for slight inconsistencies in the size of sample injections. If the external standard technique is preferred, the internal standard can be eliminated).

4.5. Purified GC grade helium, hydrogen and air.

5. Procedure.

5.1. Cleaning of equipment. All glassware used for the laboratory analysis should be properly cleaned and free of organics which could interfere in the analysis.

5.2. Calibration of personal pumps. Each pump must be calibrated with a representative charcoal tube in the line.

5.3. Collection and shipping of samples.

5.3.1. Immediately before sampling, break the ends of the tube to provide an opening at least one-half the internal diameter of the tube (2 mm).

5.3.2. The smaller section of the charcoal is used as the backup and should be placed nearest the sampling pump.

5.3.3. The charcoal tube should be placed in a vertical position during sampling to minimize channeling through the charcoal.

5.3.4 Air being sampled should not be passed through any hose or tubing before entering the charcoal tube.

5.3.5. A sample size of 10 liters is recommended. Sample at a flow rate of approximately 0.2 liters per minute. The flow rate should be known with an accuracy of at least ±5 percent.

5.3.6. The charcoal tubes should be capped with the supplied plastic caps immediately after sampling.

5.3.7. Submit at least one blank tube (a charcoal tube subjected to the same handling procedures, without having any air drawn through it) with each set of samples.

5.3.8. Take necessary shipping and packing precautions to minimize breakage of samples.

5.4. Analysis of samples.

5.4.1. Preparation of samples. In preparation for analysis, each charcoal tube is scored with a file in front of the first section of charcoal and broken open. The glass wool is removed and discarded. The charcoal in the first (larger) section is transferred to a 2-ml vial. The separating section of foam is removed and discarded; the second section is transferred to another capped vial. These two sections are analyzed separately.

5.4.2. Desorption of samples. Prior to analysis, 1.0 mL of desorbing solution is pipetted into each sample container. The desorbing solution consists of 0.05 μL internal standard per mL of carbon disulfide. The sample vials are capped as soon as the solvent is added. Desorption should be done for 30 minutes with occasional shaking.

5.4.3. GC conditions. Typical operating conditions for the gas chromatograph are:

1.30 mL/min (60 psig) helium carrier gas flow.

2.30 mL/min (40 psig) hydrogen gas flow to detector.

3.240 mL/min (40 psig) air flow to detector.

4.150 ° C injector temperature.

5.250 ° C detector temperature.

6.100 ° C column temperature.

5.4.4. Injection size. 1 μ L.

5.4.5. Measurement of area. The peak areas are measured by an electronic integrator or some other suitable form of area measurement.

5.4.6. An internal standard procedure is used. The integrator is calibrated to report results in ppm for a 10 liter air sample after correction for desorption efficiency.

5.5. Determination of desorption efficiency.

5.5.1. Importance of determination. The desorption efficiency of a particular compound can vary from one laboratory to another and from one lot of chemical to another. Thus, it is necessary to determine, at least once, the percentage of the specific compound that is removed in the desorption process, provided the same batch of charcoal is used.

5.5.2. Procedure for determining desorption efficiency. The reference portion of the charcoal tube is removed. To the remaining portion, amounts representing 0.5X, 1X, and 2X and (X represents target concentration) based on a 10 L air sample are injected into several tubes at each level. Dilutions of benzene with carbon disulfide are made to allow injection of measurable quantities. These tubes are then allowed to equilibrate at least overnight. Following equilibration they are analyzed following the same procedure as the samples. Desorption efficiency is determined by dividing the amount of benzene found by amount spiked on the tube.

6. Calibration and standards. A series of standards varying in concentration over the range of interest is prepared and analyzed under the same GC conditions that will be used on the samples. A calibration curve is prepared by plotting concentration (µg/mL) versus peak area.

7. Calculations. Benzene air concentration can be calculated from the following equation:

$$mg/m^3 = (A)(B)/(C)(D)$$

Where: A = mg/mL benzene, obtained from the calibration curve

B = desorption volume (1 mL)

C = Liters of air sampled

D = desorption efficiency

The concentration in $mg/m^3$ can be converted to ppm (at 25°C and 760 mm) with following equation:

$$ppm = (mg/m^3)(24.46)/(78.11)$$

Where: 24.46 = molar volume of an ideal gas

25 ° C and 760 mm

78.11 = molecular weight of benzene

8. Backup Data.

8.1 Detection limit—Air Samples.

The detection limit for the analytical procedure is 1.28 ng with a coefficient of variation of 0.023 at this level. This would be equivalent to an air concentration of 0.04 ppm for a 10 L air sample. This amount provided a chromatographic peak that could be identifiable in the presence of possible interferences. The detection limit data were obtained by making 1 µL injections of a 1.283 µg/mL standard.

| Injection | Area Count | |
|---|---|---|
| 1 | 655.4 | |
| 2 | 617.5 | |
| 3 | 662.0 | X̄=640.2 |
| 4 | 641.1 | SD=14.9 |
| 5 | 636.4 | CV=0.023 |
| 6 | 629.2 | |

8.2. Pooled coefficient of variation—Air Samples. The pooled coefficient of variation for the analytical procedure was determined by 1 µL replicate injections of analytical standards. The standards were 16.04, 32.08, and 64.16 µg/mL, which are equivalent to 0.5, 1.0, and 2.0 ppm for a 10 L air sample respectively.

| Injection | Area Counts 0.5 ppm | 1.0 ppm | 2.0 ppm |
|---|---|---|---|
| 1 | 3996.5 | 8130.2 | 16481 |
| 2 | 4059.4 | 8235.6 | 16493 |
| 3 | 4052.0 | 8307.9 | 16535 |
| 4 | 4027.2 | 8263.2 | 16609 |
| 5 | 4046.8 | 8291.1 | 16552 |
| 6 | 4137.9 | 8288.8 | 16618 |
| X̄= | 4053.3 | 8254.0 | 16548.3 |
| SD= | 47.2 | 62.5 | 57.1 |
| CV= | 0.0116 | 0.0076 | 0.0034 |
| C̄V̄=0.008 | ........ | ........ | ........ |

8.3. Storage data—Air Samples

Samples were generated at 1.03 ppm benzene at 80% relative humidity, 22 ° C, and 643 mm. All samples were taken for 50 minutes at 0.2 L/min. Six samples were analyzed immediately and the rest of the samples were divided into two groups by fifteen samples each. One group was stored at refrigerated temperature of -25 ° C, and the other group was stored at ambient temperature (approximately 23 ° C). These samples were analyzed over a period of fifteen days. The results are tabulated below.

**PERCENT RECOVERY**

| Day analyzed | Refrigerated | | | Ambient | | |
|---|---|---|---|---|---|---|
| 0 | 97.4 | 98.7 | 98.9 | 97.4 | 98.7 | 98.9 |
| 0 | 97.1 | 100.6 | 100.9 | 97.1 | 100.6 | 100.9 |
| 2 | 95.8 | 96.4 | 95.4 | 95.4 | 96.6 | 96.9 |
| 5 | 93.9 | 93.7 | 92.4 | 92.4 | 94.3 | 94.1 |
| 9 | 93.6 | 95.5 | 94.6 | 95.2 | 95.6 | 96.6 |
| 13 | 94.3 | 95.3 | 93.7 | 91.0 | 95.0 | 94.6 |
| 15 | 96.8 | 95.8 | 94.2 | 92.9 | 96.3 | 95.9 |

8.4. Desorption data.

Samples were prepared by injecting liquid benzene onto the A section of charcoal tubes. Samples were prepared that would be equivalent to 0.5, 1.0, and 2.0 ppm for a 10 L air sample.

**PERCENT RECOVERY**

| Sample | 0.5 ppm | 1.0 ppm | 2.0 ppm |
|---|---|---|---|
| 1 | 99.4 | 98.8 | 99.5 |
| 2 | 99.5 | 98.7 | 99.7 |
| 3 | 99.2 | 98.6 | 99.8 |
| 4 | 99.4 | 99.1 | 100.0 |
| 5 | 99.2 | 99.0 | 99.7 |
| 6 | 99.8 | 99.1 | 99.9 |
| X̄= | 99.4 | 98.9 | 99.8 |
| SD= | 0.22 | 0.21 | 0.18 |
| CV= | 0.0022 | 0.0021 | 0.0018 |
| X̄=99.4 | ........ | ........ | ........ |

8.5. Carbon disulfide.

Carbon disulfide from a number of sources was analyzed for benzene contamination. The results are given in the following table. The benzene contaminant can be removed with the procedures given in section 4.1.

| Sample | µg Benzene/ mL | ppm equivalent (for 10 L air sample) |
|---|---|---|
| Aldrich Lot 83017 | 4.20 | 0.13 |
| Baker Lot 720364 | 1.01 | 0.03 |
| Baker Lot 822351 | 1.01 | 0.03 |
| Malinkrodt Lot WEMP | 1.74 | 0.05 |
| Malinkrodt Lot WDSJ | 5.65 | 0.18 |
| Malinkrodt Lot WHGA | 2.90 | 0.09 |
| Treated $CS_2$ | ........ | ........ |

## II. OSHA LABORATORY METHOD NO. 12 FOR BULK SAMPLES

Analyte: Benzene.

Matrix: Bulk Samples.

Procedure: Bulk Samples are analyzed directly by high performance liquid chromatography (HPLC).

Detection limits: 0.01% by volume.

1. Principle of the method.

1.1. An aliquot of the bulk sample to be analyzed is injected into a liquid chromatograph.

1.2. The peak area for benzene is determined and compared to areas obtained from standards.

2. Advantages and disadvantages of the method.

2.1. The analytical procedure is quick, sensitive, and reproducible.

2.2. Reanalysis of samples is possible.

2.3. Interferences can be circumvented by proper selection of HPLC parameters.

2.4. Samples must be free of any particulates that may clog the capillary tubing in the liquid chromatograph. This may require distilling the sample or clarifying with a clarification kit.

3. Apparatus.

3.1. Liquid chromatograph equipped with a UV detector.

3.2. HPLC Column that will separate benzene from other components in the bulk sample being analyzed. The column used for validation studies was a Waters uBondapack C18, 30 cm x 3.9 mm.

3.3. A clarification kit to remove any particulates in the bulk if necessary.

3.4. A micro-distillation apparatus to distill any samples if necessary.

3.5. An electronic integrator or some other suitable method of measuring peak areas.

3.6. Microliter syringes—10 μL syringe and other convenient sizes for making standards. 10 μL syringe for sample injections.

3.7. Volumetric flasks, 5 mL and other convenient sizes for preparing standards and making dilutions.

4. Reagents.

4.1. Benzene, reagent grade.

4.2. HPLC grade water, methyl alcohol, and isopropyl alcohol.

5. Collection and shipment of samples.

5.1. Samples should be transported in glass containers with Teflon-lined caps.

5.2. Samples should not be put in the same container used for air samples.

6. Analysis of samples.

6.1. Sample preparation.

If necessary, the samples are distilled or clarified. Samples are analyzed undiluted. If the benzene concentration is out of the working range, suitable dilutions are made with isopropyl alcohol.

6.2. HPLC conditions.

The typical operating conditions for the high performance liquid chromatograph are:

1. Mobile phase—Methyl alcohol/water, 50/50

1. Analytical wavelength—254 nm

3. Injection size—10 μL

6.3. Measurement of peak area and calibration.

Peak areas are measured by an integrator or other suitable means. The integrator is calibrated to report results % in benzene by volume.

7. Calculations.

Since the integrator is programmed to report results in % benzene by volume in an undiluted sample, the following equation is used:

**% Benzene by Volume = A x B**

Where: A = % by volume on report

B = Dilution Factor

(B = 1 for undiluted sample)

8. Backup Data.

8.1. Detection limit—Bulk Samples.

The detection limit for the analytical procedure for bulk samples is 0.88 μg, with a coefficient of variation of 0.019 at this level. This amount provided a chromatographic peak that could be identifiable in the presence of possible interferences. The detection limit date were obtained by making 10 μL injections of a 0.10% by volume standard.

| Injection | Area Count | |
|---|---|---|
| 1 | 45386 | |
| 2 | 44214 | |
| 3 | 43822 | X̄=44040.1 |
| 4 | 44062 | SD=852.5 |
| 6 | 42724 | CV=0.019 |

8.2. Pooled coefficient of variation—Bulk Samples.

The pooled coefficient of variation for analytical procedure was determined by 50 μL replicate injections of analytical standards. The standards were 0.01, 0.02, 0.04, 0.10, 1.0, and 2.0% benzene by volume.

**AREA COUNT (PERCENT)**

| Injection No. | 0.01 | 0.02 | 0.04 | 0.10 | 1.0 | 2.0 |
|---|---|---|---|---|---|---|
| 1 | 45386 | 84737 | 166097 | 448497 | 4395380 | 9339150 |
| 2 | 44241 | 84300 | 170832 | 441299 | 4590800 | 9484900 |
| 3 | 43822 | 83835 | 164160 | 443719 | 4593200 | 9557580 |
| 4 | 44062 | 84381 | 164445 | 444842 | 4642350 | 9677060 |
| 5 | 44006 | 83012 | 168398 | 442564 | 4646430 | 9766240 |
| 6 | 42724 | 81957 | 173002 | 443975 | 4646260 | .......... |
| X̄= | 44040.1 | 83703.6 | 167872 | 444149 | 4585767 | 9564986 |
| SD= | 852.5 | 1042.2 | 3589.8 | 2459.1 | 96839.3 | 166233 |
| CV= | 0.0194 | 0.0125 | 0.0213 | 0.0055 | 0.0211 | 0.0174 |
| C̄V̄= | 0.017 | | | | | |

[52 FR 34562, Sept. 11, 1987, as amended at 54 FR 24334, June 7, 1989; 61 FR 5508, Feb. 13, 1996; 63 FR 1289, Jan. 8, 1998; 63 FR 20099, Apr. 23, 1998]

## §1926.1129 Coke oven emissions.

**Note:** The requirements applicable to construction work under this section are identical to those set forth at §1910.1029 of this chapter.

[61 FR 31434, June 20, 1996]

[Section 1910.1029 is printed below—CCH.]

## §1910.1029 Coke oven emissions.

**1910.1029(a)** ***Scope and application.*** This section applies to the control of employee exposure to coke oven emissions, except that this section shall not apply to working conditions with regard to which other Federal agencies exercise statutory authority to prescribe or enforce standards affecting occupational safety and health.

**1910.1029(b)** ***Definitions.*** For the purpose of this section:

*Authorized person* means any person specifically authorized by the employer whose duties require the person to enter a regulated area, or any person entering such an area as a designated representative of employees for the purpose of exercising the opportunity to observe monitoring and measuring procedures under paragraph (n) of this section.

*Beehive oven* means a coke oven in which the products of carbonization other than coke are not recovered, but are released into the ambient air.

*Coke oven* means a retort in which coke is produced by the destructive distillation or carbonization of coal.

*Coke oven battery* means a structure containing a number of slot-type coke ovens.

*Coke oven emissions* means the benzene-soluble fraction of total particulate matter present during the destructive distillation or carbonization of coal for the production of coke.

*Director* means the Director, National Institute for Occupational Safety and Health, U.S. Department of Health, Education, and Welfare, or his or her designee.

*Emergency* means any occurence such as, but not limited to, equipment failure which is likely to, or does, result in any massive release of coke oven emissions.

*Existing coke oven battery* means a battery in operation or under construction on January 20, 1977, and which is not a rehabilitated coke oven battery.

*Rehabilitated coke oven battery* means a battery which is rebuilt, overhauled, renovated, or restored such as from the pad up, after January 20, 1977.

*Secretary* means the Secretary of Labor, U.S. Department of Labor, or his or her designee.

*Stage charging* means a procedure by which a predetermined volume of coal in each larry car hopper is introduced into an oven such that no more than two hoppers are discharging simultaneously.

*Sequential charging* means a procedure, usually automatically timed, by which a predetermined volume of coal in each larry car hopper is introduced into an oven such that no more than two hoppers commence or finish discharging simultaneously although, at some point, all hoppers are discharging simultaneously.

*Pipeline charging* means any apparatus used to introduce coal into an oven which uses a pipe or duct permanently mounted onto an oven and through which coal is charged.

*Green plush* means coke which when removed from the oven results in emissions due to the presence of unvolatilized coal.

**1910.1029(c)** ***Permissible exposure limit.*** The employer shall assure that no employee in the regulated area is exposed to coke oven emissions at concentrations greater than 150 micrograms per cubic meter of air (150 $\mu g/m^3$), averaged over any 8-hour period.

**1910.1029(d)** ***Regulated areas.*** (1) The employer shall establish regulated areas and shall limit access to them to authorized persons.

**1910.1029(d)(2)** The employer shall establish the following as regulated areas:

**1910.1029(d)(2)(i)** The coke oven battery including topside and its machinery, pushside and its machinery, coke side and its machinery, and the battery ends; the wharf; and the screening station;

**1910.1029(d)(2)(ii)** The beehive oven and its machinery.

**1910.1029(e)** ***Exposure monitoring and measurement***— (1) *Monitoring program.* (i) Each employer who has a place of employment where coke oven emissions are present shall monitor employees employed in the regulated area to measure their exposure to coke oven emissions.

**1910.1029(e)(1)(ii)** The employer shall obtain measurements which are representative of each employee's exposure to coke oven emissions over an eight-hour period. All measurements shall determine exposure without regard to the use of respiratory protection.

**1910.1029(e)(1)(iii)** The employer shall collect fullshift (for at least seven continuous hours) personal samples, including at least one sample during each shift for each battery and each job classification within the regulated areas including at least the following job classifications:

**1910.1029(e)(1)(iii)*(a)*** Lidman;

**1910.1029(e)(1)(iii)*(b)*** Tar chaser;

**1910.1029(e)(1)(iii)*(c)*** Larry car operator;

**1910.1029(e)(1)(iii)*(d)*** Luterman;

**1910.1029(e)(1)(iii)*(e)*** Machine operator, coke side;

**1910.1029(e)(1)(iii)*(f)*** Benchman, coke side;

**1910.1029(e)(1)(iii)*(g)*** Benchman, pusher side;

**1910.1029(e)(1)(iii)*(h)*** Heater;

**1910.1029(e)(1)(iii)*(i)*** Quenching car operator;

**1910.1029(e)(1)(iii)*(j)*** Pusher machine operator;

**1910.1029(e)(1)(iii)*(k)*** Screening station operator;

**1910.1029(e)(1)(iii)*(l)*** Wharfman;

**1910.1029(e)(1)(iii)*(m)*** Oven patcher;

**1910.1029(e)(1)(iii)*(n)*** Oven repairman;

**1910.1029(e)(1)(iii)*(o)*** Spellman; and

**1910.1029(e)(1)(iii)*(p)*** Maintenance personnel.

**1910.1029(e)(1)(iv)** The employer shall repeat the monitoring and measurements required by this paragraph (e)(1) at least every three months.

**1910.1029(e)(2)** *Redetermination.* Whenever there has been a production, process, or control change which may result in new or additional exposure to coke oven emissions, or whenever the employer has any other reason to suspect an increase in employee exposure, the employer shall repeat the monitoring and measurements required by paragraph (e)(1) of this section for those employees affected by such change or increase.

**1910.1029(e)(3)** *Employee notification.* (i) The employer shall notify each employee in writing of the exposure measurements which represent that employe's exposure within five working days after the receipt of the results of measurements required by paragraphs (e)(1) and (e)(2) of this section.

**1910.1029(e)(3)(ii)** Whenever such results indicate that the representative employee exposure exceeds the permissible exposure limit, the employer shall, in such notification, inform each employee of that fact and of the corrective action being taken to reduce exposure to or below the permissible exposure limit.

**1910.1029(e)(4)** *Accuracy of measurement.* The employer shall use a method of monitoring and measurement which has an accuracy (with a confidence level of 95%) of not less than plus or minus 35% for concentrations of coke oven emissions greater than or equal to 150 $\mu g/m^3$.

**1910.1029(f)** ***Methods of compliance.*** The employer shall control employee exposure to coke oven emmissions by the use of engineering controls, work practices and respiratory protection as follows:

**1910.1029(f)(1)** *Priority of compliance methods*—(i) *Existing coke oven batteries. (a)* The employer shall institute the engineering and work practice controls listed in paragraphs (f)(2), (f)(3) and (f)(4) of this section in existing coke oven batteries at the earliest possible time, but not later than January 20, 1980, except to the extent that the employer can establish that such controls are not feasible. In determining the earliest possible time for institution of engineering and work practice controls, the requirement, effective August 27, 1971, to implement feasible administrative or engineering controls to reduce exposures to coal tar pitch volatiles, shall be considered. Wherever the engineering and work practice controls which can be instituted are not sufficient to reduce employee exposures to or below the permissible exposure limit, the employer shall nonetheless use them to reduce exposures to the lowest level achievable by these controls and shall supplement them by the use of respiratory protection which complies with the requirements of paragraph (g) of this section.

**1910.1029(f)(1)(i)*(b)*** The engineering and work practice controls required under paragraphs (f)(2), (f)(3) and (f)(4) of this section are minimum requirements generally applicable to all existing coke oven batteries. If, after implementing all controls required by paragraphs (f)(2), (f)(3) and (f)(4) of this section, or after January 20, 1980, whichever is sooner, employee exposures still exceed the permissible exposure limit, employers shall implement any other engineering and work practice controls necessary to reduce exposure to or below the permissible exposure limit except to the extent that the employer can establish that such controls are not feasible. Whenever the engineering and work practice controls which can be instituted are not sufficient to reduce employee exposures to or below the permissible exposure limit, the employer shall nonetheless use them to reduce exposures to the lowest level achievable by these controls and shall supplement them by the use of respiratory protection which complies with the requirements of paragraph (g) of this section.

**1910.1029(f)(1)(ii)** *New or rehabilitated coke oven batteries. (a)* The employer shall institute the best available engineering and work practice controls on all new or rehabilitated coke oven batteries to reduce and maintain employee exposures at or below the permissible exposure limit, except to the extent that the employer can establish that such controls are not feasible. Wher ever the engineering and work practice controls which can be instituted are not sufficient to reduce employee exposures to or below the permissible exposure limit, the employer shall nonetheless use them to reduce exposures to the lowest level achievable by these controls and shall supplement them by the use of respiratory protection which complies with the requirements of paragraph (g) of this section.

**1910.1029(f)(1)(ii)*(b)*** If, after implementing all the engineering and work practice controls required by paragraph (f)(1)(ii)*(a)* of this section, employee exposures still exceed the permissible exposure limit, the employer shall implement any other engineering and work practice controls necessary to reduce exposure to or below the permissible exposure limit except to the extent that the employer can establish that such controls are not feasible. Wherever the engineering and work practice controls which can be instituted are not sufficient to reduce employee exposures to or below the permissible exposure limit, the employer shall nonetheless use them to reduce exposures to the lowest level achievable by these controls and shall supplement them by the use of respiratory protection which complies with the requirements of paragraph (g) of this section.

**1910.1029(f)(1)(iii)** *Beehive ovens. (a)* The employer shall institute engineering and work practice controls on all beehive ovens at the earliest possible time to reduce and maintain employee exposures at or below the permissible exposure limit, except to the extent that the employer can establish that such controls are not feasible. In determining the earliest possible time for institution of engineering and work practice controls, the requirement, effective August 27, 1971, to implement feasible administrative or engineering controls to reduce exposures to coal tar pitch volatiles, shall be considered. Wherever the engineering and work practice controls which can be instituted are not sufficient to reduce employee exposures to or below the permissible exposure limit, the employer shall nonetheless use them to reduce exposures to the lowest level achievable by these controls and shall supplement them by the use of respiratory protection which complies with the requirements of paragraph (g) of this section.

**1910.1029(f)(1)(iii)*(b)*** If, after implementing all engineering and work practice controls required by paragraph (f)(1)(iii)*(a)* of this section, employee exposures still exceed the permissible exposure limit, the employer shall implement any other engineering and work practice controls necessary to reduce exposures to or below the permissible exposure limit except to the extent that the employer can establish that such controls are not feasible. Whenever the engineering and work practice controls which can be instituted are not sufficient to reduce employee exposures to or below the permissible exposure limit, the employer shall nonetheless use them to reduce exposures to the lowest level achievable by these controls and shall supplement them by the use of respiratory protection which complies with the requirements of paragraph (g) of this section.

**1910.1029(f)(2)** *Engineering controls*—(i) *Charging.* The employer shall equip and operate existing coke oven batteries with all of the following engineering controls to control coke oven emissions during charging operations:

**1910.1029(f)(2)(i)*(a)*** One of the following methods of charging:

**1910.1029(f)(2)(i)*(a)(1)*** Stage charging as described in paragraph (f)(3)(i)*(b)* of this section; or

**1910.1029(f)(2)(i)*(a)(2)*** Sequential charging as described in paragraph (f)(3)(i)*(b)* of this section except that paragraph (f)(3)(i)*(b)(3)(iv)* of this section does not apply to sequential charging; or

**1910.1029(f)(2)(i)*(a)(3)*** Pipeline charging or other forms of enclosed charging in accordance with paragraph (f)(2)(i) of this section, except that paragraphs (f)(2)(i)*(b)*, *(d)*, *(e)*, *(f)* and *(h)* of this section do not apply;

**1910.1029(f)(2)(i)*(b)*** Drafting from two or more points in the oven being charged, through the use of double collector mains, or a fixed or moveable jumper pipe system to another oven, to effectively remove the gases from the oven to the collector mains;

**1910.1029(f)(2)(i)*(c)*** Aspiration systems designed and operated to provide sufficient negative pressure and flow volume to effectively move the gases evolved during charging into the collector mains, including sufficient steam pressure, and steam jets of sufficient diameter;

**1910.1029(f)(2)(i)*(d)*** Mechanical volumetric controls on each larry car hopper to provide the proper amount of coal to be charged through each charging hole so that the tunnel head will be sufficient to permit the gases to move from the oven into the collector mains;

**1910.1029(f)(2)(i)*(e)*** Devices to facilitate the rapid and continuous flow of coal into the oven being charged, such as stainless steel liners, coal vibrators or pneumatic shells;

**1910.1029(f)(2)(i)*(f)*** Individually operated larry car drop sleeves and slide gates designed and maintained so that the gases are effectively removed from the oven into the collector mains;

**1910.1029(f)(2)(i)*(g)*** Mechanized gooseneck and standpipe cleaners;

**1910.1029(f)(2)(i)*(h)*** Air seals on the pusher machine leveler bars to control air infiltration during charging; and

**1910.1029(f)(2)(i)*(i)*** Roof carbon cutters or a compressed air system or both on the pusher machine rams to remove roof carbon.

**1910.1029(f)(2)(ii)** *Coking.* The employer shall equip and operate existing coke oven batteries with all of the following engineering controls to control coke oven emissions during coking operations;

**1910.1029(f)(2)(ii)*(a)*** A pressure control system on each battery to obtain uniform collector main pressure;

**1910.1029(f)(2)(ii)*(b)*** Ready access to door repair facilities capable of prompt and efficient repair of doors, door sealing edges and all door parts;

**1910.1029(f)(2)(ii)*(c)*** An adequate number of spare doors available for replacement purposes;

**1910.1029(f)(2)(ii)*(d)*** Chuck door gaskets to control chuck door emissions until such door is repaired, or replaced; and

**1910.1029(f)(2)(ii)*(e)*** Heat shields on door machines.

**1910.1029(f)(3)** *Work practice controls*—(i) *Charging.* The employer shall operate existing coke oven batteries with all of the following work practices to control coke oven emissions during the charging operation:

**1910.1029(f)(3)(i)*(a)*** Establishment and implementation of a detailed, written inspection and cleaning procedure for each battery consisting of at least the following elements:

**1910.1029(f)(3)(i)*(a)(1)*** Prompt and effective repair or replacement of all engineering controls;

**1910.1029(f)(3)(i)*(a)(2)*** Inspection and cleaning of goosenecks and standpipes prior to each charge to a specified minimum diameter sufficient to effectively move the evolved gases from the oven to the collector mains;

**1910.1029(f)(3)(i)*(a)(3)*** Inspection for roof carbon build-up prior to each charge and removal of roof carbon as necessary to provide an adequate gas channel so that the gases are effectively moved from the oven into the collector mains;

**1910.1029(f)(3)(i)*(a)(4)*** Inspection of the steam aspiration system prior to each charge so that sufficient pressure and volume is maintained to effectively move the gases from the oven to the collector mains;

**1910.1029(f)(3)(i)*(a)(5)*** Inspection of steam nozzles and liquor sprays prior to each charge and cleaning as necessary so that the steam nozzles and liquor sprays are clean;

**1910.1029(f)(3)(i)*(a)(6)*** Inspection of standpipe caps prior to each charge and cleaning and luting or both as necessary so that the gases are effectively moved from the oven to the collector mains; and

**1910.1029(f)(3)(i)*(a)(7)*** Inspection of charging holes and lids for cracks, warpage and other defects prior to each charge and removal of carbon to prevent emissions, and application of luting material to standpipe and charging hole lids where necessary to obtain a proper seal.

**1910.1029(f)(3)(i)*(b)*** Establishment and implementation of a detailed written charging procedure, designed and operated to eliminate emissions during charging for each battery, consisting of at least the following elements:

**1910.1029(f)(3)(i)(b)(1)** Larry car hoppers filled with coal to a predetermined level in accordance with the mechanical volumetric controls required under paragraph (f)(2)(i)*(d)* of this section so as to maintain a sufficient gas passage in the oven to be charged;

**1910.1029(f)(3)(i)(b)(2)** The larry car aligned over the oven to be charged, so that the drop sleeves fit tightly over the charging holes; and

**1910.1029(f)(3)(i)(b)(3)** The oven charged in accordance with the following sequence of requirements:

**1910.1029(f)(3)(i)(b)(3)(i)** The aspiration system turned on;

**1910.1029(f)(3)(i)(b)(3)(ii)** Coal charged through the outermost hoppers, either individually or together depending on the capacity of the aspiration system to collect the gases involved;

**1910.1029(f)(3)(i)(b)(3)(iii)** The charging holes used under paragraph (f)(3)(i)*(b)(3)(ii)* of this section relidded or otherwise sealed off to prevent leakage of coke oven emissions;

**1910.1029(f)(3)(i)(b)(3)(iv)** If four hoppers are used, the third hopper discharged and relidded or otherwise sealed off to prevent leakage of coke oven emissions;

**1910.1029(f)(3)(i)(b)(3)(v)** The final hopper discharged until the gas channel at the top of the oven is blocked and then the chuck door opened and the coal leveled;

**1910.1029(f)(3)(i)(b)(3)(vi)** When the coal from the final hopper is discharged and the leveling operation complete, the charging hole relidded or otherwise sealed off to prevent leakage of coke oven emissions; and

**1910.1029(f)(3)(i)(b)(3)(vii)** The aspiration system turned off only after the charging holes have been closed.

**1910.1029(f)(3)(i)(c)** Establishment and implementation of a detailed written charging procedure, designed and operated to eliminate emissions during charging of each pipeline or enclosed charged battery.

**1910.1029(f)(3)(ii)** *Coking.* The employer shall operate existing coke oven batteries pursuant to a detailed written procedure established and implemented for the control of coke oven emissions during coking, consisting of at least the following elements:

**1910.1029(f)(3)(ii)(a)** Checking oven back pressure controls to maintain uniform pressure conditions in the collecting main;

**1910.1029(f)(3)(ii)(b)** Repair, replacement and adjustment of oven doors and chuck doors and replacement of door jambs so as to provide a continuous metal-to-metal fit;

**1910.1029(f)(3)(ii)(c)** Cleaning of oven doors, chuck doors and door jambs each coking cycle so as to provide an effective seal;

**1910.1029(f)(3)(ii)(d)** An inspection system and corrective action program to control door emissions to the maximum extent possible; and

**1910.1029(f)(3)(ii)(e)** Luting of doors that are sealed by luting each coking cycle and reluting, replacing or adjusting as necessary to control leakage.

**1910.1029(f)(3)(iii)** *Pushing.* The employer shall operate existing coke oven batteries with the following work practices to control coke oven emissions during pushing operations:

**1910.1029(f)(3)(iii)(a)** Coke and coal spillage quenched as soon as practicable and not shoveled into a heated oven; and

**1910.1029(f)(3)(iii)(b)** A detailed written procedure for each battery established and implemented for the control of emissions during pushing consisting of the following elements:

**1910.1029(f)(3)(iii)(b)(1)** Dampering off the ovens and removal of charging hole lids to effectively control coke oven emissions during the push;

**1910.1029(f)(3)(iii)(b)(2)** Heating of the coal charge uniformly for a sufficient period so as to obtain proper coking including preventing green pushes;

**1910.1029(f)(3)(iii)(b)(3)** Prevention of green pushes to the maximum extent possible;

**1910.1029(f)(3)(iii)(b)(4)** Inspection, adjustment and correction of heating flue temperatures and defective flues at least weekly and after any green push, so as to prevent green pushes;

**1910.1029(f)(3)(iii)(b)(5)** Cleaning of heating flues and related equipment to prevent green pushes, at least weekly and after any green push.

**1910.1029(f)(3)(iv)** *Maintenance and repair.* The employer shall operate existing coke oven batteries pursuant to a detailed written procedure of maintenance and repair established and implemented for the effective control of coke oven emissions consisting of the following elements:

**1910.1029(f)(3)(iv)(a)** Regular inspection of all controls, including goosenecks, standpipes, standpipe caps, charging hold lids and castings, jumper pipes and air seals for cracks, misalignment or other defects and prompt implementation of the necessary repairs as soon as possible;

**1910.1029(f)(3)(iv)(b)** Maintaining the regulated area in a neat, orderly condition free of coal and coke spillage and debris;

**1910.1029(f)(3)(iv)(c)** Regular inspection of the damper system, aspiration system and collector main for cracks or leakage, and prompt implementation of the necessary repairs;

**1910.1029(f)(3)(iv)(d)** Regular inspection of the heating system and prompt implementation of the necessary repairs;

**1910.1029(f)(3)(iv)(e)** Prevention of miscellaneous fugitive topside emissions;

**1910.1029(f)(3)(iv)(f)** Regular inspection and patching of oven brickwork;

**1910.1029(f)(3)(iv)(g)** Maintenance of battery equipment and controls in good working order;

**1910.1029(f)(3)(iv)(h)** Maintenance and repair of coke oven doors, chuck doors, door jambs and seals; and

**1910.1029(f)(3)(iv)(i)** Repairs instituted and completed as soon as possible, including temporary repair measures instituted and completed where necessary, including but not limited to:

**1910.1029(f)(3)(iv)(i)(1)** Prevention of miscellaneous fugitive topside emissions; and

**1910.1029(f)(3)(iv)(i)(2)** Chuck door gaskets, which shall be installed prior to the start of the next coking cycle.

**1910.1029(f)(4)** *Filtered air.* (i) The employer shall provided positive-pressure, temperature controlled filtered air for larry car, pusher machine, door machine, and quench car cabs.

**1910.1029(f)(4)(ii)** The employer shall provide standby pulpits on the battery topside, at the wharf, and at ther screening station, equipped with positive-pressure, temperature controlled filtered air.

**1910.1029(f)(5)** *Emergencies.* Whenever an emergency occurs, the next coking cycle may not begin until the cause of the emergency is determined and corrected, unless the employer can establish that it is necessary to initiate the next coking cycle in order to determine the cause of the emergency.

**1910.1029(f)(6)** *Compliance program.* (i) Each employer shall establish and implement a written program to reduce exposures solely by means of the engineering and work practice controls required in paragraph (f) of this section.

**1910.1029(f)(6)(ii)** The written program shall include at least the following:

**1910.1029(f)(6)(ii)(a)** A description of each coke oven operation by battery, including work force and operating crew, coking time, operating procedures and maintenance practices;

**1910.1029(f)(6)(ii)(b)** Engineering plans and other studies used to determine the controls for the coke battery;

**1910.1029(f)(6)(ii)(c)** A report of the technology considered in meeting the permissible exposure limit;

**1910.1029(f)(6)(ii)(d)** Monitoring data obtained in accordance with paragraph (e) of this section;

**1910.1029(f)(6)(ii)(e)** A detailed schedule for the implementation of the engineering and work practice controls required in paragraph (f) of this section; and

**1910.1029(f)(6)(ii)(f)** Other relevant information.

**1910.1029(f)(6)(iii)** If, after implementing all controls required by paragraph (f)(2)-(f)(4) of this section, or after January 20, 1980, whichever is sooner, or after completion of a new or rehabilitated battery the permissible exposure limit is still exceeded, the employer shall develop a detailed written program and schedule for the implementation of any additional engineering controls and work practices necessary to reduce exposure to or below the permissible exposure limit.

**1910.1029(f)(6)(iv)** Written plans for such programs shall be submitted, upon request, to the Secretary and the Director, and shall be available at the worksite for examination and copying by the Secretary, the Director, and the authorized employee representative. The plans required under paragraph (f)(6) of this section shall be revised and updated at least every six months to reflect the current status of the program.

**1910.1029(f)(7)** *Training in compliance procedures.* The employer shall incorporate all written procedures and schedules required under this paragraph (f) in the information and training program required under paragraph (k) of this section and, where appropriate, post in the regulated area.

**1910.1029(g)** ***Respiratory protection***— (1) *General.* For employees who use respirators required by this section, the employer must provide respirators that comply with the requirements of this paragraph. Compliance with the permissible exposure limit may not be achieved by the use of respirators except during:

**1910.1029(g)(1)(i)** Periods necessary to install or implement feasible engineering and work-practice controls.

**1910.1029(g)(1)(ii)** Work operations, such as maintenance and repair activity, for which engineering and work-practice controls are technologically not feasible.

**1910.1029(g)(1)(iii)** Work operations for which feasible engineering and work-practice controls are not yet sufficient to reduce employee exposure to or below the permissible exposure limit.

**1910.1029(g)(1)(iv)** Emergencies.

**1910.1029(g)(2)** *Respirator program.* The employer must implement a respiratory protection program in accordance with 29 CFR 1910.134 (b) through (d) (except (d)(1)(iii)), and (f) through (m).

**1910.1029(g)(3)** *Respirator selection.* The employer must select appropriate respirators or combination of respirators from Table I of this section.

## TABLE I—RESPIRATORY PROTECTION FOR COKE OVEN EMISSIONS

| Airborne concentration of coke oven emissions | Required respirator |
|---|---|
| (a) Any concentration | (1) A Type C supplied air respirator operated in pressure demand or other positive pressure or continuous flow mode; or |
| | (2) A powered air-purifying particulate filter respirator for dust and mist or |
| | (3) A powered air-purifying particulate filter respirator or combination chemical cartridge and particulate filter respirator for coke oven emissions. |
| (b) Concentrations not greater than 1500 $ug/m_3$. | (1) Any particulate filter respirator for dust and mist except single-use respirator; or |
| | (2) Any particulate filter respirator or combination chemical cartridge and particulate filter respirator for coke oven emissions; or |
| | (3) Any respirator listed in paragraph (g)(3)(a) of this section. |

**1910.1029(h)** ***Protective clothing and equipment***—(1) *Provision and use.* The employer shall provide and assure the use of appropriate protective clothing and equipment, such as but not limited to:

**1910.1029(h)(1)(i)** Flame resistant jacket and pants;

**1910.1029(h)(1)(ii)** Flame resistant gloves;

**1910.1029(h)(1)(iii)** Face shields or vented goggles which comply with §1910.133(a)(2) of this part;

**1910.1029(h)(1)(iv)** Footwear providing insulation from hot surfaces for footwear;

**1910.1029(h)(1)(v)** Safety shoes which comply with §1910.136 of this part; and

**1910.1029(h)(1)(vi)** Protective helmets which comply with §1910.135 of this part.

**1910.1029(h)(2)** *Cleaning and replacement.* (i) The employer shall provide the protective clothing required by paragraphs (h)(1) (i) and (ii) of this section in a clean and dry condition at least weekly.

**1910.1029(h)(2)(ii)** The employer shall clean, launder, or dispose of protective clothing required by paragraphs (h)(1) (i) and (ii) of this section.

**1910.1029(h)(2)(iii)** The employer shall repair or replace the protective clothing and equipment as needed to maintain their effectiveness.

**1910.1029(h)(2)(iv)** The employer shall assure that all protective clothing is removed at the completion of a work shift only in change rooms prescribed in paragraph (i)(1) of this section.

**1910.1029(h)(2)(v)** The employer shall assure that contaminated protective clothing which is to be cleaned, laundered, or disposed of, is placed in a closable container in the change room.

**1910.1029(h)(2)(vi)** The employer shall inform any person who cleans or launders protective clothing required by this section, of the potentially harmful effects of exposure to coke oven emissions.

**1910.1029(i)** ***Hygiene facilities and practices***— (1) *Change rooms.* The employer shall provide clean change rooms equipped with storage facilities for street clothes and separate storage facilities for protective clothing and equipment whenever employees are required to wear protective clothing and equipment in accordance with paragraph (h)(1) of this section.

**1910.1029(i)(2)** *Showers.* (i) The employer shall assure that employees working in the regulated area shower at the end of the work shift.

**1910.1029(i)(2)(ii)** The employer shall provide shower facilities in accordance with §1910.141(d)(3) of this part.

**1910.1029(i)(3)** *Lunchrooms.* The employer shall provide lunchroom facilities which have a temperature controlled, positive pressure, filtered air supply, and which are readily accessible to employees working in the regulated area.

**1910.1029(i)(4)** *Lavatories.* (i) The employer shall assure that employees working in the regulated area wash their hands and face prior to eating.

**1910.1029(i)(4)(ii)** The employer shall provide lavatory facilities in accordance with §1910.141(d)(1) and (2) of this part.

**1910.1029(i)(5)** *Prohibition of activities in the regulated area.* (i) The employer shall assure that in the regulated area, food or beverages are not present or consumed, smoking products are not present or used, and cosmetics are not applied, except that these activities may be conducted in the lunchrooms, change rooms and showers required under paragraphs (i)(1)-(i)(3) of this section.

**1910.1029(i)(5)(ii)** Drinking water may be consumed in the regulated area.

**1910.1029(j)** ***Medical surveillance***—(1) *General requirements.* (i) Each employer shall institute a medical surveillance program for all employees who are employed in a regulated area at least 30 days per year.

**1910.1029(j)(1)(ii)** This program shall provide each employee covered under paragraph ( j)(1)(i) of this section with an opportunity for medical examinations in accordance with this paragraph ( j).

**1910.1029(j)(1)(iii)** The employer shall inform any employee who refuses any required medical examination of the possible health consequences of such refusal and shall obtain a signed statement from the employee indicating that the employee understands the risk involved in the refusal to be examined.

**1910.1029(j)(1)(iv)** The employer shall assure that all medical examinations and procedures are performed by or under the supervision of a licensed physician, and are provided without cost to the employee.

**1910.1029(j)(2)** *Initial examinations.* At the time of initial assignment to a regulated area or upon the institution of the medical surveillance program, the employer shall provide a medical examination for employees covered under paragraph ( j)(1)(i) of this section including at least the following elements:

**1910.1029(j)(2)(i)** A work history and medical history which shall include smoking history and the presence and degree of respiratory symptoms, such as breathlessness, cough, sputum production, and wheezing;

**1910.1029(j)(2)(ii)** A 14″ × 17″ posterior-anterior chest x-ray and International Labour Office UICC/Cincinnati (ILO U/C) rating;

**1910.1029(j)(2)(iii)** Pulmonary function tests including forced vital capacity (FVC) and forced expiratory volume at one second (FEV 1.0) with recording of type of equipment used;

**1910.1029(j)(2)(iv)** Weight;

**1910.1029(j)(2)(v)** A skin examination;

**1910.1029(j)(2)(vi)** Urinalysis for sugar, albumin, and hematuria; and

**1910.1029(j)(2)(vii)** A urinary cytology examination.

**1910.1029(j)(3)** *Periodic examinations.* (i) The employer shall provide the examinations specified in paragraphs (j)(2) (i)-(vi) of this section at least annually for employees covered under paragraph (j)(1)(i) of this section.

**1910.1029(j)(3)(ii)** The employer shall provide the examinations specified in paragraphs (j)(2)(i) and (j)(2)(iii) through (vii) of this section at least semi-annually for employees 45 years of age or older or with five (5) or more years employment in the regulated area.

**1910.1029(j)(3)(iii)** Whenever an employee who is 45 years of age or older or with five (5) or more years employment in the regulated area transfers or is transferred from employment in a regulated area, the employer shall continue to provide the examinations specified in paragraphs (j)(2)(i) and (j)(2)(iii) through (vii) of this section semi-annually, as long as that employee is employed by the same employer or a successor employer.

**1910.1029(j)(3)(iv)** The employer shall provide the x-ray specified in paragraph (j)(2)(ii) of this section at least annually for employees covered under paragraph (j)(3) of this section.

**1910.1029(j)(3)(v)** Whenever an employee has not taken the examinations specified in paragraphs ( j)(3) (i)-(iii) of this section with the six (6) months preceding the termination of employment the employer shall provide such examinations to the employee upon termination of employment.

**1910.1029(j)(4)** *Information provided to the physician.* The employer shall provide the following information to the examining physician:

**1910.1029(j)(4)(i)** A copy of this regulation and its Appendixes;

**1910.1029(j)(4)(ii)** A description of the affected employee's duties as they relate to the employee's exposure;

**1910.1029(j)(4)(iii)** The employee's exposure level or estimated exposure level;

**1910.1029(j)(4)(iv)** A description of any personal protective equipment used or to be used; and

**1910.1029(j)(4)(v)** Information from previous medical examinations of the affected employee which is not readily available to the examining physician.

**1910.1029(j)(5)** *Physician's written opinion.* (i) The employer shall obtain a written opinion from the examining physician which shall include:

**1910.1029(j)(5)(i)*(a)*** The results of the medical examinations;

**1910.1029(j)(5)(i)*(b)*** The physician's opinion as to whether the employee has any detected medical conditions which would place the employee at increased risk of material impairment of the employee's health from exposure to coke oven emissions;

**1910.1029(j)(5)(i)*(c)*** Any recommended limitations upon the employee's exposure to coke oven emissions or upon the use of protective clothing or equipment such as respirators; and

**1910.1029(j)(5)(i)*(d)*** A statement that the employee has been informed by the physician of the results of the medical examination and any medical conditions which require further explanation or treatment.

**1910.1029(j)(5)(ii)** The employer shall instruct the physician not to reveal in the written opinion specific findings or diagnoses unrelated to occupational exposure.

**1910.1029(j)(5)(iii)** The employer shall provide a copy of the written opinion to the affected employee.

**1910.1029(k)** ***Employee information and training***— (1) *Training program.* (i) The employer shall institute a training program for employees who are employed in the regulated area and shall assure their participation.

**1910.1029(k)(1)(ii)** The training program shall be provided as of January 27, 1977 for employees who are employed in the regulated area at that time or at the time of initial assignment to a regulated area.

**1910.1029(k)(1)(iii)** The training program shall be provided at least annually for all employees who are employed in the regulated area, except that training regarding the occupational safety and health hazards associated with exposure to coke oven emissions and the purpose, proper use, and limitations of respiratory protective devices shall be provided at least quarterly until January 20, 1978.

**1910.1029(k)(1)(iv)** The training program shall include informing each employee of:

**1910.1029(k)(1)(iv)*(a)*** The information contained in the substance information sheet for coke oven emissions (Appendix A);

**1910.1029(k)(1)(iv)*(b)*** The purpose, proper use, and limitations of respiratory protective devices required in accordance with paragraph (g) of this section;

**1910.1029(k)(1)(iv)*(c)*** The purpose for and a description of the medical surveillance program required by paragraph ( j) of this section including information on the occupational safety and health hazards associated with exposure to coke oven emissions;

**1910.1029(k)(1)(iv)*(d)*** A review of all written procedures and schedules required under paragraph (f ) of this section; and

**1910.1029(k)(1)(iv)*(e)*** A review of this standard.

**1910.1029(k)(2)** *Access to training materials.* (i) The employer shall make a copy of this standard and its appendixes readily available to all employees who are employed in the regulated area.

**1910.1029(k)(2)(ii)** The employer shall provide upon request all materials relating to the employee information and training program to the Secretary and the Director.

**1910.1029(l)** ***Precautionary signs and labels***— (1) *General.* (i) The employer may use labels or signs required by other statutes, regulations or ordinances in addition to, or in combination with, signs and labels required by this paragraph.

**1910.1029(l)(1)(ii)** The employer shall assure that no statement appears on or near any sign required by this paragraph which contradicts or detracts from the effects of the required sign.

**1910.1029(l)(1)(iii)** The employer shall assure that signs required by this paragraph are illuminated and cleaned as necessary so that the legend is readily visible.

**1910.1029(l)(2)** *Signs.* (i) The employer shall post signs in the regulated area bearing the legends:

**DANGER**

**CANCER HAZARD**

**AUTHORIZED PERSONNEL ONLY**

**NO SMOKING OR EATING**

**1910.1029(l)(2)(ii)** In addition, not later than January 20, 1978, the employer shall post signs in the areas where the permissible exposure limit is exceeded bearing the legend:

**DANGER**

**RESPIRATOR REQUIRED**

**1910.1029(l)(3)** *Labels.* The employer shall apply precautionary labels to all containers of protective clothing contaminated with coke oven emissions bearing the legend:

**CAUTION**

**CLOTHING CONTAMINATED WITH COKE EMISSIONS**

**DO NOT REMOVE DUST BY BLOWING OR SHAKING**

**1910.1029(m)** ***Recordkeeping***— (1) *Exposure measurements.* The employer shall establish and maintain an accurate record of all measurements taken to monitor employee exposure to coke oven emissions required in paragraph (e) of this section.

**1910.1029(m)(1)(i)** This record shall include:

**1910.1029(m)(1)(i)*(a)*** Name, social security number, and job classification of the employees monitored;

**1910.1029(m)(1)(i)*(b)*** The date(s), number, duration and results of each of the samples taken, including a description of the sampling procedure used to determine representative employee exposure where applicable;

**1910.1029(m)(1)(i)*(c)*** The type of respiratory protective devices worn, if any;

**1910.1029(m)(1)(i)*(d)*** A description of the sampling and analytical methods used and evidence of their accuracy; and

**1910.1029(m)(1)(i)*(e)*** The environmental variables that could affect the measurement of employee exposure.

**1910.1029(m)(1)(ii)** The employer shall maintain this record for at lest 40 years or for the duration of employment plus 20 years, whichever is longer.

**1910.1029(m)(2)** *Medical surveillance.* The employer shall establish and maintain an accurate record for each employee subject to medical surveillance as required by paragraph ( j) of this section.

**1910.1029(m)(2)(i)** The record shall include:

**1910.1029(m)(2)(i)*(a)*** The name, social security number, and description of duties of the employee;

**1910.1029(m)(2)(i)*(b)*** A copy of the physician's written opinion;

**1910.1029(m)(2)(i)*(c)*** The signed statement of any refusal to take a medical examination under paragraph ( j)(1)(ii) of this section; and

**1910.1029(m)(2)(i)*(d)*** Any employee medical complaints related to exposure to coke oven emissions.

**1910.1029(m)(2)(ii)** The employer shall keep, or assure that the examining physician keeps, the following medical records:

**1910.1029(m)(2)(ii)*(a)*** A copy of the medical examination results including medical and work history required under paragraph ( j)(2) of this section;

**1910.1029(m)(2)(ii)*(b)*** A description of the laboratory procedures used and a copy of any standards or guidelines used to interpret the test results;

**1910.1029(m)(2)(ii)*(c)*** The initial x-ray;

**1910.1029(m)(2)(ii)*(d)*** The x-rays for the most recent five (5) years;

**1910.1029(m)(2)(ii)*(e)*** Any x-ray with a demonstrated abnormality and all subsequent x-rays;

**1910.1029(m)(2)(ii)*(f)*** The initial cytologic examination slide and written description;

**1910.1029(m)(2)(ii)*(g)*** The cytologic examination slide and written description for the most recent 10 years; and

**1910.1029(m)(2)(ii)*(h)*** Any cytologic examination slides with demonstrated atypia, if such atypia persists for 3 years, and all subsequent slides and written descriptions.

**1910.1029(m)(2)(iii)** The employer shall maintain medical records required under paragraph (m)(2) of this section for at least 40 years, or for the duration of employment plus 20 years, whichever is longer.

**1910.1029(m)(3)** *Availability.* (i) The employer shall make available upon request all records required to be maintained by paragraph (m) of this section to the Secretary and the Director for examination and copying.

**1910.1029(m)(3)(ii)** Employee exposure measurement records and employee medical records required by this paragraph shall be provided upon request to employees, designated representatives, and the Assistant Secretary in accordance with 29 CFR 1910.20(a)-(e) and (g)-(i).

**1910.1029(m)(4)** *Transfer of records.* (i) Whenever the employer ceases to do business, the successor employer shall receive and retain all records required to be maintained by paragraph (m) of this section.

**1910.1029(m)(4)(ii)** Whenever the employer ceases to do business and there is no successor employer to receive and retain the records for the prescribed period, these records shall be transmitted by registered mail to the Director.

**1910.1029(m)(4)(iii)** At the expiration of the retention period for the records required to be maintained under paragraphs (m)(1) and (m)(2) of this section, the employer shall transmit these records by registered mail to the Director or shall continue to retain such records.

**1910.1029(m)(4)(iv)** The employer shall also comply with any additional requirements involving transfer of records set forth in 29 CFR 1910.20(h).

**1910.1029(n)** ***Observation of monitoring***— (1) *Employee observation.* The employer shall provide affected employees or their representatives an opportunity to observe any measuring or monitoring of employee exposure to coke oven emissions conducted pursuant to paragraph (e) of this section.

**1910.1029(n)(2)** *Observation procedures.* (i) Whenever observation of the measuring or monitoring of employee exposure to coke oven emissions requires entry into an area where the ues of protective clothing or equipment is required, the employer shall provide the observer with and assure the use of such equipment and shall require the observer to comply with all other applicable safety and health procedures.

**1910.1029(n)(2)(ii)** Without interfering with the measurement, observers shall be entitled to:

**1910.1029(n)(2)(ii)*(a)*** An Explanation of the measurement procedures;

**1910.1029(n)(2)(ii)*(b)*** Observe all steps related to the measurement of coke oven emissions performed at the place of exposure; and

**1910.1029(n)(2)(ii)*(c)*** Record the results obtained.

**1910.1029(o)** ***Effective date.*** This standard shall become effective January 20, 1977.

**1910.1029(p)** ***Appendices.*** The information contained in the appendixes to this section is not intended, by itself, to create any additional obligations not otherwise imposed or to detract from any existing obligation.

[39 FR 23502, June 27, 1974, as amended at 63 FR 33468, June 18, 1998]

## APPENDIX A TO §1910.1029—COKE OVEN EMISSIONS SUBSTANCE INFORMATION SHEET

### I. SUBSTANCE IDENTIFICATION

A. *Substance:* Coke Oven Emissions

B. *Definition:* The benzene-soluble fraction of total particulate matter present during the destructive distillation or carbonization of coal for the production of coke.

C. *Permissible Exposure Limit:* 150 micrograms per cubic meter of air determined as an average over an 8-hour period.

D. *Regulated areas:* Only employees authorized by your employer should enter a regulated area. The employer is required to designate the following areas as regulated areas: the coke oven battery, including topside and its machinery, pushside and its machinery, cokeside and its machinery, and the battery ends; the screening station; and the wharf; and the beehive ovens and their machinery.

### II. HEALTH HAZARD DATA

Exposure to coke oven emissions is a cause of lung cancer, and kidney cancer, in humans. Although there have not been an excess number of skin cancer cases in humans, repeated skin contact with coke oven emissions should be avoided.

### III. PROTECTIVE CLOTHING AND EQUIPMENT

A. *Respirators:* Respirators will be provided by your employer for routine use if your employer is in the process of implementing engineering and work practice controls or where engineering and work practice controls are not feasible or insufficient to reduce exposure to or below the PEL. You must wear respirators for non-routine activities or in emergency situations where you are likely to be exposed to levels of coke oven emissions in excess of the permissible exposure limit. Until January 20, 1978, the routine wearing of respirators is voluntary. Until that date, if you choose not to wear a respirator you do not have to do so. You must still have your respirator with you and you must still wear it if you are near visible emissions. Since how well your respirator fits your face is very important, your employer is required to conduct fit tests to make sure the respirator seals properly when you wear it. These tests are simple and rapid and will be explained to you during your training sessions.

B. *Protective clothing:* Your employer is required to provide, and you must wear, appropriate, clean, protective clothing and equipment to protect your body from repeated skin contact with coke oven emissions and from the heat generated during the coking process. This clothing should include such items as jacket and pants and flame resistant gloves. Protective equipment should include face shield or vented goggles, protective helmets and safety shoes, insulated from hot surfaces where appropriate.

## IV. HYGIENE FACILITIES AND PRACTICES

You must not eat, drink, smoke, chew gum or tobacco, or apply cosmetics in the regulated area, except that drinking water is permitted. Your employer is required to provide lunchrooms and other areas for these purposes.

Your employer is required to provide showers, washing facilities, and change rooms. If you work in a regulated area, you must wash your face, and hands before eating. You must shower at the end of the work shift. Do not take used protective clothing out of the change rooms without your employer's permission. Your employer is required to provide for laundering or cleaning of your protective clothing.

## V. SIGNS AND LABELS

Your employer is required to post warning signs and labels for your protection. Signs must be posted in regulated areas. The signs must warn that a cancer hazard is present, that only authorized employees may enter the area, and that no smoking or eating is allowed. In regulated areas where coke oven emissions are above the permissible exposure limit, the signs should also warn that respirators must be worn.

## VI. MEDICAL EXAMINATIONS

If you work in a regulated area at least 30 days per year, your employer is required to provide you with a medical examination every year. The medical examination must include a medical history, a chest x-ray, pulmonary function test, weight comparison, skin examination, a urinalysis, and a urine cytology exam for early detection of urinary cancer. The urine cytology exam is only included in the initial exam until you are either 45 years or older, or have 5 or more years employment in the regulated areas when the medical exams including this test, but excepting the x-ray exam, are to be given every six months; under these conditions, you are to be given an x-ray exam at least once a year. The examining physician will provide a written opinion to your employer containing the results of the medical exams. You should also receive a copy of this opinion.

## VII. OBSERVATION OF MONITORING

Your employer is required to monitor your exposure to coke oven emissions and you are entitled to observe the monitoring procedure. You are entitled to receive an explanation of the measurement procedure, observe the steps taken in the measurement procedure, and to record the results obtained. When the monitoring procedure is taking place in an area where respirators or personal protective clothing and equipment are required to be worn, you must also be provided with and must wear the protective clothing and equipment.

## VIII. ACCESS TO RECORDS

You or your representative are entitled to records of your exposure to coke oven emissions upon request to your employer. Your medical examination records can be furnished to your physician upon request to your employer.

## IX. TRAINING AND EDUCATION

Additional information on all of these items plus training as to hazards of coke oven emissions and the engineering and work practice controls associated with your job will also be provided by your employer.

[39 FR 23502, June 27, 1974, as amended at 63 FR 33468, June 18, 1998]

# APPENDIX B TO §1910.1029—INDUSTRIAL HYGIENE AND MEDICAL SURVEILLANCE GUIDELINES

## I. INDUSTRIAL HYGIENE GUIDELINES

A. *Sampling* (Benzene-Soluble Fraction Total Particulate Matter).

Samples collected should be full shift (at least 7-hour) samples. Sampling should be done using a personal sampling pump with pulsation damper at a flow rate of 2 liters per minute. Samples should be collected on 0.8 micrometer pore size silver membrane filters (37 mm diameter) preceded by Gelman glass fiber type A-E filters encased in three-piece plastic (polystyrene) field monitor cassettes. The cassette face cap should be on and the plug removed. The rotameter should be checked every hour to ensure that proper flow rates are maintained.

A minimum of three full-shift samples should be collected for each job classification on each battery, at least one from each shift. If disparate results are obtained for particular job classification, sampling should be repeated. It is advisable to sample each shift on more than one day to account for environmental variables (wind, precipitation, etc.) which may affect sampling. Differences in exposures among different work shifts may indicate a need to improve work practices on a particular shift. Sampling results from different shifts for each job classification should not be averaged. Multiple samples from same shift on each battery may be used to calculate an average exposure for a particular job classification.

B. *Analysis.*

1. All extraction glassware is cleaned with dichromic acid cleaning solution, rinsed with tap water, then dionized water, acetone, and allowed to dry completely. The glassware is rinsed with nanograde benzene before use. The Teflon cups are cleaned with benzene then with acetone.

2. Pre-weigh the 2 ml Teflon cups to one hundredth of a milligram (0.01 mg) on a autobalance AD 2 Tare weight of the cups is about 50 mg.

3. Place the silver membrane filter and glass fiber filter into a 15 ml test tube.

4. Extract with 5 ml of benzene for five minutes in an ultrasonic cleaner.

5. Filter the extract in 15 ml medium glass fritted funnels.

6. Rinse test tube and filters with two 1.5 ml aliquots of benzene and filter through the fritted glass funnel.

7. Collect the extract and two rinses in a 10 ml Kontes graduated evaporative concentrator.

8. Evaporate down to 1 ml while rinsing the sides with benzene.

9. Pipet 0.5 ml into the Teflon cup and evaporate to dryness in a vacuum oven at 40 °C for 3 hours.

10. Weigh the Teflon cup and the weight gain is due to the benzene soluble residue in half the Sample.

## II. MEDICAL SURVEILLANCE GUIDELINES

A. *General.* The minimum requirements for the medical examination for coke oven workers are given in paragraph (j) of the standard. The initial examination is to be provided to all coke oven workers who work at least 30 days in the regulated area. The examination includes a 14" X 17" posterior-anterior chest x-ray reading and a ILO/UC rating to assure some standardization of x-ray reading, pulmonary function tests (FVC and FEV 1.0), weight, urinalysis, skin examination, and a urinary cytologic examination. These tests are needed to serve as the baseline for comparing the employee's future test results. Periodic exams include all the elements of the initial exam, except that the urine cytologic test is to be performed only on those employees who are 45 years or older or who have worked for 5 or more years in the regulated area; periodic exams, with the exception of x-rays, are to be performed semiannually for this group instead of annually; for this group, x-rays will continue to be given at least annually. The examination contents are minimum requirements; additional tests such as lateral and oblique x-rays or additional pulmonary function tests may be performed if deemed necessary.

B. *Pulmonary function tests.*

Pulmonary function tests should be performed in a manner which minimizes subject and operator bias. There has been shown to be learning effects with regard to the results obtained from certain tests, such as FEV 1.0. Best results can be obtained by multiple trials for each subject. The best of three trials or the average of the last three of five trials may be used in obtaining reliable results. The type of equipment used (manufacturer, model, etc.) should be recorded with the results as reliability and accuracy varies and such information may be important in the evaluation of test results. Care should be exercised to obtain the best possible testing equipment.

[41 FR 46784, Oct. 22, 1976, as amended at 42 FR 3304, Jan. 18, 1977; 45 FR 35283, May 23, 1980; 50 FR 37353, 37354, Sept. 13, 1985; 54 FR 24334, June 7, 1989; 61 FR 5508, Feb. 13, 1996; 63 FR 1290, Jan. 8, 1998; 63 FR 33468, June 18, 1998]

## §1926.1144 1,2-dibromo-3-chloropropane.

**Note:** The requirements applicable to construction work under this section are identical to those set forth at §1910.1044 of this chapter.

[61 FR 31434, June 20, 1996]

[Section 1910.1044 is printed below—CCH.]

## §1910.1044 1,2-dibromo-3-chloropropane.

**1910.1044(a)** ***Scope and application.*** (1) This section applies to occupational exposure to 1,2-dibromo-3-chloropropane (DBCP).

**1910.1044(a)(2)** This section does not apply to:

**1910.1044(a)(2)(i)** Exposure to DBCP which results solely from the application and use of DBCP as a pesticide; or

**1910.1044(a)(2)(ii)** The storage, transportation, distribution or sale of DBCP in intact containers sealed in such a manner as to prevent exposure to DBCP vapors or liquid, except for the requirements of paragraphs (i), (n) and (o) of this section.

**1910.1044(b)** ***Definitions.*** *Authorized person* means any person required by his duties to be present in regulated areas and authorized to do so by his employer, by this section, or by the Act. *Authorized person* also includes any person entering such areas as a designated representative of employees exercising an opportunity to observe employee exposure monitoring.

*DBCP* means 1,2-dibromo-3-chloropropane, Chemical Abstracts Service Registry Number 96-12-8, and includes all forms of DBCP.

*Director* means the Director, National Institute for Occupational Safety and Health, U.S. Department of Health and Human Services, or designee.

*Emergency* means any occurrence such as, but not limited to equipment failure, rupture of containers, or failure of control equipment which may, or does, result in an unexpected release of DBCP.

*OSHA Area Office* means the Area Office of the Occupational Safety and Health Administration having jurisdiction over the geographic area where the affected workplace is located.

*Assistant Secretary* means the Assistant Secretary of Labor for Occupational Safety and Health, U.S. Department of Labor, or designee.

**1910.1044(c)** ***Permissible exposure limit***— (1) *Inhalation.* The employer shall assure that no employee is exposed to an airborne concentration of DBCP in excess of 1 part DBCP per billion parts of air (ppb) as an 8-hour time-weighted average.

**1910.1044(c)(2)** *Dermal and eye exposure.* The employer shall assure that no employee is exposed to eye or skin contact with DBCP.

**1910.1044(d)** ***Notification of use.*** Within ten (10) days following the introduction of DBCP into the workplace, every employer who has a workplace where DBCP is present, shall report the following information to the nearest OSHA Area Office for each such workplace;

**1910.1044(d)(1)** The address and location of the workplace:

**1910.1044(d)(2)** A brief description of each process or operation which may result in employee exposure to DBCP;

**1910.1044(d)(3)** The number of employees engaged in each process or operation who may be exposed to DBCP and an estimate of the frequency and degree of exposure that occurs; and

**1910.1044(d)(4)** A brief description of the employer's safety and health program as it relates to limitation of employee exposure to DBCP.

**1910.1044(e)** ***Regulated areas.*** (1) The employer shall establish, within each place of employment, regulated areas wherever DBCP concentrations are in excess of the permissible exposure limit.

**1910.1044(e)(2)** The employer shall limit access to regulated areas to authorized persons.

**1910.1044(f)** ***Exposure monitoring***— (1) *General.* (i) Determinations of airborne exposure levels shall be made from air samples that are representative of each employee's exposure to DBCP over an 8-hour period.

**1910.1044(f)(1)(ii)** For the purposes of this paragraph, employee exposure is that exposure which would occur if the employee were not using a respirator.

**1910.1044(f)(2)** *Initial.* Each employer who has a place of employment in which DBCP is present, shall monitor each workplace and work operation to accurately determine the airborne concentrations of DBCP to which employees may be exposed.

**1910.1044(f)(3)** *Frequency.* (i) If the monitoring required by this section reveals employee exposures to be below the permissible exposure limit, the employer shall repeat these measurements at least quarterly.

**1910.1044(f)(3)(ii)** If the monitoring required by this section reveals employee exposures to be in excess of the permissible exposure limit, the employer shall repeat these measurements for each such employee at least monthly. The employer shall continue monthly monitoring until at least two consecutive measurements, taken at least seven (7) days apart, are below the permissible exposure limit. Thereafter the employer shall monitor at least quarterly.

**1910.1044(f)(4)** *Additional.* Whenever there has been a production, process, control, or personnel change which may result in any new or additional exposure to DBCP, or whenever the employer has any reason to suspect new or additional exposures to DBCP, the employer shall monitor the employees potentially affected by such change for the purpose of redetermining their exposure.

**1910.1044(f)(5)** *Employee notification.* (i) Within five (5) working days after the receipt of monitoring results, the employer shall notify each employee in writing of the measurements which represent the employee's exposure.

**1910.1044(f)(5)(ii)** Whenever the results indicate that employee exposure exceeds the permissible exposure limit, the employer shall include in the written notice a statement that the permissible exposure limit was exceeded and a description of the corrective action being taken to reduce exposure to or below the permissible exposure limit.

**1910.1044(f)(6)** *Accuracy of measurement.* The employer shall use a method of measurement which has an accuracy, to a confidence level of 95 percent, of not less than plus or minus 25 percent for concentrations of DBCP at or above the permissible exposure limit.

**1910.1044(g)** ***Methods of compliance***— (1) *Priority of compliance methods.* The employer shall institute engineering and work practice controls to reduce and maintain employee exposures to DBCP at or below the permissible exposure limit, except to the extent that the employer establishes that such controls are not feasible. Where feasible engineering and work practice controls are not sufficient to reduce employee exposures to within the permissible exposure limit, the employer shall nonetheless use them to reduce exposures to the lowest level achievable by these controls, and shall supplement them by use of respiratory protection.

**1910.1044(g)(2)** *Compliance program.* (i) The employer shall establish and implement a written program to reduce employee exposures to DBCP to or below the permissible exposure limit solely by means of engineering and work practice controls as required by paragraph (g)(1) of this section.

**1910.1044(g)(2)(ii)** The written program shall include a detailed schedule for development and implementation of the engineering and work practice controls. These plans shall be revised at least every six months to reflect the current status of the program.

**1910.1044(g)(2)(iii)** Written plans for these compliance programs shall be submitted upon request to the Assistant Secretary and the Director, and shall be available at the worksite for examination and copying by the Assistant Secretary, the Director, and any affected employee or designated representative of employees.

**1910.1044(g)(2)(iv)** The employer shall institute and maintain at least the controls described in his most recent written compliance program.

**1910.1044(h)** ***Respiratory protection***— (1) *General.* For employees who are required to use respirators by this section, the employer must provide respirators that comply with the requirements of this paragraph. Respirators must be used during:

**1910.1044(h)(1)(i)** Periods necessary to install or implement feasible engineering and work-practice controls.

**1910.1044(h)(1)(ii)** Maintenance and repair activities for which engineering and work-practice controls are not feasible.

**1910.1044(h)(1)(iii)** Work operations for which feasible engineering and work-practice controls are not yet sufficient to reduce employee exposure to or below the permissible exposure limit.

**1910.1044(h)(1)(iv)** Emergencies.

**1910.1044(h)(2)** *Respirator program.* The employer must implement a respiratory protection program in accordance with 29 CFR 1910.134 (b) through (d) (except (d)(1)(iii)), and (f) through (m).

**1910.1044(h)(3)** *Respirator selection.* The employer must select the appropriate respirator from Table 1 of this section.

## TABLE 1—RESPIRATORY PROTECTION FOR DBCP

| Airborne concentration of DBCP or condition of use | Respirator type |
|---|---|
| (a) Less than or equal to 10 ppb. | (1) Any supplied-air respirator; or (2) any self-contained breathing apparatus. |
| (b) Less than or equal to 50 ppb. | (1) Any supplied-air respirator with full facepiece, helmet, or hood; or (2) any self-contained breathing apparatus with full facepiece. |
| (c) Less than or equal to 1,000 ppb. | (1) A Type C supplied-air respirator operated in pressure-demand or other positive pressure or continuous flow mode. |
| (d) Less than or equal to 2,000 ppb. | (1) A Type C supplied-air respirator with full facepiece operated in pressure-demand or other positive pressure mode, or with full facepiece, helmet, or hood operated in continuous flow mode. |
| (e) Greater than 2,000 ppb or entry and escape from unknown concentrations. | (1) A combination respirator which includes a Type C supplied-air respirator with full facepiece operated in pressure-demand or other positive pressure or continuous flow mode and an auxiliary self-contained breathing apparatus operated in pressure-demand or positive pressure mode; or (2) a self-contained breathing apparatus with full facepiece operated in pressure-demand or other positive pressure mode. |
| (f) Firefighting | (1) A self-contained breathing apparatus with full facepiece operated in pressure-demand or other positive pressure mode. |

**1910.1044(i)** ***Emergency situations***— (1) *Written plans.* (i) A written plan for emergency situations shall be developed for each workplace in which DBCP is present.

**1910.1044(i)(1)(ii)** Appropriate portions of the plan shall be implemented in the event of an emergency.

**1910.1044(i)(2)** Employees engaged in correcting emergency conditions shall be equipped as required in paragraphs (h) and (j) of this section until the emergency is abated.

**1910.1044(i)(3)** *Evacuation.* Employees not engaged in correcting the emergency shall be removed and restricted from the area and normal operations in the affected area shall not be resumed until the emergency is abated.

**1910.1044(i)(4)** *Alerting employees.* Where there is a possibility of employee exposure to DBCP due to the occurrence of an emergency, a general alarm shall be installed and maintained to promptly alert employees of such occurrences.

**1910.1044(i)(5)** *Medical surveillance.* For any employee exposed to DBCP in an emergency situation, the employer shall provide medical surveillance in accordance with paragraph (m)(6) of this section.

**1910.1044(i)(6)** *Exposure monitoring.* (i) Following an emergency, the employer shall conduct monitoring which complies with paragraph (f) of this section.

**1910.1044(i)(6)(ii)** In workplaces not normally subject to periodic monitoring, the employer may terminate monitoring when two consecutive measurements indicate exposures below the permissible exposure limit.

**1910.1044(j)** ***Protective clothing and equipments***— (1) *Provision and use.* Where there is any possibility of eye or dermal contact with liquid or solid DBCP, the employer shall provide, at no cost to the employee, and assure that the employee wears impermeable protective clothing and equipment to protect the area of the body which may come in contact with DBCP. Eye and face protection shall meet the requirements of §1910.133 of this part.

**1910.1044(j)(2)** *Removal and storage.* (i) The employer shall assure that employees remove DBCP contaminated work clothing only in change rooms provided in accordance with paragraph (l) (1) of this section.

**1910.1044(j)(2)(ii)** The employer shall assure that employees promptly remove any protective clothing and equipment which becomes contaminated with DBCP-containing liquids and solids. This clothing shall not be reworn until the DBCP has been removed from the clothing or equipment.

**1910.1044(j)(2)(iii)** The employer shall assure that no employee takes DBCP contaminated protective devices and work clothing out of the change room, except those employees authorized to do so for the purpose of laundering, maintenance, of disposal.

**1910.1044(j)(2)(iv)** DBCP-contaminated protective devices and work clothing shall be placed and stored in closed containers which prevent dispersion of the DBCP outside the container.

**1910.1044(j)(2)(v)** Containers of DBCP contaminated protective devices or work clothing which are to be taken out of change rooms or the workplace for cleaning, maintenance or disposal, shall bear labels in accordance with paragraph (o)(3) of this section.

**1910.1044(j)(3)** *Cleaning and replacement.* (i) The employer shall clean, launder, repair, or replace protective clothing and equipment required by this paragraph to maintain their effectiveness. The employer shall provide clean protective clothing and equipment at least daily to each affected employee.

**1910.1044(j)(3)(ii)** The employer shall inform any person who launders or clean DBCP-contaminated protective clothing or equipment of the potentially harmful effects of exposure to DBCP.

**1910.1044(j)(3)(iii)** The employer shall prohibit the removal of DBCP from protective clothing and equipment by blowing or shaking.

**1910.1044(k)** ***Housekeeping***— (1) *Surfaces.* (i) All workplace surfaces shall be maintained free of visible accumulations of DBCP.

**1910.1044(k)(1)(ii)** Dry sweeping and the use of compressed air for the cleaning of floors and other surfaces is prohibited where DBCP dusts or liquids are present.

**1910.1044(k)(1)(iii)** Where vacuuming methods are selected to clean floors and other surfaces, either portable units or a permanent system may be used.

**1910.1044(k)(1)(iii)*(a)*** If a portable unit is selected, the exhaust shall be attached to the general workplace exhaust ventilation system or collected within the vacuum unit, equipped with high efficiency filters or other appropriate means of contaminant removal, so that DBCP is not reintroduced into the workplace air; and

**1910.1044(k)(1)(iii)*(b)*** Portable vacuum units used to collect DBCP may not be used for other cleaning purposes and shall be labeled as prescribed by paragraph (o)(3) of this section.

**1910.1044(k)(1)(iv)** Cleaning of floors and other surfaces contaminated with DBCP-containing dusts shall not be performed by washing down with a hose, unless a fine spray has first been laid down.

**1910.1044(k)(2)** *Liquids.* Where DBCP is present in a liquid form, or as a resultant vapor, all containers or vessels containing DBCP shall be enclosed to the maximum extent feasible and tightly covered when not in use.

**1910.1044(k)(3)** *Waste disposal.* DBCP waste scrap, debris, containers or equipment, shall be disposed of in sealed bags or other closed containers which prevent dispersion of DBCP outside the container.

**1910.1044(l)** ***Hygiene facilities and practices***— (1) *Change rooms.* The employer shall provide clean change rooms equipped with storage facilities for street clothes and separate storage facilities for protective clothing and equipment whenever employees are required to wear protective clothing and equipment in accordance with paragraphs (h) and (j) of this section.

**1910.1044(l)(2)** *Showers.* (i) The employer shall assure that employees working in the regulated area shower at the end of the work shift.

**1910.1044(l)(2)(ii)** The employer shall assure that employees whose skin becomes contaminated with DBCP-containing liquids or solids immediately wash or shower to remove any DBCP from the skin.

**1910.1044(l)(2)(iii)** The employer shall provide shower facilities in accordance with 29 CFR 1910.141(d)(3).

**1910.1044(l)(3)** *Lunchrooms.* The employer shall provide lunchroom facilities which have a temperature controlled, positive pressure, filtered air supply, and which are readily accessible to employees working in regulated areas.

**1910.1044(l)(4)** *Lavatories.* (i) The employer shall assure that employees working in the regulated area remove protective clothing and wash their hands and face prior to eating.

**1910.1044(l)(4)(ii)** The employer shall provide a sufficient number of lavatory facilities which comply with 29 CFR 1910.141(d)(1) and (2).

**1910.1044(l)(5)** *Prohibition of activities in regulated areas.* The employer shall assure that, in regulated areas, food or beverages are not present or consumed, smoking products and implements are not present or used, and cosmetics are not present or applied.

**1910.1044(m)** ***Medical surveillance***— (1) *General.* (i) The employer shall make available a medical surveillance program for employees who work in regulated areas and employees who are subjected to DBCP exposures in an emergency situation.

**1910.1044(m)(1)(ii)** All medical examinations and procedures shall be performed by or under the supervision of a licensed physician, and shall be provided without cost to the employee.

**1910.1044(m)(2)** *Frequency and content.* At the time of initial assignment, and annually thereafter, the employer shall provide a medical examination for employees who work in regulated areas, which includes at least the following:

**1910.1044(m)(2)(i)** A medical and occupational history including reproductive history.

**1910.1044(m)(2)(ii)** A physical examination, including examination of the genito-urinary tract, testicle size and body habitus, including a determination of sperm count.

**1910.1044(m)(2)(iii)** A serum specimen shall be obtained and the following determinations made by radioimmunoassay techniques utilizing National Institutes of Health (NIH) specific antigen or one of equivalent sensitivity:

**1910.1044(m)(2)(iii)*(a)*** Serum follicle stimulating hormone (FSH);

**1910.1044(m)(2)(iii)*(b)*** Serum luteinizing hormone (LH); and

**1910.1044(m)(2)(iii)*(c)*** Serum total estrogen (females).

**1910.1044(m)(2)(iv)** Any other tests deemed appropriate by the examining physician.

**1910.1044(m)(3)** *Additional examinations.* If the employee for any reason develops signs or symptoms commonly associated with exposure to DBCP, the employer shall provide the employee with a medical examination which shall include those elements considered appropriate by the examining physician.

**1910.1044(m)(4)** *Information provided to the physician.* The employer shall provide the following information to the examining physician:

**1910.1044(m)(4)(i)** A copy of this regulation and its appendices;

**1910.1044(m)(4)(ii)** A description of the affected employee's duties as they relate to the employee's exposure;

**1910.1044(m)(4)(iii)** The level of DBCP to which the employee is exposed; and

**1910.1044(m)(4)(iv)** A description of any personal protective equipment used or to be used.

**1910.1044(m)(5)** *Physician's written opinion.* (i) For each examination under this section, the employer shall obtain and provide the employee with a written opinion from the examining physician which shall include:

**1910.1044(m)(5)(i)*(a)*** The results of the medical tests performed;

**1910.1044(m)(5)(i)*(b)*** The physician's opinion as to whether the employee has any detected medical condition which would place the employee at an increased risk of material impairment of health from exposure to DBCP; and

**1910.1044(m)(5)(i)*(c)*** Any recommended limitations upon the employee's exposure to DBCP or upon the use of protective clothing and equipment such as respirators.

**1910.1044(m)(5)(ii)** The employer shall instruct the physician not to reveal in the written opinion specific findings or diagnoses unrelated to occupational exposure.

**1910.1044(m)(6)** *Emergency situations.* If the employee is exposed to DBCP in an emergency situation, the employer shall provide the employee with a sperm count test as soon as practicable, or, if the employee has been vasectionized or is unable to produce a semen specimen, the hormone tests contained in paragraph (m)(2)(iii) of this section. The employer shall provide these same tests three months later.

**1910.1044(n)** ***Employee information and training***—(1) *Training program.* (i) The employer shall institute a training program for all employees who may be exposed to DBCP and shall assure their participation in such training program.

**1910.1044(n)(1)(ii)** The employer shall assure that each employee is informed of the following:

**1910.1044(n)(1)(ii)*(a)*** The information contained in Appendix A;

**1910.1044(n)(1)(ii)*(b)*** The quantity, location, manner of use, release or storage of DBCP and the specific nature of operations which could result in exposure to DBCP as well as any necessary protective steps;

**1910.1044(n)(1)(ii)*(c)*** The purpose, proper use, and limitations of respirators;

**1910.1044(n)(1)(ii)*(d)*** The purpose and description of the medical surveillance program required by paragraph (m) of this section; and

**1910.1044(n)(1)(ii)*(e)*** A review of this standard, including appendices.

**1910.1044(n)(2)** *Access to training materials.* (i) The employer shall make a copy of this standard and its appendices readily available to all affected employees.

**1910.1044(n)(2)(ii)** The employer shall provide, upon request, all materials relating to the employee information and training program to the Assistant Secretary and the Director.

**1910.1044(o)** ***Signs and labels***— (1) *General.* (i) The employer may use labels or signs required by other statutes, regulations, or ordinances in addition to or in combination with, signs and labels required by this paragraph.

**1910.1044(o)(1)(ii)** The employer shall assure that no statement appears on or near any sign or label required by this paragraph which contradicts or detracts from the required sign or label.

**1910.1044(o)(2)** *Signs.* (i) The employer shall post signs to clearly indicate all regulated areas. These signs shall bear the legend:

**DANGER**

**1,2-Dibromo-3-chloropropane**

**(Insert appropriate trade or common names)**

**CANCER HAZARD**

**AUTHORIZED PERSONNEL ONLY**

**RESPIRATOR REQUIRED**

**1910.1044(o)(3)** *Labels.* (i) The employer shall assure that precautionary labels are affixed to all containers of DBCP and of products containing DBCP in the workplace, and that the labels remain affixed when the DBCP or products containing DBCP are sold, distributed, or otherwise leave the employer's workplace. Where DBCP or products containing DBCP are sold, distributed or otherwise leave the employer's workplace bearing appropriate labels required by EPA under the regulations in 40 CFR Part 162, the labels required by this paragraph need not be affixed.

**1910.1044(o)(3)(ii)** The employer shall assure that the precautionary labels required by this paragraph are readily visible and legible. The labels shall bear the following legend:

**DANGER**

**1,2-Dibromo-3-chloropropane**

**CANCER HAZARD**

**1910.1044(p)** ***Recordkeeping***— (1) *Exposure monitoring.* (i) The employer shall establish and maintain an accurate record of all monitoring required by paragraph (f) of this section.

**1910.1044(p)(1)(ii)** This record shall include:

**1910.1044(p)(1)(ii)*(a)*** The dates, number, duration and results of each of the samples taken, including a description of the sampling procedure used to determine representative employee exposure;

**1910.1044(p)(1)(ii)*(b)*** A description of the sampling and analytical methods used;

**1910.1044(p)(1)(ii)*(c)*** Type of respiratory protective devices worn, if any; and

**1910.1044(p)(1)(ii)*(d)*** Name, social security number, and job classification of the employee monitored and of all other employees whose exposure the measurement is intended to represent.

**1910.1044(p)(1)(iii)** The employer shall maintain this record for at least 40 years or the duration of employment plus 20 years, whichever is longer.

**1910.1044(p)(2)** *Medical surveillance.* (i) The employer shall establish and maintain an accurate record for each employee subject to medical surveillance required by paragraph (m) of this section.

**1910.1044(p)(2)(ii)** This record shall include:

**1910.1044(p)(2)(ii)*(a)*** The name and social security number of the employee;

**1910.1044(p)(2)(ii)*(b)*** A copy of the physician's written opinion;

**1910.1044(p)(2)(ii)*(c)*** Any employee medical complaints related to exposure to DBCP;

**1910.1044(p)(2)(ii)*(d)*** A copy of the information provided the physician as required by paragraphs (m)(4)(ii) through (m)(4)(iv) of this section; and

**1910.1044(p)(2)(ii)*(e)*** A copy of the employee's medical and work history.

**1910.1044(p)(2)(iii)** The employer shall maintain this record for at least 40 years or the duration of employment plus 20 years, whichever is longer.

**1910.1044(p)(3)** *Availability.* (i) The employer shall assure that all records required to be maintained by this section be made available upon request to the Assistant Secretary and the Director for examination and copying.

**1910.1044(p)(3)(ii)** Employee exposure monitoring records and employee medical records required by this paragraph shall be provided upon request to employees, designated representatives, and the Assistant Secretary in accordance with 29 CFR 1910.20(a) through (e) and (g) through (i).

**1910.1044(p)(4)** *Transfer of records.* (i) If the employer ceases to do business, the successor employer shall receive and retain all records required to be maintained by paragraph (p) of this section for the prescribed period.

**1910.1044(p)(4)(ii)** If the employer ceases to do business and there is no successor employer to receive and retain the records for the prescribed period, the employer shall transmit these records by mail to the Director.

**10.1044(p)(4)(iii)** At the expiration of the retention period for the records required to be maintained under paragraph (p) of this section, the employer shall transmit these records by mail to the Director.

**1910.1044(p)(4)(iv)** The employer shall also comply with any additional requirements involving transfer of records set forth in 29 CFR 1910.20(h).

**1910.1044(q)** ***Observation of monitoring***— (1) Employee observation. The employer shall provide affected employees, or their designated representatives, with an opportunity to observe any monitoring of employee exposure to DBCP required by this section.

**1910.1044(q)(2)** *Observation procedures.* (i) Whenever observation of the measuring or monitoring of employee exposure to DBCP requires entry into an area where the use of protective clothing or equipment is required, the employer shall provide the observer with personal protective clothing or equipment required to be worn by employees working in the area, assure the use of such clothing and equipment, and require the observer to comply with all other applicable safety and health procedures.

**1910.1044(q)(2)(ii)** Without interfering with the monitoring or measurement, observers shall be entitled to:

**1910.1044(q)(2)(ii)*(a)*** Receive an explanation of the measurement procedures;

**1910.1044(q)(2)(ii)*(b)*** Observe all steps related to the measurement of airborne concentrations of DBCP performed at the place of exposure; and

**1910.1044(q)(2)(ii)*(c)*** Record the results obtained.

**1910.1044(r)** ***Appendices.*** The information contained in the appendices is not intended, by itself, to create any additional obligations not otherwise imposed or to detract from any existing obligation.

## APPENDIX A TO §1910.1044— SUBSTANCE SAFETY DATA SHEET FOR DBCP

### I. SUBSTANCE IDENTIFICATION

A. Synonyms and trades names: DBCP; Dibromochloropropane; Fumazone (Dow Chemical Company TM); Nemafume; Nemagon (Shell Chemical Co. TM); Nemaset; BBC 12; and OS 1879.

B. Permissible exposure:

1. *Airborne.* 1 part DBCP vapor per billion parts of air (1 ppb); time-weighted average (TWA) for an 8-hour workday.

2. *Dermal.* Eye contact and skin contact with DBCP are prohibited.

C. Appearance and odor: Technical grade DBCP is a dense yellow or amber liquid with a pungent odor. It may also appear in granular form, or blended in varying concentrations with other liquids.

D. Uses: DBCP is used to control nematodes, very small wormlike plant parasites, on crops including cotton, soybeans, fruits, nuts, vegetables and ornamentals.

### II. HEALTH HAZARD DATA

A. Routes of entry: Employees may be exposed:

1. Through inhalation (breathing);

2. Through ingestion (swallowing);

3. Skin contact; and

4. Eye contact.

B. Effects of exposure:

1. *Acute exposure.* DBCP may cause drowsiness, irritation of the eyes, nose, throat and skin, nausea and vomiting. In addition, overexposure may cause damage to the lungs, liver or kidneys.

2. *Chronic exposure.* Prolonged or repeated exposure to DBCP has been shown to cause sterility in humans. It also has been shown to produce cancer and sterility in laboratory animals and has been determined to constitute an increased risk of cancer in man.

3. *Reporting Signs and Symptoms.* If you develop any of the above signs or symptoms that you think are caused by exposure to DBCP, you should inform your employer.

### III. EMERGENCY FIRST AID PROCEDURES

A. *Eye exposure.* If DBCP liquid or dust containing DBCP gets into your eyes, wash your eyes immediately with large amounts of water, lifting the lower and upper lids occasionally. Get medical attention immediately. Contact lenses should not be worn when working with DBCP.

B. *Skin exposure.* If DBCP liquids or dusts containing DBCP get on your skin, immediately wash using soap or mild detergent and water. If DBCP liquids or dusts containing DBCP penetrate through your clothing, remove the clothing immediately and wash. If irritation is present after washing get medical attention.

C. *Breathing.* If you or any person breathe in large amounts of DBCP, move the exposed person to fresh air at once. If breathing has stopped, perform artificial respiration. Do not use mouth-to-mouth. Keep the affected person warm and at rest. Get medical attention as soon as possible.

D. *Swallowing.* When DBCP has been swallowed and the person is conscious, give the person large amounts of water immediately. After the water has been swallowed, try to get the person to vomit by having him touch the back of his throat with his finger. Do not make an unconscious person vomit. Get medical attention immediately.

E. *Rescue.* Notify someone. Put into effect the established emergency rescue procedures. Know the locations of the emergency rescue equipment before the need arises.

## IV. RESPIRATORS AND PROTECTIVE CLOTHING

A. *Respirators.* You may be required to wear a respirator in emergencies and while your employer is in the process of reducing DBCP exposures through engineering controls. If respirators are worn, they must have a National Institute for Occupational Safety and Health (NIOSH) approval label (Older respirators may have a Bureau of Mines Approval label). For effective protection, a respirator must fit your face and head snugly. The respirator should not be loosened or removed in work situations where its use is required. DBCP does not have a detectable odor except at 1,000 times or more above the permissible exposure limit. If you can smell DBCP while wearing a respirator, the respirator is not working correctly; go immediately to fresh air. If you experience difficulty breathing while wearing a respirator, tell your employer.

B. *Protective clothing.* When working with DBCP you must wear for your protection impermeable work clothing provided by your employer. (Standard rubber and neoprene protective clothing do not offer adequate protection).

DBCP must never be allowed to remain on the skin. Clothing and shoes must not be allowed to become contaminated with DBCP, and if they do, they must be promptly removed and not worn again until completely free of DBCP. Turn in impermeable clothing that has developed leaks for repair or replacement.

C. *Eye protection.* You must wear splash-proof safety goggles where there is any possibility of DBCP liquid or dust contacting your eyes.

## V. PRECAUTIONS FOR SAFE USE, HANDLING, AND STORAGE

A. DBCP must be stored in tightly closed containers in a cool, well-ventilated area.

B. If your work clothing may have become contaminated with DBCP, or liquids or dusts containing DBCP, you must change into uncontaminated clothing before leaving the work premises.

C. You must promptly remove any protective clothing that becomes contaminated with DBCP. This clothing must not be reworn until the DBCP is removed from the clothing.

D. If your skin becomes contaminated with DBCP, you must immediately and thoroughly wash or shower with soap or mild detergent and water to remove any DBCP from your skin.

E. You must not keep food, beverages, cosmetics, or smoking materials, nor eat or smoke, in regulated areas.

F. If you work in a regulated area, you must wash your hands thoroughly with soap or mild detergent and water, before eating, smoking or using toilet facilities.

G. If you work in a regulated area, you must remove any protective equipment or clothing before leaving the regulated area.

H. Ask your supervisor where DBCP is used in your work area and for any additional safety and health rules.

## VI. ACCESS TO INFORMATION

A. Each year, your employer is required to inform you of the information contained in this Substance Safety Data Sheet for DBCP. In addition, your employer must instruct you in the safe use of DBCP, emergency procedures, and the correct use of protective equipment.

B. Your employer is required to determine whether you are being exposed to DBCP. You or your representative have the right to observe employee exposure measurements and to record the result obtained. Your employer is required to inform you of your exposure. If your employer determines that you are being overexposed, he is required to inform you of the actions which are being taken to reduce your exposure.

C. Your employer is required to keep records of your exposure and medical examinations. Your employer is required to keep exposure and medical data for at least 40 years or the duration of your employment plus 20 years, whichever is longer.

D. Your employer is required to release exposure and medical records to you, your physician, or other individual designated by you upon your written request.

# APPENDIX B TO §1910.1044—SUBSTANCE TECHNICAL GUIDELINES FOR DBCP

## I. PHYSICAL AND CHEMICAL DATA

A. Substance Identification

1. Synonyms: 1,2-dibromo-3-chloropropane; DBCP, Fumazone; Nemafume; Nemagon; Nemaset; BBC 12; OS 1879. DBCP is also included in agricultural pesticides and fumigants which include the phrase "Nema—" in their name.

2. Formula: C3H5Br2 C1.

3. Molecular Weight: 236.

B. Physical Data:

1. Boiling point (760 mm HG): 195C (383F)

2. Specific gravity (water=1): 2.093.

3. Vapor density (air=1 at boiling point of DBCP): Data not available.

4. Melting point: 6C (43F).

5. Vapor pressure at 20C (68F): 0.8 mm Hg

6. Solubility in water: 1000 ppm.

7. Evaporation rate (Butyl Acetate=1): very much less than 1.

8. Appearance and odor: Dense yellow or amber liquid with a pungent odor at high concentrations. Any detectable odor of DBCP indicates overexposure.

## II. FIRE EXPLOSION AND REACTIVITY HAZARD DATA

A. Fire

1. Flash point: 170F (77C)

2. Autoignition temperature: Data not available.

3. Flammable limits in air, percent by volume: Data not available.

4. Extinguishing media: Carbon dioxide, dry chemical.

5. Special fire-fighting procedures: Do not use a solid stream of water since a stream will scatter and spread the fire. Use water spray to cool containers exposed to a fire.

6. Unusual fire and explosion hazards: None known.

7. For purposes of complying with the requirements of §1910.106, liquid DBCP is classified as a Class III A combustible liquid.

8. For the purpose of complying with §1910.309, the classification of hazardous locations as described in article 500 of the National Electrical Code for DBCP shall be Class I, Group D.

9. For the purpose of compliance with §1910.157, DBCP is classified as a Class B fire hazard.

10. For the purpose of compliance with §1910.178, locations classified as hazardous locations due to the presence of DBCP shall be Class I, Group D.

11. Sources of ignition are prohibited where DBCP presents a fire or explosion hazard.

B. Reactivity

1. Conditions contributing to instability: None known.

2. Incompatibilities: Reacts with chemically active metals, such as aluminum, magnesium and tin alloys.

3. Hazardous decomposition products: Toxic gases and vapors (such as HBr, HCl and carbon monoxide) may be released in a fire involving DBCP.

4. Special precautions: DBCP will attack some rubber materials and coatings.

## III. SPILL, LEAK AND DISPOSAL PROCEDURES

A. If DBCP is spilled or leaked, the following steps should be taken:

1. The area should be evacuated at once and re-entered only after thorough ventilation.

2. Ventilate area of spill or leak.

3. If in liquid form, collect for reclamation or absorb in paper, vermiculite, dry sand, earth or similar material.

4. If in solid form, collect spilled material in the most convenient and safe manner for reclamation or for disposal.

B. Persons not wearing protective equipment must be restricted from areas of spills or leaks until cleanup has been completed.

C. Waste Disposal Methods:

1. For small quantities of liquid DBCP, absorb on paper towels, remove to a safe place (such as a fume hood) and burn the paper. Large quantities can be reclaimed or collected and atomized in a suitable combustion chamber equipped with an appropriate effluent gas cleaning device. If liquid DBCP is absorbed in vermiculite, dry sand, earth or similar material and placed in sealed containers it may be disposed of in a Stateapproved sanitary landfill.

2. If in solid form, for small quantities, place on paper towels, remove to a safe place (such as a fume hood) and burn. Large quantities may be reclaimed. However, if this is not practical, dissolve in a flammable solvent (such as alcohol) and atomize in a suitable combustion chamber equipped with an appropriate effluent gas cleaning device. DBCP in solid form may also be disposed in a state-approved sanitary landfill.

## IV. MONITORING AND MEASUREMENT PROCEDURES

A. Exposure above the permissible exposure limit.

1. *Eight Hour Exposure Evaluation:* Measurements taken for the purpose of determining employee exposure under this section are best taken so that the average 8-hour exposure may be determined from a single 8-hour sample or two (2) 4-hour samples. Air samples should be taken in the employee's breathing zone (air that would most nearly represent that inhaled by the employee).

2. *Monitoring Techniques:* The sampling and analysis under this section may be performed by collecting the DBCP vapor on petroleum based charcoal absorption tubes with subsequent chemical analyses. The method of measurement chosen should determine the concentration of airborne DBCP at the permissible exposure limit to an accuracy of plus or minus 25 percent. If charcoal tubes are used, a total volume of 10 liters should be collected at a flow rate of 50 cc. per minute for each tube. Analyze the resultant samples as you would samples of halogenated solvent.

B. Since many of the duties relating to employee protection are dependent on the results of monitoring and measuring procedures, employers should assure that the evaluation of employee exposures is performed by a competent industrial hygienist or other technically qualified person.

## V. PROTECTIVE CLOTHING

Employees should be required to wear appropriate protective clothing to prevent any possibility of skin contact with DBCP. Because DBCP is absorbed through the skin, it is important to prevent skin contact with both liquid and solid forms of DBCP. Protective clothing should include impermeable coveralls or similar fullbody work clothing, gloves, headcoverings, and workshoes or shoe coverings. Standard rubber and neoprene gloves do not offer adequate protection and should not be relied upon to keep DBCP off the skin. DBCP should never be allowed to remain on the skin. Clothing and shoes should not be allowed to become contaminated with the material, and if they do, they should be promptly removed and not worn again until completely free of the material. Any protective clothing which has developed leaks or is otherwise found to be defective should be repaired or replaced. Employees should also be required to wear splash-proof safety goggles where there is any possibility of DBCP contacting the eyes.

## VI. HOUSEKEEPING AND HYGIENE FACILITIES

1. The workplace must be kept clean, orderly and in a sanitary condition;

2. Dry sweeping and the use of compressed air is unsafe for the cleaning of floors and other surfaces where DBCP dust or liquids are found. To minimize the contamination of air with dust, vacuuming with either portable or permanent systems must be used. If a portable unit is selected, the exhaust must be attached to the general workplace exhaust ventilation system, or collected within the vacuum unit equipped with high efficiency filters or other appropriate means of contamination removal and not used for other purposes. Units used to collect DBCP must be labeled.

3. Adequate washing facilities with hot and cold water must be provided, and maintained in a sanitary condition. Suitable cleansing agents should also be provided to assure the effective removal of DBCP from the skin.

4. Change or dressing rooms with individual clothes storage facilities must be provided to prevent the contamination of street clothes with DBCP. Because of the hazardous nature of DBCP, contaminated protective clothing must be stored in closed containers for cleaning or disposal.

## VII. MISCELLANEOUS PRECAUTIONS

A. Store DBCP in tightly closed containers in a cool, well ventilated area.

B. Use of supplied-air suits or other impervious clothing (such as acid suits) may be necessary to prevent skin contact with DBCP. Supplied-air suits should be selected, used, and maintained under the supervision of persons knowlegeable in the limitations and potential life-endangering characteristics of supplied-air suits.

C. The use of air-conditioned suits may be necessary in warmer climates.

D. Advise employees of all areas and operations where exposure to DBCP could occur.

## VIII. COMMON OPERATIONS

Common operations in which exposure to DBCP is likely to occur are: during its production; and during its formulation into pesticides and fumigants.

# APPENDIX C TO §1910.1044—MEDICAL SURVEILLANCE GUIDELINES FOR DBCP

## I. ROUTE OF ENTRY

Inhalation; skin absorption

## II. TOXICOLOGY

Recent data collected on workers involved in the manufacture and formulation of DBCP has shown that DBCP can cause sterility at very low levels of exposure. This finding is supported by studies showing that DBCP causes sterility in animals. Chronic exposure to DBCP resulted in pronounced necrotic action on the parenchymatous organs (i.e., liver, kidney, spleen) and on the testicles of rats at concentrations as low as 5 ppm. Rats that were chronically exposed to DBCP also showed changes in the composition of the blood, showing low RBC, hemoglobin, and WBC, and high reticulocyte levels as well as functional hepatic disturbance, manifesting itself in a long prothrombin time. Reznik et al. noted a single dose of 100 mg produced profound depression of the nervous system of rats. Their condition gradually improved. Acute exposure also resulted in the destruction of the sex gland activity of male rats as well as causing changes in the estrous cycle in female rats. Animal studies have also associated DBCP with an increased incidence of carcinoma. Olson, et al. orally administered DBCP to rats and mice 5 times per week at experimentally predetermined maximally tolerated doses and at half those doses. As early as ten weeks after initiation of treatment, DBCP induced a high incidence of squamous cell carcinomas of the stomach with metastases in both species. DBCP also induced mammary adenocarcinomas in the female rats at both dose levels.

## III. SIGNS AND SYMPTOMS

A. Inhalation: Nausea, eye irritation, conjunctivitis, respiratory irritation, pulmonary congestion or edema, CNS depression with apathy, sluggishness, and ataxia.

B. Dermal: Erythema or inflammation and dermatitis on repeated exposure.

## IV. SPECIAL TESTS

A. *Semen analysis:* The following information excerpted from the document "Evaluation of Testicular Function", submitted by the Corporate Medical Department of the Shell Oil Company (exhibit 39-3), may be useful to physicians conducting the medical surveillance program;

In performing semen analyses certain minimal but specific criteria should be met:

1. It is recommended that a minimum of three valid semen analyses be obtained in order to make a determination of an individual's average sperm count.

2. A period of sexual abstinence is necessary prior to the collection of each masturbatory sample. It is recommended that intercourse or masturbation be performed 48 hours before the actual specimen collection. A period of 48 hours of abstinence would follow; then the masturbatory sample would be collected.

3. Each semen specimen should be collected in a clean, widemouthed, glass jar (not necessarily pre-sterilized) in a manner designated by the examining physician. Any part of the seminal fluid exam should be initialed *only after liquifaction* is complete, i.e., 30 to 45 minutes after collection.

4. Semen volume should be measured to the nearest 1/10 of a cubic centimeter.

5. Sperm density should be determined using routine techniques involving the use of a white cell pipette and a hemocytometer chamber. The immobilizing fluid most effective and most easily obtained for this process is distilled water.

6. Thin, dry smears of the semen should be made for a morphologic classification of the sperm forms and should be stained with either hematoxalin or the more difficult, yet more precise, Papanicolaou technique. Also of importance to record is obvious sperm agglutination, pyospermia, delayed liquifaction (greater than 30 minutes), and hyperviscosity. In addition, pH, using nitrazine paper, should be determined.

7. A total morphology evaluation should include percentages of the following:

a. Normal (oval) forms,

b. Tapered forms,

c. Amorphous forms (include large and small sperm shapes),

d. Duplicated (either heads or tails) forms, and

e. Immature forms.

8. Each sample should be evaluated for sperm *viability* (percent viable sperm moving at the time of examination) as well as sperm *motility* (subjective characterization of "purposeful forward sperm progression" of the majority of those viable sperm analyzed) within two hours after collection, ideally by the same or equally qualified examiner.

B. *Serum determinations:* The following serum determinations should be performed by radioimmuno-assay techniques using National Institutes of Health (NIH) specific antigen or antigen preparations of equivalent sensitivity:

1. Serum follicle stimulating hormone (FSH);

2. Serum luteinizing hormone (LH); and

3. Serum total estrogen (females only).

## V. TREATMENT

Remove from exposure immediately, give oxygen or artificial resuscitation if indicated. Contaminated clothing and shoes should be removed immediately. Flush eyes and wash contaminated skin. If swallowed and the person is conscious, induce vomiting. Recovery from mild exposures is usually rapid and complete.

## VI. SURVEILLANCE AND PREVENTIVE CONSIDERATIONS

A. *Other considerations.* DBCP can cause both acute and chronic effects. It is important that the physician become familiar with the operating conditions in which exposure to DBCP occurs. Those with respiratory disorders may not tolerate the wearing of negative pressure respirators.

B. *Surveillance and screening.* Medical histories and laboratory examinations are required for each employee subject to exposure to DBCP. The employer should screen employees for history of certain medical conditions (listed below) which might place the employee at increased risk from exposure.

1. *Liver disease.* The primary site of biotransformation and detoxification of DBCP is the liver. Liver dysfunctions likely to inhibit the conjugation reactions will tend to promote the toxic actions of DBCP. These precautions should be considered before exposing persons with impaired liver function to DBCP.

2. *Renal disease.* Because DBCP has been associated with injury to the kidney it is important that special consideration be given to those with possible impairment of renal function.

3. *Skin desease.* DBCP can penetrate the skin and can cause erythema on prolonged exposure. Persons with pre-existing skin disorders may be more susceptible to the effects of DBCP.

4. *Blood dyscrasias.* DBCP has been shown to decrease the content of erythrocytes, hemoglobin, and leukocytes in the blood, as well as increase the prothrombin time. Persons with existing blood disorders may be more susceptible to the effects of DBCP.

5. *Reproductive disorders.* Animal studies have associated DBCP with various effects on the reproductive organs. Among these effects are atrophy of the testicles and changes in the estrous cycle. Persons with pre-existing reproductive disorders may be at increased risk to these effects of DBCP.

## REFERENCES

1. Reznik, Ya. B. and Sprinchan, G. K.: Experimental Data on the Gonadotoxic effect of Nemagon, *Gig. Sanit.*, (6), 1975, pp. 101-102, (translated from Russian).

2. Faydysh, E. V., Rakhmatullaev, N. N. and Varshavskii, V. A.: The Cytotoxic Action of Nemagon in a Subacute Experiment, *Med. Zh. Uzbekistana*, (No. 1), 1970, pp. 64-65, (translated from Russian).

3. Rakhmatullaev, N. N.: Hygienic Characteristics of the Nematocide Nemagon in Relation to Water Pollution Control, *Hyg. Sanit.*, 36(3), 1971, pp. 344-348, (translated from Russian).

4. Olson, W. A. *et al.:* Induction of Stomach Cancer in Rats and Mice by Halogenated Aliphatic Fumigants, *Journal of the National Cancer Institute*, (51), 1973, pp. 1993-1995.

5. Torkelson, T. R. *et al.:* Toxicologic Investigations of 1,2-Dibromo-3-chloropropane, *Toxicology and Applied Pharmacology, 3, 1961 pp. 545-559.*

[43 FR 11527, Mar. 17, 1978, as amended at 45 FR 35283, May 23, 1980; 49 FR 18295, Apr. 30, 1984; 54 FR 24334, June 7, 1989; 58 FR 35310, June 30, 1993; 61 FR 5508, Feb. 13, 1996; 63 FR 1291, Jan. 8, 1998]

### §1926.1145 Acrylonitrile

**Note:** The requirements applicable to construction work under this section are identical to those set forth at §1910.1045 of this chapter.

[61 FR 31434, June 20, 1996]

[Section 1910.1045 is printed below—CCH.]

### §1910.1045 Acrylonitrile.

1910.1045(a) ***Scope and application.*** (1) This section applies to all occupational exposures to acrylonitrile (AN), Chemical Abstracts Service Registry No. 000107131, except as provided in paragraphs (a)(2) and (a)(3) of this section.

1910.1045(a)(2) This section does not apply to exposures which result solely from the processing, use, and handling of the following materials:

1910.1045(a)(2)(i) ABS resins, SAN resins, nitrile barrier resins, solid nitrile elastomers, and acrylic and modacrylic fibers, when these listed materials are in the form of finished polymers, and products fabricated from such finished polymers;

**1910.1045(a)(2)(ii)** Materials made from and/or containing AN for which objective data is reasonably relied upon to demonstrate that the material is not capable of releasing AN in airborne concentrations in excess of 1 ppm as an eight (8)-hour time-weighted average, under the expected conditions of processing, use, and handling which will cause the greatest possible release; and

**1910.1045(a)(2)(iii)** Solid materials made from and/or containing AN which will not be heated above 170 °F during handling, use, or processing.

**1910.1045(a)(3)** An employer relying upon exemption under paragraph (a)(2)(ii) shall maintain records of the objective data supporting that exemption, and of the basis of the employer's reliance on the data, as provided in paragraph (q) of this section.

**1910.1045(b)** *Definitions. Acrylonitrile* or *AN* means acrylonitrile monomer, chemical formula $CH_2$=CHCN.

*Action level* means a concentration of AN of 1 ppm as an eight (8)-hour time-weighted average.

*Assistant Secretary* means the Assistant Secretary of Labor for Occupational Safety and Health, U.S. Department of Labor, or designee.

*Authorized person* means any person specifically authorized by the employer whose duties require the person to enter a regulated area, or any person entering such an area as a designated representative of employees for the purpose of exercising the opportunity to observe monitoring procedures under paragraph (r) of this section.

*Decontamination* means treatment of materials and surfaces by water washdown, ventilation, or other means, to assure that the materials will not expose employees to airborne concentrations of AN above 1 means the Director, National Institute for Occupational Safety and Health, U.S. Department of Health and Human Services, or designee.

*Emergency* means any occurrence such as, but not limited to, equipment failure, rupture of containers, or failure of control equipment, which results in an unexpected massive release of AN.

*Liquid AN* means AN monomer in liquid form, and liquid or semiliquid polymer intermediates, including slurries, suspensions, emulsions, and solutions, produced during the polymerization of AN.

*OSHA Area Office* means the Area Office of the Occupational Safety and Health Administration having jurisdiction over the geographic area where the affected workplace is located.

**1910.1045(c)** ***Permissible exposure limits***—(1) *Inhalation.* (i) *Time weighted average limit (TWA).* The employer shall assure that no employee is exposed to an airborne concentration of acrylonitrile in excess of two (2) parts acrylonitrile per million parts of air (2 ppm) as an eight (8)-hour time-weighted average.

**1910.1045(c)(1)(ii)** *Ceiling limit.* The employer shall assure that no employee is exposed to an airborne concentration of acrylonitrile in excess of ten (10) ppm as averaged over any fifteen (15)-minute period during the work day.

**1910.1045(c)(2)** *Dermal and eye exposure.* The employer shall assure that no employee is exposed to skin contact or eye contact with liquid AN.

**1910.1045(d)** ***Notification of regulated areas and emergencies***— (1) *Regulated areas.* Within thirty (30) days following the establishment of a regulated area pursuant to paragraph (f) of this section, the employer shall report the following information to the OSHA Area Office:

**1910.1045(d)(1)(i)** The address and location of each establishment which has one or more regulated areas;

**1910.1045(d)(1)(ii)** The locations, within the establishment, of each regulated area;

**1910.1045(d)(1)(iii)** A brief description of each process or operation which results in employee exposure to AN in regulated areas; and

**1910.1045(d)(1)(iv)** The number of employees engaged in each process or operation within each regulated area which results in exposure to AN, and an estimate of the frequency and degree of exposure that occurs.

Whenever there has been a significant change in the information required to be reported by this paragraph, the employer shall promptly provide the new information to the OSHA Area Office.

**1910.1045(d)(2)** *Emergencies.* Emergencies, and the facts obtainable at that time, shall be reported within seventy-two (72) hours of the initial occurrence to the OSHA Area Office. Upon request of the OSHA Area Office; the employer shall submit additional information in writing relevant to the nature and extent of employee exposures and measures taken to prevent future emergencies of a similar nature.

**1910.1045(e)** ***Exposure monitoring***— (1) *General.* (i) Determinations of airborne exposure levels shall be made from air samples that are representative of each employee's exposure to AN over an eight (8)-hour period.

**1910.1045(e)(1)(ii)** For the purposes of this section, employee exposure is that exposure which would occur if the employee were not using a respirator.

**1910.1045(e)(2)** *Initial monitoring.* Each employer who has a place of employment in which AN is present shall monitor each such workplace and work operation to accurately determine the airborne concentrations of AN to which employees may be exposed.

**1910.1045(e)(3)** *Frequency.* (i) If the monitoring required by this section reveals employee exposure to be below the action level, the employer may discontinue monitoring for that employee.

**1910.1045(e)(3)(ii)** If the monitoring required by this section reveals employee exposure to be at or above the action level but below the permissible exposure limits, the employer shall repeat such monitoring for each such employee at least quarterly. The employer shall continue these quarterly measurements until at least two consecutive measurements taken at least seven (7) days apart, are below the action level, and thereafter the employer may discontinue monitoring for that employee.

**1910.1045(e)(3)(iii)** If the monitoring required by this section reveals employee exposure to be in excess of the permissible exposure limits, the employer shall repeat these determinations for each such employee at least monthly. The employer shall continue these monthly measurements until at least two consecutive measurements, taken at least seven (7) days apart, are below the permissible exposure limits, and thereafter the employer shall monitor at least quarterly.

**1910.1045(e)(4)** *Additional monitoring.* Whenever there has been a production, process, control, or personnel change which may result in new or additional exposures to AN, or whenever the employer has any other reason to suspect a change which may result in new or additional exposures to AN, additional monitoring which complies with this paragraph shall be conducted.

**1910.1045(e)(5)** *Employee notification.* (i) Within five (5) working days after the receipt of the results of monitoring required by this paragraph, the employer shall notify each employee in writing of the results which represent that employee's exposure.

**1910.1045(e)(5)(ii)** Whenever the results indicate that the representative employee exposure exceeds the permissible exposure limits, the employer shall include in the written notice a statement that the permissible exposure limits were exceeded and a description of the corrective action being taken to reduce exposure to or below the permissible exposure limits.

**1910.1045(e)(6)** *Accuracy of measurement.* The method of measurement of employee exposures shall be accurate to a confidence level of 95 percent, to within plus or minus 35 percent for concentrations of AN at or above the permissible exposure limits, and plus or minus 50 percent for concentrations of AN below the permissible exposure limits.

**1910.1045(f)** ***Regulated areas.*** (1) The employer shall establish regulated areas where AN concentrations are in excess of the permissible exposure limits.

**1910.1045(f)(2)** Regulated areas shall be demarcated and segregated from the rest of the workplace, in any manner that minimizes the number of persons who will be exposed to AN.

**1910.1045(f)(3)** Access to regulated areas shall be limited to authorized persons or to persons otherwise authorized by the act or regulations issued pursuant thereto.

**1910.1045(f)(4)** The employer shall assure that food or beverages are not present or consumed, tobacco products are not present or used, and cosmetics are not applied in the regulated area.

**1910.1045(g)** ***Methods of compliance***— (1) *Engineering and work practice controls.* (i) By November 2, 1980, the employer shall institute engineering and work practice controls to reduce and maintain employee exposures to AN, to or below the permissible exposure limits, except to the extent that the employer establishes that such controls are not feasible.

**1910.1045(g)(1)(ii)** Wherever the engineering and work practice controls which can be instituted are not sufficient to reduce employee exposures to or below the permissible exposure limits, the employer shall nonetheless use them to reduce exposures to the lowest levels achievable by these controls, and shall supplement them by the use of respiratory protection which complies with the requirements of paragraph (h) of this section.

**1910.1045(g)(2)** *Compliance program.* (i) The employer shall establish and implement a written program to reduce employee exposures to or below the permissible exposure limits solely by means of engineering and work practice controls, as required by paragraph (g)(1) of this section.

**1910.1045(g)(2)(ii)** Written plans for these compliance programs shall include at least the following:

**1910.1045(g)(2)(ii)(A)** A description of each operation or process resulting in employee exposure to AN above the permissible exposure limits;

**1910.1045(g)(2)(ii)(B)** An outline of the nature of the engineering controls and work practices to be applied to the operation or process in question;

**1910.1045(g)(2)(ii)(C)** A report of the technology considered in meeting the permissible exposure limits;

**1910.1045(g)(2)(ii)(D)** A schedule for implementation of engineering and work practice controls for the operation or process, which shall project completion no later than November 2, 1980; and

**1910.1045(g)(2)(ii)(E)** Other relevant information.

**1910.1045(g)(2)(iii)** The employer shall complete the steps set forth in the compliance program by the dates in the schedule.

**1910.1045(g)(2)(iv)** Written plans shall be submitted upon request to the Assistant Secretary and the Director, and shall be available at the worksite for examination and copying by the Assistant Secretary, the Director, or any affected employee or representative.

**1910.1045(g)(2)(v)** The plans required by this paragraph shall be revised and updated at least every six (6) months to reflect the current status of the program.

**1910.1045(h)** ***Respiratory protection***— (1) *General.* For employees who use respirators required by this section, the employer must provide respirators that comply with the requirements of this paragraph. Respirators must be used during:

**1910.1045(h)(1)(i)** Periods necessary to install or implement feasible engineering and work-practice controls.

**1910.1045(h)(1)(ii)** Work operations, such as maintenance and repair activities or reactor cleaning, for which the employer establishes that engineering and work-practice controls are not feasible.

**1910.1045(h)(1)(iii)** Work operations for which feasible engineering and work-practice controls are not yet sufficient to reduce employee exposure to or below the permissible exposure limits.

**1910.1045(h)(1)(iv)** Emergencies.

**1910.1045(h)(2)** *Respirator program.* (i) The employer must implement a respiratory protection program in accordance with 29 CFR 1910.134(b) through (d) (except (d)(1)(iii), (d)(3)(iii)(B)(1), and (2)), and (f) through (m).

**1910.1045(h)(2)(ii)** If air-purifying respirators (chemical-cartridge or chemical-canister types) are used:

**1910.1045(h)(2)(ii)(A)** The air-purifying canister or cartridge must be replaced prior to the expiration of its service life or at the completion of each shift, whichever occurs first.

**1910.1045(h)(2)(ii)(B)** A label must be attached to the cartridge or canister to indicate the date and time at which it is first installed on the respirator.

**1910.1045(h)(3)** *Respirator selection.* The employer must select the appropriate respirator from Table I of this section.

## TABLE I—RESPIRATORY PROTECTION FOR ACRYLONITRILE (AN)

| Concentration of AN or condition of use | Respirator type |
|---|---|
| (a) Less than or equal to 20 ppm. | (1) Chemical cartridge respirator with organic vapor cartridge(s) and half-mask facepiece; or<br>(2) Supplied air respirator with half-mask facepiece. |
| (b) Less than or equal to 100 ppm or maximum use concentration (MUC) of cartridges or canisters, whichever is lower. | (1) Full facepiece respirator with (A) organic vapor cartridges, (B) organic vapor gas mask chin-style, or (C) organic vapor gas mask canister, front-or back-mounted;<br>(2) Supplied air respirator with full facepiece; or<br>(3) Self-contained breathing apparatus with full facepiece. |
| (c) Less than or equal to 4,000 ppm. | (1) Supplied air respirator operated in the positive pressure mode with full facepiece, helmet, suit, or hood. |
| (d) Greater than 4,000 ppm or unknown concentration. | (1) Supplied air and auxiliary self-contained breathing apparatus with full facepiece in positive pressure mode; or<br>(2) Self-contained breathing apparatus with full facepiece in positive pressure mode. |
| (e) Firefighting | Self-contained breathing apparatus with full facepiece in positive pressure mode. |
| (f) Escape | (1) Any organic vapor respirator, or<br>(2) Any self-contained breathing apparatus. |

**1910.1045(i)** ***Emergency situations***— (1) *Written plans.* (i) A written plan for emergency situations shall be developed for each workplace where liquid AN is present. Appropriate portions of the plan shall be implemented in the event of an emergency.

**1910.1045(i)(1)(ii)** The plan shall specifically provide that employees engaged in correcting emergency conditions shall be equipped as required in paragraph (h) of this section until the emergency is abated.

**1910.1045(i)(1)(iii)** Employees not engaged in correcting the emergency shall be evacuated from the area and shall not be permitted to return until the emergency is abated.

**1910.1045(i)(2)** *Alerting employees.* Where there is the possibility of employee exposure to AN in excess of the ceiling limit, a general alarm shall be installed and used to promptly alert employees of such occurrences.

**1910.1045(j)** ***Protective clothing and equipment***— (1) *Provision and use.* Where eye or skin contact with liquid AN may occur, the employer shall provide at no cost to the employee, and assure that employees wear, impermeable protective clothing or other equipment to protect any area of the body which may come in contact with liquid AN. The provision of §§1910.132 and 1910.133 shall be complied with.

1910.1045(j)(2) *Cleaning and replacement.* (i) The employer shall clean, launder, maintain, or replace protective clothing and equipment required by this section as needed to maintain their effectiveness.

1910.1045(j)(2)(ii) The employer shall assure that impermeable protective clothing which contacts or is likely to have contacted liquid AN shall be decontaminated before being removed by the employee.

1910.1045(j)(2)(iii) The employer shall assure that an employee whose nonimpermeable clothing becomes wetted with liquid AN shall immediately remove that clothing and proceed to shower. The clothing shall be decontaminated before it is removed from the regulated area.

1910.1045(j)(2)(iv) The employer shall assure that no employee removes protective clothing or equipment from the change room, except for those employees authorized to do so for the purpose of laundering, maintenance, or disposal.

1910.1045(j)(2)(v) The employer shall inform any person who launders or cleans protective clothing or equipment of the potentially harmful effects of exposure to AN.

1910.1045(k) ***Housekeeping.*** (1) All surfaces shall be maintained free of visible accumulations of liquid AN.

1910.1045(k)(2) For operations involving liquid AN, the employer shall institute a program for detecting leaks and spills of liquid AN, including regular visual inspections.

1910.1045(k)(3) Where spills of liquid AN are detected, the employer shall assure that surfaces contacted by the liquid AN are decontaminated. Employees not engaged in decontamination activities shall leave the area of the spill, and shall not be permitted in the area until decontamination is completed.

1910.1045(l) ***Waste disposal.*** AN waste, scrap, debris, bags, containers, or equipment shall be decontaminated before being incorporated in the general waste disposal system.

1910.1045(m) ***Hygiene facilities and practices.*** (1) Where employees are exposed to airborne concentrations of AN above the permissible exposure limits, or where employees are required to wear protective clothing or equipment pursuant to paragraph (j) of this section, the facilities required by 29 CFR 1910.141, including clean change rooms and shower facilities, shall be provided by the employer for the use of those employees, and the employer shall assure that the employees use the facilities provided.

1910.1045(m)(2) The employer shall assure that employees wearing protective clothing or equipment for protection from skin contact with liquid AN shall shower at the end of the work shift.

1910.1045(m)(3) The employer shall assure that, in the event of skin or eye exposure to liquid AN, the affected employee shall shower immediately to minimize the danger of skin absorption.

1910.1045(m)(4) The employer shall assure that employees working in the regulated area wash their hands and faces prior to eating.

1910.1045(n) ***Medical surveillance***— (1) *General.* (i) The employer shall institute a program of medical surveillance for each employee who is or will be exposed to AN at or above the action level, without regard to the use of respirators. The employer shall provide each such employee with an opportunity for medical examinations and tests in accordance with this paragraph.

1910.1045(n)(1)(ii) The employer shall assure that all medical examinations and procedures are performed by or under the supervision of a licensed physician, and that they shall be provided without cost to the employee.

1910.1045(n)(2) *Initial examinations.* At the time of initial assignment, or upon institution of the medical surveillance program, the employer shall provide each affected employee an opportunity for a medical examination, including at least the following elements:

1910.1045(n)(2)(i) A work history and medical history with special attention to skin, respiratory, and gastrointestinal systems, and those nonspecific symptoms, such as headache, nausea, vomiting, dizziness, weakness, or other central nervous system dysfunctions that may be associated with acute or with chronic exposure to AN;

1910.1045(n)(2)(ii) A complete physical examination giving particular attention to the peripheral and central nervous system, gastrointestinal system, respiratory system, skin, and thyroid;

1910.1045(n)(2)(iii) A 14- by 17-inch posteroanterior chest X-ray; and

1910.1045(n)(2)(iv) Further tests of the intestinal tract, including fecal occult blood screening, for all workers 40 years of age or older, and for any other affected employees for whom, in the opinion of the physician, such testing is appropriate.

1910.1045(n)(3) *Periodic examinations.* (i) The employer shall provide the examinations specified in paragraph (n)(2) of this section at least annually for all employees specified in paragraph (n)(1) of this section.

1910.1045(n)(3)(ii) If an employee has not had the examination specified in paragraph (n)(2) of this section within 6 months preceding termination of employment, the employer shall make such examination available to the employee prior to such termination.

1910.1045(n)(4) *Additional examinations.* If the employee for any reason develops signs or symptoms which may be associated with exposure to AN, the employer shall provide an appropriate examination and emergency medical treatment.

1910.1045(n)(5) *Information provided to the physician.* The employer shall provide the following information to the examining physician:

1910.1045(n)(5)(i) A copy of this standard and its appendixes;

1910.1045(n)(5)(ii) A description of the affected employee's duties as they relate to the employee's exposure;

1910.1045(n)(5)(iii) The employee's representative exposure level;

1910.1045(n)(5)(iv) The employee's anticipated or estimated exposure level (for preplacement examinations or in cases of exposure due to an emergency);

1910.1045(n)(5)(v) A description of any personal protective equipment used or to be used; and

1910.1045(n)(5)(vi) Information from previous medical examinations of the affected employee, which is not otherwise available to the examining physician.

1910.1045(n)(6) *Physician's written opinion.* (i) The employer shall obtain a written opinion from the examining physician which shall include:

1910.1045(n)(6)(i)(A) The results of the medical examination and test performed;

1910.1045(n)(6)(i)(B) The physician's opinion as to whether the employee has any detected medical condition(s) which would place the employee at an increased risk of material impairment of the employee's health from exposure to AN;

1910.1045(n)(6)(i)(C) Any recommended limitations upon the employee's exposure to AN or upon the use of protective clothing and equipment such as respirators; and

1910.1045(n)(6)(i)(D) A statement that the employee has been informed by the physician of the results of the medical examination and any medical conditions which require further examination or treatment.

1910.1045(n)(6)(ii) The employer shall instruct the physician not to reveal in the written opinion specific findings or diagnoses unrelated to occupational exposure to AN.

1910.1045(n)(6)(iii) The employer shall provide a copy of the written opinion to the affected employee.

1910.1045(o) ***Employee information and training***— (1) *Training program.* (i) By January 2, 1979, the employer shall institute a training program for and assure the participation of all employees exposed to AN above the action level, all employees whose exposures are maintained below the action level by engineering and work practice controls, and all employees subject to potential skin or eye contact with liquid AN.

1910.1045(o)(1)(ii) Training shall be provided at the time of initial assignment, or upon institution of the training program, and at least annually thereafter, and the employer shall assure that each employee is informed of the following:

1910.1045(o)(1)(ii)(A) The information contained in appendixes A and B;

1910.1045(o)(1)(ii)(B) The quantity, location, manner of use, release, or storage of AN, and the specific nature of operations which could result in exposure to AN, as well as any necessary protective steps;

**1910.1045(o)(1)(ii)(C)** The purpose, proper use, and limitations of respirators and protective clothing;

**1910.1045(o)(1)(ii)(D)** The purpose and a description of the medical surveillance program required by paragraph (n) of this section;

**1910.1045(o)(1)(ii)(E)** The emergency procedures developed, as required by paragraph (i) of this section;

**1910.1045(o)(1)(ii)(F)** Engineering and work practice controls, their function, and the employee's relationship to these controls; and

**1910.1045(o)(1)(ii)(G)** A review of this standard.

**1910.1045(o)(2)** *Access to training materials.* (i) The employer shall make a copy of this standard and its appendixes readily available to all affected employees.

**1910.1045(o)(2)(ii)** The employer shall provide, upon request, all materials relating to the employee information and training program to the Assistant Secretary and the Director.

**1910.1045(p)** ***Signs and labels***— (1) *General.* (i) The employer may use labels or signs required by other statutes, regulations, or ordinances in addition to, or in combination with, signs and labels required by this paragraph.

**1910.1045(p)(1)(ii)** The employer shall assure that no statement appears on or near any sign or label required by this paragraph which contradicts or detracts from the required sign or label.

**1910.1045(p)(2)** *Signs.* (i) The employer shall post signs to clearly indicate all workplaces where AN concentrations exceed the permissible exposure limits. The signs shall bear the following legend:

**DANGER**

**ACRYLONITRILE (AN)**

**CANCER HAZARD**

**AUTHORIZED PERSONNEL ONLY**

**RESPIRATORS MAY BE**

**REQUIRED**

**1910.1045(p)(2)(ii)** The employer shall assure that signs required by this paragraph are illuminated and cleaned as necessary so that the legend is readily visible.

**1910.1045(p)(3)** *Labels.* (i) The employer shall assure that precautionary labels are affixed to all containers of liquid AN and AN-based materials not exempted under paragraph (a)(2) of this standard. The employer shall assure that the labels remain affixed when the materials are sold, distributed, or otherwise leave the employer's workplace.

**1910.1045(p)(3)(ii)** The employer shall assure that the precautionary labels required by this paragraph are readily visible and legible. The labels shall bear the following legend:

**DANGER**

**CONTAINS ACRYLONITRILE (AN)**

**CANCER HAZARD**

**1910.1045(q)** ***Recordkeeping***— (1) *Objective data for exempted operations.* (i) Where the processing, use, and handling of materials made from or containing AN are exempted pursuant to paragraph (a)(2)(ii) of this section, the employer shall establish and maintain an accurate record of objective data reasonably relied upon in support of the exemption.

**1910.1045(q)(1)(ii)** This record shall include at least the following information:

**1910.1045(q)(1)(ii)(A)** The material qualifying for exemption;

**1910.1045(q)(1)(ii)(B)** The source of the objective data;

**1910.1045(q)(1)(ii)(C)** The testing protocol, results of testing, and/or analysis of the material for the release of AN;

**1910.1045(q)(1)(ii)(D)** A description of the operation exempted and how the data supports the exemption; and

**1910.1045(q)(1)(ii)(E)** Other data relevant to the operations, materials, and processing covered by the exemption.

**1910.1045(q)(1)(iii)** The employer shall maintain this record for the duration of the employer's reliance upon such objective data.

**1910.1045(q)(2)** *Exposure monitoring.* (i) The employer shall establish and maintain an accurate record of all monitoring required by paragraph (e) of this section.

**1910.1045(q)(2)(ii)** This record shall include:

**1910.1045(q)(2)(ii)(A)** The dates, number, duration, and results of each of the samples taken, including a description of the sampling procedure used to determine representative employee exposure;

**1910.1045(q)(2)(ii)(B)** A description of the sampling and analytical methods used and the data relied upon to establish that the methods used meet the accuracy and precision requirements of paragraph (e)(6) of this section;

**1910.1045(q)(2)(ii)(C)** Type of respiratory protective devices worn, if any; and

**1910.1045(q)(2)(ii)(D)** Name, social security number, and job classification of the employee monitored and of all other employees whose exposure the measurement is intended to represent.

**1910.1045(q)(2)(iii)** The employer shall maintain this record for at least forty (40) years, or for the duration of employment plus twenty (20) years, whichever is longer.

**1910.1045(q)(3)** *Medical surveillance.* (i) The employer shall establish and maintain an accurate record for each employee subject to medical surveillance as required by paragraph (n) of this section.

**1910.1045(q)(3)(ii)** This record shall include:

**1910.1045(q)(3)(ii)(A)** A copy of the physician's written opinions;

**1910.1045(q)(3)(ii)(B)** Any employee medical complaints related to exposure to AN;

**1910.1045(q)(3)(ii)(C)** A copy of the information provided to the physician as required by paragraph (n)(5) of this section; and

**1910.1045(q)(3)(ii)(D)** A copy of the employee's medical and work history.

**1910.1045(q)(3)(iii)** The employer shall assure that this record be maintained for at least forty (40) years, or for the duration of employment plus twenty (20) years, whichever is longer.

**1910.1045(q)(4)** *Availability.* (i) The employer shall make all records required to be maintained by this section available, upon request, to the Assistant Secretary and the Director for examination and copying.

**1910.1045(q)(4)(ii)** Records required by paragraphs (q)(1) through (q)(3) of this section shall be provided upon request to employees, designated representatives, and the Assistant Secretary in accordance with 29 CFR 1910.20(a) through (e) and (q) through (i). Records required by paragraph (q)(1) shall be provided in the same manner as exposure monitoring records.

**1910.1045(q)(5)** *Transfer of records.* (i) Whenever the employer ceases to do business, the successor employer shall receive and retain all records required to be maintained by this section for the prescribed period.

**1910.1045(q)(5)(ii)** Whenever the employer ceases to do business and there is no successor employer to receive and retain the records for the prescribed period, these records shall be transmitted to the Director.

**1910.1045(q)(5)(iii)** At the expiration of the retention period for the records required to be maintained pursuant to this section, the employer shall notify the Director at least 3 months prior to the disposal of the records, and shall transmit them to the Director upon request.

**1910.1045(q)(5)(iv)** The employer shall also comply with any additional requirements involving transfer of records set forth in 29 CFR 1910.20(h).

**1910.1045(r)** ***Observation of monitoring***— (1) *Employee observation.* The employer shall provide affected employees, or their designated representatives, an opportunity to observe any monitoring of employee exposure to AN conducted pursuant to paragraph (e) of this section.

**1910.1045(r)(2)** *Observation procedures.* (i) Whenever observation of the monitoring of employee exposure to AN requires entry into an area where the use of protective clothing or equipment is required, the employer shall provide the observer with personal protective clothing and equipment required to be worn by employees working in the area, assure the use of such clothing and equipment, and require the observer to comply with all other applicable safety and health procedures.

**1910.1045(r)(2)(ii)** Without interfering with the monitoring, observers shall be entitled:

**1910.1045(r)(2)(ii)(A)** To receive an explanation of the measurement procedures;

**1910.1045(r)(2)(ii)(B)** To observe all steps related to the measurement of airborne concentrations of AN performed at the place of exposure; and

**1910.1045(r)(2)(ii)(C)** To record the results obtained.

**1910.1045(s)** ***Effective date:*** (1) This section shall become effective November 2, 1978.

**1910.1045(s)(2)** Monitoring and medical surveillance conducted since January 17, 1978, under the provisions of the emergency temporary standard may be relied upon by the employer to meet the initial monitoring and initial medical surveillance requirements of this section.

**1910.1045(s)(3)** Training programs must be implemented by January 2, 1979.

**1910.1045(s)(4)** Engineering and work practice controls required by paragraph (g) of this section shall be implemented no later than November 2, 1980.

**1910.1045(t)** ***Appendixes.*** The information contained in the appendixes is not intended, by itself, to create any additional obligation not otherwise imposed, or to detract from any obligation.

## APPENDIX A TO §1910.1045—SUBSTANCE SAFETY DATA SHEET FOR ACRYLONITRILE

### I. SUBSTANCE IDENTIFICATION

A. Substance: Acrylonitrile ($CH_2$ CHCN).

B. Synonyms: Propenenitrile; vinyl cyanide; cyanoethylene; AN; VCN; acylon; carbacryl; fumigrian; ventox.

C. Acrylonitrile can be found as a liquid or vapor, and can also be found in polymer resins, rubbers, plastics, polyols, and other polymers having acrylonitrile as a raw or intermediate material.

D. AN is used in the manufacture of acrylic and modiacrylic fibers, acrylic plastics and resins, speciality polymers, nitrile rubbers, and other organic chemicals. It has also been used as a fumigant.

E. Appearance and odor: Colorless to pale yellow liquid with a pungent odor which can only be detected at concentrations above the permissible exposure level, in a range of 13-19 parts AN per million parts of air (13-19 ppm).

F. Permissible exposure: Exposure may not exceed either:

1. Two parts AN per million parts of air (2 ppm) averaged over the 8-hour workday; or

2. Ten parts AN per million parts of air (10 ppm) averaged over any 15-minute period in the workday.

3. In addition, skin and eye contact with liquid AN is prohibited.

### II. HEALTH HAZARD DATA

A. Acrylonitrile can affect your body if you inhale the vapor (breathing), if it comes in contact with your eyes or skin, or if you swallow it. It may enter your body through your skin.

B. Effects of overexposure: 1. Short-term exposure: Acrylonitrile can cause eye irritation, nausea, vomiting, headache, sneezing, weakness, and light-headedness. At high concentrations, the effects of exposure may go on to loss of consciousness and death. When acrylonitrile is held in contact with the skin after being absorbed into shoe leather or clothing, it may produce blisters following several hours of no apparent effect. Unless the shoes or clothing are removed immediately and the area washed, blistering will occur. Usually there is no pain or inflammation associated with blister formation.

2. Long-term exposure: Acrylonitrile has been shown to cause cancer in laboratory animals and has been associated with higher incidences of cancer in humans. Repeated or prolonged exposure of the skin to acrylonitrile may produce irritation and dermatitis.

3. Reporting signs and symptoms: You should inform your employer if you develop any signs or symptoms and suspect they are caused by exposure to acrylonitrile.

### III. EMERGENCY FIRST AID PROCEDURES

A. Eye exposure: If acrylonitrile gets into your eyes, wash your eyes immediately with large amounts of water, lifting the lower and upper lids occasionally. Get medical attention immediately. Contact lenses should not be worn when working with this chemical.

B. Skin exposure: If acrylonitrile gets on your skin, immediately wash the contaminated skin with water. If acrylonitrile soaks through your clothing, especially your shoes, remove the clothing immediately and wash the skin with water. If symptoms occur after washing, get medical attention immediately. Thoroughly wash the clothing before reusing. Contaminated leather shoes or other leather articles should be discarded.

C. Inhalation: If you or any other person breathes in large amounts of acrylonitrile, move the exposed person to fresh air at once. If breathing has stopped, perform artificial respiration. Keep the affected person warm and at rest. Get medical attention as soon as possible.

D. Swallowing. When acrylonitrile has been swallowed, give the person large quantities of water immediately. After the water has been swallowed, try to get the person to vomit by having him touch the back of his throat with his finger. Do not make an unconscious person vomit. Get medical attention immediately.

E. Rescue: Move the affected person from the hazardous exposure. If the exposed person has been overcome, notify someone else and put into effect the established emergency procedures. Do not become a casualty yourself. Understand your emergency rescue procedures and know the location of the emergency equipment before the need arises.

F. Special first aid procedures: First aid kits containing an adequate supply (at least two dozen) of amyl nitrite pearls, each containing 0.3 ml, should be maintained at each site where acrylonitrile is used. When a person is suspected of receiving an overexposure to acrylonitrile, immediately remove that person from the contaminated area using established rescue procedures. Contaminated clothing must be removed and the acrylonitrile washed from the skin immediately. Artificial respiration should be started at once if breathing has stopped. If the person is unconscious, amyl nitrite may be used as an antidote by a properly trained individual in accordance with established emergency procedures. Medical aid should be obtained immediately.

### IV. RESPIRATORS AND PROTECTIVE CLOTHING

A. Respirators. You may be required to wear a respirator for nonroutine activities, in emergencies, while your employer is in the process of reducing acrylonitrile exposures through engineering controls, and in areas where engineering controls are not feasible. If respirators are worn, they must have a label issued by the National Institute for Occupational Safety and Health under the provisions of 42 CFR part 84 stating that the respirators have been approved for use with organic vapors. For effective protection, respirators must fit your face and head snugly. Respirators must not be loosened or removed in work situations where their use is required.

Acrylonitrile does not have a detectable odor except at levels above the permissible exposure limits. Do not depend on odor to warn you when a respirator cartridge or canister is exhausted. Cartridges or canisters must be changed daily or before the end-of-service-life, whichever comes first. Reuse of these may allow acrylonitrille to gradually filter through the cartridge and cause exposures which you cannot detect by odor. If you can smell acrylonitrile while wearing a respirator, proceed immediately to fresh air. If you experience difficulty breathing while wearing a respirator, tell your employer.

B. Supplied-air suits: In some work situations, the wearing of supplied-air suits may be necessary. Your employer must instruct you in their proper use and operation.

C. Protective clothing: You must wear impervious clothing, gloves, face shield, or other appropriate protective clothing to prevent skin contact with liquid acrylonitrile. Where protective clothing is required, your employer is required to provide clean garments to you as necessary to assume that the clothing protects you adequately.

Replace or repair impervious clothing that has developed leaks.

Acrylonitrile should never be allowed to remain on the skin. Clothing and shoes which are not impervious to acrylonitrile should not be allowed to become contaminated with acrylonitrile, and if they do the clothing and shoes should be promptly removed and decontaminated. The clothing should be laundered or discarded after the AN is removed. Once acrylonitrile penetrates shoes or other leather articles, they should not be worn again.

D. Eye protection: You must wear splashproof safety goggles in areas where liquid acrylonitrile may contact your eyes. In addition, contact lenses should not be worn in areas where eye contact with acrylonitrile can occur.

## V. PRECAUTIONS FOR SAFE USE, HANDLING, AND STORAGE

A. Acrylonitrile is a flammable liquid, and its vapors can easily form explosive mixtures in air.

B. Acrylonitrile must be stored in tightly closed containers in a cool, well-ventilated area, away from heat, sparks, flames, strong oxidizers (especially bromine), strong bases, copper, copper alloys, ammonia, and amines.

C. Sources of ignition such as smoking and open flames are prohibited wherever acrylonitrile is handled, used, or stored in a manner that could create a potential fire or explosion hazard.

D. You should use non-sparking tools when opening or closing metal containers of acrylonitrile, and containers must be bonded and grounded when pouring or transferring liquid acrylonitrile.

E. You must immediately remove any non-impervious clothing that becomes wetted with acrylonitrile, and this clothing must not be reworn until the acrylonitrile is removed from the clothing.

F. Impervious clothing wet with liquid acrylonitrile can be easily ignited. This clothing must be washed down with water before you remove it.

G. If your skin becomes wet with liquid acrylonitrile, you must promptly and thoroughly wash or shower with soap or mild detergent to remove any acrylonitrile from your skin.

H. You must not keep food, beverages, or smoking materials, nor are you permitted to eat or smoke in regulated areas where acrylonitrile concentrations are above the permissible exposure limits.

I. If you contact liquid acrylonitrile, you must wash your hands thoroughly with soap or mild detergent and water before eating, smoking, or using toilet facilities.

J. Fire extinguishers and quick drenching facilities must be readily available, and you should know where they are and how to operate them.

K. Ask your supervisor where acrylonitrile is used in your work area and for any additional plant safety and health rules.

## VI. ACCESS TO INFORMATION

A. Each year, your employer is required to inform you of the information contained in this Substance Safety Data Sheet for acrylonitrile. In addition, you employer must instruct you in the proper work practices for using acrylonitrile, emergency procedures, and the correct use of protective equipment.

B. Your employer is required to determine whether you are being exposed to acrylonitrile. You or your representative has the right to observe employee measurements and to record the results obtained. Your employer is required to inform you of your exposure. If your employer determines that you are being overexposed, he or she is required to inform you of the actions which are being taken to reduce your exposure to within permissible exposure limits.

C. Your employer is required to keep records of your exposures and medical examinations. These records must be kept by the employer for at least forty (40) years or for the period of your employment plus twenty (20) years, whichever is longer.

D. Your employer is required to release your exposure and medical records to you or your representative upon your request.

# APPENDIX B TO §1910.1045—SUBSTANCE TECHNICAL GUIDELINES FOR ACRYLONITRILE

## I. PHYSICAL AND CHEMICAL DATA

A. Substance identification: 1. Synonyms: AN; VCN; vinyl cyanide; propenenitrile; cyanoethylene; Acrylon; Carbacryl; Fumigrain; Ventox.

2. Formula: CH2 = CHCN.

3. Molecular weight: 53.1.

B. Physical data: 1. Boiling point (760 mm Hg): 77.3 ° C (171 ° F);

2. Specific gravity (water = 1): 0.81 (at 20 °C or 68 °F);

3. Vapor density (air = 1 at boiling point of acrylonitrile): 1.83;

4. Melting point: -83 °C (-117 °F);

5. Vapor pressure (@20 °F): 83 mm Hg;

6. Solubility in water, percent by weight @20 °C (68 °F): 7.35;

7. Evaporation rate (Butyl Acetate = 1): 4.54; and

8. Appearance and odor: Colorless to pale yellow liquid with a pungent odor at concentrations above the permissible exposure level. Any detectable odor of acrylonitrile may indicate overexposure.

## II. FIRE, EXPLOSION, AND REACTIVITY HAZARD DATA

A. Fire: 1. Flash point: -1 °C (30 °F) (closed cup).

2. Autoignition temperature: 481 °C (898 °F).

3. Flammable limits air, percent by volume: Lower: 3, Upper: 17.

4. Extinguishing media: Alcohol foam, carbon dioxide, and dry chemical.

5. Special fire-fighting procedures: Do not use a solid stream of water, since the stream will scatter and spread the fire. Use water to cool containers exposed to a fire.

6. Unusual fire and explosion hazards: Acrylonitrile is a flammable liquid. Its vapors can easily form explosive mixtures with air. All ignition sources must be controlled where acrylonitrile is handled, used, or stored in a manner that could create a potential fire or explosion hazard. Acrylonitrile vapors are heavier than air and may travel along the ground and be ignited by open flames or sparks at locations remote from the site at which acrylonitrile is being handled.

7. For purposes of compliance with the requirements of 29 CFR 1910.106, acrylonitrile is classified as a class IB flammable liquid. For example, 7,500 ppm, approximately one-fourth of the lower flammable limit, would be considered to pose a potential fire and explosion hazard.

8. For purposes of compliance with 29 CFR 1910.157, acrylonitrile is classified as a Class B fire hazard.

9. For purpose of compliance with 29 CFR 1919.309, locations classified as hazardous due to the presence of acrylonitrile shall be Class I, Group D.

B. Reactivity:

1. Conditions contributing to instability: Acrylonitrile will polymerize when hot, and the additional heat liberated by the polymerization may cause containers to explode. Pure AN may self-polymerize, with a rapid build-up of pressure, resulting in an explosion hazard. Inhibitors are added to the commercial product to prevent self-polymerization.

2. Incompatibilities: Contact with strong oxidizers (especially bromine) and strong bases may cause fires and explosions. Contact with copper, copper alloys, ammonia, and amines may start serious decomposition.

3. Hazardous decompostion products: Toxic gases and vapors (such as hydrogen cyanide, oxides of nitrogen, and carbon monoxide) may be released in a fire involving acrylonitrile and certain polymers made from acrylonitrile.

4. Special precautions: Liquid acrylonitrile will attack some forms of plastics, rubbers, and coatings.

## III. SPILL, LEAK, AND DISPOSAL PROCEDURES

A. If acrylonitrile is spilled or leaked, the following steps should be taken:

1. Remove all ignition sources.

2. The area should be evacuated at once and re-entered only after the area has been thoroughly ventilated and washed down with water.

3. If liquid acrylonitrile or polymer intermediate, collect for reclamation or absorb in paper, vermiculite, dry sand, earth, or similar material, or wash down with water into process sewer system.

B. Persons not wearing protective equipment should be restricted from areas of spills or leaks until clean-up has been completed.

C. Waste disposal methods: Waste material shall be disposed of in a manner that is not hazardous to employees or to the general population. Spills of acrylonitrile and flushing of such spills shall be channeled for appropriate treatment or collection for disposal. They shall not be channeled directly into the sanitary sewer system. In selecting the method of waste disposal, applicable local, State, and Federal regulations should be consulted.

## IV. MONITORING AND MEASUREMENT PROCEDURES

A. Exposure above the Permissible Exposure Limit:

1. Eight-hour exposure evaluation: Measurements taken for the purpose of determining employee exposure under this section are best taken so that the average 8-hour exposure may be determined from a single 8-hour sample or two (2) 4-hour samples. Air samples should be taken in the employee's breathing zone (air that would most nearly represent that inhaled by the employee.)

2. Ceiling evaluation: Measurements taken for the purpose of determining employee exposure under this section must be taken during periods of maximum expected airborne concentrations of acrylonitrile in the employee's breathing zone. A minimum of three (3) measurements should be taken on one work shift. The average of all measurements taken is an estimate of the employee's ceiling exposure.

3. Monitoring techniques: The sampling and analysis under this section may be performed by collecting the acrylonitrile vapor on charcoal adsorption tubes or other composition adsorption tubes, with subsequent chemical analysis. Sampling and analysis may also be performed by instruments such as real-time continuous monitoring systems, portable direct-reading instruments, or passive dosimeters. Analysis of resultant samples should be by gas chromatograph.

Appendix D lists methods of sampling and analysis which have been tested by NIOSH and OSHA for use with acrylonitrile. NIOSH and OSHA have validated modifications of NIOSH Method S-156 (See Appendix D) under laboratory conditions for concentrations below 1 ppm. The employer has the obligation of selecting a monitoring method which meets the accuracy and precision requirements of the standard under his unique field conditions. The standard requires that methods of monitoring must be accurate, to a 95-percent confidence level, to ±35-percent for concentrations of AN at or above 2 ppm, and to ±-50-percent for concentrations below 2 ppm. In addition to the methods described in Appendix D, there are numerous other methods available for monitoring for AN in the workplace. Details on these other methods have been submitted by various companies to the rulemaking record, and are available at the OSHA Docket Office.

B. Since many of the duties relating to employee exposure are dependent on the results of monitoring and measuring procedures, employers shall assure that the evaluation of employee exposures is performed by a competent industrial hygienist or other technically qualified person.

## V. PROTECTIVE CLOTHING

Employees shall be provided with and required to wear appropriate protective clothing to prevent any possibility of skin contact with liquid AN. Because acrylonitrile is absorbed through the skin, it is important to prevent skin contact with liquid AN. Protective clothing shall include impermeable coveralls or similar full-body work clothing, gloves, headcoverings, as appropriate to protect areas of the body which may come in contact with liquid AN.

Employers should ascertain that the protective garmets are impermeable to acrylonitrile. Non-impermeable clothing and shoes should not be allowed to become contaminated with liquid AN. If permeable clothing does become contaminated, it should be promptly removed, placed in a regulated area for removal of the AN, and not worn again until the AN is removed. If leather footwear or other leather garments become wet from acrylonitrile, they should be replaced and not worn again, due to the ability of leather to absorb acrylonitrile and hold it against the skin. Since there is no pain associated with the blistering which may result from skin contact with liquid AN, it is essential that the employee be informed of this hazard so that he or she can be protected.

Any protective clothing which has developed leaks or is otherwise found to be defective shall be repaired or replaced. Clean protective clothing shall be provided to the employee as necessary to assure its protectiveness. Whenever impervious clothing becomes wet with liquid AN, it shall be washed down with water before being removed by the employee. Employees are also required to wear splash-proof safety goggles where there is any possibility of acrylonitrile contacting the eyes.

## VI. HOUSEKEEPING AND HYGIENE FACILITIES

For purposes of complying with 29 CFR 1910.141, the following items should be emphasized:

A. The workplace should be kept clean, orderly, and in a sanitary condition. The employer is required to institute a leak and spill detection program for operations involving liquid AN in order to detect sources of fugitive AN emissions.

B. Dry sweeping and the use of compressed air is unsafe for the cleaning of floors and other surfaces where liquid AN may be found.

C. Adequate washing facilities with hot and cold water are to be provided, and maintained in a sanitary condition. Suitable cleansing agents are also to be provided to assure the effective removal of acrylonitrile from the skin.

D. Change or dressing rooms with individual clothes storage facilities must be provided to prevent the contamination of street clothes with acrylonitrile. Because of the hazardous nature of acrylonitrile, contaminated protective clothing should be placed in a regulated area designated by the employer for removal of the AN before the clothing is laundered or disposed of.

## VII. MISCELLANEOUS PRECAUTIONS

A. Store acrylonitrile in tightly-closed containers in a cool, well-ventilated area and take necessary precautions to avoid any explosion hazard.

B. High exposures to acrylonitrile can occur when transferring the liquid from one container to another.

C. Non-sparking tools must be used to open and close metal acrylonitrile containers. These containers must be effectively grounded and bonded prior to pouring.

D. Never store uninhibited acrylonitrile.

E. Acrylonitrile vapors are not inhibited. They may form polymers and clog vents of storage tanks.

F. Use of supplied-air suits or other impervious coverings may be necessary to prevent skin contact with and provide respiratory protection from acrylonitrile where the concentration of acrylonitrile is unknown or is above the ceiling limit. Supplied-air suits should be selected, used, and maintained under the immediate supervision of persons knowledgeable in the limitations and potential life-endangering characteristics of supplied-air suits.

G. Employers shall advise employees of all areas and operations where exposure to acrylonitrile could occur.

## VIII. COMMON OPERATIONS

Common operations in which exposure to acrylonitrile is likely to occur include the following: Manufacture of the acrylonitrile monomer; synthesis of acrylic fibers, ABS, SAN, and nitrile barrier plastics and resins, nitrile rubber, surface coatings, specialty chemicals, use as a chemical intermediate, use as a fumigant and in the cyanoethylation of cotton.

# APPENDIX C TO §1910.1045— MEDICAL SURVEILLANCE GUIDELINES FOR ACRYLONITRILE

## I. ROUTE OF ENTRY

Inhalation; skin absorption; ingestion.

## II. TOXICOLOGY

Acrylonitrile vapor is an asphyxiant due to inhibitory action on metabolic enzyme systems. Animals exposed to 75 or 100 ppm for 7 hours have shown signs of anoxia; in some animals which died at the higher level, cyanomethemoglobin was found in the blood. Two human fatalities from accidental poisioning have been reported; one was caused by inhalation of an unknown concentration of the vapor, and the other was thought to be caused by skin absorption or inhalation. Most cases of intoxication from industrial exposure have been mild, with rapid onset of eye irritation, headache, sneezing, and nausea. Weakness, lightheadedness, and vomiting may also occur. Exposure to high concentrations may produce profound weakness, asphyxia, and death. The vapor is a severe eye irritant. Prolonged skin contract with the liquid may result in absorption with systemic effects, and in the formation of large blisters after a latent period of several hours. Although there is usually little or no pain or inflammation, the affected skin resembles a second-degree thermal burn. Solutions spilled on exposed skin, or on areas covered only by a light layer of clothing, evaporate rapidly, leaving no irritation, or, at the most, mild transient redness. Repeated spills on exposed skin may result in dermatitis due to solvent effects.

Results after 1 year of a planned 2-year animal study on the effects of exposure to acrylonitrile have indicated that rats ingesting as little as 35 ppm in their drinking water develop tumors of the central nervous system. The interim results of this study have been supported by a similar study being conducted by the same laboratory, involving exposure of rats by inhalation of acrylonitrile vapor, which has shown similar types of tumors in animals exposed to 80 ppm.

In addition, the preliminary results of an epidemiological study being performed by duPont on a cohort of workers in their Camden, S.C. acrylic fiber plant indicate a statistically significant increase in the incidence of colon and lung cancers among employees exposed to acrylonitrile.

### III. SIGNS AND SYMPTOMS OF ACUTE OVEREXPOSURE

Asphyxia and death can occur from exposure to high concentrations of acrylonitrile. Symptoms of overexposure include eye irritation, headache, sneezing, nausea and vomiting, weakness, and light-headedness. Prolonged skin contact can cause blisters on the skin with appearance of a second-degree burn, but with little or no pain. Repeated skin contact may produce scaling dermatits.

### IV. TREATMENT OF ACUTE OVEREXPOSURE

Remove employee from exposure. Immediately flush eyes with water and wash skin with soap or mild detergent and water. If AN has been swallowed, and person is conscious, induce vomiting. Give artificial resuscitation if indicated. More severe cases, such as those associated with loss of consciousness, may be treated by the intravenous administration of sodium nitrite, followed by sodium thiosulfate, although this is not as effective for acrylonitrile poisoning as for inorganic cyanide poisoning.

### V. SURVEILLANCE AND PREVENTIVE CONSIDERATIONS

A. As noted above, exposure to acrylonitrile has been linked to increased incidence of cancers of the colon and lung in employees of the duPont acrylic fiber plant in Camden, S.C. In addition, the animal testing of acrylonitrile has resulted in the development of cancers of the central nervous system in rats exposed by either inhalation or ingestion. The physician should be aware of the findings of these studies in evaluating the health of employees exposed to acrylonitrile.

Most reported acute effects of occupational exposure to acrylonitrile are due to its ability to cause tissue anoxia and asphyxia. The effects are similar to those caused by hydrogen cyanide. Liquid acrylonitrile can be absorbed through the skin upon prolonged contact. The liquid readily penetrates leather, and will produce burns of the feet if footwear contaminated with acrylonitrile is not removed.

It is important for the physician to become familiar with the operating conditions in which exposure to acrylonitrile may occur. Those employees with skin diseases may not tolerate the wearing of whatever protective clothing may be necessary to protect them from exposure. In addition, those with chronic respiratory disease may not tolerate the wearing of negative-pressure respirators.

B. Surveillance and screening. Medical histories and laboratory examinations are required for each employee subject to exposure to acrylonitrile above the action level. The employer must screen employees for history of certain medical conditions which might place the employee at increased risk from exposure.

1. *Central nervous system dysfunction.* Acute effects of exposure to acrylonitrile generally involve the central nervous system. Symptoms of acrylonitrile exposure include headache, nausea, dizziness, and general weakness. The animal studies cited above suggest possible carcinogenic effects of acrylonitrile on the central nervous system, since rats exposed by either inhalation or ingestion have developed similar CNS tumors.

2. *Respiratory disease.* The du Pont data indicate an increased risk of lung cancer among employees exposed to acrylonitrile.

3. *Gastrointestinal disease.* The du Pont data indicate an increased risk of cancer of the colon among employees exposed to acrylonitrile. In addition, the animal studies show possible tumor production in the stomachs of the rats in the ingestion study.

4. *Skin disease.* Acrylonitrile can cause skin burns when prolonged skin contact with the liquid occurs. In addition, repeated skin contact with the liquid can cause dermatitis.

5. *General.* The purpose of the medical procedures outlined in the standard is to establish a baseline for future health monitoring. Persons unusually susceptible to the effects of anoxia or those with anemia would be expected to be at increased risk. In addition to emphasis on the CNS, respiratory and gastro-intestinal systems, the cardiovascular system, liver, and kidney function should also be stressed.

## APPENDIX D TO §1910.1045—SAMPLING AND ANALYTICAL METHODS FOR ACRYLONITRILE

There are many methods available for monitoring employee exposures to acrylonitrile. Most of these involve the use of charcoal tubes and sampling pumps, with analysis by gas chromatograph. The essential differences between the charcoal tube methods include, among others, the use of different desorbing solvents, the use of different lots of charcoal, and the use of different equipment for analysis of the samples.

Besides charcoal, considerable work has been performed on methods using porous polymer sampling tubes and passive dosimeters. In addition, there are several portable gas analyzers and monitoring units available on the open market.

This appendix contains details for the methods which have been tested at OSHA Analytical Laboratory in Salt Lake City, and NIOSH in Cincinnati. Each is a variation on NIOSH Method S-156, which is also included for reference. This does not indicate that these methods are the only ones which will be satisfactory. There also may be workplace situations in which these methods are not adequate, due to such factors as high humidity. Copies of the other methods available to OSHA are available in the rulemaking record, and may be obtained from the OSHA Docket Office. These include, the Union Carbide, Monsanto, Dow Chemical and Dow Badische methods, as well as NISOH Method P & CAM 127.

Employers who note problems with sample breakthrough should try larger charcoal tubes. Tubes of larger capacity are available, and are often used for sampling vinyl chloride. In addition, lower flow rates and shorter sampling times should be beneficial in minimizing breakthrough problems.

Whatever method the employer chooses, he must assure himself of the method's accuracy and precision under the unique conditions present in his workplace.

### NIOSH METHOD S-156 (UNMODIFIED)

Analyte: Acrylonitrile.

Matrix: Air.

Procedure: Absorption on charcoal, desorption with methanol, GC.

1. *Principle of the method* (Reference 11.1).

1.1 A known volume of air is drawn through a charcoal tube to trap the organic vapors present.

1.2 The charcoal in the tube is transferred to a small, stoppered sample container, and the analyte is desorbed with methanol.

1.3 An aliquot of the desorbed sample is injected into a gas chromatograph.

1.4 The area of the resulting peak is determined and compared with areas obtained for standards.

2. *Range and sensitivity.*

2.1 This method was validated over the range of 17.5-70.0 mg/cu m at an atmospheric temperature and pressure of 22 °C and 760 MM Hg, using a 20-liter sample. Under the conditions of sample size (20-liters) the probable useful range of this method is 4.5-135 mg-cu m. The method is capable of measuring much smaller amounts if the desorption efficiency is adequate. Desorption efficiency must be determined over the range used.

2.2 The upper limit of the range of the method is dependent on the adsorptive capacity of the charcoal tube. This capacity varies with the concentrations of acrylonitrile and other substances in the air. The first section of the charcoal tube was found to hold at least 3.97 mg of acrylonitrile when a test atmosphere containing 92.0 mg/cu m of acrylonitrile in air was sampled 0.18 liter per minute for 240 minutes; at that time the concentration of acrylonitrile in the effluent was less than 5 percent of that in the influent. (The charcoal tube consists of two sections of activated charcoal separated by a section of urethane foam. See section 6.2.) If a particular atmosphere is suspected of containing a large amount of contaminant, a smaller sampling volume should be taken.

3. *Interference.*

3.1 When the amount of water in the air is so great that condensation actually occurs in the tube, organic vapors will not be trapped efficiently. Preliminary experiments using toluene indicate that high humidity severely decreases the breakthrough volume.

3.2 When interfering compounds are known or suspected to be present in the air, such information, including their suspected identities, should be transmitted with the sample.

3.3 It must be emphasized that any compound which has the same retention time as the analyte at the operating conditions described in this method is an interference. Retention time data on a single column cannot be considered proof of chemical identity.

3.4 If the possibility of interference exists, separation conditions (column packing, temperature, etc.) must be changed to circumvent the problem.

4. *Precision and accuracy.*

4.1 The Coefficient of Variation (CVT) for the total analytical and sampling method in the range of 17.5-70.0 mg/cu m was 0.073. This value corresponds to a 3.3 mg/cu m standard deviation at the (previous) OSHA standard level (20 ppm). Statistical information and details of the validation and experimental test procedures can be found in Reference 11.2.

4.2 On the average the concentrations obtained at the 20 ppm level using the overall sampling and analytical method were 6.0 percent lower than the "true" concentrations for a limited number of laboratory experiments. Any difference between the "found" and "true" concentrations may not represent a bias in the sampling and analytical method, but rather a random variation from the experimentally determined "true" concentration. Therefore, no recovery correction should be applied to the final result in section 10.5.

5. *Advantages and disadvantages of the method.*

5.1 The sampling device is small, portable, and involves no liquids. Interferences are minimal, and most of those which do occur can be eliminated by altering chromatographic conditions. The tubes are analyzed by means of a quick, instrumental method.

The method can also be used for the simultaneous analysis of two or more substances suspected to be present in the same sample by simply changing gas chromatographic conditions.

5.2 One disadvantage of the method is that the amount of sample which can be taken is limited by the number of milligrams that the tube will hold before overloading. When the sample value obtained for the backup section of the charcoal tube exceeds 25 percent of that found on the front section, the possibility of sample loss exists.

5.3 Furthermore, the precision of the method is limited by the reproducibility of the pressure drop across the tubes. This drop will affect the flow rate and cause the volume to be imprecise, because the pump is usually calibrated for one tube only.

6. *Apparatus.*

6.1 A calibrated personal sampling pump whose flow can be determined within ±5 percent at the recommended flow rate. (Reference 11.3).

6.2 Charcoal tubes: Glass tubes with both ends flame sealed, 7 cm long with a 6-mm O.D. and a 4-mm I.D., containing 2 sections of 20/40 mesh activated charcoal separated by a 2-mm portion of urethane foam. The activated charcoals prepared from coconut shells and is fired at 600 °C prior to packing. The adsorbing section contains 100 mg of charcoal, the backup section 50 mg. A 3-mm portion of urethane foam is placed between the outlet end of the tube and the backup section. A plug of silicated glass wool is placed in front of the adsorbing section. The pressure drop across the tube must be less than 1 inch of mercury at a flow rate of 1 liter per minute.

6.3 Gas chromatograph equipped with a flame ionization detector.

6.4 Column (4-ft × 1/4-in stainless steel) packed with 50/80 mesh Poropak, type Q.

6.5 An electronic integrator or some other suitable method for measuring peak areas.

6.6 Two-milliliter sample containers with glass stoppers or Teflon-lined caps. If an automatic sample injector is used, the associated vials may be used.

6.7 Microliter syringes: 10-microliter and other convenient sizes for making standards.

6.8 Pipets: 1.0-ml delivery pipets.

6.9 Volumetric flask: 10-ml or convenient sizes for making standard solutions.

7. *Reagents.*

7.1 Chromatographic quality methanol.

7.2 Acrylonitrile, reagent grade.

7.3 Hexane, reagent grade.

7.4 Purified nitrogen.

7.5 Prepurified hydrogen.

7.6 Filtered compressed air.

8. *Procedure.*

8.1 Cleaning of equipment. All glassware used for the laboratory analysis should be detergent washed and thoroughly rinsed with tap water and distilled water.

8.2 Calibration of personal pumps. Each personal pump must be calibrated with a representative charcoal tube in the line. This will minimize errors associated with uncertainties in the sample volume collected.

8.3 Collection and shipping of samples.

8.3.1 Immediately before sampling, break the ends of the tube to provide an opening at least one-half the internal diameter of the tube (2 mm).

8.3.2 The smaller section of charcoal is used as a backup and should be positioned nearest the sampling pump.

8.3.3 The charcoal tube should be placed in a vertical direction during sampling to minimize channeling through the charcoal.

8.3.4 Air being sampled should not be passed through any hose or tubing before entering the charcoal tube.

8.3.5 A maximum sample size of 20 liters is recommended. Sample at a flow of 0.20 liter per minute or less. The flow rate should be known with an accuracy of at least ±5 percent.

8.3.6 The temperature and pressure of the atmosphere being sampled should be recorded. If pressure reading is not available, record the elevation.

8.3.7 The charcoal tubes should be capped with the supplied plastic caps immediately after sampling. Under no circumstances should rubber caps be used.

8.3.8 With each batch of 10 samples submit one tube from the same lot of tubes which was used for sample collection and which is subjected to exactly the same handling as the samples except that no air is drawn through it. Label this as a blank.

8.3.9 Capped tubes should be packed tightly and padded before they are shipped to minimize tube breakage during shipping.

8.3.10 A sample of the bulk material should be submitted to the laboratory in a glass container with a Teflon-lined cap. This sample should not be transported in the same container as the charcoal tubes.

8.4 *Analysis of samples.*

8.4.1 Preparation of samples. In preparation for analysis, each charcoal tube is scored with a file in front of the first section of charcoal and broken open. The glass wool is removed and discarded. The charcoal in the first (larger) section is transferred to a 2-ml stoppered sample container. The separating section of foam is removed and discarded; the second section is transferred to another stoppered container. These two sections are analyzed separately.

8.4.2 Desorption of samples. Prior to analysis, 1.0 ml of methanol is pipetted into each sample container. Desorption should be done for

30 minutes. Tests indicate that this is adequate if the sample is agitated occasionally during this period. If an automatic sample injector is used, the sample vials should be capped as soon as the solvent is added to minimize volatilization.

8.4.3 GC conditions. The typical operating conditions for the gas chromatograph are:

1. 50 ml/min (60 psig) nitrogen carrier gas flow.
2. 65 ml/min (24 psig) hydrogen gas flow to detector.
3. 500 ml/min (50 psig) air flow to detector.
4. 235 °C injector temperature.
5. 255 °C manifold temperature (detector).
6. 155 °C column temperature.

8.4.4 Injection. The first step in the analysis is the injection of the sample into the gas chromatograph. To eliminate difficulties arising from blowback or distillation within the syringe needle, one should employ the solvent flush injection technique. The 10-microliter syringe is first flushed with solvent several times to wet the barrel and plunger. Three microliters of solvent are drawn into the syringe to increase the accuracy and reproducibility of the injected sample volume. The needle is removed from the solvent, and the plunger is pulled back about 0.2 microliter to separate the solvent flush from the sample with a pocket of air to be used as a marker. The needle is then immersed in the sample, and a 5-microliter aliquot is withdrawn, taking into consideration the volume of the needle, since the sample in the needle will be completely injected. After the needle is removed from the sample and prior to injection, the plunger is pulled back 1.2 microliters to minimize evaporation of the sample from the tip of the needle. Observe that the sample occupies 4.9-5.0 microliters in the barrel of the syringe. Duplicate injections of each sample and standard should be made. No more than a 3 percent difference in area is to be expected. An automatic sample injector can be used if it is shown to give reproducibility at least as good as the solvent flush method.

8.4.5 Measurement of area. The area of the sample peak is measured by an electronic integrator or some other suitable form of area measurement, and preliminary results are read from a standard curve prepared as discussed below.

8.5 *Determination of desorption efficiency.*

8.5.1 Importance of determination. The desorption efficiency of a particular compound can vary from one laboratory to another and also from one batch of charcoal to another. Thus, it is necessary to determine at least once the percentage of the specific compound that is removed in the desorption process, provided the same batch of charcoal is used.

8.5.2 Procedure for determining desorption efficiency. Activated charcoal equivalent to the amount in the first section of the sampling tube (100 mg) is measured into a 2.5 in, 4-mm I.D. glass tube, flame sealed at one end. This charcoal must be from the same batch as that used in obtaining the samples and can be obtained from unused charcoal tubes. The open end is capped with Parafilm. A known amount of hexane solution of acrylonitrile containing 0.239 g/ml is injected directly into the activated charcoal with a microliter syringe, and tube is capped with more Parafilm. When using an automatic sample injector, the sample injector vials, capped with Teflon-faced septa, may be used in place of the glass tube.

The amount injected is equivalent to that present in a 20liter air sample at the selected level.

Six tubes at each of three levels (0.5X, 1X, and 2X of the standard) are prepared in this manner and allowed to stand for at least overnight to assure complete adsorption of the analyte onto the charcoal. These tubes are referred to as the sample. A parallel blank tube should be treated in the same manner except that no sample is added to it. The sample and blank tubes are desorbed and analyzed in exactly the same manner as the sampling tube described in section 8.4.

Two or three standards are prepared by injecting the same volume of compound into 1.0 ml of methanol with the same syringe used in the preparation of the samples. These are analyzed with the samples.

The desorption efficiency (D.E.) equals the average weight in mg recovered from the tube divided by the weight in mg added to the tube, or

$$\text{D.E.} = \frac{\text{Average weight recovered (mg)}}{\text{weight added (mg)}}$$

The desorption efficiency is dependent on the amount of analyte collected on the charcoal. Plot the desorption efficiency versus weight of analyte found. This curve is used in section 10.4 to correct for adsorption losses.

9. *Calibration and standards.*

It is convenient to express concentration of standards in terms of mg/1.0 ml methanol, because samples are desorbed in this amount of methanol. The density of the analyte is used to convert mg into microliters for easy measurement with a microliter syringe. A series of standards, varying in concentration over the range of interest, is prepared and analyzed under the same GC conditions and during the same time period as the unknown samples. Curves are established by plotting concentration in mg/1.0 ml versus peak area.

**Note:** Since no internal standard is used in the method, standard solutions must be analyzed at the same time that the sample analysis is done. This will minimize the effect of known day-to-day variations and variations during the same day of the FID response.

10. *Calculations.*

10.1 Read the weight, in mg, corresponding to each peak area from the standard curve. No volume corrections are needed, because the standard curve is based on mg/1.0 ml methanol and the volume of sample injected is identical to the volume of the standards injected.

10.2 Corrections for the bank must be made for each sample.

**mg = mg sample-mg blank**

Where:

mg sample = mg found in front section of sample tube.

mg sample = mg found in front section of blank tube.

A similar procedure is followed for the backup sections.

10.3 Add the weights found in the front and backup sections to get the total weight in the sample.

10.4 Read the desorption efficiency from the curve (see sec. 8.5.2) for the amount found in the front section. Divide the total weight by this desorption efficiency to obtain the corrected mg/sample.

**Corrected mg/sample = Total weight/D.E.**

10.5 The concentration of the analyte in the air sampled can be expressed in mg/cu m.

**mg/cu m = Corrected mg (section 10.4) ×**
**1,000 (liter/cu m)/air volume sampled (liter)**

10.6 Another method of expressing concentration is ppm.

**ppm = m mg/cu × 24.45/M.W. × 760/PX T. + 273/298**

Where:

*P* = Pressure (mm Hg) of air sampled.

*T* = Temperature (° C) of air sampled.

24.45 = Molar volume (liter/mole) at 25° C and 760 mm Hg.

*M.W.* = Molecular weight (g/mole) of analyte.

760 = Standard pressure (mm Hg).

298 = Standard temperature (° K).

11. *References.*

11.1 White, L. D. et al., "A Convenient Optimized Method for the Analysis of Selected Solvent Vapors in the Industrial Atmosphere," *Amer. Ind. Hyg. Assoc. J., 31:* 225 (1970).

11.2 Documentation of NIOSH Validation Tests, NIOSH Contract No. CDC-99-74-45.

11.3 Final Report, NIOSH Contract HSM-99-71-31, "Personal Sampler Pump for Charcoal Tubes," September 15, 1972.

## NIOSH Modification of NIOSH Method S-156

The NIOSH recommended method for low levels for acrylonitrile is a modification of method S-156. It differs in the following respects:

(1) Samples are desorbed using 1 ml of 1 percent acetone in CS2 rather than methanol.

(2) The analytical column and conditions are:

Column: 20 percent SP-1000 on 80/100 Supelcoport 10 feet × 1/8 inch S.S.

Conditions:

Injector temperature: 200 °C.

Detector temperature: 100 °C.

Column temperature: 85 °C.

Helium flow: 25 ml/min.

Air flow: 450 ml/min.

Hydrogen flow: 55 ml/min.

(3) A 2 μl injection of the desorbed analyte is used.

(4) A sampling rate of 100 ml/min is recommended.

## OSHA Laboratory Modification of NIOSH Method S-156

Analyte: Acrylonitrile.

Matrix: Air.

Procedure: Adsorption on charcoal, desorption with methanol, GC.

1. *Principle of the Method* (Reference 1).

1.1 A known volume of air is drawn through a charcoal tube to trap the organic vapors present.

1.2 The charcoal in the tube is transferred to a small, stoppered sample vial, and the analyte is desorbed with methanol.

1.3 An aliquot of the desorbed sample is injected into a gas chromatograph.

1.4 The area of the resulting peak is determined and compared with areas obtained for standards.

2. *Advantages and disadvantages of the method.*

2.1 The sampling device is small, portable, and involves no liquids. Interferences are minimal, and most of those which do occur can be eliminated by altering chromatographic conditions. The tubes are analyzed by means of a quick, instrumental method.

2.2 This method may not be adequate for the simultaneous analysis of two or more substances.

2.3 The amount of sample which can be taken is limited by the number of milligrams that the tube will hold before overloading. When the sample value obtained for the backup section of the charcoal tube exceeds 25 percent of that found on the front section, the possibility of sample loss exists.

2.4 The precision of the method is limited by the reproducibility of the pressure drop across the tubes. This drop will affect the flow rate and cause the volume to be imprecise, because the pump is usually calibrated for one tube only.

3. *Apparatus.*

3.1 A calibrated personal sampling pump whose flow can be determined within ±5 percent at the recommended flow rate.

3.2 Charcoal tubes: Glass tube with both ends flame sealed, 7 cm long with a 6-mm O.D. and a 4-mm I.D., containing 2 sections of 20/40 mesh activated charcoal separated by a 2-mm portion of urethane foam. The activated charcoal is prepared from coconut shells and is fired at 600 °C prior to packing. The adsorbing section contains 100 mg of charcoal, the back-up section 50 mg. A 3-mm portion of urethane foam is placed between the outlet end of the tube and the back-up section. A plug of sililated glass wool is placed in front of the adsorbing section. The pressure drop across the tube must be less than one inch of mercury at a flow rate of 1 liter per minute.

3.3 Gas chromatograph equipped with a nitrogen phosphorus detector.

3.4 Column (10-ft x 1/8″-in stainless steel) packed with 100/120 Supelcoport coated with 10 percent SP 1000.

3.5 An electronic integrator or some other suitable method for measuring peak area.

3.6 Two-milliliter sample vials with Teflon-lined caps

3.7 Microliter syringes: 10-microliter, and other convenient sizes for making standards.

3.8 Pipets: 1.0-ml delivery pipets.

3.9 Volumetric flasks: convenient sizes for making standard solutions.

4. *Reagents.*

4.1 Chromatographic quality methanol.

4.2 Acrylonitrile, reagent grade.

4.3 Filtered compressed air.

4.4 Purified hydrogen.

4.5 Purified helium.

5. *Procedure.*

5.1 Cleaning of equipment. All glassware used for the laboratory analysis should be properly cleaned and free of organics which could interfere in the analysis.

5.2 Calibration of personal pumps. Each pump must be calibrated with a representative charcoal tube in the line.

5.3 Collection and shipping of samples.

5.3.1 Immediately before sampling, break the ends of the tube to provide an opening at least one-half the internal diameter of the tube (2 mm).

5.3.2 The smaller section of the charcoal is used as the backup and should be placed nearest the sampling pump.

5.3.3 The charcoal should be placed in a vertical position during sampling to minimize channeling through the charcoal.

5.3.4 Air being sampled should not be passed through any hose or tubing before entering the charcoal tube.

5.3.5 A sample size of 20 liters is recommended. Sample at a flow rate of approximately 0.2 liters per minute. The flow rate should be known with an accuracy of at least ±5 percent.

5.3.6 The temperature and pressure of the atmosphere being sampled should be recorded.

5.3.7 The charcoal tubes should be capped with the supplied plastic caps immediately after sampling. Rubber caps should not be used.

5.3.8 Submit at least one blank tube (a charcoal tube subjected to the same handling procedures, without having any air drawn through it) with each set of samples.

5.3.9. Take necessary shipping and packing precautions to minimize breakage of samples.

5.4 Analysis of samples.

5.4.1 Preparation of samples. In preparation for analysis, each charcoal tube is scored with a file in front of the first section of charcoal and broken open. The glass wool is removed and discarded. The charcoal in the first (larger) section is transferred to a 2-ml vial. The separating section of foam is removed and discarded; the section is transferred to another capped vial. These two sections are analyzed separately.

5.4.2 Desorption of samples. Prior to analysis, 1.0 ml of methanol is pipetted into each sample container. Desorption should be done for 30 minutes in an ultrasonic bath. The sample vials are recapped as soon as the solvent is added.

5.4.3 GC conditions. The typical operating conditions for the gas chromatograph are:

1. 30 ml/min (60 psig) helium carrier gas flow.
2. 3.0 ml/min (30 psig) hydrogen gas flow to detector.
3. 50 ml/min (60 psig) air flow to detector.
4. 200 °C injector temperature.

5. 200 °C dejector temperature.

6. 100 °C column temperature.

5.4.4 Injection. Solvent flush technique or equivalent.

5.4.5 Measurement of area. The area of the sample peak is measured by an electronic integator or some other suitable form of area measurement, and preliminary results are read from a standard curve prepared as discussed below.

5.5 Determination of desorption efficiency.

5.5.1 Importance of determination. The desorption efficiency of a particular compound can vary from one laboratory to another and also from one batch of charcoal to another. Thus, it is necessary to determine, at least once, the percentage of the specific compound that is removed in the desorption process, provided the same batch of charcoal is used.

5.5.2 Procedure for determining desorption efficiency. The reference portion of the charcoal tube is removed. To the remaining portion, amounts representing 0.5X, 1X, and 2X (X represents TLV) based on a 20 l air sample are injected onto several tubes at each level. Dilutions of acrylonitrile with methanol are made to allow injection of measurable quantities. These tubes are then allowed to equilibrate at least overnight. Following equilibration they are analyzed following the same procedure as the samples A curve of the desorption efficiency amt recovered/amt added is plotted versus amount of analyte found. This curve is used to correct for adsorption losses.

6. *Calibration and standards.*

A series of standards, varying in concentration over the range of interest, is prepared and analyzed under the same GC conditions and during the same time period as the unknown samples. Curves are prepared by plotting concentration versus peak area.

**Note:** Since no internal standard is used in the method, standard solutions must be analyzed at the same time that the sample analysis is done. This will minimize the effect of known day-to-day variations and variations during the same day of the NPD response. Multiple injections are necessary.

7. *Calculations.*

Read the weight, corresponding to each peak area from the standard curve, correct for the blank, correct for the desorption efficiency, and make necessary air volume corrections.

8. *Reference.* NIOSH Method S-156.

[43 FR 45809, Oct. 3, 1978, as amended at 45 FR 35283, May 23, 1980; 54 FR 24334, June 7, 1989; 58 FR 35310, June 30, 1993; 61 FR 5508, Feb. 13, 1996; 63 FR 1291, Jan. 8, 1998; 63 FR 20099, Apr. 23, 1998]

## §1926.1147 Ethylene oxide.

**Note:** The requirements applicable to construction work under this section are identical to those set forth at §1910.1047 of this chapter.

[61 FR 31434, June 20, 1996]

[Section 1910.1047 is printed below—CCH.]

## §1910.1047 Ethylene oxide.

**1910.1047(a)** ***Scope and application.***(1) This section applies to all occupational exposures to ethylene oxide (EtO), Chemical Abstracts Service Registry No. 75-21-8, except as provided in paragraph (a)(2) of this section.

**1910.1047(a)(2)** This section does not apply to the processing, use, or handling of products containing EtO where objective data are reasonably relied upon that demonstrate that the product is not capable of releasing EtO in airborne concentrations at or above the action level, and may not reasonably be foreseen to release EtO in excess of the excursion limit, under the expected conditions of processing, use, or handling that will cause the greatest possible release.

**1910.1047(a)(3)** Where products containing EtO are exempted under paragraph (a)(2) of this section, the employer shall maintain records of the objective data supporting that exemption and the basis for the employer's reliance on the data, as provided in paragraph (k)(1) of this section.

**1910.1047(b)** ***Definitions:*** For the purpose of this section, the following definitions shall apply:

*Action level* means a concentration of airborne EtO of 0.5 ppm calculated as an eight (8)-hour time-weighted average.

*Assistant Secretary* means the Assistant Secretary of Labor for Occupational Safety and Health, U.S. Department of Labor, or designee.

*Authorized person* means any person specifically authorized by the employer whose duties require the person to enter a regulated area, or any person entering such an area as a designated representative of employees for the purpose of exercising the right to observe monitoring and measuring procedures under paragraph (l) of this section, or any other person authorized by the Act or regulations issued under the Act.

*Director* means the Director of the National Institute for Occupational Safety and Health, U.S. Department of Health and Human Services, or designee.

*Emergency* means any occurrence such as, but not limited to, equipment failure, rupture of containers, or failure of control equipment that is likely to or does result in an unexpected significant release of EtO.

*Employee exposure* means exposure to airborne EtO which would occur if the employee were not using respiratory protective equipment.

*Ethylene oxide* or *EtO* means the three-membered ring organic compound with chemical formula $C_2H_4O$.

**1910.1047(c)** ***Permissible exposure limits***—(1) *8-hour time weighted average (TWA).* The employer shall ensure that no employee is exposed to an airborne concentration of EtO in excess of one (1) part EtO per million parts of air (1 ppm) as an 8-hour time-weighted average (8-hour TWA).

**1910.1047(c)(2)** *Excursion limit.* The employer shall ensure that no employee is exposed to an airborne concentration of EtO in excess of 5 parts of EtO per million parts of air (5 ppm) as averaged over a sampling period of fifteen (15) minutes.

**1910.1047(d)** ***Exposure monitoring***—(1) *General.* (i) Determinations of employee exposure shall be made from breathing zone air samples that are representative of the 8-hour TWA and 15-minute short-term exposures of each employee.

**1910.1047(d)(1)(ii)** Representative 8-hour TWA employee exposure shall be determined on the basis of one or more samples representing fullshift exposure for each shift for each job classification in each work area. Representative 15-minute short-term employee exposures shall be determined on the basis of one or more samples representing 15-minute exposures associated with operations that are most likely to produce exposures above the excursion limit for each shift for each job classification in each work area.

**1910.1047(d)(1)(iii)** Where the employer can document that exposure levels are equivalent for similar operations in different work shifts, the employer need only determine representative employee exposure for that operation during one shift.

**1910.1047(d)(2)** *Initial monitoring.* (i) Each employer who has a workplace or work operation covered by this standard, except as provided for in paragraph (a)(2) or (d)(2)(ii) of this section, shall perform initial monitoring to determine accurately the airborne concentrations of EtO to which employees may be exposed.

**1910.1047(d)(2)(ii)** Where the employer has monitored after June 15, 1983 and the monitoring satisfies all other requirements of this section, the employer may rely on such earlier monitoring results to satisfy the requirements of paragraph (d)(2)(i) of this section.

**1910.1047(d)(2)(iii)** Where the employer has previously monitored for the excursion limit and the monitoring satisfies all other requirements of this sections, the employer may rely on such earlier monitoring results to satisfy the requirements of paragraph (d)(2)(i) of this section.

**1910.1047(d)(3)** *Monitoring frequency (periodic monitoring).* (i) If the monitoring required by paragraph (d)(2) of this section reveals employee exposure at or above the action level but at or below the 8-hour TWA, the employer shall repeat such monitoring for each such employee at least every 6 months.

**1910.1047(d)(3)(ii)** If the monitoring required by paragraph (d)(2)(i) of this section reveals employee exposure above the 8-hour TWA, the employer shall repeat such monitoring for each such employee at least every 3 months.

**1910.1047(d)(3)(iii)** The employer may alter the monitoring schedule from quarterly to semiannually for any employee for whom two consecutive measurements taken at least 7 days apart indicate that the employee's exposure has decreased to or below the 8-hour TWA.

**1910.1047(d)(3)(iv)** If the monitoring required by paragraph (d)(2)(i) of this section reveals employee exposure above the 15 minute excursion limit, the employer shall repeat such monitoring for each such employee at least every 3 months, and more often as necessary to evaluate exposure the employee's short-term exposures.

**1910.1047(d)(4)** *Termination of monitoring.* (i) If the initial monitoring required by paragraph (d)(2)(i) of this section reveals employee exposure to be below the action level, the employer may discontinue TWA monitoring for those employees whose exposures are represented by the initial monitoring.

**1910.1047(d)(4)(ii)** If the periodic monitoring required by paragraph (d)(3) of this section reveals that employee exposures, as indicated by at least two consecutive measurements taken at least 7 days apart, are below the action level, the employer may discontinue TWA monitoring for those employees whose exposures are represented by such monitoring.

**1910.1047(d)(4)(iii)** If the initial monitoring required by paragraph (d)(2)(1) of this section reveals employee exposure to be at or below the excursion limit, the employer may discontinue excursion limit monitoring for those employees whose exposures are represented by the initial monitoring.

**1910.1047(d)(4)(iv)** If the periodic monitoring required by paragraph (d)(3) of this section reveals that employee exposures, as indicated by at least two consecutive measurements taken at least 7 days apart, are at or below the excursion limit, the employer may discontinue excursion limit monitoring for those employees whose exposures are represented by such monitoring.

**1910.1047(d)(5)** *Additional monitoring.* Notwithstanding the provisions of paragraph (d)(4) of this section, the employer shall institute the exposure monitoring required under paragraphs (d)(2)(i) and (d)(3) of this section whenever there has been a change in the production, process, control equipment, personnel or work practices that may result in new or additional exposures to EtO or when the employer has any reason to suspect that a change may result in new or additional exposures.

**1910.1047(d)(6)** *Accuracy of monitoring.* (i) Monitoring shall be accurate, to a confidence level of 95 percent, to within plus or minus 25 percent for airborne concentrations of EtO at the 1 ppm TWA and to within plus or minus 35 percent for airborne concentrations of EtO at the action level of 0.5 ppm.

**1910.1047(d)(6)(ii)** Monitoring shall be accurate, to a confidence level of 95 percent, to within plus or minus 35 percent for airborne concentrations of EtO at the excursion limit.

**1910.1047(d)(7)** *Employee notification of monitoring results.* (i) The employer shall, within 15 working days after the receipt of the results of any monitoring performed under this standard, notify the affected employee of these results in writing either individually or by posting of results in an appropriate location that is accessible to affected employees.

**1910.1047(d)(7)(ii)** The written notification required by paragraph (d)(7)(i) of this section shall contain the corrective action being taken by the employer to reduce employee exposure to or below the TWA and/or excursion limit, wherever monitoring results indicated that the TWA and/or excursion limit has been exceeded.

**1910.1047(e)** ***Regulated areas.*** (1) The employer shall establish a regulated area wherever occupational exposure to airborne concentrations of EtO may exceed the TWA or wherever the EtO concentration exceeds or can reasonably be expected to exceed the excursion limit.

**1910.1047(e)(2)** Access to regulated areas shall be limited to authorized persons.

**1910.1047(e)(3)** Regulated areas shall be demarcated in any manner that minimizes the number of employees within the regulated area.

**1910.1047(f)** ***Methods of compliance.*** (1) *Engineering controls and work practices.* (i) The employer shall institute engineering controls and work practices to reduce and maintain employee exposure to or below the TWA and to or below the excursion limit, except to the extent that such controls are not feasible.

**1910.1047(f)(1)(ii)** Wherever the feasible engineering controls and work practices that can be instituted are not sufficient to reduce employee exposure to or below the TWA and to or below the excursion limit, the employer shall use them to reduce employee exposure to the lowest levels achievable by these controls and shall supplement them by the use of respiratory protection that complies with the requirements of paragraph (g) of this section.

**1910.1047(f)(1)(iii)** Engineering controls are generally infeasible for the following operations: collection of quality assurance sampling from sterilized materials removal of biological indicators from sterilized materials: loading and unloading of tank cars; changing of ethylene oxide tanks on sterilizers; and vessel cleaning. For these operations, engineering controls are required only where the Assistant Secretary demonstrates that such controls are feasible.

**1910.1047(f)(2)** *Compliance program.* (i) Where the TWA or excursion limit is exceeded, the employer shall establish and implement a written program to reduce exposure to or below the TWA and to or below the excursion limit by means of engineering and work practice controls, as required by paragraph (f)(1) of this section, and by the use of respiratory protection where required or permitted under this section.

**1910.1047(f)(2)(ii)** The compliance program shall include a schedule for periodic leak detection surveys and a written plan for emergency situations, as specified in paragraph (h)(i) of this section.

**1910.1047(f)(2)(iii)** Written plans for a program required in paragraph (f)(2) shall be developed and furnished upon request for examination and copying to the Assistant Secretary, the Director, affected employees and designated employee representatives. Such plans shall be reviewed at least every 12 months, and shall be updated as necessary to reflect significant changes in the status of the employer's compliance program.

**1910.1047(f)(2)(iv)** The employer shall not implement a schedule of employee rotation as a means of compliance with the TWA or excursion limit.

**1910.1047(g)** ***Respiratory protection and personal protective equipment***— (1) *General.* For employees who use respirators required by this section, the employer must provide respirators that comply with the requirements of this paragraph. Respirators must be used during:

**1910.1047(g)(1)(i)** Periods necessary to install or implement feasible engineering and work-practice controls.

**1910.1047(g)(1)(ii)** Work operations, such as maintenance and repair activities and vessel cleaning, for which engineering and workpractice controls are not feasible.

**1910.1047(g)(1)(iii)** Work operations for which feasible engineering and work-practice controls are not yet sufficient to reduce employee exposure to or below the TWA.

**1910.1047(g)(1)(iv)** Emergencies.

**1910.1047(g)(2)** *Respirator program.* The employer must implement a respiratory protection program in accordance with 29 CFR 1910.134 (b) through (d) (except (d)(1)(iii)), and (f) through (m).

**1910.1047(g)(3)** *Respirator selection.* The employer must select the appropriate respirator from Table 1 of this section.

### TABLE 1—MINIMUM REQUIREMENTS FOR RESPIRATORY PROTECTION FOR AIRBORNE EtO

| Condition of use or concentration of airborne EtO (ppm) | Minimum required respirator |
|---|---|
| Equal to or less than 50 | (a) Full facepiece respirator with EtO approved canister, front-or back-mounted. |
| Equal to or less than 2,000 | (a) Positive-pressure supplied air respirator, equipped with full facepiece, hood, or helmet, or |
| | (b) Continuous-flow supplied air respirator (positive pressure) equipped with hood, helmet or suit. |
| Concentration above 2,000 or unknown concentration (such as in emergencies) | (a) Positive-pressure self-contained breathing apparatus (SCBA), equipped with full facepiece, or |
| | (b) Positive-pressure full facepiece supplied air respirator equipped with an auxiliary positive-pressure self-contained breathing apparatus. |
| Firefighting | (a) Positive pressure self-contained breathing apparatus equipped with full facepiece. |
| Escape | (a) Any respirator described above. |

**Note.** Respirators approved for use in higher concentrations are permitted to be used in lower concentrations.

**1910.1047(g)(4)** *Protective clothing and equipment.* When employees could have eye or skin contact with EtO or EtO solutions, the employer must select and provide, at no cost to the employee, appropriate protective clothing or other equipment in accordance with 29 CFR 1910.132 and 1910.133 to protect any area of the employee's body that may come in contact with the EtO or EtO solution, and must ensure that the employee wears the protective clothing and equipment provided.

**1910.1047(h)** ***Emergency situations***— (1) *Written plan.* (i) A written plan for emergency situations shall be developed for each workplace where there is a possibility of an emergency. Appropriate portions of the plan shall be implemented in the event of an emergency.

**1910.1047(h)(1)(ii)** The plan shall specifically provide that employees engaged in correcting emergency conditions shall be equipped with respiratory protection as required by paragraph (g) of this section until the emergency is abated.

**1910.1047(h)(1)(iii)** The plan shall include the elements prescribed in 29 CFR 1910.38, "Employee emergency plans and fire prevention plans."

**1910.1047(h)(2)** *Alerting employees.* Where there is the possibility of employee exposure to EtO due to an emergency, means shall be developed to alert potentially affected employees of such occurrences promptly. Affected employees shall be immediately evacuated from the area in the event that an emergency occurs.

**1910.1047(i)** ***Medical Surveillance***— (1) *General*—(i) *Employees covered.* (A) The employer shall institute a medical surveillance program for all employees who are or may be exposed to EtO at or above the action level, without regard to the use of respirators, for at least 30 days a year.

**1910.1047(i)(1)(i)(B)** The employer shall make available medical examinations and consultations to all employees who have been exposed to EtO in an emergency situation.

**1910.1047(i)(1)(ii)** *Examination by a physician.* The employer shall ensure that all medical examinations and procedures are performed by or under the supervision of a licensed physician, and are provided without cost to the employee, without loss of pay, and at a reasonable time and place.

**1910.1047(i)(2)** *Medical examinations and consultations*—(i) *Frequency.* The employer shall make available medical examinations and consultations to each employee covered under paragraph (i)(1)(i) of this section on the following schedules:

**1910.1047(i)(2)(i)(A)** Prior to assignment of the employee to an area where exposure may be at or above the action level for at least 30 days a year.

**1910.1047(i)(2)(i)(B)** At least annually each employee exposed at or above the action level for at least 30 days in the past year.

**1910.1047(i)(2)(i)(C)** At termination of employment or reassignment to an area where exposure to EtO is not at or above the action level for at least 30 days a year.

**1910.1047(i)(2)(i)(D)** As medically appropriate for any employee exposed during an emergency.

**1910.1047(i)(2)(i)(E)** As soon as possible, upon notification by an employee either (1) that the employee has developed signs or symptoms indicating possible overexposure to EtO, or (2) that the employee desires medical advice concerning the effects of current or past exposure to EtO on the employee's ability to produce a healthy child.

**1910.1047(i)(2)(i)(F)** If the examining physician determines that any of the examinations should be provided more frequently than specified, the employer shall provide such examinations to affected employees at the frequencies recommended by the physician.

**1910.1047(i)(2)(ii)** *Content.* (A) Medical examinations made available pursuant to paragraphs (i)(2)(i)(A)-(D) of this section shall include:

**1910.1047(i)(2)(ii)(A)*(1)*** A medical and work history with special emphasis directed to symptoms related to the pulmonary, hematologic, neurologic, and reproductive systems and to the eyes and skin.

**1910.1047(i)(2)(ii)(A)*(2)*** A physical examination with particular emphasis given to the pulmonary, hematologic, neurologic, and reproductive systems and to the eyes and skin.

**1910.1047(i)(2)(ii)(A)*(3)*** A complete blood count to include at least a white cell count (including differential cell count), red cell count, hematocrit, and hemoglobin.

**1910.1047(i)(2)(ii)(A)*(4)*** Any laboratory or other test which the examining physician deems necessary by sound medical practice.

**1910.1047(i)(2)(ii)(B)** The content of medical examinations or consultation made available pursuant to paragraph (i)(2)(i)(E) of this section shall be determined by the examining physician, and shall include pregnancy testing or laboratory evaluation of fertility, if requested by the employee and deemed appropriate by the physician.

**1910.1047(i)(3)** *Information provided to the physician.* The employer shall provide the following information to the examining physician:

**1910.1047(i)(3)(i)** A copy of this standard and Appendices A, B, and C.

**1910.1047(i)(3)(ii)** A description of the affected employee's duties as they relate to the employee's exposure.

**1910.1047(i)(3)(iii)** The employee's representative exposure level or anticipated exposure level.

**1910.1047(i)(3)(iv)** A description of any personal protective and respiratory equipment used or to be used.

**1910.1047(i)(3)(v)** Information from previous medical examinations of the affected employee that is not otherwise available to the examining physician.

**1910.1047(i)(4)** *Physician's written opinion.* (i) The employer shall obtain a written opinion from the examining physician. This written opinion shall contain the results of the medical examination and shall include:

**1910.1047(i)(4)(i)(A)** The physician's opinion as to whether the employee has any detected medical conditions that would place the employee at an increased risk of material health impairment from exposure to EtO;

**1910.1047(i)(4)(i)(B)** Any recommended limitations on the employee or upon the use of personal protective equipment such as clothing or respirators; and

**1910.1047(i)(4)(i)(C)** A statement that the employee has been informed by the physician of the results of the medical examination and of any medical conditions resulting from EtO exposure that require further explanation or treatment.

**1910.1047(i)(4)(ii)** The employer shall instruct the physician not to reveal in the written opinion given to the employer specific findings or diagnoses unrelated to occupational exposure to EtO.

1910.1047(i)(4)(iii) The employer shall provide a copy of the physician's written opinion to the affected employee within 15 days from its receipt.

1910.1047(j) ***Communication of EtO hazards to employees***—(1) *Signs and labels.* (i) The employer shall post and maintain legible signs demarcating regulated areas and entrances or accessways to regulated areas that bear the following legend:

**DANGER**

**ETHYLENE OXIDE**

**CANCER HAZARD AND REPRODUCTIVE HAZARD**

**AUTHORIZED PERSONNEL ONLY**

**RESPIRATORS AND PROTECTIVE CLOTHING MAY BE REQUIRED**

**TO BE WORN IN THIS AREA**

1910.1047(j)(1)(ii) The employer shall ensure that precautionary labels are affixed to all containers of EtO whose contents are capable of causing employee exposure at or above the action level or whose contents may reasonably be foreseen to cause employee exposure above the excursion limit, and that the labels remain affixed when the containers of EtO leave the workplace. For the purpose of this paragraph, reaction vessels, storage tanks, and pipes or piping systems are not considered to be containers. The labels shall comply with the requirements of 29 CFR 1910.1200(f) of OSHA's Hazard Communication standard, and shall include the following legend:

**(A) DANGER**

**CONTAINS ETHYLENE OXIDE**

**CANCER HAZARD AND REPRODUCTIVE HAZARD;**

and

1910.1047(j)(1)(ii)(B) A warning statement against breathing airborne concentrations of EtO.

1910.1047(j)(1)(iii) The labeling requirements under this section do not apply where EtO is used as a pesticide, as such term is defined in the Federal Insecticide. Fungicide, and Rodenticide Act (7 U.S.C. 136 et seq.), when it is labeled pursuant to that Act and regulations issued under that Act by the Environmental Protection Agency.

1910.1047(j)(2) *Material safety data sheets.* Employers who are manufacturers or importers of EtO shall comply with the requirements regarding development of material safety data sheets as specified in 29 CFR 1910.1200(g) of OSHA's Hazard Communication standard.

1910.1047(j)(3) *Information and training.* (i) The employer shall provide employees who are potentially exposed to EtO at or above the action level or above the excursion limit with information and training on EtO at the time of initial assignment and at least annually thereafter.

1910.1047(j)(3)(ii) Employees shall be informed of the following:

1910.1047(j)(3)(ii)(A) The requirements of this section with an explanation of its contents, including Appendices A and B;

1910.1047(j)(3)(ii)(B) Any operations in their work area where EtO is present;

1910.1047(j)(3)(ii)(C) The location and availability of the written EtO final rule; and

1910.1047(j)(3)(ii)(D) The medical surveillance program required by paragraph (i) of this section with an explanation of the information in Appendix C.

1910.1047(j)(3)(iii) Employee training shall include at least:

1910.1047(j)(3)(iii)(A) Methods and observations that may be used to detect the presence or release of EtO in the work area (such as monitoring conducted by the employer, continuous monitoring devices, etc.);

1910.1047(j)(3)(iii)(B) The physical and health hazards of EtO;

1910.1047(j)(3)(iii)(C) The measures employees can take to protect themselves from hazards associated with EtO exposure, including specific procedures the employer has implemented to protect employees from exposure to EtO, such as work practices, emergency procedures, and personal protective equipment to be used; and

1910.1047(j)(3)(iii)(D) The details of the hazard communication program developed by the employer, including an explanation of the labeling system and how employees can obtain and use the appropriate hazard information.

1910.1047(k) ***Recordkeeping***—(1) *Objective data for exempted operations.* (i) Where the processing, use, or handling of products made from or containing EtO are exempted from other requirements of this section under paragraph (a)(2) of this section, or where objective data have been relied on in lieu of initial monitoring under paragraph (d)(2)(ii) of this section, the employer shall establish and maintain an accurate record of objective data reasonably relied upon in support of the exemption.

1910.1047(k)(1)(ii) This record shall include at least the following information:

1910.1047(k)(1)(ii)(A) The product qualifying for exemption;

1910.1047(k)(1)(ii)(B) The source of the objective data;

1910.1047(k)(1)(ii)(C) The testing protocol, results of testing, and/or analysis of the material for the release of EtO;

1910.1047(k)(1)(ii)(D) A description of the operation exempted and how the data support the exemption; and

1910.1047(k)(1)(ii)(E) Other data relevant to the operations, materials, processing, or employee exposures covered by the exemption.

1910.1047(k)(1)(iii) The employer shall maintain this record for the duration of the employer's reliance upon such objective data.

1910.1047(k)(2) *Exposure measurements.* (i) The employer shall keep an accurate record of all measurements taken to monitor employee exposure to EtO as prescribed in paragraph (d) of this section.

1910.1047(k)(2)(ii) This record shall include at least the following information:

1910.1047(k)(2)(ii)(A) The date of measurement;

1910.1047(k)(2)(ii)(B) The operation involving exposure to EtO which is being monitored;

1910.1047(k)(2)(ii)(C) Sampling and analytical methods used and evidence of their accuracy;

1910.1047(k)(2)(ii)(D) Number, duration, and results of samples taken;

1910.1047(k)(2)(ii)(E) Type of protective devices worn, if any; and

1910.1047(k)(2)(ii)(F) Name, social security number and exposure of the employees whose exposures are represented.

1910.1047(k)(2)(iii) The employer shall maintain this record for at least thirty (30) years, in accordance with 29 CFR 1910.20.

1910.1047(k)(3) *Medical surveillance.* (i) The employer shall establish and maintain an accurate record for each employee subject to medical surveillance by paragraph (i)(1)(i) of this section, in accordance with 29 CFR 1910.20.

1910.1047(k)(3)(ii) The record shall include at least the following information:

1910.1047(k)(3)(ii)(A) The name and social security number of the employee;

1910.1047(k)(3)(ii)(B) Physicians' written opinions;

1910.1047(k)(3)(ii)(C) Any employee medical complaints related to exposure to EtO; and

1910.1047(k)(3)(ii)(D) A copy of the information provided to the physician as required by paragraph (i)(3) of this section.

1910.1047(k)(3)(iii) The employer shall ensure that this record is maintained for the duration of employment plus thirty (30) years, in accordance with 29 CFR 1910.20.

1910.1047(k)(4) *Availability.* (i) The employer, upon written request, shall make all records required to be maintained by this section available to the Assistant Secretary and the Director for examination and copying.

1910.1047(k)(4)(ii) The employer, upon request, shall make any exemption and exposure records required by paragraphs (k) (1) and (2) of this section available for examination and copying to affected employees, former employees, designated representatives and the

Assistant Secretary, in accordance with 29 CFR 1910.20(a) through (e) and (g) through (i).

**1910.1047(k)(4)(iii)** The employer, upon request, shall make employee medical records required by paragraph (k)(3) of this section available for examination and copying to the subject employee, anyone having the specific written consent of the subject employee, and the Assistant Secretary, in accordance with 29 CFR 1910.20.

**1910.1047(k)(5)** *Transfer of records.* (i) The employer shall comply with the requirements concerning transfer of records set forth in 29 CFR 1910.20(h).

**1910.1047(k)(5)(ii)** Whenever the employer ceases to do business and there is no successor employer to receive and retain the records for the prescribed period, the employer shall notify the Director at least 90 days prior to disposal and transmit them to the Director.

**1910.1047(l)** ***Observation of monitoring***— (1) *Employee observation.* The employer shall provide affected employees or their designated representatives an opportunity to observe any monitoring of employee exposure to EtO conducted in accordance with paragraph (d) of this section.

**1910.1047(l)(2)** *Observation procedures.* When observation of the monitoring of employee exposure to EtO requires entry into an area where the use of protective clothing or equipment is required, the observer shall be provided with and be required to use such clothing and equipment and shall comply with all other applicable safety and health procedures.

**1910.1047(m)** ***Dates***— (1)(i) *Effective date.* The paragraphs contained in this section shall become effective August 21, 1984, except for paragraphs (a)(2), (d), (e), (f)(2), (g)(3), (h), (i), and (j) which shall become effective on March 12, 1985.

**1910.1047(m)(1)(ii)** The requirements in this section which pertain only to or are triggered by the excursion limit shall become effective June 6, 1988, except for the excursion limit provisions in paragraphs (a)(2), (d), (f)(2), (g)(3) and (j) of this section which shall become effective August 25, 1988.

**1910.1047(m)(2)** *Start-up dates.* (i) The start-up date for the requirements in those paragraphs that were effective on August 21, 1984, including institution of work practice controls specified in paragraph (f)(1), shall be February 19, 1985, except as provided for in paragraph (m)(2)(ii), and the start-up date for paragraphs (a)(2), (d), (e), (f)(2), (g)(3), (h), (i), and (j) of this section shall be September 9, 1985.

**1910.1047(m)(2)(ii)** Engineering controls specified by paragraph (f)(1) of this section shall be implemented by August 21, 1985.

**1910.1047(m)(2)(iii)** Compliance with the requirements in this section which pertain only to or are triggered by the excursion limit shall be by September 6, 1988, except for compliance with the excursion limit provisions of paragraphs (a)(2), (d), (f)(2), (g)(3), and (j) of this section, which shall be by October 6, 1988, and implementation of engineering controls specified for compliance with the excursion limit, which shall be by December 6, 1988.

**1910.1047(m)(3)** *Labeling.* (i) Paragraph (j)(1)(ii)(A) of this section as amended is effective January 9, 1986.

**1910.1047(m)(3)(ii)** Paragraph (j)(1)(iii) of this is effective October 11, 1985.

**1910.1047(n)** ***Appendices.*** The information contained in the appendices is not intended by itself to create any additional obligations not otherwise imposed or to detract from any existing obligation.

## APPENDIX A TO §1910.1047— SUBSTANCE SAFETY DATA SHEET FOR ETHYLENE OXIDE (NON-MANDATORY)

### I. SUBSTANCE IDENTIFICATION

A. Substance: Ethylene oxide ($C_2 H_4O$).

B. Synonyms: dihydrooxirene, dimethylene oxide, EO, 1,2-epoxyethane, EtO, ETO, oxacyclopropane, oxane, oxidoethane, alpha/beta-oxidoethane, oxiran, oxirane.

C. Ethylene oxide can be found as a liquid or vapor.

D. EtO is used in the manufacture of ethylene glycol, surfactants, ethanolamines, glycol ethers, and other organic chemicals. EtO is also used as a sterilant and fumigant.

E. Appearance and odor: Colorless liquid below 10.7 °C (51.3 °F) or colorless gas with ether-like odor detected at approximately 700 parts EtO per million parts of air (700 ppm).

F. Permissible Exposure: Exposure may not exceed 1 part EtO per million parts of air averaged over the 8-hour workday.

### II. HEALTH HAZARD DATA

A. Ethylene oxide can cause bodily harm if you inhale the vapor, if it comes into contact with your eyes or skin, or if you swallow it.

B. Effects of overexposure;

1. Ethylene oxide in liquid form can cause eye irritation and injury to the cornea, frostbite, and severe irritation and blistering of the skin upon prolonged or confined contact. Ingestion of EtO can cause gastric irritation and liver injury. Acute effects from inhalation of EtO vapors include respiratory irritation and lung injury, headache, nausea, vomiting, diarrhea, shortness of breath, and cyaonosis (blue or purple coloring of skin). Exposure has also been associated with the occurrence of cancer, reproductive effects, mutagenic changes, neurotoxicity, and sensitization.

1. EtO has been shown to cause cancer in laboratory animals and has been associated with higher incidences of cancer in humans. Adverse reproductive effects and chromosome damage may also occur from EtO exposure.

a. Reporting signs and symptoms: You should inform your employer if you develop any signs or symptoms and suspect that they are caused by exposure to EtO.

### III. EMERGENCY FIRST AID PROCEDURES

A. Eye exposure: If EtO gets into your eyes, wash your eyes immediately with large amounts of water, lifting the lower and upper eyelids. Get medical attention immediately. Contact lenses should not be worn when working with this chemical.

B. Skin exposure: If EtO gets on your skin, immediately wash the contaminated skin with water. If EtO soaks through your clothing, especially your shoes, remove the clothing immediately and wash the skin with water using an emergency deluge shower. Get medical attention immediately. Thoroughly wash contaminated clothing before reusing. Contaminated leather shoes or other leather articles should not be reused and should be discarded.

C. Inhalation: If large amounts of EtO are inhaled, the exposed person must be moved to fresh air at once. If breathing has stopped, perform cardiopulmonary resuscitation. Keep the affected person warm and at rest. Get medical attention immediately.

D. Swallowing: When EtO has been swallowed, give the person large quantities of water immediately. After the water has been swallowed, try to get the person to vomit by having him or her touch the back of the throat with his or her finger. Do not make an unconscious person vomit. Get medical attention immediately.

E. Rescue: Move the affected person from the hazardous exposure. If the exposed person has been overcome, attempt rescue only after notifying at least one other person of the emergency and putting into effect established emergency procedures. Do not become a casualty yourself. Understand your emergency rescue procedures and know the location of the emergency equipment before the need arises.

### IV. RESPIRATORS AND PROTECTIVE CLOTHING

A. Respirators. You may be required to wear a respirator for nonroutine activities, in emergencies, while your employer is in the process of reducing EtO exposures through engineering controls, and in areas where engineering controls are not feasible. As of the effective date of this standard, only air-supplied, positive-pressure, full-facepiece respirators are approved for protection against EtO. If air-purifying respirators are worn in the future, they must have a label issued by the National Institute for Occupational Safety and Health under the provisions of 42 CFR part 84 stating that the respirators have been approved for use with ethylene oxide. For effective protection, respirators must fit your face and head snugly. Respirators must not be loosened or removed in work situations where their use is required.

EtO does not have a detectable odor except at levels well above the permissible exposure limits. If you can smell EtO while wearing a respirator, proceed immediately to fresh air. If you experience difficulty breathing while wearing a respirator, tell your employer.

B. Protective clothing: You may be required to wear impermeable clothing, gloves, a face shield, or other appropriate protective clothing to prevent skin contact with liquid EtO or EtO-containing solutions. Where protective clothing is required, your employer must provide clean garments to you as necessary to assure that the clothing protects you adequately.

Replace or repair protective clothing that has become torn or otherwise damaged.

EtO must never be allowed to remain on the skin. Clothing and shoes which are not impermeable to EtO should not be allowed to become contaminated with EtO, and if they do, the clothing should be promptly removed and decontaminated. Contaminated leather shoes should be discarded. Once EtO penetrates shoes or other leather articles, they should not be worn again.

C. Eye protection: You must wear splashproof safety goggles in areas where liquid EtO or EtO-containing solutions may contact your eyes. In addition, contact lenses should not be worn in areas where eye contact with EtO can occur.

## V. PRECAUTIONS FOR SAFE USE, HANDLING, AND STORAGE

A. EtO is a flammable liquid, and its vapors can easily form explosive mixtures in air.

B. EtO must be stored in tighly closed containers in a cool, well-ventilated area, away from heat, sparks, flames, strong oxidizers, alkalines, and acids, strong bases, acetylide-forming metals such as cooper, silver, mercury and their alloys.

C. Sources of ignition such as smoking material, open flames and some electrical devices are prohibited wherever EtO is handled, used, or stored in a manner that could create a potential fire or explosion hazard.

D. You should use non-sparking tools when opening or closing metal containers of EtO, and containers must be bonded and grounded in the rare instances in which liquid EtO is poured or transferred.

E. Impermeable clothing wet with liquid EtO or EtO-containing solutions may be easily ignited. If your are wearing impermeable clothing and are splashed with liquid EtO or EtO-containing solution, you should immediately remove the clothing while under an emergency deluge shower.

F. If your skin comes into contact with liquid EtO or EtO-containing solutions, you should immediately remove the EtO using an emergency deluge shower.

G. You should not keep food, beverages, or smoking materials in regulated areas where employee exposures are above the permissible exposure limits.

H. Fire extinguishers and emergency deluge showers for quick drenching should be readily available, and you should know where they are and how to operate them.

I. Ask your supervisor where EtO is used in your work area and for any additional plant safety and health rules.

## VI. ACCESS TO INFORMATION

A. Each year, your employer is required to inform you of the information contained in this standard and appendices for EtO. In addition, your employer must instruct you in the proper work practices for using EtO emergency procedures, and the correct use of protective equipment.

B. Your employer is required to determine whether you are being exposed to EtO. You or your representative has the right to observe employee measurements and to record the results obtained. Your employer is required to inform you of your exposure. If your employer determine that you are being overexposed, he or she is required to inform you of the actions which are being taken to reduce your exposure to within permissible exposure limits.

C. Your employer is required to keep records of your exposures and medical examinations. These exposure records must be kept by the employer for at least thirty (30) years. Medical records must be kept for the period of your employment plus thirty (30) years.

D. Your employer is required to release your exposure and medical records to your physician or designated representative upon your written request.

## VII. STERILANT USE OF ETO IN HOSPITALS AND HEALTH CARE FACILITIES

This section of Appendix A, for informational purposes, sets forth EPA's recommendations for modifications in workplace design and practice in hospitals and health care facilities for which the Environmental Protection Agency has registered EtO for uses as a sterilant or fumigant under the Federal Insecticide, Funigicide, and Rodenticide Act, 7 U.S.C. 136 *et seq.* These new recommendations, published in the FEDERAL REGISTER by EPA at 49 FR 15268, as modified in today's REGISTER, are intended to help reduce the exposure of hospital and health care workers to EtO to 1 ppm. EPA's recommended workplace design and workplace practice are as follows:

### 1. Workplace Design

a. *Installation of gas line hand valves.* Hand valves must be installed on the gas supply line at the connection to the supply cylinders to minimize leakage during cylinder change.

b. *Installation of capture boxes.* Sterilizer operations result in a gas/water discharge at the completion of the process. This discharge is routinely piped to a floor drain which is generally located in an equipment or an adjacent room. When the floor drain is not in the same room as the sterilizer and workers are not normally present, all that is necessary is that the room be well ventilated.

The installation of a "capture box" will be required for those work place layouts where the floor drain is located in the same room as the sterilizer or in a room where workers are normally present. A "capture box" is a piece of equipment that totally encloses the floor drain where the discharge from the sterilizer is pumped. The "capture box" is to be vented directly to a non-recirculating or dedicated ventilation system. Sufficient air intake should be allowed at the bottom of the box to handle the volume of air that is ventilated from the top of the box. The "capture box" can be made of metal, plastic, wood or other equivalent material. The box is intended to reduce levels of EtO discharged into the work room atmosphere. The use of a "capture box" is not required if: (1) The vacuum pump discharge floor drain is located in a well ventilated equipment or other room where workers are not normally present or (2) the water sealed vacuum pump discharges directly to a closed sealed sewer line (check local plumbing codes).

If it is impractical to install a vented "capture box" and a well ventilated equipment or other room is not feasible, a box that can be sealed over the floor drain may be used if: (1) The floor drain is located in a room where workers are not normally present and EtO cannot leak into an occupied area, and (2) the sterilizer in use is less than 12 cubic feet in capacity (check local plumbing codes).

c. *Ventilation of aeration units* i. *Existing aeration units.* Existing units must be vented to a non-recirculating or dedicated system or vented to an equipment or other room where workers are not normally present and which is well ventilated. Aerator units must be positioned as close as possible to the sterilizer to minimize the exposure from the off-gassing of sterilized items.

ii. *Installation of new aerator units (where none exist).* New aerator units must be vented as described above for existing aerators. Aerators must be in place by July 1, 1986.

d. *Ventilation during cylinder change.* Workers may be exposed to short but relatively high levels of EtO during the change of gas cylinders. To reduce exposure from this route, users must select one of three alternatives designed to draw off gas that may be released when the line from the sterilizer to the cylinder is disconnected:

i. Location of cylinders in a well ventilated equipment room or other room where workers are not normally present.

ii. Installation of a flexible hose (at least 4" in diameter) to a non-recirculating or dedicated ventilation system and located in the area of cylinder change in such a way that the hose can be positioned at the point where the sterilizer gas line is disconnected from the cylinder.

iii. Installation of a hood that is part of a nonrecirculating or dedicated system and positioned no more than one foot above the point where the change of cylinders takes place.

e. *Ventilation of sterilizer door area.* One of the major sources of exposure to EtO occurs when the sterilizer door is opened following the completion of the sterilization process. In order to reduce this avenue of exposure, a hood or metal canopy closed on each end must be installed over the sterilizer door. The hood or metal canopy must be connected to a nonrecirculating or dedicated ventilation system or one that exhausts gases to a well ventilated equipment or other room where workers are not normally present. A hood or canopy over the sterilizer door is required for use even with those sterilizers that have a purge cycle and must be in place by July 1, 1986.

f. *Ventilation of sterilizer relief valve.* Sterilizers are typically equipped with a safety relief device to release gas in case of increased pressure in the sterilizer. Generally, such relief devices are used on pressure vessels. Although these pressure relief devices are rarely opened for hospital and health care sterilizers, it is suggested that they be designed to exhaust vapor from the sterilizer by one of the following methods:

i. Through a pipe connected to the outlet of the relief valve ventilated directly outdoors at a point high enough to be away from passers by, and not near any windows that open, or near any air conditioning or ventilation air intakes.

ii. Through a connection to an existing or new nonrecirculating or dedicated ventilation system.

iii. Through a connection to a well ventilated equipment or other room where workers are not normally present.

g. *Ventilation systems.* Each hospital and health care facility affected by this notice that uses EtO for the sterilization of equipment and supplies must have a ventilation system which enables compliance with the requirements of section (b) through (f) in the manner described in these sections and within the timeframes allowed. Thus, each affected hospital and health care facility must have or install a non-recirculating or dedicated ventilation equipment or other room where workers are not normally present in which to vent EtO.

h. *Installation of alarm systems.* An audible and visual indicator alarm system must be installed to alert personnel of ventilation system failures, i.e., when the ventilation fan motor is not working.

### 2. Workplace Practices

All the workplace practices discussed in this unit must be permanently posted near the door of each sterilizer prior to use by any operator.

a. *Changing of supply line filters.* Filters in the sterilizer liquid line must be changed when necessary, by the following procedure:

i. Close the cylinder valve and the hose valve.

ii. Disconnect the cylinder hose (piping) from the cylinder.

iii. Open the hose valve and bleed slowly into a proper ventilating system at or near the in-use supply cylinders.

iv. Vacate the area until the line is empty.

v. Change the filter.

vi. Reconnect the lines and reverse the value position.

vii. Check hoses, filters, and valves for leaks with a fluorocarbon leak detector (for those sterilizers using the 88 percent chlorofluorocarbon, 12 percent ethylene oxide mixture (12/88)).

b. *Restricted access area.* i. Areas involving use of EtO must be designated as restricted access areas. They must be identified with signs or floor marks near the sterilizer door, aerator, vacuum pump floor drain discharge, and in-use cylinder storage.

ii. All personnel must be excluded from the restricted area when certain operations are in progress, such as discharging a vacuum pump, emptying a sterilizer liquid line, or venting a nonpurge sterilizer with the door ajar or other operations where EtO might be released directly into the face of workers.

c. *Door opening procedures.* i. *Sterilizers with purge cycles.* A load treated in a sterilizer equipped with a purge cycle should be removed immediately upon completion of the cycle (provided no time is lost opening the door after cycle is completed). If this is not done, the purge cycle should be repeated before opening door.

ii. *Sterilizers without purge cycles.* For a load treated in a sterilizer not equipped with a purge cycle, the sterilizer door must be ajar 6 " for 15 minutes, and then fully opened for at least another 15 minutes before removing the treated load. The length of time of the second period should be established by peak monitoring for one hour after the two 15-minute periods suggested. If the level is above 10 ppm time-weighted average for 8 hours, more time should be added to the second waiting period (door wide open). However, in no case may the second period be shortened to less than 15 minutes.

d. *Chamber unloading procedures.* i. Procedures for unloading the chamber must include the use of baskets or rolling carts, or baskets and rolling tables to transfer treated loads quickly, thus avoiding excessive contact with treated articles, and reducing the duration of exposures.

ii. If rolling carts are used, they should be pulled not pushed by the sterilizer operators to avoid offgassing exposure.

e. *Maintenance.* A written log should be instituted and maintained documenting the date of each leak detection and any maintenance procedures undertaken. This is a suggested use practice and is not required.

i. *Leak detection.* Sterilizer door gaskets, cylinder and vacuum piping, hoses, filters, and valves must be checked for leaks under full pressure with a Fluorocarbon leak detector (for 12/88 systems only) every two weeks by maintenance personnel. Also, the cylinder piping connections must be checked after changing cylinders. Particular attention in leak detection should be given to the automatic solenoid valves that control the flow of EtO to the sterilizer. Specifically, a check should be made at the EtO gasline entrance port to the sterilizer, while the sterilizer door is open and the solenoid valves are in a closed position.

ii. *Maintenance procedures.* Sterilizer/areator door gaskets, valves, and fittings must be replaced when necessary as determined by maintenance personnel in their bi-weekly checks; in addition, visual inspection of the door gaskets for cracks, debris, and other foreign substances should be conducted daily by the operator.

## APPENDIX B TO §1910.1047— SUBSTANCE TECHNICAL GUIDELINES FOR ETHYLENE OXIDE (NON-MANDATORY)

### I. PHYSICAL AND CHEMICAL DATA

A. Substance identification:

1. Synonyms: dihydrooxirene, dimethylene oxide, EO, 1,2epoxyethane, EtO ETO oxacyclopropane, oxane, oxidoethane, alpha/beta-oxidoethane, oxiran, oxirane.

2. Formula: ($C_2$ $H_4$ O).

3. Molecular weight: 44.06

B. Physical data:

1. Boiling point (760 mm Hg): 10.70 ° C (51.3° F);

2. Specific gravity (water = 1): 0.87 (at 20 ° C or 68 ° F)

3. Vapor density (air = 1): 1.49;

4. Vapor pressure (at 20° C); 1,095 mm Hg;

5. Solubility in water: complete;

6. Appearance and odor: colorless liquid; gas at temperature above 10.7° F or 51.3° C with ether-like odor above 700 ppm.

### II. FIRE, EXPLOSION, AND REACTIVITY HAZARD DATA

A. Fire:

1. Flash point: less than 0 ° F (open cup);

2. Stability: decomposes violently at temperatures above 800° F;

3. Flammable limits in air, percent by volume: Lower: 3, Upper: 100;

4. Extinguishing media: Carbon dioxide for small fires, polymer or alcohol foams for large fires;

5. Special fire fighting procedures: Dilution of ethylene oxide with 23 volumes of water renders it non-flammable;

6. Unusual fire and explosion hazards: Vapors of EtO will burn without the presence of air or other oxidizers. EtO vapors are heavier than air and may travel along the ground and be ignited by open flames or sparks at locations remote from the site at which EtO is being used.

7. For purposes of compliance with the requirements of 29 CFR 1910.106, EtO is classified as a flammable gas. For example, 7,500 ppm, approximately one-fourth of the lower flammable limit, would be considered to pose a potential fire and explosion hazard.

8. For purposes of compliance with 29 CFR 1910.155, EtO is classified as a Class B fire hazard.

9. For purpose of compliance with 29 CFR 1919.307, locations classified as hazardous due to the presence of EtO shall be Class I.

B. Reactivity:

1. Conditions contributing to instability: EtO will polymerize violently if contaminated with aqueous alkalies, amines, mineral acids, metal chlorides, or metal oxides. Violent decomposition will also occur at temperatures above 800 ° F;

2. Incompatabilities: Alkalines and acids;

3. Hazardous decomposition products: Carbon monoxide and carbon dioxide.

## III. SPILL, LEAK, AND DISPOSAL PROCEDURES

A. If EtO is spilled or leaked, the following steps should be taken:

1. Remove all ignition sources.

2. The area should be evacuated at once and re-entered only after the area has been thoroughly ventilated and washed down with water.

B. Persons not wearing appropriate protective equipment should be restricted from areas of spills or leaks until cleanup has been completed.

C. Waste disposal methods: Waste material should be disposed of in a manner that is not hazardous to employees or to the general population. In selecting the method of waste disposal, applicable local, State, and Federal regulations should be consulted.

## IV. MONITORING AND MEASUREMENT PROCEDURES

A. Exposure above the Permissible Exposure Limit:

1. Eight-hour exposure evaluation: Measurements taken for the purpose of determining employee exposure under this section are best taken with consecutive samples covering the full shift. Air samples should be taken in the employee's breathing zone (air that would most nearly represent that inhaled by the employee.)

2. Monitoring techniques: The sampling and analysis under this section may be performed by collection of the EtO vapor on charcoal adsorption tubes or other composition adsorption tubes, with subsequent chemical analysis. Sampling and analysis may also be performed by instruments such as real-time continuous monitoring systems, portable direct reading instruments, or passive dosimeters as long as measurements taken using these methods accurately evaluate the concentration of EtO in employees' breathing zones.

Appendix D describes the validated method of sampling and analysis which has been tested by OSHA for use with EtO. Other available methods are also described in Appendix D. The employer has the obligation of selecting a monitoring method which meets the accuracy and precision requirements of the standard under his unique field conditions. The standard requires that the method of monitoring should be accurate, to a 95 percent confidence level, to plus or minus 25 percent for concentrations of EtO at 1 ppm, and to plus or minus 35 percent for concentrations at 0.5 ppm. In addition to the method described in Appendix D, there are numerous other methods available for monitoring for EtO in the workplace. Details on these other methods have been submitted by various companies to the rulemaking record, and are available at the OSHA Docket Office.

B. Since many of the duties relating to employee exposure are dependent on the results of measurement procedures, employers should assure that the evaluation of employee exposures is performed by a technically qualified person.

## V. PROTECTIVE CLOTHING AND EQUIPMENT

Employees should be provided with and be required to wear appropriate protective clothing wherever there is significant potential for skin contact with liquid EtO or EtO-containing solutions. Protective clothing shall include impermeable coveralls or similar full-body work clothing, gloves, and head coverings, as appropriate to protect areas of the body which may come in contact with liquid EtO or EtO-containing solutions.

Employers should ascertain that the protective garments are impermeable to EtO. Permeable clothing, including items made of rubber, and leather shoes should not be allowed to become contaminated with liquid EtO. If permeable clothing does become contaminated, it should be immediately removed, while the employer is under an emergency deluge shower. If leather footwear or other leather garments become wet from EtO they should be discarded and not be worn again, because leather absorbs EtO and holds it against the skin.

Any protective clothing that has been damaged or is otherwise found to be defective should be repaired or replaced. Clean protective clothing should be provided to the employee as necessary to assure employee protection. Whenever impermeable clothing becomes wet with liquid EtO, it should be washed down with water before being removed by the employee. Employees are also required to wear splash-proof safety goggles where there is any possibility of EtO contacting the eyes.

## VI. MISCELLANEOUS PRECAUTIONS

A. Store EtO in tightly closed containers in a cool, wellventilated area and take all necessary precautions to avoid any explosion hazard.

B. Non-sparking tools must be used to open and close metal containers. These containers must be effectively grounded and bonded.

C. Do not incinerate EtO cartridges, tanks or other containers.

D. Employers should advise employees of all areas and operations where exposure to EtO occur.

## VII. COMMON OPERATIONS

Common operations in which exposure to EtO is likely to occur include the following: Manufacture of EtO, surfactants, ethanolamines, glycol ethers, and specialty chemicals, and use as a sterilant in the hospital, health product and spice industries.

# APPENDIX C TO §1910.1047— MEDICAL SURVEILLANCE GUIDELINES FOR ETHYLENE OXIDE (Non-Mandatory)

## I. ROUTE OF ENTRY

Inhalation.

## II. TOXICOLOGY

Clinical evidence of adverse effects associated with the exposure to EtO is present in the form of increased incidence of cancer in laboratory animals (leukemia, stomach, brain), mutation in offspring in animals, and resorptions and spontaneous abortions in animals and human populations respectively. Findings in humans and experimental animals exposed to airborne concentrations of EtO also indicate damage to the genetic material (DNA). These include hemoglobin alkylation, unsecheduled DNA synthesis, sister chromatid exchange chromosomal aberration, and functional sperm abnormalities.

Ethylene oxide in liquid form can cause eye irritation and injury to the cornea, frostbite, severe irritation, and blistering of the skin upon prolonged or confined contact. Ingestion of EtO can cause gastric irritation and liver injury. Other effects from inhalation of EtO vapors include respiratory irritation and lung injury, headache, nausea, vomiting, diarrhea, dyspnea and cyanosis.

## III. SIGNS AND SYMPTOMS OF ACUTE OVEREXPOSURE

The early effects of acute overexposure to EtO are nausea and vomiting, headache, and irritation of the eyes and respiratory passages. The patient may notice a "peculiar taste" in the mouth. Delayed effects can include pulmonary edema, drowsiness, weakness, and incoordination. Studies suggest that blood cell changes, an increase in chromosomal aberrations, and spontaneous abortion may also be causally related to acute overexposure to EtO.

Skin contact with liquid or gaseous EtO causes characteristic burns and possibly even an allergic-type sensitization. The edema and erythema occurring from skin contact with EtO progress to vesiculation with a tendency to coalesce into blebs with desquamation. Healing occurs within three weeks, but there may be a residual brown pigmentation. A 40-80% solution is extremely dangerous, causing extensive blistering after only brief contact. Pure liquid EtO causes frostbite because of rapid evaporation. In contrast, the eye is relatively insensitive to EtO, but there may be some irritation of the cornea.

Most reported acute effects of occupational exposure to EtO are due to contact with EtO in liquid phase. The liquid readily penetrates rubber and leather, and will produce blistering if clothing or footwear contaminated with EtO are not removed.

## IV. SURVEILLANCE AND PREVENTIVE CONSIDERATIONS

As noted above, exposure to EtO has been linked to an increased risk of cancer and reproductive effects including decreased male fertility, fetotoxicity, and spontaneous abortion. EtO workers are more likely to have chromosomal damage than similar groups not exposed to EtO. At the present, limited studies of chronic effects in humans resulting from exposure to EtO suggest a causal association with leukemia. Animal studies indicate leukemia and cancers at other sites (brain, stomach) as well. The physician should be aware of the findings of these studies in evaluating the health of employees exposed to EtO.

Adequate screening tests to determine an employee's potential for developing serious chronic diseases, such as cancer, from exposure to EtO do not presently exist. Laboratory tests may, however, give evidence to suggest that an employee is potentially overexposed to EtO. It is important for the physician to become familiar with the operating conditions in which exposure to EtO is likely to occur. The physician also must become familiar with the signs and symptoms that indicate a worker is receiving otherwise unrecognized and unacceptable exposure to EtO. These elements are especially important in evaluating the medical and work histories and in conducting the physical exam. When an unacceptable exposure in an active employee is identified by the physician, measures taken by the employer to lower exposure should also lower the risk of serious long-term consequences.

The employer is required to institute a medical surveillance program for all employees who are or will be exposed to EtO at or above the action level (0.5 ppm) for at least 30 days per year, without regard to respirator use. All examinations and procedures must be performed by or under the supervision of a licensed physician at a reasonable time and place for the employee and at no cost to the employee.

Although broad latitude in prescribing specific tests to be included in the medical surveillance program is extended to the examining physician, OSHA requires inclusion of the following elements in the routine examination:

(i) Medical and work histories with special emphasis directed to symptoms related to the pulmonary, hematologic, neurologic, and reproductive systems and to the eyes and skin.

(ii) Physical examination with particular emphasis given to the pulmonary, hematologic, neurologic, and reproductive systems and to the eyes and skin.

(iii) Complete blood count to include at least a white cell count (including differential cell count), red cell count, hematocrit, and hemoglobin.

(iv) Any laboratory or other test which the examining physician deems necessary by sound medical practice.

If requested by the employee, the medical examinations shall include pregnancy testing or laboratory evaluation of fertility as deemed appropriate by the physician.

In certain cases, to provide sound medical advice to the employer and the employee, the physician must evaluate situations not directly related to EtO. For example, employees with skin diseases may be unable to tolerate wearing protective clothing. In addition those with chronic respiratory diseases may not tolerate the wearing of negative pressure (air purifying) respirators. Additional tests and procedures that will help the physician determine which employees are medically unable to wear such respirators should include: An evaluation of cardiovascular function, a baseline chest x-ray to be repeated at five year intervals, and a pulmonary function test to be repeated every three years. The pulmonary function test should include measurement of the employee's forced vital capacity (FVC), forced expiratory volume at one second (FEV1), as well as calculation of the ratios of FEV1 to FVC, and measured FVC and measured FEV1 to expected values corrected for variation due to age, sex, race, and height.

The employer is required to make the prescribed tests available at least annually to employees who are or will be exposed at or above the action level, for 30 or more days per year; more often than specified if recommended by the examining physician; and upon the employee's termination of employment or reassignment to another work area. While little is known about the long term consequences of high short-term exposures, it appears prudent to monitor such affected employees closely in light of existing health data. The employer shall provide physician recommended examinations to any employee exposed to EtO in emergency conditions. Likewise, the employer shall make available medical consultations including physician recommended exams to employees who believe they are suffering signs or symptoms of exposure to EtO.

The employer is required to provide the physician with the following informatin: a copy of this standard and its appendices; a description of the affected employee's duties as they relate to the employee exposure level; and information from the employee's previous medical examinations which is not readily available to the examining physician. Making this information available to the physician will aid in the evaluation of the employee's health in relation to assigned duties and fitness to wear personal protective equipment, when required.

The employer is required to obtain a written opinion from the examining physician containing the results of the medical examinations; the physician's opinion as to whether the employee has any detected medical conditions which would place the employee at increased risk of material impairment of his or her health from exposure to EtO; any recommended restrictions upon the employee's exposure to EtO, or upon the use of protective clothing or equipment such as respirators; and a statement that the employee has been informed by the physician of the results of the medical examination and of any medical conditions which require further explanation or treatment. This written opinion must not reveal specific findings or diagnoses unrelated to occupational exposure to EtO, and a copy of the opinion must be provided to the affected employee.

The purpose in requiring the examining physician to supply the employer with a written opinion is to provide the employer with a medical basis to aid in the determination of initial placement of employees and to assess the employee's ability to use protective clothing and equipment.

## APPENDIX D TO §1910.1047—SAMPLING AND ANALYTICAL METHODS FOR ETHYLENE OXIDE (NON-MANDATORY)

A number of methods are available for monitoring employee exposures to EtO. Most of these involve the use of charcoal tubes and sampling pumps, followed by analysis of the samples by gas chromatograph. The essential differences between the charcoal tube methods include, among others, the use of different desorbing solvents, the use of different lots of charcoal, and the use of different equipment for analysis of the samples.

Besides charcoal, methods using passive dosimeters, gas sampling bags, impingers, and detector tubes have been utilized for determination of EtO exposure. In addition, there are several commercially available portable gas analyzers and monitoring units.

This appendix contains details for the method which has been tested at the OSHA Analytical Laboratory in Salt Lake City. Inclusion of this method in the appendix does not mean that this method is the only one which will be satisfactory. Copies of descriptions of other methods available are available in the rulemaking record, and may be obtained from the OSHA Docket Office. These include the Union Carbide, Dow Chemical, 3M, and DuPont methods, as well as NIOSH Method S-286. These methods are briefly described at the end of this appendix.

Employers who note problems with sample breakthrough using the OSHA or other charcoal methods should try larger charcoal tubes. Tubes of larger capacity are available. In addition, lower flow rates and shorter sampling times should be beneficial in minimizing breakthrough problems. Whatever method the employer chooses, he must assure himself of the method's accuracy and precision under the unique conditions present in his workplace.

### ETHYLENE OXIDE

Method No.: 30.

Matrix: Air.

Target Concentration: 1.0 ppm (1.8 mg/m$^3$).

Procedure: Samples are collected on two charcoal tubes in series and desorbed with 1% $CS_2$ in benzene. The samples are derivatized with HBr and treated with sodium carbonate. Analysis is done by gas chromatography with an electron capture detector.

Recommended Air Volume and Sampling Rate: 1 liter and 0.05 Lpm.

Detection Limit of the Overall Procedure: 13.3 ppb (0.024 mg/m$^3$) (Based on 1.0 liter air sample).

Reliable Quantitation Limit: 52.2 ppb (0.094 mg/m$^3$) (Based on 1.0 liter air sample).

Standard Error of Estimate: 6.59% (See Backup Section 4.6).

Special Requirements: Samples must be analyzed within 15 days of sampling date.

Status of Method: The sampling and analytical method has been subjected to the established evaluation procedures of the Organic Method Evaluations Branch.

Date: August 1981.

Chemist: Wayne D. Potter.

## ORGANIC SOLVENTS BRANCH, OSHA ANALYTICAL LABORATORY, SALT LAKE CITY, UTAH

1. General Discussion.

1.1 Background.

1.1.1 History of Procedure.

Ethylene oxide samples analyzed at the OSHA Laboratory have normally been collected on activated charcoal and desorbed with carbon disulfide. The analysis is performed with a gas chromatograph equipped with a FID (Flame ionization detector) as described in NIOSH Method S286 (Ref. 5.1). This method is based on a PEL of 50 ppm and has a detection limit of about 1 ppm.

Recent studies have prompted the need for a method to analyze and detect ethylene oxide at very low concentrations.

Several attempts were made to form an ultraviolet (UV) sensitive derivative with ethylene oxide for analysis with HPLC. Among those tested that gave no detectable product were: panisidine, methylimidazole, aniline, and 2,3,6-trichlorobenzoic acid. Each was tested with catalysts such as triethylamine, aluminum chloride, methylene chloride and sulfuric acid but no detectable derivative was produced.

The next derivatization attempt was to react ethylene oxide with HBr to form 2-bromoethanol. This reaction was successful. An ECD (electron capture detector) gave a very good response for 2bromoethanol due to the presence of bromine. The use of carbon disulfide as the desorbing solvent gave too large a response and masked the 2-bromoethanol. Several other solvents were tested for both their response on the ECD and their ability to desorb ethylene oxide from the charcoal. Among those tested were toluene, xylene, ethyl benzene, hexane, cyclohexane and benzene. Benzene was the only solvent tested that gave a suitable response on the ECD and a high desorption. It was found that the desorption efficiency was improved by using 1% $CS_2$ with the benzene. The carbon disulfide did not significantly improve the recovery with the other solvents. SKC Lot 120 was used in all tests done with activated charcoal.

1.1.2 Physical Properties (Ref. 5.2-5.4).

Synonyms: Oxirane; dimethylene oxide, 1,2-epoxy-ethane; oxane; $C_2H_4O$; ETO;

Molecular Weight: 44.06

Boiling Point: 10.7° C (51.3°)

Melting Point: -111° C

Description: Colorless, flammable gas

Vapor Pressure: 1095 mm. at 20° C

Odor: Ether-like odor

Lower Explosive Limits: 3.0% (by volume)

Flash Point (TOC): Below 0° F

Molecular Structure: $CH_2$—$CH_2$

1.2 Limit Defining Parameters.

1.2.1 Detection Limit of the Analytical Procedure.

The detection limit of the analytical procedure is 12.0 picograms of ethylene oxide per injection. This is the amount of analyte which will give a peak whose height is five times the height of the baseline noise. (See Backup Data Section 4.1).

1.2.2 Detection Limit of the Overall Procedure.

The detection limit of the overall procedure is 24.0 ng of ethylene oxide per sample.

This is the amount of analyte spiked on the sampling device which allows recovery of an amount of analyte equivalent to the detection limit of the analytical procedure. (See Backup Data Section 4.2).

1.2.3 Reliable Quantitation Limit.

The reliable quantitation limit is 94.0 nanograms of ethylene oxide per sample. This is the smallest amount of analyte which can be quantitated within the requirements of 75% recovery and 95% confidence limits. (See Backup Data Section 4.2).

It must be recognized that the reliable quantitation limit and detection limits reported in the method are based upon optimization of the instrument for the smallest possible amount of analyte. When the target concentration of an analyte is exceptionally higher than these limits, they may not be attainable at the routine operating parameters. In this case, the limits reported on analysis reports will be based on the operating parameters used during the analysis of the samples.

1.2.4 Sensitivity.

The sensitivity of the analytical procedure over a concentration range representing 0.5 to 2 times the target concentration based on the recommended air volume is 34105 area units per µg/mL. The sensitivity is determined by the slope of the calibration curve (See Backup Data Section 4.3).

The sensitivity will vary somewhat with the particular instrument used in the analysis.

1.2.5 Recovery.

The recovery of analyte from the collection medium must be 75% or greater. The average recovery from spiked samples over the range of 0.5 to 2 times the target concentration is 88.0% (See Backup Section 4.4). At lower concentrations the recovery appears to be non-linear.

1.2.6 Precision (Analytical Method Only).

The pooled coefficient of variation obtained from replicate determination of analytical standards at 0.5X, 1X and 2X the target concentration is 0.036 (See Backup Data Section 4.5).

1.2.7 Precision (Overall Procedure).

The overall procedure must provide results at the target concentration that are 25% of better at the 95% confidence level. The precision at the 95% confidence level for the 15 day storage test is plus or minus 12.9% (See Backup Data Section 4.6).

This includes an additional plus or minus 5% for sampling error.

1.3 Advantages.

1.3.1 The sampling procedure is convenient.

1.3.2 The analytical procedure is very sensitive and reproducible.

1.3.3 Reanalysis of samples is possible.

1.3.4 Samples are stable for at least 15 days at room temperature.

1.3.5 Interferences are reduced by the longer GC retention time of the new derivative.

1.4 Disadvantages.

1.4.1 Two tubes in series must be used because of possible breakthrough and migration.

1.4.2 The precision of the sampling rate may be limited by the reproducibility of the pressure drop across the tubes. The pumps are usually calibrated for one tube only.

1.4.3 The use of benzene as the desorption solvent increases the hazards of analysis because of the potential carcinogenic effects of benzene.

1.4.4 After repeated injections there can be a buildup of residue formed on the electron capture detector which decreases sensitivity.

1.4.5 Recovery from the charcoal tubes appears to be nonlinear at low concentrations.

2. Sampling Procedure.

2.1 Apparatus.

2.1.1 A calibrated personal sampling pump whose flow can be determined within plus or minus 5% of the recommended flow.

2.1.2 SKC Lot 120 Charcoal tubes: glass tube with both ends flame sealed, 7 cm long with a 6 mm O.D. and a 4-mm I.D., containing 2 sections of coconut shell charcoal separated by a 2-mm portion of urethane foam. The adsorbing section contains 100 mg of charcoal, the backup section 50 mg. A 3-mm portion of urethane foam is placed between the outlet end of the tube and the backup section. A plug of silylated glass wool is placed in front of the adsorbing section.

2.2 Reagents.

2.2.1 None required.

2.3 Sampling Technique.

2.3.1 Immediately before sampling, break the ends of the charcoal tubes. All tubes must be from the same lot.

2.3.2 Connect two tubes in series to the sampling pump with a short section of flexible tubing. A minimum amount of tubing is used to connect the two sampling tubes together. The tube closer to the pump is used as a backup. This tube should be identified as the backup tube.

2.3.3 The tubes should be placed in a vertical position during sampling to minimize channeling.

2.3.4 Air being sampled should not pass through any hose or tubing before entering the charcoal tubes.

2.3.5 Seal the charcoal tubes with plastic caps immediately after sampling. Also, seal each sample with OSHA seals lengthwise.

2.3.6 With each batch of samples, submit at least one blank tube from the same lot used for samples. This tube should be subjected to exactly the same handling as the samples (break, seal, transport) except that no air is drawn through it.

2.3.7 Transport the samples (and corresponding paperwork) to the lab for analysis.

2.3.8 If bulk samples are submitted for analysis, they shoud be transported in glass containers with Teflon-lined caps. These samples must be mailed separately from the container used for the charcoal tubes.

2.4 Breakthrough.

2.4.1 The breakthrough (5% breakthrough) volume for a 3.0 mg/m ethylene oxide sample stream at approximately 85% relative humidity, 22° C and 633 mm is 2.6 liters sampled at 0.05 liters per minute. This is equivalent to 7.8 $\mu$ g of ethylene oxide. Upon saturation of the tube it appeared that the water may be displacing ethylene oxide during sampling.

2.5 Desorption Efficiency.

2.5.1 The desorption efficiency, from liquid injection onto charcoal tubes, averaged 88.0% from 0.5 to 2.0 x the target concentration for a 1.0 liter air sample. At lower ranges it appears that the desorption efficiency is non-linear (See Backup Data Section 4.2).

2.5.2 The desorption efficiency may vary from one laboratory to another and also from one lot of charcoal to another. Thus, it is necessary to determine the desorption efficiency for a particular lot of charcoal.

2.6 Recommended Air Volume and Sampling Rate.

2.6.1 The recommended air volume is 1.0 liter.

2.6.2 The recommended maximum sampling rate is 0.05 Lpm.

2.7 Interferences.

2.7.1 Ethylene glycol and Freon 12 at target concentration levels did not interfere with the collection of ethylene oxide.

2.7.2 Suspected interferences should be listed on the sample data sheets.

2.7.3 The relative humidity may affect the sampling procedure.

2.8 Safety Precautions.

2.8.1 Attach the sampling equipment to the employee so that it does not interfere with work performance.

2.8.2 Wear safety glasses when breaking the ends of the sampling tubes.

2.8.3 If possible, place the sampling tubes in a holder so the sharp end is not exposed while sampling.

3. Analytical Method.

3.1 Apparatus.

3.1.1 Gas chromatograph equipped with a linearized electron capture detector.

3.1.2 GC column capable of separating the derivative of ethylene oxide (2-bromoethanol) from any interferences and the 1% $CS_2$ in benzene solvent. The column used for validation studies was: 10 ft x 1/8 inch stainless steel 20% SP-2100, .1% Carbowax 1500 on 100/120 Supelcoport.

3.1.3 An electronic integrator or some other suitable method of measuring peak areas.

3.1.4 Two milliliter vials with Teflon-lined caps.

3.1.5 Gas tight syringe—500 $\mu$L or other convenient sizes for preparing standards.

3.1.6 Microliter syringes—10 $\mu$ L or other convenient sizes for diluting standards and 1 $\mu$ L for sample injections.

3.1.7 Pipets for dispensing the 1% $CS_2$ in benzene solvent. The Glenco 1 mL dispenser is adequate and convenient.

3.1.8 Volumetric flasks—5 mL and other convenient sizes for preparing standards.

3.1.9 Disposable Pasteur pipets.

3.2 Reagents.

3.2.1 Benzene, reagent grade.

3.2.2 Carbon Disulfide, reagent grade.

3.2.3 Ethylene oxide, 99.7% pure.

3.2.4 Hydrobromic Acid, 48% reagent grade.

3.2.5 Sodium Carbonate, anhydrous, reagent grade.

3.2.6 Desorbing reagent, 99% Benzene/1% $CS_2$.

3.3 Sample Preparation.

3.3.1 The front and back sections of each sample are transferred to separate 2-mL vials.

3.3.2 Each sample is desorbed with 1.0 mL of desorbing reagent.

3.3.3 The vials are sealed immediately and allowed to desorb for one hour with occasional shaking.

3.3.4 Desorbing reagent is drawn off the charcoal with a disposable pipet and put into clean 2-mL vials.

3.3.5 One drop of HBr is added to each vial. Vials are resealed and HBr is mixed well with the desorbing reagent.

3.3.6 About 0.15 gram of sodium carbonate is carefully added to each vial. Vials are again resealed and mixed well.

3.4 Standard Preparation.

3.4.1 Standards are prepared by injecting the pure ethylene oxide gas into the desorbing reagent.

3.4.2 A range of standards are prepared to make a calibration curve. A concentration of 1.0 $\mu$ L of ethylene oxide gas per 1 mL desorbing reagent is equivalent to 1.0 ppm air concentration (all gas volumes at 25 ° C and 760 mm) for the recommended 1 liter air sample. This amount is uncorrected for desorption efficiency (See Backup Data Section 4.2. for desorption efficiency corrections).

3.4.3 One drop of HBr per mL of standard is added and mixed well.

3.4.4 About 0.15 grams of sodium carbonate is carefully added for each drop of HBr (A small reaction will occur).

3.5 Analysis.

3.5.1 GC Conditions.

Nitrogen flow rate—10mL/min.

Injector Temperature—250° C

Detector Temperature—300° C

Column Temperature—100° C

Injection size—0.8 μ L

Elution time—3.9 minutes

3.5.2 Peak areas are measured by an integrator or other suitable means.

3.5.3 The integrator results are in area units and a calibration curve is set up with concentration vs. area units.

3.6 Interferences.

3.6.1 Any compound having the same retention time of 2-bromoethanol is a potential interference. Possible interferences should be listed on the sample data sheets.

3.6.2 GC parameters may be changed to circumvent interferences.

3.6.3 There are usually trace contaminants in benzene. These contaminants, however, posed no problem of interference.

3.6.4 Retention time data on a single column is not considered proof of chemical identity. Samples over the 1.0 ppm target level should be confirmed by GC/Mass Spec or other suitable means.

3.7 Calculations

3.7.1 The concentration in μg/mL for a sample is determined by comparing the area of a particular sample to the calibration curve, which has been prepared from analytical standards.

3.7.2 The amount of analyte in each sample is corrected for desorption efficiency by use of a desorption curve.

3.7.3 Analytical results (A) from the two tubes that compose a particular air sample are added together.

3.7.4 The concentration for a sample is calculated by the following equation:

$$\text{ETO, mg/m}^3 = \frac{AXB}{C}$$

where:

A = mg/mL

B = desorption volume in milliliters

C = air volume in liters.

3.7.5 To convert mg/m$^3$ to parts per million (ppm) the following relationship is used:

$$\text{ETO, ppm} = \frac{\text{mg/m}^3 \times 24.45}{44.05}$$

where:

mg/m$^3$ = results from 3.7.4

24.45 = molar volume at 25° C and 760mm Hg

44.05 = molecular weight of ETO.

3.8 Safety Precautions

3.8.1 Ethylene oxide and benzene are potential carcinogens and care must be exercised when working with these compounds.

3.8.2 All work done with the solvents (preparation of standards, desorption of samples, etc.) should be done in a hood.

3.8.3 Avoid any skin contact with all of the solvents.

3.8.4 Wear safety glasses at all times.

3.8.5 Avoid skin contact with HBr because it is highly toxic and a strong irritant to eyes and skin.

4. Backup Data.

4.1 Detection Limit Data. The detection limit was determined by injecting 0.8 μL of a 0.015 μg/mL standard of ethylene oxide into 1% $CS_2$ in benzene. The detection limit of the analytical procedure is taken to be 1.20×10$^{-5}$ μg per injection. This is equivalent to 8.3 ppb (0.015 mg/m$^3$) for the recommended air volume.

4.2 Desorption Efficiency.

Ethylene oxide was spiked onto charcoal tubes and the following recovery data was obtained.

| Amount spiked (μg) | Amount recovered (μg) | Percent recovery |
|---|---|---|
| 4.5 | 4.32 | 96.0 |
| 3.0 | 2.61 | 87.0 |
| 2.25 | 2.025 | 90.0 |
| 1.5 | 1.365 | 91.0 |
| 1.5 | 1.38 | 92.0 |
| .75 | .6525 | 87.0 |
| .375 | .315 | 84.0 |
| .375 | .312 | 83.2 |
| .1875 | .151 | 80.5 |
| .094 | .070 | 74.5 |

At lower amounts the recovery appears to be non-linear.

4.3 Sensitivity Data.

The following data was used to determine the calibration curve.

| Injection | 0.5 × .75 μg/mL | 1 × 1.5 μg/mL | 2 × 3.0 μg/mL |
|---|---|---|---|
| 1 | 30904 | 59567 | 111778 |
| 2 | 30987 | 62914 | 106016 |
| 3 | 32555 | 58578 | 106122 |
| 4 | 32242 | 57173 | 109716 |
| X | 31672 | 59558 | 108408 |

Slope = 34.105.

4.4 Recovery.

The recovery was determined by spiking ethylene oxide onto lot 120 charcoal tubes and desorbing with 1% $CS_2$ in Benzene. Recoveries were done at 0.5, 1.0, and 2.0 X the target concentration (1 ppm) for the recommended air volume.

**PERCENT RECOVERY**

| Sample | 0.5× | 1.0× | 2.0× |
|---|---|---|---|
| 1 | 88.7 | 95.0 | 91.7 |
| 2 | 83.8 | 95.0 | 87.3 |
| 3 | 84.2 | 91.0 | 86.0 |
| 4 | 88.0 | 91.0 | 83.0 |
| 5 | 88.0 | 86.0 | 85.0 |
| X | 86.5 | 90.5 | 87.0 |

Weighted Average=88.2.

4.5 Precision of the Analytical Procedure.

The following data was used to determine the precision of the analytical method:

| Concentration | 0.5 × .75 μg/mL | 1 × 1.5 μg/mL | 2 × 3.0 μg/mL |
|---|---|---|---|
| Injection | .7421 | 1.4899 | 3.1184 |
| | .7441 | 1.5826 | 3.0447 |
| | .7831 | 1.4628 | 2.9149 |
| | .7753 | 1.4244 | 2.9185 |
| Average | .7612 | 1.4899 | 2.9991 |
| Standard Deviation | .0211 | .0674 | .0998 |
| CV | .0277 | .0452 | .0333 |

$$CV = \frac{3(.0277)^2 + 3(.0452)^2 + 3(.0333)^2}{3 + 3 + 3}$$

CV+0.036

4.6 Storage Data.

Samples were generated at 1.5 $mg/m^3$ ethylene oxide at 85% relative humidity, 22° C and 633 mm. All samples were taken for 20 minutes at 0.05 Lpm. Six samples were analyzed as soon as possible and fifteen samples were stored at refrigerated temperature (5 ° C) and fifteen samples were stored at ambient temperature (23 ° C). These stored samples were analyzed over a period of nineteen days.

**PERCENT RECOVERY**

| Day analyzed | Refrigerated | Ambient |
|---|---|---|
| 1 | 87.0 | 87.0 |
| 1 | 93.0 | 93.0 |
| 1 | 94.0 | 94.0 |
| 1 | 92.0 | 92.0 |
| 4 | 92.0 | 91.0 |
| 4 | 93.0 | 88.0 |
| 4 | 91.0 | 89.0 |
| 6 | 92.0 | ........ |
| 6 | 92.0 | ........ |
| 8 | ........ | 92.0 |
| 8 | ........ | 86.0 |
| 10 | 91.7 | ........ |
| 10 | 95.5 | ........ |
| 10 | 95.7 | ........ |
| 11 | ........ | 90.0 |
| 11 | ........ | 82.0 |
| 13 | 78.0 | ........ |
| 13 | 81.4 | ........ |
| 13 | 82.4 | ........ |
| 14 | ........ | 78.5 |
| 14 | ........ | 72.1 |
| 18 | 66.0 | ........ |
| 18 | 68.0 | ........ |
| 19 | ........ | 64.0 |
| 19 | ........ | 77.0 |

4.7 Breakthrough Data.

Breakthrough studies were done at 2 ppm (3.6 $mg/m^3$) at approximately 85% relative humidity at 22° C (ambient temperature). Two charcoal tubes were used in series. The backup tube was changed every 10 minutes and analyzed for breakthrough. The flow rate was 0.050 Lpm.

| Tube No. | Time (minutes) | Percent breakthrough |
|---|---|---|
| 1 | 10 | (1) |
| 2 | 20 | (1) |
| 3 | 30 | (1) |
| 4 | 40 | 1.23 |
| 5 | 50 | 3.46 |
| 6 | 60 | 18.71 |
| 7 | 70 | 39.2 |
| 8 | 80 | 53.3 |
| 9 | 90 | 72.0 |
| 10 | 100 | 96.0 |
| 11 | 110 | 113.0 |
| 12 | 120 | 133.9 |

1 None.

The 5% breakthrough volume was reached when 2.6 liters of test atmosphere were drawn through the charcoal tubes.

5. References.

5.1 "NIOSH Manual of Analytical Methods," 2nd ed. NIOSH: Cincinnati, 1977; Method S286.

5.2 "IARC Monographs on the Evaluation of Carcinogenic Risk of Chemicals to Man," International Agency for Research on Cancer: Lyon, 1976; Vol. II, p. 157.

5.3 Sax., N.I. "Dangerous Properties of Industrial Materials," 4th ed.; Van Nostrand Reinhold Company. New York, 1975; p. 741.

5.4 "The Condensed Chemical Dictionary", 9th ed.; Hawley, G.G., ed.; Van Nostrand Reinhold Company, New York, 1977; p. 361.

## Summary of Other Sampling Procedures

OSHA believes that served other types of monitoring equipment and techniques exist for monitoring time-weighted averages. Considerable research and method development is currently being performed, which will lead to improvements and a wider variety of monitoring techniques. A combination of monitoring procedures can be used. There probably is no one best method for monitoring personal exposure to ethylene oxide in all cases. There are advantages, disadvantages, and limitations to each method. The method of choice will depend on the need and requirements. Some commonly used methods include the use of charcoal tubes, passive dosimeters, Tedler gas sampling bags, detector tubes, photoionization detection units, infrared detection units and gas chromatographs. A number of these methods are described below.

### A. CHARCOAL TUBE SAMPLING PROCEDURES

*Qazi-Ketcham method* (Ex. 11-133)—This method consists of collecting EtO on Columbia JXC activated carbon, desorbing the EtO with carbon disulfide and analyzing by gas chromatography with flame ionization detection. Union Carbide has recently updated and revalidated this monitoring procedures. This method is capable of determining both eight-hour time-weighted average exposures and short-term exposures. The method was validated to 0.5 ppm. Like other charcoal collecting procedures, the method requires considerable analytical expertise.

*ASTM-proposed method*—The Ethylene Oxide Industry Council (EOIC) has contracted with Clayton Environmental Consultants, Inc. to conduct a collaborative study for the proposed method. The ASTM-Proposed method is similar to the method published by Qazi and Ketcham is the November 1977 American Industrial Hygiene Association Journal, and to the method of Pilney and Coyne, presented at the 1979 American Industrial Hygiene Conference. After the air to be sampled is drawn through an activated charcoal tube, the ethylene oxide is desorbed from the tube using carbon disulfide and is quantitated by gas chromatography utilizing a flame ionization detector. The ASTM-proposed method specifies a large two-section charcoal tube, shipment in dry ice, storage at less than -5° C, and analysis within three weeks to prevent migration and sample loss. Two types of charcoal tubes are being tested—Pittsburgh Coconut-Based (PCB) and Columbia JXC charcoal. This collaborative study will give an indication of the inter- and intralaboratory precision and accuracy of the ASTM-proposed method. Several laboratories have considerable expertise using the Qazi-Ketcham and Dow methods.

B. Passive Monitors—Ethylene oxide diffuses into the monitor and is collected in the sampling media. The DuPont Pro-Tek badge collects EtO in an absorbing solution, which is analyzed colorimetrically to determine the amount of EtO present. The 3M 350 badge collects the EtO on chemically treated charcoal. Other passive monitors are currently being developed and tested. Both 3M and DuPont have submitted data indicating their dosimeters meet the precision and accuracy requirements of the proposed ethylene oxide standard. Both presented laboratory validation data to 0.2 ppm (Exs. 11-65, 4-20, 108, 109, 130).

C. Tedlar Gas Sampling Bags-Samples are collected by drawing a known volume of air into a Tedlar gas sampling bag. The ethylene oxide concentration is often determined on-site using a portable gas chromatograph or portable infrared spectometer.

D. Detector tubes—A known volume of air is drawn through a detector tube using a small hand pump. The concentration of EtO is related to the length of stain developed in the tube. Detector tubes are economical, easy to use, and give an immediate readout. Unfortunately, partly because they are nonspecific, their accuracy is often questionable. Since the sample is taken over a short period of time, they may be useful for determining the source of leaks.

E. Direct Reading Instruments—There are numerous types of direct reading instruments, each having its own strengths and weaknesses (Exs. 135B, 135C, 107, 11-78, 11-153). Many are relatively new, offering greater sensitivity and specificity. Popular ethylene oxide direct reading instruments include infrared detection units, photoionization detection units, and gas chromatographs.

Portable infrared analyzers provide an immediate, continuous indication of a concentration value; making them particularly useful for locating high concentration pockets, in leak detection and in ambient air monitoring. In infrared detection units, the amount of infrared light absorbed by the gas being analyzed at selected infrared wavelengths is related to the concentration of a particular component. Various models have either fixed or variable infrared filters, differing cell pathlengths, and microcomputer controls for greater sensitivity, automation, and interference elimination.

A fairly recent detection system is photoionization detection. The molecules are ionized by high energy ultraviolet light. The resulting current is measured. Since different substances have different ionization potentials, other organic compounds may be ionized. The lower the lamp energy, the better the selectivity. As a continuous monitor, photoionization detection can be useful for locating high concentration pockets, in leak detection, and continuous ambient air monitoring. Both portable and stationary gas chromatographs are available with various types of detectors, including photoionization detectors. A gas chromatograph with a photoionization detector retains the photionization sensitivity, but minimizes or eliminates interferences. For several GC/PID units, the sensitivity is in the 0.1-0.2 ppm EtO range. The GC/PID with microprocessors can sample up to 20 sample points sequentially, calculate and record data, and activate alarms or ventilation systems. Many are quite flexible and can be configured to meet the specific analysis needs for the workplace.

DuPont presented their laboratory validation data of the accuracy of the Qazi-Ketcham charcoal tube, the PCB charcoal tube, Miran 103 IR analyzer, 3M #3550 monitor and the Du Pont C-70 badge. Quoting Elbert V. Kring:

We also believe that OSHA's proposed accuracy in this standard is appropriate. At plus or minus 25 percent at one part per million, and plus or minus 35 percent below that. And, our data indicates there's only one monitoring method, right now, that we've tested thoroughly, that meets that accuracy requirements. That is the Du Pont Pro-Tek badge***. We also believe that this kind of data should be confirmed by another independent laboratory, using the same type dynamic chamber testing (Tr. 1470)

Additional data by an independent laboratory following their exact protocol was not submitted. However, information was submitted on comparisons and precision and accuracy of those monitoring procedures which indicate far better precision and accuracy of those monitoring procedures than that obtained by Du Pont (Ex. 4-20, 130, 11-68, 11-133, 130, 135A).

The accuracy of any method depends to a large degree upon the skills and experience of those who not only collect the samples but also those who analyze the samples. Even for methods that are collaboratively tested, some laboratories are closer to the true values than others. Some laboratories may meet the precision and accuracy requirements of the method; others may consistently far exceed them for the same method.

[49 FR 25796, June 22, 1984, as amended at 50 FR 9801, Mar. 12, 1985; 50 FR 41494, Oct. 11, 1985; 51 FR 25053, July 10, 1986; 53 FR 11436, 11437, Apr. 6, 1988; 53 FR 27960, July 26, 1988; 54 FR 24334, June 7, 1989; 61 FR 5508, Feb. 13, 1996; 63 FR 1292, Jan. 8, 1998]

## §1926.1148 Formaldehyde.

**Note:** The requirements applicable to construction work under this section are identical to those set forth at §1910.1048 of this chapter.

[61 FR 31434, June 20, 1996]

[Section 1910.1048 is printed below—CCH.]

## §1910.1048 Formaldehyde.

**1910.1048(a)** ***Scope and application.*** This standard applies to all occupational exposures to formaldehyde, i.e. from formaldehyde gas, its solutions, and materials that release formaldehyde.

**1910.1048(b)** ***Definitions.*** For purposes of this standard, the following definitions shall apply:

*Action level* means a concentration of 0.5 part formaldehyde per million parts of air (0.5 ppm) calculated as an eight (8)-hour time-weighted average (TWA) concentration.

*Assistant Secretary* means the Assistant Secretary of Labor for the Occupational Safety and Health Administration, U.S. Department of Labor, or designee.

*Authorized person* means any person required by work duties to be present in regulated areas, or authorized to do so by the employer, by this section, or by the OSH Act of 1970.

*Director* means the Director of the National Institute for Occupational Safety and Health, U.S. Department of Health and Human Services, or designee.

*Emergency* is any occurrence, such as but not limited to equipment failure, rupture of containers, or failure of control equipment that results in an uncontrolled release of a significant amount of formaldehyde.

*Employee exposure* means the exposure to airborne formaldehyde which would occur without corrections for protection provided by any respirator that is in use.

*Formaldehyde* means the chemical substance, HCHO, Chemical Abstracts Service Registry No. 50-00-0.

**1910.1048(c)** ***Permissible Exposure Limit (PEL)***— (1) *TWA:* The employer shall assure that no employee is exposed to an airborne concentration of formaldehyde which exceeds 0.75 parts formaldehyde per million parts of air (0.75 ppm) as an 8hour TWA.

**1910.1048(c)(2)** *Short Term Exposure Limit (STEL):* The employer shall assure that no employee is exposed to an airborne concentration of formaldehyde which exceeds two parts formaldehyde per million parts of air (2 ppm) as a 15-minute STEL.

**1910.1048(d)** ***Exposure monitoring***— (1) *General.* (i) Each employer who has a workplace covered by this standard shall monitor employees to determine their exposure to formaldehyde.

**1910.1048(d)(1)(ii)** *Exception.* Where the employer documents, using objective data, that the presence of formaldehyde or formaldehydereleasing products in the workplace cannot result in airborne concentrations of formaldehyde that would cause any employee to be exposed at or above the action level or the STEL under foreseeable conditions of use, the employer will not be required to measure employee exposure to formaldehyde.

**1910.1048(d)(1)(iii)** When an employee's exposure is determined from representative sampling, the measurements used shall be representative of the employee's full shift or short-term exposure to formaldehyde, as appropriate.

**1910.1048(d)(1)(iv)** Representative samples for each job classification in each work area shall be taken for each shift unless the employer can document with objective data that exposure levels for a given job classification are equivalent for different work shifts.

**1910.1048(d)(2)** *Initial monitoring.* The employer shall identify all employees who may be exposed at or above the action level or at or above the STEL and accurately determine the exposure of each employee so identified.

**1910.1048(d)(2)(i)** Unless the employer chooses to measure the exposure of each employee potentially exposed to formaldehyde, the employer shall develop a representative sampling strategy and measure sufficient exposures within each job classification for each workshift to correctly characterize and not underestimate the exposure of any employee within each exposure group.

**1910.1048(d)(2)(ii)** The initial monitoring process shall be repeated each time there is a change in production, equipment, process, personnel, or control measures which may result in new or additional exposure to formaldehyde.

**1910.1048(d)(2)(iii)** If the employer receives reports of signs or symptoms of respiratory or dermal conditions associated with formaldehyde exposure, the employer shall promptly monitor the affected employee's exposure.

**1910.1048(d)(3)** *Periodic monitoring.* (i) The employer shall periodically measure and accurately determine exposure to formaldehyde for employees shown by the initial monitoring to be exposed at or above the action level or at or above the STEL.

**1910.1048(d)(3)(ii)** If the last monitoring results reveal employee exposure at or above the action level, the employer shall repeat monitoring of the employees at least every 6 months.

**1910.1048(d)(3)(iii)** If the last monitoring results reveal employee exposure at or above the STEL, the employer shall repeat monitoring of the employees at least once a year under worst conditions.

**1910.1048(d)(4)** *Termination of monitoring.* The employer may discontinue periodic monitoring for employees if results from two consecutive sampling periods taken at least 7 days apart show that employee exposure is below the action level and the STEL. The results must be statistically representative and consistent with the employer's knowledge of the job and work operation.

**1910.1048(d)(5)** *Accuracy of monitoring.* Monitoring shall be accurate, at the 95 percent confidence level, to within plus or minus 25 percent for airborne concentrations of formaldehyde at the TWA and the STEL and to within plus or minus 35 percent for airborne concentrations of formaldehyde at the action level.

**1910.1048(d)(6)** *Employee notification of monitoring results.* Within 15 days of receiving the results of exposure monitoring conducted under this standard, the employer shall notify the affected employees of these results. Notification shall be in writing, either by distributing copies of the results to the employees or by posting the results. If the employee exposure is over either PEL, the employer shall develop and implement a written plan to reduce employee exposure to or below both PELs, and give written notice to employees. The written notice shall contain a description of the corrective action being taken by the employer to decrease exposure.

**1910.1048(d)(7)** *Observation of monitoring.* (i) The employer shall provide affected employees or their designated representatives an opportunity to observe any monitoring of employee exposure to formaldehyde required by this standard.

**1910.1048(d)(7)(ii)** When observation of the monitoring of employee exposure to formaldehyde requires entry into an area where the use of protective clothing or equipment is required, the employer shall provide the clothing and equipment to the observer, require the observer to use such clothing and equipment, and assure that the observer complies with all other applicable safety and health procedures.

**1910.1048(e)** ***Regulated areas.*** (1) The employer shall establish regulated areas where the concentration of airborne formaldehyde exceeds either the TWA or the STEL and post all entrances and accessways with signs bearing the following information:

**DANGER**
**FORMALDEHYDE**
**IRRITANT AND POTENTIAL CANCER HAZARD**
**AUTHORIZED PERSONNEL ONLY**

**1910.1048(e)(2)** The employer shall limit access to regulated areas to authorized persons who have been trained to recognize the hazards of formaldehyde.

**1910.1048(e)(3)** An employer at a multiemployer worksite who establishes a regulated area shall communicate the access restrictions and locations of these areas to other employers with work operations at that worksite.

**1910.1048(f)** ***Methods of compliance***— (1) *Engineering controls and work practices.* The employer shall institute engineering and work practice controls to reduce and maintain employee exposures to formaldehyde at or below the TWA and the STEL.

**1910.1048(f)(2)** *Exception.* Whenever the employer has established that feasible engineering and work practice controls cannot reduce employee exposure to or below either of the PELs, the employer shall apply these controls to reduce employee exposures to the extent feasible and shall supplement them with respirators which satisfy this standard.

**1910.1048(g)** ***Respiratory protection***— (1) *General.* For employees who use respirators required by this section, the employer must provide respirators that comply with the requirements of this paragraph. Respirators must be used during:

**1910.1048(g)(1)(i)** Periods necessary to install or implement feasible engineering and work-practice controls.

**1910.1048(g)(1)(ii)** Work operations, such as maintenance and repair activities or vessel cleaning, for which the employer establishes that engineering and work-practice controls are not feasible.

**1910.1048(g)(1)(iii)** Work operations for which feasible engineering and work-practice controls are not yet sufficient to reduce employee exposure to or below the PELs.

**1910.1048(g)(1)(iv)** Emergencies.

**1910.1048(g)(2)** *Respirator program.* (i) The employer must implement a respiratory protection program in accordance with 29 CFR 1910.134(b) through (d) (except (d)(1)(iii), (d)(3)(iii)(B)(1), and (2)), and (f) through (m).

**1910.1048(g)(2)(ii)** If air-purifying chemical-cartridge respirators are used, the employer must:

**1910.1048(g)(2)(ii)(A)** Replace the cartridge after three (3) hours of use or at the end of the workshift, whichever occurs first, unless the cartridge contains a NIOSH-approved end-of-service-life indicator (ESLI) to show when breakthrough occurs.

**1910.1048(g)(2)(ii)(B)** Unless the canister contains a NIOSH-approved ESLI to show when breakthrough occurs, replace canisters used in atmospheres up to 7.5 ppm (10xPEL) every four (4) hours and industrial-sized canisters used in atmospheres up to 75 ppm (100xPEL) every two (2) hours, or at the end of the workshift, whichever occurs first.

**1910.1048(g)(3)** *Respirator selection.* (i) The employer must select appropriate respirators from Table 1 in this section.

**TABLE 1—MINIMUM REQUIREMENTS FOR RESPIRATORY PROTECTION AGAINST FORMALDEHYDE**

| Condition of use or formaldehyde concentration (ppm) | Minimum required respirator[1] |
|---|---|
| Up to 7.5 ppm. (10×PEL) | Full facepiece with cartridges or canisters specifically approved for protection against formaldehyde.[2] |
| Up to 75 ppm. (100×PEL) | Full-face mask with chin style or chest or back mounted type, with industrial size canister specifically approved for protection against formaldehyde. Type C supplied air respirator, demand type, or continuous flow type, with full facepiece, hood, or helmet. |
| Above 75 ppm or unknown. (emergencies). (100×PEL). | Self-contained breathing apparatus (SCBA) with positive pressure full facepiece. Combination supplied-air, full facepiece positive pressure respirator with auxiliary self-contained air supply. |
| Firefighting | SCBA with positive pressure in full facepiece. |
| Escape | SCBA in demand or pressure demand mode. Full-face mask with chin style or front or back mounted type industrial size canister specifically approved for protection against formaldehyde. |

[1] Respirators specified for use at higher concentrations may be used at lower concentrations.

[2] A half-mask respirator with cartridges specifically approved for protection against formaldehyde can be substituted for the full facepiece respirator providing that effective gas-proof goggles are provided and used in combination with the half-mask respirator.

**1910.1048(g)(3)(ii)** The employer must provide a powered air-purifying respirator adequate to protect against formaldehyde exposure to any employee who has difficulty using a negative-pressure respirator.

**1910.1048(h)** ***Protective equipment and clothing.*** Employers shall comply with the provisions of 29 CFR 1910.132 and 29 CFR 1910.133. When protective equipment or clothing is provided under these provisions, the employer shall provide these protective devices at no cost to the employee and assure that the employee wears them.

**1910.1048(h)(1)** *Selection.* The employer shall select protective clothing and equipment based upon the form of formaldehyde to be encountered, the conditions of use, and the hazard to be prevented.

**1910.1048(h)(1)(i)** All contact of the eyes and skin with liquids containing 1 percent or more formaldehyde shall be prevented by the use of chemical protective clothing made of material impervious to formaldehyde and the use of other personal protective equipment, such as goggles and face shields, as appropriate to the operation.

**1910.1048(h)(1)(ii)** Contact with irritating or sensitizing materials shall be prevented to the extent necessary to eliminate the hazard.

**1910.1048(h)(1)(iii)** Where a face shield is worn, chemical safety goggles are also required if there is a danger of formaldehyde reaching the area of the eye.

**1910.1048(h)(1)(iv)** Full body protection shall be worn for entry into areas where concentrations exceed 100 ppm and for emergency reentry into areas of unknown concentration.

**1910.1048(h)(2)** *Maintenance of protective equipment and clothing.* (i) The employer shall assure that protective equipment and clothing that has become contaminated with formaldehyde is cleaned or laundered before its reuse.

**1910.1048(h)(2)(ii)** When ventilating formaldehyde-contaminated clothing and equipment, the employer shall establish a storage area so that employee exposure is minimized. Containers for contaminated clothing and equipment and storage areas shall have labels and signs containing the following information:

**DANGER**

**FORMALDEHYDE-CONTAMINATED [CLOTHING] EQUIPMENT**

**AVOID INHALATION AND SKIN CONTACT**

**1910.1048(h)(2)(iii)** The employer shall assure that only persons trained to recognize the hazards of formaldehyde remove the contaminated material from the storage area for purposes of cleaning, laundering, or disposal.

**1910.1048(h)(2)(iv)** The employer shall assure that no employee takes home equipment or clothing that is contaminated with formaldehyde.

**1910.1048(h)(2)(v)** The employer shall repair or replace all required protective clothing and equipment for each affected employee as necessary to assure its effectiveness.

**1910.1048(h)(2)(vi)** The employer shall inform any person who launders, cleans, or repairs such clothing or equipment of formaldehyde's potentially harmful effects and of procedures to safely handle the clothing and equipment.

**1910.1048(i)** ***Hygiene protection.*** (1) The employer shall provide change rooms, as described in 29 CFR 1910.141 for employees who are required to change from work clothing into protective clothing to prevent skin contact with formaldehyde.

**1910.1048(i)(2)** If employees' skin may become spashed with solutions containing 1 percent or greater formaldehyde, for example, because of equipment failure or improper work practices, the employer shall provide conveniently located quick drench showers and assure that affected employees use these facilities immediately.

**1910.1048(i)(3)** If there is any possibility that an employee's eyes may be splashed with solutions containing 0.1 percent or greater formaldehyde, the employer shall provide acceptable eyewash facilities within the immediate work area for emergency use.

**1910.1048(j)** ***Housekeeping.*** For operations involving formaldehyde liquids or gas, the employer shall conduct a program to detect leaks and spills, including regular visual inspections.

**1910.1048(j)(1)** Preventative maintenance of equipment, including surveys for leaks, shall be undertaken at regular intervals.

**1910.1048(j)(2)** In work areas where spillage may occur, the employer shall make provisions to contain the spill, to decontaminate the work area, and to dispose of the waste.

**1910.1048(j)(3)** The employer shall assure that all leaks are repaired and spills are cleaned promptly by employees wearing suitable protective equipment and trained in proper methods for cleanup and decontamination.

**1910.1048(j)(4)** Formaldehyde-contaminated waste and debris resulting from leaks or spills shall be placed for disposal in sealed containers bearing a label warning of formaldehyde's presence and of the hazards associated with formaldehyde.

**1910.1048(k)** ***Emergencies.*** For each workplace where there is the possibility of an emergency involving formaldehyde, the employer shall assure appropriate procedures are adopted to minimize injury and loss of life. Appropriate procedures shall be implemented in the event of an emergency.

**1910.1048(l)** ***Medical surveillance***— (1) *Employees covered.* (i) The employer shall institute medical surveillance programs for all employees exposed to formaldehyde at concentrations at or exceeding the action level or exceeding the STEL.

**1910.1048(l)(1)(ii)** The employer shall make medical surveillance available for employees who develop signs and symptoms of over exposure to formaldehyde and for all employees exposed to formaldehyde in emergencies. When determining whether an employee may be experiencing signs and symptoms of possible overexposure to formaldehyde, the employer may rely on the evidence that signs and symptoms associated with formaldehyde exposure will occur only in exceptional circumstances when airborne exposure is less than 0.1 ppm and when formaldehyde is present in material in concentrations less than 0.1 percent.

**1910.1048(l)(2)** *Examination by a physician.* All medical procedures, including administration of medical disease questionnaires, shall be performed by or under the supervision of a licensed physician and shall be provided without cost to the employee, without loss of pay, and at a reasonable time and place.

**1910.1048(l)(3)** *Medical disease questionnaire.* The employer shall make the following medical surveillance available to employees prior to assignment to a job where formaldehyde exposure is at or above the action level or above the STEL and annually thereafter. The employer shall also make the following medical surveillance available promptly upon determining that an employee is experiencing signs and symptoms indicative of possible overexposure to formaldehyde.

**1910.1048(l)(3)(i)** Administration of a medical disease questionnaire, such as in appendix D, which is designed to elicit information on work history, smoking history, any evidence of eye, nose, or throat irritation; chronic airway problems or hyperreactive airway disease: allergic skin conditions or dermatitis; and upper or lower respiratory problems.

**1910.1048(l)(3)(ii)** A determination by the physician, based on evaluation of the medical disease questionnaire, of whether a medical examination is necessary for employees not required to wear respirators to reduce exposure to formaldehyde.

**1910.1048(l)(4)** *Medical examinations.* Medical examinations shall be given to any employee who the physician feels, based on information in the medical disease questionnaire, may be at increased risk from exposure to formaldehyde and at the time of initial assignment and at least annually thereafter to all employees required to wear a respirator to reduce exposure to formaldehyde. The medical examination shall include:

**1910.1048(l)(4)(i)** A physical examination with emphasis on evidence of irritation or sensitization of the skin and respiratory system, shortness of breath, or irritation of the eyes.

**1910.1048(l)(4)(ii)** Laboratory examinations for respirator wearers consisting of baseline and annual pulmonary function tests. As a minimum, these tests shall consist of forced vital capacity (FVC), forced expiratory volume in one second ($FEV_1$), and forced expiratory flow (FEF).

**1910.1048(l)(4)(iii)** Any other test which the examining physician deems necessary to complete the written opinion.

**1910.1048(l)(4)(iv)** Counseling of employees having medical conditions that would be directly or indirectly aggravated by exposure to formaldehyde on the increased risk of impairment of their health.

**1910.1048(l)(5)** *Examinations for employees exposed in an emergency.* The employer shall make medical examinations available as soon as possible to all employees who have been exposed to formaldehyde in an emergency.

**1910.1048(l)(5)(i)** The examination shall include a medical and work history with emphasis on any evidence of upper or lower respiratory problems, allergic conditions, skin reaction or hypersensitivity, and any evidence of eye, nose, or throat irritation.

**1910.1048(l)(5)(ii)** Other examinations shall consist of those elements considered appropriate by the examining physician.

**1910.1048(l)(6)** *Information provided to the physician.* The employer shall provide the following information to the examining physician:

**1910.1048(l)(6)(i)** A copy of this standard and appendix A, C, D, and E;

**1910.1048(l)(6)(ii)** A description of the affected employee's job duties as they relate to the employee's exposure to formaldehyde;

**1910.1048(l)(6)(iii)** The representative exposure level for the employee's job assignment;

**1910.1048(l)(6)(iv)** Information concerning any personal protective equipment and respiratory protection used or to be used by the employee; and

**1910.1048(l)(6)(v)** Information from previous medical examinations of the affected employee within the control of the employer.

**1910.1048(l)(6)(vi)** In the event of a nonroutine examination because of an emergency, the employer shall provide to the physician as soon as possible: A description of how the emergency occurred and the exposure the victim may have received.

**1910.1048(l)(7)** *Physician's written opinion.* (i) For each examination required under this standard, the employer shall obtain a written opinion from the examining physician. This written opinion shall contain the results of the medical examination except that it shall not reveal specific findings or diagnoses unrelated to occupational exposure to formaldehyde. The written opinion shall include:

**1910.1048(l)(7)(i)(A)** The physician's opinion as to whether the employee has any medical condition that would place the employee at an increased risk of material impairment of health from exposure to formaldehyde;

**1910.1048(l)(7)(i)(B)** Any recommended limitations on the employee's exposure or changes in the use of personal protective equipment, including respirators;

**1910.1048(l)(7)(i)(C)** A statement that the employee has been informed by the physician of any medical conditions which would be aggravated by exposure to formaldehyde, whether these conditions may have resulted from past formaldehyde exposure or from exposure in an emergency, and whether there is a need for further examination or treatment.

**1910.1048(l)(7)(ii)** The employer shall provide for retention of the results of the medical examination and tests conducted by the physician.

**1910.1048(l)(7)(iii)** The employer shall provide a copy of the physician's written opinion to the affected employee within 15 days of its receipt.

**1910.1048(l)(8)** *Medical removal.* (i) The provisions of paragraph (l)(8) apply when an employee reports significant irritation of the mucosa of the eyes or the upper airways, respiratory sensitization, dermal irritation, or dermal sensitization attributed to workplace formaldehyde exposure. Medical removal provisions do not apply in the case of dermal irritation or dermal sensitization when the product suspected of causing the dermal condition contains less than 0.05% formaldehyde.

**1910.1048(l)(8)(ii)** An employee's report of signs or symptoms of possible overexposure to formaldehyde shall be evaluated by a physician selected by the employer pursuant to paragraph (l)(3). If the physician determines that a medical examination is not necessary under paragraph (l)(3)(ii), there shall be a two-week evaluation and remediation period to permit the employer to ascertain whether the signs or symptoms subside untreated or with the use of creams, gloves, first aid treatment or personal protective equipment. Industrial hygiene measures that limit the employee's exposure to formaldehyde may also be implemented during this period. The employee shall be referred immediately to a physician prior to expiration of the two-week period if the signs or symptoms worsen. Earnings, seniority and benefits may not be altered during the two-week period by virtue of the report.

**1910.1048(l)(8)(iii)** If the signs or symptoms have not subsided or been remedied by the end of the two-week period, or earlier if signs or symptoms warrant, the employee shall be examined by a physician selected by the employer. The physician shall presume, absent contrary evidence, that observed dermal irritation or dermal sensitization are not attributable to formaldehyde when products to which the affected employee is exposed contain less than 0.1% formaldehyde.

**1910.1048(l)(8)(iv)** Medical examinations shall be conducted in compliance with the requirements of paragraph (l)(5) (i) and (ii). Additional guidelines for conducting medical exams are contained in appendix C.

**1910.1048(l)(8)(v)** If the physician finds that significant irritation of the mucosa of the eyes or of the upper airways, respiratory sensitization, dermal irritation, or dermal sensitization result from workplace formaldehyde exposure and recommends restrictions or removal, the employer shall promptly comply with the restrictions or recommendation of removal. In the event of a recommendation of removal, the employer shall remove the effected employee from the current formaldehyde exposure and if possible, transfer the employee to work having no or significantly less exposure to formaldehyde.

**1910.1048(l)(8)(vi)** When an employee is removed pursuant to paragraph (l)(8)(v), the employer shall transfer the employee to comparable work for which the employee is qualified or can be trained in a short period (up to 6 months), where the formaldehyde exposures are as low as possible, but not higher than the action level. The employeer shall maintain the employee's current earnings, seniority, and other benefits. If there is no such work available, the employer shall maintain the employee's current earnings, seniority and other benefits until such work becomes available, until the employee is determined to be unable to return to workplace formaldehyde exposure, until the employee is determined to be able to return to the original job status, or for six months, whichever comes first.

**1910.1048(l)(8)(vii)** The employer shall arrange for a follow-up medical examination to take place within six months after the employee is removed pursuant to this paragraph. This examination shall determine if the employee can return to the original job status, or if the removal is to be permanent. The physician shall make a decision within six months of the date the employee was removed as to whether the employee can be returned to the original job status, or if the removal is to be permanent.

**1910.1048(l)(8)(viii)** An employer's obligation to provide earnings, seniority and other benefits to a removed employee may be reduced to the extent that the employee receives compensation for earnings lost during the period of removal either from a publicly or employer-funded compensation program or from employment with another employer made possible by virtue of the employee's removal.

**1910.1048(l)(8)(ix)** In making determinations of the formaldehyde content of materials under this paragraph the employer may rely on objective data.

**1910.1048(l)(9)** *Multiple physician review.* (i) After the employer selects the initial physician who conducts any medical examination or consultation to determine whether medical removal or restriction is appropriate, the employee may designate a second physician to review any findings, determinations or recommendations of the initial physician and to conduct such examinations, consultations, and laboratory tests as the second physician deems necessary and appropriate to evaluate the effects of formaldehyde exposure and to facilitate this review.

**1910.1048(l)(9)(ii)** The employer shall promptly notify an employee of the right to seek a second medical opinion after each occasion that an initial physician conducts a medical examination or consultation for the purpose of medical removal or restriction.

**1910.1048(l)(9)(iii)** The employer may condition its participation in, and payment for, the multiple physician review mechanism upon the employee doing the following within fifteen (15) days after receipt of the notification of the right to seek a second medical opinion, or receipt of the initial physician's written opinion, whichever is later;

**1910.1048(l)(9)(iii)(A)** The employee informs the employer of the intention to seek a second medical opinion, and

**1910.1048(l)(9)(iii)(B)** The employee initiates steps to make an appointment with a second physician.

**1910.1048(l)(9)(iv)** If the findings, determinations or recommendations of the second physician differ from those of the initial physician, then the employer and the employee shall assure that efforts are made for the two physicians to resolve the disagreement. If the two physicians are unable to quickly resolve their disagreement, then the employer and the employee through their respective physicians shall designate a third physician who shall be a specialist in the field at issue:

**1910.1048(l)(9)(iv)(A)** To review the findings, determinations or recommendations of the prior physicians; and

**1910.1048(l)(9)(iv)(B)** To conduct such examinations, consultations, laboratory tests and discussions with the prior physicians as the third physician deems necessary to resolve the disagreement of the prior physicians.

**1910.1048(l)(9)(v)** In the alternative, the employer and the employee or authorized employee representative may jointly designate such third physician.

**1910.1048(l)(9)(vi)** The employer shall act consistent with the findings, determinations and recommendations of the third physician, unless the employer and the employee reach an agreement which is otherwise consistent with the recommendations of at least one of the three physicians.

**1910.1048(m)** ***Hazard communication***— (1) *General.* Communication of the hazards associated with formaldehyde in the workplace shall be governed by the requirements of paragraph (m). The definitions of 29 CFR 1910.1200(c) shall apply under this paragraph.

**1910.1048(m)(1)(i)** The following shall be subject to the hazard communication requirements of this paragraph: Formaldehyde gas, all mixtures or solutions composed of greater than 0.1 percent formaldehyde, and materials capable of releasing formaldehyde into the air, under reasonably foreseeable conditions of use, at concentrations reaching or exceeding 0.1 ppm.

**1910.1048(m)(1)(ii)** As a minimum, specific health hazards that the employer shall address are: Cancer, irritation and sensitization of the skin and respiratory system, eye and throat irritation, and acute toxicity.

**1910.1048(m)(2)** Manufacturers and importers who produce or import formaldehyde or formaldehyde-containing products shall provide downstream employers using or handling these products with an objective determination through the required labels and MSDSs if these items may constitute a health hazard within the meaning of 29 CFR 1910.1200(d) under normal conditions of use.

**1910.1048(m)(3)** *Labels.* (i) The employer shall assure that hazard warning labels complying with the requirements of 29 CFR 1910.1200(f) are affixed to all containers of materials listed in paragraph (m)(1)(i), except to the extent that 29 CFR 1910.1200(f) is inconsistent with this paragraph.

**1910.1048(m)(3)(ii)** *Information on labels.* As a minimum, for all materials listed in paragraph (m)(1)(i) capable of releasing formaldehyde at levels of 0.1 ppm to 0.5 ppm, labels shall identify that the product contains formaldehyde; list the name and address of the responsible party; and state that physical and health hazard information is readily available from the employer and from material safety data sheets.

**1910.1048(m)(3)(iii)** For materials listed in paragraph (m)(1)(i) capable of releasing formaldehyde at levels above 0.5 ppm, labels shall appropriately address all hazards as defined in 29 CFR 1910.1200(d) and 29 CFR 1910.1200 appendices A and B, including respiratory sensitization, and shall contain the words "Potential Cancer Hazard."

**1910.1048(m)(3)(iv)** In making the determinations of anticipated levels of formaldehyde release, the employer may rely on objective data indicating the extent of potential formaldehyde release under reasonably foreseeable conditions of use.

**1910.1048(m)(3)(v)** *Substitute warning labels.* The employer may use warning labels required by other statutes, regulations, or ordinances which impart the same information as the warning statements required by this paragraph.

**1910.1048(m)(4)** *Material safety data sheets.* (i) Any employer who uses formaldehyde-containing materials listed in paragraph (m)(1)(i) shall comply with the requirements of 29 CFR 1910.1200(g) with regard to the development and updating of material safety data sheets.

**1910.1048(m)(4)(ii)** Manufacturers, importers, and distributors of formaldehyde-containing materials listed in paragraph (m)(1)(i) shall assure that material safety data sheets and updated information are provided to all employers purchasing such materials at the time of the initial shipment and at the time of the first shipment after a material safety data sheet is updated.

**1910.1048(m)(5)** *Written hazard communication program.* The employer shall develop, implement, and maintain at the workplace, a written hazard communication program for formaldehyde exposures in the workplace, which at a minimum describes how the requirements specified in this paragraph for labels and other forms of warning and material safety data sheets, and paragraph (n) for employee information and training, will be met. Employers in multi-employer workplaces shall comply with the requirements of 29 CFR 1910.1200(e)(2).

**1910.1048(n)** ***Employee information and training***— (1) *Participation.* The employer shall assure that all employees who are assigned to workplaces where there is exposure to formaldehyde participate in a training program, except that where the employer can show, using objective data, that employees are not exposed to formaldehyde at or above 0.1 ppm, the employer is not required to provide training.

**1910.1048(n)(2)** *Frequency.* Employers shall provide such information and training to employees at the time of initial assignment, and whenever a new exposure to formaldehyde is introduced into the work area. The training shall be repeated at least annually.

**1910.1048(n)(3)** *Training program.* The training program shall be conducted in a manner which the employee is able to understand and shall include:

**1910.1048(n)(3)(i)** A discussion of the contents of this regulation and the contents of the Material Safety Data Sheet.

**1910.1048(n)(3)(ii)** The purpose for and a description of the medical surveillance program required by this standard, including:

**1910.1048(n)(3)(ii)(A)** A description of the potential health hazards associated with exposure to formaldehyde and a description of the signs and symptoms of exposure to formaldehyde.

**1910.1048(n)(3)(ii)(B)** Instructions to immediately report to the employer the development of any adverse signs or symptoms that the employee suspects is attributable to formaldehyde exposure.

**1910.1048(n)(3)(iii)** Description of operations in the work area where formaldehyde is present and an explanation of the safe work practices appropriate for limiting exposure to formaldehyde in each job;

**1910.1048(n)(3)(iv)** The purpose for, proper use of, and limitations of personal protective clothing and equipment;

**1910.1048(n)(3)(v)** Instructions for the handling of spills, emergencies, and clean-up procedures;

**1910.1048(n)(3)(vi)** An explanation of the importance of engineering and work practice controls for employee protection and any necessary instruction in the use of these controls; and

**1910.1048(n)(3)(vii)** A review of emergency procedures including the specific duties or assignments of each employee in the event of an emergency.

**1910.1048(n)(4)** *Access to training materials.* (i) The employer shall inform all affected employees of the location of written training materials and shall make these materials readily available, without cost, to the affected employees.

**1910.1048(n)(4)(ii)** The employer shall provide, upon request, all training materials relating to the employee training program to the Assistant Secretary and the Director.

**1910.1048(o)** ***Recordkeeping***— (1) *Exposure measurements.* The employer shall establish and maintain an accurate record of all measurements taken to monitor employee exposure to formaldehyde. This record shall include:

**1910.1048(o)(1)(i)** The date of measurement;

**1910.1048(o)(1)(ii)** The operation being monitored;

**1910.1048(o)(1)(iii)** The methods of sampling and analysis and evidence of their accuracy and precision;

**1910.1048(o)(1)(iv)** The number, durations, time, and results of samples taken;

**1910.1048(o)(1)(v)** The types of protective devices worn; and

**1910.1048(o)(1)(vi)** The names, job classifications, social security numbers, and exposure estimates of the employees whose exposures are represented by the actual monitoring results.

**1910.1048(o)(2)** *Exposure determinations.* Where the employer has determined that no monitoring is required under this standard, the employer shall maintain a record of the objective data relied upon to support the determination that no employee is exposed to formaldehyde at or above the action level.

**1910.1048(o)(3)** *Medical surveillance.* The employer shall establish and maintain an accurate record for each employee subject to medical surveillance under this standard. This record shall include:

**1910.1048(o)(3)(i)** The name and social security number of the employee;

**1910.1048(o)(3)(ii)** The physician's written opinion;

**1910.1048(o)(3)(iii)** A list of any employee health complaints that may be related to exposure to formaldehyde; and

**1910.1048(o)(3)(iv)** A copy of the medical examination results, including medical disease questionnaires and results of any medical tests required by the standard or mandated by the examining physician.

**1910.1048(o)(4)** *Respirator fit testing.* (i) The employer shall establish and maintain accurate records for employees subject to negative pressure respirator fit testing required by this standard.

**1910.1048(o)(4)(ii)** This record shall include:

**1910.1048(o)(4)(ii)(A)** A copy of the protocol selected for respirator fit testing.

**1910.1048(o)(4)(ii)(B)** A copy of the results of any fit testing performed.

**1910.1048(o)(4)(ii)(C)** The size and manufacturer of the types of respirators available for selection.

**1910.1048(o)(4)(ii)(D)** The date of the most recent fit testing, the name and social security number of each tested employee, and the respirator type and facepiece selected.

**1910.1048(o)(5)** *Record retention.* The employer shall retain records required by this standard for at least the following periods:

**1910.1048(o)(5)(i)** Exposure records and determinations shall be kept for at least 30 years.

**1910.1048(o)(5)(ii)** Medical records shall be kept for the duration of employment plus 30 years.

**1910.1048(o)(5)(iii)** Respirator fit testing records shall be kept until replaced by a more recent record.

**1910.1048(o)(6)** *Availability of records.* (i) Upon request, the employer shall make all records maintained as a requirement of this standard available for examination and copying to the Assistant Secretary and the Director.

**1910.1048(o)(6)(ii)** The employer shall make employee exposure records, including estimates made from representative monitoring and available upon request for examination, and copying to the subject employee, or former employee, and employee representatives in accordance with 29 CFR 1910.20(a)-(e) and (g)-(i).

**1910.1048(o)(6)(iii)** Employee medical records required by this standard shall be provided upon request for examination and coying, to the subject employee or former employee or to anyone having the specific written consent of the subject employee or former employee in accordance with 29 CFR 1910.20(a)-(e) and (g)-(i).

**1910.1048(p)** ***Dates***— (1) *Effective dates*—(i) *General.* This section shall become effective February 2, 1988, except as noted below.

**1910.1048(p)(1)(ii)** *Laboratories.* This standard shall become effective for anatomy, histology, and pathology laboratories February 2, 1988, except as noted in the start-up date section. For all other laboratories, paragraphs (a) and (c) of this standard shall become effective February 2, 1988, and paragraphs (b) and (d)-(o) of this standard shall become effective on September 1, 1988 except as noted in the start-up date section.

**1910.1048(p)(2)** *Start-up dates*—(i) *Exposure determinations.* Initial monitoring or objective determinations that no monitoring is required by the standard shall be completed by 6 months after the effective date of the standard.

**1910.1048(p)(2)(ii)** *Medical surveillance.* The initial medical surveillance of all eligible employees shall be completed by 6 months after the effective date of the standard.

**1910.1048(p)(2)(iii)** *Emergencies.* The emergency procedures required by this standard shall be implemented by 6 months after the effective date of the standard.

**1910.1048(p)(2)(iv)** *Respiratory protection.* Respiratory protection as required in this standard shall be provided as soon as possible and no later than 9 months after the effective date of the standard.

**1910.1048(p)(2)(v)** *Engineering and work practice controls.* Engineering and work practice controls required by this standard shall be implemented as soon as possible, but no later than one year after the effective date of the standard.

**1910.1048(p)(2)(vi)** *Employee training.* Written materials for employee training shall be updated as soon as possible, but no later than 2 months after the effective date of the standard.

**1910.1048(p)(3)** *Start-up dates of amended paragraphs*—(i) *Respiratory protection.* Respiratory protection required to meet the amended PEL of 0.75 ppm TWA shall be provided as soon as possible but no later than September 24, 1992.

**1910.1048(p)(3)(ii)** *Engineering and work practice controls.* Engineering and work practice controls required to meet the amended PEL of 0.75 ppm TWA shall be implemented as soon as possible, but no later than June 26, 1993.

**1910.1048(p)(3)(iii)** *Medical removal protection.* The medical removal protection provisions including the multiple physician review mechanism shall be implemented no later than December 28, 1992.

**1910.1048(p)(3)(iv)** *Hazard communication.* The labeling provisions contained in amended paragraph (m) of this standard shall be implemented no later than December 28, 1992. Labeling of containers of formaldehyde products shall continue to comply with the provisions of 29 CFR 1910.1200(e)-(j) until that time.

**1910.1048(p)(3)(v)** *Training.* The periodic training mandated for all employees exposed to formaldehyde between 0.1 ppm and 0.5 ppm shall begin no later than August 25, 1992.

## APPENDIX A TO §1910.1048—SUBSTANCE TECHNICAL GUIDELINES FOR FORMALIN

The following Substance Technical Guideline for Formalin provides information on uninhibited formalin solution (37% formaldehyde, no methanol stabilizer). It is designed to inform employees at the production level of their rights and duties under the formaldehyde standard whether their job title defines them as workers or supervisors. Much of the information provided is general; however, some information is specific for formalin. When employee exposure to formaldehyde is from resins capable of releasing formaldehyde, the resin itself and other impurities or decomposition products may also be toxic, and employers should include this information as well when informing employees of the hazards associated with the materials they handle. The precise hazards associated with exposure to formaldehyde depend both on the form (solid, liquid, or gas) of the material and the concentration of formaldehyde present. For example, 37-50 percent solutions of formaldehyde present a much greater hazard to the skin and eyes from spills or splashes than solutions containing less than 1 percent formaldehyde. Individual Substance Technical Guidelines used by the employer for training employees should be modified to properly give information on the material actually being used.

### Substance Identification

*Chemical Name:* Formaldehyde

*Chemical Family:* Aldehyde

*Chemical Formula:* HCHO

*Molecular Weight:* 30.03

*Chemical Abstracts Service Number (CAS Number):* 50-00-0

*Synonyms:* Formalin; Formic Aldehyde; Paraform; Formol; Formalin (Methanol-free); Fyde; Formalith; Methanal; Methyl Aldehyde; Methylene Glycol; Methylene Oxide; Tetraoxymethalene; Oxomethane; Oxymethylene

### Components and Contaminants

*Percent:* 37.0 Formaldehyde

*Percent:* 63.0 Water

(**Note**—Inhibited solutions contain methanol.)

*Other Contaminants:* Formic acid (alcohol free)

*Exposure Limits:*

OSHA TWA—0.75 ppm

OSHA STEL—2 ppm

### Physical Data

*Description:* Colorless liquid, pungent odor

*Boiling point:* 214 °F (101 °C)

*Specific Gravity:* 1.08 ($H_2O$=1 @ 20 °C)

*pH:* 2.8-4.0

*Solubility in Water:* Miscible

*Solvent Solubility:* Soluble in alcohol and acetone

*Vapor Density:* 1.04 (Air=1 @ 20 °C)

*Odor Threshold:* 0.8-1 ppm

### Fire and Explosion Hazard

Moderate fire and explosion hazard when exposed to heat or flame.

The flash point of 37% formaldehyde solutions is above normal room temperature, but the explosion range is very wide, from 7 to 73% by volume in air.

Reaction of formaldehyde with nitrogen dioxide, nitromethane, perchloric acid and aniline, or peroxyformic acid yields explosive compounds.

*Flash Point:* 185 °F (85 °C) closed cup

*Lower Explosion Limit:* 7%

*Upper Explosion Limit:* 73%

*Autoignition Temperature:* 806 °F (430 °C)

*Flammability Class (OSHA):* III A

*Extinguishing Media:* Use dry chemical, "alcohol foam", carbon dioxide, or water in flooding amounts as fog. Solid streams may not be effective. Cool fire-exposed containers with water from side until well after fire is out.

Use of water spray to flush spills can also dilute the spill to produce nonflammable mixtures. Water runoff, however, should be contained for treatment.

*National Fire Protection Association Section 325M Designation:*

*Health:* 2—Materials hazardous to health, but areas may be entered with full-faced mask self-contained breathing apparatus which provides eye protection.

*Flammability:* 2—Materials which must be moderately heated before ignition will occur. Water spray may be used to extinguish the fire because the material can be cooled below its flash point.

*Reactivity:* D—Materials which (in themselves) are normally stable even under fire exposure conditions and which are not reactive with water. Normal fire fighting procedures may be used.

### Reactivity

*Stability:* Formaldehyde solutions may self-polymerize to form paraformaldehyde which precipitates.

*Incompatibility (Materials to Avoid):* Strong oxidizing agents, caustics, strong alkalies, isocyanates, anhydrides, oxides, and inorganic acids. Formaldehyde reacts with hydrochloric acid to form the potent carcinogen, bis-chloromethyl ether. Formaldehyde reacts with nitrogen dioxide, nitromethane, perchloric acid and aniline, or peroxyformic acid to yield explosive compounds. A violent reaction occurs when formaldehyde is mixed with strong oxidizers.

Hazardous Combustion or Decomposition Products: Oxygen from the air can oxidize formaldehyde to formic acid, especially when heated. Formic acid is corrosive.

### Health Hazard Data

#### Acute Effects of Exposure

*Ingestion (Swallowing):* Liquids containing 10 to 40% formaldehyde cause severe irritation and inflammation of the mouth, throat, and stomach. Severe stomach pains will follow ingestion with possible loss of consciousness and death. Ingestion of dilute formaldehyde solutions (0.03-0.04%) may cause discomfort in the stomach and pharynx.

*Inhalation (Breathing):* Formaldehyde is highly irritating to the upper respiratory tract and eyes. Concentrations of 0.5 to 2.0 ppm may irritate the eyes, nose, and throat of some individuals. Concentrations of 3 to 5 ppm also cause tearing of the eyes and are intolerable to some persons. Concentrations of 10 to 20 ppm cause difficulty in breathing, burning of the nose and throat, cough, and heavy tearing of the eyes, and 25 to 30 ppm causes severe respiratory tract injury leading to pulmonary edema and pneumonitis. A concentration of 100 ppm is immediately dangerous to life and health. Deaths from accidental exposure to high concentrations of formaldehyde have been reported.

*Skin (Dermal):* Formalin is a severe skin irritant and a sensitizer. Contact with formalin causes white discoloration, smarting, drying, cracking, and scaling. Prolonged and repeated contact can cause numbness and a hardening or tanning of the skin. Previously exposed persons may react to future exposure with an allergic eczematous dermatitis or hives.

*Eye Contact:* Formaldehyde solutions splashed in the eye can cause injuries ranging from transient discomfort to severe, permanent corneal clouding and loss of vision. The severity of the effect depends on the concentration of formaldehyde in the solution and whether or not the eyes are flushed with water immediately after the accident.

**Note.**—The perception of formaldehyde by odor and eye irritation becomes less sensitive with time as one adapts to formaldehyde. This can lead to overexposure if a worker is relying on formaldehyde's warning properties to alert him or her to the potential for exposure.

*Acute Animal Toxicity:*

*Oral,* rats: LD50=800 mg/kg

*Oral,* mouse: LD50=42 mg/kg

*Inhalation,* rats: LCLo=250 mg/kg

*Inhalation,* mouse: LCLo=900 mg/kg

*Inhalation,* rats: LC50=590 mg/kg

### Chronic Effects of Exposure

*Carcinogenicity:* Formaldehyde has the potential to cause cancer in humans. Repeated and prolonged exposure increases the risk. Various animal experiments have conclusively shown formaldehyde to be a carcinogen in rats. In humans, formaldehyde exposure has been associated with cancers of the lung, nasopharynx and oropharynx, and nasal passages.

*Mutagenicity:* Formaldehyde is genotoxic in several *in vitro* test systems showing properties of both an initiator and a promoter.

*Toxicity:* Prolonged or repeated exposure to formaldehyde may result in respiratory impairment. Rats exposed to formaldehyde at 2 ppm developed benign nasal tumors and changes of the cell structure in the nose as well as inflamed mucous membranes of the nose. Structural changes in the epithelial cells in the human nose have also been observed. Some persons have developed asthma or bronchitis following exposure to formaldehyde, most often as the result of an accidental spill involving a single exposure to a high concentration of formaldehyde.

### Emergency and First Aid Procedures

*Ingestion (Swallowing):* If the victim is conscious, dilute, inactivate, or absorb the ingested formaldehyde by giving milk, activated charcoal, or water. Any organic material will inactivate formaldehyde. Keep affected person warm and at rest. Get medical attention immediately. If vomiting occurs, keep head lower than hips.

*Inhalation (Breathing):* Remove the victim from the exposure area to fresh air immediately. Where the formaldehyde concentration may be very high, each rescuer must put on a selfcontained breathing apparatus before attempting to remove the victim, and medical personnel should be informed of the formaldehyde exposure immediately. If breathing has stopped, give artificial respiration. Keep the affected person warm and at rest. Qualified first-aid or medical personnel should administer oxygen, if available, and maintain the patient's airways and blood pressure until the victim can be transported to a medical facility. If exposure results in a highly irritated upper respiratory tract and coughing continues for more than 10 minutes, the worker should be hospitalized for observation and treatment.

*Skin Contact:* Remove contaminated clothing (including shoes) immediately. Wash the affected area of your body with soap or mild detergent and large amounts of water until no evidence of the chemical remains (at least 15 to 20 minutes). If there are chemical burns, get first aid to cover the area with sterile, dry dressing, and bandages. Get medical attention if you experience appreciable eye or respiratory irritation.

*Eye Contact:* Wash the eyes immediately with large amounts of water occasionally lifting lower and upper lids, until no evidence of chemical remains (at least 15 to 20 minutes). In case of burns, apply sterile bandages loosely without medication. Get medical attention immediately. If you have experienced appreciable eye irritation from a splash or excessive exposure, you should be referred promptly to an opthamologist for evaluation.

### Emergency Procedures

*Emergencies:* If you work in an area where a large amount of formaldehyde could be released in an accident or from equipment failure, your employer must develop procedures to be followed in event of an emergency. You should be trained in your specific duties in the event of an emergency, and it is important that you clearly understand these duties. Emergency equipment must be accessible and you should be trained to use any equipment that you might need. Formaldehyde contaminated equipment must be cleaned before reuse.

If a spill of appreciable quantity occurs, leave the area quickly unless you have specific emergency duties. Do not touch spilled material. Designated persons may stop the leak and shut off ignition sources if these procedures can be done without risk. Designated persons should isolate the hazard area and deny entry except for necessary people protected by suitable protective clothing and respirators adequate for the exposure. Use water spray to reduce vapors. Do not smoke, and prohibit all flames or flares in the hazard area.

*Special Firefighting Procedures:* Learn procedures and responsibilities in the event of a fire in your workplace. Become familiar with the appropriate equipment and supplies and their location. In firefighting, withdraw immediately in case of rising sound from venting safety device or any discoloration of storage tank due to fire.

### Spill, Leak, and Disposal Procedures

*Occupational Spill:* For small containers, place the leaking container in a well ventilated area. Take up small spills with absorbent material and place the waste into properly labeled containers for later disposal. For larger spills, dike the spill to minimize contamination and facilitate salvage or disposal. You may be able to neutralize the spill with sodium hydroxide or sodium sulfite. Your employer must comply with EPA rules regarding the clean-up of toxic waste and notify state and local authorities, if required. If the spill is greater than 1,000 lb/day, it is reportable under EPA's Superfund legislation.

*Waste Disposal:* Your employer must dispose of waste containing formaldehyde in accordance with applicable local, state, and Federal law and in a manner that minimizes exposure of employees at the site and of the clean-up crew.

### Monitoring and Measurement Procedures

*Monitoring Requirements:* If your exposure to formaldehyde exceeds the 0.5 ppm action level or the 2 ppm STEL, your employer must monitor your exposure. Your employer need not measure every exposure if a "high exposure" employee can be identified. This person usually spends the greatest amount of time nearest the process equipment. If you are a "representative employee", you will be asked to wear a sampling device to collect formaldehyde. This device may be a passive badge, a sorbent tube attached to a pump, or an impinger containing liquid. You should perform your work as usual, but inform the person who is conducting the monitoring of any difficulties you are having wearing the device.

*Evaluation of 8-hour Exposure:* Measurements taken for the purpose of determining time-weighted average (TWA) exposures are best taken with samples covering the full shift. Samples collected must be taken from the employee's breathing zone air.

*Short-term Exposure Evaluation:* If there are tasks that involve brief but intense exposure to formaldehyde, employee exposure must be measured to assure compliance with the STEL. Sample collections are for brief periods, only 15 minutes, but several samples may be needed to identify the peak exposure.

*Monitoring Techniques:* OSHA's only requirement for selecting a method for sampling and analysis is that the methods used accurately evaluate the concentration of formaldehyde in employees' breathing zones. Sampling and analysis may be performed by collection of formaldehyde on liquid or solid sorbents with subsequent chemical analysis. Sampling and analysis may also be performed by passive diffusion monitors and shortterm exposure may be measured by instruments such as real-time continuous monitoring systems and portable direct reading instruments.

*Notification of Results:* Your employer must inform you of the results of exposure monitoring representative of your job. You may be informed in writing, but posting the results where you have ready access to them constitutes compliance with the standard.

### Protective Equipment and Clothing

[Material impervious to formaldehyde is needed if the employee handles formaldehyde solutions of 1% or more. Other employees may also require protective clothing or equipment to prevent dermatitis.]

*Respiratory Protection:* Use NIOSH-approved full facepiece negative pressure respirators equipped with approved cartridges or canisters within the use limitations of these devices. (Present restrictions on cartridges and canisters do not permit them to be used for a full workshift.) In all other situations, use positive pressure respirators such as the positive-pressure air purifying respirator or the self-contained breathing apparatus (SCBA). If you use a negative pressure respirator, your employer must provide you with fit testing of the respirator at least once a year in accordance with the procedures outlined in Appendix E.

*Protective Gloves:* Wear protective (impervious) gloves provided by your employer, at no cost, to prevent contact with formalin. Your employer should select these gloves based on the results of permeation testing and in accordance with the ACGIH Guidelines for Selection of Chemical Protective Clothing.

*Eye Protection:* If you might be splashed in the eyes with formalin, it is essential that you wear goggles or some other type of complete protection for the eye. You may also need a face shield if your face is likely to be splashed with formalin, but you must not substitute face shields for eye protection. (This section pertains to formaldehyde solutions of 1% or more.)

*Other Protective Equipment:* You must wear protective (impervious) clothing and equipment provided by your employer at no cost to prevent repeated or prolonged contact with formaldehyde liquids. If you are required to change into wholebody chemical protective clothing, your employer must provide a change room for your privacy and for storage of your normal clothing.

If you are splashed with formaldehyde, use the emergency showers and eyewash fountains provided by your employer immediately to prevent serious injury. Report the incident to your supervisor and obtain necessary medical support.

### ENTRY INTO AN IDLH ATMOSHPERE

Enter areas where the formaldehyde concentration might be 100 ppm or more only with complete body protection including a self-contained breathing apparatus with a full facepiece operated in a positive pressure mode or a supplied air respirator with full facepiece and operated in a positive pressure mode. This equipment is essential to protect your life and health under such extreme conditions.

### Engineering Controls

Ventilation is the most widely applied engineering control method for reducing the concentration of airborne substances in the breathing zones of workers. There are two distinct types of ventilation.

*Local Exhaust:* Local exhaust ventilation is designed to capture airborne contaminants as near to the point of generation as possible. To protect you, the direction of contaminant flow must always be toward the local exhaust system inlet and away from you.

*General (Mechanical):* General dilution ventilation involves continuous introduction of fresh air into the workroom to mix with the contaminated air and lower your breathing zone concentration of formaldehyde. Effectiveness depends on the number of air changes per hour. Where devices emitting formaldehyde are spread out over a large area, general dilution ventilation may be the only practical method of control.

*Work Practices:* Work practices and administrative procedures are an important part of a control system. If you are asked to perform a task in a certain manner to limit your exposure to formaldehyde, it is extremely important that you follow these procedures.

### *Medical Surveillance*

Medical surveillance helps to protect employees' health. You are encouraged strongly to participate in the medical surveillance program.

Your employer must make a medical surveillance program available at no expense to you and at a reasonable time and place if you are exposed to formaldehyde at concentrations above 0.5 ppm as an 8-hour average or 2 ppm over any 15-minute period. You will be offered medical surveillance at the time of your initial assignment and once a year afterward as long as your exposure is at least 0.5 ppm (TWA) or 2 ppm (STEL). Even if your exposure is below these levels, you should inform your employer if you have signs and symptoms that you suspect, through your training, are related to your formaldehyde exposure because you may need medical surveillance to determine if your health is being impaired by your exposure.

The surveillance plan includes:

(a) A medical disease questionnaire.

(b) A physical examination if the physician determines this is necessary.

If you are required to wear a respirator, your employer must offer you a physical examination and a pulmonary function test every year.

The physician must collect all information needed to determine if you are at increased risk from your exposure to formaldehyde. At the physician's discretion, the medical examination may include other tests, such as a chest x-ray, to make this determination.

After a medical examination the physician will provide your employer with a written opinion which includes any special protective measures recommended and any restrictions on your exposure. The physician must inform you of any medical conditions you have which would be aggravated by exposure to formaldehyde.

All records from your medical examinations, including disease surveys, must be retained at your employer's expense.

### EMERGENCIES

If you are exposed to formaldehyde in an emergency and develop signs or symptoms associated with acute toxicity from formaldehyde exposure, your employer must provide you with a medical examination as soon as possible. This medical examination will include all steps necessary to stabilize your health. You may be kept in the hospital for observation if your symptoms are severe to ensure that any delayed effects are recognized and treated.

## APPENDIX B TO §1910.1048—SAMPLING STRATEGY AND ANALYTICAL METHODS FOR FORMALDEHYDE

To protect the health of employees, exposure measurements must be unbiased and representative of employee exposure. The proper measurement of employee exposure requires more than a token commitment on the part of the employer. OSHA's mandatory requirements establish a baseline; under the best of circumstances all questions regarding employee exposure will be answered. Many employers, however, will wish to conduct more extensive monitoring before undertaking expensive commitments, such as engineering controls, to assure that the modifications are truly necessary. The following sampling strategy, which was developed at NIOSH by Nelson A. Leidel, Kenneth A. Busch, and Jeremiah R. Lynch and described in NIOSH publication No. 77-173 (Occupational Exposure Sampling Strategy Manual) will assist the employer in developing a strategy for determining the exposure of his or her employees.

There is no one correct way to determine employee exposure. Obviously, measuring the exposure of every employee exposed to formaldehyde will provide the most information on any given day. Where few employees are exposed, this may be a practical solution. For most employers, however, use of the following strategy will give just as much information at less cost.

Exposure data collected on a single day will not automatically guarantee the employer that his or her workplace is always in compliance with the formaldehyde standard. This does not imply, however, that it is impossible for an employer to be sure that his or her worksite is in compliance with the standard. Indeed, a properly designed sampling strategy showing that all employees are exposed below the PELs, at least with a 95 percent certainty, is compelling evidence that the exposure limits are being achieved provided that measurements are conducted using valid sampling strategy and approved analytical methods.

There are two PELs, the TWA concentration and the STEL. Most employers will find that one of these two limits is more critical in the control of their operations, and OSHA expects that the employer will concentrate monitoring efforts on the critical component. If the more difficult exposure is controlled, this information, along with calculations to support the assumptions, should be adequate to show that the other exposure limit is also being achieved.

### *Sampling Strategy*

### Determination of the Need for Exposure Measurements

The employer must determine whether employees may be exposed to concentrations in excess of the action level. This determination becomes the first step in an employee exposure monitoring program that minimizes employer sampling burdens while providing adequate employee protection. If employees may be exposed above the action level, the employer must measure exposure. Otherwise, an objective determination that employee exposure is low provides adequate evidence that exposure potential has been examined.

The employer should examine all available relevant information, *eg.* insurance company and trade association data and information from suppliers or exposure data collected from similar operations. The employer may also use previouslyconducted sampling including area monitoring. The employer must make a determination relevant to each operation although this need not be on a separate piece of paper. If the employer can demonstrate conclusively that no employee is exposed above the action level or the STEL through the use of objective data, the employer need proceed no further on employee exposure monitoring until such time that conditions have changed and the determination is no longer valid.

If the employer cannot determine that employee exposure is less than the action level and the STEL, employee exposure monitoring will have to be conducted.

### Workplace Material Survey

The primary purpose of a survey of raw material is to determine if formaldehyde is being used in the work environment and if so, the conditions under which formaldehyde is being used.

The first step is to tabulate all situations where formaldehyde is used in a manner such that it may be released into the workplace atmosphere or contaminate the skin. This information should be available through analysis of company records and information on the MSDSs available through provisions of this standard and the Hazard Communication standard.

If there is an indication from materials handling records and accompanying MSDSs that formaldehyde is being used in the following types of processes or work operations, there may be a potential for releasing formaldehyde into the workplace atmosphere:

(1) Any operation that involves grinding, sanding, sawing, cutting, crushing, screening, sieving, or any other manipulation of material that generates formaldehyde-bearing dust

(2) Any processes where there have been employee complaints or symptoms indicative of exposure to formaldehyde

(3) Any liquid or spray process involving formaldehyde

(4) Any process that uses formaldehyde in preserved tissue

(5) Any process that involves the heating of a formaldehydebearing resin.

Processes and work operations that use formaldehyde in these manners will probably require further investigation at the worksite to determine the extent of employee monitoring that should be conducted.

### Workplace Observations

To this point, the only intention has been to provide an indication as to the existence of potentially exposed employees. With this information, a visit to the workplace is needed to observe work operations, to identify potential health hazards, and to determine whether any employees may be exposed to hazardous concentrations of formaldehyde.

In many circumstances, sources of formaldehyde can be identified through the sense of smell. However, this method of detection should be used with caution because of olfactory fatigue.

Employee location in relation to source of formaldehyde is important in determining if an employee may be significantly exposed to formaldehyde. In most instances, the closer a worker is to the source, the higher the probability that a significant exposure will occur.

Other characteristics should be considered. Certain high temperature operations give rise to higher evaporation rates. Locations of open doors and windows provide natural ventilation that tend to dilute formaldehyde emissions. General room ventilation also provides a measure of control.

## Calculation of Potential Exposure Concentrations

By knowing the ventilation rate in a workplace and the quantity of formaldehyde generated, the employer may be able to determine by calculation if the PELs might be exceeded. To account for poor mixing of formaldehyde into the entire room, locations of fans and proximity of employees to the work operation, the employer must include a safety factor. If an employee is relatively close to a source, particularly if he or she is located downwind, a safety factor of 100 may be necessary. For other situations, a factor of 10 may be acceptable. If the employer can demonstrate through such calculations that employee exposure does not exceed the action level or the STEL, the employer may use this information as objective data to demonstrate compliance with the standard.

## Sampling Strategy

Once the employer determines that there is a possibility of substantial employee exposure to formaldehyde, the employer is obligated to measure employee exposure.

The next step is selection of a maximum risk employee. When there are different processes where employees may be exposed to formaldehyde, a maximum risk employee should be selected for each work operation.

Selection of the maximum risk employee requires professional judgment. The best procedure for selecting the maximum risk employee is to observe employees and select the person closest to the source of formaldehyde. Employee mobility may affect this selection; *eg.* if the closest employee is mobile in his tasks, he may not be the maximum risk employee. Air movement patterns and differences in work habits will also affect selection of the maximum risk employee.

When many employees perform essentially the same task, a maximum risk employee cannot be selected. In this circumstance, it is necessary to resort to random sampling of the group of workers. The objective is to select a subgroup of adequate size so that there is a high probability that the random sample will contain at least one worker with high exposure if one exists. The number of persons in the group influences the number that need to be sampled to ensure that at least one individual from the highest 10 percent exposure group is contained in the sample. For example, to have 90 percent confidence in the results, if the group size is 10, nine should be sampled; for 50, only 18 need to be sampled.

If measurement shows exposure to formaldehyde at or above the action level or the STEL, the employer needs to identify all other employees who may be exposed at or above the action level or STEL and measure or otherwise accurately characterize the exposure of these employees.

Whether representative monitoring or random sampling are conducted, the purpose remains the same—to determine if the exposure of any employee is above the action level. If the exposure of the most exposed employee is less than the action level and the STEL, regardless of how the employee is identified, then it is reasonable to assume that measurements of exposure of the other employees in that operation would be below the action level and the STEL.

## Exposure Measurements

There is no "best" measurement strategy for all situations. Some elements to consider in developing a strategy are:

(1) Availability and cost of sampling equipment

(2) Availability and cost of analytic facilities

(3) Availability and cost of personnel to take samples

(4) Location of employees and work operations

(5) Intraday and interday variations in the process

(6) Precision and accuracy of sampling and analytic methods, and

(7) Number of samples needed.

Samples taken for determining compliance with the STEL differ from those that measure the TWA concentration in important ways. STEL samples are best taken in a nonrandom fashion using all available knowledge relating to the area, the individual, and the process to obtain samples during periods of maximum expected concentrations. At least three measurements on a shift are generally needed to spot gross errors or mistakes; however, only the highest value represents the STEL.

If an operation remains constant throughout the workshift, a much greater number of samples would need to be taken over the 32 discrete nonoverlapping periods in an 8-hour workshift to verify compliance with a STEL. If employee exposure is truly uniform throughout the workshift, however, an employer in compliance with the 1 ppm TWA would be in compliance with the 2 ppm STEL, and this determination can probably be made using objective data.

## Need to Repeat the Monitoring Strategy

Interday and intraday fluctuations in employee exposure are mostly influenced by the physical processes that generate formaldehyde and the work habits of the employee. Hence, in-plant process variations influence the employer's determination of whether or not additional controls need to be imposed. Measurements that employee exposure is low on a day that is not representative of worst conditions may not provide sufficient information to determine whether or not additional engineering controls should be installed to achieve the PELs.

The person responsible for conducting sampling must be aware of systematic changes which will negate the validity of the sampling results. Systematic changes in formaldehyde exposure concentration for an employee can occur due to:

(1) The employee changing patterns of movement in the workplace

(2) Closing of plant doors and windows

(3) Changes in ventilation from season to season

(4) Decreases in ventilation efficiency or abrupt failure of engineering control equipment

(5) Changes in the production process or work habits of the employee.

Any of these changes, if they may result in additional exposure that reaches the next level of action (*i.e.* 0.5 or 1.0 ppm as an 8-hr average or 2 ppm over 15 minutes) require the employer to perform additional monitoring to reassess employee exposure.

A number of methods are suitable for measuring employee exposure to formaldehyde or for characterizing emissions within the worksite. The preamble to this standard describes some methods that have been widely used or subjected to validation testing. A detailed analytical procedure derived from the OSHA Method 52 for acrolein and formaldehyde is presented below for informational purposes.

Inclusion of OSHA's method in this appendix in no way implies that it is the only acceptable way to measure employee exposure to formaldehyde. Other methods that are free from significant interferences and that can determine formaldehyde at the permissible exposure limits within ±25 percent of the "true" value at the 95 percent confidence level are also acceptable. Where applicable, the method should also be capable of measuring formaldehyde at the action level to ±35 percent of the "true" value with a 95 percent confidence level. OSHA encourages employers to choose methods that will be best for their individual needs. The employer must exercise caution, however, in choosing an appropriate method since some techniques suffer from interferences that are likely to be present in workplaces of certain industry sectors where formaldehyde is used.

## *OSHA's Analytical Laboratory Method*

*Method No:* 52

*Matrix:* Air

*Target Concentration:* 1 ppm (1.2 mg/m$^3$)

*Procedures:* Air samples are collected by drawing known volumes of air through sampling tubes containing XAD-2 adsorbent which have been coated with 2-(hydroxymethyl) piperidine. The samples are desorbed with toluene and then analyzed by gas chromatography using a nitrogen selective detector.

*Recommended Sampling Rate and Air Volumes:* 0.1 L/min and 24 L

*Reliable Quantitation Limit:*16 ppb (20 μg/m$^3$)

*Standard Error of Estimate at the Target Concentration:* 7.3%

*Status of the Method:* A sampling and analytical method that has been subjected to the established evaluation procedures of the Organic Methods Evaluation Branch.

*Date:* March 1985

## 1. General Discussion

1.1 *Background:* The current OSHA method for collecting acrolein vapor recommends the use of activated 13X molecular sieves. The samples must be stored in an ice bath during and after sampling and also they must be analyzed within 48 hours of collection. The current OSHA method for collecting formaldehyde vapor recommends the use of bubblers containing 10% methanol in water as the trapping solution.

This work was undertaken to resolve the sample stability problems associated with acrolein and also to eliminate the need to use bubblers to sample formaldehyde. A goal of this work was to develop and/or to evaluate a common sampling and analytical procedure for acrolein and formaldehyde.

NIOSH has developed independent methodologies for acrolein and formaldehyde which recommend the use of reagent-coated adsorbent tubes to collect the aldehydes as stable derivatives. The formaldehyde sampling tubes contain Chromosorb 102 adsorbent coated with N-benzylethanolamine (BEA) which reacts with formaldehyde vapor to form a stable oxazolidine compound. The acrolein sampling tubes contain XAD-2 adsorbent coated with 2(hydroxymethyl)piperidine (2-HMP) which reacts with acrolein vapor to form a different, stable oxazolidine derivative. Acrolein does not appear to react with BEA to give a suitable reaction product. Therefore, the formaldehyde procedure cannot provide a common method for both aldehydes. However, formaldehyde does react with 2-HMP to form a very suitable reaction product. It is the quantitative reaction of acrolein and formaldehyde with 2-HMP that provides the basis for this evaluation.

This sampling and analytical procedure is very similar to the method recommended by NIOSH for acrolein. Some changes in the NIOSH methodology were necessary to permit the simultaneous determination of both aldehydes and also to accommodate OSHA laboratory equipment and analytical techniques.

1.2 *Limit-defining parameters:* The analyte air concentrations reported in this method are based on the recommended air volume for each analyte collected separately and a desorption volume of 1 mL. The amounts are presented as acrolein and/or formaldehyde, even though the derivatives are the actual species analyzed.

1.2.1 *Detection limits of the analytical procedure:* The detection limit of the analytical procedure was 386 pg per injection for formaldehyde. This was the amount of analyte which gave a peak whose height was about five times the height of the peak given by the residual formaldehyde derivative in a typical blank front section of the recommended sampling tube.

1.2.2 *Detection limits of the overall procedure:* The detection limits of the overall procedure were 482 ng per sample (16 ppb or 20 μg/m$^3$ for formaldehyde). This was the amount of analyte spiked on the sampling device which allowed recoveries approximately equal to the detection limit of the analytical procedure.

1.2.3 *Reliable quantitation limits:* The reliable quantitation limit was 482 ng per sample (16 ppb or 20 μg/m$^3$) for formaldehyde. These were the smallest amounts of analyte which could be quantitated within the limits of a recovery of at least 75% and a precision (±1.96 SD) of ± 25% or better.

The reliable quantitation limit and detection limits reported in the method are based upon optimization of the instrument for the smallest possible amount of analyte. When the target concentration of an analyte is exceptionally higher than these limits, they may not be attainable at the routine operating parameters.

1.2.4 *Sensitivity:* The sensitivity of the analytical procedure over concentration ranges representing 0.4 to 2 times the target concentration, based on the recommended air volumes, was 7,589 area units per μg/mL for formaldehyde. This value was determined from the slope of the calibration curve. The sensitivity may vary with the particular instrument used in the analysis.

1.2.5 *Recovery:* The recovery of formaldehyde from samples used in an 18-day storage test remained above 92% when the samples were stored at ambient temperature. These values were determined from regression lines which were calculated from the storage data. The recovery of the analyte from the collection device must be at least 75% following storage.

1.2.6 *Precision (analytical method only):* The pooled coefficient of variation obtained from replicate determinations of analytical standards over the range of 0.4 to 2 times the target concentration was 0.0052 for formaldehyde (Section 4.3).

1.2.7 *Precision (overall procedure):* The precision at the 95% confidence level for the ambient temperature storage tests was ±14.3% for formaldehyde. These values each include an additional ±5% for sampling error. The overall procedure must provide results at the target concentrations that are ±25% at the 95% confidence level.

1.2.8 *Reproducibility:* Samples collected from controlled test atmospheres and a draft copy of this procedure were given to a chemist unassociated with this evaluation. The formaldehyde samples were analyzed following 15 days storage. The average recovery was 96.3% and the standard deviation was 1.7%.

1.3 *Advantages:*

1.3.1 The sampling and analytical procedures permit the simultaneous determination of acrolein and formaldehyde.

1.3.2 Samples are stable following storage at ambient temperature for at least 18 days.

1.4 *Disadvantages:* None.

## 2. Sampling Procedure

2.1 *Apparatus:*

2.1.1 Samples are collected by use of a personal sampling pump that can be calibrated to within ±5% of the recommended 0.1 L/min sampling rate with the sampling tube in line.

2.1.2 Samples are collected with laboratory prepared sampling tubes. The sampling tube is constructed of silane treated glass and is about 8-cm long. The ID is 4 mm and the OD is 6 mm. One end of the tube is tapered so that a glass wool end plug will hold the contents of the tube in place during sampling. The other end of the sampling tube is open to its full 4-mm ID to facilitate packing of the tube. Both ends of the tube are firepolished for safety. The tube is packed with a 75-mg backup section, located nearest the tapered end and a 150-mg sampling section of pretreated XAD-2 adsorbent which has been coated with 2-HMP. The two sections of coated adsorbent are separated and retained with small plugs of silanized glass wool. Following packing, the sampling tubes are sealed with two 7/32 inch OD plastic end caps. Instructions for the pretreatment and the coating of XAD-2 adsorbent are presented in Section 4 of this method.

2.1.3 Sampling tubes, similar to those recommended in this method, are marketed by Supelco, Inc. These tubes were not available when this work was initiated; therefore, they were not evaluated.

2.2 *Reagents:* None required.

2.3 *Technique:*

2.3.1 Properly label the sampling tube before sampling and then remove the plastic end caps.

2.3.2 Attach the sampling tube to the pump using a section of flexible plastic tubing such that the large, front section of the sampling tube is exposed directly to the atmosphere. Do not place any tubing ahead of the sampling tube. The sampling tube should be attached in the worker's breathing zone in a vertical manner such that it does not impede work performance.

2.3.3 After sampling for the appropriate time, remove the sampling tube from the pump and then seal the tube with plastic end caps.

2.3.4 Include at least one blank for each sampling set. The blank should be handled in the same manner as the samples with the exception that air is not drawn through it.

2.3.5 List any potential interferences on the sample data sheet.

2.4 *Breakthrough:*

2.4.1 Breakthrough was defined as the relative amount of analyte found on a backup sample in relation to the total amount of analyte collected on the sampling train.

2.4.2 For formaldehyde collected from test atmospheres containing 6 times the PEL, the average 5% breakthrough air volume was 41 L. The sampling rate was 0.1 L/min and the average mass of formaldehyde collected was 250 μg.

2.5 *Desorption Efficiency:* No desorption efficiency corrections are necessary to compute air sample results because analytical standards are prepared using coated adsorbent. Desorption efficiencies were determined, however, to investigate the recoveries of the analytes from the sampling device. The average recovery over the range of 0.4 to 2 times the target concentration, based on the recommended air volumes, was 96.2% for formaldehyde. Desorption efficiencies were essentially constant over the ranges studied.

2.6 *Recommended Air Volume and Sampling Rate:*

2.6.1 The recommended air volume for formaldehyde is 24 L.

2.6.2 The recommended sampling rate is 0.1 L/min.

2.7 *Interferences:*

2.7.1 Any collected substance that is capable of reacting 2-HMP and thereby depleting the derivatizing agent is a potential interference. Chemicals which contain a carbonyl group, such as acetone, may be capable or reacting with 2-HMP.

2.7.2 There are no other known interferences to the sampling method.

2.8 *Safety Precautions:*

2.8.1 Attach the sampling equipment to the worker in such a manner that it well not interfere with work performance or safety.

2.8.2 Follow all safety practices that apply to the work area being sampled.

## 3. Analytical Procedure

3.1 *Apparatus:*

3.1.1 A gas chromatograph (GC), equipped with a nitrogen selective detector. A Hewlett-Packard Model 5840A GC fitted with a nitrogen-phosphorus flame ionization detector (NPD) was used for this evaluation. Injections were performed using a Hewlett-Packard Model 7671A automatic sampler.

3.1.2 A GC column capable of resolving the analytes from any interference. A 6 ft × 1/4 in OD (2mm ID) glass GC column containing 10% UCON 50-HB-5100 + 2% KOH on 80/100 mesh Chromosorb W-AW was used for the evaluation. Injections were performed on-column.

3.1.3 Vials, glass 2-mL with Teflon-lined caps.

3.1.4 Volumetric flasks, pipets, and syringes for preparing standards, making dilutions, and performing injections.

3.2 *Reagents:*

3.2.1 Toluene and dimethylformamide. Burdick and Jackson solvents were used in this evaluation.

3.2.2 Helium, hydrogen, and air, GC grade.

3.2.3 Formaldehyde, 37%, by weight, in water. Aldrich Chemical, ACS Reagent Grade formaldehyde was used in this evaluation.

3.2.4 Amberlite XAD-2 adsorbent coated with 2-(hydroxymethyl—piperidine (2-HMP), 10% by weight (Section 4).

3.2.5 Desorbing solution with internal standard. This solution was prepared by adding 20 μL of dimethylformamide to 100 mL of toluene.

3.3 *Standard preparation:*

3.3.1 *Formaldehyde:* Prepare stock standards by diluting known volumes of 37% formaldehyde solution with methanol. A procedure to determine the formaldehyde content of these standards is presented in Section 4. A standard containing 7.7 mg/mL formaldehyde was prepared by diluting 1 mL of the 37% reagent to 50 mL with methanol.

3.3.2 It is recommended that analytical standards be prepared about 16 hours before the air samples are to be analyzed in order to ensure the complete reaction of the analytes with 2-HMP. However, rate studies have shown the reaction to be greater than 95% complete after 4 hours. Therefore, one or two standards can be analyzed after this reduced time if sample results are outside the concentration range of the prepared standards.

3.3.3 Place 150-mg portions of coated XAD-2 adsorbent, from the same lot number as used to collect the air samples, into each of several glass 2-mL vials. Seal each vial with a Teflon-lined cap.

3.3.4 Prepare fresh analytical standards each day by injecting appropriate amounts of the diluted analyte directly onto 150-mg portions of coated adsorbent. It is permissible to inject both acrolein and formaldehyde on the same adsorbent portion. Allow the standards to stand at room temperature. A standard, approximately the target levels, was prepared by injecting 11 μL of the acrolein and 12 μL of the formaldehyde stock standards onto a single coated XAD-2 adsorbent portion.

3.3.5 Prepare a sufficient number of standards to generate the calibration curves. Analytical standard concentrations should bracket sample concentrations. Thus, if samples are not in the concentration range of the prepared standards, additional standards must be prepared to determine detector response.

3.3.7 Desorb the standards in the same manner as the samples following the 16-hour reaction time.

3.4 *Sample preparation:*

3.4.1 Transfer the 150-mg section of the sampling tube to a 2-mL vial. Place the 75-mg section in a separate vial. If the glass wool plugs contain a significant number of adsorbent beads, place them with the appropriate sampling tube section. Discard the glass wool plugs if they do not contain a significant number of adsorbent beads.

3.4.2 Add 1 mL of desorbing solution to each vial.

3.4.3 Seal the vials with Teflon-lined caps and then allow them to desorb for one hour. Shake the vials by hand with vigorous force several times during the desorption time.

3.4.4 Save the used sampling tubes to be cleaned and recycled.

3.5 *Analysis:*

3.5.1 GC *Conditions*

*Column Temperature:*

Bi-level temperature program—First level: 100 to 140 ° C at 4 ° C/min following completion of the first level.

Second level: 140 to 180 ° C at 20 ° C/min following completion of the first level.

Isothermal period: Hold column at 180 ° C until the recorder pen returns to baseline (usually about 25 min after injection).

*Injector temperature:* 180 ° C

*Helium flow rate:* 30 mL/min (detector response will be reduced if nitrogen is substituted for helium carrier gas).

*Injection volume:* 0.8 μL

*GC column:* Six-ft × 1/4-in OD (2 mm ID) glass GC column containing 10% UCON 50-HB-5100 +2% KOH on 80/100 Chromosorb W-AW.

*NPD conditions:*

Hydrogen flow rate: 3 mL/min

Air flow rate: 50 mL/min

Detector temperature: 275 ° C

3.5.2 *Chromatogram:* For an example of a typical chromatogram, see Figure 4.11 in OSHA Method 52.

3.5.3 Use a suitable method, such as electronic integration, to measure detector response.

3.5.4 Use an internal standard method to prepare the calibration curve with several standard solutions of different concentrations. Prepare the calibration curve daily. Program the integrator to report results in g/mL.

3.5.5 Bracket sample concentrations with standards.

3.6 *Interferences (Analytical)*

3.6.1 Any compound with the same general retention time as the analytes and which also gives a detector response is a potential interference. Possible interferences should be reported to the laboratory with submitted samples by the industrial hygienist.

3.6.2 GC parameters (temperature, column, etc.) may be changed to circumvent interferences.

3.6.3 A useful means of structure designation is GC/MS. It is recommended this procedure be used to confirm samples whenever possible.

3.6.4 The coated adsorbent usually contains a very small amount of residual formaldehyde derivative (Section 4.8).

3.7 *Calculations:*

3.7.1 Results are obtained by use of calibration curves. Calibration curves are prepared by plotting detector response against concentration for each standard. The best line through the data points is determined by curve fitting.

3.7.2 The concentration, in µg/mL, for a particular sample is determined by comparing its detector response to the calibration curve. If either of the analytes is found on the backup section, it is added to the amount found on the front section. Blank corrections should be performed before adding the results together.

3.7.3 The acrolein and/or formaldehyde air concentration can be expressed using the following equation:

$$\mathbf{mg/m^3 = (A)(B)/C}$$

where A = µg/mL from 3.7.2, B = desorption volume, and C = L of air sampled.

No desorption efficiency corrections are required.

3.7.4 The following equation can be used to convert results in mg/$m^3$ to ppm.

$$\mathbf{ppm = (mg/m^3)(24.45)/MW}$$

where mg/$m^3$ = result from 3.7.3, 24.45 = molar volume of an ideal gas at 760 mm Hg and 25 ° C, MW = molecular weight (30.0).

## 4. Backup Data

4.1 Backup data on detection limits, reliable quantitation limits, sensitivity and precision of the analytical method, breakthrough, desorption efficiency, storage, reproducibility, and generation of test atmospheres are available in OSHA Method 52, developed by the Organics Methods Evaluation Branch, OSHA Analytical Laboratory, Salt Lake City, Utah.

4.2 *Procedure to Coat XAD-2 Adsorbent with 2-HMP:*

4.2.1 *Apparatus:* Soxhlet extraction apparatus, rotary evaporation apparatus, vacuum dessicator, 1-L vacuum flask, 1-L round-bottomed evaporative flask, 1-L Erlenmeyer flask, 250-mL Buchner funnel with a coarse fritted disc, etc.

4.2.2 *Reagents:*

4.2.2.1 Methanol, isooctane, and toluene.

4.2.2.2 2-(Hydroxymethyl)piperidine.

4.2.2.3 Amberlite XAD-2 non-ionic polymeric adsorbent, 20 to 60 mesh, Aldrich Chemical XAD-2 was used in this evaluation.

4.2.3 *Procedure:* Weigh 125 g of crude XAD-2 adsorbent into a 1-L Erlenmeyer flask. Add about 200 mL of water to the flask and then swirl the mixture to wash the adsorbent. Discard any adsorbent that floats to the top of the water and then filter the mixture using a fritted Buchner funnel. Air dry the adsorbent for 2 minutes. Transfer the adsorbent back to the Erlenmeyer flask and then add about 200 mL of methanol to the flask. Swirl and then filter the mixture as before. Transfer the washed adsorbent back to the Erlenmeyer flask and then add about 200 mL of methanol to the flask. Swirl and then filter the mixture as before. Transfer the washed adsorbent to a 1-L roundbottomed evaporative flask, add 13 g of 2-HMP and then 200 mL of methanol, swirl the mixture and then allow it to stand for one hour. Remove the methanol at about 40 ° C and reduced pressure using a rotary evaporation apparatus. Transfer the coated adsorbent to a suitable container and store it in a vacuum desiccator at room temperature overnight. Transfer the coated adsorbent to a Soxhlet extractor and then extract the material with toluene for about 24 hours. Discard the contaminated toluene, add methanol in its place and then continue the Soxhlet extraction for an additional 4 hours. Transfer the adsorbent to a weighted 1-L round-bottom evaporative flask and remove the methanol using the rotary evaporation apparatus. Determine the weight of the adsorbent and then add an amount of 2HMP, which is 10% by weight of the adsorbent. Add 200 mL of methanol and then swirl the mixture. Allow the mixture to stand for one hour. Remove the methanol by rotary evaporation. Transfer the coated adsorbent to a suitable container and store it in a vacuum desiccator until all traces of solvents are gone. Typically, this will take 2-3 days. The coated adsorbent should be protected from contamination. XAD-2 adsorbent treated in this manner will probably not contain residual acrolein derivative. However, this adsorbent will often contain residual formaldehyde derivative levels of about 0.1 µg per 150 mg of adsorbent. If the blank values for a batch of coated adsorbent are too high, then the batch should be returned to the Soxhlet extractor, extracted with toluene again and then recoated. This process can be repeated until the desired blank levels are attained.

The coated adsorbent is now ready to be packed into sampling tubes. The sampling tubes should be stored in a sealed container to prevent contamination. Sampling tubes should be stored in the dark at room temperature. The sampling tubes should be segregated by coated adsorbent lot number. A sufficient amount of each lot number of coated adsorbent should be retained to prepare analytical standards for use with air samples from that lot number.

4.3 *A Procedure to Determine Formaldehyde by Acid Titration:* Standardize the 0.1 N HCl solution using sodium carbonate and methyl orange indicator.

Place 50 mL of 0.1 M sodium sulfite and three drops of thymophthalein indicator into a 250-mL Erlenmeyer flask. Titrate the contents of the flask to a colorless endpoint with 0.1 N HCl (usually one or two drops is sufficient). Transfer 10 mL of the formaldehyde/methanol solution (prepared in 3.3.1) into the same flask and titrate the mixture with 0.1 N HCl, again, to a colorless endpoint. The formaldehyde concentration of the standard may be calculated by the following equation:

$$\text{Formaldehyde, mg/mL} = \frac{\text{acid titer} \times \text{acid normality} \times 30.0}{\text{mL of sample}}$$

This method is based on the quantitative liberation of sodium hydroxide when formaldehyde reacts with sodium sulfite to form the formaldehyde-bisulfite addition product. The volume of sample may be varied depending on the formaldehyde content but the solution to be titrated must contain excess sodium sulfite. Formaldehyde solutions containing substantial amounts of acid or base must be neutralized before analysis.

## APPENDIX C TO §1910.1048—MEDICAL SURVEILLANCE—FORMALDEHYDE

### I. Health Hazards

The occupational health hazards of formaldehyde are primarily due to its toxic effects after inhalation, after direct contact with the skin or eyes by formaldehyde in liquid or vapor form, and after ingestion.

### II. Toxicology

#### A. Acute Effects of Exposure

1. *Inhalation (breathing):* Formaldehyde is highly irritating to the upper airways. The concentration of formaldehyde that is immediately dangerous to life and health is 100 ppm. Concentrations above 50 ppm can cause severe pulmonary reactions within minutes. These include pulmonary edema, pneumonia, and bronchial irritation which can result in death. Concentrations above 5 ppm readily cause lower airway irritation characterized by cough, chest tightness and wheezing. There is some controversy regarding whether formaldehyde gas is a pulmonary sensitizer which can cause occupational asthma in a previously normal individual. Formaldehyde can produce symptoms of bronchial asthma in humans. The mechanism may be either sensitization of the individual by exposure to formaldehyde or direct irritation by formaldehyde in persons with pre-

existing asthma. Upper airway irritation is the most common respiratory effect reported by workers and can occur over a wide range of concentrations, most frequently above 1 ppm. However, airway irritation has occurred in some workers with exposures to formaldehyde as low as 0.1 ppm. Symptoms of upper airway irritation include dry or sore throat, itching and burning sensations of the nose, and nasal congestion. Tolerance to this level of exposure may develop within 1-2 hours. This tolerance can permit workers remaining in an environment of gradually increasing formaldehyde concentrations to be unaware of their increasingly hazardous exposure.

2. *Eye contact:* Concentrations of formaldehyde between 0.05 ppm and 0.5 ppm produce a sensation of irritation in the eyes with burning, itching, redness, and tearing. Increased rate of blinking and eye closure generally protects the eye from damage at these low levels, but these protective mechanisms may interfere with some workers' work abilities. Tolerance can occur in workers continuously exposed to concentrations of formaldehyde in this range. Accidental splash injuries of human eyes to aqueous solutions of formaldehyde (formalin) have resulted in a wide range of ocular injuries including corneal opacities and blindness. The severity of the reactions have been directly dependent on the concentration of formaldehyde in solution and the amount of time lapsed before emergency and medical intervention.

3. *Skin contact:* Exposure to formaldehyde solutions can cause irritation of the skin and allergic contact dermatitis. These skin diseases and disorders can occur at levels well below those encountered by many formaldehyde workers. Symptoms include erythema, edema, and vesiculation or hives. Exposure to liquid formalin or formaldehyde vapor can provoke skin reactions in sensitized individuals even when airborne concentrations of formaldehyde are well below 1 ppm.

4. *Ingestion:* Ingestion of as little as 30 ml of a 37 percent solution of formaldehyde (formalin) can result in death. Gastrointestinal toxicity after ingestion is most severe in the stomach and results in symptoms which can include nausea, vomiting, and servere abdominal pain. Diverse damage to other organ systems including the liver, kidney, spleen, pancreas, brain, and central nervous systems can occur from the acute response to ingestion of formaldehyde.

### B. Chronic Effects of Exposure

Long term exposure to formaldehyde has been shown to be associated with an increased risk of cancer of the nose and accessory sinuses, nasopharyngeal and oropharyngeal cancer, and lung cancer in humans. Animal experiments provide conclusive evidence of a causal relationship between nasal cancer in rats and formaldehyde exposure. Concordant evidence of carcinogenicity includes DNA binding, genotoxicity in short-term tests, and cytotoxic changes in the cells of the target organ suggesting both preneoplastic changes and a dose-rate effect. Formaldehyde is a complete carcinogen and appears to exert an effect on at least two stages of the carcinogenic process.

## *III. Surveillance considerations*

### A. History

1. *Medical and occupational history:* Along with its acute irritative effects, formaldehyde can cause allergic sensitization and cancer. One of the goals of the work history should be to elicit information on any prior or additional exposure to formaldehyde in either the occupational or the nonoccupational setting.

2. *Respiratory history:* As noted above, formaldehyde has recognized properties as an airway irritant and has been reported by some authors as a cause of occupational asthma. In addition, formaldehyde has been associated with cancer of the entire respiratory system of humans. For these reasons, it is appropriate to include a comprehensive review of the respiratory system in the medical history. Components of this history might include questions regarding dyspnea on exertion, shortness of breath, chronic airway complaints, hyperreactive airway disease, rhinitis, bronchitis, bronchiolitis, asthma, emphysema, respiratory allergic reaction, or other preexisting pulmonary disease.

In addition, generalized airway hypersensitivity can result from exposures to a single sensitizing agent. The examiner should, therefore, elicit any prior history of exposure to pulmonary irritants, and any short- or long-term effects of that exposure.

Smoking is known to decrease mucociliary clearance of materials deposited during respiration in the nose and upper airways. This may increase a worker's exposure to inhaled materials such as formaldehyde vapor. In addition, smoking is a potential confounding factor in the investigation of any chronic respiratory disease, including cancer. For these reasons, a complete smoking history should be obtained.

3. *Skin Disorders:* Because of the dermal irritant and sensitizing effects of formaldehyde, a history of skin disorders should be obtained. Such a history might include the existence of skin irritation, previously documented skin sensitivity, and other dermatologic disorders. Previous exposure to formaldehyde and other dermal sensitizers should be recorded.

4. *History of atopic or allergic diseases:* Since formaldehyde can cause allergic sensitization of the skin and airways, it might be useful to identify individuals with prior allergen sensitization. A history of atopic disease and allergies to formaldehyde or any other substances should also be obtained. It is not definitely known at this time whether atopic diseases and allergies to formaldehyde or any other substances should also be obtained. Also it is not definitely known at this time whether atopic individuals have a greater propensity to develop formaldehyde sensitivity than the general population, but identification of these individuals may be useful for ongoing surveillance.

5. *Use of disease questionnaires:* Comparison of the results from previous years with present results provides the best method for detecting a general deterioration in health when toxic signs and symptoms are measured subjectively. In this way recall bias does not affect the results of the analysis. Consequently, OSHA has determined that the findings of the medical and work histories should be kept in a standardized form for comparison of the year-to-year results.

### B. Physical Examination

1. *Mucosa of eyes and airways:* Because of the irritant effects of formaldehyde, the examining physician should be alert to evidence of this irritation. A speculum examination of the nasal mucosa may be helpful in assessing possible irritation and cytotoxic changes, as may be indirect inspection of the posterior pharynx by mirror.

2. *Pulmonary system:* A conventional respiratory examination, including inspection of the thorax and auscultation and percussion of the lung fields should be performed as part of the periodic medical examination. Although routine pulmonary function testing is only required by the standard once every year for persons who are exposed over the TWA concentration limit, these tests have an obvious value in investigating possible respiratory dysfunction and should be used wherever deemed appropriate by the physician. In cases of alleged formaldehydeinduced airway disease, other possible causes of pulmonary disfunction (including exposures to other substances) should be ruled out. A chest radiograph may be useful in these circumstances. In cases of suspected airway hypersensitivity or allergy, it may be appropriate to use bronchial challenge testing with formaldehyde or methacholine to determine the nature of the disorder. Such testing should be performed by or under the supervision of a physician experienced in the procedures involved.

3. *Skin:* The physician should be alert to evidence of dermal irritation of sensitization, including reddening and inflammation, urticaria, blistering, scaling, formation of skin fissures, or other symptoms. Since the integrity of the skin barrier is compromised by other dermal diseases, the presence of such disease should be noted. Skin sensitivity testing carries with it some risk of inducing sensitivity, and therefore, skin testing for formaldehyde sensitivity should not be used as a routine screening test. Sensitivity testing may be indicated in the investigation of a suspected existing sensitivity. Guidelines for such testing have been prepared by the North American Contact Dermatitis Group.

### C. Additional Examinations or Tests

The physician may deem it necessary to perform other medical examinations or tests as indicated. The standard provides a mechanism whereby these additional investigations are covered under the standard for occupational exposure to formaldehyde.

### D. Emergencies

The examination of workers exposed in an emergency should be directed at the organ systems most likely to be affected. Much of the content of the examination will be similar to the periodic examination unless the patient has received a severe acute exposure requiring immediate attention to prevent serious consequences. If a severe overexposure requiring medical intervention or hospitalization has occurred, the physician must be alert to the possibility of delayed symptoms. Followup nonroutine examinations may be necessary to assure the patient's well-being.

## E. Employer Obligations

The employer is required to provide the physician with the following information: A copy of this standard and appendices A, C, D, and E; a description of the affected employee's duties as they relate to his or her exposure concentration; an estimate of the employee's exposure including duration (*e.g.* 15 hr/wk, three 8-hour shifts, full-time); a description of any personal protective equipment, including respirators, used by the employee; and the results of any previous medical determinations for the affected employee related to formaldehyde exposure to the extent that this information is within the employer's control.

## F. Physician's Obligations

The standard requires the employer to obtain a written statement from the physician. This statement must contain the physician's opinion as to whether the employee has any medical condition which would place him or her at increased risk of impaired health from exposure to formaldehyde or use of respirators, as appropriate. The physician must also state his opinion regarding any restrictions that should be placed on the employee's exposure to formaldehyde or upon the use of protective clothing or equipment such as respirators. If the employee wears a respirator as a result of his or her exposure to formaldehyde, the physician's opinion must also contain a statement regarding the suitability of the employee to wear the type of respirator assigned. Finally, the physician must inform the employer that the employee has been told the results of the medical examination and of any medical conditions which require further explanation or treatment. This written opinion is not to contain any information on specific findings or diagnoses unrelated to occupational exposure to formaldehyde.

The purpose in requiring the examining physician to supply the employer with a written opinion is to provide the employer with a medical basis to assist the employer in placing employees initially, in assuring that their health is not being impaired by formaldehyde, and to assess the employee's ability to use any required protective equipment.

## APPENDIX D TO §1910.1048—NONMANDATORY MEDICAL DISEASE QUESTIONNAIRE

### NONMANDATORY MEDICAL DISEASE QUESTIONNAIRE

#### A. Identification

Plant Name ______ Date ______
Employee Name ______ S.S. # ______
Job Title ______ Birthdate: ______
Age: ______ Sex: ______
Height: ______ Weight: ______

#### B. Medical History

1. Have you ever been in the hospital as a patient? Yes ☐ No ☐
   If yes, what kind of problem were you having? ______
2. Have you ever had any kind of operation? Yes ☐ No ☐
   If yes, what kind? ______
3. Do you take any kind of medicine regularly? Yes ☐ No ☐
   If yes, what kind? ______
4. Are you allergic to any drugs, foods, or chemicals? Yes ☐ No ☐
   If yes, what kind of allergy is it? ______
   What causes the allergy? ______
5. Have you ever been told that you have asthma, hayfever, or sinusitis? Yes ☐ No ☐
6. Have you ever been told that you have emphysema, bronchitis, or any other respiratory problems? Yes ☐ No ☐
7. Have you ever been told you had hepatitis? Yes ☐ No ☐
8. Have you ever been told that you had cirrhosis? Yes ☐ No ☐
9. Have you ever been told that you had cancer? Yes ☐ No ☐
10. Have you ever had arthritis or joint pain? Yes ☐ No ☐
11. Have you ever been told that you had high blood pressure? Yes ☐ No ☐
12. Have you ever had a heart attack or heart trouble? Yes ☐ No ☐

#### B-1. Medical History Update

1. Have you been in the hospital as a patient any time within the past year? Yes ☐ No ☐
   If so, for what condition? ______
2. Have you been under the care of a physician during the past year? Yes ☐ No ☐
   If so, for what condition? ______
3. Is there any change in your breathing since last year? Yes ☐ No ☐
   Better? ______ Worse? ______ No change? ______
   If change, do you know why? ______
4. Is your general health different this year from last year? Yes ☐ No ☐
   If different, in what way? ______
5. Have you in the past year or are you now taking any medication on a regular basis? Yes ☐ No ☐
   Name Rx ______
   Condition being treated ______

#### C. Occupational History

1. How long have you worked for your present employer? ______
2. What jobs have you held with this employer? Include job title and length of time in each job. ______
3. In each of these jobs, how many hours a day were you exposed to chemicals? ______
4. What chemicals have you worked with most of the time? ______
5. Have you ever noticed any type of skin rash you feel was related to your work? Yes ☐ No ☐
6. Have you ever noticed that any kind of chemical makes you cough? Yes ☐ No ☐;
   Wheeze? Yes ☐ No ☐
   Become short of breath or cause your chest to become tight? Yes ☐ No ☐
7. Are you exposed to any dust or chemicals at home? Yes ☐ No ☐
   If yes, explain: ______
8. In other jobs, have you ever had exposure to: Wood dust? Yes ☐ No ☐
   Nickel or chromium? Yes ☐ No ☐ Silica (foundry, sand blasting)? Yes ☐ No ☐
   Arsenic or asbestos? Yes ☐ No ☐ Organic solvents? Yes ☐ No ☐ Urethane foams? Yes ☐ No ☐

#### C-1. Occupational History Update

1. Are you working on the same job this year as you were last year? Yes ☐ No ☐
   If not, how has your job changed? ______
2. What chemicals are you exposed to on your job? ______
3. How many hours a day are you exposed to chemicals? ______
4. Have you noticed any skin rash within the past year you feel was related to your work? Yes ☐ No ☐
   If so, explain circumstances: ______

5. Have you noticed that any chemical makes you cough, be short of breath, or wheeze? Yes ❑ No ❑
If so, can you identify it? ____________

*D. Miscellaneous*

1. Do you smoke? Yes ❑ No ❑
If so, how much and for how long? Pipe ____________
Cigars ____________ Cigarettes ____________
2. Do you drink alcohol in any form? Yes ❑ No ❑
If so, how much, how long, and how often? ____________
3. Do you wear glasses or contact lenses? Yes ❑ No ❑
4. Do you get any physical exercise other than that required to do your job? Yes ❑ No ❑
If so, explain: ____________
5. Do you have any hobbies or "side jobs" that require you to use chemicals, such as furniture stripping, sand blasting, insulation or manufacture of urethane foam, furniture, etc? Yes ❑ No ❑
If so, please describe, giving type of business or hobby, chemicals used and length of exposures.
____________

*E. Symptoms Questionnaire*

1. Do you ever have any shortness of breath? Yes ❑ No ❑
If yes, do you have to rest after climbing several flights of stairs? Yes ❑ No ❑
If yes, if you walk on the level with people your own age, do you walk slower than they do? Yes ❑ No ❑
If yes, if you walk slower than a normal pace, do you have to limit the distance that you walk? Yes ❑ No ❑
If yes, do you have to stop and rest while bathing or dressing? Yes ❑ No ❑
2. Do you cough as much as three months out of the year? Yes ❑ No ❑
If yes, have you had this cough for more than two years? Yes ❑ No ❑
If yes, do you ever cough anything up from chest? Yes ❑ No ❑
3. Do you ever have a feeling of smothering, unable to take a deep breath, or tightness in your chest? Yes ❑ No ❑
If yes, do you notice that this on any particular day of the week? Yes ❑ No ❑
If yes, what day or the week? Yes ❑ No ❑
If yes, do you notice that this occurs at any particular place? Yes ❑ No ❑
If yes, do you notice that this is worse after you have returned to work after being off for several days? Yes ❑ No ❑
4. Have you ever noticed any wheezing in your chest? Yes ❑ No ❑
If yes, is this only with colds or other infections? Yes ❑ No ❑
Is this caused by exposure to any kind of dust or other material? Yes ❑ No ❑
If yes, what kind? ____________
5. Have you noticed any burning, tearing, or redness of your eyes when you are at work? Yes ❑ No ❑
If so, explain circumstances: ____________
6. Have you noticed any sore or burning throat or itchy or burning nose when you are at work? Yes ❑ No ❑
If so, explain circumstances: ____________
7. Have you noticed any stuffiness or dryness of your nose? Yes ❑ No ❑
8. Do you ever have swelling of the eyelids or face? Yes ❑ No ❑
9. Have you ever been jaundiced? Yes ❑ No ❑
If yes, was this accompanied by any pain? Yes ❑ No ❑
10. Have you ever had a tendency to bruise easily or bleed excessively? Yes ❑ No ❑
11. Do you have frequent headaches that are not relieved by aspirin or tylenol? Yes ❑ No ❑
If yes, do they occur at any particular time of the day or week? Yes ❑ No ❑
If yes, when do they occur? ____________
12. Do you have frequent episodes of nervousness or irritability? Yes ❑ No ❑
13. Do you tend to have trouble concentrating or remembering? Yes ❑ No ❑
14. Do you ever feel dizzy, light-headed, excessively drowsy or like you have been drugged? Yes ❑ No ❑
15. Does your vision ever become blurred? Yes ❑ No ❑
16. Do you have numbness or tingling of the hands or feet or other parts of your body? Yes ❑ No ❑
17. Have you ever had chronic weakness or fatigue? Yes ❑ No ❑
18. Have you ever had any swelling of your feet or ankles to the point where you could not wear your shoes? Yes ❑ No ❑
19. Are you bothered by heartburn or indigestion? Yes ❑ No ❑
20. Do you ever have itching, dryness, or peeling and scaling of the hands? Yes ❑ No ❑
21. Do you ever have a burning sensation in the hands, or reddening of the skin? Yes ❑ No ❑
22. Do you ever have cracking or bleeding of the skin on your hands? Yes ❑ No ❑
23. Are you under a physician's care? Yes ❑ No ❑
If yes, for what are you being treated? ____________
24. Do you have any physical complaints today? Yes ❑ No ❑
If yes, explain? ____________
25. Do you have other health conditions not covered by these questions? Yes ❑ No ❑
If yes, explain: ____________

[57 FR 22310, May 27, 1992; 57 FR 27161, June 18, 1992; 61 FR 5508, Feb. 13, 1996; 63 FR 1292, Jan. 8, 1998; 63 FR 20099, Apr. 23, 1998]

## §1926.1152 Methylene chloride.

**Note:** The requirements applicable to construction employment under this section are identical to those set forth at 29 CFR 1910.1052.

[62 FR 1619, Jan. 10, 1997]

[Section 1910.1052 is printed below—CCH.]

## §1910.1052 Methylene Chloride.

This occupational health standard establishes requirements for employers to control occupational exposure to methylene chloride (MC). Employees exposed to MC are at increased risk of developing cancer, adverse effects on the heart, central nervous system and liver, and skin or eye irritation. Exposure may occur through inhalation, by absorption through the skin, or through contact with the skin. MC is a solvent which is used in many different types of work activities, such as paint stripping, polyurethane foam manufacturing, and cleaning and degreasing. Under the requirements of paragraph (d) of this section, each covered employer must make an initial determination of each employee's exposure to MC. If the employer determines that employees are exposed below the action level, the only other provisions of this section that apply are that a record must be made of the determination, the employees must receive information and training under paragraph (l) of this section and, where appropriate, employees must be protected from contact with liquid MC under paragraph (h) of this section. The provisions of the MC standard are as follows:

**1910.1052(a)** ***Scope and application.*** This section applies to all occupational exposures to methylene chloride (MC), Chemical Abstracts Service Registry Number 75-09-2, in general industry, construction and shipyard employment.

**1910.1052(b)** ***Definitions.*** For the purposes of this section, the following definitions shall apply:

*Action level* means a concentration of airborne MC of 12.5 parts per million (ppm) calculated as an eight (8)-hour time-weighted average (TWA).

*Assistant Secretary* means the Assistant Secretary of Labor for Occupational Safety and Health, U.S. Department of Labor, or designee.

*Authorized person* means any person specifically authorized by the employer and required by work duties to be present in regulated areas, or any person entering such an area as a designated representative of employees for the purpose of exercising the right to observe monitoring and measuring procedures under paragraph (d) of this section, or any other person authorized by the OSH Act or regulations issued under the Act.

*Director* means the Director of the National Institute for Occupational Safety and Health, U.S. Department of Health and Human Services, or designee.

*Emergency* means any occurrence, such as, but not limited to, equipment failure, rupture of containers, or failure of control equipment, which results, or is likely to result in an uncontrolled release of MC. If an incidental release of MC can be controlled by employees such as maintenance personnel at the time of release and in accordance with the leak/spill provisions required by paragraph (f) of this section, it is not considered an emergency as defined by this standard.

*Employee exposure* means exposure to airborne MC which occurs or would occur if the employee were not using respiratory protection.

*Methylene chloride* (MC) means an organic compound with chemical formula, $CH_2 Cl_2$. Its Chemical Abstracts Service Registry Number is 75-09-2. Its molecular weight is 84.9 g/mole.

*Physician or other licensed health care professional* is an individual whose legally permitted scope of practice (i.e., license, registration, or certification) allows him or her to independently provide or be delegated the responsibility to provide some or all of the health care services required by paragraph (j) of this section.

*Regulated area* means an area, demarcated by the employer, where an employee's exposure to airborne concentrations of MC exceeds or can reasonably be expected to exceed either the 8-hour TWA PEL or the STEL.

*Symptom* means central nervous system effects such as headaches, disorientation, dizziness, fatigue, and decreased attention span; skin effects such as chapping, erythema, cracked skin, or skin burns; and cardiac effects such as chest pain or shortness of breath.

*This section* means this methylene chloride standard.

1910.1052(c) ***Permissible exposure limits (PELs)***—(1) *Eight-hour time weighted average (TWA) PEL.* The employer shall ensure that no employee is exposed to an airborne concentration of MC in excess of twenty-five parts of MC per million parts of air (25 ppm) as an 8-hour TWA.

1910.1052(c)(2) *Short-term exposure limit (STEL).* The employer shall ensure that no employee is exposed to an airborne concentration of MC in excess of one hundred and twenty-five parts of MC per million parts of air (125 ppm) as determined over a sampling period of fifteen minutes.

1910.1052(d) ***Exposure monitoring***—(1) *Characterization of employee exposure.* (i) Where MC is present in the workplace, the employer shall determine each employee's exposure by either:

1910.1052(d)(1)(i)(A) Taking a personal breathing zone air sample of each employee's exposure; or

1910.1052(d)(1)(i)(B) Taking personal breathing zone air samples that are representative of each employee's exposure.

1910.1052(d)(1)(ii) *Representative samples.* The employer may consider personal breathing zone air samples to be representative of employee exposures when they are taken as follows:

1910.1052(d)(1)(ii)(A) *8-hour TWA PEL.* The employer has taken one or more personal breathing zone air samples for at least one employee in each job classification in a work area during every work shift, and the employee sampled is expected to have the highest MC exposure.

1910.1052(d)(1)(ii)(B) *Short-term exposure limits.* The employer has taken one or more personal breathing zone air samples which indicate the highest likely 15-minute exposures during such operations for at least one employee in each job classification in the work area during every work shift, and the employee sampled is expected to have the highest MC exposure.

1910.1052(d)(1)(ii)(C) *Exception.* Personal breathing zone air samples taken during one work shift may be used to represent employee exposures on other work shifts where the employer can document that the tasks performed and conditions in the workplace are similar across shifts.

1910.1052(d)(1)(iii) *Accuracy of monitoring.* The employer shall ensure that the methods used to perform exposure monitoring produce results that are accurate to a confidence level of 95 percent, and are:

1910.1052(d)(1)(iii)(A) Within plus or minus 25 percent for airborne concentrations of MC above the 8-hour TWA PEL or the STEL; or

1910.1052(d)(1)(iii)(B) Within plus or minus 35 percent for airborne concentrations of MC at or above the action level but at or below the 8-hour TWA PEL.

1910.1052(d)(2) *Initial determination.* Each employer whose employees are exposed to MC shall perform initial exposure monitoring to determine each affected employee's exposure, except under the following conditions:

1910.1052(d)(2)(i) Where objective data demonstrate that MC cannot be released in the workplace in airborne concentrations at or above the action level or above the STEL. The objective data shall represent the highest MC exposures likely to occur under reasonably foreseeable conditions of processing, use, or handling. The employer shall document the objective data exemption as specified in paragraph (m) of this section;

1910.1052(d)(2)(ii) Where the employer has performed exposure monitoring within 12 months prior to April 10, 1997 and that exposure monitoring meets all other requirements of this section, and was conducted under conditions substantially equivalent to existing conditions; or

1910.1052(d)(2)(iii) Where employees are exposed to MC on fewer than 30 days per year (e.g., on a construction site), and the employer has measurements by direct-reading instruments which give immediate results (such as a detector tube) and which provide sufficient information regarding employee exposures to determine what control measures are necessary to reduce exposures to acceptable levels.

1910.1052(d)(3) *Periodic monitoring.* Where the initial determination shows employee exposures at or above the action level or above the STEL, the employer shall establish an exposure monitoring program for periodic monitoring of employee exposure to MC in accordance with Table 1:

## TABLE 1—INITIAL DETERMINATION EXPOSURE SCENARIOS AND THEIR ASSOCIATED MONITORING FREQUENCIES

| Exposure scenario | Required monitoring activity |
|---|---|
| Below the action level and at or below the STEL. | No 8-hour TWA or STEL monitoring required. |
| Below the action level and above the STEL. | No 8-hour TWA monitoring required; monitor STEL exposures every three months. |
| At or above the action level, at or below the TWA, and at or below the STEL. | Monitor 8-hour TWA exposures every six months. |
| At or above the action level, at or below the TWA, and above the STEL. | Monitor 8-hour TWA exposures every six months and monitor STEL exposures every three months. |
| Above the TWA and at or below the STEL. | Monitor 8-hour TWA exposures every three months. In addition, without regard to the last sentence of the note to paragraph (d)(3), the following employers must monitor STEL exposures every three months until either the date by which they must achieve the 8-hour TWA PEL under paragraph (n) of this section or the date by which they in fact achieve the 8-hour TWA PEL, whichever comes first: employers engaged in polyurethane foam manufacturing; foam fabrication; furniture refinishing; general aviation aircraft stripping; product formulation; use of MC-based adhesives for boat building and repair, recreational vehicle manufacture, van conversion, or upholstery; and use of MC in construction work for restoration and preservation of buildings, painting and paint removal, cabinet making, or floor refinishing and resurfacing. |
| Above the TWA and above the STEL. | Monitor 8-hour TWA exposures and STEL exposures every three months. |

[**Note to paragraph (d)(3):** The employer may decrease the frequency of 8-hour TWA exposure monitoring to every six months when at least two consecutive measurements taken at least seven days apart show exposures to be at or below the 8-hour TWA PEL. The employer may discontinue the periodic 8-hour TWA monitoring for employees where at least two consecutive measurements taken at least seven days apart are below the action level. The employer may discontinue the periodic STEL monitoring for employees where at least two consecutive measurements taken at least 7 days apart are at or below the STEL.]

**1910.1052(d)(4)** *Additional monitoring.* (i) The employer shall perform exposure monitoring when a change in workplace conditions indicates that employee exposure may have increased. Examples of situations that may require additional monitoring include changes in production, process, control equipment, or work practices, or a leak, rupture, or other breakdown.

**1910.1052(d)(4)(ii)** Where exposure monitoring is performed due to a spill, leak, rupture or equipment breakdown, the employer shall clean-up the MC and perform the appropriate repairs before monitoring.

**1910.1052(d)(5)** *Employee notification of monitoring results.* (i) The employer shall, within 15 working days after the receipt of the results of any monitoring performed under this section, notify each affected employee of these results in writing, either individually or by posting of results in an appropriate location that is accessible to affected employees.

**1910.1052(d)(5)(ii)** Whenever monitoring results indicate that employee exposure is above the 8-hour TWA PEL or the STEL, the employer shall describe in the written notification the corrective action being taken to reduce employee exposure to or below the 8-hour TWA PEL or STEL and the schedule for completion of this action.

**1910.1052(d)(6)** *Observation of monitoring*—(i) *Employee observation.* The employer shall provide affected employees or their designated representatives an opportunity to observe any monitoring of employee exposure to MC conducted in accordance with this section.

**1910.1052(d)(6)(ii)** *Observation procedures.* When observation of the monitoring of employee exposure to MC requires entry into an area where the use of protective clothing or equipment is required, the employer shall provide, at no cost to the observer(s), and the observer(s) shall be required to use such clothing and equipment and shall comply with all other applicable safety and health procedures.

**1910.1052(e)** ***Regulated areas.*** (1) The employer shall establish a regulated area wherever an employee's exposure to airborne concentrations of MC exceeds or can reasonably be expected to exceed either the 8-hour TWA PEL or the STEL.

**1910.1052(e)(2)** The employer shall limit access to regulated areas to authorized persons.

**1910.1052(e)(3)** The employer shall supply a respirator, selected in accordance with paragraph (h)(3) of this section, to each person who enters a regulated area and shall require each affected employee to use that respirator whenever MC exposures are likely to exceed the 8-hour TWA PEL or STEL.

[**Note to paragraph (e)(3):** An employer who has implemented all feasible engineering, work practice and administrative controls (as required in paragraph (f) of this section), and who has established a regulated area (as required by paragraph (e)(1) of this section) where MC exposure can be reliably predicted to exceed the 8-hour TWA PEL or the STEL only on certain days (for example, because of work or process schedule) would need to have affected employees use respirators in that regulated area only on those days.]

**1910.1052(e)(4)** The employer shall ensure that, within a regulated area, employees do not engage in non-work activities which may increase dermal or oral MC exposure.

**1910.1052(e)(5)** The employer shall ensure that while employees are wearing respirators, they do not engage in activities (such as taking medication or chewing gum or tobacco) which interfere with respirator seal or performance.

**1910.1052(e)(6)** The employer shall demarcate regulated areas from the rest of the workplace in any manner that adequately establishes and alerts employees to the boundaries of the area and minimizes the number of authorized employees exposed to MC within the regulated area.

**1910.1052(e)(7)** An employer at a multi-employer worksite who establishes a regulated area shall communicate the access restrictions and locations of these areas to all other employers with work operations at that worksite.

**1910.1052(f)** ***Methods of compliance***—(1) *Engineering and work practice controls.* The employer shall institute and maintain the effectiveness of engineering controls and work practices to reduce employee exposure to or below the PELs except to the extent that the employer can demonstrate that such controls are not feasible. Wherever the feasible engineering controls and work practices which can be instituted are not sufficient to reduce employee exposure to or below the 8-TWA PEL or STEL, the employer shall use them to reduce employee exposure to the lowest levels achievable by these controls and shall supplement them by the use of respiratory protection that complies with the requirements of paragraph (g) of this section.

**1910.1052(f)(2)** *Prohibition of rotation.* The employer shall not implement a schedule of employee rotation as a means of compliance with the PELs.

**1910.1052(f)(3)** *Leak and spill detection.* (i) The employer shall implement procedures to detect leaks of MC in the workplace. In work areas where spills may occur, the employer shall make provisions to contain any spills and to safely dispose of any MC-contaminated waste materials.

**1910.1052(f)(3)(ii)** The employer shall ensure that all incidental leaks are repaired and that incidental spills are cleaned promptly by employees who use the appropriate personal protective equipment and are trained in proper methods of cleanup.

[**Note to paragraph (f)(3)(ii):** See Appendix A of this section for examples of procedures that satisfy this requirement. Employers covered by this standard may also be subject to the hazardous waste and emergency response provisions contained in 29 CFR 1910.120(q).]

**1910.1052(g)** ***Respiratory protection***—(1) *General.* For employees who use respirators required by this section, the employer must provide respirators that comply with the requirements of this paragraph. Respirators must be used during:

**1910.1052(g)(1)(i)** Periods when an employee's exposure to MC exceeds the 8-hour TWA PEL, or STEL (for example, when an employee is using MC in a regulated area).

**1910.1052(g)(1)(ii)** Periods necessary to install or implement feasible engineering and work-practice controls.

**1910.1052(g)(1)(iii)** A few work operations, such as some maintenance operations and repair activities, for which the employer demonstrates that engineering and work-practice controls are infeasible.

**1910.1052(g)(1)(iv)** Work operations for which feasible engineering and work-practice controls are not sufficient to reduce employee exposures to or below the PELs.

**1910.1052(g)(1)(v)** Emergencies.

**1910.1052(g)(2)** *Respirator program.* (i) The employer must implement a respiratory protection program in accordance with 29 CFR 1910.134 (b) through (m) (except (d)(1)(iii) and (d)(3)(iii)(B) (1) and (2)).

**1910.1052(g)(2)(ii)** Employers who provide employees with gas masks with organic-vapor canisters for the purpose of emergency escape must replace the canisters after any emergency use and before the gas masks are returned to service.

**1910.1052(g)(3)** *Respirator selection.* The employer must select appropriate atmosphere-supplying respirators from Table 2 of this section.

### TABLE 2—MINIMUM REQUIREMENTS FOR RESPIRATORY PROTECTION FOR AIRBORNE METHYLENE CHLORIDE

| Methylene chloride airborne concentration (ppm) or condition of use | Minimum respirator required [1] |
|---|---|
| Up to 625 ppm (25×PEL) | (1) Continuous flow supplied-air respirator, hood or helmet. |
| Up to 1250 ppm (50×8-TWA PEL) | (1) Full facepiece supplied-air respirator operated in negative pressure (demand) mode.<br>(2) Full facepiece self-contained breathing apparatus (SCBA) operated in negative pressure (demand) mode. |
| Up to 5000 ppm (200×8-TWA PEL) | (1) Continuous flow supplied-air respirator, full facepiece.<br>(2) Pressure demand supplied-air respirator, full facepiece.<br>(3) Positive pressure full facepiece SCBA. |
| Unknown concentration, or above 5000 ppm (Greater than 200×8-TWA PEL) | (1) Positive pressure full facepiece SCBA.<br>(2) Full facepiece pressure demand supplied-air respirator with an auxiliary self-contained air supply. |
| Fire fighting | Positive pressure full facepiece SCBA. |
| Emergency escape | (1) Any continuous flow or pressure demand SCBA.<br>(2) Gas mask with organic vapor canister. |

[1] Respirators assigned for higher airborne concentrations may be used at lower concentrations.

**1910.1052(g)(4)** *Medical evaluation.* Before having an employee use a supplied-air respirator in the negative-pressure mode, or a gas mask with an organic-vapor canister for emergency escape, the employer must:

**1910.1052(g)(4)(i)** Have a physician or other licensed health-care professional (PLHCP) evaluate the employee's ability to use such respiratory protection.

**1910.1052(g)(4)(ii)** Ensure that the PLHCP provides their findings in a written opinion to the employee and the employer.

**1910.1052(h)** ***Protective Work Clothing and Equipment.*** (1) Where needed to prevent MC-induced skin or eye irritation, the employer shall provide clean protective clothing and equipment which is resistant to MC, at no cost to the employee, and shall ensure that each affected employee uses it. Eye and face protection shall meet the requirements of 29 CFR 1910.133 or 29 CFR 1915.153, as applicable.

**1910.1052(h)(2)** The employer shall clean, launder, repair and replace all protective clothing and equipment required by this paragraph as needed to maintain their effectiveness.

**1910.1052(h)(3)** The employer shall be responsible for the safe disposal of such clothing and equipment.

**[Note to paragraph (h)(4):** See Appendix A for examples of disposal procedures that will satisfy this requirement.]

**1910.1052(i)** ***Hygiene facilities.*** (1) If it is reasonably foreseeable that employees' skin may contact solutions containing 0.1 percent or greater MC (for example, through splashes, spills or improper work practices), the employer shall provide conveniently located washing facilities capable of removing the MC, and shall ensure that affected employees use these facilities as needed.

**1910.1052(i)(2)** If it is reasonably foreseeable that an employee's eyes may contact solutions containing 0.1 percent or greater MC (for example through splashes, spills or improper work practices), the employer shall provide appropriate eyewash facilities within the immediate work area for emergency use, and shall ensure that affected employees use those facilities when necessary.

**1910.1052(j)** ***Medical surveillance***—(1) *Affected employees.* The employer shall make medical surveillance available for employees who are or may be exposed to MC as follows:

**1910.1052(j)(1)(i)** At or above the action level on 30 or more days per year, or above the 8- hour TWA PEL or the STEL on 10 or more days per year;

**1910.1052(j)(1)(ii)** Above the 8-TWA PEL or STEL for any time period where an employee has been identified by a physician or other licensed health care professional as being at risk from cardiac disease or from some other serious MC-related health condition and such employee requests inclusion in the medical surveillance program;

**1910.1052(j)(1)(iii)** During an emergency.

**1910.1052(j)(2)** *Costs.* The employer shall provide all required medical surveillance at no cost to affected employees, without loss of pay and at a reasonable time and place.

**1910.1052(j)(3)** *Medical personnel.* The employer shall ensure that all medical surveillance procedures are performed by a physician or other licensed health care professional, as defined in paragraph (b) of this section.

**1910.1052(j)(4)** *Frequency of medical surveillance.* The employer shall make medical surveillance available to each affected employee as follows:

**1910.1052(j)(4)(i)** *Initial surveillance.* The employer shall provide initial medical surveillance under the schedule provided by paragraph (n)(2)(iii) of this section, or before the time of initial assignment of the employee, whichever is later. The employer need not provide the initial surveillance if medical records show that an affected employee has been provided with medical surveillance that complies with this section within 12 months before April 10, 1997.

**1910.1052(j)(4)(ii)** *Periodic medical surveillance.* The employer shall update the medical and work history for each affected employee annually. The employer shall provide periodic physical examinations, including appropriate laboratory surveillance, as follows:

**1910.1052(j)(4)(ii)(A)** For employees 45 years of age or older, within 12 months of the initial surveillance or any subsequent medical surveillance; and

**1910.1052(j)(4)(ii)(B)** For employees younger than 45 years of age, within 36 months of the initial surveillance or any subsequent medical surveillance.

**1910.1052(j)(4)(iii)** *Termination of employment or reassignment.* When an employee leaves the employer's workplace, or is reassigned to an area where exposure to MC is consistently at or below the action level and STEL, medical surveillance shall be made available if six months or more have elapsed since the last medical surveillance.

**1910.1052(j)(4)(iv)** *Additional surveillance.* The employer shall provide additional medical surveillance at frequencies other than those listed above when recommended in the written medical opinion. (For example, the physician or other licensed health care professional may determine an examination is warranted in less than 36 months for employees younger than 45 years of age based upon evaluation of the results of the annual medical and work history.)

**1910.1052(j)(5)** *Content of medical surveillance*—(i) *Medical and work history.* The comprehensive medical and work history shall emphasize neurological symptoms, skin conditions, history of hematologic or liver disease, signs or symptoms suggestive of heart disease (angina, coronary artery disease), risk factors for cardiac disease, MC exposures, and work practices and personal protective equipment used during such exposures.

**[Note to paragraph (j)(5)(i):** See Appendix B of this section for an example of a medical and work history format that would satisfy this requirement.]

**1910.1052(j)(5)(ii)** *Physical examination.* Where physical examinations are provided as required above, the physician or other licensed health care professional shall accord particular attention to the lungs, cardiovascular system (including blood pressure and pulse), liver, nervous system, and skin. The physician or other licensed health care professional shall determine the extent and nature of the physical examination based on the health status of the employee and analysis of the medical and work history.

**1910.1052(j)(5)(iii)** *Laboratory surveillance.* The physician or other licensed health care professional shall determine the extent of any required laboratory surveillance based on the employee's observed health status and the medical and work history.

**[Note to paragraph (j)(5)(iii):** See Appendix B of this section for information regarding medical tests. Laboratory surveillance may include before- and after-shift carboxyhemoglobin determinations, resting ECG, hematocrit, liver function tests and cholesterol levels.]

**1910.1052(j)(5)(iv)** *Other information or reports.* The medical surveillance shall also include any other information or reports the physician or other licensed health care professional determines are necessary to assess the employee's health in relation to MC exposure.

**1910.1052(j)(6)** *Content of emergency medical surveillance.* The employer shall ensure that medical surveillance made available when an employee has been exposed to MC in emergency situations includes, at a minimum:

**1910.1052(j)(6)(i)** Appropriate emergency treatment and decontamination of the exposed employee;

**1910.1052(j)(6)(ii)** Comprehensive physical examination with special emphasis on the nervous system, cardiovascular system, lungs, liver and skin, including blood pressure and pulse;

**1910.1052(j)(6)(iii)** Updated medical and work history, as appropriate for the medical condition of the employee; and

**1910.1052(j)(6)(iv)** Laboratory surveillance, as indicated by the employee's health status.

**[Note to paragraph (j)(6)(iv):** See Appendix B for examples of tests which may be appropriate.]

**1910.1052(j)(7)** *Additional examinations and referrals.* Where the physician or other licensed health care professional determines it is necessary, the scope of the medical examination shall be expanded and the appropriate additional medical surveillance, such as referrals for consultation or examination, shall be provided.

**1910.1052(j)(8)** *Information provided to the physician or other licensed health care professional.* The employer shall provide the following information to a physician or other licensed health care professional who is involved in the diagnosis of MC-induced health effects:

**1910.1052(j)(8)(i)** A copy of this section including its applicable appendices;

**1910.1052(j)(8)(ii)** A description of the affected employee's past, current and anticipated future duties as they relate to the employee's MC exposure;

**1910.1052(j)(8)(iii)** The employee's former or current exposure levels or, for employees not yet occupationally exposed to MC, the employee's anticipated exposure levels and the frequency and exposure levels anticipated to be associated with emergencies;

**1910.1052(j)(8)(iv)** A description of any personal protective equipment, such as respirators, used or to be used; and

**1910.1052(j)(8)(v)** Information from previous employment-related medical surveillance of the affected employee which is not otherwise available to the physician or other licensed health care professional.

**1910.1052(j)(9)** *Written medical opinions.* (i) For each physical examination required by this section, the employer shall ensure that the physician or other licensed health care professional provides to the employer and to the affected employee a written opinion regarding the results of that examination within 15 days of completion of the evaluation of medical and laboratory findings, but not more than 30 days after the examination. The written medical opinion shall be limited to the following information:

**1910.1052(j)(9)(i)(A)** The physician or other licensed health care professional's opinion concerning whether exposure to MC may contribute to or aggravate the employee's existing cardiac, hepatic, neurological (including stroke) or dermal disease or whether the employee has any other medical condition(s) that would place the employee's health at increased risk of material impairment from exposure to MC.

**1910.1052(j)(9)(i)(B)** Any recommended limitations upon the employee's exposure to MC, including removal from MC exposure, or upon the employee's use of respirators, protective clothing, or other protective equipment.

**1910.1052(j)(9)(i)(C)** A statement that the employee has been informed by the physician or other licensed health care professional that MC is a potential occupational carcinogen, of risk factors for heart disease, and the potential for exacerbation of underlying heart disease by exposure to MC through its metabolism to carbon monoxide; and

**1910.1052(j)(9)(i)(D)** A statement that the employee has been informed by the physician or other licensed health care professional of the results of the medical examination and any medical conditions resulting from MC exposure which require further explanation or treatment.

**1910.1052(j)(9)(ii)** The employer shall instruct the physician or other licensed health care professional not to reveal to the employer, orally or in the written opinion, any specific records, findings, and diagnoses that have no bearing on occupational exposure to MC.

**[Note to paragraph (j)(9)(ii):** The written medical opinion may also include information and opinions generated to comply with other OSHA health standards.]

**1910.1052(j)(10)** *Medical presumption.* For purposes of this paragraph (j) of this section, the physician or other licensed health care professional shall presume, unless medical evidence indicates to the contrary, that a medical condition is unlikely to require medical removal from MC exposure if the employee is not exposed to MC above the 8-hour TWA PEL. If the physician or other licensed health care professional recommends removal for an employee exposed below the 8-hour TWA PEL, the physician or other licensed health care professional shall cite specific medical evidence, sufficient to rebut the presumption that exposure below the 8-hour TWA PEL is unlikely to require removal, to support the recommendation. If such evidence is cited by the physician or other licensed health care professional, the employer must remove the employee. If such evidence is not cited by the physician or other licensed health care professional, the employer is not required to remove the employee.

**1910.1052(j)(11)** *Medical Removal Protection (MRP).* (i) Temporary medical removal and return of an employee.

**1910.1052(j)(11)(i)(A)** Except as provided in paragraph (j)(10) of this section, when a medical determination recommends removal because the employee's exposure to MC may contribute to or aggravate the employee's existing cardiac, hepatic, neurological (including stroke), or skin disease, the employer must provide medical removal protection benefits to the employee and either:

**1910.1052(j)(11)(i)(A)*(1)*** Transfer the employee to comparable work where methylene chloride exposure is below the action level; or

**1910.1052(j)(11)(i)(A)*(2)*** Remove the employee from MC exposure.

**1910.1052(j)(11)(i)(B)** If comparable work is not available and the employer is able to demonstrate that removal and the costs of extending MRP benefits to an additional employee, considering feasibility in relation to the size of the employer's business and the other requirements of this standard, make further reliance on MRP an inappropriate remedy, the employer may retain the additional employee in the existing job until transfer or removal becomes appropriate, provided:

**1910.1052(j)(11)(i)(B)*(1)*** The employer ensures that the employee receives additional medical surveillance, including a physical examination at least every 60 days until transfer or removal occurs; and

**1910.1052(j)(11)(i)(B)*(2)*** The employer or PLHCP informs the employee of the risk to the employee's health from continued MC exposure.

**1910.1052(j)(11)(i)(C)** The employer shall maintain in effect any job-related protective measures or limitations, other than removal, for as long as a medical determination recommends them to be necessary.

**1910.1052(j)(11)(ii)** End of MRP benefits and return of the employee to former job status.

**1910.1052(j)(11)(ii)(A)** The employer may cease providing MRP benefits at the earliest of the following:

**1910.1052(j)(11)(ii)(A)*(1)*** Six months;

**1910.1052(j)(11)(ii)(A)*(2)*** Return of the employee to the employee's former job status following receipt of a medical determination concluding that the employee's exposure to MC no longer will aggravate any cardiac, hepatic, neurological (including stroke), or dermal disease;

**1910.1052(j)(11)(ii)(A)*(3)*** Receipt of a medical determination concluding that the employee can never return to MC exposure.

**1910.1052(j)(11)(ii)(B)** For the purposes of this paragraph (j), the requirement that an employer return an employee to the employee's former job status is not intended to expand upon or restrict any rights an employee has or would have had, absent temporary medical removal, to a specific job classification or position under the terms of a collective bargaining agreement.

**1910.1052(j)(12)** *Medical removal protection benefits.* (i) For purposes of this paragraph (j), the term medical removal protection benefits means that, for each removal, an employer must maintain for up to six months the earnings, seniority, and other employment rights and benefits of the employee as though the employee had not been removed from MC exposure or transferred to a comparable job.

**1910.1052(j)(12)(ii)** During the period of time that an employee is removed from exposure to MC, the employer may condition the provision of medical removal protection benefits upon the employee's participation in follow-up medical surveillance made available pursuant to this section.

**1910.1052(j)(12)(iii)** If a removed employee files a workers' compensation claim for a MC-related disability, the employer shall continue the MRP benefits required by this paragraph until either the claim is resolved or the 6-month period for payment f MRP benefits has passed, whichever occurs first. To the extent the employee is entitled to indemnity payments for earnings lost during the period of removal, the employer's obligation to provide medical removal protection benefits to the employee shall be reduced by the amount of such indemnity payments.

**1910.1052(j)(12)(iv)** The employer's obligation to provide medical removal protection benefits to a removed employee shall be reduced to the extent that the employee receives compensation for earnings lost during the period of removal from either a publicly or an employer-funded compensation program, or receives income from employment with another employer made possible by virtue of the employee's removal.

**1910.1052(j)(13)** *Voluntary removal or restriction of an employee.* Where an employer, although not required by this section to do so, removes an employee from exposure to MC or otherwise places any limitation on an employee due to the effects of MC exposure on the employee's medical condition, the employer shall provide medical removal protection benefits to the employee equal to those required by paragraph (j)(12) of this section.

**1910.1052(j)(14)** *Multiple health care professional review mechanism.* (i) If the employer selects the initial physician or licensed health care professional (PLHCP) to conduct any medical examination or consultation provided to an employee under this paragraph (j)(11), the employer shall notify the employee of the right to seek a second medical opinion each time the employer provides the employee with a copy of the written opinion of that PLHCP.

**1910.1052(j)(14)(ii)** If the employee does not agree with the opinion of the employer-selected PLHCP, notifies the employer of that fact, and takes steps to make an appointment with a second PLHCP within 15 days of receiving a copy of the written opinion of the initial PLHCP, the employer shall pay for the PLHCP chosen by the employee to perform at least the following:

**1910.1052(j)(14)(ii)(A)** Review any findings, determinations or recommendations of the initial PLHCP; and

**1910.1052(j)(14)(ii)(B)** Conduct such examinations, consultations, and laboratory tests as the PLHCP deems necessary to facilitate this review.

**1910.1052(j)(14)(iii)** If the findings, determinations or recommendations of the second PLHCP differ from those of the initial PLHCP, then the employer and the employee shall instruct the two health care professionals to resolve the disagreement.

**1910.1052(j)(14)(iv)** If the two health care professionals are unable to resolve their disagreement within 15 days, then those two health care professionals shall jointly designate a PLHCP who is a specialist in the field at issue. The employer shall pay for the specialist to perform at least the following:

**1910.1052(j)(14)(iv)(A)** Review the findings, determinations, and recommendations of the first two PLHCPs; and

**1910.1052(j)(14)(iv)(B)** Conduct such examinations, consultations, laboratory tests and discussions with the prior PLHCPs as the specialist deems necessary to resolve the disagreements of the prior health care professionals.

**1910.1052(j)(14)(v)** The written opinion of the specialist shall be the definitive medical determination. The employer shall act consistent with the definitive medical determination, unless the employer and employee agree that the written opinion of one of the other two PLHCPs shall be the definitive medical determination.

**1910.1052(j)(14)(vi)** The employer and the employee or authorized employee representative may agree upon the use of any expeditious alternate health care professional determination mechanism in lieu of the multiple health care professional review mechanism provided by this paragraph so long as the alternate mechanism otherwise satisfies the requirements contained in this paragraph.

**1910.1052(k)** ***Hazard communication.*** The employer shall communicate the following hazards associated with MC on labels and in material safety data sheets in accordance with the requirements of the Hazard Communication Standard, 29 CFR 1910.1200, 29 CFR 1915.1200, or 29 CFR 1926.59, as appropiate: cancer, cardiac effects (including elevation of carboxyhemoglobin), central nervous system effects, liver effects, and skin and eye irritation.

**1910.1052(l)** ***Employee information and training.*** (1) The employer shall provide information and training for each affected employee prior to or at the time of initial assignment to a job involving potential exposure to MC.

**1910.1052(l)(2)** The employer shall ensure that information and training is presented in a manner that is understandable to the employees.

**1910.1052(l)(3)** In addition to the information required under the Hazard Communication Standard at 29 CFR 1910.1200, 29 CFR 1915.1200, or 29 CFR 1926.59, as appropriate:

**1910.1052(l)(3)(i)** The employer shall inform each affected employee of the requirements of this section and information available in its appendices, as well as how to access or obtain a copy of it in the workplace;

**1910.1052(l)(3)(ii)** Wherever an employee's exposure to airborne concentrations of MC exceeds or can reasonably be expected to exceed the action level, the employer shall inform each affected employee of the quantity, location, manner of use, release, and storage of MC and the specific operations in the workplace that could result in exposure to MC, particularly noting where exposures may be above the 8-hour TWA PEL or STEL;

**1910.1052(l)(4)** The employer shall train each affected employee as required under the Hazard Communication standard at 29 CFR 1910.1200, 29 CFR 1915.1200, or 29 CFR 1926.59, as appropriate.

**1910.1052(l)(5)** The employer shall re-train each affected employee as necessary to ensure that each employee exposed above the action level or the STEL maintains the requisite understanding of the principles of safe use and handling of MC in the workplace.

**1910.1052(l)(6)** Whenever there are workplace changes, such as modifications of tasks or procedures or the institution of new tasks or procedures, which increase employee exposure, and where those exposures exceed or can reasonably be expected to exceed the action level, the employer shall update the training as necessary to ensure that each affected employee has the requisite proficiency.

**1910.1052(l)(7)** An employer whose employees are exposed to MC at a multi-employer worksite shall notify the other employers with work operations at that site in accordance with the requirements of the Hazard Communication Standard, 29 CFR 1910.1200, 29 CFR 1915.1200, or 29 CFR 1926.59, as appropiate.

1910.1052(l)(8) The employer shall provide to the Assistant Secretary or the Director, upon request, all available materials relating to employee information and training.

1910.1052(m) ***Recordkeeping***—(1) *Objective data.* (i) Where an employer seeks to demonstrate that initial monitoring is unnecessary through reasonable reliance on objective data showing that any materials in the workplace containing MC will not release MC at levels which exceed the action level or the STEL under foreseeable conditions of exposure, the employer shall establish and maintain an accurate record of the objective data relied upon in support of the exemption.

1910.1052(m)(1)(ii) This record shall include at least the following information:

1910.1052(m)(1)(ii)(A) The MC-containing material in question;

1910.1052(m)(1)(ii)(B) The source of the objective data;

1910.1052(m)(1)(ii)(C) The testing protocol, results of testing, and/or analysis of the material for the release of MC;

1910.1052(m)(1)(ii)(D) A description of the operation exempted under paragraph (d)(2)(i) of this section and how the data support the exemption; and

1910.1052(m)(1)(ii)(E) Other data relevant to the operations, materials, processing, or employee exposures covered by the exemption.

1910.1052(m)(1)(iii) The employer shall maintain this record for the duration of the employer's reliance upon such objective data.

1910.1052(m)(2) *Exposure measurements.* (i) The employer shall establish and keep an accurate record of all measurements taken to monitor employee exposure to MC as prescribed in paragraph (d) of this section.

1910.1052(m)(2)(ii) Where the employer has 20 or more employees, this record shall include at least the following information:

1910.1052(m)(2)(ii)(A) The date of measurement for each sample taken;

1910.1052(m)(2)(ii)(B) The operation involving exposure to MC which is being monitored;

1910.1052(m)(2)(ii)(C) Sampling and analytical methods used and evidence of their accuracy;

1910.1052(m)(2)(ii)(D) Number, duration, and results of samples taken;

1910.1052(m)(2)(ii)(E) Type of personal protective equipment, such as respiratory protective devices, worn, if any; and

1910.1052(m)(2)(ii)(F) Name, social security number, job classification and exposure of all of the employees represented by monitoring, indicating which employees were actually monitored.

1910.1052(m)(2)(iii) Where the employer has fewer than 20 employees, the record shall include at least the following information:

1910.1052(m)(2)(iii)(A) The date of measurement for each sample taken;

1910.1052(m)(2)(iii)(B) Number, duration, and results of samples taken; and

1910.1052(m)(2)(iii)(C) Name, social security number, job classification and exposure of all of the employees represented by monitoring, indicating which employees were actually monitored.

1910.1052(m)(2)(iv) The employer shall maintain this record for at least thirty (30) years, in accordance with 29 CFR 1910.1020.

1910.1052(m)(3) *Medical surveillance.* (i) The employer shall establish and maintain an accurate record for each employee subject to medical surveillance under paragraph (j) of this section.

1910.1052(m)(3)(ii) The record shall include at least the following information:

1910.1052(m)(3)(ii)(A) The name, social security number and description of the duties of the employee;

1910.1052(m)(3)(ii)(B) Written medical opinions; and

1910.1052(m)(3)(ii)(C) Any employee medical conditions related to exposure to MC.

1910.1052(m)(3)(iii) The employer shall ensure that this record is maintained for the duration of employment plus thirty (30) years, in accordance with 29 CFR 1910.1020.

1910.1052(m)(4) *Availability.* (i) The employer, upon written request, shall make all records required to be maintained by this section available to the Assistant Secretary and the Director for examination and copying in accordance with 29 CFR 1910.1020.

**[Note to paragraph (m)(4)(i):** All records required to be maintained by this section may be kept in the most administratively convenient form (for example, electronic or computer records would satisfy this requirement).]

1910.1052(m)(4)(ii) The employer, upon request, shall make any employee exposure and objective data records required by this section available for examination and copying by affected employees, former employees, and designated representatives in accordance with 29 CFR 1910.1020.

1910.1052(m)(4)(iii) The employer, upon request, shall make employee medical records required to be kept by this section available for examination and copying by the subject employee and by anyone having the specific written consent of the subject employee in accordance with 29 CFR 1910.1020.

1910.1052(m)(5) *Transfer of records.* The employer shall comply with the requirements concerning transfer of records set forth in 29 CFR 1910.1020(h).

1910.1052(n) ***Dates***—(1) *Effective date.* This section shall become effective April 10, 1997.

1910.1052(n)(2) *Start-up dates.* (i) Initial monitoring required by paragraph (d)(2) of this section shall be completed according to the following schedule:

1910.1052(n)(2)(i)(A) For employers with fewer than 20 employees, within 300 days after the effective date of this section.

1910.1052(n)(2)(i)(B) For polyurethane foam manufacturers with 20 to 99 employees, within 255 days after the effective date of this section.

1910.1052(n)(2)(i)(C) For all other employers, within 150 days after the effective date of this section.

1910.1052(n)(2)(ii) Engineering controls required under paragraph (f)(1) of this section shall be implemented according to the following schedule:

1910.1052(n)(2)(ii)(A) For employers with fewer than 20 employees: within three (3) years after the effective date of this section.

1910.1052(n)(2)(ii)(B) For employers with fewer than 150 employees engaged in foam fabrication; for employers with fewer than 50 employees engaged in furniture refinishing, general aviation aircraft stripping, and product formulation; for employers with fewer than 50 employees using MC-based adhesives for boat building and repair, recreational vehicle manufacture, van conversion, and upholstering; for employers with fewer than 50 employees using MC in construction work for restoration and preservation of buildings, painting and paint removal, cabinet making and/or floor refinishing and resurfacing: within three (3) years after the effective date of this section.

1910.1052(n)(2)(ii)(C) For employers engaged in polyurethane foam manufacturing with 20 employees or more: within thirty (30) months after the effective date of this section.

1910.1052(n)(2)(ii)(D) For employers with 150 or more employees engaged in foam fabrication; for employers with 50 or more employees engaged in furniture refinishing, general aviation aircraft stripping, and product formulation; for employers with 50 or more employees using MC-based adhesives in boat building and repair, recreational vehicle manufacture, van conversion and upholstering; and for employers with 50 or more employees using MC in construction work for restoration and preservation of buildings, painting and paint removal, cabinet making and/or floor refinishing and resurfacing: within two (2) years after the effective date of this section.

1910.1052(n)(2)(ii)(E) For all other employers: within one (1) year after the effective date of this section.

1910.1052(n)(2)(iii) Employers identified in paragraphs (n)(2)(ii)(B), (C), and (D) of this section shall comply with the requirements listed below in this subparagraph by the dates indicated:

1910.1052(n)(2)(iii)(A) Use of respiratory protection whenever an employee's exposure to MC exceeds or can reasonably be expected to exceed the 8-hour TWA PEL, in accordance with paragraphs (c)(1), (e)(3), (f)(1) and (g)(1) of this section: by the applicable dates set out in paragraphs (n)(2)(ii)(B), (C) and (D) of this section for the installation of engineering controls.

**1910.1052(n)(2)(iii)(B)** Use of respiratory protection whenever an employee's exposure to MC exceeds or can reasonably be expected to exceed the STEL in accordance with paragraphs (e)(3), (f)(1), and (g)(1) of this section: by the applicable dates indicated in paragraph (n)(2)(iv) of this section.

**1910.1052(n)(2)(iii)(C)** Implementation of work practices (such as leak and spill detection, cleanup and enclosure of containers) required by paragraph (f)(1) of this section: by the applicable dates indicated in paragraph (n)(2)(iv) of this section.

**1910.1052(n)(2)(iii)(D)** Notification of corrective action under paragraph (d)(5)(ii) of this section: no later than (90) days before the compliance date applicable to such corrective action.

**1910.1052(n)(2)(iv)** Unless otherwise specified in this paragraph (n), all other requirements of this section shall be complied with according to the following schedule:

**1910.1052(n)(2)(iv)(A)** For employers with fewer than 20 employees, within one (1) year after the effective date of this section.

**1910.1052(n)(2)(iv)(B)** For employers engaged in polyurethane foam manufacturing with 20 to 99 employees, within 270 days after the effective date of this section.

**1910.1052(n)(2)(iv)(C)** For all other employers, within 255 days after the effective date of this section.

**1910.1052(n)(3)** *Transitional dates.* The exposure limits for MC specified in 29 CFR 1910.1000 (1996), Table Z-2, shall remain in effect until the start-up dates for the exposure limits specified in paragraph (n) of this section, or if the exposure limits in this section are stayed or vacated.

**1910.1052(o)** ***Appendices.*** The information contained in the appendices does not, by itself, create any additional obligations not otherwise imposed or detract from any existing obligation.

**[Note to paragraph (o):** The requirement of 29 CFR 1910.1052(g)(1) to use respiratory protection whenever an employee's exposure to methylene chloride exceeds or can reasonably be expected to exceed the 8-hour TWA PEL is hereby stayed until August 31, 1998 for employers engaged in polyurethane foam manufacturing; foam fabrication; furniture refinishing; general aviation aircraft stripping; formulation of products containing methylene chloride; boat building and repair; recreational vehicle manufacture; van conversion; upholstery; and use of methylene chloride in construction work for restoration and preservation of buildings, painting and paint removal, cabinet making and/or floor refinishing and resurfacing.

The requirement of 29 CFR 1910.1052(f)(1) to implement engineering controls to achieve the 8-hour TWA PEL and STEL is hereby stayed until December 10, 1998 for employers with more than 100 employees engaged in polyurethane foam manufacturing and for employers with more than 20 employees engaged in foam fabrication; furniture refinishing; general aviation aircraft stripping; formulation of products containing methylene chloride; boat building and repair; recreational vehicle manufacture; van conversion; upholstery; and use of methylene chloride in construction work for restoration and preservation of buildings, painting and paint removal, cabinet making and/or floor refinishing and resurfacing.]

## APPENDIX A TO SECTION 1910.1052—SUBSTANCE SAFETY DATA SHEET AND TECHNICAL GUIDELINES FOR METHYLENE CHLORIDE

### I. SUBSTANCE IDENTIFICATION

A. Substance: Methylene chloride ($CH_2$ $Cl_2$).

B. Synonyms: MC, Dichloromethane (DCM); Methylene dichloride; Methylene bichloride; Methane dichloride; CAS: 75-09-2; NCI-C50102.

C. Physical data:

1. Molecular weight: 84.9.

2. Boiling point (760 mm Hg): 39.8 °C (104 °F).

3. Specific gravity (water = 1): 1.3.

4. Vapor density (air = 1 at boiling point): 2.9.

5. Vapor pressure at 20 °C (68 °F): 350 mm Hg.

6. Solubility in water, g/100 g water at 20 °C (68 °F) = 1.32.

7. Appearance and odor: colorless liquid with a chloroform-like odor.

D. Uses:

MC is used as a solvent, especially where high volatility is required. It is a good solvent for oils, fats, waxes, resins, bitumen, rubber and cellulose acetate and is a useful paint stripper and degreaser. It is used in paint removers, in propellant mixtures for aerosol containers, as a solvent for plastics, as a degreasing agent, as an extracting agent in the pharmaceutical industry and as a blowing agent in polyurethane foams. Its solvent property is sometimes increased by mixing with methanol, petroleum naphtha or tetrachloroethylene.

E. Appearance and odor:

MC is a clear colorless liquid with a chloroform-like odor. It is slightly soluble in water and completely miscible with most organic solvents.

F. Permissible exposure:

Exposure may not exceed 25 parts MC per million parts of air (25 ppm) as an eight-hour time-weighted average (8-hour TWA PEL) or 125 parts of MC per million parts of air (125 ppm) averaged over a 15-minute period (STEL).

### II. HEALTH HAZARD DATA

A. MC can affect the body if it is inhaled or if the liquid comes in contact with the eyes or skin. It can also affect the body if it is swallowed.

B. Effects of overexposure:

1. Short-term Exposure:

MC is an anesthetic. Inhaling the vapor may cause mental confusion, light-headedness, nausea, vomiting, and headache. Continued exposure may cause increased light-headedness, staggering, unconsciousness, and even death. High vapor concentrations may also cause irritation of the eyes and respiratory tract. Exposure to MC may make the symptoms of angina (chest pains) worse. Skin exposure to liquid MC may cause irritation. If liquid MC remains on the skin, it may cause skin burns. Splashes of the liquid into the eyes may cause irritation.

2. Long-term (chronic) exposure:

The best evidence that MC causes cancer is from laboratory studies in which rats, mice and hamsters inhaled MC 6 hours per day, 5 days per week for 2 years. MC exposure produced lung and liver tumors in mice and mammary tumors in rats. No carcinogenic effects of MC were found in hamsters.

There are also some human epidemiological studies which show an association between occupational exposure to MC and increases in biliary (bile duct) cancer and a type of brain cancer. Other epidemiological studies have not observed a relationship between MC exposure and cancer. OSHA interprets these results to mean that there is suggestive (but not absolute) evidence that MC is a human carcinogen.

C. Reporting signs and symptoms:

You should inform your employer if you develop any signs or symptoms and suspect that they are caused by exposure to MC.

D. Warning Properties:

1. Odor Threshold:

Different authors have reported varying odor thresholds for MC. Kirk-Othmer and Sax both reported 25 to 50 ppm; Summer and May both reported 150 ppm; Spector reports 320 ppm. Patty, however, states that since one can become adapted to the odor, MC should not be considered to have adequate warning properties.

2. Eye Irritation Level:

Kirk-Othmer reports that "MC vapor is seriously damaging to the eyes." Sax agrees with Kirk-Othmer's statement. The ACGIH Documentation of TLVs states that irritation of the eyes has been observed in workers exposed to concentrations up to 5000 ppm.

3. Evaluation of Warning Properties:

Since a wide range of MC odor thresholds are reported (25-320 ppm), and human adaptation to the odor occurs, MC is considered to be a material with poor warning properties.

## III. EMERGENCY FIRST AID PROCEDURES

In the event of emergency, institute first aid procedures and send for first aid or medical assistance.

A. Eye and Skin Exposures:

If there is a potential for liquid MC to come in contact with eye or skin, face shields and skin protective equipment must be provided and used. If liquid MC comes in contact with the eye, get medical attention. Contact lenses should not be worn when working with this chemical.

B. Breathing:

If a person breathes in large amounts of MC, move the exposed person to fresh air at once. If breathing has stopped, perform cardiopulmorary resuscitation. Keep the affected person warm and at rest. Get medical attention as soon as possible.

C. Rescue:

Move the affected person from the hazardous exposure immediately. If the exposed person has been overcome, notify someone else and put into effect the established emergency rescue procedures. Understand the facility's emergency rescue procedures and know the locations of rescue equipment before the need arises. Do not become a casualty yourself.

## IV. RESPIRATORS, PROTECTIVE CLOTHING, AND EYE PROTECTION

A. Respirators:

Good industrial hygiene practices recommend that engineering controls be used to reduce environmental concentrations to the permissible exposure level. However, there are some exceptions where respirators may be used to control exposure. Respirators may be used when engineering and work practice controls are not feasible, when such controls are in the process of being installed, or when these controls fail and need to be supplemented. Respirators may also be used for operations which require entry into tanks or closed vessels, and in emergency situations.

If the use of respirators is necessary, the only respirators permitted are those that have been approved by the Mine Safety and Health Administration (MSHA) or the National Institute for Occupational Safety and Health (NIOSH). Supplied-air respirators are *required* because air-purifying respirators do not provide adequate respiratory protection against MC.

In addition to respirator selection, a complete written respiratory protection program should be instituted which includes regular training, maintenance, inspection, cleaning, and evaluation. If you can smell MC while wearing a respirator, proceed immediately to fresh air. If you experience difficulty in breathing while wearing a respirator, tell your employer.

B. Protective Clothing:

Employees must be provided with and required to use impervious clothing, gloves, face shields (eight-inch minimum), and other appropriate protective clothing necessary to prevent repeated or prolonged skin contact with liquid MC or contact with vessels containing liquid MC. Any clothing which becomes wet with liquid MC should be removed immediately and not reworn until the employer has ensured that the protective clothing is fit for reuse. Contaminated protective clothing should be placed in a regulated area designated by the employer for removal of MC before the clothing is laundered or disposed of. Clothing and equipment should remain in the regulated area until all of the MC contamination has evaporated; clothing and equipment should then be laundered or disposed of as appropriate.

C. Eye Protection:

Employees should be provided with and required to use splash-proof safety goggles where liquid MC may contact the eyes.

## V. HOUSEKEEPING AND HYGIENE FACILITIES

For purposes of complying with 29 CFR 1910.141, the following items should be emphasized:

A. The workplace should be kept clean, orderly, and in a sanitary condition. The employer should institute a leak and spill detection program for operations involving liquid MC in order to detect sources of fugitive MC emissions.

B. Emergency drench showers and eyewash facilities are recommended. These should be maintained in a sanitary condition. Suitable cleansing agents should also be provided to assure the effective removal of MC from the skin.

C. Because of the hazardous nature of MC, contaminated protective clothing should be placed in a regulated area designated by the employer for removal of MC before the clothing is laundered or disposed of.

## VI. PRECAUTIONS FOR SAFE USE, HANDLING, AND STORAGE

A. Fire and Explosion Hazards:

MC has no flash point in a conventional closed tester, but it forms flammable vapor-air mixtures at approximately 100 °C (212 °F), or higher. It has a lower explosion limit of 12%, and an upper explosion limit of 19% in air. It has an autoignition temperature of 556.1 °C (1033 °F), and a boiling point of 39.8 °C (104 °F). It is heavier than water with a specific gravity of 1.3. It is slightly soluble in water.

B. Reactivity Hazards:

Conditions contributing to the instability of MC are heat and moisture. Contact with strong oxidizers, caustics, and chemically active metals such as aluminum or magnesium powder, sodium and potassium may cause fires and explosions. Special precautions: Liquid MC will attack some forms of plastics, rubber, and coatings.

C. Toxicity:

Liquid MC is painful and irritating if splashed in the eyes or if confined on the skin by gloves, clothing, or shoes. Vapors in high concentrations may cause narcosis and death. Prolonged exposure to vapors may cause cancer or exacerbate cardiac disease.

D. Storage:

Protect against physical damage. Because of its corrosive properties, and its high vapor pressure, MC should be stored in plain, galvanized or lead lined, mild steel containers in a cool, dry, well ventilated area away from direct sunlight, heat source and acute fire hazards.

E. Piping Material:

All piping and valves at the loading or unloading station should be of material that is resistant to MC and should be carefully inspected prior to connection to the transport vehicle and periodically during the operation.

F. Usual Shipping Containers:

Glass bottles, 5- and 55-gallon steel drums, tank cars, and tank trucks.

**Note:** This section addresses MC exposure in marine terminal and longshore employment only where leaking or broken packages allow MC exposure that is not addressed through compliance with 29 CFR parts 1917 and 1918, respectively.

G. Electrical Equipment:

Electrical installations in Class I hazardous locations as defined in Article 500 of the National Electrical Code, should be installed according to Article 501 of the code; and electrical equipment should be suitable for use in atmospheres containing MC vapors. See Flammable and Combustible Liquids Code (NFPA No. 325M), Chemical Safety Data Sheet SD-86 (Manufacturing Chemists' Association, Inc.).

H. Fire Fighting:

When involved in fire, MC emits highly toxic and irritating fumes such as phosgene, hydrogen chloride and carbon monoxide. Wear breathing apparatus and use water spray to keep fire-exposed containers cool. Water spray may be used to flush spills away from exposures. Extinguishing media are dry chemical, carbon dioxide, foam. For purposes of compliance with 29 CFR 1910.307, locations classified as hazardous due to the presence of MC shall be Class I.

I. Spills and Leaks:

Persons not wearing protective equipment and clothing should be restricted from areas of spills or leaks until cleanup has been completed. If MC has spilled or leaked, the following steps should be taken:

1. Remove all ignition sources.

2. Ventilate area of spill or leak.

3. Collect for reclamation or absorb in vermiculite, dry sand, earth, or a similar material.

J. Methods of Waste Disposal:

Small spills should be absorbed onto sand and taken to a safe area for atmospheric evaporation. Incineration is the preferred method for disposal of large quantities by mixing with a combustible solvent and spraying into an incinerator equipped with acid scrubbers to remove hydrogen chloride gases formed. Complete combustion will convert carbon monoxide to carbon dioxide. Care should be taken for the presence of phosgene.

K. You should not keep food, beverage, or smoking materials, or eat or smoke in regulated areas where MC concentrations are above the permissible exposure limits.

L. Portable heating units should not be used in confined areas where MC is used.

M. Ask your supervisor where MC is used in your work area and for any additional plant safety and health rules.

## VII. MEDICAL REQUIREMENTS

Your employer is required to offer you the opportunity to participate in a medical surveillance program if you are exposed to MC at concentrations at or above the action level (12.5 ppm 8-hour TWA) for more than 30 days a year or at concentrations exceeding the PELs (25 ppm 8-hour TWA or 125 ppm 15-minute STEL) for more than 10 days a year. If you are exposed to MC at concentrations over either of the PELs, your employer will also be required to have a physician or other licensed health care professional ensure that you are able to wear the respirator that you are assigned. Your employer must provide all medical examinations relating to your MC exposure at a reasonable time and place and at no cost to you.

## VIII. MONITORING AND MEASUREMENT PROCEDURES

A. Exposure above the Permissible Exposure Limit:

1. Eight-hour exposure evaluation: Measurements taken for the purpose of determining employee exposure under this section are best taken with consecutive samples covering the full shift. Air samples must be taken in the employee's breathing zone.

2. Monitoring techniques: The sampling and analysis under this section may be performed by collection of the MC vapor on two charcoal adsorption tubes in series or other composition adsorption tubes, with subsequent chemical analysis. Sampling and analysis may also be performed by instruments such as real-time continuous monitoring systems, portable direct reading instruments, or passive dosimeters as long as measurements taken using these methods accurately evaluate the concentration of MC in employees" breathing zones.

OSHA method 80 is an example of a validated method of sampling and analysis of MC. Copies of this method are available from OSHA or can be downloaded from the Internet at http://www.osha.gov. The employer has the obligation of selecting a monitoring method which meets the accuracy and precision requirements of the standard under his or her unique field conditions. The standard requires that the method of monitoring must be accurate, to a 95 percent confidence level, to plus or minus 25 percent for concentrations of MC at or above 25 ppm, and to plus or minus 35 percent for concentrations at or below 25 ppm. In addition to OSHA method 80, there are numerous other methods available for monitoring for MC in the workplace.

B. Since many of the duties relating to employee exposure are dependent on the results of measurement procedures, employers must assure that the evaluation of employee exposure is performed by a technically qualified person.

## IX. OBSERVATION OF MONITORING

Your employer is required to perform measurements that are representative of your exposure to MC and you or your designated representative are entitled to observe the monitoring procedure. You are entitled to observe the steps taken in the measurement procedure, and to record the results obtained. When the monitoring procedure is taking place in an area where respirators or personal protective clothing and equipment are required to be worn, you or your representative must also be provided with, and must wear, protective clothing and equipment.

## X. ACCESS TO INFORMATION

A. Your employer is required to inform you of the information contained in this Appendix. In addition, your employer must instruct you in the proper work practices for using MC, emergency procedures, and the correct use of protective equipment.

B. Your employer is required to determine whether you are being exposed to MC. You or your representative has the right to observe employee measurements and to record the results obtained. Your employer is required to inform you of your exposure. If your employer determines that you are being over exposed, he or she is required to inform you of the actions which are being taken to reduce your exposure to within permissible exposure limits.

C. Your employer is required to keep records of your exposures and medical examinations. These records must be kept by the employer for at least thirty (30) years.

D. Your employer is required to release your exposure and medical records to you or your representative upon your request.

E. Your employee is required to provide labels and material safety data sheets (MSDS) for all materials, mixtures or solutions composed of greater than 0.1 percent MC. An example of a label that would satisfy these requirements would be:

**DANGER**
**CONTAINS METHYLENE CHLORIDE**
**POTENTIAL CANCER HAZARD**

May worsen heart disease because methylene chloride is converted to carbon monoxide in the body.

May cause dizziness, headache, irritation of the throat and lungs, loss of consciousness and death at high concentrations (for example, if used in a poorly ventilated room).

*Avoid Skin Contact.* Contact with liquid causes skin and eye irritation.

## XI. COMMON OPERATIONS AND CONTROLS

The following list includes some common operations in which exposure to MC may occur and control methods which may be effective in each case:

| Operations | Controls |
|---|---|
| Use as solvent in paint and varnish removers; manufacture of aerosols; cold cleaning and ultrasonic cleaning; and as a solvent in furniture stripping | General dilution ventilation; local exhaust ventilation; personal protective equipment; substitution. |
| Use as solvent in vapor degreasing | Process enclosure; local exhaust ventilation; chilling coils; substitution. |
| Use as a secondary refrigerant in air conditioning and scientific testing | General dilution ventilation; local exhaust ventilation; personal protective equipment. |

# APPENDIX B TO SECTION 1910.105[sic1910.1052]—MEDICAL SURVEILLANCE FOR METHYLENE CHLORIDE

## I. PRIMARY ROUTE OF ENTRY

Inhalation.

## II. TOXICOLOGY

Methylene Chloride (MC) is primarily an inhalation hazard. The principal acute hazardous effects are the depressant action on the central nervous system, possible cardiac toxicity and possible liver toxicity. The range of CNS effects are from decreased eye/hand coordination and decreased performance in vigilance tasks to narcosis and even death of individuals exposed at very high doses. Cardiac toxicity is due to the metabolism of MC to carbon monoxide, and the effects of carbon monoxide on heart tissue. Carbon monoxide displaces oxygen in the blood, decreases the oxygen available to heart tissue, increasing the risk of damage to the heart, which may result in heart attacks in susceptible individuals. Susceptible individuals include persons with heart disease and those with risk factors for heart disease.

Elevated liver enzymes and irritation to the respiratory passages and eyes have also been reported for both humans and experimental animals exposed to MC vapors.

MC is metabolized to carbon monoxide and carbon dioxide via two separate pathways. Through the first pathway, MC is metabolized to carbon monoxide as an end-product via the P-450 mixed function oxidase pathway located in the microsomal fraction of the cell. This biotransformation of MC to carbon monoxide occurs through the process of microsomal oxidative dechlorination which takes place primarily in the liver. The amount of conversion to carbon monoxide is significant as measured by the concentration of carboxyhemoglobin, up to 12% measured in the blood following occupational exposure of up to 610 ppm. Through the second pathway, MC is metabolized to carbon dioxide as an end product (with formaldehyde and formic acid as metabolic intermediates) via the glutathione dependent enzyme found in the cytosolic fraction of the liver cell. Metabolites along this pathway are believed to be associated with the carcinogenic activity of MC.

MC has been tested for carcinogenicity in several laboratory rodents. These rodent studies indicate that there is clear evidence that MC is carcinogenic to male and female mice and female rats. Based on epidemiologic studies, OSHA has concluded that there is suggestive evidence of increased cancer risk in MC-related worker populations. The epidemiological evidence is consistent with the finding of excess cancer in the experimental animal studies. NIOSH regards MC as a potential occupational carcinogen and the International Agency for Research Cancer (IARC) classifies MC as an animal carcinogen. OSHA considers MC as a suspected human carcinogen.

## III. MEDICAL SIGNS AND SYMPTOMS OF ACUTE EXPOSURE

Skin exposure to liquid MC may cause irritation or skin burns. Liquid MC can also be irritating to the eyes. MC is also absorbed through the skin and may contribute to the MC exposure by inhalation.

At high concentrations in air, MC may cause nausea, vomiting, lightheadedness, numbness of the extremities, changes in blood enzyme levels, and breathing problems, leading to bronchitis and pulmonary edema, unconsciousness and even death.

At lower concentrations in air, MC may cause irritation to the skin, eye, and respiratory tract and occasionally headache and nausea. Perhaps the greatest problem from exposure to low concentrations of MC is the CNS effects on coordination and alertness that may cause unsafe operations of machinery and equipment, leading to self-injury or accidents.

Low levels and short duration exposures do not seem to produce permanent disability, but chronic exposures to MC have been demonstrated to produce liver toxicity in animals, and therefore, the evidence is suggestive for liver toxicity in humans after chronic exposure.

Chronic exposure to MC may also cause cancer.

## IV. SURVEILLANCE AND PREVENTIVE CONSIDERATIONS

As discussed above, MC is classified as a suspect or potential human carcinogen. It is a central nervous system (CNS) depressant and a skin, eye and respiratory tract irritant. At extremely high concentrations, MC has caused liver damage in animals.

MC principally affects the CNS, where it acts as a narcotic. The observation of the symptoms characteristic of CNS depression, along with a physical examination, provides the best detection of early neurological disorders. Since exposure to MC also increases the carboxyhemoglobin level in the blood, ambient carbon monoxide levels would have an additive effect on that carboxyhemoglobin level. Based on such information, a periodic post-shift carboxyhemoglobin test as an index of the presence of carbon monoxide in the blood is recommended, but not required, for medical surveillance.

Based on the animal evidence and three epidemiologic studies previously mentioned, OSHA concludes that MC is a suspect human carcinogen. The medical surveillance program is designed to observe exposed workers on a regular basis. While the medical surveillance program cannot detect MC-induced cancer at a preneoplastic stage, OSHA anticipates that, as in the past, early detection and treatments of cancers leading to enhanced survival rates will continue to evolve.

### A. Medical and Occupational History:

The medical and occupational work history plays an important role in the initial evaluation of workers exposed to MC. It is therefore extremely important for the examining physician or other licensed health care professional to evaluate the MC-exposed worker carefully and completely and to focus the examination on MC's potentially associated health hazards. The medical evaluation must include an annual detailed work and medical history with special emphasis on cardiac history and neurological symptoms.

An important goal of the medical history is to elicit information from the worker regarding potential signs or symptoms associated with increased levels of carboxyhemoglobin due to the presence of carbon monoxide in the blood. Physicians or other licensed health care professionals should ensure that the smoking history of all MC exposed employees is known. Exposure to MC may cause a significant increase in carboxyhemoglobin level in all exposed persons. However, smokers as well as workers with anemia or heart disease and those concurrently exposed to carbon monoxide are at especially high risk of toxic effects because of an already reduced oxygen carrying capacity of the blood.

A comprehensive or interim medical and work history should also include occurrence of headache, dizziness, fatigue, chest pain, shortness of breath, pain in the limbs, and irritation of the skin and eyes.

In addition, it is important for the physician or other licensed health care professional to become familiar with the operating conditions in which exposure to MC is likely to occur. The physician or other licensed health care professional also must become familiar with the signs and symptoms that may indicate that a worker is receiving otherwise unrecognized and exceptionally high exposure levels of MC.

An example of a medical and work history that would satisfy the requirement for a comprehensive or interim work history is represented by the following:

The following is a list of recommended questions and issues for the self-administered questionnaire for methylene chloride exposure.

## QUESTIONNAIRE FOR METHYLENE CHLORIDE EXPOSURE

### I. Demographic Information

1. Name
2. Social Security Number
3. Date
4. Date of Birth
5. Age
6. Present occupation
7. Sex
8. Race

### II. Occupational History

1. Have you ever worked with methylene chloride, dichloromethane, methylene dichloride, or $CH_2$ $Cl_2$ (all are different names for the same chemical)? Please list which on the occupational history form if you have not already.

2. If you have worked in any of the following industries and have not listed them on the occupational history form, please do so.

Furniture stripping

Polyurethane foam manufacturing

Chemical manufacturing or formulation

Pharmaceutical manufacturing

Any industry in which you used solvents to clean and degrease equipment or parts

Construction, especially painting and refinishing

Aerosol manufacturing

Any industry in which you used aerosol adhesives

3. If you have not listed hobbies or household projects on the occupational history form, especially furniture refinishing, spray painting, or paint stripping, please do so.

### III. Medical History

#### A. General

1. Do you consider yourself to be in good health? If no, state reason(s).

2. Do you or have you ever had:

a. Persistent thirst

b. Frequent urination (three times or more at night)

c. Dermatitis or irritated skin

d. Non-healing wounds

3. What prescription or non-prescription medications do you take, and for what reasons?

4. Are you allergic to any medications, and what type of reaction do you have?

#### B. Respiratory

1. Do you have or have you ever had any chest illnesses or diseases? Explain.

2. Do you have or have you ever had any of the following:

a. Asthma

b. Wheezing

c. Shortness of breath

3. Have you ever had an abnormal chest X-ray? If so, when, where, and what were the findings?

4. Have you ever had difficulty using a respirator or breathing apparatus? Explain.

5. Do any chest or lung diseases run in your family? Explain.

6. Have you ever smoked cigarettes, cigars, or a pipe? Age started:

7. Do you now smoke?

8. If you have stopped smoking completely, how old were you when you stopped?

9. On the average of the entire time you smoked, how many packs of cigarettes, cigars, or bowls of tobacco did you smoke per day?

#### C. Cardiovascular

1. Have you ever been diagnosed with any of the following: Which of the following apply to you now or did apply to you at some time in the past, even if the problem is controlled by medication? Please explain any yes answers (i.e., when problem was diagnosed, length of time on medication).

a. High cholesterol or triglyceride level

b. Hypertension (high blood pressure)

c. Diabetes

d. Family history of heart attack, stroke, or blocked arteries

2. Have you ever had chest pain? If so, answer the next five questions.

a. What was the quality of the pain (i.e., crushing, stabbing, squeezing)?

b. Did the pain go anywhere (i.e., into jaw, left arm)?

c. What brought the pain out?

d. How long did it last?

e. What made the pain go away?

3. Have you ever had heart disease, a heart attack, stroke, aneurysm, or blocked arteries anywhere in you body? Explain (when, treatment).

4. Have you ever had bypass surgery for blocked arteries in your heart or anywhere else? Explain.

5. Have you ever had any other procedures done to open up a blocked artery (balloon angioplasty, carotid endarterectomy, clot-dissolving drug)?

6. Do you have or have you ever had (explain each):

a. Heart murmur

b. Irregular heartbeat

c. Shortness of breath while lying flat

d. Congestive heart failure

e. Ankle swelling

f. Recurrent pain anywhere below the waist while walking

7. Have you ever had an electrocardiogram (EKG)? When?

8. Have you ever had an abnormal EKG? If so, when, where, and what were the findings?

9. Do any heart diseases, high blood pressure, diabetes, high cholesterol, or high triglycerides run in your family? Explain.

#### D. Hepatobiliary and Pancreas

1. Do you now or have you ever drunk alcoholic beverages? Age started:______ Age stopped:______ .

2. Average numbers per week:

a. Beers:______ , ounces in usual container:

b. Glasses of wine:______ , ounces per glass:

c. Drinks:______ , ounces in usual container:

3. Do you have or have you ever had (explain each):

a. Hepatitis (infectious, autoimmune, drug-induced, or chemical)

b. Jaundice

c. Elevated liver enzymes or elevated bilirubin

d. Liver disease or cancer

#### E. Central Nervous System

1. Do you or have you ever had (explain each):

a. Headache

b. Dizziness

c. Fainting

d. Loss of consciousness

e. Garbled speech

f. Lack of balance

g. Mental/psychiatric illness

h. Forgetfulness

#### F. Hematologic

1. Do you have, or have you ever had (explain each):

a. Anemia

b. Sickle cell disease or trait

c. Glucose-6-phosphate dehydrogenase deficiency

d. Bleeding tendency disorder

2. If not already mentioned previously, have you ever had a reaction to sulfa drugs or to drugs used to prevent or treat malaria? What was the drug? Describe the reaction.

### B. Physical Examination

The complete physical examination, when coupled with the medical and occupational history, assists the physician or other licensed health care professional in detecting pre-existing conditions that might place the employee at increased risk, and establishes a baseline for future health monitoring. These examinations should include:

1. Clinical impressions of the nervous system, cardiovascular function and pulmonary function, with additional tests conducted where indicated or determined by the examining physician or other licensed health care professional to be necessary.

2. An evaluation of the advisability of the worker using a respirator, because the use of certain respirators places an additional burden on the cardiopulmonary system. It is necessary for the attending physician or other licensed health care professional to evaluate the cardiopulmonary function of these workers, in order to inform the employer in a written medical opinion of the worker's ability or fitness to work in an area requiring the use of certain types of respiratory protective equip-

ment. The presence of facial hair or scars that might interfere with the worker's ability to wear certain types of respirators should also be noted during the examination and in the written medical opinion.

Because of the importance of lung function to workers required to wear certain types of respirators to protect themselves from MC exposure, these workers must receive an assessment of pulmonary function before they begin to wear a negative pressure respirator and at least annually thereafter. The recommended pulmonary function tests include measurement of the employee's forced vital capacity (FVC), forced expiratory volume at one second (FEV1), as well as calculation of the ratios of FEV1 to FVC, and the ratios of measured FVC and measured FEV1 to expected respective values corrected for variation due to age, sex, race, and height. Pulmonary function evaluation must be conducted by a physician or other licensed health care professional experienced in pulmonary function tests.

The following is a summary of the elements of a physical exam which would fulfill the requirements under the MC standard:

## PHYSICAL EXAM

### *I. Skin and appendages*

1. Irritated or broken skin
2. Jaundice
3. Clubbing cyanosis, edema
4. Capillary refill time
5. Pallor

### *II. Head*

1. Facial deformities
2. Scars
3. Hair growth

### *III. Eyes*

1. Scleral icterus
2. Corneal arcus
3. Pupillary size and response
4. Fundoscopic exam

### *IV. Chest*

1. Standard exam

### *V. Heart*

1. Standard exam
2. Jugular vein distension
3. Peripheral pulses

### *VI. Abdomen*

1. Liver span

### *VII. Nervous System*

1. Complete standard neurologic exam

### *VIII. Laboratory*

1. Hemoglobin and hematocrit
2. Alanine aminotransferase (ALT, SGPT)
3. Post-shift carboxyhemoglobin

### *IX. Studies*

1. Pulmonary function testing
2. Electrocardiogram

An evaluation of the oxygen carrying capacity of the blood of employees (for example by measured red blood cell volume) is considered useful, especially for workers acutely exposed to MC.

It is also recommended, but not required, that end of shift carboxyhemoglobin levels be determined periodically, and any level above 3% for non-smokers and above 10% for smokers should prompt an investigation of the worker and his workplace. This test is recommended because MC is metabolized to CO, which combines strongly with hemoglobin, resulting in a reduced capacity of the blood to transport oxygen in the body. This is of particular concern for cigarette smokers because they already have a diminished hemoglobin capacity due to the presence of CO in cigarette smoke.

## C. Additional Examinations and Referrals

1. Examination by a Specialist

When a worker examination reveals unexplained symptoms or signs (i.e. in the physical examination or in the laboratory tests), follow-up medical examinations are necessary to assure that MC exposure is not adversely affecting the worker's health. When the examining physician or other licensed health care professional finds it necessary, additional tests should be included to determine the nature of the medical problem and the underlying cause. Where relevant, the worker should be sent to a specialist for further testing and treatment as deemed necessary.

The final rule requires additional investigations to be covered and it also permits physicians or other licensed health care professionals to add appropriate or necessary tests to improve the diagnosis of disease should such tests become available in the future.

2. Emergencies

The examination of workers exposed to MC in an emergency should be directed at the organ systems most likely to be affected. If the worker has received a severe acute exposure, hospitalization may be required to assure proper medical intervention. It is not possible to precisely define "severe," but the physician or other licensed health care professional's judgement should not merely rest on hospitalization. If the worker has suffered significant conjunctival, oral, or nasal irritation, respiratory distress, or discomfort, the physician or other licensed health care professional should instigate appropriate follow-up procedures. These include attention to the eyes, lungs and the neurological system. The frequency of follow-up examinations should be determined by the attending physician or other licensed health care professional. This testing permits the early identification essential to proper medical management of such workers.

## D. Employer Obligations

The employer is required to provide the responsible physician or other licensed health care professional and any specialists involved in a diagnosis with the following information: a copy of the MC standard including relevant appendices, a description of the affected employee's duties as they relate to his or her exposure to MC; an estimate of the employee's exposure including duration (e.g., 15hr/wk, three 8-hour shifts/wk, full time); a description of any personal protective equipment used by the employee, including respirators; and the results of any previous medical determinations for the affected employee related to MC exposure to the extent that this information is within the employer's control.

## E. Physicians' or Other Licensed Health Care Professionals' Obligations

The standard requires the employer to ensure that the physician or other licensed health care professional provides a written statement to the employee and the employer. This statement should contain the physician's or licensed health care professional's opinion as to whether the employee has any medical condition placing him or her at increased risk of impaired health from exposure to MC or use of respirators, as appropriate. The physician or other licensed health care professional should also state his or her opinion regarding any restrictions that should be placed on the employee's exposure to MC or upon the use of protective clothing or equipment such as respirators. If the employee wears a respirator as a result of his or her exposure to MC, the physician or other licensed health care professional's opinion should also contain a statement regarding the suitability of the employee to wear the type of respirator assigned. Furthermore, the employee should be informed by the physician or other licensed health care professional about the cancer risk of MC and about risk factors for heart disease, and the potential for exacerbation of underlying heart disease by exposure to MC through its metabolism to carbon monoxide. Finally, the physician or other licensed health care professional should inform the employer that the employee has been told the results of the medical examination and of any medical conditions which require further explanation or treatment. This written opinion must not contain any information on specific findings or diagnosis unrelated to employee's occupational exposures.

The purpose in requiring the examining physician or other licensed health care professional to supply the employer with a written opinion is to provide the employer with a medical basis to assist the employer in placing employees initially, in assuring that their health is not being impaired by exposure to MC, and to assess the employee's ability to use any required protective equipment.

**APPENDIX C TO SECTION 1910.1052—QUESTIONS AND ANSWERS—METHYLENE CHLORIDE CONTROL IN FURNITURE STRIPPING**

CDC
CENTERS FOR DISEASE CONTROL AND PREVENTION

NIOSH

—Questions and Answers—
Methylene Chloride Control in Furniture Stripping

U.S. DEPARTMENT OF HEALTH AND HUMAN SERVICES
Public Health Service
Centers for Disease Control and Prevention
National Institute for Occupational Safety and Health
September 1993

Q's & A's

DISCLAIMER

Mention of any company or product does not constitute endorsement by the Centers for Disease Control and Prevention, National Institute for Occupational Safety and Health.

This document is in the public domain and may be freely copied or reprinted.

Copies of this and other NIOSH documents are available from

Publications Dissemination, DSDTT
National Institute for Occupational Safety and Health
4676 Columbia Parkway
Cincinnati, OH 45226
FAX (513) 533-8573

For information about other occupational safety and health problems, call 1-800-35-NIOSH

DHHS (NIOSH) Publication NO. 93-133

# Q's & A's

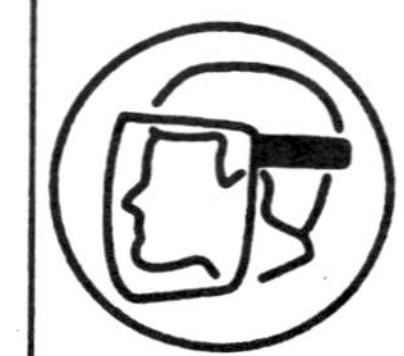

## Introduction

This Pamphlet answers commonly asked questions about the hazards from exposure to methylene chloride. It also describes approaches to controlling methylene chloride exposure during the most common furniture stripping processes. Although these approaches were developed and field tested by NIOSH, each setting requires custom installation because of the different air flow interferences at each site.

## What is the Stripping Solution Base?

The most common active ingredient in paint removers is a chemical called methylene chloride. Methylene chloride is present in the paint remover to penetrate, blister, and finally lift the old finish. Other chemicals in paint removers work to accelerate the stripping process, to retard evaporation, and to act as thickening agents. These other ingredients may include: methanol, toluene, acetone, or paraffin.[1]

## Is Methylene Chloride Bad for Me?

Exposure to methylene chloride may cause short-term health effects or long-term health effects.

### Short-Term (acute) Health Effects

Exposure to high levels of paint removers over short periods of time can cause irritation to the skin, eyes, mucous membranes, and respiratory tract. Other symptoms of high exposure are dizziness headache, and lack of coordination. The occurrence of any of these symptoms indicates that you are being exposed to high levels of the methylene chloride. At the onset of any of these symptoms, you should leave the work area, get some fresh air, and determine why the levels were high.

A portion of inhaled methylene chloride is converted by the body to carbon monoxide, which can lower the blood's ability to carry oxygen. When the solvent is used properly, however, the levels of carbon monoxide should not be hazardous. Individuals with cardiovascular or pulmonary health problems should check with their physician before using the paint stripper. Individuals experiencing severe symptoms such as shortness of breath or chest pains should obtain proper medical care immediately.[2]

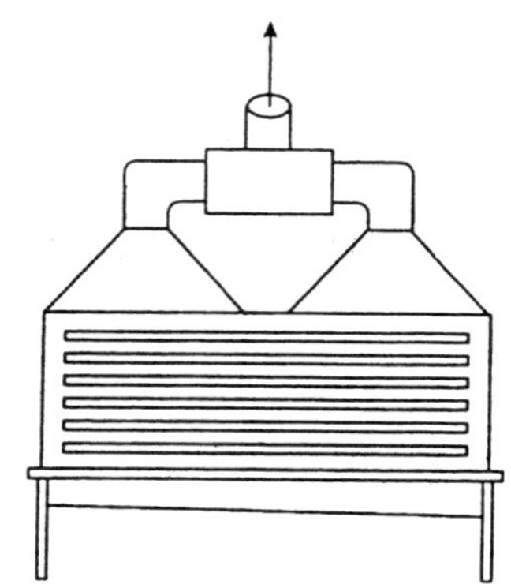

Figure 1 — Slot Hood

### Long-term (Chronic) Health Effects

Methylene chloride has been shown to cause cancer in certain laboratory animal tests. The available human studies do not provide the necessary information to determine whether methylene chloride causes cancer in humans. However, as a result of the animal studies, methylene chloride is

## Q's & A's

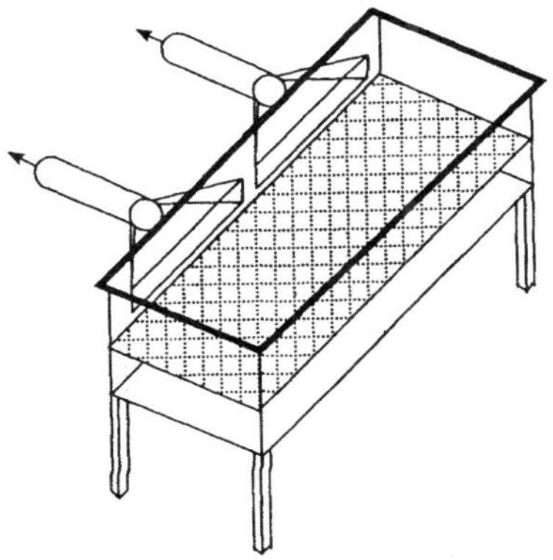



Figure 2 — Downdraft Hood

considered a potential occupational carcinogen. There is also considerable indirect evidence to suggest that workers exposed to methylene chloride may be at increased risk of developing ischemic heart disease. Therefore, it is prudent to minimize exposures to solvent vapors.[3]

### What Do Federal Agencies Say About Methylene Chloride?

In 1991, the Occupational Safety and Health Administration published a Notice of Proposed Rulemaking for methylene chloride. The proposed standard would establish an eight-hour time-weighted average exposure limit of 25 parts per million (ppm), as well as a short-term exposure limit of 125 ppm determined from a 15 minute sampling period. That is a sharp reduction from the current limit of 500 ppm. The proposed standard would also set a 12.5 ppm action level (a level that would trigger periodic exposure monitoring and medical surveillance provisions.[4]

The National Institute for Occupational Safety and Health recommends that methylene chloride be regarded as a "potential occupational carcinogen." NIOSH further recommends that occupational exposure to methylene chloride be controlled to the lowest feasible limit. This recommendation was based on the observation of cancers and tumors in both rats and mice exposed to methylene chloride in air.[5]

### How Can I Be Exposed to Methylene Chloride while Stripping Furniture?

Methylene chloride can be inhaled when vapors are in the air. Inhalation of the methylene chloride vapors is generally the most important source of exposure. Methylene chloride evaporates quicker than most chemicals. The odor threshold of methylene chloride is 300 ppm.[6] Therefore, once you smell methylene chloride, you are being over-exposed. Pouring, moving, or stirring the chemical will increase the rate of evaporation.

Methylene chloride can be absorbed through the skin either by directly touching the chemical or through your gloves. Methylene chloride can be swallowed if it gets on your hands, clothes, or beard, or if food or drinks become contaminated.

### How Can Breathing Exposures be Reduced?

**Install a Local Exhaust Ventilation System**

Local exhaust ventilation can be used to control exposures. Local exhaust ventilation systems capture contaminated air from the source before it spreads into the workers' breathing zone.[7] If engineering controls are not effective, only a self-contained breathing apparatus equipped with a full facepiece and operated in a positive-pressure mode or a supplied-air respirator affords the necessary level of protection. Air-purifying respirators such as organic vapor cartridges can only be used for escape situations.[8]

A local exhaust system consists of the following: a hood, a fan, ductwork, and a replacement air system.[9,10,11] Two processes are commonly used in furniture stripping: flow-over and dip tanks. For flow-over systems there are two common local exhaust controls for methylene chloride — a slot hood and a downdraft hood. A slot hood of different design is most often used for dip tanks. (See Figures 1, 2, and 3)

The hood is made of sheet metal and connected to the tank. All designs require a centrifugal fan to exhaust the fumes, ductwork connecting the hood and the fan, and a replacement air system to bring conditioned air into the building to replace the air exhausted.

In constructing or designing a slot or downdraft hood, use the following data:

## Q's & A's

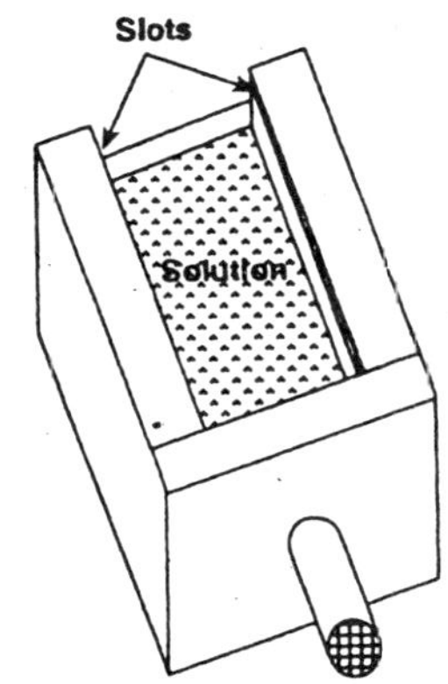



Figure 3 — Slot Hood for Dip Tank

*Slot hood (Figure 1)*

- At least 2200 cfm per 8' X 4' tank
- 1 - 2 inch slots
- Slot velocity - 1000 fpm
- 3 - 5 slots
- Plenum at least 1 foot deep

*Downdraft hood (Figure 2)*

- At least 1600 cfm per 8' X 4' tank
- Plenum at least 9" deep

*Slot hood for Dip Tank (Figure 3)*

- At least 2900 cfm per 8' X 4' tank
- 3/4" slot that runs the length of the front and back of the tank
- Slot velocity — 3200 fpm
- Plenum on the sides of the tank should be 6" deep by 36" long.
- 12" duct leads from the center of the front plenum to the fan

**Safe work practices**

Workers can lower exposures by decreasing their access to the methylene chloride.[12]

1) Turn on dip tank control system several minutes before entering the stripping area.

2) Avoid unnecessary transferring or moving of stripping solution.

3) Keep face out of the air stream between the solution-covered furniture and the exhaust system.

4) Keep face out of vapor zone above the stripping solution and dip tank.

5) Retrieve dropped items with a long handled tool.

6) Keep the solution-recycling system off when not in use. Cover reservoir for recycling system.

7) Cover dip tank when not in use.

8) Provide adequate ventilation for rinse area.

## Q's & A's

### How Can Skin Exposures be Reduced?

Skin exposures can be reduced by wearing gloves whenever you are in contact with the stripping solution.[13]

1) Two gloves should be worn. The inner glove should be made from polyethylene/ethylene vinyl alcohol (e.g. Silver Shield®, or 4H®). This material, however, does not provide good physical resistance against tears, so an outer glove made from nitrile or neoprene should be worn.

2) Shoulder-length gloves will be more protective.

3) Change gloves before the break-through time occurs. Rotate several pairs of gloves throughout the day. Let the gloves dry in a warm well ventilated area at least over night before reuse.

4) Keep gloves clean by rinsing often. Keep gloves in good condition. Inspect the gloves before use for pin-holes, cracks, thin spots, and stiffer than normal or sticky surfaces.

5) Wear a face shield or goggles to protect face and eyes.

### What Other Problems Occur?

#### Stripping Solution Temperature

Most manufacturers of stripping solution recommend controlling the solution to a temperature of 70°F. This temperature is required for the wax in the solution to form a vapor barrier on top of the solution to keep the solution from evaporating too quickly. If the temperature is too high, the wax will not form the vapor barrier. If it is too cold, the wax will solidify and separate from the solvent causing increased evaporation. Use a belt heater to heat the solution to the correct temperature. Call your solution manufacturer for the correct temperature for your solution.[14]

#### Make-Up Air

Air will enter a building in an amount to equal the amount of air exhausted whether or not provision is made for this replacement. If a local exhaust system is added a make-up or replacement air system must be added to replace the air removed. Without a replacement air system, air will enter the building through cracks causing uncontrollable eddy currents. If the building perimeter is tightly sealed, it will prevent the air from entering and severely decrease the amount exhausted from the ventilation system. This will cause the building to be under negative pressure and decrease the performance of the exhaust system.[15]

#### Dilution Ventilation

With general or dilution ventilation, uncontaminated air is moved through the workroom by means of fans or open windows, which dilutes the pollutants in the air. Dilution ventilation does not provide effective protection to other workers and does not confine the methylene chloride vapors to one area.[16]

#### Phosgene Poisoning from Use of Kerosene Heaters

Do not use kerosene heaters or other open flame heaters while stripping furniture. Use of kerosene heaters in connection with methylene chloride can create lethal or dangerous concentrations of phosgene. Methylene chloride vapor is mixed with the air used for the combustion of kerosene in kerosene stoves. The vapor thus passes through the flames, coming into close contact with carbon monoxide at high temperatures. Any chlorine formed by decomposition may, under these conditions, react with carbon monoxide and form phosgene.[17]

REFERENCES

[1] Halogenated Solvents Industry Alliance and Consumer Product Safety Commission [1990]. Stripping Paint from Wood (Pamphlet for consumers on how to strip furniture and precautions to take). Washington DC: Consumer Product Safety Commission.

[2] *Ibid.*

[3] NIOSH [1992]. NIOSH Testimony on Occupational Safety and Health Administration's proposed rule on occupational exposure to methylene chloride, September 21, 1992, OSHA Docket No. H-71. NIOSH policy statements. Cincinnati, OH: U.S. Department of Health and Human Services, Public Health Service, Centers for Disease Control, National Institute for Occupational Safety and Health.

[4] 56 Fed. Reg. 57036 [1991]. Occupational Safety and Health Administration: Proposed rule on occupational exposure to methylene chloride.

[5] NIOSH [1992].

[6] Kirk, R.E. and P.F. Othmer, Eds. [1978]. Encyclopedia of Chemical Technology, 3rd Ed., Vol. 5:690. New York: John Wiley & Sons, Inc.

[7] ACGIH [1988]. Industrial Ventilation: A Manual of Recommended Practice. 20th Ed. Cincinnati, OH: American Conference of Governmental Industrial Hygienists.

[8] NIOSH [1992].

[9] Fairfield, C.L. and A.A. Beasley [1991]. In-depth Survey Report at the Association for Retarded Citizens, Meadowlands, PA. The Control of Methylene Chloride During Furniture Stripping. Cincinnati, OH: U.S. Department of Health and Human Services, Public Health Service, Centers for Disease Control, National Institute for Occupational Safety and Health.

[10] Fairfield, C.L. [1991]. In-depth Survey Report at the J.M. Murray Center, Cortland, NY. The Control of Methylene Chloride During Furniture Stripping. Cincinnati, OH: U.S. Department of Health and Human Services, Public Health Service, Centers for Disease Control, National Institute for Occupational Safety and Health.

[11] Hall, R.M., K.F. Martinez, and P.A. Jensen [1992]. In-depth Survey Report at Tri-County Furniture Stripping and Refinishing, Cincinnati, OH. The Control of Methylene Chloride During Furtniture Stripping. Cincinnati, OH: U.S. Department of Health and Human Services, Public Health Service, Centers for Disease Control, National Institute for Occupational Safety and Health.

[12] Fairfield, C.L. and A.A. Beasley [1991]. In-depth Survey Report at the Association for Retarded Citizens, Meadowlands, PA. The Control of Methylene Chloride During Furniture Stripping. Cincinnati, OH: U.S. Department of Health and Human Services, Public Health Service, Centers for Disease Control, National Institute for Occupational Safety and Health.

[13] Roder, M. [1991]. Memorandum of March 11, 1991 from Michael Roder of the Division of Safety Research to Cheryl L. Fairfield of the Division of Physical Sciences and Engineering, National Institute for Occupational Safety and Health, Centers for Disease Control, Public Health Service, U.S. Department of Health and Human Services.

[14] Kwick Kleen Industrial Solvents, Inc. [1981]. Operations Manual, Kwick Kleen Industrial Solvents, Inc., Vincennes, IN.

[15] ACGIH [1988].

[16] *Ibid.*

[17] Gerritsen, W.B. and C.H. Buschmann [1960]. Phosgene Poisoning Caused by the Use of Chemical Paint Removers containing Methylene Chloride in Ill-Ventilated Rooms Heated by Kerosene Stoves. British Journal of Industrial Medicine *17*:187.

Q's & A's

**Where Should I go for More Information?**

The NIOSH 800-number is a toll-free technical information service that provides convenient public access to NIOSH and its information resources. Callers may request information about any aspect of occupational safety and health.

**1-800-35-NIOSH**
(1-800-356-4674)

U.S. GOVERNMENT PRINTING OFFICE: 1993: 550-147/80026

[62 FR 1601, Jan. 10, 1997, as amended at 62 FR 42667, Aug. 8, 1997; 62 FR 54383, Oct. 20, 1997; 62 FR 66277, Dec. 18, 1997; 63 FR 1295, Jan. 8, 1998; 63 FR 20099, Apr. 23, 1998; 63 FR 50729, Sept. 22, 1998]

# Occupational Safety and Health Act, Sec. 5. Duties "General Duty Clause"



[The following is commonly referred to as the "General Duty Clause"—CCH.]

Sec. 5. (a) Each employer—

(1) shall furnish to each of his employees employment and a place of employment which are free from recognized hazards that are causing or are likely to cause death or serious physical harm to his employees;

(2) shall comply with occupational safety and health standards promulgated under this Act.

[The following is not part of the "General Duty Clause" but is provided for convenience—CCH.]

(b) Each employee shall comply with occupational safety and health standards and all rules, regulations, and orders issued pursuant to this Act which are applicable to his own actions and conduct.

# 29 CFR Part 1903— Inspections, Citations, and Proposed Penalties

**AUTHORITY:** Sections 8 and 9 of the Occupational Safety and Health Act of 1970 (29 U.S.C. 657, 658); 5 U.S.C. 553; Secretary of Labor's Order No. 1-90 (55 FR 9033) or 6-96 (62 FR 111), as applicable.

**SOURCE:** 1903.7 also issued under 5 U.S.C. 553.

**SOURCE:** 36 FR 17850, Sept. 4, 1971, unless otherwise noted.

## §1903.1 Purpose and scope.

The Williams-Steiger Occupational Safety and Health Act of 1970 (84 Stat. 1590 et seq., 29 U.S.C. 651 et seq.) requires, in part, that every employer covered under the Act furnish to his employees employment and a place of employment which are free from recognized hazards that are causing or are likely to cause death or serious physical harm to his employees. The Act also requires that employers comply with occupational safety and health standards promulgated under the Act, and that employees comply with standards, rules, regulations and orders issued under the Act which are applicable to their own actions and conduct. The Act authorizes the Department of Labor to conduct inspections, and to issue citations and proposed penalties for alleged violations. The Act, under section 20(b), also authorizes the Secretary of Health, Education, and Welfare to conduct inspections and to question employers and employees in connection with research and other related activities. The Act contains provisions for adjudication of violations, periods prescribed for the abatement of violations, and proposed penalties by the Occupational Safety and Health Review Commission, if contested by an employer or by an employee or authorized representative of employees, and for judicial review. The purpose of this Part 1903 is to prescribe rules and to set forth general policies for enforcement of the inspection, citation, and proposed penalty provisions of the Act. In situations where this Part 1903 sets forth general enforcement policies rather than substantive or procedural rules, such policies may be modified in specific circumstances where the Secretary or his designee determines that an alternative course of action would better serve the objectives of the Act.

## §1903.2 Posting of notice; availability of Act, regulations and applicable standards.

**1903.2(a)(1)** Each employer shall post and keep posted a notice or notices, to be furnished by the Occupational Safety and Health Administration, U.S. Department of Labor, informing employees of the protections and obligations provided for in the Act, and that for assistance and information, including copies of the Act and of specific safety and health standards, employees should contact the employer or the nearest office of the Department of Labor. Such notice or notices shall be posted by the employer in each establishment in a conspicuous place or places where notices to employees are customarily posted. Each employer shall take steps to insure that such notices are not altered, defaced, or covered by other material.

**1903.2(a)(2)** Where a State has an approved poster informing employees of their protections and obligations as defined in §1952.10 of this chapter, such poster, when posted by employers covered by the State plan, shall constitute compliance with the posting requirements of section 8(c)(1) of the Act. Employers whose operations are not within the issues covered by the State plan must comply with paragraph (a)(1) of this section.

**1903.2(a)(3)** Reproductions or facsimiles of such Federal or State posters shall constitute compliance with the posting requirements of section 8(c)(1) of the Act where such reproductions or facsimiles are at least 8 1/2 inches by 14 inches, and the printing size is at least 10 pt. Whenever the size of the poster increases, the size of the print shall also increase accordingly. The caption or heading on the poster shall be in large type, generally not less than 36 pt.

**1903.2(b)** *Establishment* means a single physical location where business is conducted or where services or industrial operations are performed. (For example: A factory, mill, store, hotel, restaurant, movie theatre, farm, ranch, bank, sales office, warehouse, or central administrative office.) Where distinctly separate activities are performed at a single physical location (such as contract construction activities from the same physical location as a lumber yard), each activity shall be treated as a separate physical establishment, and a separate notice or notices shall be posted in each such establishment, to the extent that such notices have been furnished by the Occupational Safety and Health Administration, U.S. Department of Labor. Where employers are engaged in activities which are physically dispersed, such as agriculture, construction, transportation, communications, and electric, gas and sanitary services, the notice or notices required by this section shall be posted at the location to which employees report each day. Where employees do not usually work at, or report to, a single establishment, such as longshoremen, traveling salesmen, technicians, engineers, etc., such notice or notices shall be posted at the location from which the employees operate to carry out their activities. In all cases, such notice or notices shall be posted in accordance with the requirements of paragraph (a) of this section.

**1903.2(c)** Copies of the Act, all regulations published in this chapter and all applicable standards will be available at all Area Offices of the Occupational Safety and Health Administration, U.S. Department of Labor. If an employer has obtained copies of these materials, he shall make them available upon request to any employee or his authorized representative for review in the establishment where the employee is employed on the same day the request is made or at the earliest time mutually convenient to the employee or his authorized representative and the employer.

**1903.2(d)** Any employer failing to comply with the provisions of this section shall be subject to citation and penalty in accordance with the provisions of section 17 of the Act.

[36 FR 17850, Sept. 4, 1971, as amended at 39 FR 39036, Nov. 5, 1974]

## §1903.3 Authority for inspection.

**1903.3(a)** Compliance Safety and Health Officers of the Department of Labor are authorized to enter without delay and at reasonable times any factory, plant, establishment, construction site, or other area, workplace or environment where work is performed by an employee of an employer; to inspect and investigate during regular working hours and at other reasonable times, and within reasonable limits and in a reasonable manner, any such place of employment, and all pertinent conditions, structures, machines, apparatus, devices, equipment and materials therein; to question privately any employer, owner, operator, agent or employee; and to review records required by the Act and regulations published in this chapter, and other records which are directly related to the purpose of the inspection. Representatives of the Secretary of Health, Education, and Welfare are authorized to make inspections and to question employers and employees in order to carry out the functions of the Secretary of Health, Education, and Welfare under the Act. Inspections conducted by Department of Labor Compliance Safety and Health Officers and representatives of the Secretary of Health, Education, and Welfare under section 8 of the Act and pursuant to this Part 1903 shall not affect the authority of any State to conduct inspections in accordance with agreements and plans under section 18 of the Act.

**1903.3(b)** Prior to inspecting areas containing information which is classified by an agency of the United States Government in the interest of national security, Compliance Safety and Health Officers shall have obtained the appropriate security clearance.

## §1903.4 Objection to inspection.

**1903.4(a)** Upon a refusal to permit the Compliance Safety and Health Officer, in exercise of his official duties, to enter without delay and at reasonable times any place of employment or any place therein, to inspect, to review records, or to question any employer, owner, operator, agent, or employee, in accordance with §1903.3 or to permit a representative of employees to accompany the Compliance Safety and Health Officer during the physical inspection of any workplace in accordance with §1903.8, the Safety and Health Officer shall terminate the inspection or confine the inspection to other areas, conditions, structures, machines, apparatus, devices, equipment, materials, records, or interviews concerning which no objection is raised. The Compliance Safety and Health Officer shall endeavor to ascertain the reason for such refusal, and shall immediately report the refusal and the reason therefor to the Area Director. The Area Director shall consult with the Regional Solicitor, who shall take appropriate action, including compulsory process, if necessary.

**1903.4(b)** Compulsory process shall be sought in advance of an attempted inspection or investigation if, in the judgment of the Area Director and the Regional Solicitor, circumstances exist which make such preinspection process desirable or necessary. Some examples of circumstances in which it may be desirable or necessary to seek compulsory process in advance of an attempt to inspect or investigate include (but are not limited to):

**1903.4(b)(1)** When the employer's past practice either implicitly or explicitly puts the Secretary on notice that a warrantless inspection will not be allowed;

**1903.4(b)(2)** When an inspection is scheduled far from the local office and procuring a warrant prior to leaving to conduct the inspection would avoid, in case of refusal of entry, the expenditure of significant time and resources to return to the office, obtain a warrant and return to the worksite;

**1903.4(b)(3)** When an inspection includes the use of special equipment or when the presence of an expert or experts is needed in order to properly conduct the inspection, and procuring a warrant prior to an attempt to inspect would alleviate the difficulties or costs encountered in coordinating the availability of such equipment or expert.

**1903.4(c)** With the approval of the Regional Administrator and the Regional Solicitor, compulsory process may also be obtained by the Area Director or his designee.

**1903.4(d)** For purposes of this section, the term compulsory process shall mean the institution of any appropriate action, including *ex parte* application for an inspection warrant or its equivalent. *Ex parte* inspection warrants shall be in the preferred form of compulsory process in all circumstances where compulsory process is relied upon to seek entry to a workplace under this section.

[45 FR 65923, Oct. 3, 1980]

### §1903.5 Entry not a waiver.

Any permission to enter, inspect, review records, or question any person, shall not imply or be conditioned upon a waiver of any cause of action, citation, or penalty under the Act. Compliance Safety and Health Officers are not authorized to grant any such waiver.

### §1903.6 Advance notice of inspections.

**1903.6(a)** Advance notice of inspections may not be given, except in the following situations:

**1903.6(a)(1)** In cases of apparent imminent danger, to enable the employer to abate the danger as quickly as possible;

**1903.6(a)(2)** In circumstances where the inspection can most effectively be conducted after regular business hours or where special preparations are necessary for an inspection;

**1903.6(a)(3)** Where necessary to assure the presence of representatives of the employer and employees or the appropriate personnel needed to aid in the inspection; and

**1903.6(a)(4)** In other circumstances where the Area Director determines that the giving of advance notice would enhance the probability of an effective and thorough inspection.

**1903.6(b)** In the situations described in paragraph (a) of this section, advance notice of inspections may be given only if authorized by the Area Director, except that in cases of apparent imminent danger, advance notice may be given by the Compliance Safety and Health Officer without such authorization if the Area Director is not immediately available. When advance notice is given, it shall be the employer's responsibility promptly to notify the authorized representative of employees of the inspection, if the identity of such representative is known to the employer. (See §1903.8(b) as to situations where there is no authorized representative of employees.) Upon the request of the employer, the Compliance Safety and Health Officer will inform the authorized representative of employees of the inspection, provided that the employer furnishes the Compliance Safety and Health Officer with the identity of such representative and with such other information as is necessary to enable him promptly to inform such representative of the inspection. An employer who fails to comply with his obligation under this paragraph promptly to inform the authorized representative of employees of the inspection or to furnish such information as is necessary to enable the Compliance Safety and Health Officer promptly to inform such representative of the inspection, may be subject to citation and penalty under section 17(c) of the Act. Advance notice in any of the situations described in paragraph (a) of this section shall not be given more than 24 hours before the inspection is scheduled to be conducted, except in apparent imminent danger situations and in other unusual circumstances.

**1903.6(c)** The Act provides in section 17(f) that any person who gives advance notice of any inspection to be conducted under the Act, without authority from the Secretary or his designees, shall, upon conviction, be punished by fine of not more than $1,000 or by imprisonment for not more than 6 months, or by both.

### §1903.7 Conduct of inspections.

**1903.7(a)** Subject to the provisions of §1903.3, inspections shall take place at such times and in such places of employment as the Area Director or the Compliance Safety and Health Officer may direct. At the beginning of an inspection, Compliance Safety and Health Officers shall present their credentials to the owner, operator, or agent in charge at the establishment; explain the nature and purpose of the inspection; and indicate generally the scope of the inspection and the records specified in §1903.3 which they wish to review. However, such designation of records shall not preclude access to additional records specified in §1903.3.

**1903.7(b)** Compliance Safety and Health Officers shall have authority to take environmental samples and to take or obtain photographs related to the purpose of the inspection, employ other reasonable investigative techniques, and question privately any employer, owner, operator, agent or employee of an establishment. (See §1903.9 on trade secrets.) As used herein, the term "employ other reasonable investigative techniques" includes, but is not limited to, the use of devices to measure employee exposures and the attachment of personal sampling equipment such as dosimeters, pumps, badges and other similar devices to employees in order to monitor their exposures.

[§1903.7(b) amended at 47 FR 55478 December 10, 1982, effective January 10, 1983.]

**1903.7(c)** In taking photographs and samples, Compliance Safety and Health Officers shall take reasonable precautions to insure that such actions with flash, spark-producing, or other equipment would not be hazardous. Compliance Safety and Health Officers shall comply with all employer safety and health rules and practices at the establishment being inspected, and they shall wear and use appropriate protective clothing and equipment.

**1903.7(d)** The conduct of inspections shall be such as to preclude unreasonable disruption of the operations of the employer's establishment.

**1903.7(e)** At the conclusion of an inspection, the Compliance Safety and Health Officer shall confer with the employer or his representative and informally advise him of any apparent safety or health violations disclosed by the inspection. During such conference, the employer shall be afforded an opportunity to bring to the attention of the Compliance Safety and Health Officer any pertinent information regarding conditions in the workplace.

**1903.7(f)** Inspections shall be conducted in accordance with the requirements of this part.

[36 FR 17850, Sept. 14, 1971, as amended at 47 FR 6533, Feb. 12, 1982; 47 FR 55481, Dec. 10, 1982]

### §1903.8 Representatives of employers and employees.

**1903.8(a)** Compliance Safety and Health Officers shall be in charge of inspections and questioning of persons. A representative of the employer and a representative authorized by his employees shall be given an opportunity to accompany the Compliance Safety and Health Officer during the physical inspection of any workplace for the purpose of aiding such inspection. A Compliance Safety and Health Officer may permit additional employer representatives and additional representatives authorized by employees to accompany him where he determines that such additional representatives will further aid the inspection. A different employer and employee representative may accompany the Compliance Safety and Health Officer during each different phase of an inspection if this will not interfere with the conduct of the inspection.

**1903.8(b)** Compliance Safety and Health Officers shall have authority to resolve all disputes as to who is the representative authorized by the employer and employees for the purpose of this section. If there is no authorized representative of employees, or if the Compliance Safety and Health Officer is unable to determine with reasonable certainty who is such representative, he shall consult with a reasonable number of employees concerning matters of safety and health in the workplace.

**1903.8(c)** The representative(s) authorized by employees shall be an employee(s) of the employer. However, if in the judgment of the Compliance Safety and Health Officer, good cause has been shown why accompaniment by a third party who is not an employee of the employer (such as an industrial hygienist or a safety engineer) is reasonably necessary to the conduct of an effective and thorough physical inspection of the workplace, such third party may accompany the Compliance Safety and Health Officer during the inspection.

**1903.8(d)** Compliance Safety and Health Officers are authorized to deny the right of accompaniment under this section to any person whose conduct interferes with a fair and orderly inspection. The right of accompaniment in areas containing trade secrets shall be subject to the provisions of §1903.9(d). With regard to information classified by an agency of the U.S. Government in the interest of national security, only persons authorized to have access to such information may accompany a Compliance Safety and Health Officer in areas containing such information.

### §1903.9 Trade secrets.

**1903.9(a)** Section 15 of the Act provides: "All information reported to or otherwise obtained by the Secretary or his representative in connection with any inspection or proceeding under this Act which contains or which might reveal a trade secret referred to in section 1905 of title 18 of the United States Code shall be considered confidential for the purpose of that section, except that such information may be disclosed to other officers or employees concerned with carrying out this Act or when relevant in any proceeding under this Act. In any such proceeding the Secretary, the Commission, or the court shall issue such orders as may be appropriate to protect the confidentiality of trade secrets." Section 15 of the Act is considered a statute within the meaning of section 552(b)(3) of title 5 of the United States Code, which exempts from the disclosure requirements matters that are "specifically exempted from disclosure by statute."

**1903.9(b)** Section 1905 of title 18 of the United States Code provides: "Whoever, being an officer or employee of the United States or of any department or agency thereof, publishes, divulges, discloses, or makes known in any manner or to any extent not authorized by law any information coming to him in the course of his employment or official duties or by reason of any examination or investigation made by, or return, report or record made to or filed with, such department or agency or officer or employee thereof, which information concerns or relates to the trade secrets, processes, operations, style of work, or apparatus, or to the identity, confidential statistical data, amount or source of any income, profits, losses, or expenditures of any person, firm, partnership, corporation, or association; or permits any income return or copy thereof or any book containing any abstract or particulars thereof to be seen or examined by any person except as provided by law; shall be fined not more than $1,000, or imprisoned not more than 1 year, or both; and shall be removed from office or employment."

**1903.9(c)** At the commencement of an inspection, the employer may identify areas in the establishment which contain or which might reveal a trade secret. If the Compliance Safety and Health Officer has no clear reason to question such identification, information obtained in such areas, including all negatives and prints of photographs, and environmental samples, shall be labeled "confidential—trade secret" and shall not be disclosed except in accordance with the provisions of section 15 of the Act.

**1903.9(d)** Upon the request of an employer, any authorized representative of employees under §1903.8 in an area containing trade secrets shall be an employee in that area or an employee authorized by the employer to enter that area. Where there is no such representative or employee, the Compliance Safety and Health Officer shall consult with a reasonable number of employees who work in that area concerning matters of safety and health.

### §1903.10 Consultation with employees.

Compliance Safety and Health Officers may consult with employees concerning matters of occupational safety and health to the extent they deem necessary for the conduct of an effective and thorough inspection. During the course of an inspection, any employee shall be afforded an opportunity to bring any violation of the Act which he has reason to believe exists in the workplace to the attention of the Compliance Safety and Health Officer.

### §1903.11 Complaints by employees.

**1903.11(a)** Any employee or representative of employees who believe that a violation of the Act exists in any workplace where such employee is employed may request an inspection of such workplace by giving notice of the alleged violation to the Area Director or to a Compliance Safety and Health Officer. Any such notice shall be reduced to writing, shall set forth with reasonable particularity the grounds for the notice, and shall be signed by the employee or representative of employees. A copy shall be provided the employer or his agent by the Area Director or Compliance Safety Health Officer no later than at the time of inspection, except that, upon the request of the person giving such notice, his name and the names of individual employees referred to therein shall not appear in such copy or on any record published, released, or made available by the Department of Labor.

**1903.11(b)** If upon receipt of such notification the Area Director determines that the complaint meets the requirements set forth in paragraph (a) of this section, and that there are reasonable grounds to believe that the alleged violation exists, he shall cause an inspection to be made as soon as practicable, to determine if such alleged violation exists. Inspections under this section shall not be limited to matters referred to in the complaint.

**1903.11(c)** Prior to or during any inspection of a workplace, any employee or representative of employees employed in such workplace may notify the Compliance Safety and Health Officer, in writing, of any violation of the Act which they have reason to believe exists in such workplace. Any such notice shall comply with the requirements of paragraph (a) of this section.

**1903.11(d)** Section 11(c)(1) of the Act provides: "No person shall discharge or in any manner discriminate against any employee because such employee has filed any complaint or instituted or caused to be instituted any proceeding under or related to this Act or has testified or is about to testify in any such proceeding or because of the exercise by such employee on behalf of himself or others of any right afforded by this Act."

(Approved by the Office of Management and Budget under control number 1218-0064)

[36 FR 17850, Sept. 4, 1973, as amended at 54 FR 24333, June 7, 1989]

### §1903.12 Inspection not warranted; informal review.

**1903.12(a)** If the Area Director determines that an inspection is not warranted because there are no reasonable grounds to believe that a violation or danger exists with respect to a complaint under §1903.11, he shall notify the complaining party in writing of such determination. The complaining party may obtain review of such determination by submitting a written statement of position with the Assistant Regional Director and, at the same time, providing the employer with a copy of such statement by certified mail. The employer may submit an opposing written statement of position with the Assistant Regional Director and, at the same time, provide the complaining party with a copy of such statement by certified mail. Upon the request of the complaining party or the employer, the Assistant Regional Director at his discretion, may hold an informal conference in which the complaining party and the employer may orally present their views. After considering all written and oral views presented, the Assistant Regional Director shall affirm, modify, or reverse the determination of the Area Director and furnish the complaining party and the employer and written notification of his decision and the reasons therefor. The decision of the Assistant Regional Director shall be final and not subject to further review.

**1903.12(b)** If the Area Director determines that an inspection is not warranted because the requirements of §1903.11(a) have not been met, he shall notify the complaining party in writing of such determination. Such determination shall be without prejudice to the filing of a new complaint meeting the requirements of §1903.11(a).

### §1903.13 Imminent danger.

Whenever and as soon as a Compliance Safety and Health Officer concludes on the basis of an inspection that conditions or practices exist in any place of employment which could reasonably be expected to cause death or serious physical harm immediately or before the imminence of such danger can be eliminated through the enforcement procedures otherwise provided by the Act, he shall inform the affected employees and employers of the danger and that he is recommending a civil action to restrain such conditions or practices and for other appropriate relief in accordance with the provisions of section 13(a) of the Act. Appropriate citations and notices of proposed penalties may be issued with respect to an imminent danger even though, after being informed of such danger by the Compliance Safety and Health Officer, the employer immediately eliminates the imminence of the danger and initiates steps to abate such danger.

### §1903.14 Citations; notices of de minimis violations; policy regarding employee rescue activities.

**1903.14(a)** The Area Director shall review the inspection report of the Compliance Safety and Health Officer. If, on the basis of the report the Area Director believes that the employer has violated a requirement of section 5 of the Act, of any standard, rule or order promulgated pursuant to section 6 of the Act, or of any substantive rule published in this chapter, he shall, if appropriate, consult with the Regional Solicitor, and he shall issue to the employer either a citation or a notice of de minimis violations which have no direct or immediate relationship to safety or health. An appropriate citation or notice of de minimis violations shall be issued even though after being informed of an alleged violation by the Compliance Safety and Health Officer, the employer immediately abates, or initiates steps to abate, such alleged violation. Any citation or notice of de minimis violations shall be issued with reasonable promptness after termination of the inspection. No citation may be issued under this section after the expiration of 6 months following the occurrence of any alleged violation.

**1903.14(b)** Any citation shall describe with particularity the nature of the alleged violation, including a reference to the provision(s) of the Act, standard, rule, regulation, or order alleged to have been violated. Any citation shall also fix a reasonable time or times for the abatement of the alleged violation.

**1903.14(c)** If a citation or notice of de minimis violations is issued for a violation alleged in a request for inspection under §1903.11(a) or a notification of violation under §1903.11(c), a copy of the citation or notice of de minimis violations shall also be sent to the employee or representative of employees who made such request or notification.

**1903.14(d)** After an inspection, if the Area Director determines that a citation is not warranted with respect to a danger or violation alleged to exist in a request for inspection under §1903.11(a) or a notification of violation under §1903.11(c), the informal review procedures prescribed in §1903.12(a) shall be applicable. After considering all views presented, the Assistant Regional Director shall affirm the determination of the Area Director, order a reinspection, or issue a citation if he believes that the inspection disclosed a violation. The Assistant Regional Director shall furnish the complaining party and the employer with written notification of his determination and the reasons therefor. The determination of the Assistant Regional Director shall be final and not subject to review.

**1903.14(e)** Every citation shall state that the issuance of a citation does not constitute a finding that a violation of the Act has occurred unless there is a failure to contest as provided for in the Act or, if contested, unless the citation is affirmed by the Review Commission.

**1903.14(f)** No citation may be issued to an employer because of a rescue activity undertaken by an employee of that employer with respect to an individual in imminent danger unless:

**1903.14(f)(1)(i)** Such employee is designated or assigned by the employer to have responsibility to perform or assist in rescue operations, and

**1903.14(f)(1)(ii)** the employer fails to provide protection of the safety and health of such employee, including failing to provide appropriate training and rescue equipment; or

**1903.14(f)(2)(i)** Such employee is directed by the employer to perform rescue activities in the course of carrying out the employee's job duties, and

**1903.14(f)(2)(ii)** The employer fails to provide protection of the safety and health of such employee, including failing to provide appropriate training and rescue equipment; or

**1903.14(f)(3)(i)** Such employee is employed in a workplace that requires the employee to carry out duties that are directly related to a workplace operation where the likelihood of life-threatening accidents is foreseeable, such as a workplace operation where employees are located in confined spaces or trenches, handle hazardous waste, respond to emergency situations, perform excavations, or perform construction over water; and

**1903.14(f)(3)(ii)** Such employee has not been designated or assigned to perform or assist in rescue operations and voluntarily elects to rescue such an individual; and

**1903.14(f)(3)(iii)** The employer has failed to instruct employees not designated or assigned to perform or assist in rescue operations of the arrangements for rescue, not to attempt rescue, and of the hazards of attempting rescue without adequate training or equipment.

**1903.14(f)(4)** For purposes of this policy, the term "imminent danger" means the existence of any condition or practice that could reasonably be expected to cause death or serious physical harm before such condition or practice can be abated.

[36 FR 17850, Sept. 4, 1971, as amended at 59 FR 66613, Dec. 27, 1994]

## §1903.14a Petitions for modification of abatement date.

**1903.14a(a)** An employer may file a petition for modification of abatement date when he has made a good faith effort to comply with the abatement requirements of a citation, but such abatement has not been completed because of factors beyond his reasonable control.

**1903.14a(b)** A petition for modification of abatement date shall be in writing and shall include the following information:

**1903.14a(b)(1)** All steps taken by the employer, and the dates of such action, in an effort to achieve compliance during the prescribed abatement period.

**1903.14a(b)(2)** The specific additional abatement time necessary in order to achieve compliance.

**1903.14a(b)(3)** The reasons such additional time is necessary, including the unavailability of professional or technical personnel or of materials and equipment, or because necessary construction or alteration of facilities cannot be completed by the original abatement date.

**1903.14a(b)(4)** All available interim steps being taken to safeguard the employees against the cited hazard during the abatement period.

**1903.14a(b)(5)** A certification that a copy of the petition has been posted and, if appropriate, served on the authorized representative of affected employees, in accordance with subsection (c)(1) of this section and a certification of the date upon which such posting and service was made.

**1903.14a(c)** A petition for modification of abatement date shall be filed with the Area Director of the United States Department of Labor who issued the citation no later than the close of the next working day following the date on which abatement was originally required. A later-filed petition shall be accompanied by the employer's statement of exceptional circumstances explaining the delay.

**1903.14a(c)(1)** A copy of such petition shall be posted in a conspicuous place where all affected employees will have notice thereof or near such location where the violation occurred. The petition shall remain posted for a period of ten (10) working days. Where affected employees are represented by an authorized representative, said representative shall be served with a copy of such petition.

**1903.14a(c)(2)** Affected employees or their representatives may file an objection in writing to such petition with the aforesaid Area Director. Failure to file such objection within ten (10) working days of the date of posting of such petition or of service upon an authorized representative shall constitute a waiver of any further right to object to said petition.

**1903.14a(c)(3)** The Secretary or his duly authorized agent shall have the authority to approve any petition for modification of abatement date filed pursuant to paragraphs (b) and (c) of this section. Such uncontested petitions shall become final orders pursuant to sections 10(a) and (c) of the Act.

**1903.14a(c)(4)** The Secretary or his authorized representative shall not exercise his approval power until the expiration of fifteen (15) working days from the date the petition was posted or served pursuant to paragraphs (c)(1) and (2) of this section by the employer.

**1903.14a(d)** Where any petition is objected to by the Secretary or affected employees, the petition, citation, and any objections shall be forwarded to the Commission within three (3) working days after the expiration of the fifteen (15) day period set out in paragraph (c)(4) of this section.

[40 FR 6334, Feb. 11, 1975; 40 FR 11351, Mar. 11, 1975]

## §1903.15 Proposed penalties.

**1903.15(a)** After, or concurrent with, the issuance of a citation, and within a reasonable time after the termination of the inspection, the Area Director shall notify the employer by certified mail or by personal service by the Compliance Safety and Health Officer of the proposed penalty under section 17 of the Act, or that no penalty is being proposed. Any notice of proposed penalty shall state that the proposed penalty shall be deemed to be the final order of the Review Commission and not subject to review by any court or agency unless, within 15 working days from the date of receipt of such notice, the employer notifies the Area Director in writing that he intends to contest the citation or the notification of proposed penalty before the Review Commission.

**1903.15(b)** The Area Director shall determine the amount of any proposed penalty, giving due consideration to the appropriateness of the penalty with respect to the size of the business of the employer being charged, the gravity of the violation, the good faith of the employer, and the history of previous violations, in accordance with the provisions of section 17 of the Act.

**1903.15(c)** Appropriate penalties may be proposed with respect to an alleged violation even though after being informed of such alleged violation by the Compliance Safety and Health Officer, the employer immediately abates, or initiates steps to abate, such alleged violation. Penalties shall not be proposed for de minimis violations which have no direct or immediate relationship to safety or health.

## §1903.16 Posting of citations.

**1903.16(a)** Upon receipt of any citation under the Act, the employer shall immediately post such citation, or a copy thereof, unedited, at or near each place an alleged violation referred to in the citation occurred, except as provided below. Where, because of the nature of the employer's operations, it is not practicable to post the citation at or near each place of alleged violation, such citation shall be posted, unedited, in a prominent place where it will be readily observable by all affected employees. For example, where employers are engaged in activities which are physically dispersed (see §1903.2(b)), the citation may be posted at the location to which employees report each day. Where employees do not primarily work at or report to a single location (see §1903.2(b)), the citation may be posted at the location from which the employees operate to carry out their activities. The employer shall take steps to ensure that the citation is not altered, defaced, or covered by other material. Notices of de minimis violations need not be posted.

**1903.16(b)** Each citation, or a copy thereof, shall remain posted until the violation has been abated, or for 3 working days, whichever is later. The filing by the employer of a notice of intention to contest under §1903.17 shall not affect his posting responsibility under this section unless and until the Review Commission issues a final order vacating the citation.

**1903.16(c)** An employer to whom a citation has been issued may post a notice in the same location where such citation is posted indicating that the citation is being contested before the Review Commission, and such notice may explain the reasons for such contest. The employer may also indicate that specified steps have been taken to abate the violation.

**1903.16(d)** Any employer failing to comply with the provisions of paragraphs (a) and (b) of this section shall be subject to citation and penalty in accordance with the provisions of section 17 of the Act.

## §1903.17 Employer and employee contests before the Review Commission.

**1903.17(a)** Any employer to whom a citation or notice of proposed penalty has been issued may, under section 10(a) of the Act, notify the Area Director in writing that he intends to contest such citation or proposed penalty before the Review Commission. Such notice of intention to contest shall be postmarked within 15 working days of the receipt by the employer of the notice of proposed penalty. Every notice of intention to contest shall specify whether it is directed to the citation or to the proposed penalty, or both. The Area Director shall immediately transmit such notice to the Review Commission in accordance with the rules of procedure prescribed by the Commission.

**1903.17(b)** Any employee or representative of employees of an employer to whom a citation has been issued may, under section 10(c) of the Act, file a written notice with the Area Director alleging that the period of time fixed in the citation for the abatement of the violation is unreasonable. Such notice shall be postmarked within 15 working days of the receipt by the employer of the notice of proposed penalty or notice that no penalty is proposed. The Area Director shall immediately transmit such notice to the Review Commission in accordance with the rules of procedure prescribed by the Commission.

## §1903.18 Failure to correct a violation for which a citation has been issued.

**1903.18(a)** If an inspection discloses that an employer has failed to correct an alleged violation for which a citation has been issued within the period permitted for its correction, the Area Director shall, if appropriate, consult with the Regional Solicitor, and he shall notify the employer by certified mail or by personal service by the Compliance Safety and Health Officer of such failure and of the additional penalty proposed under section 17(d) of the Act by reason of such failure. The period for the correction of a violation for which a citation has been issued shall not begin to run until the entry of a final order of the Review Commission in the case of any review proceedings initiated by the employer in good faith and not solely for delay or avoidance of penalties.

**1903.18(b)** Any employer receiving a notification of failure to correct a violation and of proposed additional penalty may, under section 10(b) of the Act, notify the Area Director in writing that he intends to contest such notification or proposed additional penalty before the Review Commission. Such notice of intention to contest shall be postmarked within 15 working days of the receipt by the employer of the notification of failure to correct a violation and of proposed additional penalty. The Area Director shall immediately transmit such notice to the Review Commission in accordance with the rules of procedure prescribed by the Commission.

**1903.18(c)** Each notification of failure to correct a violation and of proposed additional penalty shall state that it shall be deemed to be the final order of the Review Commission and not subject to review by any court or agency unless, within 15 working days from the date of receipt of such notification, the employer notifies the Area Director in writing that he intends to contest the notification or the proposed additional penalty before the Review Commission.

## §1903.19 Abatement verification.

***Purpose.*** OSHA's inspections are intended to result in the abatement of violations of the Occupational Safety and Health Act of 1970 (the OSH Act). This section sets forth the procedures OSHA will use to ensure abatement. These procedures are tailored to the nature of the violation and the employer's abatement actions.

**1903.19(a)** ***Scope and application.*** This section applies to employers who receive a citation for a violation of the Occupational Safety and Health Act.

**1903.19(b)** ***Definitions.*** (1) *Abatement* means action by an employer to comply with a cited standard or regulation or to eliminate a recognized hazard identified by OSHA during an inspection.

**1903.19(b)(2)** *Abatement date* means:

**1903.19(b)(2)(i)** For an uncontested citation item, the later of:

**1903.19(b)(2)(i)(A)** The date in the citation for abatement of the violation;

**1903.19(b)(2)(i)(B)** The date approved by OSHA or established in litigation as a result of a petition for modification of the abatement date (PMA); or

**1903.19(b)(2)(i)(C)** The date established in a citation by an informal settlement agreement.

**1903.19(b)(2)(ii)** For a contested citation item for which the Occupational Safety and Health Review Commission (OSHRC) has issued a final order affirming the violation, the later of:

**1903.19(b)(2)(ii)(A)** The date identified in the final order for abatement; or

**1903.19(b)(2)(ii)(B)** The date computed by adding the period allowed in the citation for abatement to the final order date;

**1903.19(b)(2)(ii)(C)** The date established by a formal settlement agreement.

**1903.19(b)(3)** *Affected employees* means those employees who are exposed to the hazard(s) identified as violation(s) in a citation.

**1903.19(b)(4)** *Final order date* means:

**1903.19(b)(4)(i)** For an uncontested citation item, the fifteenth working day after the employer's receipt of the citation;

**1903.19(b)(4)(ii)** For a contested citation item:

**1903.19(b)(4)(ii)(A)** The thirtieth day after the date on which a decision or order of a commission administrative law judge has been docketed with the commission, unless a member of the commission has directed review; or

**1903.19(b)(4)(ii)(B)** Where review has been directed, the thirtieth day after the date on which the Commission issues its decision or order disposing of all or pertinent part of a case; or

**1903.19(b)(4)(ii)(C)** The date on which a federal appeals court issues a decision affirming the violation in a case in which a final order of OSHRC has been stayed.

**1903.19(b)(5)** *Movable equipment* means a hand-held or non-hand-held machine or device, powered or unpowered, that is used to do work and is moved within or between worksites.

**1903.19(c)** ***Abatement certification.*** (1) Within 10 calendar days after the abatement date, the employer must certify to OSHA (the Agency) that each cited violation has been abated, except as provided in paragraph (c)(2) of this section.

**1903.19(c)(2)** The employer is not required to certify abatement if the OSHA Compliance Officer, during the on-site portion of the inspection:

**1903.19(c)(2)(i)** Observes, within 24 hours after a violation is identified, that abatement has occurred; and

**1903.19(c)(2)(ii)** Notes in the citation that abatement has occurred.

**1903.19(c)(3)** The employer's certification that abatement is complete must include, for each cited violation, in addition to the information required by paragraph (h) of this section, the date and method of abatement and a statement that affected employees and their representatives have been informed of the abatement.

**NOTE TO PARAGRAPH (c):** Appendix A contains a sample Abatement Certification Letter.

**1903.19(d)** ***Abatement documentation.*** (1) The employer must submit to the Agency, along with the information on abatement certification required by paragraph (c)(3) of this section, documents demonstrating that abatement is complete for each willful or repeat violation and for any serious violation for which the Agency indicates in the citation that such abatement documentation is required.

**1903.19(d)(2)** Documents demonstrating that abatement is complete may include, but are not limited to, evidence of the purchase or repair of equipment, photographic or video evidence of abatement, or other written records.

**1903.19(e)** ***Abatement plans.*** (1) The Agency may require an employer to submit an abatement plan for each cited violation (except an other-than-serious violation) when the time permitted for abatement is more than 90 calendar days. If an abatement plan is required, the citation must so indicate.

**1903.19(e)(2)** The employer must submit an abatement plan for each cited violation within 25 calendar days from the final order date when the citation indicates that such a plan is required. The abatement plan must identify the violation and the steps to be taken to achieve abatement, including a schedule for completing abatement and, where necessary, how employees will be protected from exposure to the violative condition in the interim until abatement is complete.

**NOTE TO PARAGRAPH (e):** Appendix B contains a Sample Abatement Plan form.

**1903.19(f)** ***Progress reports.*** (1) An employer who is required to submit an abatement plan may also be required to submit periodic progress reports for each cited violation. The citation must indicate:

**1903.19(f)(1)(i)** That periodic progress reports are required and the citation items for which they are required;

**1903.19(f)(1)(ii)** The date on which an initial progress report must be submitted, which may be no sooner than 30 calendar days after submission of an abatement plan;

**1903.19(f)(1)(iii)** Whether additional progress reports are required; and

**1903.19(f)(1)(iv)** The date(s) on which additional progress reports must be submitted.

**1903.19(f)(2)** For each violation, the progress report must identify, in a single sentence if possible, the action taken to achieve abatement and the date the action was taken.

**NOTE TO PARAGRAPH (f):** Appendix B contains a Sample Progress Report Form.

**1903.19(g)** ***Employee notification.*** (1) The employer must inform affected employees and their representative(s) about abatement activities covered by this section by posting a copy of each document submitted to the Agency or a summary of the document near the place where the violation occurred.

**1903.19(g)(2)** Where such posting does not effectively inform employees and their representatives about abatement activities (for example, for employers who have mobile work operations), the employer must:

**1903.19(g)(2)(i)** Post each document or a summary of the document in a location where it will be readily observable by affected employees and their representatives; or

**1903.19(g)(2)(ii)** Take other steps to communicate fully to affected employees and their representatives about abatement activities.

**1903.19(g)(3)** The employer must inform employees and their representatives of their right to examine and copy all abatement documents submitted to the Agency.

**1903.19(g)(3)(i)** An employee or an employee representative must submit a request to examine and copy abatement documents within 3 working days of receiving notice that the documents have been submitted.

**1903.19(g)(3)(ii)** The employer must comply with an employee's or employee representative's request to examine and copy abatement documents within 5 working days of receiving the request.

**1903.19(g)(4)** The employer must ensure that notice to employees and employee representatives is provided at the same time or before the information is provided to the Agency and that abatement documents are:

**1903.19(g)(4)(i)** Not altered, defaced, or covered by other material; and

**1903.19(g)(4)(ii)** Remain posted for three working days after submission to the Agency.

**1903.19(h)** ***Transmitting abatement documents.*** (1) The employer must include, in each submission required by this section, the following information:

**1903.19(h)(1)(i)** The employer's name and address;

**1903.19(h)(1)(ii)** The inspection number to which the submission relates;

**1903.19(h)(1)(iii)** The citation and item numbers to which the submission relates;

**1903.19(h)(1)(iv)** A statement that the information submitted is accurate; and

**1903.19(h)(1)(v)** The signature of the employer or the employer's authorized representative.

**1903.19(h)(2)** The date of postmark is the date of submission for mailed documents. For documents transmitted by other means, the date the Agency receives the document is the date of submission.

**1903.19(i)** ***Movable equipment.*** (1) For serious, repeat, and willful violations involving movable equipment, the employer must attach a warning tag or a copy of the citation to the operating controls or to the cited component of equipment that is moved within the worksite or between worksites.

**NOTE TO PARAGRAPH (i)(1):** Attaching a copy of the citation to the equipment is deemed by OSHA to meet the tagging requirement of paragraph (i)(1) of this section as well as the posting requirement of 29 CFR 1903.16.

**1903.19(i)(2)** The employer must use a warning tag that properly warns employees about the nature of the violation involving the equipment and identifies the location of the citation issued.

**NOTE TO PARAGRAPH (i)(2):** Non-Mandatory Appendix C contains a sample tag that employers may use to meet this requirement.

**1903.19(i)(3)** If the violation has not already been abated, a warning tag or copy of the citation must be attached to the equipment:

**1903.19(i)(3)(i)** For hand-held equipment, immediately after the employer receives the citation; or

**1903.19(i)(3)(ii)** For non-hand-held equipment, prior to moving the equipment within or between worksites.

**1903.19(i)(4)** For the construction industry, a tag that is designed and used in accordance with 29 CFR 1926.20(b)(3) and 29 CFR 1926.200(h) is deemed by OSHA to meet the requirements of this section when the information required by paragraph (i)(2) is included on the tag.

**1903.19(i)(5)** The employer must assure that the tag or copy of the citation attached to movable equipment is not altered, defaced, or covered by other material.

**1903.19(i)(6)** The employer must assure that the tag or copy of the citation attached to movable equipment remains attached until:

**1903.19(i)(6)(i)** The violation has been abated and all abatement verification documents required by this regulation have been submitted to the Agency;

**1903.19(i)(6)(ii)** The cited equipment has been permanently removed from service or is no longer within the employer's control; or

**1903.19(i)(6)(iii)** The Commission issues a final order vacating the citation.

## APPENDICES TO §1903.19— ABATEMENT VERIFICATION

**NOTE:** Appendices A through C provide information and nonmandatory guidelines to assist employers and employees in complying with the appropriate requirements of this section.

## APPENDIX A TO SECTION 1903.19— SAMPLE ABATEMENT-CERTIFICATION LETTER (NONMANDATORY)

(Name), Area Director

U.S. Department of Labor—OSHA

Address of the Area Office (on the citation)

[Company's Name]

[Company's Address]

The hazard referenced in Inspection Number [insert 9-digit #] for violation identified as:

Citation [insert #] and item [insert #] was corrected on [insert date] by:

______________________________

Citation [insert #] and item [insert #] was corrected on [insert date] by:

______________________________

Citation [insert #] and item [insert #] was corrected on [insert date] by:

Citation [insert #] and item [insert #] was corrected on [insert date] by:

Citation [insert #] and item [insert #] was corrected on [insert date] by:

Citation [insert #] and item [insert #] was corrected on [insert date] by:

Citation [insert #] and item [insert #] was corrected on insert date by:

Citation [insert #] and item [insert #] was corrected on [insert date] by:

I attest that the information contained in this document is accurate.

Signature

Typed or Printed Name

### APPENDIX B TO SECTION 1903.19— SAMPLE ABATEMENT PLAN OR PROGRESS REPORT (NONMANDATORY)

(Name), Area Director

U. S. Department of Labor—OSHA

Address of Area Office (on the citation)

[Company's Name]

[Company's Address]

Check one:

Abatement Plan [ ]

Progress Report [ ]

Inspection Number ______________________

Page _____ of _____

Citation Number(s)* ______________________

Item Number(s)* ______________________

| Action | Proposed Completion Date (for abatement plans only) | Completion Date (for progress reports only) |
|---|---|---|
| 1. .............................. | .............................. | .......................... |
| 2. .............................. | .............................. | .......................... |
| 3. .............................. | .............................. | .......................... |
| 4. .............................. | .............................. | .......................... |
| 5. .............................. | .............................. | .......................... |
| 6. .............................. | .............................. | .......................... |
| 7. .............................. | .............................. | .......................... |

Date required for final abatement: ______________________

I attest that the information contained in this document is accurate.

Signature

Typed or Printed Name

Name of primary point of contact for questions: [optional]

Telephone number: ______________________

*Abatement plans or progress reports for more than one citation item may be combined in a single abatement plan or progress report if the abatement actions, proposed completion dates, and actual completion dates (for progress reports only) are the same for each of the citation items.

### APPENDIX C TO SECTION 1903.19— Sample Warning Tag (Nonmandatory)

**WARNING:**

EQUIPMENT HAZARD

CITED BY OSHA

EQUIPMENT CITED:

HAZARD CITED:

FOR DETAILED INFORMATION

SEE OSHA CITATION POSTED AT:

BACKGROUND COLOR—ORANGE
MESSAGE COLOR—BLACK

[62 FR 15337, Mar. 31, 1997]

### §1903.20 Informal conferences.

At the request of an affected employer, employee, or representative of employees, the Assistant Regional Director may hold an informal conference for the purpose of discussing any issues raised by an inspection, citation, notice of proposed penalty, or notice of intention to contest. The settlement of any issue at such conference shall be subject to the rules of procedure prescribed by the Review Commission. If the conference is requested by the employer, an affected employee or his representative shall be afforded an opportunity to participate, at the discretion of the Regional Administrator. If the conference is requested by an employee or representative of employees, the employer shall be afforded an opportunity to participate, at the discretion of the Regional Administrator. Any party may be represented by counsel at such conference. No such conference or request for such conference shall operate as a stay of any 15-working-day period for filing a notice of intention to contest as prescribed in §1903.17.

[36 FR 17850, Sept. 4, 1971. Redesignated at 62 FR 15337, Mar. 31, 1997]

### §1903.21 State administration.

Nothing in this Part 1903 shall preempt the authority of any State to conduct inspections, to initiate enforcement proceedings or otherwise to implement the applicable provisions of State law with respect to State occupational safety and health standards in accordance with agreements and plans under section 18 of the Act and Parts 1901 and 1902 of this chapter.

[36 FR 17850, Sept. 4, 1971. Redesignated at 62 FR 15337, Mar. 31, 1997]

### §1903.22 Definitions.

**1903.22(a)** *Act* means the Williams-Steiger Occupational Safety and Health Act of 1970. (84 Stat. 1590 *et seq.*, 29 U.S.C. 651 *et seq.*)

**1903.22(b)** The definitions and interpretations contained in section 3 of the Act shall be applicable to such terms when used in this Part 1903.

**1903.22(c)** *Working days* means Mondays through Fridays but shall not include Saturdays, Sundays, or Federal holidays. In computing 15 working days, the day of receipt of any notice shall not be included, and the last day of the 15 working days shall be included.

**1903.22(d)** *Compliance Safety and Health Officer* means a person authorized by the Occupational Safety and Health Administration, U.S. Department of Labor, to conduct inspections.

**1903.22(e)** *Area Director* means the employee or officer regularly or temporarily in charge of an Area Office of the Occupational Safety and Health Administration, U.S. Department of Labor, or any other person or persons who are authorized to act for such employee or officer. The latter authorizations may include general delegations of the authority of an Area Director under this part to a Compliance Safety and Health Officer or delegations to such an officer for more limited purposes, such as the exercise of the Area Director's duties under §1903.14(a). The term also includes any employee or officer exercising supervisory responsibilities over an Area Director. A supervisory employee or officer is considered to exercise concurrent authority with the Area Director.

**1903.22(f)** *Assistant Regional Director* means the employee or officer regularly or temporarily in charge of a Region of the Occupational Safety and Health Administration, U.S. Department of Labor, or any other person or persons who are specifically designated to act for such employee or officer in his absence. The term also includes any employee or officer in the Occupational Safety and Health Administration exercising supervisory responsibilities over the Assistant Regional Director. Such supervisory employee or officer is considered to exercise concurrent authority with the Assistant Regional Director. No delegation of authority under this paragraph shall adversely affect the procedures for independent informal review of investigative determinations prescribed under §1903.12 of this part.

**1903.22(g)** *Inspection* means any inspection of an employer's factory, plant, establishment, construction site, or other area, workplace or environment where work is performed by an employee of an employer, and includes any inspection conducted pursuant to a complaint filed under §1903.11(a) and (c), any reinspection, followup inspection, accident investigation or other inspection conducted under section 8(a) of the Act.

[36 FR 17850, Sept. 4, 1971, as amended at 38 FR 22624, Aug. 23, 1973. Redesignated at 62 FR 15337, Mar. 31, 1997]

# 29 CFR Part 1904—Recording and Reporting Occupational Injuries and Illnesses



**Authority:** 29 U.S.C. 657, 658, 660, 666, 669, 673, Secretary of Labor's Order No. 3—2000 (65 FR 50017), and 5 U.S.C. 533.

## Subpart A—Purpose

### §1904.0 Purpose.

The purpose of this rule (Part 1904) is to require employers to record and report work-related fatalities, injuries and illnesses.

**Note to §1904.0:** Recording or reporting a work-related injury, illness, or fatality does not mean that the employer or employee was at fault, that an OSHA rule has been violated, or that the employee is eligible for workers' compensation or other benefits.

[66 FR 5915, Jan. 19, 2001]

## Subpart B—Scope

**Note to Subpart B:** All employers covered by the Occupational Safety and Health Act (OSH Act) are covered by these Part 1904 regulations. However, most employers do not have to keep OSHA injury and illness records unless OSHA or the Bureau of Labor Statistics (BLS) informs them in writing that they must keep records. For example, employers with 10 or fewer employees and business establishments in certain industry classifications are partially exempt from keeping OSHA injury and illness records.

### §1904.1 Partial exemption for employers with 10 or fewer employees.

**1904.1(a)** ***Basic requirement.*** (1) If your company had ten (10) or fewer employees at all times during the last calendar year, you do not need to keep OSHA injury and illness records unless OSHA or the BLS informs you in writing that you must keep records under §1904.41 or §1904.42. However, as required by §1904.39, all employers covered by the OSH Act must report to OSHA any workplace incident that results in a fatality or the hospitalization of three or more employees.

**1904.1(a)(2)** If your company had more than ten (10) employees at any time during the last calendar year, you must keep OSHA injury and illness records unless your establishment is classified as a partially exempt industry under §1904.2.

**1904.1(b)** ***Implementation.*** (1) *Is the partial exemption for size based on the size of my entire company or on the size of an individual business establishment?* The partial exemption for size is based on the number of employees in the entire company.

**1904.1(b)(2)** *How do I determine the size of my company to find out if I qualify for the partial exemption for size?* To determine if you are exempt because of size, you need to determine your company's peak employment during the last calendar year. If you had no more than 10 employees at any time in the last calendar year, your company qualifies for the partial exemption for size.

[66 FR 5915, Jan. 19, 2001]

### §1904.2 Partial exemption for establishments in certain industries.

**1904.2(a)** ***Basic requirement.*** (1) If your business establishment is classified in a specific low hazard retail, service, finance, insurance or real estate industry listed in Appendix A to this Subpart B, you do not need to keep OSHA injury and illness records unless the government asks you to keep the records under §1904.41 or §1904.42. However, all employers must report to OSHA any workplace incident that results in a fatality or the hospitalization of three or more employees (see §1904.39).

**1904.2(a)(2)** If one or more of your company's establishments are classified in a non-exempt industry, you must keep OSHA injury and illness records for all of such establishments unless your company is partially exempted because of size under §1904.1.

**1904.2(b)** ***Implementation.*** (1) *Does the partial industry classification exemption apply only to business establishments in the retail, services, finance, insurance or real estate industries (SICs 52-89)?* Yes, business establishments classified in agriculture; mining; construction; manufacturing; transportation; communication, electric, gas and sanitary services; or wholesale trade are not eligible for the partial industry classification exemption.

**1904.2(b)(2)** *Is the partial industry classification exemption based on the industry classification of my entire company or on the classification of individual business establishments operated by my company?* The partial industry classification exemption applies to individual business establishments. If a company has several business establishments engaged in different classes of business activities, some of the company's establishments may be required to keep records, while others may be exempt.

**1904.2(b)(3)** *How do I determine the Standard Industrial Classification code for my company or for individual establishments?* You determine your Standard Industrial Classification (SIC) code by using the Standard Industrial Classification Manual, Executive Office of the President, Office of Management and Budget. You may contact your nearest OSHA office or State agency for help in determining your SIC.

[66 FR 5915, Jan. 19, 2001]

### §1904.3 Keeping records for more than one agency.

If you create records to comply with another government agency's injury and illness recordkeeping requirements, OSHA will consider those records as meeting OSHA's Part 1904 recordkeeping requirements if OSHA accepts the other agency's records under a memorandum of understanding with that agency, or if the other agency's records contain the same information as this Part 1904 requires you to record. You may contact your nearest OSHA office or State agency for help in determining whether your records meet OSHA's requirements.

[66 FR 5915, Jan. 19, 2001]

## Non-Mandatory Appendix A to Subpart B—Partially Exempt Industries

Employers are not required to keep OSHA injury and illness records for any establishment classified in the following Standard Industrial Classification (SIC) codes, unless they are asked in writing to do so by OSHA, the Bureau of Labor Statistics ( BLS), or a state agency operating under the authority of OSHA or the BLS. All employers, including those partially exempted by reason of company size or industry classification, must report to OSHA any workplace incident that results in a fatality or the hospitalization of three or more employees (see §1904.39).

| SIC code | Industry description |
|---|---|
| 525 | Hardware Stores |
| 542 | Meat and Fish Markets |
| 544 | Candy, Nut, and Confectionery Stores |
| 545 | Dairy Products Stores |
| 546 | Retail Bakeries |
| 549 | Miscellaneous Food Stores |
| 551 | New and Used Car Dealers |
| 552 | Used Car Dealers |
| 554 | Gasoline Service Stations |
| 557 | Motorcycle Dealers |
| 56 | Apparel and Accessory Stores |
| 573 | Radio, Television, & Computer Stores |
| 58 | Eating and Drinking Places |
| 591 | Drug Stores and Proprietary Stores |
| 592 | Liquor Stores |
| 594 | Miscellaneous Shopping Goods Stores |
| 599 | Retail Stores, Not Elsewhere Classified |
| 60 | Depository Institutions (banks & savings institutions) |
| 61 | Nondepository |
| 62 | Security and Commodity Brokers |
| 63 | Insurance Carriers |
| 64 | Insurance Agents, Brokers & Services |
| 653 | Real Estate Agents and Managers |
| 654 | Title Abstract Offices |
| 67 | Holding and Other Investment Offices |
| 722 | Photographic Studios, Portrait |
| 723 | Beauty Shops |
| 724 | Barber Shops |
| 725 | Shoe Repair and Shoeshine Parlors. |

| SIC code | Industry description |
|---|---|
| 726 | Funeral Service and Crematories. |
| 729 | Miscellaneous Personal Services. |
| 731 | Advertising Services. |
| 732 | Credit Reporting and Collection Services. |
| 733 | Mailing, Reproduction, & Stenographic Services. |
| 737 | Computer and Data Processing Services. |
| 738 | Miscellaneous Business Services. |
| 764 | Reupholstery and Furniture Repair. |
| 78 | Motion Picture. |
| 791 | Dance Studios, Schools, and Halls. |
| 792 | Producers, Orchestras, Entertainers. |
| 793 | Bowling Centers. |
| 801 | Offices & Clinics Of Medical Doctors. |
| 802 | Offices and Clinics Of Dentists. |
| 803 | Offices Of Osteopathic. |
| 804 | Offices Of Other Health Practitioners. |
| 807 | Medical and Dental Laboratories. |
| 809 | Health and Allied Services, Not Elsewhere Classified. |
| 81 | Legal Services. |
| 82 | Educational Services (schools, colleges, universities and libraries). |
| 832 | Individual and Family Services. |
| 835 | Child Day Care Services. |
| 839 | Social Services, Not Elsewhere Classified. |
| 841 | Museums and Art Galleries. |
| 86 | Membership Organizations. |
| 87 | Engineering, Accounting, Research, Management, and Related Services. |
| 899 | Services, not elsewhere classified. |

## Subpart C—Recordkeeping Forms and Recording Criteria

**Note to Subpart C:** This Subpart describes the work-related injuries and illnesses that an employer must enter into the OSHA records and explains the OSHA forms that employers must use to record work-related fatalities, injuries, and illnesses.

### §1904.4 Recording criteria.

**1904.4(a)** ***Basic requirement.*** Each employer required by this Part to keep records of fatalities, injuries, and illnesses must record each fatality, injury and illness that:

**1904.4(a)(1)** Is work-related; and

**1904.4(a)(2)** Is a new case; and

**1904.4(a)(3)** Meets one or more of the general recording criteria of §1904.7 or the application to specific cases of §1904.8 through §1904.12.

**1904.4(b)** ***Implementation.*** (1) *What sections of this rule describe recording criteria for recording work-related injuries and illnesses?* The table below indicates which sections of the rule address each topic.

**1904.4(b)(1)(i)** Determination of work-relatedness. See §1904.5.

**1904.4(b)(1)(ii)** Determination of a new case. See §1904.6.

**1904.4(b)(1)(iii)** General recording criteria. See §1904.7.

**1904.4(b)(1)(iv)** Additional criteria. (Needlestick and sharps injury cases, tuberculosis cases, hearing loss cases, medical removal cases, and musculoskeletal disorder cases). See §1904.8 through §1904.12.

**1904.4(b)(2)** *How do I decide whether a particular injury or illness is recordable?* The decision tree for recording work-related injuries and illnesses below shows the steps involved in making this determination.

[66 FR 5915, Jan. 19, 2001]

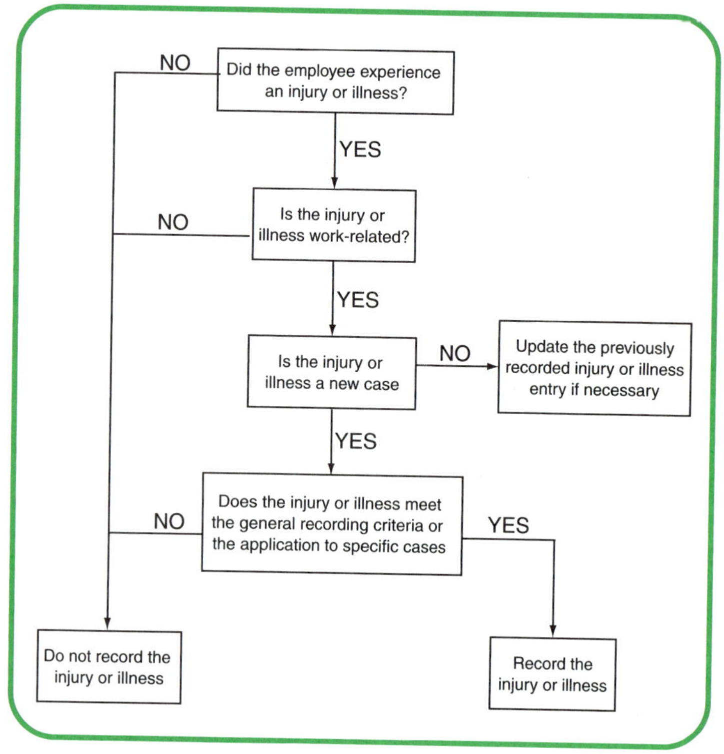


### §1904.5 Determination of work-relatedness.

**1904.5(a)** ***Basic requirement.*** You must consider an injury or illness to be work-related if an event or exposure in the work environment either caused or contributed to the resulting condition or significantly aggravated a pre-existing injury or illness. Work-relatedness is presumed for injuries and illnesses resulting from events or exposures occurring in the work environment, unless an exception in §1904.5(b)(2) specifically applies.

**1904.5(b)** ***Implementation.*** (1) *What is the "work environment"?* OSHA defines the work environment as "the establishment and other locations where one or more employees are working or are present as a condition of their employment. The work environment includes not only physical locations, but also the equipment or materials used by the employee during the course of his or her work."

**1904.5(b)(2)** *Are there situations where an injury or illness occurs in the work environment and is not considered work-related?* Yes, an injury or illness occurring in the work environment that falls under one of the following exceptions is not work-related, and therefore is not recordable.

| 1904.5(b)(2) | You are not required to record injuries and illnesses if . . . |
|---|---|
| (i) ............. | At the time of the injury or illness, the employee was present in the work environment as a member of the general public rather than as an employee. |
| (ii) ............ | The injury or illness involves signs or symptoms that surface at work but result solely from a non-work-related event or exposure that occurs outside the work environment. |
| (iii) ........... | The injury or illness results solely from voluntary participation in a wellness program or in a medical, fitness, or recreational activity such as blood donation, physical examination, flu shot, exercise class, racquetball, or baseball. |
| (iv) ........... | The injury or illness is solely the result of an employee eating, drinking, or preparing food or drink for personal consumption (whether boughton the employer's premises or brought in). For example, if the employee is injured by choking on a sandwich while in the employer's establishment, the case would not be considered work-related. **Note:** If the employee is made ill by ingesting food contaminated by workplace contaminants (such as lead), or gets food poisoning from food supplied by the employer, the case would be considered work-related. |
| (v) ............ | The injury or illness is solely the result of an employee doing personal tasks (unrelated to their employment) at the establishment outside of the employee's assigned working hours. |
| (vi) ........... | The injury or illness is solely the result of personal grooming, self medication for a non work-related condition, or is intentionally self inflicted. |
| (vii) .......... | The injury or illness is caused by a motor vehicle accident and occurs on a company parking lot or company access road while the employee is commuting to or from work. |
| (viii) ......... | The illness is the common cold or flu (Note: contagious diseases such as tuberculosis, brucellosis, hepatitis A, or plague are considered work-related if the employee isinfected at work). |
| (ix) ........... | The illness is a mental illness. Mental illness will not be considered work-related unless the employee voluntarily provides the employer with an opinion from a physician or other licensed health care professional with appropriate training and experience (psychiatrist, psychologist, psychiatric nurse practitioner, etc.) stating that the employee has a mental illness that is work-related. |

**1904.5(b)(3)** *How do I handle a case if it is not obvious whether the precipitating event or exposure occurred in the work environment or occurred away from work?* In these situations, you must evaluate the employee's work duties and environment to decide whether or not one or more events or exposures in the work environment either caused or contributed to the resulting condition or significantly aggravated a pre-existing condition.

**1904.5(b)(4)** *How do I know if an event or exposure in the work environment "significantly aggravated" a preexisting injury or illness?* A preexisting injury or illness has been significantly aggravated, for purposes of OSHA injury and illness recordkeeping, when an event or exposure in the work environment results in any of the following:

**1904.5(b)(4)(i)** Death, provided that the preexisting injury or illness would likely not have resulted in death but for the occupational event or exposure.

**1904.5(b)(4)(ii)** Loss of consciousness, provided that the preexisting injury or illness would likely not have resulted in loss of consciousness but for the occupational event or exposure.

**1904.5(b)(4)(iii)** One or more days away from work, or days of restricted work, or days of job transfer that otherwise would not have occurred but for the occupational event or exposure.

**1904.5(b)(4)(iv)** Medical treatment in a case where no medical treatment was needed for the injury or illness before the workplace event or exposure, or a change in medical treatment was necessitated by the workplace event or exposure.

**1904.5(b)(5)** *Which injuries and illnesses are considered pre-existing conditions?* An injury or illness is a preexisting condition if it resulted solely from a non-work-related event or exposure that occured outside the work environment.

**1904.5(b)(6)** *How do I decide whether an injury or illness is work-related if the employee is on travel status at the time the injury or illness occurs?* Injuries and illnesses that occur while an employee is on travel status are work-related if, at the time of the injury or illness, the employee was engaged in work activities "in the interest of the employer." Examples of such activities include travel to and from customer contacts, conducting job tasks, and entertaining or being entertained to transact, discuss, or promote business (work-related entertainment includes only entertainment activities being engaged in at the direction of the employer).

Injuries or illnesses that occur when the employee is on travel status do not have to be recorded if they meet one of the exceptions listed below.

| 1904.5 (b)(6) | If the employee has . . . | You may use the following to determine if an injury or illness is work related |
|---|---|---|
| (i) | checked into a hotel or motel for one or more days. | When a traveling employee checks into a hotel, motel, or into a other temporary residence, he or she establishes a "home away from home." You must evaluate the employee's activities after he or she checks into the hotel, motel, or other temporary residence for their work-relatedness in the same manner as you evaluate the activities of a non traveling employee. When the employee checks into the temporary residence, he or she is considered to have left the work environment. When the employee begins work each day, he or she re-enters the work environment. If the employee has established a "home away from home" and is reporting to a fixed worksite each day, you also do not consider injuries or illnesses work-related if they occur while the employee is commuting between the temporary residence and the job location. |
| (ii) | taken a detour for personal reasons. | Injuries or illnesses are not considered work-related if they occur while the employee is on a personal detour from a reasonably direct route of travel (e.g., has taken a side trip for personal reasons). |

**1904.5(b)(7)** *How do I decide if a case is work-related when the employee is working at home?* Injuries and illnesses that occur while an employee is working at home, including work in a home office, will be considered work-related if the injury or illness occurs while the employee is performing work for pay or compensation in the home, and the injury or illness is directly related to the performance of work rather than to the general home environment or setting. For example, if an employee drops a box of work documents and injures his or her foot, the case is considered work-related. If an employee's fingernail is punctured by a needle from a sewing machine used to perform garment work at home, becomes infected and requires medical treatment, the injury is considered work-related. If an employee is injured because he or she trips on the family dog while rushing to answer a work phone call, the case is not considered work-related. If an employee working at home is electrocuted because of faulty home wiring, the injury is not considered work-related.

[66 FR 5915, Jan. 19, 2001]

## §1904.6 Determination of new cases.

**1904.6(a)** ***Basic requirement.*** You must consider an injury or illness to be a "new case" if:

**1904.6(a)(1)** The employee has not previously experienced a recorded injury or illness of the same type that affects the same part of the body, or

**1904.6(a)(2)** The employee previously experienced a recorded injury or illness of the same type that affected the same part of the body but had recovered completely (all signs and symptoms had disappeared) from the previous injury or illness and an event or exposure in the work environment caused the signs or symptoms to reappear.

**1904.6(b)** ***Implementation.*** (1) *When an employee experiences the signs or symptoms of a chronic work-related illness, do I need to consider each recurrence of signs or symptoms to be a new case?* No, for occupational illnesses where the signs or symptoms may recur or continue in the absence of an exposure in the workplace, the case must only be recorded once. Examples may include occupational cancer, asbestosis, byssinosis and silicosis.

**1904.6(b)(2)** *When an employee experiences the signs or symptoms of an injury or illness as a result of an event or exposure in the workplace, such as an episode of occupational asthma, must I treat the episode as a new case?* Yes, because the episode or recurrence was caused by an event or exposure in the workplace, the incident must be treated as a new case.

**1904.6(b)(3)** *May I rely on a physician or other licensed health care professional to determine whether a case is a new case or a recurrence of an old case?* You are not required to seek the advice of a physician or other licensed health care professional. However, if you do seek such advice, you must follow the physician or other licensed health care professional's recommendation about whether the case is a new case or a recurrence. If you receive recommendations from two or more physicians or other licensed health care professionals, you must make a decision as to which recommendation is the most authoritative (best documented, best reasoned, or most authoritative), and record the case based upon that recommendation.

[66 FR 5915, Jan. 19, 2001]

## §1904.7 General recording criteria.

**1904.7(a)** ***Basic requirement.*** You must consider an injury or illness to meet the general recording criteria, and therefore to be recordable, if it results in any of the following: death, days away from work, restricted work or transfer to another job, medical treatment beyond first aid, or loss of consciousness. You must also consider a case to meet the general recording criteria if it involves a significant injury or illness diagnosed by a physician or other licensed health care professional, even if it does not result in death, days away from work, restricted work or job transfer, medical treatment beyond first aid, or loss of consciousness.

**1904.7(b)** ***Implementation.*** (1) *How do I decide if a case meets one or more of the general recording criteria?* A work-related injury or illness must be recorded if it results in one or more of the following:

**1904.7(b)(1)(i)** Death. See §1904.7(b)(2).

**1904.7(b)(1)(ii)** Days away from work. See §1904.7(b)(3).

**1904.7(b)(1)(iii)** Restricted work or transfer to another job. See §1904.7(b)(4).

**1904.7(b)(1)(iv)** Medical treatment beyond first aid. See §1904.7(b)(5).

**1904.7(b)(1)(v)** Loss of consciousness. See §1904.7(b)(6).

**1904.7(b)(1)(vi)** A significant injury or illness diagnosed by a physician or other licensed health care professional. See §1904.7(b)(7).

**1904.7(b)(2)** *How do I record a work-related injury or illness that results in the employee's death?* You must record an injury or illness that results in death by entering a check mark on the OSHA 300 Log in the space for cases resulting in death. You must also report any work-related fatality to OSHA within eight (8) hours, as required by §1904.39.

**1904.7(b)(3)** *How do I record a work-related injury or illness that results in days away from work?* When an injury or illness involves one or more days away from work, you must record the injury or illness on the OSHA 300 Log with a check mark in the space for cases involving days away and an entry of the number of calendar days away from work in the number of days column. If the employee is out for an extended period of time, you must enter an estimate of the days that the employee will be away, and update the day count when the actual number of days is known.

**1904.7(b)(3)(i)** *Do I count the day on which the injury occurred or the illness began?* No, you begin counting days away on the day after the injury occurred or the illness began.

**1904.7(b)(3)(ii)** *How do I record an injury or illness when a physician or other licensed health care professional recommends that the worker stay at home but the employee comes to work anyway?* You must record these injuries and illnesses on the OSHA 300 Log using the check box for cases with days away from work and enter the number of calendar days away recommended by the physician or other licensed health care professional. If a physician or other licensed health care professional recommends days away, you should encourage your employee to follow that recommendation. However, the days away must be recorded whether the injured or ill employee follows the physician or licensed health care professional's recommendation or not. If you receive recommendations from two or more physicians or other licensed health care professionals, you may make a decision as to which recommendation is the most authoritative, and record the case based upon that recommendation.

**1904.7(b)(3)(iii)** *How do I handle a case when a physician or other licensed health care professional recommends that the worker return to work but the employee stays at home anyway?* In this situation, you must end the count of days away from work on the date the physician or other licensed health care professional recommends that the employee return to work.

**1904.7(b)(3)(iv)** *How do I count weekends, holidays, or other days the employee would not have worked anyway?* You must count the number of calendar days the employee was unable to work as a result of the injury or illness, regardless of whether or not the employee was scheduled to work on those day(s). Weekend days, holidays, vacation days or other days off are included in the total number of days recorded if the employee would not have been able to work on those days because of a work-related injury or illness.

**1904.7(b)(3)(v)** *How do I record a case in which a worker is injured or becomes ill on a Friday and reports to work on a Monday, and was not scheduled to work on the weekend?* You need to record this case only if you receive information from a physician or other licensed health care professional indicating that the employee should not have worked, or should have performed only restricted work, during the weekend. If so, you must record the injury or illness as a case with days away from work or restricted work, and enter the day counts, as appropriate.

**1904.7(b)(3)(vi)** *How do I record a case in which a worker is injured or becomes ill on the day before scheduled time off such as a holiday, a planned vacation, or a temporary plant closing?* You need to record a case of this type only if you receive information from a physician or other licensed health care professional indicating that the employee should not have worked, or should have performed only restricted work, during the scheduled time off. If so, you must record the injury or illness as a case with days away from work or restricted work, and enter the day counts, as appropriate.

**1904.7(b)(3)(vii)** *Is there a limit to the number of days away from work I must count?* Yes, you may "cap" the total days away at 180 calendar days. You are not required to keep track of the number of calendar days away from work if the injury or illness resulted in more than 180 calendar days away from work and/or days of job transfer or restriction. In such a case, entering 180 in the total days away column will be considered adequate.

**1904.7(b)(3)(viii)** *May I stop counting days if an employee who is away from work because of an injury or illness retires or leaves my company?* Yes, if the employee leaves your company for some reason unrelated to the injury or illness, such as retirement, a plant closing, or to take another job, you may stop counting days away from work or days of restriction/job transfer. If the employee leaves your company because of the injury or illness, you must estimate the total number of days away or days of restriction/job transfer and enter the day count on the 300 Log.

**1904.7(b)(3)(ix)** *If a case occurs in one year but results in days away during the next calendar year, do I record the case in both years?* No, you only record the injury or illness once. You must enter the number of calendar days away for the injury or illness on the OSHA 300 Log for the year in which the injury or illness occurred. If the employee is still away from work because of the injury or illness when you prepare the annual summary, estimate the total number of calendar days you expect the employee to be away from work, use this number to calculate the total for the annual summary, and then update the initial log entry later when the day count is known or reaches the 180-day cap.

**1904.7(b)(4)** *How do I record a work-related injury or illness that results in restricted work or job transfer?* When an injury or illness involves restricted work or job transfer but does not involve death or days away from work, you must record the injury or illness on the OSHA 300 Log by placing a check mark in the space for job transfer or restriction and an entry of the number of restricted or transferred days in the restricted workdays column.

**1904.7(b)(4)(i)** *How do I decide if the injury or illness resulted in restricted work?* Restricted work occurs when, as the result of a work-related injury or illness:

**1904.7(b)(4)(i)(A)** You keep the employee from performing one or more of the routine functions of his or her job, or from working the full workday that he or she would otherwise have been scheduled to work; or

**1904.7(b)(4)(i)(B)** A physician or other licensed health care professional recommends that the employee not perform one or more of the routine functions of his or her job, or not work the full workday that he or she would otherwise have been scheduled to work.

**1904.7(b)(4)(ii)** *What is meant by "routine functions"?* For recordkeeping purposes, an employee's routine functions are those work activities the employee regularly performs at least once per week.

**1904.7(b)(4)(iii)** *Do I have to record restricted work or job transfer if it applies only to the day on which the injury occurred or the illness began?* No, you do not have to record restricted work or job transfers if you, or the physician or other licensed health care professional, impose the restriction or transfer only for the day on which the injury occurred or the illness began.

**1904.7(b)(4)(iv)** *If you or a physician or other licensed health care professional recommends a work restriction, is the injury or illness automatically recordable as a "restricted work" case?* No, a recommended work restriction is recordable only if it affects one or more of the employee's routine job functions. To determine whether this is the case, you must evaluate the restriction in light of the routine functions of the injured or ill employee's job. If the restriction from you or the physician or other licensed health care professional keeps the employee from performing one or more of his or her routine job functions, or from working the full workday the injured or ill employee would otherwise have worked, the employee's work has been restricted and you must record the case.

**1904.7(b)(4)(v)** *How do I record a case where the worker works only for a partial work shift because of a work-related injury or illness?* A partial day of work is recorded as a day of job transfer or restriction for recordkeeping purposes, except for the day on which the injury occurred or the illness began.

**1904.7(b)(4)(vi)** *If the injured or ill worker produces fewer goods or services than he or she would have produced prior to the injury or illness but otherwise performs all of the routine functions of his or her work, is the case considered a restricted work case?* No, the case is considered restricted work only if the worker does not perform all of the routine functions of his or her job or does not work the full shift that he or she would otherwise have worked.

**1904.7(b)(4)(vii)** *How do I handle vague restrictions from a physician or other licensed health care professional, such as that the employee engage only in "light duty" or "take it easy for a week"?* If you are not clear about the physician or other licensed health care professional's recommendation, you may ask that person whether the employee can do all of his or her routine job functions and work all of his or her normally assigned work shift. If the answer to both of these questions is "Yes," then the case does not involve a work restriction and does not have to be recorded as such. If the answer to one or both of these questions is "No," the case involves restricted work and must be recorded as a restricted work case. If you are unable to obtain this additional information from the physician or other licensed health care professional who recommended the restriction, record the injury or illness as a case involving restricted work.

**1904.7(b)(4)(viii)** *What do I do if a physician or other licensed health care professional recommends a job restriction meeting OSHA's definition, but the employee does all of his or her routine job functions anyway?* You must record the injury or illness on the OSHA 300 Log as a restricted work case. If a physician or other licensed health care professional recommends a job restriction, you should ensure that the employee complies with that restriction. If you receive recommendations from two or more physicians or other licensed health care professionals, you may make a decision as to which recommendation is the most authoritative, and record the case based upon that recommendation.

**1904.7(b)(4)(ix)** *How do I decide if an injury or illness involved a transfer to another job?* If you assign an injured or ill employee to a job other than his or her regular job for part of the day, the case involves transfer to another job. Note: This does not include the day on which the injury or illness occurred.

**1904.7(b)(4)(x)** *Are transfers to another job recorded in the same way as restricted work cases?* Yes, both job transfer and restricted work cases are recorded in the same box on the OSHA 300 Log. For example, if you assign, or a physician or other licensed health care professional recommends that you assign, an injured or ill worker to his or her routine job duties for part of the day and to another job for the rest of the day, the injury or illness involves a job transfer. You must record an injury or illness that involves a job transfer by placing a check in the box for job transfer.

**1904.7(b)(4)(xi)** *How do I count days of job transfer or restriction?* You count days of job transfer or restriction in the same way you count days away from work, using §1904.7(b)(3)(i) to (viii), above. The only difference is that, if you permanently assign the injured or ill employee to a job that has been modified or permanently changed in a manner that eliminates the routine functions the employee was restricted from performing, you may stop the day count when the modification or change is made permanent. You must count at least one day of restricted work or job transfer for such cases.

**1904.7(b)(5)** *How do I record an injury or illness that involves medical treatment beyond first aid?* If a work-related injury or illness results in medical treatment beyond first aid, you must record it on the OSHA 300 Log. If the injury or illness did not involve death, one or more days away from work, one or more days of restricted work, or one or more days of job transfer, you enter a check mark in the box for cases where the employee received medical treatment but remained at work and was not transferred or restricted.

**1904.7(b)(5)(i)** *What is the definition of medical treatment?* "Medical treatment" means the management and care of a patient to combat disease or disorder. For the purposes of Part 1904, medical treatment does not include:

**1904.7(b)(5)(i)(A)** Visits to a physician or other licensed health care professional solely for observation or counseling;

**1904.7(b)(5)(i)(B)** The conduct of diagnostic procedures, such as x-rays and blood tests, including the administration of prescription medications used solely for diagnostic purposes (e.g., eye drops to dilate pupils); or

**1904.7(b)(5)(i)(C)** "First aid" as defined in paragraph (b)(5)(ii) of this section.

**1904.7(b)(5)(ii)** *What is "first aid"?* For the purposes of Part 1904, "first aid" means the following:

**1904.7(b)(5)(ii)(A)** Using a non-prescription medication at nonprescription strength (for medications available in both prescription and non-prescription form, a recommendation by a physician or other licensed health care professional to use a non-prescription medication at prescription strength is considered medical treatment for recordkeeping purposes);

**1904.7(b)(5)(ii)(B)** Administering tetanus immunizations (other immunizations, such as Hepatitis B vaccine or rabies vaccine, are considered medical treatment);

**1904.7(b)(5)(ii)(C)** Cleaning, flushing or soaking wounds on the surface of the skin;

**1904.7(b)(5)(ii)(D)** Using wound coverings such as bandages, Band-Aids™, gauze pads, etc.; or using butterfly bandages or Steri-Strips™ (other wound closing devices such as sutures, staples, etc., are considered medical treatment);

**1904.7(b)(5)(ii)(E)** Using hot or cold therapy;

**1904.7(b)(5)(ii)(F)** Using any non-rigid means of support, such as elastic bandages, wraps, non-rigid back belts, etc. (devices with rigid stays or other systems designed to immobilize parts of the body are considered medical treatment for recordkeeping purposes);

**1904.7(b)(5)(ii)(G)** Using temporary immobilization devices while transporting an accident victim (e.g., splints, slings, neck collars, back boards, etc.).

**1904.7(b)(5)(ii)(H)** Drilling of a fingernail or toenail to relieve pressure, or draining fluid from a blister;

**1904.7(b)(5)(ii)(I)** Using eye patches;

**1904.7(b)(5)(ii)(J)** Removing foreign bodies from the eye using only irrigation or a cotton swab;

**1904.7(b)(5)(ii)(K)** Removing splinters or foreign material from areas other than the eye by irrigation, tweezers, cotton swabs or other simple means;

**1904.7(b)(5)(ii)(L)** Using finger guards;

**1904.7(b)(5)(ii)(M)** Using massages (physical therapy or chiropractic treatment are considered medical treatment for recordkeeping purposes); or

**1904.7(b)(5)(ii)(N)** Drinking fluids for relief of heat stress.

**1904.7(b)(5)(iii)** *Are any other procedures included in first aid?* No, this is a complete list of all treatments considered first aid for Part 1904 purposes.

**1904.7(b)(5)(iv)** *Does the professional status of the person providing the treatment have any effect on what is considered first aid or medical treatment?* No, OSHA considers the treatments listed in §1904.7(b)(5)(ii) of this Part to be first aid regardless of the professional status of the person providing the treatment. Even when these treatments are provided by a physician or other licensed health care professional, they are considered first aid for the purposes of Part 1904. Similarly, OSHA considers treatment beyond first aid to be medical treatment even when it is provided by someone other than a physician or other licensed health care professional.

**1904.7(b)(5)(v)** *What if a physician or other licensed health care professional recommends medical treatment but the employee does not follow the recommendation?* If a physician or other licensed health care professional recommends medical treatment, you should encourage the injured or ill employee to follow that recommendation. However, you must record the case even if the injured or ill employee does not follow the physician or other licensed health care professional's recommendation.

**1904.7(b)(6)** *Is every work-related injury or illness case involving a loss of consciousness recordable?* Yes, you must record a work-related injury or illness if the worker becomes unconscious, regardless of the length of time the employee remains unconscious.

**1904.7(b)(7)** *What is a "significant" diagnosed injury or illness that is recordable under the general criteria even if it does not result in death, days away from work, restricted work or job transfer, medical treatment beyond first aid, or loss of consciousness?* Work-related cases involving cancer, chronic irreversible disease, a fractured or cracked bone, or a punctured eardrum must always be recorded under the general criteria at the time of diagnosis by a physician or other licensed health care professional.

**Note to §1904.7:** OSHA believes that most significant injuries and illnesses will result in one of the criteria listed in §1904.7(a): death, days away from work, restricted work or job transfer, medical treatment beyond first aid, or loss of consciousness. However, there are some significant injuries, such as a punctured eardrum or a fractured toe or rib, for which neither medical treatment nor work restrictions may be recommended. In addition, there are some significant progressive diseases, such as byssinosis, silicosis, and some types of cancer, for which medical treatment or work restrictions may not be recommended at the time of diagnosis but are likely to be recommended as the disease progresses. OSHA believes that cancer, chronic irreversible diseases, fractured or cracked bones, and punctured eardrums are generally considered significant injuries and illnesses, and must be recorded at the initial diagnosis even if medical treatment or work restrictions are not recommended, or are postponed, in a particular case.

[66 FR 5915, Jan. 19, 2001]

## §1904.8 Recording criteria for needlestick and sharps injuries.

**1904.8(a)** ***Basic requirement.*** You must record all work-related needlestick injuries and cuts from sharp objects that are contaminated with another person's blood or other potentially infectious material (as defined by 29 CFR 1910.1030). You must enter the case on the OSHA 300 Log as an injury. To protect the employee's privacy, you may not enter the employee's name on the OSHA 300 Log (see the requirements for privacy cases in paragraphs 1904.29(b)(6) through 1904.29(b)(9)).

**1904.8(b)** ***Implementation.*** (1) *What does "other potentially infectious material" mean?* The term "other potentially infectious materials" is defined in the OSHA Bloodborne Pathogens standard at §1910.1030(b). These materials include:

**1904.8(b)(1)(i)** Human bodily fluids, tissues and organs, and

**1904.8(b)(1)(ii)** Other materials infected with the HIV or hepatitis B (HBV) virus such as laboratory cultures or tissues from experimental animals.

**1904.8(b)(2)** *Does this mean that I must record all cuts, lacerations, punctures, and scratches?* No, you need to record cuts, lacerations, punctures, and scratches only if they are work-related and involve contamination with another person's blood or other potentially infectious material. If the cut, laceration, or scratch involves a clean object, or a contaminant other than blood or other potentially infectious material, you need to record the case only if it meets one or more of the recording criteria in §1904.7.

**1904.8(b)(3)** *If I record an injury and the employee is later diagnosed with an infectious bloodborne disease, do I need to update the OSHA 300 Log?* Yes, you must update the classification of the case on the OSHA 300 Log if the case results in death, days away from work, restricted work, or job transfer. You must also update the description to identify the infectious disease and change the classification of the case from an injury to an illness.

**1904.8(b)(4)** *What if one of my employees is splashed or exposed to blood or other potentially infectious material without being cut or scratched?* Do I need to record this incident? You need to record such an incident on the OSHA 300 Log as an illness if:

**1904.8(b)(4)(i)** It results in the diagnosis of a bloodborne illness, such as HIV, hepatitis B, or hepatitis C; or

**1904.8(b)(4)(ii)** It meets one or more of the recording criteria in §1904.7.

[66 FR 5915, Jan. 19, 2001]

## §1904.9 Recording criteria for cases involving medical removal under OSHA standards.

**1904.9(a)** ***Basic requirement.*** If an employee is medically removed under the medical surveillance requirements of an OSHA standard, you must record the case on the OSHA 300 Log.

**1904.9(b)** ***Implementation.*** (1) *How do I classify medical removal cases on the OSHA 300 Log?* You must enter each medical removal case on the OSHA 300 Log as either a case involving days away from work or a case involving restricted work activity, depending on how you decide to comply with the medical removal requirement. If the medical removal is the result of a chemical exposure, you must enter the case on the OSHA 300 Log by checking the "poisoning" column.

**1904.9(b)(2)** *Do all of OSHA's standards have medical removal provisions?* No, some OSHA standards, such as the standards covering bloodborne pathogens and noise, do not have medical removal provisions. Many OSHA standards that cover specific chemical substances have medical removal provisions. These standards include, but are not limited to, lead, cadmium, methylene chloride, formaldehyde, and benzene.

**1904.9(b)(3)** *Do I have to record a case where I voluntarily removed the employee from exposure before the medical removal criteria in an OSHA standard are met?* No, if the case involves voluntary medical removal before the medical removal levels required by an OSHA standard, you do not need to record the case on the OSHA 300 Log.

[66 FR 5915, Jan. 19, 2001]

## §1904.10 Recording criteria for cases involving occupational hearing loss.

**1904.10(a)** ***Basic requirement.*** If an employee's hearing test (audiogram) reveals that a Standard Threshold Shift (STS) has occurred, you must record the case on the OSHA 300 Log by checking the "hearing loss" column.

**1904.10(b)** ***Implementation.*** (1) *What is a Standard Threshold Shift?* A Standard Threshold Shift, or STS, is defined in the occupational noise exposure standard at 29 CFR 1910.95(c)(10)(i) as a change in hearing threshold, relative to the most recent audiogram for that employee, of an average of 10 decibels (dB) or more at 2000, 3000, and 4000 hertz in one or both ears.

**1904.10(b)(2)** *How do I determine whether an STS has occurred?* If the employee has never previously experienced a recordable hearing loss, you must compare the employee's current audiogram with that employee's baseline audiogram. If the employee has previously experienced a recordable hearing loss, you must compare the employee's current audiogram with the employee's revised baseline audiogram (the audiogram reflecting the employee's previous recordable hearing loss case).

**1904.10(b)(3)** *May I adjust the audiogram results to reflect the effects of aging on hearing?* Yes, when comparing audiogram results, you may adjust the results for the employee's age when the audiogram was taken using Tables F-1 or F-2, as appropriate, in Appendix F of 29 CFR 1910.95.

**1904.10(b)(4)** *Do I have to record the hearing loss if I am going to retest the employee's hearing?* No, if you retest the employee's hearing within 30 days of the first test, and the retest does not confirm the STS, you are not required to record the hearing loss case on the OSHA 300 Log. If the retest confirms the STS, you must record the hearing loss illness within seven (7) calendar days of the retest.

**1904.10(b)(5)** *Are there any special rules for determining whether a hearing loss case is work-related?* Yes, hearing loss is presumed to be work-related if the employee is exposed to noise in the workplace at an 8-hour time-weighted average of 85 dBA or greater, or to a total noise dose of 50 percent, as defined in 29 CFR 1910.95. For hearing loss cases where the employee is not exposed to this level of noise, you must use the rules in §1904.5 to determine if the hearing loss is work-related.

**1904.10(b)(6)** *If a physician or other licensed health care professional determines the hearing loss is not work-related, do I still need to record the case?* If a physician or other licensed health care professional determines that the hearing loss is not work-related or has not been significantly aggravated by occupational noise exposure, you are not required to consider the case work-related or to record the case on the OSHA 300 Log.

**1904.10(c)** *Recording criteria for calendar year 2002.* From January 1, 2002 until December 31, 2002, you are required to record a work-related hearing loss averaging 25dB or more at 2000, 3000, and 4000 hertz in either ear on the OSHA 300 Log. You must use the employee's original baseline audiogram for comparison. You may make a correction for presbycusis (aging) by using the tables in Appendix F of 29 CFR 1910.95. The requirement of §1904.37(b)(1) that States with OSHA-approved state plans must have the same requirements for determining which injuries and illnesses are recordable and how they are recorded shall not preclude the states from retaining their existing criteria with regard to this section during calendar year 2002.

**Note to §1904.10:** Paragraphs (a) and (b) of this section are effective on January 1, 2003. Paragraph (c) of this section applies from January 1, 2002 until December 31, 2002.

[66 FR 5915, Jan. 19, 2001]

### §1904.11 Recording criteria for work-related tuberculosis cases.

**1904.11(a)** ***Basic requirement.*** If any of your employees has been occupationally exposed to anyone with a known case of active tuberculosis (TB), and that employee subsequently develops a tuberculosis infection, as evidenced by a positive skin test or diagnosis by a physician or other licensed health care professional, you must record the case on the OSHA 300 Log by checking the "respiratory condition" column.

**1904.11(b)** ***Implementation.*** (1) *Do I have to record, on the Log, a positive TB skin test result obtained at a pre-employment physical?* No, you do not have to record it because the employee was not occupationally exposed to a known case of active tuberculosis in your workplace.

**1904.11(b)(2)** *May I line-out or erase a recorded TB case if I obtain evidence that the case was not caused by occupational exposure?* Yes, you may line-out or erase the case from the Log under the following circumstances:

**1904.11(b)(2)(i)** The worker is living in a household with a person who has been diagnosed with active TB;

**1904.11(b)(2)(ii)** The Public Health Department has identified the worker as a contact of an individual with a case of active TB unrelated to the workplace; or

**1904.11(b)(2)(iii)** A medical investigation shows that the employee's infection was caused by exposure to TB away from work, or proves that the case was not related to the workplace TB exposure.

[66 FR 5915, Jan. 19, 2001]

### §1904.12 Recording criteria for cases involving work-related musculoskeletal disorders.

**1904.12(a)** ***Basic requirement.*** If any of your employees experiences a recordable work-related musculoskeletal disorder (MSD), you must record it on the OSHA 300 Log by checking the "musculoskeletal disorder" column.

**1904.12(b)** ***Implementation.*** (1) *What is a "musculoskeletal disorder" or MSD?* Musculoskeletal disorders (MSDs) are disorders of the muscles, nerves, tendons, ligaments, joints, cartilage and spinal discs. MSDs do not include disorders caused by slips, trips, falls, motor vehicle accidents, or other similar accidents. Examples of MSDs include: Carpal tunnel syndrome, Rotator cuff syndrome, De Quervain's disease, Trigger finger, Tarsal tunnel syndrome, Sciatica, Epicondylitis, Tendinitis, Raynaud's phenomenon, Carpet layers knee, Herniated spinal disc, and Low back pain.

**1904.12(b)(2)** *How do I decide which musculoskeletal disorders to record?* There are no special criteria for determining which musculoskeletal disorders to record. An MSD case is recorded using the same process you would use for any other injury or illness. If a musculoskeletal disorder is work-related, and is a new case, and meets one or more of the general recording criteria, you must record the musculoskeletal disorder. The following table will guide you to the appropriate section of the rule for guidance on recording MSD cases.

**1904.12(b)(2)(i)** Determining if the MSD is work-related. See §1904.5.

**1904.12(b)(2)(ii)** Determining if the MSD is a new case. See §1904.6.

**1904.12(b)(2)(iii)** Determining if the MSD meets one or more of the general recording criteria:

**1904.12(b)(2)(iii)(A)** Days away from work, see §1904.7(b)(3).

**1904.12(b)(2)(iii)(B)** Restricted work or transfer to another job, or see §1904.7(b)(4).

**1904.12(b)(2)(iii)(C)** Medical treatment beyond first aid. See §1904.7(b)(5).

**1904.12(b)(3)** *If a work-related MSD case involves only subjective symptoms like pain or tingling, do I have to record it as a musculoskeletal disorder?* The symptoms of an MSD are treated the same as symptoms for any other injury or illness. If an employee has pain, tingling, burning, numbness or any other subjective symptom of an MSD, and the symptoms are work-related, and the case is a new case that meets the recording criteria, you must record the case on the OSHA 300 Log as a musculoskeletal disorder.

**Note to §1904.12:** This section is effective January 1, 2003. From January 1, 2002 until December 31, 2002, you are required to record work-related injuries and illnesses involving muscles, nerves, tendons, ligaments, joints, cartilage and spinal discs in accordance with the requirements applicable to any injury or illness under §1904.5, §1904.6, §1904.7, and §1904.29. For entry (M) on the OSHA 300 Log, you must check either entry for "injury" or "all other illnesses."

[66 FR 5915, Jan. 19, 2001]

### §1904.13—1904.28 [Reserved]

[66 FR 5915, Jan. 19, 2001]

### §1904.29 Forms

**1904.29(a)** ***Basic requirement.*** You must use OSHA 300, 300-A, and 301 forms, or equivalent forms, for recordable injuries and illnesses. The OSHA 300 form is called the Log of Work-Related Injuries and Illnesses, the 300-A is the Summary of Work-Related Injuries and Illnesses, and the OSHA 301 form is called the Injury and Illness Incident Report.

**1904.29(b)** ***Implementation.*** (1) *What do I need to do to complete the OSHA 300 Log?* You must enter information about your business at the top of the OSHA 300 Log, enter a one or two line description for each recordable injury or illness, and summarize this information on the OSHA 300-A at the end of the year.

**1904.29(b)(2)** *What do I need to do to complete the OSHA 301 Incident Report?* You must complete an OSHA 301 Incident Report form, or an equivalent form, for each recordable injury or illness entered on the OSHA 300 Log.

**1904.29(b)(3)** *How quickly must each injury or illness be recorded?* You must enter each recordable injury or illness on the OSHA 300 Log and 301 Incident Report within seven (7) calendar days of receiving information that a recordable injury or illness has occurred.

**1904.29(b)(4)** *What is an equivalent form?* An equivalent form is one that has the same information, is as readable and understandable, and is completed using the same instructions as the OSHA form it replaces. Many employers use an insurance form instead of the OSHA 301 Incident Report, or supplement an insurance form by adding any additional information required by OSHA.

**1904.29(b)(5)** *May I keep my records on a computer?* Yes, if the computer can produce equivalent forms when they are needed, as described under §§1904.35 and 1904.40, you may keep your records using the computer system.

**1904.29(b)(6)** *Are there situations where I do not put the employee's name on the forms for privacy reasons?* Yes, if you have a "privacy concern case," you may not enter the employee's name on the OSHA 300 Log. Instead, enter "privacy case" in the space normally used for the employee's name. This will protect the privacy of the injured or ill employee when another employee, a former employee, or an authorized employee representative is provided access to the OSHA 300 Log under §1904.35(b)(2). You must keep a separate, confidential list of the case numbers and employee names for your privacy concern cases so you can update the cases and provide the information to the government if asked to do so.

**1904.29(b)(7)** *How do I determine if an injury or illness is a privacy concern case?* You must consider the following injuries or illnesses to be privacy concern cases:

**1904.29(b)(7)(i)** An injury or illness to an intimate body part or the reproductive system;

**1904.29(b)(7)(ii)** An injury or illness resulting from a sexual assault;

**1904.29(b)(7)(iii)** Mental illnesses;

**1904.29(b)(7)(iv)** HIV infection, hepatitis, or tuberculosis; (**Note:** The first sentence of this §1904.29(b)(7)(iv) is effective on January 1, 2002. The second sentence is effective beginning on January 1, 2003.)

**1904.29(b)(7)(v)** Needlestick injuries and cuts from sharp objects that are contaminated with another person's blood or other potentially infectious material (see §1904.8 for definitions); and

**1904.29(b)(7)(vi)** Other illnesses, if the employee independently and voluntarily requests that his or her name not be entered on the log. Musculoskeletal disorders (MSDs) are not considered privacy concern cases.

**1904.29(b)(8)** *May I classify any other types of injuries and illnesses as privacy concern cases?* No, this is a complete list of all injuries and illnesses considered privacy concern cases for Part 1904 purposes.

**1904.29(b)(9)** *If I have removed the employee's name, but still believe that the employee may be identified from the information on the forms, is there anything else that I can do to further protect the employee's privacy?* Yes, if you have a reasonable basis to believe that information describing the privacy concern case may be personally identifiable even though the employee's name has been omitted, you may use discretion in describing the injury or illness on both the OSHA 300 and 301 forms. You must enter enough information to identify the cause of the incident and the general severity of the injury or illness, but you do not need to include details of an intimate or private nature. For example, a sexual assault case could be described as "injury from assault," or an injury to a reproductive organ could be described as "lower abdominal injury."

**1904.29(b)(10)** *What must I do to protect employee privacy if I wish to provide access to the OSHA Forms 300 and 301 to persons other than government representatives, employees, former employees or authorized representatives?* If you decide to voluntarily disclose the Forms to persons other than government representatives, employees, former employees or authorized representatives (as required by §§1904.35 and 1904.40), you must remove or hide the employees' names and other personally identifying information, except for the following cases. You may disclose the Forms with personally identifying information only:

**1904.29(b)(10)(i)** to an auditor or consultant hired by the employer to evaluate the safety and health program;

**1904.29(b)(10)(ii)** to the extent necessary for processing a claim for workers' compensation or other insurance benefits; or

**1904.29(b)(10)(iii)** to a public health authority or law enforcement agency for uses and disclosures for which consent, an authorization, or opportunity to agree or object is not required under Department of Health and Human Services Standards for Privacy of Individually Identifiable Health Information, 45 CFR 164.512.

[66 FR 5915, Jan. 19, 2001]

## Subpart D—Other OSHA Injury and Illness Recordkeeping Requirements

### §1904.30 Multiple business establishments.

**1904.30(a)** ***Basic requirement.*** You must keep a separate OSHA 300 Log for each establishment that is expected to be in operation for one year or longer.

**1904.30(b)** ***Implementation.*** (1) *Do I need to keep OSHA injury and illness records for short-term establishments (i.e., establishments that will exist for less than a year)?* Yes, however, you do not have to keep a separate OSHA 300 Log for each such establishment. You may keep one OSHA 300 Log that covers all of your short-term establishments. You may also include the short-term establishments' recordable injuries and illnesses on an OSHA 300 Log that covers short-term establishments for individual company divisions or geographic regions.

**1904.30(b)(2)** *May I keep the records for all of my establishments at my headquarters location or at some other central location?* Yes, you may keep the records for an establishment at your headquarters or other central location if you can:

**1904.30(b)(2)(i)** Transmit information about the injuries and illnesses from the establishment to the central location within seven (7) calendar days of receiving information that a recordable injury or illness has occurred; and

**1904.30(b)(2)(ii)** Produce and send the records from the central location to the establishment within the time frames required by §1904.35 and §1904.40 when you are required to provide records to a government representative, employees, former employees or employee representatives.

**1904.30(b)(3)** *Some of my employees work at several different locations or do not work at any of my establishments at all. How do I record cases for these employees?* You must link each of your employees with one of your establishments, for recordkeeping purposes. You must record the injury and illness on the OSHA 300 Log of the injured or ill employee's establishment, or on an OSHA 300 Log that covers that employee's short-term establishment.

**1904.30(b)(4)** *How do I record an injury or illness when an employee of one of my establishments is injured or becomes ill while visiting or working at another of my establishments, or while working away from any of my establishments?* If the injury or illness occurs at one of your establishments, you must record the injury or illness on the OSHA 300 Log of the establishment at which the injury or illness occurred. If the employee is injured or becomes ill and is not at one of your establishments, you must record the case on the OSHA 300 Log at the establishment at which the employee normally works.

[66 FR 5915, Jan. 19, 2001]

### §1904.31 Covered employees.

**1904.31(a)** ***Basic requirement.*** You must record on the OSHA 300 Log the recordable injuries and illnesses of all employees on your payroll, whether they are labor, executive, hourly, salary, part-time, seasonal, or migrant workers. You also must record the recordable injuries and illnesses that occur to employees who are not on your payroll if you supervise these employees on a day-to-day basis. If your business is organized as a sole proprietorship or partnership, the owner or partners are not considered employees for recordkeeping purposes.

**1904.31(b)** ***Implementation.*** (1) *If a self-employed person is injured or becomes ill while doing work at my business, do I need to record the injury or illness?* No, self-employed individuals are not covered by the OSH Act or this regulation.

**1904.31(b)(2)** *If I obtain employees from a temporary help service, employee leasing service, or personnel supply service, do I have to record an injury or illness occurring to one of those employees?* You must record these injuries and illnesses if you supervise these employees on a day-to-day basis.

**1904.31(b)(3)** *If an employee in my establishment is a contractor's employee, must I record an injury or illness occurring to that employee?* If the contractor's employee is under the day-to-day supervision of the contractor, the contractor is responsible for recording the injury or illness. If you supervise the contractor employee's work on a day-to-day basis, you must record the injury or illness.

**1904.31(b)(4)** *Must the personnel supply service, temporary help service, employee leasing service, or contractor also record the injuries or illnesses occurring to temporary, leased or contract employees that I supervise on a day-to-day basis?* No, you and the temporary help service, employee leasing service, personnel supply service, or contractor should coordinate your efforts to make sure that each injury and illness is recorded only once: either on your OSHA 300 Log (if you provide day-to-day supervision) or on the other employer's OSHA 300 Log (if that company provides day-to-day supervision).

[66 FR 5915, Jan. 19, 2001]

### §1904.32 Annual summary.

**1904.32(a)** ***Basic requirement.*** At the end of each calendar year, you must:

**1904.32(a)(1)** Review the OSHA 300 Log to verify that the entries are complete and accurate, and correct any deficiencies identified;

**1904.32(a)(2)** Create an annual summary of injuries and illnesses recorded on the OSHA 300 Log;

**1904.32(a)(3)** Certify the summary; and

**1904.32(a)(4)** Post the annual summary.

**1904.32(b)** ***Implementation.*** (1) *How extensively do I have to review the OSHA 300 Log entries at the end of the year?* You must review the entries as extensively as necessary to make sure that they are complete and correct.

**1904.32(b)(2)** *How do I complete the annual summary?* You must:

**1904.32(b)(2)(i)** Total the columns on the OSHA 300 Log (if you had no recordable cases, enter zeros for each column total); and

**1904.32(b)(2)(ii)** Enter the calendar year covered, the company's name, establishment name, establishment address, annual average number of employees covered by the OSHA 300 Log, and the total hours worked by all employees covered by the OSHA 300 Log.

**1904.32(b)(2)(iii)** If you are using an equivalent form other than the OSHA 300-A summary form, as permitted under §1904.6(b)(4), the summary you use must also include the employee access and employer penalty statements found on the OSHA 300-A Summary form.

**1904.32(b)(3)** *How do I certify the annual summary?* A company executive must certify that he or she has examined the OSHA 300 Log and that he or she reasonably believes, based on his or her knowledge of the process by which the information was recorded, that the annual summary is correct and complete.

**1904.32(b)(4)** *Who is considered a company executive?* The company executive who certifies the log must be one of the following persons:

**1904.32(b)(4)(i)** An owner of the company (only if the company is a sole proprietorship or partnership);

**1904.32(b)(4)(ii)** An officer of the corporation;

**1904.32(b)(4)(iii)** The highest ranking company official working at the establishment; or

**1904.32(b)(4)(iv)** The immediate supervisor of the highest ranking company official working at the establishment.

**1904.32(b)(5)** *How do I post the annual summary?* You must post a copy of the annual summary in each establishment in a conspicuous place or places where notices to employees are customarily posted. You must ensure that the posted annual summary is not altered, defaced or covered by other material.

**1904.32(b)(6)** *When do I have to post the annual summary?* You must post the summary no later than February 1 of the year following the year covered by the records and keep the posting in place until April 30.

[66 FR 5915, Jan. 19, 2001]

### §1904.33 Retention and updating.

**1904.33(a)** ***Basic requirement.*** You must save the OSHA 300 Log, the privacy case list (if one exists), the annual summary, and the OSHA 301 Incident Report forms for five (5) years following the end of the calendar year that these records cover.

**1904.33(b)** ***Implementation.*** (1) *Do I have to update the OSHA 300 Log during the five-year storage period?* Yes, during the storage period, you must update your stored OSHA 300 Logs to include newly discovered recordable injuries or illnesses and to show any changes that have occurred in the classification of previously recorded injuries and illnesses. If the description or outcome of a case changes, you must remove or line out the original entry and enter the new information.

**1904.33(b)(2)** *Do I have to update the annual summary?* No, you are not required to update the annual summary, but you may do so if you wish.

**1904.33(b)(3)** *Do I have to update the OSHA 301 Incident Reports?* No, you are n t required to update the OSHA 301 Incident Reports, but you may do so if you wish.

[66 FR 5915, Jan. 19, 2001]

### §1904.34 Change in business ownership.

If your business changes ownership, you are responsible for recording and reporting work-related injuries and illnesses only for that period of the year during which you owned the establishment. You must transfer the Part 1904 records to the new owner. The new owner must save all records of the establishment kept by the prior owner, as required by §1904.33 of this Part, but need not update or correct the records of the prior owner.

[66 FR 5915, Jan. 19, 2001]

### §1904.35 Employee involvement.

**1904.35(a)** ***Basic requirement.*** Your employees and their representatives must be involved in the recordkeeping system in several ways.

**1904.35(a)(1)** You must inform each employee of how he or she is to report an injury or illness to you.

**1904.35(a)(2)** You must provide limited access to your injury and illness records for your employees and their representatives.

**1904.35(b)** ***Implementation.*** (1) *What must I do to make sure that employees report work-related injuries and illnesses to me?*

**1904.35(b)(1)(i)** You must set up a way for employees to report work-related injuries and illnesses promptly; and

**1904.35(b)(1)(ii)** You must tell each employee how to report work-related injuries and illnesses to you.

**1904.35(b)(2)** *Do I have to give my employees and their representatives access to the OSHA injury and illness records?* Yes, your employees, former employees, their personal representatives, and their authorized employee representatives have the right to access the OSHA injury and illness records, with some limitations, as discussed below.

**1904.35(b)(2)(i)** *Who is an authorized employee representative?* An authorized employee representative is an authorized collective bargaining agent of employees.

**1904.35(b)(2)(ii)** *Who is a "personal representative" of an employee or former employee?* A personal representative is:

**1904.35(b)(2)(ii)(A)** Any person that the employee or former employee designates as such, in writing; or

**1904.35(b)(2)(ii)(B)** The legal representative of a deceased or legally incapacitated employee or former employee.

**1904.35(b)(2)(iii)** *If an employee or representative asks for access to the OSHA 300 Log, when do I have to provide it?* When an employee, former employee, personal representative, or authorized employee representative asks for copies of your current or stored OSHA 300 Log(s) for an establishment the employee or former employee has worked in, you must give the requester a copy of the relevant OSHA 300 Log(s) by the end of the next business day.

**1904.35(b)(2)(iv)** *May I remove the names of the employees or any other information from the OSHA 300 Log before I give copies to an employee, former employee, or employee representative?* No, you must leave the names on the 300 Log. However, to protect the privacy of injured and ill employees, you may not record the employee's name on the OSHA 300 Log for certain "privacy concern cases," as specified in paragraphs 1904.29(b)(6) through 1904.29(b)(9).

**1904.35(b)(2)(v)** *If an employee or representative asks for access to the OSHA 301 Incident Report, when do I have to provide it?*

**1904.35(b)(2)(v)(A)** When an employee, former employee, or personal representative asks for a copy of the OSHA 301 Incident Report describing an injury or illness to that employee or former employee, you must give the requester a copy of the OSHA 301 Incident Report containing that information by the end of the next business day.

**1904.35(b)(2)(v)(B)** When an authorized employee representative asks for a copies of the OSHA 301 Incident Reports for an establishment where the agent represents employees under a collective bargaining agreement, you must give copies of those forms to the authorized employee representative within 7 calendar days. You are only required to give the authorized employee representative information from the OSHA 301 Incident Report section titled "Tell us about the case." You must remove all other information from the copy of the OSHA 301 Incident Report or the equivalent substitute form that you give to the authorized employee representative.

**1904.35(b)(2)(vi)** *May I charge for the copies?* No, you may not charge for these copies the first time they are provided. However, if one of the designated persons asks for additional copies, you may assess a reasonable charge for retrieving and copying the records.

[66 FR 5915, Jan. 19, 2001]

### §1904.36 Prohibition against discrimination.

Section 11(c) of the Act prohibits you from discriminating against an employee for reporting a work-related fatality, injury or illness. That provision of the Act also protects the employee who files a safety and health complaint, asks for access to the Part 1904 records, or otherwise exercises any rights afforded by the OSH Act.

[66 FR 5915, Jan. 19, 2001]

### §1904.37 State recordkeeping regulations.

**1904.37(a)** ***Basic requirement.*** Some States operate their own OSHA programs, under the authority of a State Plan approved by OSHA. States operating OSHA-approved State Plans must have occupational injury and illness recording and reporting requirements that are substantially identical to the requirements in this Part (see 29 CFR 1902.3(k), 29 CFR 1952.4 and 29 CFR 1956.10(i)).

**1904.37(b)** ***Implementation.*** (1) State-Plan States must have the same requirements as Federal OSHA for determining which injuries and illnesses are recordable and how they are recorded.

**1904.37(b)(2)** For other Part 1904 provisions (for example, industry exemptions, reporting of fatalities and hospitalizations, record retention, or employee involvement), State-Plan State requirements may be more stringent than or supplemental to the Federal requirements, but because of the unique nature of the national recordkeeping program, States must consult with and obtain approval of any such requirements.

**1904.37(b)(3)** Although State and local government employees are not covered Federally, all State-Plan States must provide coverage, and must develop injury and illness statistics, for these workers. State Plan recording and reporting requirements for State and local government entities may differ from those for the private sector but must meet the requirements of paragraphs 1904.37(b)(1) and (b)(2).

**1904.37(b)(4)** A State-Plan State may not issue a variance to a private sector employer and must recognize all variances issued by Federal OSHA.

**1904.37(b)(5)** A State Plan State may only grant an injury and illness recording and reporting variance to a State or local government employer within the State after obtaining approval to grant the variance from Federal OSHA.

[66 FR 5915, Jan. 19, 2001]

## §1904.38 Variances from the recordkeeping rule.

**1904.38(a)** ***Basic requirement.*** If you wish to keep records in a different manner from the manner prescribed by the Part 1904 regulations, you may submit a variance petition to the Assistant Secretary of Labor for Occupational Safety and Health, U.S. Department of Labor, Washington, DC 20210. You can obtain a variance only if you can show that your alternative recordkeeping system:

**1904.38(a)(1)** Collects the same information as this Part requires;

**1904.38(a)(2)** Meets the purposes of the Act; and

**1904.38(a)(3)** Does not interfere with the administration of the Act.

**1904.38(b)** ***Implementation.*** (1) *What do I need to include in my variance petition?* You must include the following items in your petition:

**1904.38(b)(1)(i)** Your name and address;

**1904.38(b)(1)(ii)** A list of the State(s) where the variance would be used;

**1904.38(b)(1)(iii)** The address(es) of the business establishment(s) involved;

**1904.38(b)(1)(iv)** A description of why you are seeking a variance;

**1904.38(b)(1)(v)** A description of the different recordkeeping procedures you propose to use;

**1904.38(b)(1)(vi)** A description of how your proposed procedures will collect the same information as would be collected by this Part and achieve the purpose of the Act; and

**1904.38(b)(1)(vii)** A statement that you have informed your employees of the petition by giving them or their authorized representative a copy of the petition and by posting a statement summarizing the petition in the same way as notices are posted under §1903.2(a).

**1904.38(b)(2)** *How will the Assistant Secretary handle my variance petition?* The Assistant Secretary will take the following steps to process your variance petition.

**1904.38(b)(2)(i)** The Assistant Secretary will offer your employees and their authorized representatives an opportunity to submit written data, views, and arguments about your variance petition.

**1904.38(b)(2)(ii)** The Assistant Secretary may allow the public to comment on your variance petition by publishing the petition in the **Federal Register**. If the petition is published, the notice will establish a public comment period and may include a schedule for a public meeting on the petition.

**1904.38(b)(2)(iii)** After reviewing your variance petition and any comments from your employees and the public, the Assistant Secretary will decide whether or not your proposed recordkeeping procedures will meet the purposes of the Act, will not otherwise interfere with the Act, and will provide the same information as the Part 1904 regulations provide. If your procedures meet these criteria, the Assistant Secretary may grant the variance subject to such conditions as he or she finds appropriate.

**1904.38(b)(2)(iv)** If the Assistant Secretary grants your variance petition, OSHA will publish a notice in the **Federal Register** to announce the variance. The notice will include the practices the variance allows you to use, any conditions that apply, and the reasons for allowing the variance.

**1904.38(b)(3)** *If I apply for a variance, may I use my proposed recordkeeping procedures while the Assistant Secretary is processing the variance petition?* No, alternative recordkeeping practices are only allowed after the variance is approved. You must comply with the Part 1904 regulations while the Assistant Secretary is reviewing your variance petition.

**1904.38(b)(4)** *If I have already been cited by OSHA for not following the Part 1904 regulations, will my variance petition have any effect on the citation and penalty?* No, in addition, the Assistant Secretary may elect not to review your variance petition if it includes an element for which you have been cited and the citation is still under review by a court, an Administrative Law Judge (ALJ), or the OSH Review Commission.

**1904.38(b)(5)** *If I receive a variance, may the Assistant Secretary revoke the variance at a later date?* Yes, the Assistant Secretary may revoke your variance if he or she has good cause. The procedures revoking a variance will follow the same process as OSHA uses for reviewing variance petitions, as outlined in paragraph 1904.38(b)(2). Except in cases of willfulness or where necessary for public safety, the Assistant Secretary will:

**1904.38(b)(5)(i)** Notify you in writing of the facts or conduct that may warrant revocation of your variance; and

**1904.38(b)(5)(ii)** Provide you, your employees, and authorized employee representatives with an opportunity to participate in the revocation procedures.

[66 FR 5915, Jan. 19, 2001]

## Subpart E—Reporting Fatality, Injury and Illness Information to the Government

## §1904.39 Reporting fatalities and multiple hospitalization incidents to OSHA.

**1904.39(a)** ***Basic requirement.*** Within eight (8) hours after the death of any employee from a work-related incident or the in-patient hospitalization of three or more employees as a result of a work-related incident, you must orally report the fatality/multiple hospitalization by telephone or in person to the Area Office of the Occupational Safety and Health Administration (OSHA), U.S. Department of Labor, that is nearest to the site of the incident. You may also use the OSHA toll-free central telephone number, 1-800-321-OSHA (1-800-321-6742).

**1904.39(b)** ***Implementation.*** (1) *If the Area Office is closed, may I report the incident by leaving a message on OSHA's answering machine, faxing the area office, or sending an e-mail?* No, if you can't talk to a person at the Area Office, you must report the fatality or multiple hospitalization incident using the 800 number.

**1904.39(b)(2)** *What information do I need to give to OSHA about the incident?* You must give OSHA the following information for each fatality or multiple hospitalization incident:

**1904.39(b)(2)(i)** The establishment name;

**1904.39(b)(2)(ii)** The location of the incident;

**1904.39(b)(2)(iii)** The time of the incident;

**1904.39(b)(2)(iv)** The number of fatalities or hospitalized employees;

**1904.39(b)(2)(v)** The names of any injured employees;

**1904.39(b)(2)(vi)** Your contact person and his or her phone number; and

**1904.39(b)(2)(vii)** A brief description of the incident.

**1904.39(b)(3)** *Do I have to report every fatality or multiple hospitalization incident resulting from a motor vehicle accident?* No, you do not have to report all of these incidents. If the motor vehicle accident occurs on a public street or highway, and does not occur in a construction work zone, you do not have to report the incident to OSHA. However, these injuries must be recorded on your OSHA injury and illness records, if you are required to keep such records.

**1904.39(b)(4)** *Do I have to report a fatality or multiple hospitalization incident that occurs on a commercial or public transportation system?* No, you do not have to call OSHA to report a fatality or multiple hospitalization incident if it involves a commercial airplane, train, subway or bus accident. However, these injuries must be recorded on your OSHA injury and illness records, if you are required to keep such records.

**1904.39(b)(5)** *Do I have to report a fatality caused by a heart attack at work?* Yes, your local OSHA Area Office director will decide whether to investigate the incident, depending on the circumstances of the heart attack.

**1904.39(b)(6)** *Do I have to report a fatality or hospitalization that occurs long after the incident?* No, you must only report each fatality or multiple hospitalization incident that occurs within thirty (30) days of an incident.

**1904.39(b)(7)** *What if I don't learn about an incident right away?* If you do not learn of a reportable incident at the time it occurs and the incident would otherwise be reportable under paragraphs (a) and (b) of this section, you must make the report within eight (8) hours of the time the incident is reported to you or to any of your agent(s) or employee(s).

[66 FR 5915, Jan. 19, 2001]

## §1904.40 Providing records to government representatives.

**1904.40(a)** ***Basic requirement.*** When an authorized government representative asks for the records you keep under Part 1904, you must provide copies of the records within four (4) business hours.

**1904.40(b) *Implementation.*** (1) *What government representatives have the right to get copies of my Part 1904 records?* The government representatives authorized to receive the records are:

**1904.40(b)(1)(i)** A representative of the Secretary of Labor conducting an inspection or investigation under the Act;

**1904.40(b)(1)(ii)** A representative of the Secretary of Health and Human Services (including the National Institute for Occupational Safety and Health—NIOSH) conducting an investigation under section 20(b) of the Act, or

**1904.40(b)(1)(iii)** A representative of a State agency responsible for administering a State plan approved under section 18 of the Act.

**1904.40(b)(2)** *Do I have to produce the records within four (4) hours if my records are kept at a location in a different time zone?* OSHA will consider your response to be timely if you give the records to the government representative within four (4) business hours of the request. If you maintain the records at a location in a different time zone, you may use the business hours of the establishment at which the records are located when calculating the deadline.

[66 FR 5915, Jan. 19, 2001]

### §1904.41 Annual OSHA injury and illness survey of ten or more employers.

**1904.41(a) *Basic requirement.*** If you receive OSHA's annual survey form, you must fill it out and send it to OSHA or OSHA's designee, as stated on the survey form. You must report the following information for the year described on the form:

**1904.41(a)(1)** the number of workers you employed;

**1904.41(a)(2)** the number of hours worked by your employees; and

**1904.41(a)(3)** the requested information from the records that you keep under Part 1904.

**1904.41(b) *Implementation.*** (1) *Does every employer have to send data to OSHA?* No, each year, OSHA sends injury and illness survey forms to employers in certain industries. In any year, some employers will receive an OSHA survey form and others will not. You do not have to send injury and illness data to OSHA unless you receive a survey form.

**1904.41(b)(2)** *How quickly do I need to respond to an OSHA survey form?* You must send the survey reports to OSHA, or OSHA's designee, by mail or other means described in the survey form, within 30 calendar days, or by the date stated in the survey form, whichever is later.

**1904.41(b)(3)** *Do I have to respond to an OSHA survey form if I am normally exempt from keeping OSHA injury and illness records?* Yes, even if you are exempt from keeping injury and illness records under §1904.1 to §1904.3, OSHA may inform you in writing that it will be collecting injury and illness information from you in the following year. If you receive such a letter, you must keep the injury and illness records required by §1904.5 to §1904.15 and make a survey report for the year covered by the survey.

**1904.41(b)(4)** *Do I have to answer the OSHA survey form if I am located in a State-Plan State?* Yes, all employers who receive survey forms must respond to the survey, even those in State-Plan States.

**1904.41(b)(5)** *Does this section affect OSHA's authority to inspect my workplace?* No, nothing in this section affects OSHA's statutory authority to investigate conditions related to occupational safety and health.

[66 FR 5915, Jan. 19, 2001]

### §1904.42 Requests from the Bureau of Labor Statistics for data.

**1904.42(a) *Basic requirement.*** If you receive a Survey of Occupational Injuries and Illnesses Form from the Bureau of Labor Statistics (BLS), or a BLS designee, you must promptly complete the form and return it following the instructions contained on the survey form.

**1904.42(b) *Implementation.*** (1) *Does every employer have to send data to the BLS?* No, each year, the BLS sends injury and illness survey forms to randomly selected employers and uses the information to create the Nation's occupational injury and illness statistics. In any year, some employers will receive a BLS survey form and others will not. You do not have to send injury and illness data to the BLS unless you receive a survey form.

**1904.42(b)(2)** *If I get a survey form from the BLS, what do I have to do?* If you receive a Survey of Occupational Injuries and Illnesses Form from the Bureau of Labor Statistics (BLS), or a BLS designee, you must promptly complete the form and return it, following the instructions contained on the survey form.

**1904.42(b)(3)** *Do I have to respond to a BLS survey form if I am normally exempt from keeping OSHA injury and illness records?* Yes, even if you are exempt from keeping injury and illness records under §1904.1 to §1904.3, the BLS may inform you in writing that it will be collecting injury and illness information from you in the coming year. If you receive such a letter, you must keep the injury and illness records required by §1904.5 to §1904.15 and make a survey report for the year covered by the survey.

**1904.42(b)(4)** *Do I have to answer the BLS survey form if I am located in a State-Plan State?* Yes, all employers who receive a survey form must respond to the survey, even those in State-Plan States.

[66 FR 5915, Jan. 19, 2001]

## Subpart F—Transition From the Former Rule

### §1904.43 Summary and posting of the 2001 data.

**1904.43(a) *Basic requirement.*** If you were required to keep OSHA 200 Logs in 2001, you must post a 2000 annual summary from the OSHA 200 Log of occupational injuries and illnesses for each establishment.

**1904.43(b) *Implementation.*** (1) *What do I have to include in the summary?*

**1904.43(b)(1)(i)** You must include a copy of the totals from the 2001 OSHA 200 Log and the following information from that form:

**1904.43(b)(1)(i)(A)** The calendar year covered;

**1904.43(b)(1)(i)(B)** Your company name;

**1904.43(b)(1)(i)(C)** The name and address of the establishment; and

**1904.43(b)(1)(i)(D)** The certification signature, title and date.

**1904.43(b)(1)(ii)** If no injuries or illnesses occurred at your establishment in 2001, you must enter zeros on the totals line and post the 2001 summary.

**1904.43(b)(2)** *When am I required to summarize and post the 2001 information?*

**1904.43(b)(2)(i)** You must complete the summary by February 1, 2002; and

**1904.43(b)(2)(ii)** You must post a copy of the summary in each establishment in a conspicuous place or places where notices to employees are customarily posted. You must ensure that the summary is not altered, defaced or covered by other material.

**1904.43(b)(3)** You must post the 2001 summary from February 1, 2002 to March 1, 2002.

[66 FR 5915, Jan. 19, 2001]

### §1904.44 Retention and updating of old forms.

You must save your copies of the OSHA 200 and 101 forms for five years following the year to which they relate and continue to provide access to the data as though these forms were the OSHA 300 and 301 forms. You are not required to update your old 200 and 101 forms.

[66 FR 5915, Jan. 19, 2001]

### §1904.45 OMB control numbers under the Paperwork Reduction Act

The following sections each contain a collection of information requirement which has been approved by the Office of Management and Budget under the control number listed

| 29 CFR citation | OMB Control No. |
|---|---|
| 1904.4-35 | 1218-0176 |
| 1904.39-41 | 1218-0176 |
| 1904.42 | 1220-0045 |
| 1904.43-44 | 1218-0176 |

[66 FR 5915, Jan. 19, 2001]

## Subpart G—Definitions

### §1904.46 Definitions

*The Act.* The Act means the Occupational Safety and Health Act of 1970 (29 U.S.C. 651 *et seq.*). The definitions contained in section 3 of the Act

(29 U.S.C. 652) and related interpretations apply to such terms when used in this Part 1904.

*Establishment.* An establishment is a single physical location where business is conducted or where services or industrial operations are performed. For activities where employees do not work at a single physical location, such as construction; transportation; communications, electric, gas and sanitary services; and similar operations, the establishment is represented by main or branch offices, terminals, stations, etc. that either supervise such activities or are the base from which personnel carry out these activities.

**1904.46(1)** *Can one business location include two or more establishments?* Normally, one business location has only one establishment. Under limited conditions, the employer may consider two or more separate businesses that share a single location to be separate establishments. An employer may divide one location into two or more establishments only when:

**1904.46(1)(i)** Each of the establishments represents a distinctly separate business;

**1904.46(1)(ii)** Each business is engaged in a different economic activity;

**1904.46(1)(iii)** No one industry description in the Standard Industrial Classification Manual (1987) applies to the joint activities of the establishments; and

**1904.46(1)(iv)** Separate reports are routinely prepared for each establishment on the number of employees, their wages and salaries, sales or receipts, and other business information. For example, if an employer operates a construction company at the same location as a lumber yard, the employer may consider each business to be a separate establishment.

**1904.46(2)** *Can an establishment include more than one physical location?* Yes, but only under certain conditions. An employer may combine two or more physical locations into a single establishment only when:

**1904.46(2)(i)** The employer operates the locations as a single business operation under common management;

**1904.46(2)(ii)** The locations are all located in close proximity to each other; and

**1904.46(2)(iii)** The employer keeps one set of business records for the locations, such as records on the number of employees, their wages and salaries, sales or receipts, and other kinds of business information. For example, one manufacturing establishment might include the main plant, a warehouse a few blocks away, and an administrative services building across the street.

**1904.46(3)** *If an employee telecommutes from home, is his or her home considered a separate establishment?* No, for employees who telecommute from home, the employee's home is not a business establishment and a separate 300 Log is not required. Employees who telecommute must be linked to one of your establishments under §1904.30(b)(3).

*Injury or illness.* An injury or illness is an abnormal condition or disorder. Injuries include cases such as, but not limited to, a cut, fracture, sprain, or amputation. Illnesses include both acute and chronic illnesses, such as, but not limited to, a skin disease, respiratory disorder, or poisoning. (Note: Injuries and illnesses are recordable only if they are new, work-related cases that meet one or more of the Part 1904 recording criteria.)

*Physician or Other Licensed Health Care Professional.* A physician or other licensed health care professional is an individual whose legally permitted scope of practice (i.e., license, registration, or certification) allows him or her to independently perform, or be delegated the responsibility to perform, the activities described by this regulation.

*You.* "You" means an employer as defined in Section 3 of the Occupational Safety and Health Act of 1970 (29 U.S.C. 652).

[66 FR 5915, Jan. 19, 2001]

**AUTHORITY**: Secs. 4, 6, 8, Occupational Safety and Health Act of 1970 (29 U.S.C. 653, 655, 657); Secretary of Labor's Order 8-76 (41 FR 25059); and 29 CFR Part 1911.

SOURCE: 45 FR 5282, Jan. 22, 1980, unless otherwise noted.

## GENERAL

### §1990.101 Scope.

This part establishes criteria and procedures for the identification, classification, and regulation of potential occupational carcinogens found in each workplace in the United States regulated by the Occupational Safety and Health Act of 1970 (the Act). The procedures contained in this part supplement the procedural regulations in other parts of this chapter. In the event of a conflict, the procedures contained in this part shall govern the identification, classification, and regulation of potential occupational carcinogens. This part may be referred to as "The OSHA Cancer Policy."

### §1990.102 Purpose.

The Act provides, among other things, that the Secretary, in promulgating standards dealing with toxic materials or harmful physical agents under this subsection, shall set the standard which most adequately assures, to the extent feasible, on the basis of the best available evidence, that no employee will suffer material impairment of health or functional capacity even if such employee has regular exposure to the hazard dealt with by such standard for the period of his or her working life. Development of standards under this section shall be based upon research, demonstrations, experiments, and such other information as may be appropriate. In addition to the attainment of the highest degree of health and safety protection for the employee, other considerations shall be the latest available scientific data in the field, the feasibility of the standards, and experience gained under this and other health and safety laws. Whenever practicable, the standard promulgated shall be expressed in terms of objective criteria and of the performance desired (section 6(b)(5)).

It is the purpose of the regulations of this part to carry out the intent of the Act with respect to the identification, classification, and regulation of potential occupational carcinogens.

### §1990.103 Definitions.

Terms used in this part shall have the meanings set forth in the Act. In addition, as used in this part, the following terms shall have the meanings set forth below:

*Act* means the Occupational Safety and Health Act of 1970 (Pub. L. 91-596, 84 Stat. 1590 *et seq.*, 29 U.S.C. 551 *et seq.* ).

*Administrator of EPA* means the Administrator of the United States Environmental Protection Agency, or designee.

*Chairperson of CPSC* means the Chairman of the United States Consumer Product Safety Commission, or designee.

*Commissioner of FDA* means the Commissioner of the Food and Drug Administration, United States Department of Health and Human Services, or designee.

*Director of NCI* means the Director of the National Cancer Institute, United States Department of Health and Human Services, or designee.

*Director of NIEHS* means the Director of the National Institute of Environmental Health Sciences, United States Department of Health and Human Services, or designee.

*Director of NIOSH* means the Director of the National Institute for Occupational Safety and Health, United States Department of Health and Human Services, or designee.

*Mutagenesis* means the induction of heritable changes in the genetic material of either somatic or germinal cells.

*Positive results in short-term tests* means positive results in assays for two or more of the following types of effect:

(1) The induction of DNA damage and/or repair;

(2) Mutagenesis in bacteria, yeast, *Neurospora* or *Drosophila melanogaster;*

(3) Mutagenesis in mammalian somatic cells;

(4) Mutagenesis in mammalian germinal cells; or

(5) Neoplastic transformation of mammalian cells in culture.

*Potential occupational carcinogen* means any substance, or combination or mixture of substances, which causes an increased incidence of benign and/or malignant neoplasms, or a substantial decrease in the latency period between exposure and onset of neoplasms in humans or in one or more experimental mammalian species as the result of any oral, respiratory or dermal exposure, or any other exposure which results in the induction of tumors at a site other than the site of administration. This definition also includes any substance which is metabolized into one or more potential occupational carcinogens by mammals.

*Secretary of HHS* means the Secretary of the United States Department of Health and Human Services, or designee.

### §1990.104 Scientific review panel.

**1990.104(a)** ***General.*** At any time, the Secretary may request the Director of NCI, the Director of NIEHS and/or the Director of NIOSH to convene a scientific review panel ("the panel") to provide recommendations to the Secretary in the identification, classification, or regulation of any potential occupational carcinogen.

**1990.104(b)** ***Membership.*** The panel will consist of individuals chosen by the respective Director(s). The panel will consist of individuals who are appropriately qualified in the disciplines relevant to the issues to be considered, and who are employed by the United States. The panel does not constitute an advisory committee within the meaning of section 6(b) or 7(b) of the Act, or the Federal Advisory Committee Act (Pub. L. 92-463, 86 Stat. 770).

**1990.104(c)** ***Report.*** The Secretary shall request that the panel submit a report of its evaluation within ninety (90) days after the appointment of the members of the panel. The Secretary shall place a copy of the report in the record of any relevant rulemaking undertaken pursuant to this part and allow an appropriate time for public review and comment. If a panel is not established or fails to file a timely report, or if the Secretary determines that it is necessary to proceed without waiting for the panel's report, the Secretary may proceed in making any determination without such report.

**1990.104(d)** ***Other aid and assistance.*** Nothing herein precludes the Secretary from obtaining advice or other aid from any person or organization including NCI, NIEHS, and NIOSH.

### §1990.105 Advisory committees.

The Secretary may appoint an Advisory Committee, pursuant to sections 6(b) and 7 of the Act, and 29 CFR part 1912, concerning any potential occupational carcinogen. The Secretary shall require the Advisory Committee to submit its recommendations to assist the Secretary in standard setting no later than ninety (90) days from the date of the Advisory Committee's appointment, unless extended by the Secretary for exceptional circumstances. If an Advisory Committee fails to file a timely report, the Secretary may proceed in standard setting activities without such a report.

### §1990.106 Amendments to this policy.

**1990.106(a)** ***Initiation of review of this policy.*** —(1) *Secretary's request.* No later than every three (3) years from the effective date of this part, or from the last general review, the Secretary shall request the Director of NCI, the Director of NIEHS and/or the Director of NIOSH, to review this part and render their opinions on whether significant scientific or technical advances made since the effective date of this part warrant any amendment to this part. The request shall ask that the answer be provided to the Secretary within one hundred twenty (120) days.

**1990.106(a)(2)** *Recommendations by the institutes.* At any time, the Director of NCI, the Director of NIEHS and/or the Director of NIOSH may submit recommendations to the Secretary for amendments to this part whenever any of them believes that scientific or technical advances justify such amendments.

**1990.106(a)(3)** *Petitions from the public.* (i) Any interested person may petition the Secretary concerning amendments to this part based upon substantial new issues or substantial new evidence.

**1990.106(a)(3)(ii)** For the purposes of this part, substantial new evidence is evidence which differs significantly from that presented in establishing this part, including amendments.

**1990.106(a)(3)(iii)** For the purposes of this part, substantial new issues are issues which differ significantly from those upon which the Secretary has reached a conclusion in the rulemaking establishing this part (including the conclusions reached in the preamble).

**1990.106(a)(3)(iv)** Each petition to amend this part shall contain at least the following information:

**1990.106(a)(3)(iv)(A)** Name and address of petitioner;

**1990.106(a)(3)(iv)(B)** The provisions which the petitioner believes are inappropriate;

**1990.106(a)(3)(iv)(C)** All data, views and arguments relied upon by the petitioner; and

**1990.106(a)(3)(iv)(D)** A detailed statement and analysis as to why the petitioner believes that the data, views and arguments presented by petitioner:

**1990.106(a)(3)(iv)(D)*(i)*** Constitute substantial new issues or substantial new evidence; and

**1990.106(a)(3)(iv)(D)*(ii)*** Are so significant as to warrant amendment of this part.

**1990.106(b)** ***Response to recommendations and petitions.*** —(1) *By the institutes.* Whenever any Director recommends an amendment to this part, the Secretary shall, within one hundred twenty (120) days after receipt of the recommendation, publish in the FEDERAL REGISTER, a notice which:

**1990.106(b)(1)(i)** States the reasons why the Secretary has determined not to commence a rulemaking proceeding to amend this part, in whole or in part, at that time; or

**1990.106(b)(1)(ii)** Commences a rulemaking proceeding to consider amending this part accordingly; or

**1990.106(b)(1)(iii)** Appoints an Advisory Committee as provided for by §1990.105 of this part and sections 6(b) and 7 of the Act.

**1990.106(b)(2)** *By the public.* Within ninety (90) days, or as soon thereafter as possible, after receipt of a petition pursuant to section 1990.106(a)(3), the Secretary shall:

**1990.106(b)(2)(i)** Refer the petition to the Director of NCI, the Director of NIEHS and/or the Director of NIOSH, in which case the provisions of §1990.106(a)(1) and (b)(1) are applicable; or

**1990.106(b)(2)(ii)** Appoint an advisory committee;

**1990.106(b)(2)(iii)** Deny the petition, briefly giving the reasons therefor; or

**1990.106(b)(2)(iv)** Commence a rulemaking proceeding to consider amending this part accordingly.

**1990.106(b)(3)** *On the Secretary's motion.* At any time, the Secretary may, on his own motion, commence a rulemaking proceeding to amend this part.

[45 FR 5282, Jan. 22, 1980; 45 FR 43405, June 27, 1980]

## THE OSHA CANCER POLICY

### §1990.111 General statement of regulatory policy.

**1990.111(a)** This part establishes the criteria and procedures under which substances will be regulated by OSHA as potential occupational carcinogens. Although the conclusive identification of "carcinogens" is a complex matter "on the frontiers of science," ( *IUD v. Hodgson* 499 F.2d 467, 474 (D.C. Cir. 1974)), responsible health regulatory policy requires that criteria should be specified for the identification of substances which should be regulated as posing potential cancer risks to workers.

**1990.111(b)** The criteria established by this part are based on an extensive review of scientific data and opinions. The part provides for amending these criteria in light of new scientific developments. Decisions as to whether any particular substance meets the criteria or not will be consistent with the policies and procedures established by this part and will be based upon scientific evaluation of the evidence on that substance.

**1990.111(c)** This part applies to individual substances, groups of substances, or combinations or mixtures of substances which may be found in workplaces in the United States. In individual rulemaking proceedings under this part, the identity and range of substances and mixtures to be covered by the standard will be specified and the appropriateness of applying the available evidence to the range of substances and mixtures proposed for regulation will be subject to scientific and policy review.

**1990.111(d)** Potential occupational carcinogens will be identified and classified on the basis of human epidemiological studies and/or experimental carcinogenesis bioassays in mammals. Positive results in short term tests will also be used as concordant evidence.

**1990.111(e)** Potential occupational carcinogens will be classified and regulated in accordance with the policy. The scientific evidence as to whether individual substances meet these criteria will be considered in individual rulemakings. The issues which may be considered in these rulemakings will be limited as specified herein.

**1990.111(f)** This policy provides for the classification of potential occupational carcinogens into two categories depending on the nature and extent of the available scientific evidence. The two categories of potential occupational carcinogens may be regulated differently.

**1990.111(g)** The policy establishes a procedure for setting priorities and making them public.

**1990.111(h)** Worker exposure to Category I Potential Carcinogens will be reduced primarily through the use of engineering and work practice controls.

**1990.111(i)** Worker exposure to Category II Potential Carcinogens will be reduced as appropriate and consistent with the statutory requirements on a case-by-case basis in the rulemaking proceedings on individual substances. Any permissible exposure level so established shall be met primarily through engineering and work practice controls.

**1990.111(j)** The assessment of cancer risk to workers resulting from exposure to a potential occupational carcinogen will be made on the basis of available data. Because of the uncertainties and serious consequences to workers if the estimated risk is understated, cautious and prudent assumptions will be utilized to perform risk assessments.

**1990.111(k)** Where the Secretary determines that one or more suitable substitutes exist for certain uses of Category I Potential Carcinogens that are less hazardous to humans, a no occupational exposure level shall be set for those uses, to be achieved solely through the use of engineering and work practice controls to encourage substitution. In determining whether a substitute is suitable, the Secretary will consider the technological and economic feasibility of the introduction of the substitute, including its relative effectiveness and other relevant factors, such as regulatory requirements and the time needed for an orderly transition to the substitute.

[45 FR 5282, Jan. 22, 1980, as amended at 46 FR 5881, Jan. 21, 1981]

### §1990.112 Classification of potential carcinogens.

The following criteria for identification, classification and regulation of potential occupational carcinogens will be applied, unless the Secretary considers evidence under the provisions of §§1990.143, 1990.144 and 1990.145 and determines that such evidence warrants an exception to these criteria.

**1990.112(a)** ***Category I Potential Carcinogens.*** A substance shall be identified, classified, and regulated as a Category I Potential Carcinogen if, upon scientific evaluation, the Secretary determines that the substance meets the definition of a potential occupational carcinogen in (1) humans, or (2) in a single mammalian species in a long-term bioassay where the results are in concordance with some other scientifically evaluated evidence of a potential carcinogenic hazard, or (3) in a single mammalian species in an adequately conducted long-term bioassay, in appropriate circumstances where the Secretary determines the requirement for concordance is not necessary. Evidence of concordance is any of the following: positive results from independent testing in the same or other species, positive results in short-term tests, or induction of tumors at injection or implantation sites.

**1990.112(b)** ***Category II Potential Carcinogens.*** A substance shall be identified, classified, and regulated as a Category II Potential Carcinogen if, upon scientific evaluation, the Secretary determines that:

**1990.112(b)(i)** The substance meets the criteria set forth in §1990.112(a), but the evidence is found by the Secretary to be only "suggestive"; or

**1990.112(b)(ii)** The substance meets the criteria set forth in §1990.112(a) in a single mammalian species without evidence of concordance.

## PRIORITY SETTING

### §1990.121 Candidate list of potential occupational carcinogens.

**1990.121(a)** ***Contents.*** The Secretary shall prepare a list of substances (the "Candidate List") which are reported to be present in any American workplace and which, on the basis of a brief scientific review of available data, may be considered candidates for further scientific review and possible regulation as Category I Potential Carcinogens or Category II Potential Carcinogens. For the purposes of this paragraph, "available data" means:

**1990.121(a)(1)** The data submitted by any person;

**1990.121(a)(2)** Any data referred to by the Secretary of HHS or by the Director of NIOSH, either in the latest list entitled "Suspected Carcinogens" or any other communication;

**1990.121(a)(3)** Literature referred to in U.S. Public Health Service, Publication No. 149;

**1990.121(a)(4)** Data summarized and reviewed in Monographs of the International Agency for Research on Cancer (IARC) of the World Health Organization;

**1990.121(a)(5)** The Toxic Substances Control Act Inventory of Chemical Substances, published by the Administrator of EPA;

**1990.121(a)(6)** The Secretary of HHS's Annual Report to the President and the Congress as required by the Community Mental Health Centers Extension Act of 1978, sec. 404(a)(9), 42 U.S.C. sec. 285.

**1990.121(a)(7)** Any other relevant data of which the Secretary has actual knowledge.

**1990.121(b)** ***Tentative classification.*** The Secretary may tentatively designate substances on the Candidate List as candidates for classification as Category I Potential Carcinogens or as Category II Potential Carcinogens, or may list substances without a tentative designation, based on the brief scientific review of available data for the purpose of initiating a more extensive scientific review.

**1990.121(c)** ***No legal rights established.*** The inclusion or exclusion of any substance from the Candidate List shall not be subject to judicial review nor be the basis of any legal action, nor shall the exclusion of any substance from the list prevent the regulation of that substance as a potential occupational carcinogen. The inclusion of a substance on the Candidate List and its possible tentative designation as a Category I Potential Carcinogen or a Category II Potential Carcinogen therein do not reflect a final scientific determination that the substance is, in fact, a Category I Potential Carcinogen or a Category II Potential Carcinogen. It is a policy determination based on the brief scientific review that the Secretary should conduct a thorough review of all relevant scientific data concerning the substance.

EFFECTIVE DATE NOTE: Paragraphs (a) and (b) of §1990.121 were stayed at 48 FR 243, Jan. 4, 1983, in order to evaluate the impact of publishing the Candidate Lists and Priority List and to reconsider the criteria used in establishing the lists (see also 47 FR 187, Jan. 5, 1982).

### §1990.122 Response to petitions.

Whenever the Secretary receives any information submitted in writing by any interested person concerning the inclusion or omission of any substance from the Candidate List, the Secretary shall briefly review the information and any other available data, as defined in §1990.121(a). The results of the Secretary's review shall be transmitted to the petitioner, together with a short statement of the Secretary's reasons therefor, and made public upon request.

EFFECTIVE DATE NOTE: Section 1990.122 was stayed at 48 FR 243, Jan. 4, 1983, in order to evaluate the impact of publishing the Candidate List and Priority Lists and to reconsider the criteria used in establishing the lists (see also 47 FR 187, Jan. 5, 1982).

### §1990.131 Priority lists for regulating potential occupational carcinogens.

The Secretary shall establish two priority lists for regulating potential occupational carcinogens. One list should include approximately ten (10) candidates for rulemaking as Category I Potential Carcinogens; the other approximately ten (10) candidates for rulemaking as Category II Potential Carcinogens. The order of placement of substances on these lists will not reflect the Secretary's determination of the exact order in which these substances should be regulated in rulemaking proceedings but rather a policy determination that the Secretary plans to address some or all of these substances prior to proceeding with a thorough scientific review of data concerning other substances on the Candidate List. The inclusion or exclusion of any substance on these lists shall not be subject to judicial review or be the basis for any legal action. The Secretary may regulate a potential occupational carcinogen which has not been placed on these lists. The inclusion of a substance on either of these lists does not reflect a final scientific determination that the substance is, in fact, a Category I Potential Carcinogen or a Category II Potential Carcinogen.

EFFECTIVE DATE NOTE: Section 1990.131 was stayed at 48 FR 243, Jan. 4, 1983, in order to evaluate the impact of publishing the Candidate List and Priority Lists and to reconsider the criteria used in establishing the lists (see also 47 FR 187, Jan. 5, 1982).

### §1990.132 Factors to be considered.

**1990.132(a)** The setting of priorities is a complex matter which requires subjective and policy judgments. It is not appropriate to establish a rigid formula or to assign predetermined weight to each factor. The identification of some of the elements is to guide the OSHA staff and inform the public on the development of priorities. It is not intended to create any legal rights with respect to the setting of priorities.

**1990.132(b)** Some factors which may be taken into account in setting priorities for regulating potential occupational carcinogens, when such data are available, are:

**1990.132(b)(1)** The estimated number of workers exposed;

**1990.132(b)(2)** The estimated levels of human exposure;

**1990.132(b)(3)** The levels of exposure to the substance which have been reported to cause an increased incidence of neoplasms in exposed humans, animals or both;

**1990.132(b)(4)** The extent to which regulatory action could reduce not only risks of contracting cancer but also other occupational and environmental health hazards;

**1990.132(b)(5)** Whether the molecular structure of the substance is similar to the molecular structure of another substance which meets the definition of a potential occupational carcinogen;

**1990.132(b)(6)** Whether there are substitutes that pose a lower risk of cancer or other serious human health problems, or available evidence otherwise suggests that the social and economic costs of regulation would be small; and

**1990.132(b)(7)** OSHA will also consider its responsibilities for dealing with other health and safety hazards and will consider the actions being taken or planned by other governmental agencies in dealing with the same or similar health and safety hazards.

### §1990.133 Publication.

**1990.133(a)** The Secretary shall publish the Candidate List in the FEDERAL REGISTER at least annually.

**1990.133(b)** The Secretary shall publish the Priority Lists in the FEDERAL REGISTER at least every six months and may seek public comment thereon.

**1990.133(c)** The Secretary may periodically publish in the FEDERAL REGISTER a notice requesting information concerning the classification and establishment of priorities for substances on the Candidate List together with a brief statement describing the type of information being sought.

EFFECTIVE DATE NOTE: Section 1990.133 was stayed at 48 FR 243, Jan. 4, 1983, in order to evaluate the impact of publishing the Candidate List and Priority Lists and to reconsider the criteria used in establishing the lists (see also 47 FR 187, Jan. 5, 1982).

## REGULATION OF POTENTIAL OCCUPATIONAL CARCINOGENS

### §1990.141 Advance notice of proposed rulemaking.

**1990.141(a)** Within thirty (30) days after OSHA initiates a study concerning the economic and/or technological feasibility of specific standards that might be applied in the regulation of a potential occupational carcinogen, the Secretary will normally publish, in the FEDERAL REGISTER, a notice which includes at least the following:

**1990.141(a)(1)** The name of the substance(s),

**1990.141(a)(2)** The scope of the study, including where possible,

**1990.141(a)(2)(i)** Affected industries,

**1990.141(a)(2)(ii)** Levels of exposure being studied,

**1990.141(a)(2)(iii)** The anticipated completion date of the study;

**1990.141(a)(3)** A brief summary of the available data on health effects;

**1990.141(a)(4)** An estimate of when the Secretary anticipates the issuance of a proposal;

**1990.141(a)(5)** An invitation to interested parties to provide relevant information;

**1990.141(a)(6)** A statement that persons wishing to provide OSHA with their own study should complete it within 30 days after the anticipated proposal date; and

**1990.141(a)(7)** A statement of the procedural requirements that must be met before substantial new issues or substantial new evidence will be considered in the proceeding pursuant to §1990.145.

**1990.141(b)** Where the Secretary determines to discontinue a feasibility study, the Secretary should publish, within 30 days, a notice in the FED ERAL REGISTER so indicating.

## §1990.142 Initiation of a rulemaking.

Where the Secretary decides to regulate a potential occupational carcinogen, the Secretary shall initiate a rulemaking proceeding in accordance with one of the following procedures, as appropriate.

**1990.142(a)** ***Notice of proposed rulemakings (section 6(b) of the Act)***—(1) *General.* The Secretary may issue a notice of proposed rulemaking in the FEDERAL REGISTER, pursuant to section 6(b) of the Act and part 1911 of this chapter. The notice shall provide for no more than a sixty (60) day comment period, and may provide for a hearing, which shall be scheduled for no later than one hundred (100) days after publication of the Notice of Proposed Rulemaking. The commencement of the hearing may be postponed once, for no more than thirty (30) days, for good cause shown.

**1990.142(a)(2)** *Provisions of the proposed standard for Category I Potential Carcinogens.* Whenever the Secretary issues a notice of proposed rulemaking to regulate a substance as a Category I Potential Carcinogen:

**1990.142(a)(2)(i)** The proposed standard shall contain at least provisions for scope and application, definitions, notification of use, a permissible exposure limit, monitoring, regulated areas, methods of compliance including the development of a compliance plan, respiratory protection, protective clothing and equipment, housekeeping, waste disposal, hygiene facilities, medical surveillance, employee information and training, signs and labels, recordkeeping, and employee observation of monitoring as set forth in §1990.151, unless the Secretary explains why any or all such provisions are not appropriate;

**1990.142(a)(2)(ii)** The model standard set forth in §1990.151 shall be used as a guideline, and

**1990.142(a)(2)(iii)** The permissible exposure limit shall be achieved primarily through engineering and work practice controls except that if a suitable substitute is available for one or more uses no occupational exposure shall be permitted for those uses.

**1990.142(a)(3)** *Provisions of the proposed standard for Category II Potential Carcinogens.* Whenever the Secretary issues a Notice of Proposed Rulemaking to regulate a substance as a Category II Potential Carcinogen:

**1990.142(a)(3)(i)** The proposed standard shall contain at least provisions for scope and application, definitions, notification of use, monitoring, respiratory protection, protective clothing and equipment, housekeeping, waste disposal, medical surveillance, employee information and training, recordkeeping and employee observation of monitoring as set forth in §1990.151, unless the Secretary explains why any or all such provisions are not appropriate; and

**1990.142(a)(3)(ii)** The model standard set forth in §1990.151 shall be used as a guideline; and

**1990.142(a)(3)(iii)** Worker exposure to Category II Potential Carcinogens will be reduced as appropriate and consistent with the statutory requirements on a case-by-case basis in the individual rulemaking proceedings. Any permissible exposure level so established shall be met primarily through engineering and work practice controls.

**1990.142(b)** ***Emergency temporary standards*** *(Section 6(c) of the Act)*—(1) *General.* The Secretary may issue an Emergency Temporary Standard (ETS) for a Category I Potential Carcinogen in accordance with section 6(c) of the Act.

**1990.142(b)(2)** *Provisions of the ETS.* (i) The ETS shall contain at least provisions for scope and application, definitions, notification of use, a permissible exposure limit, monitoring, methods of compliance including the development of a compliance plan, respiratory protection, protective clothing and equipment, housekeeping, waste disposal, medical surveillance, employee information and training, signs and labels, recordkeeping and employee observation of monitoring, unless the Secretary explains why any or all such provisions are not appropriate.

**1990.142(b)(2)(ii)** The model standard set forth in §1990.152 shall be used as a guideline.

**1990.142(b)(2)(iii)** The permissible exposure limit shall be achieved through any practicable combination of engineering controls, work practice controls and respiratory protection.

[45 FR 5282, Jan. 22, 1980, as amended at 46 FR 5881, Jan. 21, 1981]

## §1990.143 General provisions for the use of human and animal data.

Human and animal data which are scientifically evaluated to be positive evidence for carcinogenicity including the following policies shall be uniformly relied upon for the identification of potential occupational carcinogens. Arguments challenging the following provisions or their application to specific substances will be considered in individual rulemaking proceedings only if the evidence presented in support of the arguments meets the criteria for consideration specified in §1990.144 or §1990.145.

**1990.143(a)** ***Positive human studies.*** Positive results obtained in one or more human epidemiologic studies will be used to establish the qualitative inference of carcinogenic hazards to workers.

**1990.143(b)** ***Positive animal studies.*** Positive results obtained in one or more experimental studies conducted in one or more mammalian species will be used to establish the qualitative inference of carcinogenic hazard to workers. Arguments that positive results obtained in mammalian species should not be relied upon will be considered only if evidence is presented which meets the criteria for consideration specified in §1990.144(c) or 1990.144(f).

**1990.143(c)** ***Non-positive human studies.*** Positive results in human or mammalian studies generally will be used for the qualitative identification of potential occupational carcinogens, even where non-positive results from human studies exist. Such non-positive results will be considered by the Secretary only if the studies or results meet the criteria set forth in §1990.144(a).

**1990.143(d)** ***Non-positive animal studies.*** Positive results in one or more mammalian studies will be used for the qualitative identification of potential occupational carcinogens, even where non-positive studies exist in other mammalian species. Where non-positive and positive results exist in studies in the same species, the non-positive results will be evaluated.

**1990.143(e)** ***Spontaneous tumors.*** Positive results in human or mammalian studies for the induction or acceleration of induction of tumors of a type which occurs "spontaneously" in unexposed individuals will be used for the qualititative identification of potential occupational carcinogens.

**1990.143(f)** ***Routes of exposure.*** (1) Positive results in studies in which mammals are exposed via the oral, respiratory or dermal routes will be used for the qualitative identification of potential occupational carcinogens, whether tumors are induced at the site of application or distant sites.

**1990.143(f)(2)** Positive results in studies in which mammals are exposed via any route of exposure and in which tumors are induced at sites distant from the site of administration will be used for the qualitative identification of potential occupational carcinogens.

**1990.143(f)(3)(i)** Positive results in mammalian studies in which tumors are induced only at the site of administration, in which a substance or mixture of substances is administered by routes other than oral, respiratory or dermal, will be used as "concordant" evidence that a substance is a potential occupational carcinogen.

**1990.143(f)(3)(ii)** Arguments that such studies should not be relied upon will be considered only if evidence which meets the criteria set forth in §1990.144(b) is provided.

**1990.143(g)** ***Use of high doses in animal testing.*** Positive results for carcinogenicity obtained in mammals exposed to high doses of a substance will be used to establish the qualitative inference of carcinogenic hazard to workers. Arguments that such studies should not be relied upon will be considered only if evidence which meets the criteria set forth in §1990.144(d) is provided.

**1990.143(h)** ***"Threshold" or "No-effect" Levels.*** No determination will be made that a "threshold" or "no-effect" level of exposure can be established for a human population exposed to carcinogens in general, or to any specific substance.

**1990.143(i)** ***Benign tumors.*** Results based on the induction of benign or malignant tumors, or both, will be used to establish a qualitative inference of carcinogenic hazard to workers. Arguments that substances that induce benign tumors do not present a carcinogenic risk to workers will be considered only if evidence that meets the criteria set forth in §1990.144(e) is provided.

**1990.143(j)** ***Statistical evaluation.*** Statistical evaluation will be used in the determination of whether results in human, animal or short-term studies provide positive evidence for carcinogenicity, but will not be the exclusive means for such evaluation.

**1990.143(k)** ***Carcinogenicity of metabolites.*** A substance which is metabolized by mammals to yield one or more potential occupational carcinogens will itself be identified and classified as a potential occupational carcinogen, whether or not there is direct evidence that it induces tumors in humans or experimental animals. Evidence for such metabolism will normally be derived from *in vivo* studies in mammals. In appropriate circumstances, evidence may be derived from *in vitro* studies of mammalian tissues or fractions thereof. Arguments that evidence from *in vitro* metabolic studies in mammals is not relevant to the inference of carcinogenic hazard to humans will be considered only if such evidence meets the criteria set forth in §1990.144(c).

[45 FR 5282, Jan. 22, 1980; 45 FR 43405, June 27, 1980]

## §1990.144 Criteria for consideration of arguments on certain issues.

Arguments on the following issues will be considered by the Secretary in identifying or classifying any substance pursuant to this part, if evidence for the specific substance subject to the rulemaking conforms to the following criteria. Such arguments and evidence will be evaluated based upon scientific and policy judgments.

**1990.144(a)** ***Non-positive results obtained in human epidemiologic studies.*** Non-positive results obtained in human epidemiologic studies regarding the substance subject to the rulemaking or to a similar or closely related substance will be considered by the Secretary only if they meet the following criteria:

*Criteria.* (i) The epidemiologic study involved at least 20 years' exposure of a group of subjects to the substance and at least 30 years' observation of the subjects after initial exposure;

(ii) Documented reasons are provided for predicting the site(s) at which the substance would induce cancer if it were carcinogenic in humans; and

(iii) The group of exposed subjects was large enough for an increase in cancer incidence of 50% above that in unexposed controls to have been detected at any of the predicted sites.

Arguments that non-positive results obtained in human epidemiologic studies should be used to establish numerical upper limits on potential risks to humans exposed to specific levels of a substance will be considered only if criteria (i) and (ii) are met and, in addition:

(iv) Specific data on the level of exposure of the group of workers are provided, based either on direct measurements made periodically throughout the period of exposure, or upon other data which provide reliable evidence of the magnitude of exposure.

**1990.144(b)** ***Tumors induced at site of administration.*** Arguments that tumors at the site of administration should not be considered will be considered only if:

**1990.144(b)(i)** The route of administration is not oral, respiratory or dermal; and

**1990.144(b)(ii)** Evidence is provided which establishes that induction of local tumors is related to the physical configuration or formulation of the material administered (e.g., crystalline form or dimensions of a solid material, or matrix of an impregnated implant) and that tumors are not induced when the same material is administered in a different configuration or formula.

**1990.144(c)** ***Metabolic differences.*** Arguments that differences in metabolic profiles can be used to demonstrate that a chemical found positive in an experimental study in a mammalian species would pose no potential carcinogenic risk to exposed workers will be considered by the Secretary only if the evidence presented for the specific substance subject to the rulemaking meets the following criteria:

*Criteria* . (i) A complete metabolic profile, including identities of trace metabolites, is presented for the experimental animal species;

(ii) A complete metabolic profile, including identities of trace metabolites, is available for a human population group representative of those who are occupationally exposed;

(iii) Documented evidence is provided for ascribing the carcinogenic activity of the substance in the test animal species to metabolite(s) produced only in that species and not in humans; and

(iv) Documented evidence is provided to show that other metabolites produced also in humans have been adequately tested and have not been shown to be carcinogenic.

**1990.144(d)** ***Use of high doses in animal testing.*** Arguments that positive results obtained in carcinogenesis bioassays with experimental animals subjected to high doses of a substance are not relevant to potential carcinogenic risks to exposed workers will be considered by the Secretary only if the evidence for the specific substance subject to the rulemaking meets the following criteria:

*Criteria.* (i) Documented evidence is presented to show that the substance in question is metabolized by the experimental animal species exposed at the dose levels used in the bioassay(s) to metabolic products which include one or more that are not produced in the same species at lower doses.

(ii) Documented evidence is presented to show that the metabolite(s) produced only at high doses in the experimental animal species are the ultimate carcinogen(s) and that the metabolites produced at low doses are not also carcinogenic; and

(iii) Documented evidence is presented to show that the metabolite(s) produced only at high doses in the experimental animal species are not produced in humans exposed to low doses.

**1990.144(e)** ***Benign tumors.*** The Secretary will consider evidence that the substance subject to the rulemaking proceeding is capable only of inducing benign tumors in humans or experimental animals provided that the evidence for the specific substance meets the following criteria:

*Criteria* (i) Data are available from at least two well-conducted biosassays in each of two species of mammals (or from equivalent evidence in more than two species);

(ii) Each of the bioassays to be considered has been conducted for the full lifetime of the experimental animals;

(iii) The relevant tissue slides are made available to OSHA or its designee and the diagnoses of the tumors as benign are made by at least one qualified pathologist who has personally examined each of the slides and who provides specific diagnostic criteria and descriptions and

(iv) All of the induced tumors must be shown to belong to a type which is known not to progress to malignancy or to be at a benign stage when observed. In the latter case, data must be presented to show that multiple sections of the affected organ(s) were adequately examined to search for invasion of the tumor cells into adjacent tissue, and that multiple sections of other organs were adequately examined to search for tumor metastases.

**1990.144(f)** ***Indirect mechanisms.*** The Secretary will consider evidence that positive results obtained in a carcinogenesis bioassay with experimental animals are not relevant to a determination of a carcinogenic risk to exposed workers, if the evidence demonstrates that the mechanism by which the observed tumor incidence is effected is indirect and would not occur if humans were exposed. As examples, evidence will be considered that a substance causes a carcinogenic effect by augmenting caloric intake or that the carcinogenic effect from exposure to a substance is demonstrated to be the result of the presence of a carcinogenic virus and it is demonstrated that, in either case, the effect would not take place in the absence of the particular carcinogenic virus or the augmented caloric intake.

[45 FR 5282, Jan. 22, 1980, as amended at 46 FR 5881, Jan. 21, 1981]

## §1990.145 Consideration of substantial new issues or substantial new evidence.

**1990.145(a)** ***Substantial new issues.*** Notwithstanding any other provision of this part, the Secretary will consider in a rulemaking proceeding on a specific substance any substantial new issues upon which the Secretary did not reach a conclusion in the rulemaking proceeding(s) underlying this part including conclusions presented in the preamble.

**1990.145(b)** ***Substantial new evidence.*** Notwithstanding any other provision of this part, the Secretary will consider in a rulemaking proceeding on a specific substance any arguments, data or views which he determines are based upon substantial new evidence which may warrant the amendment of one or more provisions of this part. For the purposes of this part, "substantial new evidence" is evidence directly relevant to any provision of this part and is based upon data, views or arguments which differ significantly from those presented in establishing this part, including amendments thereto.

**1990.145(c)** ***Petitions for consideration of substantial new evidence.*** —(1) *Petition.* Any interested person may file a written petition with the Secretary to consider "substantial new evidence" or one or more "substantial new issues" which contains the information specified in paragraph (c)(2) of this section. The Secretary shall treat such a petition as a request to amend this part, as well as a petition to consider "substantial new evidence".

**1990.145(c)(2)** *Contents.* Each petition for consideration of "substantial new evidence" or one or more "substantial new issues" shall contain at least the following information:

**1990.145(c)(2)(i)** Name and address of the petitioner;

**1990.145(c)(2)(ii)** All of the data, views and arguments that the petitioner would like the Secretary to consider;

**1990.145(c)(2)(iii)** The provision or provisions that petitioner believes are inappropriate or should be added to this part in light of the new data, views, and arguments;

**1990.145(c)(2)(iv)** A statement which demonstrates that the data, views, and arguments relied upon by petitioners are directly relevant to the substance or class of substances that is the subject of a rulemaking or an Advance Notice of Proposed Rulemaking;

**1990.145(c)(2)(v)** A detailed statement and analysis as to why the petitioner believes that the data, views, and arguments presented by the petitioner:

**1990.145(c)(2)(v)(A)** Differ significantly from those presented in the proceeding(s) which establish this part;

**1990.145(c)(2)(v)(B)** Are so substantial as to warrant amendment of this part; and

**1990.145(c)(2)(v)(C)** Constitute a new issue or new evidence within the meaning of paragraphs (a) and (b) of this section.

**1990.145(c)(3)** *Deadline for petitions.* (i) Petitions which comply with paragraph (c) of this section, shall be filed in accordance with the schedule set forth in the Advance Notice of Proposed Rulemaking.

**1990.145(c)(3)(ii)** In extraordinary cases the Secretary may consider evidence submitted after the deadline if the petitioner establishes that the evidence relied upon was not available and could not have reasonably been available in whole or substantial part by the deadline and that it is being submitted at the earliest possible time.

**1990.145(d)** ***Secretary's response.*** (1) The Secretary shall respond to petitions under this paragraph in accordance with §1990.106.

**1990.145(d)(2)** Whenever the Secretary determines that the "substantial new issue" or the "substantial new evidence" submitted under this paragraph is sufficient to initiate a proceeding to amend this part, the Secretary shall:

**1990.145(d)(2)(i)** Issue a notice to consider amendment to this part and not proceed on the rulemaking concerning the individual substance until completion of the amendment proceeding; or

**1990.145(d)(2)(ii)** Issue a notice to consider amendment to this part and consolidate it with the proceeding on the individual substance.

## §1990.146 Issues to be considered in the rulemaking.

Except as provided in §1990.145, after issuance of the advance notice of rulemaking, the proceedings for individual substances under this part shall be limited to consideration of the following issues:

**1990.146(a)** Whether the substance, group of substances or combination of substances subject to the proposed rulemaking is appropriately considered in a single proceeding;

**1990.146(b)** Whether the substance or group of substances subject to the rulemaking meets the definition of a potential occupational carcinogen set forth in §1990.103, including whether the scientific studies are reliable;

**1990.146(c)** Whether the available data can appropriately be applied to the substance, group of substances or combination of substances covered by the rulemaking;

**1990.146(d)** Whether information, data, and views that are submitted in accordance with §1990.144 are sufficient to warrant an exception to this part;

**1990.146(e)** Whether the data, views and arguments that are submitted in accordance with §1990.145 are sufficient to warrant amendment of this part;

**1990.146(f)** Whether the potential occupational carcinogen meets the criteria for a Category I Potential Carcinogen or a Category II Potential Carcinogen;

**1990.146(g)** The environmental impact arising from regulation of the substance;

**1990.146(h)** Any issues required by statute or executive order;

**1990.146(i)** The determination of the level to control exposures to Category I Potential Carcinogens primarily through the use of engineering and work practice controls including technological and economic considerations.

**1990.146(j)** The determination of the appropriate employee exposure level, consistent with the Act's requirements, for Category II Potential Carcinogens;

**1990.146(k)** Whether suitable substitutes are available for one or more uses of Category I Potential Carcinogens and; if so, the no occupational exposure level to be achieved solely with engineering and work practice controls and other issues relevant to substitution; and

**1990.146(l)** Whether the provisions of the proposal and of §§1990.151 and 1990.152 (model standards) are appropriate, except as limited by §1990.142 and whether additional regulatory provisions may be appropriate.

[45 FR 5282, Jan. 22, 1980, as amended at 46 FR 5881, Jan. 21, 1981]

## §1990.147 Final action.

**1990.147(a)** Within one hundred twenty (120) days from the last day of any hearing or ninety (90) days from the close of any post hearing comment period, whichever occurs first, the Secretary shall publish in the FEDERAL REGISTER:

**1990.147(a)(1)** A final standard based upon the record in the proceeding; or

**1990.147(a)(2)** A statement that no final standard will be issued, and the reasons therefor, or

**1990.147(a)(3)** A statement that the Secretary intends to issue a final rule, but that he is unable to do so at the present time, including:

**1990.147(a)(3)(i)** The reasons therefor; and

**1990.147(a)(3)(ii)** The date by which the standard will be published, which may not exceed one hundred twenty (120) days thereafter.

**1990.147(a)(3)(iii)** The Secretary may issue no more than one such notice, unless the Secretary determines that (A) new evidence which was unavailable during the rulemaking proceeding has just become available; (B) the evidence is so important that a final rule could not reasonably be issued without this evidence, and; (C) the record is reopened for receipt of comments and/or a hearing on this evidence. This paragraph does not require the Secretary to consider any evidence which is submitted after the dates established for the submission of evidence.

**1990.147(b)** The failure of the Secretary to comply with the required timeframes shall not be a basis to set aside any standard or to require the issuance of a new proposal on any individual substance.

**1990.147(c)** The final standard shall state whether the substance or group of substances subject to the rulemaking is classified as a Category I Potential Carcinogen or as a Category II Potential Carcinogen. If the classification differs from that in the notice of proposed rulemaking, the Secretary shall explain the reasons for the change in classification in the preamble to the final standard.

**1990.147(d)** If the substance is classified as a Category I Potential Carcinogen, the final standard shall conform to the provisions of §1990.142(a)(2)(iii). If the final standard contains other provisions that substantially differ from the proposed provisions, the Secretary shall explain the reasons for the changes in the preamble to the final standard.

**1990.147(e)** If the substance is classified as a Category II potential carcinogen, the final standard shall conform to the provisions of §1990.142(a)(3)(iii). If the final standard contains other provisions that substantially differ from the proposed provisions, the Secretary shall explain the reasons for the changes in the preamble to the final standard.

**1990.147(f)** If the substance is classified as a Category II potential carcinogen, the Secretary shall notify the applicable federal and state agencies, including the Administrator of EPA, the Director of NCI, the Director of NIEHS, the Director of NIOSH, the Commissioner of FDA and the Chairperson of CPSC of such determination and request that the applicable agencies engage in, or stimulate, further research pursuant to their legislative authority, to develop new and additional scientific data.

**1990.147(g)** If, after a rulemaking, the Secretary determines that the substance under consideration should not be classified as a Category I potential carcinogen or a Category II potential carcinogen, the Secretary shall publish a notice of this determination in the FEDERAL REGISTER, together with the reasons therefor.

## MODEL STANDARDS

### §1990.151 Model standard pursuant to section 6(b) of the Act.

Occupational Exposure to ________________
Permanent Standard (insert section number of standard)

**1990.151(a)** ***Scope and application*** —(1) *General.* This section applies to all occupational exposures to ________________ or to (specify those uses or classes of uses of ________________ [Chemical Abstracts Service Registry Number 0000] which are covered by the standard, including, where appropriate, the type of exposure to be regulated by the standard) except as provided in paragraph (a)(2).

**1990.151(a)(2)** *Exemptions.* This section does not apply to (insert those uses or classes of uses of ________________ which are exempted from compliance with the standard, including, where appropriate,

**1990.151(a)(2)(i)** Workplaces where exposure to ________________ results from solid or liquid mixtures containing a specified percentage of ________________ or less;

**1990.151(a)(2)(ii)** Workplaces where another Federal agency is exercising statutory authority to prescribe or enforce standards or regulations affecting occupational exposure to ________________; or

**1990.151(a)(2)(iii)** Workplaces which are appropriately addressed in a separate standard).

**1990.151(b)** ***Definitions.***

________________ means (definition of the substance, group of substances, *or combination of substances, to be regulated* ).

*Action level* means an airborne concentration of ________________ of (insert appropriate level of exposure).

**NOTE:** Where appropriate, consider an action level as a limitation on requirements for periodic monitoring (para. (e)(3)), medical surveillance (para. (n)), training (para. (o)), labels (para. (p)(3)), and other provisions.

*Assistant Secretary* means the Assistant Secretary of Labor for Occupational Safety and Health, U.S. Department of Labor, or designee.

*Authorized person* means any person specifically authorized by the employer whose duties require the person to enter regulated areas or any person entering such an area as a designated representative of employees for the purpose of exercising the opportunity to observe monitoring procedures under paragraph (r) of this section.

*Director* means the Director, National Institute for Occupational Safety and Health, U.S. Department of Health, and Health Services, or designee.

*Emergency* means in any occurrence such as, but not limited to, equipment failure, rupture of containers, or failure of control equipment which may result in a release of ________________ which is (insert appropriate quantitative or qualitative level of release which constitutes an emergency).

*OSHA Area Office* means the Area Office of the Occupational Safety and Health Administration having jurisdiction over the geographic area where the affected workplace is located.

**1990.151(c)** ***Permissible exposure limits provisions*** —(1) *Inhalation* —(i) *Time weighted average limit (TWA).* Within (insert appropriate time period) of the effective date of this section, the employer shall assure that no employee is exposed to an airborne concentration of ________________ in excess of: (insert appropriate exposure limit or when it is determined by the Secretary that there are available suitable substitutes for uses or classes of uses that are less hazardous to humans, the proposal shall permit no occupational exposure) as an eight (8)-hour-time-weighted average.

(Where the Secretary finds that suitable substitutes for ________________ may exist, the determination of the ________________ level shall include consideration of the availability, practicability, relative degree of hazard, and economic consequences of the substitutes.)

**1990.151(c)(1)(ii)** *Ceiling limit (If appropriate).* Within (insert appropriate time period) of the effective date of this section, the employer shall assure that no employee is exposed to an airborne concentration of ________________ in excess of: (insert exposure limit) as averaged over any: (insert appropriate time period) during the working day.

**1990.151(c)(2)** *Dermal and eye exposure.* (As appropriate.) (i) Within (insert appropriate time period) of the effective date of this section, the employer shall (If eye exposure to ________________ does not create a risk of cancer, insert exposure level or criteria which will prevent other adverse health effects of eye exposure to ________________ if any. If eye exposure creates a risk of cancer, insert exposure level or criteria which represents the level of eye exposure to ________________).

**1990.151(c)(2)(ii)** Within (insert appropriate time period) of the effective date of this section, the employer shall (If skin exposure to ________________ does not create a risk of cancer, insert exposure level or criteria which will prevent other adverse health effects of skin exposure to ________________ if any. If skin exposure creates a risk of cancer, insert exposure level or criteria which represents the level of skin exposure to ________________).

**1990.151(d)** ***Notification of use and emergencies***— (1) *Use.* Within (insert appropriate time period and additional information requirements if appropriate), of the effective date of this standard or within thirty days of the introduction of ________________ into the workplace, every employer who has a place of employment in which ________________, is present shall report the address and location of each place of employment to the OSHA Area Office and an estimate of the number of employees exposed.

**1990.151(d)(2)** *Emergencies.* Emergencies, and the facts obtainable at that time, shall be reported within (insert appropriate number) hours of, or during the first federal working day after, the time the employer becomes aware of the emergency to the OSHA Area Office, whichever is longer. Upon request of the OSHA Area Office, the employer shall submit additional information in writing relevant to the nature and extent of employee exposures and measures taken to prevent future emergencies of a similar nature.

**1990.151(e)** ***Exposure monitoring***— (1) *General.* (i) Determinations of airborne exposure levels shall be made from air samples that are representative of each employee's exposure to ________________, over an eight (8) hour period. (Modify the time period as appropriate to be practical in the relevant industries yet reasonably representative of full shift exposures.) Monitoring of exposure levels required under this paragraph shall be made as follows: [insert method or alternative methods to be used to meet the requirements of this paragraph].

**1990.151(e)(1)(ii)** For the purpose of this section, employee exposure is that exposure which would occur if the employee were not using a respirator.

**1990.151(e)(2)** *Initial monitoring.* Each employer who has one or more workplaces where (specify the types of workplaces subject to the monitoring requirement) shall, within (insert appropriate period) of the effective date of this section (insert requirements for initial monitoring, as appropriate).

**1990.151(e)(3)** *Frequency.* (Insert, if appropriate, provisions prescribing the minimum frequency at which monitoring must be repeated, the conditions under which such frequency must be increased or may be reduced, and conditions under which such routine monitoring may be discontinued (for example, where the action level is not exceeded). Where appropriate, specify different frequency requirements for certain types of workplaces where, for example, exposure levels are subject to greater or less variability.)

**1990.151(e)(4)** *Additional monitoring.* (Insert, if appropriate, provisions for monitoring, in addition to the requirements (if any) of paragraph (e)(3). This may include a production, process, control or personnel change which might result in new or additional exposure to ________________, or whenever the employer has any other reason to suspect a change which might result in new or additional exposures to ________________.)

**1990.151(e)(5)** *Employee notification.* (i) Within (insert appropriate period) after the receipt of monitoring results, the employer shall notify each employee in writing of the results which represent that employee's exposure.

**1990.151(e)(5)(ii)** Whenever the results indicate that the representative employee exposure exceeds the permissible exposure limits, the employer shall include in the written notice a statement that the permissible exposure limits were exceeded and a description of the corrective action being taken to reduce exposure to or below the permissible exposure limits.

**1990.151(e)(6)** *Accuracy of measurement.* (Insert requirements for accuracy of methods of measurement or detection used to comply with the paragraph).

**1990.151(f)** ***Regulated areas*** —(1) Within (insert appropriate time period) of the effective date of this section, the employer shall, where practicable, establish regulated areas where ________________, concentrations are in excess of the permissible exposure limits.

**1990.151(f)(2)** Regulated areas shall be demarcated and segregated from the rest of the workplace, in any manner that minimizes the number of persons who will be exposed to ________________.

**1990.151(f)(3)** Access to regulated areas shall be limited to authorized persons or to persons otherwise authorized by the Act or regulations issued pursuant thereto.

**1990.151(f)(4)** The employer shall assure that in the regulated area, food or beverages are not present or consumed, smoking products are not present or used, and cosmetics are not applied (except that these activities may be conducted in the lunchroom, change rooms and showers required under paragraphs (m)(1) through (m)(3) of this section).

**1990.151(g)** ***Methods of compliance*** —(1) *Engineering and work practice controls.* (i) The employer shall institute engineering or work practice controls to reduce and maintain employee exposures to ________ to or below the permissible exposure limits, except to the extent that the employer establishes that such controls are not feasible.

**1990.151(g)(1)(ii)** Engineering and work practice controls shall be implemented to reduce exposures even if they will not be sufficient to reduce exposures to or below the permissible exposure limits.

**1990.151(g)(2)** *Compliance program.* (i) Within (insert appropriate period) of the effective date of this section, the employer shall establish and implement a written program to reduce exposures to or below the permissible exposure limits by means of engineering and work practice controls, as required by paragraph (g)(1) of this section.

**1990.151(g)(2)(ii)** Written plans for these compliance programs shall include at least the following:

**1990.151(g)(2)(ii)(A)** A description of each operation or process resulting in employee exposure to ________________;

**1990.151(g)(2)(ii)(B)** Engineering plans and other studies contemplated or used to determine the controls for each process;

**1990.151(g)(2)(ii)(C)** A report of the technology considered or to be considered in meeting the permissible exposure limits;

**1990.151(g)(2)(ii)(D)** A detailed schedule for the implementation of engineering or work practice controls; and

**1990.151(g)(2)(ii)(E)** Other relevant information reasonably requested by OSHA.

**1990.151(g)(2)(iii)** Written plans for such a program shall be submitted, upon request, to the Assistant Secretary and the Director, and shall be available at the worksite for examination and copying by the Assistant Secretary, the Director, or any affected employee or designated representative.

**1990.151(g)(2)(iv)** The plans required by this paragraph shall be revised and updated periodically to reflect the current status of the program.

**1990.151(h)** ***Respiratory protection*** —(1) *General.* The employer shall assure that respirators are used where required pursuant to this section to reduce employee exposures to or below the permissible exposure limits and in emergencies. Compliance with the permissible exposure limits may not be achieved by the use of respirators except:

**1990.151(h)(1)(i)** During the time period necessary to install or implement feasible engineering and work practice controls; or

**1990.151(h)(1)(ii)** In work operations in which the employer establishes that engineering and work practice controls are not feasible; or

**1990.151(h)(1)(iii)** In work situations where feasible engineering and work practice controls are not yet sufficient to reduce exposure to or below the permissible exposure limits; or

**1990.151(h)(1)(iv)** In emergencies.

**1990.151(h)(2)** *Respirator selection.* (i) Where respiratory protection is required under this section, the employer shall select and provide at no cost to the employee, the appropriate type of respirator from Table 1 below and shall assure that the employee wears the respirator provided.

**TABLE 1—RESPIRATORY PROTECTION FOR**

________________________

(The table will contain a listing of the appropriate type of respirator for various conditions of exposure to ________________).

**1990.151(h)(2)(ii)** The employer shall select respirators from those approved by the National Institute for Occupational Safety and Health under the provisions of 30 CFR part 11.

**1990.151(h)(3)** *Respirator program.* (i) The employer shall institute a respiratory protection program in accordance with 29 CFR 1910.134(b), (d), (e), and (f).

**1990.151(h)(3)(ii)** Employees who wear respirators shall be allowed to wash their face and respirator facepiece to prevent potential skin irritation associated with respirator use.

**1990.151(h)(3)(iii)** The employer shall assure that the respirator issued to each employee is properly fitted (as appropriate, indicate the requirement for a qualitative or quantitative respirator fit testing program).

**1990.151(i)** ***Emergency situations*** —(1) *Written plans.* (i) A written plan for emergency situations shall be developed for each workplace where ________________ is present. Appropriate portions of the plan shall be implemented in the event of an emergency.

**1990.151(i)(1)(ii)** The plan shall specifically provide that employees engaged in correcting emergency conditions shall be equipped with respirators as required in paragraph (h) of this section and other necessary personal protective equipment as required in paragraph (j) until the emergency is abated.

**1990.151(i)(2)** *Alerting employees* —(i) *Alarms.* Where there is the possibility of employee exposure to ________________, due to the occurrence of an emergency, a general alarm shall be installed and maintained to promptly alert employees of such occurrences.

**1990.151(i)(2)(ii)** *Evacuation.* Employees not engaged in correcting the emergency shall be restricted from the area and shall not be permitted to return until the emergency is abated.

**1990.151(j)** ***Protective clothing and equipment*** —(1) *Provision and use.* Where employees are exposed to eye or skin contact with ________________ (insert criteria which trigger this requirement as appropriate), the employer shall, within (insert appropriate time period) of the effective date of this section provide at no cost to such employees, and assure that such employees wear, appropriate protective clothing or other equipment in accordance with 29 CFR 1910.132 and 1910.133 to protect the area of the body which may come in contact with ________________.

**1990.151(j)(2)** *Cleaning and replacement.* (i) The employer shall clean, launder, maintain, or replace protective clothing and equipment required to maintain their effectiveness.

**1990.151(k)** ***Housekeeping*** —(1) *General.* The employer shall, within appropriate time period of the effective date of this section, implement a housekeeping program to minimize accumulation of ________________.

**1990.151(k)(2)** *Specific provisions.* The program shall include (insert appropriate elements).

**1990.151(k)(2)(i)** Periodic scheduling of routine housekeeping.

**1990.151(k)(2)(ii)** Provision for periodic cleaning of dust collection systems.

**1990.151(k)(2)(iii)** Provision for maintaining clean surfaces.

**1990.151(k)(2)(iv)** Provision for assigning personnel to housekeeping procedures; and the

**1990.151(k)(2)(v)** Provision for informing employees about the housekeeping program.

**1990.151(l)** ***Waste disposal*** —(1) *General.* The employer shall assure that no waste material containing ________________ is dispersed into the workplace, to the extent practicable.

**1990.151(l)(2)** The employer shall label, or otherwise inform employees who may contact waste material containing ________________, the contents of such waste material.

**1990.151(l)(3)** (Insert specific disposal methods, as appropriate.)

**1990.151(m)** ***Hygiene facilities and practices.*** Where employees are exposed to airborne concentrations of ________________ in excess of the permissible exposure limits specified in paragraph (c)(1), or where employees are required to wear protective clothing or equipment pursuant to paragraph (j) of this section, or where otherwise found to be appropriate, the following facilities shall be provided by the employer for the use of those employees and the employer shall assure that the employees use the facilities provided.

[Specify appropriate hygiene facilities and practices such as]:

**1990.151(m)(1)** *Change rooms.* The employer shall provide clean change rooms in accordance with 29 CFR 1910.141(e).

**1990.151(m)(2)** *Showers.* (i) The employer shall provide shower facilities in accordance with 29 CFR 1910.141(d)(3).

**1990.151(m)(2)(ii)** The employer shall assure that employees exposed to ________________ shower at the end of the work shift.

**1990.151(m)(3)** Lunchrooms. (if appropriate or other suitable requirements depending on the circumstances). Whenever food or beverages are consumed in the workplace, the employer shall provide lunchroom facilities which have a temperature controlled, positive pressure, filtered air supply, and which are readily accessible to employees exposed to ________________.

**1990.151(n)** ***Medical surveillance***—(1) *General.* (i) The employer shall institute a program of medical surveillance for (specify the types of employees subject to the medical surveillance requirement, for example, by specifying the level, duration, and frequency of exposure to ________________ which make medical surveillance appropriate for individual employees). The employer shall provide each such employee with an opportunity for medical examinations and tests in accordance with this paragraph.

**1990.151(n)(1)(ii)** The employer shall assure that all medical examinations and procedures are performed by or under the supervision of a licensed physician, and shall be provided without cost to the employee.

**1990.151(n)(2)** *Initial examinations.* Within (insert appropriate time period) of the effective date of this section or thereafter at the time of initial assignment, the employer shall provide each employee specified in paragraph (n)(1) of this section an opportunity for a medical examination, including at least the following elements:

**1990.151(n)(2)(i)** A work history and a medical history which shall include: (insert specific areas to be covered pertinent to the health hazards posed by ________________).

**1990.151(n)(2)(ii)** A physical examination which shall include: (insert specific tests, procedures, etc., pertinent to the health hazards posed by ________________.) Where appropriate, provide that the examining physician shall conduct such additional examinations and tests as are needed according to his professional judgment).

**NOTE:** Where appropriate, require or permit different medical protocols, or different frequencies of medical examinations, for separate sub-populations of employees covered under paragraph (n)(1).

**1990.151(n)(3)** *Periodic examinations.* (i) The employer shall provide the examinations specified below in this subparagraph at least (insert appropriate time) for all employees specified in paragraph (n)(1) of this section: (insert appropriate medical protocol for periodic examinations).

**1990.151(n)(3)(ii)** If an employee has not had the examinations prescribed in paragraph (n)(3)(i) of this section within (insert appropriate time period) prior to termination of employment, the employer shall make such examination available to the employee upon such termination.

**1990.151(n)(4)** *Additional examinations.* If the employee for any reason develops signs or symptoms commonly associated with exposure to ________________, the employer shall provide appropriate examination and emergency medical treatment.

**1990.151(n)(5)** *Information provided to the physician.* The employer shall provide the following information to the examining physician:

**1990.151(n)(5)(i)** A copy of this standard and its appendices:

**1990.151(n)(5)(ii)** A description of the affected employee's duties as they relate to the employee's exposure;

**1990.151(n)(5)(iii)** The employee's actual or representative exposure level;

**1990.151(n)(5)(iv)** The employee's anticipated or estimated exposure level (for preplacement examinations or in cases of exposure due to an emergency);

**1990.151(n)(5)(v)** A description of any personal protective equipment used or to be used; and

**1990.151(n)(5)(vi)** The names and addresses of physicians who, under the sponsorship of the employer, provided previous medical examinations of the affected employee, if such records are not otherwise available to the examining physician.

**1990.151(n)(6)** *Physician's written opinion.* (i) The employer shall obtain a written opinion from the examining physician which shall include:

**1990.151(n)(6)(i)(A)** The physician's certification that he has received the information from the employer required under the paragraph (n)(5) and has performed all medical examinations and tests which are in his opinion appropriate under this standard;

**1990.151(n)(6)(i)(B)** The physician's opinion as to whether the employee has any detected medical condition which would place the employee at an increased risk of material impairment of the employee's health from exposure to ________________;

**1990.151(n)(6)(i)(C)** Any recommended limitations upon the employee's exposure to ________________ or upon the use of protective clothing and equipment such as respirators; and

**1990.151(n)(6)(i)(D)** A statement that the employee has been informed by the physician of the results of the medical examination and any medical conditions which require further examination or treatment.

**1990.151(n)(6)(ii)** The employer shall instruct the physician not to reveal in the written opinion specific findings or diagnoses unrelated to occupational exposure to ________________;

**1990.151(n)(6)(iii)** The employer shall provide a copy of the written opinion to the affected employee.

**1990.151(o)** ***Employee Information and Training***—(1) *Training Program.* (i) Within (insert appropriate time period) from the effective date of this section, the employer shall institute a training program for all employees who (specify the employees subject to the training requirement), and shall assure their participation in the training program.

**1990.151(o)(1)(ii)** The training program shall be provided at the time of initial assignment, or upon institution of the training program, and at least (insert appropriate time period) thereafter, and the employer shall assure that each employee is informed of the following:

**NOTE:** Specify, as appropriate, some or all of the following information, or any other appropriate information. Where appropriate, require training programs with different contents, or different frequencies, for separate subpopulations of the employees specified in paragraph (o)(1).

**1990.151(o)(1)(ii)(A)** The information contained in the Appendices;

**1990.151(o)(1)(ii)(B)** The quantity, location, manner of use, release or storage of ________________ and the specific nature of operations which could result in exposure to ________________, as well as any necessary protective steps;

**1990.151(o)(1)(ii)(C)** The purpose, proper use, and limitations of respirators;

**1990.151(o)(1)(ii)(D)** The purpose and a description of the medical surveillance program required by paragraph (n) of this section;

**1990.151(o)(1)(ii)(E)** The emergency procedures developed, as required by paragraph (i) of of this section;

**1990.151(o)(1)(ii)(F)** The engineering and work practice controls, their function and the employee's relationship thereto; and

**1990.151(o)(1)(ii)(G)** A review of this standard.

**1990.151(o)(2)** *Access to training materials.* (i) The employer shall make a copy of this standard and its appendices readily available to all affected employees.

**1990.151(o)(2)(ii)** The employer shall provide, upon request, all materials relating to the employee information and training program to the Assistant Secretary and the Director.

**1990.151(p)** ***Signs and labels***—(1) *General.* (i) The employer may use labels or signs required by other statutes, regulations, or ordinances in addition to, or in combination with, signs and labels required by this paragraph.

**1990.151(p)(1)(ii)** The employer shall assure that no statement appears on or near any sign or label, required by this paragraph, which contradicts or detracts from the meaning of the required sign or label.

**1990.151(p)(2)** *Signs.* (i) The employer shall post signs to clearly indicate all workplaces (Specify as appropriate the description of the area to be signposted such as "where employees are exposed to ________________," or "where exposures exceed the action level," or "where exposures exceed the PEL," or "which are regulated areas"). The signs shall bear the following legend:

**DANGER**

________________________________

**(insert appropriate trade or common names)**

**CANCER HAZARD**

**AUTHORIZED PERSONNEL ONLY**

**1990.151(p)(2)(ii)** The employer shall assure that signs required by this paragraph are illuminated and cleaned as necessary so that the legend is readily visible.

**1990.151(p)(2)(iii)** Where airborne concentrations of ________________ exceed the permissible exposure limits, the signs shall bear the additional legend: "Respirator Required" or "Respirator May Be Required" as appropriate.

**1990.151(p)(3)** *Labels.* (i) The employer shall assure that precautionary labels are affixed to all containers of ________________ and of products containing ________________ (specify if appropriate suitable modifications), and that the labels remain affixed when the ________________ or products containing ________________ are sold, distributed or otherwise leave the employer's workplace.

**1990.151(p)(3)(ii)** The employer shall assure that the precautionary labels required by this paragraph are readily visible and legible. The labels shall bear the following legend:

**DANGER**

**CONTAINS ________________**

**CANCER HAZARD**

**NOTE:** Utilize the clause "POTENTIAL CANCER HAZARD" if it is appropriate to include a signs and labels provision for a Category II potential carcinogen.

**1990.151(q)** ***Recordkeeping*** —(1) *Exposure monitoring.* (i) The employer shall establish and maintain an accurate record of all monitoring required by paragraph (e) of this section.

**1990.151(q)(1)(ii)** This record shall include:

**1990.151(q)(1)(ii)(A)** The dates, number, duration, and results of each of the samples taken, including a description of the sampling procedure used to determine representative employees exposure;

**1990.151(q)(1)(ii)(B)** A description of the sampling and analytical methods used;

**1990.151(q)(1)(ii)(C)** Type of respiratory protective devices worn, if any; and

**1990.151(q)(1)(ii)(D)** Name, social security number and job classification of the employees monitored and of all other employees whose exposure the measurement is intended to represent.

**1990.151(q)(1)(iii)** The employer shall maintain this record for (insert appropriate period) or for the duration of employment plus (insert appropriate period) whichever is longer.

**1990.151(q)(2)** *Medical surveillance.* (i) The employer shall establish and maintain an accurate record of each employee subject to medical surveillance as required by paragraph (n) of this section.

**1990.151(q)(2)(ii)** This record shall include:

**1990.151(q)(2)(ii)(A)** A copy of the physicians' written opinions or a written explanation of the absence of any such opinion or employee refusal to take the medical examination:

**1990.151(q)(2)(ii)(B)** Any employees medical complaints related to exposure to ________________;

**1990.151(q)(2)(ii)(C)** A copy of the information provided to the physician as required by paragraphs (n)(5)(ii) through (v) of this section unless it is systematically retained elsewhere by the employer for the period of time specified in paragraph (q)(2)(ii); and

**1990.151(q)(2)(ii)(D)** A copy of the employee's work history.

**1990.151(q)(2)(iii)** The employer shall assure that this record be maintained for (insert appropriate period) or for the duration of employment plus (insert appropriate period) whichever is longer.

**1990.151(q)(3)** *Availability.* (i) The employer shall assure that all records required to be maintained by this section be made available upon request to the Assistant Secretary and the Director for examination and copying.

**1990.151(q)(3)(ii)** Employee exposure measurement records and employee medical records required by this section shall be provided upon request to employees, designated representatives, and the Assistant Secretary in accordance with 29 CFR 1910.20(a) through (e) and (g) through (i).

**1990.151(q)(4)** *Transfer of records.* (i) Whenever the employer ceases to do business, the successor employer shall receive and retain all records required to be maintained by this section.

**1990.151(q)(4)(ii)** Whenever the employer ceases to do business and there is no successor employer to receive and retain the records for the prescribed period, these records shall be transmitted to the Director.

**1990.151(q)(4)(iii)** At the expiration of the retention period for the records required to be maintained pursuant to this section, the employer shall transmit these records to the Director.

**1990.151(q)(4)(iv)** The employer shall also comply with any additional requirements involving transfer of records set forth in 29 CFR 1910.20(h).

**NOTE:** Include other recordkeeping requirements if appropriate.

**1990.151(r)** ***Observation of monitoring*** —(1) *Employee observation.* The employer shall provide affected employees, or their designated representatives, an opportunity to observe any monitoring of employee exposure to ________________ conducted pursuant to paragraph (e) of this section.

**1990.151(r)(2)** *Observation procedures.* (i) Whenever observation of the monitoring of employee exposure to ________________ requires entry into an area where the use of protective clothing or equipment is required, the employer shall provide the observer with personal protective clothing or equipment required to be worn by employees working in the area, assure the use of such clothing and equipment, and require the observer to comply with all other applicable safety and health procedures.

**1990.151(r)(2)(ii)** Without interfering with the monitoring, observers shall be entitled to:

**1990.151(r)(2)(ii)(A)** Receive an explanation of the measurement procedures;

**1990.151(r)(2)(ii)(B)** Observe all steps related to the measurement of airborne concentrations of ________________ performed at the place of exposure; and

**1990.151(r)(2)(ii)(C)** Record the results obtained and receive results supplied by the laboratory.

**1990.151(s)** ***Effective date.*** This section shall become effective (insert effective date).

**1990.151(t)** ***Appendices.*** The information contained in the appendices is not intended, by itself, to create any additional obligations not otherwise imposed or to detract from any existing obligation. (In normal circumstances three appendices will be included in each standard, an "Appendix A—Substance Safety Data Sheet," an "Appendix B—Substance Technical Guidelines." and an "Appendix C—Medical Surveillance Guidelines." Insert additional appendices or delete any of the suggested appendices as appropriate).

[45 FR 5282, Jan. 22, 1980; 45 FR 43405-43406, June 27, 1980, as amended at 46 FR 5881, Jan. 21, 1981]

## §1990.152 Model emergency temporary standard pursuant to section 6(c) of the Act.

**Occupational Exposure to ________________**

**Emergency Temporary Standard**
**(insert section number of standard.)**

**1990.152(a)** ***Scope and Application*** —(1) *General.* This section applies to all occupational exposures to ________________, or to (specify the uses of classes of uses of ________________ [Chemical Abstracts Service Registry Number 00000], which are covered by the standard, including, where appropriate, the type of exposure to be regulated by the standard) except as provided in paragraph (a)(2).

**1990.152(a)(2)** *Exemption.* This section does not apply to (insert those uses or classes of uses of ________________ which are exempted from compliance with the standard, including, where appropriate,

**1990.152(a)(2)(i)** Workplaces where exposure to ________________ results from solid or liquid mixtures containing a specified percentage of ________________ or less;

**1990.152(a)(2)(ii)** Workplaces where another Federal agency is exercising statutory authority to prescribe or enforce standards or regulations affecting occupational exposure to ________________ or

**1990.152(a)(2)(iii)** Workplaces which are appropriately addressed in a separate standard.

**1990.152(b)** ***Definitions.***

________________ means (definition of the substance, group of substances, or combination of substances, to be regulated).

*Action level* means an airborne concentration of ________________ of (insert appropriate level of exposure).

NOTE: Where appropriate, consider an action level as a limitation on requirements for periodic monitoring (para. (e)(3)), medical surveillance (para. (n)), training (para. (o)), and other provisions.

*Assistant Secretary* means the Assistant Secretary of Labor for Occupational Safety and Health, U.S. Department of Labor, or designee.

*Authorized person* means any person specifically authorized by the employer whose duties require the person to enter a regulated area or any person entering such an area as a designated representative of employees exercising the opportunity to observe monitoring procedures under paragraph (r) of this section.

*Director* means the Director, National Institute for Occupational Safety and Health, U.S. Department of Health, Education and Welfare, or designee.

*Emergency* means any occurrence such as, but not limited to, equipment failure, rupture of containers, or failure of control equipment which may result in a release of ________________ which is (insert appropriate quantitative or qualitative level of release which constitutes an emergency).

*OSHA Area Office* means the Area Office of the Occupational Safety and Health Administration having jurisdiction over the geographic area where the affected workplace is located.

**1990.152(c)** ***Permissible exposure limits*** —(1) *Inhalation* —(i) *Time-weighted average limit (TWA).* Within (insert appropriate time) from the effective date of this emergency temporary standard, the employer shall assure that no employee is exposed to an airborne concentration of ________________ in excess of: (insert appropriate exposure limit representing a level that can be complied with immediately) as an eight (8)-hour-time-weighted average.

**1990.152(c)(1)(ii)** *Ceiling limit (If appropriate.)* The employer shall assure that no employee is exposed to an airborne concentration of ________________ in excess of: (insert appropriate exposure limit representing a level that can be complied with immediately) as averaged over any: (insert appropriate time period) during the working day.

**1990.152(c)(2)** *Dermal and eye exposure.* (As appropriate.) (i) Within (insert appropriate time period) of the effective date of this section, the employer shall (If eye exposure to ________________ does not create a risk of cancer, insert exposure level or criteria which will prevent other adverse effects of eye exposure to ________________, if any. If eye exposure creates a risk of cancer, insert exposure level or criteria which represent the level of eye exposure to ________________.)

**1990.152(c)(2)(ii)** Within (insert appropriate time period) of the effective date of this section, the employer shall (If skin exposure to ________________ does not create a risk of cancer, insert exposure level or criteria which will prevent other adverse health affects of skin exposure to ________________ if any. If skin exposure creates a risk of cancer, insert exposure level or criteria which represents the level of skin exposure to ________________).

**1990.152(d)** ***Notification of Use.*** Within (insert appropriate time and omit specific categories of information if appropriate) of the effective date of this section, or within fifteen (15) days following the introduction of ________________ into the workplace, every employer shall report the following information to the nearest OSHA Area Office for each such workplace.

**1990.152(d)(1)** The address and location of each workplace in which ________________ is present;

**1990.152(d)(2)** A brief description of each process or operation which may result in employee exposure to ________________;

**1990.152(d)(3)** The number of employees engaged in each process or operation who may be exposed to ________________ and an estimate of the frequency and degree of exposure that occurs; and

**1990.152(d)(4)** A brief description of the employer's safety and health program as it relates to limitation of employee exposure to ________________;

**1990.152(e)** ***Exposure monitoring*** —(1) *General.* (i) Determinations of airborne exposure levels shall be made from air samples that are representative of each employee's exposure to ________________ over an eight (8) hour period. (Modify the time period as appropriate to be practical in the relevant industries yet reasonably representative of full shift exposures.) Monitoring of exposure levels required under this paragraph shall be made as follows: [insert method or alternative methods to be used to meet the requirements of this paragraph].

**1990.152(e)(1)(ii)** For the purposes of this section, employee exposure is the exposure which would occur if the employee were not using a respirator.

**1990.152(e)(2)** *Initial monitoring.* Each employer who has one or more workplaces where (specify the types of workplaces subject to the monitoring requirement), shall within (insert appropriate period) of the effective date of this section (insert requirements for initial monitoring, as appropriate).

**1990.152(e)(3)** *Frequency.* (Insert, if appropriate, provisions prescribing the minimum frequency at which monitoring must be repeated, the conditions under which such frequency must be increased, or may be reduced, and conditions under which such routine monitoring may be discontinued (for example where the action level is not exceeded). Where appropriate, specify different frequency requirements for certain types of workplaces where, for example, exposure levels are subject to greater or less variability.)

**1990.152(e)(4)** *Additional monitoring.* (Insert, if appropriate, provisions for monitoring, in addition to the requirements (if any) of paragraph (e)(3). This may include a production, process, control or personnel change which might result in new or additional exposure to ________________ or whenever the employer has any other reason to suspect a change which might result in new or additional exposures to ________________.)

**1990.152(e)(5)** *Employee notification.* (i) Within (insert appropriate period) after the receipt of monitoring results, the employer shall notify each employee in writing of the results which represent that employee's exposure.

**1990.152(e)(5)(ii)** Whenever the results indicate that the representative employee exposure exceeds the permissible exposure limits, the employer shall include in the written notice a statement that permissible exposure limits were exceeded and a description of the corrective action being taken to reduce exposure to or below the permissible exposure limits.

**1990.152(e)(6)** *Accuracy of measurement.* (Insert requirements for accuracy of methods of measurement or detection used to comply with the paragraph.)

**1990.152(f)** [Reserved]

**1990.152(g)** ***Methods of compliance*** —(1) *General.* (i) Employee exposures to ________________ shall be controlled to or below the permissible exposure limits by any practicable combination of engineering controls, work practices and personal protective devices and equipment, during the effective period of this emergency temporary standard.

**NOTE:** Where engineering controls or work practices can reduce employee exposures to ________________ it is recommended that they be implemented where practicable, even where they do not themselves reduce exposures to, or below the permissible exposure limits. Work practices which can be implemented by the employer to help reduce employee exposures to ________________ include limiting access to work areas to authorized personnel, prohibiting smoking and consumption of food and beverages in work areas, and establishing good maintenance and housekeeping practices, including the prompt clean-up of spills and repair of leaks.)

**1990.152(g)(2)** *Engineering and work practice control plan.* (i) Within (insert appropriate time period) of the effective date of this emergency temporary standard, the employer shall develop a written plan describing proposed means to reduce employee exposures to the lowest feasible level by means of engineering and work practice controls (which will be eventually required by a permanent standard for occupational exposure to ________________, as provided for by §1990.151(g) of this subpart).

**1990.152(g)(2)(ii)** Written plans required by this paragraph shall be submitted, upon request, to the Assistant Secretary and the Director and shall be available at the worksite for examination and copying by the Assistant Secretary, the Director, and any affected employee or designated representative.

**1990.152(h)** ***Respiratory protection*** —(1) *Required use.* The employer shall assure that respirators are used where required pursuant to this section to reduce employee exposures to within the permissible exposure limits and in emergencies.

**1990.152(h)(2)** *Respirator selection.* (i) Where respiratory protection is required under this section, the employer shall select and provide at no cost to the employee, the appropriate respirator from Table 1 below and shall assure that the employee wears the respirator provided.

**TABLE 1.—RESPIRATORY PROTECTION FOR**

________________________________

(The table will contain a listing of the appropriate type of respirator for various conditions of exposure to ________________.)

**1990.152(h)(2)(ii)** The employer shall select respirators from those approved by the National Institute for Occupational Safety and Health under the provisions of 30 CFR part 11.

**1990.152(h)(3)** *Respirator program.* (i) The employer shall institute a respirator protection program in accordance with 29 CFR 1910.134(b), (d), (e) and (f).

**1990.152(h)(3)(ii)** Employees who wear respirators shall be allowed to wash their face and respirator face piece to prevent potential skin irritation associated with respirator use.

**1990.152(h)(3)(iii)** The employer shall assure that the respirator issued to each employee is properly fitted (as appropriate, indicate the requirement for a qualitative or quantitative respirator fit testing program.)

**1990.152(i)** [Reserved]

**1990.152(j)** ***Protective clothing and equipment*** —(1) *Provision and use.* Where employees are exposed to eye or skin contact with ________________ (insert criteria which trigger this requirement as appropriate), the employer shall within (insert appropriate time period) of the effective date of this standard provide, at no cost to the employees, and assure that employees wear, appropriate protective clothing or other equipment in accordance with 29 CFR 1910.132 and 1910.133 to protect the area of the body which may come in contact with ________________.

**1990.152(j)(2)** *Cleaning and replacement.* (i) The employer shall clean, launder, maintain, or replace protective clothing and equipment required by this paragraph, as needed to maintain their effectiveness.

**1990.152(k)** ***Housekeeping*** —(1) *General.* The employer shall, within (insert appropriate time period) of the effective date of this section, implement a housekeeping program to minimize accumulations of ________________.

**1990.152(k)(2)** *Specific provisions.* The program shall include (insert appropriate elements):

**1990.152(k)(2)(i)** Periodic scheduling of routine housekeeping procedures;

**1990.152(k)(2)(ii)** Provision for periodic cleaning of dust collection systems;

**1990.152(k)(2)(iii)** Provision for maintaining clean surfaces;

**1990.152(k)(2)(iv)** Provision for assigning personnel to housekeeping procedures; and

**1990.152(k)(2)(v)** Provision for informing employees about the housekeeping program.

**1990.152(l)** ***Waste disposal*** —(1) *General.* The employer shall assure that no waste material containing ________________ is dispersed into the workplace, to the extent practicable.

**1990.152(l)(2)** The employer shall label, or otherwise inform employees who may contact waste material containing ________________ of the contents of such waste material.

**1990.152(l)(3)** (Insert specific disposal methods, as appropriate.)

**1990.152(m)** [Reserved]

**1990.152(n)** ***Medical surveillance*** —(1) *General.* (i) The employer shall institute a program of medical surveillance for (specify the types of employees subject to the medical surveillance requirement, for example, by specifying the level, duration, and frequency of exposure to ________________ which make medical surveillance appropriate for individual employees). The employer shall provide each such employee with an opportunity for medical examinations and tests in accordance with this paragraph.

**1990.152(n)(1)(ii)** The employer shall assure that all medical examinations and procedures are performed by or under the supervision of a licensed physician, and shall be provided without cost to the employee.

**1990.152(n)(2)** *Initial examinations.* Within (insert appropriate time period) of the effective date of this section, or thereafter at the time of initial assignment, the employer shall provide each employee specified in paragraph (n)(1) of this section an opportunity for a medical examination, including at least the following elements:

**1990.152(n)(2)(i)** A work history and a medical history which shall include (insert specific areas to be covered pertinent to the health hazards posed by ________________).

**1990.152(n)(2)(ii)** A physical examination which shall include: (insert specific tests, procedures, etc., pertinent to the health hazards posed by ________________. Where appropriate, provide that the examining physician shall conduct such additional examinations and tests as are needed according to his professional judgement).

**NOTE:** Where appropriate, require or permit different medical protocols, or different frequencies of medical examinations, for separate sub-populations of employees covered under paragraph (n)(1).

**1990.152(n)(3)** *Periodic examinations.* (If appropriate insert appropriate medical protocol and time.)

**1990.152(n)(4)** *Additional examinations.* If the employee for any reason develops signs or symptoms commonly associated with exposure to ________________, the employer shall provide an appropriate examination and emergency medical treatment.

**1990.152(n)(5)** *Information provided to the physician.* The employer shall provide the following information to the examining physician:

**1990.152(n)(5)(i)** A copy of this emergency temporary standard and its appendices;

**1990.152(n)(5)(ii)** A description of the affected employee's duties as they relate to the employee's exposure;

**1990.152(n)(5)(iii)** The employee's actual or representative exposure level;

**1990.152(n)(5)(iv)** The employee's anticipated or estimated exposure level (for preplacement examinations or in cases of exposures due to an emergency);

**1990.152(n)(5)(v)** A description of any personal protective equipment used or to be used; and

**1990.152(n)(5)(vi)** The names and addresses of physicians who, under the sponsorship of the employer, provided previous medical examinations of the affected employee, if such records are not otherwise available to the examining physician.

**1990.152(n)(6)** *Physician's written opinion.* (i) The employer shall obtain a written opinion from the examining physician which shall include:

**1990.152(n)(6)(i)(A)** The results of the medical tests performed;

**1990.152(n)(6)(i)(B)** The physician's opinion as to whether the employee has any detected medical condition which would place the employee at an increased risk of material impairment of the employee's health from exposure to ________________;

**1990.152(n)(6)(i)(C)** Any recommended limitations upon the employee's exposure to ________________ or upon the use of protective clothing and equipment such as respirators; and

**1990.152(n)(6)(i)(D)** A statement that the employee has been informed by the physician of the results of the medical examination and any medical conditions which require further examination or treatment.

**1990.152(n)(6)(ii)** The employer shall instruct the physician not to reveal in the written opinion specific findings or diagnoses unrelated to occupational exposure to ________________;

**1990.152(n)(6)(iii)** The employer shall provide a copy of the written opinion to the affected employee.

**1990.152(o)** ***Employee information and training*** —(1) *Training program.* (i) Within (insert appropriate time period) from the effective date of this standard, the employer shall institute a training program for all employees who (specify the employees subject to the training requirement), and shall assure their participation in the training program.

**1990.152(o)(1)(ii)** The employer shall assure that each employee is informed of the following:

**1990.152(o)(1)(ii)(A)** The information contained in the Appendices;

**1990.152(o)(1)(ii)(B)** The quantity, location, manner of use, release, or storage of ________________ and the specific nature of operations which could result in exposure to ________________, as well as any necessary protective steps;

**1990.152(o)(1)(ii)(C)** The purpose, proper use, and limitations of respirators;

**1990.152(o)(1)(ii)(D)** The purpose and description of the medical surveillance program required by paragraph (n) of this section; and

**1990.152(o)(1)(ii)(E)** A review of this standard.

**1990.152(o)(2)** *Access to training materials.* (i) The employer shall make a copy of this standard and its appendices readily available to all affected employees.

**1990.152(o)(2)(ii)** The employer shall provide, upon request, all materials relating to the employee information and training program to the Assistant Secretary and the Director.

**1990.152(p)** ***Signs and labels.*** (include a signs or a signs and labels provision if it is appropriate for the duration of the ETS)—(1) *General.* (i) The employer may use labels or signs required by other statutes, regulations, or ordinances in addition to, or in combination with, signs and labels required by this paragraph.

**1990.152(p)(1)(ii)** The employer shall assure that no statement appears on or near any sign or label, required by this paragraph, which contradicts or detracts from the meaning of the required sign or label.

**1990.152(p)(2)** *Signs.* (i) The employer shall post signs to clearly indicate all workplaces (specify as appropriate the description of the area to be signposted such as "where employees are exposed to ________," or "where exposures exceed the PEL," or "which are regulated areas"). The signs shall bear the following legend:

**DANGER**

________________________________________

**(insert appropriate trade or common names)**

**CANCER HAZARD**

**AUTHORIZED PERSONNEL ONLY**

**1990.152(p)(2)(ii)** The employer shall assure that signs required by this paragraph are illuminated and cleaned as necessary so that the legend is readily visible.

**1990.152(p)(2)(iii)** Where airborne concentrations of ________ exceed the permissible exposure limits, the signs shall bear the additional legend: ("Respirator Required" or "Respirator may be Required" as appropriate).

**1990.152(p)(3)** *Labels.* (i) The employer shall assure that precautionary labels are affixed to all containers of ________ and of products containing ________ (specify if appropriate suitable modifications), and that the labels remain affixed when ________ or products containing ________ are sold, distributed or otherwise leave the employer's workplace.

**1990.152(p)(3)(ii)** The employer shall assure that the precautionary labels required by this paragraph are readily visible and legible. The labels shall bear the following legend:

**DANGER**

**CONTAINS** ________________________

**CANCER HAZARD**

**1990.152(q)** ***Recordkeeping***—(1) *Exposure monitoring.* (i) The employer shall establish and maintain an accurate record of all monitoring required by paragraph (e) of this section.

**1990.152(q)(1)(ii)** This record shall include:

**1990.152(q)(1)(ii)(A)** The dates, number, duration, and results of each of the samples taken, including a description of the sampling procedures used to determine representative employee exposure;

**1990.152(q)(1)(ii)(B)** A description of the sampling and analytical methods used;

**1990.152(q)(1)(ii)(C)** Type of respiratory protective devices worn, if any; and

**1990.152(q)(1)(ii)(D)** Name, social security number, and job classification of the employee monitored and of all other employees whose exposure the measurement is intended to represent.

**1990.152(q)(1)(iii)** The employer shall maintain this record for the effective period of this emergency temporary standard, and for any additional period required by the permanent standard.

**1990.152(q)(2)** *Medical surveillance.* (i) The employer shall establish and maintain an accurate record for each employee subject to medical surveillance as required by paragraph (n) of this section.

**1990.152(q)(2)(ii)** This record shall include:

**1990.152(q)(2)(ii)(A)** A copy of the physicians' written opinions or a written explanation of the absence of any such opinion or employee refusal to take the medical examination;

**1990.152(q)(2)(ii)(B)** Any employee medical complaints related to exposure to ________;

**1990.152(q)(2)(ii)(C)** A copy of the information provided to the physician as required by paragraphs (n)(5)(ii)—(iv) of this section unless it is systematically retained elsewhere by the employer for the period of time specified in paragraph (q)(2)(iii); and

**1990.152(q)(2)(ii)(D)** A copy of the employee's work history. (1) The employer shall assure that employee exposure measurement records, as required by this section, be made available upon request to the Assistant Secretary and the Director for examination and copying.

**1990.152(q)(2)(iii)** The employer shall assure that this record be maintained for the effective period of this emergency temporary standard, and for any additional period required by the permanent standard.

**1990.152(q)(3)** *Availability.* (i) The employer shall assure that all records required to be maintained by this section be made available upon request, to the Assistant Secretary and the Director for examination and copying.

**1990.152(q)(3)(ii)** Employee exposure measurement records and employee medical records required by this section shall be provided upon request to employees, designated representatives, and the Assistant Secretary in accordance with 29 CFR 1910.20 (a) through (e) and (g) through (i).

**1990.152(r)** ***Observation of monitoring.*** (1) Employee observation. The employer shall provide affected employees, or their designated representatives, an opportunity to observe any monitoring of employee exposure to ________ conducted pursuant to paragraph (e) of this section.

**1990.152(r)(2)** *Observation procedures.* (i) Whenever observation of the monitoring of employee exposure to ________ requires entry into an area where the use of protective clothing or equipment is required, the employer shall provide the observer with personal protective clothing or equipment required to be worn by employees working in the area, assure the use of such clothing and equipment, and require the observer to comply with all other applicable safety and health procedures.

**1990.152(r)(2)(ii)** Without interfering with the monitoring, observers shall be entitled to:

**1990.152(r)(2)(ii)(A)** Receive an explanation of measurement procedures;

**1990.152(r)(2)(ii)(B)** Observe all steps related to the measurement of airborne concentrations of ________ performed at the place of exposure; and

**1990.152(r)(2)(ii)(C)** Record the results obtained and receive results supplied by the laboratory.

**1990.152(s)** ***Effective date.*** This section shall become effective (insert effective date).

**1990.152(t)** ***Appendices.*** The information contained in the appis not intended, itself, to create any additional obligations not otherwise imposed or to detract from any existing obligation. (In normal circumstances three appendices will be included in each standard, an "Appendix A—Substance Safety Data Sheet," an "Appendix B—Substance Technical Guidelines," and an "Appendix C—Medical Surveillance Guidelines." Insert additional appendices or delete any of the suggested appendices as appropriate.)

[45 FR 5282, Jan. 22, 1980; 45 FR 43406-43407, June 27, 1980, as amended at 46 FR 5882, Jan. 21, 1981]

# Subject Index For 29 CFR 1926, Sec. 5, 1903, 1904, and 1990

| Subject term, *Section No.* | Page |
|---|---|



| Subject term, *Section No.* | Page |
|---|---|

Subject term, *Section No.* — Page

U.S. Department of Labor
Occupational Safety and Health Administration

# An Overview: Recording Work-Related Injuries and Illnesses

*The Occupational Safety and Health (OSH) Act of 1970 requires certain employers to prepare and maintain records of work-related injuries and illnesses. Use these definitions when you classify cases on the Log. OSHA's recordkeeping regulation (see 29 CFR Part 1904) provides more information about the definitions below.*

The *Log of Work-Related Injuries and Illnesses* (Form 300) is used to classify work-related injuries and illnesses and to note the extent and severity of each case. When an incident occurs, use the *Log* to record specific details about what happened and how it happened. The *Summary* — a separate form (Form 300A) — shows the totals for the year in each category. At the end of the year, post the *Summary* in a visible location so that your employees are aware of the injuries and illnesses occurring in their workplace.

Employers must keep a *Log* for each establishment or site. If you have more than one establishment, you must keep a separate *Log* and *Summary* for each physical location that is expected to be in operation for one year or longer.

Note that your employees have the right to review your injury and illness records. For more information, see 29 Code of Federal Regulations Part 1904.35, *Employee Involvement.*

Cases listed on the *Log of Work-Related Injuries and Illnesses* are not necessarily eligible for workers' compensation or other insurance benefits. Listing a case on the *Log* does not mean that the employer or worker was at fault or that an OSHA standard was violated.

### When is an injury or illness considered work-related?

An injury or illness is considered work-related if an event or exposure in the work environment caused or contributed to the condition or significantly aggravated a preexisting condition. Work-relatedness is presumed for injuries and illnesses resulting from events or exposures occurring in the workplace, unless an exception specifically applies. See 29 CFR Part 1904.5(b)(2) for the exceptions. The work environment includes the establishment and other locations where one or more employees are working or are present as a condition of their employment. See 29 CFR Part 1904.5(b)(1).

### Which work-related injuries and illnesses should you record?

Record those work-related injuries and illnesses that result in:

- ▼ death,
- ▼ loss of consciousness,
- ▼ days away from work,
- ▼ restricted work activity or job transfer, or
- ▼ medical treatment beyond first aid.

You must also record work-related injuries and illnesses that are significant (as defined below) or meet any of the additional criteria listed below.

You must record any significant work-related injury or illness that is diagnosed by a physician or other licensed health care professional. You must record any work-related case involving cancer, chronic irreversible disease, a fractured or cracked bone, or a punctured eardrum. See 29 CFR 1904.7.

### What are the additional criteria?

You must record the following conditions when they are work-related:

- ▼ any needlestick injury or cut from a sharp object that is contaminated with another person's blood or other potentially infectious material;
- ▼ any case requiring an employee to be medically removed under the requirements of an OSHA health standard;
- ▼ tuberculosis infection as evidenced by a positive skin test or diagnosis by a physician or other licensed health care professional after exposure to a known case of active tuberculosis.

### What is medical treatment?

Medical treatment includes managing and caring for a patient for the purpose of combating disease or disorder. The following are not considered medical treatments and are NOT recordable:

- ▼ visits to a doctor or health care professional solely for observation or counseling;
- ▼ diagnostic procedures, including administering prescription medications that are used solely for diagnostic purposes; and
- ▼ any procedure that can be labeled first aid. (*See below for more information about first aid.*)

### What do you need to do?

1. Within 7 calendar days after you receive information about a case, decide if the case is recordable under the OSHA recordkeeping requirements.
2. Determine whether the incident is a new case or a recurrence of an existing one.
3. Establish whether the case was work-related.
4. If the case is recordable, decide which form you will fill out as the injury and illness incident report.

   You may use *OSHA's 301: Injury and Illness Incident Report* or an equivalent form. Some state workers compensation, insurance, or other reports may be acceptable substitutes, as long as they provide the same information as the OSHA 301.

#### How to work with the Log

1. Identify the employee involved unless it is a privacy concern case as described below.
2. Identify when and where the case occurred.
3. Describe the case, as specifically as you can.
4. Classify the seriousness of the case by recording the **most serious outcome** associated with the case, with column J (Other recordable cases) being the least serious and column G (Death) being the most serious.
5. Identify whether the case is an injury or illness. If the case is an injury, check the injury category. If the case is an illness, check the appropriate illness category.

U.S. Department of Labor
Occupational Safety and Health Administration

### What is first aid?

If the incident required only the following types of treatment, consider it first aid. Do NOT record the case if it involves only:

- ▼ using non-prescription medications at non-prescription strength;
- ▼ administering tetanus immunizations;
- ▼ cleaning, flushing, or soaking wounds on the skin surface;
- ▼ using wound coverings, such as bandages, BandAids™, gauze pads, etc., or using SteriStrips™ or butterfly bandages.
- ▼ using hot or cold therapy;
- ▼ using any totally non-rigid means of support, such as elastic bandages, wraps, non-rigid back belts, etc.;
- ▼ using temporary immobilization devices while transporting an accident victim (splints, slings, neck collars, or back boards).
- ▼ drilling a fingernail or toenail to relieve pressure, or draining fluids from blisters;
- ▼ using eye patches;
- ▼ using simple irrigation or a cotton swab to remove foreign bodies not embedded in or adhered to the eye;
- ▼ using irrigation, tweezers, cotton swab or other simple means to remove splinters or foreign material from areas other than the eye;
- ▼ using finger guards;
- ▼ using massages;
- ▼ drinking fluids to relieve heat stress

#### How do you decide if the case involved restricted work?

Restricted work activity occurs when, as the result of a work-related injury or illness, an employer or health care professional keeps, or recommends keeping, an employee from doing the routine functions of his or her job or from working the full workday that the employee would have been scheduled to work before the injury or illness occurred.

#### How do you count the number of days of restricted work activity or the number of days away from work?

Count the number of calendar days the employee was on restricted work activity or was away from work as a result of the recordable injury or illness. Do not count the day on which the injury or illness occurred in this number. Begin counting days from the day after the incident occurs. If a single injury or illness involved both days away from work and days of restricted work activity, enter the total number of days for each. You may stop counting days of restricted work activity or days away from work once the total of either or the combination of both reaches 180 days.

#### Under what circumstances should you NOT enter the employee's name on the OSHA Form 300?

You must consider the following types of injuries or illnesses to be privacy concern cases:

- ▼ an injury or illness to an intimate body part or to the reproductive system,
- ▼ an injury or illness resulting from a sexual assault,
- ▼ a mental illness,
- ▼ a case of HIV infection, hepatitis, or tuberculosis,
- ▼ a needlestick injury or cut from a sharp object that is contaminated with blood or other potentially infectious material (see 29 CFR Part 1904.8 for definition), and
- ▼ other illnesses, if the employee independently and voluntarily requests that his or her name not be entered on the log.

You must not enter the employee's name on the OSHA 300 *Log* for these cases. Instead, enter "privacy case" in the space normally used for the employee's name. You must keep a separate, confidential list of the case numbers and employee names for the establishment's privacy concern cases so that you can update the cases and provide information to the government if asked to do so.

If you have a reasonable basis to believe that information describing the privacy concern case may be personally identifiable even though the employee's name has been omitted, you may use discretion in describing the injury or illness on both the OSHA 300 and 301 forms. You must enter enough information to identify the cause of the incident and the general severity of the injury or illness, but you do not need to include details of an intimate or private nature.

#### What if the outcome changes after you record the case?

If the outcome or extent of an injury or illness changes after you have recorded the case, simply draw a line through the original entry or, if you wish, delete or white-out the original entry. Then write the new entry where it belongs. Remember, you need to record the most serious outcome for each case.

### Classifying injuries

An injury is any wound or damage to the body resulting from an event in the work environment.

***Examples:*** Cut, puncture, laceration, abrasion, fracture, bruise, contusion, chipped tooth, amputation, insect bite, electrocution, or a thermal, chemical, electrical, or radiation burn. Sprain and strain injuries to muscles, joints, and connective tissues are classified as injuries when they result from a slip, trip, fall or other similar accidents.

**U.S. Department of Labor**
Occupational Safety and Health Administration

## Classifying illnesses

### Skin diseases or disorders

Skin diseases or disorders are illnesses involving the worker's skin that are caused by work exposure to chemicals, plants, or other substances.

*Examples:* Contact dermatitis, eczema, or rash caused by primary irritants and sensitizers or poisonous plants; oil acne; friction blisters, chrome ulcers; inflammation of the skin.

### Respiratory conditions

Respiratory conditions are illnesses associated with breathing hazardous biological agents, chemicals, dust, gases, vapors, or fumes at work.

*Examples:* Silicosis, asbestosis, pneumonitis, pharyngitis, rhinitis or acute congestion; farmer's lung, beryllium disease, tuberculosis, occupational asthma, reactive airways dysfunction syndrome (RADS), chronic obstructive pulmonary disease (COPD), hypersensitivity pneumonitis, toxic inhalation injury, such as metal fume fever, chronic obstructive bronchitis, and other pneumoconioses.

### Poisoning

Poisoning includes disorders evidenced by abnormal concentrations of toxic substances in blood, other tissues, other bodily fluids, or the breath that are caused by the ingestion or absorption of toxic substances into the body.

*Examples:* Poisoning by lead, mercury, cadmium, arsenic, or other metals; poisoning by carbon monoxide, hydrogen sulfide, or other gases; poisoning by benzene, benzol, carbon tetrachloride, or other organic solvents; poisoning by insecticide sprays, such as parathion or lead arsenate; poisoning by other chemicals, such as formaldehyde.

### All other illnesses

All other occupational illnesses.

*Examples:* Heatstroke, sunstroke, heat exhaustion, heat stress and other effects of environmental heat; freezing, frostbite, and other effects of exposure to low temperatures; decompression sickness; effects of ionizing radiation (isotopes, x-rays, radium); effects of nonionizing radiation (welding flash, ultra-violet rays, lasers); anthrax; bloodborne pathogenic diseases, such as AIDS, HIV, hepatitis B or hepatitis C; brucellosis; malignant or benign tumors; histoplasmosis; coccidioidomycosis.

## When must you post the Summary?

You must post the *Summary* only — not the *Log* — by February 1 of the year following the year covered by the form and keep it posted until April 30 of that year.

## How long must you keep the Log and Summary on file?

You must keep the *Log* and *Summary* for 5 years following the year to which they pertain.

## Do you have to send these forms to OSHA at the end of the year?

No. You do not have to send the completed forms to OSHA unless specifically asked to do so.

## How can we help you?

If you have a question about how to fill out the *Log*,

- ❑ **visit us online at www.osha.gov** or
- ❑ **call your local OSHA office.**

U.S. Department of Labor
Occupational Safety and Health Administration

**Optional**

# Calculating Injury and Illness Incidence Rates

### What is an incidence rate?

An incidence rate is the number of recordable injuries and illnesses occurring among a given number of full-time workers (usually 100 full-time workers) over a given period of time (usually one year). To evaluate your firm's injury and illness experience over time or to compare your firm's experience with that of your industry as a whole, you need to compute your incidence rate. Because a specific number of workers and a specific period of time are involved, these rates can help you identify problems in your workplace and/or progress you may have made in preventing work-related injuries and illnesses.

### How do you calculate an incidence rate?

You can compute an occupational injury and illness incidence rate for all recordable cases or for cases that involved days away from work for your firm quickly and easily. The formula requires that you follow instructions in paragraph (a) below for the total recordable cases or those in paragraph (b) for cases that involved days away from work, *and* for both rates the instructions in paragraph (c).

(a) *To find out the total number of recordable injuries and illnesses that occurred during the year,* count the number of line entries on your OSHA Form 300, or refer to the OSHA Form 300A and sum the entries for columns (G), (H), (I), and (J).

(b) *To find out the number of injuries and illnesses that involved days away from work,* count the number of line entries on your OSHA Form 300 that received a check mark in column (H), or refer to the entry for column (H) on the OSHA Form 300A.

(c) *The number of hours all employees actually worked during the year.* Refer to OSHA Form 300A and optional worksheet to calculate this number.

You can compute the incidence rate for all recordable cases of injuries and illnesses using the following formula:

*Total number of injuries and illnesses ÷ Number of hours worked by all employees* X *200,000 hours = Total recordable case rate*

(The 200,000 figure in the formula represents the number of hours 100 employees working 40 hours per week, 50 weeks per year would work, and provides the standard base for calculating incidence rates.)

You can compute the incidence rate for recordable cases involving days away from work, days of restricted work activity or job transfer (DART) using the following formula:

*(Number of entries in column H + Number of entries in column I) ÷ Number of hours worked by all employees* X *200,000 hours = DART incidence rate*

You can use the same formula to calculate incidence rates for other variables such as cases involving restricted work activity (column (I) on Form 300A), cases involving skin disorders (column (M-2) on Form 300A), etc. Just substitute the appropriate total for these cases, from Form 300A, into the formula in place of the total number of injuries and illnesses.

### What can I compare my incidence rate to?

The Bureau of Labor Statistics (BLS) conducts a survey of occupational injuries and illnesses each year and publishes incidence rate data by various classifications (e.g., by industry by employer size, etc.). You can obtain these published data at www.bls.gov or by ca ling a BLS Regional Office.

### Worksheet

Total number of recordable injuries and illnesses in your establishment

[ ] ÷ [ ] X 200,000 = [ ]

Hours worked by all your employees

Total recordable cases incidence rate

Total number of recordable injuries and illnesses with a checkmark in column H or column I

[ ] ÷ [ ] X 200,000 = [ ]

Hours worked by all your employees

DART incidence rate

U.S. Department of Labor
Occupational Safety and Health Administration

# How to Fill Out the Log

The *Log of Work-Related Injuries and Illnesses* is used to classify work-related injuries and illnesses and to note the extent and severity of each case. When an incident occurs, use the *Log* to record specific details about what happened and how it happened.

If your company has more than one establishment or site, you must keep separate records for each physical location that is expected to remain in operation for one year or longer.

We have given you several copies of the *Log* in this package. If you need more than we provided, you may photocopy and use as many as you need.

The *Summary* — a separate form — shows the work-related injury and illness totals for the year in each category. At the end of the year, count the number of incidents in each category and transfer the totals from the *Log* to the *Summary*. Then post the *Summary* in a visible location so that your employees are aware of injuries and illnesses occurring in their workplace.

***You don't post the Log. You post only the Summary at the end of the year.***

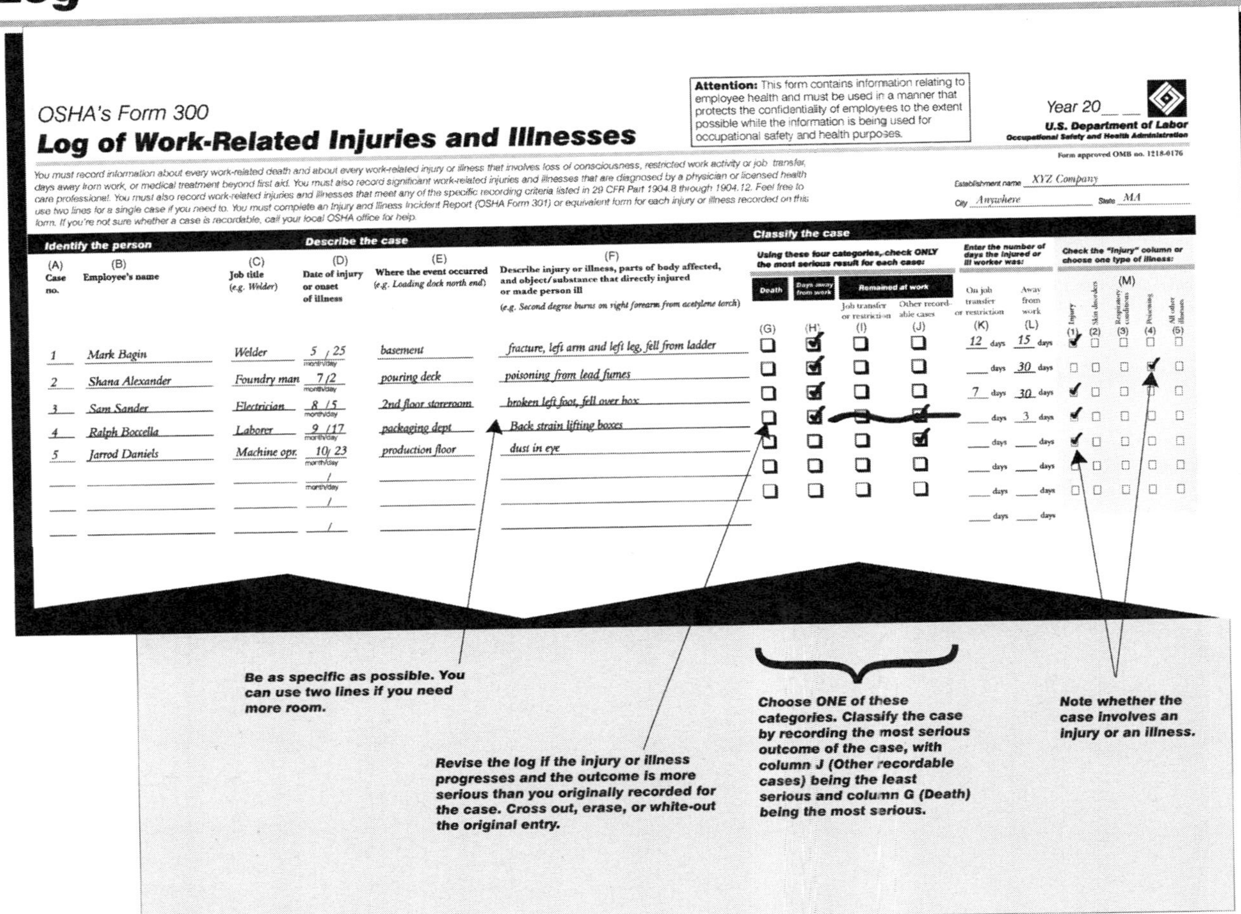

OSHA's Form 300

**Log of Work-Related Injuries and Illnesses**

**Attention:** This form contains information relating to employee health and must be used in a manner that protects the confidentiality of employees to the extent possible while the information is being used for occupational safety and health purposes.

Year 20__ U.S. Department of Labor, Occupational Safety and Health Administration. Form approved OMB no. 1218-0176

You must record information about every work-related death and about every work-related injury or illness that involves loss of consciousness, restricted work activity or job transfer, days away from work, or medical treatment beyond first aid. You must also record significant work-related injuries and illnesses that are diagnosed by a physician or licensed health care professional. You must also record work-related injuries and illnesses that meet any of the specific recording criteria listed in 29 CFR Part 1904.8 through 1904.12. Feel free to use two lines for a single case if you need to. You must complete an Injury and Illness Incident Report (OSHA Form 301) or equivalent form for each injury or illness recorded on this form. If you're not sure whether a case is recordable, call your local OSHA office for help.

Establishment name XYZ Company
City Anywhere State MA

| (A) Case no. | (B) Employee's name | (C) Job title (e.g. Welder) | (D) Date of injury or onset of illness (month/day) | (E) Where the event occurred (e.g. Loading dock north end) | (F) Describe injury or illness, parts of body affected, and object/substance that directly injured or made person ill (e.g. Second degree burns on right forearm from acetylene torch) | (G) Death | (H) Days away from work | (I) Remained at work: Job transfer or restriction | (J) Remained at work: Other recordable cases | (K) On job transfer or restriction (days) | (L) Away from work (days) | (M)(1) Injury | (2) Skin disorder | (3) Respiratory condition | (4) Poisoning | (5) All other illnesses |
|---|---|---|---|---|---|---|---|---|---|---|---|---|---|---|---|---|
| 1 | Mark Bagin | Welder | 5/25 | basement | fracture, left arm and left leg, fell from ladder | ☐ | ☑ | ☐ | ☐ | 12 | 15 | ✓ | ☐ | ☐ | ☐ | ☐ |
| 2 | Shana Alexander | Foundry man | 7/2 | pouring deck | poisoning from lead fumes | ☐ | ☑ | ☐ | ☐ | | 30 | ☐ | ☐ | ☐ | ✓ | ☐ |
| 3 | Sam Sander | Electrician | 8/5 | 2nd floor storeroom | broken left foot, fell over box | ☐ | ☑ | ☐ | ☐ | 7 | 30 | ✓ | ☐ | ☐ | ☐ | ☐ |
| 4 | Ralph Boccella | Laborer | 9/17 | packaging dept | Back strain lifting boxes | ☐ | ☑ | ~~☐~~ | ~~☑~~ | | 3 | ✓ | ☐ | ☐ | ☐ | ☐ |
| 5 | Jarrod Daniels | Machine opr. | 10/23 | production floor | dust in eye | ☐ | ☐ | ☐ | ☑ | | | ✓ | ☐ | ☐ | ☐ | ☐ |

OSHA's Form 300

# Log of Work-Related Injuries and Illnesses

**Attention:** This form contains information relating to employee health and must be used in a manner that protects the confidentiality of employees to the extent possible while the information is being used for occupational safety and health purposes.

Year 20__ __

**U.S. Department of Labor**
**Occupational Safety and Health Administration**

Form approved OMB no. 1218-0176

*You must record information about every work-related death and about every work-related injury or illness that involves loss of consciousness, restricted work activity or job transfer, days away from work, or medical treatment beyond first aid. You must also record significant work-related injuries and illnesses that are diagnosed by a physician or licensed health care professional. You must also record work-related injuries and illnesses that meet any of the specific recording criteria listed in 29 CFR Part 1904.8 through 1904.12. Feel free to use two lines for a single case if you need to. You must complete an Injury and Illness Incident Report (OSHA Form 301) or equivalent form for each injury or illness recorded on this form. If you're not sure whether a case is recordable, call your local OSHA office for help.*

*Establishment name* ______________________

*City* ______________ *State* ______

| **Identify the person** | | | **Describe the case** | | | **Classify the case** | | | | | | | | | | |
|---|---|---|---|---|---|---|---|---|---|---|---|---|---|---|---|---|
| | | | | | | **Using these four categories, check ONLY the most serious result for each case:** | | | | **Enter the number of days the injured or ill worker was:** | | **Check the "Injury" column or choose one type of illness:** | | | | |
| | | | | | | **Death** | **Days away from work** | **Remained at work** | | | | (M) | | | | |
| (A) Case no. | (B) Employee's name | (C) Job title (*e.g., Welder*) | (D) Date of injury or onset of illness | (E) Where the event occurred (*e.g., Loading dock north end*) | (F) Describe injury or illness, parts of body affected, and object/substance that directly injured or made person ill (*e.g., Second degree burns on right forearm from acetylene torch*) | (G) | (H) | Job transfer or restriction (I) | Other record-able cases (J) | On job transfer or restriction (K) | Away from work (L) | Injury (1) | Skin disorder (2) | Respiratory condition (3) | Poisoning (4) | All other illnesses (5) |
| ___ | ___ | ___ | ___/___ month/day | ___ | ___ | ☐ | ☐ | ☐ | ☐ | ___ days | ___ days | ☐ | ☐ | ☐ | ☐ | ☐ |
| ___ | ___ | ___ | ___/___ month/day | ___ | ___ | ☐ | ☐ | ☐ | ☐ | ___ days | ___ days | ☐ | ☐ | ☐ | ☐ | ☐ |
| ___ | ___ | ___ | ___/___ month/day | ___ | ___ | ☐ | ☐ | ☐ | ☐ | ___ days | ___ days | ☐ | ☐ | ☐ | ☐ | ☐ |
| ___ | ___ | ___ | ___/___ month/day | ___ | ___ | ☐ | ☐ | ☐ | ☐ | ___ days | ___ days | ☐ | ☐ | ☐ | ☐ | ☐ |
| ___ | ___ | ___ | ___/___ month/day | ___ | ___ | ☐ | ☐ | ☐ | ☐ | ___ days | ___ days | ☐ | ☐ | ☐ | ☐ | ☐ |
| ___ | ___ | ___ | ___/___ month/day | ___ | ___ | ☐ | ☐ | ☐ | ☐ | ___ days | ___ days | ☐ | ☐ | ☐ | ☐ | ☐ |
| ___ | ___ | ___ | ___/___ month/day | ___ | ___ | ☐ | ☐ | ☐ | ☐ | ___ days | ___ days | ☐ | ☐ | ☐ | ☐ | ☐ |
| ___ | ___ | ___ | ___/___ month/day | ___ | ___ | ☐ | ☐ | ☐ | ☐ | ___ days | ___ days | ☐ | ☐ | ☐ | ☐ | ☐ |
| ___ | ___ | ___ | ___/___ month/day | ___ | ___ | ☐ | ☐ | ☐ | ☐ | ___ days | ___ days | ☐ | ☐ | ☐ | ☐ | ☐ |
| ___ | ___ | ___ | ___/___ month/day | ___ | ___ | ☐ | ☐ | ☐ | ☐ | ___ days | ___ days | ☐ | ☐ | ☐ | ☐ | ☐ |
| ___ | ___ | ___ | ___/___ month/day | ___ | ___ | ☐ | ☐ | ☐ | ☐ | ___ days | ___ days | ☐ | ☐ | ☐ | ☐ | ☐ |
| ___ | ___ | ___ | ___/___ month/day | ___ | ___ | ☐ | ☐ | ☐ | ☐ | ___ days | ___ days | ☐ | ☐ | ☐ | ☐ | ☐ |
| ___ | ___ | ___ | ___/___ month/day | ___ | ___ | ☐ | ☐ | ☐ | ☐ | ___ days | ___ days | ☐ | ☐ | ☐ | ☐ | ☐ |
| | | | | | **Page totals** ➤ | ___ | ___ | ___ | ___ | ___ | ___ | Injury (1) | Skin disorder (2) | Respiratory condition (3) | Poisoning (4) | All other illnesses (5) |

*Be sure to transfer these totals to the Summary page (Form 300A) before you post it.*

Page ___ of ___

Public reporting burden for this collection of information is estimated to average 14 minutes per response, including time to review the instructions, search and gather the data needed, and complete and review the collection of information. Persons are not required to respond to the collection of information unless it displays a currently valid OMB control number. If you have any comments about these estimates or any other aspects of this data collection, contact: US Department of Labor, OSHA Office of Statistics, Room N-3644, 200 Constitution Avenue, NW, Washington, DC 20210. Do not send the completed forms to this office.

Year 20__ __

U.S. Department of Labor
Occupational Safety and Health Administration

Form approved OMB no. 1218-0176

OSHA's Form 300A

# Summary of Work-Related Injuries and Illnesses

*All establishments covered by Part 1904 must complete this Summary page, even if no work-related injuries or illnesses occurred during the year. Remember to review the Log to verify that the entries are complete and accurate before completing this summary.*

*Using the Log, count the individual entries you made for each category. Then write the totals below, making sure you've added the entries from every page of the Log. If you had no cases, write "0."*

*Employees, former employees, and their representatives have the right to review the OSHA Form 300 in its entirety. They also have limited access to the OSHA Form 301 or its equivalent. See 29 CFR Part 1904.35, in OSHA's recordkeeping rule, for further details on the access provisions for these forms.*

## Number of Cases

| Total number of deaths | Total number of cases with days away from work | Total number of cases with job transfer or restriction | Total number of other recordable cases |
|---|---|---|---|
| ____ (G) | ____ (H) | ____ (I) | ____ (J) |

## Number of Days

| Total number of days of job transfer or restriction | Total number of days away from work |
|---|---|
| ____ (K) | ____ (L) |

## Injury and Illness Types

Total number of . . .
(M)

(1) Injuries ____

(2) Skin disorders ____

(3) Respiratory conditions ____

(4) Poisonings ____

(5) All other illnesses ____

**Post this Summary page from February 1 to April 30 of the year following the year covered by the form.**

Public reporting burden for this collection of information is estimated to average 50 minutes per response, including time to review the instructions, search and gather the data needed, and complete and review the collection of information. Persons are not required to respond to the collection of information unless it displays a currently valid OMB control number. If you have any comments about these estimates or any other aspects of this data collection, contact: US Department of Labor, OSHA Office of Statistics, Room N-3644, 200 Constitution Avenue, NW, Washington, DC 20210. Do not send the completed forms to this office.

## Establishment information

**Your establishment name** ____________________

Street ____________________

City ____________ State ____ ZIP ______

Industry description (*e.g., Manufacture of motor truck trailers*)

____________________

Standard Industrial Classification (SIC), if known (*e.g., SIC 3715*)

__ __ __ __

## Employment information

(*If you don't have these figures, see the Worksheet on the back of this page to estimate.*)

Annual average number of employees ________

Total hours worked by all employees last year ________

## Sign here

**Knowingly falsifying this document may result in a fine.**

I certify that I have examined this document and that to the best of my knowledge the entries are true, accurate, and complete.

____________________ Company executive ____________ Title

( ) - ____________ Phone / / ____ Date

U.S. Department of Labor
Occupational Safety and Health Administration

**Optional**

# Worksheet to Help You Fill Out the Summary

*At the end of the year, OSHA requires you to enter the average number of employees and the total hours worked by your employees on the summary. If you don't have these figures, you can use the information on this page to estimate the numbers you will need to enter on the Summary page at the end of the year.*

### How to figure the average number of employees who worked for your establishment during the year:

❶ **Add** the total number of employees your establishment paid in all pay periods during the year. Include all employees: full-time, part-time, temporary, seasonal, salaried, and hourly.

The number of employees paid in all pay periods = ❶ ________

❷ **Count** the number of pay periods your establishment had during the year. Be sure to include any pay periods when you had no employees.

The number of pay periods during the year = ❷ ________

❸ **Divide** the number of employees by the number of pay periods.

❶ / ❷ ________ = ❸ ________

❹ **Round the answer** to the next highest whole number. Write the rounded number in the blank marked *Annual average number of employees.*

The number rounded = ❹ ________

For example, Acme Construction figured its average employment this way:

| For pay period... | Acme paid this number of employees... |
|---|---|
| 1 | 10 |
| 2 | 0 |
| 3 | 15 |
| 4 | 30 |
| 5 | 40 |
| ▼ | ▼ |
| 24 | 20 |
| 25 | 15 |
| 26 | +10 |
| | 830 |

Number of employees paid = 830 ❶

Number of pay periods = 26 ❷

$\frac{830}{26}$ = 31.92 ❸

31.92 rounds to 32 ❹

32 is the annual average number of employees

### How to figure the total hours worked by all employees:

Include hours worked by salaried, hourly, part-time and seasonal workers, as well as hours worked by other workers subject to day to day supervision by your establishment (e.g., temporary help services workers).

Do not include vacation, sick leave, holidays, or any other non-work time, even if employees were paid for it. If your establishment keeps records of only the hours paid or if you have employees who are not paid by the hour, please estimate the hours that the employees actually worked.

If this number isn't available, you can use this optional worksheet to estimate it.

### Optional Worksheet

________ **Find** the number of full-time employees in your establishment for the year.

X ________ **Multiply** by the number of work hours for a full-time employee in a year.

________ This is the number of full-time hours worked.

\+ ________ **Add** the number of any overtime hours as well as the hours worked by other employees (part-time, temporary, seasonal)

________ **Round** the answer to the next highest whole number. Write the rounded number in the blank marked *Total hours worked by all employees last year.*

OSHA's Form 301

# Injury and Illness Incident Report

**Attention:** This form contains information relating to employee health and must be used in a manner that protects the confidentiality of employees to the extent possible while the information is being used for occupational safety and health purposes.

**U.S. Department of Labor**
**Occupational Safety and Health Administration**

Form approved OMB no. 1218-0176

This *Injury and Illness Incident Report* is one of the first forms you must fill out when a recordable work-related injury or illness has occurred. Together with the *Log of Work-Related Injuries and Illnesses* and the accompanying *Summary*, these forms help the employer and OSHA develop a picture of the extent and severity of work-related incidents.

Within 7 calendar days after you receive information that a recordable work-related injury or illness has occurred, you must fill out this form or an equivalent. Some state workers' compensation, insurance, or other reports may be acceptable substitutes. To be considered an equivalent form, any substitute must contain all the information asked for on this form.

According to Public Law 91-596 and 29 CFR 1904, OSHA's recordkeeping rule, you must keep this form on file for 5 years following the year to which it pertains.

If you need additional copies of this form, you may photocopy and use as many as you need.

Completed by ______________________

Title ______________________

Phone (____) ____-______ Date ___/___/___

## Information about the employee

1) Full name ______________________

2) Street ______________________

City ______________ State ______ ZIP ______

3) Date of birth ___/___/___

4) Date hired ___/___/___

5) ☐ Male
☐ Female

## Information about the physician or other health care professional

6) Name of physician or other health care professional ______________________

7) If treatment was given away from the worksite, where was it given?

Facility ______________________

Street ______________________

City ______________ State ______ ZIP ______

8) Was employee treated in an emergency room?
☐ Yes
☐ No

9) Was employee hospitalized overnight as an in-patient?
☐ Yes
☐ No

## Information about the case

10) Case number from the *Log* ______________ *(Transfer the case number from the Log after you record the case.)*

11) Date of injury or illness ___/___/___

12) Time employee began work ______________ AM / PM

13) Time of event ______________ AM / PM ☐ Check if time cannot be determined

14) ***What was the employee doing just before the incident occurred?*** Describe the activity, as well as the tools, equipment, or material the employee was using. Be specific. *Examples:* "climbing a ladder while carrying roofing materials"; "spraying chlorine from hand sprayer"; "daily computer key-entry."

15) ***What happened?*** Tell us how the injury occurred. *Examples:* "When ladder slipped on wet floor, worker fell 20 feet"; "Worker was sprayed with chlorine when gasket broke during replacement"; "Worker developed soreness in wrist over time."

16) ***What was the injury or illness?*** Tell us the part of the body that was affected and how it was affected; be more specific than "hurt," "pain," or sore." *Examples:* "strained back"; "chemical burn, hand"; "carpal tunnel syndrome."

17) ***What object or substance directly harmed the employee?*** *Examples:* "concrete floor"; "chlorine"; "radial arm saw." *If this question does not apply to the incident, leave it blank.*

18) ***If the employee died, when did death occur?*** Date of death ___/___/___

Public reporting burden for this collection of information is estimated to average 22 minutes per response, including time for reviewing instructions, searching existing data sources, gathering and maintaining the data needed, and completing and reviewing the collection of information. Persons are not required to respond to the collection of information unless it displays a current valid OMB control number. If you have any comments about this estimate or any other aspects of this data collection, including suggestions for reducing this burden, contact: US Department of Labor, OSHA Office of Statistics, Room N-3644, 200 Constitution Avenue, NW, Washington, DC 20210. Do not send the completed forms to this office.

NOTES